DECORATIONS
UNITED STATES ARMY
1862–1926

War Department
Office of the Adjutant General

— ▪ — ▪ —

AMERICAN DECORATIONS
A List of Awards of the
Congressional Medal of Honor
the
Distinguished-Service Cross
and the
Distinguished-Service Medal

AWARDED UNDER AUTHORITY OF THE
CONGRESS OF THE UNITED STATES

1862–1926

Compiled in the Office of the Adjutant General of the Army
and Published by Order of the Secretary of War

HERITAGE BOOKS
2014

HERITAGE BOOKS

AN IMPRINT OF HERITAGE BOOKS, INC.

Books, CDs, and more—Worldwide

For our listing of thousands of titles see our website
at
www.HeritageBooks.com

A Facsimile Reprint
Published 2014 by
HERITAGE BOOKS, INC.
Publishing Division
5810 Ruatan Street
Berwyn Heights, Md. 20740

Originally published
United States
Government Printing Office
Washington
1927

WAR DEPARTMENT
Document No. 18a
Office of the Adjutant General

International Standard Book Numbers
Paperbound: 978-0-7884-5542-1
Clothbound: 978-0-7884-9004-0

TABLE OF CONTENTS

EXTRACTS FROM STATUTES AUTHORIZING THE CONGRESSIONAL MEDAL OF HONOR, THE DISTINGUISHED-SERVICE CROSS, AND THE DISTINGUISHED-SERVICE MEDAL

A resolution to provide for the presentation of "medals of honor" to the enlisted men of the Army and Volunteer Forces who have distinguished, or may distinguish, themselves in battle during the present rebellion.

Resolved by the Senate and House of Representatives of the United States of America in Congress assembled, That the President of the United States be, and he is hereby, authorized to cause two thousand "medals of honor" to be prepared with suitable emblematic devices, and to direct that the same be presented, in the name of Congress, to such noncommissioned officers and privates as shall most distinguish themselves by their gallantry in action, and other soldier-like qualities, during the present insurrection. And that the sum of ten thousand dollars be, and the same is hereby, appropriated out of any money in the Treasury not otherwise appropriated, for the purpose of carrying this resolution into effect.

Approved, July 12, 1862.

An act making appropriations for sundry civil expenses of the Government for the year ending June thirty, eighteen hundred and sixty-four, and for the year ending the 30(th) of June, 1863, and for other purposes.

*   *   *   *   *   *   *

SEC. 6. *And be it further enacted,* That the President cause to be struck from the dies recently prepared at the United States mint for that purpose, "medals of honor" additional to those authorized by the act (resolution) of July twelfth, eighteen hundred and sixty-two, and present the same to such officers, noncommissioned officers, and privates as have most distinguished, or who may hereafter most distinguish, themselves in action; and the sum of twenty thousand dollars is hereby appropriated out of any money in the Treasury not otherwise appropriated, to defray the expenses of the same.

*   *   *   *   *   *   *

Approved, March 3, 1863.

[Extract from the act making appropriations for the support of the Army for the fiscal year ending June 30, 1919, approved July 9, 1918]

Medals of honor, distinguished-service crosses, and distinguished-service medals.—That the provisions of existing law relating to the award of medals of honor to officers, noncommissioned officers, and privates of the Army be, and they hereby are, amended so that the President is authorized to present, in the name of the Congress, a medal of honor only to each person who, while an officer or enlisted man of the Army, shall hereafter, in action involving actual conflict with an enemy, distinguish himself conspicuously by gallantry and intrepidity at the risk of his life above and beyond the call of duty.

That the President be, and he is hereby, further authorized to present, but not in the name of Congress, a distinguished-service cross of appropriate design and a ribbon, together with rosette or other device, to be worn in lieu thereof, to any person who, while serving in any capacity with the Army of the United States since the sixth day of April, nineteen hundred and seventeen, has distinguished, or who shall hereafter distinguish, himself or herself by extraordinary heroism in connection with military operations against an armed enemy.

That the President be, and he is hereby, further authorized to present, but not in the name of Congress, a distinguished-service medal of appropriate design and a ribbon, together with a rosette or other device, to be worn in lieu thereof, to any person who, while serving in any capacity with the Army of the United States since the sixth day of April nineteen hundred and seventeen, has distinguished, or who hereafter shall distinguish, himself or herself by exceptionally meritorious service to the Government in a duty of great responsibility; and said distinguished-service medal shall also be issued to all enlisted men of the Army to whom the certificate of merit has been granted up to and including the date of the passage of this act under the provisions of previously existing law, in lieu of such certificates of merit, and after the passage of this act the award of the certificate of merit for distinguished service shall cease; and additional pay heretofore authorized by law for holders of the certificate of merit shall not be paid to them beyond the date of the award of the distinguished-service medal in lieu thereof as aforesaid.

That each enlisted man of the Army to whom there has been or shall be awarded a medal of honor, a distinguished-service cross, or a distinguished-service medal shall, for each such award, be entitled to additional pay at the rate of $2 per month from the date of the distinguished act or service on which the award is based, and each bar, or other suitable device, in lieu of a medal of honor, a distinguished-service cross, or a distinguished-service medal, as hereinafter provided for, shall entitle him to further additional pay at the rate of $2 per month from the date of the distinguished act or service for which the bar is awarded, and said additional pay shall continue throughout his active service, whether such service shall or shall not be continuous; but when the award is in lieu of the certificate of merit, as provided for in section three hereof, the additional pay shall begin with the date of the award.

That no more than one medal of honor or one distinguished-service cross or one distinguished-service medal shall be issued to any one person; but for each succeeding deed or act sufficient to justify the award of a medal of honor or a distinguished-service cross, or a distinguished-service medal, respectively, the President may award a suitable bar, or other suitable device, to be worn as he shall direct; and for each other citation of an officer or enlisted man for gallantry in action published in orders issued from the headquarters of a force commanded by a general officer he shall be entitled to wear, as the President shall direct, a silver star three-sixteenths of an inch in diameter.

That the Secretary of War be, and he is hereby, authorized to expend from the appropriations for contingent expenses of his department from time to time so much as may be necessary to defray the cost of the medals of honor, distinguished-service crosses, distinguished-service medals, bars, rosettes, and other devices hereinbefore provided for.

That whenever a medal, cross, bar, ribbon, rosette, or other device presented under the provisions of this act shall have been lost, destroyed, or rendered unfit for use, without fault or neglect on the part of the person to whom it was awarded, such medal, cross, bar, ribbon, rosette, or device shall be replaced without charge therefor.

That, except as otherwise prescribed herein, no medals of honor, distinguished-service cross, distinguished-service medal, or bar, or other suitable device in lieu of either of said medals or of said cross, shall be issued to any person after more than three years from the date of the act justifying the award thereof, nor unless a specific statement or report distinctly setting forth the distinguished service and suggesting or recommending official recognition thereof shall have been made at the time of the distinguished service or within two years thereafter, nor unless it shall appear from official records in the War Department that such person has so distinguished himself as to entitle him thereto; but in case an individual who shall distinguish himself dies before the making of the award to which he may be entitled, the award may, nevertheless, be made and the medal or cross or the bar or other emblem or device presented within three years from the date of the act justifying the award thereof to such representative of the deceased as the President may designate; but no medal, cross, bar, or other device, hereinbefore authorized, shall be awarded or presented to any individual whose entire service subsequently to the time he distinguished himself shall not have been honorable; but in cases of officers and enlisted men now in the Army for whom the award of the medal of honor has been recommended in full compliance with then existing regulations but on account of services which, though insufficient fully to justify the award of the medal of honor, appear to have been such as to justify the award of the distinguished-service cross or distinguished-service medal hereinbefore provided for, such cases may be considered and acted upon under the provisions of this act authorizing the award of the distinguished-service cross and distinguished-service medal, notwithstanding that said services may have been rendered more than three years before said cases shall have been considered as authorized by this act, but all consideration of and action upon any of said cases shall be based exclusively upon official records now on file in the War Department; and in the cases of officers and enlisted men now in the Army who have been mentioned in orders, now a part of official records, for extraordinary heroism or especially meritorious services, such as to justify the award of the distinguished-service cross or the distinguished-service medal hereinbefore provided for, such cases may be considered and acted on under the provisions of this act, notwithstanding that said act or services may have been rendered more than three years before said cases shall have been considered as authorized by this act, but all consideration of and action upon any said cases shall be based exclusively upon official records of the War Department.

That the President be, and he is hereby, authorized to delegate, under such conditions, regulations, and limitations as he shall prescribe, to the commanding general of a separate army or higher unit in the field, the power conferred upon him by this act to award the medal of honor, the distinguished-service cross, and the distinguished-service medal; and he is further authorized to make from time to time any and all rules, regulations, and orders which he shall deem necessary to carry into effect the provisions of this act and to execute the full purpose and intention thereof.

That the President is authorized, under regulations to be prescribed by him, to confer such medals and decorations as may be authorized in the military service of the United States upon officers and enlisted men of the military forces of the countries concurrently engaged with the United States in the present war.

EXPLANATORY NOTES

The number following name is the Army serial number.

(*) Indicates posthumous award.

R—Residence at entry into service.

B—Place of birth.

NR—No record found in War Department.

G. O. No.—, W. D.—War Department general orders in which citation is published.

The rank in each instance is that held at time of act or service for which the decoration was awarded.

Errors or omissions should be immediately reported to The Adjutant General of the Army by those concerned.

The Congressional Medal of Honor

1

ALPHABETICAL LIST OF AWARDS OF THE CONGRESSIONAL MEDAL OF HONOR

[Awarded for conspicuous gallantry and intrepidity above and beyond the call of duty in action with the enemy]

ACHESAY _____
   Winter of 1872–73.
   R—NR.
   B—Arizona Territory.

Sergeant, Indian Scouts.
Gallant conduct during campaigns and engagements with Apaches.

ADAMS, JAMES F _____
   At Nineveh, Va., Nov. 12, 1864.
   R—NR.
   B—Cabell County, Va.

Private, Company D, 1st West Virginia Cavalry.
Capture of State flag of 14th Virginia Cavalry (C. S. A.).

ADAMS, JOHN G. B _____
   At Fredericksburg, Va., Dec. 13, 1862.
   R—NR.
   B—Groveland, Mass.

Second lieutenant, Company I, 19th Massachusetts Infantry.
Seized the two colors from the hands of a corporal and a lieutenant as they fell mortally wounded, and with a color in each hand advanced across the field to a point where the regiment was reformed on those colors.

ADKISON, JOSEPH B. (1315019) _____
   Near Bellicourt, France, Sept. 29, 1918.
   R—Memphis, Tenn.
   B—Atoka, Tenn.
   G. O. No. 59, W. D., 1919.

Sergeant, Company C, 119th Infantry, 30th Division.
When murderous machine-gun fire at a range of 50 yards had made it impossible for his platoon to advance, and had caused the platoon to take cover, Sergeant Adkison alone, with the greatest intrepidity, rushed across the 50 yards of open ground directly into the face of the hostile machine gun, kicked the gun from the parapet into the enemy trench, and at the point of the bayonet captured the three men manning the gun. The gallantry and quick decision of this soldier enabled the platoon to resume its advance.

ALBEE, GEORGE E _____
   At Brazos River, Tex., Oct. 28, 1869.
   R—Owatonna, Minn.
   B—Lisbon, N. H.

First lieutenant, 41st U. S. Infantry.
Attacked with 2 men a force of 11 Indians, drove them from the hills, and reconnoitered the country beyond.

ALBER, FREDERICK _____
   At Spotsylvania, Va., May 12, 1864.
   R—NR.
   B—Germany.

Private, Company A, 17th Michigan Infantry.
Bravely rescued a lieutenant of his regiment who had been captured by a party of Confederates by shooting down one, knocking over another with the butt of his musket, and taking them both prisoners.

ALBERT, CHRISTIAN _____
   At Vicksburg, Miss., May 22, 1863.
   R—NR.
   B—Cincinnati, Ohio.

Private, Company G, 47th Ohio Infantry.
Gallantry in the charge of the "volunteer storming party."

ALLEN, ABNER P _____
   At Petersburg, Va., Apr. 2, 1865.
   R—NR.
   B—Woodford County, Ill.

Corporal, Company K, 39th Illinois Infantry.
Gallantry as color bearer in the assault on Fort Gregg.

ALLEN, JAMES _____
   At South Mountain, Md., Sept. 14, 1862.
   R—NR.
   B—NR.

Private, Company F, 16th New York Infantry.
Capture of flag of 16th Georgia Infantry (C. S. A.).

ALLEN, NATHANIEL M _____
   At Gettysburg, Pa., July 2, 1863.
   R—Boston, Mass.
   B—Boston, Mass.

Corporal, Company B, 1st Massachusetts Infantry.
When his regiment was falling back, this soldier, bearing the national color returned in the face of the enemy's fire, pulled the regimental flag from under the body of its bearer, who had fallen, saved the flag from capture, and brought both colors off the field.

ALLEN, WILLIAM _____
   At Turret Mountain, Ariz., Mar. 27, 1873.
   R—Lansingburg, N. Y.
   B—Brightstown, N. Y.

First sergeant, Company I, 23d U. S. Infantry.
Gallantry in action.

ALLEX, JAKE (1387815) _____
   At Chipilly Ridge, France, Aug. 9, 1918.
   R—Chicago, Ill.
   B—Servia.
   G. O. No. 44, W. D., 1919.

Corporal, Company H, 131st Infantry, 33d Division.
At a critical point in the action, when all the officers with his platoon had become casualties, Corporal Allex took command of the platoon and led it forward until the advance was stopped by fire from a machine-gun nest. He then advanced alone for about 30 yards in the face of intense fire and attacked the nest. With his bayonet he killed five of the enemy, when it was broken, used the butt of his rifle, capturing 15 prisoners.

ALLWORTH, EDWARD C._____ | Captain, 60th Infantry, 5th Division.
At Clery-le-Petit, France, Nov. 5, 1918. | While his company was crossing the Meuse River and canal at a bridgehead opposite Clery-le-Petit, the bridge over the canal was destroyed by shell fire and Captain Allworth's command became separated, part of it being on the east bank of the canal and the remainder on the west bank. Seeing his advance units making slow headway up the steep slope ahead, this officer mounted the canal bank and called for his men to follow. Plunging in he swam across the canal under fire from the enemy, followed by his men. Inspiring his men by his example of gallantry, he led them up the slope, joining his hard-pressed platoons in front. By his personal leadership he forced the enemy back for more than a kilometer, overcoming machine-gun nests and capturing 100 prisoners, whose number exceeded that of the men in his command. The exceptional courage and leadership displayed by Captain Allworth made possible the reestablishment of a bridgehead over the canal and the successful advance of other troops.
R—Oregon.
B—Crawford, Wash.
G O. No. 16, W. D., 1919.

AMES, ADELBERT_____ | First lieutenant, 5th U. S. Artillery.
At Bull Run, Va., July 21, 1861. | Remained upon the field in command of a section of Griffin's Battery, directing its fire after being severely wounded and refusing to leave the field until too weak to sit upon the caisson where he had been placed by men of his command.
R—NR.
B—East Thomaston, Me.

AMMERMAN, ROBERT W._____ | Private, Company B, 148th Pennsylvania Infantry.
At Spotsylvania, Va., May 12, 1864. | Capture of battle flag of 8th North Carolina (C. S. A.), being one of the foremost in the assault.
R—Center County, Pa.
B—Center County, Pa.

ANDERS, FRANK L._____ | Corporal, Company B, 1st North Dakota Volunteer Infantry.
At San Miguelde Mayumo, Luzon, P. I., May 13, 1899. | With 11 other scouts, without waiting for the supporting battalion to aid them or to get into a position to do so, charged over a distance of about 150 yards and completely routed about 300 of the enemy who were in line and in a position that could only be carried by a frontal attack.
R—Fargo, N. Dak.
B—Fort Lincoln, Dak. T.

ANDERSON, BRUCE_____ | Private, Company K, 142d New York Infantry.
At Fort Fisher, N. C., Jan. 15, 1865. | Voluntarily advanced with the head of the column and cut down the palisading.
R—Mexico.
B—NR.

ANDERSON, CHARLES W._____ | Private, Company K, 1st New York (Lincoln) Cavalry.
At Waynesboro, Va., Mar. 2, 1865. | Capture of unknown Confederate flag.
R—NR.
B—New Orleans, La.

ANDERSON, EVERETT W._____ | Sergeant, Company M, 15th Pennsylvania Cavalry.
At Crosbys Creek, Tenn., Jan. 14, 1864. | Captured, single handed, a Confederate general during a charge upon the enemy.
R—NR.
B—Chester County, Pa.

ANDERSON, FREDERICK C._____ | Private, Company A, 18th Massachusetts Infantry.
At Weldon Railroad, Va., Aug. 21, 1864. | Capture of battle flag of 27th South Carolina (C. S. A.) and the color bearer.
R—NR.
B—Boston, Mass.

ANDERSON, JAMES_____ | Private, Company M, 6th U. S. Cavalry.
At Wichita River, Tex., Oct. 5, 1870. | Gallantry during the pursuit and fight with Indians.
R—NR.
B—Canada East.

ANDERSON, JOHANNES S. (1389034)_____ | First sergeant, Company B, 132d Infantry, 33d Division.
At Consenvoye, France, Oct. 8, 1918. | While his company was being held up by intense artillery and machine-gun fire, Sergeant Anderson, without aid, voluntarily left the company and worked his way to the rear of the nest that was offering the most stubborn resistance. His advance was made through an open area and under constant hostile fire, but the mission was successfully accomplished, and he not only silenced the gun and captured it, but also brought back with him 23 prisoners.
R—Chicago, Ill.
B—Finland.
G. O. No. 16, W. D., 1919.

ANDERSON, MARION T._____ | Captain, Company D, 51st Indiana Infantry.
At Nashville, Tenn., Dec. 16, 1864. | Led his regiment over five lines of the enemy's works, where he fell, severely wounded.
R—NR.
B—Decatur County, Ind.

ANDERSON, PETER_____ | Private, Company B, 31st Wisconsin Infantry.
At Bentonville, N. C., Mar. 19, 1865. | Entirely unassisted, brought from the field an abandoned piece of artillery and saved the gun from falling into the hands of the enemy.
R—NR.
B—LaFayette County, Wis.

ANDERSON, THOMAS_____ | Corporal, Company I, 1st West Virginia Cavalry.
At Appomattox Station, Va., Apr. 8, 1865. | Capture of Confederate flag.
R—NR.
B—Washington County, Pa.

APPLE, ANDREW O._____ | Corporal, Company I, 12th West Virginia Infantry.
At Petersburg, Va., Apr. 2, 1865. | Conspicuous gallantry as color bearer in the assault on Fort Gregg.
R—NR.
B—Northampton, Pa.

APPLETON, WILLIAM H_____
   At Petersburg, Va., June 15, 1864.
   At New Market Heights, Va., Sept.
     29, 1864.
   R—NR.
   B—NR.

First lieutenant, Company H, 4th U. S. Colored Troops.
The first man of the Eighteenth Corps to enter the enemy's works.
Valiant service in a desperate assault, inspiring the Union troops by his example of steady courage.

ARCHER, JAMES W_____
   At Corinth, Miss., Oct. 4, 1862.
   R—NR.
   B—Edgar, Ill.

First lieutenant and adjutant, 59th Indiana Infantry.
Voluntarily took command of another regiment, with the consent of one or more of his seniors, who were present, rallied the command, and led it in the assault.

ARCHER, LESTER_____
   At Fort Harrison, Va., Sept. 29, 1864.
   R—NR.
   B—Fort Anne, N. Y.

Sergeant, Company E, 96th New York Infantry.
Gallantry in placing the colors of his regiment on the fort.

ARCHINAL, WILLIAM_____
   At Vicksburg, Miss., May 22, 1863.
   R—NR.
   B—Germany.

Corporal, Company I, 30th Ohio Infantry.
Gallantry in the charge of the "volunteer storming party."

ARMSTRONG, CLINTON L_____
   At Vicksburg, Miss., May 22, 1863.
   R—NR.
   B—Franklin, Ind.

Private, Company D, 83d Indiana Infantry.
Gallantry in the charge of the "volunteer storming party."

ARNOLD, ABRAHAM K_____
   At Davenport Bridge, Va., May 10,
     1864.
   R—NR.
   B—Pennsylvania.

Captain, 5th U. S. Cavalry.
By a gallant charge against a superior force of the enemy, extricated his command from a perilous position to which it had been ordered.

ASTON, EDGAR R_____
   At San Carlos, Ariz., May 30, 1868.
   R—NR.
   B—Clermont County, Ohio.

Private, Company L, 8th U. S. Cavalry.
Bravery in action with Indians.

AUSTIN, WILLIAM G_____
   At Wounded Knee Creek, S. Dak.,
     Dec. 29, 1890.
   R—New York, N. Y
   B—Galveston, Tex.

Sergeant, Company E, 7th U. S. Cavalry.
While the Indians were concealed in a ravine, assisted men on the skirmish line, directing their fire, etc., and using every effort to dislodge the enemy.

AVERY, WILLIAM B_____
   At Tranters Creek, N. C., June 5,
     1862.
   R—NR.
   B—NR.

Lieutenant, 1st New York Marine Artillery.
Handled his battery with greatest coolness amidst the hottest fire.

AYERS, DAVID_____
   At Vicksburg, Miss., May 22, 1863.
   R—Upper Sandusky, Ohio.
   B—Kalida, Ohio.

Sergeant, Company A, 57th Ohio Infantry.
Gallantry in the charge of the "volunteer storming party."

AYERS, JAMES F_____
   At Sappa Creek, Kans., Apr. 23,
     1875.
   R—NR.
   B—Collinstown, Va.

Private, Company H, 6th U. S. Cavalry.
Rapid pursuit, gallantry, energy, and enterprise in an engagement with Indians.

AYRES, JOHN G. K_____
   At Vicksburg, Miss., May 22, 1863.
   R—NR.
   B—NR.

Private, Company H, 8th Missouri Infantry.
Gallantry in the charge of the "volunteer storming party."

BABCOCK, JOHN B_____
   At Spring Creek, Nebr., May 16,
     1869.
   R—Stonington, Conn.
   B—New Orleans, La.

First lieutenant, 5th U. S. Cavalry.
While serving with a scouting column this officer's troop was attacked by a vastly superior force of Indians. Advancing to high ground he dismounted his men, remaining mounted himself to encourage them, and there fought the Indians until relieved, his horse being wounded.

BABCOCK, WILLIAM J_____
   At Petersburg, Va., Apr. 2, 1865.
   R—NR.
   B—Griswold, Conn.

Sergeant, Company E, 2d Rhode Island Infantry.
Planted the flag upon the parapet while the enemy still occupied the line; was the first of his regiment to enter the works.

BACON, ELIJAH W_____
   At Gettysburg, Pa., July 3, 1863.
   R—Berlin, Conn.
   B—Burlington, Conn.

Private, Company F, 14th Connecticut Infantry.
Capture of flag of 16th North Carolina (C. S. A.).

Baesel, Albert E.
Near Ivoiry, France, Sept. 27, 1918.
R— Berea, Ohio.
B—Berea, Ohio.
G. O. No. 43, W. D., 1922.

Second lieutenant, 148th Infantry, 37th Division.
Upon hearing that a squad leader of his platoon had been severely wounded while attempting to capture an enemy machine-gun nest about 200 yards in advance of the assault line and somewhat to the right, Lieutenant Baesel requested permission to go to the rescue of the wounded corporal. After thrice repeating his request and permission having been reluctantly given, due to the heavy artillery, rifle, and machine-gun fire and heavy deluge of gas in which the company was at the time, accompanied by a volunteer he worked his way forward, and in spite of a heavy direct machine-gun fire succeeded in reaching the wounded man, whom he just succeeded in placing upon his shoulders when both were instantly killed by enemy fire.
Posthumously awarded. Medal presented to widow, Mrs. Albert E. Baesel.

Bailey, James E.
Winter of 1872–73.
R—NR.
B—Dexter, Me.

Sergeant, Company E, 5th U. S. Cavalry.
Gallant conduct during the campaigns and engagements with Apaches.

Baird, Absalom.
At Jonesboro, Ga., Sept. 1, 1864.
R—NR.
B—Washington, Pa.

Brigadier general, U. S. Volunteers.
Voluntarily led a detached brigade in an assault upon the enemy's works.

Baird, George W.
At Bear Paw Mountain, Mont., Sept. 30, 1877.
R—Milford, Conn.
B—Connecticut.

First lieutenant and adjutant, 5th U. S. Infantry.
Most distinguished gallantry in action with the Nez Percé Indians.

Baker, Edward L., Jr.
At Santiago, Cuba, July 1, 1898.
R—NR.
B—Laramie County, Wyo.

Sergeant major, 10th U. S. Cavalry.
Left cover and, under fire, rescued a wounded comrade from drowning.

Baker, John.
At Cedar Creek, etc., Mont., October, 1876, to January, 1877.
R—Brooklyn, N. Y.
B—Germany.

Musician, Company D, 5th U. S. Infantry.
Gallantry in engagements.

Baldwin, Frank D.
(2 medals of honor.)
At Peach Tree Creek, Ga., July 20, 1864.

At McClellans Creek, Tex., Nov. 8, 1874.
R—Constantine, Mich.
B—Michigan.

Captain, Company D, 19th Michigan Infantry.

Led his company in a countercharge, under a galling fire ahead of his own men, and singly entered the enemy's line, capturing and bringing back two commissioned officers, fully armed, besides a guidon of a Georgia regiment.
First lieutenant, 5th U. S. Infantry.
Rescued, with two companies, two white girls by a voluntary attack upon Indians whose superior numbers and strong position would have warranted delay for reinforcements, but which delay would have permitted the Indians to escape and kill their captives.

Ballen, Frederick.
At Vicksburg, Miss., May 3, 1863.
R—NR.
B—Germany.

Private, Company B, 47th Ohio Infantry.
Was one of a party that volunteered and attempted to run the enemy's batteries with a steam tug and two barges loaded with subsistence stores.

Bancroft, Neil.
At Little Big Horn, Mont., June 25, 1876.
R—NR.
B—Oswego, N. Y.

Private, Company A, 7th U. S. Cavalry.
Brought water for the wounded under a most galling fire.

Banks, George L.
At Missionary Ridge, Tenn., Nov. 25, 1863.
R—NR.
B—NR.

Sergeant, Company C, 15th Indiana Infantry.
As color bearer, led his regiment in the assault, and, though wounded, carried the flag forward to the enemy's works, where he was again wounded. In a brigade of eight regiments this flag was the first planted on the parapet.

Barber, James A.
At Petersburg, Va., Apr. 2, 1865.
R—Westerly, R. I.
B—Westerly, R. I.

Corporal, Company G, 1st Rhode Island Light Artillery.
Was one of a detachment of 20 picked artillerymen who voluntarily accompanied an infantry assaulting party, and who turned upon the enemy the guns captured in the assault.

Barger, Charles D. (2205271)
Near Bois-de-Bantheville, France, Oct. 31, 1918.
R—RR 1, Stotts City, Mo.
B—Mount Vernon, Mo.
G. O. No. 20, W. D., 1919.

Private, first class, Company L, 354th Infantry, 89th Division.
Learning that two daylight patrols had been caught out in No Man's Land and were unable to return, Private Barger and another stretcher bearer upon their own initiative made two trips 500 yards beyond our lines, under constant machine-gun fire, and rescued two wounded officers.

Barkeley, David B. (1488756)
Near Pouilly, France, Nov. 9, 1918.
R—San Antonio, Tex.
B—Laredo, Tex.
G. O. No. 26, W. D., 1919.

Private, Company E, 356th Infantry, 89th Division.
When information was desired as to the enemy's position on the opposite side of the River Meuse, Private Barkeley, with another soldier, volunteered without hesitation and swam the river to reconnoiter the exact location. He succeeded in reaching the opposite bank, despite the evident determination of the enemy to prevent a crossing. Having obtained his information, he again entered the water for his return, but before his goal was reached, he was seized with cramps and drowned.
Posthumously awarded. Medal presented to mother, Mrs. Antonio Barkeley.

BARKER, NATHANIEL C.
At Spotsylvania, Va., May 12, 1864.
R—NR.
B—Piermont, N. H.

Sergeant, Company E, 11th New Hampshire Infantry.
Six color bearers of the regiment having been killed, he voluntarily took both flags of the regiment and carried them through the remainder of the battle.

BARKLEY, JOHN L. (2214317).
Near Cunel, France, Oct. 7, 1918.
R—Blairstown, Mo.
B—Blairstown, Mo.
G. O. No. 44, W. D., 1919.

Private, first class, Company K, 4th Infantry, 3d Division.
Private Barkley, who was stationed in an observation post half a kilometer from the German line, on his own initiative repaired a captured enemy machine gun and mounted it in a disabled French tank near his post. Shortly afterwards, when the enemy launched a counterattack against our forces, Private Barkley got into the tank, waited under the hostile barrage until the enemy line was abreast of him, and then opened fire, completely breaking up the counterattack and killing and wounding a large number of the enemy. Five minutes later an enemy 77-millimeter gun opened fire on the tank point-blank. One shell struck the drive wheel of the tank, but this soldier nevertheless remained in the tank and after the barrage ceased broke up a second enemy counter attack, thereby enabling our forces to gain and hold Hill 25.

BARNES, WILLIAM C.
At Fort Apache, Ariz., Sept. 11, 1881.
R—Washington, D. C.
B—San Francisco, Calif.

Private, first class, Signal Corps, U. S. Army.
Bravery in action.

BARNES, WILLIAM H.
At Chapins Farm, near Richmond, Va., Sept. 29, 1864.
R—NR.
B—St. Marys County, Md.

Private, Company C, 38th U. S. Colored Troops.
Among the first to enter the enemy's works, although wounded.

BARNUM, HENRY A.
At Chattanooga, Tenn., Nov. 23, 1863.
R—NR.
B—NR.

Colonel, 149th New York Infantry.
Although suffering severely from wounds, he led his regiment, inciting the men to greater action by word and example until again severely wounded.

BARRELL, CHARLES L.
Near Camden, S. C., April, 1865.
R—NR.
B—NR.

First lieutenant, Company C, 102d U. S. Colored Troops.
Hazardous service in marching through the enemy's country to bring relief to his command.

BARRETT, RICHARD.
At Sycamore Canyon, Ariz., May 23, 1872.
R—NR.
B—Ireland.

First sergeant, Company A, 1st U. S. Cavalry.
Conspicuous gallantry in a charge upon the Tonto Apaches.

BARRICK, JESSE.
Near Duck River, Tenn., May 26–June 2, 1863.
R—Columbiana, Ohio.
B—NR.

Corporal, Company H, 3d Minnesota Infantry.
While on a scout captured single handed two desperate Confederate guerrilla officers, both of whom were together and well armed at the time.

BARRINGER, WILLIAM H.
At Vicksburg, Miss., May 22, 1863.
R—NR.
B—NR.

Private, Company F, 4th Virginia Infantry.
Gallantry in the charge of the "volunteer storming party."

BARRY, AUGUSTUS.
1863 to 1865.
R—NR.
B—Ireland.

Sergeant major, 16th U. S. Infantry.
Gallantry in various actions during the rebellion.

BART, FRANK J. (38512).
Near Médéah Ferme, France, Oct. 3, 1918.
R—Newark, N. J.
B—New York, N. Y.
G. O. No. 16, W. D., 1919.

Private, Company C, 9th Infantry, 2d Division.
Private Bart, being on duty as a company runner, when the advance was held up by machine-gun fire voluntarily picked up an automatic rifle, ran out ahead of the line, and silenced a hostile machine-gun nest, killing the German gunners. The advance then continued, and when it was again hindered shortly afterwards by another machine-gun nest this courageous soldier repeated his bold exploit by putting the second machine gun out of action.

BATCHELDER, RICHARD N.
Between Catlett and Fairfax Stations, Va., Oct. 13–15, 1863.
R—Manchester, N. H.
B—NR.

Lieutenant colonel and chief quartermaster, 2d Corps.
Being ordered to move his trains by a continuous day-and-night march, and without the usual military escort, armed his teamsters and personally commanded them, successfully fighting against heavy odds and bringing his trains through without the loss of a wagon.

BATES, DELAVAN.
At Cemetery Hill, Va., July 30, 1864.
R—Oswego County, N. Y.
B—NR.

Colonel, 30th U. S. Colored Troops.
Gallantry in action where he fell, shot through the face, at the head of his regiment.

BATES, NORMAN F.
At Columbus, Ga., Apr. 16, 1865.
R—NR.
B—Vermont.

Sergeant, Company E, 4th Iowa Cavalry.
Capture of flag and bearer.

BATSON, MATTHEW A.
At Calamba, Luzon, P. I., July 26, 1899.
R—Carbondale, Ill.
B—Anna, Ill.

First lieutenant, 4th U. S. Cavalry.
Swam the San Juan River, in the face of the enemy's fire and drove him from his entrenchments.

BAYBUTT, PHILIP_____
    At Luray, Va., Sept. 24, 1864.
    R—NR.
    B—England.

Private, Company A, 2d Massachusetts Cavalry.
Capture of flag.

BEATTIE, ALEXANDER M_____
    At Cold Harbor, Va., June 5, 1864.
    R—NR.
    B—NR.

Captain, Company F, 3d Vermont Infantry.
Removed, under a hot fire, a wounded member of his command to a place of safety.

BEATY, POWHATAN_____
    At Chapins Farm, near Richmond,
    Va., Sept. 29, 1864.
    R—NR.
    B—Richmond, Va.

First Sergeant, Company G, 5th U. S. Colored Troops.
Took command of his company, all the officers having been killed or wounded, and gallantly led it.

BEAUFORD, CLAY_____
    Winter of 1872–73.
    R—NR.
    B—Washington County, Md.

First sergeant, Company B, 5th U. S. Cavalry.
Gallant conduct during the campaigns and engagements with Apaches.

BEAUFORT, JEAN J._____
    At Port Hudson, La., about May 20,
    1863.
    R—New Orleans, La.
    B—France.

Corporal, Company A, 2d Louisiana Infantry.
Volunteered to go within the enemy's lines and at the head of a party of eight destroyed a signal station, thereby greatly aiding in the operations against Port Hudson that immediately followed.

BEAUMONT, EUGENE B_____
    At Harpeth River, Tenn., Dec. 17,
    1864.

    At Selma, Ala., Apr. 2, 1865_____
    R—Pennsylvania.
    B—Luzerne County, Pa.

Major and assistant adjutant general, Cavalry Corps, Army of the Mississippi.
Obtained permission from the corps commander to advance upon the enemy's position with the 4th U. S. Cavalry, of which he was a lieutenant; led an attack upon a battery, dispersed the enemy, and captured the guns.
Charged, at the head of his regiment, into the second and last line of the enemy's works.

BEBB, EDWARD J_____
    At Columbus, Ga., Apr. 16, 1865.
    R—NR.
    B—Butler County, Ohio.

Private, Company D, 4th Iowa Cavalry.
Capture of flag.

BECKWITH, WALLACE A_____
    At Fredericksburg, Va., Dec. 13,
    1862.
    R—NR.
    B—New London, Conn.

Private, Company F, 21st Connecticut Infantry.
Gallantly responded to a call for volunteers to man a battery, serving with great heroism until the termination of the engagement.

BEDDOWS, RICHARD_____
    At Spotsylvania Courthouse, Va.,
    May 18, 1864.
    R—NR.
    B—England.

Private, 34th New York Battery.
Brought his guidon off in safety under a heavy fire of musketry after he had lost it by his horse becoming furious from the bursting of a shell.

BEEBE, WILLIAM S_____
    At Cane River Crossing, La., Apr.
    23, 1864.
    R—Thompson, Conn.
    B—Ithaca, N. Y.

First lieutenant, Ordnance Department, U. S. Army.
Voluntarily led a successful assault on a fortified position.

BEECH, JOHN P_____
    At Spotsylvania Courthouse, Va.,
    May 12, 1864.
    R—NR.
    B—England.

Sergeant, Company B, 4th New Jersey Infantry.
Voluntarily assisted in working the guns of a battery, all the members of which had been killed or wounded.

BEGLEY, TERRENCE_____
    At Cold Harbor, Va., June 3, 1864.
    R—NR.
    B—Ireland.

Sergeant, Company D, 7th New York Heavy Artillery.
Shot a Confederate color bearer, rushed forward and seized his colors, and, although exposed to heavy fire, regained the lines in safety.

BELCHER, THOMAS_____
    At Chapins Farm, near Richmond,
    Va., Sept. 29, 1864.
    R—NR.
    B—Bangor, Me.

Private, Company I, 9th Maine Infantry.
Took a guidon from the hands of the bearer, mortally wounded, and advanced with it nearer to the battery than any other man.

BELL, DENNIS_____
    At Tayabacoa, Cuba, June 30, 1898.
    R—Washington, D. C.
    B—Washington, D. C.

Private, Troop H, 10th U. S. Cavalry.
Voluntarily went ashore in the face of the enemy and aided in the rescue of his wounded comrades; this after several previous attempts at rescue had been frustrated.

BELL, HARRY_____
    Near Porac, Luzon, P. I., Oct. 17,
    1899.
    R—NR.
    B—NR.

Captain, 36th Infantry, U. S. Volunteers.
Led a successful charge against a superior force, capturing and dispersing the enemy and relieving other members of his regiment from a perilous position.

BELL, J. FRANKLIN_____
    Near Porac, Luzon, P. I., Sept. 9,
    1899.
    R—Shelbyville, Ky.
    B—Shelbyville, Ky.
    Distinguished-service cross and
    distinguished-service medal also
    awarded.

Colonel, 36th Infantry, U. S. Volunteers.
While in advance of his regiment charged seven insurgents with his pistol and compelled the surrender of the captain and two privates under a close fire from the remaining insurgents concealed in a bamboo thicket.

BELL, JAMES.
At Big Horn, Mont., July 9, 1875.
R—NR.
B—Ireland.

Private, Company E, 7th U. S. Infantry.
Carried dispatches to General Crook at the imminent risk of his life.

BELL, JAMES B.
At Missionary Ridge, Tenn., Nov. 25, 1863.
R—NR.
B—NR.

Sergeant, Company H, 11th Ohio Infantry.
Though severely wounded, was first of his regiment on the summit of the ridge, planted his colors inside the enemy's works, and did not leave the field until after he had been wounded five times.

BENEDICT, GEORGE G.
At Gettysburg, Pa., July 3, 1863.
R—NR.
B—Burlington, Vt.

Second lieutenant, Company C, 12th Vermont Infantry.
Passed through a murderous fire of grape and canister in delivering orders and re-formed the crowded lines.

BENJAMIN, JOHN F.
At Sailors Creek, Va., Apr. 6, 1865.
R—NR.
B—Orange City, N. Y.

Corporal, Company M, 2d New York Cavalry.
Capture of battle flag of 9th Virginia Infantry (C. S. A.).

BENJAMIN, SAMUEL N.
From Bull Run to Spotsylvania, Va., July, 1861, to May, 1864.
R—New York, N. Y.
B—New York, N. Y.

First lieutenant, 2d U. S. Artillery.
Particularly distinguished services as an artillery officer.

BENNETT, ORREN.
At Sailors Creek, Va., Apr. 6, 1865.
R—Towanda, Pa.
B—Bradford County, Pa.

Private, Company D, 141st Pennsylvania Infantry.
Capture of flag.

BENNETT, ORSON W.
At Honey Hill, S. C., Nov. 30, 1864.
B—NR.
B—NR.

First lieutenant, Company A, 102d U. S. Colored Troops.
After several unsuccessful efforts to recover three pieces of abandoned artillery, this officer gallantly led a small force fully 100 yards in advance of the Union lines and brought in the guns, preventing their capture.

BENSINGER, WILLIAM.
Georgia, April, 1862.
R—NR.
B—Ohio.

Private, Company G, 21st Ohio Infantry.
One of 22 men (including two civilians) who, by direction of General Mitchel (or Buell), penetrated nearly 200 miles south into the enemy's territory and captured a railroad train at Big Shanty, Ga., in an attempt to destroy the bridges and track between Chattanooga and Atlanta.

BENYAURD, WILLIAM H. H.
At Five Forks, Va., Apr. 1, 1865.
R—Pennsylvania.
B—Philadelphia, Pa.

First lieutenant, Engineers.
With one companion voluntarily advanced in a reconnoissance beyond the skirmishers, where he was exposed to imminent peril; also, in the same battle, rode to the front with the commanding general to encourage wavering troops to resume the advance, which they did successfully.

BERG, GEORGE.
At El Caney, Cuba, July 1, 1898.
R—NR.
B—Wayne County, Ill.

Private Company C, 17th U. S. Infantry.
Gallantly assisted in the rescue of the wounded from in front of the lines and while under heavy fire of the enemy.

BERGENDAHL, FREDERICK.
At Staked Plains, Tex., Dec. 8, 1874.
R—NR.
B—Sweden.

Private, Band, 4th U. S. Cavalry.
Gallantry in a long chase after Indians.

BERTRAM, HEINRICH.
Arizona, 1868.
R—NR.
B—Germany.

Corporal, Company B, 8th U. S. Cavalry.
Bravery in scouts and actions with Indians.

BESSEY, CHARLES A.
Near Elkhorn Creek, Wyo., Jan. 13, 1877.
R—NR.
B—Reading, Mass.

Corporal, Company A, 3d U. S. Cavalry.
While scouting with 4 men and attacked in ambush by 14 hostile Indians, held his ground, 2 of his men being wounded, and kept up the fight until himself wounded in the side, and then went to the assistance of his wounded comrades.

BETTS, CHARLES M.
At Greensboro, N. C., Apr. 19, 1865.
R—Philadelphia, Pa.
B—Bucks County, Pa.

Lieutenant colonel, 15th Pennsylvania Cavalry.
With a force of but 75 men, while on a scouting expedition, by a judicious disposition of his men, surprised and captured an entire battalion of the enemy's cavalry.

BEYER, HILLARY.
At Antietam, Md., Sept. 17, 1862.
R—NR.
B—NR.

Second lieutenant, Company H, 90th Pennsylvania Infantry.
After his command had been forced to fall back, remained alone on the line of battle, caring for his wounded comrades and carrying one of them to a place of safety.

BICKFORD, HENRY H.
At Waynesboro, Va., Mar. 2, 1865.
R—NR.
B—Michigan.

Corporal, Company E, 8th New York Cavalry.
Recapture of flag.

BICKFORD, MATTHEW
    At Vicksburg, Miss., May 22, 1863.
    R—NR.
    B—NR.
	Corporal, Company G, 8th Missouri Infantry.
	Gallantry in the charge of the "volunteer storming party."

BICKHAM, CHARLES G.
    At Bayong, near Lake Lanao, Mindando, P. I., May 2, 1902.
    R—Dayton, Ohio.
    B—Dayton, Ohio.
	First Lieutenant, 27th U. S. Infantry.
	Crossed a fire-swept field, in close range of the enemy, and brought a wounded soldier to a place of shelter.

BIEGER, CHARLES
    At Ivy Farm, Miss., Feb. 22, 1864.
    R—NR.
    B—Germany.
	Private, Company D, 4th Missouri Cavalry.
	Voluntarily risked his life by taking a horse, under heavy fire, beyond the line of battle for the rescue of his captain, whose horse had been killed in a charge and who was surrounded by the enemy's skirmishers.

BIEGLER, GEORGE W.
    Near Loac, Luzon, P. I., Oct. 21, 1900.
    R—NR.
    B—NR.
	Captain, 28th Infantry, U. S. Volunteers.
	With but 19 men resisted and at close quarters defeated 300 of the enemy.

BINGHAM, HENRY H.
    At Wilderness, Va., May 6, 1864.
    R—NR.
    B—Philadelphia, Pa.
	Captain, Company G, 140th Pennsylvania Infantry.
	Rallied and led into action a portion of the troops who had given way under the fierce assaults of the enemy.

BIRDSALL, HORATIO L.
    At Columbus, Ga., Apr. 16, 1865.
    R—NR.
    B—Monroe County, N. Y.
	Sergeant, Company B, 3d Iowa Cavalry.
	Capture of flag and bearer.

BIRKHIMER, WILLIAM E.
    At San Miguel de Mayumo, Luzon, P. I., May 13, 1899.
    R—Iowa.
    B—Somerset, Ohio.
	Captain, 3d U. S. Artillery.
	With 12 men charged and routed 300 of the enemy.

BISHOP, DANIEL
    At Turret Mountain, Ariz., Mar. 25, 1873.
    R—NR.
    B—Monroe County, Ohio.
	Sergeant, Company A, 5th U. S. Cavalry.
	Gallantry in engagements.

BISHOP, FRANCIS A.
    At Spotsylvania, Va., May 12, 1864.
    R—NR.
    B—Bradford County, Pa.
	Private, Company C, 57th Pennsylvania Infantry.
	Capture of flag.

BLACK, JOHN C.
    At Prairie Grove, Ark., Dec. 7, 1862.
    R—NR.
    B—Holmes, Miss.
	Lieutenant colonel, 37th Illinois Infantry.
	Gallantly charged the position of the enemy at the head of his regiment, after two other regiments had been repulsed and driven down the hill, and captured a battery; was severely wounded.

BLACK, WILLIAM P.
    At Pea Ridge, Ark., Mar. 7, 1862.
    R—NR.
    B—Woodford, Ky.
	Captain, Company K, 37th Illinois Infantry.
	Single-handed confronted the enemy, firing a rifle at them and thus checking their advance within 100 yards of the lines.

BLACKMAR, WILMON W.
    At Five Forks, Va., Apr. 1, 1865.
    R—NR.
    B—Bristol, Pa.
	Lieutenant, Company H, 1st West Virginia Cavalry.
	At a critical stage of the battle, without orders, led a successful advance upon the enemy.

*BLACKWELL, ROBERT L. (1316563)
    Near St. Souplet, France, Oct. 11, 1918.
    R—Hurdle Mills, N. C.
    B—Person, N. Dak.
    G. O. No. 13, W. D., 1919.
	Private, Company K, 119th Infantry, 30th Division.
	When his platoon was almost surrounded by the enemy and his platoon commander asked for volunteers to carry a message calling for reinforcements, Private Blackwell volunteered for this mission, well knowing the extreme danger connected with it. In attempting to get through the heavy shell and machine-gun fire this gallant soldier was killed.
	Posthumously awarded. Medal presented to father, James B. Blackwell.

BLACKWOOD, WILLIAM R. D.
    At Petersburg, Va., Apr. 2, 1865.
    R—NR.
    B—NR.
	Surgeon, 48th Pennsylvania Infantry.
	Removed severely wounded officers and soldiers from the field while under a heavy fire from the enemy, exposing himself beyond the call of duty, thus furnishing an example of most distinguished gallantry.

BLAIR, JAMES
    Winter of 1872–73.
    R—NR.
    B—Schuyler County, Pa.
	First sergeant, Company I, 1st U. S. Cavalry.
	Gallant conduct during the campaigns and engagements with Apaches.

BLANQUET
    Winter of 1872–73.
    R—NR.
    B—NR.
	Indian scout.
	Gallant conduct during the campaigns and engagements with Apaches.

BLASDEL, THOMAS A.................... Private, Company H, 83d Indiana Infantry.
At Vicksburg, Miss., May 22, 1863. Gallantry in the charge of the "volunteer storming party."
R—NR.
B—Dearborn County, Ind.

*BLECKLEY, ERWIN R................... Second lieutenant, 130th Field Artillery, observer 50th Aero Squadron, Air
Service.
Near Binarville, France, Oct. 6, 1918. Lieutenant Bleckley, with his pilot, First Lieut. Harold E. Goettler, Air
R—Wichita, Kans. Service, left the airdrome late in the afternoon on their second trip to drop
B—Wichita, Kans. supplies to a battalion of the 77th Division, which had been cut off by the
G. O. No. 56, W. D., 1922. enemy in the Argonne Forest. Having been subjected on the first trip to
violent fire from the enemy, they attempted on the second trip to come still
lower in order to get the packages even more precisely on the designated spot.
In the course of his mission the plane was brought down by enemy rifle and
machine-gun fire from the ground, resulting in fatal wounds to Lieutenant
Bleckley, who died before he could be taken to a hospital. In attempting
and performing this mission, Lieutenant Bleckley showed the highest possible
contempt of personal danger, devotion to duty, courage, and valor.
Posthumously awarded. Medal presented to father, E. E. Bleckley.

BLICKENSDERFER, MILTON............. Corporal, Company E, 126th Ohio Infantry.
At Petersburg, Va., Apr. 3, 1865. Capture of flag.
R—NR.
B—Lancaster, Pa.

BLISS, GEORGE N...................... Captain, Company C, 1st Rhode Island Cavalry.
At Waynesboro, Va., Sept. 28, 1864. While in command of the provost guard in the village, he saw the Union lines
R—NR. returning before the attack of a greatly superior force of the enemy, mustered
B—NR. his guard, and, without orders, joined in the defense and charged the enemy
without support. He received three saber wounds, his horse was shot, and
he was taken prisoner.

BLISS, ZENAS R....................... Colonel, 7th Rhode Island Infantry.
At Fredericksburg, Va., Dec. 13, This officer, to encourage his regiment which had never before been in action,
1862. and which had been ordered to lie down to protect itself from the enemy's
R—NR. fire, arose to his feet, advanced in front of the line, and himself fired several
B—NR. shots at the enemy at short range, being fully exposed to their fire at the time.

BLODGETT, WELLS H.................... First lieutenant, Company D, 37th Illinois Infantry.
At Newtonia, Mo., Sept. 30, 1862. With a single orderly, captured an armed picket of eight men and marched them
R—NR. in prisoners.
B—Downers Grove, Ill.

BLUCHER, CHARLES.................... Corporal, Company H, 188th Pennsylvania Infantry.
At Fort Harrison, near Richmond, Planted first national colors on the fortifications.
Va., Sept. 29, 1864.
R—NR.
B—Germany.

BLUNT, JOHN W....................... First lieutenant, Company K, 6th New York Cavalry.
At Cedar Creek, Va., Oct. 19, 1864. Voluntarily led a charge across a narrow bridge over the creek, against the lines
R—Chatham Four Corners, N. Y. of the enemy.
B—Columbia County, N. Y.

BOEHLER, OTTO....................... Private, Company I, 1st North Dakota Volunteer Infantry.
Near San Isidro, P. I., May 16, 1899. With 21 other scouts charged across a burning bridge, under heavy fire, and
R—Wahpeton, N. Dak. completely routed 600 of the enemy who were entrenched in a strongly fortified
B—Germany. position.

BOEHM, PETER M...................... Second lieutenant, Company K, 15th New York Cavalry.
At Dinwiddie Courthouse, Va., While acting as aide to General Custer took a flag from the hands of a color
Mar. 31, 1865. bearer, rode in front of a line that was being driven back and, under a heavy
R—Brooklyn, N. Y. fire, rallied the men, re-formed the line, and repulsed the charge.
B—NR.

BONEBRAKE, HENRY G................. Lieutenant, Company G, 17th Pennsylvania Cavalry.
At Five Forks, Va., Apr. 1, 1865. Capture of flag.
R—NR.
B—Waynesboro, Pa.

BONNAFFON, SYLVESTER, Jr........... First lieutenant, Company G, 99th Pennsylvania Infantry.
At Boydton Plank Road, Va., Oct. Checked the rout and rallied the troops of his command in the face of a terrible
27, 1864. fire of musketry; was severely wounded.
R—NR.
B—NR.

BOODY, ROBERT M.................... Sergeant, Company B, 40th New York Infantry.
At Williamsburg, Va., May 5, 1862. This soldier, then a corporal, at great personal risk, voluntarily saved the lives
of and brought from the battlefield two wounded comrades.
At Chancellorsville, Va. May 2, 1863. Voluntarily, and at great personal risk, brought from the field of battle and
R—NR. saved the life of Capt. George B. Carse, Company C, 40th New York Vol-
B—Lemington, Me. unteer Infantry.

BOON, HUGH P........................ Captain, Company B, 1st West Virginia Cavalry.
At Sailors Creek, Va., Apr. 6, 1865. Capture of flag.
R—NR.
B—NR.

BOQUET, NICHOLAS_____
At Wilsons Creek, Mo., Aug. 10, 1861.
R—NR.
B—NR.

Private, Company D, 1st Iowa Infantry.
Voluntarily left the line of battle, and, exposing himself to imminent danger from a heavy fire of the enemy, assisted in capturing a riderless horse at large between the lines and, hitching him to a disabled gun, saved the gun from capture.

BOSS, ORLANDO_____
At Cold Harbor,Va., June 3, 1864.
R—NR.
B—Fitchburg, Mass.

Corporal, Company F, 25th Massachusetts Infantry.
Rescued his lieutenant, who was lying between the lines mortally wounded; this under a heavy fire of the enemy.

BOURKE, JOHN G_____
At Stone River, Tenn., Dec. 31, 1862, and Jan. 1, 1863.
R—NR.
B—Philadelphia, Pa.

Private, Company E, 15th Pennsylvania Cavalry.
Gallantry in action.

BOURY, RICHARD_____
At Charlottesville,Va., Mar. 5, 1865.
R—NR.
B—Monroe County, Ohio.

Sergeant, Company C, 1st West Virginia Cavalry.
Capture of flag.

BOUTWELL, JOHN W_____
At Petersburg, Va., Apr. 2, 1865.
R—NR.
B—Hanover, N. H.

Private, Company B, 18th New Hampshire Infantry.
Brought off from the picket line, under heavy fire, a comrade who had been shot through both legs.

BOWDEN, SAMUEL_____
At Wichita River, Tex., Oct. 5, 1870.
R—NR.
B—Salem, Mass.

Corporal, Company M, 6th U. S. Cavalry.
Gallantry in pursuit of and fight with Indians.

BOWEN, CHESTER B_____
At Winchester, Va., Sept. 19, 1864.
R—NR.
B—Nunda, N. Y.

Corporal, Company I, 19th New York Cavalry (1st New York Dragoons).
Capture of flag.

BOWEN, EMMER_____
At Vicksburg, Miss., May 22, 1863.
R—NR.
B—Erie County, N. Y.

Private, Company C, 127th Illinois Infantry.
Gallantry in the charge of the "volunteer storming party."

BOWMAN, ALONZO_____
At Cibicu Creek, Ariz., Aug. 30, 1881.
R—NR.
B—Washington, Me.

Sergeant, Company D, 6th U. S. Cavalry.
Conspicuous and extraordinary bravery in attacking mutinous scouts.

BOX, THOMAS J_____
At Resaca, Ga., May 14, 1864.
R—Bedford, Ind.
B—NR.

Captain, Company D, 27th Indiana Infantry.
Capture of flag of the 38th Alabama Infantry (C. S. A.).

BOYNE, THOMAS_____
At Mimbres Mountains, N. Mex., May 29, 1879.
At Cuchillo Negro, N. Mex., Sept. 27, 1879.
R—NR.
B—Prince Georges County, Md.

Sergeant, Company C, 9th U. S. Cavalry.
Bravery in action.

BOYNTON, HENRY V_____
At Missionary Ridge, Tenn., Nov. 25, 1863.
R—NR.
B—NR.

Lieutenant colonel, 35th Ohio Infantry.
Led his regiment in the face of a severe fire of the enemy; was severely wounded.

BRADBURY, SANFORD_____
At Hell Canyon, Ariz., July 3, 1869.
R—NR.
B—Sussex County, N. J.

First sergeant, Company L, 8th U. S. Cavalry.
Conspicuous gallantry in action.

BRADLEY, THOMAS W_____
At Chancellorsville, Va., May 3, 1863.
R—NR
B—England.

Sergeant, Company H, 124th New York Infantry.
Volunteered in response to a call and alone, in the face of a heavy fire of musketry and canister, went and procured ammunition for the use of his comrades.

BRADY, JAMES_____
At Chapins Farm, near Richmond, Va., Sept. 29, 1864.
R—NR.
B—Boston, Mass.

Private, Company F, 10th New Hampshire Infantry.
Capture of flag.

BRANAGAN, EDWARD_____
At Red River, Tex., Sept. 29, 1872.
R—NR.
B—Ireland.

Private, Company F, 4th U. S. Cavalry.
Gallantry in action.

BRANDLE, JOSEPH E _____
At Lenoire, Tenn., Nov. 16, 1863.
R—NR.
B—Seneca County, Ohio.

Private, Company C, 17th Michigan Infantry.
While color bearer of his regiment, having been twice wounded and the sight of one eye destroyed, still held to the colors until ordered to the rear by his regimental commander.

BRANNIGAN, FELIX _____
At Chancellorsville, Va., May 2, 1863.
R—NR.
B—Ireland.

Private, Company A, 74th New York Infantry.
Volunteered on a dangerous service and brought in valuable information.

BRANT, ABRAM B _____
At Little Big Horn, Mont., June 25, 1876.
R—NR.
B—New York, N. Y.

Private, Company D, 7th U. S. Cavalry.
Brought water for the wounded under a most galling fire.

BRANT, WILLIAM _____
At Petersburg, Va., Apr. 3, 1865.
R—NR.
B—Elizabeth, N. J.

Lieutenant, Company B, 1st New Jersey Veteran Battalion.
Capture of battle flag of 46th North Carolina (C. S. A.).

BRAS, EDGAR A _____
Spanish Fort, Ala., Apr. 8, 1865.
R—NR.
B—Jefferson County, Iowa.

Sergeant, Company K, 8th Iowa Infantry.
Capture of flag.

BRATLING, FRANK _____
Near Fort Selden, N. Mex., July 8 to 11, 1873.
R—NR.
B—Germany.

Corporal, Company C, 8th U. S. Cavalry.
Services against hostile Indians.

BREST, LEWIS F _____
At Sailors Creek, Va., Apr. 6, 1865.
R—NR.
B—Mercer, Pa.

Private, Company D, 57th Pennsylvania Infantry.
Capture of flag.

BRETT, LLOYD M _____
At O'Fallons Creek, Mont., Apr. 1, 1880.
R—Malden, Mass.
B—Maine.
Distinguished-service medal also awarded.

Second lieutenant, 2d U. S. Cavalry.
Fearless exposure and dashing bravery in cutting off the Indians' pony herd, thereby greatly crippling the hostiles.

BREWER, WILLIAM J _____
At Appomattox campaign, Va., Apr. 4, 1865.
R—NR.
B—Putnam County, N. Y.

Private, Company C, 2d New York Cavalry.
Capture of engineer flag, Army of Northern Virginia.

BREWSTER, ANDRE W _____
At Tientsin, China, July 13, 1900.
R—Philadelphia, Pa.
B—Hoboken, N. J.
Distinguished-service medal also awarded.

Captain, 9th U. S. Infantry.
While under fire rescued two of his men from drowning.

BREYER, CHARLES _____
At Rappahannock Station, Va., Aug. 23, 1862.
R—NR.
B—NR.

Sergeant, Company I, 90th Pennsylvania Infantry.
Voluntarily, and at great personal risk, picked up an unexploded shell and threw it away, thus doubtless saving the life of a comrade whose arm had been taken off by the same shell.

BRIGGS, ELIJAH A _____
At Petersburg, Va., Apr. 3, 1865.
R—Salisbury, Conn.
B—Salisbury, Conn.

Corporal, Company B, 2d Connecticut Heavy Artillery.
Capture of battle flag.

BRINGLE, ANDREW _____
At Sailors Creek, Va., Apr. 6, 1865.
R—NR.
B—Buffalo, N. Y.

Corporal, Company F, 10th New York Cavalry.
Charged the enemy and assisted Sergeant Norton in capturing a fieldpiece and two prisoners.

BROGAN, JAMES _____
At Simon Valley, Ariz., Dec. 14, 1877.
R—NR.
B—Ireland.

Sergeant, Company G, 6th U. S. Cavalry.
Engaged single handed two renegade Indians until his horse was shot under him and then pursued them so long as he was able.

BRONNER, AUGUST F _____
At White Oak Swamp, Va., June 30, 1862.
At Malvern Hill, Va., July 1, 1862.
R—NR.
B—Germany.

Private, Company C, 1st New York Artillery.
Continued to fight after being severely wounded.

BRONSON, JAMES H_____ First sergeant, Company D, 5th U. S. Colored Troops.
   At Chapins Farm, near Richmond,   Took command of his company, all the officers having been killed or wounded,
     Va., Sept. 29, 1864.   and gallantly led it.
     R—NR.
     B—Indiana County, Pa.

BROOKIN, OSCAR_____ Private, Company C, 17th U. S. Infantry.
   At El Caney, Cuba, July 1, 1898.   Gallantly assisted in the rescue of the wounded from in front of the lines and
     R—Green County, Ohio.   under heavy fire from the enemy.
     B—Byron, Wis.

BROPHY, JAMES_____ Private, Company B, 8th U. S. Cavalry.
   Arizona, 1868.   Bravery in scouts and actions with Indians.
     R—NR.
     B—Ireland.

BROSNAN, JOHN_____ Sergeant, Company E, 164th New York Infantry.
   At Petersburg, Va., June 17, 1864.   Rescued a wounded comrade who lay exposed to the enemy's fire, receiving a
     B—NR.   severe wound in the effort.
     B—Ireland.

BROUSE, CHARLES W_____ Captain, Company K, 100th Indiana Infantry.
   At Missionary Ridge, Tenn., Nov.   To encourage his men whom he had ordered to lie down while under severe
     25, 1863.   fire, and who were partially protected by slight earthworks, himself refused
     R—NR.   to lie down, but walked along the top of the works until he fell severely
     B—NR.   wounded.

BROWN, BENJAMIN_____ Sergeant, Company C, 24th U. S. Infantry.
   Arizona, May 11, 1889.   Although shot in the abdomen, in a fight between a paymaster's escort and
     R—NR.   robbers, did not leave the field until again wounded through both arms.
     B—Spotsylvania County, Va.

BROWN, CHARLES_____ Sergeant, Company C, 50th Pennsylvania Infantry.
   At Weldon Railroad, Va., Aug. 19,   Capture of flag of 47th Virginia Infantry (C. S. A.).
     1864.
     R—NR.
     B—Schuylkill County, Pa.

BROWN, EDWARD, Jr_____ Corporal, Company G, 62d New York Infantry.
   At Fredericksburg and Salem   Severely wounded while carrying the colors, he continued at his post, under
     Heights, Va., May 3-4, 1863.   fire, until ordered to the rear.
     R—New York, N. Y.
     B—New York, N. Y.

BROWN, HENRI LE FEVRE_____ Sergeant, Company B, 72d New York Infantry.
   At Wilderness, Va., May 6, 1864.   Voluntarily and under a heavy fire from the enemy, three times crossed the
     R—Ellicott, N. Y.   field of battle with a load of ammunition in a blanket on his back, thus sup-
     B—Jamestown, N. Y.   plying the Federal forces, whose ammunition had nearly all been expended,
     and enabling them to hold their position until reinforcements arrived, when
     the enemy were driven from their position.

BROWN, JAMES_____ Sergeant, Company F, 5th U. S. Cavalry.
   At Davidson Canyon, near Camp   In command of a detachment of four men defeated a superior force.
     Crittenden, Ariz., Aug. 27, 1872.
     R—NR.
     B—Wexford, Ireland.

BROWN, JEREMIAH Z_____ Captain, Company K, 148th Pennsylvania Infantry.
   At Petersburg, Va., Oct. 27, 1864.   With 100 selected volunteers, assaulted and captured the works of the enemy,
     R—Rimmersburg, Pa.   together with a number of officers and men.
     B—Clarion County, Pa.

BROWN, JOHN H_____ First sergeant, Company A, 47th Ohio Infantry.
   At Vicksburg, Miss., May 19, 1863.   Voluntarily carried a verbal message from Col. A. C. Parry to Gen. Hugh
     R—NR.   Ewing through a terrific fire and in plain view of the enemy.
     B—Boston, Mass.

BROWN, JOHN H_____ Captain, Company D, 12th Kentucky Infantry.
   At Franklin, Tenn., Nov. 30, 1864.   Capture of flag.
     R—NR.
     B—NR.

BROWN, LORENZO D_____ Private, Company A, 7th U. S. Infantry.
   At Big Hole, Mont., Aug. 9, 1877.   After having been severely wounded in right shoulder continued to do duty in
     R—Lincoln, Ind.   a most courageous manner.
     B—Davidson County, N. C.

BROWN, MORRIS, Jr_____ Captain, Company A, 126th New York Infantry.
   At Gettysburg, Pa., July 3, 1863.   Capture of flag.
     R—NR.
     B—NR.

BROWN, ROBERT B_____ Private, Company A, 15th Ohio Infantry.
   At Missionary Ridge, Tenn., Nov.   Capture of flag.
     25, 1863.
     R—NR.
     B—Muskingum County, Ohio.

BROWN, URIAH_____
   At Vicksburg, Miss., May 22, 1863.
   R—NR.
   B—Miami County, Ohio.

Private, Company G, 30th Ohio Infantry.
Gallantry in the charge of the "volunteer storming party."

BROWN, WILSON_____
   Georgia, April, 1862.
   R—NR.
   B—Logan County, Ohio.

Private, Company F, 21st Ohio Infantry.
One of 22 men (including two civilians) who, by direction of General Mitchel (or Buell), penetrated nearly 200 miles south into the enemy's territory and captured a railroad train at Big Shanty, Ga., in an attempt to destroy the bridges and track between Chattanooga and Atlanta.

BROWNELL, FRANCIS E_____
   ————, May 24, 1861.
   R—NR.
   B—NR.

Private, Company A, 11th New York Infantry.
Killed the murderer of Colonel Ellsworth at the Marshall House, Alexandria, Va.

BRUNER, LOUIS J_____
   At Walkers Ford, Tenn., Dec. 2, 1863.
   R—NR.
   B—Monroe County, Ind.

Private, Company H, 5th Indiana Cavalry.
Voluntarily passed through the enemy's lines under fire and conveyed to a battalion, then in a perilous position and liable to capture, information which enabled it to reach a point of safety.

BRUSH, GEORGE W_____
   At Ashepoo River, S. C., May 24, 1864.
   R—NR.
   B—NR.

Lieutenant, Company B, 34th U. S. Colored Troops.
Voluntarily commanded a boat crew, which went to the rescue of a large number of Union soldiers on board the stranded steamer Boston, and with great gallantry succeeded in conveying them to shore, being exposed during the entire time to a heavy fire from a Confederate battery.

BRUTON, CHRISTOPHER C_____
   At Waynesboro, Va., Mar. 2, 1865.
   R—NR.
   B—NR.

Captain, Company C, 22d New York Cavalry.
Capture of General Early's headquarters flag, Confederate national standard.

BRYAN, WILLIAM C_____
   On Powder River, Wyo., Mar. 17, 1876.
   R—Zanesville, Ohio.
   B—Zanesville, Ohio.

Hospital steward, U. S. Army.
Accompanied a detachment of cavalry in a charge on a village of hostile Indians and fought through the engagements, having his horse killed under him. He continued to fight on foot, and under severe fire and without assistance conveyed two wounded comrades to places of safety, saving them from capture.

BRYANT, ANDREW S_____
   At New Bern, N. C., May 23, 1863.
   R—NR.
   B—NR.

Sergeant, Company A, 46th Massachusetts Infantry.
By his courage and judicious disposition of his guard of 16 men, stationed in a small earthwork at the head of the bridge, held in check and repulsed for a half hour a fierce attack of a strong force of the enemy, thus probably saving the city (New Bern) from capture.

BUCHANAN, GEORGE A_____
   At Chapins Farm, near Richmond, Va., Sept. 29, 1864.
   R—Ontario County, N. Y
   B—New York.

Private, Company G, 148th New York Infantry.
Took position in advance of the skirmish line and drove the enemy's cannoneers from their guns; was mortally wounded.

BUCK, F. CLARENCE_____
   At Chapins Farm, near Richmond, Va., Sept. 29, 1864.
   R—Windsor, Conn.
   B—Hartford, Conn.

Corporal, Company A, 21st Connecticut Infantry.
Although wounded, refused to leave the field until the fight closed.

BUCKINGHAM, DAVID E_____
   At Rowanty Creek, Va., Feb. 5, 1865.
   R—NR.
   B—NR.

First lieutenant, Company E, 4th Delaware Infantry.
Swam the partly frozen creek, under fire, in the attempt to capture a crossing.

BUCKLES, ABRAM J_____
   At Wilderness, Va., May 5, 1864.
   R—Muncie, Ind.
   B—Delaware County, Ind.

Sergeant, Company E, 19th Indiana Infantry.
Though suffering from an open wound, carried the regimental colors until again wounded.

BUCKLEY, DENIS_____
   At Peach Tree Creek, Ga., July 20, 1864.
   R—NR.
   B—Canada.

Private, Company G, 136th New York Infantry.
Capture of flag of 31st Mississippi (C. S. A.).

BUCKLEY, JOHN C_____
   At Vicksburg, Miss., May 22, 1863.
   R—NR.
   B—Fayette County, Va.

Sergeant, Company G, 4th Virginia Infantry.
Gallantry in the charge of the "volunteer storming party."

BUCKLYN, JOHN K_____
   At Chancellorsville, Va., May 3, 1863.
   R—NR.
   B—NR.

First lieutenant, Battery E, 1st Rhode Island Light Artillery.
Though himself wounded, gallantly fought his section of the battery under a fierce fire from the enemy until his ammunition was all expended, many of the cannoneers and most of the horses killed or wounded, and the enemy within 25 yards of the guns, when, disabling one piece, he brought off the other in safety.

BUFFINGTON, JOHN E_____
   At Petersburg, Va., Apr. 2, 1865.
   R—NR.
   B—Carroll County, Md.

Sergeant, Company C, 6th Maryland Infantry.
Was the first enlisted man of the 3d Division to mount the parapet of the enemy's line.

BUFFUM, ROBERT.
   Georgia, April, 1862.
   R—NR.
   B—Salem, Mass.

Private, Company H, 21st Ohio Infantry.
One of 22 men (including two civilians) who, by direction of General Mitchel (or Buell), penetrated nearly 200 miles south into the enemy's territory and captured a railroad train at Big Shanty, Ga., in an attempt to destroy the bridges and track between Chattanooga and Atlanta.

BUHRMAN, HENRY G.
   At Vicksburg, Miss., May 22, 1863.
   R—NR.
   B—Cincinnati, Ohio.

Private, Company H, 54th Ohio Infantry.
Gallantry in the charge of the "volunteer storming party."

BUMGARNER, WILLIAM.
   At Petersburg, Va., Apr. 2, 1862.
   R—NR.
   B—NP.

Sergeant, Company A, 4th Virginia Infantry.
Gallantry in the charge of the "volunteer storming party."

BURBANK, JAMES H.
   At Blackwater, near Franklin, Va., Oct. 3, 1862.
   R—NR.
   B—Holland.

Sergeant, Company K, 4th Rhode Island Infantry.
Gallantry in action while on detached service on board a gunboat.

BURGER, JOSEPH.
   At Nolensville, Tenn., Feb. 15, 1863.
   R—NR.
   B—Austria.

Private, Company H, 2d Minnesota Infantry.
Was one of a detachment of 16 men who heroically defended a wagon train against the attack of 125 cavalry, repulsed the attack, and saved the train.

BURK, MICHAEL.
   At Spotsylvania, Va., May 12, 1864.
   R—NR.
   B—Ireland.

Private, Company D, 125th New York Infantry.
Capture of flag, seizing it as his regiment advanced over the enemy's works.

BURK, THOMAS.
   At Wilderness, Va., May 6, 1864.
   R—NR.
   B—Lewis County, N. Y.

Sergeant, Company H, 97th New York Infantry.
At the risk of his own life, went back while the rebels were still firing and, finding Colonel Wheelock unable to move, alone and unaided carried him off the field of battle.

BURKARD, OSCAR.
   At Leech Lake, Minn., Oct. 5, 1898.
   R—Hay Creek, Minn.
   B—Germany.

Private, Hospital Corps, U. S. Army.
Bravery in action against hostile Indians.

BURKE, DANIEL W.
   At Shepherdstown Ford, Va., Sept. 20, 1862.
   R—NR.
   B—New Haven, Conn.

First sergeant, Company B, 2d U. S. Infantry.
Voluntarily attempted to spike a gun in the face of the enemy.

BURKE, PATRICK J.
   Arizona, 1868.
   R—NR.
   B—Ireland.

Farrier, Company B, 8th U. S. Cavalry.
Bravery in scouts and actions with Indians.

BURKE, RICHARD.
   At Cedar Creek, etc., Mont., October, 1876, to January, 1877.
   R—NR.
   B—Ireland.

Private, Company G, 5th U. S. Infantry.
Gallantry in engagements.

BURKE, THOMAS.
   At Hanover Courthouse, Va., June 30, 1863.
   R—NR.
   B—Ireland.

Private, Company A, 5th New York Cavalry.
Capture of battle flag.

BURNETT, GEORGE R.
   At Cuchillo Negro Mountains, N. Mex., Aug. 16, 1881.
   R—Spring Mills, Pa.
   B—Lower Providence Township, Pa.

Second lieutenant, 9th U. S. Cavalry.
Saved the life of a dismounted soldier, who was in imminent danger of being cut off, by alone galloping quickly to his assistance under a heavy fire and escorting him to a place of safety, his horse being twice shot in this action.

BURNS, JAMES M.
   At New Market, Va., May 15, 1864.
   R—Jefferson County, Ohio.
   B—Jefferson County, Ohio.

Sergeant, Company B, 1st West Virginia Infantry.
Under a heavy fire of musketry, rallied a few men to the support of the colors, in danger of capture, and bore them to a place of safety. One of his comrades having been severely wounded in the effort, Sergeant Burns went back a hundred yards in the face of the enemy's fire and carried the wounded man from the field.

BURRITT, WILLIAM W.
   At Vicksburg, Miss., Apr. 27, 1863.
   R—NR.
   B—Campbell, N. Y.

Private, Company G, 113th Illinois Infantry.
Voluntarily acted as a fireman on a steam tug which ran the blockade and passed the batteries under a heavy fire.

BUTLER, EDMOND.
   At Wolf Mountain, Mont., Jan. 8, 1877.
   R—Brooklyn, N. Y.
   B—Ireland.

Captain, 5th U. S. Infantry.
Most distinguished gallantry in action with hostile Indians.

BUTTERFIELD, DANIEL............
    At Gaines Mill, Va., June 27, 1862.
    R—NR.
    B—NR.

Brigadier general, U. S. Volunteers.
Seized the colors of the 83d Pennsylvania Volunteers at a critical moment and, under a galling fire of the enemy, encouraged the depleted ranks to renewed exertion.

BUTTERFIELD, FRANK G............
    At Salem Heights, Va., May 4, 1863.
    R—NR.
    B—Rockingham, Vt.

First lieutenant, Company C, 6th Vermont Infantry.
Took command of the skirmish line and covered the movement of his regiment out of a precarious position.

BUZZARD, ULYSSES G............
    At El Caney, Cuba, July 1, 1898.
    R—NR.
    B—Armstrong, Pa.

Private, Company C, 17th U. S. Infantry.
Gallantly assisted in the rescue of the wounded from in front of the lines and under heavy fire from the enemy.

BYRNE, BERNARD A............
    At Bobong, Negros, P. I., July 19, 1899.
    R—Washington, D. C.
    B—Newport Barracks, Va.

Captain, 6th U. S. Infantry.
Most distinguished gallantry in rallying his men on the bridge after the line had been broken and pushed back.

BYRNE, DENIS............
    At Cedar Creek, etc., Mont., October, 1876, to January, 1877.
    R—NR.
    B—Ireland.

Sergeant, Company G, 5th U. S. Infantry.
Gallantry in engagements.

CABLE, JOSEPH A............
    At Cedar Creek, etc., Mont., October, 1876, to January, 1877.
    R—NR.
    B—Cape Girardeau, Mo.

Private, Company I, 5th U. S. Infantry.
Gallantry in actions.

CADWALLADER, ABEL G............
    At Hatchers Run and Dabneys Mills, Va., Feb. 6, 1865.
    R—NR.
    B—Baltimore, Md.

Corporal, Company H, 1st Maryland Infantry.
Gallantly planted the colors on the enemy's works in advance of the arrival of his regiment.

CADWELL, LUMAN L............
    At Alabama Bayou, La., Sept. 20, 1864.
    R—NR.
    B—Broome, N. Y.

Sergeant, Company B, 2d New York Veteran Cavalry.
Swam the bayou under fire of the enemy and captured and brought off a boat by means of which the command crossed and routed the enemy.

CALDWELL, DANIEL............
    At Hatchers Run, Va., Feb. 6, 1865.
    R—NR.
    B—Montgomery County, Pa.

Sergeant, Company H, 13th Pennsylvania Cavalry.
In a mounted charge, dashed into center of the enemy's line and captured their colors.

CALKIN, IVERS S............
    At Sailors Creek, Va., Apr. 6, 1865.
    R—Willsborough, N. Y.
    B—Essex County, N. Y.

First sergeant, Company M, 2d New York Cavalry.
Capture of flag of 18th Virginia Infantry (C. S. A.).

CALL, DONALD M............
    Near Varennes, France, Sept. 26, 1918.
    R—Larchmont Manor, N. Y.
    B—New York, N. Y.
    G. O. No. 13, W. D., 1919.

Corporal, 344th Battalion, Tank Corps, United States Army.
During an operation against enemy machine-gun nests west of Varennes, Corporal Call was in a tank with an officer, when half of the turret was knocked off by a direct artillery hit. Choked by gas from the high-explosive shell, he left the tank and took cover in a shell hole 30 yards away. Seeing that the officer did not follow, and thinking that he might be alive, Corporal Call returned to the tank under intense machine-gun and shell fire and carried the officer over a mile under machine-gun and sniper fire to safety.

CALLAHAN, JOHN H............
    At Fort Blakely, Ala., Apr. 9, 1865.
    R—NR.
    B—Shelby County, Ky.

Private, Company B, 122d Illinois Infantry.
Capture of flag.

CALLAN, THOMAS J............
    At Little Big Horn, Mont., June 25 and 26, 1876.
    R—NR.
    B—Ireland.

Private, Company B, 7th U. S. Cavalry.
Volunteered and succeeded in obtaining water for the wounded of the command; also displayed conspicuously good conduct in assisting to drive away the Indians.

CALVERT, JAMES S............
    At Cedar Creek, etc., Mont., October, 1876, to January, 1877.
    R—NR.
    B—Athens County, Ohio.

Private, Company C, 5th U. S. Infantry.
Gallantry in actions.

CAMP, CARLTON N............
    At Petersburg, Va., Apr. 2, 1865.
    R—Hanover, N. H.
    B—Hanover, N. H.

Private, Company B, 18th New Hampshire Infantry.
Brought off from the picket line, under heavy fire, a comrade who had been shot through both legs.

CAMPBELL, JAMES A _____
   At Woodstock, Va., Jan. 22, 1865.

   At Amelia Courthouse, Va., Apr. 5, 1865.
   R—NR.
   B—New York, N. Y.

Private, Company A, 2d New York Cavalry.
While his command was retreating before superior numbers, he voluntarily rushed back with one companion and rescued his commanding officer, who had been unhorsed and left behind.
Captured two battle flags.

CAMPBELL, WILLIAM _____
   At Vicksburg, Miss., May 22, 1863.
   R—NR.
   B—Ireland.

Private, Company I, 30th Ohio Infantry.
Gallantry in the charge of the "volunteer storming party."

CANFIELD, HETH _____
   At Little Blue, Nebr., May 15, 1870.
   R—NR.
   B—New Meddford, Conn.

Private, Company C, 2d U. S. Cavalry.
Gallantry in action.

CANTRELL, CHARLES P _____
   At Santiago, Cuba, July 1, 1898.
   R—Nashville, Tenn.
   B—Smithville, Tenn.

Private, Company F, 10th U. S. Infantry.
Gallantly assisted in the rescue of the wounded from in front of the lines and under heavy fire from the enemy.

CAPEHART, CHARLES E _____
   At Monterey Mountain, Pa., July 4, 1863.
   R—NR.
   B—NR.

Major, 1st West Virginia Cavalry.
While commanding the regiment, charged down the mountain side at midnight, in a heavy rain, upon the enemy's fleeing wagon train. Many wagons were captured and destroyed and many prisoners taken.

CAPEHART, HENRY _____
   At Greenbrier River, W. Va., May 22, 1864.
   R—NR.
   B—NR.

Colonel, 1st West Virginia Cavalry.
Saved, under fire, the life of a drowning soldier.

CAPRON, HORACE, Jr _____
   At Chickahominy and Ashland, Va., June, 1862.
   R—NR.
   B—Laurel, Md.

Sergeant, Company G, 8th Illinois Cavalry.
Gallantry in action.

CAREY, HUGH _____
   At Gettysburg, Pa., July 2, 1863.
   R—NR.
   B—Ireland.

Sergeant, Company E, 82d New York Infantry.
Captured the flag of the 7th Virginia Infantry (C. S. A.), being twice wounded in the effort.

CAREY, JAMES L _____
   At Appomattox Courthouse, Va., Apr. 9, 1865.
   R—NR.
   B—Onondaga County, N. Y.

Sergeant, Company G, 10th New York Cavalry.
Daring bravery and urging the men forward in a charge.

CARLISLE, CASPER _____
   At Gettysburg, Pa., July 2, 1863.
   R—NR.
   B—Allegheny County, Pa.

Private, Company F, Independent Pennsylvania Light Artillery.
Saved a gun of his battery under heavy musketry fire, most of the horses being killed and the drivers wounded.

CARMAN, WARREN _____
   At Waynesboro, Va., Mar. 2, 1865.
   R—NR.
   B—Seneca County, N. Y.

Private, Company H, 1st New York (Lincoln) Cavalry.
Capture of flag and several prisoners.

CARMIN, ISAAC H _____
   At Vicksburg, Miss., May 22, 1863.
   R—NR.
   B—Monmouth County, N. J.

Corporal, Company A, 48th Ohio Infantry.
Saved his regimental flag; also seized and threw a shell, with burning fuse, from among his comrades.

CARNEY, WILLIAM H _____
   At Fort Wagner, S. C., July 18, 1863.
   R—NR.
   B—New Bedford, Mass.

Sergeant, Company C, 54th Massachusetts Colored Infantry.
When the color sergeant was shot down, this soldier grasped the flag, led the way to the parapet, and planted the colors thereon. When the troops fell back he brought off the flag, under a fierce fire in which he was twice severely wounded.

CARPENTER, LOUIS H _____
   At Indian campaigns, Kansas and Colorado, September and October, 1868.
   R—Philadelphia, Pa.
   B—Glassboro, N. J.

Captain, Company H, 10th U. S. Cavalry.
Was gallant and meritorious throughout the campaigns, especially in the combat of Oct. 15 and in the forced march on Sept. 23, 24, and 25 to the relief of Forsyth's Scouts, who were known to be in danger of annihilation by largely superior forces of Indians.

CARR, EUGENE A _____
   At Pea Ridge, Ark., Mar. 7, 1862.
   R—NR.
   B—NR.

Colonel, 3d Illinois Cavalry.
Directed the deployment of his command and held his ground, under a brisk fire of shot and shell in which he was several times wounded.

CARR, FRANKLIN _____
   At Nashville, Tenn., Dec. 16, 1864.
   R—NR.
   B—Stark County, Ohio.

Corporal, Company D, 124th Ohio Infantry.
Recapture of United States guidon from a rebel battery.

CARR, JOHN.................
At Chiricahua Mountain, Ariz.,
Oct. 29, 1869.
R—NR.
B—Columbus, Ohio.

Private, Company G, 8th U. S. Cavalry.
Gallantry in action.

CARROLL, THOMAS...............
Arizona, August to October, 1868.
R—NR.
B—Ireland.

Private, Company L, 8th U. S. Cavalry.
Bravery in scouts and actions with Indians.

CARSON, ANTHONY J..............
At Catubig, Samar, P. I., Apr. 15-19,
1900.
R—Malden, Mass.
B—Boston, Mass.

Corporal, Company H, 43d Infantry, U. S. Volunteers.
Assumed command of a detachment of the company which had survived an
overwhelming attack of the enemy, and by his bravery and untiring efforts
and the exercise of extraordinary good judgment in the handling of his men
successfully withstood for two days the attacks of a large force of the enemy,
thereby saving the lives of the survivors and protecting the wounded until
relief came.

CARSON, WILLIAM J..............
At Chickamauga, Ga., Sept. 19,
1863.
R—NR.
B—Washington County, Pa.

Musician, Company E, 1st Battalion, 15th U. S. Infantry.
Most distinguished gallantry in battle.

CART, JACOB...................
At Fredericksburg, Va., Dec. 13,
1862.
R—NR.
B—Carlisle, Pa.

Private, Company A, 7th Pennsylvania Reserve Corps.
Capture of flag of 19th Georgia Infantry (C. S. A.), wresting it from the hands
of the color bearer.

CARTER, GEORGE...............
Arizona, August to October, 1868.
R—NR.
B—Ireland.

Private, Company B, 8th U. S. Cavalry.
Bravery in scouts and actions with Indians.

CARTER, JOHN J...............
At Antietam, Md., Sept. 17, 1862.
R—NR.
B—NR.

Second lieutenant, Company B, 33d New York Infantry.
While in command of a detached company, seeing his regiment thrown into
confusion by a charge of the enemy, without orders made a countercharge
upon the attacking column and checked the assault. Penetrated within the
enemy's lines at night and obtained valuable information.

CARTER, JOSEPH F.............
At Fort Stedman, Va., Mar. 25, 1865.
R—Baltimore, Md.
B—NR.

Captain, Company D, 3d Maryland Infantry.
Captured the colors of the 51st Virginia Infantry (C. S. A.).

CARTER, MASON...............
At Bear Paw Mountain, Mont.,
Sept. 30, 1877.
R—Augusta, Ga.
B—Augusta, Ga.

First lieutenant, 5th U. S. Infantry.
Led a charge under a galling fire, in which he inflicted great loss upon the enemy.

CARTER, ROBERT G............
On Brazos River, Tex., Oct. 10, 1871.
R—Bradford, Mass.
B—Bridgeport, Me.

Second lieutenant, 4th U. S. Cavalry.
Held the left of the line with a few men during the charge of a large body of
Indians, after the right of the line had retreated, and by delivering a rapid
fire succeeded in checking the enemy until other troops came to the rescue.

CARTER, WILLIAM H...........
At Cibicu, Ariz., Aug. 30, 1881.
R—New York, N. Y.
B—Nashville, Tenn.
Distinguished-service medal also
awarded.

First lieutenant, 6th U. S. Cavalry.
Rescued, with the voluntary assistance of two soldiers, the wounded from
under a heavy fire.

CARUANA, ORLANDO E.........
At New Bern, N. C., Mar. 14, 1862.

At South Mountain, Md., Sept. 14,
1862.
R—NR.
B—Ca Valletta, Malta.

Private, Company K, 51st New York Infantry.
Brought off the wounded color sergeant and the colors under a heavy fire of the
enemy.
Was one of four soldiers who volunteered to determine the position of the
enemy. While so engaged was fired upon and his three companions killed,
but he escaped and rejoined his command in safety.

CASEY, DAVID................
At Cold Harbor, Va., June 3, 1884.
R—NR.
B—Ireland.

Private, Company C, 25th Massachusetts Infantry.
Two color bearers having been shot dead one after the other, the last one far
in advance of his regiment and close to the enemy's lines, this soldier rushed
forward, and, under a galling fire, after removing the dead body of the bearer
therefrom, secured the flag and returned with it to the Union lines.

CASEY, HENRY...............
At Vicksburg, Miss., Apr. 22, 1863.
R—NR.
B—Fayette County, Pa.

Private, Company C, 20th Ohio Infantry.
Voluntarily served as one of the crew of a transport that passed the forts under
a heavy fire.

CASEY, JAMES S..............
At Wolf Mountain, Mont., Jan. 8,
1877.
R—New York, N. Y.
B—Philadelphia, Pa.

Captain, 5th U. S. Infantry.
Led his command in a successful charge against superior numbers of the enemy
strongly posted.

CATLIN, ISAAC S.
    At Petersburg, Va., July 30, 1864.
    R—Oswego, N. Y.
    B—New York.

Colonel, 109th New York Infantry.
In a heroic effort to rally the disorganized troops was disabled by a severe wound. While being carried from the field he recovered somewhat and bravely started to return to his command, when he received a second wound, which necessitated amputation of his right leg.

CAWETZKA, CHARLES.
    Near Sariaya, Luzon, P. I., Aug. 23, 1900.
    R—Wayne, Mich.
    B—Detroit, Mich.

Private, Company F, 30th Infantry, U. S. Volunteers.
Single-handed, he defended a disabled comrade against a greatly superior force of the enemy.

CAYER, OVILA.
    At Weldon Railroad, Va., Aug. 19, 1864.
    R—NR.
    B—NR.

Sergeant, Company A, 14th U. S. Infantry.
Commanded the regiment, all the officers being disabled.

CECIL, JOSEPHUS S.
    At Bud-Dajo, Jolo, P. I. Mar. 7, 1906.
    R—New River, Tenn.
    B—New River, Tenn.

First lieutenant, 19th U. S. Infantry.
While at the head of the column about to assault the first cotta under a superior fire at short range personally carried to a sheltered position a wounded man and the body of one who was killed beside him.

CHAMBERLAIN, JOSHUA L.
    At Gettysburg, Pa., July 2, 1863.
    R—Brunswick, Me.
    B—NR.

Colonel, 20th Maine Infantry.
During heroism and great tenacity in holding his position on the Little Round Top against repeated assaults, and carrying the advance position on the Great Round Top.

CHAMBERLAIN, ORVILLE T.
    At Chickamauga, Ga., Sept. 20, 1863.
    R—NR.
    B—Kosciusko County, Ind.

Second lieutenant, Company G, 74th Indiana Infantry.
While exposed to a galling fire, went in search of another regiment, found its location, procured ammunition from the men thereof, and returned with the ammunition to his own company.

CHAMBERS, JOSEPH B.
    At Petersburg, Va., Mar. 25, 1865.
    R—East Brook, Pa.
    B—Beaver County, Pa.

Private, Company F, 100th Pennsylvania Infantry.
Capture of colors of 1st Virginia Infantry (C. S. A.).

CHANDLER, HENRY F.
    At Petersburg, Va., June 17, 1864.
    R—NR.
    B—Andover, Mass.

Sergeant, Company E, 59th Massachusetts Infantry.
Though seriously wounded in a bayonet charge and directed to go to the rear he declined to do so, but remained with his regiment and helped to carry the breastworks.

CHANDLER, STEPHEN E.
    At Amelia Springs, Va., Apr. 5, 1865.
    R—NR.
    B—Michigan.

Quartermaster sergeant, Company A, 24th New York Cavalry.
Under severe fire of the enemy and of the troops in retreat, went between the lines to the assistance of a wounded and helpless comrade, and rescued him from death or capture.

CHAPIN, ALARIC B.
    At Fort Fisher, N. C., Jan. 15, 1865.
    R—Pamelia, N. Y.
    B—Ogdensburg, N. Y.

Private, Company G, 142d New York Infantry.
Voluntarily advanced with the head of the column and cut down the palisading.

CHAPMAN, JOHN.
    At Sailors Creek, Va., Apr. 6, 1865.
    R—St. John, New Brunswick.
    B—St. John, New Brunswick.

Private, Company B, 1st Maine Heavy Artillery.
Capture of flag.

CHASE, JOHN F.
    At Chancellorsville, Va., May 3, 1863.
    R—NR.
    B—Chelsea, Me.

Private, 5th Battery, Maine Light Artillery.
Nearly all the officers and men of the battery having been killed or wounded, this soldier with a comrade continued to fire his gun after the other guns had ceased. The piece was then dragged off by the two, the horses having been shot, and its capture by the enemy was prevented.

CHEEVER, BENJAMIN H., Jr.
    At White River, S. Dak., Jan. 1, 1891.
    R—Washington, D. C.
    B—Washington, D. C.

First lieutenant, 6th U. S. Cavalry.
Headed the advance across White River, partly frozen, in a spirited movement to the effective assistance of Troop K, 6th U. S. Cavalry.

CHILD, BENJAMIN H.
    At Antietam, Md., Sept. 17, 1862.
    R—NR.
    B—Providence, R. I.

Corporal, Battery A, 1st Rhode Island Light Artillery.
Was wounded and taken to the rear insensible, but when partially recovered insisted on returning to the battery and resumed command of his piece, so remaining until the close of the battle.

*CHILES, MARCELLUS H.
    Near Le Champy Bas, France, Nov. 3, 1918.
    R—Denver, Colo.
    B—Eureka Springs, Ark.
    G. O. No. 20, W. D., 1919.

Captain, 356th Infantry, 89th Division.
When his battalion, of which he had just taken command, was halted by machine-gun fire from the front and left flank, he picked up the rifle of a dead soldier and, calling on his men to follow, led the advance across a stream, waist deep, in the face of the machine-gun fire. Upon reaching the opposite bank this gallant officer was seriously wounded in the abdomen by a sniper, but before permitting himself to be evacuated he made complete arrangements for turning over his command to the next senior officer, and under the inspiration of his fearless leadership his battalion reached its objective. Captain Chiles died shortly after reaching the hospital.
Posthumously awarded. Medal presented to father, John Horne Chiles.

CHIQUITO.
    Winter of 1872-73.
    R—NR.
    B—NR.

Indian scout.
Gallant conduct during the campaigns and engagements with Apaches.

CHISMAN, WILLIAM W...............
At Vicksburg, Miss., May 22, 1863.
R—NR.
B—Dearborn County, Ind.

Private, Company I, 83d Indiana Infantry.
Gallantry in the charge of the "volunteer storming party."

CHRISTIANCY, JAMES I...............
At Hawes Shops, Va., May 28, 1864.
R—NR.
B—NR.

First lieutenant, Company D, 9th Michigan Cavalry.
While acting as aid, voluntarily led a part of the line into the fight, and was twice wounded.

CHURCH, JAMES ROBB...............
At Las Guasimas, Cuba, June 24, 1898.
R—Washington, D. C.
B—Chicago, Ill.

Assistant surgeon, 1st U. S. Volunteer Cavalry.
In addition to performing gallantly the duties pertaining to his position, voluntarily and unaided carried several seriously wounded men from the firing line to a secure position in the rear, in each instance being subjected to a very heavy fire and great exposure and danger.

CHURCHILL, SAMUEL J...............
At Nashville, Tenn., Dec. 15, 1864.
R—NR.
B—Rutland County, Vt.

Corporal, Company G, 2d Illinois Light Artillery.
When the fire of the enemy's batteries compelled the men of his detachment for a short time to seek shelter, he stood manfully at his post and for some minutes worked his gun alone.

CILLEY, CLINTON A...............
At Chickamauga, Ga., Sept. 20, 1863.
R—Farmington, N. H.
B—Rockingham County, N. H.

Captain, Company C, 2d Minnesota Infantry.
Seized the colors of a retreating regiment and led it into the thick of the attack.

CLANCY, JAMES T...............
At Vaughn Road, Va., Oct. 1, 1864.
R—NR.
B—Albany, N. Y.

Sergeant, Company C, 1st New Jersey Cavalry.
Shot the Confederate General Dunovant dead during a charge, thus confusing the enemy and greatly aiding in his repulse.

CLANCY, JOHN E...............
At Wounded Knee Creek, S. Dak., Dec. 29, 1890.
R—NR.
B—New York, N. Y.

Musician, Company E, 1st U. S. Artillery.
Twice voluntarily rescued wounded comrades under fire of the enemy.

CLAPP, ALBERT A...............
At Sailors Creek, Va., Apr. 6, 1865.
R—NR.
B—Pompey, N. Y.

First sergeant, Company G, 2d Ohio Cavalry.
Capture of battle flag of the 8th Florida Infantry (C. S. A.).

CLARK, CHARLES A...............
At Brooks Ford, Va., May 4, 1863.
R—NR.
B—NR.

Lieutenant and adjutant, 6th Maine Infantry.
Having voluntarily taken command of his regiment in the absence of its commander, at great personal risk and with remarkable presence of mind and fertility of resource led the command down an exceedingly precipitous embankment to the Rappahannock River and by his gallantry, coolness, and good judgment in the face of the enemy saved the command from capture or destruction.

CLARK, HARRISON...............
At Gettysburg, Pa., July 2, 1863.
R—Chatham, N. Y.
B—Chatham, N. Y.

Corporal, Company E, 125th New York Infantry.
Seized the colors and advanced with them after the color bearer had been shot.

CLARK, JAMES G...............
At Petersburg, Va., June 18, 1864.
R—NR.
B—Germantown, Pa.

Private, Company F, 88th Pennsylvania Infantry.
Distinguished bravery in action; was severely wounded.

CLARK, JOHN W...............
Near Warrenton, Va., July 28, 1863.
R—NR.
B—NR.

First lieutenant and regimental quartermaster, 6th Vermont Infantry.
Defended the division train against a vastly superior force of the enemy; he was severely wounded, but remained in the saddle for 20 hours afterwards, until he had brought his train through in safety.

CLARK, WILFRED...............
At Big Hole, Mont., Aug. 9, 1877; at Camas Meadows, Idaho, Aug. 20, 1877.
R—NR.
B—Philadelphia, Pa.

Private, Company L, 2d U. S. Cavalry.
Conspicuous gallantry; especial skill as sharpshooter.

CLARK, WILLIAM A...............
At Nolensville, Tenn., Feb. 15, 1863.
R—NR.
B—Pennsylvania.

Corporal, Company H, 2d Minnesota Infantry.
Was one of a detachment of 16 men who heroically defended a wagon train against the attack of 125 cavalry, repulsed the attack, and saved the train.

CLARKE, DAYTON P...............
At Spotsylvania, Va., May 12, 1864.
R—Hermon, N. Y.
B—Hermon, N. Y.

Captain, Company F, 2d Vermont Infantry.
Distinguished conduct in a desperate hand-to-hand fight while commanding the regiment.

CLARKE, POWHATAN H...............
At Pinito Mountains, Sonora, Mexico, May 3, 1886.
R—Baltimore, Md.
B—Alexandria, La.

Second lieutenant, 10th U. S. Cavalry.
Rushed forward to the rescue of a soldier who was severely wounded and lay, disabled, exposed to the enemy's fire, and carried him to a place of safety.

CLAUSEN, CHARLES H.
At Spotsylvania, Va., May 12, 1864.
R—Philadelphia, Pa.
B—Philadelphia, Pa.

First lieutenant, Company H, 61st Pennsylvania Infantry.
Although severely wounded, he led the regiment against the enemy, under a terrific fire, and saved a battery from capture.

CLAY, CECIL.
At Fort Harrison, Va., Sept. 29, 1864.
R—NR.
B—Philadelphia, Pa.

Captain, Company K, 58th Pennsylvania Infantry.
Led his regiment in the charge, carrying the colors of another regiment, and when severely wounded in the right arm, incurring loss of same, he shifted the colors to the left hand, which also became disabled by a gunshot wound.

CLEVELAND, CHARLES F.
At Antietam, Md., Sept. 17, 1862.
R—NR.
B—Hartford, N. Y.

Private, Company C, 26th New York Infantry.
Voluntarily took and carried the colors into action after the color bearer had been shot.

CLOPP, JOHN E.
At Gettysburg, Pa., July 3, 1863.
R—Philadelphia, Pa.
B—Philadelphia, Pa.

Private, Company F, 71st Pennsylvania Infantry.
Capture of flag of 9th Virginia Infantry (C. S. A.), wresting it from the color bearer.

CLUTE, GEORGE W.
At Bentonville, N. C., Mar. 19, 1865.
R—NR.
B—Marathon, Mich.

Corporal, Company I, 14th Michigan Infantry.
In a charge, captured the flag of the 40th North Carolina (C. S. A.), the flag being taken in a personal encounter with an officer who carried and defended it.

COATES, JEFFERSON.
At Gettysburg, Pa., July 1, 1863.
R—NR.
B—Grant County, Wis.

Sergeant, Company H, 7th Wisconsin Infantry.
Unsurpassed courage in battle, where he had both eyes shot out.

COCKLEY, DAVID L.
At Waynesboro, Ga., Dec. 4, 1864.
R—NR.
B—NR.

First lieutenant, Company L, 10th Ohio Cavalry.
While acting as aid-de-camp to a general officer he three times asked permission to join his regiment in a proposed charge upon the enemy, and in response to the last request, having obtained such permission, joined his regiment and fought bravely at its head throughout the action.

COEY, JAMES.
At Hatchers Run, Va., Feb. 6, 1865.
R—NR.
B—NR.

Major, 147th New York Infantry.
Seized the regimental colors at a critical moment and by a prompt advance on the enemy caused the entire brigade to follow him; and, after being himself severely wounded, he caused himself to be lifted into the saddle and a second time rallied the line in an attempt to check the enemy.

COFFEY, ROBERT J.
At Banks Ford, Va., May 4, 1863.
R—NR.
B—St. John, New Brunswick.

Sergeant, Company K, 4th Vermont Infantry.
Single-handed captured two officers and five privates of the 8th Louisiana Regiment (C. S. A.).

COHN, ABRAHAM.
At Wilderness, Va., May 6, 1864.

At the mine, Petersburg, Va., July 30, 1864.
R—NR.
B—Prussia.

Sergeant major, 6th New Hampshire Infantry.
Rallied and formed, under heavy fire, disorganized and fleeing troops of different regiments.
Bravely and coolly carried orders to the advanced line under severe fire.

COLBY, CARLOS W.
At Vicksburg, Miss., May 22, 1863.
R—Madison County, Ill.
B—Merrimack, N. H.

Sergeant, Company G, 97th Illinois Infantry.
Was a member of a volunteer storming party which made a most gallant assault upon the enemy's works.

COLE, GABRIEL.
At Winchester, Va., Sept. 19, 1864.
R—New Salem, Mich.
B—Chenango County, N. Y.

Corporal, Company I, 5th Michigan Cavalry.
Capture of flag.

COLLINS, HARRISON.
At Richland Creek, Tenn., Dec. 24, 1864.
R—NR.
B—Hawkins County, Tenn.

Corporal, Company A, 1st Tennessee Cavalry.
Capture of flag of Chalmers's Division (C. S. A.).

COLLINS, THOMAS D.
At Resaca, Ga., May 15, 1864.
R—NR.
B—Neversink, N. Y.

Sergeant, Company H, 143d New York Infantry.
Captured a regimental flag of the enemy.

COLLIS, CHARLES H. T.
At Fredericksburg, Va., Dec. 13, 1862.
R—NR.
B—NR.

Colonel, 114th Pennsylvania Infantry.
Gallantly led his regiment in battle at a critical moment.

COLWELL, OLIVER.
At Nashville, Tenn., Dec. 16, 1864.
R—NR.
B—Champaign County, Ohio.

First lieutenant, Company G, 95th Ohio Infantry.
Capture of flag.

*COLYER, WILBUR E (154550)_____
  Near Verdun, France, Oct. 9, 1918.
  R—South Ozone, Long Island, N. Y.
  B—Brooklyn, N. Y.
  G. O. No. 20, W. D., 1919.

Sergeant, Company A, 1st Engineers, 1st Division.
Volunteering with two other soldiers to locate machine-gun nests, Sergeant Colyer advanced on the hostile positions to a point where he was half surrounded by the nests, which were in ambush. He killed the gunner of one gun with a captured German grenade and then turned this gun on the other nests, silencing all of them before he returned to his platoon. He was later killed in action.
Posthumously awarded. Medal presented to father, William H. Colyer.

COMFORT, JOHN W_____
  At Staked Plain, Tex., Nov. 5, 1874.
  R—NR.
  B—Philadelphia, Pa.

Corporal, Company A, 4th U. S. Cavalry.
Ran down and killed an Indian, etc.

COMPSON, HARTWELL B_____
  At Waynesboro, Va., Mar. 2, 1865.
  R—NR.
  B—Seneca, N. Y.

Major, 8th New York Cavalry.
Capture of flag belonging to General Early's headquarters.

CONAWAY, JOHN W_____
  At Vicksburg, Miss., May 22, 1863.
  R—NR.
  B—Dearborn County, Ind.

Private, Company C, 83d Indiana Infantry.
Gallantry in the charge of the "volunteer storming party."

CONBOY, MARTIN_____
  At Williamsburg, Va., May 5, 1862.
  R—New York, N. Y.
  B—NR.

Sergeant, Company B, 37th New York Infantry.
Took command of the company in action, the captain having been wounded, the other commissioned officers being absent, and handled it with skill and bravery.

CONDON, CLARENCE M_____
  Near Calulut, Luzon, P. I., Nov. 5,
  1899.
  R—NR.
  B—South Brooksville, Me.

Sergeant, Battery G, 3d U. S. Artillery.
While in command of a detachment of 4 men, charged and routed 40 entrenched insurgents, inflicting on them heavy loss.

CONGDON, JAMES_____
  Sergeant, Company E, 8th New
    York Cavalry.

See James Madison, true name.

CONNELL, TRUSTRIM_____
  At Sailors Creek, Va., Apr. 6, 1865.
  R—Fort Kennedy, Pa.
  B—Lancaster, Pa.

Corporal, Company I, 138th Pennsylvania Infantry.
Capture of flag.

CONNER, RICHARD_____
  At Bull Run, Va., Aug. 30, 1862.
  R—NR.
  B—Philadelphia, Pa.

Private, Company F, 6th New Jersey Infantry.
The flag of his regiment having been abandoned during retreat, he voluntarily returned with a single companion under a heavy fire and secured and brought off the flag, his companion being killed.

CONNOR, JOHN_____
  At Wichita River, Tex., July 12, 1870.
  R—NR.
  B—Ireland.

Corporal, Company H, 6th U. S. Cavalry.
Gallantry in action.

CONNORS, JAMES_____
  At Fishers Hill, Va., Sept. 22, 1864.
  R—NR.
  B—Ireland.

Private, Company E, 43d New York Infantry.
Capture of flag.

COOK, JOHN_____
  At Antietam, Md., Sept. 17, 1862.
  R—Cincinnati, Ohio.
  B—Hamilton County, Ohio.

Bugler, Battery B, 4th U. S. Artillery.
Volunteered at the age of 15 years to act as a cannoneer, and as such volunteer served a gun under a terrific fire of the enemy.

COOK, JOHN H_____
  At Pleasant Hill, La., Apr. 9, 1864.
  R—Quincy, Ill.
  B—England.

Sergeant, Company A, 119th Illinois Infantry.
During an attack by the enemy, voluntarily left the brigade quartermaster, with whom he had been detailed as a clerk, rejoined his command, and, acting as first lieutenant, led the line farther toward the charging enemy.

COOKE, WALTER H_____
  At Bull Run, Va., July 21, 1861.
  R—NR.
  B—NR.

Captain, Company K, 4th Pennsylvania Infantry Militia.
Voluntarily served as an aide on the staff of Col. David Hunter and participated in the battle, his term of service having expired on the previous day.

COONROD, AQUILLA_____
  At Cedar Creek, etc., Mont., October, 1876, to January, 1877.
  R—Bryan, Ohio.
  B—Williams County, Ohio.

Sergeant, Company C, 5th U. S. Infantry.
Gallantry in actions.

COPP, CHARLES D_____
  At Fredericksburg, Va., Dec. 13, 1862.
  R—Nashua, N. H.
  B—NR.

Second lieutenant, Company C, 9th New Hampshire Infantry.
Seized the regimental colors, the color bearer having been shot down, and, waiving them, rallied the regiment under a heavy fire.

CORCORAN, JOHN_____
  At Petersburg, Va., Apr. 2, 1865.
  R—NR.
  B—Pawtucket, R. I.

Private, Company G, 1st Rhode Island Light Artillery.
Was one of a detachment of 20 picked artillerymen who voluntarily accompanied an infantry assaulting party, and who turned upon the enemy the guns captured in the assault.

CORCORAN, MICHAEL.................. | Corporal, Company E, 8th U. S. Cavalry.
At Agua Fria River, Ariz., Aug. 25, 1869. | Gallantry in action.
R—NR.
B—Philadelphia, Pa.

CORLISS, GEORGE W................... | Captain, Company C, 5th Connecticut Infantry.
At Cedar Mountains, Va., Aug. 9, 1862. | Seized a fallen flag of the regiment, the color bearer having been killed, carried it forward in the face of a severe fire, and, though himself shot down and permanently disabled, planted the staff in the earth and kept the flag flying.
R—NR.
B—NR.

CORLISS, STEPHEN P................... | First lieutenant, Company F, 4th New York Heavy Artillery.
At South Side Railroad, Va., Apr. 2, 1865. | Raised the fallen colors and, rushing forward in advance of the troops, placed them on the enemy's works.
R—NR.
B—NR.

CORSON, JOSEPH K.................... | Assistant surgeon, 6th Pennsylvania Reserves (35th Pennsylvania Volunteers).
Near Bristoe Station, Va., Oct. 14, 1863. | With one companion returned in the face of the enemy's heavy artillery fire and removed to a place of safety a severely wounded soldier who had been left behind as the regiment fell back.
R—Philadelphia, Pa.
B—NR.

CO-RUX-TE-CHOD-ISH (MAD BEAR)... | Sergeant, Pawnee Scouts, U. S. Army.
At Republican River, Kans., July 8, 1869. | Ran out from the command in pursuit of a dismounted Indian; was shot down and badly wounded by a bullet from his own command.
R—NR.
B—Nebraska.

COSGRIFF, RICHARD H................. | Private, Company L, 4th Iowa Cavalry.
At Columbus, Ga., Apr. 16, 1865. | Capture of flag in a personal encounter with its bearer.
R—NR.
B—Dunkirk County, N. Y.

COSGROVE, THOMAS.................. | Private, Company F, 40th Massachusetts Infantry.
At Drurys Bluff, Va., May 15, 1864. | Individually demanded and received the surrender of seven armed Confederates concealed in a cellar, disarming and marching them in as prisoners of war.
R—NR.
B—Ireland.

*COSTIN, HENRY G. (1285528).......... | Private, Company H, 115th Infantry, 29th Division.
Near Bois-de-Consenvoye, France, Oct. 8, 1918. | When the advance of his platoon had been held up by machine-gun fire and a request was made for an automatic rifle team to charge the nest, Private Costin was the first to volunteer. Advancing with his team, under terrific fire of enemy artillery, machine guns, and trench mortars, he continued after all his comrades had become casualties and he himself had been seriously wounded. He operated his rifle until he collapsed. His act resulted in the capture of about 100 prisoners and several machine guns. He succumbed from the effects of his wounds shortly after the accomplishment of his heroic deed.
R—Baltimore, Md.
B—Baltimore, Md.
G. O. No. 34, W. D., 1919. | Posthumously awarded. Medal presented to widow, Mrs. Hythron Costin.

COUGHLIN, JOHN..................... | Lieutenant colonel, 10th New Hampshire Infantry.
At Swifts Creek, Va., May 9, 1864. | During a sudden night attack upon Burnham's Brigade, resulting in much confusion, this officer, without waiting for orders, led his regiment forward and interposed a line of battle between the advancing enemy and Hunt's Battery, repulsing the attack and saving the guns.
R—Manchester, N. H.
B—NR.

COX, ROBERT M..................... | Corporal, Company K, 55th Illinois Infantry.
At Vicksburg, Miss., May 22, 1863. | Bravely defended the colors planted on the outward parapet of Fort Hill.
R—Prairie City, Ill.
B—Guernsey County, Ohio.

COYNE, JOHN N..................... | Sergeant, Company B, 70th New York Infantry.
At Williamsburg, Va., May 5, 1862. | Capture of a flag after a severe hand-to-hand contest; was mentioned in orders for his gallantry.
R—NR.
B—NR.

CRAIG, SAMUEL H.................... | Sergeant, Company D, 4th U. S. Cavalry.
At Santa Cruz Mountains, Mexico, May 15, 1886. | Conspicuous gallantry during an attack on a hostile Apache Indian camp; seriously wounded.
R—NR.
B—New Market, N. H.

CRANDALL, CHARLES................. | Private, Company B, 8th U. S. Cavalry.
Arizona, August to October, 1868. | Bravery in scouts and actions with Indians.
R—Worcester, Mass.
B—Worcester, Mass.

CRANSTON, WILLIAM W.............. | Private, Company A, 66th Ohio Infantry.
At Chancellorsville, Va., May 2, 1863. | One of a party of four who voluntarily brought in a wounded Confederate officer from within the enemy's line in the face of a constant fire.
R—NR.
B—Champaign County, Ohio.

CREED, JOHN....................... | Private, Company D, 23d Illinois Infantry.
At Fishers Hill, Va., Sept. 22, 1864. | Capture of flag.
R—NR.
B—Ireland.

CRIST, JOHN_____ | Sergeant, Company L, 8th U. S. Cavalry.
   Arizona, Nov. 26, 1869. | Gallantry in action.
   R—NR. |
   B—Baltimore, Md. |

CRISWELL, BENJAMIN C_____ | Sergeant, Company B, 7th U. S. Cavalry.
   At Little Big Horn River, Mont., | Rescued the body of Lieutenant Hodgson from within the enemy's lines;
     June 25, 1876. | brought up ammunition and encouraged the men in the most exposed positions
   R—NR. | under heavy fire.
   B—Marshall County, W. Va. |

CROCKER, HENRY H_____ | Captain, Company F, 2d Massachusetts Cavalry.
   At Cedar Creek, Va., Oct. 19, 1864. | Voluntarily led a charge, which resulted in the capture of 14 prisoners and in
   R—NR. | which he himself was wounded.
   B—NR. |

CROCKER, ULRIC L_____ | Private, Company M, 6th Michigan Cavalry.
   At Cedar Creek, Va., Oct. 19, 1864. | Capture of flag of 18th Georgia (C. S. A.).
   R—NR. |
   B—Ohio. |

CROFT, JAMES E_____ | Private, 12th Battery, Wisconsin Light Artillery.
   At Allatoona, Ga., Oct, 5, 1864. | Took the place of a gunner who had been shot down and inspired his comrades
   R—Janesville, Wis. | by his bravery and effective gunnery, which contributed largely to the defeat
   B—England. | of the enemy.

CROSIER, WILLIAM H. H_____ | Sergeant, Company G, 149th New York Infantry.
   At Peach Tree Creek, Ga., July 20, | Severely wounded and ambushed by the enemy, he stripped the colors from the
     1864. | staff and brought them back into line.
   R—NR. |
   B—Skaneateles, N. Y. |

CROSS, JAMES E_____ | Corporal, Company K, 12th New York Infantry.
   At Blackburns Ford, Va., July 18, | With a companion, refused to retreat when the part of the regiment to which
     1861. | he was attached was driven back in dis rder, but remained upon the skir-
   R—NR. | mish line for some time thereafter, firing upon the enemy.
   B—Darien, N. Y. |

CROWLEY, MICHAEL_____ | Private, Company A, 22d New York Cavalry.
   At Waynesboro, Va., Mar. 2, 1865. | Capture of flag.
   R—NR. |
   B—Rochester, N. Y. |

CRUSE, THOMAS_____ | Second lieutenant, 6th U. S. Cavalry.
   At Big Dry Fork, Ariz., July 17, 1882. | Gallantly charged hostile Indians, and with his carbine compelled a party of
   R—Owensboro, Ky. | them to keep under cover of their breastworks, thus being enabled to recover a
   B—Owensboro, Ky. | severely wounded soldier.

CUBBERLY, WILLIAM G_____ | Private, Company L, 8th U. S. Cavalry.
   At San Carlos, Ariz., May 30, 1868. | Gallantry in action.
   R—NR. |
   B—Butler County, Ohio. |

CUKELA, LOUIS_____ | Sergeant, 66th Company, 5th Regiment, U. S. Marine Corps.
   Near Villers-Cotterets, France, July | When his company, advancing through a wood, met with strong resistance from
     18, 1918. | an enemy strong point, Sergeant Cukela crawled out from the flank and made
   R—Minneapolis, Minn. | his way toward the German lines in the face of heavy fire, disregarding the
   B—Austria. | warnings of his comrades. He succeeded in getting behind the enemy posi-
   G. O. No. 34, W. D., 1919. | tion and rushed a machine-gun emplacement, killing or driving off the crew
   | with his bayonet. With German hand grenades he then bombed out the re-
   | maining portion of the strong point, capturing four men and two damaged
   | machine guns.

CULLEN, THOMAS_____ | Corporal, Company I, 82d New York Infantry.
   At Bristoe Station Va., Oct. 14, 1863. | Capture of flag of 22d or 28th North Carolina (C. S. A.).
   R—NR. |
   B—Ireland. |

CUMMINGS, AMOS J_____ | Sergeant major, 26th New Jersey Infantry.
   At Salem Heights, Va., May 4, 1863. | Rendered great assistance in the heat of the action in rescuing a part of one of
   R—NR. | the field batteries from an extremely dangerous and exposed position.
   B—NR. . |

CUMMINS, ANDREW J_____ | Sergeant, Company F, 10th U. S. Infantry.
   At Santiago, Cuba, July 1, 1898. | Gallantly assisted in the rescue of the wounded from in front of the lines and
   R—NR. | under heavy fire of the enemy.
   B—Alexandria, Ind. |

CUMPSTON, JAMES M_____ | Private, Company D, 91st Ohio Infantry.
   At Shenandoah Valley campaign, | Capture of flag.
     August to November, 1864. |
   R—NR. |
   B—Gallia County, Ohio. |

CUNNINGHAM, CHARLES_____ | Corporal, Company B, 7th U. S. Cavalry.
   At Little Big Horn River, Mont., | Declined to leave the line when wounded in the neck during heavy fire and
     June 25, 1876. | fought bravely all next day.
   R—NR. |
   B—Hudson, N. Y. |

CUNNINGHAM, FRANCIS M _____ | First sergeant, Company H, 1st West Virginia Cavalry.
  At Sailors Creek, Va., Apr. 6, 1865. | Capture of battle flag of 12th Virginia Infantry (C. S. A.).
  R—NR. |
  B—Somerset County, Pa. |

CUNNINGHAM, JAMES S _____ | Private, Company D, 8th Missouri Infantry.
  At Vicksburg, Miss., May 22, 1863. | Gallantry in the charge of the "volunteer storming party."
  R—NR. |
  B—NR. |

CURRAN, RICHARD _____ | Assistant surgeon, 33d New York Infantry.
  At Antietam, Md., Sept. 17, 1862. | Voluntarily exposed himself to great danger by going to the fighting line,
  R—NR. | there succoring the wounded and helpless and conducting them to the field
  B—NR. | hospital.

CURTIS, JOHN C _____ | Sergeant major, 9th Connecticut Infantry.
  At Baton Rouge, La., Aug. 5, 1862. | Voluntarily sought the line of battle and alone and unaided captured two
  R—NR. | prisoners, driving them before him to regimental headquarters at the point
  B—Bridgeport, Conn. | of the bayonet.

CURTIS, JOSIAH M _____ | Second lieutenant, Company I, 12th West Virginia Infantry.
  At Petersburg, Va., Apr. 2, 1865. | Seized the colors of his regiment after two color bearers had fallen, bore them
  R—NR. | gallantly, and was among the first to gain a foothold, with his flag, inside
  B—NR. | the enemy's works.

CURTIS, N. MARTIN _____ | Brigadier general, U. S. Volunteers.
  At Fort Fisher, N. C., Jan. 15, 1865. | The first man to pass through the stockade, he personally led each assault on
  R—De Peyster, N. Y. | the traverses and was four times wounded.
  B—NR. |

CUSTER, THOMAS W _____ | Second lieutenant, Company B, 6th Michigan Cavalry.
  (2 medals of honor.) |
  At Namozine Church, Va., Apr. 2, | Capture of flag.
  1865. |
  At Sailors Creek, Va., Apr. 6, 1865. | Leaped his horse over the enemy's works and captured two stands of colors,
  R—NR. | having his horse shot under him and receiving a severe wound.
  B—NR. |

CUTCHEON, BYRON M _____ | Major, 20th Michigan Infantry.
  At Horseshoe Bend, Ky., May 10, | Distinguished gallantry in leading his regiment in a charge on a house occupied
  1863. | by the enemy.
  R—NR. |
  B—NR. |

CUTTS, JAMES M _____ | Captain, 11th U. S. Infantry.
  At Wilderness, Spotsylvania, Peters- | Gallantry in actions.
  burg, Va., 1864. |
  R—Illinois. |
  B—Washington, D. C. |

DAILY, CHARLES _____ | Private, Company B, 8th U. S. Cavalry.
  Arizona, August to October, 1868. | Bravery in scouts and actions with Indians
  R—NR. |
  B—Ireland. |

DANIELS, JAMES T _____ | Sergeant, Company L, 4th U. S. Cavalry.
  Arizona, Mar. 7, 1890. | Untiring energy and cool gallantry under fire in an engagement with Apache
  R—NR. | Indians.
  B—Richland County, Ill. |

DARROUGH, JOHN S _____ | Sergeant, Company F, 113th Illinois Infantry.
  At Eastport, Miss., Oct. 10, 1864. | Saved the life of a captain.
  R—NR. |
  B—Kentucky. |

DAVIDSIZER, JOHN A _____ | Sergeant, Company A, 1st Pennsylvania Cavalry.
  At Paines Crossroads, Va., Apr. 5, | Capture of flag.
  1865. |
  R—Lewiston, Pa. |
  B—Milford, Pa. |

DAVIDSON, ANDREW _____ | First lieutenant, Company H, 30th U. S. Colored Troops.
  At the mine, Petersburg, Va., July | One of the first to enter the enemy's works, where, after his colonel, major, and
  30, 1864. | one-third the company officers had fallen, he gallantly assisted in rallying and
  R—NR. | saving the remnant of the command.
  B—NR. |

DAVIDSON, ANDREW _____ | Assistant surgeon, 47th Ohio Infantry.
  At Vicksburg, Miss., May 3, 1863. | Voluntarily attempted to run the enemy's batteries.
  R—Middlebury, Vt. |
  B—NR. |

DAVIS, CHARLES C _____ | Major, 7th Pennsylvania Cavalry.
  At Shelbyville, Tenn., June 27, 1863. | Led one of the most desperate and successful charges of the war.
  R—NR. |
  B—Harrisburg, Pa. |

DAVIS, CHARLES P.
Near San Isidro, P. I., May 16, 1899.
R—Valley City, N. Dak.
B—Long Prairie, Minn.
Private, Company G, 1st North Dakota Volunteer Infantry.
With 21 other scouts charged across a burning bridge, under heavy fire, and completely routed 600 of the enemy who were intrenched in a strongly fortified position.

DAVIS, FREEMAN.
At Missionary Ridge, Tenn., Nov. 25, 1863.
R—NR.
B—Newcomerstown, Ohio.
Sergeant, Company B, 80th Ohio Infantry.
This soldier, while his regiment was falling back, seeing the two color bearers shot down, under a severe fire and at imminent peril recovered both the flags and saved them from capture.

DAVIS, GEORGE E.
At Monocacy, Md., July 9, 1864.
R—Burlington, Vt.
B—Dunstable, Mass.
First lieutenant, Company D, 10th Vermont Infantry.
While in command of a small force, held the approaches to the two bridges against repeated assaults of superior numbers, thereby materially delaying Early's advance on Washington.

DAVIS, HARRY.
At Atlanta, Ga., July 28, 1864.
R—NR.
B—Franklin County, Ohio.
Private, Company G, 46th Ohio Infantry.
Capture of flag of 30th Louisiana Infantry (C. S. A.).

DAVIS, JOHN.
At Culloden, Ga., Apr. —, 1865.
R—Indianapolis, Ind.
B—Carroll, Ky.
Private, Company F, 17th Indiana Mounted Infantry.
Capture of flag of Worrill Grays (C. S. A.).

DAVIS, JOSEPH.
At Franklin, Tenn., Nov. 30, 1864.
R—NR.
B—Wales.
Corporal, Company C, 104th Ohio Infantry.
Capture of flag.

DAVIS, MARTIN K.
At Vicksburg, Miss., May 22, 1863.
R—NR.
B—Marion, Ill.
Sergeant, Company H, 116th Illinois Infantry.
Gallantry in the charge of the "volunteer storming party."

DAVIS, THOMAS.
At Sailors Creek, Va., Apr. 6, 1865.
R—NR.
B—Wales.
Private, Company C, 2d New York Heavy Artillery.
Capture of flag.

DAWSON, MICHAEL.
At Sappa Creek, Kans., Apr. 23, 1875.
R—NR.
B—Boston, Mass.
Trumpeter, Company H, 6th U. S. Cavalry.
Gallantry in action.

DAY, CHARLES.
At Hatchers Run, Va., Feb. 6, 1865.
R—NR.
B—Otsego County, N. Y.
Private, Company K, 210th Pennsylvania Infantry.
Seized the colors of another regiment of the brigade, the regiment having been thrown into confusion and the color bearer killed, and bore said colors throughout the remainder of the engagement.

DAY, DAVID F.
At Vicksburg, Miss., May 22, 1863.
R—NR.
B—Dallasburg, Ohio.
Private, Company D, 57th Ohio Infantry.
Gallantry in the charge of the "volunteer storming party."

DAY, MATTHIAS W.
At Las Animas Canyon, N. Mex., Sept. 18, 1879.
R—Oberlin, Ohio.
B—Mansfield, Ohio.
Second lieutenant, 9th U. S. Cavalry.
Advanced alone into the enemy's lines and carried off a wounded soldier of his command under a hot fire and after he had been ordered to retreat.

DAY, WILLIAM L.
1872-73.
R—NR.
B—Barron County, Ky.
First sergeant, Company E, 5th U. S. Cavalry.
Gallantry during the different campaigns and fights with Apaches.

DEANE, JOHN M.
At Fort Stedman, Va., Mar. 25, 1865.
R—NR.
B—NR.
Major, 29th Massachusetts Infantry.
This officer, observing an abandoned gun within Fort Haskell, called for volunteers, and, under a heavy fire, worked the gun until the enemy's advancing line was routed.

DE ARMOND, WILLIAM.
At Upper Washita, Tex., Sept. 9-11, 1874.
R—NR.
B—Butler County, Ohio.
Sergeant, Company I, 5th U. S. Infantry.
Gallantry in action.

DEARY, GEORGE.
At Apache Creek, Ariz., Apr. 2, 1874.
R—NR.
B—Philadelphia, Pa.
Sergeant, Company L, 5th U. S. Cavalry.
Gallantry in action.

DE CASTRO, JOSEPH H.
At Gettysburg, Pa., July 3, 1863.
R—NR.
B—Boston, Mass.
Corporal, Company I, 19th Massachusetts Infantry.
Capture of flag of 19th Virginia (C. S. A.).

De Cesnola, Louis P_____
    At Aldie, Va., June 17, 1863.
    R—NR.
    B—NR.

Colonel, 4th New York Cavalry.
Was present, in arrest, when, seeing his regiment fall back, he rallied his men, accompanied them, without arms, in a second charge, and in recognition of his gallantry was released from arrest. He continued in the action at the head of his regiment until he was desperately wounded and taken prisoner.

Deetline, Frederick_____
    At Little Big Horn, Mont., June 25, 1876.
    R—Baltimore, Md.
    B—Germany.

Private, Company D, 7th U. S. Cavalry.
Voluntarily brought water to the wounded under fire.

De Lacey, Patrick_____
    At Wilderness, Va., May 6, 1864.
    R—NR.
    B—Luzerne County, Pa.

First sergeant, Company A, 143d Pennsylvania Infantry.
Running ahead of the line, under a concentrated fire, he shot the color bearer of a Confederate regiment on the works, thus contributing to the success of the attack.

Deland, Frederick N_____
    At Port Hudson, La., May 27, 1863.
    R—NR.
    B—NR.

Private, Company B, 49th Massachusetts Infantry.
Volunteered in response to a call and, under a heavy fire from the enemy, advanced and assisted in filling with fascines a ditch which presented a serious obstacle to the troops attempting to take the works of the enemy by assault.

Delaney, John C_____
    At Dabneys Mills, Va., Feb. 6, 1865.
    R—NR.
    B—Honesdale, Pa.

Sergeant, Company I, 107th Pennsylvania Infantry.
Sprang between the lines and brought out a wounded comrade about to be burned in the brush.

De Lavie, Hiram H_____
    At Five Forks, Va., Apr. 1, 1865.
    R—Allegheny County, Pa.
    B—Stark County, Ohio.

Sergeant, Company 1, 11th Pennsylvania Infantry.
Capture of flag.

Denny, John_____
    At Las Animas Canyon, N. Mex., Sept. 18, 1879.
    R—Elmira, N. Y.
    B—Big Flats, N. Y.

Sergeant, Troop B, 9th U. S. Cavalry.
Removed a wounded comrade, under a heavy fire, to a place of safety.

De Puy, Charles H_____
    At Petersburg, Va., July 30, 1864.
    R—St. Louis, Mo.
    B—Sherman, Mich.

First sergeant, Company H, 1st Michigan Sharpshooters.
Being an old artillerist, aided General Bartlett in working the guns of the dismantled fort.

De Swan, John F_____
    At Santiago, Cuba, July 1, 1898.
    R—Philadelphia, Pa.
    B—Philadelphia, Pa.

Private, Company H, 21st U. S. Infantry.
Gallantly assisted in the rescue of the wounded from in front of the lines and under heavy fire from the enemy.

De Witt, Richard W_____
    At Vicksburg, Miss., May 22, 1863.
    R—Oxford, Ohio.
    B—Butler County, Ohio.

Corporal, Company D, 47th Ohio Infantry.
Gallantry in the charge of the "volunteer storming party."

Dickens, Charles H_____
    At Chiricahua Mountains, Ariz., Oct. 20, 1869.
    R—NR.
    B—Ireland.

Corporal, Company G, 8th U. S. Cavalry.
Gallantry in action.

Dickey, William D_____
    At Petersburg, Va., June 17, 1864.
    R—Newburgh, N. Y.
    B—NR.

Captain, Battery M, 15th New York Heavy Artillery.
Refused to leave the field, remaining in command after being wounded by a piece of shell, and led his command in the assault on the enemy's works on the following day.

Dickie, David_____
    At Vicksburg, Miss., May 22, 1863.
    R—NR.
    B—Scotland.

Sergeant, Company A, 97th Illinois Infantry.
Was a member of a volunteer storming party which made a most gallant assault upon the enemy's works.

*Dilboy, George (68595)_____
    Near Belleau, France, July 18, 1918.
    R—Keene, N. H.
    B—Greece.
    G. O. No. 13, W. D., 1919.

Private, first class, Company H, 103d Infantry, 26th Division.
After his platoon had gained its objective along a railroad embankment, Private Dilboy, accompanying his platoon leader to reconnoiter the ground beyond, was suddenly fired upon by an enemy machine gun from 100 yards. From a standing position on the railroad track, fully exposed to view, he opened fire at once, but, failing to silence the gun, rushed forward with his bayonet, fixed, through a wheat field toward the gun emplacement, falling within 25 yards of the gun with his right leg nearly severed above the knee and with several bullet holes in his body. With undaunted courage he continued to fire into the emplacement from a prone position, killing two of the enemy and dispersing the rest of the crew.
Posthumously awarded. Medal presented to father, Antone Dilboy.

Dilger, Hubert_____
    At Chancellorsville, Va., May 2, 1863.
    R—NR.
    B—NR.

Captain, Battery I, 1st Ohio Light Artillery.
Fought his guns until the enemy were upon him, then with one gun hauled in the road by hand he formed the rear guard and kept the enemy at bay by the rapidity of his fire and was the last man in the retreat.

DILLON, MICHAEL A.
At Williamsburg, Va., May 5, 1862.
At Oak Grove, Va., June 25, 1862.
R—NR.
B—Chelmsford, Mass.

Private, Company G, 2d New Hampshire Infantry.
Bravery in repulsing the enemy's charge on a battery.
Crawled outside the lines and brought in important information.

DOCKUM, WARREN C.
At Sailors Creek, Va., Apr. 6, 1865.
R—NR.
B—Clintonville, N. Y.

Private, Company H, 121st New York Infantry.
Capture of flag of Savannah Guards (C. S. A.), after two other men had been killed in the effort.

DODD, ROBERT F.
At Petersburg, Va., July 30, 1864.
R—NR.
B—Canada.

Private, Company E, 27th Michigan Infantry.
While acting as orderly, voluntarily assisted to carry off the wounded from the ground in front of the Crater while exposed to a heavy fire.

DODDS, EDWARD E.
At Ashbys Gap, Va., July 19, 1864.
R—Rochester, N. Y.
B—Canada.

Sergeant, Company C, 21st New York Cavalry.
At great personal risk rescued his wounded captain and carried him from the field to a place of safety.

DODGE, FRANCIS S.
Near White River Agency, Colo., Sept. 29, 1879.
R—Danvers, Mass.
B—Massachusetts.

Captain, Troop D, 9th U. S. Cavalry.
With a force of 40 men rode all night to the relief of a command that had been defeated and was besieged by an overwhelming force of Indians, reached the field at daylight, joined in the action and fought for three days.

DOHERTY, THOMAS M.
At Santiago, Cuba, July 1, 1898.
R—Newcastle, Me.
B—Ireland.

Corporal, Company H, 21st U. S. Infantry.
Gallantly assisted in the rescue of the wounded from in front of the lines and while under heavy fire from the enemy.

DOLLOFF, CHARLES W.
At Petersburg, Va., Apr. 2, 1865.
R—NR.
B—Parishville, N. Y.

Corporal, Company K, 1st Vermont Infantry.
Capture of flag.

DONAHUE, JOHN L.
At Chiricahua Mountains, Ariz., Oct. 20, 1869.
R—NR.
B—Baltimore County, Md.

Private, Company G, 8th U. S. Cavalry.
Gallantry in action.

DONALDSON, JOHN.
At Appomattox Courthouse, Va., Apr. 9, 1865.
R—NR.
B—Butler County, Pa.

Sergeant, Company L, 4th Pennsylvania Cavalry.
Capture of flag of 14th Virginia Cavalry (C. S. A.).

DONALDSON, MICHAEL A (89868).
At Sommerance - Landres - et - St. Georges Road, France, Oct. 14, 1918.
R—Haverstraw, N. Y.
B—Haverstraw, N. Y.
G. O. No. 9, W. D., 1923.

Sergeant, Company I, 165th Infantry, 42d Division.
The advance of his regiment having been checked by intense machine-gun fire of the enemy, who were entrenched on the crest of a hill before Landres-et, St. Georges, his company retired to a sunken road to reorganize their position, leaving several of their number wounded near the enemy lines. Of his own volition, in broad daylight and under direct observation of the enemy and with utter disregard for his own safety, he advanced to the crest of the hill, rescued one of his wounded comrades, and returned under withering fire to his own lines, repeating his splendidly heroic act until he had brought in all the men, six in number.

DONAVAN, CORNELIUS.
At Agua Fria River, Ariz., Aug. 25, 1869.
R—NR.
B—Ireland.

Sergeant, Company E, 8th U. S. Cavalry.
Gallantry in action.

DONELLY, JOHN S.
At Cedar Creek, etc., Mont., October, 1876, to Jan. 8, 1877.
R—Buffalo, N. Y.
B—Ireland.

Private, Company G, 5th U. S. Infantry.
Gallantry in actions.

DONOGHUE, TIMOTHY.
At Fredericksburg, Va., Dec. 13, 1862.
R—NR.
B—Ireland.

Private, Company B, 69th New York Infantry.
Voluntarily carried a wounded officer off the field from between the lines; while doing this he was himself wounded.

DONOVAN, WILLIAM JOSEPH.
Near Landres-et-St. Georges, France, Oct. 14-15, 1918.
R—Buffalo, N. Y.
B—Buffalo, N. Y.
G. O. No. 56, W. D., 1922.
Distinguished-service medal and distinguished-service cross also awarded.

Lieutenant colonel, 165th Infantry, 42d Division.
Colonel Donovan personally led the assaulting wave in an attack upon a very strongly organized position, and when our troops were suffering heavy casualties he encouraged all near him by his example, moving among his men in exposed positions, reorganizing decimated platoons, and accompanying them forward in attacks. When he was wounded in the leg by a machine-gun bullet, he refused to be evacuated and continued with his unit until it withdrew to a less exposed position.

DOODY, PATRICK_____ Corporal, Company E, 164th New York Infantry.
  At Cold Harbor, Va., June 7, 1864.   After making a successful personal reconnaissance, he gallantly led the skir-
  R—NR.   mishers in a night attack, charging the enemy, and thus enabling the pioneers
  B—Ireland.   to put up works.

DORE, GEORGE H_____ Sergeant, Company D, 126th New York Infantry.
  At Gettysburg, Pa., July 3, 1863.   The colors being struck down by a shell as the enemy were charging, this sol-
  R—NR.   dier rushed out and seized it, exposing himself to the fire of both sides.
  B—NR.

DORLEY, AUGUST_____ Private, Company B, 1st Louisiana Cavalry.
  At Mount Pleasant, Ala., Apr. 11,   Capture of flag.
  1865.
  R—NR.
  B—Germany.

DORSEY, DANIEL_____ Corporal, Company H, 33d Ohio Infantry.
  Georgia, April, 1862.   One of 22 men (including two civilians) who by direction of General Mitchel
  R—NR.   (or Buell) penetrated nearly 200 miles south into the enemy's territory and
  B—Fairfield County, Ohio.   captured a railroad train at Big Shanty, Ga., in an attempt to destroy the
                                           bridges and track between Chattanooga and Atlanta.

DORSEY, DECATUR_____ Sergeant, Company B, 39th U. S. Colored Troops.
  At Petersburg, Va., July 30, 1864.   Planted his colors on the Confederate works in advance of his regiment, and
  R—Baltimore County, Md.   when the regiment was driven back to the Union works he carried the colors
  B—Howard County, Md.   there and bravely rallied the men.

DOUGALL, ALLAN H_____ First lieutenant and adjutant, 88th Indiana Infantry.
  At Bentonville, N. C., Mar. 19, 1865. In the face of a galling fire from the enemy he voluntarily returned to where
  R—NR.   the color bearer had fallen wounded and saved the flag of his regiment from
  B—Scotland.   capture.

DOUGHERTY, MICHAEL_____ Private, Company B, 13th Pennsylvania Cavalry.
  At Jefferson, Va., Oct. 12, 1863.   At the head of a detachment of his company dashed across an open field, exposed
  R—NR.   to a deadly fire from the enemy, and succeeded in dislodging them from an
  B—NR.   unoccupied house, which he and his comrades defended for several hours
                                           against repeated attacks, thus preventing the enemy from flanking the posi-
                                           tion of the Union forces.

DOUGHERTY, WILLIAM_____ Blacksmith, Company B, 8th U. S. Cavalry.
  Arizona, August to October, 1868.   Bravery in scouts and actions with Indians.
  R—NR.
  B—Detroit, Mich.

DOW, GEORGE P_____ Sergeant, Company C, 7th New Hampshire Infantry.
  Near Richmond, Va., Oct. —, 1864.   Gallantry while in command of his company during a reconnaissance toward
  R—NR.   Richmond.
  B—Atkinson, N. H.

DOWLING, JAMES_____ Corporal, Company B, 8th U. S. Calvary.
  Arizona, August to October, 1868.   Bravery in scouts and actions with Indians.
  R—Cleveland, Ohio.
  B—Ireland.

DOWNEY, WILLIAM_____ Private, Company B, 4th Massachusetts Cavalry.
  At Ashepoo River, S. C., May 24,   Volunteered as a member of a boat crew which went to the rescue of a large
  1864.   number of Union soldiers on board the stranded steamer Boston, and with
  R—NR.   great gallantry assisted in conveying them to shore, being exposed during
  B—Ireland.   the entire time to a heavy fire from a Confederate battery.

DOWNS, HENRY W_____ Sergeant, Company I, 8th Vermont Infantry.
  At Winchester, Va., Sept. 19, 1864.   With one comrade voluntarily crossed an open field, exposed to a raking fire,
  R—NR.   and returned with a supply of ammunition, successfully repeating the at-
  B—Jamaica, Vt.   tempt a short time thereafter.

DOWNS, WILLIS H_____ Private, Company H, 1st North Dakota Volunteer Infantry.
  At San Miguel de Mayumo, Luzon,   With 11 other scouts, without waiting for the supporting battalion to aid them
  P. I., May 13, 1899.   or to get into a position to do so, charged over a distance of about 150 yards
  R—Jamestown, N. Dak.   and completely routed about 300 of the enemy who were in line and in a posi-
  B—Mount Carmel, Conn.   tion that could only be carried by a frontal attack.

DOZIER, JAMES C_____ First lieutenant, Company G, 118th Infantry, 30th Division.
  Near Montbrehain, France, Oct. 8,   In command of two platoons, Lieutenant Dozier was painfully wounded in
  1918.   the shoulder early in the attack, but he continued to lead his men, displaying
  R—Rock Hill, S. C.   the highest bravery and skill. When his command was held up by heavy
  B—Marion, S. C.   machine-gun fire, he disposed his men in the best cover available and with
  G. O. No. 16, W. D., 1919.   a soldier continued forward to attack a machine-gun nest. Creeping up to
                                           the position in the face of intense fire, he killed the entire crew with hand
                                           grenades and his pistol and a little later captured a number of Germans, who
                                           had taken refuge in a dugout near by.

DRAKE, JAMES M_____ Second lieutenant, Company D, 9th New Jersey Infantry.
  At Bermuda Hundred, Va., May 6,   Commanded the skirmish line in the advance and held his position all day
  1864.   and during the night.
  R—Elizabeth, N. J.
  B—Union County, N. J.

DRURY, JAMES_____
At Weldon Railroad, Va., June 23, 1864.
R—Chester, Vt.
B—Ireland.

Sergeant, Company C, 4th Vermont Infantry.
Saved the colors of his regiment when it was surrounded by a much larger force of the enemy and after the greater part of the regiment had been killed or captured.

DUFFEY, JOHN_____
At Ashepoo River, S. C., May 24, 1864.
R—NR.
B—New Bedford, Mass.

Private, Company B, 4th Massachusetts Cavalry.
Volunteered as a member of a boat crew which went to the rescue of a large number of Union soldiers on board the stranded steamer Boston, and with great gallantry assisted in conveying them to shore, being exposed during the entire time to a heavy fire from a Confederate battery.

DUNLAVY, JAMES_____
At Osage, Kans., Oct. 25, 1864.
R—Davis County, Iowa.
B—Decatur County, Ind.

Private, Company D, 3d Iowa Cavalry.
Gallantry in capturing General Marmaduke.

*DUNN, PARKER F. (2943321)_____
Near Grand-Pre, France, Oct. 23, 1918.
R—Albany, N. Y.
B—Albany, N. Y.
G. O. No. 49, W. D., 1922.

Private, first class, Company A, 312th Infantry, 78th Division.
When his battalion commander found it necessary to send a message to a company in the attacking line and hesitated to order a runner to make the trip because of the extreme danger involved, Private Dunn, a member of the intelligence section, volunteered for the mission. After advancing but a short distance across a field swept by artillery and machine-gun fire, he was wounded, but continued on and fell wounded a second time. Still undaunted, he persistently attempted to carry out his mission until he was killed by a machine-gun bullet before reaching the advance line.
Posthumously awarded. Medal presented to father, James C. Dunn.

DUNNE, JAMES_____
At Vicksburg, Miss., May 22, 1863.
R—NR.
B—Detroit, Mich.

Corporal, Chicago Mercantile Battery, Illinois Light Artillery.
Carried with others by hand a cannon up to and fired it through an embrasure of the enemy's works.

DU PONT, HENRY A_____
At Cedar Creek, Va., Oct. 19, 1864.
R—Wilmington, Del.
B—Delaware.

Captain, 5th U. S. Artillery.
By his distinguished gallantry, and voluntary exposure to the enemy's fire at a critical moment, when the Union line had been broken, encouraged his men to stand to their guns, checked the advance of the enemy, and brought off most of his pieces.

DURHAM, JAMES R_____
At Winchester, Va., June 14, 1863.
R—Clarksburg, W. Va.
B—NR.

Second lieutenant, Company E, 12th West Virginia Infantry.
Led his command over the stone wall, where he was wounded.

DURHAM, JOHN S_____
At Perryville, Ky., Oct. 8, 1862.
R—NR.
B—NR.

Sergeant, Company F, 1st Wisconsin Infantry.
Seized the flag of his regiment when the color sergeant was shot and advanced with the flag midway between the lines, amid a shower of shot, shell, and bullets, until stopped by his commanding officer.

ECKES, JOHN N_____
At Vicksburg, Miss., May 22, 1863.
R—Weston, Va.
B—Lewis County, Va.

Private, Company E, 47th Ohio Infantry.
Gallantry in the charge of the "volunteer storming party."

EDDY, SAMUEL E_____
At Sailors Creek, Va., Apr. 6, 1865.
R—Chesterfield, Mass.
B—Vermont.

Private, Company D, 37th Massachusetts Infantry.
Saved the life of the adjutant of his regiment by voluntarily going beyond the line and there killing one of the enemy then in the act of firing upon the wounded officer. Was assailed by several of the enemy, run through the body with a bayonet, and pinned to the ground, but while so situated he shot and killed his assailant.

EDGERTON, NATHAN H_____
At Chapins Farm, Va., Sept. 29, 1864.
R—Philadelphia, Pa.
B—NR.

Lieutenant and adjutant, 6th U. S. Colored Troops.
Took up the flag after three color bearers had been shot down and bore it forward, though himself wounded.

EDWARDS, DANIEL R. (106546)_____
Near Soissons, France, July 18, 1918.
R—Bruceville, Tex.
B—Moorville, Tex.
G. O. No. 14, W. D., 1923.
Distinguished-service cross also awarded.

Private, first class, Company C, 3d Machine Gun Battalion, 1st Division.
Reporting for duty from hospital where he had been for several weeks under treatment for numerous and serious wounds and although suffering intense pain from a shattered arm, he crawled alone into an enemy trench for the purpose of capturing or killing enemy soldiers known to be concealed therein. He killed four of the men and took the remaining four men prisoners; while conducting them to the rear one of the enemy was killed by a high explosive enemy shell which also completely shattered one of Private Edwards's legs, causing him to be immediately evacuated to the hospital. The bravery of Private Edwards, now a tradition in his battalion because of his previous gallant acts, again caused the morale of his comrades to be raised to a high pitch.

EDWARDS, DAVID_____
At Five Forks, Va., Apr. 1, 1865.
R—NR.
B—Wales.

Private, Company H, 146th New York Infantry.
Capture of flag.

EDWARDS, WILLIAM D_____
At Big Hole, Mont., Aug. 9, 1877.
R—NR.
B—Brooklyn, N. Y.

First sergeant, Company F, 7th U. S. Infantry.
Bravery in action.

EGGERS, ALAN LOUIS (1212557)_____
 Near Le Catelet, France, Sept. 29,
  1918.
 R—Summit, N. J.
 B—Saranac Lake, N. Y.
 G. O. No. 20, W. D., 1919.

Sergeant, Machine Gun Company, 107th Infantry, 27th Division.
Becoming separated from their platoon by a smoke barrage, Sergeant Eggers, Sergt. John C. Latham, and Corpl. Thomas E. O'Shea took cover in a shell hole well within the enemy's lines. Upon hearing a call for help from an American tank, which had become disabled 30 yards from them, the three soldiers left their shelter and started toward the tank, under heavy fire from German machine guns and trench mortars. In crossing the fire-swept area Corporal O'Shea was mortally wounded, but his companions, undeterred, proceeded to the tank, rescued a wounded officer, and assisted two wounded soldiers to cover in a sap of a near-by trench. Sergeant Eggers and Sergeant Latham then returned to the tank in the face of the violent fire, dismounted a Hotchkiss gun, and took it back to where the wounded men were, keeping off the enemy all day by effective use of the gun and later bringing it, with the wounded men, back to our lines under cover of darkness.

ELDRIDGE, GEORGE H_____
 At Wichita River, Tex., July 12,
  1870.
 R—NR.
 B—Sacketts Harbor, N. Y

Sergeant, Company C, 6th U. S. Cavalry.
Gallantry in action.

ELLIOTT, ALEXANDER_____
 At Paines Crossroads, Va., Apr. 5,
  1865.
 R—North Sewickley, Pa.
 B—Beaver County, Pa.

Sergeant, Company A, 1st Pennsylvania Cavalry.
Capture of flag.

ELLIOTT, RUSSELL C_____
 At Natchitoches, La., Apr. 19, 1864.
 R—NR.
 B—Concord, N. H.

Sergeant, Company B, 3d Massachusetts Cavalry.
Seeing a Confederate officer in advance of his command, charged on him alone and unaided and captured him.

ELLIS, HORACE_____
 At Weldon Railroad, Virginia, Aug.
  21, 1864.
 R—Chippewa Falls, Wis.
 B—Mercer County, Pa.

Private, Company A, 7th Wisconsin Infantry.
Capture of flag of 16th Mississippi (C. S. A.).

ELLIS, MICHAEL B. (56976)_____
 Near Exermont, France, Oct. 5, 1918.
 R—East St. Louis, Ill.
 B—St. Louis, Mo.
 G. O. No. 74, W. D., 1919.

Sergeant, Company C, 28th Infantry, 1st Division.
During the entire day's engagement he operated far in advance of the first wave of his company, voluntarily undertaking most dangerous missions and single handed attacking and reducing machine-gun nests. Flanking one emplacement, he killed 2 of the enemy with rifle fire and captured 17 others. Later he single handed advanced under heavy fire and captured 27 prisoners, including 2 officers and six machine guns, which had been holding up the advance of the company. The captured officers indicated the locations of four other machine guns, and he in turn captured these, together with their crews, at all times showing marked heroism and fearlessness.

ELLIS, WILLIAM_____
 At Dardanelles, Ark., Jan. 14, 1865.
 R—Little Rock, Ark.
 B—England.

First sergeant, Company K, 3d Wisconsin Cavalry.
Remained at his post after receiving three wounds, and only retired, by his commanding officer's orders, after being wounded the fourth time.

ELLSWORTH, THOMAS F_____
 At Honey Hill, S. C., Nov. 30, 1864.
 R—NR.
 B—Ipswich, Mass.

Captain, Company B, 55th Massachusetts Infantry.
Under a heavy fire carried his wounded commanding officer from the field.

ELSATSOOSU_____
 Winter of 1872–73.
 R—NR.
 B—NR.

Corporal, Indian Scouts.
Gallantry in campaigns and actions with Indians.

ELSON, JAMES M_____
 At Vicksburg, Miss., May 22, 1863.
 R—Shellsburg, Iowa.
 B—Coshocton, Ohio.

Sergeant, Company C, 9th Iowa Infantry.
Carried the colors in advance of his regiment and was shot down while attempting to plant them on the enemy's works.

ELWOOD, EDWIN L_____
 At Chiricahua Mountains, Ariz.,
  Oct. 20, 1869.
 R—NR.
 B—St. Louis, Mo.

Private, Company G, 8th U. S. Cavalry.
Gallantry in action.

EMBLER, ANDREW H_____
 At Boydton Plank Road, Va., Oct.
  27, 1864.
 R—NR.
 B—New York, N. Y

Captain, Company D, 59th New York Infantry.
Charged at the head of two regiments, which drove the enemy's main body, gained the crest of the hill near the Burgess house and forced a barricade on the Boydton road.

EMMET, ROBERT TEMPLE_____
 At Las Animas Canon, N. Mex.,
  Sept. 18, 1879.
 R—New York, N. Y.
 B—New York, N. Y.

Second lieutenant, 9th U. S. Cavalry.
Distinguished gallantry in action against hostile Indians.

ENDERLIN, RICHARD............
At Gettysburg, Pa., July 1-3, 1863.
R—Chillicothe, Ohio.
B—Germany.

Musician, Company B, 73d Ohio Infantry.
Voluntarily took a rifle and served as a soldier in the ranks during the first and second days of the battle. Voluntarily and at his own imminent peril went into the enemy's lines at night and, under a sharp fire, rescued a wounded comrade.

ENGLE, JAMES E............
At Bermuda Hundred, Va., May 18, 1864.
R—NR.
B—Chester, Pa.

Sergeant, Company I, 97th Pennsylvania Infantry.
Responded to a call for volunteers to carry ammunition to the regiment on the picket line and under a heavy fire from the enemy assisted in carrying a box of ammunition to the front and remained to distribute the same.

ENGLISH, EDMUND............
At Wilderness, Va., May 6, 1864.
R—NR.
B—NR.

First sergeant, Company C, 2d New Jersey Infantry.
During a rout and while under orders to retreat, seized the colors, rallied the men, and drove the enemy back.

ENNIS, CHARLES D............
At Petersburg, Va., Apr. 2, 1865.
R—NR.
B—Stonington, Conn.

Private, Company G, 1st Rhode Island Light Artillery.
Was one of a detachment of 20 picked artillerymen who voluntarily accompanied an Infantry assaulting party and who turned upon the enemy the guns captured in the assault.

EPPS, JOSEPH L............
At Vigan, Luzon, P. I., Dec. 4, 1899.
R—Oolagah, Indian Territory.
B—Jamestown, Mo.

Private, Company B, 33d Infantry, U. S. Volunteers.
Discovered a party of insurgents inside a wall, climbed to the top of the wall, covered them with his gun, and forced them to stack arms and surrender.

ESTES, LEWELLYN G............
At Flint River, Ga., Aug. 30, 1864.
R—NR.
B—Oldtown, Me.

Captain and assistant adjutant general, Volunteers.
Voluntarily led troops in a charge over a burning bridge.

EVANS, CORON D............
At Sailors Creek, Va., Apr. 6, 1865.
R—Jefferson County, Ind.
B—Jefferson County, Ind.

Private, Company A, 3d Indiana Cavalry.
Capture of flag of 26th Virginia Infantry (C. S. A.).

EVANS, IRA H............
At Hatchers Run, Va., Apr. 2, 1865.
R—NR.
B—NR.

Captain, Company B, 116th U. S. Colored Troops.
Voluntarily passed out between the lines, under a heavy fire from the enemy, and obtained important information.

EVANS, JAMES R............
At Wilderness, Va., May 5, 1864.
R—New York, N. Y.
B—New York, N. Y.

Private, Company H, 62d New York Infantry.
Went out in front of the line under a fierce fire and, in the face of the rapidly advancing enemy, rescued the regimental flag with which the color bearer had fallen.

EVANS, THOMAS............
At Piedmont, Va., June 5, 1864.
R—NR.
B—Wales.

Private, Company D, 54th Pennsylvania Infantry.
Capture of flag of 45th Virginia (C. S. A.).

EVANS, WILLIAM............
At Big Horn, Mont., July 9, 1876.
R—St. Louis, Mo.
B—Ireland.

Private, Company E, 7th U. S. Infantry.
Carried dispatches to Brigadier General Crook through a country occupied by Sioux.

EVERSON, ADELBERT............
At Five Forks, Va., Apr. 1, 1865.
R—NR.
B—Cicero, N. Y.

Private, Company D, 185th New York Infantry.
Capture of flag.

EWING, JOHN C............
At Petersburg, Va., Apr. 2, 1865.
R—NR.
B—Westmoreland County, Pa.

Private, Company E, 211th Pennsylvania Infantry.
Capture of flag.

FACTOR, POMPEY............
At Pecos River, Tex., Apr. 25, 1875.
R—NR.
B—Arkansas.

Private, Indian Scouts.
Gallantry in action.

FALCONER, JOHN A............
At Fort Sanders, Knoxville, Tenn., Nov. 20, 1863.
R—NR.
B—Wachtenaw, Mich.

Corporal, Company A, 17th Michigan Infantry.
Conducted the "burning party" of his regiment at the time a charge was made on the enemy's picket line, and burned the house which had sheltered the enemy's sharpshooters, thus insuring success to a hazardous enterprise.

FALCOTT, HENRY............
Arizona, August to October, 1868.
R—NR.
B—France.

Sergeant, Company L, 8th U. S. Cavalry.
Bravery in scouts and actions with Indians.

FALL, CHARLES S............
At Spotsylvania Courthouse, Va., May 12, 1864.
R—NR.
B—Noble County, Ind.

Sergeant, Company E, 26th Michigan Infantry.
Was one of the first to mount the Confederate works, where he bayoneted two of the enemy and captured a Confederate flag, but threw it away to continue the pursuit of the enemy.

FALLON, THOMAS T.   Private, Company K, 37th New York Infantry.
  At Williamsburg, Va., May 5, 1862.   Assisted in driving rebel skirmishers to their main line.
  At Fair Oaks, Va., May 30–31, 1862. Participated in action, though excused from duty because of disability.
  At Big Shanty, Ga., June 14–15, 1864. In a charge with his company, was the first man on the enemy's works.
  R—Freehold, N. J.
  B—Ireland.

FALLS, BENJAMIN F.   Color sergeant, Company A, 19th Massachusetts Infantry.
  At Gettysburg, Pa., July 3, 1863.   Capture of flag.
  R—NR.
  B—Portsmouth, N. H.

FANNING, NICHOLAS.   Private, Company B, 4th Iowa Cavalry.
  At Selma, Ala., Apr. 2, 1865.   Capture of silk Confederate States flag and two staff officers.
  R—NR.
  B—Carroll County, Ind.

FARNSWORTH, HERBERT E.   Sergeant major, 10th New York Cavalry.
  At Trevilian Station, Va., June 11,   Voluntarily carried a message which stopped the firing of a Union battery into
    1864.   his regiment, in which service he crossed a ridge in plain view and swept by
  R—NR.   the fire of both armies.
  B—Cattaraugus County, N. Y.

FARQUHAR, JOHN M.   Sergeant major, 89th Illinois Infantry.
  At Stone River, Tenn., Dec. 31, When a break occurred on the extreme right wing of the Army of the Cumber-
    1862.   land, this soldier rallied fugitives from other commands, and deployed his
  R—NR.   own regiment, thereby checking the Confederate advance until a new line was
  B—Scotland.   established.

FARREN, DANIEL.   Private, Company B, 8th United States Cavalry.
  Arizona, August to October, 1868.   Bravery in scouts and actions with Indians.
  R—NR.
  B—Ireland.

FASNACHT, CHARLES H.   Sergeant, Company A, 99th Pennsylvania Infantry.
  At Spotsylvania, Va., May 12, 1864. Capture of flag of 2d Louisiana Tigers (C. S. A.) in a hand-to-hand contest.
  R—NR.
  B—Lancaster County, Pa.

FASSETT, JOHN B.   Captain, Company F, 23d Pennsylvania Infantry.
  At Gettysburg, Pa., July 2, 1863.   While acting as an aide, voluntarily led a regiment to the relief of a battery and
  R—NR.   recaptured its guns from the enemy.
  B—NR.

FEASTER, MOSHEIM.   Private, Company E, 7th U. S. Cavalry.
  At Wounded Knee Creek, S. Dak., Extraordinary gallantry.
    Dec. 29, 1890.
  R—Schellburg, Pa.
  B—Schellburg, Pa.

FEGAN, JAMES.   Sergeant, Company H, 3d U. S. Infantry.
  At Plum Creek, Kans., Mar. —,   While in charge of a powder train en route from Fort Harker to Fort Dodge,
    1868.   Kans., was attacked by a party of desperadoes, who attempted to rescue a
  R—NR.   deserter in his charge and to fire the train. Sergeant Fegan, single handed,
  B—Ireland.   repelled the attacking party, wounding two of them, and brought his train
          through in safety.

FERGUSON, ARTHUR M.   First lieutenant, 36th Infantry, U. S. Volunteers.
  Near Porac, Luzon, P. I., Sept. 28, Charged alone a body of the enemy and captured a captain.
    1899.
  R—Burlington, Kans.
  B—Coffey County, Kans.
  Distinguished-service cross also
    awarded.

FERNALD, ALBERT E.   First lieutenant, Company F, 20th Maine Infantry.
  At Five Forks, Va., Apr. 1, 1865.   Capture of flag.
  R—Winterport, Me.
  B—Winterport, Me.

FERRARI, GEORGE.   Corporal, Company D, 8th U. S. Cavalry.
  At Red Creek, Ariz., Sept. 23, 1869. Gallantry in action.
  R—Montgomery County, Ohio.
  B—New York, N. Y.

FERRIER, DANIEL T.   Sergeant, Company K, 2d Indiana Cavalry.
  At Varnells Station, Ga., May 9, While his regiment was retreating, voluntarily gave up his horse to his brigade
    1864.   commander who had been unhorsed and was in danger of capture, thereby
  R—NR.   enabling him to rejoin and rally the disorganized troops. Sergeant Ferrier
  B—NR.   himself was captured and confined in Confederate prisons, from which he
          escaped, and, after great hardship, rejoined the Union lines.

FERRIS, EUGENE W.   First lieutenant and adjutant, 30th Massachusetts Infantry.
  At Berryville, Va., Apr. 1, 1865.   Accompanied only by an orderly, outside the lines of the Army, he gallantly
  R—Lowell, Mass.   resisted an attack of five of Mosby's cavalry, mortally wounded the leader of
  B—NR.   the party, seized his horse and pistols, wounded three more, and though
          wounded himself escaped.

FESQ, FRANK_____
  At Petersburg, Va., Apr. 2, 1865.
  R—Newark, N. J.
  B—NR.
> Private, Company A, 40th New Jersey Infantry.
> Capture of flag of 18th North Carolina (C. S. A.) within the enemy's works.

FICHTER, HERMANN_____
  At Whetstone Mountains, Ariz.,
   May 5, 1871.
  R—NR.
  B—Germany.
> Private, Company F, 3d U. S. Cavalry.
> Gallantry in action.

FINKENBINER, HENRY S._____
  At Dingles Mill, S. C., Apr. 9, 1865.
  R—NR.
  B—North Industry, Ohio.
> Private, Company D, 107th Ohio Infantry.
> While on the advance skirmish line and within direct and close fire of the enemy's artillery, crossed the mill race on a burning bridge and ascertained the enemy's position.

FISHER, JOHN H._____
  At Vicksburg, Miss., May 22, 1863.
  R—Chicago, Ill.
  B—Monmouth, Pa.
> First lieutenant, Company B, 55th Illinois Infantry.
> Gallantry in the charge of the "volunteer storming party."

FISHER, JOSEPH._____
  At Petersburg, Va., Apr. 2, 1865.
  R—Philadelphia, Pa.
  B—Philadelphia, Pa.
> Corporal, Company C, 61st Pennsylvania Infantry.
> Carried the colors 50 yards in advance of his regiment, and after being painfully wounded attempted to crawl into the enemy's works in an endeavor to plant his flag thereon.

FLANAGAN, AUGUSTIN_____
  At Chapins Farm, near Richmond,
   Va., Sept. 29, 1864.
  R—Chest Springs, Pa.
  B—Cambria County, Pa.
> Sergeant, Company A, 55th Pennsylvania Infantry.
> Gallantry in the charge on the enemy's works; rushing forward with the colors and calling upon the men to follow him; was severely wounded.

FLANNIGAN, JAMES_____
  At Nolensville, Tenn., Feb. 15, 1863.
  R—NR.
  B—New York.
> Private, Company H, 2d Minnesota Infantry.
> Was one of a detachment of 16 men who heroically defended a wagon train against the attack of 125 cavalry, repulsed the attack, and saved the trains.

FLEETWOOD, CHRISTIAN A._____
  At Chapins Farm, near Richmond,
   Va., Sept. 29, 1864.
  R—NR.
  B—Baltimore, Md.
> Sergeant major, 4th U. S. Colored Troops.
> Seized the colors, after two color bearers had been shot down, and bore them nobly through the fight.

FLYNN, CHRISTOPHER_____
  At Gettysburg, Pa., July 3, 1863.
  R—NR.
  B—Ireland.
> Corporal, Company K, 14th Connecticut Infantry.
> Capture of flag of 52d North Carolina (C. S. A.).

FLYNN, JAMES E._____
  At Vicksburg, Miss., May 22, 1863.
  R—NR.
  B—Pittsfield, Ill.
> Sergeant, Company G, 6th Missouri Infantry.
> Gallantry in the charge of the "volunteer storming party."

FOLEY, JOHN H._____
  At Loupe Fork, Platte River, Nebr.,
   Apr. 26, 1872.
  R—NR.
  B—Ireland.
> Sergeant, Company B, 3d U. S. Cavalry.
> Gallantry in action.

FOLLETT, JOSEPH L._____
  At New Madrid, Mo., Mar. 3, 1862.
  At Stone River, Tenn., Dec. 31, 1862.
  R—NR.
  B—Newark, N. J.
> Sergeant, Company G, 1st Missouri Light Artillery.
> Remained on duty though severely wounded.
> While procuring ammunition from the supply train was captured, but made his escape, secured the ammunition, and in less than an hour from the time of his capture had the batteries supplied.

FOLLY, WILLIAM H._____
  Arizona, August to October, 1868.
  R—NR.
  B—Bergen County, N. J.
> Private, Company B, 8th U. S. Cavalry.
> Bravery in scouts and actions with Indians.

FORAN, NICHOLAS_____
  Arizona, August to October, 1868.
  R—NR.
  B—Ireland.
> Private, Company L, 8th U. S. Cavalry.
> Bravery in scouts and actions with Indians.

FORCE, MANNING F._____
  At Atlanta, Ga., July 22, 1864.
  R—Cincinnati, Ohio.
  B—NR.
> Brigadier general, U. S. Volunteers.
> Charged upon the enemy's works, and after their capture defended his position against assaults of the enemy until he was severely wounded.

FORD, GEORGE W._____
  At Sailors Creek, Va., Apr. 6, 1865.
  R—NR.
  B—Ireland.
> First lieutenant, Company E, 88th New York Infantry.
> Capture of flag.

FORMAN, ALEXANDER A._____
  At Fair Oaks, Va., May 31, 1862.
  R—Jonesville, Mich.
  B—Scipio, Mich.
> Corporal, Company E, 7th Michigan Infantry.
> Although wounded, he continued fighting until, fainting from loss of blood, he was carried off the field.

FORREST, ARTHUR J. (2178726).
Near Remonville, France, Nov. 1, 1918.
R—Hannibal, Mo.
B—St. Louis, Mo.
G. O. No. 50, W. D., 1919.

Sergeant, Company D, 354th Infantry, 89th Division.
When the advance of his company was stopped by bursts of fire from a nest of six enemy machine guns, without being discovered, he worked his way single handed to a point within 50 yards of the machine-gun nest. Charging, single handed, he drove out the enemy in disorder, thereby protecting the advance platoon from annihilating fire, and permitting the resumption of the advance of his company.

FORSYTH, THOMAS H.
At Powder River, Wyo., Nov. 25, 1876.
R—NR.
B—Hartford, Conn.

First sergeant, Company M, 4th U. S. Cavalry.
Though dangerously wounded, he maintained his ground with a small party against a largely superior force after his commanding officer had been shot down during a sudden attack and rescued that officer and a comrade from the enemy.

FOSTER, GARY EVANS (1311059).
Near Montbrehain, France, Oct. 8, 1918.
R—Inman, S. C.
B—Spartanburg, S. C.
G. O. No. 16, W. D., 1919.

Sergeant, Company F, 118th Infantry, 30th Division.
When his company was held up by violent machine-gun fire from a sunken road, Sergeant Foster with an officer went forward to attack the hostile machine-gun nests. The officer was wounded, but Sergeant Foster continued on alone in the face of the heavy fire and by effective use of hand grenades and his pistol killed several of the enemy and captured 18.

FOSTER, WILLIAM.
At Red River, Tex., Sept. 29, 1872.
R—NR.
B—England.

Sergeant, Company F, 4th U. S. Cavalry.
Gallantry in action.

FOURNIA, FRANK O.
At Santiago, Cuba, July 1, 1898.
R—Plattsburg, N. Y.
B—Rome, N. Y.

Private, Company H, 21st U. S. Infantry.
Gallantly assisted in the rescue of the wounded from in front of the lines and while under heavy fire of the enemy.

FOUT, FREDERICK W.
Near Harpers Ferry, W. Va., Sept. 15, 1862.
R—Indianapolis, Ind.
B—Germany.

Second lieutenant, 15th Battery Indiana Light Artillery.
Voluntarily gathered the men of the battery together, remanned the guns, which had been ordered abandoned by an officer, opened fire, and kept up the same on the enemy until after the surrender.

FOX, HENRY.
Near Jackson, Tenn., Dec. 23, 1862.
R—Lincoln, Ill.
B—Germany.

Sergeant, Company H, 106th Illinois Infantry.
When his command was surrounded by a greatly superior force, voluntarily left the shelter of the breastworks, crossed an open railway trestle under a concentrated fire from the enemy, made his way out and secured reinforcements for the relief of his command.

FOX, HENRY M.
At Winchester, Va., Sept. 19, 1864.
R—NR.
B—Trumbull, Ohio.

Sergeant, Company M, 5th Michigan Cavalry.
Capture of flag.

FOX, NICHOLAS.
At Port Hudson, La., June 14, 1863.
R—Greenwich, Conn.
B—NR.

Private, Company H, 28th Connecticut Infantry.
Made two trips across an open space in the face of the enemy's concentrated fire, and secured water for the sick and wounded.

FOX, WILLIAM R.
At Petersburg, Va., Apr. 2, 1865.
R—NR.
B—Philadelphia, Pa.

Private, Company A, 95th Pennsylvania Infantry.
Bravely assisted in the capture of one of the enemy's guns; with the first troops to enter the city, captured the flag of the Confederate customhouse.

FRANTZ, JOSEPH.
At Vicksburg, Miss., May 22, 1863.
R—Osgood, Ind.
B—France.

Private, Company E, 83d Indiana Infantry.
Gallantry in the charge of the "volunteer storming party."

FRASER, WILLIAM W.
At Vicksburg, Miss., May 22, 1863.
R—NR.
B—Scotland.

Private, Company I, 97th Illinois Infantry.
Gallantry in the charge of the "volunteer storming party."

FREEMAN, ARCHIBALD.
At Spotsylvania, Va., May 12, 1864.
R—NR.
B—Goshen, N. Y.

Private, Company E, 124th New York Infantry.
Capture of flag of 17th Louisiana (C. S. A.).

FREEMAN, HENRY B.
At Stone River, Tenn., Dec. 31, 1862.
R—Mount Vernon, Ohio.
B—Mount Vernon, Ohio.

First lieutenant, 18th U. S. Infantry.
Voluntarily went to the front and picked up and carried to a place of safety, under a heavy fire from the enemy, an acting field officer who had been wounded and was about to fall into the enemy's hands.

FREEMAN, WILLIAM H.
At Fort Fisher, N. C., Jan. 15, 1865.
R—NR.
B—Troy, N. Y.

Private, Company B, 169th New York Infantry.
Volunteered to carry the brigade flag after the bearer was wounded.

FREEMEYER, CHRISTOPHER.
At Cedar Creek, etc., Mont., Oct. 21, 1876, to Jan. 8, 1877.
R—Chicago, Ill.
B—Germany.

Private, Company D, 5th U. S. Infantry.
Gallantry in actions.

FRENCH, SAMUEL S.
  At Fair Oaks, Va., May 31, 1862.
  R—Gifford, Mich.
  B—Erie County, N. Y.
Private, Company E, 7th Michigan Infantry.
Continued fighting, although wounded, until he fainted from loss of blood.

FREY, FRANZ.
  At Vicksburg, Miss., May 22, 1863.
  R—Cleveland, Ohio.
  B—Switzerland.
Corporal, Company H, 37th Ohio Infantry.
Gallantry in the charge of the "volunteer storming party."

FRICK, JACOB G.
  At Fredericksburg, Va., Dec. 13,
    1862.
  At Chancellorsville, Va., May 3,
    1863.
  R—NR.
  B—NR.
Colonel, 129th Pennsylvania Infantry.
Seized the colors and led the command through a terrible fire of cannon and
  musketry.
In a hand-to-hand fight, recaptured the colors of his regiment.

FRIZZELL, HENRY F.
  At Vicksburg, Miss., May 22, 1863.
  R—NR.
  B—Madison County, Mo.
Private, Company B, 6th Missouri Infantry.
Gallantry in the charge of the "volunteer storming party."

FUGER, FREDERICK.
  At Gettysburg, Pa., July 3, 1863.
  R—NR.
  B—Germany.
Sergeant, Battery A, 4th U. S. Artillery.
All the officers of his battery having been killed or wounded and five of its
  guns disabled in Pickett's assault, he succeeded to the command and fought
  the remaining gun with most distinguished gallantry until the battery was
  ordered withdrawn.

FUNK, JESSE N. (2187583)
  Near Bois-de-Bantheville, France,
    Oct. 31, 1918.
  R—Calhan, Colo.
  B—New Hampton, Mo.
  G. O. No. 20, W.D., 1919.
Private, first class, Company L, 354th Infantry, 89th Division.
Learning that two daylight patrols had been caught out in No Man's Land and
  were unable to return, Private Funk and another stretcher bearer, upon their
  own initiative, made two trips 500 yards beyond our lines under constant
  machine-gun fire and rescued two wounded officers.

FUNK, WEST.
  At Appomattox Courthouse, Va.,
    Apr. 9, 1865.
  R—Philadelphia, Pa.
  B—Boston, Mass.
Major, 121st Pennsylvania Infantry.
Capture of flag of 46th Virginia Infantry (C. S. A.).

FUNSTON, FREDERICK.
  At Rio Grande de la Pampanga,
    Luzon, P. I., Apr. 27, 1899.
  R—Iola, Kans.
  B—Springfield, Ohio.
Colonel, 20th Kansas Volunteer Infantry.
Crossed the river on a raft and by his skill and daring enabled the general com-
  manding to carry the enemy's entrenched position on the north bank of the
  river and to drive him with great loss from the important strategic position
  of Calumpit.

FURLONG, HAROLD A.
  Near Bantheville, France, Nov. 1,
    1918.
  R—Detroit, Mich.
  B—Pontiac, Mich.
  G. O. No. 16, W. D., 1919.
First lieutenant, 353d Infantry, 89th Division.
Immediately after the opening of the attack in the Bois-de-Bantheville, when
  his company was held up by severe machine-gun fire from the front, which
  killed his company commander and several soldiers, Lieutenant Furlong
  moved out in advance of the line with great courage and coolness, crossing an
  open space several hundred yards wide. Taking up a position behind the
  line of machine guns, he closed in on them, one at a time, killing a number
  of the enemy with his rifle, putting four machine-gun nests out of action, and
  driving 20 German prisoners into our lines.

FURMAN, CHESTER S.
  At Gettysburg, Pa., July 2, 1863.
  R—NR.
  B—Columbia, Pa.
Corporal, Company A, 6th Pennsylvania Reserves.
Was one of six volunteers who charged upon a log house near the Devil's Den,
  where a squad of the enemy's sharpshooters were sheltered, and compelled
  their surrender.

FURNESS, FRANK.
  At Trevilian Station, Va., June 12,
    1864.
  R—Philadelphia, Pa.
  B—NR.
Captain, Company F, 6th Pennsylvania Cavalry.
Voluntarily carried a box of ammunition across an open space swept by the
  enemy's fire to the relief of an outpost whose ammunition had become almost
  exhausted, but which was thus enabled to hold its important position.

GAFFNEY, FRANK (1214882)
  Near Ronssoy, France, Sept. 29, 1918.
  R—Niagara Falls, N. Y.
  B—Buffalo, N. Y.
  G. O. No. 20, W. D., 1919.
Private, first class, Company G, 108th Infantry, 27th Division.
Private Gaffney, an automatic rifleman, pushing forward alone with his gun,
  after all the other members of his squad had been killed, discovered several
  Germans placing a heavy machine gun in position. He killed the crew, cap-
  tured the gun, bombed several dugouts and, after killing four more of the
  enemy with his pistol, held the position until reinforcements came up, when
  80 prisoners were captured.

GAGE, RICHARD J.
  At Elk River, Tenn., July 2, 1863.
  R—Ottawa, Ill.
  B—Grafton County, N. H.
Private, Company D, 104th Illinois Infantry.
Voluntarily joined a small party that, under a heavy fire, captured a stockade
  and saved the bridge.

GALLOWAY, GEORGE N.
  At Alsops Farm, Va., May 8, 1864.
  R—NR.
  B—Philadelphia, Pa.
Private, Company G, 95th Pennsylvania Infantry.
Voluntarily held an important position under heavy fire.

GALLOWAY, JOHN
At Farmville, Va., Apr. 7, 1865.
R—NR.
B—Philadelphia, Pa.

Commissary sergeant, 8th Pennsylvania Cavalry.
His regiment being surprised and nearly overwhelmed, he dashed forward under a heavy fire, reached the right of the regiment, where the danger was greatest, rallied the men, and prevented a disaster that was imminent.

GALT, STERLING A
At Bamban, Luzon, P. I., Nov. 9, 1899.
R—Pawneytown, Md.
B—Pawneytown, Md.

Artificer, Company F, 36th Infantry, U. S. Volunteers.
Distinguished bravery and conspicuous gallantry in action against insurgents.

GARDINER, JAMES
At Chapins Farm, near Richmond, Va., Sept. 29, 1864.
R—NR.
B—Gloucester, Va.

Private, Ccompany I, 36th U. S. Colored Troops.
Rushed in advance of his brigade, shot a rebel officer who was on the parapet rallying his men, and then ran him through with his bayonet.

GARDINER, PETER W
At Sappa Creek, Kans., Apr. 23, 1875.
R—NR.
B—Carlisle, N. Y.

Private, Company H, 6th U. S. Cavalry.
Gallantry in action.

GARDNER, CHARLES
Arizona, August to October, 1868.
R—NR.
B—Bavaria.

Private, Company B, 8th U. S. Cavalry.
Bravery in actions and scouts against Indians.

GARDNER, CHARLES N
At Five Forks, Va., Apr. 1, 1865.
R—NR.
B—South Scituate, Mass.

Private, Company E, 32d Massachusetts Infantry.
Capture of flag.

GARDNER, ROBERT J
At Petersburg, Va., Apr. 2, 1865.
R—Berkshire County, Mass.
B—Livingston, N. Y.

Sergeant, Company K, 34th Massachusetts Infantry.
Was among the first to enter Fort Gregg, clearing his way by using his musket on the heads of the enemy.

GARLAND, HARRY
At Little Muddy Creek, Mont., May 7, 1877.
At Camas Meadows, Idaho, Aug. 29, 1877.
R—NR.
B—Boston, Mass.

Corporal, Company L, 2d U. S. Cavalry.
Gallantry in action with hostile Sioux.

Having been wounded in the hip so as to be unable to stand, he still continued to direct the men under his charge until the enemy withdrew.

GARLINGTON, ERNEST A
At Wounded Knee Creek, S. Dak., Dec. 29, 1890.
R—Athens, Ga.
B—South Carolina.

First Lieutenant, 7th U. S. Cavalry.
Distinguished gallantry.

GARRETT, WILLIAM
At Nashville, Tenn., Dec. 16, 1864.
R—NR.
B—England.

Sergeant, Company G, 41st Ohio Infantry.
With several companions dashed forward, the first to enter the enemy's works, taking possession of four pieces of artillery and capturing a flag.

ºGASSON, RICHARD
At Chapins Farm, near Richmond, Va., Sept. 29, 1864.
R—NR.
B—Ireland.

Sergeant, Company K, 47th New York Infantry.
Fell dead while planting the colors of his regiment on the enemy's works.

GATES, GEORGE
At Picacho Mountain, Ariz., June 4, 1869.
R—NR.
B—Delaware County, Ohio.

Bugler, Company F, 8th U. S. Cavalry.
Killed an Indian warrior and captured his arms.

GAUJOT, ANTOINE A
At San Mateo, P. I., Dec. 19, 1899.
R—Williamson, W. Va.
B—Keweenaw, Mich.

Corporal, Company M, 27th Infantry, U. S. Volunteers.
Attempted, under a heavy fire of the enemy, to swim a river for the purpose of obtaining and returning with a canoe.

GAUJOT, JULIEN E
At Agua Prieta, Mexico, April 13, 1911.
R—Williamson, W. Va.
B—Keweenaw, Mich.

Captain, Troop K, 1st U. S. Cavalry.
Crossed the field of fire to obtain the permission of the rebel commander to receive the surrender of the surrounded forces of Mexican Federals and escort such forces, together with five Americans held as prisoners, to the American line.

GAUNT, JOHN C
At Franklin, Tenn., Nov. 30, 1864.
R—Damascoville, Ohio.
B—Columbiana County, Ohio.

Private, Company G, 104th Ohio Infantry.
Capture of flag.

GAUSE, ISAAC
Near Berryville, Va., Sept. 13, 1864.
R—NR.
B—Trumbull County, Ohio.

Corporal, Company E, 2d Ohio Cavalry.
Capture of the colors of the 8th South Carolina Infantry while engaged in a reconnaissance along the Berryville and Winchester Pike.

GAY, THOMAS H.
  Arizona, August to October, 1868.
  R—N R.
  B—Prince Edward Island.

Private, Company B, 8th U. S. Cavalry.
Bravery in actions and scouts against Indians.

GAYLORD, LEVI B
  At Fort Stedman, Va., Mar. 25, 1865.
  R—N R.
  B—Boston, Mass.

Sergeant, Company A, 29th Massachusetts Infantry.
Voluntarily assisted in working an abandoned gun, while exposed to a heavy
  fire, until the enemy's advancing line was routed by a charge on its left flank.

GEDEON, LOUIS
  At Fort Anita, Cebu, P. I., Feb. 4,
    1900.
  R—Pittsburgh, Pa.
  B—Pittsburgh, Pa.

Private, Company G, 19th U. S. Infantry.
Single handed, defended his mortally wounded captain from an overwhelming
  force of the enemy.

GEIGER, GEORGE
  At Little Big Horn River, Mont.,
    June 25, 1876.
  R—N R.
  B—Cincinnati, Ohio.

Sergeant, Company H, 7th U. S. Cavalry.
With three comrades during the entire engagement courageously held a position
  that secured water for the command.

GEORGIAN, JOHN
  At Chiricahua Mountains, Ariz.,
    Oct. 20, 1869.
  R—N R.
  B—Germany.

Private, Company G, 8th U. S. Cavalry.
Bravery in action.

GERBER, FREDERICK W
  1839-1871.
  R—N R.
  B—Germany.

Sergeant major, Battalion U. S. Engineers.
Distinguished gallantry in many actions and in recognition of long, faithful,
  and meritorious services covering a period of 32 years.

GERE, THOMAS P
  At Nashville, Tenn., Dec. 16, 1864.
  R—N R.
  B—Chemung County, N. Y.

First lieutenant and adjutant, 5th Minnesota Infantry.
Capture of flag of 4th Mississippi (C. S. A.).

GESCHWIND, NICHOLAS
  At Vicksburg, Miss., May 22, 1863.
  R—Pleasant Hill, Ill.
  B—France.

Captain, Company F, 116th Illinois Infantry.
Gallantry in the charge of the "volunteer storming party."

GIBBS, WESLEY
  At Petersburg, Va., Apr. 2, 1865.
  R—Salisbury, Conn.
  B—Sharon, Conn.

Sergeant, Company B, 2d Connecticut Heavy Artillery.
Capture of flag.

GIBSON, EDWARD H
  At San Mateo, P. I., Dec. 19, 1899.
  R—Boston, Mass.
  B—Boston, Mass.

Sergeant, Company M, 27th Infantry, U. S. Volunteers.
Attempted under a heavy fire of the enemy to swim a river for the purpose of
  obtaining and returning with a canoe.

GIFFORD, BENJAMIN
  At Sailors Creek, Va., Apr. 6, 1865.
  R—N R.
  B—German Flatts, N. Y.

Private, Company H, 121st New York Infantry.
Capture of flag.

GIFFORD, DAVID L
  At Ashepoo River, S. C., May 24,
    1864.
  R—N R.
  B—Dartmouth, Mass.

Private, Company B, 4th Massachusetts Cavalry.
Volunteered as a member of a boat crew which went to the rescue of a large
  number of Union soldiers on board the stranded steamer Boston and with
  great gallantry assisted in conveying them to shore, being exposed during
  the entire time to a heavy fire from a Confederate battery.

GILLENWATER, JAMES R
  Near Porac, Luzon, P. I., Feb. 7,
    1902.
  R—Rye Cove, Va.
  B—Rye Cove, Va.

Corporal, Company A, 36th Infantry, U. S. Volunteers.
While on a scout drove off a superior force of insurgents and with the assistance
  of one comrade brought from the field of action the bodies of two comrades,
  one killed and the other severely wounded.

GILLESPIE, GEORGE L
  Near Bethesda Church, Va., May
    31, 1864.
  R—Chattanooga, Tenn.
  B—Tennessee.

First lieutenant, Corps of Engineers, U. S. Army.
Exposed himself to great danger by voluntarily making his way through the
  enemy's lines to communicate with General Sheridan. While rendering
  this service he was captured; but escaped; again came in contact with the
  enemy, was again ordered to surrender, but escaped by dashing away under
  fire.

GILLIGAN, EDWARD L
  At Gettysburg, Pa., July 1, 1863.
  R—Philadelphia, Pa.
  B—Philadelphia, Pa.

First sergeant, Company E, 88th Pennsylvania Infantry.
Assisted in the capture of a Confederate flag by knocking down the color sergeant.

GILMORE, JOHN C
  At Salem Heights, Va., May 3, 1863.
  R—N R.
  B—N R.

Major, 16th New York Infantry.
Seized the colors of his regiment and gallantly rallied his men under a very severe
  fire.

GINLEY, PATRICK At Reams Station, Va., Aug. 25, 1864. R—NR. B—Ireland.	Private, Company G, 1st New York Light Artillery. The command having been driven from the works, he, having been left alone between the opposing lines, crept back into the works, put three charges of canister in one of the guns, and fired the piece directly into a body of the enemy about to seize the works; he then rejoined his command, took the colors, and ran toward the enemy, followed by the command, which recaptured the works and guns.
GION, JOSEPH At Chancellorsville, Va., May 2, 1863. R—NR. B—NR.	Private, Company A, 74th New York Infantry. Voluntarily and under heavy fire advanced toward the enemy's lines and secured valuable information.
GIVEN, JOHN J. At Wichita River, Tex., July 12, 1870. R—NR. B—Davis County, Texas.	Corporal, Company K, 6th U. S. Cavalry. Bravery in action.
GLAVINSKI, ALBERT At Powder River, Mont., Mar. 17, 1876. R—NR. B—Germany.	Blacksmith, Company M, 3d U. S. Cavalry. Gallantry in action.
GLOVER, T. B. At Mizpah Creek, Mont., Apr. 10, 1879; at Pumpkin Creek, Mont., Feb. 10, 1880. R—NR. B—New York, N. Y.	Sergeant, Troop B, 2d U. S. Cavalry. While in charge of small scouting parties, fought, charged, surrounded, and captured war parties of Sioux Indians.
GLYNN, MICHAEL At Whetstone Mountains, Ariz., July 13, 1872. R—NR. B—Ireland.	Private, Company F, 5th U. S. Cavalry. Drove off, single handed, eight hostile Indians, killing and wounding five.
GODFREY, EDWARD S. At Bear Paw Mountains, Mont., Sept. 30, 1877. R—Ohio. B—Ottawa, Ohio.	Captain, 7th U. S. Cavalry. Led his command into action when he was severely wounded.
GODLEY, LEONIDAS M. At Vicksburg, Miss., May 22, 1863. R—NR. B—Mason County, Va.	First sergeant, Company E, 22d Iowa Infantry. Led his company in the assault on the enemy's works and gained the parapet, there receiving three very severe wounds. He lay all day in the sun, was taken prisoner, and had his leg amputated without anesthetics.
GOETTEL, PHILIP At Ringgold, Ga., Nov. 27, 1863. R—Syracuse, N. Y. B—Syracuse, N. Y.	Private, Company B, 149th New York Infantry. Capture of flag and a battery guidon.
*GOETTLER, HAROLD ERNEST Near Binarville, France, Oct. 6, 1918. R—Chicago, Ill. B—Chicago, Ill. G. O. No. 56, W. D. 1922.	First lieutenant, pilot, 50th Aero Squadron, Air Service. Lieutenant Goettler, with his observer, Second Lieut. Erwin R. Bleckley, 130th Field Artillery, left the airdrome late in the afternoon on their second trip to drop supplies to a battalion of the 77th Division which had been cut off by the enemy in the Argonne Forest. Having been subjected on the first trip to violent fire from the enemy, they attempted on the second trip to come still lower in order to get the packages even more precisely on the designated spot. In the course of this mission the plane was brought down by enemy rifle and machine-gun fire from the ground, resulting in the instant death of Lieutenant Goettler. In attempting and performing this mission Lieutenant Goettler showed the highest possible contempt of personal danger, devotion to duty, courage, and valor. Posthumously awarded. Medal presented to mother, Mrs. Gertrude Goettler.
GOHEEN, CHARLES A. At Waynesboro, Va., Mar. 2, 1865. R—NR. B—Groveland, N. Y.	First sergeant, Company G, 8th New York Cavalry. Capture of flag.
GOLDEN, PATRICK Arizona, August to October, 1868. R—NR. B—Ireland.	Sergeant, Company B, 8th U. S. Cavalry. Bravery in actions and scouts against Indians.
GOLDIN, THEODORE W. At Little Big Horn, Mont., June 26, 1876. R—NR. B—Green County, Wis.	Private, Troop G, 7th U. S. Cavalry. One of a party of volunteers who, under a heavy fire from the Indians, went for and brought water to the wounded.
GOLDSBERY, ANDREW E. At Vicksburg, Miss., May 22, 1863. R—NR. B—St. Charles, Ill.	Private, Company E, 127th Illinois Infantry. Gallantry in the charge of the "volunteer storming party."

GOODALL, FRANCIS H.
At Fredericksburg, Va., Dec. 13, 1862.
R—Bath, N. H.
B—Bath, N. H.

First sergeant, Company G, 11th New Hampshire Infantry.
With the assistance of another soldier brought a wounded comrade into the lines, under heavy fire.

GOODMAN, DAVID.
At Lyry Creek, Ariz., Oct. 14, 1869.
R—NR.
B—Paxton, Mass.

Private, Company L, 8th U. S. Cavalry.
Bravery in action.

GOODMAN, WILLIAM E.
At Chancellorsville, Va., May 3, 1863.
R—NR.
B—NR.

First lieutenant, Company D, 147th Pennsylvania Infantry.
Rescued the colors of the 107th Ohio Volunteers from the enemy.

GOODRICH, EDWIN.
Near Cedar Creek, Va., Nov. —, 1864.
R—NR.
B—New York, N. Y.

First lieutenant, Company D, 9th New York Cavalry.
While the command was falling back, he returned, and in the face of the enemy rescued a sergeant from under his fallen horse.

GOULD, CHARLES G.
At Petersburg, Va., Apr. 2, 1865.
R—Windham, Vt.
B—Windham, Vt.

Captain, Company H, 5th Vermont Infantry.
Among the first to mount the enemy's works in the assault, he received a serious bayonet wound in the face, was struck several times with clubbed muskets, but bravely stood his ground, and with his sword killed the man who bayoneted him.

GOULD, NEWTON T.
At Vicksburg, Miss., May 22, 1863.
R—NR.
B—Elk Grove, Ill.

Private, Company G, 113th Illinois Infantry.
Gallantry in the charge of the "volunteer storming party."

GOURAUD, GEORGE E.
At Honey Hill, S. C., Nov. 30, 1864.
R—NR.
B—New York, N. Y.

Captain and aide-de-camp, U. S. Volunteers.
While under severe fire of the enemy, which drove back the command, rendered valuable assistance in rallying the men.

GRACE, PETER.
At Wilderness, Va., May 5, 1864.
R—Berkshire, Mass.
B—Berkshire, Mass.

Sergeant, Company G, 83d Pennsylvania Infantry.
Single handed, rescued a comrade from two Confederate guards, knocking down one and compelling surrender of the other.

GRAHAM, THOMAS N.
At Missionary Ridge, Tenn., Nov. 25, 1863.
R—NR.
B—NR.

Second lieutenant, Company G, 15th Indiana Infantry.
Seized the colors from the color bearer, who had been wounded, and, exposed to a terrible fire, carried them forward, planting them on the enemy's breastworks.

GRANT, GABRIEL.
At Fair Oaks, Va., June 1, 1862.
R—NR.
B—NR.

Surgeon, U. S. Volunteers.
Removed severely wounded officers and soldiers from the field while under a heavy fire from the enemy, exposing himself beyond the call of duty, thus furnishing an example of most distinguished gallantry.

GRANT, GEORGE.
At Fort Phil. Kearney to Fort C. F. Smith, Dakota Territory, February, 1867.
R—NR.
B—Raleigh, Tenn.

Sergeant, Company E, 18th U. S. Infantry.
Bravery, energy, and perseverance, involving much suffering and privation through attacks by hostile Indians, deep snows, etc., while voluntarily carrying dispatches.

GRANT, LEWIS A.
At Salem Heights, Va., May 3, 1863.
R—NR.
B—Vermont.

Colonel, 5th Vermont Infantry.
Personal gallantry and intrepidity displayed in the management of his brigade and in leading it in the assault in which he was wounded.

GRAUL, WILLIAM.
At Fort Harrison, near Richmond, Va., Sept. 29, 1864.
R—Reading, Pa.
B—Reading, Pa.

Corporal, Company I, 188th Pennsylvania Infantry.
First to plant the colors of his State on the fortifications.

GRAVES, THOMAS J.
At El Caney, Cuba, July 1, 1898.
R—Millville, Ind.
B—Milton, Ind.

Private, Company C, 17th U. S. Infantry.
Gallantly assisted in the rescue of the wounded from in front of the lines and under heavy fire from the enemy.

GRAY, JOHN.
At Port Republic, Va., June 9, 1862.
R—NR.
B—Scotland.

Private, Company B, 5th Ohio Infantry.
Mounted an artillery horse of the enemy and captured a brass 6-pound piece in the face of the enemy's fire and brought it to the rear.

GRAY, ROBERT A.
At Drurys Bluff, Va., May 16, 1864.
R—NR.
B—Philadelphia, Pa.

Sergeant, Company C, 21st Connecticut Infantry.
While retreating with his regiment, which had been repulsed, he voluntarily returned, in face of the enemy's fire, to a former position and rescued a wounded officer of his company who was unable to walk.

102444°—27——4

GREAVES, CLINTON. At Florida Mountains, N. Mex., Jan. 24, 1877. R—Prince Georges County, Md. B—Madison County, Va.	Corporal, Company C, 9th U. S. Cavalry. Gallantry in a hand-to-hand fight.
GREBE, M. R. WILLIAM. At Jonesboro, Ga., Aug. 31, 1864. R—NR. B—NR.	Captain, Company F, 4th Missouri Cavalry. While acting as aid and carrying orders across a most dangerous part of the battlefield, being hindered by a Confederate advance, seized a rifle, took a place in the ranks, and was conspicuous in repulsing the enemy.
GREEN, FRANCIS C. Arizona, 1868 and 1869. R—NR. B—Mount Vernon, Ind.	Sergeant, Company K, 8th U. S. Cavalry. Bravery in actions.
GREEN, GEORGE. At Missionary Ridge, Tenn., Nov. 25, 1863. R—NR. B—NR.	Corporal, Company H, 11th Ohio Infantry. Scaled the enemy's works and in a hand-to-hand fight captured a flag.
GREEN, JOHN. At the Lava Beds, Calif., Jan. 17, 1873. R—Ohio. B—Germany.	Major, 1st U. S. Cavalry. In order to reassure his command, this officer, in the most fearless manner and exposed to very great danger, walked in front of the line; the command, thus encouraged, advanced over the lava upon the Indians who were concealed among the rocks.
GREENAWALT, ABRAHAM. At Franklin, Tenn., Nov. 30, 1864. R—Salem, Ohio. B—Montgomery County, Pa.	Private, Company G, 104th Ohio Infantry. Capture of corps headquarters flag (C. S. A.).
GREENE, OLIVER D. At Antietam, Md., Sept. 17, 1862. R—Scott, N. Y. B—New York.	Major and assistant adjutant general, U. S. Army. Formed the columns under heavy fire and put them into position.
GREER, ALLEN J. Near Majada, Laguna Province, P. I., July 2, 1901. R—Memphis, Tenn. B—Memphis, Tenn.	Second lieutenant, 4th U. S. Infantry. Charged alone an insurgent outpost with his pistol, killing one, wounding two, and capturing three insurgents with their rifles and equipments.
GREGG, JOSEPH O. Near the Richmond & Petersburg Ry., Virginia., June 16, 1864. R—NR. B—NR.	Private, Company F, 133d Ohio Infantry. Voluntarily returned to the breastworks which his regiment had been forced to abandon to notify three missing companies that the regiment was falling back; found the enemy already in the works, refused a demand to surrender, returning to his command, under a concentrated fire, several bullets passing through his hat and clothing.
GREGORY, EARL D. (1290053) At Bois-de-Consenvoye, north of Verdun, France, Oct. 8, 1918. R—Chase City, Va. B—Chase City, Va. G. O. No. 34, W. D., 1919.	Sergeant, Headquarters Company, 116th Infantry, 29th Division. With the remark "I will get them," Sergt. Gregory seized a rifle and a trench-mortar shell, which he used as a hand grenade, left his detachment of the trench-mortar platoon, and, advancing ahead of the infantry, captured a machine gun and three of the enemy. Advancing still farther from the machine-gun nest, he captured a 7.5-centimeter mountain howitzer and, entering a dugout in the immediate vicinity, single-handed captured 19 of the enemy.
GREIG, THEODORE W. At Antietam, Md., Sept. 17, 1862. R—Staten Island, N. Y. B—New York.	Second lieutenant, Company C, 61st New York Infantry. A Confederate regiment, having planted its battle flag slightly in advance of the regiment, this officer rushed forward and seized it, and, although shot through the neck, retained the flag and brought it within the Union lines.
GRESHAM, JOHN C. At Wounded Knee Creek, S. Dak., Dec. 29, 1890. R—Lancaster Courthouse, Va. B—Virginia.	First lieutenant, 7th U. S. Cavalry. Voluntarily led a party into a ravine to dislodge Indians concealed therein.
GRESSER, IGNATZ. At Antietam, Md., Sept. 17, 1862. R—NR. B—NR.	Corporal, Company D, 128th Pennsylvania Infantry. While exposed to the fire of the enemy, carried from the field a wounded comrade.
GRIBBEN, JAMES H. At Sailors Creek, Va., Apr. 6, 1865. R—NR. B—Ireland.	Lieutenant, Company C, 2d New York Cavalry. Capture of flag of 12th Virginia Infantry (C. S. A.).
GRIMES, EDWARD P. At Milk River, Colo., Sept. 29 to Oct. 5, 1879. R—NR. B—Dover, N. H.	Sergeant, Company F, 5th U. S. Cavalry. The command being almost out of ammunition and surrounded on three sides by the enemy, he voluntarily brought up a supply under heavy fire at almost point blank.

GRIMSHAW, SAMUEL............
    At Atlanta, Ga., Aug. 6, 1864.
    R—NR.
    B—Jefferson County, Ohio.

Private, Company B, 52d Ohio Infantry.
Saved the lives of some of his comrades, and greatly imperiled his own, by picking up and throwing away a lighted shell which had fallen in the midst of the company.

GRINDLAY, JAMES G.............
    At Five Forks, Va., Apr. 1, 1865.
    R—Utica, N. Y.
    B—NR.

Colonel, 146th New York Infantry.
The first to enter the enemy's works, where he captured two flags.

GROVE, WILLIAM R.............
    Near Porac, Luzon, P. I., Sept. 9, 1899.
    R—Denver, Colo.
    B—Montezuma, Iowa.
    Distinguished-service medal also awarded.

Lieutenant colonel, 36th Infantry, U. S. Volunteers.
In advance of his regiment, rushed to the assistance of his colonel, charging, pistol in hand, seven insurgents, and compelling surrender of all not killed or wounded.

GRUEB, GEORGE.............
    At Chapins Farm, near Richmond, Va., Sept. 29, 1864.
    R—NR.
    B—Germany.

Private, Company E, 158th New York Infantry.
Gallantry in advancing to the ditch of the enemy's works.

GUERIN, FITZ W.............
    At Grand Gulf, Miss., Apr. 28–29, 1863.
    R—NR.
    B—New York, N. Y.

Private, Battery A, 1st Missouri Light Artillery.
With two comrades voluntarily took position on board the steamer Cheeseman, in charge of all the guns and ammunition of the battery, and remained in charge of the same for a considerable time while the steamer was unmanageable and subjected to a heavy fire from the enemy.

GUINN, THOMAS.............
    At Vicksburg, Miss., May 22, 1863.
    R—NR.
    B—Clinton County, Ohio.

Private, Company D, 47th Ohio Infantry.
Gallantry in the charge of the "volunteer storming party."

GUMPERTZ, SYDNEY G (1388848)........
    In the Bois de Forges, France, Sept. 26, 1918.
    R—Chicago, Ill.
    B—San Raphael, Calif.
    G. O. No. 16, W. D., 1919.

First sergeant, Company E, 132d Infantry, 33d Division.
When the advancing line was held up by machine-gun fire, Sergeant Gumpertz left the platoon of which he was in command and started with two other soldiers through a heavy barrage toward the machine-gun nest. His two companions soon became casualties from bursting shells, but Sergeant Gumpertz continued on alone in the face of direct fire from the machine gun, jumped into the nest, and silenced the gun, capturing nine of the crew.

GUNTHER, JACOB.............
    Arizona, 1868 and 1869.
    R—NR.
    B—Schuylkill County, Pa.

Corporal, Company E, 8th U. S. Cavalry.
Bravery in actions and scouts against Indians.

GWYNNE, NATHANIEL.............
    At Petersburg, Va., July 30, 1864.
    R—Fairmount, Mo.
    B—Champaign County, Ohio.

Private, Company H, 13th Ohio Cavalry.
When about entering upon the charge, this soldier, then but 15 years old, was cautioned not to go in, as he had not been mustered. He indignantly protested and participated in the charge, his left arm being crushed by a shell and amputated soon afterwards.

HACK, JOHN.............
    At Vicksburg, Miss., May 3, 1863.
    R—NR.
    B—Germany.

Private, Company B, 47th Ohio Infantry.
Was one of a party which volunteered and attempted to run the enemy's batteries with a steam tug and two barges loaded with subsistence stores.

HACK, LESTER G.............
    At Petersburg, Va., Apr. 2, 1865.
    R—Salisbury, Vt.
    B—Bolton, N. Y.

Sergeant, Company F, 5th Vermont Infantry.
Capture of flag of 23d Tennessee Infantry (C. S. A.).

HADDOO, JOHN.............
    At Cedar Creek, etc., Mont., October, 1876, to Jan. 8, 1877.
    R—NR.
    B—Boston, Mass.

Corporal, Company B, 5th U. S. Infantry.
Gallantry in actions.

HADLEY, CORNELIUS M.............
    At Siege of Knoxville, Tenn., Nov. 20, 1863.
    R—NR.
    B—Oswego, N. Y.

Sergeant, Company F, 9th Michigan Cavalry.
With one companion, voluntarily carried through the enemy's lines important dispatches from General Grant to General Burnside, then besieged within Knoxville, and brought back replies, his comrade's horse being killed and the man taken prisoner.

HADLEY, OSGOOD T.............
    Near Pegram House, Va., Sept. 30, 1864.
    R—NR.
    B—Nashua, N. H.

Corporal, Company E, 6th New Hampshire Veteran Infantry.
As color bearer of his regiment he defended his colors with great personal gallantry and brought them safely out of the action.

HAGERTY, ASEL.............
    At Sailors Creek, Va., Apr. 6, 1865.
    R—NR.
    B—Canada.

Private, Company A, 61st New York Infantry.
Capture of flag.

HAIGHT, JOHN H. At Williamsburg, Va., May 5, 1862. At Bristol Station, Va., Aug. 27, 1862. At Manassas, Va., Aug. 29 and 30, 1862. R—NR. B—NR.	Sergeant, Company G, 72d New York Infantry. Voluntarily carried a severely wounded comrade off the field in the face of a large force of the enemy; in doing so was himself severely wounded and taken prisoner. Went into the fight, although severely disabled. Volunteered to search the woods for the wounded.
HAIGHT, SIDNEY. At Petersburg, Va., July 30, 1864. R—NR. B—Reading, Mich.	Corporal, Company E, 1st Michigan Sharpshooters. Instead of retreating, remained in the captured works, regardless of his personal safety and exposed to the firing, which he boldly and deliberately returned until the enemy was close upon him.
HALL, FRANCIS B. At Salem Heights, Va., May 3, 1863. R—NR. B—NR.	Chaplain, 16th New York Infantry. Voluntarily exposed himself to a heavy fire during the thickest of the fight and carried wounded men to the rear for treatment and attendance.
HALL, H. SEYMOUR. At Gaines Mill, Va., June 27, 1862. At Rappahannock Station, Va., Nov. 7, 1863. R—NR. B—NR.	Second lieutenant, Company G, 27th New York Infantry, and captain, Company F, 121st New York Infantry. Although wounded, he remained on duty and participated in the battle with his company. While acting as aid, rendered gallant and prompt assistance in re-forming the regiments inside the enemy's works.
HALL, JOHN. Arizona, August to October, 1868. R—NR. B—Logan County, Ill.	Private, Company B, 8th U. S. Cavalry. Bravery in scouts and actions with Indians.
HALL, NEWTON H. At Franklin, Tenn., Nov. 30, 1864. R—NR. B—Portage County, Ohio.	Corporal, Company I, 104th Ohio Infantry. Capture of flag, believed to have belonged to Stewart's Corps (C. S. A.).
*HALL, THOMAS LEE (1311234). Near Montbrehain, France, Oct. 8, 1918. R—Fort Mill, S. C. B—Fort Mill, S. C. G. O. No. 50, W. D., 1919.	Sergeant, Company G, 118th Infantry, 30th Division. Having overcome two machine-gun nests under his skillful leadership, Sergeant Hall's platoon was stopped 800 yards from its final objective by machine-gun fire of particular intensity. Ordering his men to take cover in a sunken road, he advanced alone on the enemy machine-gun post and killed five members of the crew with his bayonet and thereby made possible the further advance of the line. While attacking another machine-gun nest later in the day this gallant soldier was mortally wounded. Posthumously awarded. Medal presented to father, William L. Hall.
HALL, WILLIAM P. Near camp on White River, Colo., Oct. 20, 1879. R—Huntsville, Mo. B—Randolph County, Mo.	First lieutenant, 5th U. S. Cavalry. With a reconnoitering party of three men, was attacked by 35 Indians and several times exposed himself to draw the fire of the enemy, giving his small party opportunity to reply with much effect.
HALLOCK, NATHAN M. At Bristoe Station, Va., June 15, 1863. R—NR. B—Orange County, N. Y.	Private, Company K, 124th New York Infantry. At imminent peril saved from death or capture a disabled officer of his company by carrying him, under a hot musketry fire, to a place of safety.
HAMILTON, FRANK. At Agua Fria River, Ariz., Aug. 25, 1869. R—NR. B—Ireland.	Private, Company E, 8th U. S. Cavalry. Gallantry in action.
HAMILTON, MATHEW H. At Wounded Knee Creek, S. Dak., Dec. 29, 1890. R—New York, N. Y. B—Australia.	Private, Company G, 7th U. S. Cavalry. Bravery in action.
HAMMEL, HENRY A. At Grand Gulf, Miss., Apr. 28–29, 1863. R—NR. B—NR.	Sergeant, Battery A, 1st Missouri Light Artillery. With two comrades voluntarily took position on board the steamer Cheeseman, in charge of all the guns and ammunition of the battery, and remained in charge of the same for considerable time while the steamer was unmanageable and subjected to a heavy fire from the enemy.
HANEY, MILTON L. At Atlanta, Ga., July 22, 1864. R—NR. B—Ohio.	Chaplain, 55th Illinois Infantry. Voluntarily carried a musket in the ranks of his regiment and rendered heroic service in retaking the Federal works which had been captured by the enemy.

HANFORD, EDWARD R................
    At Woodstock, Va., Oct. 9, 1864.
    R—NR.
    B—Allegany County, N. Y.

Private, Company H, 2d U. S. Cavalry.
Capture of flag of 32d Battalion Virginia Cavalry (C. S. A.).

HANKS, JOSEPH................
    At Vicksburg, Miss., May 22, 1863.
    R—Chillicothe, Ohio.
    B—Chillicothe, Ohio.

Private, Company E, 37th Ohio Infantry.
Voluntarily and under fire went to the rescue of a wounded comrade lying between the lines, gave him water, and brought him off the field.

HANLEY, RICHARD P................
    At Little Big Horn River, Mont., June 25, 1876.
    R—NR.
    B—Boston, Mass.

Sergeant, Company C, 7th U. S. Cavalry.
Recaptured, single handed, and without orders, within the enemy's lines and under a galling fire lasting some 20 minutes, a stampeded pack mule loaded with ammunition.

HANNA, MARCUS A................
    At Port Hudson, La., July 4, 1863.
    R—NR.
    B—Bristol, Me.

Sergeant, Company B, 50th Massachusetts Infantry.
Voluntarily exposed himself to a heavy fire to get water for comrades in rifle pits.

HANNA, MILTON................
    At Nolensville, Tenn., Feb. 15, 1863.
    R—NR.
    B—Ohio.

Corporal, Company H, 2d Minnesota Infantry.
Was one of a detachment of 16 men who heroically defended a wagon train against the attack of 125 cavalry, repulsed the attack, and saved the train.

HANSCOM, MOSES C................
    At Bristol Station, Va., Oct. 14, 1863.
    R—NR.
    B—Danville, Me.

Corporal, Company F, 19th Maine Infantry.
Capture of flag of 26th North Carolina (C. S. A.).

HAPEMAN, DOUGLAS................
    At Peach Tree Creek, Ga., July 20, 1864.
    R—Ottawa, Ill.
    B—NR.

Lieutenant colonel, 104th Illinois Infantry.
With conspicuous coolness and bravery rallied his men under a severe attack, re-formed the broken ranks, and repulsed the attack.

HARBOURNE, JOHN H................
    At Petersburg, Va., June 17, 1864.
    R—Boston, Mass.
    B—England.

Private, Company K, 29th Massachusetts Infantry.
Capture of flag.

HARDAWAY, BENJAMIN F................
    At El Caney, Cuba, July 1, 1898.
    R—NR.
    B—Benleyville, Ky.

First lieutenant, 17th U. S. Infantry.
Gallantly assisted in the rescue of the wounded from in front of the lines and under heavy fire from the enemy.

HARDENBERGH, HENRY M................
    At Deep Run, Va., Aug. 16, 1864.
    R—NR.
    B—Noble County, Ind.

Private, Company G, 39th Illinois Infantry.
Capture of flag.

HARDING, MOSHER A................
    At Chiricahua Mountains, Ariz., Oct. 20, 1869.
    R—NR.
    B—Canada West.

Blacksmith, Company G, 8th U. S. Cavalry.
Gallantry in action.

HARING, ABRAM P................
    At Bachelors Creek, N. C., Feb. 1, 1864.
    R—NR.
    B—New York, N. Y.

First lieutenant, Company G, 132d New York Infantry.
With a command of 11 men, on picket, resisted the attack of an overwhelming force of the enemy.

HARMON, AMZI D................
    At Petersburg, Va., Apr. 2, 1865.
    R—NR.
    B—Westmoreland County, Pa.

Corporal, Company K, 211th Pennsylvania Infantry.
Capture of flag.

HARRINGTON, EPHRAIM W................
    At Fredericksburg, Va., May 3, 1863.
    R—NR.
    B—Waterford, Me.

Sergeant, Company G, 2d Vermont Infantry.
Carried the colors to the top of the heights and almost to the muzzle of the enemy's guns.

HARRINGTON, JOHN................
    At Washita River, Tex., Sept. 12, 1874.
    R—NR.
    B—Detroit, Mich.

Private, Company H, 6th U. S. Cavalry.
While carrying dispatches was attacked by 125 hostile Indians, whom he and his comrades fought throughout the day.

HARRIS, CHARLES D................
    At Red Creek, Ariz., Sept. 23, 1869.
    R—NR.
    B—Albion, N. Y.

Sergeant, Company D, 8th U. S. Cavalry.
Gallantry in action.

HARRIS, DAVID W.............. | Private, Company A, 7th U. S. Cavalry.
At Little Big Horn River, Mont., June 25, 1876. | Brought water to the wounded, at great danger to his life, under a most galling fire from the enemy.
R—NR.
B—Indianapolis, Ind.

HARRIS, GEORGE W.............. | Private, Company B, 148th Pennsylvania Infantry.
At Spotsylvania, Va., May 12, 1864. | Capture of flag, wresting it from the color bearer and shooting an officer who attempted to regain it.
R—Bellefonte, Pa.
B—Schuylkill, Pa.

HARRIS, JAMES H.............. | Sergeant, Company B, 38th U. S. Colored Troops.
At New Market Heights, Va., Sept. 29, 1864. | Gallantry in the assault.
R—NR.
B—St. Marys County, Md.

HARRIS, MOSES............. | First lieutenant, 1st U. S. Cavalry.
At Smithfield, Va., Aug. 28, 1864. | In an attack upon a largely superior force, his personal gallantry was so conspicuous as to inspire the men to extraordinary efforts, resulting in complete rout of the enemy.
R—NR.
B—Andover, N. H.

HARRIS, SAMPSON............. | Private, Company K, 30th Ohio Infantry.
At Vicksburg, Miss., May 22, 1863. | Gallantry in the charge of the "volunteer storming party."
R—Olive, Ohio.
B—Noble County, Ohio.

HARRIS, WILLIAM M............. | Private, Company D, 7th U. S. Cavalry.
At Little Big Horn River, Mont., June 25, 1876. | Voluntarily brought water to the wounded under fire of the enemy.
R—NR.
B—Madison County, Ky.

HART, JOHN W............. | Sergeant, Company D, 6th Pennsylvania Reserves.
At Gettysburg, Pa., July 2, 1863. | Was one of six volunteers who charged upon a log house near the Devil's Den, where a squad of the enemy's sharpshooters were sheltered and compelled their surrender.
R—NR.
B—Germany.

HART, WILLIAM E............. | Private, Company B, 8th New York Cavalry.
At Shenandoah Valley, Va., 1864 and 1865. | Gallant conduct and services as scout in connection with capture of the guerrilla Harry Gilmore, and other daring acts.
R—Pittsford, N. Y.
B—Rushville, N. Y.

HARTRANFT, JOHN F............. | Colonel, 4th Pennsylvania Militia.
At Bull Run, Va., July 21, 1861. | Voluntarily served as an aid and participated in the battle after expiration of his term of service, distinguishing himself in rallying several regiments which had been thrown into confusion.
R—NR.
B—NR.

HARTZOG, JOSHUA B............. | Private, Company E, 1st U. S. Artillery.
At Wounded Knee Creek, S. Dak., Dec. 29, 1890. | Went to the rescue of the commanding officer who had fallen severely wounded, picked him up, and carried him out of range of the hostile guns.
R—NR.
B—Paulding County, Ohio.

HARVEY, HARRY............. | Corporal, Company A, 22d New York Cavalry.
At Waynesboro, Va., Mar. 2, 1865. | Capture of flag and bearer, with two other prisoners.
R—NR.
B—England.

HASKELL, FRANK W............. | Sergeant major, 3d Maine Infantry.
At Fair Oaks, Va., June 1, 1862. | Assumed command of a portion of the left wing of his regiment, all the company officers present having been killed or disabled, led it gallantly across a stream and contributed most effectively to the success of the action.
R—NR.
B—NR.

HASKELL, MARCUS M............. | Sergeant, Company C, 35th Massachusetts Infantry.
At Antietam, Md., Sept. 17, 1862. | Although wounded and exposed to a heavy fire from the enemy, at the risk of his own life he rescued a badly wounded comrade and succeeded in conveying him to a place of safety.
R—Chelsea, Mass.
B—Chelsea, Mass.

HASTINGS, SMITH H............. | Captain, Company M, 5th Michigan Cavalry.
At Newbys Crossroads, Va., July 24, 1863. | While in command of a squadron in rear guard of a cavalry division, then retiring before the advance of a corps of infantry, was attacked by the enemy and, orders having been given to abandon the guns of a section of field artillery with the rear guard that were in imminent danger of capture, he disregarded the orders received and aided in repelling the attack and saving the guns.
R—NR.
B—NR.

HATCH, JOHN P............. | Brigadier general, U. S. Volunteers.
At South Mountain, Md., Sept. 14, 1862. | Was severely wounded while leading one of his brigades in the attack under a heavy fire from the enemy.
R—New York, N. Y.
B—New York, N. Y.

HATLER, M. WALDO (2199881)_____
  Near Pouilly, France, Nov. 8, 1918.
  R—Neosho, Mo.
  B—Bolivar, Mo.
  G. O. No. 74, W. D., 1919.

Sergeant, Company B, 356th Infantry, 89th Division.
When volunteers were called for to secure information as to the enemy's position on the opposite bank of the Meuse River, Sergeant Hatler was the first to offer his services for this dangerous mission. Swimming across the river, he succeeded in reaching the German lines, after another soldier, who had started with him, had been seized with cramps and drowned in midstream. Alone he carefully and courageously reconnoitered the enemy's positions, which were held in force, and again successfully swam the river, bringing back information of great value.

HAUPT, PAUL_____
  At Hell Canyon, Ariz., July 3, 1869.
  R—N R.
  B—Prussia.

Corporal, Company L, 8th U. S. Cavalry.
Gallantry in action.

HAVRON, JOHN H_____
  At Petersburg, Va., Apr. 2, 1865.
  R—Providence, R. I.
  B—Ireland.

Sergeant, Company G, 1st Rhode Island Light Artillery.
Was one of a detachment of 20 picked artillerymen who voluntarily accompanied an infantry assaulting party and who turned upon the enemy the guns captured in the assault.

HAWKINS, GARDNER C_____
  At Petersburg, Va., Apr. 2, 1865.
  R—Woodstock, Vt.
  B—Pomfret, Vt.

First lieutenant, Company E, 3d Vermont Infantry.
When the lines were wavering from the well-directed fire of the enemy, this officer, acting adjutant of the regiment, sprang forward, and with encouraging words cheered the soldiers on and, although dangerously wounded, refused to leave the field until the enemy's works were taken.

HAWKINS, MARTIN J_____
  Georgia, April, 1862.
  R—N R.
  B—Mercer County, Pa.

Corporal, Company A, 33d Ohio Infantry.
One of 22 men (including two civilians) who, by direction of General Mitchel (or Buell), penetrated nearly 200 miles south into the enemy's territory and captured a railroad train at Big Shanty, Ga., in an attempt to destroy the bridges and track between Chattanooga and Atlanta.

HAWKINS, THOMAS_____
  At Deep Bottom, Va., July 21, 1864.
  R—Philadelphia, Pa.
  B—Cincinnati, Ohio.

Sergeant major, 6th United States Colored Troops.
Rescue of regimental colors.

HAWTHORN, HARRIS S_____
  At Sailors Creek, Va., Apr. 6, 1865.
  R—N R.
  B—Salem, N. Y.

Corporal, Company F, 121st New York Infantry.
Captured the Confederate Gen. G. W. Custis Lee.

HAWTHORNE, HARRY L_____
  At Wounded Knee Creek, S. Dak.,
    Dec. 29, 1890.
  R—Kentucky.
  B—Minnesota.

Second lieutenant, 2d U. S. Artillery.
Distinguished conduct in battle with hostile Indians.

HAY, FRED S_____
  At Upper Washita, Tex., Sept. 9,
    1874.
  R—N R.
  B—Scotland.

Sergeant, Company I, 5th U. S. Infantry.
Gallantry in action.

HAYES, WEBB C_____
  At Vigan, P. I., Dec. 4, 1899.
  R—Fremont, Ohio.
  B—Cincinnati, Ohio.

Lieutenant colonel, 31st Infantry, U. S. Volunteers.
Pushed through the enemy's lines alone, during the night, from the beach to the beleaguered force at Vigan, and returned the following morning to report the condition of affairs to the Navy and secure assistance.

HAYNES, ASBURY F_____
  At Sailors Creek, Va., Apr. 6, 1865.
  R—Maine.
  B—Edinburgh, Me.

Corporal, Company F, 17th Maine Infantry.
Capture of flag.

HAYS, GEORGE PRICE_____
  Near Greves Farm, France, July
    14-15, 1918.
  R—Okarche, Okla.
  B—China.
  G. O. No. 34, W. D., 1919.

First lieutenant, 10th Field Artillery, 3d Division.
At the very outset of the unprecedented artillery bombardment by the enemy of July 14-15, 1918, his line of communication was destroyed beyond repair. Despite the hazard attached to the mission of runner, he immediately set out to establish contact with the neighboring post of command and further established liaison with two French batteries, visiting their position so frequently that he was mainly responsible for the accurate fire therefrom. While thus engaged, seven horses were shot under him and he was severely wounded. His activity under most severe fire was an important factor in checking the advance of the enemy.

HAYS, JOHN H_____
  At Columbus, Ga., Apr. 16, 1865.
  R—N R.
  B—Jefferson County, Ohio.

Private, Company F, 4th Iowa Cavalry.
Capture of flag and bearer, Austin's Battery (C. S. A.).

HEALEY, GEORGE W_____
  At Newnan, Ga., July 29, 1864.
  R—N R.
  B—Dubuque, Iowa.

Private, Company E, 5th Iowa Cavalry.
When nearly surrounded by the enemy, captured a Confederate soldier, and, with the aid of a comrade who joined him later, captured four other Confederate soldiers, disarmed the five prisoners, and brought them all into the Union lines.

HEARD, JOHN W.
   At Mouth of Manimani River, west of Bahia Honda, Cuba, July 23, 1898.
   R—Mississippi.
   B—Mississippi.

First lieutenant, 3d U. S. Cavalry.
After two men had been shot down by Spaniards while transmitting orders to the engine room on the Wanderer, the ship having become disabled, this officer took the position held by them and personally transmitted the orders, remaining at his post until the ship was out of danger.

HEARTERY, RICHARD
   At Cibicu, Ariz., Aug. 30, 1881.
   R—NR.
   B—Ireland.

Private, Company D, 6th U. S. Cavalry.
Bravery in action.

HEDGES, JOSEPH
   Near Harpeth River, Tenn., Dec. 17, 1864.
   R—Ohio.
   B—Ohio.

First lieutenant, 4th U. S. Cavalry.
At the head of his regiment charged a field battery with strong infantry supports, broke the enemy's line and, with other mounted troops, captured three guns and many prisoners.

HEERMANCE, WILLIAM L
   At Chancellorsville, Va., Apr. 30, 1863.
   R—Kinderhook, N. Y.
   B—NR.

Captain, Company C, 6th New York Cavalry.
Took command of the regiment as its senior officer when surrounded by Stuart's Cavalry. The regiment cut its way through the enemy's line and escaped, but Captain Heermance was desperately wounded, left for dead on the field, and was taken prisoner.

HEISE, CLAMOR
   Arizona, August to October, 1863.
   R—NR.
   B—Germany.

Private, Company B, 8th U. S. Cavalry.
Bravery in scouts and actions with Indians.

HELLER, HENRY
   At Chancellorsville, Va., May 2, 1863.
   R—Urbana, Ohio.
   B—NR.

Sergeant, Company A, 66th Ohio Infantry.
One of a party of four who, under heavy fire, voluntarily brought into the Union lines a wounded Confederate officer from whom was obtained valuable information concerning the position of the enemy.

HELMS, DAVID H
   At Vicksburg, Miss., May 22, 1863.
   R—NR.
   B—Dearborn, Ind.

Private, Company B, 83d Indiana Infantry.
Gallantry in the charge of the "volunteer storming party."

HENDERSON, JOSEPH
   At Patian Island, P. I., July 2, 1909.
   R—Leavenworth, Kans.
   B—Leavenworth, Kans.

Sergeant, Troop B, 6th U. S. Cavalry.
While in action against hostile Moros, voluntarily advanced alone, in the face of a heavy fire, to within about 15 yards of the hostile position and refastened to a tree a block and tackle used in checking the recoil of a mountain gun.

HENRY, GUY V
   At Cold Harbor, Va., June 1, 1864.
   R—Reading, Pa.
   B—NR.

Colonel, 40th Massachusetts Infantry.
Led the assaults of his brigade upon the enemy's works, where he had two horses shot under him.

HENRY, JAMES
   At Vicksburg, Miss., May 22, 1863.
   R—NR.
   B—Sunfish, Ohio.

Sergeant, Company B, 113th Illinois Infantry.
Gallantry in the charge of the "volunteer storming party."

HENRY, JOHN
   First sergeant, Troop I, 3d U. S. Cavalry.

See John H. Shingle, true name.

HENRY, WILLIAM W
   At Cedar Creek, Va., Oct. 19, 1864.
   R—Waterbury, Vt.
   B—NR.

Colonel, 10th Vermont Infantry.
Though suffering from severe wounds, rejoined his regiment and led it in a brilliant charge, recapturing the guns of an abandoned battery.

HERINGTON, PITT B
   Near Kenesaw Mountain, Ga., June 15, 1864.
   R—NR.
   B—Michigan.

Private, Company E, 11th Iowa Infantry.
With one companion and under a fierce fire of the enemy at close range, went to the rescue of a wounded comrade who had fallen between the lines and carried him to a place of safety.

*HERIOT, JAMES D. (1311750)
   At Vaux-Andigny, France, Oct. 12, 1918.
   R—Providence, S. C.
   B—Providence, S. C.
   G. O. No. 13, W. D., 1919.

Corporal, Company I, 118th Infantry, 30th Division.
Corporal Heriot, with four other soldiers, organized a combat group and attacked an enemy machine-gun nest which had been inflicting heavy casualties on his company. In the advance two of his men were killed, and because of the heavy fire from all sides the remaining two sought shelter. Unmindful of the hazard attached to his mission, Corporal Heriot, with fixed bayonet, alone charged the machine gun, making his way through the fire for a distance of 30 yards and forcing the enemy to surrender. During this exploit he received several wounds in the arm, and later in the same day, while charging another nest, he was killed.
Posthumously awarded. Medal presented to mother, Mrs. Carrie C. Heriot.

HERRON, FRANCIS J
   At Pea Ridge, Ark., Mar. 7, 1862.
   R—Pittsburgh, Pa.
   B—NR.

Lieutenant colonel, 9th Iowa Infantry.
Was foremost in leading his men, rallying them to repeated acts of daring, until himself disabled and taken prisoner.

HERRON, LEANDER ....
    Near Fort Dodge, Kans., Sept. 2,
    1868.
    R—NR.
    B—Bucks County, Pa.

Corporal, Company A, 3d U. S. Infantry.
While detailed as mail courier from the fort, voluntarily went to the assistance of a party of four enlisted men, who were attacked by about 50 Indians at some distance from the fort and remained with them until the party was relieved.

HESSELTINE, FRANCIS S ....
    At Matagorda Bay, Tex., Dec. 29-30,
    1863.
    R—NR.
    B—NR.

Colonel, 13th Maine Infantry.
In command of a detachment of 100 men, conducted a reconnaissance for two days, baffling and beating back an attacking force of more than a thousand Confederate cavalry, and regained his transport without loss.

HEYL, CHARLES H ....
    Near Fort Hartsuff, Nebr., Apr. 28,
    1876.
    R—Camden, N. J.
    B—Philadelphia, Pa.

Second lieutenant, 23d U. S. Infantry.
Voluntarily, and with most conspicuous gallantry, charged with three men upon six Indians who were intrenched upon a hillside.

HIBSON, JOSEPH C ....
    Near Fort Wagner, S. C., July 13,
    1863.
    Near Fort Wagner, S. C., July 14,
    1863.
    Near Fort Wagner, S. C., July 18,
    1863.
    R—New York, N. Y.
    B—England.

Private, Company C, 48th New York Infantry.
While voluntarily performing picket duty under fire, was attacked and his surrender demanded, but he killed his assailant.
Responded to a call for a volunteer to reconnoiter the enemy's position, and went within the enemy's lines under fire and was exposed to great danger.
Voluntarily exposed himself with great gallantry during an assault, and received three wounds that permanently disabled him for active service.

HICKEY, DENNIS W ....
    At Stony Creek Bridge, Va., June
    29, 1864.
    R—NR.
    B—Troy, N. Y.

Sergeant, Company E, 2d New York Cavalry.
With a detachment of three men, tore up the bridge at Stony Creek, being the last man on the bridge and covering the retreat until he was shot down.

HICKOK, NATHAN E ....
    At Chapin's Farm, near Richmond,
    Va., Sept. 29, 1864.
    R—NR.
    B—Danbury, Conn.

Corporal, Company A, 8th Connecticut Infantry.
Capture of flag.

HIGBY, CHARLES ....
    At Appomattox campaign, Virginia,
    Mar. 29 to Apr. 9, 1865.
    R—NR.
    B—Pittsburgh, Pa.

Private, Company F, 1st Pennsylvania Cavalry.
Capture of flag.

HIGGINS, THOMAS J ....
    At Vicksburg, Miss., May 22, 1863.
    R—NR.
    B—Canada.

Sergeant, Company D, 99th Illinois Infantry.
When his regiment fell back in the assault, repulsed, this soldier continued to advance and planted the flag on the parapet, where he was captured by the enemy.

HIGGINS, THOMAS P ....
    Arizona, August to October, 1868.
    R—NR.
    B—Ireland.

Private, Company B, 8th U. S. Cavalry.
Bravery in scouts and actions with Indians.

HIGH, FRANK C ....
    Near San Isidro, P. I., May 16, 1899.
    R—Picard, Calif.
    B—Yolo County, Calif.

Private, Company G, 2d Oregon Volunteer Infantry
With 21 other scouts charged across a burning bridge, under heavy fire, and completely routed 600 of the enemy who were intrenched in a strongly fortified position.

HIGHLAND, PATRICK ....
    At Petersburg, Va., Apr. 2, 1865.
    R—Chicago, Ill.
    B—Ireland.

Corporal, Company D, 23d Illinois Infantry.
Conspicuous gallantry as color bearer in the assault on Fort Gregg.

HILL, EDWARD ....
    At Cold Harbor, Va., June 1, 1864.
    R—Detroit, Mich.
    B—Liberty, N. Y.

Captain, Company K, 16th Michigan Infantry.
Led the brigade skirmish line in a desperate charge on the enemy's masked batteries to the muzzles of the guns, where he was severely wounded.

HILL, FRANK E ....
    At Date Creek, Ariz., Sept. 8, 1872.
    R—NR.
    B—Mayfield, Wis.

Sergeant, Company E, 5th U. S. Cavalry.
Secured the person of a hostile Apache chief, although while holding the chief he was severely wounded in the back by another Indian.

HILL, HENRY ....
    At The Wilderness, Va., May 6,
    1864.
    R—NR.
    B—Schuylkill County, Pa.

Corporal, Company C, 50th Pennsylvania Infantry.
This soldier, with one companion, would not retire when his regiment fell back in confusion after an unsuccessful charge, but instead advanced and continued firing upon the enemy until the regiment reformed and regained its position.

HILL, JAMES ....
    At Petersburg, Va., July 30, 1864.
    R—NR.
    B—Lyons, N. Y.

Sergeant, Company C, 14th New York Heavy Artillery.
Capture of flag, shooting a Confederate officer who was rallying his men with the colors in his hand.

HILL, JAMES_____
  At Champion Hill, Miss., May 16, 1863.
  R—Cascade, Iowa.
  B—England.

First lieutenant, Company I, 21st Iowa Infantry.
By skillful and brave management captured three of the enemy's pickets.

HILL, JAMES M_____
  At Turret Mountain, Ariz., Mar. 25, 1873.
  R—NR.
  B—Washington County, Pa.

First sergeant, Company A, 5th U. S. Cavalry.
Gallantry in action.

HILL, RALYN (1381313)_____
  Near Donnevoux, France, Oct. 7, 1918.
  R—Oregon, Ill.
  B—Lindenwood, Ill.
  G. O. No. 34, W. D., 1919.

Corporal, Company H, 129th Infantry, 33d Division.
Seeing a French airplane fall out of control on the enemy side of the Meuse River with its pilot injured, Corporal Hill voluntarily dashed across the foot bridge to the side of the wounded man and, taking him on his back, started back to his lines. During the entire exploit he was subjected to murderous fire of enemy machine guns and artillery, but he successfully accomplished his mission and brought his man to a place of safety, a distance of several hundred yards.

HILLIKER, BENJAMIN F_____
  At Mechanicsburg, Miss., June 4, 1863.
  R—NR.
  B—New York.

Musician, Company A, 8th Wisconsin Infantry.
Exchanging his drum for a rifle, he voluntarily joined the skirmish line under fire, and advancing with it, received a severe wound, which resulted in his permanent disability.

HILLOCK, MARVIN C_____
  At Wounded Knee Creek, S. Dak., Dec. 29, 1890.
  R—Lead City, S. Dak.
  B—Michigan.

Private, Company B, 7th U. S. Cavalry.
Distinguished bravery.

HILLS, WILLIAM G_____
  At North Fork, Va., Sept. 26, 1864.
  R—NR.
  B—NR.

Private, Company E, 9th New York Cavalry.
Voluntarily carried a severely wounded comrade out of a heavy fire of the enemy.

HILTON, ALFRED B_____
  At Chapin's Farm, near Richmond, Va., Sept. 29, 1864.
  R—NR.
  B—Harford County, Md.

Sergeant, Company H, 4th U. S. Colored Troops.
When the regimental color bearer fell, this soldier seized the color and carried it forward, together with the national standard, until disabled at the enemy's inner line.

HILTON, RICHMOND H (1312381)_____
  At Brancourt, France, Oct. 11, 1918.
  R—Westville, S. C.
  B—Westville, S. C.
  G. O. No. 16, W. D., 1919.

Sergeant, Company M, 118th Infantry, 30th Division.
While Sergeant Hilton's company was advancing through the village of Brancourt it was held up by intense enfilading fire from a machine gun. Discovering that this fire came from a machine-gun nest among shell holes at the edge of the town, Sergeant Hilton, accompanied by a few other soldiers, but well in advance of them, pressed on toward this position, firing with his rifle until his ammunition was exhausted, and then with his pistol, killing 6 of the enemy and capturing 10. In the course of this daring exploit he received a wound from a bursting shell, which resulted in the loss of his arm.

HIMMELSBACK, MICHAEL_____
  At Little Blue, Nebr., May 15, 1870.
  R—NR.
  B—Allegheny County, Pa.

Private, Company C, 2d U. S. Cavalry.
Gallantry in action.

HINCKS, WILLIAM B_____
  At Gettysburg, Pa., July 3, 1863.
  R—Bridgeport, Conn.
  B—Bucksport, Me.

Sergeant major, 14th Connecticut Infantry.
Capture of flag of 14th Tennessee Infantry (C. S. A.).

HINEMANN, LEHMANN_____
  Winter of 1872–73.
  R—NR.
  B—Germany.

Sergeant, Company L, 1st U. S. Cavalry.
Gallant conduct in campaigns and engagements with Apaches.

HOBDAY, GEORGE_____
  At Wounded Knee Creek, S. Dak., Dec. 29, 1890.
  R—NR.
  B—Pulaski County, Ill.

Private, Company A, 7th U. S. Cavalry.
Conspicuous and gallant conduct in battle.

HODGES, ADDISON J_____
  At Vicksburg, Miss., May 3, 1863.
  R—Adrian, Mich.
  B—Hillsdale, Mich.

Private, Company B, 47th Ohio Infantry.
Was one of a party that volunteered and attempted to run the enemy's batteries with a steam tug and two barges loaded with subsistence stores.

HOFFMAN, CHARLES F_____ | Gunnery sergeant, 49th Company, 5th Regiment U. S. Marine Corps, 2d Division.
Near Chateau-Thierry, France, June 6, 1918.
R—Brooklyn, N. Y.
B—New York, N. Y.
G. O. No. 34, W. D., 1919.

Immediately after the company to which he belonged had reached its objective on Hill 142, several hostile counterattacks were launched against the line before the new position had been consolidated. Sergeant Hoffman was attempting to organize a position on the north slope of the hill when he saw 12 of the enemy, armed with five light machine guns, crawling toward his group. Giving the alarm, he rushed the hostile detachment, bayoneted the two leaders, and forced the others to flee, abandoning their guns. His quick action, initiative, and courage drove the enemy from a position from which they could have swept the hill with machine-gun fire and forced the withdrawal of our troops.

HOFFMAN, HENRY_____
At Sailors Creek, Va., Apr. 6, 1865.
R—N R.
B—Germany.

Corporal, Company M, 2d Ohio Cavalry.
Capture of flag.

HOFFMAN, THOMAS W_____
At Petersburg, Va., Apr. 2, 1865.
R—N R.
B—Perrysburg, Pa.

Captain, Company A, 208th Pennsylvania Infantry.
Prevented a retreat of his regiment during the battle.

HOGAN, FRANKLIN_____
Front of Petersburg, Va., July 30, 1864.
R—Howard, Pa.
B—Center County, Pa.

Corporal, Company A, 45th Pennsylvania Infantry.
Capture of flag of 6th Virginia Infantry (C. S. A.).

HOGAN, HENRY_____
(2 medals of honor.)
At Cedar Creek, etc., Mont., October, 1876, to Jan. 8, 1877.
At Bear Paw Mountain, Mont., Sept. 30, 1877.
R—N R.
B—Ireland.

First sergeant, Company G, 5th U. S. Infantry.
Gallantry in actions.
Carried Lieutenant Romeyn, who was severely wounded, off the field of battle under heavy fire.

HOGARTY, WILLIAM P_____
At Antietam, Md., Sept. 17, 1862; at Fredericksburg, Va., Dec. 13, 1862.
R—N R.
B—New York, N. Y.

Private, Company D, 23d New York Infantry.
Distinguished gallantry in actions while attached to Battery B, 4th U. S. Artillery; lost his left arm at Fredericksburg.

HOLCOMB, DANIEL I_____
At Brentwood Hills, Tenn., Dec. 16, 1864.
R—N R.
B—Hartford, Ohio.

Private, Company A, 41st Ohio Infantry.
Capture of Confederate guidon.

HOLDEN, HENRY_____
At Little Big Horn River, Mont., June 25, 1876.
R—N R.
B—England.

Private, Company D, 7th U. S. Cavalry.
Brought up ammunition under a galling fire from the enemy.

HOLDERMAN, NELSON M_____
Northeast of Binarville, in the forest of Argonne, France, Oct. 2-8, 1918.
R—Santa Ana, Calif.
B—Trumbell, Nebr.
G. O. No. 11, W. D., 1921.

Captain, 307th Infantry, 77th Division.
Captain Holderman commanded a company of a battalion which was cut off and surrounded by the enemy. He was wounded on Oct. 4, on Oct. 5, and again on Oct. 7, but throughout the entire period, suffering great pain and subjected to fire of every character, he continued personally to lead and encourage the officers and men under his command with unflinching courage and with distinguished success. On Oct. 6, in a wounded condition, he rushed through enemy machine-gun and shell fire and carried two wounded men to a place of safety.

HOLEHOUSE, JAMES_____
At Maryes Heights, Va., May 3, 1863.
R—N R.
B—England.

Private, Company B, 7th Massachusetts Infantry.
With one companion voluntarily and with conspicuous daring advanced beyond his regiment, which had been broken in the assault, and halted beneath the crest. Following the example of these two men, the colors were brought to the summit, the regiment was advanced and the position held.

HOLLAND, DAVID_____
At Cedar Creek, etc., Mont., October, 1876, to Jan. 8, 1877.
R—N R.
B—Dearborn, Mich.

Corporal, Company A, 5th U. S. Infantry.
Gallantry in actions.

HOLLAND, LEMUEL F_____
At Elk River, Tenn., July 2, 1863.
R—N R.
B—Burlington, Ohio.

Corporal, Company D, 104th Illinois Infantry.
Voluntarily joined a small party that, under a heavy fire, captured a stockade and saved the bridge.

HOLLAND, MILTON M_____
At Chapins Farm, near Richmond, Va., Sept. 29, 1864.
R—Athens, Ohio.
B—Austin, Tex.

Sergeant major, 5th U. S. Colored Troops.
Took command of Company C, after all the officers had been killed or wounded, and gallantly led it.

HOLMES, LOVILO N_____
At Nolensville, Tenn., Feb. 15, 1863.
R—NR.
B—Cattaraugus County, N. Y.

First sergeant, Company H, 2d Minnesota Infantry.
Was one of a detachment of 16 men who heroically defended a wagon train against the attack of 125 cavalry, repulsed the attack, and saved the train.

HOLMES, WILLIAM T_____
At Sailors Creek, Va., Apr. 6, 1865.
R—Indianapolis, Ind.
B—Vermilion County, Ill.

Private, Company A, 3d Indiana Cavalry.
Capture of flag of 27th Virginia Infantry (C. S. A.).

HOLTON, CHARLES M_____
At Falling Waters, Va., July 14, 1863.
R--NR.
B—Potter, N. Y.

First sergeant, Company A, 7th Michigan Cavalry.
Capture of flag of 55th Virginia Infantry (C. S. A.).

HOLTON, EDWARD A_____
At Lees Mills, Va., Apr. 16, 1862.
R—NR.
B—Westminster, Vt.

First sergeant, Company I, 6th Vermont Infantry.
Rescued the colors of his regiment under heavy fire, the color bearer having been shot down while the troops were in retreat.

HOMAN, CONRAD_____
Near Petersburg, Va., July 30, 1864.
R—NR.
B—Roxbury, Mass.

Color sergeant, Company A, 29th Massachusetts Infantry.
Fought his way through the enemy's lines with the regimental colors, the rest of the color guard being killed or captured.

*HOOKER, GEORGE_____
At Tonto Creek, Ariz., Jan. 22, 1873.
R—NR.
B—Frederick, Md.

Private, Company K, 5th U. S. Cavalry.
Gallantry in action in which he was killed.

HOOKER, GEORGE W_____
At South Mountain, Md., Sept. 14, 1862.
R—Boston, Mass.
B--Salem, N. Y.

First lieutenant, Company E, 4th Vermont Infantry.
Rode alone, in advance of his regiment, into the enemy's lines, and before his own men came up received the surrender of the major of a Confederate regiment, together with the colors and 116 men.

HOOPER, WILLIAM B_____
At Chamberlains Creek, Va., Mar. 31, 1865.
R—NR.
B—Willimantic, Conn.

Corporal, Company L, 1st New Jersey Cavalry.
With the assistance of a comrade headed off the advance of the enemy, shooting two of his color bearers; also posted himself between the enemy and the led horses of his own command, thus saving the herd from capture.

HOOVER, SAMUEL_____
At Santa Maria Mountains, Ariz., May 6, 1873.
R--NR.
B—Dauphin County, Pa.

Bugler, Company A, 1st U. S. Cavalry.
Gallantry in action; also services as trailer in May, 1872.

HOPKINS, CHARLES F_____
At Gaines Mill, Va., June 27, 1862.
R—NR.
B—Warren County, N. J.

Corporal, Company I, 1st New Jersey Infantry.
Voluntarily carried a wounded comrade, under heavy fire, to a place of safety; though twice wounded in the act, he continued in action until again severely wounded.

HORAN, THOMAS_____
At Gettysburg, Pa., July 2, 1863.
R—Dunkirk, N. Y.
B—NR.

Sergeant, Company E, 72d New York Infantry.
In a charge of his regiment this soldier captured the regimental flag of the 8th Florida Infantry (C. S. A.).

HORNADAY, SIMPSON_____
At Sappa Creek, Kans., Apr. 23, 1875.
R—NR.
B—Hendricks County, Ind.

Private, Company H, 6th U. S. Cavalry.
Gallantry in action.

HORNE, SAMUEL B_____
At Fort Harrison, Va., Sept. 29, 1864.
R—Winsted, Conn.
B—England.

Captain, Company H, 11th Connecticut Infantry.
While acting as an aid and carrying an important message, was severely wounded and his horse killed, but delivered the order and rejoined his general.

HORSFALL, WILLIAM H_____
At Corinth, Miss., May 21, 1862.
R—NR.
B—NR.

Drummer, Company G, 1st Kentucky Infantry.
Saved the life of a wounded officer lying between the lines.

HOTTENSTINE, SOLOMON J_____
At Petersburg and Norfolk Railroad, Va., Aug. 19, 1864.
R—Philadelphia, Pa.
B—Leigh County, Pa.

Private, Company C, 107th Pennsylvania Infantry.
Capture of flag belonging to a North Carolina regiment.

HOUGH, IRA_____
At Cedar Creek, Va., Oct. 19, 1864.
R—Henry County, Ind.
B—Henry County, Ind.

Private, Company E, 8th Indiana Infantry.
Capture of flag.

HOUGHTON, CHARLES H.
At Petersburg, Va., July 30, 1864; Mar. 25, 1865.
R—Ogdensburg, N. Y.
B—NR.

Captain, Company L, 14th New York Artillery.
In the Union assault at the crater and in the Confederate assault repelled at Fort Haskell, displayed most conspicuous gallantry and repeatedly exposed himself voluntarily to great danger, was three times wounded, and suffered the loss of a leg.

HOUGHTON, GEORGE L.
At Elk River, Tenn., July 2, 1863.
R—NR.
B—Canada.

Private, Company D, 104th Illinois Infantry.
Voluntarily joined a small party that, under a heavy fire, captured a stockade and saved the bridge.

HOULTON, WILLIAM.
At Sailors Creek, Va., Apr. 6, 1865.
R—NR.
B—Clymer, N. Y.

Commissary sergeant, 1st West Virginia Cavalry.
Capture of flag.

HOWARD, HENDERSON C.
At Glendale, Va., June 30, 1862.
R—NR.
B—NR.

Corporal, Company B, 11th Pennsylvania Reserves.
While pursuing one of the enemy's sharpshooters, encountered two others, whom he bayoneted in hand-to-hand encounters; was three times wounded in action.

HOWARD, HIRAM R.
At Missionary Ridge, Tenn., Nov. 25, 1863.
R—NR.
B—NR.

Private, Company H, 11th Ohio Infantry.
Scaled the enemy's works and in a hand-to-hand fight captured a flag.

HOWARD, JAMES.
At Battery Gregg, near Petersburg, Va., Apr. 2, 1865.
R—NR.
B—Newton, N. J.

Sergeant, Company K, 158th New York Infantry.
Carried the colors in advance of the line of battle, the flagstaff being shot off while he was planting it on the parapet of the fort.

HOWARD, OLIVER O.
At Fair Oaks, Va., June 1, 1862.
R—Maine.
B—Maine.

Brigadier general, U. S. Volunteers.
Led the 61st New York Infantry in a charge in which he was twice severely wounded in the right arm, necessitating amputation.

HOWARD, SQUIRE E.
At Bayou Teche, La., Jan. 14, 1863.
R—Townshend, Vt.
B—Jamaica, Vt.

First sergeant, Company H, 8th Vermont Infantry.
Voluntarily carried an important message through the heavy fire of the enemy.

HOWE, ORION P.
At Vicksburg, Miss., May 19, 1863.
R—Waukegan, Ill.
B—Portage County, Ohio.

Musician, Company C, 55th Illinois Infantry.
A drummer boy, 14 years of age, and severely wounded and exposed to a heavy fire from the enemy, he persistently remained upon the field of battle until he had reported to Gen. W. T. Sherman the necessity of supplying cartridges for the use of troops under command of Colonel Malmborg.

HOWE, WILLIAM H.
At Fort Stedman, Va., Mar. 25, 1865.
R—NR.
B—Haverhill, Mass.

Sergeant, Company K, 29th Massachusetts Infantry.
Served an abandoned gun under heavy fire.

HOWZE, ROBERT L.
At White River, S. Dak., Jan. 1, 1891.
R—Texas.
B—Texas.
Distinguished-service medal also awarded.

Second lieutenant, Company K, 6th U. S. Cavalry.
Bravery in action.

HUBBARD, THOMAS.
At Little Blue, Nebr., May 15, 1870.
R—NR.
B—Philadelphia, Pa.

Private, Company C, 2d U. S. Cavalry.
Gallantry in action.

HUBBELL, WILLIAM S.
At Fort Harrison, Va., Sept. 30, 1864.
R—North Stonington, Conn.
B—NR.

Captain, Company A, 21st Connecticut Infantry.
Led out a small flanking party and by a dash and at great risk captured a large number of prisoners.

HUDSON, AARON R.
At Culloden, Ga., Apr. —, 1865.
R—La Porte County, Ind.
B—Madison County, Ky.

Private, Company C, 17th Indiana Mounted Infantry.
Capture of flag of Worrill Grays (C. S. A.).

HUFF, JAMES W.
Winter of 1872–73.
R—NR.
B—Washington, Pa.

Private, Company L, 1st U. S. Cavalry.
Gallant conduct in campaigns and engagements with Apaches.

HUGGINS, ELI L.
At O'Fallons Creek, Mont., Apr. 1, 1880.
R—Minnesota.
B—Illinois.

Captain, 2d U. S. Cavalry.
Surprised the Indians in their strong position and fought them until dark with great boldness.

HUGHEY, JOHN........................
    At Sailors Creek, Va., Apr. 6, 1865.
    R—NR.
    B—Highland, Ohio.

Corporal, Company L, 2d Ohio Cavalry.
Capture of flag of 38th Virginia Infantry (C. S. A.).

HUGHS, OLIVER........................
    At Weldon Railroad, Va., June 24,
    1864.
    R—NR.
    B—Fentress County, Tenn.

Corporal, Company C, 12th Kentucky Infantry.
Capture of flag of 11th South Carolina (C. S. A.).

HUIDEKOPER, HENRY S................
    At Gettysburg, Pa., July 1, 1863.
    R—NR.
    B—NR.

Lieutenant colonel, 150th Pennsylvania Infantry.
While engaged in repelling an attack of the enemy, received a severe wound of the right arm, but instead of retiring remained at the front in command of the regiment.

HUMPHREY, CHARLES F...............
    At Clearwater, Idaho, July 11, 1877.
    R—NR.
    B—New York.

First lieutenant, 4th U. S. Artillery.
Voluntarily and successfully conducted, in the face of a withering fire, a party which recovered possession of an abandoned howitzer and two Gatling guns lying between the lines a few yards from the Indians.

HUNT, FRED O.......................
    At Cedar Creek, etc., Mont.,
    October, 1876, to January 8, 1877.
    R—NR.
    B—New Orleans, La.

Private, Company A, 5th U. S. Infantry.
Gallantry in actions.

HUNT, LEWIS T......................
    At Vicksburg, Miss., May 22, 1863.
    R—NR.
    B—Montgomery County, Ind.

Private, Company H, 6th Missouri Infantry.
Gallantry in the charge of the "volunteer storming party."

HUNTER, CHARLES A.................
    At Petersburg, Va., Apr. 2, 1865.
    R—Spencer, Mass.
    B—Spencer, Mass.

Sergeant, Company E, 34th Massachusetts Infantry.
In the assault on Fort Gregg, bore the regimental flag bravely and was among the foremost to enter the work.

HUNTERSON, JOHN C.................
    On the Peninsula, Va., June 5, 1862.
    R—Philadelphia, Pa.
    B—Philadelphia, Pa.

Private, Company B, 3d Pennsylvania Cavalry.
While under fire, between the lines of the two armies, voluntarily gave up his own horse to an engineer officer whom he was accompanying on a reconnaissance and whose horse had been killed, thus enabling the officer to escape with valuable papers in his possession.

HUNTSMAN, JOHN A.................
    At Bamban, Luzon, P. I., Nov. 9,
    1899.
    R—Lawrence, Kans.
    B—Oskaloosa County, Iowa.

Sergeant, Company E, 36th Infantry, U. S. Volunteers.
For distinguished bravery and conspicuous gallantry in action against insurgents.

HUTCHINSON, RUFUS D..............
    At Little Big Horn River, Mont.,
    June 25, 1876.
    R—NR.
    B—Butlerville, Ohio.

Sergeant, Company B, 7th U. S. Cavalry.
Guarded and carried the wounded, brought water for the same, and posted and directed the men in his charge under galling fire from the enemy.

HYATT, THEODORE...................
    At Vicksburg, Miss., May 22, 1863.
    R—Gardner, Ill.
    B—Pennsylvania.

First sergeant, Company D, 127th Illinois Infantry.
Gallantry in the charge of the "volunteer storming party."

HYDE, HENRY J.....................
    Winter of 1872–73.
    R—NR.
    B—Bangor, Me.

Sergeant, Company M, 1st U. S. Cavalry.
Gallant conduct in campaigns and engagements with Apaches.

HYDE, THOMAS W...................
    At Antietam, Md., Sept. 17, 1862.
    R—NR.
    B—NR.

Major, 7th Maine Infantry.
Led his regiment in an assault on a strong body of the enemy's infantry and kept up the fight until the greater part of his men had been killed or wounded, bringing the remainder safely out of the fight.

HYMER, SAMUEL.....................
    At Buzzard's Roost Gap, Ga., Oct.
    13, 1864.
    R—NR.
    B—NR.

Captain, Company D, 115th Illinois Infantry.
With only 41 men under his command, defended and held a blockhouse against the attack of Hood's Division for nearly 10 hours, thus checking the advance of the enemy and insuring the safety of the balance of the regiment, as well as that of the 8th Kentucky Infantry, then stationed at Ringgold, Ga.

ILGENFRITZ, CHARLES H.............
    At Fort Sedgwick, Va., Apr. 2, 1865.
    R—Pennsylvania.
    B—York County, Pa.

Sergeant, Company E, 207th Pennsylvania Infantry.
The color bearer falling, pierced by seven balls, he immediately sprang forward and grasped the colors, planting them upon the enemy's forts amid a murderous fire of grape, canister, and musketry from the enemy.

IMMELL, LORENZO D................
    At Wilsons Creek, Mo., Aug. 10,
    1861.
    R—NR.
    B—Ross, Ohio.

Corporal, Company F, 2d U. S. Artillery.
Bravery in action.

INGALLS, LEWIS J. ....
At Boutte Station, La., Sept. 4, 1862.
R—NR.
B—Boston, Mass.

Private, Company K, 8th Vermont Infantry.
A railroad train guarded by about 60 men on flat cars having been sidetracked by a misplaced switch into an ambuscade of guerillas who were rapidly shooting down the unprotected guards, this soldier, under a severe fire in which he was wounded, ran to another switch and, opening it, enabled the train and the surviving guards to escape.

INSCHO, LEONIDAS H. ....
At South Mountain, Md., Sept. 14, 1862.
R—Charleston, W. Va.
B—Chatham, Ohio.

Corporal, Company E, 12th Ohio Infantry.
Alone and unaided and with his left hand disabled, captured a Confederate captain and four men.

IRSCH, FRANCIS ....
At Gettysburg, Pa., July 1, 1863.
R—New York, N. Y.
B—NR.

Captain, Company D, 45th New York, Infantry.
Gallantry in flanking the enemy and capturing a number of prisoners and in holding a part of the town against heavy odds while the Army was rallying on Cemetery Hill.

IRWIN, BERNARD J. D. ....
Apache Pass, Ariz., Feb.13-14, 1861.
R—New York.
B—Ireland.

Assistant surgeon, U. S. Army.
Voluntarily took command of troops and attacked and defeated the hostile Indians he met on the way.

IRWIN, PATRICK ....
At Jonesboro, Ga., Sept. 1, 1864.
R—NR.
B—Ireland.

First sergeant, Company H, 14th Michigan Infantry.
In a charge by the 14th Michigan Infantry against the intrenched enemy was the first man over the line of works of the enemy, and demanded and received the surrender of a Confederate general officer and his command.

JACKSON, FREDERICK R. ....
At James Island, S. C., June 16, 1862.
R—New Haven, Conn.
B—New Haven, Conn.

First sergeant, Company F, 7th Connecticut Infantry.
Having his left arm shot away in a charge on the enemy, he continued on duty, taking part in a second and a third charge until he fell exhausted from the loss of blood.

JACKSON, JAMES ....
At Camas Meadows, Idaho, Aug. 20, 1877.
R—NR.
B—New Jersey.

Captain, 1st United States Cavalry.
Dismounted from his horse in the face of a heavy fire from pursuing Indians, and with the assistance of one or two of the men of his command secured and carried to a place of safety the body of his trumpeter, who had been shot and killed.

JACOBSON, EUGENE P. ....
At Chancellorsville, Va., May 2, 1863.
R—New York, N. Y.
B—NR.

Sergeant major, 74th New York Infantry.
Bravery in conducting a scouting party in front of the enemy.

JAMES, ISAAC ....
At Petersburg, Va., Apr. 2, 1865.
R—NR.
B—Montgomery County, Ohio.

Private, Company H, 110th Ohio Infantry.
Capture of flag.

JAMES, JOHN ....
At Upper Washita, Tex., Sept. 9 to 11, 1874.
R—NR.
B—England.

Corporal, 5th U. S. Infantry.
Gallantry in action.

JAMES, MILES ....
At Chapins Farm, near Richmond, Va., Sept. 30, 1864.
R—Norfolk, Va.
B—Princess Anne County, Va.

Corporal, Company B, 36th U. S. Colored Troops.
Having had his arm mutilated, making immediate amputation necessary, he loaded and discharged his piece with one hand and urged his men forward; this within 30 yards of the enemy's works.

JAMIESON, WALTER ....
At Petersburg, Va., July 30, 1864.

At Fort Harrison, Va., Sept. 29, 1864.
R—NR.
B—France.

First sergeant, Company B, 139th New York Infantry
Voluntarily went between the lines under a heavy fire to the assistance of a wounded and helpless officer, whom he carried within the Union lines.
Seized the regimental color, the color bearer and guard having been shot down, and, rushing forward, planted it upon the fort in full view of the entire brigade.

JANSON, ERNEST AUGUST ....
Service rendered under the name Charles F. Hoffman.

Gunnery sergeant, 49th Company, 5th Regiment U. S. Marine Corps, 2d Division.

JARDINE, JAMES ....
At Vicksburg, Miss., May 22, 1863.
R—Hamilton County, Ohio.
B—Scotland.

Sergeant, Company F, 54th Ohio Infantry.
Gallantry in the charge of the "volunteer storming party."

JARVIS, FREDERICK ....
At Chiricahua Mountains, Ariz., Oct. 20, 1869.
R—NR.
B—Essex County, N. Y.

Sergeant, Company G, 1st U. S. Cavalry.
Gallantry in action.

JELLISON, BENJAMIN H. ....
At Gettysburg, Pa., July 3, 1863.
R—Newburyport, Mass.
B—Newburyport, Mass.

Sergeant, Company C, 19th Massachusetts Infantry.
Capture of flag of 57th Virginia Infantry (C. S. A.).

JENNINGS, JAMES T_____   Private, Company K, 56th Pennsylvania Infantry.
    At Weldon Railroad, Va., Aug. 20,   Capture of flag of 55th North Carolina Infantry (C. S. A.).
      1864.
    R—Bucks County, Pa.
    B—England.

JENSEN, GOTFRED_____   Private, Company D, 1st North Dakota Volunteer Infantry.
    At San Miguel de Mayumo, Luzon,   With 11 other scouts, without waiting for the supporting battalion to aid them
      P. I., May 13, 1899.   or to get into a position to do so, charged over a distance of about 150 yards and
    R—Devils Lake, N. Dak.   completely routed about 300 of the enemy, who were in line and in a position
    B—Denmark.   that could only be carried by a frontal attack.

JETTER, BERNHARD_____   Sergeant, Company K, 7th U. S. Cavalry.
    At Sioux campaign, December, 1890.   Distinguished bravery.
    R—NR.
    B—Germany.

JEWETT, ERASTUS W_____   First lieutenant, Company A, 9th Vermont Infantry.
    At Newport Barracks, N. C., Feb.   By long and persistent resistance and burning the bridges kept a superior force
      2, 1864.   of the enemy at a distance and thus covered the retreat of the garrison.
    R—St. Albans, Vt.
    B—St. Albans, Vt.

JIM_____   Sergeant, Indian Scouts.
    Winter of 1872–73.   Gallantry during campaigns and actions with Apaches.
    R—NR.
    B—Arizona Territory.

JOHN, WILLIAM_____   Private, Company E, 37th Ohio Infantry.
    At Vicksburg, Miss., May 22, 1863.   Gallantry in the charge of the "volunteer storming party."
    R—Chillicothe, Ohio.
    B—Germany.

JOHNDRO, FRANKLIN_____   Private, Company A, 118th New York Infantry.
    At Chapins Farm, near Richmond,   Capture of 40 prisoners.
      Va., Sept. 30, 1864.
    R—NR.
    B—Highgate Falls, Vt.

JOHNS, ELISHA_____   Corporal, Company B, 113th Illinois Infantry.
    At Vicksburg, Miss., May 22, 1863.   Gallantry in the charge of the "volunteer storming party."
    R—NR.
    B—Clinton, Ohio.

JOHNS, HENRY T_____   Private, Company C, 49th Massachusetts Militia Infantry.
    At Port Hudson, La., May 27, 1863.   Volunteered in response to a call and took part in the movement that was made
    R—NR.   upon the enemy's works under a heavy fire therefrom in advance of the gen-
    B—NR.   eral assault.

JOHNSON, ANDREW_____   Private, Company G, 116th Illinois Infantry.
    At Vicksburg, Miss., May 22, 1863.   Gallantry in the charge of the "volunteer storming party."
    R—Assumption, Ill.
    B—Delaware County, Ohio.

JOHNSON, FOLLETT_____   Corporal, Company H, 60th New York Infantry.
    At New Hope Church, Ga., May 27,   Voluntarily exposed himself to the fire of a Confederate sharpshooter, thus
      1864.   drawing fire upon himself and enabling his comrade to shoot the sharpshooter.
    R—NR.
    B—St. Lawrence, N. Y.

JOHNSON, HENRY_____   Sergeant, Company D, 9th U. S. Cavalry.
    At Milk River, Colo., Oct. 2–5, 1879.   Voluntarily left fortified shelter and under heavy fire at close range made the
    R—NR.   rounds of the pits to instruct the guards; fought his way to the creek and back
    B—Boynton, Va.   to bring water to the wounded.

JOHNSON, JOHN_____   Private, Company D, 2d Wisconsin Infantry.
    At Fredericksburg, Va., Dec. 13,   Conspicuous gallantry in battle in which he was severely wounded.
      1862.
    R—Rochester, Minn.
    B—Norway.

JOHNSON, JOSEPH E_____   First lieutenant, Company A, 58th Pennsylvania Infantry.
    At Fort Harrison, Va., Sept. 29,   Though twice severely wounded while advancing in the assault, he disregarded
      1864.   his injuries and was among the first to enter the fort, where he was wounded
    R—NR.   for the third time.
    B—NR.

JOHNSON, RUEL M_____   Major, 100th Indiana Infantry.
    At Chattanooga, Tenn., Nov. 25,   While in command of the regiment bravely exposed himself to the fire of the
      1863.   enemy, encouraging and cheering his men.
    R—NR.
    B—NR.

JOHNSON, SAMUEL_____   Private, Company G, 9th Pennsylvania Reserves.
    At Antietam, Md., Sept. 17, 1862.   Individual bravery and daring in capturing from the enemy 2 colors, receiv-
    R—NR.   ing in the act a severe wound.
    B—Fayette County, Pa.

JOHNSON, WALLACE W.
At Gettysburg, Pa., July 2, 1863.
R—Waverly, N. Y.
B—Newfield, N. Y.

Sergeant, Company G, 6th Pennsylvania Reserves.
With 5 other volunteers gallantly charged on a number of the enemy's sharpshooters concealed in a log house, captured them, and brought them into the Union lines.

JOHNSTON, DAVID.
At Vicksburg, Miss., May 22, 1863.
R—NR.
B—NR.

Private, Company K, 8th Missouri Infantry.
Gallantry in the charge of the "volunteer storming party."

JOHNSTON, EDWARD.
At Cedar Creek, etc., Mont., October 1876, to January, 1877.
R—Buffalo, N. Y.
B—Pen Yan, N. Y.

Corporal, Company C, 5th U. S. Infantry.
Gallantry in actions.

JOHNSTON, GORDON.
At Mount Bud-Dajo, Jolo, P. I., Mar. 7, 1906.
R—Birmingham, Ala.
B—Charlotte, N. C.
Distinguished-service cross and distinguished - service medal also awarded.

First lieutenant, U. S. Signal Corps.
Voluntarily took part in and was dangerously wounded during an assault on the enemy's works.

JOHNSTON, HAROLD I. (2202872)
Near Pouilly, France, Nov. 9, 1918.
R—Chicago, Ill.
B—Kendell, Kans.
G. O. No. 20, W. D., 1919.

Private, first class, Company A, 356th Infantry, 89th Division.
When information was desired as to the enemy's position on the opposite side of the River Meuse, Sergeant Johnston, with another soldier, volunteered without hesitation and swam the river to reconnoiter the exact location of the enemy. He succeeded in reaching the opposite bank, despite the evident determination of the enemy to prevent a crossing. Having obtained his information, he again entered the water for his return. This was accomplished after a severe struggle which so exhausted him that he had to be assisted from the water, after which he rendered his report of the exploit.

JOHNSTON, WILLIE.
—————
R—St. Johnsbury, Vt.
B—Morristown, N. Y.

Musician, Company D, 3d Vermont Infantry.
Date and place of act not of record in the War Department.

JONES, DAVID.
At Vicksburg, Miss., May 22, 1863.
R—NR.
B—Fayette County, Ohio.

Private, Company I, 54th Ohio Infantry.
Gallantry in the charge of the "volunteer storming party."

JONES, WILLIAM.
At Spotsylvania, Va., May 12, 1864.
R—New York, N. Y.
B—Ireland.

First sergeant, Company A, 73d New York Infantry.
Capture of flag of 65th Virginia Infantry (C. S. A.).

JONES, WILLIAM H.
At Little Muddy Creek, Mont., May 7, 1877; at Camas Meadows, Idaho, Aug. 20, 1877.
R—NR.
B—Davidson County, N. C.

Farrier, Company L, 2d U. S. Cavalry.
Gallantry in actions.

JORDAN, ABSALOM.
At Sailors Creek, Va., Apr. 6, 1865.
R—North Madison, Ind.
B—Brown County, Ohio.

Corporal, Company A, 3d Indiana Cavalry.
Capture of flag.

JORDAN, GEORGE.
At Fort Tularosa, N. Mex., May 14, 1880.
At Carrizo Canyon, N. Mex., Aug. 12, 1881.
R—NR.
B—Williamson County, Tenn.

Sergeant, Company K, 9th U. S. Cavalry.
While commanding a detachment of 25 men, repulsed a force of more than 100 Indians.
While commanding the right of a detachment of 19 men, he stubbornly held his ground in an extremely exposed position and gallantly forced back a much superior number of the enemy, preventing them from surrounding the command.

JOSSELYN, SIMEON T.
At Missionary Ridge, Tenn., Nov. 25, 1863.
R—NR.
B—NR.

First lieutenant, Company C, 13th Illinois Infantry.
While commanding his company, deployed as skirmishers, came upon a large body of the enemy, taking a number of them prisoners. Lieutenant Josselyn himself shot their color bearer, seized the colors and brought them back to his regiment.

JUDGE, FRANCIS W.
At Fort Sanders, Knoxville, Tenn., Nov. 29, 1863.
R—NR.
B—England.

First sergeant, Company K, 79th New York Infantry.
The color bearer of the 51st Georgia Infantry (C. S. A.), having planted his flag upon the side of the work, Sergeant Judge leaped from his position of safety, sprang upon the parapet, and in the face of a concentrated fire seized the flag and returned with it in safety to the fort.

KAISER, JOHN.
At Richmond, Va., June 27, 1862.
R—NR.
B—Germany.

Sergeant, Company E, 2d U. S. Artillery.
Gallant and meritorious service during the seven days' battles before Richmond, Va.

KALTENBACH, LUTHER.
At Nashville, Tenn., Dec. 16, 1864.
R—Honey Creek, Iowa.
B—Germany.

Corporal, Company F, 12th Iowa Infantry.
Capture of flag, supposed to be of 5th Mississippi Infantry (C. S. A.).

KANE, JOHN.
At Petersburg, Va., Apr. 2, 1865.
R—NR.
B—Ireland.

Corporal, Company K, 100th New York Infantry.
Gallantry as color bearer in the assault on Fort Gregg.

KAPPESSER, PETER.
At Lookout Mountain, Tenn., Nov. 24, 1863.
R—NR.
B—Germany.

Private, Company B, 149th New York Infantry.
Capture of Confederate flag (Bragg's army).

KARNES, JAMES E. (1307595).
Near Estrees, France, Oct. 8, 1918.
R—Knoxville, Tenn.
B—Arlington, Tenn.
G. O. No. 50, W. D., 1919.

Sergeant, Company D, 117th Infantry, 30th Division.
During an advance, his company was held up by a machine gun, which was enfilading the line. Accompanied by another soldier, he advanced against this position and succeeded in reducing the nest by killing three and capturing seven of the enemy and their guns.

KARPELES, LEOPOLD.
At Wilderness, Va., May 6, 1864.
R—Springfield, Mass.
B—Hungary.

Sergeant, Company E, 57th Massachusetts Infantry.
While color bearer, rallied the retreating troops and induced them to check the enemy's advance.

KATZ, PHILLIP C. (2263512).
Near Eclisfontaine, France, Sept. 26, 1918.
R—San Francisco, Calif.
B—San Francisco, Calif.
G. O. No. 16, W. D., 1919.

Sergeant, Company C, 363d Infantry, 91st Division.
After his company had withdrawn for a distance of 200 yards on a line with the units on its flanks, Sergeant Katz learned that one of his comrades had been left wounded in an exposed position at the point from which the withdrawal had taken place. Voluntarily crossing an area swept by heavy machine-gun fire, he advanced to where the wounded soldier lay and carried him to a place of safety.

KAUFMAN, BENJAMIN (1709789).
In the Forest of Argonne, France, Oct. 4, 1918.
R—Brooklyn, N. Y.
B—Buffalo, N. Y.
G. O. No. 50, W. D., 1919.

First sergeant, Company K, 308th Infantry, 77th Division.
He took out a patrol for the purpose of attacking an enemy machine gun which had checked the advance of his company. Before reaching the gun he became separated from his patrol and a machine-gun bullet shattered his right arm. Without hesitation he advanced on the gun alone, throwing grenades with his left hand and charging with an empty pistol, taking one prisoner and scattering the crew, bringing the gun and prisoner back to the first-aid station.

KAUSS, AUGUST.
At Five Forks, Va., Apr. 1, 1865.
R—NR.
B—Germany.

Corporal, Company H, 15th New York Heavy Artillery.
Capture of battle flag.

KAY, JOHN.
Arizona, Oct. 21, 1868.
R—NR.
B—England.

Private, Company L, 8th U. S. Cavalry.
Brought a comrade, severely wounded, from under the fire of a large party of the enemy.

KEATING, DANIEL.
Wichita River, Tex., Oct. 5, 1870.
R—NR.
B—Ireland.

Corporal, Company M, 6th U. S. Cavalry.
Gallantry in action and in pursuit of Indians.

KEELE, JOSEPH.
At North Anna River, Va., May 23, 1864.
R—Staten Island, N. Y.
B—Ireland.

Sergeant major, 182d New York Infantry.
Voluntarily and at the risk of his life carried orders to the brigade commander, which resulted in saving the works his regiment was defending.

KEEN, JOSEPH S.
Near Chattahoochee River, Ga., Oct. 1, 1864.
R—NR.
B—England.

Sergeant, Company D, 13th Michigan Infantry.
While an escaped prisoner of war within the enemy's lines witnessed an important movement of the enemy, and at great personal risk made his way through the enemy's lines and brought news of the movement to Sherman's army.

KEENAN, BARTHOLOMEW T.
At Chiricahua Mountains, Ariz., Oct. 20, 1869.
R—NR.
B—Brooklyn, N. Y.

Trumpeter, Company G, 1st U. S. Cavalry.
Gallantry in action.

KEENAN, JOHN.
Arizona, August to October, 1868.
R—NR.
B—Ireland.

Private, Company B, 8th U. S. Cavalry.
Bravery in actions and scouts.

KEENE, JOSEPH.
At Fredericksburg, Va., Dec. 13, 1862.
R—NR.
B—England.

Private, Company B, 26th New York Infantry.
Voluntarily seized the colors after several color bearers had been shot down and led the regiment in the charge.

KELLER, WILLIAM
    At Santiago, Cuba, July 1, 1898.
    R—Buffalo, N. Y.
    B—Buffalo, N. Y.

Private, Company F, 10th U. S. Infantry.
Gallantly assisted in the rescue of the wounded from in front of the lines and under heavy fire of the enemy.

KELLEY, ANDREW J
    At Knoxville, Tenn., Nov. 20, 1863.
    R—NR.
    B—La Grange County, Ind.

Private, Company E, 17th Michigan Infantry.
Having voluntarily accompanied a small party to destroy buildings within the enemy's lines whence sharpshooters had been firing, disregarded an order to retire, remained and completed the firing of the buildings, thus insuring their total destruction; this at the eminent risk of his life from the fire of the advancing enemy.

KELLEY, CHARLES
    At Chiricahua Mountains, Ariz., Oct. 20, 1869.
    R—NR.
    B—Ireland.

Private, Company G, 1st U. S. Cavalry.
Gallantry in action.

KELLEY, GEORGE V
    At Franklin, Tenn., Nov. 30, 1864.
    R—NR.
    B—NR.

Captain, Company A, 104th Ohio Infantry.
Capture of flag supposed to be of Cheatham's Corps.

KELLEY, LEVERETT M
    At Missionary Ridge, Tenn., Nov. 25, 1863.
    R—Rutland, Ill.
    B—Schenectady, N. Y.

Sergeant, Company A, 36th Illinois Infantry.
Sprang over the works just captured from the enemy, and calling upon his comrades to follow, rushed forward in the face of a deadly fire and was among the first over the works on the summit, where he compelled the surrender of a Confederate officer and received his sword.

KELLY, ALEXANDER
    At Chapins Farm, near Richmond, Va., Sept. 29, 1864.
    R—NR.
    B—Pennsylvania.

First sergeant, Company F, 6th U. S. Colored Troops.
Gallantly seized the colors, which had fallen near the enemy's lines of abatis, raised them and rallied the men at a time of confusion and in a place of the greatest danger.

KELLY, DANIEL
    At Waynesboro, Va., Mar. 2, 1865.
    R—NR.
    B—Groveland, N. Y.

Sergeant, Company G, 8th New York Cavalry.
Capture of flag.

KELLY, JOHN JOSEPH
    At Blanc Mont Ridge, France, Oct. 3, 1918.
    R—Chicago, Ill.
    B—Chicago, Ill.
    G. O. No. 16, W. D., 1919.

Private, 78th Company, 6th Regiment, U. S. Marine Corps, 2d Division.
Private Kelly ran through our own barrage 100 yards in advance of the front line and attacked an enemy machine-gun nest, killing the gunner with a grenade, shooting another member of the crew with his pistol, and returning through the barrage with eight prisoners.

KELLY, JOHN J. H
    At Upper Washita, Tex., Sept. 9, 1874.
    R—NR.
    B—Schuyler County, Ill.

Corporal, Company I, 5th U. S. Infantry.
Gallantry in action.

KELLY, THOMAS
    At Front Royal, Va., Aug. 16, 1864.
    R—NR.
    B—Ireland.

Private, Company A, 6th New York Cavalry.
Capture of flag.

KELLY, THOMAS
    At Upper Washita, Tex., Sept. 9, 1874.
    R—NR.
    B—NR.

Private, Company I, 5th U. S. Infantry.
Gallantry in action.

KELLY, THOMAS
    At Santiago, Cuba, July 1, 1898.
    R—New York, N. Y.
    B—Ireland.

Private, Company H, 21st U. S. Infantry.
Gallantly assisted in the rescue of the wounded from in front of the lines and while under heavy fire from the enemy.

KELSAY
    Winter of 1872–73.
    R—NR.
    B—NR.

Indian scout.
Gallantry during the campaigns against Apaches.

KEMP, JOSEPH
    At Wilderness, Va., May 6, 1864.
    R—Sault Ste. Marie, Mich.
    B—Lima, Ohio.

First sergeant, Company D, 5th Michigan Infantry.
Capture of flag of 31st North Carolina (C. S. A.) in a personal encounter.

KENDALL, WILLIAM W
    At Black River Bridge, Miss., May 17, 1863.
    R—Dubois County, Ind.
    B—Dubois County, Ind.

First sergeant, Company A, 49th Indiana Infantry.
Voluntarily led the company in a charge and was the first to enter the enemy's works, taking a number of prisoners.

KENNEDY, JOHN.
    At Trevilian Station, Va., June 11, 1864.
    R—NR.
    B—Ireland.

Private, Company M, 2d U. S. Artillery.
Remained at his gun, resisting with its implements the advancing cavalry, and thus secured the retreat of his detachment.

KENNEDY, JOHN T.
    At Patian Island, P. I., July 4, 1909.
    R—Orangeburg, S. C.
    B—Hendersonville, S. C.
    Distinguished-service medal also awarded.

Second lieutenant, 6th U. S. Cavalry.
While in action against hostile Moros he entered with a few enlisted men the mouth of a cave occupied by a desperate enemy, this act having been ordered after he had volunteered several times. In this action Lieutenant Kennedy was severely wounded.

KENNEDY, PHILIP.
    At Cedar Creek, etc., Mont., Oct. 21, 1876, to Jan. 8, 1877.
    R—NR.
    B—Ireland.

Private, Company C, 5th U. S. Infantry.
Gallantry in actions.

KENYON, JOHN S.
    At Trenton, N. C., May 15, 1862.
    R—NR.
    B—NR.

Sergeant, Company D, 3d New York Cavalry.
Voluntarily left a retiring column, returned in face of the enemy's fire, helped a wounded man upon a horse, and so enabled him to escape capture or death.

KENYON, SAMUEL P.
    At Sailors Creek, Va., Apr. 6, 1865.
    R—Oriskany Falls, N. Y.
    B—Ira, N. Y.

Private, Company B, 24th New York Cavalry.
Capture of battle flag.

KEOUGH, JOHN.
    At Sailors Creek, Va., Apr. 6, 1865.
    R—Albany, N. Y.
    B—Ireland.

Corporal, Company E, 67th Pennsylvania Infantry.
Capture of battle flag of 50th Georgia Infantry (C. S. A.).

KEPHART, JAMES.
    At Vicksburg, Miss., May 19, 1863.
    R—NR.
    B—Venango County, Pa.

Private, Company C, 13th U. S. Infantry.
Voluntarily and at the risk of his life, under a severe fire of the enemy, aided and assisted to the rear an officer who had been severely wounded and left on the field.

KERR, JOHN B.
    At White River, S. Dak., Jan. 1, 1891.
    R—Hutchison Station, Ky.
    B—Fayette County, Ky.

Captain, 6th U. S. Cavalry.
Distinguished services.

KERR, THOMAS R.
    At Moorfield, W. Va., Aug. 7, 1864.
    R—Pittsburgh, Pa.
    B—NR.

Captain, Company C, 14th Pennsylvania Cavalry.
After being most desperately wounded, he captured the colors of a Confederate regiment.

KERRIGAN, THOMAS.
    At Wichita River, Tex., July 12, 1870.
    R—NR.
    B—Ireland.

Sergeant, Company H, 6th U. S. Cavalry.
Gallantry in action.

KIGGINS, JOHN.
    At Lookout Mountain, Tenn., Nov. 24, 1863.
    R—NR.
    B—Syracuse, N. Y.

Sergeant, Company D, 149th New York Infantry.
Waved the colors to save the lives of the men who were being fired upon by their own batteries, and thereby drew upon himself a concentrated fire from the enemy.

KILBOURNE, CHARLES E.
    At Paco Bridge, P. I., Feb. 5, 1899.
    R—Portland, Oreg.
    B—Fort Myer, Va.
    Distinguished-service cross and distinguished-service medal also awarded.

First lieutenant, U. S. Volunteer Signal Corps.
Within a range of 250 yards of the enemy and in the face of a rapid fire climbed a telegraph pole at the east end of the bridge and in full view of the enemy coolly and carefully repaired a broken telegraph wire, thereby reestablishing telegraphic communication to the front.

KILMARTIN, JOHN.
    At Whetstone Mountains, Ariz., May 5, 1871.
    R—NR.
    B—Canada.

Private, Company F, 3d U. S. Cavalry.
Gallantry in action.

KIMBALL, JOSEPH.
    At Sailors Creek, Va., Apr. 6, 1865.
    R—NR.
    B—Littleton, N. H.

Private, Company B, 2d West Virginia Cavalry.
Capture of flag of 6th North Carolina Infantry (C. S. A.).

KINDIG, JOHN M.
    At Spotsylvania, Va., May 12, 1864.
    R—NR.
    B—East Liberty, Pa.

Corporal, Company A, 63d Pennsylvania Infantry.
Capture of flag of 28th North Carolina Infantry (C. S. A.).

KING, HORATIO C _____ *_____
    Near Dinwiddie Courthouse, Va.,
    Mar. 31, 1865.
    R—NR.
    B—NR.

Major and quartermaster, U. S. Volunteers.
While serving as a volunteer aid carried orders to the reserve brigade and participated with it in the charge which repulsed the enemy.

KING, RUFUS, Jr _____
    At White Oak Swamp Bridge, Va.,
    June 30, 1862.
    R—New York.
    B—New York.

First lieutenant, 4th U. S. Artillery.
This officer, when his captain was wounded, succeeded to the command of two batteries while engaged against a superior force of the enemy and fought his guns most gallantly until compelled to retire.

KINNE, JOHN B _____
    Near San Isidro, P. I., May 16, 1899.
    R—Fargo, N. Dak.
    B—Beloit, Wis.

Private, Company B, 1st North Dakota Infantry.
With 21 other scouts charged across a burning bridge under heavy fire and completely routed 600 of the enemy who were entrenched in a strongly fortified position.

KINSEY, JOHN _____
    At Spotsylvania, Va., May 18, 1864.
    R—NR.
    B—Lancaster County, Pa.

Corporal, Company B, 45th Pennsylvania Infantry.
Seized the colors, the color bearer having been shot, and with great gallantry succeeded in saving them from capture.

KIRBY, DENNIS T _____
    At Vicksburg, Miss., May 22, 1863.
    R—St. Louis, Mo.
    B—NR.

Major, 8th Missouri Infantry.
Seized the colors when the color bearer was killed and bore them himself in the assault.

KIRK, JOHN _____
    At Wichita River, Tex., July 12,
    1870.
    R—NR.
    B—York, Pa.

First sergeant, Company L, 6th U. S. Cavalry.
Gallantry in action.

KIRK, JONATHAN C _____
    At North Anna River, Va., May 23,
    1864.
    R—Wilmington, Ohio.
    B—Clinton County, Ohio.

Captain, Company F, 20th Indiana Infantry.
Volunteered for dangerous service and single handed captured 13 armed Confederate soldiers and marched them to the rear.

KIRKWOOD, JOHN A _____
    At Slim Buttes, Dak. Territory,
    Sept. 9, 1876.
    R—NR.
    B—Allegheny City, Pa.

Sergeant, Company M, 3d U. S. Cavalry.
Bravely endeavored to dislodge some Sioux Indians secreted in a ravine.

KITCHEN, GEORGE K _____
    At Upper Washita, Tex., Sept. 9,
    1874.
    R—NR.
    B—Lebanon County, Pa.

Sergeant, Company H, 6th U. S. Cavalry.
Gallantry in action.

KLINE, HARRY _____
    At Sailors Creek, Va., Apr. 6, 1865.
    R—Syracuse, N. Y.
    B—Germany.

Private, Company E, 40th New York Infantry.
Capture of battle flag.

KLOTH, CHARLES H _____
    At Vicksburg, Miss., May 22, 1863.
    R—NR.
    B—Europe.

Private, Chicago Mercantile Battery, Illinois Light Artillery.
Carried with others by hand a cannon up to and fired it through an embrasure of the enemy's works.

KNAAK, ALBERT _____
    Arizona, Aug. to Oct., 1868.
    R—NR.
    B—Switzerland.

Private, Company B, 8th U. S. Cavalry.
Bravery in actions and scouts.

KNIGHT, CHARLES H _____
    At Petersburg, Va., July 30, 1864.
    R—Keene, N. H.
    B—Keene, N. H.

Corporal, Company I, 9th New Hampshire Infantry
In company with a sergeant was the first to enter the exploded mine; was wounded but took several prisoners to the Federal lines.

KNIGHT, JOSEPH F _____
    At White River, S. Dak., Jan. 1, 1891.
    R—NR.
    B—Danville, Ill.

Sergeant, Company F, 6th U. S. Cavalry.
Led the advance in a spirited movement to the assistance of Troop K, 6th U. S. Cavalry.

KNIGHT, WILLIAM _____
    Georgia, April, 1862.
    R—NR.
    B—Wayne County, Ohio.

Private, Company E, 21st Ohio Infantry.
One of 22 men (including two civilians) who, by direction of General Mitchel (or Buell), penetrated nearly 200 miles south into the enemy's territory and captured a railroad train at Big Shanty, Ga., in an attempt to destroy the bridges and track between Chattanooga and Atlanta.

KNOWLES, ABIATHER J _____
    At Bull Run, Va., July 21, 1861.
    R—NR.
    B—NR.

Private, Company D, 2d Maine Infantry.
Removed dead and wounded under heavy fire.

KNOX, EDWARD M._____ | Second lieutenant, 15th New York Battery.
At Gettysburg, Pa., July 2, 1863. | Held his ground with the battery after the other batteries had fallen back until
R—New York, N. Y. | compelled to draw his piece off by hand; he was severely wounded.
B—NR. |

KNOX, JOHN W._____ | Corporal, Company I, 5th U. S. Infantry.
At Upper Washita, Tex., Sept. 9, | Gallantry in action.
1874. |
R—NR. |
B—Burlington, Iowa. |

*KOCAK, MATEJ_____ | Sergeant, 66th Company, 5th Regiment, U. S. Marine Corps, 2d Division.
Near Soissons, France, July 18, 1918. | When the advance of his battalion was checked by a hidden machine-gun nest,
R—New York, N. Y. | he went forward alone, unprotected by covering fire from his own men, and
B—Austria. | worked in between the German positions in the face of fire from enemy cover-
G. O. No. 34, W. D., 1919. | ing detachments. Locating the machine-gun nest, he rushed it and with his
| bayonet drove off the crew. Shortly after this he organized 25 French colonial
| soldiers who had become separated from their company and led them in attack-
| ing another machine-gun nest, which was also put out of action.
| Posthumously awarded. Medal presented to sister-in-law, Mrs. Julia Kocak.

KOELPIN, WILLIAM_____ | Sergeant, Company I, 5th U. S. Infantry.
At Upper Washita, Tex., Sept. 9, | Gallantry in action.
1874. |
R—New York, N. Y. |
B—Prussia. |

KOOGLE, JACOB_____ | First lieutenant, Company G, 7th Maryland Infantry.
At Five Forks, Va., Apr. 1, 1865. | Capture of battle flag.
R—NR. |
B—Frederick, Md. |

KOSOHA_____ | Indian scout.
Winter of 1872–73. | Gallantry during the campaigns against Apaches.
R—NR. |
B—NR. |

KOUNTZ, JOHN S._____ | Musician, Company G, 37th Ohio Infantry.
At Missionary Ridge, Tenn., Nov. | Seized a musket and joined in the charge, in which he was severely wounded.
25, 1863. |
R—Maumee, Ohio. |
B—Maumee, Ohio. |

KRAMER, THEODORE L._____ | Private, Company G, 188th Pennsylvania Infantry.
At Chapins Farm, near Richmond, | Took one of the first prisoners, a captain.
Va., Sept. 29, 1864. |
R—Danville, Pa. |
B—Luzerne County, Pa. |

KREHER, WENDELIN_____ | First sergeant, Company C, 5th U. S. Infantry.
At Cedar Creek, etc., Mont., Oct. | Gallantry in actions.
21, 1876, to Jan. 8, 1877. |
R—NR. |
B—Prussia. |

KRETSINGER, GEORGE_____ | Private, Chicago Mercantile Battery, Illinois Light Artillery.
At Vicksburg, Miss., May 22, 1863. | Carried with others by hand a cannon up to and fired it through an embrasure
R—Chicago, Ill. | of the enemy's works.
B—Herkimer County, N. Y. |

KUDER, ANDREW_____ | Second lieutenant, Company G, 8th New York Cavalry.
At Waynesboro, Va., Mar. 2, 1865. | Capture of flag.
R—NR. |
B—Groveland, N. Y. |

KUDER, JEREMIAH_____ | Lieutenant, Company A, 74th Indiana Infantry.
At Jonesboro, Ga., Sept. 1, 1864. | Capture of flags of 8th and 19th Arkansas (C. S. A.).
R—Warsaw, Ind. |
B—Seneca County, Ohio. |

KYLE, JOHN_____ | Corporal, Company M, 5th U. S. Cavalry.
Near Republican River, Kans., | This soldier and two others were attacked by eight Indians, but beat them off
July 8, 1869. | and badly wounded two of them.
R—NR. |
B—Cincinnati, Ohio. |

LABILL, JOSEPH S._____ | Private, Company C, 6th Missouri Infantry.
At Vicksburg, Miss., May 22, 1863. | Gallantry in the charge of the "volunteer storming party."
R—Vandalia, Ill. |
B—France. |

LADD, GEORGE_____ | Private, Company H, 22d New York Cavalry.
At Waynesboro, Va., Mar. 2, 1865. | Captured a standard bearer, his flag, horse, and equipments.
R—NR. |
B—Carmullus, N. Y. |

LAING, WILLIAM_____
    At Chapins Farm, near Richmond,
    Va., Sept. 29, 1864.
    R—NR.
    B—Hempstead, N. Y.

Sergeant, Company F, 158th New York Infantry.
Was among the first to scale the parapet.

LANDIS, JAMES P_____
    At Paines Crossroads, Va., Apr. 5,
    1865.
    R—NR.
    B—Mifflin County, Pa.

Chief bugler, 1st Pennsylvania Cavalry.
Capture of flag.

LANE, MORGAN D_____
    Near Jetersville, Va., Apr. 6, 1865.
    R—NR.
    B—Monroe, N. Y.

Private, Signal Corps, U. S. Army.
Capture of flag of gunboat Nansemond.

LANFARE, AARON S_____
    At Sailors Creek, Va., Apr. 6, 1865.
    R—NR.
    B—Bradford, Conn.

First lieutenant, Company B, 1st Connecticut Cavalry
Capture of flag of 11th Florida Infantry (C. S. A.).

LANGBEIN, J. C. JULIUS_____
    At Camden, N. C., Apr. 19, 1862.
    R—NR.
    B—NR.

Musician, Company B, 9th New York Infantry.
A drummer boy, 15 years of age, he voluntarily and under a heavy fire went to the aid of a wounded officer, procured medical assistance for him, and aided in carrying him to a place of safety.

LARIMER, SMITH_____
    At Sailors Creek, Va., Apr. 6, 1865.
    R—Columbus, Ohio.
    B—Richland County, Ohio.

Corporal, Company G, 2d Ohio Cavalry.
Capture of flag of General Kershaw's headquarters.

LARKIN, DAVID_____
    At Red River, Tex., Sept. 29, 1872.
    R—NR.
    B—Ireland.

Farrier, Company F, 4th U. S. Cavalry.
Gallantry in action.

LARRABEE, JAMES W_____
    At Vicksburg, Miss., May 22, 1863.
    R —Mendota, Ill.
    B—Rensselaer County, N. Y.

Corporal, Company I, 55th Illinois Infantry.
Gallantry in the charge of the "volunteer storming party."

LATHAM, JOHN CRIDLAND (1212528)_____
    Near Le Catelet, France, Sept. 29,
    1918.
    R—Rutherford, N. J.
    B—England.
    G. O. No. 20, W. D., 1919.

Sergeant, Machine Gun Company, 107th Infantry, 27th Division.
Becoming separated from their platoon by a smoke barrage, Sergeant Latham, Sergeant Alan L. Eggers, and Corporal Thomas E. O'Shea took cover in a shell hole well within the enemy's lines. Upon hearing a call for help from an American tank which had become disabled 30 yards from them, the three soldiers left their shelter and started toward the tank under heavy fire from German machine guns and trench mortars. In crossing the fire-swept area, Corporal O'Shea was mortally wounded, but his companions, undeterred, proceeded to the tank, rescued a wounded officer, and assisted two wounded soldiers to cover in the sap of a near-by trench. Sergeant Latham and Sergeant Eggers then returned to the tank in the face of the violent fire, dismounted a Hotchkiss gun, and took it back to where the wounded men were, keeping off the enemy all day by effective use of the gun and later bringing it with the wounded men back to our lines under cover of darkness.

LAWRENCE, JAMES_____
    Arizona, August to October, 1868.
    R—NR.
    B—Scotland.

Private, Company B, 8th U. S. Cavalry.
Bravery in actions and scouts against Indians.

LAWSON, GAINES_____
    At Minnville, Tenn., Oct. 3, 1863.
    R—NR.
    B—Hawkins County, Tenn.

First sergeant, Company D, 4th East Tennessee Infantry.
Went to the aid of a wounded comrade between the lines and carried him to a place of safety.

LAWTON, HENRY W_____
    At Atlanta, Ga., Aug. 3, 1864.
    R—NR.
    B—Ohio.

Captain, Company A, 30th Indiana Infantry.
Led a charge of skirmishers against the enemy's rifle pits and stubbornly and successfully resisted two determined attacks of the enemy to retake the works.

LAWTON, JOHN S_____
    At Milk River, Colo., Sept. 29, 1879.
    R—NR.
    B—Bristol, R. I.

Sergeant, Company D, 5th U. S. Cavalry.
Coolness and steadiness under fire; volunteered to accompany a small detachment on a very dangerous mission.

LAWTON, LOUIS B_____
    At Tientsin, China, July 13, 1900.
    R—Auburn, N. Y.
    B—Independence, Iowa.

First lieutenant, 9th U. S. Infantry.
Carried a message and guided reinforcements across a wide and fire-swept space, during which he was thrice wounded.

LEAHY, CORNELIUS J_____
    Near Porac, Luzon, P. I., Sept. 3,
    1899.
    R—San Francisco, Calif.
    B—Ireland.

Private, Company A, 36th Infantry, U. S. Volunteers.
Distinguished gallantry in action in driving off a superior force and with the assistance of one comrade brought from the field of action the bodies of two comrades, one killed and the other severely wounded, this while on a scout.

LEE, FITZ. At Tayabacoa, Cuba, June 30, 1898. R—Dinwiddie County, Va. B—Dinwiddie County, Va.	Private, Troop M, 10th U. S. Cavalry. Voluntarily went ashore in the face of the enemy and aided in the rescue of his wounded comrades, this after several previous attempts had been frustrated.
LEMERT, MILO (1315827). Near Bellicourt, France, Sept. 29, 1918. R—Crossville, Tenn. B—Marshalltown, Iowa. G. O. No. 59, W. D., 1919.	First, sergeant Company G, 119th Infantry, 30th Division. Seeing that the left flank of his company was held up, he located the enemy machine-gun emplacement, which had been causing heavy casualties. In the face of heavy fire he rushed it single handed, killing the entire crew with grenades. Continuing along the enemy trench in advance of the company, he reached another emplacement, which he also charged, silencing the gun with grenades. A third machine-gun emplacement opened upon him from the left and with similar skill and bravery he destroyed this also. Later, in company with another sergeant, he attacked a fourth machine-gun nest, being killed as he reached the parapet of the emplacement. His courageous action in destroying in turn four enemy machine-gun nests prevented many casualties among his company and very materially aided in achieving the objective. Posthumously awarded. Medal presented to widow, Mrs. Nellie V. Lemert.
LENIHAN, JAMES. At Clear Creek, Ariz., Jan. 2, 1873. R—NR. B—Ireland.	Private, Company K, 5th U. S. Cavalry. Gallantry in action.
LEONARD, EDWIN. Near Petersburg, Va., June 18, 1864. R—Agawam, Mass. B—Agawam, Mass.	Sergeant, Company I, 37th Massachusetts Infantry. Voluntarily exposed himself to the fire of a Union brigade to stop their firing on the Union skirmish line.
LEONARD, PATRICK. Near Fort Hartsuff, Nebr., Apr. 28, 1876. R—NR. B—Ireland.	Corporal, Company A, 23d U. S. Infantry. Gallantry in charge on hostile Sioux.
LEONARD, PATRICK. At Little Blue, Nebr., May 15, 1870. R—NR. B—Ireland.	Sergeant, Company C, 2d U. S. Cavalry. Gallantry in action.
LEONARD, WILLIAM. At Muddy Creek, Mont., May 7, 1877. R—NR. B—Ypsilanti, Mich.	Private, Company L, 2d U. S. Cavalry. Bravery in action.
LEONARD, WILLIAM E. At Deep Bottom, Va., Aug. 16, 1864. R—Pennsylvania. B—Greene County, Pa.	Private, Company F, 85th Pennsylvania Infantry. Capture of battle flag.
LESLIE, FRANK. At Front Royal, Va., Aug. 15, 1864. R—NR. B—England.	Private, Company B, 4th New York Cavalry. Capture of colors of 3d Virginia Infantry (C. S. A.).
LEVY, BENJAMIN. At Glendale, Va., June 30, 1862. R—NR. B—NR.	Private, Company B, 40th New York Infantry. This soldier, a drummer boy, took the gun of a sick comrade, went into the fight, and when the color bearers were shot down, carried the colors and saved them from capture.
LEWIS, DE WITT CLINTON. At Secessionville, S. C., June 16, 1862. R—NR. B—West Chester, Pa.	Captain, Company F, 97th Pennsylvania Infantry. While retiring with his men before a heavy fire of canister shot at short range, returned in the face of the enemy's fire and rescued an exhausted private of his company who but for this timely action would have lost his life by drowning in the morass through which the troops were retiring.
LEWIS, HENRY. At Vicksburg, Miss., May 3, 1863. R—NR. B—Wayne County, Mich.	Corporal, Company B, 47th Ohio Infantry. For being one of a party that volunteered and attempted to run the enemy's batteries with a steam tug and two barges loaded with subsistence stores.
LEWIS, SAMUEL E. At Petersburg, Va., Apr. 2, 1865. R—Coventry, R. I. B—Coventry, R. I.	Corporal, Company G, 1st Rhode Island Light Artillery. Was one of a detachment of 20 picked artillerymen who voluntarily accompanied an Infantry assaulting party and who turned upon the enemy the guns captured in the assault.
LEWIS, WILLIAM B. At Bluff Station, Wyo., Jan. 20 to 22, 1877. R—NR. B—Boston, Mass.	Sergeant, Company B, 3d U. S. Cavalry. Bravery in skirmish.
LIBAIRE, ADOLPHE. At Antietam, Md., Sept. 17, 1862. R—NR. B—NR.	Captain, Company E, 9th New York Infantry. In the advance on the enemy and after his color bearer and the entire color guard of 8 men had been shot down, this officer seized the regimental flag and with conspicuous gallantry carried it to the extreme front, urging the line forward.

LILLEY, JOHN
At Petersburg, Va., Apr. 2, 1865.
R—NR.
B—Mifflin County, Pa.

Private, Company F, 205th Pennsylvania Infantry.
Capture of battle flag.

LITTLE, HENRY F. W.
Near Richmond, Va., Sept. —, 1864.
R—New Hampshire.
B—Manchester, N. H.

Sergeant, Company D, 7th New Hampshire Infantry.
Gallantry on the skirmish line.

LITTLE, THOMAS.
Arizona, August to October, 1868.
R—NR.
B—West Indies.

Bugler, Company B, 8th U. S. Cavalry.
Bravery in actions and scouts against Indians.

LITTLEFIELD, GEORGE H.
At Fort Fisher, Va., Mar. 25, 1865.
R—NR.
B—Skowhegan, Me.

Corporal, Company G, 1st Maine Infantry.
The color sergeant having been wounded, this soldier picked up the flag and bore it to the front, to the great encouragement of the charging column.

LIVINGSTON, JOSIAH O.
At Newport Barracks, N. C., Feb. 2, 1864.
R—NR.
B—Walden, Vt.

First lieutenant and adjutant, 9th Vermont Infantry.
When, after desperate resistance, the small garrison had been driven back to the river by a vastly superior force, this officer, while a small force held back the enemy, personally fired the railroad bridge and, although wounded himself, assisted a wounded officer over the burning structure.

LOCKE, LEWIS.
At Fames Cross Roads, Va., Apr. 5, 1865.
R—Jersey City, N. J.
B—Clintonville, N. Y.

Private, Company A, 1st New Jersey Cavalry.
Capture of a Confederate flag.

*LOGAN, JOHN A.
At San Jacinto, P. I., Nov. 11, 1899.
R—Youngstown, Ohio.
B—Carbondale, Ill.

Major, 33d Infantry, U. S. Volunteers.
For most distinguished gallantry in leading his battalion upon the intrench-ments of the enemy, on which occasion he fell mortally wounded.

LOHNES, FRANCIS W.
At Gilmans Ranch, Nebr., May 12, 1865.
R—NR.
B—Oneida County, N. Y.

Private, Company H, 1st Nebraska Veteran Cavalry.
Gallantry in defending Government property against Indians.

LOMAN, BERGER (1389565)
Near Consenvoye, France, Oct. 9, 1918.
R—Chicago, Ill.
B—Norway.
G. O. No. 16, W. D., 1919.

Private, Company H, 132d Infantry, 33d Division.
When his company had reached a point within 100 yards of its objective, to which it was advancing under terrific machine-gun fire, Private Loman voluntarily and unaided made his way forward after all others had taken shelter from the direct fire of an enemy machine gun. He crawled to a flank position of the gun and, after killing or capturing the entire crew, turned the machine gun on the retreating enemy.

LONERGAN, JOHN.
At Gettysburg, Pa., July 2, 1863.
R—Burlington, Vt.
B—Ireland.

Captain, Company A, 13th Vermont Infantry.
Gallantry in the recapture of four guns and the capture of two additional guns from the enemy; also the capture of a number of prisoners.

LONG, OSCAR F.
At Bear Paw Mountain, Mont., Sept. 30, 1877.
R—Utica, N. Y.
B—New York.

Second lieutenant, 5th U. S. Infantry.
Having been directed to order a troop of cavalry to advance, and finding both its officers killed, he voluntarily assumed command, and under a heavy fire from the Indians advanced the troop to its proper position.

LONGFELLOW, RICHARD M.
Near San Isidro, P. I., May 16, 1899.
R—Mandan, N. Dak.
B—Logan County, Ill.

Private, Company A, 1st North Dakota Volunteer Infantry.
With 21 other scouts charged across a burning bridge under heavy fire and completely routed 600 of the enemy who were intrenched in a strongly fortified position.

LONGSHORE, WILLIAM H.
At Vicksburg, Miss., May 22, 1863.
R—NR.
B—Muskingum County, Ohio.

Private, Company D, 30th Ohio Infantry.
Gallantry in the charge of the "volunteer storming party."

LONSWAY, JOSEPH.
At Murfrees Station, Va., Oct. 16, 1864.
R—NR.
B—Clayton, N. Y.

Private, Company D, 20th New York Cavalry.
Volunteered to swim Blackwater River to get a large flat used as a ferry on other side; succeeded in getting the boat safely across, making it possible for a detachment to cross the river and take possession of the enemy's breast-work.

LORD, WILLIAM.
At Drurys Bluff, Va., May 16, 1864.
R—NR.
B—England.

Musician, Company C, 40th Massachusetts Infantry.
Went to the assistance of a wounded officer lying helpless between the lines, and under fire from both sides removed him to a place of safety.

LORISH, ANDREW J.
At Winchester, Va., Sept. 19, 1864.
R—NR.
B—Steuben, N. Y.

Commissary sergeant, 19th New York Cavalry (1st New York Dragoons).
Capture of Confederate flag.

LOVE, GEORGE M.
At Cedar Creek, Va., Oct. 19, 1864.
R—NR.
B—NR.

Colonel 116th New York Infantry.
Capture of battle flag of 2d South Carolina (C. S. A.).

LOVERING, GEORGE M.
At Port Hudson, La., June 14, 1863.
R—NR.
B—NR.

First sergeant, Company I, 4th Massachusetts Infantry.
During a momentary confusion in the ranks caused by other troops rushing upon the regiment, this soldier, with coolness and determination, rendered efficient aid in preventing a panic among the troops.

LOWER, CYRUS B.
At Wilderness, Va., May 7, 1864.
R—NR.
B—Lawrence County, Pa.

Private, Company K, 13th Pennsylvania Reserves.
Gallant services and soldierly qualities in voluntarily rejoining his command after having been wounded.

LOWER, ROBERT A.
At Vicksburg, Miss., May 22, 1863.
R—NR.
B—Illinois.

Private, Company K, 55th Illinois Infantry.
Gallantry in the charge of the "volunteer storming party."

LOWTHERS, JAMES.
At Sappa Creek, Kans., Apr. 23, 1875.
R—NR.
B—Boston, Mass.

Private, Company H, 6th U. S. Cavalry.
Bravery in action.

LOYD, GEORGE.
At Wounded Knee Creek, S. Dak., Dec. 29, 1890.
R—NR.
B—Ireland.

Sergeant, Company I, 7th U. S. Cavalry.
Bravery, especially after having been severely wounded through the lung.

LOYD, GEORGE.
At Petersburg, Va., Apr. 2, 1865.
R—NR.
B—Muskingum County, Ohio.

Private, Company A, 122d Ohio Infantry.
Capture of division flag of General Heth.

LUCAS, GEORGE W.
At Benton, Ark., July 25, 1864.
R—NR.
B—Adams County, Ill.

Private, Company C, 3d Missouri Cavalry.
Pursued and killed Confederate Brig. Gen. George M. Holt, Arkansas Militia, capturing his arms and horse.

LUCE, MOSES A.
At Laurel Hill, Va., May 10, 1864.
R—NR.
B—NR.

Sergeant, Company E, 4th Michigan Infantry.
Voluntarily returned in the face of the advancing enemy to the assistance of a wounded and helpless comrade, and carried him, at imminent peril, to a place of safety.

LUDGATE, WILLIAM.
At Farmville, Va., Apr. 7, 1865.
R—New York, N. Y.
B—England.

Captain, Company G, 59th New York Veteran Infantry.
Gallantry and promptness in rallying his men and advancing with a small detachment to save a bridge about to be fired by the enemy.

LUDWIG, CARL.
At Petersburg, Va., June 18, 1864.
R—NR.
B—France.

Private, 34th New York Battery.
As gunner of his piece, inflicted singly a great loss upon the enemy and distinguished himself in the removal of the piece while under a heavy fire.

*LUKE, FRANK, Jr.
Near Murvaux, France, Sept, 29, 1918.
R—Phoenix, Ariz.
B—Phoenix, Ariz.
G. O. No. 59, W. D., 1919.
Distinguished-service cross also awarded.

Second lieutenant, 27th Aero Squadron, 1st Pursuit Group, Air Service.
After having previously destroyed a number of enemy aircraft within 17 days, he voluntarily started on a patrol after German observation balloons. Though pursued by eight German planes which were protecting the enemy balloon line, he unhesitatingly attacked and shot down in flames three German balloons, being himself under heavy fire from ground batteries and the hostile planes. Severely wounded, he descended to within 50 meters of the ground, and flying at this low altitude near the town of Murvaux opened fire upon enemy troops, killing six and wounding as many more. Forced to make a landing and surrounded on all sides by the enemy, who called upon him to surrender, he drew his automatic pistol and defended himself gallantly until he fell dead from a wound in the chest.
Posthumously awarded. Medal presented to father, Frank Luke, sr.

LUNT, ALPHONSO M.
At Opequan Creek, Va., Sept. 19, 1864.
R—Cambridge, Mass.
B—Berwick, Me.

Sergeant, Company F, 38th Massachusetts Infantry.
Carried his flag to the most advanced position where, left almost alone close to the enemy's lines, he refused their demand to surrender, withdrew at great personal peril, and saved his flag.

LUTES, FRANKLIN W.
At Petersburg, Va., Mar. 31, 1865.
R—NR.
B—Oneida County, N. Y.

Corporal, Company D, 111th New York Infantry.
Capture of flag of 41st Alabama Infantry (C. S. A.), together with the color bearer and one of the color guard.

LUTHER, JAMES H.
At Fredericksburg, Va., May 3, 1863.
R—NR.
B—Dighton, Mass.

Private, Company D, 7th Massachusetts Infantry.
Among the first to jump into the enemy's rifle pits, he himself captured and brought out three prisoners.

LUTY, GOTLIEB_____
    At Chancellorsville, Va., May 3, 1863.
    R—West Manchester, Pa.
    B—Allegheny County, Pa.

Corporal, Company A, 74th New York Infantry.
Bravely advanced to the enemy's line under heavy fire and brought back valuable information.

LYMAN, JOEL H_____
    At Winchester, Va., Sept. 19, 1864.
    R—East Randolph, N. Y.
    B—Cattaraugus, N. Y.

Quartermaster sergeant, Company B, 9th New York Cavalry.
In an attempt to capture a Confederate flag he captured one of the enemy's officers and brought him within the lines.

LYON, EDWARD E_____
    At San Miguel de Mayomo, Luzon, P. I., May 13, 1899.
    R—Amboy, Wash.
    B—Hixton, Wis.

Private, Company B, 2d Oregon Volunteer Infantry.
With 11 other scouts, without waiting for the supporting battalion to aid them or to get into a position to do so, charged over a distance of about 150 yards and completely routed about 300 of the enemy, who were in line and in a position that could only be carried by a frontal attack.

LYON, FREDERICK A_____
    At Cedar Creek, Va., Oct. 19, 1864.
    R—NR.
    B—Williamsburg, Mass.

Corporal, Company A, 1st Vermont Cavalry.
With one companion, captured the flag of a Confederate regiment, 3 officers, and an ambulance with its mules and driver.

LYTLE, LEONIDAS S_____
    Near Fort Selden, N. Mex., July 8 to 11, 1873.
    R—NR.
    B—Warren County, Pa.

Sergeant, Company C, 8th U. S. Cavalry.
Services against hostile Indians.

LYTTON, JEPTHA L_____
    Near Fort Hartsuff, Nebr., Apr. 28, 1876.
    R—NR.
    B—Lawrence County, Ind.

Corporal, Company A, 23d U. S. Infantry.
Gallantry in charge on hostile Sioux.

MACARTHUR, ARTHUR, Jr_____
    At Missionary Ridge, Tenn., Nov. 25, 1863.
    R—NR.
    B—Springfield, Mass.

First lieutenant and adjutant, 24th Wisconsin Infantry.
Seized the colors of his regiment at a critical moment and planted them on the captured works on the crest of Missionary Ridge.

McADAMS, PETER_____
    At Salem Heights, Va., May 3, 1863.
    R—Philadelphia, Pa.
    B—Ireland.

Corporal, Company A, 98th Pennsylvania Infantry.
Went 250 yards in front of his regiment toward the position of the enemy and under fire brought within the lines a wounded and unconscious comrade.

McALWEE, BENJAMIN F_____
    At Petersburg, Va., July 20, 1864.
    R—NR.
    B—Washington, D. C.

Sergeant, Company D, 3d Maryland Infantry.
Picked up a shell with burning fuse and threw it over the parapet into the ditch, where it exploded; by this act he probably saved the lives of comrades at the great peril of his own.

McANALLY, CHARLES_____
    At Spotsylvania, Va., May 12, 1864.
    R—NR.
    B—Ireland.

Lieutenant, Company D, 69th Pennsylvania Infantry.
In a hand-to-hand encounter with the enemy captured a flag, was wounded in the act, but continued on duty until he received a second wound.

McBRIDE, BERNARD_____
    Arizona, August to October, 1868.
    R—NR.
    B—Brooklyn, N. Y.

Private, Company B, 8th U. S. Cavalry.
Bravery in actions and scouts against Indians.

McBRYAR, WILLIAM_____
    Arizona, Mar. 7, 1890.
    R—New York, N. Y.
    B—Elizabethtown, N. C.

Sergeant, Company K, 10th U. S. Cavalry.
Bravery in action with Apache Indians.

McCABE, WILLIAM_____
    Near Red River, Tex., Sept. 26-28, 1874.
    R—NR.
    B—Ireland.

Private, Company E, 4th U. S. Cavalry.
Gallantry in attack on a large party of Cheyennes.

McCAMMON, WILLIAM W_____
    At Corinth, Miss., Oct. 3, 1862.
    R—NR.
    B—NR.

First lieutenant, Company E, 24th Missouri Infantry.
While on duty as provost marshal, voluntarily assumed command of his company, then under fire, and so continued in command until the repulse and retreat of the enemy on the following day, the loss to this company during the battle being very great.

McCANN, BERNARD_____
    At Cedar Creek, etc., Mont., Oct. 21, 1876, to Jan. 8, 1877.
    R—NR.
    B—Ireland.

Private, Company F, 22d U. S. Infantry.
Gallantry in actions.

McCARREN, BERNARD_____
    At Gettysburg, Pa., July 3, 1863.
    R—NR.
    B—Ireland.

Private, Company C, 1st Delaware Infantry.
Capture of flag.

McCARTHY, MICHAEL.
    At White Bird Cannon, Idaho, June
      1876, to Jan. 8, 1877.
    R—NR.
    B—St. Johns, Newfoundland.

First sergeant, Troop H, 1st U. S. Cavalry.
Was detailed with 6 men to hold a commanding position, and held it with great gallantry until the troops fell back. He then fought his way through the Indians, rejoined a portion of his command, and continued the fight in retreat. He had two horses shot under him and was captured, but escaped and reported for duty after three days' hiding and wandering in the mountains.

McCAUSLIN, JOSEPH.
    At Petersburg, Va., Apr. 2, 1865.
    R—Ohio County, W. Va.
    B—Ohio County, W. Va.

Private, Company D, 12th West Virigia Infantry.
Conspicuous gallantry as color bearer in the assault on Fort Gregg.

McCLEARY, CHARLES H.
    At Nashville, Tenn., Dec. 16, 1864.
    R—NR.
    B—Sandusky County, Ohio.

First lieutenant, Company C, 72d Ohio Infantry.
Capture of flag of 4th Florida Infantry (C. S. A.).

McCLELLAND, JAMES M.
    At Vicksburg, Miss., May 22, 1863.
    R—Ohio.
    B—Harrison County, Ohio.

Private, Company B, 30th Ohio Infantry.
Gallantry in the charge of the "volunteer storming party."

McCLERNAND, EDWARD J.
    At Bear Paw Mountain, Mont.,
      Sept. 30, 1877.
    R—Springfield, Ill.
    B—Jacksonville, Ill.

Second lieutenant, 2d U. S. Cavalry.
Gallantly attacked a band of hostiles and conducted the combat with excellent skill and boldness.

McCONNELL, JAMES.
    At Vigan, Luzon, P. I., Dec. 4, 1899.
    R—Detroit, Mich.
    B—Syracuse, N. Y.

Private, Company B, 33d Infantry, U. S. Volunteers.
Fought for hours lying between two dead comrades, notwithstanding his hat was pierced, his clothing plowed through by bullets, and his face cut and bruised by flying gravel.

McCONNELL, SAMUEL.
    At Fort Blakely, Ala., Apr. 9, 1865.
    R—NR.
    B—Belmont County, Ohio.

Captain, Company H,119th Illinois Infantry.
Capture of flag.

McCORMICK, MICHAEL.
    At Cedar Creek, etc., Mont., Oct.
      21, 1876, to Jan. 8, 1877.
    R—NR.
    B—Rutland, Vt.

Private, Company G, 5th U. S. Infantry.
Gallantry in actions.

McCORNACK, ANDREW.
    At Vicksburg, Miss., May 22, 1863.
    R—NR.
    B—Kane, Ill.

Private, Company I, 127th Illinois Infantry.
Gallantry in the charge of the "volunteer storming party."

McDONALD, FRANKLIN M.
    Near Fort Griffin, Tex., Aug. 5, 1872.
    R—NR.
    B—Bowling Green, Ky.

Private, Company G, 11th U. S. Infantry.
Gallantry in defeating Indians who attacked the mail.

McDONALD, GEORGE E.
    At Fort Stedman, Va., Mar. 25,
      1865.
    R—NR.
    B—Warwick, R. I.

Private, Company L, 1st Connecticut Heavy Artillery.
Capture of flag.

McDONALD, JAMES.
    Arizona, August to October, 1868.
    R—NR.
    B—Scotland.

Corporal, Company B, 8th U. S. Cavalry.
Bravery in actions and scouts against Indians.

McDONALD, JOHN WADE.
    At Pittsburg Landing, Tenn., Apr.
      6, 1862.
    R—NR.
    B—Lancaster, Ohio.

Private, Company E, 20th Illinois Infantry.
Was severely wounded while endeavoring, at the risk of his life, to carry to a place of safety a wounded and helpless comrade.

McDONALD, ROBERT.
    At Wolf Mountain, Mont., Jan. 8,
      1877.
    R—Fort Sumner, N. Mex.
    B—New York.

First lieutenant, 5th U. S. Infantry.
Led his command in a successful charge against superior numbers of hostile Indians, strongly posted.

McELHINNY, SAMUEL O.
    At Sailors Creek, Va., Apr. 6, 1865.
    R—NR.
    B—Meigs County, Ohio.

Private, Company A, 2d West Virginia Cavalry.
Capture of flag.

McENROE, PATRICK H.
    At Winchester, Va., Sept. 19, 1864.
    R—New York.
    B—Ireland.

Sergeant, Company D, 6th New York Cavalry.
Capture of colors of 36th Virginia Infantry (C. S. A.).

McFALL, DANIEL
At Spotsylvania, Va., May 12, 1864.
R—NR.
B—Niagara County, N. Y.

Sergeant, Company E, 17th Michigan Infantry.
Captured the colonel commanding the Confederate brigade that charged the Union batteries; on the same day rescued an officer of his regiment from the enemy.

McGANN, MICHAEL A.
At Rosebud River, Mont., June 17, 1876.
R—NR.
B—Ireland.

First sergeant, Company F, 3d U. S. Cavalry.
Gallantry in action.

McGAR, OWEN
At Cedar Creek, etc., Mont., Oct. 21, 1876, to Jan. 8, 1877.
R—Pawtucket, R. I.
B—North Attleboro, Mass.

Private, Company C, 5th U. S. Infantry.
Gallantry in actions.

McGINN, EDWARD
At Vicksburg, Miss., May 22, 1863.
R—NR.
B—New York, N. Y.

Private, Company F, 54th Ohio Infantry.
Gallantry in the charge of the "volunteer storming party."

McGONAGLE, WILSON
At Vicksburg, Miss., May 22, 1863.
R—NR.
B—Jefferson County, Ohio.

Private, Company B, 30th Ohio Infantry.
Gallantry in the charge of the "volunteer storming party."

McGONNIGLE, ANDREW J.
At Cedar Creek, Va., Oct. 19, 1864.
R—Cumberland, Md.
B—New York, N. Y.

Captain and assistant quartermaster, U. S. Volunteers.
While acting chief quartermaster of General Sheridan's forces operating in the Shenandoah Valley was severely wounded while voluntarily leading a brigade of infantry and was commended for the greatest gallantry by General Sheridan.

McGOUGH, OWEN
At Bull Run, Va., July 21, 1861.
R—NR.
B—Ireland.

Corporal, Battery D, 5th U. S. Artillery.
Through his personal exertions under a heavy fire, one of the guns of his battery was brought off the field; all the other guns were lost.

McGRATH, HUGH J.
At Calamba, Luzon, P. I., July 26, 1899.
R—Eau Claire, Wis.
B—Fon du Lac, Wis.

Captain, 4th U. S. Cavalry.
Swam the San Juan River in the face of the enemy's fire and drove him from his entrenchments.

McGRAW, THOMAS
At Petersburg, Va., Apr. 2, 1865.
R—Chicago, Ill.
B—Ireland.

Sergeant, Company B, 23d Illinois Infantry.
One of the three soldiers most conspicuous for gallantry in the final assault.

McGUIRE, PATRICK
At Vicksburg, Miss., May 22, 1863.
R—NR.
B—Ireland.

Private, Chicago Mercantile Battery, Illinois Light Artillery.
Carried with others by hand a cannon up to and fired it through an embrasure of the enemy's works.

McHALE, ALEXANDER U.
At Spotsylvania Courthouse, Va., May 12, 1864.
R—NR.
B—Ireland.

Corporal, Company C, 26th Michigan Infantry.
Captured a Confederate color in a charge, threw the flag over in front of the works, and continued in the charge upon the enemy.

McHUGH, JOHN
At Cedar Creek, etc., Mont., Oct. 21, 1876, to Jan. 8, 1877.
R—NR.
B—Syracuse, N. Y.

Private, Company A, 5th U. S. Infantry.
Gallantry in actions.

McKAY, CHARLES W.
At Dug Gap, Ga., May 8, 1864.
R—NR.
B—Mansfield, N. Y.

Sergeant, Company C, 154th New York Infantry.
Voluntarily risked his life in rescuing under the fire of the enemy a wounded comrade who was lying between the lines.

McKEE, GEORGE
At Petersburg, Va., Apr. 2, 1865.
R—NR.
B—Ireland.

Color sergeant, Company D, 89th New York Infantry.
Gallantry as color bearer in the assault on Fort Gregg.

McKEEN, NINEVEH S.
At Stone River, Tenn., Dec. 30, 1862.
At Liberty Gap, Tenn., June 25, 1863.
R—NR.
B—Clark County, Ill.

First lieutenant, Company H, 21st Illinois Infantry.
Conspicuous in the charge, where he was three times wounded.

Captured colors of 8th Arkansas Infantry (C. S. A.).

McKEEVER, MICHAEL
At Burnt Ordinary, Va., Jan. 19, 1863.
R—Philadelphia, Pa.
B—Ireland.

Private, Company K, 5th Pennsylvania Cavalry.
Was one of a small scouting party that charged and routed a mounted force of the enemy six times their number. He led the charge in a most gallant and distinguished manner, going far beyond the call of duty.

McKinley, Daniel_____ | Private, Company B, 8th U. S. Cavalry.
Arizona, August to October, 1868. | Bravery in actions and scouts against Indians.
R—NR.
B—Boston, Mass.

McKown, Nathaniel A_____ | Sergeant, Company B, 58th Pennsylvania Infantry.
At Chapins farm, near Richmond, | Capture of flag.
Va., Sept. 29, 1864.
R—NR.
B—Susquehanna County, Pa.

McLennon, John_____ | Musician, Company A, 7th U.S. Infantry.
At Big Hole, Mont., Aug. 9, 1877. | Gallantry in action.
R—NR.
B—Fort Belknap, Tex.

McLoughlin, Michael_____ | Sergeant, Company A, 5th U. S. Infantry.
At Cedar Creek, etc., Mont., Oct. 21, | Gallantry in action.
1876, to Jan. 8, 1877.
R—NR.
B—Ireland.

McMahon, Martin T_____ | Captain and aid-de-camp, U. S. Volunteers.
At White Oak Swamp, Va., June 30, | Under fire of the enemy successfully destroyed a valuable train that had been
1862. | abandoned and prevented it from falling into the hands of the enemy.
R—NR.
B—NR.

McMasters, Henry A_____ | Corporal, Company A, 4th U. S. Cavalry.
At Red River, Tex., Sept. 29, 1872. | Gallantry in action.
R—NR.
B—Augusta, Me.

McMillan, Albert W_____ | Sergeant, Company E, 7th U. S. Cavalry.
At Wounded Knee Creek, S. Dak., | While engaged with Indians concealed in a ravine, he assisted the men on the
Dec. 29, 1890. | skirmish line, directed their fire, encouraged them by example, and used
R—Baltimore, Md. | every effort to dislodge the enemy.
B—Baltimore, Md.

McMillen, Francis M_____ | Sergeant, Company C, 110th Ohio Infantry.
At Petersburg, Va., Apr. 2, 1865. | Capture of flag.
R—NR.
B—Bracken County, Ky.

McMurtry, George G_____ | Captain, 308th Infantry, 77th Division.
At Charlevaux, in the Forest D'Ar- | Commanded a battalion which was cut off and surrounded by the enemy and,
gonne, France, Oct. 2–8, 1918. | although wounded in the knee by shrapnel on October 4 and suffering great
R—New York, N. Y. | pain, he continued throughout the entire period to encourage his officers and
B—Pittsburgh, Pa. | men with a resistless optimism that contributed largely toward preventing
G. O. No. 118, W. D., 1918. | panic and disorder among the troops, who were without food, cut off from
| communication with our lines. On October 4, during a heavy barrage, he
| personally directed and supervised the moving of the wounded to shelter
| before himself seeking shelter. On October 6 he was again wounded in the
| shoulder by a German grenade, but continued personally to organize and
| direct the defense against the German attack on the position until the attack
| was defeated. He continued to direct and command his troops, refusing
| relief, and personally led his men out of the position after assistance arrived
| before permitting himself to be taken to the hospital on October 8. During
| this period the successful defense of the position was due largely to his
| efforts.

McNally, James_____ | First sergeant, Company E, 8th U. S. Cavalry.
Arizona, 1868 and 1869. | Bravery in actions and scouts against Indians.
R—NR.
B—Ireland.

McNamara, William_____ | First sergeant, Company F, 4th U. S. Cavalry.
At Red River, Tex., Sept. 29, 1872. | Gallantry in action.
R—NR.
B—Ireland.

McPhelan, Robert_____ | Sergeant, Company E, 5th U. S. Infantry.
At Cedar Creek, etc., Mont., Oct. 21, | Gallantry in actions.
1876, to Jan. 8, 1877.
R—NR.
B—Ireland.

McVeagh, Charles H_____ | Private, Company B, 8th U. S. Cavalry.
Arizona, August to October, 1868. | Bravery in actions and scouts against Indians.
R—NR.
B—New York, N. Y.

McVean, John P_____ | Corporal, Company D, 49th New York Infantry.
At Fredericksburg Heights, Va., | Shot a Confederate color bearer and seized the flag; also approached, alone,
May 4, 1863. | a barn between the lines and demanded and received the surrender of a num-
R—NR. | ber of the enemy therein.
B—Canada.

McWHORTER, WALTER F.
At Sailors Creek, Va., Apr. 6, 1865.
R—Harrison County, Va.
B—Lewis County, Va.

Commissary sergeant, Company E, 3d West Virginia Cavalry.
Capture of flag of 6th Tennessee Infantry (C. S. A.).

MACHOL.
Arizona, 1872 and 1873.
R—NR.
B—NR.

Private, Indian Scouts.
Gallant services in campaigns against the Apache Indians.

MACLAY, WILLIAM P.
At Hilongas, Leyte, P. I., May 6, 1900.
R—Altoona, Pa.
B—Spruce Creek, Pa.

Private, Company A, 43d Infantry, U. S. Volunteers.
Charged an occupied bastion, saving the life of an officer in a hand-to-hand combat and destroying the enemy.

MADDEN, MICHAEL.
At Masons Island, Md., Sept. 3, 1861.
R—NR.
B—NR.

Private, Company K, 42d New York Infantry.
Assisted a wounded comrade to the river bank and, under fire of the enemy, swam with him across a branch of the Potomac to the Union lines.

MADISON, JAMES.
At Waynesboro, Va., Mar. 2, 1865.
R—NR.
B—Niagara, N. Y.
Service rendered under name of James Congdon.

Sergeant, Company E, 8th New York Cavalry.
Recapture of General Crook's headquarters flag.

MAGEE, WILLIAM.
At Murfreesboro, Tenn., Dec. 5, 1864.
R—NR.
B—Newark, N. J.

Drummer, Company C, 33d New Jersey Infantry.
In a charge was among the first to reach a battery of the enemy and, with one or two others, mounted the artillery horses and took two guns into the Union lines.

MAHERS, HERBERT.
At Seneca Mountain, Ariz., Aug. 25, 1869.
R—NR.
B—Canada.

Private, Company, F, 8th U. S. Cavalry.
Gallantry in action.

MAHONEY, GREGORY.
Near Red River, Tex., Sept. 26–28, 1874.
R—NR.
B—South Wales.

Private, Company E, 4th U. S. Cavalry.
Gallantry in attack on a large party of Cheyennes.

MAHONEY, JEREMIAH.
At Fort Sanders, Knoxville, Tenn., Nov. 29, 1863.
R—NR.
B—NR.

Sergeant, Company A, 29th Massachusetts Infantry.
Capture of flag of 17th Mississippi Infantry (C. S. A.).

MALLON, GEORGE H.
In the Bois-de-Forges, France, Sept. 26, 1918.
R—Minneapolis, Minn.
B—Ogden, Kans.
G. O. No. 16, W. D., 1919.

Captain, 132d Infantry, 33d Division.
Becoming separated from the balance of his company because of a fog, Captain Mallon, with nine soldiers, pushed forward and attacked nine active hostile machine guns, capturing all of them without the loss of a man. Continuing on through the woods, he led his men in attacking a battery of four 155-millimeter howitzers, which were in action, rushing the position and capturing the battery and its crew. In this encounter Captain Mallon personally attacked one of the enemy with his fists. Later, when the party came upon two more machine guns, this officer sent men to the flanks while he rushed forward directly in the face of the fire and silenced the guns, being the first one of the party to reach the nest. The exceptional gallantry and determination displayed by Captain Mallon resulted in the capture of 100 prisoners, 11 machine guns, four 155-millimeter howitzers and one antiaircraft gun.

MANDY, HARRY J.
At Front Royal, Va., Aug. 15, 1864.
R—New York, N. Y.
B—England.

First sergeant, Company B, 4th New York Cavalry.
Capture of flag of 3d Virginia Infantry (C. S. A.).

MANGAM, RICHARD C.
At Hatchers Run, Va., Apr. 2, 1865.
R—NR.
B—Ireland.

Private, Company H, 148th New York Infantry.
Capture of flag of 8th Mississippi Infantry (C. S. A.).

MANNING, JOSEPH S.
At Fort Sanders, Knoxville, Tenn., Nov. 29, 1863.
R—NR.
B—Ipswich, Mass.

Private, Company K, 29th Massachusetts Infantry.
Capture of flag of 16th or 18th Georgia Infantry (C. S. A.).

MANNING, SIDNEY E. (97184)............
Near Breuvannes, France, July 28, 1918.
R—Flomaton, Ala.
B—Butler County, Ala.
G. O. No. 44, W. D., 1919.

Corporal, Company G, 167th Infantry, 42d Division.
When his platoon commander and platoon sergeant had both become casualties soon after the beginning of an assault on strongly fortified heights overlooking the Ourcq River, Corporal Manning took command of his platoon, which was near the center of the attacking line. Though himself severely wounded he led forward the 35 men remaining in the platoon and finally succeeded in gaining a foothold on the enemy's position, during which time he had received more wounds and all but seven of his men had fallen. Directing the consolidation of the position, he held off a large body of the enemy only 50 yards away by fire from his automatic rifle. He declined to take cover until the line had been entirely consolidated with the line of the platoon on the flank, when he dragged himself to shelter, suffering from nine wounds in all parts of the body.

MARLAND, WILLIAM............
At Grand Coteau, La., Nov. 3, 1863.
R—NR.
B—NR.

First lieutenant, 2d Independent Battery, Massachusetts Light Artillery.
After having been surrounded by the enemy's cavalry, his support having surrendered, he ordered a charge and saved the section of the battery that was under his command.

MARQUETTE, CHARLES...........
At Petersburg, Va., Apr. 2, 1865.
R—NR.
B—Lebanon County, Pa.

Sergeant, Company F, 93d Pennsylvania Infantry.
Capture of flag.

MARSH, ALBERT.............
At Spotsylvania, Va., May 12, 1864.
R—Randolph, N. Y.
B—Cattaraugus County, N. Y.

Sergeant, Company B, 64th New York Infantry.
Capture of flag.

MARSH, CHARLES H..........
At Back Creek Valley, Va., July 31, 1864.
R—NR.
B—Milford, Conn.

Private, Company D, 1st Connecticut Cavalry.
Capture of flag and its bearer.

MARSH, GEORGE.............
At Elk River, Tenn., July 2, 1863.
R—NR.
B—La Salle County, Ill.

Sergeant, Company D, 104th Illinois Infantry.
Voluntarily led a small party and, under a heavy fire, captured a stockade and saved the bridge.

MARTIN, GEORGE.............
At Millerstown, Pa., July —, 1863.
R—NR.
B—Germany.
Service rendered under name of Martin Schwenk.

Sergeant, Company B, 6th U. S. Cavalry.
Bravery in an attempt to carry a communication through the enemy's lines; also rescued an officer from the hands of the enemy.

MARTIN, PATRICK............
At Castle Dome and Santa Maria Mountains, Ariz., June and July, 1873.
R—NR.
B—Ireland.

Sergeant, Company G, 5th U. S. Cavalry.
Gallant services in operations of Capt. James Burns, 5th U. S. Cavalry.

MARTIN, SYLVESTER H............
At Weldon Railroad, Va., Aug. 19, 1864.
R—NR.
B—Chester County, Pa.

Lieutenant, Company K, 88th Pennsylvania Infantry.
Gallantly made a most dangerous reconnaissance, discovering the position of the enemy and enabling the division to repulse an attack made in strong force.

MASON, ELIHU H............
Georgia, April, 1862.
R—NR.
B—Wayne County, Ind.

Sergeant, Company K, 21st Ohio Infantry.
One of 22 men (including two civilians) who by direction of General Mitchel (or Buell) penetrated nearly 200 miles south into the enemy's territory and captured a railroad train at Big Shanty, Ga., in an attempt to destroy the bridges and track between Chattanooga and Atlanta.

MATHEWS, GEORGE W............
Near Labao, Luzon, P. I., Oct. 29, 1899.
R—Worcester, Mass.
B—Worcester, Mass.

Assistant surgeon, 36th Infantry, U. S. Volunteers.
While in attendance upon the wounded and under a severe fire from the enemy, seized a carbine and beat off an attack upon wounded officers and men under his charge.

MATHEWS, WILLIAM H............
At Petersburg, Va., July 30, 1864.
R—England.
B—NR.
Service rendered under name of Henry Sivel.

First sergeant, Company E, 2d Maryland Veteran Infantry.
Finding himself among a squad of Confederates, he fired into them, killing one, and was himself wounded, but succeeded in bringing in a sergeant and two men of the 17th South Carolina Regiment as prisoners.

MATTHEWS, DAVID A............
Arizona, 1868 and 1869.
R—NR.
B—Boston, Mass.

Corporal, Company E, 8th U. S. Cavalry.
Bravery in actions and scouts against Indians.

MATTHEWS, JOHN C............
At Petersburg, Va., Apr. 2, 1865.
R—NR.
B—Westmoreland County, Pa.

Corporal, Company A, 61st Pennsylvania Infantry.
Voluntarily took the colors, whose bearer had been disabled, and, although himself severely wounded, carried the same until the enemy's works were taken.

MATTHEWS, MILTON_____
   At Petersburg, Va., Apr. 2, 1865.
   R—Pittsburgh, Pa.
   B—Pittsburgh, Pa.

Private, Company C, 61st Pennsylvania Infantry.
Capture of flag of 7th Tennessee Infantry (C. S. A.).

MATTINGLY, HENRY B_____
   At Jonesboro, Ga., Sept. 1, 1864.
   R—NR.
   B—Marion County, Ky.

Private, Company B, 10th Kentucky Infantry.
Capture of flags of 6th and 7th Arkansas Infantry (C. S. A.).

MATTOCKS, CHARLES P_____
   At Sailors Creek, Va., Apr. 6, 1865.
   R—NR.
   B—NR.

Major, 17th Maine Infantry.
Displayed extraordinary gallantry in leading a charge of his regiment which resulted in the capture of a large number of prisoners and a stand of colors.

MAUS, MARION P_____
   At Sierra Madre Mountains, Mexico, Jan. 11, 1886.
   R—Maryland.
   B—Burnt Mills, Md.

First lieutenant, 1st U. S. Infantry.
Most distinguished gallantry in action with hostile Apaches led by Geronimo and Natchez.

MAXHAM, LOWELL M_____
   At Fredericksburg, Va., May 3, 1863.
   R—NR.
   B—Carver, Mass.

Corporal, Company F, 7th Massachusetts Infantry.
Though severely wounded and in face of a deadly fire from the enemy at short range, he rushed bravely forward and was among the first to enter the enemy's works on the crest of Maryes Heights and helped to plant his regimental colors there.

MAY, JOHN_____
   At Wichita River, Tex., July 12, 1870.
   R—NR.
   B—Germany.

Sergeant, Company L, 6th U. S. Cavalry.
Gallantry in action.

MAY, WILLIAM_____
   At Nashville, Tenn., Dec. 16, 1864.
   R—NR.
   B—Pennsylvania.

Private, Company H, 32d Iowa Infantry.
Ran ahead of his regiment over the enemy's works and captured from its bearer the flag of Bonanchad's Confederate battery.

MAYBERRY, JOHN B_____
   At Gettysburg, Pa., July 3, 1863.
   R—NR.
   B—Smyrna, Del.

Private, Company F, 1st Delaware Infantry.
Capture of flag.

MAYES, WILLIAM B_____
   Near Kenesaw Mountain, Ga., June 15, 1864.
   R—NR.
   B—Marion County, Ohio.

Private, Company K, 11th Iowa Infantry.
With one companion and under a fierce fire from the enemy at short range went to the rescue of a wounded comrade who had fallen between the lines and carried him to a place of safety.

MAYNARD, GEORGE H_____
   At Fredericksburg, Va., Dec. 13, 1862.
   R—NR.
   B—NR.

Private, Company D, 13th Massachusetts Infantry.
A wounded and helpless comrade, having been left on the skirmish line, this soldier voluntarily returned to the front under a severe fire and carried the wounded man to a place of safety.

MAYS, ISAIAH_____
   Arizona, May 11, 1889.
   R—Fort Grant, Ariz.
   B—Carters Bridge, Va.

Corporal, Company B, 24th U. S. Infantry.
Gallantry in the fight between Paymaster Wham's escort and robbers.

MEACH, GEORGE E_____
   At Winchester, Va., Sept. 19, 1864.
   R—NR.
   B—New York.

Ferrier, Company I, 6th New York Cavalry.
Capture of flag.

MEAGHER, THOMAS_____
   At Chapins Farm, near Richmond, Va., Sept. 29, 1864.
   R—Brooklyn, N. Y.
   B—Scotland.

First Sergeant, Company G, 158th New York Infantry.
Led a section of his men on the enemy's works, receiving a wound while scaling a parapet.

MEAHER, NICHOLAS_____
   At Chiricahua Mountains, Ariz., Oct. 20, 1869.
   R—NR.
   B—Perry County, Ohio.

Corporal, Company G, 1st U. S. Cavalry.
Gallantry in action.

MEARS, GEORGE W_____
   At Gettysburg, Pa., July 2, 1863.
   R—Bloomsburgh, Pa.
   B—Bloomsburgh, Pa.

Sergeant, Company A, 6th Pennsylvania Reserves.
With five volunteers he gallantly charged on a number of the enemy's sharpshooters concealed in a log house, captured them, and brought them into the Union lines.

MECHLIN, HENRY W. B_____
   At Little Big Horn, Mont., June 25, 1876.
   R—NR.
   B—Westmoreland County, Pa.

Blacksmith, Company H, 7th U. S. Cavalry.
With three comrades during the entire engagement courageously held a position that secured water for the command.

MENTER, JOHN W. At Sailors Creek, Va., Apr. 6, 1865. R—Detroit, Mich. B—Palmer, N. Y.	Sergeant, Company D, 5th Michigan Infantry. Capture of flag.
MERRIAM, HENRY C. At Fort Blakely, Ala., Apr. 9, 1865. R—Houlton, Me. B—Houlton, Me.	Lieutenant colonel, 73d U. S. Colored Troops. Volunteered to attack the enemy's works in advance of orders and, upon permission being given, made a most gallant assault.
MERRIFIELD, JAMES K. At Franklin, Tenn., Nov. 30, 1864. R—NR. B—Pennsylvania.	Corporal, Company C, 88th Illinois Infantry. Captured two battle flags from the enemy and returned with them to his own lines.
MERRILL, AUGUSTUS. At Petersburg, Va., Apr. 2, 1865. R—NR. B—Byron, Me.	Captain, Company B, 1st Maine Veteran Infantry. With 6 men captured 69 Confederate prisoners and recaptured several soldiers who had fallen into the enemy's hands.
MERRILL, GEORGE. At Fort Fisher, N. C., Jan. 15, 1865. R—NR. B—Queensberry, N. Y.	Private, Company I, 142d New York Infantry. Voluntarily advanced with the head of the column and cut down the palisading.
MERRILL, JOHN. At Milk River, Colo., Sept. 29, 1879. R—NR. B—New York, N. Y.	Sergeant, Company F, 5th U. S. Cavalry Though painfully wounded, he remained on duty and rendered gallant and valuable service.
MERRITT, JOHN G. At Bull Run, Va., July 21, 1861. R—NR. B—NR.	Sergeant, Company K, 1st Minnesota Infantry. Gallantry in action; was wounded in advance of his regiment.
ᵒMESTROVITCH, JAMES I. (1243675) At Fismette, France, Aug. 10, 1918. R—Pittsburgh, Pa. B—Montenegro. G. O. No. 20, W. D., 1919.	Sergeant, Company C, 111th Infantry, 28th Division. Seeing his company commander lying wounded 30 yards in front of the line after his company had withdrawn to a sheltered position behind a stone wall, Sergeant Mestrovitch voluntarily left cover and crawled through heavy machine-gun and shell fire to where the officer lay. He took the officer upon his back and crawled back to a place of safety, where he administered first-aid treatment, his exceptional heroism saving the officer's life. Posthumously awarded. Medal presented to mother, Mrs. Mary I. Mestrovitch.
MEYER, HENRY C. At Petersburg, Va., June 17, 1864. R—Dobbs Ferry, N. Y. B—Hamburg, N. Y.	Captain, Company D, 24th New York Cavalry. During an assault and in the face of a heavy fire rendered heroic assistance to a wounded and helpless officer, thereby saving his life and in the performance of this gallant act sustained a severe wound.
MILES, L. WARDLAW. Near Revillon, France, Sept. 14, 1918. R—Princeton, N. J. B—Baltimore, Md. G. O. No. 44, W. D., 1919.	Captain, 308th Infantry, 77th Division. Volunteered to lead his company in a hazardous attack on a commanding trench position near the Aisne Canal, which other troops had previously attempted to take without success. His company immediately met with intense machine-gun fire, against which it had no artillery assistance, but Captain Miles preceded the first wave and assisted in cutting a passage through the enemy's wire entanglements. In so doing he was wounded five times by machine-gun bullets, both legs and one arm being fractured, whereupon he ordered himself placed on a stretcher and had himself carried forward to the enemy trench in order that he might encourage and direct his company, which by this time had suffered numerous casualties. Under the inspiration of this officer's indomitable spirit his men held the hostile position and consolidated the front line after an action lasting two hours, at the conclusion of which Captain Miles was carried to the aid station against his will.
MILES, NELSON A. At Chancellorsville, Va., May 2 and 3, 1863. R—Roxbury, Mass. B—Westminster, Mass.	Colonel, 61st New York Infantry. Distinguished gallantry while holding with his command an advanced position against repeated assaults by a strong force of the enemy; was severely wounded.
MILLER, ARCHIE. At Patian Island, P. I., July 2, 1909. R—St. Louis, Mo. B—Fort Sheridan, Ill.	First lieutenant, 6th U. S. Cavalry. While in action against hostile Moros, when the machine-gun detachment, having been driven from its position by a heavy fire, one member being killed, did, with the assistance of an enlisted man, place the machine gun in advance of its former position at a distance of about 20 yards from the enemy, in accomplishing which he was obliged to splice a piece of timber to one leg of the gun tripod, all the while being under a heavy fire, and the gun tripod being several times struck by bullets.
MILLER, DANIEL H. At Whetstone Mountains, Ariz., May 5, 1871. R—NR. B—Fairfield County, Ohio.	Private, Company F, 3d U. S. Cavalry. Gallantry in action.

MILLER, FRANK\
    At Sailors Creek, Va., Apr. 6, 1865.\
    R—Jamaica, N. Y.\
    B—New York.

Private, Company M, 2d New York Cavalry.\
Capture of flag of 25th Battalion Virginia Infantry (C. S. A.); was taken prisoner, but successfully retained his trophy until recaptured.

MILLER, GEORGE\
    At Cedar Creek, etc., Mont., Oct. 21, 1876, to Jan. 8, 1877.\
    R—NR.\
    B—Brooklyn, N. Y.

Corporal, Company H, 5th U. S. Infantry.\
Gallantry in actions.

MILLER, GEORGE W\
    Arizona, August to October, 1868.\
    R—NR.\
    B—Philadelphia, Pa.

Private, Company B, 8th U. S. Cavalry.\
Bravery in actions and scouts against Indians.

MILLER, HENRY A\
    At Fort Blakely, Ala., Apr. 9, 1865.\
    R—Decatur, Ill.\
    B—Germany.

Captain, Company B, 8th Illinois Infantry.\
Capture of flag.

MILLER, JACOB C\
    At Vicksburg, Miss., May 22, 1863.\
    R—NR.\
    B—Bellevue, Ohio.

Private, Company G, 113th Illinois Infantry.\
Gallantry in the charge of the "volunteer storming party."

MILLER, JAMES P\
    At Selma, Ala., Apr. 2, 1865.\
    R—Henry County, Iowa.\
    B—Franklin, Ohio.

Private, Company D, 4th Iowa Cavalry.\
Capture of standard of 12th Mississippi Cavalry (C. S. A.).

MILLER, JOHN\
    At Waynesboro, Va., Mar. 2, 1865.\
    R—Rochester, N. Y.\
    B—Germany.

Private, Company H, 8th New York Cavalry.\
Capture of flag.

MILLER, JOHN\
    At Gettysburg, Pa., July 3, 1863.\
    R—NR.\
    B—NR.

Corporal, Company G, 8th Ohio Infantry.\
Capture of two flags.

*MILLER, OSCAR F\
    Near Gesnes, France, Sept. 28, 1918.\
    R—Los Angeles, Calif.\
    B—Franklin County, Ark.\
    G. O. No. 16, W. D., 1919.

Major, 361st Infantry, 91st Division.\
After two days of intense physical and mental strain, during which Major Miller had led his battalion in the front line of the advance through the forest of Argonne, the enemy was met in a prepared position south of Gesnes. Though almost exhausted, he energetically reorganized his battalion and ordered an attack. Upon reaching open ground, the advancing line began to waver in the face of machine-gun fire from the front and flanks and direct artillery fire. Personally leading his command group forward between his front-line companies, Major Miller inspired his men by his personal courage, and they again pressed on toward the hostile position. As this officer led the renewed attack he was shot in the right leg, but he nevertheless staggered forward at the head of his command. Soon afterwards he was again shot in the right arm, but he continued the charge, personally cheering his troops on through the heavy machine-gun fire. Just before the objective was reached he received a wound in the abdomen, which forced him to the ground, but he continued to urge his men on, telling them to push on to the next ridge and leave him where he lay. He died from his wounds a few days later.\
Posthumously awarded. Medal presented to widow, Mrs. Anna M. Miller.

MILLER, WILLIAM E\
    At Gettysburg, Pa., July 3, 1863.\
    R—NR.\
    B—NR.

Captain, Company H, 3d Pennsylvania Cavalry.\
Without orders led a charge of his squadron upon the flank of the enemy, checked his attack, and cut off and dispersed the rear of his column.

MILLS, ALBERT L\
    Near Santiago, Cuba, July 1, 1898.\
    R—NR.\
    B—NR.

Captain and assistant adjutant general, U. S. Volunteers.\
Distinguished gallantry in encouraging those near him by his bravery and coolness after being shot through the head and entirely without sight.

MILLS, FRANK W\
    At Sandy Cross Roads, N. C., Sept. 4, 1862.\
    R—NR.\
    B—Middletown, N. Y.

Sergeant, Company C, 1st New York Mounted Rifles.\
While scouting, this soldier, in command of an advance of but three or four men, came upon the enemy, and charged them without orders, the rest of the troop following, the whole force of the enemy, 120 men, being captured.

MINDIL, GEORGE W\
    At Williamsburg, Va., May 5, 1862.\
    R—Philadelphia, Pa.\
    B—NR.

Captain, Company I, 61st Pennsylvania Infantry.\
As aid-de-camp led the charge with a part of a regiment, pierced the enemy's center, silenced some of his artillery, and, getting in his rear, caused him to abandon his position.

MITCHELL, ALEXANDER H\
    At Spotsylvania, Va., May 12, 1864.\
    R—Hamilton, Pa.\
    B—Perrysville, Pa.

First lieutenant, Company A, 105th Pennsylvania Infantry.\
Capture of flag of 18th North Carolina Infantry (C. S. A.), in a personal encounter with the color bearer.

MITCHELL, JOHN. — First sergeant, Company I, 5th U. S. Infantry.
At Upper Washita, Tex., Sept. 9 to 11, 1874. — Gallantry in engagement with Indians.
R—NR.
B—Ireland.

MITCHELL, JOHN J. — Corporal, Company L, 8th U. S. Cavalry.
At Hell Canyon, Ariz., July 3, 1869. — Gallantry in action.
R—NR.
B—Ireland.

MITCHELL, THEODORE. — Private, Company C, 61st Pennsylvania Infantry.
At Petersburg, Va., Apr. 2, 1865. — Capture of flag.
R—Pittsburgh, Pa.
B—Tarentum, Pa.

MOFFITT, JOHN H. — Corporal, Company C, 16th New York Infantry.
At Gaines Mill, Va., June 27, 1862. — Voluntarily took up the regimental colors after several color bearers had been shot down and carried them until himself wounded.
R—NR.
B—NR.

MOLBONE, ARCHIBALD. — Sergeant, Company G, 1st Rhode Island Light Artillery.
At Petersburg, Va., Apr. 2, 1865. — Was one of a detachment of 20 picked artillerymen who voluntarily accompanied an Infantry assaulting party and who turned upon the enemy the guns captured in the assault.
R—NR.
B—Coventry, R. I.

MONAGHAN, PATRICK. — Corporal, Company F, 48th Pennsylvania Infantry.
At Petersburg, Va., June 17, 1864. — Recapture of colors of 7th New York Heavy Artillery.
R—Minersville, Pa.
B—Ireland.

MONTROSE, CHARLES H. — Private, Company I, 5th U. S. Infantry.
At Cedar Creek, etc., Mont., Oct. 21, 1876, to Jan. 8, 1877. — Gallantry in actions.
R—St. Louis, Mo.
B—St. Paul, Minn.

MOORE, DANIEL B. — Corporal, Company E, 11th Wisconsin Infantry.
At Fort Blakely, Ala., Apr. 9, 1865. — At the risk of his own life saved the life of an officer who had been shot down and overpowered by superior numbers.
R—NR.
B—Iowa County, Wis.

MOORE, GEORGE G. — Private, Company D, 11th West Virginia Infantry.
At Fishers Hill, Va., Sept. 22, 1864. — Capture of flag.
R—NR.
B—Tylor, Va.

MOORE, WILBUR F. — Private, Company C, 117th Illinois Infantry.
At Nashville, Tenn., Dec. 16, 1864. — Captured flag of a Confederate battery while far in advance of the Union lines.
R—NR.
B—St. Clair County, Ill.

MOQUIN, GEORGE. — Corporal, Company F, 5th U. S. Cavalry.
At Milk River, Colo., Sept. 29 to Oct. 5, 1879. — Gallantry in action.
R—NR.
B—New York, N. Y.

MORAN, JOHN. — Private, Company F, 8th U. S. Cavalry.
At Seneca Mountain, Ariz., Aug. 25, 1869. — Gallantry in action.
R—NR.
B—Ireland.

MORAN, JOHN E. — Captain, Company L, 37th Infantry, U. S. Volunteers.
Near Mabitac, Laguna, Luzon, P. I., Sept. 17, 1900. — After the attacking party had become demoralized, fearlessly led a small body of troops under a severe fire and through water waist deep in the attack against the enemy.
R—NR.
B—NR.

MORELOCK, STERLING (2661521). — Private, Company M, 28th Infantry, 1st Division.
Near Exermont, France, Oct. 4, 1918. — While his company was being held up by heavy enemy fire Private Morelock, with three other men who were acting as runners at company headquarters, voluntarily led them as a patrol in advance of his company's front line through an intense rifle, artillery, and machine-gun fire and penetrated a woods which formed the German front line. Encountering a series of five hostile machine-gun nests, containing from one to five machine guns each, with his patrol he cleaned them all out, gained and held complete mastery of the situation until the arrival of his company commander with reinforcements, even though his entire party had become casualties. He rendered first aid to the injured and evacuated them by using as stretcher bearers 10 German prisoners whom he had captured. Soon thereafter his company commander was wounded and while dressing his wound Private Morelock was very severely wounded in the hip, which forced his evacuation. His heroic action and devotion to duty were an inspiration to the entire regiment.
R—Oquawka, Ill.
B—Silver Run, Md.
G. O. No. 43, W. D., 1922.

MOREY, DELANO_____
At McDowell, Va., May 8, 1862.
R—Hardin County, Ohio.
B—Licking County, Ohio.

Private, Company B, 82d Ohio Infantry.
After the charge of the command had been repulsed, he rushed forward alone with an empty gun and captured two of the enemy's sharpshooters.

MORFORD, JEROME_____
At Vicksburg, Miss., May 22, 1863.
R—Bridgers Corner, Ill.
B—Mercer County, Pa.

Private, Company K, 55th Illinois Infantry.
Gallantry in the charge of the "volunteer storming party."

MORGAN, GEORGE H_____
At Big Dry Fork, Ariz., July 17, 1882.
R—Minneapolis, Minn.
B—Canada.

Second lieutenant, 3d U. S. Cavalry.
Gallantly held his ground at a critical moment and fired upon the advancing enemy (hostile Indians) until himself disabled by a shot.

MORGAN, LEWIS_____
At Spotsylvania, Va., May 12, 1864.
R—Delaware County, Ohio.
B—Delaware County, Ohio.

Private, Company I, 4th Ohio Infantry.
Capture of flag from the enemy's works.

MORGAN, RICHARD H_____
At Columbus, Ga., Apri. 16, 1865.
R—NR.
B—Dubois County, Ind.

Corporal, Company A, 4th Iowa Cavalry.
Capture of flag inside the enemy's works, contesting for its possession with its bearer.

MORIARTY, JOHN_____
Arizona, 1868 and 1869.
R—NR.
B—England.

Sergeant, Company E, 8th U. S. Cavalry.
Bravery in actions and scouts against Indians.

MORRILL, WALTER G_____
At Rappahannock Station, Va., Nov. 7, 1863.
R—Brownville, Me.
B—Brownville, Me.

Captain, Company B, 20th Maine Infantry.
Learning that an assault was to be made upon the enemy's works by other troops, this officer voluntarily joined the storming party with about 50 men of his regiment, and by his dash and gallantry rendered effective service in the assault.

MORRIS, JAMES L_____
Near Fort Selden, N. Mex., July 8 to 11, 1873.
R—NR.
B—Ireland.

First sergeant, Company C, 8th U. S. Cavalry.
Services against hostile Indians.

MORRIS, WILLIAM _____
At Sailors Creek, Va., Apr. 6, 1865.
R—Philadelphia, Pa.
B—Philadelphia, Pa.

Sergeant, Company C, 1st New York (Lincoln) Cavalry.
Capture of flag of 40th Virginia Infantry (C. S. A.).

MORRIS, WILLIAM W_____
At Upper Washita, Tex., Sept. 9 to 11, 1874.
R—NR.
B—Stewart County, Tenn.

Corporal, Company H, 6th U. S. Cavalry.
Gallantry in engagement with Indians.

MORRISON, FRANCIS_____
At Bermuda Hundred, Va., June 17, 1864.
R—Drakestown, Pa.
B—Fayette County, Pa.

Private, Company H, 85th Pennsylvania Infantry.
Voluntarily exposed himself to a heavy fire to bring off a wounded comrade.

MORSE, BENJAMIN_____
At Spotsylvania, Va., May 12, 1864.
R—NR.
B—Livingston, N. Y.

Private, Company C, 3d Michigan Infantry.
Capture of colors of 4th Georgia Battery (C. S. A.).

MORSE, CHARLES E_____
At Wilderness, Va., May 5, 1864.
R—NR.
B—France.

Sergeant, Company I, 62d New York Infantry.
Voluntarily rushed back into the enemy's lines, took the colors from the color sergeant, who was mortally wounded, and, although himself wounded, carried them through the fight.

MOSHER, LOUIS C_____
At Bagsak Mountain, Jolo, P. I., June 11, 1913.
R—Brockton, Mass.
B—Westport, Mass.

Second lieutenant, Philippine Scouts.
Voluntary entered a cleared space within about 20 yards of the Moro trenches under a furious fire from them and carried a wounded soldier of his company to safety at the risk of his own life.

MOSTOLLER, JOHN W_____
At Lynchburg, Va., June 18, 1864.
R—NR.
B—Somerset County, Pa.

Private, Company B, 54th Pennsylvania Infantry.
Voluntarily led a charge on a Confederate battery (the officers of the company being disabled) and compelled its hasty removal.

MOTT, JOHN_____
At Whetstone Mountains, Ariz., May 5, 1871.
R—NR.
B—Scotland.

Sergeant, Company F, 3d U. S. Cavalry.
Gallantry in action.

MOYLAN, MYLES_____
At Bear Paw Mountain, Mont., Sept. 30, 1877.
R—Essex, Mass.
B—Ireland.

Captain, 7th U. S. Cavalry.
Gallantly led his command in action against Nez Percés Indians until himself severely wounded.

MULHOLLAND, ST. CLAIR A_____
At Chancellorsville, Va., May 4 and 5, 1863.
R—NR.
B—NR.

Major, 116th Pennsylvania Infantry.
In command of the picket line held the enemy in check all night to cover the retreat of the Army.

MUNDELL, WALTER L_____
At Sailors Creek, Va., Apr. 6, 1865.
R—Dallas, Mich.
B—Marshall, Va.

Corporal, Company E, 5th Michigan Infantry.
Capture of flag.

MUNSELL, HARVEY M_____
At Gettysburg, Pa., July 1–3, 1863.
R—Venango County, Pa.
B—Steuben County, N. Y.

Sergeant, Company A, 99th Pennsylvania Infantry.
Gallant and courageous conduct as color bearer. (This noncommissioned officer carried the colors of his regiment through 13 engagements.)

MURPHY, CHARLES J_____
At Bull Run, Va., July 21, 1861.
R—NR.
B—NR.

First lieutenant and quartermaster, 38th New York Infantry.
Took a rifle and voluntarily fought with his regiment in the ranks; when the regiment was forced back, voluntarily remained on the field caring for the wounded, and was there taken prisoner.

MURPHY, DANIEL_____
At Hatchers Run, Va., Oct. 27, 1864.
R—NR.
B—Philadelphia, Pa.

Sergeant, Company F, 19th Massachusetts Infantry.
Capture of flag of 47th North Carolina Infantry (C. S. A.).

MURPHY, DENNIS J. F_____
At Corinth, Miss., Oct. 3, 1862.
R—NR.
B—Ireland.

Sergeant, Company F, 14th Wisconsin Infantry.
Although wounded three times, carried the colors throughout the conflict.

MURPHY, EDWARD_____
At Chiricahua Mountains, Ariz., Oct. 20, 1869.
R—NR.
B—Ireland.

Private, Company G, 1st U. S. Cavalry.
Gallantry in action.

MURPHY, EDWARD F_____
At Milk River, Colo., Sept. 29, 1879.
R—NR.
B—Wayne County, Pa.

Corporal, Company D, 5th U. S. Cavalry.
Gallantry in action.

MURPHY, JAMES T_____
At Petersburg, Va., Mar. 25, 1865.
R—NR.
B—Canada.

Private, Company L, 1st Connecticut Artillery.
A piece of artillery having been silenced by the enemy, this soldier voluntarily assisted in working the piece, conducting himself throughout the engagement in a gallant and fearless manner.

MURPHY, JEREMIAH_____
At Powder River, Mont., Mar. 17, 1876.
R—NR.
B—Ireland.

Private, Company M, 3d U. S. Cavalry.
Bravery in action with Sioux.

MURPHY, JOHN P_____
At Antietam, Md., Sept. 17, 1862.
R—NR.
B—Ireland.

Private, Company K, 5th Ohio Infantry.
Capture of flag of 13th Alabama Infantry (C. S. A.).

MURPHY, MICHAEL C_____
At North Anna River, Va., May 24, 1864.
R—NR.
B—NR.

Lieutenant colonel, 170th New York Infantry.
This officer, commanding the regiment, kept it on the field exposed to the fire of the enemy for three hours without being able to fire one shot in return because of the ammunition being exhausted.

MURPHY, PHILIP_____
At Seneca Mountain, Ariz., Aug. 25, 1869.
R—NR.
B—Ireland.

Corporal, Company F, 8th U. S. Cavalry.
Gallantry in action.

MURPHY, ROBINSON B_____
At Atlanta, Ga., July 28, 1864.
R—NR.
B—Illinois.

Musician, Company A, 127th Illinois Infantry.
Being orderly to the brigade commander, he voluntarily led two regiments as reinforcements into line of battle, where he had his horse shot under him.

MURPHY, THOMAS_____
At Seneca Mountain, Ariz., Aug. 25, 1869.
R—NR.
B—Ireland.

Corporal, Company F, 8th U. S. Cavalry.
Gallantry in action.

MURPHY, THOMAS_____
   At Chapins Farm, near Richmond,
     Va., Sept. 30, 1864.
   R—NR.
   B—New York, N. Y.

Corporal, Company K, 158th New York Infantry.
Capture of flag.

MURPHY, THOMAS C_____
   At Vicksburg, Miss., May 22, 1863.
   R—NR.
   B—Ireland.

Corporal, Company I, 31st Illinois Infantry.
Voluntarily crossed the line of heavy fire of Union and Confederate forces,
carrying a message to stop the firing of one Union regiment on another.

MURPHY, THOMAS J_____
   At Five Forks, Va., Apr. 1, 1865.
   R—New York, N. Y.
   B—Ireland.

First sergeant, Company G, 146th New York Infantry.
Capture of flag.

MURRAY, THOMAS_____
   At Little Big Horn, Mont., June 25,
     1876.
   R—NR.
   B—Ireland.

Sergeant, Company B, 7th U. S. Cavalry.
Brought up the pack train, and on the second day the rations, under a heavy
fire from the enemy.

MYERS, FRED_____
   At White River, S. Dak., Jan. 1,
     1891.
   R—Washington, D. C.
   B—Germany.

Sergeant, Company K, 6th U. S. Cavalry.
With five men repelled a superior force of the enemy and held his position
against their repeated efforts to recapture it.

MYERS, GEORGE S_____
   At Chickamauga, Tenn., Sept. 19,
     1863.
   R—NR.
   B—Fairfield, Ohio.

Private, Company F, 101st Ohio Infantry.
Saved the regimental colors by greatest personal devotion and bravery.

MYERS, WILLIAM H_____
   At Appomattox Court House, Va.,
     Apr. 9, 1865.
   R—Baltimore, Md.
   B—Philadelphia, Pa.

Private, Company A, 1st Maryland Cavalry.
Gallantry in action; was five times wounded.

NANNASADDIE_____
   1872-1873.
   R—NR.
   B—NR.

Indian scout.
Gallant conduct in campaigns and engagements with Apaches.

NANTAJE_____
   1872-1873.
   R—NR.
   B—NR.

Indian scout.
Gallant conduct in campaigns and engagements with Apaches.

NASH, HENRY_____
   At Vicksburg, Miss., May 3, 1863.
   R—NR.
   B—Lanawee, Mich.

Corporal, Company B, 47th Ohio Infantry.
Was one of a party that volunteered and attempted to run the enemy's batteries
with a steam tug and two barges loaded with subsistence stores.

NASH, JAMES J_____
   At Santiago, Cuba, July 1, 1898.
   R—Louisville, Ky.
   B—Louisville, Ky.

Private, Company F, 10th U. S. Infantry.
Gallantly assisted in the rescue of the wounded from in front of the lines and
under heavy fire from the enemy.

NEAHR, ZACHARIAH C_____
   At Fort Fisher, N. C., Jan. 16, 1865.
   R—NR.
   B—Canajoharie, N. Y.

Private, Company K, 142d New York Infantry.
Voluntarily advanced with the head of the column and cut down the palisading.

NEAL, SOLON D_____
   At Wichita River, Tex., July 12,
     1870.
   R—NR.
   B—Hanover, N. H.

Private, Company L, 6th U. S. Cavalry.
Gallantry in action.

NEDER, ADAM_____
   Sioux campaign, December, 1890.
   R—NR.
   B—Bavaria.

Corporal, Company A, 7th U. S. Cavalry.
Distinguished bravery.

NEE, GEORGE H_____
   At Santiago, Cuba, July 1, 1898.
   R—Boston, Mass.
   B—Boston, Mass.

Private, Company H, 21st U. S. Infantry.
Gallantly assisted in the rescue of the wounded from in front of the lines and
under heavy fire from the enemy.

NEIBAUR, THOMAS C (98595)_____
    Near Landres-et-St. Georges, France,
      Oct. 16, 1918.
    R—Sugar City, Idaho.
    B—Sharon, Idaho.
    G. O. No. 118, W. D., 1918.

Private, Company M, 167th Infantry, 42d Division.
On the afternoon of Oct. 16, 1918, when the Cote-de-Chatillon had just been gained after bitter fighting and the summit of that strong bulwark in the Kriemhilde Stellung was being organized, Private Neibaur was sent out on patrol with his automatic rifle squad to enfilade enemy machine-gun nests. As he gained the ridge he set up his automatic rifle and was directly thereafter wounded in both legs by fire from a hostile machine gun on his flank. The advance wave of the enemy troops, counterattacking, had about gained the ridge, and although practically cut off and surrounded, the remainder of his detachment being killed or wounded, this gallant soldier kept his automatic rifle in operation to such effect that by his own efforts and by fire from the skirmish line of his company, at least 100 yards in his rear, the attack was checked. The enemy wave being halted and lying prone, four of the enemy attacked Private Neibaur at close quarters. These he killed. He then moved alone among the enemy lying on the ground about him, in the midst of the fire from his own lines, and by coolness and gallantry captured 11 prisoners at the point of his pistol and, although painfully wounded, brought them back to our lines. The counterattack in full force was arrested to a large extent by the single efforts of this soldier, whose heroic exploits took place against the sky line in full view of his entire battalion.

NEILON, FREDERICK S_____
    At Upper Washita, Tex., Sept. 9–11,
      1874.
    R—NR.
    B—Boston, Mass.
    Service rendered under name of
      Frank Singleton.

Sergeant, Company A, 6th U. S. Cavalry.
Gallantry in action.

NEVILLE, EDWIN M_____
    At Sailors Creek, Va., Apr. 6, 1865.
    R—NR.
    B—NR.

Captain, Company C, 1st Connecticut Cavalry.
Capture of flag.

NEWMAN, HENRY_____
    At Whetstone Mountains, Ariz.,
      July 13, 1872.
    R—NR.
    B—Germany.

First sergeant, Company F, 5th U. S. Cavalry.
Gallantry in action.

NEWMAN, MARCELLUS J_____
    At Resaca, Ga., May 14, 1864.
    R—NR.
    B—Washington, Ill.

Private, Company B, 111th Illinois Infantry.
Voluntarily returned, in the face of a severe fire from the enemy, and rescued a wounded comrade who had been left behind as the regiment fell back.

NEWMAN, WILLIAM H_____
    Near Amelia Springs, Va., Apr. 6,
      1865.
    R—NR.
    B—Orange County, N. Y.

Lieutenant, Company B, 86th New York Infantry.
Capture of flag.

NICHOLS, HENRY C_____
    At Fort Blakely, Ala., Apr. 9, 1865.
    R—NR.
    B—Brandon, Vt.

Captain, Company E, 73d U. S. Colored Troops.
Voluntarily made a reconnaissance in advance of the line held by his regiment and, under a heavy fire, obtained information of great value

NIHILL, JOHN_____
    At Whetstone Mountains, Ariz.,
      July 13, 1872.
    R—Brooklyn, N. Y.
    B—Ireland.

Private, Company F, 5th U. S. Cavalry
Gallantry in action.

NISPEROS, JOSE B_____
    At Lapurap, Basilan, P. I., Sept. 24,
      1911.
    R—San Fernandos Union, P. I.
    B—San Fernandos Union, P. I.

Private, 34th Company, Philippine Scouts.
Having been badly wounded (his left arm was broken and lacerated and he had received several spear wounds in the body so that he could not stand) continued to fire his rifle with one hand until the enemy was repulsed, thereby aiding materially in preventing the annihilation of his party and the mutilation of their bodies.

NIVEN, ROBERT_____
    At Waynesboro, Va., Mar. 2, 1865.
    R—NR.
    B—Harlem, N. Y.

Second lieutenant, Company H, 8th New York Cavalry.
Capture of two flags.

NOLAN, JOHN J_____
    At Georgia Landing, La., Oct. 27,
      1862.
    R—NR.
    B—Ireland.

Sergeant, Company K, 8th New Hampshire Infantry.
Although prostrated by a cannon shot, refused to give up the flag which he was carrying as color bearer of his regiment and continued to carry it at the head of the regiment throughout the engagement.

NOLAN, JOSEPH A_____
    At Labao, Luzon, P. I., May 29, 1900.
    R—South Bend, Ind.
    B—Elkhart, Ind.

Artificer, Company B, 45th Infantry, U. S. Volunteers.
Voluntarily left shelter and at great personal risk passed the enemy's lines and brought relief to besieged comrades.

NOLAN, RICHARD J.
    At White Clay Creek, S. Dak., Dec.
    30, 1890.
    R—NR.
    B—Ireland.
Farrier, Company I, 7th U. S. Cavalry.
Bravery.

NOLL, CONRAD.
    At Spotsylvania, Va., May 12, 1864.
    R—Ann Arbor, Mich.
    B—Germany.
Sergeant, Company D, 20th Michigan Infantry.
Seized the colors, the color bearer having been shot down, and gallantly fought his way out with them, though the enemy were on the left flank and rear.

NORTH, JASPER N.
    At Vicksburg, Miss., May 22, 1863.
    R—NR.
    B—NR.
Private, Company D, 4th Virginia Infantry.
Gallantry in the charge of the "volunteer storming party."

NORTON, ELLIOTT M.
    At Sailors Creek, Va., Apr. 6, 1865.
    R—Cooper, Mich.
    B—Connecticut.
Second lieutenant, Company H, 6th Michigan Cavalry.
Capture of two flags.

NORTON, JOHN R.
    At Sailors Creek, Va., Apr. 6, 1865.
    R—NR.
    B—Ontario County, N. Y.
Lieutenant, Company M, 1st New York (Lincoln) Cavalry.
Capture of flag.

NORTON, LLEWELLYN P.
    At Sailors Creek, Va., Apr. 6, 1865.
    R—NR.
    B—Scott, N. Y.
Sergeant, Company L, 10th New York Cavalry.
Charged the enemy and, with the assistance of Corporal Bringle, captured a fieldpiece with two prisoners.

NOYES, WILLIAM W.
    At Spotsylvania, Va., May 12, 1864.
    R—Montpelier, Vt.
    R—Montpelier, Vt.
Private, Company F, 2d Vermont Infantry.
Standing upon the top of the breastworks, deliberately took aim and fired no less than 15 shots into the enemy's lines, but a few yards away.

NUTTING, LEE.
    At Todds Tavern, Va., May 8, 1864.
    R—NR.
    B—Orange County, N. Y.
Captain, Company C, 61st New York Infantry.
Led the regiment in the charge at a critical moment under a murderous fire until he fell desperately wounded.

O'BEIRNE, JAMES R.
    At Fair Oaks, Va., May 31 and June
    1, 1862.
    R—NR.
    B—NR.
Captain, Company C, 37th New York Infantry.
Gallantly maintained the line of battle until ordered to fall back.

O'BRIEN, HENRY D.
    At Gettysburg, Pa., July 3, 1863.
    R—St. Anthony Falls, Minn.
    B—Maine.
Corporal, Company E, 1st Minnesota Infantry.
Taking up the colors where they had fallen, he rushed ahead of his regiment, close to the muzzles of the enemy's guns, and engaged in the desperate struggle in which the enemy was defeated, and, though severely wounded, he held the colors until wounded a second time.

O'BRIEN, PETER.
    At Waynesboro, Va., Mar. 2, 1865.
    R—NR.
    B—Ireland.
Private, Company A, 1st New York (Lincoln) Cavalry.
Capture of flag and of a Confederate officer with his horse and equipments.

O'CALLAGHAN, JOHN.
    Arizona, August to October, 1868.
    R—NR.
    B—New York, N. Y.
Sergeant, Company B, 8th U. S. Cavalry.
Bravery in actions and scouts against Indians.

O'CONNOR, ALBERT.
    At Gravelly Run, Va., Mar. 31 and
    Apr. 1, 1865.
    R—NR.
    B—Canada.
Sergeant, Company A, 7th Wisconsin Infantry.
On Mar. 31, 1865, with a comrade recaptured a Union officer from a detachment of nine Confederates, capturing three of the detachment and dispersing the remainder, and on April 1, 1865, seized a stand of Confederate colors, killing a Confederate officer in a hand-to-hand contest over the colors and retaining the colors until surrounded by Confederates and compelled to relinquish them.

O'CONNOR, TIMOTHY.
    R—NR.
    B—Ireland.
Private, Company E, 1st U. S. Cavalry.
Date and place of act not of record in the War Department.

O'DEA, JOHN.
    At Vicksburg, Miss., May 22, 1863.
    R—NR.
    B—NR.
Private, Company D, 8th Missouri Infantry.
Gallantry in the charge of the "volunteer storming party."

O'DONNELL, MENOMEN.
    At Vicksburg, Miss., May 22, 1863.

    At Fort DeRussey, La., Mar. 14,
    1864.
    R—Illinois.
    B—NR.
First lieutenant, Company A, 11th Missouri Infantry.
Voluntarily joined the color guard in the assault on the enemy's works when he saw indications of wavering and caused the colors of his regiment to be planted on the parapet.
Voluntarily placed himself in the ranks of an assaulting column (being then on staff duty) and rode with it into the enemy's works, being the only mounted officer present; was twice wounded in battle.

OLIVER, CHARLES At Petersburg, Va., Mar. 25, 1865. R—NR. B—Allegheny County, Pa.	Sergeant, Company M, 100th Pennsylvania Infantry. Capture of flag of 31st Georgia Infantry (C. S. A.).
OLIVER, FRANCIS At Chiricahua Mountains, Ariz., Oct. 20, 1869. R—NR. B—Baltimore, Md.	First sergeant, Company G, 1st U. S. Cavalry. Bravery in action.
OLIVER, PAUL A At Resaca, Ga., May 15, 1864. R—New York, N. Y. B—NR.	Captain, Company D, 12th New York Infantry. While acting as aid, assisted in preventing a disaster caused by Union troops firing into each other.
O'NEILL, RICHARD W. (89741) On the Ourcq River, France, July 30, 1918. R—New York, N. Y. B—New York, N. Y. G. O. No. 30, W. D., 1921.	Sergeant, Company D, 165th Infantry, 42d Division. In advance of an assaulting line, he attacked a detachment of about 25 of the enemy. In the ensuing hand-to-hand encounter he sustained pistol wounds, but heroically continued in the advance, during which he received additional wounds; but, with great physical effort, he remained in active command of his detachment. Being again wounded he was forced by weakness and loss of blood to be evacuated, but insisted upon being taken first to the battalion commander in order to transmit to him valuable information relative to enemy positions and the disposition of our men.
O'NEILL, STEPHEN At Chancellorsville, Va., May 1, 1863. R—NR. B—St. Johns, New Brunswick.	Corporal, Company E, 7th U. S. Infantry. Took up the colors from the hands of the color bearer who had been shot down and bore them through the remainder of the battle.
O'NEILL, WILLIAM At Red River, Tex., Sept. 29, 1872. R—NR. B—Tariffville, Conn.	Corporal, Company I, 4th U. S. Cavalry. Bravery in action.
OPEL, JOHN N At Wilderness, Va., May 5, 1864. R—NR. B—NR.	Private, Company G, 7th Indiana Infantry. Capture of flag of 50th Virginia Infantry (C. S. A.).
ORBANSKY, DAVID At Shiloh, Tenn., Vicksburg, Miss., etc., 1862 and 1863. R—NR. B—Prussia.	Private, Company B, 58th Ohio Infantry. Gallantry in actions.
O'REGAN, MICHAEL Arizona, August to October, 1868. R—NR. B—Fall River, Mass.	Private, Company B, 8th U. S. Cavalry. Bravery in actions and scouts against Indians.
ORR, CHARLES A At Hatchers Run, Va., Oct. 27, 1864. R—Bennington, N. Y. B—Holland, N. Y.	Private, Company G, 187th New York Infantry. This soldier and two others, voluntarily and under fire, rescued several wounded and helpless soldiers.
ORR, MOSES Winter of 1872–73. R—NR. B—Ireland.	Private, Company A, 1st U. S. Cavalry. Gallantry in campaigns and engagements with Apaches.
ORR, ROBERT L At Petersburg, Va., Apr. 2, 1865. R—Philadelphia, Pa. B—NR.	Major, 61st Pennsylvania Infantry. Carried the colors, at the head of the column, in the assault after two color bearers had been shot down.
ORTH, JACOB G At Antietam, Md., Sept. 17, 1862. R—NR. B—Philadelphia, Pa.	Corporal, Company D, 28th Pennsylvania Infantry. Capture of flag supposed to be of 7th South Carolina Infantry (C. S. A.).
OSBORNE, WILLIAM Winter of 1872–73. R—NR. B—Boston, Mass.	Sergeant, Company M, 1st U. S. Cavalry. Gallantry in campaigns and engagements with Apaches.
OSBORNE, WILLIAM H At Malvern Hill, Va., July 1, 1862. R—NR. B—Scituate, Mass.	Private, Company C, 29th Massachusetts Infantry. Although wounded and carried to the rear, he secured a rifle and voluntarily returned to the front, where, failing to find his own regiment, he joined another and fought with it until again severely wounded and taken prisoner.

*O'SHEA, THOMAS E. (1212577)_____
  Near Le Catelet, France, Sept. 29, 1918.
  R—Summit, N. J.
  B—New York, N. Y.
  G. O. No. 20, W. D., 1919.

Corporal, Machine Gun Company, 107th Infantry, 27th Division.
Becoming separated from their platoon by a smoke barrage, Corporal O'Shea, with two other soldiers, took cover in a shell hole well within the enemy's lines. Upon hearing a call for help from an American tank, which had become disabled 30 yards from them, the three soldiers left their shelter and started toward the tank under heavy fire from German machine guns and trench mortars. In crossing the fire-swept area Corporal O'Shea was mortally wounded and died of his wounds shortly afterwards.
Posthumously awarded. Medal presented to father, Thomas E. O'Shea.

OSS, ALBERT_____
  At Chancellorsville, Va., May 3, 1863.
  R—Newark, N. J.
  B—Belgium.

Private, Company B, 11th New Jersey Infantry.
Remained in the rifle pits after the others had retreated, firing constantly and contesting the ground step by step.

O'SULLIVAN, JOHN_____
  At Staked Plains, Tex., Dec. 8, 1874.
  R—New York, N. Y.
  B—Ireland.

Private, Company I, 4th U. S. Cavalry.
Gallantry in a long chase after Indians.

OVERTURF, JACOB H_____
  At Vicksburg, Miss., May 22, 1863.
  R—Holton, Ind.
  B—Jefferson County, Ind.

Private, Company K, 83d Indiana Infantry.
Gallantry in the charge of the "volunteer storming party."

PACKARD, LORON F_____
  At Raccoon Ford, Va., Nov. 27, 1863.
  R—Cuba, N. Y.
  B—Cattaraugus County, N. Y.

Private, Company E, 5th New York Cavalry.
After his command had retreated, this soldier, voluntarily and alone, returned to the assistance of a comrade and rescued him from the hands of 3 armed Confederates.

PAINE, ADAM_____
  At Staked Plains, Tex., Sept. 20, 1874.
  R—NR.
  B—Florida.

Private, Indian Scouts.
Gallantry in action.

PALMER, GEORGE H_____
  At Lexington, Mo., Sept. 20, 1861.
  R—NR.
  B—NR.

Musician, 1st Illinois Cavalry.
Volunteered to fight in the trenches and also led a charge which resulted in the recapture of a Union hospital, together with Confederate sharpshooters then occupying the same.

PALMER, JOHN G_____
  At Fredericksburg, Va., Dec. 13, 1862.
  R—Montville, Conn.
  B—Montville, Conn.

Corporal, Company F, 21st Connecticut Infantry.
Volunteered to assist as gunner of a battery upon which the enemy was concentrating its fire, and fought with the battery until the close of the engagement.

PALMER, WILLIAM J_____
  At Red Hill, Ala., Jan. 14, 1865.
  R—NR.
  B—NR.

Colonel, 15th Pennsylvania Cavalry.
With less than 200 men attacked and defeated a superior force of the enemy, capturing their fieldpiece and about 100 prisoners without the loss of a man.

PARKER, JAMES_____
  At Vigan, Luzon, P. I., Dec. 4, 1899.
  R—Newark, N. J.
  B—Newark, N. J.
  Distinguished-service medal also awarded.

Lieutenant colonel, 45th Infantry, U. S. Volunteers.
While in command of a small garrison repulsed a savage night attack by overwhelming numbers of the enemy, fighting at close quarters in the dark for several hours.

PARKER, THOMAS_____
  At Petersburg, Va., Apr. 2, 1865.
  At Sailors Creek, Va., Apr. 6, 1865.
  R—NR.
  B—England.

Corporal, Company B, 2d Rhode Island Infantry.
Planted the first color on the enemy's works.
Carried the regimental colors over the creek after the regiment had broken and been repulsed.

PARKS, JAMES W_____
  At Nashville, Tenn., Dec. 16, 1864.
  R—NR.
  B—Lawrence County, Ohio.

Corporal, Company F, 11th Missouri Infantry.
Capture of flag.

PARKS, JEREMIAH_____
  At Cedar Creek, Va., Oct. 19, 1864.
  R—Orangeville, N. J.
  B—Orangeville, N. J.

Private, Company A, 9th New York Cavalry.
Capture of flag.

PARNELL, WILLIAM R_____
  At White Bird Canyon, Idaho, June 17, 1877.
  R—New York.
  B—Ireland.

First lieutenant, 1st U. S. Cavalry.
With a few men, in the face of a heavy fire from pursuing Indians and at imminent peril, returned and rescued a soldier whose horse had been killed and who had been left behind in the retreat.

PARROTT, JACOB_____
  Georgia, April, 1862.
  R—Ohio.
  B—Fairfield County, Ohio.

Private, Company K, 33d Ohio Infantry.
One of 22 men (including two civilians) who by direction of General Mitchel (or Buell) penetrated nearly 200 miles south into the enemy's territory and captured a railroad train at Big Shanty, Ga., in an attempt to destroy the bridges and track between Chattanooga and Atlanta.

PARSONS, JOEL_____
  At Vicksburg, Miss., May 22, 1863.
  R—NR.
  B—NR.

Private, Company B, 4th Virginia Infantry.
Gallantry in the charge of the "volunteer storming party."

PATTERSON, JOHN H_____
  At Wilderness, Va., May 5, 1864.
  R—New York.
  B—New York.

First lieutenant, 11th U. S. Infantry.
Under the heavy fire of the advancing enemy, picked up and carried several hundred yards to a place of safety a wounded officer of his regiment who was helpless and would otherwise have been burned in the forest.

PATTERSON, JOHN T_____
  At Winchester, Va., June 14, 1863.
  R—NR.
  B—Morgan County, Ohio.

Principal musician, 122d Ohio Infantry.
With one companion voluntarily went in front of the Union line under a heavy fire from the enemy and carried back a helpless wounded comrade, thus saving him from death or capture.

PAUL, WILLIAM H_____
  At Antietam, Md., Sept. 17, 1862.
  R—NR.
  B—Philadelphia, Pa.

Private, Company E, 90th Pennsylvania Infantry.
Under a most withering and concentrated fire, voluntarily picked up the colors of his regiment, when the bearer and two of the color guard had been killed, and bore them aloft throughout the entire battle.

PAY, BYRON E_____
  At Nolensville, Tenn., Feb. 15, 1863.
  R—NR.
  B—New York.

Private, Company H, 2d Minnesota Infantry.
Was one of a detachment of 16 men who heroically defended a wagon train against the attack of 125 cavalry, repulsed the attack, and saved the train.

PAYNE, IRVIN C_____
  At Sailors Creek, Va., Apr. 6, 1865.
  R—NR.
  B—Wayne County, Pa.

Corporal, Company M, 2d New York Cavalry.
Capture of Virginia State colors.

PAYNE, ISAAC_____
  At Pecos River, Tex., Apr. 25, 1875.
  R—NR.
  B—Mexico.

Trumpeter, Indian Scouts.
Gallantry in action.

PAYNE, THOMAS H. L_____
  At Fort Blakely, Ala., Apr. 9, 1865.
  R—NR.
  B—Maine.

First lieutenant, Company E, 37th Illinois Infantry.
While acting regimental quartermaster, learning of an expected assault, requested assignment to a company that had no commissioned officers present; was so assigned, and was one of the first to lead his men into the enemy's works.

PEARSALL, PLATT_____
  At Vicksburg, Miss., May 22, 1863.
  R—NR.
  B—Meigs County, Ohio.

Corporal, Company C, 30th Ohio Infantry.
Gallantry in the charge of the "volunteer storming party."

PEARSON, ALFRED L_____
  At Lewis' Farm, Va., Mar. 29, 1865.
  R—Pittsburgh, Pa.
  B—Pittsburgh, Pa.

Colonel, 155th Pennsylvania Infantry.
Seeing a brigade forced back by the enemy, he seized his regimental color, called on his men to follow him, and advanced upon the enemy under a severe fire. The whole line took up the advance, the lost ground was regained, and the enemy was repulsed.

PECK, ARCHIE A. (1704658)_____
  In the Argonne Forest, France, Oct. 6, 1918.
  R—Hornell, N. Y.
  B—Tyrone, N. Y.
  G. O. No. 16, W. D., 1919.

Private, Company A, 307th Infantry, 77th Division.
While engaged with two other soldiers on patrol duty, he and his comrades were subjected to the direct fire of an enemy machine gun, at which time both his companions were wounded. Returning to his company, he obtained another soldier to accompany him to assist in bringing in the wounded men. His assistant was killed in the exploit, but he continued on, twice returning and safely bringing in both men, being under terrific machine-gun fire during the entire journey.

PECK, CASSIUS_____
  Near Blackburn's Ford, Va., Sept. 19, 1862.
  R—NR.
  B—Brookfield, Vt.

Private, Company F, 1st U. S. Sharpshooters.
Took command of such soldiers as he could get and attacked and captured a Confederate battery of four guns. Also, while on a reconnaissance, overtook and captured a Confederate soldier.

PECK, THEODORE S_____
  At Newport Barracks, N. C., Feb. 2, 1864.
  R—NR.
  B—Burlington, Vt.

First lieutenant, Company H, 9th Vermont Infantry.
By long and persistent resistance and burning the bridges, kept a superior force of the enemy at bay and covered the retreat of the garrison.

PEIRSOL, JAMES K_____
  At Paines Crossroads, Va., Apr. 5, 1865.
  R—NR.
  B—Beaver County, Pa.

Sergeant, Company F, 13th Ohio Cavalry.
Capture of flag.

PENGALLY, EDWARD_____
  At Chiricahua Mountains, Ariz., Oct. 20, 1869.
  R—NR.
  B—England.

Private, Company G, 8th U. S. Cavalry.
Gallantry in action.

PENNSYL, JOSIAH............... | Sergeant, Company M, 6th U. S. Cavalry.
At Upper Washita, Tex., Sept. 11, 1871. | Gallantry in action.
R—N R.
B—Frederick County, Md.

PENNYPACKER, GALUSHA............... | Colonel, 97th Pennsylvania Infantry.
At Fort Fisher, N. C., Jan. 15, 1865. | Gallantly led the charge over a traverse and planted the colors of one of his
R—West Chester, Pa. | regiments thereon; was severely wounded.
B—NR.

PENTZER, PATRICK H............... | Captain, Company C, 97th Illinois Infantry.
At Blakely, Ala., Apr. 9, 1865. | Among the first to enter the enemy's intrenchments, he received the surrender
R—N R. | of a Confederate general officer and his headquarters flag.
B—Marion County, Mo.

*PERKINS, MICHAEL J. (60527)......... | Private, first class, Company D, 101st Infantry, 26th Division.
At Belleu Bois, France, Oct. 27, 1918. | He, voluntarily and alone, crawled to a German "pill box" machine-gun emplace-
R—Boston, Mass. | ment, from which grenades were being thrown at his platoon. Awaiting his
B—Boston, Mass. | opportunity, when the door was again opened and another grenade thrown,
G. O. No. 34, W D., 1919. | he threw a bomb inside, bursting the door open, and then, drawing his trench
 | knife, rushed into the emplacement. In a hand-to-hand struggle he killed
 | or wounded several of the occupants and captured about 25 prisoners, at the
 | same time silencing seven machine guns.
 | Posthumously awarded. Medal presented to father, Michael Perkins.

PESCH, JOSEPH............... | Private, Battery A, 1st Missouri Light Artillery.
At Grand Gulf, Miss., Apr. 28-29, 1863. | With two comrades voluntarily took position on board the steamer Cheeseman,
R—N R. | in charge of all the guns and ammunition of the battery, and remained in
B—Prussia. | charge of the same, although the steamer became unmanageable and was
 | exposed for some time to a heavy fire from the enemy.

PETERS, HENRY C............... | Private, Company B, 47th Ohio Infantry.
At Vicksburg, Miss., May 3, 1863. | For being one of a party that volunteered and attempted to run the enemy's
R—Michigan. | batteries with a steam tug and two barges loaded with subsistence stores.
B—Monroe County, Mich.

PETTY, PHILIP............... | Sergeant, Company A, 136th Pennsylvania Infantry.
At Fredericksburg, Va., Dec. 13, 1862. | Took up the colors as they fell out of the hands of the wounded color bearer
R—N R. | and carried them forward in the charge.
B—N R.

PFISTERER, HERMAN............... | Musician, Company H, 21st U. S. Infantry.
At Santiago, Cuba, July 1, 1898. | Gallantly assisted in the rescue of the wounded from in front of the lines and
R—New York, N. Y. | under heavy fire from the enemy.
B—Brooklyn, N. Y.

PHELPS, CHARLES E............... | Colonel, 7th Maryland Infantry.
At Laurel Hill, Va., May 8, 1864. | Rode to the head of the assaulting column, then much broken by severe losses
R—Baltimore, Md. | and faltering under the close fire of artillery, placed himself conspicuously
B—NR. | in front of the troops, and gallantly rallied and led them to within a few feet
 | of the enemy's works, where he was severely wounded and captured.

PHIFE, LEWIS............... | Sergeant, Company B, 8th U. S. Cavalry.
Arizona, August to October, 1868. | Bravery in actions and scouts against Indians.
R—Marion County, Oreg.
B—Des Moines County, Iowa.

PHILIPSEN, WILHELM O............... | Blacksmith, Troop D, 5th U. S. Cavalry.
At Milk Creek, Colo., Sept. 29, 1879. | With nine others voluntarily attacked and captured a strong position held by
R—N R. | Indians.
B—Germany.

PHILLIPS, JOSIAH............... | Private, Company E, 148th Pennsylvania Infantry.
At Sutherland Station, Va., Apr. 2, 1865. | Capture of flag.
R—N R.
B—Wyoming County, N. Y.

PHILLIPS, SAMUEL D............... | Private, Company H, 2d U. S. Cavalry.
At Muddy Creek, Mont., May 7, 1877. | Gallantry in action.
R—N R.
B—Butler County, Ohio.

PFISTERER, FREDERICK............... | First lieutenant, 18th U. S. Infantry.
At Stone River, Tenn., Dec. 31, 1862. | Voluntarily conveyed, under a heavy fire, information to the commander of a
R—Medina County, Ohio. | battalion of Regular troops by which the battalion was saved from capture
B—Germany. | or annihilation.

PHOENIX, EDWIN............... | Corporal, Company E, 4th U. S. Cavalry.
Near Red River, Tex., Sept. 26-28, 1874. | Gallantry in action.
R—NR.
B—St. Louis, Mo.

PICKLE, ALONZO H.
At Deep Bottom, Va., Aug. 14, 1864.
R—Dover, Minn.
B—Canada.

Sergeant, Company B, 1st Minnesota Infantry.
At the risk of his life, voluntarily went to the assistance of a wounded officer lying close to the enemy's lines and, under fire, carried him to a place of safety.

PIERCE, CHARLES H.
Near San Isidro, Luzon, P. I., Oct. 19, 1899.
R—Delaware City, Del.
B—Cecil County, Md.

Private, Company I, 22d U. S. Infantry.
Held a bridge against a superior force of the enemy and fought, though severely wounded, until the main body came up to cross.

PIKE, EDWARD M.
At Cache River, Ark., July 7, 1862.
R—Bloomington, Ill.
B—Casco, Me.

First sergeant, Company A, 33d Illinois Infantry.
While the troops were falling back before a superior force, this soldier, assisted by 1 companion, and while under severe fire at close range, saved a cannon from capture by the enemy.

*PIKE, EMORY J.
Near Vandieres, France, Sept. 15, 1918.
R—Cuba.
B—Columbus City, Iowa.
G. O. No. 16, W. D., 1919.

Lieutenant colonel, division machine-gun officer, 82d Division.
Having gone forward to reconnoiter new machine-gun positions, Colonel Pike offered his assistance in reorganizing advance Infantry units which had become disorganized during a heavy artillery shelling. He succeeded in locating only about 20 men, but with these he advanced and when later joined by several Infantry platoons rendered inestimable service in establishing outposts, encouraging all by his cheeriness, in spite of the extreme danger of the situation. When a shell had wounded one of the men in the outpost, Colonel Pike immediately went to his aid and was severely wounded himself when another shell burst in the same place. While waiting to be brought to the rear, Colonel Pike continued in command, still retaining his jovial manner of encouragement, directing the reorganization until the position could be held. The entire operation was carried on under terrific bombardment, and the example of courage and devotion to duty, as set by Colonel Pike, established the highest standard of morale and confidence to all under his charge. The wounds he received were the cause of his death.
Posthumously awarded. Medal presented to daughter, Miss Martha Agnes Pike.

PINGREE, SAMUEL E.
At Lees Mills, Va., Apr. 16, 1862.
R—Hartford, Vt.
B—Salisbury, N. H.

Captain, Company F, 3d Vermont Infantry.
Gallantly led his company across a wide, deep creek, drove the enemy from the rifle pits, which were within 2 yards of the farther bank, and remained at the head of his men until a second time severely wounded.

PINKHAM, CHARLES H.
At Fort Stedman, Va., Mar. 25, 1865.
R—NR.
B—Grafton, Mass.

Sergeant major, 57th Massachusetts Infantry.
Captured the flag of the 57th North Carolina Infantry (C. S. A.), and saved his own colors by tearing them from the staff while the enemy was in the camp.

PINN, ROBERT
At Chapins farm, near Richmond, Va., Sept. 29, 1864.
R—Massillon, Ohio.
B—Stark County, Ohio.

First sergeant, Company I, 5th U. S. Colored Troops.
Took command of his company after all the officers had been killed or wounded and gallantly led it in battle.

PIPES, JAMES M.
At Gettysburg, Pa., July 2, 1863.

At Reams Station, Va., Aug. 25, 1864.
R—NR.
B—Green County, Pa.

Captain, Company A, 140th Pennsylvania Infantry.
While a sergeant and retiring with his company before the rapid advance of the enemy, he and a companion stopped and carried to a place of safety a wounded and helpless comrade; in this act both he and his companion were severely wounded.
While commanding a skirmish line, voluntarily assisted in checking a flank movement of the enemy, and while so doing was severely wounded, suffering the loss of an arm.

PITMAN, GEORGE J.
At Sailors Creek, Va., Apr. 6, 1865.
R—Philadelphia, Pa.
B—Recklestown, N. J.

Sergeant, Company C, 1st New York (Lincoln) Cavalry.
Capture of flag of the Sumter Heavy Artillery (C. S. A.).

PITTINGER, WILLIAM
Georgia, April, 1862.
R—NR.
B—NR.

Sergeant, Company G, 2d Ohio Infantry.
One of 22 men (including 2 civilians) who by direction of General Mitchel (or Buell) penetrated nearly 200 miles south into the enemy's territory and captured a railroad train at Big Shanty, Ga., in an attempt to destroy the bridges and track between Chattanooga and Atlanta.

PLANT, HENRY E.
At Bentonville, N. C., Mar. 19, 1865.
R—Cockery, Mich.
B—Oswego County, N. Y.

Corporal, Company F, 14th Michigan Infantry.
Rushed into the midst of the enemy and rescued the colors, the color bearer having fallen mortally wounded.

PLATT, GEORGE C.
At Fairfield, Pa., July 3, 1863.
R—NR.
B—Ireland.

Private, Troop H, 6th U. S. Cavalry.
Seized the regimental flag upon the death of the standard bearer in a hand-to-hand fight and prevented it from falling into the hands of the enemy.

PLATTEN, FREDERICK
At Sappa Creek, Kans., Apr. 23, 1875.
R—NR.
B—Ireland.

Sergeant, Company H, 6th U. S. Cavalry.
Gallantry in action.

PLIMLEY, WILLIAM_____
 At Hatchers Run, Va., Apr. 2, 1865.
 R—Catskill, N. Y.
 B—Catskill, N. Y.

First lieutenant, Company F, 120th New York Infantry.
While acting as aid to a general officer voluntarily accompanied a regiment in an assault on the enemy's works and acted as leader of the movement which resulted in the rout of the enemy and the capture of a large number of prisoners.

PLOWMAN, GEORGE H_____
 At Petersburg, Va., June 17, 1864.
 R—NR.
 B—England.

Sergeant major, 3d Maryland Infantry.
Recaptured the colors of the 2d Pennsylvania Provisional Artillery.

PLUNKETT, THOMAS_____
 At Fredericksburg, Va., Dec. 11, 1862.
 R—West Boylston, Mass.
 B—Ireland.

Sergeant, Company E, 21st Massachusetts Infantry.
Seized the colors of his regiment, the color bearer having been shot down, and bore them to the front where both his arms were carried off by a shell.

POLOND, ALFRED_____
 At Santiago, Cuba, July 1, 1898.
 R—Lapeer, Mich.
 B—Lapeer, Mich.

Private, Company F, 10th U. S. Infantry.
Gallantly assisted in the rescue of the wounded from in front of the lines and while under heavy fire of the enemy.

POND, GEORGE F_____
 At Drywood, Kans., May 15, 1864.
 R—NR.
 B—Lake County, Ill.

Private, Company C, 3d Wisconsin Cavalry.
With two companions attacked a greatly superior force of guerillas, routed them, and rescued several prisoners.

POND, JAMES B_____
 At Baxter Springs, Kans., Oct. 6, 1863.
 R—NR.
 B—Allegany, N. Y.

First lieutenant, Company C, 3d Wisconsin Cavalry.
While in command of two companies of Cavalry, was surprised and attacked by several times his own number of guerillas, but gallantly rallied his men, and after a severe struggle drove the enemy outside the fortifications. Lieutenant Pond then went outside the works and, alone and unaided, fired a howitzer three times, throwing the enemy into confusion and causing him to retire.

POPE, THOMAS A. (1387320)_____
 At Hamel, France, July 4, 1918.
 R—Chicago, Ill.
 B—Chicago, Ill.
 G. O. No. 44, W. D., 1919.

Corporal, Company E, 131st Infantry, 33d Division.
His company was advancing behind the tanks when it was halted by hostile machine-gun fire. Going forward alone he rushed a machine-gun nest, killed several of the crew with his bayonet, and, standing astride of his gun, held off the others until reinforcements arrived and captured them.

POPPE, JOHN A_____
 At Milk River, Colo., Sept. 29 to Oct. 5, 1879.
 R—NR.
 B—Cincinnati, Ohio.

Sergeant, Company F, 5th U.S. Cavalry.
Gallantry in action.

PORTER, AMBROSE_____
 At Tallahatchie River, Miss., Aug. 7, 1864.
 R—NR.
 B—Allegany County, Md.

Commissary sergeant, Company D, 12th Missouri Cavalry.
Was one of four volunteers who swam the river under a brisk fire of the enemy's sharpshooters and brought over a ferryboat by means of which the troops crossed and dislodged the enemy from a strong position.

PORTER, HORACE_____
 At Chickamauga, Ga., Sept. 20, 1863.
 R—Pennsylvania.
 B—Pennsylvania.

Captain, Ordnance Department, U. S. Army.
While acting as a volunteer aid, at a critical moment when the lines were broken, rallied enough fugitives to hold the ground under heavy fire long enough to effect the escape of wagon trains and batteries.

PORTER, JOHN R_____
 Georgia, April, 1862.
 R—NR.
 B—Delaware County, Ohio.

Private, Company G, 21st Ohio Infantry.
One of 22 men (including 2 civilians) who by direction of General Mitchel (or Buell) penetrated nearly 200 miles south into the enemy's territory and captured a railroad train at Big Shanty, Ga., in an attempt to destroy the bridges and track between Chattanooga and Atlanta.

PORTER, SAMUEL_____
 At Wichita River, Tex., July 12, 1870.
 R—NR.
 B—Montgomery County, Md.

Farrier, Company L, 6th U. S. Cavalry.
Gallantry in action.

PORTER, WILLIAM_____
 At Sailors Creek, Va., Apr. 6, 1865.
 R—NR.
 B—New York, N. Y.

Sergeant, Company H, 1st New Jersey Cavalry.
Among the first to check the enemy's countercharge.

POST, PHILIP SIDNEY_____
 At Nashville, Tenn., Dec. 15-16, 1864.
 R—Galesburg, Ill.
 B—NR.

Colonel, 59th Illinois Infantry.
Led his brigade in an attack upon a strong position under a terrific fire of grape, canister, and musketry; was struck down by a grapeshot after he reached the enemy's works.

POSTLES, JAMES PARKE_____
 At Gettysburg, Pa., July 2, 1863.
 R—Wilmington, Del.
 B—NR.

Captain, Company A, 1st Delaware Infantry.
Voluntarily delivered an order in the face of heavy fire of the enemy.

POTTER, GEORGE W.   Private, Company G, 1st Rhode Island Light Artillery.
  At Petersburg, Va., Apr. 2, 1865.   Was one of a detachment of 20 picked artillerymen who voluntarily accom-
  R—Coventry, R. I.   panied an Infantry assaulting party, and who turned upon the enemy the
  B—Coventry, R. I.   guns captured in the assault.

POTTER, NORMAN F.   First sergeant, Company E, 149th New York Infantry.
  At Lookout Mountain, Tenn., Nov.   Capture of flag (Bragg's army).
    24, 1863.
  R—Pompey, N. Y.
  B—Pompey, N. Y.

POWELL, WILLIAM H.   Major, 2d West Virginia Cavalry.
  At Sinking Creek Valley, Va., Nov. Distinguished services in raid, where, with 20 men, he charged and captured
    26, 1862.   the enemy's camp, 500 strong, without the loss of man or gun.
  R—NR.
  B—NR.

POWER, ALBERT   Private, Company A, 3d Iowa Cavalry.
  At Pea Ridge, Ark., Mar. 7, 1862.   Under a heavy fire and at great personal risk went to the aid of a dismounted
  R—Davis County, Iowa.   comrade who was surrounded by the enemy, took him up on his own horse,
  B—Guernsey County, Ohio.   and carried him to a place of safety.

POWERS, THOMAS   Corporal, Company G, 1st U. S. Cavalry.
  At Chiricahua Mountains, Ariz.,   Gallantry in action.
    Oct. 20, 1869.
  R—NR.
  B—New York, N. Y.

POWERS, WESLEY J.   Corporal, Company F, 147th Illinois Infantry.
  At Oostanaula, Ga., Apr. 3, 1865.   Voluntarily swam the river under heavy fire and secured a ferryboat, by means
  R—Virgil, Ill.   of which the command crossed.
  B—Canada.

PRATT, JAMES   Blacksmith, Company I, 4th U. S. Cavalry.
  At Red River, Tex., Sept. 29, 1872. Gallantry in action.
  R—Bellefontaine, Ohio.
  B—Bellefontaine, Ohio.

PRENTICE, JOSEPH R.   Private, Company E, 19th U. S. Infantry.
  At Stone River, Tenn., Dec. 31, 1862. Voluntarily rescued the body of his commanding officer, who had fallen mor-
  R—NR.   tally wounded.
  B—Lancaster, Ohio.

PRESTON, NOBLE D.   First lieutenant and commissary, 10th New York Cavalry.
  At Trevilian Station, Va., June 11, Voluntarily led a charge in which he was severely wounded.
    1864.
  R—Fulton, N. Y.
  B—NR.

*PRUITT, JOHN H.   Corporal, 78th Company, 6th Regiment, U. S. Marine Corps, 2d Division.
  At Blanc Mont Ridge, France, Oct.   Corporal Pruitt single handed attacked two machine guns, capturing them,
    3, 1918.   and killing two of the enemy. He then captured 40 prisoners in a dugout
  R—Tucson, Ariz.   near by. This gallant soldier was killed soon afterward by shell fire while
  B—Sadeville, Ark.   he was sniping at the enemy.
  G. O. No. 62, W. D., 1919.   Posthumously awarded. Medal presented to mother, Mrs. Belle Pruitt.

PURCELL, HIRAM W.   Sergeant, Company G, 104th Pennsylvania Infantry.
  At Fair Oaks, Va., May 31, 1862.   While carrying the regimental colors on the retreat he returned to face the
  R—NR.   advancing enemy, flag in hand, and saved the other color, which would
  B—Bucks County, Pa.   otherwise have been captured.

PURMAN, JAMES J.   Lieutenant, Company A, 140th Pennsylvania Infantry.
  At Gettysburg, Pa., July 2, 1863.   Voluntarily assisted a wounded comrade to a place of apparent safety while the
  R—NR.   enemy were in close proximity; he received the fire of the enemy and a
  B—NR.   wound which resulted in the amputation of his left leg.

PUTNAM, EDGAR P.   Sergeant, Company D, 9th New York Cavalry.
  At Crumps Creek, Va., May 27, 1864. With a small force on a reconnaissance drove off a strong body of the enemy,
  R—Stockton, N. Y.   charged into another force of the enemy's cavalry and stampeded them,
  B—Stockton, N. Y.   taking 27 prisoners.

PUTNAM, WINTHROP D.   Corporal, Company A, 77th Illinois Infantry.
  At Vicksburg, Miss., May 22, 1863. Carried, with others, by hand, a cannon up to and fired it through an embrasure
  R—NR.   of the enemy's works.
  B—Southbridge, Mass.

PYM, JAMES   Private, Company B, 7th U. S. Cavalry.
  At Little Big Horn River, Mont.,   Voluntarily went for water and secured the same under heavy fire.
    June 25, 1876.
  R—NR.
  B—England.

QUAY, MATTHEW S.   Colonel, 134th Pennsylvania Infantry.
  At Fredericksburg, Va., Dec. 13,   Although out of service, he voluntarily resumed duty on the eve of battle and
    1862.   took a conspicuous part in the charge on the heights.
  R—NR.
  B—NR.

QUINLAN, JAMES_____
At Savage Station, Va., June 29, 1862.
R—NR.
B—NR.

Major, 88th New York Infantry.
Led his regiment on the enemy's battery, silenced the guns, held the position against overwhelming numbers, and covered the retreat of the 2d Army Corps.

QUINN, ALEXANDER M_____
At Santiago, Cuba, July 1, 1898.
R—Philadelphia, Pa.
B—Passaic, N. J.

Sergeant, Company A, 13th U. S. Infantry.
Gallantly assisted in the rescue of the wounded from in front of the lines and under heavy fire from the enemy.

QUINN, PETER H_____
At San Miguel de Mayumo, Luzon,
P. I., May 13, 1899.
R—San Francisco, Calif.
B—San Francisco, Calif.

Private, Company L, 4th U. S. Cavalry.
With 11 other scouts, without waiting for the supporting battalion to aid them or to get into a position to do so, charged over a distance of about 150 yards and completely routed about 300 of the enemy who were in line and in a position that could only be carried by a frontal attack.

RAERICK, JOHN_____
At Lyry Creek, Ariz., Oct. 14, 1869.
R—NR.
B—Germany.

Private, Company L, 8th U. S. Cavalry.
Gallantry in action with Indians.

RAFFERTY, PETER_____
At Malvern Hill, Va., July 1, 1862.
R—New York, N. Y.
B—Ireland.

Private, Company B, 69th New York Infantry.
Having been wounded and directed to the rear, declined to go, but continued in action, receiving several additional wounds, which resulted in his capture by the enemy and his total disability for military service.

RAGNAR, THEODORE_____
At White Clay Creek, S. Dak., Dec.
30, 1890.
R—NR.
B—Sweden.

First sergeant, Company K, 7th U. S. Cavalry.
Bravery.

RAND, CHARLES F_____
At Blackburns Ford, Va., July 18,
1861.
R—Batavia, N. Y.
B—Batavia, N. Y.

Private, Company K, 12th New York Infantry.
Remained in action when a part of his regiment broke in disorder, joined another company, and fought with it through the remainder of the engagement.

RANKIN, WILLIAM_____
At Red River, Tex., Sept. 29, 1872.
R—NR.
B—Lewistown, Pa.

Private, Company F, 4th U. S. Cavalry.
Gallantry in action with Indians.

RANNEY, GEORGE E_____
At Resaca, Ga., May 14, 1864.
R—NR.
B—New York, N. Y.

Assistant surgeon, 2d Michigan Cavalry.
At great personal risk went to the aid of a wounded soldier lying under heavy fire between the lines, and with the aid of an orderly carried him to a place of safety.

RANNEY, MYRON H_____
At Bull Run, Va., Aug. 30, 1862.
R—NR.
B—Franklinville, N. Y.

Private, Company G, 13th New York Infantry.
Picked up the colors and carried them off the field after the color bearer had been shot down; was himself wounded.

RANSBOTTOM, ALFRED_____
At Franklin, Tenn., Nov. 30, 1864.
R—NR.
B—Delaware County, Ohio.

First sergeant, Company K, 97th Ohio Infantry.
Capture of flag in a hand-to-hand fight with a Confederate color bearer.

RATCLIFF, EDWARD_____
At Chapins Farm, near Richmond,
Va., Sept. 29, 1864.
R—NR.
B—James City County, Va.

First sergeant, Company C, 38th U. S. Colored Troops.
Commanded and gallantly led his company after the commanding officer had been killed; was the first enlisted man to enter the enemy's works.

RAUB, JACOB F_____
At Hatchers Run, Va., Feb. 5, 1865.
R—Weaversville, Pa.
B—NR.

Assistant surgeon, 210th Pennsylvania Infantry.
Discovering a flank movement by the enemy, apprised the commanding general at great peril, and though a noncombatant voluntarily participated with the troops in repelling this attack.

RAY, CHARLES W_____
Near San Isidro, Luzon, P. I., Oct.
19, 1899.
R—Delta, Iowa.
B—Yancey County, N. C.

Sergeant, Company I, 22d U. S. Infantry.
Captured a bridge with the detachment he commanded and held it against a superior force of the enemy, thereby enabling an army to come up and cross.

RAYMOND, WILLIAM H_____
At Gettysburg, Pa., July 3, 1863.
R—Penfield, N. Y.
B—Penfield, N. Y.

Corporal, Company A, 108th New York Infantry.
Voluntarily and under a severe fire brought a box of ammunition to his comrades on the skirmish line.

READ, MORTON A_____
At Appomattox Station, Va., Apr.
8, 1865.
R—NR.
B—Brockport, N. Y.

Lieutenant, Company D, 8th New York Cavalry.
Capture of flag of 1st Texas Infantry (C. S. A.).

REBMANN, GEORGE F.
  At Blakely, Ala., Apr. 9, 1865.
  R—NR.
  B—Schuyler, Ill.

Sergeant, Company B, 119th Illinois Infantry.
Capture of flag.

REDDICK, WILLIAM H.
  Georgia, April, 1862.
  R—Ohio.
  B—Adams County, Ohio.

Corporal, Company B, 33d Ohio Infantry.
One of 22 men (including two civilians) who by direction of General Mitchel (or Buell) penetrated nearly 200 miles south into the enemy's territory and captured a railroad train at Big Shanty, Ga., in an attempt to destroy the bridges and tracks between Chattanooga and Atlanta.

REED, AXEL H.
  At Chickamauga, Ga., Sept. 19, 1863.

  At Missionary Ridge, Tenn., Nov. 25, 1863.
  R—NR.
  B—Maine.

Sergeant, Company K, 2d Minnesota Infantry.
While in arrest, left his place in the rear and voluntarily went to the line of battle, secured a rifle, and fought gallantly during the two days' battle; was released from arrest in recognition of his bravery.
Commanded his company and gallantly led it, being among the first to enter the enemy's works; was severely wounded, losing an arm, but declined a discharge and remained in active service to the end of the war.

REED, CHARLES W.
  At Gettysburg, Pa., July 2, 1863.
  R—NR.
  B—Charlestown, Mass.

Bugler, 9th Independent Battery, Massachusetts Light Artillery.
Rescued his wounded captain from between the lines.

REED, GEORGE W.
  At Weldon Railroad, Va., Aug. 21, 1864.
  R—Johnstown, Pa.
  B—Cambria County, Pa.

Private, Company E, 11th Pennsylvania Infantry.
Capture of flag of 24th North Carolina Volunteers (C. S. A.).

REED, JAMES C.
  Arizona, Apr. 29, 1868.
  R—NR.
  B—Ireland.

Private, Company A. 8th U. S. Cavalry.
Defended his position (with three others) against a party of 17 hostile Indians under heavy fire at close quarters, the entire party except himself being severely wounded.

REED, WILLIAM.
  At Vicksburg, Miss., May 22, 1863.
  R—NR.
  B—NR.

Private, Company H, 8th Missouri Infantry.
Gallantry in the charge of the "volunteer storming party."

REEDER, CHARLES A.
  At Battery Gregg, near Petersburg, Va., Apr. 2, 1865.
  R—NR.
  B—Harrison, W. Va.

Private, Company G, 12th West Virginia Infantry.
Capture of flag.

REGAN, PATRICK.
  At the Bois-de-Consenvoye, France, Oct. 8, 1918.
  R—Los Angeles, Calif.
  B—Middleboro, Mass.
  G. O. No. 50, W. D., 1919.

Second lieutenant, 115th Infantry, 29th Division.
While leading his platoon against a strong enemy machine-gun nest which had held up the advance of two companies, Lieutenant Regan divided his men into three groups, sending one group to either flank, and he himself attacking with an automatic rifle team from the front. Two of the team were killed outright, while Lieutenant Regan and the third man were seriously wounded, the latter unable to advance. Although severely wounded, Lieutenant Regan dashed with empty pistol into the machine-gun nest, capturing 30 Austrian gunners and four machine guns. This gallant deed permitted the companies to advance, avoiding a terrific enemy fire. Despite his wounds, he continued to lead his platoon forward until ordered to the rear by his commanding officer.

REID, ROBERT.
  At Petersburg, Va., June 17, 1864.
  R—Pottsville, Pa.
  B—Scotland.

Private, Company G, 48th Pennsylvania Infantry.
Capture of flag of 44th Tennessee Infantry (C. S. A.).

REIGLE, DANIEL P.
  At Cedar Creek, Va., Oct., 19, 1864.
  R—NR.
  B—Adams County, Pa.

Corporal, Company F, 87th Pennsylvania Infantry.
Capture of flag.

REISINGER, J. MONROE.
  At Gettysburg, Pa., July 1, 1863.
  R—Meadville, Pa.
  B—Beaver County, Pa.

Corporal, Company H, 150th Pennsylvania Infantry..
Specially brave and meritorious conduct in the face of the enemy.

RENNINGER, LOUIS.
  At Vicksburg, Miss., May 22, 1863.
  R—NR.
  B—Liverpool, Ohio.

Corporal, Company H, 37th Ohio Infantry.
Gallantry in the charge of the "volunteer storming party."

RESSLER, NORMAN W.
  At El Caney, Cuba, July 1, 1898.
  R—Dalmatia, Pa.
  B—Dalmatia, Pa.

Corporal, Company D, 7th U. S. Infantry.
Gallantly assisted in the rescue of the wounded from in front of the lines and under heavy fire of the enemy.

REYNOLDS, GEORGE.
  At Winchester, Va., Sept. 19, 1864.
  R—NR.
  B—Ireland.

Private, Company M, 9th New York Cavalry.
Capture of Virginia State flag.

RHODES, JULIUS D.
At Thoroughfare Gap, Va., Aug. 28, 1862.
At Bull Run, Va., Aug. 30, 1862.
R—NR.
B—Monroe County, Mich.

Private, Company F, 5th New York Cavalry.
After having had his horse shot under him in the fight, he voluntarily joined the 105th New York Volunteers and was conspicuous in the advance on the enemy's lines.
Gallantry in the advance on the skirmish line, where he was wounded.

RHODES, SYLVESTER D.
At Fishers Hill, Va., Sept. 22, 1864.
R—Wilkes-Barre, Pa.
B—Plains, Pa.

Sergeant, Company D, 61st Pennsylvania Infantry.
Was on the skirmish line which drove the enemy from the first intrenchment and was the first man to enter the breastworks, capturing one of the guns and turning it upon the enemy.

RICE, EDMUND
At Gettysburg, Pa., July 3, 1863.
R—Boston, Mass.
B—Brighton, Mass.

Major, 19th Massachusetts Infantry.
Conspicuous bravery on the third day of the battle in the countercharge against Pickett's division where he fell severely wounded within the enemy's lines.

RICH, CARLOS H.
At Wilderness, Va., May 5, 1864.
R—Northfield, Mass.
B—Canada.

First sergeant, Company K, 4th Vermont Infantry.
Saved the life of an officer.

RICHARDSON, WILLIAM R.
At Sailors Creek, Va., Apr. 6, 1865.
R—Washington, Ohio.
B—Cleveland, Ohio.

Private, Company A, 2d Ohio Cavalry.
Having been captured and taken to the rear, made his escape, rejoined the Union lines, and furnished information of great importance as to the enemy's position and the approaches thereto.

RICHEY, WILLIAM E.
At Chickamauga, Ga., Sept. 19, 1863.
R—NR.
B—Athens County, Ohio.

Corporal, Company A, 15th Ohio Infantry.
While on the extreme front, between the lines of the combatants single-handed he captured a Confederate major who was armed and mounted.

RICHMAN, SAMUEL
Arizona, 1868–1869.
R—NR.
B—Cleveland, Ohio.

Private, Company E, 8th U. S. Cavalry.
Bravery in actions with Indians.

RICHMOND, JAMES
At Gettysburg, Pa., July 3, 1863.
R—Toledo, Ohio.
B—Maine.

Private, Company F, 8th Ohio Infantry.
Capture of flag.

RICKSECKER, JOHN H.
At Franklin, Tenn., Nov. 30, 1864.
R—NR.
B—Springfield, Ohio.

Private, Company D, 104th Ohio Infantry.
Capture of flag of 16th Alabama Artillery (C. S. A.).

RIDDELL, RUDOLPH
At Sailors Creek, Va., Apr. 6, 1865.
R—Hamilton, N. Y.
B—Hamilton, N. Y.

Lieutenant, Company I, 61st New York Infantry.
Capture of flag.

RILEY, THOMAS
At Fort Blakely, Ala., Apr. 4, 1865.
R—NR.
B—Ireland.

Private, Company D, 1st Louisiana Cavalry.
Capture of flag.

RIPLEY, WILLIAM Y. W.
At Malvern Hill, Va., July 1, 1862.
R—Rutland, Vt.
B—NR.

Lieutenant colonel, 1st U. S. Sharpshooters.
At a critical moment brought up two regiments, which he led against the enemy himself, being severely wounded.

ROACH, HAMPTON M.
At Milk River, Colo., Sept. 29 to Oct. 5, 1879.
R—NR.
B—Concord, La.

Corporal, Company F, 5th U. S. Cavalry.
Erected breastworks under fire; also kept the command supplied with water three consecutive nights while exposed to fire from ambushed Indians at close range.

ROBB, GEORGE S.
Near Sechault, France, Sept. 29–30, 1918.
R—Salina, Kans.
B—Assaria, Kans.
G. O. No. 16, W. D., 1919.

First lieutenant, 369th Infantry, 93d Division.
While leading his platoon in the assault on Sechault Lieutenant Robb was severely wounded by machine-gun fire, but rather than go to the rear for proper treatment he remained with his platoon until ordered to the dressing station by his commanding officer. Returning within 45 minutes, he remained on duty throughout the entire night, inspecting his lines and establishing outposts. Early the next morning he was again wounded, once again displaying his remarkable devotion to duty by remaining in command of his platoon. Later the same day a bursting shell added two more wounds, the same shell killing his commanding officer and two officers of his company. He then assumed command of the company and organized its position in the trenches. Displaying wonderful courage and tenacity at the critical times, he was the only officer of his battalion who advanced beyond the town, and by clearing machine-gun and sniping posts contributed largely to the aid of his battalion in holding their objective. His example of bravery and fortitude and his eagerness to continue with his mission despite severe wounds set before the enlisted men of his command a most wonderful standard of morale and self-sacrifice.

ROBBINS, AUGUSTUS J_____
At Spotsylvania, Va., May 12, 1864.
R—NR.
B—Grafton, Vt.

Second lieutenant, Company B, 2d Vermont Infantry.
While voluntarily serving as a staff officer successfully withdrew a regiment across and around a severely exposed position to the rest of the command; was severely wounded.

ROBBINS, MARCUS M_____
At Sappa Creek, Kans., Apr. 23, 1875.
R—NR.
B—Elba, Wis.

Private, Company H, 6th U. S. Cavalry.
Gallantry in engagement with Indians.

ROBERTS, CHARLES D_____
At El Caney, Cuba, July 1, 1898.
R—Cheyenne Agency, S. Dak.
B—Fort D. A. Russell, Wyo.
Distinguished-service medal also awarded.

Second lieutenant, 17th U. S. Infantry.
Gallantly assisted in the rescue of the wounded from in front of the lines and under heavy fire of the enemy.

*ROBERTS, HAROLD W. (1013943)_____
In the Montrebeau Woods, France, Oct. 4, 1918.
R—San Francisco, Calif.
B—San Francisco, Calif.
G. O. No. 16, W. D., 1919.

Corporal, Company A, 344th Battalion, Tank Corps.
Corporal Roberts, a tank driver, was moving his tank into a clump of bushes to afford protection to another tank which had become disabled. The tank slid into a shell hole, 10 feet deep, filled with water, and was immediately submerged. Knowing that only one of the two men in the tank could escape, Corporal Roberts said to the gunner. "Well, only one of us can get out, and out you go," whereupon he pushed his companion through the back door of the tank and was himself drowned.
Posthumously awarded. Medal presented to father, John A. Roberts.

ROBERTS, OTIS O_____
At Rappahannock Station, Va., Nov. 7, 1863.
R—Dexter, Me.
B—Sangerville, Me.

Sergeant, Company H, 6th Maine Infantry.
Capture of flag of 8th Louisiana Infantry (C. S. A.), in a hand-to-hand struggle with the color bearer.

ROBERTSON, MARCUS W_____
Near San Isidro, P. I., May 16, 1899.
R—Hood River, Oreg.
B—Flintville, Wis.

Private, Company B, 2d Oregon Volunteer Infantry.
With 21 other scouts charged across a burning bridge, under heavy fire, and completely routed 600 of the enemy who were intrenched in a strongly fortified position.

ROBERTSON, ROBERT S_____
At Corbins Bridge, Va., May 8, 1864.
R—Argyle, N. Y.
B—Argyle, N. Y.

First lieutenant, Company K, 93d New York Infantry.
While acting as aid-de-camp to a general officer, seeing a regiment break to the rear, he seized its colors, rode with them to the front in the face of the advancing enemy, and rallied the retreating regiment.

ROBERTSON, SAMUEL_____
Georgia, April, 1862.
R—NR.
B—Muskingum County, Ohio.

Private, Company G, 33d Ohio Infantry.
One of 22 men (including two civilians) who, by direction of General Mitchel (or Buell) penetrated nearly 200 miles south into the enemy's territory and captured a railroad train at Big Shanty, Ga., in an attempt to destroy the bridges and track between Chattanooga and Atlanta.

ROBIE, GEORGE F_____
Before Richmond, Va., Sept. —, 1864.
R—NR.
B—Candia, N. H.

Sergeant, Company D, 7th New Hampshire Infantry.
Gallantry on the skirmish line.

ROBINSON, ELBRIDGE_____
At Winchester, Va., June 14, 1863.
R—NR.
B—Morgan County, Ohio.

Private, Company C, 122d Ohio Infantry.
With one companion, voluntarily went in front of the Union line, under a heavy fire from the enemy, and carried back a helpless, wounded comrade, thus saving him from death or capture.

ROBINSON, JAMES H_____
Arkansas.
R—Victor, Mich.
B—Oakland County, Mich.

Private, Company B, 3d Michigan Cavalry (Civil War).
Successfully defended himself, single handed, against seven guerrillas, killing the leader and driving off the remainder of the party.

ROBINSON, JOHN_____
At Gettysburg, Pa., July 3, 1863.
R—NR.
B—Ireland.

Private, Company I, 19th Massachusetts Infantry.
Capture of flag of 57th Virginia Infantry (C. S. A.).

ROBINSON, JOHN C_____
At Laurel Hill, Va., May 8, 1864.
R—Binghamton, N. Y.
B—Binghamton, N. Y.

Brigadier general, U. S. Volunteers.
Placed himself at the head of the leading brigade in a charge upon the enemy's breastworks; was severely wounded.

ROBINSON, JOSEPH_____
At Rosebud River, Mont., June 17, 1876.
R—NR.
B—Ireland.

First sergeant, Company D, 3d U. S. Cavalry.
Discharged his duties while in charge of the skirmish line under fire with judgment and great coolness and brought up the led horses at a critical moment.

ROBINSON, THOMAS_____
At Spotsylvania, Va., May 12, 1864.
R—Tamaqua, Pa.
B—Ireland.

Private, Company H, 81st Pennsylvania Infantry.
Capture of flag in a hand-to-hand conflict.

ROCHE, DAVID.......................... First sergeant, Company A, 5th U. S. Infantry.
  At Cedar Creek, etc., Mont., Oct. Gallantry in actions.
    21, 1876, to Jan. 8, 1877.
  R—NR.
  B—Ireland.

ROCK, FREDERICK...................... Private, Company A, 37th Ohio Infantry.
  At Vicksburg, Miss., May 22, 1863. Gallantry in the charge of the "volunteer storming party."
  R—Cleveland, Ohio.
  B—Germany.

ROCKEFELLER, CHARLES M........... Lieutenant, Company A, 178th New York Infantry.
  At Fort Blakely, Ala., Apr. 9, 1865. Voluntarily and alone, under a heavy fire, obtained valuable information which
  R—NR.   a reconnoitering party of 25 men had previously attempted and failed to
  B—NR.   obtain, suffering severe loss in the attempt. The information obtained by
                                    him was made the basis of the orders for the assault that followed. He also
                                    advanced with a few followers, under the fire of both sides, and captured 300
                                    of the enemy who would otherwise have escaped.

RODENBOUGH, THEOPHILUS F......... Captain, 2d U. S. Cavalry.
  At Trevilian Station, Va., June 11, Handled the regiment with great skill and valor; was severely wounded.
    1864.
  R—Pennsylvania.
  B—Pennsylvania.

RODENBURG, HENRY................... Private, Company A, 5th U. S. Infantry.
  At Cedar Creek, etc., Mont., Oct. Gallantry in actions.
    21, 1876, to Jan. 8, 1877.
  R—NR.
  B—Germany.

ROGAN, PATRICK....................... Sergeant, Company A, 7th U. S. Infantry.
  At Big Hole, Mont., Aug. 9, 1877. Verified and reported the company while subjected to a galling fire from the
  R—NR.   enemy.
  B—Ireland.

ROHM, FERDINAND F.................. Chief bugler, 16th Pennsylvania Cavalry.
  At Reams Station, Va., Aug. 25, 1864. While his regiment was retiring under fire voluntarily remained behind to
  R—Juniata County, Pa.   succor a wounded officer who was in great danger, secured assistance, and
  B—Juniata County, Pa.   removed the officer to a place of safety.

ROMEYN, HENRY....................... First lieutenant, 5th U. S. Infantry.
  At Bear Paw Mountain, Mont., Led his command into close range of the enemy, there maintained his position,
    Sept. 30, 1877.   and vigorously prosecuted the fight until he was severely wounded.
  R—Michigan.
  B—Galen, N. Y.

ROOD, OLIVER P....................... Private, Company B, 20th Indiana Infantry.
  At Gettysburg, Pa., July 3, 1863. Capture of flag of 21st North Carolina Infantry (C. S. A.).
  R—NR.
  B—Frankfort County, Ky.

ROONEY, EDWARD..................... Private, Company D, 5th U. S. Infantry.
  At Cedar Creek, etc., Mont., Oct. Gallantry in actions.
    21, 1876, to Jan. 8, 1877.
  R—Poughkeepsie, N. Y.
  B—Poughkeepsie, N. Y.

ROOSEVELT, GEORGE W.............. First sergeant, Company K, 26th Pennsylvania Infantry.
  At Bull Run, Va., Aug. 30, 1862. Recaptured the colors, which had been seized by the enemy.
  At Gettysburg, Pa., July 2, 1863. Captured a Confederate color bearer and color, in which effort he was severely
  R—Chester, Pa.   wounded.
  B—Chester, Pa.

ROSS, FRANK F......................... Private, Company H, 1st North Dakota Volunteer Infantry.
  Near San Isidro, P. I., May 16, 1899. With 21 other scouts charged across a burning bridge under heavy fire and
  R—Langdon, N. Dak.   completely routed 600 of the enemy who were intrenched in a strongly fortified
  B—Avon, Ill.   position.

ROSS, MARION A....................... Sergeant major, 2d Ohio Infantry.
  Georgia, April, 1862. One of 22 men (including two civilians) who by direction of General Mitchel (or
  R—NR.   Buell) penetrated nearly 200 miles south into the enemy's territory and
  B—NR.   captured a railroad train at Big Shanty, Ga., in an attempt to destroy the
                                      bridges and track between Chattanooga and Atlanta.

ROSSBACH, VALENTINE.............. Sergeant, 34th New York Battery.
  At Spotsylvania, Va., May 12, 1864. Encouraged his cannoneers to hold a very dangerous position, and when all
  R—NR.   depended on several good shots it was from his piece that the most effective
  B—Germany.   were delivered, causing the enemy's fire to cease and thereby relieving the
                                      critical position of the Federal troops.

ROTH, PETER.......................... Private, Company A, 6th U. S. Cavalry.
  At Washita River, Tex., Sept. 12, Gallantry in action with Indians.
    1874.
  R—NR.
  B—Germany.

ROUGHT, STEPHEN
  At Wilderness, Va., May 6, 1864.
  R—Crampton, Pa.
  B—Bradford County, Pa.

Sergeant, Company A, 141st Pennsylvania Infantry.
Capture of flag of 13th North Carolina Infantry (C. S. A.).

ROUNDS, LEWIS A.
  At Spotsylvania, Va., May 12, 1864.
  R—Huron County, Ohio.
  B—Cattaraugus County, N. Y.

Private, Company D, 8th Ohio Infantry.
Capture of flag.

ROUSH, J. LEVI.
  At Gettysburg, Pa., July 2, 1863.
  R—NR.
  B—Bedford County, Pa.

Corporal, Company D, 6th Pennsylvania Reserves.
Was one of six volunteers who charged upon a log house near the Devil's Den, where a squad of the enemy's sharpshooters were sheltered, and compelled their surrender.

ROWALT, JOHN F.
  At Lyry Creek, Ariz., Oct. 14, 1869.
  R—Belleville, Ohio.
  B—Belleville, Ohio.

Private, Company L, 8th U. S. Cavalry.
Gallantry in action with Indians.

ROWAND, ARCHIBALD H., Jr.
  Winter of 1864-65.
  R—NR.
  B—Philadelphia, Pa.

Private, Company K, 1st West Virginia Cavalry.
Was one of two men who succeeded in getting through the enemy's lines with dispatches to General Grant.

ROWDY.
  Arizona, Mar. 7, 1890.
  R—NR.
  B—Arizona.

Sergeant, Company A, Indian Scouts.
Bravery in action with Apache Indians.

ROWE, HENRY W.
  At Petersburg, Va., June 17, 1864.
  R—Candia, N. H.
  B—Candia, N. H.

Private, Company I, 11th New Hampshire Infantry.
Capture of flag.

ROY, STANISLAUS.
  At Little Big Horn, Mont., June 25, 1876.
  R—NR.
  B—France.

Sergeant, Company A, 7th U. S. Cavalry.
Brought water to the wounded at great danger to life and under a most galling fire of the enemy.

RUNDLE, CHARLES W.
  At Vicksburg, Miss., May 22, 1863.
  R—NR.
  B—Cincinnati, Ohio.

Private, Company A, 116th Illinois Infantry.
Gallantry in the charge of the "volunteer storming party"

RUSSELL, CHARLES L.
  At Spotsylvania, Va., May 12, 1864.
  R—Malone, N. Y.
  B—Malone, N. Y.

Corporal, Company H, 93d New York Infantry.
Capture of flag of 42d Virginia Infantry (C. S. A.).

RUSSELL, JAMES.
  At Chiricahua Mountains, Ariz., Oct. 20, 1869.
  R—NR.
  B—New York, N. Y.

Private, Company G, 1st U. S. Cavalry.
Gallantry in action with Indians.

RUSSELL, MILTON.
  At Stone River, Tenn., Dec. 29, 1862.
  R—NR.
  B—Hendricks County, Ind.

Captain, Company A, 51st Indiana Infantry.
Was the first man to cross Stone River and, in the face of a galling fire from the concealed skirmishers of the enemy, led his men up the hillside, driving the opposing skirmishers before them.

RUTHERFORD, JOHN T.
  At Yellow Tavern, Va., May 11, 1864.
  At Hanovertown, Va., May 27, 1864.
  R—Canton, N. Y.
  B—NR.

First lieutenant, Company L, 9th New York Cavalry.
Made a successful charge, by which 90 prisoners were captured.

In a gallant dash on a superior force of the enemy and in a personal encounter captured his opponent.

RUTTER, JAMES M.
  At Gettysburg, Pa., July 1, 1863.
  R—NR.
  B—Wilkesbarre, Pa.

Sergeant, Company C, 143d Pennsylvania Infantry.
At great risk of his life went to the assistance of a wounded comrade, and while under fire removed him to a place of safety.

RYAN, DAVID.
  At Cedar Creek, etc., Mont., Oct. 21, 1876, to Jan. 8, 1877.
  R—NR.
  B—Ireland.

Private, Company G, 5th U. S. Infantry.
Gallantry in actions.

RYAN, DENNIS.
  At Gageby Creek, Indian Territory, Dec. 2, 1874.
  R—NR.
  B—Ireland.

First sergeant, Company I, 6th U. S. Cavalry
Courage while in command of a detachment.

RYAN, PETER J. _____
    At Winchester, Va., Sept. 19, 1864.
    R—NR.
    B—Ireland.

Private, Company D, 11th Indiana Infantry.
With one companion captured 14 Confederates in the severest part of the battle.

SACRISTE, LOUIS J. _____
    At Chancellorsville, Va., May 3, 1863.
    At Auburn, Va. Oct. 14, 1863.
    R—Philadelphia, Pa.
    B—Philadelphia, Pa.

First lieutenant, Company D, 116th Pennsylvania Infantry.
Saved from capture a gun of the 5th Maine Battery.

Voluntarily carried orders which resulted in saving from destruction or capture the picket line of the First Division, Second Army Corps.

SAGE, WILLIAM H. _____
    Near Zapote River, Luzon, P. I., June 13, 1899.
    R—Binghamton, N. Y.
    B—Centerville, N. Y.

Captain, 23d U. S. Infantry.
With nine men volunteered to hold an advanced position and held it against a terrific fire of the enemy, estimated at 1,000 strong. Taking a rifle from a wounded man, and cartridges from the belts of others, Captain Sage himself killed five of the enemy.

SAGELHURST, JOHN C. _____
    At Hatchers Run, Va., Feb. 6, 1865.
    R—Buffalo, N. Y.
    B—Buffalo, N. Y.

Sergeant, Company B, 1st New Jersey Cavalry.
Under a heavy fire from the enemy carried off the field a commissioned officer who was severely wounded and also led a charge on the enemy's rifle pits.

SALE, ALBERT. _____
    At Santa Maria River, Ariz., June 29, 1869.
    R—NR.
    B—Broom County, N. Y.

Private, Company F, 8th U. S. Cavalry.
Gallantry in killing an Indian warrior and capturing pony and effects.

SAMPLER, SAMUEL M. (1490609) _____
    Near St. Etienne, France, Oct. 8, 1918.
    R—Altus, Okla.
    B—Decatur, Tex.
    G. O. No. 59, W. D., 1919.

Corporal, Company H, 142d Infantry, 36th Division.
His company having suffered severe casualties during an advance under machine-gun fire, was finally stopped. Corporal Sampler detected the position of the enemy machine guns on an elevation. Armed with German hand grenades, which he had picked up, he left the line and rushed forward in the face of heavy fire until he was near the hostile nest, where he grenaded the position. His third grenade landed among the enemy, killing 2, silencing the machine guns, and causing the surrender of 28 Germans, whom he sent to the rear as prisoners. As a result of his act the company was immediately enabled to resume the advance.

SANCRAINTE, CHARLES F. _____
    At Atlanta, Ga., July 22, 1864.
    R—NR.
    B—Monroe, Mich.

Private, Company B, 15th Michigan Infantry.
Voluntarily scaled the enemy's breastworks and signaled to his commanding officer to charge; also in single combat captured the colors of the 5th Texas regiment.

SANDLIN, WILLIE (2078103) _____
    At Bois-de-Forges, France, Sept. 26, 1918.
    R—Hyden, Ky.
    B—Jackson, Ky.
    G. O. No. 16, W. D., 1919.

Sergeant, Company A, 132d Infantry, 33d Division.
He showed conspicuous gallantry in action by advancing alone directly on a machine-gun nest which was holding up the line with its fire. He killed the crew with a grenade and enabled the line to advance. Later in the day he attacked alone and put out of action two other machine-gun nests, setting a splendid example of bravery and coolness to his men.

SANDS, WILLIAM. _____
    At Dabneys Mills, Va., Feb. 6-7, 1865.
    R—Reading, Pa.
    B—Reading, Pa.

First sergeant, Company G, 88th Pennsylvania Infantry.
Grasped the enemy's colors in the face of a deadly fire and brought them inside the lines.

SANFORD, JACOB. _____
    At Vicksburg, Miss., May 22, 1863.
    R—NR.
    B—Fulton County, Ill.

Private, 55th Illinois Infantry.
Gallantry in the charge of the "volunteer storming party."

SARGENT, JACKSON. _____
    At Petersburg, Va., Apr. 2, 1865.
    R—Stowe, Vt.
    B—Stowe, Vt.

Sergeant, Company D, 5th Vermont Infantry.
First to scale the enemy's works and plant the colors thereon.

SARTWELL, HENRY. _____
    At Chancellorsville, Va., May 3, 1863.
    R—NR.
    B—Ticonderoga, N. Y.

Sergeant, Company D, 123d New York Infantry.
Was severely wounded by a gunshot in his left arm, went half a mile to the rear, but insisted on returning to his company and continued to fight bravely until he became exhausted from the loss of blood and was compelled to retire from the field.

SAVACOOL, EDWIN F. _____
    At Sailors Creek, Va., Apr. 6, 1865.
    R—Marshall, Mich.
    B—Jackson, Mich.

Captain, Company K, 1st New York (Lincoln) Cavalry.
Capture of flag.

*SAWELSON, WILLIAM (1752909) _____
    At Grand-Pre, France, Oct. 26, 1918.
    R—Harrison, N. J.
    B—Newark, N. J.
    G. O. No. 16, W. D., 1919.

Sergeant, Company M, 312th Infantry, 78th Division.
Hearing a wounded man in a shell hole some distance away calling for water, Sergeant Sawelson, upon his own initiative, left shelter and crawled through heavy machine-gun fire to where the man lay, giving him what water he had in his canteen. He then went back to his own shell hole, obtained more water, and was returning to the wounded man when he was killed by a machine-gun bullet.
Posthumously awarded. Medal presented to father, Jacob L. Sawelson.

SAXTON, RUFUS_____ | Brigadier general, U. S. Volunteers.
  At Harpers Ferry, Va., May 26 to 30, 1862. | Distinguished gallantry and good conduct in the defense.
  R—Massachusetts.
  B—Greenfield, Mass.

SCANLAN, PATRICK_____ | Private, Company A, 4th Massachusetts Cavalry.
  Ashepoo River, S. C., May 24, 1864. | Volunteered as a member of a boat crew which went to the rescue of a large
  R—N R. | number of Union soldiers on board the stranded steamer Boston and with
  B—Ireland. | great gallantry assisted in conveying them to shore, being exposed during the entire time to a heavy fire from a Confederate battery.

SCHAFFNER, DWITE H_____ | First lieutenant, 306th Infantry, 77th Division.
  Near St. Hubert's Pavillion, Boureuilles, France, Sept. 28, 1918. | In command of Company K, 306th Infantry, he led his men in an attack on St. Hubert's Pavillion through terrific enemy machine-gun, rifle, and artillery
  R—Falls Creek, Pa. | fire and drove the enemy from a strongly held intrenched position after hand-
  B—Arroya, Pa. | to-hand fighting. His bravery and contempt for danger inspired his men,
  G. O. No. 15, W. D., 1923. | enabling them to hold fast in the face of three determined enemy counterattacks. His company's position being exposed to enemy fire from both flanks, he made three efforts to locate an enemy machine gun which had caused heavy casualties in his company. On his third reconnaissance he discovered the gun position and personally silenced the gun, killing or wounding the crew thereof. The third counterattack made by the enemy was initiated by the appearance of a small detachment advancing well in advance of the enemy attacking wave, calling as they advanced "Kamerad." When almost within reach of the American front line the enemy attacking wave behind them appeared, attacking vigorously with pistols, rifles, and hand grenades, causing heavy casualties in the American platoon holding the advanced position. Lieutenant Schaffner mounted the parapet of the trench and used his pistol and grenades with great gallantry and effect, killing a number of enemy soldiers, finally reaching the enemy officer leading the attacking forces, a captain, shooting and mortally wounding the latter with his pistol, and dragging the captured officer back to the company's trench, securing from him valuable information as to the enemy's strength and position. The information so secured enabled Lieutenant Schaffner to maintain for five hours the advanced position of his company despite the fact that it was surrounded on three sides by strong enemy forces. The undaunted bravery, gallant soldierly conduct, and leadership displayed by Lieutenant Schaffner undoubtedly saved the survivors of the company from death or capture.

SCHEIBNER, MARTIN E_____ | Private, Company G, 90th Pennsylvania Infantry.
  At Mine Run, Va., Nov. 27, 1863. | Voluntarily extinguished the burning fuse of a shell which had been thrown into
  R—N R. | the lines of the regiment by the enemy.
  B—Germany.

SCHENCK, BENJAMIN W_____ | Private, Company D, 116th Illinois Infantry.
  At Vicksburg, Miss., May 22, 1863. | Gallantry in the charge of the "volunteer storming party."
  R—Maroa, Ill.
  B—Butler County, Ohio.

SCHILLER, JOHN_____ | Private, Company E, 158th New York Infantry.
  At Chapins Farm, near Richmond, Va., Sept. 29, 1864. | Advanced to the ditch of the enemy's works.
  R—N R.
  B—Germany.

SCHLACHTER, PHILIPP_____ | Private, Company F, 73d New York Infantry.
  At Spotsylvania, Va., May 12, 1864. | Capture of flag of 15th Louisiana Infantry (C. S. A.).
  R—N R.
  B—Germany.

SCHMAL, GEORGE W_____ | Blacksmith, Company M, 24th New York Cavalry.
  At Paines Crossroads, Va., Apr. 5, 1865. | Capture of flag.
  R—Buffalo, N. Y.
  B—Germany.

SCHMAUCH, ANDREW_____ | Private, Company A, 30th Ohio Infantry.
  At Vicksburg, Miss., May 22, 1863. | Gallantry in the charge of the "volunteer storming party."
  R—N R.
  B—Germany.

SCHMIDT, CONRAD_____ | First sergeant, Company K, 2d U. S. Cavalry.
  At Winchester, Va., Sept. 19, 1864. | Went to the assistance of his regimental commander whose horse had been killed
  R—N R. | under him in a charge, mounted the officer behind him, under a heavy fire
  B—Germany. | from the enemy, and returned him to his command.

SCHMIDT, WILLIAM_____ | Private, Company G, 37th Ohio Infantry.
  At Missionary Ridge, Tenn., Nov. 25, 1863. | Rescued a wounded comrade under terrific fire.
  R—Maumee, Ohio.
  B—Tiffin, Ohio.

SCHNEIDER, GEORGE_____ | Sergeant, Company A, 3d Maryland Veteran Infantry.
  At Petersburg, Va., July 30, 1864. | After the color sergeant had been shot down, seized the colors and planted them
  R—N R. | on the enemy's works during the charge.
  B—Baltimore, Md.

SCHNELL, CHRISTIAN
  At Vicksburg, Miss., May 22, 1863.
  R—Wapakoneta, Ohio.
  B—Virginia.
Corporal, Company C, 37th Ohio Infantry.
Gallantry in the charge of the "volunteer storming party."

SCHNITZER, JOHN
  At Horseshoe Canyon, N. Mex.,
    Apr. 23, 1882.
  R—NR.
  B—Bavaria.
Wagoner, Troop G, 4th U. S. Cavalry.
Assisted, under a heavy fire, to rescue a wounded comrade.

SCHOFIELD, JOHN M
  At Wilsons Creek, Mo., Aug. 10,
    1861.
  R—NR.
  B—NR.
Major, 1st Missouri Infantry.
Was conspicuously gallant in leading a regiment in a successful charge against
the enemy.

SCHOONMAKER, JAMES M
  At Winchester, Va., Sept. 19, 1864.
  R—NR.
  B—NR.
Colonel, 14th Pennsylvania Cavalry.
At a critical period, gallantly led a cavalry charge against the left of the enemy's
line of battle, drove the enemy out of his works, and captured many prisoners.

SCHORN, CHARLES
  At Appomattox, Va., Apr. 8, 1865.
  R—NR.
  B—Germany.
Chief bugler, Company M, 1st West Virginia Cavalry.
Capture of flag of the Sumter Flying Artillery (C. S. A.).

SCHOU, JULIUS
  Sioux campaign, 1876.
  R—NR.
  B—Denmark.
Corporal, Company I, 22d U. S. Infantry.
Carried dispatches to Fort Buford.

SCHROEDER, HENRY F
  At Carig, P. I., Sept. 14, 1900.
  R—Chicago, Ill.
  B—Chicago, Ill.
Sergeant, Company L, 16th U. S. Infantry.
With 22 men defeated 400 insurgents, killing 36 and wounding 90.

SCHROETER, CHARLES
  At Chiricahua Mountains, Ariz.,
    Oct. 20, 1869.
  R—NR.
  B—Germany.
Private, Company G, 8th U. S. Cavalry.
Gallantry in action.

SCHUBERT, MARTIN
  At Fredericksburg, Va., Dec. 13,
    1862.
  R—NR.
  B—Germany.
Private, Company E, 26th New York Infantry.
Relinquished a furlough granted for wounds, entered the battle, where he picked
up the colors after several bearers had been killed or wounded, and carried
them until himself again wounded.

SCHWAN, THEODORE
  At Peebles Farm, Va., Oct. 1, 1864.
  R—New York.
  B—Germany.
First lieutenant, 10th U. S. Infantry.
At the imminent risk of his own life, while his regiment was falling back before
a superior force of the enemy, he dragged a wounded and helpless officer to
the rear, thus saving him from death or capture.

SCHWENK, MARTIN
  Sergeant, Company B, 6th Cavalry.
See George Martin, true name.

SCOFIELD, DAVID H
  At Cedar Creek, Va., Oct. 19, 1864.
  R—NR.
  B—Mamaroneck, N. Y.
Quartermaster sergeant, Company K, 5th New York Cavalry.
Capture of flag of 13th Virginia Infantry (C. S. A.).

SCOTT, ALEXANDER
  At Monocacy, Md., July 9, 1864.
  R—Winooski, Vt.
  B—Canada.
Corporal, Company D, 10th Vermont Infantry.
Under a very heavy fire of the enemy saved the national flag of his regiment
from capture.

SCOTT, GEORGE
  At Little Big Horn, Mont., June
    25–26, 1876.
  R—NR.
  B—Lancaster County, Ky.
Private, Company D, 7th U. S. Cavalry.
Voluntarily brought water to the wounded under fire.

SCOTT, JOHN M
  Georgia, April, 1862.
  R—NR.
  B—Stark County, Ohio.
Sergeant, Company F, 21st Ohio Infantry.
One of 22 men (including two civilians) who by direction of General Mitchel
(or Buell) penetrated nearly 200 miles south into the enemy's territory and
captured a railroad train at Big Shanty Ga., in an attempt to destroy the
bridges and tracks between Chattanooga and Atlanta.

SCOTT, JOHN WALLACE
  At Five Forks, Va., Apr. 1, 1865.
  R—NR.
  B—NR.
Captain, Company D, 157th Pennsylvania Infantry.
Capture of flag.

SCOTT, JULIAN A
  At Lees Mills, Va., Apr. 16, 1862.
  R—NR.
  B—Johnson, Vt.
Drummer, Company E, 3d Vermont Infantry.
Crossed the creek under a terrific fire of musketry several times to assist in
bringing off the wounded.

SCOTT, ROBERT B_____
    At Chiricahua Mountains, Ariz.,
      Oct. 20, 1869.
    R—NR.
    B—Washington County, N. Y.

> Private, Company G, 8th U. S. Cavalry.
> Gallantry in action.

SEAMAN, ELISHA B_____
    At Chancellorsville, Va., May 2,
      1863.
    R—Logan County, Ohio.
    B—Logan County, Ohio.

> Private, Company A, 66th Ohio Infantry.
> Was one of a party of four who voluntarily brought into the Union lines, under fire, a wounded Confederate officer from whom was obtained valuable information concerning the enemy.

SEARS, CYRUS_____
    At Iuka, Miss., Sept. 19, 1862.
    R—Bucyrus, Ohio.
    B—NR.

> First lieutenant, 11th Battery, Ohio Light Artillery.
> Although severely wounded, fought his battery until the cannoneers and horses were nearly all killed or wounded.

SEAVER, THOMAS O_____
    At Spotsylvania Courthouse, Va.,
      May 10, 1864.
    R—NR.
    B—NR.

> Colonel, 3d Vermont Infantry.
> At the head of three regiments and under a most galling fire attacked and occupied the enemy's works.

SEIBERT, LLOYD M (2266821)_____
    Near Epinonville, France, Sept. 26,
      1918.
    R—Salinas, Calif.
    B—Caladonia, Mich.
    G. O. No. 16, W. D., 1919.

> Sergeant, Company F, 364th Infantry, 91st Division.
> Suffering from illness, Sergeant Seibert remained with his platoon and led his men with the highest courage and leadership under heavy shell and machine-gun fire. With two other soldiers he charged a machine-gun emplacement in advance of their company, he himself killing one of the enemy with a shot-gun and captured two others. In this encounter he was wounded, but he nevertheless continued in action, and when a withdrawal was ordered he returned with the last unit, assisting a wounded comrade. Later in the evening he volunteered and carried in wounded until he fainted from exhaustion.

SEITZINGER, JAMES M_____
    At Cold Harbor, Va., June 3, 1864.
    R—Worcester, Pa.
    B—Germany.

> Private, Company G, 116th Pennsylvania Infantry.
> When the color bearer was shot down, this soldier seized the colors and bore them gallantly in a charge against the enemy.

SELLERS, ALFRED J_____
    At Gettysburg, Pa., July 1, 1863.
    R—NR.
    B—NR.

> Major, 90th Pennsylvania Infantry.
> Voluntarily led the regiment under a withering fire to a position from which the enemy was repulsed.

SESTON, CHARLES H_____
    At Winchester, Va., Sept. 19, 1864.
    R—New Albany, Ind.
    B—New Albany, Ind.

> Sergeant, Company I, 11th Indiana Infantry.
> Gallant and meritorious service in carrying the regimental colors.

SEWARD, GRIFFIN_____
    At Chiricahua Mountains, Ariz.,
      Oct. 20, 1869.
    R—NR.
    B—Dover, Del.

> Wagoner, Company G, 8th U. S. Cavalry.
> Gallantry in action.

SEWELL, WILLIAM J_____
    At Chancellorsville, Va., May 3,
      1863.
    R—NR.
    B—NR.

> Colonel, 5th New Jersey Infantry.
> Assuming command of a brigade, he rallied around his colors a mass of men from other regiments and fought these troops with great brilliancy through several hours of desperate conflict, remaining in command though wounded and inspiring them by his presence and the gallantry of his personal example.

SHAFFER, WILLIAM_____
    Arizona, August to October, 1868.
    R—NR.
    B—Germany.

> Private, Company B, 8th U. S. Cavalry.
> Bravery in scouts and actions with Indians.

SHAFTER, WILLIAM R_____
    At Fair Oaks, Va., May 31, 1862.
    R—Galesburg, Mich.
    B—Galesburg, Mich.

> First lieutenant, Company I, 7th Michigan Infantry.
> Remained to the close of the battle, although severely wounded.

SHAHAN, EMISIRE_____
    At Sailors Creek, Va., Apr. 6, 1865.
    R—NR.
    B—Preston County, W. Va.

> Corporal, Company A, 1st West Virginia Cavalry.
> Capture of flag of 76th Georgia Infantry (C. S. A.).

SHALER, ALEXANDER_____
    At Marye's Heights, Va., May 3,
      1863.
    R—NR.
    B—NR.

> Colonel, 65th New York Infantry.
> At a most critical moment, the head of the charging column being about to be crushed by the severe fire of the enemy's artillery and infantry, he pushed forward with a supporting column, pierced the enemy's works, and turned their flank.

SHAMBAUGH, CHARLES_____
    At Charles City Crossroads, Va.,
      June 30, 1862.
    R—NR.
    B—Prussia.

> Corporal, Company B, 11th Pennsylvania Reserves.
> Capture of flag.

SHANES, JOHN_____
At Carters Farm, Va., July 20, 1864.
R—NR.
B—Monongalia County, W. Va.

Private, Company K, 14th West Virginia Infantry.
Charged upon a Confederate fieldpiece in advance of his comrades and by his individual exertions silenced the piece.

SHAPLAND, JOHN_____
At Elk River, Tenn., July 2, 1863.
R—Ottawa, Ill.
B—England.

Private Company D, 104th Illinois Infantry.
Voluntarily joined a small party that, under a heavy fire, captured a stockade and saved the bridge.

SHARPLESS, EDWARD C_____
At Upper Washita, Tex., Sept. 9-11, 1874.
R—NR.
B—Marion County, Ohio.

Corporal, Company H, 6th U. S. Cavalry.
While carrying dispatches was attacked by 125 hostile Indians, whom he (and a comrade) fought throughout the day.

SHAW, GEORGE C_____
At Fort Pitacus, Lake Lanao, Mindanao, P. I., May 4, 1903.
R—Washington, D. C.
B—Pontiac, Mich.

First lieutenant, 27th U. S. Infantry.
For distinguished gallantry in leading the assault and, under a heavy fire from the enemy, maintaining alone his position on the parapet after the first three men who followed him there had been killed or wounded, until a foothold was gained by others and the capture of the place assured.

SHAW, THOMAS_____
At Carizo Canyon, N. Mex., Aug. 12, 1881.
R—NR.
B—Covington, Ky.

Sergeant, Company K, 9th U. S. Cavalry.
Forced the enemy back after stubbornly holding his ground in an extremely exposed position and prevented the enemy's superior numbers from surrounding his command.

SHEA, JOSEPH H_____
At Chapins Farm, near Richmond, Va., Sept. 29, 1864.
R—NR.
B—Baltimore, Md.

Private, Company K, 92d New York Infantry.
Gallantry in bringing wounded from the field under heavy fire.

SHEERIN, JOHN_____
Near Fort Selden, N. Mex., July 8-11, 1873.
R—NR.
B—Camden County, N. J.

Blacksmith, Company C, 8th U. S. Cavalry.
Service against hostile Indians.

SHELLENBERGER, JOHN S_____
At Deep Run, Va., Aug. 16, 1864.
R—NR.
B—NR.

Corporal, Company B, 85th Pennsylvania Infantry.
Capture of flag.

SHELTON, GEORGE M_____
At La Paz, Leyte, P. I., Apr. 26, 1900.
R—Bellington, Tex.
B—Brownwood, Tex.

Private, Company I, 23d U. S. Infantry.
Advanced alone under heavy fire of the enemy and rescued a wounded comrade.

SHEPARD, IRWIN_____
At Knoxville, Tenn., Nov. 20, 1863.
R—Chelsea, Mich.
B—Skaneateles, N. Y.

Corporal, Company E, 17th Michigan Infantry.
Having voluntarily accompanied a small party to destroy buildings within the enemy's lines, whence sharpshooters had been firing, disregarded an order to retire, remained and completed the firing of the buildings, thus insuring their total destruction, this at the imminent risk of his life from the fire of the advancing enemy.

SHEPHERD, WARREN J_____
At El Caney, Cuba, July 1, 1898.
R—Westover, Pa.
B—Cherry Tree, Pa.

Corporal, Company D, 17th U. S. Infantry.
Gallantly assisted in the rescue of the wounded from in front of the lines and under heavy fire from the enemy.

SHEPHERD, WILLIAM_____
At Sailors Creek, Va., Apr. 6, 1865.
R—Dillsboro, Ind.
B—Dearborn County, Ind.

Private, Company A, 3d Indiana Cavalry.
Capture of flag.

SHEPPARD, CHARLES_____
At Cedar Creek, etc., Mont., Oct. 21, 1876, to Jan. 8, 1877.
R—St. Louis, Mo.
B—Rocky Hill, Conn.

Private, Company A, 5th U. S. Infantry.
Bravery in action with Sioux.

SHERMAN, MARSHALL_____
At Gettysburg, Pa., July 3, 1863.
R—St. Paul, Minn.
B—Burlington, Vt.

Private, Company C, 1st Minnesota Infantry.
Capture of flag of 58th Virginia Infantry (C. S. A.).

SHIEL, JOHN_____
At Fredericksburg, Va., Dec. 13, 1862.
R—NR.
B—NR.

Corporal, Company E, 90th Pennsylvania Infantry.
Carried a dangerously wounded comrade into the Union lines, thereby preventing his capture by the enemy.

SHIELDS, BERNARD_____
At Appomattox, Va., Apr. 8, 1865.
R—NR.
B—Ireland.

Private, Company E, 2d West Virginia Cavalry.
Capture of flag of the Washington Artillery (C. S. A.).

SHIELS, GEORGE F_____ | Surgeon, U. S. Volunteers.
At Tuliahan River, P. I., Mar. 25, 1899. | Voluntarily exposed himself to the fire of the enemy and went with four men to the relief of two native Filipinos lying wounded about 150 yards in front of the lines and personally carried one of them to a place of safety.
R—California.
B—California.

SHILLING, JOHN_____ | First sergeant, Company H, 3d Delaware Infantry.
At Weldon Railroad, Va., Aug. 21, 1864. | Capture of flag.
R—NR.
B—NR.

SHINGLE, JOHN H_____ | First sergeant, Troop I, 3d U. S. Cavalry.
At Rosebud River, Mont., June 17, 1876. | Gallantry in action.
R—NR.
B—Philadelphia, Pa.
Service rendered under name of John Henry.

SHIPLEY, ROBERT F_____ | Sergeant, Company A, 140th New York Infantry.
At Five Forks, Va., Apr. 1, 1865. | Capture of flag.
R—NR.
B—Wayne, N. Y.

SHOEMAKER, LEVI_____ | Sergeant, Company A, 1st West Virginia Cavalry.
At Nineveh, Va., Nov. 12, 1864. | Capture of flag of 22d Virginia Cavalry (C. S. A.).
R—NR.
B—Monongalia County, W. Va.

SHOPP, GEORGE J_____ | Private, Company E, 191st Pennsylvania Infantry.
At Five Forks, Va., Apr. 1, 1865. | Capture of flag.
R—Reading, Pa.
B—Equinunk, Pa.

SHUBERT, FRANK_____ | Sergeant, Company E, 43d New York Infantry.
At Petersburg, Va., Apr. 2, 1865. | Capture of two markers.
R—NR.
B—Germany.

SICKLES, DANIEL E_____ | Major general, U. S. Volunteers.
At Gettysburg, Pa., July 2, 1863. | Displayed most conspicuous gallantry on the field, vigorously contesting the advance of the enemy and continuing to encourage his troops after being himself severely wounded.
R—New York.
B—New York.

SICKLES, WILLIAM H_____ | Sergeant, Company B, 7th Wisconsin Infantry.
At Gravelly Run, Va., Mar. 31, 1865. | With a comrade attempted capture of a stand of Confederate colors and detachment of nine Confederates, actually taking prisoners three members of the detachment, dispersing the remainder, and recapturing a Union officer who was a prisoner in hands of the detachment.
R—Columbia County, Wis.
B—Danube, N. Y.

SIDMAN, GEORGE D_____ | Private, Company C, 16th Michigan Infantry.
At Gaines Mills, Va., June 27, 1862. | Distinguished bravery in battle.
R—Owosso, Mich.
B—Monroe, N. Y.

SIMMONS, JOHN_____ | Private, Company D, 2d New York Heavy Artillery.
At Sailors Creek, Va., Apr. 6, 1865. | Capture of flag.
R—Liberty, N. Y.
B—Bethel, N. Y.

SIMMONS, WILLIAM T_____ | Lieutenant, Company C, 11th Missouri Infantry.
At Nashville, Tenn., Dec. 16, 1864. | Capture of flag of 34th Alabama Infantry (C. S. A.).
R—NR.
B—Green County, Ill.

SIMONDS, WILLIAM EDGAR_____ | Sergeant major, 25th Connecticut Infantry.
At Irish Bend, La., Apr. 14, 1863. | Displayed great gallantry, under a heavy fire from the enemy, in calling in the skirmishers and assisting in forming the line of battle.
R—NR.
B—NR.

SIMONS, CHARLES J_____ | Sergeant, Company A, 9th New Hampshire Infantry.
At Petersburg, Va., July 30, 1864. | Was one of the first in the exploded mine, captured a number of prisoners, and was himself captured, but escaped.
R—NR.
B—India.

SINGLETON, FRANK_____ | See Frederick S. Neilon, true name.
Sergeant, Company A, 6th U. S. Cavalry.

SIVEL, HENRY_____ | See William H. Mathews, true name.
First sergeant, Company E, 2d Maryland Veteran Infantry.

SKELLIE, EBENEZER_____ | Corporal, Company D, 112th New York Infantry.
At Chapins Farm, near Richmond, Sept. 29, 1864. | Took the colors of his regiment, the color bearer having fallen, and carried them through the first charge; also, in the second charge, after all the color guard had been killed or wounded he carried the colors up to the enemy's works, where he fell wounded.
R—Mina, N. Y.
B—Mina, N. Y.

SKINKER, ALEXANDER R.
At Cheppy, France, Sept. 26, 1918.
R—St. Louis, Mo.
B—St. Louis, Mo.
G. O. No. 13, W. D., 1919.

Captain, 138th Infantry, 35th Division.
Unwilling to sacrifice his men when his company was held up by terrific machine-gun fire from iron pill boxes in the Hindenburg line, Captain Skinker personally led an automatic rifleman and a carrier in an attack on the machine guns. The carrier was killed instantly, but Captain Skinker seized the ammunition and continued through an opening in the barbed wire, feeding the automatic rifle until he, too, was killed.
Posthumously awarded. Medal presented to widow, Mrs. Caroline F. Skinker.

SKINNER, JOHN O.
At Lava Beds, Oreg., Jan. 17, 1873.
R—Maryland.
B—Maryland.

Contract surgeon, U. S. Army.
Rescued a wounded soldier who lay under a close and heavy fire during the assault on the Modoc stronghold after two soldiers had unsuccessfully attempted to make the rescue and both had been wounded in doing so.

SLACK, CLAYTON K (2055341)
Near Consenvoye, France, Oct. 8, 1918.
R—Madison, Wis.
B—Plover, Wis.
G. O. No. 16, W. D., 1919.

Private, Company D, 124th Machine Gun Battalion, 33d Division.
Observing German soldiers under cover 50 yards away on the left flank, Private Slack, upon his own initiative, rushed them with his rifle and, single handed, captured 10 prisoners and two heavy-type machine guns, thus saving his company and neighboring organizations from heavy casualties.

SLADEN, JOSEPH A.
At Resaca, Ga., May 14, 1864.
R—NR.
B—England.

Private, Company A, 33d Massachusetts Infantry.
While detailed as clerk at headquarters, voluntarily engaged in action at a critical moment and by personal example inspired the troops to repel the enemy.

SLAGLE, OSCAR.
At Elk River, Tenn., July 2, 1863.
R—NR.
B—Fulton County, Ohio.

Private, Company D, 104th Illinois Infantry.
Voluntarily joined a small party that, under a heavy fire, captured a stockade and saved the bridge.

SLAVENS, SAMUEL.
Georgia, April, 1862.
R—NR.
B—Pike County, Ohio.

Private, Company E, 33d Ohio Infantry.
One of 22 men (including two civilians) who by direction of General Mitchel (or Buell) penetrated nearly 200 miles south into the enemy's territory and captured a railroad train at Big Shanty, Ga., in an attempt to destroy the bridges and tracks between Chattanooga and Atlanta.

SLETTELAND, THOMAS.
Near Paete, Luzon, P. I., April 12, 1899.
R—Grafton, N. Dak.
B—Norway.

Private, Company C, 1st North Dakota Infantry.
Single-handed and alone defended his dead and wounded comrades against a greatly superior force of the enemy.

SLOAN, ANDREW J.
At Nashville, Tenn., Dec. 16, 1864.
R—NR.
B—Bedford County, Pa.

Private, Company H, 12th Iowa Infantry.
Capture of flag.

SLUSHER, HENRY C.
Near Moorefield, W. Va., Sept. 11, 1863.
R—NR.
B—Washington County, Pa.

Private, Company F, 22d Pennsylvania Cavalry.
Voluntarily crossed a branch of the Potomac River under fire to rescue a wounded comrade held prisoner by the enemy. Was wounded and taken prisoner in the attempt.

SMALLEY, REUBEN.
At Vicksburg, Miss., May 22, 1863.
R—NR.
B—Redding, N. Y.

Private, Company F, 83d Indiana Infantry.
Gallantry in the charge of the "volunteer storming party."

SMALLEY, REUBEN S.
At Elk River, Tenn., July 2, 1863.
R—NR.
B—Washington County, Pa.

Private, Company D, 104th Illinois Infantry.
Voluntarily joined a small party that, under a heavy fire, captured a stockade and saved the bridge.

SMITH, ALONZO.
At Hatchers Run, Va., Oct. 27, 1864.
R—NR.
B—Niagara County, N. Y.

Sergeant, Company C, 7th Michigan Infantry.
Capture of flag of 26th North Carolina Infantry (C. S. A.).

SMITH, ANDREW J.
At Chiricahua Mountains, Ariz., Oct. 20, 1869.
R—Baltimore, Md.
B—Baltimore, Md.

Sergeant, Company G, 8th U. S. Cavalry.
Gallantry in action.

SMITH, CHARLES E.
At Wichita River, Tex., July 12, 1870.
R—NR.
B—Auburn, N. Y.

Corporal, Company H, 6th U. S. Cavalry.
Gallantry in action.

SMITH, CHARLES H.
At St. Mary's Church, Va., June 24, 1864.
R—Maine.
B—Hollis, Me.

Colonel, 1st Maine Cavalry.
Remained in the fight to the close, although severely wounded.

SMITH, CORNELIUS C_____ | Corporal, Company K, 6th U. S. Cavalry.
Near White River, S. Dak., Jan. 1, 1891.
R—Helena, Mont.
B—Tucson, Ariz.

With four men of his troop drove off a superior force of the enemy and held his position against their repeated efforts to recapture it, and subsequently pursued them a great distance.

SMITH, DAVID L_____ | Sergeant, Battery E, 1st New York Light Artillery.
At Warwick Court House, Va., Apr. 6, 1862.
R—Bath, N. Y.
B—NR.

This soldier, when a shell struck an ammunition chest, exploding a number of cartridges and setting fire to the packing tow, procured water and extinguished the fire, thus preventing the explosion of the remaining ammunition.

SMITH, FRANCIS M_____ | First lieutenant and adjutant, 1st Maryland Infantry.
At Dabney's Mill, Va., Feb. 6, 1865.
R—Frederick, Md.
B—Frederick, Md.

Voluntarily remained with the body of his regimental commander under a heavy fire after the brigade had retired and brought the body off the field.

*SMITH, FRED E_____ | Lieutenant colonel, 308th Infantry, 77th Division.
Near Binarville, France, Sept. 29, 1918.
R—Bartlett, N. Dak.
B—Rockford, Ill.
G. O. No. 49, W. D., 1922.

When communication from the forward regimental post of command to the battalion leading the advance had been interrupted temporarily by the infiltration of small parties of the enemy armed with machine guns, Lieutenant Colonel Smith personally led a party of 2 other officers and 10 soldiers, and went forward to reestablish runner posts and carry ammunition to the front line. The guide became confused and the party strayed to the left flank beyond the outposts of supporting troops, suddenly coming under fire from a group of enemy machine guns only 50 yards away. Shouting to the other members of his party to take cover, this officer, in disregard of his own danger, drew his pistol and opened fire on the German gun crew. About this time he fell, severely wounded in the side, but regaining his footing, he continued to fire on the enemy until most of the men in his party were out of danger. Refusing first-aid treatment he then made his way in plain view of the enemy to a hand-grenade dump and returned under continued heavy machine-gun fire for the purpose of making another attack on the enemy emplacements. As he was attempting to ascertain the exact location of the nearest nest, he again fell, mortally wounded.

Posthumously awarded. Medal presented to widow, Mrs. Clara R. Smith.

SMITH, GEORGE W_____ | Private, Company M, 6th U. S. Cavalry.
At Washita River, Tex., Sept. 12, 1874.
R—NR.
B—Greenfield, N. Y.

Gallantry in action.

SMITH, HENRY I_____ | First lieutenant, Company B, 7th Iowa Infantry.
At Black River, N. C., Mar. 15, 1865.
R—NR.
B—England.

Voluntarily and under fire rescued a comrade from death by drowning

SMITH, JAMES_____ | Private, Company I, 2d Ohio Infantry.
Georgia, April, 1862.
R—NR.
B—NR.

One of 22 men (including two civilians) who by direction of General Mitchel (or Buell) penetrated nearly 200 miles south into the enemy's territory and captured a railroad train at Big Shanty, Ga., in an attempt to destroy the bridges and track between Chattanooga and Atlanta.

SMITH, JOSEPH S_____ | Lieutenant colonel and commissary of subsistence, 2d Army Corps.
At Hatchers Run, Va., Oct. 27, 1864.
R—Maine.
B—Wiscasset, Me.

Led a part of a brigade, saved two pieces of artillery, captured a flag, and secured a number of prisoners.

SMITH, OTIS W_____ | Private, Company G, 95th Ohio Infantry.
At Nashville, Tenn., Dec. 16, 1864.
R—NR.
B—Logan County, Ohio.

Capture of flag of 6th Florida Infantry (C. S. A.).

SMITH, OTTO_____ | Private, Company K, 8th U. S. Cavalry.
Arizona, 1868 and 1869.
R—NR.
B—Baltimore, Md.

Bravery in scouts and actions with Indians.

SMITH, RICHARD_____ | Private, Company B, 95th New York Infantry.
Weldon Railroad, Va., Aug. 21, 1864.
R—NR.
B—Rockland, N. Y.

Captured 2 officers and 20 men of Hagood's brigade while they were endeavoring to make their way back through the woods.

SMITH, ROBERT_____ | Private, Company M, 3d U. S. Infantry.
At Slim Buttes, Mont., Sept. 9, 1876.
R—NR.
B—Philadelphia, Pa.

Special bravery in endeavoring to dislodge Indians secreted in a ravine.

SMITH, S. RODMOND_____ | Captain, Company C, 4th Delaware Infantry.
At Rowanty Creek, Va., Feb. 5, 1865.
R—Wilmington, Del.
B—Delaware.

Swam the partly frozen creek under fire to establish a crossing.

SMITH, THADDEUS S_____ | Corporal, Company E, 6th Pennsylvania Reserve Infantry.
At Gettysburg, Pa., July 2, 1863.
R—NR.
B—Franklin County, Pa.

Was one of six volunteers who charged upon a log house near the Devil's Den, where a squad of the enemy's sharpshooters were sheltered, and compelled their surrender.

SMITH, THEODORE F.
At Chiricahua Mountains, Ariz.,
Oct. 20, 1869.
R—NR.
B—Rahway, N. J.

Private, Company G, 1st U. S. Cavalry.
Gallantry in action.

SMITH, THOMAS.
At Chiricahua Mountains, Ariz.,
Oct. 20, 1869.
R—NR.
B—Boston, Mass.

Private, Company G, 1st U. S. Cavalry.
Gallantry in action.

SMITH, THOMAS J.
At Chiricahua Mountains, Ariz.,
Oct. 20, 1869.
R—NR.
B—England.

Private, Company G, 1st U. S. Cavalry.
Gallantry in action.

SMITH, WILLIAM.
At Chiricahua Mountains, Ariz.,
Oct. 20, 1869.
R—NR.
B—Bath, Me.

Private, Company G, 8th U. S. Cavalry.
Gallantry in action.

SMITH, WILLIAM H.
At Chiricahua Mountains, Ariz.,
Oct. 20, 1869.
R—NR.
B—Lapeer County, Mich.

Private, Company G, 1st U. S. Cavalry.
Gallantry in action.

SMITH, WILSON.
At Washington, N. C., Sept. 6, 1862.
R—Madison, N. Y.
B—Madison, N. Y.

Corporal, Battery H, 3d New York Light Artillery.
Took command of a gun (the lieutenant in charge having disappeared) and fired the same so rapidly and effectively that the enemy were repulsed, although for a time a hand-to-hand conflict was had over the gun.

SNEDDEN, JAMES.
At Piedmont, Va., June 5, 1864.
R—NR.
B—Scotland.

Musician, 54th Pennsylvania Infantry.
Left his place in the rear, took the rifle of a disabled soldier, and fought through the remainder of the action.

SNOW, ELMER A.
At Rosebud Creek, Mont., June 17, 1876.
R—NR.
B—Hardwick, Mass.

Trumpeter, Company M, 3d U. S. Cavalry.
Bravery in action; was wounded in both arms.

SOUTHARD, DAVID.
At Sailors Creek, Va. Apr. 6, 1865.
R—NR.
B—Ocean County, N. J.

Sergeant, Company C, 1st New Jersey Cavalry.
Capture of flag; and was the first man over the works in the charge.

SOVA, JOSEPH E.
At Appomattox campaign, Virginia, Mar. 29 to Apr. 9, 1865.
R—NR.
B—Chili, N. Y.

Saddler, Company H, 8th New York Cavalry.
Capture of flag.

SOWERS, MICHAEL.
At Stony Creek Station, Va., Dec. 1, 1864.
R—NR.
B—Allegheny County, Pa.

Private, Company L, 4th Pennsylvania Cavalry.
His horse having been shot from under him, he voluntarily and on foot participated in the cavalry charge made upon one of the forts, conducting himself throughout with great personal bravery.

SPALDING, EDWARD B.
At Pittsburg Landing, Tenn., Apr. 6, 1862.
R—NR.
B—Ogle County, Ill.

Sergeant, Company E, 52d Illinois Infantry.
Although twice wounded, and thereby crippled for life, he remained fighting in open ground to the close of the battle.

SPENCE, ORIZOBA.
At Chiricahua Mountains, Ariz.,
Oct. 20, 1869.
R—Tionesta, Pa.
B—Forest County, Pa.

Private, Company G, 8th U. S. Cavalry.
Gallantry in action.

SPERRY, WILLIAM J.
At Petersburg, Va., Apr. 2, 1865.
R—Vermont.
B—Cavendish, Vt.

Major, 6th Vermont Infantry.
With the assistance of a few men, captured two pieces of artillery and turned them upon the enemy.

SPILLANE, TIMOTHY.
At Hatchers Run, Va., Feb. 5-7, 1865.
R—Erie, Pa.
B—Ireland.

Private, Company C, 16th Pennsylvania Cavalry.
Gallantry and good conduct in action; bravery in a charge and reluctance to leave the field after being twice wounded.

SPRAGUE, BENONA_____ | Corporal, Company F, 116th Illinois Infantry.
At Vicksburg, Miss., May 22, 1863 | Gallantry in the charge of the "volunteer storming party."
R—Cheneys Grove, Ill. |
B—Onondaga County, N. Y. |

SPRAGUE, JOHN W_____ | Colonel, 63d Ohio Infantry.
At Decatur, Ga., July 22, 1864. | With a small command defeated an overwhelming force of the enemy and
R—NR. | saved the trains of the corps.
B—NR. |

SPRINGER, GEORGE_____ | Private, Company G, 1st U. S. Cavalry.
At Chiricahua Mountains, Ariz., | Gallantry in action.
Oct. 20, 1869. |
R—NR. |
B—York County, Pa. |

SPURLING, ANDREW B_____ | Lieutenant colonel, 2d Maine Cavalry.
At Evergreen, Ala., Mar. 23, 1865. | Advanced alone in the darkness beyond the picket line, came upon three of
R—NR. | the enemy, fired upon them (his fire being returned), wounded two, and
B—Cranberry Isles, Me. | captured the whole party.

STACEY, CHARLES_____ | Private, Company D, 55th Ohio Infantry.
At Gettysburg, Pa., July 2, 1863. | Voluntarily took an advanced position on the skirmish line for the purpose of
R—NR. | ascertaining the location of Confederate sharpshooters, and under heavy
B—England. | fire held the position thus taken until the company of which he was a mem-
| ber went back to the main line.

STAHEL, JULIUS_____ | Major general, U. S. Volunteers.
At Piedmont, Va., June 5, 1864. | Led his division into action until he was severely wounded.
R—NR. |
B—Hungary. |

STANCE, EMANUEL_____ | Sergeant, Company F, 9th U. S. Cavalry.
At Kickapoo Springs, Tex., May 20, | Gallantry on scout after Indians.
1870. |
R—NR. |
B—Carroll County, La. |

STANLEY, DAVID S_____ | Major general, U. S. Volunteers.
At Franklin, Tenn., Nov. 30, 1864. | At a critical moment rode to the front of one of his brigades, reestablished its
R—Washington, D. C. | lines, and gallantly led it in a successful assault.
B—Fort Sully, S. Dak. |

STANLEY, EBEN_____ | Private, Company A, 5th U. S. Cavalry.
Near Turret Mountain, Ariz., Mar. | Gallantry in action.
25 and 27, 1873. |
R—NR. |
B—Decatur County, Iowa. |

STANLEY, EDWARD_____ | Corporal, Company F, 8th U. S. Cavalry.
At Seneca Mountain, Ariz., Aug. 26, | Gallantry in action.
1869. |
R—NR. |
B—New York, N. Y. |

STARKINS, JOHN H_____ | Sergeant, 34th New York Battery.
At Campbell Station, Tenn., Nov. | Brought off his piece without losing a man.
16, 1863. |
R—NR. |
B—NR. |

STAUFFER, RUDOLPH_____ | First sergeant, Company K, 5th U. S. Cavalry.
Near Camp Hualpai, Ariz., 1872. | Gallantry on scouts after Indians.
R—NR. |
B—Switzerland. |

STEELE, JOHN W_____ | Major and aid-de-camp, U. S. Volunteers.
At Spring Hill, Tenn., Nov. 29, 1864. | During a night attack of the enemy upon the wagon and ammunition train of
R—NR. | this officer's corps, he gathered up a force of stragglers and others, assumed
B—NR. | command of it, though himself a staff officer, and attacked and dispersed
| the enemy's forces, thus saving the train.

STEINER, CHRISTIAN_____ | Saddler, Company G, 8th U. S. Cavalry.
At Chiricahua Mountains, Ariz., | Gallantry in action.
Oct. 20, 1869. |
R—NR. |
B—Germany. |

STEINMETZ, WILLIAM_____ | Private, Company G, 83d Indiana Infantry.
At Vicksburg, Miss., May 22, 1863. | Gallantry in the charge of the "volunteer storming party."
R—NR. |
B—Newport, Ky. |

STEPHENS, WILLIAM G_____ | Private, Chicago Mercantile Battery, Illinois Light Artillery.
At Vicksburg, Miss., May 22, 1863. | Carried with others by hand a cannon up to and fired it through an embrasure
R—NR. | of the enemy's works.
B—New York, N. Y. |

STERLING, JOHN T.
    At Winchester, Va., Sept. 19, 1864.
    R—Marion County, Ind.
    B—Edgar County, Ill.

Private, Company D, 11th Indiana Infantry.
With one companion captured 14 of the enemy in the severest part of the battle.

STEVENS, HAZARD.
    At Fort Huger, Va., Apr. 19, 1863.
    R—Olympia, Washington Territory.
    B—Newport, R. I.

Captain and assistant adjutant general, U. S. Volunteers.
Gallantly led a party that assaulted and captured the fort.

STEVENS, THOMAS W.
    At Little Big Horn, Mont., June 25-26, 1876.
    R—NR.
    B—Madison County, Ky.

Private, Company D, 7th U. S. Cavalry.
Voluntarily brought water to the wounded under fire.

STEWART, BENJAMIN F.
    At Big Horn River, Mont., July 9, 1876.
    R—NR.
    B—Norfolk, Va.

Private, Company E, 7th U. S. Infantry.
Carried dispatches to General Crook at imminent risk of his life.

STEWART, GEORGE E.
    At Passi, Island of Panay, Philippine Islands, Nov. 26, 1899.
    R—New York, N. Y.
    B—New South Wales.

Second lieutenant, 19th U. S. Infantry.
While crossing a river in the face of the enemy, this officer plunged in and at the imminent risk of his own life saved from drowning an enlisted man of his regiment.

STEWART, GEORGE W.
    At Paine's Cross Roads, Va., Apr. 5, 1865.
    R—NR.
    B—Salem, N. J.

First sergeant, Company E, 1st New Jersey Cavalry.
Capture of flag.

STEWART, JOSEPH.
    At Five Forks, Va., Apr. 1, 1865.
    R—NR.
    B—Ireland.

Private, Company G, 1st Maryland Infantry.
Capture of flag.

STICKELS, JOSEPH.
    At Fort Blakely, Ala., Apr. 9, 1865.
    R—Bethany, Ohio.
    B—Butler County, Ohio.

Sergeant, Company A, 83d Ohio Infantry.
Capture of flag.

STICKOFFER, JULIUS H.
    At Cienaga Springs, Utah, Nov. 11, 1868.
    R—NR.
    B—Switzerland.

Saddler, Company L, 8th U. S. Cavalry.
Gallantry in action.

STOCKMAN, GEORGE H.
    At Vicksburg, Miss., May 22, 1863.
    R—NR.
    B—Germany.

First lieutenant, Company C, 6th Missouri Infantry.
Gallantry in the charge of the "volunteer storming party."

STOKES, ALONZO.
    At Wichita River, Tex., July 12, 1870.
    R—NR.
    B—Logan County, Ohio.

First sergeant, Company H, 6th U. S. Cavalry.
Gallantry in action.

STOKES, GEORGE.
    At Nashville, Tenn., Dec. 16, 1864.
    R—Jerseyville, Ill.
    B—England.

Private, Company C, 122d Illinois Infantry.
Capture of flag.

STOLTZ, FRANK.
    At Vicksburg, Miss., May 22, 1863.
    R—NR.
    B—Dearborn County, Ind.

Private, Company G, 83d Indiana Infantry.
Gallantry in the charge of the "volunteer storming party."

STOREY, JOHN H. R.
    At Dallas, Ga., May 28, 1864.
    R—NR.
    B—Philadelphia, Pa.

Sergeant, Company F, 109th Pennsylvania Infantry.
While bringing in a wounded comrade, under a destructive fire, he was himself wounded in the right leg, which was amputated on the same day.

STRAUB, PAUL F.
    At Alos, Zambales, Luzon, P. I., Dec. 21, 1899.
    R—Iowa.
    B—Germany.

Surgeon, 36th Infantry, U. S. Volunteers.
Voluntarily exposed himself to a hot fire from the enemy in repelling with pistol fire an insurgent attack and at great risk of his own life went under fire to the rescue of a wounded officer and carried him to a place of safety.

STRAUSBAUGH, BERNARD A.
    At Petersburg, Va., June 17, 1864.
    R—Warfordsburg, Pa.
    B—Adams County, Pa.

First sergeant, Company A, 3d Maryland Infantry.
Recaptured the colors of 2d Pennsylvania Provisional Artillery.

STRAYER, WILLIAM H. At Loupe Fork, Platte River, Nebr., Apr. 26, 1872. R—N R. B—Maytown, Pa.	Private, Company B, 3d U. S. Cavalry. Gallantry in action.
STEEILE, CHRISTIAN. At Paine's Cross Roads, Va., Apr. 5, 1865. R—Jersey City, N. J. B—Germany.	Private, Company I, 1st New Jersey Cavalry. Capture of flag.
STRIVSON, BENONI. Arizona, August to October, 1868. R—N R. B—Overton, Tenn.	Private, Company B, 8th U. S. Cavalry. Bravery in scouts and actions with Indians.
STRONG, JAMES N. At Port Hudson, La., May 27, 1863. R—N R. B—N R.	Sergeant, Company C, 49th Massachusetts Militia. Volunteered in response to a call and took part in the movement that was made upon the enemy's works under a heavy fire therefrom in advance of the general assault.
STURGEON, JAMES K. At Kenesaw Mountain, Ga., June 15, 1864. R—N R. B—Perry County, Ohio.	Private, Company F, 46th Ohio Infantry. Advanced beyond the lines, and in an encounter with three Confederates shot two and took the other prisoner.
SULLIVAN, THOMAS. At Chiricahua Mountains, Ariz., Oct. 20, 1869. R—N R. B—Covington, Ky.	Private, Company G, 1st U. S. Cavalry. Gallantry in action against Indians concealed in a ravine.
SULLIVAN, THOMAS. At Wounded Knee Creek, S. Dak., Dec. 29, 1890. R—Newark, N. J. B—Ireland.	Private, Company E, 7th U. S. Cavalry. Conspicuous bravery in action against Indians concealed in a ravine.
SUMMERS, JAMES C. At Vicksburg, Miss., May 22, 1863. R—N R. B—N R.	Private, Company H, 4th West Virginia Infantry. Gallantry in the charge of the "volunteer storming party."
SUMNER, JAMES. At Chiricahua Mountains, Ariz., Oct. 20, 1869. R—Chicago, Ill. B—England.	Private, Company G, 1st U. S. Cavalry. Gallantry in action.
SURLES, WILLIAM H. At Perryville, Ky., Oct. 8, 1862. R—N R. B—N R.	Private, Company G, 2d Ohio Infantry. In the hottest part of the fight he stepped in front of his colonel to shield him from the enemy's fire.
SUTHERLAND, JOHN A. At Arizona, August to October, 1868. R—Montgomery County, Ind. B—Monroe County, Ind.	Corporal, Company L, 8th U. S. Cavalry. Bravery in scouts and actions with Indians.
SWAN, CHARLES A. At Selma, Ala., Apr. 2, 1865. R—N R. B—Sweden.	Private, Company K, 4th Iowa Cavalry. Capture of flag (supposed to be 11th Mississippi) and bearer.
SWAP, JACOB E. At The Wilderness, Va., May 5, 1864. R—Springs, Pa. B—Calnehoose, N. Y.	Private, Company H, 83d Pennsylvania Infantry. Although assigned to other duty, he voluntarily joined his regiment in a charge and fought with it until severely wounded.
SWAYNE, WAGER. At Corinth, Miss., Oct. 4, 1862. R—N R. B—N R.	Lieutenant colonel, 43d Ohio Infantry. Conspicuous gallantry in restoring order at a critical moment and leading his regiment in a charge.
SWEATT, JOSEPH S. G. At Carsville, Va., May 15, 1863. R—N R. B—N R.	Private, Company C, 6th Massachusetts Infantry. When ordered to retreat this soldier turned and rushed back to the front, in the face of the heavy fire of the enemy, in an endeavor to rescue his wounded comrades, remaining by them until overpowered and taken prisoner.
SWEENEY, JAMES. At Cedar Creek, Va., Oct. 19, 1864. R—N R. B—England.	Private, Company A, 1st Vermont Cavalry. With one companion captured the State flag of a North Carolina regiment, together with three officers and an ambulance with its mules and driver

SWEGHEIMER, JACOB.........................
    At Vicksburg, Miss., May 22, 1863.
    R—NR.
    B—Germany.

Private, Company I, 54th Ohio Infantry.
Gallantry in the charge of the "volunteer storming party."

SWIFT, FREDERIC W.........................
    At Lenoir Station, Tenn., Nov. 16, 1863.
    R—Michigan.
    B—NR.

Lieutenant colonel, 17th Michigan Infantry.
Gallantly seized the colors and rallied the regiment after three color bearers had been shot and the regiment, having become demoralized, was in imminent danger of capture.

SWIFT, HARLAN J..........................
    At Petersburg, Va., July 30, 1864.
    R—New York.
    B—New Hudson, N. Y.

Second lieutenant, Company H, 2d New York Militia Regiment.
Having advanced with his regiment and captured the enemy's line, saw four of the enemy retiring toward their second line of works. He advanced upon them alone, compelled their surrender, and regained his regiment with the four prisoners.

SYPE, PETER.............................
    At Vicksburg, Miss., May 3, 1863.
    R—NR.
    B—Monroe County, Mich.

Private, Company B, 47th Ohio Infantry.
Was one of a party that volunteered and attempted to run the enemy's batteries with a steam tug and two barges loaded with subsistence stores.

TABOR, WILLIAM L. S....................
    At the seige of Port Hudson, La., July, 1863.
    R—NR.
    B—NR.

Private, Company K, 15th New Hampshire Infantry.
Voluntarily exposed himself to the enemy only a few feet away to render valuable services for the protection of his comrades.

TAGGART, CHARLES A....................
    At Sailors Creek, Va., Apr. 6, 1865.
    R—NR.
    B—Blandford, Mass.

Private, Company B, 37th Massachusetts Infantry.
Capture of flag.

TALLEY, EDWARD R. (1309598).........
    Near Ponchaux, France, Oct. 7, 1918.
    R—Russellville, Tenn.
    B—Russellville, Tenn.
    G. O. No. 50, W. D., 1919.

Sergeant, Company L, 117th Infantry, 30th Division.
Undeterred by seeing several comrades killed in attempting to put a hostile machine-gun nest out of action, Sergeant Talley attacked the position single handed. Armed only with a rifle, he rushed the nest in the face of intense enemy fire, killed or wounded at least six of the crew, and silenced the gun. When the enemy attempted to bring forward another gun and ammunition he drove them back by effective fire from his rifle.

TANNER, CHARLES B....................
    At Antietam, Md., Sept. 17, 1862.
    R—NR.
    B—Pennsylvania.

Second lieutenant, Company H, 1st Delaware Infantry.
Carried off the regimental colors, which had fallen within 20 yards of the enemy's lines, the color guard of nine men having all been killed or wounded; was himself three times wounded.

TAYLOR, ANTHONY.......................
    At Chickamauga, Ga., Sept. 20, 1863.
    R—NR.
    B—Burlington, N. J.

First lieutenant, Company A, 15th Pennsylvania Cavalry.
Held out to the last with a small force against the advance of superior numbers of the enemy.

TAYLOR, BERNARD.......................
    Near Sunset Pass, Ariz., Nov. 1, 1874.
    R—NR.
    B—St. Louis, Mo.

Sergeant, Company A, 5th U. S. Cavalry.
Bravery in rescuing Lieutenant King, 5th U. S. Cavalry, from Indians.

TAYLOR, CHARLES........................
    At Big Dry Wash, Ariz., July 17, 1882.
    R—NR.
    B—Baltimore, Md.

First sergeant, Company D, 3d U. S. Cavalry.
Gallantry in action.

TAYLOR, FORRESTER L...................
    At Chancellorsville, Va., May 3, 1863.
    R—NR.
    B—Philadelphia, Pa.

Captain, Company H, 23d New Jersey Infantry.
At great risk voluntarily saved the lives of and brought from the battlefield two wounded comrades.

TAYLOR, HENRY H.......................
    At Vicksburg, Miss., June 25, 1863.
    R—NR.
    B—Jo Daviess County, Ill.

Sergeant, Company C, 45th Illinois Infantry.
Was the first to plant the Union colors upon the enemy's works.

TAYLOR, JOSEPH........................
    At Weldon Railroad, Va., Aug. 18, 1864.
    R—NR.
    B—England.

Private, Company E, 7th Rhode Island Infantry.
While acting as an orderly to a general officer on the field and alone, encountered a picket of three of the enemy and compelled their surrender.

TAYLOR, RICHARD.......................
    At Cedar Creek, Va., Oct. 19, 1864.
    R—NR.
    B—Madison County, Ala.

Private, Company E, 18th Indiana Infantry.
Capture of flag.

TAYLOR, WILBUR N......................
    Arizona, 1868 and 1869.
    R—NR.
    B—Hamden, Me.

Corporal, Company K, 8th U. S. Cavalry.
Bravery in actions with Indians.

TAYLOR, WILLIAM............................
    At Front Royal, Va., May 23, 1862.

    At Weldon Railroad, Va., Aug. 19. 1864.
    R—NR.
    B—Washington, D. C.

Sergeant, Company H, and second lieutenant, Company M, 1st Maryland Infantry.
Then a sergeant, he was painfully wounded while obeying an order to burn a bridge, but, persevering in the attempt, he burned the bridge and prevented its use by the enemy.
Then a lieutenant, he voluntarily took the place of a disabled officer and undertook a hazardous reconnaissance beyond the lines of the army; was taken prisoner in the attempt.

TEA, RICHARD L............................
    At Sappa Creek, Kans., Apr. 23, 1875.
    R—NR.
    B—Philadelphia, Pa.

Sergeant, Company H, 6th U. S. Cavalry.
Gallantry in action.

TERRY, JOHN D............................
    At Newbern, N. C., Mar. 14, 1862.
    R—Boston, Mass.
    B—Montville, Me.

Sergeant, Company E, 23d Massachusetts Infantry.
In the thickest of the fight, where he lost his leg by a shot, still encouraged the men until carried off the field.

THACKRAH, BENJAMIN............................
    Near Fort Gates, Fla., Apr. 1, 1864.
    R—Johnsonville, N. Y.
    B—Scotland.

Private, Company H, 115th New York Infantry.
Was a volunteer in the surprise and capture of the enemy's picket.

THATCHER, CHARLES M............................
    At Petersburg, Va., July 30, 1864.
    R—NR.
    B—Coldwater, Mich.

Private, Company B, 1st Michigan Sharpshooters.
Instead of retreating or surrendering when the works were captured, regardless of his personal safety continued to return the enemy's fire until he was captured.

THAXTER, SIDNEY W............................
    At Hatchers Run, Va., Oct. 27, 1864.
    R—Maine.
    B—NR.

Major, 1st Maine Cavalry.
Voluntarily remained and participated in the battle with conspicuous gallantry, although his term of service had expired and he had been ordered home to be mustered out.

THOMAS, CHARLES L............................
    At Powder River Expedition, Dakota Territory, Sept. 12-17, 1865.
    R—NR.
    B—Philadelphia, Pa.

Sergeant, Company E, 11th Ohio Cavalry.
Carried a message through a country infested with hostile Indians and saved the life of a comrade en route.

THOMAS, HAMPTON S............................
    At Amelia Springs, Va., Apr. 5, 1865.
    R—Pennsylvania.
    B—NR.

Major, 1st Pennsylvania Veteran Cavalry.
Conspicuous gallantry in the capture of a field battery and a number of battle flags and in the destruction of the enemy's wagon train.

THOMAS, STEPHEN............................
    At Cedar Creek, Va., Oct. 19, 1864.
    R—Montpelier, Vt.
    B—NR.

Colonel, 8th Vermont Infantry.
Distinguished conduct in a desperate hand-to-hand encounter, in which the advance of the enemy was checked.

THOMPKINS, GEORGE W............................
    At Petersburg, Va., Mar. 25, 1865.
    R—NR.
    B—Orange County, N. Y.

Corporal, Company F, 124th New York Infantry.
Capture of flag of 59th Alabama Infantry (C. S. A.) from an officer who, with colors in hand, was rallying his men.

THOMPKINS, WILLIAM H............................
    At Tayabacoa, Cuba, June 30, 1898.
    R—Paterson, N. J.
    B—Paterson, N. J.

Private, Troop G, 10th U. S. Cavalry.
Voluntarily went ashore in the face of the enemy and aided in the rescue of his wounded comrades; this after several previous attempts at rescue had been frustrated.

THOMPSON, ALLEN............................
    At White Oak Road, Va., Apr. 1, 1865.
    R—Sandy Creek, N. Y.
    B—New York, N. Y.

Private, Company I, 4th New York Heavy Artillery.
Made a hazardous reconnaissance through timber and slashings preceding the Union line of battle, signaling the troops and leading them through the obstructions.

THOMPSON, CHARLES A............................
    At Spotsylvania, Va., May 12, 1864.
    R—NR.
    B—Perrysburg, Ohio.

Sergeant, Company D, 17th Michigan Infantry.
After the regiment was surrounded and all resistance seemed useless, fought single handed for the colors and refused to give them up until he had appealed to his superior officers.

THOMPSON, FREEMAN C............................
    At Petersburg, Va., Apr. 2, 1865.
    R—NR.
    B—Monroe County, Ohio.

Corporal, Company F, 116th Ohio Infantry.
Was twice knocked from the parapet of Fort Gregg by blows from the enemy's muskets but at the third attempt fought his way into the works.

THOMPSON, GEORGE W............................
    At Little Blue, Nebr., May 15, 1870.
    R—NR.
    B—Victory, N. Y.

Private, Company C, 2d U. S. Cavalry.
Gallantry in action.

THOMPSON, JAMES............................
    At White Oak Road, Va., Apr. 1, 1865.
    R—Sandy Creek, N. Y.
    B—Sandy Creek, N. Y.

Private, Company K, 4th New York Heavy Artillery.
Made a hazardous reconnaissance through timber and slashings, preceding the Union line of battle, signaling the troops and leading them through the obstructions.

THOMPSON, JAMES B.
  At Gettysburg, Pa., July 3, 1863.
  R—Perrysville, Pa.
  B—Juniata County, Pa.

Sergeant, Company G, 1st Pennsylvania Rifles.
Capture of flag of 15th Georgia Infantry (C. S. A.).

THOMPSON, J. HARRY.
  At New Bern, N. C., Mar. 14, 1862.
  R—NR.
  B—NR.

Surgeon, U. S. Volunteers.
Voluntarily reconnoitered the enemy's position and carried orders under the hottest fire.

THOMPSON, JOHN.
  At Chiricahua Mountains, Ariz., Oct. 20, 1869.
  R—New York, N. Y.
  B—Scotland.

Sergeant, Company G, 1st U. S. Cavalry.
Bravery in action with Indians.

THOMPSON, JOHN.
  At Hatchers Run, Va., Feb. 6, 1865.
  R—Baltimore, Md.
  B—Denmark.

Corporal, Company C, 1st Maryland Infantry.
As color bearer with most conspicuous gallantry preceded his regiment in the assault and planted his flag upon the enemy's works.

THOMPSON, JOSEPH H.
  Near Apremont, France, Oct. 1, 1918.
  R—Beaver Falls, Pa.
  B—Ireland.
  G. O. No. 21, W. D., 1925.

Major, 110th Infantry, 28th Division.
Counterattacked by two regiments of the enemy, Major Thompson encouraged his battalion in the front line by constantly braving the hazardous fire of machine guns and artillery. His courage was mainly responsible for the heavy repulse of the enemy. Later in the action, when the advance of his assaulting companies was held up by fire from a hostile machine-gun nest and all but one of the six assaulting tanks were disabled, Major Thompson, with great gallantry and coolness, rushed forward on foot three separate times in advance of the assaulting line, under heavy machine-gun and antitank-gun fire, and led the one remaining tank to within a few yards of the enemy machine-gun nest, which succeeded in reducing it, thereby making it possible for the infantry to advance.

THOMPSON, PETER.
  At Little Big Horn Mont., June 25, 1876.
  R—NR.
  B—Scotland.

Private, Company C, 7th U. S. Cavalry.
After having voluntarily brought water to the wounded, in which effort he was shot through the head, he made two more successful trips for the same purpose, notwithstanding remonstrances of his sergeant.

THOMPSON, THOMAS.
  At Chancellorsville, Va., May 2, 1863.
  R—NR.
  B—Champaign County, Ohio.

Sergeant, Company A, 66th Ohio Infantry.
One of a party of four who voluntarily brought into the Union lines, under fire, a wounded Confederate officer from whom was obtained valuable information concerning the enemy.

THOMPSON, WILLIAM P.
  At Wilderness, Va., May 6, 1864.
  R—NR.
  B—Brooklyn, N. Y.

Sergeant, Company G, 20th Indiana Infantry.
Capture of flag of 55th Virginia Infantry (C. S. A.).

THOMSON, CLIFFORD.
  At Chancellorsville, Va., May 2, 1863.
  R—New York, N. Y.
  B—NR.

First lieutenant, Company A, 1st New York Cavalry.
Volunteered to ascertain the character of approaching troops; rode up so closely as to distinguish the features of the enemy, and as he wheeled to return they opened fire with musketry, the Union troops returning same. Under a terrific fire from both sides Lieutenant Thomson rode back unhurt to the Federal lines, averting a disaster to the Army by his heroic act.

THORN, WALTER.
  At Dutch Gap Canal, Va., Jan. 1, 1865.
  R—NR.
  B—New York, N. Y.

Second lieutenant, Company G, 116th U. S. Colored Troops.
After the fuse to the mined bulkhead had been lit this officer, learning that the picket guard had not been withdrawn, mounted the bulkhead and at great personal peril warned the guard of its danger.

TIBBETS, ANDREW W.
  At Columbus, Ga., Apr. 16, 1865.
  R—Appanoose County, Iowa.
  B—Clark County, Ind.

Private, Company I, 3d Iowa Cavalry.
Capture of flag and bearer, Austin's Battery (C. S. A.).

TILTON, HENRY R.
  At Bear Paw Mountain, Mont., Sept. 30, 1877.
  R—Jersey City, N. J.
  B—Barnegat, N. J.

Major and surgeon, U. S. Army.
Fearlessly risked his life and displayed great gallantry in rescuing and protecting the wounded men.

TILTON, WILLIAM.
  At Richmond campaign, Virginia, 1864.
  R—NR.
  B—St. Albans, Vt.

Sergeant, Company C, 7th New Hampshire Infantry.
Gallant conduct in the field.

TINKHAM, EUGENE M.
  At Cold Harbor, Va., June 3, 1864.
  R—NR.
  B—NR.

Corporal, Company H, 148th New York Infantry.
Though himself wounded, voluntarily left the rifle pits, crept out between the lines and, exposed to the severe fire of the enemy's guns at close range, brought within the lines two wounded and helpless comrades.

TITUS, CALVIN PEARL_____ | Musician, Company E, 14th U. S. Infantry.
At Pekin, China, Apr. 14, 1900. | Gallant and daring conduct in the presence of his colonel and other officers and
R—NR. | enlisted men of his regiment; was first to scale the wall of the city.
B—Vinton, Iowa. |

TITUS, CHARLES_____ | Sergeant, Company H, 1st New Jersey Cavalry.
At Sailors Creek, Va., Apr. 6, 1865. | Was among the first to check the enemy's countercharge.
R—New Brunswick, N. J. |
B—Millstone, N. J. |

TOBAN, JAMES W_____ | Sergeant, Company C, 9th Michigan Cavalry.
At Aiken, S. C., Feb. 11, 1865. | Voluntarily and at great personal risk returned, in the face of the advance of
R—NR. | the enemy, and rescued from impending death or capture Maj. William C.
B—Northfield, Mich. | Stevens, 9th Michigan Cavalry, who had been thrown from his horse.

TOBIE, EDWARD P_____ | Sergeant major, 1st Maine Cavalry.
Appomattox campaign, Virginia, | Though severely wounded at Sailors Creek, April 6, and at Farmville, Apr. 7,
Mar. 29 to Apr. 9, 1865. | refused to go to the hospital, but remained with his regiment, performed the
R—Lewiston, Me. | full duties of adjutant upon the wounding of that officer, and was present for
B—Lewiston, Me. | duty at Appomattox.

TOBIN, JOHN M_____ | First lieutenant and adjutant, 9th Massachusetts Infantry.
At Malvern Hill, Va., July 1, 1862. | Voluntarily took command of the 9th Massachusetts while adjutant, bravely
R—NR. | fighting from 3 p. m. until dusk, rallying and re-forming the regiment under
B—NR. | fire; twice picked up the regimental flag, the color bearer having been shot
 | down, and placed it in worthy hands.

TOFFEY, JOHN J_____ | First lieutenant, Company G, 33d New Jersey Infantry.
At Chattanooga, Tenn., Nov. 23, | Although excused from duty on account of sickness, went to the front in com-
1863. | mand of a storming party, and with conspicuous gallantry participated in
R—Hudson, N. J. | the assault on Missionary Ridge; was here wounded and permanently dis-
B—Duchess, N. Y. | abled.

TOLAN, FRANK_____ | Private, Company D, 7th U. S. Cavalry.
At Little Big Horn, Mont., June 25, | Voluntarily brought water to the wounded under fire.
1876. |
R—NR. |
B—Malone, N. Y. |

TOMPKINS, AARON B_____ | Sergeant, Company G, 1st New Jersey Cavalry.
At Sailors Creek, Va., Apr. 5, 1865. | Charged into the enemy's ranks and captured a battle flag, having a horse shot
R—NR. | under him and his cheeks and shoulders cut with a saber.
B—Orange County, N. J. |

TOMPKINS, CHARLES H_____ | First lieutenant, 2d U. S. Cavalry.
At Fairfax, Va., June 1, 1861. | Twice charged through the enemy's lines and, taking a carbine from an enlisted
R—Brooklyn, N. Y. | man, shot the enemy's captain.
B—Fort Monroe, Va. |

TOOHEY, THOMAS_____ | Sergeant, Company F, 24th Wisconsin Infantry.
At Franklin, Tenn., Nov. 30, 1864. | Gallantry in action; voluntarily assisting in working guns of battery near right
R—NR. | of the regiment after nearly every man had left them, the fire of the enemy
B—New York, N. Y. | being hotter at this than at any other point on the line.

TOOMER, WILLIAM_____ | Sergeant, Company G, 127th Illinois Infantry.
At Vicksburg, Miss., May 22, 1863. | Gallantry in the charge of the "volunteer storming party."
R—NR. |
B—Ireland. |

TORGLER, ERNST_____ | Sergeant, Company G, 37th Ohio Infantry.
At Ezra Chapel, Ga., July 28, 1864. | At great hazard of his life he saved his commanding officer, then badly wounded,
R—NR. | from capture.
B—Germany. |

TOY, FREDERICK E_____ | First sergeant, Company G, 7th U. S. Cavalry.
At Wounded Knee Creek, S. Dak., | Bravery.
Dec. 29, 1890. |
R—NR. |
B—Buffalo, N. Y. |

TOZIER, ANDREW J_____ | Sergeant, Company I, 20th Maine Infantry.
At Gettysburg, Pa., July 2, 1863. | At the crisis of the engagement this soldier, a color bearer, stood alone in an
R—NR. | advanced position, the regiment having been borne back, and defended his
B—NR. | colors with musket and ammunition picked up at his feet.

TRACY, AMASA S_____ | Lieutenant colonel, 2d Vermont Infantry.
At Cedar Creek, Va., Oct. 19, 1864. | Took command of and led the brigade in the assault on the enemy's works.
R—Middlebury, Vt. |
B—Maine. |

TRACY, BENJAMIN F_____ | Colonel, 109th New York Infantry.
At Wilderness, Va., May 6, 1864. | Seized the colors and led the regiment when other regiments had retired and
R—NR. | then re-formed his line and held it.
B—New York. |

TRACY, CHARLES H.
   At Spotsylvania, Va., May 12, 1864.

   At Petersburg, Va., Apr. 2, 1865.
   R—Springfield, Mass.
   B—Jewett City, Conn.

Sergeant, Company A, 37th Massachusetts Infantry.
At the risk of his own life, assisted in carrying to a place of safety a wounded and helpless officer.
Advanced with the pioneers, and, under heavy fire, assisted in removing two lines of chevaux-de-frise; was twice wounded but advanced to the third line, where he was again severely wounded, losing a leg.

TRACY, JOHN.
   At Chiricahua Mountains, Ariz., Oct. 20, 1869.
   R—St. Paul, Minn.
   B—Ireland.

Private, Company G, 8th U. S. Cavalry.
Bravery in action with Indians.

TRACY, WILLIAM G.
   At Chancellorsville, Va., May 2, 1863.
   R—NR.
   B—Onondaga, N. Y.

Second lieutenant Company I, 122d New York Infantry.
Having been sent outside the lines to obtain certain information of great importance and having succeeded in his mission, was surprised upon his return by a large force of the enemy, regaining the Union lines only after greatly imperiling his life.

TRAUTMAN, JACOB.
   At Wounded Knee Creek, S. Dak., Dec. 29, 1890.
   R—NR.
   B—Germany.

First sergeant, Company I, 7th U. S. Cavalry.
Killed a hostile Indian at close quarters, and, although entitled to retirement from service, remained to the close of the campaign.

TRAYNOR, ANDREW.
   At Masons Hill, Va., Mar. 16, 1864.
   R—Rome, N. Y.
   B—Newark, N. J.

Corporal, Company D, 1st Michigan Cavalry.
Having been surprised and captured by a detachment of guerrillas, this soldier, with other prisoners, seized the arms of the guard over them, killed two of the guerrillas, and enabled all the prisoners to escape.

TREAT, HOWELL B.
   At Buzzard's Roost, Ga., May 11, 1864.
   R—Painesville, Ohio.
   B—Painesville, Ohio.

Sergeant, Company I, 52d Ohio Infantry.
Risked his life in saving a wounded comrade.

TREMAIN, HENRY E.
   At Resaca, Ga., May 15, 1864.
   R—New York, N. Y.
   B—New York, N. Y.

Major and aid-de-camp, U. S. Volunteers.
Voluntarily rode between the lines while two brigades of Union troops were firing into each other and stopped the firing.

TREMBLEY, WILLIAM B.
   At Calumpit, Luzon, P. I., Apr. 27, 1899.
   R—Kansas City, Kans.
   B—Johnson, Kans.

Private, Company B, 20th Kansas Volunteer Infantry.
Swam the Rio Grande de Pampanga in face of the enemy's fire and fastened a rope to the occupied trenches, thereby enabling the crossing of the river and the driving of the enemy from his fortified position.

TRIBE, JOHN.
   At Waterloo Bridge, Va., Aug. 25, 1862.
   R—Oswego, N. Y.
   B—Tioga County, N. Y.

Private, Company G, 5th New York Cavalry.
Voluntarily assisted in the burning and destruction of the bridge under heavy fire of the enemy.

TROGDEN, HOWELL G.
   At Vicksburg, Miss., May 22, 1863.
   R—NR.
   B—NR.

Private, Company B, 8th Missouri Infantry.
Gallantry in the charge of the "volunteer storming party."

TRUELL, EDWIN M.
   Near Atlanta, Ga., July 21, 1864.
   R—Manston, Wis.
   B—Lowell, Mass.

Private, Company E, 12th Wisconsin Infantry.
Although severely wounded in a charge, he remained with the regiment until again severely wounded, losing his leg.

TUCKER, ALLEN.
   At Petersburg, Va., Apr. 2, 1865.
   R—NR.
   B—Lyme, Conn.

Sergeant, Company F, 10th Connecticut Infantry.
Gallantry as color bearer in the assault on Fort Gregg.

TUCKER, JACOB R.
   At Petersburg, Va., Apr. 1, 1865.
   R—Baltimore, Md.
   B—Chester County, Pa.

Corporal, Company G, 4th Maryland Infantry.
Was one of the three soldiers most conspicuous in the final assault.

TURNER, HAROLD L. (1490302)
   Near St. Etienne, France, Oct. 8, 1918.
   R—Seminole, Okla.
   B—Aurora, Mo.
   G. O. No. 59, W. D., 1919.

Corporal, Company F, 142d Infantry, 36th Division.
After his platoon had started the attack Corporal Turner assisted in organizing a platoon consisting of the battalion scouts, runners, and a detachment of Signal Corps. As second in command of this platoon he fearlessly led them forward through heavy enemy fire, continually encouraging the men. Later he encountered deadly machine-gun fire which reduced the strength of his command to but four men, and these were obliged to take shelter. The enemy machine-gun emplacement, 25 yards distant, kept up a continual fire from four machine guns. After the fire had shifted momentarily Corporal Turner rushed forward with fixed bayonet and charged the position alone, capturing the strong point with a complement of 50 Germans and 4 machine guns. His remarkable display of courage and fearlessness was instrumental in destroying the strong point, the fire from which had blocked the advance of his company.

TURNER, WILLIAM B
Near Roussoy, France, Sept. 27, 1918.
R—Garden City, N. Y.
B—Boston, Mass.
G. O. No. 81, W. D., 1919.

First lieutenant, 105th Infantry, 27th Division.
He led a small group of men to the attack, under terrific artillery and machine-gun fire, after they had become separated from the rest of the company in the darkness. Single handed he rushed an enemy machine gun which had suddenly opened fire on his group and killed the crew with his pistol. He then pressed forward to another machine-gun post 25 yards away and had killed one gunner himself by the time the remainder of his detachment arrived and put the gun out of action. With the utmost bravery he continued to lead his men over three lines of hostile trenches, cleaning up each one as they advanced, regardless of the fact that he had been wounded three times, and killed several of the enemy in hand-to-hand encounters. After his pistol ammunition was exhausted, this gallant officer seized the rifle of a dead soldier, bayoneted several members of a machine-gun crew, and shot the other. Upon reaching the fourth-line trench, which was his objective, Lieutenant Turner captured it with the nine men remaining in his group and resisted a hostile counterattack until he was finally surrounded and killed. Posthumously awarded. Medal presented to mother, Mrs. William Turner.

TURPIN, JAMES H
Arizona, 1872–1874.
R—NR.
B—Easton, Mass.

First sergeant, Company L, 5th U. S. Cavalry.
Gallantry in actions with Apaches.

TWEEDALE, JOHN
At Stone River, Tenn., Dec. 31, 1862, and Jan. 1, 1863.
R—NR.
B—Philadelphia, Pa.

Private, Company B, 15th Pennsylvania Cavalry.
Gallantry in action.

TWOMBLY, VOLTAIRE P
At Fort Donelson, Tenn., Feb. 15, 1862.
R—NR.
B—Van Buren, Iowa.

Corporal, Company F, 2d Iowa Infantry.
Took the colors after three of the color guard had fallen, and although almost instantly knocked down by a spent ball, immediately arose and bore the colors to the end of the engagement.

TYRRELL, GEORGE WILLIAM
At Resaca, Ga., May 14, 1864.
R—NR.
B—Ireland.

Corporal, Company H, 5th Ohio Infantry.
Capture of flag.

UHRI, GEORGE
At White Oak Swamp Bridge, Va., June 30, 1862.
R—NR.
B—Germany.

Sergeant, Light Battery F, 5th U. S. Artillery.
Was one of a party of three who, under heavy fire of the advancing enemy, voluntarily secured and saved from capture a field gun belonging to another battery, and which had been deserted by its officers and men.

URELL, M. EMMET
At Bristoe Station, Va., Oct. 14, 1863.
R—NR.
B—Ireland.

Private, Company E, 82d New York Infantry.
Gallantry in action while detailed as color bearer; was severely wounded.

VALE, JOHN
At Nolensville, Tenn., Feb. 15, 1863.
R—Rochester, Minn.
B—England.

Private, Company H, 2d Minnesota Infantry.
Was one of a detachment of 16 men who heroically defended a wagon train against the attack of 125 cavalry, repulsed the attack, and saved the train.

VANCE, WILSON
At Stone River, Tenn., Dec. 31, 1862.
R—NR.
B—Hancock County, Ohio.

Private, Company B, 21st Ohio Infantry.
Voluntarily and under a heavy fire, while his command was falling back, rescued a wounded and helpless comrade from death or capture.

VANDERSLICE, JOHN M
At Hatchers Run, Va., Feb. 6, 1865.
R—NR.
B—Philadelphia, Pa.

Private, Company D, 8th Pennsylvania Cavalry.
Was the first man to reach the enemy's rifle pits, which were taken in the charge.

VAN IERSEL, LUDOVICUS M. M. (40858)
At Mouzon, France, November 9, 1918.
R—Glen Rock, N. J.
B—Holland.
G. O. No. 34, W. D., 1919.

Sergeant, Company M, 9th Infantry, 2d Division.
While a member of the reconnaissance patrol, sent out at night to ascertain the condition of a damaged bridge, Sergeant Van Iersel volunteered to lead a party across the bridge in the face of heavy machine-gun and rifle fire from a range of only 75 yards. Crawling alone along the débris of the ruined bridge he came upon a trap, which gave away and precipitated him into the water. In spite of the swift current he succeeded in swimming across the stream and found a lodging place among the timbers on the opposite bank. Disregarding the enemy fire, he made a careful investigation of the hostile position by which the bridge was defended and then returned to the other bank of the river, reporting this valuable information to the battalion commander.

VAN MATRE, JOSEPH
At Petersburg, Va., Apr. 2, 1865.
R—NR.
B—Mason County, Va.

Private, Company G, 116th Ohio Infantry.
In the assault on Fort Gregg this soldier climbed upon the parapet and fired down into the fort as fast as the loaded guns could be passed up to him by comrades.

VAN SCHAICK, LOUIS J
Near Nasugbu, Batangas, P. I., Nov. 23, 1901.
R—Cobleskill, N. Y.
B—Cobleskill, N. Y.

First lieutenant, 4th U. S. Infantry.
While in pursuit of a band of insurgents was the first of his detachment to emerge from a canyon, and seeing a column of insurgents and fearing they might turn and dispatch his men as they emerged one by one from the canyon, galloped forward and closed with the insurgents, thereby throwing them into confusion until the arrival of others of the detachment.

VANWINKLE, EDWARD............
At Chapins Farm, near Richmond, Va., Sept. 29, 1864.
R—NR.
B—NR.

Corporal, Company C, 148th New York Infantry.
Took position in advance of the skirmish line and drove the enemy's cannoneers from their guns.

VARNUM, CHARLES A............
At White Clay Creek, S. Dak., Dec. 30, 1890.
R—Florida.
B—Troy, N. Y.

Captain, Company B, 7th U. S. Cavalry.
While executing an order to withdraw, seeing that a continuance of the movement would expose another troop of his regiment to being cut off and surrounded, he disregarded orders to retire, placed himself in front of his men, led a charge upon the advancing Indians, regained a commanding position that had just been vacated, and thus insured a safe withdrawal of both detachments without further loss.

VEAL, CHARLES....................
At Chapins Farm, near Richmond, Va., Sept. 29, 1864.
R—Portsmouth, Va.
B—Portsmouth, Va.

Private, Company D, 4th U. S. Colored Troops.
Seized the national colors, after two color bearers had been shot down, close to the enemy's works, and bore them through the remainder of the battle.

VEALE, MOSES....................
At Wauhatchie, Tenn., Oct. 28, 1863.
R—Philadelphia, Pa.
B—NR.

Captain, Company F, 109th Pennsylvania Infantry.
Gallantry in action; manifesting throughout the engagement coolness, zeal, judgment, and courage.

VEAZEY, WHEELOCK G.............
At Gettysburg, Pa., July 3, 1863.
R—NR.
B—NR.

Colonel, 16th Vermont Infantry.
Rapidly assembled his regiment and charged the enemy's flank; changed front under heavy fire, and charged and destroyed a Confederate brigade, all this with new troops in their first battle.

VERNAY, JAMES D.................
At Vicksburg, Miss., Apr. 22, 1863.
R—NR.
B—Lacon, Ill.

Second lieutenant, Company B, 11th Illinois Infantry.
Served gallantly as a volunteer with the crew of the steamer Horizon that, under a heavy fire, passed the Confederate batteries.

VEUVE, ERNEST..................
At Staked Plains, Tex., Nov. 3, 1874.
R—NR.
B—Switzerland.

Farrier, Company A, 4th U. S. Cavalry.
Gallant manner in which he faced a desperate Indian.

VIFQUAIN, VICTOR................
At Fort Blakely, Ala., Apr. 9, 1865.
R—NR.
B—NR.

Lieutenant colonel, 97th Illinois Infantry.
Capture of flag.

VILLEPIGUE, JOHN C (1312401)........
At Vaux-Andigny, France, Oct. 15, 1918.
R—Camden, S. C.
B—Camden, S. C.
G. O. No. 16, W. D., 1919.

Corporal, Company M, 118th Infantry, 30th Division.
Having been sent out with two other soldiers to scout through the village of Vaux-Andigny, he met with strong resistance from enemy machine-gun fire, which killed one of his men and wounded the other. Continuing his advance without aid 500 yards in advance of his platoon and in the face of machine-gun and artillery fire, he encountered four of the enemy in a dugout, whom he attacked and killed with a hand granade. Crawling forward to a point 150 yards in advance of his first encounter, he rushed a machine-gun nest, killing four and capturing six of the enemy and taking two light machine guns. After being joined by his platoon he was severley wounded in the arm.

VOIT, OTTO.....................
At Little Big Horn, Mont., June 25, 1876.
R—NR.
B—Germany.

Saddler, Company H, 7th U. S. Cavalry.
Bravery in action.

VOKES, LEROY H.................
At Loupe Fork, Platte River, Nebr., Apr. 26, 1872.
R—NR.
B—Lake County, Ill.

First sergeant, Company B, 3d U. S. Cavalry.
Gallantry in action.

VON MEDEM, RUDOLPH..............
1872 and 1873.
R—NR.
B—Germany.

Sergeant, Company A, 5th U. S. Cavalry.
Gallantry in actions and campaigns.

VON SCHLICK, ROBERT H...........
At Tientsin, China, July 13, 1900.
R—San Francisco, Calif.
B—Germany.

Private, Company C, 9th U. S. Infantry.
Although previously wounded while carrying a wounded comrade to a place of safety, rejoined his command, which partly occupied an exposed position upon a dike, remaining there after his command had been withdrawn, singly keeping up the fire, and obliviously presenting himself as a conspicuous target until he was literally shot off this position by the enemy.

VON VEGESACK, ERNEST............
At Gaines Mills, Va., June 27, 1862.
R—NR.
B—NR.

Major and aid-de-camp, U. S. Volunteers.
While voluntarily serving as aid-de-camp, successfully and advantageously changed the position of troops under fire.

WAALER, REIDAR (1209189)
Near Roussoy, France, Sept. 27, 1918.
R—New York, N. Y.
B—Norway.
G. O. No. 20, W. D., 1919.

Sergeant, Company A, 105th Machine-Gun Battalion, 27th Division.
In the face of heavy artillery and machine-gun fire, he crawled forward in a burning British tank, in which some of the crew were imprisoned, and succeeded in rescuing two men. Although the tank was then burning fiercely and contained ammunition which was likely to explode at any time, this soldier immediately returned to the tank and, entering it, made a search for the other occupants, remaining until he satisfied himself that there were no more living men in the tank.

WAGEMAN, JOHN H
At Petersburg, Va., June 17, 1864.
R—Amelia, Ohio.
B—Clermont County, Ohio.

Private, Company I, 60th Ohio Infantry.
Remained with the command after being severely wounded until he had fired all the cartridges in his possession, when he had to be carried from the field.

WAGNER, JOHN W
At Vicksburg, Miss., May 22, 1863.
R—NR.
B—NR.

Corporal, Company F, 8th Missouri Infantry.
Gallantry in the charge of the "volunteer storming party."

WAINWRIGHT, JOHN
At Fort Fisher, N. C., Jan. 15, 1865.
R—NR.
B—Onondaga County, N. Y.

First lieutenant, Company F, 97th Pennsylvania Infantry.
Gallant and meritorious conduct, where, as first lieutenant, he commanded the regiment.

WALKER, ALLEN
Texas, Dec. 30, 1891.
R—NR.
B—Patroit, Ind.

Private, Company C, 3d U. S. Cavalry.
While carrying dispatches, he attacked a party of three armed men and secured papers valuable to the United States.

WALKER, FRANK O
Near Taal, Luzon, P. I., Jan. 18, 1900.
R—Burlington, Mass.
B—South Boston, Mass.

Private, Company F, 46th Infantry, U. S. Volunteers.
Under heavy fire of the enemy in rescuing a dying comrade who was sinking beneath the water.

WALKER, JAMES C
At Missionary Ridge, Tenn., Nov. 25, 1863.
R—Springfield, Ohio.
B—Clark County, Ohio.

Private, Company K, 31st Ohio Infantry.
After 2 color bearers had fallen, seized the flag and carried it forward, assisting in the capture of a battery. Shortly thereafter he captured the flag of the 41st Alabama and the color bearer.

WALKER, JOHN
At Red Creek, Ariz., Sept. 23, 1869.
R—NR.
B—France.

Private, Company D, 8th U. S. Cavalry.
Gallantry in action with Indians.

WALL, JERRY
At Gettysburg, Pa., July 3, 1863.
R—NR.
B—Geneva, N. Y.

Private, Company B, 126th New York Infantry.
Capture of flag.

WALLACE, GEORGE W
At Tinuba, Luzon, P. I., March 4, 1900.
R—Denver, Colo.
B—Fort Riley, Kans.

Second lieutenant, 9th U. S. Infantry.
With another officer and a native Filipino, was shot at from ambush, the other officer falling severely wounded. Lieutenant Wallace fired in the direction of the enemy, put them to rout, removed the wounded officer from the path, returned to the town, a mile distant, and summoned assistance from his command.

WALLACE, WILLIAM
At Cedar Creek, etc., Mont., Oct. 21, 1876, to Jan. 8, 1877.
R—NR.
B—Ireland.

Sergeant, Company C, 5th U. S. Infantry.
Gallantry in actions.

WALLAR, FRANCIS A
At Gettysburg, Pa., July 1, 1863.
R—NR.
B—Gurney, Ohio.

Corporal, Company I, 6th Wisconsin Infantry.
Capture of flag of 2d Mississippi Infantry (C. S. A.).

WALLEY, AUGUSTUS
At Cuchillo Negro Mountains, N. Mex., Aug. 16, 1881.
R—NR.
B—Reistertown, Md.

Private, Company I, 9th U. S. Cavalry.
Bravery in action with hostile Apaches.

WALLING, WILLIAM H
At Fort Fisher, N. C., Dec. 25, 1864.
R—NR.
B—NR.

Captain, Company C, 142d New York Infantry.
During the bombardment of the fort by the fleet, captured and brought off the flag of the fort, the flagstaff having been shot down.

WALSH, JOHN
At Cedar Creek, Va., Oct. 19, 1864.
R—NR.
B—Ireland.

Corporal, Company D, 5th New York Cavalry.
Recaptured the flag of the 15th New Jersey Infantry.

WALTON, GEORGE W
At Fort Hell, Petersburg, Va., Aug. 29, 1864.
R—NR.
B—Chester, Pa.

Private, Company C, 97th Pennsylvania Infantry.
Went outside the trenches, under heavy fire at short range, and rescued a comrade who had been wounded and thrown out of the trench by an exploding shell.

WAMBSGAN, MARTIN_____
    At Cedar Creek, Va., Oct. 19, 1864.
    R—Cayuga County, N. Y.
    B—Germany.

Private, Company D, 90th New York Infantry.
While the enemy were in close proximity, this soldier sprang forward and bore off in safety the regimental colors, the color bearer having fallen on the field of battle.

WANTON, GEORGE H_____
    At Tayabacoa, Cuba, June 30, 1898.
    R—Paterson, N. J.
    B—Paterson, N. J.

Private, Troop M, 10th U. S. Cavalry.
Voluntarily went ashore in the face of the enemy and aided in the rescue of his wounded comrades; this after several previous attempts at rescue had been frustrated.

WARD, CALVIN JOHN (1307698)_____
    Near Estrees, France, Oct. 8, 1918.
    R—Morristown, Tenn.
    B—Green County, Tenn.
    G. O. No. 16, W. D., 1919.

Private, Company D, 117th Infantry, 30th Division.
During an advance Private Ward's company was held up by a machine gun, which was enfilading the line. Accompanied by a noncommissioned officer, he advanced against this post and succeeded in reducing the nest by killing three and capturing seven of the enemy and their guns.

WARD, CHARLES H_____
    At Chiricahua Mountains, Ariz.,
      Oct. 20, 1869.
    R—Philadelphia, Pa.
    B—England.

Private, Company G, 1st U. S. Cavalry.
Gallantry in action with Indians.

WARD, JAMES_____
    At Wounded Knee Creek, S. Dak.,
      Dec. 29, 1890.
    R—Boston, Mass.
    B—Quincy, Mass.

Sergeant, Company B, 7th U. S. Cavalry.
Continued to fight after being severely wounded

WARD, JOHN_____
    At Pecos River, Tex., Apr. 25, 1875.
    R—NR.
    B—Arkansas.

Sergeant, Indian Scouts.
Gallantry in action with Indians.

WARD, NELSON W_____
    At Staunton River Bridge, Va.,
      June 25, 1864.
    R—NR.
    B—Columbiana County, Ohio.

Private, Company M, 11th Pennsylvania Cavalry.
Voluntarily took part in a charge; went alone in front of his regiment under a heavy fire to secure the body of his captain, who had been killed in the action.

WARD, THOMAS J_____
    At Vicksburg, Miss., May 22, 1863.
    R—NR.
    B—Romney, Va.

Private, Company C, 116th Illinois Infantry.
Gallantry in the charge of the "volunteer storming party."

WARD, WILLIAM H_____
    At Vicksburg, Miss., May 3, 1863.
    R—NR.
    B—NR.

Captain, Company B, 47th Ohio Infantry.
Voluntarily commanded the expedition which, under cover of darkness, attempted to run the enemy's batteries.

WARDEN, JOHN_____
    At Vicksburg, Miss., May 22, 1863.
    R—Lemont, Ill.
    B—Cook County, Ill.

Corporal, company E, 55th Illinois Infantry.
Gallantry in the charge of the "volunteer storming party."

WARFEL, HENRY C_____
    At Paines Crossroads, Va., Apr. 5,
      1865.
    R—NR.
    B—Huntington, Pa.

Private, Company A, 1st Pennsylvania Cavalry.
Capture of Virginia State colors.

WARREN, FRANCIS E_____
    At Port Hudson, La., May 27, 1863.
    R—NR.
    B—NR.

Corporal, Company C, 49th Massachusetts Infantry.
Volunteered in response to a call, and took part in the movement that was made upon the enemy's works under a heavy fire therefrom, in advance of the general assault.

WARRINGTON, LEWIS_____
    At Muchague Valley, Tex., Dec. 8,
      1874.
    R—Washington, D. C.
    B—Washington, D. C.

First lieutenant, 4th U. S. Cavalry.
Gallantry in a combat with five Indians.

WATSON, JAMES C_____
    At Wichita River, Tex., July 12,
      1870.
    R—NR.
    B—Cochecton, N. Y.

Corporal, Company L, 6th U. S. Cavalry.
Gallantry in action.

WATSON, JOSEPH_____
    Near Picacho Mountain, Ariz.,
      June 4, 1869.
    R—St. Joseph, Mich.
    B—Union City, Mich.

Private, Company F, 8th U. S. Cavalry.
Killed an Indian warrior and captured his arms.

WEAHER, ANDREW J_____
    Arizona, August to October, 1868.
    R—NR.
    B—Philadelphia, Pa.

Private, Company B, 8th U. S. Cavalry.
Bravery in actions and scouts against Indians.

WEAVER, AMOS.
    Between Calubus and Malalong, P. I., Nov. 5, 1899.
    R—NR.
    B—NR.

Sergeant, Company F, 36th Infantry, U. S. Volunteers.
Alone and unaided, charged a body of 15 insurgents, dislodging them, killing 4 and wounding several.

WEBB, ALEXANDER S.
    At Gettysburg, Pa., July 3, 1863.
    R—New York, N. Y.
    B—New York, N. Y.

Brigadier general, U. S. Volunteers.
Distinguished personal gallantry in leading his men forward at a critical period in the contest.

WEBB, JAMES.
    At Bull Run, Va., Aug. 30, 1862.
    R—NR.
    B—Brooklyn, N. Y.

Private, Company F, 5th New York Infantry.
Under heavy fire voluntarily carried information to a battery commander that enabled him to save his guns from capture. Was severely wounded, but refused to go to hospital, and participated in the remainder of the campaign.

WEBBER, ALANSON P.
    At Kenesaw Mountain, Ga., June 27, 1864.
    R—Illinois.
    B—Greene County, N. Y.

Musician, 86th Illinois Infantry.
Voluntarily joined in a charge against the enemy, which was repulsed, and by his rapid firing in the face of the enemy enabled many of the wounded to return to the Federal lines; with others, held the advance of the enemy while temporary works were being constructed.

WEEKS, JOHN H.
    At Spotsylvania, Va., May 12, 1864.
    R—Hartwick Seminary, New York.
    B—Hampton, Conn.

Private, Company H, 152d New York Infantry.
Capture of flag and color bearer.

WEINERT, PAUL H.
    At Wounded Knee Creek, S. Dak., Dec. 29, 1890.
    R—Baltimore, Md.
    B—Germany.

Corporal, Company E, 1st U. S. Artillery.
Taking the place of his commanding officer, who had fallen severely wounded, he gallantly served his piece, after each fire advancing it to a better position.

WEIR, HENRY C.
    At St. Marys Church, Va., June 24, 1864.
    R—NR.
    B—West Point, N. Y.

Captain and assistant adjutant general, U. S. Volunteers.
The division being hard pressed and falling back, this officer dismounted, gave his horse to a wounded officer, and thus enabled him to escape. Afterwards, on foot, Captain Weir rallied and took command of some stragglers and helped to repel the last charge of the enemy.

WEISS, ENOCH R.
    At Chiricahua Mountains, Ariz., Oct. 20, 1869.
    R—NR.
    B—Kosciusko County, Ind.

Private, Company G, 1st U. S. Cavalry.
Gallantry in action with Indians.

WELBORN, IRA C.
    At Santiago, Cuba, July 2, 1898.
    R—Mico, Miss.
    B—Mico, Miss.
    Distinguished-service medal also awarded.

Second lieutenant, 9th U. S. Infantry.
Voluntarily left shelter and went, under fire, to the aid of a private of his company who was wounded.

WELCH, CHARLES H.
    At Little Big Horn, Mont., June 25–26, 1876.
    R—NR.
    B—New York, N. Y.

Sergeant, Company D, 7th U. S. Cavalry.
Voluntarily brought water to the wounded, under fire.

WELCH, GEORGE W.
    At Nashville, Tenn., Dec. 16, 1864.
    R—NR.
    B—Brown County, Iowa.

Private, Company A, 11th Missouri Infantry.
Capture of flag.

WELCH, MICHAEL.
    At Wichita River, Tex., Oct. 5, 1870.
    R—NR.
    B—Poughkeepsie, N. Y.

Sergeant, Company M, 6th U. S. Cavalry.
Gallantry in action.

WELCH, RICHARD.
    At Petersburg, Va., Apr. 2, 1865.
    R—NR.
    B—Ireland.

Corporal, Company E, 37th Massachusetts Infantry.
Capture of flag.

WELCH, STEPHEN.
    At Dug Gap, Ga., May 8, 1864.
    R—NR.
    B—Groton, N. Y.

Sergeant, Company C, 154th New York Infantry.
Risked his life in rescuing a wounded comrade under fire of the enemy.

WELD, SETH L.
    At La Paz, Leyte, P. I., Dec. 5, 1906.
    R—Altamont, N. C.
    B—Sandy Hook, Md.

Corporal, Company L, 8th U. S. Infantry.
With his right arm cut open with a bolo, went to the assistance of a wounded constabulary officer and a fellow soldier who were surrounded by about 40 Pulajanes, and, using his disabled rifle as a club, beat back the assailants and rescued his party.

WELLS, HENRY S.
At Chapin's Farm, near Richmond, Va., Sept. 29, 1864.
R—NR.
B—NR.

Private, Company C, 148th New York Infantry.
With two comrades, took position in advance of the skirmish line, within short distance of the enemy's gunners, and drove them from their guns.

WELLS, THOMAS M.
At Cedar Creek, Va., Oct. 19, 1864.
R—Dekalb, N. Y.
B—Ireland.

Chief bugler, 6th New York Cavalry.
Capture of colors of 44th Georgia Infantry (C. S. A.).

WELLS, WILLIAM.
At Gettysburg, Pa., July 3, 1863.
R—Waterbury, Vt.
B—NR.

Major, 1st Vermont Cavalry.
Led the second battalion of his regiment in a daring charge.

WELSH, EDWARD.
At Vicksburg, Miss., May 22, 1863.
R—Cincinnati, Ohio.
B—Ireland.

Private, Company D, 54th Ohio Infantry.
Gallantry in the charge of the "volunteer storming party."

WELSH, JAMES.
At Petersburg, Va., July 30, 1864.
R—NR.
B—Ireland.

Private, Company E, 4th Rhode Island Infantry.
Bore off the regimental colors after the color sergeant had been wounded and the color corporal bearing the colors killed, thereby saving the colors from capture.

WENDE, BRUNO.
At El Caney, Cuba, July 1, 1898.
R—Canton, Ohio.
B—Germany.

Private, Company C, 17th U. S. Infantry.
Gallantly assisted in the rescue of the wounded from in front of the lines and under heavy fire from the enemy.

WEST, CHESTER H. (2263711).
Near Bois-de-Cheppy, France, Sept. 26, 1918.
R—Los Banos, Calif.
B—Fort Collins, Colo.
G. O. No. 34, W. D., 1919.

First sergeant, Company D, 363d Infantry, 91st Division.
While making his way through a thick fog with his automatic rifle section, his advance was halted by direct and unusual machine-gun fire from two guns. Without aid, he at once dashed through the fire and, attacking the nest, killed two of the gunners, one of whom was an officer. This prompt and decisive hand-to-hand encounter on his part enabled his company to advance farther without the loss of a man.

WEST, FRANK.
At Big Dry Wash, Ariz., July 17, 1882.
R—Mohawk, N. Y.
B—Mohawk, N. Y.

First lieutenant, 6th U. S. Cavalry.
Rallied his command and led it in the advance against the enemy's fortified position.

WESTERHOLD, WILLIAM.
At Spotsylvania, Va., May 12, 1864.
R—NR.
B—Prussia.

Sergeant, Company G, 52d New York Infantry.
Capture of flag of 23d Virginia Infantry (C. S. A.) and its bearer.

WESTON, JOHN F.
Near Wetumpka, Ala., Apr. 13, 1865.
R—NR.
N—NR.

Major, 4th Kentucky Cavalry.
This officer, with a small detachment, while en route to destroy steamboats loaded with supplies for the enemy, was stopped by an unfordable river, but with 5 of his men swam the river, captured two leaky canoes, and ferried his men across. He then encountered and defeated the enemy, and on reaching Wetumpka found the steamers anchored in midstream. By a ruse obtained possession of a boat, with which he reached the steamers and demanded and received their surrender.

WETHERBY, JOHN C.
Near Imus, Luzon, P. I., Nov. 20, 1899.
R—Martinsville, Ind.
B—Morgan County, Ind.

Private, Company L, 4th U. S. Infantry.
While carrying important orders on the battle field, was desperately wounded and, being unable to walk, crawled far enough to deliver his orders.

WHEATON, LOYD.
At Fort Blakely, Ala., Apr. 9, 1865.
R—NR
B—NR.

Lieutenant colonel, 8th Illinois Infantry.
Led the right wing of his regiment and, springing through an embrasure, was the first to enter the enemy's works, against a strong fire of artillery and infantry.

WHEELER, DANIEL D.
At Salem Heights, Va., May 3, 1863.
R—NR.
B—Cavendish, Vt.

First lieutenant, Company G, 4th Vermont Infantry.
Distinguished bravery in action where he was wounded and had a horse shot under him.

WHEELER, HENRY W.
At Bull Run, Va., July 21, 1861.
R—NR.
B—NR.

Private, Company A, 2d Maine Infantry.
Voluntarily accompanied his commanding officer and assisted in removing the dead and wounded from the field under a heavy fire of artillery and musketry.

WHERRY, WILLIAM M.
At Wilsons Creek, Mo., Aug. 10, 1861.
R—NR.
B—NR.

First lieutenant, Company D, 3d U. S. Reserve Missouri Infantry.
Displayed conspicuous coolness and heroism in rallying troops that were recoiling under heavy fire.

WHITAKER, EDWARD W_____
  At Reams Station, Va., June 29, 1864.
  R—NR.
  B—NR.

Captain, Company E, 1st Connecticut Cavalry.
While acting as an aid voluntarily carried dispatches from the commanding general to General Meade, forcing his way with a single troop of Cavalry, through an Infantry division of the enemy in the most distinguished manner, though he lost half his escort.

WHITE, ADAM_____
  At Hatchers Run, Va., Apr. 2, 1865.
  R—Parkersburg, W. Va.
  B—Switzerland.

Corporal, Company G, 11th West Virginia Infantry.
Capture of flag.

WHITE, EDWARD_____
  At Calumpit, Luzon, P. I., Apr. 27, 1899.
  R—Kansas City, Kans.
  B—Seneca, Kans.

Private, Company D, 20th Kansas Volunteer Infantry.
Swam the Rio Grande de Pampanga in face of the enemy's fire and fastened a rope to the occupied trenches, thereby enabling the crossing of the river and the driving of the enemy from his fortified position.

WHITE, J. HENRY_____
  At Rappahannock Station, Va., Aug. 23, 1862.
  R—NR.
  B—Philadelphia, Pa.

Private, Company A, 90th Pennsylvania Infantry.
At the imminent risk of his life, crawled to a near-by spring within the enemy's range and exposed to constant fire filled a large number of canteens, and returned in safety to the relief of his comrades who were suffering for want of water.

WHITE, PATRICK H_____
  At Vicksburg, Miss., May 22, 1863.
  R—Chicago, Ill.
  B—NR.

Captain, Chicago Mercantile Battery, Illinois Light Artillery.
Carried with others by hand a cannon up to and fired it through an embrasure of the enemy's works.

WHITEHEAD, JOHN M_____
  At Stone River, Tenn., Dec. 31, 1862.
  R—Westville, Ind.
  B—NR.

Chaplain, 15th Indiana Infantry.
Went to the front during a desperate contest and unaided carried to the rear several wounded and helpless soldiers.

WHITEHEAD, PATTON G_____
  At Cedar Creek, etc., Mont., Oct. 21, 1876, to Jan. 8, 1877.
  R—NR.
  B—Russell County, Va.

Private, Company C, 5th U. S. Infantry.
Gallantry in actions.

WHITMAN, FRANK M_____
  At Antietam, Md., Sept. 17, 1862.

  At Spotsylvania, Va., May 18, 1864.
  R—Ayersville, Mass.
  B—Woodstock, Me.

Private, Company G, 35th Massachusetts Infantry.
Was among the last to leave the field and was instrumental in saving the lives of several of his comrades at the imminent risk of his own.
Was foremost in line in the assault, where he lost a leg.

WHITMORE, JOHN_____
  At Blakely, Ala., Apr. 9, 1865.
  R—NR.
  B—Brown County, Ill.

Private, Company F, 119th Illinois Infantry.
Capture of flag.

WHITNEY, WILLIAM G_____
  At Chickamauga, Ga., Sept. 20, 1863.
  R—NR.
  B—Allen, Mich.

Sergeant, Company B, 11th Michigan Infantry.
As the enemy were about to charge, this officer went outside the temporary Union works among the dead and wounded enemy and at great exposure to himself cut off and removed their cartridge boxes, bringing the same within the Union lines, the ammunition being used with good effect in again repulsing the attack.

WHITTIER, EDWARD N_____
  At Fishers Hill, Va., Sept. 22, 1864.
  R—NR.
  B—Portland, Me.

First lieutenant, 5th Battery, Maine Light Artillery.
While acting as assistant adjutant general, Artillery brigade, 6th Army Corps, went over the enemy's works, mounted, with the assaulting column, to gain quicker possession of the guns and to turn them upon the enemy.

WHITTLESEY, CHARLES W_____
  Northeast of Binarville, in the Forest D'Argonne, France, Oct. 2–7, 1918.
  R—Pittsfield, Mass.
  B—Florence, Wis.
  G. O. No. 118, W. D., 1918.

Major, 308th Infantry, 77th Division.
Although cut off for five days from the remainder of his division, Major Whittlesey maintained his position, which he had reached under orders received for an advance, and held his command, consisting originally of 463 officers and men of the 308th Infantry and of Company K of the 307th Infantry, together in the face of superior numbers of the enemy during the five days. Major Whittlesey and his command were thus cut off, and no rations or other supplies reached him, in spite of determined efforts which were made by his division. On the fourth day Major Whittlesey received from the enemy a written proposition to surrender, which he treated with contempt, although he was at that time out of rations and had suffered a loss of about 50 per cent in killed and wounded of his command and was surrounded by the enemy.

*WICKERSHAM, J. HUNTER_____
  Near Limey, France, Sept. 12, 1918.
  R—Denver, Colo.
  B—New York, N. Y.
  G. O. No. 16, W. D., 1919.

Second lieutenant, 353d Infantry, 89th Division.
Advancing with his platoon during the St. Mihiel offensive, he was severely wounded in four places by the bursting of a high-explosive shell. Before receiving any aid for himself he dressed the wounds of his orderly, who was wounded at the same time. He then ordered and accompanied the further advance of his platoon, although weakened by the loss of blood. His right hand and arm being disabled by wounds, he continued to fire his revolver with his left hand until, exhausted by loss of blood, he fell and died from his wounds before aid could be administered.
Posthumously awarded. Medal presented to mother, Mrs. M. E. Damon.

WIDICK, ANDREW J.
 At Vicksburg, Miss., May 22, 1863.
 R—NR.
 B—Macon County, Ill.

Private, Company B, 116th Illinois Infantry.
Gallantry in the charge of the "volunteer storming party."

WIDMER, JACOB.
 At Milk Creek, Colo., Sept. 29, 1879.
 R—NR.
 B—Germany.

First sergeant, Company D, 5th U. S. Cavalry.
Volunteered to accompany a small detachment on a very dangerous mission.

WILCOX, WILLIAM H.
 At Spotsylvania, Va., May 12, 1864.
 R—Lempster, N. H.
 B—Lempster, N. H.

Sergeant, Company G, 9th New Hampshire Infantry.
Took command of his company, deployed as skirmishers, after the officers in command of the skirmish line had both been wounded, conducting himself gallantly; afterwards, becoming separated from his command, he asked and obtained permission to fight in another company.

WILDER, WILBER E.
 At Horseshoe Canyon, N. Mex.,
  Apr 23, 1882.
 R—Detroit, Mich.
 B—Atlas, Mich.

First lieutenant, 4th U. S. Cavalry.
Assisted, under a heavy fire, to rescue a wounded comrade.

WILEY, JAMES.
 At Gettysburg, Pa., July 3, 1863.
 R—NR.
 B—Ohio.

Sergeant, Company B, 59th New York Infantry.
Capture of flag of a Georgia regiment.

WILHELM, GEORGE.
 At Champion Hill, or Bakers Creek,
  Miss., May 16, 1863.
 R—Lancaster, Ohio.
 B—NR.

Captain, Company F, 56th Ohio Infantry.
Having been badly wounded in the breast and captured, he made a prisoner of his captor and brought him into camp.

WILKENS, HENRY.
 At Little Muddy Creek, Mont.,
  May 7, 1877; at Camas Meadows,
  Idaho, Aug. 20, 1877.
 R—NR.
 B—Germany.

First sergeant, Company L, 2d U. S. Cavalry.
Bravery in actions with Indians.

WILKINS, LEANDER A.
 At Petersburg, Va., July 30, 1864.
 R—NR.
 B—Lancaster, N. H.

Sergeant, Company H, 9th New Hampshire Infantry.
Recaptured the colors of 21st Massachusetts Infantry in a hand-to-hand encounter.

WILLCOX, ORLANDO B.
 At Bull Run, Va., July 21, 1861.
 R—Detroit, Mich.
 B—Detroit, Mich.

Colonel, 1st Michigan Infantry.
Led repeated charges until wounded and taken prisoner.

WILLIAMS, ELLWOOD N.
 At Shiloh, Tenn., Apr. 6, 1862.
 R—Havanna, Ill.
 B—Philadelphia, Pa.

Private, Company A, 28th Illinois Infantry.
A box of ammunition having been abandoned between the lines, this soldier voluntarily went forward with one companion, under a heavy fire from both armies, secured the box, and delivered it within the line of his regiment, his companion being mortally wounded.

WILLIAMS, GEORGE C.
 At Gaines Mill, Va., June 27, 1862.
 R—NR.
 B—England.

Quartermaster sergeant, 1st Battalion, 14th U. S. Infantry.
While on duty with the wagon train as quartermaster sergeant he voluntarily left his place of safety in the rear, joined a company, and fought with distinguished gallantry through the action.

WILLIAMS, LE ROY.
 At Cold Harbor, Va., June 3, 1864.
 R—NR.
 B—Oswego, N. Y.

Sergeant, Company G, 8th New York Heavy Artillery.
Voluntarily exposed himself to the fire of the enemy's sharpshooters and located the body of his colonel who had been killed close to the enemy's lines. Under cover of darkness, with 4 companions, he recovered the body and brought it within the Union lines, having approached within a few feet of the Confederate pickets while so engaged.

WILLIAMS, MOSES.
 At foothills of the Cuchillo Negro
  Mountains, N. Mex., Aug. 16,
  1881.
 R—NR.
 B—Carroll County, Pa.

First sergeant, Company I, 9th U. S. Cavalry.
Rallied a detachment, skillfully conducted a running fight of three or four hours, and by his coolness, bravery, and unflinching devotion to duty in standing by his commanding officer in an exposed position under a heavy fire from a large party of Indians saved the lives of at least three of his comrades.

WILLIAMS, WILLIAM H.
 At Peach Tree Creek, Ga., July 20,
  1864.
 R—Miami County, Ohio.
 B—Hancock County, Ohio.

Private, Company C, 82d Ohio Infantry.
Voluntarily went beyond the lines to observe the enemy; also aided a wounded comrade.

WILLIAMSON, JAMES A.
 At Chickasaw Bayou, Miss., Dec.
  29, 1862.
 R—NR.
 B—NR.

Colonel, 4th Iowa Infantry.
Led his regiment against a superior force, strongly entrenched, and held his ground when all support had been withdrawn.

WILLISTON, EDWARD B. 　At Trevilian Station, Va., June 12, 1864. 　R—San Francisco, Calif. 　B—Norwich, Vt.	First lieutenant, 2d U. S. Artillery. Distinguished gallantry.
WILLS, HENRY. 　Near Fort Selden, N. Mex., July 8 to 11, 1873. 　R—NR. 　B—Gracon, Va.	Private, Company C, 8th U. S. Cavalry. Services against hostile Indians.
WILSON, ARTHUR H. 　At Patian Island, P. I., July 4, 1909. 　R—Springfield, Ill. 　B—Springfield, Ill.	Second lieutenant, 6th U. S. Cavalry. While in action against hostile Moros, when, it being necessary to secure a mountain gun in position by rope and tackle, voluntarily with the assistance of an enlisted man, carried the rope forward and fastened it, being all the time under heavy fire of the enemy at short range.
WILSON, BENJAMIN. 　At Wichita River, Tex., Oct. 5, 1870. 　R—NR. 　B—Pittsburgh, Pa.	Private, Company M, 6th U. S. Cavalry. Gallantry in action.
WILSON, CHARLES. 　At Cedar Creek, etc., Mont., Oct. 21, 1876, to Jan. 8, 1877. 　R—Beardstown, Ill. 　B—Petersburg, Ill.	Corporal, Company H, 5th U. S. Infantry. Gallantry in actions.
WILSON, CHARLES E. 　At Sailors Creek, Va., Apr. 6, 1865. 　R—NR. 　B—Bucks County, Pa.	Sergeant, Company A, 1st New Jersey Cavalry. Charged the enemy's works, colors in hand, and had two horses shot under him.
WILSON, CHRISTOPHER W. 　At Spotsylvania, Va., May 12, 1864. 　R—West Meriden, Conn. 　B—Ireland.	Private, Company E, 73d New York Infantry. Took the flag from the wounded color bearer and carried it in the charge over the Confederate works, in which charge he also captured the colors of the 56th Virginia (C. S. A.), bringing off both flags in safety.
WILSON, FRANCIS A. 　At Petersburg, Va., Apr. 2, 1865. 　R—Philadelphia, Pa. 　B—Philadelphia, Pa.	Corporal, Company B, 95th Pennsylvania Infantry. Was among the first to penetrate the enemy's lines and himself captured a gun of the two batteries captured.
WILSON, JOHN. 　At Chamberlains Creek, Va., Mar. 31, 1865. 　R—Jersey City, N. J. 　B—England.	Sergeant, Company L, 1st New Jersey Cavalry. With the assistance of one comrade, headed off the advance of the enemy, shooting two of his color bearers; also posted himself between the enemy and the led horses of his own command, thus saving the herd from capture.
WILSON, JOHN A. 　Georgia, April, 1862. 　R—NR. 　B—Columbus, Ohio.	Private, Company C, 21st Ohio Infantry. One of 22 men (including two civilians) who by direction of General Mitchel (or Buell) penetrated nearly 200 miles south into the enemy's territory and captured a railroad train at Big Shanty, Ga., in an attempt to destroy the bridges and track between Chattanooga and Atlanta.
WILSON, JOHN M. 　At Malvern Hill, Va., Aug. 6, 1862. 　R—Washington Territory. 　B—Washington, D. C.	First lieutenant, U. S. Engineers. Remained on duty, while suffering from an acute illness and very weak, and participated in the action of that date. A few days previous he had been transferred to a staff corps, but preferred to remain until the close of the campaign, taking part in several actions.
WILSON, MILDEN H. 　At Big Hole, Mont., Aug. 9, 1877. 　R—Newark, Ohio. 　B—Huron County, Ohio.	Sergeant, Company I, 7th U. S. Infantry. Gallantry in forming company from line of skirmishers and deploying again under a galling fire, and in carrying dispatches at the imminent risk of his life.
WILSON, WILLIAM. 　(2 medals of honor.) 　At Colorado Valley, Tex., Mar., 28, 1872. 　At Red River, Tex., Sept. 29, 1872. 　R—Philadelphia, Pa. 　B—Philadelphia, Pa.	Sergeant, Company I, 4th U. S. Cavalry. Gallantry in pursuit of a band of cattle thieves from New Mexico. Distinguished conduct in action with Indians.
WILSON, WILLIAM O. 　Sioux campaign, 1890. 　R—Dakota. 　B—Hagerstown, Md.	Corporal, Company I, 9th U. S. Cavalry. Bravery.
WINDOLPH, CHARLES. 　At Little Big Horn, Mont., June 25-26, 1876. 　R—Brooklyn, N. Y. 　B—Germany.	Private, Company H, 7th U. S. Cavalry. With three comrades, during the entire engagement, courageously held a position that secured water for the command.

WINDUS, CLARON A._____ | Bugler, Company L, 6th U. S. Cavalry.
At Wichita River, Tex., July 12, 1870. | Gallantry in action.
R—NR.
B—Janesville, Wis.

WINEGAR, WILLIAM W._____ | Lieutenant, Company B, 19th New York Cavalry (1st New York Dragoons).
At Five Forks, Va., Apr. 1, 1865. | Capture of battle flag.
R—NR.
B—Springport, N. Y.

WINTERBOTTOM, WILLIAM_____ | Sergeant, Company A, 6th U. S. Cavalry.
At Wichita River, Tex. July 12, | Gallantry in action.
1870.
R—NR.
B—England.

WISNER, LEWIS S._____ | First lieutenant, Company K, 124th New York Infantry.
At Spotsylvania, Va., May 12, 1864. | While serving as an engineer officer voluntarily exposed himself to the enemy's
R—NR. | fire.
B—Wallkill, N. Y.

WITCOME, JOSEPH_____ | Private, Company B, 8th U. S. Cavalry.
Arizona, August to October, 1868. | Bravery in actions and scouts against Indians.
R—NR.
B—Mechanicsburg, Pa.

WITHINGTON, WILLIAM H_____ | Captain, Company B, 1st Michigan Infantry.
At Bull Run, Va., July 21, 1861. | Remained on the field under heavy fire to succor his superior officer.
R—NR.
B—NR.

*WOLD, NELS (2140528) _____ | Private, Company I, 138th Infantry, 35th Division.
Near Cheppy, France, Sept. 26, 1918. | He rendered most gallant service in aiding the advance of his company, which
R—Minnewaukan, N. Dak. | had been held up by machine-gun nests, advancing, with one other soldier,
B—Winger, Minn. | and silencing the guns, bringing with him, upon his return, 11 prisoners.
G. O. No. 16, W. D., 1919. | Later the same day he jumped from a trench and rescued a comrade who was
 | about to be shot by a German officer, killing the officer during the exploit.
 | His actions were entirely voluntary, and it was while attempting to rush a
 | fifth machine-gun nest that he was killed. The advance of his company was
 | mainly due to his great courage and devotion to duty.
 | Posthumously awarded. Medal presented to sister, Mrs. G. H. Dale.

WOLLAM, JOHN_____ | Private, Company C, 33d Ohio Infantry.
Georgia, April, 1862. | One of 22 men (including two civilians) who by direction of General Mitchel
R—NR. | (or Buell) penetrated nearly 200 miles south into the enemy's territory and
B—Hamilton, Ohio. | captured a railroad train at Big Shanty, Ga., in an attempt to destroy the
 | bridges and track between Chattanooga and Atlanta.

WOOD, H. CLAY_____ | First lieutenant, 1st U. S. Infantry.
At Wilsons Creek, Mo., Aug. 10, 1861. | Distinguished gallantry.
R—Winthrop, Me.
B—Winthrop, Me.

WOOD, LEONARD_____ | Assistant surgeon, U. S. Army.
In Apache campaign, summer of 1886. | Voluntarily carried dispatches through a region infested with hostile Indians,
R—Massachusetts. | making a journey of 70 miles in one night and walking 30 miles the next day.
B—Winchester, N. H. | Also for several weeks, while in close pursuit of Geronimo's band and con-
Distinguished-service medal also | stantly expecting an encounter, commanded a detachment of Infantry, which
awarded. | was then without an officer, and to the command of which he was assigned
 | upon his own request.

WOOD, MARK_____ | Private, Company C, 21st Ohio Infantry.
Georgia, April, 1862. | One of 22 men (including two civilians) who by direction of General Mitchel
R—NR. | (or Buell) penetrated nearly 200 miles south into the enemy's territory and
B—England. | captured a railroad train at Big Shanty, Ga., in an attempt to destroy the
 | bridges and track between Chattanooga and Atlanta.

WOOD, RICHARD H_____ | Captain, Company A, 97th Illinois Infantry.
At Vicksburg, Miss., May 22, 1863. | Led the "volunteer storming party," which made a most gallant assault upon
R—Illinois. | the enemy's works.
B—NR.

WOODALL, ZACHARIAH_____ | Sergeant, Company I, 6th U. S. Cavalry.
At Washita River, Tex., Sept. 12, | While in command of 5 men and carrying dispatches was attacked by 125
1874. | Indians, whom he with his command fought throughout the day, he being
R—NR. | severely wounded.
B—Alexandria, Va.

WOODBURY, ERI D_____ | Sergeant, Company E, 1st Vermont Cavalry.
At Cedar Creek, Va., Oct. 19, 1864. | Capture of flag of 12th North Carolina Infantry (C. S. A.)
R—St. Johnsbury, Vt.
B—Francistown, N. H.

102444°—27——9

WOODFILL, SAMUEL_____
   At Cunel, France, Oct. 12, 1918.
   R—Bryantsburg, Ind.
   B—Jefferson County, Ind.
   G. O. No. 16, W. D., 1919.

First lieutenant, 60th Infantry, 5th Division.
While he was leading his company against the enemy, his line came under heavy machine-gun fire, which threatened to hold up the advance. Followed by two soldiers at 25 yards, this officer went out ahead of his first line toward a machine-gun nest and worked his way around its flank, leaving the two soldiers in front. When he got within 10 yards of the gun it ceased firing, and four of the enemy appeared, three of whom were shot by Lieutenant Woodfill. The fourth, an officer, rushed at Lieutenant Woodfill, who attempted to club the officer with his rifle. After a hand-to-hand struggle, Lieutenant Woodfill killed the officer with his pistol. His company thereupon continued to advance, until shortly afterwards another machine-gun nest was encountered. Calling on his men to follow, Lieutenant Woodfill rushed ahead of his line in the face of heavy fire from the nest, and when several of the enemy appeared above the nest he shot them, capturing three other members of the crew and silencing the gun. A few minutes later this officer for the third time demonstrated conspicuous daring by charging another machine-gun position, killing five men in one machine-gun pit with his rifle. He then drew his revolver and started to jump into the pit, when two other gunners only a few yards away turned their gun on him. Failing to kill them with his revolver, he grabbed a pick lying near by and killed both of them. Inspired by the exceptional courage displayed by this officer, his men pressed on to their objective under severe shell and machine-gun fire.

WOODRUFF, ALONZO_____
   At Hatchers Run, Va., Oct. 27, 1864.
   R—Ionia, Mich.
   B—Ionia, Mich.

Sergeant, Company I, 1st U. S. Sharpshooters.
Went to the assistance of a wounded and overpowered comrade, and in a hand-to-hand encounter effected his rescue.

WOODRUFF, CARLE A._____
   At Newbys Crossroads, Va., July 24, 1863.
   R—Washington, D. C.
   B—Buffalo, N. Y.

First lieutenant, 2d U. S. Artillery.
While in command of a section of a battery constituting a portion of the rear guard of a division then retiring before the advance of a corps of Infantry was attacked by the enemy and ordered to abandon his guns. Lieutenant Woodruff disregarded the orders received and aided in repelling the attack and saving the guns.

WOODS, BRENT_____
   New Mexico, Aug. 19, 1881.
   R—Louisville, Ky.
   B—Pulaski, Ky.

Sergeant, Company B, 9th U. S. Cavalry.
Saved the lives of his comrades and citizens of the detachment.

WOODS, DANIEL A._____
   At Sailors Creek, Va., Apr. 6, 1865.
   R—NR.
   B—Ohio County, W. Va.

Private, Company K, 1st Virginia Cavalry.
Capture of flag of 18th Florida Infantry (C. S. A.).

WOODWARD, EVAN M._____
   At Fredericksburg, Va., Dec. 13, 1862.
   R—NR.
   B—NR.

First lieutenant and adjutant, 2d Pennsylvania Reserve Infantry.
Advanced between the lines, demanded and received the surrender of the 19th Georgia Infantry and captured their battle flag.

WORTICK, JOSEPH_____
   At Vicksburg, Miss., May 22, 1863.
   R—NR.
   B—Fayette County, Pa.

Private, Company A, 8th Missouri Infantry.
Gallantry in the charge of the "volunteer storming party."

WORTMAN, GEORGE G._____
   Arizona, August to October, 1868.
   R—NR.
   B—Monckton, New Brunswick.

Sergeant, Company B, 8th U. S. Cavalry.
Bravery in actions and scouts against Indians.

WRAY, WILLIAM J._____
   At Fort Stevens, D. C., July 12, 1864.
   R—NR.
   B—Philadelphia, Pa.

Sergeant, Company K, 1st Veteran Reserve Corps.
Rallied the company at a critical moment during a change of position under fire.

WRIGHT, ALBERT D._____
   At Petersburg, Va., July 30, 1864.
   R—NR.
   B—NR.

Captain, Company G, 43d U. S. Colored Troops.
Advanced beyond the enemy's lines, capturing a stand of colors and its color guard; was severely wounded.

WRIGHT, ROBERT_____
   At Chapel House Farm, Va., Oct. 1, 1864.
   R—Woodstock, Conn.
   B—Ireland.

Private, Company G, 14th U. S. Infantry.
Gallantry in action.

WRIGHT, SAMUEL_____
   At Nolensville, Tenn., Feb. 15, 1863.
   R—NR.
   B—Indiana.

Corporal, Company H, 2d Minnesota Infantry.
Was one of a detachment of 16 men who heroically defended a wagon train against the attack of 125 cavalry, repulsed the attack, and saved the train.

WRIGHT, SAMUEL C._____
   At Antietam, Md., Sept. 17, 1862.
   R—Plympton, Mass.
   B—Plympton, Mass.

Private, Company E, 29th Massachusetts Infantry.
Voluntarily advanced under a destructive fire and removed a fence which would have impeded a contemplated charge.

YEAGER, JACOB F.
    At Buzzards Roost, Ga., May 11, 1864.
    R—Tiffin, Ohio.
    B—Lehigh County, Pa.

Private, Company H, 101st Ohio Infantry.
Seized a shell with fuze burning that had fallen in the ranks of his company and threw it into a stream, thereby probably saving his comrades from injury.

YORK, ALVIN C. (1910421)
    Near Chatel-Chehery, France, Oct. 8, 1918.
    R—Pall Mall, Tenn.
    B—Fentress County, Tenn.
    G. O. No. 59, W. D., 1919.

Corporal, Company G, 328th Infantry, 82d Division.
After his platoon had suffered heavy casualties and three other noncommissioned officers had become casualties Corporal York assumed command. Fearlessly leading seven men, he charged with great daring, a machine-gun nest which was pouring deadly and incessant fire upon his platoon. In this heroic feat the machine-gun nest was taken, together with 4 officers and 128 men and several guns.

YOUNG, ANDREW J.
    At Paines Cross Roads, Va., Apr. 5, 1865.
    R—Carmichaelstown, Pa.
    B—Greene County, Pa.

Sergeant, Company F, 1st Pennsylvania Cavalry.
Capture of flag.

YOUNG, CALVARY M.
    At Osage, Kans., Oct. 25, 1864.
    R—NR.
    B—Washington County, Ohio.

Sergeant, Company L, 3d Iowa Cavalry.
Gallantry in capturing General Cabell.

YOUNG, JAMES M.
    At the Wilderness, Va., May 6, 1864.
    R—Chautauqua County, N. Y.
    B—Chautauqua County, N. Y.

Private, Company B, 72d New York Infantry.
With 2 companions, voluntarily went forward in the forest to reconnoiter the enemy's position; was fired upon and 1 of his companions disabled. Private Young took the wounded man upon his back and, under fire, carried him within the Union lines.

YOUNGS, BENJAMIN F.
    At Petersburg, Va., June 17, 1864.
    R—Canada.
    B—Canada.

Corporal, Company I, 1st Michigan Sharpshooters.
Capture of flag of 35th North Carolina Infantry (C. S. A.).

YOUNKER, JOHN L.
    At Cedar Mountain, Va., Aug. 9, 1862.
    R—NR.
    B—Germany.

Private, Company A, 12th U. S. Infantry.
Voluntarily carried an order, at great risk of life in the face of a fire of grape and canister; in doing this he was wounded.

YOUNT, JOHN P.
    At Whetstone Mountains, Ariz., May 5, 1871.
    R—NR.
    B—Putnam County, Ind.

Private, Company F, 3d U. S. Cavalry.
Gallantry in action with Indians.

ZIEGNER, HERMANN.
    At Wounded Knee Creek and White Clay Creek, S. Dak., Dec. 29-30, 1890.
    R—NR.
    B—Germany.

Private, Company E, 7th U. S. Cavalry.
Conspicuous bravery.

The Distinguished-Service Cross

125

ALPHABETICAL LIST OF AWARDS OF THE DISTINGUISHED-SERVICE CROSS IN NATIONAL GROUPS

AMERICANS

[Awarded for extraordinary heroism in action under the provisions of the act of Congress approved July 9, 1918]

AMODT, MORRIS H. G. (2061663)........
Near Heurne, Belgium, Nov. 3, 1918.
R—Chicago, Ill.
B—St. Paul, Minn.
G. O. No. 44, W. D., 1919.

Sergeant, Company K, 148th Infantry, 37th Division.
He advanced alone through violent artillery fire to reconnoiter the new position to be occupied by his company beyond the l'Escaut River. He made the reconnaissance and returned with valuable information for his company commander, but was wounded while advancing to the new position with his company.

AAMOT, ARTHUR (84000).................
Near Juvigny, France, Aug. 29, 1918.
R—Saco, Mont.
B—Shelly, Minn
G. O. No. 44, W. D., 1919.

Sergeant, Company D, 126th Infantry, 32d Division.
Sergeant Aamot had sought cover in a shell hole, after a difficult advance in the face of heavy machine-gun fire, when he observed distress signals from a tank near by on which concentrated artillery and machine-gun fire was being directed by the enemy. Leaving his shelter, Sergeant Aamot proceeded through the fire to the tank, where he found a wounded man, whom he courageously carried to safety.

AARONSON, JULIUS (1811631) ....
Near Apremont, France, Oct. 7, 1918.
—Lansford, Pa.
—Russia.
O. No. 35, W. D., 1919.

Private, Company G, 109th Infantry, 28th Division.
When his company was suddenly fired upon by enemy machine guns during an advance and forced to seek shelter Private Aaronson remained in the open under a continuous shower of machine-gun bullets, caring for eight wounded men, dressing their wounds and securing their evacuation.
Oak-leaf cluster.
For the following act of extraordinary heroism in action near Apremont, France, on the same date, Private Aaronson is awarded an oak-leaf cluster to be worn with the distinguished-service cross. Having become separated from his company and wounded by a bullet which pierced his helmet, he advanced alone on a machine-gun nest across an open field in broad daylight, killed the gunner and captured two of the crew, whom he pressed into the service of carrying the wounded.

GEORGE C. (2216261)...........
Near Fey-en-Haye, France, Sept. 12, 1918.
R—Norman, Okla.
B—Cleveland County, Okla.
G. O. No. 123, W. D., 1918.

Sergeant, Company A, 357th Infantry, 90th Division.
He saved the life of a soldier who was directly under the fire of an enemy machine gun by rushing the gun, killing the gunner, and capturing the gun. His gallant conduct inspired the men of his platoon to continue the advance.

ABBOTT, ROBERT L. (748894)...........
At Chateau-Thierry, France, May 31 to June 4, 1918.
R—Sherman, Tex.
B—Springdale, Ark.
G. O. No. 132, W. D., 1918.

Corporal, Company B, 3d Supply Train, 3d Division.
Wounded in the hand by a bursting shell, he voluntarily drove a motor cycle, carrying messages and information to and from French and other headquarters. He was without sleep for 36 hours and constantly passed through hostile machine-gun and shell fire.

ABLE, HERBERT A....................
Near Ville-en-Woevre, France, Nov. 9, 1918.
R—Memphis, Tenn.
B—Memphis, Tenn.
G. O. No. 32, W. D., 1919.

First lieutenant, 324th Infantry, 81st Division.
He voluntarily went through an intense machine-gun barrage at great personal risk in order to rescue a wounded soldier.

ABEND, LOUIS (58629)..................
Near Cantigny, France, May 28-30, 1918.
R—New York, N. Y.
B—New York, N. Y.
G. O. No. 4, W. D., 1923.

Corporal, Company M, 28th Infantry, 1st Division.
When all the officers of his platoon had become casualties under a heavy artillery and machine-gun fire, Corporal Abend, displaying great bravery and initiative, voluntarily took command, effected a reorganization of the platoon which was being rapidly depleted, and held his men so well in hand that they completely repulsed two powerful counterattacks launched by the enemy.

ABERNATHY, CHARLES V..............
Near Thiaucourt, France, Sept. 14, 1918.
R—Palatka, Fla.
B—Shelby, N. C.
G. O. No. 70, W. D., 1919.

Second lieutenant, 6th Infantry, 5th Division.
Commanding the regimental pioneer platoon, he led it and the Stokes mortar platoon as infantry, and overcame a machine-gun nest, capturing several machine guns and disposing of the crew. He continued to advance under heavy shell and machine-gun fire until he fell wounded in the head, hip, and leg.

ABERNETHY, THOMAS J.
    Near Vourbin, France, July 15, 1918.
    R—West Pembroke, Me.
    B—Perry, Me.
    G. O. No. 121, W. D., 1918.

Second lieutenant, 147th Aero Squadron, Air Service.
Lieutenant Abernethy, while on patrol duty, attacked an enemy plane at close range, firing 100 rounds at a distance of from 50 to 200 yards. He followed the German ship down and saw it fall out of control, and as he turned he found five enemy planes diving at him. Without hesitation, he took the offensive and fired 200 rounds into enemy ships at not more than 15 to 20 yards. He observed tracer bullets entering the bodies of the enemy aircraft, but owing to the violence of the combat he did not have time to observe whether any of his force were shot down. Fighting vigorously, he succeeded in dispersing the enemy ships and making a safe landing within his own lines, although his own engine and plane were almost shot to pieces.

ABRAMS, ROLAND H. (1781249).
    Near Crepion, France, Nov. 3, 1918.
    R—Baltimore, Md.
    B—Baltimore, Md.
    G. O. No. 21, W. D., 1919.

Private, Company K, 313th Infantry, 79th Division.
While carrying a message from battalion to regimental headquarters, Private Abrams was seriously injured in the lungs and eyes by concentrated mustard gas, but he continued on his mission and reported back to battalion headquarters before seeking medical attention.

*ACHENBACH, MAX (1692125).
    Near Fleville, France, Oct. 5, 1918.
    R—Revere, Mass.
    B—Boston, Mass.
    G. O. No. 64, W. D., 1919.

Private, Company A, 2d Machine Gun Battalion, 1st Division.
When his gun squad had received orders to withdraw to a better position because of the intense fire to which they were subjected, the retreat was delayed because the gunner had been wounded. Private Achenbach rushed to the gun and endeavored to remove it, but while thus engaged he received severe wounds in the arms, legs, and stomach, which shortly after caused his death.
Posthumously awarded.   Medal presented to sister, Miss Emile Achenbach.

*ACHESON, WILLIAM CHALMERS.
    Near St. Juvin, France, Oct. 14, 1918.
    R—Pittsburgh, Pa.
    B—La Junta, Colo.
    G. O. No. 44, W. D., 1919.

Second lieutenant, 320th Machine Gun Battalion, 82d Division.
Seeing a flank position exposed by the nonarrival of an infantry regiment Lieutenant Acheson promptly moved his four guns to the position and held off a strong attack by the enemy. During the action 30 prisoners were taken, but nearly all his platoon had been killed or wounded. Lieutenant Acheson personally operated a gun, and, although wounded, poured a most effective fire in the ranks of the enemy, continuing until he died from loss of blood.
Posthumously awarded.   Medal presented to father, Rev. T. H. Acheson.

ACKERS, DEANE E.
    Near Soissons, France, July 22, 1918.
    R—Abilene, Kans.
    B—Abilene, Kans.
    G. O. No. 46, W. D., 1919.

Second lieutenant, 16th Infantry, 1st Division.
He assumed command of several detachments which were nearly surrounded by superior forces of the enemy, and with singular gallantry and leadership fought his way through the enemy lines back to our positions.

*ACKLEY, FRANCIS (2383027).
    Near Cunel, France, Oct. 14, 1918.
    R—Elmira, N. Y.
    B—Westfield, Pa.
    G. O. No. 37, W. D., 1919.

Corporal, Company D, 60th Infantry, 5th Division.
After his company had suffered severe losses from an enemy machine gun, Corporal Ackley volunteered to silence it single-handed. Advancing from the flank, under heavy sniping fire, he surprised the crew, killed the three gunners with his pistol, and then turned the machine gun on the enemy, covering the advance of his detachment to the position and inflicting severe losses on the hostile troops.
Posthumously awarded.   Medal presented to mother, Mrs. Rose Bailey.

ACKLEY, GEORGE E. (1530988).
    Near Montfaucon, France, Sept. 27, 1918.
    R—Pomeroy, Ohio.
    B—Pomeroy, Ohio.
    G. O. No. 59, W. D., 1919.

Sergeant, Company L, 148th Infantry, 37th Division.
While leading his platoon he stormed and destroyed two machine-gun nests. Later he again displayed utter disregard for his personal safety when he extricated his platoon from a perilous position, forcing a passage through the enemy and rejoining the remainder of the company.

ACKLEY, JAMES (49505).
    Near Chateau-Thierry, France, June 6, 1918.
    R—Siegel, Pa.
    B—Siegel, Pa.
    G. O. No. 87, W. D., 1919.

Private, Company A, 23d Infantry, 2d Division.
Even after he had been painfully wounded, Private Ackley remained on duty during the attack, performing his mission as runner until ordered to the dressing station by his commanding officer.

ACKLEY, WARD M.
    Near Very, France, Sept. 27, 1918.
    R—Portland, Oreg.
    B—Oakwood, Ohio.
    G. O. No. 44, W. D., 1919.

First lieutenant, 363d Infantry, 91st Division.
Exposing himself to heavy machine-gun and artillery fire in leading his platoon forward, Lieutenant Ackley himself captured a machine-gun nest, killing seven of the enemy with his automatic pistol.

ADAIR, RUSSEL K. (731659).
    Near Fontaines, France, Nov. 7, 1918.
    R—Tolesboro, Ky.
    B—Tolesboro, Ky.
    G. O. No. 37, W. D., 1919.

Private, Company B, 6th Infantry, 5th Division.
Private Adair, accompanied by three other soldiers, volunteered and went out under heavy machine-gun and artillery fire to rescue a wounded comrade. Failing in the first attempt, they again tried, and this time succeeded in bringing the wounded man to shelter.

ADAIR, WILLIAM R. (2239467).
    Near Montfaucon, France, Oct. 24, 1918.
    R—Fort Gibson, Okla.
    B—Tahlequah, Okla.
    G. O. No. 37, W. D., 1919.

Sergeant, Company C, 315th Field Signal Battalion, 90th Division.
After being severely gassed, he stayed at his post and ran his telephone lines through a terrific artillery barrage. He remained on duty, though he was blinded and could hardly talk, until his organization was relieved.

ADAMS, EDWARD (39001)----------------
Near Medeah Ferme, France, Oct. 5, 1918.
R—Petersburg, Va.
B—Petersburg, Va.
G. O. No. 70, W. D., 1919.

Private, Company E, 9th Infantry, 2d Division.
After all the other runners in his company had become casualties, he carried numerous messages through heavy barrage and maintained communication with battalion headquarters at a critical moment in the operations.

ADAMS, FRANK H.----------------------
Near Dormans, France, July 15, 1918.
R—Vincennes, Ind.
B—Petersburg, Ind.
G. O. No. 101, W. D., 1918.

Lieutenant colonel, 38th Infantry, 3d Division.
He was conspicuous for gallantry in action when, with courage and forcefulness and without regard to his personal safety, he voluntarily organized detachments of units other than his own and led them into effective combat.

ADAMS, FREDERICK W.------------------
Near Soissons, France, July 22, 1918.
R—Kansas City, Mo.
B—Kansas City, Mo.
G. O. No. 44, W. D., 1919.

First lieutenant, 16th Infantry, 1st Division.
During the violent fighting of July 22, 1918, he distinguished himself by his courage, judgment, and efficient leadership. After the strength of the regiment had been seriously reduced by losses, he took command of a large number of the remaining troops, disposed them in effective positions, walking up and down the lines under constant fire from the enemy, and by his example of coolness and bravery inspired his men to hold the positions they had gained.

ADAMS, HARRY J. (2177024)-----------
At Bouillonville, France, Sept. 12-13, 1918.
R—Sweetwater, Tex.
B—Girard, Ill.
G. O. 20, W. D., 1919.

Sergeant, Company K, 353d Infantry, 89th Division.
He followed a retreating German into a house in the town of Bouillonville and, ascertaining that the enemy had entered a dugout, fired the remaining two shots in his pistol through the door and ordered the surrender of the occupants. By his bravery, coolness, and confidence he captured single handed approximately 300 prisoners, including 7 officers.

ADAMS, JAMES P.---------------------
Near Blanc Mont, France, Oct 3, 1918.
R—North Augusta, S. C.
B—North Augusta, S. C.
G. O. No. 35, W. D., 1919.

First lieutenant, 6th Regiment, U. S. Marine Corps, 2d Division.
Voluntarily leading four soldiers through a heavy barrage, he attacked and killed a machine-gun crew which was enfilading his company's first line. His willingness, fearlessness, and great courage made possible the cleaning out of many more machine guns which were holding up the advance of his company.

ADAMS, JAMES S. (2410107)-----------
At Mon Plaisir Farm, north of Thiaucourt, France, Sept. 22, 1918.
R—Arlington, N. J.
B—England.
G. O. No. 127, W. D., 1918.

Private, Company M, 310th Infantry, 78th Division.
Disregarding his own personal safety, he went to an open field, swept by heavy machine-gun fire, to the assistance of an officer who had been wounded during the withdrawal of his company from a raid. He bandaged the officer's wound and carried him to shelter, thereby saving the officer's life.

*ADAMS, JOHN C.---------------------
At Crezancy, France, July 15, 1918.
R—Portland, Oreg.
B—Gaithersburg, Md.
G. O. No. 32, W. D., 1919.

Captain, 30th Infantry, 3d Division.
Captain Adams was gassed to such an extent that he vomited several times in his gas mask and had to lie on top of his dugout under heavy shell fire to get sufficient air. He refused to leave his post for medical treatment and remained to direct the movements of his company during the entire day. He was killed in action July 25, 1918.
Posthumously awarded. Medal presented to mother, Mrs. Katherine R. Adams.

ADAMS, JOHN ORA-------------------
Near Medeah Ferme, France, Oct. 3, 1918.
R—Kalispel, Mont.
B—Amery, Wis.
G. O. No. 15, W. D., 1919.

Second lieutenant, 9th Infantry, 2d Division.
He remained on duty after receiving two shrapnel wounds in the arm, and continued to lead platoon to its objective. He directed the consolidation of his position and the reorganization of his platoon before finally reporting to the aid station, eight hours after being wounded.

ADAMS, QUINCY (1896536)-----------
Near Xon Hill, France, Sept. 13, 1918.
R—Limestone, Me.
B—Limestone, Me.
G. O. No. 71, W. D., 1919.

Corporal, Company C, 320th Machine Gun Battalion, 82d Division.
Facing intense machine-gun fire, he went forward with another soldier for 200 yards and rescued a wounded infantryman who had fallen when the patrol he was with had been forced back by hostile fire. Corporal Adams showed utter disregard for personal danger.

ADAMS, ROLAND LEE----------------
Near Sommerance, France, Oct. 16-18, 1918.
R—Auburn, Ala.
B—Pine Apple, Ala.
G. O. No. 37, W. D., 1919.

First lieutenant, 327th Infantry, 82d Division.
During an attack he led his company through a heavy artillery and machine-gun fire. When he had advanced more than a kilometer in front of the other troops in the vicinity he found his flank exposed to terrific fire, which made it necessary to draw back the right wing for connection with nearest division. He personally placed each group in position and was at all times exposed to sniper and machine-gun fire. Even after being seriously gassed he volunteered for duty in the front line.

*ADAMS, SAMUEL T.----------------
Near Exermont, France, Sept. 29, 1918.
R—Kenneth, Mo.
B—Bowling Green, Ky.
G. O. No. 70, W. D., 1919.

First lieutenant, 140th Infantry, 35th Division.
After all the other officers of his company had become casualties, Lieutenant Adams reorganized his company and led it brilliantly in the assault on the town of Exermont. He was killed later during the consolidation of the new position.
Posthumously awarded. Medal presented to mother, Mrs. Sallie C. Adams.

ADAMSKI, JIOZEF (1829142)---------
Near Bois-des-Ogons, France, Oct. 9, 1918.
R—Pittsburgh, Pa.
B—Poland.
G. O. No. 44, W. D., 1919.

Corporal, Company C, 320th Infantry, 80th Division.
Rushing ahead of his advancing lines, exposed to heavy enemy fire, Corporal Adamski discovered a trail which was not being covered by enemy fire and through which men could pass in safety. He returned with this valuable information to his company commander, his action permitting his company to safely pass through the zone.

ADELHELM, HUGO C. (1393402)_____
Near Consenvoye, France, Oct. 8, 1918.
R—Chicago, Ill.
B—Chicago, Ill.
G. O. No. 93, W. D., 1919.

Sergeant, first class, Company C, 108th Engineers, 33d Division.
While a member of a working party engaged in building a bridge across the Meuse River, Sergeant Adelhelm, with another soldier, volunteered to cross the river in order to handle guy ropes, though there were two enemy machine-gun nests on the opposite bank within 100 meters of the bridge site. Though his gas mask was rendered useless in crossing the river, he remained for more than two hours under a bombardment of gas and high-explosive shells and machine-gun fire until the work was completed.

ADELSPERGER, EARL (261841)_____
Near Gesnes, France, Oct. 9, 1918.
R—Carey, Ohio.
B—Tiffin, Ohio.
G. O. No. 64, W. D., 1919.

Sergeant, Company C, 125th Infantry, 32d Division.
In charge of a reconnaissance patrol, he led his command far into enemy lines, until he encountered intense flanking machine-gun fire. Having guided his men to cover, he continued to advance alone until he discovered the source of the enemy fire, after which he returned to his men and started back to our lines. On the return journey he stopped and assisted a badly wounded man, picking him up and carrying him to safety, at the same time guiding his patrol in safety to the lines.

ADKISSON, SAMUEL P._____
Near Septsarges, France, Oct. 10, 1918.
R—Los Angeles, Calif.
B—Louisville, Ky.
G. O. No. 44, W. D., 1919.

Second lieutenant, 39th Infantry, 4th Division.
Leading his platoon through an unusually heavy barrage, Lieutenant Adkisson filled a gap on his right flank which was until then exposed. From this point he attacked and captured several machine guns and 20 prisoners. During an attack he was badly gassed and his platoon reduced in strength to six men, but he held his position under a murderous cross fire of artillery and machine guns until relieved three days later.

ADLER, JULIUS O._____
At St. Juvin, France, Oct. 14, 1918.
R—New York, N. Y.
B—Chattanooga, Tenn.
G. O. No. 44, W. D., 1919.

Major, 306th Infantry, 77th Division.
Accompanied by another officer, Major Adler was supervising the work of clearing the enemy from St. Juvin when they suddenly came upon a party of the enemy numbering 150. Firing on the enemy with his pistol, Major Adler ran toward the party, calling on them to surrender. His bravery and good marksmanship resulted in the capture of 50 Germans, and the remainder fled.

ADLER, NICK (2357310)_____
Near St. Gilles, south of Fismes, France, Aug. 8–20, 1918.
R—Eau Claire, Wis.
B—Eau Claire, Wis.
G. O. No. 139, W. D., 1918.

Wagoner, Supply Company, 120th Field Artillery, 32d Division.
Through the operations near the Vesle River, covering a period of 12 days, he delivered hot meals to the firing battery at great personal risk, due to enemy shell fire. The battery position and all roads leading to them were subject to frequent enemy fire, but this soldier on every occasion delivered the meals to the battery without delay.

ADREAN, CHARLES H. (1209634)_____
East of Ronssoy, France, Sept. 29, 1918.
R—Utica, N. Y.
B—Utica, N. Y.
G. O. No. 16, W. D., 1919.

First sergeant, Company A, 107th Infantry, 27th Division.
While commanding part of his company he was wounded in the head, but continued to direct his men, reorganizing a detachment of soldiers and establishing a line of defense in a trench. Later, while going to the assistance of some members of his command who had pushed far to the front, he was again hit in the shoulder and severely wounded. His heroic and voluntary disregard of self in order to save his comrades set a splendid example to all ranks. He has since died of the wounds received in this action.
Posthumously awarded. Medal presented to widow, Mrs. Charles H. Adrean.

ADSIT, HENRY_____
Near Le Catelet, France, Sept. 29, 1918.
R—Buffalo, N. Y.
B—St. Louis, Mo.
G. O. No. 37, W. D., 1919.

Captain, Machine Gun Company, 107th Infantry, 27th Division.
While leading a platoon of heavy machine guns through a smoke screen and under terrific fire, he suddenly became pocketed in the midst of enemy machine-gun strongholds. He personally went forward and, with the aid of bombs and the effective use of his pistol, made possible the holding of the position until a defense was organized.

AFFATATO, EPIFANIO (2450088)_____
Near Ronssoy, France, Sept. 29, 1918.
R—Brooklyn, N. Y.
B—Italy.
G. O. No. 46, W. D., 1919.

Private, Company C, 107th Infantry, 27th Division.
After being severely wounded by flying shrapnel, Private Affatato took shelter in a shell hole somewhat in advance of his company, from which he had become separated in the fog and smoke. He saved the lives of four of his wounded comrades who were occupying the shell hole by throwing live grenades, which had been tossed into the shell hole by members of his own company in the rear, into the enemy's lines.

AGHABABIAN, VARTAN (1523283)_____
Northwest of Montfaucon, France, Sept. 28, 1918.
R—Springfield, Ohio.
B—Armenia.
G. O. No. 15, W. D., 1923.

Private, first class, medical detachment, 146th Infantry, 37th Division.
He voluntarily accompanied the first attack wave of the 146th Infantry, seeking out the wounded under terrific enemy machine-gun and artillery fire, carrying them to places of safety and applying first aid until he himself was seriously wounded. This soldier's heroic conduct and devotion to his comrades greatly inspired the men of his regiment.

AHEARN, TIMOTHY (64112)_____
Near Verdun, France, Oct. 27, 1918.
R—New Haven, Conn.
B—New Haven, Conn.
G. O. No. 44, W. D., 1919.

Corporal, Company C, 102d Infantry, 26th Division.
After all of the officers and sergeants had become casualties, Corporal Ahearn took command of his company, leading it through the remainder of the day's action with great bravery and ability. Later in the day he went to the rescue of a wounded officer and succeeded in bringing him to a place of safety through terrific machine-gun fire.

AIBNER, AUGUST (560341)_____
Near Bois-du-Fays, France, Oct. 4, 1918.
R—Thermopolis, Wyo.
B—Cold Creek, Colo.
G. O. No. 2, W. D., 1920.

Sergeant, Company M, 58th Infantry, 4th Division.
Sergeant Aibner advanced in the midst of an enemy barrage and rescued two of his comrades who were lying wounded in advance of our front line and under heavy enemy fire.

AIELLO, ANTONIO (106825)_____
    Near Vierzy, France, July 19, 1918.
    R—Philadelphia, Pa.
    B—Italy.
    G. O. No. 132, W. D., 1918.

Private, Company A, 4th Machine Gun Battalion, 2d Division.
He voluntarily left the safety of the trench, advanced nearly a hundred yards in the open under heavy artillery fire, and carried back to safety a severely wounded marine.

AIRD, WILLIAM A_____
    Between the Meuse River and the Argonne Forest, France, Sept. 26 to Oct. 4, 1918.
    R—Gardena, Calif.
    B—Australia.
    G. O. No. 139, W. D., 1919.

Major, 348th Machine Gun Battalion, 91st Division.
Major Aird repeatedly reorganized the Infantry on the left flank of his division sector and by his sound tactics and good judgment kept the attack progressing against snipers, machine-gun nests, and artillery. He personally on several occasions went forward to reconnoiter and then led the attack against the discovered positions. He captured three enemy 77-millimeter guns, and by sighting through the bore turned their fire on hostile emplacements and machine-gun nests, destroying many such nests. His fearlessness, courage, and intitiative were an inspiration and example to those under him, and to his efforts much of the splendid progress on this portion of the field was due.

AKERS, EDGAR W_____
    Near Binarville, France, Sept. 28, 1918.
    R—Seattle, Wash.
    B—New York, N. Y.
    G. O. No. 20, W. D., 1919.

Second lieutenant, 308th Infantry, 77th Division.
During the advance in the Argonne Forest, France, Lieutenant Akers, having been severely wounded, led his platoon in a successful assault on two machine-gun nests, thereby aiding in the advance of his battalion sergeant.

AKINS, BENNIE A. (410274)_____
    Near Munster, Alsace, Sept. 12 and 13, 1918.
    R—Madison, Ga.
    B—Union Point, Ga.
    G. O. No. 23, W. D., 1919.

Corporal, Company A, 52d Infantry, 6th Division.
In repulsing a raid on our trenches, Corporal Akins seized an automatic rifle and pursued the Germans across No Man's Land in the face of converging fire of several enemy machine guns.

ALBERT, RALPH F. (69822)_____
    Near Trugny, France, July 23, 1918.
    R—Houlton, Me.
    B—Houlton, Me.
    G. O. No. 10, W. D., 1920.

Sergeant, Headquarters Company, 103d Infantry, 26th Division.
Although wounded by the explosion of his gun, Sergeant Albert remained on duty throughout the action in the Belleau Wood. Exposing himself to withering machine-gun fire, he carried a wounded comrade to a place of safety.

ALBRECHT, GEORGE W. (1408457)_____
    Near Septsarges, France, Oct. 24, 1918.
    R—Watertown, Ill.
    B—Shellville, Calif.
    G. O. No. 37, W. D., 1919.

Sergeant, Company G, 5th Ammunition Train, 5th Division.
When an enemy shell struck some pyrotechnics stored in the ammunition dump of his organization, he directed and assisted in the removal of inflammable material and placing the fire under control. Through his coolness and courage the destruction of a large quantity of near-by ammunition was avoided.

ALBRIGHT, FRED C_____
    Near Xammes, France, Sept. 13, 1918.
    R—Garland, Kans.
    B—Kalamazoo, Mich.
    G. O. No. 95, W. D., 1919.

Captain, 353d Infantry, 89th Division.
When the battalion on the left of his own met with such heavy fire, as it was attempting to take up its position, that it was forced to withdraw, leaving many wounded men behind, Captain Albright, with fearless disregard for his own safety, went to the adjoining area, and, under continued heavy artillery fire, cared for all the wounded who had been left there.

*ALDRICH, PERRY H_____
    Near St. Mihiel, France, Oct. 29, 1918.
    R—Essex Junction, Vt.
    B—West Kill, N. Y.
    G. O. No. 13, W. D., 1919.

First lieutenant, observer, 135th Aero Squadron, Air Service.
He, as an observer, with First Lieut. E. C. Landen, volunteered and went on an important mission for the corps commander without the usual protection. Forced to fly at an altitude of 1,000 meters because of poor visibility, soon after crossing the lines they encountered an enemy Rumpler plane and forced it to the ground. On returning they attacked another Rumpler and drove it off. After completing their mission and seeing an enemy observation tower on Lake Lachaussee, they reentered enemy territory and fired upon it. Immediately attacked by seven enemy planes (Fokker type), a combat followed, in which Lieutenant Aldrich was mortally wounded.
Posthumously awarded. Medal presented to father, Rev. Leonard Aldrich.

ALDRIDGE, JOSEPH S., Jr. (2414730)_____
    Near Vieville-en-Haye, France, Sept. 24-25, 1918.
    R—Elizabeth, N. J.
    B—Easton, Pa.
    G. O. No. 26, W. D., 1919.

Private, first class, Company B, 311th Infantry, 78th Division.
On the night of Sept. 24, 1918, Private Aldridge repeatedly carried messages between his company and battalion headquarters through a heavy barrage. He also took the place of a wounded litter bearer and assisted in bringing in wounded under shell fire.

ALE, JOHN H_____
    North of Flirey, France, Sept. 12, 1918.
    R—Muncie, Ind.
    B—Benton County, Ind.
    G. O. No. 128, W. D., 1918.

First lieutenant, 355th Infantry, 89th Division.
After having been badly wounded early in the action, losing his right hand and being wounded in both legs and the chest, he returned to his platoon and addressed the men, telling them he was unable to go with them, but that he had confidence in their ability, to go ahead without him and urged them to sustain the high reputation of the platoon, company, and battalion, thereby inspiring his men with his own personal courage to advance.

*ALEKNO, FRANK (49744)_____
    Maujouy Farm, Senoncourt, France, Apr. 21, 1918.
    R—Lawrence, Mass.
    B—Russia.
    G. O. No. 88, W. D., 1918.

Private, Company B, 23d Infantry, 2d Division.
While a member of a patrol of three men on Apr. 21, 1918, he attacked a hostile patrol of seven men, and although fatally wounded continued in action until the hostile patrol was driven back and the officer commanding it, with a noncommissioned officer, was killed. Although mortally wounded he carried a message for assistance to a point 200 yards away. Died Apr. 21, 1918.
Posthumously awarded. Medal presented to uncle, Anthony Kanopa.

ALEXANDER, ARTHUR H_____
Between Friauville and Lamorville, France, Sept. 4, 1918.
R—Wellesley, Mass.
B—Decatur, Ill.
G. O. No. 121, W. D., 1918.

First lieutenant, 96th Aero Squadron, Air Service.
While on a bombing expedition with other planes from his squadron he engaged in a running fight over hostile territory with a superior number of enemy battle planes from Friauville to Lamorville, France, he was seriously wounded in the abdomen by a machine-gun bullet, and his observer was shot through the legs. Although weak from pain and loss of blood, Lieutenant Alexander piloted his plane back to his own airdrome and concealed the fact of his injury until after his observer had been cared for.

ALEXANDER, LEON R. (2262236)_____
Near Eclisfontaine, France, Sept. 27 to Oct. 1, 1918.
R—National City, Calif.
B—National City, Calif.
G. O. No. 70, W. D., 1919.

Private, first class, Company B, 348th Machine Gun Battalion, 91st Division.
Private Alexander repeatedly carried messages between his company and battalion posts of command through heavy enemy artillery and snipers' fire.

*ALEXANDER, MEARL C_____
At Chateau-Thierry, France, June 6, 1918.
R—Birmingham, Ala.
B—Cleveland, Ohio.
G. O. No. 110, W. D., 1918.

Corporal, Headquarters Company, 5th Regiment, U. S. Marine Corps, 2d Division.
Killed in action at Chateau-Thierry, France, June 6, 1918, he gave the supreme proof of that extraordinary heroism which will serve as an example to hitherto untried troops.
Posthumously awarded. Medal presented to sister, Mrs. Anna Dean.

ALEXANDER, ROBERT_____
Near Grand Pre, France, Oct. 11, 1918.
R—Fortress Monroe, Va.
B—Baltimore, Md.
G. O. No. 35, W. D., 1919.

Major general, 77th Division.
During the advance in the Argonne Forest, and at a time when his forces were fatigued by the stress of battle and a long period of active front-line service, Major General Alexander visited the units in the front line, cheering and encouraging them to greater efforts. Unmindful of the severe fire to which he was subjected, he continued until he had inspected each group, his utter disregard of danger and inspiring example resulting in the crossing of the Aire and the capture of Grand Pre and St. Juvin.

ALEXANDER, STIRLING CAMPBELL_____
In the region of Landres-et-St. Georges, France, Oct. 6, 1918.
R—Philadelphia, Pa.
B—Philadelphia, Pa.
G. O. No. 138, W. D., 1918.

First lieutenant, pilot, 99th Aero Squadron, Air Service.
He, with Lieutenant Atwater, observer, on a photographic mission, was forced back by seven enemy pursuit planes. A few minutes later he returned over the lines and while deep in enemy territory was cut off by 12 enemy planes (Pfals scouts). He maneuvered his plane to give battle and so effectively managed the machine that he, with his observer, was able to destroy one and force the others to withdraw. With his observer severely wounded, he managed to bring his plane safely back to his own aerodrome with his mission completed.

ALEXANDER, THOMAS L_____
Near Chatel-Chehery, France, Oct. 8, 1918.
R—Charlotte, N. C.
B—Charlotte, N. C.
G. O. No. 37, W. D., 1919.

First lieutenant, 327th Infantry, 82d Division.
Leading the first attack wave, he was painfully wounded in the mouth. He continued on through the heavy fire for a distance of 10,000 yards until his objective was reached. Organizing his position and consolidating his men, he remained in command, though very weak from exhaustion and loss of blood, refusing treatment until relieved.

ALFONTE, DALLAS R_____
Near Exermont, France, Oct. 4, 1918.
R—Ingalls, Ind.
B—Ingalls, Ind.
G. O. No. 11, W. D., 1921.

Captain, 18th Infantry, 1st Division.
Although painfully wounded by a machine-gun bullet on Oct. 4, 1918, Captain Alfonte refused to be evacuated and remained on duty in an exposed position on Hill 240 until his regiment was relieved.

ALLAMONG, ISAAC F. (1289021)_____
Near Marlbrouke, France, Oct. 8, 1918.
R—Winchester, Va.
B—Stephenson, Va.
G. O. No. 44, W. D., 1919.

Corporal, Company I, 116th Infantry, 29th Division.
He displayed exceptional daring in capturing single-handed 3 guns and 20 prisoners.

ALLEN, ABE L. (56781)_____
Near Cantigny, France, May 22, 1918.
R—Leesville, La.
B—Nona, Tex.
G. O. No. 99, W. D., 1918.

Corporal, Company B, 28th Infantry, 1st Division.
During a heavy bombardment of the front line near Cantigny, France, May 22, 1918, although severely injured by the explosion of a shell which buried two comrades, he promptly and courageously dug them out with his hands and took them to shelter, being subjected all the time to severe fire of shell and shrapnel.

ALLEN, CHARLES B_____
Near Baulny, France, Sept. 26–29, 1918.
R—St. Louis, Mo.
B—St. Louis, Mo.
G. O. No. 71, W. D., 1919.

Second lieutenant, 137th Infantry, 35th Division.
Though suffering from the effects of gas, he refused to be evacuated, and upon his company commander being called upon to assume command of the battalion Lieutenant Allen displayed marked bravery and skill in leading the advance of his company. Seriously wounded, he again refused to be evacuated, remaining on duty until his command was withdrawn.

ALLEN, CHARLES W. (221668)_____
Near Bantheville, France, Oct. 23, 1918.
R—Drummond, Okla.
B—Drummond, Okla.
G. O. No. 46, W. D., 1919.

Sergeant, Company E, 357th Infantry, 90th Division.
During a fight between his company and a superior force of the enemy Sergeant Allen observed a machine gun in action on the flank of his platoon. He charged the emplacement and captured the crew of six men.

ALLEN, CLARENCE E_____
Near Crezancy, France, July 15, 1918.
R—Salt Lake City, Utah.
B—Salt Lake City, Utah.
G. O. No. 64, W. D., 1919.

First lieutenant, 30th Infantry, 3d Division.
Lieutenant Allen displayed exceptional courage, bravery, and self-sacrifice by moving about the woods in which his platoon was quartered during a heavy bombardment, placing his men in safe dugouts and rendering aid to wounded men under an intense shelling of high explosives and gas shells. He was killed while in the execution of this mission.
Posthumously awarded. Medal presented to father, Clarence E. Allen.

ALLEN, EDWARD (550550)
Near Moulin, France, July 15, 1918.
R—Philadelphia, Pa.
B—Philadelphia, Pa.
G. O. No. 5, W. D., 1920.

Sergeant, Company E, 38th Infantry, 3d Division.
At a critical time in the last German offensive he voluntarily made a reconnaissance of the enemy position at a time when his company was almost surrounded. He was mortally wounded as he returned to our lines, but lived to give information which was of vital importance in repelling the enemy attacks.
Posthumously awarded. Medal presented to father, Charles Allen.

ALLEN, FRED (56769)
Near Cantigny, France, May 28-30, 1918.
R—Chicago, Ill.
B—Chicago, Ill.
G. O. No. 87, W. D., 1919.

Sergeant, Company B, 28th Infantry, 1st Division.
During the attack and defense of Cantigny, Sergeant Allen established an automatic rifle post 75 yards in front of our lines and under heavy machine-gun and shell fire of the enemy. He had previously rendered invaluable aid in fearlessly crawling from shell hole to shell hole to aid wounded comrades.

ALLEN, GARDNER PHILIP
Near Thiaucourt, France, Oct. 9, 1918.
R—Flint, Mich.
B—Green Bay, Wis.
G. O. No. 145, W. D., 1918.

First lieutenant, Coast Artillery Corps, observer, 8th Aero Squadron, Air Service.
He, as an observer, with First Lieutenant Edward Russell Moore, pilot, took advantage of a short period of fair weather during generally unfavorable atmospheric conditions to undertake a photographic mission behind the German lines. Accompanied by two protecting planes, they had just commenced their mission when they were attacked by eight enemy planes, which followed them throughout their course, firing at the photographic plane. Lieutenant Moore, pilot, with both flying wires cut by bullets, a landing wire shot away, his elevators riddled with bullets, and both wings punctured, continued on the prescribed course, although it made him an easy target. Lieutenant Allen was thus enabled in the midst of the attack to take pictures of the exact territory assigned, and he made no attempt to protect the plane with his machine guns. Displaying entire disregard for personal danger and steadfast devotion to duty, these two officers successfully accomplished their mission.

ALLEN, JAMES
At the entrance of the harbor of Santiago, Cuba, June 2-5, 1898.
R—La Porte, Ind.
B—Indiana.
G. O. No. 14, W. D., 1925.

Lieutenant colonel, Chief Signal Officer, U. S. Volunteers.
Colonel Allen, by his persistent and untiring efforts on an unarmed transport, the Adria, and under fire of the Spanish batteries, succeeded in raising and severing two submarine cables used by the enemy.

ALLEN, JOSEPH E (1287266)
Near Brabrant, France, Oct. 8, 1918.
R—Richmond, Va.
B—Richmond, Va.
G. O. No. 37, W. D., 1919.

Corporal, Company B, 116th Infantry, 29th Division.
Corporal Allen, in company with four other soldiers, attacked and captured eight machine guns, together with their crews, in the face of determined resistance.

ALLEN, LESLIE (2267602)
Near Eclisfontaine, France, Sept. 28, 1918.
R—Hawthorne, Calif.
B—Averyville, Ill.
G. O. No. 37, W. D., 1919.

Corporal, Company K, 364th Infantry, 91st Division.
Responding to a call for volunteers, Corporal Allen, with five others, advanced 400 yards beyond their front to bring in wounded comrades. They succeeded in rescuing seven of their men and also in bringing in the dead body of a lieutenant while exposed to terrific machine-gun fire.

ALLEN, OLIVER
In the Forest of Argonne, France, Oct. 4-11, 1918.
R—New York, N. Y.
B—England.
G. O. No. 21, W. D., 1919.

Captain, 16th Infantry, 1st Division.
Captain Allen remained in command of his company after he had been wounded and, after the battalion commander had been wounded, took command of the battalion and led it forward under heavy fire from artillery and machine guns, taking and holding all objectives.

ALLEN, WILLIAM Y. (1346096)
Near Medeah Ferme, France, Oct. 3, 1918.
R—Atlanta, Ga.
B—Burke County, Ga.
G. O. No. 21, W. D., 1919.

Private, Company F, 9th Infantry, 2d Division.
Private Allen, together with four other men, charged a machine-gun nest containing three heavy machine guns, and captured the 3 guns and 20 prisoners.

ALLEY, ARVLE H. (1289728)
Near Samogneux, France, Oct. 12, and 15, 1918.
R—Radford, Va.
B—Snowville, Va.
G. O. No. 37, W. D., 1919.

Sergeant, Company M, 116th Infantry, 29th Division.
On Oct. 12 he repeatedly exposed himself while aiding wounded comrades under terrific bombardment. On Oct. 15 he fearlessly entered the wood and drove back enemy detachments before they could place machine guns on the flank of his battalion.

ALLISON, CARL OSCAR (1788002)
In the Argonne Forest, Nov. 2, 1918.
R—Washington, D. C.
B—Washington, D. C.
G. O. No. 46, W. D., 1919.

Sergeant, Company C, 312th Machine Gun Battalion, 79th Division.
In the face of direct machine-gun fire not more than 40 yards distant, he mounted a machine gun and succeeded in knocking out one of the enemy guns and taking 25 prisoners. Although seriously wounded during this most gallant exploit, he remained at his gun and ably assisted the advancing Infantry until weakness and loss of blood forced him to go to the rear.

ALLMAN, FRANK (2467794)
At Beaumont, France, Nov. 5, 1918.
R—Big Stone Gap, Va.
B—Virginia City, Va.
G. O. No. 35, W. D., 1919.

Corporal, Company A, 305th Engineers, 80th Division.
Corporal Allman and a comrade were severely wounded by the explosion of a shell. He administered first aid to his companion, himself refusing medical attention. He then carried the wounded man through the heavily shelled town to a dressing station. Although again wounded by machine-gun fire, he continued to assist the man, refusing medical attention until his comrade had been attended to.

ALMON, EARL_____
Near Fleville, France, Oct. 4, 1918.
R—Warm Springs, Ark.
B—Princeton, Ind.
G. O. No. 44, W. D., 1919.

First lieutenant, 16th Infantry, 1st Division.
After the battalion commander and all the company officers had been killed or wounded, Lieutenant Almon, battalion adjutant, took command, and although wounded by high-explosive fire, reorganized the battalion under violent artillery and machine-gun fire and continued the advance. Although twice counterattacked, he reached and held his objective.

ALOE, ALFRED_____
At Mozambique, Ilocos Norte, Philippine Islands, Sept. 22, 1900.
R—Philadelphia, Pa.
B—St. Louis, Mo.
G. O. No. 10, W. D., 1924.

First lieutenant, 12th Infantry, U. S. Army.
In the face of a heavy fire from insurgent forces, with great gallantry he advanced to within 75 yards of the enemy's line and rescued the dead body, arms, and equipment of a soldier of his command and dressed the serious wounds of another.

ALONZO, EUGENE (2284326)_____
Near Eclisfontaine, France, Sept. 27, 1918, and Oct. 4, 1918.
R—Los Angeles, Calif.
B—Philippine Islands.
G. O. No. 46, W. D., 1919.

Private, Machine Gun Company, 364th Infantry, 91st Division.
On Sept. 27 he, with two other soldiers, volunteered and went 300 yards beyond our outposts' lines, through heavy shell fire, to bring in a wounded private of his regiment. On Oct. 4 he remained in the open, under heavy shrapnel and high-explosive fire, giving first aid to our wounded men until he was wounded by shrapnel.

ALSUP, JULIAN W_____
At Blanc Mont, France, Oct. 3, 1918.
R—Nashville, Tenn.
B—Nashville, Tenn.
G. O. No. 37, W. D., 1919.

Private, 78th Company, 6th Regiment, U. S. Marine Corps, 2d Division.
When the advance of their company was held up by enfilading fire from a hostile machine-gun nest, Private Alsup, with three other soldiers, volunteered and made a flank attack on the nest with bombs and rifles, killing 3 members of the crew and capturing 25 others, together with three machine guns.

ALT, WALTER F. (1735515)_____
Near Bois-de-Loges, France, Nov. 1, 1918.
R—Buffalo, N. Y.
B—Buffalo, N. Y.
G. O. No. 49, W. D., 1922.

First sergeant, Company G, 309th Infantry, 78th Division.
With another soldier, Sergeant Alt went to the rescue of an officer and several men of his company who had been severely wounded and were lying in some bushes more than 100 yards to the front. Crossing and recrossing five times an open area which was under a heavy high-explosive shell fire and swept by machine-gun fire, he brought in the officer and wounded men of the party. The soldier who accompanied Sergeant Alt was instantly killed by machine-gun fire in the attempt to effect the rescue.

AMBRUNN, WILLIAM C. (551441)_____
Near Mezy, France, July 15, 1918.
R—Newark, N. J.
B—Bridgeport, Conn.
G. O. No. 32, W. D., 1919.

Corporal, Company H, 38th Infantry, 3d Division.
He remained at his post, bombing incoming German boats with hand grenades, although wounded during the battle.

*AMES, OLIVER, Jr_____
Near Villers-sur-Fere, France, July 27–28, 1918.
R—Boston, Mass.
B—Boston, Mass.
G. O. No. 120, W. D., 1918.

Second lieutenant, 165th Infantry, 42d Division.
During the fighting at Meurcy Farm, near Villers-sur-Fere, France, July 27–28, 1918, his heroic leadership was an inspiration to his command. He fought gallantly until on the last day he was killed while going forward voluntarily through machine-gun and sniper's fire to the assistance of his battalion commander.
Posthumously awarded. Medal presented to widow, Mrs. Oliver Ames.

*AMES, PATRICK (89024)_____
Near Landres-et-St.Georges, France, Oct. 14–18, 1918.
R—New York, N. Y.
B—Ireland.
G. O. No. 74, W. D., 1919.

Corporal, Company M, 165th Infantry, 42d Division.
Under direct fire from enemy machine guns, Corporal Ames made four trips across open ground, carrying messages during the attack on Landres-et-St. Georges. On the night of Oct. 17 he accompanied a patrol sent out to penetrate the enemy's line and showed exceptional coolness in covering the retirement of the patrol under heavy shell and machine-gun fire. On the following night, being in charge of another similar patrol, this soldier was mortally wounded, but he again displayed superior courage and leadership in withdrawing his men without further casualties.
Posthumously awarded. Medal presented to father, Michael Ames.

AMMONS, GEORGE H. (1378745)_____
Near Remonville, France, Oct. 31, 1918.
R—Chicago, Ill.
B—Central City, Iowa.
G. O. No. 37, W. D., 1919.

Sergeant, Battery A, 124th Field Artillery, 33d Division.
While in charge of the limbers and horses of a platoon sent to the front-line Infantry trenches, he, although himself wounded, took the place of a driver who had fallen from his horse. He refused medical attention until all the pieces were in position and the limbers and horses taken to a place of safety.

*AMORY, THOMAS D_____
Near Verdun, France, Oct. 2, 1918.
R—Wilmington, Del.
B—Duluth, Minn.
G. O. No. 37, W. D., 1919.

Second lieutenant, 26th Infantry, 1st Division.
Lieutenant Amory took out a patrol of 64 men, penetrating the enemy lines for the purpose of reconnoitering terrain over which an advance was to be made on the following morning. When his patrol was fired on by machine guns from all sides, this officer led three of his men forward to clear the machine-gun nests, placing the rest of his men under cover. He succeeded in overcoming one of these nests and killing the crew, but as he was advancing on another gun, located in a house about 10 yards away, he was killed by a machine-gun bullet, his last words being, "We'll take that nest or die trying."
Posthumously awarded. Medal presented to father, Edward J. Amory.

ANDERSON, ALEXANDER E_____
At Bois Colas near Sergines-et-Nesles, France, July 29, 1918.
R—New York, N. Y.
B—New York, N. Y.
G. O. No. 9, W. D., 1923.
Distinguished-service medal also awarded.

Major, 165th Infantry, 42d Division.
During a counterattack by the enemy Major Anderson, with great courage and disregard for his own safety, gathered together a small number of men and with them rushed to the support of a thinly held line. Exposing himself to concentrated machine-gun fire, he exhorted his men to stand fast, his example of courage and contempt for the heavy enemy fire greatly encouraging the men engaged and resulting in the complete repulse of the enemy forces.

ANDERSON, CARTER LEONARD.
Near St. Etienne, France, Oct. 3,
1918.
R—Cleveland, Ohio.
B—Cleveland, Ohio.
G. O. No. 53, W. D., 1920.

Private, 76th Company, 6th Regiment, U. S. Marine Corps, 2d Division.
After his platoon had been halted by machine-gun fire, Private Anderson exposed himself to heavy enemy fire to attack the enemy position. In spite of the enemy fire, he advanced and by his automatic rifle delivered an effective fire on the enemy. He was later severely wounded while defending his position against an enemy counterattack.

ANDERSON, CHARLES L. (1869638).
Near Haudiomont, France.
R—Mayville, N. Y.
B—Mayville, N. Y.
G. O. No. 32, W. D., 1919.

Sergeant, first class, 306th Field Signal Battalion, 81st Division.
He worked incessantly during a very heavy enemy barrage, keeping up the lines of communication between the regiment and battalions in the field. He was often buried in débris and knocked down by shell explosions, and was both wounded and gassed, but bravely continued his work.

ANDERSON, EMORY L. (54626).
Near Ploisy, France, July 19, 1918.
R—Nashville, Ga.
B—Jeff Davis County, Ga.
G. O. No. 87, W. D., 1919.

Private, Company K, 26th Infantry, 1st Division.
Disregarding a painful wound in the shoulder, Private Anderson kept his automatic rifle in action and remained in the advance until ordered to the rear.

*ANDERSON, ERNEST H. (2273765).
Near Moulin de Guenoville, France,
Sept. 26, 1918.
R—Anaconda, Mont.
B—Anaconda, Mont.
G. O. No. 44, W. D., 1919.

Private, first class, Company F, 1st Gas Regiment.
Private Anderson, with three other soldiers, advanced nearly 200 yards over an open hillside exposed to machine-gun fire and carried two wounded men to the protection of a near-by trench. Private Anderson has since been killed in action.
Posthumously awarded. Medal presented to mother, Mrs. Christine Anderson.

*ANDERSON, FLETCHER D. (935146).
North of Cierges, France, Oct. 5,
1918.
R—Jamestown, N. Dak.
B—Forman, N. Dak.
G. O. No. 10, W. D., 1920.

Private, medical detachment, 7th Infantry, 3d Division.
While attending wounded under fire of artillery and machine guns, Private Anderson, although himself wounded, continued to render first aid to the wounded while subject to fire, until mortally wounded by machine-gun fire.
Posthumously awarded. Medal presented to father, James N. Anderson.

ANDERSON, FRANK E. (1244634).
Service rendered under the name
Frank E. Andrea.

Sergeant, Company G, 111th Infantry, 28th Division.

ANDERSON, HARRY N. (107149).
Near Greves Farm, France, July
15, 1918.
R—Wataga, Ill.
B—Wataga, Ill.
G. O. No. 44, W. D., 1919.

Sergeant, Battery E, 10th Field Artillery, 3d Division.
He displayed notable courage in continuing to direct the fire of his piece under terrific bombardment after being twice wounded, continuing on duty until he was ordered to the rear.

ANDERSON, JOEL (R-43872).
Near Fleville, France, Oct. 4, 1918.
R—San Francisco, Calif.
B—Sweden.
G. O. No. 35, W. D., 1920.

Sergeant, Company K, 16th Infantry, 1st Division.
After all the company officers had been killed or wounded, he assisted in organizing his company under heavy fire and led a portion in the attack. When the advance was halted by heavy enfilade machine-gun fire, he led a platoon to the attack of the machine-gun nest and captured the guns and their crews, thus permitting the continuation of the advance.

*ANDERSON, LANE S.
In attack on the Hindenburg line,
Sept. 27, 1918.
R—Charleston, W. Va.
B—Richmond, Va.
G. O. No. 10, W. D., 1920.

Second lieutenant, 106th Infantry, 27th Division.
Shortly after reaching his objective his detachment was attacked and almost encircled by the enemy, who were covering the position with sweeping machine-gun fire. In the face of this fire Lieutenant Anderson, with a grenade in each hand, jumped up from the cover of his position and attacked the enemy by bombing. He had advanced but a short distance when he was struck several times by machine-gun shots. He was dragged back to his trench and died shortly after.
Posthumously awarded. Medal presented to widow, Mrs. Julia L. Anderson.

ANDERSON, LEONARD (1308354).
Near Premont, France, Oct. 8, 1918.
R—Hillsboro, Tenn.
B—Coffee County, Tenn.
G. O. No. 81, W. D., 1919.

Corporal, Company G, 117th Infantry, 30th Division.
Wounded by shell fire, he led an automatic-rifle team forward under intense enemy fire to knock out an enemy machine-gun position which had held up the advance of his company. He refused to be evacuated until the final objective had been reached.

ANDERSON, OLIVER (2261510).
Near Steenbrugge, Belgium, Oct.
31, 1918.
R—Sand Creek, Mont.
B—Grafton, N. Dak.
G. O. No. 37, W. D., 1919.

Sergeant, Company L, 362d Infantry, 91st Division.
Sergeant Anderson, with two other soldiers, attacked a strong machine-gun position from which a destructive fire had been poured into his platoon and the platoon of the flank company, wounding his lieutenant, the platoon sergeant, and many others. They drove the machine gunners from the position, thereby enabling the line to continue the advance.

ANDERSON, PAUL H. (2810466).
Near Le Grand Carre Farm, France,
Nov. 1, 1918.
R—El Paso, Tex.
B—El Paso, Tex.
G. O. No. 46, W. D., 1919.

Private, Company G, 360th Infantry, 90th Division.
After being severely wounded early in the combat, Private Anderson went on two missions for his battalion commander, which necessitated his passing through heavy machine-gun and shell fire, not mentioning the fact that he had been wounded.

ANDERSON, RICHARD C. (263992).
Near Sergy, France, July 31, 1918.
R—Manistique, Mich.
B—Manistique, Mich.
G. O. No. 81, W. D., 1919.

Private, Company M, 125th Infantry, 32d Division.
Assisted by another soldier, Private Anderson rescued a wounded comrade from within 100 feet of the enemy line, dragging him back to safety through annihilating machine-gun fire.

*ANDERSON, ROBERT B.
 At Cantigny, France, May 28-30,
  1918.
 R—Wilson, N. C.
 B—Wilson, N. C.
 G. O. No. 99, W. D., 1918.

First lieutenant, 28th Infantry, 1st Division.
In the attack and defense at Cantigny, France, May 28–30, 1918, he showed
 utter disregard for his personal safety in leading his command forward in spite
 of artillery and machine-gun fire. While directing the security of his men
 after the advance and in order to make certain that they were protected
 first, he himself was killed.
Posthumously awarded. Medal presented to father, Dr. W. S. Anderson.

*ANDERSON, THOMAS B.
 Near Courmont, France, July 30,
  1918, and near Baslieux, France,
  Sept. 5, 1918.
 R—Latrobe, Pa.
 B—Latrobe, Pa.
 G. O. No. 1, W. D., 1926.

Major, 110th Infantry, 28th Division.
Leading his battalion in an attack, he refused to be evacuated when wounded
 and gassed until the objective had been gained and the position consolidated.
 His courage was an inspiration to his men. Five weeks later he was killed
 while leading a patrol across open ground, swept by heavy fire, against an
 enemy machine-gun nest.
Posthumously awarded. Medal presented to widow, Mrs. Ruth Anderson.

ANDERSON, WALTER N.
 Near Very, France, Sept. 26, 1918.
 R—Berkeley, Calif.
 B—Marionette, Wis.
 G. O. No. 37, W. D., 1919.

First lieutenant, 363d Infantry, 91st Division.
With the aid of an enlisted man, he attacked a nest of enemy snipers and succeeded in killing two, wounding one, and taking the remaining two as
 prisoners.

ANDERSON, WILLIAM A. (2087317).
 At Chipilly Ridge, France, Aug. 9,
  1918.
 R—Chicago, Ill.
 B—Chicago, Ill.
 G. O. No. 128, W. D., 1918.

Private, Company B, 131st Infantry, 33d Division.
He rendered service as stretcher bearer under heavy shell fire, continuing on
 duty for 48 hours, until complete exhaustion compelled him to be evacuated.

ANDERSON, WILLIAM L. (550049).
 During the Meuse-Argonne offen-
  sive, Oct. 1 to 16, 1918.
 R—Cartersville, Va.
 B—Cartersville, Va.
 G. O. No. 16, W. D., 1920.

Mechanic, Company C, 38th Infantry, 3d Division.
Mechanic Anderson acted as company runner and carried numerous messages
 on routes exposed to artillery and machine-gun fire. When on the point of
 collapse from exhaustion and ordered to the rear, he refused to leave his
 company. His individual efforts in maintaining communication regardless
 of the great dangers to which he was exposed were of great value to his company commander.

*ANDES, JAMES COWAN.
 Near Soissons, France, July 19, 1918.
 R—New York, N. Y.
 B—Knoxville, Tenn.
 G. O. No. 15, W. D., 1919.

Second lieutenant, 16th Infantry, 1st Division.
He fearlessly led his platoon in the face of heavy machine-gun fire to a cave in
 which several hundred Germans had taken shelter. Rather than subject
 any of his men to such extreme danger, he entered the cave alone and
 demanded the surrender of the enemy and was killed as a result of this
 heroic act.
Posthumously awarded. Medal presented to father, George S. Andes.

ANDES, ROY B. (1785940).
 At Bois de Beuge, north of Mont-
  faucon, France, Sept. 29, 1918.
 R—East Petersburg, Pa.
 B—Sporting Hill, Pa.
 G. O. No. 15, W. D., 1923.

Private, first class, Headquarters Company, 316th Infantry, 79th Division.
Voluntarily accompanying Lieutenant Goetz, under intense enemy machine-
 gun and artillery fire, he made his way over rough and broken terrain to the
 advanced position held by a provisional battalion of his regiment with a
 message to the battalion commander to immediately withdraw his forces to
 a position in the rear, as an American barrage was about to fall, and a short
 time later did fall, on the woods which was formerly occupied by the battalion. The heroic conduct of Private Andes and the officer he accompanied
 saved the lives of many of his comrades.

ANDRE, CHARLES H.
 Near St. Juvin, France, Oct. 14-15,
  1918.
 R—Detroit, Mich.
 B—Ithaca, Mich.
 G. O. No. 71, W. D., 1919.

First lieutenant, 305th Machine Gun Battalion, 77th Division.
Coming face to face with a large number of the enemy while he was on a reconnoissance patrol, he opened fire with his revolver and continued to advance,
 demoralizing the enemy and proving instrumental in the capture of 50
 prisoners. During a counterattack he voluntarily went forward with a
 machine gun in the face of heavy fire, and operated it to such good effect as
 to break two waves of the advancing enemy.

*ANDREA, FRANK E. (1244634).
 Near Crezancy, France, July 16, 1918.
 R—Pittsburgh, Pa.
 B—Sioux City, Iowa.
 G. O. No. 102, W. D., 1918

Sergeant, Company G, 111th Infantry, 28th Division.
Sergeant Andrea was told by a runner that an enemy patrol had captured two
 ambulances containing American wounded on the road east of his position.
 He organized a relief party, personally commanded it, drove the enemy to
 rout, recovered the ambulance and the wounded men, and brought them
 back to our lines.
Posthumously awarded. Medal presented to widow, Mrs. Josephine Anderson.

*ANDRES, JOHN, Jr. (553061).
 Near Cierges, France, Oct. 4, 1918.
 R—Forks, N. Y.
 B—Cheektowago, N. Y.
 G. O. No. 2, W. D., 1920.

Private, Company C, 8th Machine Gun Battalion, 3d Division.
As the company runner he repeatedly carried important messages across areas
 swept by artillery and machine-gun fire. Upon returning after delivering a
 message he picked up an abandoned machine gun of a wounded comrade
 and bravely bore it forward into the attack 500 meters until killed by enemy
 machine-gun fire.
Posthumously awarded. Medal presented to father, John Andres, sr.

ANDREW, FLYNN LAMBERT ANTHONY.
 Near Landres-et-St. Georges, France,
  Oct. 30, 1918.
 R—Denver, Colo.
 B—Denver, Colo.
 G. O. No. 126, W. D., 1919.

First lieutenant, observer, 104th Aero Squadron, Air Service.
Unable to complete a photographic mission, owing to motor trouble, Lieutenant
 Andrew, with his pilot, made a reconnaissance behind the German lines.
 They dispersed a battalion of enemy troops, and, although twice attacked
 by enemy patrols, drove them off and in each case brought down one enemy
 plane. They remained in the air until their motor failed completely.

ANDREWS, CHARLES F_____
At Paranas, Samar, Philippine Islands, Mar. 11, 1900.
R—San Francisco, Calif.
B—Fort Adams, R. I.
G. O. No. 13, W. D., 1924.

Second lieutenant, 43d Infantry, United States Volunteers.
During an attack by a greatly superior number of insurgents on the town which he was occupying with a small detachment, Lieutenant Andrews displayed exceptional courage, initiative, and highest qualities of leadership in successfully repulsing a desperate attack of the enemy, most of whom were armed with bolos.

ANDREWS, MYRON MORRIS_____
Near Soissons, France, July 19, 1918.
R—West Hartford, Conn.
B—West Hartford, Conn.
G. O. No. 100, W. D., 1918.

First lieutenant, 26th Infantry, 1st Division.
Besides inspiring his men and by his conduct in the fighting near Soissons, France, he promptly disposed his company to cover a battalion front in a critical situation on July 19, 1918, and by fearless exposure under fire successfully directed the operations of the command.

*ANDREWS, SAM E_____
Near Montfaucon, France, Sept. 26, 1918.
R—Ozark, Ala.
B—Ozark, Ala.
G. O. No. 70, W. D., 1919.

First lieutenant, 145th Infantry, 37th Division.
Lieutenant Andrews displayed brilliant courage and leadership in leading his platoon against and capturing a strong enemy machine-gun nest. In this exploit he was killed, but his notable coolness and determination furnished an inspiration to his men.
Posthumously awarded. Medal presented to mother, Mrs. Lettie Andrews.

ANDREWS, WALTER GRESHAM_____
Near Vendhuile, France, Sept. 29-30, 1918.
R—Buffalo, N. Y.
B—Evanston, Ill.
G. O. No. 72, W. D., 1920.

Captain, 107th Infantry, 27th Division.
Although severely wounded in the arm on the morning of Sept. 29, Captain Andrews gallantly led his company throughout the attack on the Hindenburg line. After the advance he made a personal reconnaissance under heavy shell and machine-gun fire and organized a section of a trench within 20 yards of the enemy line. This position was held against enemy grenade and machine-gun fire until his company was relieved on the afternoon of Sept. 30.

*ANDRYKOWSKI, VICTOR (30812)_____
Near Courmont and St. Martin, France, July 31 to Aug. 3, 1918.
R—Saginaw, Mich.
B—Poland.
G. O. No. 20, W. D., 1919.

Private, Company G, 125th Infantry, 32d Division.
Throughout the battle to force passage of the Ourcq River and capture the heights beyond, Private Andrykowski, a stretcher bearer, worked day and night, evacuating wounded under heavy artillery and machine-gun fire. On August 3, under violent shell fire, opposite Mont St. Martin, he made repeated trips between the firing line and the dressing station until he was killed by a shell.
Posthumously awarded. Medal presented to father, Francis Andrykowski.

ANGELL, HOWARD M. (2260613)_____
Near Gesnes, France, Sept. 29, 1918.
R—Bingham, Utah.
B—Salt Lake City, Utah.
G. O. No. 27, W. D., 1919.

Sergeant, Company F, 362d Infantry, 91st Division.
He was wounded during the advance of his regiment on Gesnes, but refusing medical treatment he continued in command of his section until next morning, when he was ordered to the dressing station by his battalion commander.

ANGELLY, HENRY M (2220218)_____
During the St. Mihiel offensive, Sept. 14, 1918.
R—Boswell, Okla.
B—Gasville, Okla.
G. O. No. 72, W. D., 1920.

Private, Company D, 358th Infantry, 90th Division.
Private Angelly, with four other men, volunteered to cross a valley to the woods opposite and silence machine guns which had held up the advance of his company. In the face of heavy enemy fire this small group accomplished its mission, thus enabling the company to cross the valley without further loss. Private Angelly was severely wounded in the performance of this act.

ANGELO, JOSEPH T. (243496)_____
Near Cheppy, France, Sept. 26, 1918.
R—Pennsgrove, N. J.
B—Lotterman, Pa.
G. O. No. 44, W. D., 1919.

Private, first class, Headquarters Company, 1st Brigade Tank Corps.
Within 40 meters of the German machine guns he carried his wounded commanding officer into a shell hole and remained with him under continuous shell fire for over an hour, except when he twice carried orders to passing tanks.

*ANGIER, ALBERT E_____
Near Revillon, France, Sept. 14, 1918.
R—Waban, Mass.
B—Waban, Mass.
G. O. No. 32, W. D., 1919.

First lieutenant, 308th Infantry, 77th Division.
Although wounded, he continued to lead his men in an attack. By his gallant example he urged them forward through enemy wire to their objective. Even when mortally wounded he continued to direct the consolidation of his position, refusing medical attention in favor of others who had a better chance to live than himself.
Posthumously awarded. Medal presented to father, George M. Angier.

ANKUDOVITCH, WILLIAM D. (1253167)___
Near Courville, France, Aug. 29, 1918.
R—West Hazleton, Pa.
B—Harwood, Pa.
G. O. No. 78, W. D., 1919.

Corporal, Battery A, 109th Field Artillery, 28th Division.
When the battery position was being subjected to concentrated enemy fire a shell burst near Corporal Ankudovitch's gun, wounding him and four other members of the gun crew. Disregarding his own injuries, Corporal Ankudovitch assisted in caring for the other men and carrying them to the rear until he fell, exhausted from loss of blood.

*ANTES, JAY LE R. (2693)_____
At Cantigny, France, May 28-29, 1918.
R—Norristown, Pa.
B—Norristown, Pa.
G. O. No. 15, W. D., 1921.

Private, medical detachment, 28th Infantry, 1st Division.
Private Antes fearlessly exposed himself to barrage and machine-gun fire to perform his duty as a stretcher bearer. In order that the suffering of wounded might be relieved and lives saved, with unselfish heroism he left the security of the trench to go to wounded in a machine-gun emplacement, and while performing this noble duty was killed.
Posthumously awarded. Medal presented to mother, Mrs. Catherine Antes.

ANTHONY, CLEM (281234)_____
Near Juvigny, north of Soissons, France, Aug. 30, 1918.
R—Newaygo, Mich.
B—Kansas City, Kans.
G. O. No. 20, W. D., 1919.

Private, Company L, 128th Infantry, 32d Division.
When a retirement had been ordered, he was the last to leave his post, fearlessly exposing himself to fire from machine guns and snipers to bring in a wounded soldier, together with his automatic rifle and ammunition. Throughout the engagement his conduct under fire furnished an example of coolness and courage to his comrades.

ANTHONY, GEORGE W. (2256998)........
Near Rembercourt, France, Oct. 9–Nov. 11, 1918.
R—Blackfoot, Idaho.
B—Portage, Utah.
G. O. No. 70, W. D., 1919.

Private, Company C, 56th Infantry, 7th Division.
As company and platoon runner he worked tirelessly and unceasingly without regard to personal safety, carrying messages both day and night under violent machine gun and artillery fire.

°ANTHONY, HAROLD B. (2260112)........
At Bois-de-Vary, France, Sept. 26, 1918.
R—Miles City, Mont.
B—Spencer, Ind.
G. O. No. 59, W. D., 1919.

Supply sergeant, Company D, 362d Infantry, 91st Division.
Sergeant Anthony, while leading a small detachment operating on the flank of his company, suddenly came under heavy machine-gun fire. Alone, he crawled up close to the machine gun, killed the gunner, and captured four prisoners. Again, at Eclisfontaine, France, Sept. 29, 1918, the company was held up by machine-gun fire from front and flank. Sergeant Anthony spotted the machine-gun nest. While attempting to reach an automatic squad to point out the hostile gun he was killed by the machine-gun fire.
Posthumously awarded. Medal presented to father, Rev. A. H. Anthony.

ANTHONY, ROY C. (2178410)..........
Near Remonville, France, Nov. 1, 1918.
R—Currington, Mo.
B—Boydsville, Mo.
G. O. No. 44, W. D., 1919.

Sergeant, Company B, 354th Infantry, 89th Division.
Leading his platoon against perilous fire, he showed great courage in advancing and breaking down resistance of the enemy machine guns and artillery. Although twice wounded, he refused treatment until the company was relieved. His efforts were mainly responsible for the successful gaining of all objectives.

°ARCHER, JOSEPH D. (1781968)........
Near Montfaucon, France, Oct. 11, 1918.
R—Philadelphia, Pa.
B—Philadelphia, Pa.
G. O. No. 13, W. D., 1919.

Private, first class, Company D, 117th Ammunition Train, 42d Division.
Private Archer, on duty at the ammunition dump of the 42d Division when it was violently bombarded by the enemy, volunteered to assist another soldier, who was wounded. He was killed shortly before reaching the dressing station.
Posthumoulsy awarded. Medal presented to father, Joseph J. Archer.

°ARKMAN, FRANK (1429432)..........
Near Bois-de-la-Naza, France, Oct. 5, 1918.
R—Bellingham, Minn.
B—Glenwood Springs, Colo.
G. O. No. 59, W. D., 1919.

Private, Company L, 305th Infantry, 77th Division.
With utter disregard for his personal safety he went forward with three other soldiers, in the face of heavy machine-gun and grenade fire, and brought back five seriously wounded men to a first-aid station. He displayed bravery, coolness, and good judgment in effecting the rescue.
Posthumously awarded. Cross on display at the Smithsonian Institution, Washington, D. C., pending delivery to next of kin when and if located.

°ARMIJO, MARCUS B. (251354)..........
Near Fismes, France, Aug. 5, 1918.
R—El Paso, Tex.
B—Rincon, N. Mex.
G. O. No. 116, W. D., 1918.

Private, Company C, 125th Infantry, 32d Division.
While his company was under a heavy barrage fire, Private Armijo was hit by shell and both his legs blown off. Private Armijo lifted himself up on his elbow and rolled and smoked cigarettes. By this display of nerve he conveyed to his comrades an unconquerable spirit of fearlessness, pluck, and will power.
Posthumously awarded. Medal presented to widow, Mrs. Maria Armijo.

°ARMISTEAD, JOSEPH G. (730778)........
Near Thiaucourt, France, Sept. 14, 1918.
R—San Antonio, Tex.
B—Pembroke, Ky.
G. O. No. 87, W. D., 1919.

Corporal, Headquarters Company, 6th Infantry, 5th Division.
After his Stokes mortar had been destroyed by enemy shell fire and the officer in charge severely wounded, Corporal Armistead formed his squad as riflemen and led them forward against a machine-gun nest which was firing on our line from the flank. In attempting this bold feat Corporal Armistead was killed.
Posthumously awarded. Medal presented to sister, Mrs. Mary E. Burrus.

ARMSTRONG, EDWARD V. (58494)........
Near Seicheprey, France, Mar. 28, 1918.
R—Marianna, Pa.
B—Irwin, Pa.
G. O. No. 126, W. D., 1919.

Private, Company L, 28th Infantry, 1st Division.
He was a member of a patrol consisting of an officer and four men who, with great daring, entered a dangerous portion of the enemy trenches, where they surrounded a party nearly double their own strength, captured a greater number than themselves, drove off an enemy rescuing party, and made their way back to our lines with four prisoners, from whom valuable information was obtained.

ARMSTRONG, RODNEY M..............
Near Rembercourt and Charey, France, Nov. 4, 1918.
R—Topeka, Kans.
B—Topeka, Kans.
G. O. No. 7, W. D., 1919.

First lieutenant, pilot, 163th Aero Squadron, Air Service.
As pilot of a De Haviland 4 plane, he flew an infantry contact mission over the lines of the 7th Division, Nov. 4, 1918. Owing to low clouds and rain, he crossed the line at 1,000 feet in order to enable his observer to locate the position more accurately. While on the enemy's side he was wounded by an explosive bullet. In spite of his wound and weakness, he continued his mission, coming down to within 500 feet of the enemy's machine guns and troops until his observer had signaled him that the mission was completed.

ARMSTRONG, THOMAS (1208024)........
East of Ronssoy, France, Sept. 29, 1918.
R—Brooklyn, N. Y.
B—Ireland.
G. O. No. 142, W. D., 1918.

Sergeant, Company H, 106th Infantry, 27th Division.
During the operations against the Hindenburg line he alone attacked and drove back an enemy patrol. Later, when his captain was wounded, he remained with him and killed two Germans who attacked them.

ARNOLD, ALBERT C..............
Near St. Juvin, France, Oct. 10–11, 1918.
R—Jacksonville, Fla.
B—Jefferson County, Ky.
G. O. No. 46, W. D., 1919.

First lieutenant, 326th Infantry, 82d Division.
On the night of Oct. 10, Lieutenant Arnold was painfully wounded while reconnoitering the enemy's positions, but continued with his mission after receiving first aid from an accompanying soldier. Early in the morning of Oct. 11, with the assistance of one soldier he silenced a machine gun which was enfilading our line. He was again wounded while accomplishing this mission, but continued his efforts until another machine gun had been put out of action through his personal direction. He remained on duty with his men until he became so weak from loss of blood and exposure to gas that he collapsed and was carried from the field.

ARNOLD, ALFRED C._____
Near Medeah Ferme, France, Oct. 4–9, 1918.
R—Middletown, Conn.
B—St. Johnsburg, Vt.
G. O. No. 44, W. D., 1919.

Lieutenant colonel, 9th Infantry, 2d Division.
This officer displayed the most inspiring personal bravery and cool judgment under massed counterattacks, heavy machine-gun fire, and intensive artillery barrage. Performing many gallant acts beyond those in the line of his duty, he held his line, maintained liaison under difficult conditions with the unit on his right, and at a critical time repelled a serious counterattack.
Oak-leaf cluster.
In addition to the distinguished-service cross, Lieutenant Colonel Arnold is awarded an oak-leaf cluster for the following act of extraordinary heroism in action near Thiaucourt, France, Sept. 12, 1918: At a critical moment in the advance he went through a barrage and stopped the assaulting lines of a neighboring unit which had failed to halt on their objective and were in danger from their own barrage. His coolness in walking up and down the line under heavy enemy bombardment inspired confidence and restored order in a wavering line.

ARNOLD, DEWEY G. (57713)_____
Near Nonsard, France, Sept. 12, 1918.
R—Roebuck, S. C.
B—Cherokee, S. C.
G. O. No. 37, W. D., 1919.

Corporal, Company G, 28th Infantry, 1st Division.
Accompanied by another soldier, he attacked and destroyed an enemy machine-gun nest, using only his rifle and bayonet.

*ARNOLD, HOWARD W._____
At Villers-sur-Fere, France, July 28, 1918.
R—New York, N. Y.
B—Elberon, N. J.
G. O. No. 15, W. D., 1923.

First lieutenant, 165th Infantry, 42d Division.
With complete disregard for his own safety he left the shelter of a shallow trench and in full view of the enemy crossed a field swept by a hail of enemy machine-gun and rifle fire to warn his company commander of a strong enemy force which threatened to isolate the companies holding the advanced position. Although warned by his company commander of his danger he calmly returned to his platoon's position under the hottest enemy fire. As he was about to enter a place of shelter he was killed by enemy fire.
Posthumously awarded. Medal presented to father, Oscar M. Arnold.

ARNOLD, WALTER F. (43957)_____
Near Exermont, France, Oct. 4, 1918.
R—Philadelphia, Pa.
B—Newport News, Va.
G. O. No. 24, W. D., 1920.

Private, Company I, 16th Infantry, 1st Division.
Private Arnold, while passing through an enemy machine-gun barrage, located an enemy machine-gun position. He single-handedly rushed the position, captured the gun, and forced the crew to surrender.

ARPIN, EDMUND P._____
Near Gesnes, France, Oct. 7, 1918.
R—Grand Rapids, Wis.
B—Grand Rapids, Wis.
G. O. No. 47, W. D., 1921.

First lieutenant, 128th Infantry, 32d Division.
He volunteered to lead and led a platoon of 41 men in an attack on Hill 269. Although all but four became casualties, this small group, under the leadership of Lieutenant Arpin, continued on its mission, took the hill, and held it for some time without hope of reinforcements.

ARRANTS, WILLIAM R._____
In the Bois-de-la-Cote, Lemont, France, Sept. 28, 1918, and near Nantillois, France, Oct. 5, 1918.
R—Decatur, Tenn.
B—Decatur, Tenn.
G. O. No. 140, W. D., 1918.

First lieutenant, Medical Corps, attached to 317th Infantry, 80th Division.
He with his battalion aid unit accompanied his battalion into action in the Bois-de-la-Cote, Lemont, and promptly opened his aid station within 100 yards of the front line, where he worked all night under continuous fire giving aid to the wounded. When there was a shortage of stretcher bearers he assisted in bringing in the wounded. Under intense fire he undertook to locate the ambulance dressing station and personally directed the evacuation of wounded to it. In the attack from the Bois-du-Fays, Oct. 5, he again went with the attacking troops and opened a first-aid station in an old cellar with no cover. Under an intense barrage of shrapnel and high-explosive shells he performed the most devoted service in attending the wounded, working continuously for nine hours until after his unit had been ordered to retire.

*ARSENAULT, LUCIAN L. (67164)_____
Near Torcy, France, July 18, 1918.
R—Mexico, Me.
B—Canada.
G. O. No. 9, W. D., 1923.

Private, Company B, 103d Infantry, 26th Division.
When he declined to leave his platoon for treatment, though severely wounded in the leg by machine-gun fire, continuing with the advance of his platoon, he was again wounded and again refused to go to the rear but continued with the assaulting line until he received a third wound, which caused instantaneous death. His high courage, unselfish devotion to duty, and utter disregard for his own safety raised the morale of his company to a high pitch and spurred them on to great endeavors.
Posthumously awarded. Medal presented to father, Leon Arsenault.

ARSENAULT, THOMAS (1679188)_____
Near Bazoches, France, Aug. 27, 1918.
R—Newcomb, N. J.
B—Canada.
G. O. No. 87, W. D., 1919.

Private, first class, Company G, 306th Infantry, 77th Division.
With an utter disregard for his personal safety, Private Arsenault rescued a wounded officer and carried him across an area swept by a withering machine-gun fire to a dressing station, preventing the capture of a wounded man by the enemy.

ARTHUR, DOGAN H.
In the St. Mihiel salient, Sept. 12, 1918.
R—Highland Park, Mich.
B—Union, S. C.
G. O. No. 126, W. D., 1919.

Second lieutenant, pilot, 12th Aero Squadron, Air Service.
Lieutenant Arthur, pilot, and Second Lieut. Howard T. Fleeson, observer, executed a difficult mission of Infantry contact patrol without protection of accompanying battle planes on the first day of the St. Mihiel offensive. After being driven back twice by a patrol of nine enemy planes, they courageously made a third attempt in the face of a third attack by the same planes, found the American lines, and after being shot down, but falling uninjured in friendly territory, communicated their valuable information to headquarters.
Oak-leaf cluster.
A bronze oak leaf is awarded Lieutenant Arthur for the following acts of extraordinary heroism in action Oct. 18–30, 1918: On Oct. 18, 1918, while on Artillery reglage, he and his observer were attacked by four enemy planes. His observer's guns were jammed, but Lieutenant Arthur, with splendid courage and coolness, outmaneuvered the hostile aircraft and escaped, although they followed his plane to within 25 meters of the ground, badly damaging it by machine-gun fire. On Oct. 30, 1918, his was one of a formation of nine planes which were to take photographs in German territory. Before the lines were reached six planes dropped out, but the remaining three entered the German lines, although they observed several large formations of enemy planes in the near vicinity. When they were 12 kilometers within the German lines, they were attacked by 18 enemy Fokkers. Regardless of his own safety, Lieutenant Arthur engaged these planes in order to allow his companions to escape, and turned toward his own lines only when he saw them shot down. Then he fought his way home, and in the fight which ensued his observer shot down two enemy planes.

ASCHER, OSCAR (1698293).
In the Argonne Forest, France, Oct. 6, 1918.
R—New York, N. Y.
B—Bohemia.
G. O. No. 64, W. D., 1919.

Corporal, Company K, 305th Infantry, 77th Division.
Engaged as messenger, Corporal Ascher made repeated trips to the most advanced positions, each time under the severest of machine-gun fire. On one occasion he volunteered and carried a message to a platoon sergeant when the latter was actually engaged in charging the enemy. Corporal Ascher accomplished this hazardous mission by verbally delivering the instructions contained in the message, thereby materially aiding in the success of the attack.

*ASELTON, ERNEST K.
Near St. Etienne, France, Oct. 8, 1918.
R—St. Paul, Minn.
B—Portland, Mich.
G. O. No. 70, W. D., 1919.

Private, 76th Company, 6th Regiment, U. S. Marine Corps, 2d Division.
He volunteered and, under extremely heavy shell and machine-gun fire, established liaison for his company, bringing reinforcements to the line at a critical time, and thereby assisting materially in repelling a hostile counterattack. He was killed later during this attack.
Posthumously awarded. Medal presented to father, Isaac Aselton.

ASH, HAROLD JAMES (9589).
Near Bois-de-Montrebeau, France, Oct. 4, 1918.
R—New York, N. Y.
B—New York, N. Y.
G. O. No. 46, W. D., 1919.

Sergeant, 345th Battalion, Tank Corps.
Driving his tank in the face of a 77-millimeter gun, Sergeant Ash continued with his mission until his tank was destroyed. He remained with the tank until a machine-gun nest was destroyed, and then accompanied the tank commander on foot through severe fire, killing two snipers with his pistol, while the commander was disabling machine and antitank guns, after which he returned to his lines.

ASHBURN, ISAAC S.
Near Fey-en-Haye, France, during the attack on the St. Mihiel salient, Sept. 12, 1918.
R—Greenville, Tex.
B—Farmersville, Tex.
G. O. No. 128, W. D., 1918.

Major, 358th Infantry, 90th Division.
After being practically paralyzed for more than an hour from a wound in the neck, he resumed command of his battalion and continued to lead it with exceptional daring and effect until he was incapacitated by a second wound two days later.

ASHCRAFT, EUGENE M. (1565755).
Near Exermont, France, Oct. 6, 1918.
R—Newcastle, Ind.
B—Williamstown, Ky.
G. O. No. 44, W. D., 1919.

Private, Company E, 28th Infantry, 1st Division.
Responding to a call for volunteers, he proceeded 400 yards ahead of his platoon to ascertain the location of the enemy. The mission was accomplished through an extremely heavy fire, but Private Ashcraft, after obtaining his information, successfully returned over the same ground, and made his report to the platoon commander.

ATCHAVIT, CALVIN (2806696).
Near Fey-en-Haye, France, Sept. 12, 1918.
B—Walters, Okla.
R—Walters, Okla.
G. O. No. 87, W. D., 1919.

Private, Company A, 357th Infantry, 90th Division.
During the attack of his company, though he had been severely wounded in his right arm, Private Atchavit shot and killed one of the enemy and captured another.

ATKINS, MARVIN L.
Near St. Souplet, France, Oct. 18, 1918.
B—Lakeville, Conn.
B—Lakeville, Conn.
G. O. No. 44, W. D., 1919.

First lieutenant, 105th Infantry, 27th Division.
Continuing in action after being gassed, Lieutenant Atkins displayed exceptional personal bravery when the advance of his platoon was checked by heavy machine-gun fire in seizing the gun of a wounded soldier and attacking a machine gun, which he silenced and captured. Under the inspiration of this fearless act his company overcame several other machine-gun emplacements and reached the objective.

ATKINS, MOSES D.
Near St. Mihiel, France, Sept. 12–13, 1918.
R—Leavenworth, Kans.
B—Galveston, Tex.
G. O. No. 70, W. D., 1919.

Captain, 353d Infantry, 89th Division.
Though wounded at the outset of the attack, he continued to lead his company with skill and entire disregard of danger, until during the attack on Thiaucourt he was wounded a second time so severely that he was unable to proceed.

ATKINSON, JOSEPH T. (1537922)_____
Near Heuvel, Belgium, Nov. 2, 1918.
R—Cleveland, Ohio.
B—Freeport, Pa.
G. O. No. 37, W. D., 1919.

Private, Company B, 112th Engineers, 37th Division.
Private Atkinson, with two other soldiers, crossed the Scheldt River, after two attempts, and succeeded in stretching a line for a bridge across the stream. They were discovered and fired upon by the enemy, but they continued at work driving stakes, and made a second trip across the river to obtain wire, despite the fact that a violent artillery barrage had been laid down on their position.

ATKINSON, RALPH (98612)_____
Near Landres-et-St. Georges, France,
Oct. 16, 1918.
R—Camden, Ark.
B—Montgomery, Ala.
G. O. No. 131, W. D., 1918.

Sergeant, Headquarters Company, 167th Infantry, 42d Division.
During the attack on the Cote-de-Chatillon, Sergeant Atkinson, in command of the Stokes Mortar Platoon, together with three other soldiers, was advancing with the first wave of the assault, when, on nearing the objective, he discovered about 250 of the enemy forming for a counterattack. At this juncture he and his party advanced with a Stokes mortar, under heavy fire, to a position where he could get a fair field of fire, set up the mortar, and opened a murderous fire on the approaching enemy, dispersing them in every direction. His quick action, good judgment, and leadership undoubtedly not only broke up the enemy counterattack but inflicted severe losses on the enemy. He showed extraordinary heroism and courage at a critical time.

ATWATER, BENJAMIN L_____
Near Landres-et-St. Georges, France,
Oct. 5, 1918.
R—Redbank, N. J.
B—Redbank, N. J.
G. O. No. 1, W. D., 1919.

First lieutenant, observer, 99th Aero Squadron, Air Service.
He started on a photographic mission with Lieutenant Alexander, pilot, over the enemy lines. Forced back by seven enemy pursuit planes, he determined to complete his mission and recrossed the line eight minutes later. A large group of enemy pursuit machines again attacked his plane. Disregarding his wound, he operated his machine gun with such effect that the nearest of the enemy planes was put down out of control.

*ATWOOD, JOHN BAIRD_____
North of Montfaucon, France,
Sept. 28, 1918.
R—Pittsburgh, Pa.
B—Pittsburgh, Pa.
G. O. No. 15, W. D., 1923.

Major, 316th Infantry, 79th Division.
The elements of the leading company of his battalion being stopped by terrific enemy machine-gun and artillery fire from the Bois-de-Beuge and the sunken Nantillois-Cunel Road and in danger of disorganization, Major Atwood then rushed personally into a position of the most extreme danger in the first wave, called to his men to follow, turned the retreat into an advance, and fell dead from a rifle wound in the face as he personally directed fire on an enemy machine-gun nest.
Posthumously awarded. Medal presented to mother, Mrs. M. Atwood.

AUBER, JOHN J._____
Near Samogneux, France, Nov. 1,
1918.
R—Elm Grove, W. Va.
B—Wheeling, W. Va.
G. O. No. 37, W. D., 1919.

Private, Company E, 314th Infantry, 79th Division.
While standing in the entrance of his dugout he saw a grenade, with fuse burning, rolling into the dugout where his comrades were sleeping. He picked up the grenade and attempted to throw it away, but it exploded in his hand, blowing off the hand and forearm.

*AUER, CHARLES (118503)_____
At Chateau-Thierry, France, June
6, 1918.
R—Bandon, Oreg.
B—Willamina, Oreg.
G. O. No. 110, W. D., 1918.

Corporal, 20th Company, 5th Regiment, U. S. Marine Corps, 2d Division.
Killed in action at Chateau-Thierry, France, June 6, 1918, he gave the supreme proof of that extraordinary heroism which will serve as an example to hitherto untried troops.
Posthumously awarded. Medal presented to father, John F. Auer.

AUSTERMANN, RICHARD W_____
Near Fismes, France, Aug. 3, 1918.
R—Loyal, Wis.
B—Waukesha, Wis.
G. O. No. 143, W. D., 1918.

Second lieutenant, 128th Infantry, 32d Division.
He collected several groups of disorganized men from different companies, organized a patrol, and, advancing across a creek, so deployed his men as to pour a cross fire on enemy machine-gun nests. Observing other nests then out of range, he led a volunteer squad and cleaned out 3 more machine-gun nests. He continued with the squad and put out of action three more nests.

AUSTIN, CLAUDE W_____
South of Dun-sur-Meuse, France,
Oct. 5, 1918.
R—Effingham, Ill.
B—Effingham, Ill.
G. O. No. 44, W. D., 1919.

First lieutenant, 130th Infantry, 33d Division.
On the evening of Oct. 5 a shell struck an old building in front of a dugout occupied by 1 of his machine-gun teams, wounding 2 of the men who were just coming out and hurling them to the bottom of the steps. Fire from the building spread to the framework of the dugout, which contained a quantity of grenades and high explosives. He unhesitatingly ran to the rescue of the 2 men and dragged them out, 1 at a time, but they died a short time later. He then entered the dugout, bringing out 5 unwounded men, undoubtedly saving their lives, for the dugout was totally destroyed a short time later. The entire exploit was carried on under sniping fire from across the river as well as machine-gun and artillery fire from three sides of the salient.

*AUSTIN, EDWIN (274176)_____
Near Roncheres, France, July 30,
1918.
R—Shawano, Wis.
B—Tavining, Mich.
G. O. No. 66, W. D., 1919.

Private, Company F, 127th Infantry, 32d Division.
He volunteered to go out in advance of our front lines and bring back wounded who had been left there when his company was withdrawn. He made 2 trips, under heavy fire, bringing back wounded with the aid of another soldier, but was killed by machine-gun fire when he went out for the third time.
Posthumously awarded. Medal presented to father, George Austin.

*AUSTIN, FRANCIS R_____
Near Haumont, France, Nov. 11,
1918.
R—Boston, Mass.
B—Boston, Mass.
G. O. No. 37, W. D., 1919.

First lieutenant, 109th Infantry, 28th Division.
He led a platoon of machine guns and two 1-pounder guns with their crews under cover of a fog within the enemy's wire and attacked at close range a strong point held by 25 men and 10 machine guns. After this position had been reduced, concentrated machine-gun fire from the ranks forced Lieutenant Austin and his party to withdraw. Exposing himself, in order to place his men under cover, he was mortally wounded, but he directed the dressing of the wounds of his men and their evacuation before he would accept any aid for himself. He died a few hours later.
Posthumously awarded. Medal presented to father, Francis B. Austin.

*AUSTIN, JAMES B _____
Near Cierges, France, Oct. 8, 1918.
R—Chicago, Ill.
B—Kansas City, Mo.
G. O. No. 126, W. D., 1919.

Captain, 38th Infantry, 3d Division.
Captain Austin continued for several hours to command his company after he had been shot through the body and in the leg. He sent back numerous reports to his regimental commander during this period, but never mentioned the fact that he was severely wounded.
Posthumously awarded. Medal presented to widow, Mrs. James B. Austin.

AUSTIN, JOHN C. (98724) _____
Near Landres-et-St. Georges, France, Oct. 16, 1918.
R—Sylacauga, Ala.
B—Talladega, Ala.
G. O. No. 81, W. D. 1919.

Corporal, Headquarters Company, 167th Infantry, 42d Division.
Corporal Austin volunteered and crawled forward more than 50 yards in the open under heavy rifle, machine-gun, and shell fire from the enemy in order to observe the fire effect of a Stokes mortar. He returned with valuable information which assisted in breaking up a hostile counterattack. A few minutes later with another soldier he went to the assistance of a comrade, who had fallen wounded in an open field 50 yards away, and carried him to shelter under heavy enemy machine-gun fire.

*AUSTIN, RAYMOND B _____
Near Fleville, France, Oct. 7, 1918.
R—Delaware, Ohio.
B—Delaware, Ohio.
G. O. No. 56, W. D., 1922.

Major, 6th Field Artillery, 1st Division.
In order to obtain first-hand information of the enemy's disposition and render the maximum support to the attacking Infantry, Major Austin, with absolute disregard for his own personal safety, made repeated trips to the Infantry front line during the attack of the 1st Division, Oct. 7, 1918. He continued, until mortally wounded, to send to the Artillery information of the utmost value concerning the enemy's disposition.
Posthumously awarded. Medal presented to father, Cyrus B. Austin.

AVERY, CHARLES D _____
Near Cantigny, France, May 27, 1918.
R—Emporia, Kans.
B—Emporia, Kans.
G. O. No. 64, W. D., 1919.

Second lieutenant, 28th Infantry, 1st Division.
After a two-hour barrage, which caused many casualties in our forces, the enemy raided a sector occupied by our troops. During the attack lieutenant Avery exhibited unusual courage in holding together his handful of men, after one-third had become casualties, and distributing ammunition to the remaining men, which finally stopped the attack. Two prisoners were taken during the battle. He was severely wounded about the head and later buried in a trench where he remained for three and one-half hours before being dug out.

AVERY, WALTER L _____
North of Chateau-Thierry, France, July 25, 1918.
R—Columbus, Ohio.
B—Columbus, Ohio.
G. O. No. 121, W. D., 1918.

First lieutenant, 95th Aero Squadron, Air Service.
While on his first patrol over the enemy's lines he attacked an enemy two-seater biplane. While thus occupied he was vigorously attacked by another enemy plane, but by a quick turn, skillful maneuvering, and accurate shooting he drove the second plane to the American side of the lines, where it crashed into the woods. Lieutenant Avery's motor had been badly damaged by bullets, but he made a successful landing back of our lines, where he learned that the enemy pilot who had been made a prisoner was a German ace credited with 16 victories. Lieutenant Avery's conduct was especially commendable because his plane had been seriously damaged at the beginning of the combat.

AWBERY, CLARENCE (731663) _____
Near Fontaines, France, Nov. 7, 1918.
R—Elk Creek, Ky.
B—Henry County, Ky.
G. O. No. 37, W. D., 1919.

Private, Company B, 6th Infantry, 5th Division.
Private Awbrey, accompanied by three other soldiers, volunteered and went out under heavy machine-gun and artillery fire to rescue a wounded comrade. Failing in the first attempt, they again tried, and this time succeeded in bringing the wounded man to shelter.

AWL, FRANCIS A _____
During the Meuse-Argonne offensive, Sept. 26–29, 1918.
R—Harrisburg, Pa.
B—Harrisburg, Pa.
G. O. No. 38, W. D., 1922.

Captain, 315th Infantry, 79th Division.
On Sept. 26, 1918, when a portion of his company had been pocketed on a ridge and subjected to enfilade fire of the enemy's machine guns and snipers, Captain Awl, being unable to obtain volunteers, took a rifle and, with absolute disregard for his own safety, crawled into the open on the exposed flank and, by engaging the enemy, enabled his men to withdraw, thereby saving a great loss of life and permitting the unit to reform for a further advance. Later he repeatedly led successful attacks against superior numbers and machine-gun nests. While in an attack on Sept. 29 he was severely wounded and refused to be evacuated until forced to do so.

*AXTON, ANDREW P _____
In the Bois-de-Belleau, France, June 6, 1918.
R—W. Brownsville, Pa.
B—Uniontown, Pa.
G. O. No. 126, W. D., 1918.

Private, 82d Company, 6th Regiment, U. S. Marine Corps, 2d Division.
In the Bois-de-Belleau, France, June 6, 1918, he was conspicuous for his bravery and coolness in advancing with an automatic rifle on a strongly defended machine-gun position. He was killed in the performance of this duty.
Posthumously awarded. Medal presented to mother, Mrs. Nell P. Axton.

AYERS, JOHN W. (1284342) _____
Near Sivry, France, Oct. 18, 1918.
R—Easton, Md.
B—Easton, Md.
G. O. No. 16, W. D., 1919.

Corporal, Company C, 115th Infantry, 29th Division.
During several engagements in the vicinity of Sivry he upon his own initiative went forward and located enemy machine-gun nests. On another occasion his platoon having lost connection with his company during a heavy enemy artillery fire, he reconnoitered the position and established liaison with his company.

AYLWARD, WILLIAM B. (63592) _____
Near Epieds, France, July 23, 1918.
R—Waterbury, Conn.
B—Westfield, Mass.
G. O. No. 74, W. D., 1919.

Corporal, Company A, 102d Infantry, 26th Division.
Corporal Aylward maintained liaison between the platoons of his company, and after his platoon commander and sergeant had been shot down he took command of the platoon, remaining in command until only 2 men and himself were left alive and unwounded. Although slightly gassed, he remained on duty, rendering first aid and carrying wounded to the first-aid station, until he became so overcome from the effects of the gas that he had to be evacuated.

AYOTTE, EDWARD E. (69840)_____
Near Bouresches, France, July 20, 1918.
R—Houlton, Me.
B—Houlton, Me.
G. O. No. 87, W. D., 1919.

Private, Headquarters Company, 103d Infantry, 26th Division.
Under the deadly fire of the enemy's artillery and machine guns, Private Ayotte administered first aid to many wounded. He carried a wounded officer some distance to safety, after which he returned, ceasing in his attention to the wounded only after all had received aid.

BABCOCK, PHILIP R_____
Near Fismes, France, Aug. 11, 1918.
R—Lynn, Mass
B—Lyme, Conn.
G. O. No. 44, W. D., 1919.

First lieutenant, pilot, 88th Aero Squadron, Air Service.
Louis G. Bernheimer, first lieutenant, pilot; John W. Jordan, second lieutenant, 7th Field Artillery, observer; Roger W. Hitchcock, second lieutenant, pilot; James S. D. Burns, deceased, second lieutenant, 165th Infantry, observer; Joel H. McClendon, deceased, first lieutenant, pilot; Charles W. Plummer, deceased, second lieutenant, 101st Field Artillery, observer; Philip R. Babcock, first lieutenant, pilot; and Joseph A. Palmer, second lieutenant, 15th Field Artillery, observer. All of these men were attached to the 88th Aero Squadron, Air Service.
For extraordinary heroism in action near Fismes, France, Aug. 11, 1918. Under the protection of three pursuit planes, each carrying a pilot and an observer, Lieutenants Bernheimer and Jordan, in charge of a photo plane, carried out successfully a hazardous photographic mission over the enemy's lines to the River Aisne. The four American ships were attacked by 12 enemy battle planes. Lieutenant Bernheimer, by coolly and skillfully maneuvering his ship, and Lieutenant Jordan, by accurate operation of his machine gun, in spite of wounds in the shoulder and leg, aided materially in the victory which came to the American ships, and returned safely with 36 valuable photographs. The pursuit plane operated by Lieutenants Hitchcock and Burns was disabled while these two officers were fighting effectively. Lieutenant Burns was mortally wounded and his body jammed the controls. After a headlong fall of 2,500 meters, Lieutenant Hitchcock succeeded in regaining control of this plane and piloted it back to his airdrome. Lieutenants McClendon and Plummer were shot down and killed after a vigorous combat with five of the enemy's planes. Lieutenants Babcock and Palmer, by gallant and skillful fighting, aided in driving off the German planes and were materially responsible for the successful execution of the photographic mission.

BABST, JULIUS J_____
At Chateau-Thierry, France, June 6-7, 1918.
R—Denver, Colo.
B—Naperville, Ill.
G. O. No. 46, W. D., 1919.

First lieutenant, chaplain, 23d Infantry, 2d Division.
Chaplain Babst displayed exceptional bravery and devotion to duty by repeatedly going out from the first-aid station of his battalion to care for the wounded and voluntarily exposed himself to terrific artillery and machine-gun fire to administer the last sacraments to the dying. At imminent risk to his own life he worked to improve the conditions at the aid station and fearlessly conducted burial services under fire.
Oak-leaf cluster.
For the following acts of extraordinary heroism in action near St. Etienne, Oct. 3–9, 1918, Chaplain Babst is awarded an oak-leaf cluster, to be worn with the distinguished-service cross: He showed magnificent courage in caring for the wounded under heavy fire, having personally administered to over 50 officers and men, also assuring their evacuation. He showed remarkable devotion to duty by refusing an opportunity to attend chaplain's school, preferring to accompany his regiment into battle, where he labored unceasingly for 7 days, during which time he performed many acts of bravery.

*BACHMAN, JOHN A_____
Near Jaulny, France, Sept. 26, 1918.
R—Buffalo, N. Y.
B—Austin, Pa.
G. O. No. 44, W. D., 1919.

Second lieutenant, 308th Machine Gun Battalion, 78th Division.
During an early morning raid he attempted to place two guns in position, when the enemy opened a terrific barrage. He was ordered to shelter on the slope of the hill, and, after his men had taken refuge there, he went back to determine whether or not all of his men had found shelter. In passing through the heavy barrage he was hit by a shell and instantly killed.
Posthumously awarded. Medal presented to father, F. A. Bachman.

BACKLEY, EDWARD J. (2059313)_____
Near Berzy-le-Sec, France, July 20, 1918.
R—Chicago, Ill.
B—Chicago, Ill.
G. O. No. 117, W. D., 1918.

Private, Company D, 28th Infantry, 1st Division.
He showed exceptional courage and devotion to duty in unhesitatingly advancing against the intense fire of a machine gun and assisting in capturing the gun and crew.

*BACKSTROM, ROBERT E. (1386523)_____
At Chipilly Ridge, France, Aug. 10, 1918.
R—Chicago, Ill.
B—Chicago Heights, Ill.
G. O. No. 38, W. D., 1922.

Sergeant, Company B, 131st Infantry, 33d Division.
When the outpost along the sunken Bray Corbie Road was repeatedly shelled and driven back by snipers, Sergeant Backstrom voluntarily went out alone, with utter disregard for his own safety, held this position all day under heavy sniper and machine-gun fire and repeated shelling on his position.
Posthumously awarded. Medal presented to father, Martin Backstrom.

BACKUS, DAVID H_____
Near Etain, France, Sept. 26, 1918.
R—St. Paul, Minn.
B—St. Paul, Minn.
G. O. No. 138, W. D., 1918.

First lieutenant, pilot, 49th Aero Squadron, Air Service.
He was one of a patrol of five monoplanes that were attacked by nine enemy planes (Fokker type) in a superior position. The American patrol leader, seeing the futility of giving combat, turned toward our lines with the enemy in close pursuit. One of our patrol, however, fell behind, and the enemy planes dove upon him. Lieutenant Backus, although beyond danger, seeing the predicament of his comrade, turned, and alone attacked the enemy destroying one and dispersing the others.
Oak-leaf cluster.
A bronze oak leaf for extraordinary heroism in action in the region of Landreville, France, Oct. 23, 1918. A patrol of American monoplane planes attacked an enemy formation of superior number. Flying rear position, he maneuvered above the attack to prevent other enemy planes from assisting their companions. In the midst of the combat he saw three planes escaping from battle. He immediately gave chase and attacked and shot down all three of the enemy.

BACON, BENJAMIN R. (1901194)_____
  South of Champigneulle, France,
    Oct. 16, 1918.
  R—Philadelphia, Pa.
  B—Philadelphia, Pa.
  G. O. No. 46, W. D., 1919.

Sergeant, Company D, 326th Infantry, 82d Division.
Twice wounded by machine-gun bullets, he continued to lead his platoon through heavy artillery and machine-gun fire, penetrating the enemy lines and silencing several machine guns.

BADHAM, WILLIAM T_____
  Near Buzancy, France, Oct. 23, 1918.
  R—Birmingham, Ala.
  B—Birmingham, Ala.
  G. O. No. 7, W. D., 1919.

Second lieutenant, observer, 91st Aero Squadron, Air Service.
This officer gave proof of exceptional bravery while on a photographic mission 25 kilometers within the enemy lines. His plane was attacked by a formation of 30 enemy aircraft; by skillful work with his machine gun he successfully repelled the attack and destroyed two German planes. At the same time he manipulated his camera and obtained photographs of great military value.

BAER, PAUL FRANK_____
  Northeast of Reims, France, Mar.
    11 and 16, 1918.
  R—Fort Wayne, Ind.
  B—Fort Wayne, Ind.
  G. O. No. 128, W. D., 1918.

First lieutenant, pilot, 103d Aero Squadron, Air Service.
On March 11, 1918, he attacked, alone, a group of seven enemy pursuit machines, destroying one, which fell near the French lines northeast of Reims, France. On Mar. 16, 1918, he attacked two enemy two-seaters, one of which fell in flames in approximately the same region.
Oak-leaf cluster.
He was awarded a bronze oak leaf for the following acts of extraordinary heroism in action: He brought down enemy planes on Apr. 5, 12, and 23, 1918, and on May 8, 1918, he destroyed two German machines, and on May 21, 1918, he destroyed his eighth enemy plane.

BAGBY, RALPH B_____
  Near Tailly, France, Nov. 2, 1918.
  R—New Haven, Mo.
  B—New Haven, Mo.
  G. O. No. 44, W. D., 1919.

First lieutenant, Field Artillery, observer, 88th Aero Squadron, Air Service.
Lieutenant Bagby, with First Lieut. Louis G. Bernheimer, pilot, on their own initiative went on a reconnaissance mission, flying 50 kilometers behind the German lines, securing valuable information as to the condition of the bridges across the Meuse River and enemy activity in the back areas, and also harassing enemy troops.

*BAILEY, ALFRED G. (550543)_____
  Near Moulins, France, July 15, 1918.
  R—Eli, Okla.
  B—Hulbert, Okla.
  G. O. No. 32, W. D., 1919.

Sergeant, Company E, 38th Infantry, 3d Division.
Sergeant Bailey, unaided, killed two enemy machine gunners and captured a third, together with his machine gun.
Posthumously awarded. Medal presented to father, George F. Bailey.

BAILEY, EARL WALLACE (549202)_____
  Near Jaulgonne, France, July 22,
    1918.
  R—Cardiff, N. Y.
  B—Cardiff, N. Y.
  G. O. No. 19, W. D., 1920.

Private, Machine Gun Company, 38th Infantry, 3d Division.
Private Bailey crawled forward through severe machine-gun and rifle fire and killed two members of an enemy machine-gun crew. He immediately turned the captured gun around and with it opened an effective fire on the enemy, thus enabling a company of infantry to advance.

BAILEY, ERNEST O. (2155676)_____
  Near Premont, France, Oct. 9, 1918,
    and near Molain, France, Oct. 17,
    1918.
  R—Roosevelt, Minn.
  B—Jones, Okla.
  G. O. No. 81, W. D., 1919.

Corporal, Company E, 117th Infantry, 30th Division.
When his company was held up by machine-gun fire, he carried a message across open ground to a tank commander, whose tank brought the needed support for reducing the nests. Later, in company with an officer, he braved intense shell fire to rescue a wounded soldier.

BAILEY, GEORGE W _____

  Near St. Etienne, France, Oct. 4,
    1918.
  R—Ogdensburg, N. Y.
  B—Ogdensburg, N. Y.
  G. O. No. 35, W. D., 1919.

Pharmacist's mate, third class, U. S. Navy, attached to 5th Regiment, U. S. Marine Corps, 2d Division.
He voluntarily went out in front of the most advanced positions of our troops in order to render first aid to a number of wounded soldiers. He continued the work until all the wounded had been given first aid and evacuated.

BAILEY, HENRY S_____
  During the Argonne-Meuse Offensive, France, Sept. 26, 1918.
  R—Berkeley, Calif.
  B—Long Beach, Calif.
  G. O. No. 37, W. D., 1919.

First lieutenant 363d Infantry, 91st Division.
Commanding a small detachment which was being held up by machine-gun fire, Lieutenant Bailey, with one soldier, proceeded to force the enemy's withdrawal. Working ahead under terrific fire, although wounded, he made his way to the right flank of the enemy's position, and within 15 minutes silenced the fire.

BAILEY, IVAN Y. (2257549)_____
  Near Gesnes, France, Oct. 10, 1918.
  R—Fort Shaw, Mont.
  B—Canton, S. Dak.
  G. O. No. 37, W. D., 1919.

Private, Company B, 361st Infantry, 91st Division.
While on a liaison patrol Private Bailey and Corporal Carl G. Theobald attacked and captured a hostile machine-gun nest and its entire crew. Private Bailey then took the prisoners across No Man's Land to our lines under machine gun fire.

BAILEY, JESSE M. (3000)_____
  Near Chateau-Thierry, France, July
    18–24, 1918.
  R—Providence, R. I.
  B—Watertown, N. Y.
  G. O. No. 125, W. D., 1918.

Private, sanitary detachment, 103d Machine Gun Battalion, 26th Division.
He gave an inspiring example of courage and coolness in treating wounded for two days without rest or food and under intense artillery and machine-gun fire. While maintaining a dressing station in a crater, an enemy shell struck in the center of the hole, wounding nearly all in the crater and severely injuring one man, who was buried in the earth and débris. Upon regaining consciousness Private Bailey treated all of the wounded men, helped them to the first-aid station, returned and dug out the man who had been buried and left for dead and succeeded in resuscitating him.

*BAILEY, ROBERT M_____
  Near Verdun, France, October 12,
    1918.
  R—Anderson, S. C.
  B—Anderson, S. C.
  G. O. No. 44, W. D., 1919.

Second lieutenant, 114th Infantry, 29th Division.
Leading his platoon against an enemy position, Lieutenant Bailey was fatally wounded but refused to leave until his position was organized and a counterattack repulsed.
Posthumously awarded. Medal presented to mother, Mrs. Alice C. Bailey.

BAILEY, THOMAS_____ | First lieutenant, 111th Infantry, 28th Division.
Near Fismes, France, Sept. 4, 1918. | When the advance was held up, owing to lack of information, and no man
R—Philadelphia, Pa. | volunteered for a reconnaissance mission because of the hazard attached
B—Philadelphia, Pa. | thereto, Lieutenant Bailey undertook the mission. Crawling on his belly
G. O. No. 50, W. D., 1919. | 100 yards across an open space and then traversing 200 yards of woods infested
 | by the enemy, he gained and returned with information of the greatest value,
 | making possible a subsequent and successful attack.

BAILEY, THOMAS F. (2178317)_____ | Corporal, Company A, 354th Infantry, 89th Division.
Near Barricourt, France, Nov. 1, | Corporal Bailey, seeing that the advance of his platoon was stopped by machine-
1918. | gun fire from the right front, took three men with him, and, with a rush in
R—Williamstown, Mo. | face of a withering fire, charged the machine gun and captured it and the
B—Lewiston, Mo. | crew of five prisoners, thereby saving his platoon from the destructive fire
G. O. No. 4, W. D., 1923. | of the gun and enabling it to immediately continue to advance.

BAILEY, WALTER J. (214742)_____ | Private, first class, Machine Gun Company, 325th Infantry, 82 Division.
Near St. Juvin, France, Oct. 12, | Securing a captured German machine gun, Private Bailey operated it against
1918. | the enemy from an exposed position until he was wounded and rendered
R—Lansing, Iowa. | unconscious by an enemy shell.
B—Oquawka, Ill. |
G. O. No. 78, W. D., 1919. |

BAIN, EDGAR H._____ | Captain, 119th Infantry, 30th Division.
Near Busigny, France, Oct.9, 1918. | Advancing under heavy fire, with orders to pass through the front-line com-
R—Goldsboro, N. C. | pany, he found the troops he was to relieve 1,000 yards from their position,
B—Goldsboro, N. C. | falling back. Rallying them, he personally led the troops in advance under
G. O. No. 81, W. D., 1919. | terrific fire, assaulting and capturing the assigned objective.

BAINBRIDGE, ROGER J. (52418)_____ | Corporal, Company A, 26th Infantry, 1st Division.
Near Soissons, France, July 19, 1918. | While acting as liaison corporal near Soissons, France, July 19, 1918, he was
R—Edmond, Okla. | severely wounded, but nevertheless continued in action and killed three of
B—Galesburg, Ill. | the enemy before being ordered to a dressing station for treatment.
G. O. No. 132, W. D., 1918. |

*BAIR, HOWARD A._____ | Second lieutenant, 354th Infantry, 89th Division.
Near Barricourt, France, Nov. 2, | Calling on his platoon to follow, he pushed forward and attacked enemy ma-
1918. | chine-gun nests. After killing two of the enemy he himself was killed by a
R—Columbus, Ohio. | hand grenade while accepting the surrender of another of the enemy.
B—Sterling, Ohio. | Posthumously awarded. Medal presented to mother, Mrs. Francis Bair.
G. O. No. 44, W. D., 1919. |

BAKER, DOUGLAS B._____ | First lieutenant, 30th Infantry, 3d Division.
Near Bois-de-Beuge and Bois-de-la- | During the period Oct. 9-15, 1918, he made frequent trips through heavy shell,
Pultiere, France, Oct. 9-15, 1918. | gas, and machine-gun fire to repair broken telephone and telegraph wires,
R—Melrose, Mass. | and when they could not longer be repaired he personally carried messages
B—Springfield, Mass. | through the shell-swept area. On Oct. 15 he personally reconnoitered the
G. O. No. 32, W. D., 1919. | Bois-de-la-Pultiere under heavy machine-gun and shell fire in an endeavor to
 | find a suitable location for his regimental post of command.

*BAKER, EMERY L. (58640)_____ | Private, first class, Company M, 28th Infantry, 1st Division.
Near Cantigny, France, May 28, | While acting as platoon runner, he passed through three violent artillery bar-
1918. | rages with coolness and apparent contempt for danger and repeatedly carried
R—New Bloomfield, Mo. | ammunition to his comrades under fire.
B—Greeley, Colo. | Posthumously awarded. Medal presented to mother, Mrs. Augusta Baker.
G. O. No. 139, W. D., 1918. |

BAKER, HARRY J._____ | Sergeant, 66th Company, 5th Regiment, U. S. Marine Corps, 2d Division.
Near St. Etienne, France, Oct. 4, | He disregarded his own safety by going out under a heavy shell and machine-
1918. | gun fire to carry a wounded comrade to a place of safety.
R—Denver, Colo. |
B—Mobile, Ala. |
G. O. No. 35, W. D., 1919. |

BAKER, JESSE (1763574)_____ | Private, Company A, 312th Infantry, 78th Division.
Near Grand Pre, France, Oct. 18, | He carried a message from his platoon leader to the leader of an adjoining pla-
1918. | toon, crossing an area swept by an intense machine-gun fire. While waiting
R—Phillipsburg, N. J. | for a reply, he was seriously wounded, but returned with an answer to his
B—Phillipsburg, N. J. | platoon leader, remaining on duty until ordered evacuated by his command-
G. O. No. 37, W. D., 1919. | ing officer.

BAKER, JOHN (2716416)_____ | Private, Company I, 368th Infantry, 92d Division.
Near Binarville, France, Sept. 28, | Although severely wounded in the right hand, losing two fingers, Private
1918. | Baker, a runner, continued 300 yards through heavy enemy machine-gun
R—Philadelphia, Pa. | fire to the forward battalion and delivered his message alone, having been
B—Cheriton, Va. | deserted by an unwounded fellow runner.
G. O. No. 81, W. D., 1919. |

BAKER, JOHN M. (2339046)_____ | Corporal, Company G, 4th Infantry, 3d Division.
Near Roncheres, France, July 29, | He led a patrol through heavy machine-gun fire in an attack on an enemy nest.
1918. | Seeing all the members of his patrol lying about, either killed or wounded,
R—Raleigh, N. C. | he courageously continued to fire, killing a sniper who had been inflicting
B—Raleigh, N. C. | severe losses.
G. O. No. 103, W. D., 1919. | Oak-leaf cluster.
 | For the following act of extraordinary heroism in action near Cunel, France,
 | Oct. 14, 1918, Corporal Baker is awarded an oak-leaf cluster, to be worn with
 | the distinguished-service cross: After his platoon commander had been
 | wounded, Corporal Baker took command, and, after being wounded him-
 | self, refused to go for treatment, remaining to lead his platoon for two days
 | until relieved.

BAKER, JOHN T. (915250)
Near Brieulles, France, Oct. 18 to Nov. 3-4, 1918.
R—Bancroft, Iowa.
B—Hastings, Pa.
G. O. No. 37, W. D., 1919.

Sergeant, first class, Company F, 7th Engineers, 5th Division.
On Oct. 18 Sergeant Baker, with a detail of 19 men, carrying wire for wiring in outposts, followed the Infantry through the Bois-des-Rappes, where he employed his men as Infantry to assist in holding the captured position. Later they wired in four outposts in direct view of and under heavy fire from the enemy. On Nov. 3 and 4 they succeeded several times in laying footbridges across the Meuse under heavy fire, allowing the Infantry to cross to the east bank.

BAKER, JOSEPH M.
At Belleau Wood, France, June 6, 1918.
R—Logansport, Ind.
B—Logansport, Ind.
G. O. No. 60, W. D., 1920.

Private, 67th Company, 5th Regiment, U. S. Marine Corps, 2d Division.
When his platoon was suffering casualties from the fire from a hidden machine gun, Private Baker exposed himself to heavy fire to take up a position on the flank of the enemy gun. He attacked and killed the gunner by rifle fire and then rushed the gun, killing the crew with his bayonet.

BAKER, McLAURIN (1330208)
Near Mazinghein, France, Oct. 18, 1918.
R—Lamar, S. C.
B—Lamar, S. C.
G. O. No. 35, W. D., 1919.

Sergeant, Company C, 105th Field Signal Battalion, 30th Division.
During the fighting around Mazinghein, Sergeant Baker, while attached to the 120th Infantry, was painfully wounded by shrapnel, which necessitated his going to the first-aid station for treatment. Realizing that his services were greatly needed at the line, he refused to be evacuated, but remained in action until the troops were withdrawn.

BALCH, JOHN H.
In the Bois-de-Belleau, France, June 6-8, 1918; near Vierzy, France, July 19, 1918; and near St. Etienne-a-Arnes, France, Oct. 5, 1918.
R—Kansas City, Mo.
B—Edgerton, Kans.
G. O. No. 70, W. D., 1919.

Pharmacist's mate, first class, U. S. Navy, attached to 6th Regiment, U. S. Marine Corps, 2d Division.
During the attack in the Bois de Belleau he displayed conspicuous coolness under shell fire in evacuating wounded men. During the action near Vierzy he worked unceasingly for 16 hours giving assistance to the wounded on a field torn by high-explosive shells and covered by direct machine-gun fire. Near St. Etienne-a-Arnes he again gave proof of excellent judgment and courage in establishing an advance dressing station under violent shell and machine-gun fire, thereby saving many lives which would otherwise have been lost.

BALD, EDWARD
Near Somme-Py, France, Oct. 2-10, 1918.
R—Camden, N. J.
B—Philadelphia, Pa.
G. O. No. 21, W. D., 1919.

Corporal, 15th Company, 6th Machine Gun Battalion, U. S. Marine Corps, 2d Division.
He maneuvered his machine gun squad independently of the platoon, going forward under intense enemy machine-gun and artillery fire and concentrations of gas. On one occasion he led his squad, regardless of personal danger, in the rear of the German positions and laid down a flanking fire against a portion of the enemy line, facilitating its capture, together with a number of the enemy.

BALDRIDGE, ROBERT L. (1540532)
Near Heurne, Belgium, Nov. 4, 1918.
R—Toledo, Ohio.
B—Dexter, Mo.
G. O. No. 59, W. D., 1919.

Private, 148th Ambulance Company, 112th Sanitary Train, 37th Division.
With two other soldiers he volunteered to rescue two wounded men who had been lying in an exposed position on the opposite bank of the Scheldt River for two days. Making two trips across the stream in the face of heavy machine-gun and shell fire, he and his companions succeeded in carrying both the wounded men to shelter.

BALDRIDGE, TROY J. (45903)
South of Soissons, France, July 18, 1918.
R—Woodlawn, Ill.
B—Madison County, Mo.
G. O. No. 53, W. D., 1920.

Corporal, Company A, 18th Infantry, 1st Division.
Corporal Baldridge voluntarily led four men a distance of about 400 yards in advance of our line and attacked a superior force of the enemy who were attempting to man a machine gun in a disabled French tank. Due to his bold attack, the enemy was driven off and the tank retaken. His company was thereby enabled to continue the advance with slight loss.

BALDWIN, GEOFFREY P.
In the Bois-des-Rappes, France, Oct. 15, 1918.
R—Battle Creek, Mich.
B—Madison Barracks, N. Y.
G. O. No. 22, W. D., 1920.

Major, 60th Infantry, 5th Division.
He personally led his small detachment through woods that were infested with machine-gun nests and attained his objective on the northern edge of the Bois-des-Rappes and held same regardless of the danger due to his position in advance of the line and being enveloped by the enemy on both flanks. In utter disregard of his own life, he personally charged a machine-gun nest and killed the gunner.

BALDWIN, MOSES S. (97121)
Near Landres-et-St. Georges, France, Oct. 15, 1918.
R—Midland City, Ala.
B—Dale County, Ala.
G. O. No. 131, W. D., 1918.

Corporal, Company G, 167th Infantry, 42nd Division.
In an attack on the Cote-de-Chatillon, disregarding all personal danger, he repeatedly went over shell-swept areas under heavy machine-gun fire to give first-aid treatment to the wounded and carry them to shelter.

BALDWIN, THOMAS (40366)
Near Soissons, France, July 18, 1918.
R—Ironton, Ohio.
B—Ironton, Ohio.
G. O. No. 35, W. D., 1919.

Private, first class, Company K, 9th Infantry, 2d Division.
After all the other runners of his platoon had been either killed or wounded by machine-gun and shell fire and he himself had been seriously wounded, he refused evacuation, continuing with his mission. After making three trips through the heavy barrage, he guided platoons to their designated objectives, after which he was ordered to a dressing station.

*BALDWIN, WILLIAM W.
Near Meurcy Farm on the Ourcq River, France, July 30, 1918.
R—Chicago, Ill.
B—Burlington, Iowa.
G. O. No. 4, W. D., 1923.

First lieutenant, 165th Infantry, 42d Division.
When the advance of the assault company which he was commanding was held up by a very heavy artillery and machine-gun fire, Lieutenant Baldwin, realizing the emergency, went to the front and personally participated in the capture of the machine-gun nests. Due to his personal leadership and outstanding courage, his company overcame the resistance, crossed the Ourcq River, and captured Meurcy Farm, when he was killed.
Posthumously awarded. Medal presented to father, W. W. Baldwin.

BALL, ERNEST B._____
  Near St. Etienne, France, Oct. 3, 1918.
  R—Ukiah, Calif.
  B—Boonville, Calif.
  G. O. No. 46, W. D., 1919.

Pharmacist's mate, second class, U. S. Navy, attached to 18th Company, 5th Regiment, U. S. Marine Corps, 2d Division.
He continually exposed himself to severe machine-gun and artillery fire while dressing and carrying wounded soldiers belonging to the unit to which he was attached.

BALL, ERNEST W. (2257088)_____
  Near Eclisfontaine, France, Sept. 29, 1918.
  R—Rigby, Idaho.
  B—Sandy, Utah.
  G. O. No. 44, W. D., 1919.

Private, Company H, 361st Infantry, 91st Division.
While his company was being harassed by enemy snipers hidden in imitation tanks, Private Ball, without aid, went forward and succeeded in killing one and capturing another.

BALL, RALPH (44380)_____
  Near Fleville, France, Oct. 9, 1918.
  R—Milbrook, N. Y.
  B—Milbrook, N. Y.
  G. O. No. 23, W. D., 1919.

Corporal, Company M, 16th Infantry, 1st Division.
Although severely wounded, Corporal Ball led his section through a terrific enemy barrage and advanced until his men had safely passed the bombed area before he would allow himself to be evacuated.

*BALL, WILLIAM R. (2810709)_____
  Near Fey-en-Haye, France, Sept. 9–17, 1918.
  R—Lindsay, Okla.
  B—Fleetwood, Okla.
  G. O. No. 128, W. D., 1918.

Corporal, Company G, 357th Infantry, 90th Division.
Becoming separated from his patrol, Corporal Ball, with another soldier, attacked an enemy patrol and drove it off, though the number of their opponents was estimated at 50. He did excellent work with his platoon in the advance of Sept. 12 north of Fey-en-Haye, in rushing machine-gun nests. On Sept. 17 he was a member of an outpost attacked by a larger body of Germans. Though wounded, he remained at his post.
Posthumously awarded. Medal presented to father, William E. Ball.

*BALLARD, BLACKBURN W. (154334)____
  Near Verdun, France, Oct. 9, 1918.
  R—Colusa, Calif.
  B—Colusa, Calif.
  G. O. No. 44, W. D., 1919.

Corporal, Company A, 1st Engineers, 1st Division.
Advancing alone ahead of his squad, in the face of unusual machine-gun fire, he set out to attack a machine-gun nest. He so inspired his men that they came to his assistance, and under his direction the stronghold was taken, together with many prisoners. While guarding the collected prisoners he was killed.
Posthumously awarded. Medal presented to father, R. B. Ballard.

*BALLARD, FREDERICK E. (109677)_____
  Near Marcheville, France, Sept. 26, 1918.
  R—Ludlow, Vt.
  B—Ludlow, Vt.
  G. O. No. 21, W. D., 1919.

Private, Company C, 102d Machine Gun Battalion, 26th Division.
He displayed remarkable courage and coolness during this engagement. When apparently trapped in an enemy trench near a machine-gun emplacement, he worked his way out under the wire entanglements in plain view of the enemy, and, returning with hand grenades, assisted in bombing out the machine-gun nest and capturing some of the men who were defending it. Later he accompanied a detachment and assisted in mopping up the town, driving out the enemy and taking several prisoners. While thus engaged he was struck by an exploding shell and killed.
Posthumously awarded. Medal presented to mother, Mrs. Jessie E. Ballard.

BALLARD, WALTER D. (41872)_____
  South of Soissons, France, July 21, 1918.
  R—Redfield, Iowa.
  B—Red Oak, Iowa.
  G. O. No. 117, W. D., 1918.

Private, Company B, 16th Infantry, 1st Division.
Displaying exceptional initiative and bravery throughout the operations south of Soissons, France, July 18 to 22, 1918, he, with extraordinary heroism on July 21, 1918, with two companions, captured two machine guns that were causing heavy losses to his company.

BALLESTERO, FRED V. (2265095)_____
  Near Very, France, Sept. 26, 1918.
  R—Towle, Calif.
  B—San Francisco, Calif.
  G. O. No. 64, W. D., 1919.

Private, Company A, 363d Infantry, 91st Division.
At the very outset of the action Private Ballestero took six prisoners single handed and under fire. After his company had been stopped by heavy firing, he and another soldier went ahead on a reconnaissance mission and encountered heavy machine-gun fire. Sending his companion back with information and for help, he alone pressed on, capturing a machine gun and its entire crew.

BALLING, JOSEPH P. (1749662)_____
  Near Grand Pré, France, Nov. 1, 1918.
  R—Buffalo, N. Y.
  B—Buffalo, N. Y.
  G. O. No. 64, W. D., 1919.

Corporal, Company M, 311th Infantry, 78th Division.
Assigned to the duty of carrying rations, water, and ammunition to the front line, Corporal Balling led his details over ground under incessant gas attacks and terrific shell fire. Although his force was greatly decreased through casualties, he maintained complete control and succeeded in supplying the troops with necessities. After being helplessly wounded he directed the movements for the safety of his detail and the care of the wounded.

BALMAYNE, COLIN B. (51680)_____
  Service rendered under the name, Colin B. Joe.

Sergeant, Company K, 23d Infantry, 2d Division.

BANAHAN, RAYMOND T. (1286340)_____
  Near Bois-de-Consenvoye, France, Oct. 10, 1918.
  R—Baltimore, Md.
  B—Baltimore, Md.
  G. O. No. 37, W. D., 1919.

Sergeant, Company L, 115th Infantry, 29th Division.
He went through heavy artillery fire to the side of a wounded comrade who was exposed to the enemy and in a helpless condition. Taking his wounded comrade with him, he returned to safety, his whole journey being made through an artillery barrage.

BANE, THOMAS P. (1307266)_____
  Near Busigny, France, Oct. 9, 1918.
  R—Knoxville, Tenn.
  B—South Boston, Va.
  G. O. No. 78, W. D., 1919.

Corporal, Company C, 117th Infantry, 30th Division.
Corporal Bane, while leading his squad in the advance with his company, was wounded by a machine-gun bullet in the head. Despite his wound he continued in the advance until the objective was reached and the position consolidated. Corporal Bane on the day previous, in company with two companions, rushed a near-by machine-gun nest, killing five of the enemy and capturing the remainder.

BANISTER, MORRIS A. (1210112). Near Mount Kemmel, Belgium, on the night of Aug. 17, 1918. R—Watertown, N. Y. B—Watertown, N. Y. G. O. No. 9, W. D., 1923.

Private, first class, Company C, 107th Infantry, 27th Division. While engaged with a working party in the repair of communicating trenches, he voluntarily went to the rescue of a British ration party which had been struck by a high-explosive shell. Under concentrated machine gun and rifle fire and continuous shell fire he assisted in carrying the British wounded for a distance of 125 yards while fully exposed to enemy fire and utterly without regard for his own safety.

BANK, CARL (280908). Near Juvigny, north of Soissons, France, Aug. 28, 1918. R—Lowell, Mich. B—Auburn, Ind. G. O. No. 139, W. D., 1918.

Sergeant, Company K, 126th Infantry, 32d Division. Though he was severely wounded by machine-gun fire, he continued to lead his platoon throughout the attack until his company re-formed and an emergency no longer existed.

***BANKS, LEONARD S. (2213782).** In the Foret-de-Fere, France, July 23, 1918. R—Farnum, Nebr. B—Webster County, Nebr. G. O. No. 32, W. D., 1919.

Private, Company G, 4th Infantry, 3d Division. Badly wounded while on patrol, Private Banks returned to his company to get assistance for wounded comrades. He then volunteered and led the first-aid men through heavy gas and shell bombardment to the place where his wounded comrades were. Posthumously awarded. Medal presented to father, Swan Banks.

BANN, EDWARD (2664718). In the Bois-des-Ogons, France, October 4, 1918. R—Pittsburgh, Pa. B—Pittsburgh, Pa. G. O. No. 7, W. D., 1919.

Private, Company M, 318th Infantry, 80th Division. He was acting as stretcher bearer with another soldier, who was shot by a sniper. Going out under fire from the sniper, he captured the latter with the aid of another man. While taking his prisoner to the rear he found a wounded man, whom he carried to the aid station under heavy fire, while his companion went on with the prisoner. Upon returning from the aid station he continued his work of rescuing the wounded.

BARBER, HENRY A., Jr. Near Moulins, France, July 14–15, 1918. R—Cambridge, Md. B—Fort Reno, Okla. G. O. No. 37, W. D., 1919.

First lieutenant, 9th Machine Gun Battalion, 3d Division. Seeing his right flank badly exposed to the enemy's advance across the Marne, Lieutenant Barber changed the position of two of his guns to meet this emergency, performing this during terrific enemy fire. He then ran a distance of 150 yards in the open to stop the fire of our infantry on our troops. Going forward to the aid of a wounded soldier, Lieutenant Barber administered first aid and was carrying the man to safety when the latter died. Picking up the one remaining undamaged gun, he opened fire on the enemy, who were crossing the river, sinking one boat, killing many, and causing the others to abandon their boats.

BARBER, THOMAS M. Near Cantigny, France, May 28–30, 1918. R—Charleston, W. Va. B—Charleston, W. Va. G. O. No. 99, W. D., 1918.

First lieutenant, Medical Corps, attached to 28th Infantry, 1st Division. He repeatedly demonstrated heroic self-sacrifice by caring for wounded under enemy fire and with apparent contempt for his own safety. When his aid station had been destroyed by shell fire, he promptly moved into a shell hole near by and continued his faithful work.

BARBIER, ALEX J. (1595630). Near Bantheville, France, Oct. 22, 1918. R—Whitecastle, La. B—Paincourtville, La. G. O. No. 3, W. D., 1921.

Private, Headquarters Company, 356th Infantry, 89th Division. While engaged in a raid on enemy positions, he was painfully wounded in the hand, but refused to go to the rear, remaining on duty with his platoon and taking an active part in the action for 24 hours after being wounded.

***BARBOUR, WILLIAM C. (1315081).** Near Busigny, France, Oct. 9, 1918. R—Smithfield, N. C. B—Smithfield, N. C. G. O. No. 44, W. D., 1919.

Private, first class, Company C, 119th Infantry, 30th Division. During the operations near Busigny on Oct. 9, he, with one other soldier, voluntarily left his place of comparative safety and advanced into the open in the face of close-range machine-gun fire to rescue a severely wounded comrade. He received a severe wound while engaged in this self-appointed task, from which he later died. Posthumously awarded. Medal presented to mother, Mrs. Mandy Barbour.

BARCSYKOWSKI, FRANK JOHN. Near Vierzy, France, July 19, 1918. R—Buffalo, N. Y. B—Buffalo, N. Y. G. O. No. 71, W. D., 1919.

Private, 55th Company, 5th Regiment, U. S. Marine Corps, 2d Division. He displayed exceptional bravery in charging three machine guns with the aid of a small detachment of his comrades, killing the crews and capturing the guns, which were immediately turned on the Germans, thereby opening the line for the advance of his company, which had been held up by the enemy's fire. Oak-leaf cluster. For the following act of extraordinary heroism in action in the Bois-de-Belleau, France, June 11, 1918, Private Barcsykowski is awarded an oak-leaf cluster, to be worn with the distinguished-service cross: When all the other members of their group had been killed or wounded by fire from an enemy machine gun Private Barcsykowski and another soldier charged this gun and killed the entire crew.

BARD, FRANKLIN C. (1899616). Near St. Juvin, France, Oct. 14–18, 1918. R—Adams, Mass. B—Adams, Mass. G. O. No. 78, W. D., 1919.

Sergeant, Company K, 325th Infantry, 82d Division. After all the officers of his company had become casualties, Sergeant Bard reorganized the company with notable bravery and skill and led it forward to its objective. Retaining command next day, though handicapped by numerous casualties, he kept his men well in hand and successfully led them in all advances which were ordered, inspiring them by his courage and coolness.

*BARDMAN, BARNEY (1704820)_____
   Near Grand Pre, France, Oct. 15,
    1918.
   R—Brooklyn, N. Y.
   B—Russia.
   G. O. No. 142, W. D., 1918.

Private, first class, Company B, 307th Infantry, 77th Division.
He, acting as scout for his platoon, was attempting to effect a crossing over the Aire River under heavy machine-gun and sniper fire. When he was struck by a bullet and mortally wounded he called out a warning to the other five members of the platoon to take cover, thereby saving many lives.
Posthumously awarded. Medal forwarded to Samuel Bardman, cousin, for delivery to the next of kin.

BARFIELD, HARRY M._____
   Near Mezy, France, July 15, 1918.
   R—Macon, Ga.
   B—Macon, Ga.
   G. O. No. 64, W. D., 1919.

Second lieutenant, 38th Infantry, 3d Division.
During the battle of the Marne, near Mezy, Lieutenant Barfield rallied the men of a company who were falling back in disorder, formed them, and stopped what otherwise would have been a rout. This in the face of heavy machine-gun and artillery fire. At the same time he managed and controlled the fire of his four machine guns so as to deliver an annihilating fire on the enemy.

BARKALOW, JAMES W. (238115)_____
   Near Exermont, France, Oct. 9,
    1918.
   R—South Amboy, N. J.
   B—South Amboy, N. J.
   G. O. No. 9, W. D., 1923.

Private, Company C, 18th Infantry, 1st Division.
After all, except himself and one other man, of two squads had been either killed or wounded, Private Barkalow gathered together seven men and continued to advance in the face of heavy shell and machine-gun fire, capturing an enemy machine gun which had been causing heavy losses in our lines, killing two of the operators and taking the third prisoner. He then continued his advance to a trench where he and his men captured 35 additional prisoners. Sending these prisoners to the rear under proper guard he again advanced and captured a trench mortar and 5 German prisoners, all these acts being performed under heavy fire.

BARKER, MANDEVILLE J., Jr._____
   Near Baslieux, France, Sept. 15,
    1918.
   R—Garrettsville, Ohio.
   B—Rochester, N. Y.
   G. O. No. 140, W. D., 1918.

Secretary, Y. M. C. A., on duty with 108th Machine Gun Battalion, 28th Division.
Mr. Barker showed a fearless disregard of his own safety by crawling out in front of the line under heavy enemy machine-gun and sniper fire to aid wounded soldiers, whom he carried back to shelter after dressing their wounds. He also administered aid to a wounded German within 20 yards of the enemy lines and brought him in a prisoner.

BARKSDALE, ALFRED D._____
   Near Samogneux, France, Oct. 8,
    1918; near Molleville, France, Oct.
    12, 1918; and in the Bois-de-la-
    Grande Montagne, France, Oct.15,
    1918.
   R—Lynchburg, Va.
   B—Houston, Va.
   G. O. No. 44, W. D., 1919.

Captain, 116th Infantry, 29th Division.
Commanding a support company during the attack of Oct. 8, Captain Barksdale discovered that his battalion had advanced ahead of the unit on the right flank, and was suffering heavy losses from machine-gun fire. Without orders he attacked and captured the guns, taking many prisoners. On Oct. 12 he worked for over an hour, exposed to a terrific bombardment, binding the wounds of his men. On Oct. 15 he advanced alone into a thick wood and, with the aid of his pistol, put out of action a destructive machine gun which was pouring such a deadly fire that his men could not raise their heads.

BARLOW, ALFRED M._____
   Near Heurne, Belgium, Nov. 3, 1918.
   R—Gallipolis, Ohio.
   B—Gallipolis, Ohio.
   G. O. No. 37, W. D., 1919.

First lieutenant, 148th Infantry, 37th Division.
Although suffering from a painful shrapnel wound in the leg, he led his company, with excellent leadership and command, over the river, and not until he had received wounds in both legs would he give his consent to be taken to a dressing station.

BARLOW, CLYDE (554644)_____
   Near Cierges, France, Oct. 9, 1918.
   R—Town Bluff, Tex.
   B—Town Bluff, Tex.
   G. O. No. 2, W. D., 1920.

Private, Company A, 9th Machine Gun Battalion, 3d Division.
Private Barlow exposed himself to artillery and direct machine-gun fire while going 300 yards in advance of our front lines to render first aid to a wounded soldier, whom he carried to shelter with the assistance of a comrade.

*BARLOW, FRANCIS A._____
   Near Ourcq River, France, July 31,
    1918.
   R—Cheboygan, Mich.
   B—Edmore, Mich.
   G. O. No. 46, W. D., 1919.

First lieutenant, 125th Infantry, 32d Division.
Lieutenant Barlow was severely wounded in the passage of the Ourcq River and the capture of the heights beyond, but continued in command of his company. When another officer of his company was wounded he attempted to carry him to a place of safety, but was physically unable to do so. Although repeatedly urged to go to the rear, he continually refused and remained in command of the company until it was ordered to withdraw.
Posthumously awarded. Medal presented to widow, Mrs. Eva A. Barlow.

BARNARD, COURTNEY H._____
   At Cheppy, France, Sept. 26, 1918.
   R—Albany, N. Y.
   B—Albany, N. Y.
   G. O. No. 4, W. D., 1923.

Captain, 345th Battalion, Tank Corps.
During the attack on Cheppy, he personally reconnoitered on foot a passage through an extensive mine field while under heavy artillery and machine-gun fire, and later led his company through this mine field and destroyed the machine-gun nests which were holding up the advance of our Infantry. Captain Barnard was exposed to terrific machine-gun and artillery fire throughout the entire action, and by his coolness and courage set a splendid example to his men, which produced a noticeable determination on their part, which materially aided in the success of the attack.

BARNES, HARRY C., Jr._____
   In the Bois-Brule, near St. Die, in
    the Vosges, France, on July 21,
    1918.
   R—Tulsa, Okla.
   B—Guthrie, Okla.
   G. O. No. 24, W. D., 1920.

First lieutenant, 6th Infantry, 5th Division.
He showed good judgment, dash, bravery, and determination in leading his patrol into the enemy's lines in the Bois-Brule, near St. Die, in the Vosges, on July 21, 1918. Although seriously wounded, he continued to direct his men, and succeeded in driving off the enemy, at the same time wounding four of the German patrol with his own revolver.

BARNES, JULIUS L. (62418)_____
   Near Vaux, France, July 15, 1918.
   R—Stow, Mass.
   B—Everett, Mass.
   G. O. No. 39, W. D., 1920.

Private, Company M, 101st Infantry, 26th Division.
When two members of a small group were severely wounded during a counterattack against the enemy, Private Barnes advanced twice across the open under machine-gun fire and brought each man back to cover.

BARNES, RAYMOND (106163)_____ | Private, Company B, 3d Machine Gun Battalion, 1st Division.
Near Berzy-le-Sec, France, July 18, 1918. | On July 18, 1918, near Berzy-le-Sec, France, he was severely wounded by a shell, but as soon as he regained consciousness he went forward, rejoined former position in squad, and fought with it until ordered to an aid station by his platoon commander.
R—Taylorsville, N. C. |
B—Alexander County, N. C. |
G. O. No. 109, W. D., 1918. |

BARNES, ROBERT (1403375)_____ | Sergeant, Company L, 370th Infantry, 93d Division.
Near Mont de Singes, France, Sept. 19, 1918. | After starting on a mission to reconnoiter the front lines of the enemy he received wounds which proved serious. Disregarding advice to return, he continued, collecting valuable information, which was submitted to his company commander through a member of his patrol party.
R—Danville, Ill. |
B—Danville, Ill. |
G. O. No. 37, W. D., 1919. |

BARNETT, CECIL E. (1306573)_____ | Private, Machine Gun Company, 117th Infantry, 30th Division.
Near Gusigny, France, Oct. 8–9, 1918 | After repeatedly carrying messages under heavy fire, he was painfully wounded while acting as a connecting file, but he continued on duty until he was ordered to the dressing station. He then insisted on being permitted to rejoin his company rather than be sent to the hospital.
R—Pittsburg Landing, Tenn. |
B—Pittsburg Landing, Tenn. |
G. O. No. 37, W. D., 1919. |

BARNETT, LELAND M._____ | First lieutenant, 148th Infantry, 37th Division.
Near Ivoiry, France, Sept. 27, 1918. | Having become separated from his battalion headquarters, Lieutenant Barnett, battalion adjutant, on his own initiative undertook, under heavy artillery and machine gun fire, to locate machine-gun nests which were hindering the advance. He ignored the warnings of his orderly as to the danger of this work and continued at it until he was killed.
R—Norwood, Ohio. | Posthumously awarded. Medal presented to widow, Mrs. Leland M. Barnett.
B—Columbus, Ohio. |
G. O. No. 126, W. D., 1919. |

BARNHART, FRANK A._____ | Sergeant, Headquarters Company, 5th Regiment, U. S. Marine Corps, 2d Division.
Near Somme-Py, France, Oct. 4, 1918. | He left his trench several times and helped to carry wounded soldiers from the field through machine gun and artillery fire.
R—Roberton, Ohio. |
B—Ellis City, Pa. |
G. O. No. 46, W. D., 1919. |

BARNHART, WALTER I. (1830737)____ | Sergeant, Company I, 320th Infantry, 80th Division.
Near Bois d'en dela, France, Sept. 26, 1918. | Accompanied by an officer, Sergeant Barnhart advanced against an enemy machine-gun nest which was holding up the advance of his company. When they had reached a point within 50 yards of the gun they were fired upon, and the officer was severely wounded. Sergeant Barnhart immediately picked the officer up and brought him through a hail of machine-gun bullets to a position of safety.
R—Saltdenville, Pa. |
B—Ligonier Valley, Pa. |
G. O. No. 44, W. D., 1919. |

BARNWELL, FRANK H._____ | Second lieutenant, 26th Infantry, 1st Division.
At Soissons, France, July 18, 1918. | He distinguished himself by heroic and inspiring leadership at Soissons, France, July 18, 1918, directing his platoon with unusual effectiveness until he fell wounded.
R—Memphis, Tenn. |
B—Yazoo City, Miss. |
G. O. No. 132, W. D., 1918. |

BARR, JOSEPH T. (1550805)._____ | Sergeant, Battery D, 76th Field Artillery, 3d Division.
Near Chateau-Thierry, France, during the night of July 14 and 15, 1918. | While in command of a "roving" gun section, Sergeant Barr took up a position about 500 yards in rear of the front line and executed a fire mission while exposed to heavy enemy bombardment. Although wounded, he refused to go to the rear until his mission was completed and the gun returned to the battery.
R—South Orange, N. J. |
B—Avondale, N. J. |
G. O. No. 16, W. D., 1920. |

BARRETT, HERBERT W._____ | Second lieutenant, 9th Infantry, 2d Division.
Near Blanc Mont, France, Oct. 3, 1918. | Reorganizing his company after the other officers had become casualties, Lieutenant Barrett led them in an attack, capturing a machine-gun nest, capturing or killing the crew. Under heavy fire, he rescued two of his men who had been wounded and buried by a high-explosive shell. He was wounded while administering first aid to one of his men under machine-gun fire.
R—Somerville, Mass. |
B—Somerville, Mass. |
G. O. No. 87, W. D., 1919. |

*BARROW, WILLIAM LAWRENCE_____ | Private, 16th Company, 5th Regiment, U. S. Marine Corps, 2d Division.
Near Chateau-Thierry, France, June 23, 1918. | After making several successful journeys over an area swept by artillery, machine gun, and rifle fire in his capacity as runner, he was caught in a heavy barrage and seriously wounded. By a superhuman effort he made his way to his objective, and after being dragged into safety insisted on personally delivering his message. He then fell exhausted from loss of blood.
R—North Tonawanda, N. Y. | Posthumously awarded. Medal presented to mother, Mrs. Della Wires.
B—North Tonawanda, N. Y. |
G. O. No. 44, W. D., 1919. |

BARROW, CHARLES L._____ | First lieutenant, 149th Machine Gun Battalion, 42d Division.
Northwest of Chateau-Thierry, France, July 30–31, 1918 | Displaying the greatest courage and disregard of personal danger, at all times throughout the critical period of 48 hours, July 30–31, 1918, near Sergy, he particularly distinguished himself when leading his platoon into position in the face of fire on the crest of Hill 212 and when presiding over the reorganization of the position thus won.
R—Austin, Tex. |
B—Houston, Tex. |
G. O. No. 4, W. D., 1927. |

BARROWS, ALBERT EDWARD_____
Near Vierzy, France, July 19, 1918.
R—Lynn, Mass.
B—Lynn, Mass.
G. O. No. 126, W. D., 1919.

Private, 55th Company, 5th Regiment, U. S. Marine Corps, 2d Division.
He displayed exceptional bravery in charging three machine guns with the aid of a small detachment of his comrades, killing the crews and capturing the guns, which were immediately turned on the Germans, thereby opening the line for the advance of his company, which had been held up by the enemy's fire.
Oak-leaf cluster.
For the following act of extraordinary heroism in action in the Bois-de-Belleau, France, June 11, 1918, Private Barrows was awarded an oak-leaf cluster: After the remainder of their group had become casualties from machine-gun fire, Private Barrows and a comrade displayed exceptional courage in charging and killing the crew of the hostile machine gun which had held up the advance.

BARROWS, CLAYTON E. (554136)_____
Near Le Rocq Farm, France, July 14–15, 1918.
R—Springfield, Mass.
B—Burlington, Vt.
G. O. No. 98, W. D., 1919.

Private, Company D, 8th Machine Gun Battalion, 3d Division.
On duty as a runner at battalion headquarters Private Barrows was sent with a message to his company commander during a terrific enemy bombardment. Passing for 2½ kilometers over ground where gas and high-explosive shells were constantly falling, he was forced to tear off his gas mask in order to find his way, but he succeeded in delivering his message, falling exhausted just as his mission was completed.

*BARRY, EDWARD (2088934)_____
Near Bois-du-Barricourt, France, Nov. 1, 1918.
R—Chicago, Ill.
B—Chicago, Ill.
G. O. No. 126, W. D., 1919.

Corporal, Company K, 354th Infantry, 89th Division.
After receiving a dangerous wound in the chest from a shell fragment, Corporal Barry insisted on going forward with his combat group. He pushed forward through intense machine-gun fire until again hit by a machine-gun bullet, which caused his death.
Posthumously awarded. Medal presented to father, J. T. Barry.

*BARRY, EDWARD W. (1751055)_____
Near Grand Pre, France, Oct. 25, 1918.
R—Palmyra, N. Y.
B—Farmington, N. Y.
G. O. No. 37, W. D. 1919.

Private, Company I, 311th Infantry, 78th Division.
Acting as stretcher bearer, under heavy machine-gun fire, he was wounded, but, disregarding his own injuries, he went to the aid of another wounded man, and while attending him was killed by shrapnel.
Posthumously awarded. Medal presented to mother, Mrs. Cora Barry.

BARRY, HERBERT E. (2852029)_____
Near Fey-en-Haye, France, Sept. 12, 1918.
R—Brainerd, Minn.
B—Iron Mountain, Mich.
G. O. No. 128, W. D., 1918.

Private, Company D, 359th Infantry, 90th Division.
Private Barry, while carrying a message through a heavy barrage, was knocked down by shell explosions several times, but with heroic devotion to duty he continued on and delivered his message. On the return trip he was rendered unconscious for three hours by a shell. Upon being revived he immediately reported to battalion headquarters for duty.

BARRY, JOHN (57885)_____
Near Soissons, France, July 19, 1918.
R—New York, N. Y.
B—New York, N. Y
G. O. No. 68, W. D., 1920.

Sergeant, Company H, 28th Infantry, 1st Division.
After his platoon had been halted by the fire from a concealed machine-gun nest, Sergeant Barry, alone, exposed to heavy enemy fire, advanced and attacked the enemy position. He, single handed, put the machine-gun nest out of action and thus enabled his platoon to continue the advance.

BARRY, WILLIAM H._____
Near Exermont, France, Oct. 5, 1918.
R—Langley, Wash.
B—Mexico.
G. O. No. 103, W. D., 1919.

Second lieutenant, 28th Infantry, 1st Division.
Assuming command of his company after his company commander and a major portion of the company became casualties, he reorganized his company and personally led it forward in the attack, successfully attaining his objective in the face of intense machine-gun and artillery fire. He constantly exposed himself to enemy fire in order to encourage and insure the protection of his men.

BARTELS, HERMAN B. (1383967)_____
At Marcheville, France, Nov. 10, 1918.
R—Effingham, Ill.
B—Effingham, Ill.
G. O. No. 23, W. D., 1919.

Corporal, Company F, 130th Infantry, 33d Division.
While maintaining liaison with an adjacent company during a raid, Corporal Bartels personally cleared out two dugouts, taking a number of prisoners. He was later wounded, but continued in action until he fell.

BARTH, FREDERICK (1697302)_____
Near Villers-devant-Muzon, France, Nov. 8, 1918.
R—Brooklyn, N. Y.
B—New York, N. Y.
G. O. No. 24, W. D., 1920.

Corporal, Company C, 305th Infantry, 77th Division.
Corporal Barth with an officer penetrated about 3 kilometers into the enemy lines. Due to the fearlessness of Corporal Barth, the patrol evaded the challenge of an enemy sentry and returned to our lines with valuable information. Previously, on Sept. 2, 1918, near Bazoches, Corporal Barth swam the Vesle and fixed a rope for the crossing of a patrol. The patrol penetrated the enemy position and reconnoitered same, during which time Corporal Barth was wounded while going, under enemy fire, to the assistance of a wounded comrade whom he helped to our lines after obtaining special information of the enemy. The return of the patrol was made possible only by the heroic efforts of Corporal Barth.

BARTHOLF, HERBERT B._____
Near Ancerville, France, Oct. 30, 1918, and near Baalon, France, Nov. 4, 1918.
R—Glencoe, Ill.
B—Chicago, Ill.
G. O. No. 35, W. D., 1919

First lieutenant, pilot, 103d Aero Squadron, Air Service.
On Oct. 30, in the region of Ancerville, Lieutenant Bartholf, with one other pilot, engaged five enemy planes. Outnumbered, he did not hesitate to attack, and, although subjected to the severe fire of five enemy planes, he succeeded in destroying one. On Nov. 4, in the region of Baalon, while on a bombing expedition, he encountered an enemy patrol of eight machines (Fokker type). He immediately dived into their formation and, despite the severe fire to which he was subjected, continued a spirited combat with one of the enemy until it crashed to the ground.

BARTLETT, ELMER E., Jr. (2405377)_____
  Near Medeah Ferme France, Oct.
    9, 1918.
  R—Florence, N. J.
  B—Florence, N. J.
  G. O. No. 23, W. D., 1919.

Private, Company C, 2d Engineers, 2d Division.
Crawling forward under heavy machine-gun fire, he assisted in bringing a wounded comrade to safety.

BARTLETT, GEORGE W. (54347)_____
  Near Soissons, France, July 19, 1918.
  R—Honeoye, N. Y.
  B—Rochester, N. Y.
  G. O. No. 44, W. D., 1919.

Private, first class, Company I, 26th Infantry, 1st Division.
Acting in the capacity of battalion runner, Private Bartlett volunteered and carried an important attack order from his regimental headquarters to the front-line battalion through a terrific artillery bombardment. He was twice knocked down while in the performance of this mission, but successfully delivered his message on time. Although in a dazed condition upon reaching the forward line, he again volunteered to carry a message across an area swept by enemy machine-gun fire.

BARTLETT, MURRAY_____
  In Villers-Tournelle, France, May,
    1918, and in the Chazelle Ravine,
    south of Soissons, France, July 22,
    1918.
  B—Geneva, N. Y.
  R—Poughkeepsie, N. Y.
  G. O. No. 15, W. D., 1923.

Secretary, Y. M. C. A., acting assistant chaplain, 18th Infantry, 1st Division.
Voluntarily assuming the duties of chaplain, 18th Infantry, he displayed conspicuous bravery in caring for the wounded and burying the dead of his regiment under intense enemy fire, working constantly with the advanced elements of the command until July 22, 1918, when he was seriously wounded while in close proximity to the front line. His cheerful, heroic energy and indifference to personal danger exerted a profound effect upon the morale of the men of his regiment and inspired them to many deeds of gallantry and supreme devotion to duty.

*BARTO, TOM F. (568752)_____
  On the Vesle River, near Ville-
    Savoye, France, Aug. 11, 1918.
  R—Bellingham, Wash.
  B—Neosho, Mo.
  G. O. No. 129, W. D., 1918.

Corporal, Company D, 4th Engineers, 4th Division.
He volunteered to go into Ville-Savoye at a time when it was under a heavy bombardment to rescue a wounded officer.
Posthumously awarded. Medal presented to father, Alexander Barto.

BARTON, CHARLES R. (69290)_____
  Near Belleau Wood, France, July
    20, 1918.
  R—Houlton, Me.
  B—Houlton, Me.
  G. O. No. 37, W. D., 1919.

Sergeant, Company L, 103d Infantry, 26th Division.
Wounded in the leg by a machine-gun bullet, he dressed his wound and continued to lead his section in a successful attack on two machine-gun nests. He remained on duty for 14 hours, until weakness forced him to be evacuated.

BARTON, HARRY D. (125468)_____
  Near Fleville, France, Oct. 4, 1918.
  R—Youngstown, Ohio.
  B—Cincinnati, Ohio.
  G. O. No. 37, W. D., 1919.

Corporal, Battery A, 6th Field Artillery, 1st Division.
Being in charge of one of the sections of drivers of his platoon when it was caught in an enemy barrage, Corporal Barton unlimbered his gun and caisson, removed his horses and drivers to a place of safety, and returning to his platoon acted as runner, passing four times over heavily shelled areas.

BARTON, JESSE M._____
  Near Becquigny, France, Oct. 17,
    1918.
  R—Barton, Ohio.
  B—Barton, Ohio.
  G. O. No. 68, W. D., 1920.

Second lieutenant, 118th Infantry, 30th Division.
After his superior officer had been wounded, he assumed command of and personally led the advance of his unit until he was struck by an enemy shell and severely wounded. Although suffering intense pain and almost unconscious, he refused to be evacuated until after he had given instructions to the platoon sergeant to continue the advance. His gallant conduct was an inspiring example to the men of his platoon.

BARTON, THOMAS D._____
  Near St. Etienne, France, Oct. 8–10,
    1918.
  R—Amarillo, Tex.
  B—Kilgore, Tex.
  G. O. No. 81, W. D., 1919.

Captain, 142d Infantry, 36th Division.
Captain Barton advanced his company against a strongly fortified enemy position, and succeeded in capturing the enemy works, together with 20 machine guns and 90 prisoners. After he had lost all his company officers and sustained many casualties in his command, Captain Barton again moved forward through an intense barrage and established the most advanced position of the first day of the battle.

BASCOM, ROBERT (80766)_____
  Near Badricourt, Alsace, France,
    July 19, 1918.
  R—Portland, Oreg.
  B—Rochester, N. Y.
  G. O. No. 22, W. D., 1920.

Private, Company E, 127th Infantry, 32d Division.
During an enemy raid on a platoon sector Private Bascom, although wounded three times by shell fragments, carried a message through artillery barrages to the company headquarters and returned with reinforcements. He did not accept first aid until after the raid had been repulsed.

*BASS, URBANE F._____
  Near Monthois, France, Oct. 1–6,
    1918.
  R—Fredericksburg, Va.
  B—Richmond, Va.
  G. O. No. 13, W. D., 1919.

First lieutenant, Medical Corps, attached to 372d Infantry, 93d Division.
During the attack on Monthois he administered first aid in the open under prolonged and intense shell fire until he was severely wounded and carried from the field.
Posthumously awarded. Medal presented to widow, Mrs. Maude L. Bass.

*BASSETT, REXFORD O. (2249609)_____
  Near Les Huit Chemins, France,
    Sept. 26, 1918.
  R—Enid, Okla.
  B—Denver, Colo.
  G. O. No. 72, W. D., 1920.

Corporal, Company D, 358th Infantry, 90th Division.
Corporal Bassett volunteered to rush an enemy machine gun which had concentrated its fire on an opening in some barbed-wire entanglement through which his company was endeavoring to advance. He gallantly attacked the pill box and temporarily silenced one of the fire openings when he was mortally wounded.
Posthumously awarded. Medal presented to father, Clarence W. Bassett.

BASSETT, WALDO S. (40552)_____
  South of Soissons, France, July 18,
    1918.
  R—Franklin, Mass.
  B—Natick, Mass.
  G. O. No. 117, W. D., 1918

Corporal, Company L, 9th Infantry, 2d Division.
He volunteered to carry messages through intense shell and machine-gun fire after all of the runners had been killed or wounded. He was wounded while carrying a message, but continued on in spite of his injuries until it was delivered, and refused to go to the rear for treatment until ordered to do so by his company commander.

BASSI, JOSEPH (562059).
Near Bois-du-Fays, France, Oct. 4-7, 1918.
R—Memphis, Tenn.
B—Italy.
G. O. No. 71, W. D., 1919.

Private, Company I, 59th Infantry, 4th Division.
Showing marked personal courage, he repeatedly crossed ground swept by heavy artillery and machine-gun fire to deliver important messages. He volunteered for dangerous missions, his example being an inspiration to other runners.

BASSMAN, BARNETT (1897412).
Near St. Juvin, France, Oct. 16, 1918.
R—New York, N. Y.
B—Russia.
G. O. No. 46, W. D., 1919.

Private, Company A, 325th Infantry, 82d Division.
When many squads of his company had broken up and the men scattered, Private Bassman with great courage collected 15 men who had become separated from their squads and organized them into a provisional platoon, which he successfully led in the attack, thereby contributing materially to the success of his company.

BASTON, ALBERT P.
Near Chateau-Thierry, France, June 6, 1918.
R—St. Louis Park, Minn.
B—St. Louis Park, Minn.
G. O. No. 99, W. D., 1918.

First lieutenant, 5th Regiment, United States Marine Corps, 2d Division.
Although shot in both legs while leading his platoon through the woods at Hill 142, near Chateau-Thierry, France, on June 6, 1918, he refused treatment until after he had personally assured himself that every man in his platoon was under cover and in good firing position.

BATCHELDER, HAROLD W.
Near Bois d'Aigremont, France, July 15, 1918.
R—Hardwick, Vt.
B—South Ryegate, Vt.
G. O. No. 46, W. D., 1919.

First lieutenant, 30th Infantry, 3d Division.
When it seemed impossible for a runner to get through the violent barrage, he volunteered and carried an important message to regimental headquarters, returning with an answer.

\TEMAN, CHARLES W.
Near Thiaucourt, France, Sept. 12-15, 1918.
R—Booneville, Ind.
B—Booneville, Ind.
G. O. No. 46, W. D., 1919.

Pharmacist's mate, third class, U. S. Navy, attached to 6th Machine Gun Battalion, United States Marine Corps, 2d Division.
Rendering first aid under heavy artillery and machine-gun fire, he showed utter disregard for his own personal safety, venturing through shelled areas to the assistance of the wounded. He obtained most valuable information for the guidance of stretcher bearers.

BATEMAN, HENRY (1705363).
Near St. Pierremont, France, Nov. 4, 1918.
R—Buffalo, N. Y.
B—Salamanca, N. Y.
G. O. No. 37, W. D., 1919.

Corporal, Headquarters Company, 307th Infantry, 77th Division.
After passing through a heavily bombarded area, he learned that a soldier of his platoon had been wounded and had fallen in the shelled area. He at once volunteered and went back for him, assisted in bringing him to a place of safety, and later helped to carry him through another shelled area to the first-aid station.

BATEMAN, HENRY E.
Near Verdun, France, Oct. 12-13, 1918.
R—Easton, Md.
B—Easton, Md.
G. O. No. 44, W. D., 1919.

First lieutenant, 114th Infantry, 29th Division.
After all the battalion runners had become casualties, he volunteered and carried important messages to the rear through violent artillery and machine-gun fire. Next day this officer went through a wood occupied by enemy machine-gun nests and snipers and established liaison with three companies in an advanced position.

BATES, BRET V.
Near Chaudron Farm, France, Oct. 1, 1918.
R—Wheaton, Minn.
B—Springfield, Nebr.
G. O. No. 70, W. D., 1919.

First lieutenant, Medical Corps, 139th Ambulance Company, 110th Sanitary Train, 35th Division.
When the sanitary detachment with which he was working was ordered to fall back under intense artillery and machine-gun fire, Lieutenant Bates, upon his own volition and contrary to the advice of others, refused to leave the wounded men who had not been evacuated, but remained with them throughout the day, ministering to them under the most violent fire, in utter disregard for his own safety. When night came he secured litter bearers and succeeded in taking all the wounded to safety.

BATES, CHARLES E. H.
Near Marcheville, France, Sept. 26, 1918.
R—Alameda, Calif.
B—Alameda, Calif.
G. O. No. 46, W. D., 1919.

Second lieutenant, 103d Infantry, 26th Division.
He displayed the highest quality of courage and leadership in leading his platoon through to its objective under a heavy barrage of machine-gun and artillery fire, without flank support. He held his objective under murderous artillery and machine-gun fire until relieved.

*BATES, PAUL A. (546660).
Near Crezancy, France, July 15, 1918, and near Cunel, France, Oct. 10, 1918.
R—Williamstown, Pa.
B—Williamstown, Pa.
G. O. No. 32, W. D., 1919.

Private, Company F, 30th Infantry, 3d Division.
During a terrific bombardment on the morning of July 15 Private Bates carried in and dressed the wounded at great personal risk to himself. On Oct. 10, near Cunel, although fatally wounded, he continued to command his men in attack on machine guns until the end.
Posthumously awarded. Medal presented to sister, Mrs. Vida Viola Snell.

BATLEY, HAROLD (1681167).
Near Badonvilliers, France, June 24, 1918.
R—Lawrence, Mass.
B—Dover, N. H.
G. O. No. 123, W. D., 1919.

Private, Company C, 308th Infantry, 77th Division.
Private Batley, after two patrols had failed, volunteered and went alone to the grouped combat through the barrage and brought back information of the highest value.

BATSON, GEORGE WELLS (1309986).
Near Bellicourt, France, Sept. 26, 1918.
R—Greenville, S. C.
B—Greenville, S. C.
G. O. No. 35, W. D., 1919.

Corporal, Company A, 118th Infantry, 30th Division.
With absolute disregard for his personal safety, he went 300 yards beyond the front line, in full view of the enemy and under heavy machine-gun fire, and brought back a wounded soldier.

BATTA, FRANK................................
Near Fismes and Fismette, France, Aug. 10, 1918.
R—Chillicothe, Mo.
B—Mooresville, Mo.
G. O. No. 126, W. D., 1919.

Second lieutenant, 111th Infantry, 28th Division.
By bravely taking an exposed position at great risk to himself, he successfully maneuvered his command across a railroad track which was enfiladed by machine-gun and sniper fire. While doing so he was wounded in the neck, yet he fearlessly led his troops in a successful assault. His exceptional courage and initiative inspired his men to a victorious attack.

BATTEN, HAROLD A. (109326)............
At Marcheville, France, Sept. 26, 1918.
R—South Boston, Mass.
B—England.
G. O. No. 15, W. D., 1919.

Private, first class, Company A, 102d Machine Gun Battalion, 26th Division.
While the squad to which he belonged was proceeding to its objective, all the other members were either killed or wounded. After procuring assistance for the wounded soldiers, he immediately attached himself to another squad of his section and remained on duty with it throughout the day.

BAUCOM, BYRNE V........................
In the Chateau-Thierry and St. Mihiel salients, France, June and July, 1918; Sept. 12–16, 1918.
R—Milford, Tex.
B—Milford, Tex.
G. O. No. 64, W. D., 1919.

Second lieutenant, observer, 1st Aero Squadron, Air Service.
Lieutenant Baucom, with First Lieut. William P. Erwin, pilot, by a long period of faithful and heroic operations, set an inspiring example of courage and devotion to duty to his entire squadron. Throughout the Chateau-Thierry actions in June and July, 1918, he flew under the worst weather conditions and successfully carried out his missions in the face of heavy odds. In the St. Mihiel sector, Sept. 12–16, 1918, he repeated his previous courageous work. He flew as low as 50 feet from the ground behind the enemy's lines, harassing German troops with machine guns and rifles. He twice drove off enemy planes which were attempting to destroy an American observation balloon. On Sept. 12–13, 1918, he flew at extremely low altitudes and carried out infantry contact patrols successfully. Again on Sept. 12 he attacked a German battery, forced the crew to abandon it, shot off his horse a German officer who was trying to escape, drove the cannoneers to their dugouts, and kept them there until the Infantry could come up and capture them.
Oak-leaf cluster.
For the following act of extraordinary heroism in action near Sedan, France, Nov. 5, 1918, Lieutenant Baucom is awarded an oak-leaf cluster to be worn with the distinguished-service cross awarded him Oct. 1, 1918. With atmospheric conditions such that flying was nearly impossible he voluntarily undertook a flight as observer to locate the position of enemy troops and machine-gun nests which had been holding up our advance and causing severe casualties. Forced to fly at a very low altitude and subject to almost constant antiaircraft and rifle fire, he obtained the information that was vital to the success of our operations and dropped the message at division headquarters. He then penetrated far into the enemy lines and opening fire upon enemy crews routed them from a series of machine-gun nests. When his machine was finally shot down, he succeeded in operating the gun and beat off an attack by the enemy in force. Armed only with revolvers and German grenades which they found in an enemy emplacement, he and his observer then worked their way back to the American lines with valuable information, repeatedly subjected to enemy fire on their way.

BAUERNFIEND, JOHN R. (1812747)......
Near Verdun, France, Nov. 5, 1918.
R—Baltimore, Md.
B—Pittsburgh, Pa.
G. O. No. 44, W. D., 1919.

Private, Company B, 310th Machine Gun Battalion, 79th Division.
With two other soldiers, he voluntarily left a place of safety, went forward 40 meters under machine-gun fire in plain view of the enemy, and rescued another soldier who had been blinded by a machine-gun bullet, and was helplessly staggering about.

*BAUGHN, ROBERT O. (43098)............
In the Argonne Forest, France, Oct. 4, 1918.
R—Calhoun, Ky.
B—Calhoun, Ky.
G. O. No. 44, W. D., 1919.

Sergeant, Company G, 16th Infantry, 1st Division.
During offensive operations he carried important messages across fire-swept territory, continuing with this work until seriously wounded.
Posthumously awarded. Medal presented to mother, Mrs. Martha F. Guy.

BAUME, JOHN............................

Near St. Etienne, France, Oct. 3–5, 1918.
R—Rochester, N. Y.
B—England.
G. O. No. 87, W. D., 1919.

Pharmacist's mate, first class, U. S. Navy, attached to 1st Battalion, 5th Regiment, U. S. Marine Corps, 2d Division.
Pharmacist's Mate Baume gave aid to the wounded under shell and machine-gun fire and went forward several times during the advance to locate advanced dressing stations.

BAXTER, ALBERT F......................
In the Bois-de-Barricourt, France, Nov. 1, 1918.
R—San Jose, Calif.
B—New York, N. Y.
G. O. No. 95, W. D., 1919.

Captain, 353d Infantry, 89th Division.
When his company was held up by machine-gun nests on three separate occasions Captain Baxter moved forward to a position in advance of his leading elements and, with rifle grenades, put the machine-gun nests out of action. On each of these occasions it was Captain Baxter's act that enabled his company to continue its advance. Throughout the attack his leadership and initiative were largely responsible for the success of his company.

BAXTER, GEORGE K. (1210168)..........
Near Ronssoy, France, Sept. 29, 1918.
R—Watertown, N. Y.
B—Mechanicville, N. Y.
G. O. No. 16, W. D., 1920.

Private, first class, Company C, 107th Infantry, 27th Division.
Private Baxter was a scout for a Lewis gun squad during the attack. After all members of his squad, except the corporal, had been killed he ran 30 yards, picked up the Lewis gun, and opened fire on a group of 20 of the enemy who were attempting to turn the flank of platoon. He was exposed to heavy machine-gun fire, but his own fire was so effective that his unit was able to continue the advance.

BAXTER, STUART A. _____
Near Montdidier, France, June 5, 1918; near Soissons, France, July 21, 1918; and near Verdun, France, Oct. 4, 1918.
R—Detroit, Mich.
B—Brooklyn, N. Y.
G. O. No. 21, W. D., 1919.

Second lieutenant, 26th Infantry, 1st Division.
On June 5 Lieutenant Baxter led a patrol across a heavily shelled area and established liaison with an adjoining battalion. On July 21, while his platoon was being held up by terrific machine-gun fire, he crawled forward to dress the wounds of his men, and he so encouraged and rallied his men that further advance was made possible. On Oct. 4, though suffering from wounds, he advanced with his company in the face of most destructive fire of machine guns and artillery until further advance was impossible.

BAXTER, WILLIAM V. (1711191) _____
Near Revillon, France, Sept. 8, 1918, and in the Argonne offensive, Sept. 28, 1918.
R—Red Hook, N. Y.
B—Dutchess County, N. Y.
G. O. No. 35, W. D., 1919.

Private, medical detachment, 308th Infantry, 77th Division.
On Sept. 8 Private Baxter went to the aid of wounded comrades, despite the deadly fire of rifles and machine guns, and after administering to them in a shell hole he carried the men one at a time to safety. On Sept. 28, after being painfully wounded, he refused to go to the rear until he had rendered first aid to a more seriously wounded comrade.

BAY, ROLAND W. (50004) _____
Near St. Etienne-a-Arnes, France, Oct. 3, 1918.
R—Salem, Ill.
B—Vandalia, Ill.
G. O. No. 46, W. D., 1919.

Corporal, Company C, 23d Infantry, 2d Division.
He went out 50 yards in front of the line under heavy machine-gun fire and brought back a wounded soldier. Later he carried a message 200 yards through a heavy barrage to battalion headquarters. He also volunteered to bring up ammunition under heavy fire and was wounded while performing that mission.

*BAYLY, HARRY E. (2175201) _____
Near Remonville, France, Oct. 22, 1918.
R—Topeka, Kans.
B—Jennings, Kans.
G. O. No. 66, W. D., 1919.

Sergeant, Headquarters Company, 353d Infantry, 89th Division.
When the advance was checked by severe machine-gun fire, Sergeant Bayly, who was in charge of a 1-pounder section, volunteered to open fire on the nest with a Stokes mortar. Holding the mortar between his legs he put it in action immediately and silenced the machine-gun nest, allowing the advance to continue. He was killed by shellfire several days later.
Posthumously awarded. Medal presented to father, Allyn R. Bayly.

BEACH, WILLIAM B. (568500) _____
Near St. Thibaut, France, Aug. 6 and 8, 1918.
R—Kingman, Ariz.
B—Kingman, Ariz.
G. O. No. 46, W. D., 1919.

Sergeant, first class, Company C, 4th Engineers, 4th Division.
Being a member of a covering detachment sent out to protect a detail which was constructing a bridge over the Vesle River, he voluntarily left his squad and fought his way alone down the river in order to locate an enemy machine-gun nest. The flashes from his automatic rifle drew fire from the enemy, and he was forced to jump into the river for protection. Swimming back to his squad, he organized a detail and led it in a successful attack on the hostile position. Two nights later, after the bridge had been destroyed, this soldier, with three others, volunteered to rebuild it. Under continuous fire from the enemy, he swam the river several times and set the posts for the bridge, thereby making possible the infantry attack on the following morning.

BEAL, HAROLD V. (562801) _____
Near Chery-Chartreuve, France, Aug. 13, 1918.
R—Oak Ridge, Mo.
B—Oak Ridge, Mo.
G. O. No. 89, W. D., 1919.

Corporal, Battery A, 13th Field Artillery, 4th Division.
Corporal Beal displayed unusual courage in repairing shattered telephone lines during a heavy barrage under direct observation by the enemy. He was repeatedly knocked down by concussion of shells and he was painfully wounded in the shoulder by a bursting shell, but he continued at his work until it was completed without seeking medical aid.

BEAN, FRANCIS A. (1209669) _____
Near St. Souplet, France, Oct. 18, 1918.
R—Utica, N. Y.
B—Utica, N. Y.
G. O. No. 70, W. D., 1919.

Sergeant, Company A, 107th Infantry, 27th Division.
His company having been stopped by heavy enemy machine-gun fire, Sergeant Bean and two other soldiers worked their way into the enemy position, putting out of action two hostile machine guns. They then proceeded toward a dugout near by under heavy fire and, upon encountering a German in the act of throwing a grenade at them, Sergeant Bean shot him. Reaching the entrance to the dugout they forced the surrender of the occupants, who numbered 35, including 3 officers.

BEAN, RUFUS (541927) _____
Near Cunel, France, Oct. 11, 1918.
R—Poplar Bluff, Mo.
B—Ridgeway, Ill.
G. O. No. 95, W. D., 1919.

Sergeant, Company G, 7th Infantry, 3d Division.
Though he had been so severely gassed as to be incapacitated for duty, and was also suffering from a painful rupture, Sergeant Bean remained in command of his platoon, which had been reduced to two squads, and led it under heavy artillery and machine-gun fire in an attack on an enemy machine-gun position, killing the gunners and capturing the gun. With conspicuous bravery he directed the consolidation of the position in the captured trench under continuous machine-gun and artillery barrage.

*BEANE, JAMES D. _____
Near Bantheville, France, Oct. 29, 1918.
R—Concord, Mass.
B—Concord, Mass.
G. O. No. 46, W. D., 1919.

First lieutenant, 22d Aero Squadron, Air Service.
When Lieutenant Beane's patrol was attacked by eight enemy planes (type Fokker) he dived into their midst in order to divert their attention from the other machines of his group and shot down one of the Fokkers in flames. Four other Fokkers then joined in the battle, one of which was also destroyed by this officer.
Posthumously awarded. Medal presented to foster father, Wilfrid Wheeler.

BEAR, ABSALOM F. (303756) _____

Near St. Etienne, France, Oct. 4, 1918.
R—Iberia, Mo.
B—Iberia, Mo.
G. O. No. 35, W. D., 1919.

Hospital apprentice, first class, U. S. Navy, attached to 5th Regiment, U. S. Marine Corps, 2d Division.
During a heavy bombardment he went to an advanced observation post, dressed the wounds of a comrade, and conducted him to the rear.

BEARD, CORNELIUS------------------- | First lieutenant, 101st Engineers, 26th Division.
Near Chavignon, France, Mar. 17, 1918. | On Mar. 17, 1918, at the front near Chavignon, France, he was knocked down by a shell explosion, which caused him to lose consciousness. Upon regaining consciousness he searched for and found some of his men. During two hours he assisted Sergeant Reed and Corporal Belanger, of his detachment, back to the trenches, part of the time under fire of a German aviator and of German artillery. His energy, self-sacrifice, and spirit throughout the operation were of the highest order.
R—Boston, Mass.
B—Boston, Mass.
G. O. No. 129, W. D., 1918.

BEARD, EDWIN L. (52149)------------- | Private, first class, Company M, 23d Infantry, 2d Division.
Near Chateau-Thierry, France, June 6, 1918. | After his platoon was practically wiped out and had been withdrawn near Chateau-Thierry, France, on June 6, 1918, he continued forward to his objective, and remained throughout the night under heavy fire in hope of keeping the ground gained until reinforcements came up.
R—Redfield, S. Dak.
B—Fayetteville, Mo.
G. O. No. 99, W. D., 1918.

BEARSS, HIRAM I---------------------- | Colonel, U. S. Marine Corps, attached to 102d Infantry, 26th Division.
At Marcheville and Riaville, France, Sept. 26, 1918. | His indomitable courage and leadership led to the complete success of the attack by two battalions of his regiment on Marcheville and Riaville. During the attacks these two towns changed hands four times, finally remaining in our possession until the troops were ordered to withdraw. Under terrific machine-gun and artillery fire he was the first to enter Marcheville, where he directed operations. Later, upon finding his party completely surrounded, he personally assisted in fighting the enemy off with pistol and hand grenades.
R—Peru, Ind.
B—Peru, Ind.
G. O. No. 143, W. D., 1918.
Distinguished-service medal also awarded.

*BEASLEY, SHADWORTH O------------- | Major, Medical Corps, attached to 76th Artillery, 3d Division.
Near Petit, Bordeaux Woods, France, July 14–16, 1918. | During the entire action Major Beasley braved the danger of continuous shell-fire by constantly searching for wounded and administering treatment.
R—San Francisco, Calif. | Posthumously awarded. Medal forwarded to Rufus L. Rigdon for delivery to the next of kin.
B—San Francisco, Calif.
G. O. No. 44, W. D., 1919.

BEATO, JOHN (1367822)-------------- | Corporal, Company H, 131st Infantry, 33d Division.
Near Chipilly Ridge, France, Aug. 9–10, 1918. | He volunteered and led a patrol of 8 men, which located an enemy nest, attacked it, and brought back 40 prisoners, among them 2 wounded officers. Throughout the fight he showed marked personal bravery and ability in leading ration parties through heavy enemy barrages.
R—Chicago, Ill.
B—Chicago, Ill.
G. O. No. 71, W. D., 1919.

BEATON, STANLEY (3164)------------- | Sergeant, 101st Ambulance Company, 101st Sanitary Train, 26th Division.
At Wadonville, France, Sept. 25, 1918. | Sergeant Beaton established a dressing station outpost under extremely heavy shellfire and cared for his wounded companions in the open under fire from enemy snipers and artillery.
R—Brookline, Mass.
B—Brookline, Mass.
G. O. No. 142, W. D., 1918.

BEATTIE, MORSE N. (180811)---------- | Private, first class, sanitary detachment, 126th Infantry, 32d Division.
Near Cierges, France, July 31, 1918. | Voluntarily leaving shelter, he crossed an open field, subjected to heavy machine-gun and artillery fire, to give first aid to two wounded soldiers. His heroic action saved the lives of the wounded men.
R—Kalamazoo, Mich.
B—Wayland, Mich.
G. O. No. 71, W. D., 1919.

BEATTY, GEORGE S------------------- | Second lieutenant, 7th Infantry, 3d Division.
Near Le Rocq Ferme, France, July 15, 1918. | Having remained at battalion headquarters, after the relief of his battalion, when the German barrage preceding the second battle of the Marne opened, Lieutenant Beatty, realizing the gravity of the situation, voluntarily went out through heavy destructive fire on a reconnaissance of the front lines and obtained information which could not be secured in any other manner. He encouraged the troops by his disregard for personal danger, and gave directions for the defense of the positions. It being necessary for him to remove his gas mask in order to accomplish this mission, he was seriously burned by mustard gas.
R—Clinton, N. C.
B—Ivanhoe, N. C.
G. O. No. 89, W. D., 1919.

BEATTY, GEORGE W. (935602) -------- | Sergeant, medical detachment, 306th Infantry, 77th Division.
At St. Juvin, France, Sept. 15, 1918. | He went forward to dress the wounds of an officer who could not be brought in because of the exceedingly heavy machine-gun fire, his bravery being instrumental in saving the officer's life. Throughout the entire day this soldier worked tirelessly at the dressing station under heavy shell fire until he was completely exhausted, showing a persistent devotion to duty.
R—Denver, Colo.
B—Pittsburgh, Pa.
G. O. No. 21, W. D., 1919.

BEATY, LESLIE (1315887)------------- | Private, first class, Company G, 119th Infantry, 30th Division.
Near Ribeauville, France, Oct. 18, 1918. | When enemy machine-gun fire was holding up the advance of the line, Private Beaty carried ammunition forward and aided a comrade to knock an enemy nest out of action and kill a sniper who was inflicting heavy losses on our troops.
R—Jamestown, Tenn.
B—Tennessee.
G. O. No. 50, W. D., 1919.

BEAUCHAMP, FELIX------------- ---- | Captain, 5th Regiment, U. S. Marine Corps, 2d Division.
Near St. Etienne, France, Oct. 3–4, 1918. | He took command after his company commander had been evacuated, and, despite severe wounds, participated in many engagements, continuing until additional wounds forced his withdrawal from the field.
R—Alaska.
B—Nashville, Tenn.
G. O. No. 37, W. D., 1919.

BEAUDETTE, JOSEPH A. (49881)_____
   Near Landres-et-St. Georges, France,
     Nov. 1 and 3-6, 1918.
   R—Saginaw, Mich.
   B—Saginaw, Mich.
   G. O. No. 126, W. D., 1919.

First sergeant, Company C, 23d Infantry, 2d Division.
Single handed, Sergeant Beaudette attacked and captured a German machine-gun nest, killing with his pistol the seven members of the crew. Two days later, after all the officers of his company had become casualties, he assumed command, and for three days led the company in the advance from Posse to Beaumont, displaying exceptional qualities of courage and leadership during severe fighting.

*BEAUVAIS, WALTON U_____
   Near Bois-d'Harville, France,
     Nov. 10, 1918.
   R—Tottenville, N. Y.
   B—Chicago, Ill.
   G. O. No. 64, W. D., 1919.

Second lieutenant, 131st Infantry, 33d Division.
In command of the left of the assaulting wave when it was held up by machine-gun fire, he placed himself in an exposed position, where he could command a view of the enemy position, and shot the enemy machine gunner, thus allowing the resumption of the advance. He exposed himself continually to heavy fire, setting an example of courage and coolness. He was mortally wounded later in the performance of duty.
Posthumously awarded. Medal presented to mother, Mrs. U. F. Beauvais.

BECK, ALBERT (546040)_____
   Near Jaulgonne, France, July 23-26,
     1918.
   R—Magnet, Ind.
   B—Perry County, Ind.
   G. O. No. 64, W. D., 1919.

Sergeant, Company D, 30th Infantry, 3d Division.
During the exceptionally heavy bombardment of the enemy from July 23-26 Sergeant Beck volunteered and carried messages after wire communication had been destroyed and runners wounded by the heavy firing. He also aided in guiding parties in bringing food and ammunition to the front line.

BECK, CHARLES L. (3091584)_____
   Near Romagne, France, Oct. 14,
     1918.
   R—Westphalia, Ind.
   B—Brownstown, Ind.
   G. O. No. 21, W. D., 1919.

Private, Company M, 126th Infantry, 32d Division.
In an attack on Cote Dame Marie the 126th Infantry was held up, owing to intense machine-gun fire and grenades. Private Beck volunteered as a member of a combat patrol which cut through the enemy lines, captured 10 machine guns, killed or captured 15 of the enemy, and forced a large number to surrender, clearing the Cote Dame Marie of the enemy, thus enabling the regiment to continue their advance.

BECK, JOHN I. (541731)_____
   At Fossoy, France, July 15, 1918.
   R—McIntyre, Ga.
   B—Wilkinson County, Ga.
   G. O. No. 89, W. D., 1919.

Corporal, Company F, 7th Infantry, 3d Division.
After the remainder of his platoon had become casualties, Corporal Beck, with five other soldiers, succeeded in holding his platoon position against a flank attack by the enemy, inspiring his men by his courage to pour a deadly fire into the ranks of the approaching Germans. Though he was wounded by machine-gun bullet, he refused to leave his post until he was finally ordered to be evacuated.

BECKER, EDWARD (284943)_____
   At Juvigny, France, Aug. 30, 1918,
     and in the Argonne Forest, France,
     Oct. 8, 1918.
   R—Sparta, Wis.
   B—Sparta, Wis.
   G. O. No. 98, W. D., 1919.

First sergeant, Company L, 128th Infantry, 32d Division.
During the attack on Juvigny, when all the officers in his company had been wounded, Sergeant Becker immediately took command of the company and led them in the attack. After the engagement he supervised and assisted in the work of clearing the field of wounded, working under incessant machine-gun and artillery fire. In the Argonne Woods, Oct. 8, when the advance of his organization was held up by cleverly concealed enemy machine guns, Sergeant Becker, with an officer, exposed himself to the enemy fire while reconnoitering the enemy positions. As a result of their observations the company was successfully disposed with only three casualties.

*BECKER, FRED H_____
   Near Vierzy, France, July 18, 1918.
   R—Waterloo, Iowa.
   B—Waterloo, Iowa.
   G. O. No. 116, W. D., 1918.

Second lieutenant, Infantry, attached to 5th Regiment, U. S. Marine Corps.
Lieutenant Becker went forward in advance of his platoon and destroyed a machine-gun nest, thereby preventing the death or injury of many men of his command. His self-sacrificing courage permitted his platoon to advance, but, as he completed the performance of this noble work, he himself was killed.
Posthumously awarded. Medal presented to father, J. B. Becker.

BECKWITH, BRYAN (1315641)_____
   Near Ypres, Belgium, Aug. 25, 1918.
   R—Fayetteville, N. C.
   B—Lilesville, N. C.
   G. O. No. 44, W. D., 1919.

First sergeant, 119th Infantry, 30th Division.
At imminent peril to his own life, Sergeant Beckwith and two companions extinguished a fire in an ammunition dump, caused by a bursting shell, thereby preventing the explosion of the dump and saving the lives of a large number of men who were in the vicinity.

BEDOLFE, HAROLD (568256)_____
   Near St. Thibaut, France, Aug. 5,
     1918.
   R—Oakland, Calif.
   B—San Francisco, Calif.
   G. O. No. 5, W. D., 1920.

Sergeant, first class, Company B, 4th Engineers, 4th Division.
He went forward, exposed to intense rifle, machine-gun, and artillery fire, and carried on his back, with the assistance of another soldier, a badly wounded comrade to a shell hole, thus saving the life of the wounded soldier. In the performance of this act Sergeant Bedolfe was severely wounded.

BEEBE, DAVID C_____
   Near St. Mihiel, France, Sept. 13,
     1918.
   R—Syracuse, N. Y.
   B—Syracuse, N. Y.
   G. O. No. 124, W. D., 1918.

Second lieutenant, pilot, 50th Aero Squadron, Air Service.
With Second Lieut. Franklin D. Bellows, observer, he executed a reconnaissance mission early in the morning of the second day of the St. Mihiel offensive in spite of the clouds, high wind, mist, flying at an altitude of only 300 meters and without protection of accompanying battle planes. Although subjected to severe fire from ground batteries, they penetrated 8 kilometers behind the German lines. His motor was badly damaged and his observer, Lieutenant Bellows, was mortally wounded. Despite these conditions he succeeded in bringing the disabled machine safely to his lines.

BEEBE, LEWIS C_____
   Near Crezancy, France, July 15,
     1918.
   R—Eugene, Oreg.
   B—Ashton, Iowa.
   G. O. No. 32, W. D., 1919.

Second lieutenant, 30th Infantry, 3d Division.
During the terrific artillery bombardment of the German offensive of July 15, 1918, Lieutenant Beebe carried a wounded man 300 yards to a dressing station. In order to maintain the liaison Lieutenant Beebe made repeated trips through the heavy shelling, repairing the wires, and reestablishing communication.

BEEBY, ALBERT E. (106762)............
  Near Vierzy, France, July 19, 1918.
  R—Hill City, Kans.
  B—Hill City, Kans.
  G. O. No. 15, W. D., 1919.

Corporal, Company A, 4th Machine Gun Battalion, 2d Division
He voluntarily left the safety of the trench, advanced about 100 yards in the open, under heavy artillery fire, and carried to safety a severely wounded marine.

BEECHER, HARRISON S................
  Near Gesnes, France, Sept. 29, 1918.
  R—Tacoma, Wash.
  B—Forestville, Ky.
  G. O. No. 28, W. D., 1921.

Captain, 347th Machine Gun Battalion, 91st Division.
Although suffering from a gunshot wound in the leg, he continued to lead his company in the attack. Later he was again wounded but continued in active command until exhausted and ordered by his superior officer to be evacuated. His fearless conduct and example were an inspiration to his officers and men.

BEERS, J. CLYDE (1240530)............
  Near Sergy, France, July 29, 1918.
  R—Indiana, Pa.
  B—Dayton, Pa.
  G. O. No. 53, W. D., 1920.

Mechanic, Company F, 110th Infantry, 28th Division.
After an unsuccessful attack on the Bois-de-Grimpettes, Mechanic Beers went out alone in front of our line, in plain view of the enemy, under heavy machine-gun fire from the front and flank, and gathered up the Chauchat rifles and Musette bags of ammunition which had been abandoned by wounded men. He made several trips, distributing the badly needed equipment to the advanced elements of our line.

*BEGLEY, WILLIAM (1709131)..........
  Near Charlevaux, France, Oct. 3–6, 1918.
  R—Brooklyn, N. Y.
  B—Brooklyn, N. Y.
  G. O. No. 87, W. D., 1919.

Private, Company G, 308th Infantry, 77th Division.
When his battalion was surrounded in the Argonne Forest, Oct. 3–7, Private Begley took charge of his squad, after the corporal had been killed, and, despite the fact that he was wounded in the arm by a machine-gun bullet, encouraged his men through all the attacks of the four days until he was killed, Oct. 6.
Posthumously awarded. Medal presented to mother, Mrs. Margaret Begley.

BEHAN, JAMES P. (562804)............
  Near Chery-Chartreuve, France, Aug. 13, 1918.
  R—New Orleans, La.
  B—New Orleans, La.
  G. O. No. 89, W. D., 1919.

Private, Battery A, 13th Field Artillery, 4th Division.
Private Behan displayed unusual courage in repairing shattered telephone lines during a heavy barrage, under direct observation by the enemy. He was repeatedly knocked down by concussion of shells, and his helmet was smashed by a bursting shell, but he continued at his work until it was completed without seeking medical aid.

BEHRENDT, AUGUST F................
  Near Exermont, France, Oct. 9, 1918.
  R—Kansas City, Mo.
  B—Germany.
  G. O. No. 44, W. D., 1919.

Captain, 16th Infantry, 1st Division.
On three different occasions Captain Behrendt reorganized his command and, placing himself in the lead, advanced against machine-gun nests, each time accomplishing his mission, despite severe losses. He led a patrol against a nest which was firing point blank on his troops, continuing until half of his patrol had been killed or wounded, and rushing the nest and capturing the gun and crew.

*BEIFUS, MARTIN (1710290)............
  Near Serval, France, on or about Sept. 9, 1918.
  R—Brooklyn, N. Y.
  B—Brooklyn, N. Y.
  G. O. No. 21, W. D., 1925.

Sergeant, Company M, 308th Infantry, 77th Division.
During the advance of his platoon he went out alone, and with a chauchat rifle and grenades drove the enemy out of a trench which was later occupied by our troops. Mortally wounded, he continued to encourage and direct his men in the work of consolidating the position, refusing to be evacuated till this work had been accomplished.
Posthumously awarded. Medal presented to sister, Mrs. M. L. Lorance.

BEINLICH, HARRY F. (1215693)........
  Near St. Souplet, France, Oct. 15, 1918.
  R—Elmira, N. Y.
  B—Sayre, Pa.
  G. O. No. 37, W. D., 1919.

Corporal, Company L, 108th Infantry, 27th Division.
Accompanied by an officer and three other soldiers, he made a reconnaissance of the River La Selle, the journey being made under constant and heavy machine-gun fire. To secure the desired information it was necessary to wade the stream for the entire distance.

BEIRD, ROY H.....................
  At Blanc Mont, France, Oct. 3, 1918.
  R—Bluffs, Ill.
  B—Illinois.
  G. O. No. 37, W. D., 1919.

Private, 78th Company, 6th Regiment, U. S. Marine Corps, 2d Division.
When the advance of their company was held up by enfilading fire from a hostile machine-gun nest, Private Beird, with three other soldiers, volunteered and made a flank attack on the nest with bombs and rifles, killing 3 members of the crew and capturing 25 others, together with 3 machine guns.

BELANGER, EDWARD A. (39258)........
  At Vaux, France, July 1, 1918.
  R—Chicopee Falls, Mass.
  B—North Adams, Mass.
  G. O. No. 99, W. D., 1918.

Private, Company F, 9th Infantry, 2d Division.
At Vaux, July 1, 1918, he bravely attacked eight of the enemy, killing four and capturing four.

BELEFANT, ABRAHAM (1703911)........
  Near St. Pierremont, France, Nov. 4, 1918.
  R—Brooklyn, N. Y.
  B—New York, N. Y.
  G. O. No. 37, W. D., 1919.

Sergeant, Headquarters Company, 307th Infantry, 77th Division.
After passing through a heavily bombarded area, he learned that a soldier of his platoon had been wounded and had fallen in the shelled area. He at once volunteered and went back for him, assisted in bringing him to a place of safety, and later helped to carry him through another shelled area to the first-aid station.

*BELFRY, EARL....................
  In the capture of Bouresches, France, June 6, 1918.
  R—New York, N. Y.
  B—Chicago, Ill.
  G. O. No. 32, W. D., 1918.

Private, 96th Company, 6th Regiment, U. S. Marine Corps, 2d Division.
He showed exceptional courage in the capture of Bouresches, France, on June 6, 1918, entering the town after being wounded and taking a leading part in forcing the machine guns of the enemy to evacuate.
Posthumously awarded. Medal presented to widow, Mrs. Marguerite F. Belfry-White.

BELK, EDD (2383775)...................
   Near Cunel, France, Oct. 12, 1918.
   R—Iberia, Mo.
   B—Iberia, Mo.
   G. O. No. 37, W. D., 1919.

Private, first class, Company G, 60th Infantry, 5th Division.
Although seriously wounded and ordered to the rear, he continued in the advance with his company through an intense barrage of artillery and machine-gun fire. Later in the day he had his wound dressed and was tagged for evacuation, but returned to his company and continued on active duty with his organization.

*BELKO, JOHN G. (1230395)..............
   Near Montblainville, France, Sept. 27, 1918.
   R—Donora, Pa.
   B—Plymouth, Mass.
   G. O. No. 98, W. D., 1919.

Private, first class, Company A, 110th Infantry, 28th Division.
Private Belko made several trips under heavy fire, carrying wounded comrades to shelter. On the same day he advanced alone 60 yards ahead of the line under heavy machine-gun fire and drove off about 20 of the enemy, who had been harassing his company with rifle grenade fire.
Posthumously awarded. Medal presented to father, John Belko.

*BELL, ALBERT H., Jr..................
   Near les Franquettes Farm, France, July 23, 1918.
   R—Greensburg, Pa.
   B—Greensburg, Pa.
   G. O. No. 32, W. D., 1919.

Second lieutenant, 4th Infantry, 3d Division.
While acting as battalion gas officer Lieutenant Bell volunteered and took charge of a squad of men and captured a machine gun and eight prisoners. He was killed while attempting to take a second machine gun.
Posthumously awarded. Medal presented to father, A. H. Bell, sr

BELL, BLAKE (2337289)................
   Near Cunel, France, Oct. 20, 1918.
   R—Kinde, Mich.
   B—Kinde, Mich.
   G. O. No. 44, W. D., 1919.

Private, Headquarters Company, 4th Infantry, 3d Division.
When all superiors of the platoons had become casualties, Private Bell assumed command and efficiently conducted the operations. When his guns had been put out of action he personally salvaged the parts and resumed fire. When relieved he safely conducted the remnants of the platoon from the line.

BELL, CHARLES J. (2893884)............

   Near Tulgas, Russia, Nov. 12, 1918.
   R—Louisville, Ky.
   B—Louisville, Ky.
   G. O. No. 108, W. D., 1919.

Private, Company B, 339th Infantry, 85th Division (detachment in north Russia).
After the blockhouse in which he and several other comrades were stationed had been hit by a high-explosive shell, killing two and wounding five, and himself had been so severely wounded as to be blinded in one eye, he continued to remain at his post and fired his Lewis gun until relieved. This continued under heavy shell fire.

BELL, FRANK (2220227)................
   Near Fey-en-Haye, France, Sept. 12, 1918.
   R—Vian, Okla.
   B—Sallisaw, Okla.
   G. O. No. 95, W. D., 1919.

Private, Company D, 358th Infantry, 90th Division.
Private Bell saved the lives of 30 of his comrades by coming out of a dugout into which the Germans were throwing grenades, shooting the leader and dispersing the remainder of the party.

BELL, FRANK E. (568528)..............
   Service rendered under the name, James Manning.

Corporal, Company C, 4th Engineers, 4th Division.

BELL, FRANK J. (283221)..............
   Near Gesnes, France, Oct. 6, 1918.
   R—Chicago, Ill.
   B—White Earth, Minn.
   G. O. No. 95, W. D., 1919.

Private, Company A, 128th Infantry, 32d Division.
Private Bell, while acting as runner, repeatedly volunteered to take the place of other runners who had become exhausted, and delivered messages under severe enemy artillery and machine-gun fire. On the night of Oct. 6, while on an important mission, he received a painful wound in the hand. He nevertheless continued on and delivered his message. On his return to battalion headquarters he refused to be evacuated, dressed the wound himself, and continued in the capacity of runner.

*BELL, GEORGE (2168986)..............
   Near Lesseau, France, Sept. 4, 1918.
   R—Athens, Ala.
   B—Athens, Ala.
   G. O. No. 139, W. D., 1918.

Private, Company E, 366th Infantry, 92d Division.
Although he was severely wounded he remained at his post and continued to fight a superior enemy force which had attempted to enter our lines, thereby preventing the success of an enemy raid in force.
Posthumously awarded. Medal presented to father, Albert Bell.

BELL, GLENN A. (2248895)............
   Near Villers-devant-Dun, France, Nov. 2, 1918.
   R—Barry, Tex.
   B—Corsicana, Tex.
   G. O. No. 46, W. D., 1919.

Corporal, Company D, 359th Infantry, 90th Division.
He was wounded in the arm by machine-gun fire, but in spite of his wound continued to lead his squad, and assisted in taking several machine-gun nests.

*BELL, JAMES FRANKLIN..............
   Against Spanish forces at Manila, Philippine Islands, Aug. 10, 1898.
   R—Shelbyville, Ky.
   B—Shelbyville, Ky.
   G. O. No. 3, W. D., 1925.
   Medal of honor and distinguished-service medal also awarded.

Major, Engineer officer, U. S. Volunteers.
Major Bell, with utter disregard for his personal safety, conducted a bold and daring reconnaissance of the creek in front of Fort San Antonio de Abad held by Spanish forces, and ascertained not only that it was fordable, but the exact width of the ford at the beach, and swimming in the bay to a point from which he could examine the Spanish line from the rear, secured information which facilitated the planning of the successful attack of Aug. 13, 1898, on Manila, Philippine Islands.
Posthumously awarded. Medal presented to widow, Mrs. James Franklin Bell.

BELL, JOE..........................
   Near Chateau-Thierry, France, June 23, 1918.
   R—Jackson, Miss.
   B—Mississippi.
   G. O. No. 35, W. D., 1919.

Sergeant, 16th Company, 5th Regiment, U. S. Marine Corps, 2d Division.
After becoming separated from his own platoon, he attached himself to another platoon of the company, and, learning that all the runners who had attempted to carry important messages had been killed or wounded, immediately volunteered and made several trips across an open area subjected to a continuous and intense barrage of artillery and machine-gun fire. He performed this important mission despite the fact that he was almost exhausted when he joined the platoon.

BELL, JOHN A. (2780031) _____ | Sergeant, Machine Gun Company, 363d Infantry, 91st Division.
Near Waereghem, Belgium, Oct. 31, 1918.
R—Pasadena, Calif.
B—Champaign, Ill.
G. O. No. 74, W. D., 1919.

Sergeant Bell showed great devotion to duty and extreme bravery under fire when he refused to leave the field until ordered to do so after his arm had been so badly wounded that amputation was necessary.

BELL, WILLIAM B. (263269) _____
Near Cierges, France, July 31, 1918.
R—Keno, Mich.
B—Pittsburgh, Pa.
G. O. No. 81, W. D., 1919.

Corporal, Company I, 125th Infantry, 32d Division.
When his company was held up by severe machine-gun fire from the right flank, Corporal Bell voluntarily went out in front of our lines and carried a wounded comrade to cover and administered first aid.

*BELL, WILLIAM Z. (1351191) _____
East of Grand Pre, France, Oct. 15, 1918.
R—Malone, Fla.
B—Chipley, Fla.
G. O. No. 21, W. D., 1925.

Private, Company C, 168th Infantry, 42d Division.
Private Bell, serving as stretcher bearer during two attacks, with exceptional bravery and disregard of danger, exposed himself during an intense artillery bombardment to assist a wounded soldier in imminent need of first aid and was killed at his work by an enemy shell.
Posthumously awarded. Medal presented to widow, Mrs. Zula Bell.

*BELLOWS, FRANKLIN B. _____
Near St. Mihiel, France, Sept. 13, 1918.
R—Wilmette, Ill.
B—Evanston, Ill.
G. O. No. 124, W. D., 1918.

Second lieutenant, observer, 50th Aero Squadron, Air Service.
Second Lieutenant Bellows, with Second Lieut. David C. Beebe, pilot, executed a reconnaissance mission early in the morning of the second day of the St. Mihiel offensive, in spite of low clouds, high winds, and mist, flying at an altitude of only 300 meters and without protection of accompanying battle planes. Although subjected to severe fire from ground batteries, they penetrated 8 kilometers beyond the German lines. Lieutenant Beebe's motor was badly damaged, and Lieutenant Bellows was mortally wounded and died just after the disabled machine landed safely in friendly territory.
Posthumously awarded. Medal presented to father, John A. Bellows.

*BELOUNGEA, WILLIAM A. (263947) _____
Northeast of Chateau-Thierry, France, July 31, 1918.
R—Manistique, Mich.
B—Epoufette, Mich.
G. O. No. 78, W. D., 1919.

Corporal, Company M, 125th Infantry, 32d Division.
With the assistance of another soldier, Corporal Beloungea dragged a wounded comrade to his own trench, a distance of 150 yards, through an intense barrage of machine-gun and artillery fire.
Posthumously awarded. Medal presented to sister, Mrs. Frank Sly.

BELT, BILLIE W. (2195748) _____
At Bouillonville, France, Sept. 24, 1918.
R—Windsor, Mo.
B—Lincoln, Mo.
G. O. No. 37, W. D., 1919.

Corporal, Company B, 314th Motor Supply Train, 89th Division.
When an enemy shell struck a truck loaded with gasoline, killing two men and wounding several others, Corporal Belt rushed to the burning truck, in spite of the danger from the exploding bedons of gasoline, pulled one of the men from beneath the burning truck, and extinguished the flames on his clothing. He then returned and attempted to rescue another man, but was unable to do so.

BELZER, WILLIAM E. _____
Near Jaulny, France, Sept. 12–13, 1918.
R—Helena, Mont.
B—Ackley, Iowa.
G. O. No. 128, W. D., 1918.

Second lieutenant, observer, observation group, Air Service, attached to 4th Army Corps.
On Sept. 12 Lieutenant Belzer, observer, and First Lieut. Wallace Coleman, pilot, while on an artillery surveillance mission, were attacked by an enemy plane. They waited until the enemy was at close range and then fired 50 rounds directly into the vital parts of the enemy machine, which was seen to disappear out of control. The next day Lieutenants Belzer and Coleman, while on a reconnaissance mission, were attacked by seven enemy aircraft. They unhesitatingly opened fire, but owing to their guns being jammed were forced to withdraw to our lines where, clearing the jam, they returned to finish the mission. Their guns again jammed, and they were driven back by a large patrol of enemy planes. After skillful maneuvering they succeeded in putting one gun into use and returned a third time, only to be driven back. Undaunted, they returned the fourth time and accomplished their mission, transmitting valuable information to the Infantry headquarters.

*BENDER, JOHN F. (2357247) _____
Near les Franquette Farm, France, July 23, 1918.
R—Mount Carmel, Pa.
B—Mount Carmel, Pa.
G. O. No. 32, W. D., 1919.

Private, Company B, 4th Infantry, 3d Division.
He crawled forward and continued to fire into a machine-gun nest until he was killed.
Posthumously awarded. Medal presented to mother, Mrs. Anna V. Bender.

*BENEFIELD, CORBETT (1490084) _____
Near St. Etienne, France, Oct. 8, 1918.
R—Caddo, Okla.
B—Birmingham, Ala.
G. O. No. 66, W. D., 1919.

Corporal, Company E, 142d Infantry, 36th Division.
Wounded in the arm by a machine-gun bullet while leading his squad through enemy entanglements, Corporal Benefield rallied his men and led them in an attack on the machine gun which was enfilading the line, and captured the gun with its entire crew. Continuing on, despite his wound, he was killed shortly afterwards while leading his squad under a heavy enemy bombardment.
Posthumously awarded. Medal presented to father, Tom Benefield.

BENELL, OTTO E. _____
Near Thiaucourt, France, Sept. 12, 1918.
R—Fort Collins, Colo.
B—Pueblo, Colo.
G. O. No. 140, W. D., 1918.

Second lieutenant, 135th Aero Squadron, Air Service.
He went out on a two-hour counterattack artillery adjustment under adverse weather conditions. Soon after he suffered an accident to his right hand, which made it useless. In spite of this injury, he continued to operate his wireless key with his left hand, directing the fire of the batteries on concentration behind the enemy lines.

BENJAMIN, RAY N.
Near Blanc Mont, France, Oct. 6, 1918.
R—Gray Court, S. C.
B—Quarry, S. C.
G. O. No. 32, W. D., 1919.

First lieutenant, 2d Engineers, 2d Division.
While commanding a detachment of wire cutters, working in advance of the Infantry, he was painfully wounded by a shell fragment, but he refused to leave his men until his mission was accomplished and the advance of the Infantry assured.

BENJAMIN, WILLIAM (1706230).
At Chateau-Diable, near Fismes, France, Aug. 27, 1918.
R—New York, N. Y.
B—England.
G. O. No. 128, W. D., 1918.

Sergeant, Company H, 307th Infantry, 77th Division.
Although severely wounded in the first minutes of a three-hour engagement, he continued to lead his platoon with entire disregard of personal safety, and although suffering intensely from his wounds, he refused to be evacuated until the action was over and he had found cover for his men.

*BENNETT, CHARLES S. (43931).
South of Soissons, France, July 19 and 20, 1918.
R—Tulsa, Okla.
B—St. Paul, Minn.
G. O. No 4, W. D., 1923.

Corporal, Company K, 16th Infantry, 1st Division.
During the night of July 19, while in charge of an automatic rifle squad on out-post duty, Corporal Bennett was taken prisoner by an enemy patrol. He effected his escape by means of securing a German Luger pistol from his captor and taking single-handed 30 of the enemy as prisoners, all of whom he marched to his own lines. After rejoining his company, he was directed to the rear, but declined to go, and remained with his comrades until he was mortally wounded about 5 p. m., July 20, 1918. After being wounded he persistently refused to be evacuated from the field until forced to do so by his platoon commander.
Posthumously awarded. Medal presented to brother, J. C. Bennett.

BENNETT, HARRY L., Jr.
At Soissons, France, July 18, 1918.
R—Houston, Tex.
B—Houston, Tex.
G. O. No. 56, W. D., 1922.

Captain, 26th Infantry, 1st Division.
Being in command of a regimental train and having been instructed to deliver medical supplies at all costs to the advance first-aid station, and after all means of transportation had failed, he secured a wheel litter which he packed with medical supplies, then filled his arms and started forward under heavy artillery bombardment. He delivered the supplies as directed after having passed through the intense hostile counter barrage and aided in saving the lives of many wounded men. His personal courage and utter disregard of danger gave proof of his high soldierly qualities. At St. Baussant on Sept. 12, 1918, he personally and gallantly led a charge on an enemy machine gun which held up the advance of his battalion. He shot and killed the gunner and killed and captured many of the enemy. His splendid valor and coolness were an inspiration to his men and materially assisted in the success of the operation.

BENNETT, JACK (736647).
Near Louppy, France, Nov. 10, 1918.
R—Jasper, Mo.
B—Waxahatchie, Tex.
G. O. No. 37, W. D., 1919.

Sergeant, Company K, 11th Infantry, 5th Division.
He led a patrol in a flank attack on a machine gun nest which was holding up the advance, and though half of his party was killed or wounded, he succeeded in putting the gun out of action. Coming under the fire of another machine gun, he was forced to take cover until after dark, when he returned to our lines, bringing with him the wounded men.

BENNING, FRED G. (45097).
South of Exermont, France, Oct. 9, 1918.
R—Norfolk, Nebr.
B—Norfolk, Nebr.
G. O. No. 5, W. D., 1920.

Corporal, Machine Gun Company, 16th Infantry, 1st Division.
After his platoon commander had been killed and two senior noncommissioned officers disabled, he took command of the platoon and, by his able leadership and courage, conducted it through heavy fire to its assigned position on Hill 240.

BENOIT, HENRY N. (2293659).
Near Gesnes, France, Sept. 26 to Oct. 4, 1918.
R—Ekalaka, Mont.
B—Verdigris, Nebr.
G. O. No. 64, W. D., 1919.

Private, first class, Company D, 361st Infantry, 91st Division.
During eight days of action while acting in the capacity of runner between his company and battalion headquarters Private Benoit was constantly subjected to heavy shell fire, but performed his mission without thought of personal danger, carrying the many messages promptly and successfully.

BENSON, ANDREW A. (180727).
Near Bantheville, France, Nov. 1, 1918.
R—Bertrand, Nebr.
B—Bertrand, Nebr.
G. O. No. 37, W. D., 1919.

Private, medical detachment, 1st Gas Regiment.
Severely wounded by shell fire, Private Benson continued to give first aid to the wounded until struck the second time. After receiving the second wound he remained on duty, giving directions for the care of other wounded.

*BENTLEY, RICHARD E. (1215694).
Near St. Souplet, France, Oct. 15, 1918.
R—Horseheads, N. Y.
B—Big Flats, N. Y.
G. O. No. 37, W. D., 1919.

Corporal, Company L, 108th Infantry, 27th Division.
Accompanied by an officer and three other soldiers, he made a reconnaissance of the River La Selle, the journey being made under constant and heavy machine-gun fire. To secure the desired information it was necessary to wade the stream for the entire distance.
Posthumously awarded. Medal presented to mother, Mrs. William Bentley.

BENTON, HARWOOD O.
Near Montrebeau Woods, France, Sept. 29, 1918.
R—Oberlin, Kans.
B—Oberlin, Kans.
G. O. No. 46, W. D., 1919.

Second lieutenant, 137th Infantry, 35th Division.
When the advance of his company had been checked and forced back into the woods, Lieutenant Benton, although himself wounded, went into an open field and, under heavy machine-gun fire, rescued two wounded comrades. He remained in action despite his wounds for three days, when he was ordered to the hospital by the battalion commander.

BENZ, CEDRIC CHARLES................
West of Chateau-Thiery, France, July 1, 1918.
R—Pittsburgh, Pa.
B—Pittsburgh, Pa.
G. O. No. 100, W. D., 1918.

First lieutenant, 111th Infantry, 28th Division.
While trying to assist a wounded companion in the attack on Hill 204, west of Chateau-Thierry, France, July 1, 1918, he discovered a party of Germans, and with the aid of two wounded soldiers boldly rushed them and made 38 prisoners.

BERG, JOHN N. (2472741)...............
Near Sommauthe, France, Nov. 4, 1918.
R—Crosby, Pa.
B—Sweden.
G. O. No. 37, W. D., 1919.

Corporal, Company C, 317th Infantry, 80th Division.
He led his squad under heavy machine-gun fire in an attack on a machine-gun nest, capturing two machine guns, killing the gunners, and driving off the remainder of the crews. With his squad he held the position for one hour, until the arrival of the rest of his company.

BERG, JOSEPH (97327)..................
Near Croix Rouge Farm, northeast of Chateau-Thierry, France, July 27, 1918.
R—Mount Vernon, Wash.
B—Belgium.
G. O. No. 102, W. D., 1918.

Private, Company G, 167th Infantry, 42d Division.
When his company was in action near Hill 212, Private Berg was posted as lookout while his company was intrenching. He observed the enemy bringing forward machine guns through the wheat fields to place them in position. Waiting until they were within close range, he exposed himself to heavy machine gun and artillery fire and succeeded in killing or disabling the crews of three machine guns, thus saving his company from heavy casualties.

BERGASSE, HERMAN J. (1707565).......
Near Binarville, France, Sept. 28, 1918.
R—New York, N. Y.
B—New York, N. Y.
G. O. No. 35, W. D., 1919.

First sergeant, Company A, 308th Infantry, 77th Division.
Assuming command of the command after his commanding officer had become a casualty, Sergeant Bergasse led a formidable attack on an enemy machine-gun emplacement, silencing two guns in the nest and permitting the further advance of his battalion.

*BERGEN, WILLIAM J. (91226)..........
Near Villers-sur-Fere, France, July 28, 1918.
R—Bronx, N. Y.
B—New York, N. Y.
G. O. No. 88, W. D., 1918.

Private, Company K, 165th Infantry, 42d Division.
On duty as a litter bearer in action near Villers-sur-Fere, France, July 28, 1918, he was killed while going into heavy machine-gun and shell fire to rescue the wounded. He had worked tirelessly and fearlessly throughout the attack on the enemy north of the River Ourcq.
Posthumously awarded. Medal presented to widow, Mrs. William J. Bergen.

BERGSTEIN, ALFRED M.................
Near Exermont, France, Oct. 8, 1918.
R—Pottsville, Pa.
B—Philadelphia, Pa.
G. O. No. 46, W. D., 1919.

First lieutenant, Medical Corps, attached to 18th Infantry, 1st Division.
Under heavy shell fire, Lieutenant Bergstein cared for the wounded, although he had been severely wounded and was suffering great pain. He refused to be evacuated until all the wounded had been treated.

BERKELEY, THEODORE I................
Near Mont St. Pere, France, July 23, 1918.
R—Morristown, N. J.
B—New York, N. Y.
G. O. No. 32, W. D., 1919.

Second lieutenant, 4th Infantry, 3d Division.
While in an open field swept by heavy machine-gun and rifle fire he was ordered by his battalion commander to seek cover in the woods, but seeing a wounded man farther to the front, he went to his aid and brought him to a place of safety.

BERKLEY, GEORGE (736604)............
Near Cunel, France, Oct. 14–18, 1918.
R—Golden Pond, Ky.
B—Trigg County, Ky.
G. O. No. 37, W. D., 1919.

First sergeant, Company K, 11th Infantry, 5th Division.
After all the officers of his company had been killed or wounded he successfully led his men until compelled, through wounds, to leave the field.

*BERKOMPAS, OLIUS (263169)..........
Near Romagne, France, Oct. 11, 1918.
R—Rudyard, Mich.
B—West Olive, Mich.
G. O. No. 64, W. D., 1919.

Bugler, Company I, 125th Infantry, 32d Division.
In the attack on Hill 258 he volunteered to carry messages from his company in the attacking line to the battalion post of command. In order to reach the post of command it was necessary to cross an open area of about 500 yards in width, subjected to intense machine-gun fire and under direct observation of the enemy. He was killed while engaged in this mission.
Posthumously awarded. Medal presented to father, Tate Berkompas.

BERLANDER, ALBERT M. (168973).......
Near Attigny, France, Oct. 15, 1918.
R—San Francisco, Calif.
B—England.
G. O. No. 27, W. D., 1920.

Sergeant, first class, Company D, 2d Engineers, 2d Division.
While making a reconnaissance of the Aisne River and the Ardennes Canal, in advance of the line of American outposts, Sergeant Berlander was wounded by a sniper's bullet. He called to his comrades in time to warn them and ordered them not to come to his assistance. By lying still until darkness came he was able to continue his reconnaissance and return with valuable information.

BERNHEIMER, LOUIS G
Near Fismes, France, Aug. 11, 1918.
R—New York, N. Y.
B—New York, N. Y.
G. O. No. 44, W. D., 1919.

First lieutenant, pilot, 88th Aero Squadron, Air Service.
Louis G. Bernheimer, first lieutenant, pilot; John W. Jordan, second lieutenant, 7th Field Artillery, observer; Roger W. Hitchcock, second lieutenant, pilot; James S. D. Burns, deceased, second lieutenant, 165th Infantry, observer; Joel H. McClendon, deceased, first lieutenant, pilot; Charles W. Plummer, deceased, second lieutenant, 101st Field Artillery, observer; Philip R. Babcock, first lieutenant, pilot; and Joseph A. Palmer, second lieutenant, 15th Field Artillery, observer. All of these men were attached to the 88th Aero Squadron, Air. Service.
For extraordinary heroism in action near Fismes, France, Aug. 11, 1918. Under the protection of three pursuit planes, each carrying a pilot and an observer, Lieutenants Bernheimer and Jordan, in charge of a photo plane, carried out successfully a hazardous photographic mission over the enemy's line to the River Aisne. The 4 American ships were attacked by 12 enemy battle planes. Lieutenant Bernheimer, by cooly and skillfully maneuvering his ship, and Lieutenant Jordan, by accurate operation of his machine gun, in spite of wounds in the shoulder and leg, aided materially in the victory which came to the American ships, and returned safely with 36 valuable photographs. The pursuit plane operated by Lieutenants Hitchcock and Burns was disabled while these two officers were fighting effectively. Lieutenant Burns was mortally wounded and his body jammed the controls. After a headlong fall of 2,500 meters, Lieutenant Hitchcock succeeded in regaining control of this plane and piloted it back to his airdrome. Lieutenants McClendon and Plummer were shot down and killed after a vigorous combat with 5 of the enemy's planes. Lieutenants Babcock and Palmer, by gallant and skillful fighting, aided in driving off the German planes and were materially responsible for the successful execution of the photographic mission.
Oak-leaf cluster.
Lieutenant Bernheimer is also awarded an oak-leaf cluster for the following act of extraordinary heroism in action near Tailly, France, Nov. 2, 1918: Lieutenant Bernheimer and First Lieut. Ralph P. Bagby, observer, on their own initiative, went on a reconnaissance mission, flying 15 kilometers behind the German lines, securing valuable information as to the condition of the bridges across the Meuse River and enemy activity in the back areas, and harassing enemy troops.

BERNIER, OLIVER D
Near Chateau-Thierry, France, June 6, 1918.
R—Syracuse, N. Y.
B—Syracuse, N. Y.
G. O. No. 23, W. D., 1919.

Second lieutenant, 5th Regiment, U. S. Marine Corps, 2d Division.
Exposing himself to very heavy concentrated machine-gun and rifle fire, Lieutenant Bernier rushed ahead and broke down a strong wire fence, thereby preventing a delay in his progress and consequent exposure of his men to fire.

BERNSTEIN, DAVID
Near Blanc Mont, France, Oct. 5, 1918.
R—New York, N. Y.
B—New York, N. Y.
G. O. No. 46, W. D., 1919.

Corporal, 43d Company, 5th Regiment, U. S. Marine Corps, 2d Division.
Learning that a number of wounded soldiers were lying in no man's land, he immediately volunteered to help carry them in. He made several trips over an area constantly shelled and subjected to machine-gun and rifle fire.

*BERRY, BENJAMIN I
Near Mont Blanc, France, Oct. 4–6, 1918.
R—Carrizozo, N. Mex.
B—McKenzie, Tenn.
G. O. No. 20, W. D., 1919.

Second lieutenant, 5th Machine Gun Battalion, 2d Division.
Upon hearing that his company commander had been killed and that the second in command was wounded, Lieutenant Berry went immediately to the front line and took command. On Oct. 5, 1918, he was wounded in the head and the surgeon ordered him evacuated. Lieutenant Berry removed the evacuation tag and went to the front line, where he remained for 24 hours. Although in a weakened condition, he personally guided the company after they had been relieved.
Posthumously awarded. Medal presented to brother, R. E. Berry.

BERRY, BENJAMIN S
In the Bois-de-Belleau, France, northwest of Chateau-Thierry, France, June 6, 1918.
R—Chester, Pa.
B—Chester, Pa.
G. O. No. 99, W. D., 1918.

Major, 5th Regiment, U. S. Marine Corps, 2d Division.
He led his men in a gallant attack across open ground and into the Bois-de-Belleau, France, northwest of Chateau-Thierry, on the afternoon of June 6, 1918, inspiring them to deeds of valor by his example. When he reached the edge of the woods he fell, severely wounded. Nevertheless he arose and made a final dash of 30 yards across, through a storm of bullets, and reached again the first wave of his command, before yielding to exhaustion from his injury.

*BERRY, ERNEST (2248627)
During the St. Mihiel offensive, France, Sept. 14, 1918.
R—Kiefer, Okla.
B—Union Springs, Ala.
G. O. No. 72, W. D., 1920.

Private, Company D, 358th Infantry, 90th Division.
Private Berry, with four other men, volunteered to cross a valley to the woods opposite and silence machine guns which had held up the advance of his company. In the face of heavy enemy fire this small group accomplished its mission, thus enabling the company to cross the valley without further loss. He was slightly wounded in the performance of this act and killed in action a few days thereafter.
Posthumously awarded. Medal presented to sister, Mrs. Belle Walker.

*BERRY, STANLEY H. (306727)
Near Chateau-Thierry, France, July 14–15, 1918.
R—Philadelphia, Pa.
B—Germantown, Pa.
G. O. No. 43, W. D., 1922.

Private, Battery D, 76th Field Artillery, 3d Division.
During an intensive hostile bombardment, when all telephone lines had been cut off by shell fire, Private Berry voluntarily went out and continued to repair the forward lines until killed. His heroic conduct and self-sacrifice were an inspiration to all his comrades at a most trying time.
Posthumously awarded. Medal presented to mother, Mrs. Mary Berry.

BERRY, THOMAS A. (2248629)
Near Vilcey, France, Sept. 12, 1918.
R—Drumright, Okla.
B—Sallisaw, Okla.
G. O. No. 98, W. D., 1919.

Private, Company E, 357th Infantry, 90th Division.
Private Berry was a member of a patrol cleaning up a trench, when an enemy grenade was thrown into the midst of the group. With notable presence of mind and entire disregard for his own safety, Private Berry seized the grenade and threw it over the parapet, where it exploded an instant later, thereby saving the lives of his comrades.

BERRY, WAYNE R. (1459841)
Near Charpentry, France, Sept. 29, 1918.
R—Speed, Mo.
B—Speed, Mo.
G. O. No. 70, W. D., 1919.

Private, Company B, 140th Infantry, 35th Division.
Although seriously wounded, Private Berry, disregarding the danger, advanced alone on a reconnaissance under heavy machine-gun fire and brought back important information of the enemy.

BERRY, WILLIAM (2386003)
Near Lion-devant-Dun, France, Nov. 7, 1918.
R—Lockport, N. Y.
B—Lockport, N. Y.
G. O. No. 37, W. D., 1919.

Private, Company A, 61st Infantry, 5th Division.
In the attack on the town of Lion-devant-Dun, Private Berry's company was halted by heavy machine-gun fire. Advancing ahead and urging his comrades to follow, he succeeded in gaining a foothold in the town, until he was felled by the murderous fire of the enemy guns.

BERRYHILL, JOHN W. (1320092)
Near Bellicourt, France, Sept. 29, 1918.
R—Charlotte, N. C.
B—Charlotte, N. C.
G. O. No. 37, W. D., 1919.

Private, first class, Company D, 120th Infantry, 30th Division.
With eight other soldiers, comprising the company headquarters detachment, he assisted his company commander in cleaning out enemy dugouts along a canal and capturing 242 prisoners.

BERWICK, ELWYN L. (563217)
Near Chery-Chartreuve, France, Aug. 13, 1918.
R—Alameda, Calif.
B—Oakland, Calif.
G. O. No. 27, W. D., 1920.

Corporal, Battery C, 13th Field Artillery, 4th Division.
When an enemy shell struck his battery position, setting fire to a powder dump and killing or wounding 30 men, Corporal Berwick, though himself wounded, went into the burning dump at imminent risk to his life and assisted in extinguishing the flames. He then assisted in removing the other men before securing aid for himself. Refusing to be evacuated, he reported back to his battery with one arm in a sling and resumed his place as gunner.

BESS, ROY A. (2205709)
Near Beaufort, France, Nov. 4, 1918.
R—St. Louis, Mo.
B—Lutesville, Mo.
G. O. No. 37, W. D., 1919.

Private, Company L, 355th Infantry, 89th Division.
Although wounded by machine-gun fire, he refused first aid and continued in the engagement for two days without treatment.

*BESSINGER, EDWARD (129736)
Near Chateau-Thierry, France, June 1–July 1, 1918, and near Thiaucourt, France, Sept. 17, 1918.
R—Chicago, Ill.
B—Chicago, Ill.
G. O. No. 37, W. D., 1919.

Corporal, Headquarters Company, 15th Field Artillery, 2d Division.
Near Chateau-Thierry he repeatedly exposed himself to heavy shell and gas bombardments in order to maintain telephone communication between the Infantry and Artillery posts of command. Near Thiaucourt, on Sept. 17, he accompanied the first wave of Infantry, carrying a projector, and, in spite of the heavy shellfire, kept the Artillery informed of the progress of the attack. He was killed near Somme-Py, France, on Oct. 7, while in the faithful performance of his duty.
Posthumously awarded. Medal presented to mother, Mrs. Mabel Hansen.

BEST, EDWARD G. (1697579)
In the Forest of Argonne, France, Oct. 3, 1918.
R—New York, N. Y.
B—Boston, Mass.
G. O. No. 78, W. D., 1919.

Private, Company E, 305th Infantry, 77th Division.
During an attack on a series of strong enemy machine-gun nests Private Best took charge of company liaison and voluntarily carried messages to all the platoons of the company, exposing himself fearlessly to sweeping machine-gun fire.

BEVAN, STANLEY (1197031)
At Brieulles, France, Nov. 3–4, 1918.
R—Frostburg, Md.
B—Frostburg, Md.
G. O. No. 35, W. D., 1919.

Sergeant, Company D, 15th Machine Gun Battalion, 5th Division.
When the footbridges over the Meuse River were destroyed by artillery fire, Sergeant Bevan volunteered and assisted in repairing the damage under violent machine-gun fire. On the night of November 4, while leading his platoon across the footbridge, part of it was blown away and he fell into the water, but pulling himself out, he continued to lead his men, regardless of the cold and extreme fatigue. Later he exposed himself to the enemy fire while carrying a wounded man to a place of safety.

BIBLE, PAUL (42729)
South of Sedan, France, Nov. 7, 1918.
R—Foley, Minn.
B—Alberta, Minn.
G. O. No. 35, W. D., 1920.

Corporal, Company E, 16th Infantry, 1st Division.
Corporal Bible, aided by a comrade, advanced under heavy fire upon an enemy machine-gun position which was causing severe losses to their company. Disregarding personal danger, they silenced the machine gun and thus enabled their company to continue the advance with few losses.

*BICKFORD, ERNEST E. (43389)
Near Soissons, France, July 18, 1918.
R—North English, Iowa.
B—Humboldt, Nebr.
G. O. No. 15, W. D., 1919.

Corporal, Company H, 16th Infantry, 1st Division.
Without assistance he attacked an enemy machine gun which was located in a tree and dislodged the gun, but was himself killed while performing this courageous duty.
Posthumously awarded. Medal presented to father, Jesse Bickford.

BICKNELL, LEROY E. (555306)
During the Argonne-Meuse operations, France, Oct. 9–28, 1918.
R—Westford, Mass.
B—Westford, Mass.
G. O. No. 26, W. D., 1919.

Sergeant, Company D, 9th Machine Gun Battalion, 3d Division.
With no infantry support, his platoon withstood an enemy attack for two days, during which time two of his men and one gun were captured. He planned and carried out a counterattack, using in part captured enemy guns, and succeeded in releasing his own men and capturing about 50 prisoners.

BIDDLE, CHARLES JOHN_____
   In the region of Danvillers, France,
     Sept. 26, 1918.
   R—Andalusia, Pa.
   B—Andalusia, Pa.
   G. O. No. 60, W. D., 1920.

Captain, 13th Aero Squadron, Air Service.
During an engagement between 11 Spads and 12 enemy Fokkers, Captain Biddle, perceiving a comrade in distress from the attack of two planes, dived upon them and by his fire forced them to withdraw. His prompt action saved the life of his comrade, who was in imminent danger of being shot to the ground.

BIEMUELLER, ORIGINES P. (1261250)_____
   Near Fismes, France, Aug. 10-13,
     1918.
   R—Philadelphia, Pa.
   B—Philadelphia, Pa.
   G. O. No. 15, W. D., 1919.

Wagoner, 110th Ambulance Company, 103d Sanitary Train, 28th Division.
Because of the destruction from shell fire of 10 of the 13 ambulances of his company, he worked for 72 hours, 48 of them without rest, driving through a shell-swept and gas-infested area, and thereby making possible the evacuation of the wounded.

BIERYTA, MICHAEL (2086881)_____
   Near Bois d'Harville, France, Nov.
     10, 1918.
   R—Chicago, Ill.
   B—Russia.
   G. O. No. 71, W. D., 1919.

Private, Company M, 131st Infantry, 33d Division.
Under terrific machine-gun fire, he advanced through 40 feet of wire entanglements, hacking his way with his bayonet, so that his platoon could pass through to their objective. He was mortally wounded by enemy fire as he finished his work.
Posthumously awarded. Medal presented to sister, Wiktavia Laysienska.

BIGONEY, PHILIP W. (1706251)_____
   Near Chateau-Diable, France, Aug.
     27, 1918, and Sept. 8, 1918.
   R—Brooklyn, N. Y.
   B—Sayville, N. Y.
   G. O. No. 64, W. D., 1919.

First sergeant, Company H, 307th Infantry, 77th Division.
With utter disregard for his own safety, he dressed the wounds of many of his comrades under the intense machine-gun and rifle fire from the enemy's lines. On Sept. 8 he rescued a wounded officer from a heavy barrage, carried him to a place of safety, and dressed his wounds.

BILITZKI, JOHN (1213407)_____
   East of Ronssoy, France, Sept. 29,
     1918.
   R—Buffalo, N. Y.
   B—Pembina, N. Dak.
   G. O. No. 26, W. D., 1919.

Sergeant, Company A, 108th Infantry, 27th Division.
During the operations against the Hindenburg line, Sergeant Bilitzki, although twice wounded, refused to leave the field, but remained with his platoon, exhibiting magnificent courage and bravery, until he was wounded a third time. His devotion to duty set a splendid example to the men of his company.

BILLINGSLEY, EARL (2387705)_____
   Near Dun-sur-Meuse, France, Nov.
     5, 1918.
   R—Anniston, Ala.
   B—Tecumseh, Ala.
   G. O. No. 37, W. D., 1919.

Sergeant, Company H, 61st Infantry, 5th Division.
He voluntarily went forward alone against an enemy machine-gun nest, which was holding up the advance of his line, wounding and capturing one prisoner and putting the remaining occupant to flight.

BILLIS, GUST (2041307)_____
   Near Bois-d'Ormont, France, Oct.
     12, 1918.
   R—Menomonie, Wis.
   B—Greece.
   G. O. No. 32, W. D., 1919.

Sergeant, Company A, 113th Infantry, 29th Division.
With disregard for his personal safety, he saved the life of an officer of his company by attacking and killing two Germans who were about to strike the officer in the back. Later the same day he attacked, of his own accord, several machine-gun nests, always returning with prisoners, machine guns, or both.

BILLMAN, FRED E. (570919)_____
   At Sergy, France, July 29-30, 1918.
   R—Wind Gap, Pa.
   B—New Tripoli, Pa.
   G. O. No. 35, W. D., 1919.

Private, medical detachment, 47th Infantry, 4th Division.
He displayed conspicuous bravery by administering first aid to wounded soldiers in areas swept by shell and machine-gun fire.

*BILLS, ANTHONY C. (126780)_____
   Near Cantigny, France, May 28-31,
     1918.
   R—Hartford, Conn.
   B—Hastings, Pa.
   G. O. No. 99, W. D., 1918.

Corporal, Headquarters Company, 7th Field Artillery, 1st Division.
He voluntarily and constantly was exposed to shellfire to repair important telephone lines. During the performance of this work he lost an arm and was otherwise seriously injured.
Posthumously awarded. Medal presented to father, Anthony Bills.

BINGHAM, JOHN P. (1210303)_____
   Near Ronssoy, France, Sept. 29,
     1918.
   R—New York, N. Y.
   B—East Orange, N. J.
   G. O. No. 20, W. D., 1919.

Corporal, Company D, 107th Infantry, 27th Division.
During the operations against the Hindenburg line Corporal Bingham left shelter and went forward, crawling on his hands and knees under heavy machine-gun fire to the aid of a wounded officer and a wounded soldier. With the assistance of another soldier he succeeded in dragging and carrying them back to the shelter of a trench.

BINKLEY, DAVID V. (101873)_____
   At Hill 212, near Sergy, northeast
     of Chateau-Thierry, France, July
     28, 1918.
   R—Ames, Iowa.
   B—Lancaster, Pa.
   G. O. No. 117, W. D., 1918.

Private, Company I, 168th Infantry, 42d Division.
He sought and obtained permission to go out in front of our lines and recover his corporal, who was lying severely wounded in the open. He crossed an open area that was swept for more than 50 yards by enemy machine guns, reached the corporal and carried him safely back into our lines. Later he was wounded, but refused to go to the aid station until his company had won its objective.

*BIRCH, ALBERT E_____
   Near Bois-de-Bantheville, France,
     Nov. 1, 1918.
   R—Lawrence, Kans.
   B—England.
   G. O. No. 37, W. D., 1919.

Second lieutenant, 342d Machine Gun Battalion, 89th Division.
Although suffering from a wound received during the action of Nov. 1, Lieutenant Birch refused to go to the rear for treatment, but continued on duty with his platoon throughout a very critical period. He remained on duty until the morning of Nov. 11, when he was killed.
Posthumously awarded. Medal presented to father, Charles E. Birch.

BIRCH, ERNEST W. (126023)_____
   In the Ansauville sector, France,
     on or about Mar. 3, 1918.
   R—Lankershim, Calif.
   B—Rye, Calif.
   G. O. No. 129, W. D., 1918.

Corporal, Battery D, 6th Field Artillery, 1st Division.
With conspicuous bravery, he voluntarily left his dugout under intense enemy bombardment and, without assistance, rescued a comrade who was lying outside, wounded and exposed to enemy fire.

BIRCH, ROBERT I. (42660) _____ Sergeant, Company E, 16th Infantry, 1st Division.
Near Sedan, France, Nov. 6–7, 1918. He voluntarily led a small group against an enemy machine-gun nest which
R—Albee, S. Dak. was impeding the progress of his company. Although severely wounded in
B—Albee, S. Dak. the attack, he succeeded in silencing the gun nest and remained until the
G. O. No. 44, W. D., 1919. mission was completed.

BIRCHFIELD, KENNETH (2382541) _____ Corporal, Company B, 60th Infantry, 5th Division.
Near Cunel, France, Oct. 14, 1918. He advanced alone upon a machine gun which was holding up the platoon,
R—Henderson, W. Va. 150 yards in advance of his company, killed the gunner with the butt of his
B—Henderson, W. Va. rifle, and forced two other gunners to surrender.
G. O. No. 37, W. D., 1919.

BIRD, FELIX (1389424) _____ Private, Company C, 132d Infantry, 33d Division.
Near Consenvoye, France, Oct. 9, Advancing alone against a dugout, Private Bird captured 49 of the enemy and
1918. killed 1 officer who attempted to escape.
R—Chicago, Ill.
B—Chicago, Ill.
G. O. No. 71, W. D., 1919.

BIRD, FRANCIS M _____ Pharmacist's mate, first class, U. S. Navy, attached to 5th Regiment, U. S.
Marine Corps, 2d Division.
Near Suippes, France, Oct. 3–7, 1918. He showed great courage in caring for and evacuating the wounded under
R—Salt Lake City, Utah. heavy shellfire, and, at one time, he alone brought a wounded man from the
B—Amon, Idaho. field after two litter bearers had been killed.
G. O. No. 35, W. D., 1919.

BIRD, HOBART M. (139033) _____ Sergeant, Battery A, 147th Field Artillery, 32d Division.
Near St. Gilles, France, Aug. 12, 1918. After being painfully wounded by an exploding shell, Sergeant Bird, with no
R—Portland, Oreg. thought of his own wound, assisted a more severely wounded comrade to
B—Viento, Oreg. the first-aid station, and then walked a distance of 1½ kilometers over a
G. O. No. 46, W. D. 1919. heavily shelled road in quest of ambulance and stretchers.

BIRMINGHAM, DANIEL J. _____ First lieutenant, 28th Infantry, 1st Division.
Near Soissons, France, July 18, 1918, Although twice wounded, he refused to be evacuated until the objective had
and near Exermont, France, Oct. been gained and the position consolidated. Again, in the attack on Exermont,
4, 1918. Oct. 4, he continued in command of his battalion, after suffering a dangerous
R—New York, N. Y. wound in the hip, until all the objectives had been taken and the positions
B—Ireland. consolidated.
G. O. No. 44, W. D., 1919.

*BIRNEY, KNOX B. _____ First lieutenant, 6th Engineers, 3d Division.
At Claire-Chenes Woods, France, He, on his own initiative, took 12 men from his platoon and charged a number of
Oct. 20, 1918. machine-gun nests which had been holding up the advance for two hours.
R—Philadelphia, Pa. Attacking them across open ground, he cleaned out the nests and captured 42
B—Philadelphia, Pa. prisoners. In attacking other nests in the vicinity he lost his life. His
G. O. No. 142, W. D., 1918. courageous act made it possible for the attacking troops to gain and hold the
woods with a minimum number of casualties.
Posthumously awarded. Medal presented to father, Herman H. Birney.

BISCHOFF, CLIFFORD E. _____ First lieutenant, 128th Infantry, 32d Division.
Near Juvigny, France, Aug. 29– Rendered unconscious by a bursting shell, which wounded his company com-
Sept. 2, 1918. mander, Lieutenant Bischoff took command of his company as soon as he
R—Superior, Wis. regained consciousness and later assumed command of the battalion when the
B—Superior, Wis. battalion commander was gassed. After being relieved of command he
G. O. No. 32, W. D., 1919. went forward on a reconnaissance with one soldier, and the two of them alone
captured 75 of the enemy.

*BISER, JOHN L. (1284110) _____ Private, Company B, 115th Infantry, 29th Division.
Near Verdun, France, Oct. 10, 1918. While under intense machine-gun and artillery fire he disregarded his personal
R—Hagerstown, Md. safety, administered first-aid to a wounded comrade near him, and was
B—Adams Grove, Pa. instantly killed by a shell.
G. O. No. 139, W. D., 1918. Posthumously awarded. Medal presented to widow, Mrs. Minnie Biser.

BISHOP, GEORGE O. (551180) _____ Private, first class, Company G, 38th Infantry, 3d Division.
Near Mezy, France, July 15, 1918. Against the advice of his companions he advanced through intense artillery and
R—Salem, Va. machine-gun fire against an enemy machine gun which was maintaining a
B—Copper Hill, Va. damaging fire on his company. Single-handed he killed the crew of this gun,
G. O. No. 46, W. D., 1919. returning to our lines with the captured gun.

BISHOP, RALPH L. _____ Second lieutenant, 102d Infantry, 26th Division.
Near Chemin-des-Dames, France, He was in command of a working party of about 30 men on the night of Feb. 28,
Feb. 28, 1918. 1918. He encountered a heavy barrage of the enemy, which protected the
R—New Haven, Conn. advance of enemy assault troops. With coolness and courage he immediately
B—New Haven, Conn. placed his men in shell holes, fought off the enemy, and twice walked through
G. O. No. 126, W. D., 1918. the enemy's and our own barrage to recover the remains of one of his party and
to collect his own men.

BISSELL, CLAYTON L. _____ First lieutenant, 148th Aero Squadron, Air Service.
In the vicinity of Jenlain, France, While a member of a flight he was attacked by greatly superior numbers of
Oct. 28, 1918. enemy planes. Lieutenant Bissell, observing an American plane attacked by
R—Kane, Pa. eight of the enemy, dived into their midst, destroying one plane, whereupon
B—Kane, Pa. he was set upon by three enemy Fokkers, one of which he shot down, driving
G. O. No. 14, W. D., 1923. the remaining planes to their own lines. His own plane was so badly crippled
as to be beyond repair. The outstanding bravery displayed by Lieutenant
Bissell greatly inspired the members of his squadron.

BIWAN, JOSEPH (2302111)............
At St. Gilles, near Fismes, France, Aug. 4–5, 1918.
R—Sheboygan, Wis.
B—Sheboygan, Wis.
G. O. No. 139, W. D., 1918.

Private, first class, Headquarters Company, 120th Field Artillery, 32d Division. Throughout two days he maintained the telephone lines running into battalion headquarters, making frequent repairs of the lines amid falling walls and heavy bombardment by both gas and high-explosives shells.

BJORNSTAD, ALFRED W..............
South of Manila, Philippine Islands, Aug. 13, 1898.
R—St. Paul, Minn.
B—St. Paul, Minn.
G. O. No. 126, W. D., 1919.
Distinguished-service medal also awarded.

Captain, 13th Minnesota Volunteer Infantry.
Though wounded, he commanded a firing line at a critical stage of the combat and continued to command and encourage his men after having been wounded a second time, until he was compelled to be removed from the field.

BLACK, FREDERICK W...............
Near Soissons, France, July 18–22, 1918.
R—Huntington, Pa.
B—Lenox, Iowa.
G. O. No. 117, W. D., 1918.

Captain, Medical Corps, attached to 28th Infantry, 1st Division.
He went over the top to the attack in the first wave and was wounded on the morning of the first day. Disregarding his wound, he pressed on with the attacking troops and crossed and recrossed the sector immediately behind the most advanced wave, rendering first aid to wounded and placing them in shell holes. He worked unceasingly without sleep or rest and was again wounded on the fourth day by shell fire. Though twice wounded, he steadily refused to be evacuated until the evening of the fourth day, when he was exhausted and suffering from his wounds.

BLACK, WILLIAM A. (732379)...........
During Meuse offensive, France, Nov. 3–4, 1918.
R—Clarks, La.
B—Clarks, La.
G. O. No. 37, W. D., 1919.

Private, Company E, 6th Infantry, 5th Division.
After three runners had been wounded in an attempt to deliver an important message, he volunteered and delivered the message, twice crossing a valley swept by machine-gun and artillery fire and wading a river filled with ice and slush.

BLACKBURN, RAYMOND G. (1708112)....
Near Binarville, France, Oct. 2, 1918.
R—Yonkers, N. Y.
B—Yonkers, N. Y.
G. O. No. 35, W. D., 1919.

Sergeant, Company C, 308th Infantry, 77th Division.
He volunteered and led a reconnaissance patrol, and while returning to his company commander with his information one of the patrol became detached and was in danger of being captured by the enemy. Realizing his comrade's predicament, he rushed to his aid and rescued him, killing two of the enemy and dispersing the others.

BLACKBURN, WALDEN E. (1858690)......
Near Manheulles, France, Nov. 9, 1918.
R—Bellbuckle, Tenn.
B—Beech Grove, Tenn.
G. O. No. 32, W. D., 1919.

Corporal, Company C, 324th Infantry, 81st Division.
He, after having part of his right breast torn away, remained on duty directing his squad for 36 hours, when he fainted from exhaustion and was carried from the field.

*BLACKHAM, HENRY RYSDYK...........
Near Brabant, France, Oct. 11, 1918.
R—Jersey City, N. J.
B—Jersey City, N. J.
G. O. No. 20, W. D., 1919.

Second lieutenant, 116th Infantry, 29th Division.
Although severely wounded by machine-gun fire, he refused to go to the rear and continued to lead his company until he was killed.
Posthumously awarded. Medal presented to mother, Mrs. Clara A. Blackham.

BLACKINTON, GEORGE W.............
Near Xammes, France, Sept. 12–13, 1918.
R—Detroit, Mich.
B—Flint, Mich.
G. O. No. 37, W. D., 1919.

Major, 353d Infantry, 89th Division.
Having moved his battalion to an advanced position in accordance with orders, he found himself without support on either flank and no supporting machine guns or artillery 2 kilometers in advance of our main front line. In spite of his perilous situation, this officer, with the utmost coolness and good judgment, set to work intrenching and consolidating the position, determined to hold it at all costs, though his battalion was subjected to artillery and machine-gun fire and was threatened by counterattack by the enemy in force. .

BLAIR, GEORGE A..................
Near Premont, France, Oct. 8, 1918.
R—Knoxville, Tenn.
B—Wise, Va.
G. O. No. 46, W. D., 1919.

Captain, 117th Infantry, 30th Division.
During the advance from Geneve to Premont he was seriously wounded by machine-gun fire. Despite his condition he insisted on remaining with his company, and allowed himself to be evacuated only after his objective had been reached, his position reorganized, and liaison established with flanking units.

*BLAIR, JOSEPH E. (72121)...........
During the action of Apr. 12, 1918, at Bois Brule, near Apremont, France.
R—Holyoke, Mass.
B—Dublin, N. H.
G. O. No. 88, W. D., 1918.

Private, Company E, 104th Infantry, 26th Division.
During action of Apr. 12, 1918, he displayed exceptional coolness and devotion to duty in declining to seek cover during bombardment and continuing at his post in exposed position awaiting attack of enemy. Killed in action Apr. 13, 1918.
Posthumously awarded. Medal presented to mother, Mrs. Rosie Blair.

*BLAIR, TRACY S. (2187066)...........
Near Barricourt, France, Nov. 1–2, 1918.
R—Buffalo, Kans.
B—Cotter, Iowa.
G. O. No. 70, W. D., 1919.

Corporal, Company E, 353d Infantry, 89th Division.
After his platoon had reached its objective, he voluntarily accompanied his platoon commander on a reconnaissance patrol of the enemy's positions. They came upon a large body of German troops without being discovered, and Corporal Blair, under fire of artillery and machine guns, went back and brought up two platoons, which drove off the hostile force and captured a number of prisoners. The following day, while advancing in the face of severe machine-gun fire, he was fatally wounded.
Posthumously awarded. Medal presented to mother, Mrs. Mary C. Blair.

BLAKE, ARTHUR DAVID (1723612)_____
  Near St. Remy, France, Sept. 12,
    1918.
  R—Gary, Ind.
  B—Jackson, Tenn.
  G. O. No. 24, W. D., 1920.

Private, first class, Company B, 103d Machine Gun Battalion, 26th Division.
During an enemy attack, Private Blake, although not on duty, first discovered the presence of the enemy and gave the alarm. During the action four of the enemy attempted to flank one of the machine-gun positions. Private Blake killed one of the flanking group with his pistol and captured the other three unaided.

BLAKE, CHARLES RAYMOND_____
  Near Lassigny, France, Aug. 9, 1918.
  R—Westerly, R. I.
  B—Westerly, R. I.
  G. O. No. 20, W. D., 1919.

First lieutenant, pilot, Air Service, U. S. Army, attached to French Army.
Lieutenant Blake, with Second Lieut. Earle W. Porter, observer, while on a reconnaissance expedition at a low altitude far beyond the enemy lines, was attacked by five German battle planes. His observer was wounded at the beginning of the combat, but he maneuvered his plane so skillfully that the observer was able to shoot down one of their adversaries. By more skillful maneuvering he enabled his observer to fight off the remaining planes and returned safely to friendly territory.

BLAKE, ROBERT_____
  Near Bois-de-Belleau, France, June
    6, 1918.
  B—Berkeley, Calif.
  B—Seattle, Wash.
  G. O. No. 46, W. D., 1919.

First lieutenant, 5th Regiment, U. S. Marine Corps, 2d Division.
When the line was temporarily held up, he volunteered and maintained liaison with the 49th Company, continually crossing and recrossing an open field swept by intense machine-gun fire. Later in the engagement he established liaison with the French unit on the left flank, crossing a wheat field under heavy machine-gun and sniping fire, and returned with valuable information.

BLAKEMAN, CHESTER W. (38840)_____
  Near the Meuse River, France,
    Nov. 1–4, 1918.
  R—Horse Cave, Ky.
  B—Horse Cave, Ky.
  G. O. No. 70, W. D., 1919.

Sergeant, Company D, 9th Infantry, 2d Division.
After being wounded by machine-gun fire, he treated his wound himself and continued to lead his men under heavy shell and machine-gun fire until he was again wounded and ordered to a first-aid station.

²BLAKNEE, FAUN (107438)_____
  Near Somme-Py, France, Oct. 5,
    1918.
  R—Bellaire, Ohio.
  B—Businessburg, Ohio.
  G. O. No. 27, W. D., 1919.

First sergeant, Company B, 5th Machine Gun Battalion, 2d Division.
He volunteered to carry an important message across an area swept by machine-gun fire. He arrived at the company post of command with the message, and fell dead from a wound he received while in the execution of his mission.
Posthumously awarded. Medal presented to sister, Mrs. Pearl Creamer.

BLALOCK, ROBERT (53141)_____
  Near Verdun, France, Oct. 7, 1918.
  R—Krebs, Okla.
  B—Ada, Okla.
  G. O. No. 46, W. D., 1919.

Sergeant, Company D, 26th Infantry, 1st Division.
He led a patrol of 10 men against a strong enemy machine-gun position, flanking the strong point and attacking it from the rear with admirable judgment. After expending all his ammunition, this soldier continued to fight with two captured Luger pistols and himself killed eight of the enemy, in spite of being wounded. Nine machine-gun nests were wiped out as a result of this attack, and the position was organized for defense with the captured guns.

BLANCHARD, EDGAR (1880391)_____
  Near Bellicourt, France, Sept. 29,
    1918.
  R—Fayetteville, N. C.
  B—Fayetteville, N. C.
  G. O. No. 81, W. D., 1919.

Private, Company G, 120th Infantry, 30th Division.
He displayed marked personal bravery, capturing single-handed seven Germans whom he came upon in a trench and dugout. While taking the prisoners to the rear, he met a wounded soldier, and preferring to return to the firing line, turned the prisoners over to the wounded man and rejoined his squad.

BLANCHARD, HAROLD_____
  During the Meuse-Argonne offen-
    sive, France, Oct. 7–21, 1918.
  R—Boston, Mass.
  B—Boston, Mass.
  G. O. No. 37, W. D., 1919.

Major, 327th Infantry, 82d Division.
During 14 days of severe fighting he was constantly on duty with his battalion, although suffering severely from bronchitis, the result of being gassed. He personally took command of a company, after all the officers had become casualties, and led them through a heavy artillery barrage and machine-gun fire, gaining his objective. Immediately after his battalion was relieved he collapsed from the severe strain.

BLANCHARD, WALTER H. (213198)_____
  Near Varennes, France, Sept. 26,
    1918.
  R—Haverhill, Mass.
  B—Gloucester, Mass.
  G. O. No. 81, W. D., 1919.

Corporal, Company B, 344th Battalion, Tank Corps.
Corporal Blanchard, in company with an officer, crawled forward under heavy fire at the risk of his own life and dragged back a wounded man who was lying about 150 meters in front of our trenches.

BLANCHETTE, EDWARD W. (71066)_____
  Near Verdun, France, Oct. 16, 1918.
  R—Millbury, Mass.
  B—Millbury, Mass.
  G. O. No. 21, W. D., 1919.

Corporal, Company A, 104th Infantry, 26th Division.
Although wounded and ordered to the rear, he continued to lead his platoon after his sergeant had been killed and continued in command until he dropped from exhaustion.

²BLANCHFIELD, JOHN_____
  At Chateau-Thierry, France, June
    6, 1918.
  R—Brooklyn, N. Y.
  B—Ireland.
  G. O. No. 5, W. D., 1920.

Captain, 5th Regiment, U. S. Marine Corps, 2d Division.
He demonstrated exceptional ability in organizing his line at Chateau-Thierry, France, June 6, 1918, and showed heroic leadership in holding it under violent attack. His company successfully repelled two assaults by superior forces, in the second of which he was mortally wounded.
Posthumously awarded. Medal presented to widow, Mrs. John Blanchfield.

BLANKENSHIP, JOHN C. (244442)_____
  Near Fossoy, France, July 15, 1918.
  R—Ottawa, Ill.
  B—Niantic, Ill.
  G. O. No. 44, W. D., 1919.

Corporal, Company C, 5th Field Signal Battalion, 3d Division.
During the intense artillery bombardment preparatory to the great German offensive of July 15 he voluntarily led a medical officer to the aid of wounded men, following broken wire through woods. He guided the party on their return over the same route, although suffering from a severe wound.

BLAUROCK, OSCAR (2338476)_____
Near Nesles, France, July 15, 1918.
R—Brooklyn, N. Y.
B—New York, N. Y.
G. O. No. 32, W. D., 1919.

Private, first class, Company D, 4th Infantry, 3d Division.
During a heavy shell and gas bombardment he made repeated trips with messages to the various platoons, at the same time volunteering and assisting in the removal of the wounded to a place of safety.

BLEASDALE, REDWALD H. (544469)_____
Near Mezy, France, July 15, 1918.
R—Janesville, Wis.
B—Janesville, Wis.
G. O. No. 32, W. D., 1919.

Private, Headquarters Company, 30th Infantry, 3d Division.
He remained with his gun during a heavy bombardment until his gun pit was blown in and then removed his gun to another position and continued the fire under heavy machine-gun fire. Later, in the same action, he volunteered and went to reconnoiter a small woods believed to be occupied by enemy troops. There he killed several Germans single handed and returned to our lines with an American soldier.

BLEASDALE, VICTOR F_____
Near Blanc Mont, France, Oct. 8, 1918.
R—Janesville, Wis.
B—Janesville, Wis.
G. O. No. 21, W. D., 1919.

First lieutenant, 6th Machine Gun Battalion, U. S. Marine Corps, 2d Division.
On several occasions, regardless of his personal safety, he led his machine-gun platoon through heavy machine-gun and artillery fire. When the infantry company which he was supporting was halted by the fire of 2 enemy Maxims he formed his platoon as infantry and assaulted and captured both the enemy guns.

*BLEAU, HOMER J. (560708)_____
Near Brieulles, France, Sept. 29, 1918.
R—Munising, Mich.
B—Canada.
G. O. No. 89, W. D., 1919.

Sergeant, Company A, 59th Infantry, 4th Division.
When his company was held up by heavy artillery and machine-gun fire Sergeant Bleau displayed exceptional bravery and devotion to duty in leading his platoon across an open field in an attack upon an enemy machine-gun nest. Even after receiving a wound from the effects of which he died next morning he remained with his men, encouraging them on and inspiring them by his fortitude.
Posthumously awarded. Medal presented to brother, Charles Bleau.

BLEAZARD, ORSON D., Jr. (1113348)____
In the Bois-des-Rappes, France, Oct. 22, 1918.
R—Taylorsville, Utah.
B—Kamas, Utah.
G. O. No. 27, W. D., 1920.

Private, Company C, 9th Field Signal Battalion, 5th Division.
Although he was almost exhausted from gas and fatigue, Private Bleazard remained on duty throughout the day and night, laying telephone lines from the regimental relay station to the front line through heavy barrage fire from artillery and machine guns.

*BLESSING, GEORGE (550602)_____
Near Moulins, France, July 14–15, 1918.
R—Pittsfield, Mass.
B—New York, N. Y.
G. O. No. 10, W. D., 1920.

Corporal, Company E, 38th Infantry, 3d Division.
Although exposed to severe artillery fire, this noncommissioned officer kept the automatic rifle teams under his command in action, thus causing very heavy casualties to the enemy, who were attempting to cross the Marne River in boats. The stubborn resistance of this unit prevented the enemy from gaining a foothold at a critical point of our lines. While encouraging his men to greater efforts he was killed by shell fire.
Posthumously awarded. Medal presented to father, George Blessing.

BLEWETT, CHARLES H. (2225348)_____
At Moulins, France, July 14–20, 1918.
R—Richardson, Tex.
B—Richardson, Tex.
G. O. No. 44, W. D., 1919.

Private, Company A, 9th Machine Gun Battalion, 3d Division.
After being wounded in the arm by shrapnel Private Blewett continued on duty with his platoon until it was relieved four days later. He then joined another platoon going back to the line and remained in action until the condition of his wound necessitated his evacuation.

BLOCK, SAMUEL M. (1351878)_____
At Marcheville, France, Sept. 26, 1918.
R—Tampa, Fla.
B—Vanndale, Ark.
G. O. No. 15, W. D., 1919.

Private, Company A, 102d Infantry, 26th Division.
After several other runners had failed he volunteered and was successful in carrying a message through an intense machine-gun and artillery barrage.

BLOHM, JOHN (1697164)_____
Near St. Thibaut, France, Sept. 2, 1918.
R—New York, N. Y.
B—Germany.
G. O. No. 99, W. D., 1918.

Sergeant, Company B, 305th Infantry, 77th Division.
From a shell hole, in which he had taken shelter while returning from a successful daylight patrol across the Vesle River, Sergeant Blohm saw a corporal of his patrol dragging himself through the grass and bleeding profusely from a wound in the neck. He unhesitatingly left his shelter, carried the corporal behind a tree near the river bank, dressed his wound, and using boughs from a fallen tree as improvised raft towed the injured man across the river and carried him 200 yards over an open field to the American outpost line, all under continuous rifle and machine-gun fire.

BLOMBERG, HENRY S_____
Near Juvigny, north of Soissons, France, Aug. 30, 1918.
R—Superior, Wis.
B—Superior, Wis.
G. O. No. 143, W. D., 1918.

First lieutenant, 127th Infantry, 32d Division.
Inspiring his men by his own personal bravery, he vigorously led his company forward in the face of heavy machine-gun and artillery fire, capturing the heights overlooking Juvingy with many prisoners. After reaching the objective he repeatedly exposed himself to hostile fire time after time in reorganizing the line. During the defense of the position won he personally set up and operated a captured German machine gun against the enemy while under terrific fire.

BLOMGREN, ERNEST W. (1209144)_____
Near Ronssoy, France, Sept. 27, 1918.
R—New York, N. Y.
B—New York, N. Y.
G. O. No. 64, W. D., 1919.

Private, sanitary detachment, 106th Infantry, 27th Division.
During the operations against the Hindenburg line, east of Ronssoy, on Sept. 27, 1918, Private Blomgren displayed unusual courage and bravery by going forward through the terrific shell and machine-gun fire to rescue wounded comrades.

BLOND, PERCY S. (1520648) _____
  Near Montfaucon, France, Sept. 26–
    Oct. 1, 1918.
  R—Washington, Pa.
  B—Washington, Pa.
  G. O. No. 139, W. D., 1918.

First sergeant, Company C, 146th Infantry, 37th Division.
Crossing an exposed area under heavy shell and machine-gun fire he went forward and rescued a wounded comrade, carrying him 200 yards up a steep slope. On another occasion, during a severe artillery and machine-gun bombardment, he crept alone to an advanced post and carried back another wounded soldier. During the 5 days' action he gave first aid treatment to 20 members of his company, inspiring everyone by his valiant conduct in ministering to the wounded.

BLOOD, ROBERT O _____
  Near Bouresches, France, July
    20–23, 1918.
  R—Concord, N. H.
  B—Enfield, N. H.
  G. O. No. 125, W. D., 1918.

First lieutenant, Medical Corps, attached to 103d Infantry, 26th Division.
He remained with his battalion during the entire advance, working untiringly under heavy enemy fire at all times, superintending the evacuation of the wounded and caring for them in the most dangerous and exposed positions. On July 22 he established his dressing station in an advanced position that was constantly under shell fire, and many times left his station to go into the front lines to treat the wounded.

BLOOMBERG, SAM (2407217) _____
  Near Grand Pre, France, Oct. 18–23,
    1918.
  R—Newark, N. J.
  B—Newark, N. J.
  G. O. No. 44, W. D., 1919.

Private, first class, Company B, 312th Infantry, 78th Division.
During the period of 5 days that his company was occupying an advanced and isolated position Private Bloomberg volunteered and carried messages to his company headquarters after seeing 2 other runners wounded in the attempt to cross through the sweeping barrage. He was at all times under constant fire and observation of enemy machine gunners, but he succeeded in establishing liaison during the entire operations.

BLOSSOM, LYNN (280016) _____
  Near Juvigny, France, Aug. 30, 1918.
  R—Liberty, Mich.
  B—Liberty, Mich.
  G. O. No. 64, W. D., 1919.

Private, Company I, 128th Infantry, 32d Division.
While engaged as runner during an attack he maintained liaison with adjoining units throughout a most intense fire of artillery and machine guns, continuing his work until wounded by machine-gun fire.

BLUME, FERDINAND F. (1825965) _____
  On the west bank of the Meuse,
    France, Sept. 26, 1918.
  R—Millvale, Pa.
  B—Millvale, Pa.
  G. O. No. 44, W. D., 1919.

Corporal, Company C, 319th Infantry, 80th Division.
While his platoon was being held up by wire and other obstacles and the fire of the enemy threatened to annihilate it he made his way through the wire to the German trenches, from which position he bombed the enemy from their trenches. He thus saved the lives of many of his comrades and enabled them to take the trenches with a minimum of casualties.

BLUME, LEO H (1763718) _____
  Near Talma Farm, France, Oct. 18,
    1918.
  R—Rensselaer, N. Y.
  B—Albany, N. Y.
  G. O. No. 64, W. D., 1919.

Private, first class, Company C, 312th Infantry, 78th Division.
He volunteered to carry a message from his platoon across a zone of 200 yards swept by heavy artillery and machine-gun fire. He was seriously wounded while making the attempt to perform the mission, and remained under this terrific fire for over 2 hours before it was possible to rescue him.

*BLUMENTHAL, A. LABEL (17453) _____
  Near Cierges, France, Aug. 2, 1918.
  R—Chicago, Ill.
  B—Lincoln, Nebr.
  G. O. No. 70, W. D., 1919.

Private, medical detachment, 128th Infantry, 32d Division.
As he was dressing wounded men and carrying them into shell holes for protection a heavy barrage was put down in the field where he was working, but he nevertheless refused to seek cover, ministering to the wounded and reassuring them until he was mortally wounded by a bursting shell.
Posthumously awarded. Medal presented to widow, Mrs. Lena L. Blumenthal.

BLUST, PAUL E. (1592747) _____
  Near Medeah Ferme, France, Oct. 9,
    1918.
  R—New Orleans, La.
  B—New Orleans, La.
  G. O. No. 23, W. D., 1919.

Private, Company C, 2d Engineers, 2d Division.
Crawling forward under heavy machine-gun fire, he assisted in bringing a wounded comrade to safety.

BLY, ROBERT _____
  Near Cote-de-Chatillon, France,
    Oct. 15, 1918.
  R—Fort Scott, Kans.
  B—Loup City, Nebr.
  G. O. No. 13, W. D., 1919.

First lieutenant, 168th Infantry, 42d Division.
Leading his own and another company by a flanking movement around Hill 288 in the face of terrific machine-gun fire, Lieutenant Bly, with remarkable courage and skill, reached the enemy's line and effected the capture of a strongly fortified and entrenched position on the crest of the hill, together with numerous machine guns, 92 prisoners, and 1 minenwerfer. He personally took charge of the minenwerfer and turned it on the enemy, firing all their available ammunition. During these operations this officer himself killed or captured two complete machine-gun crews. Later in the day he again led his company forward and captured Hill 242, together with another minenwerfer, under circumstances which required the greatest determination and courage.

BLYNN, JOHN M. (10675) _____
  Near Somme-Py, France, Oct. 2–9,
    1918.
  R—Philadelphia, Pa.
  B—Philadelphia, Pa.
  G. O. No. 145, W. D., 1918.

Private, first class, section No. 554, Ambulance Service.
Throughout the attack north of Somme-Py he worked day and night, repeatedly driving over roads under constant shell fire to the advance dressing stations, and when necessary driving to points still farther to the front. On Oct. 3, in front of the advanced Infantry post, his ambulance was wrecked by a bursting shell. Securing another car, he evacuated the wounded.

BLYTHE, WILLIAM JESSE _____
  East of Belleau, France, July 21, 1918.
  R—Chino, Calif.
  B—Lawrence, Mass.
  G. O. No. 125, W. D., 1918.

First lieutenant, 104th Infantry, 26th Division.
Lieutenant Blythe, with two enlisted men, charged a machine-gun nest, captured two machine guns, and killed or captured 12 of the enemy.

BOAL, THEODORE D._____
Near Montblainville, France, Sept. 27, 1918.
R—Boalsburg, Pa.
B—Iowa City, Iowa.
G. O. No. 56, W. D., 1922.

Captain, aide-de-camp to the commanding general, 28th Division.
While serving as aide to the commanding general, 28th Division, he voluntarily exposed himself to great danger by repeatedly crossing an elevation swept by extremely heavy fire from rifles and machine guns in order to carry information to some 37-millimeter guns that were enabled to neutralize the machine guns of the enemy which were enfilading the entire front line of the division. His actions were an important factor in the destruction of the hostile machine guns and contributed materially to the success of the attack on the enemy position.

*BOARDMAN, GUY W. (2280987)_____
Near Courchamps, France, July 19, 1918.
R—Hughson, Calif.
B—Kenton, Ohio.
G. O. No. 108, W. D., 1919.

Private, Company A, 59th Infantry, 4th Division.
Though he had been wounded in the ankle, Private Boardman crawled out from a shell hole under heavy machine-gun fire and made several trips to a small stream 100 yards away for the purpose of filling the canteens of his wounded comrades, until he was ordered to the rear for medical aid. He was later killed in action while charging an enemy machine-gun nest.
Posthumously awarded. Medal presented to father, William Boardman.

BOARDMAN, WALTER J. (67103)_____
At Belleau Wood, France, July 18, 1918.
R—Manchester, N. H.
B—Manchester, N. H.
G. O. No. 9, W. D., 1923.

Private, first class, Company B, 103d Infantry, 26th Division.
He openly and fearlessly exposed himself to severe hostile fire and the fire of enemy snipers while rushing to capture an enemy ammunition carrier who was bringing up ammunition to one of the enemy's machine-gun squads. Capturing the ammunition carrier, he threw many bombs into the machine-gun positions, killing or driving out the crews.

BOAS, ROSS H._____
Near Soissons, France, July 19, 1918.
R—Harrisburg, Pa.
B—Harrisburg, Pa.
G. O. No. 35, W. D., 1919.

Second lieutenant, 1st Engineers, 1st Division.
After being wounded Lieutenant Boas continued to lead two platoons of engineers, acting as Infantry in the protection of the flank of the brigade, for three days, exposed to terrific machine-gun and artillery fire throughout the attack, during which time more than two-thirds of his detachment were lost.

BOBB, LEWIS C. (1237251)_____
Near Apremont, France, Oct. 2, 1918.
R—Williamsport, Pa.
B—Elkhart, Ind.
G. O. No. 87, W. D., 1919.

Sergeant, Company K, 109th Infantry, 28th Division.
Sergeant Bobb was a member of a reconnaissance patrol consisting of an officer and four soldiers, which was stopped and in danger of being surrounded by enemy machine gunners and snipers. Risking his own life to save his comrades, he dashed from cover to draw the enemy fire, calling on the others to run. Crossing an open space for 150 yards under fire, he returned the fire with his pistol, and upon reaching a sheltered position continued to keep the enemy down by his fire, while the other members of the patrol succeeded in escaping.

*BOBO, JOHN (42643)_____
Near Soissons, France, July 18–19, 1918.
R—Bonanza, Ark.
B—Fort Smith, Ark.
G. O. No. 37, W. D., 1919.

Sergeant, Company E, 16th Infantry, 1st Division.
When his company's advance was seriously threatened by the terrific fire from a machine-gun nest, Sergeant Bobo personally killed the machine-gun crew and captured the gun. Later, in the same action, he led a party of two squads against an enemy strongpoint, capturing 2 officers, 125 men, and 12 machine guns that were delivering a sweeping fire, threatening the success of the entire operation. He was killed shortly after the completion of this extraordinary feat.
Posthumously awarded. Medal presented to mother, Mrs. Lulu Bobo.

BOBRYK, JOSEPH (1907052)_____
Near Chatel-Chehery, France, Oct. 7, 1918.
R—Dickson, Pa.
B—Russia.
G. O. No. 189, W. D., 1919.

Private, Company G, 327th Infantry, 82d Division.
After his company had suffered heavy casualties in reaching its objective and consolidating its position, Private Bobryk volunteered, and, single handed, carried numerous wounded men to the dressing station, crossing the Aire River in so doing and passing each time through terrific artillery and machine-gun fire. Even after being painfully gassed, he continued this work without thought for his own personal safety.

BOEHLE, WILLIAM E. (2177788)_____
Near Crezancy, France, July 15–16, 1918.
R—O'Fallon, Mo.
B—Darden, Mo.
G. O. No. 32, W. D., 1919.

Private, Company A, 30th Infantry, 3d Division.
After his company had withdrawn from their position he voluntarily returned to the former position and throughout the night of July 15 assisted in evacuating the wounded.

BOGAN, HENRY S._____
Near Thiaucourt, France, Sept. 15, 1918.
R—Franklin, Ky.
B—Franklin, Ky.
G. O. No. 37, W. D., 1919.

Sergeant, 78th Company, 6th Regiment U. S. Marine Corps, 2d Division.
He led a small detachment in an attack on a machine gun which was holding up the advance, capturing the gun and five of its crew. He then continued the advance, entered the hostile trenches, and cleared them for a distance of 150 yards, remaining all day in this advanced position under continuous artillery and machine-gun fire.
Oak-leaf cluster.
Sergeant Bogan is also awarded an oak-leaf cluster, to be worn with his distinguished-service cross, for the following act of extraordinary heroism in action near Blanc Mont, France, Oct. 3, 1918: During the attack on Blanc Mont, Sergeant Bogan, without aid, captured three machine-gun nests, and, after being wounded, took 30 prisoners. He himself escorted these prisoners to the rear rather than have the line weakened by taking men for this duty.

BOGGS, ERNEST H. (1491288)_____
Near St. Etienne, France, Oct. 8, 1918.
R—Pilot Point, Tex.
B—Cooke County, Tex.
G. O. No. 37, W. D., 1919.

Corporal, Company M, 142d Infantry, 36th Division.
After his company had made an attack and had taken up a new position, he rendered great assistance in the reorganization of the new position. He was seriously wounded while in the performance of this work, but refused to go to the rear because it might weaken the position.

BOGGS, JOHN C.
Near Soissons, France, July 21, 1918.
R—Richmond, Va.
B—Norfolk, Va.
G. O. No. 15, W. D., 1919.

Second lieutenant, 2d Machine Gun Battalion, 1st Division.
He displayed exceptional personal bravery and initiative by volunteering to take charge of a machine gun and crew, protecting an exposed flank with them and dispersing an enemy counterattack.

BOHAN, WILLIAM J. (552258)
Near Mezy, France, July 15, 1918.
R—Newburgh, N. Y.
B—Newburgh, N. Y.
G. O. No. 32, W. D., 1919.

Sergeant, Company L, 38th Infantry, 3d Division.
During the intense enemy artillery preparation prior to the German offensive of July 15 he voluntarily left the shelter of a trench to aid a wounded comrade to a place of safety.

BOLACK, WILLIAM F. (70729)
Near Verdun, France, Oct. 25, 1918.
R—Springfield, Mass.
B—Springfield, Mass.
G. O. No. 23, W. D., 1919.

Mechanic, Machine Gun Company, 104th Infantry, 26th Division.
While taking a train of machine-gun carts to the relief of his company in the front line Mechanic Bolack was caught in a terrific bombardment, his train scattered, several of his mules killed, and he himself wounded. He had his wounds dressed at a near-by station, and, refusing to be evacuated, passed through the bombardment three times while reorganizing his train and carrying out his mission.

BOLEN, JACOB (1815674)
Near Verdun, France, Nov. 2, 1918.
R—Philadelphia, Pa.
B—Philadelphia, Pa.
G. O. No. 46, W. D., 1919.

Private, Company C, 314th Infantry, 79th Division.
Although suffering from a painful shell-fragment wound in the head, he remained on duty with his platoon, exposing himself to machine-gun and sniper fire while acting as outpost. Advancing alone at daylight, he reconnoitered what appeared to be a machine-gun position, returning with information which enabled his outguard to better their location.

BOLIN, HERALD E. (2262490)
At Waereghem, Belgium, Oct. 31, 1918.
R—Wenatchee, Wash.
B—Stanbury, Mo.
G. O. No. 46, W. D., 1919.

Battalion sergeant major, Headquarters Company, 363d Infantry, 91st Division.
Seeing a wounded soldier lying in an exposed position, he started to go to the former's assistance, and as he did so was knocked down by a bullet which struck him in the hip. He, nevertheless, continued on in the face of the dangerous fire and succeeded in moving his wounded comrade to shelter before attending to his own wound. Although he was suffering intense pain, he refused to go to the rear, but remained constantly at his post under artillery and machine-gun fire, having been on strenuous duty and without sleep for two days.

BOLLES, FRANK C.
During the attack on Jaro, Panay, Philippine Islands, Feb. 12, 1899.
R—Rolla, Mo.
B—Elgin, Ill.
G. O. No. 14, W. D., 1923.
Distinguished-service medal also awarded.

Second lieutenant, 18th Infantry, U. S. Army.
He exhibited conspicuous bravery and skill in handling his detachment and directing the fire of his piece. Even after he was seriously wounded in the leg he continued to encourage his men and could scarcely be prevailed upon to desist from attempting mounting his horse when so crippled as to be unable to do so.
Oak-leaf cluster.
Colonel Bolles was awarded an oak-leaf cluster for the following act of extraordinary heroism in action near Septsarges, France, Sept. 26, and near Bois-du-Fays, France, Sept. 28, 1918, while serving as colonel, 39th Infantry, 4th Division. On Sept. 26 Colonel Bolles personally directed the assaulting battalion of his regiment when the line was temporarily held up by hostile fire, leading the attacking troops forward to their objective. After reaching the objective, terrific hostile fire caused many casualties, and the line was beginning to waver, when Colonel Bolles assisted in the reorganization of the line, and by his personal example of courage and fearlessness encouraged his men to hold, in the face of the withering machine-gun and artillery fire, until the flank division had advanced abreast. On Sept. 28, he rallied his men under the sweeping fire of machine guns, minenwerfers, and artillery, and although painfully wounded, personally assisted in the reorganization of the positions.

BOLLING, ALEXANDER R.
In Bois-des-Nesles, France, July 14–15, 1918.
R—Philadelphia, Pa.
B—Philadelphia, Pa.
G. O. No. 32, W. D., 1919.

Second lieutenant, 4th Infantry, 3d Division.
While in command of three widely separated platoons in the Bois-des-Nesles, on the night of July 14, 1918, Lieutenant Bolling continually exposed himself to very heavy gas and shell fire by going from one platoon to another.

BOLLINGER, ERNEST V. (1587207)
Near Soissons, France, July 19, 1918.
R—Little Rock, Ark.
B—McAlister, Okla.
G. O. No. 117, W. D., 1918.

Private, Company F, 28th Infantry, 1st Division.
He advanced on machine-gun snipers on the Paris-Soissons road, showing exceptional bravery by reaching the machine gun and killing the snipers with hand grenades and automatic rifle.

BOLT, BERNARD H. (58348)
At Seicheprey, France, Mar. 28, 1918.
R—Birmingham, Ala.
B—South Bethlehem, Pa.
G. O. No. 129, W. D., 1918.

Private, Company K, 28th Infantry, 1st Division.
He was a member of a patrol consisting of an officer and four men, who with great daring entered a dangerous portion of the enemy trenches, where they surrounded a party nearly double their own strength, captured a greater number than themselves, drove off an enemy rescuing party, and made their way back to our lines with four prisoners, from whom valuable information was obtained. He died from wounds received in this expedition. Posthumously awarded. Medal presented to father, Carl Bolt.

BOLTON, ARTIE EARL
In the Bois-de-Grande-Montagne, France, Oct. 16, 1918.
R—Wingina, Va.
B—Norwood, Va.
G. O. No. 44, W. D., 1919.

First lieutenant, 115th Infantry, 29th Division.
Having been ordered to take up his position on the final objective, Lieutenant Bolton made a personal reconnaissance of his company front line, during which time he was subjected to the artillery fire of both friendly and enemy guns and machine guns directed on his position. He again went out on the same mission and captured two prisoners who were carrying a machine gun.

BONACK, PAUL J. (275139)_____
Near Juvigny, France, Aug. 30, 1918.
R—Three Lakes, Wis.
B—Three Lakes, Wis.
G. O. No. 98, W. D., 1919.

Sergeant, Company L, 127th Infantry, 32d Division.
When his company was stopped by a concealed machine-gun nest, Sergeant Bonack ascertained its position and courageously attacked it, single handed, upon his own initiative, killing the crew and enabling his company to continue the advance.

BONAVANTURA, FERDINANDO (2405383)__
Near St. Juvin, France, Oct. 19, 1918.
R—Burlington, N. J.
B—Italy.
G. O. No. 35, W. D., 1919.

Private, Company B, 309th Infantry, 78th Division.
Private Bonavantura, armed with an automatic rifle, captured a machine gun single-handed under heavy flanking fire from machine guns. The gun was supported by a squad of infantry, whom he forced to flee. Later he led a detail far into the enemy lines, encouraging his men by his fearless example.

BONDAY, ROBERT_____
Near St. Etienne, France, Oct. 4-6, 1918.
R—Mount Clemens, Mich.
B—Michigan.
G. O. No. 37, W. D., 1919.

Private, 45th Company, 5th Regiment, U. S. Marine Corps, 2d Division.
Private Bonday, as a runner, displayed exceptional courage in carrying messages for three days under shell and machine-gun fire.

BONGARDT, CHARLES F_____
At Vaux, France, July 1, 1918.
R—Omaha, Nebr.
B—Omaha, Nebr.
G. O. No. 101, W. D., 1918.

Second lieutenant, 17th Field Artillery, 2d Division.
While serving as a telephone officer, he crossed an open field in full view of the enemy and under constant bombardment three times to repair telephone lines vitally necessary to keep six batteries in operation.

BONNALIE, ALLAN F_____
Near Bruges, Belgium, Aug. 13, 1918.
R—San Francisco, Calif.
B—Denver, Colo.
G. O. No. 99, W. D., 1918.

First lieutenant, Air Service, U. S. Army, attached to Royal Air Forces, British Army.
On Aug. 13, 1918, this officer led 2 other machines on a long photographic reconnaissance. In spite of the presence of numerous enemy aircraft, they were able to take all the photographs required, but were attacked by 6 Fokker biplanes. During the combat Lieutenant Bonnalie saw that one of his accompanying machines was in difficulty and that an enemy airplane was nearly on its tail. He at once broke off combat with the enemy with whom he was engaged and dived to the assistance of the machine in trouble. He drove off the enemy plane, regardless of the bullets which were ripping up his own machine. Eventually, however, his tail planes and his elevator wires were shot away and his machine began to fall in side slips. Lieutenant Bonnalie managed to keep his machine facing toward the British lines by means of the rudder control, while his observer and the third machine drove off the enemy aircraft, which was still attacking. In its damaged condition Lieutenant Bonnalie's machine was tail heavy, and he therefore had his observer leave his cockpit and lie out along the cowl in front of the pilot. In this manner he recrossed the British trenches at a low altitude and righted his machine sufficiently to avoid a fatal crash. Had it not been for the gallantry of Lieutenant Bonnalie the injured machine to whose assistance he went would have fallen into enemy territory, as pilot had been wounded and its observer killed. Lieutenant Bonnalie's own machine was riddled with bullets and it was a marvelous performance to bring it safely to the ground.

BONNEVILLE, MARION SPENCER_____
Near Bouresches, France, June 6, 1918.
R—Chicago, Ill.
B—Snow Hill, Md.
G. O. No. 49, W. D., 1922.

Private, Headquarters Company, 6th Regiment, U. S. Marine Corps, 2d Division.
For extraordinary heroism in action on June 6, 1918, when he volunteered and assisted in taking a truck load of ammunition and material into Bouresches, France, over a road swept by artillery and machine-gun fire, thereby relieving a critical situation.

BONNEY, TIMOTHY D_____
Near Belleau, France, July 20, 1918.
R—Mexico, Me.
B—Mexico, Me.
G. O. No. 9, W. D., 1923.

First lieutenant, 103d Infantry, 26th Division.
With complete disregard for his own safety, he rescued two wounded soldiers under an extremely heavy barrage of machine-gun fire by which they were surrounded. Unassisted, he carried both men through a hail of fire to the nearest dressing station, his great courage and devotion to duty undoubtedly saving their lives.

BOOHER, ARTHUR J. (322991)_____
In the Suchan Valley, Siberia, during June and July, 1919.
R—Pequot, Minn.
B—Hayfield, Minn.
G. O. No. 133, W. D., 1919.

Corporal, Company D, 31st Infantry.
He distinguished himself on numerous occasions by his capabilities and extraordinary daring as a patrol leader while under fire, notably at Kazanka, Siberia, July 1, 1919, and in the lower Suchan Valley, Siberia, July 5, 1919.

BOONE, JOEL THOMPSON_____
In the Bois-de-Belleau, France, June 9-10 and 25, 1918.
R—Philadelphia, Pa.
B—St. Clair, Pa.
G. O. No. 137, W. D., 1918.

Passed assistant surgeon, U. S. Navy, attached to 6th Regiment, U. S. Marine Corps, 2d Division.
On two successive days the regimental aid station in which he was working was struck by heavy shells and in each case demolished. Ten men were killed and a number of wounded were badly hurt by falling timbers and stone. Under these harassing conditions this officer continued without cessation his treatment of the wounded, superintending their evacuation, and setting an inspiring example of heroism to the officers and men serving under him. On June 25, 1918, Surgeon Boone followed the attack by one battalion against enemy machine-gun positions in the Bois-de-Belleau, establishing advanced dressing stations under continuous shell fire.

BOONE, LEWIS W. (1312417)_____
Near Vaux Andigny, France, Oct. 11, 1918.
R—Westville, S. C.
B—Kershaw County, S. C.
G. O. No. 81, W. D., 1919.

Private, Company M, 118th Infantry, 30th Division.
On duty as a company runner, he carried an important message through an artillery and machine-gun barrage to battalion headquarters. Starting back through the barrage to the front lines he was wounded, but believing he might be needed at the front attempted to make his way back to his company, displaying unusual fortitude and devotion to duty.

BOONE, RAYMOND W.
In advance on Bouresches, France, June 6, 1918.
R—Elsinore, Calif.
B—Parsons, Kans.
G. O. No. 119, W. D., 1918.

Corporal, 79th Company, 6th Regiment, U. S. Marine Corps, 2d Division.
After receiving three severe wounds he continued in the advance on Bouresches, France, on June 6, 1918. Having been sent to the rear, he returned close to the advanced lines, where he assisted in bringing in the wounded.

*BOONE, WILLIAM EWING.
Near Soissons, France, July 18–19, 1918.
R—Kansas City, Mo.
B—Los Angeles County, Calif.
G. O. No. 59, W. D., 1919.

Second lieutenant, 26th Infantry, 1st Division.
Displaying valorous leadership throughout two days of attack near Soissons, France, July 18–19, 1918, he was killed while charging enemy machine guns at the head of his platoon.
Posthumously awarded. Medal presented to father, Howard C. Boone.

BOOP, LAWRENCE S. (2659181).
Near Bois-de-Brieulles, France, Sept. 29, 1918.
R—Girard, Ohio.
B—Glenn Grove, Pa.
G. O. No. 39, W. D., 1920.

Private, Company A, 59th Infantry, 4th Division.
After all communication with the company on the left had been broken by an intense machine-gun and artillery fire, he volunteered and reestablished liaison with the flank company, successfully performing this mission by going a distance of over 300 yards through a terrific artillery and machine-gun fire.

BOOTH, CHARLES W. (1383973).
Near Marcheville, France, Nov. 10, 1918.
R—Rockford, Ill.
B—Alexandria, Minn.
G. O. No. 44, W. D., 1919.

Private, Company F, 130th Infantry, 33d Division.
After being wounded twice by sniper's fire, he continued to crawl forward until he located and killed the sniper who was picking off our men. His act saved many lives.

BOOTH, JAMES O. (1827490).
Near Sivry-sur-Meuse, France, Sept. 26–28, 1918.
R—Glassport, Pa.
B—DuBois, Pa.
G. O. No. 140, W. D., 1918.

Cook, Company I, 319th Infantry, 80th Division.
He displayed exceptional courage when, under heavy shell fire and in an exposed position, he constantly made coffee for the battalion and carried it to the lines. On another occasion he assisted in the evacuation of the wounded, carrying them over a half mile under severe fire to the battalion first-aid station.

BOOTZ, HENRY A.
In the Luneville sector, France, Mar. 20 and 21, 1918, and near Villers-sur-Fere, France, July 28, 1918.
R—New York, N. Y.
B—Germany.
G. O. No. 55, W. D., 1920.

First lieutenant, 165th Infantry, 42d Division.
While conducting a raid in the Luneville sector, Lieutenant Bootz exposed himself to heavy enemy fire to carry a severely wounded soldier to a place of safety. He later reorganized a patrol to search for missing members of his raiding party. Near Meurcy Farm, July 28, this officer gallantly led his company until he was severely wounded. He then continued to direct the operations of his organization until the position was consolidated.

BORDEN, HORACE L.
Near Cunel France, Oct. 29, 1918.
R—Newport, R. I.
B—Newport, R. I.
G. O. No. 44, W. D., 1919.

Second lieutenant, Signal Corps, U. S. Army, 90th Aero Squadron, Air Service.
While carrying out a difficult contact mission without the protection of friendly planes he was attacked by three hostile machines, which he succeeded in driving off. He secured the information he sought, but while attempting to fire a signal rocket it exploded, setting the machine on fire. Lieutenant Borden crawled back on the fuselage of the machine and extinguished the flames with his bare hands. Although suffering great pain, he refused to be sent to the rear for treatment, but remained on duty with his squadron.

BORDKAS, GUS (2834009).
Near Barricourt, France, Nov. 1, 1918.
R—Kansas City, Mo.
B—Kansas City, Mo.
G. O. No. 44, W. D., 1919.

Private, medical detachment, 354th Infantry, 89th Division.
He showed great bravery and extraordinary heroism by dressing and evacuating the wounded under a terrific shell fire. While advancing with the front echelon, he saw a comrade fall when a withering machine-gun fire was encountered and, without hesitation, ran to his rescue, carrying the wounded man to the shelter of a shell hole.

BORDVICK, MONRED A. (2060138).
Near Consenvoye, France, Oct. 8, 1918.
R—Chicago, Ill.
B—Norway.
G. O. No. 71, W. D., 1919.

Sergeant, Company C, 132d Infantry, 33d Division.
In charge of a flank patrol of four men during an attack, he entered a village occupied by the enemy in force and captured 42 prisoners and 3 machine guns which were holding up the advance of the battalion from the left flank. He displayed marked courage and ability as a leader.

BORETZ, HARRY (1211226).
At Abre Guernon, Belgium, Oct. 18, 1918.
R—New York, N. Y.
B—Russia.
G. O. No. 16, W. D., 1923.

Corporal, Company H, 107th Infantry, 27th Division.
Although severely wounded and ordered to dressing station, he returned to the front lines, rejoined his company and voluntarily joined a patrol which attacked and routed enemy machine-gun crews. The officer in command of the patrol having been killed, Corporal Boretz assumed command thereof and directed the capture of an enemy machine-gun crew of four men.

BORKUS, THOMAS (154423).
Near Exermont, France, Oct. 9, 1918.
R—Chicago, Ill.
B—Russia.
G. O. No. 32, W. D., 1919.

Corporal, Company B, 1st Engineers, 1st Division.
Upon his own initiative, Corporal Borkus, with another soldier, displayed notable courage in attacking two machine guns which were hindering the advance. Undaunted by the heavy machine-gun fire, they poured a deadly rifle fire upon the enemy gunners and forced them to flee toward our attacking troops, who captured them.

BORST, RALPH P. (544706).
Near Crezancy, France, July 15, 1918.
R—Syracuse, N. Y.
B—Frankfort, Ky.
G. O. No. 32, W. D., 1919.

Regimental supply sergeant, 30th Infantry, 3d Division.
Hearing that the wounded at a dressing station had no means of evacuation, he took four escort wagons over shell-swept territory and carried the men to an ambulance station. He then established ambulance service between this point and the battalion dressing station.

BORTON, EDWARD W. (2267407)_____
  Near Eclisfontaine, France, Oct. 4,
    1918.
  R—Taft, Calif.
  B—Emporia, Kans.
  G. O. No. 46, W. D., 1919.

Private, Machine Gun Company, 364th Infantry, 91st Division.
After obtaining permission to go to the aid of wounded soldiers, he remained in the open under heavy shrapnel and high-explosive fire, giving first aid to our wounded men until he was wounded by shrapnel.

BOS, LAMBERT_____
  Near Blanc Mont Ridge, France,
    Oct. 3, 1918.
  R—Granite, Idaho.
  B—El Paso, Tex.
  G. O. No. 35, W. D., 1919.

Private, 78th Company, 6th Regiment, U. S. Marine Corps, 2d Division.
Private Bos, with two other volunteers, flanked a machine-gun nest, and, after one of his comrades had been wounded, captured 14 men and 2 machine guns. Later he aided in the capture of 40 other prisoners in a dugout.

BOSONE, PETER P. (2260628)_____
  During the Argonne offensive,
    France, Sept. 26–Oct. 12, 1918.
  R—Bingham, Utah.
  B—Castle Gate, Utah.
  G. O. No. 2, W. D., 1919.

Sergeant, Company F, 362d Infantry, 91st Division.
He was knocked unconscious by shell fire, but after recovering he immediately continued in action, thus setting a good example of devotion to duty to his men.

BOSTON, WILLIE (1931412)_____
  Near Ardeuil, France, Sept. 29, 1918.
  R—Roopville, Ga.
  B—Roopville, Ga.
  G. O. No. 46, W. D., 1919.

Private, Machine Gun Company, 371st Infantry, 93d Division.
With three other soldiers, Private Boston crawled 200 yards ahead of our lines, under violent machine-gun fire, and rescued an officer who was lying mortally wounded in a shell hole.

BOTELLE, GEORGE W. (1682967)_____
  Near Charlevaux Mill, France, Oct.
    4, 1918.
  R—Waterbury, Conn.
  B—Oakville, Conn.
  G. O. No. 71, W. D., 1919.

Private, Company C, 308th Infantry, 77th Division.
He repeatedly carried messages over ground swept by intense enemy fire. When his battalion had been surrounded and several other runners had been killed or wounded in the attempt, he volunteered to carry a message through the enemy lines to the regimental post of command, being severely wounded in the performance of this mission.

BOTHWELL, EUGENE (541632)_____
  Near Fossoy, France, July 15, 1918.
  R—Tonawanda, N. Y.
  B—North Tonawanda, N. Y.
  G. O. No. 44, W. D., 1919.

Private, Company K, 7th Infantry, 3d Division.
Through the heavy artillery preparation of the enemy of July 15, he volunteered and carried a message over an entirely strange route. He successfully completed his mission and returned with the answer.

BOTSFORD, NORMAN L._____
  Near Montfaucon, France, Sept. 28,
    1918.
  R—Pittsburgh, Pa.
  B—Pittsburgh, Pa.
  G. O. No. 56, W. D., 1922.

First lieutenant, 316th Infantry, 79th Division.
Although suffering from a painful shell wound while leading his platoon, Lieutenant Botsford declined to receive medical attention. He courageously kept on in the advance, bravely leading his men under extremely heavy shell and machine-gun fire until he collapsed. His devotion to duty and example in his wounded condition was a great inspiration to his men.

BOUCHARD, JOSEPH A. (1408465)_____
  Near Septsarges, France, Oct. 24,
    1918.
  R—Detroit, Mich.
  B—Chassell, Mich.
  G. O. No. 37, W. D., 1919.

Sergeant, Company G, 5th Ammunition Train, 5th Division.
When an enemy shell struck some pyrotechnics stored in the ammunition dump of his organization he directed and assisted in the removal of the inflammable material and placed the fire under control. Through his coolness and courage the destruction of a large quantity of near-by ammunition was avoided.

BOUCHER, ADELARDE, (62505)_____
  In the Belieu Bois, Oct. 23-26, 1918.
  R—Fitchburg, Mass.
  B—Canada.
  G. O. No. 64, W. D., 1919.

Private, first class, Headquarters Company, 101st Infantry, 26th Division.
During three days of intense shelling and concentrated machine-gun fire, Private Boucher acted as guide and liaison runner for the several units of his regiment time and again, both day and night, passing through the heavy fire, keeping contact with all adjacent units. He played a very strong part in maintaining the excellent liaison and aided materially in the efficient distribution of ammunition throughout the entire attack.

BOUGHAN, JOSEPH F. (60153)_____
  North of Verdun, France, Oct. 27,
    1918.
  R—Newton, Mass.
  B—Newton, Mass.
  G. O. No. 21, W. D., 1919.

Private, Company C, 101st Infantry, 26th Division.
While advancing with the first wave, Private Boughan, with another soldier, attacked a machine-gun nest and killed two of the crew. He accomplished this feat only after a hand-to-hand encounter, in which he was severely wounded.

BOUGIE, JAMES E. (1207891)_____
  Near Ronssoy, France, Sept. 29,
    1918.
  R—Brooklyn, N. Y.
  B—Brooklyn, N. Y.
  G. O. No. 37, W. D., 1919.

Private, sanitary detachment, 106th Infantry, 27th Division.
During operations against the Hindenburg line he went forward under a heavy shell and machine-gun fire and brought in wounded comrades, continuing his work even after he himself had been wounded.

BOURDON, WILLIAM R._____
  Near the Bois-des-Forges, France,
    Sept. 26, 1918.
  R—Minneapolis, Minn.
  B—Minneapolis, Minn.
  G. O. No. 46, W. D., 1919.

First lieutenant, 124th Machine Gun Battalion, 33d Division.
Lieutenant Bourdon was advancing with his platoon, when it came under heavy enemy machine-gun fire from a small clump of woods. Ordering his men to take cover, he went forward alone, located the machine gun, and killed the gunner.

*BOURLAND, WILLIAM F._____
  Near Verdun, France, Oct. 8–9, 1918.
  R—Rock Springs, Tex.
  B—Valley Springs, Tex.
  G. O. No. 27, W. D., 1919.

First lieutenant, 1st Engineers, 1st Division.
On Oct. 8, leading his company to the assault of a hill, he captured many prisoners and machine guns. On Oct. 9, while defending the hill, the enemy launched a counterattack of greatly superior numbers, but, in the face of great danger, Lieutenant Bourland proceeded to an outpost, and by skillful direction he contributed greatly to the successful defense of the hill. He was killed while in command at this outpost.
Posthumously awarded. Medal presented to mother, Mrs. Jessie C. Bourland Newsom.

*BOURNE, RUSSEL K. (136664)_____
  Near Samogneux, France, Oct. 24, 1918.
  R—Manville, R. I.
  B—Hartford, Conn.
  G. O. No. 21, W. D., 1919.

Corporal, Battery C, 103d Field Artillery, 26th Division.
After his piece had received two direct hits he refused to seek safety, and helped to carry a wounded comrade across a terrifically shelled area. While in the performance of this task he was instantly killed.
Posthumously awarded. Medal presented to father, Howard P. Bourne.

BOUSTEAD, GEORGE R. (102480)_____
  Northeast of Chateau - Thierry, France, July 28, 1918.
  R—Woodbine, Iowa.
  B—Woodbine, Iowa.
  G. O. No. 108, W. D., 1918.

Corporal, Company M, 168th Infantry, 42d Division.
Corporal Boustead distinguished himself northeast of Chateau-Thierry, France, on July 28, 1918, when, as the leader of a squad of four men, he raided an enemy machine-gun nest held by 12 Germans. As a result of this daring work 1 of the enemy was killed, the other 11 captured, and their four machine guns turned upon the retreating foe.

*BOUTON, ARTHUR E._____
  Near Soissons, France, July 18, 1918.
  R—Trumansburg, N. Y.
  B—Trumansburg, N. Y.
  G. O. No. 116, W. D., 1918.

Major, 9th Infantry, 2d Division.
His exhibition of dash and courage in leading an assaulting line against enemy machine-gun nests under terrific artillery fire and the successful protection of his left flank, which became exposed when liaison was broken, aided materially the success of the whole attack. He was killed by shellfire while leading his battalion in the assault.
Posthumously awarded. Medal presented to father, Edwin P. Bouton.

BOWER, GEORGE_____
  Near Thiaucourt, France, Sept. 15, 1918.
  R—Torresdale, Pa.
  B—Pennsylvania.
  G. O. No. 44, W. D., 1919.

Second lieutenant, 6th Machine Gun Battalion, U. S. Marine Corps, 2d Division.
Aiding an infantry platoon which had been forced to withdraw because of heavy machine-gun and artillery fire, Lieutenant Bower, while suffering from severe wounds, kept his guns in position, consolidating his location and preventing the danger of an enemy counterattack.

*BOWER, JAMES R. (2288499)_____
  Near Gesnes, France, Sept. 29, 1918.
  R—Pasadena, Calif.
  B—Decatur, Ill.
  G. O. No. 13, W. D., 1919.

Private, Company L, 362d Infantry, 91st Division.
He was with his company commander and three other soldiers firing at Germans in trees, when he observed another group of the enemy about to open fire on his party. He directed attention toward the enemy in order to warn his company commander and other soldiers. In so doing he drew the first shots from the Germans and was killed.
Posthumously awarded. Medal presented to father, J. M. Bower.

BOWERS, LLOYD G._____
  Near Gironville, and Chatel-Chehery, France, Aug. 14–29, and Oct. 27, 1918.
  R—Birmingham, Ala.
  B—Alabama.
  G. O. No. 3, W. D., 1919.

First lieutenant, 3d Balloon Squadron, Air Service.
On Aug. 14 this officer's balloon was attacked by four enemy chase machines, and though urged to jump he remained at his post and secured information of great value. On Aug. 29 he was attacked by enemy planes using incendiary bullets, but would not leave his post before his balloon caught fire; he insisted at once upon reascending, although he knew that the enemy was constantly patrolling the air. On Oct. 27, near Chatel-Chehery, while regulating artillery fire, he was attacked by several enemy planes, and his balloon was perforated by incendiary bullets. He remained in the air and carried out his observation. His extreme courage and devotion to duty furnished a splendid example to the officers and men of his command.

BOWES, DAVID M._____
  Near Varennes, France, Sept. 26, 1918.
  R—Bath, N. Y.
  B—Bath, N. Y.
  G. O. No. 20, W. D., 1919.

Second lieutenant, 304th Brigade, Tank Corps.
He crawled forward under heavy fire at the risk of his own life and rescued a wounded soldier, who was lying about 150 meters in front of the trenches occupied by the advanced Infantry.

BOWLES, ELVER J. (1956164)_____
  Near Brabant-sur-Meuse, France, Oct. 23, 1918.
  R—Columbus, Ohio.
  B—Columbus, Ohio.
  G. O. No. 37, W. D., 1919.

Sergeant, 308th Trench Mortar Battery, 158th Field Artillery Brigade, 83d Division.
During an offensive action in the Bossois-Bois, he remained in the open under direct fire of machine guns and artillery, assisting another soldier in operating a trench mortar for 57 minutes, firing 230 bombs. Repeatedly knocked down by concussion from exploding shells and bombs, he remained at his post until exhausted.

*BOWLES, WHITNEY (1225334)_____
  Near Le Catelet, France, Sept. 29, 1918.
  R—Forest Hills, N. Y.
  B—Greenwich, Conn.
  G. O. No. 16, W. D., 1920

Sergeant, Company L, 107th Infantry, 27th Division.
After the advance had ceased Sergeant Bowles exposed himself to intense machine-gun fire in order to place the remains of his platoon in shell hole, and organize his platoon for defense. Later learning that an officer of another company lay severely wounded a short distance in front of our lines, Sergeant Bowles unhesitatingly advanced into heavy machine-gun fire toward the officer to rescue him. He was killed while attempting the rescue.
Posthumously awarded. Medal presented to widow, Mrs. Whitney Bowles.

BOWMAN, ALVIN LESTER_____
Near the Meuse River, France, Nov. 3, 4, and 10, 1918.
R—Falls City, Oreg.
B—Philomath, Oreg.
G. O. No. 126, W. D., 1919.

Pharmacist's mate, second class, U. S. Navy, attached to 51st Company, Fifth Regiment, U. S. Marine Corps, 2d Division.
He displayed exceptional coolness and bravery under intense artillery and machine-gun fire, dressing wounded and carrying them to safety. On the night of Nov. 10, under violent machine-gun and shell fire, he carried three wounded men across the Meuse River to a point where they could be reached by stretcher bearers, exposing himself without thought of personal danger.

BOWMAN, SAMUEL A_____
Near Fleville, France, Oct. 4, 1918.
R—Springfield, Ohio.
B—Springfield, Ohio.
G. O. No. 15, W. D., 1919.

Second lieutenant, Field Artillery, observer, 12th Aero Squadron, Air Service.
He displayed remarkable bravery and devotion to duty while on an Infantry contact-patrol mission. In the performance of this duty the poor visibility necessitated flying at an altitude of less than 100 meters in order to distinguish front lines, and heavy machine-gun fire was encountered from the enemy positions in Fleville. The plane was pierced many times and he was severely wounded, but, in spite of this fact, he continued on his mission until the front line was located, after which he wrote and dropped clear and accurate messages to division and corps command posts, giving valuable and timely information.

BOWMAN, SILAS E. (1308160)_____
Near Premont, France, Oct. 7-9, 1918.
R—Johnson City, Tenn.
B—Washington County, Tenn.
G. O. No. 44, W. D., 1919.

Private, Company F, 117th Infantry, 30th Division.
For three successive days he carried messages over ground swept by machine-gun fire and heavy shelling. He disregarded personal safety, and was instrumental in establishing liaison with units in the vicinity.

*BOWMAN, WILLIAM H. (2020842)_____
Near Tulgas, Russia, Nov. 12, 1918, and on Mar. 1, 1919.
R—Detroit, Mich.
B—Penn Laird, Va.
G. O. No. 108, W. D., 1919.

Sergeant, Company B, 339th Infantry, 85th Division (detachment in north Russia).
During the engagement at Tulgas, rather than order any of his men to take the risk, he personally delivered a message over a road torn and swept by machine-gun and shell fire. On Mar. 1, 1919, when knee-deep in snow, and after he had been exposed for almost three hours to a temperature below zero and to enemy fire, he was mortally wounded while passing down the firing line in an heroic effort to keep up the spirits of his men.
Posthumously awarded. Medal presented to widow, Mrs. William H. Bowman.

BOYATT, CHARLES R. (1388370)_____
Near Bois-d'Harville, France, Nov. 10, 1918.
R—Chicago, Ill.
B—New York, N. Y.
G. O. No. 17, W. D., 1924.

Corporal, Company L, 131st Infantry, 33d Division.
Advancing through a clearing in the woods subjected to heavy fire he, with another soldier, flanked, and destroyed an enemy machine-gun nest, capturing prisoners, and allowing their company to resume the advance.

BOYD, CLYDE M. (92171)_____
Near Suippes, France, July 14-15, 1918.
R—Payne, Ohio.
B—Niniville, Ohio.
G. O. No. 21, W. D., 1919.

Private, first class, Headquarters Company, 166th Infantry, 42d Division.
Private Boyd's position as 37-millimeter gunner was subjected to an all-night shelling, so intense that, although wounded, he would not allow his comrades to carry him to safety until the bombardment slackened. He remained for three hours after being wounded, and when examined it was found that he had received 22 wounds.

*BOYD, JOSEPH A. (540794)_____
During the Meuse-Argonne offensive, France, Oct. 12 and 14, 1918.
R—Scipio, Okla.
B—Francis, Okla.
G. O. No. 10, W. D., 1920.

Sergeant, Company B, 7th Infantry, 3d Division.
He returned to his company on Oct. 11, and due to his poor physical condition on account of wounds received in June he was directed to remain with the kitchen. Upon hearing of the heavy casualties of officers and noncommissioned officers in his company he voluntarily joined his company and showed great initiative in leading a platoon in the attack of Oct. 12. On Oct. 14, while gallantly leading his platoon in attack, he was killed by shell fire.
Posthumously awarded. Medal presented to mother, Mrs. M. E. Boyd.

BOYD, LAYTON A. (950454)_____
Near Medeah Ferme, France, Oct. 9, 1918.
R—Nowata, Okla.
B—Lenapah, Okla.
G. O. No. 23, W. D., 1919.

Private, medical detachment, 2d Engineers, 2d Division.
Exposed to enemy sniper and machine-gun fire, Private Boyd went in front of our lines to administer aid to a wounded officer and also to wounded enemy troops. He also crawled to within 50 feet of an enemy machine gun and assisted in bringing a wounded comrade to safety.

BOYD, LOGAN W_____
Near Bellicourt, France, Sept. 29, 1918.
R—Knoxville, Tenn.
B—Stanford, Ky.
G. O. No. 43, W. D., 1922.

Second lieutenant, 117th Infantry, 30th Division.
Assuming command of his company after all other company officers had become casualties or were lost in the fog, Lieutenant Boyd, although wounded in the arm by shell fragments, skillfully led his company forward under a heavy hostile artillery fire until their objective was reached and the position consolidated, when he consented to go to the rear for treatment. During the advance through the heavy fog Lieutenant Boyd, by his personal leadership, kept the company intact, thereby making possible the capture of 100 prisoners by his organization.

*BOYD, RICHARD H_____
Near La Haie Menneresse, France, Oct. 17, 1918.
R—Knoxville, Tenn.
B—Richmond, Ky.
G. O. No. 64, W. D., 1919.

Second lieutenant, 117th Infantry, 30th Division.
After his platoon had suffered severe casualties, and his ammunition entirely exhausted, Lieutenant Boyd went about the town, under annihilating fire, to collect the stragglers. With the few he managed to collect he attacked and put out of action three enemy machine-gun nests, after which he reported to the front line. Here he voluntarily exposed himself in assisting wounded from the face of murderous fire, and while standing in full view of the enemy, directing his men to seek shelter, he was killed by a shell explosion.
Posthumously awarded. Medal presented to mother, Mrs. Ben S. Boyd.

BOYD, THEODORE E.

Near Conflans, France, Sept. 14, 1918.
R—Carthage, Tenn.
B—Ashland City, Tenn.
G. O. No. 20, W. D., 1919.

Second lieutenant, 7th Field Artillery, observer, attached to 88th Aero Squadron, Air Service.
This officer, being detailed for the protection of a photographic mission with five other planes, proceeded on his mission, when three of the escorting planes failed to join the formation. While flying near Conflans the formation engaged in combat with five enemy pursuit planes. Wounded in both legs, the left foot, and the right elbow, he displayed exceptional tenacity and courage by continuing to fire his guns until the enemy were put to flight.

BOYD, WILLIAM C.

Near Ponchaux, France, Oct. 7, 1918.
R—Knoxville, Tenn.
B—Crab Orchard Springs, Ky.
G. O. No. 4, W. D., 1923.

First lieutenant, 117th Infantry, 30th Division.
During the attack of his company, displaying exceptional bravery and disregard of personal danger, Lieutenant Boyd went forward in the face of a heavy machine-gun fire, and in full view of the enemy carried a badly wounded soldier to safety in a near-by shell hole. His gallant conduct was an excellent example and an inspiration to his company.

BOYKIN, SAMUEL V. (1203478)

East of Ronssoy, France, Sept. 29, 1918.
R—New York, N. Y.
B—Richmond, Va.
G. O. No. 20, W. D., 1919.

Sergeant, Company B, 105th Infantry, 27th Division.
During the operation against the Hindenburg line Sergeant Boykin, with an officer and two other sergeants, occupied an outpost position in advance of the line, which was attacked by a superior force of the enemy. Sergeant Boykin assisted in repulsing this attack and in killing 10 Germans, capturing 5, and driving off the others. The bravery and determination displayed by this group was an inspiration to all who witnessed it.

BOYLE, J. EDWARD

Near Varennes, France, Sept. 26, 1918.
R—Beaver Falls, Pa.
B—New Brighton, Pa.
G. O. No. 64, W. D., 1919.

Captain, 110th Infantry, 28th Division.
While bringing his company into position he was severely wounded, being struck in 11 places by fragments of high-explosive shells. Seeing that his being wounded and the enemy fire had caused some disorganization of his command, he assembled his platoon and section leaders, assigned them their missions, and, although bleeding profusely, set them an example of coolness and bravery. Inspired by his bravery, the company moved forward, getting into action at a critical period of the fight.

BOYLE, JAMES B.

Near Verdun, France, Oct. 8, 1918.
R—Baltimore, Md.
B—Baltimore, Md.
G. O. No. 2, W. D., 1919.

First lieutenant, 115th Infantry, 29th Division.
During an offensive of his organization on the edge of Consenvoye Wood he led a flanking attack on the enemy, and by skillful handling of his platoon captured two machine guns and opened a way for an advance which resulted in clearing the wood of the enemy and greatly assisted in obtaining our objective. Later he was severely wounded while leading a wire-carrying party through a heavy artillery barrage, refusing first aid until a soldier wounded at the same time had been attended to.

*BOYLE, JUNIUS I.

In the Bois Dommartin, northwest of Thiaucourt, France, Oct. 11, 1918.
R—Baltimore, Md.
B—Gaithersburg, Md.
G. O. No. 68, W. D., 1920.

First lieutenant, 147th Infantry, 37th Division.
He voluntarily and alone made a reconnaissance of the Bois Dommartin, a strongly fortified enemy position, and returned with valuable information. Later, accompanied by another soldier, he again made a reconnaissance, and while returning with valuable information was attacked by a superior enemy force and killed.
Posthumously awarded. Medal presented to mother, Mrs. M. A. Boyle.

BOYLE, WILLIAM J. (41319)

Near Medeah Ferme, France, Oct. 5, 1918.
R—New York, N. Y.
B—Ireland.
G. O. No. 46, W. D., 1919.

Corporal, Machine Gun Company, 9th Infantry, 2d Division.
He saved the lives of many of his comrades by killing two enemy machine gunners and putting the guns out of action.

BOYSEN, ERNEST J.

At Champigneulle, France, Nov. 1, 1918.
R—Harlan, Iowa.
B—Atlantic, Iowa.
G. O. No. 81, W. D., 1919.

Second lieutenant, 305th Infantry, 77th Division.
When his platoon was held up by fire from enemy machine guns and snipers, Lieutenant Boysen went forward in advance of his platoon in disregard of personal danger, and with a rifle brought down three enemy snipers and drove off the enemy machine gunners, thereby enabling his platoon to resume its advance.

*BOZARTH, LOUIS (2240361)

Near Madeleine Farm, France, Nov. 8, 1918.
R—El Paso, Tex.
B—Kirksville, Mo.
G. O. No. 60, W. D., 1920.

Corporal, Company F, 315th Supply Train, 90th Division.
While driving a supply train truck he was severely wounded in the left leg by fragment of an enemy high-explosive shell and the truck was badly damaged by shellfire. Assisted by his helper, he made temporary repairs and drove the truck to its destination. He later died as a result of the wounds received.
Posthumously awarded. Medal presented to mother, Mrs. Mary Frances Hamilton.

*BOZENHART, ERNEST G. (1543504)

Near Ivoiry, France, Sept. 29, 1918.
R—Toledo, Ohio.
B—Toledo, Ohio.
G. O. No. 37, W. D., 1919.

Private, medical detachment, 147th Infantry, 37th Division.
Making his way through heavy artillery and machine-gun fire, he rendered valuable medical treatment to the wounded and assisted in bringing the men to safety and forwarding them to a first-aid station. In the performance of his duties he was shortly afterwards killed.
Posthumously awarded. Medal presented to father, George Bozenhart.

BRACEY, JAMES E. (154431)

North of Exermont, France, Oct. 9, 1918.
R—Rowland, N. C.
B—Robeson County, N. C.
G. O. No. 72, W. D., 1920.

Sergeant, first class, Company A, 1st Engineers, 1st Division.
During the attack on Hill 269, when his group came under direct machine-gun fire, Sergeant Bracey skillfully advanced his men and then alone he rushed and captured the enemy gunner. His gallant act enabled other members of his group to close in on the enemy without loss, capturing the gun and forcing five of the enemy to surrender.

BRACKETT, ALBERT C. (67541) _____
At Marcheville, France, Sept. 26, 1918.
R—South Paris, Me.
B—Portland, Me.
G. O. No. 142, W. D., 1918.

Sergeant, Company D, 103d Infantry, 26th Division.
Although he was severely wounded, he insisted upon continuing his duties. Under severe fire from snipers, machine guns, and artillery he repeatedly stood up in the open, offering himself as a target for the snipers in order to locate their positions.

BRADBURY, ARTHUR W. _____
Near Gesnes, France, Sept. 29, 1918.
R—Seattle, Wash.
B—Santa Rosa, Calif.
G. O. No. 59, W. D., 1919.

Captain, 362d Infantry, 91st Division.
Severely wounded while his battalion was attacking Gesnes, he refused to be evacuated. Though unable to walk, he remained in active command of his battalion during a critical period of the engagement, until the objective had been gained, his own battalion and the one on his left reorganized, the occupied position consolidated, and food and water procured for his men.

*BRADEN, CHARLES. _____
Near the mouth of the Big Horn River, Mont., Aug. 11, 1873.
R—East Saginaw, Mich.
B—Detroit, Mich.
G. O. No. 7, W. D., 1925.

Second lieutenant, 7th Cavalry, U. S. Army.
Lieutenant Braden, with 20 men, having been attacked by nearly 200 Indians, although severely wounded in the encounter, by his personal gallantry and splendid leadership so inspired his small command as to enable it to repulse the attack by overwhelmingly superior numbers
Posthumously awarded. Medal presented to widow, Mrs. Charles Braden.

BRADFIELD, WALTER E. (1489844) _____
Near St. Etienne, France, Oct. 8, 1918.
R—Sapulpa, Okla.
B—Muskogee, Okla.
G. O. No. 66, W. D., 1919.

Corporal, Company H, 142d Infantry, 36th Division.
While advancing with his company he was wounded in the hip by a sniper who was inflicting severe losses on the company. Crawling forward, he killed this sniper and continued in action even after receiving a second wound. He refused to go to the rear until he was ordered to do so by his company commander.

BRADFORD, JOSEPH W. (44365) _____
In the Argonne-Meuse offensive, France, Oct. 4, 1918.
R—Brockton, Mass.
B—North Carver, Mass.
G. O. No. 46, W. D., 1919.

Sergeant, Company M, 16th Infantry, 1st Division.
Consolidating remnants of other platoons with his own, he advanced against violent machine-gun fire of the enemy. He directed an attack against two nests, which were reduced, and his objective gained. While reconnoitering in front of his objective he was severely wounded.

BRADLEY, JOSEPH L. (40617) _____
Near Medeah Ferme, France, Oct. 8, 1918.
R—Marshall, N. C.
B—Marshall, N. C.
G. O. No. 78, W. D., 1919.

Private, first class, Company L, 9th Infantry, 2d Division.
Without regard for his own safety, Private Bradley worked unceasingly as a stretcher bearer, caring for the wounded of other companies as well as of those of his own, and inspiring others to greater efforts by his example of courage and endurance.

BRADLEY, MANLEY (1817652) _____
Near Nantillois, France, Oct. 5, 1918.
R—Montebello, Va.
B—Montebello, Va.
G. O. No. 46, W. D., 1919.

Sergeant, Company D, 317th Infantry, 80th Division.
He was wounded in the head while leading his platoon across a valley swept by machine-gun fire, but he continued to lead his men on to their objective, refusing to report to the dressing station until he was ordered to do so.

BRADLEY, PAUL W. (88900) _____
At Ancerviller, France, Apr. 26, 1918; near Meurcy Farm, France, July 29, 1918; and near Landres-et-St. Georges, France, Oct. 14, 1918.
R—Short Hills, N. J.
B—Geneva, Ill.
G. O. No. 78, W. D., 1919.

Private, first class, Machine Gun Company, 165th Infantry, 42d Division.
When an enemy shell struck the gun position of his squad near Ancerviller, France, severely wounding him, Private Bradley coolly removed the gun to a place of safety and returned for the tripod, being wounded for a second time in so doing. Near Meurcy Farm this soldier assumed leadership of his squad after his corporal was severely wounded in the arm by a machine-gun bullet until the objective had been taken and the line firmly established. During the advance on the enemy position near Landres-et-St. Georges, Private Bradley again displayed conspicuous coolness and courage in taking charge of his section, after the section sergeant had been wounded, and directing the placing and firing of the guns.

BRADLEY, ROE (1311167) _____
Near Brancourt, France, Oct. 8, 1918.
R—Glendale, S. C.
B—Marion, N. C.
G. O. No. 78, W. D., 1919.

Private, Company F, 118th Infantry, 30th Division.
When a party of 25 of the enemy threatened the advance by machine-gun fire from a sunken road, Private Bradley, who was ahead of the front line, quickly got his automatic rifle into action, and by well-directed enfilading fire killed a large number of the enemy, capturing the remainder. His timely act prevented an interruption of the attack.

*BRADSHAW, HOWARD W. (2385949) _____
At Cunel, France, Oct. 14, 1918.
R—Akron, Ohio.
B—Brockwayville, Pa.
G. O. No. 20, W. D., 1919.

Sergeant, Company A, 61st Infantry, 5th Division.
His company being left without officers, he reorganized the company under severe shellfire. With absolute disregard for his personal safety, he led the company against machine-gun emplacements until he was killed.
Posthumously awarded. Medal presented to father, Charles L. Bradshaw.

BRADSNYDER, HENRY (65120) _____
Near Verdun, France, Oct. 23, 1918.
R—Thompsonville, Conn.
B—Torrington, Conn.
G. O. No. 44, W. D. 1919.

Corporal, Company G, 102d Infantry, 26th Division.
Accompanied by his lieutenant, Corporal Bradsnyder charged two machine-gun nests which were causing heavy losses in our ranks. To reach the positions it was necessary to pass through an intense machine-gun and artillery fire. The first gun was successfully reached, a German officer and one of the crew killed, and the gun captured. In the dash for the second gun the lieutenant was severely wounded. Corporal Bradsnyder gave him first-aid treatment and remained with him, protecting him from death or capture until aid arrived some time later.

BRADY, CLIFFORD W. (2805988) _____
Near Bantheville, France, Oct. 23-24, 1918.
R—Enid. Okla.
B—Henry, Ill.
G. O. No. 46, W. D., 1919.

Private, Company K, 357th Infantry, 90th Division.
On many occasions, while his battalion was holding a position, Private Brady crawled far in advance of the outposts and, with glasses and compass, located enemy machine-gun positions. His work was done under most hazardous conditions and heaviest fire, but he succeeded after all other attempts had proved fatal.

BRADY, DALTON E_____
  Near Cunel, France, Oct. 14, 1918.
  R—French Creek, W. Va.
  B—French Creek, W. Va.
  G. O. No. 37, W. D., 1919.

Captain, 60th Infantry, 5th Division.
In the face of heavy machine gun and shell fire, Captain Brady left a sheltered position to go to the rescue of First Sergeant Kenneth Romaine, who was lying wounded some distance away in a shell hole, upon which the enemy was directing a heavy machine-gun fire. With utter disregard for his own personal safety, he carried the wounded sergeant through the terrific machine gun and shell fire to a place of safety.

BRADY, FRANCIS M_____
  Near Cunel, France, Oct. 12, 1918.
  R—Yonkers, N. Y.
  B—Yonkers, N. Y.
  G. O. No. 26, W. D., 1919.

First lieutenant, 9th Machine Gun Battalion, 3d Division.
He led his platoon over 300 meters of open ground, attacking and capturing five enemy machine guns, with their officers and crews, thereby saving many lives and establishing liaison with the troops on his right. He recrossed the open ground to report his location. Despite intense artillery fire, he held the captured position for 48 hours, and with a leader personally silenced two enemy machine guns which were enfilading the troops advancing to his support.

BRADY, JOHN J_____
  Near Chateau-Thierry, France, June 6–7, 1918.
  R—New York, N. Y.
  B—New York, N. Y.
  G. O. No. 46, W. D., 1919.

Chaplain, U. S. Navy, attached to 5th Regiment, U. S. Marine Corps, 2d Division.
He made two complete tours of the front line under severe fire, carried on his duties as chaplain with untiring service, and ministered to the men of the regiment under unusually trying circumstances. He continually exposed himself to carry cigarettes to men of the line, who had no opportunity to get them otherwise.

BRAGG, JAMES W. (754936)_____
  Near Binarville, France, Oct. 2–7, 1918.
  R—Webster County, W. Va.
  B—Wayneville, W. Va.
  G. O. No. 20, W. D., 1919.

Private, medical detachment, 308th Infantry, 77th Division.
He was on duty with a detachment of his regiment which was cut off and surrounded by the enemy in the Argonne Forest, France, for five days. Though he was without food throughout this period, he continued to render first aid to the wounded, exposing himself to heavy shell and machine gun fire at the risk of his life until he was completely exhausted.

BRAMBLE, EDWIN D. (241077)_____
  At Marcheville, France, Sept. 26, 1918.
  R—Mapleton, Iowa.
  B—Mapleton, Iowa.
  G. O. No. 139, W. D., 1918.

Private, first class, Headquarters Company, 102d Infantry, 26th Division.
He performed valuable service in maintaining communication by voluntarily repairing telephone lines under a violent artillery bombardment. While so engaged he was seriously wounded.

BRANDON, CLYDE_____
  Near St. Etienne, France, Oct. 3–9, 1918.
  R—St. Lake City, Utah.
  B—Denver, Colo.
  G. O. No. 78, W. D., 1919.

Private, 82d Company, 6th Regiment, U. S. Marine Corps, 2d Division.
For six days and nights Private Brandon, a battalion scout, worked unceasingly in supplying his battalion commander with accurate information, repeatedly volunteering for hazardous reconnaissances over fire-swept terrain, penetrating the enemy's lines without hesitation to observe hostile positions, and also establishing liaison under conditions of exceptional difficulty.

*BRANDT, ARTHUR F. (100943)_____
  Northeast of Verdun, France, Oct. 16, 1918.
  R—Pottsville, Iowa.
  B—Doon, Iowa.
  G. O. No. 32, W. D., 1919.

Corporal, Company E, 168th Infantry, 42d Division.
After his company had been in action three days during the attack on the Cote-de-Chatillon and was to be relieved, Corporal Brandt volunteered to guide the company to a position of security in the rear which he had selected. While the relief was being made under shell fire, this soldier and four others were severely wounded by a bursting shell. Realizing that this wound would prove fatal, Corporal Brandt, while being carried on a stretcher, indicated the route to be taken by the company, being wounded in the face and scarcely able to talk. Through his extraordinary fortitude and will power the company was able to reach its position over difficult terrain and under enemy fire. He died of his wounds the next day.
Posthumously awarded. Medal presented to father, William F. Brandt.

BRANDT, LENNO H. (43704)_____
  Near Fleville, France, Oct. 4, 1918.
  R—Boone, Iowa.
  B—Peoria, Ill.
  G. O. No. 35, W. D. 1920.

Corporal, Company I, 16th Infantry, 1st Division.
Corporal Brandt, with one companion, advanced ahead of his organization, exposed to heavy fire, and silenced an enemy machine-gun nest which had halted his company.

BRANSON, WALTER W. (42850)_____
  Near Soissons, France, July 19, 1918.
  R—Evansville, Ind.
  B—Castlewood, Va.
  G. O. No. 15, W. D., 1919.

Corporal, Company F, 16th Infantry, 1st Division.
Upon finding six men who during the advance had become separated from their companies, he voluntarily organized them into a detachment, led an attack upon two machine guns, killed seven of the crew, and captured five.

BRANTLEY, ROBERT CLINE (244899)_____
  Near Malancourt Woods, France, Sept. 26, 1918.
  R—Monterey, Calif.
  B—Salisbury, N. C.
  G. O. No. 87, W. D., 1919.

Sergeant, Company D, 1st Gas Regiment.
After his detachment had been ordered to the rear, Sergeant Brantley remained to administer first aid to a wounded comrade, bringing him to safety through withering machine-gun fire.

BRAUN, GUSTAV J_____
  Near Sergy, France, July 29–30, 1918.
  R—Indianapolis, Ind.
  B—Buffalo, N. Y.
  G. O. No. 46, W. D., 1919.

Captain, 47th Infantry, 4th Division.
No medical officer or first-aid men being present, Captain Braun, then first lieutenant and battalion liaison officer, established a first-aid station and worked throughout the day and night dressing the wounded. On both days he repeatedly went out himself in the most intense shell fire and carried wounded men to shelter. When the water supply was exhausted, he made several trips through heavy machine-gun fire and filled canteens at water holes and a creek in front of the line.

BRAUN, PAUL (52990)................... | Sergeant, Company C, 26th Infantry, 1st Division.
  In the battle near Soissons, France, | He led his platoon against a machine gun, captured the gun, and killed its crew.
    July 21, 1918.
  R—Philadelphia, Pa.
  B—Germany.
  G. O. No. 132, W. D., 1918.

BRAUNGARDT, LAFAYETTE (1039102)..... | Private, Battery F, 10th Field Artillery, 3d Division.
  Near Greves Farm, France, July 15, | Responding to a call for volunteers, Private Braungardt, with eight other
    1918. | soldiers, manned two guns of a French battery which had been deserted by
  R—St. Louis, Mo. | the French during the unprecedented fire, after many casualties had been
  B—Urich, Mo. | inflicted on their forces. For two hours he remained at this post and poured
  G. O. No. 46, W. D., 1919. | an effective fire into the ranks of the enemy.

BRAUTIGAM, GEORGE F................. | Private, 49th Company, 5th Regiment, U. S. Marine Corps, 2d Division.
  In the Bois-de-Belleau, France, June | After carrying messages all night under intense artillery fire he volunteered,
    23, 1918. | with another soldier, and carried a wounded officer through the shell fire to a
  R—Cincinnati, Ohio. | dressing station 1 kilometer away.
  B—Ohio.
  G. O. No. 70, W. D., 1919.

BREAKEY, JOHN W. (1387830).......... | Sergeant, Company H, 131st Infantry, 33d Division.
  At Chipilly Ridge, France, Aug. 9, | After being shot through both legs he gallantly continued to perform his duty,
    1918. | charging one machine-gun nest after another, until the objective was reached.
  R—Chicago, Ill.
  B—Chicago, Ill.
  G. O. No. 128, W. D., 1918.

*BRECKENRIDGE, ROBERT M. (1967624). | Private, first class, Company H, 365th Infantry, 92d Division.
  At Ferme de Bel-Air, France, Oct. | Although severely wounded in the leg from shell fire, he, an automatic rifleman,
    29, 1918. | continued in action, crawled forward for a distance of 100 yards to a position
  R—Hennessey, Okla. | where he obtained a better field of fire, and assisted preventing any enemy
  B—Hennessey, Okla. | party from taking a position on the company flank. In spite of his wound,
  G. O. No. 31, W. D., 1919. | he continued to use his weapon with great courage and skill until he was killed
   | by enemy machine-gun fire.
   | Posthumously awarded. Medal presented to mother, Mrs. Amelia Wilson.

BRECKINRIDGE, LUCIEN S.............. | Captain, 308th Infantry, 77th Division.
  Near Grand Pre, France, Oct. 14, | All the bridges over the Meuse River having been destroyed by artillery fire,
    1918. | Captain Breckenridge, who had been ordered to cross the river with his
  R—New York, N. Y. | battalion, personally reconnoitered the banks of the river in utter disregard
  B—Washington, D. C. | for his own safety until he found a ford. He then led his command across
  G. O. No. 78, W. D., 1919. | the stream under intense machine gun and artillery fire and established a
   | position on the heights of the opposite bank.

BREEDEN, ELDON...................... | First lieutenant, 357th Infantry, 90th Division.
  Near Fey-en-Haye, France, Sept. | Though he had been wounded in the side by a machine-gun bullet, Lieutenant
    12, 1918. | Breeden refused medical aid until he had led his platoon to its objective.
  R—Medford, Okla.
  B—Corydon, Ind.
  G. O. No. 78, W. D., 1919.

BREEN, VINCENT C................... | Captain, 101st Infantry, 26th Division.
  In the Bois Belleau north of Verdun, | During the attack made to retake the woods lost by the retirement of our units,
    France, Oct. 27, 1918. | Captain Breen was severely wounded in the arm. After receiving first aid
  R—Boston, Mass. | he again led his company forward through heavy fire until wounded a second
  B—Boston, Mass. | time, this time in the shoulder. It was largely due to his courage and initia-
  G. O. No. 39, W. D., 1920. | tive that his company was able to advance to take its objective.

BREESE, CLINTON S.................. | Second lieutenant, observer, 12th Aero Squadron, Air Service.
  Near Argonne, France, Nov. 2, 1918. | While on an infantry contact mission, Lieutenant Breese and his pilot were
  R—Waukesha, Wis. | attacked by four enemy planes and driven back, but, realizing the importance
  B—Waukesha, Wis. | of their mission, deliberately returned and attacked the four planes, sending
  G. O. No. 32, W. D., 1919. | one to the earth and driving the others away. Unmindful of the damaged
   | condition of their plane and of their own danger, they then flew for an hour
   | within 100 meters of the ground, through a continuous heavy machine-gun
   | fire, until they had accurately located our front-line positions.

BREGGER, THOMAS (543321).......... | Private, first class, medical detachment, 7th Infantry, 3d Division.
  Near La Tuilerie Farm, France, | He displayed exceptional courage while caring for the wounded soldiers of his
    July 22, 1918, and near Le Charmel, | battalion while under intense machine-gun and artillery fire. His efforts were
    France, July 25, 1918. | unceasing. For a long period he was the only Hospital Corps man on duty,
  R—Ithaca, N. Y. | and when the battalion was relieved he voluntarily remained on duty, stay-
  B—Quincy, Ill. | ing with the wounded throughout the night under the severe artillery fire.
  G. O. No. 64, W. D., 1919.

BREKKE, OLAF (1424692)............. | Private, Company C, 58th Infantry, 4th Division.
  Near Nantillois, France, Oct. 2-5, | Though wounded in the chest by shrapnel, he refused to be evacuated, con-
    1918. | tinuing his duties as runner for three days, till his organization was relieved.
  R—South Fergus Falls, Minn. | He showed marked personal heroism in performing dangerous missions,
  B—Grafton, N. Dak. | exposing himself to heavy artillery and machine-gun fire.
  G. O. No. 81, W. D., 1919.

*BREMER, HERMAN F. (1317082)....... | Sergeant, Machine Gun Company, 119th Infantry, 30th Division.
  Near Bellicourt, France, Sept. 29, | Sergeant Bremer displayed coolness, excellent judgment, and efficient leader-
    1918. | ship in keeping his platoon intact, while advancing with the regiment. Ex-
  R—Charleston, S. C. | posed to fire from all sides, he set his guns and engaged the enemy. While
  B—Charleston, S. C. | leading his men to a new position, he was instantly killed.
  G. O. No. 21, W. D., 1919. | Posthumously awarded. Medal presented to sister, Mrs. Eleanor Bremer
   | Brown.

BREMNER, FRANK M. (2061802) _____
  Near Chipilly Ridge, France, Aug.
    9, 1918.
  R—Chicago, Ill.
  B—Chicago, Ill.
  G. O. No. 71, W. D., 1919.

Private, Company G, 131st Infantry, 33d Division.
When an enemy machine gun was holding up our advance he worked out alone in advance of our front lines and, getting in rear of the hostile position, captured the enemy gun and its crew. While advancing, Private Bremner showed utter disregard of the heavy artillery and machine-gun fire to which he was subjected.

BRENNAN, ELMER W. (261954) _____
  Near Cierges, northeast of Chateau-
    Thierry, France, July 31, 1918.
  R—Detroit, Mich.
  B—Detroit, Mich.
  G. O. No. 139, W. D., 1918.

Corporal, Company D, 125th Infantry, 32d Division.
During the heavy shelling in the Bois-de-Grimpettes he rendered himself conspicuous by exposing himself to great personal danger in order to give aid to wounded companions, frequently searching the woods for wounded soldiers. He aided seven comrades who had been badly wounded to places of safety. That afternoon he conducted a liaison officer from the position his company occupied forward through a barrage to the town of Cierges. When this officer had been wounded and gassed, Corporal Brennan assisted him to reach his destination and deliver his message.

BRENNAN, HUGH F. (1898616) _____
  Near Fleville, France, Oct. 8–13, 1918.
  R—Pittsburgh, Pa.
  B—Pittsburgh, Pa.
  G. O. No. 78, W. D., 1919.

Sergeant, Company K, 328th Infantry, 82d Division.
After being severely wounded in the shoulder by a machine-gun bullet, Sergeant Brennan refused to be evacuated but remained in command of his company, to which he had succeeded after all the officers had become casualties. For four days he led his men in the operations against the enemy, though he was suffering severely, until his wound became infected, and he was evacuated against his protest.

BRENNAN, MATTHEW (89359) _____
  Near Landres-et-St. Georges, France,
    Oct. 15, 1918.
  R—Hoboken, N. J.
  B—Hoboken, N. J.
  G. O. No. 78, W. D., 1919.

Corporal, Company B, 165th Infantry, 42d Division.
After his platoon commander, platoon sergeant, and all other noncommissioned officers had become casualties, Corporal Brennan assumed command of his platoon, reorganized it under trying conditions, and continued the advance under heavy machine-gun and artillery fire, inspiring his men by his coolness and leadership.

BRENNER, JACOB P. _____
  Near Molleville Farm, France, Oct.
    12, 1918.
  R—Youngstown, Ohio.
  B—Warren, Ohio.
  G. O. No. 71, W. D., 1919.

Second lieutenant, 322d Field Artillery, 83d Division.
As executive officer at a battery and responsible for the prompt delivery of the barrage to repel a counterattack by the enemy, he remained under heavy fire at his post after being severely wounded and successfully carried out his mission. His example of heroism was an inspiration to his men.

BRENSTUHL, JOHN G. (95139) _____
  Near St. Georges, France, Oct. 15,
    1918.
  R—Lancaster, Ohio.
  B—Columbus, Ohio.
  G. O. No. 23, W. D., 1919.

Private, Company L, 166th Infantry, 42d Division.
Seeing the only other company runner killed while delivering a message of vital importance, Private Brenstuhl crawled from shell hole to shell hole during a rain of machine-gun bullets, took the message from the dead man, and completed the mission.

BRERETON, LEWIS H. _____
  Over Thiaucourt, France, Sept. 12,
    1918.
  R—Allegheny, Pa.
  B—Allegheny, Pa.
  G. O. No. 15, W. D., 1919.

Major, pilot, corps observation wing, Air Service.
He, together with an observer, voluntarily and pursuant to a request for a special mission, left his airdrome and crossed the enemy's lines over Lironville and proceeded to Thiaucourt. In spite of poor visibility, which forced them to fly at a very low altitude, and in spite of intense and accurate antiaircraft fire, they maintained their flight along their course and obtained valuable information. Over Thiaucourt they were suddenly attacked by four enemy monoplane Fokkers. Maneuvering his machine so that his observer could obtain a good field of fire, he entered into combat. His observer's guns becoming jammed, he withdrew until the jam was cleared, when he returned to the combat. His observer then becoming wounded, he coolly made a landing within friendly lines, although followed down by the enemy to within 25 meters of the ground. By this act he made himself an inspiration and example to all the members of his command.

BRESLIN, JAMES E. _____
  Near Cote-de-Chatillon, France, Oct.
    15, 1918.
  R—Malden, Mass.
  B—Boston, Mass.
  G. O. No. 13, W. D., 1919.

Second lieutenant, 168th Infantry, 42d Division.
He was in charge of a combat liaison platoon during the offensive operations at Cote-de-Chatillon. When the leading companies were held up by intense concentrated machine-gun fire, he courageously led his platoon forward and penetrated the enemy's lines for a depth of 1 kilometer, his command being reduced by heavy casualties to only 12 men. In severe hand-to-hand fighting he captured 2 machine-gun nests and 40 prisoners, and obtained valuable information regarding the hostile positions, which enabled the leading companies to continue the advance.

BRESNAHAN, THOMAS F. _____
  Near Mezy, France, July 15, 1918.
  R—Fitchburg, Mass.
  B—Fitchburg, Mass.
  G. O. No. 23, W. D., 1919.

First lieutenant, 38th Infantry, 3d Division.
While acting as battalion signal officer Lieutenant Bresnahan organized a detachment of orderlies, runners, and casuals and attacked a German patrol, which was completely routed.

BRETT, SERENO E. _____
  Near Richecourt, France, Sept. 12,
    1918.
  R—Corvallis, Oreg.
  B—Portland, Oreg.
  G. O. No. 15, W. D., 1919.
  Distinguished-service medal also
    awarded.

Major, 326th Battalion, Tank Corps.
On the opening day of the St. Mihiel offensive he led his battalion on foot from Richecourt to the Boise Quart de Reserve in the face of heavy machine-gun and artillery fire; by his coolness and courage setting an example to the entire battalion.

BREWER, GUY S.
Near St. Mihiel, France, Sept. 12, 1918.
R—Des Moines, Iowa.
B—State Center, Iowa.
G. O. No. 26, W. D., 1919.

Major, 168th Infantry, 42d Division.
He personally led the assaulting wave of his battalion at St. Mihiel, continuing to the enemy's wire, despite the fact that he was wounded by a shell fragment. While directing his men through the wire entanglements, his right arm was shattered by a machine-gun bullet, but he remained on the field for more than an hour directing the disposition of his forces and giving careful directions to the succeeding commander.

BREWER, JOHN B. (40255).
Near Soissons, France, July 18, 1918.
R—Middletown, Ohio.
B—Martin County, Ky.
G. O. No. 35, W. D., 1919.

Sergeant, Company K, 9th Infantry, 2d Division.
Leading his platoon in attack, he encountered heavy machine-gun and shell-fire, but he continued to press on, despite a severe wound which he received early in the fight. While his line was being held up by machine-gun fire he encouraged his men, despite a second wound, which shortly after compelled his removal from the field.

BREWER, LOUIS M. (280431).
Near Romagne, France, Oct. 9, 1918.
R—Detroit, Mich.
B—Woodbury, Ill.
G. O. No. 126, W. D., 1919.

Corporal, Company H, 126th Infantry, 32d Division.
When his platoon sergeant was severely wounded during an attack on enemy machine-gun nests, Corporal Brewer took command of the platoon and led it forward. In so doing he was himself wounded, but he refused to seek medical aid until the objective had been reached and the platoon reorganized.

BREWSTER, HUGH.
In the region of Hageville, France, Sept. 14, 1918.
R—Fort Worth, Tex.
B—Bankston, Ala.
G. O. No. 123, W. D., 1918.

First lieutenant, 49th Aero Squadron, Air Service.
With First Lieut. Hugh L. Fontaine, he attacked nine enemy monoplanes (Fokkers) at an altitude of 4,000 meters. He dived into the midst of the enemy formation without consideration for his personal safety, subjecting himself to great danger. By the suddenness and extreme vehemence of his attack, the machines were driven into confusion. Although greatly outnumbered, he and Lieutenant Fontaine succeeded in shooting down two of the enemy.

BRICE, ARTHUR T., Jr.
Near Bois d'Aigremont, France, July 15, 1918.
R—Washington, D. C.
B—Washington, D. C.
G. O. No. 64, W. D., 1919.

First lieutenant, 7th Infantry, 3d Division.
With a total disregard for his own danger, Lieutenant Brice went into the Bois d'Aigremont under a heavy artillery fire and led scattered troops to the new line of resistance near Fossoy, where they were successfully employed against the enemy's advance. Later, he successfully led his company into Le Charmel in the face of a heavy artillery and machine-gun fire and cleared that town of the enemy.

BRICKLEY, DAVID J.
Near Verdun, France, Oct. 23-24, 1918.
R—Dorchester, Mass.
B—Boston, Mass.
G. O. No. 46, W. D., 1919.

First lieutenant, 101st Infantry, 26th Division.
Stubbornly resisting three strong enemy counterattacks, Lieutenant Brickley, without aid, went forward and by effective machine-gun fire drove the enemy from and captured a strong pill box which had been raising havoc in our ranks.

BRIDENSTINE, LESLIE M. (2273688).
At Audenarde, Belgium, Nov. 1, 1918.
R—Holtville, Calif.
B—El Centro, Calif.
G. O. No. 1, W. D., 1919.

Sergeant, Company F, 316th Engineers, 91st Division.
He volunteered to accompany an officer and three other soldiers on a reconnaissance patrol of the city of Audenarde. Entering under heavy shellfire, the party reconnoitered the city for seven hours while it was still being patrolled by the enemy, advancing 2 kilometers in front of our own outposts and beyond those of the enemy.

BRIDGES, ALVIN O. (1320097).
Near Bellicourt, France, Sept. 29, 1918.
R—Jonesboro, N. C.
B—Lee County, N. C.
G. O. No. 37, W. D., 1919.

Private, first class, Company D, 120th Infantry, 30th Division.
With eight other soldiers, comprising the company headquarters detachment, he assisted his company commander in clearing out enemy dugouts along a canal and capturing 242 prisoners.

BRIDGFORD, JOHN V.
South of Mouzon, France, Nov. 11, 1918.
R—New York, N. Y.
B—New York, N. Y.
G. O. No. 27, W. D., 1920.

Private, 51st Company, 5th Regiment, U. S. Marine Corps, 2d Division.
Private Bridgford delivered an important message from his company to regimental headquarters. In the performance of this mission, exposed to heavy machine-gun and artillery fire, and in full view of enemy machine gunners, he swam the Meuse River. In spite of the great danger, he delivered the message as directed.

BRIGANDO, WILLIAM J. (65548).
Near Verdun, France, Oct. 27, 1918.
R—Meriden, Conn.
B—Meriden, Conn.
G. O. No. 103, W. D., 1919.

Private, first class, Company I, 102d Infantry, 26th Division.
During the attack on Hill 360, north of Verdun, on Oct. 27, Private Brigando, in charge of a squad, made a desperate attempt to silence a machine gun which was holding up the advance of the entire company. The nature of the terrain made this attack more than hazardous, but Private Brigando nevertheless persisted until the entire squad, himself included, had become casualties.

BRIGGS, CHARLES A. (1711811).
In the Foret d'Argonne, north of La Harazee, France, Sept. 29, 1918.
R—Binghamton, N. Y.
B—Buffalo, N. Y.
G. O. No. 128, W. D., 1918.

Sergeant, Company D, 306th Machine Gun Battalion, 77th Division.
Knowing that his commanding officers and three soldiers had been shot down and reported killed, he volunteered, obtained permission, and passed into a zone of heavy and continuous machine-gun fire to where his comrades lay, to render first aid and to rescue them if alive, but unfortunately he found his comrades dead.

BRIGHAM, GEORGE N. (558268).
At St. Thibaut, France, Aug. 10, 1918.
R—Rockville, Conn.
B—Ellington, Conn.
G. O. No. 15, W. D., 1919.

Corporal, Company I, 47th Infantry, 4th Division.
Accompanied by another soldier, he penetrated the enemy's lines and patrolled a sector from the north bank of the River Vesle to the town of Bazoches. These two men entered an enemy dugout and killed two Germans, at the same time locating a machine-gun emplacement. Corporal Brigham, though wounded, completed his mission before obtaining first aid.

*BRIMER, FRANK M. (2262207).
Near Eclisfontaine, France, Sept. 27–30, 1918.
R—Los Angeles, Calif.
B—Wamego, Kans.
G. O. No. 129, W. D., 1918.

First sergeant, Company B, 348th Machine Gun Battalion, 91st Division. This soldier exemplified in the highest degree the spirit of bravery, devotion to duty, and self-sacrifice. Though he had been badly gassed during the action of Sept. 27, he maintained liaison between his company and the battalion post of command, at one time carrying an important message from the front to the artillery. On Sept. 29, though still suffering from the effects of gas, he refused to leave his company, and on a cold, rainy night brought up fresh ammunition over a rough, unfamiliar road in the dark through heavy shell-fire. He took part in the action on Tronsal Farm Hill Sept. 30, assisting in the direction and control of fire of the machine guns of his company. In the evening of Sept. 30 he was killed while taking a wounded soldier to the rear. Posthumously awarded. Medal presented to mother, Mrs. Ida M. Brimer.

BRINDA, JOHN (2181349).
Near Flirey, France, Sept. 12, 1918.
R—Oasis, Nebr.
B—Mount Vernon, Ill.
G. O. No. 128, W. D., 1918.

Sergeant, Company B, 355th Infantry, 89th Division. Without waiting orders he went forward against a concealed enemy machine gun which was holding up his platoon, killed the gunner, and captured four men, thereby enabling his platoon to continue the advance.

BRINK, HERBERT M. (1217163).
Near Montzeville, France, Sept. 14, 1918.
R—New York, N. Y.
B—New York, N. Y.
G. O. No. 32, W. D., 1919.

Mechanic, Battery B, 104th Field Artillery, 27th Division. When a continuous bombardment had set fire to the camouflage covering of a large ammunition dump of 75-millimeter shells and exploded nine of the shells, he, utterly disregarding his personal safety, left a sheltered position and ran to the dump and, with the aid of three other men, extinguished the fire, not only saving the ammunition but also preventing the exact location of the dump by the enemy.

BRINKLEY, AMIEL W.
Near Beaurevoir, France, Oct. 7, 1918.
R—Memphis, Tenn.
B—Memphis, Tenn.
G. O. No. 37, W. D., 1919.

Captain, 117th Infantry, 30th Division. While commanding his company in action Captain Brinkley was wounded by shellfire, which fractured his jawbone in two places. He continued in command of his company, reorganizing it sufficiently to advance, when he was forced to withdraw because of the loss of speech.

BRISON, CHARLES W. (2382260).
Near Cunel, France, Oct. 12, 1918.
R—Reading, Pa.,
B—Leesport, Pa.
G. O. No. 64, W. D., 1919.

Mechanic, Company A, 60th Infantry, 5th Division. While his company was occupying a position on the forward slope of a hill subjected to incessant sniping, machine-gun and artillery fire, Mechanic Brison voluntarily covered a large area searching for the wounded and administering first aid to them. He also carried important messages between the different units of his company.

BRITT, CHARLES (3182478).
Near Remilly, France, Nov. 10, 1918.
R—Cornwall-on-Hudson, N. Y.
B—Pine Plains, N. Y.
G. O. No. 46, W. D., 1919.

Corporal, Company H, 307th Infantry, 77th Division. While accompanying a patrol he swam the Meuse River to repair a footbridge. His exploit was accomplished under most severe fire of enemy machine guns and artillery, but his act enabled the patrol to cross the river and return with information of the utmost value.

BRITTAIN, WILLIAM S.
Near Juvigny, France, Aug. 31, 1918.
R—Flint, Mich.
B—Michigan.
G. O. No. 20, W. D., 1919.

Second lieutenant, 125th Infantry, 32d Division. Lieutenant Brittain by his aggressive spirit and action when out beyond the main Infantry line pushed forward unsupported with his platoon and captured 2 German officers, 94 men, and 8 machine guns. From this point of action Lieutenant Brittain with his platoon proceeded in a direction from which firing could be heard and by his quick decision captured 2 enemy field guns, caliber 105, and 6 trench mortars, and immediately thereafter attacked and captured 30 prisoners and 3 additional heavy machine guns.

*BRITTON, JOE (1453175).
At Varennes, France, Sept. 26, 1918.
R—St. Louis, Mo.
B—St. Louis, Mo.
G. O. No. 13, W. D., 1919.

Sergeant, Company I, 138th Infantry, 35th Division. This soldier was in command of a platoon which became separated from the rest of the company in a heavy fog. Entering Varennes in advance of any other troops Sergeant Britton, with his command, occupied the southern edge of the town in the face of heavy machine-gun fire and took 24 prisoners. When he saw that the odds were hopelessly against him he successfully withdrew, with only one casualty. Later, after being gassed, he manifested exceptional courage and endurance by remaining on duty with his company and advancing in the attack. Posthumously awarded. Medal presented to father, Richard L. Britton.

*BROADFOOT, JOSIAH (9280).
Near La Forge Farm, France, Sept. 27, 1918.
R—Westerly, R. I.
B—Westerly, R. I.
G. O. No. 89, W. D., 1919.

Corporal, Company B, 344th Battalion, Tank Corps. Corporal Broadfoot volunteered to drive a tank, and, his services being accepted, he attacked enemy machine-gun nests until his tank was put out of action, destroying several enemy nests which had held up our Infantry. He then continued the advance on foot until he was fatally wounded. Posthumously awarded. Medal presented to mother, Mrs. Annie H. Broadfoot.

*BROADHEAD, JOSHUA K. (1280).
Near Seicheprey, France, Apr. 20, 1918.
R—Providence, R. I.
B—Johnstown, R. I.
G. O. No. 88, W. D., 1918.

Sergeant, Battery A, 103d Field Artillery, 26th Division. He displayed exceptional bravery and devotion to duty during the action of Apr. 20, 1918, when, although wounded early in the engagement, he refused to leave his section, remaining on duty and keeping his gun in action until killed later in the day. Posthumously awarded. Medal presented to father, Firth Broadhead.

BROBERG, CARL J.
Near St. Etienne, France, Oct. 9, 1918.
R—Aneta, N. Dak.
B—Aneta, N. Dak.
G. O. No. 44, W. D., 1919.

Private, 75th Company, 6th Regiment, U. S. Marine Corps, 2d Division. When many members of his company had been killed and wounded, he went out onto the field under heavy machine-gun fire and administered first aid to several wounded officers and soldiers and carried them to shelter in shell holes.

BROCK, EDWARD J. (2265853)_____
Near Wortegen, Belgium, Nov. 3, 1918.
R—Los Angeles, Calif.
B—Chicago, Ill.
G. O. No. 37, W. D., 1919.

Sergeant, Supply Company, 364th Infantry, 91st Division.
While his wagon train was being heavily shelled he was severely wounded and his horse killed. He refused, however, to be evacuated, and mounting another horse led the train to a location where it was safely parked.

BROCKI, MIECZYSLAW (41911)_____
South of Soissons, France, July 18–22, 1918.
R—Chicago, Ill.
B—Russia.
G. O. No. 14, W. D., 1920.

Corporal, Company B, 16th Infantry, 1st Division.
Displaying exceptional initiative and bravery throughout the operations south of Soissons, France, July 18 to 22, 1918, he with extraordinary heroism on July 21, 1918, with 2 companions, captured 2 machine guns that were causing heavy losses in his company.
Oak-leaf cluster.
For the following act of extraordinary heroism in action at Soissons, France, July 21, 1918, an oak-leaf cluster was awarded Corporal Brocki to be worn with the distinguished-service cross: Regardless of his personal safety, Corporal Brocki successfully led his platoon in action against the enemy, although greatly outnumbered. When in danger of being surrounded he killed with clubbed rifle 6 of the enemy and wounded 3 and fought his way out to a wounded comrade and rescued him.

BRODNICKI, VALERYAN J. (2369044)_____
Near Romanovka, Siberia, June 25, 1919.
R—Chicago, Ill.
B—Chicago, Ill.
G. O. No. 27, W. D., 1920.

Corporal, Company A, 31st Infantry.
When his platoon was completely surrounded by the enemy, and although already twice wounded, he volunteered to carry a message through the enemy's lines to the nearest American troops at Novo-Nezhino, 6 miles away. He succeeded in getting through the enemy's line and reached the nearest American forces and returned with reenforcements for his platoon.

BROGDEN, RONALD_____
Near Thiaucourt, France, Sept. 15, 1918.
R—Goldsboro, N. C.
B—Goldsboro, N. C.
G. O. No. 37., W. D., 1919.

Pharmacist's mate, third class, U. S. Navy, attached to 6th Regiment, U. S. Marine Corps, 2d Division.
He displayed exceptional courage and devotion to duty by going through heavy artillery and machine-gun fire to the aid of a wounded officer belonging to another organization. After giving first-aid treatment to the officer he carried him back to shelter.

BROOKS, ARTHUR R_____
Over Mars-la-Tour, France, Sept. 14, 1918.
R—Framingham, Mass.
B—Framingham, Mass.
G. O. No. 123, W. D., 1918.

Second lieutenant, 22d Aero Squadron, Air Service.
When his patrol was attacked by 12 enemy Fokkers over Mars-la-Tour, 8 miles within the enemy lines, he alone fought bravely and relentlessly with 8 of them, pursuing the fight from 5,000 meters to within a few meters of the ground, and though his right rudder control was out and his plane riddled with bullets, he destroyed 2 Fokkers, 1 falling out of control and the other bursting into flames.

BROOKS, CHARLES W_____
In the Bois-de-Belleau, France, June 8, 1918.
R—Wheaton, Ill.
B—Wyanet, Ill.
G. O. No. 110, W. D., 1918.

Corporal, 83d Company, 6th Regiment, U. S. Marine Corps, 2d Division.
In the Bois-de-Belleau, France, on June 8, 1918, he displayed great courage and disregard for personal safety in repeatedly going through heavy machine-gun fire with messages.

BROOKS, EDWARD H_____
At Montfaucon, France, Oct. 5, 1918.
R—Concord, N. H.
B—Concord, N. H.
G. O. No. 27, W. D., 1920.

Captain, 76th Field Artillery, 3d Division.
Captain Brooks exposed himself to heavy and accurate artillery fire directed on an ammunition train while driving a loaded ammunition truck to safety, the driver of which had been killed by the enemy fire. This truck was attached to a burning truck, and the prompt action of Captain Brooks averted a possible explosion which would have caused serious losses.

BROOKS, EDWARD P_____
At Pont-Maugis, France, Nov. 7, 1918.
R—Westbrook, Me.
B—Westbrook, Me.
G. O. No. 78, W. D., 1919.

First lieutenant, 1st Engineers, 1st Division.
Exposing himself to intense machine-gun and artillery fire, Lieutenant Brooks personally reconnoitered the ground over which an attack was to be made. He then skillfully led his men in a successful attack on the village, which was defended by machine guns and a 77-millimeter gun firing on the advancing troops. Lieutenant Brooks coolly led his platoon in the face of this fire until they were near enough to put the enemy gunners to rout by rifle fire.

BROOKS, ELBERT E_____
At Bouresches, France, June 6, 1918.
R—Memphis, Tenn.
B—Tennessee.
G. O. No. 119, W. D., 1918.

Private, 79th Company, 6th Regiment, U. S. Marine Corps, 2d Division.
He was conspicuous for heroic action in placing his body in front of his platoon leader while under heavy machine-gun fire, in order that he might dress the officer's wounds. He was shot twice in the hip while shielding the body of his leader.

BROOKS, FLOYD A. (40373)_____
Near Champagne, France, Oct. 13, 1918.
R—Newport, Ky.
B—Covington, Ky.
G. O. No. 46, W. D., 1919.

Sergeant, Company K, 9th Infantry, 2d Division.
After his lieutenant had been seriously wounded, Sergeant Brooks took command of the platoon and led it with marked ability for seven days. Later, in the Argonne-Meuse engagement, he personally led his platoon against a machine gun which was holding up the advance of our line and destroyed the gun.

BROOKS, ORA B. (3355974)_____
During attack on Hill 272, north of Exermont, France, Oct. 8 to 12, 1918.
R—Findlay, Ohio.
B—Findlay, Ohio.
G. O. No. 27, W. D., 1920.

Private, first class, Company C, 16th Infantry, 1st Division.
Private Brooks, as a runner, carried messages through artillery and machine-gun fire. He was of great value to his company commander, as he was at all times performing tasks involving great hazard. On one occasion he advanced beyond the lines and located an enemy machine gun, returned to our lines, secured a squad, and gallantly assisted in the capture of the gun and crew.

BROOKSHIRE, ALBERT B. (1305450)......
    Near La Salle River, France, Oct. 16–19, 1918.
    R—Paris, Tenn.
    B—Paris, Tenn.
    G. O. No. 81, W. D., 1919.

Sergeant, Company B, 113th Machine Gun Battalion, 30th Division.
In addition to performing his duties as gas noncommissioned officer, he on five occasions volunteered and successfully delivered important messages, passing through intense enemy fire, from his company to the advanced positions in the front line.

BROOKSHIRE, WILSON D. (1320832)......
    Near Bellicourt, France, Sept. 29, 1918.
    R—Taylorsville, N. C.
    B—Taylorsville, N. C.
    G. O. No. 37, W. D., 1919.

Private, Company G, 120th Infantry, 30th Division.
Private Brookshire, with one other soldier, attacked a machine-gun post which was causing much damage. They captured the post, taking prisoner one officer and eight men and put the machine gun out of action.

*BROOMFIELD, HUGH D. G...........
    Near Cunel, France, Oct. 21, 1918.
    R—Gladstone, Oreg.
    B—Hudson, Ill.
    G. O. No. 37, W. D., 1919.

First lieutenant, pilot, 90th Aero Squadron, Air Service.
Responding to an urgent request for a plane to penetrate the enemy lines to ascertain whether or not the enemy was preparing a counterattack, he immediately volunteered for the mission. Obliged to fly at a very low altitude on account of the unfavorable weather conditions, he was under terrific fire of the enemy at all times, but by skillful dodging he managed to cross the enemy lines.
Posthumously awarded. Medal presented to father, Thomas F. Broomfield.

BROPHY, WILLIAM E. (243403)...........
    Near Argonne Forest, France, Oct. 6, 1918.
    R—McAdoo, Pa.
    B—Philadelphia, Pa.
    G. O. No. 46, W. D., 1919.

Corporal, Company A, 345th Battalion, Tank Corps.
Although wounded in the arm, Corporal Brophy insisted upon returning to his tank and taking part in the counterattack. He remained with the tank, doing very effective work until the attack had been repulsed.

BROSNAHAN, DANIEL T. (1903134).......
    Near St. Juvin, France, Oct. 16, 1918.
    R—Springfield, Mass.
    B—Springfield, Mass.
    G. O. No. 59, W. D., 1919.

Private, Company M, 326th Infantry, 82d Division.
On three different occasions he advanced under heavy machine-gun and artillery fire to a point considerably in advance of our front lines and rescued wounded comrades, bringing them safely back to our lines.

*BROTHERTON, WILLIAM E.............
    Near Fere-en-Tardenois, France, Aug. 1, 1918.
    R—Chicago, Ill.
    B—Guthrie, Ill.
    G. O. No. 7, W. D., 1919.

Second lieutenant, 147th Aero Squadron, Air Service.
An enemy Rumpler plane being reported over the airdrome, he, with another officer, ascended, and soon encountered six Fokker planes that were protecting another Fokker serving as a decoy. Disregarding the enemy's superiority in numbers, he maneuvered so as to secure the advantage of the sun and dived on the decoy plane; pouring in destruction fire, he killed the pilot and crashed the machine to the ground.
Posthumously awarded. Medal presented to father, C. J. Brotherton.

BROWN, ALBERT B. (2304957)...........
    Near St. Giles, south of Fismes, France, Aug. 7 and 14, 1918.
    R—Milwaukee, Wis.
    B—Sandusky, Ohio.
    G. O. No. 20, W. D., 1919.

Sergeant, medical detachment, 121st Field Artillery, 32d Division.
When a bombardment was laid down on the batteries of his regiment Sergeant Brown, on his own initiative, rushed his detachment to the assistance of the wounded, administering first aid and evacuating the wounded in spite of continued enemy shelling. On another occasion, when he was severely gassed and blinded by the explosion of a mustard-gas shell, he made his way to a telephone and summoned medical assistance for the wounded, exposing himself to the increased danger from the effect of gas in order that the other wounded might be cared for.

*BROWN, BAYARD.................
    Near Soissons, France, July 20–22, and near Verdun, France, Oct. 9, 1918.
    R—Genoa, Ill.
    B—Genoa, Ill.
    G. O. No. 44, W. D., 1919.

Second lieutenant, 26th Infantry, 1st Division.
He took command of his battalion at Soissons, after all his senior officers had been killed or wounded, organized for a counterattack, and held his command all day, although he was dangerously wounded. In the fight for Hill 212 in the Argonne he took command of his company, after his company commander had been killed, and led it forward until mortally wounded within 50 yards of the enemy position.
Posthumously awarded. Medal presented to father, D. S. Brown.

BROWN, BERLIN WESLEY (1309673)....
    Near Busigny, France, Oct. 18, 1918.
    R—Tellico Plains, Tenn.
    B—Towns County, Ga.
    G. O. No. 133, W. D., 1918.

Sergeant, Company M, 117th Infantry, 30th Division.
When his platoon had been driven back by a concentrated machine-gun barrage and his platoon commander had been seriously wounded and had fallen on the field, Sergeant Brown and another soldier volunteered and brought the officer back to our lines.

BROWN, BILL (97125)...................
    Near Landres-et-St. Georges, France, Oct. 16, 1918.
    R—Ozark, Ala.
    B—Dale County, Ala.
    G. O. No. 131, W. D., 1918.

Sergeant, Company G, 167th Infantry, 42d Division.
During the attack on the Cote-de-Chatillon, after having been severely wounded and gassed, he refused to go to the hospital, realizing that his presence with his platoon, which had suffered heavy casualties, would greatly assist in the attack. He reorganized his platoon and personally led it in the attack, later consolidating his positions, thereby setting an example of utter disregard for danger and inspiring his men by his remarkable courage and devotion to duty.

BROWN, CLIFFORD E. (3486042)...........
    Near Verdun, France, Oct. 23, 1918.
    R—Akron, Ohio.
    B—Frazee, Minn.
    G. O. No. 37, W. D., 1919.

Private, Company A, 110th Machine Gun Battalion, 29th Division.
He voluntarily went into a sector that was under an intense barrage of enemy artillery, machine-gun, and gas-shell fire and assisted in bringing a wounded comrade to the first-aid station.

BROWN, CLIFFORD R. (1680666) _____
Near Binarville, France, Oct. 2-7, 1918.
R—Panama, N. Y.
B—Fluvannia, N. Y.
G. O. No. 37, W. D., 1919.

Private, Company C, 308th Infantry, 77th Division.
During the time when his company was isolated in the Argonne Forest and cut off from communication with friendly troops Private Brown, together with another soldier, volunteered to carry a message through the German lines, although he was aware that several unsuccessful attempts had been previously made by patrols, the members of which were either killed, wounded, or driven back. By his courage and determination he succeeded in delivering the message and brought relief to his battalion.

BROWN, DEWEY S. (1320297) _____
Near Bellicourt, France, Sept. 29, 1918.
R—Oxford, N. C.
B—Statesville, N. C.
G. O. No. 81, W. D., 1919.

Sergeant, Company E, 120th Infantry, 30th Division.
Wounded twice at the start of an advance, he remained in command of his platoon, carrying it through to a position near its objective, when he was wounded a third time and forced to retire. His personal courage was an inspiration to the men under him.

*BROWN, DILMUS _____
In the Bois-de-Belleau, France, June 11, 1918.
R—Comer, Ga.
B—Georgia.
G. O. No. 71, W. D., 1919.

Private, 55th Company, 5th Regiment, U. S. Marine Corps, 2d Division.
After all the other members of his squad had become casualties Private Brown, single-handed, charged and captured a hostile machine gun.
Posthumously awarded. Medal presented to father, John W. Brown.

BROWN, EDWARD B. (42739) _____
South of Sedan, France, Nov. 7, 1918.
R—Kansas, Ill.
B—Kansas, Ill.
G. O. No. 35, W. D., 1920.

Private, Company E, 18th Infantry, 1st Division.
Private Brown, aided by his corporal, advanced upon a machine-gun position which was causing severe losses to their company. Disregarding personal danger, they silenced the machine gun, thus enabling their company to continue the advance with few losses.

*BROWN, FRANCIS J. (1561206) _____
Near Chateau-Thierry, France, July 14-15, 1918.
R—Englewood, N. J.
B—New York, N. Y.
G. O. No. 44, W. D., 1919.

Sergeant, Battery F, 76th Field Artillery, 3d Division.
He kept in repair the telephone lines which were constantly being broken by the heavy shelling, remaining at this hazardous task until killed by an enemy shell.
Posthumously awarded. Medal presented to father, Thomas Brown.

*BROWN, FRANK (545204) _____
Near Mezy, France, July 15, 1918.
R—New York, N. Y.
B—Russia.
G. O. No. 32, W. D., 1919.

Corporal, Company A, 30th Infantry, 3d Division.
Having been detailed to carry rations to another company of his regiment, and learning upon his arrival there that an attack was imminent, Corporal Brown volunteered and remained with the company commander. He continued on duty long after he had been mortally wounded, and was killed in action later in the day.
Posthumously awarded. Medal presented to aunt, Mrs. M. Mosberg.

*BROWN, FREDERICK H., Jr. (1211433) ___
Near Ronssoy, France, Sept. 29, 1918.
R—Englewood, N. J.
B—Buffalo, N. Y.
G. O. No. 32, W. D., 1919.

Sergeant, Company I, 107th Infantry, 27th Division.
On two occasions he averted heavy casualties in his platoon by going forward and, single-handed, destroying machine-gun nests with hand grenades. At the time of his death he had brought his platoon to the farthest point of advance.
Posthumously awarded. Medal presented to father, Frederick H. Brown.

BROWN, GEORGE (560709) _____
Near Bois-du-Fays, France, Oct. 4-5, 1918.
R—Moundsville, W. Va.
B—Gleneaston, W. Va.
G. O. No. 71, W. D., 1919.

Private, Headquarters Company, 59th Infantry, 4th Division.
As a battalion runner, he repeatedly exposed himself to intense artillery and machine-gun fire, crossing open spaces in view of the enemy to deliver important messages. He aided largely in maintaining liaison and his courage was an inspiration to those near him.

BROWN, GEORGE L. (1781205) _____
Near Montfaucon, France, Sept. 27, 1918.
R—Baltimore, Md.
B—Whitehaven, Pa.
G. O. No. 37, W. D., 1919.

Corporal, Company K, 313th Infantry, 79th Division.
He crawled ahead of his platoon and located and killed a sniper who had wounded him and several others. Although seriously wounded, he remained in command of his platoon until he was ordered to the rear, when he insisted on going back without assistance, though he was so weak he could hardly walk.

BROWN, GEORGE V. (2311980) _____
Near Fleville, France, Oct. 4, 1918.
R—Marysville, Calif.
B—Denver, Colo.
G. O. No. 81, W. D., 1919.

Sergeant, Company I, 16th Infantry, 1st Division.
During an attack Sergeant Brown advanced with five other soldiers into the enemy's line, and although surrounded by six German machine guns, held his position until support reached him. Although badly gassed, he refused to be evacuated.

*BROWN, HAROLD (1634272) _____
Near Charlevaux, France, Oct. 4, 1918.
R—Bakersfield, Calif.
B—Bakersfield, Calif.
G. O. No. 74, W. D., 1919.

Private, Company D, 308th Infantry, 77th Division.
When the first two battalions of his regiment had been surrounded by the enemy, Private Brown volunteered to accompany a patrol for the purpose of establishing liaison with the forward troops, knowing from the fate of previous patrols that the mission would probably prove fatal. He was killed as the patrol was attempting unsuccessfully to reach the forward battalions.
Posthumously awarded. Medal presented to widow, Mrs. Grace G. Brown.

*BROWN, HARRY A. (1210115) _____
Near Ronssoy, France, Sept. 29, 1918.
R—Watertown, N. Y.
B—Brattleboro, Vt.
G. O. No. 9, W. D., 1923.

Private, first class, Company C, 107th Infantry, 27th Division.
He voluntarily assisted in putting out of action an enemy pyrotechnic expert whose activities were hampering and disclosing the position of the company about to assault the enemy lines. This act was performed with utter disregard of personal danger in the face of intense machine-gun and rifle fire and necessitated crawling over rough and broken terrain under observation of enemy machine gunners and snipers. His courageous action doubtless saved his company heavy losses and greatly strengthened the morale of his organization.
Posthumously awarded. Medal presented to brother, Earl H. Brown.

BROWN, HERBERT A. (40400) _____
Near Soissons, France, July 18, 1918.
R—Lauraville, Md.
B—Brooklyn, N. Y.
G. O. No. 35, W. D., 1919.

Corporal, Company K, 9th Infantry, 2d Division.
When his company was being swept by a withering machine-gun fire from a hidden nest, he voluntarily made his way around the flank of the emplacement and attacked the crew. After causing the crew to flee in disorder, he manned the gun and poured a heavy fire into their retreating ranks.

BROWN, JAMES _____
Near Ville - devant - Chaumont, France, Nov. 10, 1918.
R—Quechee, Vt.
B—Belmont, Mass.
G. O. No. 9, W. D., 1923.

Second lieutenant, 104th Infantry, 26th Division.
In command of the 2d Battalion of his regiment and directed to take the town of Ville-devant-Chaumont he led his battalion in person, and by a process of slow infiltration successfully carried out his mission. His entire disregard for his own safety, his coolness under heavy enemy fire, and his gallant leadership proving an inspiration to his men, enabling them in the face of concentrated machine-gun fire to take the town, making possible the advance of the troops of his division on both sides.

BROWN, JAMES E. (69546) _____
Near Torcy and Belleau, France, July 18–20, 1918.
R—Revere, Mass.
B—Holyoke, Mass.
G. O. No. 37, W. D., 1919.

Corporal, Company M, 103d Infantry, 26th Division.
He displayed exceptional bravery as a member of a patrol of six men which entered Belleau from the rear and captured four prisoners. He also worked untiringly in the evacuation of wounded at Torcy in the face of constant and intense shell fire. Later, when his platoon commander and most of the commissioned officers had been incapacitated, he took command of his platoon and skillfully led it in the advance, keeping his command together under heavy shell and machine-gun fire until he was struck four times.

*BROWN, JAMES FINLAY _____
At Villers-devant-Mouzon, France, Nov. 7, 1918.
R—Brooklyn, N. Y.
B—Brooklyn, N. Y.
G. O. No. 21, W. D., 1919.

First lieutenant, 302d Engineers, 77th Division.
He displayed remarkable bravery in reconnoitering sites for footbridges across the Meuse River and later directing the construction of these bridges under heavy machine-gun fire. While so engaged, he was killed by a sniper.
Posthumously awarded. Medal presented to widow, Mrs. Olivine K. Brown

BROWN, JAMES R. (1261273) _____
Near Fismes, France, Aug. 10–13, 1918.
R—Philadelphia, Pa.
B—Philadelphia, Pa.
G. O. No. 15, W. D., 1919.

Private, 110th Ambulance Company, 103d Sanitary Train, 28th Division.
Because of the destruction from shell fire of 10 of the 13 ambulances of his company, he worked for 72 hours, 48 of them without rest, driving through a shell-swept and gas-infested area, and thereby making possible the evacuation of the wounded.

*BROWN, JOHN (793900) _____
Near Remonville, France, Nov. 5, 1918.
R—New York, N. Y.
B—England.
G. O. No. 44, W. D., 1919.

Private, medical detachment, 11th Infantry, 5th Division.
While making his way to the side of a wounded comrade he was seriously wounded, but he refused to permit the litter bearers to take him to a dressing station until those wounded about him were first evacuated.
Posthumously awarded. Medal presented to uncle, James Brown.

BROWN, JOSEPH FRANCE (1678312) _____
Near the Forest of Argonne, France, Sept. 27, 1918.
R—Hogansburg, N. Y.
B—Hogansburg, N. Y.
G. O. No. 78, W. D., 1919.

Private, Company K, 306th Infantry, 77th Division.
During an attack on the trenches held by his company, Private Brown found an automatic rifle which had been abandoned by a wounded soldier. Though he was unfamiliar with the operation of the weapon, Private Brown opened fire on the enemy with it, killing two of them, and thereby making possible the escape of three of his comrades who had been captured by the enemy. The remainder of the hostile force was driven off.

BROWN, JOSEPH J _____
In the Bois-de-Belleau, France, June 20, 1918.
R—Philadelphia, Pa.
B—Beverly, N. J.
G. O. No. 99, W. D., 1918.

First lieutenant, 7th Infantry, 3d Division.
He went out into no man's land in the face of violent German machine-gun fire to carry back into his own trenches a severely wounded soldier.

BROWN, JOSHUA D. (1284010) _____
Near Verdun, France, Oct. 14–15, 1918.
R—Hagerstown, Md.
B—Emmitsburg, Md.
G. O. No. 2, W. D., 1919.

Sergeant, Company B, 115th Infantry, 29th Division.
On Oct. 14, while commanding a platoon in the Bois-de-Consenvoye, north of Verdun, he was wounded. He refused to be sent to the hospital and continued in command of his platoon, doing excellent work until Oct. 16, when he was again severely wounded and carried from the field.

BROWN, LELAND (2386239) _____
Near Cote St. Germain, France, Nov. 6, 1918.
R—Crossville, Ill.
B—White County, Ill.
G. O. No. 35, W. D., 1919.

Private, first class, Company B, 61st Infantry, 5th Division.
He attacked a machine-gun nest single-handed, and in the face of heavy fire reduced the nest, capturing one prisoner. Later in the same day he patrolled alone under heavy fire in advance of his company and attacked another machine-gun position, capturing the gun and four prisoners.

BROWN, LESTER (737186) _____
Near Joully, France, Nov. 10, 1918.
R—Murfreesboro, Tenn.
B—Donalds Chapel, Tenn.
G. O. No. 50, W. D., 1919.

Private, Company M, 11th Infantry, 5th Division.
After a small patrol had failed to silence a machine gun, Private Brown rushed forward with an automatic rifle through a heavy machine-gun fire and was wounded.

BROWN, MITCHELL H_____
Near Beffu - et - le - Morthomme, France, Oct. 23, 1918.
R—Rockwall, Tex.
B—Rockwall, Tex.
G. O. No. 20, W. D., 1919.

Second lieutenant, observer, 50th Aero Squadron, Air Service.
Lieutenant Brown, observer, piloted by Lieutenant Phillips, while on a reconnaissance for the 78th Division, attacked an enemy balloon and forced it to descend. They were in turn attacked by three enemy planes (Fokker type). The incendiary bullets from the enemy's machine set the signal rockets in Lieutenant Brown's cockpit afire. Disregarding the flames, he continued to fire, destroying one enemy plane and forcing the others to retire. He then used the extinguisher handed him by his pilot and put out the flames. They successfully completed their mission and secured valuable information.

BROWN, PAUL FRANCIS_____
Near Eclisfontaine, France, Sept. 26-27, 1918.
R—Minneapolis, Minn.
B—Lake City, Minn.
G. O. No. 35, W. D., 1920.

Captain, Medical Corps, attached to 361st Infantry, 91st Division.
Captain Brown voluntarily advanced in front of our lines for the purpose of rescuing the wounded left in advance of the new lines by the retirement of a unit of the regiment. Due to his efforts 14 wounded Americans were brought safely back to our lines.

BROWN, ROY A. (2168841)_____
Near Lesseau, France, Sept. 4, 1918.
R—Decatur, Ala.
B—Decatur, Ala.
G. O. No. 139, W. D., 1918.

Sergeant, Company E, 366th Infantry, 92d Division.
He was a member of a combat group which was attacked by 20 of an enemy raiding party, advancing under a heavy barrage and using liquid fire. The sergeant in charge of the group was killed and several others, including Sergeant Brown, were wounded. Nevertheless, this soldier, with three others, fearlessly resisted the enemy until they were driven off.

BROWN, RUSSELL A. (1209742)_____
Near St. Souplet, France, Oct. 18, 1918.
R—Morristown, N. J.
B—Morristown, N. J.
G. O. No. 87, W. D., 1919.

Private, first class, Company A, 107th Infantry, 27th Division.
When the advance of his battalion was checked by heavy machine-gun fire, Private Brown, with two other soldiers, went forward under heavy fire to reconnoiter the enemy positions. By effective rifle fire they drove the gunners from two machine-gun nests into a dugout near by, which they captured, together with 35 prisoners, including 3 officers.

BROWN, SAMUEL A., JR._____
Near Ronssoy, France, Sept. 29, 1918.
R—Jamestown, N. Y.
B—Jamestown, N. Y.
G. O. No. 21, W. D., 1919.

Second lieutenant, 108th Infantry, 27th Division.
Advancing with his platoon through heavy fog and dense smoke, and in the face of terrific fire, which inflicted heavy casualties on his forces, Lieutenant Brown reached the wire in front of the main Hindenburg line, and after reconnoitering for gaps, assaulted the position and effected a foothold. Having been reinforced by another platoon, he organized a small force, and by bombing and trench fighting captured over a hundred prisoners. Repeated attacks throughout the day were repelled by his small force. He also succeeded in taking four field pieces, a large number of machine guns, antitank rifles, and other military property, at the same time keeping in subjection the prisoners he had taken.

BROWN, SAMUEL R. (1864673)_____
Near Moranville and Grimaucourt, France, Nov. 9, 1918.
R—Macclesfield, N. C.
B—Macclesfield, N. C.
G. O. No. 32, W. D., 1919.

Sergeant, Company F, 322d Infantry, 81st Division.
After having been wounded in the afternoon of Nov. 9 he had his wound dressed and returned to his platoon through very heavy enemy artillery and machine-gun fire. When his platoon was relieved, he returned to the former position through enemy artillery fire to the rescue of a wounded man and assisted him to the rear.

BROWN, VINCIL E. (2382491)_____
Near St. Mihiel, France, Sept. 16, 1918.
R—Norman, Okla.
B—Jamesport, Mo.
G. O. No. 37, W. D., 1919.

First sergeant, Company B, 60th Infantry, 5th Division.
Seeing a comrade lying wounded and exposed to great danger of machine-gun and shell fire, Sergeant Brown went to his aid and, after rescuing him, carried him through the sweeping barrage to the first-aid station, a distance of one-half a kilometer.

*BROWN, WALTER B. (238783)_____
Near Gesnes, France, Oct. 9, 1918.
R—Brainard, Minn.
B—Holdingford, N. Dak.
G. O. No. 64, W. D., 1919.

Private, Company K, 125th Infantry, 32d Division.
Exposing himself to the greatest danger, he constantly carried messages from the company to the platoons occupying the front lines. The journey necessitated his crossing an area swept by intense artillery and withering machine-gun fire, but he successfully maintained liaison during a very critical period of the attack.
Posthumously awarded. Medal presented to father, Andrew Brown.

BROWN, WILLIAM J. (199111)_____
At Riaville, France, Sept. 26, 1918.
R—Roxbury, Mass.
B—Boston, Mass.
G. O. No. 15, W. D., 1919.

Corporal, 101st Field Signal Battalion, 26th Division.
At a critical time when the need for a barrage was imperative and telephone communication impossible, he voluntarily carried a message to the artillery across an open field which was subjected to intense artillery, machine-gun, and rifle fire.

*BROWNE, JENNINGS B. (2398042)_____
Near Fossoy, France, July 14-15, 1918.
R—Columbus, Ga.
B—Newborn, Ga.
G. O. No. 10, W. D., 1920.

Bugler, machine gun company, 7th Infantry, 3d Division.
Bugler Brown exposed himself to heavy fire while carrying numerous messages during the heavy artillery preparation preceding the last German offensive. He was killed while acting as guide to a front-line platoon.
Posthumously awarded. Medal presented to mother, Mrs. S. J. Brown.

BROWNVILLE, CHARLES G. (2903)_____
East of Belleau, France, July 20 and 22, 1918.
R—Needham, Mass.
B—Wollaston, Mass.
G. O. No. 125, W. D., 1918.

Private, medical detachment, 103d Infantry, 26th Division.
He was conspicuous for his unfailing assistance to wounded under heavy fire of machine guns, and his absolute fearlessness.

+BROXUP, JOHN.
Near St. Étienne, France, Oct. 4, 1918.
R—Buffalo, N. Y.
B—Rochester, N. Y.
G. O. No. 35, W. D., 1919.

Private, 49th Company, 5th Regiment, U. S. Marine Corps, 2d Division.
Private Broxup succeeded in bringing a wounded officer back to our lines when his company was forced back to a new position by superior numbers.
Posthumously awarded. Medal presented to widow, Mrs. Grace Broxup.

BRUCE, ANDREW D.
Near, Vierzy, France, July 18-19, and near Blanc Mont, France, Oct. 3-4, 1918.
R—San Antonio, Tex.
B—St. Louis, Mo.
G. O. No. 44, W. D., 1919.

Major, 4th Machine Gun Battalion, 2d Division.
On the night of July 18-19 he made a personal reconnaissance ahead of his troops, through heavy flanking machine-gun fire. He pushed forward to the outpost lines through heavy artillery and machine-gun fire to keep in touch with all his platoon. On Oct. 3-4 he made a personal reconnaissance on the left flank of his division through heavy shell fire and continual sniping, and gained information which enabled him to well place his battalion and cover an exposed flank.

BRUCE, JOHN S.
Near Trugny, France, July 23, 1918.
R—Franklin, N. H.
B—Canada.
G. O. No. 78, W. D., 1919.

Second lieutenant, 102d Infantry, 26th Division.
Despite the fact that he had been wounded Lieutenant Bruce continued in command of his machine-gun platoon, firing the gun himself when the crew was depleted, until he received a second wound.

BRUCE, WILL (42150).
South of Soissons, France, July 18, 1918.
R—Spartanburg, S. C.
B—Pickens, S. C.
G. O. No. 24, W. D., 1920.

Private, Company C, 16th Infantry, 1st Division.
Private Bruce with three others advanced in front of our lines and silenced an enemy machine-gun post, killing the crew of four. The post was causing heavy casualties on the left flank of the company. On the following day with a comrade he advanced beyond the lines and silenced two machine guns, forcing the crews to surrender. In the latter operation he was wounded.

BRUMMETT, JAMES R.
Near Thiaucourt, France, Sept. 12-16, 1918.
R—Dime Box, Tex.
B—Wilmer, Tex.
G. O. No. 46, W. D., 1919.

Private, 81st Company, 6th Machine Gun Battalion, U. S. Marine Corps, 2d Division.
By effective use of an automatic rifle Private Brummett defended the left flank of his platoon, thereby preventing the enemy from reaching the rear of his lines. On several other occasions he volunteered and carried messages through terrific bombardment.

BRUNDRETT, CHARLES E. (64374).
Near Seicheprey, France, Apr. 20, 1918.
R—New Haven, Conn.
B—West Springfield, Mass.
G. O. No. 87, W. D., 1919.

Private, Company D, 102d Infantry, 26th Division.
Although surrounded on all sides by the enemy, and with his ammunition entirely exhausted, Private Brundrett offered a most stubborn resistance to the enemy's attack, fighting his way through their ranks with his rifle and bayonet to the support platoon, where he again took up the fight.

BRUNNER, HOWARD V. (112783).
Near Sergy, France, July 29-30, 1918.
R—Bethlehem, Pa.
B—Bethlehem, Pa.
G. O. No. 71, W. D., 1919.

Private, first class, Company B, 149th Machine Gun Battalion, 42d Division.
As platoon runner he showed marked heroism in volunteering for dangerous missions, repeatedly carrying important messages through zones swept by intense artillery and machine-gun fire. He remained on duty after being wounded and until his command was relieved.

BRYAN, CLAUDE (483006).
Near Preny Ridge, France, Nov. 10, 1918.
R—Chicago, Ill.
B—Chicago, Ill.
G. O. No. 44, W. D., 1919.

Corporal, Company I, 56th Infantry, 7th Division.
After being wounded in the foot by a machine-gun bullet he refused to go to the rear when ordered to do so, but remained with the company until the whole line was relieved.

+BRYANT, HOMER E. (1311604).
Near St.-Martin-Riviere, France, Oct. 11, 1918.
R—Salem, S. C.
B—Piedmont, S. C.
G. O. No. 74, W. D., 1919.

Private, Company H, 118th Infantry, 30th Division.
Hearing a call for help from a man lying beyond the front line, Private Bryant, a stretcher bearer, unhesitatingly went to his assistance, although the spot was under heavy fire from enemy machine guns and snipers. As he was approaching the wounded man he was instantly killed by an enemy sniper.
Posthumously awarded. Medal presented to father, William H. Bryant.

BRYANT, WILLIAM E. (2220212).
During the St. Mihiel offensive, France, Sept. 14, 1918.
R—Boswell, Okla.
B—Bowling Green, Ky.
G. O. No. 72, W. D., 1920.

Sergeant, Company D, 358th Infantry, 90th Division.
Sergeant Bryant, with four other men, volunteered to cross a valley to the woods opposite and silence machine guns which had held up the advance of his company. In the face of heavy enemy fire this small group accomplished its mission, thus enabling the company to cross the valley without further loss.

BRYSON, JULIUS J. (1310635).
Near Bellicourt, France, Sept. 27, 1918.
R—Webster, N. C.
B—Webster, N. C.
G. O. No. 64, W. D., 1919.

First sergeant, Company D, 118th Infantry, 30th Division.
Although wounded very severely in the knee by shrapnel, Sergeant Bryson remained in charge of his platoon for more than 24 hours, during a critical period of the operations. Due to his excellent example of courage, leadership, and skill in handling them, his platoon successfully repelled a number of enemy attacks during this period of time.

BRYSON, SAMUEL R.
In the Champagne sector, France, Sept. 29, 1918.
R—Mauch Chunk, Pa.
B—Mauch Chunk, Pa.
G. O. No. 46, W. D., 1919.

First lieutenant, 371st Infantry, 93d Division.
After being seriously wounded Lieutenant Bryson remained in command of his platoon, never hesitating in his attempts to gain his objective in the face of the greatest hazards.

BRYSON, WILLIAM (2870626).
Near Verdun, France, Nov. 7, 1918.
R—Decatur, Ark.
B—Cassville, Mo.
G. O. No. 37, W. D., 1919.

Private, Company I, 315th Infantry, 79th Division.
He volunteered to reconnoiter an enemy trench to determine its exact location and to ascertain whether or not it was protected by wire. His mission was accomplished under heavy fire, and the information he obtained proved of the greatest assistance and value.

*BUB, ELROY (544477)_____
 Near Mezy, France, July 15, 1918.
 R—Milwaukee, Wis.
 B—Milwaukee, Wis.
 G. O. No. 32, W. D., 1919.

Private, Headquarters Company, 30th Infantry, 3d Division.
He successfully carried messages through terrific artillery and machine-gun fire and was wounded while performing the mission.
Posthumously awarded. Medal presented to mother, Mrs. Bertha Bub.

BUCHANAN, ALFRED B. (2229446) _____
 Near Le Grand Carre Farm, France, Nov. 1, 1918.
 R—Bryan, Tex.
 B—Bryan, Tex.
 G. O. No. 46, W. D., 1919.

Sergeant, Company G, 360th Infantry, 90th Division.
After being severely wounded early in the action Sergeant Buchanan had his wound dressed and started to rejoin his platoon, but passed through a gap in our lines and reached the German lines instead. He returned from there, located his platoon, and led it with marked courage and coolness until he was severely wounded the second time.

*BUCHANAN, MAX C._____
 In the assault at Cantigny, France, May 28-31, 1918.
 R—Boston, Mass.
 B—Cabot, Vt.
 G. O. No. 99, W. D., 1918.

Second lieutenant, 28th Infantry, 2d Division.
He brilliantly led his platoon in the assault at Cantigny, France, reached his objective, consolidated his position successfully under heavy fire, continually walked up and down his line to instruct and encourage his men, until he was killed by an enemy shell.
Posthumously awarded. Medal presented to father, W. H. Buchanan.

BUCHANAN, ROBERT (419002)_____
 Near Metzeral, Alsace, Sept. 16-17, 1918.
 R—Maitland, W. Va.
 B—Tazewell, Va.
 G. O. No. 32, W. D., 1919.

Private, first class, Company B, 54th Infantry, 6th Division.
On the night of Sept. 16-17 he led a patrol into an entirely unfamiliar sector of the enemy positions and, without artillery support, captured four prisoners and secured much valuable information; despite the heavy hostile grenade, rifle, automatic rifle, and machine-gun fire he cut through the enemy's electrified wire and overcame a sentinel without losing a man. Three of the prisoners were captured by him when he alone stormed a barricaded dugout and disarmed three Germans.

BUCK, BEAUMONT B._____
 During the attack on Berzy-le-Sec, France, July 21, 1918.
 R—Hillsboro, Tex.
 B—Mayhew, Miss.
 G. O. No. 20, W. D., 1919.

Brigadier general, 2d Infantry Brigade, 1st Division.
Before and during the attack on Berzy-le-Sec, France, July 21, 1918, he displayed conspicuous gallantry and heroic leadership of his command. When most of the officers of his brigade had fallen, General Buck, with contempt of personal danger and in spite of heavy artillery bombardment and machine-gun fire, traversed the front of his advancing forces, gave correct directions to his organization commanders, and led the first wave of the culminating attack which stormed and captured the town.

BUCK, BENJAMIN (300943)_____
 Near Romagne, France, Oct. 14, 1918.
 R—Fond du Lac, Wis.
 B—Shawano, Wis.
 G. O. No. 98, W. D., 1919.

Corporal, Company G, 128th Infantry, 32d Division.
Corporal Buck advanced alone beyond the front lines, through a terrific barrage and in the face of unusually active machine-gun fire, to the rescue of a wounded comrade. While making his way through the wood he came upon two other members of the company whom he utilized as a patrol attacking and capturing an enemy machine-gun nest, together with six prisoners. He then went to the rescue of the wounded soldier, forcing the prisoners to carry the man to the rear.

BUCK, OSCAR L._____
 In the attack on Landres-et-St. Georges line, France, Oct. 15, 1918.
 R—Detroit, Mich.
 B—Big Rapids, Mich.
 G. O. No. 19, W. D., 1920.

Captain, 165th Infantry, 42d Division.
Captain Buck led his company ably and efficiently in the attack. Although severely wounded, he refused to go to the rear, but continued to direct his men under terrific artillery and machine-gun fire until he was exhausted.

BUCK, ROBERT (1314609)_____
 Near Bellicourt, France, Sept. 29, 1918.
 R—Goldsboro, N. C.
 B—Goldsboro, N. C.
 G. N. No. 35, W. D., 1919.

Private, Company A, 119th Infantry, 30th Division.
Although seriously wounded in the arm by machine-gun fire early in the engagement, Private Buck for three hours continued on duty as an automatic rifle carrier and did not go to the rear until his company had been reorganized.

BUCKENDAHL, EMIL (1419977)_____
 Near Gesnes, France, Oct. 5, 1918.
 R—Pierce, Nebr.
 B—Pierce, Nebr.
 G. O. No. 66, W. D., 1919.

Private, Company F, 127th Infantry, 32d Division.
Private Buckendahl, a litter bearer, on his own inititative, went out from a position of shelter to an exposed flank under intense machine-gun fire, and carried back to safety a wounded soldier who had been left in the field.

BUCKLEY, HAROLD R._____
 Near Perles, France, Aug. 10, 1918.
 R—Agawam, Mass.
 B—Westfield, Mass.
 G. O. No. 138, W. D., 1918.

First lieutenant, pilot, 95th Aero Squadron, Air Service.
He was on a patrol protecting a French biplace observation machine, when they were suddenly set upon by six enemy planes. He attacked and destroyed the nearest, and the remainder fled into their own territory. He then carried on with his mission until he had escorted the allied plane safely to its own aerodrome.
Oak-leaf cluster.
A bronze oak leaf for extraordinary heroism in action near Neuville, France, and Boureuilles, France, Sept. 16-27, 1918. He dived through a violent and heavy antiaircraft and machine-gun fire and set on fire an enemy balloon that was being lowered to its nest. On the next day, while leading a patrol, he met and sent down in flames an enemy plane while it was engaged in reglage work.

BUDD, ARTHUR D._____
 Near Grand Pre, France, Oct. 28, 1918.
 R—Meriden, Conn.
 B—Meriden, Conn.
 G. O. No. 35, W. D., 1919.

Lieutenant colonel, 311th Infantry, 78th Division.
After our troops had established a new line and before the position had been consolidated, the enemy put down an extraordinarily heavy barrage on the position. Colonel Budd went through this barrage from one end of the line to the other to prepare for the expected counterattack before returning to his post of command.

BUDD, JOHN O. (936877)_____
Near Fossoy, France, July 15, 1918.
R—Minden, Nebr.
B—Tanerdale, N. Y.
G. O. No. 44, W. D., 1919.

Private, medical detachment, 7th Infantry, 3d Division.
Working throughout the heavy enemy artillery fire of July 15, which preceded the German offensive, he aided the wounded and evacuated 12 comrades from an exposed position.

BUDD, KENNETH P_____
Near Ville-Savoye, France, Aug. 16, 1918.
R—New York, N. Y.
B—New York, N. Y.
G. O. No. 32, W. D., 1919.

Major, 308th Infantry, 77th Division.
Although his post of command was subjected to continuous and concentrated gas attacks, and despite the fact that he was severely gassed during the bombardment, he refused to be evacuated, remaining for three days to personally superintend the relief of his battalion and the removal to the rear of men who had been gassed.

*BUDDE, GEORGE WILLIAM_____
Near Villemontry, France, Nov. 11, 1918.
R—Cincinnati, Ohio.
B—Cincinnati, Ohio.
G. O. No. 32, W. D., 1919.

Private, 17th Company, 5th Regiment, U. S. Marine Corps, 2d Division.
Upon his own initiative, Private Budde advanced in front of the line to determine whether a certain machine-gun position was hostile or friendly and was killed by a machine-gun bullet.
Posthumously awarded. Medal presented to mother, Mrs. Elizabeth Budde.

BUELL, RALPH POLK_____
In an attack on Ronssoy, France, Sept. 29, 1918.
R—Bayside, N. Y.
B—Washington, D. C.
G. O. No. 16, W. D., 1920.

First lieutenant, 107th Infantry, 27th Division.
Lieutenant Buell led his company in attack exposed to heavy artillery fire. When confronted by a strong enemy machine-gun position, firing pointblank on his advancing unit, he led the dash which resulted in the capture of the trench. He fell wounded 30 yards in advance of his men.

BUFFALO, JOSEPH A. (2220710)_____
Near Fey-en-Haye, France, Sept. 12, 1918.
R—Bixby, Okla.
B—Marion County, Ark.
G. O. No. 128, W. D., 1918.

Private, Company F, 358th Infantry, 90th Division.
Although he was seriously wounded early in action, he remained in the fight throughout the day, leading small parties of men against machine-gun emplacements, killing two of the enemy himself, and refusing to be evacuated till late at night, unable to fight further.

BUFORD, DAVID L_____

In the Bois-de-Belleau, France, June 13, 1918.
R—Frankston, Tex.
B—Taylor, Ark.
G. O. No. 70, W. D., 1919.

Gunnery sergeant, 55th Company, 5th Regiment, U. S. Marine Corps, 2d Division.
After being wounded Sergeant Buford, with exceptional courage, continued to lead his section forward against a machine-gun nest, and captured it.

BUFORD, EDWARD, Jr_____
In the region of Commercy-St. Mihiel, France, May 22, 1918.
R—Nashville, Tenn.
B—Nashville, Tenn.
G. O. No. 129, W. D., 1918.

First lieutenant, 95th Aero Squadron, Air Service.
While on barrage patrol against German photographic machines in the region of Commercy-St. Mihiel, France, he engaged in combat, alone, five German biplane machines, attacking one or more of them in three separate combats in 25 minutes. One of the machines he shot down and the others he drove off, thus fulfilling his mission against heavy odds.

BULKLEY, STANLEY_____
Near Ronssoy, France, Sept. 29, 1918.
R—New York, N. Y.
B—Minneapolis, Minn.
G. O. No. 13, W. D., 1923.

Captain, 105th Infantry, 27th Division.
Commanding the 3d Battalion of his regiment, and having led them forward to a position covering the left flank of the division, he discovered a small unit of another battalion located in a shell hole some distance in advance of his position. Finding this unit entirely surrounded by the enemy and in imminent danger of total annihilation Captain Bulkley, though suffering from a severe head wound, rushed forward to their assistance, firing a Lewis gun, breaking the enemy's attack, and killing and wounding several of their machine gunners. This action was in full view of the enemy and performed with utter disregard of heavy enemy machine-gun and rifle fire, and served as an inspiration to the members of the American fighting forces engaged in that operation.

BULLION, GEORGE (77242)_____
Near Cierges, France, Oct. 2, 1918.
R—Centralia, Wash.
B—Denison, Tex.
G. O. No. 59, W. D., 1919.

Private, first class, Company C, 125th Infantry, 32d Division.
While our troops were endeavoring to establish a line 600 meters in front of the town of Cierges, heavy artillery and enfilading machine-gun fire from enemy guns threatened to hinder the operation. When it became necessary to establish liaison with adjoining units he volunteered and undertook the mission, crossing and recrossing the area under heavy fire. He continued with his work, although weak from exhaustion and lack of food, until the line was established and the crisis passed.

*BULLOCK, BENJAMIN, 3d_____
Near Nantillois, France, Sept. 28–29, 1918.
R—Ardmore, Pa.
B—Bayhead, N. J.
G. O. No. 35, W. D., 1919.

First lieutenant, 3d Battalion, 315th Infantry, 79th Division
On the afternoon of Sept. 28 Lieutenant Bullock displayed great bravery and fearlessness by assisting two wounded men to a place of safety while under heavy sniper and artillery fire. On the morning of the 29th of September he again demonstrated great bravery by advancing alone into a wood and killing a sniper. On the afternoon of the 29th of September, while carrying a message to the regimental post of command, he was killed by a high-explosive shell.
Posthumously awarded. Medal presented to father, Benjamin Bullock, jr.

*BUMA, RAYMOND (556232)_____
Near Cuisy, France, Sept. 26, 1918.
R—Whitinsville, Mass.
B—Holland.
G. O. No. 44, W. D., 1919.

Corporal, Machine Gun Company, 39th Infantry, 4th Division.
After all his squad members had become casualties, Corporal Buma alone continued to operate his gun, and after his ammunition was exhausted he ran from shell hole to shell hole picking up ammunition and carrying it back to his gun, resuming fire on the enemy, which was very instrumental in the success of the attack. He was killed in action shortly afterwards.
Posthumously awarded. Medal presented to father, Minne Buma.

BUMP, ARTHUR L.
Near Preny, France, Nov. 2, 1918.
R—New London, Ohio.
B—Eau Claire, Wis.
G. O. No. 37, W. D., 1919.

Colonel, 56th Infantry, 7th Division.
While his regiment was being subjected to an intense enemy bombardment he visited every platoon in the front line and so encouraged and inspired his men by his bravery that they successfully met and repulsed every counter-attack made upon them.

BUNCH, HENRY E.
Near the Bois-de-Chatillon, France, Oct. 13–15, 1918.
R—Camilla, Ga.
B—Clarks Hill, S. C.
G. O. No. 13, W. D., 1919.

Captain, Medical Corps, attached to 168th Infantry, 42d Division.
During the advance of his regiment in the Verdun sector he established aid stations at points as far advanced as possible and supervised them throughout the combat, working continuously, tirelessly, and fearlessly without food or rest. On Oct. 14 this officer went out in advance of the front line to reconnoiter a site for an aid station and an ambulance route. Seeing a wounded officer lying about 300 meters from the enemy's line, he went to his rescue and carried him through terrific machine-gun and rifle fire to a shell hole, where he administered aid, in entire disregard of his own safety.

BUNGE, ROBERT C.
Near Montfaucon, France, Sept. 26, 1918.
R—Cincinnati, Ohio.
B—New York, N. Y.
G. O. No. 43, W. D., 1922.

Captain, 148th Infantry, 37th Division.
While in command of a combat liaison group operating between the 37th and 91st Divisions, and under heavy hostile artillery fire, Captain Bunge, although painfully wounded by a shell fragment and burned with gas, courageously remained in command of his company, maintained contact with the enemy, and directed the company movements. When the attack was continued on Sept. 27 and his company was acting in the same capacity, while passing through a terrible hostile artillery barrage he received a serious fracture of the skull from enemy shell fragments, and refusing to be evacuated he tenaciously continued with his group. Later, on the same day, while leading his company, he was again seriously wounded by shell fire, which necessitated his evacuation.

BUNYARD, CLARENCE H. (1305272)
Near Bellicourt, France, Sept. 29, 1918.
R—Memphis, Tenn.
B—Canton, Miss.
G. O. No. 59, W. D., 1919.

Corporal, Company A, 113th Machine Gun Battalion, 30th Division.
Though wounded soon after the opening of the attack, he continued to lead his squad until he was incapacitated by a second wound.

BUONOMO, ANTHONY (2451073)
Near Chevieres, France, Oct. 19, 1918.
R—New York, N. Y.
B—Italy.
G. O. No. 44, W. D., 1919.

Private, first class, Company F, 310th Infantry, 78th Division.
Private Buonomo was voluntarily acting as guide on a reconnaissance with an officer, when the latter was severely wounded by a bursting shell. Having himself been struck by a shell fragment, he disregarded his own injuries, and immediately bandaged the officer's wound and assisted him to the dressing station, 800 meters away, across an open field swept by shell fire. He then volunteered to return to division headquarters through intense artillery fire to report that the reconnaissance had not yet been completed.

BURBANK, FRANK J.
West of Bouresches, France, July 20, 1918.
R—Livermore Falls, Me.
B—Livermore Falls, Me.
G. O. No. 11, W. D., 1921.

Second lieutenant, 103d Infantry, 26th Division.
After his company had suffered severe casualties Lieutenant Burbank reorganized his company under heavy machine-gun and artillery fire. He then led his organization in the attack on Hill 190. He was the first to enter the enemy trenches and personally captured a machine gun, with its crew. Due to his gallantry, the objective was reached and held.

BURCH, ALBERT S.
Near St. Juvin, France, Oct. 14, 1918.
R—Atlanta, Ga.
B—Baltimore, Md.
G. O. No. 81, W. D., 1919.

First lieutenant, 326th Infantry, 82d Division.
While leading his men against determined enemy resistance, Lieutenant Burch was severely wounded in the arm by four machine-gun bullets. Although suffering intense pain, he continued to press on until ordered to the rear. On his way to the dressing station he endeavored to carry a more severely wounded officer, and, although greatly weakened, he struggled with his burden until the arrival of litter bearers.

BURCHFIELD, JOSEPH H. (2579782)
Throughout the operations south of Soissons, France, July 18–22, 1918.
R—Cleveland, Ohio.
B—Youngstown, Ohio.
G. O. No. 99, W. D., 1918.

Private, medical detachment, 16th Infantry, 1st Division.
During the entire operation he repeatedly exposed himself to heavy enemy fire in order to dress and evacuate the wounded. On July 22 he went through a heavy enemy barrage to render first aid to the wounded in the front line and to evacuate them to the rear and was himself wounded while engaged in this work.

BURCHILL, GEORGE H. (1375803)
Near Very, France, Sept. 26, 1918.
R—Chicago, Ill.
B—Chicago, Ill.
G. O. No. 44, W. D., 1919.

Private, first class, Battery C, 122d Field Artillery, 33d Division.
Though suffering from illness, he volunteered and performed valiant service as a telephone operator under heavy shell fire. Later he went out alone through shell fire to repair the telephone line, which had been broken in several places by shells.

BURDEN, EDWARD F. (2672130)
Near St. Juvin, France, Oct. 11, 1918.
R—Long Island City, N. Y.
B—New York, N. Y.
G. O. No. 78, W. D., 1919.

Private, sanitary detachment, 326th Infantry, 82d Division.
Under heavy machine-gun and shell fire, Private Burden crossed and recrossed the Aire River five times, administering first aid to 40 wounded soldiers until he collapsed from exhaustion.

BURDETT, WILLIAM C.
Near Medeah Ferme, France, Oct. 3, 1918.
R—Mortel, Tenn.
B—Nashville, Tenn.
G. O. No. 21, W. D., 1919.

Captain, 9th Infantry, 2d Division.
Throughout five days of the most bitter fighting he displayed most exceptional valor and coolness in leading his men through intense machine-gun and barrage fire. He fell, seriously wounded, while at the head of his men.

BURDICK, HOWARD------------------
Northwest of Cambrai, France, Sept. 28, 1918.
R—Brooklyn, N. Y.
B—New York, N. Y.
G. O. No. 38, W. D., 1921.

Second lieutenant, 17th Aero Squadron, Air Service.
Attacked by two Fokker biplanes, he outmaneuvered both machines, shot one into flames and routed the other one. Later, seeing three Fokkers attacking an American aviator, he at once dove into the combat to his assistance, shooting down one and driving off the other two. His quick and unhesitating attack, single-handed, on the three Fokkers saved the life of his fellow pilot.

BURGARD, JOHN C------------------
Near Epinonville, France, Sept. 27, 1918.
R—Portland, Oreg.
B—Portland, Oreg.
G. O. No. 70, W. D., 1919.

First lieutenant, 362d Infantry, 91st Division.
On duty as battalion liaison officer, Lieutenant Burgard was establishing the battalion post of command at daybreak when he suddenly discovered a party of the enemy placing machine guns so as to fire upon the position from the flank. Firing a rifle to give the alarm, Lieutenant Burgard advanced toward the enemy, followed by the battalion headquarters group, whom he led in a vigorous attack on the hostile force, capturing 21 of the enemy, 1 machine gun, and 2 light machine rifles, with but 1 casualty among his own men.

BURGER, VALENTINE JOSEPH----------
Near Hill 360, over the region of the Meuse, France, Oct. 27, 1918.
R—Leonia, N. J.
B—Brooklyn, N. Y.
G. O. No. 39, W. D., 1920.

Second lieutenant, observer, 3d Observation Group, Air Service.
He, with his pilot, flying at an altitude of less than 50 meters, within close range of numerous machine guns and light artillery pieces firing continually on them, staked the American advance lines and helped silence enemy machine-gun nests which were holding up the advance of the Infantry at this point. Although the plane was riddled with over 300 bullet holes and the pilot severely wounded, he gathered valuable and accurate information and assisted his pilot to a safe landing within reach of the post of command and delivered his valuable information.
Oak-leaf cluster.
For the following acts of extraordinary heroism in action in Europe this man was awarded an oak-leaf cluster: On the morning of Nov. 1, 1918, during the progress of an important attack, Lieutenant Burger volunteered on a mission to fly through a heavy fog in order to locate the then advanced infantry of the attack. In accomplishing this mission it was necessary to fly at a very low altitude and through the American barrage, which was being fired during the flight. He penetrated several kilometers into the enemy's lines, being subjected to heavy machine-gun fire from the ground, which struck his plane many times, obtained information of the disposition of the enemy artillery, infantry, and our own front line. He returned through the fire with the information, which was the first authentic data to reach the division commander.

BURGESS, FREDERICK V------------
Near St. Mihiel, France, Sept. 13, 1918.
R—Burlington, Vt.
B—Burlington, Vt.
G. O. No. 35, W. D., 1919.

First lieutenant, 15th Machine Gun Battalion, 5th Division.
After being painfully wounded by a machine-gun bullet, in a particularly intense barrage of machine-gun and shell fire, he remained with his platoon, visiting his guns and directing their fire throughout a determined counterattack, refusing to be evacuated until the attack was over.

BURGH, DAVID T------------------
East of Ronssoy, France, Sept. 29, 1918.
R—Warren, Me.
B—Scotland.
G. O. No. 137, W. D., 1918.

First lieutenant, chaplain, 105th Infantry, 27th Division.
During the operations against the Hindenburg line he displayed remarkable devotion to duty and courage in caring for the wounded under heavy shell and machine-gun fire. The splendid example set by this officer was an inspiration to the combat troops.

BURGIN, JOHN C. (913946)------------
Near Romagne, France, Oct. 14, 1918.
R—Bond, Ky.
B—Richmond, Ky.
G. O. No. 37, W. D., 1919.

Sergeant, Company A, 7th Engineers, 5th Division.
Seriously wounded while advancing with his platoon under terrific shell and machine-gun fire, he refused treatment and led his men on to the objective.

BURK, WALTER S------------------
Near Villers-devant-Dun, France, Nov. 2, 1918.
R—Troy, N. Y.
B—Ridley Park, Pa.
G. O. No. 44, W. D., 1919.

First lieutenant, 359th Infantry, 90th Division.
Lieutenant Burk refused to leave his platoon after being wounded. He led his men in the advance under heavy machine-gun fire and held all the positions taken until relieved the following morning.

*BURKE, CAMPBELL------------------
Near Gesnes, France, Oct. 9, 1918.
R—Beatrice, Ky.
B—Beatrice, Ky.
G. O. No. 20, W. D., 1919.

Captain, 361st Infantry, 91st Division.
The battalion which Captain Burke commanded was ordered to attack a position on Hill 255 under terrific machine-gun and artillery fire. His coolness and personal example contributed largely to the success of the battalion and enabled it to capture substantially the entire objective. He was severely wounded in this engagement.
Posthumously awarded. Medal presented to father, T. B. Burke.

*BURKE, CHARLES H. (57347)----------
At Bois-de-Remieres, Seicheprey, France, Mar. 16, 1918.
R—Wellsville, Ohio.
B—West Bridgeport, Pa.
G. O. No. 88, W. D., 1918.

Corporal, Company E, 28th Infantry, 1st Division.
While on patrol duty and being severely wounded he refused to leave his platoon leader, who had also been severely wounded, and stayed by his side during intense bombardment and assisted in driving off an enemy patrol. Died Mar. 17, 1918, of wounds received in action.
Posthumously awarded. Medal presented to father, Samuel Burke.

BURKE, JACKSON D. (56301)----------
At Cantigny, France, May 28-30, 1918.
R—Beatrice, Ky.
B—Pike County, Ky.
G. O. No. 99, W. D., 1918.

Sergeant, Headquarters Company, 28th Infantry, 1st Division.
On May 28-30, 1918, at Cantigny, France, he showed exceptional energy, bravery, and loyalty to duty. At one period of the fight it was necessary to send a message of great importance to the regimental commander. It was considered impossible for a runner to reach regimental headquarters because of the intensity of the enemy fire. He nevertheless volunteered to carry the message, and by crawling several hundred yards through machine-gun fire he successfully executed his mission.

BURKE, JOHN J._____
Near Villers-sur-Fere, France, July 27-28, 1918.
R—Jersey City, N. J.
B—Elizabeth, N. J.
G. O. No. 44, W. D., 1919.

Second lieutenant, 165th Infantry, 42d Division.
Lieutenant Burke was instructed by his regimental commander to take four men and locate the position of the assaulting battalion. Upon leaving regimental headquarters he was serverely wounded, but continued on his mission in the face of unusually heavy artillery and machine-gun fire. He succeeded in locating the battalion only after four hours' search, constantly under fire, whereupon he returned and reported to his regimental commander.

BURKE, JOHN T. (976787)_____
Service rendered under the name, John F. O'Rourke.

Private, medical detachment, 9th Infantry, 2d Division.

BURKE, WALTER F. (2412662)_____
Near Vieville-en-Haye, France, Sept. 25-26, 1918.
R—Orange, N. J.
B—Orange, N. J.
G. O. No. 44, W. D., 1919.

Private, first class, medical detachment, 311th Infantry, 78th Division.
During an extreme shelling he cared for the wounded, although exposed at all times to the hazard of the rain of shells. He was stunned by the concussion of a high-explosive shell, which killed men on both sides of him, but he continued until ordered to the aid post. He volunteered and returned to the lines to relieve a comrade who had fallen from exhaustion.

"BURKS, CHARLES R. (101895)_____
Near Sergy, northeast of Chateau-Thierry, France, July 30, 1918.
R—Malvern, Iowa.
B—Strahn, Iowa.
G. O. No. 116, W. D., 1918.

Private, Company I, 165th Infantry, 42d Division.
During the midday attack on Sergy, after all the runners had been exhausted and many men had been killed or wounded, Private Burks volunteered to take a message to a neighboring unit through violent bombardment and machine-gun fire. He was killed by a machine-gun bullet while on his way with the message.
Posthumously awarded. Medal presented to father, John H. Burks.

*BURKS, JAMES B._____
At Etraye Ridge, France, Oct. 23, 1918.
R—Newport News, Va.
B—Lawyer, Va.
G. O. No. 78, W. D., 1919.

Second lieutenant, 113th Infantry, 29th Division.
Having gone out from his position under heavy machine-gun fire in an effort to establish a liaison with the unit on his right, Lieutenant Burks encountered an enemy patrol. In the combat which followed he was killed after he had killed several of his adversaries.
Posthumously awarded. Medal presented to father, R. H. Burke.

BURLEIGH, NELSON (262223)_____
Near Cierges, northeast of Chateau-Thierry, France, July 31, 1918.
R—Flint, Mich.
B—Saco, Me.
G. O. No. 117, W. D., 1918.

Private, Company E, 125th Infantry, 32d Division.
Although severely wounded, he crawled to an exposed and dangerous place where a comrade lay seriously injured and rendered first aid, thereby saving his comrade's life.

'BURNES, JOHN F._____
In the attack on Bois-de-Belleau, France, June 12, 1918.
R—Binghamton, N. Y.
B—Binghamton, N. Y.
G. O. No. 99, W. D., 1918.

Captain, 5th Regiment, U. S. Marine Corps, 2d Division.
In the attack on Bois-de-Belleau June 12, 1918, he was badly wounded, but completed the disposition of his platoon under violent fire. The injuries which he sustained in the performance of this self-sacrificing duty later caused his death.
Posthumously awarded. Medal presented to sister, Mrs. Jacob Keigler.

BURNETT, CLIFTON (45910)_____
Near Montrefagne, France, Oct. 9, 1918.
R—Pilgrim, Tex.
B—Pilgrim, Tex.
G. O. No. 81, W. D., 1919.

Sergeant, headquarters 1st Infantry Brigade, 1st Division.
On Oct. 9 Sergeant Burnett volunteered and established liaison between battalion commanders and brigade headquarters under heavy artillery and machine-gun fire. On the same day he twice volunteered and carried messages and acted as guide across fields subjected to shell and machine-gun fire. He repeatedly repaired telephone wires when no linemen were available, and continued this work through shell fire and gas bombardment, although four of the men working with him were wounded.

BURNS, EDWARD N. (1243650)_____
Near Fismette, France, Aug. 10, 1918.
R—Philadelphia, Pa.
B—Philadelphia, Pa.
G. O. No. 49, W. D., 1922.

Sergeant, Company B, 111th Infantry, 28th Division.
When the attack of his company was held up by fire from a hostile strong point, Sergeant Burns with two other men voluntarily cut their way through enemy wire entanglements under heavy fire, reached their objective and engaged the enemy in hand-to-hand combat. During this latter action six of the enemy were killed, and the attacking line was enabled to advance to the new position. At further risk of his life, Sergeant Burns carried back one of his comrades who had been fatally wounded in the action.

BURNS, FORREST (53085)_____
Near Soissons, France, July 18-19, 1918.
R—Richmond, Ky.
B—Richmond, Ky.
G. O. No. 72, W. D., 1920.

First sergeant, Company I, 28th Infantry, 1st Division.
When the advance of his platoon had been halted by heavy machine-gun fire from the front, Sergeant Burns, with three others, advanced through heavy machine-gun fire and attacked the enemy position. His group succeeded in capturing two enemy machine guns, thereby enabling his organization to advance with slight loss. After the platoon commander had been killed, he directed the advance of his platoon until severely wounded.

*BURNS, HAROLD W._____

Near Brabant-sur-Meuse, France, Oct. 23, 1918.
R—Gary, Ind.
B—Canada.
G. O. No. 50, W. D., 1919.

Second lieutenant, 308th Trench Mortar Battery, 158th Field Artillery Brigade, 83d Division.
In the open, under direct fire from the enemy machine guns and artillery, Lieutenant Burns went from gun to gun of his platoon, encouraging his men to continued effort. Gassed, he refused to be evacuated, but remained in command of his platoon until after the action, when he gave first aid to the wounded.
Posthumously awarded. Medal presented to mother, Mrs. L. P. Goodwin.

*BURNS, JAMES S. D.
    Near Fismes, France, Aug. 11, 1918.
    R—New York, N. Y.
    B—New York, N. Y.
    G. O. No. 44, W. D., 1919.

Second lieutenant, 165th Infantry, observer, 88th Aero Squadron, Air Service. Louis G. Bernheimer, first lieutenant, pilot; John W. Jordan, second lieutenant, 7th Field Artillery, observer; Roger W. Hitchcock, second lieutenant, pilot; James S. D. Burns, deceased, second lieutenant, 165th Infantry, observer; Joel H. McClendon, deceased, first lieutenant, pilot; Charles W. Plummer, deceased, second lieutenant, 101st Field Artillery, observer; Philip R. Babcock, first lieutenant, pilot; and Joseph A. Palmer, second lieutenant, 15th Field Artillery, observer. All of these men were attached to the 88th Aero Squadron, Air Service.
For extraordinary heroism in action near Fismes, France, Aug. 11, 1918. Under the protection of three pursuit planes, each carrying a pilot and an observer, Lieutenants Bernheimer and Jordan, in charge of a photo plane, carried out successfully a hazardous photographic mission over the enemy's lines to the River Aisne. The four American ships were attacked by 12 enemy battle planes. Lieutenant Bernheimer, by coolly and skilfully maneuvering his ship, and Lieutenant Jordan, by accurate operation of his machine gun, in spite of wounds in the shoulder and leg, aided materially in the victory which came to the American ships, and returned safely with 36 valuable photographs. The pursuit plane operated by Lieutenants Hitchcock and Burns was disabled while these two officers were fighting effectively. Lieutenant Burns was mortally wounded and his body jammed the controls. After a headlong fall of 2,500 meters, Lieutenant Hitchcock succeeded in regaining control of this plane and piloted it back to his airdrome. Lieutenants McClendon and Plummer were shot down and killed after a vigorous combat with five of the enemy's planes. Lieutenants Babcock and Palmer, by gallant and skillful fighting, aided in driving off the German planes and were materially responsible for the successful execution of the photographic mission.
Posthumously awarded. Medal presented to father, Z. James Burns.

BURNS, JOSEPH W. (89500)
    Near Landres-et-St. Georges, France,
    Oct. 15, 1918.
    R—Brooklyn, N. Y.
    B—Brooklyn, N. Y.
    G. O. No. 37, W. D., 1919.

Sergeant, Company C, 165th Infantry, 42d Division.
Assisted by another soldier, he voluntarily went to the aid of a comrade who was lying in front of his lines, in full view of the enemy. After administering first-aid, they succeeded in bringing the wounded man to safety.

*BURNS, MYRON D (2395250)
    Near Fossoy, France, July 15, 1918.
    R—Shamrock, Okla.
    B—Eldred, Pa.
    G. O. No. 44, W. D., 1919.

Private, Company F, 7th Infantry, 3d Division.
Although suffering intense agony from severe wounds, he killed eight of the enemy with his rifle and bayonet, and then crawled about for two days before being picked up. He died shortly after from his wounds.
Posthumously awarded. Medal presented to mother, Mrs. Jennie Douglas.

BURNS, THOMAS V. (114861)
    Near Vierzy, France, July 18, 1918.
    R—Scranton, Pa.
    B—Scranton, Pa.
    G. O. No. 117, W. D., 1918.

Private, Company E, 9th Infantry, 2d Division.
He fearlessly sprang to the assistance of a French officer and helped him, under fire, removed a wounded French soldier from a burning tank which had been struck by a shell. Afterwards seeing a gap opening in the firing line, he collected four men, dashed forward and captured five machine guns, with which he held the line until the arrival of reinforcements. He then rejoined his platoon, where he rendered valuable service during the remainder of the battle.

BURR, GEORGE E. (252092)
    Near Cierges, France, Aug. 2, 1918.
    R—Milwaukee, Wis.
    B—Hill City, S. Dak.
    G. O. No. 147, W. D. 1918.

Sergeant, first class, Company C, 107th Field Signal Battalion, 32d Division.
He, in charge of a detachment, strung wire far in advance of the front lines, working through a heavy artillery fire to the point where the regimental post of command was to be situated, 100 yards from the enemy line. Here he was ordered to leave one man at the instrument. While the rest of the detachment returned to the rear he himself volunteered and remained alone in this dangerous position.

BURR, JOHN G. (1383200)
    Near Riaville, France, Nov. 9, 1918.
    R—Effingham, Ill.
    B—Effingham, Ill.
    G. O. No. 70, W. D., 1919.

Mechanic, Company A, 130th Infantry, 33d Division.
As he was administering first aid to a wounded comrade during a raid, they were attacked by several of the enemy. Undaunted by this superior force, Mechanic Burr succeeded in killing four and driving off the others, thereby setting a conspicuous example of courage and coolness.

BURRELL, REUBEN (1794075)
    In the Champagne sector, France,
    Sept. 30, 1918.
    R—Conshohocken, Pa.
    B—New Kent County, Va.
    G. O. No. 46, W. D., 1919.

Private, Machine Gun Company, 371st Infantry, 93d Division.
Private Burrell, although painfully wounded in the knee, refused to be evacuated, stating that if he went to the rear there would not be enough left for his group to function.

BURROUGHS, FRANK ALBERT (2193111).
    Near la-Haie-Menneresse, France,
    Oct. 17, 1918.
    R—Watauga, S. Dak.
    B—Des Moines, Iowa.
    G. O. No. 50, W. D., 1919.

Sergeant, Machine Gun Company, 118th Infantry, 30th Division.
When his platoon commander was wounded and all the noncommissioned officers had become casualties, under a heavy enemy barrage which fell upon his company, killing or wounding more than a third of the men, Sergeant Burroughs reorganized the platoon under heavy shellfire, directed the evacuation of the wounded, and then led the one remaining gun team forward, displaying remarkable coolness and initiative.

BURT, BYRON T., Jr.
    Near Griscourt, France, Aug. 4–11;
    near Sommedieue, France, Sept.
    16; and near Avocourt, France,
    Oct. 1, 1918.
    R—Port Henry, N. Y.
    B—Italy.
    G. O. No. 46, W. D., 1919.

First lieutenant, observer, balloon section, Air Service, 1st Army.
On each of these occasions Lieutenant Burt remained with his balloon, making important observations of the enemy's positions and directing our artillery fire until his balloon was set on fire by incendiary bullets from enemy aircraft. On one occasion he refused to jump until his companion, a student observer, was safely away.

*BURTON, EDWARD A_____ | First lieutenant, 128th Infantry, 32d Division.
Near Cierges, France, Aug. 1, 1918. | He was mortally wounded while carrying a wounded man from a position
R—Reedsburg, Wis. | exposed to artillery and machine-gun fire, but regardless of his own suffering,
B—Hillsboro, Wis. | he persisted in his task until he had placed the wounded man in a place of
G. O. No. 44, W. D., 1919. | safety. He died while being evacuated.
| Posthumously awarded. Medal presented to mother, Mrs. Charles Burton.

BURTON, MILTON G. (2273452)_____ | Sergeant, Company E, 316th Engineers, 91st Division.
Near Eclisfontaine, France, Sept. | While attached to an infantry unit, Sergeant Burton evidenced great bravery
28, 1918. | in bringing in four severely wounded men from the front lines to the dressing
R—Los Angeles, Calif. | station. After reaching the dressing station he immediately administered
B—Iowa. | first aid to the men, being continually exposed to the machine-gun and
G. O. No. 46, W. D., 1919. | sniper fire.

BUSCH, GEORGE L. (2178732)_____ | Sergeant, Company D, 354th Infantry, 89th Division.
Near Remonville, France, Nov. 1, | His company was waiting at its first objective for the barrage to advance,
1918. | when five enemy machine guns opened fire on it from a point in front of
R—Troy, Mo. | the barrage. Realizing the gravity of the situation, Sergeant Busch led a
B—Troy, Mo. | combat group from his platoon with exceptional skill and bravery through
G. O. No. 70, W. D., 1919. | the barrage to the flank of the enemy position and silenced the machine
| guns, capturing prisoners from their crews. While returning through the
| barrage, he was knocked down by concussion from a bursting shell, but he
| immediately arose and led his men back to the platoon.

BUSCHING, GEORGE A. (2149481)_____ | Private, Company G, 118th Infantry, 30th Division.
Near Brancourt, France, Oct. 8, 1918. | He observed a severely wounded soldier about 100 yards from his post on a
R—Plainfield, Iowa. | sunken road heavily shelled by artillery and machine-gun enfilading fire.
B—Guttenberg, Iowa. | He voluntarily went out and carried this soldier to a place of safety.
G. O. No. 133, W. D., 1918. |

BUSCHMANN, JEROME (39521)_____ | Sergeant, Company G, 9th Infantry, 2d Division.
South of Soissons, France, July 18, | Jerome Buschmann, sergeant; John Rockwell, private; William F. Rockwell,
1918. | private; Alfred Shimanoski, private; and Watzlaw Viniarsky, private,
R—Webb City, Mo. | all of Company G, 9th Infantry. For extraordinary heroism in action south
B—St. Charles, Mo. | of Soissons, France, July 18, 1918. They conspicuously distinguished them-
G. O. No. 20, W. D., 1919. | selves by attacking a party of more than 60 Germans and, in an intense and
| desperate hand-to-hand fight, succeeded in killing 22 men and capturing 40
| men and 5 machine guns.

*BUSEY, CHARLES BOWEN_____ | Second lieutenant, 310th Infantry, 78th Division.
In the Bois-des-Loges, France, Nov. | While on duty as instructor at the school at Langres, France, Lieutenant
1, 1918. | Busey was sent to the 78th Division for a week of observation work, where
R—Urbana, Ill. | on his own request, he was attached to a front-line battalion; and again, on
B—Urbana, Ill. | his own request, was assigned to duty with a company. During the attack
G. O. No. 43, W. D., 1922. | on the enemy strong point in the Bois-des-Loges, Lieutenant Busey unhesi-
| tatingly and with the utmost gallantry led a patrol of four men through a
| heavy artillery and machine-gun fire toward the position of a machine-gun
| nest which was holding up the company's advance, when he was killed by a
| hostile hand grenade within a few yards of his objective.
| Posthumously awarded. Medal presented to widow, Mrs. Louise Carter Busey.

*BUSH, ALDEN (261832)_____ | Corporal, Company C, 125th Infantry, 32d Division.
At Cierges, northeast of Chateau- | During the attack on and capture of the village of Cierges, northeast of Chateau-
Thierry, France, Aug. 1, 1918. | Thierry, France, Aug. 1, 1918, Corporal Bush was fatally wounded. In spite
R—Detroit, Mich. | of his wound he struggled forward, urging on and inspiring his men, and
B—Rockford, Mich. | keeping up with the attacking wave until he fell.
G. O. No. 102, W. D., 1918. | Posthumously awarded. Medal presented to father, R. E. Bush.

BUSH, GARRET (52475)_____ | Corporal, Company A, 26th Infantry, 1st Division.
Near Soissons, France, July 19–21, | During the fighting near Soissons, France, July 19–21, 1918, he repeatedly
1918. | passed through shell and machine-gun fire to locate dangerous enemy po-
R—Sheyenne, N. Dak. | sitions
B—Holland. |
G. O. No. 132, W. D., 1918. |

BUSH, HERMAN L. (109458)_____ | First sergeant, Company B, 102d Machine Gun Battalion, 26th Division.
Near Verdun, France, Oct. 25, 1918. | Sergeant Bush, learning that an officer was lying wounded in both legs in a
R—Dorchester, Mass. | zone of heavy machine-gun fire, immediately left a position of shelter, went
B—Boston, Mass. | to his aid, and succeeded in bringing the officer back to a place of safety.
G. O. No. 21, W. D., 1919. |

*BUSHNELL, THEODORE K._____ | Second lieutenant, 2d Machine Gun Battalion, 1st Division.
Near Fleville, France, Oct. 5, 1918. | He showed exceptional bravery by remaining with his platoon after being
R—Denver, Colo. | wounded. He refused evacuation until he received a second wound, the
B—Denver, Colo. | nature of which demanded his immediate return to the rear.
G. O. No. 44, W. D., 1919. | Posthumously awarded. Medal presented to father, George A. Bushnell.

BUSK, JOSEPH R._____ | Second lieutenant, 38th Infantry, 3d Division.
East of Chateau-Thierry, France, | Despite the coldness of the water, the swiftness of the current, and the presence
June 17, 1918. | of the enemy on the opposite bank, Lieutenant Busk completed a personal
R—New York, N. Y. | reconnaissance of the enemy's position by swimming the River Marne,
B—New York, N. Y. | after which he took a patrol across the river in boats and obtained valuable
G. O. No. 126, W. D., 1919. | information regarding the movements of the enemy.

*BUTCHER, GEORGE S._____
   Near Verdun, France, Oct. 27, 1918.
   R—Upper Montclair, N. J.
   B—Chatham, N. J.
   G. O. No. 15, W. D., 1919.

Captain, 111th Machine Gun Battalion, 29th Division.
Hearing a call for help from a neighboring platoon of another company, whose men were all casualties, he quickly made his way there, manned the guns, and kept up a steady fire until he himself was killed by a shell. His action was purely voluntary, but realizing the necessity of opening fire immediately, he disregarded his own safety in order to protect others, displaying the most heroic self-sacrifice.
Posthumously awarded. Medal presented to father, Charles R. Butcher.

BUTCHER, ORA LEE (2205393)_____
   In the Bois de Mort Mare, near Flirey, France, Sept. 12, 1918.
   R—Pattonsburg, Mo.
   B—Pattonsburg, Mo.
   G. O. No. 139, W. D., 1918.

Private, Company M, 356th Infantry, 89th Division.
He was on duty as an observer at battalion headquarters, twice volunteering to carry important messages from his battalion commander to company commanders. In so doing he passed through heavy barrages.

BUTCHER, THOMAS W (2224707)_____
   Near Villers-devant-Dun, France, Nov. 2, 1918.
   R—Fort Worth, Tex.
   B—Waxahachie, Tex.
   G. O. No. 46, W. D., 1919.

Corporal, Company C, 359th Infantry, 90th Division.
Having been wounded in the back by a machine-gun bullet, Corporal Butcher led his squad through three bands of machine-gun fire, capturing three guns, and capturing or killing all of the crews.

BUTLER, CHARLES (2860948)_____
   Near Ardeuil, France, Sept. 29, 1918.
   R—New Orleans, La.
   B—McComb City, Miss.
   G. O. No. 46, W. D., 1919.

Private, Machine Gun Company, 371st Infantry, 93d Division.
With three other soldiers, Private Butler crawled 200 yards ahead of our lines under violent machine-gun fire and rescued an officer who was lying mortally wounded in a shell hole.

BUTLER, EMORY L. (1316455)_____
   Near Bellicourt, France, Sept. 29, 1918.
   R—Rowan County, N. C.
   B—Rowan County, N. C.
   G. O. No. 81, W. D., 1919.

Corporal, Company K, 119th Infantry, 30th Division.
Becoming separated from his platoon during the advance, he continued 500 yards beyond the objective, and, although there were several enemy machine guns near him, he went to a dugout and forced the 35 occupants to come out and surrender. He was soon joined by other members of his platoon and aided in cleaning out other near-by dugouts, displaying absolute disregard of danger.

BUTLER, JAMES S (68302)_____
   Near Verdun, France, Nov. 3, 1918.
   R—Keene, N. H.
   B—Stoddard, N. H.
   G. O. No. 21, W. D., 1919.

Sergeant, Company G, 103d Infantry, 26th Division.
While leading a daylight patrol into the Bois Moirey to ascertain the enemy's position, Sergeant Butler volunteered and advanced alone into a machine-gun nest to draw fire. He went forward until fired upon by enemy machine guns and snipers. He then crawled back and reported the position of the enemy to his battalion commander.

BUTLER, LAWRENCE DONALD_____
   Near Romanovka, Siberia, June 25, 1919.
   R—San Francisco, Calif.
   B—Plano, Tex.
   G. O. No. 133, W. D., 1919.

Second lieutenant, 31st Infantry.
Although twice wounded, once severely, early in the action, and after over 50 per cent of the detachment were casualties and the detachment completely surrounded by the enemy, he continued courageously to direct the men, and by his heroism, bearing, and skill so inspired the few survivors that they were enabled to completely repulse greatly superior numbers of the enemy.

*BUTLER, RICHARD (109870)_____
   Near Marcheville, France, Sept. 26, 1918.
   R—New Haven, Conn.
   B—New Haven, Conn.
   G. O. No. 133, W. D., 1918.

Private, Company D, 102d Machine Gun Battalion, 26th Division.
He volunteered to accompany a party whose mission was to bomb a hostile machine-gun emplacement. Under heavy shell fire he approached to within 30 feet of the emplacement when he was fired upon from loopholes in a stone wall. Working his way behind the wall, this courageous soldier enfiladed the enemy with rifle fire and effected their capture. While he was disarming prisoners he was shot and mortally wounded.
Posthumously awarded. Medal presented to mother, Mrs. Johanna Butler.

BUTLER, WILLIAM (104464)_____
   Near Maison-de-Champagne, France, Aug. 18, 1918.
   R—New York, N. Y.
   B—White Plains, Md.
   G. O. No. 37, W. D., 1919.

Sergeant, Company I, 369th Infantry, 93d Division.
He broke up a German raiding party which had succeeded in entering our trenches and capturing some of our men. With an automatic rifle he killed four of the raiding party and captured or put to flight the remainder of the invaders.

BUTTERFIELD, CLARK (1196687)_____
   Near Cunel, France, Oct. 14, 1918.
   R—Minneapolis, Minn.
   B—St. Paul, Minn.
   G. O. No. 44, W. D., 1919.

Sergeant, Company B, 13th Machine Gun Battalion, 5th Division.
Leaving his shelter in a shallow machine-gun emplacement and accompanying an officer, Sergeant Butterfield ventured forth through a most intense fire to the aid of a wounded officer, and assisted in carrying him a distance of 170 yards to safety.

BUTTERFIELD, OLIN J. (108225)_____
   Near St. Etienne, France, Oct. 3, 1918.
   R—Denver, Colo.
   B—Terre Haute, Ind.
   G. O. No. 23, W. D., 1919.

Corporal, 77th Company, 6th Machine Gun Battalion, U. S. Marine Corps, 2d Division.
When our advance Infantry was forced to withdraw, Corporal Butterfield's machine-gun crew refused to withdraw, but calmly set up their machine gun. The gun was upset by a bursting hand grenade, which also injured Corporal Butterfield and another member of the squad. Despite their injuries, they immediately reset the gun and opened fire on the advancing Germans when 20 feet distant, causing the Germans to break and retreat in disorder.

BUTTS, EDMUND L.
In the Bois d'Aigremont, near Cre-
zancy, France, July 14-18, 1918.
R—Stillwater, Minn.
B—Stillwater, Minn.
G. O. No. 116, W. D., 1919.

Colonel, 30th Infantry, 3d Division.
On repeated occasions during the intense enemy bombardment preceding the
second battle of the Marne and on the following day Colonel Butts went to
exposed positions under heavy shellfire for the purpose of making personal
reconnaissances, securing information of great value. The personal courage
and determination displayed by him inspired his regiment to withstand
successfully the principal shock of the German attack and drive the enemy
back across the Marne by the brilliant counterattacks which he planned.

*BYAM, OLIVER P.
Near Cunel Heights, France, Oct.
11, 1918.
R—Gooding, Idaho.
B—Iowa.
G. O. No. 89, W. D., 1919.

Second lieutenant, 7th Infantry, 3d Division.
Upon his own initiative Lieutenant Byam moved his machine-gun platoon
through heavy artillery and machine-gun fire 400 meters in advance of the
front line, and from there opened fire on the enemy, who was holding up our
advance, displaying exceptional bravery in holding this position against
several hostile attacks. This officer was later killed by machine-gun fire
while leading a patrol to the enemy's line.
Posthumously awarded. Medal presented to father, Oliver L. Byam.

*BYINGTON, RUSSELL P. (1205046).
East of Ronssoy, France, Sept. 29,
1918.
R—Ossining, N. Y.
B—Croton, N. Y.
G. O. No. 16, W. D., 1919.

Private, first class, Company I, 105th Infantry, 27th Division.
During the operations against the Hindenburg line, he was wounded early in
the action, but continued to advance with his company and declined to
go to the rear for medical treatment. Later in the engagement he was killed
by a machine-gun bullet. His gallantry and bravery and absolute disregard
for his personal safety was a splendid example to all ranks.
Posthumously awarded. Medal presented to father, Dr. C. P. Byington.

BYRAM, GEORGE L.
At Las Guasimas, Cuba, June 24,
1898.
R—Mobile, Ala.
B—Shuqualak, Miss.
G. O. No. 71, W. D., 1919.

First lieutenant, 1st Cavalry, U. S. Army.
For extraordinary heroism in an engagement with an armed enemy at Las
Guasimas, Cuba, June 24, 1898.

BYRD, DANIEL B.
Near Escaufourt, France, Oct. 10,
1918.
R—Fayetteville, N. C.
B—Fayetteville, N. C.
G. O. No. 37, W. D., 1919.

First lieutenant, 119th Infantry, 30th Division.
Leading a small detachment under heavy fire, while the regiment was making
an advance, he encountered stiff resistance which threatened to cut his
detachment from the main line. By his utter disregard of the great danger,
and the prompt placing of his automatic rifles he made it possible for his
detachment to return to the lines. He was wounded by shrapnel but he
remained with the men until ordered to the rear by his commanding officer.

BYRD, MACK C. (156766).
Near Bois-de-Belleau, France, June
3, 1918.
R—Elkin, N. C.
B—Roaring River, N. C.
G. O. No. 23, W. D., 1919.

First sergeant, Company D, 2d Engineers, 2d Division.
Although badly wounded and suffering intense pain, Sergeant Byrd refused
evacuation, remaining and assisting his commanding officer throughout the
operations.

BYRD, WOODIE E. (128080).
Near Samogneux, France, Oct. 15,
1918.
R—Port Norfolk, Va.
B—Hartford County, N. C.
G. O. No. 37, W. D., 1919.

Bugler, Company E, 116th Infantry, 29th Division.
He displayed notable bravery in successfully carrying messages to the right
flank of his company after 4 other soldiers had been killed or wounded in
attempting to carry out this mission.

BYRNE, JAMES J. (1284614).
Near Bois de Consenvoye, France,
Oct. 10, 1918.
R—Baltimore, Md.
B—Midland, Md.
G. O. No. 37, W. D., 1919.

Private, Company D, 115th Infantry, 29th Division.
While the advance of his platoon was being held up by machine-gun fire from
a tree Private Byrne made his way through heavy and constant fire to a
position from which he was able to kill the gunner and rout the remainder
of the enemy. His valiant action made possible the further advance of his
platoon without serious loss.

BYRNS, ROBERT A.
Near Venduil, France, Sept. 29, 1918.
R—Staten Island, N. Y.
B—Maryland.
G. O. No. 20, W. D., 1919.

First lieutenant, 107th Infantry, 27th Division.
Lieutenant Byrns, although himself wounded, reorganized his company after
the captain was killed and led it forward in the face of intense machine-gun
fire until he was wounded a second time. His splendid courage and gallant
conduct set an inspiring example to all ranks.

BYRON, THOMAS F. (6526).
Near St.-Hilaire, France, Sept. 22,
1918.
R—Waterbury, Conn.
B—Waterbury, Conn.
G. O. No. 26, W. D., 1919.

Sergeant, Company C, 102d Infantry, 26th Division.
Assisted by another soldier, Sergeant Byron rushed a machine-gun nest, which
had been firing on their patrol. They succeeded in killing the crew.

BYRUM, JOHN C. (1320291).
Near Bellicourt, France, Sept. 29,
1918.
R—Edenton, N. C.
B—Edenton, N. C.
G. O. No. 50, W. D., 1919.

First sergeant, Company E, 120th Infantry, 30th Division.
Although he was wounded at the very start of the attack, Sergeant Byrum
continued with the advance, reorganizing scattered units and leading them
back to the line. Later his arm was shot off, but he steadfastly refused evacuation until loss of blood so weakened him that he was taken to the rear.

CABLE, ROBERT B. (1309715)_____
  Near Montbrehain and Busigny, France, Oct. 7-17, 1918.
  R—Tellico Plains, Tenn.
  B—Carter County, Tenn.
  G. O. No. 46, W. D., 1919.

First sergeant, Company M, 117th Infantry, 30th Division.
For repeated acts of extraordinary heroism near Montbrehain and Busigny, France. Leading 2 platoons of his company, after the officers had become casualties, Sergeant Cable effectively cleared the ground on the right flank of the company of machine-gun nests, capturing 2 guns. Later in the day he took command of the company, when no officers remained with it, and continued to be in charge for a week, in which time he led his men in 6 attacks, inspiring them by his fearlessness. On Oct. 9 he led an attack on the town of Busigny, charging across an open field in the face of heavy machine gun fire from the houses of the village and clearing the town of the enemy. This gallant soldier was later wounded while leading 2 platoons against an enemy machine-gun nest.

CADDLE, JAMES J. (1680035)_____
  Near Ville-Savoye, France, Aug. 23-25, 1918.
  R—Churchville, N. Y.
  B—New York, N. Y.
  G. O. No. 78, W. D., 1919.

Private, Company B, 308th Infantry, 77th Division.
Private Caddle, a battalion runner, displayed exceptional bravery in carrying numerous messages under heavy artillery fire to the front-line positions, crossing the Vesle River and proceeding for more than a kilometer in plain view of the enemy, over terrain which was continually bombarded with gas and high-explosive shells.

*CAGLE, THOMAS G. (1309440)_____
  Near Ponchaux, France, Oct. 6, 1918.
  R—Lenoir City, Tenn.
  B—Lenoir City, Tenn.
  G. O. No. 14, W. D., 1925.

Private, first class, Company L, 117th Infantry, 30th Division.
When part of the line had been halted by heavy fire from 3 machine-gun nests Private Cagle and Corpl. George W. Spears, armed only with rifles and bayonets, rushed the nearest hostile position and, of the crew of 6, killed 3 and put the remainder to flight. Being unable to advance on 2 other guns because of their heavy fire, these 2 soldiers then opened fire with their rifles and forced the remainder of the crew of approximately 12 to abandon the position after 2 of their number had been killed and 2 wounded. Private Cagle was wounded in this action, but he declined to be evacuated and shortly afterwards was killed.
Posthumously awarded. Medal presented to widow, Mrs. Addie Cagle.

CAHILL, HARRY F._____
  Near Soissons, France, July 18-22, 1918.
  R—New York, N. Y.
  B—New York, N. Y.
  G. O. No. 120, W. D., 1918.

First lieutenant, 18th Infantry, 1st Division.
He was at all times regardless of personal safety and commanded successively a platoon, a company, and a battalion, carrying again and again his command through heavy fire to all assigned objectives by sheer leadership and personal example. With a very small force he successfully organized and held a wide front under intense bombardment and against the pressure of enemy infantry.

CAHILL, WILLIAM J. (1684623)_____
  Near Bois-de-la-Cote-Lemont, France, Oct. 3, and the Bois-du-Fays, France, Oct. 9, 1918.
  R—Manchester, N. H.
  B—Ireland.
  G. O. No. 46, W. D., 1919.

Private, Company D, 59th Infantry, 4th Division.
On Oct. 3, while acting in the capacity of company runner, he carried messages to two platoons of his company through a heavy fire from guns and snipers. He successfully delivered the messages after crawling for a distance of 400 yards. On Oct. 9, in company with one other runner, he delivered messages to a platoon which was engaged in combat liaison duty in the Bois-du-Fays, passing through a severe artillery fire while in the execution of this mission.

CAIN, CHARLES (89855)_____
  Near Landres-et-St. Georges, France, Oct. 15, 1918.
  R—New York, N. Y.
  B—Waltham, Mass.
  G. O. No. 95, W. D., 1919.

Corporal, Company D, 165th Infantry, 42d Division.
Volunteering for the mission, Corporal Cain exposed himself in the open to heavy shell and machine-gun fire to obtain ammunition for his company after all on hand had been exhausted. He made repeated trips over the battlefield to gather ammunition from the bodies of the dead until his entire company had been supplied.

CAIN, JAMES S. (107626)_____
  Near Medeah Ferme, France, Oct. 4-6, 1918.
  R—Troy, N. Y.
  B—Troy, N. Y.
  G. O. No. 27, W. D., 1919.

Sergeant, Company C, 5th Machine Gun Battalion, 2d Division.
Accompanied by another soldier, he left the shelter of his trench, under heavy shellfire, to render assistance to soldiers buried by the explosion of a shell. Shortly after he left cover again to go to the assistance of other members of his section wounded by shellfire. On Oct. 6 he was wounded by machine-gun fire in the performance of his duty.

CAIN, LYLE B. (2400011)_____
  Near Fismes, France, Aug. 10, 1918.
  R—Wenatchee, Wash.
  B—Burlington, Colo.
  G. O. No. 32, W. D., 1919.

Private, Company K, 38th Infantry, 3d Division.
With one other soldier he volunteered and went to the rescue of a wounded man from another regiment and returned through heavy machine-gun and shell-fire, bringing the wounded man to his own trench.

CAIN, ROBERT S._____
  Near Fismette, France, Aug. 10-12, 1918.
  R—Pittsburgh, Pa.
  B—Scotland.
  G. O. No. 37, W. D., 1919.

Captain, 111th Infantry, 28th Division.
Armed with an automatic rifle, he personally led the advance elements of the line in driving the enemy from the forest north of the Vesle River, thereby maintaining liaison at great personal risk.

CALBI, CARMEN (1709580)_____
  Near Grand Pre, France, Oct. 14, 1918.
  R—New York, N. Y.
  B—Italy.
  G. O. No. 27, W. D., 1920.

Sergeant, Company I, 308th Infantry, 77th Division.
Sergeant Calbi, with two others, made a flank attack upon an enemy machine-gun nest. He rushed through enemy machine-gun fire and captured the gun.

*CALDEIRA, JOSEPH R. (1632972)_____
  In Bois de Cunel, near Madeline Farm, France, Oct. 9, 1918.
  R—Hayward, Calif.
  B—Castro Valley, Calif.
  G. O. No. 49, W. D., 1922.

Private, medical detachment, 30th Infantry, 3d Division.
Subjected to heavy shell, machine-gun, and rifle fire, he displayed exceptional courage in evacuating wounded from the front line to battalion aid station, which he continued until instantly killed by a shell.
Posthumously awarded. Medal presented to father, Joe S. Caldeira.

CALDWELL, EDGAR N._____
Near St. Mihiel, France, Sept. 12, 1918.
R—Fort Atkinson, Wis.
B—Glasgow, Wis.
G. O. No. 66, W. D., 1919.

Captain, 16th Infantry, 1st Division.
When his company met with enemy machine-gun fire of such intensity that the success of the operation was threatened Captain Caldwell, disregarding personal danger, walked up and down the front line, designating targets to his men and encouraging them. He then led an automatic-rifle squad, proceeding 200 yards ahead of the line, and captured an enemy machine gun.

CALDWELL, GEORGE S. (57790)_____
Near Soissons, France, July 18, 1918.
R—McKeesport, Pa.
B—Herminie, Pa.
G. O. No. 145, W. D., 1918.

Private, Company G, 28th Infantry, 1st Division.
In order to stop artillery fire, which was causing heavy losses in our ranks, he, with another soldier, rushed 300 yards to the front, attacked a machine-gun strong point and a 77-millimeter artillery gun, captured the position and the gun, killed 2, and captured 13 of the enemy.

CALDWELL, GEORGE W. (1551561)_____
Near la Trinite Ferme, France, July 15, 1918.
R—Lake George, N. Y.
B—Lake George, N. Y.
G. O. No. 44, W. D., 1919.

Sergeant, medical detachment, 76th Field Artillery, 3d Division.
Although suffering from a severe gassing, received after his mask had been shot away by a fragment of a shell, he continued through the heavy shelling to administer aid to the wounded.

*CALHOUN, GROVER W. (105179)_____
Near Soissons, France, July 19, 1918.
R—Morgan, Ga.
B—Edison, Ga.
G. O. No. 99, W. D., 1918.

Private, Company B, 3d Machine Gun Battalion, 1st Division.
He distinguished himself near Soissons, France, by exceptional coolness and heroic handling of his gun. While under intense fire he inflicted heavy casualties on the enemy, thereby materially aiding the advance, and was killed on July 19, 1918, while seeking an advantageous position during the advance of that date.
Posthumously awarded. Medal presented to mother, Mrs. Sallie Calhoun.

CALLAHAN, WILLIAM (1708546)_____
Near Revillon, France, Sept. 9, 1918.
R—Elmhurst, N. Y.
B—Ireland.
G. O. No. 35, W. D., 1919.

Sergeant, Company E, 308th Infantry, 77th Division.
In order to clean out an enemy machine-gun nest which was holding up the advance of his company, Sergeant Callahan volunteered and, with an officer, crawled through the enemy wire into his lines, killed two of the enemy, and, although their position was discovered and the area was swept by machine-gun fire, he remained with the officer, killed an enemy machine gunner, and drove another away with his gun, and finally returned with information concerning the enemy positions.

CALLARD, ARTHUR (49465)_____
Near St. Etienne-a-Arnes, France, Oct. 3, 1918.
R—Fall River, Mass.
B—Fall River, Mass.
G. O. No. 13, W. D., 1919.

Private, Company A, 23d Infantry, 2d Division.
While on duty as a company runner he carried a message through two barrages to regimental commanders. By his act many casualties were avoided.

CALLEN, NATHANIEL ERNEST_____
Near Molain, France, Oct. 17, 1918.
R—Athens, Tenn.
B—Knoxville, Tenn.
G. O. No. 35, W. D., 1919.

Major, 117th Infantry, 30th Division.
While leading his battalion into advanced positions, Major Callen made a personal reconnaissance of the territory ahead of his troops in order to locate strongly held machine-gun nests which were holding up the advance of his battalion. On several occasions he personally superintended the cleaning out of machine-gun nests. Throughout the engagement he was subjected to continuous machine gun, sniper, trench mortar, and artillery fire, but continued his work, setting an excellent example of courage and bravery.

*CALLEWAERT, ALBERIS (2101504)_____
Near Chezy, July 18, 1918; near Les Pres Farm, Aug. 4, 1918; near Bois-du-Fays, France, Sept. 28, 1918.
R—St. Paul, Minn.
B—Belgium.
G. O. No. 64, W. D., 1919.

Private, Headquarters Company, 58th Infantry, 4th Division.
Facing heavy fire, he carried ammunition from regimental headquarters to the companies of the assaulting battalion, returning with prisoners. In a later engagement he carried and laid wire while under heavy fire from snipers, machine guns, and artillery, thus maintaining telephonic communication with the front-line companies. Subsequently, while endeavoring to establish telephonic communications, he was killed while carrying wire across ground swept by machine guns and artillery.
Posthumously awarded. Medal presented to father, Constant Callewaert.

CALVIN, HARRY LESLIE_____
Near Tigny, France, July 21, 1918.
R—Brooklyn, N. Y.
B—Youngstown, Ohio.
G. O. No. 87, W. D., 1919.

Captain, 12th Field Artillery, 2d Division.
With utter disregard for personal danger, he passed for 200 yards under intense artillery and machine-gun fire to rescue a wounded officer. Finding the wounded officer could only be moved on a stretcher, he placed him in a shell hole and started back for one. He was severely wounded in the head, falling unconscious. Recovering a half hour later, he tried to go back to rescue the wounded officer, but again fell senseless.

CAMELL, HARVEY E. (2160829)_____
Near Brieulles, France, Oct. 10, 1918.
R—Cohasset, Minn.
B—Duluth, Minn.
G. O. No. 32, W. D., 1919.

Private, Company M, 132d Infantry, 33d Division.
After seeing several other runners fail in the attempt to get through the barrage, Private Camell volunteered and carried a message through the violent barrage to his battalion commander. During the entire action of Oct. 6–13 he performed most valiant service in maintaining liaison between his company and battalion headquarters.

CAMERON, CHARLES (106182)_____
Near Soissons, France, July 19, 1918.
R—Youngstown, Ohio.
B—New Castle, Pa.
G. O. No. 99, W. D., 1918.

Private, first class, Company B, 3d Machine Gun Battalion, 1st Division.
When the Infantry of which he was a part was held up by a trench occupied by Germans near Soissons, France, July 19, 1918, he voluntarily ran around the trench to its rear, shot and killed one of the enemy, and captured the remainder.

CAMPBELL, ALEXANDER (1735613)......
At Rembercourt, France, Sept. 23, 1918.
R—Lockport, N. Y.
B—Lockport, N. Y.
G. O. No. 71, W. D., 1919.

Sergeant, Company I, 309th Infantry, 78th Division.
A shell landed in an observation post occupied by Sergeant Campbell and two other soldiers, seriously wounding all three. Though he himself had been struck in seven places, this soldier placed both his companions under shelter and then walked through the barrage to company headquarters, where he sent stretcher bearers to the assistance of his wounded comrades, before securing first aid for himself.

CAMPBELL, DOUGLAS..................
East of Flirey, France, May 19, 1918
R—Cambridge, Mass.
B—San Francisco, Calif.
G. O. No. 121, W. D., 1918.

First lieutenant, 94th Aero Squadron, Air Service.
He attacked an enemy biplane at an altitude of 4,500 meters, east of Flirey, France. He rushed to the attack, but after shooting a few rounds his gun jammed. Undeterred by this accident, he maneuvered so as to protect himself, corrected the jam in midair, and returned to the assault. After a short, violent action, the enemy plane took fire and crashed to the earth.
Oak-leaf clusters (4).
One bronze oak leaf is awarded to Lieutenant Campbell for each of the following acts of extraordinary heroism in action: On May 27, 1918, he encountered three enemy monoplanes at an altitude of 3,000 meters over Montsec, France. Despite the superior strength of the enemy, he promptly attacked, and fighting a brilliant battle, shot down one German machine, which fell in three pieces, and drove the other two well within the enemy lines. On May 28, 1918, he saw six German Albatross aeroplanes flying toward him at an altitude of 2,000 meters, near Bois Rata, France. Regardless of personal danger, he immediately attacked, and by skillful maneuvering and accurate operation of his machine gun he brought one plane down in flames and drove the other five back into their own lines. On May 13, 1918, he took the offensive against two German planes at an altitude of 2,500 meters over Lironville, France, shot down one of them, and pursued the other far behind the German lines. On June 5, 1918, accompanied by another pilot, he attacked two enemy battle planes at an altitude of 5,700 meters over Epley, France. After a spirited combat he was shot through the back by a machine-gun bullet, but in spite of his injury he kept on fighting until he had forced one of the enemy planes to the ground, where it was destroyed by artillery fire, and had driven the other plane back into its own territory.

*CAMPBELL, GEORGE A..............
Near St. Mihiel, France, Sept. 12, 1918.
R—Woburn, Mass.
B—Charlottstown, Conn.
G. O. No. 44, W. D., 1919.

Captain, 18th Infantry, 1st Division.
He displayed exceptional bravery when with three men he preceded his battalion into le Jolie Bois and captured three machine guns and 20 prisoners which had been maintaining a heavy fire upon our lines.
Posthumously awarded. Medal presented to mother, Mrs. Annie Campbell.

CAMPBELL, HARRY W. (913943)........
At Romagne, France, Oct. 14, 1918.
R—New York, N. Y.
B—West Alexander, Ohio.
G. O. No. 37, W. D., 1919.

Sergeant, Company A, 7th Engineers, 5th Division.
Sergeant Campbell, on patrol with two other soldiers, captured a machine gun and 13 prisoners, killing two others who tried to escape.

CAMPBELL, JAMES E. (1747255)........
Near Grand Pre, France, Nov. 1, 1918.
R—Oswego, N. Y.
B—Oswego, N. Y.
G. O. No. 35, W. D., 1919.

Private, first class, Company K, 311th Infantry, 78th Division.
After all the regular company runners had become casualties, Private Campbell volunteered as a runner, carrying a call for reinforcements through the enemy barrage and guiding the reinforcing troops back to his company. He then aided in first aid work until his company was relieved.

CAMPBELL, JOHN A................
Near Blanc Mont Ridge, France, Oct. 3, 1918.
R—Brooklyn, N. Y.
B—Lehighton, Pa.
G. O. No. 21, W. D., 1919.

Second lieutenant, 9th Infantry, 2d Division.
While in an advanced position flanked by machine guns and under heavy artillery fire, he carried a message to regimental headquarters by the shortest and most direct route through woods occupied by the enemy.

CAMPBELL, MARTIN H. (2337312)........
Near Le Charmel, France, July 24, 1918.
R—Flint, Mich.
B—Cross Plains, Wis.
G. O. No. 64, W. D., 1919.

Band sergeant, Headquarters Company, 4th Infantry, 3d Division.
Sergeant Campbell frequently requested and was granted permission to render aid to the wounded. Making his way into an open field swept by a withering machine-gun fire, he rendered first aid to many of the wounded. While performing this highly meritorious work he himself was wounded by grenade fire.

CAMPBELL, ROBERT L................
Near Binarville, France, Sept. 27, 1918.
R—Greensboro, N. C.
B—Athens, Ga.
G. O. No. 27, W. D., 1919.

First lieutenant, 368th Infantry, 92d Division.
In the afternoon of Sept. 27, 1918, Lieutenant Campbell saw a runner fall wounded in the middle of a field swept by heavy machine-gun fire. At imminent peril to his own life, and in full view of the enemy, he crossed the field and carried the wounded soldier to shelter.

CAMPBELL, WILLIAM E. (119363)........
Near Blanc Mont, France, Oct. 3–5, 1918.
R—Tuscaloosa, Ala.
B—Mobile, Ala.
G. O. No. 46, W. D., 1919.

Sergeant, 43d Company, 5th Regiment, United States Marine Corps, 2d Division.
On the 3d and 4th of October, while detailed on statistical work, he voluntarily assisted in giving first aid to the wounded. On Oct. 5, when the enemy advanced within 300 yards of the dressing station, he took up a position in the lines helping in defense. Although twice wounded, he remained in action under heavy fire until the enemy had been repulsed.

CAMPBELL, WILLIS M. (2108668)........
Near Brieulles-sur-Meuse, France, Sept. 29, 1918.
R—New Castle, Pa.
B—New Castle, Pa.
G. O. No. 50, W. D., 1919.

Sergeant, Company B, 59th Infantry, 4th Division.
He made his way forward, in the face of annihilating fire, to the aid of a wounded comrade who was lying exposed to this great hazard and carried him across an open field to safety.

CAMPITELLI, DONATO (542921)..........
On Hill 299, France, Oct. 21, 1918.
R—Gibbstown, N. J.
B—Italy.
G. O. No. 39, W. D., 1920.

Corporal, Company L, 7th Infantry, 3d Division.
Shortly after the attack started he discovered a machine gun to his left. He worked his way toward it and without aid captured it with its crew and garrison of 14 men.

CANAVAN, PATRICK (5853)...........
Near St. Etienne-a-Arnes, France, Oct. 3-9, 1918.
R—Campello, Mass.
B—Ireland.
G. O. No. 35, W. D., 1919.

Private, first class, medical detachment, 23d Infantry, 2d Division.
He gave aid to the wounded under severe shell and machine-gun fire and was wounded twice before he left the field. He refused the aid of stretcher bearers and walked alone to the ambulance station.

CANNEY, JAMES H. (322055)...........
In the lower Suchan Valley, Siberia, July 5, 1919.
R—Dorchester, Mass.
B—Charlestown, Mass.
G. O. No. 133, W. D., 1919.

Sergeant, Company D, 31st Infantry.
While in command of a patrol of 3 men, he charged a body of about 50 of the enemy and put them to flight, killing 1 of the enemy.

CANNON, CLARENCE F. (1309393).......
Near Ponchaux, France, Oct. 7, 1918.
R—Lenoir City, Tenn.
B—Concord, Tenn.
G. O. No. 37, W. D., 1919.

Sergeant, Company L, 117th Infantry, 30th Division.
After being severely wounded by machine-gun fire he continued to lead his platoon in attack until he was ordered to the rear by his comanding officer.

CAPEN, RALPH A. (1783758)...........
Near Montfaucon, France, Sept. 29 and Oct. 1, 1918.
R—Mattoon, Ill.
B—Eastport, Me.
G. O. No. 37, W. D., 1919.

First sergeant, Company A, 311th Machine Gun Battalion, 79th Division.
On Sept. 29 Sergeant Capen voluntarily left his place of safety and rescued a wounded comrade and brought him to a place of shelter. On Oct. 1 he volunteered to assist in carrying rations to his company, making repeated trips through heavy artillery fire and each time successfully accomplished his mission.

CAPEZIO, JOHN (38848).............
Near Beaumont, France, Nov. 4, 1918.
R—Newark, N. J.
B—Italy.
G. O. No. 73, W. D., 1919.

Private, Company D, 9th Infantry, 2d Division.
In the face of heavy machine-gun and shell fire Private Capezio led a squad in a flank attack on an enemy machine-gun nest which was holding up his company and endangering the success of its mission. Through his coolness and bravery under fire, the enemy machine guns and the entire crew were captured.

CAPPADOCIA, LOUIS (550861)..........
Near Moulins, France, July 15, 1918.
R—New York, N. Y.
B—Italy.
G. O. No. 19, W. D., 1920.

Sergeant, Company F, 38th Infantry, 3d Division.
Sergeant Cappadocia led his platoon successfully against enemy machine guns that were harrassing the company from the flank. After being wounded in the chin by a machine-gun bullet he refused to be evacuated and went forward with his platoon in the counterattack.

CAPPEL, MARVIN............
Near Medeah Ferme, France, Oct. 3, 1918.
R—Alexandria, La.
B—Evergreen, La.
G. O. No. 21, W. D., 1919.

Captain, Medical Corps, attached to 9th Infantry, 2d Division.
He visited the front lines continually, both night and day, supervising the evacuation of the wounded, personally directing the work of the stretcher bearers, and on several occasions, when the fighting was most severe, he ran forward under intense artillery and machine-gun fire and personally gave first aid and carried in the wounded.

CAPPS, ELIJAH A. (1320859)..........
Near Bellicourt, France, Sept. 29, 1918.
R—Princeton, N. C.
B—Princeton, N. C.
G. O. No. 37, W. D., 1919.

Private, Company G, 120th Infantry, 30th Division.
In the face of heavy machine-gun fire, Private Capps, with 2 other soldiers, attacked and put out of action an enemy machine-gun post, capturing a German officer and 3 soldiers.

CAPWELL, ANDREW W. (110387).......
In the Bois de la Brigade de Marines (Bois-de-Belleau), France, July 19, 1918.
R—Chepachet, R. I.
B—Ashland, R. I.
G. O. No. 87, W. D., 1919.

Private, Company C, 103d Machine Gun Batalion, 26th Division.
Private Capwell made 19 trips from the wood across a field swept by heavy machine-gun fire, carrying to shelter 18 soldiers and 1 officer who had been wounded. He himself was wounded next morning by a shell-shocked soldier whom he was trying to aid.

CARAGEORGE, SOCRATES (106840).......
Near Vierzy, France, July 19, 1918.
R—Philadelphia, Pa.
B—Greece.
G. O. No. 99, W. D., 1918.

Corporal, Company A, 4th Machine Gun Battalion, 2d Division.
He voluntarily left shelter and went 300 yards across an open field through a heavy barrage and procured signal rockets, with which he successfully directed the fire of our artillery.

CARBARY, JAMES (118242)...........
During the attack on Bois-de-Belleau, France, June 12, 1918.
R—Kansas City, Mo.
B—Muscatine, Iowa.
G. O. No. 99, W. D., 1918.

Gunnery sergeant, 47th Company, 5th Regiment, U. S. Marine Corps, 2d Division.
He voluntarily made two trips in the open under terrific fire to within 50 yards of a machine gun and rescued, successfully, two wounded men.

CARBAUGH, CHARLES F. (558049).......
Southeast of Bazoches, France, Aug. 9, 1918.
R—Stephens City, Va.
B—Stephens City, Va.
G. O. No. 21, W. D., 1919.

Private, Company F, 47th Infantry, 4th Division.
He was sent as a runner to direct a platoon of his company to assemble and return to its position. He displayed unusual leadership in performing his mission by himself taking command of the disorganized unit, getting it well in hand, and leading it back under a hostile shelling without losses and without confusion.

*CARDER, CYRIL_____ | Second lieutenant, 16th Infantry, 1st Division.
Near Soissons, France, July 21, 1918. | Having been wounded in the back and arm, Lieutenant Carder refused to be
R—New York, N. Y. | evacuated, but continued to lead his platoon forward in the face of intense
B—Corning, N. Y. | machine-gun fire, repeatedly exposing himself, with total disregard for per-
G. O. No. 89, W. D., 1919. | sonal safety, until he was killed by machine-gun fire.
 | Posthumously awarded. Medal presented to father, Frederick Carder.

CARDWELL, HENRY WARREN (1307522). | Corporal, Company D, 117th Infantry, 30th Division.
Near Molain, France, Oct. 17, 1918. | Having been separated from their company in a smoke barrage, Corporal Card-
R—Normandy, Tenn. | well and Private Carl Lee found themselves face to face with a party of the
B—Madera, Calif. | enemy. Private Lee brought his automatic rifle to his shoulder and at-
G. O. No. 37, W. D., 1919. | tempted to fire, but the gun was jammed and would not shoot. Seeing
 | themselves covered by the gun and not knowing its condition, the Germans
 | threw up their hands, and while Private Lee kept the rifle at his shoulder
 | Corporal Cardwell rounded up the Germans and disarmed them. Their
 | ruse resulted in the capture of 12 of the enemy, comprising three machine-gun
 | crews.

CARDWELL, MARION H_____ | First lieutenant, 58th Infantry, 4th Division.
Near Chevillon, France, July 18, | After his company had failed in an attack on Hill 208 with the loss of 2 officers
1918. | and 65 men, Lieutenant Cardwell reorganized the remaining men of his
R—Dayton, Ohio. | organization and personally led them in a second attack on the same objective.
B—Princeton, Ky. | The advance was made in the face of heavy machine-gun and trench mortar
G. O. No. 10, W. D., 1920. | fire, but due to the example and individual bravery of Lieutenant Cardwell,
 | objective was taken and held.

CAREY, EDDIE (68658)_____ | Corporal, Company H, 103d Infantry, 26th Division.
Near Bois-de-St. Remy, France, Sept. | When his platoon was forced to halt by enemy machine-gunfire, Corporal Carey
12, 1918. | crawled forward with an automatic rifle under machine-gun fire, opened fire
R—Waterville, Me. | on the enemy's position, killed two of the Germans, and captured the gun.
B—Waterville, Me. |
G. O. No. 26, W. D., 1919. |

CARGIN, GEORGE I. (1210383)_____ | Private, Company D, 107th Infantry, 27th Division.
Near Ronssoy, France, Sept. 29, 1918. | During the operations against the Hindenburg line, he, with four other soldiers,
R—Stalker, Pa. | left shelter and went forward into an open field, under heavy shell and
B—Stalker, Pa. | machine-gun fire, and succeeded in bandaging and carrying back to our lines
G. O. No. 20, W. D., 1919. | two wounded comrades.

CARHART, JOSEPH B_____ | Second lieutenant, 5th Regiment, U. S. Marine Corps, 2d Division.
Near Vierzy, France, July 18, 1918. | He displayed exceptional bravery in charging 3 machine guns with a small
R—Weehawken, N. J. | detachment of his men, killing the gun crews and capturing the guns, which
B—Guttenberg, N. J. | were immediately turned on the Germans, thereby opening the line for the
G. O. No. 3, W. D., 1925. | advance of his company, which had been held up by the enemy's fire.

*CARKENER, STUART (656044)_____ | Corporal, Headquarters Company, 76th Field Artillery, 3d Division.
Near Roncheres, France, July 30, | Despite the fierce shelling to which he was subjected, he remained at his forward
1918. | observation post for many hours, until killed by enemy shell fire.
R—Kansas City, Mo. | Posthumously awarded. Medal presented to father, G. S. Carkener.
B—Boulder, Colo. |
G. O. No. 14, W. D., 1925. |

CARLEY, VICTOR A. (2785507)_____ | Private, Company D, 361st Infantry, 91st Division.
Near Gesnes, France, Oct. 3, 1918. | He voluntarily and unhesitatingly left shelter under heavy shell fire and,
R—Seattle, Wash. | without thought of personal danger, rendered first aid and carried a wounded
B—Cincinnati, Ohio. | comrade to a place of safety.
G. O. No. 20, W. D., 1919. |

CARLISLE, ROBERT GUNN_____ | Second lieutenant, 308th Infantry, 77th Division.
Near St. Juvin, France, Oct. 14, 1918. | After his platoon had suffered very heavy casualties Lieutenant Carlisle led a
R—Aberdeen, Miss. | group of eight men on a reconnaissance along the Aire River. Encountering
B—Egypt, Miss. | enemy machine-gun fire, he gallantly led his group in the attack and com-
G. O. No. 3, W. D., 1921. | pletely silenced the enemy fire. Due, in part, to his heroism, his organization
 | was able to cross the Aire River on the following day

CARLISLE, THOMAS W. (1315229)____ | Sergeant, Company D, 119th Infantry, 30th Division.
Near Bellicourt, France, Sept. 29, | He volunteered with two comrades, and went in advance of our lines, under
1918, and near St. Souplet, France, | heavy machine-gun fire, and rescued a wounded soldier. Later, when his
Oct. 12, 1918. | platoon had been reduced to four men, he inspired them by his personal
R—Goldsboro, N. C. | courage to hold their position till reinforcements arrived.
B—Edgecombe County, N. C. |
G. O. No. 81, W. D., 1919. |

CARLSON, CHARLES G. (2067825)____ | Private, Headquarters Company, 129th Infantry, 33d Division.
Near Consenvoye, France, Oct. 17, | Though he was seriously wounded, he succeeded in reaching his destination
1918. | with an important message, his route being under heavy shell fire through
R—Chicago, Ill. | ravines filled with gas.
B—Chicago, Ill. |
G. O. No. 64, W. D., 1919. |

CARLSON, EARNEST A. (1113300)____ | Corporal, Company C, 9th Field Signal Battalion, 5th Division.
In the Bois-des-Rappes, France, Oct. | Although he was almost exhausted from gas and fatigue, he remained on duty
22, 1918. | throughout the day and night, laying telephone lines from the regimental
R—Houston, Tex. | relay station to the front lines through a heavy barrage fire from artillery
B—El Campo, Tex. | and machine guns.
G. O. No. 37, W. D., 1919. |

CARLSON, EMIL A. (283225)------------
  Near Cierges, France, Aug. 1, 1918,
    and near Gesnes, France, Oct. 14,
    1918.
  R—Merrillan, Wis.
  B—Negaunee, Mich.
  G. O. No. 95, W. D., 1919.

Private, first class, Company A, 128th Infantry, 32d Division.
For repeated acts of extraordinary heroism in action near Cierges, France, and near Gesnes, France. During an attack near Cierges, Private Carlson, assisted by 2 comrades, repeatedly exposed himself to enemy machine-gun and artillery fire in order to assist the wounded to the first-aid station. On Oct. 14, in the Meuse-Argonne offensive, when the advance of his company was held up by enemy machine-gun fire, he advanced in attacking a machine-gun nest until all the members of the squad were killed except himself. He then pushed on alone and killed one of the enemy. His rifle became jammed and useless, he continued to advance and succeeded in capturing the two guns, together with the crew, consisting of 8 men.

CARLSON, GUSTAVE H. (71281)---------
  Near Verdun, France, Oct. 15, 1918.
  R—Framingham, Mass.
  B—Framingham, Mass.
  G. O. No. 17, W. D., 1924.

Sergeant, Company B, 104th Infantry, 26th Division.
He showed extraordinary courage and bravery in going beyond our front line under heavy machine-gun fire and bringing back 2 wounded comrades.

CARLSON, SWEN (1388564)--------------
  In the Meuse-Argonne, France, Sept.
    26, 1918.
  R—Chicago, Ill.
  B—Chicago, Ill.
  G. O. No. 71, W. D., 1919.

Private, Company M, 131st Infantry, 33d Division.
On his own initiative, he crawled out with 3 other soldiers across an open field for 200 yards, subject the while to intense artillery and machine-gun fire, to flank 3 machine-gun replacements which were holding up the advance. With his comrades, Private Carlson killed 7 of the enemy and captured 23 prisoners.

CARLSON, WALTER C------------------
  Near Geneve, France, Oct. 9, 1918.
  R—Chicago, Ill.
  B—Chicago, Ill.
  G. O. No. 133, W. D., 1918.

Second lieutenant, 117th Infantry, 30th Division.
Remaining on duty after being wounded in the shoulder by a shell fragment, he aided the advance of his battalion by leading his platoon in flanking attacks on machine-gun nests. He advanced with his men for 400 yards across a field heavily swept by machine-gun fire to a railroad embankment and held the position for more than an hour; reinforcements were prevented from reaching him by the intense fire. By his courage and determination in maintaining this position, he protected the flank of his battalion and made possible its further advance. On Oct. 17, near La Selle River, this officer was knocked down and wounded by a bursting shell; when he recovered he moved forward with his platoon until weakness compelled his evacuation.

CARNAHAN, HARRY F. (1245941)-------
  Near Fismette, France, Aug. 11,
    1918.
  R—Birdville, Pa.
  B—Leechburg, Pa.
  G. O. No. 50, W. D., 1919.

Private, Company M, 111th Infantry, 28th Division.
Although the bridge crossing the Vesle was being heavily shelled and many men were killed in attempting to cross it, Private Carnahan made repeated trips, each time carrying a badly wounded man. He was finally wounded, but refused evacuation, volunteering for duty the following day and providing medical attention for his wounded comrades by venturing across the bridge through the murderous fire.

CARNER, FRANK W. (2110693)----------
  South of Sedan, France, Nov. 7, 1918.
  R—Cleveland, Ohio.
  B—Cleveland, Ohio.
  G. O. No. 68, W. D., 1920.

Corporal, Company E, 16th Infantry, 1st Division.
Although wounded, Corporal Carner refused to be evacuated and continued to lead his squad forward. Finding another squad without a leader, he combined it with his own, led them both to the attack under heavy fire, and succeeded in driving the enemy from a strong position. His action enabled his company to advance to its objective.

CARNEY, THOMAS J. (45109)-----------
  South of Soissons, France, July 18 to
    22, 1918.
  R—Brooklyn, N. Y.
  B—New York, N. Y.
  G. O. No. 55, W. D., 1920.

Corporal, Machine Gun Company, 16th Infantry, 1st Division.
After his platoon commander and platoon sergeant had both been wounded, he took command of the platoon, reorganized it under heavy fire, and successfully led the attack of his platoon on all objectives. This noncommissioned officer had previously performed gallant service while commanding a machine-gun section during a heavy enemy bombardment at Cantigny, France, May 28, 1918.

*CARPENTER, FRANK B. (1680323)-----
  Near Moulin de Charlavaux, France,
    Oct. 5, 1918.
  R—Lockport, N. Y.
  B—Medina, N. Y.
  G. O. No. 44, W. D., 1919.

Corporal, Company C, 307th Infantry, 77th Division.
While advancing with his platoon in the Argonne Forest, Corporal Carpenter located a machine-gun sniper who was directing fire on his squad. Ordering his men to take cover, he drew the sniper's attention to himself by fire from his rifle. His gallant efforts prevented heavy casualties among his squad, but he himself was killed.
Posthumously awarded. Medal presented to mother, Mrs. Mary Carpenter.

*CARPENTER, JAMES B. (558226)-------
  Near Bazoches, France, Aug. 9, 1918.
  R—Avant, Okla.
  B—Cameron, Okla.
  G. O. No. 21, W. D., 1919.

Private, Company H, 47th Infantry, 4th Division.
He responded to a call for volunteers to destroy a hostile machine gun, the approach to which was covered by fire from three other machine guns. With seven other soldiers he went forward and skillfully and boldly accomplished the mission. This courageous soldier has since been killed in action.
Posthumously awarded. Medal presented to sister, Mrs. Lena Woods.

CARPENTIER, GEORGE R--------------
  Near Cheveuges, France, Nov. 7,
    1918.
  R—Washington, D. C.
  B—France.
  G. O. No. 126, W. D., 1919.

First lieutenant, chaplain, 166th Infantry, 42d Division.
Volunteering for the service, he accompanied a patrol as interpreter and later, when our troops encountered stubborn resistance and sustained heavy casualties, he established a dressing station and, under heavy shell fire, administered to the wounded and dying, continuing this service after he himself had been wounded twice.

*CARR, JOHN M. (1306503)------------
  Near Montbrehain, France, Oct. 8,
    1918.
  R—Knoxville, Tenn.
  B—Knoxville, Tenn.
  G. O. No. 37, W. D., 1919.

Sergeant, Machine Gun Company, 117th Infantry, 30th Division.
While leading his section in an assault upon a hostile machine-gun nest Sergeant Carr fell mortally wounded, but he inspired his men by urging them on and giving detailed instructions to the soldier whom he placed in command to succeed himself.
Posthumously awarded. Medal presented to father, Andrew J. Carr.

CARR, WARNER W.
Near Vaux, France, June 30, 1918.
R—Fowler, Ind.
B—Oxford, Ind.
G. O. No. 7, W. D., 1925.

Captain, 9th Infantry, 2d Division.
Preparatory to an attack upon Vaux, and in broad daylight, he voluntarily crawled out into no man's land to a point close to the enemy lines, remaining there several hours, under fire from machine guns and snipers, while he made sketches of the town and its defenses. The information he obtained was of the utmost importance in planning the attack, which was made the following day with marked success.

CARRIER, WILLIAM H.
Near Tuilerie Farm, France, Nov. 5, 1918.
R—Glastonbury, Conn.
B—Glastonbury, Conn.
G. O. No. 37, W. D., 1919.

First lieutenant, 9th Infantry, 2d Division.
While on his way to investigate the situation on the flank of his battalion, he received word that the flank company was hard pressed and without officers. He fearlessly made his way across an open field swept by continuous machine-gun fire, and although four guides were hit and his own clothing pierced, reached the company and saved the situation.

CARRIGAN, ALFRED H., Jr.
Near St. Etienne, France, Oct. 8, 1918.
R—Wichita Falls, Tex.
B—Wichita Falls, Tex.
G. O. No. 20, W. D., 1919.

First lieutenant, 142d Infantry, 36th Division.
After he had led his men through the wire and obtained cover, Lieutenant Carrigan saw one of his men wounded and entangled in the wire. He left cover under heavy fire to bring this man to safety. As he was helping him out of the wire he was shot in the neck by a machine-gun bullet.
Posthumously awarded. Medal presented to father, A. H. Carrigan.

CARROLE, ALICK (42450).
In the Argonne Forest, France, Oct. 9, 1918.
R—Watertown, Mass.
B—Russia.
G. O. No. 46, W. D., 1919.

Private, first class, Company D, 16th Infantry, 1st Division.
While his platoon was being seriously menaced by hostile machine-gun fire Private Carrole, the only survivor of his squad, crawled forward, and with his rifle killed two gunners and captured the remaining six men. His act enabled the further advance of his platoon.

CARROLL, CLARENCE E. (741520).
Near Ampiersbach, France, Sept. 28, 1918.
R—Winnsboro, La.
B—Winnsboro, La.
G. O. No. 32, W. D., 1919.

Corporal, Company E, 52d Infantry, 6th Division.
Although badly wounded, he continued in action against an attempted raid by the enemy until he was blinded by the explosion of a hand grenade. His determination was evidenced by the remark, "I can't see, you give it to them," which he made when passing his rifle to a sergeant of his company.

CARROLL, DANIEL B.
Near Bois-de-Cheppy, France, Sept. 26–28, 1918.
R—Santa Cruz, Calif.
B—Australia.
G. O. No. 39, W. D., 1920.

First lieutenant, 364th Infantry, 91st Division.
Although wounded in the arm in the attack of Sept. 26, Lieutenant Carroll gallantly led his platoon forward, under heavy artillery and machine-gun fire, through the Bois-de-Cheppy. Later, while leading his platoon in an attack near Neuve Grange Farm, he continued on until severely wounded a second time.

CARROLL, GEORGE A. (2152536).
Near St. Juvin, France, Oct. 11, 1918.
R—Rock Island, Ill.
B—Ireland.
G. O. No. 81, W. D., 1919.

Corporal, Machine Gun Company, 326th Infantry, 82d Division.
Seeing an officer lying wounded and unable to return, Corporal Carroll went forth in full view of the enemy and, under terrific machine-gun and shell fire, assisted the officer to safety. He returned to his post and helped in the operation of his gun until ordered to withdraw.

CARROLL, GEORGE C.
Near Fort de Marr, France, Sept. 26, 1918.
R—Garrett, Ind.
B—Garrett, Ind.
G. O. No. 46, W. D., 1919.

First lieutenant, 2d Balloon Squadron, Air Service.
Lieutenant Carroll had ascended in a balloon to a height of 1 kilometer on a réglage mission when he was attacked by enemy planes, but he refused to leave his post, and fired on the planes with his pistol while incendiary bullets were striking his basket and balloon. He was finally forced to jump when his balloon burst into flames, but he reascended as soon as a new balloon could be inflated. On three other occasions he also gave proof of exceptional courage by remaining in his balloon in the face of aeroplane attacks, jumping only when his balloon took fire, and immediately reascending when a new balloon could be inflated.

CARROLL, PATRICK J. (1700228).
Near Bazoches, France, Aug. 15, 1918.
R—New York, N. Y.
B—Ireland.
G. O. No. 81, W. D., 1919.

Corporal, Company F, 306th Infantry, 77th Division.
Corporal Carroll led a patrol of 5 men to the rescue of his company commander, who was lying concealed within 20 yards of an enemy machine-gun nest. He advanced through the intense machine-gun fire to the enemy's position and, although wounded in 9 places, returned to our lines with important information.

CARROLL, ROBERT E. (53831).
Near Soissons, France, July 19, 1918.
R—Cedartown, Ga.
B—Polk County, Ga.
G. O. No. 132, W. D., 1918.

Sergeant, Company G, 26th Infantry, 1st Division.
His platoon having been held up by machine-gun fire from an enemy dugout near Soissons, France, July 19, 1918, he crawled to the door of the dugout, killed the crew, and captured the gun.

CARROLL, THOMAS A. (42882).
Near Cantigny, France, on or about June 3, 1918.
R—Cincinnati, Ohio.
B—St. Louis, Mo.
G. O. No. 129, W. D., 1918.

Corporal, Company F, 16th Infantry, 1st Division.
While a member of a patrol which was rushed by a greatly superior hostile patrol, he opened fire on the enemy at 15 yards, and although severely wounded displayed marked courage in covering the retirement of his patrol.

CARROLL, THOMAS P. (2452643).
Near Grand Pre, France, Oct. 18, 1918.
R—New York, N. Y.
B—Ireland.
G. O. No. 37, W. D., 1919.

Private, first class, Company D, 312th Infantry, 78th Division.
Accompanying a patrol, Private Carroll voluntarily advanced, in the face of annihilating machine-gun fire, in attacking a machine-gun nest which was halting the progress of his platoon. He forced the enemy to retreat and then captured the gun.

CARROLL, TROY C. (1572832)_____
Near Brancourt, France, Oct. 8, 1918.
R—Elkhart, Ind.
B—Anderson, Ind.
G. O. No. 32, W. D., 1919.

Corporal, Company A, 301st Battalion, Tank Corps.
Corporal Carroll was a gunner in a tank which was struck 4 times by shells, which killed or wounded the entire crew. Disregarding his own wounds in the back and leg, he walked 3 miles to secure assistance, guided stretcher bearers back to the disabled tank, and assisted in evacuating the wounded until he was completely exhausted.

CARROLL, WILLIAM M., Jr_____
Near Nantillois, France, Sept. 29, 1918.
R—Rutherford, N. J.
B—Rutherford, N. J.
G. O. No. 37, W. D., 1919.

Captain, 315th Infantry, 79th Division.
Captain Carroll, with a sergeant of his company, outflanked a machine-gun nest which was holding up their advance, shot one German noncommissioned officer who tried to escape, and captured two prisoners, the other occupants fleeing. The reduction of this machine-gun nest made it possible for the flank of the battalion to advance.

*CARSON, BEN C. (3130335)_____
Near Moranville, France, Nov. 9, 1918.
R—Seattle, Wash.
B—Martin, Tenn.
G. O. No. 32, W. D., 1919.

Mechanic, Machine Gun Company, 322d Infantry, 81st Division.
Although suffering acutely from a wound, he continued with the advance, and after setting up his gun, preparing to open fire, he received a second wound, which caused his death.
Posthumously awarded. Medal presented to mother, Mrs. Eva S. Carson.

*CARSON, JOSEPH C. (2806618)_____
Near Les Huit Chemins, France, Sept. 26, 1918.
R—Tulsa, Okla.
B—Canada.
G. O. No. 72, W. D., 1920.

Sergeant, Company D, 358th Infantry, 90th Division.
Sergeant Carson, with two other men, volunteered to rush an enemy machine gun which had concentrated its fire on an opening in some barbed-wire entanglement through which his company was endeavoring to advance. He had successfully silenced its fire when he was killed by the fire of other enemy machine guns.
Posthumously awarded. Medal presented to widow, Mrs. Ruth V. Brown Carson.

CARSON, LESTER (1245729)_____
At Fismette, France, Aug. 11, 1918.
R—Clearfield, Pa.
B—Clearfield, Pa.
G. O. No. 99, W. D., 1918.

Private, Company L, 111th Infantry, 28th Division.
After a runner had been killed trying to carry a message from Fismette to Fismes, Private Carson volunteered and successfully delivered a duplicate message over the same route through heavy artillery and machine-gun fire.

*CARTER, BUCK A. (1316101)_____
Near Bellicourt, France, Sept. 29, 1918.
R—Ingold, N. C.
B—Sampson County, N. C.
G. O. No. 87, W. D., 1919.

Private, Company H, 119th Infantry, 30th Division.
Wounded in the hand, he continued in the advance, operating his Lewis gun effectively. He aided in the capture of two enemy machine-gun posts, inspiring those serving with him by his personal fortitude. He was killed later in performance of duty.
Posthumously awarded. Medal presented to father, Lewis Carter.

*CARTER, CARL C. (540599)_____
In Belleau Wood, France, June 21, 1918.
R—Fresno, Calif.
B—Moundsville, W. Va.
G. O. No. 16, W. D., 1920.

Sergeant, Company A, 7th Infantry, 3d Division.
Sergeant Carter fearlessly led his platoon in an attack until shot down by machine-gun bullet, after which he rose to his feet and urged his men forward and then fell dead.
Posthumously awarded. Medal presented to half brother, Russell K. Price.

CARTER, CARL H. (3510054)_____
Near Rembercourt, France, Oct. 9–Nov. 11, 1918.
R—Claremore, Okla.
B—Amo, Ind.
G. O. No. 133, W. D., 1919.

Private, Company C, 56th Infantry, 7th Division.
As a battalion runner he worked unceasingly, without regard for personal safety, carrying messages night and day under fire from enemy artillery and machine guns.

CARTER, CLARY (551513)_____
Near Mezy, France, July 15, 1918.
R—Naulakla, Va.
B—Caroline County, Va.
G. O. No. 21, W. D., 1925.

Private, Company H, 38th Infantry, 3d Division.
He held a post where the Germans made repeated attempts to cross the River Marne in boats. He continued to fire his automatic rifle into the boatloads of Germans after being wounded.

CARTER, EDWARD J. (1709698)_____
Near Grand Pre, France, Oct. 14, 1918.
R—Binghamton, N. Y.
B—Rutland, Vt.
G. O. No. 35, W. D., 1919.

Sergeant, Company I, 308th Infantry, 77th Division.
When his company was halted by machine-gun fire which threatened to wipe out the entire company, Sergeant Carter led a patrol and charged the nest, and was successful not only in cleaning out the stronghold but in enabling his company to command a more favorable position.

CARTER, ELIOT A._____
Near Bois-de-St. Remy, France, Sept. 12, 1918.
R—Nashua, N. H.
B—West Newton, Mass.
G. O. No. 44, W. D., 1919.

Second lieutenant, 103d Infantry, 26th Division.
Advancing against greatly superior numbers of the enemy, Lieutenant Carter was painfully wounded. He refused evacuation until his objective was reached, and during the combat captured or killed more than 60 of the enemy.

CARTER, FRANK_____
Near St. Juvin, France, Oct. 11, 1918.
R—Atlanta, Ga.
B—Atlanta, Ga.
G. O. No. 81, W. D., 1919.

First lieutenant, 326th Infantry, 82d Division.
Leading two platoons across the Aire River, Lieutenant Carter brought his command to the aid of an assault company which had been stopped by withering enemy fire. As senior officer with the troops who had crossed the river he constantly exposed himself in organizing units and directing fire. He continued in action after being wounded, being evacuated only after he had dropped from exhaustion and after he had rescued three wounded enlisted men and one officer.

CARTER, FRANKLIN W._____
Near Villers-sur-Fere, France, July 28, 1918.
R—Pittsburgh, Pa.
B—Warrenton, Va.
G. O. No. 64, W. D., 1919.

Second lieutenant, 165th Infantry, 42d Division.
When all the men in his platoon had become casualties he operated the one remaining machine gun with the aid of two volunteers from a line company. Even after he himself had been severely wounded, he remained at his post until the Infantry, having effected a crossing of the Ourcq, were firmly established. Refusing assistance to the dressing station, he tried to crawl back, but dropped exhausted.

CARTER, HUGH C. (1286762)_____
Near Verdun, France, Oct. 11–17, 1918.
R—Baltimore, Md.
B—Lent, Va.
G. O. No. 1, W. D., 1919.

Private, first class, sanitary detachment, 115th Infantry, 29th Division.
On Oct. 11 he dressed and treated wounded men on the front line under shellfire continuously for two hours. On Oct. 16 he carried a wounded officer on his back under shellfire to the first-aid station. On Oct. 17 he directed litter bearers to the front line and helped to evacuate the wounded. All during the drive he went back and forth to the dressing station for bandages and medicine for the wounded. Daily during the attacks he would search the woods for wounded men.

CARTER, JAMES (122867)_____
At Bouresches, France, June 6, 1918.
R—South Wilmington, Ill.
B—South Wilmington, Ill.
G. O. No. 119, W. D., 1918.

Private, 96th Company, 6th Regiment, U. S. Marine Corps, 2d Division.
After having been wounded in the capture of Bouresches, France, on June 6, 1918, he displayed remarkable energy and courage in fearlessly attacking superior numbers of the enemy and materially aiding in their defeat.

CARTER, JOHN C. (1312964)_____
Near Vaux-Andigny, France, Oct. 5–17, 1918.
R—Columbia, S. C.
B—Silver, S. C.
G. O. No. 87, W. D., 1919.

Private, medical detachment, 118th Infantry, 30th Division.
Private Carter displayed notable bravery in administering aid to wounded men and carrying them to the aid station under heavy fire. He also assisted in maintaining liaison to the flanks and rear of his company under continuous fire, volunteering and carrying a message under especially hazardous conditions and during an enemy counterattack. During this engagement Private Carter was wounded, but he declined to leave his post until ordered to do so by an officer.

CARTER, MICHAEL (2256223)_____
Near Gesnes, France, Sept. 28, 1918.
R—Los Angeles, Calif.
B—Ireland.
G. O. No. 15, W. D., 1919.

Corporal, Headquarters Company, 361st Infantry, 91st Division.
While attached to the signal section of the attacking battalion, he repeatedly spliced telephone wires in the midst of heavy artillery and machine-gun fire during the attack, displaying at all times exceptional coolness and personal bravery and aiding materially in maintaining communication between battalion and regimental command posts.

CARTER, PAUL D._____
During the Argonne-Meuse offensive, France, Oct. 9, 1918.
R—Knoxville, Tenn.
B—Knoxville, Tenn.
G. O. No. 37, W. D., 1919.

Second lieutenant, 28th Infantry, 1st Division.
Rendered helpless by the explosion of a gas shell, he refused evacuation, but after regaining consciousness returned to his command and was instrumental in repulsing a strong enemy counterattack.

CARTER, ROBERT G._____
Near Thiaucourt, France, Sept. 12–14, 1918.
R—Chevy Chase, Md.
B—Washington, D. C.
G. O. No. 46, W. D., 1919.

Second lieutenant, 6th Infantry, 5th Division.
Commanding a platoon of Stokes mortars, he showed marked bravery and leadership, capturing many prisoners and directing the organization of captured positions with utter disregard of his personal danger. On Sept. 14 he directed the operations of his mortars under an intense artillery and machine-gun fire, until the last one was smashed by shell fire, and having been twice wounded and unable to continue forward, he called his section leaders and ordered them to continue the advance with the Infantry before he would allow himself to be taken to the rear.

*CARTER, THOMAS E._____
Near Tuilerie Farm, France, Nov. 4, 1918.
R—Andover, Mass.
B—Pembroke, Me.
G. O. No. 37, W. D., 1919.

Second lieutenant, 9th Infantry, 2d Division.
Lieutenant Carter showed extraordinary heroism when he led his company through intense machine-gun and artillery fire against superior numbers. He made an important gain, but was killed by machine-gun fire during the engagement.
Posthumously awarded. Medal presented to mother, Mrs. George M. Carter.

CARTER, WILLIAM C. (2418011)_____
Near Grand Pre, France, Nov. 1, 1918.
R—Auburn, Ill.
B—Logan County, Ky.
G. O. No. 44, W. D., 1919.

Private, Company F, 311th Infantry, 78th Division.
While the advance of his company was being held up by hostile machine-gun fire, he worked his way around the enemy's flank, and although exposed to sniper fire he charged the nest and by the effective use of his Chaucat rifle captured that nest and the one on the right. His action made possible the further advance of his company and the capture of 47 more prisoners.

CARTON, CHARLES A. (128334)_____
At Givry, France, Oct. 15, 1918.
R—Canada.
B—Ireland.
G. O. No. 13, W. D., 1919.

Corporal, Headquarters Company, 12th Field Artillery, 2d Division.
He displayed conspicuous courage and gallantry under fire by organizing a detachment, leading it out in front of our line under heavy fire, and bringing back the body of an officer who had been killed by a sniper.

CARTONA, CHARLES (65624)_____
Near Verdun, France, Oct. 26, 1918.
R—Terryville, Conn.
B—Russia.
G. O. No. 64, W. D., 1919.

Private, Company I, 102d Infantry, 26th Division.
During the operations in the Belleu Bois, he went forward at the risk of personal danger, and succeeded in rescuing a wounded comrade and brought him back to safety, being subject the whole journey to intense and deadly artillery and machine-gun fire.

CARTY, JAMES F. (65866)_____
Near Bois-de-St. Remy, France, Sept. 12, 1918.
R—Wallingford, Conn.
B—Rockville Center, N. Y.
G. O. No. 26, W. D., 1919.

Private, Company K, 102d Infantry, 26th Division.
When Infantry advance had been held up by machine-gun fire, the strength of which could not be determined, Private Carty and another soldier scouted far beyond their lines into enemy territory, and, after cutting telephone cables, crept up on the nests from the rear. The entire personnel of the guns, consisting of 1 officer and 39 men, was taken prisoner by Private Carty, after which he marched them into our lines.

CARVER, JOHN (1316155)................
Near Bellicourt, France, Sept. 29, 1918.
R—Maggie, N. C.
B—Maggie, N. C.
G. O. No. 81, W. D., 1919.

Corporal, Company H, 119th Infantry, 30th Division.
With another soldier, he attacked and demolished two enemy machine-gun posts 200 yards in advance of our lines. He then stood guard at the entrance of a dugout while the other soldier entered it and brought out 75 German soldiers and 3 officers, who were taken back to the line as prisoners.

CARVER, PAUL M................
Near Soissons, France, July 18, 1918.
R—Dexter, Me.
B—Wolfeboro, N. H.
G. O. No. 78, W. D., 1919.

Second lieutenant, 16th Infantry, 1st Division.
Lieutenant Carver displayed exceptional qualities of courage and leadership in conducting his platoon through heavy shell and machine-gun fire to its objective. He was later severely wounded while rallying his men and consolidating the new position.

CARVO, JOSEPH H. (2268262)................
Near Sergy, France, July 29–30, 1918.
R—Yakima, Wash.
B—Crookston, Minn.
G. O. No. 71, W. D., 1919.

Private, first class, Company I, 47th Infantry, 4th Division.
Acting as runner, he carried messages repeatedly over open ground swept by terrific machine-gun fire, aiding materially in the maintenance of liaison.

CASAGA, SAMUEL E. (1390334)................
At St. Maurice, France, Nov. 4, 1918.
R—Chicago, Ill.
B—Chicago, Ill.
G. O. No. 59, W. D., 1919.

Sergeant, Company A, 132d Infantry, 33d Division.
He was a member of a patrol which was stopped on the edge of a wood by enemy machine-gun fire. While his comrades returned the fire he crawled to the flank of the enemy's position, disregarding the machine-gun fire, and, single handed, captured a prisoner, whom he brought back to our lines.

CASE, ARCHIE B. (1216127)................
East of Ronssoy, France, Sept. 29, 1918.
R—Rochester, N. Y.
B—Rochester, N. Y.
G. O. No. 20, W. D., 1919.

Private, sanitary detachment, 108th Infantry, 27th Division.
During the operations against the Hindenburg line he repeatedly left shelter and went forward into the open, under heavy shell and machine-gun fire, and succeeded in bandaging and carrying back to our lines many wounded soldiers.

CASERTA, VINCENZO (1383927)................
Near Marcheville, France, Nov. 10, 1918.
R—Rend, Ill.
B—Italy.
G. O. No. 23, W. D., 1919.

Corporal, Company F, 130th Infantry, 33d Division.
Although wounded early in the attack, he continued to lead his squad to its objective and refused to leave the field until ordered to do so by an officer.

CASEY, CHARLES J................
In the Salient-du-Feys, France, Mar. 9, 1918.
R—Red Oak, Iowa.
B—Elmira, N. Y.
G. O. No. 126, W. D., 1919.

Captain, 168th Infantry, 42d Division.
He displayed notable gallantry on Mar. 9, 1918, in leading a command of untried men in company with French troops in a successful raid on enemy trenches in the Salient-du-Feys, France. By his heroic conduct he inspired both his own men and the men of our ally participating in the operation.

*CASEY, GEORGE A. (1697879)................
Near Barricade Pavillion, France, Sept. 27, 1918.
R—Cold Spring, N. Y.
B—Cold Spring, N. Y.
G. O. No. 89, W. D., 1919.

Sergeant, Company G, 305th Infantry, 77th Division.
Although he had been mortally wounded by enemy shellfire, which caused heavy casualties in his platoon, sergeant Casey reorganized the platoon and directed the placing of outposts so that the position could be held, refusing to have his own wounds dressed until the other wounded men were evacuated. Posthumously awarded. Medal presented to mother, Mrs. Margaret Casey.

CASEY, JOHN................
At Chateau-Thierry, France, June 6, 1918.
R—West Lynn, Mass.
B—Ireland.
G. O. No. 110, W. D., 1918.

Sergeant, 49th Company, 6th Regiment, U. S. Marine Corps, 2d Division.
Although wounded during a counterattack he remained with his group, refusing to go to the rear or to accept medical attention until assured that the enemy had retired and that his men were properly dug in.

CASEY, JOHN L. (65625)................
Near Chateau-Thierry, France, July 23, 1918.
R—Fairfield, Conn.
B—Yonkers, N. Y.
G. O. No. 125, W. D., 1918.

Corporal, Company I, 102d Infantry, 26th Division.
After seeing 3 runners shot down while trying to cross an open field through violent machine-gun fire, to establish liaison with a regiment 500 yards away on his left, Corporal Casey undertook the same mission and successfully accomplished it.

CASEY, JOSEPH W. (60820)................
East of Epieds, France, July 23, 1918.
R—Lawrence, Mass.
B—Canada.
G. O. No. 30, W. D., 1921.

Sergeant, Company F, 101st Infantry, 26th Division.
Leading his platoon under a heavy machine-gun barrage and through the fire of snipers from trees, he attacked 2 German machine-gun nests, captured their guns, and killed their crews. Sergeant Casey then saw 3 Germans crawling toward his men to open fire on them. He dashed forward, attacked them single-handed, and killed them all.
Oak-leaf cluster.
This man was awarded an oak-leaf cluster for extraordinary heroism at Belleu Bois, north of Verdun, France, Oct. 25, 1918. While leading his patrol in advance of our lines Sergeant Casey, encountering an enemy machine-gun nest, rushed in advance of his patrol and captured the position.

CASSADY, THOMAS G................
Near Fismes, May 29, 1918, and near Epieds, France, June 5, 1918.
R—Spencer, Ind.
B—Freedom, Ind.
G. O. No. 138, W. D., 1918.

First lieutenant, 28th Aero Squadron, Air Service.
On May 29, 1918, he, single-handed, attacked an Lvg German plane, which crashed near Fismes. On June 5, 1918, as patrol leader of 5 Spads, while being attacked by 12 German Fokkers, he brought down 1 of the enemy planes near Epieds, and by his dash and courage broke the enemy formation.
Oak-leaf cluster.
A bronze oak leaf to be worn on the distinguished-service cross is awarded for the following act of extraordinary heroism in action: On Aug. 15, 1918, near St. Maire, while acting as protection for a Salmson he was attacked by 7 Fokkers, 2 of which he brought down and enabled the Salmson to accomplish its mission and return safely.

CASSELO, ANGELO (732149)------------
Near Fontaine, France, Nov. 8, 1918.
R—West Paterson, N. J.
B—Italy.
G. O. No. 37, W. D., 1919.

Private, Company D, 6th Infantry, 5th Division.
While engaged as scout he put to flight an enemy machine-gun crew unaided and maintained his distance ahead of his company in the face of flanking machine-gun fire which had prevented others from gaining ground.

CASSIDY, EUGENE B. (1241866)--------
Near the Vesle River, France, Aug. 26, 1918.
R—New Derry, Pa.
B—Morell, Pa.
G. O. No. 98, W. D., 1919.

Sergeant, Company M, 110th Infantry, 28th Division.
With utter disregard for his own safety, Sergeant Cassidy voluntarily left shelter and, going 100 yards in advance of the line under heavy machine-gun and shell fire, rescued a wounded officer.

CASSIDY, HENRY K.------------
Near the Oureq River, France, July 28, 1918.
R—Wichita, Kans.
B—Fort Scott, Kans.
G. O. No. 81, W. D., 1919.

First lieutenant, 165th Infantry, 42d Division.
After his battalion commander had been killed and he himself so severely wounded that he was unable to walk without assistance, Lieutenant Cassidy (then battalion adjutant) remained on duty for 3 days, despite the fact that he had been ordered to the rear, and assisted the new battalion commander in re-forming the battalion. His remarkable fortitude and courage furnished an inspiration to the members of the battalion and aided materially in the attack.

CASSIDY, JOSEPH D. (1168061)----------
Near Romagne, France, Nov. 1, 1918.
R—Pekin, Ill.
B—Argyle, Ill.
G. O. No. 37, W. D., 1919.

Sergeant, Battery C, 124th Field Artillery, 33d Division.
After 3 members of his gun crew had been wounded during heavy enemy shell fire he alone continued to keep his gun in action. Later, after reorganizing his section, he administered first-aid treatment to the wounded men.

⁴CASSIDY, JOSEPH J. (1282963) ---------
In the Bois d'Etrayes, near Verdun, France, Oct. 23, 1918.
R—Princeton, N. J.
B—Ireland.
G. O. No. 3, W. D., 1919.

Private, first class, Company C, 111th Machine Gun Battalion, 29th Division.
In spite of being very ill and near exhaustion, with his gun crew under almost continuous shell fire for more than 24 hours, he remained, directing and encouraging the men of his depleted squad. He was killed by shell fire while he was faithfully engaged in keeping his machine gun in action.
Posthumously awarded. Medal presented to father, Joseph J. Cassidy.

CASSINGHAM, LEROY (2809082)----------
Near Fey-en-Haye, France, Sept. 12, 1918.
R—Osage, Okla.
B—Bosworth, Mo.
G. O. No. 81, W. D., 1919.

Private, Company M, 358th Infantry, 90th Division.
During the attack of his company between Fey-en-Haye and Vilcey on Sept. 12 Private Cassingham became separated from his company but continued to advance alone. He entered an enemy dugout and, single-handed, captured 13 prisoners, including 1 German major.

CASTLEMAN, JAMES P.------------
At Columbus, N. Mex., Mar. 9, 1916.
R—Louisville, Ky.
B—New Castle, Ky.
G. O. No. 2, W. D., 1920.

First lieutenant, 13th Cavalry, U. S. Army.
On Mar. 9, 1916, during the attack on the town of Columbus, N. Mex., by Mexicans under Villa, Lieutenant Castleman, by his gallant conduct and the successful and effective disposition of his troops in defense, saved the lives of many civilians and prevented the loss of much property in the town.

CASTLEMAN, JOHN R.------------
Near Romagne, France, Oct. 5, 1918.
R—Berryville, Va.
B—Berryville, Va.
G. O. No. 37, W. D., 1919.

First lieutenant, 99th Aero Squadron, Air Service.
In spite of being attacked by 7 enemy planes (type Fokker) and later by 5 (type Pfalz), Lieutenant Castleman successfully accomplished a photographic mission 6 kilometers behind the German lines, without protection, and also destroyed 2 of the enemy planes.

CASTLEMAN, LAWRENCE A. (2940920)--
Near Grand Pre, France, Oct. 25, 1918.
R—Pawnee, Ill.
B—Rolla, Mo.
G. O. No. 37, W. D., 1919.

Private, first class, Company K, 311th Infantry, 78th Division.
Despite the fact that the remainder of his company had been forced to evacuate a height, Private Castleman remained at his post and by exceptional handling of his automatic rifle stopped an enemy advance and made it possible for his company to again occupy the position.

CASTURA, MICHAEL (261828)----------
At Cierges and Mont St. Martin, northeast of Chateau-Thierry, France, Aug. 1–3, 1918.
R—Hazelton, Pa.
B—Eckley, Pa.
G. O. No. 20, W. D., 1919.

Sergeant, Company C, 125th Infantry, 32d Division.
For repeated acts of extraordinary heroism in action at Cierges and Mont St. Martin, northeast of Chateau-Thierry, France. During the attack on Cierges Sergeant Castura took command of his platoon when the commanding officer had been evacuated and led it successfully through the barrage to its objective. On Aug. 3, when 1 of his men had been left wounded on the field and no first-aid men were present, this soldier dashed through a terrific barrage and carried the wounded man to shelter.

CASWELL, GEORGE D. (1224730)-------
Near Ronssoy, France, Sept. 27, 1918.
R—Troy, N. Y.
B—Troy, N. Y.
G. O. No. 32, W. D., 1919.

Corporal, Company M, 105th Infantry, 27th Division.
He braved the perils of exacting machine-gun fire when he ventured out to rescue a wounded comrade. He completed this mission and returned for another comrade, who was lying wounded still farther forward. He also successfully brought this man to safety. That evening he led a detail through the murderous fire to replenish the supply of hand grenades. While returning to the lines he was seriously wounded when the box which he was carrying exploded.

CATALANO, SOLOMON (1697260)--------
Near Bazoches, France, Sept. 2, 1918.
R—New York, N. Y.
B—Italy.
G. O. No. 35, W. D., 1920.

Corporal, Company C, 305th Infantry, 77th Division.
As a member of a small patrol, Corporal Catalano crossed the Vesle River to reconnoiter the enemy's position. The patrol, having accomplished its mission, was attacked by the enemy and subjected to heavy machine-gun fire. In covering the withdrawal Corporal Catalano attacked, single-handed, and drove off 6 of the enemy. By his deed he enabled the others to return to our lines with the valuable information obtained. In the encounter he was severely wounded in the throat.

CATES, CLIFTON B ....
Near Chateau-Thierry, France,
June, 6, 1918.
R—Tiptonville, Tenn.
B—Tiptonville, Tenn.
G. O. No. 81, W. D., 1919.

First lieutenant, 6th Regiment, U. S. Marine Corps, 2d Division.
While advancing with his company on the town of Bouresches their progress
was greatly hindered by withering machine-gun and artillery fire of the enemy, which caused many casualties, one of whom was his commanding
officer. Taking command, Lieutenant Cates led them on to the objective,
despite the fact that he was rendered temporarily unconscious by a bullet
striking his helmet and that this was his first engagement. Exposing himself
to the extreme hazard, he reorganized his position with but a handful of men.
Oak-leaf cluster.
For the following act of extraordinary heroism in action near Bois-de-Belleau,
France, June 13-14, 1918, Lieutenant Cates is awarded an oak-leaf cluster,
to be worn with the distinguished-service cross: During the night a severe
gas attack made it necessary to evacuate practically the entire personnel of 2
companies, including officers. Lieutenant Cates, suffering painfully from
wounds, refused evacuation, remaining and rendering valuable assistance to
another company.

CATHCART, JAMES O ....
Near Gesnes, France, Oct. 14-19,
1918.
R—Detroit, Mich.
B—Canada.
G. O. No. 71, W. D., 1919.

Major, 126th Infantry, 32d Division.
Finding 2 companies of another battalion badly disorganized, he effected a
reorganization and ordered them to advance, personally assuming command
of these troops, of his own battalion, which had been in support, and of another
battalion. He conducted the successful attack on Cote Dame Marie, exhibiting extraordinary initiative and bravery and showing utter disregard of his
own personal danger.

CATHCART, WILBUR (1211949) ....
Near Lake Dickebusch, Belgium,
Aug. 22, 1918.
R—Elmhurst, N. Y.
B—Baltimore, Md.
G. O. No. 9, W. D., 1923.

Private, first class, Company L, 107th Infantry, 27th Division.
All wire communications between Company L and battalion headquarters
having been destroyed by shell fire, Private Cathcart voluntarily and with
utter disregard for his personal safety carried an important message to his
battalion command through terrific machine-gun and artillery fire and under
observation of the enemy. His hazardous mission successfully accomplished,
he returned through terrific machine-gun fire to his company, his conduct
having been a splendid example of bravery and devotion to the men of his
company.

*CATHER, GROSVENOR P ....
Near Cantigny, France, May 27,
1918.
R—Bladen, Nebr.
B—Webster County, Nebr.
G. O. No. 44, W. D., 1919.

Second lieutenant, 26th Infantry, 1st Division.
During a strong enemy attack Lieutenant Cather mounted the parapet of his
trench, and, although exposed to withering machine-gun fire, he so skillfully
directed the fire of his automatic rifles that the attack was repulsed. In this
action he fell mortally wounded.
Posthumously awarded. Medal presented to widow, Mrs. Grosvenor P. Cather.

CATTUS, JOHN C ....
Near Bois-de-Cunel, France, Oct.
10-11, 1918.
R—New York, N. Y.
B—New York, N. Y.
G. O. No. 32, W. D., 1919.

First lieutenant, 30th Infantry, 3d Division.
Although severely wounded while advancing with his company and assisting
in taking enemy trenches, he returned to the post of command through heavy
machine-gun and shell fire with valuable information of the troops in the line.

CAULDER, LAWRENCE E. (1311769) ......
Near Brancourt, France, Oct. 8, 1918.
R—Chesterfield, S. C.
B—Cheraw, S. C.
G. O. No. 78, W. D., 1919.

Private, Company I, 118th Infantry, 30th Division.
With another soldier, Private Caulder crawled through intense artillery and
machine-gun fire, 50 yards in advance of their platoon, for the purpose of
sniping the enemy machine gunners who were holding up the platoon. His
companion was killed, but Private Caulder remained at his post and kept
up an effective rifle fire on the enemy nest until the tanks came up and
destroyed it.

*CAUSLAND, HARRY L. (3133469) ....
Near Bantheville, France, Oct. 24,
1918.
R—Anacortes, Wash.
B—Quemes Island, Wash.
G. O. No. 44, W. D., 1919.

Private, Company I, 357th Infantry, 90th Division.
He was acting as ammunition carrier for a machine gun which was supporting
a raid on the enemy by the Infantry. While this gun was being operated
from a shell hole its ammunition became exhausted, and the corporal called
back to the carriers, who were in shell holes 25 yards behind, to bring up more
ammunition. Though the gun position was under heavy enemy fire, Private
Causland called out, "I'll take it," and rushed forward with two boxes of
ammunition. Just as he reached the emplacement with the ammunition
he was killed instantly by a machine-gun bullet.
Posthumously awarded. Medal presented to father, Frank Causland.

CAVANAUGH, JAMES A. (1202061) ......
Near Mount Kemmel, Belgium,
Aug. 29, 1918.
R—New York, N. Y.
B—Pittston, Pa.
G. O. No. 23, W. D., 1919.

Corporal, Company D, 102d Engineers, 27th Division.
After several runners, sent back through a heavy barrage for reinforcements
and ammunition, had failed to return, Corporal Cavanaugh, who was on
duty with the Infantry, volunteered for this mission and successfully accomplished it.

CAVANAUGH, LEONARD (48254) ....
Near Exermont, France, Oct. 4,
1918.
R—West Duluth, Minn.
B—Stillwater, Minn.
G. O. No. 53, W. D., 1920.

Corporal, Company L, 18th Infantry, 1st Division.
After all the officers of the company had become casualties, Corporal Cavanaugh
took command of and reorganized the company under heavy fire and led it
in the attack on Hill 240. It was due to his courage and initiative that his
company reached all objectives.

CAVANAUGH, THOMAS J. (1243911) ......
At Fismette, France, Aug. 11-12,
1918.
R—Pittsburgh, Pa.
B—Ireland.
G. O. No. 116, W. D., 1918.

Sergeant, Company D, 111th Infantry, 28th Division.
After he had been wounded by shrapnel he refused to go to the rear, but directed
the operations of his platoon in resisting enemy attacks for an hour and a
half, when he finally collapsed. The following day he returned and unhesitatingly selected a position of great danger to himself in order to direct
machine-gun fire upon enemy snipers. By his courageous exposure he thus
made possible the evacuation of 25 wounded men across an exposed area and
over the Vesle River.

CAVENAUGH, HARRY LaTOUR_____
In the Argonne-Meuse offensive, France, Sept. 26–Oct. 5, 1918.
R—Guthrie, Okla.
B—Fort Douglas, Utah.
G. O. No. 64, W. D., 1919.

Colonel, 363d Infantry, 91st Division.
During the entire operations of the Argonne-Meuse offensive Colonel Cavenaugh personally commanded his regiment after all the officers of his staff had been evacuated because of sickness or wounds. He continually exposed himself to terrific enemy fire while leading and organizing broken units and making reconnaissances of the front lines. On Sept. 26 he personally led his command out of Cheppy Woods and thereafter kept his post of command in close contact with his fighting units.

CAVENEE, CLAUDE E. (2178440)_____
Near Remonville, France, Nov. 1, 1918.
R—Monroe City, Mo.
B—Rice County, Kans.
G. O. No. 95, W. D., 1919.

Sergeant, Company B, 354th Infantry, 89th Division.
After his platoon commander had been wounded Sergeant Cavenee took command of his platoon, which was under heavy machine-gun and artillery fire, and by the force of his own example of bravery led his men in an attack, successfully overcoming the enemy's resistance in spite of unfavorable odds.

CAYER, ALBERT J. (2315048)_____
Near Mezy, France, July 15, 1918.
R—Brooklyn, N. Y.
B—Superior, Wis.
G. O. No. 23, W. D., 1919.

Private, Company B, 38th Infantry, 3d Division.
During the intense enemy artillery preparation just prior to the German offensive of July 15, 1918, he voluntarily made several trips through the heaviest shelling to bring wounded comrades from the field.

CAYWOOD, HUGH T. (1467557)_____
Near Cheppy, France, Sept. 26, 1918.
R—Eureka, Kans.
B—Eureka, Kans.
G. O. No. 59, W. D., 1919.

Sergeant, first class, Company A, 110th Engineers, 35th Division.
While a member of a platoon of wire cutters he, with the assistance of a comrade, attacked and captured an enemy machine-gun nest that was holding up the advance. One officer, six men, and two guns were taken in the face of intense machine-gun fire.

CECILIA, LOUIS (1391098)_____
Near Consenvoye, France, Oct. 8, 1918.
R—Chicago, Ill.
B—Italy.
G. O. No. 37, W. D., 1919.

Private, Company G, 132d Infantry, 33d Division.
While his company was being held up by machine-gun fire, he crawled to a point within 10 yards of the nest and bombed out the enemy, where they came under fire from our guns and were killed. During the exploit Private Cecilia was wounded by enemy bombs.

CELLAR, CHESTER M. (1213529)_____
East of Ronssoy, France, Sept. 29, 1918.
R—Rochester, N. Y.
B—Waverly, Kans.
G. O. No. 20, W. D., 1919.

Corporal, Company A, 108th Infantry, 27th Division.
After the other members of his automatic-rifle squad had been killed or wounded in an assault against an enemy machine-gun nest, Corporal Cellar operated his gun alone, holding the fire of the machine gun until reenforcements arrived and put it out of action. His great courage and gallantry set an inspiring example to all his comrades.

CERBIN, STANLEY F. (47469)_____
Near Exermont, France, Oct. 4 and 5, 1918.
R—Chicago, Ill.
B—Chicago, Ill.
G. O. No. 39, W. D., 1920.

Sergeant, Company H, 18th Infantry, 1st Division.
On Oct. 4, 1918, Sergeant Cerbin led a small group through heavy fire and captured an enemy machine-gun nest with its crew of 6 men. The following day he again assisted in the capture of a machine-gun nest. Sergeant Cerbin later, after all officers had become casualties, took command of the company and led it forward to its objective.

CEPAGLIA, PHILIP (1708116)_____
Near Binarville, France, Oct. 2–8, 1918.
R—New York, N. Y.
B—Italy.
G. O. No. 20, W. D., 1919.

Private, Company C, 308th Infantry, 77th Division.
Private Cepaglia was on duty as a battalion runner during the period of 6 days in which his own and another battalion were surrounded by the enemy in the Argonne Forest, France, and cut off from communication with friendly troops. Although he was without food and toward the end of the period almost exhausted, this soldier carried messages to all parts of the position. Constantly under heavy fire from machine guns and trench mortars, he showed an utter disregard for his own personal safety.

CHADWICK, HARRY R._____
Near Bois-du-Fays, France, Oct. 11, 1918.
R—Chicago, Ill.
B—Hamilton, Ohio.
G. O. No. 71, W. D., 1919.

Captain, 132d Infantry, 33d Division.
Although wounded while placing his machine guns in position preparatory to an attack, he remained on duty for several hours, constantly exposing himself to enemy fire as he moved along the front line to encourage his men. He remained on duty until exhausted from loss of blood

CHAFIN, MARILE (542435)_____
Near Fossoy, France, July 15, 1918.
R—Williamson, W. Va.
B—Williamson, W. Va.
G. O. No. 44, W. D., 1919.

Corporal, Company I, 7th Infantry, 3d Division.
While delivering a message, he was severely wounded by the explosion of a shell, but continued with his mission and returned with an answer before reporting for treatment.

CHAMBERLAIN, ISRAEL J. (369047)_____
In Bois Bossois, France, Oct. 9, 1918.
R—Hartland, Vt.
B—Plattsburg, N. Y.
G. O. No. 44, W. D., 1919.

Private, first class, Company B, 116th Infantry, 29th Division.
Private Chamberlain went through an open country under heavy machine-gun fire to ascertain whether friendly troops were ahead of his regiment, after unknown soldiers had been observed; he was urged by the French troops on the flank not to make the return trip, as certain death seemed sure to be the outcome, but without hesitation, returned with information which resulted in the wounding of 1 of the enemy, the killing of 2, and the capture of 37, including 1 officer.

CHAMBERLAIN, MAX C. (56599)_____
South of Soissons, France, July 19, 1918.
R—Provo, Utah.
B—Salt Lake City, Utah.
G. O. No. 39, W. D., 1920.

Private, first class, Company A, 28th Infantry, 1st Division.
Although twice wounded by machine-gun fire, he continued on in the assault wave and assisted in driving the enemy from their positions. The courage of this noncommissioned officer was a material factor in the capture of a number of prisoners.

CHAMBERLAIN, WARD B.
Near Merval, France, Sept. 15, 1918.
R—New York, N. Y.
B—New York, N. Y.
G. O. No. 46, W. D., 1919.

First lieutenant, 307th Infantry, 77th Division.
While leading his company in attack through terrific shell fire Lieutenant Chamberlin was severely wounded in the right hand, this wound rendering his entire right arm useless. Despite his weakness from loss of blood, he refused to be evacuated until forced to do so.

CHAMBERS, CHARLES C.
In Bois-de-Septsarges, near Montfaucon, France, Sept. 26, 1918.
R—Cleveland, Ohio.
B—Galena, Ill.
G. O. No. 49, W. D., 1922.

Major, 135th Machine Gun Battalion, 37th Division.
While voluntarily going forward on a mission of establishing liaison between a front-line unit of his own division and the division on the right, Major Chambers encountered a large number of men falling back in confusion, badly disorganized and without leaders, as a result of a heavy artillery fire and machine-gun fire from pill boxes in the woods and from a strong point on the heights beyond. With the greatest energy, courage, and leadership, at a most critical time and under a heavy fire, he reorganized the scattered troops, put them in trenches, and later led them forward, overcoming a stubborn resistance from machine guns, drove the enemy from his position, reestablished the front line and accomplished his liaison mission. By his calmness, decision, and courage he inspired great confidence among the scattered and confused troops.

CHAMBERS, REED M.
Over the region of Epinonville, France, Sept. 29, 1918.
R—Memphis, Tenn.
B—Onaga, Kans.
G. O. No. 14, W. D., 1920.

First lieutenant, 94th Aero Squadron, Air Service.
While on a mission, Lieutenant Chambers, accompanied by another machine piloted by First Lieut. Samuel Kaye, jr., encountered a formation of six enemy machines (Fokker type) at an altitude of 3,000 feet. Despite numerical superiority of the enemy, Lieutenant Chambers and Lieutenant Kaye immediately attacked and succeeded in destroying one and forced the remaining five to retreat into their own lines.
Oak-leaf clusters (3).
A bronze oak-leaf cluster was awarded to Lieutenant Chambers for each of the following acts of extraordinary heroism in action: First oak-leaf cluster: Near Montfaucon and Vilosnes-sur-Meuse, France, Oct. 2, 1918, Lieutenant Chambers, while on a mission, at an altitude of 2,000 feet, encountered an enemy two-seater (Halberstadt type). He immediately attacked and, after a brief combat, succeeded in shooting it down. Second oak-leaf cluster: Near Montfaucon and Vilosnes-sur-Meuse, France, Oct. 2, 1918, at 7.40 o'clock, Lieutenant Chambers saw four enemy machines (Fokker type) attacking another American machine (Spad type). He immediately went to its rescue, and, after a few minutes of fierce combat, he succeeded in shooting down one. Third oak-leaf cluster: Near the Bois-de-la-Cote-Lemonte, France, Oct. 22, 1918, while on a voluntary patrol, Captain Chambers encountered five enemy planes (Fokker type) harassing our infantry at an altitude of 300 meters. Attacking them without hesitation, he shot down two of them and drove the others off.

CHAMPENY, ARTHUR S.
Near St. Mihiel, France, Sept. 12, 1918.
R—Lyons, Kans.
B—Briggsville, Wis.
G. O. No. 37, W. D., 1919.

First lieutenant, 356th Infantry, 89th Division.
Assisting the battalion commander, who had been severely wounded in the early fighting, he maintained the liaison personnel, making many journeys himself through heavy shelling. When the battalion commander had been evacuated, he assumed command, and moved the battalion to its new position.

CHAMPION, HERBERT O. (1328377).
At Proven, Belgium, July 16, 1918.
R—Mooresboro, N. C.
B—Cleveland County, N. C.
G. O. No. 145, W. D., 1918.

Private, first class, sanitary detachment, 105th Engineers, 30th Division.
When an enemy airplane dropped a bomb in the camp of his organization, killing one soldier and wounding seven, including himself, he administered first aid to the other wounded, helped carry them to the dressing station, and there gave further assistance in dressing and evacuating the wounded men, never mentioning his own serious injuries until he knew that all the others had been cared for.

*CHANDLER, HENRY E.
Near Thiaucourt, France, Sept. 15, 1918.
R—Ruston, La.
B—Ruston, La.
G. O. No. 37, W. D., 1919.

First lieutenant, 6th Regiment, U. S. Marine Corps, 2d Division.
He fearlessly exposed himself to severe artillery and machine-gun fire and located machine-gun nests and sniper posts harassing his company and hindering its advance. He then led his platoon forward in the face of heavy fire and destroyed the nests.
Posthumously awarded. Medal presented to widow, Mrs. Irene M. Chandler.

CHANDLER, ISAAC (263881).
Near Cierges, France, July 31, 1918.
R—Menominee, Mich.
B—Menominee, Mich.
G. O. No. 46, W. D., 1919.

Corporal, Company L, 125th Infantry, 32d Division.
While his company was leading in a battalion attack, Corporal Chandler persisted in carrying messages to the battalion commander and carrying rations to the line under intensive machine-gun fire from the front and flanks, after he had been very severely wounded.

CHANEY, EDWARD (1817402).
Near Sommauthe, France, Nov. 4, 1918.
R—Smith, Va.
B—Spencer, Va.
G. O. No. 37, W. D., 1919.

Private, Company C, 317th Infantry, 80th Division.
He crawled in front of the line under heavy machine-gun fire and carried a wounded soldier to safety.

CHAPIN, IVORY H. (937039).
At Wadonville, France, Sept. 25-26, 1918.
R—Hurley, N. Mex.
B—Arlington, Okla.
G. O. No. 139, W. D., 1918.

Private, 101st Ambulance Company, 101st Sanitary Train, 26th Division.
He assisted in establishing a dressing station in a dugout under a heavy shell-fire. When it was destroyed by a shell he worked unceasingly in the open under fire from enemy machine guns and snipers, caring for the wounded. He remained at his post for several hours after his station had been ordered closed, permitting neither his own exhaustion nor the enemy fire to deter him from aiding the wounded.

CHAPIN, WILBUR M. (2192850) .... Private, first class, Machine Gun Company, 7th Infantry, 3d Division.
Near Fossoy, France, night of July During the German offensive of July 15, Private Chapin, while under heavy
14-15, 1918, and near Cunel, artillery and machine-gun fire, carried three messages to front-line platoons.
France, Oct. 11, 1918. On Oct. 11, in the Meuse-Argonne offensive, after all his squad were killed
R—Peever, S. Dak. by a shell which buried the gun, Private Chapin dug the gun out and moved
B—Luverne, Minn. it to a new position, where he resumed firing on the enemy and assisted in
G. O. No. 16, W. D., 1920. repulsing a counterattack.

*CHAPMAN, CHARLES W., Jr .... Second lieutenant, 94th Aero Squadron, Air Service.
In the region of Autrepierre, France, While on patrol duty, he courageously attacked a group of four monoplanes
May 3, 1918. and one biplane and succeeded in bringing one down before he himself was
R—Waterloo, Iowa. shot down in flames.
B—Dubuque, Iowa. Posthumously awarded. Medal presented to father, C. W. Chapman.
G. O. No. 101, W. D., 1918.

CHAPMAN, ELDRIDGE G., Jr .... Captain, 5th Machine Gun Battalion, attached to 9th Infantry, 2d Division.
Near Thiaucourt, France, Sept. 12, During a heavy enemy counterattack he remained constantly in front of his
1918. company, directing its fire and encouraging its efforts. His bravery was
R—Denver, Colo. mainly responsible in preventing the enemy's advance and the taking of its
B—Denver, Colo. position.
G. O. No. 44, W. D., 1919.

*CHAPPEL, CHARLES F .... First lieutenant, 339th Infantry, 85th Division (detachment in north Russia).
Near Kadish, Russia, Sept. 27, 1918. Lieutenant Chappel led his men in attack against a superior force of the enemy.
R—Toledo, Ohio. Disregarding all personal danger, he personally advanced upon enemy ma-
B—Dayton, Ohio. chine guns, then in action, thereby assisting in building up a firing line of the
G. O. No. 19, W. D., 1920. attacking force. He was killed while making this dash.
Posthumously awarded. Medal presented to widow, Mrs. Zada H. Chappel.

CHAPPELL, RALPH A. (1268697) .... Private, Company I, 30th Infantry, 3d Division.
Near Crezancy, France, July 15, 1918. Although badly wounded by shell fire, during the heavy artillery fire of the
R—Lincoln, Nebr. enemy's offensive, Private Chappell remained at his post in the front line,
B—Annawan, Ill. until ordered to the rear by his commanding officer.
G. O. No. 44, W. D., 1919.

CHARLES, JOSEPH (1721431) .... Private, Company L, 307th Infantry, 77th Division.
In the Forest of Argonne, France, When his company was held up by barbed wire during an attack on hostile
Oct. 2, 1918. machine guns, he displayed marked courage in crawling over the wire to the
R—Kingston, N. Y. dugout occupied by the enemy and capturing, single handed, 20 prisoners and
B—High Falls, N. Y. a machine gun.
G. O. No. 20, W. D., 1919.

CHARTIER, ERNEST J. (2216108) .... First sergeant, Machine Gun Company, 357th Infantry, 90th Division.
Near St. Marie Farm, France, Sept. He volunteered to go forward with the patrol of three on two occasions to destroy
14, 1918. machine-gun nests which were holding up our advance. Both missions were
R—Buffalo, N. Dak. successful, the enemy emplacements being reduced and the advance resumed.
B—Belle Prairie, Minn. Sergeant Chartier exposed himself voluntarily to the fire of the enemy guns
G. O. No. 87, W. D., 1919. that they might be located by the flash.

CHARTIER, PEARL D. (2176261) .... Private, Company H, 140th Infantry, 35th Division.
Near Charpentry, France, Sept. 27, Private Chartier voluntarily went forward in the face of intense artillery and
1918. machine-gun fire and cut gaps through wire entanglements in order to facili-
R—Concordia, Kans. tate the advance of his battalion.
B—Clyde, Kans.
G. O. No. 70, W. D., 1919.

CHASE, JOHN W. (2396788) .... Corporal, Company L, 30th Infantry, 3d Division.
Near Jaulgonne, France, July 15, For repeated acts of extraordinary heroism in action near Jaulgonne, France.
and July 23-24, 1918. On July 15 Corporal Chase made two trips through a violent barrage to locate
R—Fort Lupton, Colo. missing comrades. On July 23 he volunteered and found a first-aid station,
B—Jewell City, Kans. to which he carried many of the wounded. On July 24, after his company
G. O. No. 35, W. D., 1920. had been relieved, he voluntarily remained behind until all wounded were
evacuated, during which time he was exposed to unusual shellfire.

CHASE, ROY WESLEY .... Corporal, 80th Company, 6th Regiment, U. S. Marine Corps, 2d Division.
In the Bois-de-Belleau, France, He assumed command of his platoon in the attack on enemy machine-gun
June 6, 1918. positions in the Bois-de-Belleau, France, on June 6, 1918, during which he and
R—Chicago, Ill. his men captured two machine guns and killed their crews. He did not
B—Chicago, Ill. retire from the action until all of his men had been killed or wounded.
G. O. No. 126, W. D., 1918.

CHATMAN, GROVER M .... Private, 75th Company, 6th Regiment, U. S. Marine Corps, 2d Division.
Near Thiaucourt, France, Sept. 15, While his platoon was occupying an outpost line in close proximity to the
1918. enemy, he voluntarily crossed an open space swept by machine-gun fire and
R—San Antonio, Tex. overcame three snipers who had been harrassing his platoon.
B—Austin, Tex.
G. O. No. 37, W. D., 1919.

CHAVIE, JOSEPH A. (275360) .... Private, first class, Company M, 127th Infantry, 32d Division.
Near Terny-Sorny, north of Soissons, Under heavy shell fire, when movement in the open was extremely hazardous,
France, Sept. 2, 1918. he made eight trips from the front lines to battalion headquarters with im-
R—Milwaukee, Wis. portant messages. His courage, high sense of duty, and coolness under fire
B—Calumet, Mich. were an example of heroism and devotion to duty which inspired his com-
G. O. No. 128, W. D., 1918. rades.

CHEEVERS, EARL J. (1165622).
Near Bois-de-Forges, France, Sept. 26, 1918.
R—Chicago, Ill.
B—Chicago, Ill.
G. O. No. 81, W. D., 1919.

Sergeant, Headquarters Company, 132d Infantry, 33d Division.
While engaged in maintaining a line of communication, Sergeant Cheevers saw 4 of the enemy enter a dugout during an attack. Armed with only a pistol, he followed, and upon reaching the dugout he ordered the men to come out. When they refused he entered, routed out and captured 12 prisoners.

ELLIS, WALTER L. (26417).
Near Fismes, France, and between the Ourcq and Vesle Rivers, July 31 to Aug. 6, 1918.
R—Hookset, Mich.
B—Shelby, Mich.
G. O. No. 100, W. D., 1918.

Sergeant, Headquarters Company, 125th Infantry, 32d Division.
He displayed exceptional courage and skill in maintaining observation posts under heavy artillery fire. On his own initiative, accompanied only by his brother, he reconnoitered the advanced positions of his brigade, passing through three heavily shelled areas, subjected also to the fire of enemy snipers and machine gunners, obtaining information of great importance to his brigade commander.

KNEY, HENRY A. (110250).
Near Chateau-Thierry, France, July 18-24, 1918.
R—Hookset, N. H.
B—Goffstown, N. H.
G. O. No. 125, W. D., 1918.

Private, Company B, 103d Machine Gun Battalion, 26th Division.
When his unit advanced on Belleau and Givry he was seriously wounded in the left arm, but continued to carry his heavy gun with his right arm through 1,200 meters of enemy fire and took up his position with his squad. When a shell struck his position, severely wounding a comrade and half burying him, the gun, and the remainder of the squad, he extricated himself, assisted his companions to recover the gun, and remained at his post until ordered to a dressing station by his platoon commander.

CHENOWETH, CHARLES E.
In the forest of Argonne, France, Sept. 29-30, 1918.
R—Lima, Ohio.
B—St. Johns, Ohio.
G. O. No. 20, W. D., 1919.

Captain, 363d Infantry, 91st Division.
At the time when troops on the left had retired, Captain Chenoweth, with his company, covered the left flank of his division and thus prevented an attack by the enemy upon its flank. After being severely wounded, he remained at his post until he had issued the necessary orders for holding the position he had seized.

CHERRY, CLAUD E. (559081).
Near St. Thibaut, France, Aug. 7, 1918.
R—Joilet, Ill.
B—Basco, Ill.
G. O. No. 15, W. D., 1919.

Sergeant, Company B, 11th Machine Gun Battalion, 4th Division.
He commanded the third platoon of his company during the engagement near St. Thibaut. On Aug. 7, 1918, he crossed the Vesle River and took up a position in front of his own Infantry on terrain constantly swept by heavy artillery, machine-gun, and sniper fire and directed his guns so skillfully as to silence a machine-gun nest and make possible the Infantry advance. His conspicuous courage was an inspiration to his men. This gallant soldier was killed on Aug. 9, 1918, by a fragment from an aerial bomb. Posthumously awarded. Medal presented to mother, Mrs. Sarah Demarest.

CHESNEY, ANTHONY (52420).
Near Soissons, France, July 19, 1918.
R—Kulpmont, Pa.
B—Greenbridge, Pa.
G. O. No. 132, W. D., 1918.

Corporal, Company A, 26th Infantry, 1st Division.
After being twice wounded and unable to advance, near Soissons, France, July 19, 1918, he took over an automatic rifle and used it effectively until ordered to the rear by an officer.
Posthumously awarded. Medal presented to brother, Zigmont Chesney.

CHILDERS, JOHN W. (2178442).
Near Remonville, France, Nov. 1, 1918.
R—La Grange, Mo.
B—La Grange, Mo.
G. O. No. 44, W. D., 1919.

Private, first class, Company B, 354th Infantry, 89th Division.
After all his superiors had become casualties, he assumed command of the platoon and, reorganizing the scattered groups, he led them forward against great resistance and gained his objective.

CHILDS, HOWARD JAMES.
In the Bois-de-Belleau, France, June 6 and 8, 1918.
R—Granville, Ill.
B—Granville, Ill.
G. O. No. 70, W. D., 1919.

Corporal, 83d Company, 6th Regiment, U. S. Marine Corps, 2d Division.
Howard J. Childs, Joseph A. Durgis, and Allen Benjamin Tilghman, corporals, and Herman L. McLeod, private, Company K, 6th Regiment, United States Marine Corps. These four men were prominent in the attack on enemy machine-gun positions in the Bois-de-Belleau on June 6 and 8, 1918; were foremost in their company at all times, and acquitted themselves with such distinction that they were an example for the remainder of their command.

CHILES, WALTER K. (154525).
Near Soissons, France, July 20, 1918.
R—Ensley, Ala.
B—Anniston, Ala.
G. O. No. 87, W. D., 1919.

Sergeant, Company B, 1st Engineers, 1st Division.
Sergeant Chiles displayed exceptional courage after being wounded, by remaining in command of his platoon under intense bombardment until he received a second severe wound.

CHINN, ANTHONY J. (736116).
Near Brandeville, France, Nov. 8, 1918.
R—High Cliff, Wis.
B—Italy.
G. O. No. 37, W. D., 1919.

Sergeant, Company H, 11th Infantry, 5th Division.
He volunteered to establish liaison between the battalions of his regiment. He crossed a valley under heavy artillery fire and climbed for over an hour on a hill exposed to steady machine-gun fire, accomplished his mission, and returned by the same route.

CHISHOLM, RAYMOND C. (243374).
Near Varennes, France, Sept. 26, 1918.
R—Springfield, N. J.
B—Brooklyn, N. Y.
G. O. No. 78, W. D., 1919.

Sergeant, Company A, 345th Battalion, Tank Corps.
After his tank had been hit by a shell, Sergeant Chisholm ordered his driver ahead, although fatally wounded. He continued to take an important part in the action until he dropped dead at his post.
Posthumously awarded. Medal presented to father, Robert M. Chisholm.

CHITWOOD, WARREN A. (2783841).
Near Eclisfontaine, France, Sept. 1, 1918.
R—San Francisco, Calif.
B—Weston, W. Va.
G. O. No. 44, W. D., 1919.

Private, Company K, 364th Infantry, 91st Division.
Although his right thumb had been shot off, he remained at his post as carrier of an automatic-rifle squad throughout the engagement.

CHRISTENBERRY, CURN (98227)_____
At Souain, France, July 16, 1918.
R—Landersville, Ala.
B—Landersville, Ala.
G. O. No. 101, W. D., 1918.

Private, Company L, 167th Infantry, 42d Division.
After having been wounded he remained courageously at his post under heavy shell fire, and not only afforded an inspiring example by that fortitude but rescued comrades who had been buried when a shell caved in their trench at Souain, France, July 16, 1918.

CHRISTENSEN, LEROY C_____
Near Blanc Mont Ridge, France, Oct. 4, 1918.
R—St. Paul, Minn.
B—Gibbon, Minn.
G. O. No. 81, W. D., 1919.

Private, 43d Company, 5th Regiment, U. S. Marine Corps, 2d Division.
Upon seeing his captain wounded and lying in an exposed position, Private Christensen left shelter and rendered him first aid. He then carried the officer from the area where he had been lying, which was subjected to an intense machine-gun and artillery barrage.

CHRISTENSEN, WALTER_____
Near Reims, France, July 15, 1918.
R—Gowen, Mich.
B—Greenville, Mich.
G. O. No. 99, W. D., 1918.

First lieutenant, 166th Infantry, 42d Division.
On July 15, 1918, near Reims, France, immediately following the bombardment of his position, he conducted himself with heroic disregard of personal safety in speedily placing his men in such advantageous posts and spurring them to vigorous action by seizing a rifle and fighting himself on the parapet of the trench that the German charge at that point was instantly repulsed. Although wounded he remained on duty until the enemy was defeated and his men were safe.

CHRISTENSON, WALTER T. (56299)_____
Near Sedan, France, Nov. 7, 1918.
R—Bridgman, Mich.
B—Chicago, Ill.
G. O. No. 44, W. D., 1919.

Private, Machine Gun Company, 28th Infantry, 1st Division.
After his platoon commander and the second in command had became casualties, Private Christenson took charge of the platoon, reorganized it, and led it forward, set up his machine guns in the open under the direct fire of enemy machine guns and artillery, and successfully silenced four enemy machine guns which were impeding the advance of the Infantry. Again he volunteered and carried a message from his company commander to the battalion post of command and brought back an answer without loss of time through a terrific enemy fire. After his company was relieved he returned to the former position with a detail of men and brought back all the wounded of his platoon, personally seeing that they were properly cared for and evacuated.

CHRISTIANSEN, HANS P. (56558)_____
At Cantigny, France, May 27–28, 1918.
R—St. Paul, Minn.
B—Denmark.
G. O. No. 99, W. D., 1918.

First Sergeant, Company A, 28th Infantry, 1st Division.
He refused to receive treatment, although sick, and walked up and down the line encouraging his men and exposing himself to shell and machine-gun fire.

CHRISTIANSEN, HENRY_____
At Saulx, France, Sept. 25–26, 1918.
R—Chicago, Ill.
B—Chicago, Ill.
G. O. No. 16, W. D., 1919.

First lieutenant, 101st Ambulance Company, 101st Sanitary Train, 26th Division.
He established and operated a dressing station in an advanced position under constant, heavy bombardment by the enemy. When word was received that our troops were withdrawing and permission had been given to move his station to a safer position, he declined to withdraw, but continued his work of ministering to the wounded.

CHRISTIANSON, ENOCH (1707575)_____
Near Binarville, France, Oct. 1, 1918.
R—New York, N. Y.
B—Norway.
G. O. No. 32, W. D., 1919.

Private, first class, Company A, 308th Infantry, 77th Division.
When the advance of his platoon had been checked by enemy machine-gun fire, Private Christianson deliberately exposed himself to sniper fire in order to locate the position of the sniper who had caused several casualties in his platoon.

CHRISTMAN, CLARENCE R_____
South of Soissons, France, July 22, 1918.
R—New York, N. Y.
B—Amsterdam, N. Y.
G. O. No. 99, W. D., 1918.

Second lieutenant, 2d Machine Gun Battalion, 1st Division.
By leading one of his machine guns in advance of the Infantry through a wheat field south of Soissons, France, July 22, 1918, he succeeded in obtaining a position of advantage, from which he cleaned out an enemy trench and forced out of action a hostile machine gun which had been holding up the advance of our Infantry.

CHRISTOPHER, JOHN C_____
Near Sergy, France, July 28, 1918.
R—Red Oak, Iowa.
B—Red Oak, Iowa.
G. O. No. 99, W. D., 1918.

First lieutenant, 168th Infantry, 42d Division.
He led his platoon against the Prussian Guards on Hill No. 212, near Sergy, France, on July 28, 1918. So courageous was he and so skillful in directing the attack that 13 of the enemy's best troops were captured at their guns and 6 machine guns were taken and turned on the foe.

CHURCH, JOHN H_____
At Cantigny, France, May 28–31, 1918.
R—New York, N. Y.
B—Gleniron, Pa.
G. O. No. 99, W. D., 1918.

Second lieutenant, 28th Infantry, 1st Division.
Knocked down and rendered unconscious by the explosion of a shell early in the attack on Cantigny, France, May 28–31, 1918, he staggered forward as soon as he regained consciousness and insisted upon resuming command, thereby giving a striking example of fortitude to his men.

CHYKO, JOHN (1782520)_____
Near Moirey, France, Nov. 10, 1918.
R—Beaver Valley, Pa.
B—Shenandoah, Pa.
G. O. No. 37, W. D., 1919.

Corporal, Company E, 314th Infantry, 79th Division.
Taking command of a platoon in the absence of officers and sergeants, he led a successful assault. With a small detachment of his men he wiped out several machine-gun nests, thereby aiding in the advance of his battalion. On the same evening, after establishing listening posts, he patroled with three men 1 kilometer from the front line, thus establishing security of the battalion position for the night.

CINAMON, ARCHIE (1306540)_____
Near Bellicourt, France, Sept. 29, 1918.
R—Rogersville, Tenn.
B—Hawkins County, Tenn.
G. O. No. 37, W. D., 1919.

Private, Machine Gun Company, 117th Infantry, 30th Division.
Private Cinamon, a runner, carried many messages under heavy fire. At one time when his platoon was held up by machine-gun fire he, with another soldier, refused to take cover, but delivered effective rifle fire until the machine gun was silenced, thereby enabling his platoon to continue its advance.

CISEK, JULIAN (48445)
Near Exermont, France, Oct. 4, 1918.
R—Bessemer, Mich.
B—Austria.
G. O. No. 60. W. D., 1920.

Private, first class, Company M, 18th Infantry, 1st Division.
After his company had suffered very heavy casualties, Private Cisek reorganized a group of about 35 men, under heavy artillery and machine-gun fire, and led them forward in the attack, a distance of about 2 kilometers, to the objective. Upon the capture of the objective he assisted in the consolidation of the position taken.

*CLABBY, JOHN L. (60916)
Near Epieds, France, July 23, 1918.
R—Lawrence, Mass.
B—Providence, R. I.
G. O. No. 125, W. D., 1918.

Sergeant, Company F, 101st Infantry, 26th Division.
During the advance by his platoon upon machine-gun nests in Trugny Woods he observed a German machine gun on his right flank. He charged it single handed in the face of its fire, killed the gunners, and destroyed the gun. Posthumously awarded. Medal presented to mother, Mrs. Bridget Clabby.

CLAFLIN, JAMES A. (5011)
Near Thiaucourt, France, Sept. 13, 1918.
R—Philadelphia, Pa.
B—Brooklyn, N. Y.
G. O. No. 37, W. D., 1919.

Sergeant, medical detachment, 5th Machine Gun Battalion, 2d Division.
In a territory swept by the direct fire of two German batteries, he displayed great courage and devotion to duty in giving first aid to the wounded and in superintending their removal to a place of safety.

CLAPP, DAVID O. (56810)
During the defense of Cantigny, France, May 28, 1918.
R—Ada, Okla.
B—Sulphur Springs, Tex.
G. O. No. 99, W. D., 1918.

Private, first class, Company B, 28th Infantry, 1st Division.
Although wounded several times, he stayed with his automatic rifle and assisted in the reconsolidation of his platoon. He was ordered to the aid station, but after receiving first aid insisted on returning to his post, where he remained working until forced to be evacuated.

CLAPP, KENNETH SMITH
Near Luneville, France, June 13, 1918.
R—Fort Wayne, Ind.
B—Fort Wayne, Ind.
G. O. No. 132, W. D., 1918.

Second lieutenant, 27th Aero Squadron, Air Service.
Outnumbered and handicapped by his presence far behind the German lines, he and three flying companions fought brilliantly a large group of enemy planes, bringing down or putting to flight all in the attacking party, while performing an important mission near Luneville, France, June 13, 1918.

CLARK, ARTHUR I. (2253790)
Near Esnes, France, Sept. 26, 1918.
R—Moore, Mont
B—Newton, Iowa.
G. O. No. 98, W. D., 1919.

Sergeant, Company C, 39th Infantry, 4th Division.
Sergeant Clark was in command of one platoon of his company, which was held up by intense enemy machine-gun fire. Accompanied by two other soldiers, he voluntarily made an attack on one of the nests under heavy fire, firing a rifle grenade into it and forcing its surrender. He then advanced on another machine-gun nest and captured it, taking seven prisoners from both nests. His platoon having been forced to fall back by machine-gun fire from the rear, he reorganized it and led it in a successful attack on 75 of the enemy whom he discovered near by.

CLARK, CHALMERS
Near St. Etienne, France, Oct. 3-9, 1918.
R—Detroit, Mich.
B—Exeter, Mo
G. O. No. 78, W. D., 1919.

Private, 83d Company, 6th Regiment, U. S. Marine Corps, 2d Division.
Displaying remarkable devotion to duty, Private Clark remained on duty as a battalion runner for six days and nights, almost without rest, continually risking his life in crossing fields swept by machine-gun and shellfire on liaison and reconnaissance missions, for which he volunteered. Each night he organized and guided carrying parties, bringing food and water to the men in the front lines.

CLARK, GEORGE E. (1375938)
Near Epinonville, France, Oct. 2, 1918.
R—Chicago, Ill.
B—Chicago, Ill.
G. O. No. 78, W. D., 1919.

Sergeant, Battery D, 122d Field Artillery, 33d Division.
When his battery echelon was bombed, Sergeant Clark, with great courage and presence of mind, conducted his men to shelter, and then took charge of rescuing and treating the wounded, until he was himself severely wounded by an exploding bomb, necessitating the amputation of one of his arms.

CLARK, GUY H.
Near Blanc Mont Ridge, France, Oct. 5, 1918.
R—St. Paul, Minn.
B—White Bear Lake, Minn.
G. O. No. 46, W. D., 1919.

Corporal, 43d Company, 5th Regiment, U. S. Marine Corps, 2d Division.
Learning that a number of wounded soldiers were lying in no man's land, he immediately volunteered to help carry them in. He made several trips over an area which was constantly shelled and subjected to machine-gun and rifle fire.

CLARK, HAROLD S. (2952)
Near Belleau and Givry, France, July 18-21, 1918.
R—Ware, Mass.
B—Ware, Mass.
G. O. No. 15, W. D., 1923.

Private, first class, medical detachment, 104th Infantry, 26th Division.
While a member of the medical detachment, 104th Infantry, he exhibited rare courage and devotion to duty, repeatedly exposing himself to a concentration of hostile machine-gun fire while applying first aid and carrying wounded men from the front lines to the dressing stations. His complete disregard for his own safety and his devotion to his comrades greatly inspired the men of his battalion.

CLARK, HARRY G. (1285262)
In the Bois-de-Montagne, France, Oct. 15, 1918.
R—Cumberland, Md.
B—Dawsonville, Md.
G. O. No. 37, W. D., 1919.

Sergeant, Company G, 115th Infantry, 29th Division.
He remained for two days by himself in a sniper's post in advance of the front line, killing 12 enemy scouts. When all of the officers of his company had been incapacitated, this soldier took command and steadied his men by his own coolness and courage. He remained on duty until he was severely wounded while leading a combat liaison patrol.

CLARK, JAMES L. (118290)
At Bois-de-Belleau, France, June 6, 1918.
R—Seaman, Ohio.
B—Ohio.
G. O. No. 99, W. D., 191

Private, 47th Company, 5th Regiment, U. S. Marine Corps, 2d Division.
Although wounded in the attack on Bois-de-Belleau, France, he crossed through enemy territory to convey a message.

*CLARK, JAMES PAUL (1214559)_____
Near Ronssoy, France, Sept. 29, 1918.
R—Medina, N. Y.
B—Pierre, S. Dak.
G. O. No. 15, W. D., 1919.

Corporal, Company F, 108th Infantry, 27th Division.
He displayed unusual courage and leadership in taking command of his company after all the officers had been killed and leading it into effective combat. Posthumously awarded. Medal presented to father, James Clark.

CLARK, JOSEPH (1563168)_____
North of Cierges, France, Oct. 4, 1918.
R—Cameron, W. Va.
B—Cameron, W. Va.
G. O. No. 10, W. D., 1920.

Private, Company C, 8th Machine Gun Battalion, 3d Division.
Private Clark, during an attack under terrific shell and machine-gun fire, carried messages from company to platoon headquarters, thus maintaining the necessary communication required in a successful operation.

CLARK, JOHN F. (550617)_____
Near Moulins, France, July 15, 1918.
R—Bloomington, Ill.
B—Bloomington, Ill.
G. O. No. 10, W. D., 1920.

Private, Company E, 38th Infantry, 3d Division.
He carried numerous messages from company headquarters to the front-line platoons, when it was almost impossible to maintain liaison, particularly exposing himself to intense enemy fire. Through his efforts communication was maintained at a critical time in the defensive operations against the last enemy offensive.

CLARK, MERL E. (100408)_____
At the Cote-de-Chatillon, east of Grand Pre, France, Oct. 16, 1918.
R—Webster City, Iowa.
B—Webster City, Iowa.
G. O. No. 1, W. D., 1919.

Sergeant, Company C, 168th Infantry, 42d Division.
After leading his platoon in a resolute assault across open ground swept by machine-gun fire, he saw his left held up by a machine-gun nest. Taking 4 soldiers, he flanked the enemy position, killed 4 Germans, and captured 2 prisoners and 2 heavy machine guns, his own detachment suffering no casualties. He executed this movement with exceptional skill, daring, and promptness, and in less than 10 minutes cleared the ground for the advance of two companies.

CLARK, ORRIE A. (1679806)_____
On Hill 273, in the Forest of Argonne, France, Oct. 5, 1918.
R—Geneva, N. Y.
B—Phelps, N. Y.
G. O. No. 74, W. D., 1919.

Private, first class, Company F, 307th Infantry, 77th Division.
When an officer called for volunteers to cut an opening in a thick barbed-wire entanglement, Private Clark unhesitatingly responded and, under the enemy's rifle fire and cross fire from two machine guns, worked for more than an hour at cutting the wire, desisting only when ordered to do so by his officers. He then returned to his organization through the enemy's artillery barrage.
Oak-leaf cluster.
For the following act of extraordinary heroism in action near Fismes, France, Sept. 5, 1918, Private Clark is awarded an oak-leaf cluster, to be worn with the distinguished-service cross: He volunteered and carried a message from his platoon commander to a squad leader through continuous shell and machine-gun fire and was thereby the means of saving the lives of 6 men of the squad.

CLARK, PATRICK J. (107637)_____
Near Thiaucourt, France, Sept. 13, 1918.
R—Bridgeport, Conn.
B—Scotland.
G. O. No. 46, W. D., 1919.

Private, first class, Company C, 5th Machine Gun Battalion, 2d Division.
On two occasions he unhesitatingly went through heavy machine-gun fire carrying messages. Later, when volunteers were called for to take a message through a violent barrage he was the first to respond.

CLARK, ROBERT P_____
Near Cantigny, France, July 4, 1918.
R—Orono, Me.
B—Bethlehem, Pa.
G. O. No. 46, W. D., 1919.

Second lieutenant, 16th Infantry, 1st Division.
Accompanied by 5 men, he rushed a trench manned by a greatly superior number of the enemy and fought until the entire enemy garrison had been killed or wounded. Four of his patrol were killed and 1 wounded in the encounter. He bandaged the wounded man and returned to our lines, carrying a wounded German prisoner.

CLARK, WILLIAM L. (1320230)_____
Near Bellicourt, France, Sept. 29, 1918.
R—Knoxville, Tenn.
B—Claiborne County, Tenn.
G. O. No. 37, W. D., 1919.

Private, first class, Company D, 120th Infantry, 30th Division.
With 8 other soldiers, comprising the company headquarters detachment, he assisted his company commander in cleaning out enemy dugouts along a canal and capturing 242 prisoners.

CLARKE, LEO GEORGE_____
Near Remonville, France, Nov. 5–10, 1918.
R—Waukon, Iowa.
B—Waukon, Iowa.
G. O. No. 37, W. D., 1919.

Second lieutenant, 11th Infantry, 5th Division.
He set an example of bravery and self-sacrifice to his men during the period Nov. 5–9, 1918. On Nov. 10, while assembling his company, he discovered a wounded man lying in a place exposed to machine-gun fire, and regardless of his own danger carried him to a place of safety.

CLARKE, SHELDON V._____
Near Raulecourt, France, Aug. 28 and Sept. 26, 1918.
R—Williamsport, Pa.
B—Williamsport, Pa.
G. O. No. 46, W. D., 1919.

First lieutenant, 9th Balloon Company, Air Service.
While making a general surveillance of enemy territory on Aug. 28 and Sept. 26 he was attacked by enemy planes. On both occasions he remained at his post and directed the fire by telephone until his balloon had been set on fire. On Aug. 28 he assisted a passenger to descend and did not jump himself until the other's parachute had opened. On both occasions he reascended as soon as another balloon could be obtained.

CLARKSTON, SAMUEL_____
Near Beaumont, France, Nov. 10–11, 1918.
R—Kyle, Ohio.
B—Leslie County, Ky.
G. O. No. 37, W. D., 1919.

Gunnery sergeant, Machine Gun Company, 5th Regiment, U. S. Marine Corps, 2d Division.
Leading the one remaining machine-gun crew across the River Meuse, Sergeant Clarkston, under trying conditions, established a stronghold from which he inflicted severe losses on the enemy.

CLART, EMMETT S. (106187)
Near Soissons, France, July 19, 1918.
R—Odum, Ga.
B—Odum, Ga.
G. O. No. 108, W. D., 1918.

Private, Company B, 3d Machine Gun Battalion, 1st Division.
Shocked and bruised by a shell near Soissons, France, July 19, 1918, he was taken to an aid station and put with the wounded who were to be evacuated, but when Infantry reenforcements passed by he joined them, participated in their attack, and fought effectively with rifle and bayonet. He took 5 prisoners and was ordered to escort them to the rear. On his way back he ascertained the location of his company commander, and, after disposing of his prisoners, he reported back to his organization and asked for further duty.

CLASBY, DANIEL J. (60837)
Near Bois-de-St.-Remy, France, Sept. 12, 1918.
R—Waltham, Mass.
B—Woburn, Mass.
G. O. No. 26, W. D., 1919.

Private, Company F, 101st Infantry, 26th Division.
Accompanying two other soldiers, Private Clasby rushed forward in advance of his lines, exposed to heavy machine-gun fire, and captured two machine guns and six of the enemy who were manning the position.

CLAUSON, OSCAR (2293182)
Near the Scheldt River, Belgium, Oct. 31, 1918.
R—Havre, Mont.
B—Sweden.
G. O. No. 46, W. D., 1919.

Private, Company F, 362d Infantry, 91st Division.
When the advance of the front line was held up by fire from a machine-gun nest 300 yards to the front, Private Clauson, with two others, crossed the open field in the face of fire from enemy artillery, machine guns, and snipers. Charging the nest they killed two of the crew, wounded two others, and captured five, together with the machine gun.

CLAY, FRED (263425)
Near Sergy, France, July 31, 1918.
R—Saginaw, Mich.
B—Saginaw, Mich.
G. O. No. 64, W. D., 1919.

Corporal, Company K, 125th Infantry, 32d Division.
After all runners had become casualties in attempting to carry out their missions Corporal Clay voluntarily assumed the duties of company runner. His efforts were materially responsible for success gained during the day, during the course of which he repeatedly exposed himself to direct enemy artillery and machine-gun fire.

*CLAY, HENRY ROBINSON, Jr.
Near Sains-les-Marquion, France, Sept. 4, 1918.
R—Fort Worth, Tex.
B—Plattsburg, Mo.
G. O. No. 60, W. D., 1920.

First lieutenant, 148th Aero Squadron, Air Service.
In an action wherein Lieutenant Clay's patrol was outnumbered two to one, he attacked the group and shot down the enemy aircraft in flames. He continued in the combat and later attacked two enemy aircraft which were pursuing a plane of his patrol and succeeded in shooting one enemy aircraft down. Again, on Sept. 27, 1918, near Cambrai, France, with one other pilot, Lieutenant Clay observed five enemy planes approaching our lines and, although hopelessly outnumbered, immediately attacked and singled out a plane which was seen to crash to the ground. He was immediately attacked by the other enemy planes and compelled to fight his way back to our lines.
Posthumously awarded. Medal presented to father, Henry R. Clay.

CLELAND, JOHN R. D.
Near Soissons, France, July 21, 1918.
R—Jacksonville, Fla.
B—Waycross, Ga.
G. O. No. 15, W. D., 1919.

First lieutenant, 28th Infantry, 1st Division.
Although he was wounded before and in the attack upon Berzy-le-Sec, he declined an opportunity to be evacuated and led his platoon to its final objective, which he consolidated and held.

CLEMENT, JOSEPH T.
Near La Ferte-Milon, France, July 18, 1918.
R—Dunnellon, Fla.
B—Charleston, S. C.
G. O. No. 43, W. D., 1922.

Major, 39th Infantry, 4th Division.
He reported for duty from the hospital as his regiment was preparing to advance to the attack. Soon thereafter, when it was of urgent necessity that important orders of the regimental commander be carried to the officer commanding the assault battalion, he volunteered for the dangerous mission. Accompanied by two men he worked his way forward through a heavy artillery shell fire, located the assault units, and delivered the attack order. Endeavoring to locate the commander of the assault battalion, he fearlessly went into a heavy artillery barrage and continued this important and hazardous task until he was wounded by shell fire.

CLEMENTSON, HARRY B. (3130713)
Near Carrefour-de-Meurrissons, France, Sept. 27, 1918.
R—Eagle Bend, Minn.
B—Long Prairie, Minn.
G. O. No. 44, W. D., 1919.

Private, Company A, 305th Infantry, 77th Division.
After his company had taken shelter from the enfilading machine-gun and trench-mortar fire, Private Clementson, accompanied by two other soldiers, crawled out, in the face of a machine-gun barrage, to the aid of wounded comrades, thus saving the lives of at least two of his companions.

CLERMONT, JOSEPH R. (1903154)
Near St. Juvin, France, Oct. 16, 1918.
R—Fall River, Mass.
B—Canada.
G. O. No. 50, W. D., 1919.

Private, Company M, 326th Infantry, 82d Division.
With another soldier Private Clermont advanced several hundred yards ahead of the front line under heavy artillery and machine-gun fire and rescued a wounded comrade.

*CLEVELAND, VICTOR A. (1384760)
In the Bois-de-Chaume, France, Oct. 12, 1918.
R—Louisville, Ill.
B—Louisville, Ill.
G. O. No. 20, W. D., 1919.

Corporal, Company L, 130th Infantry, 33d Division.
Corporal Cleveland led a detail to the rescue of an officer who had been caught in a heavy barrage of gas and high-explosive shells and seriously wounded. In his efforts to get the wounded officer to an aid station this courageous soldier was killed.
Posthumously awarded. Medal presented to mother, Mrs. Ella Steele.

CLEVERLY, IRVING N. (1212822)
Near Ronssoy, France, Sept. 28, 1918.
R—Brooklyn, N. Y.
B—Brooklyn, N. Y.
G. O. No. 15, W. D., 1923.

Private, sanitary detachment, 107th Infantry, 27th Division.
He voluntarily accompanied Sergeants John W. Schwengler and Harry W. Greene, 107th Infantry, advanced 250 yards in front of the company's position in the face of intense enemy machine-gun fire at short range, rescued a badly wounded soldier who had lain exposed to this fire for 24 hours, carried the wounded man back to comparative shelter, and dressed his wounds, undoubtedly saving the life of the wounded man. This gallant and heroic act was an object lesson of soldierly conduct and greatly inspired the men of the 107th Infantry.

CLINCY, WILL (2169151)_____
Near Frapelle, France, Sept. 4, 1918.
R—Birmingham, Ala.
B—Marion, Ala.
G. O. No. 15, W. D., 1919.

Private, first class, Company F, 366th Infantry, 92d Division.
He showed exceptional bravery during an enemy raid, his team mate on an automatic rifle having been mortally wounded, and although he was himself severely wounded he continued to serve his weapon alone until the raid was driven back.

CLINE, FLOYD (541122)_____
At La Tuilerie Farm, near La Charmel France, during the Aisne-Marne offensive, July 23, 1918.
R—Pineville, W. Va.
B—Long Branch, W. Va.
G. O. No. 9, W. D., 1923.

Private, Company C, 7th Infantry, 3d Division.
When no stretcher bearers were available Private Cline voluntarily crossed an open space of 200 yards swept by heavy enemy machine-gun fire directly in front of his company's lines and brought back to safety a severely wounded comrade who was calling for help.

*CLINE, JACOB F. (1272934)_____
Near Verdun, France, Oct. 11, 1918.
R—Union Hill, N. J.
B—West Hoboken, N. J.
G. O. No. 32, W. D., 1919.

Bugler, Company D, 111th Machine Gun Battalion, 29th Division.
He voluntarily left cover to carry a litter through an intense bombardment after the litter bearers had been wounded. He was killed while on this duty.
Posthumously awarded. Medal presented to father, Jacob F. Cline.

CLINE, JESSE L. (3167473)_____
Near Lion-devant-Dun, France, Nov. 7, 1918.
R—Justice, W. Va.
B—Mingo County, W. Va.
G. O. No. 44, W. D., 1919.

Private, Company A, 61st Infantry, 5th Division.
Although entirely unacquainted with the personnel of his company, to which he had just been assigned, Private Cline volunteered and led a party of 20 men in a successful attack on a strong enemy machine-gun position.

CLOONAN, JOHN J. (67179)_____
At Belleau Wood, France, July 20, 1918.
R—Manchester, N. H.
B—Ireland.
G. O. No. 9, W. D., 1923.

Private, first class, Company D, 103d Infantry, 26th Division.
For extraordinary heroism in action at Belleau Wood, France, July 20, 1918, while acting as an automatic-rifle man. His assistant having been wounded, Private Cloonan advanced under a heavy concentration of enemy machine-gun fire and artillery fire and volunteered to destroy, single handed, two enemy machine guns which were directing a severe fire against the flank of his company. With great bravery and utter disregard for his own safety, he successfully accomplished his mission; his act undoubtedly saved the lives of many of his comrades and also permitted the further advance of his company.

CLOSE, HARRY S. (1207119)_____
Near Ronssoy, France, Sept. 27, 1918.
R—Brooklyn, N. Y.
B—Brooklyn, N. Y.
G. O. No. 37, W. D., 1919.

Corporal, Company D, 106th Infantry, 27th Division.
During operations against the Hindenburg line he, single handed, attacked a group of 13 of the enemy. By hard fighting he succeeded in killing 3 and taking the remainder as prisoners, marching them to the rear under heavy fire of machine guns and shells. When returning to his command he was wounded.

*CLOWE, EDWARD P. (89617)_____
South of Landres-et-St. Georges, France, Oct. 15, 1918.
R—New York, N. Y.
B—New York, N. Y.
G. O. No. 72, W. D., 1920.

Sergeant, Company C, 165th Infantry, 42d Division.
Sergeant Clowe led his section in the attack through heavy enemy machine-gun fire on a strongly held enemy position. His organization was compelled to halt, due to the intensity of the enemy fire. For four hours Sergeant Clowe held an advanced position until he was ordered to withdraw. After assisting in the evacuation of the wounded, he, with two others, acted as rear guard. This noncommissioned officer continued to resist superior numbers of the enemy until he fell mortally wounded and his death followed within the enemy lines.
Posthumously awarded. Medal presented to father, Edward J. Clowe.

COAKLEY, JOHN L. (131756)_____
Near Somme-Py, France, Oct. 9, 1918.
R—Kansas City, Kans.
B—St. Louis, Mo.
G. O. No. 37, W. D., 1919.

Corporal, Battery B, 17th Field Artillery, 2d Division.
Although seriously wounded and still exposed to enemy shell fire, he refused treatment until three other members of his squad had been removed and attended to.

COATS, LORENZA C. (3509173)_____
Near Rembercourt, France, Oct. 9–Nov. 11, 1918.
R—Winnsboro, Tex.
B—Winnsboro, Tex.
G. O. No. 78, W. D., 1919.

Private, Company A, 56th Infantry, 7th Division.
As company and platoon runner, Private Coats worked tirelessly, being on duty almost constantly both day and night, and carrying numerous messages under enemy machine-gun and artillery fire.

COCHRAN, CARLISLE C. (545136)_____
Near Crezancy, France, July 15, 1918.
R—Huntersville, N. C.
B—Verona, Ky.
G. O. No. 16, W. D., 1920.

Sergeant, medical detachment, 30th Infantry, 3d Division.
Sergeant Cochran, though severely injured in one foot early in the morning, persevered in the work of rendering first aid and assistance to the wounded, exposed to heavy shell fire, until it became necessary for him to be evacuated later in the afternoon.

COCHRAN, JOHN B. (2216257)_____
Near Fey-en-Haye, France, Sept. 12, 1918.
R—Gildford, Mont.
B—Lamont, Okla.
G. O. No. 128, W. D., 1918.

Sergeant, Company A, 357th Infantry, 90th Division.
Although severely wounded he led his section forward and captured three machine guns.

*COCHRAN, WILLIAM B. (2385037)_____
In the Bois-des-Rappes, France, Oct. 14, 1918.
R—Gulfport, Miss.
B—Gulfport, Miss.
G. O. No. 70, W. D., 1919.

Sergeant, Company A, 61st Infantry, 5th Division.
Disregarding his own personal safety, he assisted in reorganizing his company under heavy machine-gun and artillery fire and leading it against machine-gun nests which were holding up the advance of his battalion. He was killed in action during this engagement.
Posthumously awarded. Medal presented to brother, David V. Cochran.

COCHRANE, ROBERT S._____
Near St. Etienne, France, Oct. 3-4, 1918.
R—Richburg, S. C.
B—Due West, S. C.
G. O. No. 15, W. D., 1919.

Chief pharmacist's mate, U. S. Navy, attached to 6th Machine Gun Battalion, United States Marine Corps, 2d Division.
He continued to dress wounded when the area in which he was working was swept by machine-gun fire. He was an example of coolness to all during 48 hours of continuous shell fire, never hesitating to expose himself to danger when assistance was needed.

CODAY, WILLIAM C. (3510370)_____
Near Rembercourt, France, Oct. 9-Nov. 11, 1918.
R—Stonebluff, Okla.
B—Hartsguld, Mo.
G. O. No. 78, W. D., 1919.

Private, Company C, 56th Infantry, 7th Division.
As company and platoon runner Private Coday worked tirelessly, being on duty almost constantly both day and night, and carrying numerous messages under enemy machine-gun and artillery fire.

COFF, JOSEPH J. (1452466)_____
Near Very, France, Sept. 26, 1918.
R—St. Louis, Mo.
B—St. Louis, Mo.
G. O. No. 46, W. D., 1919.

Bugler, Company F, 138th Infantry, 35th Division.
Bugler Coff was a member of a liaison group who worked their way 1,000 yards in advance of their first wave. Surrounded by enemy machine guns, Bugler Coff, accompanied by two more soldiers, silenced two machine guns, and took 23 prisoners. Seriously gassed, he persisted in remaining in the fight until exhausted.

*COFFEY, JOSEPH E. (42227)_____
Near Soissons, France, July 20, 1918.
R—Erie, Pa.
B—Bradford, Pa.
G. O. No. 43, W. D., 1922.

Sergeant, Company C, 16th Infantry, 1st Division.
After all the officers of his company had been killed or wounded and his company commander had taken command of the battalion, Sergeant Coffey unhesitatingly took command of and skillfully led the company forward in the face of a heavy hostile fire and successfully took their objective. On Oct. 9, 1918, in the Argonne Forest, after being severely wounded, he refused to be evacuated and continued to advance with his platoon. Displaying the greatest courage and gallantry, he voluntarily attempted to save the life of his company commander, who had been severely wounded by enemy machine-gun fire. It was during this heroic effort that Sergt. Coffey was killed.
Posthumously awarded. Medal presented to mother, Mrs. Emma P. Coffey.

*COFFMAN, RALPH L. (794687)_____
Near Brieulles, France, Nov. 4, 1918.
R—Marceline, Mo.
B—Rutledge, Mo.
G. O. No. 21, W. D., 1919.

Sergeant, Company B, 15th Machine Gun Battalion, 5th Division.
When his advance had been held up by an enemy machine gun, and having been advised that the crew manning the gun would die rather than surrender, Sergeant Coffman alone attacked the gun. His attempt proved fatal, for he was killed before reaching the gun. His action, however, enabled his platoon to overcome the resistance without further serious loss.
Posthumously awarded. Medal presented to father, J. M. Coffman.

COGSWELL, JULIUS C._____
In the bombardment of La Cense Farm, France, June 6, 1918.
R—Charleston, S. C.
B—Sullivans Island, S. C.
G. O. No. 126, W. D., 1918.

First lieutenant, 6th Regiment, U. S. Marine Corps, 2d Division.
Having been previously wounded in the bombardment of La Cense Farm, France, he refused to be evacuated, and handled his platoon with marked bravery and skill in an assault on a formidable machine-gun position until seriously wounded on June 6, 1918.

COGSWELL, THEODORE L._____
At Landres-et-St. Georges, France, Nov. 1, 1918.
R—Washington, D. C.
B—Washington, D. C.
G. O. No. 14, W. D., 1923.

First lieutenant, 319th Infantry, 80th Division.
In the advance of that day his company was halted by intense machine-gun fire while passing through barbed-wire entanglements. In the face of this fire Lieutenant Cogswell voluntarily crawled down a slope in front of the enemy's position and in full view thereof, discovered a sunken road at the foot of the slope, crawled back to his company, and led them to the advanced position. Immediately upon reaching the new position he again voluntarily sought a favorable forward position, crawling a distance of 75 yards toward the enemy, when he was severely wounded. The outstanding bravery and devotion to duty displayed by Lieutenant Cogswell served to incite the men of his company to heroic endeavors, enabling them to assist in the capture of the town of Landres-et-St. Georges, together with 209 prisoners, 9 field pieces, and 15 machine guns.

COHAN, ABRAHAM (3177)_____
Near Verdun, France, Nov. 9, 1918.
R—Needham, Mass.
B—Russia.
G. O. No. 21, W. D., 1919.

Private, sanitary detachment, 103d Infantry, 26th Division.
After three others had failed in the attempt and were wounded, Private Cohan went out under terrific machine-gun fire and gave first aid to a wounded soldier.

COHEE, ORA J._____
Near Rembercourt, France, Nov. 1-2, 1918.
R—South Bend, Ind.
B—Logansport, Ind.
G. O. No. 35, W. D., 1919.

First lieutenant, chaplain, 34th Infantry, 7th Division.
He worked untiringly, under constant artillery fire, for two days in charge of the stretcher bearers, personally assisting in carrying wounded men to safety.

COHEN, FRANK J. (1249864)_____
Near Baslieux, France, Sept. 5, 1918.
R—Pittsburgh, Pa.
B—Pittsburgh, Pa.
G. O. No. 71, W. D., 1919.

Private, Headquarters Company, 107th Field Artillery, 28th Division.
On duty with an officer at an observation post far in advance of the Infantry line, he exposed himself to heavy machine-gun fire to obtain stretcher and bandages when the officer was hit by a machine-gun bullet. The officer's wound having proved fatal, Private Cohen the following day volunteered to go out and assist in bringing back the body, being constantly subjected to enemy fire.

*COHN, DAVID H._____
Near Spitaals Bosschen, Belgium, Oct. 31, 1918.
R—Spokane, Wash.
B—Portland, Oreg.
G. O. No. 50, W. D., 1919.

First lieutenant, 363d Infantry, 91st Division.
When his company met with determined resistance from enemy machine guns and snipers, Lieutenant Cohn, being the only officer remaining with his company, undertook a dangerous reconnaissance himself rather than assign the mission to others, and while so engaged was mortally wounded.
Posthumously awarded. Medal presented to father, Hyman Cohn.

COHN, EUGENE S ....
Near Exmorieux Farm, France, Oct.
2, 1918.
R—Spokane, Wash.
B—Portland, Oreg.
G. O. No. 46, W. D., 1919.

Captain, 364th Infantry, 91st Division.
After being painfully wounded by shrapnel Captain Cohn refused to go to the rear and remained on duty with his company in the front line without medical attention for 54 hours.

COHN, HERBERT ARNOLD ....
East of Montfaucon, France, Sept.
26, 1918.
R—San Francisco, Calif.
B—Carson City, Nev.
G. O. No. 3, W. D., 1921.

Second lieutenant, 39th Infantry, 4th Division.
Lieutenant Cohn led the assault wave forward through heavy machine gun fire in an attack against a strongly defended enemy position. He was forced to pass through barbed-wire entanglements before entering the enemy trenches. While exposing himself to intense machine-gun fire, he was badly wounded, but his command, inspired by his gallant example, gained and held the objective sought.

COHOON, WILLIAM M.(1224828) ....
Near Ronssoy, France, Sept. 27,
1918.
R—Wappingers Falls, N. Y.
B—Wappingers Falls, N. Y.
G. O. No. 13, W. D., 1923.

Private, Company M, 105th Infantry, 27th Division.
In the face of concentrated fire from hostile machine guns, he, together with Corpl. George D. Caswell, made his way over rough and broken terrain under observation of the enemy, rescued a severely wounded comrade, carried him to a place of safety, and returned to a point still farther in advance of his own lines and carried another member of his company to shelter. This action, showing indomitable bravery and devotion to duty, undoubtedly saved the lives of the wounded men and set a splendid example of courage and self-sacrifice to other members of the command.

COLBURN, ALVIN ....
At Vaux, France, July 1-2, 1918.
R—Washington, D. C.
B—Dedham, Mass.
G. O. No. 130, W. D., 1919.

Captain, 9th Infantry, 2d Division.
After undergoing a severe bombardment from trench mortars and 77's, which caused numerous casualties in his company, Captain Colburn led his company (Company H) in an attack and succeeded in capturing 100 prisoners and 13 machine guns. He constantly exposed himself to enemy fire while leading his command toward its objective. His gallant conduct and able leadership gave his men the confidence necessary to accomplish their mission and to repel a strong counterattack in the darkness of the early morning of July 2, 1918.

COLE, ALAN RAMSEY ....
Near Soissons, France, July 18, 1918.
R—Portland, Me.
B—Bath, Me.
G. O. No. 46, W. D., 1919.

Second lieutenant, 16th Infantry, 1st Division.
Although wounded early in the morning, he continued to lead his platoon in the front wave, personally silencing machine-gun nests, displaying wonderful courage, leadership, and devotion to duty during the entire operations. He remained with his platoon until ordered to the rear by his battalion commander.

COLE, ARTHUR CLEMENCE ....
Near Landres-et-St. Georges, France,
Nov. 1, 1918.
R—Providence, R. I.
B—East Providence, R. I.
G. O. No. 35, W. D., 1920.

First lieutenant, 23d Infantry, 2d Division.
Lieutenant Cole exposed himself to heavy machine-gun fire in order to lead a small detachment against a strongly held enemy position. Armed with a rifle, he attacked an enemy machine-gun position from the flank, capturing same. He then led his unit against other machine-gun nests, which resulted in the capture of some guns and about 40 of the enemy crews.

COLE, CHARLES EDWARD (2809360) ......
Near Bourrut, France, Nov. 1, 1918.
R—Dallas, Tex.
B—New York, N. Y.
G. O. No. 46, W. D., 1919.

Private, first class, Company F, 359th Infantry, 90th Division.
Under heavy machine-gun fire, Private Cole organized 2 broken platoons and guided them into position. Through two days of strong operations he acted as runner, carrying messages to the front line and returning with valuable information, at all times exposed to severe shell and machine-gun fire.

*COLE, EDWARD B. ....
In the Bois-de-Belleau, France,
June 10, 1918.
R—Hingham, Mass.
B—Boston, Mass.
G. O. No. 119, W. D., 1918.

Major, 6th Regiment, U. S. Marine Corps, 2d Division.
In the Bois-de-Belleau, France, on June 10, 1918, his unusual heroism in leading his company under heavy fire enabled it to fight with exceptional effectiveness. He personally worked fearlessly until he was mortally wounded.
Posthumously awarded. Medal presented to widow, Mrs. Mary E. Cole.

COLE, JAMES E. ....
Near Beaumont, France, Nov. 10,
1918.
R—Port Jervis, N. J.
B—Montague, N. J.
G. O. No. 32, W. D., 1919.

Private, 75th Company, 6th Machine Gun Battalion, U. S. Marine Corps, 2d Division.
He was painfully wounded in the foot by a bursting shell which killed or wounded all the members of his gun crew, but as soon as he had obtained first-aid treatment he immediately returned to his comrades and worked all night under heavy shell fire carrying wounded to the dressing station.

*COLEBANK, PHILIP R. ....
Near Ivoiry, France, Sept. 29, 1918.
R—Cincinnati, Ohio.
B—Denton, Tex.
G. O. No. 133, W. D., 1918.

First lieutenant, 147th Infantry, 37th Division.
This officer, with 2 soldiers, went out in the face of heavy machine-gun and artillery fire to bring in a wounded soldier. As they reached the wounded man a shell burst, killing him instantly.
Posthumously awarded. Medal presented to widow, Mrs. Emma R. Colebank.

COLEMAN, CARROLL J. (1764771) ....
Near Grand Pre, France, Oct. 17
and Nov. 1, 1918.
R—Jamaica, Vt.
B—Jamaica, Vt.
G. O. No. 70, W. D., 1919.

Private, first class, Company H, 311th Infantry, 78th Division.
With another soldier Private Coleman went out 25 yards in advance of the front line under severe shell and machine-gun fire to shelter a wounded comrade. After administering first aid, they carried him 4 kilometers to the dressing station across fields exposed to heavy hostile fire. On Nov. 1 he went through heavy artillery and machine-gun fire and assisted in carrying his commanding officer, who had been wounded, to the aid station. Upon his return, though nearly exhausted, he volunteered to go back with an ammunition detail.

COLEMAN, RUFUS M. (2220909)_____
    Near Verdun, France, Oct. 8-24, 1918.
    R—Weleetka, Okla.
    B—Muldrow, Okla.
    G. O. No. 2, W. D., 1919.

Private, Company B, 115th Infantry, 29th Division.
In the Verdun sector, east of the Meuse, he volunteered on several occasions during a heavy barrage to take messages to the battalion commander. He at all times disregarded his personal safety, and his splendid work was an inspiration to all those associated with him.

COLEMAN, WALLACE A_____
    Near Jaulny, France, Sept. 12-13, 1918.
    R—Racine, Wis.
    B—Columbus, Ind.
    G. O. No. 128, W. D., 1918.

First lieutenant, pilot, observation group, Air Service, 4th Army Corps.
On Sept. 12 Lieutenant Coleman, pilot, and second Lieutenant William Belzer, observer, while on an artillery surveillance mission, were attacked by an enemy plane. They waited until the enemy was at close range and then fired 50 rounds directly into the vital parts of the enemy machine, which was seen to disappear out of control. The next day Lieutenants Belzer and Coleman, while on a reconnaissance mission, were attacked by seven enemy aircraft. They unhesitatingly opened fire, but, owing to their guns being jammed, were forced to withdraw to our lines, where, clearing the jam, they returned to finish the mission. Their guns again jammed, and they were driven back by a large patrol of enemy planes. After skillful maneuvering they succeeded in putting one gun into use and returned a third time, only to be driven back. Undaunted, they returned the fourth time and accomplished their mission, transmitting valuable information to the Infantry headquarters.

COLEMAN, WILLIAM O_____
    Near Soissons, France, July 20, 1918.
    R—Chappell, S. C.
    B—Chappell, S. C.
    G. O. No. 60, W. D., 1920.

First lieutenant, Cavalry, aid-de-camp, 1st Infantry Brigade, 1st Division.
His liaison group having suffered many casualties, Lieutenant Coleman exposed himself to heavy fire on two different occasions in traversing the front line of the 18th Infantry during the attack. He returned each time with valuable information. Later, while on a mission to the front line, he was severely wounded, which necessitated the amputation of his left arm. Previously at Seicheprey, France, Mar. 1, 1918, while on a mission to obtain identifications of enemy participants in a raid, he went out in advance of our lines, under heavy enemy fire, to assist in the capture of an enemy raider.

COLFLESH, ROBERT W. (2395261)_____
    Near Fossoy, France, July 14, 1918.
    R—Des Moines, Iowa.
    B—Des Moines, Iowa.
    G. O. No. 59, W. D., 1919.

Corporal, Company M, 7th Infantry, 3d Division.
After his men had been caught in an intense artillery shelling, Corporal Colflesh, although wounded, refused to seek shelter until all his men had taken cover. While aiding the last man into a trench, he received a second wound.

COLLETTE, JOE (1422945)_____
    Near Sedan, France, Nov. 7, 1918.
    R—Elk River, Minn.
    B—Dayton, Minn.
    G. O. No. 44, W. D., 1919.

Private first class, Company L, 166th Infantry, 42d Division.
After his company had taken up a position and were waiting for orders, Private Collette, in the absence of the company runners, volunteered and carried all messages to and from the battalion post of command, some 800 yards distant. His route lay over a steep hillside, subjected to a heavy concentration of artillery, machine-gun, and sniper fire. He continued to carry messages after learning that the enemy had the exact range of the post of command and a shell struck the building, killing several of the men. Having had nothing to eat for 36 hours, he finally fell, completely exhausted.

COLLEY, DWIGHT T._____
    Near Bois-d'Haumont, France, Oct. 16, 1918.
    R—Nayatt, R. I.
    B—Barrington, R. I.
    G. O. No. 46, W. D., 1919.

Second lieutenant, 104th Infantry, 26th Division.
On Oct. 16 Lieutenant Colley led his company to the enemy's trenches despite the failure of supporting tanks to advance. After the order to withdraw had been given, he remained on the field, personally superintending the removal of every wounded man. He crawled along the ground for a long distance under the close-range fire of enemy machine guns in order to make sure that no wounded men had been left behind.

COLLEY, THOMAS M._____
    Near Stonne, France, Nov. 6, 1918.
    R—Mart, Tex.
    B—Texas.
    G. O. No. 35, W. D., 1919.

First lieutenant, 308th Infantry, 77th Division.
Though wounded, he voluntarily went through shell fire and gave first aid to wounded members of his platoon, thereby receiving additional wounds.

COLLIER, CLIVE C. (2808730)_____
    Near Villers-devant-Dun, France, Nov. 1-2, 1918.
    R—Waxahachie, Tex.
    B—Petersburg, Tenn.
    G. O. No. 46, W. D., 1919.

Corporal, Company D, 359th Infantry, 90th Division.
He courageously led his squad through severe machine-gun and artillery fire and drove off several machine-gun crews. Next day, after being wounded by fire from a heavy Maxim gun, he continued the advance, captured the gun, killed one of the crew, and took two prisoners.

COLLIER, JAMES W. (125842)_____
    Near Fleville, France, Oct. 6, 1918.
    R—Hot Springs, Ark.
    B—Jamestown, Tenn.
    G. O. No. 35, W. D., 1920.

Private, Headquarters Company, 6th Field Artillery, 1st Division.
Private Collier voluntarily went forward and made his way to the enemy front lines to locate hostile artillery firing at short range on our batteries. While on this mission, he was seriously wounded by machine-gun fire.

COLLINGE, PERCY T. (2284355)_____
    Near Very, France, Oct. 1-2, 1918.
    R—Los Angeles, Calif.
    B—Canada.
    G. O. No. 71, W. D., 1919.

Sergeant, first class, medical detachment, 316th Engineers, 91st Division.
He courageously and skillfully directed his men in giving aid to wounded soldiers under heavy shellfire. After being wounded he continued on duty until weariness forced him to go to the hospital. Within two days he was again with his detachment.

COLLINS, CORNELIUS P. (2309730)_____
    Near Meurcy Farm, north of Chateau-Thierry, France, July 29, 1918.
    R—Falmouth, Mass.
    B—Woods Hole, Mass.
    G. O. No. 4, W. D., 1923.

Private, Company C, 165th Infantry, 42d Division.
When the advance of his platoon was held up by enemy fire from a machine-gun nest on Hill 212, Private Collins with several others charged the nest, but were repulsed with heavy losses, all but three of the party being killed or wounded. He and two comrades immediately volunteered and made the second assault, which was successful. His courageous and gallant actions were an inspiration to his company. He was severely wounded on July 30.

*COLLINS, EMMETT E. (99699)
Near the Ourcq River, France, July 28, 1918.
R—Des Moines, Iowa.
B—Knoxville, Iowa.
G. O. No. 81, W. D., 1919.

Sergeant, Machine Gun Company, 168th Infantry, 42d Division.
After being wounded Sergeant Collins voluntarily returned to his company as soon as he had received first aid and fought courageously until he was killed. Posthumously awarded. Medal presented to mother, Mrs. Fanny Collins.

COLLINS, IRUM Q. (2039257)
Near Verdun, France, Oct. 12–17, 1918.
R—Sterling, Mich.
B—Marlette, Mich.
G. O. No. 130, W. D., 1918.

Private, Company D, 114th Infantry, 29th Division.
He displayed exceptional bravery carrying messages as a runner through barrage fire and gassed areas with heroic devotion to duty until he was finally overcome by gas.

COLLINS, JAMES H. (1710117)
West of St. Juvin, France, Oct. 16, 1918.
R—Sag Harbor, N. Y.
B—Sag Harbor, N. Y.
G. O. No. 20, W. D., 1919.

Private, Company L, 308th Infantry, 77th Division.
Private Collins, with another soldier, volunteered to cross a level open space for 600 yards, swept by converging machine-gun fire, to deliver a message to the front line, undeterred by the knowledge that six other soldiers had been wounded in a similar attempt. Crawling from one shell hole to another he succeeded in reaching the front line and delivering the message.

COLLINS, JAMES P. (1750872)
Near Grand Pre, France, Oct. 23, 1918.
R—Newark, N. J.
B—Newark, N. J.
G. O. No. 35, W. D., 1919.

First sergeant, Company D, 312th Infantry, 78th Division.
Upon his own initiative he crossed an open field heavily swept by machine-gun fire, and assisted by two other soldiers, carried to shelter his company commander, who had been mortally wounded.

COLLINS, PATRICK (48684)
Near Sergy, France, July 31, 1918.
R—El Paso, Tex.
B—Ireland.
G. O. No. 23, W. D., 1919.

Sergeant, Company M, 167th Infantry, 42d Division.
Being informed that a wounded man was lying in No Man's Land, Sergeant Collins immediately volunteered and with Pvt. William A. Pitts, of Company M, of the same regiment, went to his aid. The intense fire of the enemy necessitated crawling the entire distance. While on the return trip the wounded man was hit by a machine-gun bullet and instantly killed, but these two men brought in the dead body, crawling with great difficulty over the shell-torn ground.

*COLLINS, PETER, Jr. (1210325)
Near Ronssoy, France, Sept. 29, 1918.
R—Mount Vernon, N. Y.
B—Mount Vernon, N. Y.
G. O. No. 49, W. D., 1922.

Corporal, Company D, 107th Infantry, 27th Division.
When his platoon was checked at the wire in front of a strongly manned enemy trench, acting on his own initiative, he led his Lewis-gun squad around the left flank of the platoon under terrific hand grenade, rifle, and machine-gun fire from the enemy. Leading his squad, he jumped into a German trench and, with his Lewis gun, inflicted heavy casualties, silenced an enemy machine gun, and compelled many of the enemy to surrender. Then, by his rifle fire, he prevented the enemy from advancing in the trench from the left. While in the act of mopping up this trench Corporal Collins was killed. Posthumously awarded. Medal presented to father, Peter Collins, sr.

COLLINS, ROBERT L. (1698435)
Near Bois-de-la-Naza, France, Oct. 5, 1918.
R—Brewster, N. Y.
B—Lake Mahopac, N. Y.
G. O. No. 59, W. D., 1919.

Sergeant, Company L, 305th Infantry, 77th Division.
In the face of heavy machine-gun fire and grenades, he went forward with three other soldiers and brought back five seriously wounded men to a point where they could be given first-aid treatment. With utter disregard for his personal safety, he displayed coolness and good judgment effecting the rescue.

COLLINS, WILBUR M.
Near Chatel-Chehery, France, Oct. 8, 1918.
R—Macon, Ga.
B—Cochran, Ga.
G. O. No. 37, W. D., 1919.

Captain, 327th Infantry, 82d Division.
After the barrage had failed to fall on time, he led his platoon into the face of machine-gun fire, personally capturing one gun and turning it on the enemy, causing them to flee in disorder. Having reached his objective, he organized his positions under a heavy artillery barrage.

COLONNA, THOMAS (1751384)
At Grand Pre, France, Oct. 17, 1918.
R—Jersey City, N. J.
B—Italy.
G. O. No. 35, W. D., 1919.

Private, Company F, 312th Infantry, 78th Division.
Upon being wounded by a shell, he refused to go to the rear, but remained on the firing line against the advice of his commanding officer, dressing wounds of two comrades.

COLTON, JAMES STANLEY
Near Fismes, France, Aug. 4–6, 1918.
R—Pasadena, Calif.
B—Wapello, Iowa.
G. O. No. 99, W. D., 1918.

Second lieutenant, 116th Engineers, attached to Company F, 107th Engineers, 32d Division.
From Aug. 4 to 6, 1918, he successfully carried out a reconnaissance for the location of possible bridge sites across the River Vesle, near Fismes, France. He was constantly under heavy shellfire and was frequently harassed by fire from hidden machine-gun nests in the town. Nevertheless, he passed beyond our farthest lines and secured the desired information. He was wounded before his mission was accomplished, but refused to return to his battalion until he had made his reconnaissance and had been relieved by another detail.

COLVILLE, GEORGE, Jr. (2173873)
Near Remonville, France, Nov. 1, 1918.
R—Marceline, Mo.
B—Pleasanton, Kans.
G. O. No. 95, W. D., 1919.

Private, first class, Machine Gun Company, 354th Infantry, 89th Division.
Private Colville was a member of a machine-gun crew firing at close range from a shell hole in an open field when their gun became disabled. Thereupon he and two other soldiers advanced with pistols upon the enemy machine-gun nest at which they had been firing and captured it with three guns and nine prisoners. Putting one of the captured guns into immediate action against the enemy, they enabled the Infantry to advance with a minimum of casualties.

*COLVIN, DAVID P_____
In the Bois-de-Belleau, France, June 13, 1918.
R—Greensburg, Pa.
B—Greensburg, Pa.
G. O. No. 71, W. D., 1919.

Private, 18th Company, 5th Regiment, U. S. Marine Corps, 2d Division.
With another soldier Private Colvin advanced under the fire of an enemy machine gun, killed four of the crew, and captured the gun.
Posthumously awarded. Medal presented to mother, Mrs. Catherine Quigley.

*COMBS, HANON FIELDS_____
In France against the enemy during the months of July, August, and October, 1918.
R—Typo, Ky.
B—Hazard, Ky.
G. O. No. 19, W. D., 1920.

Captain, 38th Infantry, 3d Division.
On July 22, 1918, while at the head of his company, which was advancing in the direction of Le Charmel, Department of Aisne, France, against the enemy, he noticed that after passing a certain spot on the road his company was subject to the machine-gun fire of a sniper. Organizing a small party he led them in capturing the machine gun and killing the sniper. On Aug. 7, 1918, at the Vesle River, he personally silenced a machine-gun sniper that had been keeping up a harassing fire on his command. This was done in the face of machine-gun fire. On Oct. 9, 1918, near Cierges, Department of Meuse, Captain Combs was wounded in action by shrapnel in the back. He refused to go back to the first-aid station. Oct. 22, 1918, he was killed by a sniper while establishing a new post command.
Posthumously awarded. Medal presented to father, William G. Combs.

*COMBS, STEVE (2337876)_____
Near Cunel, France, Oct. 16, 1918.
R—Tallega, Ky.
B—Owsley County, Ky.
G. O. No. 64, W. D., 1919.

Corporal, Company A, 4th Infantry, 3d Division.
After all his superior officers and noncommissioned officers were killed or wounded, Corporal Combs took command of the platoon and pushed forward, capturing a machine gun and its crew. Although being mortally wounded a short time afterwards, he constantly called to his companions to continue the advance until he lost consciousness.
Posthumously awarded. Medal presented to sister, Mary Palmer.

COMERFORD, JOHN T_____
Near Bois-de-Belleau, north of Verdun, France, Oct. 28, 1918.
R—Brookline, Mass.
B—Brookline, Mass.
G. O. No. 56, W. D., 1922.

Captain, Machine Gun Company, 101st Infantry, 26th Division.
Following five days' combat, during which his company made three attacks and repulsed four counterattacks in which his company was well-nigh exhausted by uninterrupted fighting, the enemy placed a barrage of minenwerfer, machine-gun and artillery fire on a slightly entrenched front line, causing the Infantry to fall back, leaving a gap in the line. Captain Comerford volunteered to reestablish the line, gathered a group of 10 men, organized them, and led them into the gap, encountered an enemy patrol coming through, charged and drove them out, reestablished the line, and held it under a heavy machine-gun fire until reinforcements arrived. During this action he and a majority of his men were wounded, and some of the latter killed, but their heroic action prevented the enemy from inflicting heavy losses by flanking fire.

COMFORT, CHARLES W., Jr_____
At Seicheprey, France, Apr. 20, 1918.
R—New Haven, Conn.
B—Philadelphia, Pa.
G. O. No. 15, W. D., 1919.

First lieutenant, Medical Corps, attached to 102d Infantry, 26th Division.
He administered first aid for 36 hours without rest or relief to numerous wounded in the open, almost constantly under heavy artillery fire, and assisted in their evacuation, thereby setting an example of heroic performance of duties under the most trying circumstances.
Oak-leaf cluster.
A bronze oak leaf is awarded to him for the following act of extraordinary heroism: On Sept. 26, 1918, near Marcheville, France, he displayed the highest courage and devotion to duty, being continually present on the front line, administering first aid to the wounded under violent artillery and machine-gun fire.

COMFORT, WILLIS E_____
Near Soissons, France, July 18, 1918.
R—Kit Carson, Colo.
B—Onaga, Kans.
G. O. No. 15, W. D., 1919.

Captain, 16th Infantry, 1st Division.
After being severely injured he refused to be evacuated, but energetically led his company forward to its objective and maintained it there until he was mortally wounded.
Posthumously awarded. Medal presented to mother, Mrs. L. L. Comfort.

*COMINA, LOUIS (1629839)_____
Near Varennes, France, Sept. 26, 1918.
R—Chrysotile, Ariz.
B—Italy.
G. O. No. 126, W. D., 1919.

Private, Company C, 110th Infantry, 28th Division.
With 2 other soldiers, Private Comina voluntarily went forward under heavy artillery and machine-gun fire and silenced an enemy machine-gun nest, killing 4 of the crew and bringing back 11 prisoners. He was killed in action eight days later.
Posthumously awarded. Medal presented to mother, Mrs. Redenta Comina.

*COMPTON, LETCHER C_____
Near Charpentry, France, Sept. 27 and 28, 1918.
R—Kirkwood, Mo.
B—St. Louis, Mo.
G. O. No. 60, W. D., 1920.

Second lieutenant, intelligence officer, 1st Battalion, 140th Infantry, 35th Division.
Lieutenant Compton exposed himself to heavy machine-gun fire to advance ahead of his battalion for the purpose of locating accurately the enemy's positions. His repeated reconnaissances secured valuable information for his battalion commander. This officer was killed while making a reconnaissance south of Exermont.
Posthumously awarded. Medal presented to father, Richard J. Compton.

COMSTOCK, GEORGE E. (2021117)_____
Near Kadish, Russia, Dec. 30-31, 1918.
R—Detroit, Mich.
B—Detroit, Mich.
G. O. No. 14, W. D., 1920.

First sergeant, Company E, 339th Infantry, 85th Division (detachment in north Russia).
Sergeant Comstock went forward 200 yards in advance of our lines and guided back a man who had been blinded by shellfire and was stumbling around in full view of the enemy, who were sniping at him. On two other occasions he went forward in advance of our lines, exposing himself to heavy enemy fire, and returned with a wounded man.

CONATY, CHARLES C_____
Near Crezancy, France, July 16, 1918.
R—Taunton, Mass.
B—Taunton, Mass.
G. O. No. 99, W. D., 1918.

First lieutenant, chaplain, 11th Infantry, 28th Division.
Without regard for his personal safety Chaplain Conaty, under intense shellfire, following the attack of his troops from Crezancy to the Marne River, attended the wounded and throughout the night searched and assisted in carrying wounded to the dressing station.

CONDIT, GEORGE W. (2177534)_____
Near Barricourt, France, Nov. 2–3, 1918.
R—Troy, Kans.
B—Robinson, Kans.
G. O. No. 66, W. D., 1919

Private, first class, medical detachment, 353d Infantry, 89th Division.
Undeterred by seeing another first-aid man killed in attempting to reach a wounded soldier who was lying in the open a hundred yards from a wood, Private Condit fearlessly exposed himself to fire from enemy snipers and machine guns and succeeded in carrying the wounded soldier to shelter. Throughout the attack of Nov. 2–3 he worked under severe machine-gun fire without cover in dressing wounds after all other first-aid men had become casualties.

CONDIT, PHILLIP H_____
Near Cunel, France, Oct. 11, 1918.
R—East Orange, N. J.
B—East Orange, N. J.
G. O. No. 89, W. D., 1919.

First lieutenant, 7th Infantry, 3d Division.
With marked bravery, Lieutenant Condit led two platoons under withering machine-gun fire in an attack on a machine-gun nest, and succeeded in killing the entire enemy crew, including an officer. His company commander having been wounded, he organized his company's sector in a shallow enemy trench, which had been captured, and for two days held this position against repeated hostile counterattacks, inspiring his men by his courage and cheerful bearing in the face of hardships.

*CONE, BEN_____
In the Bois-de-Belleau, France, June 6, 1918.
R—Detroit, Mich.
B—Detroit, Mich.
G. O. No. 126, W. D., 1918.

Corporal, 82d Company, 6th Regiment, U. S. Marine Corps, 2d Division.
He showed exceptional heroism and coolness by advancing with an automatic rifle on a strongly defended enemy machine gun which he knew it was necessary to silence. He was killed while fearlessly going forward in this endeavor.
Posthumously awarded. Medal presented to mother, Mrs. Bertha Cone.

CONE, HERBERT A_____
Service rendered under the name, Herbert A. Cohn.

Second lieutenant, 39th Infantry, 4th Division.

CONKLIN, MATTHEW E_____
Near St. Juvin, France, Oct. 16, 1918.
R—Syracuse, N. Y.
B—Syracuse, N. Y.
G. O. No. 44, W. D., 1919.

Second lieutenant, 310th Infantry, 78th Division.
Lieutenant Conklin was wounded while leading his company in advance, but continued to lead his men across an open area under a heavy machine-gun and artillery fire, wading the Agran River, and took his position on a hillside, where he directed the digging in of his company before he would be evacuated.

*CONKLING, JOSEPH W_____
During the advance from the Sommerance-St. Juvin Road toward Landres-et-St. Georges, France, Oct. 11, 1918.
R—Atlanta, Ga.
B—Lewisville, Tex.
G. O. No. 3, W. D., 1921.

Captain, 327th Infantry, 82d Division.
When the entire line was held up by direct artillery fire and concentrated machine-gun fire he crawled out on the open crest of the hill for a distance of 200 yards, alone, for the purpose of reconnoitering and spotting enemy emplacements. Though the fire was constant and direct, he reached his objective and returned, seriously wounded several times by machine-gun bullets, which later caused his death.
Posthumously awarded. Medal presented to widow, Mrs. Joseph W. Conkling,

CONLIN, JOHN J. (1706995)_____
Near Merval, France, Sept. 14, 1918.
R—Brooklyn, N. Y.
B—Brooklyn, N. Y.
G. O. No. 10, W. D., 1920.

Sergeant, Company L, 307th Infantry, 77th Division.
Although wounded in the head by machine-gun bullet during an attack, Sergeant Conlin continued to lead his platoon in the advance and organized his position after the objective had been captured. He refused to be evacuated until loss of blood prevented his continuing with his company.

CONN, JEROME W. (254461)_____
At Le Charmel, France, July 28, 1918.
R—Toledo, Ohio.
B—Berkey, Ohio.
G. O. No. 10, W. D., 1920.

Private, Battery A, 76th Field Artillery, 3d Division.
While delivering an important message Private Conn was wounded in the arm and head by shell fragments. Neglectful of self, he completed his mission, delivered his message, and fell unconscious.

CONN, ROBBINS L_____
Near Revillon, France, Sept. 10, 1918.
R—New York, N. Y.
B—Chicago, Ill.
G. O. No. 35, W. D., 1919.

First lieutenant, 308th Infantry, 77th Division.
Lieutenant Conn volunteered and, with two soldiers, went on a patrol for the purpose of capturing prisoners. They crawled forward to within a few yards of the enemy lines, overpowered two sentries, and succeeded in delivering them to the battalion commander, despite the fact that the enemy put down a heavy barrage of rifle fire and rifle grenades.

*CONNELL, ANDREW F. (272)_____
At Saulx, France, Sept. 26, 1918.
R—Rochester, N. H.
B—Rochester, N. H.
G. O. No. 133, W. D., 1918.

Sergeant, 101st Ambulance Company, 101st Sanitary Train, 26th Division.
He labored unceasingly throughout the engagement, treating and evacuating the wounded soldiers in the advanced areas. He made repeated trips through an intense barrage and was again returning to duty at the front after a trip through exceptionally heavy fire when he was killed by an exploding shell.
Posthumously awarded. Medal presented to father, Joseph Connell.

CONNELLY, FRANCIS J. (1783359)_____
Near Bellicourt, France, Sept. 29, 1918.
R—Conshohocken, Pa.
B—Conshohocken, Pa.
G. O. No. 50, W. D., 1919.

Sergeant, Company A, 301st Battalion, Tank Corps.
Sergeant Connelly was on duty as gunner in a tank, whose track was broken by a direct hit from an enemy shell. Because of the heavy machine-gun fire it was impossible to repair the track, but Sergeant Connelly, accompanied by another soldier, left the tank, picked up some rifles, and, crawling through the trenches and brush to the rear of the machine-gun position, killed 4 of the enemy crew. They then returned to the tank and assisted in repairing the track under heavy shellfire.

CONNELLY, MICHAEL F. (264411)_____
Near Romagne, France, Oct. 10–13, 1918.
R—Lansing, Mich.
B—Boston, Mass.
G. O. No. 37, W. D., 1919.

Private, Headquarters Company, 125th Infantry, 32d Division.
Private Connelly, acting as battalion runner, repeatedly crossed the valley between Hill 258 and La Cote Dame Marie, a distance of 500 yards, swept continually by machine-gun and sniper fire.

CONNETTE, FRED (262518).
Near the Bois-les-Jomblets, north-
east of Chateau-Thierry, France,
July 31, 1918.
R—Detroit, Mich.
B—Canada.
G. O. No. 117, W. D., 1918.

Private, Company F, 125th Infantry, 32d Division.
While advancing with the first wave under heavy machine-gun fire and artillery
barrage, he was severely wounded. Knowing there was a machine-gun nest
directly in front of him which was decimating the company by its fire, he
disregarded his injuries, went forward, and killed the machine gunner and
captured the machine gun.

CONNOLLY, PATRICK A.
Near San Jose, Batangas, Luzon,
Philippine Islands, Dec. 23, 1901.
R—Springfield, Mass.
B—Ireland.
G. O. No. 17, W. D., 1924.

First lieutenant, 21st Infantry, U. S. Army.
While leading his detachment of 7 men in an attack against a band of 23 insurgent bolomen, which resulted in a desperate hand-to-hand encounter, he
displayed personal bravery, courage, and leadership qualities of the highest
order, and despite a severe wound on the head received early in the engagement he heroically continued the fight until all the enemy were killed.

°CONNORS, JOHN (551429).
Near Chateau-Thierry, France, July
15, 1918.
R—Pawtucket, R. I.
B—Pawtucket, R. I.
G. O. No. 88, W. D., 1918.

Corporal, Company H, 38th Infantry, 3d Division.
On the river bank near Chateau-Thierry, France, July 15, 1918, he commanded
a squad that kept 2 machine guns in operation to prevent Germans landing
until all in the group were killed. He was the last to fall, being shot as he
was in the act of throwing a hand grenade into a boat filled with the enemy.
Posthumously awarded. Medal presented to father, John Connors.

CONOVER, HARVEY.
Near Consenvoye, France, Oct. 27,
1918.
R—Hinsdale, Ill.
B—Chicago, Ill.
G. O. No. 37, W. D., 1919.

First lieutenant, pilot, 3d Observation Group, Air Service.
Flying at an altitude of less than 50 meters over enemy artillery and machine
guns, which were constantly firing on him, he and his observer staked the
American front lines and gave valuable information and assistance to the
advancing Infantry. Although suffering from 2 severe wounds, and with
a seriously damaged plane, he delivered a harassing fire on 6 enemy machine-
gun nests which were checking the advance of the ground troops, and successfully drove off the crews of 4 guns and silenced the other 2. He then
made a safe landing and forwarded his information to division headquarters
before seeking medical aid.

CONOVER, HOWARD R. (2404064).
Near Grand Pre, France, Oct. 20,
1918.
R—Trenton, N. J.
B—Trenton, N. J.
G. O. No. 46, W. D., 1919.

Private, Company A, 312th Infantry, 78th Division.
Although painfully wounded in the hand, he went to the assistance of a wounded
comrade who was lying helpless in an exposed position. He carried him from
the front, a distance of 400 yards, on his back, and when forced to relinquish
his burden because of exhaustion, he informed and directed stretcher bearers,
thus assuring the safety of his comrade.

*CONRAD, ROBERT Y.
Near Samogneux, France, Oct. 8,
1918.
R—Winchester, Va.
B—Winchester, Va.
G. O. No. 32, W. D., 1919.

Captain, 116th Infantry, 29th Division.
Captain Conrad led his company in assault, capturing many prisoners and
machine guns. He continually inspired his men by utter disregard of danger
and was mortally wounded while leading a charge on a machine-gun nest.
Posthumously awarded. Medal presented to widow, Mrs. Robert Y. Conrad.

CONROY, LAWRENCE (1737034).
Near Grand Pre, France, Oct. 23,
1918.
R—Newark, N. J.
B—Newark, N. J.
G. O. No. 81, W. D., 1919.

Mechanic, Company I, 312th Infantry, 78th Division.
Seeing a comrade lying wounded in front of our lines, Mechanic Conroy voluntarily made his way forward and rescued the man from direct machine-gun
fire. After all platoon leaders had become casualties he assumed command
and very creditably directed the action throughout the entire attack.

CONSIDINE, ALBERT J. (62602).
Near Vaux, France, July 13, 1918.
R—Newton, Mass.
B—Newton, Mass.
G. O. No. 30, W. D., 1921.

Corporal, Headquarters Company, 101st Infantry, 26th Division.
Leaving a place of safety he voluntarily dashed through a dense enemy barrage
to the rescue of comrades who were entombed in a signal station which had
been demolished by shellfire. Despite the continuing enemy fire, he dug
away the ruins, rescued the wounded, and assisted them to first aid.

CONSIDINE, FRANCIS (1707613).
Near La Harazee, France, Sept.
26, 1918.
R—New York, N. Y.
B—Somerville, Mass.
G. O. No. 35, W. D., 1919.

Private first class, Company A, 308th Infantry, 77th Division.
As acting corporal, Private Considine was in charge of a group which ran upon
an enemy machine-gun nest in a swamp. Although wounded in one foot by
a machine-gun bullet and in the other foot by a grenade, he continued to hold
his post and encouraged his men until assistance came.

CONTENT, HAROLD A.
Near Bois de Consenvoye, France,
Oct. 23, 1918.
R—New York, N. Y.
B—New York, N. Y.
G. O. No. 15, W. D., 1923.

Captain, field director, American Red Cross.
Although lame from causes existing prior to the World War and suffering from
gas poison and painfully wounded by shellfire, he continued to assist in the
removal of wounded soldiers and while so engaged encountered a severely
wounded runner, who was unable to continue on his mission of carrying an
important message from the commanding general, 58th Infantry Brigade, to
the commanding officer, 116th Infantry. Captain Content took the message,
and though suffering greatly himself, passed through heavy machine-gun and
shell fire and gassed woods for a distance of 500 yards and delivered the message.
Volunteering to carry the answer to the message, he again passed through
the gassed woods under a heavy machine-gun and shell fire and delivered in
person the message to the 58th Brigade command post. He then returned to
the task of succoring the wounded where he remained on duty until ordered
to the rear.

CONWAY, JAMES (2721014).
Near Nantillois, France, Sept. 29,
1918.
R—New Bedford, Mass.
B—Ireland.
G. O. No. 66, W. D., 1919.

Private, Company C, 58th Infantry, 4th Division.
Private Conway, a company runner, repeatedly volunteered for the most dangerous missions, carrying messages through enemy machine-gun and shell
fire on numerous occasions. Several days later, when his eardrum was broken
by concussion from a bursting shell, he refused to go to the rear for treatment,
but remained on duty until his company was relieved.

COOK, EVERETT RICHARD_____
Near Damvillers, France, Sept. 26, 1918.
R—Memphis, Tenn.
B—Indianapolis, Ind.
G. O. No. 13, W. D., 1919.

Captain, pilot, 91st Aero Squadron, Air Service.
While on a photographic mission in the vicinity of Damvillers, which necessitated a penetration of 20 kilometers within the enemy lines, Captain Cook was attacked by seven enemy pursuit planes and his plane was riddled with bullets. In spite of the attack he continued on his mission, turning only for our lines when his observer had secured photographs of great military value. In the combat one enemy aircraft was destroyed.

COOK, FRANK B., Jr_____
Near Ville-Savoye, France, Aug. 11, 1918.
R—Oakland, Calif.
B—Oakland, Calif.
G. O. No. 143, W. D., 1918.

Second lieutenant, 4th Engineers, 4th Division.
He directed the construction of an artillery bridge on the Vesle River under constant machine-gun and shell fire, setting a splendid example to the members of his command by his disregard of danger. On the morning of Aug. 11, he was wounded while personally looking after the safety of an outguard during a heavy bombardment.

*COOK, FRED A_____
Near St. Etienne-a-Arnes, France, Oct. 3–9, 1918.
R—Post Mills, Vt.
B—Strafford, Vt.
G. O. No. 37, W. D., 1919.

Major, 23d Infantry, 2d Division.
He led his battalion in an attack, although exposed to machine-gun fire from both flanks and front, steadying and encouraging his men by his fearless example. He was instantly killed while directing the reduction of a strongly intrenched machine-gun position.
Posthumously awarded. Medal presented to widow, Mrs. Fred A. Cook.

COOK, HARVEY WEIR_____
Near the Bois-de-Dole, France, Aug. 1, 1918.
R—Anderson, Ind.
B—Wilkinson, Ind.
G. O. No. 44, W. D., 1919.

First lieutenant, 94th Aero Squadron, Air Service.
Sighting six enemy monoplace planes, at an altitude of 3,500 meters, he attacked them, despite their numerical superiority, shooting down one and driving off the others.
Oak-leaf cluster.
For the following act of extraordinary heroism in action near Crepion, France, Oct. 30, 1918, Lieutenant Cook is awarded an oak-leaf cluster, to be worn with the distinguished-service cross: He attacked three enemy biplanes at an altitude of 1,660 meters. After a few minutes of severe fighting his guns jammed, but after clearing the jam he returned to the attack, shot down one of his adversaries in flames and forced the other two to retire to their own lines.

COOK, HOWARD C_____
Near Thiaucourt, France, Sept. 15, 1918.
R—Chicago, Ill.
B—Chicago, Ill.
G. O. No. 87, W. D., 1919.

Private, 95th Company, 6th Regiment, U. S. Marine Corps, 2d Division.
Private Cook repeatedly volunteered and carried messages for his battalion commander through severe machine-gun and artillery fire. He also exposed himself in an open field for several hours under fire in order to locate enemy snipers and machine-gun nests.

COOK, LLOYD H_____
Near the Bois Claire-Chenes, France, Oct. 20, 1918.
R—Chelsea, Mass.
B—Chelsea, Mass.
G. O. No. 98, W. D., 1919.

Captain, 7th Machine Gun Battalion, 3d Division.
After marching all night to his company's position for the initial attack, Captain Cook personally led the advance upon the enemy, inspiring his men by his bravery and determination, frequently going ahead of his company to reconnoiter its position. Even after being wounded in the leg by machine-gun fire he continued forward until he received three more wounds.

COOK, ROBERT P. (1320798)_____
Near Bellicourt, France, Sept. 29, 1918.
R—Altamahaw, N. C.
B—Reidsville, N. C.
G. O. No. 37, W. D., 1919.

Sergeant, Company G, 120th Infantry, 30th Division.
When his platoon was held up by machine-gun fire during an advance, although suffering from a painful machine-gun bullet wound in the hand, he personally killed the gunner and put the gun out of action, thus permitting the further advance of his platoon.

COOK, ROBERT R. (1386272)_____
At Bois-de-Chaume, France, Oct. 11, 1918.
R—Caledonia, Mich.
B—Caledonia, Mich.
G. O. No. 35, W. D., 1919.

Sergeant, Company A, 131st Infantry, 33d Division.
He crawled out in front of the lines some hundred yards to locate enemy snipers. While in this perilous position, he fired upon and put out of action a group of enemy machine gunners, thus exposing his position, and drawing enemy sniper fire. Having in his possession a number of asphyxiating grenades which emit a dense white smoke, he hurled one of them at the sniper's position, and, under cover of this improvised smoke screen, walked back to the lines.

COOK, WALTER_____
Near Blanc Mont, France, Oct. 6, 1918.
R—Priceburg, Pa.
B—Austria.
G. O. No. 46, W. D., 1919.

Gunnery sergeant, 43d Company, 5th Regiment, U. S. Marine Corps, 2d Division.
Without regard for his own personal safety, he rescued two men who were buried with dirt by the explosion of a German ammunition dump, and refused to find cover for himself until every man of his command had found a place of safety.

COOKE, JAMES H_____
At Santiago, Cuba, July 2, 1898.
R—New York, N. Y.
B—Ireland.
G. O. No. 32, W. D., 1919.

Corporal, Company B, 3d Infantry, U. S. Army.
For gallantry in action in the trenches at Santiago, Cuba, on July 2, 1898.

COOKSEY, THOMAS LARKIN (2239613)_____
Near Montigny - devant - Sassey, France, Nov. 5, 1918.
R—Lamesa, Tex.
B—Midlothian, Tex.
G. O. No. 37, W. D., 1919.

Sergeant, 315th Train Headquarters and Military Police, 90th Division.
During a very heavy attack in the vicinity of his post, where artillery fire and aircraft machine-gun fire had created a most confusing situation, he calmly directed traffic, aided wounded, and removed obstructions, thereby preventing wild disorder. He also assisted the drivers of ammunition trucks in getting their machines to places of safety.

COOLAHAN, WILLIAM T. (1257762)........
Near Fismes, France, Aug. 9, 1918.
R—Philadelphia, Pa.
B—Philadelphia, Pa.
G. O. No. 60, W. D., 1920.

Private, first class, Company B, 109th Machine Gun Battalion, 28th Division.
Private Coolahan exposed himself to heavy machine-gun fire to stand in the middle of the Vesle River and assist the members of his section across the stream. His individual courage was an important factor in enabling his organization to cross the stream and take up a position to cover the crossing of Infantry units.

COOLIDGE, EDMUND (9984)..............
Near Belleau Bois, France, Oct. 23, 1918.
R—Concord, Mass.
B—Concord, Mass.
G. O. No. 71, W. D., 1919.

Private, first class, Headquarters Company, 101st Infantry, 26th Division.
He left shelter and exposed himself to intense machine-gun fire when he saw a soldier lying wounded in advance of our lines. He reached the wounded man, despite the enemy fire, and dragged him back to a place of safety.

°COOLIDGE, HAMILTON.................
Near Grand Pre, France, Oct. 27, 1918.
R—Miami, Fla.
B—Chestnut Hill, Mass.
G. O. No. 37, W. D., 1919.

Captain, 94th Aero Squadron, Air Service.
Leading a protection patrol, he went to the assistance of two observation planes which were being attacked by six German machines. Observing this maneuver, the enemy sent up a terrific barrage from antiaircraft guns on the ground. Disregarding the extreme danger, he dived straight into the barrage, and his plane was struck and sent down in flames.
Posthumously awarded. Medal presented to father, J. R. Coolidge.

COONEY, JAMES M. (2805663)..........
Near Fey-en-Haye, France, Sept. 12, 1918.
R—Shawnee, Okla.
B—Norman, Okla.
G. O. No. 87, W. D., 1919.

Corporal, Company A, 357th Infantry, 90th Division.
Corporal Cooney, single handed, captured 43 Germans at one time, and later assisted in the capture of many other prisoners and machine guns.

COONEY, MICHAEL (89552)...........
Near Villers-sur-Fere, Aisne, France, Aug. 1, 1918.
R—New York, N. Y.
B—Ireland.
G. O. No. 26, W. D., 1919.

Corporal, Company C, 165th Infantry, 42d Division.
He carried a wounded soldier 150 yards to safety through heavy machine-gun fire. Then seeing his platoon about to advance, he returned under fire to the place where he had picked up the wounded man, secured his own rifle, and returned to join the advance.

COOPER, EDWIN H...................
Near Torcy, France, July 18, 1918, and near Bouresches, France, July 20, 1918.
R—Atlantic City, N. J.
B—Wilmington, Del.
G. O. No. 24, W. D., 1920.

Captain, photographic section, Signal Corps, 26th Division.
On July 18, 1918, he advanced fearlessly under enemy fire to an exposed position in a shell hole in front of the attacking troops in order to carry out a photographic mission. While in this position he went to the rescue of a wounded man and carried him to the shelter of a shell hole about 100 yards to the rear. Later he assisted in the evacuation of enemy prisoners. On July 20 he again advanced to a forward position in order to secure pictures of the attacking troops. His gallant conduct stimulated the morale of the advancing troops.

COOPER, EVERETT E. (1289814)........
Near Samogneux, France, Oct. 12, 1918.
R—Cambria, Va.
B—Montgomery County, Va.
G. O. No. 37, W. D., 1919.

Private, Company M, 116th Infantry, 29th Division.
He left a safe place, went through a terrific barrage to help a wounded comrade, and brought him back under heavy machine-gun fire. He also carried important messages through the barrage.

*COOPER, JAMES A...................
Between Berzy-le-Sec and Soissons, France, July 18-19, 1918.
R—Hale Center, Tex.
B—Troupe, Tex.
G. O. No. 99, W. D., 1918.

Second lieutenant, 3d Machine Gun Battalion, 1st Division.
His leadership under fire was distinguished by heroic conduct and the prompt utilization of every advantage. Through his fearlessness, devotion to duty, and while leading the men forward on the second day he was killed.
Posthumously awarded. Medal presented to father, J. T. Cooper.

COOPER, OSCAR M...................
Near St. Etienne, France, Oct. 4-6, 1918.
R—Sedro Woolley, Wash.
B—Osceola, Wash.
G. O. No. 37, W. D., 1919.

Private, 20th Company, 5th Regiment, U. S. Marine Corps, 2d Division.
Private Cooper, a runner, displayed exceptional courage in carrying messages for 3 days under shell and machine-gun fire.

COOPER, THOMAS (1874651)..........
At Trieres Farm, France, Sept. 30 to Oct. 2, 1918.
R—Florence, S. C.
B—Darlington, S. C.
G. O. No. 81, W. D., 1919.

Sergeant, Company K, 371st Infantry, 93d Division.
Wounded in an attack on Trieres Farm, France, Sept. 30, Sergeant Cooper remained on duty with his company and commanded his platoon until evacuated on Oct. 2, 1918.

*COOPER, WILLIAM N. (1316775)........
Near St. Souplet, France, Oct. 10, 1918.
R—Knoxville, Tenn.
B—Knoxville, Tenn.
G. O. No. 87, W. D., 1919.

Private, Company L, 119th Infantry, 30th Division.
He left shelter to advance under heavy machine-gun fire and rescued a wounded soldier, carrying him back to safety. He was killed in action the following day.
Posthumously awarded. Medal presented to mother, Mrs. Mary Cooper.

COPE, ONAL M. (156607)............
Near Vaux, France, July 1-4, 1918.
R—Arrowsmith, Ill.
B—Arrowsmith, Ill.
G. O. No. 23, W. D., 1919.

Corporal, Company C, 2d Engineers, 2d Division.
Acting as runner during the entire action, Corporal Cope volunteered and carried messages, making 8 trips one night, exposed at all times to high explosives and gas shells and machine-gun fire.

COPE, TOBE C.
At Trieres Farm, France, Sept. 30, 1918.
R—Fort Thomas, Ky.
B—Sparta, Tenn.
G. O. No. 21, W. D., 1919.

Major, 371st Infantry, 93d Division.
Wounded in the arm, Major Cope remained on duty throughout the engagement, led his battalion, and encouraged his men by his gallant example, and refused to be evacuated.
Posthumously awarded. Medal presented to widow, Mrs. Lillian Cope.

COPELAND, FRANCIS T. (1209958)
Near Ronssoy, France, Sept. 29, 1918.
R—New York, N. Y.
B—East Orange, N. J.
G. O. No. 19, W. D., 1920.

Mechanic, Company B, 107th Infantry, 27th Division.
Mechanic Copeland, with a companion, left the protection of a trench, and in the face of heavy machine-gun and grenade fire went in advance of our lines to rescue a wounded comrade. They were exposed to heavy fire from the time they left the trench. Mechanic Copeland's companion was killed as they were returning to the trench, but he, however, struggled on and succeeded in dragging the wounded man to safety.

COPELAND, SAMUEL
Against Filipino insurgents at Naguilian, Luzon, Philippine Islands, Dec. 7, 1899.
R—Washington, D. C.
B—Lynchburg, Va.
G. O. No. 3, W. D., 1925.

Private, Company A, 24th Infantry, U. S. Army.
When the command of which he was a member was held up in the crossing of the Rio Grande de Cagayan by rifle fire from a well-intrenched enemy, and being without boats or rafts with which to cross, Private Copeland, with 5 other members of his company, volunteered to swim the river. Displaying great gallantry and with utter disregard for his life, he swam the river in the face of heavy rifle fire, returned on a raft, secured arms and ammunition, crossed a second time, and took part in an attack which drove a superior force of the enemy from their trenches and the town occupied by them, thereby making possible the further advance of his company.

CORAM, CLAUDE A. (1306542)
Near Bellicourt and Nauroy, France, Sept. 29, 1918; and near Premont, France, Oct. 9, 1918.
R—Knoxville, Tenn.
B—Knoxville, Tenn.
G. O. No. 78, W. D., 1919.

Private, first class, Machine Gun Company, 117th Infantry, 30th Division.
When his platoon was held up by enemy artillery fire Private Coram, a runner, succeeded in passing through the severe fire and establishing liaison with the Infantry near by. On Oct. 9 he again showed unusual coolness in carrying a message through a wood containing numerous enemy snipers and machine guns.

CORBETT, MURL
Near Belleau Wood, France, June 6–25, 1918.
R—Edwardsville, Ill.
B—Pierce County, Mo.
G. O. No. 9, W. D., 1923.

First sergeant, 49th Company, 5th Regiment, U. S. Marine Corps, 2d Division.
Finding himself one of a few noncommissioned officers left alive after a desperate attack by the enemy, he organized a defensive position under heavy fire, withstanding numerous counterattacks. On the night of June 12 he led a patrol of three men into the enemy's lines, secured valuable information, and although wounded in the eye assisted in carrying from no man's land men more seriously wounded than himself, refusing evacuation until receiving peremptory orders from his company commander. Realizing his battalion was hard pressed for noncommissioned officers, he left the hospital without permission, returned to the front lines, and brought up reinforcements under heavy shell fire at a critical time. On Oct. 4, 1918, on Blanc Mont Ridge, Champagne, France, as battalion intelligence officer, he fearlessly exposed himself to enemy fire to secure important information; severely wounded by a bursting shell, and incapacitated by gas, he refused aid from comrades in order that the attack might not be delayed.

CORNELL, ELMO (53940)
Near Soissons, France, July 19, 1918.
R—Harbor Springs, Mich.
B—St. Johns, Mich.
G. O. No. 132, W. D., 1918.

Corporal, Company G, 26th Infantry, 1st Division.
He led an automatic-rifle squad near Soissons, France, July 19, 1918, until all his men had been killed or wounded, and then alone, from a farther advanced position, silenced the machine gun which had decimated his command.

CORNELL, PERCY DURYEA
Near St. Etienne, France, Oct. 4, 1918.
R—Galveston, Tex.
B—New York, N. Y.
G. O. No. 60, W. D., 1920.

Captain, 5th Regiment, U. S. Marine Corps, 2d Division.
Captain Cornell courageously led his company through heavy artillery and machine-gun fire in the attack on a strongly defended enemy position. His company held the exposed flank, and later, when the battalion withdrew, he skillfully covered the other shifting units while exposed to heavy enemy fire.

CORNELL, THOMAS L.
Near Soissons, France, July 20–21, 1918.
R—East Orange, N. J.
B—New Haven, Conn.
G. O. No. 132, W. D., 1918.

First lieutenant, 26th Infantry, 1st Division.
He showed complete disregard of self in placing his men to the best advantage under machine-gun and artillery fire near Soissons, France, July 20–21, 1918, and in the last of the fighting rendered invaluable assistance under fire in the reorganization of the battalion of which his command was a part.

CORNELL, WALTER R.
At Chateau-Thierry, France, June 6, 1918.
R—Cornell, Ill.
B—Cornell, Ill.
G. O. No. 119, W. D., 1918.

Marine gunner, 73d Company, 6th Regiment, U. S. Marine Corps, 2d Division.
Killed in action at Chateau-Thierry, France, June 6, 1918, he gave the supreme proof of that extraordinary heroism which will serve as an example to hitherto untried troops.
Posthumously awarded. Medal presented to father, Henry M. Cornell.

CORNWELL, ARTHUR B.
Near Exermont, France, Oct. 4, 1918.
R—Saginaw, Mich.
B—Saginaw, Mich.
G. O. No. 46, W. D., 1920.

First lieutenant, 18th Infantry, 1st Division.
When his battalion commander was missing Lieutenant Cornwell assumed command of the Second Battalion and personally led it in the attack on Hill 240, which was strongly defended by the enemy, who, by their machine-gun fire, inflicted heavy casualties upon the attacking troops. The position was taken and the units on both flanks, previously held up by the fire from this position, were enabled to continue their advance.

*COSGROVE, JOHN D.
  Near Charpentry, France, Sept.
  26-27, 1918.
  R—St. Louis, Mo.
  B—St. Louis, Mo.
  G. O. No. 16, W. D., 1919.

Second lieutenant, 139th Infantry, 35th Division.
As battalion intelligence officer, he repeatedly went in front of his own and adjoining battalions to secure information which he conveyed to regimental headquarters over fields swept by artillery and machine-gun fire. When the advance of his battalion was checked by destructive hostile fire, this officer, disregarding personal danger, conducted a personal reconnaissance, locating many machine guns and strongholds. He was killed while voluntarily leading and placing troops in advantageous positions to reduce these machine-gun nests.
Posthumously awarded. Medal presented to mother, Mrs. Sarah Agnes Nace.

COSTIANES, NICK (246618).
  Northeast of Chateau-Thierry,
  France, July 28, 1918.
  R—Greenville, Pa.
  B—Greece.
  G. O. No. 99, W. D., 1918.

Private, Company M, 168th Infantry, 42d Division.
He distinguished himself northeast of Chateau-Thierry, France, on July 28, 1918, when with four other men he raided an enemy machine-gun nest held by 12 Germans. As a result of daring and presence of mind, one of the enemy was killed, the other 11 captured, and their four machine guns turned upon the retreating foe.

COSTNER, OLEY (1309445).
  Near Ponchaux, France, Oct. 7,
  1918.
  R—London, Tenn.
  B—London, Tenn.
  G. O. No. 81, W. D., 1919.

Private, Company L, 117th Infantry, 30th Division.
When his company commander, for whom he was orderly, was wounded, he carried the wounded officer for 75 yards, under heavy fire, to a shell hole. When this location was fired upon by a machine gun, he crawled out, against his captain's advice, attacked the enemy position, and brought back the gun. Though ordered by his company commander to leave him and save himself from the heavy fire to which they were subjected, he remained at his post until he could effect the officer's evacuation to the rear.

COSTON, TONY.
  Against insurgent forces near Caridad, Leyte, Philippine Islands,
  Apr. 9, 1901.
  R—Webber, Kans.
  B—Atherton, Mo.
  G. O. No. 21, W. D., 1925.

Private, Company D, 44th Infantry, U. S. Volunteers.
Private Coston, accompanied by five other men, with utter disregard for his personal safety, gallantly ascended a steep, narrow, rocky trail under a direct fire of both cannon and rifles, and a hail of huge bowlders hurled down upon him, and succeeded in driving the insurgents from their strongly entrenched position and captured 12 cannon and a quantity of supplies.

COTTEN, ORVIL L. (1330179).
  Near Bellicourt, France, Sept. 27,
  1918.
  R—Memphis, Tenn.
  B—Falkner, Miss.
  G. O. No. 21, W. D., 1919.

Corporal, Company C, 105th Field Signal Battalion, 30th Division.
In order to maintain communication between two regiments of infantry, and, after an assisting detachment had suffered severe casualties, Corporal Cotten alone kept the line in repair, working under constant heavy shellfire. Although badly gassed, he refused evacuation, requesting and obtaining permission to continue his work.

COTTON, JOHN W.
  Near Soissons, France, July 20, 1918.
  R—Hollywood, Calif.
  B—Phoenix, Ariz.
  G. O. No. 98, W. D., 1919.

Captain, 2d Machine Gun Battalion, 1st Division.
When the advancing Infantry line had been held up by heavy artillery and machine-gun fire and all the officers had become casualties, Captain Cotton sent what remained of his machine-gun company to protect the flanks, and then voluntarily reorganized the assaulting line, breaking up a hostile counterattack which had been launched from the hill in front of them. Twice during the action he personally led tanks forward in advance of the line and reduced hostile machine-gun nests. As he was leading the attacking waves forward in the face of heavy fire, he was seriously wounded in the head by a machine-gun bullet, but he pushed on until he was forced to stop by loss of blood.

COUCHOT, OTTO V. (45798).
  Near Soissons, France, July 18, 1918.
  R—Dayton, Ohio.
  B—Versailles, Ohio.
  G. O. No. 3, W. D., 1921.

Sergeant, Company A, 18th Infantry, 1st Division.
Sergeant Couchot voluntarily proceeded with four other men the distance of about 400 yards in advance of our line and attacked a superior force of the enemy who were attempting to man a machine gun in a disabled French tank. Due to this bold attack the enemy was driven off and the tank retaken. Company A, 18th Infantry, was thereby enabled to continue the advance with slight loss.

COUGHLIN, WILLIAM C. (1754113).
  At Grand Pre, France, Oct. 23, 1918.
  R—Aurora, N. Y.
  B—Aurora, N. Y.
  G. O. No. 35, W. D., 1919.

Private, Company L, 312th Infantry, 78th Division.
He volunteered and maintained liaison with a company which had been cut off from the rest of the battalion, making several trips across open ground for 150 yards under intense machine-gun fire.

COURTER, JAMES L. (199132).
  At Marcheville, France, Sept. 26,
  1918.
  R—Thornton, Pa.
  B—Lansdowne, Pa.
  G. O. No. 15, W. D., 1919.

Corporal, Company C, 101st Field Signal Battalion, 26th Division.
When telephone communication to the rear had been cut and its reestablishment was impossible because of the intensity of the bombardment, he voluntarily carried important messages to the rear through a violent artillery barrage and machine-gun and rifle fire.

COURTNEY, ARTHUR M. (1495063).
  Near La Fontaine au Croncq Farm,
  France, Nov. 4, 1918.
  R—Friday, Tex.
  B—Brush Prairie, Tex.
  G. O. No. 78, W. D., 1919.

Private, Company D, 9th Infantry, 2d Division.
Private Courtney, a stretcher bearer, displayed exceptional gallantry in continuing to remove wounded from a field swept by machine-gun fire of such intensity that five of his associates were killed and two others wounded.

COURTNEY, GERALD.
  At Wadonville, France, Sept. 25,
  1918.
  R—Boston, Mass.
  B—Boston, Mass.
  G. O. No. 137, W. D., 1918.

Second lieutenant, 102d Machine Gun Battalion, 26th Division.
He was wounded while conducting his platoon into position preparatory to laying a barrage for a raid. With utter disregard for his personal safety he remained on duty for more than an hour, satisfying himself that all of his guns were properly laid and adjusted for firing. After his wounds were dressed he returned to duty and remained with his platoon until it was relieved.

COURTNEY, JAMES_____
Near Blanc Mont, France, Oct. 5, 1918.
R—Covington, Ky.
B—Bracken County, Ky.
G. O. No. 46, W. D., 1919.

Sergeant, 43d Company, 5th Regiment, U. S. Marine Corps, 2d Division.
Upon learning that a number of soldiers were lying wounded in no man's land, he immediately volunteered and made trips over an area swept by machine-gun and rifle fire until all the wounded had been carried to shelter.

COURTNEY, JOHN T. (73504)_____
At St. Agnant, France, Apr. 10, 1918.
R—Waltham, Mass.
B—Aurora, Ill.
G. O. No. 109, W. D., 1918.

Sergeant, Company L, 104th Infantry, 26th Division.
In the action of Apr. 10, 1918, he displayed courage, coolness, and the spirit of self-sacrifice, when he obtained permission to leave shelter and went through a shell-swept area to bring in wounded, carrying one wounded man more than 50 yards under heavy shellfire.

COUSINS, JOHN W_____
Near Conflans, France, Nov. 2, 1918.
R—New Haven, Conn.
B—New Haven, Conn.
G. O. No. 15, W. D., 1919.

First lieutenant, Infantry, observer, 91st Aero Squadron, Air Service.
In the course of a photographic mission of a particularly dangerous character he and his pilot were attacked by a superior number of enemy pursuit planes. During the combat that ensued, with remarkable coolness and excellent shooting he destroyed one of the attacking machines. Notwithstanding that the enemy aircraft continued to attack and harass them Lieutenant Cousins and his pilot reached all their objectives and returned to our lines with photographs of great military importance.

COVERDELL, VERN A_____
Near St. Etienne, France, Oct. 4, 1918.
R—Atwood, Colo.
B—Allendale, Mo.
G. O. No. 46, W. D., 1919.

Second lieutenant, 5th Regiment, U. S. Marine Corps, 2d Division.
After being wounded he dressed his own wounds and those of three men near him under heavy artillery and machine-gun fire. He reorganized his platoon, strengthened his position, and made a written report before allowing himself to be evacuated.

*COVERT, SAMUEL J. (1523288)_____
Near Montfaucon, France, Sept. 28, 1918.
R—Londonville, Ohio.
B—Perrysville, Ohio.
G. O. No. 50, W. D., 1919.

Private, sanitary detachment, 146th Infantry, 37th Division.
Voluntarily leaving cover, he went through intense machine-gun and artillery fire to the assistance of a wounded soldier and was himself killed while administering first aid to the latter.
Posthumously awarded. Medal presented to father, J. W. Covert.

*COWAN, JACK (2806615)_____
Near Vilcey, France, during the offensive against the St. Mihiel salient, Sept. 12, 1918.
R—Tulsa, Okla.
B—Dallas, Tex.
G. O. No. 129, W. D., 1918.

Private, first class, Machine Gun Company, 358th Infantry, 90th Division.
As a runner, he made four trips through a barrage, carrying important messages from the company command post. At another time he recovered a machine gun, the crew of which had been knocked out by a shell, and carried it unaided to a position where it was put in action. This brave soldier was killed while in the faithful performance of his duties.
Posthumously awarded. Medal presented to sister, Miss Willie Cowan.

COWIE, JAMES (40392)_____
Near St. Georges, France, Nov. 3, 1918.
R—Du Bois, Pa.
B—Adrian, Pa.
G. O. No. 35, W. D., 1919.

Sergeant, Company K, 9th Infantry, 2d Division.
While advancing with his platoon under heavy machine-gun fire, and after providing shelter for his men, Sergeant Cowie with two of his men outflanked a gun, capturing it and 18 prisoners.

COX, AULBERT D. (2021026)_____
At Vistofka, Russia, Mar. 3, 1919.
R—Arthur, Ill.
B—Terre Haute, Ind.
G. O. No. 108, W. D., 1919.

Sergeant, Company D, 339th Infantry, 85th Division (detachment in North Russia).
Upon learning that two companies of the enemy had worked their way to the rear of the allied lines, Sergeant Cox, a patient in a hospital, voluntarily left his bed, secured a Lewis gun, and successfully held off the enemy until assistance came up. The daring act of this gallant soldier prevented serious losses from being inflicted on the allied forces.

*COX, EDGAR L. (1980555)_____
Near Bellicourt, France, Sept. 29, 1918.
R—Lebanon, Ky.
B—Mansfield, Ky.
G. O. No. 32, W. D., 1919.

Private, first class, Machine Gun Company, 120th Infantry, 30th Division.
With five other soldiers Private Cox succeeded in breaking up three machine-gun nests and capturing eight prisoners under heavy artillery and machine-gun fire. After his platoon had reached its objective, he and four others volunteered and made a reconnaissance 600 yards in front of the line to make sure that the valley beyond was clear of the enemy. Private Cox has since been killed in action.
Posthumously awarded. Medal presented to father, Mathew Cox.

COX, EDWARD J. (5480206)_____
Near Jaulgonne, France, July 23, 1918.
R—Schaghticoke, N. Y.
B—Schaghticoke, N. Y.
G. O. No. 46, W. D., 1919.

First sergeant, Company L, 30th Infantry, 3d Division.
Although suffering from severe mustard-gas burns, he led a platoon through the attack of July 23 with unquestionable initiative, coolness, and courage.

COX, GEORGE C_____
Near Hill 240, north of Exermont, France, Oct. 4-5, 1918.
R—Cullowhee, N. C.
B—Cullowhee, N. C.
G. O. No. 15, W. D., 1923.

Captain, Signal Corps, 2d Field Signal Battalion, 1st Division.
Captain Cox made a most hazardous daylight reconnaissance in front of the American lines to determine the disposition of the enemy then holding this hill. During this entire exploit he was under close-range rifle and machine-gun fire. Subsequently he accompanied the assault wave in the attack, and, with two noncommissioned officers, established and maintained, from Oct. 5-9, 1918, telephonic communication on Hill 240 under intense artillery concentrations and in spite of direct machine-gun fire. This action was vital to the combat efficiency of the command and the success of the 1st Brigade.

Cox, John J. (551705).............
Near Cunel, France, Oct. 9, 1918.
R—Vandling, Pa.
B—William Penn, Pa.
G. O. No. 19, W. D., 1920.

Sergeant, Company I, 38th Infantry, 3d Division.
Sergeant Cox bravely took command of Company I after all officers of his company had been killed or wounded. He quickly reorganized the company, and, in the face of heavy machine-gun fire, led the company forward to its objective. He consolidated his position and held the line until relieved the following day.

Cox, Leonard..................
On the Vesle River, near Bazoches, France, Sept. 2, 1918.
R—New York, N. Y.
B—New York, N. Y.
G. O. No. 15, W. D., 1919.

Second lieutenant, 305th Infantry, 77th Division.
He left St. Thibaut in broad daylight with another officer and a patrol of 10 men to reconnoiter the enemy's positions across the Vesle River. The patrol divided and Lieutenant Cox conducted his half to the chateau in Bazoches, a recognized German post. He entered the yard of the chateau, met parties of the enemy personally, killed two and wounded another, who were firing on members of his patrol; continued his observations, though fired upon by machine guns, and with great skill withdrew his patrol under fire without loss, having gained valuable information.

Cox, Lewis B.................
During the St. Mihiel offensive, France, Sept. 12-15, 1918.
R—Lexington, Va.
B—Portland, Oreg.
G. O. No. 128, W. D., 1918.

First lieutenant, 6th Infantry, 5th Division.
From the beginning of the battle till evacuated from the field severely wounded, he displayed exceptional heroism, bravery, and devotion to duty of the highest order. Especially courageous was his work on Sept. 14 in the Bois-de-Bonvaux, when, facing a murderous machine-gun fire, many of his men down, himself twice wounded, he held his ground until reinforcements came up and surrounded and captured the enemy machine-gun nest. His work was a splendid example to the entire command.

Cox, Omar Clark (1330030)..........
Near Ypres, Belgium, July 16, 1918.
R—La Follette, Tenn.
B—Heiskell, Tenn.
G. O. No. 21, W. D., 1919.

Sergeant, first class, Company A, 105th Field Signal Battalion, 30th Division.
He volunteered and assisted a British soldier to reestablish communication lines, which the heavy artillery fire had made useless. Crawling through almost direct machine-gun fire and making his way through barbed wire, he reached his point, where he remained for almost two hours. He made his way back to our lines, through an intense hand-grenade bombardment.

*Coxe, Edward G. (89745)..........
At Ferme de Meurcy, near Villers-sur-Fere, France, July 28, 1918.
R—New York, N. Y.
B—New York, N. Y.
G. O. 116, W. D., 1918.

Private, Company D, 165th Infantry, 42d Division.
He continued to care for the wounded under heavy machine-gun and artillery fire after he himself was severely injured. He has since died as a result of the wounds received in action.
Posthumously awarded. Medal presented to father, Michael J. Coxe.

*Coyle, Edward A. (43448)..........
South of Soissons, France, July 18 and 19, 1918.
R—Darien Center, N. Y.
B—Albany, N. Y.
G. O. No. 27, W. D., 1920.

First Sergeant, Company H, 16th Infantry, 1st Division.
On July 18, 1918, Sergeant Coyle, after carrying his wounded company commander to safety and dressing his wounds, took command of the company, all the officers having been either killed or wounded, and reorganized it under heavy enemy fire. On the following day he successfully led the company in the attack, attaining all objectives. He later exposed himself to heavy machine-gun fire while directing units in order to repulse an enemy counterattack. While in performance of this act he was killed by enemy fire.
Posthumously awarded. Medal presented to mother, Mrs. Josephine Sanderson.

Coyle, William J..................
Near Cheppy, France, Sept. 26, 1918.
R—Seattle, Wash.
B—Sutter Creek, Calif.
G. O. No. 70, W. D., 1919.

Captain, 363d Infantry, 91st Division.
While on duty as liaison officer, Captain Coyle observed a strongly fortified enemy trench, which was a menace to further advance. He organized a group of 14 men, and in the face of heavy machine-gun and sniper fire led them in an attack on the trench. Through his quick decision and courage the hostile position was captured with 52 prisoners, 3 heavy and 2 light machine guns.

Crabbe, George S..................
Near Cierges, France, July 31, 1918.
R—Saginaw, Mich.
B—Saginaw, Mich.
G. O. No. 64, W. D., 1919.

First lieutenant, 125th Infantry, 32d Division.
While advancing with his company he wrenched his leg severely in the crossing of the Ourcq River, but continued in the advance. Later he was severely wounded by machine-gun bullets in the left thigh, but again refused evacuation, and continued in command of his company until the objective had been reached and the position consolidated, remaining nine hours with his company after having been wounded.

Crabbe, Thomas P..................
At Blanc Mont Massif, France, Oct. 4, 1918.
R—Cleveland, Ohio.
B—Oliveburg, Ohio.
G. O. No. 21, W. D., 1919.

Private, 6th Machine Gun Battalion, U. S. Marine Corps, 2d Division.
He voluntarily left a sheltered position under intense enemy bombardment, dressed the wounds of four wounded men lying in a position exposed to intense enemy machine-gun fire, carried them one by one to a place of safety, and then went for a stretcher bearer to assist him in evacuating them.

Craddock, John E. (2274429)..........
Near Very, France, Oct. 20, 1918.
R—Riverside, Calif.
B—Mount Carroll, Ill.
G. O. No. 81, W. D., 1919.

Corporal, 316th Train Headquarters and Military Police, 91st Division.
Engaged in regulating traffic at an important distributing point, which was suddenly bombarded, Corporal Craddock, by his coolness, prevented a general stampede when drivers and working parties started to seek cover.

Craft, Urban V. (2038959)..........
At Bois-de-Grand Montagne, France, Oct. 15-18, 1918.
R—West Branch, Mich.
B—Rose City, Mich.
G. O. No. 37, W. D., 1919.

Private, Headquarters Company, 322d Field Artillery, 83d Division.
He, with utter disregard for his personal safety, constantly exposed himself to enemy fire while repairing wires and maintaining important telephone communications within his area.

*CRAIDGE, ROBERT E.
At Hill 212, near Cierges, northeast of Chateau-Thierry, France, July 31, 1918.
R—Bay City, Mich.
B—Saginaw, Mich.
G. O. No. 116, W. D., 1918.

Corporal, Company I, 125th Infantry, 32d Division.
Corporal Craidge was in charge of an advanced Chauchat rifle position while his company was exposed to severe machine-gun fire from the front and right flank. After 3 of his men had been killed and the other totally blinded by shellfire, Corporal Craidge remained at his post and kept his gun in operation until he was killed, this brave act inspiring the members of his organization who were eyewitnesses.
Posthumously awarded. Medal presented to mother, Mrs. Elinor Craidge.

CRAIG, CLAUDE H.
At Bois-des-Forges, France, Sept. 26, 1918.
R—Austin, Minn.
B—Milbank, S. Dak.
G. O. No. 56, W. D., 1922.

First lieutenant, 132d Infantry, 33d Division.
When the advance of his platoon was held up by enemy machine-gun fire, Lieutenant Craig, with utter disregard for his own safety and in front of his platoon, personally led it in the attack on the machine-gun nests, and assisted in destroying the hostile nests, killing 2 officers and 6 men. His courageous action was an inspiration to his men and made possible the advance of the attacking wave.

*CRAIG, JOHN M.
Near Soissons, France, July 18–19, 1918.
R—Garnett, Kans.
B—Mahaska County, Iowa.
G. O. No. 35, W. D., 1919.

Lieutenant colonel, headquarters 16th Infantry, 1st Division.
After the commander of the front-line battalion and most of the junior officers had been killed, he voluntarily left regimental headquarters and assumed command of the assault battalion. He led a small detachment of soldiers in a personal reconnaissance, locating and destroying machine-gun nests, thereby permitting the battalion to continue the advance, which he led personally until he was killed.
Posthumously awarded. Medal presented to widow, Mrs. Catherine C. Craig.

CRAIG, WILLIAM H. (1834369).
Near Grand Carre Farm, France, Nov. 1, 1918.
R—Castle Shannon, Pa.
B—Castle Shannon, Pa.
G. O. No. 10, W. D., 1920.

Corporal, Battery F, 313th Field Artillery, 80th Division.
While acting as battery scout he carried several important messages through intense artillery and machine-gun fire from the Infantry front lines to the battery position. On the afternoon of Nov. 1, he was severely wounded while carrying a message to an advance observation post.

CRAMER, JOHN W. (2257086).
At Audenarde, Belgium, Nov. 1, 1918.
R—Hailey, Idaho.
B—Hailey, Idaho.
G. O. No. 21, W. D., 1919.

Corporal, Company H, 361st Infantry, 91st Division.
He was a member of a patrol sent out to reconnoiter the town of Audenarde. This patrol discovered several enemy machine-gun sniper posts, located in buildings, which were enfilading the streets of the town. With another soldier, he dodged from building to building and entering one of the houses containing a machine gun, captured 2 machine gunners.

*CRAMP, TONY (113196).
North of the River Ourcq, near Villers-sur-Fere, France, July 28, 1918.
R—Fond du Lac, Wis.
B—Germany.
G. O. No. 88, W. D., 1918.

Private, Company B, 150th Machine Gun Battalion, 42d Division.
He showed extraordinary courage and ability as a leader of men in field of battle. When his section sergeant had been killed and his corporal wounded, he assumed command of his gun section and led them forward against the enemy, directing the fire with effect until killed.
Posthumously awarded. Medal presented to mother, Mrs. Victoria Cramp.

CRANDALL, DEWITT H. (1215536).
East of Roussoy, France, Sept. 29, 1918.
R—Canisteo, N. Y.
B—Campbell, N. Y.
G. O. No. 20, W. D., 1919.

Private, medical detachment, 108th Infantry, 27th Division.
During the operations against the Hindenburg line, Private Crandall, although he had been twice wounded, courageously treated the wounded, inspiring the combat troops by his example until wounded a third time.

*CRANDALL, JOSEPH B. (554827).
Near Chateau-Thierry, France, July 14–15, 1918.
R—Old Bridge, N. J.
B—Old Bridge, N. J.
G. O. No. 44, W. D., 1919.

Sergeant, Company B, 9th Machine Gun Battalion, 3d Division.
Being detached from his platoon with a machine-gun section for the purpose of making a relief, Sergeant Crandall, although severely wounded and suffering great pain, continued to direct his section until killed at his post.
Posthumously awarded. Medal presented to father, Ira C. Crandall.

*CRANDALL, ROBERT F.
Near Chateau-Thierry, France, June 18, 1918.
R—Stamford, Conn.
B—New Canaan, Conn.
G. O. No. 35, W. D., 1920.

First lieutenant, 38th Infantry, 3d Division.
Lieutenant Crandall led a platoon across the Marne River. When fired upon from three directions the patrol was forced back to the river bank. Finding some of the men missing, he alone returned to the place within the enemy lines in order to guide back the missing members of the patrol. In an encounter with a group of the enemy he captured and brought back an enemy noncommissioned officer.
Posthumously awarded. Medal presented to widow, Mrs. Agnes E. Crandall.

CRANDALL, ROBERT L. (3127323).
Near Carrefour - de - Meurrissons, France, Sept. 27, 1918.
R—Peoa, Utah.
B—Peoa, Utah.
G. O. No. 44, W. D., 1919.

Private, Company A, 305th Infantry, 77th Division.
After his command had taken shelter from the enfilading machine-gun and trench-mortar fire of the enemy, Private Crandall, with 2 other soldiers, crawled to the aid of wounded comrades, thus saving the lives of at least 2, while exposed to terrific fire of the enemy.

CRANE, JOHN A.
Near Cantigny and again in the battle of July 18, 1918, southwest of Soissons, France.
R—Pikesville, Md.
B—St. Georges, Md.
G. O. No. 15, W. D., 1923.

Lieutenant colonel, 6th Field Artillery, 1st Division.
Having on the morning of that day (July 18, 1918) advanced his batteries 3 miles and, finding that the normal liaison between Artillery and Infantry was not then operating and that observation of fire from near the batteries was impossible, went forward himself to locate our Infantry front line and to personally conduct fire on the enemy's forces. He advanced through heavy enemy artillery and machine-gun fire over a terrain easily visible to the enemy to the front line of the supported infantry at Chaudun, where he drew down a violent shelling from enemy artillery. Undeterred by this and seeing German machine-gun detachments advancing to position a short distance from him, he continued to seek a favorable point of observation. While engaged in this dangerous and self imposed mission he was desperately wounded. This inspiring action on the part of the battalion commander, locating and supporting the advance Infantry elements, permitted the supported Infantry to seize and hold a line ahead of adjacent units, and, protected by its supporting Artillery, to hold the lead in the attack of the Tenth French Army.

CRANFORD, ALBERT LEE (1310721)_____
Near Bellicourt, France, Sept. 27, 1918.
R—Concord, N. C.
B—Concord, N. C.
G. O. No. 50, W. D., 1919.

Private, Company D, 118th Infantry, 30th Division.
After all his comrades had been killed or wounded, and he himself injured by an enemy hand grenade, Private Cranford defended his post single-handed in the face of a German bombing attack until reinforcements arrived. He then continued on duty with his company, refusing to be evacuated until he was severely gassed later.

CRANFORD, RALPH F. (322881)_____
Near Sitsa, Siberia, June 26, 1919.
R—Franklin, Ind.
B—Scottsburg, Ind.
G. O. No. 133, W. D., 1919.

Sergeant, Company C, 31st Infantry.
He was in command of a platoon of his company and successfully defended the railroad and bridges against an attack by a force of partisans greatly superior in numbers. Due to Sergeant Cranford's skill in placing the members of his platoon, they were able to drive the partisans away from the strong position they had taken upon the hills without loss to the platoon.

CRAVEN, ALBERT D. (542618)_____
North of Cierges, France, Oct. 7, 1918.
R—Weiser, Idaho.
B—Weiser, Idaho.
G. O. No. 10, W. D., 1920.

Sergeant, Company K, 7th Infantry, 3d Division.
Sergeant Craven, on 3 occasions, voluntarily exposed himself to heavy machine-gun fire; crawling in advance of our lines, he rescued wounded comrades.

CRAVEN, FREDERICK P. (2309519)_____
Near Landres-et-St. Georges, France, Oct. 15, 1918.
R—Roxbury, Mass.
B—Roxbury, Mass.
G. O. No. 37, W. D., 1919.

Private, Company C, 165th Infantry, 42d Division.
After all company and battalion agents had been killed or wounded in an attempt to deliver an important message to the battalion commander, Private Craven voluntarily undertook the task. Under heavy machine-gun, sniper, and artillery fire he accomplished his mission and returned with an answer.

CRAVEN, HERMAN C. (1329839)_____
Near Premont, France, Oct. 9, 1918.
R—Eupora, Miss.
B—Eupora, Miss.
G. O. No. 31, W. D., 1919.

Private, Company G, 120th Infantry, 30th Division.
While serving as a runner he volunteered to go to an exposed position on the flank to a body of troops, deliver a message to them if they were Americans, and report back if they were Germans. Using a captured German bicycle, he rode along a road subjected to heavy fire, found that the troops were American, and delivered an important message.

CRAVEN, HOWARD (2256233)_____
Near Gesnes, France, Sept. 26, 1918.
R—Emporium, Pa.
B—Long Island, N. Y.
G. O. No. 20, W. D., 1919.

Battalion sergeant major, Headquarters Company, 361st Infantry, 91st Division.
Although wounded he remained on duty, and during the heaviest bombardment of the battalion command post reorganized and kept under control the liaison section, which was essential to the successful operation of the battalion. He constantly exposed himself to danger and rendered service of great value.

CRAVEN, WILLIAM J. (3180)_____
At Wadonville, France, Sept. 25–26, 1918.
R—Boston, Mass.
B—Boston, Mass.
G. O. No. 139, W. D., 1918.

Private, first class, 101st Ambulance Company, 101st Sanitary Train, 26th Division.
He assisted in establishing a dressing station in a dugout in an advanced position. When it was destroyed by a shell he worked unceasingly in the open under fire from enemy machine guns and snipers, caring for the wounded. He remained at his post for several hours after his station had been ordered closed, permitting neither his own exhaustion nor the enemy fire to deter him from aiding the wounded.

CRAWFORD, CLIFFORD (2335087)_____
Near Bussy Farm, France, Sept. 28–29, 1918.
R—Boston, Mass.
B—Union Springs, Ala.
G. O. No. 13, W. D., 1919.

Private, Headquarters Company, 372d Infantry, 93d Division.
He was acting as liaison agent between regimental headquarters and the battalion. Having carried a message through a heavy bombardment to the commander of a battalion which was about to make an attack he joined the first wave of the attack and dashed into the enemy's trenches. Seeing 2 of the enemy rush to a dugout he followed them and brought 10 prisoners from the dugout, killing 2 who tried to escape.

CRAWFORD, HAROLD E. (106190)_____
Between Soissons and Berzy-le-Sec, France, July 19, 1918.
R—Lorain, Ohio.
B—Avilla, Ind.
G. O. No. 99, W. D., 1918.

Private, Company B, 3d Machine Gun Battalion, 1st Division.
He bravely carried messages through shellfire and reorganized and directed Infantry units. Intrusted with a particularly important message, that he knew must be delivered, he fearlessly started through a heavy bombardment to execute his mission, and while so engaged was severely wounded.

CRAWFORD, NED (2294889)_____
At Epinonville, France, Oct. 2, 1918.
R—Hood River, Oreg.
B—Marietta, Ohio.
G. O. No. 127, W. D., 1918.

Private, first class, Company C, 316th Field Signal Battalion, 91st Division.
When the telephone station in which he was working was struck by a shell, killing 2 men and injuring 5, he disregarded personal safety and continued to operate his switchboard in an exposed position, in order that communication might be maintained until a new central could be established in a new location.

CRAWFORD, ROBERT E. (915492)_____
Near Verdun, France, Nov. 4, 1918.
R—Oklahoma City, Okla.
B—Hillsboro, Tex.
G. O. No. 37, W. D., 1919.

Corporal, Company D, 7th Engineers, 5th Division.
When 3 boats in a pontoon bridge across the Meuse River were destroyed by artillery fire he volunteered and waded into the river under heavy shellfire and, by holding up the deck until new boats were launched and placed in position, permitted the uninterrupted crossing of the Infantry.

CRAWFORD, WILLIAM B_____
At Ferme de la Riviere, France, Sept. 30, 1918.
R—Denison, Tex.
B—Corinth, Miss.
G. O. No. 46, W. D., 1919.

Captain, 370th Infantry, 93d Division.
Having been placed in command of Company L, whose task it was to lead the advance in an attack, the same undertaking having failed the day previous, Captain Crawford, in order to assure the success of the attack, personally led the advanced element of his company in the face of heavy fire. The objective was successfully carried, due to Captain Crawford's gallant conduct.

CREECH, JESSE O.
　　Various dates and places in France, 1918.
　　R—Takoma Park, D. C.
　　B—Harlan, Ky.
　　G. O. No. 19, W. D., 1926.

First lieutenant, 148th Aero Squadron, Air Service.
Being on enemy patrol on Sept. 26, 1918, at Cambrai, France, when a large number of enemy airplanes were encountered, in the fight that ensued Lieutenant Creech shot down 2 of the enemy planes and saved the commander of our patrol from being shot down. On Oct. 28, 1918, near Jenlain, France, Lieutenant Creech's flight of 5 planes was attacked by 8 Fokker biplanes. In this encounter Lieutenant Creech also shot down 2 enemy planes. On Sept. 28, 1918, south of Masnieres, France, Lieutenant Creech with his flight attacked an enemy balloon and compelled the observers to jump. Enemy troops were then attacked in close formation, causing many casualties and scattering all the troops. In all of these encounters Lieutenant Creech displayed high courage, great valor, and utter disregard of danger. He constantly went to the assistance of members of his flight and exposed himself with great fearlessness, and yet with all displayed keen judgment and tireless energy. He proved himself a leader of unusual ability, and was a constant inspiration to the members of his command.

CREPEAU, LOUIS J.
　　Near Belleau Wood, France, June 13, 1918.
　　R—Boston, Mass.
　　B—Detroit, Mich.
　　G. O. No. 50, W. D., 1919.

Private, 55th Company, 5th Regiment, U. S. Marine Corps, 2d Division.
While carrying a message in the Bois-de-Belleau he was surrounded by a detachment of Germans, who demanded his surrender. By his initiative and quick action he not only was able to return to our lines but captured four of the Germans and brought them to our lines. He then selected a different route and delivered the message.

CRESSMAN, CALVIN J. (2716337).
　　Near Moirey, France, Nov. 9, 1918.
　　R—Coopersburg, Pa.
　　B—Center Valley, Pa.
　　G. O. No. 46, W. D., 1919.

Private, Company E, 314th Infantry, 79th Division.
Private Cressman, first carrier for his automatic-rifle team, advanced with a patrol against strong machine-gun positions. Although wounded five times, he refused to be taken back and continued to load the automatic rifle in the face of heavy machine-gun fire.

CRESSWELL, JAMES A. (61857).
　　North of Verdun, France, Oct. 26, 1918.
　　R—Hingham, Mass.
　　B—Hingham, Mass.
　　G. O. No. 56, W. D., 1922.

Corporal, Company K, 101st Infantry, 26th Division.
In the desperate attacks of the 101st Infantry he captured and held the last heights of the Meuse between the 23d and 29th of October, 1918. Corporal Cresswell, having succeeded to the command of his platoon, displayed the utmost bravery and fearlessness in personally leading his platoon in repeated attacks against the enemy, during which he exposed himself with great courage. He proved himself an inspiring and successful leader in all these attacks. On Oct. 26, with utter disregard of danger, he personally led an attack, successfully wiping out a group of 4 machine-gun nests, thereby clearing dangerous terrain and allowing the unit on his right to successfully proceed with the counterattack. He went without rest and sleep to the point of exhaustion, and his personal valor and bravery gave added impetus to his men, so that under his able leadership all the new objectives during these dates were captured until the platoon itself was relieved on Oct. 29.

CRISP, CURTIS M. (1383886).
　　At Marcheville, France, Nov. 10, 1918.
　　R—Parrish, Ill.
　　B—Parrish, Ill.
　　G. O. No. 46, W. D., 1919.

First sergeant, Company F, 130th Infantry, 33d Division.
When all the company runners had been wounded during a raid by his battalion, he volunteered to establish liaison with an adjacent company. While going through a heavy barrage under sniper fire from 3 directions he was knocked unconscious by the concussion of a bursting shell. Upon recovering he succeeded in killing a sniper who was picking off our men and had wounded his company commander. Though unable to stand, Sergeant Crisp insisted upon remaining on duty with his company.

CRISTOFARO, VITTORIO (1746634).
　　Near Grand Pre, France, Oct. 23, 1918.
　　R—Syracuse, N. Y.
　　B—Italy.
　　G. O. No. 37, W. D., 1919.

Private, Company K, 312th Infantry, 78th Division.
While his company was forced to lie in the open for a period of 24 hours, because of the murderous machine-gun and artillery fire of the enemy, Private Cristofaro, occupying a position in front of the company, advanced and attacked the enemy gun nests, silencing no less than three guns and killing their crews.

CRITES, HERMAN.
　　Near Juvigny, France, Aug. 31, 1918.
　　R—Flint, Mich.
　　B—Burr Oak, Mich.
　　G. O. No. 64, W. D., 1919.

First lieutenant, 125th Infantry, 32d Division.
After his own company had reached the objective assigned to it, he made a personal reconnaissance of the front and flank and discovered that the right-flank regiment was being held up by a strong machine-gun nest. He immediately maneuvered his company to a position where an enfilading fire could be delivered, enabling the flanking regiment to attain its objective. By utilizing the German arms and ammunition his company took up the advance, when it was discovered that a wide interval had been left between the two regiments because of a terrific machine-gun fire on the flanks of the regiments. He at once directed his company in the filling of this gap, holding the position during the night, as well as capturing four heavy and two light machine guns.

CROCKER, JOHN M. (1961691).
　　Near Verdun, France, Oct. 16–17, 1918.
　　R—Hamilton, Ohio.
　　B—Hamilton, Ohio.
　　G. O. No. 37, W. D., 1919.

Corporal, Headquarters Company, 322d Field Artillery, 83d Division.
Maintaining a telephone line between infantry and artillery under a constant artillery barrage, his courage made possible the launching of artillery fire which stopped enemy counterattacks. During the action he repaired 30 breaks, his line being so badly cut that he was obliged to use enemy wire for repairing. Because of the intensity of the fire he sent his men to a place of safety, remaining alone at the hazardous post.

CROCKETT, EDWARD L. (2103705).
　　Near Samogneux, France, Oct. 15, 1918.
　　R—Florence, Ala.
　　B—Barton, Ala.
　　G. O. No. 37, W. D., 1919.

Sergeant, Company E, 116th Infantry, 29th Division.
When his company was subjected to severe machine-gun fire, Sergeant Crockett, with two other soldiers, attacked a nest of four machine guns, killing 8 of the enemy and capturing 27.

CROCKETT, OREN O. (1631317)_____
    North of Cierges, France, Oct. 4,
      1918.
    R—Plain View, N. Mex.
    B—Pearsal, Tex.
    G. O. No. 10, W. D., 1920.

Private, Company C, 8th Machine Gun Battalion, 3d Division.
Private Crockett, during an attack under terrific shell and machine-gun fire, carried messages from company to platoon headquarters, thus maintaining the necessary communication required in a successful operation.

CROFTS, JOHN A. (1979865)_____
    Near Bellicourt, France, Sept. 29,
      1918.
    R—Evansville, Ind.
    B—Center Trop, Ind.
    G. O. No. 44, W. D., 1919.

Private, Company C, 120th Infantry, 30th Division.
After being wounded in the right arm to such an extent that he could not continue his duties as stretcher-bearer, and after being ordered back for treatment, Private Crofts continued throughout the day and night under heavy shellfire to assist such wounded as were able to walk.

CROLL, GEORGE H_____
    In Chennery, France, Nov. 2, 1918.
    R—Evanston, Ill.
    B—Evanston, Ill.
    G. O. No. 35, W. D., 1919.

Private, 83d Company, 6th Regiment, U. S. Marine Corps, 2d Division.
He courageously entered three dugouts alone and captured 34 of the enemy.

CROMPTON, WILLIAM H. (5785)_____
    Near Blanc Mont Ridge, France,
      Oct. 7, 1918.
    R—Fall River, Mass.
    B—Fall River, Mass.
    G. O. No. 37, W. D., 1919.

Sergeant, medical detachment, 9th Infantry, 2d Division.
He continued attending the wounded after the first-aid station, in which he was working, was struck by a shell, which wounded him and killed one of the men.

CROMWELL, JOSEPH P_____
    Near Cunel, France, Oct. 14, 1918.
    R—Lynchburg, Va.
    B—Baltimore, Md.
    G. O. No. 49, W. D., 1922.

Captain, 11th Infantry, 5th Division.
When the assault battalion of his regiment had been held up by terrific hostile artillery and machine-gun fire, upon learning of the loss of all the company commanders, Captain Cromwell voluntarily left the supporting battalion and went forward through an almost overwhelming enemy fire to the advance position of the assault battalion, where, although wounded in the arm, he assisted the battalion commander in leading the men from a very disadvantageous position to the capture of a near-by hill held by the enemy, and later in the hostile counterattack assisted in the defense of the position.

CRONE, JOHN B_____
    Near Murvaux, France, Nov. 6, 1918.
    R—West Lebanon, Ind.
    B—West Lebanon, Ind.
    G. O. No. 71, W. D., 1919.

Second lieutenant, 60th Infantry, 5th Division.
During the operations of his company which resulted in the capture of Murvaux, Lieutenant Crone, with the aid of two soldiers, rushed a machine-gun nest, capturing the gun and gunner. Later in the day he repulsed a heavy counterattack on the hill Cote St. Germaine.

*CRONIN, RAYMOND P_____
    In the vicinity of Chateau-Thierry,
      France, June 6, 1918.
    R—Pittsburgh, Pa.
    B—Pittsburgh, Pa.
    G. O. No. 101, W. D., 1918.

Sergeant, 49th Company, 5th Regiment, U. S. Marine Corps, 2d Division.
Under heavy machine-gun fire, he attempted to establish liason with an adjoining French unit, during which he was killed.
Posthumously awarded. Medal presented to mother, Mrs. Edna A. Cronin.

CRONKHITE, LEROY G_____
    Near Binarville, France, Sept. 28–
      Oct. 1, 1918.
    R—Seattle, Wash.
    B—Argyle, Minn.
    G. O. No. 81, W. D., 1919.

Second lieutenant, 308th Infantry, 77th Division.
In the face of heavy machine-gun fire, Lieutenant Cronkhite went forward to within hand-grenade range of the enemy lines and brought back to shelter a soldier who had been severely wounded. Later in the day he went out alone and located a dangerous machine-gun nest, which was thereupon destroyed. Although wounded, Lieutenant Cronkhite refused to be evacuated until Oct. 1, when he was ordered to the hospital by the battalion commander.

*CROSBIE, SAMUEL F. (1210091)_____
    Near Ronssoy, France, Sept. 29,
      1918.
    R—Malone, N. Y.
    P—Canada.
    G. O. No. 49, W. D., 1922.

Sergeant, Company C, 107th Infantry, 27th Division.
After his company had passed the first and second lines of the enemy's resistance and when all his company officers and sergeants senior to him had been killed, Sergeant Crosbie, while under a heavy hostile machine-gun fire, which was sweeping his company's position at short range, voluntarily moved from shell hole to shell hole, reorganized a portion of the company, and was killed as he was about to lead the line forward.
Posthumously awarded. Medal presented to father, Gilbert Crosbie.

CROSS, HERBERT A. (129179)_____
    Near Thiaucourt, France, Sept. 12,
      1918.
    R—Wayne, Mich.
    B—Wayne, Mich.
    G. O. No. 37, W. D., 1919.

Corporal, Battery E, 12th Field Artillery, 2d Division.
Acting as gunner of the second piece, Corporal Cross continued in the service of his piece under heavy hostile shellfire. When the entire gun crew of the first piece was wiped out, at a word from his executive officer he assumed command of the first piece, with a hastily organized crew. He assisted in lifting aside the dead and wounded and continued in the service of the first piece during the barrage until the Infantry had attained their objective.

CROSS, JAMES_____
    Near St. Souplet, France, Oct. 15,
      1918.
    R—Helenwood, Tenn.
    B—Huntsville, Tenn.
    G. O. No. 74, W. D., 1919.

Second lieutenant, 108th Infantry, 27th Division.
Accompanied by 4 soldiers, Lieutenant Cross made a reconnaissance of the River La Salle, the journey being under constant heavy machine-gun fire. To secure the desired information it was necessary to wade the stream for the entire distance. On the following evening Lieutenant Cross tapped the line from which his regiment would launch their attack, and in the battle that followed he was severely wounded.

CROSS, PAUL (2148351)_____
    Near St. Juvin, France, Oct. 12,
      1918.
    R—Lacey, Iowa.
    B—Plane, Iowa.
    G. O. No. 46, W. D., 1919.

Private, first class, Machine Gun Company, 325th Infantry, 82d division.
After his machine-gun squad had been dispersed by a sweeping enemy fire he continued to operate his gun alone until forced to leave it by the overwhelming enemy attack. He then killed two of the enemy with his pistol, but was severely wounded in the encounter. His unusual bravery and daring contributed materially to the success of his regiment in the action.

CROSS, WILLIE (1309733)............
Near Bellicourt, France, Oct. 7, 1918.
R—Jefferson City, Tenn.
B—Jefferson City, Tenn.
G. O. No. 21, W. D., 1919.

Private, Company M, 117th Infantry, 30th Division.
After three runners had been killed in attempting to carry a message to an advance platoon through a heavy artillery and machine-gun barrage, he volunteered for the mission and carried the message through.

*CROSSEN, VERNON J............
Near Landres-et-St. Georges, France, Nov. 1–4, 1918.
R—Rochester, Nev.
B—Salt Lake City, Utah.
G. O. No. 32, W. D., 1919.

Sergeant, 18th Company, 5th Regiment, U. S. Marine Corps, 2d Division.
While he was forming an attack a shell hit in the midst of his platoon. Disregarding the heavy counterbarrage, he reorganized his command and led them in attack, continuing for three days, when he was killed.
Posthumously awarded. Medal presented to father, John W. Crossen.

CROW, JOHN H. (1280785)............
Near Verdun, France, Oct. 12–17, 1918.
R—Salem, N. J.
B—Salem, N. J.
G. O. No. 130, W. D., 1918.

Private, Company F, 114th Infantry, 29th Division.
Throughout the six-days' engagement this soldier performed heroic duty in maintaining liaison between regimental and battalion posts of command. He reorganized relay posts, beyond his required route, delivered messages when relays were gone, passing through three barrages to do so, and, although gassed, refused to quit his work.

CROWE, FRED A. (914831)............
Near Cunel, France, Oct. 14, 1918.
R—Minneapolis, Minn.
B—Sioux City, Iowa.
G. O. No. 37, W. D., 1919.

Private, Company D, 7th Engineers, 5th Division.
He crawled forward under fire from machine guns and snipers and killed two Germans who were operating a machine gun in a tree.

CROWELL, CLARENCE A. (551763)........
Near Counigis, France, July 15, 1918.
R—Hyannis, Mass.
B—Hyannis, Mass.
G. O. No. 53, W. D., 1920.

Private, first class, Company I, 38th Infantry, 3d Division.
Private, first class, Crowell carried messages from his company and battalion headquarters over routes swept by heavy artillery fire. Due to the individual gallantry of this runner, communication was maintained during the terrific bombardment which preceded the German attack on the Marne.

*CROWLEY, EDWARD J. (1705739)........
Near Glennes, France, Sept. 15, 1918.
R—New York, N. Y.
B—New York, N. Y.
G. O. No. 74, W. D., 1919.

Sergeant, Company F, 307th Infantry, 77th Division.
Leaving cover, under heavy machine-gun and shell fire, he visited the outposts in order to assure himself of their security. After accomplishing this mission he fearlessly exposed himself to rescue a wounded comrade, and in so doing lost his life.
Posthumously awarded. Medal presented to mother, Mrs. George Crowley.

*CROWLEY, JOHN J. (1212204)............
Near Ronssoy, France, Sept. 29, 1918.
R—New York, N. Y.
B—Little Falls, N. Y.
G. O. No. 37, W. D., 1919.

Sergeant, Company M, 107th Infantry, 27th Division.
When the advance of his platoon was checked by perilously heavy machine-gun fire from the direct front, he, after ordering his men to cover, advanced alone and bombed the gun out of action. After successfully leading his platoon to one of the farthermost points of the advance he was killed.
Posthumously awarded. Medal presented to mother, Mrs. Hannah Crowley.

*CROWTHER, ORLANDO C............
Near Chateau-Thierry, France, June 6, 1918.
R—Canton, Ill.
B—Arcadia, Ill.
G. O. No. 119, W. D., 1918.

First Lieutenant, 5th Regiment, U. S. Marine Corps, 2d Division.
He displayed the highest type of courage and leadership. After all the men near him had been killed or wounded, he captured one machine gun and crew unaided, and while attempting to take a second was himself killed.
Posthumously awarded. Medal presented to mother, Mrs. Lou Crowther.

CRUM, LEO J............
Near Cierges, France, July 31, 1918, and Aug. 1, 1918.
R—Kalamazoo, Mich.
B—Corunna, Mich.
G. O. No. 124, W. D., 1918.

First lieutenant, Medical Corps, attached to 126th Infantry, 32d Division.
During the attack against Cierges by his regiment he worked continuously and heroically under fire to treat and evacuate the wounded. When the house in which his first-aid station was located was struck by an enemy shell he safely evacuated all of his patients and promptly established another aid station near the front. His untiring efforts and personal bravery saved the lives of many wounded and suffering men and were a source of inspiration to the entire command.

*CRYDER, CHARLES C. (93148)........
Near Suippes, France, July 15, 1918.
R—Irwin, Ohio
B—Blue Island, Ill.
G. O. No. 27, W. D., 1919.

Corporal, Company G, 166th Infantry, 42d Division.
He remained with his commanding officer, who was overcome with gas, and assisted him to a place of safety during a severe bombardment, despite the fact that the officer repeatedly suggested that he leave him and seek safety for himself and that he had previously been instructed, in case of bombardment, to seek shelter in a concrete dugout about 1,000 yards in the rear.
Posthumously awarded. Medal presented to father, J. B. Cryder.

*CUDDY, GEORGE J. (1210851)........
Near Ronssoy, France, Sept. 29, 1918.
R—New York, N. Y.
B—New York, N. Y.
G. O. No. 71, W. D., 1919.

Private, Company F, 107th Infantry, 27th Division.
Shot through both cheeks and the throat, he refused to be evacuated and continued to advance through intense machine-gun fire. Later he fell mortally wounded by a third bullet. His heroic conduct was an inspiration to those near him.
Posthumously awarded. Medal presented to mother, Mrs. Hannah Cuddy.

*CUFF, WILLIAM (1402324)............
At Mont-de-Sanges, France, Sept. 23, 1918.
R—Chicago, Ill.
B—Kimball, W. Va.
G. O. No. 44, W. D., 1919.

Private, Machine Gun Company, 370th Infantry, 93d Division.
Private Cuff carried important messages for his regimental commander, constantly exposing himself to heavy artillery and machine-gun fire until he was killed.
Posthumously awarded. Medal presented to mother, Mrs. Emma Cuff.

CULLEN, FRED E. (567377)............
Near Bazoches, France, Aug. 7, 1918.
R—Skaneateles, N. Y.
B—Motterville, N. Y.
G. O. No. 27, W. D., 1920.

Corporal, Company D, 12th Machine Gun Battalion, 4th Division.
When two American soldiers attempted to cross an open space and were fired upon by enemy machine guns, one of the soldiers fell wounded, and the enemy concentrated their fire upon the fallen man. Under the enemy machine-gun fire Corporal Cullen rushed forward 75 yards and in spite of the heavy fire carried his wounded comrade to a place of safety.

CULLEN, MICHAEL J. (2781603)..........
Near Eclisfontaine, France, Sept. 23, 1918.
R—San Francisco, Calif.
B—Chicago, Ill.
G. O. No. 50, W. D., 1919.

Corporal, Company K, 364th Infantry, 91st Division.
With six other soldiers he responded to a call for volunteers to go 400 yards beyond the front line to bring in wounded comrades. Under terrific machine-gun and shell fire he assisted in the rescue of seven wounded men, also bringing in the dead body of a company lieutenant.

CULLEN, WILLIAM J..........
Near Binarville, France, Oct. 2–8, 1918.
R—New York, N. Y.
B—New York, N. Y.
G. O. No. 20, W. D., 1919.

First lieutenant, 308th Infantry, 77th Division.
During the advance of his regiment through the Forest of Argonne, France, Lieutenant Cullen led his company, under intense concentration of machine-gun fire, to the day's objective, steadying his men and directing the organization and entrenchment of his position. During the period in which part of the regiment was cut off by the enemy he continued to visit his posts and encourage his men under intense concentrations of trench-mortar and machine-gun fire, effectively directing the repulse of attacks on his position. On Oct. 4–6 this officer, observing friendly airplanes, left his shelter and went out into a cleared space in plain view of the enemy and under intense machine-gun fire signaled the position to the airplanes. During all this critical time when his company, as well as the battalion, was entirely without food for 5 days, he displayed coolness, good judgment, and efficiency, furnishing an inspiring example to his men. His gallantry in action contributed materially to the holding of the left flank and the successful resistance made by his battalion.

CULLISON, JESSE M..........
At Laversines, France, July 18, 1918.
R—Hampstead, Md.
B—Baltimore, Md.
G. O. No. 44, W. D., 1919.

Lieutenant colonel, 28th Infantry, 1st Division.
He distinguished himself by doing more than his duty in disposing front-line troops in effective position, fearlessly subjecting himself to danger in order to accomplish his task and thereby inspiring the officers and men of his brigade to valorous and successful attack.

CULNAN, JOHN H..........
At Chateau-Thierry, France, June 6, 1918.
R—New Orleans, La.
B—Calumet, Mich.
G. O. No. 110, W. D., 1918.

Sergeant, 49th Company, 5th Regiment, U. S. Marine Corps, 2d Division.
While assisting a wounded man to the rear he was himself wounded in the head, but carried out his mission, succeeding in bringing the other wounded man to the dressing station.

CULVER, STANLEY (1808456)..........
Near Moirey, France, Nov. 10, 1918.
R—Berwick, Pa.
B—Broadway, Pa.
G. O. No. 46, W. D., 1919.

Private, Company C, 314th Infantry, 79th Division.
During an advance Private Culver, though suffering from wounds, carried important messages under heavy shellfire from battalion to company commander, refusing to have his wounds dressed until his mission was completed. He returned after treatment was given and remained at his work as runner, not allowing himself to be evacuated.

CUMMINGS, AVERY D..........
Near Gesnes, France, Sept. 29, 1918.
R—Couer D'Alene, Idaho.
B—Mitchell, Iowa.
G. O. No. 139, W. D., 1918.

Lieutenant colonel, 181st Infantry Brigade, 91st Division.
During the attack on Gesnes he, in addition to performing his regular duties as brigade adjutant, 181st Brigade, went forward with the front line of attack, directing the organization and outposting of the front line after Gesnes and the army objective beyond it had been captured. All of the senior officers of the assaulting regiment having been killed or wounded in the attack on Gesnes, he unhesitatingly organized the scattered elements of the regiment and pushed the attack home to final success.

CUMMINGS, FRANK J. (63256)..........
At Marcheville, France, Sept. 26, 1918.
R—New Haven, Conn.
B—New Haven, Conn.
G. O. No. 15, W. D., 1919.

Sergeant, Headquarters Company, 102d Infantry, 26th Division.
He repeatedly volunteered for dangerous missions, carrying messages through violent artillery and machine-gun fires. When a small portion of his organization was cut off by the enemy, he went to their aid alone and with his pistol cleaned out a bombers' nest.

CUMMINS, FRED (1975075)..........
Near Consenvoye, France, Oct. 9, 1918.
R—Harrisburg, Ill.
B—Stoneford, Ill.
G. O. No. 44, W. D., 1919.

Private, Company F, 132d Infantry, 33d Division.
He, single-handed, captured a German machine gun, killing one of the crew and routing the others. He then turned the gun on the enemy with great effectiveness, protecting the right flank of his battalion. Later in the day he volunteered and rescued an outpost of 3 men which was surrounded by the enemy. He performed these missions with great courage, bravery, and initiative, subjected to severe enemy fire throughout the entire exploit.

*CUNNINGHAM, CHARLES E. (28088)....
In the vicinity of Hecken, Alsace, May 27, 1918.
R—Grand Rapids, Mich.
B—Grand Rapids, Mich.
G. O. No. 99, W. D., 1918.

Sergeant, Company K, 126th Infantry, 32d Division.
Although seriously wounded during a surprise attack by a German patrol in the vicinity of Hecken, Alsace, on May 27, 1918, he continued to direct his men and succeeded in driving the enemy off. He has since died of injuries sustained.
Posthumously awarded. Medal presented to mother, Mrs. Augusta Pangburn.

*CUNNINGHAM, FLOYD L..........
Near Brabant, France, Oct. 8, 1918.
R—Northfield, Minn.
B—Dale, Ky.
G. O. No. 46, W. D., 1919.

First lieutenant, 116th Infantry, 29th Division.
He displayed rare courage in voluntarily going to the assistance of a wounded comrade under heavy machine-gun and shell fire.
Posthumously awarded. Medal presented to father, George W. Cunningham.

*CUNNINGHAM, OLIVER B..........
Near Villemontoire, Chateau-Thierry, and St. Mihiel, France, July 21, 1918, to Sept. 17, 1918.
R—Evanston, Ill.
B—Chicago, Ill.
G. O. No. 37, W. D., 1919.

Captain, 15th Field Artillery, 2d Division.
During this period, Captain Cunningham, with utter disregard for his personal danger, on numerous occasions exposed himself to the enemy fire while reconnoitering and performing liaison work. On Sept. 17, while at his post as liaison officer with the most advanced Infantry unit, he was killed by an enemy shell.
Posthumously awarded. Medal presented to father, Frank S. Cunningham.

CUNNINGHAM, WILLIAM A.
Near Sommerance, France, Oct. 12, 1918.
R—Athens, Ga.
B—Nashville, Tenn.
G. O. No. 13, W. D., 1919.

Captain, 321st Machine Gun Battalion, 82d Division.
Captain Cunningham, though painfully wounded in the face by shrapnel when his battalion was seriously engaged, continued to lead his men through heavy shellfire, leading them with skill and inspiring them with courage.

ᶜCURFMAN, THOMAS D. (2660009).
Near Exermont, France, Oct. 5, 1918.
R—Steubenville, Ohio.
B—Steubenville, Ohio.
G. O. No. 74, W. D., 1919.

Sergeant, Company G, 28th Infantry, 1st Division.
When the advance of his company was retarded by machine-gun fire he took an automatic rifle from a wounded gunner and went forward alone to a position from which he opened fire and destroyed the enemy nest. He was himself mortally wounded.
Posthumously awarded. Medal presented to mother, Mrs. George Curfman.

CURLEE, WILLIAM (39478).
Near Medeah Ferme, France, Oct. 3, 1918.
R—Polkton, N. C.
B—Polkton, N. C.
G. O. No. 21, W. D., 1919.

Corporal, Company F, 9th Infantry, 2d Division.
Corporal Curlee, together with four men, charged a machine-gun nest containing three machine guns and captured the three guns and 20 prisoners.

CURNOW, EARL M. (261678).
Near Juvigny, north of Soissons, France, Aug. 29–Sept. 2, 1918.
R—Detroit, Mich.
B—Detroit, Mich.
G. O. No. 139, W. D., 1919.

Corporal, Company H, 128th Infantry, 32d Division.
Though he had been severely gassed, he remained with his company while it was in the front line. When it was in support he aided in carrying wounded across an area covered by machine-gun and artillery fire. Although this work was not required of him, he volunteered for it in spite of the danger and his own physical condition.

CURRAN, FRED F. (2780146).
Near Very, France, Sept. 26, 1918.
R—San Francisco, Calif.
B—Oakland, Calif.
G. O. No. 37, W. D., 1919.

Corporal, Company B, 363d Infantry, 91st Division.
He accompanied a lieutenant on an attack against a nest of enemy snipers. They succeeded in killing two, wounding one, and taking the remaining two prisoners.

CURRIE, WALTER (567917).
Near Ville-Savoye, France, Aug. 7, 1918, and near Brieulles, France, Sept. 29, 1918.
R—Detroit, Mich.
B—Detroit, Mich.
G. O. No. 98, W. D., 1919.

Private, Company A, 59th Infantry, 4th Division.
On the Vesle River, Aug. 7, 1918, when his company was in need of ammunition and after several men had been killed in the attempt to secure it, Private Currie volunteered and went for ammunition across an open field swept by machine-gun fire. He successfully returned with the ammunition, thereby greatly assisting his company to hold its position. He was severely wounded near Brieulles while making a gallant stand against the enemy with 12 other men, the only survivors of his platoon.

CURRY, GEORGE F. (2397837).
Near Moulins, France, July 15, 1918.
R—Capitol Heights, Md.
B—Washington, D. C.
G. O. No. 19, W. D., 1920.

Private, Company F, 38th Infantry, 3d Division.
Private Curry voluntarily took up an abandoned automatic rifle and with one man moved to an exposed flank and opened a continuous harassing fire on the advancing enemy. Although exposed to direct fire, Private Curry kept his gun in action until an Infantry platoon could take up a better defensive position.

CURTI, MIKE (3142771).
Near Cesnes, France, Oct. 4, 1918.
R—Reno, Nev.
B—Italy.
G. O. No. 66, W. D., 1919.

Private, Company F, 127th Infantry, 32d Division.
Private Curti, a litter bearer, went out alone in front of the lines several times under the severest of fire and carried back wounded men from an exposed area, from which his company had been forced to withdraw.

CURTIN, DAVID F. (62859).
In Bois Belleau, north of Verdun, France, Oct. 27, 1918.
R—Brookline, Mass.
B—Charlestown, Mass.
G. O. No. 39, W. D., 1920.

Sergeant, Machine Gun Company, 101st Infantry, 26th Division.
The attacking Infantry having been temporarily halted by machine-gun fire from the front, Sergeant Curtin, grasping the situation, led a group of men in a flank attack upon the enemy machine-gun position, which resulted in the capture of the position and 30 prisoners. His deed enabled the left of the line to advance to its objective.

CURTIS, BERNARD B. (1289643).
In the Bois-deBrabant, France, Oct. 8, 1918.
R—Hopewell, Va.
B—Mecklensburg, Va.
G. O. No. 126, W. D., 1919.

Corporal, Company G, 116th Infantry, 29th Division.
Corporal Curtis courageously continued to lead his squad against the enemy after being painfully wounded in the face by shell fragments, refusing to obtain first aid in order not to delay the advance of his squad.

CURTIS, CLYDE O. (43092).
South of Soissons France, July 18, 1918.
R—Stella, Nebr.
B—Stella, Nebr.
G. O. No. 44, W. D., 1919.

First sergeant, Company G, 16th Infantry, 1st Division.
Leading his platoon against an enemy battery in the face of direct fire, he personally killed the gunner and, with the aid of his men, either killed or wounded the entire crew, thus preventing further casualties on his troops.

CURTIS, EARL W. (2101765).
Near Brieulles-sur-Meuse, France, Sept. 29, 1918.
R—Virden, Ill.
B—Houston, Mo.
G. O. No. 46, W. D., 1919.

Private, Company B, 59th Infantry, 4th Division.
Advancing alone across open territory and exposed to extremely heavy machine-gun fire, he rescued a fellow soldier who was lying wounded beyond the front line. He accomplished his mission, even after being painfully wounded in the head during his return.

CURTIS, EDWARD P.
In the region of Stenay, France, Sept. 27, 1918.
R—Rochester, N. Y.
B—Rochester, N. Y.
G. O. No. 138, W. D., 1919.

First lieutenant, pilot, 95th Aero Squadron, Air Service.
He volunteered to perform a reconnaissance patrol of particular danger and importance, 30 kilometers within the enemy's territory. He made the entire journey through a heavy antiaircraft and machine-gun fire, and flew at an extremely low altitude to secure the desired information.

CURTIS, JOHN E.
  At Andevanne, France, Nov 1, 1918.
  R—Meridian, Miss.
  B—Beauregard, Ill.
  G. O. No. 49, W. D., 1922.

Captain, 360th Infantry, 90th Division.
After his company had reached its objective and was under a terrific hostile artillery and machine-gun fire which caused the loss of 10 killed and 25 wounded in a short time. Captain Curtiss diplayed conspicuous bravery by exposing himself to a heavy shelling while going forward and bringing to shelter one of his men who had been wounded and was unable to move.

CURTIS, NATHAN M. (2059989).
  Near Consenvoye, France, Oct. 10, 1918.
  R—Chicago, Ill.
  B—St. Joseph, Ill.
  G. O. No. 70, W. D., 1919.

Corporal, Company L, 131st Infantry, 33d Division.
Voluntarily leaving shelter, he led a patrol of 3 men across an open field under heavy shell fire and captured an enemy machine gun, killing 2 and capturing 1 of the crew.

CURTIS, ROLLIN B. (285563).
  Near Cierges, France, Aug. 1, 1918, and near Gesnes, France, Oct. 9, 1918.
  R—Baraboo, Wis.
  B—Baraboo, Wis.
  G. O. No. 126, W. D., 1919.

First sergeant, Company A, 128th Infantry, 32d Division.
During the advance near Reddy Farm, Sergeant Curtis, then a platoon commander, was painfully wounded in the leg, but refused to be evacuated until wounded a second time. During the advance the battalion commander asked for a volunteer to establish liaison with the unit on the right. Sergeant Curtis immediately offered his services for this hazardous mission. On Oct. 9, near Gesnes, when the only officer in his company was killed, he took command, leading the company through an intense enemy barrage. Although wounded and knocked down by the explosion of a shell, he continued to lead the company until relieved.

*CURTIS, WILL C. (1484824).
  Near St. Etienne, France, Oct. 8, 1918.
  R—Krum, Tex.
  B—Mart, Tex.
  G. O. No. 21, W. D., 1919.

Private, Company M, 142d Infantry, 36th Division.
While passing through a heavy enemy barrage he was mortally wounded, but continued to advance, encouraging his comrades to follow him. His example of courage and fearlessness gave confidence to his comrades and the advance was successful. He later died of his wounds.
Posthumously awarded. Medal presented to father, H. V. Curtis.

CUSHING, FREDERICK R. (127418).
  Near Exermont, France, Oct. 4, 1918.
  R—Roxbury, Mass.
  B—Cambridge, Mass.
  G. O. No. 32, W. D., 1919.

Sergeant, Battery C, 7th Field Artillery, 1st Division.
He volunteered and took his piece and gun squad forward to the infantry first lines, where he was subjected to violent bombardment. He refused to take shelter until nearly all his horses and three of his men were wounded, and then left shelter no less than five times to rescue wounded comrades, this being done under a heavy barrage.

*CUSHION, LEON J. (67780).
  At Marcheville, France, Sept. 26, 1918.
  R—East Hardwick, Vt.
  B—Greensboro Bend, Vt.
  G. O. No. 133, W. D., 1918.

Private, Company D, 103d Infantry, 26th Division.
Under terrific machine-gun, artillery, and rifle fire he displayed great courage in locating and fighting enemy machine gunners. He was killed while rushing a machine-gun nest.
Posthumously awarded. Medal presented to father, Nelson J. Cushion.

CUSTEAU, ODILON (67257).
  Near Belleau Wood, France, July 20, 1918.
  R—Concord, N. H.
  B—Canada.
  G. O. No. 125, W. D., 1918.

Corporal, Company C, 103d Infantry, 26th Division.
In the advance of the first wave, east from Belleau Woods, he cleaned out, single handed, a dugout of German machine guns.

CUTLER, MERRITT D. (1211452).
  Near Ronssoy, France, Sept. 29, 1918.
  R—Freeport, N. Y.
  B—New York, N. Y.
  G. O. No. 32, W. D., 1919.

Corporal, Company I, 107th Infantry, 27th Division.
Although suffering from wounds, he went forth under treacherous enemy fire and dragged two wounded comrades to safety. Later the same day he organized a stretcher party and brought in 3 wounded comrades under machine-gun fire, which was so severe that it had stopped the advance of neighboring troops.

*CUTTER, EDWARD B.
  Near Cunei, France, Oct. 21, 1918.
  R—Anoka, Minn.
  B—Anoka, Minn.
  G. O. No. 37, W. D., 1919.

First lieutenant, observer, 90th Aero Squadron, Air Service.
Responding to an urgent request for a plane to penetrate the enemy lines to ascertain whether or not the enemy was preparing a counterattack, Lieutenant Cutter immediately volunteered for the mission. Obliged to fly at a very low altitude on account of the unfavorable weather conditions, he was under terrific fire of the enemy at all times, but by skillful dodging he managed to cross the enemy lines. His plane was seen to suddenly lurch and crash the short distance to the ground, both he and his pilot being killed.
Posthumously awarded. Medal presented to mother, Mrs. Mary S. Cutter.

DABNEY, MILAN (2293685).
  Near Eclisfontaine and Tronsol Farm, France, Sept. 27–Oct. 1, 1918.
  R—Butte, Mont.
  B—Amador City, Calif.
  G. O. No. 87, W. D., 1919.

Private, Company B, 348th Machine Gun Battalion, 91st Division.
Throughout 5 days of action Private Dabney maintained liaison between company and battalion posts of command, repeatedly passing through the enemy's barrages and constantly subjected to enemy sniping.

DABNEY, WILLIAM C.
  Near Soissons, France, July 18–22, 1918.
  R—Louisville, Ky.
  B—Charlottesville, Va.
  G. O. No. 126, W. D., 1919.

First lieutenant, 26th Infantry, 1st Division.
He took command in going over the top 4 times in 4 successive attacks near Soissons, France, July 18–22, 1918, and by his bravery achieved success in reaching his objective.

DAKIN, HURSEY A. (847688)................ | Corporal, Company F, 1st Gas Regiment.
In the Bois-de-Jure, near Gercourt, France, Sept. 26, 1918.
R—Freewater, Oreg.
B—Stillwater, Minn.
G. O. No. 27, W. D., 1919. | He volunteered with another soldier to attack a machine-gun nest which was holding up the advance. They advanced against very heavy machine-gun fire and captured the position, killing 2 Germans and routing the remainder of the gun crew.

DALEY, PHILIP A. (154611)................ | Private, Company A, 1st Engineers, 1st Division.
Northwest of Verdun, France, Oct. 9, 1918.
R—Hollister, Calif.
B—Idria, Calif.
G. O. No. 32, W. D., 1919. | On 2 occasions, when intense machine-gun fire threatened his platoon, Private Daley voluntarily accompanied another soldier and attacked the enemy positions in the face of heavy fire, silencing the guns by effective rifle fire.

DALLAS, FRED W. (2811224)................ | Corporal, Headquarters Company, 360th Infantry, 90th Division.
Near the Bois-d'Argonne, France, Nov. 2, 1918.
R—Beaumont, Tex.
B—New Iberia, La.
G. O. No. 46, W. D., 1919. | Although his arm was shattered by a machine-gun bullet, he refused to go to the rear, but continued to advance under extremely heavy machine-gun fire until he received 2 more wounds, 1 piercing his leg and making it impossible for him to advance farther. While crawling to the rear he encountered another wounded man and assisted him to the dressing station.

*DALRYMPLE, THERON E. (154613)........ | Sergeant, first class, Company A, 1st Engineers, 1st Division.
At Bois-de-Villers, France, May 9, 1918.
R—Rochester, N. Y.
B—Mount Morris, N. Y.
G. O. No. 100, W. D., 1918. | He displayed heroic devotion to duty by rendering first-aid assistance to the wounded, by handling his platoon under shellfire with coolness and courage, and by attempting to protect a comrade while he himself was mortally wounded.
Posthumously awarded. Medal presented to father, Asa C. Dalrymple.

DALRYMPLE, WILLIAM V. (53884)........ | Private, Company G, 26th Infantry, 1st Division.
Near Soissons, France, July 19, 1918.
R—Albertville, Ala.
B—Albertville. Ala.
G. O. No. 132, W. D., 1918. | He crawled to an enemy sniper's post and killed or wounded all its occupants.

DALTON, GILBERT R. (1975888)........... | Bugler, Company M, 132d Infantry, 33d Division.
At Bois-de-Malaumont, France, Oct. 9, 1918.
R—Eldorado, Ill.
B—Galatia, Ill.
G. O. No. 59, W. D., 1919. | Bugler Dalton and an officer were making a reconnaissance of Bois-de-Malaumont. Upon entering the woods they were suddenly fired upon by machine guns. Together they rushed the machine gun. The officer was wounded and unable to take cover. Bugler Dalton ran across an open space, exposing himself to short range of machine-gun fire, and carried the officer to a position of safety.

DALY, DANIEL............................. | First sergeant, 73d Company, 6th Regiment, U. S. Marine Corps, 2d Division.
At Lucy-le-Bocage, and during the attack on Bouresches, France, June 5, 1918, June 7, 1918, and June 10, 1918.
R—Brooklyn, N. Y.
B—Glen Cove, Long Island, N. Y.
G. O. No. 101, W. D., 1918. | Sergeant Daly repeatedly performed deeds of heroism and great service on June 5, 1918. At the risk of his life he extinguished a fire in an ammunition dump at Lucy-le-Bocage. On June 7, 1918, while his position was under violent bombardment, he visited all the gun crews of his company, then posted over a wide portion of the front, to cheer his men. On June 10, 1918, he attacked an enemy machine-gun emplacement unassisted and captured it by use of hand grenades and his automatic pistol. On the same day, during the German attack on Bouresches, he brought in wounded under fire.

DALY, PAUL.............................. | First lieutenant, 18th Infantry, 1st Division.
Near Soissons, France, July 19–22, 1918.
R—New York, N. Y.
B—New York, N. Y.
G. O. No. 46, W. D., 1919. | Although wounded, he took command of the first battalion of his regiment after the previous battalion commanders had been wounded, displaying the highest courage in advancing and holding the objective of his battalion. Three days later he was again wounded, but remained with his command until he was ordered to the rear by his regimental commander.

DANIEL, CHARLES E. (1287351).......... | Private, first class, Company B, 116th Infantry, 29th Division.
In the Bois-d'Etrayes, France, Oct. 27, 1918.
R—Richmond, Va.
B—Louisa County, Va.
G. O. No. 15, W. D., 1919. | He crawled from his own trenches to within range of an enemy machine-gun nest, which had been harassing his company all day, and bombed out the gunners with hand grenades, thereby enabling his company to occupy a more advantageous position.

DANIEL, HENRY (1038067)............... | Private, Battery A, 10th Field Artillery, 3d Division.
Near St. Eugene, France, July 16, 1918.
R—Crete, Nebr.
B—Crete, Nebr.
G. O. No. 44, W. D., 1919. | He repeatedly volunteered and carried important messages 4 kilometers through heavy shellfire near the enemy lines. On one trip he found a wounded soldier and carried him through an enemy barrage to the dressing station.

DANIEL, WILLIAM H. (1995315)........ | Private, Company G, 119th Infantry, 30th Division.
Near Ribeauville, France, Oct. 18, 1918.
R—White Hall, Ill.
B—Carrollton, Ill.
G. O. No. 50, W. D., 1919. | When enemy sniper and machine-gun fire had held up his line and caused many casualties in his ranks Private Daniel went forward a distance of about 150 yards and with the aid of his machine gun put an enemy nest out of action and killed one of the snipers.

DANIELL, JOSIAH (42756)............... | Sergeant, Company E, 16th Infantry, 1st Division.
South of Soissons, France, July 18, 1918.
R—Watkinsville, Ga.
B—Watkinsville, Ga.
G. O. No. 15, W. D., 1919. | He voluntarily and alone advanced against a machine gun and captured the gun and its crew.

DANIELS, FRANCIS L. (2176745)........
    Near Tailly, France, Nov. 2, 1918.
    R—Smith Center, Kans.
    B—Franklin County, N. C.
    G. O. No. 46, W. D., 1919.

Corporal, Company H, 353d Infantry, 89th Division.
    When his company had been held up by a machine gun and a comrade killed at his side Corporal Daniels voluntarily advanced alone over an open space for 200 yards in the face of the machine-gun fire into a clump of bushes, from which he succeeded in killing the enemy gunner and capturing the gun.

DANLEY, RAYMOND (553753)...........
    At Courboin, France, night of July 14–15, 1918.
    R—Harrisburg, Pa.
    B—Steelton, Pa.
    G. O. No. 10, W. D., 1920.

Private, first class, Company B, 8th Machine Gun Battalion, 3d Division.
    When other means of liaison were impossible Private Danley volunteered to act as a runner. He repeatedly carried messages through intense artillery fire from the company P. C. to a front-line platoon.

DANTUONO, FERDINAND A. (554825)....
    Near Chateau-Thierry, France, July 14–15, 1918.
    R—Brooklyn, N. Y.
    B—Italy.
    G. O. No. 10, W. D., 1920.

Private, Company B, 9th Machine Gun Battalion, 3d Division.
    He exposed himself to intense artillery fire to carry numerous messages. Liaison was maintained at critical times through the individual efforts of this soldier.

*DANYSCH, STEVE G. (106938)........
    South of Soissons, France, July 19, 1918.
    R—Westhoff, Tex.
    B—Yorktown, Tex.
    G. O. No. 132, W. D., 1918.

Sergeant, Company B, 4th Machine Gun Battalion, 2d Division.
    Having received a severe wound in the head, he refused to be evacuated and continued to lead his platoon with great bravery until he fell unconscious. He died from his wounds several days later.
    Posthumously awarded. Medal presented to father, Florian Danysch.

DANZIG, SAMUEL V. H...........
    Near Fossoy, France, July 14–15, 1918, and near Le Charmel, France, July 26, 1918.
    R—New York, N. Y.
    B—Torrington, Conn.
    G. O. No. 98, W. D., 1919.

First lieutenant, 8th Machine Gun Battalion, 3d Division.
    Lieutenant Danzig repeatedly crossed open spaces swept by shellfire during a violent barrage to inspect the gun positions of his platoon. He established his post of command at an exposed position under artillery fire, encouraging his men and looking after their safety. Later in the day he took charge of the guns of an officer who had been wounded and placed them in new positions. On July 26 he advanced through a heavy barrage on a reconnaissance and placed four guns in support of a battalion of infantry. After the infantry had withdrawn he held this position under enemy fire from the rear and flanks until ordered to withdraw the following night.

DARCY, JAMES J. (89332) ....
    Near Baccarat, France, June 12, 1918.
    R—Brooklyn, N. Y.
    B—Ireland.
    G. O. No. 15, W. D., 1923.

Private, first class, Company B, 165th Infantry, 42d Division.
    Exposing himself to intense enemy fire and with complete disregard for his own safety, he made his way under observation of the enemy to within a short distance of their lines, rescued a wounded soldier, whom he carried a distance of 300 yards to a position of comparative safety, in broad daylight and at all times under observation of the enemy. The superb courage and soldierly devotion to duty thus displayed greatly inspired the men of the regiment.

DARGIS, JOSEPH A.............
    In the Bois-de-Belleau, France, June 6–8, 1918.
    R—Chicago, Ill.
    B—Russia.
    G. O. No. 70, W. D., 1919.

Corporal, 83d Company, 6th Regiment, U. S. Marine Corps, 2d Division.
    Howard J. Childs, Joseph A. Dargis, and Allen Benjamin Tilghman, corporals, and Herman L. McLeod, private, Company K, 6th Regiment, U. S. Marine Corps. These four men were prominent in the attack on enemy machine-gun positions in the Bois-de-Belleau on June 6 and 8, 1918, were foremost in their company at all times, and acquitted themselves with such distinction that they were an example for the remainder of their command.

DARKOSKI, WACLAW (54604)...........
    In the Argonne Forest, France, Oct. 9, 1918.
    R—Jersey City, N. J.
    B—Russia.
    G. O. No. 44, W. D., 1919.

Sergeant, Company K, 26th Infantry, 1st Division.
    During the fighting in the Argonne, Oct. 4–13, Sergeant Darkoski displayed extraordinary heroism and ability in leading his platoon against strong points under the most severe shell and machine-gun fire. On Oct. 9, in a hand-to-hand fight, he alone captured two enemy guns.

DARLING, HOMER O............
    Near Bois-du-Fays, France, Oct. 10–12, 1918.
    R—Mendon, Mass.
    B—Mendon, Mass.
    G. O. No. 59, W. D., 1919.

Second lieutenant, 132d Infantry, 33d Division.
    Exposed to heavy machine-gun fire from the front and right flank, he led his platoon forward through heavy brush, although suffering heavy casualties. During the advance he and one other member of his platoon attacked a machine-gun nest and captured three machine guns and five prisoners. In hand-to-hand fighting he personally killed five Germans and wounded others.

DASCH, CARL W. (98757)...........
    Northeast of Chateau-Thierry, France, July 26 to Aug. 1, 1918.
    R—Weiser, Idaho.
    B—Boise, Idaho.
    G. O. No. 102, W. D., 1918.

Private, Headquarters Company, 167th Infantry, 42d Division.
    During the entire period, July 26 to Aug. 1, 1918, in action northeast of Chateau-Thierry, France, he carried messages between the firing line and battalion headquarters through heavy enemy shellfire. On returning from the firing line he would pick up a severely wounded man each time and carry him through the barrage to a first-aid section. He finally became so exhausted he could not continue his work, yet he had to be ordered to report to the aid section for treatment. During the whole series of engagements he did not sleep and taxed his physical endurance to the utmost at all times, setting to his comrades an example of utter disregard of danger and exceptional devotion to duty.

DAUGHTY, JOHN E. (1383909)...........
    In the Evergreen Woods, France, Nov. 6, 1918.
    R—National City, Ill.
    B—Carrollton, Ill.
    G. O. No. 46, W. D., 1919.

Sergeant, Company F, 130th Infantry, 33d Division.
    While on a daylight patrol he displayed exceptional bravery, when with one other man he fought a large force of the enemy, killing a German officer and two machine gunners who were attempting to put their guns in action. Surrounded by the enemy, he capture a German corporal and fought his way out with his pistol, not having time to load his rifle. He marched his prisoner to the rear at the point of his bayonet, thereby running the risk of being captured himself in order to carry out orders to take at least one prisoner.

*DAUSCH, WILLIAM (1211430).
Near Roussoy, France, Sept. 29, 1918.
R—New York, N. Y.
B—New York, N. Y.
G. O. No. 32, W. D., 1919.

Sergeant, Company I, 107th Infantry, 27th Division.
During operations against the Hindenburg line, he rendered valuable assistance and demonstrated rare courage in attacking and destroying two enemy machine-gun nests by the accurate fire of his rifle. Even after being mortally wounded in the head he continued in the combat until he collapsed. Posthumously awarded. Medal presented to mother, Mrs. Katherine Dausch.

*DAVIDSON, ALEX.
Near Madeleine Farm, France, Oct. 16, 1918.
R—Boat, Ky.
B—Boat, Ky.
G. O. No. 53, W. D., 1920.

Second lieutenant, 7th Infantry, 3d Division.
While leading his platoon in an attack Lieutenant Davidson was struck down by enemy fire. He got up and started forward, urging his men on, until he was hit a second time. He refused medical attention, and, although mortally wounded, he motioned to his men to continue the attack.
Posthumously awarded. Medal presented to father, Robert Davidson.

DAVIDSON, ALEXANDER H.
Near Porac, Province of Pampanga, Island of Luzon, Philippine Islands, Oct. 17, 1899.
R—Delaware City, Del.
B—Delaware City, Del.
G. O. No. 27, W. D., 1919.

First lieutenant, 36th Infantry, U. S. Volunteers.
For extraordinary heroism in an engagement with hostile Filipino insurgents near Porac, Province of Pampanga, Island of Luzon, Philippine Islands, on Oct. 17, 1899.

*DAVIDSON, JOSEPH M.
Near La Polka Farm, France, Nov. 4, 1918.
R—St. Joseph, La.
B—St. Joseph, La.
G. O. No. 3, W. D., 1922.

First lieutenant, 318th Infantry, 80th Division.
His line was stopped 300 yards from the La Polka Farm by a terrific concentration of machine-gun fire directed from the high ground and woods on the farm. By his coolness and courage he quickly collected together a platoon, and leading them across open ground swept by machine-gun fire, he attacked the machine guns. Just before reaching the enemy, Lieutenant Davidson was killed, but his men, inspired by his coolness and devotion, fought on and captured the machine-gun nests.
Posthumously awarded. Medal presented to father, William M. Davidson.

DAVIDSON, LILBURN C.
Near Villers-sous-Preny, France, Sept. 15, 1918.
R—Jackson, Ky.
B—Jackson, Ky.
G. O. No. 87, W. D., 1919.

Captain, 359th Infantry, 90th Division.
When his own lines were being heavily shelled at night he led a patrol of 36 men to gain contact with the enemy, and, after being challenged by German outposts, led his men in a charge, under heavy fire, killing 40 of the enemy and capturing 36 prisoners, 4 trench mortars, and 4 machine guns. He established and held an important position until leapfrogged by the remainder of his company in an attack several hours later.

DAVIDSON, WILLIAM LEE.
North of Verdun, France, Oct. 12, 1918.
R—Chester, S. C.
B—Carlisle, S. C.
G. O. No. 24, W. D., 1920.

Captain, Dental Corps, attached to 114th Infantry, 29th Division.
Captain Davidson, while attending wounded under heavy fire, was himself wounded by several pieces of shell fragments. Regardless of his own wounds, he continued in his care of the wounded, refusing to be treated until his regiment was relieved from the line, when several pieces of shell were removed from his head and shoulders.

DAVIS, ABEL.
Near Consenvoye, France, Oct. 9, 1918.
R—Chicago, Ill.
B—Germany.
G. O. No. 44, W. D., 1919.
Distinguished-service medal also awarded.

Colonel, 132d Infantry, 33d Division.
Upon reaching its objective, after a difficult advance, involving two changes of direction, his regiment was subjected to a determined enemy counterattack. Disregarding the heavy shell and machine-gun fire, Colonel Davis personally assumed command, and by his fearless leadership and courage the enemy was driven back.

DAVIS, BEN G. (1323057).
Near Ypres, Belgium, Aug. 23, 1918.
R—Wilmington, N. C.
B—Wilmington, N. C.
G. O. No. 35, W. D., 1919.

Private, Company C, 115th Machine Gun Battalion, 30th Division.
When several members of his platoon were severely wounded by shellfire, Private Davis, though himself wounded, went through the bombardment to a dugout and procured assistance for his comrades, guiding a rescuing party to their assistance.

DAVIS, CHESTER A.
Near Hattonville, France, in the St. Mihiel salient, Sept. 12–13, 1918.
R—Salem, Mass.
B—Salem, Mass.
G. O. No. 126, W. D., 1918.

Major, 3d Machine Gun Battalion, 1st Division.
Sent by his brigade commander to verify the positions of the first lines of the assaulting battalions, he continued forward to ascertain the approximate location of the enemy lines. Accompanied by only two others, he overtook a hostile formation of about 60 men, wagons, animals, and machine guns, which formed a rear guard to facilitate the retreat of the enemy convoys, and by the exercise of rare gallantry and judgment captured this entire enemy force, despite its overwhelming superiority in numbers and strength.

DAVIS, CHESTER V. (732421).
Near Brieulles, France, Nov. 4, 1918.
R—Tennyson, Ind.
B—Tennyson, Ind.
G. O. No. 37, W. D., 1919.

Corporal, Company E, 6th Infantry, 5th Division.
He volunteered to carry a message to the battalion commander notifying him that our barrage was falling short. Crossing a valley swept by shell and machine-gun fire, he swam the icy Meuse River and succeeded in delivering his message in time to avoid further casualties.

DAVIS, CLARENCE A. (100707).
Near Cote-de-Chatillon, France, Oct. 14, 1918.
R—Clarence, Iowa.
B—Clinton, Iowa.
G. O. No. 23, W. D., 1919.

Corporal, Company D, 168th Infantry, 42d Division.
During the attack he made his way forward through intense artillery and machine-gun fire to rescue a wounded comrade. When he had reached a point about 25 yards in front of the enemy trenches he found that the man was dead, and he was himself so seriously wounded that he was compelled to return.

DAVIS, DUNK (1315838).
Near Bellicourt, France, Sept. 29, 1918.
R—Raeford, N. C.
B—Red Springs, N. C.
G. O. No. 50, W. D., 1919.

First sergeant, Company G, 119th Infantry, 30th Division.
He voluntarily went forward to attack enemy machine guns which were carefully concealed and raising havoc with his section of the line. He succeeded in putting both posts out of action and killing all the occupants.

DAVIS, DWIGHT F.
Between Baulny and Chaudron Farm, France, Sept. 29-30, 1918.
R—St. Louis, Mo.
B—St. Louis, Mo.
G. O. No. 9, W. D., 1922.

Major, Infantry, General Staff Corps, Assistant Chief of Staff, 35th Division.
After exposure to severe shelling and machine-gun fire for three days, during which time he displayed rare courage and devotion to duty, Major Davis, then adjutant, Sixty-ninth Infantry Brigade, voluntarily and in the face of intense enemy machine-gun and artillery fire proceeded to various points in his brigade sector, assisted in reorganizing positions, and in replacing units of the brigade, this self-imposed duty necessitating continued exposure to concentrated enemy fire. Sept. 30, 1918, learning that a strong counterattack had been launched by the enemy against Baulny Ridge and was progressing successfully, he voluntarily organized such special duty men as could be found and with them rushed forward to reinforce the line under attack, exposing himself with such coolness and great courage that his conduct inspired the troops in this crisis and enabled them to hold on in the face of vastly superior numbers.

DAVIS, EDGAR C. (2387703).
Near Dun-sur-Meuse, France, Nov.5, 1918.
R—Lewisburg, Tenn.
B—Lewisburg, Tenn.
G. O. No. 37, W. D., 1919.

Sergeant, Company H, 61st Infantry, 5th Division.
He voluntarily advanced alone against an enemy machine gun which was holding up his platoon, capturing 4 prisoners and the gun. Later he led a few men into a dugout and captured 13 prisoners. He then continued the advance until the day's objective was reached and the platoon position organized.

DAVIS, FRED C.
Near Ronssoy, France, Sept. 29-30, 1918.
R—Madrid, Iowa.
B—Polk City, Iowa.
G. O. No. 37, W. D., 1919.

First lieutenant, 108th Infantry, 27th Division.
He successfully held a trench several hundred yards in advance of the Hindenburg line, under heavy shell and machine-gun fire, with a detachment which he had organized with men from different organizations. The following morning he led his detachment still farther, cleaning up about 500 yards of enemy trenches.

*DAVIS, GUY K. (2853390).
Near Villers-devant-Dun, France, Nov. 2, 1918.
R—Utica, Ill.
B—Lawrence County, Ind.
G. O. No. 87, W. D., 1919.

Corporal, Company F, 359th Infantry, 90th Division.
After his platoon sergeant had become a casualty he assumed command and led his men with marked personal bravery and skill, advancing until he was mortally wounded. His example inspired his men with his own fighting spirit.
Posthumously awarded. Medal presented to mother, Mrs. Charity Davis.

*DAVIS, HARRY J. (2282992).
At Cunel, France, Oct. 14, 1918.
R—Camden, N. J.
B—Jersey City, N. J.
G. O. No. 44, W. D., 1919.

Corporal, Company D, 60th Infantry, 5th Division.
While the advance of his platoon was being held up by deadly enfilading fire of the enemy, he voluntarily rushed over open ground, through direct machine-gun fire, for a distance of over 100 meters, and, with the aid of his bayonet attacked the nest. He killed both of the gunners, thereby silencing the fire but during the combat he himself was severly wounded. It was due to his prompt and fearless action that further advance of his platoon was made possible.
Posthumously awarded. Medal presented to widow, Mrs. Lena Marz.

*DAVIS, HENRY (735200).
Near Romagne-sous-Montfaucon, France, Oct. 14, 1918.
R—Pomeroyton, Ky.
B—Pomeroyton, Ky.
G. O. No. 37, W. D., 1919.

Private, first class, Company H, 6th Infantry, 5th Division.
Although wounded, he courageously led several attacks against machine-gun emplacements. While reconnoitering from an exposed position in front of the lines he was wounded a second time by a sniper, but refused to go to the rear until ordered to do so.
Posthumously awarded. Medal presented to father, J. P. Davis.

DAVIS, HERMAN (2129720).
At Molleville, Farm, France, Oct. 10, 1918.
R—Big Lake, Ark.
B—Manila, Ark.
G. O. No. 46, W. D., 1919.

Private, Company I, 113th Infantry, 29th Division.
On duty as a company runner, he was accompanying the left assault platoon of his company during the advance through the woods, when it was fired on by an enemy machine gun. As soon as the gun opened fire the members of the platoon scattered and attempted to flank the gun, but Private Davis pushed on ahead, being the first to reach the nest, attacked it single handed, and killed the four enemy gunners. His gallant act enabled his platoon to continue the advance.

DAVIS, HILDRED D. (1392055).
Near Butgneville, France, Nov. 11, 1918.
R—Springfield, Ill.
B—Elkhart, Ill.
G. O. No. 46, W. D., 1919.

Bugler, Company C, 124th Machine Gun Battalion, 33d Division.
When his platoon and the infantry company to which it was attached were held up by hostile obstructions and machine-gun fire he volunteered and carried a message to the left flank of the company. A few minutes afterwards he again distinguished himself by leaving the cover of a trench, going forward under heavy machine-gun fire, and helping rescue a wounded officer.

DAVIS, HOWARD HUBBER.
In Templeux Quarries, France, Jan. 8, 1918.
R—Cleveland, Ohio.
B—Cleveland, Ohio.
G. O. No. 133, W. D., 1918.

First lieutenant, Medical Corps, attached to 12th Sherwood Foresters, British Army.
He entered a dugout which had been caved in by enemy shell fire and ministered to the wounded. Although the dugout was under heavy shellfire, he performed an operation for amputation of a leg and thereby saved a soldier's life.

*DAVIS, JOSEPH CARLTON.
Near Beaumont, France, Apr. 26, 1918.
R—Providence, R. I.
B—Providence, R. I.
G. O. No. 21, W. D., 1925.

First lieutenant, 103d Field Artillery, 26th Division.
Lieutenant Davis kept his battery in action during a heavy destructive enemy bombardment which lasted for several hours. Although almost half his detachment were killed or wounded and three of his guns put out of action, he personally directed the fire of his remaining gun. Later, while assisting a wounded officer to safety through a shell-swept area, a shell exploded near them, killing the wounded officer and knocking Lieutenant Davis to the ground. His work throughout the engagement was an inspiration to his men.
Posthumously awarded. Medal presented to mother, Mrs. Clara L. Davis.

DAVIS, LEROY (1403409)................
   At Mont-de-Sanges, France, Sept. 18, 1918.
   R—Champaign, Ill.
   B—Kansas City, Mo.
   G. O. No. 46, W. D., 1919.

Private, Company L, 370th Infantry, 93d Division.
He went out, under heavy fire, to the aid of a runner who had been wounded, applied first aid, took the messages of the wounded man, and delivered them to their destination.

DAVIS, LUTHER F. (1309122)............
   Near Busigny, France, Oct. 9, 1918.
   R—Chattanooga, Tenn.
   B—Flat Rock, Ky.
   G. O. No. 37, W. D., 1919.

Sergeant, Company K, 117th Infantry, 30th Division.
While commanding a platoon he repeatedly volunteered and went forward to draw fire from enemy machine guns, thereby locating the positions of the machine-gun posts and facilitating their destruction.

*DAVIS, MURRAY................
   Near Exermont, France, Sept. 26–29, 1918.
   R—Kansas City, Mo.
   B—Burlingame, Kans.
   G. O. No. 59, W. D., 1919.

Major, 140th Infantry, 35th Division.
He led his battalion brilliantly, and, when wounded, refused to go to the rear, but having his wound dressed on the spot continued in command of his battalion. Later he was killed while leading his command in an advance. Posthumously awarded. Medal presented to father, William B. Davis.

DAVIS, NEWELL B. (58168)............
   Near Cantigny, France, May 28, 1918.
   R—Hornell, N. Y.
   B—Greenwood, N. Y.
   G. O. No. 99, W. D., 1918.

Corporal, Company I, 28th Infantry, 1st Division.
He voluntarily left shelter and exposed himself to violent machine-gun fire in order to bring to shelter a wounded comrade. While performing this meritorious deed he was himself seriously wounded.

DAVIS, NEWMAN (735171)............
   Near Brandeville, France, Nov. 7–8, 1918.
   R—Jacksonville, Ala.
   B—Russia.
   G. O. No. 37, W. D., 1919.

Corporal, Company D, 11th Infantry, 5th Division.
After all the officers of his company had become casualties, Corporal Davis took command, displaying exceptional gallantry in leading his men

DAVIS, THOMAS H. (1799330)............
   At Binarville, France, Sept. 30, 1918.
   R—Hampton, Va.
   B—Hampton, Va.
   G. O. No. 20, W. D., 1919.

Private, first class, sanitary detachment, 368th Infantry, 92d Div.
Private Davis, with an officer and another soldier, voluntarily left shelter and crossed an open space 50 yards wide swept by shell and machine-gun fire to rescue a wounded soldier, whom they carried to a place of safety

DAVIS, WALTER E. (2338820)............
   Near Les Evaux, France, July 13, 1918.
   R—Lyndora, Pa.
   B—Springfield, Ill.
   G. O. No. 32, W. D., 1919.

Sergeant, Headquarters Company, 4th Infantry, 3d Division
After several night patrols had failed in the attempt to cross the Marne, Sergeant Davis, with three companions, crossed the river in daylight, and, in full view of the enemy, remained in enemy territory throughout the day

*DAVIS, WILLIAM D................
   Near Gesnes, France, Sept. 26–Oct. 2, 1918.
   R—Neosho, Mo.
   B—Duplain, Mich.
   G. O. No. 139, W. D., 1918.
   Distinguished-service medal also awarded.

Colonel, 361st Infantry, 91st Division.
He displayed distinguished gallantry in leading and directing the front line in the four days' advance on Gesnes and in the four following days, holding the front line under heavy shellfire. During this period his regiment was suffering heavy casualties, but he remained constantly with the front line, encouraging his men by his presence to hold out under this most dangerous and trying condition of warfare. Twice wounded he remained in command of his regiment throughout the entire action until it was finally achieved on Oct. 12.
Posthumously awarded. Medal presented to widow, Mrs. Abbie Greene Davis.

DAVIS, WILLIAM R. (72149)............
   At Bois Brule, near Apremont, France, Apr. 12, 1918.
   R—Brockton, Mass.
   B—Avon, Mass.
   G. O. No. 99, W. D., 1918.

Private, Company E, 104th Infantry, 26th Division.
For exceptional courage and devotion to duty in action on Apr. 12, 1918, remaining at his post and continuing to fire his rifle and grenades at the enemy after he was severely wounded in action.

DAVIS, WILTSHIRE C. (1288664)............
   Near Haumont, France, Oct. 11, 1918.
   R—Farmville, Va.
   B—Farmville, Va.
   G. O. No. 37, W. D., 1919.

First sergeant, Company G, 116th Infantry, 29th Division.
After the loss of all his officers, and his company was becoming disorganized, he took command and reassembled the company, bringing it to the objective at the most opportune moment. He commanded for a period of 11 days thereafter and successfully continued the operations by his leadership and exceptional courage under fire.

DAWSON, HAROLD A. (180793)............
   Near Juvigny, France, Aug. 31, 1918.
   R—Kalamazoo, Mich.
   B—Waynesville, Ohio.
   G. O. No. 71, W. D., 1919.

Sergeant, sanitary detachment, 126th Infantry, 32d Division.
He volunteered to go out into a field swept by artillery and machine-gun fire to administer first aid to wounded soldiers, constantly exposing himself to fire. His work saved the lives of many soldiers. After dark, when it was possible to remove the wounded from the field, he worked tirelessly during their evacuation to the first-aid station.

DAWSON, LEO H................
   Near Hartennes, France, July 19, 1918.
   R—Denver, Colo.
   B—Maxwell, N. Mex.
   G. O. No. 21, W. D., 1919.

First lieutenant, 94th Aero Squadron, Air Service.
While on a voluntary patrol, he encountered seven enemy monoplace planes at an altitude of 2,000 meters. After a brief engagement his guns jammed, but, after repairing the jam in the air and under heavy fire, he returned to the fight, shot down one of the enemy in flames and drove off the others.
Oak-leaf cluster.
For the following act of extraordinary heroism in action near Clery-le-Petit, France, Nov. 4, 1918, Lieutenant Dawson is awarded an oak-leaf cluster: Sighting four enemy planes (type, Rumpler), he immediately attacked, despite the numerical superiority of the enemy, and destroyed one of the group, whereupon the remaining three scattered and returned to their lines.

DAY, CLINTON (2264227)
Near St. Thibault, France, Aug. 7, 1918.
R—Fillmore, Utah.
B—Fillmore, Utah.
G. O. No. 66, W. D., 1919.

Private, first class, Company C, 58th Infantry, 4th Division.
Private Day repeatedly volunteered and carried messages from his company in the front line across an open field swept by enemy machine-gun and sniper fire to the battalion post of command. He also voluntarily made trips across this dangerous area for the purpose of filling canteens for wounded soldiers and securing stretchers.

DAY, LOUIS T. (69877)
At Apremont, France, May 20, 1918.
R—Somerville, Mass.
B—Somerville, Mass.
G. O. No. 49, W. D., 1922.

Sergeant, Headquarters Company, 103d Infantry, 26th Division.
Displaying remarkable coolness and courage, Sergeant Day volunteered and, with one other man, at about midnight and under a heavy enemy artillery and machine-gun fire crawled through and over several wire entanglements to a point well within the hostile machine-gun outpost line, where they cut several meters from the armored cable which supplied electric current to the enemy's wire entanglements which were an impediment and danger to our reconnoitering patrols.

DAYTON, ALLAN S.
Near Fismes, France, Sept. 6, 1918.
R—Pittsburgh, Pa.
B—Hackensack, N. J.
G. O. No. 1, W. D., 1919.

First lieutenant, 107th Field Artillery, 28th Division.
He led a patrol out to the Infantry lines in order to adjust the artillery fire on machine guns which were holding up the advance. It was found necessary to advance about half a mile beyond the front lines across open ground swept by machine-gun fire; but, undaunted, this officer continued on for a half hour, until he established telephone communication with his regiment. Having finished his work he helped to carry a wounded officer back through an enemy barrage safely to our lines.

DEAKINS, JESSE S. (45842)
South of Soissons, France, July 18, 1918.
R—St. Joseph, Mo.
B—Buchanan County, Mo.
G. O. No. 72, W. D., 1920.

Mechanic, Company A, 18th Infantry, 1st Division.
Mechanic Deakins voluntarily proceeded with four other men the distance of about 400 yards in advance of our line and attacked a superior force of the enemy who were attempting to man a machine gun in a disabled French tank. Due to this bold attack the enemy was driven off and the tank retaken. Company A, 18th Infantry, was thereby enabled to continue the advance with slight loss.

DEAN, JOHN J. (1240919)
Near Montblainville, France, Sept. 27, 1918.
R—Philadelphia, Pa.
B—Ireland.
G. O. No. 64, W. D., 1919.

Sergeant, Company H, 110th Infantry, 28th Division.
Although wounded in the lung by a machine-gun bullet, he refused to be evacuated. Upon his platoon reaching its objective, he consolidated the new position and repulsed a strong counterattack with heavy losses to the enemy. He then reported to his battalion commander, refusing a litter, saying that all effectives were needed in the line, and walked to the dressing station.

DEAN, THOMAS G., Jr. (1210293)
Near Ronssoy, France, Sept. 29, 1918.
R—New York, N. Y.
B—New York, N. Y.
G. O. No. 20, W. D., 1919.

Corporal, Company D, 107th Infantry, 27th Division.
During the operations against the Hindenburg line Corporal Dean, with four other soldiers, left shelter and went forward into an open field under heavy shell and machine-gun fire, and succeeded in bandaging and carrying back to our lines two wounded men.

*DEARING, VINTON ADAMS
Near Cantigny, France, May 28, 1918.
R—Cambridge, Mass.
B—Japan.
G. O. No. 99, W. D., 1918.

Second lieutenant, 28th Infantry, 1st Division.
Detailed in command of a carrying party on May 28, 1918, near Cantigny, France, he bravely proceeded under fire to execute his mission, and by his example of bravery heartened his men, who were under fire for the first time. On May 29, 1918, he took his party through heavy shelling to carry ammunition to the front lines without being ordered to do so.
Posthumously awarded. Medal presented to mother, Mrs. J. L. Dearing.

°DEASEY, HERBERT A. (2061894)
Near Chipilly Ridge, France, Aug. 9, 1918.
R—Chicago, Ill.
B—England.
G. O. No. 71, W. D., 1919.

Private, Company F, 131st Infantry, 33d Division.
Acting on his own initiative, he advanced alone against a machine-gun nest that had been causing heavy casualties among his comrades. He crawled to within a short distance of the enemy position before he was detected. He then rushed the post and bayoneted the three gunners, being himself killed in the encounter.
Posthumously awarded. Medal presented to father, James Deasey.

DE BATTISTA, GEORGE (322977)
Near Kazanka, Siberia, July 13, 1919.
R—San Francisco, Calif.
B—Malta.
G. O. No. 133, W. D., 1919.

Sergeant, Company D, 31st Infantry.
He led a patrol of eight men through enemy country to Kazanka, about 10 miles distant from camp, and returned with an accurate report of enemy activities in the town. The circumstances governing at the time required extraordinary daring and skill on the part of Sergeant De Battista in order to carry out the instructions given him.

DE BERARDINAS, PIETRO (1285622)
Near Verdun, France, Oct. 17, 1918.
R—Highlandtown, Md.
B—Italy.
G. O. No. 15, W. D., 1919.

Private, Company H, 115th Infantry, 29th Division.
In the Bois de Consenvoye, east of the Meuse, he, acting in the capacity of runner, carried three successive messages through continuous and heavy barrages of both our own and the enemy's artillery, traversing a path where two men had previously been killed by the same barrage.

DECAIRE, GEORGE (262723)
Near the Ourcq River, France, July 31, 1918.
R—Baltic, Mich.
B—Houghton, Mich.
G. O. No. 46, W. D., 1919.

Bugler, Company G, 125th Infantry, 32d Division.
Bugler Decaire, acting as runner to battalion headquarters, continued in the advance and performed his duties under extremely heavy shell and machine-gun fire after he had been very severely wounded in the knee.

DE CARL, THEODORE J. (2383577)
Near Cunel, France, Oct. 12, 1918.
R—Philadelphia, Pa.
B—New York, N. Y.
G. O. No. 37, W. D., 1919.

Corporal, Company F, 60th Infantry, 5th Division.
In order to deliver a message from the regimental headquarters to the attacking battalion with all the speed possible, he crossed an open area under continuous machine-gun fire. Although painfully wounded while traversing this open stretch of 500 meters, he successfully delivered the message.

De Carre, Alphonse
In the Bois-de-Belleau, France, June 11, 1918.
R—St. Louis, Mo.
B—Washington, D. C.
G. O. No. 81, W. D., 1919.

Captain, 5th Regiment, U. S. Marine Corps, 2d Division.
Commanding a detachment of about 60 men in the rear of the attacking battalion, Captain De Carre continued forward alone, after the advance battalion had swerved to the right. About 150 yards distant he encountered deadly enemy machine-gun fire, but he continued on, capturing two guns. He then brought up his detachment and by exceptional handling effected the capture of an entire machine-gun company, consisting of three officers and 160 men. His effort prevented the enemy from firing on our troops from the rear.

De Castro, Ralph Ellison
Near St. Mihiel, France, Sept. 12, 1918.
R—Brooklyn, N. Y.
B—Brooklyn, N. Y.
G. O. No. 116, W. D., 1918.

Second lieutenant, pilot, 1st Aero Squadron, Air Service.
Because of intense aerial activity on the opening day of the St. Mihiel offensive, Lieutenant De Castro, pilot, and First Lieutenant Arthur E. Easterbrook, observer, volunteered to fly over the enemy's lines on a photographic mission, without the usual protection of accompanying battle planes. Notwithstanding low-hanging clouds, which necessitated operation at an altitude of only 400 meters, they penetrated 4 kilometers beyond the German lines. Attacked by four enemy machines, they fought their foes, completed their photographic mission, and returned safely.

Dechert, Robert
Near Le Charmel, France, July 23–25, 1918.
R—Philadelphia, Pa.
B—Philadelphia, Pa.
G. O. No. 89, W. D., 1919.

First lieutenant, 7th Infantry, 3d Division.
When the attacking battalion had been held up by machine-gun fire, Lieutenant Dechert, who was on duty as regimental signal officer, personally carried wire across an open field in full view of the enemy and established a telephone station within 200 yards of the front line. He then went forward under heavy shell fire to report to the battalion commander, and returning to the telephone kept it in operation for 24 hours under intense artillery and machine-gun fire.

De Coppet, Andre
Near Merval, France, Sept. 14, 1918.
R—New York, N. Y.
B—New York, N. Y.
G. O. No. 21, W. D., 1919.

First lieutenant, Infantry, aid-de-camp, 77th Division.
In preparation for an attack by units of his division, Lieutenant De Coppet helped establish an observation post. Learning a wounded officer was in front, he made his way twice through intense fire from artillery and small arms to where the wounded officer lay, and assisted in carrying him back to safety.

De Cota, Joseph (59854)
Near Verdun, France, Oct. 23, 1918.
R—Charlestown, Mass.
B—Charlestown, Mass.
G. O. No. 21, W. D., 1919.

Sergeant, Company B, 101st Infantry, 26th Division.
During the advance of his battalion at Molleville Farm, Sergeant De Cota was rendered unconscious and wounded by the explosion of a shell. Recovering his senses, he quickly rejoined his platoon and led it in the attack. Although suffering from a painful wound in the arm, he remained on duty until ordered to the rear by his company commander late the next day.

Dee, Frank E. (1211440)
Near Ronssoy, France, Sept. 29, 1918.
R—Waterbury, Conn.
B—Waterbury, Conn.
G. O. No. 32, W. D., 1919.

Sergeant, Company I, 107th Infantry, 27th Division.
When the advance of his platoon had been held up by direct machine-gun fire, he, severely wounded during the advance, went out in plain view of the enemy, pulled the pin of a grenade with his teeth, and, throwing the bomb with his left arm, put the gun and its crew out of action.

Deeringer, Henry (2155793)
At Estrees, France, Oct. 8, 1918.
R—Knoxville, Iowa.
B—Formoso, Kans.
G. O. No. 46, W. D., 1919.

Private, Company B, 117th Infantry, 30th Division.
While working as a stretcher bearer, Private Deeringer was himself severely wounded, but he nevertheless succeeded in getting his patient to the dressing station, where he himself received first aid and was tagged for evacuation. Tearing the tag from his coat, he returned to the field and continued to perform his duties until afternoon, when he was hardly able to walk, and was again ordered to the rear.

*Deese, Peyton V.
Near Nesles, France, July 28, 1918.
R—Skipperville, Ala.
B—Dale County, Ala.
G. O. No. 20, W. D., 1919.

First lieutenant, 167th Infantry, 42d Division.
Although wounded, Lieutenant Deese led his platoon against enemy machine-gun nests, silencing them. He made his way through a heavy barrage, encountering and breaking an enemy counterattack. Progressing slowly and with great difficulty on account of the deadly fire, he reached a strongly fortified position of the enemy, where he captured many prisoners and machine guns, which he defended until the arrival of support. During the action he was again wounded, the effects of which caused his death.
Posthumously awarded. Medal presented to father, S. S. Deese.

Deford, August H. (58679)
At Cantigny, France, May 28, 1918.
R—Chicago, Ill.
B—New York, N. Y.
G. O. No. 99, W. D., 1918.

Corporal, Company M, 28th Infantry, 1st Division.
Although wounded three times in the attack at Cantigny, France, May 28, 1918, he showed conspicuous bravery in assisting to capture a machine gun which was causing heavy losses. Although nearly exhausted, he captured three prisoners and made them secure before accepting attention to his injuries.

*Deggs, George (3069505)
Near Preny, France, Nov. 2, 1918.
R—Benford, Tex.
B—Stryker, Tex.
G. O. No. 44, W. D., 1919.

Private, Company E, 56th Infantry, 7th Division.
During an enemy counterattack the dugout in which he and his companions were taking shelter was surrounded by a group of the enemy, who were demanding surrender. Jumping to the fire step and with the aid of a rifle, Private Deggs killed four of the enemy and caused the rest to flee in confusion.
Posthumously awarded. Medal presented to mother, Mrs. Julia Deggs.

De Lacour, Reginald B.
North of Chalons-sur-Marne, France, July 15, 1918.
R—Stratford, Conn.
B—Wichita, Kans.
G. O. No. 30, W. D., 1921.

First lieutenant, 165th Infantry, 42d Division.
After having been knocked down and severely wounded in the leg by an exploding shell at about 7 o'clock a. m., and although suffering great pain and much weakened by loss of blood, he refused to leave his platoon, but remained on duty therewith under heavy enemy fire until about 11 o'clock a. m., when relieved by another officer. His fortitude and disregard of personal danger were a source of inspiration to the men of his platoon.

DE LACY, AUBREY B
Near Haumont, France, Sept. 27, 1918.
R—New York, N. Y.
B—New York, N. Y.
G. O. No. 23, W. D., 1919.

First lieutenant, 166th Infantry, 42d Division.
Leading a patrol into the town to ascertain whether or not it was still occupied by the enemy, Lieutenant De Lacy came under heavy machine-gun fire. Against greatly superior numbers he continued forward and, entering the town, took two prisoners, from whom he gained valuable information.

*DE LAITE, DONALD K. (1550907)
Near Romagne, France, Oct. 15, 1918.
R—Portage, Me.
B—Kingman, Me.
G. O. No. 27, W. D., 1920.

Private, first class, Battery D, 76th Field Artillery, 3d Division.
As a member of a gun crew Private De Laite kept his gun in action during heavy enemy bombardment. When 3 shells exploded among the gun crew, killing or wounding all of them, Private De Laite, although fatally wounded, disregarded his own wounds and went to obtain aid for his wounded comrades.
Posthumously awarded. Medal presented to father, Fred De Laite.

DELAMBO, MIKE (543095)
At Claire Chenes, France, Oct. 20, 1918.
R—Greensburg, Pa.
B—Connellsville, Pa.
G. O. No. 27, W. D., 1920.

Private, Company M, 7th Infantry, 3d Division.
When his company was held up by machine-gun fire Private Delambo, with 2 comrades, advanced and made a flank attack on an enemy machine-gun position, capturing the gun and forcing 1 officer, 2 sergeants, and 6 privates to surrender. By his deed the company was enabled to continue the advance.

DELAND, THORNDIKE
Near Marimbois Farm, France, Nov 4, 1918.
R—Denver, Colo.
B—Cleveland, Ohio.
G. O. No. 37, W. D., 1919.

First lieutenant, 340th Field Artillery, 89th Division.
Assisted by a soldier, he went forward in advance of the Infantry to lay telephone wires. Nearing Marimbois Farm they found the place occupied by the enemy. Armed with hand grenades he advanced on a dugout, where he routed out 17 of the enemy, bringing them back to our lines in the midst of a severe shell and machine-gun fire.

*DELARIO, CHARLES E.
Near Verdun, France, Nov. 2, 1918.
R—Los Angeles, Calif.
B—Laramie, Wyo.
G. O. No. 37, W. D., 1919.

Captain, 360th Infantry, 90th Division.
Wounded while leading his company in the advance he turned over the command to another officer and went to the rear for first aid. Upon reaching the aid station he learned that his company was without officers, whereupon he immediately started back to the front through heavy machine-gun fire and was killed on the way.
Posthumously awarded. Medal presented to mother, Mrs. Anna McKee.

DELESDERNIER, LIONEL W. (66230)
Near Seicheprey, France, Apr. 20, 1918.
R—Meriden, Conn.
B—Meriden, Conn.
G. O. No. 9, W. D., 1923.

Private, first class, Company L, 102d Infantry, 26th Division.
With great daring and utter disregard for his own safety he made 3 hazardous journeys from his company command post to the battalion ammunition dump through the Bois-de-Remoirre, which was infested with enemy machine-gun nests and snipers. Twice he was driven off by pistol and grenade fire which he returned, and the third time forced to abandon further attempts when the dump was destroyed by enemy fire. Using the top of a dugout as a parapet he, with a few men of his company, engaged the enemy and with pistols and grenades alone kept the enemy from the company post of command. Finally Private Delesdernier, running through a terrific barrage of enemy fire, carried a message to the company support, accomplished his mission, and again returned through the enemy barrage to the defense of his company's headquarters.

DELEUW, CHARLES E.
Near Ville-Savoye, France, Aug. 11, 1918.
R—Riverside, Ill.
B—Jacksonville, Ill.
G. O. No. 138, W. D., 1918.

First lieutenant, 4th Engineers, 4th Division.
He was in command of a detachment of Engineers engaged in constructing an artillery bridge across the River Vesle under constant fire from machine guns and bombardment by both high-explosive and gas shells. Although he was suffering from the effects of gas this officer remained in charge of the party, directing the work and furnishing his men a splendid example of courage under fire and disregard for personal safety.

DE LOISELLE, HAROLD C.
Near Ronssoy, France, Sept. 27, 1918.
R—Brooklyn, N. Y.
B—New York, N. Y.
G. O. No. 39, W. D., 1920.

First lieutenant, 106th Infantry, 27th Division.
Lieutenant de Loiselle twice voluntarily went forward under heavy shell and machine-gun fire to reconnoiter the enemy positions. While leading the second of these patrols the officer who accompanied him was killed. Notwithstanding, with but one other man, he continued in his reconnaissance until almost surrounded by the enemy. He succeeded in outmaneuvering the enemy and returned with valuable information.

DELOTO, PETER (1737909)
Near Grand Pre, France, Oct. 25, 1918.
R—Fulton, N. Y.
B—Passaic, N. J.
G. O. No. 37, W. D., 1919.

Sergeant, Company K, 311th Infantry, 78th Division.
During the attack Sergeant Deloto single-handed captured 8 prisoners and later aided materially in the capture of 16 more. When his company had been held up by the fire of a machine gun he set out, unassisted, and succeeded in capturing it, during which exploit he was wounded in the leg.

DE MAY, JOSEPH (51984)
Near Vaux, France, July 1, 1918.
R—Lawrence, Mass.
B—Italy.
G. O. No. 99, W. D., 1918.

Private, Company L, 23d Infantry, 2d Division.
He gamely continued to fire his automatic rifle after falling wounded in both legs.

DENIG, ROBERT L
Near Medeah Ferme, France, Oct. 3, 1918.
R—Sandusky, Ohio.
B—Clinton, N. Y.
G. O. No. 21, W. D., 1919.

Major, U. S. Marine Corps, attached to 9th Infantry, 2d Division.
While directing his battalion in cleaning out woods filled with enemy machine guns and snipers, himself severely wounded, he remained on duty until his mission had been accomplished.

DENN, ANDREW (42760)_____
Near Soissons, France, July 22, 1918.
R—Albany, N. Y.
B—Albany, N. Y.
G. O. No. 44, W. D., 1919.

Corporal, Company E, 16th Infantry, 1st Division.
Reorganizing his platoon, after all officers and noncommissioned officers had become casualties, he led a charge on an enemy machine-gun nest which threatened to annihilate his platoon and halted its advance. Although wounded in the attack he continued on until he had killed the gunner and the entire crew.

DENNELLEY, JOHN HENRY (89034)_____
Near Landres-et-St. Georges, France, Oct. 15, 1918.
R—Great Neck Station, N. Y.
B—Great Neck, N. Y.
G. O. No. 126, W. D., 1919.

Sergeant, Company A, 165th Infantry, 42d Division.
When his company had been ordered to withdraw under intense machine-gun and artillery fire Sergeant Dennelley remained in the position until the other men had safely retired, when he saw to the removal of 6 wounded men.

*DENNIS, CLARENCE A._____
At Chateau-Thierry, France, June 6, 1918.
R—Hackensack, N. J.
B—Hackensack, N. J.
G. O. No. 126, W. D., 1919.

Second lieutenant, 6th Regiment, U. S. Marine Corps, 2d Division.
Killed in action at Chateau-Thierry, France, June 6, 1918. He gave the supreme proof of that extraordinary heroism which will serve as an example to hitherto untried troops.
Posthumously awarded. Medal presented to father, W. H. Dennis.

DENNIS, ERWIN A._____
Near St. Souplet, France, Oct. 17, 1918.
R—Auburn, N. Y.
B—Auburn, N. Y.
G. O. No. 37, W. D., 1919.

Second lieutenant, 108th Infantry, 27th Division.
He led a small patrol against an enemy machine-gun nest, which he successfully captured. He discovered a large enemy nest, and for three hours held a position against it until reinforced by a Vickers machine gun. This aid forced the enemy to surrender. The capture consisted of 3 officers, 145 men, 3 large Maxim guns, 7 light machine guns, and 3 antitank guns.

DENNISON, CHARLES S._____
In the Forest of Argonne, France, Sept. 27, 1918.
R—Denver, Colo.
B—Leonia, N. J.
G. O. No. 21, W. D., 1919.

Second lieutenant, 306th Infantry, 77th Division.
He was in command of a patrol sent out in the afternoon to locate a machine-gun nest. In the course of this operation he received a severe wound, but after reporting the location of the machine-gun nest at company headquarters he immediately returned to the vicinity of the machine-gun nest and spent the greater part of the night searching for a member of his patrol who was missing. Although he was suffering severe pain from his wounds, he refused to go to the first-aid station before the missing soldier was found.

*DEPUE, DAVID T._____
Near St. Georges, France, Nov. 1, 1918.
R—South Chicago, Ill.
B—Whitehall, Mich.
G. O. No. 71, W. D., 1919.

Private, first class, 76th Company, 6th Regiment, U. S. Marine Corps, 2d Division.
When his platoon was held up by barbed-wire entanglements within 30 yards of an enemy machine-gun nest, he took an automatic rifle from a dead gunner near him and, firing as he advanced, charged through the wire. He fell twice, but reached the enemy position after his ammunition was exhausted, swinging the rifle above his head as a club upon the enemy defenders. When the platoon reached the enemy nest Private Depue was found lying mortally wounded among four enemy dead.
Posthumously awarded. Medal presented to father, James Depue.

*DE RHAM, CHARLES, Jr._____
Near Bazoches, France, Sept. 3, 1918, and in the Meuse-Argonne offensive, Sept. 27, 1918.
R—Cold Spring, N. Y.
B—New York, N. Y.
G. O. No. 24, W. D., 1920.

First lieutenant, 305th Infantry, 77th Division.
On Sept. 3 Lieutenant de Rham led the first patrol of his brigade across the Vesle River, exposed to heavy enemy machine-gun fire. On Sept. 27 he led 5 attacks against enemy machine-gun positions. He was killed while making the fifth attack.
Posthumously awarded. Medal presented to widow, Mrs. Charles de Rham.

*DEROCHERS, RODOLPH (1668157)_____
Near Exermont, France, Oct. 9, 1918.
R—Fall River, Mass.
B—Canada.
G. O. No. 44, W. D., 1919.

Private, first class, Company C, 2d Machine Gun Battalion, 1st Division.
Private Derochers, a company runner, displayed exceptional bravery in carrying messages through heavy shell and machine-gun fire until killed.
Posthumously awarded. Medal presented to mother, Mrs. Cleophe Derochers.

DE ROGATIS, ALBERT (1752789)_____
Near St. Juvin, France, Oct. 16, 1918.
R—Asbury Park, N. J.
B—Italy.
G. O. No. 81, W. D., 1919.

Private, Company M, 309th Infantry, 78th Division.
When his company was held up by heavy machine-gun fire Private de Rogatis voluntarily worked his way behind an enemy machine-gun position, killed a German soldier, and captured 7 others, together with 2 machine guns.

DE ROSSELLI, PETER L. (2786122)_____
Near Gesnes, France, Oct. 4, 1918.
R—Hollywood, Calif.
B—Toledo, Ohio.
G. O. No. 103, W. D., 1919.

Private, Company F, 361st Infantry, 91st Division.
Accompanying a patrol on a reconnaissance, Private de Rosselli penetrated enemy positions, the exploit being accomplished under heavy fire. Although wounded, he returned with valuable information regarding the positions of enemy machine-gun nests and snipers' posts.

*DERRICKSON, PAUL W._____
In the advance on Cantigny, France, May 28, 1918.
R—Norfolk, Va.
B—Norfolk, Va.
G. O. No. 99, W. D., 1918.

Second lieutenant, 28th Infantry, 1st Division.
He courageously went forward with his platoon and reached the position he had been directed to take. Fearlessly walking up and down his line, he cheered and directed the work of his men until he was killed.
Posthumously awarded. Medal presented to mother, Mrs. Mary G. Derrickson.

DERRY, JOHN W. (2213873)_____
Near Nesles, France, July 15, 1918.
R—Petersburg, Ill.
B—Petersburg, Ill.
G. O. No. 32, W. D., 1919.

Private, first class, Company D, 4th Infantry, 3d Division.
He requested permission to leave the trenches and to assist the wounded. Under heavy artillery bombardment of the enemy, he aided many wounded comrades to a dressing station, returning with litters for the more seriously wounded.

*DE RUM, HOWARD D (1200961)_____
Near Roussoy, France, Sept. 29, 1918.
R—Buffalo, N. Y.
B—Buffalo, N. Y.
G. O. No. 37, W. D., 1919.

Corporal, Company C, 102d Field Signal Battalion, 27th Division.
Corporal de Rum accompanied the first attacking wave, stringing telephone lines under terrific enemy fire, even after being advised by the signal officer to seek shelter, courageously maintained communication until he was killed. Posthumously awarded. Medal presented to widow, Mrs. Helen I. de Rum.

*DESAUSSURE, EDWARD C_____
Near Sommerance, France, Oct. 18, 1918.
R—Jacksonville, Fla.
B—Atlanta, Ga.
G. O. No. 73, W. D., 1919.

First lieutenant, 328th Infantry, 82d Division.
Lieutenant Desaussure was painfully wounded by shrapnel while in command of his company. Continuing to direct its operations, while he was having his wound attended to at the dressing station, he insisted upon returning to his command immediately thereafter, and in attempting to do so was killed by a bursting shell. His conspicuous devotion to duty and self-sacrificing spirit furnished an inspiration to his men which contributed materially to the ultimate success of the attack.
Posthumously awarded. Medal presented to mother, Mrs. George R. Desaussure.

DESKINS, FRANK (542198)_____
Near Remonville, France, Nov. 1, 1918.
R—Myrtle, W. Va.
B—Myrtle, W. Va.
G. O. No. 44, W. D., 1919.

Corporal, Company D, 354th Infantry, 89th Division.
He took command of the 4 leading combat groups after 4 other leaders had been disabled, directing them with such skill and coolness that many machine guns were taken from the enemy. Four hours before the day's objective had been reached his shoulder was pierced by a machine-gun bullet, but he told no one of the wound until his line had been organized along the objective.

DE SMIDT, JOHN (1390996)_____
At Hamel, France, July 4, 1918.
R—Chicago, Ill.
B—Chicago, Ill.
G. O. No. 81, W. D., 1919.

Corporal, Company G, 132d Infantry, 33d Division.
With the assistance of an Australian soldier, Corporal de Smidt crept up on the position of an enemy machine gun, captured the gun, and forced its crew to carry it back to our lines.

DESSEZ, PAUL T_____
At Chateau-Thierry, France, June 6, 1918.
R—Washington, D. C.
B—Washington, D. C.
G. O. No. 110, W. D., 1918.

Commander, surgeon, U. S. Navy, attached to Headquarters, 5th Regiment, U. S. Marine Corps, 2d Division.
He organized the service of caring for and evacuating the wounded in a most systematic and admirable manner, constantly exposing himself to the enemy, displaying extraordinary heroism, coolness, and energy.

*DETROW, WALTER H. (2110211)_____
Near Sergy, France, Aug. 1, 1918.
R—New Springfield, Ohio.
B—New Springfield, Ohio.
G. O. No. 44, W. D., 1919.

Private, Company B, 47th Infantry, 4th Division.
After all the officers and noncommissioned officers of his platoon had been lost, Private Detrow assumed command of the platoon, successfully leading it from its critical situation to the objective, through terrific machine-gun and shellfire. He performed this gallant act without any previous instructions or orders, and acted entirely upon his own initiative.
Posthumously awarded. Medal presented to mother, Mrs. J. W. Detrow.

DETTRE, REXFORD H. (125227)_____
At Villers-Tournelle, Cantigny sector, France, May 1, 1918.
R—Bradenton, Fla.
B—Philadelphia, Pa.
G. O. No. 100, W. D., 1918.

Corporal, Headquarters Company, 6th Field Artillery, 1st Division.
He displayed distinguished bravery in twice leaving his shelter during a heavy bombardment and going to the assistance of wounded men lying exposed in the open.

*DEVALLES, JOHN B._____
Near Apremont, Toul sector, France, Apr. 10 to 13, 1918.
R—New Bedford, Mass.
B—Azores.
G. O. No. 35, W. D., 1920.

Chaplain, 104th Infantry, 26th Division.
Chaplain DeValles repeatedly exposed himself to heavy artillery and machine-gun fire in order to assist in the removal of the wounded from exposed points in advance of the lines. He worked for long periods of time with stretcher bearers in carrying wounded men to safety. Chaplain DeValles previously rendered gallant service in the Chemin des Dames sector, Mar. 11, 1918, by remaining with a group of wounded during a heavy enemy bombardment.
Posthumously awarded. Medal presented to sister, Mrs. Marie Hill.

DEVANE, DUNCAN J. (1323052)_____
Near Ypres, Belgium, Aug. 23, 1918.
R—Clarkton, N. C.
B—Clarkton, N. C.
G. O. No. 35, W. D., 1919.

Sergeant, Company C, 115th Machine Gun Battalion, 30th Division.
Upon learning that several members of his platoon had been wounded by enemy shellfire, he immediately left his dugout and went to their assistance. After carrying one man to shelter and being knocked down by a bursting shell in so doing, he returned to the shelled area and helped carry the rest of the wounded men to the dressing station, 500 yards away, across a field which was being heavily bombarded with gas and high-explosive shells.

DEVEREAUX, HAROLD J. (264010)_____
Near Sergy, France, July 31, 1918.
R—Elsie, Mich.
B—Elsie, Mich.
G. O. No. 117, W. D., 1918.

Private, Company M, 125th Infantry, 32d Division.
When his company had crossed the River Ourcq and captured the Bois Pelger, the corporal of his squad, fighting beside him, was wounded by machine-gun fire. The enemy continued to fire on the wounded man, and Private Devereaux, single handed, with the fire of his rifle, attacked the machine gun and succeeded in putting it out of action.

DEVLIN, BERT WILLIAM_____
Near Blanc Mont, France, Oct. 5, 1918.
R—Bennington, Vt.
B—Chelsea, Mass.
G. O. No. 46, W. D., 1919.

Private, first class, 43d Company, 5th Regiment, U. S. Marine Corps, 2d Division.
He demonstrated the highest degree of courage by offering his services in bringing the wounded to a place of safety from a region which was under constant shell and machine-gun fire.

DE VOS, PETER A. (2382811)_____
Near Cunel, France, Oct. 14, 1918.
R—Minneapolis, Minn.
B—Belgium.
G. O. No. 44, W. D., 1919.

Private, first class, Company C, 60th Infantry, 5th Division.
Accompanied by one other soldier, Private De Vos flanked 2 machine-gun nests, killed 7 of the enemy and captured 4 machine guns, thereby making it possible for two companies of his battalion to enter the woods and continue the advance.

DEWALT, CLYDE H. (2185154)........
    Near Pouilly, France, Nov. 10–11, 1918.
    R—Danville, Pa.
    B—Danville, Pa.
    G. O. No. 37, W. D., 1919.

Sergeant, Company K, 356th Infantry, 89th Division.
He volunteered and led a patrol against enemy machine guns which were flanking his company. He captured 2 of the guns and returned with 5 prisoners, making possible the continuance of his company's advance.

DEXTER, ALLAN L........
    At Bois Brule, near Apremont, France, Apr. 12–13, 1918.
    R—Brookline, Mass.
    B—Brookline, Mass.
    G. O. No. 99, W. D., 1918.

Second lieutenant, 104th Infantry, 26th Division.
While acting as battalion scout officer during the action of Apr. 12 and 13, 1918, he displayed conspicuous courage and devotion to duty by exposing himself constantly under heavy shellfire to secure information, continuing his work for 24 hours after being wounded and until he collapsed at dressing station, where he had been sent for treatment.

*DIAL, WALTER V........
    Near Fleville, France, Oct. 4, 1918.
    R—Huntington, W. Va.
    B—W. Va.
    G. O. No. 63, W. D., 1920.

Second lieutenant, 2d Machine Gun Battalion, 1st Division.
Lieutenant Dial displayed exceptional courage in leading his platoon in attacking and breaking up German machine-gun nests under heavy artillery and machine-gun fire. Although he was wounded, he refused to be evacuated and continued to advance until he was killed.
Posthumously awarded. Medal presented to father, A. G. Dial.

DICARLO, SALVATORE (74508)........
    Near Vierzy, France, July 18, 1918.
    R—Los Angeles, Calif.
    B—Italy.
    G. O. No. 32, W. D., 1919.

Private, first class, 4th Machine Gun Battalion, 2d Division.
Single-handed, he attacked and captured 3 enemy machine guns and 8 prisoners. Later he rendered first aid to wounded comrades and assisted them from the field under intense machine-gun and artillery fire.

DICK, HENRY J........
    Near Chipilly Ridge, France, Aug. 9, 1918.
    R—Emporia, Kans.
    B—Burton, Kans.
    G. O. No. 71, W. D., 1919.

Second lieutenant, 131st Infantry, 33d Division.
Although wounded, he rushed a machine-gun nest that was causing heavy casualties, and bayoneted 1 of the crew, shot 2, and captured 5 of the enemy. Seeing some of the enemy enter a dugout, he followed, capturing 12 more Germans and 3 machine guns. His bravery was an inspiration to his men.

DICKENS, BENJAMIN (42761)........
    In the Forest of Argonne, France, Oct. 4, 1918.
    R—Harmony, Ind.
    B—Harmony, Ind.
    G. O. No. 44, W. D., 1919.

Sergeant, Company E, 16th Infantry, 1st Division.
Although he was severely wounded early in the attack, he refused to be evacuated, and, when all of the officers had become casualties, took command of the company and led it to the objective. Shortly afterwards the enemy made a strong attack against the position, and although he was scarcely able to walk, he so successfully led his company that the superior number of the enemy was forced to withdraw.

DICKERSON, WILLIAM L. (53942)........
    Near Soissons, France, July 19, 1918.
    R—Pelzer, S. C.
    B—Anderson, S. C.
    G. O. No. 132, W. D., 1918.

Private, Company G, 26th Infantry, 1st Division.
With two other soldiers he rushed a machine-gun position near Soissons, France, July 19, 1918, killed the crew and captured the gun in order to make the advance of his platoon possible.

DICKEY, LESLIE J. (1864758)........
    Near Grimaucourt, France, Nov. 10, 1918.
    R—Vincennes, Ind.
    B—Gallatin County, Ill.
    G. O. No. 32, W. D., 1919.

First sergeant, Company F, 322d Infantry, 81st Division.
After his company had been relieved he voluntarily returned to the position which they formerly occupied and carried a wounded comrade to safety through fierce artillery fire.

*DICKINSON, CLEMENT P........
    Near Charpentry, France, Sept. 27, 1918.
    R—Clinton, Mo.
    B—Clinton, Mo.
    G. O. No. 31, W. D., 1922.

First lieutenant, 129th Machine Gun Battalion, 35th Division.
Lieutenant Dickinson voluntarily went forward alone through intense artillery and machine-gun fire to locate machine-gun positions and having accomplished this, returned and led his company forward to the attack, thus passing three times through this intense field of fire. On the following morning, having again volunteered to reconnoiter the new positions, he was instantly killed by a high-explosive shell.
Posthumously awarded. Medal presented to widow, Mrs. Emma Dickinson.

DICKINSON, DWIGHT, Jr........
    Near St. Etienne, France, Oct. 4, 1918.
    R—Washington, D. C.
    B—Jamestown, N. Y.
    G. O. No. 46, W. D., 1919.

Passed assistant surgeon, U. S. Navy, attached to 5th Regiment, U. S. Marine Corps, 2d Division.
Under terrific shell and machine-gun fire, he attended the wounded with utter disregard for his own safety. When a shell struck the dressing station which he had established in an advanced zone, he rushed to the assistance of the wounded, and through his devotion to duty many lives were saved.

*DICKOP, RAY C........
    In the attack on Fismes, France, Aug. 4, 1918.
    R—Beloit, Wis.
    B—Beloit, Wis.
    G. O. No. 117, W. D., 1918.

First lieutenant, 127th Infantry, 32d Division.
On reaching Chezelles Farm, he was shot in the head, body, and legs. Although thus fatally wounded, when orders came for another assault, he gave the command "Charge" to his company and led the assault until he fell dead.
Posthumously awarded. Medal presented to aunt, Miss Lena Schiller.

*DICKSON, HARRISON A........
    Near Chipilly Ridge, France, Aug. 9, 1918.
    R—Jacksonville, Ill.
    B—Jacksonville, Ill.
    G. O. No. 74, W. D., 1919.

First lieutenant, 131st Infantry, 33d Division.
When his company was held up by heavy machine-gun fire, he ordered his men to lie down and went out alone, facing intense fire, in an effort to capture the hostile nest. Shortly after starting forward, he was shot through the heart.
Posthumously awarded. Medal presented to mother, Mrs. Charles E. Dickson.

DIEKEMA, WILLIS A.
In the region of Metz, France, Sept. 15, 1918.
R—Chicago, Ill.
B—Holland, Mich.
G. O. No. 143, W. D., 1918.

First lieutenant, pilot, 91st Aero Squadron, Air Service.
While on a photographic mission, his formation was attacked by a superior number of enemy aircraft, and in the course of the combat his companion planes were driven off. Disregarding the fact that his machine was without protection, he continued on his mission until his observer, Lieutenant Hammond, had completed the photographs. On the return they fought their way through an enemy patrol and destroyed one of the machines.

DIENER, LOUIS.
In the Ravine de la Veux Michieux, France, Oct. 26–27, 1918.
R—Culpeper, Va.
B—Culpeper, Va.
G. O. No. 15, W. D., 1919.

Captain, Medical Corps, sanitary detachment, 112th Machine Gun Battalion, 29th Division.
Upon being notified that an enemy shell had struck a dugout occupied by the brigade radio detachment he ran to the aid of the buried men and worked tirelessly to rescue them. Despite the fact that numerous gas and high-explosive shells were falling in the vicinity, he continued his efforts until he was certain that the three men remaining in the ruined dugout were dead.

DIETER, ARTHUR (1254583).
At St. Agnan, France, July 16, 1918.
R—Dunmore, Pa.
B—Scranton, Pa.
G. O. No. 99, W. D., 1918.

Private, Company A, 103d Engineers, 28th Division.
On four different occasions during the night of July 16 he volunteered and under heavy shell and machine-gun fire successfully rescued wounded comrades.

DIETZ, ALBERT (562072).
In the Bois-du-Fays, France, Oct. 6, 1918.
R—Vincennes, Ind.
B—Vincennes, Ind.
G. O. No. 35, W. D., 1919.

Sergeant, Company I, 59th Infantry, 4th Division.
When one of his men was wounded and his clothing and bandoleer of ammunition caught fire, he cried for help, and Sergeant Dietz left a place of safety and regardless of his personal safety went through intense machine-gun fire and rescued him.

DIETZ, EDWARD W. A. (2176527).
Near Bantheville, France, Nov. 6–7, 1918.
R—Omaha, Nebr.
B—Newark, N. J.
G. O. No. 46, W. D., 1919.

Private, first class, 314th Engineers, 89th Division.
Private Dietz accompanied an officer on a reconnaissance of the bridge at Pouilly and the road from Pouilly to Inor. Successfully reaching the river, he crossed, an act which had not been done by any troops previously. Recrossing under heavy enemy fire, he made his way to Pouilly, collecting on the way most valuable information and data for engineer work. Just as they approached their destination he was wounded, the officer with him being killed. When darkness set in, he returned and supplied most valuable information regarding the reconnaissance.

DIGGINS, JOHN P.
At Marcheville, France, Sept. 26, 1918
R—Nashua, N. H.
B—Nashua, N. H.
G. O. No. 142, W. D., 1918.

Sergeant, Company D, 103d Infantry, 26th Division.
Sergeant Diggins, together with Private Ivor Grindle, climbed out of a trench in the face of severe shrapnel and machine-gun fire, proceeded 150 yards across an open space to the aid of a wounded officer, and dressed his wounds.

DIGGS, BENJAMIN W. (1309738).
Near Montbrehain, France, Oct. 7, 1918.
R—Oliver Springs, Tenn.
B—Oliver Springs, Tenn.
G. O. No. 50, W. D., 1919.

Private, first class, Company M, 117th Infantry, 30th Division.
Private Diggs volunteered and successfully carried a message through heavy shell and machine-gun fire in plain view of the enemy after one runner had been killed and two others wounded in attempting to accomplish this mission. Though he was gassed in performing this feat, he refused to seek first aid until he was wounded later in the afternoon.

DIGGS, JUNIUS (1871496).
Near Ardeuil, France, Sept. 30, 1918.
R—Chesterfield, S. C.
B—Lilesville, N. C.
G. O. No. 46, W. D., 1919.

Private, Company G, 371st Infantry, 93d Division.
After his company had been forced to withdraw from an advanced position under severe machine-gun and artillery fire, this soldier went forward and rescued wounded soldiers, working persistently until all of them had been carried to shelter.

DIGIACOMO, PASQUALE (1518469).
East of Baccarat, France, Aug. 15, 1918.
R—Akron, Ohio.
B—Italy.
G. O. No. 50, W. D., 1919.

Private, Company F, 145th Infantry, 37th Division.
He was one of four men who successfully held a small advanced post against a raid of 80 of the enemy. Two of the defenders were killed, but the staunch work of the others drove off the raiders. He engaged in a hand-to-hand encounter with the assailants with hand grenades and his rifle.

*DILBECK, ANDREW W. (3207561).
Near Pouilly, France, Nov. 10–11, 1918.
R—Crossville, Ala
B—Crossville, Ala.
G. O. No. 44, W. D., 1919.

Private, Company I, 356th Infantry, 89th Division.
Private Dilbeck accompanied Lieutenant Murphy and three other soldiers in a flank attack on three heavy machine guns. Fired on directly at 30 yards, they charged the guns, and in the hand-to-hand fight which followed this soldier and two of his comrades were killed.
Posthumously awarded. Medal presented to father, William Dilbeck.

DILE, PERCY L. (3485137).
In the St. Mihiel offensive, France, Sept. 13, 1918.
R—Lawrenceville, Ill.
B—Lawrenceville, Ill.
G. O. No. 35, W. D., 1919.

Private, Company C, 15th Machine Gun Battalion, 5th Division.
After being severely wounded he refused to be evacuated, but made several trips through intense machine-gun and shell fire, bringing up ammunition to his squad. After the gunner was killed he took his place and fired the gun until completely exhausted.

DILL, LESTER C. (557748).
At Sergy, France, Aug. 1, 1918.
R—Towanda, Pa.
B—Plainfield, Pa.
G. O. No. 145, W. D., 1918.

Private, Company B, 47th Infantry, 4th Division.
After being wounded twice while he was carrying a message he bandaged his wounds under fire and delivered his message.

DILLARD, MARQUIS L. (2176369)........
Near Barricourt, France, Nov. 1, 1918.
R—Laddonia, Mo.
B—Rush Hill, Mo.
G. O. No. 87, W. D., 1919.

Sergeant, Company A, 354th Infantry, 89th Division.
While leading a patrol in advance of our lines he was subjected to intense machine-gun and minenwerfer fire. With marked courage, he successively led his men in a charge, first upon the machine gun and then upon the minenwerfer, capturing both guns and the crews.

DILLENBECK, WILLARD (264808)........
Near Soissons, France, July 21, 1918.
R—Delavan, Wis.
B—Delavan, Wis.
G. O. No. 132, W. D., 1918.

Private, Company A, 26th Infantry, 1st Division.
He repeatedly carried messages from his company to platoon commanders near Soissons, France, July 21, 1918, in daylight across open ground in full view of the enemy and under heavy bombardment.

DILLIARD, JOHN A. (1711535)...........
Near Marcq, France, Oct. 14, 1918.
R—Brooklyn, N. Y.
B—Brooklyn, N. Y.
G. O. No. 21, W. D., 1919.

Private, Company B, 306th Machine Gun Battalion, 77th Division.
In the performance of his duties as runner, Private Dilliard was obliged to travel over a road which was under constant and heavy shellfire, but he succeeded in delivering a message to his commanding officer, which enabled the latter to so place his guns that a direct fire was made on the enemy.

DILLINGHAM, CHARLES K............
Near Nantillois, France, Oct. 8, 1918.
R—Germantown, Pa.
B—Cheraw, S. C.
G. O. No. 7, W. D., 1919.

Second lieutenant, 318th Infantry, 80th Division.
On duty as battalion intelligence officer, twice he volunteered and led a patrol through woods known to be occupied by hostile machine guns. Working his way through artillery and machine-gun fire, he succeeded in ascertaining the position of units on the right and left of his own. Throughout the action around Nantillois and the Bois-des-Ogons this officer was a constant inspiration to his men by his devotion to duty and disregard of personal safety.

DILLON, JOHN E......................
Near Beaumont, France, Nov. 10, 1918.
R—Middletown, Mo.
B—Middletown, Mo.
G. O. No. 32, W. D., 1919.

Private, 75th Company, 6th Machine Gun Battalion, U. S. Marine Corps, 2d Division.
He was painfully wounded in the foot by a bursting shell which killed or wounded all the members of his gun crew, but as soon as he had obtained first-aid treatment he immediately returned to his comrades and worked all night under heavy shellfire carrying wounded to the dressing station.

DILLON, HARRY......................
Near Soissons, France, July 22, 1918.
R—Mondovi, Wis.
B—Wisconsin.
G. O. No. 15, W. D., 1919.

Second lieutenant, 26th Infantry, 1st Division.
He carried his platoon forward in four attacks and took all objectives assigned to him.

DILLON, JOHN T. (61635).............
Near Chateau-Thierry, France, July 22, 1918.
R—New Haven, Conn.
B—Ireland.
G. O. No. 125, W. D., 1918.

Sergeant, Company C, 102d Infantry, 26th Division.
After being wounded he refused to go to the rear, but volunteered to act as a runner and repeatedly carried messages through the enemy barrage. Later the same day he voluntarily joined a platoon and fought with it in a successful attack against the enemy's line.

DILLON, RAYMOND P................
Near Mezieres, France, Nov. 3, 1918.
R—Chicago, Ill.
B—Chicago, Ill.
G. O. No. 7, W. D., 1919.

First lieutenant, pilot, 24th Aero Squadron, Air Service.
He exhibited courage in the course of a long and dangerous photographic and visual reconnaissance in the region of Mezieres with two other planes of the 24th Aero Squadron. Their formation was broken by the attack of 10 enemy pursuit planes. Five enemy planes attacked Lieutenant Dillon and his observer, who succeeded in shooting down two of these out of control. They then had a clear passage to their own lines, but turned back into Germany to assist a friendly plane with several hostile aircraft attacking it. They succeeded in shooting down one more of the enemy.

*DILWORTH, JOSEPH (573543).........
Near Montfaucon Hill, France, Sept. 26, 1918.
R—South Manchester, Conn.
B—New York, N. Y.
G. O. No. 89, W. D., 1919.

Private, Company A, 39th Infantry, 4th Division.
After his squad leader had become a casualty, he assumed command and led his men against machine-gun nests, materially assisting in the capture of two guns and prisoners. He was killed in the performance of duty.
Posthumously awarded. Medal presented to widow, Mrs. Helga Dilworth.

DION, ARTHUR J. (1901659).........
Near Fleville, France, Oct. 14, 1918.
R—Milford, Mass.
B—Pawtucket, R. I.
G. O. No. 50, W. D., 1919.

Sergeant, Company F, 326th Infantry, 82d Division.
During the advance of his battalion, and at a time when the concentrated fire of the enemy had caused numerous casualties in our ranks, Sergeant Dion not only reorganized his own platoon, but assisted in returning others after the leaders had been lost. He was selected as patrol leader to gain contact with the enemy, and, although painfully wounded and nearly exhausted from loss of blood, he continued on. After having his wound dressed he insisted on returning to complete his mission.

DION, EDWARD L..................
During the Seicheprey, France, engagement, Apr. 20, 1918.
R—Hartford, Conn.
B—Biddeford, Me.
G. O. No. 99, W. D., 1918.

Private, Company C, 102d Infantry, 26th Division.
He displayed extraordinary heroism in defending his post during the Seicheprey engagement on the morning of Apr. 20, 1918. Although completely surrounded by the enemy on several occasions, he fought them off with grenades and rifle fire, finally succeeding in driving them away, after which he carried a wounded comrade through a rain of shrapnel to a first-aid station and returned to his post.

*DIPASQUALE, AMERICO (3110554)......
Near Verdun, France, Nov. 11, 1918.
R—Philadelphia, Pa.
B—Italy.
G. O. No. 37, W. D., 1919.

Private, Company G, 315th Infantry, 79th Division.
He volunteered his services as a connecting file, and during the course of operations was obliged to cross and recross fields swept by shell and machine-gun fire. His efforts were instrumental in keeping contact with the unit on his left. While he was thus engaged he was killed.
Posthumously awarded. Medal presented to father, Signiora Felician Dipasquale.

*DI PASQUALE, FORTUNATO (1680690)___ Near Ville-Savoye, France, Aug. 23, 1918.
R—Niagara Falls, N. Y.
B—Italy.
G. O. No. 56, W. D., 1922.

Private, Company D, 308th Infantry, 77th Division.
During the attack of his company to regain ground in the outpost zone on the Vesle River, Private Di Pasquale found himself holding an important post on the left flank of the company. He advanced across a railroad track in the face of terrific machine-gun fire from the high bank beyond the railroad cut, and, undaunted by enemy fire and with great courage, climbed half way up the steep railroad embankment and aided materially to the success of his company in driving the enemy from their machine-gun emplacement. Private Di Pasquale was killed as he made this advance.
Posthumously awarded. Medal presented to father, Carlo Di Pasquale.

*DISALVO, CHARLES (2848232)_____ Near Remonville, France, Nov. 1, 1918.
R—St. Louis, Mo.
B—New York, N. Y.
G. O. No. 71, W. D., 1919.

Private, Company B, 354th Infantry, 89th Division.
When the combat group of which he was a member was held up by enemy machine guns he charged forward alone and, attacking the nest, killed one gunner and forced the rest to surrender. His heroic act enabled the advance to be resumed, though Private Disalvo had himself been mortally wounded.
Posthumously awarded. Medal presented to widow, Mrs. Rose Disalvo.

*DIXON, BEN F._____ Near Bellicourt, France, Sept. 29, 1918.
R—Asheboro, N. C.
B—Kings Mountain, N. C.
G. O. No. 5, W. D., 1924.

Captain, 120th Infantry, 30th Division.
He was severely wounded during the early part of the operations against the Hindenburg line; his company having only one officer, he remained on duty. Shortly afterwards he received a second wound, and again refused to leave his men. When he saw that the front waves of his company were getting into a barrage, he at once went forward to stop them, and while doing so he was killed.
Posthumously awarded. Medal presented to mother, Mrs. B. F. Dixon, sr.

DIXON, JOHN R. (1258639)_____ West of Varennes, France, Sept. 27, 1918.
R—Pittsburgh, Pa.
B—Pittsburgh, Pa.
G. O. No. 11, W. D., 1921.

Sergeant, Company D, 112th Infantry, 28th Division.
While advancing in the attack against a strongly held enemy position, Sergeant Dixon's platoon was suddenly fired on by several machine guns. The platoon commander was killed and the platoon forced to withdraw. Hearing the cries of a wounded comrade, he advanced through heavy machine-gun fire over 100 yards toward the enemy lines and returned with a severely wounded comrade.

DIXON, ROY T. (1386569)_____ Near Consenvoye, France, Oct. 14, 1918.
R—Chicago, Ill.
B—Chicago, Ill.
G. O. No. 78, W. D., 1919.

Corporal, Company B, 131st Infantry, 33d Division.
After five runners had been killed or wounded in attempting to reach the battalions on the flanks of his own, Corporal Dixon volunteered to lead a patrol to establish liaison with them. In so doing he encountered an enemy machine gun, which he boldly attacked and silenced, successfully accomplishing his mission.

DOBBS, LAIN (52670) _____ Near Verdun, France, Oct. 4, 1918.
R—Rapids, Ky.
B—Rapids, Ky.
G. O. No. 64, W. D., 1919.

Sergeant, Company B, 26th Infantry, 1st Division.
While attacking woods protected by machine guns, Sergeant Dobbs, in command of a small patrol, was surrounded by about 50 of the enemy. Under his direction the party succeeded in killing or wounding many of the enemy, causing the remainder to withdraw. Although affected by gas, he continued to fight on, until seriously wounded by shellfire.

DOBSON, WOODRUFF W._____ Near le Moulin de l'Homme Mort, France, Sept. 29, 1918.
R—New York, N. Y.
B—New York, N. Y.
G. O. No. 35, W. D., 1919.

First lieutenant, 308th Infantry, 77th Division.
He volunteered and reconnoitered in front of the first-line battalion to secure information regarding enemy machine guns and minenwerfers which had checked the advance of his organization. He was wounded by a sniper's bullet as he crawled back from this reconnaissance, but refused to submit to first aid until he made his report to the battalion commander and informed his men of the enemy's position.

†DOCK, FRANCIS JOSEPH_____ At Chateau-Thierry, France, June 6, 1918.
R—South Boston, Mass.
B—Belgium.
G. O. No. 110, W. D., 1918.

Corporal, 55th Company, 5th Regiment, U. S. Marine Corps, 2d Division.
Killed in action at Chateau-Thierry, France, June 6, 1918, he gave the supreme proof of that extraordinary heroism which will serve as an example to hitherto untried troops.
Posthumously awarded. Medal presented to father, Joseph Dock.

DODD, BRENDAN J._____ Near Consenvoye, France, Oct. 8, 1918.
R—Chicago, Ill.
B—Ireland.
G. O. No. 32, W. D., 1919.

Major, 132d Infantry, 33d Division.
While the attacking first wave was halted by machine-gun fire, Major Dodd crossed the line and, getting in front of it, located the direction from which the fire was coming. He then directed a flanking fire on the stronghold and so encouraged his men that the attack was renewed. His great bravery resulted in a highly successful attack, during which many of the enemy were killed and captured and a large number of our men recovered who had been taken prisoners earlier in the day.

DODDER, ALEXANDER (448061)_____ Near Landersbach, Alsace, Oct. 4, 1918.
R—Walworth, Wis.
B—Argentine, Mich.
G. O. No. 120, W. D., 1918.

Corporal, Company H, 53d Infantry, 6th Division.
He was in a detachment of 50 soldiers who were attacked by a raiding party of the enemy composed of 300 storm troops. Though severely wounded by shellfire and grenades during the combat, this courageous soldier continued to operate his automatic rifle until the enemy retreated. The fire of the rifle which he was manning alone dispersed the main body of the enemy and prevented them from capturing prisoners and gaining valuable information.

DODGE, CHARLIE M. (2940)_____ At Bois Brule, near Apremont, France, Apr. 10, 1918.
R—Springfield, Mass.
B—Melrose, Mass.
G. O. No. 99, W. D., 1918.

Private, first class, sanitary detachment, 104th Infantry, 26th Division.
He displayed conspicuous gallantry during the action of Apr. 10, 1918, in running through heavily shelled area to rescue an officer who had fallen mortally wounded and at great personal risk carrying him to dressing station.

*DODGE, ROWLAND S _____
   Near Verdun, France, Oct. 24–25, 1918.
   R—Pawtucket, R. I.
   B—Pawtucket, R. I.
   G. O. No. 37, W. D., 1919.

Second lieutenant, 101st Infantry, 26th Division.
Learning of the proposed advance in which his company was to participate, although sick in a hospital, Lieutenant Dodge secured his release and joined his command. He was at all times in advance of his front line, reconnoitering the ground, thereby facilitating the advance. Leading a counterattack, he was killed by rifle fire.
Posthumously awarded. Medal presented to mother, Mrs. Ellen A. Dodge.

D'OGOSTINO, ANTONIO (1235320) _____
   Near St. Agnan, France, July 15, 1918.
   R—Williamsport, Pa.
   B—Italy.
   G. O. No. 109, W. D., 1918.

Private, Company B, 109th Infantry, 28th Division.
He voluntarily went out alone through strange territory under heavy bombardment near St. Agnan, France, July 15, 1918, obtained important information, and on his own initiative rescued 2 wounded French soldiers exposed to enemy fire.

DOGRESS, CHRISTIAN (2106526) _____
   Near Medeah Ferme, France, Oct. 4, 1918.
   R—New Castle, Pa.
   B—New Castle, Pa.
   G. O. No. 147, W. D., 1918.

Private, Company A, 9th Infantry, 2d Division.
Though he had been wounded 3 times by machine-gun fire, he refused to go to the rear and remained in the advance of his company until the final position was reached and consolidated.

DOHERTY, JOHN (2357261) _____
   Near Les Franquettes Farm, France, July 23, 1918.
   R—New York, N. Y.
   B—Ireland.
   G. O. No. 32, W. D., 1919.

Sergeant, Company B, 4th Infantry, 3d Division.
After his company had withdrawn, Sergeant Doherty remained for 5 hours in an open field, swept by machine-gun fire, with his company commander, who was seriously wounded, carrying him to a dressing station after darkness had set in.

*DOLAN, BERNARD LEO (1735635) _____
   Near Champigneulle, France, Oct. 16, 1918.
   R—Lockport, N. Y.
   B—Lockport, N. Y.
   G. O. No. 35, W. D., 1920.

Corporal, Company I, 309th Infantry, 78th Division.
Although he was himself wounded, Corporal Dolan left the shelter of a shell hole and went out under terrific machine-gun fire to assist a wounded comrade. In so doing he received another wound, which caused his death.
Posthumously awarded. Medal presented to mother, Mrs. Ella Dolan.

DOLAN, CHARLES E. (2057515) _____
   Near Jaulny, France, Nov. 8, 1918.
   R—Zion City, Ill.
   B—Meckling, S. Dak.
   G. O. No. 44, W. D., 1919.

Private, Company D, 55th Infantry, 7th Division.
After repeated efforts had failed he carried a message from his platoon to his company commander, through an intense artillery and machine-gun barrage. He then guided a detail of stretcher bearers back through the barrage to his platoon and assisted in the evacuation of the wounded.

DOLAN, JAMES (1709086) _____
   Near Charlevaux, France, Oct. 3–7, 1918.
   R—New York, N. Y.
   B—Ireland.
   G. O. No. 32, W. D., 1919.

Corporal, Company G, 308th Infantry, 77th Division.
He was very severely wounded while in charge of his automatic rifle section, which was a unit of a surrounded battalion. After receiving first aid, he resumed his post and remained in command of his section until the battalion was relieved.

DOLCE, LOUIS C. (198798) _____
   Near Exermont, France, Oct. 8, 1918.
   R—Butte, Mont.
   B—Italy.
   G. O. No. 44, W. D., 1919.

Corporal, Company C, 2d Field Signal Battalion, 1st Division.
He volunteered and laid a telephone line to an advanced observation post under heavy artillery and machine-gun fire, working his way the entire distance of more than 1 kilometer through dense undergrowth and barbed-wire entanglements.

D'OLIVE, CHARLES R _____
   Near St. Benoit, France, Sept. 12, 1918.
   R—Memphis, Tenn.
   B—Suggsville, Ala.
   G. O. No. 123, W. D., 1918.

First lieutenant, pilot, 93d Aero Squadron, Air Service.
He, in conjunction with another American pilot, engaged and fought five enemy planes. Outnumbered and fighting against tremendous odds, he shot down three enemy planes and outfought the entire enemy formation.

*DOLL, JOHN A. (2965378) _____
   Near Olsene, Belgium, Oct. 31, 1918.
   R—York, Pa.
   B—Glen Rock, Pa.
   G. O. No. 37, W. D., 1919.

Private, first class, Company E, 145th Infantry, 37th Division.
While leading a squad forward, Private Doll suddenly found himself in the midst of an enemy barrage, but he exposed himself to the severe fire, in trying to keep his men organized and continue with the advance. He was killed while thus engaged.
Posthumously awarded. Medal presented to mother, Mrs. Lizzie Doll.

*DOLLARD, WILLIAM B. (1596) _____
   Near Soissons, France, July 22, 1918.
   R—New Bedford, Mass.
   B—New Bedford, Mass.
   G. O. No. 1, W. D., 1926.

Private, medical detachment, 1st Engineers, 1st Division.
When he learned that a colonel lay wounded in an exposed position in front of the lines, he asked and obtained permission of his company commander to go to the officer's assistance. While attempting to perform this courageous duty he was killed by machine-gun fire.
Posthumously awarded. Medal presented to sister, Mrs. Elizabeth Dollard Glynn.

DOMBROWSKI, LEON A. (542186) _____
   North of Montfaucon, France, Oct. 4 to 5, 1918.
   R—Buffalo, N. Y.
   B—Buffalo, N. Y.
   G. O. No. 53, W. D., 1920.

Corporal, Company H, 7th Infantry, 3d Division.
Corporal Dombrowski exposed himself to heavy machine-gun fire in order to crawl out in advance of our lines and drag a severely wounded noncommissioned officer to a near-by shell hole. Later under the cover of darkness he assisted this wounded noncommissioned officer to the first-aid station during an enemy bombardment.

*DOMMET, C. HARRY (1257013)_____
Near Villette, France, Sept. 5, 1918.
R—Lancaster, Pa.
B—Lancaster, Pa.
G. O. No. 13, W. D., 1923.

Private, first class, medical detachment, 108th Machine Gun Battalion, 28th Division.
For extraordinary heroism in action near Villette, France, Sept. 5, 1918, while serving as first-aid man attached to Company B. Wounded, and sent back to the dressing station, he learned that the other first-aid man attached to his company had been sent back and that the company was without medical aid, he, despite his own wounds and weakened condition, returned to his company under heavy enemy fire and continued to minister to the wounded until relieved and sent to hospital, where he died as a result of gunshot wounds on Sept. 5, 1918. His great courage and self-sacrifice were an inspiration to the men of his battalion.
Posthumously awarded. Medal presented to father, Joe Dommet.

DONAGHUE, ROBERT H._____
Northwest of Chateau-Thierry, France, in the Bois-de-Belleau, June 8, 1918.
R—Rock River, Wyo.
B—Thornfield, Mo.
G. O. No. 101, W. D., 1918.

Sergeant, 82d Company, 6th Regiment, U. S. Marine Corps, 2d Division.
He led his platoon against violent fire to destroy a machine-gun position, killed or wounded 8 Germans himself, and did not cease firing until overcome from loss of blood from his own injuries.

DONAHOE, FRANK O. (553510)_____
Near Cunel, France, Oct. 4, 1918.
R—Philadelphia, Pa.
B—Fulton, N. Y.
G. O. No. 44, W. D., 1919.

Private, Company A, 4th Infantry, 3d Division.
While his company was in support of the attacking company, communication was temporarily lost in the darkness. After all the runners had become casualties, Private Donahoe volunteered and succeeded in establishing liaison, pushing forward through the severe machine-gun and artillery fire despite the fact that he was painfully wounded soon after starting on his mission.

DONAHUE, HARRY J. (42114)_____
Near Soissons, France, July 18-20, 1918.
R—Philadelphia, Pa.
B—Philadelphia, Pa.
G. O. No. 35, W. D., 1920.

Sergeant, Company C, 16th Infantry, 1st Division.
Sergeant Donahue assisted his company commander in the reorganization of his company under heavy shellfire. Due to the loss of officers, he led a few men through heavy machine-gun fire and captured a machine-gun position which was causing losses to his company. In the performance of this act he was severely wounded in the head by a machine-gun bullet.

DONAHUE, JOE J. (1427391)_____
Near Bois-de-Consenvoye, France, Oct. 24, 1918.
R—Stewartville, Minn.
B—Stewartville, Minn.
G. O. No. 78, W. D., 1919.

Private, first class, Battery E, 323d Field Artillery, 83d Division.
Venturing over a road where three other runners had failed, Private Donahue carried a most important message over an area which was subjected to the fiercest kind of shelling. He chose this route to expedite the delivery, even though it was possible to make the journey by a longer but less dangerous route. He completed his mission despite two severe wounds he had received on the way.

DONAHUE, MICHAEL J. (46744)_____
Near Exermont, France, Oct. 4-5, 1918.
R—Springfield, Mass.
B—Ireland.
G. O. No. 35, W. D., 1920.

Sergeant, Company H, 18th Infantry, 1st Division.
On Oct. 4, assisted by a small group of his platoon, he attacked an enemy machine-gun nest and captured it. On the following day, while attacking an enemy machine-gun nest, his platoon was repulsed with many casualties. Accompanied by a comrade, he advanced through heavy machine-gun fire and brought in the wounded men to safety. He then reorganized his platoon and made a second attack on the enemy position, capturing the gun and crew.

DONAHUE, WILLIAM H._____
Near Pexonne, France, Mar. 5, 1918.
R—Minneapolis, Minn.
B—Gold Hill, Nev.
G. O. No. 50, W. D., 1919.

Lieutenant colonel, 151st Field Artillery, 42d Division.
In the action near Pexonne, France, on Mar. 5, 1918, he entered the quarry of Battery C, 151st Field Artillery, when it was under accurately-adjusted shellfire, for the purpose of aiding the officers and men of that battery, when he might with propriety have stayed away.

*DONALDSON, GLENN S. (614090)_____
Northwest of Somme-Py, France, near St. Etienne, Oct. 8, 1918.
R—Winona, Minn.
B—Winona, Minn.
G. O. No. 21, W. D., 1919.

Private, first class, section No. 606, Ambulance Service.
He showed conspicuous courage and devotion to duty in evacuating the wounded under the most trying conditions. He made repeated trips in plain view of enemy observers over roads under continuous shell fire. He was killed by a shell fragment while he was driving his ambulance over a heavily shelled road.
Posthumously awarded. Medal presented to father, A. Donaldson.

DONALDSON, JAMES HOWLAND_____
Near Cantigny, France, May 28, 1918.
R—Richmond Hill, N. Y.
B—Richmond Hill, N. Y.
G. O. No. 13, W. D., 1923.

Second lieutenant, 28th Infantry, 1st Division.
At the crucial moment of a strong enemy counterattack launched while the 28th Infantry was digging in and strengthening its newly captured position in front of Cantigny, Lieutenant Donaldson with great courage and splendid leadership, seeing one flank of a company without officer giving ground under intense fire, rushed over an area swept by concentrated machine-gun fire, steadied and reorganized the unit, and in conjunction with his own platoon led them in a successful counterattack, regaining all ground and holding for three days against persistent efforts by the enemy to dislodge him.
Oak-leaf cluster.
Lieutenant Donaldson was awarded an oak-leaf cluster for the following act of extraordinary heroism at Berzy-le-Sec, France, July 21, 1918: His captain having been killed, and after three days of continuous fighting, with many casualties, Lieutenant Donaldson reorganized his company and with such other available men as could be found led the attacking first wave on Berzy-le-Sec, displaying rare courage and splendid leadership, advancing in front of his company through a terrific enemy artillery and machine-gun barrage. Charging at the head of his men, five enemy machine-gun nests were destroyed, this officer personally killing two gunners with his pistol. Later his company was halted by terrific machine-gun fire from the front and left flank. Lieutenant Donaldson rushed to the edge of the wood, drove the enemy gunners to flight, and calling on his men to follow he led them through a treacherous muddy swamp across the river Crise in mud and water waist deep, in the face of terrific enemy artillery, machine-gun, and rifle fire. Upon reaching the river bank Lieutenant Donaldson was wounded but refused to be evacuated to the hospital, remaining on duty during several counterattacks until relieved two days later.

DONALDSON, JOHN O.............
Various dates and places in France.
R—Washington, D. C.
B—North Dakota.
G. O. No. 13, W. D., 1924.

Second lieutenant, Air Service, U. S. Army, attached to 32d Squadron, Royal Air Forces, British Expeditionary Forces.
For extraordinary heroism in action near Mont-Notre-Dame, France, on July 22, 1918, when, on patrol, he attacked a formation of 20 Fokker enemy biplanes. Singling out one of the hostile machines he engaged it from behind, firing a short burst at close range, the plane bursting into flames and crashing to the ground. On Aug. 8, 1918, he engaged 5 enemy scout planes over Licourt, France; singling out 1 and diving on it, he opened fire at close range, causing it to crash to the ground. On Aug. 9, 1918, over Licourt, France, observing a British plane being attacked by three enemy scout planes, he immediately engaged one of the enemy, firing a long burst at very close range, the enemy plane bursting into flames and crashing to the ground. On Aug. 25, 1918, over Hancourt, France, he attacked four Fokker enemy planes, diving into their midst and firing a short burst at one of them from a short range, destroying the plane, the pilot of which descended to safety in a parachute. On July 25, 1918, over Fismes, France, he drove down out of control an enemy Fokker plane; on Aug. 10, over Perrone, France, 1 Fokker biplane; and on Aug. 29, over Cambria, France, 1 Fokker biplane. In all these engagements he displayed the greatest devotion to duty and gallantry in the face of the enemy.

DONALDSON, STUART S. (1540547).....
Near Heurne, Belgium, Nov. 4, 1918.
R—Toledo, Ohio.
B—Detroit, Mich.
G. O. No. 87, W. D., 1919.

Private, 148th Ambulance Company, 112th Sanitary Train, 37th Division.
With 2 other soldiers, Private Donaldson volunteered to rescue 2 wounded men who had been lying in an exposed position on the opposite bank of the Scheldt River for 2 days. Making 2 trips across the stream, in the face of heavy machine-gun and shell fire, he and his companions succeeded in carrying both the wounded men to shelter.

DONNELLY, EDWARD (53915)...........
At Soissons, France, July 19, 1918.
R—Providence, R. I.
B—Ireland.
G. O. No. 132, W. D., 1918.

First sergeant, Company G, 26th Infantry, 1st Division.
As liaison sergeant he showed conspicuous bravery and good judgment at Soissons, France, July 19, 1918, when the loss of his captain threatened the success of the company's operations. He maintained communication and went forward on 4 attacks under heavy fire.

DONNELLY, PATRICK C. (158042).......
Near Hamel, France, Mar. 28, 1918.
R—Philadelphia, Pa.
B—Centralia, Pa.
G. O. No. 32, W. D., 1919.

Wagoner, Company B, 6th Engineers, 3d Division.
He placed himself in a most exposed position under direct observation of the enemy to cover the attempted rescue of a wounded comrade, maintaining his position until the party returned.

DONOGHUE, WALTER P. (1708284)......
Near Moulin-de-Charlevaux, in the Argonne Forest, France, Oct. 6, 1918.
R—Bronx, N. Y.
B—Bronx, N. Y.
G. O. No. 126, W. D., 1919.

Sergeant, Company D, 308th Infantry, 77th Division.
He was sent out on a patrol to investigate machine-gun fire from the left flank and to the rear of his company's position, and was wounded in the left leg by shrapnel fragments. Upon reporting back to his company commander he refused to be evacuated, but insisted in taking an active and gallant part in 4 subsequent attacks made to reach a battalion of our troops who were cut off and surrounded by a superior force of the enemy.

DONOVAN, JAMES J. (2411771)........
Near Grand Pre, France, Oct. 16-20, 1918.
R—Bayonne, N. J.
B—Bayonne, N. J.
G. O. No. 35, W. D., 1919.

Corporal, medical detachment, 312th Infantry, 78th Division.
In the face of heavy shell and machine-gun fire and continuous gas attacks, he established and maintained for four days a dressing station in a most advanced position. When an enemy attack seemed imminent, he refused to retire to safety, but remained at his post, being relieved after his comrades advanced.

DONOVAN, JOHN P. (2135351)..........
At Pouilly, France, Nov. 5 and 6, 1918.
R—Glasgow, Mo.
B—What Cheer, Iowa.
G. O. No. 3, W. D., 1921.

Private, Company L, 356th Infantry, 89th Division.
Participating in the first reconnaissance of the damaged bridges at Pouilly, with 2 others, he advanced more than 500 meters beyond the American outposts, crossing 3 branches of the Meuse River and successfully encountering the enemy.

DONOVAN, PAUL J. (1387516)........
At Chipilly Ridge, France, Aug. 9, 1918.
R—Chicago, Ill.
B—Chicago, Ill.
G. O. No. 16, W. D., 1920.

Private, Headquarters Company, 131st Infantry, 33d Division.
Private Donovan, while exposed to machine-gun and artillery fire, went forward and killed an enemy sniper. Later, while moving forward to the attack, he entered single handed a dugout and captured five of the enemy.

DONOVAN, WILLIAM JOSEPH..........
Near Villers-sur-Fere, France, July 28-31, 1918.
R—Buffalo, N. Y.
B—Buffalo, N. Y.
G. O. No. 71, W. D., 1919.
Medal of honor and distinguished-service medal also awarded.

Major, 165th Infantry, 42d Division.
He led his battalion across the River Ourcq and captured important enemy strongholds near Villers-sur-Fere, France, on July 28-31, 1918. He was in advance of the division for 4 days, all the while under shell and machine-gun fire from the enemy, who were on three sides of him, and he was repeatedly and persistently counterattacked, being wounded twice. His coolness, courage, and efficient leadership rendered possible the maintenance of this position.

DOOCY, ELMER T..............
Near Snippes, northeast of Chalons-sur-Marne, France, July 14-15, and near Sergy, northeast of Chateau-Thierry, France, July 28 and 30-31, 1918.
R—Pittsfield, Ill.
B—Pittsfield, Ill.
G. O. No. 99, W. D., 1918.

Second lieutenant, 168th Infantry, 42d Division.
After being severely wounded, with utter disregard of his own safety and comfort, he remained on duty with his platoon under heavy fire of gas and high-explosive shells. Again, on Hill 212, near Sergy, he led his platoon and that of another wounded officer forward into a machine-gun nest, under heavy fire, capturing 4 prisoners and 2 machine guns, and 2 days later, at night, near Sergy, at great risk of his own life, he bravely went out in front of a German sniper and brought back into the line a wounded corporal of his platoon.

DOODY, JOHN
Near Vierzy, France, July 19, 1918.
R—New York N. Y.
B—New York, N. Y.
G. O. No. 127, W. D., 1918.

Corporal, 55th Company, 5th Regiment, U. S. Marine Corps, 2d Division.
He displayed exceptional bravery in charging 3 machine guns with the aid of a small detachment of his comrades, killing the crews and capturing the guns, which were immediately turned on the Germans, thereby opening the line for the advance of his company, which had been held up by the enemy's fire.

DOOGS, JOHN A. (652363)
Near Medeah Ferme, France, Oct. 9, 1918.
R—St. Francis, Tex.
B—Branchville, Ind.
G. O. No. 23, W. D., 1919.

Private, Company C, 2d Engineers, 2d Division.
Crawling forward under heavy machine-gun fire, he assisted in bringing a wounded comrade to safety.

DOOLEY, JAMES (550012)
On the Vesle River, France, Aug. 7, 1918.
R—Jersey City, N. J.
B—Ireland.
G. O. No. 4, W. D., 1923.

Corporal, Company C, 38th Infantry, 3d Division.
While acting as litter bearer, Corporal Dooley displayed exceptional bravery and courage in crossing in broad daylight the river on a narrow footbridge within 100 yards of enemy machine guns, and where several men had been killed by enemy fire. He succeeded in crossing, and carried back an officer who had been mortally wounded on the north bank of the river.

DOOLEY, JOHN J. (545194)
Near Crezancy, France, July 15-16, 1918.
R—Jamaica Plain, Mass.
B—Jamaica Plain, Mass.
G. O. No. 32, W. D., 1919.

Sergeant, Company A, 30th Infantry, 3d Division.
Throughout the engagement he encouraged the men of his company by his gallant conduct. After the company was ordered to withdraw, he voluntarily returned to the position his company had held and throughout the night of July 15-16 assisted in evacuating the wounded.

*DOREMUS, HARRY B.
Near Verdun, France, Oct. 27, 1918.
R—Hackensack, N. J.
B—Hackensack, N. J.
G. O. No. 21, W. D., 1919.

Captain, 114th Infantry, 29th Division.
Having been ordered to establish liaison between his company and the support unit on the right, Captain Doremus led his detail to its objective under heavy machine-gun fire. The successful completion of his work saved a most serious situation, but in the performance of his duty he was killed.
Posthumously awarded. Medal presented to widow, Mrs. Harry B. Doremus.

DOREY, HALSTEAD
North of Montfaucon, France, Oct. 15, 1918.
R—St. Louis, Mo.
B—St. Louis, Mo.
G. O. No. 46, W. D., 1919.
Distinguished-service medal also awarded.

Colonel, 4th Infantry, 3d Division.
When his men had become almost exhausted by 12 days' continuous fighting against stubborn resistance and had suffered heavy casualties, Colonel Dorey, himself suffering from a painful wound, went forward from his post of command through a heavy enemy barrage to the front line, where he reorganized his forces and directed the attacking units for two days until he was again severely wounded. His conspicuous bravery inspired his troops to the successful assault of a strongly fortified ravine and woods, which were of vital importance, and resulted in the capture of numerous prisoners and much material.

DOREY, LEO J. (68182)
Near Bois-de-St. Remy, France, Sept. 12, 1918.
R—Burlington, Vt.
B—Burlington, Vt.
G. O. No. 64, W. D., 1919.

Private, Company F, 103d Infantry, 26th Division.
Throughout a period of extreme shelling and unusually heavy machine-gun fire, Private Dorey volunteered and carried messages repeatedly from his platoon to his company commander. He conveyed information which resulted in the capture of 2 officers and 22 men of the enemy.

DORGAN, JOHN JOSEPH (2207992)
Near Boney, France, Sept. 23, 1918.
R—Toledo, Ohio.
B—East St. Louis, Ill.
G. O. No. 37, W. D., 1919.

Private, Company C, 356th Infantry, 89th Division.
Although wounded 4 times at the start of the engagement, Private Dorgan continued with the advance of his platoon. Under heavy fire, his comrades began to waver, and Private Dorgan immediately went to the assistance of the sergeant and aided greatly in keeping the men under control.

*DORSEY, JAMES W., JR. (1270299)
At Brabant-sur-Meuse, France, Oct. 26, 1918.
R—Washington, D. C.
B—Washington, D. C.
G. O. No. 21, W. D., 1919.

Private, first class, Company B, 104th Field Signal Battalion, 29th Division.
Upon learning that a number of other soldiers had been buried in a dugout struck by an enemy shell, he immediately, of his own volition, left shelter, organized a rescuing party and went to their aid, fearlessly exposing himself to the heavy shellfire. To save others, he gave his own life.
Posthumously awarded. Medal presented to father, James W. Dorsey, sr.

DOTY, MEEL (2154315)
Near Beaurevoir, France, Oct. 6, 1918.
R—Rockwell City, Iowa.
B—Rockwell City, Iowa.
G. O. No. 37, W. D., 1919.

Corporal, Company K, 117th Infantry, 30th Division.
He volunteered and crossed an open space swept by fire from enemy machine gun and snipers to rescue wounded comrades.

DOUCETTE, GEORGE (61739)
North of Verdun, France, Oct. 23, 1918.
R—Attleboro, Mass.
B—Canada.
G. O. No. 126, W. D., 1919.

Corporal, Headquarters Company, 101st Infantry, 26th Division.
Although not required by his regular duties to do so, Corporal Doucette voluntarily went forward with the first attacking wave and displayed notable bravery in attacking machine-gun nests. After being wounded he continued to advance until he received a second wound which necessitated his evacuation.

DOUDNA, JOHN F.
Near Geanes, France, Sept. 25, 1918.
R—Lake City, Mich.
B—Richmond Center, Wis.
G. O. No. 139, W. D., 1918.

First lieutenant, Medical Corps, attached to 362d Infantry, 91st Division.
This officer was under constant shellfire with his battalion for 17 days, and though he had been painfully wounded by a machine-gun bullet, he remained at his post, rendering first aid to the wounded night and day, performing the duties of 2 other medical officers who had been incapacitated, in addition to his own. Lieutenant Doudna's utter disregard for personal danger and complete devotion to duty made possible the rapid evacuation of the wounded, thus materially keeping up the morale of the combat troops, and alleviating the suffering of the wounded.

DOUGENECK, FRANCIS J. (63410) _____
At Marcheville, France, Sept. 26, 1918.
R—Bristol, Conn.
B—Southwick, Mass.
G. O. No. 15, W. D., 1919.

Corporal, Headquarters Company, 102d Infantry, 26th Division.
When a patrol in charge of a wounded officer was entirely cut off by machine-gun and artillery fire, he displayed great bravery by voluntarily carrying a message over ground swept by machine guns and later leading a rescuing party to the position.

DOUGHERTY, NEIL F _____
Near Bayonville, France, Nov. 2, 1918.
R—Los Angeles, Calif.
B—Chicago, Ill.
G. O. No. 35, W. D., 1919.

First lieutenant, 6th Regiment, U. S. Marine Corps, 2d Division.
Displaying conspicuous leadership, he led his platoon against an enemy battery while it was in action. Through his skillful maneuvering 42 prisoners, 10 pieces of artillery, and 5 machine guns were captured.

DOUGHERTY, RAYMOND M. (514358) ____
At Bois-d'Aigremont, France, July 15, 1918.
R—Lincoln, Nebr.
B—Independence, Iowa.
G. O. No. 32, W. D., 1919.

Regimental sergeant major, Headquarters Company, 30th Infantry, 3d Division.
He constantly exposed himself to shellfire while receiving messages from runners and in giving directions to them. He also frequently removed his gas mask to make himself more clearly understood, and encourage the runners by his example.

DOUGHTY, CHARLES A. (3170629) _____
Near La Fontaine au Croncq Farm, France, Nov. 4, 1918.
R—Broadwater, Va.
B—New Inlet, Va.
G. O. No. 95, W. D., 1919.

Private, Company C, 9th Infantry, 2d Division.
As a stretcher bearer Private Doughty displayed exceptional courage in removing wounded men from a field swept by enemy machine-gun fire of such intensity that 5 other soldiers engaged in this work were killed and another wounded.

DOUGLAS, JOHN E _____
Near St. Etienne, France, Oct. 8, 1918.
R—Bessemer, Ala.
B—Nashville, Tenn.
G. O. No. 66, W. D., 1919.

First lieutenant, 142d Infantry, 36th Division.
Lieutenant Douglas was a member of a patrol consisting of himself, another officer, and 3 soldiers, which came under fire from an enemy machine-gun nest. Leading the patrol forward by short rushes to within a short distance of the enemy position, he had his companions take cover, while he continued on alone in full view of the enemy, and by accurate fire from an automatic rifle at a range of 40 yards silenced the nest, capturing 20 men and 4 machine guns.

*DOUGLAS, OTIS R. (1315115) _____
Near Bellicourt, France, Sept. 29, 1918.
R—Taylorsville, N. C.
B—Stony Point, N. C.
G. O. No. 44, W. D., 1919.

Private, Company C, 119th Infantry, 30th Division.
Hearing cries of distress from a disabled tank, he assisted an officer by advancing in the face of terrific machine-gun and shell fire to the spot. Notwithstanding the fact that the tank was subjected to point-blank fire of artillery, he succeeded in rescuing the badly wounded tank commander and removing him to a place of safety.
Posthumously awarded. Medal presented to widow, Mrs. Otis R. Douglas.

DOUGLAS, REED S. (2284309) _____
Near Bois-de-Pultier, France, Oct. 15, 1918, and near the Meuse River, Nov. 5, 1918.
R—Pittsburgh, Pa.
B—Pittsburgh, Pa.
G. O. No. 64, W. D., 1919.

Sergeant, Company I, 60th Infantry, 5th Division.
On Oct. 15 Sergeant Douglas led his platoon against a strong machine-gun nest in the open field east of the Bois-de-Pultier, capturing about 20 prisoners, with very few casualties in his own platoon. On Nov. 5 he led his platoon across the Meuse in the face of heavy machine-gun, rifle, and artillery fire and successfully cleared the heights east of the river of strong enemy machine-gun positions.

*DOUGLASS, ALLAN W _____
Near Limey, France, Sept. 12, 1918.
R—Buffalo, N. Y.
B—Plainfield, N. J.
G. O. No. 15, W. D., 1923.

First lieutenant, 113th Field Artillery, 30th Division.
During the engagement, after having been struck by a shell splinter, Lieutenant Douglas continued his work of removing dead and wounded horses, and assisted in moving the carriages to a place of safety. He was again struck and killed. By his courage and devotion to duty he inspired the men of this section to continue their work successfully.
Posthumously awarded. Medal presented to father, E. T. Douglas.

DOUGLASS, JAMES M _____
Near St. Juvin, France, Oct. 11, 1918.
R—Russellville, Ala.
B—Russellville, Ala.
G. O. No. 46, W. D., 1919.

First lieutenant, 329th Machine Gun Battalion, 82d Division.
When his platoon had become greatly disorganized through many casualties, Lieutenant Douglass, although wounded, reorganized it and led it through hazardous machine-gun and artillery fire. He remained on duty for several days, during which time he suffered acutely from his wounds, until ordered to a dressing station.

DOUGLASS, JOSEPH U. (1210860) _____
East of Ronssoy, France, Sept. 29, 1918.
R—Ridgewood, N. J.
B—Cherokee, Iowa.
G. O. No. 20, W. D., 1919.

Corporal, Company K, 107th Infantry, 27th Division.
He, with 3 other soldiers, went out into an open field under heavy shell and machine-gun fire and succeeded in carrying back to our lines four seriously wounded men.

DOUGLASS, KINGMAN _____
Near Longuyon, France, Oct. 31, 1918.
R—Oak Park, Ill.
B—Oak Park, Ill.
G. O. No. 1, W. D., 1919.

First lieutenant, pilot, 91st Aero Squadron, Air Service.
While on a photographic mission, he encountered a superior number of enemy pursuit planes. Notwithstanding the odds against him, he turned and dived on the hostile formation, destroying one plane and damaging another. He then continued on his mission and returned with photographs of great military value.

DOVELL, CHAUNCEY E _____
Near Bussy Farm, Ardeuil-et-Montfauxelles, and Trieres Farm, France, Sept. 26–30, 1918.
R—Somerset, Va.
B—Une, Va.
G. O. No. 78, W. D., 1919.

First lieutenant, Medical Corps, attached to 371st Infantry, 93d Division.
Throughout 5 days of most intense action Lieutenant Dovell worked unceasingly in caring for the wounded, disregarding a severe wound which he himself received in the neck by a shell fragment. He remained continuously on duty, giving an example of fortitude and courage to all about him.

*DOWD, MEREDITH L.
Near Dannevoux, France, Oct. 26, 1918.
R—Orange, N. J.
B—Orange, N. J.
G. O. No. 37, W. D., 1919.

Second lieutenant, 147th Aero Squadron, Air Service.
Having been unable to overtake and join a patrol, Lieutenant Dowd alone encountered four German planes, which he daringly attacked. He fought with most wonderful skill and bravery, diving into the formation and sending one of the enemy to earth. In the course of the combat his machine was disabled and crashed to the earth, killing him in the fall.
Posthumously awarded. Medal presented to widow, Mrs. Mary B. Dowd.

DOWNER, JOHN W.
Near Beaumont, France, Mar. 11, 1918.
R—Norfolk, Va.
B—Charles Town, W. Va.
G. O. No. 32, W. D., 1919.

Lieutenant colonel, 6th Field Artillery, 1st Division.
While commanding a battalion of artillery in support of an extensive raid, Colonel Downer was severely gassed. Despite his sickness and suffering from pain, he remained at his post, which was subjected to several direct hits, and directed the fire of his battalion. He rendered invaluable aid to the advancing Infantry, holding his men at their posts during the intensity of the continual gas bombardment, lasting one entire night.

DOWNEY, ERNEST L. (1448656)
Near the Bois-de-Montrebeau, France, Sept. 28, 1918.
R—Pleasanton, Kans.
B—Platte County, Mo.
G. O. No. 93, W. D., 1919.

Sergeant, Company G, 137th Infantry, 35th Division.
After being severely wounded, Sergeant Downey refused to go to the rear, but continued in the advance until the final objective was reached and his company relieved.

DOWNHAM, LEXIE (56822)
At Cantigny, France, May 28-30, 1918.
R—Anderson, Ind.
B—Perkinsville, Ind.
G. O. No. 89, W. D., 1918.

Private, first class, Company B, 28th Infantry, 1st Division.
He captured, single handed, by exercising unusual bravery, 10 of the enemy.

DOWNS, FRANK J. (559681)
Near Courchamps, France, July 19, 1918.
R—Pottsville, Pa.
B—Sunbury, Pa.
G. O. No. 27, W. D., 1920.

Sergeant, Company B, 58th Infantry, 4th Division.
During an enemy counterattack, when the units on his right and left were falling back, Sergeant Downs, promptly grasping the situation, fearlessly led his platoon forward 400 yards, broke up the enemy attack, and established a new line of resistance. His heroic conduct enabled adjoining units to advance.

DOYLE, JOHN J. (49946)
Near St. Etienne-a-Arnes, France, Oct. 4, 1918.
R—Peabody, Mass.
B—Peabody, Mass.
G. O. No. 46, W. D., 1919.

Corporal, Company C, 23d Infantry, 2d Division.
Corporal Doyle in charge of the runners, repeatedly carried important messages himself from company to battalion headquarters. He volunteered to deliver a message to an outpost through a hail of machine-gun bullets, and while performing this mission was wounded.

DOYLE, JOHN W. (1302)
Near Cierges, France, Aug. 1, 1918.
R—Breckenridge, Mich.
B—Breckenridge, Mich.
G. O. No. 9, W. D., 1923.

Sergeant, medical detachment, 125th Infantry, 32d Division.
The Infantry, forced by vastly superior numbers to seek a more advantageous position, left 21 wounded men between the lines. Sergeant Doyle, together with Lieut. Warde B. Smith and Private Krause, medical detachment, 125th Infantry, in broad daylight under concentrated machine-gun and artillery fire carried the wounded men to a place of safety, thus undoubtedly saving their lives. The indomitable courage and spirit thus displayed inspired the troops with renewed determination and courage.

DOZER, OTIS V. (2176368)
In the Bois-de-Barricourt, France, Nov. 2, 1918.
R—Cedar Vale, Kans.
B—Cedar Vale, Kans.
G. O. No. 37, W. D., 1919.

Sergeant, Company F, 353d Infantry, 89th Division.
He fearlessly exposed himself in the face of machine-gun fire for the purpose of setting an example to the men of his company. He was wounded while advancing, but continued until exhausted. His coolness and courage resulted in the capture of three machine guns and their crews.

DOZIER, CARMON (1307019)
At Estrees, France, Oct. 8, 1918.
R—Yorkville, Tenn.
B—Yorkville, Tenn.
G. O. No. 46, W. D., 1919.

Private, first class, Company B, 117th Infantry, 30th Division.
After being severely wounded by shell fire, he crawled forward, killed two enemy machine gunners, and captured their gun, thereby clearing the way for the further advance of his company.

*DOZIER, ROY C. (32162)
Near Soissons, France, July 22, 1918.
R—Hillman, Ga.
B—Hillman, Ga.
G. O. No. 26, W. D., 1919.

Private, Company C, 16th Infantry, 1st Division.
While engaged as a runner, Private Dozier displayed extreme courage and devotion to duty by carrying messages through heavy machine-gun and artillery fire and through enemy-occupied territory.
Posthumously awarded. Medal presented to mother, Mrs. C. W. Dozier.

DRAKE, CHARLES B.
At the siege and taking of Cotta Pang Pang, Jolo, Philippine Islands, Feb. 14, 1904.
R—Old Forge, Pa.
B—Old Forge, Pa.
G. O. No. 126, W. D., 1919.
Distinguished-service medal also awarded.

Captain, 14th Cavalry, U. S. Army.
His conspicuous bravery and daring were demonstrated in leading the men to the firing line and being the first over the bamboo fence and the stone wall of the cotta.

DRAKE, THOMAS D. (560684)
Near Courchamps, France, July 19, and near Brieulles, France, Sept. 29, 1918.
R—Lumberport, W. Va.
B—Preston, W. Va.
G. O. No. 10, W. D., 1920.

Sergeant, Company A, 59th Infantry, 4th Division.
After having successfully led his platoon to its objective in the attack of July 19, he gathered together groups of other companies of the battalion which had become disorganized due to heavy losses and established under intense fire a line of defense which was held until the unit was relieved. On Sept. 29 he was wounded in the hand, but refused to go to the rear and continued to perform his duties, frequently exposing himself to heavy machine-gun fire in order to control his command.

DRAPER, CHARLES L. (1210189)_____
Near Ronssoy, France, Sept. 29, 1918.
R—Carthage, N. Y.
B—Carthage, N. Y.
G. O. No. 16, W. D., 1920.

Private, first class, Company C, 107th Infantry, 27th Division.
After his company had passed beyond the first line of the enemy's resistance and at a time when hostile machine-gun fire presented the most formidable resistance to the advance, Private Draper, single-handed, rushed a hostile machine-gun position, killed both of its defenders, and captured the gun.

DRAUGHON, EDGAR S. W. (1880373)____
Near St. Quentin, France, Sept. 29 to Oct. 20, 1918.
R—Fayetteville, N. C.
B—Sampson County, N. C.
G. O. No. 35, W. D., 1919.

Private, sanitary detachment, 120th Infantry, 30th Division.
Throughout this period, Private Draughon labored unceasingly in evacuating the wounded from the front lines to the battalion aid post. On Oct. 19, with complete disregard for his personal safety, he advanced under heavy shell and machine-gun fire beyond the front line, rendered first aid to a wounded officer, and assisted him to the rear.

DRAVLAND, ALBERT B. (2143891)_____
Near Chatel-Chehery, France, Oct. 7, 1918.
R—Carbury, N. Dak.
B—Carbury, N. Dak.
G. O. No. 15, W. D., 1919.

Private, first class, Company G, 328th Infantry, 32d Division.
Acting as stretcher bearer, he displayed exceptional courage in transporting wounded from the battle field under machine-gun and artillery fire. He worked continuously for more than 48 hours without a rest supervising a detachment of stretcher bearers searching the woods for wounded.

DRAVO, CHARLES A_____
Near Sedan, France, Nov. 6–7, 1918.
R—Pittsburgh, Pa.
B—Fort McDowell, Ariz.
G. O. No. 37, W. D., 1919.

Lieutenant colonel, 165th Infantry, 42d Division.
Leading the front-line battalion of his regiment throughout the entire attack, Colonel Dravo was constantly under accurate machine-gun fire and incessant artillery fire. Having been ordered to advance, he personally formed his lines for attack, despite the fact that for 29 hours he had labored without rest or relief, and led his command forward, in the face of fiercest fire, encountering and subduing the enemy after a hand-to-hand struggle.

DREBEN, SAM (1480481)_____
Near St. Etienne, France, Oct. 8, 1918.
R—El Paso, Tex.
B—Russia.
G. O. No. 37, W. D., 1919.

First sergeant, Company A, 141st Infantry, 36th Division.
He discovered a party of German troops going to the support of a machine-gun nest situated in a pocket near where the French and American lines joined. He called for volunteers and, with the aid of about 30 men, rushed the German positions, captured 4 machine guns, killed more than 40 of the enemy, captured 2, and returned to our lines without the loss of a man.

DRECHSEL, GEORGE (475273)_____
Near Moulin de Guenoville, France, Sept. 26, 1918.
R—Chicago, Ill.
B—Chicago, Ill.
G. O. No. 44, W. D., 1919.

Private, Company F, 1st Gas Regiment.
With 3 other soldiers, he advanced nearly 200 yards, over an open hillside exposed to machine-gun fire and carried 2 wounded men to the protection of a near-by trench.

DREES, ALBERT J. (2265790)_____
In the Meuse-Argonne offensive, France, Sept. 26–29, 1918.
R—Los Angeles, Calif.
B—Uniontown, Wash.
G. O. No. 44, W. D., 1919.

Private, first class, Machine Gun Company, 364th Infantry, 91st Division.
Seriously wounded by shrapnel, he refused evacuation, but remained with his gun team for 3 days, rendering great assistance through a terrific encounter. His sufferings finally necessitated his removal to the hospital.

DRENNON, LOU H. (1383907)_____
At Marcheville, France, Nov. 10, 1918.
R—Ina, Ill.
B—Ina, Ill.
G. O. No. 32, W. D., 1919.

Sergeant, Company F, 130th Infantry, 33d Division.
During an attack on the town, Sergeant Drennon, although severely wounded, continued to lead his men until they had reached their objective, when he fell completely exhausted.

DRESBACH, IRVIN H. (94363)_____
Near Sommerance, France, Oct. 14, 1918.
R—Hallsville, Ohio.
B—Hallsville, Ohio.
G. O. No. 35, W. D., 1919.

Sergeant, Company H, 166th Infantry, 42d Division.
Taking command of the platoon and also the advance patrol, when both their leaders had become casualties, and despite the fact that he was so severely gassed that he could hardly speak, Sergeant Dresbach pushed vigorously forward, compelling the enemy to run in confusion and causing them to surrender to a near-by division.

DRESSELL, EVERETT C. (261498) _____
Near Juvigny, France, Aug. 31, 1918.
R—Flint, Mich.
B—Allegan County, Mich.
G. O. No. 3, W. D., 1921.

Private, first class, Machine Gun Company, 125th Infantry, 32d Division.
After his company had captured two enemy fieldpieces and a large quantity of ammunition, enemy artillery attempted to blow up the ammunition with incendiary shells. At great risk to his life from the explosion of shells and grenades he extinguished the fires, thereby insuring the safety of his comrades and the retention by his company of its advanced position.
Posthumously awarded. Medal presented to father, Fred Dressell.

DREW, CHARLES W_____
Near Flircy, France, Aug. 15, 1918.
R—Philadelphia, Pa.
B—Rochester, N. Y.
G. O. No. 15, W. D., 1926.

First lieutenant, 13th Aero Squadron, Air Service.
Lieutenant Drew operated one of a patrol of four machines which attacked four enemy battle planes. In the fight which followed he attacked in succession three of the enemy airships, driving one of them out of the battle. He then engaged another machine at close range and received 10 bullets in his own plane, one of which penetrated his radiator, while another pierced his helmet. In spite of this he followed the German plane to a low altitude within the enemy's lines and shot it down in flames. During the latter part of the combat he courageously refused to abandon the fight, although he had become separated from his companions and his engine had become so hot, because of the leak in his radiator, that there was imminent danger of its failing him at any moment.

DROTNING, HAROLD J. (2302303)_____
Near St. Gilles, south of Fismes, France, Aug. 4, 1918.
R—Milwaukee, Wis.
B—Stoughton, Wis.
G. O. No. 21, W. D., 1919.

Private, first class, Battery A, 120th Field Artillery, 32d Division.
When the men of his battery position had been ordered to shelter on account of enemy shelling, Private Drotning, in company with two other men, rescued a French soldier from drowning in a stream. This act was performed while the valley was filled with mustard gas.

DRUMM, CLARENCE M.
At Cantigny, France, May 28, 1918.
R—Bigelow, Kans.
B—Marshall County, Kans.
G. O. No. 99, W. D., 1918.

Second lieutenant, 28th Infantry, 1st Division.
He bravely led his platoon, through shell and machine-gun fire, to its objective, and fearlessly exposed himself by walking up and down his line to direct and encourage his men. After making certain that they were well cared for and just before it was possible for him to think of himself, he was killed.
Posthumously awarded. Medal presented to father, W. M. Drumm.

DRY, CLARENCE C. (1461547).
At Exermont, France, Sept. 28, 1918.
R—Kansas City, Mo.
B—McKinney, Tex.
G. O. No. 70, W. D., 1919.

Sergeant, Company I, 140th Infantry, 35th Division.
Volunteering to ascertain the location of an enemy machine-gun nest, Sergeant Dry walked out into the open to draw its fire, and when he was about 50 yards from the hostile position he was killed. His self-sacrificing act enabled his platoon to destroy the enemy nest.
Posthumously awarded. Medal presented to grandfather, John A. Furr.

DRYSDALE, GEORGE (98693).
Near Beuvardes, France, July 29, 1918.
R—Pratt City, Ala.
B—Pratt City, Ala.
G. O. No. 81, W. D., 1919.

Private, Headquarters Company, 167th Infantry, 42d Division.
Upon learning that his platoon commander had been wounded, Private Drysdale voluntarily left cover and went to his assistance under intense enemy machine-gun fire. With another soldier he administered first aid to the wounded officer and then carried him through the machine-gun fire to a place of safety.

DUBIE, EUGENE (67364).
Near Belleau, France, July 20, 1918.
R—Livermore Falls, Me.
B—Riley, Me.
G. O. No. 37, W. D., 1919.

Private, Company C, 103d Infantry, 26th Division.
He displayed exceptional courage and initiative in single-handed attacks on groups of hostile machine-gun crews. In the face of murderous fire he led attacks in which centers of enemy resistance were overpowered.

DUBLINSKY, MORRIS (552617).
Near Mezy, France, July 15, 1918.
R—Newark, N. J.
B—Russia.
G. O. No. 32, W. D., 1919.

Private, first class, Company M, 38th Infantry, 3d Division.
Prior to the German offensive of July 15, 1918, during an intense artillery bombardment by the enemy, he voluntarily brought in two wounded comrades through the heavy fire.
Posthumously awarded. Medal presented to widow, Mrs. Ida Dublinsky.

DUBOIS, RUSSELL L. (1711194).
Near Ville-Savoye, France, Aug. 20, 1918.
R—Mount Kisco, N. Y.
B—Annandale-on-Hudson, N. Y.
G. O. No. 35, W. D., 1919.

Private, first class, medical detachment, 308th Infantry, 77th Division.
Although suffering acutely from the effects of mustard gas, he refused to be evacuated because of the great need of medical attention among his comrades. For three days he remained at his post, and only went to the rear when ordered to do so by his commanding officer.

DU BOIS, VICTOR A. (62691).
North of Verdun, France, Oct. 23, 1918.
R—Wakefield, Mass.
B—Needham Heights, Mass.
G. O. No. 21, W. D., 1919.

Corporal, Headquarters Company, 101st Infantry, 26th Division.
When the sergeant in charge of his wire-laying detail was wounded he assumed command and, although himself wounded, continued to work and established liaison with the assaulting battalion. He then went back to the wounded sergeant, administered first aid, and brought stretcher bearers to him.

DUBORD, FRANK F. (2725538).
At Exermont, France, Oct. 6, 1918.
R—Chisholm, Me.
B—Lewiston, Me.
G. O. No. 44, W. D., 1919.

Private, Company E, 28th Infantry, 1st Division.
Responding to a call for volunteers, he proceeded 400 yards ahead of his platoon to ascertain the location of the enemy. The mission was accomplished through an extremely heavy fire, but he, after obtaining his information, returned over the same ground and made his report to the platoon commander.
Posthumously awarded. Medal presented to father, Frank Dubord.

DUCKSTEIN, ARTHUR WILLIAM.
Between Montrebeau and Exermont, France, Sept. 29, 1918.
R—New York, N. Y.
B—Philadelphia, Pa.
G. O. No. 128, W. D., 1918.

First lieutenant, pilot, 1st Aero Squadron, Air Service.
While on a special command reconnaissance to ascertain whether or not there was any concentration of enemy troops between Montrebeau and Exermont which might indicate a possible counterattack, this officer obtained information of the very greatest value. Flying over the enemy's lines at an altitude of less than 200 meters, in spite of most unfavorable atmospheric conditions, in the presence of numerous enemy aircraft, and under continuous heavy rifle and machine-gun fire from the ground, he spotted enemy troops massed for counterattack, and, although severely wounded by a machine-gun bullet from the ground, continued his mission until he had clearly and accurately located the position. He then returned and, though suffering from the pain of his wound, succeeded in writing out and dropping a clear and complete message. The counterattack, launched shortly afterwards by a fresh enemy division, was crushed, and the accurate and timely information brought back by Lieutenant Duckstein, after a very gallant flight under highly adverse conditions, was of the greatest importance in this success.

DUDDERAR, MARSHALL B. (1309120).
Near Geneva, France, Oct. 7, 1918.
R—Chattanooga, Tenn.
B—Corbin, Ky.
G. O. No. 21, W. D., 1919.

Sergeant, Company K, 117th Infantry, 30th Division.
Taking command of his company after the company commander had been wounded, Sergeant Dudderar led his men forward in the face of intense fire until further advance was impossible, when he proceeded alone for 25 yards, exposing himself in full view of the enemy in an effort to draw fire and thus locate a machine-gun nest that was causing losses. He returned with the desired information, but in the advance he was mortally wounded and died shortly afterwards.
Posthumously awarded. Medal presented to mother, Mrs. R. S. Dudderar.

DUDLEY, MACK (1865502).
Near Terny-Sorny, France, Sept. 1, 1918.
R—Dickson, Tenn.
B—Dickson, Tenn.
G. O. No. 64, W. D., 1919.

Private, first class, Company G, 128th Infantry, 32d Division.
During the preparations for attack, he acted as runner, carrying several messages through extreme machine-gun and shell fire to the different platoon leaders, maintaining liaison until the reorganization was complete.

DUDZINSKI, FRANCISZAK (263277)_____
Near Cierges, France, July 31, 1918.
R—Hamtramck, Mich.
B—Russia.
G. O. No. 14, W. D., 1925.

Private, Company I, 125th Infantry, 32d Division.
Under heavy machine-gun fire, he went out in front of the position of his unit and administered first aid to three wounded men, being himself wounded while engaged in this courageous service.

DUELL, HOLLAND S_____
Near Binarville, France, Sept. 28–29, 1918.
R—Yonkers, N. Y.
B—Syracuse, N. Y.
G. O. No. 38, W. D., 1922.

Major, 306th Field Artillery, 77th Division.
While in command of the 2d Battalion, 306th Field Artillery, he voluntarily took one of the guns of his battalion forward to a position in advance of the immediate front line of the 368th Infantry. Although subjected to heavy machine-gun fire at short range and artillery fire he continued to direct the fire of his gun, and by his example of coolness and bravery encouraged his gun detachment to remain at their gun, thereby assisting greatly in repulsing a servere counterattack of the enemy.

DUEY, ARMA (548062)_____
Near Jaulgonne, France, July 23, 1918.
R—Houtzdale, Pa.
B—Belgium.
G. O. No. 64, W. D., 1919.

Private, first class, Company L, 30th Infantry, 3d Division.
During the attack made by his company, Private Duey constantly carried messages under heaviest shellfire, insuring and maintaining liaison with all neighboring units.

DUFF, ELDON A. (240592)_____
At Belleau Wood, France, July 20, 1918.
R—Eskridge, Kans.
B—Lyndon, Kans.
G. O. No. 9, W. D., 1923.

Private, Company B, 103d Infantry, 26th Division.
In the face of terrific enemy machine-gun fire he voluntarily joined a squad of men from his company, openly exposed himself to the enemy, charged a machine-gun crew, and with his bayonet killed the gunner and all other members of the crew. His outstanding courage and devotion to duty inspired his comrades, raising their morale to a high pitch.

DUFF, PHILLIP T. (2087059)_____
Near Consenvoye, France, Oct. 9, 1918.
R—Chicago, Ill.
B—Chicago, Ill.
G. O. No. 78, W. D., 1919.

Private, Company E, 132d Infantry, 33d Division.
During an enemy counterattack, accompanied by heavy fire, Private Duff volunteered and carried an important message to the battalion commander. He later led the supporting company to its position in the line, displaying notable courage in facing machine-gun fire.

DUFFY, FRANCIS A. (2414057)_____
Near Thiaucourt and Grand Pre, France, Sept. 21 and Oct. 16, 1918.
R—Florence, N. J.
B—Florence, N. J.
G. O. No. 20, W. D., 1919.

Private, medical detachment, 310th Infantry, 78th Division.
On Sept. 21, at Thiaucourt, France, he remained in the frontline trenches under heavy artillery and machine-gun fire, caring for the wounded and displaying utter disregard for personal danger while administering first-aid treatment to 40 men. During the advance of his company from St. Juvin to Grand Pre he followed immediately behind the first wave, under heavy machine gun and shellfire, caring for the wounded as they fell.

DUFFY, FRANCIS P_____
In the village of Villers-sur-Fere, France, July 28–31, 1918.
R—New York, N. Y.
R—Canada.
G. O. No. 99, W. D., 1918.
Distinguished-service medal also awarded.

First lieutenant, chaplain, 165th Infantry, 42d Division.
He devoted himself tirelessly and unceasingly to the care of the wounded and dying in the village of Villers-sur-Fere, France, from July 28 to 31, 1918. Despite a constant and severe bombardment with shells and aerial bombs, he continued to circulate in and about the two aid stations and the hospitals, creating an atmosphere of cheerfulness and confidence by his courageous and inspiring example.

DUFFY, JOHN (1702076)_____
At Bazoches, France, Aug. 14, 1918.
R—Brooklyn, N. Y.
B—Ireland.
G. O. No. 16, W. D., 1923.

Private, first class, Company F, 306th Infantry, 77th Division.
Voluntarily joining a daylight patrol seeking information as to the strength and positions of the enemy, which was attacked about 100 yards beyond its own lines by an enemy hostile post of seven men. The enemy was immediately attacked from the rear, several of the men killed, and the survivors scattered. A moment later another enemy post was attacked and in hand-to-hand fighting, Private Duffy killed one of the enemy and was himself badly wounded. Although unable to walk and under heavy fire from near-by enemy posts, Private Duffy dragged himself to his own lines and gave valuable information as to the disposition of the enemy forces.

DUFFY, JOHN C_____
Near Landersbach, Alsace, Oct. 4, 1918.
R—New Bern, N. C.
B—New Bern, N. C.
G. O. No. 130, W. D., 1918.

Second lieutenant, 53d Infantry, 6th Division.
During an attack by a German raiding party of about 300 men he took command of a post where the 5 men manning it had been killed or wounded by liquid fire. By his coolness and fearless exposure of himself he was able to hold the post with a small reinforcement. After the raid he removed some 20 grenades which had become dangerously hot, due to the fire, and were about to explode.

DUFFY, MARK MATTHEW_____
Near Romagne, France, Nov. 1, 1918.
R—Chicago, Ill.
B—Springvale, Wis.
G. O. No. 37, W. D., 1919.

First lieutenant, Medical Corps, attached to 124th Field Artillery, 33d Division.
He displayed exceptional bravery in caring for the wounded and directing their evacuation under heavy shellfire. During the action he went under concentrated fire and rescued several wounded men.

DUFFY, OWEN F. (1899652)_____
Near St. Juvin, France, Oct. 16, 1918.
R—Wellsville, N. Y.
B—Wellsville, N. Y.
G. O. No. 78, W. D., 1919.

Private, Company K, 325th Infantry, 82d Division.
Private Duffy, with another soldier, voluntarily made several trips out into No Man's Land, under heavy fire, and carried eight wounded men to cover.

DUFFY, THOMAS J. (61226)_____
In the Bois-de-Wavrille, Troyon sector, France, Oct. 2, 1918.
R—Woburn, Mass.
B—Woburn, Mass.
G. O. No. 39, W. D., 1920.

Private, first class, Company G, 101st Infantry, 26th Division.
While a member of a raiding party which penetrated to the enemy's second line, Private Duffy located a machine gun which was about to fire on the rear of his platoon. Exposing himself to heavy machine-gun fire he crawled to a position of vantage and shot the gunner and captured the gun and two other members of the crew.

*DUGAN, DANIEL, Jr.
Near Nantillois, France, Oct. 4, 1918.
R—Orange, N. J.
B—Orange, N. J.
G. O. No. 19, W. D., 1920.

First lieutenant, 317th Infantry, 80th Division.
Lieutenant Dugan, after several attempts to gain a foothold in the Bois-des-Ogons had failed, courageously led a platoon across a ravine covered by enemy fire and gained a foothold in the woods. This attack was made in the face of heavy machine-gun fire from the front and flank. The personal leadership and courage displayed by Lieutenant Dugan were vital factors in the success of the attack.
Posthumously awarded. Medal presented to father, Daniel A. Dugan.

*DUGAN, FRANK (1943639)
In the Argonne sector, France, Oct. 5-6, 1918.
R—Cleveland, Ohio.
B—Cleveland, Ohio.
G. O. No. 44, W. D., 1919.

Private, Company A, 26th Infantry, 1st Division.
When his section leader was killed, he took command of the section and led it against a machine-gun nest, capturing three machine guns and prisoners. He was dangerously wounded in this encounter, but continued to lead his men forward until he fell exhausted from loss of blood.
Posthumously awarded. Medal presented to mother, Mrs. Bridget Dugan.

DUGAN, JOHN I. (2844913)
Near Bois-de-Bantheville, France, Oct. 23, 1918.
R—Fort Scott, Kans.
B—Anna, Kans.
G. O. No. 37, W. D., 1919.

Private, Company B, 353d Infantry, 89th Division.
Although badly wounded in the face, he refused medical attention and assisted in an attack on a machine-gun nest, capturing one gun by himself.

*DUKE, ARY A. (97417)
Near Souain, France, July 15-16, 1918.
R—Andalusia, Ala.
B—Petrey, Ala.
G. O. No. 99, W. D., 1918.

Private, Company H, 167th Infantry, 42d Division.
He displayed praiseworthy devotion to duty and courageous self-sacrifice when he remained at his post under heavy fire, receiving wounds that later caused his death.
Posthumously awarded. Medal presented to mother, Mrs. E. L. Duke.

*DULANEY, DICK (2218073)
Near Bantheville, France, Nov. 10, 1918.
R—El Reno, Okla.
B—El Reno, Okla.
G. O. No. 37, W. D., 1919.

Private, first class, medical detachment, 357th Infantry, 90th Division.
Having been assigned to a company as first-aid man, he rendered most valiant service, working constantly under terrific fire. Never considering his own safety, he was always ready to administer to the wounded, his continued exposure resulting in wounds so severe that his death followed a few hours after he had received them.
Posthumously awarded. Medal presented to father, T. H. Dulaney.

*DULEVITZ, FRED C. (65874)
Near Verdun, France, Oct. 23-27, 1918.
R—Hartford, Conn.
B—Russia.
G. O. No. 64, W. D., 1919.

Private, Company K, 102d Infantry, 26th Division.
When it became vitally important to get a message to the battalion commander Private Dulevitz volunteered for the mission, knowing that it was necessary to pass through a terrific enemy barrage. Shortly after starting on his mission and while passing through the murderous fire he was instantly killed.
Posthumously awarded. Medal presented to mother, Mrs. Ida Dulevitz.

DULY, JOHN (1099774)
Near Jaulny, France, Nov. 8, 1918.
R—Bridgeport, Conn.
B—Pittsburgh, Pa.
G. O. No. 32, W. D., 1919.

Sergeant, Company D, 55th Infantry, 7th Division.
When one of his men became entangled in the barbed wire, Sergeant Duly alone went to his rescue under heavy fire. He then went forward alone, through intense artillery and machine-gun fire, to attack a machine-gun nest. While returning to our lines he fell exhausted from fatigue and shell shock.

DUMAIS, CANDIDE (68685)
Near Bois-de-St. Remy, France, Sept. 12, 1918.
R—Waterville, Me.
B—Van Buren, Me.
G. O. No. 19, W. D., 1921.

Corporal, Company H, 103d Infantry, 26th Division.
When under heavy machine-gun fire, Corporal Dumais charged the machine gun from the flank, killed the gunner, dispersed the crew, and put the gun out of action.

DUNBAR, CHARLES T. (569264)
West of Fismes, France, Aug. 5, 1918.
R—Sidney, Nebr.
B—Jumping Branch, W. Va.
G. O. No. 145, W. D., 1918.

Corporal, Company F, 4th Engineers, 4th Division.
He was a member of a small detachment of Engineers which went out in advance of the front line of the Infantry through an enemy barrage from 77-millimeter and 1-pounder guns to construct a footbridge over the River Vesle. As soon as their operations were discovered machine-gun fire was opened up on them, but, undaunted, the party continued at work, removing the German wire entanglements and successfully completing a bridge which was of great value in subsequent operations.

DUNBECK, CHARLEY
Near St. Etienne, France, Oct. 4, 1918.
R—Anderson, Ind.
B—Lucasville, Ohio.
G. O. No. 44, W. D., 1919.

Captain, 5th Regiment, U. S. Marine Corps, 2d Division.
Although wounded in the head and in a position exposed to intense machine-gun and artillery fire, he refused to be evacuated until he had personally given instructions to the second in command for carrying on the advance.

DUNCAN, ALVIN P. (42460)
Near Sedan, France, Nov. 7, 1918.
R—Smithville, Miss.
B—Smithville, Miss.
G. O. No. 64, W. D., 1919.

Sergeant, Company D, 16th Infantry, 1st Division.
He led a daylight patrol against an enemy machine-gun emplacement which was harassing the advance of his company. Under the murderous fire from this gun his patrol was wiped out, but he pressed on alone, capturing the machine gun and gunner. He showed coolness and bravery under heavy fire, and his action materially facilitated the advance of his company.

DUNCAN, BASIL E. (1306546)
Near Busigny, France, Oct. 8-9, 1918.
R—Fordtown, Tenn.
B—Washington County, Tenn.
G. O. No. 78, W. D., 1919.

Private, first class, Machine Gun Company, 117th Infantry, 30th Division.
Private Duncan, a runner, repeatedly carried messages with great coolness and daring, undeterred by the most intense artillery and machine-gun fire.

102444°—27——18

*Duncan, Charles B.
Near Bois-de-Septsarges, France, Sept. 29, 1918.
R—Nashville, Tenn.
B—New York, N. Y.
G. O. No. 71, W. D., 1919.

Captain, 77th Field Artillery, 4th Division.
When an enemy shell landed in the ammunition dump of his battery he jumped in among the burning shells and succeeded in getting the fuses away and extinguishing the fire. Later he was mortally wounded by enemy shellfire.
Posthumously awarded. Medal presented to grandmother, Mrs. W. M. Duncan.

*Duncan, Donald F.
At Chateau-Thierry, France, June 6, 1918.
R—St. Joseph, Mo.
B—Kansas City, Mo.
G. O. No. 119, W. D., 1918.

Captain, 6th Regiment, U. S. Marine Corps, 2d Division.
Killed in action at Chateau-Thierry, France, June 6, 1918, he gave the supreme proof of that extraordinary heroism which will serve as an example to hitherto untried troops.
Posthumously awarded. Medal presented to father, John A. Duncan.

Duncan, Edward A. (1213403)
East of Ronssoy, France, Sept. 29, 1918.
R—Derby, N. Y.
B—Buffalo, N. Y.
G. O. No. 20, W. D., 1919.

First sergeant, Company A, 108th Infantry, 27th Division.
During the operations against the Hindenburg line he displayed great gallantry and courage by going forward under heavy shell and machine-gun fire and bandaging the wounded and bringing them back to our lines. Throughout the engagement he exhibited a fearless disregard of the enemy's fire and performed valuable service by organizing new squads when his company was suffering heavy casualties as a result of shell and machine-gun fire.

Duncan, Ernest (2180096)
Near Crezancy, France, July 15, 1918.
R—Paris, Mo.
B—Paris, Mo.
G. O. No. 32, W. D., 1919.

Private, Company A, 30th Infantry, 3d Division.
During the engagement he set an example to the other members of his company by his gallant conduct. After the company was ordered to withdraw he voluntarily returned to the position his company had held, and throughout the night assisted in evacuating the wounded.

Duncan, George E. (1457730)
In Montrebeau Woods, France, Sept. 28, 1918.
R—Platte City, Mo.
B—Camden Point, Mo.
G. O. No. 95, W. D., 1919.

Private, first class, Company K, 139th Infantry, 35th Division.
Upon seeing his brother killed by a bursting shell as he was leading his platoon forward, Private Duncan displayed the utmost bravery and initiative in rushing forward, taking command of the platoon, which had been depleted by casualties and was in danger of becoming disorganized, and leading it forward in the advance.

Duncan, John C. (2178736)
Near Remouville, France, Nov. 1, 1918.
R—Whiteside, Mo.
B—Silex, Mo.
G. O. No. 87, W. D., 1919.

Corporal, Company D, 354th Infantry, 89th Division.
In command of the leading group of his platoon, he encountered a machine-gun nest, containing 6 guns, about 100 feet ahead of him. After opening fire on the enemy position, the automatic gunner and carrier of his group were shot. Corporal Duncan then took the automatic rifle and running forward into the enemy fire shot and took prisoner the German machine-gun crews.

Dunigan, Patrick Richard
Near Sergy, in the valley of the Ourcq River, France, July 31, Aug. 1–4, 1918.
R—Emmett, Mich.
B—Emmett, Mich.
G. O. No. 15, W. D., 1923.

Major, chaplain, 126th Infantry, 32d Division.
While performing his duties as chaplain of the 126th Infantry, 32d Division, he repeatedly displayed exceptional qualities of personal courage and heroism in attendance upon the wounded, and dying, going from the most advanced points of the firing line to the several dressing stations, crossing and recrossing the lowlands of the Ourcq River which were under intense artillery and machine-gun fire; again crossing and recrossing them when the entire valley was drenched with gas, and against the protests of officers who warned him that he was doing so at the imminent risk of his own life. After having been severely gassed while in the performance of these errands of mercy, and tagged for evacuation, he refused to be evacuated and still persisted in carrying his errands of mercy to the wounded and dying, thereby enhancing the morale of the troops in his fearless disregard of personal danger and contributed materially to the success of the capture of the heights beyond the Ourcq River and the advance to the Vesle River at Fismes.

Dunlap, Jay (1245684)
In the Argonne Forest, France, Oct. 1, 1918.
R—East Pittsburgh, Pa.
B—Pittsburgh, Pa.
G. O. No. 50, W. D., 1919.

Corporal, Company L, 111th Infantry, 28th Division.
Knowing that two men were killed in attempting to rescue a wounded comrade who was lying far in front of the line, Corporal Dunlap volunteered and attempted the rescue. Despite the severe fire directed at this point, he successfully accomplished his mission.

*Dunlavy, Herbert D.
In Bouresches, France, June 6, 1918.
R—Houston, Tex.
B—Richmond, Tex.
G. O. No. 119, W. D., 1918.

Private, 96th Company, 6th Regiment, U. S. Marine Corps, 2d Division.
He showed conspicuous courage in capturing a machine gun unassisted during the street fighting in Bouresches, France, on the night of June 6, 1918. He was killed in the repulse of the enemy on the night of June 7–8, 1918.
Posthumously awarded. Medal presented to mother, Mrs. Hattie Hall.

Dunn, Don (1472688)
Near Charpentry, France, Sept. 29, 1918.
R—Oklahoma City, Okla.
B—Fort Sill, Okla.
G. O. No. 59, W. D., 1919.

Private, first class, 140th Ambulance Company, 110th Sanitary Train, 35th Division.
Working as a litter bearer in the advanced area, when our front line dropped back, he remained on the field alone during the night, dressed the wounds of a number of soldiers, and carried them to the protection of a shell hole, later carrying them back to the dressing station. He continually exposed himself to machine-gun and artillery fire during this work.

Dunn, George H. (1241813)
Near Cierges, France, July 30, 1918.
R—Latrobe, Pa.
B—Latrobe, Pa.
G. O. No. 55, W. D., 1920.

Sergeant, Company M, 110th Infantry, 28th Division.
Early in the attack Sergeant Dunn was wounded in the leg by a bullet and ordered to the first-aid station, but he refused to leave and continued on in the advance. He assisted in the reorganization of his platoon and led a section forward in the attack. His conduct was of great value to his organization commander during this difficult operation.

*DUNN, JAMES R. (1287216)............
Near Bois - Brabant - sur - Meuse,
France, Oct. 8, 1918.
R—Richmond, Va.
B—Portsmouth, Va.
G. O. No. 37, W. D., 1919.

Private, first class, Company B, 115th Infantry, 29th Division.
Private Dunn, with 4 other soldiers, fearlessly attacked 8 machine-gun positions
and succeeded, after stubborn resistance, in capturing both the guns and their
crews.
Posthumously awarded. Medal presented to father, James F. Dunn.

DUNN, JOSEPH H.................

Near Marcheville, France, Sept. 25–
26, 1918.
R—Rockland, Mass.
B—Rockland, Mass.
G. O. No. 14, W. D., 1925.

First lieutenant, 101st Ambulance Company, 101st Sanitary Train, 26th Division.
He assisted in establishing and maintaining an ambulance dressing station in
an advanced position, where he labored heroically, dressing and evacuating
the wounded in full view of the enemy, under constant heavy bombardment by the enemy.

DUNN, JOSEPH J. (1245747)............
In the Argonne Forest, France, Sept.
26, 1918.
R—Philadelphia, Pa.
B—Philadelphia, Pa.
G. O. No. 50, W. D., 1919.

Private, Company K, 111th Infantry, 28th Division.
Having become separated from his combat group, Private Dunn was making
his way back to our lines when he came upon a machine-gun crew of 4 of the
enemy. Rushing them with his pistol and hand grenades, he demanded
and accomplished their surrender. He returned with the captured men
and gun.

DUNNE, CHRISTOPHER C. (1287125).....
At Chipilly Ridge, France, Aug. 9,
1918.
R—Chicago, Ill.
B—Chicago, Ill.
G. O. No. 128, W. D., 1918.

Private, Company D, 131st Infantry, 33d Division.
In an attack on an enemy machine-gun nest he bayoneted the gunner and
captured 4 others of the crew. Although wounded in this action, he showed
great devotion to duty by remaining with his squad until the line was consolidated.

DUNNE, JAMES F. (126794)............
Near Very, France, Oct. 4, 1918.
R—Dorchester, Mass.
B—Boston, Mass.
G. O. No. 32, W. D., 1919.

Corporal, Headquarters Company, 7th Field Artillery, 1st Division.
After three of his operators were killed and he himself was wounded, he refused
to be evacuated, but remained at his switchboard, maintaining uninterrupted communication.

DUNNE, THOMAS J..............
During the Meuse-Argonne offensive,
France, Sept. 26–Nov. 11, 1918.
R—Stapleton, N. Y.
B—Stapleton, N. Y.
G. O. No. 14, W. D., 1923.

First lieutenant, chaplain, 306th Infantry, 77th Division.
While a crossroads was being heavily shelled by enemy artillery and after
several men had been killed and others wounded, Chaplain Dunne, utterly
disregarding his own safety, went to their assistance and ministered to them.
He constantly exposed himself to the heaviest fire in order to assist the
wounded men of his regiment, at all times displaying heroic conduct and
superb devotion to his duty. His splendid and consistent bravery and contempt for his own safety was a continuing inspiration to every man in his
regiment and served to build up a fine sense of duty and soldierly obligation
in the organization.

DUNNINGTON, WALTER G., Jr...........
Near St. Eugene, France, July 14–15,
1918.
R—Farmville, Va.
B—Farmville, Va.
G. O. No. 37, W. D., 1919.

Second lieutenant, 10th Field Artillery, 3d Division.
Wounded and gassed while directing the fire of 1 platoon of his company under
terrific bombardment, he refused to be relieved. Although the area surrounding his position was heavily saturated with gas, he removed his gas
mask in order that he could make his commands heard above the roar of
the guns.

DUNSING, CLARENCE L. A. (1283787)....
Near Molleville, France, Oct. 18,
1918.
R—Frederick, Md.
B—New Haven, Conn.
G. O. No. 44, W. D., 1919.

Corporal, Company A, 115th Infantry, 29th Division.
During a heavy bombardment Corporal Dunsing, who was on duty as gas
noncommissioned officer, went along the front line of his company and administered first-aid treatment to several seriously wounded men.

*DUPRE, HAROLD J. (40565).............
Near Medeah Ferme, France, Oct.
3, 1918.
R—Syracuse, N. Y.
B—Utica, N. Y.
G. O. No. 20, W. D., 1919.

Sergeant, Company L, 9th Infantry, 2d Division.
He gallantly led his half platoon against an enemy machine-gun nest and
captured the position, together with 4 machine guns and about 60 prisoners.
Immediately after this he fell mortally wounded.
Posthumously awarded. Medal presented to father, David Dupre.

DUPREE, GEORGE A. (1217159).........
Near Montzeville, France, Sept. 14,
1918.
R—New York, N. Y.
B—New York, N. Y.
G. O. No. 32, W. D., 1919.

Corporal, Battery B, 104th Field Artillery, 27th Division.
When a continuous bombardment had set fire to the camouflage covering of
a large ammunition dump of 75-millimeter shells and exploded 9 of the shells,
he, utterly disregarding his personal safety, left a sheltered position and
ran to the dump and with the aid of 3 other men, extinguished the fire, not
only saving the ammunition but also preventing the exact location of the
dump by the enemy.

DURHAM, JAMES E. (106355)...........
Near Ploisy, France, July 19, 1918.
R—Buffalo, Ky.
B—Buffalo, Ky.
G. O. No. 81, W. D., 1919.

Sergeant, Company D, 3d Machine Gun Battalion, 1st Division.
When the officer commanding his machine-gun platoon had been wounded,
Sergeant Durham took command of the platoon and the adjacent infantry,
whose officers had become casualties, displaying extraordinary heroism in
leading both to attack the enemy artillery while subject to direct fire from
their guns. By his conspicuous coolness and unfaltering courage at a very
critical time he inspired great confidence in the men under his control while
consolidating the positions at the final objective.

DURING, FRED..............
Near Nantillois, France, Oct. 4, 1918.
R—Lawton, Okla.
B—St. Louis, Mo.
G. O. No. 19, W. D., 1920.

First lieutenant, 4th Infantry, 3d Division.
Lieutenant During crossed an open field toward the enemy under the heaviest
of machine-gun fire, enfilading and frontal, in order to give directions to
French tanks for an attack on enemy machine guns. He was wounded in
the leg and received a second wound while returning to his company. In
spite of his wounds, he directed the attack from a shell hole until carried
from the field, due to loss of blood.

DUTTON, HUBERT WALLACE............
  Near Fismette, France, Sept. 5, 1918.
  R—Stroudsburg, Pa.
  B—Stroudsburg, Pa.
  G. O. No. 24, W. D., 1920.

First lieutenant, 109th Infantry, 28th Division.
When his battalion commander was wounded during the crossing of the Vesle River he assumed command, rallied the scattered men, and reorganized them into fighting units, inspiring them by his own brave and determined spirit. His judgment in selecting strong points and making his dispositions made possible the success of the operations. In the course of the action he encountered a hostile machine-gun nest, killing the officer in command of it with a rifle and capturing 14 prisoners. He then turned the captured gun on the enemy and expended 4,000 rounds of captured ammunition in covering an exposed flank.

DUTY, RAY (1038082)................
  During the Meuse-Argonne offensive, France, Oct. 23 to 26, 1918.
  R—Cincinnati, Ohio.
  B—Sanderville, Ohio.
  G. O. No. 10, W. D., 1920.

Corporal, Battery A, 10th Field Artillery, 3d Division.
Corporal Duty volunteered and succeeded in carrying messages between the front line and advance P. C., when other runners failed to get through under severe fire of artillery and machine guns.

DWIGGINS, DANIEL M...............
  Near Beuvardes, France, July 29, 1918.
  R—Drew, Miss.
  B—Dwiggins, Miss.
  G. O. No. 98, W. D., 1919.

First lieutenant, 167th Infantry, 42d Division.
Commanding a platoon of 37-millimeter guns, Lieutenant Dwiggins carried 2 guns to the top of a hill under intense machine-gun fire and opened effective fire at short range on the enemy machine-gun nests. Heavy shellfire was directed at his guns and one of them was put out of action. Sending his men to cover, he remained with the second gun in the face of heavy machine-gun and artillery fire. Even after being wounded in the leg by a machine-gun bullet he continued to fire his gun until it exploded, displaying remarkable disregard for personal safety.

DYE, HENRY E. (2181655)..............
  Near Juvigny, France, Aug. 29, 1918.
  R—Walworth, Nebr.
  B—Sargent, Nebr.
  G. O. No. 44, W. D., 1919.

Corporal, Company D, 126th Infantry, 32d Division.
After an advance through heavy machine-gun fire, from which his organization had suffered many casualties and he himself had been twice wounded, Corporal Dye had taken refuge in a shell hole, when he heard the cries of a wounded comrade who was lying in an exposed position. Disregarding the intense machine-gun fire, he crossed an open space, dressed the wounds of his comrade, and carried him to shelter.

DYER, HERBERT RALPH..............
  Near Premont, France, Oct. 9, and near Molain, France, Oct. 17, 1918.
  R—Columbia, Tenn.
  B—Durand, Wis.
  G. O. No. 63, W. D., 1920.

First lieutenant, 117th Infantry, 30th Division.
On Oct. 9, when his company was held up by heavy fire from numerous machine guns, Lieutenant Dyer showed extraordinary bravery in driving them out and allowing the advance of the battalion to continue. On Oct. 17, when all the officers of an adjoining company had been killed, he promptly took command under the terrific enemy fire, and although twice wounded continued to lead the two companies in the advance.

*EADS, LEE S...............
  Near Regnieville, France, Sept. 15–17, 1918.
  R—Hamilton, Mo.
  B—Pike, Tex.
  G. O. No. 20, W. D., 1919.

Captain, 60th Infantry, 5th Division.
Prompted by his great devotion to duty, he preferred to remain on duty with a detachment of his companions organized as a carrying party, desiring to see that his detail performed the work assigned to them to the very utmost. While thus engaged he received severe wounds, which proved to be the cause of his death.
Posthumously awarded. Medal presented to father, Dr. J. B. Eads.

EARL, ELMER (104395)..............
  At Ripont Swamp, France, Sept. 26, 1918.
  R—Goshen, N. Y.
  B—Goshen, N. Y.
  G. O. No. 37, W. D., 1919.

Corporal, Company K, 369th Infantry, 93d Division.
While passing through a swamp where most of the platoon was wounded, Corporal Earl dressed the wounds of several of his comrades and, after reaching the shelter of a hill beyond, returned repeatedly and assisted many of his comrades to a place of safety.

EARLE, WILLIAM J. (39077)..............
  Near Medeah Ferme, France, Oct. 8, 1918.
  R—East St. Louis, Ill.
  B—East St. Louis, Ill.
  G. O. No. 21, W. D., 1919.

Corporal, Company E, 9th Infantry, 2d Division.
When his company had been held up by a machine-gun nest, he advanced on the nest from the flank and captured it single handed.

EAST, JAMES (1707625)..............
  Near Binarville, France, Sept. 29, 1918.
  R—Quinton, Ky.
  B—Frazer, Ky.
  G. O. No. 35, W. D., 1919.

Sergeant, Company A, 308th Infantry, 77th Division.
He volunteered and guided 3 wounded men to a first-aid station through machine-gun fire. He was wounded while on this mission, but, learning that his company was to make an advance, refused to be evacuated and returned to duty, bringing important information as to the enemy positions.

EASTERBROOK, ARTHUR E..............
  Near St. Mihiel, France, Sept. 12, 1918.
  R—Fort Flagler, Wash.
  B—Amsterdam, N. Y.
  G. O. No. 116, W. D., 1919.

First lieutenant, Infantry, observer, 1st Aero Squadron, Air Service.
Because of intense aerial activity on the opening day of the St. Mihiel offensive, Lieutenant Easterbrook, observer, and Second Lieutenant Ralph E. de Castro, pilot, volunteered to fly over the enemy's lines on a photographic mission without the usual protection of accompanying battle planes. Notwithstanding the low-hanging clouds, which necessitated operating at an altitude of only 400 meters, they penetrated 4 kilometers beyond the German lines. Attacked by four enemy machines, they fought off their foes, completed their photographic mission, and returned safely.
Oak-leaf cluster.
Lieutenant Easterbrook is also awarded an oak-leaf cluster for the following acts of extraordinary heroism in action near Exermont and Varennes, France, Oct. 8, 1918: Lieutenant Easterbrook, with Lieutenant Erwin, pilot, successfully carried out a mission of locating our Infantry, despite 5 encounters with enemy planes. During these encounters he broke up a formation of 3 planes, sending 1 down out of control, killed or wounded an observer in an encounter with another formation, and sent a biplane crashing to the ground, besides driving away a formation of two planes and several single machines.

*EATON, STARR SEDGWICK_____
Near Chateau-Thierry, France, July 1, 1918.
R—Madison, Wis.
B—Milwaukee, Wis.
G. O. No. 117, W. D., 1918.

Captain, 23d Infantry, 2d Division.
Captain Eaton displayed notable coolness and courage during the attack by his company, winning a brilliant success and capturing a large number of prisoners and a quantity of enemy munitions. After obtaining his objective he personally led a small detachment against a hostile machine gun and silenced it, capturing in person the machine gunners and destroying the nest.
Posthumously awarded. Medal presented to widow, Mrs. Louise S. Eaton.

EATON, WARREN EDWIN_____
Near Bantheville, France, Oct. 10, 1918.
R—Norwich, N. Y.
B—Norwich, N. Y.
G. O. No. 46, W. D., 1919.

First lieutenant, 103d Aero Squadron, Air Service.
With 1 other pilot, Lieutenant Eaton engaged an enemy formation of 11 planes (type Fokker), though another hostile formation was directly above them. After a severe combat Lieutenant Eaton destroyed 1 of the enemy and with his companion drove down another out of control.

*EBBERT, PETER W_____
At Ville-Savoye, France, Aug. 7, 1918.
R—Glen Rock, N. J.
B—New York, N. Y.
G. O. No. 27, W. D., 1919.

Captain, 58th Infantry, 4th Division
Lieutenant Ebbert, acting as battalion supply officer, conducted numerous details of food and ammunition through the heavy enemy artillery barrage. Later in the day he volunteered for observation duty and was posted in a prominent tower, where he was killed by a direct artillery hit.
Posthumously awarded. Medal presented to widow, Mrs. Peter W. Ebbert.

EBERLIN, RALPH_____
Near Mezy, France, July 15, 1918.
R—New York, N. Y.
B—Russia.
G. O. No. 59, W. D., 1919.

First lieutenant, 38th Infantry, 3d Division.
Although severely wounded during the first attack of the Germans on the railroad line near Mezy, he remained in command of his platoon and held an exposed flank against repeated attacks of the enemy.

ECHOLS, BYRON C. (106203)_____
Northwest of Epinonville, France, Oct. 4, 1918.
R—Brilliant, Ala.
B—Armeston, Ala.
G. O. No. 10, W. D., 1920.

Corporal, Company B, 3d Machine Gun Battalion, 1st Division.
When the advance of the Infantry line was held up by enemy machine-gun fire, Corporal Echols led his machine-gun crew in advance of the Infantry line and, although exposed to heavy machine-gun and artillery fire, he acted as gunner and opened an effective fire which silenced 2 enemy machine guns and allowed the Infantry to advance. Later he again led his crew forward and while engaging an enemy machine-gun nest was severely wounded.

*ECKEL, WILLIAM H_____
Near Ponchaux, France, Oct. 7, 1918.
R—Knoxville, Tenn.
B—Knoxville, Tenn.
G. O. No. 74, W. D., 1919.

Second lieutenant, 117th Infantry, 30th Division.
When heavy fire was encountered from enemy machine-gun nests located in a railroad cut, Lieutenant Eckel led his platoon in several successful attacks on these nests, personally killing and wounding a number of the enemy with his pistol and disarming 1 of them in a hand-to-hand struggle. He then reorganized the captured position and held out against the enemy with such men as he had left, opening fire himself with an automatic rifle, which he secured from a dead soldier. While attempting to destroy some enemy machine-gun nests in front of his platoon this gallant officer was mortally wounded by a shell fragment.
Posthumously awarded. Medal presented to father, Hugh Eckel.

ECKER, FREDERICK W_____
Near Verdun, France, Oct. 10, 1918.
R—New York, N. Y.
B—New York, N. Y.
G. O. No. 2, W. D., 1919.

First lieutenant, 115th Infantry, 29th Division.
While leading his platoon in the Bois-de-Consenvoye in an attack against strong enemy machine-gun nests he was severely wounded. He continued to advance, routed the enemy from their positions, and refused to leave his platoon until it was reorganized.

*ECKWEILER, ROBERT J. (748073)_____
Near Crezancy and Chateau-Thierry, France, July 15, 1918.
R—Notch, Pa.
B—New York, N. Y.
G. O. No. 78, W. D., 1919.

Private, Company C, 3d Ammunition Train, 3d Division.
On the morning of the 15th of July Private Eckweiler, with Private McNamee, volunteered and brought up a truck for the purpose of saving the records of the 30th Infantry, which were in danger of capture. He was killed while attempting this mission.
Posthumously awarded. Medal presented to father, John A. Eckweiler.

ECONOMOS, CONSTANTINE D. (2078874)_____
During the Meuse-Argonne offensive, France, Sept. 26, 1918.
R—Chicago, Ill.
B—Greece.
G. O. No. 5, W. D., 1920.

Private, Company B, 131st Infantry, 33d Division.
With three other soldiers he charged and captured a battery of three .77 field pieces which, protected by machine guns, were firing point-blank on the position held by his company. This deed enabled his company to continue the advance.

*EDDY, HENRY LESLIE_____
At Chateau-Thierry, France, June 6, 1918.
R—New Britain, Conn.
B—New Britain, Conn.
G. O. No. 15, W. D., 1919.

Second lieutenant, Infantry, U. S. Army, attached to 82d Company, 6th Regiment, U. S. Marine Corps, 2d Division.
He gave the supreme proof of that extraordinary heroism which will serve as an example to hitherto untried troops.
Posthumously awarded. Medal presented to father, H. W. Eddy.

EDDY, WILLIAM A_____
Near Torcy, France, June 4, 1918.
R—New Rochelle, N. Y.
B—Syria.
G. O. No. 119, W. D., 1918.

Second lieutenant, intelligence officer, 6th Regiment, U. S. Marine Corps, 2d Division.
While leader of a raiding patrol, he displayed great courage and devotion to duty by fearlessly entering dangerous areas and obtaining valuable information.

EDGAR, FRED W. (546891)_____
Near Jaulgonne, France, July 24, 1918.
R—La Raysville, Pa.
B—New Hyde Park, Long Island, N. Y.
G. O. No. 32, W. D., 1919.

Corporal, Company G, 30th Infantry, 3d Division.
He placed five wounded men in a shell hole in front of the enemy's machine-gun emplacements and protected them until nightfall with his rifle fire. He then went for aid and returned with stretchers to find that three of them had left the place. After helping to carry the two remaining to the rear, he returned to make further search, in which he was unsuccessful, but returned with 1 German prisoner whom he had captured.

EDMISTON, ANDREW, Jr _____
    In the Forêt-de-Fère, France, Aug. 2, 1918.
    R—Weston, W. Va.
    B—Weston, W. Va.
    G. O. No. 9, W. D., 1923.

Second lieutenant, 39th Infantry, 4th Division.
While in command of a platoon of his regiment it was severely bombed by enemy planes and under intense artillery fire. Serious disorganization followed the bombing. Taking steps for the safety of his unit he went forward under heavy artillery fire and assisted in the reorganization of units and, though badly wounded, insisted upon the evacuation of all other wounded before accepting first aid for himself. His splendid example of courage and devotion inspired every man of his command.

EDMUNDS, EDWARD, Jr _____
    Near Verdun, France, Oct. 27, 1918.
    R—Newton, Mass.
    B—England.
    G. O. No. 44, W. D., 1919.

First lieutenant, 102d Infantry, 26th Division.
An order having been received from division headquarters for an accurate report on the strength present in the front line, Lieutenant Edmunds crawled from shell hole to shell hole in broad daylight and in plain view of the enemy, who kept him under continuous sniping fire from numerous machine guns. Going on under these conditions, he personally counted every man in the front line of the battalion that he was commanding, and made his report to the regimental commander.

EDSALL, WILLIAM A. (281508) _____
    Near Romagne, France, Oct. 14, 1918.
    R—Muskegon, Mich.
    B—Ravenna, Mich.
    G. O. No. 21, W. D., 1919.

Private, Company M, 126th Infantry, 32d Division.
In an attack on Cote Dame Marie the 126th Infantry was held up owing to intense enemy machine-gun fire and grenades. Private Edsall volunteered as a member of a combat patrol which cut through the enemy lines, captured 10 machine guns, killed and captured 15 of the enemy, and forced a large number to surrender, clearing the Cote Dame Marie of the enemy, thus enabling the regiment to continue their advance.

EDWARDS, DANIEL R. (106546) _____
    At Cantigny, France, May 28–30, 1918.
    R—Bruceville, Tex.
    B—Mooreville, Tex.
    G. O. No. 15, W. D., 1923.
    Medal of honor also awarded.

Private, Company C, 3d Machine Gun Battalion, 1st Division.
Serving as gunner of his machine-gun squad he advanced with the first assault line of the Infantry and while passing through the village of Cantigny at 5.30 a. m., May 28, carrying his machine gun upon his shoulder, he was attacked by an enemy soldier and bayoneted, receiving a severe wrist wound; the enemy soldier was killed by an infantryman. Continuing in the advance beyond Cantigny and meeting intense enemy fire, the attacking wave was halted. Private Edwards with his squad remained in an advanced position, protecting with his fire the Infantry which had fallen back to a more advantageous position and were intrenching. While thus engaged the machine gunners repulsed two determined enemy counterattacks, during which the three members of the squad accompanying Private Edwards were killed and he himself severely wounded. Despite these wounds, which he himself dressed, he remained alone in his position throughout the day, firing whenever a target offered, withstanding attacks by liquid fire and machine-gun fire; he refused to be evacuated and continued to operate his gun until nightfall, when his company was relieved. His extraordinary bravery and devotion to duty, his fortitude and undaunted determination despite his numerous and painful wounds, incited the men of his battalion to splendid endeavors and raised their morale to an extremely high pitch.

*EDWARDS, GARRETT (42866) _____
    Near Soissons, France, July 19, 1918.
    R—Kittyton, Tenn.
    B—Unicoi County, Tenn.
    G. O. No. 15, W. D., 1919.

Sergeant, Company F, 16th Infantry, 1st Division.
When the Infantry of which he was a part was held up by an enemy machine-gun nest which was inflicting heavy losses upon his platoon, he unhesitatingly went forward to ascertain its location and was killed while performing this courageous duty.
Posthumously awarded. Medal presented to father, J. B. Edwards.

*EDWARDS, GEORGE H. (1218659) _____
    Near Forges, France, Sept. 26, 1918.
    R—Brooklyn, N. Y.
    B—New York, N. Y.
    G. O. No. 37, W. D., 1919.

Private, first class, Battery C, 105th Field Artillery, 27th Division.
In the face of heavy machine-gun fire, at great personal risk, he crawled out from a position of safety and rescued a wounded soldier who was lying exposed to enemy fire. He was killed later in the advance while charging a machine gun.
Posthumously awarded. Medal presented to mother, Mrs. H. G. Edwards.

EDWARDS, HARLEY S. (2308797) _____
    Near Somme-Py, France, Oct. 4–5, 1918.
    R—Titusville, Pa.
    B—Titusville, Pa.
    G. O. No. 37, W. D., 1919.

Private, Battery E, 12th Field Artillery, 2d Division.
During a violent enemy counterbarrage, Private Edwards, with Pvt. Russell Moran, remained on duty for 14 hours repairing the telephone line from their battery position to the battalion post of command, 250 meters away. Within this period the wires were cut by shellfire more than 20 times, but these two soldiers, displaying remarkable coolness and disregard of danger, promptly mended all breaks and maintained constant communication between the battalion and the battery commanders.

EDWARDS, HUGH F _____
    Near Vaux, France, July 1, 1918.
    R—Hamilton, Kans.
    B—Hamilton, Kans.
    G. O. No. 99, W. D., 1918.

Second lieutenant, 9th Infantry, 2d Division.
He led his platoon against heavy machine-gun fire, silenced several machine guns, established his guns at his objective, repelled a counterattack on his left flank, and personally killed two of the enemy.

EDWARDS, NORMAN E. (2267278) _____
    Near Eclisfontaine, France, Sept. 28, 1918.
    R—Chicago, Ill.
    B—Chicago, Ill.
    G. O. No. 37, W. D., 1919.

Corporal, Company H, 364th Infantry, 91st Division.
Going forward alone and in the open 100 yards in front of his company, he fired a rifle grenade into a machine-gun nest. The fire of the machine guns immediately slackened, and when they were captured it was found that only three of the enemy remained to man four guns.

EDWARDS, PAUL S _____
    Near Cheppy, France, Sept. 26, 1918.
    R—Little Rock, Ark.
    B—Ithaca, Mich.
    G. O. No. 20, W. D., 1919.

First lieutenant, Signal Corps, attached to 304th Brigade, Tank Corps.
During the attack on Cheppy he displayed unusual gallantry by crossing a fire-swept area and carrying messages to tanks, coolly moving from one to another and informing the drivers of his mission, under a fire so heavy and accurate that many of the tanks were struck by bullets while he was standing by.

*EGLER, FREDERICK A. (1831686)........
Near Cunel, France, Oct. 11, 1918.
R—Pittsburgh, Pa.
B—Pittsburgh, Pa.
G. O. No. 32, W. D., 1919.

Sergeant, Company M, 320th Infantry, 80th Division.
Advancing alone far beyond his lines, Sergeant Egler encountered and attacked an enemy machine-gun emplacement, killing an officer and capturing two guns, causing the enemy, about eight in number, to flee in disorder. During the action on Nov. 1, 1918, Sergeant Egler received wounds which caused his death.
Posthumously awarded. Medal presented to widow, Mrs. Frederick A. Egler.

EHLERS, CARL H. (2780582)............
Near Eclisfontaine, France, Sept. 28, 1918.
R—Sacramento, Calif.
B—Davenport, Iowa.
G. O. No. 37, W. D., 1919.

Corporal, Company K, 364th Infantry, 91st Division.
Responding to a call for volunteers, Corporal Ehlers, with five others, advanced 400 yards beyond their front to bring in wounded comrades. They succeeded in rescuing seven of their men, also in bringing in the dead body of a lieutenant, while exposed to terrific machine-gun fire.

EICH, WERNER (94889).................
Near Sedan, France, Nov. 7, 1918.
R—Mount Washington, Ohio.
B—Cincinnati, Ohio.
G. O. No. 44, W. D., 1919.

Private, Company K, 166th Infantry, 42d Division.
Private Eich was a member of a patrol sent out to silence machine-gun nests which were holding up the battalion's advance. When the officer leading the patrol fell mortally wounded he went to his assistance in the face of heavy fire from machine guns only 100 yards away, 3 other soldiers being killed in similar attempts.

EICHELBERGER, ROBERT L............
In Siberia, June 28–July 3, 1919.
R—Urbana, Ohio.
B—Urbana, Ohio.
G. O. No. 9, W. D., 1923.
Distinguished-service medal also awarded.

Lieutenant colonel (Infantry), General Staff Corps, U. S. Army.
For extraordinary heroism in action June 28–July 3, 1919, while serving as assistant chief of staff, G-2, American Expeditionary Forces, Siberia. On July 2, 1919, after the capture, by American troops of Novitskaya, an American platoon detailed to clear hostile patrols from a commanding ridge was halted by enemy enfilading fire, seriously wounding the members of the patrol. Colonel Eichelberger, without regard to his own safety and armed with a rifle, voluntarily covered the withdrawal of the platoon. On June 28, at the imminent danger of his own life, he entered the partisan lines and effected the release of one American officer and three enlisted men in exchange for a Russian prisoner. On July 3 an American column being fired upon when debouching from a mountain pass, Colonel Eichelberger voluntarily assisted in establishing the firing line, prevented confusion, and, by his total disregard for his own safety, raised the morale of the American forces to a high pitch.

EICHELSDOERFER, ROBERT M..........
Near Mouzon, France, Nov. 7, 1918.
R—Indianapolis, Ind.
B—Grand Rapids, Mich.
G. O. No. 3, W. D., 1922.

First lieutenant, Cavalry, aid-de-camp, 2d Infantry Brigade, 1st Division.
Lieutenant Eichelsdoerfer with another officer made a most hazardous reconnaissance of the enemy position along the River Meuse and supplied valuable information of these positions. During the entire exploit they were constantly under enemy observations and heavy fire of their guns.

EICHORN, VICTOR L. (959)............
At St. Benoit, France, Sept. 16, 1918.
R—Brooklyn, N. Y.
B—Brooklyn, N. Y.
G. O. No. 131, W. D., 1918.

Sergeant, sanitary detachment, 165th Infantry, 42d Division.
While the regimental dressing station was under heavy shell fire he volunteered to lead a squad of litter bearers to rescue several wounded men of another regiment who had been caught in a heavy barrage. He succeeded in leading his squad for a distance of 3 kilometers through a constant severe bombardment under direct observation of the enemy artillery and snipers to an outpost outside of his own regimental sector. He brought in 1 wounded officer and 7 severely wounded soldiers without losing any of his own men.

EIGENAUER, JOHN E. (2383990)........
Near Bois-des-Rappes, France, Oct. 15, 1918.
R—Philadelphia, Pa.
B—Philadelphia, Pa.
G. O. No. 87, W. D., 1919.

Sergeant, Company H, 60th Infantry, 5th Division.
Advancing ahead of his platoon, Sergeant Eigenauer engaged 2 enemy machine-gun nests, killing all the gunners. He then led his platoon forward until the fire of the enemy became so dense that he was obliged to dig in.

EIKLEBERRY, GIDEON J. (210663).......
Near Bouresches, France, July 20, 1918.
R—Jeffersonville, Ill.
B—Jeffersonville, Ill.
G. O. No. 68, W. D., 1920.

Sergeant, first class, photographic section, Signal Corps, attached to 26th Division.
He fearlessly advanced to a forward position during the attack of the 104th Infantry, in order to secure pictures of the attacking troops. After being severely wounded he continued his work until ordered to the rear. His gallant conduct stimulated the morale of the advancing troops.

EKLUND, EMIL J......................
Near Bois-du-Fays, France, Oct. 5, 1918.
R—Butte, Mont.
B—Lead, S. Dak.
G. O. No. 16, W. D., 1920.

First lieutenant, 58th Infantry, 4th Division.
The courageous conduct of Lieutenant Eklund while in command of the left flank platoon of the division during a strong enemy counterattack, exposed to heavy machine-gun and rifle fire, had a great moral effect on his men. The attack was repulsed after a hand-to-hand encounter in which many casualties were suffered.

*ELIA, PASQUALE (542705)............
Near Cierges, France, Oct. 5, 1918.
R—Webster, Pa.
B—Italy.
G. O. No. 10, W. D., 1920.

Private, Company K, 7th Infantry, 3d Division.
Private Elia, on his own initiative, advanced to an exposed position of great danger, and with his automatic rifle killed 18 Germans. Due to the individual efforts of this soldier his company was able to advance. He continued his fire until he was killed.
Posthumously awarded. Medal presented to father, Charles Elia.

ELICKY, FRANK (1039122)............
Near Greves Farm, France, July 15, 1918.
R—New York, N. Y.
B—New York, N. Y.
G. O. No. 46, W. D., 1919.

Private, Battery F, 10th Field Artillery, 3d Division.
Responding to a call for volunteers, Private Elicky, with 8 other soldiers, manned 2 guns of a French battery which had been deserted by the French during the unprecedented fire, after many casualties had been inflicted on their forces. For two hours he remained at this post and poured an effective fire into the ranks of the enemy.

ELIOT, AMORY VIVIEN
Between Fismette and Bazoches, France, Sept. 2, 1918.
R—New York, N. Y.
B—New York, N. Y.
G. O. No. 2, W. D., 1920.

Second lieutenant, 307th Infantry, 77th Division.
He advanced in front of our lines, exposed to close range enemy machine-gun fire, and rescued a severely wounded soldier.

ELKINS, STEPHEN B
East of Ronssoy, France, Sept. 29, 1918.
R—Birmingham, Ala.
B—Eupora, Miss.
G. O. No. 20, W. D., 1919.

Second lieutenant, 105th Infantry, 27th Division.
During the operations against the Hindenburg line Lieutenant Elkins, with 3 sergeants, occupied an outpost position in advance of the line, which was attacked by a superior force of the enemy. He so directed his small detachment that he succeeded in repulsing the attack, killing 10 Germans, capturing 5, and putting the rest of the enemy to flight. The bravery and determination displayed by this group were an inspiration to all who witnessed them.

ELLET, MONROE (1576886)
Near Thiaucourt, France, Sept. 12, 1918.
R—Huntington, Ind.
B—Huntington, Ind.
G. O. No. 37, W. D., 1919.

Private, Battery E, 12th Field Artillery, 2d Division.
Acting as spare cannoneer, Private Ellet sprang at a word from his executive officer to act as No. 1 in the supplementary gun crew for the first piece, whose entire crew were casualties. He assisted in laying aside the dead and wounded and continued in the service of his piece until the barrage was completed.

ELLINGTON, JAMES M
Near Bellicourt, France, Sept. 29, 1918.
R—Oxford, N. C.
B—Grandville County, N. C.
G. O. No. 81, W. D., 1919.

First lieutenant, 120th Infantry, 30th Division.
Severely wounded in an attack, he refused to stop for first aid, leading his men forward under heavy fire. When, after several hours' fighting, he was ordered to the rear by his battalion commander, he returned to the front line after having his wound dressed, directing the work of reorganizing his command and consolidating the position that had been won.

ELLIOTT, CHARLES B
Near Chateau-Thierry, France, June 6–7, 1918.
R—Alexandria, Va.
B—Washington, D. C.
G. O. No. 126, W. D., 1919.

Major, 23d Infantry, 2d Division.
Leading his command in combat on June 6–7, he reformed his badly shattered units in the face of unusual and annihilating fire and directed the placing of his guns throughout a severe gas attack which severely poisoned him. While leading two companies of his battalion forward in the attack, July 19, he was wounded by artillery fire.

ELLIOTT, CHARLES G. (1099102)
Near Preny, France, Nov. 5, 1918.
R—Elmira, N. Y.
B—Springfield, Pa.
G. O. No. 32, W. D., 1919.

Sergeant, Company M, 56th Infantry, 7th Division.
While leading a patrol in front of his company sector on November 5, Sergeant Elliott and another member of his patrol were wounded when heavy machine-gun fire was encountered. However, he retained command of the patrol, took care of the wounded man, and reported to his company commander the results of the patrol before he would be evacuated.

*ELLIOTT, CLARK R
Near Soissons, France, July 21, 1918.
R—Minneapolis, Minn.
B—Morgan County, Ohio.
G. O. No. 132, W. D., 1918.

Lieutenant colonel, 26th Infantry, 1st Division.
He promptly and courageously took command of the men of the front lines at a critical stage of the engagement near Soissons, France, July 21, 1918, and while leading them forward in a successful attack was himself killed.
Posthumously awarded. Medal presented to widow, Mrs. C. R. Elliott.

ELLIOTT, EDWIN A. (556067)
North of Nantillois, France, Oct. 12, 1918.
R—Elm Mott, Tex.
B—Troy, Tex.
G. O. No. 15, W. D., 1921.

Sergeant, Company E, 39th Infantry, 4th Division.
An ammunition detail having failed several times to carry ammunition over a barraged zone, Sergeant Elliott voluntarily gathered and conducted an ammunition detail over 3 kilometers under extremely heavy artillery and machine-gun fire to the front line. In advance of his men he dragged a full box of Springfield ammunition for over a kilometer and distributed it to the front line. Later, he volunteered to carry, and carried, a message back to regimental headquarters.

ELLIOTT, ROBERT P
Near Olizy, France, Oct. 27, 1918.
R—Los Angeles, Calif.
B—Los Angeles, Calif.
G. O. No. 7, W. D., 1919.

First lieutenant, pilot, 96th Aero Squadron, Air Service.
He flew in a formation over the enemy's lines on a bombing expedition. Attacked by a superior number of enemy pursuit planes, his aileron controls were soon shot away. He continued to pilot his machine and give protection to his comrades. When his observer was seriously wounded he left the formation, at great risk to himself, and with a disabled machine made a safe landing.

ELLIS, GEORGE L. (737106)
Near Brieulles, France, Nov. 4, 1918.
R—Aurora, Mo.
B—Mattoon, Ill.
G. O. No. 37, W. D., 1919.

Sergeant, Company M, 11th Infantry, 5th Division.
While his regiment was trying to force a crossing over the Meuse River, Sergeant Ellis swam and saved the lives of two of his comrades who had become exhausted in the cold water.

ELLIS, HERBERT J
Near Montfaucon, France, Sept. 26, 1918.
R—Toledo, Ohio.
B—Toledo, Ohio.
G. O. No. 128, W. D., 1918.

First lieutenant, 3d Brigade, Tank Corps.
He set a conspicuous example of daring and disregard of personal safety throughout the advance toward Montfaucon. Standing up in the open under fire from enemy snipers and machine guns, he directed the engineers in the work of clearing a path for the tanks. Single handed he routed a sniper who was harassing the engineers engaged in this work. With a French officer he entered the Bois-de-Cuisy in advance and assisted in putting out of action seven Germans who were hindering the advance.

ELLIS, LUTHER E
In Bois-d'Ormont, France, Oct. 23, 1918.
R—Montpelier, Ind.
B—Butler, Ky.
G. O. No. 133, W. D., 1919.

Captain, 102d Infantry, 26th Division.
He personally led his company against a strongly held enemy machine-gun position. During the advance he was shot through the lung. When wounded his men halted to render first aid, but he ordered them forward. His example of gallantry contributed greatly to the success of the attack.

ELLIS, NATHANIEL WATSON_____
  Near Montbrehain, France, Oct. 7,
    1918.
  R—Tellico Plains, Tenn.
  B—Elizabeth, Tenn.
  G. O. No. 46, W. D., 1919.

First lieutenant, 117th Infantry, 30th Division.
When his company was held up by sweeping machine-gun fire, Lieutenant Ellis rushed forward alone, in the face of direct machine-gun fire, to an enemy machine-gun nest 60 yards in advance of his platoon, and by the effective use of his pistol killed 5 of the enemy and captured 26 prisoners, together with the machine gun. Although he had been seriously wounded in two places while advancing he held the position until his platoon came up.

ELLSWORTH, BRADFORD_____
  Near Le Besace, France, Nov. 5, 1918.
  R—New York, N. Y.
  B—New Hartford, Conn.
  G. O. No. 37, W. D., 1919.

Captain, 366th Infantry, 77th Division.
He displayed great courage by leading a mounted reconnaissance patrol a kilometer in advance of our lines, developed the enemy's line by drawing fire from his machine guns, and remained under this heavy machine-gun fire until the emplacements had been located.

ELLSWORTH, LEON E. (70157)_____
  Near Bouresches, France, July 20,
    1918.
  R—Jeffersonville, Vt.
  B—Cambridge, Vt.
  G. O. No. 15, W. D., 1923.

Private, Machine Gun Company, 103d Infantry, 26th Division.
The left company of the battalion engaged in attack against Hill 190 between Bouresches and Belleau was checked by intense enemy fire, the right company proceeding toward its objective, thus creating a dangerous break in the line. The battalion commander found it imperative to establish communication with the commander of the right company. Private Ellsworth volunteered to carry the message, crossing an open field in full view of the enemy machine gunners for a distance of 200 yards. Returning to the battalion post of command he was wounded in the hand and shot through the chest, the bullet passing through the body. Notwithstanding his severe wounds he reported back after accomplishing his mission and then walked to a dressing station, from which he was evacuated. The heroic act of Private Ellsworth proved an inspiration to the men of his battalion.

ELMER, CLARENCE G_____
  Near Belval-Bois-des-Dames, France,
    Nov. 2, 1918.
  R—Lanark, Ill.
  B—Lanark, Ill.
  G. O. No. 44, W. D., 1919.

Second lieutenant, 9th Infantry, 2d Division.
When he was asked to send out a squad in advance of the line under heavy machine-gun fire to enfilade enemy artillery, Lieutenant Elmer asked for and received permission to lead the party, and by his bravery and coolness succeeded in driving the enemy gunners away from their guns.

ELMES, CHESTER H_____
  Near Verdun, France, Oct. 12, 1918.
  R—Concord Junction, Mass.
  B—Concord Junction, Mass.
  G. O. No. 130, W. D., 1918.

Second lieutenant, 114th Infantry, 29th Division.
Though suffering from a painful wound in the head and ordered to the rear, he reluctantly left his platoon, and, in spite of weakness, carried a wounded soldier under heavy shell fire to a dressing station 500 yards away.

ELPERN, BENJAMIN D. (546414)_____
  At Jaulgonne, France, July 23-26,
    1918.
  R—Greensburg, Pa.
  B—Greensburg, Pa.
  G. O. No. 87, W. D., 1919.

Corporal, Company E, 30th Infantry, 3d Division.
Corporal Elpern volunteered and carried messages under hazardous circumstances when wire communication had been entirely destroyed by the intense shell fire. He also rendered timely aid in guiding parties bringing up food and ammunition.

ELSEA, ALBERT E. (1446131)_____
  Near Hilsenfirst, France, July 6, 1918.
  R—Joplin, Mo.
  B—Carterville, Mo.
  G. O. No. 16, W. D., 1920.

Sergeant, Company B, 129th Machine Gun Battalion, 35th Division.
While acting as machine-gun leader near Hilsenfirst, France, July 6, 1918, he was wounded in the face by a bursting shell, but continued to direct his men until attack ended, and then insisted on walking to a dressing station.
Oak-leaf cluster.
For the following acts of extraordinary heroism in action in Europe this man was awarded an oak-leaf cluster: During the Meuse-Argonne offensive, Oct. 12, 1918 (while he was serving as second lieutenant, Company B, 129th Machine Gun Battalion, 35th Division), he took up a position exposed to heavy machine-gun and artillery fire in order to direct the fire of his machine guns. Although knocked down twice by concussion of high-explosive shells, he continued at his position. On Oct. 15, 1918, he gathered a squad of stragglers under fire and led them, together with his own unit, to their objective.

*ELSWORTH, EDWARD, Jr_____
  At Claire-Chenes Woods, France,
    Oct. 20, 1918.
  R—New York, N. Y.
  B—New York, N. Y.
  G. O. No. 142, W. D., 1918.

First lieutenant, 6th Engineers, 3d Division.
When two machine-gun nests were holding up the advance of his company, he took 3 soldiers, and by daring and skillful maneuvering captured both guns. Later in the same day he charged another machine-gun nest and was killed.
Posthumously awarded. Medal presented to brother, Oliver B. Elsworth.

ELY, EARNEST E. (199423)_____
  In the Aire Valley, France, Sept. 26,
    Oct. 12, 1918.
  R—Eugene, Oreg.
  B—Clancey, Mont.
  G. O. No. 37, W. D., 1919.

Sergeant, first class, Headquarters Company, 1st Brigade, Tank Corps.
He worked for five days and nights under intense shell fire to establish signal communications. While engaged in this work he was gassed and carried to a dressing station unconscious. On regaining consciousness he escaped from the dressing station, returned to his post, and continued to work for 12 hours more until he was gassed for a second time.

ELY, HANSON E_____
  Near Vierzy, France, July 18, 1918.
  R—Iowa City, Iowa.
  B—Buffalo Grove, Iowa.
  G. O. No. 15, W. D., 1923.
  Distinguished-service medal also
    awarded.

Brigadier general, 3d Infantry Brigade, 2d Division.
In order that he might personally direct the attack, General Ely attempted to enter Vierzy which was not yet cleared of the enemy; he was fired on at short range by enemy machine guns in the town. He then personally organized and directed an attack which resulted in the capture of the town and in the advance of the lines well beyond. General Ely's indomitable bravery, disregard for his own safety, his devotion to his men, and his frequent presence with them in the front line, inspired them to deeds of great courage and enabled the troops to take the town, despite strong enemy resistance made by vastly superior numbers.

EMERSON, HARLOW B. (263625) _____
  Near Sergy, France, July 31, 1918.
  R—Saginaw, Mich.
  B—Saginaw, Mich.
  G. O. No. 64, W. D., 1919.

Corporal, Company K, 125th Infantry, 32d Division.
Early in the attack on Hill 212, Corporal Emerson was severely wounded but continued to lead his squad in advance until ordered to the rear. Having successfully reached the objective, he administered first aid to a number of his wounded comrades, fearlessly exposing himself to further danger in the performance of this voluntary work.

EMERSON, WILLARD INGHAM _____
  Near Grand Pre, France, Nov. 1, 1918.
  R—New York, N. Y.
  B—Ithaca, N. Y.
  G. O. No. 3, W. D., 1919.

First lieutenant, 311th Infantry, 78th Division.
Lieutenant Emerson displayed remarkable gallantry and leadership during the fighting north of Grand Pre when the line of his regiment was the pivot for the advance of the Army. He personally led his company around machine-gun nests, frequently going out with selected patrols for the purpose of bombing out enemy machine gunners. When the commanding officer of the company on his left was incapacitated he at once took command of the company and maneuvered it with his own. By his example, in undergoing hardships during the advance, he sustained the morale of his men and inspired them to valiant combat.

EMERY, DONALD (1212809) _____
  Near Dickebusch, Belgium, Aug. 22, 1918.
  R—New York, N. Y.
  B—Chelsea, Vt.
  G. O. No. 99, W. D., 1918.

Private, first class, sanitary detachment, 107th Infantry, 27th Division.
Displaying an absolute disregard of danger in caring for wounded under shell and rifle fire and a continuous cheerfulness under trying conditions, his courageous example was inspiring to his comrades.

EMERY, JOHN G _____
  Near Hill 240, Exermont sector, France, Oct. 9, 1918.
  R—Grand Rapids, Mich.
  B—Grand Rapids, Mich.
  G. O. No. 43, W. D., 1922.

Major, 18th Infantry, 1st Division.
Major Emery, with the greatest gallantry and utter disregard for his own safety, personally led the advance of his battalion against the strongly occupied position of the enemy, holding it to the attack by force of leadership, in an extremely thick fog and under heavy enemy artillery and machine-gun fire. When the fog lifted, his battalion was so close upon the enemy that the latter, though equipped with many machine guns, retired from their position. While deliberately proceeding to a vantage point which was under very heavy enemy fire, in order that he might better direct the combat, he fell severely wounded by shell fire.

*EMERY, JOSEPH W., Jr _____
  Near Vierzy, France, July 18, 1918.
  R—Quincy, Ill.
  B—Quincy, Ill.
  G. O. No. 103, W. D., 1919.

First lieutenant, 9th Infantry, 2d Division.
While attached to the regimental post of command in the rear, Lieutenant Emery voluntarily joined the assaulting battalion as a platoon leader in a company which was short of officers. When desperate hostile resistance was encountered at the outset of the attack and all the other officers of the company became casualties, he took command of the company and courageously led his men in overcoming enemy machine-gun nests. While rushing an enemy position at the head of his men this gallant officer was killed.
Posthumously awarded. Medal presented to father, Joseph W. Emery, sr.

*EMORY, GERMAN H. H _____
  Near Sommerance, France, Nov. 1, 1918.
  R—Baltimore, Md.
  B—Allegany County, Md.
  G. O. No. 27, W. D., 1919.

Major, 320th Infantry, 80th Division.
After advancing to the north slope of the Ravine Aux Pierres, through heavy machine-gun and shell fire, Major Emory's battalion was momentarily held up by a very intense machine-gun fire. He was killed while personally directing the attack and encouraging his troops, moving back and forth in front of the line in plain view of the enemy and under direct machine-gun fire.
Posthumously awarded. Medal presented to widow, Mrs. Lucy S. Emory.

ENDICOTT, BURTIS A. (2100275) _____
  Near Trugny, France, July 22, 1918.
  R—Ridgway, Ill.
  B—Ridgway, Ill.
  G. O. No. 87, W. D., 1919.

Private, Company C, 102d Infantry, 26th Division.
After 5 runners had been shot down trying to get through a heavy artillery barrage with a message to the battalion commander, he volunteered to carry the same message over the same route and succeeded.
Oak-leaf cluster.
For the following act of extraordinary heroism in action at Marcheville, France, Sept. 26, 1918, he is awarded a bronze oak-leaf cluster to be worn on the distinguished-service cross. While isolated from the rest of his platoon and under violent machine-gun and artillery fire, he was wounded in the arm by a machine-gun bullet. After receiving first aid he returned to duty and continued in the fight until again wounded.

ENDLER, FRANK (732170) _____
  Near Fontaines, France, Nov. 8, 1918.
  R—Avella, Pa.
  B—Bowhill, Pa.
  G. O. No. 37, W. D., 1919.

Private, Company D, 6th Infantry, 5th Division.
Private Endler, while preceding his platoon in an advance, met 2 German machine-gun groups, and by his rifle fire and accurate information sent back to his platoon made it possible for that flank of his company to advance. During the engagement he saved many lives by carrying valuable information to his company commander through heavy machine-gun fire.

ENGEL, GEORGE F. (274128) _____
  Near Gesnes, France, Oct. 10, 1918.
  R—Milwaukee, Wis.
  B—Milwaukee, Wis.
  G. O. No. 59, W. D., 1921.

Sergeant, Company F, 127th Infantry, 32d Division.
Sergeant Engel repeatedly advanced under fire in front of our lines, locating enemy positions. He displayed the utmost bravery and fearlessness on all occasions, as well as cool judgment. On Oct. 10 he was leading a patrol under heavy enemy fire and was wounded by an enemy machine-gun bullet. Although badly wounded he refused to go to the rear for aid, but continued the advance until he had obtained the desired information.

ENGEL, WILLIAM FREDERICK _____
  Near Blanc Mont Ridge, France, Oct. 4, 1918.
  R—Berea, Ohio.
  B—Medina, Ohio.
  G. O. No. 46, W. D., 1919.

Sergeant, 55th Company, 5th Regiment, U. S. Marine Corps, 2d Division.
After locating a machine-gun nest, he destroyed one of the guns and returned to our lines with valuable information concerning the location of the machine-gun nest.

ENGELBRECHT, FRED (43385)_____
Near Fleville, France, Oct. 4, 1918.
R—Elkton, S. Dak.
B—New Liberty, Iowa.
G. O. No. 35, W. D., 1920.

Sergeant, Company K, 16th Infantry, 1st Division.
When the attack of his company on the second objective was halted by machine-gun fire from the front, Sergeant Engelbrecht with a comrade advanced and destroyed the enemy machine-gun position, forcing 4 of the enemy to surrender.

*ENGLANDER, LEO (1708449)_____
On the Vesle River, near Ville-Savoye, France, Aug. 23, 1918.
R—New York, N. Y.
B—New York, N. Y.
G. O. No. 56, W. D., 1922

Private, Company D, 308th Infantry, 77th Division.
He volunteered to go out into no man's land to bring in a comrade from his platoon who had been seriously wounded and unable to move. Private Englander reached the man and was about to carry him to safety when he was killed by enemy machine-gun fire. His heroism was an inspiration to the members of his company.
Posthumously awarded. Medal presented to sister, Mrs Sadie Englander Liswood.

ENGLE, JOSEPH C. (1315222)_____
Near Bellicourt, France, Sept. 29, 1918.
R—Memphis, Tenn.
B—Cincinnati, Ohio.
G. O. No. 81, W. D., 1919.

Sergeant, Company D, 119th Infantry, 30th Division.
Although himself wounded, he continued to lead the advance of his platoon, and when it was held up by fire from an enemy machine-gun nest he advanced against it alone, bombed it, and thus allowed a resumption of the advance.

ENGLISH, LEE M._____
Near Montfaucon, France, Sept. 29, 1918.
R—Bedford, Ohio.
B—Freeport, Ohio.
G. O. No. 37, W. D., 1919.

First lieutenant, Dental Corps, attached to 314th Infantry, 79th Division.
Although he had received a very painful wound 2 days previous, he remained on duty at an aid station under heavy shell fire until it was completely destroyed and many of the inmates and attendants killed or wounded. He assisted in caring for these wounded and directing their evacuation and then dressed the wounded on the field until an aid station could be located farther to the rear.

*ENGLISH, MATH L._____
Near Cheppy, France, Sept. 26, 1918.
R—Coupeville, Wash.
B—Gibson, Ga.
G. O. No. 37, W. D., 1919.

Captain, 344th Battalion, Tank Corps.
During the attack on Cheppy, he dismounted from his tank and, under heavy machine-gun fire, personally supervised the cutting of a passage for his tanks through three hostile trenches.
Oak-leaf cluster.
For the following act of extraordinary heroism in action near Exermont, France, Oct. 4, 1918, Captain English is awarded an oak-leaf cluster, to be worn with the distinguished-service cross: He left his tank under heavy machine-gun and artillery fire to make a personal reconnaissance, in the course of which he was killed.
Posthumously awarded. Medal presented to widow, Mrs. Math L. English.

ENOCHS, REX PAUL._____
In the Bois-de-Bon-Vaux, France, Sept. 16-17, 1918, near the Bois-des-Rappes, France, Oct. 15-16, 1918, and in the Murvaux Valley, France, Nov. 6, 1918.
R—Pleasantville, Ind.
B—Pleasantville, Ind.
G. O. No. 78, W. D., 1919.

Second lieutenant, 60th Infantry, 5th Division.
In the Bois-de-Bon-Vaux Lieutenant Enochs displayed exceptional bravery in maintaining liaison between his battalion and adjoining units under heavy artillery and machine-gun fire. On Oct. 15-16, he was part of a small detachment which advanced to the objective several kilometers beyond the other troops, being completely cut off by the enemy. During the night Lieutenant Enochs led patrols and personally overcame enemy machine gunners. On Nov. 6, this officer maintained a liaison system in advance of our troops despite enemy machine-gun fire from the front and flanks.

ENRIGHT, HAROLD W. (2658588)_____
Near Bois-de-Brieulles, France, Sept. 28, 1918.
R—Warren, Ill.
B—Rochester, N. Y.
G. O. No. 46, W. D., 1919.

Private, Company I, 47th Infantry, 4th Division.
He charged an enemy machine gun which was inflicting heavy losses upon our troops and delaying the advance. He wounded the gunner and captured the gun, thereby enabling our advance to continue.

EPLER, CHARLES H. (560697)_____
Near Ville-Savoye, France, Aug. 8, 1918.
R—Dayton, Ohio.
B—Dayton, Ohio.
G. O. No. 95, W. D., 1919.

Private, Company A, 59th Infantry, 4th Division.
After several unsuccessful attempts to silence an enemy machine-gun nest had been made, Private Epler and another soldier volunteered to put the nest out of action. His companion was killed, but Private Epler succeeded in throwing grenades into the nest, setting fire to the ammunition boxes with which it was surrounded, killing several of the crew and stopping the fire of the gun.

EPLEY, BRUCE (734104)_____
At Frapelle, France, Aug. 17, 1918.
R—Newport, Tenn.
B—Newport, Tenn.
G. O. No. 44, W. D., 1919.

Sergeant, Company M, 6th Infantry, 5th Division.
Rather than ask another to undertake so dangerous a mission, Sergeant Epley himself cut the enemy's wire under heavy bombardment and led his section to its objective. Later, in the absence of his platoon commander, he assumed command of his platoon and directed the consolidation of its position. He then crossed an open field under heavy machine-gun fire to locate another platoon which had become lost, found the remnants of the platoon, which he reorganized and led to the objective. Though he had been seriously gassed, this soldier continued on duty throughout the 3 days' bombardment, refusing to be evacuated until he was ordered to the rear.

EPPIHEIMER, ELMER L. (1245197)_____
Near Fismes, France, Aug. 4, 1918.
R—West Chester, Pa.
B—Chester Co., Pa.
G. O. No. 50, W. D., 1919

Private, first class, Company I, 111th Infantry, 28th Division.
Volunteering as runner on a most hazardous mission across a ruined bridge, Private Eppiheimer, after successfully making the crossing, again offered to recross to inform our batteries that their barrage was falling short. When he had reached the middle of the bridge an enemy shell exploded, throwing him into the river and slightly wounding him. He completed his journey by swimming and delivered his message without delay.

ERB, DANIEL (735179)\
    Near the village of Dun-sur-Meuse,\
    France, Nov. 5, 1918.\
    R—Emaus, Pa.\
    B—Treichlersville, Pa.\
    G. O. No. 37, W. D., 1919.

Private, Company D, 11th Infantry, 5th Division.\
Having become separated from the remainder of his company, he discovered and captured, single handed, a hostile machine-gun crew. Taking his prisoners to a dugout near by, he found 48 more Germans, whom he also disarmed.

ERKENBRACK, HARRY B. (75674)\
    Near Medeah Ferme, France, Oct.\
    30, 1918.\
    R—Okanogan, Wash.\
    B—Lecky, Iowa.\
    G. O. No. 95, W. D., 1919.

Private, Company F, 9th Infantry, 2d Division.\
As a runner Private Erkenbrack proved himself a fearless soldier by maintaining liaison for his battalion commander at all times through the most severe shellfire. He was always at the front when called upon for dangerous missions, and he carried them out with the greatest gallantry.

EROMO, GILIMO (543173)\
    At Claire-Chenes, France, Oct. 24,\
    1918.\
    R—McKees Rocks, Pa.\
    B—Italy.\
    G. O. No. 27, W. D., 1920.

Private, Company M, 7th Infantry, 3d Division.\
When his company was held up by heavy machine-gun fire, Private Eromo on his own initiative made a flank attack upon a machine-gun nest, capturing same and forcing one officer, two sergeants, and six privates to surrender. By his act the company was thus enabled to continue in the advance.

ERWIN, WILLIAM P.\
    In the Chateau-Thierry and St.\
    Mihiel salients, France, June and\
    July, 1918; Sept. 12 to 15, 1918.\
    R—Chicago, Ill.\
    B—Ryan, Okla.\
    G. O. No. 70, W. D., 1919.

First lieutenant, 1st Aero Squadron, Air Service.\
Lieutenant Erwin, with Second Lieutenant Byrne E. Baucom, observer, by a long period of faithful and heroic operations, set an inspiring example of courage and devotion to duty to his entire squadron. Throughout the Chateau-Thierry actions, in June and July, 1918, he flew under the worst weather conditions and successfully carried out his missions in the face of heavy odds. In the St. Mihiel sector, Sept. 12 to 15, 1918, he repeated his previous courageous work. He flew as low as 50 feet from the ground behind the enemy's lines, harassing German troops with machine-gun fire and subjecting himself to attack from ground batteries, machine guns, and rifles. He twice drove off enemy planes which were attempting to destroy an American observation balloon. On Sept. 12 and 13, 1918, he flew at extremely low altitudes and carried out Infantry contact patrols successfully. Again, on Sept. 12, he attacked a German battery, forced the crew to abandon it, shot off his horse a German officer who was trying to escape, drove the cannoneers to their dugouts, and kept them there until the Infantry could come up and capture them.\
Oak-leaf cluster.\
For the following act of extraordinary heroism in action near Sedan, France, Nov. 5, 1918, Lieutenant Erwin was awarded an oak-leaf cluster to be worn with his distinguished-service cross: Against the advice of experienced officers, he undertook a reconnaissance flight in the face of atmospheric conditions that rendered flying most dangerous. In order that his observer might gain the necessary information, he was forced to fly at a perilously low altitude and was subject to continuous antiaircraft and rifle fire. When information gained on the flight had been dropped at division headquarters, he circled and returned over the enemy lines, although on the first reconnaissance mission his plane had been repeatedly hit by bullets. Penetrating far into enemy territory, he maneuvered most skillfully and, with shells bursting near him, flew low while his observer poured deadly fire upon machine-gun nests that had been holding up the advance of our troops. When his machine was crippled by enemy fire he displayed exceptional skill in effecting a landing upon rocky land within the enemy lines. With his observer, he beat off repeated enemy attacks and fought his way back to the American lines with information of vital importance to our troops.

ESPY, ROBERT\
    Near Courpoil, France, July 26, 1918.\
    R—Abbeville, Ala.\
    B—Abbeville, Ala.\
    G. O. No. 23, W. D., 1919.

First lieutenant, 167th Infantry, 42d Division.\
While making a reconnaissance of his position 200 yards in front of our lines he noticed the enemy preparing a counterattack. Taking an automatic rifle from a dead gunner's hands, he alone, although twice wounded, opened fire on the enemy, breaking up the attack and inflicting severe losses on their ranks.

ESSER, JOSEPH F. (1827192)\
    At Imecourt, France, Nov. 1, 1918.\
    R—Springdale, Pa.\
    B—Tarentum, Pa.\
    G. O. No. 87, W. D., 1919.

Corporal, Company H, 319th Infantry, 80th Division.\
When his company was held up by enemy machine-gun fire, Corporal Esser, having no rifle grenades, searched about with disregard for his own safety until he found two, with which he boldly attacked the enemy, causing the surrender of 200 Germans.

ESTE, J. DICKINSON\
    Near Chambley, France, Sept. 13,\
    1918.\
    R—Haverford, Pa.\
    B—Philadelphia, Pa.\
    G. O. No. 123, W. D., 1918.

First lieutenant, pilot, 13th Aero Squadron, Air Service.\
He was leading an offensive patrol of five machines when a formation of seven enemy single-seaters approached the patrol from above. Although outnumbered and in a very disadvantageous position, he did not hesitate to lead his patrol to the attack. Through the combat which followed he fought with the greatest bravery, in spite of the fact that he was himself attacked by two enemy planes, which fired at him at point-blank range from the rear and above. By his skill and courage he was able to keep his formation together, and they succeeded in shooting down three of the enemy planes, of which he himself destroyed one and drove down another out of control.

ESTEP, ISAAC (3175493)\
    Near La-Fontaine-au-Croncq Farm,\
    France, Nov. 4, 1918.\
    R—Clinchport, Va.\
    B—Scott County, Va.\
    G. O. No. 95, W. D., 1919.

Private, Company C, 9th Infantry, 2d Division.\
Being on duty as a stretcher bearer, he displayed exceptional gallantry and disregard of danger to self in removing wounded from a field so swept by machine-gun fire that the ordinary man would have felt justified in leaving them there until the storm had abated. Of the five men engaged in this work one was killed and Private Estep and one other wounded, while the clothing and equipment of all were riddled by bullets.

ETHIER, RALPH (2256264)..............
Near Gesnes, France, Sept. 26, 1918.
R—Hoquiam, Wash.
B—Spokane, Wash.
G. O. No. 13, W. D., 1919.

Sergeant, Headquarters Company, 361st Infantry, 91st Division.
He was in charge of the signal section attached to the attacking battalion; he displayed remarkable coolness and disregard for personal danger in bringing the battalion telephone line, through heavy artillery and machine-gun fire, to the ridge which was being attacked and there establishing communication with regimental headquarters.

EVANS, BENJAMIN A. (1785434)..........
North of Verdun, France, Nov. 7, 1918.
R—Philadelphia, Pa.
B—Philadelphia, Pa.
G. O. No. 39, W. D., 1920.

Corporal, Company I, 315th Infantry, 79th Division.
Corporal Evans on the night of Nov. 7, 1918, when hearing the company commander asking for volunteers to reconnoiter about 100 yards of unknown ground and locate an enemy trench, did volunteer to undertake the mission and, accompanied by Private Bryson, successfully accomplished the same and brought back valuable information. As a noncommissioned officer his example and courage aided greatly in holding his platoon together.

EVANS, CHARLES H. (574149)..........
Near the Bois-de-Brieulles, France, Sept. 26, 1918.
R—Lewistown, Mont.
B—Jamaica, Iowa.
G. O. No. 98, W. D., 1919.

Private, Company B, 39th Infantry, 4th Division.
When his company was held up by heavy enemy machine-gun fire, Private Evans and two other soldiers advanced in the face of intense fire and captured the enemy machine-gun nest, from which the fire had been coming, killing two of the enemy and capturing 3 prisoners with their machine gun.

EVANS, EDWARD F. (2365769)..........
In the vicinity of Kraefski, Siberia, June 12, 1919.
R—Logan, W. Va.
B—Williamson, W. Va.
G. O. No. 133, W. D., 1919.

Cook, Company F, 27th Infantry.
When his detachment was attacked by a superior force of the enemy he took up a position alone on the thatched roof of a shed on the flank of his platoon, from which, owing to his calmness and excellent marksmanship, he successfully broke up a flanking party of the enemy by killing and wounding seven of them, thus preventing his detachment from being cut off or from suffering serious casualties.

*EVANS, EDWIN V..........
South of Soissons, France, July 23, 1918.
R—Boulder, Colo.
B—Bedford, Iowa.
G. O. No. 44, W. D., 1919.

Second lieutenant, 16th Infantry, 1st Division.
When his battalion was forced to fall back before heavy machine-gun and artillery fire, Lieutenant Evans led 6 runners in an attack on an enemy machine-gun nest. They had progressed about 200 yards ahead of the line when this officer, who was in advance of his men, was instantly killed just as he reached the machine-gun nest.
Posthumously awarded. Medal presented to mother, Mrs. C. T. Evans.

EVANS, GEORGE R. (3210491)..........
Near Grimaucourt, France, Nov. 10, 1918.
R—Lapine, Ala.
B—Lapine, Ala.
G. O. No. 32, W. D., 1919.

Private, Company E, 322d Infantry, 81st Division.
He repeatedly carried messages through very heavy artillery and machine-gun fire and at one time successfully carried a wounded comrade through an intense barrage to a first-aid station.

EVANS, GWILYN R. (2176436)..........
In the Bois-d'Aigremont, France, July 15, 1918.
R—Lebo, Kans.
B—Arvonia, Kans.
G. O. No. 78, W. D., 1919.

Private, Company M, 30th Infantry, 3d Division.
Private Evans, a runner, frequently exposed himself to heavy artillery fire in carrying messages for his battalion commander.

EVANS, HENRY C..........
Near Chaudun, France, July 19, 1918.
R—Baltimore, Md.
B—Baltimore, Md.
G. O. No. 56, W. D., 1922.

First lieutenant, 6th Field Artillery, 1st Division.
Learning that the advance of the Infantry which his battery was supporting was meeting with stubborn resistance, Lieutenant Evans voluntarily went forward through artillery and machine-gun fire to the crest of a hill and climbed a tree overlooking the enemy position for the purpose of adjusting fire upon the enemy. Though subjected to severe fire from hostile artillery and machine guns, he courageously remained in this exposed position, and was thus able to direct the fire of his battery as to materially assist the advance of the Infantry.

*EVANS, PLUMMER (2000093)..........
Near Verdun, France, Oct. 11, 1918.
R—Soldier, Ky.
B—Soldier, Ky.
G. O. No. 32, W. D., 1919.

Private, Company D, 111th Machine Gun Battalion, 29th Division.
Disregarding his own personal safety, he volunteered to assist in aiding his wounded comrades, leaving his place of safety during an intense bombardment. While performing this meritorious work he himself was killed.
Posthumously awarded. Medal presented to father, N. E. Evans.

EVANS, RODERICK (2659414)..........
Near Hill 240, Exermont, France, Oct. 6, 1918.
R—Girard, Ohio.
B—Girard, Ohio.
G. O. No. 78, W. D., 1919.

Corporal, Company G, 28th Infantry, 1st Division.
After Corporal Evans had been wounded and sent to the dressing station he voluntarily returned to the front line and led his platoon in action until he he was wounded a second time.

EVANS, WILLIAM C. (588438)..........
Near Bazoches, France, Aug. 27–28, 1918.
R—Plainsville, Pa.
B—Nanticoke, Pa.
G. O. No. 100, W. D., 1918.

Private, first class, sanitary detachment, 306th Infantry, 77th Division.
This soldier showed extraordinary heroism and devotion to duty by attending to wounded without rest or relief, under heavy shellfire, until he fell unconscious from exhaustion. He remained at work for two days and a half under circumstances which called for the greatest determination and courage.

EVENSON, ELMER (284813)..........
Near Juvigny, France, Aug. 30, 1918.
R—Elkhorn, Wis.
B—Elkhorn, Wis.
G. O. No. 66, W. D., 1919.

Sergeant, Headquarters Company, 128th Infantry, 32d Division.
Sergeant Evenson voluntarily made five trips in front of the line under intense machine-gun and artillery fire and brought in five wounded men who had been left in an exposed position after a withdrawal of the line.

EVERHART, CHESTER H. (2018995)_____
Near Bolshieozerki, Russia, Apr. 2, 1919.
R—Detroit, Mich.
B—Detroit, Mich.
G. O. No. 16, W. D., 1920.

Private, 337th Ambulance Company, 310th Sanitary Train, 85th Division (Detachment in north Russia).
Private Everhart went forward some 200 yards in front of our lines into an area swept by artillery and machine-gun fire to assist in bringing a wounded man to a place of safety.

EVERSOLE, OLAY (1543587)_____
Near Cierges, France, Sept. 27, 1918.
R—Dayton, Ohio.
B—Dayton, Ohio.
G. O. No. 27, W. D., 1922.

Private, medical detachment, 148th Infantry, 37th Division.
During the advance of Company K, 148th Infantry, Private Eversole frequently exposed himself to great danger by carrying a number of wounded men through heavy machine-gun fire from an open field to a place of safety.

*EVERSON, LUDWIG L_____
In the Argonne Forest, France, Sept. 26, 1918.
R—Joplin, Mo.
B—Webb City, Mo.
G. O. No. 59, W. D., 1919.

First lieutenant, 129th Machine Gun Battalion, 35th Division.
Meeting with stubborn resistance from the enemy, he reorganized scattered personnel from other units, attached them to his platoon, and continued to advance. When mortally wounded, he refused all assistance, ordering his detachment forward.
Posthumously awarded. Medal presented to widow, Mrs. Mabel Everson.

*EYLER, WILLIAM H_____
Northeast of Chateau-Thierry, France, Aug. 2, 1918.
R—Paulding, Ohio.
B—Chillicothe, Ohio.
G. O. No. 132, W. D., 1918.

First lieutenant, 166th Infantry, 42d Division.
Never faltering in his advance during the attack on the hill commanding Marcuil-en-Dole, and disregarding all personal danger, he led his company forward through heavy fire until he was killed.
Posthumously awarded. Medal presented to widow, Mrs. William H. Eyler.

FAATZ, STEPHEN R. (2302308)_____
Near St. Gilles, south of Fismes, France, Aug. 4, 1918.
R—Milwaukee, Wis.
B—Hampton, Iowa.
G. O. No. 147, W. D., 1918.

Private, Battery A, 120th Field Artillery, 32d Division.
When the men of his battery position had been ordered to shelter on account of enemy shelling, he, in company with two other men, rescued a French soldier from drowning in a stream. The act was performed while the valley was filled with mustard gas.

FAGA, WILLIAM HENRY_____
Near Vierzy, France, July 19, 1918.
R—Chicago, Ill.
B—Adair, Iowa.
G. O. No. 116, W. D., 1919.

Sergeant, 76th Company, 6th Regiment, U. S. Marine Corps, 2d Division.
He attacked and captured a machine gun that was inflicting losses on the American lines. In addition, he volunteered and successfully delivered messages of great importance to his battalion commander through a machine-gun and artillery barrage.
Oak-leaf cluster.
Sergeant Faga was also awarded an oak-leaf cluster, to be worn with the distinguished-service cross, for the following act of extraordinary heroism in action near Ardennes, France, Nov. 1, 1918: Just as he was to be evacuated to the rear because of severe wounds he saw that his platoon was without a commander, both leaders having become casualties. Returning to the lines, he assumed command for the remainder of the attack, successfully accomplishing his mission.

FAHEY, JOSEPH H. (2414738)_____
Near Vieville-en-Haye, France, Sept. 26, 1918.
R—Elizabeth, N. J.
B—Medford, Mass.
G. O. No 26, W. D., 1919.

Sergeant, Company B, 311th Infantry, 78th Division.
He made three attempts to silence several machine guns which had held up his platoon. He retired only after he had been wounded and his companions killed or wounded.

FAIR, HAROLD I_____
In the Belleau Bois, north of Verdun, France, Oct. 25, 1918.
R—Brooklyn, N. Y.
B—Staten Island, N. Y.
G. O. No. 39, W. D., 1920.

First lieutenant, 101st Infantry, 26th Division.
Lieutenant Fair, with a noncommissioned officer, while in advance of our line, encountered an enemy patrol engaged in removing wounded Americans. They attacked and succeeded in putting the enemy to flight. The noncommissioned officer was sent back for reinforcements, but Lieutenant Fair continued on and captured an enemy officer. Later, with the assistance of others, he removed to safety all the wounded.

FAIRALL, GEORGE S. (1246726)_____
Near Fismes, France, Aug. 7, 1918.
R—Steelton, Pa.
B—Iowa City, Iowa.
G. O. No. 11, W. D., 1921.

Sergeant, Machine Gun Company, 112th Infantry, 28th Division.
While in command of a section of machine guns, Sergeant Fairall made a reconnaissance in advance of his gun position. Encountering a party of the enemy, he kept them engaged until he was wounded in the left arm. He refused immediate evacuation and later guided a company of infantry to a position in the front line in time to avoid a serious situation. He continued to act as runner for his organization until forced to be evacuated on account of his wounds.

FAISON, JAMES K. (1312984)_____
Near Vaux-Andigny, France, Oct. 13–17, 1918.
R—Bennettsville, S. C.
B—Mount Olive, N. C.
G. O. No. 133, W. D., 1918.

Private, medical detachment, 118th Infantry, 30th Division.
During the advance in the vicinity of Vaux-Andigny Private James K. Faison for four days and four nights worked unceasingly dressing the wounded and giving them water. On five different occasions he went out over ground swept by enemy shell and machine-gun fire to rescue the wounded, at times within 100 yards and in direct view of the enemy positions.

FALKINBURG, EDWIN A. (2411496)_____
Near Chevieres, France, Oct. 21, 1918.
R—Barnegat, N. J.
B—Barnegat, N. J.
G. O. No. 15, W. D., 1919.

Private, medical detachment, 311th Infantry, 78th Division.
He remained on duty continuously for four hours, administering aid to wounded men under heavy shellfire. Finding that he could not properly work while wearing his gas mask, he removed it, although many gas shells were bursting in his vicinity. After being gassed he continued to work for an hour until all the wounded were attended.

FALLAW, THOMAS H............
   Near Landres-et-St. Georges, France,
     Oct. 16, 1918.
   R—Opelika, Ala.
   B—Waverly, Ala.
   G. O. No. 131, W. D., 1918.

Captain, 167th Infantry, 42d Division.
In the attack on the Cote-de-Chatillon, seeing that the entire advance was being held up in an open field by heavy machine-gun fire from the edge of the woods, he personally organized a detachment and led it in a rush on the woods under heavy fire, making a daring and vigorous attack on the enemy machine-gun nests, clearing the edge of the woods, capturing prisoners, and inflicting severe losses on the enemy. Through this gallant act he gained the final objective with a minimum loss to his command and set an inspiring example of disregard for personal safety and devotion to duty.

FALLS, ROY NISEL (1286789)...........
   Service rendered under the name,
   Nisel Rafalsky.

Sergeant, sanitary detachment, 115th Infantry, 29th Division.

FANNIN, OLIVER W............
   At Nogales, Ariz., Aug. 27, 1918.
   R—Colorado, Tex.
   B—Winters, Tex.
   G. O. No. 44, W. D., 1919.

First lieutenant, 35th Infantry, 18th Division.
In an engagement with hostile Mexicans at Nogales, Ariz., Aug. 27, 1918, while commanding the guard, 35th U. S. Infantry.

FARBER, LLOYD (2175210)...........
   Near Bantheville, France, Nov. 2,
     1918.
   R—Selden, Kans.
   B—Hoxie, Kans.
   G. O. No. 37, W. D., 1919.

Corporal, Headquarters Company, 353d Infantry, 89th Division.
He displayed exceptional bravery in three times carrying messages through severe artillery barrages.

FARKAS, JOE F. (50381)...........
   Near Thiaucourt, France, Sept. 12,
     1918.
   R—Bradley, Ohio.
   B—Hungary.
   G. O. No. 126, W. D., 1919.

Sergeant, Company E, 23d Infantry, 2d Division.
As Sergeant Farkas was leading his platoon in an attack on a machine-gun nest, a bullet struck him in the eye, destroying its sight. Despite this agonizing wound he continued in advance and put the enemy machine-gun nest out of action, refusing to seek medical aid until the success of his platoon's mission had been assured, more than an hour and a half later

FARMER, JAMES B. (1289744)...........
   Near Soissons, France, Aug. 30, 1918,
     and in the Argonne sector, France,
     Oct. 8, 1918.
   R—Radford, Va.
   B—Pulaski, Va.
   G. O. No. 98, W. D., 1919.

Corporal, Company L, 128th Infantry, 32d Division.
During the attack near Soissons, on Aug. 30, Corporal Farmer led his squad through a downpour of shell and machine-gun fire. After the attack he worked in the face of machine-gun fire, assisting and carrying the wounded from the field. On October 8, while on a reconnoitering party, three of his men were killed and two wounded. He carried the two wounded men back to safety in the face of intense machine-gun fire.

FARMER, WILLIAM R............

   Near Blanc Mont Ridge, France,
     Oct. 4, 1918.
   R—Pittsburgh, Pa.
   B—New York, N. Y.
   G. O. No. 87, W. D., 1919.

Secretary Y. M. C. A., attached to 3d Battalion, 5th Regiment, U. S. Marine Corps, 2d Division.
Secretary Farmer voluntarily established an advanced dressing station under enemy machine-gun and artillery fire. He continued to render first aid until forced back by a threatened counterattack, at which time he personally assisted two seriously wounded men to the first-aid station, then returning to the line and remaining with the unit until it was relieved.

°FARNSWORTH, THOMAS H............
   Near Thiaucourt, France, Sept. 13,
     1918.
   R—Washington, D. C.
   B—Washington, D. C.
   G. O. No. 126, W. D., 1919.

First lieutenant, pilot, 96th Aero Squadron, Air Service.
After being badly wounded Lieutenant Farnsworth shot down an enemy plane and made a perfect landing. His first thoughts upon landing were for his observer, who had been thrown from his plane during the action. Lieutenant Farnsworth died shortly after being taken to the aid station. Posthumously awarded. Medal presented to father, John Farnsworth.

FARRANT, OLIVER C............
   In the attack on Tigny, France,
     July 19, 1918.
   R—St. Paul, Minn.
   B—Baltimore, Md.
   G. O. No. 132, W. D., 1918.

Sergeant 96th Company, 6th Regiment, U. S. Marine Corps, 2d Division.
He led his section with remarkable coolness and courage in the advance under heavy fire of machine guns and artillery. He was severely wounded but continued to advance, thereby setting such an inspiring example to the men of his section that they routed the enemy.

FARRELL, EDWARD J. (63636)...........
   On the Monte des Tombes position,
     France, Feb. 28, 1918.
   R—Hartford, Conn.
   B—Hartford, Conn.
   G. O. No. 126, W. D., 1919.

Private, Company A, 102d Infantry, 26th Division.
On the night of Feb. 28, 1918, while under heavy barrage fire on the Monte des Tombes position, France, this soldier twice ran through the barrage to assist a comrade who had been wounded near him in a trench and assisted in carrying the man back to a dugout where first aid could be rendered.

°FARRELL, JOHN J. (2843306)...........
   Near Remonville, France, Nov. 1,
     1918.
   R—Put in Bay, Mich.
   B—Paterson, N. J.
   G. O. No. 32, W. D., 1919.

Private, Company B, 354th Infantry, 89th Division.
When the combat group of which he was a member was held up by machine-gun fire of the enemy, Private Farrell left the group and, crawling around to the flank of the nest, charged with his bayonet. The enemy surrendered and his comrades took the gun, after which the advance continued. Private Farrell was so seriously wounded during the combat that he died before he could be removed from the field.
Posthumously awarded. Medal presented to grandfather, Frank Delvin.

FARRELL, THOMAS F............
   At Bois-de-Money, France, Oct. 8-9,
     1918.
   R—Troy, N. Y.
   B—Brunswick, N. Y.
   G. O. No. 15, W. D., 1923.

Major, 1st Engineers, 1st Division.
On October 8 when ordered to take and hold Hill 269, which was strongly held by enemy forces, Major Farrell with great skill and with undaunted courage and determination led his battalion to the attack, seized and held this vital point despite the fact that he was attacked by greatly superior numbers on three sides and nearly surrounded by strong enemy forces who showed extraordinary determination to regain this highly important position. He held the hill until reinforcements could reach him after darkness had fallen on Oct. 9, 1918. His fearless leadership, utter disregard for his own safety, and complete devotion to duty raised the morale of his battalion to a high pitch and inspired them to acts of great endeavor.

FARRELL, WILLIAM J............
  At Seicheprey, France, Apr. 20, 1918.
  R—Dorchester, Mass.
  B—Boston, Mass.
  G. O. No. 49, W. D., 1922.

First lieutenant, chaplain, 104th Infantry, 26th Division.
With great gallantry and with utter disregard for his own danger, he personally conducted an ambulance from the battalion command post to the position of a supporting battery, where he assisted in the evacuation of the wounded. At Ville-devant-Chaumont, France, Nov. 9, 1918, when informed that one of the men of his battalion had been mortally wounded, Chaplain Farrell, in spite of extremely heavy artillery and flanking machine-gun fire, made his way by running and crawling from shell hole to shell hole until he reached the dying soldier to whom he gave the last rites of his church and with whom he remained until the soldier died.

FARRINGTON, DELBERT (317910)........
  At Novitskaya, Siberia, July 2, 1919.
  R—Greensboro, N. C.
  B—Rockingham County, N. C.
  G. O. No. 133, W. D., 1919.

Sergeant, Company M, 31st Infantry.
After his platoon commander was severely wounded, he assumed command of the platoon and led it in such a skillful manner as to gain superiority of fire and drive the enemy from his position without further loss to the platoon.

*FARST, LEROY (1935403)...............
  Near Cierges, France, Oct. 8, 1918.
  R—New Madison, Ohio.
  B—Eldorado, Ohio.
  G. O. No. 32, W. D., 1919.

Private, Company K, 38th Infantry, 3d Division.
Making his way forward under unusually heavy shellfire, he crawled to within 75 yards of the enemy's lines to bring in his wounded platoon commander. In attempting this heroic mission he was killed.
Posthumously awarded. Medal presented to mother, Mrs. Jennie Farst.

*FARWELL, GEORGE W..............
  Near Gesnes, France, Sept. 28–29, 1918.
  R—Los Angeles, Calif.
  B—Tower City, N. Dak.
  G. O. No. 20, W. D., 1919.

Major, 361st Infantry, 91st Division.
He displayed exceptional personal bravery in leading his command to the capture of enemy positions near Gesnes, France, Sept. 28–29, 1918. In each of these actions his troops were subjected to heavy artillery bombardment and machine-gun fire, but due to his coolness and the inspiration of his personal leadership and bravery his battalion in each instance captured and held the positions attacked.
Posthumously awarded. Medal presented to father, G. D. Farwell.

FARWELL, WREY GILMORE............
  In the Bois-de-Belleau, France, June 6, 1918.
  R—Brooklyn, N. Y.
  B—Norfolk, Va.
  G. O. No. 7, W. D., 1919.

Lieutenant commander, Medical Corps, U. S. Navy, attached to 6th Regiment, U. S. Marine Corps, 2d Division.
He voluntarily exceeded the demand of duty by personally supervising the evacuation of his wounded commanding officer across a field under fire of machine guns and snipers.

FAY, JOHN H............
  At Chateau-Thierry, France, June 6, 1918.
  R—Philadelphia, Pa.
  B—Philadelphia, Pa.
  G. O. No. 110, W. D., 1918.

Captain, 8th Machine Gun Company, 5th Regiment, U. S. Marine Corps, 2d Division.
He displayed extraordinary heroism in the disposition of his machine gun under particularly difficult conditions. Opposed by superior forces, his utter indifference to personal danger furnished an example which inspired his men to success.

FAZLER, ROBERT M. (1751332)........
  In Grand Pre, France, Oct. 22, 1918.
  R—Newark, N. J.
  B—Newark, N. J.
  G. O. No. 44, W. D., 1919.

Private, first class, Company L, 312th Infantry, 78th Division.
After delivering a message to his battalion headquarters, and while passing through the main street of Grand Pre, which was being enfiladed by a terrific enemy fire, Private Fazler went to the rescue of a comrade who had fallen under the direct fire and observation of the enemy, notwithstanding the warning that to go to the man's aid would mean certain death to himself. He gave the wounded man first-aid treatment and successfully carried him to a place of safety.

FEATHERS, EARL H. (554567)...........
  Near Bois-de-Cunel, France, Oct. 12–15, 1918.
  R—Dover Plains, N. Y.
  B—Berlin, N. Y.
  G. O. No. 26, W. D., 1919.

Sergeant, medical detachment, 9th Machine Gun Battalion, 3d Division.
Sergeant Feathers, with utter disregard of his personal safety, voluntarily made 5 trips into no man's land under heavy artillery and machine-gun fire, bringing in wounded. On another occasion he went through a barrage in order to replenish the front-line medical supply.

FEB, ELMER E. (57565)...............
  Near Soissons, France, July 19, 1918.
  R—St. Louis, Mo.
  B—St. Louis, Mo.
  G. O. No. 117, W. D., 1918.

Private, Company F, 28th Infantry, 1st Division.
He advanced on machine-gun snipers on the Paris-Soissons road, showing exceptional bravery by reaching the machine gun and killing the snipers with hand grenades and automatic rifle.

FECHET, D'ALARY............
  Near Vierzy, France, July 18, 1918.
  R—Jacksonville, Fla.
  B—Washington, D. C.
  G. O. No. 3, W. D., 1921.

Major, 23d Infantry, 2d Division.
He personally led his battalion in the attack, during which he was severely wounded in the neck by a shell fragment. After receiving first aid he rejoined his battalion and remained in command throughout the operation. His energy and personal heroism were material factors in the successful attack made by his battalion on the strongly fortified town of Vierzy.

FEEGAL, JOHN R............
  Near Vaux, France, July 20, 1918.
  R—Meriden, Conn.
  B—Brooklyn, N. Y.
  G. O. No. 125, W. D., 1918.

First lieutenant, 102d Infantry, 26th Division.
When his company was held up by a machine gun he went ahead under fire alone and killed the machine-gun operator with his pistol, thereby enabling his company to continue the advance.

*FEGELEY, BYRON S. (543326)..........
  Near Cierges, France, Oct. 1, 1918.
  R—Hamburg, N. J.
  B—Sayre, Pa.
  G. O. No. 16, W. D., 1920.

Private, medical detachment, 7th Infantry, 3d Division.
After having been rendered unconscious for two hours by an exploding shell he, upon recovering consciousness, continued to administer first aid to the wounded. Later, while carrying a severely wounded man to safety, he was fatally wounded by a shell fragment.
Posthumously awarded. Medal presented to father, Rev. W. O. Fegeley.

FEIGLE, WILLIAM M.
Near Beaumont, France, Nov. 10, 1918.
R—Galveston, Tex.
B—Galveston, Tex.
G. O. No. 81, W. D., 1919.

Sergeant, Headquarters Company, 5th Regiment, U. S. Marine Corps, 2d Division.
While an ammunition train was passing through a town, one of the trucks was struck by a shell and set on fire. The blazing truck was abandoned and, knowing that it soon would explode and kill all those in the vicinity, Sergeant Feigle jumped on the truck and drove it to the outskirts of the town, thereby saving the lives of at least 35 people.

*FEINBERG, HIRCHE J. (2354893)
Near Cunel, France, Oct. 6, 1918.
R—New York, N. Y.
B—Russia.
G. O. No. 37, W. D., 1919.

Private, medical detachment, 4th Infantry, 3d Division.
After all the officers and noncommissioned officers of the company to which he was attached had become casualties, with the exception of the company commander, Private Feinberg voluntarily assisted in reorganizing the company and holding the men together, continually going up and down along the line under the severe enemy fire until he was killed.
Posthumously awarded. Medal presented to widow, Mrs. Senie Feinberg.

FELAND, LOGAN.
At Bois-de-Belleau, France, June 6-14, 1918.
R—Hopkinsville, Ky.
B—Kentucky.
G. O. No. 99, W. D., 1918.
Distinguished-service medal also awarded.

Colonel, 5th Regiment, U. S. Marine Corps, 2d Division.
During the operations at Bois-de-Belleau, June 6-14, 1918, he distinguished himself by his energy, courage, and disregard for personal safety in voluntarily leading troops into action through heavy artillery and machine-gun fire. His efforts contributed largely to our successes at this point.

FELITTO, CARMINE (1708328)
Near Binarville, France, Oct. 4, 1918.
R—New York, N. Y.
B—New York, N. Y.
G. O. No. 32, W. D., 1919.

Corporal, Company D 308th Infantry, 77th Division.
When his platoon leader and a small detachment of men were surrounded by the enemy and subjected to a terrific artillery and machine-gun fire, Corporal Felitto volunteered and brought a message from his lieutenant to the company commander, bravely making his way through the enemy's lines, despite the fact that he had seen other men killed while making the attempt. He brought the first message from the detachment, which had been cut off from the company for 18 hours.

FENOUILLET, CARL (1210528)
Near St. Souplet, France, Oct. 18, 1918.
R—Brooklyn, N. Y.
B—New York, N. Y.
G. O. No. 11, W. D., 1921.

Corporal, Company E, 107th Infantry, 27th Division.
Corporal Fenouillet, after his commanding officer and four sergeants had become casualties, took command of the company, rallied it by his personal heroism and exposure to fire and gallantly led it in the attack.

*FENTON, NEAL D. (2062)
Near Verdun, France, Oct. 6, 1918.
R—Hutchinson, Kans.
B—Omaha, Nebr.
G. O. No. 44, W. D., 1919.

Corporal, medical detachment, 26th Infantry, 1st Division.
He volunteered and went forward, in the face of direct concentrated machine-gun fire, to the rescue of a wounded man. He was killed while attempting this rescue.
Posthumously awarded. Medal presented to father, Clarence F. Fenton.

FENWICK, EDWARD G. (8078)
Near Abeele, France, June 7, 1918.
R—East Falls Church, Va.
B—Washington, D. C.
G. O. No. 109, W. D., 1918.

Private, section No. 517, Ambulance Service, with French Army.
On the morning of June 7, 1918, near Abeele, France, after having suffered severe injuries, with no thought of himself, he walked some distance to camp and procured assistance for the three wounded soldiers whom he had removed from his car after it had been struck by a high-explosive shell.

FERENTCHAK, MARTIN (48032)
North of Baulny, France, Oct. 1, 1918.
R—Chicago, Ill.
B—Austria.
G. O. No. 53, W. D., 1920.

Sergeant, Company K, 18th Infantry, 1st Division.
After the patrol leader had been killed, Sergeant Ferentchak assumed command of a patrol which had advanced about a kilometer in front of our lines. This patrol encountered the enemy and was exposed to heavy artillery and machine-gun fire. In spite of enemy fire, Sergeant Ferentchak carefully observed and sketched the enemy positions. The patrol suffered heavy casualties, but this noncommissioned officer led the remainder of his group back to our lines and furnished information which proved of great value to his commander in the attack which followed a few days later.

FERGUSON, ARTHUR M.
At Calumpit, Philippine Islands, Apr. 26, 1899.
R—Burlington, Kans.
B—Coffey County, Kans.
G. O. No. 126, W. D., 1919.
Medal of honor also awarded.

Corporal, Company E, 20th Kansas Volunteer Infantry.
Against an armed enemy at Calumpit, Philippine Islands, Apr. 26, 1899, while serving as corporal, 20th Kansas Volunteer Infantry. At the imminent risk of his life he voluntarily crawled through a network of iron beams underneath a bridge and, inch by inch, worked his way hand over hand across the bridge until he was underneath an insurgent's outpost, obtaining a complete description of the condition of the bridge.

FERGUSON, DOUGALD (278102)
At Cierges, northeast of Chateau-Thierry, France, Aug. 1, 1918.
R—Grand Rapids, Mich.
B—Belding, Mich.
G. O. No. 66, W. D., 1919.

Sergeant, Machine Gun Company, 126th Infantry, 32d Division.
When the infantry on his right was held up by fire from an enemy machine gun at Cierges, northeast of Chateau-Thierry, France, Aug. 1, 1918, he seized a rifle and rushed around the flank of the enemy's position, bayoneting 2 of the machine-gun crew and shooting the third, thus enabling the infantry to advance.

FERGUSON, EUGENE R. (1448642)
Near Montrebeau Woods, France, Sept. 29, 1918.
R—Minneapolis, Kans.
B—Minneapolis, Kans.
G. O. No. 78, W. D., 1919.

Corporal, Company G, 137th Infantry, 35th Division.
Seeing a comrade lying severely wounded and unable to reach our lines, Corporal Ferguson left his own shelter and, in the face of intense machine-gun and artillery fire, went out into the open and carried his comrade back, being himself severely wounded before he regained our lines.

FERGUSON, GEORGE H.
Near Romagne, France, Oct. 14, 1918.
R—Newark, Del.
B—Newark, Del.
G. O. No. 44, W. D., 1919.

Second lieutenant, 6th Infantry, 5th Division.
After being painfully wounded in the leg early in the attack, he continued forward, leading his platoon through an unusually heavy artillery and machine-gun fire. Later he left a shell hole in which he had taken refuge and administered first aid to soldiers who had fallen near him until forced to abandon this work because of exhaustion.

FERGUSON, GEORGE L. (1869671)
Near Bois-de-Manheulles, France, Nov. 9, 1918.
R—Elyria, Ohio.
B—Toledo, Ohio.
G. O. No. 32, W. D., 1919.

Sergeant, first class, 306th Field Signal Battalion, 81st Division.
While making a reconnaissance of the enemy's advanced positions he alone routed a German machine-gun squad who were setting up a machine gun along a road over which our troops were advancing. He continued the reconnaissance with the battalion commander until the latter was fatally wounded, and then assisted him to a dressing station, being subjected to heavy machine-gun fire the entire time.

*FERGUSON, HAROLD (1210566)
Near Ronssoy) France, Sept. 29, 1918.
R—Thiells, N. Y.
B—Thiells, N. Y.
G. O. No. 19, W. D., 1920.

Corporal, Company E, 107th Infantry, 27th Division.
During the attack on the Hindenburg line Corporal Ferguson assumed command of a platoon after the officers and senior noncommissioned officers had become casualties. He rapidly reorganized the platoon and fearlessly led it into effective combat through heavy shell and machine-gun fire. He was later killed after reaching his objective.
Posthumously awarded. Medal presented to father, John Ferguson.

FERGUSON, JOHN E. (369915)
Near Samogneux, France, Oct. 8–29, 1918.
R—New York, N. Y.
B—El Reno, Okla.
G. O. No. 37, W. D., 1919.

Corporal, Company H, 115th Infantry, 29th Division.
Throughout the offensive east of the Meuse, near Samogneux, he displayed exceptional bravery and endurance as a battalion runner, repeatedly carrying important messages through intense artillery and machine-gun fire after other runners had been killed in traversing the same routes. On numerous occasions he alone was responsible for the maintenance of both forward and rear liaison.

*FERGUSON, JOSEPH SIMPSON
Near Varennes, France, Sept. 26, 1918.
R—Philadelphia, Pa.
B—Philadelphia, Pa.
G. O. No. 95, W. D., 1919.

First lieutenant, 110th Infantry, 28th Division.
Lieutenant Ferguson was engaged in "mopping up" the town when he discovered a considerable force of the enemy coming from dugouts and taking up positions in the rear of the front line with machine guns and antitank guns, from which they fired upon the front line and almost immediately destroyed five tanks. Realizing at once the heavy casualties that might be caused to our troops and the impossibility of further advance by the front line with this enemy force in its rear, Lieutenant Ferguson, on his own initiative, assembled a portion of his men, skillfully conducted them to a point on the eastern edge of the town, thence toward the west, and captured the entire enemy forces, consisting of about 20 officers and over 100 men, thus enabling the front line to continue its advance.
Oak-leaf cluster.
For the following act of extraordinary heroism in action near Courmont, France, July 30, 1918, Lieutenant Ferguson is awarded an oak-leaf cluster to be worn with the distinguished-service cross: As he was reforming his platoon in a sunken road after a withdrawal he went forward under enemy machine-gun and sniper fire and carried to shelter a severely wounded soldier.
Posthumously awarded. Medal presented to widow, Mrs. Elizabeth Ferguson,

FERGUSON, LLOYD L. (736952)
Near Brandeville, France, Nov. 8–9, 1918.
R—Chatfield, Minn.
B—Chatfield, Minn.
G. O. No. 37, W. D., 1919.

Sergeant, Company L, 11th Infantry, 5th Division.
After leading his platoon against a superior number of the enemy he was wounded and taken to an aid station, where he was ordered evacuated. He refused and rejoined his company the following day and was again wounded while advancing at the head of his platoon.

FERGUSON, WILLIAM J.
Near Villemontry, France, Nov. 10, 1918.
R—Cleveland, Ohio.
B—Cleveland, Ohio.
G. O. No. 32, W. D., 1919.

Corporal, 17th Company, 5th Regiment, U. S. Marine Corps, 2d Division.
Corporal Ferguson and a companion went out ahead of the line and silenced a machine gun which threatened to hold up the advance of his company.

FERRELL, WILLIAM M. (794616)
Near Verdun, France, Nov. 5, 1918.
R—Ashland, Miss.
B—Ashland, Miss.
G. O. No. 46, W. D., 1919.

Private, first class, medical detachment, 11th Infantry, 5th Division.
While rendering first aid under terrific fire he was severely wounded. In spite of his injury, he continued to dress the wounds of a comrade, after which he helped him back to the first-aid station. Even after returning he displayed more interest in the wounds of another than he did in caring for his own wounds.

FERRENBACH, LEO C.
Near Ansauville, France, July 22, 1918.
R—St. Louis, Mo.
B—St. Louis, Mo.
G. O. No. 7, W. D., 1919.

First lieutenant, 14th Balloon Company, Air Service.
Acting as balloon observer, he was conducting an important surveillance of his sector when, at an altitude of 800 meters, successive attacks were made upon the balloon by enemy planes. This officer refused to leave his post and continued his work with strong enemy patrols hovering above him until one of the hostile machines dived and set fire to the balloon. After he had jumped in his parachute the burning balloon fell and barely missed him. He immediately reascended while enemy patrols were still in the vicinity.

FERRY, BRADFORD (1284862)
Near Montagne, France, Oct. 17, 1918.
R—Elkton, Md.
B—Elkton, Md.
G. O. No. 37, W. D., 1919.

Private, Company E, 115th Infantry, 29th Division.
After three men of his carrying party had been wounded and the others scattered by shellfire, Private Ferry organized a second detail, which he led through heavy shellfire to his company kitchen and returned after dark with rations for the men in the front line.

FESSELMEYER, WILLIAM T.
Near Grand Ballois Farm, France,
July 14-15, 1918.
R—New York, N. Y.
B—New York, N. Y.
G. O. No. 32, W. D., 1919.

Second lieutenant, 4th Infantry, 3d Division.
On the night of July 14-15, 1918, he continually exposed himself to heavy gas and shell fire while caring for the wounded until he was overcome by gas and exhaustion.

FIECHTER, WALTER.
Near St. Agnan, France, July 16,
1918.
R—Philadelphia, Pa.
B—Philadelphia, Pa.
G. O. No 71, W. D., 1919.

First lieutenant, 109th Infantry, 28th Division.
Although painfully wounded at the start of an attack, he refused to be evacuated, but continued to lead his platoon forward until ordered to withdraw. His courage was an inspiration to his command.

FIELD, JOHN HENRY, Jr.
Near Champigneulles, France, Oct.
16, 1918.
R—Nashua, N. H.
B—Nashua, N. H.
G. O. No. 56, W. D., 1922.

First lieutenant, 309th Infantry, 78th Division.
In the attack on the Bois-de-Loges, when his company commander was severely wounded, Lieutenant Field took command of the company and, although he himself was severely wounded, continued to lead his men with great coolness and aggressiveness, being the only officer with the company. In the face of extremely heavy machine-gun and artillery fire, he conducted the advance so skillfully as to capture over 60 prisoners and 5 machine guns. Although his wound had so disabled him that he had great difficulty in adjusting his gas mask, in spite of continued periods of gas shelling, he refused to be evacuated until the following morning, when his battalion was relieved by another organization.

*FIELDS, JAMES O. (2162649)
At Fresnes-en-Woevre, France, Nov.
10, 1918.
R—Livermore, Iowa.
B—Loogocite, Ind.
G. O. No. 35, W. D., 1919.

Corporal, Company D, 130th Infantry, 33d Division.
With utter disregard of his personal danger, he went in to an intense enemy barrage to rescue a wounded soldier. While accomplishing this heroic feat he was seriously wounded.
Posthumously awarded. Medal presented to mother, Mrs. Ida Fields.

FIELDS, WILLIAM E. (2390419)
Near Cunel, France, Oct. 14, 1918.
R—Uniontown, Pa.
B—Smithfield, Pa.
G. O. No. 27, W. D., 1920.

Sergeant, Company C, 14th Machine Gun Battalion, 5th Division.
Sergeant Fields placed his machine gun on a forward and exposed slope and opened an effective fire on the enemy position. The enemy concentrated machine-gun fire on his position, but by his fearless conduct he continued to fire, and succeeded in silencing 3 enemy machine guns, thus enabling the Infantry to advance. At Clery-le-Grande, Nov. 1, 1918, after being severely wounded, Sergeant Fields refused medical attention, directing the medical personnel to other wounded.

FIGGINS, CHARLES R. (2237567)
Near Eclisfontaine, France, Sept.
28, 1918.
R—Wasco, Calif.
B—Randolph, Nebr.
G. O. No. 37, W. D., 1919.

Corporal, Company K, 364th Infantry, 91st Division.
Responding to a call for volunteers, Corporal Figgins, with 5 others, advanced 400 yards beyond their front to bring in wounded comrades. They succeeded in rescuing 7 of their men, also in bringing in the dead body of a lieutenant while exposed to terrific machine-gun fire.

FIGEH, PETER (1746496)
Near Vieville-en-Haye, France, Sept.
23-24, 1918.
R—Perth Amboy, N. J.
B—Perth Amboy, N. J.
G. O. No. 23, W. D., 1919.

Private, Company D, 311th Infantry, 78th Division.
On the night of Sept. 23-24, Private Figen repeatedly carried messages between his company and battalion headquarters through a heavy barrage, until completely exhausted. On the morning of Sept. 26, he volunteered and carried an important message to battalion headquarters through a heavy machine-gun fire.

°FILLINGIN, LINIE G. (97502)
Near Pexonne, France, Mar. 5, 1918.
R—Coffee, Ala.
B—Coffee, Ala.
G. O. No. 58, W. D., 1918.

Private, Company H, 167th Infantry, 42d Division.
In the action of Mar. 5, 1918, near Pexonne, France, he displayed courageous devotion to duty by remaining at his post under heavy fire and after being wounded. Died from wounds received Mar. 6, 1918.
Posthumously awarded. Medal presented to father, G. L. Fillingin.

FILLYAW, WALTER J. (2340137)
Near Cunel, France, Oct. 5, 1918.
R—Fayetteville, N. C.
B—Fayetteville, N. C.
G. O. No. 89, W. D., 1919.

Private, medical detachment, 4th Infantry, 3d Division.
Having been wounded and ordered to the rear, Private Fillyaw nevertheless continued to administer first-aid treatment to other wounded men under constant shell fire until he was wounded a second time, when he was evacuated, despite his protests.

FINCH, ROBERT L.
Near Bussy Farm, France, Sept.
28, 1918.
R—Tempe, Ariz.
B—Tempe, Ariz.
G. O. No. 37, W. D., 1919.

First lieutenant, adjutant, 3d Battalion, 372d Infantry, 93d Division.
He voluntarily led a portion of the first attacking wave over the enemy's position in the face of intense artillery and machine-gun fire. Later he, in company with another officer, voluntarily advanced under heavy fire to the enemy's wire and cut an opening for the passage of our troops.

FINK, WILLIAM W. (1378728)
Near Remonville, France, Oct. 31,
1918.
R—Springfield, Ill.
B—Strasburg, Ill.
G. O. No. 37, W. D., 1919.

Private, Battery A, 124th Field Artillery, 33d Division.
Private Fink, a driver, was seriously wounded by shell fire while going forward to the front-line Infantry trenches with his platoon, but he remained at his post, refusing to seek medical attention until he fell from his horse exhausted.

FINKLE, BURR (91319)
Near Villers-sur-Fere, France, July
28, 1918
R—Middletown, N. Y.
B—Liberty, N. Y.
G. O. No. 108, W. D., 1918.

Private, Company K, 165th Infantry, 42d Division.
He saw six Germans about to make a prisoner of his corporal, who had been severely wounded in the ankle. He called a comrade and advanced on the Germans, killed two of them, took the other four prisoners, and returned with his corporal to our lines.

FINLEY, ARCHIE J. (263447).
Near Gesnes, France, Oct. 9, 1918.
R—Winters, Mich.
B—Marquette, Mich.
G. O. No. 78, W. D., 1919

Corporal, Company K, 125th Infantry, 32d Division.
In the fighting near Gesnes, Corporal Finley's platoon became isolated in a position far in advance of the rest of his company. In order to obtain liaison Corporal Finley twice voluntarily carried messages from his platoon to the remainder of the company. In order to do this it was necessary to cross two stretches which were entirely open to the enemy's fire. In spite of this, Corporal Finley carried out his mission successfully and completed plans for the withdrawal of the platoon that night. During the withdrawal of the platoon Corporal Finley repeatedly returned to search for wounded men whom he carried on his back to a place of safety.

*FINN, HENRY (2340153).
Near Les Franquettes Farm, France, July 23, 1918.
R—Portal, N. Dak.
B—Lesueur Center, Minn.
G. O. No. 32, W. D., 1919.

Private, medical detachment, 4th Infantry, 3d Division.
Despite the severe bombardment of machine guns, he went into an open field to administer to wounded officers and men. He was killed while rendering aid to these men.
Posthumously awarded. Medal presented to father, Michael Finn.

FINN, JOHN J. (1204570).
East of Ronssoy, France, Sept. 29, 1918.
R—New Rochelle, N. Y.
B—Ireland.
G. O. No. 14, W. D., 1925.

Mechanic, Company G, 105th Infantry, 27th Division.
During the operations against the Hindenburg line he left shelter and went forward under heavy shell and machine-gun fire and rescued five wounded soldiers. While in the performance of this gallant act he and another soldier attacked an enemy dugout, killing two of the enemy and taking one prisoner. This courageous act set splendid example to all.

*FINNEGAN, JOHN J. (90455).
At Ourcq, France, July 28, 1918.
R—New York, N. Y.
B—Ireland.
G. O. No. 16, W. D., 1923.

Corporal, Company F, 165th Infantry, 42d Division.
While acting as scout and sniper he made his way across the Ourcq River, obtained valuable information of the enemy, and returned to his own lines, at all times under intense enemy machine-gun, rifle, artillery, and sniper fire. His company concentrating their fire upon enemy positions indicated by Corporal Finnegan, silenced enemy machine guns. He performed the same mission shortly thereafter, and while so engaged for the third time within a few hours, he was mortally wounded by enemy fire and died upon the field.
Posthumously awarded. Medal presented to sister, Miss M. Finnegan.

*FINNEGAN, ROBERT (56649).
At Cantigny, France, May 27–28, 1918.
R—Pittsburgh, Pa.
B—Ridgeway, Pa.
G. O. No. 99, W. D., 1918.

Corporal, Company A, 28th Infantry, 1st Division.
Although mortally wounded he concealed that fact, encouraged his men by his example of fortitude, and continued to fire his automatic rifle until he became exhausted.
Posthumously awarded. Medal presented to father, Hugh Finnegan.

*FINNERTY, BERNARD R. (90705).
Near Auberive, France, July 16, 1918.
R—New York, N. Y.
B—Ireland.
G. O. No. 99, W. D., 1918.

Sergeant, Company H, 165th Infantry, 42d Division.
He bravely attacked a group of the enemy without assistance in a bayou near Auberive, France, July 16, 1918, and drove them out, thereby saving his unit from surprise attack. While engaged i this courageous enterprise he was killed.
Posthumously awarded. Medal presented to father, Bernard Finnerty.

FINUCANE, PETER (1702065).
Near Bazoches, France, Aug. 14, 1918.
R—New York, N. Y.
B—Ireland.
G. O. No. 14, W. D., 1923.

Corporal, Company F, 306th Infantry, 77th Division.
He voluntarily exposed himself to intense enemy machine-gun and artillery fire, crawling forward, in company with four other men of his company, in search of their wounded company commander, who had f llen a short distance in front of his company's position. After a fruitless search for the wounded officer the patrol engaged the nearest enemy post and in a fight with hand grenades destroyed it. Corporal Finucane then assisted a wounded comrade to return to his own lines. The heroic conduct of Corporal Finucane greatly encouraged the men of his company, inciting them to heroic endeavor.

*FIORENTINO, ANTHONY (1241719).
Near Magneux, France, Aug. 25, 1918.
R—Rankin, Pa.
B—Rankin, Pa.
G. O. No. 95, W. D., 1919.

Private, Company L, 110th Infantry, 28th Division.
Volunteering to locate an enemy machine-gun nest which was inflicting heavy casualties on our forces, Private Fiorentino advanced alone and by drawing the enemy fire enabled his company to destroy the nest and continue the advance. In exposing himself to the hostile fire, this gallant soldier was fatally wounded and died on the field shortly afterwards.
Posthumously awarded. Medal presented to father, Frank Fiorentino.

FIORITO, DIONIGO (40897).
Near Tuilerie Farm, France, Nov. 4, 1918.
R—New York, N. Y.
B—Steelton, Pa.
G. O. No. 46, W. D., 1919.

Private, Company M, 9th Infantry, 2d Division.
After many of the stretcher bearers had become casualties, Private Fiorito, without assistance, carried many of his wounded comrades to the rear, through heavy machine-gun and shell fire.

*FISCHER, ROBERT McCAUGHIN.
At Chateau-Thierry, France, June 6, 1918.
R—New Ulm, Minn.
B—New Ulm, Minn.
G. O. No. 110, W. D., 1918.

Corporal, 20th Company, 5th Regiment, U. S. Marine Corps, 2d Division.
Killed in action at Chateau-Thierry, France, June 6, 1918, he gave the supreme proof of that extraordinary heroism which will serve as an example to hitherto untried troops.
Posthumously awarded. Medal presented to aunt, Miss Minna Fischer.

FISHER, AARON R.
Near Lesseau, France, Sept. 3, 1918.
R—Lyles, Ind.
B—Lyles, Ind.
G. O. No. 147, W. D., 1918.

Second lieutenant, 366th Infantry, 92d Division.
He showed exceptional bravery in action when his position was raided by a superior force of the enemy by directing his men and refusing to leave his position, although he was severely wounded. He and his men continued to fight the enemy until the latter were beaten off by a counterattack.

*FISHER, FRANK J._____ Near Beauclair, France, Nov. 3-4, 1918. R—Kansas City, Kans. B—Kansas City, Kans. G. O. No. 87, W. D., 1919.	Second lieutenant, 355th Infantry, 89th Division. Largely as a result of his personal bravery he advanced our front line a distance of 2 kilometers, going out in advance and capturing two German machine-guns and killing the crews. When his line was later halted by heavy fire, he exposed himself fearlessly in passing among his men to steady them, and direct the consolidation of the position they held. While so doing he was mortally wounded. Posthumously awarded. Medal presented to father, William M. Fisher.
FISHER, GEORGE F._____ In the vicinity of Xonville, France, Sept. 16, 1918. R—Chicago, Ill. B—Harrisburg, Pa. G. O. No. 35, W. D., 1920.	Captain, 49th Aero Squadron, Air Service. While flying alone Captain Fisher encountered two enemy biplanes which were attempting to cross our lines. He attacked them and shot down one of them. He then pursued the other, forcing it to go down out of control about 10 kilo- meters within the enemy lines.
FISHER, HAROLD (69834)_____ Near Riaville, France, Sept. 25, 1918. R—Houlton, Me. B—Oakfield, Me. G. O. No. 9, W. D., 1923.	Corporal, Headquarters Company, 103d Infantry, 26th Division. While a member of the party making a raid on Riaville, after the detachment had been ordered to retire upon completion of its mission, Corporal Fisher voluntarily went forward at great personal risk of his life through extremely heavy hostile machine-gun fire and succeeded in carrying back to his own lines a helpless wounded comrade. His act was one of great devotion to duty and was an inspiration to his comrades.
FISHER, RONALD E._____ At Peruka Utig's Cotta, Philippine Islands, May 3, 1905. R—Annapolis, Md. B—Croziesville, Pa. G. O. No. 3, W. D., 1922.	Second lieutenant, 14th Cavalry, U. S. Army. In a hand-to-hand fight in a trench in the jungle, Lieutenant Fisher killed three Moros and cleared the trench.
FISHER, RUSSELL S._____ Near Dun-sur-Meuse, France, Nov. 4-5, 1918. R—Chicago, Ill. B—Chicago, Ill. G. O. No. 37, W. D., 1919.	Captain, 61st Infantry, 5th Division. After all means of conveyance across the canal had been destroyed by enemy shell fire, Captain Fisher bravely called on his company to swim; he himself leading his men into the water. The successful accomplishment of this task made it possible for him to attack and capture the height on the east side of the river and drive out the enemy, who were holding the bridgehead.
*FISKE, NEWELL R._____ Near Fossoy, France, July 15, 1918. R—Cranford, N. J. B—West Hampton, N. Y. G. O. No. 44, W. D., 1919.	Captain, 7th Infantry, 3d Division. He fearlessly led his troops in a counterattack through intense artillery fire, inspiring his men by his gallant conduct. He lost his life while in a perform- ance of this act. Posthumously awarded. Medal presented to father, H. M. Fiske.
FITTS, WILLIAM T., Jr_____ Near Ribeauville, France, Oct. 17, 1918. R—Knoxville, Tenn. B—Keysville, Va. G. O. No. 3, W. D., 1921.	Second lieutenant, 117th Infantry, 30th Division. When a friendly tank appeared suddenly out of the fog and opened fire on his organization, Lieutenant Fitts rushed in front of the tank, signaling it to cease firing. He was wounded in the chest in the performance of this act, but remained with his organization until compelled to go to the rear due to weak- ness caused from loss of blood.
*FITZGERALD, HOWARD P. (73590)_____ At Bois Brule, near Apremont, France, Apr. 10, 1918. R—Springfield, Mass. B—Springfield, Mass. G. O. No. 88, W. D., 1918.	Private, Company L, 104th Infantry, 26th Division. He displayed coolness, courage, and the spirit of self-sacrifice in action of Apr. 10, 1918, digging out a buried comrade while under heavy fire, persisting in his work until he received a mortal wound, of which he died Apr. 11, 1918. Posthumously awarded. Medal presented to father, Edward Fitzgerald.
FITZGERALD, JAMES (552557)_____ During the battle of the Marne, France, July 19, 1918. R—Elmira, N. Y. B—Elmira, N. Y. G. O. No. 9, W. D., 1923.	Sergeant, Company M, 38th Infantry, 3d Division. Sergeant Fitzgerald repeatedly volunteered for patrol duty during the period of the Germans' occupation south of the Marne, and while a member of a reconnoitering patrol, his patrol was fired upon by machine guns from a Ger- man strong point. The patrol withdrew with the exception of Sergeant Fitz- gerald, who remained in the vicinity under a heavy fire and courageously continued his reconnaissance until his mission had been accomplished, after which he killed two Germans and returned with valuable information.
FITZGERALD, ROBERT J. (9632)_____ Near Soissons, France, Sept. 3, 1918. R—Ben Avon, Pa. B—Allegheny, Pa. G. O. No. 15, W. D., 1919.	Private, section No. 625, Ambulance Service. His ambulance broke down while he was on his way to an advance post along a road then under steady machine-gun fire. In spite of the enemy fire, he attempted to repair the machine, but was unable to do so. Securing assist- ance, he repaired it under fire in full view of the enemy and continued to make repeated trips with wounded through machine-gun and artillery bombard- ment.
FITZPATRICK, MICHAEL F. (91543)_____ Near Landres-et-St. Georges, France, Oct. 14, 1918. R—Brooklyn, N. Y. B—Brooklyn, N. Y. G. O. No. 32, W. D., 1919.	Sergeant, Company L, 165th Infantry, 42d Division. After his platoon commander had been killed, Sergeant Fitzpatrick took com- mand of the platoon. Although painfully wounded in the arm early in the engagement, and constantly exposed to intense machine-gun and artillery fire and gas, he remained at his post directing and encouraging his men until his platoon was relieved late that night, when he was evacuated.
FITZSIMMONS, THOMAS E. (88483)_____ Near Landres-et-St. Georges, France, Oct. 15, 1918. R—South Orange, N. J. B—South Orange, N. J. G. O. No. 44, W. D., 1919.	Sergeant, Headquarters Company, 165th Infantry, 42d Division. Running forward to a slope just above the enemy's wire, Sergeant Fitzsimmons, although exposing himself to direct fire of all kinds, was able to conduct such an effective fire with his trench mortars that a threatened enemy counter- attack was broken up.

FLAGG, DANIEL S. (1388595) _____ Private, Company M, 131st Infantry, 33d Division.
Near Gercourt, France, Sept. 26, 1918.
R—Chicago, Ill.
B—Louisiana, Mo.
G. O. No. 87, W. D., 1919.
With 3 other soldiers he crawled across an open field, subjected to heavy machine-gun and artillery fire, for 200 yards, to flank 3 enemy emplacements which were holding up the advance. This volunteer patrol captured the machine-gun positions, killing 7 and capturing 23 of the enemy.

FLANAGAN, GORDON D. (1833380) _____ Private, Battery A, 313th Field Artillery, 80th Division.
Near Dannevoux, France, Sept. 28, 1918.
R—Red Creek, W. Va.
B—Red Creek, W. Va.
G. O. No. 24, W. D., 1920.
The fieldpiece of which Private Flanagan was gunner was moved by hand to a position exposed to observation and enemy fire, in order to deliver direct fire upon the enemy guns near Vilosnes. The officer in charge was called away early in the engagement and Private Flanagan took command during the critical period and fired 260 rounds at enemy guns. In spite of the heavy enemy fire concentrated upon his position, he delivered a very effective fire on the enemy. He ceased to fire only after his gun had been struck by shell splinters from both front and rear.

FLANAGAN, HUGH M _____ Second lieutenant, 28th Infantry, 1st Division.
At Cutry, France, July 18, 1918.
R—Fond du Lac, Wis.
B—Fond du Lac, Wis.
G. O. No. 126, W. D., 1919.
Lieutenant Flanagan personally led his platoon in an attack on several machine-gun nests which were holding up the advance of the entire battalion. At times he was 50 yards in advance of his men. His action so inspired his command that the enemy strong point was destroyed and 100 prisoners were captured. He was severely wounded in the attack.

FLANNERY, HARRY E. (2192862) _____ Sergeant, Company D, 341st Machine Gun Battalion, 89th Division.
In the Bois-de-Bantheville, France, Nov. 1, 1918.
R—Hillsview, S. Dak.
B—Hillsview, S. Dak.
G. O. No. 37, W. D., 1919.
During heavy enemy shell fire he kept excellent control over his gun section, and when severely wounded refused to be cared for until other men received first aid.

FLANNERY, WALTER R _____ First lieutenant, 7th Infantry, 3d Division.
Near Sauvigny, France, June 3, 1918.
R—Pittsburgh, Pa.
B—Pittsburgh, Pa.
G. O. No. 99, W. D., 1918.
At great peril to his life on the night of June 3, 1918, he voluntarily swam the River Marne, near Sauvigny, France, and brought back a wounded French soldier, who, having escaped from his German captors, was unable to return to his own lines.

FLEESON, HOWARD T _____ Second lieutenant, observer, 12th Aero Squadron, Air Service.
In the St. Mihiel salient, France, Sept. 12, 1918.
R—Sterling, Kans.
B—Sterling, Kans.
G. O. No. 27, W. D., 1919.
He and 2d Lieut. Dogan H. Arthur, pilot, executed a difficult mission of infantry contact patrol, without protection of accompanying battle planes, on the first day of the St. Mihiel offensive. After being driven back twice by a patrol of 9 enemy planes, they courageously made a third attempt in the face of a third attack by the same planes, found the American lines, and after being shot down, but falling uninjured in friendly territory, communicated their valuable information to headquarters.
Oak-leaf cluster.
An oak-leaf cluster is awarded Lieutenant Fleeson for the following act: On Oct. 30, 1918, at Carbuzancy, France, he accompanied a formation of 9 planes on a photographic mission in German territory. Six planes turned back before reaching the enemy line, and the remaining 3 were attacked by 18 planes when they had penetrated 12 kilometers into enemy country. After his 2 companions, whom he tried to assist, were shot down, Lieutenant Fleeson fought his way back to his own lines, destroying 2 enemy planes in the combat.

FLEET, GEORGE T _____ Captain, 26th Infantry, 1st Division.
At Berzy-le-Sec, France, July 21, 1918.
R—Douglas, Ariz.
B—Blacksburg, Va.
G. O. No. 132, W. D., 1918.
Amid showers of machine-gun bullets and artillery shells of all calibers, at Berzy-le-Sec, France, July 21, 1918, he gallantly proceeded to the front lines for vital information needed by the division commander, and accomplished his mission in spite of the great danger to which he was exposed.

FLEGEL, FRANK H. (1785713) _____ Private, Company L, 315th Infantry, 79th Division.
Near Nantillois, France, Sept. 29, 1918.
R—Philadelphia, Pa.
B—Egg Harbor, N. J.
G. O. No. 37, W. D., 1919.
When his platoon had been ordered to cover because of annihilating machine-gun and artillery fire, Private Flegel accompanied another soldier to the rescue of a comrade who was lying 300 yards distant. The journey was made through heavy and continuous fire, but Private Flegel, with his fellow soldier, succeeded in bringing their wounded comrade to safety.

*FLEISCHMAN, FRANK F. (1286125) _____ Private, Company K, 115th Infantry, 29th Division.
Near Balschwiller, Alsace, Aug. 31, 1918.
R—Back River, Md.
B—Back River, Md.
G. O. No. 102, W. D., 1918.
After a raid against enemy trenches, he volunteered to accompany his platoon leader into no man's land to rescue a missing member of the platoon who had been wounded. While engaged in this courageous duty he was mortally wounded.
Posthumously awarded. Medal presented to father, Charles Fleischman.

FLEITZ, MORRIS FREDERICK _____ Private, Headquarters Company, 6th Regiment, U. S. Marine Corps, 2d Division.
In the attack on Bois-de-Belleau, France, June 9–10, 1918.
R—Louisville, Ky.
B—Louisville, Ky.
G. O. No. 109, W. D., 1918.
He showed extraordinary heroism and faithfulness in the face of great danger, remaining on arduous duty without rest for 2 days, under constant fire, to supply his battalion with rations and ammunition on June 9–10, 1918, and in the attack on Bois-de-Belleau, France. He made 2 trips with ammunition in broad daylight and in plain view of the enemy and carried ammunition across the field under heavy shellfire.

*FLEMING, JOHN L. (1249332) _____ Sergeant, Company M, 112th Infantry, 28th Division.
Near Apremont, France, Oct. 2, 1918.
R—Grove City, Pa.
B—Wampum, Pa.
G. O. No. 11, W. D., 1921.
When the advance of his organization was held up by a strong enemy counterattack, and his company was suffering heavy casualties from enfilading fire from the enemy position on Chene Tondu Hill, Sergeant Fleming placed two automatic rifle teams in a position in advance of our front line. He directed the fire of these guns until he fell mortally wounded. His conduct was a material factor in the successful repulse of the enemy.
Posthumously awarded. Medal presented to mother, Mrs. Jennie Fleming.

*FLEMING, PATRICK F. (1454042)_____
Near Oderon, Alsace, July 12, 1918.
R—St. Louis, Mo.
B—Brooklyn, N. Y.
G. O. No. 59, W. D., 1919.

Private, Company M, 138th Infantry, 35th Division.
He bravely attempted to pick up and throw away, near Oderon, Alsace, on July 12, 1918, a live grenade that had fallen among 5 soldiers, but because of irregularities of the trench he could not reach it before it burst. He thrust his foot on it, thereby saving his companions from death or injury, but causing wounds that necessitated amputation of the foot.
Posthumously awarded. Medal presented to father, Joseph F. Fleming.

FLEMING, SAMUEL WILSON, Jr_____
Near Beaumont, France, Nov. 9, 1918.
R—Harrisburg, Pa.
B—Harrisburg, Pa.
G. O. No. 44, W. D., 1919.

Major, 315th Infantry, 79th Division.
On the night of Nov. 9, he received a serious and painful wound from a high-explosive shell, but refused to be evacuated, and continued in command of his battalion until the signing of the armistice on Nov. 11. He was exposed not only to heavy enemy fire but to severe weather conditions as well.

FLEMING, THOMAS W. (263515)_____
Near Romagne and Bantheville, France, Oct. 14, 1918.
R—Merrill, Mich.
B—Merrill, Mich.
G. O. No. 98, W. D., 1919.

Corporal, Company G, 128th Infantry, 32d Division.
When the advance was held up by fire from an enemy machine gun, Corporal Fleming, with utter disregard of personal danger, rushed out alone in the face of terrific machine-gun and shell fire, attacked the machine-gun nest, killing the gunner, capturing the 4 remaining members of the crew and bringing them back to our lines.

FLESHER, HERBERT W. (1537837)_____
Near Heuvel, Belgium, Nov. 2, 1918.
R—Cleveland, Ohio.
B—Cleveland, Ohio.
G. O. No. 37, W. D., 1919.

Sergeant, first class, Company B, 112th Engineers, 37th Division.
Sergeant Flesher, with two other soldiers, crossed the Scheldt River after two attempts and succeeded in stretching a line for the bridge across the stream. They were discovered and fired upon by the enemy, but they continued at work driving stakes and made a second trip across the river to obtain wire, despite the fact that a violent artillery barrage had been laid down on their positions.

FLETCHER, ALLEN_____
At Gesnes, France, Sept. 29, 1918.
R—Victorville, Calif.
B—Cincinnati, Ohio.
G. O. No. 27, W. D., 1919.

Captain, 362d Infantry, 91st Division.
Although he was severely wounded, he continued to lead his company in the assault on Gesnes; then, much weakened by his wound, he reorganized his company and directed its employment as a covering detachment in the withdrawal. He remained with his company until ordered to leave his post and receive medical treatment. He showed throughout the engagement a devotion to duty only exceeded by his utter disregard of personal safety.

FLETCHER, HARRY BENTON_____
In the capture of Bouresches, France, June 6, 1918.
R—Albion, Calif.
B—Greenville, Tex.
G. O. No. 119, W. D., 1918.

Corporal, 79th Company, 6th Regiment, U. S. Marine Corps, 2d Division.
After being severely wounded in the capture of Bouresches, France, June 6, 1918, he refused to go to the rear for treatment, but remained at his post and urged on his men to renewed efforts.

FLETCHER, JEFFERSON B_____
At Autry (Ardennes), France, Oct. 6, 1918.
R—New York, N. Y.
B—Chicago, Ill.
G. O. No. 137, W. D., 1918.

First lieutenant, section No. 517, Ambulance Service.
He was on his way to establish an advance aid station during a heavy bombardment when he was wounded by a shell fragment and his right eardrum broken. Two men accompanying him were killed, and the concussion of the exploding shell knocked him into a ditch full of water, where he lay for an hour. Displaying unfaltering devotion to duty, he continued on his mission, established the aid station, and remained in command of his section.

*FLETCHER, LEE C_____
At Fismette, France, Aug. 10–13, 1918.
R—Wellsburg, N. Y.
B—Gillett, Pa.
G. O. No. 37, W. D., 1919.

First lieutenant, 111th Infantry, 28th Division.
When his battalion was attacked by a greatly superior force, Lieutenant Fletcher, intelligence officer, organized a detachment and successfully defended an important position. Later, when our barrage was falling short, he voluntarily carried a message to the Artillery for the purpose of correcting the fire.
Posthumously awarded. Medal presented to father, George Fletcher.

FLING, JOHN H. (1453190)_____
At Cheppy, France, Sept. 26–27, 1918.
R—St. Louis, Mo.
B—St. Paul, Minn.
G. O. No. 37, W. D., 1919.

First sergeant, Company I, 138th Infantry, 35th Division.
When his company commander was killed and his company left without officers, Sergeant Fling took command of the company, successfully withdrew it from the midst of machine-gun nests, reorganized it, and continued the advance. He was severely wounded in the head the next day, but insisted on remaining on duty with his company and advancing in the attack, notwithstanding the fact that the advance was over a gassed area and his wounds prevented him wearing a gas mask.

FLINT, JOHN H. (88870)_____
Near Ferme de Jonchery, France, July 15, 1918, and near La Marche, France, Sept. 23, 1918.
R—Brooklyn, N. Y.
B—New York, N. Y.
G. O. No. 59, W. D., 1919.

Corporal, Machine Gun Company, 165th Infantry, 42d Division.
Stationed with the company train near a French battery of artillery, Corporal Flint, then a mechanic, left a concrete shelter and volunteered to carry a severely wounded French soldier to the dressing station. To reach the dressing station, over a kilometer away, he crossed an open field, subjected at the time to intense artillery bombardment. Later when a 150-millimeter shell burst near a shelter tent in which he was sleeping, killing one and wounding three other noncommissioned officers and hurling Corporal Flint several yards, he called assistance and supervised the care of the wounded.

FLOCKEN, JOHN BECK_____
In the capture of Bouresches, France, June 6, 1918.
R—St. Louis, Mo.
B—Peoria, Ill.
G. O. No. 126, W. D., 1918.

Private, 79th Company, 6th Regiment, U. S. Marine Corps, 2d Division.
In the capture of Bouresches, France, June 6, 1918, he was twice hit in the leg, but he dragged his automatic rifle 200 yards forward, opened fire on an enemy machine gun, and silenced it.

FLOOD, JAMES E. (1938517)_____
Near Romagne, France, Oct. 11, 1918.
R—Dennison, Ohio.
B—Dennison, Ohio.
G. O. No. 37, W. D., 1919.

Sergeant, Company F, 38th Infantry, 3d Division.
He kept his platoon advancing under heavy artillery and machine gun fire, and although seriously wounded, continued in command until the objective was reached.

FLOOD, JOHN VINCENT_____
Near Badonvillers, France, June 24, 1918.
R—New York, N. Y.
B—New York, N. Y.
G. O. No. 24, W. D., 1920.

Second lieutenant, 308th Infantry, 77th Division.
After being severely wounded he continued to direct his platoon with great courage and determination.

FLORIAN, PAUL A., Jr_____
East of Ronssoy, France, Sept. 29, 1918.
R—Troy, N. Y.
B—San Antonio, Tex.
G. O. No. 20, W. D., 1919.

Second lieutenant, 105th Infantry, 27th Division.
During the operations against the Hindenburg line Lieutenant Florian exhibited splendid courage and gallantry. After having been twice wounded he continued under heavy shell and machine gun fire to install telephone wires to an advance headquarters.

FLOYD, WILLIAM M. (2339814)_____
Near Les Evaux, France, July 13, 1918.
R—Norris City, Ill.
B—Norris City, Ill.
G. O. No. 32, W. D., 1919.

Private, Company L, 4th Infantry, 3d Division.
After seeing several patrols fail in the attempt to cross the River Marne during the night, Private Floyd, with three companions, successfully crossed in broad day light and in full view of the enemy, remaining in hostile territory throughout the day.

*FLYNN, FRANCIS J_____
At Chateau-Thierry, France, June 6, 1918.
R—Binghamton, N. Y.
B—Elmira, N. Y.
G. O. No. 110, W. D., 1918.

Gunnery sergeant, 20th Company, 5th Regiment, U. S. Marine Corps, 2d Division.
Killed in action at Chateau-Thierry, France, June 6, 1918, he gave the supreme proof of that extraordinary heroism which will serve as an example to hitherto untried troops.
Posthumously awarded. Medal presented to mother, Mrs. Elizabeth Flynn.

FLYNN, JOHN L. (2058791)_____
Near Consenvoye, France, Oct. 9, 1918.
R—Whiteside County, Ill.
B—Morrison, Ill.
G. O. No. 46, W. D., 1919.

Corporal, Company G, 131st Infantry, 33d Division.
Upon his own initiative, he advanced by short rushes under machine gun and sniper fire to a point from which he successfully bombed and silenced an enemy machine-gun sniper who was holding up the advance of his company.

FOCHT, JOHN A_____
Near Madeleine Farm, France, Oct. 27, 1918.
R—Sweetwater, Tex.
B—Sweetwater, Tex.
G. O. No. 87, W. D., 1919.

First lieutenant, 315th Engineers, 90th Division.
Knocked down and severely wounded when enemy shells hit the building where his company was at mess, he remained in the shelled area and, with the assistance of one of his men, carried out the wounded to a place of safety, where their wounds could be dressed.

FOGG, PRESTON DOANE_____
Near Champigneulles, France, Oct. 16, 1918.
R—Brighton, Mass.
B—Chelsea, Mass.
G. O. No. 35, W. D., 1919.

First lieutenant, 309th Infantry, 78th Division.
After leading his company in the attack on Champigneulles and thereafter successfully withdrawing the remnants of his command, Lieutenant Fogg, although himself wounded, carried to shelter another wounded officer who was unable to walk.

FOGO, EDWARD T_____
Near La Selle River, France, Oct. 18, 1918.
R—Wellsville, Ohio.
B—Wellsville, Ohio.
G. O. No. 37, W. D., 1919.

First lieutenant, 120th Infantry, 30th Division.
When his company had received orders to advance from the front line, he, then in command, led the company to its objective, despite severe wounds he had received prior to the start of the attack. He refused treatment until the mission was complete, when he went to the dressing station. He returned as soon as possible and remained with his company during the entire operations.

FOLEY, HARRY J. (262279)_____
Near Cierges, northeast of Chateau-Thierry, France, July 31, 1918.
R—Waterford, Mich.
B—Detroit, Mich.
G. O. No. 132, W. D., 1918.

Private, Company E, 125th Infantry, 32d Division.
After he had been wounded in both arms July 31, 1918, near Cierges, northeast of Chateau-Thierry, France, he collected ammunition from the dead and wounded who were lying on the battle field and carried it under fire to his comrades on the firing line.

FOLEY, THOMAS F_____
Near Vaux, France, July 15 to 22, 1918.
R—Worcester, Mass.
B—Worcester, Mass.
G. O. No. 125, W. D., 1918.

Captain, 101st Infantry, 26th Division.
Throughout the four days of the advance he commanded and led his battalion with exceptional bravery and judgment, thereby inspiring his men. When strong resistance was encountered he personally went forward and reconnoitered the terrain under heavy machine gun and sniper fire, and on July 15 and again on July 22, he personally led his battalion in successful attacks.

FOLLETTE, JUSTIN P_____
Near Chatel-Chehery, France, Oct. 15, 1918.
R—Jamul, Calif.
B—Libertyville, Ill.
G. O. No. 143, W. D., 1918.

First lieutenant, observer, 12th Aero Squadron, Air Service.
He volunteered under the most adverse weather conditions to stake the advance of the 82d Division. Disregarding the fact that darkness would set in before he and his pilot could complete their mission, he made observation at the extremely low altitude of 1,500 feet, amid a most terrific antiaircraft and ground machine gun fire until the necessary information was secured. On the return, due to darkness, his pilot was forced to land on a shell-torn field, whence he proceeded on foot to headquarters with valuable information

FOLLIS, CHARLES.
Near Sergy, France, July 31, 1918.
R—Sault Ste. Marie, Mich.
B—Canada.
G. O. No. 64, W. D., 1919.

First lieutenant, 125th Infantry, 32d Division.
After the capture of Hill 212, near Sergy, Lieutenant Follis personally directing the consolidation of the position. He continually exposed himself to sweeping machine-gun fire while looking after the care and evacuation of all wounded men on the field, personally making sure that all were taken to the aid station.

FOLSOM, JOHN V. (2258313).
In Bois-Communal-de-Cierges, east of Exmorieux Ferme, France, Sept. 28, 1918.
R—Cedarhill, Idaho.
B—Lamont, Wash.
G. O. No. 9, W. D., 1923.

Private, first class, Company L, 361st Infantry, 91st Division.
His company, checked by a heavy barrage of machine gun and artillery fire, with gas shells interspersed, he volunteered to carry an important message from his company commander to the commander of an artillery unit. With great gallantry he accomplished his mission in the face of intense enemy fire. He was killed in action on Oct. 10, 1918, fighting for Hill 255. His high courage and splendid soldierly qualities were important factors in the operations of his company, and inspired his comrades to great endeavors.

FOLSOM, LYNN H.
Near Fremont, France, Oct. 8-20, 1918.
R—Elizabethton, Tenn.
B—Elizabethton, Tenn.
G. O. No. 137, W. D., 1918.

First lieutenant, 117th Infantry, 30th Division.
Although he was painfully wounded on Oct. 8, he remained on duty, taking command of his company 6 days later, when he was the only officer present, and effectively reorganizing the command after its strength had been greatly reduced. Still suffering from his wound, he led his company in attack on Oct. 17 and stayed at his post for 2 days thereafter until his battalion was relieved.

*FOLZ, ALEXANDER (2194430).
Near Remonville, France, Nov. 1, 1918.
R—Rockford, Ill.
B—Russia.
G. O. No. 44, W. D., 1919.

Corporal, Company B, 354th Infantry, 89th Division.
While leading his squad in attack on a machine-gun nest, his automatic rifleman was wounded and unable to continue. He took the gun, and, firing as he advanced, put the machine-gun out of action, but he was so badly wounded during the exploit that he died from his wounds a few minutes after.
Posthumously awarded. Medal presented to widow, Mrs. Alexander Folz.

FONTAINE, HUGH L.
In the region of Hageville, France, Sept. 14, 1918.
R—Memphis, Tenn.
B—New Orleans, La.
G. O. No. 13, W. D., 1919.

First lieutenant, 49th Aero Squadron, Air Service.
He, together with First Lieutenant Hugh Brewster, attacked 9 enemy monoplanes (Fokkers) at an altitude of 4,000 meters. He dived into the midst of the enemy formation without consideration for his personal safety, subjecting himself to great danger. By the suddenness and extreme vehemence of his attack the machines were driven into confusion. Although greatly outnumbered, he and Lieutenant Brewster succeeded in shooting down 2 of the enemy.
Oak-leaf cluster.
A bronze oak leaf is awarded Lieutenant Fontaine for extraordinary heroism in action near Champigneulles, France, Oct. 10, 1918. While leading a patrol of 3 other machines he attacked 4 enemy planes in the region of Champigneulles. He succeeded in shooting down 2 of the enemy planes in flames. The first of these he shot down in the initial attack. The second he attacked while it was endeavoring to shoot down one of our planes, which had been rendered helpless by the loss of 1 of its wings. He dived on the attacking plane and shot it down in flames.

FOOKS, HERBERT C.
Near Eyne, Belgium, Nov. 4, 1918.
R—Salisbury, Md.
B—Salisbury, Md.
G. O. No. 19, W. D., 1920.

Major, 145th Infantry, 37th Division.
Although severely wounded and his jaw shattered by a machine-gun bullet, he refused to be evacuated, administered first aid himself, and continued to fearlessly direct his battalion during a strong counterattack. The personal example of this officer was a vital factor in the success of the operation.

FORBES, BURT T. (1316258).
Near Ypres, Belgium, Sept. 1, 1918.
R—Greenville, N. C.
B—Camden County, N. C.
G. O. No. 81, W. D., 1919.

Corporal, Company I, 119th Infantry, 30th Division.
While his patrol was acting as a flank guard, with orders not to fire unless absolutely necessary, he detected an enemy patrol of 8 men approaching and starting to set up a machine gun. Crawling forward alone, he charged the enemy patrol and, single-handed, killed 3 Germans and routed the other 5.

FORBIS, FRED M. (2845581).
Southeast of Remonville, France, Nov. 1, 1918.
R—Holts Summit, Mo.
B—Holts Summit, Mo.
G. O. No. 95, W. D., 1919.

Private, first class, Company D, 354th Infantry, 89th Division.
Private Forbis, a member of the leading combat group of his platoon, encountered a machine-gun nest of 6 guns. Although only 100 feet from the nest, he remained at his automatic rifle, pouring in such a sheath of bullets on the enemy that it enabled his comrades to outflank the nest. He remained at his post until severely wounded. His exceptional nerve and coolness were an inspiration to his comrades throughout the action.

*FORD, CHARLES M.
Near St. Etienne, France, Oct. 9, 1918.
R—West Plains, Mo.
B—Woodbine, Iowa.
G. O. No. 78, W. D., 1919.

Second lieutenant, 141st Infantry, 36th Division.
After all the officers of his company had been killed or wounded, Lieutenant Ford took command of the company and with about 24 men advanced beyond the main line, over extremely difficult ground, capturing 24 enemy machine guns. Lieutenant Ford established his men in a good position, practically isolated from the rest of the line, and, manning the captured guns, held the position under heavy machine-gun and shell fire for about 60 hours.
Posthumously awarded. Medal presented to mother, Mrs. Mary R. Ford.

FORD, CHRISTOPHER W.
Near Reims, France, Mar. 27, 1918, and near Armentieres, France, May 21, 1918.
R—New York, N. Y.
B—New York, N. Y.
G. O. No. 37, W. D., 1919.

Captain, 103d Aero Squadron, Air Service.
Near Reims, on Mar. 27, he, while on a patrol with 2 other pilots, led his formation in an attack on 8 enemy planes. After 20 minutes of fighting the American formation shot down 3 German machines, of which 1 was destroyed by this officer. Near Armentieres, on May 21, he again led a patrol of 6 planes in attacking 20 enemy aircraft. The attack resulted in 10 individual combats. Captain Ford shot down 1 hostile plane and, with his patrol, routed the others.

FORD, DARIS V. (568604).
On the River Vesle, east of St. Thibaut, France, Aug. 6, 1918.
R—North Platte, Nebr.
B—Lincoln, N. C.
G. O. No. 78, W. D., 1919.

Private, Company C, 4th Engineers, 4th Division.
While his company was advancing ahead of the Infantry toward the Vesle River to put in footbridges, Private Ford, acting as liaison messenger, displayed undaunted courage and utter disregard for his personal safety by time after time carrying messages through a terrific enemy barrage and heavy machine-gun fire, each time successfully accomplishing his mission.

FORD, FRANK M. (1554255)............
Northeast of Chateau-Thierry, France, July 27, 1918.
R—Covington, Ky.
B—Clay County, Ky.
G. O. No. 108, W. D., 1918.

Private, Company D, 166th Infantry, 42d Division.
After he and a comrade had located a hostile machine gun in a clump of trees 500 meters north of a chateau which their platoon was holding east of Fere-en-Tardenois, they secured the permission of their platoon commander to attempt to force the enemy to abandon this position, and advanced over open ground and in the face of fire. With their rifles they drove the enemy from their gun, killing one and wounding another.

FORE, JAMES EDWARD (1310840)........
At St. Martin, Riviere, France, Oct. 17, 1918.
R—Union, S. C.
B—Asheville, N. C.
G. O. No. 44, W. D., 1919.

Sergeant, Company E, 118th Infantry, 30th Division.
While engaged with 4 other soldiers in mopping up a village, he led his men in a flank attack on a machine-gun nest and captured the crew, numbering 18. Pushing forward, he organized a squad of stragglers and captured an entire company of Germans, including 2 officers.

FORE, WITT SAMUEL (1310827)..........
Near Brancourt, France, Oct. 8, 1918.
R—Union, S. C.
B—Asheville, N. C.
G. O. No. 50, W. D., 1919.

Sergeant, Company E, 118th Infantry, 30th Division.
Disregarding personal safety, Sergeant Fore ran forward through heavy machine-gun and shell fire to a shell hole where a wounded soldier lay mortally wounded and carried his comrade to shelter.

FOREHAND, WALTER S (1316251)........
Near Bellicourt, France, Sept. 29, 1918.
R—South Mills, N. C.
B—South Mills, N. C.
G. O. No. 78, W. D., 1919.

Sergeant, Company I, 119th Infantry, 30th Division.
Sergeant Forehand showed exceptional bravery and devotion to duty by advancing with another soldier, though separated from his platoon, in the attack by his regiment on Sept. 29, 1918. They found 4 privates, also lost in the smoke and fog, and, with this small party, proceeded toward the objective. During their advance they surprised and captured 92 Germans, including several officers, without other aid. They succeeded in getting all of the prisoners back to the military police, and then rejoined their platoon.

FOREMAN, MILTON J....................
Near Ferme de Maucourt, north-west of Beauclair, France, Nov. 4, 1918.
R—Chicago, Ill.
B—Chicago, Ill.
G. O. No. 9, W. D., 1923.
Distinguished-service medal also awarded.

Colonel, 122d Artillery, 33d Division.
When the advance of the Infantry was held up by heavy fire from hostile machine guns and artillery, which his artillery was unable to locate and neutralize, Colonel Foreman personally advanced by creeping through a heavy enemy artillery barrage to a point in the Infantry front line where he could by direct visual observation locate the position of enemy machine guns. Remaining at this post exposed to a terrific hostile bombardment, he transmitted information to the supporting artillery and directed their fire on hostile strong points until the advance of the Infantry line was effected.

FORMICA, PIETRO (567932)............
Near Ville-Savoye, France, Aug. 8, 1918.
R—Detroit, Mich.
B—Italy.
G. O. No. 95, W. D., 1919.

Private, first class, Company A, 59th Infantry, 4th Division.
After several other soldiers had been killed in attempting to carry a message across an open field under intense enemy fire, Private Formica volunteered for this perilous mission and successfully accomplished it. He continued to display marked courage in carrying messages under fire until he was wounded next day.

*FORREST, HARRY E. (1781060)..........
Near Montfaucon, France, Sept. 26, 1918.
R—Baltimore, Md.
B—Baltimore, Md.
G. O. No. 32, W. D., 1919.

Sergeant, Company I, 313th Infantry, 79th Division.
He led his platoon in an attack against an enemy machine-gun nest which was impeding the progress of his company. During the course of the exploit he was killed by fire from the nest, but his action enabled his men to accomplish the capture of 12 of the enemy, who were manning the guns in the nest. Posthumously awarded. Medal presented to mother, Mrs. Annie Forrest.

FORRESTER, ROBERT R................
Near Pont-a-Mousson, France, Sept. 13, 1918.
R—Atlanta, Ga.
B—Atlanta, Ga.
G. O. No. 103, W. D., 1919.

First lieutenant, 327th Infantry, 82d Division.
Lieutenant Forrester volunteered for duty with other organizations ordered to make a daylight raid against the enemy. His superb leadership and coolness under very trying circumstances greatly reduced the number of casualties among his troops and his disregard of personal safety greatly inspired his men.

*FORSTER, HAROLD R. (1213977)........
East of Ronssoy, France, Sept. 29, 1918.
R—Clyde, N. Y.
B—Dunkirk, N. Y.
G. O. No. 20, W. D., 1919.

Private, Company C, 108th Infantry, 27th Division.
During the operations against the Hindenburg line, when the advance of his company was held up by an enemy machine-gun nest, he crawled forward to a small shell hole, killed 4 of the German machine gunners with a Lewis gun, and put their gun out of action, thereby permitting the advance to continue. In accomplishing this courageous act he was seriously wounded. Posthumously awarded. Medal presented to father, Robert Forster.

*FORSYTH, MATTHEW W., Jr. (129182)..
Near Thiaucourt, France, Sept. 12, 1918.
R—Cheltenham, Pa.
B—Philadelphia, Pa.
G. O. No. 59, W. D., 1919.

Private, first class, Battery E, 12th Field Artillery, 2d Division.
When his gun position was subjected to a heavy enfilading fire Private Forsyth continued in the service of his piece, regardless of his personal safety, and was killed while in the performance of his duty.
Posthumously awarded. Medal presented to father, Matthew W. Forsyth, Sr.

FORT, HOWARD H. (53445)_____
    Near Soissons, France, July 19-21,
      1918.
    R—Anderson, Ind.
    B—Madison County, Ind.
    G. O. No. 15, W. D., 1923.

Sergeant, Company L, 28th Infantry, 1st Division.
As mess sergeant, Company L, 28th Infantry, he established the advance kitchen of the 3d Battalion of his regiment under direct observation of an enemy balloon and under heavy enemy shellfire. Due to congestion of traffic, the roads being blocked with ammunition trains, the delivery of rations at night was forbidden. Sergeant Fort for 3 successive days voluntarily led a ration train in daylight in constant exposure to enemy shelling from his kitchen to the troops of the battalion in the front lines, showing splendid devotion to duty and utter disregard for his own safety. His bravery and determined devotion to his comrades inspired them to great endeavors.
Oak-leaf cluster.
Sergeant Fort was awarded an oak-leaf cluster for the following act of heroism: For extraordinary heroism in action near Exermont, France, Oct. 1-12, 1918. Sergeant Fort, then mess sergeant, Company L, 28th Infantry, displayed outstanding courage and bravery and exceptional devotion to duty by voluntarily exposing himself daily to intense enemy artillery fire in conducting ration trains and delivering rations to the front-line troops. This duty was performed during 9 successive nights under intense and concentrated fire, under untold hazards and difficulties. Sergeant Fort's indomitable spirit, his devotion to his comrades, and his utter disregard for his own safety greatly inspired the men of his organization and were important factors in the successful operations of this regiment during this engagement.

*FORTH, HERMAN (2180322)_____
    In the Bois-de-Barricourt, France,
      Nov. 2, 1918.
    R—Wayne City, Ill.
    B—Wayne City, Ill.
    G. O. No. 37, W. D., 1919.

Private, medical detachment, 341st Machine-Gun Battalion, 89th Division.
In the face of enemy machine-gun fire, he went to the aid of 2 severely wounded soldiers, carried them into a shell hole, dressed their wounds, and while endeavoring to get the blankets from the packs of the wounded men was struck in the head by a machine-gun bullet and instantly killed.
Posthumously awarded. Medal presented to mother, Mrs. Dora A. Forth.

*FOSNES, ERNEST (567092)_____
    During the Aisne-Marne offensive,
      France, July 19, and August 8,
      1918.
    R—Montevideo, Minn.
    B—Montevideo, Minn.
    G. O. No. 5, W. D., 1920

Corporal, Company A, 59th Infantry, 4th Division.
On July 19 he exposed himself to intense machine-gun and artillery fire to assist in the reorganization of his company, which had become temporarily disorganized, due to heavy losses. On Aug. 8, when his platoon had become separated from the company during the attack, he exposed himself to direct machine-gun fire in order to encourage the members of his platoon in their task. He was mortally wounded a short time afterwards.
Posthumously awarded. Medal presented to father, Christopher Fosnes.

*FOSS, SAXTON C. (39287)_____
    Near Medeah Ferme, France, Oct. 8,
      1918.
    R—Somerville, Mass.
    B—Somerville, Mass.
    G. O. No. 89, W. D., 1919.

Private, Company F, 9th Infantry, 2d Division.
With exceptional courage, Private Foss voluntarily advanced to flank a machine-gun nest which was holding up the advancing battalion, and in so doing was fatally wounded.
Posthumously awarded. Medal presented to mother, Mrs. Carrie Foss.

FOSSETT, EDWARD JOSEPH (1245972)____
    During an attack in the Courbandon
      Woods, Sept. 5-6, 1918.
    R—Philadelphia, Pa.
    B—Philadelphia, Pa.
    G. O. No. 72, W. D., 1920.

Private, Company M, 111th Infantry, 28th Division.
While acting as liaison between the 109th and 111th Infantry he was severely wounded by a high-explosive shell, but refused to be evacuated until his command was relieved on Sept. 9.

FOSSIE, LESTER (1403546)_____
    At Ferme-de-la-Riviere, France, Oct.
      5, 1918.
    R—Metropolis, Ill.
    B—Metropolis, Ill.
    G. O. No 46, W. D., 1919.

Supply sergeant, Company M, 370th Infantry, 93d Division.
A messenger having been wounded by an enemy sniper in the open between the lines, Sergeant Fossie immediately went to his rescue, and brought him into the company headquarters over ground swept by machine-gun and sniper fire.

*FOSTER, HAMILTON K._____
    Near Soissons, France, July 22, 1918.
    R—New Rochelle, N. Y.
    B—New York, N. Y.
    G. O. No. 15, W. D., 1919.

Captain, 26th Infantry, 1st Division.
A courageous and inspiring leader at all times during the fighting near Soissons, France, July 22, 1918, he particularly distinguished himself for bravery and judgment by charging and capturing a machine gun that threatened his advance.
Posthumously awarded. Medal presented to father, Dr. Matthias L. Foster.

*FOSTER, WALTER L. (1866552)_____
    Near Bellicourt, France, Sept. 29,
      1918.
    R—Haw River, N. C.
    B—Alamance County, N. C.
    G. O No. 21, W. D., 1919.

Private, Company D, 119th Infantry, 30th Division.
Acting as a runner, Private Foster carried frequent messages between his platoon leader and company commander, exposed at all times to heavy enemy fire of artillery and machine guns. While performing this meritorious work he was killed by machine-gun fire.
Posthumously awarded. Medal presented to sister, Miss Nezzie Foster.

FOUREMAN, ROY B._____
    Near Brabant-sur-Meuse, France,
      Oct. 23, 1918.
    R—Greenville, Ohio.
    B—Greenville, Ohio.
    G. O. No. 21, W. D., 1919.

Second lieutenant, 308th Trench Mortar Battery, 158th Field Artillery Brigade, 83d Division.
During the offensive action in Bossois Bois, the 4 trench mortars operated by his platoon were put out of action. Under an enemy barrage Lieutenant Foureman went from gun to gun, encouraging his men to continued effort, until his last gun was out of action, when he turned his attention to assisting the wounded.

FOUST, BENJAMIN E. (1456831)_____
    Near Exermont, France, Sept. 29,
      1918.
    R—Augusta, Kans.
    B—Cass County, Mo.
    G. O. No. 59, W. D., 1919.

Mechanic, Company F, 139th Infantry, 35th Division.
After having one eye shot out, Mechanic Foust refused to avail himself of the opportunity to be evacuated to the rear, but rendered first aid to himself and continued to dress the wounds of his comrades, until a heavy concentration of gas so affected his wounded eye that he was forced to go to the rear. His work was the means of saving the lives of many of his comrades.

FOUST, JOHN W. (549232)............
Near Cunel, France, Oct. 22, 1918.
R—Lexington, N. C.
B—Lexington, N. C.
G. O. No. 44, W. D., 1919.

Corporal, Machine Gun Company, 38th Infantry, 3d Division.
After all the officers of his company had been wounded, Corporal Foust assumed command, and with great courage and bravery organized a detachment, recapturing two of his machine guns that had fallen to the enemy in a counter-attack earlier in the day.

FOWLE, JOHN G. (180303)............
Near Juvigny, France, Aug. 31, 1918.
R—Traverse City, Mich.
B—Traverse City, Mich.
G. O. No. 64, W. D., 1919.

Sergeant, sanitary detachment, 126th Infantry, 32d Division.
Under harassing machine-gun fire and in plain view of the enemy Sergeant Fowle voluntarily went forward a distance of 150 yards to dress the wounds of an officer. He returned for assistance, after which he removed the officer to a place of safety.

FOWLER, EDWARD C..................
Near Blanc Mont Ridge, France, Oct. 2–3, 1918.
R—South Boston, Mass.
B—South Boston, Mass.
G. O. No. 46, W. D., 1919.

Second lieutenant, 6th Regiment, U. S. Marine Corps, 2d Division.
On Oct. 2 he led his men into an advance trench and cleared it of the enemy without a casualty. That night he went out alone and killed the crew of a machine-gun nest with bombs. During the attack on Blanc Mont the following morning he led his men, capturing about 80 prisoners and 15 machine guns. After consolidating his position on Blanc Mont, he went out alone and, while exposed to artillery fire, sniped the crew of a machine-gun nest.

*FOWLER, LEWIS K. (1319673)..........
Near Busigny, France, Oct. 19, 1918.
R—Cardenas, N. C.
B—Wake County, N. C.
G. O. No. 32, W. D., 1919.

Private, first class, Company B, 120th Infantry, 30th Division.
He remained at his post, covering the withdrawal of his company with his automatic rifle, in order that the company might take up a better position. He was instantly killed while in the performance of this mission.
Posthumously awarded. Medal presented to mother, Mrs. John W. Fowler.

FOX, CHARLES M..................
Near Bantheville, France, Oct. 26, 1918.
R—Chicago, Ill.
B—Stinesville, Ill.
G. O. No. 66, W. D., 1919.

Captain, Medical Corps, attached to 353d Infantry, 89th Division.
Although he was suffering from the effects of gas, Captain Fox maintained his battalion dressing station under a terrific bombardment of gas and high-explosive shells, which had almost demolished his station, continuing to care for the wounded and refusing to be evacuated until blindness rendered him unable to work.

FOX, DANIEL R...................
Near St. Etienne, France, Oct. 4, 1918.
R—Pottstown, Pa.
B—Shenkel, Pa.
G. O. No. 35, W. D., 1919.

Sergeant, 17th Company, 5th Regiment, U. S. Marine Corps, 2d Division.
He volunteered and carried an important message across a heavily shelled area, returning through a barrage to report the result of his mission. Later, after being wounded, he remained on duty for four hours, carrying messages across a field swept by machine-gun fire.

FOX, FRANK I. (2302048)..............
At St. Gilles, near Fismes, France, Aug. 4, 1918.
R—Milwaukee, Wis.
B—Centerville, Iowa.
G. O. No. 139, W. D., 1918.

Corporal, Headquarters Company, 120th Field Artillery, 32d Division.
He, with other soldiers, made frequent trips to maintain telephone communication between battalion and regimental headquarters during a destructive bombardment. All other lines had been destroyed, and as this line was used by both Infantry and Artillery for communication with the rear, it was of utmost importance that it be maintained.

FOX, GEORGE F. (1709954)..........
Near St. Juvin, France, Oct. 15, 1918.
R—New York, N. Y.
B—New York, N. Y.
G. O. No. 10, W. D., 1920.

Corporal, Company F, 308th Infantry, 77th Division.
Corporal Fox exposed himself to machine-gun fire to rescue a wounded comrade who lay in an exposed position. While crawling out to bring in his comrade he was under direct enemy observation and bursts of machine-gun fire.

FOX, MATHEW S. (1224926)............
Near Consenvoye, France, Nov. 4, 1918.
R—New York, N. Y.
B—New York, N. Y.
G. O. No. 46, W. D., 1919.

Corporal, Battery F, 104th Field Artillery, 27th Division.
While the battery position was being subjected to severe bombardment of gas and high-explosive shells, Corporal Fox, in an effort to rescue two wounded comrades, extinguished a pile of burning camouflage, which was used as a cover for the ammunition and fuses. While fighting the burning camouflage the ammunition was exploded by another bursting shell.

FOX, WADE H...................
Near St. Etienne, France, Oct. 4, 1918.
R—Lost Creek, W. Va.
B—Berea, W. Va.
G. O. No. 37, W. D., 1919.

Private, 16th Company, 5th Regiment, U. S. Marine Corps, 2d Division.
Under constant shell and machine-gun fire for three days he performed his duties as runner with exceptional courage and daring, establishing efficient service between company and battalion headquarters.

FOY, RICHARD (1702105)..............
Near Bazoches, France, Aug. 14, 1918.
R—Brooklyn, N. Y.
B—Brooklyn, N. Y.
G. O. No. 16, W. D., 1923.

Private, first class, Company F, 306th Infantry, 77th Division.
He voluntarily exposed himself to intense machine-gun and artillery fire, crawling forward, in company with four other men of his company, in search of their wounded company commander who had fallen a short distance in front of his company's position. After a fruitless search for the wounded officer, the patrol engaged the nearest enemy post and in a fight with hand grenades destroyed it. Private Foy then assisted a wounded comrade to return to his own line. The heroic conduct of Private Foy greatly encouraged the men of his company, inciting them to heroic endeavor.

*FRANCIS, RAYMOND (1551161)..........
Near Montfaucon, France, Oct. 6, 1918.
R—Philadelphia, Pa.
B—Philadelphia, Pa.
G. O. No. 19, W. D., 1920.

Corporal, Battery E, 76th Field Artillery, 3d Division.
While acting as chief of section during an important firing, Corporal Francis was mortally wounded by an enemy shell which rendered all but one of the section casualties. Although suffering great pain, he directed the remaining members of the gun crew in the firing until he was relieved.
Posthumously awarded. Medal presented to mother, Mrs. Catherine Francis.

FRANCISCO, JOHN (1390657)..........
In the Bois-de-Foret, France, Oct. 12, 1918.
R—Chicago, Ill.
B—Chicago, Ill.
G. O. No. 78, W. D., 1919.

Private, Company M, 132d Infantry, 33d Division.
Private Francisco displayed remarkable heroism and leadership. During the afternoon the enemy made three strong counter-attacks, and it was during these attacks that Private Francisco gathered together fragments of squads and assumed command of them. He led them against the enemy, approaching from the rear of the right flank, and was personally responsible for the capture of four machine guns and five prisoners.

FRANK, EMANUEL (549181)........... Corporal, Machine Gun Company, 38th Infantry, 3d Division.
Near Launay, France, July 15, 1918. Corporal Frank, with an officer and another soldier of his company, attacked
R—Rockaway Beach, N. Y. a patrol of seven Germans who had captured four American soldiers, killed
B—Brooklyn, N. Y. one of the Germans and captured the others.
G. O. No. 23, W. D., 1919.

FRANK, GEORGE PERCY........... Sergeant, 82d Company, 6th Regiment, U. S. Marine Corps, 2d Division.
In the Bois-de-Belleau, France, He showed exceptional bravery and coolness in leading his platoon against
June 8, 1918. superior numbers of the enemy strongly fortified in a machine-gun nest,
R—Rochester, N. Y. which he captured and held.
B—Auburn, N. Y.
G. O. No. 110, W. D., 1918.

*FRANK, WILLIAM F........... First lieutenant, 20th Aero Squadron, Air Service.
Near Buzancy, France, Oct. 23, While flying in the rear of a formation, returning from a bombing raid, Lieu-
1918. tenant Frank's machine was attacked by three hostile planes (Fokker type).
R—Chicago, Ill. Lieutenant Frank was wounded and rendered unconscious early in the
B—Chicago, Ill. encounter, but upon recovering, he shot down a Fokker which was attacking
G. O. No. 35, W. D., 1919. the leader of the formation, and drove off two others which were pressing
him from the side.
Posthumously awarded. Medal presented to brother, Paul A. Frank.

FRANK, WILLIAM J. (544519)........... Private, Headquarters Company, 30th Infantry, 3d Division.
Near Bois-d'Aigremont, France, On the night of July 14-15 he volunteered and carried messages through
July 14-15, 1918. heavy shellfire after other runners had been killed in the attempt to perform
R—Wahpeton, N. Dak. the same mission.
B—Grafton, N. Dak.
G. O. No. 32, W. D., 1919.

FRANKLENFELD, CHARLES (329250)...... Corporal, Company H, 31st Infantry.
In the Suchan Valley, Siberia, during He distinguished himself on numerous occasions by his capabilities and
June, July, and August, 1919. extraordinary daring as a patrol leader while under fire, notably at Novo
R—Tuckahoe, N. Y. Litovsk on August 8, 1919, when he rushed the door of a hut occupied by the
B—Gettysburg, Pa. enemy, who were firing directly at him.
G. O. No. 133, W. D., 1919.

FRARY, FRANK M. (107165)........... Sergeant, Battery F, 10th Field Artillery, 3d Division.
Near Greves Farm, France, July 15, Responding to a call for volunteers, Sergeant Frary, with 8 other soldiers,
1918. manned 2 guns of a French battery which had been deserted by the French
R—Ogden, Utah. during the unprecedented fire after many casualties had been inflicted on
B—Salt Lake City, Utah. their forces. For 2 hours he remained at his post and poured an effective
G. O. No. 46, W. D., 1919. fire into the ranks of the enemy.

FRASER, DUNCAN........... First lieutenant, 16th Infantry, 1st Division.
Near Soissons, France, July 18, 1918. With four men, Lieutenant Fraser captured a machine-gun nest which was
R—Ardsley-on-Hudson, N. Y. delivering an annihilating fire upon his company and holding up its advance.
B—Pittsburgh, Pa.
G. O. No. 81, W. D., 1919.

*FRASER, HARRY L........... Captain, Quartermaster Corps, 5th Division.
Near Brandeville, France, Nov. 10, He was directing a working party which was being heavily bombarded. Hav-
1918. ing been ordered to safety, he, thinking only of the welfare of his men, went
R—El Paso, Tex. forth to see that all had found shelter, and was instantly killed.
B—St. Louis, Mo. Posthumously awarded. Medal presented to mother, Mrs. Mary E. Fraser.
G. O. No. 20, W. D., 1919.

FRASIER, LYMAN S........... Major, 26th Infantry, 1st Division.
Near Verdun, France, Oct. 7, 1918. While conducting a flanking movement to reduce the enemy defenses on Hill
R—Amsterdam, N. Y. 272, at the head of his two assaulting companies, Major Frasier met a bat-
B—Amsterdam, N. Y. talion of enemy, formed for counterattack against our advanced positions, in
G. O. No. 78, W. D., 1919. the Bois-de-Moncy. Disposing of his force with excellent judgment, Major
Frasier himself conducted an assault that routed the enemy, driving him
from the field in complete disorganization. Later in the action, when
wounded, he refused to relinquish command and continued to direct the
operations of his battalion until he had placed his troops on their final
objective.

FRATUS, GEORGE R. (2385733)........... Private, Company F, 61st Infantry, 5th Division.
At Aincreville, France, Nov. 1, 1918. When telephone communications had been cut off Private Fratus volunteered
R—Providence, R. I. to carry messages to the battalion commander. Though he was wounded
B—Hawaii. in passing through an intense artillery barrage, he succeeded in delivering
G. O. No. 95, W. D., 1919. the messages.

FRAY, JOHN P. (65276)........... Corporal, Company H, 102d Infantry, 26th Division.
At St. Hilaire, France, Sept. 18, 1918. While under heavy enemy machine-gun fire, Corporal Fray charged a machine
R—Waterbury, Conn. gun single-handed, putting it out of action, and dispersing its crew. Later,
B—Waterbury, Conn. as the raiding party withdrew, he assisted in carrying back the wounded.
G. O. No. 26, W. D., 1919.

FRAZIER, JOSEPH........... First lieutenant, 9th Infantry, U. S. Army.
At Tientsin, China, July 13, 1900. He displayed conspicuous gallantry and absolute disregard of personal safety
R—Grand Center, Mo. in rescuing, under a terrific enemy fire, the colonel of his regiment, who had
B—Thomas Hill, Mo. fallen mortally wounded.
G. O. No. 3, W. D., 1925

*FRAZIER, WALTER D........... Second lieutenant, 5th Regiment, U. S. Marine Corps, 2d Division.
At Chateau-Thierry, France, June 6, Killed in action at Chateau-Thierry, France, June 6, 1918, he gave the supreme
1918. proof of that extraordinary heroism which will serve as an example to hitherto
R—Pittsburgh, Pa. untried troops.
B—Pittsburgh, Pa. Posthumously awarded. Medal presented to mother, Mrs. W. D. Frazier.
G. O. No. 119, W. D., 1918.

*FREDERICKS, CORNELIUS C. (734175)___
    Near Frapelle, France Aug 17, 1918.
    R—Brooklyn, N. Y.
    B—New York, N. Y.
    G. O. No. 15, W. D., 1919.

Private, Company M, 6th Infantry, 5th Division.
He displayed great coolness and courage under a heavy enemy barrage when he unhesitatingly went forward to destroy enemy wire entanglements and continued this extremely hazardous work until killed.
Posthumously awarded. Medal presented to mother, Mrs. A. M. Fredericks.

FREDLUND, VICTOR (2854501)_____
    Near Preny, France, Sept. 25, 1918.
    R—Moline, Ill.
    B—Boulder, Colo.
    G. O. No. 140, W. D., 1918.

Private, Company C, 315th Engineers, 90th Division.
While withdrawing from a daylight raid with his detail he saw a wounded infantry soldier lying helpless behind a heavy machine-gun and artillery barrage. He returned through the intense fire and, finding that the wounded soldier's leg was practically severed, he tourniqueted the leg and carried him 100 yards through the barrage to a place of temporary shelter. Securing additional assistance, he took the man to the first-aid station. By his courage and efforts the wounded soldier's life was saved.

*FREE, GEORGE M. (1829131)_____
    Near Bois-des-Ogons, France, Oct. 10, 1918.
    R—Pittsburgh, Pa.
    B—Pittsburgh, Pa.
    G. O. No. 37, W. D., 1919.

Sergeant, Company C, 320th Infantry, 80th Division.
After half of his platoon and his officers had been killed or wounded he, under terrific barrage and machine-gun fire, organized a charge to attack a strong point, which was seriously menacing the whole command. In attempting this most hazardous task he was killed.
Posthumously awarded. Medal presented to brother, Conrad G. Free.

FREEHOFF, WILLIAM F._____
    Along the Marne River, July 15, 1918.
    R—Vestal Center, N. Y.
    B—Philadelphia, Pa.
    G. O. No. 56, W. D., 1922.

Captain, 38th Infantry, 3d Division.
Under a heavy hostile artillery fire and fire from low-flying airplanes, he rallied and reorganized scattered elements of his own and another company and courageously led them in the face of heavy machine-gun fire against a strong machine-gun nest, killed and wounded several of the crew and captured the gun.

FREEMAN, EDGAR H. (96367)_____
    Near Ancerviller, France, Mar. 4, 1918.
    R—Montevallo, Ala.
    B—Centerville, Ala.
    G. O. No. 126, W. D., 1919.

Corporal, Company D, 167th Infantry, 42d Division.
He conducted himself with marked bravery as a member of a patrol of 5 men which encountered an enemy patrol of 11 men, which it attacked and routed, taking 2 prisoners.

FREEMAN, PATRICK (1701019)_____
    In the Forest of Argonne, France, Sept. 27, 1918.
    R—New York, N. Y.
    B—Ireland.
    G. O. No. 20, W. D., 1919.

Sergeant, Company B, 306th Infantry, 77th Division.
He displayed exceptional courage and bravery while leading his platoon against enemy machine-gun and trench-mortar positions and putting them out of action. Although wounded, this soldier remained on duty with his platoon, killing and capturing several of the enemy and finally occupying part of the hostile trench.

FREEMAN, ROBERT L. (75083)_____
    Near Gesnes, France, Oct. 7, 1918.
    R—North Yakima, Wash.
    B—North Yakima, Wash.
    G. O. No. 47, W. D., 1921.

Sergeant, Company F, 128th Infantry, 32d Division.
1 of 4 survivors of a platoon of 41 who attacked Hill 269, he, with the 3 others, continued on their mission and held the hill for some time without hope of reinforcements.

*FREIBERG, HYMAN _____
    Near Chipilly Ridge, France, Aug. 9, 1918.
    R—New York, N. Y.
    B—New York, N. Y.
    G. O. No. 87, W. D., 1919.

Second lieutenant, 131st Infantry, 33d Division.
Although wounded early in an advance, he went forward with his men until he fell from loss of blood. He refused to be evacuated, and, while his wounds were being treated on the spot, preparatory to resuming the advance, was killed by shell fire.
Posthumously awarded. Medal presented to mother, Mrs. Jennie Freiberg.

*FREML, WESLEY_____
    Near Cantigny, France, May 30, 1918.
    R—Richmond, Calif.
    B—Vail, Iowa.
    G. O. No. 37, W. D., 1919.

First lieutenant, 26th Infantry, 1st Division.
While leading his company to the relief of a company who were holding a newly captured position, and while consolidating his new position, he was constantly subjected to perilous shelling, and even thought twice wounded by bursting shells, he refused evacuation. He successfully repulsed three strong enemy counterattacks, and while engaged in a hand-to-hand fight with two German officers he was killed.
Posthumously awarded. Medal presented to father, Wesley Freml.

FRENCH, HENRY (1316782)_____
    Near St. Souplet, France, Oct. 10, 1918.
    R—Maynardville, Tenn.
    B—Knox County, Tenn.
    G. O. No. 81, W. D., 1919.

Private, first class, Company L, 119th Infantry, 30th Division.
When his platoon was ordered to withdraw he manned a Lewis gun, the crew of which had become casualties, and by delivering a heavy fire successfully covered the withdrawal. Himself subjected to heavy enemy fire and wounded in the course of action, he remained at his post till the danger to his platoon was over.

FRESHOUR, EARNEST W. (93503)_____
    Near Ancerville, France, June 6, 1918.
    R—Marion, Ohio.
    B—Newcomerstown, Ohio.
    G. O. No. 35, W. D., 1919.

Private, Company D, 166th Infantry, 42d Division.
After all communication had been cut off and his platoon had suffered heavy casualties, Private Freshour, with another member of his platoon, volunteered and carried a message through heavy artillery and machine-gun fire to their company commander, giving him the information necessary to reinforce the position.

*FRETZ, EARL R._____
    Near Courchamps, France, July 18, 1918.
    R—Dorchester, Mass.
    B—Ottsville, Pa.
    G. O. No. 5, W. D., 1920.

First lieutenant, 12th Machine Gun Battalion, 4th Division.
After all the officers of Company E, 58th Infantry, had become casualties, he voluntarily assumed command of the infantry company in addition to his machine-gun platoon and personally led it forward to its objective. The gallantry displayed by this officer while exposed to heavy machine-gun and artillery fire was an important factor in the success of the advance.
Posthumously awarded. Medal presented to widow, Mrs. Gertrude Fretz.

FREW, STEPHEN P. (1900526)_____
   Near St. Juvin, France, Oct. 15 and 16, 1918.
   R—Punxsutawney, Pa.
   B—Punxsutawney, Pa.
   G. O. No. 9, W. D., 1923.

Private, Company A, 326th Infantry, 82d Division.
After seeing a runner killed by a shell while passing through a terrific artillery barrage, Private Frew volunteered to carry the message, well knowing that he had to cross an open space swept by machine-gun fire, 400 yards from Ravine au Pierre to battalion headquarters, in full view of the enemy. He showed a courage and bravery which was an inspiration to all his comrades.

*FREY, AMEL_____
   Near the Bois-de-Fontaine, France, May 27, 1918.
   R—Toledo, Ohio.
   B—Switzerland.
   G. O. No. 126, W. D., 1919.

Captain, 26th Infantry, 1st Division.
Captain Frey was seriously wounded while bringing his support platoons forward during a raid on his sector by the enemy, but gallantly continued to lead his men in the repulse of the raid until carried from the field by the stretcher bearers an hour later.
Posthumously awarded. Medal presented to sister, Mrs. Louisa Hofer.

FREY, CLARENCE F. (1786371)_____
   Near Verdun, France, Nov. 7, 1918.
   R—Red Lion, Pa.
   B—Red Lion, Pa.
   G. O. No. 37, W. D., 1919.

Private, Headquarters Company, 316th Infantry, 79th Division.
Acting as runner, Private Frey was sent from regimental headquarters to the front lines. On the way he was severely wounded by a fragment shell, but he continued on, despite weakness from loss of blood and dazed from shock. He delivered his message before being evacuated.

FREY, GEORGE J. (2792057)_____
   North of Exermont, France, Oct. 8, 1918.
   R—New York, N. Y.
   B—New York, N. Y.
   G. O. No. 60, W. D., 1920.

Private, first class, Company C, 16th Infantry, 1st Division.
Private Frey advanced beyond the front lines in order to locate an enemy machine-gun group who were raking the valley with their fire. Upon his approach the crew abandoned their gun and retreated into a wood. Private Frey pursued them into the shelter of the woods, firing upon them as he advanced. Upon returning he discovered a wounded comrade lying about 50 yards in advance of our lines in a place exposed to heavy fire. Private Frey went to this wounded man and carried him to a place of safety.

FRIEL, JOHN W. (2705793)_____
   Near Eyne, Belgium, Nov. 2, 1918.
   R—Philadelphia, Pa.
   B—Centerville, Md.
   G. O. No. 63, W. D., 1920.

Corporal, Company K, 145th Infantry, 37th Division.
In full view of the enemy and under heavy artillery and machine-gun fire, Corporal Friel, with two other men, swam the Escaut River and assisted in the construction of a footbridge. The construction of this bridge aided materially in the later successful operations of American troops in this vicinity.

*FRIEL, JOSEPH (1707634)_____
   Near Binarville, France, Oct. 2-5, 1918.
   R—New York, N. Y.
   B—New York, N. Y.
   G. O. No. 16, W. D., 1919.

Private, Company A, 308th Infantry, 77th Division.
He was on duty as a battalion runner during the period of six days in which his own and another battalion were surrounded by the enemy in the Argonne Forest, France, and cut off from communication with friendly troops. Although he was without food and, toward the end of the period, almost exhausted, this soldier carried messages to all parts of the position. Constantly under heavy fire from machine guns and trench mortars, he showed an utter disregard for his own personal safety. On the night of Oct. 5, 1918, he was sent to carry a message through the enemy lines to regimental headquarters. Several other attempts had been made, as this soldier knew, which had resulted in the death or capture of the runners. He made the attempt, but was killed in the performance of his mission by the enemy fire.
Posthumously awarded. Medal presented to mother, Mrs. Mary J. Friel.

*FRIERSON, MEADE, Jr_____
   Near Sergy, France, July 31, 1918.
   R—Nashville, Tenn.
   B—Columbia, Tenn.
   G. O. No. 34, W. D., 1919.

Captain, Cavalry, attached to 125th Infantry, 32d Division.
He was continually at the head of his company in the attack on Hill 212, near Sergy, and was constantly exposed to the terrific enemy fire while leading his men forward. After the objective had been gained and the men had dug in on the crest of the hill, he remained on watch the entire night, constantly patrolling his company sector under the heavy rifle, machine-gun, and artillery fire from the enemy's lines. He was later killed while on a hazardous reconnaissance in the vicinity of Juvigny, France.
Posthumously awarded. Medal presented to father, Meade Frierson, sr.

FRIES, GEORGE J., Jr. (1825534)_____
   Near Cunel, France, Oct. 11, 1918.
   R—Carrick, Pa.
   B—Pittsburgh, Pa.
   G. O. No. 35, W. D., 1919.

Private, medical detachment, 319th Infantry, 80th Division.
For two days and nights he worked incessantly as the only first-aid man with two companies in the front line. On several occasions he went out in front of our lines under heavy enemy fire to aid wounded men and to help bring them back to our line, his pack and equipment being badly torn by pieces of shrapnel.

FRITZ, ALBERT W. (283566)_____
   South of Soissons, France, July 18-23, 1918.
   R—Berlin, Wis.
   B—Berlin, Wis.
   G. O. No. 15, W. D., 1919.

Private, Company I, 16th Infantry, 1st Division.
While attached to a machine-gun company as an ammunition carrier, Private Fritz, after being twice wounded, continued to carry ammunition while exposed to heavy shell fire.

FRITZ, CLYDE A. (40409)_____
   Near Soissons, France, July 18, 1918.
   R—Kokomo, Ind.
   B—Richland Center, Wis.
   G. O. No. 37, W. D., 1919.

Corporal, Company K, 9th Infantry, 2d Division.
He accompanied a party of 10 men which attacked a ravine infested with enemy machine-gun nests. During the exploit all except Corporal Fritz and one comrade were killed or wounded, but they succeeded in silencing enough guns to make further advance possible.

FRITZ, LEONARD B. (1309759)_____
   Near Busigny, France, Oct. 18, 1918.
   R—Laurel Bloomery, Tenn.
   B—Laurel Bloomery, Tenn.
   G. O. No. 133, W. D., 1918.

Private, Company M, 117th Infantry, 30th Division.
When his platoon was held up by an enemy machine-gun post, Private Fritz, with another soldier, took their automatic rifle and rushed forward through intense fire, skillfully placed the rifle in position, and opened an effective fire.

FRIZZELL, CHARLES F._____
   Near Le Charmel, France, July 22, 1918.
   R—New York, N. Y.
   B—Nashville, Tenn.
   G. O. No. 19, W. D., 1920.

First lieutenant, 38th Infantry, 3d Division.
While making a personal reconnaissance in front of our lines he was seriously wounded, but crawled to the rear to a patrol with valuable information. He ordered that he be left behind and that the patrol return with the information he had obtained. The patrol returned and he, unable to resist by fighting, was later taken prisoner by the advancing enemy.

FROMAN, HJALMAR (2257094)_____
   Near Gesnes, France, Oct. 3, 1918.
   R—Murphy, Idaho.
   B—Sweden.
   G. O. No. 20, W. D., 1919.

Corporal, Company D, 361st Infantry, 91st Division.
He voluntarily and unhesitatingly left shelter under heavy shellfire and, without thought of personal danger, rendered first aid and carried a wounded comrade to a place of safety.

FROST, JOHN_____
   Near Verneville, France, Sept. 17, 1918.
   R—San Antonio, Tex.
   B—San Antonio, Tex.
   G. O. No. 46, W. D., 1919.

First lieutenant, 103d Aero Squadron, Air Service.
While on patrol duty with two other pilots in enemy territory Lieutenant Frost attacked an enemy formation of eight planes (type Fokker). He attacked at close range, and after a severe combat succeeded in sending one enemy down in flames. With his comrades, they destroyed in all four planes, and by repeated attacks dispersed the remainder.

FRUNDT, OSCAR C._____
   In eastern Siberia, June 12, 1919, from June 13 to June 18, 1919, and June 25, 1919.
   R—Jersey City, N. J.
   B—Jersey City, N. J.
   G. O. No. 133, W. D., 1919.

Captain, Medical Corps, U. S. Army.
For extraordinary heroism in action while in command of a hospital train in eastern Siberia on June 12, 1919, from June 13 to June 18, 1919, and on June 25, 1919, for his expeditious treatment and care of wounded and the skillful handling of a hospital train while under fire.

FRYE, JOHN G._____
   Near Blanc Mont, France, Oct. 4, 1918.
   R—Bernie, Mo.
   B—Swinton, Mo.
   G. O. No. 95, W. D., 1919.

Private, 97th Company, 6th Regiment, U. S. Marine Corps, 2d Division.
Private Frye, a platoon runner, fearlessly exposed himself in carrying important messages through a heavy enemy barrage.

*FUHRMAN, PAUL C. (554884)_____
   Near Romagne, France, Oct. 12, 1918.
   R—Brooklyn, N. Y.
   B—Harrisburg, Pa.
   G. O. No. 16, W. D., 1920.

Corporal, Company B, 9th Machine Gun Battalion, 3d Division.
Corporal Fuhrman, without regard to personal safety, operated and advanced his machine gun under intense enemy artillery and machine-gun fire. By this act he disabled and put out of action two enemy machine-gun crews which were holding up the advance of the infantry.
Posthumously awarded. Medal presented to widow, Mrs. Paul C. Fuhrman.

FULLER, ARTHUR M. (2382833)_____
   Near Cunel, France, Oct. 14, 1918.
   R—Baltimore, Md.
   B—Lauraville, Md.
   G. O. No. 44, W. D., 1919.

Supply sergeant, Company C, 60th Infantry, 5th Division.
Accompanied by one other soldier, Sergeant Fuller flanked two machine-gun nests, killed seven of the enemy, and captured four machine-guns, thereby making it possible for two companies of his battalion to enter the woods and continue the advance.

*FULLER, EDWARD C._____
   In the attack of Bois-de-Belleau, France, June 12, 1918.
   R—Hamilton, Va.
   B—Hamilton, Va.
   G. O. No. 99, W. D., 1919.

Captain, 6th Regiment, U. S. Marine Corps, 2d Division.
While fearlessly exposing himself in an artillery barrage for the purpose of getting his men into a position of security in the attack on Bois-de-Belleau, on June 12, 1918, he was killed and thereby gave his life in an effort to protect his men.
Posthumously awarded. Award made to father, Col. B. H. Fuller.

FULLER, JOSEPH M. (1253725)_____
   Near Apremont, France, Oct. 4, 1918.
   R—Wilkes-Barre, Pa.
   B—Wilkes-Barre, Pa.
   G. O. No. 64, W. D., 1919.

Sergeant, Battery D, 109th Field Artillery, 28th Division.
He left an observation post to aid in the rescue of an officer who had fallen in a field swept by artillery and machine-gun fire. After administering first aid he carried the officer to a place of safety, his prompt action saving the officer's life.

FULLER, KENNETH E._____
   Near Vaux Castille, France, July 18, 1918.
   R—Exeter, N. H.
   B—Exeter, N. H.
   G. O. No. 22, W. D., 1920.

Second lieutenant, 23d Infantry, 2d Division.
When his company was temporarily halted by heavy machine-gun fire, Second Lieutenant Fuller personally led a group of 10 men in an attack on the machine-gun position. He was killed while leading this attack, but due to his heroic example the enemy position was captured and his company was able to continue its advance.
Posthumously awarded. Medal presented to father, Arthur O. Fuller.

FULLER, LONZO L. (263176)_____
   Near Juvigny, north of Soissons, France, Sept. 1, 1918.
   R—Lansing, Mich.
   B—Midland, Mich.
   G. O. No. 20, W. D., 1919.

Private, Company H, 128th Infantry, 32d Division.
During an attack on a strong enemy position, in the face of heavy fire from artillery and machine guns, Private Fuller, a runner, worked unceasingly throughout the attack in maintaining lateral liaison between units. The entire route which he was obliged to travel was exposed to heavy fire from the enemy, and on one of his trips he succeeded in locating a machine-gun nest which had been inflicting heavy losses on our troops. Returning through a rain of bullets, he reported the exact position of the emplacement, which enabled the artillery to demolish it.

FULLER, WILLIAM H. (1449861)_____
   Near Montrebeau Woods, France, Sept. 29, 1918.
   R—Manchester, Kans.
   B—England.
   G. O. No. 71, W. D., 1919.

Corporal, Company M, 137th Infantry, 35th Division.
With another soldier he left a shell hole and advanced under heavy machine-gun fire, rescued one of our wounded soldiers, and took him to a dressing station. He showed marked heroism, the mission being undertaken against the advice of his platoon leader.

FULTON, SELMO (1464582)_____
Near Charpentry, France, Sept. 29,
1918.
R—Odessa, Mo.
B—Odessa, Mo.
G. O. No. 81, W. D., 1919.

Private, first class, Battery C, 129th Field Artillery, 35th Division.
When the rest of the gun squad was wiped out by heavy shelling and he himself
had been wounded, he continued to fire his piece single handed until another
gun squad was formed.

FUNDERBURK, MYRON F. (1311757)_____
Near Brancourt, France, Oct. 8, 1918.
R—Pageland, S. C.
B—Monroe, S. C.
G. O. No. 21, W. D., 1919.

Mechanic, Company I, 118th Infantry, 30th Division.
He was acting as a stretcher bearer for his company, which was suffering many
casualties as it advanced. While he was carrying a wounded soldier, he was
himself seriously wounded in the shoulder. He continued, under heavy
artillery fire, to evacuate the wounded until he fell from exhaustion.

FUNK, PETER_____
Near St. Etienne, France, Oct. 4,
1918.
R—Mount Healthy, Ohio.
B—Cincinnati, Ohio.
G. O. No. 37, W. D., 1919.

Private, 8th Company, 5th Regiment, U. S. Marine Corps, 2d Division.
Private Funk, together with his corporal, advanced under heavy artillery and
machine-gun fire to a forward position, where Private Funk operated the
machine gun, doing great damage to the enemy. These two remained with
their gun in a far-advanced position alone until their ammunition was
exhausted.

FUQUAY, JAMES (1402857)_____
At Guillimet Farm, France, Sept. 28,
1918.
R—Chicago, Ill.
B—Chicago, Ill.
G. O. No. 44, W. D., 1919.

Private, Company H, 370th Infantry, 93d Division.
When Private Fuquay, an automatic-rifle man, was stationed at a particularly
dangerous location his rifle became jammed, whereupon he took it apart,
remedied the trouble, and again put it into operation. While doing this he
was wounded in the left arm, but refused first aid, continuing to serve his
piece and direct the fire on the enemy positions until completely exhausted
from loss of blood.

FURBUSH, GEORGE W., Jr_____
Near Vaux, France, July 1, 1918.
R—Waltham, Mass.
B—Waltham, Mass.
G. O. No. 4, W. D. 1927.

First lieutenant, 23d Infantry, 2d Division.
Although wounded, he continued in action as leader of a platoon and by his
heroic conduct stimulated his men to success until incapacitated by a second
wound.

FUREY, JOHN PATRICK (90726)_____
Near Landres-et-St. Georges, France,
Oct. 15, 1918.
R—New York, N. Y.
B—Ireland.
G. O. No. 9, W. D., 1923.

Sergeant, Company H, 165th Infantry, 42d Division.
Although himself slightly wounded while near enemy wire entanglements,
which position was boxed in by a hostile artillery barrage and swept by
enemy machine guns, Sergeant Furey, with utter disregard for his own safety
voluntarily exposed himself to a great danger in going to the aid of a comrade,
who was severely wounded in both legs, bandaging his wounds and assisting
him to the first-aid station, all under heavy hostile fire. He then rejoined his
command near the front line. In the performance of this gallant act he was
again wounded.

FURFORO, VINACEZO (2067267)_____
During the Meuse-Argonne offen-
sive, France, Sept. 26, 1918.
R—Chicago, Ill.
B—Italy.
G. O. No. 5, W. D., 1920.

Private, Company B, 131st Infantry, 33d Division.
With 3 other soldiers he charged and captured a battery of three. 77 field pieces
which, protected by machine guns, were firing point-blank on the position
held by his company. This deed enabled his company to continue the
advance.

FURLOW, GEORGE WILLARD_____
Near Charey, France, Sept. 13, 1918.
R—Rochester, Minn.
B—Rochester, Minn.
G. O. No. 20, W. D., 1919.

First lieutenant, pilot, 103d Aero Squadron, Air Service.
Lieutenant Furlow, while leading a patrol of 3 monoplace planes at an altitude
of 400 meters, met and attacked an enemy patrol of 7 monoplace planes.
Despite numerical superiority, he destroyed 2 of the enemy's planes, and with
the aid of his companions forced the others to withdraw.
Oak-leaf cluster.
An oak leaf is awarded Lieutenant Furlow for the following act of extraordinary
heroism in action in the region of Verneville, France, Sept. 17, 1918: Lieutenant Furlow, while on patrol with 2 other companions, met and attacked
an enemy formation of 8 planes. In the course of the combat which ensued,
Lieutenant Furlow's plane was severely damaged by the enemy's fire.
Despite the damage, he continued the attack until he had destroyed 1 hostile aircraft and with his patrol forced the remainder of the enemy to retire.

FURNESS, THOMAS F_____
Near Fosse, France, Nov. 1-5, 1918.
R—Brookline, Mass.
B—Philadelphia, Pa.
G. O. No. 3, W. D., 1924.

First lieutenant, 7th Field Artillery, 1st Division.
Performing his duties as liaison officer in a most effective manner, he also took
command of Infantry platoons, after the officers had become casualties, and
led them brilliantly. On one occasion he reorganized a platoon after the
leader had been killed, and attacked a strong machine-gun position, capturing 4 guns and 15 prisoners.

FURR, WALTER E_____
Near Vierzy, France, July 19, 1918.
R—Franklin, N. C.
B—Concord, N. C.
G. O. No. 117, W. D., 1918.

Private, 82d Company, 6th Regiment, U. S. Marine Corps, 2d Division.
Unaided, Private Furr crept forward in advance of his line, searched an under-
ground tunnel, captured 5 Germans, and brought them back through heavy
machine-gun and shell fire.

FURY, WILLIAM H_____
Near Chateau-Thierry, France, June
10, 1918.
R—Mossneck, Va.
B—England.
G. O. No. 37, W. D., 1919.

Sergeant, Headquarters Detachment, 6th Regiment, U. S. Marine Corps,
2d Division.
He remained in a building subjected to heavy shell fire and mustard gas and
made coffee for wounded men until the building was almost entirely demolished by the enemy shell fire.

GABRIEL, HARRY S.
  At Molleville Farm, near Hill 378,
    Grande Montagne sector, France,
    Nov. 3, 1918.
  R—Watkins, N. Y.
  B—Watkins, N. Y.
  G. O. No. 14, W. D., 1923.

First lieutenant, 316th Infantry, 79th Division.
Between 6.30 and 9 o'clock he fought with 2 platoons through a dense thicket against an unlocated enemy, who resisted with constant artillery and machine gun fire. The 2 officers of the platoons on his left were killed, as well as his accompanying officer, Lieut. Rudolph E. Peterson. Although his men were decimated, he rallied them four separate times and took command of the men originally on his left after their officers were killed, and renewed his assaults against hidden machine-gun nests on Hill 370. His small group silenced 7 machine guns and captured several prisoners. He reached his objective—Hill 370—in spite of terrific losses, definitely located the enemy, which was the main purpose of the reconnaissance, and assisted in the capture of 7 machine-gun nests. The extraordinary courage of Lieutenant Gabriel, his coolness, and superb devotion to duty served as a constant inspiration to every man of the 316th Infantry Regiment.

GADDIS, THOMAS (1245485).
  Near La Chene Tondu, France, Oct.
    5, 1918.
  R—Oil City, Pa.
  B—England.
  G. O. No. 50, W. D., 1919.

Sergeant, Company K, 111th Infantry, 28th Division.
When his company had been held up by a sweeping fire from a machine-gun nest, Sergeant Gaddis, in charge of a patrol of 5 men, went forward to reduce the nest. The patrol was once driven back by the terrific fire, but again advanced. 15 feet from the nest 2 guns opened fire, killing 3 of the patrol, whereupon Sergeant Gaddis rushed forward alone and killed the crews of both guns with hand grenades. His action not only made possible the advance of the line, which was of extreme importance, but also saved many of his comrades at a time when his company had suffered heavy casualties.

GAFFEY, LUKE (1214669).
  Near Ronssoy, France, Sept. 28, 1918.
  R—New York, N. Y.
  B—New York, N. Y.
  G. O. No. 133, W. D., 1919.

Private, Company F, 108th Infantry, 27th Division.
He displayed rare courage in leaving shelter and going into an open field under heavy shell and machine-gun fire and rescuing wounded soldiers of another regiment.
Oak-leaf cluster.
An oak-leaf cluster, to be worn with the distinguished-service cross, is awarded Private Gaffey for the following act of extraordinary heroism in action near Ronssoy, France, Sept. 29, 1918: When all the other members of his squad had been killed or wounded, he picked up an automatic rifle and advanced alone against an enemy position.

GAGE, GEORGE H.
  At Rambucourt, France, Mar. 17,
    1918, and at Cantigny, France,
    May 28, 1918.
  R—Rochester, N. Y.
  B—Benton, N. Y.
  G. O. No. 117, W. D., 1918.

Captain, Medical Corps, attached to 28th Infantry, 1st Division.
Conspicuous for his courage in the actions at Rambucourt on Mar. 17, 1918, and at Cantigny, May 28, 1918, he gave inspiration to the officers and men of the command by his extraordinary heroism throughout the operations south of Soissons, July 18 to 22, 1918, and especially at Berzy-le-Sec, July 21, 1918, when he accompanied the first line and attended the wounded in the open under incessant machine-gun and artillery fire.

GAHRING, W. ROSS.
  At Cantigny, France, May 28–31,
    1918.
  R—Mount Vernon, Mo.
  B—La Grange, Mo.
  G. O. No. 99, W. D., 1918.

First lieutenant, 28th Infantry, 1st Division.
He was severely wounded by machine-gun fire shortly after successfully leading his platoon to its objective. Notwithstanding this, he remained on duty for nine hours, thereby setting a brave example for his men.

GAINES, JOHN P.
  Near Soissons, France, July 20, 1918.
  R—Bay City, Tex.
  B—Matagorda, Tex.
  G. O. No. 100, W. D., 1918.

Second lieutenant, 26th Infantry, 1st Division.
He stayed with his command and led it to its final objective near Soissons, France, July 20, 1918, after being wounded, directed the consolidation of his position, and yielded his post only at the command of a superior officer.

GALBRAITH, FREDERICK W., Jr.
  Near Ivoiry, France, Sept. 29, 1918.
  R—Cincinnati, Ohio.
  B—Watertown, Mass.
  G. O. No. 140, W. D., 1918.
  Distinguished-service medal also
    awarded.

Colonel, 147th Infantry, 37th Division.
When an enemy counterattack was imminent he went into the front lines under a violent artillery and machine-gun barrage, and by the coolness and certainty of his orders and the inspiring example of his personal courage reorganized his own command and took command of other units whose officers had been lost or diverted in the confusion of battle. Knocked down by a shell, he refused to be evacuated and continued to carry on the work of reorganizing his position and disposing the troops to a successful conclusion.

GALE, CARROLL M.
  North of Bois-de-Chaume, France,
    Oct. 10, 1918.
  R—Chicago, Ill.
  B—Angola, Ind.
  G. O. No. 56, W. D., 1922.

Captain, 131st Infantry, 33d Division.
Although painfully wounded while his company was advancing to the final objective, he courageously continued to lead the company and later directed the front-line operations of his battalion until the line was organized. Captain Gale was continuously under heavy shell and machine-gun fire while moving back and forth along the line, and the example of his coolness and bravery was an inspiration to his men.

*GALE, HUGH K. (1622152).
  North of Cierges, France, Oct. 4,
    1918.
  R—Liberty, N. Mex.
  B—Franklin, Ariz.
  G. O. No. 16, W. D., 1920.

Private, Company C, 8th Machine Gun Battalion, 3d Division.
When the attack on Hill 241 was held up, Private Gale led a few infantrymen and machine gunners forward under artillery and machine-gun fire to establish a new line about a kilometer in advance of our front lines. He was mortally wounded while making the dash forward.
Posthumously awarded. Medal presented to father, George H. Gale.

GALKA, TONY (545292).
  Near Crezancy, France, July 15, 1918.
  R—Barnesboro, Pa.
  B—Poland.
  G. O. No. 32, W. D., 1919.

Private, Company A, 30th Infantry, 3d Division.
After his company had withdrawn from their position he voluntarily returned to the former position and, throughout the night of July 15, assisted in evacuating the wounded.

GALLAGHER, CORNELIUS E. (2411781)....
Near Grand Pre, France, Oct. 28–29, 1918.
R—Bayonne, N. J.
B—Bayonne, N. J.
G. O. No. 35, W. D., 1919.

Sergeant, Company C, 309th Machine Gun Battalion, 78th Division.
Although painfully wounded in the shoulder, he remained at his post for 6 hours without reporting his wound. Even then he requested permission to remain, and, having obtained it, he encouraged his men to withstand a threatened counterattack. He left the field only when ordered to do so.

GALLAGHER, GEORGE (548025).........
Near Crezancy, France, July 15, and near Jaulgonne, France, July 23, 1918.
R—Brooklyn, N. Y.
B—Brooklyn, N. Y.
G. O. No. 78, W. D., 1919.

Corporal, Company L, 30th Infantry, 3d Division.
On the night of July 15, during the intense shelling which preceded the German offensive, Corporal Gallagher made 3 trips through the fire and, collecting lost troops, conducted them to their place in the line. On July 23 he led a patrol through perilous fire and established communication with troops on the right of his position.
Posthumously awarded. Medal presented to father, Hugh J. Gallagher.

GALLAGHER, JOHN M. (154803).......
Near Verdun, France, Oct. 9, 1918.
R—Ginter, Pa.
B—Houtzdale, Pa.
G. O. No. 32, W. D., 1919.

Corporal, Company C, 1st Engineers, 1st Division.
While his battalion was defending a hill captured from the enemy, Corporal Gallagher was placed on the extreme right of the line nearest to the enemy, when an enemy counterattack was launched against their position. On account of the severe casualties inflicted, orders were given to withdraw. Remaining alone at his post after the withdrawal, Corporal Gallagher valiantly resisted the attack, advancing about 30 yards, when he saw the enemy strip the body of his dead commanding officer. Later, when his company drove the enemy beyond the former position, they found the body of Corporal Gallagher lying across his rifle. In a circle facing him were the bodies of 6 Germans, whose lives he exacted during the unequal struggle.
Posthumously awarded. Medal presented to father, Thomas B. Gallagher.

GALLO, JOSEPH (241758).............
Near Vaux, France, July 1, 1918.
R—Silver Lake, N. J.
B—Italy.
G. O. No. 117, W. D., 1918.

Sergeant, first class, Company A, 2d Engineers, 2d Division.
He showed great bravery and energy and exceptional presence of mind in leading his platoon through a heavy barrage fire to reinforce a weakened section of the line. Further, after the capture of a hill which was his objective, he pursued a German officer, although exposed to heavy fire, captured him, took away his arms, and brought him back a prisoner.

GALLOWAY, JUDSON P.............
In the Chateau-Thierry sector, June 6, 1918.
R—Newburgh, N. Y.
B—Newburgh, N. Y.
G. O. No. 116, W. D., 1918.

First lieutenant, 23d Infantry, 2d Division.
Lieutenant Galloway exhibited exceptional courage and leadership when, after being mortally wounded, he continued to direct the steady advance of his platoon in the face of heavy machine-gun fire until struck a second time and killed.
Posthumously awarded. Medal presented to widow, Mrs. Jane R. Galloway.

GALOFF, FRED (284404).............
Vicinity of Breheville and Ecurey, northwest of Verdun, France, Nov. 5–11, 1918.
R—Elmwood, Wis.
B—Menomonie, Wis.
G. O. No. 3, W. D., 1922.

Sergeant, Company H, 128th Infantry, 32d Division.
He displayed remarkable leadership and courage in keeping the men together and calm under the terrific enemy artillery and machine-gun fire during the attack on Nov. 10, 1918. Although gassed and exhausted, he refused to go to the rear for medical treatment and remained with the company, assisting in giving first aid to the wounded and arranging for their evacuation, disregarding personal danger and displaying a fine spirit of sacrifice.

GAMMELL, WARREN S. (641073).......
Northwest of Somme-Py, near St. Etienne, France, Oct. 8, 1918.
R—Madison, Minn.
B—Madison, Minn.
G. O. No. 15, W. D., 1919.

Sergeant, first class, section No. 606, Ambulance Service.
He showed conspicuous courage and devotion to duty in evacuating the wounded under the most trying conditions. He made repeated trips in plain view of enemy observers over roads under continuous shell fire. He was killed by a shell fragment while riding in an ambulance to an advanced post.
Posthumously awarded. Medal presented to father, Dr. H. W. Gammell.

GANCAZ, STANLEY (56656).............
At Exermont, France, Oct. 4, 1918.
R—Jersey City, N. J.
B—Russia.
G. O. No. 103, W. D., 1919.

Private, first class, Company A, 28th Infantry, 1st Division.
When a German 77-millimeter gun, supported by numerous machine guns, broke the tank attack and held up the infantry advance, Private Gancaz, with 2 scouts, made an encircling movement amid heavy fire and put the gun out of action, capturing the entire crew. They then cleaned out the enemy dugouts in the vicinity and returned with 40 prisoners, including an officer.

GANDER, WILLIAM (2383354)...........
Near Cunel, France, Oct. 14, 1918.
R—Springfield, Ohio.
B—Russia.
G. O. No. 37, W. D., 1919.

Private, Company E, 60th Infantry, 5th Division.
Voluntarily advancing ahead of his company, he captured an enemy machine gun with 3 prisoners. The next morning, with another soldier, he again went forth and caused the surrender of several guns and 5 prisoners. His bravery in both instances greatly facilitated the advance of his company, who were meeting with resistance from the nests which he wiped out.

GANNON, JOSEPH J. (72088)..........
Bois-Brule, near Apremont, France, Apr. 12, 1918.
R—Cambridge, Mass.
B—Cambridge, Mass.
G. O. No. 107, W. D., 1918.

Private, Company E, 104th Infantry, 26th Division.
He displayed courage and self-sacrificing devotion to duty in action against the enemy on Apr. 12, 1918, voluntarily going with one comrade to an advanced post in a communication trench and with automatic rifle holding back advance of the enemy through the trench until his comrade was killed and he himself severely wounded.

GAREER, HARRY E. (1097441)..........
Near Montauville, France, Oct. 24, 1918.
R—Beaver Falls, N. Y.
B—Beaver Falls, N. Y.
G. O. No. 21, W. D., 1919.

Private, first class, Battery F, 21st Field Artillery, 5th Division.
When an enemy shell set fire to the powder dump of his battery, he crossed a shell-swept area to warn his companions of the danger from the threatened explosion of the dump. He then recrossed the shelled area to notify his officers of the conflagration and returned to the burning dump to assist in saving some of the powder.

GARCIA, AMADO (1626968)
Near Fismes, France, Aug. 26, 1918.
R—Acomita, N. Mex.
B—Acomita, N. Mex.
G. O. No. 98, W. D., 1919.

Private, first class, Company K, 110th Infantry, 28th Division.
With 2 other soldiers, Private Garcia crawled 300 yards in front of our lines, through the enemy's wire, and attacked a hostile machine-gun nest. The enemy crew opened fire on them at a range of only 10 yards and resisted stubbornly, but they succeeded in killing 3 of the crew and driving off the others with clubbed rifles. They returned to our lines under heavy fire.

GARCIA, GRAVIEL (2229509)
Near St. Juvin, France, Oct. 16, 1918.
R—Somerville, Tex.
B—Davis, Tex.
G. O. No. 46, W. D., 1919.

Private, Company C, 325th Infantry, 82d Division.
He voluntarily went out into no man's land under heavy enemy fire and administered first aid to a wounded comrade. While making his way back to our lines with the wounded man he was himself severely wounded.

GARDELLA, FRANK J., Jr. (88892)
North of the River Ourcq, near Villers-sur-Fere, France, July 28, 1918.
R—New York, N. Y.
B—New York, N. Y.
G. O. No. 99, W. D., 1918.

Sergeant, Machine Gun Company, 165th Infantry, 42d Division.
When 2 enemy airplanes flew parallel to our Infantry lines north of the River Ourcq, near Villers-sur-Fere, France, July 28, 1918, pouring machine-gun bullets into our positions and driving everyone to cover, he rushed to his machine gun and took aim at the upper of the 2 machines. Although he was constantly subject to a storm of bullets from the planes and from enemy snipers on the ground, he nevertheless coolly sighted his gun and riddled the upper plane. It collapsed and fell in flames, striking the lower one as it fell and causing it to crash to the earth also.

GARDINER, KENNETH (170093)
Near St. Eugene, France, July 17, 1918.
R—Council Bluffs, Iowa.
B—Essex, Iowa.
G. O. No. 44, W. D., 1919.

Corporal, Battery A, 10th Field Artillery, 3d Division.
Suffering from shell shock and a wound in the shoulder, he continued to carry messages over shell-swept roads until he was forced to go to the dressing station by his battery commander.

GARDNER, ALFRED W.
In the Argonne Forest, France, Oct. 3, 1918.
R—New York, N. Y.
B—Sharon Springs, N. Y.
G. O. No. 37, W. D., 1919.

First lieutenant, 305th Infantry, 77th Division.
Attacking enemy machine-gun nests, he displayed the highest courage when he led his company up a steep slope in the face of murderous fire. Before he could accomplish his objective he was killed.
Posthumously awarded. Medal presented to mother, Mrs. Mary E. Gardner.

GARDNER, ELMER W. (39586)
South of Soissons, France, July 18, 1918.
R—Binghamton, N. Y.
B—Marydale, N. Y.
G. O. No. 7, W. D., 1919.

Private, Company G, 9th Infantry, 2d Division.
While acting as a runner he was seriously wounded, but in spite of his injury he struggled forward and delivered his message.

GARDNER, GEORGE W. (2176371)
In Bois-de-Barricourt, France, Nov. 2, 1918.
R—Traer, Kans.
B—Traer, Kans.
G. O. No. 37, W. D., 1919.

Sergeant, Company F, 353d Infantry, 89th Division.
He led his platoon through shell and machine-gun fire in an attack on strong enemy positions, capturing 2 machine guns and assisting in the destruction of several others that were holding up our advance.

GARDNER, JOHN H. (1312150)
Near Ramicourt, France, Oct. 8, 1918.
R—Hartsville, S. C.
B—Hartsville, S. C.
G. O. No. 21, W. D., 1919.

Sergeant, Company L, 118th Infantry, 30th Division.
After his company commander had been wounded immediately before an attack, Sergeant Gardner took command of the company and led it throughout the action. When his company was held up by machine-gun fire, he went forward and killed 4 German machine gunners, thereby enabling his company to continue the advance. On another occasion he picked up the rifle of a wounded soldier and killed 3 of the enemy. Later, when his company was almost surrounded by hostile machine guns, his men, under his cool direction, fought their way out, reached their objective and consolidated the position.

GARDNER, MARTIN E. (1900435)
Near St. Juvin, France, Oct. 15, 1918.
R—Sharpsburg, Md.
B—Sharpsburg, Md.
G. O. No. 72, W. D., 1920.

Sergeant, Company A, 326th Infantry, 82d Division.
After his commanding officer had been severely wounded, Sergeant Gardner took command of the detachment, and although severely wounded himself, he gallantly led his group in the attack against enemy machine-gun fire. At the conclusion of this successful attack, he was shot and killed by an enemy sniper.
Posthumously awarded. Medal presented to mother, Mrs. James A. Gardner.

GAREY, ENOCH B.
In the Gerardmer defensive sector, France, Sept. 16, 1918.
R—Baltimore, Md.
B—Maryland.
G. O. No. 60, W. D., 1920.

Major, 18th Machine Gun Battalion, 6th Division.
Major Garey organized and led a combat patrol which penetrated the enemy lines and returned with several prisoners. Although exposed to machine-gun fire and later grenade fire from a superior number of the enemy, he conducted his patrol so as to accomplish its mission and returned to our lines with valuable information.

GAREY, PHILIP (1211451)
Near Ronssoy, France, Sept. 29, 1918.
R—South Orange, N. J.
B—Orange, N. J.
G. O. No. 32, W. D., 1919.

Sergeant, Company I, 107th Infantry, 27th Division.
Organizing a platoon of survivors of a battalion, he led them in attack against the enemy. Under terrific shell and machine-gun fire, he advanced against an enemy machine-gun nest and, by the effective use of hand grenades, killed or wounded the crew and destroyed the gun.

GARIEPY, THEODORE T. (261847)
East of the Bois-des-Grimpettes, near Cierges, northeast of Chateau-Thierry, France, Aug. 1, 1918.
R—Detroit, Mich.
B—Oscoda, Mich.
G. O. No. 132, W. D., 1918.

Corporal, Company C, 125th Infantry, 32d Division.
When two companies of another organization deployed in a field, a violent artillery fire was directed on them, necessitating their moving into an adjoining wood, leaving their dead and wounded on the field. Corporal Gariepy, with a party of five men, left the woods and directed the work of carrying the wounded to a safe spot, where they could be given medical attention. He directed this work under heavy fire and with an utter disregard for his own life.

GARLINGTON, CRESWELL.............
Near Merval, France, Sept. 14, 1918.
R—Ednor, Md.
B—Rock Island, Ill.
G. O. No. 21, W. D., 1919.

Lieutenant colonel, General Staff, 77th Division.
In preparation for attack by units of his division he helped establish an advance observation post. Learning a wounded officer was in front, Lieutenant Colonel Garlington made his way twice through intense fire from artillery and small arms to where the wounded officer lay and assisted in carrying him to safety.

GARNER, CORTIS H. (1330345)...........
Near Bellicourt and Nauroy, France,
Sept. 28 to Oct. 1, 1918.
R—Raleigh, N. C.
B—Wake County, N. C.
G. O. No. 37, W. D., 1919.

Private, Company C, 105th Field Signal Battalion, 30th Division.
Attached to the headquarters of the 69th Infantry Brigade as a dispatch rider, he repeatedly showed exceptional bravery throughout the operations of that brigade. During the engagement near Bellicourt he particularly distinguished himself by his prompt delivery of all messages under vigorous shellfire and bombing raids by enemy aircraft, riding day and night in all kinds of weather.

GARNER, GEORGE W. (1588420)..........
At Exermont, France, Oct. 4, 1918.
R—Lonoke, Ark.
B—Lonoke, Ark.
G. O. No. 87, W D., 1919.

Private, Company D, 28th Infantry, 1st Division.
When a German 77-millimeter gun, supported by numerous machine guns, broke the tank attack and held up the Infantry advance, Private Garner, with two scouts, made an encircling movement amid heavy fire and put the gun out of action, capturing the entire crew. They then cleaned out the enemy dugouts in the vicinity and returned with 40 prisoners, including an officer.

GARNER, JOHN B.(1932043)..............
North of Clermery, in Lorraine,
France, Aug. 16-17, 1918.
R—Cedartown, Ga.
B—Polk County, Ga.
G. O. No. 100, W. D., 1918.

Sergeant, Company F, 325th Infantry, 82d Division.
Although wounded in the face and hand, he went 50 meters up a road in the face of fire from two German machine guns and of exploding grenades to rescue a lieutenant who lay 100 meters within the German wire, so severely wounded as to be unable to move without assistance. Sergeant Garner put the officer on his back, crawled back through the enemy's wire, and from there carried him 500 meters across open ground, under fire, to safety.

GARR, CHARLES W. (156821)...........
Near St. Etienne-a-Arnes, France,
Oct. 7, 1918.
R—Spiro, Okla.
B—Spiro, Okla.
G. O. No. 23, W. D., 1919.

Corporal, Company D, 2d Engineers, 2d Division.
Advancing ahead of our infantry, he made a reconnaissance of the town of St. Etienne-a-Arnes; and, in spite of the danger, exposed to artillery and machine-gun fire of our own and enemy guns, he procured and returned with valuable information.

GARRISON, WILLIAM H. (199156).......
Near Chateau-Thierry, France, July
20-23, 1918.
R—Atlanta, Nev.
B—Pioche, Nev.
G. O. No. 125, W. D., 1918.

Private, Signal Corps platoon, 101st Infantry, 26th Division.
He displayed great personal bravery and skill in maintaining telephone lines between the regimental commander and the leading battalion for more than two days. He patrolled the line continuously and repaired it when it was cut during bombardment. Knocked down frequently by exploding shells, and once buried beneath dirt and débris, he nevertheless stuck courageously to his task, thereby making communication possible.

GARRITY, PATRICK (2087074).............
Near Remonville, France, Nov. 1,
1918.
R—Chicago, Ill.
B—Ireland.
G. O. No. 35, W. D., 1920.

Private, first class, Company C, 354th Infantry, 89th Division.
When his company was advancing across an open field an enemy machine gun opened fire upon it from the left front. Private Garrity and two other soldiers being on that flank of the company immediately advanced upon it. His two comrades were shot down, but Private Garrity advanced alone at a run and with his bayonet drove the three Germans from their guns into a nearby dugout, captured them and the gun.

GARSIDE, HENRY P., Jr. (39290).........
Near Beaumont, France, Nov. 1-5,
1918.
R—Fall River, Mass.
B—Fall River, Mass.
G. O. No. 44, W. D., 1919.

Private, Company F, 9th Infantry, 2d Division.
On duty as a company runner, he repeatedly carried messages unflinchingly through heavy enemy machine-gun and shell fire with utter disregard for personal safety.

GARST, HENRY J. (558199).............
Near Bazoches, France, Aug. 9, 1918.
R—St. Louis, Mo.
B—St. Louis, Mo.
G. O. No. 46, W. D., 1919.

Corporal, Company H, 47th Infantry, 4th Division.
Responding to a call for volunteers to destroy a hostile machine gun, Corporal Garst, with two other soldiers, boldly went forward through machine-gun fire and accomplished this mission.

*GARTHRIGHT, JOHN R. (1707274).......
In the Forest of Argonne, France,
Sept. 28-29, 1918.
R—Brooklyn, N. Y.
B—Richmond, Va.
G. O. No. 20, W. D., 1919.

Private, Company M, 307th Infantry, 77th Division.
He distinguished himself time after time, carrying in wounded under heavy shellfire until he was finally killed while engaged in this self-sacrificing work. Posthumously awarded. Medal presented to widow, Mrs. Laura Garthright.

GARVIN, FRANK W. (1210129).........
Near St. Souplet, France, Oct. 18,
1918.
R—New York, N. Y.
B—New York, N. Y.
G. O. No. 46, W. D., 1919.

Private, Company C, 107th Infantry, 27th Division.
After the advance of his company had been stopped by strong hostile machine-gun fire, Private Garvin, with three companions, advanced far ahead of the front line to attack an enemy position located in a large farmhouse. By skillful maneuvering in the broad daylight, they covered all entrances to the house and forced the surrender of the entire force of the enemy, numbering 36 men and 2 officers. During the exploit they killed two Germans who attempted to take cover in the cellar.

GASAWAY, THOMAS (731703)...........
Near Fontaines, France, Nov. 7,
1918.
R—Vanceburg, Ky.
B—Creeks Run, Ky.
G. O. No. 37, W. D., 1919.

Corporal, Company B, 6th Infantry, 5th Division.
Corporal Gasaway, accompanied by three other soldiers, volunteered and went out under heavy machine-gun and artillery fire to rescue a wounded comrade. Failing in the first attempt, they again tried, and this time succeeded in bringing the wounded man to shelter.

*GASKINS, FREDERICK O. (1311737)......
Near la Haie-Menneresse, France,
Oct. 16, 1918.
R—Chesterfield, S. C.
B—Chesterfield, S. C.
G. O. No. 133, W. D., 1918.

Corporal, Company I, 118th Infantry, 30th Division.
When the advance of his company was held up by two machine-gun nests,
he led his squad, entirely on his own initiative, in the face of intense machine-
gun fire, against an enemy post on the right flank. Followed by his men,
he rushed the position, taking it and killing two of the gun crew. He then
rushed a second post alone with his rifle, killing one of the crew. He was
himself killed before he could reach the post.
Posthumously awarded. Medal presented to father, F. W. Gaskins.

GASPAROTTO, TONY (1909797)..........
Near Bellicourt, France, Sept. 29,
1918.
R—Livingston, Ill.
B—Italy.
G. O. No. 32, W. D., 1919.

Private, Company B, 119th Infantry, 30th Division.
During the operations against Bellicourt he, without assistance, charged an
enemy position, taking and returning with 26 prisoners

GASTON, GEORGE F. (1387869)..........
At Chipilly Ridge, France, Aug. 9,
1918.
R—Findlay, Ill.
B—Lakewood, Ill.
G. O. No. 128, W. D., 1918.

Private, Company H, 131st Infantry, 33d Division.
After being severely wounded by shrapnel he showed the greatest courage by
continuing to advance on a machine-gun emplacement, keeping the gun
occupied, and thereby enabling a detachment to flank the position and
capture it.

GASTON, PAUL J. (17428).............
Bellevue Farm, near Cierges,
France, Aug. 1, 1918.
R—Rhinelander, Wis.
B—Lexington, Ky.
G. O. No. 124, W. D., 1918.

Sergeant, sanitary detachment, 121st Machine Gun Battalion, 32d Division.
During the attack on Bellevue Farm he worked energetically throughout the
engagement, which continued from 2 to 10 p. m., to give first aid to the
wounded and to carry them to the dressing station. He was under fire of
snipers continuously and frequently went into shell and machine-gun fire
to administer to wounded soldiers. His calmness and courage was a source
of inspiration to the combat troops.

*GASTON, ROBERT A. (1487711)...........
Near St. Etienne, France, Oct. 8,
1918.
R—San Antonio, Tex.
B—Fort Clark, Tex.
G. O. No. 21, W. D., 1919.

Corporal, Company F, 141st Infantry, 36th Division.
He, leading his squad, made three attempts to take an enemy's machine-gun
nest under enfilading fire. In the final attempt he was wounded, but continued to advance until he was killed by machine-gun fire.
Posthumously awarded. Medal presented to mother, Mrs. Lilly Gaston.

GATAINO, ISAAC (1558294)............
Near St. Thibaut, France, Aug. 8,
1918.
R—Chicago, Ill.
B—Greece.
G. O. No. 35, W. D., 1919.

Corporal, Company I, 47th Infantry, 4th Division.
He showed exceptional courage and judgment when patrolling the country to
the flank of his company under heavy machine-gun and artillery fire. He
obtained liaison with the flank company and brought back valuable information regarding the river to the front of our lines.

*GATES, AUSTIN (14556)..............
Near Charpentry, France, Oct. 3,
1918.
R—Drummond, Mont.
B—Philipsburg, Mont.
G. O. No. 81, W. D., 1919.

Private, Company M, 16th Infantry, 1st Division.
He went forward with three other soldiers, and, though subjected to intense
enemy fire, rescued a wounded soldier who had fallen in advance of our lines.
Posthumously awarded. Medal presented to mother, Mrs. George Duff.

GAY, JAMES H..................
Near Crezancy, France, July 15,
1918.
R—Philadelphia, Pa.
B—Philadelphia, Pa.
G. O. No. 46, W. D., 1919.

First lieutenant, 30th Infantry, 3d Division.
When his small force of about 30 men was almost entirely surrounded by greatly
superior forces of the enemy, Lieutenant Gay, refusing to surrender, cut his
way out by delivering a deadly fire from both his front and rear. He also
captured about 150 prisoners, including a major, and his remarkable gallantry
aided greatly in breaking up the German drive of July 15.

*GAY, LAWRENCE W. (132999).........
North of Verdun, France, Oct. 23,
1918.
R—Groton, Mass.
B—Groton, Mass.
G. O. No. 9, W. D., 1923.

Sergeant, Headquarters Company, 101st Field Artillery, 26th Division.
As a sergeant in charge of an artillery observation post and rocket relay station
during an intensive bombardment of gas shells, after a direct hit had been
made on his observatory he evacuated all his men, but he, himself, though
badly gassed, continued for 6 hours and until relieved to observe the enemy's
fire and relay calls for barrages. He died shortly after being evacuated thus
sacrificing his life to save the men under his command, his extraordinary
heroism and devotion to duty under a continuous bombardment of high
explosives being a great inspiration to the men of his regiment.
Posthumously awarded. Medal presented to father, Henry H. Gay.

GAY, LUTHER (733044)...............
Near the Bois-de-Chatillon, France,
Nov. 5, 1918.
R—Oak Park, Ga.
B—Wadley, Ga.
G. O. No. 98, W. D., 1919.

Private, Company L, 6th Infantry, 5th Division.
As Private Gay and another soldier were going to the rear to guide the remainder
of their company to its position they were fired on from an enemy machine-gun
nest. Attacking the nest, they killed 4 of the crew and captured 1 prisoner,
driving off the remainder.

GAYLORD, BRADLEY J...............
Between Chambley and Xammes,
France, Sept. 13, 1918.
R—Buffalo, N. Y.
B—Austria.
G. O. No. 123, W. D., 1918.

First lieutenant, pilot, 1st Day Bombardment Group, Air Service.
While leading an important mission with 2 other planes, he was attacked by
15 enemy planes. Nevertheless, he and his observer carried out the mission,
bombed the objective in a running fight, and shot down at least 1 enemy
plane.

GAYNIER, CHARLES E. (1203752).......
Near Ronssoy, France, Sept. 30,
1918.
R—Rockaway, N. Y.
B—New York, N. Y.
G. O. No. 126, W. D., 1919.

Private, Company C, 105th Infantry, 27th Division.
He exhibited exceptional bravery in voluntarily leaving shelter, going forward
under heavy shell and machine-gun fire and bringing back to our lines several
wounded comrades.

*GCALAIRTCH, AUGUST (40569)............
Near Soissons, France, July 25, 1918.
R—Newark, N. J.
B—Jersey City, N. J.
G. O. No. 21, W. D., 1919.

Private, Company L, 9th Infantry, 2d Division.
While endeavoring with his automatic rifle to silence the fire of an enemy machine-gun nest, which was holding up a portion of our line, he was killed by shell fire.
Posthumously awarded. Medal presented to mother, Mrs. Victoria Gcalairtch.

GEANEY, EDWARD J. (89730)............
Near Villers-sur-Fere, France, July 30, 1918.
R—New York, N. Y.
B—Ireland.
G. O. No. 30, W. D., 1921.

Sergeant, Company D, 165th Infantry, 42d Division.
In the assault, seeing his officer wounded, with great effort and under heavy fire he succeeded in getting the wounded officer almost to a place of safety when he sustained a machine-gun wound which broke his arm. In spite of the pain and with great fortitude he succeeded in getting the officer to a place of safety.

GEARHARD, AUGUST F............
At Norroy and Vandieres, France, Sept. 15-16, 1918, and at Chatel-Chehery and Sommerance, France, Oct. 7-20, 1918.
R—Milwaukee, Wis.
B—Milwaukee, Wis.
G. O. No. 142, W. D., 1918.

First lieutenant, chaplain, 328th Infantry, 82d Division.
During this advance of his regiment from Norroy to Vandieres he displayed exceptional bravery and fidelity to his duties by working 2 days and nights without sleeping, recovering and removing to the rear the killed and wounded, making frequent trips into no man's land under heavy shell fire for this purpose. In the operations at Chatel-Chehery and Sommerance this officer again distinguished himself by remaining constantly at the advanced aid station assisting the surgeon, administering aid to the wounded, cheering and comforting them.

*GEARY, WILLIAM J............
Near Belleau Wood, France, June 25-26, 1918.
R—Washington, D. C.
B—Harpers Ferry, W. Va.
G. O. No. 31, W. D., 1922.

Sergeant major, headquarters detachment, 4th Brigade, U. S. Marine Corps, 2d Division.
Having voluntarily and of his own initiative requested authority to join an attacking company was killed while gallantly leading his platoon forward against the enemy.
Posthumously awarded. Medal presented to mother, Mrs. Claudia E. Geary.

GEE, OTHEL J............
Near Vieville-en-Haye, France, Sept. 27, 1918.
R—Timpson, Tex.
B—Howe, Tex.
G. O. No. 37, W. D., 1919.

First lieutenant, Medical Corps, attached to 20th Field Artillery, 5th Division.
Lieutenant Gee voluntarily ran to the assistance of a wounded soldier, dressed his wounds, and carried him to shelter through severe shell fire.

GEER, CLARENCE W............
At St. Hilaire, France, Sept. 18-19, 1918.
R—Torrington, Conn.
B—New Milford, Conn.
G. O. No. 78, W. D., 1919.

First lieutenant, 102d Infantry, 26th Division.
While advancing with a raiding party, in the face of heavy machine-gun fire, he charged alone into the woods and captured 2 prisoners single-handed. After the raiding party had retired, he again went back alone into the woods, under the same heavy machine-gun fire, to locate wounded who had been left behind. His courage and exceptional bravery set an example to his men which contributed greatly to the success of the raid.

GEER, FREDERICK W. (2943018)........
Near Grand Pre, France, Oct. 26, 1918.
R—Albany, N. Y.
B—Cohoes, N. Y.
G. O. No. 35, W. D., 1919.

Private, Company M, 312th Infantry, 78th Division.
In the face of heavy machine-gun fire and on his own initiative Private Geer left a place of safety and made his way by crawling to the side of a wounded comrade, administering first-aid treatment.

GEER, PRENTICE S............
At Chateau-Thierry, France, June 6, 1918.
R—St. Paul, Minn.
B—St. Paul, Minn.
G. O. No. 110, W. D., 1918.

Corporal, 67th Company, 5th Regiment, U. S. Marine Corps, 2d Division.
At Chateau-Thierry, France, on June 6, 1918, having become isolated when the enemy counterattacked his group, he courageously charged with a bayonet and, with the assistance of his comrades, captured a machine-gun crew and repulsed the attack at that point.

GEHRIS, JACK D. (1711185)............
Near Binarville, France, Oct. 2 and 5, 1918.
R—Easton, Pa.
B—Whitehaven, Pa.
G. O. No. 35, W. D., 1919.

Private, first class, medical detachment, 308th Infantry, 77th Division.
Under a heavy enemy barrage he went to the rescue of 2 severely wounded men and carried them to a place offering scant shelter, where they were forced to remain until aid arrived the next morning. On Oct. 5, 1918, when a shell struck his first-aid station, killing 2 and wounding 5 others, he, although wounded, administered first aid to his comrades before receiving medical attention for himself.

GEIGER, FRANK F............
In the Forest of Argonne, France, Nov. 1, 1918.
R—Buffalo, N. Y.
B—Buffalo, N. Y.
G. O. No. 35, W. D., 1919.

Sergeant, 95th Company, 6th Regiment, U. S. Marine Corps, 2d Division.
He displayed great coolness and courage in leading his section against machine-gun positions. He was later severely wounded while rushing a machine-gun nest unaided.

GENEST, PAUL P. (51889)............
Near Vaux, France, July 1, 1918.
R—Lowell, Mass.
B—Canada.
G. O. No. 132, W. D., 1918.

Mechanic, Company L, 23d Infantry, 2d Division.
He picked up a live grenade thrown into his group from an enemy dugout and hurled it back into the dugout, killing and wounding a number of the enemy and saving the lives of several comrades.

GENICKE, RAYMOND J. (280280)........
Near Juvigny, France, Aug. 29, 1918.
R—Detroit, Mich.
B—Detroit, Mich.
G. O. No. 32, W. D., 1919.

Private, Company H, 128th Infantry, 32d Division.
Private Genicke accompanied an officer in advancing ahead of the first wave under intense machine-gun and shell fire to within a few feet of an enemy trench and capturing 75 prisoners. He then entered this trench and took 10 more of the enemy.

GENRICH, LUDWIG (2384249)
At Clery-le-Petit, France, Nov. 6, 1918.
R—South Bend, Ind.
B—Russia.
G. O. No. 81, W. D., 1919.

Mechanic, Company I, 60th Infantry, 5th Division.
His company having been held up by enemy machine-gun fire, Mechanic Genrich, with his platoon leader and another soldier, attacked a machine-gun nest, killing 2 of the crew and capturing 8 prisoners, thereby permitting the company to advance and establish a bridgehead.

GENTRY, CARL C. (1456423)
Near Baulny, France, Sept. 30, 1918.
R—Trenton, Mo.
B—Grundy County, Mo.
G. O. No. 95, W. D., 1919.

Private, first class, Company D, 139th Infantry, 35th Division.
Upon his own initiative, Private Gentry organized a platoon composed of disorganized men from various units, and led it through terrific machine-gun fire, stopping an advance of the enemy. In performing this courageous feat he was severely wounded.

*GENTRY, HERMAN D. (96166)
Near Pexonne, France, Mar. 5, 1918.
R—Huntsville, Ala.
B—Gurley, Ala.
G. O. No. 22, W. D., 1926.

Private, Company C, 167th Infantry, 42d Division.
He showed courageous devotion to duty by remaining at his post under heavy shellfire, after being wounded. Died Mar. 11, 1918, of wounds received in action.
Posthumously awarded. Medal presented to father, Morgan Gentry.

GEORGE, HAROLD H.
Near Bantheville, France, Oct. 27, 1918.
R—Niagara Falls, N. Y.
B—Niagara Falls, N. Y.
G. O. No. 46, W. D., 1919.

First lieutenant, 139th Aero Squadron, Air Service.
Lieutenant George displayed great courage in attacking a formation of four enemy planes (type Fokker), destroying two of them in a terrific fight and driving the other two back to their own territory.

GEORGE, REUBIN L. (2284172)
Near Brieulles, France, Sept. 29, 1918.
R—Chualar, Calif.
B—Hollister, Calif.
G. O. No. 95, W. D., 1919.

Corporal, Company A, 59th Infantry, 4th Division.
After his platoon had become badly disorganized under heavy fire and all the sergeants had been killed or wounded, Corporal George took charge of the platoon, reorganized it with great courage and initiative, and led it on in the attack against hostile machine guns. He was wounded shortly afterwards, but he remained throughout the night where he had fallen, refusing to be evacuated till all the other wounded had been cared for.

GERLING, FRANK C. (1937573)
On Hill 253, Meuse-Argonne offensive, France, Oct. 8 and 9, 1918.
R—Fort Loramie, Ohio.
B—Fort Loramie, Ohio.
G. O. No. 16, W. D., 1920.

Private, first class, Company I, 38th Infantry, 3d Division.
He carried numerous messages over a dangerous and hazardous route, which was constantly swept by machine-gun and artillery fire. Due to his efforts the company was able to maintain communication with the battalion and with the platoons in advance.

GEST, SYDNEY G.
In the Bois-de-Belleau, France, June 11, 1918.
R—Overbrook, Pa.
B—Overbrook, Pa.
G. O. No. 15, W. D., 1923.

Private, 43d Company, 5th Regiment, U. S. Marine Corps, 2d Division.
The attacking line of the 43d Company being held up at one point by the fire of an enemy heavy Maxim machine gun, Private Gest, in the face of the machine-gun fire, crawled around the flank and, undaunted by grenades and rifle fire, rushed the gun crew's nest, killing the enemy gunner and four others as they attempted to escape.

GETCHELL, CHARLES H. (146244)
Near Pannes, France, Sept. 21, 1918.
R—Minneapolis, Minn.
B—Minneapolis, Minn.
G. O. No. 128, W. D., 1918.

Mess sergeant, Battery F, 151st Field Artillery, 42d Division.
While the terrain was under heavy artillery fire from the enemy, Sergeant Getchell, observing some wounded men lying in the open without attention, ran to their aid of his own volition, administered to their wounds, and helped carry them to a place of safety, after which he again went out into the shell-swept area and made a careful reconnaissance of the fields, searching for any other casualties which he might have overlooked.

GEYER, ROBERT E. (129240)
Near Thiaucourt, France, Sept. 12, 1918.
R—South Bend, Ind.
B—South Bend, Ind.
G. O. No. 37, W. D., 1919.

Private, Battery E, 12th Field Artillery, 2d Division.
While acting as spare gunner, Private Geyer sprang at a word from his executive officer to act as No. 1 in the supplementary gun crew for the first piece, whose entire crew were casualties. He assisted in laying aside the dead and wounded and continued in the service of his piece until the barrage was completed.

GHOLSTON, JABEZ G.
Near Fontaines, France, Nov. 7, 1918.
R—Woodland, Miss.
B—Griffith, Miss.
G. O. No. 37, W. D., 1919.

Captain, 6th Infantry, 5th Division.
He personally led several of his platoons against machine guns on the flank of the attacking battalion and reduced that resistance. He was wounded by shellfire upon reaching the objective, but remained with his company until the position had been organized and an advance upon a new objective begun.

*GIBBONS, JAMES J.
At Chateau-Thierry, France, June 6, 1918.
R—Buffalo, N. Y.
B—Brooklyn, N. Y.
G. O. No. 110, W. D., 1918.

Sergeant, 45th Company, 5th Regiment, U. S. Marine Corps, 2d Division.
Killed in action at Chateau-Thierry, France, June 6, 1918, he gave the supreme proof of that extraordinary heroism which will serve as an example to hitherto untried troops.
Posthumously awarded. Medal presented to sister, Miss Evelyn Gibbons.

GIBBS, DUDLEY R. (144499)
Near Fleville, France, Oct. 28, 1918.
R—Miami, Okla.
B—Salt Lake City, Utah.
G. O. No. 26, W. D., 1919.

Private, first class, Battery E, 150th Field Artillery, 42d Division.
Acting as courier, Private Gibbs was riding through an intense shelling, and, stopping at a cry of help, observed two members of his organization lying in the road. Going to their aid, he found that one man was already dead and the other seriously wounded. He administered aid under most harassing conditions and assisted in the removal of the wounded man to a dressing station.

GIBNEY, JOHN J. (1039050)
Near Greves Farm, France, July 15, 1918.
R—Tucson, Ariz.
B—Mexico.
G. O. No. 46, W. D., 1919.

Corporal, Battery F, 10th Field Artillery, 3d Division.
Responding to a call for volunteers, Corporal Gibney, with 8 other soldiers, manned 2 guns of a French battery which had been deserted by the French during the unprecedented fire, after many casualties had been inflicted on their forces. For 2 hours he remained at this post and poured an effective fire into the ranks of the enemy.

GIBSON, CHARLES S.
Near Bayonville, France, Nov. 1, 1918.
R—Batesville, Ind.
B—Morris, Ind.
G. O. No. 35, W. D., 1919.

Private, 82d Company, 6th Regiment, U. S. Marine Corps, 2d Division.
He volunteered and went forward to reconnoiter a ravine which was infested with hostile machine-gun and artillery positions, returning with several prisoners.

GIBSON, HERBERT D.
Near St. Etienne-a-Arnes, France, Oct. 3, 1918.
R—East Poultney, Vt.
B—Schenectady, N. Y.
G. O. No. 35, W. D., 1920.

First lieutenant, 23d Infantry, 2d Division.
During the attack, when his company was debouching from a wood, his men were suddenly exposed to direct fire of a 150-millimeter gun some 200 meters to the front. Lieutenant Gibson rushed toward the enemy guns, killing two gunners before other members of his organization joined him and completed the capture of the guns.

GIBSON, RALPH (1402850)
At Beaune, France, Nov. 8, 1918.
R—Chicago, Ill.
B—Chicago, Ill.
G. O. No. 46, W. D., 1919.

Sergeant, Company H, 370th Infantry, 93d Division.
Sergeant Gibson led his platoon across a fire-swept zone in the advance of the objective, encouraging his men by his fearless example. On the same day he was placed in charge of two important reconnoitering patrols whose mission was to locate enemy machine-gun positions that were known to be in the path of the advance of the company. He accomplished the mission, constantly exposed to enemy fire, and brought back important information.

GIBSON, RAYMOND
In the Bois-de-Belleau, France, June 8, 1918.
R—Kingsport, Tenn.
B—Jonesboro, Tenn.
G. O. No. 110, W. D., 1918.

Corporal, 83d Company, 6th Regiment, U. S. Marine Corps, 2d Division.
He handled alone a Chauchat rifle with such accuracy in the face of an extremely heavy fire that his platoon was enabled to move against the enemy machine-gun platoon.

GIBSON, THOMAS R.
Near Soissons, France, July 21, 1918.
R—Chicago, Ill.
B—Chicago, Ill.
G. O. No. 128, W. D., 1918.

First lieutenant, 18th Infantry, 1st Division.
He displayed rare gallantry and leadership when, all the other officers of his company having been killed or wounded, he led his men through a zone of intense bombardment, and, although badly wounded himself, he attained his objective with only 5 men and 2 machine guns, refusing to be evacuated until his guns were effectively in action.

GIESCKE, LEWIS M. (1389290)
Near Brieulles, France, Oct. 9-12, 1918.
R—Poplar Grove, Ill.
B—Poplar Grove, Ill.
G. O. No. 44, W. D., 1919.

Private, Company I, 132d Infantry, 33d Division.
He administered first aid to many comrades under heavy shellfire and assisted them to the aid station. When his company was in need of water, he went alone under heavy machine-gun fire in direct view of the enemy and procured it. Each night he personally guided the ration detail through heavy bombardment. Toward the end of the engagement, after his platoon sergeant and the other noncommissioned officers had become casualties, Private Giescke took charge of the platoon, displaying unusual leadership.

GIGER, GEORGE J.
North of Verdun, France, Oct. 12, 1918.
R—Newark, N. J.
B—Paterson, N. J.
G. O. No. 26, W. D., 1919.

First lieutenant, 114th Infantry, 29th Division.
While leading his platoon in attack he was wounded, but refused evacuation, encouraging his men to continue, when he was again wounded. He then assisted in the removal of the wounded and refused attention until all others had received first aid.

GILBERT, FRANCIS W.
Near Fismes, France, Aug. 26, 1918.
R—Utica, N. Y.
B—Utica, N. Y.
G. O. No. 23, W. D., 1919.

First lieutenant, 307th Infantry, 77th Division.
On Aug. 26 Lieutenant Gilbert made a daylight reconnaissance of the ruins of the Tannerie, near Fismes, entered Fismes under direct observation and fire of the enemy, and continued his reconnaissance along the Rouen-Reims road under machine-gun fire, for the purpose of ascertaining whether or not the terrain was favorable for an attack on the Chateau du Diable. On Nov. 10 he voluntarily led a patrol across the River Meuse and located the enemy positions.

GILBERT, HAROLD N.
Near Mezy, France, July 10, 1918.
R—Williamsport, Pa.
B—Halifax, Pa.
G. O. No. 9, W. D., 1923.

First lieutenant, 30th Infantry, 3d Division.
With utter disregard for his own safety, exposed to heavy enemy machine gun and rifle fire, he crossed open ground under enemy observation and less than 75 yards from the enemy line and rendered first aid to wounded members of his platoon, thus saving their lives and setting an example of bravery and devotion to duty to other members of his command.

GILBERT, LESLIE T.
Near Blanc Mont Ridge, France, Oct. 4, 1918.
R—St. Paul, Minn.
B—Eau Claire, Wis.
G. O. No. 46, W. D., 1919.

Private, 55th Company, 5th Regiment, U. S. Marine Corps, 2d Division.
He volunteered to rescue a comrade from a most violent barrage. Although severely wounded while performing this courageous deed, he continued until his task was accomplished.

GILBERTIE, JOHN S. (1906527)
Near Cornay, France, Oct. 7-26, 1918.
R—Westport, Conn.
B—Italy.
G. O. No. 21, W. D., 1919.

Corporal, Company E, 327th Infantry, 82d Division.
During the entire action from Oct. 7, he carried messages from the front line to battalion and regimental headquarters, although suffering from the effects of gas and sickness. On two occasions he volunteered and led patrols into the enemy territory, obtaining and returning with information of the utmost importance and value.

GILCHRIST, EDWARD J.
Near Verdun, France, Oct. 29, 1918.
R—Richmond Hill, N. Y.
B—Brooklyn, N. Y.
G. O. No. 46, W. D., 1919.

Second lieutenant, 102d Infantry, 26th Division.
After the advance of his company had been held up by intense machine gun and rifle fire, he successfully gathered together the scattered members of his command and consolidated his position. He then exposed himself to great danger from heavy machine-gun fire by crawling far in advance of our lines and rescuing several wounded men who were lying helpless, unprotected in shell holes.

GILES, FLORAIN D.
Opposite Montdidier, France, May 27, 1918.
R—Wilmore, Ky.
B—Gratz, Ky.
G. O. No. 39, W. D., 1920.

First lieutenant, 26th Infantry, 1st Division.
During the heavy enemy bombardment which preceded an enemy raid, Lieutenant Giles was twice buried by débris, due to enemy shell fire. Although suffering from concussion, he rallied his men and led them in a counterattack, repulsing the enemy, who had attempted to enter his platoon sector.

GILFILLAN, DEAN McGREW.
Near Varennes, France, Oct. 26, 1918.
R—Ironton, Ohio.
B—Ironton, Ohio.
G. O. No. 44, W. D., 1919.

Captain, 345th Battalion, Tank Corps.
Captain Gilfillan destroyed two machine guns and inflicted heavy losses on a column of German infantry after his tank had been on fire by two direct hits by enemy artillery and he himself wounded by machine-gun fire. He left his tank only when an explosion was imminent, and was wounded a second time by shell fragments, but remained at his post until he had turned over his command to another officer.

GILKESON, SHANKLIN EBENEZER (711).
At St. Benoît, France, Sept. 16, 1918.
R—Claremore, Okla.
B—Warrensburg, Mo.
G. O. No. 131, W. D., 1918.

Sergeant, 167th Ambulance Company, 117th Sanitary Train, 42d Division.
While the regimental dressing station was under heavy shellfire he volunteered to lead a squad of litter bearers to rescue several wounded men of another regiment who had been caught in a heavy barrage. Although he was wounded himself, he succeeded in leading the squad for a distance of 3 kilometers through a constant severe bombardment under direct observation of the enemy artillery and snipers to an outpost outside of his own regimental sector. He brought in one wounded officer and seven severely wounded soldiers without losing any of his men.

GILL, JOHN HENRY (1319816).
Near Bellicourt, France, Sept. 29, 1918.
R—Henderson, N. C.
B—Franklin County, N. C.
G. O. No. 37, W. D., 1919.

Sergeant, Headquarters Company, 120th Infantry, 30th Division.
After being wounded during the attack, Sergeant Gill, with his trench mortar section men, who had become lost from other companies, and stragglers, attacked a strong machine-gun position at the junction of the tunnel and canal and was wounded the third time. During the attack he was wounded in 13 places by machine-gun bullets and shrapnel, but continued the attack with the utmost coolness and bravery.

GILL, MARTIN MATTHEW (89151).
Near Landres-et-St. Georges, France, Oct. 15, 1918.
R—New York, N. Y.
B—New York, N. Y.
G. O. No. 37, W. D., 1919.

Private, first class, Company A, 165th Infantry, 42d Division.
Displaying remarkable coolness and bravery under heavy artillery and machine gun fire, Private Gill volunteered in every way possible to assist in administering to the wounded. Going forward over a hazardous area, he brought in a wounded comrade to a place of safety.

*GILL, RAYMOND (1708394).
Near Ville-Savoye, France, Aug. 24, 1918.
R—Long Island City, N. Y.
B—New York, N. Y.
G. O. No. 32, W. D., 1919.

Sergeant, Company D, 308th Infantry, 77th Division.
During the advance of his company across the Vesle River, Sergeant Gill, disregarding severe wounds, insisted on leading a patrol to capture a sniper who was occupying a formidable position to fire on our men. While on this precarious mission he was killed.
Posthumously awarded. Medal presented to father, Robinson Gill.

GILL, WILLIAM TIGNOR, Jr.
Near Vierzy, France, July 19, 1918.
R—Washington, D. C.
B—Washington, D. C.
G. O. No. 137, W. D., 1918.

Passed assistant surgeon, U. S. Navy, attached to 6th Regiment, U. S. Marine Corps, 2d Division.
He established a forward dressing station behind the advanced lines and for 15 hours treated the wounded and directed their evacuation while subjected to intense front and flank fire and in the absence of adequate shelter. His fearlessness under these conditions saved the lives of many wounded who would otherwise have been lost to the service. He disregarded personal danger and remained in an exposed position in order to give immediate care to the unfortunate.

GILLEN, EDWARD N. (1540551).
Near Heurne, Belgium, Nov. 4, 1918.
R—Toledo, Ohio.
B—Ohio.
G. O. No. 59, W. D., 1919.

Private, 148th Ambulance Company, 112th Sanitary Train, 37th Division.
With two other soldiers he volunteered to rescue two wounded men who had been lying in an exposed position on the opposite bank of the Scheldt River for two days. Making two trips across the stream in the face of heavy machine gun and shell fire he and his companions succeeded in carrying both the wounded men to shelter.

GILLESPIE, PETER (89757).
Near Villers-sur-Fere, France, July 29, 1918.
R—New York, N. Y.
B—Philadelphia, Pa.
G. O. No. 81, W. D., 1919.

Private, Machine Gun Company, 165th Infantry, 42d Division.
Locating an enemy sniper, Private Gillespie took the rifle of a dead comrade and, with no regard to personal safety, crawled forward under heavy machine-gun fire to a position far in advance of the assaulting wave. After an exchange of shots, he killed an enemy sniper who had killed or wounded several members of his battalion, and worked his way back to his own lines through an area swept by fire from near-by heights. When all of his officers had been evacuated he rallied the survivors of his company and held them to their task after another company had been sent up as relief. He aided materially in the repulse of a strong counterattack, although suffering from the effects of gas, refusing to be evacuated till he became exhausted and was carried from the field.

*GILLESPIE, WILLIAM L. (43221).
In the Forest of Argonne, France, Oct. 4, 1918.
R—Fort Wayne, Ind.
B—Fort Wayne, Ind.
G. O. No. 37, W. D., 1919.

Private, Company G, 16th Infantry, 1st Division.
While carrying a message from the support to the firing line he was severely wounded, but he continued on his mission and delivered his message.
Posthumously awarded. Medal presented to father, William R. Gillespie.

*GILLETT, TOD F. (10381).
Near Beaumont, France, June 19, 1918.
R—Tampa, Fla.
B—Weirsdale, Fla.
G. O. No. 126, W. D., 1918.

Private, first class, Section No. 647, Ambulance Service.
On June 19, 1918, near Beaumont, France, he volunteered to proceed with his ambulance under heavy bombardment to bring wounded men to a place of safety. While returning on this perilous trip he was killed by a shell.
Posthumously awarded. Medal presented to father, M. E. Gillett.

GILLETTE, NORRIS W.
Near Montfaucon, France, Sept. 26, 1918.
R—Toledo, Ohio.
B—Toledo, Ohio.
G. O. No. 56, W. D., 1922.

First lieutenant, 148th Ambulance Company, 112th Sanitary Train, 37th Division.
While in command of a medical detachment working forward through the woods on the right flank of the 73d Brigade sector, Lieutenant Gillette encountered a large number of men in confusion who were without officers and under fire from enemy snipers located both at the edge of the woods and at a strong point on a knoll beyond it. By his remarkable courage and tact, and through the power of his inspiring heroic example, the scattered troops were reorganized into squad and platoon groups and took up a position from which they as part of a battalion later moved forward and drove the enemy from the woods and overcame the enemy's strong point.

GILLIAM, REXIE E.
At Soissons, France, July 18-22, 1918.
R—Mesilla Park, N. Mex.
B—Toledo, Tex.
G. O. No. 15, W. D., 1919.

First lieutenant, 26th Infantry, 1st Division.
Four times he led his company over the top during the battle of Soissons, France, July 18-22, 1918, and by his individual bravery inspired his men to 4 successful attacks.

GILLILAND, SAMUEL F. (1039403).
Near Laneuvelle, France, Nov. 9-10, 1918.
R—South Boston, Va.
B—South Boston, Va.
G. O. No. 46, W. D., 1919.

Private, first class, medical detachment, 11th Field Artillery, 6th Division.
He made repeated trips over a road which was under continuous shell fire, he being the only driver who would risk driving over this road. On one trip his ambulance was struck by a shell, which wounded him, killed one of his patients, and caused fresh wounds to several others, but he continued on duty, evacuating the wounded of his own and other organizations.

GILLIS, ANGUS J. (41862).
Near Soissons, France, July 20, 1918.
R—Beverly, Mass.
B—Beverly, Mass.
G. O. No. 15, W. D., 1919.

Private, first class, Company F, 1st Engineers, 1st Division.
He displayed exceptional courage when he carried messages through an area under intense shell and machine-gun bombardment after 4 men had lost their lives in the attempt.

GILLOTTI, ANGELO J. (291820).
Near Thiaucourt, France, Sept. 12, 1918.
R—Utica, N. Y.
B—Italy.
G. O. No. 59, W. D., 1919.

Private, Battery E, 12th Field Artillery, 2d Division.
While acting as spare cannoneer, he sprang at a word from his executive officer to act as No. 1 of the supplementary gun crew for the first piece, whose entire crew were casualties. He assisted in laying aside the dead and wounded and continued in the service of his piece until the barrage had been completed.

GILMER, ROBERT A.
At Trieres Farm, France, Sept. 30, 1918.
R—Anderson, S. C.
B—Anderson, S. C.
G. O. No. 62, W. D., 1919.

Second lieutenant, 371st Infantry, 93d Division.
While personally reconnoitering a position to place his platoon in the defense of Trieres Farm, he, regardless of personal danger, exposed himself in an area swept by machine-gun fire and was killed while in the performance of this mission.
Posthumously awarded. Medal presented to mother, Mrs. Martha L. Gilmer.

GILSTRAP, LEE F. (1483970).
Near St. Etienne-sur-Marne, France, Oct. 8, 1918.
R—Chandler, Okla.
B—Chandler, Okla.
G. O. No. 15, W. D., 1923.

Bugler, Company B, 142d Infantry, 36th Division.
In the assault of the 142d Infantry before the town of St. Etienne, Bugler Gilstrap voluntarily accompanied the first assaulting wave of his regiment. Under direct observation of the enemy he assisted many wounded men to shelter, with utter disregard for his own safety. Throughout the day he voluntarily carried messages from battalion to regimental headquarters through a zone swept by terrific enemy machine-gun, gas-shell, and high-explosive artillery fire. Repeatedly urged by officers and noncommissioned officers to seek shelter, he coolly declined to do so, continuing on his dangerous missions in complete contempt for the hottest fire, until badly burned by gas-shell fire and evacuated to hospital. During the course of the day he found several enemy soldiers hiding in shell holes. These men he required to carry wounded American soldiers to dressing stations. The indomitable heroism, contempt for enemy fire, and superb devotion to duty of Bugler Gilstrap served as an example of soldierly bravery and conduct to every man of his regiment and inspired them all to the greatest endeavors.

GILTS, DAN (158435).
Near Crezancy, France, July 15, 1918.
R—Mishawaka, Ind.
B—Defiance, Ohio.
G. O. No. 44, W. D., 1919.

Private, first class, Company C, 6th Engineers, 3d Division.
After being wounded, he was being evacuated when he discovered that all ammunition carriers of a near-by battery had been disabled. He insisted on being allowed to carry up ammunition until the battery was completely out of action.
Posthumously awarded. Medal presented to father, Isaac Gilts.

GINGRAS, JULES, Jr. (1717529).
Near Ville-Savoye, France, Aug. 18, 1918.
R—New York, N. Y.
B—Detroit, Mich.
G. O. No. 37, W. D., 1919.

Sergeant, Company F, 302d Engineers, 77th Division.
He voluntarily plunged into the Vesle River to rescue some soldiers who had fallen into the water with full packs while crossing a footbridge and were in danger of drowning. In order to see he removed his gas mask, and as a result was severely gassed.

GIROUX, ERNEST A.
Near Armentieres, France, May 22, 1918.
R—Somerville, Mass.
B—Roxbury, Mass.
G. O. No. 35, W. D., 1919.

First lieutenant, pilot 103d Aero Squadron, Air Service.
He, while on a patrol with four other scout planes, attacked an enemy formation of eight monoplane machines. Two of his companions were forced to retire when their guns became jammed. Despite numerical superiority, Lieutenant Giroux continued the attack, endeavoring to protect his leader, until finally forced down and killed.
Posthumously awarded. Medal presented to mother, Mrs. Arthur E. Haley.

GITZ, RALPH (1253129).
Near Courville, France, Sept. 6, 1918.
R—Hazleton, Pa.
B—Hazleton, Pa.
G. O. No. 71, W. D., 1919.

Corporal, Battery A, 109th Field Artillery, 28th Division.
Exposing himself to an artillery barrage, he occupied an exposed position to read projector signals from the Infantry front line. He volunteered for this duty and rendered valuable service, receiving and transmitting messages until he was so severely wounded that it was necessary for him to be evacuated.

GIVENS, FRED G.
Near Marcheville, France, Nov. 10, 1918.
R—Carbondale, Ill.
B—Caneyville, Ky.
G. O. No. 23, W. D., 1919.

Captain, 130th Infantry, 33d Division.
Captain Givens led the attack on Marcheville with exceptional bravery, and, although wounded while passing through a heavy barrage, would not leave his company until his objective had been reached and the plans made for its defense.

GJERSTAD, GUSTAV (2106540).
Near Mayonville, France, Nov. 1, 1918.
R—Gary, Minn.
B—Clayton County, Iowa.
G. O. No. 44, W. D., 1919.

Private, Company D, 9th Infantry, 2d Division.
While assisting a squad to flank a machine-gun nest, which was delaying his company's advance, he was twice painfully wounded. He persisted in continuing in the advance until another wound forced him to the rear for treatment.

GLADNEY, WILLIAM H. (1245981).
Near Fismes, France, Aug. 8 to 12, 1918.
R—Philadelphia, Pa.
B—Philadelphia, Pa.
G. O. No. 60, W. D., 1920.

Private, first class, Company M, 111th Infantry, 28th Division.
Private Gladney repeatedly carried messages for a period of five days from Fismes to Fismette. The route over which he traveled was exposed to enemy observation and covered by enemy sniper, machine-gun, and 77-millimeter fire. He frequently delivered messages when others failed. His gallant conduct enabled his commanding officer to maintain communication throughout the operation.

GLADSTONE, LEO.
Near Belleau Wood, France, June 6, 1918.
R—Gary, Ind.
B—Siberia.
G. O. No. 15, W. D., 1923.

Private, 51st Company, 5th Regiment, U. S. Marine Corps, 2d Division.
Although wounded in the arm by a machine-gun bullet, he assisted in carrying a badly wounded soldier of his company to a dressing station and returned to the front line, to find that the company of which he was a member had changed its position, and that another member of the company had been wounded and left behind exposed to enemy fire, whereupon Private Gladstone voluntarily advanced alone under enemy observation and intense fire to the point where the wounded man was lying, killed an enemy soldier who had captured the wounded man, and after an hour of fighting during which time Private Gladstone and the wounded man were again hit by enemy machine-gun fire, the former carried the latter back to the front line, thus undoubtedly saving the soldier from death or capture.

GLASS, HENRY B. (1289551).
North of Verdun, France, Oct. 3–15, 1918.
R—Lynchburg, Va.
B—Lynchburg, Va.
G. O. No. 87, W. D., 1919.

Corporal, Company L, 116th Infantry, 29th Division.
Corporal Glass repeatedly volunteered to carry messages through violent artillery and machine-gun fire after other runners failed to get through. After being gassed and ordered to the rear, this soldier continued on duty until he collapsed.

*GLASSBRENNER, FRED L. (1995552).
Near Bellicourt, France, Sept. 29, 1918.
R—Alton, Ill.
B—Alton, Ill.
G. O. No. 32, W. D., 1919.

Private, first class, Headquarters Company, 119th Infantry, 30th Division.
He was advancing with the 1-pounder gun section when he was wounded in the leg by a machine-gun bullet and was urged to go to the rear. He nevertheless continued to advance, and was later killed upon leaving a sheltered position to go to the assistance of another wounded soldier.
Posthumously awarded. Medal presented to father, Len Glassbrenner.

GLEASON, JAMES V. (1243326).
Near Fismette, France, Aug. 9–13, 1918.
R—Pottstown, Pa.
B—Pottstown, Pa.
G. O. No. 27, W. D., 1919.

Corporal, Company A, 111th Infantry, 28th Division.
When the enemy attacked our lines, employing liquid fire, he inspired the troops of which he formed a part to hold the post. During the succeeding 3 days and nights without food he controlled the firing line of the advanced position until badly wounded by shrapnel.

GLEASON, JOHN W.
Near Cheppy, Montrebeau Woods, and Exermont, France, Sept. 26–Oct. 4, 1918.
R—Wheaton, Ill.
B—Rutland, Mass.
G. O. No. 46, W. D., 1919.

Second lieutenant, 345th Battalion, Tank Corps.
During the action at Cheppy Lieutenant Gleason led his platoon against a strong point defended by machine and antitank guns, leaving his command on foot through a mine field under heavy rifle and machine-gun fire. Two days later, in Montrebeau Woods, he led his platoon against machine-gun nests, cleaning them out, and leading a battalion of infantry forward to occupy this position. On Oct. 4 he continued in the attack on Exermont, though he had been twice wounded and ordered to the rear.

GLENDENNING, HUGH L. (1210305).
Near Ronssoy, France, Sept. 29, 1918.
R—Jersey City, N. J.
B—Ceredo, W. Va.
G. O. No. 20, W. D., 1919.

Sergeant, Company D, 107th Infantry, 27th Division.
During the operations against the Hindenburg line Sergeant Glendenning, with 4 other soldiers, left shelter and went forward into an open field under heavy shell and machine-gun fire and succeeded in bandaging and carrying back to our lines 2 wounded men.

GLENN, CHARLES (568781).
Near Ville-Savoye, France, Aug. 11, 1918.
R—Spokane, Wash.
B—Kentucky.
G. O. No. 145, W. D., 1918.

Private, Company D, 4th Engineers, 4th Division.
Although his eyes had been burned by gas, he volunteered for duty and assisted in the construction of an artillery bridge across the Vesle River under constant machine-gun and artillery fire, setting a conspicuous example of personal bravery and devotion to duty.

GLOMSKI, FRANK (273939).
Near Terny-Sorny, north of Soissons, France, Aug. 31, 1918.
R—Eau Claire, Wis.
B—Manitowoc, Wis.
G. O. No. 128, W. D., 1918.

Sergeant, 2d Battalion, Scout Platoon, 127th Infantry, 32d Division.
When the advance of his battalion was halted by heavy machine-gun fire he went forward alone, advancing from one shell hole to another, under the heavy fire mentioned, and located the exact positions of enemy machine-gun emplacements. Wounded while accomplishing this dangerous mission, he nevertheless attempted to deliver his information to battalion headquarters. Again wounded while endeavoring to reach the post of command, he gave detailed information to another soldier, who reported it to the battalion commander.

*GLUCK, FREDERICK (90185)----------
   Near Sommerance, France, Oct. 16, 1918.
   R—New York, N. Y.
   B—New York, N. Y.
   G. O. No. 78, W. D., 1919.

Private, first class, Company E, 165th Infantry, 42d Division.
Private Gluck volunteered to leave his battalion, which was in support, and help carry wounded men from the field of action. He was killed while in the performance of this mission after having assisted in bringing in at least a dozen of his wounded comrades.
Posthumously awarded. Medal presented to father, Charles Gluck.

GLUCKSMAN, SAMUEL.----------------
   At Blanc Mont, France, Oct. 3, 1918.
   R—Chicago, Ill.
   B—Austria.
   G. O. No. 32, W. D., 1919.

Private, 78th Company, 6th Regiment, U. S. Marine Corps, 2d Division.
After capturing a prisoner, Private Glucksman forced his captive to lead him to a dugout containing 20 of the enemy, whom he also captured. Later, after being wounded, he returned to the front line after securing first-aid treatment and continued in action until his wound forced him to be evacuated.

GLYNN, CORNELIUS T. (1685561)--------
   Near Bois-du-Fays, France, Oct. 5–6, 1918.
   R—Hartford, Conn.
   B—Hartford, Conn.
   G. O. No. 71, W. D., 1919.

Corporal, Company K, 59th Infantry, 4th Division.
He showed marked bravery as battalion runner, repeatedly carrying messages through heavy artillery and machine-gun fire. He remained on duty night and day, aiding materially in maintaining liaison.

GODBEY, ARNOLD DEE.----------------
   At Chateau-Thierry, France, June 6, 1918.
   R—Chicago, Ill.
   B—St. Joseph, Mo.
   G. O. No. 110, W. D., 1918.

Corporal, 67th Company, 5th Regiment, U. S. Marine Corps, 2d Division.
He volunteered to rescue wounded men from a field swept by machine-gun fire and snipers. Upon gaining permission to perform this duty, he bravely continued the hazardous work, with the aid of other volunteers, until all had been rescued.

GODFREY, JOHN R. (1288519)----------
   At Bois-d'Etraye, France, Oct. 23, 1918.
   R—Hampton, Va.
   B—Norfolk, Va.
   G. O. No. 37, W. D., 1919.

Private, Company G, 116th Infantry, 29th Division.
In the face of a terrific enemy barrage Private Godfrey went out in advance of his lines and brought in a wounded comrade.

GOETSCH, ARTHUR J. (568732)---------
   At Ville-Savoye, France, Aug. 11, 1918.
   R—Walnut, Iowa.
   B—Davenport, Iowa.
   G. O. No. 147, W. D., 1918.

Sergeant, Company D, 4th Engineers, 4th Division.
Although his eyes had been burned by gas, he volunteered for duty, and assisted in the construction of an artillery bridge across the Vesle River, under constant machine-gun and artillery fire, setting a conspicuous example of personal bravery and devotion to duty.

GOETZ, MOWRY E.--------------------
   At Bois-de-Beuge, north of Montfaucon, France, Sept. 29, 1918.
   R—Woodlawn, Pa.
   B—Cleveland, Ohio.
   G. O. No. 15, W. D., 1923.

First lieutenant, 316th Infantry, 79th Division.
Orders having been received at regimental headquarters to withdraw the troops to the Bois-de-Beuge, and a provisional battalion of the regiment occupying an advance position in which they were surrounded by enemy forces, the order for the retirement of the battalion was not delivered to the battalion commander. Later in the afternoon it was learned that an American barrage would fall on the woods (Bois-de-Beuge) occupied by the battalion. Lieutenant Goetz volunteered to carry orders for the withdrawal of the battalion and, despite terrific enemy machine-gun and artillery fire, he made his way over rough and broken country, accompanied voluntarily by Private Roy B. Andes. They accomplished their mission, returning with the battalion to the new position as the American shells commenced to fall upon the Bois-de-Beuge.

GOLD, THOMAS B.--------------------
   Near Busigny, France, Oct. 9, 1918, and Mazinghein, France, Oct. 18–19, 1918.
   R—Lawndale, N. C.
   B—Rutherford County, N. C.
   G. O. No. 44, W. D. 1919.

First lieutenant, Medical Corps, attached to 119th Infantry, 30th Division.
During the attack of Oct. 9 he established his aid post in a roadside shrine up with the front line, where he rendered valuable assistance to the wounded. On another occasion he established alone a post close to the front line, where he again gave treatment until the heavy fire of the enemy forced him to withdraw. During the advance of Oct. 18–19 he established another frontline post under the enemy fire and thus saved the lives of many of the troops.

GOLDBERG, SAM (1445961)-----------
   Near Cheppy, France, Sept. 26, 1918.
   R—St. Louis, Mo.
   B—England.
   G. O. No. 71, W. D., 1919.

Private, Headquarters Company, 138th Infantry, 35th Division.
Displaying marked heroism, he entered an enemy dugout alone, armed only with a pistol, and compelled the surrender of 18 Germans he encountered there. Exposing himself to intense machine-gun fire, he dressed the wounds of an officer, showing utter disregard of his personal danger. Later he compelled 4 German prisoners to carry a wounded officer to the rear.

*GOLDCAMP, FRANK J. (158151)--------
   Near Bois-des-Tailloux, France, Mar. 28, 1918.
   R—Akron, Ohio.
   B—Ironton, Ohio.
   G. O. No. 22, W. D., 1926.

Private, Company B, 6th Engineers, 3d Division.
He volunteered to leave shelter to rescue his patrol leader, who was lying wounded within sight of his lines. While engaged in this heroic work he came under hostile machine-gun fire and was killed.
Posthumously awarded. Medal presented to father, Frank E. Goldcamp.

GOLDEN, WILLIAM T. (1278325)--------
   Near Bois d'Etraye, France, Oct. 23, 1918.
   R—Newark, N. J.
   B—Newark, N. J.
   G. O. No. 35, W. D., 1919.

Private, Company C, 113th Infantry, 29th Division.
After his company had been forced to retire about 100 meters on account of a heavy barrage on its position, Private Golden worked his way through heavy shell and machine-gun fire to the position formerly held and succeeded in bringing a wounded comrade to a place of safety.

GOLDSTEIN, ISAAC (2406728)-----------
   Near Talma Farm, France, Oct. 19, 1918.
   R—Newark, N. J.
   B—England.
   G. O. No. 44, W. D., 1919.

Private, Company D, 312th Infantry, 78th Division.
Disregarding the warning that to leave shelter meant almost certain death, Private Goldstein went to the aid of a wounded comrade, through an open field, exposed to murderous artillery and machine-gun fire. Bringing his comrade back to safety, he took a message from the wounded man's pocket and delivered it to its destination. Under cover of darkness he carried his comrade to a first-aid station.

GOLDTHWAITE, GEORGE E.
Near the Bois - de - Bantheville,
France, Oct. 15, 1918.
R—Marion, Ind.
B—Marion, Ind.
G. O. No. 1, W. D., 1919.

First lieutenant, pilot, 24th Aero Squadron, Air Service.
In the course of a special reconnaissance to locate a hostile concentration massing for a counterattack in the vicinity of the Bois-de-Bantheville, he and his observer flew generally at an altitude of 400 meters, at times as low as 50 meters, 5 kilometers into the enemy's lines. Antiaircraft guns riddled his plane with bullets, pierced the gasoline tank, and drenched both pilot and observer. He continued on until the enemy's concentration was located and military information of great value secured. The bravery of Lieutenant Goldthwaite saved the lives of many American soldiers and brought large losses to the enemy.

*GOLTRA, ISAAC V.
In the Chateau-Thierry sector,
France, June 6-7, 1918.
R—Blue Mound, Ill.
B—Springfield, Ill.
G. O. No. 116, W. D., 1918.

First lieutenant, 23d Infantry, 2d Division.
Lieutenant Goltra exhibited exceptional self-sacrifice and courage in the face of heavy machine-gun fire, promptly taking command of his platoon when its leader was killed, and fearlessly leading its advance. The second day of the attack he was killed while directing his platoon through a heavy German barrage.
Posthumously awarded. Medal presented to sister, Mrs. H. T. Willett.

GONZALES, BENJAMIN (748181)
Near Crezancy and Chateau-Thierry, France, July 15, 1918.
R—Watrous, N. Mex.
B—Watrous, N. Mex.
G. O. No. 26, W. D., 1919.

Wagoner, Company D, 3d Ammunition Train, 3d Division.
While on duty with the 30th Infantry he saw an officer and two soldiers killed while attempting to remove company records, which were in danger of capture. He then left his dugout, succeeded in bringing up a truck, loaded what records he could, and after assisting several wounded men into the truck drove it out under heavy shellfire.

GOODALL, ROBERT M.
At Belair Farm, near Pont-a-Mousson, France, Sept. 12, 1918.
R—Birmingham, Ala.
B—Nashville, Tenn.
G. O. No. 13, W. D., 1919.

Second lieutenant, 321st Machine Gun Battalion, 82d Division
When our forces were attacked on the right flank at Belair Farm, Lieutenant Goodall defended this flank alone with his pistol, preventing the enemy from coming up the trenches until the Infantry could form to resist the attack. Throughout a very trying and critical time this officer displayed utter disregard of personal danger, and by his remarkable personal bravery inspired confidence among his own men and cooperating troops.

GOODING, ROY H. (8706)
Near Somme-Py, France, Oct. 2-9, 1918.
R—West Haven, Conn.
B—Shelton, Conn.
G. O. No. 37, W. D., 1919.

Private, section No. 554, Ambulance Service.
He volunteered and drove an ambulance at night, relieving men who were exhausted, and evacuated wounded from the most advanced posts under shellfire. On Oct. 5 he voluntarily went out and repaired an ambulance which had been damaged by a shell, in direct view of the enemy, who shelled the place continually as he worked. On Oct. 7 he repaired another car under the same circumstances, assuring the prompt evacuation of the wounded.

GOODMAN, LIONEL (264361)
Near Verdun, France, Oct. 11-13, 1918.
R—Detroit, Mich.
B—Advance, Mich.
G. O. No. 13, W. D., 1919

Private, Headquarters Company, 125th Infantry, 32d Division.
As a runner for the Third Battalion, 125th Infantry, during the taking and holding of the line near La Tuilerie Farm, he was engaged in carrying important messages, crossing and recrossing Death Valley between Hill 258 and La Cote Dame Marie, the foremost part of the line held by the Third Battalion. The valley was swept by machine-gun fire, the terrain affording absolutely no protection, requiring a perilous dash of 500 yards across open ground before any cover was reached. It was only by a display of supreme courage that important messages reached the battalion.

GOODMAN, WALTER O. (3109805)
At Molleville Farm, France, Nov. 7, 1918.
R—Philadelphia, Pa.
B—Philadelphia, Pa.
G. O. No. 37, W. D., 1919.

Private, Company E, 315th Infantry, 79th Division.
He volunteered to carry wounded to the first-aid station, through continuous shellfire. After he became too exhausted to carry more, he continued to help those who could walk with assistance.

GOODRICH, LOUIS D. (156246)
Lucy-de-Bocage to Bouresches, France, June 9, 1918.
R—Sedalia, Mo.
B—Sedalia, Mo.
G. O. No. 99, W. D., 1918.

Private, Company A, 2d Engineers, 2d Division.
On June 9, 1918, voluntarily carried an important message from Lucy-le-Bocage to Bouresches in daylight along an open road between the lines exposed to small-arms fire.

GOODRIDGE, GEORGE L.
Near Verdun, France, Nov. 8, 1918.
R—Melrose, Mass.
B—Bradford, Mass.
G. O. No. 23, W. D., 1919.

First lieutenant, 101st Infantry, 26th Division.
Lieutenant Goodridge, with about 30 men, secured a footing in an advanced enemy trench. The attacking battalion met with stubborn resistance and fell back to their starting point, but he tenaciously held his position until relieved on Nov. 11. His coolness and courage made it possible to hold this position, 800 meters in advance of our line, under intense machine-gun and artillery fire.

GOODWILLIE, HERRICK R. (1386505)
Near Bois-de-Chaume, France, Oct. 10, 1918.
R—Oak Park, Ill.
B—Oak Park, Ill.
G. O. No. 71, W. D., 1919.

Sergeant, Company B, 131st Infantry, 33d Division.
When his company was enfiladed from the right by machine guns and could neither advance nor withdraw, he volunteered to go for assistance, after several runners had been killed in similar attempts. Crawling back through heavy machine-gun fire, he reached the trench-mortar battery, guided them forward, and himself operated one of the mortars which knocked out the enemy machine-gun nest, saving his company.

GOODWIN, ROBERT HAZEN (1308606)
Near Bellicourt, France, Sept. 26-29, 1918.
R—Bristol, Tenn.
B—Hampton, Tenn.
G. O. No. 44, W. D., 1919

Sergeant, Company H, 117th Infantry, 30th Division.
He was badly gassed on Sept. 26 and ordered to the hospital, but insisted on remaining at his post, despite the fact that he was suffering great pain. On Sept. 29 he was painfully wounded by shellfire, but again remained with his platoon, refusing to be evacuated. He efficiently performed his duties until his company was relieved.

GORDON, ERSKINE............
Near Gercourt, France, Sept. 26-27, 1918.
R—Washington, D. C.
B—Washington, D. C.
G. O. No. 7, W. D., 1919.

Captain, 319th Infantry, 80th Division.
After the assaulting companies had passed over three machine-gun nests, which, not having been destroyed, opened heavy and effective fire, he reorganized scattered elements of his own company and of two others and fearlessly exposing himself to the fire of these guns, as well as that of our own artillery, personally led an attack on three nests and captured them, with 50 prisoners.

*GORDON, FRANK P. (137189)..........
Near Seicheprey, France, Apr. 20, 1918.
R—New Haven, Conn.
B—New Haven, Conn.
G. O. No. 88, W. D., 1918.

Private, first class, Battery F, 103d Field Artillery, 26th Division.
He displayed conspicuous courage on Apr. 20, 1918, in voluntarily going to the aid of a wounded comrade, during which action he and the man he was assisting were killed.
Posthumously awarded. Medal presented to father, Patrick R. Gordon.

GORDON, HAROLD J............
Near Heurne, Belgium, Nov. 4, 1918.
R—Cleveland, Ohio.
B—Sheakleyville, Pa.
G. O. No. 9, W. D., 1923.

Captain, 148th Ambulance Company, 112th Sanitary Train, 37th Division.
Although in command of an ambulance company and not required to work so far forward, he voluntarily crossed the river and sought out wounded among the troops in the advance line. Finding two severely wounded men, he gave them first aid under withering machine-gun, rifle, and shellfire, arranged such shelter for them as he could, then sought assistance to carry them on to safety. Returning with three men of his company, the bearers placed one of the wounded on an improvised litter, while the other was carried on the shoulders of the officer until the latter was exhausted. The fire becoming more intense, the wounded man was placed on the ground and encouraged by the officer to crawl to safety, the latter crawling beside him and protecting him from the enemy's fire with his own body.

*GORDON, MILLEDGE A. (1312846)......
At Harricourt, France, Sept. 26, 1918-Oct. 17, 1918.
R—Clemson College, S. C.
B—Oconee, S. C.
G. O. No. 44, W. D., 1919.

Sergeant, machine gun company, 118th Infantry, 30th Division.
Orders for his relief having failed to reach him, Sergeant Gordon remained on duty all night, maintaining liaison between gun sections of his platoon, exposed to severe shellfire, from which he was gassed. He nevertheless stayed with his company, and while going forward on Oct. 8 he fainted from the effects of the gas and was evacuated to the rear, unconscious. Regaining consciousness while en route to the casualty clearing station, he crawled out of the ambulance and worked his way back to his company without securing treatment. Though still suffering from weakness, he persistently refused to to be evacuated and took part in subsequent engagements with his platoon, until he was killed in action.
Posthumously awarded. Medal presented to father, Dr. Thomas Gordon.

GORMAN, CARLISLE A. (1287244)........
In the Bois Brabant-sur-Meuse, France, Oct. 8, 1918.
R—Richmond, Va.
B—Richmond, Va.
G. O. No. 37, W. D., 1919.

Sergeant, Company B, 116th Infantry, 29th Division.
Sergeant Gorman, with four other soldiers, fearlessly attacked eight machine-gun positions and succeeded, after stubborn resistance, in capturing both the guns and the crews.

GORMAN, JAMES A. (50036)........
Near St. Etienne-a-Arnes, France, Oct. 3, 1918.
R—Mount Carmel, Pa.
B—Mount Carmel, Pa.
G. O. No. 35, W. D., 1919.

Corporal, Company C, 23d Infantry, 2d Division.
When he saw a soldier lying wounded about 200 yards in front of his trench, Corporal Gorman, without thought of personal danger, went out and brought him to safety, under terrific machine-gun fire from both sides and from the front.

*GOSSELIN, ALEXANDER (198539)........
Near Exermont, France, Oct. 4, 1918.
R—San Francisco, Calif.
B—Chicago, Ill.
G. O. No. 44, W. D., 1919.

Sergeant, Company B, 2d Field Signal Battalion, 1st Division.
After finishing the work of laying telephone lines on high ground, under heavy artillery fire and direct observation of the enemy, he ordered his men to shelter and remained behind to repair breaks caused by the shelling. He was killed a few minutes later by the explosion of a shell.
Posthumously awarded. Medal presented to father, Anatole Gosselin.

GOTTSCHALK, FRANK L............
Near St. Gilles, France, Aug. 3, 1918.
R—Marathon City, Wis.
B—Wausau, Wis.
G. O. No. 44, W. D., 1919.

First lieutenant, 128th Infantry, 32d Division.
After being severely wounded while leading an attack on a machine-gun nest, he continued to lead his platoon, in the face of severe machine-gun fire, remaining on duty all night upon reaching the final objective.

GOTTSCHALK, JAMES N. (3489164)........
Near Cunel, France, Oct. 14, 1918.
R—Leetonia, Ohio.
B—Cleveland, Ohio.
G. O. No. 35, W. D., 1919.

Private, Company C, 15th Machine Gun Battalion, 5th Division.
Refusing to be evacuated after being seriously wounded, he continued to carry messages through heavy machine-gun and shell fire. After becoming too weak to make trips by himself, he guided another runner and later assisted a wounded comrade to the rear.

GOULD, WILLIAM (1205208)............
Near St. Souplet, France, Oct. 18, 1918.
R—West Hebron, N. Y.
B—West Hebron, N. Y.
G. O. No. 23. W. D., 1919.

Mechanic, Company K, 105th Infantry, 27th Division.
Mechanic Gould, single-handed, attacked a heavy machine gun which was covering the retreat of the Germans and drove off the crew.

GOUMAS, GEORGE (1918987)............
Near Fleville, France, Oct. 21, 1918.
R—Marietta, Ga.
B—Greece.
G. O. No. 145, W. D., 1918.

Private, first class, Company B, 307th Field Signal Battalion, 82d Division.
This soldier was in a relay station which was struck by a shell, wounding him and five others. After assisting the other wounded men to a truck near by, he returned through the falling shells to the relay station and assisted another soldier in repairing the wires which had been cut, remaining at his post until he had made sure that the lines were in good condition and he was ordered to come in by his commanding officer.

GOURLEY, GEORGE B. (1387246)_____
At Chipilly Ridge, France, Aug. 10, 1918.
R—Chicago, Ill.
B—Chicago, Ill.
G. O. No. 128, W. D., 1918.

Sergeant, Company E, 131st Infantry, 33d Division.
He displayed qualities of leadership by taking command of his platoon and continuing the advance when his platoon commander had been killed. With one other soldier he fearlessly attacked an enemy machine-gun nest, capturing the gun and killing the crew. He then carried the captured gun with him and used it effectively against the enemy.

*GOW, KENNETH_____
Near Ronssoy, France, Sept. 29, 1918.
R—Summit, N. J.
B—Summit, N. J.
G. O. No. 1, W. D., 1926.

First lieutenant, 107th Infantry, 27th Division.
While supply officer for his company he personally took rations forward with a pack mule through continuous shell and machine-gun fire. When all officers of his company were either killed or wounded he assumed command and led it forward through heavy shell and machine-gun fire until he was killed on Oct. 17, 1918, near St. Souplet, France.
Posthumously awarded. Medal presented to father, Robert M. Gow.

GRABAU, JOHN C._____
Near Brabant-sur-Meuse, France, Nov. 3, 1918.
R—Buffalo, N. Y.
B—Walmore, N. Y.
G. O. No. 56, W. D., 1922.

Captain, Medical Corps, attached to 106th Field Artillery, 27th Division.
He displayed great coolness and courage, while under a heavy shellfire, in going to the assistance of a wounded soldier, and although twice knocked down by exploding shells and badly wounded in the arm, he continued his efforts until he ascertained the soldier was dead. Later, at the dressing station, although bleeding profusely, he refused to attend to his own wounds until he had dressed the wounds of several other men.

*GRABINSKI, ELMER C. (113180)_____
Between Sergy and Villers-sur-Fere, France, July 28, 1918.
R—Fond du Lac, Wis.
B—Miller, S. C.
G. O. No. 99, W. D., 1918.

Corporal, Company B, 150th Machine Gun Battalion, 42d Division.
He led his gun crew with courage and fearlessness in the attack north of the River Ourcq. He directed the fire of his gun with excellent effect, shooting several enemy snipers. Showing always the greatest eagerness to press forward and always disregarding his own safety, he was finally killed after leading his men forward successfully to their objective.
Posthumously awarded. Medal presented to mother, Mrs. Mathilda Grabinski.

GRADDY, WILLIAM E. (1467917)_____
Near Baulny, France, Sept. 29, 1918.
R—Couch, Mo.
B—Sharp County, Ark.
G. O. No. 59, W. D., 1919.

Private, Company B, 110th Engineers, 35th Division.
As a company runner he was sent to the rear with an important message, having to pass through the enemy counterbarrage. Though severely wounded soon after starting on the mission, he struggled on and delivered the message as he fell, fainting from loss of blood.

GRADY, JOHN J. (62106)_____
Near Vaux, France, July 20, 1918.
R—South Boston, Mass.
B—Ireland.
G. O. No. 125, W. D., 1918.

Private, Company L, 101st Infantry, 26th Division.
Private Grady, Corp. Austin J. Kelley, and Pvt. Harold E. Rounds penetrated the enemy's lines in the face of machine-gun fire, captured a machine gun and its crew, and returned with valuable information concerning the enemy's positions.

GRAHAM, EDWARD L._____
Near Bellicourt, France, Sept. 29–30, 1918.
R—Lexington, Va.
B—Lexington, Va.
G. O. No. 81, W. D., 1919.

Captain, 119th Infantry, 30th Division.
Although twice wounded while leading his men, he refused to be evacuated, inspiring his command by his personal courage and fortitude. Ordered to the rear by a superior officer, he returned to his men as soon as his wounds were dressed.

GRAHAM, JOHN D. (8594)_____
Near Somme-Py, France, Oct. 2–9, 1918.
R—Philadelphia, Pa.
B—Philadelphia, Pa.
G. O. No. 37, W. D., 1919.

Corporal, section No. 554, Ambulance Service.
He was on duty continuously during this period at the most advanced marine post, assuring the prompt evacuation of the wounded and gassed. On the night of Oct. 5, hearing that several ambulances were needed to evacuate a number of wounded who were very close to the line, Corporal Graham, accompanied by the driver, passed over a road and across a field raked by machine-gun and shell fire and evacuated the wounded that were at this place. On Oct. 6 he worked under shellfire practically the whole day at an advanced post, superintending the loading of the ambulances and the rapid evacuation of the wounded.

GRAHAM, JOSEPH D. (89271)_____
At Ourcq River, near Villers-sur-Fere, France, July 28, 1918.
R—Brooklyn, N. Y.
B—Brooklyn, N. Y.
G. O. No. 14, W. D., 1923.

Private, first class, Company B, 165th Infantry, 42d Division.
After repeated attempts had been made to rescue two wounded soldiers, who were lying within a hundred yards of the enemy lines, Corporal Graham went out alone, over ground so swept by machine-gun and rifle fire that 2 men had been killed previously in the attempt, and succeeded in bringing the 2 wounded soldiers to safety, showing by his valor and devotion to duty an excellent example to the men of the organization.

GRAHEK, MATTHEW G. (2021862)_____
Near Obozerskaya, Russia, Sept. 29, 1918.
R—Detroit, Mich.
B—Germany.
G. O. No. 16, W. D., 1920.

Sergeant, Company M, 339th Infantry, 85th Division (detachment in North Russia).
Sergeant Grahek voluntarily went forward about 150 yards in advance of our line, exposed to heavy machine-gun and artillery fire, and rescued a wounded comrade. Again, on Apr. 1, 1919, near Bolshe-Ozerka, Russia, he advanced alone with a Lewis gun against enemy snipers concealed in a ditch and routed them.

GRANGER, JOHN McCLAVE (1212111)____
Near Bony, France, Sept. 29, 1918.
R—New York, N. Y.
B—New York, N. Y.
G. O. No. 19, W. D., 1920.

Corporal, Company M, 107th Infantry, 27th Division.
He crossed an area exposed to heavy fire to deliver a message, and while in the act of delivering his message his left leg was torn off by a shell. He refused assistance and shouted words of encouragement to members of his platoon in action.

GRANT, ALFRED A._____
Near Chateau-Thierry, France, July 2, 1918.
R—Manhattan, Kans.
B—Denton, Tex.
G. O. No. 121, W. D., 1918.

First lieutenant, 27th Aero Squadron, Air Service.
With several other officers, Lieutenant Grant encountered an enemy patrol of 9 planes. During the combat he became slightly separated from the other American machines and was attacked by three of the enemy. By skillful maneuvering and good marksmanship he destroyed one machine and drove off the other two.

*GRANT, JOHN_____
At Chateau-Thierry, France, June 6, 1918.
R—Boston. Mass.
B—Mars Hill, Me.
G. O. No. 110, W. D., 1918.

First sergeant, 20th Company, 5th Regiment, U. S. Marine Corps, 2d Division. Killed in action at Chateau-Thierry, France, June 6, 1918, he gave the supreme proof of that extraordinary heroism which will serve as an example to hitherto untried troops.
Posthumously awarded. Medal presented to friend, Mrs. Rosie Brown.

*GRAUER, SHIPTON G. (1257216)_____
Near Baslieux, north of the Vesle River, France, Sept. 5, 1918.
R—Reading, Pa.
B—Buffalo, N. Y.
G. O. No. 46, W. D., 1920.

Sergeant, Company B, 108th Machine Gun Battalion, 28th Division.
The advance of his platoon having been halted by a wire entanglement, Sergeant Grauer advanced alone about 50 yards in advance of his platoon, under enemy fire, and located a gap in the enemy wire. He then signaled to his platoon which advanced. He was mortally wounded in the performance of this act, but his gallant deed enabled the others to continue the attack.
Posthumously awarded. Medal presented to father, Charles E. Grauer.

GRAVE, HENRY H._____
Near St. Baussant, France, Sept. 12, 1918.
R—Columbus, Ohio.
B—Germany.
G. O. No. 24, W. D., 1920.

Captain, 166th Infantry, 42d Division.
During the St. Mihiel offensive Captain Grave was leading an assault wave which was checked by a dense machine-gun fire. In order to locate the exact position of the enemy strongholds he mounted the parapet, thus drawing a a burst of machine-gun fire, which enabled his Stokes mortars to open fire. He then led a flanking group to the position, while his assault wave went forward and captured 34 prisoners and a machine gun, rifles, and a store of ammunition, the first prisoners to be taken during the drive.

GRAVELINE, FRED C. (20083)_____
Near Villers-devant-Dun, France, and Mouzon, France, Sept. 29–Nov. 5, 1918.
R—Cleveland, Ohio.
B—West Warren, Mass.
G. O. No. 37, W. D., 1919.

Sergeant, first class, 20th Aero Squadron, Air Service.
Volunteering to act as observer and aerial gunner because of the shortage of officer observers, he started on 17 bombing missions, successfully reaching his objective on 14 of these expeditions, shooting down 2 enemy aircraft. On 2 occasions, while flying in the rear of his formation, he drove off superior numbers of German machines.

GRAVES, JOHN C. (181245)_____
Near Nantillois, France, Oct. 5, 1918.
R—Des Moines, Iowa.
B—Omaha, Nebr.
G. O. No. 37, W. D., 1919.

Corporal, Company A, 1st Gas Regiment.
After other means of communication had failed, he voluntarily carried messages from the regimental post of command to advance positions through several enemy barrages of gas and high-explosive shells. He continued on duty, even after being wounded, until he was exhausted.

GRAVES, SIDNEY C._____
In the Bois-de-Fontaine, France, Apr. 29, 1918.
R—El Paso, Tex.
B—Fort Logan, Colo.
G. O. No. 118, W. D., 1919, and G. O. No. 15, W. D., 1923.

Major, 16th Infantry, 1st Division.
Having located an enemy machine gun in front of his position, Major (then captain) Graves, with 3 men, voluntarily crawled out to the position of the machine gun, in full view and within 100 yards of the enemy lines, shot the gunner, killed the rest of the crew with grenades, and returned with his party without a casualty.
Oak-leaf cluster.
(Major, Infantry, Assistant Chief of Staff, American Expeditionary Forces, Siberia.) For extraordinary heroism in action at Vladivostok, Siberia, Nov. 18, 1919. In answer to a call to save noncombatants entrapped in the railroad station at Vladivostok, Siberia, Major Graves fearlessly entered a zone swept by intense machine-gun and artillery fire of Russian Government and insurgent forces, entered the station, and assisted in locating 6 noncombatants. He escorted them through the attacking troops to a place of safety.

GRAY, COLA A. (109262)_____
At Marcheville, France, Sept. 26, 1918.
R—Waverly, Mass.
B—Albany, Vt.
G. O. No. 15, W. D., 1919.

First sergeant, Company A, 102d Machine Gun Battalion, 26th Division.
When all of his officers had been wounded, he took command of the company and led it throughout the attack, under terrific fire from all arms, showing high qualities of leadership and personal bravery.

GRAY, JOSEPH W_____
In Romagne, France, Oct. 18, 1918.
R—Titusville, Pa.
B—Titusville, Pa.
G. O. No. 37, W. D., 1919.

First lieutenant, 7th Engineers, 5th Division.
Although wounded, he personally supervised the construction of a bridge under severe artillery and direct machine-gun fire, thereby making it possible for the Infantry and Artillery to advance to more advantageous positions.

GRAY, THOMAS J. (102627)_____
Northeast of Chateau-Thierry, France, July 28, 1918.
R—Elliott, Iowa.
B—Lorimor, Iowa.
G. O. No. 108, W. D., 1918.

Private, Company M, 168th Infantry, 42d Division.
He distinguished himself northeast of Chateau-Thierry, France, on July 28, 1918, when, with 4 other men, he raided an enemy machine-gun nest held by 12 Germans. As a result of their daring and presence of mind, 1 of the enemy was killed, the other 11 captured, and their 4 machine guns turned upon the retreating foe.

GRAYSON, THOMAS J._____
Near Exermont, France, Oct. 2–3, 1918.
R—Biloxi, Miss.
B—Mobile, Ala.
G. O. No. 16, W. D., 1920.

First lieutenant, 28th Infantry, 1st Division.
He commanded a combat patrol which penetrated 2 kilometers into the enemy lines, and although twice attacked he maintained his position for 36 hours without supplies before returning to our lines with valuable information. The leadership and initiative displayed by Lieutenant Grayson were vital factors in the successful performance of this important mission.

*GREEN, DONALD R. (129140)_____
Near Thiaucourt, France, Sept. 12, 1918.
R—Anacostia, D. C.
B—Anacostia, D. C.
G. O. No. 15, W. D., 1919.

Sergeant, Battery E, 12th Field Artillery, 2d Division.
Sergeant Green continued the service of his piece under a galling enfilading fire. In his fortitude and devotion to duty, he set a high example to the men of his section until he and his entire crew were casualties. He was killed while serving his piece.
Posthumously awarded. Medal presented to mother, Mrs. Lillie M. Green.

GREEN, DONALD W.
Northeast of Chateau-Thierry, France, July 26, 1918.
R—Chicago, Ill.
B—Anniston, Ala.
G. O. No. 108, W. D., 1918.

Second lieutenant, 167th Infantry, 42d Division.
He set an example of heroism and devotion to duty at Coix Rouge Farm, when he continued to lead his platoon through heavy fire for five days after being wounded.

GREEN, ERNEST B. (1310736)
Near Bellicourt, France, Sept. 25, 1918.
R—Concord, N. C.
B—Concord, N. C.
G. O. No. 37, W. D., 1919.

Private, Company D, 118th Infantry, 30th Division.
Although stunned and bruised by a shell which burst in his trench, he went to the aid of a comrade outside of the trench and brought him to safety. This was in full view of the enemy and under heavy shelling.

GREEN, GARLAND (546149)
Near Mezy, France, July 15, 1918.
R—Barkers Creek, N. C.
B—Barkers Creek, N. C.
G. O. No. 32, W. D., 1919.

Private, Company D, 30th Infantry, 3d Division.
During the German artillery bombardment of July 15 he carried messages between company and battalion headquarters, and, although wounded in the arm, he refused evacuation until relieved two days later.

*GREEN, HERBERT D. (2426055)
Near Cunel, France, Oct. 10, 1918.
R—Buffalo, Ohio.
B—Cumberland, Ohio.
G. O. No. 16, W. D., 1920.

Private, Company A, 7th Infantry, 3d Division.
He voluntarily crossed an area swept by heavy machine-gun and artillery fire to carry water to distressed members of his platoon. He was killed in the performance of this self-assigned task.
Posthumously awarded. Medal presented to father, Frank Green.

GREEN, JAMES O., Jr.
Near Chateau-Thierry, France, June 6, 1918.
R—Whitewater, Wis.
B—Hillsdale, Mich.
G. O. No. 19, W. D., 1921.

Captain, 23d Infantry, 2d Division.
After 2 platoons under his command had been practically wiped out, he continued forward to his objective with 2 enlisted men and remained throughout the night under heavy fire in hope of keeping the ground gained until reinforcements came up.

GREEN, ROBERT L. (2980317)

At Tulgas, Russia, Nov. 14, 1918.
R—Lincoln, Nebr.
B—Kiowa, Kans.
G. O. No. 108, W. D., 1919.

Corporal, Company D, 339th Infantry, 85th Division (detachment in North Russia).
He led an attack against snipers concealed in houses on the edge of the village. In order to reach these buildings he was forced to cross an open space of 200 yards, in clear view of the enemy. After reaching the buildings and locating the snipers he alone charged the building in which the snipers were located and captured 14 prisoners.

*GREEN, WALLACE (734184)
At Frapelle, France, Aug. 17, 1918.
R—Eure, N. C.
B—Eure, N. C.
G. O. No. 15, W. D., 1919.

Sergeant, Company M, 6th Infantry, 5th Division.
He unhesitatingly and with great coolness and courage went forward under a heavy enemy barrage to destroy wire entanglements and continued this hazardous work until killed.
Posthumously awarded. Medal presented to mother, Mrs. Elizabeth Green.

GREEN, WILLIE (2249500)
Near Vilcey, France, Sept. 13, 1918.
R—Hastings, Okla.
B—Italy, Tex.
G. O. No. 127, W. D., 1918.

Corporal, Company B, 358th Infantry, 90th Division.
While officers were holding a conference in a dugout on our outpost line, a German patrol came across a small footbridge directly to the entrance of the dugout and threw a grenade inside. Corporal Green, without any thought of personal danger, in order to save his officers from injury, stood on the grenade and then shot two of the retreating patrol.

GREENE, DON (2213364)
Near the Bois-de-Barricourt, France, Nov. 2, 1918.
R—Eldorado, Kans.
B—North Baltimore, Ohio.
G. O. No. 81, W. D., 1919.

Private, Company H, 353d Infantry, 89th Division.
Private Greene displayed conspicuous bravery in twice carrying important messages across an area under heavy shell and machine-gun fire from the front and flanks.

GREENE, HAROLD (1211192)
Near Roussoy, France, Sept. 29, 1918.
R—Binghamton, N. Y.
B—Anoka, N. Y.
G. O. No. 32, W. D., 1919.

First sergeant, Company H, 107th Infantry, 27th Division.
When his commanding officer was severely wounded and evacuated, he took command of the company and led it into effective combat. He continued to lead the company forward through a terrific fire of artillery and machine guns for more than a mile after being severely wounded, and refused to be evacuated until he had received a second wound, which made it impossible for him to continue farther.

GREENE, HENRY W. (1210036)
South of Roussoy, France, Sept. 28, 1918.
R—Brooklyn, N. Y.
B—Covington, Ky.
G. O. No. 16, W. D., 1920.

Sergeant, Company C, 107th Infantry, 27th Division.
Sergeant Greene left the shelter of a trench and went forward some 35 yards and assisted a wounded soldier to shelter. This act was performed shortly after daybreak at a time when the location of the wounded man and the trench were being swept by hostile fire.

*GREENE, JAMES A. (53739)
Near Cantigny, France, June 7, 1918.
R—Providence, R. I.
B—Providence, R. I.
G. O. No. 44, W. D., 1919.

Corporal, Company F, 26th Infantry, 1st Division.
While in charge of an outpost he was attacked by a patrol of 12 Germans and held his ground single-handed after the other members of the outpost were either killed or wounded. He inflicted heavy casualties on the enemy patrol, repulsing it, and then held his post until relief came.
Posthumously awarded. Medal presented to father, Charles F. Greene.

GREENE, JOHN N.
In Ansauville sector, Mar. 1, 1918.
R—Staunton, Va.
B—Staunton, Va.
G. O. No. 126, W. D., 1918.

Second lieutenant, 6th Field Artillery, 1st Division.
Attacked in a dugout by a large party of Germans, he was wounded by the explosion of an enemy hand grenade. He refused to surrender when ordered to do so, but instead fought vigorously until he had wounded or driven away all of the attacking party.

GREENFIELD, WILLIAM G. (2217131)____
Near Fey-en-Haye, France, Sept. 12, 1918.
R—Camargo, Okla.
B—Cedarville, Mo.
G. O. No 128, W. D., 1918.

Sergeant, Company G, 357th Infantry, 90th Division.
Although wounded, he continued to advance with his organization until its objective was reached. He showed qualities of leadership by organizing all the men he could find and assuming command until he was relieved by officers and ordered to have his wound dressed.

GREENWAY, JOHN C.____
Near Verdun, France, Oct. 23, 1918.
R—Warren, Ariz.
B—Huntington, Ala.
G. O. No. 46, W. D., 1919.

Lieutenant colonel, 101st Infantry, 26th Division.
During a terrific enemy shelling on two of his battalions, and after both his battalion commanders had been wounded, Colonel Greenway personally directed the activities and greatly encouraged his forces by his presence. Leading them in attack, he demonstrated the utmost valor at the most critical moments, and he was the first of his command to enter the German trench which marked the objective of the day's attack.

*GREENWOOD, HARRY L. (1785631)____
Near Malancourt, France, Sept. 26, 1918.
R—Philadelphia, Pa.
B—Philadelphia, Pa.
G. O. No. 37, W. D., 1919.

Sergeant, Company K, 315th Infantry, 79th Division.
He personally led a patrol of 4 men against a machine gun which was holding up the advance, captured the gun and 9 prisoners. He inspired his men to greater effort by his excellent example, under heavy machine-gun fire, and was killed while reorganizing his platoon after a counterattack.
Posthumously awarded. Medal presented to mother, Mrs. Leah Greenwood.

GREGORY, RALPH AMHERST____
Near St. Agnan, France, July 16, 1918.
R—Scranton, Pa.
B—Hollisterville, Pa.
G. O. No. 9, W. D., 1923.

Major, 109th Infantry, 28th Division.
During an attack on Hill 200 in which his battalion participated, the French and American forces were repulsed with heavy losses owing to terrific machine-gun and artillery fire. Leaving the protection afforded by the trenches and mounting the parapet thereof, with utter disregard for his own safety and in full view of the enemy, he exhorted his men to again attack, his indifference to the intense enemy machine-gun fire to which he was subjected raising the morale of his men to so high a pitch as to enable them to renew the assault with great courage and determination, driving the enemy from their strongly held positions.

GREIST, EDWARDS HAROLD____
Near Andevanne, France, Nov. 1, 1918.
R—Chicago, Ill.
B—Indianapolis, Ind.
G. O. No. 55, W. D., 1920.

First lieutenant, 3d Observation Group, Air Service.
On the morning of Nov. 1, 1918, during the progress of an important attack, Lieutenant Greist volunteered on a mission to fly through a heavy fog in order to locate the then advanced infantry of the attack. In accomplishing this mission it was necessary to fly at a very low altitude and through the American barrage which was being fired during the flight. He penetrated several kilometers into the enemy's lines, being subjected to heavy machine-gun fire from the ground, which struck his plane many times, obtained information of the disposition of the enemy artillery, infantry, and our own front line. He returned through the fire with the first authentic information to reach the division commander.

GRESHAM, FRANK B. (1098437)____
Near the Bois-du-Fays, France, Sept. 26, 1918.
R—Augusta, Ga.
B—Edgefield, S. C.
G. O. No. 46, W. D., 1919.

Sergeant, Company G, 39th Infantry, 4th Division.
After his patrol had been twice scattered by machine-gun fire, he continued his reconnaissance, accompanied by only one other soldier, and secured the information for which he had been sent. Upon rejoining his company he was placed in command of his platoon, whose commander had been wounded, and succeeded in reorganizing it under heavy shell fire.

GREY, CHARLES G____
Near Montmedy, France, Nov. 4, 1918.
R—Chicago, Ill.
B—Chicago, Ill.
G. O. No. 46, W. D., 1919.

Captain, 213th Aero Squadron, Air Service.
While leading a patrol of 3 machines, Captain Grey observed a formation of our bombing planes hard pressed by 12 of the enemy. He attacked the leading enemy machine without hesitation, thereby attracting the enemy's fire and allowing the bombing machines to escape undamaged.

GRIBBON, JOHN J. (89749)____
Near Meurcy Ferme, France, July 30, 1918.
R—Brooklyn, N. Y.
B—New York, N. Y.
G. O. No. 81, W. D., 1919.

Corporal, Company D, 165th Infantry, 42d Division.
When his patrol had been caught under a severe cross fire by machine guns and most of the members killed or wounded, and the survivors withdrew to cover, it was discovered that a wounded man had been left at the farthest point to which the patrol had advanced, Corporal Gribbon volunteered to rescue him. Running a hundred yards over open country, swept by withering machine-gun fire, he reached the severely wounded member of his patrol and half dragged half carried his helpless comrade back with him to safety.

GRIDER, THOMAS E. (42170)____
Near Soissons, France, July 22, 1918.
R—Danville, Ill.
B—Danville, Ill.
G. O. No. 87, W. D., 1919.

Private, Company C, 16th Infantry, 1st Division.
Private Grider carried important messages through heavy shell and machine-gun fire and fought his way through territory occupied by the enemy.

*GRIFFIN, HENRY Q____
On Hill 200, near St. Agnan, France, July 16, 1918.
R—Winthrop, Mass.
B—Winthrop, Mass.
G. O. No. 99, W. D., 1918.

Second lieutenant, 109th Infantry, 28th Division.
When the first attack on Hill 200, near St. Agnan, France, was made by American troops on July 16, 1918, he led his platoon to the most advanced point reached by any detachment and was killed when within 20 yards of an enemy machine-gun emplacement.
Posthumously awarded. Medal presented to father, W. I. Griffin.

*GRIFFIN, MARTIN G____
Near St. Etienne, France, Oct. 6, 1918.
R—Fall River, Mass.
B—Ireland.
G. O. No. 37, W. D., 1919.

First lieutenant, 23d Infantry, 2d Division.
After the major and adjutant of the battalion had been killed, Lieutenant Griffin took command of the battalion and led it in an attack over difficult terrain, under heavy machine-gun and artillery fire. He gained his objective and held it.
Posthumously awarded. Medal presented to widow, Mrs. Katherine Griffin.

GRIFFIN, ROBERT A.
Near Epinonville, France, Sept. 26–29, 1918.
R—San Jose, Calif.
B—Kansas City, Mo.
G. O. No. 81, W. D., 1919.

Captain, 364th Infantry, 91st Division.
His company having suddenly become trapped in the smoke and fog under heavy enemy artillery fire, Captain Griffin displayed notable coolness in getting his men to cover in a trench near by, being the last to take shelter, though he was knocked down by bursting shells in so doing. As soon as the fire abated, he quickly redisposed his men and led them toward the enemy. On the next afternoon, during the attack on Epinonville, he led 11 soldiers in a successful charge on an enemy machine-gun nest. Later he personally assisted in carrying in wounded men of his company under heavy machine-gun fire. On the afternoon of Sept. 29 this officer was severely wounded while on a patrol, but he continued on under heavy machine-gun and artillery fire and successfully accomplished his mission.

*GRIFFIN, WILLIAM L.
At Chateau-Thierry, France, June 6, 1918.
R—Hamburg, Pa.
B—Parkersburg, W. Va.
G. O. No. 110, W. D., 1918.

Corporal, 45th Company, 5th Regiment, U. S. Marine Corps, 2d Division.
Killed in action at Chateau-Thierry, France, June 6, 1918, he gave the supreme proof of that extraordinary heroism which will serve as an example to hitherto untried troops.
Posthumously awarded. Medal presented to sister, Mrs. E. M. McConnell.

GRIFFITH, CHANCY W. (1039129).
Near Greves, France, July 15, 1918.
R—Amery, Wis.
B—Eagle Grove, Iowa.
G. O. No. 46, W. D., 1919.

Private, Battery F, 10th Field Artillery, 3d Division.
Responding to a call for volunteers, Private Griffith, with eight other soldiers manned two guns of a French battery which had been deserted by the French during the unprecedented fire, after many casualties had been inflicted on their forces. For two hours he remained at this post and poured an effective fire into the ranks of the enemy.

GRIFFITH, LUTHER C. (1320912).
Near Bellicourt, France, Sept. 29, 1918.
R—Reidsville, N. C.
B—Rockingham County, N. C.
G. O. No. 81, W. D., 1919.

Private, Company G, 120th Infantry, 30th Division.
When the other members of a Lewis gun crew had become casualties, he operated the gun single-handed, and, attacking an enemy machine-gun emplacement, killed the gunner and made the other two members of the crew prisoners.

*GRIFFITH, OSCAR C. (2564).
At Villers-Tournelle, France, Apr. 26, 1918.
R—Gadsden, Ala.
B—Etowah County, Ala.
G. O. No. 88, W. D., 1918.

Private, first class, 12th Ambulance Company, 1st Sanitary Train, 1st Division.
He displayed extraordinary heroism and devotion to duty in going to the assistance of a wounded man lying in the open and administering first aid under heavy fire. He was killed while engaged in this courageous service.
Posthumously awarded. Medal presented to father, Andrew W. Griffith.

GRIFFITH, WALTER (43226).
In the Argonne Forest, France, Oct. 4, 1918.
R—Hitchins, Ky.
B—Elliott County, Ky.
G. O. No. 44, W. D., 1919.

Private, Company G, 16th Infantry, 1st Division.
He was delivering a message to the front line when he encountered an enemy machine gun firing on our front-line troops from the rear. He crawled to the emplacement and silenced the fire from the gun by killing both gunners.

GRIFFITHS, ALFRED S.
Near Ville-Savoye, France, Aug. 19, 1918.
R—Amityville, Long Island, N. Y.
B—New York, N. Y.
G. O. No. 35, W. D., 1919.

Captain, 308th Infantry, 77th Division.
While suffering from the effects of gas, Captain Griffiths led a liaison patrol to the flanking battalion across an open slope and under direct enemy observation, exposed during the whole journey to terrific artillery and machine-gun fire. He remained on duty as battalion adjutant, after all other officers had been evacuated because of the effects of gas, although he had been rendered temporarily speechless and blind by a severe gassing.

GRIMES, EUSTIS B.
At Marcheville, France, Sept. 26, 1918.
R—Fitchburg, Mass.
B—Belmont, Mass.
G. O. No. 138, W. D., 1918.

Second lieutenant, 102d Infantry, 26th Division.
During a violent enemy bombardment he advanced with his detachment under intense machine-gun fire, wiped out a machine-gun nest, and captured the gun, which had been harassing the right of our line. He displayed exceptional bravery and resourcefulness throughout the entire engagement.

GRINDLE, IVOR (67715).
At Marcheville, France, Sept. 26, 1918.
R—Bucksport, Me.
B—Penobscot, Me.
G. O. No. 143, W. D., 1918.

Private, Company D, 103d Infantry, 26th Division.
He, with Sergt. John P. Diggins, climbed out of a trench in the face of severe shrapnel and machine-gun fire, proceeded 150 yards across an open space to the aid of a wounded officer and dressed his wounds.

GRINSTEAD, JAMES R. (1449747).
Near Buzemont, France, Sept. 26–28–30, 1918.
R—Pawhuska, Okla.
B—Ridgeway, Mo.
G. O. No. 59, W. D., 1919.

Sergeant, Company M, 137th Infantry, 35th Division.
Although wounded in the foot, while still in the departure trench, he went forward in the attack with his company. Later he led two platoons to the attack against superior numbers and in the face of withering fire. Wounded a second time, he refused to be evacuated until the regiment was relieved.

*GRISHAM, JESSE M. (2226305).
Near Fey-en-Haye, France, Sept. 12, 1918.
R—Holland, Ark.
B—Holland, Ark.
G. O. No. 129, W. D., 1918.

Corporal, Company L, 359th Infantry, 90th Division.
When the advance of his company was halted by an impassable barbed-wire entanglement, he voluntarily jumped out of a trench in the face of heavy machine-gun fire and cut sufficient paths through the wire to enable the company to continue its advance. In the performance of this self-sacrificing act this gallant soldier was killed.
Posthumously awarded. Medal presented to mother, Mrs. Mary Lockey.

GRISWOLD, AVA H. (2309699)............
Near Romagne, France, Oct. 9-12, 1918.
R—Paxton, Ill.
B—Loda, Ill.
G. O. No. 71, W. D., 1919.

First sergeant, Company C, 125th Infantry, 32d Division.
He displayed initiative and marked personal bravery in voluntarily organizing and leading patrols under heavy fire beyond the front lines of his company. He thus obtained valuable information, his activity providing against surprise attacks by the enemy at a critical period. When he could not obtain volunteers for patrols, he did not hesitate to go on dangerous reconnaissance missions by himself.

GRISWOLD, CHAUNCEY J. (1750084).....
Near Grand Pre, France, Oct. 16, 1918.
R—Dorchester, Mass.
B—Nashua, N. H.
G. O. 35, W. D., 1919.

Private, first class, medical detachment, 312th Infantry, 78th Division.
He volunteered to leave his battalion, which was in support, and went forward to the front line, where he established a dressing station. He remained with this advance station during the entire time it was in the line and gave first aid to the wounded under constant shell fire and through frequent gas attacks.

*GRISWOLD, LEVI W. (1456183)..........
In the Montrebeau Woods, France, Sept. 29, 1918.
R—Yarrow, Mo.
B—Adair County, Mo.
G. O. No. 89, W. D., 1919.

Corporal, Company C, 139th Infantry, 35th Division.
During a hostile counterattack Corporal Griswold, with absolute disregard for personal safety, crept out from a shell hole, under terrific machine-gun fire and gas bombardment, in an attempt to assist a wounded comrade to adjust his gas mask. He was killed in the performance of this self-sacrificing mission.
Posthumously awarded. Medal presented to mother Mrs. Alice McAnich.

GROBBEL, CLEMENT A. (2051645)........
Near Emtsa, Russia, Nov. 4, 1918.
R—Warren, Mich.
B—Warren, Mich.
G. O. No. 14, W. D., 1920.

Corporal, Company I, 339th Infantry, 85th Division (detachment in North Russia).
When attacked by a largely superior force, in order to deliver a more effective fire, Corporal Grobbel voluntarily left his trench and took up a position on top of the railroad bank. Although exposed to heavy machine-gun fire, he held his position and fired his Lewis gun until the enemy was repulsed. The conduct of this noncommissioned officer was an important factor in the successful defense of the position.

*GROBTUCK, SAMUEL D. (1709848)......
Near Ville-Savoye, France, Aug. 22, 1918.
R—New York, N. Y.
B—New York, N. Y.
G. O. No. 32, W. D., 1919.

Private, first class, Company K, 308th Infantry, 77th Division.
While carrying a message to his battalion commander, asking for reinforcements, he passed through the village of Ville-Savoye, filled with mustard gas, and was killed by shellfire while crossing an open field under direct observation of the enemy.
Posthumously awarded. Medal presented to father, Abraham Grobtuck.

GROENENBOON, ONNO (44231)..........
At Cantigny, France, June 2, 1918.
R—Volga, S. Dak.
B—Fairview, S. Dak.
G. O. No. 26, W. D., 1919.

Private, Company L, 16th Infantry, 1st Division.
He went forward under intense machine-gun and artillery fire and assisted in the removal of a wounded soldier over a distance of 1 kilometer.

GROFF, JOHN....................
In the Bois-de-Belleau, France, June 6, 1918.
R—Kenvil, N. J.
B—Syracuse, N. Y.
G. O. No. 110, W. D., 1918.

Gunnery sergeant, 83d Company, 6th Regiment, U. S. Marine Corps, 2d Division.
In the Bois-de-Belleau, France, on June 6, 1918, while out with a patrol to obtain information essential to his commander, he was attacked by a German patrol of superior numbers. With six men he attacked the enemy, inflicted heavy losses upon them, and drove them back into the German lines.

GROSS, CHRISTIAN..................
Near Uspanka, Siberia, June 11, 1918.
R—Chicago, Ill.
B—Chicago, Ill.
G. O. No. 133, W. D., 1919.

First lieutenant, 27th Infantry, U. S. Army.
While leading a combat patrol he was fired on by an unknown force of the enemy; he left his patrol, after giving them orders to fire on the enemy, and advanced through a fire-swept zone to an advantageous position on a hill from which he assisted greatly in forcing the retirement of the enemy.

GROSS, GEORGE (2335924)............
Near Sechault, France, Sept. 29, 1918.
R—Washington, D. C.
B—Washington, D. C.
G. O. No. 13, W. D., 1919.

Private, Company D, 372d Infantry, 93d Division.
Although he had been badly gassed, he kept his machine gun in action until he fell beside his gun.

GROVE, GLENN M. (567696)..........
Near Nantillois, France, Sept. 26, 1918.
R—Tyrone, Pa.
B—James Creek, Pa.
G. O. No. 145, W. D., 1918.

Sergeant, Company D, 11th Machine Gun Battalion, 4th Division.
He, with two officers, using captured German Maxim guns, pushed forward to a heavily shelled area, from which the other troops had withdrawn, and by their accurate and effective fire kept groups of the enemy from occupying advantageous positions. When given permission to withdraw he declined to do so, but maintained fire superiority all afternoon until it became too dark to see. His conspicuous gallantry furnished an inspiration to the other members of the command.

GROVES, JOHN H. (1899872)............
Near St. Juvin, France, Oct. 16, 1918.
R—Warehouse Point, Conn.
B—Windsor Locks, Conn.
G. O. No. 46, W. D., 1919.

Sergeant, Company L, 325th Infantry, 82d Division.
He led a patrol against a machine-gun nest which was causing many casualties in his company. He then worked his way to the rear of another machine-gun position, charged it, capturing three prisoners, two guns, and killing six of the enemy.

GROWDON, JAMES P................
West of Fismes, France, Aug. 5, 1918.
R—Portland, Oreg.
B—Iowa.
G. O. No. 21, W. D., 1919.

Captain, 4th Engineers, 4th Division.
After reconnoitering a sector of the River Vesle in advance of the front lines of the infantry for the purpose of selecting a site for a footbridge, he went with a small party of engineers through an enemy barrage from 77-millimeter and 1-pounder guns and assisted in directing the construction work. As soon as the operations were discovered, machine-gun fire was opened up on the party, but they continued at work, removing the German wire entanglements and successfully completing a bridge which was of great value in subsequent operations.

GRULKEY, FRED J. (2249897)_____
  In the Bois-d'Ormont, northeast of
    Samogneux, France, Oct. 12, 1918.
  R—El Reno, Okla.
  B—Audubon, La.
  G. O. No. 130, W. D., 1918.

Private, Company C, 113th Infantry, 29th Division.
He was seriously wounded in the chest, but he continued in the advance until
  he was again wounded, when he crawled forward with his company to its
  objective.

GRUMLEY, FRED (94604)_____
  East of Reims, France, July 15, 1918.
  R—Columbus, Ohio.
  B—Columbus, Ohio.
  G. O. No. 81, W. D., 1919.

Corporal, Company I, 166th Infantry, 42d Division.
After his rifle had been put out of commission, Corporal Grumley jumped to
  the parapet with hand grenades, with which he aided materially in compelling
  the retreat of the Germans.

GRUNDY, JOHN (3357714)_____
  Near Tuilerie Farm, France, Nov. 1,
    1918.
  R—Philadelphia, Pa.
  B—Philadelphia, Pa.
  G. O. No. 46, W. D., 1919.

Private, Company K, 9th Infantry, 2d Division.
Private Grundy made his way through heavy shell and machine-gun fire to
  his lieutenant, who had been seriously wounded, and dressed his wounds.
  He later rendered excellent service by keeping liaison between the platoons
  of his company while they were under severe machine-gun and shell fire.
  He continued to render valuable service in this way until wounded.

GUCWA, JOSEPH (53662)_____
  Near Soissons, France, July 19, 1918.
  R—Newark, N. J.
  B—Scranton, Pa.
  G. O. No. 132, W. D., 1918.

Private, Company F, 26th Infantry, 1st Division.
Severely wounded at the beginning of the offensive near Soissons, France,
  July 19, 1918, he remained with his company throughout the day, attacked
  with it again in the evening, and accompanied it, fighting bravely until it
  reached its objective.

GUDE, CHARLES J. (2846082)_____
  Near Beauclair, France, Nov. 3, 1918.
  R—Nebraska City, Nebr.
  B—Watson, Mo.
  G. O. No. 44, W. D., 1919.

Private, first class, Company D, 342d Machine Gun Battalion, 89th Division.
After being twice wounded and unconscious for about an hour, upon being
  revived Private Gude took command of his squad and later of his section,
  after the squad and section leaders had been wounded, refusing first aid until
  he was relieved and ordered to the rear by his platoon commander.

GUENTHER, ALOIS J. (1243631)_____
  West of Fismette, France, Aug. 10,
    1918.
  R—Pittsburgh, Pa.
  B—Pittsburgh, Pa.
  G. O. No. 128, W. D., 1918.

Sergeant, Company C, 111th Infantry, 28th Division.
Sergeant Guenther, with another soldier, voluntarily left a place of safety and
  crawled through heavy machine-gun and shell fire to the aid of a comrade
  who had fallen wounded during the withdrawal of their company from an
  exposed position, carrying him 75 yards across an open area to shelter.

GUENTHER, CHARLES R. (134724)_____
  At Chavignon Chemin-des-Dames,
    France, Feb. 28, 1918.
  R—Webster, Mass.
  B—Webster, Mass.
  G. O. No. 129, W. D., 1918.

Private, Battery B, 102d Field Artillery, 26th Division.
He was wounded while reporting to his post under a heavy bombardment of
  his battery on Feb. 28, 1918. He nevertheless served his gun during the whole
  duration of the barrage, and, although wounded, he displayed extraordinary
  bravery and gave a fine example of devotion to duty.

*GUESS, JOHN, Jr. (2267259)_____
  Near Eclisfontaine, France, Sept. 28,
    1918.
  R—El Monte, Calif.
  B—El Monte, Calif.
  G. O. No. 21, W. D., 1919.

Sergeant, Company H, 364th Infantry, 91st Division.
Assisted by another sergeant, and leading a combat group across an open valley
  under constant hostile fire, Sergeant Guess completed the capture of four
  machine-gun nests and three prisoners. He was seriously wounded in the
  encounter and died soon afterwards.
Posthumously awarded. Medal presented to father, Richard Guess.

GUINUP, EARL M. (1746649)_____
  Near Grand Pre, France, Oct. 23,
    1918.
  R—Fulton, N. Y.
  B—Fulton, N. Y.
  G. O. No. 35, W. D., 1919.

Private, Company K, 312th Infantry, 78th Division.
While Grand Pre was being severely bombarded by artillery Private Guinup
  volunteered to enter a section of the town to determine the advisability of
  his company's entrance therein. He successfully accomplished his mission
  through the severest barrage and returned with the necessary information
  for the entrance.

*GULBRANDSEN, ARVID W._____
  Near Bois-de-Chaume, France, Oct.
    8, 1918.
  R—Chicago, Ill.
  B—Norway.
  G. O. No. 44, W. D., 1919.

Second lieutenant, 132d Infantry, 33d Division.
Leading his platoon against an enemy machine-gun nest which was inflicting
  severe casualties on his battalion, he continued to advance alone in the face
  of the annihilating machine-gun fire after 14 of his men were killed about
  him. Before reaching his objective he was killed by machine-gun fire from
  his right flank.
Posthumously awarded. Medal presented to father, Arvid M. Gulbrandsen.

GULLION, OTTO_____
  Near Bayonville, France, Nov. 1,
    1918.
  R—Glencoe, Ky.
  B—Franklin County, Ky.
  G. O. No. 35, W. D., 1919.

Sergeant, 82d Company, 6th Regiment, U. S. Marine Corps, 2d Division.
Exposing himself to enemy fire, he advanced ahead of his platoon into a ravine
  and captured, single handed, a German officer and 4 men.

GUMBS, ESRIC C. (2459352)_____
  Near Authe and Sy, France, Nov.
    2-4, 1918.
  R—Brooklyn, N. Y.
  B—Holland.
  G. O. No. 44, W. D., 1919.

Private, first class, Company E, 310th Infantry, 78th Division.
Private Gumbs, while acting as battalion runner, repeatedly crossed areas
  subjected to heavy shell and machine-gun fire, carrying messages to and
  from flank companies. His work, performed with unhesitating devotion to
  duty, materially contributed to the successful advance of his battalion.

GUMM, JAY D. (210874)_____
  At Vadenay, north of Chalons-sur-
    Marne, France, July 15, 1918.
  R—Dallas, Tex.
  B—Mexia, Tex.
  G. O. No. 20, W. D., 1919.

Sergeant, 117th Train Headquarters and Military Police, 42d Division.
During the shelling of Vadenay he voluntarily left a place of safety and went
  through heavy fire of major-caliber shells and rescued a French soldier who
  was lying, severely wounded, in the street.

GUMP, NOAH L. (914855)
Near Brieulles, France, Nov. 5, 1918.
R—Fulton, Kans.
B—Fulton, Kans.
G. O. No. 37, W. D., 1919.

Private, Company B, 7th Engineers, 5th Division.
When 3 of the boats supporting a pontoon bridge across the Meuse River were destroyed by artillery fire, he voluntarily waded into the stream under heavy artillery and machine-gun fire and held up the deck of the bridge until new boats were launched and placed in position.

GUNCKLE, WILK (1531261)
Near Heurne, Belgium, Nov. 3, 1918.
R—Osgood, Ohio.
B—Osgood, Ohio.
G. O. No. 37, W. D., 1919.

Private, Company M, 148th Infantry, 37th Division.
He volunteered and guided ammunition carriers to advanced positions, despite the fact that he was seriously wounded in the face, which made it necessary to hold a bandage in place during the journey to and from the front. After receiving treatment at the first-aid station he returned to his duties.

*GUNDELACH, ANDRE H.
Near Buxieres, France, Sept. 12, 1918.
R—Chicago, Ill.
B—Chicago, Ill.
G. O. No. 37, W. D., 1919.

First lieutenant, pilot, 96th Aero Squadron, Air Service.
Lieutenant Gundelach, with Second Lieut. Pennington H. Way, observer, volunteered for a hazardous mission to bomb concentrations of enemy troops. They successfully bombed their objectives, but while returning were attacked by 8 enemy planes. Their plane was brought down in flames and both officers killed.
Posthumously awarded. Medal presented to uncle, C. D. Meyer.

GUNDERSON, ARTHUR J. (1386503)
On the Meuse River, France, Sept. 26, 1918.
R—Chicago, Ill.
B—Chicago, Ill.
G. O. No. 56, W. D., 1922.

Sergeant, Company B, 131st Infantry, 33d Division.
When the advance of his platoon was held up by fire from three .77 fieldpieces which were being protected by 1 machine gun, Sergeant Gunderson, with 2 comrades, advanced on this battery, which was firing point-blank, and, by short rushes and with utter disregard for personal safety, put the machine gun out of action, killed the crew, and captured the 3 fieldpieces, thus enabling his platoon to advance to its objective.

GUNN, FRED L. (2677)
At Cantigny, France, May 28-31, 1918.
R—Meridian, Miss.
B—Flora, Miss.
G. O. No. 99, W. D., 1918.

Private, first class, medical detachment, 28th Infantry, 1st Division.
At the battle of Cantigny, France, May 28-31, 1918, he repeatedly, on his own initiative, left the security of the trench to administer first aid under fire and in full view of the enemy snipers and machine gunners. His brave conduct was a noble example, and his ministration relieved suffering and saved lives.

GUNST, GERALD H.
Near Andevanne, France, Nov. 2, 1918.
R—Corpus Christi, Tex.
B—Portland, Oreg.
G. O. No. 39, W. D., 1920.

First lieutenant, 360th Infantry, 90th Division.
When several officers of a company became casualties, Lieutenant Gunst took command of an assault platoon. Although wounded in the leg while leading his platoon in the attack, he continued on until the objective was reached. He then voluntarily carried an important message to regimental headquarters, crossing an area covered by enemy machine-gun fire.

*GUNTERMAN, JAMES (1556214)
At Argenthol, France, July 25, 1918.
R—Rumsey, Ky.
B—Rumsey, Ky.
G. O. No. 16, W. D., 1920.

Private, Company M, 7th Infantry, 3d Division.
While charging an enemy machine-gun nest which was holding up the advance of his squad he was mortally wounded, but his heroic deed enabled his comrades to capture the nest.
Posthumously awarded. Medal presented to father, Cal Gunterman.

GUSTAFSON, CARL E. (156704)
Near Medeah Ferme, France, Oct. 8-9, 1918.
R—Eureka, Calif.
B—Scotia, Calif.
G. O. No. 44, W. D., 1919.

Private, first class, Company C, 2d Engineers, 2d Division.
Engaged as runner, he constantly carried messages through a sector which was under intense shell and machine-gun fire and infested with sniper fire.

GUSTAFSON, CHARLES (1409011)
Near Septsarges, France, Oct. 24, 1918.
R—Cooperstown, N. Y.
B—Sweden.
G. O. No. 37, W. D., 1919.

Private, Company G, 5th Ammunition Train, 5th Division.
When an enemy shell struck some pyrotechnics stored in the ammunition dump of his organization, he assisted in removing inflammable material and placing the fire under control. Through his coolness and courage the destruction of a large quantity of near-by ammunition was avoided.

GUSTAFSON, JOHN A.
Near Chateau-Thierry, France, June 23, 1918.
R—Canton, Ill.
B—Wetmore, Pa.
G. O. No. 44, W. D., 1919.

Sergeant, 47th Company, 3d Battalion, U. S. Marine Corps, 2d Division.
During an advance by his company he directed a charge on an enemy machine-gun emplacement, destroying 4 guns and taking 28 prisoners. Later in the attack he alone charged a machine-gun nest, killing 1 gunner, wounding another, and causing the remaining 3 to surrender.

GUTHRIE, ELMER F. (2176235)
Near Barricourt, France, Nov. 1, 1918.
R—Hutchinson, Kans.
B—Hutchinson, Kans.
G. O. No. 95, W. D., 1919.

Sergeant, Company E, 353d Infantry, 89th Division.
When his battalion encountered heavy machine-gun fire which stopped its advance, Sergeant Guthrie quickly organized a group of 10 or 12 men from different companies and led them, with exceptional dash and courage, in an attack on the enemy gun. Skillfully picking out a protected route he succeeded in flanking the nest and annihilating the crew.

GUTHRIE, MURRAY K.
Near Andevanne, France, Oct. 1, 1918.
R—Minneapolis, Minn.
B—Minneapolis, Minn.
G. O. No. 37, W. D., 1919

First lieutenant, 13th Aero Squadron, Air Service.
He was a member of an offensive patrol of 4 planes which was attacked far behind the enemy's lines by 6 German machines. One of our pilots was forced to withdraw by the failure of his machine guns, and 2 others were surrounded and overpowered. Lieutenant Guthrie fought the 6 enemy planes alone for 10 minutes and destroyed one of them.
Oak-leaf cluster (2).
For the following act of extraordinary heroism in action near Montfaucon, France, Oct. 4, 1918, Lieutenant Guthrie is awarded an oak-leaf cluster to be worn with the distinguished-service cross: When the leader of his patrol was blown to pieces by a shell, Lieutenant Guthrie took command and attacked the formation of 6 enemy planes. Although he became separated from his companions, he succeeded in destroying one of his opponents.
For the following act of extraordinary heroism in action near Fontaines, France, Nov. 4, 1918, Lieutenant Guthrie is awarded another oak-leaf cluster: As flight commander he led his formation of 6 planes to the attack of 7 enemy planes (type Fokker). Six of the enemy were destroyed, 1 of which was sent down by Lieutenant Guthrie. Immediately following this combat he attacked and drove off 4 hostile machines (type Fokker) which were about to attack one of our balloons.

GUY, LEONARD E. (572657)............
Near Nantillois, France, Sept. 27, 1918.
R—Great Falls, Mont.
B—Anoka, Minn.
G. O. No. 81, W. D., 1919.

Sergeant, Company C, 58th Infantry, 4th Division.
Sergeant Guy displayed exceptional courage in attacking single handed a machine-gun emplacement, capturing the gun and taking as prisoners 3 machine gunners.

GUYER, THOMAS L. (2426651)..........
Near Bellicourt, France, Sept. 29, 1918.
R—Erie, Pa.
B—Erie, Pa.
G. O. No. 81, W. D., 1919.

Private, Company I, 119th Infantry, 30th Division.
Although severely wounded by shellfire, he remained on duty as machine-gun ammunition carrier throughout the day, refusing to be evacuated. His example of courage inspired those serving near him.

HAAS, ROBERT K...................
Near Revillon, France, Sept. 10, 1918.
R—New York, N. Y.
B—New York, N. Y.
G. O. No. 35, W. D., 1919.

First lieutenant, 308th Infantry, 77th Division.
During the attack on Revillon Lieutenant Haas voluntarily left his shelter and went across an open slope in full observation of the enemy and under heavy machine-gun fire to the aid of a wounded soldier, bringing him back to our lines for first-aid treatment.

HABECKER, GUY M. (1786483).........
Near Verdun, France, Nov. 4–6, 1918.
R—Steelton, Pa.
B—Landisville, Pa.
G. O. No. 44, W. D., 1919.

Corporal, Company I, 316th Infantry, 79th Division.
While performing the duties of supply sergeant Corporal Habecker succeeded in getting food to his company, which was holding the outpost line. He led carrying parties through heavy shellfire, bringing hot coffee and bread to the men. On one occasion he made the journey unaided, taking with him a large quantity of bread.

HACKER, HENRY E. (377831)..........
Near Apremont, France, Oct. 1, 1918.
R—New York, N. Y.
B—New York, N. Y.
G. O. No. 64, W. D., 1919.

Private, Company A, 110th Infantry, 28th Division.
He overheard a lieutenant report to the battalion commander the location of an enemy machine-gun nest which was causing heavy casualties among our troops. Due to heavy shelling, delay in the bringing up of a 1-pounder crew was serious, and Private Hacker, realizing the need for immediately silencing the enemy machine guns, acting on his own initiative, advanced alone over a fire-swept area for about 500 yards, attacked and captured the enemy crews of 6 men and 2 machine guns, and brought them back to our lines.

HADLEY, LEE A.....................
Near Ronssoy, France, Sept. 29, 1918.
R—Lacona, N. Y.
B—Lacona, N. Y.
G. O. No. 14, W. D., 1923.

First lieutenant, Medical Corps, attached to 106th Infantry, 27th Division.
Under observation of the enemy and with complete disregard for his own safety he ran and crawled 200 yards to a shell hole where 18 of his men had been killed or wounded by shellfire. Dressing the wounds of 12 of the surviving men, he carried each of them 100 yards to a place of safety under intense enemy machine-gun and artillery fire; from this point they were evacuated by members of a sanitary detachment. Two of the wounded men were killed by enemy fire while being carried in the arms of this officer. The undaunted bravery of Lieutenant Hadley was an inspiration to every member of his organization.

HADNETT, GEORGE (1518457)..........
East of Baccarat, France, Aug. 15, 1918.
R—Youngstown, Ohio.
B—Youngstown, Ohio.
G. O. No. 100, W. D., 1918.

Corporal, Company F, 145th Infantry, 37th Division.
He was in command of a small advance post which was successfully held by 3 men and himself against a raid by 80 of the enemy. Two of his party were killed, but the staunch defense of the others drove off the raiders. He personally killed 3 of the enemy in hand-to-hand fighting.

HAEFLIGER, FRED...................
Near St. Etienne, France, Oct. 3, 1918.
R—Boardman, Wis.
B—Mount Vernon, Wis.
G. O. No. 23, W. D., 1919.

Private, Company C, 6th Machine Gun Battalion, U. S. Marine Corps, 2d Division.
When our advanced Infantry was forced to withdraw, Private Haefliger's machine-gun crew refused to withdraw, but calmly set up their machine gun. The gun was upset by a bursting hand grenade, which also injured Private Haefliger and another member of the squad. Despite their injuries, they immediately reset the gun and opened fire on the advancing Germans when 20 feet distant, causing the Germans to break and retreat in disorder.

HAGAN, JOSEPH A..................
At Bois-de-Belleau, France, June 6, 1918.
R—Richmond, Va.
B—Richmond, Va.
G. O. No. 9, W. D., 1923.

First lieutenant, 5th Regiment, U. S. Marine Corps, 2d Division.
For extraordinary heroism in action at Bois-de-Belleau, France, June 6, 1918, when he rescued a platoon sergeant of his regiment from death or capture. In the face of heavy machine-gun and rifle fire he rushed across an open space of 200 yards under full view of the intrenched enemy forces and within 25 yards of his lines placed the wounded sergeant upon his back and returned under a withering fire to his own lines. His splendid act of devotion to duty and his utter disregard for his own safety inspired the men of his regiment with renewed courage and determination.

HAGAN, LUTHER J. (1980151).........
Near Bellicourt, France, Sept. 29, 1918.
R—French Lick, Ind.
B—French Lick, Ind.
G. O. No. 81, W. D., 1919.

Private, Company K, 119th Infantry, 30th Division.
When his squad was held up by fire from a sniper's post he advanced alone against it across an open space and drove the others off, allowing a renewal of the advance. Later he braved heavy shellfire to go to the aid of wounded soldiers, dressing their wounds and assisting them back to the lines.

*HAGEMAN, AUGUST (2176759).........
Near Crezancy, France, July 15, 1918.
R—Lantry, S. Dak.
B—Vincennes, Ind.
G. O. No. 95, W. D., 1919.

Private, first class, Company M, 30th Infantry, 3d Division.
This soldier carried frequent messages through barrage fire, and later brought in wounded and carried ammunition to the front line. On July 26, near Jaulgonne, this man also carried messages through the most violent shellfire. Posthumously awarded. Medal presented to father, Frank Hageman.

HAGEMEYER, GEORGE K. (1212127)____
Near Bony, France, Sept. 29, 1918.
R—New York, N. Y.
B—New York, N. Y.
G. O. No. 44, W. D., 1919.

Private, Company M, 107th Infantry, 27th Division.
During a period that his company was held up by hazardous enemy fire Private Hagemeyer acted as stretcher bearer, and even though wounded and ordered to the rear, he continued in the work of removing the wounded. He remained at his task through the violent fire until he was wounded five times and forced to retire from loss of blood.

HAGERMAN, OTHO M. (2178271)_____
Near Barricourt, France, Nov. 1, 1918.
R—Kahoka, Mo.
B—Wayland, Mo.
G. O. No. 81, W. D., 1919.

Private, first class, Company A, 354th Infantry, 89th Division.
Seeing his platoon held up by murderous machine-gun fire, Private Hagerman advanced over an open area in the face of fierce machine-gun fire to a point within 150 yards of the enemy, from where he destroyed the machine-gun nest with rifle grenades.

HAHN, FREDERICK_____
Near Cantigny, France, May 28-30, 1918.
R—Savannah, Ga.
B—Savannah, Ga.
G. O. No. 99, W. D., 1918.

Second lieutenant, 7th Field Artillery, 1st Division.
He unhesitatingly went into heavy shellfire to supervise the repair of telephone lines and to act as runner when the further maintenance of the wires became an impossibility.

HAHN, JOSEPH C. (2178107)_____
Near Remonville, France, Nov. 1, 1918.
R—St. Louis, Mo.
B—St. Louis, Mo.
G. O. No. 95, W. D., 1919.

Corporal, Machine Gun Company, 354th Infantry, 89th Division.
Corporal Hahn displayed exceptional bravery and initiative in leading his machine-gun squad through heavy fire ahead of the Infantry front line to a point only 30 yards from 6 enemy machine guns which defended a hill. Despite the intense grenade and machine-gun fire which was directed at him he maintained such effective fire that the hostile guns were put out of action and the Infantry advance thereupon resumed.

HALE, CLAUD P. (2180240)_____
Near Remonville, France, Nov. 1, 1918.
R—Martinsburg, Mo.
B—Shamrock, Mo.
G. O. No. 64, W. D., 1919.

Private, first class, medical detachment, 354th Infantry, 89th Division.
Braving the deadly machine-gun and artillery fire, he worked unceasingly, and with no regard for his personal safety in administering to wounded men and assisting them to places of safety. He constantly went forward into the open under machine-gun fire to aid fallen comrades, and his work was the means of saving many lives.

HALEY, ROSCOE R._____
Near St. Etienne, France, Oct. 8, 1918.
R—San Antonio, Tex.
B—Belton, Tex.
G. O. No. 66, W. D., 1919.

Second lieutenant, 142d Infantry, 36th Division.
Lieutenant Haley was severely wounded in the head while leading his platoon forward, but he continued in the advance, directing the cutting of wire entanglements under heavy fire. Shortly afterwards he received another wound in the face, but he again refused to go to the rear, remaining with his platoon, encouraging his men, and maintaining organization until he became unconscious.

*HALFMANN, ANTHONY N. (113161)____
Between Sergy and Villers-sur-Fere, France, July 28, 1918.
R—Fond du Lac, Wis.
B—Peebles, Wis.
G. O. No. 102, W. D., 1918.

Sergeant, Company B, 150th Machine Gun Battalion, 42d Division.
He displayed courage, coolness, and leadership throughout the attack on the enemy positions north of the River Ourcq. The machine-gun crew which he was directing shot 7 enemy snipers from their posts. He was killed while reconnoitering in advance of our lines for an advantageous position for his guns.
Posthumously awarded. Medal presented to father, John Halfmann.

HALL, CHARLES P._____
At Vierzy, France, July 18, 1918.
R—Charleston, Miss.
B—Sardis, Miss.
G. O. No. 98, W. D., 1919.

Major, adjutant, 3d Infantry Brigade, 2d Division.
At a critical time in the battle, when information was difficult to obtain, Major Hall, brigade adjutant, volunteered to report on the fighting in Vierzy, which was then in the hands of the enemy. Accompanying a group of French tanks, he entered the town under intense fire, and during the advance went forward through machine-gun fire and carried to safety a wounded man. He assisted materially in maintaining organization among the troops and established a first-aid station at which many wounded were cared for, returning later to brigade headquarters with valuable information.

HALL, FRED L. (53459)_____
Near Soissons, France, July 19, 1918.
R—Doyon, N. Dak.
B—Harrisburg, N. Dak.
G. O. No. 132, W. D., 1918.

Private, Company E, 26th Infantry, 1st Division.
He refused to go to the rear when wounded near Soissons, France, July 19, 1918, continued to fire his automatic rifle with effectiveness, and when finally exhausted directed another man in the use of the rifle and attempted to act as ammunition carrier.

*HALL, GEORGE W. (2444063)_____
Near Pincon Farm, France, Sept. 5, 1918.
R—Chicago, Ill.
B—St. Joseph, Mich.
G. O. No. 44, W. D., 1919.

Private, Machine Gun Company, 305th Infantry, 77th Division.
After having been ordered to a place of safety he left his shelter and returned to a trench which was being severely shelled, rescued and brought back 2 wounded comrades. He was killed in action 3 days later.
Posthumously awarded. Medal presented to widow, Mrs. Dorothy G. Hall.

*HALL, HAROLD DE LA MONTE (561549).
Near Bois-de-Brieulles, France, Sept. 29, 1918.
R—Buffalo, W. Va.
B—Charleston, W. Va.
G. O. No. 37, W. D., 1919.

Private, Company A, 59th Infantry, 4th Division.
When his company was in a perilous position he volunteered and carried a message to battalion headquarters, a distance of 1,000 yards, under heavy artillery and machine-gun fire. On his return journey he was killed.
Posthumously awarded. Medal presented to mother, Mrs. May E. Amberg.

*HALL, HENRY H. (1316674)_____
Near Voormezeele, Belgium, Aug. 31, 1918.
R—Lumber Bridge, N. C.
B—Lumber Bridge, N. C.
G. O. No. 87, W. D., 1919.

Private, Company L, 119th Infantry, 30th Division.
When the carrier of a Lewis gun crew was killed, he took his place, and the ammunition becoming exhausted, volunteered to go for a new supply under heavy fire. Wounded while on this mission, he opened fire on the enemy with his rifle, engaging a hostile patrol until he was mortally wounded by a second bullet.
Posthumously awarded. Medal presented to father, Horace W. Hall.

HALL, HERBERT W.............
    Near Thiaucourt, France, Sept. 27, 1918.
    R—Boston, Mass.
    B—Cliftondale, Mass.
    G. O. No. 15, W. D., 1919.

First lieutenant, 44th Artillery, Coast Artillery Corps.
He was in charge of a trainload of ammunition being sent to two 8-inch howitzer batteries in active operation against the enemy. Finding that part of the light railway track had been destroyed by enemy shellfire, he secured a detail of men under an Engineer officer and worked with them to repair the track. When the Engineer officer was killed by an exploding shell, he assumed full charge and continued the work under heavy shellfire, showing utter disregard for personal danger and inspiring confidence in his men by his calmness, decision, and courage.

HALL, JAMES G...............
    Near Montauville, France, Sept. 12–15, 1918.
    R—Atlanta, Ga.
    B—Atlanta, Ga.
    G. O. No. 27, W. D., 1920.

First lieutenant, Medical Corps, attached to 360th Infantry, 90th Division.
In spite of severe wounds, including two broken ribs, received on the first day of the action, he continued at his post for 3 days, administering aid to the wounded throughout the combat. Numbers of lives were saved by his heroism and devotion to duty.

HALL, JAMES NORMAN..........
    At Reims, France, Mar. 27, 1918.
    R—Colfax, Iowa.
    B—Colfax, Iowa.
    G. O. No. 129, W. D., 1918.

Captain, 103d Aero Squadron, Air Service.
On Mar. 27, 1918, while leading a patrol of 3, he attacked a group of 5 enemy fighters and 3 enemy 2-seaters, himself destroying 1 and forcing down 2 others in a fight lasting more than 20 minutes.

*HALL, PERCY M...............
    Near Ronssoy, France, Sept. 29, 1918.
    R—Montclair, N. J.
    B—Montclair, N. J.
    G. O. No. 32, W. D., 1919.

First lieutenant, 107th Infantry, 27th Division.
Disregarding his extremely weak condition, he insisted on going into attack with his company. Recent illness made it nearly impossible for him to stand, still he went to all parts of the line during an enemy counter barrage and murderous machine-gun fire, maintaining his platoon formations. By giving his overcoat to a wounded man, he so exposed himself that he died shortly afterwards from the effects.
Posthumously awarded. Medal presented to sister, Mrs. Joseph A. Howe.

HALL, RAMON L...............
    East of Ronssoy, France, Sept. 29, 1918.
    R—Schenectady, N. Y.
    B—Wilkes-Barre, Pa.
    G. O. No. 20, W. D., 1919.

Second lieutenant, 105th Infantry, 27th Division.
During the operations against the Hindenburg line he left shelter, went forward under heavy shell and machine-gun fire, and succeeded in bringing back to our lines a wounded soldier. His splendid courage and gallant conduct furnished a fine example to his command.

HALL, VARNER (96363).........
    Near Ancerviller, France, Mar. 4, 1918.
    R—Birmingham, Ala.
    B—Evansville, Ind.
    G. O. No. 126, W. D., 1919.

Sergeant, Company D, 167th Infantry, 42d Division.
He conducted himself with marked bravery as a member of a patrol of 5 men which, encountering an enemy patrol of 11 men, attacked vigorously, routed the enemy patrol, and took 2 prisoners.

*HALL, WILLIAM A. (2192179)....
    Near St. Mihiel, France, Sept. 12, 1918.
    R—Winfield, Kans.
    B—Rock, Kans.
    G. O. No. 20, W. D., 1919.

Private, Company A, 353d Infantry, 89th Division.
While acting as runner between his company and battalion headquarters he made several trips through severe artillery bombardment before he was severely wounded.
Posthumously awarded. Medal presented to widow, Mrs. Essie Hall.

HALLEY, ORVEL H. (58710).....
    Near Exermont, France, Oct. 1 to 3, 1918.
    R—Macon, Mo.
    B—Macon, Mo.
    G. O. No. 9, W. D., 1923.

Corporal, Company M, 28th Infantry, 1st Division.
While in charge of two squads of a reconnoitering patrol on an exceptionally hazardous mission in the vicinity of Serieux Farm, he displayed utter fearlessness, leading his men in groups through zones swept by intense machine-gun fire, returning three times to guide those left behind. Reaching his objective, he voluntarily entered and moved about within the German lines, returning with important information for the officer in charge of the patrol. When the patrol was surrounded, Corporal Halley, although dangerously situated with his men, held on, refusing to surrender and repulsed two attacks. Later, when attacked by much larger force, he succeeded under cover of semidarkness in withdrawing under heavy fire and joined his company. His bravery, coolness, and devotion to duty were a constant inspiration to his men.

HALPHEN, DEWEY (58191)......
    At Cantigny, France, May 28–30, 1918.
    R—St. Martinsville, La.
    B—St. Martinsville, La.
    G. O. No. 99, W. D., 1918.

Private, Company I, 28th Infantry, 1st Division.
He acted as liaison agent during the fight of May 28–30, 1918, at Cantigny, France, with courageous disregard of his own safety.

HAM, SAMUEL V..............
    Near Magneux, France, Sept. 6, 1918.
    R—Warrington, Ind.
    B—Markleville, Ind.
    G. O. No. 128, W. D., 1918.
    Distinguished-service medal also awarded.

Colonel, 109th Infantry, 28th Division.
By courageously leading his firing line in the advance across the Vesle River from Magneux toward Muscourt, Colonel Ham exemplified the greatest heroism and truest leadership, instilling in his men confidence in their undertaking. Having been severely wounded and unable to move, he remained for 10 hours on the field of battle directing the attack and refused to be evacuated or receive medical attention until his men had been cared for.

HAMAK, FRANK A. (2380866)....
    Near Chaumont, south of Sedan, France, Nov. 8, 1918.
    R—Wecota, S. Dak.
    B—Loyalton, S. Dak.
    G. O. No. 126, W. D., 1919.

Private, Company E, 165th Infantry, 42d Division.
Severely wounded in the legs by machine-gun fire and unable to walk, Private Hamak refused to accept assistance from his comrades rather than expose them to danger, and crawled 500 yards to a dressing station.

HAMBLETON, JOHN A.
  At Lironville, France, May 28, 1918.
  R—Lutherville, Md.
  B—Maryland.
  G. O. No. 15, W. D., 1923.

First lieutenant, 95th Aero Squadron, Air Service.
Lieutenant Hambleton, accompanied by Lieutenant Taylor, while answering an alert to Lironville, France, encountered five enemy airplanes in the vicinity of St. Mihiel. As the American airplanes approached the enemy turned away. Lieutenant Hambleton and Lieutenant Taylor followed, and at Pont-a-Mousson again overtook the enemy, one airplane flying at 1,500 meters, two at 2,000 meters, and the remaining two airplanes at 2,500 meters. Lieutenant Hambleton attacked the lowest airplane, firing 20 rounds and driving it from the formation. An enemy bullet shot the cross section wires from his airplane, the splinters from the bullet cutting his cheek and right shoulder. Notwithstanding his wounds and the disabled condition of his airplane, he continued to fight until the enemy was driven to its own lines.
Oak-leaf cluster.
Lieutenant Hambleton was awarded an oak-leaf cluster for the following act of heroism: For extraordinary heroism near Gironville, France, June 6, 1918. After becoming separated from his patrol, on account of misty weather, Lieutenant Hambleton was flying in the region of Gironville when he encountered two enemy biplane machines. He engaged them in a sharp combat despite the fact that he was being subjected to the concentrated fire of both biplanes. He pulled away and was maneuvering for a better position when the biplanes started into Germany and he promptly chased them across the lines, thereby preventing them from performing a photographic mission. On the same date in the region of Grissolles, while leading a patrol, Lieutenant Hambleton encountered a formation of five enemy pursuit airplanes. He gallantly led the attack, and after a short and decisive fight with one enemy airplane, during which time he was subjected to the concentrated fire of the other enemy airplanes, he drove it to the ground, where it crashed. Still undaunted, he returned and, without hesitation or fear, repeatedly attacked the remainder of the enemy formation until he had forced it to retire in disorder. The rare courage and superb devotion to duty displayed by Lieutenant Hambleton greatly inspired the members of his squadron.

HAMBRICK, GORDON A. (40296)
  Near Medeah Ferme, France, Oct. 3, 1918.
  R—Georgetown, Ky.
  B—Georgetown, Ky.
  G. O. No. 21, W. D., 1919.

Sergeant, Company K, 9th Infantry, 2d Division.
Believing his platoon commander to be killed and finding himself in front of our attacking wave, he gathered a number of men, detached from their organizations, who had pushed to the front, and led them against enemy positions across ground swept by machine-gun fire. With 25 men, he attacked a strongly defended enemy position occupying the right flank of the objective, capturing 80 prisoners and 5 enemy machine guns, and held the position until the arrival of our attacking wave.

*HAMEL, ALFRED R.
  Near Soissons, France, July 18, 1918.
  R—Atlantic City, N. J.
  B—Ardsley, Pa.
  G. O. No. 132, W. D., 1918.

Captain, 26th Infantry, 1st Division.
Although severely wounded on July 18, 1918, near Soissons, France, he refused to give up, and continued courageously to lead his company until killed. Posthumously awarded. Medal presented to widow, Mrs. Edith I. Hamel.

HAMES, WILLIAM W.
  Near Bussy Farm and Sechault, France, Sept. 26-29, 1918.
  R—Jonesville, S. C.
  B—Jonesville, S. C.
  G. O. No. 13, W. D., 1919.

First lieutenant, 372d Infantry, 93d Division.
He was in charge of the 37-millimeter guns, which he skillfully employed until they were put out of action. He then joined one of the assaulting waves advancing on the enemy's position, and with the aid of two men captured a machine gun and three prisoners. Although badly gassed, he continued in action until the next day, when he collapsed.

HAMILTON, ARTHUR M. (560474)
  Near Brieulles, France, Oct. 6, 1918.
  R—Des Moines, Iowa.
  B—Clearfield, Iowa.
  G. O. No. 53, W. D., 1920.

Corporal, Company E, 58th Infantry, 4th Division.
Corporal Hamilton and a comrade, under heavy enemy fire, went to the rescue of wounded lying in advance of our lines and returned to our lines with two wounded American soldiers. In accomplishing this mission they advanced to within 75 yards of the enemy lines, over an area which the enemy raked with their fire.

HAMILTON, BRYAN (1518161)
  At Olsene, Belgium, Oct. 31, 1918.
  R—Portsmouth, Ohio.
  B—Ruggles, Ky.
  G. O. No. 44, W. D., 1919.

Sergeant, Company E, 145th Infantry, 37th Division.
He was leading a detachment forward which was caught in a heavy enemy counterbarrage. Though he was badly wounded by shellfire, he kept his men organized, and, pushing forward, dislodged the enemy from a strong position, where he was again wounded.

HAMILTON, BYRON W. (102458)
  On Hill 212, near Sergy, France, July 28, 1918.
  R—Pleasant Plain, Iowa.
  B—Washington County, Iowa.
  G. O. No. 99, W. D., 1918.

Sergeant, Company M, 168th Infantry, 42d Division.
When leading a rushing attack on machine guns on Hill No. 212, near Sergy, France, July 28, 1918, he found himself ahead of his line, so wounded as to be unable to stand on his feet. Attacked by 10 Germans of the Prussian Guards, he rose to his knees and shot 5 of them. The others fled.

HAMILTON, GEORGE W.

Near Bois-de-Belleau, Chateau-Thierry, France, June 6, 1918.
R—Washington, D. C.
B—Washington, D. C.
G. O. No. 15, W. D., 1919, and G. O. No. 15, W. D., 1923.

Captain, 5th Regiment, U. S. Marine Corps, 2d Division.

He displayed the highest type of courage and leadership when on the first day of the Chateau-Thierry battle his command was under decimating fire of machine guns from the front and both flanks. All of his officers but one and most of his noncommissioned officers having been killed or wounded, he passed up and down his front lines and, by his personal bravery, inspired his men to valiant and successful combat under especially difficult conditions.

Oak-leaf cluster.

While in command of Company B (49th Company), 5th Regiment, U. S. Marine Corps, during the attack upon Hill 142 at 3.45 a. m., June 6, 1918, he led his company to its objective. While his company was engaged in digging in, 12 enemy machine gunners stealthily approached and proceeded to set up their guns for the purpose of firing upon Captain Hamilton's company. He discovered the enemy before the guns were in position, and, in company with a noncommissioned officer, gallantly rushed upon the enemy, attacking them with a bayonet, dispatching several and driving the remaining men to flight, capturing their machine guns. The undaunted bravery and soldierly conduct displayed by Captain Hamilton served to inspire the men in his regiment with increased determination and incited them to heroic endeavors.

*HAMILTON, JAMES A. (1225269)

Near Ronssoy, France, Sept. 27, 1918.
R—Hoosick Falls, N. Y.
B—Albany, N. Y.
G. O. No. 20, W. D., 1919.

First sergeant, Company M, 105th Infantry, 27th Division.

He rallied his company after it had become disorganized under a machine-gun barrage and all the officers were killed or wounded. He led his men forward in an effective attack, and was shortly afterwards killed while moving along his line.

Posthumously awarded. Medal presented to mother, Mrs. Mary Hamilton.

HAMILTON, JOHN W.

Near Bellicourt, France, Sept. 29, 1918.
R—Pleasanton, Kans.
B—Freeman, Mo.
G. O. No. 64, W. D., 1919.

First lieutenant, 120th Infantry, 30th Division.

Assuming command of his company when two senior officers became casualties, he led his men bravely under heavy fire, continuing in command after being wounded in the arm by a shell splinter. Later when knocked down by a large piece of shell, which struck him in the leg, he again refused to leave his men and kept up the advance. He personally led a patrol of three squads into the town of Bellicourt, held by the enemy, returning with prisoners and valuable information which aided in the further advance.

*HAMILTON, LLOYD A.

At Varssenaere, Belgium, Aug. 13, 1918.
R—Pittsfield, Mass.
B—Troy, N. Y.
G. O. No. 20, W. D., 1919.

First lieutenant, 17th Aero Squadron, Air Service.

Leading a low bombing attack on a German aerodrome 30 miles behind the lines, he destroyed the hangars on the north side of the aerodrome and then attacked a row of enemy machines, flying as low as 20 feet from the ground, despite intense machine-gun fire, and setting fire to three of the German planes. He then turned and fired bursts through the windows of the château in which the German pilots were quartered, 26 of whom were afterwards reported killed.

Posthumously awarded. Medal presented to father, Rev. John A. Hamilton.

*HAMILTON, OTHO

Near St. Etienne-a-Arnes, France, Oct. 3, 1918.
R—Pomeroy, Ohio.
B—Pomeroy, Ohio.
G. O. No. 20, W. D., 1919.

First lieutenant, 23d Infantry, 2d Division.

Lieutenant Hamilton advanced his company through an area swept by machine-gun and shell fire, and by his coolness and excellent leadership brought his company into an extremely advantageous position. He then caused a platoon to be sent flanking a machine gun that had retarded his advance, and with the remainder of the company maintained a continuous fire against the enemy's position, encouraging his men by his gallant example. He was later instantly killed by shellfire while returning from a reconnaissance.

Posthumously awarded. Medal presented to widow, Mrs. Otho Hamilton.

HAMILTON, REUBEN G.

Near Marcheville, France, Sept. 25 and 26, 1918.
R—Carlisle, S. C.
B—Herbert, S. C.
G. O. No. 138, W. D., 1918.

Major, Medical Corps, headquarters ambulance section, 101st Sanitary Train, 26th Division.

He established and maintained an ambulance dressing station in an advanced and hazardous position, where he labored unceasingly treating and evacuating the wounded throughout the day, in full view of the enemy and under heavy bombardment. Knowing that our troops were withdrawing and the enemy was about to enter the town, he continued his aid to the wounded, even after permission to withdraw had been given him by his commanding officer.

*HAMM, ARTHUR E.

Near Flirey, France, Aug. 4, 1918.
R—West Hampton Beach, Long Island, N. Y.
B—Groveland, Mass.
G. O. No. 71, W. D., 1919.

Captain, 326th Infantry, 82d Division.

Though wounded at the start of a daylight raid, he continued to lead his men, reaching a depth of 1,000 yards into the enemy lines and obtaining valuable information. He displayed marked personal heroism under heavy fire, setting an example of fortitude which contributed largely to the success of the operation. He was killed in action during the St. Mihiel drive.

Posthumously awarded. Medal presented to widow, Mrs. Elizabeth C. Hamm.

HAMMOND, ALEX (2169063)

Near Lesseau, France, Sept. 4, 1918.
R—Harvest, Ala.
B—Huntsville, Ala.
G. O. No. 139, W. D., 1918.

Private, Company E, 366th Infantry, 92d Division.

Although he was severely wounded, he remained at his post and continued to fight a superior force which had attempted to enter our lines, thereby preventing the success of any enemy raid in force.

*HAMMOND, CHARLES A.

Near Sergy, France, July 31, 1918.
R—Port Huron, Mich.
B—Port Huron, Mich.
G. O. No. 117, W. D., 1918.

First lieutenant, 125th Infantry, 32d Division.

Advancing up and beyond Hill No. 212, he was shot through the arm, yet he refused to go to the rear, even for first-aid treatment, but continued to assault with his platoon until he received two more wounds, from the last of which he died.

Posthumously awarded. Medal presented to brother, John J. Hammond.

HAMMOND, CHARLES GALLOWAY (115864)
Near Vadenay, France, July 15, 1918.
R—South Roanoke, Va.
B—Botetourt County, Va.
G. O. No. 46, W. D., 1919.

Private, 117th Train Headquarters and Military Police, 42d Division.
During the heavy shelling of Vadenay on the morning of July 15, he voluntarily left a place of safety, went to the aid of a wounded French soldier, and brought him to shelter through heavy fire of large-caliber shells.

HAMMOND, JOHN (114213)
Near Landres-et-St. Georges, France, Oct. 15, 1918.
R—Brooklyn, N. Y.
B—Brooklyn, N. Y.
G. O. No. 37, W. D., 1919.

Corporal, Company C, 165th Infantry, 42d Division.
Having observed four liaison men killed in an attempt to reach headquarters with an important position sketch, Corporal Hammond volunteered to attempt the mission. He not only successfully reached his destination, but also returned to his post, the entire exploit being under most severe fire.

HAMMOND, LEONARD C
In the region of Metz, France, Sept. 15, 1918.
R—San Francisco, Calif.
B—Missoula, Mont.
G. O. No. 138, W. D., 1918.

First lieutenant, observer, 91st Aero Squadron, Air Service.
While on a photographic mission his formation was attacked by a superior number of enemy pursuit planes. Notwithstanding that the enemy planes succeeded in driving off the protecting planes, he and his pilot, Lieutenant Diekma, continued on alone. Continually harassed by enemy aircraft, they completed their photographs, and on the return fought their way through an enemy patrol and destroyed one of the machines.

HAMMOND, LEROY H
Near Ardeuil-et-Montfauxelles, France, Sept. 28 and 29, 1918.
R—Mount Pleasant, Tenn.
B—Rogersville, Ala.
G. O. No. 53, W. D., 1920.

Captain, 371st Infantry, 93d Division.
On the afternoon of Sept. 28, Captain Hammond was wounded in the arm and in the leg by a shell fragment. He refused to go to the rear, knowing if he did so it would depress the morale of his men. On the following day he led one of his machine-gun sections forward in the attack through heavy fire until severely wounded a second time. The enemy position was taken, with guns, prisoners, and much material.

HAMMOND, WILLIAM H
Near Montfaucon, France, Sept. 26-27, 1918.
R—Fresno, Calif.
B—Visalia, Calif.
G. O. No. 95, W. D., 1919.

First lieutenant, 39th Infantry, 4th Division.
He fearlessly led his platoon against a German counterattack and succeeded in breaking it up. Sighting a German patrol taking American prisoners to the rear, he led a combat patrol which routed the Germans and rescued the captured Americans. In the advance in which he took part the next day, he was severely wounded in the chest, but refusing first-aid treatment continued to urge his men forward, although unable himself to go.

HAMMONS, CHARLES C. (2240260)
Near Septsarges, France, Oct. 24, 1918.
R—Southland, Tex.
B—Springtown, Tex.
G. O. No. 37, W. D., 1919.

Corporal, Company G, 5th Ammunition Train, 5th Division.
When an enemy shell struck some pyrotechnics stored in the ammunition dump of his organization, he directed and assisted in the removal of inflammable material and placing the fire under control. Through his coolness and courage the destruction of a large quantity of near-by ammunition was avoided.

HAMPLE, HARRY E. (2159880)
Near Chipilly Ridge, France, Aug. 10-19, 1918.
R—Watkins, Iowa.
B—Watkins, Iowa.
G. O. No. 71, W. D., 1919.

Private, first class, Company C, 131st Infantry, 33d Division.
He volunteered repeatedly to carry messages over ground swept by heavy machine-gun and artillery fire. He displayed marked personal courage, accomplishing every mission he was given.

HAMPSON, EDMUND R. (108815)
Near Trugny, France, July 22, 1918.
R—Waterbury, Conn.
B—Waterbury, Conn.
G. O. No. 125, W. D., 1918.

Wagoner, Company B, 101st Machine Gun Battalion, 26th Division.
Although painfully wounded by machine-gun bullets, he courageously continued his duty of evacuating the wounded until exhausted from loss of blood.

HAMRIC, ERVIN (1553834)
Near Cunel, France, Oct. 6, 1918.
R—Sutton, W. Va.
B—Braxton County, W. Va.
G. O. No. 16, W. D., 1920.

Private, Company D, 8th Machine Gun Battalion, 3d Division.
Private Hamric carried numerous messages over routes exposed to artillery and machine-gun fire from company headquarters to the front-line platoons. The individual efforts of this soldier in delivering messages when exposed to enemy fire were of great value to his company commander.

HANBERY, JAMES W
At Chateau-Thierry, France, July 19, 1918.
R—Pittsburg, Kans.
B—Hopkinsville, Ky.
G. O. No. 31, W. D., 1922.

First lieutenant, 59th Infantry, 4th Division.
For extraordinary heroism in action at Chateau-Thierry, France, July 19, 1918, in command of the attacking unit of the assault company of his battalion. After gaining his objective, in an advance through heavy machine-gun and artillery fire, the battalion on his left having been held up by enemy machine-gun nests, his company and battalion became exposed to a grazing and flanking fire which threatened the destruction of the entire battalion. Lieutenant Hanbery reorganized the attacking line and although wounded, led a brilliant and successful attack against the enemy machine-gun nests until again wounded and rendered helpless, when he refused succor in order not to endanger the lives of his men.

HANCOCK, GLENN F. (2395312)
Near Cunel, France, Oct. 11 and 19, 1918.
R—Wakeeney, Kans.
B—New Albany, Kans.
G. O. No. 98, W. D., 1919.

Private, first class, Machine Gun Company, 7th Infantry, 3d Division.
While Private Hancock was on duty as company runner he took charge of a squad whose corporal had been wounded, and moving the gun to a favorable position quickly established fire superiority over the enemy, and thereby enabled the Infantry to overcome a group of the enemy who had been delaying the advance. While he was firing the gun a bursting shell buried him and two other soldiers, but he immediately extricated himself and resumed firing. On the night of Oct. 19 he manned another gun whose squad had been put out of action, and took charge of the crew until the unit was relieved.

HAND, LEO (2192757)
Near St. Souplet, France, Oct. 10, 1918.
R—Clear Lake, S. Dak.
B—Ayrshire, Iowa.
G. O. No. 81, W. D., 1919.

Private, Machine Gun Company, 119th Infantry, 30th Division.
When his gun team had suffered heavy casualties he continued in the advance, although severely wounded by shellfire. His bravery was an inspiration to the men with him, and by continuing the operation of the machine gun he gave support which was invaluable to the advancing Infantry.

HANDWERK, RUSSELL E. (63663)_____
 Near Soissons, France, July 19, 1918.
 R—Slatington, Pa.
 B—Slatington, Pa.
 G. O. No. 132, W. D., 1918.

Private, Company F, 26th Infantry, 1st Division.
When the line of which he was a part was halted on July 19, 1918, near Soissons,
France, by a machine gun, he advanced on it alone and put it out of action.

HANDY, EDWARD H. (1709754)_____
 At Binarville, France, Sept. 30, 1918.
 R—Washington, D. C.
 B—Alexandria County, Va.
 G. O. No. 20, W. D., 1919.

Private, first class, Company B, 368th Infantry, 92d Division.
Private Handy, with an officer and another soldier, voluntarily left shelter and
crossed an open space 50 yards wide swept by shell and machine-gun fire to
rescue a wounded soldier, whom they carried to a place of safety.

HANDY, THOMAS T._____
 In the salient du Feys, France, Mar.
 9, 1918.
 R—Emory, Va.
 R—Spring City, Tenn.
 G. O. No. 126, W. D., 1919.

Captain, 7th Field Artillery, 1st Division.
When Company D, 168th Infantry, was under severe attack in the salient du
Feys, France, he voluntarily joined it upon finding that he could do so without interfering with his normal duties, and by his coolness and conspicuous
courage aided materially in its success.

HANEY, JOHN S._____
 Near Beaumont, France, Nov. 10,
 1918.
 R—Cleveland, Okla.
 B—Cleveland, Okla.
 G. O. No. 37, W. D., 1919.

Private, 55th Company, 5th Regiment, U. S. Marine Corps, 2d Division.
He displayed exceptional courage in voluntarily advancing on and destroying
machine guns which were firing directly on his company.

HANEY, MATHIAS WILLOUGHBY_____
 Near Montfaucon Hill, France,
 Sept. 26–28, 1918.
 R—Philadelphia, Pa.
 B—Bristol, W. Va.
 G. O. No. 46, W. D., 1919.

First lieutenant, 39th Infantry, 4th Division.
Lieutenant Haney displayed exceptional skill in extricating his company from
a perilous position into which it had moved because of a dense fog, and in
so doing captured prisoners whose number exceeded that of his own command.
Taking command of his battalion the next day at a critical time he succeeded
in stopping a threatened retreat, and under heavy machine-gun and shell fire
reestablished the line. On Sept. 28, near Septsarges, France, this officer led
his battalion forward through heavy fire, advancing his line 1 kilometer and
holding it against counterattacks until he was relieved.

HANGER, CHARLES R._____
 Near Pouilly, France, Nov. 10–11,
 1918.
 R—Laddonia, Mo.
 B—Andrian County, Mo.
 G. O. No. 37, W. D., 1919.

Second lieutenant, 356th Infantry, 89th Division.
When three of the companies of his battalion had become lost in the dense fog,
during the crossing of the Meuse, he made five trips through extremely heavy
shellfire and guided them to the river crossing, thus enabling the battalion
to fulfill its mission in the operation.

HANKINS, STAYTON M._____
 Near St. Etienne, France, Oct. 8,
 1918.
 R—Quanah, Tex.
 B—Quanah, Tex.
 G. O. No. 59, W. D., 1919.

First lieutenant, 142d Infantry, 36th Division.
Although severely wounded in the leg, he continued in command of his company, remaining in action until he fell from complete exhaustion and was
evacuated.

HANLEY, GEORGE L. (540503)_____
 Near Fossoy, France, July 14 and 15,
 1918.
 R—Buffalo, N. Y.
 B—Buffalo, N. Y.
 G. O. No. 16, W. D., 1920.

Private, first class, Machine Gun Company, 7th Infantry, 3d Division.
During the intensive artillery bombardment preceding the last German offensive, Private (first class) Hanley carried seven messages between company and
platoon headquarters. He exposed himself to heavy artillery fire in order
to maintain communication between the platoons of his company.

HANLEY, JAMES MATTHEW_____
 Near Villers-sur-Fere, France, July
 28, 1918.
 R—Cleveland, Ohio.
 B—Cleveland, Ohio.
 G. O. No. 37, W. D., 1919.

First lieutenant, chaplain, 165th Infantry, 42d Division.
Leaving his place of safety, Chaplain Hanley voluntarily faced the fire of artillery and machine guns, so that he could administer to the wounded. He
disregarded the advice of his commanding officer to cease exposing himself,
but he remained, giving first aid, comforting, cheering, and hearing con-
fessions of the dying. After he had been severely wounded he was taken to
the rear.

HANLEY, JOHN J. (2670941)_____
 Near St. Juvin, France, Oct. 11, 1918.
 R—New York, N. Y.
 B—New York, N. Y.
 G. O. No. 46, W. D., 1919.

Private, Company E, 326th Infantry, 82d Division.
While on duty as a stretcher-bearer he was painfully wounded in the head, but
he nevertheless continued to evacuate the wounded, crossing and recrossing
the Aire River several times under heavy fire and refusing to secure treatment
himself until all of the other wounded had been cared for.

HANLEY, WILLIAM T._____
 Near Le Charmel, France, July
 24–25, 1918.
 R—Rumford, Me.
 B—Berlin, N. H.
 G. O. No. 32, W. D., 1919.

Second lieutenant, 30th Infantry, 3d Division.
After he had led his platoon through intense shell and machine-gun fire, Lieutenant Hanley, although twice wounded, remained at his post throughout
the encounter.

HANNA, EDWARD G. (74522)_____
 At Blanc Mont Ridge, France, Oct.
 3–4, 1918.
 R—Confidence, Calif.
 B—San Francisco, Calif.
 G. O. No. 21, W. D., 1919.

Private, Headquarters Company, 4th Machine Gun Battalion, 2d Division.
Private Hanna went with two runners in advance of the front line to reconnoiter
a position which his battalion was to occupy. In the face of heavy shell and
machine-gun fire they made their way to a point just beyond the summit of
Blanc Mont Ridge. He was seriously gassed and the other two soldiers were
wounded. He accomplished his mission and remained under heavy fire all
the night, reporting to his commanding officer next day. Though suffering
from the effect of the gas, he continued on duty for seven days.

HANNA, LEON M. (1387000)
Near Bois-de-Chaume, France, Oct.
10, 1918.
R—Chicago, Ill.
B—Cameron, Tex.
G. O. No. 64, W. D., 1919.

Private, Company D, 131st Infantry, 33d Division.
When his platoon had suffered heavy casualties and was held up by fire from a hidden machine-gun emplacement, he advanced alone under heavy fire, and locating the enemy position charged it. He bayoneted the enemy gunner and captured two of the gun crew, enabling his platoon to resume the advance.

*HANNA, MARK.
Near Stenay, France, Nov. 6-11,
1918.
R—Kansas City, Mo.
B—Chillicothe, Mo.
G. O. No. 20, W. D., 1919.

Major, 356th Infantry, 89th Division.
Major Hanna displayed extreme courage on Nov. 6 by making a daring reconnaissance of the town of Pouilly, near Stenay. This town was held in strength by the enemy, with evident indication of determination to prevent a crossing of the River Meuse at this point. He remained in this town over two hours, returning with information of great value. On the night of Nov. 10-11 he was in command of the second battalion of the 356th Infantry, and while waiting to cross the River Meuse Major Hanna's battalion was subject to terrific shell fire. During this period he walked up and down the line encouraging and steadying his men. Major Hanna was killed at the head of his command.
Posthumously awarded. Medal presented to widow, Mrs. Corinne Esther Hanna.

*HANNA, SAMUEL H. (574104)
At Bois-du-Fays, France, Sept. 30,
1918.
R—Van Nuys, Calif.
B—Waukegan, Ill.
G. O. No. 5, W. D., 1920.

Sergeant, Company B, 12th Machine Gun Battalion, 4th Division.
When Company C, 58th Infantry, was temporarily halted by heavy machine-gun fire, he exposed himself to enfilading fire in order to place his guns in position to execute a covering fire for the Infantry. With the aid of the fire from the machine guns under his command the advance was resumed. In the performance of this deed he was mortally wounded.
Posthumously awarded. Medal presented to sister, Miss Jennie Hanna.

HANSEN, ARTHUR A.
Near Bois-de-Wavrille, France,
Oct. 2, 1918.
R—Waltham, Mass.
B—Waltham, Mass.
G. O. No. 46, W. D., 1919.

Captain, 101st Infantry, 26th Division.
Despite the fact that the support elements failed to arrive at the point of departure, Captain Hansen, unable to communicate with his superiors, led his troops forward. Encountering a particularly intense enemy barrage, he demonstrated unusual ability by safely conducting his command over the shell-swept area, successfully accomplishing his mission.

*HANSEN, HERMAN L. (80982)
Near St. Etienne-a-Arnes, France,
Oct. 3, 1918.
R—Nicolaus, Calif.
B—Nicolaus, Calif.
G. O. No. 20, W. D., 1919.

Corporal, Company A, 23d Infantry, 2d Division.
He voluntarily exposed himself to heavy machine-gun fire, and with the assistance of another soldier brought in from an exposed position a soldier severely wounded. Corporal Hansen was killed after he had himself obtained first aid for the wounded man.
Posthumously awarded. Medal presented to father, Harry L. Hansen.

*HANSEN, WILLIAM.
At Chateau-Thierry, France, June
6, 1918.
R—Portland, Oreg.
B—Portland, Oreg.
G. O. No. 110, W. D., 1918.

Corporal, 20th Company, 5th Regiment, U. S. Marine Corps, 2d Division.
Killed in action at Chateau-Thierry, France, June 6, 1918, he gave the supreme proof of that extraordinary heroism which will serve as an example to hitherto untried troops.
Posthumously awarded. Medal presented to father, Benhart Hansen.

HANSON, WALTER R. (303179)
Near Gesnes, France, Oct. 7-8, 1918.
R—La Crosse, Wis.
B—La Crosse, Wis.
G. O. No. 81, W. D., 1919.

Private, first class, medical detachment, 128th Infantry, 32d Division.
When his battalion was forced to withdraw three times successively within 48 hours, Private Hanson each time went out in front of the line under intense machine-gun and shell fire and rescued wounded men.

*HANTSCHKE, EDWARD (88944)
Near Pannes, France, Sept. 12, 1918.
R—Newark, N. J.
B—Brooklyn, N. Y.
G. O. No. 74, W. D., 1919.

Private, Machine Gun Company, 165th Infantry, 42d Division.
Seeing that his platoon leader was being fired upon by a German sniper, Private Hantschke, who at the time was under cover, jumped to his feet and, stepping in front of the officer, tried to push him into a shell hole. The bullet intended for the officer struck Private Hantschke in the mouth, killing him instantly.
Posthumously awarded. Medal presented to father, Max Hantschke.

HAPSCH, JOHN P. (283568)
Near Soissons, France, July 18, 1918.
R—Hudson, Wis.
B—Hudson, Wis.
G. O. No. 72, W. D., 1920.

Corporal, Company G, 16th Infantry, 1st Division.
Knowing that the enemy had captured a wounded member of his company, Corporal Hapsch with two others advanced across dangerous ground to a barn, where they routed the enemy captors and carried back their comrade to safety.

HARBIN, ELLIOTT R. (1330351)
Near Bellicourt, France, Sept. 29,
1918.
R—Greenville, S. C.
B—Greenville County, S. C.
G. O. No. 44, W. D., 1919.

Private, first class, Company C, 105th Field Signal Battalion, 30th Division.
While assisting a party in laying a telephone line, Private Harbin was seriously wounded, but refused to be evacuated and continued his work throughout the day under heavy shellfire. He also assisted in giving first aid to the wounded under fire.

HARBISON, HARRY J. (54801)
West of Berzy-le-Sec, France, July
20, 1918.
R—Philadelphia, Pa.
B—Philadelphia, Pa.
G. O. No. 15, W. D., 1919.

Private, Company K, 26th Infantry, 1st Division.
Although severely wounded in the leg by machine-gun fire, he refused to go to the rear, bandaged his own wound, and advanced with his platoon until its final objective was reached.

HARDER, FRED J. (2267547)
Near Eclisfontaine, France, Sept 29,
1918.
R—Brawley, Calif.
B—Germany.
G. O. No. 44, W. D., 1919.

First sergeant, Company K, 364th Infantry, 91st Division.
Responding to a call for volunteers, he crossed an open space a distance of about 300 yards under terrific hostile fire of machine guns and artillery to the position held by his machine-gun company to inform them of the location of enemy machine-gun nests and snipers, who had escaped detection. The enfilading fire of the enemy threatened to wipe out his company, but the success of his mission enabled our guns to silence the fire of the enemy.

HARDIE, WILLIAM C. (549510)
Near Mezy, France, July 15, 1918.
R—Jersey City, N. J.
B—Rushville, Nebr.
G. O. No. 23, W. D., 1919.

Sergeant, Company A, 38th Infantry, 3d Division.
During the preparations for the German offensive of July 15, 1918, and through the intense artillery bombardment connected therewith, Sergeant Hardie showed utter disregard of personal danger by voluntarily rescuing wounded comrades from exposed positions.

HARDIMAN, MICHAEL J.
Near St. Etienne, France, Oct. 4, 1918.
R—Chicago, Ill.
B—Ireland.
G. O. No. 46, W. D., 1919.

Private, 18th Company, 5th Regiment, U. S. Marine Corps, 2d Division.
During the offensive operations at Blanc Mont Ridge he volunteered and went into no man's land, under intense machine-gun fire, and brought in a wounded man.

HARDIN, MELVIN N. (1312188)
Near Bellicourt, France, Sept. 23–27, 1918.
R—Greer, S. C.
B—Gaffney, S. C.
G. O. No. 59, W. D., 1919.

Private, Company L, 118th Infantry, 30th Division.
During four days of operations and under unusually adverse conditions, Private Hardin, acting as company runner, repeatedly carried messages from company headquarters to the front line, over open ground, subjected to shell and direct machine-gun fire. With practically no food or sleep, and showing absolute disregard for personal safety, he successfully performed each mission, aiding materially in the maintenance of liaison and the success of the attack.

HARDIN, WILLIAM T. (1491176)
Near St. Etienne, France, Oct. 8, 1918.
R—Rhome, Tex.
B—Arkansas.
G. O. No. 50, W. D., 1919.

Sergeant, Company L, 142d Infantry, 36th Division.
He was severely wounded in the right shoulder early in the action, but although bleeding profusely, he refused to go to the rear until the enemy's position had been captured and the line consolidated.

HARDING, STACY L.
Near Malancourt and Nantillois, France, Sept. 26 to Oct. 11, 1918.
R—Antioch, Calif.
B—Waltham, Mass.
G. O. No. 66, W. D., 1919.

Second lieutenant, 120th Field Artillery, 32d Division.
On duty as artillery information officer, Lieutenant Harding displayed the utmost courage, fortitude, and devotion to duty in maintaining liaison between the Infantry and Artillery, going repeatedly to the front line and carrying messages for both Infantry and Artillery commanders. In seeking the location of an Infantry company, under heavy machine-gun and artillery fire, he searched the woods on the Cote Dame Marie, as far as the enemy's wire, where he was killed by machine-gun fire.
Posthumously awarded. Medal presented to mother, Mrs. J. S. Harding.

HARDISON, DEWITT (1330352)
Near Bellicourt, France, Sept. 29, 1918.
R—Kenly, N. C.
B—Martin County, N. C.
G. O. No. 21, W. D., 1919.

Private, first class, Company C, 105th Field Signal Battalion, 30th Division.
Being a member of a detail to establish communication with the front line, Private Hardison was caught in an enemy barrage, during which his detail suffered many casualties. Although badly gassed, he continued to work for the entire day, always exposed to heavy artillery fire, after which he assisted in the removal of the wounded.

HARDY, HARRISON A. (549424)
Near Mezy, France, July 15, 1918.
R—Peabody, Mass.
B—Salem, Mass.
G. O. No. 44, W. D., 1919.

Private, Company A, 38th Infantry, 3d Division.
Although painfully wounded during the battle of the Marne, he continued on duty with his platoon and was active in bringing wounded men to the safety of our lines.

HARDY, ORVILLE, Jr. (549164)
Near Mezy, France, July 15, 1918.
R—Shamokin, Pa.
B—Shamokin, Pa.
G. O. No. 23, W. D., 1919.

Corporal, Machine Gun Company, 38th Infantry, 3d Division.
Corporal Hardy remained with his gun after all the other members of his gun crew were killed or wounded. He continued to fire until his ammunition was exhausted, then removed the firing mechanism and returned to our lines, fighting his way with his pistol.

HARELIS, STEVE J. (551444)
Near Mezy, France, July 15, 1918.
R—Boston, Mass.
B—Greece.
G. O. No. 23, W. D., 1919.

Private, first class, Company H, 38th Infantry, 3d Division.
During the intense enemy artillery shelling in preparation for the German offensive of July 15, 1918, he voluntarily made three attempts to bring in a wounded comrade who was exposed to enemy fire before he finally succeeded in getting him to safety.

HARGRAVE, THOMAS J.
Near Grand Pre, France, Oct. 26, 1918.
R—Rochester, N. Y.
B—Wymore, Nebr.
G. O. No. 35, W. D., 1919.

First lieutenant, 309th Machine Gun Battalion, 78th Division.
He personally made a reconnaissance of the entire front of the battalion to which he was attached, under heavy machine-gun and shell fire, and returned with very valuable information in the shortest time possible. He returned through an open field, subjected to an enemy fire. This officer later rendered valuable assistance in repelling a counterattack by his coolness and intelligent direction of the company.

HARKENRIDER, LOUIS H. (5554)
In the vicinity of Chateau-Thierry, France, June 6, 1918.
R—Griffith, Ind.
B—Huntington, Ind.
G. O. No. 101, W. D., 1918.

Private, 15th Ambulance Company, 2d Sanitary Train attached to 5th Regiment, U. S. Marine Corps, 2d Division.
On June 6, 1918, in the vicinity of Chateau-Thierry, with a corporal, he went out into an open field under heavy shell and machine-gun fire and succeeded in bandaging and carrying back to our lines a wounded comrade.

HARLIN, HENRY JOHN (1212618)
Near Ronssoy, France, Sept. 29, 1918.
R—New York, N. Y.
B—New York, N. Y.
G. O. No. 44, W. D., 1919.

Private, Machine Gun Company, 107th Infantry, 27th Division.
After an advance of more than 2,000 yards with the Infantry, Private Harlin crawled through a barbed-wire entanglement and remained the entire night under machine-gun fire within a few yards of enemy position in order to protect his comrades from a surprise attack.

HARLOW, LEON E. (128049) ....
Near Cantigny, France, May 28-30, 1918.
R—Waco, Tex.
B—Waco, Tex.
G. O. No. 99, W. D., 1918.

Corporal, Battery F, 7th Field Artillery, 1st Division.
He voluntarily left his shelter and went out into a smothering bombardment to perform exhausting labor in repairing telephone lines. He repaired five breaks in a 50-yard stretch of wire, and when the same small section was again broken in four places he carried a message through heavy shelling to the regimental post of command.

HARMAN, JOHN T. ....
Near Soissons, France, July 19, 1918.
R—New York, N. Y.
B—Staunton, Va.
G. O. No. 133, W. D., 1919.

First lieutenant, 26th Infantry, 1st Division.
He refused to be evacuated when wounded near Soissons, France, July 19, 1918, but continued forward with his command in a third successful advance until wounded seriously a second time.

HARMON, HENRY C. (1428675) ....
At Marcheville, France, Sept. 26, 1918.
R—Edgemont, S. Dak.
B—Newtown, Mo.
G. O. No. 137, W. D., 1918.

Private, Company C, 102d Infantry, 26th Division.
He carried messages across an area swept by violent machine-gun and artillery fire, showing remarkable bravery and devotion to duty on several occasions when the situation was critical and the messages were of the utmost importance.

HARMON, JOHN J. (549690) ....
In the Meuse-Argonne offensive, France, Oct. 10, 1918.
R—Weir, Kans.
B—Weir, Kans.
G. O. No. 16, W. D., 1920.

Sergeant, Company B, 38th Infantry, 3d Division.
This soldier showed exceptional courage, initiative, and devotion to duty. After all the company officers were either killed or wounded, he assumed command, and with absolute disregard of his own safety did reorganize, under intense shell and machine-gun fire, the scattering units of the company and fearlessly led them to their objective. By this action many casualties were avoided, and by his personal example of coolness and courage the men were so encouraged and stimulated that they accomplished their difficult task.

HARREL, BENJAMIN H. (1977470) ....
Near Consenvoye, France, Oct. 10, 1918.
R—Indianapolis, Ind.
B—Ripley County, Ind.
G. O. No. 71, W. D., 1919.

Sergeant, Company K, 131st Infantry, 33d Division.
Upon discovering an enemy machine-gun nest beyond the objective, on his own initiative he crawled out with 2 men, and though subjected to heavy fire flanked the position, captured the machine gun and 31 prisoners, 1 of them an officer, killing 2 of the enemy who tried to escape.

HARRELL, RAYMOND (2311808) ....
Near Bois-des-Rappes, France, Oct. 21, 1918.
R—Murphysboro, Ill.
B—Murphysboro, Ill.
G. O. No. 81, W. D., 1919.

Private, Company K, 11th Infantry, 5th Division.
While carrying a message to another battalion in the line, Private Harrell was wounded by a machine-gun bullet at the same time his companion was killed. He courageously delivered his message, despite the wound, and received an answer, which he delivered to his company and battalion commanders before being evacuated.

HARRELL, WILLIAM F ....
Near Apremont, France, Oct. 4, 1918.
R—Marion, S. C.
B—Marion, S. C.
G. O. No. 44, W. D., 1919.
Distinguished-service medal also awarded.

Lieutenant colonel, 16th Infantry, 1st Division.
Colonel Harrell showed extraordinary coolness under fire in the battle of the Argonne. When one portion of his battalion was held up by machine-gun fire he went forward to the most advanced elements, reorganized them, and carried the objective, but was severely wounded while in the execution of his mission.

*HARRIMAN, LYNN H ....
In the Humbert Plantation, France, May 27, 1918.
R—Concord, N. H.
B—Warner, N. H.
G. O. No. 88, W. D., 1918.

Second lieutenant, 101st Infantry, 26th Division.
While in command of his platoon in the Humbert Plantation on May 27, 1918, he was viciously attacked by the enemy in greatly superior numbers. He led his men with determination and great courage, and himself stopped their advance in a side trench. He was mortally wounded but continued to fight on, calling upon and encouraging his men and participating with them in driving the enemy off. Died May 31, 1918.
Posthumously awarded. Medal presented to widow, Mrs. L. H. Harriman.

HARRIMAN, SHERMAN G ....
Near Crezancy, France, July 15-16, 1918.
R—Lawrence, Mass.
B—North Conway, N. H.
G. O. No. 44, W. D., 1919.

Sergeant, first class, 2d Engineers, 2d Division.
Assisting in the removal of the wounded, Sergeant Harriman drove an ambulance throughout the entire night, continuing until several hits by the enemy completely disabled his machine. The next morning he led his men into the trenches through an intense shelling, and remained in command for 12 hours after being wounded.

*HARRINGTON, ALEXANDER (235554) .....
Near Montblainville, France, Sept. 30, 1918.
R—Philadelphia, Pa.
B—Philadelphia, Pa.
G. O. No. 56, W. D., 1922.

Sergeant, Company C, 109th Infantry, 28th Division.
Sergeant Harrington, though wounded and about to be carried to the rear, seeing two comrades who were apparently more severely wounded than himself, ordered the litter bearers to carry the other men first. Before the litter bearers returned, Sergeant Harrington was killed by a machine-gun bullet. His bravery, sacrifice, and qualities of leadership were an inspiration to the members of his command.
Posthumously awarded. Medal presented to father, Richard Harrington.

HARRINGTON, ERNEST A. S. (68111) .....
Near Bois-de-St. Remy, France, Sept. 12, 1918.
R—Windham, N. H.
B—Lowell, Mass.
G. O. No. 26, W. D., 1919.

Corporal, Company F, 103d Infantry, 26th Division.
While sniper fire was holding up the advance of a section of his platoon, Corporal Harrington rushed forward and, without aid, forced an officer and 6 men to surrender at the point of his bayonet.

HARRINGTON, JOHN H. (57711) ....
At Cantigny, France, May 29, 1918.
R—New Haven, Conn.
B—New Haven, Conn.
G. O. No. 39, W. D., 1920.

Corporal, Company G, 28th Infantry, 1st Division.
In the absence of his platoon commander, Corporal Harrington led his platoon forward under heavy machine-gun fire from a support position to the assistance of a front-line unit. Although severely wounded in the hip, he delivered to the commander of the support troops a report of the situation of the front line.

HARRIS, CHARLES (66031)_____
Near Trugny, France, July 22, 1918.
R—Meriden, Conn.
B—Rumania.
G. O. No. 16, W. D., 1923.

Mechanic, Company L, 102d Infantry, 26th Division.
He voluntarily carried a message from his company post of command to regimental headquarters, amid a storm of enemy shell and machine-gun fire, accomplishing his mission and thus preventing the division artillery from firing upon its own troops. On a previous occasion during the Aisne-Marne offensive he had carried important messages through intense enemy fire, once having encountered on such a mission 4 of the enemy, 1 of whom he killed and 3 of whom he brought in as prisoners. His outstanding bravery and devotion to duty served as an example to the men of his regiment.

ᵃHARRIS, CHARLES D_____
In Claire-Chenes Woods, France, Oct. 20, 1918.
R—Cedartown, Ga.
B—Fort Niagara, N. Y.
G. O. No. 70, W. D., 1919.

Captain, 6th Engineers, 3d Division.
While leading his company in an attack on enemy machine-gun nests he, with three of his men in advance of the remainder of the company, fearlessly attacked an enemy machine-gun nest, capturing three prisoners and two guns, turning the guns against the enemy. He was mortally wounded while operating one of the guns in an exposed position.
Posthumously awarded. Medal presented to father, Maj. Gen. P. C. Harris.

ᵇHARRIS, CLIFFORD O_____
Near Juvigny, France, Sept. 1, 1918.
R—Portland, Oreg.
B—Vancouver, Wash.
G. O. No. 11, W. D., 1921.

Second lieutenant, 128th Infantry, 32d Division.
When the advance of his platoon was halted by the effective fire of an enemy machine-gun nest, Lieutenant Harris with 2 others rushed out before the front line to attack the enemy position. While in the accomplishment of this heroic act and when within the enemy's lines he was killed.
Posthumously awarded. Medal presented to father, W. R. Harris.

HARRIS, DUNCAN G_____
In the Argonne Forest, France, Sept. 30, 1918.
R—New York, N. Y.
B—New York, N. Y.
G. O. No. 46, W. D., 1919.

Major, 305th Infantry, 77th Division.
During an attack in the Argonne Forest he fell and broke his collar bone, but continued with his battalion throughout the attack and until the Meuse was reached. For 15 days he was continually with his battalion, personally leading them against strongly fortified enemy positions, although suffering acute and continual pain from his injury and being handicapped by having his arm in a sling.

ᵃHARRIS, EDWARD C_____
Near Grimaucourt, France, Nov. 11, 1918.
R—Wendell, N. C.
B—Henderson, N. C.
G. O. No. 32, W. D., 1919.

Second lieutenant, 321st Infantry, 81st Division.
Under the fire of 3 machine guns, firing upon him from different directions, he took his gun through the enemy wire and mounted it. He would not permit his men to remain in such a dangerous position, and after being wounded severely ordered his men to leave him.
Posthumously awarded. Medal presented to father, E. W. Harris.

HARRIS, GRAHAM W. (1319104)_____
Near Bellicourt, France, Sept. 29, 1918.
R—Oxford, N. C.
B—Oxford, N. C.
G. O. No. 44, W. D., 1919.

Sergeant, Machine Gun Company, 120th Infantry, 30th Division.
Becoming separated from his platoon in the dense smoke and fog with 5 other soldiers, Sergeant Harris kept his men together and continued the advance under heavy artillery and machine-gun fire. Upon reaching the objective he made a personal reconnaissance 600 yards to the front, capturing several prisoners, and assisting in breaking up three machine-gun nests. He remained in this advanced position until he was ordered back.

HARRIS, HENRY L., Jr_____
At Peruka-Utig's Cotta, Island of Jolo, P. I., May 3, 1905.
R—Hoboken, N. J.
B—West Point, N. Y.
G. O. No 126, W. D., 1919.

First lieutenant, 22d Infantry, U. S. Army.
For extraordinary heroism in action against an armed enemy at the capture of Peruka-Utig's Cotta, Island of Jolo, P. I., May 3, 1905, while serving as first lieutenant, 22d U. S. Infantry. He gallantly and fearlessly directed the movements of the most exposed part of the firing line, and, without regard for his personal safety, was the first man to enter the cotta, a stronghold of the enemy, where it was expected that a last stubborn stand would be made by the insurgents.

ᵃHARRIS, JOB R. (107063)_____
Near Medeah Ferme, France, Oct. 8, 1918.
R—Pittsburgh, Pa.
B—Pittsburgh, Pa.
G. O. No. 142, W. D., 1918.

Sergeant, Company B, 4th Machine Gun Battalion, 2d Division.
Though he had fainted twice as a result of being gassed the previous night, he remained on duty and continued to lead his section. Placing them in action, he displayed great coolness and bravery in directing the fire of his guns until he was killed by enemy machine-gun fire.
Posthumously awarded. Medal presented to sister, Mrs. William C. Palmer.

HARRIS, MAHLON H_____
Near Grand Pre, France, Oct. 26, 1918.
R—Powell, Pa.
B—Powell, Pa.
G. O. No. 28, W. D., 1921.

First lieutenant, 312th Infantry, 78th Division.
During the assault on the citadel Lieutenant Harris gallantly led his company through heavy enemy fire. Learning that the adjoining company had met with extraordinarily heavy machine-gun fire which had killed all of its officers and was holding up the company, he, of his own volition, passed through deadly fire, reorganized this adjoining company and led it, with his own, to the objective. This officer was among the first to scale the wall of the citadel.

HARRIS, MORGAN K_____
At Apremont, France, Sept. 29, 1918.
R—Lawrence, N. Y.
B—Lebanon Springs, N. Y.
G. O. No. 98, W. D., 1919.

Second lieutenant, 109th Infantry, 28th Division.
Surrounded by a number of the enemy during a hostile counterattack and called upon to surrender, Lieutenant Harris escaped by knocking down one of his adversaries and shooting another. Crawling back to his support, he organized a group of men and led them in, completely repulsing the counterattack, displaying the utmost courage and determination.

HARRIS, ROY (569559)_____
West of Fismes, France, Aug. 5, 1918.
R—Quitman, Ga.
B—Quitman, Ga.
G. O. No. 145, W. D., 1918.

Private, Company F, 4th Engineers, 4th Division.
He was a member of a small detachment of Engineers which went out in advance of the front line of the Infantry through an enemy barrage from 77-millimeter and 1-pounder guns to construct a footbridge over the River Vesle. As soon as their operations were discovered, machine-gun fire was opened up on them, but, undaunted, the party continued at work, removing the German wire entanglements and completing a bridge which was of great value in subsequent operations.

HARRIS, WILLIAM L (60605)...........
Near St. Remy, France, Sept. 13, 1918.
R—Malden, Mass.
B—Malden, Mass.
G. O. No. 78, W. D., 1919.

Corporal, Company E, 101st Infantry, 26th Division.
While acting as a runner between company and battalion headquarters, under terrific shellfire, Corporal Harris carried messages without regard to personal danger until struck and killed by a shell.
Posthumously awarded. Medal presented to father, James G. Harris.

*HARSSELL, GEORGE J. (1282974)........
Near Verdun, France, Oct. 12, 1918.
R—Jersey City, N. J.
B—Jersey City, N. J.
G. O. No. 37, W. D., 1919.

Private, first class, Company C, 111th Machine Gun Battalion, 29th Division.
By encouraging his comrades and rendering valuable aid to the wounded, he succeeded greatly in retaining the morale of those about him during an almost continual rain of shells. Even after being severely wounded by shrapnel, he attempted to retain the spirit by singing. After being removed to a hospital he died.
Posthumously awarded. Medal presented to mother, Mrs. Nellie Harssell.

HART, CLAUDE V. (102473)...........
Near Sergy, France, July 28, 1918.
R—Cherokee, Iowa.
B—Beadle County, S. Dak.
G. O. No. 99, W. D., 1918.

Sergeant, Company M, 168th Infantry, 42d Division.
He coolly and with utter disregard of danger led his platoon against enemy machine-gun emplacements. Four of the foe were captured, together with their two machine guns, which he turned and operated on the retreating Germans until he was severely wounded.

HART, FRANK I. (1303551)...........
Near Consenvoye, France, Oct. 7-8, 1918.
R—Chicago, Ill.
B—Green Bay, Wis.
G. O. No. 30, W. D., 1921.

Sergeant, Company C, 108th Engineers, 33d Division.
For extraordinary heroism in action near Consenvoye, France, on the night of Oct. 7-8, 1918, when in command of a detail of Company C, 108th Engineers, which was engaged in the construction of a bridge across the Meuse River. In directing and supervising the construction of this bridge he exposed himself to heavy enemy artillery and machine-gun fire. During an enemy gas bombardment he exposed himself to additional danger by distributing gas masks to the members of his detail, who had lost their masks during the construction of the bridge. Due in great part to his energy and gallantry, the bridge was completed in time to enable the attack to be carried out the following morning with marked success.

HART, GEORGE C..................
Near Grand Pre, France, Oct. 25-26, 1918.
R—Elmira, N. Y.
B—Elmira, N. Y.
G. O. No. 56, W. D., 1922.

Captain, 309th Machine Gun Battalion, 78th Division.
On Oct. 25, 1918, Captain Hart volunteered to lead his company into a particularly dangerous sector on the ridge north of Grand Pre, offering their use practically as infantry. His act was of inestimable value in strengthening the morale of the Infantry battalion which had suffered very heavy casualties and had no prospect of relief. On the morning of Oct. 26 during a counterattack he not only personally directed the location of his machine guns in the outpost line, but by his splendid example of fearlessness he rallied and commanded the Infantry when no other officer was available, and was an inspiration both to the infantrymen and machine gunners in holding the ridge. He fought valiantly until very seriously wounded by enemy machine-gun fire.

HART, JACK S...................
Near Thiaucourt, France, Sept. 15, 1918.
R—Weatherford, Tex.
B—Weatherford, Tex.
G. O. No. 46, W. D., 1919.

First lieutenant, 6th Machine Gun Battalion, U. S. Marine Corps, 2d Division.
Acting as company commander, Lieutenant Hart advanced with the first wave of Infantry to locate favorable positions for his guns. He discovered an enemy machine gun which he, alone, attacked, and although severely wounded, succeeded in capturing the gun and taking the crew prisoners.

*HART, LOUIS FRANCIS (367426)........
In Houppy Bois, Molleville, France, Oct. 28, 1918.
R—Hartford, Conn.
B—Hartford, Conn.
G. O. No. 13, W. D., 1923.

Private, Company B, 101st Machine Gun Battalion, 26th Division.
Volunteering to accompany a runner who had been ordered to carry an important message from his platoon commander to the battalion commander through a zone swept by heavy artillery fire, and with disregard for his own safety, he proceeded to carry out his mission until struck and killed by an enemy shell.
Posthumously awarded. Medal presented to father, William Hart.

HART, PERCIVAL G................
During the battle of the St. Mihiel salient, France, Sept. 12-13, 1918.
R—Chicago, Ill.
B—Benicia, Calif.
G. O. No. 128, W. D., 1918.

Second lieutenant, 135th Aero Squadron, Air Service.
On Sept. 12 he made three Infantry liaison patrols and obtained valuable information of the location of our advancing line, which information he conveyed to organization commanders. Bad weather conditions necessitated flying at a very low altitude, but in spite of this and repeated attacks by superior numbers of enemy aircraft he drove them off by his accurate fire and accomplished his mission. On Sept. 13 he unhesitatingly went to the assistance of three allied planes which were being attacked by a large patrol of the enemy, and by his steady fire drove off the enemy patrol and enabled the three allied planes to return.

*HART, SAMUEL C. (1315901)...........
Near Bellicourt, France, Sept. 29, 1918.
R—Mooresville, N. C.
B—Mooresville, N. C.
G. O. No. 46, W. D., 1919.

Private, first class, Company G, 119th Infantry, 30th Division.
After having been seriously wounded in the arm, which caused much pain and loss of blood, he continued to advance, carrying a Lewis gun, and pouring an effective fire into the ranks of the enemy until he was killed in the attack.
Posthumously awarded. Medal presented to father, Samuel B. Hart.

HARTIGAN, WALTER F. L............
Near Bellicourt, France, Sept. 29, 1918.
R—Lynchburg, Va.
B—Liberty, Va.
G. O. No. 49, W. D., 1922.

Major, 118th Infantry, 30th Division.
During the attack against the Hindenburg line, while at the division forward observation post, Major Hartigan, realizing that many men were losing their direction in the heavy fog, voluntarily left his post of comparative safety, went forward under a heavy hostile artillery barrage, and in the face of machine-gun fire reorganized into detachments at this critical time several hundred stragglers of the 60th Brigade, as well as two companies of the 117th Infantry. Securing tanks to accompany them, by indefatigable efforts he directed these units forward when their assistance was of prime importance to the successful issue of the combat.

HARTLEY, PAUL LIVINGSTON (1312251)...
Near Montbrehain, France, Oct. 8, 1918.
R—Hartsville, S. C.
B—Marion County, S. C.
G. O. No. 37, W. D., 1919.

Private, Company L, 118th Infantry, 30th Division.
After practically all of the other members of his squad had become casualties he maintained an effective fire with his automatic rifle from an advanced position and thereby protected his platoon. He was twice buried by exploding shells, but each time he dug himself out and resumed firing immediately. After his ammunition was exhausted he rushed forward with his empty gun and forced 20 of the enemy to surrender.

HARTMAN, GUY L...
Near Frapelle, France, Aug. 17, 1918.
R—Farmington, N. C.
B—Farmington, N. C.
G. O. No. 32, W. D., 1919.

First lieutenant, 6th Infantry, 5th Division.
After having been painfully wounded, Lieutenant Hartman refused to go to the rear for treatment. He made his way through a heavy barrage and brought up a platoon that was stopped by heavy fire. Some time later, after having his wound dressed, he conducted his brigade commander through a heavily gassed area, after which he remained constantly on duty until relieved.

HARTMAN, WILLIAM A. (257825)...
The Vesle River front near Fismes, France, Aug. 4, 1918.
R—Milwaukee, Wis.
B—Milwaukee, Wis.
G. O. No. 99, W. D., 1918.

Sergeant, Company F, 107th Engineers, 32d Division.
He was a member of a patrol sent out from the battalion post of command at midnight, Aug. 4, 1918, to reconnoiter the Vesle River front, near Fismes, France, for the location of possible sites for pontoon bridges and of material for making structures. Despite heavy artillery and machine-gun fire that forced the patrol to scatter and separated him from the lieutenant in charge, he continued the work on his own initiative, and, acting entirely without orders, started his detail on the actual construction of rafts for the pontoon bridges. His courage, ability as a leader, and his inflexible determination made the reconnaissance a complete success.

HARTMANN, MIKE A. (1783104)...
Near Montfaucon, France, Sept. 27, 1918.
R—Brooklyn, N. Y.
B—Brooklyn, N. Y.
G. O. No. 46, W. D., 1919.

First sergeant, Company I, 314th Infantry, 79th Division.
Sergeant Hartmann advanced alone and silenced a machine-gun nest which was holding up an entire platoon.

HARTNEY, HAROLD E...
Near Fismes, France, Aug. 13, 1918.
R—Canada.
B—Canada.
G. O. No. 1, W. D., 1919.

Major, 27th Aero Squadron, 1st Pursuit Group, Air Service.
He voluntarily accompanied a reconnaissance patrol. Realizing the importance of the mission, he took command, and, although 5 enemy planes repeatedly made attempts to drive them back, he continued into enemy territory, returning later to our lines with important information. The cool judgment and determination displayed by Major Hartney furnished an inspiration to all members of his command.

*HARTUNG, JOHN A. (2178459)...
Near Remonville, France, Nov. 1, 1918.
R—Ewing, Mo.
B—Camp Point, Ill.
G. O. No. 44, W. D., 1919.

Sergeant, Company B, 354th Infantry, 89th Division.
While leading his combat group forward he encountered a machine-gun stronghold, which opened fire on his force from three sides. He charged the guns to the front, but he was killed on reaching the line. His valorous act, however, enabled his men to break the line and take the flanking guns, thereby saving many casualties in his platoon.
Posthumously awarded. Medal presented to mother, Mrs. Anna E. Hartung.

HARTWELL, ALBERT S. (9921)...
East of Reims, France, Mar. 20–23, 1918.
R—France.
B—France.
G. O. No. 99, W. D., 1918.

Private, Section No. 633, Ambulance Service.
From Mar. 20 to 23, 1918, he repeatedly drove his ambulance over a road east of Reims, France, that was under bombardment of gas and explosive shells. Upon one occasion, while cranking his car, he was knocked several yards by the explosion of a shell, yet he continued his work. Another day his car was hit by a shell and badly damaged, and he himself was severely wounded in the head and both legs. In spite of his wounds he carried a wounded man who was in the ambulance to a place of safety, made him comfortable, and then crawled to a dressing station for assistance.

HARTY, BERNARD A. (2524363)...
On Patian Island, P. I., July 4, 1909.
R—Philadelphia, Pa.
B—Philadelphia, Pa.
G. O. No. 14, W. D., 1925.

Private, Troop A, 6th Cavalry, U. S. Army.
For extraordinary heroism in action against hostile Moros on Patian Island, P. I., July 4, 1909. Private Harty, with other men, entered a cave occupied by a desperate enemy, and in the face of a heavy fire, with utter disregard for his personal safety, aided in forcing the outlaws to abandon their stronghold, which resulted in their destruction by our forces.

*HARVEY, HARRY A...
Near Chateau-Thierry, France, July 14–15, 1918.
R—Memphis, Tenn.
B—McComb, Miss.
G. O. No. 22, W. D., 1920.

Captain, 18th Field Artillery, 3d Division.
During the bombardment preceding the enemy attack across the Marne the position of Battery A, 18th Field Artillery, was subjected to heavy artillery fire for a period of 4 hours. All communications were destroyed and the ammunition dump destroyed by hostile fire. Throughout the enemy bombardment, Captain Harvey kept his battery in action, exposing himself to concentrated enemy artillery fire in order to replenish his ammunition supply, and delivered an effective fire on the enemy.
Posthumously awarded. Medal presented to widow, Mrs. Ethel C. Harvey.

HARWOOD, BENJAMIN P...
Near Chateau-Thierry, France, July 5, 1918.
R—Minneapolis, Minn.
B—Helena, Mont.
G. O. No. 7, W. D., 1919.

First lieutenant, Field Artillery, observer, 12th Aero Squadron, Air Service.
He volunteered, with another plane, to protect a photographic plane. In the course of their mission they were attacked by 7 enemy planes (Fokker type). He accepted the combat and kept the enemy engaged while the photographic plane completed its mission. His guns jammed and he himself was seriously wounded. After skillfully clearing his guns, with his plane badly damaged, he fought off the hostile planes and enabled the photographic plane to return to our lines with valuable information.

HARWOOD, FRANK M. (70621) ....
Near Blanc Mont Ridge, France, Oct. 3-9, 1918.
R—Arlington, Mass.
B—Arlington, Mass.
G. O. No. 21, W. D., 1919.

Private, Company D, 9th Infantry, 2d Division.
While acting as battalion runner, Private Harwood, regardless of personal danger, many times volunteered and carried messages under the most intense shell fire, and greatly assisted in maintaining liaison with other units.

HARWOOD, RALPH W ....
Near Verdun, France, Oct. 23-29, 1918.
R—Barre, Mass.
B—Barre, Mass.
G. O. No. 46, W. D., 1919.

Second lieutenant, 102d Infantry, 26th Division.
Throughout the action from Oct. 23 to 28, he set a splendid example of courage and heroism to his company in action against overwhelming odds. On Oct. 29, upon learning that 4 of his men were in shell holes in front of the line and near the enemy's position and were so badly wounded that they could not return without assistance, he crawled forward and rescued the 4 men, one at a time, bringing them back to our lines through a constant and terrific machine-gun fire.

HASKEW, EDWARD D. (571647) ....
Between Septsarges and Fromereville, France, Oct. 6, 1918.
R—Gladstone, N. J.
B—Wilton, Conn.
G. O. No. 128, W. D., 1918.

Wagoner, 33d Ambulance Company, 4th Sanitary Train, 4th Division.
He was on duty with his ambulance, carrying wounded from a battalion aid station. He left with 4 stretcher cases and went about 2 kilometers south of Gercourt; while ascending a hill his ambulance was struck by a shell, he receiving multiple shell wounds of hands, left thigh, and feet. Although seriously wounded, he bravely remained at his post and continued on with his ambulance along a shell-swept road to the crest of the hill near an aid station, when he turned his ambulance off the road and sought assistance for his wounded.

HASKINS, CLIFFORD R. (108817) ....
Near Trugny, France, July 22, 1918.
R—Elmwood, Conn.
B—Danbury, Conn.
G. O. No. 125, W. D., 1918.

Wagoner, Company B, 101st Machine Gun Battalion, 26th Division.
He was seriously wounded in the leg while placing injured soldiers in his ambulance. Nevertheless, he insisted on driving the machine to the dressing station and continued the work of evacuating the wounded until exhausted from loss of blood.

HASLETT, ELMER R ....
Near Montfaucon, France, Sept. 28, 1918.
R—Los Angeles, Calif.
B—Carterville, Mo.
G. O. No. 35, W. D., 1919.

Captain, 12th Aero Squadron, Air Service.
While on an artillery surveillance mission, he engaged 4 enemy planes, which were about to attack the American balloon line. He succeeded in diverting them from the balloons, but in the combat his machine gun became jammed. Driving off his nearest adversary by firing a Very pistol at him, he succeeded in clearing the jam, and, returning to the fight, he destroyed one hostile plane and dispersed the remainder of the group.

HASSARD, ROBERT J. (39107) ....
Near Vierzy, France, July 18, 1918.
R—New York, N. Y.
B—New York, N. Y.
G. O. No. 132, W. D., 1918.

Corporal Company E, 9th Infantry, 2d Division.
He led his squad in the attack with conspicuous bravery and utter disregard for his own safety. Although wounded three times, he refused to go to the rear for medical attention and remained at his post on the firing line until his company was relieved the following night. Throughout the fight he rendered valuable assistance to his platoon leader, assisted in reorganizing the company after the attack and in preparing the position for the counterattack.

HASSEBROCK, WILLIAM M. (3767172) ...
Near Brandeville, France, Nov. 8, 1918.
R—Holstein, Mo.
B—St. Louis, Mo.
G. O. No. 81, W. D., 1919.

Private, Machine Gun Company, 11th Infantry, 5th Division.
Wounded in the leg by shrapnel, Private Hassebrock continued in the advance, carrying his machine gun. Soon afterwards he was knocked down and again wounded in the same leg, but, although he was ordered to a dressing station he succeeded in joining his gun crew at the firing position, where he was wounded severely for the third time.

HASSIG, ALBERT U ....
Between Fey-en-Haye and Vilcey, France, Sept. 12, 1918.
R—Oklahoma City, Okla.
B—McKees Rocks, Pa.
G. O. No. 23, W. D., 1919.

First lieutenant, 358th Infantry, 90th Division.
Although he was severely wounded, he refused to go to a first-aid station, but continued to give great assistance in silencing enemy machine-gun nests.

HASSLER, RUDOLPH P. (2261300) ....
At Gesnes, France, Sept. 29, 1918.
R—Sumatra, Mont.
B—Davenport, Iowa.
G. O. No. 1, W. D., 1919.

Sergeant, Company K, 362d Infantry, 91st Division.
Although he was seriously wounded, he remained in command of his platoon until he was relieved next morning, displaying exceptional devotion to duty.

HATCH, BENJAMIN T., Jr. (1287967) ....
North of Samogneux, France, Oct. 15, 1918.
R—Disputanta, Va.
B—Prince George County, Va.
G. O. No. 15, W. D., 1919.

Sergeant, Company E, 116th Infantry, 29th Division.
When his company was subjected to deadly machine-gun fire from two hostile machine-gun nests he, upon his own initiative, organized an attacking group and led it in an assault on the nests, putting them out of action and making the further advance of his company possible.

HATCH, CARL T ....
Near Nantillois, France, Oct. 4, 1918.
R—Baltimore, Md.
B—St. Albans, Vt.
G. O. No. 37, W. D., 1919.

Second lieutenant, 317th Infantry, 80th Division.
Seriously wounded in both knees while leading his platoon against German machine-gun nests, Lieutenant Hatch declined to be evacuated, but remained in command of his platoon for 9 hours until it was relieved.

HATCH, SIDNEY H. (1389449) ....
Near Breuilles, France, Oct. 11, 1918.
R—Chicago, Ill.
B—River Forest, Ill.
G. O. No. 44, W. D., 1919.

Private, first class, Headquarters Company, 132d Infantry, 33d Division.
After being wounded by a shell which buried him in a small hole, he made a trip to battalion headquarters, carrying a message from his platoon, and after returning with an answer assisted in carrying ammunition until the sergeant discovered that he had been wounded and sent him to the aid station.

HATCHER, SAMUEL H. (1306980)........
Near Bellicourt, France, Sept. 28, 1918.
R—Walland, Tenn.
B—Blount County, Tenn.
G. O. No. 35, W. D., 1919.

Private, Company B, 117th Infantry, 30th Division.
During the very thickest of the fighting Private Hatcher, assisted by another soldier, attacked two enemy machine-gun positions, killing the gunners and capturing the guns, thus allowing the further advance of his company.

HATLESTAD, ANDREW O. (2193012).....
Near Montfaucon, France, Oct. 3, 1918.
R—Athboy, S. Dak.
B—Fairfax, Minn.
G. O. No. 16, W. D., 1920.

Private, Machine Gun Company, 7th Infantry, 3d Division.
He repeatedly carried messages to the front-line platoons through heavy artillery and machine-gun fire. He was killed by enemy fire while carrying a message to the front line under heavy fire.
Posthumously awarded. Medal presented to father, Ole S. Hatlestad.

HAUBRICH, ROBERT............
Near Sedan, France, Nov. 7, 1918.
R—Columbus, Ohio.
B—Germany.
G. O. No 31, W. D., 1919.

Major, 166th Infantry, 42d Division.
Personally leading his battalion, which was the assault battalion, in the advance upon Sedan, Major Haubrich exposed himself many times to intense machine-gun and heavy artillery fire in order to keep contact with the enemy, greatly encouraging all the members of his command by his gallantry. Later in the day he was wounded, but he continued on duty until two hours later, when he had his wounds dressed at the first-aid station, resuming command immediately afterwards.

HAUSMANN, WILLIAM L. (2213939)......
Near Grand Ballois Farm, France, July 14-15, 1918.
R—St. Helena, Nebr.
B—St. Helena, Nebr.
G. O. No. 32, W. D., 1919.

Private, Company A, 4th Infantry, 3d Division.
Under a heavy gas and shell bombardment he repeatedly volunteered and delivered messages over routes other than his own when the runners assigned to those routes had been killed or wounded.

HAUSS, ALBERT M. (58101)............
Near Soissons, France, July 19, 1918.
R—East St. Louis, Ill.
B—East St. Louis, Ill.
G. O. No. 117, W. D., 1918.

Private, Company I, 28th Infantry, 1st Division.
With the aid of two men he charged and captured a German machine gun.

HAVEN, CHESTER (881756)...........
Near Exermont, France, Oct. 9, 1918.
R—Falls City, Oreg.
B—Hebron, N. Dak.
G. O. No. 37, W. D., 1919.

Private, Company B, 1st Engineers, 1st Division.
Upon his own initiative, Private Haven, with another soldier, displayed notable courage in attacking two machine guns which were hindering the advance. Undaunted by the heavy machine-gun fire, they poured a deadly rifle fire upon the enemy gunners and forced them to flee toward our attacking troops who captured them.

HAVERFIELD, JAMES G. (1518700)........
Near Olsene, Belgium, Oct. 31, 1918.
R—Tiltonsville, Ohio.
B—East Liverpool, Ohio.
G. O. No. 59, W. D., 1919.

Sergeant, Company G, 145th Infantry, 37th Division.
He advanced alone under heavy machine-gun fire and killed 2 of the enemy who were delivering effective machine-gun fire on the attacking wave of his company and delaying its progress.

HAWIE, ASHAD G. (97069)..............
Near Landres-et-St. Georges, France, Oct. 16, 1918.
R—Mobile, Ala.
B—Syria.
G. O. No. 131, W. D., 1918.

Private, first class, Company F, 167th Infantry, 42d Division.
Private Hawie, as company runner, without hesitation or fear of fire from heavy artillery and machine guns, made repeated trips with messages between company and battalion headquarters. On one trip he was attacked by 2 Germans, 1 of whom he killed. Taking the other prisoner, he continued on his mission and turned his prisoner over to the battalion commander after delivering his message.

HAWKE, FREDERICK J. (1399370)........
Near St. Juvin, France, Oct. 15 1918.
R—North Andover, Mass.
B—England.
G. O. No. 71, W. D., 1919.

Sergeant, Company L, 325th Infantry, 82d Division.
When his company was temporarily held up by severe machine-gun fire, Sergeant Hawke led a patrol in an attack on an enemy machine gun, working his way behind the gun and rushing it, killing one of the crew, wounding another, and taking three prisoners.

HAWKINS, PAUL G. (1317111)............
Near Bellicourt, France, Sept. 28-29, 1918.
R—Sanford, Fla.
B—Kinston, N. C.
G. O. No. 81, W. D., 1919.

Sergeant, Machine Gun Company, 119th Infantry, 30th Division.
As a platoon runner he showed marked personal bravery, repeatedly carrying important messages over shell-swept areas and under heavy machine-gun fire, sometimes for a distance of 2 miles. He remained constantly on duty for 2 days, and when his section leader became separated from his section, took command and led it with success.

HAWKINS, WILLIAM S.............
Near St. Souplet, France, Oct. 17-18, 1918.
R—Flushing, Long Island, N. Y.
B—New York, N. Y.
G. O. No. 46, W. D., 1919.

First lieutenant, 107th Infantry, 27th Division.
Acting as liaison officer during the forcing of the La Selle River, Lieutenant Hawkins was severely wounded by an exploding shell. Undeterred by the weakened condition to which his wound reduced him, he bravely continued on, working untiringly under heavy shell and machine-gun fire for 2 days until the advance of his battalion was checked.

HAWKINSON, HOWARD E.............
Near Exermont, France, Oct. 4, 1918.
R—Syracuse, N. Y.
B—Minneapolis, Minn.
G. O. No. 9, W. D., 1923.

Captain, 28th Infantry, 1st Division.
Although suffering intense pain from a gunshot wound in the leg which had not healed, he led the regimental assault battalion, of which he was in command, in the attack on that day. Forging ahead of his lines with his staff in darkness and fog, he suddenly came upon a German detachment to the rear of which were 2 machine guns which immediately opened fire. Despite surprise, courageously exposing himself, he advanced intrepidly on the nest with utter disregard for personal safety, firing his pistol, wounding some of the crew and falling, mortally wounded, when only 10 yards from the pit from which the fire was directed.
Posthumously awarded. Medal presented to mother, Mrs. Adelaide N. Hawkinson.

*HAWKS, EMERY (262185)_____
Near Romagne, France, Oct. 8, 1918.
R—Flint, Mich.
B—Whittemore, Mich.
G. O. No. 71, W. D., 1919.

Sergeant, Company E, 125th Infantry, 32d Division.
He led his platoon after his lieutenant had been killed, and when the advance was held up by intense machine-gun fire, fearlessly exposed himself in going in advance of his men to rescue a wounded soldier, being killed while administering first-aid treatment to him.
Posthumously awarded. Medal presented to father, Charles L. Hawks.

HAWS, EDWARD H_____
Near Blanc Mont, France, Oct. 2-9, 1918.
R—Philadelphia, Pa.
B—Philadelphia, Pa.
G. O. No. 26, W. D., 1919.

Private, 96th Company, 6th Regiment, U. S. Marine Corps., 2d Division.
Throughout 8 days of fighting Private Haws fearlessly and tirelessly carried messages between his company and battalion headquarters, through heavy machine gun and artillery fire.

HAYES, CASEY H_____
At Greves Farm, France, July 15, 1918.
R—San Diego, Calif.
B—Fort Worth, Tex.
G. O. No. 44, W. D., 1919.

Major, 10th Field Artillery, 3d Division.
Major Hayes, then on duty as battalion adjutant, assumed command during the absence of the battalion commander. Communication between battalion headquarters and the batteries had been cut off by an enemy bombardment of the greatest intensity, but this officer, in entire disregard for his own safety, went to each battery position and gave orders the execution of which aided materially in stopping the German advance at a critical moment.

HAYES, EDWARD S_____
Near Chevieres, France, Oct. 19, 1918.
R—Waterbury, Conn.
B—Waterbury, Conn.
G. O. No. 72, W. D., 1920.
Distinguished-service medal also awarded.

Lieutenant colonel, Infantry, Assistant Chief of Staff, G-3, 78th Division.
Lieutenant Colonel Hayes, with a private acting as guide, voluntarily made a personal reconnaissance of the front lines and the Bois-de-Loges, near the Aire River, under heavy enemy fire, being severely wounded while making this reconnaissance.

HAYES, JOSEPH_____
Near Manheulles, France, Nov. 11, 1918.
R—Haverhill, Mass.
B—Boston, Mass.
G. O. No. 32, W. D., 1919.

Second lieutenant, 323d Infantry, 81st Division.
Although severely wounded, he continued in command of his platoon in an advance under violent artillery fire until ordered to the rear by his company commander.

*HAYES, MICHAEL J_____
At Bazoches, France, Aug. 14, 1918, and St. Juvin, France, Oct. 14, 1918.
R—Cleveland, Ohio.
B—Youngstown, Ohio.
G. O. No. 20, W. D., 1919.

First lieutenant, 306th Infantry, 77th Division.
On Aug. 14 Lieutenant Hayes led a patrol of five men in broad daylight and without any cover rescued his company commander, who had fallen wounded near a German machine-gun nest. Failing to find the wounded officer, he crawled to within 20 yards of the post, attacked it with great dash and gallantry, inflicting a number of casualties in spite of heavy fire from enemy machine guns and hand grenades. On Oct. 14 this officer led his platoon forward into the attack with energy and courage, in the face of heavy artillery and machine-gun fire. In the face of direct fire from enemy machine guns upon his platoon, disregarding his own personal safety, he went forward to reconnoiter and find cover for his men from which to continue the attack. In the performance of his courageous enterprise he was killed by machine-gun fire.
Posthumously awarded. Medal presented to mother, Mrs. Julia Hayes.

*HAYES, MORRISON (567487)_____
Near Hautevesnes, France, July 19, 1918.
R—Wellsville, N. Y.
B—Wellsville, N. Y.
G. O. No. 35, W. D., 1920.

Corporal, Company D, 12th Machine Gun Battalion, 4th Division.
Although wounded during an advance, he refused to be evacuated and led his squad forward with the Infantry, placing the gun in action in the front line. Exposed to intense fire, he maintained his gun in action until he received a second wound, which later proved fatal. When ordered to withdraw, he assisted in moving the gun back to another position, inspiring his men by his personal heroism.
Posthumously awarded. Medal presented to father, Clark R. Hayes.

HAYNES, GLENN C_____
Near Bois-de-Chatillon, France, Oct. 16, 1918.
R—Centerville, Iowa.
B—Centerville, Iowa.
G. O. No. 13, W. D., 1919.

Captain, 168th Infantry, 42d Division.
Captain Haynes, as battalion commander, distinguished himself by his coolness and leadership in the attack on Bois-de-Chatillon and Cote-de-Chatillon. When the commanders of his two front-line companies were put out of action after having obtained a footing upon the slopes, Captain Haynes personally took command of the two companies and in utter disregard for his own safety successfully led them through heavy artillery, machine-gun, and rifle fire to their objective.

HAYNES, WILLIAM G. (73016)_____
Near Givry, France, July 20, 1918.
R—Revere, Mass.
B—Revere, Mass.
G. O. No. 37, W. D., 1919.

Corporal, Company I, 104th Infantry, 26th Division.
Upon learning that a soldier lay wounded in a shell hole, he voluntarily left shelter, went through a heavy machine-gun fire, and carried the wounded soldier to the dressing station.

HAYS, FRANK K_____
In the region of Chambley, France, Sept. 13, 1918.
R—Chicago, Ill.
B—Louisville, Ky.
G. O. No. 143, W. D., 1918.

Second lieutenant, pilot, 13th Aero Squadron, Air Service.
He was one of an offensive patrol of five planes attacked by seven enemy scouts (Fokker type) that dived down on them from the clouds, catching the American patrol in a disadvantageous position. In the course of the combat which followed both of his guns were jammed. By an extraordinary effort he cleared his guns and drove off the adversary. He then observed his flight commander in a dangerous situation with two enemy planes behind him. He attacked and destroyed one and forced the other to withdraw.

*HAZLETT, CLARK S. (1391468)---------
  Near Brieulles, France, Oct. 10, 1918.
  R—Wilkinsburg, Pa.
  B—Louisville, Tenn.
  G. O. No. 56, W. D., 1922.

Private, Machine Gun Company, 132d Infantry, 33d Division.
On hearing the call for aid by several men who had been hit by shellfire, Private Hazlett rushed to their aid and, although exposed to heavy artillery and machine-gun fire, he applied first aid to the wounded men. While in the act of placing the wounded on litters, he was killed.
Posthumously awarded. Medal presented to sister, Miss Mary B. Hazlett.

HEACOX, HARRY W. (1978158)---------
  Near Bois-d'Harville, France, Nov. 10, 1918.
  R—Indianapolis, Ind.
  B—Harmony, Ind.
  G. O. No. 71, W. D., 1919.

Private, Company I, 131st Infantry, 33d Division.
After performing several dangerous missions as a company runner, he volunteered to go forward with an officer to attack a machine-gun nest that was causing heavy casualties. Though the officer was killed in the attempt, Private Heacox captured the nest, took command of the company on his own initiative, and carried it forward to its objective.

HEAD, CARROLL E. (1490120)---------
  Near Attigny, France, Oct. 14, 1918.
  R—Fort Worth, Tex.
  B—Fort Worth, Tex.
  G. O. No. 81, W. D., 1919.

Corporal, Company C, 111th Field Signal Battalion, 36th Division.
The enemy having located the advance divisional information center, the personnel was removed to a new position. Corporal Head volunteered and removed the switchboard, changing the wire connections so that communication would not be interrupted. For five or six minutes he worked under intense shellfire, several shells passing through the building as he worked. He succeeded, however, in reestablishing the line connections outside the building. Immediately after he left the building it was completely demolished by a shell.

HEALEY, JEREMIAH (1709040)---------
  Near Charlevaux, France, Oct. 3–7, 1918.
  R—New York, N. Y.
  B—Ireland.
  G. O. No. 35, W. D., 1919.

Sergeant, Company G, 308th Infantry, 77th Division.
Although wounded on the third day of the battle in the Argonne Forest, Sergeant Healey continually exposed himself to machine-gun and artillery fire while aiding and cheering his men. He also volunteered his services in an attempt to break through the enemy lines and bring aid to his organization.

*HEALY, HAROLD A.---------
  Near Le Charmel, France, July 15, 1918.
  R—Norwich, Conn.
  B—Pawtucket, R. I.
  G. O. No. 89, W. D., 1919.

Second lieutenant, 8th Machine Gun Battalion, 3d Division.
After several runners had been unsuccessful in attempts to carry an important message, Lieutenant Healy volunteered and succeeded in delivering the message through intense shellfire, returning with the reply and aiding three wounded infantrymen en route. Later, with another officer, he went through heavy enemy shell and machine-gun fire and carried to shelter two wounded Frenchmen.
Posthumously awarded. Medal presented to father, John J. Healy.

HEALY, JAMES A.---------
  Near Grand Pre, France, Oct. 30, 1918.
  R—Washington, D. C.
  B—Fort Leavenworth, Kans.
  G. O. No. 37, W. D., 1919.

First lieutenant, 147th Aero Squadron, Air Service.
Becoming separated from his patrol, Lieutenant Healy, flying at an altitude of 600 meters, discovered an enemy plane (type Halberstadt) hiding in the sun 200 meters above him, which he attacked and sent to the ground in a steep dive. He then noticed two other machines (type Fokker) which had been attempting to attack him. He succeeded in outmaneuvering them and finally shot down one of the Fokkers. He returned without a drop of gasoline in his tank.

HEARD, ELMER---------
  Near Le Grande Carre Ferme, France, Nov. 1, 1918.
  R—Weleetka, Okla.
  B—Jacksonville, Ark.
  G. O. No. 37, W. D., 1919.

Captain, 360th Infantry, 90th Division.
After all the other officers of his company had been wounded, Captain Heard remained in command, though he had himself been severely wounded and was suffering from illness, courageously leading his men throughout two days and nights of severe fighting.

HEATH, FLOYD E. (49901)---------
  Near St. Etienne-a-Arnes, France, Oct. 4, 1918.
  R—Warren, Ill.
  B—Elizabeth, Ill.
  G. O. No. 35, W. D., 1919.

Corporal, Company C, 23d Infantry, 2d Division.
Anticipating an enemy counterattack, he was ordered to take out a patrol of eight men to scour the woods to the left of his position, drive out the snipers, and establish liaison with the enemy on the left. Fearing that the size of the patrol would attract too much attention, he left the others in the trench, and going out, accomplished the mission alone, returning under heavy machine-gun fire.

HEATH, LAUREL B (3185701)---------
  At Brieulles, France, Nov. 4, 1918.
  R—Windsor, N. Y.
  B—Windsor, N. Y.
  G. O. No. 37, W. D., 1919.

Private, Company C, 2d Antiaircraft Machine Gun Battalion, Air Service.
He went through intense shellfire and assisted Second Lieutenant Samuel F. Telfair in carrying a wounded comrade to safety.

HECHTL, ALBERT LOUIS (1520622)---------
  Near Montfaucon, France, Sept. 26–Oct. 1, 1918.
  R—Canton, Ohio.
  B—Colesville, N. Y.
  G. O. No. 139, W. D., 1918.

Sergeant, Company C, 146th Infantry, 37th Division.
Throughout the five days' offensive he commanded his platoon with rare coolness and was always in the first wave of his company, facing the greatest danger. He personally took charge of a thin line of outposts on the flank and broke up a German counterattack that was forming under the protection of a barrage. On the fourth day of the drive this soldier was severely gassed, but he concealed this fact from his officers until he was exhausted.

HECKMAN, JACOB H.---------
  In the Bois-de-Belleau, France, June 25, 1918.
  R—Haverhill, Mass.
  B—Haverhill, Mass.
  G. O. No. 46, W. D., 1919.

Second lieutenant, 5th Regiment, U. S. Marine Corps, 2d Division.
With the assistance of three sergeants, he started out to destroy the final stand of the enemy in the Bois-de-Belleau, an impregnable position, where enemy guns were concealed by rocks and heavy shrubbery. Armed with only a pistol, he rushed the nest which was offering the most violent resistance and captured 1 officer and 90 men. Each of his men destroyed a nest and captured two of the enemy at each position. After effecting the complete reduction of the last element, he marched his prisoners in under a severe and harassing fire of the retreating enemy.

HEDLUND, FRITZ E. (60879)
In Trugny Woods, France, July 23, 1918.
R—Waltham, Mass.
B—Waltham, Mass.
G. O. No. 37, W. D., 1919.

Private, first class, Company F, 101st Infantry, 26th Division.
Private Hedlund, a runner, maintained liaison between his company commander and an isolated combat group. Passing through two German attacking waves under intense fire, he reached the combat group. On the return trip several Germans attempted to take him prisoner. He shot one of them, bayoneted another, and escaped from the others. He made three more trips to the combat group and kept up constant liaison.

*HEDRICK, ARLY L.
Near Bauluy, France, Sept. 28, 1918.
R—Kansas City, Mo.
B—Robinson, Ark.
G. O. No. 59, W. D., 1919.

Captain, 110th Engineers, 35th Division.
While reconnoitering for mined bridges, he was warned that a certain masonry arch was held under enemy machine-gun fire. He placed his detail under cover, advanced alone under persistent fire, exposing himself continually while removing detonaters from the mines he found, and returned across open ground to his command.
Posthumously awarded. Medal presented to widow, Mrs. Geraldine Hedrick.

*HEFFRON, JOHN J. (550019)
Near Crezancy, France, July 15-18, 1918.
R—Minooka, Pa.
B—Olyphant, Pa.
G. O. No. 35, W. D., 1920.

Private, Company C, 38th Infantry, 3d Division.
Private Heffron repeatedly carried messages to front-line platoons over routes exposed to heavy machine gun and artillery fire. Due to his individual efforts communication was maintained during a critical period.
Posthumously awarded. Medal presented to father, Michael Heffron.

*HEFFRON, WALTER R. (2176373)
Near Limey, France, Sept. 12, 1918.
R—Kingman, Kans.
B—Kingman, Kans.
G. O. No. 49, W. D., 1922.

Sergeant, Company C, 353d Infantry, 89th Division.
Prior to the attack Sergeant Heffron saved his platoon from exposure to enemy fire by going out of our trench in the face of the enemy machine-gun barrage and cutting a path through our wire. He then made his way back to his men and led them through this opening. Though the enemy machine-gun fire was intense, he carefully organized his platoon and had it in perfect formation when the command was given to advance. His example of coolness and bravery steadied his men and gave them confidence at the start of their first offensive. He was killed by a shell as he was leading his men across the first enemy trenches.
Posthumously awarded. Medal presented to mother, Mrs. William Heffron.

HEIKEN, EILERT G.
In the St. Mihiel and Argonne, France, Sept. 23, Nov. 9-10, 1918.
R—Ottawa, Kans.
B—Bushton, Kans.
G. O. No. 37, W. D., 1919.

First lieutenant, 356th Infantry, 89th Division.
On the night of Sept. 23, though wounded in the shoulder at the beginning of the attack on Dommartin Wood, Lieutenant Heiken continued until his mission was successfully accomplished. On the night of Nov. 9, with 8 men he was the first to cross the Meuse and patrol enemy lines. On the night of Nov. 10, with 20 men, he covered the crossing of his battalion until severely wounded.

HEIKKINEN, JOHN (2306006)
Near Courmont and St. Martin, France, July 31-Aug. 3, 1918.
R—Atlantic Mine, Mich.
B—Calumet, Mich.
G. O. No. 21, W. D., 1919.

Private, Company G, 125th Infantry, 32d Division.
Throughout the battle to force passage of the Ourcq River and capture the heights beyond, Private Heikkinen, a stretcher bearer, worked day and night evacuating the wounded under heavy artillery and machine-gun fire. On Aug. 3, under violent shell fire, opposite Mont St. Martin, he made repeated trips between the firing line and the dressing station until wounded.

HEIM, CARL A. (1215648)
East of Roussoy, France, Sept. 28-29, 1918.
R—Elmira, N. Y.
B—Troy, N. Y.
G. O. No. 20, W. D., 1919.

Sergeant, Company L, 108th Infantry, 27th Division.
During the operations against the Hindenburg line Sergeant Heim displayed great gallantry and leadership in reorganizing and assuming command of his company and leading it into effective combat after all the officers had been killed or wounded.

HEIMERDINGER, CHARLES
Near Landres-et-St. Georges, France, Nov. 3, 1918.
R—Chicago, Ill.
B—Chicago, Ill.
G. O. No. 46, W. D., 1919.

Second lieutenant, 23d Infantry, 2d Division.
When machine-gun nests were rendering his position untenable, Lieutenant Heimerdinger led a patrol of 12 men into the enemy's lines, reducing the number of nests and returning with 20 prisoners. During his return his patrol was fired upon and 2 of his men wounded. He then took 2 of his men and kept a fire on the enemy until both his wounded and prisoners could be brought in.

HEINTZ, VICTOR
Near Cierges, France, Sept. 29, 1918.
R—Cincinnati, Ohio.
B—Grayville, Ill.
G. O. No. 37, W. D., 1919, and G. O. No. 15, W. D., 1923.

Captain, 147th Infantry, 37th Division.
Heeding a call for help from a severely wounded soldier, Captain Heintz immediately left his place of shelter and crawled through heavy artillery and machine-gun fire to the aid of the man and carried him to a place of safety.
Oak-leaf cluster.
For extraordinary heroism in action near Cierges, France, Sept. 28, 1918. While serving as operations officer of his regiment, with complete disregard for his own safety and under terrific enemy machine-gun, rifle, sniper, and artillery fire, Captain Heintz ran and crawled several hundred yards to the post of command of the commander of the assault battalion of his regiment after several runners had been killed or wounded in attempting the same object and delivered orders from the regimental commander which served to prevent a successful enemy attack and enabled the battalions in support and reserve to reorganize their forces for defense of the regiment's position. The daring and soldierly devotion to duty displayed by Captain Heintz greatly inspired the officers and men with whom he served.

HEINY, JOHN D.
Near Charpentry, France, Sept. 27, 1918.
R—Kirksville, Mo.
B—Kirksville, Mo.
G. O. No. 126, W. D., 1919.

Captain, 139th Infantry, 35th Division.
In order to secure important information as to the position of hostile batteries, he passed through the enemy's artillery fire and was twice severely wounded. In spite of his wounds he remained on duty throughout the entire engagement and gave great assistance in the reorganization of advance positions.

HEINZ, NICK (2141018)............
Near Vilcey, France, Sept. 12, 1918.
R—Holdingford, Minn.
B—Albany, Minn.
G. O. No. 128, W. D., 1918.

Private, Company M, 358th Infantry, 90th Division.
Assisted by another soldier he outflanked a German machine-gun nest, killed the gunner, and captured the gun. Although painfully wounded in both arms, he stayed with the gun until he was relieved 12 hours later.

HEINZMANN, GROVER P............
Near Verdun, France, Oct. 12, 1918.
R—Passaic, N. J.
B—Jersey City, N. J.
G. O. No. 32, W. D., 1919.

First lieutenant, 114th Infantry, 29 Division.
After having seen several of his men killed or wounded in an attempt to deliver a message through a heavily shelled area, Lieutenant Heinzmann volunteered and carried the dispatch to its objective, then offered to return through the same barrage on any mission required by his battalion commander.

HEINZMANN, LEO (520932)............
Near Romanovka, Siberia, June 25, 1919.
R—Los Angeles, Calif.
B—Garfield, N. Y.
G. O. No. 133, W. D., 1919.

Corporal, Company A, 31st Infantry.
When his platoon was completely surrounded by the enemy he volunteered to carry a message through the enemy's lines to the nearest American troops at Novo-Nezhino, 6 miles away. Corporal Heinzmann succeeded in getting through the enemy's lines, reached the nearest American forces, and returned with reinforcements for his platoon.

HEITZ, HARRY D............
Near Apremont, France, Oct. 1, 1918.
R—Kansas City, Mo.
B—Carrollton, Mo.
G. O. No. 14, W. D., 1923.

Second lieutenant, 327th Battalion, Tank Corps.
With rare coolness, courage, energy, and initiative he led two platoons of tanks against an enemy attack upon the town of Apremont. Two of the tanks being disabled, he continued to advance, causing heavy casualties among the enemy forces. He continued for a considerable distance within the enemy lines and attacked them from the rear, thus rendering highly important help to his own forces. He continued his attack upon the enemy until his own tank was completely disabled. Discovering a wounded sergeant lying exposed to violent enemy machine-gun, rifle, and artillery fire he carried the wounded man to shelter.

*HELIKSON, FRANK (1193768)............
At St. Eugene, France, July 15, 1918.
R—Franklin, Mich.
B—Houghton, Mich.
G. O. No. 32, W. D., 1919.

Private, headquarters detachment, 10th Field Artillery, 3d Division.
Being on duty as a messenger between battalion headquarters and the battery positions when the roads were under heavy bombardment of gas and high-explosive shells, Private Helikson continued to make trips until his foot was shattered. Even after receiving this wound he completed the mission upon which he was then engaged before seeking medical treatment.
Posthumously awarded. Medal presented to father, Dan Helikson.

HELLIWELL, HAROLD H............
Near Medeah Ferme, France, Oct. 8, 1918.
R—St. Paul, Minn.
B—Minneapolis, Minn.
G. O. No. 21, W. D., 1919.

First lieutenant, 9th Infantry, 2d Division.
While acting as battalion adjutant, when the latter was wounded, he made continual reconnaissances under heavy shell and machine-gun fire. He maintained liaison at all times, and carried important messages to the flanks and rear through the enemy barrage. He assisted in organizing the battalion when attacked upon the flank in successful resistance to determined counterattack by superior forces.

*HELLMAN, CHARLES F. (1901990)............
Near St. Juvin, France, Oct. 11, 1918.
R—Somerville, Mass.
B—Cambridge, Mass.
G. O. No. 78, W. D., 1919.

Private, first class, Company G, 326th Infantry, 82d Division.
Engaged as company runner, Private Hellman was given a message for delivery to the forward platoon. Through withering machine-gun and deadly artillery fire he made his way, but fell wounded after going about 25 yards. He dragged himself forward and delivered his message, after which he started back. On the return journey he was hit the second time and killed.
Posthumously awarded. Medal presented to father, Adolph E. Hellman.

HELM, LYNN, Jr............
Near Laneuville, France, Nov. 7, 1918.
R—Los Angeles, Calif.
B—Chicago, Ill.
G. O. No. 44, W. D., 1919.

First lieutenant, 11th Field Artillery, 6th Division.
Acting as battalion telephone officer, Lieutenant Helm repeatedly went out himself under heavy fire rather than assign such hazardous missions to members of his detail, personally directing the repair and maintenance of a line which was severed 15 times in two hours within a length of only 1 kilometer.

HELMICK, DAN S............
Near Bois-de-Moncy, France, Oct. 9, 1918.
B—Minneapolis, Minn.
B—Minneapolis, Minn.
G. O. No. 44, W. D., 1919.

Captain, 1st Engineers, 1st Division.
Captain Helmick personally organized his company for the attack and directed the consolidation of the captured objective. He established his post of command in the front line, and repeatedly passed up and down the line directing his men under heavy shell fire until severely wounded.

HELMORE, JOSEPH T. (683290)............
Northeast of Chateau-Thierry, France, July 31, 1918.
R—Elsie, Mich.
B—Corunna, Mich.
G. O. No. 87, W. D., 1919.

Private, Company M, 125th Infantry, 32d Division.
Through heavy machine-gun fire and artillery barrage fire he, with another man, dragged a wounded comrade from within 100 feet of the enemy lines to his own lines, a distance of 150 yards.

HELSLEY, ALBERT B............
Near Fieville, France, Oct. 9, 1918.
R—Bowling Green, Ky.
B—Central City, Ky.
G. O. No. 4, W. D., 1923.

First lieutenant, 16th Infantry, 1st Division.
During the attack of Hill 272, he courageously led, under a heavy machine-gun fire, a section of his machine-gun platoon to a position from which they could engage a machine-gun nest about 400 yards distant that was causing heavy casualties, and although severely wounded, he nevertheless remained in command and directed the advance which resulted in taking the enemy strong point and relieving an exposed flank from a heavy fire.

*HEMINGWAY, HAROLD L.
In the vicinity of St. Remy, France, Sept. 12, 1918.
R—New Haven, Conn.
B—New Haven, Conn.
G. O. No. 19, W. D., 1921.

First lieutenant, 104th Infantry, 26th Division.
In advancing with his company under heavy shell and machine-gun fire he boldly exposed himself, and his personal conduct was a splendid example of fearlessness to all his men. His company having been held up by enemy barbed-wire entanglements, and 2 enlisted men having been wounded at the enemy's wire, Lieutenant Hemingway exposed himself to machine-gun fire to advance in front of the front line in order to rescue the two wounded members of his company.
Posthumously awarded. Medal presented to father, J. S. Hemingway.

HEMPE, JOSEPH C. (1646636).
Service rendered under the name, Joseph C. Vercruysse.

Private, Company H, 306th Infantry, 77th Division.

HENDERSHOT, FORD J. (542626).
Near Fossoy, France, July 15, 1918.
R—Bad Axe, Mich.
B—Verona, Mich.
G. O. No. 44, W. D., 1919.

Corporal, Company K, 7th Infantry, 3d Division.
After all his men had been wounded by the intense artillery shelling of the enemy during the offensive of July 15, he went to their aid, applying first aid and guiding litter bearers to the dressing station. He went out again and brought a wounded comrade in on his back under constant heavy fire.

HENDERSON, CHARLES R. (1211852).
Near Dickebusch, France, Aug. 22, 1918.
R—White Plains, N. Y.
B—Westchester, N. Y.
G. O. No. 99, W. D., 1918.

Corporal, Company L, 107th Infantry, 27th Division.
When his post was attacked by a greatly superior number of the enemy, he heroically defended it in spite of the loss of 6 of his squad and succeeded in driving off the enemy.

HENDERSON, GEORGE H. (1472613).
Near Charpentry, France, Sept. 30, 1918.
R—Newark, N. J.
B—Pittsburgh, Pa.
G. O. No. 70, W. D., 1919.

Sergeant, 140th Ambulance Company, 110th Sanitary Train, 35th Division.
While leading litter bearers he came under intense shell fire. Seeing a wounded man lying in an exposed position, Sergeant Henderson, accompanied by another soldier, left shelter to go to the wounded man's assistance. While they were proceeding under terrific fire, his companion was killed and he himself severely wounded by a bursting shell. Despite his wound, Sergeant Henderson continued on his mission, administered first aid to the wounded man, and carried him back to the shelter of a shell hole.

HENDERSON, HENRY.
Near Remonville, France, Nov. 1, 1918.
R—Council Bluffs, Iowa.
B—Topeka, Kans.
G. O. No. 26, W. D., 1919.

Second lieutenant, 354th Infantry, 89th Division.
When his company was fired upon by a battery of German 77's not more than 300 yards distant, Lieutenant Henderson led his platoon at a run through 2 machine-gun nests, which were defending that flank of the battery, and succeeded in capturing the entire battery with the aid of the bayonet.

HENDERSON, PHIL A.
In the Toul sector, France, Aug. 28, 1918.
R—Seattle, Wash.
B—Chehalis, Wash.
G. O. No. 20, W. D., 1919.

First lieutenant, Infantry, observer, 12th Aero Squadron, Air Service.
While on an unprotected reconnaissance mission with Lieut. Edward Orr, pilot, he encountered a patrol of 8 enemy pursuit planes near the American balloon lines. When Lieutenant Orr attacked the planes, which had dived at the American balloon, Lieutenant Henderson engaged the other 8 enemy machines, which were attacking from the rear. In the violent battle which followed all nine of the enemy were driven off.

HENDLER, ULLMAN C.
Near Courmont, France, July 30, 1918.
R—Philadelphia, Pa.
B—Philadelphia, Pa.
G. O. No. 98, W. D., 1919.

First lieutenant, 110th Infantry, 28th Division.
Severely wounded in the shoulder early in the attack on the Bois-de-Grimpettes, Lieutenant Hendler, though he was suffering intense pain, refused to be evacuated, but continued to lead his company forward until the entire woods had been captured and the new position consolidated.

HENDRICKS, PATRICK (1703437).
Near Badonviller, France, June 24, 1918.
R—New York, N. Y.
B—New York, N. Y.
G. O. No. 46, W. D., 1919.

Private, Company C, 308th Infantry, 77th Division.
After being wounded he continued to work his automatic rifle until it was destroyed. He then secured a rifle and continued to fight, and later assisted other wounded before having his own wound dressed.

HENDRICKS, TERRY NICHOLS.
Near Mont Blanc, France, Oct. 4, 1918.
R—Valdosta, Ga.
B—Tifton, Ga.
G. O. No. 37, W. D., 1919.

First lieutenant, 12th Field Artillery, 2d Division.
During the course of a terrific bombardment by the enemy, which forced the firing battery to take shelter, and when 2 of the gun crew were seriously wounded, Lieutenant Hendricks voluntarily left cover and without regard for his own safety carried the wounded men through heavy fire to a place of safety.

HENLEY, COURTNEY S.
North of the Sommerance-St. Juvin Road, France, Oct. 11, 1918.
R—Birmingham, Ala.
B—Birmingham, Ala.
G. O. No. 105, W. D., 1919.

Captain, 327th Infantry, 82d Division.
Captain Henley led a party of 3 enlisted men in an attack on an enemy machine-gun position which was doing considerable damage to our forces. Under intense hostile fire his attack drove the enemy gunners from the machine-gun nest.

HENNELLY, JAMES H.
Near St. Eugene, France, July 14-15, 1918.
R—Norfolk, Va.
B—Norfolk, Va.
G. O. No. 59, W. D., 1919.

Second lieutenant, 10th Field Artillery, 3d Division.
On duty with the Infantry as liaison officer, he was captured by a part of the enemy while taking a message back to the artillery. Shortly afterwards he succeeded in making his escape, and, although he had been twice wounded, he made his way through a heavy barrage and delivered his message.

*HENNESSEY, JAMES (1750394)............
At Grand Pre, France, Oct. 22, 1918.
R—Newark, N. J.
B—Newark, N. J.
G. O. No. 64, W. D., 1919.

Sergeant, Company B, 312th Infantry, 78th Division.
With all disregard for his own danger, he crawled about 30 yards through a sweeping machine-gun fire to the rescue of a wounded comrade. He was killed by a machine-gun bullet while administering aid to the wounded soldier.
Posthumously awarded. Medal presented to mother, Mrs. Matty Hennessey.

HENNESSY, PATRICK (1025038)............
Near Rembercourt, France, Nov. 1-2, 1918.
R—Newark, N. J.
B—Ireland.
G. O. No. 35, W. D., 1919.

Sergeant, Company B, 34th Infantry, 7th Division.
While in command of a platoon he was wounded, but he refused to be evacuated, and remained with his men for 30 hours under a heavy bombardment of gas and high-explosive shells, looking after them and administering first aid to the wounded.

HENRIKSEN, HANS (107336)............
Near Letanne, France, Nov. 6, 1918.
R—Chicago, Ill.
B—Denmark.
G. O. No. 26, W. D., 1919.

Private, Company A, 5th Machine Gun Battalion, 2d Division.
He went out from a place of safety, through a heavy shell fire, and helped to carry a wounded comrade to safety. Without orders, he made a second trip for a distance of 500 yards through machine-gun and artillery fire to bring in the tripod of his gun squad.

*HENRY, CLIFFORD WEST (52421)......
During the St. Mihiel offensive, Sept. 14, 1918.
R—New York, N. Y.
B—New York, N. Y.
G. O. No. 28, W. D., 1921.

Second lieutenant, 102d Infantry, 26th Division.
During the St. Mihiel offensive, although mortally wounded and suffering great pain, he gave information for the disposition of his men. He refused first aid until other wounded men had been taken care of.
Posthumously awarded. Medal presented to father, Ira Walton Henry.

HENRY, DANIEL C. (545592)............
Near Chateau-Thierry, France, July 15, 1918.
R—Glen, Pa.
B—Fairmount City, Pa.
G. O. No. 16, W. D., 1920.

Private, Company B, 30th Infantry, 3d Division.
After many other runners had been killed or wounded, Private Henry carried several messages from platoon to company headquarters. At this time the enemy had crossed the Marne River, and the route followed by Private Henry was exposed to heavy artillery and machine-gun fire.

HENRY, NORMAN (1401525)............
Near Ferme de la Riviere, France, Sept. 30, 1918.
R—Chicago, Ill.
B—West New York, N. J.
G. O. No. 37, W. D., 1919.

Sergeant, Machine Gun Company, 370th Infantry, 93d Division.
While leading his squad under heavy fire, he displayed great bravery and coolness by repairing a damaged gun under direct enemy observation. He proved of valuable assistance to the one remaining officer of his company in assembling the scattered units of his company after they had reached their objective.

HENRY, RAY (262192)..................
Near Cierges, south of Fismes, France, July 31, 1918.
R—Flint, Mich.
B—Flint, Mich.
G. O. No. 20, W. D., 1919.

Corporal, Company E, 125th Infantry, 32d Division.
In the advance up the hill of Les Jumbles Woods, in the face of a heavy machine-gun and artillery barrage, he was wounded in the shoulder, but he nevertheless continued to advance. Being unable to use his rifle, he threw it down, drew his pistol, and led his platoon forward until he received 6 more bullet wounds.

HENRY, THOMAS...................
Near Soissons, France, July 18, 1918.
R—New York, N. Y.
B—Ireland.
G. O. No. 68, W. D., 1920.

First lieutenant, 28th Infantry, 1st Division.
Lieutenant Henry gallantly led his platoon forward through heavy machine-gun fire in an attack on a strongly held enemy position. Due to his initiative and gallantry, 2 enemy machine guns were captured and 25 of the enemy forced to surrender. Later, when wounded, he refused to be evacuated until, through loss of blood, he could go no farther.

HENSLEY, HENRY G. (1098784)..........
Near Preny, France, Nov. 1, 1918.
R—Vixen, N. C.
B—Pensacola, N. C.
G. O. No. 32, W. D., 1919.

Sergeant, Company H, 56th Infantry, 7th Division.
When the position of his platoon became untenable on account of machine-gun fire from a nest in their front, Sergeant Hensley attacked the nest alone and succeeded in driving off the enemy with hand grenades.

*HENSLEY, THOMAS L. (2339279)........
Near Roncheres, France, July 29, 1918.
R—Humble, Tex.
B—Lee County, Tex.
G. O. No. 32, W. D., 1919.

Private, first class, Company H, 4th Infantry, 3d Division.
Going from one wounded comrade to another, exposed to heavy machine-gun and deadly sniper fire, he administered aid to all that he could reach before being killed in the performance of this heroic mission.
Posthumously awarded. Medal presented to mother, Mrs. Mary Hensley.

HEPBURN, WILLIAM...................
Near Verdun, France, Oct. 24, 1918.
R—Hartford, Conn.
B—Hartford, Conn.
G. O. No. 37, W. D., 1919.

First lieutenant, 102d Infantry, 26th Division.
With only a handful of men, Lieutenant Hepburn charged a machine-gun nest, killing or wounding the whole crew. He silenced the fire of 5 different guns, and, in taking the last, he was severely wounded about the head. He continued, however, until his mission was accomplished and the gun silenced.

HERBERT, THOMAS J...................

Near Chaulnes, France, Aug. 1-8, 1918.
R—Cleveland, Ohio.
B—Cleveland, Ohio.
G. O. No. 9, W. D., 1923.

First lieutenant, Air Service, U. S. Army, attached to 56th Squadron, Royal Air Forces, British Expeditionary Forces.
For extraordinary heroism in action near Chaulnes, France, Aug. 1-8, 1918, when with a formation of 6 machines he attacked 18 enemy Fokker biplanes shooting one down in flames. During the fighting Lieutenant Herbert was shot in the leg and his plane was struck in the petrol tank, necessitating skillful maneuvering to regain his own lines. As he was landing he became unconscious from loss of blood. On Aug. 4, 1918, at Cappy, France, he destroyed an enemy Pfalz scout plane at an altitude of 9,000 feet, thus saving his patrol leader, who was at the mercy of the enemy plane. On Aug. 1, he joined in the attack of the enemy aerodrome at Epinoy, the altitude at time of attack being but 200 feet; he killed two enemy mechanics by machine-gun fire and shot up hangars and billets. The bravery, skill, and determination of this officer were an inspiration to other members of his squadron.

HERITIER, ARTHUR (263210)_____
On Hill 212 near Cierges, northeast, of Chateau-Thierry, France, July 31, 1918.
R—Linwood, Mich.
B—Linwood, Mich.
G. O. No. 117, W. D., 1918.

Private, first class, Company I, 125th Infantry, 32d Division.
After one man had been killed and two others wounded in attempting to render first aid to a lieutenant who lay mortally wounded in an area that was subjected to fire from enemy machine guns and snipers, he successfully reached the lieutenant and gave him first aid, risking his own life in doing so.

HERMAN, ALBERT W. (64357)_____
Near Epieds, France, July 21, 1918.
R—New Haven, Conn.
B—Germany.
G. O. No. 26, W. D., 1919.

Sergeant, Company D, 102d Infantry, 26th Division.
By the effective use of his bayonet, Sergeant Herman killed many of the enemy, and although painfully wounded, he continued to fight until ordered to the rear by his commanding officer.

HERMLE, LEO D_____
Near the Meuse River, France, Nov. 1, 1918.
R—Oakland, Calif.
B—Hastings, Nebr.
G. O. No. 35, W. D., 1919.

First lieutenant, 6th Regiment, U. S. Marine Corps, 2d Division.
When the company on his left was checked by heavy machine-gun fire, Lieutenant Hermle led a platoon forward and surrounded a large number of the enemy, capturing 155 prisoners and 17 machine guns. Pushing on, he took the town of St. Georges and many machine-gun positions. Although he was painfully wounded, he refused to be evacuated and remained with his men for 2 days until he was ordered to the rear.

HERR, CHARLES RYMAN_____
In the Bois-des-Ogons, France, Oct. 4-6, 1918.
R—Flemington, N. J.
B—Flemington, N. J.
G. O. No. 7, W. D., 1919.

First lieutenant, 319th Infantry, 80th Division.
Suffering from the effects of mustard gas, he refused to leave his platoon, and later, when his company commander was killed, took command of the company. Under the inspiration of his personal bravery his command overcame the most determined resistance and succeeded in getting a foothold in the Bois-des-Ogons while it was under flanking fire from machine guns and artillery. He personally visited his outposts under a heavy artillery and machine-gun barrage, inspiring confidence, which enabled his men to maintain their position at a critical time.

HERREN, WILLIAM (559453)_____
Near Ville-Savoye, France, Aug. 7, 1918.
R—West Asheville, N. C.
B—Buncombe County, N. C.
G. O. No. 64, W. D., 1919.

First sergeant, Machine Gun Company, 58th Infantry, 4th Division.
He carried guns and ammunition to the front-line platoons through an intense barrage after several carrying details had failed to get through. He then volunteered to stay with the right-flank platoon, which was under heavy fire in an exposed position. During the afternoon he and one other man pushed forward with a captured machine gun and assisted materially in breaking up several hostile counterattacks during the day.

HERRICK, LESTER (1213991)_____
East of Ronssoy, France, Sept. 29, 1918.
R—Rockwell Springs, N. Y.
B—Bradford, Pa.
G. O. No. 20, W. D., 1919.

Private, Company C, 108th Infantry, 27th Division.
During the operations against the Hindenburg line he exhibited great courage and bravery by advancing alone against two enemy machine-gun positions, putting both of them out of action. In accomplishing this he was seriously wounded.

*HERRMANN, BERTRAND (1750886)_____
Near Grand Pre, France, Oct. 23, 1918.
R—Newark, N. J.
B—Newark, N. J.
G. O. No. 37, W. D., 1919.

Private, first class, Company D, 312th Infantry, 78th Division.
Private Herrmann, with his twin brother, Pvt. (First Class) Victor Herrmann, voluntarily crossed an open field heavily swept by machine-gun fire and assisted First Sergt. James P. Collins to carry to shelter their company commander, who had been mortally wounded. As they were returning to our lines, this soldier was slightly wounded, and later in the day while being evacuated to the rear he was killed by a shell.
Posthumously awarded. Medal presented to mother, Mrs. Anna Herrmann.

HERRMANN, VICTOR (1750888)_____
Near Grand Pre, France, Oct. 23, 1918.
R—Newark, N. J.
B—Newark, N. J.
G. O. No. 35, W. D., 1919.

Private, first class, Company D, 312th Infantry, 78th Division.
Private Herrmann, with his twin brother, Pvt. (First Class) Bertrand Herrmann, voluntarily crossed an open field heavily swept by machine-gun fire and assisted First Sergt. James P. Collins to carry to shelter their company commander, who had been mortally wounded.

HERSCHKOWITZ, JACK (1708138)_____
Near Binarville, France, Sept. 29, 1918.
R—New York, N. Y.
B—Rumania.
G. O. No. 13, W. D., 1919.

Private, first class, Company C, 308th Infantry, 77th Division.
In order to obtain ammunition and rations, Private Herschkowitz, with another soldier, accompanied an officer in an effort to reestablish communication between battalion and regimental headquarters. They were attacked by a small party of Germans, but drove them off, killing one. When night came they crawled unknowingly into the center of a German camp, where they lay for three hours undetected. Finally discovered, they made a dash to escape. In order to protect the officer, Private Herschkowitz deliberately drew the enemy fire to himself, allowing the officer to escape. Private Herschkowitz succeeded in getting through and delivering his message the next morning.

HERTER, EDWARD (1388652)_____
Near Bois-d' Harville, France, Nov. 10, 1918.
R—Chicago, Ill.
B—Russia.
G. O. No. 71, W. D., 1919.

Private, Company M, 131st Infantry, 33d Division.
On his own initiative he left shelter and crossed open ground swept by machine-gun fire to rescue a wounded comrade. Though himself severely wounded on the mission, he succeeded in carrying the soldier back to his own lines.

*HESS, HERMAN L_____
Near Cierges, France, Sept. 29, 1918.
R—Logan, Ohio.
B—Pomeroy, Ohio.
G. O. No. 50, W. D., 1919.

First lieutenant, 148th Infantry, 37th Division.
Accompanied by a soldier, he made two trips through heavy machine-gun fire and rescued two wounded men who had been left lying in an exposed place when the battalion took up a new position.
Posthumously awarded. Medal presented to father, Frank Hess.

HESTER, HARVEY S.
Near Vaux-Audigny, France, Oct. 10, 1918.
R—Asheville, N. C.
B—Winston-Salem, N. C.
G. O. No. 35, W. D., 1919.

First lieutenant, 120th Infantry, 30th Division.
Although severely wounded in the back by shrapnel, he led his platoon forward, covering a flank of his battalion, which was exposed to heavy enemy fire.

HEUEISEN, FRANK A. (1147301).
Near Montrebeau Woods, France, Sept. 29, 1918.
R—Great Bend, Kans.
B—Stafford County, Kans.
G. O. No. 20, W. D., 1919.

Sergeant, Company A, 137th Infantry, 35th Division.
When his company was checked and forced to withdraw into the woods, Sergeant Heueisen voluntarily went into an open field under heavy machine-gun fire and carried wounded soldiers a hundred yards to shelter, preventing their capture by the enemy.

*HEWIT, BENJAMIN H.
Near Montfaucon, France, Sept. 28–29, 1918.
R—Hollidaysburg, Pa.
B—Jamestown, N. Dak.
G. O. No. 66, W. D., 1919.

Captain, 316th Infantry, 79th Division.
He led his men into battle with such fearlessness and valor that he was at all times able to reorganize and continue forward under most difficult circumstances. Although wounded, he remained in command, always being under terrific shell and machine-gun fire, but not until he had received a second wound did he relinquish his command. While being taken from the field he received a third wound.
Posthumously awarded. Medal presented to father, Oliver H. Hewit.

*HEWITT, CHARLES W., Jr.
At Chateau-Thierry, France, June 6, 1918.
R—Philadelphia, Pa.
B—Camden, N. J.
G. O. No. 110, W. D., 1918.

Corporal, 45th Company, 5th Regiment, U. S. Marine Corps, 2d Division.
Killed in action at Chateau-Thierry, France, June 6, 1918, he gave the supreme proof of that extraordinary heroism which will serve as an example to hitherto untried troops.
Posthumously awarded. Medal presented to father, Charles W. Hewitt, sr.

HEYDENBERK, DICK (106361).
Near Ploisy, France, July 19, 1918.
R—Wayland, Mich.
B—Hopkins, Mich.
G. O. No. 126, W. D., 1918.

Private, first class, Company D, 3d Machine Gun Battalion, 1st Division.
When he was wounded near Ploisy, France, July 19, 1918, he declined medical attention until he led his platoon through steady bombardment to its final objective and had seen to the proper placing of his guns. Although weak from loss of blood he stayed at his post and effectively directed the fire of his command until wounded four times.

HICKEY, THOMAS J. (136703).
Near Samogneux, France, Oct. 24, 1918.
R—Providence, R. I.
B—Rumford, R. I.
G. O. No. 21, W. D., 1919.

Private, Battery C, 103d Field Artillery, 26th Division.
After his piece had received two direct hits and he was the only man left uninjured in his squad, he refused to seek shelter and assisted in the evacuation of the wounded. During a terrific shelling he made two trips to the aid station with a litter.

HICKOK, CHARLES H. (1375262).
Near Verdun, France, Nov. 1, 1918.
R—Poughkeepsie, N. Y.
B—Poughkeepsie, N. Y.
G. O. No. 44, W. D., 1919.

Sergeant, headquarters detachment, 122d Field Artillery, 33d Division.
Commanding an Artillery liaison detail, Sergeant Hickok succeeded, after many attempts, in laying a telephone line through a heavy enemy barrage and opening up communication between Infantry and Artillery. Just as he reached a point where his line was connected with the Infantry he was severely wounded.

HICKS, ALBERT (554577).
Service rendered under the name, Frank McBride.

First sergeant, Company A, 9th Machine Gun Battalion, 3d Division.

HICKS, CHARLES L. (731463).
Near Bois-de-Chatillon, France, Nov. 5, 1918.
R—Dawson Springs, Ky.
B—Dawson Springs, Ky.
G. O. No. 44, W. D., 1919.

Sergeant, Company A, 6th Infantry, 5th Division.
Accompanied by another soldier, Sergeant Hicks advanced against an enemy machine-gun unit, which was holding up the advance of his platoon and the company on his left. He completed the capturing of the position, killing three of the enemy and routing the remaining part of the unit, thereby enabling the platoon to advance to their objective.

HIGGINS, MARTIN J.
Near Villers-sur-Fere, France, July 30, 1918.
R—New York, N. Y.
B—New York, N. Y.
G. O. No. 99, W. D., 1918.

Private, first class, Company H, 165th Infantry, 42d Division.
He twice left shelter and went out into exposed places in front of the firing line, under heavy enemy machine-gun fire, and carried in wounded comrades.

HIGGINS, PATRICK P. (914354).
Near Cunel, France, Oct. 20, 1918.
R—Denver, Colo.
B—Denver, Colo.
G. O. No. 26, W. D., 1919.

Sergeant, Company B, 7th Engineers, 5th Division.
While making a reconnaissance within the enemy lines this soldier assisted in capturing an enemy machine-gun nest which was harassing the party. Under heavy machine-gun fire he gave first-aid treatment to a wounded soldier, and when the officer in charge of the detachment was wounded he bound up the latter's injuries and carried him 2 kilometers through heavy machine-gun and shell fire to an aid station.

HIGGINS, WESTRA (60099).
North of Verdun, France, Oct. 23, 1918.
R—Charlestown, Mass.
B—Boston, Mass.
G. O. No. 35, W. D., 1920.

Sergeant, Company C, 101st Infantry, 26th Division.
Sergeant Higgins alone attacked 2 machine-gun nests which were holding up the advance of his platoon, killing 4 of the enemy and capturing 2.

HIGGINS, WILLIAM H. (1735641).
Near Bois-de-Loges, France, Oct. 16, 1918.
R—Lockport, N. Y.
B—Lockport, N. Y.
G. O. No. 71, W. D., 1919.

Sergeant, Company I, 309th Infantry, 78th Division.
Although painfully wounded early in an attack, he refused to be evacuated, and continued to lead his platoon in the advance, displaying marked personal bravery.

*HIGGINSON, WILLIAM PAUL_____
At Chateau-Thierry, France, June 6, 1918.
R—Rochester, N. Y.
B—Philadelphia, Pa.
G. O. No. 110, W. D., 1918.

First sergeant, 20th Company, 5th Regiment, U. S. Marine Corps, 2d Division.
Killed in action at Chateau-Thierry, France, June 6, 1918, he gave the supreme proof of that extraordinary heroism which will serve as an example to hitherto untried troops.
Posthumously awarded. Medal presented to mother, Mrs. Bridget Higginson.

HIGGS, HERMAN C. (280029)_____
Near Montfaucon, France, Sept. 26, 1918.
R—Lafayette, Ind.
B—Colt, Ark.
G. O. No. 147, W. D., 1918.

Private, medical detachment, 1st Gas Regiment.
He worked continuously and heroically under withering fire from machine guns, upon several occasions voluntarily going out ahead of the first wave to administer first aid to wounded soldiers. His untiring efforts and personal bravery saved the lives of many wounded soldiers and were a source of inspiration to the combat troops.

HIGGS, JAMES ALLEN, Jr_____
Near Pont-a-Mousson, France, July 31, and Aug. 21, 1918, and near Gesnes, France, Oct. 29, 1918.
R—Raleigh, N. C.
B—Raleigh, N. C.
G. O. No. 126, W. D., 1919.

First lieutenant, 3d Balloon Squadron, Air Service.
On July 31, near Pont-a-Mousson, he was carrying on a general surveillance of his sector from his balloon with a French soldier when an enemy plane dived from a cloud and opened fire on the balloon. In imminent danger, he remained in the basket until he had helped his French comrade, after whom he himself jumped. On Aug. 21, in the same sector, he was performing an important mission, regulating artillery fire. Enemy planes attacked, and with great gallantry he remained in the basket until his assistant had jumped. On Oct. 29, near Gesnes, he was conducting a reglage from the basket with a student observer. Attacked by enemy planes, after his balloon was burning he would not quit his post until he had assisted his companion to escape. In each of the foregoing instances Lieutenant Higgs at once reascended in a new balloon.

HIGHLEY, CHARLES C._____
Near Imecourt, France, Nov. 1, 1918.
R—Conshohocken, Pa.
B—Conshohocken, Pa.
G. O. No. 81, W. D., 1919.

First lieutenant, 319th Infantry, 80th Division.
In the face of heavy machine-gun fire, Lieutenant Highley personally led his platoon, or elements thereof, against several enemy machine-gun nests, putting out of action and capturing 15 guns with 50 prisoners. Later in the day he led a squad of men in an attack on a battery of enemy field pieces seen coming out of a wood south of Sivry and succeeded in capturing the 3 pieces of artillery, together with 55 prisoners and 27 horses.

HIGHLEY, ELMER W. (915598)_____
Near Nantillois, France, Oct. 13, 1918.
R—Sturgis, S. Dak.
B—Edgemont, S. Dak.
G. O. No. 37, W. D., 1919.

Sergeant, medical detachment, 7th Engineers, 5th Division.
Although wounded, he remained on duty after his battalion had passed on and the other troops in the vicinity had been forced into dugouts, giving first aid and directing the evacuation of his wounded comrades on a heavily shelled road.

HIGLEY, GEORGE R._____
Near Consenvoye, France, Oct. 8, 1918.
R—Chicago Heights, Ill.
B—Rockford, Ill.
G. O. No. 46, W. D., 1919.

First lieutenant, 124th Machine Gun Battalion, 33d Division.
Upon their own initiative Lieutenant Higley, who was on duty as reconnaissance officer, and another officer crossed the Meuse River to reconnoiter a supply route. They were fired on by 2 enemy machine guns, but, disregarding the heavy machine-gun and shell fire, they advanced on the enemy positions and captured the 2 machine guns, together with 31 Austrian prisoners.

*HIGLEY, JAMES S._____
Near Very, France, Sept. 27, 1918.
R—Phoenix, Ariz.
B—Prescott, Ariz.
G. O. No. 39, W. D., 1920.

First lieutenant, 364th Infantry, 91st Division.
Lieutenant Higley fearlessly led his platoon, under heavy enemy fire, from Mont des Allieux through the enemy position in the Bois-de-Cheppy. The following day he again valiantly led his platoon against enemy machine-gun positions near Very and continued leading the attack until killed by enemy fire.
Posthumously awarded. Medal presented to mother, Mrs. Stephen W. Higley.

HIGSON, WILLIE (1319800)_____
Near Bellicourt, France, Sept. 29, 1918.
R—Henderson, N. C.
B—Washington, D. C.
G. O. No. 37, W. D., 1919.

Corporal, Company C, 120th Infantry, 30th Division.
He showed extraordinary heroism and courage in leading men under heavy shrapnel and enfilading machine-gun fire during the attack on the Hindenburg line. During a temporary halt he acted as runner through this fire and attempted to return after being severely wounded.

HILBURN, HERBERT S._____
Near Villers-devant-Dun, France, Nov. 2, 1918.
R—Plainview, Tex.
B—Fleming, Tex.
G. O. No. 46, W. D., 1919.

Captain, 359th Infantry, 90th Division.
Under heavy machine-gun fire, he repeatedly went to the rear of his company to rally and reorganize it, and then rushed forward to lead his men on. Having taken the town of Villers-devant-Dun and the crest beyond, he held it with only 16 men until the next morning against superior numbers of the enemy.

HILDEBRAND, HARRY (106567)_____
Near Soissons, France, July 18-24, 1918.
R—Butte, Mont.
B—Huntsville, Ala.
G. O. No. 44, W. D., 1919.

Sergeant, Company C, 3d Machine Gun Battalion, 1st Division.
He went forward beyond the front line, exposed to fire of snipers, and repaired and put into action an abandoned enemy machine gun. Later, his platoon commander being wounded and the platoon disorganized by direct artillery fire, he took command, reorganized the platoon, gathered reinforcements and protected a dangerously exposed flank of the infantry. He also voluntarily led his machine guns with the attacking battalion, rendering most efficient service until wounded.

HILDRETH, RICHARD P._____
Near Landres-et-St. Georges, France, Nov. 3, 1918.
R—Chicago, Ill.
B—Wheeling, W. Va.
G. O. No. 46, W. D., 1919.

Captain, 23d Infantry, 2d Division.
During offensive operations west of the Meuse, it was found necessary to withdraw his command to a less exposed position. When he had successfully accomplished this withdrawal, Captain Hildreth saw three of his men lying wounded 100 yards in front of his lines. Assisted by another officer, he crawled out, under heavy machine-gun and shell fire, and safely conducted the wounded to shelter.

*HILGER, JOHN
   Near Soissons, France, July 19, 1918.
   R—Kansas City, Mo.
   B—Germany.
   G. O. No. 132, W. D., 1918.

Sergeant, Company M, 26th Infantry, 1st Division.
In spite of 2 wounds received on July 19, 1918, near Soissons, France, he led his platoon against a machine-gun nest and flanked and captured it, but as this courageous and useful deed was accomplished he was struck by a machine-gun bullet and killed.
Posthumously awarded. Medal presented to sister, Mrs. John Speicher.

HILL, FRANK Y.
   Near Geneve, France, Oct. 8, 1918.
   R—Sparta, Tenn.
   B—Sparta, Tenn.
   G. O. No. 50, W. D., 1919.

First lieutenant, 117th Infantry, 30th Division.
He was wounded by shellfire while he was leading his men forward, but he nevertheless continued in the attack until his men were stopped by intense machine-gun fire and forced to take cover in shell holes. Voluntarily leaving shelter, he thereupon started back to obtain reinforcements, and after going only a short distance was seriously wounded in three places by machine-gun fire.

HILL, FRED WILLIAM
   In the Bois-de-Belleau, France, June 8, 1918.
   R—Chicago, Ill.
   B—Chicago, Ill.
   G. O. No. 119, W. D., 1918.

Corporal, Headquarters Company, 6th Regiment, U. S. Marine Corps, 2d Division.
Regardless of personal danger, he showed conspicuous bravery in carrying ammunition from the dump into the fighting line in the face of heavy machine-gun and rifle fire in the Bois-de-Belleau on June 8, 1918. Learning of the need of hand grenades, he carried them forward without waiting for orders.

HILL, GLEN (541922)
   North of Rambucourt, France, Apr. 12–13, 1918.
   R—Cedar Springs, Mich.
   B—Cedar Springs, Mich.
   G. O. No. 107, W. D., 1918.

Private, Company E, 104th Infantry, 26th Division.
Having recently been assigned to the regiment and hearing order for counterattack being given in an adjoining detachment he joined the latter and himself killed 2 of the enemy with his bayonet.

HILL, GUY H.
   Near Soissons, France, July 18–20, 1918.
   R—Plattsburg, N. Y.
   B—Norfolk, N. Y.
   G. O. No. 117, W. D., 1918.

Private, Company E, 28th Infantry, 1st Division.
When the advance was held up by an enemy machine gun, he rushed it single handed, put the gun out of commission, and took the crew prisoners.

*HILL, HENRY R.
   Near Romagne-sous-Montfaucon, France, Oct. 16, 1918.
   R—Quincy, Ill.
   B—Quincy, Ill.
   G. O. No. 35, W. D., 1919.

Major, 128th Infantry, 32d Division.
With absolute disregard for his personal safety, he led his battalion over the top personally, reached the objective, and cleaned out enemy machine-gun nests. When a group of enemy machine gunners were about to open fire on his flank, Major Hill noticed them, and, armed only with a captured pistol, he immediately went forward to engage them. Taken by surprise, 3 of the crew surrendered, but 1, remaining in the pit, turned the machine gun on him, and as Major Hill's pistol failed to work, he was instantly killed by the machine-gun fire.
Posthumously awarded. Medal presented to mother, Mrs. Cecelia R. Hill.

HILL, LLOYD G.
   At Tronsel Farm, France, Sept. 29, 1918.
   R—Spokane, Wash.
   B—Spokane, Wash.
   G. O. No. 44, W. D., 1919.

First lieutenant, 363d Infantry, 91st Division.
Receiving no response to a call for volunteers, he personally went forward under machine-gun fire and rescued a wounded member of his detachment, bringing him in and transferring him to a first-aid station.

HILL, MAURY
   Near Conflans, France, Nov. 2, 1918.
   R—St. Louis, Mo.
   B—St. Louis, Mo.
   G. O. No. 46, W. D., 1919.

Captain, pilot, 24th Aero Squadron, Air Service.
While on a photographic mission of a particularly dangerous character Captain Hill and his observer, Lieut. John W. Cousins, were attacked by superior numbers of enemy pursuit planes. During the combat which ensued his skill and coolness enabled his observer to destroy one of the enemy aircraft.

*HILL, RAYMOND C.
   Near Lachaussee, France, Sept. 13, 1918.
   R—Lewiston, Idaho.
   B—Macon County, Ill.
   G. O. No. 126, W. D., 1919.

First lieutenant, 146th Field Artillery, observer, 99th Aero Squadron, Air Service.
He, with First Lieut. Clarence C. Kahle, pilot, were directed to take photographs of the old Hindenburg line. They were accompanied by two protecting planes. After they had taken some photographs the protecting planes were driven off by hostile aircraft, but he and his pilot continued on their mission alone, until in the vicinity of Lachaussee they were attacked by an enemy formation of 9 planes. Putting up a gallant fight against these overwhelming odds, he was shot through the heart and killed, but his pilot, by his pluck, determination, skill, and courage, brought the photographs and the plane to our lines.
Posthumously awarded. Medal presented to widow, Mrs. Raymond C. Hill.

HILL, ROBERT (2386550)
   At Cote-St. Germain, France, Nov. 6, 1918.
   R—Buffalo, Minn.
   B—Buffalo, Minn.
   G. O. No. 37, W. D., 1919.

Corporal, Company C, 61st Infantry, 5th Division.
When his platoon had been stopped by heavy artillery and machine-gun fire Corporal Hill crawled forward, though wounded, and led his men to their objective before being evacuated.

HILL, SAMUEL (1829765)
   Near Bethincourt, France, Sept. 26, 1918.
   R—Apollo, Pa.
   B—Apollo, Pa.
   G. O. No. 21, W. D., 1919.

Private, first class, Company E, 320th Infantry, 80th Division.
Assisted by 3 comrades, he volunteered and went to the aid of a platoon which was held up by machine-gun fire. Although under constant fire of the enemy, he flanked the stronghold, and by effective use of his automatic rifle killed 2 officers and 3 enlisted men and captured the gun.

HILL, TERRELL WINFIELD ....
  Near Grand Pre, France, Oct. 23,
    1918.
  R—Columbus, Ga.
  B—Columbus, Ga.
  G. O. No. 15, W. D., 1921.

Second lieutenant, 312th Infantry, 78th Division.
His company commander being wounded, Lieutenant Hill assumed command of the company and led it gallantly forward in the attack. Due to the intensity of the enemy fire, the advance was halted, but this officer continued to expose himself in order to place his men in a position of shelter. In the performance of this act he was himself wounded in the right arm but continued in active command of the company until relieved the following day.

HILL, WILLIAM H. (92698) ....
  Near Haumont, France, Sept. 27,
    1918.
  R—Columbus, Ohio.
  B—Columbiana County, Ohio.
  G. O. No. 23, W. D., 1919.

Corporal, Company A, 166th Infantry, 42d Division.
Corporal Hill, while establishing liaison with a separate unit of his patrol, under heavy machine-gun fire, personally captured two prisoners, and, after delivering them to a guard, continued with his mission. He was severely wounded while performing this dangerous mission.

HILLER, WALTER S ....
  Near Bellefonte Farm, France, Nov.
    10, 1918.
  R—Pocatello, Idaho.
  B—Salt Lake City, Utah.
  G. O. No. 37, W. D., 1919.

Corporal, 55th Company, 5th Regiment, U. S. Marine Corps, 2d Division.
Under direct fire from 7 enemy machine guns, he led a detachment across the River Meuse and destroyed a machine-gun nest which occupied a dominating position 1,000 yards in advance of his company.

HILLIARD, GROVER C. (1309157) ......
  Near Beaurevoir, France, Oct. 6,
    1918.
  R—Dover, Tenn.
  B—Stewart County, Tenn.
  G. O. No. 37, W. D., 1919.

Sergeant, Company K, 117th Infantry, 30th Division.
He volunteered and crossed an open space swept by fire from enemy machine guns and snipers to rescue wounded comrades.

*HILLIG, HARRY (1995326) ....
  Near Bellicourt, France, Sept. 29,
    1918.
  R—Carrollton, Ill.
  B—Carrollton, Ill.
  G. O. No. 44, W. D., 1919.

Private, Company G, 119th Infantry, 30th Division.
Although he had been seriously wounded in the neck by a machine-gun bullet, he refused to go to the rear and continued in the advance until he was killed several hours later.
Posthumously awarded. Medal presented to sister, Mrs. Katherine Hillig-Wagener.

HINE, OTTO L. H ....
  At Chaudron Farm, France, Sept.
    29-30, 1918.
  R—Muskogee, Okla.
  B—Floyd, Tex.
  G. O. No. 81, W. D., 1919.

First lieutenant, Dental Corps, attached to 139th Infantry, 35th Division.
Upon his own initiative, Lieutenant Hine went to a dressing station in advance of the line, after the Infantry had withdrawn, and worked under heavy bombardment of gas and high explosive shells dressing the patients and directing their evacuation. That night he returned to our lines through heavy artillery and machine-gun fire to arrange for ambulances and litters. Later he made another trip to rear for the purpose of securing an artillery barrage to protect his dressing station. Through his exceptional courage and energy all the wounded men were safely evacuated.

HINES, JOHN L ....
  Near Berzy-le-Sec, France, July 21,
    1918.
  R—White Sulphur Springs, W. Va.
  B—White Sulphur Springs, W. Va.
  G. O. No. 10, W. D., 1920.
  Distinguished-service medal also
    awarded.

Brigadier general, 1st Infantry Brigade, 1st Division.
At a critical time during the battle southwest of Soissons, when liaison had been broken between the 16th Infantry and 26th Infantry, and repeated efforts to reestablish it had failed, General Hines, then in command of the 1st Infantry Brigade, personally went through terrific artillery fire to the front lines of the 16th Infantry, located its left flank, and, walking in front of the lines, encouraged the troops by his example of fearlessness and disregard of danger. He then succeeded in finding the right forward elements of the 26th Infantry and directed the linking up of the two regiments, thereby enabling the operations to be pushed forward successfully.

HINES, PAUL H ....
  At Marcheville, France, and near
    Riaville, France, Sept. 25-26, 1918.
  R—South Boston, Mass.
  B—South Boston, Mass.
  G. O. No. 139, W. D., 1918.

First lieutenant, 102d Infantry, 26th Division.
He showed great bravery and devotion to duty throughout this action. At one time he went through violent artillery bombardment and machine-gun fire to reestablish broken liaison with a battalion on the left. Later he voluntarily led a party of first-aid men across an open field swept by machine-gun fire and rescued a wounded officer after several previous attempts had failed.

HINSON, GUY R. (1329688) ....
  In Canal sector, Belgium, Aug. 27,
    1918.
  R—Charlotte, N. C.
  B—Charlotte, N. C.
  G. O. No. 145, W. D., 1918.

Sergeant, first class, Company F, 105th Engineers, 30th Division.
He was in charge of a platoon, delivering a highly concentrated gas-cloud attack against the enemy, when the cloud unexpectedly flared back. After leading his men to a place of safety, this soldier went back into the cloud four times at imminent peril to his own life, collecting and rescuing others who had been overcome. Conducting his platoon through heavy machine-gun fire, he put them in charge of another sergeant with instructions to resume their mission, while he again returned to search for gassed men, and found all but two. His excellent leadership and unusual courage prevented many casualties and at the same time effected the completion of an important mission.

*HINTZ, EDWARD E. (65518) ....
  Near Bouresches, France, July 20,
    1918.
  R—New Britain, Conn.
  B—New Britain, Conn.
  G. O. No. 126, W. D., 1919.

Corporal, Company I, 102d Infantry, 26th Division.
Although seriously wounded Corporal Hintz remained on duty and with 9 remaining men of his platoon assisted in filling in a gap between his company and the company on the right, a distance of about 200 yards, until support arrived.
Posthumously awarded. Medal presented to father, Leidwig Hintz.

HIRSCHFELDER, CHESTER J ....
  Near St. Etienne, France, Oct. 4,
    1918.
  R—Beeville, Tex.
  B—Fort Wayne, Ind.
  G. O. No. 53, W. D., 1920.

Captain, 4th Machine Gun Battalion, 2d Division.
When an enemy machine gun was inflicting heavy casualties upon his company, Captain Hirschfelder crawled forward alone across a field exposed to heavy machine-gun fire and threw hand grenades into the enemy position. His action silenced the machine gun and prevented further casualties to his company.

HIRST, SAMUEL CARROLL_____
Near Blanc Mont Ridge, France, Oct. 5, 1918.
R—Purcellville, Va.
B—Purcellville, Va.
G. O. No. 46, W. D., 1919.

Private, 55th Company, 5th Regiment, U. S. Marine Corps, 2d Division.
Together with another soldier, Private Hirst successfully completed the capture of a machine gun, destroying it and killing 2 of the crew. Fighting his way back to his own lines, he furnished valuable information concerning enemy machine-gun emplacements.

HISER, HENRY (1524099)_____
Near Avocourt, France, Sept. 26, 1918.
R—Levanna, Ohio.
B—Brown County, Ohio.
G. O. No. 50, W. D., 1922.

Private, first class, Company D, 136th Machine Gun Battalion, 37th Division.
When the advance of his platoon was held up by fire from a hostile machine-gun nest. Private Hiser advanced alone ahead of the platoon, worked his way around the flank and rear, and single handed killed the officer in command and a gunner and captured 15 prisoners, thereby enabling his platoon to advance.

*HITCHCOCK, ROGER W._____
Near Fismes, France, Aug. 11, 1918.
R—Los Angeles, Calif.
B—Nutley, N. J.
G. O. No. 44, W. D., 1919.

Second lieutenant, pilot, 88th Aero Squadron, Air Service.
Louis G. Bernheimer, first lieutenant, pilot; John W. Jordan, second lieutenant, 7th Field Artillery, observer; Roger W. Hitchcock, second lieutenant, pilot; James S. D. Burns, deceased, second lieutenant, 165th Infantry, observer; Joel H. McClendon, deceased, first lieutenant, pilot; Charles W. Plummer, deceased, second lieutenant, 101st Field Artillery, observer; Philip R. Babcock, first lieutenant, pilot; and Joseph A. Palmer, second lieutenant, 15th Field Artillery, observer. All of these men were attached to the 88th Aero Squadron, Air Service.
For extraordinary heroism in action near Fismes, France, Aug. 11, 1918. Under the protection of 3 pursuit planes, each carrying a pilot and an observer, Lieutenants Bernheimer and Jordan, in charge of a photo plane, carried out successfully a hazardous photographic mission over the enemy's lines to the River Aisne. The 4 American ships were attacked by 12 enemy battle planes. Lieutenant Bernheimer, by coolly and skillfully maneuvering his ship, and Lieutenant Jordan, by accurate operation of his machine gun, in spite of wounds in the shoulder and leg, aided materially in the victory which came to the American ships, and returned safely with 36 valuable photographs. The pursuit plane operated by Lieutenants Hitchcock and Burns was disabled while these two officers were fighting effectively. Lieutenant Burns was mortally wounded and his body jammed the controls. After a headlong fall of 2,500 meters, Lieutenant Hitchcock succeeded in regaining control of this plane and piloted it back to his airdrome. Lieutenants McClendon and Plummer were shot down and killed after a vigorous combat with 5 of the enemy's planes. Lieutenants Babcock and Palmer, by gallant and skillful fighting, aided in driving off the German planes and were materially responsible for the successful execution of the photographic mission.
Posthumously awarded. Medal presented to widow, Mrs. Alta I. Hitchcock.

HITCHENS, ERNEST L. (1781220)_____
Near Crepion, France, Nov. 7–8, 1918.
R—Baltimore, Md.
B—Salisbury, Md.
G. O. No. 37, W. D., 1919.

Sergeant, Company K, 313th Infantry, 79th Division.
While his battalion was conducting a relief, he and 4 members of his platoon were wounded by shell fire. After he had taken the 4 wounded comrades to a first-aid station, he immediately returned without treatment to himself, despite his severe suffering from 3 body wounds, and remained with his platoon until the completion of the relief.

*HIX, JAMES C. (274210)_____
Near Ronchers, France, July 30, 1918.
R—Beloit, Wis.
B—Knoxville, Tenn.
G. O. No. 66, W. D., 1919.

Private, Company F, 127th Infantry, 32d Division.
With another soldier, he volunteered to go out in advance of our lines to rescue wounded soldiers who had been left there when the company withdrew. Under heavy fire they made 2 trips, bringing back wounded men. Private Hix was wounded by machine-gun fire when he went out for the third time.
Posthumously awarded. Medal presented to brother, H. W. Hix.

HOBBS, AVIS T._____
Near Voormezeele, Belgium, Aug. 30, 1918, and near Busigny, France, Oct. 9, 1918.
R—Lebanon, Tenn.
B—Lebanon, Tenn.
G. O. No. 81, W. D., 1919.

First lieutenant, 119th Infantry, 30th Division.
Under heavy shellfire and in full view of the enemy, he volunteered for and conducted a daylight patrol of 1 man and himself to reconnoiter the best route for a raid, obtaining very valuable information. Later, when in the course of an attack our line was held up by the enemy, he led a patrol of 12 men and, under heavy fire, placed his Lewis gun so as to drive out the enemy, advancing 2,000 yards ahead of our front lines to gain information.

HOBBS, HORACE P._____
At Pala's Cotta, Island of Jolo, P. I., May 5, 1905.
R—Germantown, Pa.
B—Philadelphia, Pa.
G. O. No. 4, W. D., 1923.

First lieutenant, 17th Infantry, U. S. Army.
With his company (provisional company, 17th Infantry), he held his command in a position under heavy hostile fire at a range of from 50 to 100 yards from a strong enemy position for more than 4 hours and gained valuable information as to the enemy's position by fearlessly exposing himself in climbing a tree to make observations.

HOBSCHEID, PAUL (1386758)_____
At Chipilly Ridge, France, Aug. 9, 1918.
R—Chicago, Ill.
B—Chicago, Ill.
G. O. No. 128, W. D., 1918.

Corporal, Company C, 131st Infantry, 33d Division.
Corporal Hobscheid, under heavy fire, advanced into a hostile sniping post, found and entered a long dugout, and brought out 30 prisoners.

HOBSON, HENRY W._____
Near St. Mihiel, France, Sept. 12, 1918.
R—New York, N. Y.
B—Denver, Colo.
G. O. No. 99, W. D., 1918.

Major, 356th Infantry, 89th Division.
Within 10 minutes after the beginning of the advance, at 5 a. m., he was twice wounded—in the shoulder by a machine gun bullet and by shrapnel in the leg. Being in command of the assault battalion and realizing the importance of its operations, he continued to accompany and direct his command throughout the day, notwithstanding his wounds, which caused him great pain and difficulty of movement. At halts he had to be assisted to lie down and get up by his adjutant; nevertheless he remained on duty until the fighting of the day was over.

HODGES, COURTNEY H.
Near Brieulles, France, Nov. 2–4, 1918.
R—Perry, Ga.
B—Perry, Ga.
G. O. No. 3., W. D., 1919.

Lieutenant colonel, 6th Infantry, 5th Division.
He personally conducted a reconnaissance of the Meuse River to determine the most advantageous location for a crossing and for a bridge site. Having organized a storming party, he attacked the enemy not 100 paces distant, and, although failing, he managed to effect the crossing of the canal, after 20 hours of ceaseless struggling. His fearlessness and courage were mainly responsible for the advance of his brigade to the heights east of the Meuse.

*HOEYNCK, FRANK J.
Near Bantheville, France, Nov. 6–7, 1918.
R—Chicago, Ill.
B—Quincy, Ill.
G. O. No. 37, W. D., 1919.

Second lieutenant, 314th Engineers, 89th Division.
Lieutenant Hoeynck received orders to make a reconnaissance of the bridge at Pouilly and the road from Pouilly to Inor. He was accompanied on the expedition by a soldier of his platoon, the two being the first to cross the river at Inor. At this point they gained most valuable information. Recrossing the river, he made his way to Pouilly under machine-gun fire, collecting engineer data of the greatest importance. Just as they approached their destination Lieutenant Hoeynck was killed by machine-gun fire.
Posthumously awarded. Medal presented to father, Frederick T. Hoeynck.

*HOFF, JOHN VAN R.
At Wounded Knee, S. Dak., Dec. 29, 1890.
R—New York, N. Y.
B—New York.
G. O. No. 3, W. D., 1925.

Captain, assistant surgeon, U. S. Army.
For extraordinary heroism in action against hostile Sioux Indians at Wounded Knee, S. Dak., Dec. 29, 1890. When the Indians made a sudden treacherous attack upon the troops Captain Hoff, with utter disregard for his personal safety, attended to the dressing of the wounds of fallen soldiers.
Posthumously awarded. Medal presented to widow, Mrs. John Van R. Hoff.

HOFFMAN, CLYDE O. (2217627)
Near Fey-en-Haye, France, Sept. 14, 1918.
R—El Reno, Okla.
B—Kellersville, Ill.
G. O. No. 98, W. D., 1919.

First sergeant, Company K, 357th Infantry, 90th Division.
Sergeant Hoffman was seriously wounded when 3 enemy machine guns opened fire on his platoon at a range of only 25 yards, but he continued firing on the enemy until the enemy guns were silenced, inspiring the other members of his platoon by his coolness and courage.

HOFFMAN, EARL A. (2180850)
In the Bois-de-Bantheville, France, Oct. 24, 1918.
R—Randolph, Nebr.
B—Denison, Iowa.
G. O. No. 37, W. D., 1919.

Private, Company C, 341st Machine Gun Battalion, 89th Division.
Severely wounded while dressing the wounds of 23 of his platoon, he continued his work until he fainted from pain and was evacuated.

HOFFMAN, HENRY J. (145708)
Near Suippes, France, July 15, 1918.
R—Virginia, Minn.
B—Little Falls, Minn.
G. O. No. 35, W. D., 1919.

Private, Battery C, 151st Field Artillery, 42d Division.
After telephonic communication had been severed, he volunteered and carried a message over territory perilously swept by shell fire. He was further hindered by being obliged to wear his gas mask, but the mission was successfully accomplished.

HOFFMAN, LEONARD LAWRENCE
Near Blanc Mont, France, Oct. 5, 1918.
R—Minneapolis, Minn.
B—Minneapolis, Minn.
G. O. No. 15, W. D., 1919.

Private, 81st Company, 6th Machine Gun Battalion, U. S. Marine Corps, 2d Division.
He showed exceptional courage in volunteering and carrying an important message through a heavy machine-gun barrage, after another soldier met death in attempting to do so.

*HOFFMAN, MYRON L. (2285150)
Near Paarden Kanter, Belgium, Oct. 31, 1918.
R—San Francisco, Calif.
B—San Francisco, Calif.
G. O. No. 21, W. D., 1919.

First sergeant, Company M, 363d Infantry, 91st Division.
Sergeant Hoffman was mortally wounded by shellfire while he was returning to report to his company commander the position of the dressing station which he had located in order that the wounded might be evacuated under cover of darkness. While being carried to the rear in great pain he directed his litter bearers to go by way of the battalion post of command in order that he might deliver his reports.
Posthumously awarded. Medal presented to sister, Miss Florence A. Hoffman.

HOGAN, JOHN A. (2219962)
Near Bois-des-Rappes, France, Oct. 28–30, 1918.
R—Heavener, Okla.
B—Paris, Ark.
G. O. No. 98, W. D., 1919.

Sergeant, Company C, 358th Infantry, 90th Division.
While sergeant of a liaison platoon he volunteered to locate a machine-gun nest, and under heavy fire went out alone and destroyed it. He volunteered to go out on a dangerous reconnaissance mission and returned with valuable information of the enemy, thus enabling a successful machine-gun barrage to be laid down on the hostile positions.

HOGE, WILLIAM M., Jr.
Near Brieulles, France, Nov. 4, 1918.
R—Lexington, Mo.
B—Boonville, Mo.
G. O. No. 37, W. D., 1919.

Major, 7th Engineers, 5th Division.
After personally and voluntarily reconnoitering the site of a pontoon bridge over the Meuse, in daylight and under direct shellfire, Major Hoge commanded the movement of a train of heavy wagons, under enemy observation, to this location. He then supervised the construction of the bridge and the successful crossing of the train.

HOGGLE, JOHN (1348779)
Near Brieulles, France, Nov. 4, 1918.
R—Delmar, Ala.
B—Walker County, Ala.
G. O. No. 37, W. D., 1919.

Private, Company D, 7th Engineers, 5th Division.
When 3 of the boats supporting a pontoon bridge across the Meuse River were destroyed by artillery fire, he voluntarily waded into the stream to a depth of 4 feet, under heavy artillery and machine-gun fire, and held up the deck of the bridge until new boats were launched and placed in position.

HOLCOMB, ROY (94375)
Near Sommerance, France, Oct. 14, 1918.
R—Chillicothe, Ohio.
B—Higby, Ohio.
G. O. No. 23, W. D., 1919.

Sergeant, Company H, 166th Infantry, 42d Division.
Sergeant Holcomb remained with his platoon throughout the day, leading and directing them in action, although badly gassed and ordered to the hospital.

HOLDEN, JAMES E. (1754094)_____
Near Limey, France, Sept. 20, 1918.
R—Newark, N. J.
B—Newark, N. J.
G. O. No. 44, W. D., 1919.

Private, Headquarters Company, 312th Infantry, 78th Division.
Undaunted by heavy shellfire, Private Holden, a runner between brigade and regiment headquarters, delivered an important message, walking and crawling through a wood which was being so heavily bombarded with gas and high-explosive shells that it had been evacuated. He returned to his post after delivering his message through the same severe fire.

HOLDEN, LANSING C., Jr_____
Near Montigny, France, Oct. 23, 1918.
R—New York, N. Y.
B—New York, N. Y.
G. O. No. 46, W. D., 1919.

First lieutenant, 95th Aero Squadron, 1st Pursuit Group, Air Service.
He was ordered to attack several German balloons reported to be regulating effective artillery fire on our troops. After driving off an enemy plane encountered before reaching the balloons, he soon came upon 5 balloons in ascension 1 kilometer apart. In attacking the first, which proved to be a decoy with a basket, his gun jammed; after clearing it, he attacked the second balloon, forcing the observer to jump. His gun again jammed before he could set fire to this balloon. Moving on the third balloon at a height of only 50 meters, he set fire to it and compelled the observer to jump. He was prevented from attacking the two remaining balloons by the further jamming of his machine gun.
Oak-leaf cluster.
For the following act of extraordinary heroism in action near St. Jean de Buzy, France, Nov. 4, 1918, Lieutenant Holden is awarded an oak-leaf cluster, to be worn with the distinguished-service cross: Flying at a low altitude to evade hostile pursuit patrols, he attacked a German observation balloon in the face of antiaircraft and machine-gun fire. Although the balloon was being rapidly pulled down, he set fire to it in its nest and also caused much damage to adjacent buildings.

HOLDEN, KENNETH H._____
Near the Argonne, France, Nov. 2, 1918.
R—Michigan City, Ind.
B—Three Oaks, Mich.
G. O. No. 32, W. D., 1919.

First lieutenant, pilot, 12th Aero Squadron, Air Service.
While on an Infantry contact mission he and his observer were attacked by 4 enemy planes and driven back, but, realizing the importance of their mission, deliberately returned and attacked the 4 planes, sending 1 to the earth and driving the others away. Unmindful of the damaged condition of their plane and of their own danger, they then flew for an hour within 100 meters of the ground, through a continuous heavy machine-gun fire, until they had accurately located our front-line position.

HOLLAND, HARRY G. (1392469)_____
Near Bois-de-Chaume, France, Oct. 10, 1918.
R—Lincoln, Ill.
B—Lincoln, Ill.
G. O. No. 44, W. D., 1919.

Private, Company B, 122d Machine Gun Battalion, 33d Division.
Advancing alone against 20 of the enemy, whom he saw bringing machine guns into position to enfilade his position, he killed 10 and routed the rest, his entire exploit being under heavy shell and machine-gun fire.

HOLLAND, JAMES W. (1329250)_____
Near Bellicourt, France, Sept. 29, 1918.
R—Gastonia, N. C.
B—Gastonia, N. C.
G. O. No. 21, W. D., 1919.

Sergeant, first class, Company D, 105th Engineers, 30th Division.
While suffering from severe wounds, and still subjected to intense artillery fire, Sergeant Holland directed the evacuation of his platoon commander, and fully instructed his junior sergeant before he would allow himself to be evacuated.

HOLLAND, RICHARD W. (1384638)_____
Near Marcheville, France, Nov. 10, 1918.
R—Rend, Ill.
B—Kentucky.
G. O. No. 71, W. D., 1919.

Private, first class, Company F, 130th Infantry, 33d Division.
When his platoon was in an isolated position, exposed to heavy fire, he volunteered to carry a message through intense machine-gun and artillery fire. Though his rifle was shot from his hands, he delivered the message.

HOLLAND, SPESSERD L._____
Near Bois-de-Bantheville, France, Oct. 15, 1918.
R—Bartow, Fla.
B—Bartow, Fla.
G. O. No. 37, W. D., 1919.

First lieutenant, Coast Artillery Corps, observer, 24th Aero Squadron, Air Service.
Flying at an altitude of 400 meters, 5 kilometers within the enemy lines, he and his pilot, Lieut. George A. Goldthwaite, continued on their mission in spite of being harassed by antiaircraft, securing information of great military value.

*HOLLIDAY, HARRY_____
Near Mezy, France, July 15, 1918.
R—Traverse City, Mich.
B—Traverse City, Mich.
G. O. No. 32, W. D., 1919.

First lieutenant, 30th Infantry, 3d Division.
Although wounded during the enemy's barrage, he remained in charge of his gun squad, protecting it from the flank with pistol and hand grenades. He was again wounded by a hand grenade, but persisted in protecting the flank of the gun, though the enemy had advanced as close as the muzzle. After keeping his gun in action for 10 hours, he organized a platoon of his gunners and stragglers and fought a retiring action for over a mile. He remained with his men until ordered to the hospital on account of his wounds.
Posthumously awarded. Medal presented to mother, Mrs. G. A. Holliday.

HOLLINGSWORTH, ALEXANDER (2056074).
Near Remonville, France, Nov. 1, 1918.
R—Hendersonville, N. C.
B—Hendersonville, N. C.
G. O., No. 44, W. D., 1919.

Corporal, Company B, 354th Infantry, 89th Division.
He led his combat group against a machine-gun position through an intense machine-gun and artillery fire, and although severely wounded took part in the capture of the machine gun and crew. He refused to be evacuated until he had reported to his company commander.

HOLLIS, GEORGE G. (2184500)_____
Near Lucy, France, Nov. 4, 1918.
R—Ajo, Ariz.
B—Horace, Kans.
G. O. No. 37, W. D., 1919.

Corporal, Company E, 356th Infantry, 89th Division.
While acting as scout of a patrol group, he encountered an enemy machine-gun nest and opened fire on it, killing the gunner and capturing the nest, thereby protecting the balance of his patrol from casualties. He turned the captured guns on the enemy.

HOLLOWAY, HORACE L. (1956210)
Near Brabant-sur-Meuse, France, Oct. 23, 1918.
R—Richmond, Va.
B—Richmond, Va.
G. O. No. 37, W. D., 1919.

Sergeant, 308th Trench Mortar Battery, 83d Division.
During an offensive action in Boissois Bois he remained at his post under the most hazardous circumstances. In the open, under fire of machine guns and 77's, he kept his mortar going for 57 minutes, firing 230 bombs. Repeatedly knocked down by concussion of exploding shells, he only ceased firing when exhausted.

HOLLY, GEORGE W. (1705295)
Near Baccarat, France, June 23, 1918.
R—Mount Vernon, N. Y.
B—New York, N. Y.
G. O. No. 107, W. D., 1918.

Private, Company D, 307th Infantry, 77th Division.
He attempted to catch a hand grenade thrown into the window of his dugout by a German and did succeed in diverting it and thereby saving from death or injury a number of his comrades, but in the effort lost most of his hand.

HOLMES, ALBERT P. (1387883)
At Chipilly Ridge, France, Aug. 9, 1918.
R—Chicago, Ill.
B—Chicago, Ill.
G. O. No. 128, W. D., 1918.

Private, Company H, 131st Infantry, 33d Division.
After 6 runners had been killed or wounded in an attempt to establish liaison with battalion headquarters, he volunteered for this hazardous duty and succeeded in getting through, under heavy shellfire where others had failed.

*HOLMES, BURTON (1872566)
Near Hill 188, France, Sept. 28, 1918.
R—Pendleton, S. C.
B—Pickens, S. C.
G. O. No. 37, W. D., 1919.

Private, Company C, 371st Infantry, 93d Division.
After he had been badly wounded and his automatic rifle had been put out of commission, Private Holmes returned to his company, under extremely heavy machine-gun and shell fire, and taking another automatic rifle went back and reopened fire on the enemy. While thus engaged, he was killed. Posthumously awarded. Medal presented to uncle, Bill Holmes.

HOLMES, CHARLES (3132516)
Near Gesnes, France, Oct. 14, 1918.
R—Potter Valley, Calif.
B—Oakland, Calif.
G. O. No. 66, W. D., 1919.

Private, first class, Company B, 127th Infantry, 32d Division.
He carried to safety a wounded officer, passing over a path blocked by 2 barbed-wire obstacles, and enfiladed by enemy machine-gun fire. After successfully accomplishing this perilous mission he succeeded in working his way back to his company over the same route in the face of heavy fire.

HOLMES, CHRISTIAN R.
Near Ansauville, France, Mar. 22, 1918.
R—Cincinnati, Ohio.
B—Cincinnati, Ohio.
G. O. No. 129, W. D., 1918.

First lieutenant, 28th Infantry, 1st Division.
As leader of a patrol he displayed extraordinary heroism and daring; he cut and crawled through 12 strands of wire in front of an enemy listening post, leaped upon the sentinel, made him a prisoner, and brought him back through no man's land.

HOLMES, FELIX R.
Near Consenvoye, France, Oct. 11, 1918.
R—Cranberry, W. Va.
B—Jersey City, N. J.
G. O. No. 37, W. D., 1919.

Captain, 129th Infantry, 33d Division.
After being wounded and ordered to the aid station, Captain Holmes reported to his regimental headquarters, giving a detailed report of the forward conditions. His strength failing while attempting to reach the aid station alone, he was assisted to the place, where he had his wounds dressed and was tagged for evacuation, but instead returned to the front line through intense shell, machine-gun, and sniper fire and took command of his company.

HOLMES, FRANK D. (2312486)
Near Cierges, France, Aug. 1, 1918.
R—Chicago, Ill.
B—Chicago, Ill.
G. O. No. 64, W. D., 1919.

Private, Company C, 125th Infantry, 32d Division.
After he had seen 2 runners wounded in attempting to get through, Private Holmes voluntarily undertook the mission. He crossed an open field, about 500 meters wide, thence through the town of Cierges, constantly under direct enemy observation and subjected to withering fire, and, after delivering his message, he returned over the same route. Called upon to guide a company to a new position, he led them up in groups to avoid losses, making several trips through the murderous fire.

HOLMES, HARVEY H. (551434)
Near Mezy, France, July 15–19, 1918.
R—Wells Bridge, N. Y.
B—Norwich, N. Y.
G. O. No. 46, W. D., 1919.

Corporal, Company H, 38th Infantry, 3d Division.
He observed that a number of Germans were moving toward a clump of bushes near our front during the battle of the Marne. With a patrol of 6 men he rushed the position under heavy machine-gun fire and captured 60 prisoners.

*HOLMES, JAMES H., Jr.
Near Soissons, France, July 18–19, 1918.
R—Charleston, S. C.
B—Charleston, S. C.
G. O. No. 132, W. D., 1918.

Captain, 26th Infantry, 1st Division.
After having bravely led his company in 3 attacks in 2 days near Soissons, France, July 18–19, 1918, he was killed in a fourth attack, while charging an enemy machine gun. Posthumously awarded. Medal presented to widow, Mrs. James H. Holmes.

*HOLMES, OLIVER WENDELL (2847904)
Near Limey, France, Sept. 12, 1918.
R—Hastings, Nebr.
B—Ruskin, Nebr.
G. O. No. 28, W. D., 1921.

Private, Company G, 353d Infantry, 89th Division.
Seeing his lieutenant fall severely wounded, Private Holmes, with another stretcher bearer, rushed through heavy machine-gun fire to his assistance. When they had placed the lieutenant on their stretcher and were endeavoring to go through the heavy fire to the dressing station, Private Holmes fell mortally wounded. Posthumously awarded. Medal presented to father, Winfield S. Holmes.

*HOLROYD, CROSSLEY MONTROSE (1235327).
Near St. Agnan, France, July 16, 1918.
R—Philadelphia, Pa.
B—England.
G. O. No. 108, W. D., 1919.

Private, Company B, 109th Infantry, 28th Division.
He voluntarily left the shelter of his trench and went into machine-gun and artillery fire 3 times to rescue wounded comrades 100 yards away. He was killed in action Oct. 2, 1918. Posthumously awarded. Medal presented to father, Sam Holroyd.

HOLT, FRANK MAURICE (2302277)
Near St. Gilles, south of Fismes, France, Aug. 4, 1918.
R—Milwaukee, Wis.
B—Minneapolis, Minn.
G. O. No. 15, W. D., 1919.

Sergeant, Battery A, 120th Field Artillery, 32d Division.
When the men of his battery position had been ordered to shelter on account of enemy shelling Sergeant Holt, in company with 2 other men, rescued a French soldier from drowning in a stream. This act was performed while the valley was filled with mustard gas.

HOLT, JEFFERSON L. (5890)............ | Private, medical detachment, 2d Engineers, 2d Division.
At Lucy-le-Bocage, France, June 2–3, 1918.
R—Pecos, Tex.
B—Santo, Tex.
G. O. No. 99, W. D., 1918.

During the day and night of June 2–3, 1918, he exposed himself to severe and continuous fire beyond call of duty in order to bring aid to wounded engineers and marines.

HOLT, WALTER A. (1000557)............ | Private, Company A, 326th Infantry, 82d Division.
Near St. Juvin, France, Oct. 15–16, 1918.
R—Rochelle, Ga.
B—Eastman, Ga.
G. O. No. 72, W. D., 1920.

Private Holt volunteered to establish liaison with the unit on the right, although this mission required that he cross an area swept by heavy artillery and machine-gun fire. Although fired upon at short range by enemy snipers, he successfully completed his mission and succeeded in capturing 3 German prisoners. Later he exposed himself to heavy machine-gun fire in order to assist in the rescue of the wounded.

HOLTZ, ADOLPH (60445)............ | Private, Company D, 101st Infantry, 26th Division.
In the Trugny Woods, near Chateau-Thierry, France, July 23, 1918.
R—North Plymouth, Mass.
B—Russia.
G. O. No. 126, W. D., 1918.

Private Holtz, with 2 other men in an advanced position ahead of the battalion, charged a machine gun, killed 4 of the enemy, and drove off the rest, thereby making possible the advance of their comrades.

*HOLTZ, FRANK F. (1705126)............ | Sergeant, Company C, 307th Infantry, 77th Division.
Near Moulin de Charlavaux, France, Oct. 4, 1918.
R—Buffalo, N. Y.
B—Buffalo, N. Y.
G. O. No. 71, W. D., 1919.

His platoon held up and cut off from the remainder of the company, he volunteered to establish liaison and summon reinforcements after several runners had been killed or captured in the attempt. Passing through intense artillery and machine-gun fire, he carried word to his company commander, but was mortally wounded while returning to his platoon.
Posthumously awarded. Medal presented to mother, Mrs. Susan F. Holtz.

HOLZ, WILLIAM (47461)............ | First sergeant, Company H, 18th Infantry, 1st Division.
South of Soissons, France, July 18 and 19, 1918.
R—Lancaster, Pa.
B—Lancaster, Pa.
G. O. No. 53, W. D., 1920.

Sergeant Holz exposed himself to heavy enemy fire reorganizing a platoon of his company and in leading it forward in the attack. He was wounded by a fragment of a high-explosive shell the following day while again leading his platoon in the attack. Although directed for evacuation to the hospital he refused to go, and gathered together a small group of stragglers and led them forward to the attacking line. Later he was forced to go to the rear on account of weakness from loss of blood.

HOLZGREBE, WILLIAM O. (263527)...... | Private, Company E, 125th Infantry, 32d Division.
Near Verdun, France, Oct. 11–13, 1918.
R—Escanaba, Mich.
B—Escanaba, Mich.
G. O. No. 145, W. D., 1918.

As a runner of the 3d Battalion, 125th Infantry, during the taking and holding of the line near La Tuilerie Ferme, he was engaged in carrying important messages, crossing and recrossing Death Valley between Hill 268 and La Cote Dame Marie, the foremost part of the line held by the 3d Battalion. The valley was swept by machine-gun fire, the terrain affording absolutely no protection, requiring a perilous dash of 500 yards over open ground before any cover was reached. It was only by his display of supreme courage that important messages reached the battalion.

HOLMAN, GEORGE D. (1785188)........ | Private, Company L, 332d Infantry, 91st Division.
Near Gesnes, France, Oct. 11, 1918.
R—Valley City, N. Dak.
B—Valley City, N. Dak.
G. O. No. 137, W. D., 1918.

He was on duty at one of the posts of a double relay of runners between battalion and regimental headquarters and carried numerous messages through heavily shelled area. Three other soldiers were killed and 5 seriously wounded along his route, but with unfaltering devotion to duty he continued at his work of relaying messages until he was relieved.

HOLMAN, SIDNEY (0391461)............ | Private, Machine Gun Company, 132d Infantry, 33d Division.
In the Bois-de-Foret, France, Oct. 10, 1918.
R—Chicago, Ill.
B—Chicago, Ill.
G. O. No. 32, W. D., 1919.

After 3 runners had been killed or wounded in attempts to get through heavy shellfire with an important message from the regimental commander of the 30th Infantry to the regimental commander of the 59th Infantry, Private Holman, with Pvt. James J. Snyder, responded to a call for volunteers and succeeded in delivering the message.

HONCHAR, STEVE (2832383)............ | Private, Company A, 60th Infantry, 5th Division.
Near Brandeville, France, Nov. 7, 1918.
R—Akron, Ohio.
B—Russia.
G. O. No. 37, W. D., 1919.

When a small group of his platoon with which he was advancing was surprised by enemy machine-gun fire, Private Honchar, an automatic rifleman, selected a position and opened fire. He silenced a portion of the fire and drew the rest in his direction, thereby enabling his comrades to take cover. After being wounded three times, he called a comrade and directed his fire, after which he crawled back for first-aid treatment.

HOOD, ROBERT B............ | Captain, 12th Field Artillery, 2d Division.
Near Thiaucourt, France, Sept. 12, 1918.
R—Hutchinson, Kans.
B—Wellington, Kans.
G. O. No. 37, W. D., 1919.

While acting as executive officer, Captain Hood brought the battery into action under fire, superintended the placing of the guns and the unloading of the ammunition, and executed the fire of the battery under an intense enfilading fire. When the entire gun crew of his first piece was wiped out, he hastily formed a supplementary gun squad and succeeded in getting the first piece into action again within 4 minutes.

*HOOK, ALFRED J............ | First lieutenant, 108th Infantry, 27th Division.
East of Ronssoy, France, Sept. 29, 1918.
R—Brooklyn, N. Y.
B—Brooklyn, N. Y.
G. O. No. 139, W. D., 1918.

He exhibited great courage and gallantry in taping off the line of departure for his company under a heavy shell and machine-gun fire. Later in the attack this daring officer was killed at the head of his company.
Posthumously awarded. Medal presented to mother, Mrs. Katherine C. Hook.

HOOK, FREDERICK E.
Near St. Etienne, France, Oct. 4-5, 1918.
R—New York, N. Y.
P—Rossville, Kans.
G. O. No. 37, W. D., 1919.

Passed Assistant Surgeon, U. S. Navy, attached to 5th Regiment, U. S. Marine Corps, 2d Division.
He established an advance dressing station in an exposed position under heavy artillery and machine-gun fire. He worked fearlessly and unceasingly, giving first aid to the wounded and directing their evacuation, until ordered to move to the rear.

*HOOK, JOSEPH J.
At Pouilly, France, Nov. 5 and 6, 1918.
R—Atlanta, Ga.
B—Atlanta, Ga.
G. O. No. 3, W. D., 1921.

First lieutenant, 356th Infantry, 89th Division.
Participating in the first reconnoissance of the damaged bridges at Pouilly, with 2 others, he advanced more than 550 meters beyond the American outposts, crossing 3 branches of the Meuse River and successfully encountering the enemy.
Posthumously awarded. Medal presented to father, A. Stephens Hook.

HOOK, WILLIAM B. (568360)
Near St. Thibaut, France, Aug. 9, 1918.
R—Los Angeles, Calif.
B—Zanesville, Ohio.
G. O. No. 88, W. D., 1919.

Sergeant, Company B, 4th Engineers, 4th Division.
While a member of a party engaged in constructing a bridge across the Vesle River in advance of the Infantry, Sergeant Hook voluntarily plunged into the stream under heavy enemy machine-gun and grenade fire, swam with a line to the opposite bank, which was held by the enemy, and securely tied the end of the bridge to the opposite bank.

HOOPER, MONROE C. (1316313)
Near Busigny, France, Oct. 10, 1918.
R—Andrews, N. C.
B—Resaca, Ga.
G. O. No. 81, W. D., 1919.

Corporal, Company I, 119th Infantry, 30th Division.
Knocked down by the explosion of an explosive bullet beneath his helmet, he regained his feet and led the members of his patrol against a hostile patrol which had been encountered. Though he and his men were outnumbered nearly 5 to 1, he led the advance against the enemy, himself killing 7 Germans.

HOOVER, CHARLES S. (1956211)
Near Brabant-sur-Meuse, France, Oct. 23, 1918.
R—Columbus, Ohio.
B—Lancaster, Pa.
G. O. No. 37, W. D., 1919.

Sergeant, 308th Trench Mortar Battery, 83d Division.
Sergeant Hoover, during the offensive action in the Boissois Bois, was in charge of 2 trench mortars. Wounded by shrapnel and knocked down by the explosion of bombs, he returned to the 1 mortar that was undamaged and continued to fire until it was put out of action.

HOOVER, WILLIAM J.
Near Verdilly, France, July 2, 1918.
R—Bell Buckle, Tenn.
B—Bell Buckle, Tenn.
G. O. No. 46, W. D., 1919.

First lieutenant, 27th Aero Squadron, Air Service.
On the morning of July 2 his flight patrol encountered the famous Richthofen Circus. Lieutenant Hoover was simultaneously attacked by 3 of the enemy and cut off from his comrades. By skillful maneuvering he avoided the effects of the concentrated fire and fearlessly attacked the 3. Although his machine was seriously damaged, he killed 1 of the enemy pilots and destroyed his plane, drove down another apparently out of control, and chased the third far into its own lines. He then continued the patrol until shortage of gasoline forced him to return.

*HOPE, BEN (98503)
Northeast of Château-Thierry, France, July 28, 1918.
R—Huntsville, Ala.
B—Hazel Green, Ala.
G. O. No. 117, W. D., 1918.

Private, Company D, 167th Infantry, 42d Division.
After he had been wounded in the head he continued to advance against the enemy until he had been wounded 3 more times. He died as a result of these wounds.
Posthumously awarded. Medal presented to mother, Mrs. Lizzie Hope.

HOPE, EDWARD TWIST
At Château-Thierry, France, June 6, 1918.
R—Waterboro, S. C.
B—Japan.
G. O. No. 110, W. D., 1918.

First lieutenant, 5th Regiment, U. S. Marine Corps, 2d Division.
He displayed coolness and courage in directing his platoon in attack, during which he was badly wounded, but refused assistance until wounded men near him had been treated.

HOPKINS, GEORGE A. (280173)
Near Gorges, France, Oct. 15, 1918.
R—Burlingame, Kans.
B—Burlingame, Kans.
G. O. No. 66, W. D., 1919.

Sergeant, Company G, 126th Infantry, 32d Division.
Although he had received a wound in the head, Sergeant Hopkins returned to his platoon, which had been stopped by a machine gun 100 yards away, and was suffering many casualties. Passing to the flank of the platoon, he located the enemy gunner and shot him, taking the other 2 members of the crew prisoners. The platoon thereupon resumed its advance.

*HOPKINS, STEPHEN T.
Between Chambley and Xammes, France, Sept. 13, 1918.
R—Boston, Mass.
B—Newtonville, Mass.
G. O. No. 15, W. D., 1923.

Second lieutenant, 96th Aero Squadron, Air Service.
While acting as pilot of a flight of 3 airplanes which were attacked by 15 enemy planes, he continued on his mission and bombed his objective despite the fact that he was surrounded by greatly superior numbers of the enemy. In the flight which followed the bombing operations, Lieutenant Hopkins and his observer continued the flight until shot down and killed, thus enabling 1 airplane of the flight to return to its airdrome with valuable information. The heroic conduct and superb devotion to duty of Lieutenant Hopkins proved an inspiration to the members of his squadron.
Posthumously awarded. Medal presented to father, Edward E. Hopkins.

*HOPKINS, THOMAS
Near Wesserling, Alsace, July 29, 1918.
R—Wichita, Kans.
B—Cheyenne, Wyo.
G. O. No. 126, W. D., 1918.

Second lieutenant, 139th Infantry, 35th Division.
He left his own post of comparative safety and voluntarily went through a heavy artillery barrage to rescue a private who was wounded and entangled in barbed wire. While engaged in this self-sacrificing effort he was killed.
Posthumously awarded. Medal presented to widow, Mrs. Edna Hopkins.

HOPKINS, WILLIAM T.
Near Le Charmel, France, July 28, 1918.
R—Savannah, Ga.
B—Savannah, Ga.
G. O. No. 44, W. D., 1919.

First lieutenant, 76th Field Artillery, 3d Division.
After his commanding officer, himself, and 32 of the battery had been wounded by a bomb from an enemy plane, Lieutenant Hopkins assumed command of the battery, remaining at his post until all the wounded had been evacuated and another officer found to replace him.

HOPP, HARVEY M. (1375816)_____ Private, first class, Battery C, 122d Field Artillery, 33d Division.
Near Romagne, France, Oct. 30, Under fire from enemy artillery, machine guns, and snipers, he crawled out in
1918.   the open to within 50 meters of a hostile position, remaining there several
R—Wilmette, Ill.   hours, and returning with valuable information concerning the enemy's
B—Chicago, Ill.   movements.
G. O. No. 44, W. D., 1919.

HOPPE, HARRY E. (1286009)_____ Sergeant, Company K, 115th Infantry, 29th Division.
Near Gildwiller, Alsace, July 31, He displayed extraordinary heroism, presence of mind, and physical endurance
1918.   during an enemy raid against his small advance post near Gildwiller, in
R—Baltimore, Md.   Alsace, on July 31, 1918. Although seriously wounded from grenades, he
B—Washington, D. C.   and 1 private from his company counterattacked a greatly superior enemy,
G. O. No. 7, W. D., 1925.   personally killing at least one of them with his rifle, preventing any entry
              by the enemy into his post, and finally reentering our lines after having
              received a second severe wound.

HOPPING, FLOYD (226561)_____ Mechanic, Company L, 363d Infantry, 91st Division.
Near Very, France, Sept. 29, 1918. When his company was stopped by a wide belt of barbed wire, Mechanic Hop-
R—Los Gatos, Calif.   ping voluntarily went forward under heavy machine-gun fire and cut lanes
B—San Jose, Calif.   through the wire in several places, thereby enabling his company to pass
G. O. No. 37, W. D., 1919.   through without casualties.

*HOPTA, JOSEPH L._____ Corporal, 55th Company, 5th Regiment, U. S. Marine Corps, 2d Division.
Near Vierzy, France, July 18, 1918. He captured an enemy machine gun and its crew single handed under a heavy
R—Newark, N. J.   concentrated machine-gun fire.
B—New Brunswick, N. J.   Posthumously awarded. Medal presented to father, John Hopta.
G. O. No. 117, W. D., 1918.

*HORGAN, HARRY B._____ Corporal, Company C, 165th Infantry, 42d Division.
At Ferme de Meurcy, near Villers- After his platoon had moved from an open field to the cover of the woods,
sur-Fere, France, July 31, 1918.   Corporal Horgan returned to the field under heavy machine-gun fire to rescue
R—Congress Junction, Ariz.   a wounded comrade lying in an exposed position, and was killed by the side
B—New York, N. Y.   of the man he tried to save.
G. O. No. 102, W. D., 1918.   Posthumously awarded. Medal presented to mother, Mrs. Nellie Horgan
              Birch.

HORN, IRVIN D. (914865)_____ Private, Company D, 7th Engineers, 5th Division.
Near Cunel, France, Oct. 14, 1918. He boldly entered a hostile dugout by himself, knowing that it was occupied
R—Avon, Ill.   by the enemy, and when he was confronted by a German major and his
B—Freestone, Pa.   orderly killed the major and captured the orderly.
G. O. No. 44, W. D., 1919.

HORNE, WILLIAM T. (1206616)_____ Private, first class, Machine Gun Company, 117th Infantry, 30th Division.
Near Busigny, France, Oct. 8-10, Near Busigny, on Oct. 8 he was gassed, but he remained on duty for 2 days until
1918.   the objective had been reached.
R—Knoxville, Tenn.
B—Sullivan County, Tenn.
G. O. No. 37, W. D., 1919.

HORSEMAN, CARL (1284269)_____ Sergeant, Company C, 115th Infantry, 29th Division.
In the Bois-de-Consenvoye, France, He voluntarily left shelter and went through heavy shellfire to rescue some men
Oct. 12, 1918.   who had been wounded when a shell struck their dugout. After adminis-
R—Cambridge, Md.   tering first-aid treatment he assisted them to a dressing station.
B—Cambridge, Md.
G. O. No. 37, W. D., 1919.

HORTON, EUGENE W._____ Private, first class, Company K, 308th Infantry, 77th Division.
Near Ville-Savoye, France, Aug. 22, When his company was attacked by greatly superior numbers of the enemy
1918.   Private Horton continued to operate his automatic rifle although exposed to
R—New York, N. Y.   heavy machine-gun fire. His gallant conduct was a material factor in the
B—Brooklyn, N. Y.   successful repulse of the enemy who were endeavoring to turn the flank of
G. O. No. 11, W. D., 1921.   his organization.

*HORTON, HARDIN F._____ Second lieutenant, 131st Infantry, 33d Division.
Near Bois-de-Chaume, France, Oct. Although twice wounded, Lieutenant Horton refused to leave his command,
10, 1918.   continuing to lead it through annihilating machine-gun and perilous shell fire.
R—Ithaca, N. Y.   When a machine gun on the flank opened fire and caused many casualties
B—Bethel, N. Y.   on his forces he seized a rifle and, firing from a standing position, he was
G. O. No. 78, W. D., 1919.   wounded a third time and killed.
              Posthumously awarded. Medal presented to mother, Mrs. Jennie S. Horton.

HORTON, VAN (2168859)_____ Corporal, Company E, 366th Infantry, 92d Division.
Near Lesseau, France, Sept. 4, 1918. During a hostile attack, preceded by a heavy minnenwerfer barrage, involving
R—Athens, Ala.   the entire front of the battalion, the combat group to which this courageous
B—Athens, Ala.   soldier belonged was attacked by about 20 of the enemy, using liquid fire.
G. O. No. 143, W. D., 1918.   The sergeant in charge of the group and 4 other men having been killed,
              he fearlessly rushed to receive the attack and the persistency with which he
              fought resulted in stopping the attack and driving back the enemy.

*HOSKINS, LEONARD C._____ Second lieutenant, 52d Artillery, Coast Artillery Corps.
Near La Chappelle, France, June He gave proof of great devotion and bravery when he entered a shell-swept
28, 1918.   area in search for wounded, and was killed while conducting several of his
R—East Las Vegas, N. Mex.   men to safety.
B—East Las Vegas, N. Mex.   Posthumously awarded. Medal presented to father, D. T. Hoskins.
G. O. No. 101, W. D., 1918.

HOSP, JAMES.
Near Landersbach, Alsace, Oct. 4, 1918.
R—Ilion, N. Y.
B—Washington, D. C.
G. O. No. 130, W. D., 1918.

Private, Company H, 53d Infantry, 6th Division.
He was a member of a party of 50 attacked by an enemy raiding party consisting of about 300 storm troops. During the raid and before the barrage lifted he crossed over open ground to his post and manned it alone throughout the engagement. During the latter part of the fight he was the sole protection for a group of soldiers near him who had been flanked by the enemy.

HOSTETTER, JAY F.
Near Greves Farm, France, July 14-15, 1918.
R—Lancaster, Pa.
B—Florin, Pa.
G. O. No. 27, W. D., 1920.

Second lieutenant, 10th Field Artillery, 3d Division.
Having discovered that 2 French guns on his left had lost all the crew during the terrific German bombardment, Lieutenant Hostetter requested and obtained permission to use them. Securing volunteers from his ranks, he pressed the guns into action, and for several hours poured an effective fire into the forces of the advancing enemy. His entire exploit was subjected to the extreme danger of high-explosive and gas shells.

*HOUCHINS, LYLE C.
Near Thiaucourt, France, Sept. 15, 1918.
R—Cincinnati, Ohio.
B—Coredo, W. Va.
G. O. No. 37, W. D., 1919.

Corporal, 73d Company, 6th Regiment, U. S. Marine Corps, 2d Division.
During an enemy counterattack Corporal Houchins voluntarily left a sheltered position and, in entire disregard for his own safety, set up his gun in the open on the advancing enemy. He broke up the counterattack within 100 yards of our line.
Posthumously awarded. Medal presented to father, Robert L. Houchins.

HOUSEHOLDER, JONATHAN A. (643042).
Near Dun-sur-Meuse, France, Nov. 7, 1918.
R—Irwin, Pa.
B—Westmoreland County, Pa.
G. O. No. 81, W. D., 1919.

Private, first class, section No. 590, Ambulance Service.
Proceeding along a road with a detachment of 7 ambulances he stopped his car when fired upon by a German sniper near the roadside, and though unarmed himself attacked and disarmed the German. Then, with the aid of other ambulance drivers, he attacked a patrol of Germans, capturing 5.

HOUSTON, CLYDE (1495276).
Near Tuilerie Farm, France, Nov. 4, 1918, and near Mouzon, France, Nov. 7-11, 1918.
R—Houston Heights, Tex.
B—Austin, Tex.
G. O. No. 46, W. D., 1919.

Private, first class, Company M, 9th Infantry, 2d Division.
On Nov. 4 Private Houston showed unusual courage and daring in carrying messages under heavy machine-gun and artillery fire. On Nov. 7-11 he carried messages between his company and battalion headquarters while the enemy were endeavoring to cut off communication by machine-gun and shell fire.

*HOUSTON, SAMUEL HUMES.
Near Ville-Savoye, France, Aug. 4, 1918.
R—Baltimore, Md.
B—Baltimore, Md.
G. O. No. 62, W. D., 1919.

Major, 58th Infantry, 4th Division.
With but 15 minutes in which to prepare his battalion for attack, Major Houston on horseback galloped from flank to flank, fully exposed to deadly artillery fire, in order to make the necessary preparations for the advance. After his leading element had started the attack he was killed by an enemy shell.
Posthumously awarded. Medal presented to mother, Mrs. Mary Houston.

HOVATTER, EVERETT E. (937709).
At Medeah Ferme, France, Oct. 4, 1918.
R—Auburn, S. Dak.
B—Thornton, W. Va.
G. O. No. 37, W. D., 1919.

Private, medical detachment, 5th Machine Gun Battalion, 2d Division.
When the artillery barrage of the enemy had lowered upon the infantry to which he was attached, necessitating an alteration in their position, Private Hovatter, regardless of personal safety, remained to render aid to the wounded and to provide for their evacuation.

HOWARD, CHARLES (145424).
Near Suippes, France, July 15, 1918.
R—Highwood, Minn.
B—Highwood, Minn.
G. O. No. 35, W. D., 1919.

Private, Battery B, 151st Field Artillery, 42d Division.
After all communication by telephone had been destroyed by heavy enemy shellfire, Private Howard volunteered and carried a message to the artillery post of command, calling for a barrage at several critical points, through an area subject to terrific shell fire.

HOWARD, CHARLIE (734196).
At Frapelle, France, Aug. 17, 1918.
R—Fonde, Ky.
B—Lafollette, Tenn.
G. O. No. 15, W. D., 1919.

Corporal, Company M, 6th Infantry, 5th Division.
Corporal Howard, although severely wounded early in the action, refused to quit the fight, and shortly afterwards, when his second leader was killed, took command of and led his section forward to its objective.

*HOWARD, CHESTER R.
Near Belleau Wood, France, July 20, 1918.
R—Mount Vernon, Iowa.
B—Curtis, Nebr.
G. O. No. 125, W. D., 1918.

First lieutenant, 104th Infantry, 26th Division.
Lieutenant Howard was wounded early in the action, but set a splendid example of personal bravery by retaining command of his company and leading it through a violent artillery and machine-gun barrage until he fell exhausted after advancing 200 meters.
Posthumously awarded. Medal presented to mother, Mrs. C. C. Howard.

HOWARD, GEORGE H.
Near St. Juvin, France, Oct. 16-26, 1918.
R—Philadelphia, Pa.
B—Philadelphia, Pa.
G. O. No. 46, W. D., 1919.

Second lieutenant, 326th Infantry, 82d Division.
On the 16th of October he was severely wounded by a machine-gun bullet which struck him in the hip. After having his wound dressed at the first-aid station he refused to be evacuated but instead returned to his platoon and continued to lead it for the remaining 10 days of the engagement. Although suffering intense pain from his wound, he constantly refused to leave his command until ordered to the rear by his battalion commander.

HOWARD, HARRY (1550404).
Near Le Charmel, France, July 28, 1918.
R—Summit, N. J.
B—Madison, N. J.
G. O. No. 44, W. D., 1919.

Private, Battery B, 76th Field Artillery, 3d Division.
After he, himself, had been badly wounded and, seeing his commanding officer and 32 comrades wounded by a bomb from an enemy plane, he refused treatment, but assisted in the evacuation of the wounded.

HOWARD, HENRY (556412)............
Near Septsarges, France, Sept. 27, 1918.
R—Valley View, Ky.
B—Irvin, Ky.
G. O. No. 64, W. D., 1919.

Sergeant, Company A, 39th Infantry, 4th Division.
Although seriously wounded during a bombardment which scattered his men and caused his company and battalion to retire behind a ridge in the rear, Sergeant Howard, with about 15 men, held the advanced position under the continuous fire of machine guns, 1-pounders, and artillery until relieved the following day by another battalion. He insisted on remaining with his detachment until the commanding officer of the relieving battalion personally directed his evacuation.

HOWARD, JAMES L............
At Marcheville, France, Sept. 26, 1918.
R—Hartford, Conn.
B—Hartford, Conn.
G. O. No. 132, W. D., 1918.

Lieutenant colonel, division machine-gun officer, 101st Machine Gun Battalion, 26th Division.
He directed the machine-gun attack in person. Entering Marcheville ahead of the troops, he rendered great assistance while the town changed hands four times. When he was in a small party cut off and surrounded by the enemy and under fire from every direction, by his coolness and resourcefulness he assisted materially in aiding the party to withdraw. He effectively organized machine-gun defenses when the enemy was endeavoring to drive our troops from the town. During the entire day he was under intense artillery bombardment, machine-gun and rifle fire, and hand-to-hand conflict with the enemy.

HOWARD, ROBERT P (1386525)........
Near Chipilly Ridge, France, Aug. 9, 1918.
R—Chicago, Ill.
B—New York, N. Y.
G. O. No. 71, W. D., 1919.

Corporal, Company B, 131st Infantry, 33d Division.
He displayed extraordinary bravery when, after being wounded by machine-gun bullet and with a piece of shrapnel in his lung, he refused to be evacuated. Stating that he knew most of the noncommissioned officers in the company had been killed or wounded, he remained on duty with his men, setting an example of coolness and courage.

HOWARD, WILLIAM (444484)............
Near Charpentry, France, Oct. 3, 1918.
R—Paris, Ky.
B—Nicholas County, Ky.
G. O. No. 103, W. D., 1919.

Private, Company M, 16th Infantry, 1st Division.
With a rescue party of 3 other men, Private Howard went 400 yards in advance of our lines to rescue a wounded soldier who had lain in an exposed place before an enemy machine-gun post for five days. The mission was successfully performed in broad daylight under a sweeping fire from enemy machine guns.

HOWARD, WILLIAM HARRIS H............
South of Soissons, France, July 18–19, 1918.
R—Lockport, Ill.
B—Lockport, Ill.
G. O. No. 139, W. D., 1918.

First lieutenant, 9th Infantry, 2d Division.
He conspicuously distinguished himself by his gallant actions in leading his platoon through 2 fierce attacks. By his splendid example in facing enemy fire, his platoon fought with the same qualities and succeeded in routing the enemy until the final objective was reached. His personal disregard of life consequence to himself under terrific shellfire was noted at all times by his men along the line. He was wounded just before his objective was reached.

*HOWE, GEORGE P............
On the Tower Hamlets Spur, east of Ypres, Belgium, Sept. 28, 1917.
R—Boston, Mass.
B—Lawrence, Mass.
G. O. No. 83, W. D., 1918.

First lieutenant, Medical Corps, attached to 37th Division, British Expeditionary Forces.
Although wounded in the head early on the morning of Sept. 28, 1917, during the operation on the Tower Hamlets Spur, east of Ypres, he displayed conspicuous courage and devotion in attending to wounded under very heavy and continuous shellfire, refusing to leave and continuing at his aid post until killed by a shell.
Posthumously awarded. Medal presented to widow, Mrs. Marion C. Howe.

HOWE, HARRY R............
In the Bois-de-la Croisette, France, July 14, 1918.
R—Mansfield, Mass.
B—Mansfield, Mass.
G. O. No. 125, W. D., 1918.

First lieutenant, 101st Engineers, 26th Division.
After being badly burned on the left hand by the explosion of a mustard-gas shell he declined an opportunity to be relieved and worked for more than an hour under heavy gas and high-explosive shell bombardment, getting his men out of the area of concentrated gas. Unable to use his left hand, he remained on duty during the July offensive.

HOWE, MAURICE W............
Near Haumont, France, Sept. 22, 1918.
R—Fitchburg, Mass.
B—Merrimac, Mass.
G. O. No. 21, W. D., 1919.

Captain, 167th Infantry, 42d Division.
He commanded an early morning raid on the town of Haumont, and not only executed the raid successfully, but returned alone a second time to the town to be assured that none of his men had been left wounded. He inflicted severe losses on the enemy and took 17 prisoners.

HOWE, WILLIAM J. (1677330)............
At Cantigny, France, May 28–30, 1918.
R—New Haven, Conn.
B—Schenectady, N. Y.
G. O. No. 95, W. D., 1918.

Private, Company H, 28th Infantry, 1st Division.
Acting as company runner on May 28–30, 1918, at Cantigny, France, he showed exceptional bravery in carrying messages through heavily shelled area, where he was also a target for snipers. Although rendered unconscious by shell explosion on one of his trips, he continued on duty as soon as he regained consciousness.

HOWLAND, HARRY S............
Near Cheppy, France, Sept. 26, 1918.
R—Chicago, Ill.
B—Sweden.
G. O. No. 71, W. D., 1919.

Colonel, 138th Infantry, 35th Division.
After losing touch with his first and second line battalions, due to unusually heavy fog, Colonel Howland, with a mixed detachment (partly noncombatants), penetrated to a point immediately in front of the German line of resistance. While getting better cover for his detachment under heavy shellfire and direct machine-gun fire, Colonel Howland was wounded in the hand by a shell fragment. After three hours' delay tanks arrived, and Colonel Howland advanced with his detachment, taking the enemy position, with many prisoners. After remaining in the attack for over 7 hours he was evacuated because of the wound in his hand.

HOWLAND, SYLVESTER J. (1203613)............
East of Ronssoy, France, Sept. 29, 1918.
R—Waterford, N. Y.
B—Waterford, N. Y.
G. O. No. 20, W. D., 1919.

Private, Company B, 105th Infantry, 27th Division.
During the operations against the Hindenburg line he left shelter, went forward under heavy shell and machine-gun fire, and succeeded in rescuing a wounded soldier, thereby displaying great bravery and gallantry. In performing this act he was wounded.

HOY, HENRY (1390393) ....
Near Forges, France, Sept. 26, 1918.
R—Chicago, Ill.
B—Chicago, Ill.
G. O. No. 44, W. D., 1919.

Private, Company A, 132d Infantry, 33d Division.
Private Hoy saw a hand grenade dropped near an officer of his company, which endangered not only the officer's life but also many members of the company who were in the vicinity. Rushing to the spot, he picked up the bomb and hurled it in the direction of the enemy. It exploded in the air and the lives of his comrades were thus saved by his act.

HOYT, RUSSELL E. (72039) ....
At Bois-Brule, France, Apr. 12, 1918.
R—Cambridge, Mass.
B—Cambridge, Mass.
G. O. No. 28, W. D., 1918.

Corporal, Company E, 104th Infantry, 26th Division.
During action Apr. 12, 1918, he displayed courage and self-sacrificing devotion to duty in going to communication trench with one comrade and holding back advance of enemy through trench until killed at his post.
Posthumously awarded. Medal presented to father, William Hoyt.

HUARD, SOLOMON (2725071) ....
On Hill 272, near Fleville, France, Oct. 8, 1918.
R—Winslow, Me.
B—Canada.
G. O. No. 35, W. D., 1920.

Private, first class, Company C, 16th Infantry, 1st Division.
While carrying a message for his platoon commander Private Huard was fired upon by an enemy machine-gun nest. He dropped to the ground and opened fire on it, and in the fight which ensued killed 2 of the crew and forced the remaining 3 to surrender, sending them to the rear. After delivering his message he returned to his organization and took command of a squad, which he led throughout the action.

HUBBARD, HAROLD G. (1323006) ....
Near Ypres, Belgium, Aug. 23, 1918.
R—Wilmington, N. C.
B—Clinton, N. C.
G. O. No. 35, W. D., 1919.

Sergeant, Company C, 115th Machine-Gun Battalion, 30th Division.
During heavy hostile bombardment, he voluntarily left his dugout and went through the shellfire to the assistance of his wounded platoon commander. After taking the officer to a partially sheltered position, he ran 400 yards through the barrage to secure a litter and assisted the stretcher bearer in carrying the wounded officer to a dressing station.

HUBBARD, HENRY G. (44163) ....
Near Cantigny, France, June 2, 1918.
R—Tallahassee, Fla.
B—Sale City, Ga.
G. O. No. 35, W. D., 1919.

Private, Company L, 16th Infantry, 1st Division.
He went forward under intense machine-gun and artillery fire and assisted in the removal of a wounded soldier over a distance of 1 kilometer.

HUBBARD, WILLIS W ....
Near Limeyville, France, Sept. 11, 1918.
R—Champaign, Ill.
B—Wellington, Kans.
G. O. No. 32, W. D., 1919.

First lieutenant, acting regimental adjutant, 17th Field Artillery, 2d Division.
Although starting on a mission which would have taken him away from the place of danger, Lieutenant Hubbard went to the aid of a wounded soldier and was himself severely wounded in the head. Almost blinded by blood, he assisted the soldier to a place of safety and later insisted upon being allowed to remain with the regiment.

*HUDNALL, JAMES W. (1326800) ....
Near Bellicourt, France, Sept. 29, 1918.
R—Spray, N. C.
B—Beardstown, Ill.
G. O. No. 21, W. D., 1919.

Sergeant, Company G, 120th Infantry, 30th Division.
After being twice wounded, Sergeant Hudnall continued to lead his platoon in attack, capturing two machine guns. In later action he received additional wounds which caused his death.
Posthumously awarded. Medal presented to sister, Miss Eva Hudnall.

HUDSON, BEN S ....
Near Varennes and Baulny, France, Sept. 26-28, 1918.
R—Fredonia, Kans.
B—Piedmont, Kans.
G. O. No. 37, W. D., 1919.

Captain, 137th Infantry, 35th Division.
Captain Hudson personally led an attack on a machine-gun nest which was holding up the advance and captured it, taking 9 prisoners and 3 guns. On the two following days he led his battalion in attacks under direct shell and machine-gun fire until he was wounded in the foot. As litter bearers were carrying him to the rear they met a severely wounded soldier, whereupon Captain Hudson ordered the litter bearers to carry the wounded soldier to the dressing station while he walked.

HUDSON, DONALD ....
Near Fère-en-Tardenois, France, Aug. 1, 1918.
R—Kansas City, Mo.
B—Topeka, Kans.
G. O. No. 36, W. D., 1919.

First lieutenant, 27th Aero Squadron, Air Service.
A protection patrol of which Lieutenant Hudson was a member was attacked by a large formation of enemy planes; he was separated from the formation and forced to a low altitude by 4 enemy planes (type Fokker). He shot down 1, drove off the other 3, and started to our lines with a damaged machine, but was attacked by 2 biplane planes. He shot down both of these planes and by great perseverance and determination succeeded in reaching our lines.

HUDSON, HAROLD A. (1230189) .....
Near Bellicourt, France, Sept. 29, 1918.
R—Asheville, N. C.
B—Toledo, Iowa.
G. O. No. 44, W. D., 1919.

Sergeant, first class, Company C, 105th Field Signal Battalion, 30th Division.
Sergeant Hudson and a number of other members of a signal detachment were wounded by shellfire while proceeding through an enemy counterbarrage to the front line, but, disregarding his own injuries, this soldier administered first aid to his wounded comrades and then extended a telephone line to the advance message center, and with 5 men maintained and operated the message center.

HUEBNER, CLARENCE R ....
Near Cantigny, France, May 28-30, 1918.
R—Bushton, Kans.
B—Bushton, Kans.
G. O. No. 44, W. D., 1920.
Distinguished-service medal also awarded

Major, 28th Infantry, 1st Division
For 3 days near Cantigny, France, May 28-30, 1918, he withstood German assaults under intense bombardment, heroically exposing himself to fire constantly in order to command his battalion effectively. Although his command had lost half its officers and 30 per cent of its men, he held his position and prevented a break in the line at that point.
Oak-leaf cluster.
For the following act of extraordinary heroism in action Major Huebner was awarded an oak-leaf cluster to be worn with the distinguished-service cross. South of Soissons, France, July 18-23, 1918, he displayed great gallantry, and, after all the officers of his battalion had become casualties, he reorganized his battalion while advancing, captured his objective and again reorganized his own and another battalion, carrying the line forward. He remained continuously on duty until wounded on the second day of the action.

HUELSER, CHARLES A_____
  In front of Landres-et-St. Georges,
    France, Oct. 14, 1918.
  R—Brooklyn, N. Y.
  B—Brooklyn, N. Y.
  G. O. No. 95, W. D., 1919.

Second lieutenant, 165th Infantry, 42d Division.
Sent forward through terrific machine-gun and artillery fire to take command of a platoon in the front lines, he displayed marked personal courage and heroism, inspiring the men serving under him by his example of fearlessness. During a heavy counterattack he showed great gallantry, cooly placing his guns where they could best fire upon the advancing enemy, although subjected the while to intense enemy fire. He succeeded in breaking up the enemy attack, and although wounded by a machine-gun bullet refused to be evacuated until the enemy had withdrawn.

HUFF, CHESTER RAY (1749072)_____
  Near Thiaucourt, France, Sept. 17–
    27, 1918.
  R—Jersey City, N. J.
  B—Topsfield, Me.
  G. O. No. 44, W. D., 1919.

Private, medical detachment, 310th Infantry, 78th Division.
During the night of Sept. 17, Private Huff gave proof of great devotion to duty by leaving protection and assisting another comrade in bringing a wounded soldier to safety. Again, on the morning of Sept. 27, he left shelter and journeyed over 500 yards to the side of a wounded comrade, through shellfire, carrying the victim to the first-aid station.

HUFFMAN, CHARLES E., Jr. (42123)_____
  Near Soissons, France, July 19–20,
    1918.
  R—Gadsden, Ala.
  B—Rome, Ga.
  G. O. No. 35, W. D., 1920.

Sergeant, Company C, 16th Infantry, 1st Division.
Sergeant Huffman organized an automatic-rifle squad under heavy machine-gun fire and led it in an attack upon an enemy machine-gun nest. The guns and crew were captured. The enemy gun was firing with serious effect on the right flank of his company, causing heavy losses. Later he took charge of a bombing squad and raided an enemy machine-gun position, capturing the guns and crew. In this latter operation he was severely wounded.

HUFFSTATER, LEON DAVID_____
  At Bouresches, France, June 6, 1918.
  R—Oswego, N. Y.
  B—Richland, N. Y.
  G. O. No. 110, W. D., 1918.

Private, 97th Company, 6th Regiment, U. S. Marine Corps, 2d Division.
During the action at Bouresches, France, on June 6, 1918, he volunteered to leave shelter to bring in wounded, and upon receiving permission to do so he carried injured comrades through artillery and machine-gun fire at great risk to his own life.

HUFSTEDLER, ERIE C_____
  In the occupation of Bouresches,
    France, June 6, 1918.
  R—Birdell, Ark.
  B—Birdell, Ark.
  G. O. No. 16, W. D., 1923.

Private, 79th Company, 6th Regiment, U. S. Marine Corps, 2d Division.
After being severely wounded in the occupation of Bouresches, France, on June 6, 1918, he refused to go to the rear, but remained and assisted with the wounded, displaying great self-sacrifice and devotion to duty.

°HUGHES, FLOYD A. (1520626)_____
  Near Montfaucon, France, Sept. 26–
    30, 1918.
  R—Canton, Ohio.
  B—Canton, Ohio
  G. O. No. 130, W. D., 1918.

Mechanic, Company C, 146th Infantry, 37th Division.
This soldier was constantly on duty as a runner during the offensive west of the Meuse River, many times carrying messages through heavy machine-gun and shellfire. On Sept. 30, when the enemy was reported to be forming for a counterattack on the left flank, he volunteered to take the information to the battalion commander. Passing through heavy shellfire, he delivered the message in time to enable the battalion commander to protect the threatened flank, but he was killed by a shell on his return trip to his company. Posthumously awarded. Medal presented to widow, Mrs. Kathryn Hughes.

HUGHES, GEORGE DEWEY_____
  Near St. Etienne, France, Oct. 4,
    1918.
  R—Salt Lake City, Utah.
  B—Spanish Fork, Utah.
  G. O. No. 44, W. D., 1919.

Corporal, 20th Company, 5th Regiment, U. S. Marine Corps, 2d Division.
He conducted a patrol to the front, located machine-gun nests, and gained contact with the enemy under very severe artillery and machine-gun fire, and set an example of calmness and courage under these hazardous conditions.

HUGHES, GEORGE E. (2262208)_____
  Near Eclisfontaine, France, Sept.
    27–30, 1918.
  R—Rainbow, Calif.
  B—Providence, R. I.
  G. O. No. 128, W. D., 1918.

Sergeant, Company B, 348th Machine Gun Battalion, 91st Division.
He was seriously gassed Sept. 27, but refused to leave his post and continued working and fighting with his comrades under heavy shelling for 3 days till complete exhaustion forced him to be evacuated.

HUGHES, WILLIAM E. (1289558)_____
  In the Bultruy Bois, France, Oct. 15,
    1918.
  R—Lynchburg, Va.
  B—Amherst County, Va.
  G. O. No. 98, W. D., 1919.

Private, first class, Company L, 116th Infantry, 29th Division.
Though he had been wounded in the leg and ordered to be evacuated, Private Hughes remained with his company in the advance, displaying marked fortitude. Twenty-four hours later his leg had become so stiff that he was compelled to go to the rear. This soldier had previously displayed marked courage by remaining with his automatic rifle under heavy fire, after 2 members of his squad had become casualties, until he had silenced an enemy machine gun.

HUGHES, WILLIAM J. (2156830)_____
  Near Brancourt, France, Oct. 8, 1918.
  R—Goodhope, Ill.
  B—Flemington, N. J.
  G. O. No. 74, W. D., 1919.

Private, Company M, 118th Infantry, 30th Division.
Because his company had already suffered heavy casualties and he realized that all men possible were needed on the firing line, he refused to be evacuated when seriously wounded, and continued in the advance till the objective had been reached and consolidated 8 hours later.

°HULBERT, HENRY LEWIS_____
  At Chateau-Thierry, France, June
    6, 1918.
  R—Riverdale, Md.
  B—England.
  G. O. No. 110, W. D., 1918.

Marine Gunner, 5th Regiment, U. S. Marine Corps, 2d Division.
At Chateau-Thierry, France, June 6, 1918, he displayed extraordinary heroism during attack on the enemy's lines, during which time he constantly exposed himself to the enemy's fire without regard for personal danger, thereby assuring the delivery of supplies.
Posthumously awarded. Medal presented to widow, Mrs. Victoria C. Hulbert.

HULETT, FORREST (2038156)_____
  Near Etraye Ridge, France, Oct.
    24, 1918.
  R—Bridgeport, Mich.
  B—Wellston, Mich.
  G. O. No. 37, W. D., 1919.

Private, Company F, 116th Infantry, 29th Division.
Crawling to the aid of 2 men who had been left when his company withdrew, he was exposed to both sniper and machine-gun fire. On reaching the side of his comrades he found them both to be dead, after which he made his way safely back to our lines.

HULL, ALSON J._____
Near Cormicy, France, May 27, 1918.
R—Troy, N. Y.
B—Berlin, N. Y.
G. O. No. 30, W. D., 1921.

First lieutenant, Medical Corps, attached to 4th Battalion, South Stafford Regiment, British Expeditionary Forces.
When forced to withdraw his aid post he went back to attend a severely wounded man and succeeded in rescuing him when the enemy were only 50 yards away.

HULL, HENRY C. (1211207)_____
Near Bony, France, Sept. 29, 1918.
R—White Plains, N. Y.
B—Brooklyn, N. Y.
G. O. No. 32, W. D., 1919.

Sergeant, Company H, 107th Infantry, 27th Division.
After being severely wounded in the head, he reorganized a badly scattered line in the midst of heavy shell and machine-gun fire and led it into effective combat against the enemy. He continued to lead his men forward until loss of blood compelled him to place another in command.

HULSART, C. RAYMOND_____
At Gouzeaucourt, France, Nov. 30, 1917.
R—Plainfield, N..J.
B—Seabright, N. J.
G. O. No. 129, W. D., 1918.

Captain, 11th Railway Engineers.
When an unarmed working party under his command were subjected to a sudden German attack at Gouzeaucourt, France, Nov. 30, 1917, he displayed extraordinary heroism in going through heavy shellfire to direct the escape of his men, remaining under fire until all had left, and going back into the barrage to assist in carrying a wounded soldier of another unit, and returning into the barrage a second time to search for a wounded British soldier.

HULTZEN, LEE S._____
Near Vieville-en-Haye, France, Sept. 26, 1918.
R—Norwich, N. Y.
B—Burlington Flats, N. Y.
G. O. No. 26, W. D., 1919.

First lieutenant, 311th Infantry, 78th Division.
After reaching his objective with a platoon of about 15 men, Lieutenant Hultzen organized his platoon and held it with 3 captured German machine guns. He cleaned out a "pill box" and attacked a dozen of the enemy with practically no assistance.

HUMBIRD, JOHN A._____
At Marcheville, France, Sept. 26, 1918.
R—Sandpoint, Idaho.
B—Hudson, Wis.
G. O. No. 139, W. D., 1918.

First lieutenant, 102d Machine Gun Battalion, 26th Division.
This officer displayed indomitable courage and leadership on numerous occasions during this engagement. Leading a small group of men thrugh barbed-wire entanglements in the face of machine-gun fire and hand grenades, he cleaned out a trench, capturing a strong enemy machine-gun emplacement and its entire crew. When the town of Marcheville fell into our hands, he organized a machine-gun position under heavy machine-gun fire, locating a position for antiaircraft guns, during which time hostile battle planes were flying low and firing upon our troops. After the recapture of Marcheville by the enemy, he led his platoon in the counterattack.

HUMMER, JOSEPH H. (1256327)_____
Near Fismes, France, Sept. 5, 1918.
R—Riverside, Pa.
B—Mansdale, Pa.
G. O. No. 13, W. D., 1923.

Private, medical detachment, 108th Machine Gun Battalion, 28th Division.
While assigned to duty as first-aid man he rescued 2 wounded men despite intense hostile machine-gun and rifle fire. While so engaged he himself was badly wounded, but continued on his mission, carrying the wounded men to places of safety. The indomitable spirit of self-sacrifice and splendid devotion to duty thus displayed served as an example to the men of his organization, inspiring them to greater endeavors.

HUMPHREY, CHARLES F., Jr._____
At Angeles, Luzon, P. I., Aug. 16, 1899.
R—Washington, D. C.
B—Washington, D. C.
G. O. No. 126, W. D., 1919.

First lieutenant, 12th Infantry, U. S. Army.
For extraordinary heroism in action against hostile insurgents at Angeles, Luzon, Philippine Islands, Aug. 16, 1899, while serving as first lieutenant, 12th U. S. Infantry. While exposed to heavy enemy fire he dressed the wounds of 3 men of his company, and his courage, coolness, and good judgment were an inspiration to his command.

HUMPHREY, FRANK H. (57730)_____
Near Soissons, France, July 19, 1918.
R—Minneapolis, Minn.
B—Black River Falls, Wis.
G. O. No 117, W. D., 1918.

Mechanic, Company G, 28th Infantry, 1st Division.
He courageously charged ahead with 1 man and attacked a machine-gun nest, putting the gun out of commission and killing the gunners. He showed conspicuous gallantry in action throughout the whole attack.

HUMPHREY, RAY H._____
Near the Bois-de-Septsarges, France, Oct. 4, 1918.
R—Union, N. Y.
B—Stottville, N. Y.
G. O. No. 81, W. D., 1919.

Captain, Medical Corps, attached to 130th Infantry, 33d Division.
Having just been transferred to the 130th Infantry, Captain Humphrey was seriously wounded in the head while on his way to take up his work at the regimental aid post. He nevertheless refused to be evacuated, but continued on duty caring for the wounded at this place. He later proceeded under severe artillery fire to the battalion aid post, where he continued his work under most trying conditions.

HUNT, CHARLES H. (74528)_____
At Blanc Mont Ridge, France, Oct. 3, 1918.
R—Canada.
B—Rochester, Vt.
G. O. No. 21, W. D., 1919.

Private, Headquarters Company, 4th Machine Gun Battalion, 2d Division.
Detailed with two other soldiers to undertake a dangerous reconnaissance, he made his way to the point designated through heavy shell and machine-gun fire. Neglecting a wound in the back, he proceeded to his designation and to the dressing station, where he was tagged for evacuation. Regardless of his wound, he returned and remained on duty until the battalion was relieved on Oct. 10, 1918.

HUNT, DAVID B. (62967)_____
At Bourbelin, France, July 16, 1918.
R—Jamaica Plain, Mass.
B—Ireland.
G. O. No. 44, W. D., 1919.

Corporal, Machine Gun Company, 101st Infantry, 26th Division.
When an artillery barrage was laid down on the section of which he was a member, killing 3 and wounding 5, including the section sergeant and corporal, Corporal Hunt, severely wounded himself, carried the sergeant to the first-aid station returned to his post through the barrage, assumed leadership of the section, and successfully directed their operations until he fell from exhaustion.

HUNT, HARMON (2969885)............
At Fismette, France, Aug. 27, 1918.
R—Chauncey, W. Va.
B—Mingo County, W. Va.
G. O. No. 4, W. D., 1923.

Private, Company H, 112th Infantry, 28th Division.
When the 2 companies of the 112th Infantry holding the town of Fismette were forced to withdraw after suffering heavy casualties as a result of desperate enemy assault, Private Hunt with 7 other men volunteered to remain and keep down an enfilading fire, which was seriously interfering with the withdrawal across the Vesle River of the few remaining men of the command. When again attacked by an overwhelming force, he was one of the defenders of the crossing and made possible the retirement of about 20 of his comrades. He continued to hold the position against great odds, inflicting heavy casualties upon the enemy, and finally retired under orders, fighting his way through machine-gun, rifle, and trench-mortar fire. His exceptional disregard for personal safety, resourcefulness, and bravery were an inspiration to all his comrades.

HUNT, HENRY BOICE............
Near Remicourt, France, Oct. 3, 1918.
R—Chesterfield, S. C.
B—Chesterfield, S. C.
G. O. No. 50, W. D., 1919.

Private, Company L, 118th Infantry, 30th Division.
While the advance of his company was being held up by terrific machine-gun fire from several enemy nests and after all the members of his squad had become casualties, he made his way forward with his automatic rifle. Under a continual rain of machine-gun and shell fire, he operated his gun against the enemy until the gun jammed; whereupon he took a shovel, rushed a machine-gun post 75 yards away, and killed the gunner, thereby enabling the continuance of the advance.

HUNT, LEROY P............
Near St. Etienne, France, Oct. 3-4, 1918.
R—Berkeley, Calif.
B—Newark, N. J.
G. O. No. 35, W. D., 1919.

Captain, 5th Regiment, U. S. Marine Corps, 2d Division.
After 6 hours of severe fighting, Captain Hunt and his men succeeded in reducing a large sector of trenches and machine-gun nests and captured 300 prisoners. On Oct. 4, near St. Etienne, he constantly exposed himself to enemy fire while leading his men toward their objective. His gallant conduct gave his men confidence to completely rout superior forces concentrating for a counterattack.

*HUNTER, DANIEL AMOS............
At Chateau-Thierry, France, June 6, 1918.
R—Baltimore, Md.
B—Baltimore, Md.
G. O. No. 110, W. D., 1918.

First sergeant, 67th Company, 5th Regiment, U. S. Marine Corps, 2d Division.
During the attack at Chateau-Thierry, France, on June 6, 1918, he fearlessly exposed himself and encouraged all men near him, although he himself was wounded 3 times. He subsequently died of wounds.
Posthumously awarded. Medal presented to widow, Mrs. Ida M. Hunter.

HUNTER, DAVID, Jr............
At Trugny Woods, France, July 23, 1918.
R—Rockford, Ill.
B—Rockford, Ill.
G. O. No. 81, W. D., 1919.

Second lieutenant, 101st Infantry, 26th Division.
Under terrific machine-gun and sniper fire, Lieutenant Hunter led his battalion scouts into Trugny Woods 200 yards ahead of the assaulting waves and, crawling to within 30 feet of an enemy machine-gun nest, killed or wounded every member of the crew. He then crawled from point to point along the front under intense fire and noted the position of the enemy guns, returning with information which enabled his battalion to clear the woods.

HUNTER, FRANCIS R............
Near Cierges, France, Oct. 4, 1918.
R—Racine, Wis.
B—Racine, Wis.
G. O. No. 24, W. D., 1920.

Lieutenant colonel, 76th Field Artillery, 3d Division.
While directing the operations of his battalion in a forward position under heavy artillery fire a high-explosive shell exploded under his horse, killing the horse and severely wounding Lieutenant Colonel Hunter in the right leg. In spite of his serious injuries he insisted upon seeing his battery commanders and before being evacuated he personally charged each with his mission, urging them to put forth all possible efforts in carrying out their important mission.

HUNTER, FRANK O'D............
In the region of Ypres, Belgium, June 2, 1918.
R—Savannah, Ga.
B—Savannah, Ga.
G. O. No. 147, W. D., 1918.

First lieutenant, pilot, 103d Aero Squadron, Air Service.
Lieutenant Hunter, while on patrol, alone attacked 2 enemy biplanes, destroying 1 and forcing the other to retire. In the course of the combat he was wounded in the forehead. Despite his injuries he succeeded in returning his damaged plane to his own aerodrome.
Oak-leaf clusters (4).
A bronze oak leaf, for extraordinary heroism in action in the region of Champey, France, Sept. 13, 1918. He, accompanied by 1 other plane, attacked an enemy patrol of 6 planes. Despite numerical superiority and in a decisive combat, he destroyed 1 enemy plane and, with the aid of his companion, forced the others within their own lines.
A bronze oak leaf, for extraordinary heroism in action near Verneville, France, Sept. 17, 1918. Leading a patrol of 3 planes, he attacked an enemy formation of 8 planes. Although outnumbered, they succeeded in bringing down 4 of the enemy. Lieutenant Hunter accounted for 2 of these.
A bronze oak leaf, for extraordinary heroism in action in the region of Linydevant-Dun, France, Oct. 4, 1918. While separated from his patrol he observed an allied patrol of 7 planes (Breguets) hard pressed by an enemy formation of 10 planes (Fokker type). He attacked 2 of the enemy that were harassing a single Breguet, and in a decisive fight destroyed 1 of them. Meanwhile 5 enemy planes approached and concentrated their fire upon him. Undaunted by their superiority, he attacked and brought down a second plane.
A bronze oak leaf, for extraordinary heroism in action in the region of Bantheville, France, Oct. 6, 1918. While on patrol he encountered an enemy formation of 6 monoplanes. He immediately attacked and destroyed 1 enemy plane and forced the others to disperse in confusion.

*HUNTER, JONES W. (75160)............
At Vaux, France, July 3, 1918.
R—Charlotte, N. C.
B—Huntersville, N. C.
G. O. No. 99, W. D., 1918.

Private, first class, Company G, 9th Infantry, 2d Division.
He showed himself to be conspicuously brave when at Vaux, France, July 3, 1918, although wounded by a shell and unable to carry ammunition, for which duty he had been detailed, he went over the top with his squad and fought heroically with it until killed.
Posthumously awarded. Medal presented to half-sister, Mrs. Mack Auten.

HUNTER, ROBERT L. (1283790)........
Near Verdun, France, Oct. 24, 1918.
R—Baltimore, Md.
B—Baltimore, Md.
G. O. No. 1, W. D., 1919.

Sergeant, Company A, 115th Infantry, 29th Division.
Disregarding his own danger, and encouraging his men by word and action, he led his platoon in an attack upon a machine-gun nest and was instantly killed. His men, inspired by his heroism, continued the attack and captured the machine-gun nest.
Posthumously awarded. Medal presented to father, William J. Hunter.

HUNTER, WILLIAM L. (2105021)........
In Bois-du-Fays, France, Sept. 26–Oct. 6, 1918.
R—Vincent, Ohio.
B—Belpre, Ohio.
G. O. No. 5, W. D., 1920.

Private, Company D, 58th Infantry, 4th Division.
While on duty as a company runner, he repeatedly carried messages when exposed to artillery and machine-gun fire. Although reduced to a state of physical exhaustion, he refused to be relieved and continued to perform his duty of maintaining liaison.

HUNTLEY, JOSEPH R.....................
East of Belleau, France, July 21, 1918.
R—Winchester, Mass.
B—Boston, Mass.
G. O. No. 125, W. D., 1918.

Private, Company I, 104th Infantry, 26th Division.
Under the leadership of an officer he and Pvt. Charles E. Richardson charged a machine-gun nest, captured 2 guns, and killed or captured 12 men.

HUNTON, ORAMELL E. (641896)........
Northwest of Somme-Py, near St. Etienne, France, Oct. 4, 1918.
R—Portland, Me.
B—Winthrop, Me.
G. O. No. 133, W. D., 1918.

Private, first class, Section No. 606, Ambulance Service.
He showed conspicuous courage and devotion to duty in evacuating the wounded under the most trying conditions. He made repeated trips in plain view of enemy observers over roads under continuous shellfire. He was killed by a shell fragment while standing beside his car at an advanced post.
Posthumously awarded. Medal presented to mother, Mrs. Esther E. Barbier.

HUPMAN, LOCKERN (733855)...........
At Frapelle, France, Aug. 17, 1918.
R—Long Island City, N. Y.
B—Canada.
G. O. No. 15, W. D., 1919.

Sergeant, Company L, 6th Infantry, 5th Division.
After his platoon commander had fallen he took command of his platoon and led it to its objective through a heavy enemy barrage, although himself twice wounded.

HURDLE, WILLIE G. (2465187)..........
Near Ferme La Folie, France, Sept. 30, 1918.
R—Driver, Va.
B—Suffolk, Va.
G. O. No. 46, W. D., 1919.

Private, Machine Gun Company, 370th Infantry, 93d Division.
While under heavy enemy fire, Private Hurdle volunteered and, accompanied by another soldier, rescued a wounded comrade from an exposed position. He also performed valuable service as liaison agent, and, under very heavy fire, succeeded in cases where others had failed.

HURLEY, FRANCIS E. (60662).........
In the region of Grand Pont-Moulin Rouge, France, Feb. 23, 1918.
R—Malden, Mass.
B—Malden, Mass.
G. O. No. 129, W. D., 1918.

Corporal, Company E, 101st Infantry, 26th Division.
He took part in a daring raid into the enemy's lines in the region of Grand Pont-Moulin Rouge on the night of Feb. 23, 1918. He showed great valor and fearlessness, and particularly distinguished himself by leading the patrol into a German dugout whose occupants had refused to surrender and from which 16 prisoners were taken.

HURLEY, JOHN PATRICK...............
At Villers-sur-Fere, France, July 23, 1918.
R—Brooklyn, N. Y.
B—New York, N. Y.
G. O. No. 15, W. D., 1923.

Captain, 165th Infantry, 42d Division.
His company having reached its objective, he ordered several patrols forward to silence several enemy machine guns which were causing heavy casualties in his own and other companies of his battalion. The patrols meeting heavy casualties from the intense enemy fire, Captain Hurley himself organized a patrol and led it forward. While temporarily checked by the intense fire, he crawled forward and rescued a wounded officer, carrying him to a place of shelter. Resuming the advance of the patrol Captain Hurley was severely wounded, but continued to direct the movement of his men, who, inspired by the great courage and fortitude of their leader, pushed forward and captured the machine-gun nest, killing or capturing the crews thereof.

HURLEY, PAUL THOMAS...............
Near Vierzy, France, July 19, 1918.
R—Ardmore, Pa.
B—Ardmore, Pa.
G. O. No. 117, W. D., 1918.

Private, 55th Company, 5th Regiment, U. S. Marine Corps, 2d Division.
He displayed exceptional bravery in charging 3 machine guns with the aid of a small detachment of his comrades, killing the crews and capturing the guns, which were immediately turned on the Germans, thereby opening the line for the advance of his company, which had been held up by the enemy's fire.

HURLEY, PHILIP....................
In the Bois-de-Belleau, France, June 6, 1918.
R—Milford, Conn.
B—New York, N. Y.
G. O. No. 15, W. D., 1919.

First lieutenant, Infantry, attached to 83d Company, 6th Regiment, U. S. Marine Corps, 2d Division.
In the Bois-de-Belleau, France, on June 6, 1918, he displayed coolness, judgment, and utter fearlessness in leading his platoon to its objective under heavy fire.

HURST, WILLIAM E. (280193).........
Near Juvigny, France, Aug. 27, 1918.
R—Detroit, Mich.
B—Detroit, Mich.
G. O. No. 66, W. D., 1919.

Private, first class, Company G, 126th Infantry, 32d Division.
Upon being sent with a message under heavy machine-gun fire to a platoon commander, he found the latter lying wounded. Unassisted, he carried the officer to a shell hole, bound up his wound, and returned to his company commander with important information.

HUSBANDS, WILLIAM D. (1583445)......
Near Fismes, France, Aug. 10, 1918.
R—Arkadelphia, Ark.
B—Arkadelphia, Ark.
G. O. No. 32, W. D., 1919.

Private, Company K, 38th Infantry, 3d Division.
With one other soldier he volunteered and went to the rescue of a wounded man from another regiment, and returned through heavy machine-gun and shell fire, bringing the wounded man to his own trench.

*HUSSEY, JOHN (1899366)_____
Near St. Juvin, France, Oct. 16, 1918.
R—Westfield, Mass.
B—Ireland.
G. O. No. 74, W. D., 1919.

Sergeant, Company I, 325th Infantry, 82d Division.
After his platoon leader had became a casualty he led his platoon forward with great bravery under intense direct fire from a machine-gun nest, being killed while in the attack.
Posthumously awarded. Medal presented to father, John Hussey.

*HUSTED, CHESTER SETH_____
Near Blanc Mont, France, Oct. 5, 1918.
R—Corona, Calif.
B—Rialto, Calif.
G. O. No. 142, W. D., 1918.

Private, 81st Company, 6th Machine Gun Battalion, U. S. Marine Corps, 2d Division.
Displaying great courage and disregard for his own safety, he volunteered to carry an important message through a heavy machine-gun barrage, losing his life in attempting to carry out this mission.
Posthumously awarded. Medal presented to mother, Mrs. Emma M. Husted.

*HUTCHCRAFT, REUBEN B_____
Near Sedan, France, Nov. 7, 1918.
R—Paris, Ky.
B—Paris, Ky.
G. O. No. 37, W. D., 1919.

Captain, 166th Infantry, 42d Division.
He personally took command of a platoon of his company, which was designated as advance guard, and led his patrol to the most advanced point reached by any of our troops during the engagement. He was killed while making reconnaissance within 30 yards of enemy machine guns.
Posthumously awarded. Medal presented to mother, Mrs. R. B. Hutchcraft.

HUTCHINGS, CHARLES, Jr_____
South of Beaumont, France, Nov. 2 and 3, 1918.
R—Brewster, N. Y.
B—Brewster, N. Y.
G. O. No. 35, W. D., 1920.

First lieutenant, 9th Infantry, 2d Division.
On the night march of his regiment through the enemy lines Lieutenant Hutchings commanded the advanced party. About midnight his group was halted by heavy machine-gun and Stokes-mortar fire. Lieutenant Hutchings deployed his command and went forward alone to make a reconnaissance. Although exposed to fire from both lines, he completed his reconnaissance and captured 9 of the enemy, whom he returned to our lines, together with valuable information.

*HUTH, ALBERT A. (1827299)_____
Near Imecourt, France, Nov. 1, 1918.
R—Pittsburgh, Pa.
B—Pittsburgh, Pa.
G. O. No. 44, W. D., 1919.

Corporal, Company H, 319th Infantry, 80th Division.
He voluntarily advanced 3 times against an enemy machine-gun stronghold which was holding up his platoon's advance. He was wounded during his last attempt, but his act enabled his platoon to advance and capture 25 of the enemy who were concealed near by.
Posthumously awarded. Medal presented to mother, Mrs. Fredericka Huth.

HUTSON, GEORGE R. (1550465)_____
Near Le Charmel, France, July 28, 1918.
R—Knoxville, Tenn.
B—Tullahoma, Tenn.
G. O. No. 44, W. D., 1919.

Private, Battery B, 76th Field Artillery, 3d Division.
After he, himself, his commanding officer, and 32 comrades had been wounded by a bomb from an enemy plane, he remained at his post, assisted in laying his piece and directing fire on the enemy.

HUTTO, JOHN B. (98617)_____
Near Beuvardes, France, July 29, 1918.
R—Birmingham, Ala.
B—Oakman, Ala.
G. O. No. 81, W. D., 1919.

Sergeant, Headquarters Company, 167th Infantry, 42d Division.
Sergeant Hutto voluntarily left his 37-millimeter-gun section, which was attached to the reserve battalion, in order to accompany another section into action against two enemy machine-gun nests, which were sweeping our lines with heavy fire. When his platoon commander was wounded shortly afterwards, he, with another soldier, made his way under heavy shell and machine-gun fire to where the officer lay, and after administering first aid carried him to a place of safety.

HYATT, ROBERT W. (46030)_____
At Exermont, France, Oct. 4, 1918.
R—Benton, Ill.
B—Hamilton County, Ill.
G. O. No. 3, W. D., 1921.

Sergeant, Company B, 18th Infantry, 1st Division.
On two occasions Sergeant Hyatt exposed himself to heavy enemy fire while advancing in front of our lines in order to make observations of the enemy's activities on Hill 240. Upon his second journey to this exposed position he was severely wounded in the leg and was forced to remain in the enemy's lines until rescued the following day by our advancing troops.

HYDE, JAMES (540798)_____
In the Bois-de-Belleau, France, June 20, 1918.
R—Roxborough, Pa.
B—Ireland.
G. O. No. 99, W. D., 1918.

Sergeant, Company B, 7th Infantry, 3d Division.
He went out into the open under heavy machine-gun fire to carry in a wounded man. Being unable to pick him up, he lay down, pulled the man on his back, and crawled to his position.

*HYMAN, ERNEST (1319106)_____
Near Bellicourt, France, Sept. 29, 1918.
R—Palmyra, N. C.
B—Palmyra, N. C.
G. O. No. 32, W. D., 1919.

Private, Machine Gun Company, 120th Infantry, 30th Division.
Becoming separated from his organization in the smoke and fog, Private Hyman joined another soldier and was instrumental in breaking up 3 machine-gun nests and capturing 4 prisoners. After reaching the objective he volunteered and accompanied a reconnaissance patrol 600 yards beyond the line to the enemy. He has since been killed in action.
Posthumously awarded. Medal presented to widow, Mrs. Ernest Hyman.

*HYMAN, WILLIAM P_____
At Seringes-et-Nesles, France, July 29, 1918.
R—Iowa Falls, Iowa.
B—Minneapolis, Kans.
G. O. No. 88, W. D., 1918.

Second lieutenant, 166th Infantry, 42d Division.
After the capture of Seringes-et-Nesles, France, on July 29, 1918, by the organization of which he was a part, and while holding a portion of the town with his platoon, he found that 1 of his men was missing. Being told that the man was wounded and lay beyond a hedge at the edge of the village, he unselfishly attempted to locate the wounded man, was caught in direct machine-gun fire, and killed.
Posthumously awarded. Medal presented to father, L. T. Hyman.

HYNES, THOMAS J. (1279728)_____
Near Bois-d'Ormont, France, Oct. 12, 1918.
R—Jersey City, N. J.
B—Rosendale, N. Y.
G. O. No. 30, W. D., 1921.

Sergeant, Company B, 114th Infantry, 29th Division.
Taking command of the platoon after the platoon leader had become a casualty he gallantly led it in the attack. Although wounded in the head he continued in command until exhausted. After having his wound dressed he attempted to rejoin his organization, but was severely wounded before reaching the front lines. Sergeant Hynes refused to be littered to the dressing station until other wounded soldiers had been taken care of.

IGOU, PAUL (1309134)............
Near Beaurevoir, France, Oct. 7, 1918.
R—Chattanooga, Tenn.
B—Chattanooga, Tenn.
G. O. No. 44, W. D., 1919.

Sergeant, Company K, 117th Infantry, 30th Division.
After having been severely wounded in the knee, he remained in command of his platoon. During the destructive fire, he established his headquarters in a shell hole, and, by means of runners, maintained liaison and directed the attack. During a strong counterattack by the enemy, he skillfully commanded his platoon and repulsed the attack. He remained on duty for 24 hours after being wounded.

IHRKE, ALBERT L. J. (2023072)...........
Near Sergy, France, Aug. 1, 1918.
R—Mayville, Mich.
B—Mayville, Mich.
G. O. No. 46, W. D., 1919.

Private, Company B, 47th Infantry, 4th Division.
Private Ihrke displayed great courage and devotion to duty by remaining in an exposed position under heavy machine gun and shellfire to cover the withdrawal of his company.

IMHOF, HARRY E. (1159836)............
Near Vieville, France, Nov. 1, 1918.
R—Millvale, Pa.
B—Millvale, Pa.
G. O. No. 46, W. D., 1919.

Sergeant, Company D, 21st Machine Gun Battalion, 7th Division.
During a heavy barrage of high-explosive and gas shells he assisted an officer to give first aid to a wounded officer and 2 soldiers after the platoon had withdrawn from the position.

INGALLS, JOHN J.............
In the Bois-de-Belleau, France, June 6, 1918.
R—Olin, Iowa.
B—Bellevue, Iowa.
G. O. No. 119, W. D., 1918.

Corporal, 80th Company, 6th Regiment, U. S. Marine Corps, 2d Division.
Wounded in the assault on machine-gun positions in the Bois-de-Belleau, France, on June 6, 1918, he refused to be evacuated, but assisted in the evacuation of the wounded, thereby displaying great qualities of self-sacrifice and devotion to duty.

INGALLS, RAY L. H. (1040244)...........
Near Laneuville, France, Oct. 6–7, 1918.
R—Lubec, Me.
B—Canada.
G. O. No. 44, W. D., 1919.

Sergeant, Battery E, 11th Field Artillery, 6th Division.
Sergeant Ingalls displayed conspicuous leadership in handling his section under heavy shellfire. When he was wounded by a bursting shell, he first ascertained the damage done to his section by the shell before proceeding to the dressing station, taking another wounded man with him.

*INGERSOLL, HARRY............
Near Montfaucon, France, Sept. 26, 1918.
R—Philadelphia, Pa.
B—Philadelphia, Pa.
G. O. No. 78, W. D., 1919.

Captain, 313th Infantry, 79th Division.
He showed absolute disregard for personal danger in leading his company in an attack against an enemy position, strongly intrenched and protected by barbed-wire entanglements. Although he was killed at the enemy wire by machine-gun fire, his men, inspired by his courage, carried on the attack and took the enemy position, which had been holding up the advance. Posthumously awarded. Medal presented to father, Charles E. Ingersoll.

INGOLD, ALBERT S. (240628)...........
Near Bois-de-St. Remy, France, Sept. 12, 1918.
R—Imperial, Nebr.
B—Imperial, Nebr.
G. O. No. 10, W. D., 1920.

Private, Company H, 103d Infantry, 26th Division.
Although suffering from wounds, Private Ingold continued to advance with his company, and when the advance was held up by enemy machine-gun fire, he made his way forward and with an automatic silenced the fire of 2 enemy guns. He continued on despite his condition until the objective was reached.

INGOLD, WILLIAM J. (54103)...........
Near Cantigny, France, June 3, 1918.
R—Memphis, Tenn.
B—Altoona, Pa.
G. O. No. 71, W. D., 1919.

Sergeant, Company H, 26th Infantry, 1st Division.
While posting a listening post he encountered a hostile patrol of about 40 men; he attacked the Germans, although armed only with a pistol, and, killing an officer and 1 soldier, routed the enemy. Carrying the body of the officer, he had just returned to our lines when a raid was attempted by the Germans. Running to the scene of action, he killed 2 more Germans, aiding materially in routing the raiding party.

INGRAM, ISAAC F. (1289124)...........
Near Samogneux, France, Oct. 15, 1918.
R—Ferrum, Va.
B—Roanoke, Va.
G. O. No. 37, W. D., 1919.

Private, Company I, 116th Infantry, 29th Division.
Private Ingram advanced alone and with his automatic rifle silenced a hostile machine gun whose fire was holding up the line.

INGRAM, LEE H. (1204519)...........
East of Ronssoy, France, Sept. 29, 1918.
R—Gloversville, N. Y.
B—Hope Falls, N. Y.
G. O. No. 20, W. D., 1919.

Sergeant, Company G, 105th Infantry, 27th Division.
During the operations against the Hindenburg line he left shelter and went forward under heavy shell and machine-gun fire and rescued 5 wounded soldiers. In performing this gallant act, Sergeant Ingram and another soldier attacked an enemy dugout, killing 2 of the enemy and taking 1 prisoner.

INKS, CHARLES L. (1245501)...........
At Fismette, France, Aug. 12, 1918.
R—Pittsburgh, Pa.
B—McKeesport, Pa.
G. O. No. 98, W. D., 1919.

Sergeant, Company K, 111th Infantry, 28th Division.
Upon his own initiative, Sergeant Inks crawled 35 yards in advance of the line under the most intense machine-gun fire to where a wounded soldier lay, carrying him to shelter and administering first-aid treatment, thereby saving his life.

INMAN, LEON W............
Near Blanc Mont, France, Oct. 4, 1918.
R—Detroit, Mich.
B—Sauk Rapids, Minn.
G. O. No. 46, W. D., 1919.

Sergeant, 43d Company, 5th Regiment, U. S. Marine Corps, 2d Division.
He led his platoon in attack, setting an example by keeping several yards in advance of the men. He also assisted in dressing the wounds of members of his platoon under machine-gun and artillery fire. After being wounded, he remained with his platoon until ordered to leave by the medical officer.

INMAN, PERCY E............
Near Madeleine Farm, France, Oct. 13, 1918.
R—Bangor, Me.
B—Bangor, Me.
G. O. No. 23, W. D., 1919.

Second lieutenant, 13th Machine Gun Battalion, 5th Division.
Lieutenant Inman was seriously wounded while making a reconnaissance under a heavy fire, but returned with his report.

INNES, THOMAS (137324)........... Private, Company K, 109th Infantry, 28th Division.
  Near St. Agnan, France, July 16,   Wounded during an attack, he returned to the line after having his wound
    1918.   dressed, and though incapacitated for using a rifle, assisted in carrying
  R—Philadelphia, Pa.   wounded soldiers from the field, subjected the while to intense machine-gun
  B—Philadelphia, Pa.   and artillery fire, until he was exhausted and ordered to the rear.
  G. O. No. 71, W. D., 1919.

INSLEY, HARRY B. (1284383)....... Private, Company C, 115th Infantry, 29th Division.
  Near Sivry, France, Oct. 8, 1918.   Working his way over ground swept by machine-gun fire, he attacked an enemy
  R—Wingate, Md.   machine gun which was harassing our advance from the rear, shot 1 of the
  B—Wingate, Md.   crew, mortally wounded the gunner, and returned to his command with the
  G. O. No. 37, W. D., 1919.   enemy gun.

IRBY, SPRILEY E. (2463011)........ Private, Company H, 370th Infantry, 93d Division.
  At Beaume, France, Nov. 3, 1918.   He carried messages to the various units in his vicinity under severe enemy fire.
  R—Blackstone, Va.   He was badly wounded while in the performance of this duty.
  B—Lunenburg County, Va.
  G. O. No. 46, W. D., 1919.

IRELAND, RUTHERFORD........... Captain, 166th Infantry, 27th Division.
  Near St. Souplet, France, Oct. 18, He continued to lead his battalion in attack, although suffering great pain from
    1918.   a wound caused by shellfire. After being ordered to the dressing station,
  R—Brooklyn, N. Y.   he had his wound dressed without waiting to have the shrapnel removed,
  B—Cleveland, Ohio.   returning to his battalion, and remaining on duty for 2 days.
  G. O. No. 37, W. D., 1919.

IRONS, BENJAMIN G. (2411507)..... Private, first class, medical detachment, 311th Infantry, 78th Division.
  Near Vieville-en-Haye, France, Sept. Private Irons, with another soldier, advanced to an exposed position, and
    25-26, 1918.   while administering first aid to a wounded man, the 3 were suddenly sur-
  R—Silverton, N. J.   rounded and captured by a party of the enemy. While being taken toward
  B—Silverton, N. J.   the German lines, Private Irons and his companion attacked their captors
  G. O. No. 78, W. D., 1919.   and succeeded in freeing themselves, at the same time capturing 2 Germans,
                                      whom they brought to battalion headquarters, together with the wounded
                                      man.

IRONS, JOHN K. (736749)........ Corporal, Company K, 11th Infantry, 5th Division.
  Near Bois St. Claude, in the St.   After being wounded in the foot by a machine-gun bullet and later in the leg
    Mihiel salient, France, Sept. 12, by shrapnel, he continued in the advance of his squad all that day without
    1918.   medical attention other than his own first aid, thereby displaying exceptional
  R—Steubenville, Ohio.   courage and devotion to duty.
  B—Monaca, Pa.
  G. O. No. 123, W. D., 1918.

IRGANG, ANDREW (1079137)......... Corporal, Machine Gun Company, 120th Infantry, 30th Division.
  Near Bellicourt, France, Sept. 29, After he had become separated from the rest of the platoon, Corporal Irgang
    1918.   kept his squad together and broke up a machine-gun post, capturing the
  R—Spades, Ind.   gunners and the gun. As his own gun had become disabled, he turned the
  B—Brookville, Ind.   captured gun around and fired 1,000 rounds from it, covering the advance of
  G. O. No. 44, W. D., 1919.   the Infantry. He then continued to lead his squad forward under terrific
                                      artillery and machine-gun fire.

IRVING, LIVINGSTON GILSON........ First lieutenant, 103d Aero Squadron, Air Service.
  Near Bantheville, France, Oct. 10, Accompanied by another pilot, he attacked an enemy formation of 11 planes,
    1918.   4 of which were above him. In spite of the great odds, he dived into the lower
  R—Berkeley, Calif.   formation, and after a sharp combat destroyed one plane and with the aid of
  B—San Francisco, Calif.   his companion forced a second plane to earth.
  G. O. No. 46, W. D., 1919.

IRWIN, FRANK J. (1210102)........ Corporal, Company C, 107th Infantry, 27th Division.
  In the Mount Kemmel sector, Bel-   While engaged with an American working party between the British front and
    gium, Aug. 17, 1918.   support lines a British ration party passing near by was struck by high-
  R—New York, N. Y.   explosive hostile shellfire, badly wounding several. Corporal Irwin, with
  B—East Boston, Mass.   2 assistants, voluntarily twice crossed an area under heavy enemy shell and
  G. O. No. 43, W. D., 1922.   machine-gun fire in order to carry the wounded men to a position of safety.

ISAAC, JOSEPH (264039).......... Private, Company M, 125th Infantry, 32d Division.
  Northeast of Jaulgonne, near Sergy, Although wounded in the head, he crawled from within 100 feet of the German
    France, July 31, 1918.   line back to his own line, 150 yards distant, carrying a more severely wounded
  R—Manistique, Mich.   comrade on his back.
  B—Manistique, Mich.
  G. O. No. 117, W. D., 1918.

ISRAEL, FREDERICK............ Second lieutenant, 5th Regiment, U. S. Marine Corps, 2d Division.
  Near St. Etienne, France, Oct. 4,   He twice volunteered and carried messages to the front line along a road swept
    1918.   by machine-gun and shellfire.
  R—Philadelphia, Pa.
  B—Philadelphia, Pa.
  G. O. No. 35, W. D., 1919.

IVES, EDWIN B............... First lieutenant, 9th Infantry, 2d Division.
  Near Blanc Mont Ridge, France,   He volunteered and carried an important message from battalion to regimental
    Oct. 3, 1918.   headquarters, through heavy machine-gun and artillery fire, and returned
  R—Great Bend, Kans.   with an answer.
  B—Nevada, Mo.
  G. O. No. 21, W. D., 1919.

JACK, DANIEL L. (3449673)............
Near Scheldt River, Belgium, Oct.
31, 1918.
R—Geneva, Ind.
B—Berne, Ind.
G. O. No. 46, W. D., 1919

Private, Company F, 362d Infantry, 91st Division.
When the advance of the front line was held up by fire from a machine-gun nest 300 yards to the front, Private Jack, with 2 others, crossed the open field in the face of fire from enemy artillery, machine guns, and snipers. Charging the nest, they killed 2 of the crew, wounded 2 others, and captured 5, together with the gun.

JACKSON, BURWELL C. (42996)..........
Near Soissons, France, July 19, 1918.
R—Kinston, N. C.
B—Lenoir County, N. C.
G. O. No. 15, W. D., 1919.

Private, Company F, 16th Infantry, 1st Division.
He, alone, captured a machine gun, killed 2 of the crew, and took the remaining 3 prisoners. Later in the same day he was killed while making a similar attempt.
Posthumously awarded. Medal presented to brother, Jesse L. Jackson.

JACKSON, FRANKLIN J...............
East of Ronssoy, France, Sept. 29,
1918.
R—Brooklyn, N. Y.
B—Brooklyn, N. Y.
G. O. No. 142, W. D., 1918.

First lieutenant, 106th Infantry, 27th Division.
During the operations against the Hindenburg line he, as trench mortar officer, of his regiment, twice volunteered to go forward under heavy shell and machine-gun fire on a personal reconnaissance. While gallantly and courageously engaged in the second reconnaissance he was killed.
Posthumously awarded. Medal presented to mother, Mrs. Elizabeth S. Jackson.

JACKSON, GEORGE C. (44356)..........
Near Fleville, France, Oct. 2, 1918.
R—Fort Worth, Tex.
B—Temple, Tex.
G. O. No. 44, W. D., 1919.

Sergeant, Company M, 16th Infantry, 1st Division.
His platoon becoming disorganized by enemy artillery fire, Sergeant Jackson halted his men, reorganized them under the heavy fire, and resumed the advance. Later he borrowed an automatic rifle from one of the men in his platoon and firing it from his hip advanced on an enemy machine gun, killing 2 members of the crew and capturing another.

JACKSON, HORATIO N................
Near Montfaucon, France, Sept. 26-
29, 1918.
R—Burlington, Vt.
B—Canada.
G. O. No. 37, W. D., 1919.

Major, Medical Corps, attached to 313th Infantry, 79th Division.
Constantly working in the face of heavy machine-gun and shellfire, he was most devoted in his attention to the wounded, always present in the line of advance, directing the administering of first aid and guiding the work of litter bearers. He remained on duty until severely wounded by high-explosive shells, when he was obliged to evacuate.

JACKSON, JAMES (1388190)............
Near Consenvoye, France, Oct. 11,
1918.
R—Chicago, Ill
B—Calumet, Mich.
G. O. No. 71, W. D., 1919.

First sergeant, Company K, 131st Infantry, 33d Division.
Advancing with a few companions through the enemy barrage, he attacked a machine-gun emplacement which had been inflicting heavy casualties, and captured the enemy gun and 17 prisoners. His personal heroism was an inspiration to those with him.

JACKSON, RUFUS B................
Near Farm La Folie, France, Sept.
28, 1918.
R—Chicago, Ill.
B—Buffalo, Wyo.
G. O. No. 46, W. D., 1919.

Second lieutenant, 370th Infantry, 93d Division.
Having been ordered to use his Stokes mortars in wiping out machine-gun nests which had been resisting the advance of his company, Lieutenant Jackson made a personal reconnaissance by crawling to the enemy's lines to locate the nests. Accomplishing his purpose, he returned and directed the fire, silencing the guns.

JACKSON, WILLIAM (38425)...........
Near Blanc Mont Ridge, France,
Oct. 3-9, 1918.
R—Easton, Pa.
B—Trenton, N. J.
G. O. No. 21, W. D., 1919.

Sergeant, Company B, 9th Infantry, 2d Division.
While acting as battalion runner, Sergeant Jackson, regardless of personal danger, many times volunteered and carried messages under the most intense shellfire, thereby greatly assisting in maintaining liaison with other units.

JACOBS, WENDELL W. (2185701).......
In the Bois-de-Bantheville, France,
Oct. 30, 1918.
R—Glenwood Springs, Colo.
B—Sigourney, Iowa.
G. O. No. 37, W. D., 1919.

Private, Company C, 341st Machine Gun Battalion, 89th Division.
When 6 men of his section were wounded and his own hand was partly severed by a shell fragment, Private Jacobs had another soldier sever his hand with a pocket knife and then bandage it. While in this condition he assisted other wounded soldiers in every way possible before proceeding to the first-aid station, whence he was evacuated to the hospital.

JACOBSON, WILLIAM (50211)..........
Near Chateau-Thierry, France, June
6, 1918.
R—Chicago, Ill.
B—Chicago, Ill.
G. O. No. 99, W. D., 1918.

Private, Company D, 23d Infantry, 2d Division.
After having his nose shot off, he perseveringly continued his work throughout the night as a runner under heavy fire, in order to maintain communication.

JACOBSON, WILLIAM A. (17461).........
Near Gesnes, France, Oct. 7, 1918.
R—Viroqua, Wis.
B—Viroqua, Wis.
G. O. No. 44, W. D., 1919.

Private, medical detachment, 128th Infantry, 32d Division.
When his battalion was forced to retire under heavy artillery and machine-gun fire, Private Jacobson went out in front of the line, administering first aid and bringing in the wounded who had been left lying in exposed positions. While so engaged he received 2 wounds, the second of which caused his death before he reached the aid station.
Posthumously awarded. Medal presented to father, Jacob Jacobson.

JACQUES, LEO P. T. (43493).........
At Boise Brule, France, Apr. 10,
1918.
R—Greenfield, Mass.
B—Greenfield, Mass.
G. O. No. 99, W. D., 1918.

Sergeant, Company L, 104th Infantry, 26th Division.
During the action of Apr. 10, 1918, he displayed courage, coolness, and spirit of self-sacrifice in voluntarily going through shell-swept area to bring back wounded to a place of safety, carrying one wounded man more than 50 yards under heavy shellfire.

JAGER, HOLGER (59770)_____
North of Verdun, France, Oct. 27, 1918.
R—Boston, Mass.
B—Norway.
G. O. No. 21, W. D., 1919.

Corporal, Company A, 101st Infantry, 26th Division.
He continued to carry messages after being wounded in the back by a machine-gun bullet, until he was again wounded by a bursting shell so seriously that his evacuation was necessary.

JAMES, BENJAMIN (133168)_____
At Seicheprey, France, Apr. 21, 1918.
R—Brookline, Mass.
B—South Boston, Mass.
G. O. No. 99, W. D., 1918.

First sergeant, Battery A, 101st Field Artillery, 26th Division.
During the action of Apr. 21, 1918, when a shell struck gunpit of a battery, although seriously wounded in the chest and legs by fragments, he immediately obtained stretchers for the other men, doing everything possible for their comfort until he fell. He declined medical attention until all others had been looked after, setting a splendid example of self-sacrifice.

JAMES, DARL S_____
Near Baulny, France, Sept. 29–30, 1918.
R—Kansas City, Kans.
B—Greeley County, Nebr.
G. O. No. 59, W. D., 1919.

Captain, 110th Engineers, 35th Division.
Although severely wounded within a few moments from the start of the engagement, he refused to be evacuated, but remained in command of his company and, in addition, throughout the day assisted in reorganizing scattered elements. On Sept. 30, during two counterattacks, he supervised the resistance of his sector in spite of his weakened condition and continually inspired his command by his disregard of machine-gun and artillery fire.

JAMES, JESSE A. (2339853)_____
Near Les Evaux, France, July 13, 1918.
R—Madill, Okla.
B—Eagletown, Okla.
G. O. No. 32, W. D., 1919.

Sergeant, Company L, 4th Infantry, 3d Division.
After many attempts to get patrols across the Marne had failed, Sergeant James alone swam the river, taking with him a wire, by which a boat containing 2 of his comrades was drawn across without attracting the attention of the enemy.

JAMES, JOSEPH (1798927)_____
Near Binarville, France, Sept. 30, 1918.
R—Philadelphia, Pa.
B—Philadelphia, Pa.
G. O. No. 35, W. D., 1919.

Private, Headquarters Company, 368th Infantry, 92d Division.
He went to the aid of a wounded companion under very severe machine-gun and artillery fire and brought him to cover. He stayed with the wounded man, giving him all possible aid until assistance came, when he returned to his place with the platoon.

*JAMISON, ROLAND RAY_____
Near St. Etienne, France, Oct. 4–5, 1918.
R—Hermitage, Mo.
B—Hermitage, Mo.
G. O. No. 37, W. D., 1919.

Pharmacist's mate, first class, U. S. Navy, attached to 5th Regiment, U. S. Marine Corps, 2d Division.
Regardless of his personal danger, he repeatedly exposed himself to machine-gun and shellfire to give first aid to the wounded in the open.
Posthumously awarded. Medal presented to father, Edward W. Jamison.

*JANKOWSKI, JAN (39605)_____
Near Vaux, France, July 1, 1918.
R—Chicago, Ill.
B—Russia.
G. O. No. 99, W. D., 1918.

Supply sergeant, Company G, 9th Infantry, 2d Division.
Near Vaux, July 1, 1918, he entered a dugout, killing 2 and capturing 3 of the enemy single handed.
Posthumously awarded. Medal presented to brother, Walter Jankowski.

*JANSEN, LOUIS B_____
Near Epinonville, France, Sept. 26, 1918.
R—Chicago, Ill.
B—Chicago, Ill.
G. O. No. 20, W. D., 1919.

First lieutenant, 361st Infantry, 91st Division.
When the advance of his battalion was held up by an enemy machine-gun nest Lieutenant Jansen, accompanied by a soldier, crossed the enemy wire, took the position, killed 1 of the enemy, and captured 4 prisoners and 2 machine guns.
Posthumously awarded. Medal presented to brother, Joseph Jansen.

JANSSEN, MARTIN J. (2181249)_____
Near Flirey, France, Sept. 12, 1918.
R—Rushville, Nebr.
B—Holland.
G. O. No. 127, W. D., 1918.

Sergeant, Company A, 355th Infantry, 89th Division.
Coming up in the rear of two platoons of a battalion of the first line, Sergeant Janssen, belonging to another battalion, noticed the two platoons were held up by machine-gun fire from the front and flank and appeared to be without a leader. Fearlessly exposing himself, he ran from one end of the line to the other urging the men forward until both platoons had moved across a small gully out of danger from the machine-gun fire. His bravery and leadership thus prevented the interruption of the advance of the entire first line.

JANSSEN, ROLLA (41071)_____
Near Blanc Mont Ridge, France, Oct. 3, 1918.
R—Ashley, Ill.
B—Ashley, Ill.
G. O. No. 21, W. D., 1919.

Corporal, Headquarters Company, 9th Infantry, 2d Division.
While acting as a battalion runner, Corporal Janssen carried a message through a heavy barrage, and, although wounded, succeeded in returning with an answer. After his wound had been dressed he remained on duty throughout the engagement.

JARDINE, DAVID F. (547434)_____
In the Bois-d'Aigremont, France, July 15–26, 1918.
R—Kansas City, Mo.
B—Elmira, N. Y.
G. O. No. 21, W. D., 1925.

First sergeant, Company I, 30th Infantry, 3d Division.
When the platoon-commander was killed, Sergeant Jardine was placed in command of a platoon and, without regard for his personal safety, led it to the proper place through a violent barrage and successfully maintained the leadership of it throughout the battle.

JARVI, EINO I. (263965)_____
Near Verdun, France, Oct. 11–13, 1918.
R—Rudyard, Mich.
B—Finland.
G. O. No. 13, W. D., 1919.

Mechanic, Company M, 125th Infantry, 32d Division.
As a runner for the Third Battalion, 125th Infantry, during the taking and holding of the line near La Tuilerie Farm, he was engaged in carrying important messages, crossing and recrossing Death Valley, between Hill 258 and La Cote Dame Marie, the foremost part of the line held by the Third Battalion. The valley was swept by machine-gun fire, the terrain affording absolutely no protection, requiring a perilous dash of 500 yards across open ground before any cover was reached. It was only by a display of supreme courage that important messages reached the battalion.

JARVIS, HOMER S.
Near Nantillois, France, Sept. 26, 1918.
R—Caldwell, Idaho.
B—Xenia, Ill.
G. O. No. 138, W. D., 1918.

First lieutenant, 11th Machine Gun Battalion, 4th Division.
He, with another officer and a soldier, using captured German Maxim guns, pushed forward to a heavily shelled area from which the Infantry had withdrawn, and by their accurate and effective fire kept groups of the enemy from occupying advantageous positions. Maintaining fire superiority all afternoon, he withdrew from his dangerous position only when it became too dark to see.

*JAUSS, RAYMOND B.
Near Crezancy, France, July 15, 1918.
R—New York, N. Y.
B—Albany, N. Y.
G. O. No. 32, W. D., 1919.

First lieutenant, 30th Infantry, 3d division.
After all means of liaison had failed he carried important messages to his observation posts. He was killed by shellfire while visiting one of his observation stations near Crezancy.
Posthumously awarded. Medal presented to widow, Mrs. Harriet A. Jauss.

JAWORSKI, FRANK (569251).
West of Fismes, France, Aug. 5, 1918.
R—West Hammond, Ill.
B—Chicago, Ill.
G. O. No. 145, W. D., 1918.

Corporal, Company F, 4th Engineers, 4th Division.
He was a member of a small detachment of Engineers which went out in advance of the front line of the Infantry through an enemy barrage from 77-millimeter and 1-pounder guns to construct a footbridge over the River Vesle. As soon as their operations were discovered machine-gun fire was opened up on them, but, undaunted, the party continued at work, removing the German wire entanglements and completing a bridge which was of great value in subsequent operations.

JAY, DELANCEY KANE.
At Chateau du Diable, near Fismes, France, Aug. 27, 1918.
R—Westbury, Long Island, N. Y.
B—Switzerland.
G. O. No. 9, W. D., 1923.

Major, 307th Infantry, 77th Division.
With utter disregard of his own safety he left the shelter of his command post and personally directed the attack of his battalion against the strongly fortified enemy position in and about Chateau du Diable north of the Vesle River. From the beginning of the attack he stood on a railroad embankment within 70 meters of the enemy line, fully exposed to their observation, and under a continuous and intense fire of concealed machine guns, rifles, and artillery. From this position he continued to direct, control, and encourage his officers and men during the progress of the attack, and even after he had been wounded and until exhausted by loss of blood. He refused to be evacuated until he had given full instructions to his second in command and until all wounded enlisted men had been evacuated. His exceptional example of physical and mental courage was an inspiration to all his officers and men under the most trying and dangerous conditions.

JEFFERS, JOHN N.
Over the region of Romagne, France, Oct. 5, 1918.
R—Los Angeles, Calif.
B—Los Angeles, Calif.
G. O. No. 138, W. D., 1918.

First lieutenant, 94th Aero Squadron, Air Service.
While on patrol he encountered 10 enemy machines (Fokker type) at an altitude of 2,000 feet. Despite numerical superiority and by a display of remarkable courage and skillful maneuvering, he separated one of the planes from the formation and after a brief encounter shot it down in flames.

JEFFERS, LAMAR.
Near St. Juvin, France, Oct. 11, 1918.
R—Anniston, Ala.
B—Anniston, Ala.
G. O. No. 46, W. D., 1919.

Captain, 326th Infantry, 82d Division.
On the night of Oct. 10 Captain Jeffers reconnoitered a badly damaged bridge, and early in the morning of the 11th he supervised its repair, being continuously under an intense machine-gun fire. He later led the leading company of the battalion over this bridge and across an open and level terrain, where all of his officers and almost two-thirds of his men became casualties and he himself was seriously wounded. He continued to lead his company forward, however, until he fell, shot through the jaw with a machine-gun bullet.

JEFFERS, SOLOMON L.
Against insurgent forces, near Licuan, Luzon, P. I., Jan. 25, 1900.
R—Little Rock, Ark.
B—Ozark, Ark.
G. O. No. 7, W. D., 1925.

First lieutenant, 33d Infantry, U. S. Volunteers.
With great gallantry and fearlessness Lieutenant Jeffers led 7 men up a narrow trail against unknown numbers of the enemy, flanked their position, and killed or captured the entire number.

JEFFERSON, ALBERT G.
At Le Hamel, France, July 4, 1918.
R—Oak Park, Ill.
B—Reed City, Mich.
G. O. No. 99, W. D., 1918.

First lieutenant, 131st Infantry, 33d Division.
After being severely wounded in the breast and shoulder from shellfire he continued with and commanded his platoon until its final objective was reached and its consolidation was completed.

JEFFERY, FOREST G. (1590360).
North of Cunel, France, Oct. 20, 1918.
R—Newport, Ark.
B—Mount Olive, Ark.
G. O. No. 22, W. D., 1920.

Private, Company M, 7th Infantry, 3d Division.
At a time when his platoon was held up by machine-gun fire from the front, Private Jeffery with 2 other comrades went forward and attacked the machine-gun position from the flank, capturing 1 officer, 2 sergeants, and 6 privates, and enabling his platoon to further advance without loss of men.

*JEFFORDS, PAUL (1447380).
Near Baulny Ridge, France, Sept. 28, 1918.
R—Kansas City, Kans.
B—Kansas City, Kans.
G. O. No. 89, W. D., 1919.

Corporal, Company A, 137th Infantry, 35th Division.
After gallantly leading a section of the line in the advance, Corporal Jeffords was wounded in several places by machine-gun bullets, while he was taking position for his squad, but he refused medical attention and continued to display unusual fortitude until he died several minutes later.
Posthumously awarded. Medal presented to mother, Mrs. Ella E. Jeffords.

JEFFREY, JANE.
At Jouy-sur-Morin (Seine-et-Marne), France, July 15, 1918.
R—Dorchester, Mass.
B—England.
G. O. No. 71, W. D., 1919.

Nurse, American Red Cross Hospital No. 107.
While she was on duty at American Red Cross Hospital No. 107 Miss Jeffrey was severely wounded by an exploding bomb during an air raid. She showed utter disregard for her own safety by refusing to leave her post, though suffering great pain from her wounds. Her courageous attitude and devotion to the task of helping others was inspiring to all of her associates.

JEFFREY, ROBERT F. (1240588)...........
  At Apremont, France, Sept. 29, 1918.
  R—Homer City, Pa.
  B—Morris Run, Pa.
  G. O. No. 115, W. D., 1918.

Corporal, Headquarters Company, 110th Infantry, 28th Division.
He was a member of a section operating 37-millimeter guns which was attacked by the enemy. After removing the guns to safety he learned that the officer commanding the section had been captured, whereupon he organized a party of 5, attacked the enemy's patrol, numbering 35, and succeeded in delivering the captured officer, killing 15 of the enemy and personally capturing 2. Later in the same evening, in entire disregard for his own safety, he assisted a sergeant in organizing 75 men for a counterattack, which they launched in the face of heavy fire at close range, driving the enemy back for more than a kilometer.

JENKINS, JAMES T. (1818428)...........
  Near Nantillois, France, Oct. 5, 1918.
  R—Buena Vista, Va.
  B—Greenlee, Va.
  G. O. No. 37, W. D., 1919.

Sergeant, Company G, 317th Infantry, 80th Division.
Patrolling by himself in front of the line, he came upon a machine-gun emplacement manned by a German officer and 3 men. He wounded the officer and 1 soldier by rifle fire, captured the other 2 men, and took them, with the machine gun, to the rear.

JENKINS, JOHN M...........
  Near Cunel, France, Oct. 14, 1918.
  R—Yorkville, S. C.
  B—Yorkville, S. C.
  G. O. No. 103, W. D., 1919.

Colonel, 30th Infantry, 3d Division.
He personally led a reconnaissance patrol through the eastern and northern edges of Bois-de-la-Pultiere in order to obtain most necessary information while the area was being continuously bombarded by high-explosive and gas shells and raked by machine-gun fire. His courage and bravery was a splendid example and an inspiration to the officers and men of his command.

JENKINS, MATHEW (1402492)...........
  Near Vauxaillon, France, Sept. 20, 1918.
  R—Chicago, Ill.
  B—Lookout, La.
  G. O. No. 127, W. D., 1918.

Sergeant, Company F, 370th Infantry, 93d Division.
He was in command of a detachment and was ordered to attack the German line. After rescuing, under fire, a wounded comrade he charged with his detachment, took a fortified tunnel, and, though far in advance of our lines and without rations and ammunition, held the position for more than 36 hours, until relieved, making use of captured guns and ammunition to repel the counterattacks made upon him.

*JENKINS, PAUL B. (1784119)...........
  Near Gibercy, France, Nov. 11, 1918.
  R—Philadelphia, Pa.
  B—Franklinton, N. C.
  G. O. No. 37, W. D., 1919.

Sergeant, Headquarters Company, 315th Infantry, 79th Division.
While installing a telephone line his regiment started an attack. The enemy responded with a terrific barrage and before the communication was completed Sergeant Jenkins was in the midst of a heavy encounter. Bravely he remained at his post, endeavoring to establish telephone service, but was instantly killed by shellfire.
Posthumously awarded. Medal presented to sister, Mrs. Eunice G. Strother.

JENKINS, WADE H. (2216190)...........
  Near St. Marie Farm, France, Sept. 14, 1918.
  R—Orlando, Okla.
  B—Lake City, Mo.
  G. O. No. 87, W. D., 1919.

Private, first class, Machine Gun Company, 357th Infantry, 90th Division.
He volunteered to go forward with 3 other soldiers to reduce 2 machine-gun nests which successively held up our advance. Crawling forward under heavy fire, he showed marked personal bravery, attacking and killing occupants of the enemy emplacements.

JENKINS, WESTON C...........
  In the Forest of Argonne, France, Oct. 5, 1918.
  R—Rome, N. Y.
  B—New York, N. Y.
  G. O. No. 20, W. D., 1919.

Captain, 307th Infantry, 77th Division.
He commanded the second battalion of his regiment with conspicuous gallantry. With utter disregard for his own safety, he continued throughout the action to direct his troops personally, moving about from place to place under heavy artillery and machine-gun fire. Exposing himself to the hostile fire, he encouraged his men in their efforts to break through the enemy's line and succeeded in maintaining their aggressive spirit by his personal example of fearlessness.

*JENKS, DEAN N...........
  At Fossoy, France, July 16, 1918.
  R—Denver, Colo.
  B—Brooklyn, N. Y.
  G. O. No. 44, W. D., 1919.

First lieutenant, 7th Infantry, 3d Division.
He fearlessly led his company in an attack under a heavy bombardment, encouraging his men by his example. He was killed by shellfire while in the performance of this act.
Posthumously awarded. Medal presented to widow, Mrs. D. N. Jenks.

*JENNART, LEON (14229)...........
  Near Badricourt, Alsace, July 12, 1918.
  R—Detroit, Mich.
  B—Italy.
  G. O. No. 66, W. D., 1919.

Bugler, Battery E, 119th Field Artillery, 32d Division.
As he was returning from a reconnaissance with his battery commander, under heavy enemy artillery fire, a shell struck between them, mortally wounding the officer and throwing Bugler Jennart from his horse, which was killed. Although he was himself seriously wounded, this soldier crawled across the road to the assistance of his captain, and just before he reached the latter's side was instantly killed by another bursting shell.
Posthumously awarded. Medal presented to father, Leon Jennart, sr.

JENNINGS, EDGAR A (263921)...........
  Near Cierges, France, July 31, 1918.
  R—Lynchburg, Va.
  B—Pocahontas, Va.
  G. O. No. 71, W. D., 1919.

Sergeant, Company M, 125th Infantry, 32d Division.
He displayed marked bravery during an attack by his company, and when liaison with adjoining units had been lost volunteered repeatedly for dangerous missions, he being a sergeant at the time. He exposed himself in the open to enemy machine-gun fire to rescue wounded soldiers and reorganized the company after heavy casualties had been sustained.

JENNISON, CHARLES SUMNER...........
  Near Blanc Mont Ridge, France, Oct. 4–6, 1918.
  R—Malone, N. Y.
  B—Brushton, N. Y.
  G. O. No. 46, W. D., 1919.

Pharmacist's mate, second class, U. S. Navy, attached to 5th Regiment, U. S. Marine Corps, 2d Division.
He worked fearlessly and efficiently at caring for the wounded in an advanced dressing station exposed to heavy shell and rifle fire. He was wounded when a shell struck and partly wrecked his aid station, but he refused to be evacuated, and continued at his post for 48 hours.

JENSEN, INGEMANN (1391466)_____
Near Bois-du-Fays, France, Oct. 9, 1918.
R—Chicago, Ill.
B—Denmark
G. O. No. 44, W. D., 1919.

Private, Machine Gun Company, 132d Infantry, 33d Division.
In the action in the Bois-du-Fays on Oct. 9 he was wounded, but returned to the line as soon as he had his wound dressed. He was soon wounded the second time and sent to the first-aid station, where he was ordered to the rear, but instead he returned to the line, where he was wounded for the third time, and carried to the rear on a stretcher.

JENSEN, OTTO CARSTEN (3138240)_____
Near St. Juvin, France, Nov. 1, 1918.
R—Rock Springs, Wyo.
B—Denmark.
G. O. No. 53, W. D., 1920.

Private, Company E, 305th Infantry, 77th Division.
When his company had suffered heavy casualties and had been withdrawn, Private Jensen went out with another soldier in advance of our lines under machine-gun fire to rescue the wounded. In attempting this hazardous mission he was seriously wounded.
Posthumously awarded. Medal presented to father, Karsten Jensen.

JERABEK, JERRY J. (285397)_____
In Romagne Woods, France, Oct. 11, 1918.
R—Algoma, Wis.
B—Kewaunee, Wis.
G. O. No. 81, W. D., 1919.

Corporal, Company A, 121st Machine Gun Battalion, 32d Division.
Passing through heavy fire and though wire entanglements, he led his section to a posit on 500 meters in advance of the Infantry, where he set up his guns and effectively covered the advance. He showed marked bravery and skill in leading his men, capturing 22 prisoners without sustaining a casualty.

JERMIASON, AXEL (54697)_____
Near Cantigny, France, May 27, 1918.
R—Buford, N. Dak.
B—Norway.
G. O. No. 44, W. D., 1919.

Private, Company K, 26th Infantry, 1st Division.
He was so seriously wounded that he could not operate his automatic rifle, but refused to be evacuated, and continued with the rifle team, taking the place of first leader. Becoming very weak from the loss of blood, he was ordered to the rear by his platoon commander, but on his way back picked up a rifle and continued in the fight until the enemy had been driven back.

JEROME, JOSEPH E. (2366521)_____
At Vladivostok, Siberia, Nov. 17 and 18, 1919.
R—Oakland, Calif.
B—Oakland, Calif.
G. O. No. 13, W. D., 1923.

Private, Company B, Replacement Battalion, American Expeditionary Forces, Siberia.
He drove an automobile repeatedly through a zone swept by the fire of rifles, machine guns, and fieldpieces of contending Russian Government and insurgent troops, rescuing many noncombatants who were entrapped in the Vladivostok station between the lines, carrying them to places of safety, thus undoubtedly saving many lives.

JERRY, BARNEY (1564797)_____
Near Grimaucourt, France, Nov. 10, 1918.
R—Bigfoot Branch, Memphis, Tenn.
B—Byhalia, Miss.
G. O. No. 32, W. D., 1919.

Corporal, Company F, 322d Infantry, 81st Division.
While acting as scout 400 yards in advance of his company he opened fire on 10 Germans advancing in front of his position, killing 2, wounding 1, and causing the rest to retreat.

JERVEY, FRANK JOHNSTONE_____
Near Les Franquettes Farm, France, July 22, 1918.
R—Charleston, S. C.
B—Summerville, S. C.
G. O. No. 32, W. D., 1919.

Captain, 4th Infantry, 3d Division.
Although wounded 5 times, when his company was suddenly fired upon by machine guns, while crossing an open field, Captain Jervey remained in command of his company until he became unconscious.

JERVEY, THOMAS M_____
Near Longuyon, France, Oct. 31, 1918.
R—Charleston, S. C.
R—Summerville, S. C.
G. O. No. 13, W. D., 1919.

First lieutenant, Ordnance Department, attached to 1st Army, Observation Group, Air Service.
Assigned to the 1st Army Observation Group, Air Service, as armament officer, he volunteered as observer on a photographic mission from Ontedy to Longuyon, 25 kilometers into the enemy lines. In combat with 14 enemy aircraft which followed, one enemy aircraft was destroyed. Lieutenant Jervey, regardless of the fact that his plane was badly shot up and that his hands were badly frozen, continued on the mission, returning only upon its successful conclusion.

JEWETT, HENRY C_____
During the Argonne-Meuse offensive, France, Sept. 25–Oct. 4, 1918.
R—Buffalo, N. Y.
B—Buffalo, N. Y.
G. O. No. 20, W. D., 1919.
Distinguished-service medal also awarded

Colonel, 316th Engineers, 91st Division.
Assigned to the command of an infantry brigade, he was directed to go forward, find his brigade, and consolidate his regiments, which had become separated. He crossed territory under terrific fire and pulled his rear regiment to the aid of the regiment in the front, which was seriously engaged, thereafter commanding the movements of both regiments in a highly creditable manner.

JILLSON, HOWARD D. (1735489)_____
In Bois-de-Hailbat, northeast of Jauiny, France, Sept. 17, 1918.
R—Buffalo, N. Y.
B—Youngstown, N. Y.
G. O. No. 125, W. D., 1918.

Private, Company F, 309th Infantry, 78th Division.
Although suffering from illness, he volunteered as runner, and repeatedly carried messages across heavily shelled areas, displaying the greatest courage and coolness.

JOBES, LESLIE J_____
Near Verdun, France, Oct. 8, 1918.
R—Hoboken, N. J.
B—Pittsfield, Mass.
G. O. No. 2, W. D., 1919.

First lieutenant, 115th Infantry, 29th Division.
While in command of his platoon during an engagement of the 29th Division, Verdun sector, he displayed exceptional bravery, disregarding his own safety, and encouraged his men both by words and action. While leading his platoon, in an attack on a machine-gun nest he was instantly killed, but the attack, begun by him, continued and resulted in the machine-gun nest being captured.
Posthumously awarded. Medal presented to father, W. L. J. Jobes.

JOE, COLIN B. (51680)_____
In the Meuse-Argonne offensive, France, Nov. 1 and Nov. 5, 1918.
R—Milton, Mass.
B—Wakefield, Mass.
G. O. No. 81, W. D., 1919.

Sergeant, Company K, 23d Infantry, 2d Division.
With complete disregard of his own danger, he went forward alone, when the advance was held up by 2 machine-gun nests, and single handed reduced these positions, capturing 9 prisoners. Later single handed, he attacked the crews of 3 machine guns, being severely wounded in this action.

JOERGER, CARL F. (1737681) _____
    Near Grand Pre, France, Oct. 24,
      1918.
    R—Newark, N. J.
    B—Newark, N. J.
    G. O. No. 44, W. D., 1919.

Private, Company M, 312th Infantry, 78th Division.
Leaving his dugout, Private Joerger voluntarily crawled across a machine-gun-swept area to the aid of 2 wounded comrades. While performing this heroic task he was seriously wounded.

*JOHANSON, CARL I. (1721217) _____
    In the Forest of Argonne, France,
      Sept. 27, 1918.
    R—Brooklyn, N. Y.
    B—Sweden.
    G. O. No. 21, W. D., 1919.

Private, first class, Company B, 306th Infantry, 77th Division.
He displayed exceptional bravery in volunteering to cut the enemy's wire and thereby make it possible for his company to advance upon the enemy. In performing this invaluable service he repeatedly exposed himself to heavy fire from enemy machine guns and was severely wounded.
Posthumously awarded. Medal presented to mother, Mrs. Josefine Emanuelson.

*JOHNS, LATIMER _____
    Near Gesnes, France, Sept. 30, 1918.
    R—Randolph, Wis.
    B—Cotter, Iowa.
    G. O. No. 21, W. D., 1925.

Second lieutenant, 122d Field Artillery, 33d Division.
Lieutenant Johns was in command of a platoon in support of an assaulting battalion of Infantry. During the attack he went far ahead of the Infantry to establish an observation post, where he directed fire from his guns, thereby rendering valuable assistance to the advancing battalion. After several attempts, he went through a heavy enemy barrage and enfilading machine-gun fire, but when returning to his post he was killed.
Posthumously awarded. Medal presented to brother, Haydn Johns.

JOHNS, SAMUEL H. (2337044) _____
    Near Bussy Farm, France, Sept. 28,
      1918.
    R—Needham, Mass.
    B—Philadelphia, Pa.
    G. O. No. 13, W. D., 1919.

Private, Company L, 372d Infantry, 93d Division.
After several other runners had been killed or wounded, he volunteered to carry a message over fields swept by heavy machine-gun fire and artillery bombardment. He succeeded in delivering the message, but was severely wounded while on the return trip.

JOHNSEN, EDWIN A. (46659) _____
    Near Exermont, France, Oct. 4, 1918.
    R—Elgin, Ill.
    B—St. Charles, Ill.
    G. O. No. 35, W. D., 1920.

Sergeant, Company D, 18th Infantry, 1st Division.
Sergeant Johnsen led his section in the attack through heavy artillery fire. Although severely wounded by a high-explosive shell, he remained with his organization until he was unable to continue forward, due to the loss of blood.

JOHNSON, ABE (65165) _____
    At Marcheville, France, Sept. 26,
      1918.
    R—Waterbury, Conn.
    B—Waterbury, Conn.
    G. O. No. 21, W. D., 1919.

Private, Company G, 102d Infantry, 26th Division.
This soldier volunteered to accompany a party whose mission was to bombard a hostile machine-gun emplacement. Under heavy shellfire, he approached to within 30 feet of the emplacement, when he was fired upon through loopholes in a stone wall. Working his way behind the wall, he enfiladed the enemy with rifle fire and effected their capture with the machine gun.

JOHNSON, ALGOT (1707542) _____
    Near Ville-Savoye, France, Aug. 26,
      1918.
    R—Bronx, N. Y.
    B—Sweden.
    G. O. No. 35, W. D., 1919.

Private, Company A, 308th Infantry, 77th Division.
Under heavy fire from the enemy, Private Johnson, accompanied by 1 man, crossed the Vesle River and silenced a machine gun which was causing heavy casualties in his company. They killed one gunner and wounded the other.

JOHNSON, ALVA LEE (2240264) _____
    Near Septsarges, France, Oct. 24,
      1918.
    R—El Paso, Tex.
    B—McClellan County, Tex.
    G. O. No. 37, W. D., 1919.

Corporal, Company G, 5th Ammunition Train, 5th Division.
When an enemy shell struck some pyrotechnics stored in the ammunition dump of his organization, he directed and assisted in the removal of inflammable material and placing the fire under control. Through his coolness and courage the destruction of a large quantity of near-by ammunition was avoided.

JOHNSON, ARTHUR (2463695) _____
    Near Mont des Singes, France, Sept.
      30, 1918.
    R—Newcomer, Pa.
    B—Dayton, Tenn.
    G. O. No. 46, W. D., 1919.

Private, Headquarters Company, 370th Infantry, 93d Division.
Acting as ammunition carrier, he received a painful injury in the back from a shell fragment. While engaged in carrying ammunition he found a wounded man in an exposed position and, regardless of his own wound, carried this man under heavy shellfire to the first-aid station, a distance of more than a kilometer, returning to his work immediately afterwards.

JOHNSON, BRAINARD W. (1290596) _____
    Near Verdun, France, Oct. 24, 1918.
    R—Basic, Va.
    B—Greenville, Va.
    G. O. No. 34, W. D., 1919.

Private, sanitary detachment, 116th Infantry, 29th Division.
He repeatedly exposed himself to heavy machine-gun fire in giving first aid to the wounded and carrying them to the rear. Through his devotion to duty and disregard of danger many lives were saved.

JOHNSON, CHARLES B., Jr. (133021) _____
    North of Chateau-Thierry, France,
      July 19, 1918.
    R—Somerville, Mass.
    B—Somerville, Mass.
    G. O. No. 125, W. D., 1918.

Sergeant, Battery A, 101st Field Artillery, 26th Division.
While his battery position was under shellfire and its ammunition dump had been hit and shells were exploding in it and the crew was ordered to leave, Sergeant Johnson remained, put out 3 fires in the dump, and afterwards, under heavy fire, searched for and brought wounded to shelter.

JOHNSON, CHRISTIAN A. (2036068) _____
    Near Ronssoy, France, Sept. 27,
      1918.
    R—Rapid City, Mich.
    B—Kalkaska, Mich.
    G. O. No. 78, W. D., 1919.

Private, first class, Headquarters Company, 106th Infantry, 27th Division.
When a strong force of the enemy had cut off his company from the advance units of our troops, Private Johnson volunteered to accompany 2 officers on a hazardous patrol to ascertain the exact location of the enemy and our advance troops. They came under terrific enemy fire, by which one of the officers was killed, but Private Johnson continued forward until he was completely surrounded by the enemy. He succeeded in working his way back and made his report, which was of great value in meeting the critical situation.

JOHNSON, GEORGE S. (2827875)_____
   Near Fleville, France, Oct. 4, 1918.
   R—Rockford, Ill.
   B—Rockford, Ill.
   G. O. No. 35, W. D., 1920.

Private, Company K, 16th Infantry, 1st Division.
When his organization was halted by machine-gun fire from the front Private Johnson, accompanied by a noncommissioned officer, advanced upon and destroyed the enemy machine-gun position, forcing 4 of the enemy to surrender, and thereby insuring the further advance of his unit.

JOHNSON, GILLIS AUGUSTUS_____
   Near St. Etienne, France, Oct. 4, 1918.
   R—Fort Worth, Tex.
   B—Fort Worth, Tex.
   G. O. No. 35, W. D., 1919.

Second lieutenant, 5th Regiment, U. S. Marine Corps, 2d Division.
He volunteered and led an attack upon enemy machine-gun positions under intense machine-gun and artillery barrage, and, although severely wounded in the leg, succeeded in cleaning out several machine-gun nests, capturing guns and a number of prisoners.

JOHNSON, GUSTAVE H. (1714321)_____
   Near Chery-Chartreuve, France, Aug. 16, 1918.
   R—Brooklyn, N. Y.
   B—Sweden.
   G. O. No. 37, W. D., 1919.

Corporal, Battery C, 305th Field Artillery, 77th Division.
Corporal Johnson assisted Lieut. Arthur A. Robinson, of his battery, in rescuing the body of an officer from a burning ammunition dump which was under fire.

JOHNSON, HANNING G. (145412)_____
   Near Suippes, France, July 15, 1918.
   R—Minneapolis, Minn.
   B—Lake City, Minn.
   G. O. No. 44, W. D., 1919.

Sergeant, Battery B, 151st Field Artillery, 42d Division.
He remained in command of his gun section throughout the entire day, after having been severely wounded.

JOHNSON, HAROLD R. (135610)_____
   Near Seicheprey, France, Apr. 20, 1918.
   R—Providence, R. I.
   B—Providence, R. I.
   G. O. No. 107, W. D., 1918.

Private, Headquarters Company, 102d Field Artillery, 26th Division.
For exceptional bravery and devotion to duty on Apr. 20, 1918, when, although wounded in the arm and leg he continued, while under heavy shellfire, to repair the telephone lines, and succeeded in reestablishing communication.

JOHNSON, HENRY (1316046)_____
   Near Bellicourt, France, Sept. 29, 1918.
   R—Creston, Tenn.
   B—Morgan County, Tenn.
   G. O. No. 50, W. D., 1919.

Private, first class, Company G, 119th Infantry, 30th Division.
After his platoon had been halted by unusually heavy fire from machine-gun nests, Private Johnson made his way forward and by the effective use of hand grenades killed the occupants of the nests and made possible the continuance of the advance.

JOHNSON, JOHN (1413499)_____
   At Belieu Bois, north of Verdun, France, Oct. 27, 1918.
   R—Alexander, Iowa.
   B—Denmark.
   G. O. No. 49, W. D., 1922.

Private, Company A, 102d Infantry, 26th Division.
On three separate occasions Private Johnson displayed the utmost heroism and complete disregard for personal danger by crawling from our lines while under direct observation and machine-gun fire and, by the skillful use of natural cover which the ground afforded, brought back to safety 3 wounded comrades who were lying helpless near the enemy's position, thereby saving them from death or capture.

JOHNSON, MAURICE E (1711846)_____
   Near Binarville, France, Oct. 2–7, 1918.
   R—Buffalo, N. Y.
   B—Monroeton, Pa.
   G. O. No. 13, W. D., 1923.

Corporal, Company D, 306th Machine Gun Battalion, 77th Division.
Two platoons of his company being surrounded by strong enemy forces, Corporal Johnson continued to operate his machine gun with little rest for 5 days. Each day small enemy groups endeavored to capture the machine gun, which was stationed on the right flank of the company; on each occasion Corporal Johnson remained with his gun in the face of deadly enemy fire, delivering bursts of fire which drove the enemy raiders to cover, killing and wounding many of them. Wounded on Oct. 4 by grenade fire, he still remained at his post; he was again wounded on Oct. 5. On Oct. 7, owing to his wounds, his machine gun was temporarily out of action and the ground held by riflemen. On the afternoon of this day the enemy launched an attack against the right flank, using liquid fire, whereupon Corporal Johnson, in spite of his exhaustion due to his wounds, resumed operation of his machine gun, and with such effect as to repulse the enemy.

ᵃJOHNSON, MELVIN B. (84054)_____
   Near Gesnes, France, Oct. 14, 1918.
   R—Greve, Mont.
   B—Devils Lake, N. Dak.
   G. O. No. 78, W. D., 1919.

Corporal, Company M, 127th Infantry, 32d Division.
When his battalion was held up after suffering heavy casualties from flanking machine-gun fire, he went out alone with an automatic rifle to a position 250 yards in advance of our lines, and, although subjected to intense fire from three directions, operated his gun and so neutralized the enemy fire while his battalion re-formed. He was killed on this mission, undertaken on his own initiative.
Posthumously awarded. Medal presented to father, Peter Johnson.

ᵉJOHNSON, OSCAR E. (1897971)_____
   Near St. Juvin, France, Oct. 14–15, 1918.
   R—Jamestown, N. Y.
   B—Sheffield, Pa.
   G. O. No. 44, W. D., 1919.

Private, first class, Company C, 325th Infantry, 82d Division.
In utter disregard for his own safety, Private Johnson repeatedly carried messages through heavy fire until he received wounds which later caused his death.
Posthumously awarded. Medal presented to cousin, Mrs. John Carlson.

JOHNSON, OSCAR E. (2382607)_____
   Near Cunel, France, Oct. 14, 1918.
   R—Norwalk, Conn.
   B—Norfolk, Conn.
   G. O. No. 81, W. D., 1919.

Private, first class, Company B, 60th Infantry, 5th Division.
Private Johnson volunteered and went to the rescue of his platoon commander who had been wounded and was lying in a very dangerous position, subjected to heavy machine-gun and shellfire. He succeeded in carrying the officer to a place of safety.

JOHNSON, PAUL (275289)_____
   Near Gesnes, France, Oct. 14, 1918.
   R—Oconto, Wis.
   B—Oconto, Wis.
   G. O. No. 71, W. D., 1919.

Private, Company M, 127th Infantry, 32d Division.
When the battalion was held up by heavy machine-gun fire, he volunteered to go forward 250 yards and help in establishing a position to neutralize the enemy fire. Although wounded, he remained at his post for seven hours under heavy fire till the enemy position was taken by assault.

JOHNSON, RAGNVOLD (2256921)...........
Near Gesnes, France, Sept. 29–Oct. 1, 1918.
R—Everett, Wash.
B—Norway.
G. O. No. 139, W. D., 1918.

Cook, Company D, 361st Infantry, 91st Division.
Under heavy shellfire and badly wounded, he constantly assisted for 3 days in cooking for an entire battalion in the front line.

JOHNSON, REUBEN L. (2161307)..........
Near the Bois-de-Brieulles, France, Sept. 28, 1918.
R—Ashton, S. Dak.
B—Donovan, Ill.
G. O. No. 98, W. D., 1919.

Private, first class, Company D, 47th Infantry, 4th Division.
Although he had been painfully wounded in the back by a bursting shell, Private Johnson continued to perform his duties as a runner under heavy artillery and machine-gun fire, thereby enabling his company commander to maintain control of the company. He remained on duty until late in the night when he was ordered to the dressing station.

*JOHNSON, RICHARD (275590)...........
Near Ciergos, France, July 29, 1918.
R—Eau Claire, Wis.
B—Eau Claire, Wis.
G. O. No. 74, W. D., 1919.

Sergeant, Company E, 127th Infantry, 32d Division.
Coming unexpectedly upon a German machine gun, he threw himself upon it as it started firing, being himself killed but preventing any casualties among the members of his own platoon, the enemy gunners being made prisoners.
Posthumously awarded. Medal presented to brother, Carl Johnson.

JOHNSON, ROYAL C..........
At Montfaucon, France, Sept. 26–27, 1918.
R—Aberdeen, S. Dak.
B—Cherokee, Iowa.
G. O. No. 3, W. D., 1924.

First lieutenant, 313th Infantry, 79th Division
He constantly exposed himself to the enemy fire during the action at Montfaucon, setting an example to his men by his fearlessness. When severely wounded by shellfire he assisted 2 wounded men of his company to the rear and refused to occupy space in the ambulance until these men had been provided for.

JOHNSON, SAMUEL I..........
At Vladivostok, Siberia, on the night of Nov. 17–18, 1919.
R—Honolulu, Hawaii.
B—Russia.
G. O. No. 28, W. D., 1921.

Major, 27th Infantry.
On three successive occasions he went through a zone swept by intense fire of contending factions to the railroad station and brought out noncombatants through the continuous fire from rifles and machine guns.

JOHNSON, SAMUEL M..........
Near Bussy Farm, France, Sept. 27, 1918.
R—Athens, Ohio.
B—Trimble, Ohio.
G. O. No. 13, W. D., 1919.

Major, 372d Infantry, 93d Division.
He led his battalion with exceptional bravery and judgment through a heavy enemy barrage in an attack on a strong hostile force on the plateau south of Bussy Farm, fearlessly exposing himself to point out enemy machine-gun positions. Having attained his objective, he held his ground in spite of the fact that his command had been badly cut up and participated in the attack on the following day. In spite of the strong resistance, his battalion captured a large number of guns, an ammunition dump, and valuable material. His fearlessness, energy, and leadership inspired his men to successful attack.

JOHNSON, SILAS J (1099182)..........
In the Puvenelle sector, France, Nov. 3, 1918.
R—Northfield, Minn.
B—Grundy County, Ill.
G. O. No. 44, W. D., 1919.

Sergeant, medical detachment, 56th Infantry, 7th Division.
When the company to which he was attached withdrew from their position and the medical personnel was ordered to find a place of safety, Sergeant Johnson refused to leave the wounded. He carried a seriously wounded officer 1½ kilometers through a heavy artillery barrage to the battalion aid station.

JOHNSON, SWAN (43105)..........
Near Soissons, France, July 18, 1918.
R—Seattle, Wash.
B—Seattle, Wash.
G. O. No. 15, W. D., 1919.

Sergeant, Company G, 18th Infantry, 1st Division.
He personally reconnoitered a heavily guarded enemy position and killed one of the foe. In spite of being seriously wounded, he succeeded in returning to his patrol, informing them of the enemy's position and thereby enabling them to capture the entire enemy force.

JOHNSON, SWAN E. (1386504)..........
At Chipilly Ridge, France, Aug. 9, 1918.
R—Chicago, Ill.
B—Chicago, Ill.
G. O. No 70, W. D., 1919.

Sergeant, Company B, 131st Infantry, 33d Division.
His company having been held up by concentrated machine-gun and artillery fire, Sergeant Johnson and another soldier went forward and made a reconnaissance, locating a way forward which was protected from direct fire. Although he was badly wounded, Sergeant Johnson returned and led his company along this route, thereby enabling the entire battalion to advance.

JOHNSON, WILBUR (207347)..........
Near Les Pres Farm, France, Aug. 9, 1918.
R—Lansing, Mich.
B—Parshallburg, Mich.
G. O. No. 98, W. D., 1919.

Private, Battery C, 119th Field Artillery, 32d Division.
When an enemy shell burst at the rear end of the gun pit, wounding him and all the members of the gun crew except the chief of the section, Private Johnson concealed the fact that he had received two severe wounds in the back after he had assisted in removing his wounded comrades. He then resumed his duties and continued serving the piece for about 10 minutes until he collapsed.

JOHNSON, WILLIAM T. (1826972)..........
Near Bois-du-Fays, France, Oct. 5, 1918.
R—Waverly, Va.
B—Rosewood, Va.
G. O. No. 37, W. D., 1919.

Sergeant, Company A, 318th Infantry, 80th Division.
While leading a patrol, Sergeant Johnson encountered terrific machine-gun fire, which forced him to order his patrol to cover. He then advanced alone, working his way to the nest, which he destroyed, thereby permitting the patrol to continue its operation. Later the same day he braved the perils of an extremely heavy barrage to bring to safety a wounded comrade who was lying 300 yards in advance of the lines.

JOHNSTON, EWART..........
During the attack on Malbrouck Hill and Consenvoye Woods, north of Verdun, France, Oct. 8, 1918.
R—Winchester, Va.
B—Birmingham, Ala.
G. O. No. 15, W. D., 1921.

Captain, 116th Infantry, 29th Division.
Captain Johnston led his company through heavy machine-gun and artillery fire in the attack to his objective. Upon reaching a position scheduled for a passage of the lines he located a strong enemy position. Upon his own initiative he led his company in a bayonet attack and captured about 200 prisoners.

JOHNSTON, FRANK (1491261)............
  Near St. Étienne, France, Oct. 8,
    1918.
  R—Denton, Tex.
  B—Parker County, Tex.
  G. O. No. 37, W. D., 1919.

Corporal, Company M, 142d Infantry, 36th Division.
After his company had been thrown into confusion by running into its own artillery barrage, he reorganized a large part of the company and continued the advance. His command was again separated by a barrage of enemy artillery fire, but he continued with that portion of the company left under his control until he was twice wounded and carried to the rear.

JOHNSTON, GEORGE W. (112606)........
  Near Sergy, France, July 29 to Aug.
    1, 1918.
  R—Phillipsburg, N. J.
  B—Philadelphia, Pa.
  G. O. No. 71, W. D., 1919.

Private, first class, Company A, 149th Machine Gun Battalion, 42d Division.
When all the other runners were either wounded or exhausted, he maintained liaison by constantly carrying messages through zones swept by intense enemy fire. He often volunteered to assist stretcher bearers in removing wounded from the battle field.

JOHNSTON, GORDON...................
  At Palo, Leyte, Philippine Islands,
    Feb. 1, 1900.
  R—Birmingham, Ala.
  B—Charlotte, N. C.
  G. O. No. 13, W. D., 1924.
  Medal of honor and distinguished-
    service medal also awarded.

Second lieutenant, 43d Infantry, United States Volunteers.
While in command of a small detachment of scouts he displayed remarkable gallantry and leadership in charging a greatly superior force of intrenched insurgents in the face of cannon and rifle fire, driving the enemy from their position and capturing the town of Palo.

JOHNSTON, HAMILTON................
  Near Launay, France, July 15, 1918.
  R—Troy, N. Y.
  B—Cohoes, N. Y.
  G. O. No. 23, W. D., 1919.

Second lieutenant, 38th Infantry, 3d Division.
Lieutenant Johnson, with 2 soldiers, attacked a patrol of 7 Germans who had captured 4 American soldiers, killed 1 of the Germans, and captured the others.

JOHNSTON, HUGH (1398103)..........
  Near St. Jevin, France, Oct. 11, 1918.
  R—Brooklyn, N. Y.
  B—Forest City, Pa.
  G. O. No. 40, W. D., 1919.

Sergeant, Company D, 325th Infantry, 82d Division.
Voluntarily leaving shelter, he crawled out into the open under heavy enemy fire to the aid of a wounded soldier. While administering first aid to the latter he was himself wounded, but he nevertheless attempted to carry his comrade to safety, and in so doing he received a second wound.

JOHNSTON, JOSEPH H................
  At Beulay, France, Oct. 15, 1918.
  R—Chapel Hill, N. C.
  B—Chapel Hill, N. C.
  G. O. No. 74, W. D., 1919.

First lieutenant, 322d Infantry, 81st Division.
Lieutenant Johnston led a daylight patrol behind the German front line for the purpose of securing information as to the reported retreat of the enemy. Discovering an enemy machine gun, he led his men in an attempt to capture it, but when they were about 25 yards away the gun opened fire and this officer was mortally wounded. Upon being pulled into a trench by members of the patrol he manifested no anxiety concerning himself, but urged his men to continue their mission.
Posthumously awarded. Medal presented to mother, Mrs. C. W. Johnston.

JOHNSTON, LOUIS E. (1319169).......
  Near Mazinghien, France, Oct. 18–
    19, 1918.
  R—Davidson, N. C.
  B—Davidson, N. C.
  G. O. No. 35, W. D., 1919.

Corporal, Machine Gun Company, 120th Infantry, 30th Division.
When his platoon became separated from the battalion to which it was attached Corporal Johnston proceeded under heavy shellfire along a road with which he was unfamiliar and established liaison with his battalion.

JOHNSTON, Dr. MERCER G...........
  Near Verdun, France, Oct. 27, 1918.
  R—Baltimore, Md.
  B—Church Hill, Miss.
  G. O. No. 23, W. D., 1919.

Secretary, Young Men's Christian Association.
After volunteering and going to the front line through heavy bombardment for the purpose of burying the dead, Doctor Johnston found the litter service of the 101st Infantry badly disorganized on account of heavy casualties and intense shelling. He immediately took charge of the litter bearers, reorganized the service, took care of the slightly wounded himself, saw to the procuring and loading of ambulances, and, although badly gassed and suffering severely, refused to leave his post until all had been taken care of.

JOHNSTON, SCOTT MARTIN...........
  Near Vierzy, France, July 19, 1918.
  R—St. Paul, Minn.
  B—St. Paul, Minn.
  G. O. No. 95, W. D., 1919.

Second lieutenant, 6th Regiment, U. S. Marine Corps, 2d Division.
With a small detachment Lieutenant Johnston charged a machine-gun nest and captured a German gun which was inflicting severe losses on the American lines. Although seriously wounded, he stayed with his men until ordered to a dressing station by his company commander.

JOHNSTON, WILLIAM H..............
  Northwest of Verdun, France, Sept.
    27–30, 1918.
  R—Cincinnati, Ohio.
  B—Cincinnati, Ohio.
  G. O. No. 37, W. D., 1919.
  Distinguished-service medal also
    awarded.

Major general, 91st Division.
He repeatedly showed exceptional bravery during the Argonne-Meuse offensive, frequently visiting his front lines under heavy fire from enemy artillery, machine guns, and snipers, displaying marked coolness and inspiring the members of his command with confidence and determination.

JOINER, ARTHUR ELMER (2304082)......
  Near Le Grande Carre Ferme,
    France, Nov. 1, 1918.
  R—Granbury, Tex.
  B—Cleburne, Tex.
  G. O. No. 46, W. D., 1919.

Private, Company A, 360th Infantry, 90th Division.
Private Joiner, a battalion runner, made 4 trips to his company through intense machine-gun fire. On the fourth trip he was severely wounded, but he continued to crawl along until he intercepted another runner and gave him the message.

JOLLEY, THOMAS (3019)............
  Near Chemin-des-Dames, France,
    Mar. 6, 1918.
  R—Lawrence, Mass.
  B—England.
  G. O. No. 129, W. D., 1918.

Private, first class, medical detachment, 101st Field Artillery, 26th Division.
On Mar. 6, 1918, while the area in which he was located was being heavily shelled by the enemy he showed extraordinary valor by leaving his dugout, passing through 300 meters of heavy shellfire, and rendering aid to a wounded man at great risk of his own life.

JONA, STEPHEN, Jr. (63831)
At Marcheville, France, Sept. 26, 1918.
R—Hartford, Conn.
B—Hartford, Conn.
G. O. No. 137, W. D., 1918.

Corporal, Company B, 102d Infantry, 26th Division.
During a barrage lasting 2 hours he placed his men in the best shelter available, remaining in observation himself, and refused to take cover. He organized a platoon of men who had become separated from their commands and led them forward under a heavy fire from artillery, machine guns, and snipers. Throughout the engagement he was a source of inspiration to his men.

JONES, ALVEY (202846)
At Beaumont, France, Feb. 22, 1918, and at Missy-aux-Bois, France, July 23, 1918.
R—Morton Grove, Ill.
B—Brooklyn, N. Y.
G. O. No. 14, W. D., 1920.

Battalion sergeant major, Headquarters Company, Motor Battalion, 1st Ammunition Train, 1st Division.
Near Beaumont, Sergeant Major Jones displayed marked courage by refusing to be evacuated after being wounded, remaining on duty under severe shell fire and preventing a traffic blockade. At Missy-aux-Bois an enemy shell set fire to a pile of ammunition which he was salvaging. Disregarding danger, he managed to save a larger pile near by, extinguishing the flames. Though he was slightly wounded, he continued his work until every box of ammunition was salvaged.

JONES, ARCHIE J. (3133255)
During attack on Bois-de-Baulny, north of Epinonville, France, Sept. 28, 1918.
R—Chicago, Ill.
B—Canada.
G. O. No. 39, W. D., 1920.

Private, Company D, 364th Infantry, 91st Division.
During the attack Private Jones rushed, under enemy machine-gun fire, into the open in advance of the lines held and rendered first aid to a badly wounded officer who was lying exposed to enemy machine-gun fire.

JONES, ARTHUR CARROLL (2712837)
Near Montfaucon, France, Sept. 28, 1918.
R—Baltimore, Md.
B—Suffolk, Va.
G. O. No. 4, W. D., 1923.

Private, Company G, 313th Infantry, 79th Division.
While on his way to the dressing station for treatment, hearing a call for a volunteer to carry an important message from brigade headquarters to a regiment which was being pressed back, Private Jones volunteered and carried the message through a terrific intervening enemy barrage, and as a result of his courageous actions reinforcements arrived at a critical time.

JONES, ARTHUR H.
In the Toul sector, July 16, 1918.
R—Hayward, Calif.
B—Vallejo, Calif.
G. O. No. 121, W. D., 1918.

Second lieutenant, 147th Aero Squadron, Air Service.
Lieutenant Jones and 4 other pilots were attacked by 9 German pursuit planes. Without hesitation he dived into the leader of the enemy formation, pouring machine-gun fire into him at 100 yards. After a quick and decisive combat the enemy leader fell out of control. He then attacked 2 of the other enemy planes, which were attacking him from the rear, and succeeded in driving them off.

JONES, ARTHUR W. (181193)
Near Cambrin, France, April 9, 1918.
R—Minneapolis, Minn.
B—Hastings, Minn.
G. O. No. 126, W. D., 1918.

Corporal, Company A, 30th Engineers, attached to British Expeditionary Forces.
While returning from the front lines on the morning of Apr. 9, 1918, his platoon was subjected to a heavy shell fire, several of the men being killed or wounded, the balance taking shelter near by. Corporal Jones persisted in leaving his shelter and searching for wounded, several of whom he brought back in the midst of a barrage. He carried on the work in heroic manner for the benefit of his comrades and with disregard for his own personal safety.

*JONES, CARL O. (1311053)
Near Montbrehain, France, Oct. 8, 1918.
R—Kannapolis, N. C.
B—Cabarrus, N. C.
G. O. No. 37, W. D., 1919.

Private, Company E, 118th Infantry, 30th Division.
Crawling to the flanks of a German machine-gun nest, he covered the crew with his rifle from a distance of 30 yards and captured 12 of the enemy. This gallant soldier was subsequently killed in action.
Posthumously awarded. Medal presented to father, Sam Jones.

JONES, CHARLES G. (1850221)
Near Bois-des-Ogons, France, Oct. 4–5, 1918.
R—Hopewell, Va.
B—Buffalo, N. Y.
G. O. No. 81, W. D., 1919.

Corporal, Company F, 318th Infantry, 80th Division.
Making his way through a heavy barrage, he brought valuable information as to the enemy's position to his company commander. He then returned to the enemy's position, showing absolute disregard for his own personal danger and brought back 2 wounded men who had fallen there.

JONES, CLARENCE M. (1247107)
Near Chatel-Chehery, France, Oct. 8, 1918.
R—Meadville, Pa.
B—Meadville, Pa.
G. O. No. 98, W. D., 1919.

Sergeant, Company B, 112th Infantry, 28th Division.
Ordered to clear Hill 244 of the enemy, Sergeant Jones led a patrol of 7 men up a steep slope under enemy grenade fire by which 4 of his men were killed. Pushing on with the remaining 3, he silenced 3 machine-gun nests and 12 snipers, driving off the remainder of the Germans. He then sent one of his men back with a message and with the other 2 held the position for 2 hours until he was relieved.

JONES, CLAUDE V. (2222195)
Near Fey-en-Haye, France, Sept. 12, 1918.
R—Clarita, Okla.
B—Brownwood, Tex.
G. O. No. 81, W. D., 1919.

Corporal, Company M, 358th Infantry, 90th Division.
With the assistance of one other soldier, Corporal Jones attacked a machine-gun nest from the flank and captured the gun, together with 11 prisoners.

JONES, CLINTON
Near Landres-et-St. Georges, France, Oct. 30, 1918.
R—San Diego, Calif.
B—Ross Valley, Calif.
G. O. No. 66, W. D., 1919.

Second lieutenant, pilot, 22d Aero Squadron, Air Service.
Lieutenant Jones, while attacking 4 enemy planes (Fokker type), was in turn attacked from above and obliged to dive through a formation of 15 planes (Fokker type). His plane was riddled with bullets, but he managed to destroy one of the enemy machines.
Oak-leaf cluster.
For the following act of extraordinary heroism in action near St. Mihiel, France, Oct. 18, 1918, Lieutenant Jones is awarded an oak-leaf cluster to be worn with the distinguished-service cross: He was a member of a patrol which succeeded in hedging in a fast enemy biplace plane. Approaching the enemy plane, Lieutenant Jones signaled the enemy to give up and land. The reply was a burst of machine-gun fire, which cut his wind shield and set fire to his plane. He then closed in and shot the German pilot and sent the plane crashing to the ground. He landed in his own plane and extinguished the flames.

*JONES, DANIEL L. (1238903)............
  Near Baslieux, France, Sept. 6, 1918.
  R—Latrobe, Pa.
  B—Latrobe, Pa.
  G. O. No. 95, W. D., 1919.

Sergeant, Headquarters Company, 110th Infantry, 28th Division.
In command of a detachment of signalmen in the crossing of the Vesle River, Sergeant Jones was severely gassed while transmitting an important message to the rear. He, however, continued with his message the entire distance to the regimental headquarters, where he arrived exhausted and suffering severely from burns and gas inhalation. Although tagged at the dressing station for evacuation, he refused to be evacuated and returned to the front line, later bringing 3 other men blinded by gas to the rear. Sergeant Jones subsequently died from the injuries received.
Posthumously awarded. Medal presented to widow, Mrs. Daniel L. Jones.

JONES, ECMAN T. (53482)..............
  Near Soissons, France, July 19, 1918.
  R—Ottawa, Kans.
  B—Kansas City, Mo.
  G. O. No. 15, W. D., 1919.

Sergeant, Company E, 26th Infantry, 1st Division.
After being wounded on July 19, 1918, near Soissons, France, he refused to be evacuated, led his platoon in attack, and continued fighting until incapacitated by a second severe wound.

JONES, GEORGE W................
  In the Ravine-de-Bois-de-Caures, France, Oct. 31, 1918.
  R—Worcester, Mass.
  B—Worcester, Mass.
  G. O. No. 46, W. D., 1919.

First lieutenant, 102d Field Artillery, 26th Division.
Though himself painfully wounded by a bursting shell when his battery position was heavily bombarded by the enemy, he immediately directed the the work of rescuing wounded men from demolished dugouts and evacuating them to the rear. Having finished this work, he at once reorganized his battery and carried out orders for immediate fire on the enemy.

JONES, HARRY (1390994)............
  Near Consenvoye, France, Oct. 8, 1918.
  R—Chicago, Ill.
  B—Chicago, Ill.
  G. O. No. 64, W. D., 1919.

Corporal, Company G, 132d Infantry, 33d Division.
He showed extraordinary personal bravery when his platoon was held up by fire from a machine-gun emplacement. Crawling forward alone from his own line, he worked his way to the flank of the enemy position and then rushed it, bayoneting one German and taking two prisoners. His action enabled his platoon to advance at once.

JONES, HARVEY L. (1385596)............
  Near Dannevoux, France, Oct. 3, 1918.
  R—Peoria, Ill.
  B—Peoria, Ill.
  G. O. No. 15, W. D., 1923.

Private, first class, Company C, 123d Machine Gun Battalion, 33d Division.
Under direct observation of the enemy and under heavy enemy artillery and machine-gun fire he rescued a wounded soldier who was lying exposed to heavy fire and close to the enemy's lines. With great bravery he succeeded in carrying the wounded man to a place of comparative safety whence he was removed to the hospital.

JONES, HENRY L. (1253832)............
  Near Le Chene Tondu, France, Oct. 4, 1918.
  R—Wilkes-Barre, Pa.
  B—Wilkes-Barre, Pa.
  G. O. No. 64, W. D., 1919.

Corporal, Battery D, 109th Field Artillery, 28th Division.
He left an observation post, and, exposing himself to intense artillery and machine-gun fire, dressed the wounds of an officer who had fallen in the open. Then, with the aid of another soldier, he carried the wounded officer to a dressing station. His action saved the life of the officer.

*JONES, HERBERT J................
  Near Coullemelle, France, July 4, 1918.
  R—Dresden, Tenn.
  B—Dresden, Tenn.
  G. O. No. 87, W. D., 1919.

Second lieutenant, 6th Field Artillery, 1st Division.
During a heavy enemy bombardment he exposed himself fearlessly to go to the assistance of a wounded soldier, being killed by shell fire while engaged in this heroic action.
Posthumously awarded. Medal presented to mother, Mrs. Joseph E. Jones.

JONES, JAMES T. (1307400)............
  Near Ypres, Belgium, July 24, 1918.
  R—Knoxville, Tenn.
  B—Knoxville, Tenn.
  G. O. No. 50, W. D., 1919.

Corporal, Company C, 117th Infantry, 30th Division.
He was in charge of a detached automatic-rifle post heavily bombarded by the enemy. Two of his men were killed by shellfire, 2 others and he himself seriously wounded. Though it was his first experience under fire, he exhibited unhesitating devotion to duty by remaining at his post. Sending for assistance, he reorganized his position and gave aid and comfort to the wounded.

JONES, LEROY (1234936)............
  Near Verdun, France, Oct. 8, 1918.
  R—Wilmington, Del.
  B—Stanton, Del.
  G. O. No. 32, W. D., 1919.

Private, first class, Company E, 115th Infantry, 29th Division.
While his platoon was being held up by machine-gun fire he voluntarily left his position and, crawling through intense machine-gun fire, he single handed captured 2 machine guns, killing 4 of the enemy, and taking both crews.

JONES, PERCY H. (3170783)............
  Near Consenvoye, France, Oct. 9, 1918.
  R—Buffalo Junction, Va.
  B—Mecklenburg County, Va.
  G. O. No. 35, W. D., 1920.

Private, Company B, 131st Infantry, 33d Division.
After 2 other soldiers had been killed and 1 wounded in attempts to rescue their wounded platoon leader, Private Jones went forward under heavy machine-gun fire and carried his lieutenant to a place of safety.

JONES, ROY V. (322758)............
  Near Romanovka, Siberia, June 25, 1919.
  R—Knoxville, Iowa.
  B—Beacon, Iowa.
  G. O. No. 133, W. D., 1919.

Private, first class, Company A, 31st Infantry.
Though wounded early in action he continued to operate his automatic rifle throughout the fight.

JONES, SANDY E. (1872473)............
  Near Champagne, France, Sept. 28–29, 1918.
  R—Taft, S. C.
  B—Taft, S. C.
  G. O. No. 46, W. D., 1919.

Corporal, Company C, 371st Infantry, 93d Division.
Corporal Jones was engaged as company clerk and was left behind to care for the company records. When he learned that all the company officers had become casualties he immediately went forward, and collecting the scattered elements of the company, reorganized them under most trying and difficult conditions.

JONES, THOMAS EDWARD_____
Near Binarville, France, Sept. 27, 1918.
R—Washington, D. C.
B—Lynchburg, Va.
G. O. No. 35, W. D., 1919.

First lieutenant, Medical Corps, attached to 368th Infantry, 92d Division.
Lieutenant Jones went into an open area subjected to direct machine-gun fire to care for a wounded soldier who was being carried by another officer. While dressing the wounded runner, a machine-gun bullet passed between his arms and chest and a man was killed within a few yards of him.

JONES, WILBUR S. (92530)_____
Near Chalons-sur-Marne, France, July 15, 1918.
R—Newport, Ky.
B—Georgetown, Ohio.
G. O. No. 44, W. D., 1919.

Wagoner, Supply Company, 166th Infantry, 42d Division.
During the great German artillery bombardment of July 15 he was driving a ration cart to the front lines when he was caught in the heavy bombardment. Both his mules were killed and he was blown from his seat by a shell explosion, the same shell killing a comrade on the cart ahead of him. Catching a stray mule and borrowing another, he returned with his comrade to the company, after which he delivered his rations. After his team became frightened and ran away, he remained in the fight throughout the attack of the day. While delivering rations he was wounded, but he completed his task before he would allow his wound to be dressed.

JONES, WILLIAM (94672)_____
Near St. Baussant, France, Sept. 12, 1918.
R—Newark, Ohio.
B—Prospect, Ohio.
G. O. No. 44, W. D., 1919.

Corporal, Company I, 166th Infantry, 42d Division.
When his platoon came under heavy fire from a machine-gun nest on its flank, he took part of his squad, surrounded the nest, and captured the gun and crew.

JONES, WILLIAM (1387079)_____
Near Bethincourt, France, Sept. 26, 1918.
R—Chicago, Ill.
B—Rochester, N. Y.
G. O. No. 46, W. D., 1919.

Sergeant, Company G, 131st Infantry, 33d Division.
Upon his own initiative he advanced under concentrated rifle and machine-gun fire, which was holding up his platoon, and put out of action a nest of light machine guns on the flank, thereby permitting the platoon to continue forward.

JONES, WILLIE F. (1312983)_____
Near Brancourt, France, Oct. 7, 1918.
R—Abbeville, S. C.
B—Abbeville, S. C.
G. O. No. 133, W. D., 1918.

Private, medical detachment, 118th Infantry, 30th Division.
In the face of heavy enemy fire Private Jones, together with 3 other stretcher bearers, advanced before our front line and brought back to shelter a wounded Australian officer.

²JORDAN, CHARLES F. (1316183)_____
Near Bellicourt, France, Sept. 29, 1918.
R—Cooleemee, N. C.
B—Mocksville, N. C.
G. O. No. 87, W. D., 1919.

Private, Company H, 119th Infantry, 30th Division.
He repeatedly exposed himself to enemy fire to save his comrades, going forward in advance of our lines to attack machine-gun nests. After reducing 1 enemy nest with rifle grenades, he proceeded to attack another, and while so doing was killed by enemy snipers.
Posthumously awarded. Medal presented to father, George W. Jordan.

JORDAN, CLIFTON M. (44247)_____
Near Cantigny, France, June 2, 1918.
R—Malta, Mont.
B—Hamburg, Ark.
G. O. No. 35, W. D., 1919.

Private, Company L, 16th Infantry, 1st Division.
He went forward under intense machine-gun and artillery fire and assisted in the removal of a wounded soldier over a distance of 1 kilometer.

JORDAN, EDWARD J. (1245876)_____
Near Fismette, France, Aug. 12, 1918.
R—Philadelphia, Pa.
B—Philadelphia, Pa.
G. O. No. 50, W. D., 1919.

Corporal, Company M, 111th Infantry, 28th Division.
Having learned from a raid of the previous night that a comrade was lying wounded in front of his lines, Corporal Jordan set out to rescue him after seeing two other men killed in the attempt. Crawling in full view of the enemy, he was shot through the leg when a storm of fire was directed at him, but he struggled on and brought his man to safety.

JORDAN, JACK_____
Near St. Etienne, France, Oct. 4, 1918.
R—Tulia, Tex.
B—Rockwall, Tex.
G. O. No. 46, W. D., 1919.

Corporal, 8th Company, 5th Regiment, U. S. Marine Corps, 2d Division.
After all the other members of their gun crews had become casualties, Corporal Jordan and another soldier advanced with their gun through heavy artillery and machine-gun fire to an advanced position and put up their gun in action in support of the Infantry. They remained in this position after the Infantry had retired and until their ammunition was exhausted.

JORDAN, JAMES O. (1866912)_____
Near Busigny, France, Oct. 9, 1918.
R—Winston-Salem, N. C.
B—Surry County, N. C.
G. O. No. 37, W. D., 1919.

Private, Company C, 117th Infantry, 30th Division.
When his platoon was subjected to heavy machine-gun fire from the front and flanks, Private Jordan courageously operated his automatic rifle from an exposed position with such good effect that fire superiority was maintained until reinforcements arrived.

JORDAN, JOHN P. (181327)_____
Near Nantillois, France, Oct. 5, 1918.
R—Cleveland, Ohio.
B—Haymarket, Va.
G. O. No. 37, W. D., 1919.

Corporal, Company A, 1st Gas Regiment.
After other means of communication had failed, he voluntarily carried messages from the regimental post of command to advanced positions through several enemy barrages of gas and high-explosive shells. He continued on duty, even after being wounded, until he was exhausted.

JORDAN, JOHN W _____

Near Fismes, France, Aug. 11, 1918.
R—Indianapolis, Ind.
B—Indianapolis, Ind.
G. O. No. 44, W. D., 1919.

Second lieutenant, 7th Field Artillery, observer, 88th Aero Squadron, Air Service.
Louis G. Bernheimer, first lieutenant, pilot; John W. Jordan, second lieutenant, 7th Field Artillery, observer; Roger W. Hitchcock, second lieutenant, pilot; James S. D. Burns, deceased, second lieutenant, 165th Infantry, observer; Joel H. McClendon, deceased, first lieutenant, pilot; Charles W. Plummer, deceased, second lieutenant, 101st Field Artillery, observer; Philip R. Babcock, first lieutenant, pilot; and Joseph A. Palmer, second lieutenant, 15th Field Artillery, observer. All of these men were attached to the 88th Aero Squadron, Air Service.
For extraordinary heroism in action near Fismes, France, Aug. 11, 1918. Under the protection of 3 pursuit planes, each carrying a pilot and an observer, Lieutenants Bernheimer and Jordan, in charge of a photo plane, carried out successfully a hazardous photographic mission over the enemy's lines to the River Aisne. The 4 American ships were attacked by 12 enemy battle planes. Lieutenant Bernheimer, by coolly and skillfully maneuvering his ship, and Lieutenant Jordan, by accurate operation of his machine gun, in spite of wounds in the shoulder and leg, aided materially in the victory which came to the American ship and returned safely with 36 valuable photographs. The pursuit plane operated by Lieutenants Hitchcock and Burns was disabled while these two officers were fighting effectively. Lieutenant Burns was mortally wounded and his body jammed the controls. After a headlong fall of 2,500 meters, Lieutenant Hitchcock succeeded in regaining control of this plane and piloted it back to his airdrome. Lieutenants McClendon and Plummer were shot down and killed after a vigorous combat with 5 of the enemy's planes. Lieutenants Babcock and Palmer, by gallant and skillful fighting, aided in driving off the German planes and were materially responsible for the successful execution of the photographic mission.

JORDAN, MORTIMER H _____
Near Souain, France, July 15, 1918.
R—Birmingham, Ala.
B—Birmingham, Ala.
G. O No. 27, W. D., 1919.

Captain, 167th Infantry, 42d Division.
Seeing a private of his company wounded by shellfire, Captain Jordan left his shelter and rushed to the aid of the wounded man. After administering first aid he carried him through the terrific bombardment a distance of 150 yards to a place of safety.
Posthumously awarded. Medal presented to widow, Mrs. Mortimer H. Jordan.

JORDAN, NOLAN L. (3115109) _____
Near Malancourt, France, Sept. 26, 1918.
R—Courtland, Va.
B—Courtland, Va.
G. O. No. 46, W. D., 1919.

Private, first class, Company K, 314th Infantry, 79th Division.
Private Jordan, with another soldier of his platoon, outflanked a machine gun in advance of our line, killed 3 of the crew and captured 2 others, together with the machine gun.

JORDAN, RICHARD OAKES _____
At Blanc Mont, France, Oct. 3, 1918.
R—Minneapolis, Minn.
B—Sioux Falls, S. Dak.
G. O. No. 37, W. D., 1919.

Private, 78th Company, 6th Regiment, U. S. Marine Corps, 2d Division
When the advance of their company was held up by enfilading fire from a hostile machine-gun nest, Private Jordan, with 3 other soldiers, volunteered and made a flank attack on the nest with bombs and rifles, killing 3 members of the crew and capturing 25 others, together with 3 machine guns.

JOUBERT, JAMES M. (1205174) _____
Near St. Souplet, France, Oct. 17, 1918.
R—Glens Falls, N. Y.
B—Ticonderoga, N. Y.
G. O. No. 126, W. D., 1919.

Sergeant, Company K, 105th Infantry, 27th Division.
He exhibited great daring in advancing, single handed, against two enemy machine guns, which he put out of action.

JULEWICZ, HYLORY (1900564) _____
Near St. Juvin, France, Oct. 17, 1918.
R—Jersey City, N. J.
B—Poland.
G. O. No. 9, W. D., 1923.

Private, Company A, 326th Infantry, 82d Division.
After the first-line positions in front of Champeigneulles, France, were abandoned, his company commander discovered that important papers, messages, and maps which would give information to the enemy had been left in the old front line. Private Julewicz, without announcing his purpose, at risk of life or capture, crossed 500 yards of machine-gun and shell-swept ground in full view of the enemy and recovered the papers.

JUSTESEN, WILLIAM A _____
Near Vierzy, France, July 19, 1918.
R—Moroni, Utah.
B—Moroni, Utah.
G. O. No. 137, W. D., 1918.

Private, 55th Company, 5th Regiment, U. S. Marine Corps, 2d Division.
He displayed exceptional bravery in charging 3 machine guns with the aid of a small detachment of his comrades, killing the crews and capturing the guns, which were immediately turned on the Germans, thereby opening the line for the advance of his company, which had been held up by the enemy's fire.

JUTRAS, WILLIAM H _____
Near Riaville, France, Sept. 26, 1918.
R—Manchester, N. H.
B—Peterboro, N. H.
G. O. No. 142, W. D., 1918.

First Lieutenant, 103d Infantry, 26th Division.
When the platoon on the right flank of his company was threatened by an enfilading movement of enemy machine guns, he carried a message to the commander of that platoon through deadly machine-gun and minenwerfer bombardment. It then being necessary to establish liaison with the company on the right, in order to save this platoon from annihilation, and knowing that he faced almost certain death, this gallant officer unhesitatingly volunteered for this mission and crossed a terrain swept by converging machine-gun fire. Mortally wounded, he delivered his message in time to save his platoon.
Posthumously awarded. Medal presented to mother, Mrs. Methe H. Jutras.

KACPRZYZKI, BRONISLAW (17291) _____
Near Jaulny, France, Sept. 12, 1918.
R—Detroit, Mich.
B—Poland.
G. O. No. 74, W. D., 1919.

Private, medical detachment, 9th Infantry, 2d Division.
Private Kacprzyzki, with three other soldiers, volunteered to carry wounded men of their companies from front of our advanced positions and carried this work on under violent machine-gun fire while a counterattack was developing.

*KAHLE, CLARENCE C._____
   Near Lachaussee, France, Sept. 13,
     1918.
   R—Pittsburgh, Pa.
   B—Franklin, Pa.
   G. O. No. 123, W. D., 1918.

First lieutenant, pilot, 99th Aero Squadron, Air Service.
He, with First Lieut. Raymond C. Hill, observer, was directed to take photographs of the old Hindenburg line. They were accompanied by 2 protecting planes. After they had taken some photographs the protecting
planes were driven off by hostile aircraft, but he and his observer continued
their mission alone, until in the vicinity of Lachaussee they were attacked by
an enemy formation of 9 planes. Lieutenant Kahle put up a gallant fight,
in which his observer was shot through the heart and killed. Although
pitted against overwhelming odds, Lieutenant Kahle, by his pluck, determination, skill, and courage, brought the photographs and the plane back to
his airdrome, the enemy keeping up a constant attack upon him back to our
lines, riddling the plane with machine-gun bullets.
Posthumously awarded. Medal presented to mother, Mrs. F. L. Kahle.

KAIN, RANCY R. (279193)_____
   In the woods north of Cierges,
     France, Aug. 1, 1918.
   R—Watervliet, Mich.
   B—Cecil, Ohio.
   G. O. No. 117, W. D., 1918.

Sergeant, Company C, 126th Infantry, 32d Division.
In an attack on the woods occupied by the enemy north of Cierges to clear out
machine guns and snipers, he was in charge of a platoon. Seeing 2 machine-
gun nests and no officer being present to take command, he directed his
platoon in such a manner as to silence the machine guns, killing 4 of the
enemy. He also silenced another machine gun located in a tree, and personally brought down the operator.

KALLOCH, PARKER C., Jr._____
   North of Montrebeau Woods,
     France, Sept. 29, 1918.
   R—Portland, Me.
   B—San Francisco, Calif.
   G. O. No. 126, W. D., 1919.

Major, 137th Infantry, 35th Division.
Major Kalloch, although wounded, and scarcely able to walk, personally
organized and led an attack against a superior force at Exermont in the face
of heavy artillery and intense machine-gun fire. His extraordinary courage
and utter disregard for personal safety were an inspiration to his entire command.

KAMINSKI, MIKE (2633818)_____
   Near Romagne, France, Oct. 11, 1918.
   R—Detroit, Mich.
   B—Detroit, Mich.
   G. O. No. 64, W. D., 1919.

Private, first class, Company I, 125th Infantry, 32d Division.
He displayed exceptional courage in repeatedly crossing an area swept by
machine-gun and shellfire to carry messages to battalion headquarters, after
seeing many other soldiers killed or wounded in attempting the same mission.
On one occasion he stopped in full view of the enemy to take a message from
another runner who had been wounded.

KANALEY, JOSEPH (1272983)_____
   Near Verdun, France, Oct. 11, 1918.
   R—Orange, N. J.
   B—Orange, N. J.
   G. O. No. 37, W. D., 1919.

Private, Company D, 111th Machine Gun Battalion, 29th Division.
He ran along a road that was being heavily shelled and secured a litter, returning with it over the same route, and assisted in carrying a wounded soldier
to a first-aid station.

KANE, ALVA (2177538)_____
   Near Bantheville Woods, France,
     Sept. 25–27, 1918.
   R—Stilwell, Kans.
   B—Stilwell, Kans.
   G. O. No. 81, W. D., 1919.

Private, medical detachment, 353d Infantry, 89th Division.
After 2 medical officers attached to the battalion with which he was serving
had been gassed and evacuated, he took over and operated the first-aid station,
remaining heroically at his post, though subjected to heavy shellfire. He
remained continuously on duty until blinded by mustard gas.

*KANE, CHARLES J. (51516)_____
   Near Vaux, France, July 1, 1918.
   R—Syracuse, N. Y.
   B—Syracuse, N. Y.
   G. O. No. 102, W. D., 1918.

Private, Company I, 23d Infantry, 2d Division.
Attempting to bring his captain, who was lying wounded and exposed to fire,
to shelter, near Vaux, France, July 1, 1918, he was himself killed, thereby
sacrificing his life in an effort to rescue his commanding officer.
Posthumously awarded. Medal presented to guardian, Sister Superior, Sister
Emily.

KANE, MATTHEW JOSEPH (89287)_____
   Near Landres-et-St. Georges, France,
     Oct. 15, 1918.
   R—New York, N. Y.
   B—New York, N. Y.
   G. O. No. 37, W. D., 1919.

Private, Company A, 165th Infantry, 42d Division.
He volunteered several times to act as liaison agent after every runner had been
either killed or wounded. He was untiring in his efforts to maintain communication under heavy enemy artillery and machine-gun fire.

KANE, TONY W._____
   Near Beaumont, France, Nov. 10,
     1918.
   R—Chicago, Ill.
   B—Russia.
   G. O. No. 89, W. D., 1919.

Sergeant, 55th Company, 5th Regiment, U. S. Marine Corps, 2d Division.
He reconnoitered the position of enemy machine guns which were holding up
the advance of his company across the Meuse. After he had located them he
alone silenced the fire of 2 guns, thus making possible the continuance of
his company's advance.
Oak-leaf cluster.
Sergeant Kane is also awarded an oak-leaf cluster, to be worn with the distinguished-service cross, for the following act of extraordinary heroism in
action in the Bois-de-Belleau, France, June 11, 1918: Displaying notable bravery, he single-handed charged and captured an enemy machine gun, killing
its crew.

*KANOPSKY, FRANK (2338697)_____
   Near Roncheres, France, July 28,
     1918.
   R—Harwick, Pa.
   B—Russia.
   G. O. No. 32, W. D., 1919.

Private, Company E, 4th Infantry, 3d Division.
While acting as runner he was severely burned with mustard gas and wounded
by shrapnel. He refused to be evacuated and continued to perform his duties
under intense machine-gun fire.
Posthumously awarded. Medal presented to mother, Mrs. August Kanigo-
nawski.

*KANTZ, CLARENCE E. (1747366)_____
   Near Grand Pre, France, Oct. 26,
     1918.
   R—Camden, N. J.
   B—Clearfield, Pa.
   G. O. No. 20, W. D., 1919.

Sergeant, Company K, 311th Infantry, 78th Division.
During an enemy counterattack, Sergeant Kantz by his foresight saved the
company on the right of his platoon by reinforcing that company's flank.
This action stopped the advance of several machine gunners, who were
advancing through our lines at that point, but cost Sergeant Kantz his life.
Posthumously awarded. Medal presented to mother, Mrs. Minnie Kantz.

KAPERZYNSKI, JOE (554579)_____
    Near Moulins, France, July 15, 1918.
    R—Pittsburgh, Pa.
    B—Pittsburgh, Pa.
    G. O. No. 22, W. D., 1920.

Private, Company A, 9th Machine Gun Battalion, 3d Division.
During the German offensive July 15, 1918, Private Kaperzynski, directed the firing of his machine gun while exposed to heavy enemy artillery and machine-gun fire, inflicting heavy losses on the enemy. During the attack near Madeline Farm, he, with a companion, crawled in front of our lines and rescued a wounded American soldier.

KARCH, EMANUEL (42033)_____
    Throughout the operations south of
      Soissons, France, July 18 to 22, 1918.
    R—Angela, Mont.
    B—Russia.
    G. O. No. 117, W. D., 1918.

Private, Company B, 16th Infantry, 1st Division.
Displaying exceptional initiative and bravery throughout the operations south of Soissons, France, July 18 to 22, 1918, he, with extraordinary heroism, on July 21, 1918, with 2 companions, captured 2 machine guns that were causing heavy losses to his company.

KARDOK, JOSEPH (2337898)_____
    Near Grand Ballois Farm, France,
      July 15, 1918.
    R—Philadelphia, Pa.
    B—Russia.
    G. O. No. 32, W. D., 1919.

Private, first class, Company A, 4th Infantry, 3d Division.
After being badly gassed, he continued with his duties as runner, exposed to the extreme shelling of high explosive and gas bombs.

KARELIS, KIMON (794855)_____
    Near Vieville-en-Haye, France,
      Sept. 12-13, 1918.
    R—Milwaukee, Wis.
    B—Greece.
    G. O. No. 35, W. D., 1919.

Private, first class, Company C, 15th Machine Gun Battalion, 5th Division.
After he was severely wounded and his squad leader killed, he took charge of the squad and directed their fire with such telling effect that 3 machine guns, which had caused many casualties in our ranks, were put out of action. In the counterattack which followed he personally manned the machine gun after the other members of the crew had been killed or wounded and continued to operate it until completely exhausted.

*KARG, HOWARD M. (2405429)_____
    Near St. Juvin, France, Nov. 5, 1918.
    R—Mount Holly, N. J.
    B—Mount Holly, N. J.
    G. O. No. 37, W. D., 1919.

Sergeant, Company A, 309th Infantry, 78th Division.
Under terrific enemy bombardment he carried a soldier who was overcome by gas for a distance of 50 yards over a road exposed to the direct fire of enemy machine guns, snipers, and shellfire to a place of safety.
Posthumously awarded. Medal presented to mother, Mrs. Elizabeth Karg.

*KARKER, JACK (1210137)_____
    Near Ronssoy, France, Sept. 29, 1918.
    R—Lowville, N. Y.
    B—Cobleskill, N. Y.
    G. O. No. 19, W. D., 1920.

Corporal, Company C, 107th Infantry, 27th Division.
In the face of hostile machine-gun fire Corporal Karker ran to the assistance of a wounded comrade. He pulled the wounded man into a shell hole, keeping his own body interposed between the wounded man and the enemy's fire. While giving first aid to his wounded comrade, Corporal Karker was killed.
Posthumously awarded. Medal presented to mother, Mrs. Harry Kelso.

KARWOSKI, PAUL J. (2411207)_____
    Near Grand Pre, France, Oct. 26,
      1918.
    R—Trenton, N. J.
    B—Russia.
    G. O. No. 35, W. D., 1919.

Sergeant, Company M, 311th Infantry, 78th Division.
Acting as a scout, he obtained valuable information as to the location of enemy positions, and, single-handed, attacked a machine-gun crew, killing 1 of the enemy and taking the remaining 2 prisoners.

KAUFFMAN, ABE (128071)_____
    During the enemy counterattack on
      Cantigny, France, May 28, 1918.
    R—Philadelphia Pa.
    B—Philadelphia, Pa.
    G. O. No. 99, W. D., 1918.

Private, Battery F, 7th Field Artillery, 1st Division.
He refused to leave his gun after he had lost a finger during the enemy counterattack on Cantigny, France, May 28, 1918, but continued to perform his duties until so severely wounded as to be unable to assist in serving his piece.

KAULSKY, FRANK_____
    Near Blanc Mont Ridge, France,
      Oct. 4, 1918.
    R—Baltimore, Md.
    B—Waukesha, Wis.
    G. O. No. 98, W. D., 1919.

Private, 55th Company, 5th Regiment, U. S. Marine Corps, 2d Division.
Private Kaulsky volunteered to rescue a wounded comrade from a violent barrage, and, although he was wounded, he successfully accomplished his mission.

*KAY, IVAN E. (3128362)_____
    Near Waereghem, Belgium, Oct. 31,
      1918.
    R—Little Falls, Minn.
    B—Little Falls, Minn.
    G. O. No. 66, W. D., 1919.

Private, Company L, 363d Infantry, 91st Division.
When the progress of 2 front-line companies had been held up by intense machine-gun fire and the enemy had begun to close in on our forces Private Kay, although in an exposed position, checked the advance of the enemy with rifle grenades until he was killed. His act made possible the renewal of the advance by the 2 companies.
Posthumously awarded. Medal presented to father, George W. Kay.

KAYE, SAMUEL, Jr._____
    Over the region of Epinonville,
      France, Sept. 29, 1918.
    R—Columbus, Miss.
    B—Columbus, Miss.
    G. O. No. 138, W. D., 1918.

First lieutenant, 94th Aero Squadron, Air Service.
While on a mission he, accompanied by another machine piloted by Lieut. Reed M. Chambers, encountered a formation of 6 enemy machines (Fokker type) at an altitude of 3,000 feet. Despite numerical superiority of the enemy, Lieutenant Kaye and Lieutenant Chambers immediately attacked and succeeded in destroying 1 and forced the remaining 5 to retire into their own lines.
Oak-leaf cluster.
A bronze oak leaf is awarded to Lieutenant Kaye for the following act of extraordinary heroism in action over the region of Montfaucon and Bantheville, France, Oct. 5, 1918: He encountered a formation of 7 enemy machines (Fokker type). Regardless of their numerical superiority he immediately attacked, and by skillful maneuvering succeeded in separating 1 enemy plane from its formation and, after a short combat, shot it down in flames.

KAYLOR, FRANK E. (03423)............
  Northwest of Chateau-Thierry, France, July 27, 1918.
  R—Marion, Ohio.
  B—Logan County, Ohio.
  G. O. No. 35, W. D., 1920.

Private, Company D, 166th Infantry, 42d Division.
After he and a comrade had located a hostile machine gun in a clump of trees 500 meters north of a château, east of Fere-en-Tardenois, they volunteered to attempt to force the enemy to abandon this position, and advanced against it over open ground in the face of fire. By their rifle fire they drove the enemy crew from its gun, killing 1 and wounding another.

*KAYSER, ALFRED (540713)............
  Near Mont St. Pere, Marne, France, July 21, 1918.
  R—New York, N. Y.
  B—Morristown, Pa.
  G. O. No. 19, W. D., 1920.

Private, Company A, 7th Infantry, 3d Division.
Private Kayser, with an automatic rifle, went forward in advance of our line and opened fire on an enemy machine-gun position in order to cover the flank attack of another platoon. Although mortally wounded, he continued to fire until the platoon had completed its flanking movement.
Posthumously awarded. Medal presented to mother, Mrs. Alfred Kayser.

*KEACHIE, EDWIN S. (124004)............
  At Soissons, France, July 18–24, 1918.
  R—Detroit, Mich.
  B—Chicago, Ill.
  G. O. No. 35, W. D., 1919.

Corporal, Battery B, 5th Field Artillery, 1st Division.
He displayed unusual ability and courage by taking the place of officers who had been wounded and faithfully performing their duties while his battery position was under severe flanking fire. He continually exposed himself to violent bombardment until killed.
Posthumously awarded. Medal presented to mother Mrs. Lydia Keachie.

KEAN, ROBERT WINTHROP............
  Near Vierzy and Villemontoire, France, July 18, 1918.
  R—New York, N. Y.
  B—Elberon, N. J.
  G. O. No. 87, W. D., 1919.

First lieutenant, 15th Field Artillery, 2d Division.
He accompanied 2 successive waves of Infantry in the attack on Vierzy and Villemontoire, July 18, exposed himself with almost reckless disregard of the enemy's heavy shell and machine-gun fire, secured valuable information for the artillery as to the location of our own and the enemy's lines, and personally took command of an isolated 9th Infantry machine-gun detachment that had lost its officers by heavy fire. Lieutenant Kean on July 18 and 19 actually accompanied 3 successive waves of the 23d Infantry, the 9th Infantry, and an attack by French infantry without rest. His endurance and courage were exceptional and most inspiring upon this occasion and upon all other occasions of attack by the 2d Division.

KEANE, CHRISTOPHER W. (1388750)....
  At Hamel, France, July 4, 1918.
  R—Chicago, Ill.
  B—Chicago, Ill.
  G. O. No. 99, W. D., 1918.

Private, first class, medical detachment, 131st Infantry, 33d Division.
Throughout the engagement he displayed great gallantry and devotion to duty by treating the wounded in an area swept by machine-gun and artillery fire. When 2 stretcher bearers who were working with him were killed he impressed German prisoners into the service of carrying wounded to the aid station.

KEARNEY, JOHN J. (1209988)............
  At Marcheville, France, Sept. 26, 1918.
  R—Hartford, Conn.
  B—Kingston, N. Y.
  G. O. No. 143, W. D., 1918.

Private, Company B, 102d Infantry, 26th Division.
When liaison with the front-line companies had been completely broken and several runners had failed to reach them he successfully carried a message through an intense machine-gun and artillery barrage, returning with important information.

KEARNS, CHARLIE F. (2248734)............
  Near Fey-en-Haye, France, Sept. 12, 1918.
  R—Blackwell, Okla.
  B—Mulhall, Okla.
  G. O. No. 128, W. D., 1918.

Private, Company E, 357th Infantry, 90th Division.
He rushed machine-gun nests that were holding up the advance of his company, killed the crews with his automatic rifle, and captured the guns. He continued to render valiant service throughout the engagement until he was wounded by a shell fragment.

KEARNS, THOMAS W............
  Near Sergy, France, July 29–30, 1918.
  R—Boston, Mass.
  B—Boston, Mass.
  G. O. No. 71, W. D., 1919.

First lieutenant, 47th Infantry, 4th Division.
When a counterattack was impending he successively carried 15 wounded men across a shell-swept area, in full view of the enemy, taking them to a place of safety and preventing them from being captured by the enemy. Later he reorganized groups of stragglers and led them into combat.

KEATING, JAMES A............
  Various dates and places in France.
  R—Chicago, Ill.
  B—Chicago, Ill.
  G. O. No. 9, W. D., 1923.

First lieutenant, Air Service, with headquarters, American Air Service units, attached to 49th Squadron, British Expeditionary Forces.
For extraordinary heroism in action. On Aug. 9, 1918, he bombed Falvy Bridge over 1,000 feet, obtaining a direct hit. On returning, his formation was attacked by enemy planes and a running fight ensued. By skillfully flying with exceptional coolness he enabled his observer to shoot 2 planes down in flames. On Aug. 8, 1918, after bombing Bethencourt Bridge over 800 feet, obtaining a direct hit, he found 7 enemy planes attacking his formation from the rear. By maneuvering for position he enabled his observer to shoot 1 down in flames over Bethencourt. On July 17, 1918, he bombed Passy Bridge over 500 feet, destroying it just as a column of transport was passing. He then opened machine-gun fire on the troops in the vicinity, causing many casualties and great disorder. His exceptional courage and splendid bravery were a constant inspiration to the members of his command.

*KEATING, WILLIAM JOSEPH............
  Near Melleville Farm, north of Verdun, France, Oct. 27, 1918.
  R—Texas, Md.
  B—Texas, Md.
  G. O. No. 20, W. D., 1919.

Captain, 114th Infantry, 29th Division.
While he was in command of the machine-gun company of his regiment and every available gunner had been killed or wounded he personally manned one of the guns and kept it in operation until killed.
Posthumously awarded. Medal presented to brother, Raymond M. Keating.

KEATON, ANDY W. (2248523)............
  Near Fey-en-Haye, France, Sept. 9–12, 1918.
  R—Ozark, Ark.
  B—Ozark, Ark.
  G. O. No. 128, W. D., 1918.

Private, Company G, 357th Infantry, 90th Division.
Becoming separated from his patrol, Private Keaton, with another soldier, attacked an enemy patrol and drove it off, though the number of their opponents was estimated at 50. During the advance of Sept. 12 north of Fey-en-Haye he went to the rescue of his battalion commander and another officer, who were pocketed by the fire of a machine gun. He stalked the gun position and drove off the gunners.

KEE, SING (1702357)_____
At Mont Notre Dame, west of Fismes, France, Aug. 14-15, 1918.
R— New York, N. Y.
B—Saratoga, Calif.
G. O. No. 99, W. D., 1918.

Private, Company G, 306th Infantry, 77th Division.
Although seriously gassed during shelling by high-explosive and gas shells, he refused to be evacuated and continued, practically single-handed, by his own initiative, to operate the regimental message center relay station at Mont Notre Dame. Throughout this critical period he showed extraordinary heroism, high courage, and persistent devotion to duty, and totally disregarded all personal danger. By his determination he materially aided his regimental commander in communicating with the front line.

KEENAN, JOHN J. (480917)_____
Near Jaulny, France, Nov. 8, 1918.
R—Chicago, Ill.
B—Chicago, Ill.
G. O. No. 44, W. D., 1919.

Corporal, Company D, 55th Infantry, 7th Division.
With a detail of 7 men, he volunteered and went through severe artillery and machine-gun fire to bring in a platoon which had been cut off for 5 hours, in front of a strongly fortified enemy position. Repeated efforts to reach the platoon during the day had failed, but Corporal Keenan and his party brought in the entire platoon, including several wounded men, among them the platoon commander.

KEENAN, JOSEPH A. (1785685)_____
Near Nantillois, France, Sept. 29, 1918.
R—Philadelphia, Pa.
B—Philadelphia, Pa.
G. O. No. 37, W. D., 1919.

Corporal, Company L, 315th Infantry, 79th Division.
When his platoon had been ordered to cover because of annihilating machine-gun and artillery fire Corporal Keenan accompanied another soldier to the rescue of a comrade who was lying 300 yards distant. The journey was made through heavy and continuous fire, but Corporal Keenan, with his fellow soldier, succeeded in bringing their wounded comrade to safety.

KEENAN, WILLIS H._____
In Champagne sector, France, Sept. 26-Oct. 1, 1918.
R—Coshocton, Ohio.
B—Quaker City, Ohio.
G. O. No. 37, W. D., 1919.

First lieutenant, Medical Corps, attached to 369th Infantry, 93d Division.
Although suffering from illness, this officer remained on duty day and night throughout the engagement. When his battalion was in reserve, he voluntarily went forward to the assaulting battalions whose surgeons had been evacuated. In the attack on Sechault he exposed himself continuously to intense artillery and machine-gun fire while rendering first aid.

KEEPER, EASTER E. (3485610)_____
Near Bois-d'Harville, France, Nov. 10, 1918.
R—East Liverpool, Ohio.
B—Graysville, Ohio.
G. O. No. 64, W. D., 1919.

Private, Company L, 131st Infantry, 33d Division.
Volunteering for the service, he went out in advance of our lines, under heavy machine-gun fire, to cut lanes through wide belts of barbed wire. Despite the enemy fire, to which he was constantly exposed, he accomplished his mission, enabling the attacking waves to resume the advance.

KEIRS, ROBERT M. (554695)_____
Near Moulins, France, July 15, 1918.
R—Washington, Pa.
B—Midlothian, Md.
G. O. No. 22, W. D., 1920.

Private, Company A, 9th Machine Gun Battalion, 3d Division.
During the strong enemy attack, Private Keirs made a daylight reconnaissance under enemy machine-gun and artillery fire to locate the positions reached by the enemy in their advance across the Marne. He located accurately the enemy line, and the information he secured was of the greatest value to his platoon commander. Later he carried messages, under machine gun fire, to our artillery.

*KEISER, HARRY M._____
Near Cierges, northeast of Chateau-Thierry, France, July 31, 1918.
R—Chicago, Ill.
B—Dubuque, Iowa.
G. O. No. 132, W. D., 1918.

First lieutenant, 125th Infantry, 32d Division.
In the action to force the passage of the Ourcq River and capture the heights beyond, he distinguished himself by his conduct and personal example to his men. Under artillery fire and heavy machine-gun fire he continued to direct his platoon, even though severely wounded, in an effective manner against the enemy, until he was killed by machine-gun fire.
Posthumously awarded. Medal presented to widow, Mrs. Harry M. Keiser.

KEITH, HUBERT W._____
Northeast of Cunel, France, Oct. 22, 1918.
R—Clermont, Ga.
B—Clermont, Ga.
G. O. No. 22, W. D., 1920.

First lieutenant, 38th Infantry, 3d Division.
First Lieutenant Keith courageously led his company in the attack on Hill 299, exposed to heavy machine-gun fire. He quickly organized his position, and later repelled an enemy counterattack. Although wounded on the evening of the 22d, he continued with his company until the following day.

KEITH, MONT R. (58488)_____
Near Exermont, France, Oct. 1-12, 1918.
R—Roxbury, Mass.
B—Canada.
G. O. No. 13, W. D., 1923.

Private, Company L, 28th Infantry, 1st Division.
Private Keith voluntarily joined a patrol whose mission was to locate enemy machine guns and to secure such other information as might be of value to the attacking forces. The patrol penetrated the enemy's lines to a distance of a kilometer, locating a large number of machine-gun nests. Cut off from their own troops without food or water for 36 hours and under constant enemy fire, 1 officer and 18 of the 36 men of the patrol were either killed or wounded. Private Keith with indomitable spirit made his way to the American forces, in broad daylight under intense enemy fire, furnished valuable information of the enemy to his battalion commander, and voluntarily made his way back to the beleaguered patrol in order to lead them to their own lines; they thus escaped the American barrage which preceded the attack of Oct. 4, 1918. His undaunted courage, leadership, and devotion to duty proved an inspiration to every man of his regiment.

KELL, PORTER P. (58198)_____
South of Soissons, France, July 18-23, 1918.
R—Clayton, Ga.
B—Clayton, Ga.
G. O. No. 39, W. D., 1920.

Corporal, Company I, 28th Infantry, 1st Division.
Corporal Kell carried numerous important messages over ground swept by heavy machine-gun and artillery fire, returning with valuable information from adjoining units. Due to his gallantry, his company commander was able to keep in close touch with platoon leaders. Corporal Kell on several occasions dressed the wounded under heavy enemy fire.

KELLER, ROY L. (2178408)_____
Near Remonville, France, Nov. 1, 1918.
R—Kirksville, Mo.
B—Ewing, Mo.
G. O. No. 44, W. D., 1919.

Sergeant, Company B, 354th Infantry, 89th Division.
When his combat group was heavily fired upon, while crossing open exposed ground, by 2 machine guns, he charged the position from the flank, personally killing 1 gunner and making the capture of the other members of the crews possible. His action made possible the advance of his company.

KELLER, THEODORE_____ Sergeant, 47th Company, 5th Regiment, U. S. Marine Corps, 2d Division.
Near Barricourt, France, Nov. 1-2, Working through 3 desperate counterattacks of the enemy, Sergeant Keller
1918.   established liaison with the adjacent division, maintaining contact with his
R—Lexington, Ky.   flanks at all times, despite the hazards under which he worked.
B—Lexington, Ky.
G. O. No. 32, W. D., 1919.

KELLEY, AUSTIN J. (62102)_____ Corporal, Company L, 101st Infantry, 26th Division.
Near Vaux, France, July 20, 1918.   Corporal Kelley and Pvts. Harold E. Rounds and John J. Grady penetrated
R—Malden, Mass.   the enemy's lines in the face of machine-gun fire, captured a machine gun
B—Canada.   and its crew, and returned with valuable information concerning the enemy's
G. O. No. 125, W. D., 1918.   positions.

KELLEY, HENRY G. (1315932)_____ Private, Company G, 119th Infantry, 30th Division.
Near Bellicourt, France, Sept. 20, Voluntarily advancing alone against a machine-gun nest which was causing
1918.   heavy casualties in his platoon, he bombed the enemy position, killing 5 of
R—Andrews, N. C.   the crew and capturing the remaining 3.
B—Buford, Ga.
G. O. No. 46, W. D., 1919.

KELLEY, ORLEN O. (2185903)_____ Sergeant, Company I, 356th Infantry, 89th Division.
Near Pouilly, France, Nov. 7-8 and Accompanying a patrol into the village of Pouilly to determine the strength of
11, 1918.   the enemy, Sergeant Kelley crossed the Meuse by means of a destroyed
R—Pickering, Mo.   bridge, which, together with all approaches thereto, was subjected to an
B—Pickering, Mo.   intense harassing fire of 1-pounders, machine guns, and snipers. On Nov. 11
G. O. No. 32, W. D., 1919.   he continued a gallant fight against an enemy machine-gun nest after his
                                      company commander had been evacuated because of wounds.

KELLOG, GORDON V. (1375419)_____ Corporal, Battery A, 122d Field Artillery, 33d Division.
Near the Bois-de-Bantheville,   Having been sent on a reconnaissance, he reached the enemy lines and returned
France, Oct. 30, 1918.   with information of great value.
R—Chicago, Ill.
B—Chicago, Ill.
G. O. No. 44, W. D., 1919.

KELLY, AUGUSTINE C. (1375600)_____ Corporal, Battery B, 122d Field Artillery, 33d Division.
Near Bantheville, France, Oct. 29, He voluntarily proceeded to a point within 150 meters of the enemy, where he re-
1918.   mained for more than an hour, securing valuable information regarding the
R—Laurel, Miss.   enemy's position and activity. During all this time he was subjected to
B—New Orleans, La.   severe shell, machine-gun, and snipers' fire, being wounded by a shell while
G. O. No. 44, W. D., 1919.   returning to our lines.

KELLY, CHARLES. (562583)_____ Sergeant, Company C, 12th Machine Gun Battalion, 4th Division.
During the Meuse-Argonne offen-   Sergeant Kelly led his platoon in the attack with great bravery against strongly
sive, France, Sept. 29, 1918.   held enemy trenches. Shortly after reaching his objective he was wounded
R—Dalton, Mass.   in the throat. He refused to be evacuated, but continued to actively com-
B—Hinsdale, Mass.   mand his men until the night of Oct. 1, by which time, due to his wound, he
G. O. No. 19, W. D., 1920.   had lost the power of speech.

KELLY, FRANCIS A_____ First lieutenant, chaplain, 104th Machine Gun Battalion, 27th Division.
Near Ronssoy, France, Sept. 26-30, During the operations of his regiment against the Hindenburg line and later
1918, and east of the La Selle River, east of the La Salle River he was constantly at the front, caring for the wounded
France, Oct. 13-20, 1918.   and supervising the burial of the dead, often under heavy shell and machine-
R—Cohoes, N. Y.   gun fire. His fearless conduct afforded an inspiring example to the combat
B—Cohoes, N. Y.   troops.
G. O. No. 143, W. D., 1918.

KELLY, FRANCIS J_____ First lieutenant, 5th Regiment, U. S. Marine Corps, 2d Division.
Near St. Etienne, France, Oct. 4,   After all other officers of his company had become casualties, Lieutenant Kelly
1918.   took command, and while able to retire to a stronger position, yet he ordered
R—Brooklyn, N. Y.   an advance against greatly superior numbers, breaking the enemy's attack
B—Brooklyn, N. Y.   and taking many prisoners.
G. O. No. 37, W. D., 1919.

KELLY, JAMES P_____ First lieutenant, 18th Field Artillery, 3d Division.
Near Chateau-Thierry, France,   During the bombardment preceding the enemy attack across the Marne the
July 14-15, 1918.   position of Battery B, 18th Field Artillery, was subjected to heavy artillery
R—Exeter, N. H.   fire for a period of 4 hours. All communications were destroyed and the
B—Traverse City, Mich.   ammunition dump destroyed by hostile fire. Throughout the enemy bom-
G. O. No. 22, W. D., 1920.   bardment Lieutenant Kelly kept his battery in action, exposing himself to
                                      concentrated enemy artillery fire in order to replenish his ammunition supply,
                                      and delivered an effective fire on the enemy.

KELLY, JOHN (2191080)_____ Private, Company A, 341st Machine Gun Battalion, 89th Division.
Near Bois-de-Barricourt, France,   With his machine-gun section, Private Kelly was sent to an advanced position,
Nov. 2, 1918.   where they were subjected to fire from numerous enemy snipers and machine-
R—Storm Lake, Iowa.   gun positions, and were also mistaken for the enemy by our own Infantry
B—Humboldt, Nebr.   After the section lieutenant was wounded, he directed the sergeant to signal
G. O. No. 46, W. D., 1919.   their identity to the Infantry, and, overhearing the order, Private Kelly
                                      rose on the edge of the shell hole to send the message, but was met by bursts
                                      of automatic-rifle fire. In order to save his comrades from the danger of this
                                      fire, he ran down the hill to our own lines, convinced the Infantry of the
                                      identity of the troops ahead, and then returned to his section.

KELLY, LAWRENCE J. (60970)_____ Private, Company F, 101st Infantry, 26th Division.
Near Bois-de-St. Remy, France,   Accompanying 2 other soldiers, Private Kelly rushed forward in advance of
Sept. 12, 1918.   his lines, exposed to heavy machine-gun fire, and captured 2 machine guns
R—Woburn, Mass.   and 6 of the enemy who were manning the position.
B—Woburn, Mass.
G. O. No. 26, W. D., 1919.

KELLY, LEO P.
Near Vaux, France, June 6-7, 1918.
R—Pueblo, Colo.
B—Pueblo, Colo.
G. O. No. 64, W.D., 1919.

First lieutenant, 9th Infantry, 2d Division.
During a night attack made by his battalion, Lieutenant Kelly, the battalion adjutant, voluntarily left his dugout and went to the front-line positions through the intense artillery fire. There he successfully maintained liaison with the advance troops and the artillery in the rear by means of signals sent from his exposed position. His absolute disregard for his own safety contributed largely to the success of the engagement.

KELLY, MICHAEL A.
Near Landres-et-St. Georges, France, Oct. 15, 1918.
R—New York, N. Y.
B—Ireland.
G. O. No. 37, W. D., 1919.

Major, 165th Infantry, 42d Division.
Major Kelly, because of having a very high fever, was ordered to the rear while conducting an attack against strong resistance. He refused, however, to be relieved, remaining in command for two days, after which time he collapsed. His sickness which resulted necessitated his remaining in the hospital for a period of over a month.

KELLY, THOMAS P. (65075).
Near St. Hilaire, France, Sept. 22, 1918.
R—Waterbury, Conn.
B—Waterbury, Conn.
G. O. No. 26, W. D., 1919.

Corporal, Company G, 102d Infantry, 26th Division.
Assisted by another soldier, Corporal Kelly rushed a machine-gun nest which had been firing on their patrol. They succeeded in killing the crew.

*KELSEY, HARRY R. (1456594).
Near Charpentry, France, Sept. 27, 1918.
R—Easton, Kans.
B—Easton, Kans.
G. O. No. 59, W. D., 1919.

First sergeant, Company E, 139th Infantry, 35th Division.
He volunteered to carry an important message from his company commander to battalion headquarters and, knowing the extreme importance of the message, proceeded by the most direct route, through the heavy machine-gun fire and artillery barrage, with entire disregard for his personal safety. He was killed while performing this heroic service.
Posthumously awarded. Medal presented to father, John A. Kelsey.

*KELTY, ASHER E.
Near Crepion, France, Sept. 26, 1918.
R—Rice Lake, Wis.
B—Rice Lake, Wis.
G. O. No. 21, W. D., 1919.

First lieutenant, pilot, 91st Aero Squadron, Air Service.
In the course of a photographic mission, Lieutenant Kelty, with his observer, was obliged to penetrate a heavy enemy antiaircraft barrage, realizing that obtaining the location of the artillery objectives was of the greatest importance. When a shell struck his machine, his observer was instantly killed and his machine so badly wrecked that it plunged to the earth, thereby causing his death.
Posthumously awarded. Medal presented to father, James B. Kelty.

KEMME, CHARLES (283396).
Near Soissons, France, July 18, 1918.
R—Milwaukee, Wis.
B—Cedar Rapids, Iowa.
G. O. No. 121, W. D., 1918.

Private, Company F, 28th Infantry, 1st Division.
He displayed exceptional bravery and utter disregard of his own life by advancing on a machine-gun nest of 2 guns and 6 men, and by working his way up to the rear of the guns put them out of action in hand-to-hand combat. On July 20, 1918, he again advanced on a machine gun which was doing great damage to the men in his sector by sniping, put the gun out of action, and returned to the trenches with the gun.

KEMMERER, BIRTRUS (1457369).
Near Baulny, France, Sept. 26, 1918.
R—Carrollton, Mo.
B—Spencer, Ind.
G. O. No. 59, W. D., 1919.

Private, Company H, 139th Infantry, 35th Division.
Seeing his battalion adjutant lying wounded several yards in front of our line and in great danger from heavy machine-gun and rifle fire, Private Kemmerer with a total disregard for personal danger, went to the assistance of this officer and succeeded in carrying him to safety. This gallant soldier was himself wounded while performing this heroic act.

KEMPTON, RAYMOND D. (1280007).
Near Verdun, France, Oct. 12, 1918.
R—Hackensack, N. J.
B—Hackensack, N. J.
G. O. No. 130, W. D., 1918.

Private, Company C, 114th Infantry, 29th Division.
Although severely wounded in the arm, he refused to go to the rear, and after receiving first-aid treatment rejoined his platoon under heavy shellfire and continued to fight until incapacitated by a second severe wound.

KENASTON, HAROLD W.
Near Pouilly, France, Nov. 4-10, 1918.
R—Canton, Ohio.
B—Passaic, N. J.
G. O. No. 37, W. D., 1919.

First lieutenant, 356th Infantry, 89th Division.
On Nov. 4 he was severely gassed during a heavy shelling, but remained on duty. On Nov. 5, still suffering from the effects of the gas, he volunteered and went to a near-by town to assist in the work of supplying reserve rations for the command, working continuously for two days and two nights under terrific fire. During the night of Nov. 10 he guided an element to the pontoon crossing the Meuse, and while returning he was rendered unconscious by a shell explosion. Gaining consciousness, he voluntarily started on a search for a part of his command which had been lost in the dense fog.

KENDAL, HERBERT B. (68613).
Near Bois-de-St. Remy, France, Sept. 12, 1918.
R—Wolfeboro, N. H.
B—Boston, Mass.
G. O. No. 87, W. D., 1919.

Private, Company H, 103d Infantry, 26th Division.
While under heavy fire, he cut a path through the enemy's wire entanglements and then crawled through the opening he had made, leading a small detail against an enemy machine gun, which he flanked and put out of action. With two companions he destroyed a second emplacement, and while advancing upon a third was severely wounded by shellfire.

KENDALL, PAUL W.
At Posolskaya, Siberia, Jan. 10, 1920.
R—Sheridan, Wyo.
B—Baldwin, Kans.
G. O. No. 35, W. D., 1920.

Second lieutenant, 27th Infantry.
He was in command of a detachment of his company when attacked by an armored train at 1 a. m. The detachment under his leadership and inspired by his example attacked and disabled the armored train and caused its surrender.

KENNEDY, GEORGE M. (158881)
Near Malancourt, France, Oct. 2, 1918.
R—Galesburg, Ill.
B—Galesburg, Ill.
G. O. No. 120, W. D., 1918.

Private, Company E, 6th Engineers, 3d Division.
About 2.30 a. m. he was on guard in a corral when a shell landed, wounding some of the men and killing several animals. He immediately awakened the occupants of the camp and returned to his post, finding that another shell had struck, wounding several more men and killing more horses. He went to a wagon where a sick soldier lay helpless and succeeded in carrying him and leading a badly wounded soldier away from the spot. While thus engaged a third shell struck, wounding one man and violently throwing all three men to the ground. Private Kennedy arose and carried the helpless soldier to a safe dugout, where he was given first aid. He then returned to the corral to assist other wounded.

KENNEDY, GRANT (2339022)
Near Cunel, France, Oct. 13, 1918.
R—Imboden, Va.
B—Williamsburg, Ky.
G. O. No. 98, W. D., 1919.

Corporal, Company G, 4th Infantry, 3d Division.
When his platoon commander was wounded he reorganized the platoon while under severe fire, placing his men so as best to repel the enemy's counter-attack. He inspired his men by his own personal bravery. Wounded and sent to a hospital he returned to the front line after a week's treatment, being severely wounded while leading his men in an attack.

KENNEDY, HARRY (53064)
Near Soissons, France, July 19, 1918.
R—Fairmount, Ill.
B—Fairmount, Ill.
G. O. No. 44, W. D., 1919.

Private, Company C, 26th Infantry, 1st Division.
Armed with only his rifle and bayonet, Private Kennedy alone captured a machine gun after killing the entire crew. He then turned his captured gun on the enemy, causing the retreat of an entire platoon.

KENNEDY, JOHN J.
Near Courmont, France, July 29, 1918.
R—Scottdale, Pa.
B—Scottdale, Pa.
G. O. No. 71, W. D., 1919.

Captain, 110th Infantry, 28th Division.
Learning that 2 of his men were lying wounded in an exposed position in front of our lines, he went forward alone, facing heavy fire, to their rescue. After carrying 1 of the wounded men back to our lines, he was confronted by a German upon his return. Picking up the rifle of the wounded man he had come to rescue, he shot the German, and then completed his mission of rescue.

KENNEDY, RAY R. (2158908)
Near Ribeauville, France, Oct. 18–19, 1918.
R—Estherville, Iowa.
B—De Kalb, Ill.
G. O. No. 44, W. D., 1919.

Private, Company C, 119th Infantry, 30th Division.
On the night of Oct. 18–19, when all communication had been temporarily lost with the unit on the left of his regiment, he volunteered to attempt the establishment of liaison, fully aware of the danger necessarily incurred. While attempting to cross a heavily shelled area to reach the flank regiment he received a wound which resulted in the loss of a leg.

KENNER, ALBERT W.
Near Soissons, France, July 22, 1918.
R—Washington, D. C.
B—Holyoke, Mass.
G. O. No. 15, W. D., 1919.

Major, Medical Corps, attached to 26th Infantry, 1st Division.
Learning that his regimental commander had been mortally wounded near Soissons, France, July 22, 1918, he voluntarily went through machine-gun fire beyond the front lines in the hope of helping him. Finding his colonel dead, he recovered the body, in spite of the danger to which such action subjected him.

KENNEY, GEORGE C.
Near Jametz, France, Oct. 9, 1918.
R—Boston, Mass.
B—Canada.
G. O. No. 13, W. D., 1919.

First lieutenant, pilot, 91st Aero Squadron, Air Service.
This officer gave proof of his bravery and devotion to duty when he was attacked by a superior number of aircraft. He accepted combat, destroyed 1 plane, and drove the others off. Notwithstanding that the enemy returned and attacked again in strong numbers, he continued his mission and enabled his observer to secure information of great military value.

KENNY, FRANK H., Jr. (1211265)
Near Ronssoy, France, Sept. 29, 1918.
R—Scarsdale, N. Y.
B—Brooklyn, N. Y.
G. O. No. 32, W. D., 1919.

Private, first class, Company H, 107th Infantry, 27th Division.
When his commanding officer fell wounded he made his way through intense machine-gun fire to his first sergeant and notified him that he should assume command of the company. He then continued with the company until the advance was checked and the first sergeant severely wounded, when he made his way in search of the next in command. Failing to find him, he organized a squad of slightly wounded men and, with an automatic rifle and ammunition which he salvaged, mopped up a section of the enemy trench, and then rejoined his company in its continued advance.

KENNY, THOMAS (1204791)
Near St. Souplet, France, Oct. 17, 1918.
R—New York, N. Y.
B—Ireland.
G. O. No. 23, W. D., 1919.

Sergeant, Company H, 105th Infantry, 27th Division.
While patrolling alone in advance of the line, he discovered a German officer directing a detachment in establishing machine-gun posts. He immediately opened fire, killing 1 and forcing the others to surrender. Later, reinforced by the remainder of his squad, Sergeant Kenny captured 34 of the enemy, including 7 officers.

KENOYER, JOHN (1448399)
Near Baulny, France, Sept. 28, 1918.
R—Hutchinson, Kans.
B—Medora, Kans.
G. O. No. 95, W. D., 1919.

Private, Company E, 137th Infantry, 35th Division.
Having previously made several trips to the rear with wounded comrades under heavy shellfire, Private Kenoyer, with another litter bearer, after making three attempts, succeeded in reaching their battalion commander, who lay wounded under heavy machine-gun fire, and carried him back to safety.

KENT, WALTER J. (1222934)
Near St. Souplet, France, Oct. 17, 1918.
R—Whitestone, N. Y.
B—New York, N. Y.
G. O. No. 9, W. D., 1923.

Private, first class, sanitary detachment, 105th Infantry, 27th Division.
With splendid courage and devotion to duty and without thought for his own safety, he cared for 40 wounded men of his command, at all times exposed to terrific machine-gun and high-explosive shell fire. He continued his aid until he himself was severely wounded while attempting to obtain stretchers with which to move the more serious cases. His remarkable courage and devotion to his comrades inspired every man of his command.

KENYON, THEODORE S_____
In the Forest of Argonne, France,
Sept. 27, 1918.
R—New York, N. Y.
B—New York, N. Y.
G. O. No. 21, W. D., 1919.

First lieutenant, 306th Infantry, 77th Division.
He displayed exceptional qualities of leadership and gallantry in action while
leading his company against a counterattack of the enemy in superior numbers. Later, although 3 times wounded, he remained with his command.

KEPNER, WILLIAM E_____
Near Cunel, France, Oct. 5–6, 1918.
R—Kokomo, Ind.
B—Peru, Ind.
G. O. No. 44, W. D., 1919.

Captain, 4th Infantry, 3d Division.
While in command of a battalion, Captain Kepner personally led 1 company
of his command in an attack on a woods occupied by a company of German
machine gunners. He was the first man to enter the woods, and later, when
part of the attacking company was held up by flanking machine-gun fire, he,
with a patrol of 3 men, encircled this machine gun, and after a hard hand-to-
hand fight, put the gun out of action.

KERR, ALWYN A. (2041346)_____
Near Bois-de-Ormont, France, Oct.
13, 1918.
R—Detroit, Mich.
B—Port Huron, Mich.
G. O. No. 35, W. D., 1919.

Private, Company F, 114th Infantry, 29th Division.
While on guard duty in the front line, Private Kerr saw 4 of the enemy approaching through the thick fog. Without hesitation, he gave the alarm and charged
them, capturing the entire number after determined resistance.

KERR, MARK E. (10384)_____
Near Fleville and St. Juvin, France,
Oct. 11–13, 1918.
R—Fellows, Calif.
B—Ireland.
G. O. No. 145, W. D., 1918.

Private, first class, section No. 647, Ambulance Service.
After 36 hours of continuous firing over heavily shelled roads, he, upon his
own initiative, was the first to establish liaison with an advanced dressing
station which had been isolated by the explosion of a tank trap. To do this
he lifted his car across the mine crater with the aid of some infantrymen, and
for 8 hours thereafter drove his ambulance through a heavy bombardment
of high-explosive and gas shells between the mine crater and the dressing station. During this period his car was pierced repeatedly by shell fragments,
2 of his patients receiving additional wounds. 2 days later, when the Infantry
had made a further advance, his car was again the first to establish liaison
with an advanced dressing station on the outskirts of St. Juvin.

KERR, ROLAND M. (1246005)_____
Near Fismette, France, Aug. 11,
1918.
R—Philadelphia, Pa.
B—Philadelphia, Pa.
G. O. No. 64, W. D., 1919.

Corporal, Company M, 111th Infantry, 28th Division.
He volunteered to carry a severely wounded soldier to a dressing station across
a bridge subjected to intense enemy fire. With 3 other soldiers he started
across the bridge carrying the wounded man. 1 stretcher bearer was killed
and another wounded, but Corporal Kerr continued with the wounded soldier
and after reaching the dressing station returned and rescued the stretcher
bearer wounded on the bridge.

KERWIN, JOSEPH N. (2273726)_____
At Audenarde, Belgium, Nov. 1,
1918.
R—Salt Creek, Wyo.
B—Denver, Colo.
G. O. No. 1, W. D., 1919.

Sergeant, first class, Company F, 316th Engineers, 91st Division.
He volunteered to accompany an officer and 3 other soldiers on a reconnaissance
patrol of the city of Audenarde. Entering under heavy shell fire, the party
reconnoitered the city for 7 hours while it was still being patrolled by the
enemy, and advanced 2 kilometers in front of our own outposts and beyond
those of the enemy.

*KESKE, CLARENCE E. (275379)_____
Near Gesnes, France, Oct. 14, 1918.
R—Beaver Dam, Wis.
B—Oak Grove, Wis.
G. O. No. 56, W. D., 1922.

Private, Company M, 127th Infantry, 32d Division.
When his battalion was held up and suffering heavy losses from flanking fire
of 4 enemy machine-gun nests, Private Keske volunteered and advanced to
a position 250 yards ahead of the line and helped to establish a position neutral-
izing the enemy fire, which enabled the battalion to re-form. He remained
in this advanced position, inspiring by his courageous example his comrades
to maintain the position, although under terrific machine-gun fire from 3
sides and hand grenades, until he was killed.
Posthumously awarded. Medal presented to father, Charles Keske.

*KESSLER, EDWARD M. (2176452)_____
Near Flirey, France, Sept. 12, 1918.
R—Cheney, Kans.
B—Maize, Kans.
G. O. No. 87, W. D., 1919.

Corporal, Company F, 353d Infantry, 89th Division.
When his platoon was held up by heavy machine-gun fire in front of the Bois-
de-Mort Mare, Corporal Kessler advanced with his squad and succeeded in
putting the machine guns out of action without losing a man. Farther on
in the woods 2 more machine guns were encountered, firing from a dugout.
This soldier went out alone and silenced the guns with hand grenades, thereby
facilitating the advance of the entire company.
Posthumously awarded. Medal presented to father, Rev. C. M. Kessler.

KESSLER, HENRY O. (1708151)_____
Near Ville-Savoye, France, Aug. 23,
1918.
R—Brooklyn, N. Y.
B—Brooklyn, N. Y.
G. O. No. 32, W. D., 1919.

Private, Company C, 308th Infantry, 77th Division.
He was the first to respond to a call for volunteers to rescue a wounded soldier
who had fallen severely wounded while on a patrol. Crawling forward
through intense machine-gun and artillery fire, he assisted in the rescue, being
severely wounded while engaged in the undertaking.

KETCHEM, HAROLD (2185885)_____
Near Bellicourt, France, Sept. 29,
1918.
R—Alexandria, S. Dak.
B—Elnora, Ind.
G. O. No. 37, W. D., 1919.

Private, Machine Gun Company, 117th Infantry, 30th Division.
Private Ketchem, a runner, carried many messages under heavy fire. At one
time when his platoon was held up by machine-gun fire he, with another
soldier, refused to take cover, but delivered effective rifle fire until the machine
gun was silenced, thereby enabling his platoon to continue its advance.

*KIAH, MARTIN J. (1551116)_____
Near Soissons, France, July 20, 1918.
R—Bay City, Mich.
B—Bay City, Mich.
G. O. No. 124, W. D., 1918.

Corporal, Company C, 1st Engineers, 1st Division.
When volunteers were called for his by company commander, Corporal Kiah
volunteered and rescued wounded comrades from a barrage. Although
wounded in the performance of these heroic deeds, he continued until killed
by shell fire.
Posthumously awarded. Medal presented to father, George Kiah.

KIBLER, JOHN T. _____
Near St. Etienne-a-Arnes, France,
Oct. 3–4, 1918.
R—Chestertown, Md.
B—Chestertown, Md.
G. O. No. 49, W. D., 1922.

First lieutenant, 23d Infantry, 2d Division.
Although severely gassed by a gas shell which burst in the trench beside him Lieutenant Kibler retained command of his company and led it forward in a difficult night attack, penetrating the enemy's line to a depth of 2 kilometers. Throughout the night he and his men were subjected to a deadly cross fire from enemy machine guns and at daybreak a strong counterattack was launched against him. Heroically leading a small force against the flank of the attacking party, Lieutenant Kibler succeeded in routing a greatly superior force. He courageously continued to lead his men until so weakened by the effects of the gas that he lost consciousness.

KIDD, CHARLES M. (1786710)_____
Near Verdun, France, Nov. 10, 1918.
R—New Freedom, Pa.
L—Baltimore County, Md.
G. O. No. 37, W. D., 1919.

Corporal, Company E, 316th Infantry, 79th Division.
While his company was being held up by machine-gun fire, Corporal Kidd led his squad, in spite of a severe leg wound, against the machine gun, killing the gunner and clearing the way for the advance of his company.

KIDD, CLIFFORD C. (2183276)_____
Near Remonville, France, Nov. 1,
1918.
R—Denver, Col.
B—Creston, Iowa.
G. O. No. 95, W. D., 1919.

Private, first class, Machine Gun Company, 354th Infantry, 89th Division.
Private Kidd displayed exceptional bravery in carrying his machine gun through heavy fire ahead of the Infantry front line to a point only 30 yards from 6 enemy machine guns which defended a hill. Despite the intense grenade and machine-gun fire which was directed at him, he maintained such effective fire that the hostile guns were put out of action and the infantry advance thereupon resumed.

*KIDDER, HUGH P._____
Near Blanc Mont, France, Oct. 2–3,
1918.
R—Minneapolis, Minn.
B—Waukon, Iowa.
G. O. No. 37, W. D., 1919.

Second lieutenant, 6th Regiment, U. S. Marine Corps, 2d Division.
On the morning of Oct. 2 he led a small patrol into enemy trenches and captured 2 strong machine-gun positions which were menacing his company. On Oct. 3 he, with his platoon, attacked and captured 4 machine-gun nests and many prisoners, after which he went to the aid of 2 of his wounded men. While attempting to better his position in the face of heavy machine-gun and artillery fire he was killed.
Posthumously awarded. Medal presented to mother, Mrs. Kate Kidder.

KIELPINSKI, VINCENT P. (274795)_____
Near Juvigny, France, Aug. 31, 1918.
R—Milwaukee, Wis.
B—Chicago, Ill.
G. O. No. 98, W. D., 1919.

Sergeant, Company K, 127th Infantry, 32d Division.
When his company had become disorganized and his company commander gassed, Sergeant Kielpinski carried orders under intense fire and assisted in reorganizing the company. Though he was wounded next day and ordered to the hospital he returned from the dressing station to his company and continued in action.

KILBOURNE, CHARLES E._____
Near Thiaucourt, France, Sept. 12,
1918.
R—Portland, Oreg.
B—Fort Myer, Va.
G. O. No. 143, W. D., 1918.
Medal of honor and distinguished-
service medal also awarded.

Colonel (Coast Artillery Corps), General Staff Corps, 89th Division.
As Chief of Staff, he exposed himself to artillery and machine-gun fire during the advance of his division, exercising cool judgment and strong determination in reorganizing the lines and getting troops forward to the objective.

KILBY, ROBERT E. L. (40436)_____
Near St. Mihiel, France, Sept. 14,
1918.
R—Laurel Bloomery, Tenn.
B—Ashe County, N. C.
G. O. No. 46, W. D., 1919.

Private, first class, Company K, 9th Infantry, 2d Division.
Private Kilby volunteered to go with his company commander to reconnoiter a German trench before a contemplated advance. They encountered a German officer with 7 men in the trench. Private Kilby successfully cleaned the trench and saved his captain's life by his coolness and exceptional courage.

KILCHER, ELMER J. (2162875)_____
At Fresnes-en-Woevre, France, Nov.
10, 1918.
R—Waucoma, Iowa.
B—Waucoma, Iowa.
G. O. No. 23, W. D., 1919.

Private, first class, Company D, 130th Infantry, 33d Division.
Private Kilcher voluntarily returned through the enemy's barrage after a raid to rescue another soldier who had been wounded, and was himself wounded as a result of his self-sacrificing effort.

KILFOYLE, FRANK J. (1458113)_____
At Varennes, France, Sept. 26, 1918.
R—St. Joseph, Mo.
B—St. Joseph, Mo.
G. O. No. 95, W. D., 1919.

Sergeant, Company M, 139th Infantry, 35th Division.
Under heavy machine-gun and artillery fire Sergeant Kilfoyle led an automatic rifle squad into Varennes and captured an enemy machine-gun nest, which had been inflicting heavy casualties on our forces, killing or wounding all the members of the crew, including a German major. Two days later he again displayed skillful leadership in organizing his platoon against an enemy counterattack, which was successfully repulsed.

KILMER, EVERETT A. (2662358)_____
Near Charpentry, France, Oct. 3,
1918.
R—Seneca, Ill.
B—Verona, Ill.
G. O. No. 81, W. D., 1919.

Private, Company M, 16th Infantry, 1st Division.
He voluntarily accompanied 3 other soldiers and went forward under heavy enemy fire and rescued a wounded comrade who had fallen in advance of our lines.

KILPATRICK, LLOYD (1160164)_____
Near Vieville-en-Haye, France, Nov.
1, 1918.
R—Freemansburg, Pa.
B—Northampton, Pa.
G. O. No. 37, W. D., 1919.

Private, first class, Company C, 21st Machine Gun Battalion, 7th Division.
He acted as a runner during offensive operations and under severest barrages and carried important messages to and from the front lines.

KILROY, JOSEPH F. (1785632)_____
Near Verdun, France, Nov. 4, 1918.
R—Philadelphia, Pa.
B—Philadelphia, Pa.
G. O. No. 37, W. D., 1919.

First sergeant, Company K, 315th Infantry, 79th Division.
He led a patrol of 5 men to flank a machine-gun nest, but heavy machine-gun fire caused the loss of the other members of the patrol. He picked up the automatic rifle of a fallen comrade and by his effective and severe fire rendered valuable aid in breaking up a local counterattack which was forming.

KILROY, LAWRENCE B. (2019020) ....
Near Kadish, Russia, Sept. 27-28, 1918.
R—Detroit, Mich.
B—Detroit, Mich.
G. O. No. 78, W. D., 1919.

Private, 337th Ambulance Company, attached to medical detachment, 339th Infantry, 85th Division (detachment in North Russia).
Acting as stretcher bearer to 2 companies of Infantry in action against the Bolsheviks, Private Kilroy, for 2 days and nights made his way through swamps and forests to administer first aid and carry wounded to the dressing station. His work at all times was accomplished under sweeping machine-gun and intense artillery fire, making it necessary for him to crawl on his hands and knees for long distances.

KIMBALL, ROY EDMUND (193999) ....
At Marcheville, France, Sept. 26, 1918.
R—Somerville, Mass.
B—Deering, Me.
G. O. No. 87, W. D., 1919.

Private, first class, 101st Field Signal Battalion, 26th Division.
Private Kimball displayed remarkable coolness and courage under violent bombardment when he voluntarily repaired telephone lines and rendered great assistance in maintaining communication. Although wounded, he continued his work until ordered evacuated by his commanding officer.

KIMBALL, WALTER G. ....
Near St. Mihiel, France, Sept. 12, 1918.
R—New York, N. Y.
B—Canton, N. Y.
G. O. No. 37, W. D., 1919.

First lieutenant, 9th Infantry, 2d Division.
While advancing in the first wave of the advance of Sept. 12, he was forced to deploy his platoon over a wide sector in thickly wooded territory in order to maintain contact and liaison with the unit on the flank. The crews of several machine-gun nests were routed, and in one instance he captured a machine-gun crew assisted by only two of his men.

*KIMMELL, HARRY L. ....
South of Soissons, France, July 19, 1918.
R—Washington, D. C.
B—Washington, D. C.
G. O. No. 3, W. D., 1921.

Captain, 16th Infantry, 1st Division.
When his company was halted by machine-gun fire from the front, Captain Kimmell led a platoon through heavy machine-gun fire and captured the enemy position, forcing its defenders to surrender. His gallantry enabled the entire battalion to continue the advance.
Oak-leaf cluster.
Captain Kimmell was awarded an oak-leaf cluster for the following act of extraordinary heroism north of Fleville, France, Oct. 9, 1918: He personally led 2 platoons of his company against a strongly held enemy position. He fell mortally wounded while leading this advance, but other members of his command, inspired by his gallantry, successfully assaulted the enemy position.
Posthumously awarded. Medal presented to father, Commander Harry L. Kimmell, U. S. Navy.

KINCAID, JAMES LESLIE ....
Near Ronssoy, France, Sept. 27, 1918.
R—Syracuse, N. Y.
B—Syracuse, N. Y.
G. O. No. 59, W. D., 1921.

Lieutenant colonel, Judge Advocate General's Department.
Because of a shortage of officers and after additional officers had been requested, Colonel Kincaid volunteered to command one of the battalions of the 106th Infantry. He commanded the battalion throughout the engagement of Sept. 27, 1918, with great courage and forcefulness and without regard to his personal safety, thereby setting a splendid example to all ranks.

KINDLEY, FIELD E. ....
Near Bourlon Wood, France, Sept. 24, 1918.
R—Coffeyville, Kans.
B—Pea Ridge, Ark.
G. O. No. 7, W. D., 1919.

First lieutenant, 148th Aero Squadron, Air Service.
He attacked a formation of 7 hostile planes (type, Fokker) and sent 1 crashing to the ground.
Oak-leaf cluster.
A bronze oak leaf is awarded to Lieutenant Kindley for the following act of extraordinary heroism in action near Marcoing, France, Sept. 27, 1918: Flying at a low altitude, this officer bombed the railway at Marcoing and drove down an enemy balloon. He then attacked German troops at a low altitude and silenced a hostile machine gun, after which he shot down in flames an enemy plane (type, Halberstadt) which had attacked him. He has so far destroyed 7½ enemy aircraft and driven down 3 out of control.

KING, DELANCEY ....
Near Ronssoy, France, Sept. 29, 1918.
R—Buffalo, N. Y.
B—Lockport, N. Y.
G. O. No 15, W. D., 1919.

First lieutenant, 108th Infantry, 27th Division.
He was wounded early in the engagement, but continued to lead his men until he received a second wound. His gallantry under shell and machine-gun fire and his disregard for his own safety furnished a splendid example to all ranks.

KING, EDWARD E. (44351) ....
Near Fleville, France, Oct. 4, 1918.
R—Dayton, Ohio.
B—Dayton, Ohio.
G. O. No. 50, W. D., 1919.

First sergeant, Company M, 16th Infantry, 1st Division.
Assuming command of his company after the company commander had been wounded, Sergeant King reinforced it with remnants of other units, leading the entire command through perilous machine-gun and artillery fire to his objective. After having consolidated his position, he led a volunteer patrol against several machine-gun nests which were harassing the position. Under violent fire from these guns he cleaned out the troublesome nests, with small losses to his forces.

KING, EDWARD L. ....
Near Imus, Cavite Province, P. I., Oct. 6, 1899.
R—Bridgewater, Mass.
B—Bridgewater, Mass.
G. O. No. 126, W. D., 1919.
Distinguished-service medal also awarded.

Captain, 11th Cavalry, U. S. Army.
While serving as captain, 11th U. S. Cavalry, his great personal bravery in disarming a hostile Filipino saved the life of a brother officer.

KING, FRED T. (143131) ....
Near Bulson, France, Nov. 8-9, 1918.
R—Danville, Ill.
B—Danville, Ill.
G. O. No. 71, W. D., 1919.

Sergeant, Headquarters Company, 149th Field Artillery, 42d Division.
Placed in charge of a telephone detail, he refused to be evacuated when wounded by a splinter from a shell which killed 5 of his men and wounded 12 others. He remained on duty until the regiment left the line, inspiring his men by his example of fortitude.

KING, GEORGE (42508)_____
Near the Argonne Forest, France, Oct. 9, 1918.
R—New York, N. Y.
B—Scotland.
G. O. No. 44, W. D., 1919.

Private, Company D, 16th Infantry, 1st Division.
He was a member of a reconnoitering patrol which encountered heavy fire from hostile machine guns. Private King advanced alone and killed the gunners, enabling his patrol to complete its mission.

*KING, HAROLD J_____
Near Fismes, France, Aug. 4, 1918.
R—Manistee, Mich.
B—Manistee, Mich.
G. O. No. 59, W. D., 1919.

First lieutenant, 126th Infantry, 32d Division.
Advancing in the face of terrific machine-gun fire, Lieutenant King rescued a wounded man of his platoon, the exploit being accomplished in broad daylight. After carrying the man 150 yards, in full view of the enemy, he stopped at request made by the dying man. During the advance from the Ourcq to the Vesle he demonstrated most admirable courage and fearlessness. While accompanying a reconnaissance patrol on the night of Oct. 10 he was killed by sniper fire.
Posthumously awarded. Medal presented to father, Dr. James A. King.

KING, HENRY M. (1241143)_____
At Apremont, France, Sept. 29, 1918.
R—Greensburg, Pa.
B—Greensburg, Pa.
G. O. No. 98, W. D., 1919.

Sergeant, Company I, 110th Infantry, 28th Division.
During a determined enemy counterattack, Sergeant King, volunteered with another soldier to locate the source of enemy fire which was inflicting heavy casualties on his company. Going forward under fire and discovering 7 of the enemy, they opened fire, killing 2, wounding 2, and capturing a machine gun. Their courageous feat materially aided his platoon to continue the advance.

KING, JAMES NORMAN_____
Near Thiaucourt, France, Sept. 16, 1918.
R—Lima, Ohio.
B—Fairgrove, Mich.
G. O. No. 127, W. D., 1918.

First lieutenant, chaplain, 310th Infantry, 78th Division.
He gave proof of unhesitating devotion and energy far beyond all call of his duty as battalion chaplain. He was continually on the outpost lines searching for and burying the dead and circulating among the men in the trenches. During the entire period the line was subjected to extremely heavy shelling from the enemy, yet he allowed nothing to interfere with his having burial services for the dead that were brought to the burial ground located within the shelled area. He was sent to the rear, but he prevailed upon the surgeon to allow him to return to the front and continue his work.

KING, JAMES PRYN_____
Near Consenvoye, France, Oct. 10, 1918.
R—Hillsboro, Ill.
B—Dixon, Mo.
G. O. No. 19, W. D., 1920.

First lieutenant, 122d Machine Gun Battalion, 33d Division.
While in command of a machine-gun platoon attached to the 131st Infantry, Lieutenant King, by personal reconnaissance established machine-gun and automatic-rifle outposts. During a threatened counterattack he exposed himself to point-blank fire and walked from outpost to outpost reassuring his men.

*KING, JESSE L. (2288851)_____
Near Gesnes, France, Sept. 26, 1918.
R—Laramie, Wyo.
B—Blencoe, Iowa.
G. O. No. 20, W. D., 1919.

Corporal, Company D, 361st Infantry, 91st Division.
Corporal King, together with 2 other soldiers, captured 3 enemy machine guns and 26 prisoners.
Posthumously awarded. Medal presented to sister, Mrs. Olive Jacob.

KING, OGDEN DOREMUS_____
Near the Bois-de-Belleau, France, June 9-10, 1918.
R—Albemarle, N. C.
B—Albemarle, N. C.
G. O. No. 137, W. D., 1918.

Assistant surgeon, U. S. Navy, attached to 6th Machine Gun Battalion, U. S. Marine Corps, 2d Division.
On 2 successive days the regimental aid station in which he was working was struck by heavy shells and in each case demolished. 10 men were killed and a number of wounded were badly hurt by falling timbers and stone. Under these harassing conditions this officer continued without cessation his treatment of the wounded, assisting in their evacuation and setting an inspiring example of devotion and courage to the officers and men serving under him.

KING, RICHARD E. (2187488)_____
Near Marimbois Farm, France, Nov. 4, 1918.
R—Tucson, Ariz.
B—Perry, Okla.
G. O. No. 37, W. D., 1919.

Private, Headquarters Company, 340th Field Artillery, 89th Division.
Assisting an officer in establishing communication with the advancing Infantry, Private King reached Marimbois Farm, where he found the place occupied by the enemy. Armed with hand grenades, he advanced on a dugout, where he routed out 17 of the enemy, bringing them back to our lines in the midst of severe shell and machine-gun fire.

KINGSBURY, CARL O_____
On the morning of Apr. 13, 1918.
R—New Castle, Pa.
B—Greenville, Pa.
G. O. No. 129, W. D., 1918.

Hospital apprentice, first class, U. S. Navy, attached to Headquarters 6th Regiment, U. S. Marine Corps, 2d Division.
He displayed commendable action in the immediate care and attention incident to the evacuation of more than 100 casualties following the gas-shell bombardment on the morning of Apr. 13, 1918, after he himself had been subjected to gas. He worked incessantly, disregarding his own symptoms, until he had to be evacuated.

KINKADE, BERTE L. (57247)_____
At Exermont, France, Oct. 4, 1918.
R—St. Joseph, Mo.
B—Wheaton, Minn.
G. O. No. 71, W. D., 1919.

Corporal, Company D, 28th Infantry, 1st Division.
When a German 77-millimeter gun, supported by numerous machine guns, broke the tank attack and held up the Infantry advance, Corporal Kinkade, with 2 scouts, made an encircling movement amid heavy fire and put the gun out of action, capturing the entire crew. They cleaned out the enemy dugouts in the vicinity and returned with 40 prisoners, including an officer.

KINNEER, ISAAC G. (105248)_____
South of Soissons, France, July 18, 1918.
R—Mount Pleasant, Iowa.
B—Veedersburg, Ind.
G. O. No. 72, W. D., 1920.

Supply sergeant, Company A, 2d Machine Gun Battalion, 1st Division.
Sergeant Kinneer took command of his platoon after his platoon leader had been wounded, and led one of his sections through our own barrage in order to take up a position where he could fire upon two 77-millimeter fieldpieces which were firing point blank on our troops. The fire of his gun was so effective that the guns were captured and the crews were forced to surrender

°KINNEY, CLAIR A.
    Near Doulcon, France, Oct. 4, 1918.
    R—Endicott, Wash.
    B—Endicott, Wash.
    G. O. No. 37, W. D., 1919.

First lieutenant, 49th Aero Squadron, Air Service.
With a patrol of six other machines, Lieutenant Kinney attacked 17 enemy planes (type Fokker). Diving into the midst of the enemy formation, he fired into one of the German planes and pursued it until it crashed to the ground, though he was wounded by another Fokker which attacked him from the rear. After maneuvering to escape his pursuer, he immediately attacked another enemy plane directly in front of him and forced it to the ground. In so doing he was fired upon from behind by another Fokker, several bullets striking him in the body and another setting fire to his gas tank. He succeeded in making a safe landing. This gallant officer has since died of his wounds. Posthumously awarded. Medal presented to mother, Mrs. M. P. Kinney.

KINNEY, MARTIN H. (2237426).
    Near Pout-a-Mousson, France, Sept. 26, 1918.
    R—Fort Worth, Tex.
    B—Fort Worth, Tex.
    G. O. No. 128, W. D., 1918.

Sergeant, Company E, 315th Engineers, 90th Division.
After receiving orders to withdraw from position, he saw an officer lying wounded and helpless about 15 yards in advance of the line. With utter disregard for his own safety he crawled through heavy enemy fire from the front and flanks to the aid of the officer and dragged him back about 50 yards to the shelter of a small mound, where he dressed the officer's wound, and then carried him through the barrage to a first-aid station.

KINSLEY, WILBERT E.
    East of Cunel, Verdun sector, France, Oct. 7, 1918.
    R—Elmira, N. Y.
    B—Somerville, Mass.
    G. O. No. 15, W. D., 1919.

Second lieutenant, pilot, 3d Observation Group, Air Service.
While staking the advanced lines of the 80th Division he was attacked by 8 enemy machines (Fokker type), which dived out of a near-by cloud bank. Although attacked simultaneously by the enemy planes, he placed his airplane in such a position that his observer, Second Lieut. William O. Lowe, U. S. Marine Corps, was able to shoot down and crash 1 enemy plane and disable a second so badly that it was forced to land a few kilometers inside the German lines. Later, on the same mission, he was again attacked by a patrol of 5 enemy scout machines, and in a running fight he drove these off and successfully completed his mission.

KIRK, EDGAR.
    Near Soissons, France, July 19, 1918.
    R—Princeton, Ind.
    B—Kirksville, Ind.
    G. O. No. 15, W. D., 1919.

Second lieutenant, 2d Machine Gun Battalion, 1st Division.
At a critical period in the attack south of Soissons when the Infantry was held up by a German battery, he pushed forward with one of his machine guns in the face of direct machine gun and artillery fire and by cool and courageous direction of this gun put the battery out of action, thereby causing its capture and permitting the Infantry to advance.

KIRK, HENRY S. (1203467).
    East of Ronssoy, France, Sept. 29, 1918.
    R—New York, N. Y.
    B—Dorset, Vt.
    G. O. No. 20, W. D., 1919.

Sergeant, Company B, 105th Infantry, 27th Division.
During the operations against the Hindenburg line Sergeant Kirk, with an officer and 2 other sergeants, occupied an outpost position in advance of the line which was attacked by a superior force of the enemy. Sergeant Kirk assisted in repulsing this attack and in killing 10 Germans, capturing 5, and driving off the others. The bravery and determination displayed by this group was an inspiration to all who witnessed it.

KIRK, RICHARD M. (2257820).
    At Audenarde, Belgium, Nov. 1, 1918.
    R—Seattle, Wash.
    B—Philadelphia, Pa.
    G. O. No. 21, W. D., 1919.

Sergeant, Company H, 361st Infantry, 91st Division.
He was a member of a patrol sent out to reconnoiter the town of Audenarde. This patrol discovered several enemy machine-gun sniper posts, located in buildings, which were enfilading the streets of the town. Taking another soldier with him and dodging from building to building, he entered one of these houses and captured 2 machine gunners.

KIRKPATRICK, CLIFFORD (10395).
    At Sommerance, France, Oct. 13–15, 1918.
    R—North Leominster, Mass.
    B—Fitchburg, Mass.
    G. O. No. 145, W. D., 1918.

Private, first class, Section No. 647, Ambulance Service.
While he was loading his ambulance at an advanced dressing station he was struck in the back by a shell fragment and rendered unconscious. Upon recovering he drove his car through heavy shellfire back to the field hospital and made repeated trips until he was relieved. Two days later, arriving at a point on the road near an advanced post, he saw a shell strike in the middle of an infantry detachment moving up to the line. He stopped his car and, despite continuous and intense shelling, loaded on 3 wounded soldiers, carried them to the nearest aid station, returned, and continued the work of evacuation until relieved.

KIRSCHENBAUM, EDWARD (550986).
    Near Moulins, France, July 15, 1918.
    R—New York, N. Y.
    B—New York, N. Y.
    G. O. No. 22, W. D., 1920.

Private, Company F, 38th Infantry, 3d Division.
Private Kirschenbaum carried numerous messages through heavy artillery and machine-gun fire from the front-line companies to battalion headquarters. Due to his efforts communications were maintained during the period when his organization was successfully defending its position against powerful enemy attack. On one of his trips he captured 3 enemy prisoners, who had infiltrated through our lines.

KITE, GEORGE J. (1288359).
    Near Verdun, France, Oct. 24, 1918.
    R—Grove Hill, Va.
    B—Grove Hill, Va.
    G. O. No. 21, W. D., 1919.

Private, Company B, 102d Machine Gun Battalion, 26th Division.
After several runners had been killed in the attempt, Sergeant Kite volunteered and delivered an important message under a hail of fire.

KJELLBERG, JOHN.
    Near Trugny, France, July 23, 1918.
    R—New York, N. Y.
    B—Sweden.
    G. O. No. 126, W. D., 1919.

Second lieutenant, 102d Infantry, 26th Division.
After being wounded he refused to be evacuated, but gallantly continued in command of his platoon, firing the machine gun himself when the crew was depleted until he was wounded a second time.

KJOSE, CLARENCE OLANDO (2186117).
At Claire-Chenes Woods, France, Oct. 15, 1918.
R—Akron, Iowa.
B—Sioux City, Iowa.
G. O. No. 98, W. D., 1919.

Corporal, Machine Gun Company, 7th Infantry, 3d Division.
Advancing with 2 squads in front of the company to which he was attached, after being twice stopped by enfilading machine-gun fire and losing half his detachment, Corporal Kjose continued to lead his squad through the woods, and put out of action an enemy machine gun which had been holding up the advance of the Infantry. This soldier had been wounded twice within the 3 preceding days, but continued in action, refusing to be evacuated.

*KLANSKA, FREDERIC (2338833).
Near Le Charmel, France, July 26, 1918.
R—St. Paul, Minn.
B—De Sota, Minn.
G. O. No. 32, W. D., 1919.

Private, Company F, 4th Infantry, 3d Division.
Private Klanska, at great personal risk, voluntarily left his shelter and went 100 yards into an open field swept by heavy machine-gun fire and brought in a wounded man on his back
Posthumously awarded. Medal presented to father, Henry J. Klanska.

KLAPETZKY, JOSEPH W. (1789740).
Near Montfaucon, France, Sept. 29, 1918.
R—Syracuse, N. Y.
B—Syracuse, N. Y.
G. O. No. 37, W. D., 1919.

Corporal, 314th Ambulance Company, 304th Sanitary Train, 79th Division.
He went through heavy shellfire into a burning dressing station which he knew contained a large quantity of ammunition, in order to rescue 2 wounded German prisoners. He succeeded in carrying them to safety.

KLAVITER, EMIL O. (2259450).
Near Gesnes, France, Sept. 29, 1918.
R—Reno, Nev.
B—Princeton, Wis.
G. O. No. 64, W. D., 1919.

Private, first class, Machine Gun Company, 362d Infantry, 91st Division.
Private Klaviter, a runner and signalman, while carrying a message to the attacking battalion, encountered an enemy machine-gun nest. Single-handed he killed 2 of the German machine gunners and captured 6 others, together with their gun, ammunition, and belts. The prisoners were then turned over to some other soldiers and forced to carry a wounded officer to the rear, while Private Klaviter continued on his mission.

*KLEIBER, WALTER J. (107175).
Near Greves Farm, France, July 14, 1918.
R—Whiting, Ind.
B—Chicago, Ill.
G. O. No. 44, W. D., 1919.

Private, Battery E, 10th Field Artillery, 3d Division.
He was acting as telephone operator at a gun in a detached position when all the crew became casualties. With another soldier he courageously continued to fire the piece under the heaviest bombardment until it was struck by a shell and he was killed.
Posthumously awarded. Medal presented to father, Reinholdt T. Kleiber.

KLEIN, IRVING (1707558).
Near Binarville, France, Sept. 29, 1918, and Charlevaux, France, Oct. 3–7, 1918.
R—New York, N. Y.
B—Hungary.
G. O. No. 35, W. D., 1919.

Corporal, Company A, 308th Infantry, 77th Division.
On Sept. 29, after locating the position of 3 enemy machine guns, he succeeded in silencing 1, took up a position against the other 2 under intense shellfire, and sent back information to his company commander which made it possible to clean out the entire nest. On Oct. 3, although wounded seriously, he continued to assist his men in repulsing the attack of an enemy patrol.

KLICK, ALBERT W. (68563).
Near Bois-de-St. Remy, France, Sept. 12, 1918.
R—Fairfield, Me.
B—Germany.
G. O. No. 26, W. D., 1919.

Sergeant, Company H, 103d Infantry, 26th Division.
With the aid of 6 comrades, Sergeant Klick attacked and put out of action a machine gun which was checking the advance of his company. Later he captured, without aid, about 20 prisoners, and while advancing against another nest he was twice wounded. Although in severe pain, he declined the use of a litter, walking 3 kilometers to a dressing station.

KLIER, GEORGE J. (1830246).
Near Verdun, France, Oct. 11, 1918.
R—Pittsburgh, Pa.
B—Pittsburgh, Pa.
G. O. No. 21, W. D., 1919.

Private, first class, Company G, 320th Infantry, 80th Division.
While his platoon was being forced back he remained to bind up the wounds of a comrade, although he himself was suffering from a painful wound. He then refused to be taken to the rear until all the others had been evacuated.

KLINE, DAVID (2180319).
Near Remonville and Barricourt, France, Nov. 1–2, 1918.
R—Omaha, Nebr.
B—Omaha, Nebr.
G. O. No. 13, W. D., 1919.

Private, first class, medical detachment, 341st Machine Gun Battalion, 89th Division.
Attached to a company immediately behind the assault battalion of Infantry, he worked unceasingly, giving first-aid treatment to the wounded in his vicinity in an area constantly swept by intense fire from all arms. Displaying the highest devotion to duty and disregard for his own safety, he frequently worked in plain view of the enemy.

KLINE, ORVAL.
Northeast of Nantillois, France, Oct. 12, 1918.
R—Bloomington, Md.
B—Martinsburg, W. Va.
G. O. No. 138, W. D., 1918.

Second lieutenant, 11th Machine Gun Battalion, 4th Division.
While the Infantry was falling back 200 meters to take cover from heavy artillery and machine-gun fire he, with his platoon sergeant, stayed at their 1 remaining machine gun, which they continued to operate for 45 minutes, until the Infantry position was reestablished. They not only successfully covered withdrawal of the Infantry, but also captured a German machine gun and 3 prisoners.

*KLINE, ROBERT J. (279460).
Near Gesnes, northwest of Verdun, France, Oct. 5, 1918.
R—Saranac, Mich.
B—Lowell, Mich.
G. O. No. 16, W. D., 1919.

Sergeant, Company D, 126th Infantry, 32d Division.
Picking up a light Maxim gun and ammunition left by the retreating enemy, he advanced in front of his company's line and supported it with enfilading fire so effectively that he was soon made the target for many German guns. With the utmost coolness and entire disregard for his own safety he continued to give appreciable support with his captured gun until he was killed by the hostile machine-gun fire. His example of bravery and audacity was an inspiration to the whole command.
Posthumously awarded. Medal presented to sister, Miss Ruth Kline.

KLING, ISAIAH MILLER (542647).
Near Cunel, France, Oct. 20, 1918.
R—Gettysburg, Pa.
B—Byerstown, Pa.
G. O. No. 22, W. D., 1920.

Corporal, Company K, 7th Infantry, 3d Division.
Corporal Kling courageously led a detachment of 20 men against a machine-gun position on the right flank of the company, which was causing heavy losses. During the attack all others of Corporal Kling's detail became casualties, but he continued on alone. Gaining an advanced position in a shell hole, he kept the enemy machine gun under continuous fire until reinforcements arrived and captured it.

KLINGE, WALTER (1224744)_____
Near St. Souplet, France, Oct. 17, 1918.
R—Brooklyn, N. Y.
B—Hoboken, N. J.
G. O. No. 32, W. D., 1919.

Private, first class, Company M, 105th Infantry, 27th Division.
When sent out as a scout with a small patrol consisting of an officer and 2 men, Private Klinge courageously went ahead alone, killed 2 enemy scouts whom he encountered, and drove the gunners away from 2 machine guns. When the patrol came up the capture of the guns was completed with their assistance.

*KLINGER, WALTER W. (1278414)_____
At Bois-d'Etrayes, France, Oct. 23, 1918.
R—Newark, N. J.
B—Newark, N. J.
G. O. No. 13, W. D., 1919.

Sergeant, Company B, 113th Infantry, 29th Division.
Two of his companions having been killed, he advanced alone upon 1 of the enemy's strongest machine-gun nests and destroyed it with hand grenades. He was later killed while administering first aid to a wounded soldier.
Posthumously awarded. Medal presented to mother, Mrs. John Rummell.

KMIOTEK, ALOYSIUS (2338361)_____
Near Grand Ballois Farm, France, July 15, 1918.
R—New Castle, Pa.
B—Austria.
G. O. No. 32, W. D., 1919.

Private, Company A, 4th Infantry, 3d Division.
Although badly wounded, he continued to perform his duties as runner, and before he would accept relief notified adjoining relays that his post was disabled.

KNAPP, CHARLES H. (2386369)_____
Northwest of Aincreville, France, Nov. 2, 1918.
R—New Brunswick, N. J.
B—Buffalo, N. Y.
G. O. No. 35, W. D., 1920.

Private, first class, Company B, 61st Infantry, 5th Division.
Private Knapp took command of Company B, after all the officers were killed or wounded, and led his company to the attack. When held up by machine-gun fire from enemy nest, he led out a patrol against it, capturing the guns and killing the crew. Again on Nov. 6 at Hill Cote St. Germaine he assumed command of the company and repulsed a strong enemy counterattack.

KNAUFF, RALPH E. (1250787)_____
Near Mont St. Martin, France, Aug. 19, 1918.
R—Renovo, Pa.
B—Williamsport, Pa.
G. O. No. 15, W. D., 1919.

Private, Battery D, 107th Field Artillery, 28th Division.
Seeing another soldier fall seriously wounded, Private Knauff ran to his assistance and under heavy shellfire carried him to safety.

KNESS, KARL F._____
Near St. Etienne, France, Oct. 4, 1918.
R—Wichita, Kans.
B—Wichita, Kans.
G. O. No. 35, W. D., 1919.

Private, 17th Company, 5th Regiment, U. S. Marine Corps, 2d Division.
He volunteered and assisted a wounded comrade to get to the rear, going through an area swept by terrific machine-gun and artillery fire for a distance of more than a kilometer. He carried the wounded man the greater part of the distance.

KNIGHT, HENRY (546170)_____
Near Crezancy, France, July 15, 1918.
R—Easton, Me.
B—Fort Fairfield, Me.
G. O. No. 44, W. D., 1919.

Private, Company D, 30th Infantry, 3d Division.
After his company had been relieved he remained to guide the new organization through the intense artillery and machine-gun fire, and for 3 days carried food and water to the wounded, who could not be removed during the bombardment.

KNIGHT, PAUL R._____
Near Binarville, France, Oct. 3–5, 1918.
R—Hibernia, N. Y.
B—New York, N. Y.
G. O. No. 20, W. D., 1919.

First lieutenant, 308th Infantry, 77th Division.
Although he had been twice wounded, he led his company in 4 attempts to cut through a heavy barbed-wire entanglement to capture Hill 205 in the Forest of Argonne, France, in order to reach 2 battalions of his regiment which had been cut off by the enemy.

KNOKE, EUGENE F. (2261810)_____
Near Gesnes, France, Sept. 29, 1918.
R—Glasston, Mont.
B—Deshler, Ohio.
G. O. No. 3, W. D., 1919.

Private, Company M, 362d Infantry, 91st Division.
He performed his duties as company runner with the utmost fearlessness, crossing fire-swept fields on two occasions to carry important messages to neighboring units.

KNOTTS, HOWARD C._____
Near Arieux, France, Sept. 17, 1918.
R—Carlinville, Ill.
B—Girard, Ill.
G. O. No. 19, W. D., 1921.

Second lieutenant, 17th Aero Squadron, Air Service.
During a patrol flight 5 American planes were attacked by 20 enemy Fokkers. During the combat, when Lieutenant Knotts saw one of his comrades attacked by 7 enemy planes and in imminent danger of being shot down, he, although himself engaged with the enemy, went to the assistance of his comrade and attacked 2 of his immediate pursuers. In the fight which ensued he shot 1 of the enemy down in flames and forced the other out of control. His prompt act enabled his comrade to escape destruction, although his comrade's plane was so disabled that he made the allied lines with difficulty, crashing as he landed.

KNOWLES, JAMES, Jr._____
Near Montfaucon, France, Oct. 9, 1918.
R—Cambridge, Mass.
B—Cincinnati, Ohio.
G. O. No. 127, W. D., 1918.

First lieutenant, 95th Aero Squadron, Air Service.
While on a voluntary patrol over the enemy's lines he observed 3 enemy Fokkers attacking 1 of our balloons. He unhesitatingly attacked, and in a bitter combat that lasted for 5 minutes he succeeded in bringing 1 of the enemy planes down in flames and driving off the others.

*KNOWLTON, RAYMOND F. (134079)_____
North of Verdun, France, Oct. 21, 1918.
R—Danvers, Mass.
B—Gaysville, Vt.
G. O. No. 16, W. D., 1923.

Private, first class, Battery E, 101st Field Artillery, 26th Division.
With great heroism and devotion to duty and with utter disregard for his personal safety, he rushed forward to the aid of a wounded soldier who was lying exposed to terrific enemy fire. While so engaged he was hit by enemy shellfire and mortally wounded, dying shortly thereafter. The superb devotion to duty displayed by Private Knowlton greatly inspired and encouraged the men of his battery.
Posthumously awarded. Medal presented to father, Ralph L. Knowlton.

KNOX, THOMAS T_____
    Against Spanish forces at Las Guasi-
      mas, Cuba, June 24, 1898.
    R—Nashville, Tenn.
    B—Roane County, Tenn.
    G. O. No. 34, W. D., 1924.

Captain, 1st Cavalry, U. S. Army.
Though severely wounded, he refused to leave the firing line, but continued to lead his troop until exhausted from excessive loss of blood. His great fortitude and fearless conduct was an inspiring example to his men.

KNOX, WILLIAM L. (65965)_____
    Near Seicheprey, France, Apr. 20,
      1918.
    R—South Coventry, Conn.
    B—Bridgeton, Me.
    G. O. No. 14, W. D., 1923.

Sergeant, Company L, 102d Infantry, 26th Division.
His platoon having been relieved from its position in the front lines, Sergeant Knox remained with the relieving platoon, which was without a commissioned officer, organized and maintained its position during a severe attack by enemy infantry, bombers, and machine-gun fire. Though his platoon was entirely surrounded and cut off from support, and he himself twice wounded by shrapnel and grenades, he led his men against the enemy, broke through their lines, and established contact with the units on his right. His coolness, leadership, and outstanding courage under intense enemy fire greatly inspired his men and spurred them on to great endeavors in the face of vastly superior numbers.

KOBERNAT, JAMES F. (1429404)_____
    In the Forest of Argonne, France,
      Oct. 11, 1918.
    R—Hill City, Minn.
    B—Hurley, Wis.
    G. O. No. 142, W. D., 1918.

Private, Company M, 307th Infantry, 77th Division.
He succeeded in establishing an automatic rifle post under heavy machine-gun fire. In the face of this heavy fire he continually advanced until he was killed.
Posthumously awarded. Medal presented to brother, Joseph B. Kobernat.

KOCH, ARTHUR H. (2850524)_____
    Near Fey-en-Haye, France, Sept.
      12, 1918.
    R—St. Paul, Minn.
    B—Brownsville, Minn.
    G. O. No. 98, W. D., 1919.

Corporal, Company A, 357th Infantry, 90th Division.
With the help of only one other soldier Corporal Koch successfully attacked a German machine-gun nest from the flank, killing 7 Germans, wounding 1, and capturing 3 machine guns.

KOCHANIK, JOHN (274833)_____
    Near Fismes, France, Aug. 4, 1918.
    R—Milwaukee, Wis.
    B—Austria.
    G. O. No. 95, W. D., 1919.

Corporal, Company K, 127th Infantry, 32d Division.
Ordering his squad to take cover, Corporal Kochanik single handed rushed an enemy machine gun and killed the 2 Germans manning it. As he was turning the captured gun on the enemy the courageous soldier was killed by a sniper.
Posthumously awarded. Medal presented to brother, Walter Kochanik.

KOCHENSPARGER, JAMES A. (93879)____
    Northeast of Chateau-Thierry,
      France, July 29, 1918.
    R—Circleville, Ohio.
    B—Pickaway County, Ohio.
    G. O. No. 99, W. D., 1918.

Sergeant, Company F, 166th Infantry, 42d Division.
He was killed on July 29, 1918, when establishing an outpost beyond the edge of Seringes-et-Nesles, which had just been captured by our forces. Throughout the attack he led his section of automatic riflemen with the greatest gallantry, giving an example of courage and bravery that was an inspiration to the men of his command.
Posthumously awarded. Medal presented to father, Charles Kochensparger.

KOEHLI, FRED_____
    Near Montfaucon, France, Sept. 27,
      1918.
    R—Alliance, Ohio.
    B—Alliance, Ohio.
    G. O. No. 68, W. D., 1920.

First lieutenant, 146th Infantry, 37th Division.
Lieutenant Koehli, with 2 noncommissioned officers, advanced 200 yards beyond the objective of the patrol in the face of heavy machine-gun fire and captured three 77-millimeter fieldpieces and 2 light machine guns.

KOEHLER, KURT H. A. (155137)_____
    Near Verdun, France, Oct. 9, 1918.
    R—Hillsdale, Oreg.
    B—Germany.
    G. O. No. 37, W. D., 1919.

Private, Company A, 1st Engineers, 1st Division.
Operating without assistance a machine gun which he secured by crawling out ahead of our lines, he successfully resisted a greatly superior force of the enemy, killing several and causing the rest to retreat. When wounds in the shoulder made it impossible for him to further operate the gun he rendered it unserviceable with a pick handle before retiring and reporting to his company commander.

KOEPPEL, OSCAR_____
    Near Seringes, France, July 27, 1918.
    R—Linden Heights, Ohio.
    B—Westville, Ohio.
    G. O. No. 26, W. D., 1919.

Captain, 166th Infantry, 42d Division.
After being severely wounded while leading his men through an intense barrage, Captain Koeppel refused to leave his company until they had been established on the front line and all orders and instructions turned over to the next in command.

KOERFER, FRANK P. (1335894)_____
    Near Gercourt, France, Sept. 26,
      1918.
    R—Chicago, Ill.
    B—Milwaukee, Wis.
    G. O. No. 46, W. D., 1919.

Corporal, Headquarters Company, 131st Infantry, 33d Division.
Under heavy machine-gun fire he crept up to a church and captured 4 of the enemy, who were operating machine guns from the building.

KOFMEHL, WILLIAM H_____
    North of Verdun, France, Oct. 21,
      1918.
    R—Rapid City, S. Dak.
    B—Farley, Iowa.
    G. O. No. 59, W. D., 1919.

Second lieutenant, 15th Machine Gun Battalion, 5th Division.
After his platoon had suffered heavy casualties in the Bois-des-Rappes from machine-gun fire he located the enemy guns, and, rallying a handful of his men, charged the enemy positions, capturing 37 prisoners. After getting his own machine guns in position Lieutenant Kofmehl, seeing that one of his gunners had been wounded, operated the gun himself, setting an excellent example to his men.

KOGLER, JOHN M. (2071)_____
    Near Soissons, France, July 18, 1918.
    R—Baltimore, Md.
    B—Baltimore, Md.
    G. O. No. 15, W. D., 1919.

Private, medical detachment, 26th Infantry, 1st Division.
In spite of the fact that he had been severely wounded himself near Soissons, France, July 18, 1918, he refused to be relieved and continued to treat wounded under fire for 2 days.

KOHN, MARIAN (553826)............... | Corporal, Company B, 8th Machine Gun Battalion, 3d Division.
During the Argonne-Meuse offen- | Seeing his platoon commander and platoon sergeant buried by a shell explosion,
sive, France, Oct. 5-6, 1918. | Corporal Kohn left his trench and, in the face of perilous machine-gun and
R—Toledo, Ohio. | shell fire, single handed rescued his comrades from the caved-in trench.
B—Russia.
G. O. No. 44, W. D., 1919.

KOHN, MAT A. (2705642)............... | Corporal, Company K, 145th Infantry, 37th Division.
Near Heurne, Belgium, Nov. 4, 1918. | Corporal Kohn went through heavy shell and machine-gun fire for a distance
R—Wabasha, Minn. | of 100 yards and carried a wounded comrade to safety.
B—Wabasha, Minn.
G. O. No. 59, W. D., 1919.

KOIJANE, FRANK A. (1389510).......... | Sergeant, Company G, 132d Infantry, 33d Division.
At Hamel, France, July 4, 1918. | While digging in at his final objective he came under fire from a hostile machine
R—Chicago, Ill. | gun in a sunken road 200 yards to the right front. With Lieutenant Yagle
B—Chicago, Ill. | and 2 Australian soldiers, he rushed the position and captured the gun and
G. O. No. 44, W. D., 1919. | 8 prisoners.

KOLEMAN, NORMAN (297349).......... | Sergeant, Battery C, 119th Field Artillery, 32d Division.
Near Les Pres Farm, France, Aug. | Sergeant Koleman had just returned with his gun crew to their dugout after
9, 1918. | maintaining fire for 12 hours under heavy bombardment when a shell of
R—Lansing, Mich. | large caliber struck directly over the dugout, killing or wounding the entire
B—Gibsonburg, Ohio. | crew. Regardless of the fact that he himself had been wounded in 9
G. O. No. 98, W. D., 1919. | places, Sergeant Koleman immediately walked and crawled to the nearest
| shelter to secure aid for his men.

KOLONOCZYK, WASYL (1210213)......... | Private, Company C, 107th Infantry, 27th Division.
Near St. Souplet, France, Oct. 18, | Private Kolonoczyk, under heavy shell and machine-gun fire, left the shelter
1918. | of his trench and, going forward under a thick smoke screen, single handed
R—Cohoes, N. Y. | captured between 30 and 40 German prisoners. His conspicuous gallantry
B—Russia. | and bravery upon this occasion showed a heroic disregard for his own safety,
G. O. No. 46, W. D., 1919. | which was a splendid example for all.
| Oak-leaf cluster.
| Private Kolonoczyk is also awarded an oak-leaf cluster for the following act of
| extraordinary heroism in action near St. Souplet, France, Oct. 18, 1918: After
| the advance of his company had been stopped by strong hostile machine-
| gun fire, he, with 3 companions, advanced far ahead of the front line to
| attack an enemy position located in a farmhouse. By skillful maneu-
| vering in the broad daylight, they covered all entrances to the house and
| forced the surrender of the entire force of the enemy, numbering 36 men and
| 2 officers. During the exploit they killed 2 Germans who attempted to take
| cover in the cellar.

*KOLWYCK, OREN C. (1312511).......... | Private, Company M, 118th Infantry, 30th Division.
Near Villeret, France, Sept. 27, 1918. | He displayed marked bravery as battalion runner, fearlessly exposing himself
R—Humboldt, Tenn. | to heavy fire to deliver important messages. Volunteering to deliver a mes-
B—Trenton, Tenn. | sage which necessitated his crossing a barrage of high explosive, shrapnel,
G. O. No. 87, W. D., 1919. | and gas shells, he was mortally wounded.
| Posthumously awarded. Medal presented to father, William H. Kolwyck.

*KOON, CARTER R. (155143)........... | Private, Company B, 1st Engineers, 1st Division.
South of Soissons, France, July 20, | He volunteered and obtained the permission of his company commander to
1918. | pass through an area then under heavy shellfire for the purpose of recovering
R—Seattle, Wash. | wounded comrades. He made 2 trips successfully, but on his third trip he
B—Tacoma, Wash. | was mortally wounded.
G. O. No. 15, W. D., 1919. | Posthumously awarded. Medal presented to mother, Mrs. Fannie H. White.

KOON, ETHEN S.................... | Second lieutenant, 119th Infantry, 30th Division.
Near Ypres, Belgium, Aug. 31, 1918. | Ignoring his severe wound, suffered in the advance of his platoon against the
R—Asheville, N. C. | enemy south of Ypres, he remained with his men until all the wounded had
B—Columbia, S. C. | been evacuated and personally directed the reorganization of his position
G. O. No. 44, W. D., 1919. | until ordered to the rear by his commanding officer.

*KOPP, HENRY (57929).............. | Private, first class, Company H, 28th Infantry, 1st Division.
Near St. Mihiel, France, Sept. 12, | He displayed excellent courage in capturing a machine gun, which he turned
1918. | upon an adjoining trench, forcing 20 of the enemy to surrender.
R—Brooklyn, N. Y. | Posthumously awarded. Medal presented to father, Henry Kopp.
B—Brooklyn, N. Y.
G. O. No. 46, W. D., 1919.

KORGIS, HERCULES E. (51858).......... | Sergeant, Company L, 23d Infantry, 2d Division.
Near Mont Blanc, Oct. 3-9, 1918, | While leading a small patrol in an endeavor to establish liaison with the French
and near Letanne, France, Nov. | troops on the right, during an attack, Sergeant Korgis's patrol was subjected
5, 1918. | to short-range machine-gun fire from 2 enemy guns. The fire halted
R—Worcester, Mass. | their advance, whereupon Sergeant Korgis designated a few men to fire upon
B—Turkey. | the enemy from the front, while he led the remainder in a flank attack upon
G. O. No. 46, W. D., 1920. | the enemy, charging the post, killing the enemy crew of 8 with grenades,
| and capturing their guns. On Nov. 5, when an attack section became dis-
| organized due to surprise fire from enemy machine guns, Sergeant Korgis
| fearlessly exposed himself to this fire and rushed to the "panic stricken"
| section and conducted them to cover, reorganized them, and subsequently
| led them in an attack upon the enemy position.

KORMAN, FRANK A.................. | Private, 16th Company, 5th Regiment, U. S. Marine Corps, 2d Division.
Near St. Etienne, France, Oct. 4-6, | When volunteers were called for to rescue another member of his company
1918. | who was severely wounded, he responded and in plain view of the enemy
R—Roxbury, Mass. | carried his wounded comrade to shelter through heavy machine-gun fire.
B—Roxbury, Mass.
G. O. No. 37, W. D., 1919.

KORN, WALTER S. (1981322)............
Near Bellicourt, France, Sept. 29, 1918.
R—Leetonia, Ohio.
B—Holmesville, Ohio.
G. O. No. 37, W. D., 1919.

Sergeant, Company G, 120th Infantry, 30th Division.
He continued to lead his platoon in attack on the Hindenberg line after he had received a wound from shrapnel. He was later knocked down by a rock thrown by a shell explosion, twice more wounded by shrapnel, but continued to lead his platoon until he received a severe wound, which necessitated his evacuation. He personally captured 2 prisoners in the attack.

KORTH, HERMAN (2305645)............
Near Juvigny, France, north of Soissons, Aug. 31, 1918.
R—Fond du Lac, Wis.
B—Germany.
G. O. No. 20, W. D., 1919.

Sergeant, Company D, 121st Machine Gun Battalion, 32d Division.
Under heavy fire from machine guns and artillery, he crawled to the crest of a hill, setting stakes to line our artillery on enemy machine-gun emplacements. He remained in observation in this perilous position for half an hour, signaling back when our own troops were endangered by the fire of the batteries.

*KORZYSKO, GEORGE (1389643)............
Near Forges, France, Sept. 26, 1918.
R—Chicago, Ill.
B—Russia.
G. O. No. 37, W. D., 1919.

Private, Company H, 132d Infantry, 33d Division.
During the action in Forges Wood, he, single handed, attacked and captured several machine guns, killing the gunners with hand grenades. It was while he was thus engaged that he was killed.
Posthumously awarded. Medal presented to brother, Mike Korzysko.

KOS, MAX S. (2004446)............
Near St. Thibaut, France, Aug. 8–9, 1918.
R—Indianapolis, Ind.
B—Fort Wayne, Ind.
G. O. No. 15, W. D., 1919.

Private, Company K, 47th Infantry, 4th Division.
He volunteered to patrol the valley along the railroad tracks north of St. Thibaut for the purpose of locating machine-gun nests. He was wounded early in the morning, but remained in the valley until the next night, securing the information for which he was sent, and killing 2 Germans.

KOSTAK, FRANK J. (1387719)............
At Chipilly Ridge, France, Aug. 9, 1918.
R—Chicago, Ill.
B—Chicago, Ill.
G. O. No. 128, W. D., 1918.

Private, Company G, 131st Infantry, 33d Division.
Single handed, Private Kostak with great gallantry attacked a machine-gun position, capturing 2 machine guns and 7 prisoners.

KOUTS, WILLIAM M. (2257105)............
Near Gesnes, France, Sept. 26, 1918.
R—Camas, Wash.
B—Vancouver, Wash.
G. O. No. 37, W. D., 1919.

Sergeant, Company D, 361st Infantry, 91st Division.
Sergeant Kouts, together with 2 other soldiers, captured 3 enemy machine guns and 26 prisoners.

KOWALKOWSKI, LEO (2180133)............
Near Baulny, France, Sept. 28, 1918.
R—St. Louis, Mo.
B—St. Louis, Mo.
G. O. No. 95, W. D., 1919.

Private, Company E, 137th Infantry, 35th Division.
Having previously made several trips to the rear with wounded comrades under heavy shellfire, Private Kowalkowski, with another litter bearer, after making three attempts, succeeded in reaching their battalion commander, who lay wounded under heavy machine-gun fire, and carried him back to safety.

KOWASKI, LOUIS L. (56777)............
Near Cantigny, France, May 28–30, 1918.
R—Indiana Harbor, Ind.
B—Chicago, Ill.
G. O. No. 109, W. D., 1918.

Corporal, Company B, 28th Infantry, 1st Division.
He captured an enemy machine gun and aided materially in breaking up a counterattack by using it against his foes. In company with his captain he led the way to a dugout which he had located in capturing the machine gun and assisted in taking 9 prisoners.

KOZIKOWSKI, STANISLAW (1708153)......
Near Binarville, France, Oct. 2–7, 1918.
R—Brooklyn, N. Y.
B—Poland.
G. O. No. 37, W. D., 1919.

Private, Company C, 308th Infantry, 77th Division.
During the time when his company was isolated in the Argonne Forest and cut off from communication with friendly troops, Private Kozikowski, together with another soldier, volunteered to carry a message through the German lines, although he was aware that several unsuccessful attempts had been previously made by patrols, the members of which were either killed, wounded, or driven back. By his courage and determination he succeeded in delivering the message and brought relief to his battalion.

*KOZLOSKI, JOHN (2337870)............
Near Grand Ballois Farm, France, July 14–15, 1918.
R—Baltimore, Md.
B—Baltimore, Md.
G. O. No. 32, W. D., 1919.

Private, Company A, 4th Infantry, 3d Division.
Private Kozloski repeatedly volunteered and delivered messages over routes where all previous runners had either been killed or wounded.
Posthumously awarded. Medal presented to father, Peter Kozloski.

*KRAFT, WILLIAM E. (1215872)............
Near Ronssoy, France, Sept. 29, 1918.
R—Weehawken, N. J.
B—Weehawken, N. J.
G. O. No. 46, W. D., 1920.

Sergeant, Company M, 108th Infantry, 27th Division.
In the attack on the Hindenburg line, after all the officers of his company became casualties, he took command and led his company in the attack through machine-gun and artillery fire. Shortly after returning from a personal reconnaissance of the enemy machine-gun position, he again took up the advance, and was leading the attack when killed by an enemy sniper. The heroism displayed by this noncommissioned officer was an important factor in the success of the attack of his company.
Posthumously awarded. Medal presented to father, Charles C. Kraft.

KRAMER, HENRY G. (1210300)............
Near Ronssoy, France, Sept. 29, 1918.
R—New York, N. Y.
B—New York, N. Y.
G. O. No. 20, W. D., 1919.

Corporal, Company D, 107th Infantry, 27th Division.
During the operations against the Hindenburg line Corporal Kramer, with 4 other soldiers, left shelter and went forward into an open field under heavy shell and machine-gun fire and succeeded in bandaging and carrying back to our lines two wounded men.

KRAMPS, CLARENCE O. (1038528)............
Near St. Eugene, France, July 14–15, 1918.
R—Rosedale, Kans.
B—St. Louis, Mo.
G. O. No. 32, W. D., 1919.

Corporal, Battery C, 10th Field Artillery, 3d Division.
He remained on duty as chief of section after being very severely wounded. Later in the action, when one of his gun crew was wounded, he was lifted to the seat and continued to fire the piece in addition to his other duties until forcibly taken from the seat and sent to an aid station. Here it was found that he had been wounded in 4 places.

KRAUSE, JOSEPH (1786771)............
Near Consenvoye Woods, France, Nov. 6, 1918.
R—New Rochelle, N. Y.
B—New Rochelle, N. Y.
G. O. No. 9, W. D., 1923.

Sergeant, sanitary detachment, 316th Infantry, 79th Division.
During the Meuse-Argonne, while Hill 378, north of Verdun, was being stormed by the 316th Infantry, Sergeant Krause, himself in the foremost rank, rushed forward in the face of heavy fire from a machine-gun nest and rescued successively 3 wounded soldiers who had fallen in the attempt to take the hill bringing each in turn to safety. His actions were in disregard of his own safety and were an inspiration to his comrades.

KRAUSE, WILLIAM (1332)..............
Near Cierges, France, Aug. 1, 1918.
R—Saginaw, Mich.
B—Galecia, Mich.
G. O. No. 9, W. D., 1923.

Private, medical detachment, 125th Infantry, 32d Division.
The battalion to which he was attached, forced by vastly superior numbers to take up a more advantageous position, left 21 wounded men well in advance of their second position. Private Krause, together with Lieut. Warde B. Smith, Medical Corps, and Sergt. John W. Doyle, medical detachment, under observation of the enemy and in the face of intense enemy machine-gun fire carried the wounded men to a place of safety. The indomitable spirit and extreme devotion to his comrades inspired the men of his organization with renewed courage and determination.

KRAUSE, WILLIAM H. (1213849)........
Near Roussoy, France, Sept. 29, 1918.
R—Syracuse, N. Y.
B—Syracuse, N. Y.
G. O. No. 20, W. D., 1919.

First sergeant, Company C, 108th Infantry, 27th Division.
He displayed great gallantry during the operations against the Hindenburg line. A smoke barrage was put down by the enemy between his company and the company on the left. The company commander having been wounded, Sergeant Krause sent a soldier to establish liaison with the company on the left. When this soldier was severely wounded and unable to accomplish his mission, Sergeant Krause went himself and succeeded in reaching the company. While returning to his own company Sergeant Krause met a party of Germans attempting to break through the gap between the two companies. In a personal encounter he killed a German officer and forced the rest of the party to withdraw.

KRAYER, NICHOLAS (1210537)..........
Near Roussoy, France, Sept. 29, 1918.
R—New Providence, N. J.
B—New York, N. Y.
G. O. No. 68, W. D., 1920.

Corporal, Company E, 107th Infantry, 27th Division.
During the attack on the Hindenburg line, after all officers in his company had been killed, Corporal Krayer exposed himself to heavy machine-gun fire to reorganize his command. He went from shell hole to shell hole, and by his courageous example inspired his men to continue the advance.

KREGER, EDWARD A.................
Between Los Banos and Bay Laguna, P. I., Mar. 10, 1900.
R—Cherokee, Iowa.
B—Keota, Iowa.
G. O. 108, W. D., 1919.
Distinguished-service medal also awarded.

Captain, 39th Infantry, U. S. Volunteers.
For extraordinary heroism in an engagement with an armed enemy between Los Banos and Bay Laguna, P. I., on Mar. 10, 1900.

*KREIS, JACOB (2024430)................
At St. Thibaut, France, Aug. 10, 1918.
R—Sheboygan, Wis.
B—Russia.
G. O. No. 147, W. D., 1918.

Private, Company I, 47th Infantry, 4th Division.
Accompanied by another soldier, he penetrated the enemy's lines and patrolled a sector from the north bank of the River Vesle to the town of Bazoches. These two men entered an enemy dugout and killed two Germans, at the same time locating a machine-gun emplacement.
Posthumously awarded. Medal presented to Henry Gross, administrator, for delivery to the next of kin.

KREITZER, DAVID I. (1246736)..........
Near Fismes, France, Aug. 7-8, 1918.
R—Mechanicsburg, Pa.
B—Pennsylvania.
G. O. No. 11, W. D., 1921.

Private, Machine Gun Company, 112th Infantry, 28th Division.
After being severely wounded in the right shoulder, Private Kreitzer refused immediate evacuation and continued to perform his duty as company runner. Upon several occasions, he exposed himself to heavy enemy fire in order to carry messages from company headquarters to his platoon. His gallant conduct aided materially in keeping up the morale and courage of his comrades.

KRESS, JOHN A....................
Near Umatilla, Oreg., July 8, 1878.
R—Elmira, N. Y.
B—Tioga County, Pa.
G. O. No. 7, W. D., 1925.

Captain, Ordnance Department, U. S. Army.
Captain Kress volunteered, though not charged with combat duties, to organize and lead an expedition against bands of hostile Piute-Bannock Indians and to prevent their crossing the Columbia River. Captain Kress seized a river boat and equipped it as a gunboat by his gallant and fearless leadership in patrolling the Columbia River was successful in five attacks upon the Indians, both on land and from the river; succeeded in capturing and killing their horses, destroyed their boats, arms, and ammunition, and camp equipage, thus frustrating their plan of spreading the war among the Indians to the North.

KREUZMAN, WILLIAM A..............
Near Bayonville, France, Nov. 1, 1918.
R—Batesville, Ind.
B—Georgetown, Ky.
G. O. No. 44, W. D., 1919.

Private, 82d Company, 6th Regiment, U. S. Marine Corps, 2d Division.
Private Kreuzman volunteered and went forward to reconnoiter a ravine which was infested with hostile machine guns, returning with several prisoners.

*KRIECHBAUM, PHILIP E...........
Near Apremont, France, Oct. 3, 1918.
R—Chambersburg, Pa.
B—Chambersburg, Pa.
G. O. No. 11, W. D., 1921.

Second lieutenant, 112th Infantry, 28th Division.
When his company was halted by heavy machine-gun fire, Lieutenant Kriechbaum advanced ahead of his company, thus personally leading them to the attack. He was killed by enemy machine-gun fire while some distance ahead of the first wave. His initiative and heroism were material factors in this operation.
Posthumously awarded. Medal presented to mother, Mrs. Lucy E. Kriechbaum.

KRIECHBAUM, ROY R.
Near Fismette, France, Aug. 9, 1918.
R—Chambersburg, Pa.
B—Chambersburg, Pa.
G. O. No. 11, W. D., 1921.

Captain, 112th Infantry, 28th Division.
When the town of Fismette was held both by the enemy and our forces, Captain Kriechbaum voluntarily exposed himself to heavy close-range machine-gun fire in order to rescue a wounded soldier. While in the performance of this heroic act, he was severely wounded in both legs.

KRIGBAUM, WILLIAM L.
Near Bois-de-Chaume, France, Oct. 9, 1918.
R—Decatur, Ill.
B—Decatur, Ill.
G. O. No. 64, W. D., 1919.

Captain, 124th Machine Gun Battalion, 33d Division.
When the battalion to which Captain Krigbaum was attached had reached its objective north of Bois-de-Chaume, it was subject to two counterattacks. The right flank of the battalion was left exposed and all the gun crews on that flank were either killed or wounded and the guns put out of action. At the most critical time of this emergency, Captain Krigbaum alone mounted a captured German machine gun and so successfully operated it against the enemy that the counterattack was stopped and the flank of the battalion saved from serious losses.

KROGER, CHESTER F.
Near Soissons, France, July 19, 1918.
R—Cincinnati, Ohio.
B—Newport, Ky.
G. O. No. 72, W. D., 1920.

Second lieutenant, 26th Infantry, 1st Division.
While directing the attack of his platoon against strong enemy resistance beyond the Paris-Soissons Highway, Lieutenant Kroger was severely wounded in the head by a machine-gun bullet. He refused immediate evacuation, and although staggering from the loss of blood he assisted in the reorganization of the various units of the 2d Battalion and continued with the advance until he fell from exhaustion.

KROMER, CHESTER H. (280958)
Near Juvigny, France, Aug. 30, 1918.
R—Grand Rapids, Mich.
B—Grand Rapids, Mich.
G. O. No. 66, W. D., 1919.

Corporal, Company K, 128th Infantry, 32d Division.
Corporal Kromer voluntarily made 4 trips in front of the line under machine-gun and artillery fire and brought in 4 wounded men who had been left in an exposed position after a withdrawal of the line.

KROTOSHINSKY, ABRAHAM (1706855)
In Argonne Forest, France, Oct. 6, 1918.
R—Bronx, N. Y.
B—Poland.
G. O. No. 139, W. D., 1918.

Private, Company K, 307th Infantry, 77th Division.
He was on liaison duty with a battalion of the 308th Infantry which was surrounded by the enemy north of the Forest de la Buinonne in Argonne Forest. After patrols and runners had been repeatedly shot down while attempting to carry back word of the battalion's position and condition, he volunteered for the mission and successfully accomplished it.

KRUEGER, ARTHUR (2067288)
Near Consenvoye, France, Oct. 9, 1918.
R—Chicago, Ill.
B—Chicago, Ill.
G. O. No. 126, W. D., 1919.

Private, Company B, 131st Infantry, 33d Division.
While his platoon was halted by murderous fire of the enemy, he crawled from a shell hole in which he was seeking shelter and made his way forward to the aid of a wounded comrade. On the way he was wounded, but bravely continued on, until he had dressed the wounds of his comrade. He then insisted on walking to the dressing station for treatment of his own wounds.

KRUGER, ANTHONY J. (1709796)
Near Wilhelmplatz, France, Sept. 29, 1918.
R—Patchogue, N. Y.
B—Maspeth, N. Y.
G. O. No. 35, W. D., 1919

Sergeant, Company K, 308th Infantry, 77th Division.
He was ordered to take his platoon and capture a machine gun which was holding up the advance of the company and causing many casualties. Armed with an automatic pistol, he, without hesitation and with utter disregard for his personal safety, charged the machine gun, stopping only when he was rendered unconscious by two bullet wounds in the neck.

KRUM, JAMES EDWARD (1387247)
Near Hamel, France, July 4, 1918.
R—Chicago, Ill.
B—Chicago, Ill.
G. O. No. 14, W. D., 1920.

Sergeant, Company E, 131st Infantry, 33d Division.
Although severely wounded in the right arm at the beginning of the engagement, he continued forward as squad leader, exhibiting great gallantry and setting an inspiring example to his men. After his wound had been dressed he insisted upon returning to duty with his platoon.

KRZYKWA, ALBERT S. (281563)
Near Romagne, France, Oct. 14, 1918.
R—Alto, Mich.
B—Grand Rapids, Mich.
G. O. No. 21, W. D., 1919.

Private, Company M, 126th Infantry, 32d Division.
In an attack on Cote Dame Marie the 126th Infantry was held up, owing to intense enemy machine-gun fire and grenades. Private Krzykwa volunteered as a member of a combat patrol which cut through the enemy lines, capturing 10 machine guns, killed and captured 15 of the enemy, and forced a large number to surrender, clearing that part of the Cote Dame Marie of the enemy and enabling the regiment to continue their advance.

KUBACKI, STEVE (1099757)
Near Jaulny, France, Nov. 10, 1918.
R—Milwaukee, Wis.
B—Germany.
G. O. No. 32, W. D., 1919.

Sergeant, Company D, 55th Infantry, 7th Division.
After leading his platoon to its objective and having consolidated a defense position, Sergeant Kubacki alone went forward under heavy shellfire to reconnoiter the enemy's position. Finding the area clear, he moved his platoon 300 yards forward to a more favorable position, which he held despite the fierce fire of the enemy.

KUDER, HOWARD F. (107146)
Near Greves Farm, France, July 14–15, 1918.
R—Philadelphia, Pa.
B—India.
G. O. No. 44, W. D., 1919.

Sergeant, Battery F, 10th Field Artillery, 3d Division.
He volunteered to carry messages after communication had been broken off. Although wounded, he refused to be relieved, and continued making trips to the batteries throughout the heavy bombardment without thought for personal safety.

KUHLMAN, ALFRED H. (2039948)
Near Haumont, France, Oct. 11, 1918.
R—Rogers City, Mich.
B—Rogers City, Mich.
G. O. No. 27, W. D., 1919.

Private, 116th Ambulance Company, 103d Sanitary Train, 28th Division.
As a stretcher bearer he gave proof of great courage and high sense of duty by helping transport a wounded soldier to a dressing station under heavy fire, by which 3 other stretcher bearers were killed or seriously wounded. He was wounded himself, but he nevertheless returned to the shell-swept area and assisted in rescuing a wounded officer and 6 wounded soldiers.

KUHLMAN, GEORGE WINFIELD _____
    Near Fismes, France, Aug. 5–6, 1918.
    R—Glidden, Wis.
    B—Algoma, Wis.
    G. O. No. 99, W. D., 1918.

Second lieutenant, 107th Engineers, 32d Division.
He was sent on the night of Aug. 5–6, 1918, to make a reconnaissance of all possible means of crossing the River Vesle, near Fismes, France. It had been reported that the Germans had all retreated from the south bank of the river, but he found that such was not the case; they were there in force. Nevertheless, such was his bravery and determination that he crossed into and through the German lines, made a full reconnaissance, and returned with his report.

KUKOSKI, JOHN _____
    At Chateau-Thierry, France, June 6, 1918.
    R—Buffalo, N. Y.
    B—Russia.
    G. O. No. 110, W. D., 1918.

Private, 49th Company, 5th Regiment, U. S. Marine Corps, 2d Division.
Alone, he charged a machine gun and with the utmost bravery captured it and its crew, including an officer.

KUNIEAWSKY, BEN (105711) _____
    Near Exermont, France, Oct. 4, 1918.
    R—Dickinson, N. Dak.
    B—Russia.
    G. O. No. 19, W. D., 1920.

Private, first class, Company C, 2d Machine Gun Battalion, 1st Division.
Private Kunieawsky, with a companion, went forward in front of our lines to flank and locate the enemy machine-gun nests, whose fire was halting the advance of our Infantry. They were exposed to heavy machine-gun fire but pushed forward. His companion was mortally wounded, but Private Kunieawsky continued and located the enemy position and returned with information which enabled our forces to put nests out of action.

KUNZIE, HARRY K. (280706) _____
    Near Cierges, France, Aug. 1, 1918.
    R—Big Rapids, Mich.
    B—Michigan.
    G. O. No. 74, W. D., 1919.

Corporal, Company I, 126th Infantry, 32d Division.
He crawled out in advance of his company and, single handed, killed the 6 Germans forming the crew of a machine gun. He then continued forward to silence another gun, being mortally wounded in the attempt. Posthumously awarded. Medal presented to father, William G. Kunzle.

KURLE, CHRISTIAN (3137463) _____
    Near Oches, France, Nov. 4, 1918.
    R—Forsyth, Mont.
    B—New York, N. Y.
    G. O. No. 46, W. D., 1919.

Private, Company H, 307th Infantry, 77th Division.
Exposing himself to heavy machine-gun fire, Private Kurle crossed an open field 300 yards wide and rescued a severely wounded comrade.

KWASIGROCH, PAUL J _____
    Near Bois-de-Chatillon, France, Nov. 5–6, 1918.
    R—Milwaukee, Wis.
    B—Linden Station, Wis.
    G. O. No. 32, W. D., 1919.

Second lieutenant, 6th Infantry, 5th Division.
Despite severe leg wounds, he remained on duty with his company throughout the entire action, refusing treatment until his objective had been reached and the remainder of his company reorganized and a liaison established. On the following day, after first aid had been given, he returned immediately to duty, and during the remaining advances commanded in a most skillful and courageous manner, he being the only officer left with the organization.

KYLE, ERNEST C. (880907) _____
    Near Haumont, France, Oct. 11, 1918.
    R—Portland, Oreg.
    B—Acton, Kans.
    G. O. No. 27, W. D., 1919.

Private, 116th Ambulance Company, 104th Sanitary Train, 29th Division.
As a stretcher bearer he gave proof of great courage and high sense of duty by helping transport a wounded soldier to a dressing station under heavy enemy fire, by which 3 other stretcher bearers were killed or seriously wounded. He repeatedly returned to the shell-swept area and assisted in rescuing the wounded.

LA BUHN, EDMUND F _____
    Near Brabant, France, Oct. 8–9, 1918.
    R—Detroit, Mich.
    B—Detroit, Mich.
    G. O. No. 37, W. D., 1919.

Second lieutenant, 116th Infantry, 29th Division.
Becoming detached from his regular organization on Oct. 8, Lieutenant La Buhn, accompanied by 5 soldiers, attached himself to another company and voluntarily took part in the offensive operations. Leading his men, he charged a machine-gun nest and captured several guns and 4 prisoners. He continued in action with this company until wounded on Oct. 9, 1918.

LACOSSE, LAWRENCE J. (1211766) _____
    East of Ronssoy, France, Sept. 29, 1918.
    R—Malone, N. Y.
    B—Burk, Vt.
    G. O. No. 20, W. D., 1919.

Private, Company K, 107th Infantry, 27th Division.
Private Lacosse, with 3 other soldiers, went out into an open field under heavy shell and machine-gun fire and succeeded in carrying back to our lines 4 seriously wounded men.

LA CROIX, ORIE H. (1683636) _____
    Near Binarville, France, Oct. 1, 1918.
    R—Bridgeport, Conn.
    B—Canada.
    G. O. No. 87, W. D., 1919.

Corporal, Company A, 308th Infantry, 77th Division.
When his company commander and first sergeant had been wounded, he rallied the company and continued the advance, fearlessly exposing himself to hostile fire and inspiring the men with him by his courage.

LA FORD, RUSSELL E. (1215451) _____
    East of Ronssoy, France, Sept. 29, 1918.
    R—North Tonawanda, N. Y.
    B—Vermilion, Mich.
    G. O. No. 20, W. D., 1919.

Private, Company K, 108th Infantry, 27th Division.
During the operations against the Hindenburg line he left shelter and went out into the open under heavy shell and machine-gun fire and succeeded in bandaging and carrying back to our lines a wounded officer.

LAGASSEY, NAPOLEON F. (51686) _____
    Near St. Etienne-a-Arnes, France, Oct. 3–9, 1918.
    R—North Oxford, Mass.
    B—Brooklyn, N. Y.
    G. O. No. 37, W. D., 1919.

Private, first class, Company K, 23d Infantry, 2d Division.
While carrying a message he was wounded in both legs and in the face, but delivered his message to the proper place before he collapsed from loss of blood.

LAGEAISE, STANLEY, Jr. (56046) _____
    At Cantigny, France, May 28–30, 1918.
    R—Duluth, Minn.
    B—Willow River, Minn.
    G. O. No. 99, W. D., 1918.

Corporal, Headquarters Company, 28th Infantry, 1st Division.
During the 3 days of fighting at Cantigny, France, May 28–30, 1918, he worked bravely without thought of himself to maintain lines in working condition. He was almost constantly under heavy fire, but fearlessly went into it whenever necessary and thereby aided materially in the success of the regiment's enterprise.

LA GROW, ELGIA (2852473) ....
Near Fey-en-Haye, France, Sept. 12, 1918.
R—Gladstone, Colo.
B—Morrison, Colo.
G. O. No. 81, W. D., 1919.

Private, Company A, 357th Infantry, 90th Division.
Private La Grow displayed an admirable quality of courage in always being the first to venture forth in an endeavor to wipe out obstacles in his company's advance. He captured without aid 3 enemy machine guns on different occasions.

LAIMINGER, ALFRED S. (240047) ....
Near Bois-de-St. Remy, France, Sept. 12, 1918.
R—Sopris, Colo.
B—St. Louis, Mo.
G. O. No. 87, W. D., 1919.

Private, Company H, 103d Infantry, 26th Division.
When his company was held up by heavy machine-gun fire, he circled through the woods and opened fire with his automatic rifle upon the enemy machine-gun emplacement, forcing the enemy crew to retire, leaving their gun in position.

LAIRD, FRED L. (1453036) ....
At Hilsenfirst, Alsace, July 6, 1918.
R—Hamburg, Ill.
B—Hamburg, Ill.
G. O. No. 55, W. D., 1920.

Private, first class, Company H, 138th Infantry, 35th Division.
While acting as rear guard for a raiding party he saw 2 of the enemy who had escaped the attention of the raiding party preparing to cut off the retreat of the party. In spite of the enemy machine-gun and artillery fire he remained behind to cover the retreat of the party and by his rifle fire killed 2 of the enemy who were preparing to cut off the retreat of the party. A few minutes later, with a grenade, he killed 1 of the enemy who had attacked 2 unarmed litter bearers. By his prompt deed he saved the lives of these 2 men.

*LAIT, HENRY A. (68386) ....
Near St. Remy, France, Sept. 12, 1918.
R—Old Town, Me.
B—Bangor, Me.
G. O. No. 27, W. D., 1919.

Private, first class, Company G, 103d Infantry, 26th Division.
Private Lait, with 2 other comrades, advanced into the open and fired an automatic rifle on an enemy machine-gun emplacement, thereby drawing the machine-gun fire to themselves and enabling the platoon, which had been exposed to an enfilading fire, to flank the gun and put it out of action. Private Lait was killed by a machine-gun bullet during the action. Posthumously awarded. Medal presented to father, Hyman Lait.

LA JENNESSEE, JOSEPH P. (2382994) ....
Near Cunel, France, Oct. 14, 1918.
R—Mahnomen, Minn.
B—Mahnomen, Minn.
G. O. No. 37, W. D., 1919.

Sergeant, Company D, 60th Infantry, 5th Division.
He retained the command of his platoon after he had received a severe gunshot wound in the leg, maintained the organization of his platoon under heavy fire, and directed it in the overcoming of several machine-gun positions. He consolidated his position on the line held by the company and remained on post 36 hours until ordered evacuated on account of his wound.

LAKE, CLARENCE W. (1655959) ....
Near Gesnes, France, Oct. 5, 1918.
R—Topsfield, Mass.
B—Swampscott, Mass.
G. O. No. 66, W. D., 1919.

Corporal, Company F, 127th Infantry, 32d Division.
With 2 other soldiers Corporal Lake advanced ahead of their company and rushed an enemy machine-gun nest from the flank, capturing 14 members of the crew and 2 machine guns, and thereby enabling the advance to continue. Carrying the captured guns with them to the objective, they later used them successfully in repelling a hostile counterattack.

LAKE, JAMES (1981635) ....
Near Bellicourt, France, Sept. 29, 1918.
R—Kingston, Ky.
B—Richmond, Ky.
G. O. No. 37, W. D., 1919.

Private, Company B, 126th Infantry, 30th Division.
With 8 other soldiers, comprising the company headquarters detachment, Private Lake assisted his company commander in cleaning out enemy dugouts along a canal and capturing 242 prisoners.

LAMB, EARL (1391257) ....
Near Consenvoye, France, Oct. 9, 1918.
R—Chicago, Ill.
B—Chicago, Ill.
G. O. No. 46, W. D., 1919.

Corporal, Company F, 132d Infantry, 33d Division.
When the advance of his platoon was stopped by an enemy machine gun, he charged the gun from the flank, wounded 1 of the gunners, and captured the other 2 members of the gun crew, together with the gun. Remaining in an advanced position under fire throughout the day, he used the captured machine gun in breaking up an enemy counterattack.

LAMB, JOHN R. (252120) ....
Near Juvigny, north of Soissons, France, Aug. 28–Sept. 4, 1918.
R—Rockford, Mich.
B—Rockford, Mich.
G. O. No. 20, W. D., 1919.

Sergeant, Company C, 107th Field Signal Battalion, 32d Division.
When heavy shelling and machine-gun fire destroyed telephone communication, Sergeant Lamb, without orders and upon his own initiative, went out and repeatedly patrolled the line, making repairs and reestablishing communication. Thereafter he continued to patrol the lines, constantly improving connections and placing the line in protected places. He worked indefatigably day and night during a period of seven days, and refused to rest or seek a place of safety while he could find work to do. When he saw the body of his brother who had been killed in action, he did not falter, but remained on duty as before. He was finally forced to go to a hospital by reason of complete exhaustion. By his bravery and devotion to duty he so distinguished himself as to become the object of admiration by brother officers and soldiers.

LAMB, ROBERT J ....
Near Bellicourt, France, Sept. 29, 1918.
R—Fayetteville, N. C.
B—Fayetteville, N. C.
G. O. No. 81, W. D., 1919.

Captain, 119th Infantry, 30th Division.
In command of a company he, with 2 other men, rushed a machine-gun post which was holding up the advance, killing the German crew. Later, separated from part of his command, owing to a dense smoke screen, he found himself with a few men in front of 3 German machine-gun nests. Leading the attack, he captured the enemy positions with 25 prisoners.

LAMBERT, JOHN H ....
Near Stenay, France, Oct. 30, 1918.
R—Cambridge, Mass.
B—Salem, N. J.
G. O. No. 13, W. D., 1919.

First lieutenant, 91st Aero Squadron, Air Service.
While on a photographic mission in the vicinity of Stenay, his work being seriously interfered with by the fire of a formation of enemy planes, he temporarily discontinued his mission, attacked the formation, and dispersed it, destroying 1 plane and seriously damaging another. He then returned to his objective, completed his mission, and returned with information of great military value.

*LAMBING, FLOYD C. (2470475)_____
Near Nantillois, France, Oct. 10, 1918.
R—Apollo, Pa.
B—Salina, Pa.
G. O. No. 37, W. D., 1919.

Private, Company A, 320th Infantry, 80th Division.
Private Lambing, when his company was suddenly pulled back, went forward through both friendly and enemy shellfire to an advanced post and directed the occupants of the outpost to safety, although he was killed in the act. Posthumously awarded. Medal presented to mother, Mrs. Frank J. Lambing.

LAMM, JOHNNIE (1320902)_____
Near Bellicourt, France, Sept. 29, 1918.
R—Lucama, N. C.
B—Lucama, N. C.
G. O. No. 37, W. D., 1919.

Private, Company G, 120th Infantry, 30th Division.
In the face of heavy machine-gun fire, Private Lamm, with 2 other soldiers, attacked and put out of action an enemy machine-gun post, capturing a German officer and 3 soldiers.

LA MORDER, HENRY C. (92126)_____
Near Buzancy, France, Oct. 16, 1918.
R—Akron, Ohio.
B—Salisbury, Vt.
G. O. No. 23, W. D., 1919.

Sergeant, Headquarters Company, 166th Infantry, 42d Division.
While his platoon was taking shelter from the withering machine-gun fire, Sergeant La Morder heard cries of a wounded comrade in a near-by shell hole. Braving the deadly machine-gun fire, he went to the soldier's assistance, bound up his wounds, and, when attempting to carry him to safety, was himself severely wounded. He started to crawl to the dressing station, but became exhausted after going a very short distance.

LAMPMAN, BRYAN (431676)_____
Near Moulin de Guenoville, France, Sept. 26, 1918.
R—Quincy, Mich.
B—Quincy, Mich.
G. O. No. 35, W. D., 1920.

Corporal, Company F, 1st Gas Regiment.
Corporal Lampman, with 3 other soldiers, advanced nearly 200 yards over an open hillside exposed to machine-gun fire and carried 2 wounded men to the protection of a near-by trench.

*LAMSON, DWIGHT F. (2189236)_____
Near Limey, France, Sept. 12, 1918.
R—Little River, Kans.
B—Goodland, Ind.
G. O. No. 89, W. D., 1919.

Private, Company G, 353d Infantry, 89th Division.
Seeing his lieutenant fall severely wounded, Private Lamson, with another stretcher bearer, rushed through severe machine-gun fire to his assistance. When they had placed the lieutenant on their stretcher and were endeavoring to go through the heavy fire to the dressing station, Private Lamson fell mortally wounded.
Posthumously awarded. Medal presented to father, Charles E. Lamson.

LANDES, WILLIAM S_____
North of Samogneux, France, Oct. 11–25, 1918.
R—Lansdowne, Pa.
B—Sheldon, Ill.
G. O. No. 15, W. D., 1919.

First lieutenant, 113th Infantry, 29th Division.
Acting as regimental munitions officer, he was wounded early in the action, but remained on duty for two weeks, supervising the distribution of ammunition in the front-line battalions. Through his untiring efforts and utter disregard of personal danger under heavy shellfire an adequate supply of ammunition to the battalions in the line was constantly maintained. After the attack on the Bois d'Etrayes, Oct. 24, when liaison with the advance battalion had been broken by the intense bombardment, this officer volunteered to proceed to the position occupied by this battalion and returned with valuable information as to the location of the front line.

LANDON, HAROLD M_____
Near Sechault, France, Sept. 29, 1918.
R—New York, N. Y.
B—New York, N. Y.
G. O. No. 37, W. D., 1919.

First lieutenant, 369th Infantry, 93d Division.
Lieutenant Landon, on duty as assistant liaison officer, personally carried an order to the assault battalion in order to insure its delivery, passing through heavy fire. The battalion commander being killed just as he arrived, Lieutenant Landon gave the order to the next senior, and then waited to see its execution. When the assaulting line wavered under a terrific enemy barrage, this officer jumped ahead of the line and led the first wave 1,000 meters to the objective, assisting in consolidating the new position before he returned to regimental headquarters.

LANDEY, GERARD P. (106254)_____
Near Soissons, France, July 19, 1918.
R—Dutch Town, La.
B—Dutch Town, La.
G. O. No. 99, W. D., 1918.

Sergeant, Company B, 3d Machine Gun Battalion, 1st Division.
When his platoon commander was incapacitated by wounds, July 19, 1918, near Soissons, France, he displayed instant initiative, effectively took command of his platoon, and directed its movements with marked ability and courage during the remaining three days of the advance.

LANDSTREET, ROBERT S_____
Near Bois-de-Consenvoye and Bois-de-Grande-Montague, France, Oct. 8–16, 1918.
R—Baltimore, Md.
B—Baltimore, Md.
G. O. No. 37, W. D., 1919.

First lieutenant, 115th Infantry, 29th Division.
On Oct. 8 he led his platoon through machine-gun and rifle fire in an advance which resulted in the capture of 360 prisoners and 12 machine guns. On the morning of Oct. 16 he volunteered, with 1 sergeant, and straightened out the line of an adjacent unit. His movements were under constant machine-gun fire, and so close to the enemy that he, with his sergeant, captured 2 prisoners while accomplishing their mission.

LANE, ELMER L. (72414)_____
At Bois Brule, France, Apr. 13, 1918.
R—West Somerville, Mass.
B—Woonsocket, R. I.
G. O. No. 99, W. D., 1918.

Private, Company F, 104th Infantry, 26th Division.
For coolness and gallantry in action on Apr. 13, 1918. Standing on parapet of trench in order to see advancing enemy through heavy fog, he continued, under heavy fire, to throw hand grenades at enemy until severely wounded, thus preventing enemy from penetrating line in vicinity of his post.

LANE, HERSCHEL V_____
Near Vierzy, France, July 20, 1918.
R—Minneapolis, Minn.
B—Omaha, Nebr.
G. O. No. 137, W. D., 1918.

Private, 77th Company, 6th Machine Gun Battalion, U. S. Marine Corps, 2d Division.
He volunteered and successfully carried messages from Vierzy to the front line near Tigny, through heavy artillery and machine-gun fire, after 2 others who had been detailed for the duty had failed to get through.

LANE, LESLIE M. (64075)_____
Near Seicheprey, France, Apr. 4–5, 1918.
R—Poughquag, N. Y.
B—Kent, Conn.
G. O. No. 129, W. D., 1918.

Private, first class, Company C, 102d Infantry, 26th Division.
On the night of Apr. 4–5, 1918, he was carrying rations to the men in the front trenches. He encountered a large enemy platoon who demanded his surrender. He refused to surrender, drew his pistol, and killed the enemy platoon commander, causing the enemy raiding party to retire. During the encounter he was severely wounded by hand grenades. By his quickness of action he undoubtedly saved the lives of the men in our advanced listening post.

LANERGAN, JOHN F. (600143)_____
    Near Verdun, France, Oct. 23, 1918.
    R—Dorchester, Mass.
    B—Roxbury, Mass.
    G. O. No. 46, W. D., 1919.

Private, Company B, 101st Infantry, 26th Division.
While engaged as runner during the attack on Houppy Bois on Oct. 23, he made repeated trips over an area swept by machine-gun and artillery fire. On Oct. 27, after all his superior officers had become casualties, he assembled scattered combat groups, and, after reorganizing them, led them in a successful counterattack against the enemy.

LANGDON, RUSSELL C_____
    Near Fismes, France, Aug. 5, 1918.
    R—Brooklyn, N. Y.
    B—Brooklyn, N. Y.
    G. O. No. 46, W. D., 1920.
    Distinguished-service medal also
        awarded.

Colonel, 127th Infantry, 32d Division.
After a patrol had reported to him that a bridge crossing of the Vesle could not be located due to heavy enemy machine-gun and rifle fire, Colonel Langdon personally led a patrol through an area covered by enemy shell and machine-gun fire, to the Vesle River and selected a suitable spot for the building of a bridge and gave instructions for the selection of material for construction.

LANGE, CARL M. (105472)_____
    Near Fleville, France, Oct. 5, 1918.
    R—Hartington, Nebr.
    B—Wall Lake, Iowa.
    G. O. No. 44, W. D., 1919.

Private, first class, Company B, 2d Machine Gun Battalion, 1st Division.
Seeing that his first line was being held up by machine-gun fire from the woods, Private Lange, with another soldier, voluntarily made his way through a terrific enemy barrage and entered the woods, cleared out 3 machine guns, killed several of the crew, and captured about 20 prisoners. Failing in his attempt to communicate the success of his mission to the attacking wave, he himself went back, and, finding all his officers had become casualties, assisted in organizing a small force and leading it to the objective.

LANGEMAK, FRITZHOF G. (284261)_____
    Near Exermont, France, Oct. 4, 1918.
    R—Sturgeon Bay, Wis.
    B—Milwaukee, Wis.
    G. O. No. 9, W. D., 1923.

Corporal, Company M, 28th Infantry, 1st Division.
In the absence of any commissioned officer he assumed command of the platoon of which he was a member when it was in serious danger because of lack of leadership. With utter disregard for personal safety, subjected to intense machine-gun fire, he coolly reorganized the men about him, most of whom were replacements who had never been in action. Although lost and without maps he led his men in vigorous attacks against the enemy forces and well placed machine-gun nests, eventually reaching his objectives where he reestablished liaison with his flanks and continued the advance. By his fearlessness and aggressiveness he prevented the enemy from gaining advantage of a gap which was starting to open in our attacking line, and by his initiative aided the advance of the troops on his flanks.

LANGFORD, JULIUS A. (1319446)_____
    Near St. Souplet, France, Oct. 17–19,
        1918.
    R—Swepsonville, N. C.
    B—Alamance County, N. C.
    G. O. No. 81, W. D., 1919.

Private, Company A, 120th Infantry, 30th Division.
Being a company runner, he displayed marked bravery, repeatedly crossing heavily-shelled areas and exposing himself to machine-gun fire to deliver important messages, enabling his company to maintain adequate liaison.

LANGHAM, GEORGE W. (2311110)_____
    Near Juvigny, north of Soissons,
        France, Aug. 29–Sept. 2, 1918.
    R—Roaring Spring, Pa.
    B—Puzzletown, Pa.
    G. O. No. 20, W. D., 1919.

Private, Company H, 128th Infantry, 32d Division.
Though he had been severely gassed, he remained on duty with his company while it was in the front line. Later, when it was in support, he voluntarily aided in the work of carrying wounded across an area covered by artillery and machine-gun fire.

LANGLEY, JOHN H. (2405820)_____
    Near Chevieres, France, Oct. 21,
        1918.
    R—Newfield, N. J.
    B—Pottstown, N. J.
    G. O. No. 27, W. D., 1919.

Private, medical detachment, attached to Company G, 311th Infantry, 78th Division.
He remained on duty continuously for 4 hours administering aid to wounded men under heavy shellfire. Finding that he could not properly work while wearing his gas mask, he removed it, though many gas shells were bursting in his vicinity. After being gassed, he continued to work for an hour until all the wounded were attended.

LANGSTON, LUTHER J. (1528479)_____
    Near Ivoiry, France, Sept. 27, 1918.
    R—Covington, Ohio.
    B—Covington, Ohio.
    G. O. No. 11, W. D., 1921.

First sergeant, Company A, 148th Infantry, 37th Division.
When his platoon was halted by the fire of concealed enemy machine guns, Sergeant Langston dashed ahead of his organization and, alone, captured the enemy machine gun, forcing 5 of the enemy to surrender. This act of heroism enabled his organization to resume the advance.

*LANIGHAN, MATTHEW S. (1735646)_____
    Near St. Juvin, France, Oct. 16,
        1918.
    R—Lockport, N. Y.
    B—Lockport, N. Y.
    G. O. No. 87, W. D., 1919.

Sergeant, Company I, 309th Infantry, 78th Division.
Although wounded, he refused to be evacuated and led his men with marked personal courage, capturing several enemy machine guns and prisoners. He was killed while organizing his platoon for a further advance.
Posthumously awarded. Medal presented to father, Edward Lanighan.

LAPEAN, FRED W. (543305)_____
    During the Meuse-Argonne offen-
        sive, north of Cierges, France,
        Sept. 30, 1918.
    R—Holyoke, Mass.
    B—Turners Falls, Mass.
    G. O. No. 47, W. D., 1921.

Sergeant, medical detachment, 7th Infantry, 3d Division.
Under observation of the enemy and subjected to heavy machine-gun fire, he, of his own initiative, worked his way from shell hole to shell hole in order to give first aid to 3 wounded men who were lying in an open field next to the Cierges-Romagne Road.

LARKIN, MICHAEL (58203)_____
    Near Soissons, France, July 18,
        1918.
    R—Jersey City, N. J.
    B—New York, N. Y.
    G. O. No. 72, W. D., 1920.

Corporal, Company I, 28th Infantry, 1st Division.
When his platoon had been halted by heavy machine-gun fire from the front, Corporal Larkin, with 3 others, pushed forward through heavy fire and attacked the enemy machine-gun nest. Two enemy machine guns were captured and their crews forced to surrender. He then reorganized the enemy position and assisted in the repulse of a strong enemy counterattack. He served with distinction until a serious wound forced his evacuation.

LARNER, GORMAN DE FREEST_____
    In the region of Champeny, France,
      Sept. 13, 1918.
    R—Washington, D. C.
    B—Washington, D. C.
    G. O. No. 145, W. D., 1918.

First lieutenant, pilot, 103d Aero Squadron, Air Service.
He attacked an enemy patrol of 6 machines (Fokker type) and fought against odds until he had destroyed 1 and forced the others to retire.
Oak-leaf cluster.
A bronze oak leaf, for extraordinary heroism in action in the region of Montfaucon, France, Oct. 4, 1918. While leading a patrol of 4 monoplanes, he led his patrol in an attack on an enemy formation of 7 planes. By skillfully maneuvering he crushed 1 of the enemy machines and with the aid of his patrol forced the remainder of the enemy formation to withdraw.

LARSON, COLONEL (282986)_____
    Near Juvigny, France, Sept. 1–3,
      1918.
    R—Neillsville, Wis.
    B—Lewis, Wis.
    G. O. No. 98, W. D., 1919.

Corporal, Headquarters Company, 128th Infantry, 32d Division.
Corporal Larson constantly patrolled the telephone lines in front of Juvigny, repairing the lines under heavy artillery and machine-gun fire. Near Ecurey, when the enemy artillery fire was so intense as to make telephonic communications impossible, Corporal Larson, on duty as a lineman, volunteered as a runner, and served as such under extremely heavy fire until completely exhausted.

LARSON, JAMES A. (1783289)_____
    Near Malancourt, France, Sept. 26,
      1918.
    R—Blossburg, Pa.
    B—Schenectady, N. Y.
    G. O. No. 46, W. D., 1919.

Corporal, Company K, 314th Infantry, 79th Division.
He, with another soldier from his platoon, outflanked a machine gun in advance of our line, killed 3 of the crew and captured 2 others, together with the machine gun.

LARSON, JULIUS D. (915376)_____
    Near Brieulles, France, Nov. 3, 1918.
    R—Chaseburg, Wis.
    B—Chaseburg, Wis.
    G. O. No. 37, W. D., 1919.

Private, first class, Company F, 7th Engineers, 5th Division.
With indomitable courage and bravery he rendered most valuable assistance in the construction of a pontoon bridge across the Meuse River and 2 other bridges across the Canal Est. At both places the work was done in the face of heavy machine-gun fire.

*LASHIWER, HYMAN (40917)_____
    Near Jaulny, France, Sept. 12, 1918.
    R—New York, N. Y.
    B—Russia.
    G. O. No. 37, W. D., 1919.

Private, first class, Company M, 9th Infantry, 2d Division.
Private Lashiwer, with 3 other soldiers, volunteered to carry wounded men of other companies from in front of our advanced positions and carried this work on under violent machine-gun fire while a counterattack was developing. Posthumously awarded. Medal presented to sister, Mrs. Eva Kessler.

LATHAM, DAVIDSON U. N. (1408986)___
    Near Septsarges, France, Oct. 24,
      1918.
    R—Gastonia, N. C.
    B—Lexington, N. C.
    G. O. No. 37, W. D., 1919.

Wagoner, Company G, 5th Ammunition Train, 5th Division.
When an enemy shell struck some pyrotechnics stored in the ammunition dump of his organization, he assisted in removing inflammable material and placing the fire under control. Through his coolness and courage the destruction of a large quantity of near-by ammunition was avoided.

ʹLAUBER, CLARENCE E. (3091038)_____
    Near Pouilly, France, Nov. 10–11,
      1918.
    R—Wauseon, Ohio.
    B—Archbold, Ohio.
    G. O. No. 44, W. D., 1919.

Private, Company I, 356th Infantry, 89th Division.
Private Lauber accompanied Lieutenant Murphy and 3 other soldiers in a flank attack on 3 heavy machine guns. Fired on directly at 30 yards, they charged the guns, and in the hand-to-hand fight which followed this soldier and 2 of his comrades were killed.
Posthumously awarded. Medal presented to sister, Mrs. Elsie Lauber Porter.

LAUGHLIN, JOSEPH H. (56681)_____
    Near Cantigny, France, May 28–30,
      1918.
    R—Concord, N. C.
    B—Concord, N. C.
    G. O. No. 98, W. D., 1919.

Private, Company A, 28th Infantry, 1st Division.
In command of the battalion runners, Private Laughlin volunteered to carry messages through the unusually heavy fire. During a very critical period of the fighting he twice went through a machine-gun barrage to the front line to obtain information when no word from that source had been received for a long period.

LAUNCELOT, MARC V. (38292)_____
    Near Medeah Ferme, France, Oct.
      3, 1918.
    R—Bridgeport, Conn.
    B—El Paso, Tex.
    G. O. No. 21, W. D., 1919.

Sergeant, Company B, 9th Infantry, 2d Division.
During the absence of his platoon commander, Sergeant Launcelot took command of the platoon, led an attack, and captured 7 machine guns in strong position. He was later seriously wounded.

*LAUTENSLAGER, EARL W. (1213623)___
    East of Ronssoy, France, Sept. 29,
      1918.
    R—Geneseo, N. Y.
    B—Fayette, N. Y.
    G. O. No. 16, W. D., 1919.

Private, Company B, 108th Infantry, 27th Division.
During the operations against the Hindenburg line, when his company was held up by an enemy machine-gun nest, he volunteered to cross an open field in front of his company in order to ascertain the exact location of the enemy's position. While engaged in this enterprise he was killed by a bursting shell. His heroic self-sacrifice was a splendid example to the men of his company.
Posthumously awarded. Medal presented to father, George J. Lautenslager.

LAVERY, JAMES FLAVIAN (1167691)_____
    At Nogales, Ariz., and Nogales,
      Sonora, Mexico, Aug. 27, 1918.
    R—New York, N. Y.
    B—Lavery, Pa.
    G. O. No. 9, W. D., 1923.

Private, first class, Quartermaster Corps, U. S. Army.
During an engagement with hostile Mexicans at Nogales, Ariz., and Nogales, Sonora, Mexico, on the 27th of August, 1918, Private Lavery, braving the heaviest fire, repeatedly entered the zone of fire with his motor truck and carried wounded men to places of safety, thereby saving the lives of several soldiers.

*LAVIOLETTE, HENRY J. (72416)_____
    At Chateau-Thierry, France, July
      20–23, 1918.
    R—Marlboro, Mass.
    B—Marlboro, Mass.
    G. O. No. 116, W. D., 1918.

Private, Company K, 104th Infantry, 26th Division.
Private Laviolette, acting as a runner, carried messages through heavy artillery fire with absolute fearlessness until killed.
Posthumously awarded. Medal presented to father, Midard Laviolette.

*Lavoie, Leo J. (42517)_____
   In the Argonne Forest, France, Oct. 9, 1918.
   R—Lunenburg, Mass.
   B—Fitchburg, Mass.
   G. O. No. 44, W. D., 1919.

Corporal, Company D, 16th Infantry, 1st Division.
He led his squad against an enemy machine gun which was causing severe losses in his company. His journey was made through a very difficult terrain and under deadly fire, but his mission was accomplished without the loss of a man. Later in the day he was killed while leading his squad.
Posthumously awarded. Medal presented to mother, Mrs. Margaret Lavoie.

Lawhorne, Dewie H. (1320851)_____
   Near Bellicourt, France, Sept. 29, 1918.
   R—Draper, N. C.
   B—Ambrose County, Va.
   G. O. No. 37, W. D., 1919.

Corporal, Company G, 120th Infantry, 30th Division.
In the face of heavy machine-gun fire, Private Lawhorne, with 2 other soldiers, attacked and put out of action an enemy machine-gun post, capturing a German officer and 3 soldiers.

Lawless, Edward R. (556118)_____
   Near Troesnes, France, July 18, 1918.
   R—Boston, Mass.
   B—Leominster, Mass.
   G. O. No. 44, W. D., 1919.

Sergeant major, Headquarters Company, 39th Infantry, 4th Division.
When it had become necessary to send an urgent message to the battalion base company, Sergeant Major Lawless, though under fire for the first time, voluntarily took the message across an open field, a distance of 500 yards. It seemed almost impossible to get through the murderous fire, but, knowing the importance of the message, Sergeant Major Lawless ventured through rather than take the longer yet safer route. He completed his mission, returning over the same course.

Lawless, James B. (1764959)_____
   Near Thiaucourt, France, Sept. 22, 1918.
   R—Newport, R. I.
   B—Newport, R. I.
   G. O. No. 35, W. D., 1919.

Sergeant, Machine Gun Company, 310th Infantry, 78th Division.
During a raid Sergeant Lawless bayoneted the men at 2 enemy machine guns which were firing upon our raiding party and put several others out of action with hand grenades.

Lawlor, Joseph William_____
   At St. Martin-Reviere, France, Oct. 17, 1918.
   R—Montclair, N. J.
   B—Montclair, N. J.
   G. O. No. 37, W. D., 1919.

First lieutenant, 118th Infantry, 30th Division.
Being the only officer left with the company, Lieutenant Lawlor was ordered to clear a village of the enemy. Most of his command became separated from him in a dense fog, but with his 1 remaining squad he proceeded to accomplish his mission. Lieutenant Lawlor and 1 soldier came upon 100 Germans operating machine guns from behind a hedge and succeeded in capturing 50 of the enemy. This officer then pursued the others, exchanging shots with a German officer as he ran.

Lawrance, Jackson S_____
   At Binarville, France, Sept. 30, 1918.
   R—Philadelphia, Pa.
   B—Philadelphia, Pa.
   G. O. No. 20, W. D., 1919.

Major, Medical Corps, attached to 368th Infantry, 92d Division.
Major Lawrance, with 2 soldiers, voluntarily left shelter and crossed an open space 50 yards wide, swept by shell and machine-gun fire, to rescue a wounded soldier, whom they carried to a place of safety.

Lawrence, Henry S. (1257212)_____
   Near Basheux, France, Sept. 5, 1918.
   R—Peckville, Pa.
   B—Scranton, Pa.
   G. O. No. 128, W. D., 1918.

Sergeant, Company B, 108th Machine Gun Battalion, 28th Division.
When the advance of the platoon commanded by him was held up by machine-gun fire from the front and flanks, Sergeant Lawrence took a rifle and bayonet and, accompanied by another soldier, crawled under the enemy wire in the face of severe fire, drove off the crews of several enemy machine guns and returned with 2 prisoners, thereby enabling his platoon to rush on. Subsequently, under heavy machine-gun and sniper fire, he went along the line of his gun emplacements cheering and encouraging his men, his fearlessness affording an inspiration to them throughout the engagement.

Lawrence, Hull F. (1736600)_____
   Near Grand Pre, France, Oct. 23, 1918.
   R—Newark, N. J.
   B—Newark, N. J.
   G. O. No. 37, W. D., 1919.

Private, Company K, 312th Infantry, 78th Division.
While his company was seeking shelter from a continuous rain of machine-gun bullets, Private Lawrence voluntarily carried messages from his company commander, who was lying wounded under enemy observation, to battalion headquarters. He worked under heavy bombardment at all times, but showed remarkable courage and devotion to duty while thus engaged.

Lawrence, Vivian S., Jr. (1842279)___
   Near Gercourt, France, Sept. 29, 1918.
   R—Churchland, Va.
   B—Churchland, Va.
   G. O. No. 13, W. D., 1919.

Corporal, 319th Ambulance Company, 305th Sanitary Train, 80th Division.
While he was passing along the roads leading to Septsarges, near Gercourt, in an ambulance, a large shell fell among a group of Infantry soldiers near by, severely wounding 5 of them. He stopped the ambulance and rendered efficient first aid, through concentrated shellfire. He then loaded the patients in the ambulance and removed them from the shelled area.

*Lawson, Bradley (1980806)_____
   Near Bellicourt, France, Sept. 29, 1918.
   R—Kildav, Ky.
   B—Jellico, Ky.
   G. O. No. 37, W. D., 1919.

Private, first class, Machine Gun Company, 120th Infantry, 30th Division.
Having been wounded by a bursting shell immediately after the opening of the attack, he refused to go to the rear, but remained with his corporal, who had been severely wounded by the same shell. For 2 hours, under an intense enemy barrage, he continued to minister to his wounded comrade until another shell burst near by, by which he was fatally wounded.
Posthumously awarded. Medal presented to father, Bud Lawson.

Lawson, Fred M. (542413)_____
   At Fossoy, France, July 14–15, 1918.
   R—Rosehill, Ky.
   B—Mercer County, Ky.
   G. O. No. 87, W. D., 1919.

Corporal, Company L, 7th Infantry, 3d Division.
When it was reported that the enemy had crossed the Marne River, Corporal Lawson twice led a patrol across his platoon front through heavy shellfire and at the risk of his life. On July 15, 16, and 17 he worked day and night unceasingly until forced to stop from complete exhaustion, displaying at all times the highest coolness and judgment and rendering services of the greatest value.

LAWSON, WALTER R._____
Near St. Mihiel, France, July 30 and Sept. 13, 1918.
R—Birmingham, Ala.
B—Georgia.
G. O. No. 21, W. D., 1919.

Captain, observer, 91st Aero Squadron, Air Service.
This officer showed rare courage on a reconnaissance far over the enemy lines when he continued on his mission after being seriously wounded by anti-aircraft fire. On Sept. 13, although he was still convalescing from his wound, he volunteered for a particularly dangerous mission requiring a flight of 75 kilometers within the enemy lines. Because of weather conditions he was forced to fly at a dangerously low altitude and was repeatedly fired on by anti-aircraft and machine-gun guns. He successfully accomplished his mission and returned with important information.

LAY, ARTHUR (38906)_____
Near the Meuse River, France, Nov. 3-4, 1918.
R—New York, N. Y.
B—Albany, N. Y.
G. O. No. 46, W. D., 1919.

Mechanic, Company D, 9th Infantry, 2d Division.
While passing through the German lines at night, carrying messages, Mechanic Lay captured 8 of the enemy, including 2 officers, and marched them back 4 kilometers. The next day he carried messages through artillery and machine-gun fire and fearlessly exposed himself to danger in guiding companies into position. While returning from a mission he carried a severely wounded comrade through a machine-gun barrage to the dressing station.

LAYER, JOHN L. (1038310)_____
Near St. Eugene, France, July 14, 1918.
R—Pittsburgh, Pa.
B—Chicago, Ill.
G. O. No. 44, W. D., 1919.

Private, Battery B, 10th Field Artillery, 3d Division.
He continued on duty repairing breaks in telephone line, even after being severely wounded and almost unconscious.

LEACH, GEORGE E._____
Near Pexonne, France, Mar. 5, 1918.
R—Minneapolis, Minn.
B—Cedar Rapids, Iowa.
G. O. No. 126, W. D., 1919.
Distinguished-service medal also awarded.

Colonel, 151st Field Artillery, 42d Division.
Near Pexonne, France, on Mar. 5, 1918, he entered the quarry of Battery C, 151st Field Artillery, then subjected to particularly accurate artillery bombardment, for the purpose of directing and encouraging the officers and men of that battery, when he might with propriety have sent his directions by messenger.

LEACH, JOHN A. (552565)_____
Near La Franquette Farm, France, July 22, 1918.
R—Westminster, Vt.
B—Westminster, Vt.
G. O. No. 22, W. D., 1920.

Sergeant, Company H, 38th Infantry, 3d Division.
While his unit was in close contact with the enemy, a spy circulated a report that the company had been ordered to withdraw. Those who thus retired were shot down by the enemy. During the disorder that followed, Sergeant Leach fearlessly reorganized the platoon under close-range enemy fire. Due to his heroic example, the men of his platoon held their ground and repulsed the strong enemy attack which followed.

LEACH, JOHN W._____
Near Bellicourt, France, Sept. 29, 1918.
R—Knoxville, Tenn.
B—Wytheville, Va.
G. O. No. 46, W. D., 1919.

First lieutenant, 117th Infantry, 30th Division.
About 6.30 in the morning of Sept. 29 Lieutenant Leach received a serious wound which rendered his right arm useless. Notwithstanding his suffering and weakness from loss of blood he continued to lead his platoon toward its objective until after 1 o'clock of that day, when he became so exhausted and weak that he was ordered to the aid station by his commanding officer and from there was evacuated to the hospital.

LEARY, EDMUND M._____
Near Stenay, France, Nov. 10, 1918.
R—North Whitefield, Me.
B—Cambridge, Mass.
G. O. No. 46, W. D., 1919.

Colonel, 358th Infantry, 90th Division.
Under heavy machine-gun and artillery fire, Colonel Leary personally led 2 sections of machine guns and 2 platoons of Infantry to the assistance of 1 of his battalions in order to protect its flank.

LEAVELL, JOHN H._____
At Audenarde, Belgium, Nov. 1, 1918.
R—Salt Lake City, Utah.
B—Georgetown, Tex.
G. O. No. 37, W. D., 1919.

Captain, 316th Engineers, 91st Division.
He led a patrol on a reconnaissance of the city of Audenarde at a time when it was still occupied by enemy patrols and snipers, obtaining important data on destroyed bridges and locating materials necessary in repairing them. While on this errand Captain Leavell and his men encountered a German patrol 3 times their number, and in the combat which followed several of the enemy were killed or wounded and a spy was captured.

LE CLAIR, ARTHUR H. (42518)_____
In the Argonne Forest, France, Oct. 12, 1918.
R—Gardner, Mass.
B—Fitchburg, Mass.
G. O. No. 87, W. D., 1919.

Corporal, Company D, 16th Infantry, 1st Division.
When ordered on an exploiting party 1½ kilometers in advance of our lines, he first reconnoitered the route and then led his men forward under heavy fire from the enemy artillery and machine guns. He held the position he established, under heavy fire, until relieved, 14 hours later.

*LEDWELL, HARVEY M. (2337850)_____
Near Charmel, France, July 26, 1918.
R—Greensboro, N. C.
B—Clay County, Ind.
G. O. No. 126, W. D., 1919.

Sergeant, Company A, 4th Infantry, 3d Division.
Although very seriously wounded, he refused aid of his men, who stopped to assist him, ordering them forward and directing their attack until they had passed beyond hearing distance.
Posthumously awarded. Medal presented to niece, Mrs. Gertrude Holland.

LEE, ALFRED P. (73537)_____
At Apremont, France, Apr. 10, 1918.
R—Northampton, Mass.
B—Ware, Mass.
G. O. No. 99, W. D., 1918.

Private, Company L, 104th Infantry, 26th Division.
During the action of Apr. 10, 1918, he displayed courage, coolness, and the spirit of self-sacrifice in voluntarily going through shell-swept area to bring back wounded to a place of safety, carrying 1 wounded man more than 50 yards under heavy shell fire.

LEE, ARTHUR TRUMBULL_____
Near Tronsol Farm, France, Sept. 28, and at Leeuwken, Belgium, Oct. 31, 1918.
R—Walla Walla, Wash.
B—Seattle, Wash.
G. O. No. 46, W. D., 1919.

First lieutenant, 364th Infantry, 91st Division.
On the afternoon of September 28 he advanced with his platoon to Tronsol Farm under heavy artillery and machine-gun fire and captured 7 machine guns and much ammunition. Forced to retire, he returned with his capture and platoon in good order. On Oct. 31, while attempting to locate machine-gun nests which were holding up his advance, he was so severely wounded that in spite of his desire to continue he was unable to do so.

LEE, CARL (2154320)_____
Near Molain, France, Oct. 17, 1918.
R—Osage, Iowa.
B—Norway.
G. O. No. 44, W. D., 1919.

Private, Company D, 117th Infantry, 30th Division.
Having become separated from their company in a smoke barrage, Private Lee and Corpl. Henry W. Cardwell found themselves face to face with a party of the enemy. Private Lee brought his automatic rifle to his shoulder and attempted to fire, but the gun was jammed and would not shoot. Seeing themselves covered by the gun, and not knowing its condition, the Germans threw up their hands, and while Private Lee kept the rifle at his shoulder, Corporal Cardwell rounded up the Germans and disarmed them. Their ruse resulted in the capture of 12 of the enemy, comprising 3 machine-gun crews.

LEE, CHRIS (54614)_____
Near Cantigny, France, May 27, 1918.
R—Canada.
B—Pope County, Minn.
G. O. No. 19, W. D., 1920.

Private, first class, Company K, 26th Infantry, 1st Division.
During an enemy raid Private Lee exposed himself to heavy artillery, machine-gun, and rifle fire to go to the company dump for ammunition. Although wounded twice when en route to the dump and 3 times more while returning, he persisted and delivered to his squad the needed ammunition and then acted as loader for an automatic rifle team until the attack was repulsed.

LEE, CHRISTOPHER F._____
Near Verdun, France, Oct. 25, 1918.
R—Boston, Mass.
B—Boston, Mass.
G. O. No. 37, W. D., 1919.

Major, 102d Infantry, 26th Division.
When his battalion had met with heavy artillery, machine-gun, and rifle resistance and his troops were on the verge of becoming disorganized, he took the leadership of the front-line platoon and charged the enemy. In this act he was so seriously wounded that he had to be evacuated.

LEE, EDWARD BROOKE_____
Near Balschwiller, Alsace, Aug. 31, 1918.
R—Silver Spring, Md.
B—Washington, D. C.
G. O. No. 3, W. D., 1922.

Captain, 115th Infantry, 29th Division.
For extraordinary heroism in command of a raiding party against the enemy trenches near Balschwiller on the morning of Aug. 31, 1918, leading the advance through the enemy wire, being the last to quit the enemy trenches, personally assisting in carrying the wounded back through the enemy counter-barrage, and remaining all day of Aug. 31 in a shell hole in no man's land in order to assist all wounded to return to our line.

LEE, EVERETT D. (1205388)_____
Near Ronssoy, France, Sept. 27, 1918.
R—Saratoga Springs, N. Y.
B—Saratoga Springs, N. Y.
G. O. No. 13, W. D., 1923.

Sergeant, Company L, 105th Infantry, 27th Division.
When the advance of his company was checked by concentrated machine-gun and rifle fire Sergeant Lee advanced alone, sought out enemy machine-gun nests, and by the use of hand grenades and with complete disregard for his own safety, killed and wounded members of enemy machine-gun crews, causing them to abandon the guns, and thus permitting the infantry to again advance. Sergeant Lee's courage and devotion to duty raised the morale of his company to a high pitch.

LEE, JAMES A. (1210429)_____
At St. Souplet, France, Oct. 18, 1918.
R—Norwood, N. Y.
B—Peekskill, N. Y.
G. O. No. 39, W. D., 1920.

Private, Company D, 107th Infantry, 27th Division.
The advance of his organization having been temporarily halted by machine-gun fire from a fortified house, Private Lee went forward as a scout to locate the entrance to the house. After discovering the entrance, he led a detachment in an assault, and at the point of the bayonet compelled about 35 of the enemy to surrender. A few minutes later, while advancing alone toward an outhouse, he effected the capture of more of the enemy.

LEE, JOHN B_____
Near Mezieres, France, Nov. 3, 1918.
R—Newark, N. J.
B—Newark, N. J.
G. O. No. 13, W. D., 1919.

Second lieutenant, observer, 24th Aero Squadron, Air Service.
He exhibited extreme courage in the course of a long and dangerous photographic and visual reconnaissance in the region of Mezieres with 2 other planes of the 24th Aero Squadron. Their formation was broken by the attack of 10 enemy pursuit planes, 5 enemy planes attacking Lieutenant Lee and his pilot. With remarkable coolness he succeeded in shooting down 2 of the planes. They then had a clear passage to their own lines, but turned back into Germany to assist a friendly plane with several hostile aircraft attacking it. They succeeded in shooting down 1 more of the enemy. Lieutenant Lee and pilot returned to our lines with information and photographs of great military value.

*LEE, JOHN C_____
Near Chipilly Ridge, France, Aug. 9, 1918.
R—Chicago, Ill.
B—Chicago, Ill.
G. O. No. 74, W. D., 1919.

Second lieutenant, 131st Infantry, 33d Division.
When his platoon was held up by fire from a machine-gun nest he advanced alone against the enemy position, and, although mortally wounded, attacked and killed the crew, falling dead among the bodies of the Germans.
Posthumously awarded. Medal presented to mother, Mrs. Mary A. Lee.

LEE, ORUM (1521885)_____
Near Montfaucon, France, Sept. 27, 1918.
R—Orrville, Ohio.
B—Reedy, W. Va.
G. O. No. 16, W. D., 1920.

Sergeant, Company H, 146th Infantry, 37th Division.
Sergeant Lee, with an officer and noncommissioned officer, advanced 200 yards beyond the objective of the patrol in the face of heavy machine-gun fire and captured 3 77-millimeter fieldpieces and 2 light machine guns.

LEE, ROBERT S. (1200983)_____
East of Ronssoy, France, Sept. 29, 1918.
R—Albany, N. Y.
B—Louisville, Ky.
G. O. No. 74, W. D., 1919.

Sergeant, Company C, 102d Field Signal Battalion, 27th Division.
Sergeant Lee was a member of an advanced regimental signal party which went over the top with the third wave in the attack against the Hindenburg line. The officer in charge and 3 privates were wounded, and 2 sergeants, 2 corporals, and 2 runners were killed. Sergeant Lee, in the face of terrific shell and machine-gun fire, fearlessly and courageously strung telephone lines and patrolled same in such a manner as to insure telephone communications with the battalion to which he was attached. His bravery and extreme devotion to duty was a splendid inspiration to all his comrades.

LEE, THEODORE F. (2260957)_____
Near Gesnes, France, Sept. 29, 1918.
R—Ibapah, Utah.
B—Oakley, Idaho.
G. O. No. 26, W. D., 1919.

Private, Company H, 362d Infantry, 91st Division.
When the advance of his battalion was held up by a machine-gun nest on a high ridge south of Gesnes, Private Lee and another soldier charged the emplacement, killing part of the crew and routing the others, capturing 3 heavy machine guns and 1 light Maxim gun, which they turned on the retreating Germans.

LEE, WILL H._____
Near Blanc Mont, France, Oct. 4, 1918.
R—Fort Worth, Tex.
B—Yoakum, Tex.
G. O. No. 46, W. D., 1919.

Corporal, 43d Company, 5th Regiment, U. S. Marine Corps, 2d Division.
During an advance of his company he volunteered to carry a wounded comrade to a place of safety through hostile machine-gun fire. He later brought a wounded soldier from no man's land through a heavy machine-gun and artillery barrage.

LEE, WILLIAM F._____
At Cheveuges, France, Nov. 7, 1918.
R—Amherst, Mass.
B—Fort Warren, Mass.
G. O. No. 49, W. D., 1922.

Major, 28th Infantry, 1st Division.
While his battalion was engaged with the enemy on the heights near Sedan he left his advance command post, went forward in the face of heavy enemy machine-gun and artillery fire, where he personally supervised the successful reduction of enemy machine-gun positions and strong points. Later in the engagement, when his battalion was under a heavy frontal and flanking fire of machine guns, rifles, and direct artillery fire from high ground, he again left his advance command post, and with the utmost disregard for his personal safety advanced into the assault waves, traversed them, encouraging his men, and assisted the company officers in bringing such effective fire upon the enemy as to enable the battalion to advance upon their objective in the minimum time and with the least possible loss of life. His gallant and courageous conduct was an inspiration to his command and in a large measure responsible for the success of the 28th Infantry in the operations against Sedan.

LEEB, JOSEPH S. (2293711)_____
Near Gesnes, France, Oct. 3, 1918.
R—Chicago, Ill.
B—Chicago, Ill.
G. O. No. 20, W. D., 1919.

Private, Company D, 361st Infantry, 91st Division.
He voluntarily and unhesitatingly left shelter under heavy shellfire and without thought of personal danger rendered first aid and carried a wounded comrade to a place of safety.

*LEEKER, GARRETT (550897)_____
Near Romagne, France, Oct. 9, 1918.
R—Jersey City, N. J.
B—Jersey City, N. J.
G. O. No. 27, W. D., 1920.

Private, first class, Company F, 38th Infantry, 3d Division.
In the attack on Hill 253 Private Leeker took command of his platoon after the platoon leader and the sergeant had become casualties. He reorganized the platoon under fire and fearlessly led it to its objective. He fell mortally wounded while leading his men in the attack.
Posthumously awarded. Medal presented to widow, Mrs. Sadie Leeker.

*LEEPER, DAN C_____
Near Villers-devant-Dun, France, Nov. 2, 1918.
R—Denison, Tex.
B—Denison, Tex.
G. O. No. 37, W. D., 1919.

Captain, 359th Infantry, 90th Division.
In the face of intense machine-gun fire, Captain Leeper led his company forward to its objective, capturing a machine-gun nest and making possible the capture of a strong enemy position on a hill. He was severely wounded during these operations, but his courage was an inspiration to his men in consolidating their position.
Posthumously awarded. Medal presented to widow, Mrs. Leonora O. Leeper.

LEGENDRE, JAMES HENNEN_____
On the Lucy-Torcy Road, France, June 6, 1918.
R—New York, N. Y.
B—New Orleans, La.
G. O. No. 101, W. D., 1918.

Second lieutenant, 5th Regiment, U. S. Marine Corps, 2d Division.
He displayed exceptional bravery in organizing and leading a party of volunteers through heavy machine-gun fire for the purpose of securing 2 wounded men on the Lucy-Torcy Road June 6, 1918.

LEGGE, BARNWELL R_____
Near Verdun, France, Oct. 5, 1918.
R—Charleston, S. C.
B—Charleston, S. C.
G. O. No. 87, W. D., 1919.
Distinguished-service medal also awarded.

Major, 26th Infantry, 1st Division.
Personally leading an attack against a strong enemy position, he inspired his men by his courage, cutting his way through entanglements and directing the attacks against 3 different strong points.

LEGNOSKY, JOHN (560215)_____
In the Bois-de-Malaumont, east of Breuilles, France, Oct. 4, 1918.
R—Hadley, Pa.
B—Leisenring, Pa.
G. O. No. 16, W. D., 1920.

First sergeant, Company L, 58th Infantry, 4th Division.
Although painfully wounded in the foot on Oct. 4, First Sergeant Legnosky remained on duty with his company. After his company commander had been killed he assumed command of the company and efficiently led it in action until the unit was released on Oct. 7, 1918. He repeatedly exposed himself to heavy fire in order to better control his men.

*LEHY, HOWARD C. (1746071)_____
Near Bois-de-Grande-Fontaine, France, Sept. 26, 1918.
R—Oakhurst, N. J.
B—Oakhurst, N. J.
G. O. No. 37, W. D., 1919.

Sergeant, Company B, 311th Infantry, 78th Division.
After his patrol had disposed of 2 sniper posts with rifle grenades he held his group in position and covered the withdrawal of his platoon from heavy enfilading machine-gun fire of the enemy.
Posthumously awarded. Medal presented to father, John Lehy.

*LEIBOULT, EDWARD N_____
Near St. Juvin, France, Oct. 11, 1918.
R—Fulton, N. Y.
B—Palermo, N. Y.
G. O. No. 87, W. D., 1919.

First lieutenant, 326th Infantry, 82d Division.
Under terrific fire he led his platoon across the Aire River, formed them on ground devoid of cover, and, though he had suffered 50 per cent casualties, led the survivors in a charge against the enemy, falling mortally wounded at the head of his men.
Posthumously awarded. Medal presented to widow, Mrs. Della M. Leiboult.

LEITER, WILSON H. (1245908)_____
Near Fismette, France, Aug. 11, 1918.
R—Harrisburg, Pa.
B—Harrisburg, Pa.
G. O. No. 50, W. D., 1919.

Private, Company M, 111th Infantry, 28th Division.
While his company was withdrawing to a place of safety Private Leiter stopped in a most exposed position and picking up a badly wounded man he continued in the withdrawal until the point of assembly was reached, at which time he brought him to a dressing station.

*LEITNER, ALOYSIUS_____

At Bois-de-Belleau, France, June 12, 1918.
R—Holstein, Wis.
B—Charlesburg, Wis.
G. O. No. 99, W. D., 1918.

Private, Headquarters Company, 5th Regiment, U. S. Marine Corps, 2d Division.
On June 12, 1918, in the attack on Bois-de-Belleau, although seriously wounded he displayed extraordinary heroism in assisting to capture 3 officers and 169 men of the enemy forces, after which he continued forward and aided in taking 6 more prisoners, who were operating a machine gun. The wounds received while performing these valiant deeds have since proved fatal.
Posthumously awarded. Medal presented to father, Joseph Leitner.

LEITZELL, WILBUR F._____

Near Apremont, France, Oct. 1, 1918.
R—State College, Pa.
B—Scottdale, Pa.
G. O. No. 72, W. D., 1920.

Captain, 107th Machine Gun Battalion, 28th Division.
Captain Leitzell exposed himself to heavy fire in order to place his machine guns in action against an enemy counterattack. Due to his initiative and gallantry the enemy attack was repulsed without the aid of supporting Infantry. Later, the commander of arriving Infantry support being wounded, Captain Leitzell took command of the Infantry and led them to their positions. While in the performance of this act he was seriously wounded.

LEMASTER, FRANK (155210)_____

Near Bois-de-Moncy, France, Oct. 9, 1918.
R—Morgan County, Ky.
B—Morgan County, Ky.
G. O. No. 37, W. D., 1919.

Private, Company C, 1st Engineers, 1st Division.
Remaining alone on an outpost 100 yards in advance of his detachment, Private Lemaster, by the efficient and effective use of his rifle, stopped a German counterattack, personally killing 12 of the enemy during the exploit.

LE MASTERS, CHARLES A. (2193675)____

Near Beauclair, France, Nov. 4–11, 1918.
R—Litchfield, Nebr.
B—Ord, Nebr.
G. O. No. 37, W. D., 1919.

Corporal, Company C, 314th Field Signal Battalion, 89th Division.
From Nov. 4 to 11, while continually under heavy shellfire, Corporal Le Masters laid and maintained lines of communication within his area with utter disregard for his personal safety.

LE MAY, JOSEPH J (2784906)_____

Near Eclisfontaine, France, Sept. 28, 1918.
R—Seattle, Wash.
B—New York, N. Y.
G. O. No. 44, W. D., 1919.

Private, Company K, 364th Infantry, 91st Division.
Responding to a call for volunteers, Private Le May, with 5 others, advanced 400 yards beyond their front to bring in wounded comrades. They succeeded in rescuing 7 of their men and also in bringing in the dead body of a lieutenant while exposed to terrific machine-gun fire.

LEMIEUX, WILLIAM (554737)_____

At Mezy, France, July 15, 1918.
R—Taunton, Mass.
B—Canada.
G. O. No. 93, W. D., 1919.

Sergeant, Company A, 9th Machine Gun Battalion, 3d Division.
Though the two Infantry platoons to which he was attached had been reduced to but 15 men, Sergeant Lemieux stuck to his position with his 2 guns, and, in order to obtain a better field of fire, placed them on top of a railroad embankment in plain view of the enemy under intense machine-gun and shell fire. From here he maintained a deadly fire upon enemy boats crossing the Marne River, sinking several, killing the occupants of others, and forcing several to turn back. He also wiped out several enemy platoons which were preparing to cross the river. His guns were twice buried by bursting shells, but each time he calmly cleared the guns and immediately resumed firing. His gallant stand contributed materially to the successful defense of the sector.

*LEMMA, SAMUEL (542066)_____

In the Belleau Wood, France, June 13, 1918.
R—Canandaigua, N. Y.
B—Italy.
G. O. No. 22, W. D., 1920.

Private, medical detachment, 7th Infantry, 3d Division.
After being severely wounded in the leg, Private Lemma exposed himself to heavy machine-gun fire in order to give medical attention to other wounded. His activities after being wounded caused a serious hemorrhage, which resulted in his death from loss of blood.
Posthumously awarded. Medal presented to mother, Mrs. Mary Lemma.

LEMMON, CHARLIE E. (44337)_____

Near Fleville, France, Oct. 9, 1918.
R—Augusta, Ga.
B—Charleston, S. C.
G. O. No. 74, W. D., 1919.

Sergeant, Company M, 16th Infantry, 1st Division.
He alone charged an enemy machine gun, capturing the gun and 2 German soldiers. He was seriously wounded, while in the successful execution of this mission. His timely capture of the gun made it possible for his company to advance and organize their objective with few casualties.

LEMMON, FRED L_____

Near Charpentry, France, Sept. 27–28, 1918.
R—Hutchinson, Kans.
B—Clyde, Ohio.
G. O. No. 59, W. D., 1919.

Lieutenant colonel, 140th Infantry, 35th Division.
Wounded severely in the chest, he remained in command of his battalion for 24 hours, until no longer able to walk. He showed a great personal courage and skill in leading his battalion against heavy shell and machine-gun fire, refusing to be evacuated until helpless from loss of blood.

LEMON, DWIGHT E. (3107539)_____

Near Verdun, France, Nov. 8, 1918.
R—Indianapolis, Ind.
B—Alvin, Ill.
G. O. No. 44, W. D., 1919.

Private, Company B, 310th Machine Gun Battalion, 79th Division.
When his comrades, many of whom were wounded, were suffering from want of water, he took their canteens and went 200 meters to the rear under heavy artillery and machine-gun fire; here he filled the canteens at a spring and returned through intense fire, under direct observation of machine gunners and snipers.

LENAHAN, EDWARD J. (553832)_____

Near Montfaucon, France, Oct. 7, 1918.
R—Savannah, Ga.
B—Savannah, Ga.
G. O. No. 95, W. D., 1919.

Corporal, Company B, 8th Machine Gun Battalion, 3d Division.
Corporal Lenahan, a runner, repeatedly carried messages through heavy barrages. On one occasion, when an enemy shell struck a dugout occupied by himself and several comrades, burying 2 of them, though he was suffering from the concussion, he immediately set to work digging out the imprisoned men, disregarding the intense shell and machine-gun fire to which he was subjected.

*LENAHAN, JOHN C. (1710242)_____
Near Grand Pre, France, Oct. 14,
1918.
R—Brooklyn, N. Y.
B—New York, N. Y.
G. O. No. 32, W. D., 1919.

Private, Company M, 308th Infantry, 77th Division.
When his company was ordered to take a position along the river bank, under heavy machine-gun and artillery fire, Private Lenahan, acting first sergeant, made his way from flank to flank, supervising the disposition of the troops. Despite serious wounds received, he completed his mission and reported to his company commander, dying shortly after from the effects of his wounds. Posthumously awarded. Medal presented to mother, Mrs. Lucy Lenahan.

LENNOX, HERBERT K. (54717)_____
Near Cantigny, France, May 27,
1918.
R—Pottsville, Pa.
B—Pottsville, Pa.
G. O. No. 44, W. D., 1919.

Private, Company K, 26th Infantry, 1st Division.
Engaged as gunner of an automatic rifle team, Private Lennox was so severely wounded that he was unable to withdraw from the advancing enemy. Concealing himself until the assaulting wave had passed over him, he opened fire on the enemy from the rear, completely discouraging their attack and forcing them to retire.

LEONARD, HARRY D. (42049)_____
South of Soissons, France, July 19–22, 1918.
R—Kings Mills, Ohio.
B—Waynesville, Ohio.
G. O. No. 117, W. D., 1918.

Private, Company B, 16th Infantry, 1st Division.
Severely wounded on July 19, 1918, he continued to go forward with his company until the operations were over. On July 20, in spite of his wounds, he carried a message through a heavy barrage.

LEONARD, HARRY W. (93220)_____
Near Chateau-Thierry, France, July 15–18, and July 28, 1918.
R—London, Ohio.
B—Madison County, Ohio.
G. O. No. 26, W. D., 1919.

Private, first class, Headquarters Company, 166th Infantry, 42d Division.
During the terrific struggle of July 15–18, 1918, he continually volunteered and carried messages through territory swept by high explosives, shrapnel and gas. On July 28 he left his shelter to assist his company commander, who, had been seriously wounded, carrying him through a rain of shells to a place of safety.

LEONARD, JOHN W _____
Near Romagne, France, Oct. 14, 1918.
R—Toledo, Ohio.
B—Toledo, Ohio.
G. O. No. 37, W. D., 1919.

Major, 6th Infantry, 5th Division.
Major Leonard personally led the assaulting wave in an attack under severe shell and machine-gun fire from the front and flanks. Upon reaching the objective, he directed the organization of the position and by his example of fearlessness rallied his men and kept his line intact.

LEONARD, MELVIN H_____
Near Soissons, France, July 18, 1918.
R—Boston, Mass.
B—Middleboro, Mass.
G. O. No. 46, W. D., 1919.

First lieutenant, 9th Infantry, 2d Division.
Although impeded by a very painful wound, he continued to lead his company, resisting a first and second attack of the enemy, though barely able to walk. Upon gaining his objective he supervised the consolidation of the position under severe shellfire and remained until his company was relieved.

LEPLEY, JAMES B. (102463)_____
Near Souain, to the northeast of Chalons-sur-Marne, France, July 14–15, 1918.
R—Red Oak, Iowa.
B—Pittsburgh, Pa.
G. O. No. 99, W. D., 1918.

Sergeant, Company M, 168th Infantry, 42d Division.
Near Souain, to the northeast of Chalons-sur-Marne, France, on the night of July 14–15, 1918, he left his trench and returned to the woods, through a smothering fire of gas, high explosive, and shrapnel, to search for 2 men from his platoon who were missing. He found them lost in the woods and guided them back to the platoon. On July 28, 1918, near Sergy, France, he led his platoon forward in the face of a heavy machine-gun fire and captured 6 machine guns and 13 prisoners from the Prussian Guards.

LESESNE, FRANCIS K_____
Near Ardeuil, France, Sept. 29, 1918.
R—Charleston, S. C.
B—Charleston, S. C.
G. O. No. 46, W. D., 1919.

Captain, 371st Infantry, 93d Division.
Painfully wounded in the arm by shellfire, Captain Lesesne nevertheless remained with his company until his organization two days later had gone into a reserve position, and he was ordered evacuated.

LESH, WILLIAM THEODORE (1915792)____
Near Sommerance, France, Oct. 11, 1918.
R—Scranton, Pa.
B—Scranton, Pa.
G. O. No. 87, W. D., 1919.

Sergeant, Battery A, 321st Field Artillery, 82d Division.
Wounded by shellfire, he refused to be evacuated, but after receiving first aid continued to command his section, which was under heavy fire, encouraging his men by his own bravery.

LETZING, JOHN LAWRENCE_____
North of Montfaucon, France, Sept. 29, 1918.
R—Roxbury, Mass.
B—Boston, Mass.
G. O. No. 3, W. D., 1921.

Second lieutenant, 148th Infantry, 37th Division.
During the attack Lieutenant Letzing exposed himself to heavy artillery, rifle, and machine-gun fire to lead tanks against enemy machine-gun positions. After the tanks had been withdrawn he walked up and down the firing line of his platoon and encouraged his men to greater efforts.

LEVAN, SIMPSON (156177)_____
Between Moneaux and Vaux, France, July 1–3, 1918.
R—Linn, Kans.
B—Linn, Kans.
G. O. No. 132, W. D., 1918.

Corporal, Company A, 2d Engineers, 2d Division.
Although wounded in the head and left leg by a high-explosive shell, which fact he concealed from his officers, Corporal Levan continued forward with his company. While thus wounded he led his platoon under heavy fire from Moneaux to Vaux during three days of hard and constant fighting, effectively discharging his duties until exhausted.

LEVAS, JAMES (106256)_____
Between Berzy-Le-Sec and Soissons, France, July 19, 1918.
R—New York, N. Y.
B—Turkey.
G. O. No. 99, W. D., 1918.

Sergeant, Company B, 3d Machine Gun Battalion, 1st Division.
He took charge of his platoon after his commander was killed. Soon afterwards he himself was wounded, but he dressed his own wound and continued forward. In a later advance directed by him he was severely wounded, but placed his gun in position, looked after the security of his men, and reported those facts personally to his commanding officer before permitting himself to be taken to a dressing station.

LEVENSON, ABE (173052)_____
Near Croix Rouge farm, northeast of Chateau-Thierry, France, July 27, 1918.
R—Pittsburgh, Pa.
B—Pittsburgh, Pa.
G. O. No. 102, W. D., 1918.

Private, Company G, 167th Infantry, 42d Division.
When his company was in action near Hill 212, Private Levenson was posted as lookout while his company was intrenching. He observed the enemy bringing forward machine guns through the wheat fields to place them in position. Waiting until they were within close range, he exposed himself to heavy machine-gun and artillery fire and succeeded in killing or disabling the crew of 2 machine guns, thus saving his company from heavy casualties.

LEVESQUE, ADELORD (1550809)_____
   Near Montfaucon, France, Oct. 5, 1918, and near Cierges, France, Oct. 18, 1918.
   R—Canada.
   B—Waterville, Me.
   G. O. No. 22, W. D., 1920.

Sergeant, Battery D, 76th Field Artillery, 3d Division.
On Oct. 5, when 4 men of his gun crew were wounded and he himself hit by a shell fragment, he kept the gun in action until a new gun crew was organized. On Oct. 18, when an enemy shell burst at his gun, killing or wounding all the gun crew but himself, he again kept to his gun carrying out the mission assigned to him.

LEVINE, ABEL J. (1211232)_____
   Near Bony, France, Sept. 29, 1918.
   R—Mount Vernon, N. Y.
   B—Russia.
   G. O. No. 44, W. D., 1919.

Corporal, Company H, 107th Infantry, 27th Division.
After his platoon had suffered heavy casualties and all the sergeants had been wounded, Corporal Levine collected the remaining effectives in his own and other units, formed a platoon, and continued the advance. When his rifle was rendered useless, he killed several of the enemy with his pistol. He was wounded shortly afterwards, but he refused assistance until his men had been cared for and evacuated.

*LEVINE, JACOB (1697649)_____
   Near St. Juvin, France, Nov. 1, 1918.
   R—New York, N. Y.
   B—Russia.
   G. O. No. 50, W. D., 1919.

Private, Company E, 305th Infantry, 77th Division.
While his company was being attacked from three sides and the terrific enemy fire had caused many casualties in the ranks, he volunteered and carried a message to the left flank. After he had advanced about 10 yards he was killed by a deluge of machine-gun bullets.
Posthumously awarded. Medal presented to father, Nathan Levine.

LEVIS, WILLIAM EDWARD_____
   Near Bois-d'Aigremont, France, July 15, 1918.
   R—Alton, Ill.
   B—Alton, Ill.
   G. O. No. 32, W. D., 1919.

Second lieutenant, 30th Infantry, 3d Division.
On several occasions he volunteered and went through the intense artillery bombardment of the enemy to perform important missions. He led a detail of 3 men to secure rockets, and on the journey 2 were killed and the other wounded. Undaunted, he alone pushed on, arriving at the ammunition dump just as an enemy shell exploded and destroyed it.

LEVY, JOSEPH (1746080)_____
   At Bois-de-Grande Fontaine, France, Sept. 26, 1918.
   R—Seabright, N. J.
   B—Seabright, N. J.
   G. O. No. 14, W. D., 1923.

Supply sergeant, Company B, 311th Infantry, 78th Division.
With utter disregard for his own safety, he repeatedly led details through heavy enemy artillery and machine-gun fire, carrying rations to the hungry and exhausted men in the advanced positions. Again and again he carried wounded men under the hottest enemy fire to dressing stations with complete contempt for personal danger. The consistently heroic conduct of Sergeant Levy throughout his combat service served as an example of soldierly conduct to the men of his company.

LEVY, REUBEN M. (2263302)_____
   Near Very, France, Sept. 26, 1918.
   R—Placerville, Calif.
   B—Vallejo, Calif.
   G. O. No. 72, W. D., 1920.

First sergeant, Company B, 363d Infantry, 91st Division.
After the advance of his platoon had been held up by machine-gun fire, Sergeant Levy, with 1 other man, attacked 1 machine gun and put it out of action. This act resulted in the enemy abandoning 2 other machine guns and permitted the advance of his platoon.

LEWANDOWSKI, FELIX (551730)_____
   During attack on Hill 253 in the Argonne, France, Oct. 8 and 9, 1918.
   R—Trenton, N. J.
   B—Russia.
   G. O. No. 60, W. D., 1920.

Private, first class, Company I, 38th Infantry, 3d Division.
Private Lewandowski carried numerous messages from company to battalion headquarters over a route which was constantly swept by heavy machine-gun and artillery fire. His efforts in maintaining communication were of the greatest value to his commanding officer.

LEWIS, BERNARD (1799493)_____
   Near Binarville, France, Sept. 30, 1918.
   R—Washington, D. C.
   B—Washington, D. C.
   G. O. No. 127, W. D., 1918.

Private, Company A, 368th Infantry, 92d Division.
During an attack on Binarville he volunteered to go down the road that leads into the village to rescue a wounded soldier of his company. To accomplish his mission he was compelled to go under heavy machine-gun and shell fire. In total disregard of personal danger he brought the wounded man safely to our lines.

*LEWIS, CHARLES (2388698)_____
   In the Bois-des-Rappes, France, Oct. 12, 1918, and near Cunel, France, Oct. 13, 1918.
   R—San Francisco, Calif.
   B—Toledo, Ohio.
   G. O. No. 20, W. D., 1919.

Sergeant, Company M, 61st Infantry, 5th Division.
He assisted, on Oct. 12, in carrying his company commander, who had been severely wounded, to a first-aid station under heavy shellfire. On Oct. 13 near Cunel, he took charge of his company and successfully reached the objective and held it until he was killed by shellfire.
Posthumously awarded. Medal presented to sister, Mrs. Carrie Runkel.

LEWIS, CONVERSE R._____
   On the Buluan River, island of Mindanao, P. I., June 14, 1904.
   R—New Orleans, La.
   B—Fort Custer, Mont.
   G. O. No. 111, W. D., 1919.

Second lieutenant, 23d Infantry, U. S. Army.
For extraordinary heroism in action against hostile Moros on the Buluan River, island of Mindanao, June 14, 1904.

LEWIS, EVAN E._____
   Near Marcheville, France, Sept. 26, 1918.
   R—Worthing, S. Dak.
   B—Worthing, S. Dak.
   G. O. No. 23, W. D., 1919.

Major, 102d Infantry, 26th Division.
Being second in command of the assaulting troops, he displayed great bravery and rare initiative. While under terrific artillery and machine-gun fire he reorganized scattered units, established and organized positions in depth, set up liaison from front to rear, and in hand-to-hand fighting personally led his men, inspiring in them a confidence and tenacity of purpose that were productive of success.
Oak-leaf cluster.
He is also awarded an oak-leaf cluster to be worn with the distinguished-service cross for the following act of extraordinary heroism in action near Beaumont, France, Nov. 10, 1918: Commanding his regiment, he personally led the advance of his front line, under a heavy artillery and machine-gun fire, and gained the absolute confidence of his troops by his example of courage and coolness.

*LEWIS, FRANK A. (263396)_____
Near Romange, France, Oct. 9 to 11, 1918.
R—Chavies, Ala.
B—Cherokee, Ala.
G. O. No. 64, W. D., 1919.

Private, first class, Company I, 125th Infantry, 32d Division.
During the period from Oct. 9–11, 1918, he repeatedly volunteered and carried messages from his company in the front line to his battalion post of command, crossing and recrossing a fire-swept valley while engaged in this work. After being dangerously wounded on his last trip, he bent all his energies to attracting the attention of another runner who would deliver his message, but the delay in securing first aid resulted in his death. His work during these days was not only an inspiration to his comrades but also of inestimable value in the success of the regimental attack.
Posthumously awarded. Medal presented to father, Joseph W. Lewis.

*LEWIS, FRANK N_____
Near Cunel, France, Oct. 5–6, 1918.
R—Memphis, Tenn.
B—Cismont, Va.
G. O. No. 37, W. D., 1919.

Captain, 4th Infantry, 3d Division.
Although severely wounded in both arms and both legs during the advance of his company on the night of Oct. 5–6, Captain Lewis continued in command, and by his bravery and courage contributed materially to the complete success of the attack.
Posthumously awarded. Medal presented to mother, Mrs. T. W. Lewis.

LEWIS, HAROLD A. (1277011)_____
Near Ravine de la Reine, north of Samogneux, France, Oct. 10, 1918.
R—Jersey City, N. J.
B—Jersey City, N. J.
G. O. No. 130, W. D., 1918.

Sergeant, Company K, 113th Infantry, 29th Division.
After his platoon commander had been wounded, Sergeant Lewis, although twice wounded himself, continued to lead the platoon until its objective was reached. By his bravery and persistency the platoon, greatly outnumbered, succeeded in overcoming the enemy.

*LEWIS, HARRY J. (549264)_____
Near Jaulgonne, France, July 23, 1918.
R—Martville, N. Y.
B—Rochester, N. Y.
G. O. No. 22, W. D., 1920.

Private, Machine Gun Company, 38th Infantry, 3d Division.
Private Lewis repeatedly volunteered and carried messages across a zone swept by artillery, machine-gun, and rifle fire. He fearlessly and efficiently performed his duties until killed.
Posthumously awarded. Medal presented to brother, Merton Lewis.

LEWIS, JAMES (284458)_____
Near Exermont, France, Oct. 4–11, 1918.
R—Hay River, Wis.
B—Boyceville, Wis.
G. O. No. 98, W. D., 1919.

Private, Company I, 28th Infantry, 1st Division.
He volunteered repeatedly for dangerous missions, and, a private at the time, took command of the platoon when all the noncommissioned officers had become casualties. In leading his men forward he inspired them by his personal bravery, ably directing the consolidation of the objective when taken. Placed in permanent command of his platoon by his company commander, he led an attack with marked disregard of personal danger and later led ration and water details through heavily gassed and shelled areas to obtain supplies for his men.

LEWIS, MADISON H_____
Near Ville-Savoye, France, Aug. 18, 1918.
R—New York, N. Y.
B—New York, N. Y.
G. O. No. 98, W. D., 1919.

Captain, 302d Engineers, 77th Division.
Under enemy fire, high explosive and gas, Captain Lewis voluntarily plunged into the Vesle River to rescue some soldiers who had fallen into the water with full packs while crossing a footbridge and were in danger of drowning. In order to see he removed his gas mask, and as a result was severely gassed.

LEWIS, MORRISON F. (1249842)_____
Near Baslieux, France, Sept. 5, 1918.
R—New Castle, Pa.
B—Jeanette, Pa.
G. O. No. 71, W. D., 1919.

Corporal, Headquarters Company, 107th Field Artillery, 28th Division.
He directed a detail running a telephone line to an advanced observation post under heavy artillery and machine-gun fire, and when shellfire rendered the maintenance of the line impossible, directed his detail in the evacuation of wounded infantrymen. Subjecting himself to intense enemy fire he carried to the rear the body of an officer who had been killed by enemy machine-gun fire.

LEWIS, ROBERT D. (1749676)_____
Near Grand Pre, France, Oct. 27, 1918.
R—Buffalo, N. Y.
B—Elmira, N. Y.
G. O. No. 35, W. D., 1919.

Corporal, Company M, 311th Infantry, 78th Division.
After his company had reached its objective, Corporal Lewis rendered valuable assistance in organizing positions on ground swept by enemy fire. Alone, he flanked a machine-gun position and captured two prisoners. While patroling between the outposts he was wounded by machine-gun fire.

LEWIS, ROBERT F. (1315851)_____
Near Bellicourt, France, Sept. 29, 1918.
R—Bolton, N. C.
B—Wilmington, N. C.
G. O. No. 46, W. D., 1919.

Corporal, Company G, 119th Infantry, 30th Division.
His section having been stopped by a concealed machine gun, Corporal Lewis, on his own initiative, crawled forward alone over ground swept by machine-gun fire. Attacking the nest with bombs and firing at it with his rifle, he killed the entire crew, numbering 7, and thereby cleared the way for the further advance of his section.

LEWIS, STACY A. (105252)_____
Near Soissons, France, July 22, 1918.
R—Kildare, Okla.
B—Silverdale, Kans.
G. O. No. 15, W. D., 1919.

Sergeant, Company A, 2d Machine Gun Battalion, 1st Division.
He voluntarily organized a machine-gun crew, moved forward in front of the Infantry under heavy machine-gun and shell fire, killed an entire machine-gun crew, and captured the gun.

LEWIS, WILLIAM F_____
Near Laksamana Usap's Cotta, island of Jolo, Philippine Islands, Jan. 7, 1905.
R—Kinston, N. C.
B—Chester Springs, Va.
G. O. No. 4, W. D., 1923.

Captain, Medical Corps, U. S. Army.
During the attack on the cotta, Captain Lewis showed great gallantry and courage when he went into the line of fire to the assistance of a badly wounded soldier, whom he rescued and carried to the dressing station, where the man later died.

LEWIS, WILLIAM PENN, Jr............
Near Verdun, France, Oct. 16, 1918.
R—Baltimore, Md.
B—Baltimore, Md.
G. O. No. 46, W. D., 1920.

First lieutenant, 115th Infantry, 29th Division.
Lieutenant Lewis, although ill at the time, led his platoon in an attack in the Bois-de-Grande Montagne against numerous enemy machine-gun nests and quickly dispatched the enemy troops therein. Finally he advanced alone and captured an enemy sniper from a tree and brought him into our lines. He remained with his company until Oct. 18, when he collapsed with shell shock and exhaustion.

LIBERMAN, LOUIS (125221)............
At Villers-Tournelle, Cantigny sector, France, May 1, 1918.
R—New York, N. Y.
B—Russia.
G. O. No. 100, W. D., 1918.

Corporal, Headquarters Company, 6th Field Artillery, 1st Division.
He displayed distinguished bravery in twice leaving his shelter during a heavy bombardment and going to the assistance of wounded men lying exposed in the open.

LICKLITER, JOHN D. (57133)...........
Near Berzy-le-Sec, France, July 20, 1918.
R—Martinsburg, W. Va.
B—Martinsburg, W. Va.
G. O. No. 117, W. D., 1918.

Sergeant, Company D, 28th Infantry, 1st Division.
He gave proof of unhesitating devotion and energy in brilliantly leading four men against a machine gun which was delivering intense fire and capturing the gun.

LIDDELL, CHARLES J. (1489634).........
Near Attigny, France, Oct. 14, 1918.
R—Marietta, Okla.
B—Thackerville, Okla.
G. O. No. 50, W. D., 1919.

Sergeant, Company D, 142d Infantry, 36th Division.
Sergeant Liddell was engaged on a reconnaissance with another soldier when the latter was wounded. Being unable to carry his comrade back to the dressing station, he placed him in a sheltered position and made his way to regimental headquarters, 4 kilometers to the rear, to secure an ambulance. The ambulance being driven back by enemy artillery fire, Sergeant Liddell made another trip to the rear and returned with a side car, in which he succeeded in rescuing the wounded man, despite heavy machine-gun fire.

LIDWELL, EDWARD J. (1387913).........
Near Bois-de-Chaume, France, Oct. 9, 1918.
R—Chicago, Ill.
B—Canada.
G. O. No. 126, W. D., 1919.

Private, Company H, 131st Infantry, 33d Division.
Advancing single handed against a machine gun, Private Lidwell put it out of action, killing its crew of 3 and preventing an enfilading fire on the company, thus saving many lives.

LIEBERMAN, NATHAN (1386775)..........
At Chipilly Ridge, France, Aug. 9, 1918.
R—Chicago, Ill.
B—Chicago, Ill.
G. O. No. 123, W. D., 1918.

Corporal, Company C, 131st Infantry, 33d Division.
He displayed unusual gallantry in rushing a machine-gun nest whose fire was checking the advance. With the assistance of men in his squad he put the machine gun out of action and took 4 prisoners.

LIEBESKIND, HARRY (2353)............
In the Bois-de-Caures, north of Verdun, France, Nov. 8, 1918.
R—Waterbury, Conn.
B—Brooklyn, N. Y.
G. O. No. 3, W. D., 1921.

Private, first class, medical detachment, 102d Infantry, 26th Division.
Private Liebeskind exposed himself to heavy enemy fire to accompany a medical officer to a badly wounded soldier. The patient was carried to a shell hole, and later Private Liebeskind exposed himself to heavy fire in order to cut a passage through a wire entanglement. This act made it possible to evacuate the patient to the rear.

LIENHARD, JACOB............
Near St. Etienne, France, Oct. 4, 1918.
R—Plymouth, Wis.
B—Plymouth, Wis.
G. O. No. 46, W. D., 1919.

Second lieutenant, 5th Regiment, U. S. Marine Corps, 2d Division.
He led his men in an attack on a strongly held enemy position through heavy machine-gun and shellfire, and although severely wounded continued to lead and encourage his men.

*LIETZAN, JOSEPH (126340)............
At Couellemelle, France, Apr. 27, 1918.
R—Hammond, Ind.
B—Hammond, Ill.
G. O. No. 88, W. D., 1918.

Private, Battery E, 6th Field Artillery, 1st Division.
Under a heavy bombardment, he voluntarily went to the assistance of other soldiers who had been buried in a dugout by enemy shellfire, and was killed while engaged in this heroic act.
Posthumously awarded. Medal presented to mother, Mrs. Anna Lietzan.

LIEUALLEN, FRED ADCOOK............
Near Sergy, France, July 28 to 31, 1918.
R—Portland, Oreg.
B—Umatilla County, Oreg.
G. O. No. 53, W. D., 1920.

Captain, Medical Corps, attached to 47th Infantry, 4th Division.
Captain Lieuallen operated a dressing station for 2 days under unusually heavy enemy fire. Our front line was for a time bent back by the enemy, thus exposing his position to capture by the enemy; he refused to leave his dressing station, and continued to attend to the needs of 100 wounded men until the lost ground was retaken by our troops. This officer performed gallant service also at St. Thibaut, France, Aug. 6 to 12, 1918, while maintaining a dressing station with the advanced elements under heavy enemy fire.

LIGGETT, HARRY B............
Near Bois-de-Chaume, France, Oct. 10, 1918.
R—Freeport, Ill.
B—Broadhead, Wis.
G. O. No. 44, W. D., 1919.

Second lieutenant, 122d Machine Gun Battalion, 33d Division.
Leading his platoon, under heavy shell and machine-gun fire, Lieutenant Liggett launched an attack on 2 enemy machine-gun nests. Accompanied by 1 soldier, he silenced the fire from 1 nest with rifle fire, and directed the fire of his platoon so that the other nest was destroyed. He was severely wounded in this action.

LIGHT, JOHN C. (1467652)............
Near Cheppy, France, Sept. 26, 1918.
R—Topeka, Kans.
B—Plain View, Kans.
G. O. No. 59, W. D., 1919.

Corporal, Company A, 110th Engineers, 35th Division.
He fearlessly attacked a machine-gun position and single handed killed 1 gunner and captured another. Later, when his entire squad was wounded he refused to be evacuated, although severely wounded himself, and remained at his post until his entire squad had been cared for.

LIGHT, LAVERN (2417804)...............
Near Grand Pre, France, Nov. 1, 1918.
R—Cooperstown, N. Y.
B—Exeter, N. Y.
G. O. No. 93, W. D., 1919.

Corporal, Company E, 311th Infantry, 78th Division.
While his company and 2 others were fighting to prevent being cut off and annihilated by the enemy, Corporal Light voluntarily crept out 10 yards in advance of the line and killed 5 of the enemy who were attempting to execute an encircling movement with machine guns. Remaining in his exposed position with cool tenacity he contributed materially toward reducing the hostile forces and making possible a rush by his company.

LIGHT, WILBUR S. (2220716)...........
Near Fey-en-Haye, France, Sept. 12–14, 1918.
R—Durant, Okla.
B—Fayetteville, W. Va.
G. O. No. 128, W. D., 1918.

Corporal, Company F, 358th Infantry, 90th Division.
His daring and bravery were conspicuous. He charged a German machine gun single handed, killing the gunner and putting 3 others of the crew to flight. During this action he killed 6 of the enemy. He showed rare leadership on numerous occasions in forming groups and leading them against machine-gun nests.

LIGHTNER, BLAKE...................
At Courmont, France, July 29–31, 1918, and near Montblainville, France, Sept. 27 to Oct. 3, 1918.
R—Rosebud, Pa.
B—Irvona, Pa.
G. O. No. 98, W. D., 1919.

Second lieutenant, 110th Infantry, 28th Division.
Lieutenant Lightner voluntarily established an advance observation post at Courmont. During his work he was knocked down by the concussion of an exploding shell, but remained at his post. Throughout the action in the Argonne he repeatedly exposed himself while leading his men. At Montblainville, although wounded by a shell splinter, he continued in action and succeeded in putting into operation German machine guns against the enemy, greatly assisting in repulsing their counterattack. He refused to be evacuated until ordered to the rear.

LIGON, LOUIS LUCIUS..............
Near Brancourt, France, Oct. 8, 1918.
R—Anderson, S. C.
B—Piedmont, S. C.
G. O. No. 50, W. D., 1919.

Captain, 118th Infantry, 30th Division.
Commanding a battalion which had been caught in a terrific barrage, Captain Ligon pushed forward and led all his command until the barrage had lifted. Although he was severely gassed, which rendered him nearly speechless and caused much suffering, he remained with his troops for 8 days, leaving his post only when ordered to do so by his commanding officer.

LIGSUKIS, FRANK (1899406)..........
Near St. Juvin, France, Oct. 16, 1918.
R—Hartford, Conn.
B—Russia.
G. O. No. 81, W. D., 1919.

Mechanic, Company I, 325th Infantry, 82d Division.
With another soldier, Mechanic Ligsukis voluntarily made several trips into No Man's Land under heavy enemy fire and carried to safety 8 wounded comrades who had been lying helpless and exposed to this fire.

LILJEBERG, RAGNAR (2087137).........
Near Chipilly Ridge, France, Aug. 9–11, 1918.
R—Chicago, Ill.
B—Sweden.
G. O. No. 64, W. D., 1919.

Private, Company D, 131st Infantry, 33d Division.
Being on duty as a runner, he carried messages under heavy shell and machine-gun fire. Owing to casualties he did the work of 6 runners, proving himself to be a man of unusual gallantry and devotion to duty.

LILLARD, DAVID W.............
Near Ponchaux, France, Oct. 7, 1918.
R—Etowah, Tenn.
B—Decatur, Tenn.
G. O. No. 81, W. D., 1919.

Captain, 117th Infantry, 30th Division.
Severely wounded in the side when an enemy machine-gun bullet struck and exploded two clips of shells in his magazine pouch, Captain Lillard struggled to his feet and directed the further advance of his company. For 6 hours he remained in command of his company, issuing orders from a shell hole under the most intense fire. During part of this period he was practically unconscious and was suffering severe pain, but he nevertheless successfully accomplished the organization of his company's position.

*LIMON, JOE (2268298).............
Near St. Thibaut, France, Aug. 10, 1918.
R—Seattle, Wash.
B—Spain.
G. O. No. 32, W. D., 1919.

Private, Company M, 47th Infantry, 4th Division.
Wounded in a scouting mission, Private Limon remained on observation until he had gained valuable information. After having his wound dressed, he returned to duty and made several trips to the flank regiments, each time bringing back valuable information for our own troops and of the enemy. He later voluntarily located a minnenwerfer and heavy mortar emplacement from which the enemy was firing on neighboring troops.
Posthumously awarded. Medal presented to father, Erminio Limon.

LINBERG, ALBERT W...........
Near Evermont, France, Oct. 8, 1918.
R—Hillsboro, Ill.
B—Sweden.
G. O. No. 34, W. D., 1924.

Captain, Medical Corps, attached to 18th Infantry, 1st Division.
Crawling out over shell-torn ground and in the face of direct machine-gun fire, he rescued a soldier whose leg had been shattered by shellfire. By performing an amputation while constantly exposed to sniping and machine-gun fire he made it possible to evacuate the soldier to safety a few hours later, when the infantry moved forward.

LINDAHL, LUTHER E. (553271).........
Near Bois-de-Brieulles, France, Sept. 28, 1918.
R—Sheffield, Pa.
B—Sheffield, Pa.
G. O. No. 46, W. D., 1919.

Sergeant, Company I, 47th Infantry, 4th Division.
Sergeant Lindahl charged an enemy machine gun which was inflicting heavy losses upon our troops and delaying the advance. He wounded the gunner and captured the gun, thereby enabling our advance to continue.

LINDEN, HARRY (1709302)............
Near Ville-Savoye, France, Aug. 16, 1918.
R—Brooklyn, N. Y.
B—Brooklyn, N. Y.
G. O. No. 32, W. D., 1919.

Sergeant, Company H, 308th Infantry, 77th Division.
After all his company officers and first sergeant had been evacuated because of gas, Sergeant Linden assumed command of the company, which was then occupying an extremely precarious position, exposed to an unusually heavy shell and gas bombardment. He remained in command until the company was relieved, and the following night, despite his sufferings from the effects of gas helped carry up ammunition under intense enemy artillery fire.

LINDGREN, EVERETTE E.
Near St. Etienne, France, Oct. 4, 1918.
R—Detroit, Mich.
B—Adrian, Minn.
G. O. No. 35, W. D., 1919.

Second lieutenant, 5th Regiment, U. S. Marine Corps, 2d Division.
During an attack on a strong enemy position, under terrific machine-gun and artillery fire, he led his platoon to the support of the platoon operating on his left, rallying men from another company, who had become separated from their organization, to his support. Although severely wounded, he remained in action until the position was consolidated.

LINDHOLM, REUBEN P.
In the Argonne Forest, France, Oct. 4, 1918.
R—Bayshore, Long Island, N. Y.
B—New York, N. Y.
G. O. No. 19, W. D., 1920.

Second lieutenant, 307th Infantry, 77th Division.
Although wounded severely in the arm by a machine-gun bullet, he continued to lead his company. After making the disposition of his company he reported in person to his commanding officer. He then returned to his company and continued in command until relieved.

LINDIE, ELMER H. (68223).
Near Bois-de-St. Remy, France, Sept. 12, 1918.
R—Monson, Me.
B—Monson, Me.
G. O. No. 26, W. D., 1919.

Private, Company F, 103d Infantry, 26th Division.
Under heavy grenade and rifle fire he crawled forward from shell hole to shell hole until he reached a flank position of an enemy machine-gun nest, from which point he killed a gunner and caused the rest to surrender to his comrades.

LINDQUIST, EDWARD N. (46198).
South of Soissons, France, July 19, 1918.
R—Dunnegan, Mo.
B—Leonard, Kans.
G. O. No. 72, W. D., 1920.

Sergeant, Company B, 18th Infantry, 1st Division.
His platoon having been halted by heavy machine-gun fire, Sergeant Lindquist went forward through heavy machine-gun fire and, single handed, attacked an enemy sniper who, armed with a light machine gun, was inflicting many casualties on his organization. He killed the sniper and thus enabled his organization to continue the advance.

LINDSAY, GRANT S. (95901).
Near Sedan, France, Nov. 7, 1918.
R—Lancaster, Ohio.
B—Fairfield County, Ohio.
G. O. No. 81, W. D., 1919.

Sergeant, Company L, 166th Infantry, 42d Division.
Sergeant Lindsay was in command of his platoon on the right flank of the assaulting wave, when hidden concentrations of machine guns were encountered in front and on the right flank. Skillfully maneuvering his combat groups, he led them with remarkable courage and coolness through the heavy enemy machine-gun fire and broke up the hostile counterattack which had been started. In so doing he personally advanced ahead of his men under heavy fire from machine guns and snipers and succeeded in locating some of the machine-gun nests.

LINDSAY, ROBERT O.
Near Bantheville, France, Oct. 27, 1918.
R—Madison, N. C.
B—Madison, N. C.
G. O. No. 46, W. D., 1919.

First lieutenant, 139th Aero Squadron, Air Service.
In company with two other planes, Lieutenant Lindsay attacked 3 enemy planes (Fokker type) at an altitude of 3,000 meters, and after a sharp fight brought down 1 of them. While engaged with the 2 remaining machines 8 more planes (Fokker type) came at him from straight ahead. He flew straight through their formation, gained an advantageous position, and brought down another plane before he withdrew from the combat.

LINDSAY, ROBERT W. (1386578).
Near Consenvoye, France, Oct. 10, 1918.
R—Morris, Ill.
B—Morris, Ill.
G. O. No. 71, W. D., 1919.

Private, Company B, 131st Infantry, 33d Division.
He volunteered to carry messages over ground swept by heavy fire after 2 other runners had been wounded. He delivered important messages working to maintain liaison after he had himself been wounded.

*LINDSEY, CLINTON S.
In the Bois-de-Belleau, France, June 6, 1918.
R—San Marcos, Tex.
B—Bertram, Tex.
G. O. No. 110, W. D., 1918.

Private, 82d Company, 6th Regiment, U. S. Marine Corps, 2d Division.
He displayed commendable gallantry when he voluntarily sought permission to leave shelter and, at great exposure to himself, rescued a helpless wounded officer from an open field.
Posthumously awarded. Medal presented to father, Felix W. Lindsey.

LINDSEY, CLYDE H. (3489185).
Near Bois-de-Brieulles, France, Sept. 29, 1918.
R—Cambridge, Ill.
B—Maywood, Mo.
G. O. No. 46, W. D., 1919.

Private, Company A, 59th Infantry, 4th Division.
Disregarding personal safety, Private Lindsey, in the performance of his duties as runner, carried repeated messages across a greatly exposed area, which was subjected to fierce artillery and machine-gun fire. He succeeded after another runner had been killed in the attempt.

LINDSTEN, ALBIN (2156105).
Near Bellicourt, France, Sept. 29, 1918.
R—Whitebear, Minn.
B—Whitebear, Minn.
G. O. No. 32, W. D., 1919.

Private, Company H, 117th Infantry, 30th Division.
Responding to a call for volunteers by his company commander to locate the source of machine-gun fire which had cut off the headquarters platoon from the rest of the company, Private Lindsten crawled through the barrage, ascertained the position of the guns, and led the platoon to safety. He then crawled back through the barrage, located the rest of the company, and guided the platoon to them.

LINER, IRVING L. (1708384).
Near Binarville, France, Oct. 2–7, 1918.
R—Brooklyn, N. Y.
B—New York, N. Y.
G. O. No. 21, W. D., 1919.

Private, Company D, 308th Infantry, 77th Division.
He was a battalion runner, when his battalion was surrounded by the enemy in the forest of Argonne and cut off from communication with friendly troops. He carried messages with great cheerfulness under conditions of stress and under heavy machine-gun and shellfire, at a time when he was exhausted by exposure and hunger, being without food for 5 days.

LINGO, LOVICK P.
At Cornay, France, Oct. 9–10, 1918.
R—Milledgeville, Ga.
B—McIntyre, Ga.
G. O. No. 81, W. D., 1919

First lieutenant, 328th Infantry, 82d Division.
Lieutenant Lingo was with an attacking party which, after driving off the enemy, was counterattacked and surrounded. Upon being called on to surrender, he refused and, despite the fact that 10 men had been shot down in trying to get away, fought his way out and, though wounded, reached his own lines. Later, when Lieutenant Lingo learned that his company was without officers, he returned and remained with it for several days until weakness from his wound forced his evacuation.

*LINIGER, WILLIAM (2305455)_____
  Near Romagne, France, Oct. 9, 1919.
  R—Muskegon, Mich.
  B—Muscatine, Iowa.
  G. O. No. 71, W. D., 1919.

Sergeant, Company I, 126th Infantry, 32d Division.
After all the officers of his company had been killed or wounded, an attack by his platoon was ordered on an enemy machine-gun nest. Receiving a mortal wound in the attack which captured the enemy nest, Sergeant Liniger struggled on until the objective had been reached, encouraging his men and setting them an example by his bravery.
Posthumously awarded. Medal presented to mother, Mrs. Elsie Liniger

LINK, OLLIE R. (1316897)_____
  Near St. Souplet, France, Oct. 9-10, 1918.
  R—Buies Creek, N. C.
  B—Person County, N. C.
  G. O. No. 37, W. D., 1919.

Cook, Company M, 119th Infantry, 30th Division.
Hearing that the casualties in his company were very heavy, he left his place in the kitchen and joined his comrades on the front line. From there he advanced alone a distance of 200 yards and located 2 machine-gun nests, the journey being done in the face of heavy enemy fire.

*LINSKEY, WILLIAM F. (1387408)_____
  At Hamel, France, July 4, 1918.
  R—Chicago, Ill.
  B—Chicago, Ill.
  G. O. No. 99, W. D., 1918.

Private, Company E, 131st Infantry, 33d Division.
He was severely wounded in the right arm by shrapnel at the beginning of the battle. Nevertheless he carried his automatic rifle forward and used it effectively in the assault on the village.
Posthumously awarded. Medal presented to father, Patrick Linskey.

LINTON, CLYDE W. (152364)_____
  Near Suippes, France, July 15, 1918.
  R—Akron, Mich.
  B—Unionville, Mich.
  G. O. No. 21, W. D., 1919.

Corporal, Battery E, 42d Artillery, Coast Artillery Corps.
He displayed remarkable courage and devotion to duty as a lineman in charge of exposed telephone lines between his battery commander's station and the firing battery during the German attack at Suippes. At great personal risk he repaired the lines as they were broken by shellfire. In performing this duty he was severely wounded.

*LINTON, FREDERICK M_____
  Near Marcheville and Riaville, France, Sept. 25-26, 1918.
  R—Roslindale, Mass.
  B—Boston, Mass.
  G. O. No. 44, W. D., 1919.

First Lieutenant, 51st Infantry Brigade, 26th Division.
Lieutenant Linton, while acting as liaison officer with brigade headquarters, volunteered to carry a message from the front lines to the rear through a terrific barrage and murderous machine-gun fire. After successfully accomplishing this mission he returned with a platoon of reinforcements across an open field through the same heavy fire. When the town of Marcheville fell into the enemy's hands, he volunteered to lead a platoon in the counterattack and was wounded while in command. He retained command and held his ground with the platoon until he received his second and fatal wound.
Posthumously awarded. Medal presented to widow, Mrs. Catherine S. Linton.

*LIPPE, OSCAR P. (1707011)_____
  Near Vesle River, France, Aug. 29, 1918.
  R—New York, N. Y.
  B—New York, N. Y.
  G. O. No. 21, W. D., 1919.

Sergeant, Company L, 307th Infantry, 77th Division.
He saw a wounded comrade 50 yards away and immediately started to his assistance across an open space covered by machine-gun and sniper fire. He was killed attempting to reach his comrade.
Posthumously awarded. Medal presented to mother, Mrs. Therese Radley Lippe.

*LIPPITT, ALEXANDER F_____
  In the Champagne sector, north of Chalons, France, July 15, 1918.
  R—Providence, R. I.
  B—Providence, R. I.
  G. O. No. 53, W. D., 1920.

First lieutenant, 166th Infantry, 42d Division.
During a powerful enemy attack, Lieutenant Lippitt led his platoon through heavy artillery and machine-gun fire in a counterattack against the enemy, which had gained a foothold in our line. The enemy was repulsed and the line reestablished. He assisted in the reorganization and defense of the position against 2 enemy assaults. The gallantry of this officer was a great aid to his command at a time of unusual danger. He was killed a few days later while advancing to an attack.
Posthumously awarded. Medal presented to father, Charles W. Lippitt.

LIPSCOMB, ABNER E. (1488021)_____
  Near St. Etienne, France, Oct. 8, 1918.
  R—Brenham, Tex.
  B—Brenham, Tex.
  G. O. No. 37, W. D., 1919.

Sergeant, Company I, 141st Infantry, 36th Division.
After all the officers of his company were either killed or wounded, Sergeant Lipscomb took command of his company, and, although twice wounded, continued to advance. He led the company with courage and skill, held difficult positions in the exposed salient occupied by the first battalion, and established a flank guard.

LISA, JAMES R_____
  Near St. Souplet, France, Oct. 18, 1918.
  R—Randalls Island, N. Y.
  B—Calumet, Mich.
  G. O. No. 23, W. D., 1919.

First lieutenant, 105th Infantry, 27th Division.
After his battalion had been compelled to withdraw because of enfilading fire, Lieutenant Lisa displayed marked bravery in going forward and attending wounded men, whose evacuation was impossible because of the intense fire.

*LISTER, JOHN M. (2264913)_____
  Near Waereghem, Belgium, Oct. 31, 1918.
  R—San Francisco, Calif.
  B—Austria.
  G. O. No. 20, W. D., 1919.

Corporal, Company K, 363d Infantry, 91st Division.
When the progress of 2 front-line companies had been stopped by the intense fire of enemy machine guns and the enemy began to close in on these troops, he voluntarily assembled a squad of automatic riflemen and grenadiers and went forward to an exposed position, where he directed the fire of his squad until killed.
Posthumously awarded. Medal presented to father, Michael Lister.

*LITCHFIELD, JOHN RUSSELL_____
  Near Thiaucourt, France, Sept. 15, 1918.
  R—Blackwell, Okla.
  B—Flanagan, Ill.
  G. O. No. 27, W. D., 1919.

Pharmacist's mate, third class, U. S. Navy, attached to 74th Company, 6th Regiment, U. S. Marine Corps, 2d Division.
He displayed exceptional bravery in giving first aid to the wounded under shellfire. He was killed while taking a wounded soldier out of a trench to the rear.
Posthumously awarded. Medal presented to mother, Mrs. Martha D. Litchfield.

LITTAUER, KENNETH P.
Near Conflans, France, Sept. 14, 1918, and near Doulcon, France, Oct. 30, 1918.
R—Branford, Conn.
B—Newark, N. J.
G. O. No. 37, W. D., 1919.

Captain, 88th Aero Squadron, Air Service.
He volunteered on a mission to protect a photographic plane for another squadron on Sept. 14 and continued toward the objective at Conflans even after 3 other protecting planes had failed to start. In an encounter with 5 enemy pursuit planes he completely protected the photographic plane by skillful maneuvering, although his observer was wounded and his machine seriously damaged. On Oct. 30, while on duty as Chief of Air Service of the 3d Army Corps, he volunteered and made an important reconnaissance of enemy machine-gun emplacements at a low altitude near Doulcon.

LITTLEFIELD, FRANK C. (1910940).
At Cornay, France, Oct. 10, 1918.
R—Winterport, Me.
B—Winterport, Me.
G. O. No. 87, W. D., 1919.

Corporal, Company K, 328th Infantry, 82d Division.
Corporal Littlefield was a member of a detachment which had been surrounded by the enemy and had fought for several hours against overwhelming odds. When the officer in charge was about to surrender to the enemy, this soldier refused to do so, and, dashing through severe machine-gun and rifle fire, succeeded in making his escape, though he was painfully wounded in the face in so doing. Reporting to his company without delay, he remained on duty throughout the entire operation.

LITTLEJOHN, KENNETH S.
At Claire-Chenes Woods, France, Oct. 20, 1918.
R—Nogales, Ariz.
B—Montclair, N. J.
G. O. No. 137, W. D., 1918.

Captain, 6th Engineers, 3d Division.
He reorganized 3 Engineer companies after they had retired from the woods, and by his personal example of daring and bravery successfully led his men against enemy machine guns. His gallant action resulted in the capture of of the Claire-Chenes Woods.

LIVERMORE, RUSSELL B.
Near the Bois-de-Belleau, France, July 18, 1918.
R—Yonkers, N. Y.
B—Yonkers, N. Y.
G. O. No. 81, W. D., 1919.

First lieutenant, 104th Infantry, 26th Division.
As his battalion was advancing across an open field it came under fire from a hostile machine gun located in a strong enfilading position in a ravine. Hastily gathering a group of men from his platoon, Lieutenant Livermore led them in a charge on the nest and put it out of action, capturing 11 prisoners and saving his battalion many casualties.

LLEWELLYN, FRANK A.
East of St. Die, France, Aug. 17, 1918.
R—Chicago, Ill.
B—Minneapolis, Minn.
G. O. No. 121, W. D., 1918.

First lieutenant, 99th Aero Squadron, Air Service.
Lieutenant Llewellyn, acting as pilot, and Lieutenant Neel, acting as observer, carried on successful liaison with the Infantry during the attack on Frapelle. They flew over the enemy lines at an altitude of only 400 meters, firing on and disconcerting the enemy, and thereby giving courage and confidence to the American forces. Despite heavy fire from 15 antiaircraft machine guns and several batteries of antiaircraft artillery, they performed their work efficiently. Their aeroplane was struck by a number of machine-gun bullets, one of which cut the rudder and elevator control wires and caused the rudder to jam. The broken control wire was held and operated by Lieutenant Neel, under direction of Lieutenant Llewellyn. Running the machine together in this manner they continued their liaison work until the plane began to become unmanageable, when, in spite of its damaged condition, they brought it back to their airdrome.

LLOYD, DALE W. (56128).
Near Exermont, France, Oct. 5–11, 1918.
R—Marengo, Ohio.
B—Sparta, Ohio.
G. O. No. 44, W. D., 1919.

Private, Headquarters Company, 28th Infantry, 1st Division.
Advancing with front-line units, he established observation posts under heavy and continuous shell and machine-gun fire. Although suffering from wounds, he remained at his post and rendered valuable reports regarding the progress of the battle.

LLOYD, WILFRED (275108).
Near Fismes, France, Aug. 4, 1918.
R—Beloit, Wis.
B—Roscoe, Ill.
G. O. No. 128, W. D., 1918.

Private, first class, Company L, 127th Infantry, 32d Division.
Wounded while advancing under machine-gun fire, he showed highest devotion to duty by returning to his company as soon as he had obtained first-aid treatment. He crawled to a road, secured a pistol to replace the one he had lost when he was wounded, and joined a group which attacked a machine-gun nest, capturing the position and the machine gun.

LOCKARD, DENNIS H. (198629).
Near Exermont, France, Oct. 5, 1918.
R—Muncy, Pa.
B—Muncy, Pa.
G. O. No. 44, W. D., 1919.

Sergeant, Company C, 2d Field Signal Battalion, 1st Division.
Leading a detail through an intense bombardment, Sergeant Lockard kept in operation the telephone lines of regimental headquarters, thereby enabling his regimental commander to keep in touch with elements on the firing line.

*LOCKE, KARL W.
At Chateau-Thierry, France, June 6, 1918.
R—Cleveland, Ohio.
B—Oberlin, Ohio.
G. O. No. 110, W. D., 1918.

Corporal, 51st Company, 5th Regiment, U. S. Marine Corps, 2d Division.
Killed in action at Chateau-Thierry, France, June 6, 1918, he gave the supreme proof of that extraordinary heroism which will serve as an example to hitherto untried troops.
Posthumously awarded. Medal presented to mother, Mrs. Elizabeth Locke.

LOCKE, RAYMOND I. (144600).
At Varennes, France, Sept. 26, 1918.
R—Clinton, Mo.
B—Clinton, Mo.
G. O. No. 98, W. D., 1919.

Sergeant, Company A, 129th Machine Gun Battalion, 35th Division.
Accompanied by another soldier, Sergeant Locke went through heavy fire, and by skillful maneuvering located and captured an enemy 77-millimeter gun with part of its crew, which had been checking the advance of our troops.

*LODER, JAMES C.
Near Soissons, France, July 18, 1918.
R—Chattanooga, Tenn.
B—Chattanooga, Tenn.
G. O. No. 132, W. D., 1918.

Second lieutenant, 26th Infantry, 1st Division.
On July 18, 1918, he gallantly inspired his platoon to 3 vigorous and successful advances against machine-gun fire near Soissons, France, in the last of which he was killed.
Posthumously awarded. Medal presented to widow, Mrs. James C. Loder.

LOESCHER, FRANK B. (2230212)..........
    At Le Grand Carre Ferme, France,
      Nov. 1, 1918.
    R—Sealy, Tex.
    B—Sealy, Tex.
    G. O. No. 46, W. D., 1919.

Sergeant, Company K, 360th Infantry, 90th Division.
Although wounded in the arm early in the attack, Sergeant Loescher continued to lead his platoon in the attack on a strong enemy position, and by the use of rifles, rifle grenades, and hand grenades, and after capturing several machine-gun nests, including one dugout containing 70 Germans, forced the entire strong point to yield, including two German companies armed with machine guns, which had been holding up the advance of the whole company.

*LOGUE, THOMAS (297562)..........
    Near Montfaucon, France, Sept. 29,
      1918.
    R—Detroit, Mich.
    B—Ireland.
    G. O. No 66, W. D., 1919.

Private, Battery D, 119th Field Artillery, 32d Division.
Private Logue worked ceaselessly under heavy enemy shellfire in carrying wounded comrades to the first-aid station until he was himself severely wounded.
Posthumously awarded. Medal presented to sister, Miss Annie Logue.

LOHMANN, LEWIS E. (1040424)..........
    Near Laneuville, France, Nov. 7,
      1918.
    R—Pekin, Ill.
    B—Canton, Ill.
    G. O. No. 44, W. D., 1919.

Corporal, headquarters 3d Battalion, 11th Field Artillery, 6th Division.
Corporal Lohmann displayed fearless devotion to duty in maintaining telephone lines while his battalion was in action near Laneuville. On one occasion he voluntarily accompanied an officer and, under heavy fire, repaired a telephone line which was severed 15 times in 2 hours within a length of only 1 kilometer.

LOKEN, CLARENCE (2814509)..........
    Near Beaufort, France, Nov. 4, 1918.
    R—Morris, Wis.
    B—Morris, Wis.
    G. O. No. 37, W. D., 1919.

Private, Company M, 356th Infantry, 89th Division.
He was engaged as a runner and made several trips, through heavy bombardment, from the front line to the support battalion. Even after his equipment had been riddled by shell fragments, he continued maintaining liaison between these 2 points.

LOMONACO, FRANK (3191095)..........
    In the Meuse-Argonne offensive,
      France, Nov. 4-5, 1918.
    R—Brooklyn, N. Y.
    B—Italy.
    G. O. No. 37, W. D., 1919.

Private, first class, Company K, 315th Infantry, 79th Division.
In the performance of his duties as runner he made repeated journeys across areas swept by machine-gun fire, but he never hesitated in the execution of his mission. On the firing line he was of most valuable assistance in destroying a counterattack.

LONADIER, JULES (1597382)..........
    Near Bayonville, France, Nov. 1-7,
      1918.
    R—Trichell, La.
    B—Trichell, La.
    G. O. No. 87, W. D., 1919.

Private, Company L, 23d Infantry, 2d Division.
Private Lonadier, a company runner, displayed exceptional courage in repeatedly passing through heavy enemy fire and delivering all messages intrusted to him in a prompt and efficient manner.

LONG, ARTHUR S. (44521)..........
    Near Hill 272, France, Oct. 9, 1918.
    R—Roberts, Mont.
    B—Wardner, Idaho.
    G. O. No. 71, W. D., 1919.

Private, Company D, 16th Infantry, 1st Division.
Facing direct fire from a 77-millimeter gun which was enfilading his company, he advanced against the gun with an automatic rifleman. Attacking the German gun position, he captured the crew, making it possible for Private Long's company to hold the ground it had gained.

*LONG, FRANK S..........
    Near Fleville, France, Oct. 5, 1918.
    R—Burlington, Iowa.
    B—Burlington, Iowa.
    G. O. No. 95, W. D., 1919.

First lieutenant, 110th Infantry, 28th Division.
Having been wounded in the side by shrapnel while caring for wounded men of his platoon, Lieutenant Long refused to be evacuated, but returned from the dressing station to his command. While withdrawing his platoon to a better position under a heavy barrage he was instantly killed by shellfire. His courage and self-sacrifice furnished a splendid inspiration to his men.
Posthumously awarded. Medal presented to father, Col. F. S. Long.

LONG, MILTON E..........
    At Cunel Heights, France, October,
      12, 1918.
    R—Columbus, Ga.
    B—Columbus, Ga.
    G. O. No. 98, W. D., 1919.

Second lieutenant, 7th Infantry, 3d Division.
After being severely wounded by a machine-gun bullet, Lieutenant Long displayed exceptional courage and determination by remaining with his platoon, moving it forward and clearing the woods of enemy machine guns, 3 of which he captured with their entire crews.

LONGFIELD, SIMON E. (50169)..........
    Near Bois-de-Clerembauts, France,
      June, 7, 1918.
    R—Hartford, Conn.
    B—Ireland.
    G. O. No. 108, W. D., 1919.

Corporal, Company D, 23d Infantry, 2d Division.
Even after having been seriously wounded, Corporal Longfield continued with his duties as runner throughout the whole night, after which he was ordered to the rear because of weakness caused by loss of blood.

LONGOWSKI, JOSEPH (561582)..........
    Near Bois-du-Fays, France, Oct. 4,
      1918.
    R—Winona, Minn.
    B—Winona, Minn.
    G. O. No. 71, W. D., 1919.

Private, Company L, 59th Infantry, 4th Division.
Under heavy fire, while performing a mission as battalion runner, he encountered an enemy patrol of 4 men and, forcing them to surrender, brought them to the rear.

LOOMIS, CASEY V..........
    Near Thiaucourt, France, Sept. 15,
      1918.
    R—Milwaukee, Wis.
    B—Walhalla, N. Dak.
    G. O. No. 37, W. D., 1919.

Corporal, 73d Company, 6th Regiment, U. S. Marine Corps, 2d Division.
During an enemy counterattack he voluntarily left a sheltered position and, in entire disregard for his own safety, set up his gun in the open under heavy enemy fire. By securing enfilading fire on the advancing enemy he broke up the counterattack within 100 yards of our line.

LOOMIS, JOHN H. (1738027)............. | Corporal, Company F, 311th Infantry, 78th Division.
Near Grand Pre, France, Oct. 28, 1918. R—Attica, N. Y. B—Attica, N. Y. G. O. No. 44, W. D., 1919. | Corporal Loomis volunteered to lead a 3-man patrol on a reconnaissance of enemy positions. After one of his men had been wounded, he continued on with the other one and gained the information sought. Being wounded on the return trip by fire from machine guns, he was forced to remain under cover until night, when he located his other wounded man and returned under cover of darkness with valuable information.

LOOMIS, JOHN S.......................... | First lieutenant, 132d Machine Gun Battalion, 36th Division.
Near St. Etienne, France, Oct. 10, 1918. R—Dallas, Tex. B—Dallas, Tex. G. O. No. 81, W. D., 1919. | During an attack to straighten the front of the 71st Brigade, Lieutenant Loomis, in command of a machine-gun platoon, upon arriving at the objective, discovered that all of the Infantry officers in the vicinity had fallen, thus leaving their troops in confusion. Although under a heavy barrage of high-explosive and gas shells, machine-gun and rifle fire, Lieutenant Loomis removed his gas mask in order that his voice might carry, and, with great coolness, reorganized the Infantry forces, thus enabling them to hold the ground they had gained. Lieutenant Loomis remained in command until he was so severely gassed that he had to be carried from the field.

LORD, ERNEST G. (794736)............. | Private, Company B, 15th Machine Gun Battalion, 5th Division.
At Frapelle, France, Aug. 17, 1918. R—Toledo, Ohio. B—Washington, D. C. G. O. No. 15, W. D., 1919. | Private Lord, although wounded severely early in the action and suffering great pain, refused to quit his gun squadron, but advanced with it until its objective was reached.

*LORING, DAVID WORTH................. | Second lieutenant, 115th Machine Gun Battalion, 30th Division.
Near Ypres, Belgium, Aug. 23, 1918. R—Wilmington, N. C. B—Sumter, S. C. G. O. No. 32, W. D., 1919. | When his gun positions were rendered untenable by shellfire and his men ordered to seek shelter in dugouts, Lieutenant Loring left a place of safety for the purpose of seeing that all of his men were under cover and was mortally wounded by a shell, dying on his way to the hospital.
 | Posthumously awarded. Medal presented to widow, Mrs. Viola Shaw Loring.

LOSCO, PATRICK (114893).............. | Private, first class, Company H, 9th Infantry, 2d Division.
Near Beaumont, France, Nov. 5, 1918. R—Brooklyn, N. Y. B—Brooklyn, N. Y. G. O. No. 16, W. D., 1920. | As a runner he showed absolute disregard for personal danger in carrying messages across areas swept by artillery and machine-gun fire, maintaining liaison between his company and battalion commander.

LOTSPIECH, ORR V..................... | Private, 83d Company, 6th Regiment, U. S. Marine Corps, 2d Division.
Near Vierzy, France, July 19, 1918. R—Booneville, Mo. B—Sheldon, Mo. G. O. No. 46, W. D., 1919. | While passing through an unusual barrage of artillery fire and machine guns he had his leg broken by a bullet. Undaunted, he dragged himself to the post of command and reported to his company commander that his mission had been completed, thus saving the dispatch of another runner.

LOTT, JOHN H. (2440260)............. | Corporal, Company C, 109th Infantry, 28th Division.
Near St. Agnan, France, July 16, 1918. R—Waycross, Ga. B—Waycross, Ga. G. O. No. 109, W. D., 1918. | Two different times, near St. Agnan, France, July 16, 1918, he preceded his platoon into enemy wire under fire, cut a path for it, and returned and led it through the gaps thus made.

LOUCKS, CLIFFORD C. (1537953)........ | Private, Company B, 112th Engineers, 37th Division.
Near Heuvel, Belgium, Nov. 2, 1918. R—East Cleveland, Ohio. B—Collinwood, Ohio. G. O. No. 37, W. D., 1919. | Private Loucks, with 2 other soldiers, crossed the Scheldt River after 2 attempts and succeeded in stretching a line for a bridge across the stream. They were discovered and fired upon by the enemy, but they continued at work driving stakes and made a second trip across the river to obtain wire, despite the fact that a violent artillery barrage had been laid down on their position.

LOUGH, MAXON S...................... | Major, 38th Infantry, 3d Division.
Near Romagne, France Oct. 9, 1918. R—Fargo, N. Dak. B—Fargo, N. Dak. G. O. No. 37, W. D., 1919. | Although severely wounded, Major Lough continued to lead his battalion to his objective and refused to be evacuated until his battalion was relieved.

*LOUGHLIN, JOSEPH J................... | Captain, 322d Infantry, 81st Division.
Near Moranville, France, Nov. 9, 1918. R—Wilmington, N. C. B—Swansboro, N. C. G. O. No. 32, W. D., 1919. | At the sacrifice of his own life he went forward through heavy machine-gun fire to locate a nest of machine guns which was holding up the advance of the regiment. He located the machine guns so that the 1-pounders could silence them, but was killed by the enemy machine-gun fire.
 | Posthumously awarded. Medal presented to widow, Mrs. Eleanor K. Loughlin.

*LOVE, CHARLES J. (1678783)........... | Private, Company K, 59th Infantry, 4th Division.
Near Bois-du-Fays, France, Oct. 5, 1918. R—Boonville, N. Y. B—Niles, N. Y. G. O. No. 71, W. D., 1919. | Volunteering for a dangerous liaison mission, he went out alone, crossing an open space for 400 yards, which was swept by heavy machine-gun fire. Going far in advance of our lines, he obtained the desired information and brought back a German prisoner. He was killed later in performance of duty by shellfire.
 | Posthumously awarded. Medal presented to mother, Mrs. Etta M. Love.

LOVELACE, DAVID H. (1319176)........ | Private, Machine Gun Company, 120th Infantry, 30th Division.
Near Bellicourt, France, Sept. 29, 1918. R—Jonesville, N. C. B—Grayson County, Va. G. O. No. 44, W. D., 1919. | His left arm having been rendered useless by a shrapnel wound, Private Lovelace continued to carry ammunition with his other arm until the objective was reached, when, against his protests, he was ordered to the rear for medical treatment.

LOWE, JOHN C. (2266478)_____
  Near Leeuwken, Belgium, Oct. 31
    1918.
  R—Long Beach, Calif.
  B—Hot Springs, Ark.
  G. O. No. 64, W. D., 1919.

Corporal, Company C, 364th Infantry, 91st Division.
With 3 other soldiers, Corporal Lowe skillfully worked his way under heavy fire to the flank of a machine-gun nest which was holding up the advance of his platoon with enfilading fire, killed 6 Germans, and captured 3 machine guns.

LOWE, THURMAN (741107)_____
  Near Munster, Alsace, Sept. 12–13,
    1918.
  R—Boma, Tenn.
  B—Boma, Tenn.
  G. O. No. 23, W. D., 1919.

Sergeant, Company A, 52d Infantry, 6th Division.
In repulsing a raid on our trenches, Sergeant Lowe seized an automatic rifle and pursued the Germans across no man's land in the face of converging fire of several enemy machine guns.

LOWE, WILLIAM O_____

  East of Cunel, Verdun sector,
    France, Oct. 7, 1918.
  R—Knoxville, Tenn.
  B—Athens, Tenn.
  G. O. No. 15, W. D., 1919.

Second lieutenant, U. S. Marine Corps, observer, attached to 3d Observation Group, 90th Aero Squadron, Air Service.
While staking the advance lines of the 80th Division he was suddenly attacked by a formation of 8 enemy machines (Fokker type), which dived out of a near-by cloud bank. Although greatly outnumbered, he succeeded in shooting down 1 out of control and disabled a second so that it was forced to land. Later, on the same mission, he was again attacked by a patrol of 5 enemy scout machines, and in a running fight he drove these off and successfully completed his mission.

*LOWRY, FRANCIS B_____

  Near Crepion, France, Sept. 26, 1918.
  R—Denver, Colo.
  B—Denver, Colo.
  G. O. No. 21, W. D., 1919.

Second lieutenant, Coast Artillery Corps, observer, 91st Aero Squadron, Air Service.
On Sept. 26, while on a very important photographic mission, Lieutenant Lowry, with Lieutenant Kelty, pilot, realized the importance of the mission and chose to continue their course through a harassing antiaircraft barrage. A shell made a direct hit on the plane, brought it down in fragments, and instantly killed Lieutenant Lowry.
Posthumously awarded. Medal presented to father, Walter B. Lowry.

LOYD, LOUIS H. (2189256)_____
  Near Beaufort, France, Nov. 4, 1918.
  R—Oakville, Mo.
  B—Annapolis, Mo.
  G. O. No. 37, W. D., 1919.

Private, first class, Company F, 356th Infantry, 89th Division.
Despite the fact that severe wounds made it impossible for him to take along his automatic rifle, he continued in the advance of the company. Again wounded, he refused evacuation, continuing with his comrades until the objective was reached.

LOYD, SAMUEL A. (1289500)_____
  North of Verdun, France, Oct. 15
    and 23, 1918.
  R—Lynchburg, Va.
  B—Lynchburg, Va.
  G. O. No. 37, W. D., 1919.

Sergeant, Company L, 116th Infantry, 29th Division.
After being severely gassed, Sergeant Loyd continued to advance with his company in spite of suffering great pain. On Oct. 23, when his company was under heavy flanking fire from machine guns, he rescued several wounded men and administered first-aid treatment.

*LUBECK, TONY (58210)_____
  Near Cantigny, France, May 28,
    1918.
  R—Chicago, Ill.
  B—Chicago, Ill.
  G. O. No. 99, W. D., 1918.

Private, first class, Company I, 28th Infantry, 1st Division.
While engaged on a mission to kill an enemy sniper, who was concealed in a wheat field and was inflicting severe losses upon American forces, near Cantigny, France, May 28, 1918, and after succeeding in his mission, he himself was killed.
Posthumously awarded. Medal presented to mother, Mrs. Mary Lubeck.

LUCIA, ARTHUR A. (2340082)_____
  Near Le Charmel, France, July 27,
    1918.
  R—Brooklyn, N. Y.
  B—Brooklyn, N. Y.
  G. O. No. 71, W. D., 1919.

Private, first class, Company M, 4th Infantry, 3d Division.
He volunteered to carry an important message through a heavy artillery barrage after several other runners had reported that they could not get through. He succeeded in his mission.

LUEBKE, ALVIN J_____
  At Roche, France, Oct. 27, 1918.
  R—Forestville, Wis.
  B—Chicago, Ill.
  G. O. No. 44, W. D., 1919.

First lieutenant, 142d Infantry, 36th Division.
Regaining consciousness while being carried to the rear after he had been wounded by an exploding shell, he returned immediately to his company, which he later led 500 meters into no man's land for the purpose of organizing his position preparatory to an attack the next morning.

LUKAZ, JOSEPH (2338585)_____
  Near Cunel, France, Oct. 13 1918.
  R—South Fork, Pa.
  B—Russia.
  G. O. No. 64, W. D., 1919.

Private, Company D, 4th Infantry, 3d Division.
While defending an outpost, after all his comrades had been wounded, Private Lukaz was attacked by a group of 8 Germans, 1 an officer. With his automatic rifle he disabled all but the officer and 2 men, who took cover in a shell hole. Running to their hiding place, he killed all the enemy with hand grenades and returned to hold his portion of the line intact.

*LUKE, FRANK, Jr_____
  Near St. Mihiel, France, Sept. 12–15,
    1918.
  R—Phoenix, Ariz.
  B—Phoenix, Ariz.
  G. O. No. 34, W. D., 1919.
  Medal of honor also awarded.

Second lieutenant, pilot, 27th Aero Squadron, Air Service.
By skill, determination, and bravery, and in the face of heavy enemy fire, he successfully destroyed 8 enemy observation balloons in 4 days.
Oak-leaf cluster.
He is also awarded an oak-leaf cluster for the following act of extraordinary heroism in action near Etain, France, Sept. 18, 1918: Immediately after destroying 2 enemy observation balloons he was attacked by a large formation of German planes (type Fokker). He turned to attack 2 which were directly behind him and shot them down. Sighting an enemy biplane, although his gasoline was nearly gone, he attacked and destroyed this machine also.
Posthumously awarded. Medal presented to father, Frank Luke, sr.

*LUKENS, ALAN W.
Near La Madeleine Ferme, north of Montfaucon, France, Sept. 29, 1918.
R—Haverford, Pa.
B—Elizabeth, N. J.
G. O. No. 15, W. D., 1923.

Captain, 316th Infantry, 79th Division.
The attacking forces of his regiment having been checked in their advance by strongly held enemy positions, a temporary withdrawal of the advance elements was ordered in order that the supporting artillery might lay down a barrage in the woods occupied by the extreme front-line elements. Upon learning that the order for withdrawal had apparently failed to reach a few of his men who still remained in the woods, Captain Lukens himself made his way to the woods, under a hail of enemy machine-gun fire, found his men, and directed them to places of comparative safety, and, while so engaged, was himself killed by enemy machine-gun fire. His superb devotion to the men of his command served to incite them to heroic endeavors.
Posthumously awarded. Medal presented to father, Lewis N. Lukens.

*LUKINS, FRED T.
At Chateau-Thierry, France, June 6, 1918.
R—Indianapolis, Ind.
B—Indianapolis, Ind.
G. O. No. 110, W. D., 1918.

Sergeant, 20th Company, 5th Regiment, U. S. Marine Corps, 2d Division.
Killed in action at Chateau-Thierry, France, June 6, 1918, he gave the supreme proof of that extraordinary heroism which will serve as an example to hitherto untried troops.
Posthumously awarded. Medal presented to mother, Mrs. Margaret Lukins.

LULOFF, ZALME.
Near Blanc Mont, France, Oct. 4, 1918.
R—Staten Island, N. Y.
B—Russia.
G. O. No. 46, W. D., 1919.

Private, 43d Company, 5th Regiment, U. S. Marine Corps, 2d Division.
Disregarding his personal safety, Private Luloff showed marked bravery by carrying messages through heavy machine-gun fire, and after his captain was wounded he rendered valuable assistance in establishing and maintaining liaison between the platoons of this company.

LUMLEY, ZODA D.
In the St. Mihiel offensive, Sept. 14, 1918, and near Baalon, France, Nov. 10-11, 1918.
R—Kampsville, Ill.
B—Kampsville, Ill.
G. O. No. 87, W. D., 1919.

First lieutenant, 357th Infantry, 90th Division.
In the St. Mihiel offensive he showed marked personal courage, advancing his first-aid station with the front line. Near Baalon he maintained a dressing station far to the front, under terrific artillery and machine-gun fire, showing absolute indifference to personal safety that he might aid the men on the firing line during heavy gas bombardments.

LUMPKIN, LAURENCE M. (236436).
Near Exermont, France, Oct. 4, 1918.
R—Danville, Va.
B—Danville, Va.
G. O. No. 44, W. D., 1919.

Sergeant, Quartermaster Corps, Pack Train No. 1, 1st Division.
He conducted his pack train, loaded with wire, through the enemy's counter barrage, over high, open ground, under direct observation of the enemy. Five of his 10 mules were killed, but he managed to make a second trip and bring up enough wire to finish the line.

LUND, CARL S. (109549).
Near Mouilly, France, Sept. 12, 1918.
R—Somerville, Mass.
B—Cranston, R. I.
G. O. No. 23, W. D., 1919.

Private, Company B, 102d Machine Gun Battalion, 26th Division.
At great risk of his own life from machine-gun fire at close range, Private Lund volunteered and went to the aid of a wounded comrade, bandaged his wounds, and helped him to a place of safety.

LUNDEGARD, AXEL C. (18023).
Near Verdun, France, Oct. 4, 1918.
R—Chicago, Ill.
B—Chicago, Ill.
G. O. No. 44, W. D., 1919.

Private, medical detachment, 26th Infantry, 1st Division.
He went forward alone and unarmed into the enemy's territory to rescue a wounded man who had been injured in the advance patrol fighting. He succeeded in his mission despite very heavy machine-gun and artillery fire.

*LUNSFORD, BEDFORD B. (1309812).
Near Bellicourt, France, Oct. 7, 1918.
R—Mount Vernon, Tenn.
B—Union County, Ga.
G. O. No. 133, W. D., 1918.

Corporal, Company M, 117th Infantry, 30th Division.
When the line was held up by enemy machine-gun fire he strapped an automatic rifle to his shoulder and advanced in the face of machine-gun fire. Firing as he went forward, he killed 4 of the enemy before he fell, nearly riddled with bullets.
Posthumously awarded. Medal presented to father, Thomas Lunsford.

LUNSFORD, EMMET E. (523801).
Near Romanovka, Siberia, June 25, 1919.
R—Claremore, Okla.
B—Edinburgh, Mo.
G. O. No. 133, W. D., 1919.

Private, first class, Company A, 31st Infantry.
Though wounded early in action he continued to operate his automatic rifle throughout the fight.

LUNSFORD, JESSE A. (1320936).
Near Bellicourt, France, Sept. 29, 1918.
R—Timberlake, N. C.
B—Person, N. C.
G. O. No. 32, W. D., 1919.

Corporal, Company G, 120th Infantry, 30th Division.
He attacked, single handed, a machine-gun post from which a destructive fire was being directed against his company. While he was approaching the nest the machine-gun shot the butt off his rifle and cut a hole in his breeches, but he succeeded in getting close enough to the nest to throw 4 hand grenades into it and then killed the gunner with his bayonet.

LUSK, JAMES G.
At Montrebeau Woods, near Charpentry, France, Oct. 4, 1918.
R—Greenville, Miss.
B—Greenville, Miss.
G. O. No. 53, W. D., 1920.

First lieutenant, 2d Machine Gun Battalion, attached to 16th Infantry, 1st Division.
After all the other officers in the attack had been killed or wounded, Lieutenant Lusk took command of the fragments of the battalion, reorganized them under heavy artillery and machine-gun fire, and personally led them forward in the attack, as a result of which 6 machine guns and a number of prisoners were taken. The capture effected, the advance of other troops was insured.

LUTZ, JOHN D. (52124).
Near St. Etienne-a-Arnes, France, Oct. 3-9, 1918.
R—Allston, Mass.
B—Cambridge, Mass.
G. O. No. 35, W. D., 1919.

Corporal, Company M, 23d Infantry, 2d Division.
He fearlessly exposed himself to a close-range enfilading fire of a German automatic rifle. He continued to lead his men throughout the battle after being wounded.

LUZENSKI, WILLIAM (280411)_____
Northeast of Chateau - Thierry, France, Aug. 4, 1918.
R—Detroit, Mich.
B—Detroit, Mich.
G. O. No. 117, W. D., 1918.

Sergeant, Company H, 126th Infantry, 32d Division.
Shortly after the assault was launched he was wounded by shellfire. In spite of his wounds he gallantly led his platoon, gained his objective, and remained in command of his platoon until ordered to the rear by his commanding officer.

LUZI, LUZIUS (4089)_____
Near Chateau - Thierry, France, June 6, 1918.
R—Salesville, Mont.
B—Switzerland.
G. O. No. 4, W. D., 1926.

Private, Company M, 23d Infantry, 2d Division.
He fearlessly and frequently passed through heavy machine-gun fire while performing his duty as a runner, after being twice wounded.

LUZOW, GOTTLIEB (45959)_____
Near Exermont, France, Oct. 10, 1918.
R—Chicago, Ill.
B—Russia.
G. O. No. 35, W. D., 1920.

Private, first class, Company A, 18th Infantry, 1st Division.
His platoon being held up by enemy machine-gun fire, Private Luzow, with 1 other soldier, advanced upon the machine gun with its crew. Their heroic action enabled his platoon to continue the advance with small loss.

LYERLY, WILLIAM B. (1320094)_____
Near Bellicourt, France, Sept. 29, 1918.
R—Mount Vernon, N. C.
B—Mount Vernon, N. C.
G. O. No. 37, W. D., 1919.

Private, Company D, 120th Infantry, 30th Division.
With 8 other soldiers, comprising the company headquarters detachment, he assisted his company commander in cleaning out enemy dugouts along a canal and capturing 242 prisoners.

LYNCH, ANDREW B. (1240277)_____
Near Apremont, France, Sept. 29, 1918.
R—Philadelphia, Pa.
B—Philadelphia, Pa.
G. O. No. 20, W. D., 1919.

Sergeant, Headquarters Company, 110th Infantry, 28th Division.
He was on duty with a section operating 37-millimeter guns. Under orders he moved the guns to the rear, and then, learning that his commanding officer had been taken prisoner, he, with another soldier, organized a party of 5, attacked the enemy patrol, numbering 35, and killed 15 of them, he personally rescuing his commanding officer and capturing 3 prisoners. Immediately afterwards he took command of 75 men and launched a counterattack, driving the enemy before them for over a kilometer. His conduct exemplified the greatest courage, judgment, and leadership.

*LYNCH, EDMUND W._____
At Fismette, France, Aug. 10, 1918.
R—Sharon Hill, Pa.
B—Philadelphia, Pa.
G. O. No. 37, W. D., 1919.

Captain, 111th Infantry, 28th Division.
Seeing two of his platoons being cut off by the enemy, Captain Lynch alone went to their rescue and engaged the enemy with his automatic pistol, killing several. He saved his platoons, but in so doing sacrificed his own life.
Posthumously awarded. Medal presented to widow, Mrs. Alice W. Lynch.

LYNCH, HENRY PETER (1204251)_____
Near Ronssoy, France, Sept. 29, 1918.
R—White Plains, N. Y.
B—White Plains, N. Y.
G. O. No. 46, W. D., 1919.

Corporal, Company E, 105th Infantry, 27th Division.
Having been trapped in a shell hole with 9 other men 50 yards in front of the line, Corporal Lynch crawled back to his company's position through heavy enemy machine-gun fire and reported their predicament. He then organized a bombing party, led it against the hostile machine gun, and put it out of action, thereby enabling his men to return safely to our line.

LYNCH, JOSEPH J. (89809)_____
Near Meurcy Ferme, France, July 30, 1918.
R—White Plains, N. Y.
B—White Plains, N. Y.
G. O. No. 59, W. D., 1919.

Sergeant, Company D, 165th Infantry, 42d Division.
Sergeant Lynch voluntarily went to the rescue of his lieutenant, who was severely wounded and lying in an exposed position. He succeeded in carrying the wounded officer to safety through the intense machine-gun and rifle fire.

LYNCH, RONALD D. (731749)_____
Near Fontaines, France, Nov. 7, 1918.
R—Elbridge, Tenn.
B—Hickman, Ky.
G. O. No. 37, W. D., 1919.

Private, Company B, 6th Infantry, 2d Division.
Private Lynch, accompanied by 3 other soldiers, volunteered and went out under heavy machine-gun and artillery fire to rescue a wounded comrade. Failing in the first attempt, they again tried and this time succeeded in bringing the wounded man to shelter.

LYNG, ARTHUR E._____
Near St. Etienne, France, Oct. 4, 1918.
R—Lowell, Mass.
B—New Haven, Conn.
G. O. No. 37, W. D., 1919.

Gunnery sergeant, 49th Company, 5th Regiment, U. S. Marine Corps, 2d Division.
While engaged in scouting he discovered the enemy forming for a surprise attack against an unprotected portion of his lines. He quickly organized a sufficient force to destroy the enemy's plans and accomplished the capture of 10 of the raiding party and 6 machine guns.

LYNK, HARRY E. (1207814)_____
Near Ronssoy, France, Sept. 29, 1918.
R—Brooklyn, N. Y.
B—Kingston, N. Y.
G. O. No. 23, W. D., 1919.

Sergeant, Company G, 106th Infantry, 27th Division.
While suffering from severe wounds he organized several small groups from other companies, consolidated them, and led them into effective combat, continuing with this splendid example of courage and fearlessness until wounded a second time.

*LYON, JOHN._____
Near Samogneux, France, Oct. 15, 1918.
R—Ballston, Va.
B—Ballston, Va.
G. O. No. 3, W. D., 1924.

Second lieutenant, 116th Infantry, 29th Division.
During the attack on the Bois-de-la-Grande-Montagne, Lieutenant Lyon left a place of comparative safety to cross an open space, exposed to direct observation and fire from the enemy, to attempt a rescue of a wounded officer. He and 2 men who accompanied him were killed in this attempt.
Posthumously awarded. Medal presented to father, Frank Lyon.

*LYONS, DOUGLAS M._____
Near Verdun, France, Oct. 12, 1918.
R—Fort Henry, Tenn.
B—Stewart County, Tenn.
G. O. No. 87, W. D., 1919.

Second lieutenant, 114th Infantry, 29th Division.
Inspiring his men by his fearlessness, Lieutenant Lyons led an attack against a strongly fortified enemy position, and in so doing was killed.
Posthumously awarded. Medal presented to father, Thomas M. Lyons.

LYONS, WALTER F. (136724)
Near Samogneux, France, Oct. 24, 1918.
R—North Attleboro, Mass.
B—Providence, R. I.
G. O. No. 21, W. D., 1919.

Private, first class, Battery C, 103d Field Artillery, 26th Division.
He went through a heavily shelled area to obtain medical aid for his wounded comrades, and, failing to do this, returned to the position to aid in evacuating them. While carrying a wounded man to a place of safety, 2 men who were assisting him were instantly killed; nevertheless, he continued at his task until the position was evacuated.

LYSTER, WAYNE G.
Near St. Etienne, France, Oct. 4–6, 1918.
R—Aldan, Pa.
B—Cardington, Pa.
G. O. No. 103, W. D., 1919.

Private, Headquarters Company, 5th Regiment, U. S. Marine Corps, 2d Division.
Private Lyster, a runner, displayed exceptional courage in volunteering to carry important messages over terrain constantly swept by machine-gun and shell fire.

MAAS, CHARLES S. (97085)
Near Croix Rouge Farm, France, July 26, 1918.
R—Selma, Ala.
B—Selma, Ala.
G. O. No. 46, W. D., 1919.

Corporal, Company F, 167th Infantry, 42d Division.
He voluntarily went out under a terrific machine-gun fire to the rescue of his commanding officer, who had been wounded.

MAASS, CHARLES (2334882)
Near St. Mihiel, France, Sept. 15–16, 1918.
R—Swarthmore, Pa.
B—Jersey City, N. J.
G. O. No. 37, W. D., 1919.

Private, Company L, 60th Infantry, 5th Division.
After all the other company runners had been evacuated because of wounds Private Maass, although himself wounded, refused to go to the rear and remained on duty, carrying messages through heavy shell fire and maintaining liaison with all four platoons until his company was relieved.

MABE, JAMES G. (1314736)
Near Bellicourt, France, Sept. 29, 1918.
R—Madison, N. C.
B—Madison, N. C.
G. O. No. 35, W. D., 1919.

Private, Company A, 119th Infantry, 30th Division.
Having been severely wounded in the shoulder by shrapnel early in the attack, Private Mabe refused to leave his platoon and, after losing his rifle, armed himself with grenades and cleaned out numerous enemy dugouts. Not until his company had taken its position for the night did he go to the rear.

McADAMS, HOWARD K.
Near Brieulles, France, Nov. 4–5, 1918.
R—Eveleth, Minn.
B—Duluth, Minn.
G. O. No. 37, W. D., 1919.

Captain, 7th Engineers, 5th Division.
Although severely wounded, Captain McAdams remained on duty directing the construction of a pontoon bridge across the Meuse River under heavy shell and machine-gun fire.

*McAFEE, JOHN W. (2184365)
Near Pouilly, France, Nov. 8, 1918.
R—Gallatin, Mo.
B—Gallatin, Mo.
G. O. No. 37, W. D., 1919.

Corporal, Company D, 356th Infantry, 89th Division.
While a member of a patrol sent out to reconnoiter the banks of the Meuse River, and when all means of crossing the river had been destroyed, Corporal McAfee, with another soldier, volunteered to swim across, though the other side was held in force by the enemy. Before reaching the opposite bank, he was seized with a cramp, caused by the extremely low temperature of the water, and was drowned.
Posthumously awarded. Medal presented to father, Oliver P. McAfee.

McALEXANDER, ULYSSES G.
Near Jaulgonne, France, July 22, 1918.
R—McPherson, Kans.
B—Dundas, Minn.
G. O. No. 37, W. D., 1919.
Distinguished-service medal also awarded.

Colonel, 38th Infantry, 3d Division.
As a colonel, commanding the 38th Infantry, he displayed exceptional gallantry when his regiment, attacking without support on either flank, was stopped by severe machine-gun and artillery fire, by going ahead of the most advanced elements of his command, and in full view of the enemy, leading his men by force of his own example to the successful assault of Jaulgonne and the adjoining heights. Later in the day, when progress was again checked, he personally reconnoitered to within 50 yards of hostile machine-gun nests, and through information thus obtained was enabled to hold an advanced position with both flanks exposed, for more than 36 hours.

*McANDREW, EDWARD W. (562392)
Near the Bois-des-Ogons, France, Sept. 30, 1918.
R—Chicago, Ill.
B—Vincennes, Ind.
G. O. No. 78, W. D., 1919.

Sergeant, Company B, 12th Machine Gun Battalion, 4th Division.
Exposing himself fearlessly to enfilading machine-gun fire from the enemy, Sergeant McAndrew directed the placing of the guns of his section in such positions as to protect the advance of the Infantry and in so doing was fatally wounded. Despite the fact that one-half of his body was paralyzed as a result of his injury, he insisted upon remaining in command of his section until the action was over. He died in a field hospital shortly after being evacuated.
Posthumously awarded. Medal presented to mother, Mrs. Charles Weyl.

McARDLE, ALBERT H. (3559)
East of Belleau Woods, France, July 18, 1918.
R—East Boston, Mass.
B—East Boston, Mass.
G. O. No. 125, W. D., 1918.

Private, 103d Ambulance Company, 101st Sanitary Train, 26th Division.
While giving first aid to a wounded soldier who had fallen in a wheat field in range of enemy fire he received 2 machine-gun bullets through the thigh. In spite of his injury he continued at work until the soldier's wounds had been properly dressed and endeavored to carry him out of danger, but fell exhausted from loss of blood.

MACAROVSKY, HERMAN (39895)
Near Blanc Mont Ridge, France, Oct. 3, 1918.
R—St. Paul, Minn.
B—Russia.
G. O. No. 21, W. D., 1919.

Sergeant, Company H, 9th Infantry, 2d Division.
All of his superior officers having been killed or wounded, Sergeant Macarovsky took command of and organized his company, under heavy shell fire. He then led it to the attack, captured or killed many Germans, and put several machine-gun nests out of action.

MacArthur, Douglas_____
In the Salient du Feys, France, Mar. 9, 1918.
R—Milwaukee, Wis.
B—Little Rock Barracks, Ark.
G. O. No. 27, W. D., 1919.
Distinguished-service medal also awarded.

Brigadier general, then colonel, chief of staff, 42d Division.
When Company D, 168th Infantry, was under severe attack in the Salient du Feys, France, he voluntarily joined it, upon finding that he could do so without interfering with his normal duties, and by his coolness and conspicuous courage aided materially in its success.
Oak-leaf cluster.
An oak-leaf cluster is awarded General MacArthur for the following acts of distinguished service: As brigade commander General MacArthur personally led his men and by the skillful maneuvering of his brigade made possible the capture of Hills 288, 242, and the Cote-de-Chatillon, France, Oct. 14, 15, and 16, 1918. He displayed indomitable resolution and great courage in rallying broken lines and in reforming attacks, thereby making victory possible. On a field where courage was the rule, his courage was the dominant feature.

ªMacArthur, John_____
Near Luneville, France, June 13, 1918.
R—Buffalo, N. Y.
B—Columbia, Pa.
G. O. No. 101, W. D., 1918.

Second lieutenant, 27th Aero Squadron, Air Service.
Outnumbered and handicapped by his presence far behind the German lines, he and 3 flying companions fought brilliantly a large group of enemy planes, bringing down or putting to flight all in the attacking party, while performing an important mission near Luneville, France, on June 13, 1918.
Posthumously awarded. Medal presented to father, Charles P. MacArthur.

McArty, Arnot L. (2100809)_____
Near Bois-de-la-Cote-Lamont, France, Oct. 3, 1918, and near Bois-du-Fays, France, Oct. 9, 1918.
R—Enfield, Ill.
B—Enfield, Ill.
G. O. No. 46, W. D., 1919.

Private, Company F, 59th Infantry, 4th Division.
On Oct. 3, while acting in the capacity of company runner, he carried messages to 2 platoons of his company through a heavy fire of machine guns and snipers. He successfully delivered the messages, after crawling for a distance of 400 yards. On Oct. 9, in company with one other runner, he delivered messages to a platoon which was engaged in combat liaison duty in the Bois-du-Fays, passing through a severe artillery fire while in the execution of this mission.

Macauley, Malcolm A. (73327)_____
Near Bouresches, France, July 20, 1918.
R—Springfield, Mass.
B—Roxbury, Mass.
G. O. No. 81, W. D., 1919.

Private, Company K, 104th Infantry, 26th Division.
On July 20 Private Macauley repeatedly carried messages through a field swept by a terrific machine-gun fire. At one time he crawled to a brook through this machine-gun fire and returned with water to 2 wounded men, giving them first aid and assisting them to the rear.

McAuliffe, Daniel (44515)_____
In the Argonne Forest, France, Oct. 4, 1918.
R—Butte, Mont.
B—Butte, Mont.
G. O. No. 23, W. D., 1919.

Corporal, Company M, 16th Infantry, 1st Division.
Leading his squad on enemy machine-gun nests which had been inflicting severe casualties on his platoon, Corporal McAuliffe opened an effective bombing attack on the nests, and, although severely wounded, he remained in command until the strong point was reduced.

McAuliffe, Michael J. (89784)_____
Near Landres-et-St. Georges, France, Oct. 15, 1918.
R—Brooklyn, N. Y.
B—Ireland.
G. O. No. 71, W. D., 1919.

Sergeant, Company D, 165th Infantry, 42d Division.
Volunteering for the mission, he exposed himself in the open to heavy shell and machine-gun fire to obtain ammunition for his company after all on hand had been exhausted. He made repeated trips over the battle field, gathering ammunition from the bodies of the dead until his entire company had been supplied.

MacBeth, Edwin (1243698)_____
Near Fismette, France, Aug. 10, 1918.
R—Pittsburgh, Pa.
B—Homestead, Pa.
G. O. No. 37, W. D., 1919.

Sergeant, Company C, 111th Infantry, 28th Division.
Sergeant MacBeth and another soldier voluntarily went through heavy machine-gun fire to carry an important message to an advance unit. Attracted by the cries of a wounded soldier while they were returning, they went to his assistance and were endeavoring to rescue him when Sergeant MacBeth's companion was fatally wounded. Being unable to bring in the 2 wounded men by himself, this soldier returned to the line and secured assistance.

MacBrayne, Winfred C_____
Near Fismes, France, Aug. 26, 1918.
R—Lowell, Mass.
B—Lowell Mass.
G. O. No. 1, W. D., 1919.

First lieutenant, Field Artillery, observer, 2d Balloon Squadron, Air Service.
While he was conducting an aerial reconnoissance and general surveillance from his balloon, he was repeatedly attacked by hostile airplanes, but continued his mission despite the proximity of strong enemy air patrols, against which he had no defense. When an enemy Fokker emerged from low-hanging clouds, firing at the balloon with incendiary bullets, Lieutenant MacBrayne remained in the basket until his companion, who was making his first ascension, had safely jumped. He leaped when the balloon was nearer the ground, and his parachute opened as he crossed into the woods. He insisted upon reascending immediately, thereby displaying conspicuous coolness and determination.

McBreen, Leo (2057707)_____
In the Meuse-Argonne offensive, France, Sept. 26, 1918.
R—Geneva, Ill.
B—Venice, Ill.
G. O. No. 71, W. D., 1919.

Private, Company M, 131st Infantry, 33d Division.
With 3 other soldiers he crawled across an open field for 200 yards, subjected the while to intense artillery and machine-gun fire, to execute a flank attack upon 3 machine-gun emplacements which were holding up our advance. The enemy positions were silenced, 7 of the crews being killed and 23 made prisoners.

McBride, Emmett (2387938)_____
Near Dun-sur-Meuse, France, Nov. 5, 1918.
R—Browns Spring, Mo.
B—Christian County, Mo.
G. O. No. 98, W. D., 1919.

Sergeant, Company I, 61st Infantry, 5th Division.
At a critical time, when the advance across the Meuse was being held up by enemy fire, Sergeant McBride displayed exceptional bravery in crossing a damaged pontoon bridge under terrific artillery and machine-gun fire, clearing the bridgehead of the enemy and protecting it. After crossing the canal he displayed great bravery in flanking a machine-gun nest single-handed, killing 2 of the gunners, and thereby enabling his company to advance.

McBRIDE, FRANK (554577)_____
At Paroy, France, July 14–15, 1918.
R—New York, N. Y.
B—New York, N. Y.
G. O. No. 44, W. D., 1919.

First sergeant, Company A, 9th Machine Gun Battalion, 3d Division.
Upon his own initiative, Sergeant McBride left his dugout under heavy shell fire and proceeded 200 yards to a house where an officer and another soldier were lying severely wounded. After administering first aid to them he remained with them until they were evacuated.

McCAIN, ARTHUR J. (919026)_____
Near Montfaucon, France, Sept. 28–30, 1918.
R—Watertown, S. Dak.
B—Hamlin County, S. Dak.
G. O. No. 37, W. D., 1919.

Private, first class, headquarters detachment, 79th Division.
While acting as a division observer Private McCain remained constantly on duty for several days in a building on the heights of Montfaucon. While in the building portions of it were destroyed by direct artillery fire, and hospital shelling was such that other observers located in the same building withdrew to a safer place. Private McCain, however, remained constantly at his post of duty and obtained important information.

*McCALL, ANDY (1403221)_____
Near Ferme de la Riviere, France, Sept. 30, 1918.
R—Houston, Tex.
B—Duke, Tex.
G. O. No. 37, W. D., 1919.

Private, Machine Gun Company, 370th Infantry, 93d Division.
Although relieved on the previous night, he willingly agreed to substitute for a sick comrade, returning the following day to his dangerous post as advance machine gunner. During a heavy shelling of his position he was killed.
Posthumously awarded. Medal presented to mother, Mrs. Sylvia McCall.

*McCALL, HOWARD C._____
Near Chezy, France, July 19, 1918.
R—Philadelphia, Pa.
B—Philadelphia, Pa.
G. O. No. 37, W. D., 1919.

Captain, 59th Infantry, 4th Division.
After his company had suffered heavy losses in taking its immediate objective, he placed himself at the head of his command and led his men forward in the face of violent shell and machine-gun fire, until he fell mortally wounded, cheering his men on with his last words.
Posthumously awarded. Medal presented to mother, Mrs. J. B. McCall.

McCALLISTER, JOSEPH A. (1706239)____
At Chateau-Diable, near Fismes, France, Aug. 27, 1918.
R—Brooklyn, N. Y.
B—Brooklyn, N. Y.
G. O. No. 127, W. D., 1918.

Corporal, Company H, 307th Infantry, 77th Division.
He personally led his squad in an attempt to capture an enemy machine gun, and after all of his men had been wounded and he himself severely wounded he withdrew, and collecting other men advanced three times to the attack, finally capturing the gun and driving off or killing its crew.

McCANN, KENNETH M. (1201080)_____
Near Mount Kemmel, Belgium, Aug. 29–31, 1918.
R—New York, N. Y.
B—New York, N. Y.
G. O. No. 37, W. D., 1919.

Corporal, Company C, 102d Field Signal Battalion, 27th Division.
Corporal McCann, a signalman, worked continuously for 72 hours without relief through repeated gas bombardments. When the forward lines were cut by shellfire, he personally directed the running of a new line under a heavy shell and machine-gun fire.

McCANN, WALTER J. (2946)_____
At Bois Brule, France, Apr. 10, 1918.
R—Springfield, Mass.
B—Springfield, Mass.
G. O. No. 107, W. D., 1918.

Private, first class, medical detachment, 104th Infantry, 26th Division.
During the action of Apr. 10, 1918, he displayed conspicuous gallantry by leaving shelter and running through a heavily shelled area to pick up a wounded soldier and carry him unaided and at great personal risk to a dressing station.

McCARTHY, CHARLES A. (105474)_____
Near Fleville, France, Oct. 5, 1918.
R—Webster City, Iowa.
B—Blairsburg, Iowa.
G. O. No. 44, W. D., 1919.

Private, Company B, 2d Machine Gun Battalion, 1st Division.
With the assistance of 1 other soldier, Private McCarthy entered a wood where 3 machine guns were holding up our attack, and under unusually heavy artillery and machine-gun fire knocked one of the guns out and rushed the second single handed. With the assistance of the other soldier he then succeeded in capturing about 20 prisoners who were in or near the machine-gun nest.

McCARTHY, JAMES J. (1390530)_____
In Bois-du-Fays, France, Oct. 10, 1918.
R—Chicago, Ill.
B—Chicago, Ill.
G. O. No. 5, W. D., 1920.

Corporal, Headquarters Company, 132d Infantry, 33d Division.
After being severely wounded during an enemy barrage, and although bleeding profusely, he remained in line and reorganized a shattered detachment of the 39th Infantry and held the position in the line.

McCARTHY, MICHAEL P. (1392087)_____
Near Butgneville, France, Nov. 11, 1918.
R—Girard, Ill.
B—Girard, Ill.
G. O. No. 46, W. D., 1919.

Sergeant, Company C, 124th Machine Gun Battalion, 33d Division.
When violent machine-gun fire had forced his company to take shelter, Sergeant McCarthy, with another soldier, braved the direct and short-range fire by voluntarily going forward and rescuing a wounded officer and carrying him back to a place of safety.

McCARTHY, WILLIAM (89662)_____
At Ferme de Meurcy, near Villers-sur-Fere, France, July 29, 1918.
R—New York, N. Y.
B—Ireland.
G. O. No. 117, W. D., 1918.

Private, Company C, 165th Infantry, 42d Division.
After having been wounded and ordered from the field he went out into an area that was under accurate enemy machine-gun fire and the fire of our own artillery barrage, gave first aid to a wounded comrade, and carried him back to the dressing station.

McCARTHY, WILLIAM H. (51181)_____
Near Vaux, France, July 2, 1918.
R—Lynn, Mass.
B—Lynn, Mass.
G. O. No. 19, W. D., 1920.

Private, Company H, 23d Infantry, 2d Division.
During an artillery bombardment Private McCarthy was wounded and rendered unconscious by the explosion of a large shell. Upon recovering consciousness he gave no thought to his own wound but devoted himself to the rescue of other wounded and carried them through enemy fire to a place of shelter.

McCAULEY, PHILIP J._____
Near Pexonne, France, Mar. 5, 1918.
R—St. Paul, Minn.
B—Lafayette, Ind.
G. O. No. 126, W. D., 1919.

Captain, 151st Field Artillery, 42d Division.
He displayed great presence of mind, promptness, and unusual courage in conducting the operations of Battery C, 151st Field Artillery, under exceptionally difficult conditions, due to accurately adjusted shellfire near Pexonne, France, Mar. 5, 1918. His fortitude aided materially in the success of the command.

McCELVEY, GEORGE C.
  Near St. Thibaut, France, Aug. 7–9, 1918.
  R—Mount Carmel, S. C.
  B—Mount Carmel, S. C.
  G. O. No. 37, W. D., 1919.

Captain, 47th Infantry, 4th Division.
He stood in the swift current of the Vesle River and helped the men of 3 platoons across. He was pulled into the river twice by drowning men, but each time succeeded in bringing them ashore. On succeeding days he was conspicuously present in places of danger, setting a splendid example to his command.

McCLELLAN, ARNO S.
  At Sergy, France. Aug. 1, 1918.
  R—Harveys, Pa.
  B—Scott County, Ind.
  G. O. No. 70, W. D., 1919.

Second lieutenant, 47th Infantry, 4th Division.
This officer fearlessly led his platoon in locating and successfully attacking German machine guns, thereby facilitating the advance of his company. He also led a combat patrol in front of his position for the purpose of driving out hostile snipers. Later, when his company was forced to retire to a more sheltered zone, Lieutenant McClellan, with 1 soldier, remained in an exposed position and rendered valuable service by covering the withdrawal with accurate fire from an automatic rifle.

McCLELLAND, HELEN G.
  France, Aug. 17, 1917.
  R—Fredericktown, Ohio.
  B—Austinburg, Ohio.
  G. O. No. 17, W. D., 1926.

Reserve nurse, Army Nurse Corps.
For extraordinary heroism in action while on duty with the surgical team at British Casualty Clearing Station No. 61, British area, France, Aug. 17, 1917. She occupied the same tent with Miss Beatrice MacDonald, another reserve nurse, cared for her when wounded, stopped the hemorrhage from her wounds under fire caused by bombs from German aeroplanes.

McCLELLAND, LEE R. (1870896)
  Near Ardeuil, France, Sept. 30, 1918.
  R—Asheville, N. C.
  B—Laurinburg, N. C.
  G. O. No. 46, W. D., 1919.

Sergeant, medical detachment, 371st Infantry, 93d Division.
While administering first-aid treatment to wounded soldiers on the field, Sergeant McClelland received a painful wound on the leg, but without mentioning his injury he remained on duty, caring for the wounded under shellfire, until the regiment was relieved.

McCLEMENS, FREDERICK W. (2311117)
  Near Romagne, France, Oct. 14, 1918.
  R—Carnegie, Pa.
  B—Pittsburgh, Pa.
  G. O. No. 21, W. D., 1919.

Private, Company M, 126th Infantry, 32d Division.
When the advance of his regiment was held up by enemy machine-gun fire and grenades, Private McClemens volunteered as a member of a combat patrol which cut through the enemy lines, captured 10 machine guns, killed and captured 15 Germans, and forced a large number to surrender to other troops, clearing that part of the Cote Dame Marie of the enemy, enabling the advance of the regiment to continue.

*McCLENDON, JOEL H.
  Near Fismes, France, Aug. 11, 1918.
  R—Farmers Branch, Tex.
  B—Ferris, Tex.
  G. O. No. 44, W. D., 1919.

First lieutenant, pilot, 88th Aero Squadron, Air Service.
Louis G. Bernheimer, first lieutenant, pilot; John W. Jordan, second lieutenant, 7th Field Artillery, observer; Roger W. Hitchcock, second lieutenant, pilot; James S. D. Burns, deceased, second lieutenant, 165th Infantry, observer; Joel H. McClendon, deceased, first lieutenant, pilot; Charles W. Plummer, deceased, second lieutenant, 101st Field Artillery, observer; Philip R. Babcock, first lieutenant, pilot; and Joseph A. Palmer, second lieutenant, 15th Field Artillery, observer. All of these men were attached to the 88th Aero Squadron, Air Service.
For extraordinary heroism in action near Fismes, France, Aug. 11, 1918. Under the protection of 3 pursuit planes, each carrying a pilot and an observer, Lieutenants Bernheimer and Jordan, in charge of a photo plane, carried out successfully a hazardous photographic mission over the enemy's lines to the River Aisne. The 4 American ships were attacked by 12 enemy battle planes. Lieutenant Bernheimer, by coolly and skillfully maneuvering his ship, and Lieutenant Jordan, by accurate operation of his machine gun, in spite of wounds in the shoulder and leg, aided materially in the victory which came to the American ships, and returned safely with 36 valuable photographs. The pursuit plane operated by Lieutenants Hitchcock and Burns was disabled while these two officers were fighting effectively. Lieutenant Burns was mortally wounded and his body jammed the controls. After a headlong fall of 2,560 meters, Lieutenant Hitchcock succeeded in regaining control of this plane and piloted it back to his airdrome. Lieutenants McClendon and Plummer were shot down and killed after a vigorous combat with 5 of the enemy's planes. Lieutenants Babcock and Palmer, by gallant and skillful fighting, aided in driving off the German planes and were materially responsible for the successful execution of the photographic mission.
Posthumously awarded. Medal presented to father, J. W. McClendon.

*McCLOUD, JAMES M.
  Near Soissons, France, July 19, 1918.
  R—Burbank, Calif.
  B—Santa Ana, Calif.
  G. O. No. 132, W. D., 1918.

Major, 26th Infantry, 1st Division.
After being wounded twice during an advance by his battalion on a machine-gun nest near Soissons, France, July 19, 1918, he continued in the attack until killed.
Posthumously awarded. Medal presented to widow, Mrs. Dolly McCloud.

McCLUER, EDWIN ALEXANDER
  Near Jonville, France, Sept. 14, 1918.
  R—Urbanna, Va.
  B—Norfolk, Va.
  G. O. No. 37, W. D., 1919.

Second lieutenant, 344th Battalion, Tank Corps.
Commanding a reconnaissance patrol of 3 tanks, he put to rout a company of German infantry, captured 5 pieces of artillery, and destroyed 8 machine guns. His action was 8 kilometers in advance of our front lines.
Oak-leaf cluster.
For the following act of extraordinary heroism in action near Bois-de-Montrebeau, France, Sept. 28, 1918, an oak-leaf cluster, worn with the distinguished-service cross, is awarded Lieutenant McCluer: In the attack on the woods he led his tank patrol on foot through dense wooded territory and in the face of intense fire. He was 2 kilometers in advance of the Infantry front line during this exploit.

McCLURE, HERBERT C. (1465017)
  Near Cheppy, France, Sept. 27, 1918.
  R—Independence, Mo.
  B—Independence, Mo.
  G. O. No. 98, W. D., 1919.

Sergeant, Battery E, 129th Field Artillery, 35th Division.
Sergeant McClure, although wounded by shellfire while manning a captured German gun with a detail of cannoneers, remained at his post until relieved, when he returned to his battery, keeping his gun in action until ordered to the field hospital by his battery commander.

McCLURE, LOWE A._____
In the Bois-de-Pultiere, France, Oct. 14, 1918, and during the Meuse offensive, France, Nov. 5, 1918.
R—Carson City, Nev.
B—Carson City, Nev.
G. O. No. 37, W. D., 1919.

Lieutenant colonel, 61st Infantry, 5th Division.
Lieutenant Colonel McClure, commanding the 3d Battalion, organized his companies and directed their disposition while constantly under fire. At the crossing of the Meuse, Nov. 5, he organized his battalion for the attack on Hill 292, and under direct enemy fire made a reconnaissance which enabled him to capture Hill 292, with more than 100 prisoners and 50 machine guns, without casualties.

McCLURE, ROBERT B._____
At Belieu Bois, France, Oct. 27, 1918.
R—Piedmont, Calif.
B—Rome, Ga.
G. O. No. 28, W. D., 1921.

Second lieutenant, 102d Infantry, 26th Division.
After being wounded he continued to lead his men until he was again wounded by enemy machine-gun fire in the foot and knee. Despite his wounds he reorganized his command and established a line of resistance, retaining active command until relieved by another officer several hours later.

*McCLUSKEY, ROSS (446177)_____
Near Landersbach, Gerardmer sector, Alsace, Oct. 4, 1918.
R—Eau Claire, Wis.
B—Eau Claire, Wis.
G. O. No. 130, W. D., 1918.

Corporal, Company H, 53d Infantry, 6th Division.
He was a member of a party of 50 attacked by an enemy raiding party consisting of about 300 storm troops. Although fatally wounded en route to his post from his dugout, he did not falter, and, despite a heavy bombardment, took up his position and continued to fight until the enemy was repulsed. He imbued his men with such fighting spirit that, although greatly outnumbered, they fought until the enemy was decisively beaten.
Posthumously awarded. Medal presented to mother, Mrs. Gertrude McCluskey.

McCOLLUM, JOSEPH (567527)_____
Near Bois-de-Roi, France, July 23, 1918.
R—Springfield, Mass.
B—Ludlow, Mass.
G. O. No. 71, W. D., 1919.

Wagoner, Company B, 10th Machine Gun Battalion, 4th Division.
On duty as a runner, he volunteered to reestablish liaison with the French unit to which his company was attached, after several officers and men had failed in the attempt. He performed the mission successfully, although exposed to heavy fire. Though knocked down and temporarily stunned by the explosion of a shell, he accomplished a second dangerous mission, remaining on duty until ordered to the rear.

*McCOMB, ROY E. (2183787)_____
In the Forest de Jaulnay, France, Nov. 4, 1918.
R—Cameron, Mo.
B—Welda, Kans
G. O. No. 37, W. D., 1919.

Sergeant, Machine Gun Company, 356th Infantry, 89th Division.
He led a section of machine guns with great courage under heavy enemy artillery and machine-gun fire, silencing 3 hostile machine-gun nests. Repeatedly exposing himself in order to get the maximum fire effect, this soldier was fatally wounded and died the same day.
Posthumously awarded. Medal presented to widow, Mrs. Ruth A. McComb.

McCOMBS, SHERMAN B. (1452462)_____
Near Baulny, France, Sept. 28, 1918.
R—St. Louis, Mo.
B—Albia, Iowa.
G. O. No. 95, W. D., 1919.

Mechanic, Company F, 138th Infantry, 35th Division.
When volunteers were called for to ascertain the location of the regiment on the left of his own, Mechanic McCombs immediately responded, and passing through direct machine-gun and artillery fire accomplished this dangerous mission with complete success.

*McCONNELL, JAMES_____
Near Les Franquettes Farm, France, July 23, 1918.
R—Syracuse, N. Y.
B—Syracuse, N. Y.
G. O. No. 87, W. D., 1919.

First lieutenant, 4th Infantry, 3d Division.
He continued in command of his platoon after having part of his face shot away by a machine-gun bullet. He later died of his wound.
Posthumously awarded. Medal presented to widow, Mrs. Susan S. McConnell.

*McCONNELL, WILLIAM O. (53124)_____
Near Soissons, France, July 18, 1918.
R—Princeton, N. J.
B—St. Paul, Minn.
G. O. No. 32, W. D., 1918.

Private, first class, Company I, 26th Infantry, 1st Division.
After being wounded on July 18, 1918, near Soissons, France, he continued to act as a runner for his company under fire during its 3-day advance until killed, July 21, 1918.
Posthumously awarded. Medal presented to mother, Mrs. W. W. McConnell.

McCORMACK, ALPHONSUS L._____
At Molleville Farm, France, Oct. 15, 1918.
R—Boston, Mass.
B—Boston, Mass.
G. O. No. 20, W. D., 1919.

Second lieutenant, 116th Infantry, 29th Division.
He continued in the advance after being wounded by machine-gun fire, taking command of his company and leading it until weakness from his wound necessitated his evacuation. His gallantry inspired his men to successful combat.

McCORMICK, CHRISTIE F._____
Near Consenvoye, France, Oct. 9, 1918.
R—Algona, Iowa.
B—Ottumwa, Kans.
G. O. No. 59, W. D., 1919.

Captain, 132d Infantry, 33d Division.
Surrounded by the enemy and unable to communicate with the rest of his regiment, he, with only 5 men, maintained an advanced position against a counterattack by picked enemy storm troops, remaining in this perilous place throughout the night under terrific fire of enemy artillery and machine guns until the arrival of supporting troops next day.

McCORMICK, CLARK T. (52037)_____
Near the Cote-de-Chatillon, France, Nov. 1, 1918.
R—Saginaw, Mich.
B—Coleman, Mich.
G. O. No. 98, W. D., 1919.

Sergeant, Company L, 23d Infantry, 2d Division.
Sergeant McCormick displayed exceptional bravery in voluntarily leading his platoon into a thick wood, capturing 12 prisoners and 2 machine guns, which had been causing us heavy losses.

McCORMICK, LEIGHTON (127311)_____
Near Servillers, France, May 5, 1918.
R—Wilmington, Del.
B—Wilmington, Del.
G. O. No. 46, W. D., 1920.

Sergeant, Battery B, 7th Field Artillery, 1st Division.
During a heavy enemy bombardment, a member of the battery was severely wounded. Sergeant McCormick, while going to the assistance of this wounded soldier was himself wounded by a shell splinter which passed through his leg above the knee. In spite of his wound, he assisted his comrade to a place of safety and then resumed command of his section, continuing in same until ordered to the rear.

McCOWIN, ELMER (104360)_____
At Ripont Swamp, France, Sept. 26, 1918.
R—New York, N. Y.
B—Richmond, Va.
G. O. No. 37, W. D., 1919.

Private, Company K, 369th Infantry, 93d Division.
While passing through a swamp where most of the platoon was wounded, Private McCowin dressed the wounds of several of his comrades, and after reaching the shelter of a hill beyond, returned repeatedly and assisted many of his comrades to a place of safety. He also carried messages through shell and machine-gun fire after being severely gassed.

McCOY, CHARLES A. (203088)_____
Near Verdun, France, Oct. 5, 1918.
R—Freeport, Ill.
B—Argyle, Wis.
G. O. No. 44, W. D., 1919.

Corporal, Company I, 26th Infantry, 1st Division.
Even though he was suffering from wounds, he refused evacuation, and after all superiors of his company had become casualties, he assumed command, reorganizing the forces and consolidating the position, working in the face of murderous machine-gun fire.

McCOY, CHARLES TICE_____
Near St. Etienne, France, Oct. 4-6, 1918.
R—Louisburg, Kans.
B—Logan, Iowa.
G. O. No. 37, W. D., 1919.

Private, Headquarters Company, 5th Regiment, U. S. Marine Corps, 2d Division.
Engaged as a runner, Private McCoy demonstrated the highest courage in carrying messages over hazardous territory under constant shell and machine-gun fire.

*McCOY, JAMES_____
At Chateau-Thierry, France, June 6, 1918.
R—Fall River, Mass.
B—Fall River, Mass.
G. O. No. 119, W. D., 1918.

Captain, 5th Regiment, U. S. Marine Corps, 2d Division.
Killed in action at Chateau-Thierry, France, June 6, 1918, he gave the supreme proof of that extraordinary heroism which will serve as an example to hitherto untried troops.
Posthumously awarded. Medal presented to niece, Miss Mabel Davel.

McCOY, PETE (1931743)_____
Near Bellicourt, France, Sept. 29, 1918.
R—Morell, Ky.
B—Thomas, Ky.
G. O. No. 50, W. D., 1919.

Private, Company B, 120th Infantry, 30th Division.
Unexpectedly encountering 7 of the enemy, Private McCoy, single handed, killed them all with his bayonet and a hand grenade. As a result of this feat he captured 4 hostile machine-gun emplacements and took 17 prisoners out of a dugout near by. Upon advancing farther he found a wounded officer, whom he sent to the rear in charge of another soldier, and continued on to the objective.

McCRACKEN, LYNN (105475)_____
Near Fleville, France, Oct. 4, 1918.
R—Manilla, Iowa.
B—Manilla, Iowa.
G. O. No. 16, W. D., 1920.

Private, first class, Company B, 2d Machine Gun Battalion, 1st Division.
During a heavy enemy counterattack Private McCracken conveyed important information through a heavy enemy artillery barrage to an advanced machine-gun platoon of his company. By dashing from shell hole to shell hole, exposed to heavy fire, he accomplished his mission in ample time.

McCRUDDEN, JAMES F. (2450857)_____
Near Grand Pre, France, Oct., 26, 1918.
R—Yonkers, N. Y.
B—Yonkers, N. Y.
G. O. No. 37, W. D., 1919.

Private, Company M, 312th Infantry, 78th Division.
Acting on his own initiative, after all runners had failed to deliver a message, Private McCrudden carried a message through an intense machine-gun fire for the captain of his company. Upon reaching the company he found that the captain had been killed and immediately returned through the same sweeping fire to report the fact.

McCULLOUGH, CLARE A. (246695)_____
Near Gesnes, France, Oct. 7, 1918.
R—Meadville, Pa.
B—Meadville, Pa.
G. O. No. 47, W. D., 1921.

Corporal, Company F, 128th Infantry, 32d Division.
One of four survivors of a platoon of 41 who attacked Hill 269, he, with the 3 others, continued on their mission and held the hill for some time without hope of reinforcements.

McCUNN, WALTER T. (2412654)_____
Near Thiaucourt, France, Sept. 29, 1918.
R—East Orange, N. J.
B—Green Bay, Wis.
G. O. No. 37, W. D., 1919.

Private, first class, medical detachment, 303d Engineers, 78th Division.
He was administering first aid to 2 wounded comrades in no man's land, under heavy shellfire, when an exploding shell killed both of the wounded and rendered him unconscious for more than an hour. Upon his recovery he remained at his post, administering treatment to others.

McDADE, WILLIAM J. (1306965)_____
Near Geneve, France, Oct. 8, 1918.
R—Statesville, N. C.
B—Caldwell County, N. C.
G. O. No. 46, W. D., 1919.

Sergeant, Company B, 117th Infantry, 30th Division.
While advancing with his platoon on the morning of Oct. 8, Sergeant McDade was seriously wounded in the hip, but insisted upon remaining with his platoon. He was again wounded twice by machine-gun fire, but continued to the objective, where he materially aided in consolidating the position. He was then ordered to the aid station by his commanding officer and was later evacuated to the hospital.

McDANIEL, JOHN R. (2809107)_____
Near Vilcey, France, Sept. 12, 1918.
R—Pawhuska, Okla.
B—Nowata, Okla.
G. O. No. 98, W. D., 1919.

Corporal, Company M, 358th Infantry, 90th Division.
When his group had been halted by fire from enemy snipers, Corporal McDaniel voluntarily exposed himself by standing in the open under fire, shot down three of the snipers from one tree and two from another.

*McDANIEL, LEE B. (2175601)_____
In Bois-de-Bantheville, France, Oct. 23, 1918.
R—Alma, Kans.
B—McCune, Kans.
G. O. No. 20, W. D., 1919.

Sergeant, Company A, 353d Infantry, 89th Division.
He led his platoon with great bravery and coolness against cleverly concealed machine guns until he fell severely wounded.
Posthumously awarded. Medal presented to widow, Mrs. Della McDaniel.

McDERMOTT, CLEVELAND W._____
Near Bantheville, France, Oct. 18, 1918.
R—Syracuse, N. Y.
B—Denver, Colo.
G. O. No. 1, W. D., 1919.

Second lieutenant, 147th Aero Squadron, Air Service.
In starting on a patrol mission he was delayed by motor trouble. Unable to overtake the other machines, he continued on alone. Sighting an enemy plane (Fokker), he immediately gave chase, and, despite its effort to escape, he succeeded in shooting it down. Six Fokkers then attacked him, and, though he was outnumbered and blinded by the sun, he shot down one of them and scattered the others. In the midst of this combat his motor stopped, and he was forced to glide into friendly territory.

McDermott, Francis P. (8144)_____
Near Fismes, France, Aug. 8, 1918.
R—Houtzdale, Pa.
B—Houtzdale, Pa.
G. O. No. 26, W. D., 1919.

Sergeant, first class, Section No. 524, Ambulance Service.
He remained on duty for 30 hours, guiding and directing the ambulances to and from the most advanced points. He volunteered and acted as stretcher bearer in full view of the enemy and under heavy machine-gun fire. At one point he cleared the road of débris under machine-gun fire and guided ambulances to the wounded.

ᵃMcDermott, Morgan B._____
Near Cunel, France, Oct. 20, 1918.
R—Tucson, Ariz.
B—Butte, Mont.
G. O. No. 37, W. D., 1919.

First lieutenant, 7th Engineers, 5th Division.
Accompanied by a soldier, he made a reconnaissance within the German lines and captured a machine gun. Under heavy machine-gun fire he gave first-aid treatment to a wounded soldier and continued on his mission until he was himself severely wounded.
Posthumously awarded. Medal presented to father, William McDermott.

MacDermut, Charles K._____
Near Moirey, France, Nov. 10, 1918.
R—Leonia, N. J.
B—Leonia, N. J.
G. O. No. 46, W. D., 1919.

Captain, 314th Infantry, 79th Division.
Although seriously wounded by three machine-gun bullets while reconnoitering, Captain MacDermut continued to direct his troops and refused to receive medical treatment or to allow stretcher bearers to come to the exposed position until night.

McDevitt, James A._____
Near Cuisy, France, Sept. 15, and Oct. 5–6, 1918.
R—Cincinnati, Ohio.
B—Newport, Ky.
G. O. No. 15, W. D., 1919.

First lieutenant, 281st Aero Squadron, Air Service.
On Sept. 15, 1918, while performing an important mission, his balloon was attacked and riddled by an enemy plane firing incendiary bullets. He stuck to his post and gathered valuable information. On Oct. 5 he was again attacked by several planes and the basket was set afire by incendiary bullets. While descending, he was fired upon and his parachute was hit many times. He, nevertheless, insisted upon returning to the air. On Oct. 6 he was attacked and his balloon was riddled with bullets. Again on the same day he was attacked by several enemy planes; he remained with his balloon until it came down in flames; he then resumed his post in a new balloon.

MacDonald, Beatrice Mary_____
France, Aug. 17, 1917.
R—New York, N. Y.
B—Canada.
G. O. No. 17, W. D., 1926.

Reserve nurse, Army Nurse Corps.
For extraordinary heroism while on duty with the surgical team at the British Casualty Clearing Station No. 61, British area, France, Aug. 17, 1917. During a German night air raid she continued at her post of duty caring for the sick and wounded until seriously wounded by a German bomb, thereby losing one eye.

ᵃMacDonald, Douglas (542071)_____
Near Cunel, France, Oct. 11, 1918.
R—Oliver Mills, Pa.
B—Oliver Mills, Pa.
G. O. No. 89, W. D., 1919.

Private, Company G, 7th Infantry, 3d Division.
While acting as a litter bearer Private MacDonald was painfully wounded in the hand by a machine-gun bullet, but refused to go to a dressing station, continuing to administer first aid to wounded under constant machine-gun and shell fire. On one occasion when a comrade had been buried by an exploding shell Private MacDonald rushed 200 yards into the open and worked under heavy fire at rescuing him. He was later killed by shell fire while giving aid to a wounded man.
Posthumously awarded. Medal presented to father, Carrick MacDonald.

McDonald, George G. (1762843)_____
Near Grand-Pre, France, Nov. 1, 1918.
R—Buffalo, N. Y.
B—Buffalo, N. Y.
G. O. No. 32, W. D., 1919

Corporal, Company E, 311th Infantry, 78th Division.
While accompanying a separated detachment, Corporal McDonald voluntarily entered a dense woods, infested with snipers and machine guns, to locate the main body of his company. The success of this mission enabled the lost detachment to occupy their position with the company. During the entire action of his company Corporal McDonald was forever eager, in his capacity as runner, to assume the most hazardous risks to maintain liaison between units.

McDonald, James (263604)_____
Near Sergy, France, July 31, 1918.
R—Saginaw, Mich.
B—Saginaw, Mich.
G. O. No. 64, W. D., 1919.

First sergeant, Company K, 125th Infantry, 32d Division.
After his platoon commander had fallen wounded at the beginning of the advance, and the platoon was becoming demoralized under intense machine-gun and artillery fire, Sergeant McDonald assumed command, steadied his men, and led them in a successful attack in a further advance of 150 yards, gaining an important objective and strengthening the defense of the captured position of Hill 212.

McDonald, John B._____
Near Épinonville and Gesnes, France, Sept. 26–30, 1918.
R—Athens, Ala.
B—Athens, Ala.
G. O. No. 44, W. D., 1919.
Distinguished-service medal also awarded.

Brigadier general, 181st Infantry Brigade, 91st Division.
He was almost continuously present with the leading elements of his brigade, inspiring his troops by his personal bravery and energy under fire. Near Epinonville, when his command was about to make an attack and was under heavy artillery fire, this officer, by his disregard for personal safety, steadied his men and stimulated them to successful assault on the ridge forming part of the German main line of resistance. Two days later, with one flank exposed by the withdrawal of the unit on the right, he led his brigade in the attack on and capture of Gesnes.

McDonald, Robert (1317009)_____
Near Souplet, France, Oct. 9–10, 1918.
R—Regle, N. C.
B—Grape Creek, N. C.
G. O. No. 37, W. D., 1919.

Private, first class, Company M, 119th Infantry, 30th Division.
He showed exceptional bravery and courage by going forward alone on many occasions to gain information of the enemy. He remained on duty with his company after being wounded until ordered to the rear for treatment.

McDonald, Robert M. (1311723)_____
Near Brancourt, France, Oct. 8, 1918.
R—Cheraw, S. C.
B—Cheraw, S. C.
G. O. No. 59, W. D., 1919.

Sergeant, Company I, 118th Infantry, 30th Division.
On the morning of Oct. 8, near the village of Brancourt, France, he alone charged an enemy machine-gun nest that was causing many casualties in his platoon and temporarily holding up the advance. He killed the gunner and leader, put the gun out of action, and thus enabled his platoon to advance.

MacDONNELL, JOHN L. (1212145)........
   At Guillemont Farm, near Bony,
     France, Sept. 28, 1918.
   R—New York, N. Y.
   B—Bay Shore, N. Y.
   G. O. No. 56, W. D., 1922.

Sergeant, Company M, 107th Infantry, 27th Division.
When the advance of his platoon was held up by machine-gun fire which was harassing his position, Sergeant MacDonnell voluntarily led a detachment of his platoon over the top and destroyed the machine-gun nest. His courageous actions were an inspiration to all his comrades.

McDONOUGH, JOHN F. (2357286)........
   Near Nantillois, France, Sept. 30,
     1918.
   R—Boston, Mass.
   B—Boston, Mass.
   G. O. No. 87, W. D., 1919.

Corporal, Company B, 4th Infantry, 3d Division.
Upon being wounded in the arm, going through a heavy artillery barrage, Corporal McDonough refused to go to the rear for first aid, but dressed his wound himself, remaining in command of his platoon section until he was killed by shellfire.
Posthumously awarded. Medal presented to widow, Mrs. Bridget McDonough.

MacDOUGALL, ALLAN J...............
   Near Revillon, France, Sept. 9, 1918.
   R—Detroit, Mich.
   B—Canada.
   G. O. No. 35, W. D., 1919.

Captain, 308th Infantry, 77th Division.
Captain MacDougall voluntarily assumed command of a patrol of 3 men to locate enemy lines and gun positions. Crawling through withering machine-gun fire to within 20 yards of the enemy lines, he encountered 2 Germans on outpost, whom he killed. Remaining exposed to the enemy for an hour, Captain MacDougall killed a machine gunner who attempted to take a position in front of him. His entire mission was harassed by perilous machine-gun fire and a constant hand-grenade bombardment.
Oak-leaf cluster.
For the following act of extraordinary heroism in action in the Argonne Forest, France, Oct. 4, 1918, Captain MacDougall is awarded an oak-leaf cluster, to be worn with the distinguished-service cross: Leading 3 companies to the aid of a surrounded battalion, he was rendered unconscious by a shell burst, but, upon recovery, refused treatment until properly relieved.

McDOUGALL, HARRY O...............
   Near Dun-sur-Meuse, France, Oct.
     23, 1918.
   R—Pocatello, Idaho.
   B—Malad City, Idaho.
   G. O. No. 1, W. D., 1919.

First lieutenant, pilot, 96th Aero Squadron, Air Service.
He, with Second Lieut. Elmore K. McKay, observer, while on a bombing mission, displayed exceptional courage by leaving a comparatively secure position in the center of the formation during a combat with 5 enemy planes and going to the protection of 2 other officers whose plane had been disabled and forced out of the formation. Lieutenant McDougall skillfully maneuvered his machine so as to enable Lieutenant McKay to shoot down 1 of the adversaries and fight off the others.

McDOWELL, EDGAR E. (1311232)........
   Near Montbrehain, France, Oct. 8
     1918.
   R—Hendersonville, N. C.
   B—Hendersonville, N. C.
   G. O. No. 46, W. D., 1919.

Private, Company F, 118th Infantry, 30th Division.
When the second wave of his company was confronted by 2 enemy machine-gun posts, which had been passed over by the first wave, Private McDowell, from a prone position, sniped at these posts and then rushed 1 of them. In so doing he was wounded in the wrist, but he continued on and succeeded in killing 2 Germans and capturing 4 others. The other post, containing 20 of the enemy, surrendered shortly afterwards.

McDOWELL, ELLIOTT E...............
   Near Amblimont, France, Nov. 8,
     1918.
   R—Cambridge, Mass.
   B—Dorchester, Mass.
   G. O. No. 138, W. D., 1918.

Second lieutenant, 305th Infantry, 77th Division.
He displayed unusual courage and determination in making a patrol under particularly hazardous circumstances, accompanied by only 1 soldier. Crossing the Meuse River, the east bank of which and the ridge east of it were known to be strongly held by the enemy, they proceeded through numerous machine-gun nests over the ridge, penetrating 3 kilometers into the hostile position and returning with important information concerning the enemy occupation.

MACE, JOHN H. (1461283)...............
   Near Exermont, France, Sept. 28,
     1918.
   R—Liberty, Mo.
   B—Liberty, Mo.
   G. O. No. 59, W. D., 1919.

Sergeant, Company H, 140th Infantry, 35th Division.
He volunteered to lead a detachment to attack a machine-gun nest which was holding up the advance of his battalion. Although severely wounded, he carried the position, killing the enemy gun crew and capturing the machine gun.

MacELLIGOTT, GEORGE H...............
   Near Mezy, France, July 15, 1918.
   R—West Somerville, Mass.
   B—West Townsend, Mass.
   G. O. No. 32, W. D., 1919.

First lieutenant, 30th Infantry, 3d Division.
Although mortally wounded, he remained in command of his platoon under direct view of the enemy and through a terrific bombardment until he died.
Posthumously awarded. Medal presented to mother, Mrs. S. Louise Mac-Elligott.

MacELROY, GEORGE L. (1236800)......
   Near Monthurel, France, July 17,
     1918.
   R—Philadelphia, Pa.
   B—Philadelphia, Pa.
   G. O. No. 72, W. D., 1920.

Bugler, Company H, 109th Infantry, 28th Division.
During the progress of a rather severe attack Bugler MacElroy delivered an important message from his company commander to the regimental headquarters. In order to perform this mission he was compelled to cross areas swept by heavy artillery and machine-gun fire. Due to his individual gallantry, communication was established at a critical time in the operation of this regiment.

McELWAIN, HARRY E. (543311).........
   Near Fossoy, France, July 15, 1918.
   R—Unionville, Pa.
   B—Fawngrove, Pa.
   G. O. No. 44, W. D., 1919.

Sergeant, medical detachment, 7th Infantry, 3d Division.
During an intense artillery preparation by the enemy he voluntarily went out about 1,000 yards, through this heavy shellfire, to administer first aid to 5 wounded men.

McENTEE, EUGENE (78382)...........
   Near Verdun, France, Oct. 2, 1918.
   R—Portland, Oreg.
   B—Montpelier, Idaho.
   G. O. No. 19, W. D., 1921.

Corporal, Headquarters Company, 26th Infantry, 1st Division.
In charge of maintaining telephone communication while advancing with a patrol, he showed marked personal bravery, and after being shot in the ankle refused to be evacuated and advanced for 1,300 meters under heavy machine-gun fire, repairing telephone lines and making it possible to send valuable information to the rear.

McEWEN, GLEN O. (74459)
Near Moulin de Guenoville, France, Sept. 26, 1918.
R—Spokane, Wash.
B—Nashville, Tenn.
G. O. No. 44, W. D., 1919.

Sergeant, Machine Gun Company, 161st Infantry, 41st Division, attached to 1st Gas Regiment.
Sergeant McEwen, with 3 other soldiers, advanced nearly 200 yards over an open hillside exposed to machine-gun fire and carried 2 wounded men to the protection of a near-by trench.

McFADDEN, JAMES E. (551132)
Near Mezy, France, July 15, 1918.
R—Cornplanter, Pa.
B—Venango County, Pa.
G. O. No. 24, W. D., 1920.

Corporal, Company G, 38th Infantry, 3d Division.
Corporal McFadden, in command of a squad, defended his position on the banks of the Marne against powerful enemy attacks until all of his men had been killed or wounded and he himself wounded. He later reported to his platoon commander and acted as liaison agent, refusing to go to the rear to have his wound dressed.

McFARLAND, GEORGE W. (1829414)
Near Brieulles, France, Sept. 28, 1918.
R—Latrobe, Pa.
B—Latrobe, Pa.
G. O. No. 59, W. D., 1919.

Corporal, Company D, 320th Infantry, 80th Division.
When the Germans counterattacked with a superior number in the Bois-de-Donovan, Corporal McFarland went from post to post under intense machine-gun and artillery fire, collecting all available rifle grenades in the platoon. Although he was seriously wounded by a machine-gun bullet, he continued on duty until his platoon was relieved several hours later.

MacFARLAND, JAY W.
During the St. Mihiel offensive, France, Sept. 12, 1918.
R—Cleveland, Ohio.
B—Belmont, Ohio.
G. O. No. 35, W. D., 1920.

First lieutenant, 16th Infantry, 1st Division.
Perceiving a gap which was increasing between 2 attacking companies, Lieutenant MacFarland promptly led 3 squads to fill the interval. His command was met at once by severe machine-gun fire and suffered severe casualties. He ordered his men to keep down and alone went forward to locate the enemy guns. In this act he was severely wounded, but he successfully directed the attack on the guns, which were captured before he was evacuated to the rear.

*MacFARLAND, JAMES
In the Bossois Bois, France, Oct. 12–17, 1918.
R—Burlington, N. J.
B—Burlington, N. J.
G. O. No. 20, W. D., 1919.

First lieutenant, 113th Infantry, 29th Division.
Throughout the 5 days of our attack in the Bossois Bois Lieutenant MacFarland had an advance dressing station in the woods, under constant shellfire, without protection. He repeatedly exposed himself to shellfire while going to the aid of wounded. His gallant example assisted greatly in keeping up the morale of the troops with whom he came in contact. He died from the effects of wounds received while giving aid to the wounded.
Posthumously awarded. Medal presented to widow, Mrs. James MacFarland.

*McFARLING, GEORGE (262245)
Near Cierges, northeast of Chateau-Thierry, France, July 31, 1918.
R—Flint, Mich.
B—Alpena, Mich.
G. O. No. 116, W. D., 1918.

Private, Company E, 125th Infantry, 32d Division.
Although he was himself severely wounded, Private McFarling crawled over to an exposed and dangerous place to render first aid to a seriously wounded comrade and while doing so received a fatal wound.
Posthumously awarded. Medal presented to mother, Mrs. Ada Jane McFarling.

McGAINEY, HUGH P. (1285511)
Near Verdun, France, Oct. 8–15, 1918.
R—Baltimore, Md.
B—Baltimore, Md.
G. O. No. 3, W. D., 1919.

Sergeant, Company H, 115th Infantry, 29th Division.
In the Bois-de-Consenvoye, east of the Meuse, he, in command of his platoon, led his men under heavy machine-gun fire and captured approximately 500 prisoners, 3 field pieces, and many machine guns. On Oct. 15 he voluntarily exposed himself to warn his men against gas and was wounded by shrapnel. He refused to go to the hospital until ordered to do so by a medical officer.

*McGARRY, PATRICK L. (1253447)
Near Fismes, France, Sept. 5, 1918.
R—Duryea, Pa.
B—Pittston, Pa.
G. O. No. 64, W. D., 1919.

Private, Battery B, 109th Field Artillery, 28th Division.
Seeing a wounded comrade lying in an open field swept by an enemy barrage he showed marked personal bravery in going to the rescue. Forced to expose himself to enemy fire to aid his companions, he himself was mortally wounded.
Posthumously awarded. Medal presented to father, Owen McGarry.

McGAY, GEORGE H.
Near Thiaucourt, France, Sept. 12, 1918.
R—New York, N. Y.
B—New York, N. Y.
G. O. No. 98, W. D., 1919.

Second lieutenant, 23d Infantry, 2d Division.
Organizing a group of 20 men who had become separated from their organizations, Lieutenant McGay, under machine-gun fire, attacked a strongly intrenched position defended by a greatly superior number of the enemy, killing 3 of the hostile force and capturing 25, together with 4 machine guns.

*McGEARY, JOHN (90532)
Near Villers-sur-Fere, France, Aug. 1, 1918.
R—New York, N. Y.
B—Ireland.
G. O. No. 88, W. D., 1918.

Private, Company G, 165th Infantry, 42d Division.
He left his shelter and went out into heavy shell and machine-gun fire to rescue a wounded comrade, receiving fatal wounds in the attempt.
Posthumously awarded. Medal presented to sister, Miss Mary Anna McGeary.

McGEE, EDWARD (2213042)
In the Bois-de-Barricourt, France, Nov. 1–2, 1918.
R—Logan, Kans.
B—Tipton, Kans.
G. O. No. 37, W. D., 1919.

Private, Company M, 353d Infantry, 89th Division.
When volunteers were called for to maintain liaison with the assaulting battalion during heavy shell and machine-gun fire, he volunteered and successfully carried out 5 such missions.

McGEE, LAWRENCE T. (1257894)
At Fismes, France, Aug. 12, 1918.
R—Allentown, Pa.
B—Allentown, Pa.
G. O. No. 3, W. D., 1922.

Private, Company C, 109th Machine Gun Battalion, 28th Division.
Seeing a wounded soldier in an ambulance which had broken down while crossing a small bridge, in plain view of the enemy and under heavy artillery fire, he went to his rescue, and, assisted by another soldier, carried out the wounded man under intense shell fire 1½ miles to a dressing station.

McGILL, DON R.
Near Brabant, France, Oct. 23, 1918.
R—Nelsonville, Ohio.
B—Nelsonville, Ohio.
G. O. No. 32, W. D., 1919.

Captain, Field Artillery, 308th Trench Mortar Battery, 83d Division.
Due to the untiring energy and determination of Captain McGill, eight 6-inch mortars and ammunition were transported to within 800 meters of the enemy lines, greatly aiding in the preparatory artillery bombardment. Although 15 of the 55 men engaged were killed, 13 wounded, and 4 gassed, the mortars were kept in action until the last one was destroyed by enemy fire.

McGINNIS, GEORGE E.
At Fismette, France, Aug. 9–10, 1918.
R—Philadelphia, Pa.
B—Norristown, Pa.
G. O. No. 15, W. D., 1919.

Captain, 110th Ambulance Company, 103d Sanitary Train, 28th Division.
During the night of the 9th of August Captain McGinnis, with complete disregard of his personal safety, made a reconnaissance under fire and located a line of evacuation for ambulances from Fismette and on the morning of the 10th of August, under shellfire, he personally repaired the bridge between Fismes and Fismette, thereby making possible the evacuation of 28 wounded men.

McGINNIS, WILLIAM H. (567996)
Near Chery-Chartreuve, France, Aug. 10, 1918.
R—Beckley, W. Va.
B—Beckley, W. Va.
G. O. No. 50, W. D., 1919.

Corporal, Company D, 12th Machine Gun Battalion, 4th Division.
An incendiary shell exploded near a large ammunition dump near which his company was resting, wounding several of his comrades, and setting fire to a portion of the dump. While a second explosion was imminent, Corporal McGinnis rushed into the flames and dragged a wounded man to safety.

McGINTY, JOHN J. (243468)
At Varennes, France, Sept. 26, 1918.
R—Carbondale, Pa.
B—Carbondale, Pa.
G. O. No. 46, W. D., 1919.

Sergeant, first class, Company B, 344th Battalion, Tank Corps.
Gathering several scattered infantrymen, Sergeant McGinty led them into the town of Varennes ahead of the tanks and captured a number of prisoners. He then withdrew to the outskirts of the town to direct the advance of several tanks which had arrived. Returning to the attack on foot, he continued forward until a wound compelled him to retire.

*McGLINCHEY, WILLIAM J. (1696992)
Near Carrefour-de-Meurrussons, France, Sept. 28, 1918.
R—Brooklyn, N. Y.
B—Brooklyn, N. Y.
G. O. No. 44, W. D., 1919.

Sergeant, Company A, 305th Infantry, 77th Division.
While his platoon was being heavily bombarded he left his place of safety and quiet to rescue a wounded comrade. In attempting this valiant deed Sergeant McGlinchey lost his own life.
Posthumously awarded. Medal presented to widow, Mrs. Frances M. McGlinchey.

*McGLUE, JOHN R. (1209998)
Near Ronssoy, France, Sept. 29, 1918.
R—Brooklyn, N. Y.
B—Lynn, Mass.
G. O. No. 16, W. D., 1920.

Private, first class, Company B, 107th Infantry, 27th Division.
Private, First Class, McGlue, with Mechanic Copeland, left the protection of a trench and, in the face of heavy machine-gun and grenade fire, went in advance of our lines to rescue a wounded comrade. They were exposed to heavy fire from the time they left the trench. Private McGlue was killed as he and his companion were returning to the trench with the succored wounded comrade.
Posthumously awarded. Medal presented to father, H. R. McGlue.

McGOWAN, FRANCIS (60991)
Near Bois-de-St. Remy, France, Sept. 12, 1918.
R—Waltham, Mass.
B—Waltham, Mass.
G. O. No. 26, W. D., 1919.

Sergeant, Company F, 101st Infantry, 26th Division.
Accompanying two other soldiers, Sergeant McGowan rushed foreward in advance of his lines, exposed to heavy machine-gun fire, and attacked an enemy machine-gun stronghold which was halting the progress of his platoon. He succeeded in capturing 2 guns and 6 of the crew who were manning them.

McGRATH, HENRY JOHN
On the high seas, Oct. 12, 1918.
R—Jacksonville, Fla.
B—Brookline, Mass.
G. O. No. 71, W. D., 1919.

Second lieutenant, Quartermaster Corps, with Army Transport Service.
For extraordinary heroism in action between the United States Army chartered transport Amphion and an enemy submarine on the high seas on Oct. 12, 1918. In the face of heavy enemy shellfire he took charge of and directed the laying of fire hose along the deck and extinguished a fire which had been started by an exploding shell. Exposing himself to exploding shells and without regard for his personal safety, he carried a wounded seaman across the shell-swept deck to a place of safety. During the entire engagement, which lasted 1 hour and 20 minutes, this officer displayed great coolness, going from place to place about the ship and encouraging the crew at a time when encouragement was sorely needed. His coolness and the effective manner with which he gave orders inspired everyone and greatly aided the escape of the Amphion.

McGRAW, JOE W. (1517987)
Near Heurne, Belgium, Nov. 4, 1918.
R—McGraw, Ohio.
B—McGraw, Ohio.
G. O. No. 59., W. D., 1919.

Private, Company D, 145th Infantry, 37th Division.
He displayed exceptional personal bravery when, with one other soldier, he went to the aid of a comrade who had been attacked and wounded by a patrol of 8 Germans, putting the patrol to flight and rescuing the wounded man.

McGUIRE, EARL R. (13510)
Near Chateau-Thierry, France, July 18, 1918.
R—Greenfield, Mass.
B—Greenfield, Mass.
G. O. No. 125, W. D., 1918.

Sergeant, Company L, 104th Infantry, 26th Division.
After being severely wounded in the head he struggled to his feet and led his platoon forward, instilling courage and confidence in his men.

MacGUIRE, EDWARD A.
Near the Bois-de-Chatillon, France, Nov. 5–9, 1918.
R—New York, N. Y.
B—New York, N. Y
G. O. No. 44, W. D., 1919.

First lieutenant, 6th Infantry, 5th Division.
Having developed a hernia in crossing the Meuse River, Lieutenant MacGuire displayed remarkable fortitude and devotion to duty by remaining with his company and leading it into action, reaching his objective under severe machine-gun and shellfire. He continued with his command until a double hernia developed from strain caused by 2 forced marches into advanced positions and he was ordered to the rear.

McGUIRE, JAMES (1750405)................
Near Talma Farm, France, Oct. 23, 1918.
R—Newark, N. J.
B—Ireland.
G. O. No. 37, W. D., 1919.

Sergeant, Company B, 312th Infantry, 78th Division.
Without regard for his own danger, Sergeant McGuire went to the rescue of a wounded comrade who was lying seriously wounded 100 yards in front of our lines. He carried him safely to the aid station, crossing and recrossing an area swept by intense artillery and machine-gun fire and under the direct observation of the enemy.

McGUIRE, JAMES, Jr. (1902288).........
Near St. Juvin, France, Oct. 17, 1918.
R—New Haven, Conn.
B—Ireland.
G. O. No. 46, W. D., 1919.

Private, Company H, 326th Infantry, 82d Division.
With exceptional gallantry, he voluntarily went out into an area swept by heavy machine-gun fire to rescue a wounded sergeant, and succeeded in carrying the latter a considerable distance through heavy fire to a first-aid station.

McGUIRE, LEO F. (10388).................
Near Seicheprey, France, Apr. 19, 1918.
R—Tulsa, Okla.
B—Elgin, Kans.
G. O. No. 129, W. D., 1918.

Private, first class, section No. 647, Ambulance Service.
He was on duty as driver of an ambulance at an advanced post on Apr. 19, 1918. During Apr. 19 and 20 he made several trips to and from a dressing station reached by an exposed road in daylight, for the purpose of bringing back wounded. On one of these trips the ambulance was blown from the road by the explosion of a shell and he was knocked unconscious by the shock. On recovering consciousness he returned on foot. Although suffering from an injury in the back and not yet recovered from the shock, he wished to return to duty the afternoon of the same day, but was not permitted to do so by the medical officers until the afternoon of the following day.

McGUIRE, MAURICE J.................
Near St. Agnan, France, July 16, 1918.
R—Scranton, Pa.
B—Scranton, Pa.
G. O. No. 71, W. D., 1919.

First lieutenant, 109th Infantry, 28th Division.
Although painfully wounded shortly after his platoon began an attack, Lieutenant McGuire refused to be evacuated until his command was ordered to withdraw. He then saw that their position was firmly held before going to the rear for treatment. His personal heroism was an inspiration to his men.

McGUIRK, HARRY (1284895)............
Near Haumont, France, Oct. 11, 1918.
R—Principio Furnace, Md.
B—North East, Md.
G. O. No. 7 W. D., 1919.

Corporal, 116th Ambulance Company, 104th Sanitary Train, 29th Division.
He worked for 4 days, fearlessly exposing himself to heavy enemy fire, in administering first aid and directing the evacuation of the wounded. By his conspicuous bravery and untiring energy he was an example to his men.

*McGUIRL, BERNARD (62960)...........
Near Verdun, France, Oct. 23, 1918.
R—Fitchburg, Mass.
B—Fitchburg, Mass.
G. O. No. 21, W. D., 1919.

Corporal, Headquarters Company, 101st Infantry, 26th Division.
While leading a squad of ammunition carriers he was severely wounded. After a tourniquet had been applied to his wounds he ordered the second in command to continue with the work, refusing the aid of his comrades until their mission had been completed. While on his way to a hospital he died of his wounds.
Posthumously awarded. Medal presented to mother, Mrs. Rose McGuirl.

*McHENRY, JOHN, Jr.................
Near Vierzy, France, July 19, 1918, and before Blanc Mont, Oct. 3, 1918.
R—Baltimore, Md.
B—Pikesville, Md.
G. O. No. 60, W. D., 1920.

First lieutenant, 6th Regiment, U. S. Marine Corps, 2d Division.
Lieutenant McHenry led his platoon through heavy machine-gun and artillery fire until he fell severely wounded. On Oct. 3, 1918, while leading his platoon in the attack on the strongly fortified enemy position before Blanc Mont, he fell mortally wounded.
Posthumously awarded. Medal presented to father, John McHenry.

McINTYRE, DONALD R.................
Near Thiaucourt, France, Sept. 18, 1918.
R—Springfield, Mass.
B—Manchester, N. H.
G. O. No. 37, W. D., 1919.

Second lieutenant, Company E, 310th Infantry, 78th Division.
Seeing one of his men lying wounded in a shell hole in front of his main line, Lieutenant McIntyre passed through an intense barrage to his aid. Having bandaged his wounds, he brought back his man to a place of safety.

McINTYRE, EUGENE (1396538).........
Near Hill 281, France, Oct. 3, 1918.
R—Chicago, Ill.
B—Johnstown, Pa.
G. O. No. 71, W. D., 1919.

Corporal, 130th Ambulance Company, 108th Sanitary Train, 33d Division.
After being severely wounded by the explosion of an enemy shell in the dressing station, he remained on duty, assisting in caring for the wounded until he fainted from loss of blood.

McINTYRE, JAMES B.................
Near Villers-sur-Fere, France, July 28–Aug. 3, 1918.
R—North Adams, Mass.
B—North Adams, Mass.
G. O. No. 35, W. D., 1919.

First lieutenant, 165th Infantry, 42d Division.
He organized his platoon into a carrying party and on two occasions brought up ammunition and supplies to the battalion in the front line, through heavy machine-gun, rifle, and shell fire. He was knocked down several times and once thrown into the Ourcq River, but successfully carried out his mission.

McINTYRE, JOHN (1467559)............
Near Cheppy, France, Sept. 26, 1918.
R—Topeka, Kans.
B—Nokomis, Ill.
G. O. No. 59, W. D., 1919.

Sergeant, Company A, 110th Engineers, 35th Division.
While a member of a platoon of wire cutters, he, with another sergeant, attacked and helped to capture an enemy machine-gun nest that was holding up our advance. One officer, 6 men, and 2 guns were taken in the face of intense machine-gun fire.

McINTYRE, WILLIAM.................
Near Vierzy, France, July 19, 1918.
R—Chicago, Ill.
B—Scottsville, N. Y.
G. O. No. 126, W. D., 1919.

Private, 55th Company, 5th Regiment, U. S. Marine Corps, 2d Division.
Corporal Montag and Privates McIntyre, Messinger, and Wood captured a machine gun which was holding up the 55th Company of Marines, killing the entire crew. To accomplish this hazardous and daring work it was necessary for them to expose themselves to the fire of this gun. Even though Corporal Montag and Privates McIntyre and Messinger were wounded during the advance, the party continued forward and succeeded.

MacIsaac, Donald (161412)..........
At Gouzeaucourt, France, Nov. 30, 1917.
R—Kew Gardens, Long Island, N. Y.
B—Chicago, Ill.
G. O. No. 129, W. D., 1918.

Sergeant, Company B, 11th Railway Engineers.
When the unarmed working party of which he was a member was unexpectedly attacked at Gouzeaucourt, France, Nov. 30, 1917, he displayed extraordinary heroism by declining to take advantage of shelter, in going back into the barrage to assist American soldiers of another unit, and returning into the barrage a second time to search for wounded British soldiers.

*Mack, Peter F. (255882)...........
Near Montrebeau Woods, France, Sept. 28, 1918.
R—Ottawa, Ill.
B—Chicago, Ill.
G. O. No. 60, W. D., 1920.

Private, Company K, 140th Infantry, 35th Division.
After being mortally wounded during the advance, Private Mack refused to go to the rear. He went to the place assigned to him in the line and attempted to dig in. His fearlessness and fortitude were a great inspiration to his comrades.
Posthumously awarded. Medal presented to brother, Michael J. Mack.

Mack, Walter C. (2711175)..........
Near Eyne, Belgium, Nov. 2, 1918.
R—Philadelphia, Pa.
B—Philadelphia, Pa.
G. O. No. 37, W. D., 1919.

Private, Company B, 135th Machine Gun Battalion, 37th Division.
In the face of intense machine-gun fire, he voluntarily swam the Scheldt River to obtain information regarding the enemy. His successful return with the desired information enabled his company commander to so place his guns that they could be fired with great advantage.

Mack, William............................
On the Vesle River, near Bazoches, France, Sept. 2, 1918.
R—New York, N. Y.
B—Cleveland, Ohio.
G. O. No. 15, W. D., 1919.

First lieutenant, 305th Infantry, 77th Division.
Lieutenant Mack volunteered to leave St. Thibaut in broad daylight with another officer and a patrol of 10 men to reconnoiter the enemy's lines. Upon reaching the Vesle River Lieutenant Mack swam across it and arranged the rope by means of which the remainder of the patrol crossed the stream. He divided the patrol and, taking 5 men with him, advanced on the village of Bazoches, which was occupied by the enemy. He attacked enemy hiding places in an old house in which he encountered 4 Germans. Although under machine-gun fire, he gained valuable information, having actually penetrated the enemy's advanced posts, and with great skill withdrew his patrol. Lieutenant Mack and 4 of his men were wounded, 2 mortally.

MacKall, Murray R..................
West of Fismes, France, Aug. 4-5, 1918.
R—San Francisco, Calif.
B—East Liverpool, Ohio.
G. O. No. 147, W. D., 1918.

Captain, 4th Engineers, 4th Division.
He reconnoitered a section of the River Vesle in advance of the front line of Infantry under continuous fire from machine gun and 1 pounders. Proceeding alone for about 1 kilometer along the stream, despite the fact that German machine guns were located near the opposite bank, he continued his reconnaissance and selected several suitable sites, one of which was used the next night. Captain MacKall guided the working party through the enemy's barrage.

Mackay, Donald S...................
Near Sergy, France, July 26-30, 1918.
R—Blue Hill, Me.
B—St. Albans, Vt.
G. O. No. 74, W. D., 1919.

First lieutenant, 168th Infantry, 42d Division.
In an effort to locate enemy machine-gun emplacements, Lieutenant Mackay constantly exposed himself to enemy fire and, while so doing, was severely wounded. During the entire five days of operations he led a scout group forward, locating nests that had been stubbornly resisting the progress of our troops, and supplying the artillery with most valuable information, resulting in the destruction of the nests.

*McKay, Albert G. (1865857)........
Near Montbrehain, France, Oct. 8, 1918.
R—Gastonia, N. C.
B—Mooresville, N. C.
G. O. No. 87, W. D., 1919.

Corporal, Company C, 105th Engineers, 30th Division.
Corporal McKay, a runner, passed unfalteringly through heavy enemy shell fire to inform platoon leaders of the location of cover from the advanced enemy counterbarrage, continuing to expose himself until all were protected, thereby preventing many casualties. As he was returning from this mission he was badly wounded and died shortly afterwards.
Posthumously awarded. Medal presented to mother, Mrs. Belle Branton McKay.

McKay, Elmore K...................
Near Dun-sur-Meuse, France, Oct. 23, 1918.
R—Washington, D. C.
B—Washington, D. C.
G. O. No. 1, W. D., 1919.

Second lieutenant, observer, 96th Aero Squadron, Air Service.
He, with First Lieutenant Harry O. McDougall, pilot, while on a bombing mission displayed exceptional courage by leaving a comparatively secure position in the center of the formation during a combat with 5 enemy planes and going to the protection of 2 other officers whose planes had been disabled and forced out of the formation. While his pilot skillfully maneuvered the machine, he shot down one of the adversaries and fought off the others, thereby saving the lives of the officers in the disabled American planes.

McKay, James R...................
Near Doulcon, France, Oct. 4, 1918.
R—Wheaton, Ill.
B—Grinnell, Iowa.
G. O. No. 46, W. D., 1919.

First lieutenant, 49th Aero Squadron, Air Service.
When a patrol of 7 planes attacked a group of 17 enemy planes (Fokker type) he remained above to protect from that direction. Without regard to his own danger he attacked alone 5 more enemy planes which dived into the combat, and pressing the attack succeeded in breaking up their formation and shooting down 1 of the enemy planes.

McKay, John W. (2213386)..........
Northwest of Bantheville, France, Nov. 1, 1918.
R—Independence, Kans.
B—Thayer, Kans.
G. O. No. 27, W. D., 1919.

Corporal, Company M, 353d Infantry, 89th Division.
Immediately after the beginning of the attack in the Bois-de-Bantheville, France, when his company was held up by a strong machine-gun nest and his company commander and several others were killed by its fire, Corporal McKay, accompanied by an officer and with great gallantry and coolness attacked and captured the machine-gun nest of 4 guns, killing or wounding a number of the crew.

McKendry, Stewart J. (132500).....
Near Vaux, France, July 1, 1918.
R—Philadelphia, Pa.
B—Ireland.
G. O. No. 101, W. D., 1918.

Private, Battery E, 17th Field Artillery, 2d Division.
He performed his duty of telephone-line repairman with great bravery and promptness in spite of intense bombardment of the area where he had to work, near Vaux, July 1, 1918, going fearlessly and without waiting to be ordered when communication was broken.

McKenna, Herbert F. (91192)........
Near Villers-sur-Fere, France, July 28, 1918.
R—New York, N. Y.
B—New York, N. Y.
G. O. No. 99, W. D., 1918.

Sergeant, Company K, 165th Infantry, 42d Division.
At the beginning of the attack against the enemy positions on the north bank of of the River Ourcq he was wounded in the arm by a machine-gun bullet, yet he continued in the advance and took charge of his platoon when its commander was killed. When the first attack was over, he received first aid and then returned to his company, where he assumed the duties of first sergeant in addition to his duties as a platoon commander.

*McKenna, James A., Jr............
Near Villers-sur-Fere, France, July 28, 1918.
R—New York, N. Y.
B—New York, N. Y.
G. O. No. 99, W. D., 1918.

Major, 165th Infantry, 42d Division.
He was killed while successfully leading a most difficult and trying attack across the River Ourcq and against the strongly prepared positions on the heights beyond.
Posthumously awarded. Medal presented to father, James A. McKenna.

McKenna, Patrick (51932)...........
Near Vaux, France, July 1, 1918.
R—St. Paul, Minn.
B—Fairmont, Minn.
G. O. No. 102, W. D., 1918.

Corporal, Company L, 23d Infantry, 2d Division.
After being seriously wounded near Vaux, France, July 1, 1918, he charged into thick woods held by the enemy in face of a barrage of hand grenades and killed 3 single handed.

McKeogh, Arthur F..............
Near Binarville, France, Sept. 29, 1918.
R—New York, N. Y.
B—Troy, N. Y.
G. O. No. 15, W. D., 1921.

First lieutenant, 308th Infantry, 77th Division.
In order to obtain ammunition and rations, Lieutenant McKeogh, accompanied by 2 enlisted men, attempted to reestablish communication between battalion and regimental headquarters. When night came they crawled unknowingly into the center of a German camp, where they lay over 3 hours undetected. Finally discovered, they made a dash to escape, and Lieutenant McKeogh, in order to protect his men, deliberately drew the enemy fire upon himself. He succeeded, however, in getting through the enemy lines, delivered his message, and effected the reestablishment of communication.

McKernan, William T. (65743)........
Near Verdun, France, Oct. 24–29, 1918.
R—Willimantic, Conn.
B—Hartford, Conn.
G. O. No. 46, W. D., 1919.

Sergeant, Company K, 102d Infantry, 26th Division.
After all the company officers had been killed or wounded, he took command of the company and led it in effective attack against the enemy on the 25th of October and again on the 27th. After the attack on the enemy of Oct. 28, he secured a box of rifle grenades, and while a protective barrage was laid down with these he crawled out in advance of the line and brought in several wounded comrades.

McKey, Harold, G................
At Romagne, France, Oct. 13–14, 1918.
R—Chicago, Ill.
B—Chicago, Ill.
G. O. No. 98, W. D., 1919.

First lieutenant, 128th Infantry, 32d Division.
When his company commander was wounded Lieutenant McKey took command of the company. Although wounded and weakened by the loss of blood, he refused to be evacuated, but continued in command of his men for 32 hours under severe machine-gun and artillery fire until the town of Romagne was taken, remaining in action until ordered to the rear.

*McKibbin, James M................
Near Chevieres, France, Oct. 14, 1918.
R—Buck Valley, Pa.
B—Buck Valley, Pa.
G. O. No. 21, W. D., 1919.

Captain, Medical Corps, attached to 306th Machine Gun Battalion, 77th Division.
During a very heavy artillery barrage, which lasted for approximately 2 hours, Captain McKibbin displayed great coolness and courage in dressing and administering first aid to the wounded. Informed that a sergeant had been wounded and was lying between our lines and the enemy's line, he went to administer first aid to him. While in the performance of these duties under intense fire, he was wounded by machine-gun fire and later died from the effects of the wound.
Posthumously awarded. Medal presented to widow, Mrs. Mary McKibbin.

McKiddy, Zona (1319188)..........
Near Bellicourt, France, Sept. 29, 1918.
R—Knoxville, Tenn.
B—Meadow Creek, Ky.
G. O. No. 35, W. D., 1919.

Private, Machine Gun Company, 120th Infantry, 30th Division.
When his platoon had suffered heavy casualties and the runners had been killed by heavy artillery fire, Private McKiddy volunteered to carry a message, calling for reinforcements. Making his way through a dense smoke barrage, he succeeded in reaching company headquarters and returning, despite the intense bombardment.

*McKimmey, John C............
Near St. Etienne, France, Oct. 8, 1918.
R—Comanche, Tex.
B—Goldthwaite, Tex.
G. O. No. 78, W. D., 1919.

Second lieutenant, 141st Infantry, 36th Division.
His organization harassed and suffering severe losses from enemy machine-gun fire, Lieutenant McKimmey, accompanied by 2 sergeants, left shelter and advanced across an open area, exposed to enemy fire, in order to attack their position. He lost his life while making the attempt.
Posthumously awarded. Medal presented to father, R. L. McKimmey.

Mackin, Elton Edward............
Near Blanc Mont, France, Oct. 4, 1918.
R—Lewiston, N. Y.
B—Lewiston, N. Y.
G. O. No. 37, W. D., 1919.

Private, 67th Company, 5th Regiment, U. S. Marine Corps, 2d Division.
As a runner he carried messages over territory which was subject to constant shellfire, exhibiting singular courage and devotion to duty.

McKinley, Earl M..............
Near Nantillois, France, Sept. 26, 1918.
R—East Liverpool, Ohio.
B—East Liverpool, Ohio.
G. O. No. 142, W. D., 1918.

First lieutenant, 11th Machine Gun Battalion, 4th Division.
He with another officer and a soldier, using captured German Maxim guns, pushed forward to a heavily shelled area from which other troops had withdrawn and by their accurate and effective fire kept groups of the enemy from occupying advantageous positions, maintaining fire superiority all the afternoon. He withdrew from his dangerous position only when it became too dark to see.

*McKinlock, George A., Jr
  At Berzy-le-Sec, France, July 21,
    1918.
  R—Lake Forest, Ill.
  B—Chicago, Ill.
  G. O. No. 132, W. D., 1918.

Second lieutenant, 2d Infantry Brigade, 1st Division.
He showed noble disregard of self and devotion to duty by traversing the front lines for information necessary in connection with his work as intelligence officer, and while fearlessly performing this work was killed.
Posthumously awarded. Medal presented to mother, Mrs. George A. McKinlock.

McKinney, Darel Jesse
  In the Bois-de-Belleau, France,
    June 8, 1918.
  R—Milwaukee, Wis.
  B—Milwaukee, Wis.
  G. O. No. 110, W. D., 1918.

Sergeant, 83d Company, 6th Regiment, U. S. Marine Corps, 2d Division.
In the Bois-de-Belleau, on June 8, 1918, although severely wounded he refused to go to the rear for treatment. Despite his wounds he continued to lead his platoon to the attack, inflicting great losses upon the enemy.

McKinney, Loater Lloyd (1311144)
  Near Montbrehain, France, Oct. 8,
    1918.
  R—Spartanburg, S. C.
  B—Dobson, N. C.
  G. O. No. 35, W. D., 1919.

Private, Company F, 118th Infantry, 30th Division.
Accompanying 2 comrades, he attacked with hand grenades an enemy machine gun stronghold containing at least 40 Germans and 4 machine guns and forced the enemy to surrender.

McLain, Alexis M. (1317178)
  Near St. Souplet, France, Oct. 10,
    1918.
  R—Winston-Salem, N. C.
  B—Alexandria, N. C.
  G. O. No 81, W. D., 1919.

Private, Company K, 119th Infantry, 30th Division.
After 1 soldier had been killed and another wounded in the attempt, he carried a message under heavy fire to company headquarters, bringing up reinforcements which saved his platoon.

McLain, Charles L.
  On the Marne River, France, July
    15, 1918, and at Apremont, France,
    Sept. 29, 1918.
  R—Indiana, Pa.
  B—Indiana, Pa.
  G. O. No. 143, W. D., 1918.

Captain, 110th Infantry, 28th Division.
He was an observer with the French when the enemy attack on the Marne River was started July 15, 1918. All the officers of an Infantry company having been killed or wounded, he voluntarily reorganized the remainder of the company and successfully fought his way through the enemy, upon 2 occasions being surrounded. In this operation he was badly gassed. At Apremont, Sept. 29, when his own company had reached its objective, he, finding that another company was without officers, voluntarily assumed command of it and led the first wave. In so doing he was wounded, but he continued in action until the objective was reached.

*McLaughlin, Edward J.
  At Etraye Ridge, France, Oct. 23,
    1918.
  R—Newark, N. J.
  B—Newark, N. J.
  G. O. No. 89, W. D., 1919.

First Lieutenant, 113th Infantry, 29th Division.
Assuming command of his company after the death of his commanding officer, Lieutenant McLaughlin displayed remarkable courage and coolness in leading his men to the summit of the ridge until he was severely wounded. He was killed shortly afterwards by shellfire as he was being evacuated in an ambulance.
Posthumously awarded. Medal presented to widow, Mrs. Mary McLaughlin.

*McLaughlin, Edward R. (2431613)
  Near Cunel, France, Oct. 18, 1918.
  R—Wampum, Pa.
  B—Wampum, Pa.
  G. O. No. 22, W. D., 1920.

Private, Company B, 7th Infantry, 3d Division.
The platoon of which Private McLaughlin was a member was cut off from the rest of the company by an intense artillery barrage. It was imperative to communicate with company headquarters. He delivered a message through extremely heavy fire. While on the return trip with the reply he was hit by an enemy shell and instantly killed.
Posthumously awarded. Medal presented to father, John W. McLaughlin.

McLaughlin, Edwin W. (1211454)
  Near Ronssoy, France, Sept. 29,
    1918.
  R—Middletown, N. Y.
  B—Brooklyn, N. Y.
  G. O. No. 32, W. D., 1919.

Mechanic, Company I, 107th Infantry, 27th Division.
While the rest of his company was being held up by intensive machine-gun fire of the enemy, he advanced alone and put the guns out of action. On several other occasions he volunteered and accompanied patrols in attack against enemy nests, each time proving himself of the greatest assistance, successfully accomplishing his mission despite great hazards.

McLaughlin, John (91850)
  Near Sedan, France, Nov. 7, 1918.
  R—Wards Island, N. Y.
  B—Ireland.
  G. O. No. 37, W. D., 1919.

Corporal, Company M, 165th Infantry, 42d Division.
After 10 of the 16 of his patrol had been wounded and others scattered by machine-gun fire, Corporal McLaughlin, with 2 other soldiers, continued on his mission. He located 6 guns and returned with valuable information which was turned over to the troops relieving his regiment. This mission was carried out under continuous machine-gun fire.

McLawhon, Lewis B. (49243)
  Near Chateau-Thierry, France, June
    6, 1918.
  R—Winterville, N. C.
  B—Winterville, N. C.
  G. O. No. 102, W. D., 1918.

Saddler, Machine Gun Company, 23d Infantry, 2d Division.
While attached to the headquarters of a machine-gun company of the 23d Infantry, near Chateau-Thierry, France, on June 6, 1918, he made 8 trips as a runner to and from advance platoons. He showed heroic coolness in the face of machine-gun fire and absolute fearlessness in the execution of his work.

McLean, Stephen (60026)
  Near Mouilly, France, Sept. 12, 1918.
  R—West Roxbury, Mass.
  B—Dedham, Mass.
  G. O. No. 27, W. D., 1920.

Private, Company B, 101st Infantry, 26th Division.
When the advance of his platoon was belated by heavy machine-gun fire from the front, Private McLean exposed himself to heavy fire, charged and captured the enemy position. Single handed he forced 9 of the enemy to surrender and enabled his platoon to continue the advance.

McLelland, William D.
  Near Nantillois and Montfaucon,
    France, Sept. 29–Oct. 1, 1918.
  R—Mooresville, N. C.
  B—Mooresville, N. C.
  G. O. No. 126, W. D., 1919.

First lieutenant, 314th Ambulance Company, 304th Sanitary Train, 79th Division.
Lieutenant McLelland, near Nantillois, displayed untiring energy in bringing in the wounded while continually subjected to machine-gun and shrapnel fire. It was necessary to move the dressing station to some abandoned German dugouts because of the heavy fire, and during the bombardment this station was set on fire and 6 men killed, but Lieutenant McLelland, by his coolness and courage, enabled the speedy evacuation of the wounded.

McLENDON, PRESTON ALEXANDER.....
Near Blanc Mont, France, Oct. 3–4, 1918.
R—New York, N. Y.
B—Wadesboro, N. C.
G. O. No. 37, W. D., 1919.

Lieutenant (junior grade), assistant surgeon, U. S. Navy, attached to 1st Battalion, 5th Regiment, U. S. Marine Corps, 2d Division.
During heavy action he continually pushed his dressing station to more advantageous positions. Although in great danger because of a severe shelling, he dressed his patients in an exposed position, using his dugout for the seriously wounded.

McLENNAN, DONALD J..............
Near St. Etienne, France, Oct. 8, 1918.
R—Wausau, Wis.
B—Rib Lake, Wis.
G. O. No. 50, W. D., 1919.

First lieutenant, 142d Infantry, 36th Division.
Leading a patrol of 10 men, with orders to gain contact with the enemy, Lieutenant McLennan was suddenly subjected to terrific machine-gun fire, which wounded two of his men. He quickly ordered his men to shelter, and in order to insure their safety he ran across an open space for a distance of 30 yards to draw the enemy's fire. Rejoining his patrol by passing through the same deadly fire, he ordered them to withdraw, covering their withdrawal and reporting back to his company commander with his 2 wounded men and much valuable information.

McLEOD, HERMAN L...............
In the Bois-de-Belleau, France, June 6 and 8, 1918.
R—Paulding, Ohio.
B—Cincinnati, Ohio.
G. O. No. 70, W. D., 1919.

Private, 83d Company, 6th Regiment, U. S. Marine Corps, 2d Division.
Howard J. Childs, Joseph A. Dargis, and Allen Benjamin Tilghman, corporals, and Herman L. McLeod, private, 83d Company, 6th Regiment, U. S. Marine Corps. These 4 men were prominent in the attack on enemy machine-gun positions in the Bois-de-Belleau on June 6 and 8, 1918, were foremost in their company at all times, and acquitted themselves with such distinction that they were an example for the remainder of their command.

*McLEOD, LAMAR Y...............
Near St. Juvin, France, Oct. 11, 1918.
R—Mobile, Ala.
B—Grove Hill, Ala.
G. O. No. 9, W. D., 1923.

Captain, 325th Infantry, 82d Division.
Captain McLeod at a critical time during an enemy counterattack assumed command of an improvised platoon and by fearlessly exposing himself to encourage his men repulsed the attack and held the position, although it had been previously evacuated by a company under a very severe counterattack. In the performance of this hazardous duty Captain McLeod lost his life.
Posthumously awarded. Medal presented to brother, Lucius B. McLeod.

McLEOD, MARION F.............
Near Manheulles, France, Nov. 11, 1918.
R—Columbia, S. C.
B—Lynchburg, S. C.
G. O. No. 32, W. D., 1919.

First lieutenant, 323d Infantry, 81st Division.
While advancing with his platoon, under perilous shellfire, he was severely wounded. Refusing aid, he remained, and, while his platoon was suffering heavy casualties, he succeeded in holding his platoon under control and advancing it.

MacLEOD, NORMAN D.............
At Marcheville, France, Sept. 26, 1918.
R—Providence, R. I.
B—East Providence, R. I.
G. O. No. 138, W. D., 1918.

Captain, 103d Field Artillery, 26th Division.
While acting as artillery liaison officer he displayed remarkable courage and judgment under terrific artillery and machine-gun fire. In addition to his duties as liaison officer he volunteered and took personal command of a detachment of infantrymen who were without officers, and by his personal bravery and resourcefulness successfully withstood a violent counterattack by the enemy.

McLOUD, PAUL...............
At Gouzeaucourt, France, Nov. 30, 1917.
R—Albany, N. Y.
B—Montezuma, N. Y.
G. O. No. 129, W. D., 1918.

First lieutenant, 11th Railway Engineers.
He displayed extraordinary heroism at Gouzeaucourt, France, Nov. 30, 1917, in remaining under shellfire until the escape of his men, who had been caught unarmed by the German attack, was assured. He then assisted in leading troops to the trenches, directing the procurement and distribution of ammunition, and displaying coolness and judgment while continually under fire.

McLOUGHLIN, COMERFORD...........
At Ripont, France, Sept. 26, 1918.
R—Rye, N. Y.
B—New York, N. Y.
G. O. No. 15, W. D., 1923.

First lieutenant, 369th Infantry, 93d Division.
In command of a company of his regiment during the assault on the enemy's position he voluntarily exposed himself to a concentration of enemy machine-gun and artillery fire, made his way with great difficulty over rough and broken ground, and rescued his wounded battalion commander and his battalion adjutant and several wounded enlisted men, all of whom he carried to a dressing station, thus undoubtedly saving their lives. The undaunted courage and devotion to duty displayed by Lieutenant McLoughlin inspired the men of his regiment to great endeavors.

McLOUGHLIN, JOHN J. (91270).........
Near Villers-sur-Fere, France, July 28, 1918.
R—Brooklyn, N. Y.
B—Minneapolis, Minn.
G. O. No. 108, W. D., 1918.

Corporal, Company K, 165th Infantry, 42d Division.
He killed 4 of the enemy, took 1 prisoner, and held the position by himself until support arrived. Prior to this he had left shelter to give first aid to a comrade who was lying severely wounded in the open.

McLOUGHLIN, WILLIAM (106076)........
Near Berzy-le-Sec, France, July 21, 1918.
R—Anaconda, Mont.
B—Brooklyn, N. Y.
G. O. No. 15, W. D., 1919.

Private, Company A, 3d Machine Battalion, 1st Division.
He advanced against a machine gun and, single handed, killed or captured the entire crew.

McMANAWAY, HERMAN BLAIR (1310006)
Near Hargicourt, France, Sept. 26, 1918.
R—Greenville, S. C.
B—Greer, S. C.
G. O. No. 64, W. D., 1919.

Corporal, Company A, 118th Infantry, 30th Division.
Volunteering to act as stretcher bearer, he assisted in evacuating the wounded during a severe gas shelling. Realizing the presence of strong gas, he unhesitatingly took off his mask and placed it on a wounded man whose mask had been shot away, and in so doing he was badly gassed. It then became necessary to evacuate him, but his heroic and timely act saved the life of his comrade.

McMANUS, WALTER P. (1212082)............
Near Le Catelet, France, Sept. 29, 1918.
R—Brooklyn, N. Y.
B—Worcester, Mass.
G. O. No. 14, W. D., 1923.

Corporal, Company M, 107th Infantry, 27th Division.
The platoon which he commanded being driven to shelter by intense enemy artillery and machine-gun fire and many of its members wounded, Sergeant McManus, with complete disregard for his own safety and disregarding this heavy fire, advanced in plain view of the enemy, rescued a wounded man of his platoon, and although he himself was severely wounded applied first aid and brought his wounded comrade back to a place of comparative safety, thereby saving the latter's life.

McMASTER, GEORGE H............
In the crossing of the Rio Grande de Cagayan, P. I., Dec. 7, 1899.
R—Columbia, S. C.
B—Columbia, S. C.
G. O. No. 43, W. D., 1922.

First lieutenant, 24th Infantry, U. S. Army.
While in command of Company H, 24th Infantry, which was held up in the crossing of the Rio Grande de Cagayan, P. I., Dec. 7, 1899, by rifle fire from a well-intrenched enemy on the opposite bank, and being without rafts or boats with which to cross, Lieutenant McMaster volunteered to swim the river. Displaying great gallantry and utmost disregard of his own life, with a party of 5 men he swam the river in the face of a heavy rifle fire and with reinforcements which later joined him drove the enemy from their trenches and then through and out of the town, thereby making possible the further advance of his command.

*McMANUS, HERBERT W. (1712678).....
Near Carrefour-des-Meurrussons, France, Sept. 27, 1918.
R—Collins Center, N. Y.
B—Collins Center, N. Y.
G. O. No. 70, W. D., 1919.

Private, Company A, 305th Infantry, 77th Division.
When his platoon encountered intense fire from hostile trench mortars and machine guns, he took up an exposed position on the flank and, with his automatic rifle, covered the withdrawal of the platoon to a protected position, sacrificing his life in so doing.
Posthumously awarded. Medal presented to father, George Mackmer.

McMORRIS, WILLIAM R............
In the Argonne Forest, France, Oct. 4, 1918.
R—Bay City, Mich.
B—Bay City, Mich.
G. O. No. 44, W. D., 1919.

Captain, 16th Infantry, 1st Division.
He directed the advance of his company through an intense artillery barrage and against heavy machine-gun fire. When the battalion commander had been wounded, he assumed command of the battalion and led it to its objective, after which he consolidated the new position. While personally resisting an enemy counterattack, he was wounded.

McMUNN, RICHARD L............
Near Chateau-et-Ferme-de-Aulnois, France, Nov. 7, 1918.
R—Olney, Ill.
B—Neoga, Ill.
G. O. No. 44, W. D., 1919.

Second lieutenant, 130th Infantry, 33d Division.
Leading his platoon against a strong machine-gun emplacement, he cut his way through 2 bands of barbed wire and succeeded in reaching the stronghold. Although suffering from severe wounds which he received during the raid, he remained on duty until the action was over.

McMURRY, ORA R............
Near Romagne, France, Oct. 4, 1918.
R—Evansville, Wis.
B—Lake Preston, S. Dak.

First lieutenant, 49th Aero Squadron, Air Service.
He was a member of a patrol of 7 machines which attacked 17 enemy Fokkers. After shooting down 1 of the enemy, this officer returned to the fight and shot down another.
Oak-leaf cluster.
For the following act of extraordinary heroism in action near Tages La Croix aux Bois, France, Oct. 30, 1918, Lieutenant McMurry is awarded an oak-leaf cluster, to be worn with his distinguished-service cross: After becoming separated from his patrol because of motor trouble, this officer encountered and attacked 5 enemy planes (type Fokker) and succeeded in shooting down 1 of them.

MacNAIR, HUGH W. (9527)............
At Ostel (Aisne), France, Oct. 5, 1918.
R—Houghton, Mich.
B—Houghton, Mich.
G. O. No. 20, W. D., 1919.

Private, first class, section No. 622, Ambulance Service.
Having just been relieved after 48 hours of strenuous duty, he volunteered to drive an ambulance to an advanced regimental post under constant and intense fire. While engaged in this dangerous mission, he suffered a wound which necessitated amputation of his right leg.

*McNAMARA, JOHN P. (89925)............
Near Landres-et-St. Georges, France, Oct. 14–15, 1918.
R—Staten Island, N. Y.
B—Staten Island, N. Y.
G. O. No. 16, W. D., 1923.

Private, first class, Company D, 165th Infantry, 42d Division.
Under direct and heavy enemy machine-gun and artillery fire, Private McNamara repeatedly carried messages across open ground during the attack upon Landres-et-St. Georges. Accompanying a patrol Oct. 14, he was wounded in arm and leg, but continued on with the patrol, which penetrated the enemy lines, where he engaged in hand-to-hand fighting. While thus engaged he was mortally wounded and carried by his comrades to his own lines, where he died two days later. His splendid courage and devotion to duty inspired the men of his patrol to great endeavors.
Posthumously awarded. Medal presented to mother, Mrs. Annie McNamara.

MacNAMEE, FRANK A., Jr............
In the Belleau Wood, France, July 19, 1918.
R—Albany, N. Y.
B—Albany, N. Y.
G. O. No. 125, W. D., 1918.

First lieutenant, 101st Field Artillery, 26th Division.
By his utter disregard of danger he inspired great confidence in his men during a critical period by three times going into a heavily shelled area to help rescue wounded.

*McNAMEE, WILLIAM J. (748096)......
Near Crezancy, and Chateau-Thierry, France, July 15, 1918.
R—New York, N. Y.
B—Brooklyn, N. Y.
G. O. No. 35, W. D., 1920.

Private, Company C, 3d Ammunition Train, 3d Division.
On the morning of the 15th of July, Private McNamee, with Private Eckweiler, volunteered and brought up a truck for the purpose of saving the records of the 30th Infantry, which were in danger of capture. He was killed while attempting this mission.
Posthumously awarded. Medal presented to father, David McNamee.

MacNAUGHTON, HENRY DARIUS......
Near Molain, France, Oct. 17, 1918.
R—Grand Rapids, Mich.
B—Ada, Mich.
G. O. No. 81, W. D., 1919.

Captain, 117th Infantry, 30th Division.
Advancing for 100 yards under heavy shell and machine-gun fire, he went with a soldier to the aid of a wounded man who had had a leg blown off by an antitank shell. Showing utter disregard for personal danger, Captain MacNaughton dressed the wound of the soldier, who had fallen near the tank, which was still being shelled, and then carried the wounded man to safety

McNEECE, JOHN H. (60422)_____
 In the Houppy Bois, north of Verdun, France, Oct. 23, 1918.
 R—Holbrook, Mass.
 B—Pittsfield, Mass.
 G. O. No. 46, W. D., 1919.

Private, Company D, 101st Infantry, 26th Division.
On duty as a runner, he repeatedly carried messages under the heaviest shell and machine-gun fire. He also performed heroic service in carrying wounded soldiers to shelter and administering first aid. Though he was almost exhausted, he voluntarily went to the rear under heavy artillery fire and procured food, which he brought back and distributed among his comrades in the shell hole.

McNEIL, EDWARD H. (64248)_____
 Near Bouresches, France, July 20, 1918.
 R—Mexico, Me.
 B—Bangor, Me.
 G. O. No. 125, W. D., 1918.

Private, Company B, 103d Infantry, 26th Division.
Upon reaching his objective, Hill 190, in front of Bouresches, he found he was the only man of his squad left. Going forward, he entered an enemy machine-gun position and at the point of his bayonet captured 3 machine guns and 5 prisoners.

McNERNEY, EDWARD J. (73255)_____
 Near Epieds, France, July 22, 1918.
 R—Lowell, Mass.
 B—Lowell, Mass.
 G O. No. 74, W. D., 1919.

Sergeant, Company K, 104th Infantry, 26th Division.
Sergeant McNerney, with 1 other soldier, went to the rescue of a wounded comrade through a severe machine-gun fire at a direct range of only 350 yards and carried the wounded man to safety.

McNICHOLAS, THOMAS G._____
 At Boise-de-Consenvoye, France, Oct. 26, 1918.
 R—Cockeysville, Md.
 B—Cockeysville, Md.
 G. O. No. 15, W. D., 1923.

Captain, 115th Infantry, 29th Division.
Upon learning that 1 of the officers had been wounded by enemy fire and lying in an exposed position, Captain McNicholas left his post of command at battalion headquarters and with complete disregard for his own safety made his way through concentrated enemy machine-gun, rifle, and artillery fire, found the wounded officer and carried him to a place of shelter within his own lines. The bravery and devotion to his brother officer displayed by Captain McNicholas greatly inspired the men of his battalion.

MacNIDER, HANFORD_____
 Near Medeah Ferme, France, Oct. 3–9, 1918.
 R—Mason City, Iowa.
 B—Mason City, Iowa.
 G. O. No. 44, W. D., 1919.

Captain, 9th Infantry, 2d Division.
He voluntarily joined an attacking battalion on Oct. 3, and accompanied it to its final objectives. During the second attack on the same day he acted as a runner through heavy artillery and machine-gun fire. He visited the lines both night and day, where the fighting was most severe. When higher authority could not be reached, he assumed responsibilities and gave the necessary orders to stabilize serious situations. When new and untried troops took up the attack, he joined their forward elements, determined the enemy points of resistance, by personal reconnaissance, uncovered enemy machine-gun nests, and supervised their destruction.
Oak-leaf cluster.
Captain MacNider is also awarded an oak-leaf cluster, to be worn with distinguished-service cross, for the following act of extraordinary heroism in action near Remeneauville, France, Sept. 12, 1918: On duty as regimental adjutant, while carrying instructions to the assaulting lines, he found the line unable to advance and being disorganized by a heavy machine-gun fire. Running forward in the face of the fire, this officer captured a German machine gun, drove off the crew, reorganized the line on that flank, and thereby enabled the advance to continue.

McNULTY, CLARENCE J. (17469)_____
 Near Gesnes, France, Oct. 7, 1918.
 R—Chicago, Ill.
 B—Chicago, Ill.
 G. O. No. 46, W. D., 1919.

Private, medical detachment, 128th Infantry, 32d Division.
When his battalion was forced to retire under heavy artillery and machine-gun fire, Private McNulty, accompanied by Private William A. Jacobson, went out in front of the battalion, administering first aid and bringing in the wounded, who had been left lying in exposed positions. While they were carrying back a wounded soldier, Private Jacobson was wounded, whereupon Private McNulty alone carried the wounded man to the dressing station and then immediately returned to assist Private Jacobson.

McNULTY, HERMAN L_____
 Near Remonville, France, Nov. 1, 1918.
 R—Huntington, W. Va.
 B—Washington, D. C.
 G. O. No. 87, W. D., 1919.

First lieutenant, 354th Infantry, 89th Division.
At the head of his company following close upon the barrage, he inspired his men by his personal valor, and when wounded by a machine-gun bullet in the leg refused to be evacuated. After his wound had been bound up he continued with his company in the advance, remaining on duty till the objective had been reached and consolidated.

McNULTY, JOHN_____
 Between Blanc Mont and St. Etienne, France, Oct. 4, 1918.
 R—Revere, Mass.
 B—England.
 G. O. No. 20, W. D., 1919.

Marine gunner, 66th Company, 6th Machine Gun Battalion, U. S. Marine Corps, 2d Division.
Although he was severely wounded during an enemy counterattack, he voluntarily remained on the firing line under heavy artillery and machine-gun fire, operating a machine-gun, the crew of which had all been killed or wounded. By staying at his post until the enemy was repulsed and he was ordered to the rear by his commanding officer, he furnished an inspiring example to the other members of the company.

MacPHERSON, HENRY B. (1681718)_____
 At St. Juvin, France, Oct. 16, 1918.
 R—Abington, Mass.
 B—Tewksbury, Mass.
 G. O. No. 21, W. D., 1919.

Corporal, Company C, 306th Infantry, 77th Division.
He volunteered repeatedly during the attack on St. Juvin to carry messages through a severe enemy barrage. Throughout the action this soldier showed entire disregard for personal danger and a devotion to duty far beyond the scope of his position, accomplishing several important missions with success.

McPIKE, LESLIE ALBERT (554075)_____
 In the Bois-de-Foret, France, Oct. 22, 1918.
 R—Bedford, Ind.
 B—Bedford, Ind.
 G. O. No. 89, W. D., 1919.

Sergeant, Company C, 8th Machine Gun Battalion, 3d Division.
Having been sent with his section to defend a difficult position, Sergeant McPike succeeded in breaking up a hostile counterattack, though his ammunition was exhausted in so doing. Shortly afterwards, when another counterattack was made against him, he and his men held off the enemy with their pistols, though at one time the Germans had closed in from 3 sides. Their courageous stand checked the enemy until fresh ammunition could be brought up, whereupon the hostile attack was completely repulsed.

McRae, Duncan K. (2261709)............
Near Gesnes, France, Oct. 11, 1918.
R—Helena, Mont.
B—Helena, Mont.
G. O. No. 1, W. D., 1919.

Sergeant, Company M, 362d Infantry, 91st Division.
Sergeant McRae took out a patrol for the purpose of ascertaining the position of the enemy and the location of machine guns. Three of his men were killed, but he continued on over difficult terrain and returned with information of the highest value in subsequent operations.

McSorley, James (273900)............
Near Roncheres, France, July 30, 1918.
R—Eau Claire, Wis.
B—Eau Claire, Wis.
G. O. No. 66, W. D., 1919.

Sergeant, Company E, 127th Infantry, 32d Division.
Painfully wounded in the right arm, Sergeant McSorley continued to lead his platoon in the face of terrific machine-gun fire, remaining with his men and directing the advance until he was ordered to the rear.

McSweeney, Daniel S. (2705993)......
Near Ivoiry, France, Sept. 27, 1918.
R—Pittsburgh, Pa.
B—Pittsburgh, Pa.
G. O. No. 14, W. D., 1923.

Sergeant, Company B, 148th Infantry, 37th Division.
Leaving a place of shelter, he voluntarily crawled about 400 yards in advance of the front-line elements of his battalion and attempted to rescue a wounded officer. He then crossed an area swept by intense enemy machine-gun fire and attempted to capture an enemy machine gun which was causing heavy casualties in his company. While so engaged he was severely wounded by enemy fire.

*McVey, Joseph E. (58743)............
Near Soissons, France, July 18–21, 1918.
R—Kansas City, Mo.
B—Wray, Colo.
G. O. No. 21, W. D., 1919.

Private, Company M, 28th Infantry, 1st Division.
After his platoon commander had been wounded, Private McVey took command of the platoon and displayed exceptional initiative, good judgment, and devotion to duty. He alone captured a machine gun which was holding up the advance and continued to lead his men forward until reaching their final objective; he himself was killed.
Posthumously awarded. Medal presented to sister, Miss Elizabeth Hoy.

MacVicar, Ian D. (547453)............
Near Crezancy, France, July 15, 1918.
R—Norfolk, Conn.
B—Bengies, Md.
G. O. No. 81, W. D., 1919.

Sergeant, Company I, 30th Infantry, 3d Division.
Sergeant MacVicar conducted a party of ammunition carriers to the front line during the most violent part of the shell fire near Crezancy on July 15. He also conducted a reconnoitering patrol through terrific machine-gun and shell fire that same day.

McVickar, Lansing................
Near Very, France, Oct. 4, 1918.
R—Cambridge, Mass.
B—New London, Conn.
G. O. No. 44, W. D., 1919.

First lieutenant, 1st Battalion, headquarters 7th Field Artillery, 1st Division.
Lieutenant McVickar volunteered and took forward a gun to the aid of the Infantry under most hazardous circumstances. Despite the loss of 2 horses and the wounding of several of his men, he continued until he encountered an enemy barrage, from which it was necessary to take cover. He exposed himself to the barrages on 5 different occasions to bring in wounded men.

McVicker, Franklin D (38137)........
At Vaux, France, July 1, 1918.
R—Portage, Pa.
B—Somerset, Pa.
G. O. No. 99, W. D., 1918.

Private, first class, Company A, 9th Infantry, 2d Division.
Acting as stretcher bearer, although wounded twice, he continued to make trips through an intense barrage to carry back wounded.

MacWilliam, Alexander (1815486)....
Near Nantillois, France, Oct. 4–5, 1918.
R—Erie, Pa.
B—Scotland.
G. O. No. 35, W. D., 1919.

First sergeant Company B, 313th Machine Gun Battalion, 80th Division.
Concealing the fact that he was severely wounded, he remained on duty until the afternoon of the following day. While in this condition, he went to the aid of a wounded comrade and brought him to a place of safety, his route being subjected to a concentrated artillery bombardment.

Madden, David (2338718)............
Near Cunel, France, Oct. 7, 1918.
R—Philadelphia, Pa.
B—Boston, Mass.
G. O. No. 44, W. D., 1919.

Sergeant, Company E, 4th Infantry, 3d Division.
Due to the lifting of a heavy fog, Company E was caught on an open hillside within 100 yards of the enemy's line and were forced to remain in the cover of shell holes for the entire day. Hearing a wounded man moaning in great pain, Sergeant Madden left his shelter and went to his rescue, carrying him through the terrific machine-gun and artillery fire to a shell hole, where he administered all the aid and comfort possible, and then returned to his post through the same intense fire.

Madden, Joseph A. (1335)............
Near Cierges and Fismes, France, July 31 to Aug. 4, 1918.
R—Manistee, Mich.
B—Manistee, Mich.
G. O. No. 124, W. D., 1918.

Private, medical detachment, 125th Infantry, 32d Division.
Throughout the advance by Company D, to which he was attached, his conduct in treating the wounded under fire afforded an inspiring example of devotion to duty. He was in the front lines at all times, administering relief not only to men of his company but also to the wounded of other organizations in his vicinity. He voluntarily searched the woods and fields to give first aid to the wounded. In spite of heavy and continuous shell fire, he continued working untiringly day and night until exhausted.

*Madden, Robert A. (2004650)........
Near Sergy, France, July 29–30, 1918.
B—Indianapolis, Ind.
R—Indianapolis, Ind.
G. O. No. 74, W. D., 1919.

Private, Company I, 47th Infantry, 4th Division.
Passing through heavy machine-gun and artillery fire, he maintained liaison with adjacent units, displaying marked heroism in his work. He was mortally wounded in the performance of duty.
Posthumously awarded. Medal presented to father, Robert N. Madden.

Maddox, John (2273686)............
At Audenarde, Belgium, Nov. 1, 1918.
R—Calexico, Calif.
B—Gove, Kans.
G. O. No. 1, W. D., 1919.

Sergeant, first class, Company F, 316th Engineers, 91st Division.
He volunteered to accompany an officer and 3 other soldiers on a reconnaissance patrol of the city of Audenarde. Entering under heavy shellfire, the party reconnoitered the city for 7 hours while it was still being patrolled by the enemy and advanced 2 kilometers in front of our outposts and beyond those of the enemy.

MADER, THOMAS OTTO (1253128)_____
Near Varennes, France, Oct. 2, 1918.
R—Audenried, Pa.
B—Audenried, Pa.
G. O. No. 20, W. D., 1919.

First sergeant, Battery A, 109th Field Artillery, 28th Division.
He displayed great coolness and bravery in helping to guide sections of his battery over a road swept by enemy shell fire, during which 8 men were wounded and 10 horses killed, including one which he himself rode. The driver of a swing team, having difficulty in controlling 1 of the horses of a section, was assigned to another horse and his place taken by Sergeant Mader, who guided the section until he was so severely wounded that he was unable to control. In spite of his wounds, he directed the carriages to places of safety, and, disregarding personal safety, requested the medical officer to first give attention to the other wounded. Sergeant Mader's conduct was an inspiration to the men of his battery.

MADORE, JOHN J. (558115)_____
Near Bazoches, France, Aug. 9, 1918.
R—Malden, Mass.
B—Canada.
G. O. No. 15, W. D., 1919.

Private, Company G, 47th Infantry, 4th Division.
He volunteered to carry a message to an advance squad through heavy machine-gun fire. After delivering the message and administering first-aid treatment to the wounded men in the squad he crawled up to the nearest enemy machine gun and put it out of action with a hand grenade.

*MADSEN, EDMUND TERNER_____
At Chateau-Thierry, France, June 6, 1918.
R—Indianapolis, Ind.
B—Denmark.
G. O. No. 110, W. D., 1918.

First sergeant, 47th Company, 5th Regiment, U. S. Marine Corps, 2d Division
Killed in action at Chateau-Thierry, France, June 6, 1918, he gave the supreme proof of that extraordinary heroism which will serve as an example to hitherto untried troops.
Posthumously awarded. Medal presented to mother, Mrs. Johanne Madsen.

MADSEN, HOWARD E. (1284515)_____
In the Bois-de-Consenvoye, France, Oct. 22, 1918.
R—Baltimore, Md.
B—Baltimore, Md.
G. O. No. 37, W. D., 1919.

Sergeant, Company D, 115th Infantry, 29th Division.
When his platoon was fired on at close range by a machine gun, Sergeant Madsen went forward and killed 2 of the enemy with grenades, routing the other members of the crew.

MAGUIRE, JOHN T._____
Near Exermont, France, Oct. 4, 1918.
R—St. Louis, Mo.
B—St. Louis, Mo.
G. O. No. 15, W. D., 1921.

First lieutenant, 18th Infantry, 1st Division.
Although severely wounded, Lieutenant Maguire refused to be evacuated and continued to lead his platoon until again wounded and forced by exhaustion, due to loss of blood, to be evacuated.

MAGUIRE, SIDNEY CLIFFORD (1386609)__
At Chipilly Ridge, France, Aug. 9, 1918.
R—Chicago, Ill.
B—Chicago, Ill.
G. O. No. 128, W. D., 1918.

Sergeant, Company B, 131st Infantry, 33d Division.
Although wounded early in the engagement, he showed great devotion to duty by continuing at his post as platoon leader for 2 days, relinquishing command only when forced to do so by the condition of his wound.

*MAHAFEY, EMORY (2683)_____ _____
At Cantigny, France, May 28–29, 1918.
R—Atlanta, Ga.
B—Lawrenceville, Ga.
G. O. No. 116, W. D., 1918.

Private, first class, medical detachment, 28th Infantry, 1st Division.
He did more than his duty under violent fire in the open to relieve sufferings of the wounded. On his way to a machine-gun emplacement where men who had been injured there he stopped to give first aid to Pvt. Jay Ler. Antes, who lay mortally wounded and exposed to machine-gun fire, and while performing this heroic act was killed.
Posthumously awarded. Medal presented to mother, Mrs. Victoria Mahafey.

MAHAR, DANIEL H. (1205545)_____
Near St. Souplet, France, Oct. 18, 1918.
R—Newark, N. J.
B—Brooklyn, N. Y.
G. O. No. 23, W. D., 1919.

Corporal, Company L, 105th Infantry, 27th Division.
He courageously led several attacks on enemy machine-gun nests. Later in the day he attacked, single-handed, 2 enemy snipers, killing 1 and driving off the other.

MAHONEY, CORNELIUS J. (2021686)_____
Near Kadish, Russia, Oct. 16, 1918.
R—Detroit, Mich.
B—Hartland, Mich.
G. O. No. 16, W. D., 1920.

Sergeant, Company K, 339th Infantry, 85th Division (detachment in north Russia).
After 2 unsuccessful attempts by patrols to locate the right flank of the enemy, Sergeant Mahoney went out alone behind the enemy lines at great personal risk and selected an excellent position from which to deliver a flank attack. The attack delivered by 60 men from the position selected was highly successful.

*MAHONEY, JAMES (2383800)_____
Near Cunel, France, Oct. 15, 1918.
R—Waterbury, Conn.
B—Waterbury, Conn.
G. O. No. 27, W. D., 1920.

Private, Company G, 60th Infantry, 5th Division.
Private Mahoney upon 3 occasions exposed himself to heavy machine-gun fire in attempting to work through the enemy lines to carry a message to a group which, due to enemy infiltration, had become isolated. Although mortally wounded, he insisted upon being carried to regimental headquarters in order to make a report of the information gained.
Posthumously awarded. Medal presented to father, James Mahoney.

MAIER, CARL J. (2261161)_____
At Bois-de-Cheppy, near Meuse, France, Sept. 26, 1918.
R—Glendive, Mont.
B—Minnow, S. Dak.
G. O. No. 32, W. D., 1919.

Private, first class, Company I, 362d Infantry, 91st Division.
Working with a patrol in an attack on an enemy machine gun, he crawled upon the emplacement and without assistance killed 3 enemy gunners and captured their machine gun.

MAILS, MARK W. (1521521)_____
Near Cierges, France, Sept. 28, 1918.
R—Akron, Ohio.
B—Abilene, Kans.
G. O. No. 44, W. D., 1919.

Private, first class, Company F, 146th Infantry, 37th Division.
After his platoon had withdrawn about 50 yards to an established line, a wounded comrade was seen lying ahead in the position they formerly occupied. The enemy had just launched a strong counterattack, but Private Mails, with another soldier, volunteered to go to the assistance of the wounded man. In the face of terrific fire of enemy artillery and machine guns and the fire of their own comrades, who were resisting the attack, Private Mails succeeded in bringing his man to a place of safety.

MAIN, CHARLES L. (552263)............
 Near Launay, France, July 15, 1918.
 R—Rouseville, Pa.
 B—Titusville, Pa.
 G. O. No. 37, W. D., 1919.

Sergeant, Company L, 38th Infantry, 3d Division.
Sergeant Main, with an officer and another soldier of his company, attacked a patrol of 7 Germans who had captured 4 American soldiers, killed 1 of the Germans, and captured the others.

MAIN, WILLIAM B. (42525)............
 In the Forest of Argonne, France, Oct. 9, 1918.
 R—Rouseville, Pa.
 B—Caneville, Pa.
 G. O. No. 59, W. D., 1919.

Corporal, Company D, 16th Infantry, 1st Division.
Accompanied by another soldier, he advanced on a German 77-millimeter gun which was enfilading his company, and with an automatic rifle caused such heavy casualties among the enemy crew that they were forced to withdraw.

MAJOR, IRA E. (1625474)............
 Near Rembercourt, France, Nov. 1 and 2, 1918.
 R—Greenwood, S. C.
 B—Glendale, S. C.
 G. O. No. 22, W. D., 1920.

Corporal, Company E, 34th Infantry, 7th Division.
Although suffering from 5 wounds, 1 of which caused the loss of a hand, Corporal Major refused to be evacuated. He remained in command of his squad throughout the day and night and crawled into position the following morning to aid in the repulse of a German attack. His conduct furnished a great inspiration to the men of his company.

MAKOS, SOTEREOS N. (53667)............
 Near Soissons, France, July 19, 1918.
 R—Worcester, Mass.
 B—Greece.
 G. O. No. 132, W. D., 1918.

Private, Company F, 26th Infantry, 1st Division.
He was wounded 3 times, but continued with the advance and kept up the operation of his automatic rifle despite his injuries until the objective had been attained.

MALCOLMSON, BRUCE K. (2267250)......
 Near Cheppy, France, Sept. 26, 1918.
 R—South Pasadena, Calif.
 B—Pasadena, Calif.
 G. O. No. 37, W. D., 1919.

First sergeant, Company H, 364th Infantry, 91st Division.
Accompanied by another soldier he advanced against 2 machine guns and killed 2 of the mounters. This made possible the capture of 55 men who were in a trench 100 yards farther forward. Later in the same day he assisted an officer in leading a platoon which captured approximately 100 of the enemy.

*MALICHIS, CONSTANTINE (547492)......
 Near Bois-d' Aigremont, France, July 15, 1918.
 R—New Orleans, La.
 B—Greece.
 G. O. No. 32, W. D., 1919.

Corporal, Company I, 30th Infantry, 3d Division.
He volunteered and carried the message from his company post of command to the battalion post of command through the thickest of the German barrage. He was killed in action the same day.
Posthumously awarded. Medal presented to father, Alexander Malichis.

MALLAN, JOHN C. (1387913)............
 Near Chipilly Ridge, France, Aug. 9, 1918.
 R—Chicago, Ill.
 B—Chicago, Ill.
 G. O. No. 71, W. D., 1919.

Private, Company H, 131st Infantry, 33d Division.
During an attack he worked out far ahead of our lines and personally killed 4 Germans and brought back 3 prisoners, 1 of them an officer. Later, he formed one of a raiding party and displayed marked skill and bravery aiding in the capture of 14 prisoners. Both these missions were carried out under heavy artillery and machine-gun fire.

MALONE, CLAYTON (2176872)............
 Near Bois-de-Barricourt, France, Nov. 1, 1918.
 R—Liberal, Kans.
 B—Herington, Kans.
 G. O. No. 71, W. D., 1919.

Sergeant, Company I, 353d Infantry, 89th Division.
He led his platoon with marked bravery and ability in an attack which resulted in the capture of 15 machine guns and 70 prisoners, more than 40 of the enemy being killed or wounded. Throughout the attack he showed great heroism, killing 2 of the enemy in hand-to-hand encounters.

MALONE, FRANCIS P. (109795)............
 Near Trugny, France, July 22-23, 1918.
 R—New Haven, Conn.
 B—New Haven, Conn.
 G. O. No. 37, W. D., 1919.

Corporal, Company D, 102d Machine Gun Battalion, 26th Division.
After all the other members of his squad had become casualties, he volunteered and led 2 other squads on a machine-gun nest, which he had located and put it out of action with the first burst of shot. Later he helped dress wounded men of his squad and evacuated them under heavy artillery and machine-gun fire. Although severely gassed the next day, he continued in the advance.

MALONE, PAUL B............
 South of Soissons, France, July 18-19, 1918.
 R—West Point, N. Y.
 B—New York.
 G. O. No. 15, W. D., 1921.
 Distinguished-service medal also awarded.

Colonel, 23d Infantry, 2d Division.
During the 2 days which his regiment was engaged with the enemy, Colonel Malone frequently visited the advanced troops. On the evening of July 18, after the regiment had suffered severe losses, he assisted in the reorganization of a battalion for the attack on Vierzy. On the morning of July 19 he made a personal reconnaissance of the front lines, under heavy fire, in order to ascertain the enemy position, which was of vital importance.

MALONEY, WILLIAM E............
 Near Remonville, France, Nov. 1, 1918.
 R—New York, N. Y.
 B—New York, N. Y.
 G. O. No. 44, W. D., 1919.

Second lieutenant, 354th Infantry, 89th Division.
Leading an assault platoon, he encountered a nest of 6 enemy guns, which was pouring out a deadly fire from three directions. Rushing the guns in the immediate front, he captured them, as well as taking the crews as prisoners; and while thus engaged he was knocked unconscious by a bursting shell. Regaining consciousness when picked up by first-aid men, he returned immediately and rejoined his platoon and directed operations until the objectives were reached.

MANCE, STEPHEN M. (1386580)............
 At Chipilly Ridge, France, Aug. 10, 1918.
 R—Chicago, Ill.
 B—Joliet, Ill.
 G. O. No. 128, W. D., 1918.

Corporal, Company B, 131st Infantry, 33d Division.
Sent out alone to locate the position of snipers, and coming upon a machine-gun nest, he boldly attacked it single-handed, capturing the gun, wounding 1 of the crew, and taking 3 prisoners.

MANCO, ARTIE G. (57328)............
 Near Nonsard, France, Sept. 12, 1918.
 R—Richardsville, Ky.
 B—Warren County, Ky.
 G. O. No. 55, W. D., 1920.

Sergeant, Company E, 28th Infantry, 1st Division.
When the advance of his company was halted by the fire from 4 machine guns from the front, Sergeant Manco led a patrol of 4 men, through heavy fire, to the flank and rear of the enemy position. He then attacked the enemy and captured the guns and crew.

MANDERS, FRANK (3746337)............
　At Grand Pre, France, Nov. 1, 1918.
　R—Rockland, Wis.
　B—Kaukauna, Wis.
　G. O. No. 56, W. D., 1922.

Private, Company I, 312th Infantry, 78th Division.
After requesting that he be permitted to lead a detail for the purpose of recovering the body of a wounded comrade whom he had been forced to leave about 500 yards in front of their lines earlier in the day when on patrol duty, Private Manders led the detail [of 4 men under heavy artillery and machine-gun fire on its dangerous mission. After proceeding about 200 yards the detail felt that the expedition was hopeless on account of the extremely heavy fire. Private Manders, however, refused to give up and insisted on the detail going forward. By his own absolute disregard of personal safety and by his own example he so inspired the members of the detail that they went forward with him under the gravest danger for over 300 yards, recovered the wounded comrade, and made their way back to their own lines.

MANGIARACINA, FRANK (39439)..........
　Near Beaumont, France, Nov. 1–5, 1918.
　R—Brooklyn, N. Y.
　B—Italy.
　G. O. No. 44, W. D., 1919.

Private, Company F, 9th Infantry, 2d Division.
On duty as a company runner, he repeatedly carried messages through heavy machine-gun and shell fire with utter disregard for personal safety.

MANIER, WILL R. Jr...................
　In the Claire-Chenes Woods, France, Oct. 21, 1918.
　R—Nashville, Tenn.
　B—Nashville, Tenn.
　G. O. No. 98, W. D., 1919.

Captain, 5th Infantry Brigade, 3d Division.
Captain Manier, brigade liaison officer, was at the command post of the assaulting battalion when word was received that the enemy had penetrated our line and was closing in on the command post. No reserves being available, Captain Manier assisted in hastily organizing a force of runners, a signalman and others, and himself taking a rifle led these men under heavy machine-gun and rifle fire in a counterattack on the advancing enemy. He succeeded in stopping the enemy and holding the position until the line was reestablished. As a result of his courage and initiative a large number of the enemy were captured and many killed or wounded, and the success of future operations assured.

*MANN, ALLEN R. (1212830)...........
　Near Ronssoy, France, Sept. 29, 1918.
　R—New York, N. Y.
　B—Shawnee, Ohio.
　G. O. No. 49, W. D., 1922.

Private, sanitary detachment, 107th Infantry, 27th Division.
With utter disregard for his own personal safety, Private Mann repeatedly went out into no man's land, in plain view of the enemy, and administered first aid to the wounded. While engaged in this work he was caught in a heavy enfilading fire from the enemy and was killed.
Posthumously awarded. Medal presented to father, E. C. Mann.

MANNING, JAMES (568528).............
　Near St. Thibaut, France, Aug. 8, 1918.
　R—Youngstown, Ohio.
　B—Youngstown, Ohio.
　G. O. No. 46, W. D., 1919.

Corporal, Company C, 4th Engineers, 4th Division.
He was one of 4 men who volunteered and swam the Vesle River for the purpose of doing work on the opposite bank necessary in the construction of a footbridge. With another soldier he succeeded in felling a large tree in the face of heavy machine-gun fire and 1-pounder fire after the remainder of the platoon had withdrawn.

MANNING, JAMES EUGENE.............
　Near Thiaucourt, France, Sept. 15, 1918.
　R—Canton, Ohio.
　B—Allentown, Pa.
　G. O. No. 37, W. D., 1919.

Hospital apprentice, first class, U. S. Navy, attached to 1st Battalion, 6th Regiment, U. S. Marine Corps, 2d Division.
While he was attending a wounded man his dressing station was struck by a shell from which his patient received two additional wounds. He dressed the man's new wounds and while so doing was himself struck in the back and knocked down by the explosion of another shell. He remained at his post, however, until he had finished dressing his patient's injuries and then removed him from the dressing station, which very soon was completely destroyed by a third shell.

MANNING, JAMES F., Jr...............
　Near Doulcon, France, Oct. 4, 1918.
　R—Purcellville, Va.
　B—Washington, D. C.
　G. O. No. 7, W. D., 1919.

First lieutenant, pilot, 49th Aero Squadron, Air Service.
While leading a patrol of 7 planes he accepted combat with 17 German machines (type, Fokker) at an altitude of 1,200 meters. Through his courageous leadership and skillful maneuver of his patrol 7 of the enemy planes were shot down.

MANNING, JOHN C. (54617)...........
　Near Paris-Soissons Road, France, July 18–19, 1918.
　R—Kingsville, Tex.
　B—Jena, La.
　G. O. No. 59, W. D., 1919.

Corporal, Company K, 26th Infantry, 1st Division.
He voluntarily left his shelter and went to the rescue of 2 wounded French soldiers who were in a burning tank. The rescue was made under an intense fire from the enemy in plain view and while they were using all efforts to complete the destruction of the tank and prevent the rescue of its inmates. Corporal Manning succeeded in forcing open a door of the burning tank, dragging out the wounded men, and bringing them to a place of shelter, from which they were later evacuated.

MANNING, JOHN R. (2848637)..........
　In the Bois-de-Bantheville, France, Nov. 1, 1918.
　R—Needham, Mass.
　B—Wellesley, Mass.
　G. O. No. 44, W. D., 1919.

Private, Company D, 342d Machine Gun Battalion, 89th Division.
Disregarding painful injuries in the face, head, and wrist, he continued on duty as gunner throughout the action until he was relieved and ordered to an aid station by his platoon commander, thereby affording an inspiring example of courage to his comrades.

*MANNING, WILLIAM SINKLER.........
　Near Verdun, France, Nov. 5, 1918.
　R—Washington, D. C.
　B—Sumter County, S. C.
　G. O. No. 37, W. D., 1919.

Major, 316th Infantry, 79th Division.
Leading his command in the face of extremely heavy artillery and machine-gun fire, he displayed remarkable bravery and coolness in reorganizing his battalion after severe losses had been inflicted on them. By continuous encouragement and daring he directed operations to the successful gaining of his objective. During the operations he was instantly killed by a machine-gun bullet.
Posthumously awarded. Medal presented to widow, Mrs. Barbara B. Manning.

MANNION, JOSEPH F. (1284404)........
　Near Verdun, France, Oct. 11, 1918.
　R—Baltimore, Md.
　B—Baltimore, Md.
　G. O. No. 37, W. D., 1919.

Sergeant, Company C, 115th Infantry, 29th Division.
He volunteered and left his place of safety, making his way 100 yards in advance of our lines to the aid of wounded men. At the time the enemy was delivering terrific machine-gun and artillery fire, but he continued and assisted one comrade to a dressing station. He returned and helped the other men to places of safety.

MANNION, MAURICE (1236408)_____
At Apremont, France, Sept. 29, 1918.
R—Carbondale, Pa.
B—Carbondale, Pa.
G. O. No. 98, W. D., 1919.

Sergeant, Company F, 109th Infantry, 28th Division.
During a hostile attack Sergeant Mannion displayed exceptional courage and initiative in leading a patrol to the left flank of his platoon and driving off a superior number of the enemy who were attempting to encircle the flank, killing 10 of them, capturing 8, and putting the remainder to flight.

MANSFIELD, HARRY (280667)_____
At Romagne, France, Oct. 14, 1918.
R—Mecosta, Mich.
B—Mecosta, Mich.
G. O. No. 98, W. D., 1919.

Sergeant, Company I, 126th Infantry, 32d Division.
His company having been held up by a machine-gun nest, Sergeant Mansfield, with another soldier, crawled 200 yards ahead of his company and reduced the machine-gun nest, killing 3 of the enemy and capturing 18.

MANSFIELD, JAMES R. (1783727)_____
Near Montfaucon, France, Sept. 29, 1918.
R—Haydenville, Mass.
B—Haydenville, Mass.
G. O. No. 46, W. D., 1919.

Sergeant, medical detachment, 311th Machine Gun Battalion, 79th Division.
Administering first aid in a most exposed position, Sergeant Mansfield rendered most valuable assistance not only to casualties from his own command, but also to those from other organizations. He continued with his work, although wounded, and refused to leave for the rear until ordered to do so by his commanding officer.

MANSFIELD, THAROLD B. (2193652)_____
Near Beauclair, France, Nov. 4-11, 1918.
R—Bay City, Mich.
B—Flint, Mich.
G. O. No. 37, W. D., 1919.

Corporal, Company C, 314th Field Signal Battalion, 89th Division.
From the 4th to the 11th of November, while continually under heavy shellfire, he laid and maintained lines of communication within his area with utter disregard for his personal safety.

MANTON, WALTER W._____
At Soissons, France, July 18, 1918.
R—Detroit, Mich.
B—Detroit, Mich.
G. O. No. 46, W. D., 1919.

Captain, Medical Corps, attached to 26th Infantry, 1st Division.
Accompanying his battalion in the attack, he was with the second wave when he sustained a compound fracture of the right forearm from a bursting shell. He nevertheless refused to go to the rear, but remained on duty until the final objective was reached in the afternoon, attending the wounded and directing their evacuation.

*MANWARING, CLYDE F. (735498)_____
Near Louppy, France, Nov. 9, 1918.
R—Akron, Ohio.
B—Clymer, N. Y.
G. O. No. 89, W. D., 1919.

Sergeant, Company E, 11th Infantry, 5th Division.
His platoon having been held up by enemy machine-gun fire, Sergeant Manwaring took 3 other soldiers and advanced against the machine guns. Two of the men accompanying him were wounded, but he continued on with the other one and captured 2 of the hostile guns, killing 1 of the crew and driving off the remainder.
Posthumously awarded. Medal presented to widow, Mrs. Clyde F. Manwaring.

MANZI, NICHOLAS (2414276)_____
Near Jaulny, France, Oct. 3, 1918.
R—Camden, N. J.
B—Italy.
G. O. No. 126, W. D., 1919.

Private, medical detachment, 309th Infantry, 78th Division.
During a heavy bombardment of our front lines, Private Manzi went 50 yards in advance of our positions to an automatic-rifle post to dress the wounds of 3 of the crew. He then assisted the wounded men, one by one, to reach a place of safety. All this time the line was under steady machine-gun fire as well as bombardment.

MARAGLIA, BATISTA (1681474)_____
In the Argonne Forest, France, Oct. 1, 1918.
R—Stoughton, Mass.
B—Italy.
G. O. No. 44, W. D., 1919.

Private, Company L, 305th Infantry, 77th Division.
Seeing a runner of his platoon lying helpless from a broken leg in front of an enemy machine gun, Private Maraglia volunteered and went to his aid, making his way through direct fire for a distance of 75 yards and returning with his wounded comrade.

MARCELLA, RICHARD (558037)_____
Near Bazoches, France, Aug. 9, 1918.
R—New York, N. Y.
B—New York, N. Y.
G. O. No. 46, W. D., 1919.

Bugler, Machine Gun Company, 47th Infantry, 4th Division.
Responding to a call for volunteers to destroy a hostile machine gun, Bugler Marcella, with two other soldiers, boldly went forward through machine-gun fire and accomplished this mission.

MARCH, PEYTON C._____
Before Manila, P. I., Aug. 13, 1898.
R—Easton, Pa.
B—Easton, Pa.
G. O. No. 39, W. D., 1920.
Distinguished-service medal also awarded.

First lieutenant, Astor Battery, U. S. Army.
He gallantly led a charge on the enemy's breastworks, volunteers having been called for by the brigadier general commanding.

*MARCHANT, JOHN R._____
East of the Meuse River and north of Consenvoye, France, Oct. 11, 1918.
R—Chicago, Ill.
B—Chicago, Ill.
G. O. No. 49, W. D., 1922.

First lieutenant, 131st Infantry, 33d Division.
While commanding his company in an attack in which he was severely wounded he displayed the highest type of courage and leadership by leading his men until the objective was reached, and then holding it against several severe counterattacks until he was killed. His example of bravery and devotion to duty so inspired his men that although temporarily without a company commander they successfully repulsed the counterattacks and held the line intact at a critical point.
Posthumously awarded. Medal presented to sister, Miss Annabelle Marchant.

MARCINIAK, JOHN (545327)_____
Near Crezancy, France, July 15, 1918.
R—Schenectady, N. Y.
B—Russia.
G. O. No. 32, W. D., 1919.

Corporal, Company A, 30th Infantry, 3d Division.
After his company had been ordered to withdraw, he returned to the scene of the struggle and throughout the whole night worked untiringly in the evacuation of the wounded, exposed to the terrific bombardment of the enemy.

MAREK, FRANK S. (281493)_____
At Romagne, France, Oct., 14, 1918.
R—Grand Rapids, Mich.
B—Grand Rapids, Mich.
G. O. No. 98, W. D., 1919.

Corporal, Company M, 126th Infantry, 32d Division.
With another soldier Corporal Marek crawled 200 yards ahead of his company and reduced a machine-gun nest, which had been holding up the advance, killing 3 of the enemy and capturing 18.

MARGOLIN, HARRY (2337979)_____ | Private, Company A, 4th Infantry, 3d Division.
  Near Grand Ballois Farm, France, | Under a heavy gas and shell bombardment he repeatedly volunteered and
    July 14–15, 1918. | delivered messages over routes other than his own when the runners assigned
  R—Brooklyn, N. Y. | to those routes had been killed or wounded.
  B—New York, N. Y. |
  G. O. No. 32, W. D., 1919. |

MARINO, CHARLES (73640)_____ | Private, Company L, 104th Infantry, 26th Division.
  At Bois Brule, near Apremont, | He displayed coolness, courage, and the spirit of self-sacrifice during the action
    France, Apr. 10, 1918. | of Apr. 10, 1918, in voluntarily going through shell-swept area to bring back a
  R—Greenfield, Mass. | wounded noncommissioned officer to a dressing station.
  B—Italy. |
  G. O. No. 99, W. D., 1918. |

MARK, ROY C. (2297894)_____ | Private, Company I, 125th Infantry, 32d Division.
  Near Cierges, France, July 31, 1918. | Though he was twice wounded during the attack on Hill 212, he displayed re-
  R—McMillan, Mich. | markable bravery and devotion to duty by continuing in action and advanc-
  B—McMillian, Mich. | ing with the attack. He continued to fight until he was ordered to the rear
  G. O. No. 20, W. D., 1919. | on account of his wounds.

MARKHAM, RALPH H. (2267783)_____ | Private, Company K, 364th Infantry, 91st Division.
  Near Eclisfontaine, France, Sept. | On duty as a scout, Private Markham repeatedly located organizations on the
    28, 1918. | flanks, displaying exceptional daring under fire. Though he had been with-
  R—McKittrick, Calif. | out water or food for 24 hours he went forward under heavy fire and secured
  B—Newville, Calif. | information which enabled his company commander to re-form the line and
  G. O. No. 98, W. D., 1919. | deliver a concentrated fire on the enemy.

*MARKLEY, GEORGE_____ | Sergeant, 47th Company, 5th Regiment, U. S. Marine Corps, 2d Division.
  Near Somme-Py, France, Oct. 4, | He volunteered to ascertain the position of enemy machine guns enfilading his
    1918. | company. Under heavy shell and machine-gun fire he explored the enemy's
  R—Detroit, Mich. | lines and secured valuable information. Later, when his company was
  B—Battle Creek, Mich. | ordered to attack, he preceded the line as scout and was killed.
  G. O. No. 13, W. D., 1919. | Posthumously awarded. Medal presented to father, W. B. Markley.

MARKOE, STEPHEN C_____ | Second lieutenant, 18th Infantry, 1st Division.
  Near Seicheprey, France, Mar. 1, | While occupying a combat position in the Bois Carre with his platoon of Com-
    1918. | pany I, 18th Infantry, the position was smothered under an intense
  R—Penllyn, Pa. | bombardment of enemy artillery and machine-gun fire. Following a rolling
  B—Philadelphia, Pa. | barrage, he was attacked by a vastly superior force of enemy storm troops.
  G. O. No. 14, W. D., 1923. | The company commander having been killed, Lieutenant Markoe led the
 | counterattack against the raiding forces and killed in close combat the
 | leader of the raiding party. The action of Lieutenant Markoe stopped the
 | enemy in this part of the sector and was the principal factor which led to
 | the repulse of the enemy raid. Despite the fact that he had been severely
 | wounded early in the action, he refused to relinquish command until the
 | enemy was decisively ejected with great loss from the position. Lieutenant
 | Markoe displayed exceptional leadership and extraordinary heroism in this
 | action.

MARKS, ERWIN J. (543342)_____ | Private, medical detachment, 7th Infantry, 3d Division.
  Near Le Charmel, France, July 22, | A patrol of 5 men had advanced about 500 yards in front of our lines. While
    1918. | crossing a small clearing the patrol was fired upon by a concealed machine
  R—Brooklyn, N. Y. | gun, which killed 2 and wounded 3. On his own initiative, exposed to
  B—England. | heavy machine-gun fire, Private Marks went forward and brought in the
  G. O. No. 16, W. D., 1920. | wounded men 1 at a time.

MARKS, SAMUEL J _____ | First lieutenant, Medical Corps, attached to 314th Infantry, 79th Division.
  Near Malancourt, France, Sept. | Lieutenant Marks advanced with the foremost elements of his battalion,
    27–29, 1918. | dressing and evacuating the wounded under machine-gun fire for a period
  R—Philipsburg, Pa. | of 12 hours. On Sept. 29, when his aid station was shelled, several patients
  B—Philipsburg, Pa. | and attendants being killed and wounded, this officer, though himself
  G. O. No. 31, W. D., 1919. | wounded, remained at his post caring for patients who had received fresh
 | wounds and assisted in their evacuation.

*MARKS, WILLOUGHBY R_____ | First lieutenant, 61st Infantry, 5th Division.
  Near Cunel, France, Oct. 12, 1918. | While in command of Company C, 61st Infantry, he was severely wounded but
  R—Apalachicola, Fla. | continued to lead his company, refusing to be evacuated until the objective
  B—Columbus, Ga. | was reached and his lines reorganized. About to be evacuated, he learned
  G. O. No. 15, W. D., 1923. | that an officer of his battalion was mortally wounded and lying exposed to
 | terrific enemy fire in front of the lines. With utter disregard for his own
 | safety he rushed forward to rescue his fellow officer, and in the attempt was
 | struck by enemy high-explosive shellfire and mortally wounded, dying a few
 | minutes later. His undaunted courage and devotion to duty served as a
 | splendid example of soldierly conduct to the men of his command.
 | Posthumously awarded. Medal presented to mother, Mrs. Annie Ryan
 | Marks.

MARKUS, NORBERT W_____ | Second lieutenant, 3d Machine Gun Battalion, 1st Division.
  Near Soissons, France, July 19, 1918. | After the entire personnel of the machine-gun squad under his command had
  R—Quincy, Ill. | been killed or disabled and when he himself was severely wounded near
  B—Quincy, Ill. | Soissons, France, July 19, 1918, he kept up the operation of his gun and re-
  G. O. No. 126, W. D., 1918. | fused to be taken to the rear when relieved until he had been carried to his
 | company commander and had given the latter valuable information.

MARLIN, FRANK (54038)_____ | Private, Company G, 26th Infantry, 1st Division.
  Near Soissons, France, July 19, 1918. | With 2 other soldiers he rushed a machine-gun position near Soissons, France,
  R—Gatesville, Tex. | July 19, 1918, killed the crew, and captured the gun in order to make the
  B—Christiana, Tenn. | advance of his platoon possible.
  G. O. No. 132, W. D., 1918. |

*MARLIN, JESSE (83455)_____
At Juvigny, France, Aug. 31, 1918.
R—Billings, Mont.
B—Roanoke, Va.
G. O. No. 142, W. D., 1918.

Corporal, Company B, 127th Infantry, 32d Division.
He was one of a party of 3 officers and 2 men who, armed with 1 German machine gun and 3 German rifles, attacked a machine-gun nest held by 70 Germans. Under terrific fire from the enemy, who laid down an artillery barrage upon their position, they concentrated their rifle fire so effectively that 32 Germans surrendered within an hour. After the prisoners had been brought in, Corporal Marlin, with a private, established another machine gun in an advanced position and kept up a concentrated fire on the Germans until he was wounded in the body 5 times by machine-gun bullets.
Posthumously awarded. Medal presented to mother, Mrs. R. P. Marlin.

MARLIN, WILLIAM L._____
At Heurne, Belgium, Nov. 1-2, 1918.
R—Covington, Ohio.
B—Covington, Ohio.
G. O. No. 38, W. D., 1922.

Major, 148th Infantry, 37th Division.
While commanding the 3d Battalion, 148th Infantry, Major Marlin displayed exceptional qualities of personal courage and leadership in forcing the crossing of the Escaut River, establishing a bridgehead on the right bank of the river, and maintaining his position against repeated and vigorous counterattacks, all under heavy artillery and aeroplane fire. Major Marlin exposed himself fearlessly and audaciously and without regard for danger, thereby greatly enhancing the morale of the troops and contributing materially to the success of this operation. His personal bravery in this act was markedly conspicuous and outstanding.

MARLOWE, FRED MARION_____
At Mim. St. Georges, France, Nov. 1, 1918.
R—Greensburg, Ind.
B—Williamstown, Ind.
G. O. No. 37, W. D., 1919.

Sergeant, 74th Company, 6th Regiment, U. S. Marine Corps, 2d Division.
He had just taken command of his platoon, owing to the senior platoon sergeant having been wounded, when the advancing line was held up by a concentration of enemy machine-gun fire. Taking 2 other soldiers with him, he rushed a German machine-gun nest and put it out of action. The capture of this nest compelled the surrender of the remaining machine gunners in the vicinity, and the line was again able to advance; 80 prisoners and 9 machine guns were captured through this bold exploit.

MARONEY, THOMAS F. (1708059)____
Near Badonvillers, France, June 24, 1918.
R—Brooklyn, N. Y.
B—Brooklyn, N. Y.
G. O. No. 37, W. D., 1919.

Corporal, Company C, 308th Infantry, 77th Division.
Although wounded while bringing up ammunition for his automatic rifle team, Corporal Maroney stayed with his men, encouraging and directing them.

MARONEY, WILLIAM (107705)_____
Near Chateau-Thierry, France, June 6, 1918, near Thiaucourt, France, Sept. 13, 1918, and at Medeah Ferme, France, Oct. 4-9, 1918.
R—Pensacola, Fla.
B—Syracuse, N. Y.
G. O. No. 46, W. D., 1919.

Corporal, Company C, 5th Machine Gun Battalion, 2d Division.
Near Chateau-Thierry, France, June 6, Corporal Maroney took charge of 2 squads after his lieutenant and sergeant had been disabled and fearlessly led them forward to their objective. Near Thiaucourt, France, Sept. 13, he displayed great bravery in leading his platoon through heavy machine-gun fire, at one time personally reconnoitering a machine-gun position before allowing his men to proceed. At Medeah Farm, France Oct. 4-9, 1918, he again furnished an inspiring example of fearless leadership and dauntless courage under heavy artillery and machine-gun fire.

MARQUETTE, LOUIS F. (71285)_____
Near Verdun, France, Oct. 15, 1918.
R—Springfield, Mass.
B—Springfield, Mass.
G. O. No. 23, W. D., 1919.

Corporal, Company B, 104th Infantry, 26th Division.
He showed extraordinary courage and bravery in going beyond our front line, under heavy machine-gun fire, and bringing back 2 wounded comrades.

*MARQUIS, OLIVER (1569378)_____
Near Sedan, France, Nov. 7, 1918.
R—Poseyville, Ind.
B—Mount Vernon, Ind.
G. O. No. 78, W. D., 1919.

Private, Company K, 166th Infantry, 42d Division.
Private Marquis was a member of a patrol sent out to silence machine-gun nests which were holding up the battalion's advance. When the officer leading the patrol fell mortally wounded, this soldier attempted to go to the officer's assistance, despite heavy fire from machine guns only 100 yards away, and was himself killed.
Posthumously awarded. Medal presented to mother, Mrs. Lucinda Marquis.

*MARRA, JOHN (44509)_____
Near Charpentry, France, Oct. 3, 1918.
R—New York, N. Y.
B—Italy.
G. O. No. 87, W. D., 1919.

Private, Company M, 16th Infantry, 1st Division.
In broad daylight and subjected to heavy fire from the enemy, he led a patrol of 4 men in advance of our lines to rescue a severely wounded soldier. His courage was an inspiration to the men serving with him.
Posthumously awarded. Medal presented to father, Tony Marra.

MARRIOTT, OWEN R. (1193712)____
Near Courbon, France, July 15, 1918.
R—Proctor, Mo.
B—Morgan County, Mo.
G. O. No. 44, W. D., 1919.

Corporal, Headquarters Company, 10th Field Artillery, 3d Division.
Corporal Marriott, a member of the regimental telephone detail, when it became impossible to maintain telephone communications, volunteered and carried messages under heavy shell fire, in spite of having been wounded in the knee.

*MARSH, ARTHUR D._____
Near Verdun, France, Oct. 12, 1918.
R—Newark, N. J.
B—England.
G. O. No. 44, W. D., 1919.

Captain, 113th Infantry, 29th Division.
Refusing to relinquish command of his company, even though suffering from illness, he led them up a road under a most terrific bombardment, and while assisting a wounded man to safety he was killed.
Posthumously awarded. Medal presented to widow, Mrs. Arthur D. Marsh.

MARSH, ELMER M. (43856)_____
Near Fleville, France, Oct. 3 and 4, 1918.
R—Chickasha, Okla.
B—Everton, Mo.
G. O. No. 35, W. D., 1920.

Sergeant, Company K, 16th Infantry, 1st Division.
On Oct. 4, after all officers had been killed or wounded, he took command of and led his platoon in the attack, during which he was severely wounded. The objective was taken and held, due in a large measure to his gallant efforts. On the previous day he conducted a reconnaissance of the enemy position and located machine-gun positions. The information obtained proved valuable in the subsequent attack of Oct. 4, 1918.

°MARSH, HARRY H.............
  Near Cunel, France, Oct. 14, 1918.
  R—Paxton, Ill.
  B—Burnetts Creek, Ind.
  G. O. No. 32, W. D., 1919.

First lieutenant, 30th Infantry, 3d Division.
Lieutenant Marsh, with a force of 50 men, took a line of trenches, at the same time capturing prisoners greatly in excess of the members of his own command. It was due to his gallant example that this feat was accomplished. He was killed by machine-gun fire as the trench was taken.
Posthumously awarded. Medal presented to widow, Mrs. Nellie D. Marsh.

MARSH, JOHN (2261295).............
  Near Eclisfontaine, France, Oct. 1, 1918.
  R—Vida, Mont.
  B—Lanark, Ill.
  G. O. No. 27, W. D., 1919.

First sergeant, Company K, 362d Infantry, 91st Division.
He was painfully wounded by a shell fragment, but refused to go to the rear. Remaining with his company under heavy shell fire, he continued to perform his duties.

MARSH, JOHN C. (1289360).............
  At Molleville Farm, France, Oct. 15, 1918.
  R—Charlottesville, Va.
  B—Charlottesville, Va.
  G. O. No. 37, W. D., 1919.

Private, first class, Company K, 116th Infantry, 29th Division.
After losing his ammunition carrier, Private Marsh advanced unaided and effectively operated his automatic rifle until wounded in the hand. He continued to fire until his ammunition was exhausted, and refused to go to the rear until ordered to do so.

MARSHALL, ALLAN J. (2262896).........
  Near Spitaals-Bosschen, Belgium, Oct. 31, 1918.
  R—San Francisco, Calif.
  B—San Francisco, Calif.
  G. O. No. 44, W. D., 1919.

Sergeant, Machine Gun Company, 363d Infantry, 91st Division.
Although suffering acute pain from a severe injury, he led his section through a continuous rain of machine-gun and sniper fire beyond the Infantry front line, where he encountered the enemy.

MARSHALL, HARRY F. (548161).........
  Near Crezancy, France, July 15, 1918.
  R—South Brewer, Me.
  B—Westfield, Mass.
  G. O. No. 2, W. D., 1920.

Private, Company L, 30th Infantry, 3d Division.
He displayed coolness and bravery in carrying numerous messages under shell fire, thereby keeping up liaison between adjacent units.

°MARSHALL, LEROY F. (155281).........
  Near Verdun, France, Oct. 9, 1918.
  R—Simla, Colo.
  B—Leavenworth, Kans.
  G. O. No. 46, W. D., 1919.

Corporal, Company A, 1st Engineers, 1st Division.
He voluntarily went forward to silence a sniper who was pouring a dangerous fire into his position. He crept through brush for a distance of 300 yards and then crossed a machine-gun-swept area, killing the sniper with his rifle.
Posthumously awarded. Medal presented to father, Henry L. Marshall.

MARSHALL, RALPH WILLIAM.............
  In the Bois-de-Belleau, France, June 6 and 8, 1918.
  R—West Chicago, Ill.
  B—Elgin, Ill.
  G. O. No. 110, W. D., 1918.

Second lieutenant, 6th Regiment, U. S. Marine Corps, 2d Division.
He demonstrated conspicuous bravery and coolness in fearlessly exposing himself to heavy fire from machine guns, rifles, and hand grenades in order that he might procure accurate information regarding the movements of the enemy.

MARSHALL, ROBERT E. (2180142).........
  Near Mezy, France, July 15, 1918.
  R—Graniteville, Mo.
  B—Knob Lick, Mo.
  G. O. No. 32, W. D., 1919.

Private, Company A, 30th Infantry, 3d Division.
After being wounded in the head and subjected to an intense barrage, he remained at his post for three hours, operating his automatic rifle, until ordered to the rear.

MARSHALL, ROBERT G.............
  Near the Bois-du-Fays, France, Oct. 4, 1918.
  R—Minneapolis, Minn.
  B—Minneapolis, Minn.
  G. O. No. 66, W. D., 1919.

First lieutenant, 58th Infantry, 4th Division.
When his company's advance was stopped by heavy enfilading machine-gun fire, Lieutenant Marshall took 7 soldiers and rushed the enemy nest, killing 6 of the enemy and capturing 30, including a captain. Lieutenant Marshall accomplished this daring feat without any of his own men becoming casualties.

°MARTELL, JUDSON G.............
  Near Cunel, France, Oct. 14, 1918.
  R—West Somerville, Mass.
  B—West Somerville, Mass.
  G. O. No. 20, W. D., 1919.

First lieutenant, 60th Infantry, 5th Division.
Although seriously wounded, he continued to direct his command under heavy machine-gun and sniping fire and maintained organization under heavy demoralizing circumstances until he was killed by a sniper's bullet.
Posthumously awarded. Medal presented to mother, Mrs. Edward A. Martell.

MARTENS, JOHN C. (525957).............
  Near Kazanka, Siberia, July 3, 1919.
  R—Anaheim, Calif.
  B—Canada.
  G. O. No. 39, W. D., 1920.

Private, Company C, 31st Infantry.
He carried a message across a field under fairly heavy fire. He then volunteered to go to break up a sniper's nest, and went alone to an exposed and dangerous point where he could fire at the nest and broke it up, killing one of the snipers and driving the rest away.

MARTIE, JOHN E.............
  Near Fleville, France, Oct. 4, 1918.
  R—California, Mo.
  B—California, Mo.
  G. O. No. 3, W. D., 1921.

Captain, 16th Infantry, 1st Division.
Captain Martie exposed himself to heavy artillery and machine-gun fire while leading his company forward in an attack against strongly held enemy positions. Reaching his objective, he organized his position for defense and held the same against enemy counterattacks.

°MARTIN, CECIL N. (2100547).............
  Near Sergy, France, July 29–30, 1918.
  R—Lawrenceville, Ill.
  B—Carmi, Ill.
  G. O. No. 74, W. D., 1919.

Private, Company I, 47th Infantry, 4th Division.
Exposing himself to heavy enemy machine-gun and artillery fire, he repeatedly carried messages from his company commander to the battalion post of command. He was killed in the performance of this hazardous duty.
Posthumously awarded. Medal presented to mother, Mrs. Ettie Bryant.

MARTIN, CLAUDE A.............
  Near Vaux, France, July 1, 1918.
  R—Welsh, La.
  B—Lafayette, La.
  G. O. No. 99, W. D., 1918.

Captain, Medical Corps, attached to 23d Infantry, 2d Division.
He operated a battalion dressing station near Vaux, France, July 1, 1918, and, although the station was practically destroyed by shell fire, he bravely and successfully treated the wounded and directed their safe evacuation.

*MARTIN, DANIEL J._____
Near Juvigny, north of Soissons, France, Sept. 1, 1918.
R—Waukesha, Wis.
B—Genesee, Wis.
G. O. No. 143, W. D., 1918.

Captain, 123th Infantry, 32d Division.
On numerous occasions he personally headed every forward movement of his command, displaying superb courage in his absolute disregard for personal safety. In the attack on Juvigny the battalion commanded by this officer was called upon to execute a turning movement and effect junction with the French troops on the flank. Under the personal direction of Captain Martin this turning movement was completed across a hill strongly held by the enemy, its success being due to his initiative. During this engagement he was gassed and taken from the field unconscious.
Posthumously awarded. Medal presented to widow, Mrs. Hattie E. Martin.

MARTIN, EARL J. (65521)_____
Near Verdun, France, Oct. 27, 1918.
R—New Britain, Conn.
B—White River Junction, Vt.
G. O. No. 21, W. D., 1919.

Corporal, Company I, 102d Infantry, 26th Division.
He led his platoon over the top until further advance was impossible. He then remained in observation. Upon being grenaded by two of the enemy, he shot them with his pistol and, moving forward, gained entrance to an unused "pill box." A few hours later he ventured out into enemy trenches, entered a near-by dugout, and disposed of 6 occupants, then retired to his former position in the "pill box," returning to his own lines after dark.

MARTIN, EDWARD_____
Near Courmont, France, July 29, 1918.
R—Waynesburg, Pa.
B—Tenmile, Pa.
G. O. No. 71, W. D., 1919.

Lieutenant colonel, 110th Infantry, 28th Division.
In command of an inexperienced battalion, he led an attack against a strongly held position, and advancing with the front line, raised the morale of officers and men by his coolness under heavy fire and utter disregard for personal danger.
Oak-leaf cluster.
For the following act of extraordinary heroism in action Lieut. Col. Edward Martin is awarded one oak-leaf cluster to be worn with the distinguished-service cross: For extraordinary heroism in action near Courmont, France, July 30, 1918. Although painfully wounded when regimental headquarters was destroyed by shell fire, he went with a battalion commander and directed the successful attack against a strong enemy position, remaining in command of the regiment until its relief.

MARTIN, HARRY H._____
Near Brueil, France, July 18, 1918.
R—Emporia, Kans.
B—Emporia, Kans.
G. O. No. 27, W. D., 1920.

Captain, 28th Infantry, 1st Division.
Captain Martin, although severely wounded in the right side, before reaching the first objective, refused to be evacuated and continued to direct his company in the attack through heavy artillery and machine-gun fire to the second and third objectives.

MARTIN, HUGH B. (1317775)_____
Near Busigny, France, Oct. 10, 1918.
R—Kinston, N. C.
B—Greenville County, S. C.
G. O. No. 81, W. D., 1919.

Corporal, Machine Gun Company, 119th Infantry, 30th Division.
When a battalion of Infantry was held up by heavy machine-gun fire, he rushed his section forward to a position 300 yards in advance of our front lines, engaged and silenced the enemy, and allowed a renewal of the advance. He displayed marked personal bravery under terrific enemy fire.

MARTINEZ, AUGUSTIN, (2346993)_____
Near Pouilly, France, Nov. 10–11, 1918.
R—Aztec, N. Mex.
B—Parkview, N. Mex.
G. O. No. 37, W. D., 1919.

Corporal, Company I, 356th Infantry, 89th Division.
He accompanied Lieut. John H. Murphy, of his regiment, and 3 other soldiers in a flank attack on 3 heavy machine guns. Fired on directly at 30 yards, they charged the guns, and met hand-to-hand resistance, but repulsed the enemy, capturing the guns. Corporal Martinez followed the fleeing Germans until they were lost in the fog.

MARTINEZ, LAURIANO (1626989)_____
Near Fismes, France, Aug. 26, 1918.
R—Colmor, N. Mex.
B—Chacon, N. Mex.
G. O. No. 98, W. D., 1919.

Private, Company K, 110th Infantry, 28th Division.
With 2 other soldiers Private Martinez crawled 300 yards in front of our line through the enemy's wire and attacked a hostile machine-gun nest. The enemy crew opened fire on them at a range of only 10 yards and resisted stubbornly, but they succeeded in killing 3 of the crew and driving off the others with clubbed rifles. They returned to our lines under heavy fire.

MARTINSON, ALFRED, (273955)_____
In the Carspach Woods, near Badricourt, Alsace, France, July 19, 1918.
R—Eleva, Wis.
B—Blair, Wis.
G. O. No. 63, W. D., 1920.

Private, first class, Company E, 127th Infantry, 32d Division.
During an enemy raid on the sector held by the 3d Platoon of Company E, 127th Infantry, which was preceded by a terrific and accurate barrage of high-explosive shells and machine-gun fire, and although surrounded by a superior enemy force, Private Martinson refused to surrender, but instead jumped from his trench and with his automatic rifle repulsed the enemy.

MARTZ, ALVEY C. (1239863)_____
Near Conde-en-Brie, France, July 15, 1918.
R—Glencoe, Pa.
B—Mount Pleasant, Pa.
G. O. No. 98, W. D., 1919.

Sergeant, Company C, 110th Infantry, 28th Division.
Under violent shell and machine-gun fire, Sergeant Martz assisted in reorganizing the remnants of his shattered company, which was surrounded by the enemy, and held the position until his group was again cut to pieces. With an officer and 2 other soldiers he then succeeded in fighting his way from within the enemy's lines to his regiment, killing a large number of the enemy with his pistol.

MARTZ, FORREST L. (1565239)_____
In the Bois-du-Fays, near Brieulles, France, Oct. 6, 1918.
R—Tipton, Ind.
B—Tipton, Ind.
G. O. No. 53, W. D., 1920.

Private, first class, Company C, 12th Machine Gun Battalion, 4th Division.
Private Martz and a comrade, under heavy enemy fire, went to the rescue of wounded lying in advance of our lines and returned to our lines with 2 wounded American soldiers. In accomplishing this mission they advanced to within 75 yards of the enemy lines over an area which the enemy raked with their fire.

MARX, Robert S._____
In front of Baalons, France, Nov. 10, 1918.
R—Cincinnati, Ohio.
B—Cincinnati, Ohio.
G. O. No. 14, W. D., 1923.

Captain, 357th Infantry, 90th Division.
Having been sent to make a reconnaissance and if found necessary to take command of the 3d Battalion, 357th Infantry, the advance of which had just been checked with severe losses, he displayed the highest quality of courage and leadership in the face of a murderous artillery and machine-gun fire by immediately reorganizing the battalion and after a personal reconnaissance directing the assault line, which resulted in the taking of the enemy position. During the attack Captain Marx was severely wounded. His brave example greatly inspired his men.

MASCIARELLI, GIACOMO (3110935)_____
  Near Malancourt, France, Sept. 26, 1918.
  R—Philadelphia, Pa.
  B—Italy.
  G. O. No. 37, W. D., 1919.

Private, Company L, 315th Infantry, 79th Division.
Private Masciarelli alone charged a machine-gun nest which was holding up the advance of his platoon. With a flanking fire, he killed 1 member of the crew and caused the rest to surrender. His prisoners consisted of 1 noncommissioned officer and 7 privates.

MASCORELLA, SAMUEL (343479)_____
  Near Mezy, France, July 15, 1918.
  R—Erie, Pa.
  B—Buffalo, N. Y.
  G. O. No. 23, W. D., 1919.

Private, Headquarters Company, 38th Infantry, 3d Division.
He volunteered and carried a message to headquarters after 2 runners had been killed while attempting to get through a barrage. He returned through the barrage with an answer to the message.

MASLOSKY, JOHN (2413727)_____
  Near Grand Pre, France, Oct. 26, 1918.
  R—Elizabeth, N. J.
  B—Russia.
  G. O. No. 39, W. D., 1920.

Private, Company M, 311th Infantry, 78th Division.
He displayed exemplary devotion to duty in attacking machine-gun nests without aid and capturing many prisoners. For several hours he worked in advance of the company, and, although believed to have been lost, he later returned, bringing with him many prisoners.

MASLOWSKI, VINCENT (741464)_____
  Near Munster, Vosges Front, Lorraine, Sept. 14, 1918.
  R—Racine, Wis.
  B—Russia.
  G. O. No. 27, W. D., 1919.

Private, first class, Company D, 52d Infantry, 6th Division.
Private Maslowski, with other men of his squad, was on duty in a firing trench, in combat with the enemy. A grenade thrown by 1 of the men struck the parapet and fell back into the trench. Private Maslowski seized the grenade and threw it from the trench just as it exploded, thereby saving his comrades and himself from injury and possible death.

MASON, CLAUDE H_____
  Near St. Etienne, France, Oct. 8–10, 1918.
  R—El Paso, Tex.
  B—Wadestown, W. Va.
  G. O. No. 87, W. D., 1919.

First lieutenant, 141st Infantry, 36th Division.
On October 8 Lieutenant Mason followed the wave of attack under heavy shell and sniper fire, and maintained an aid station with no protection, near the front lines, giving first aid to the wounded and evacuating the injured from a heavily shelled area for a period of 3 days.

*MASON, EDWARD G. (757108)_____
  Near Jaulny, France, Nov. 10, 1918.
  R—Detroit, Mich.
  B—Detroit, Mich.
  G. O. No. 32, W. D., 1919.

First sergeant, Company D, 55th Infantry, 7th Division.
First Sergeant Mason continuously walked up and down his company sector, caring for the wounded and encouraging the men during an intense shell and machine-gun barrage. He was mortally wounded while going to the aid of two wounded comrades and died a short time afterwards.
Posthumously awarded. Cross on display at the Smithsonian Institution, Washington, D. C., pending delivery to next of kin, when, and if located.

MASON, FRANCIS W_____
  Near St. Georges, France, Oct. 22, 1918.
  R—Salem, Oreg.
  B—Wichita, Kans.
  G. O. No. 7, W. D., 1919.

Second lieutenant, 328th Infantry, 82d Division.
He led a patrol of 40 men through a woods in order to envelope the enemy's position. Advancing under heavy shellfire, this officer was severely wounded but, displaying excellent leadership and unusual bravery, he continued the advance and succeeded in occupying the woods.

MASON, ZELNA (2075)_____
  Near Soissons, France, July 19, 1918.
  R—Memphis, Tenn.
  B—Hickory, Ky.
  G. O. No. 132, W. D., 1918.

Private, first class, medical detachment, 26th Infantry, 1st Division.
Although twice wounded near Soissons, France, July 19, 1918, he refused an offer to be carried to the rear and continued to render aid to the wounded under fire.

MASSICOTTE, AUGUST J. (294363)_____
  Near Medeah Ferme, France, Oct. 3–9, 1918.
  R—Franklin, N. H.
  B—Franklin, N. H.
  G. O. No. 21, W. D., 1919.

Private, Headquarters Company, 9th Infantry, 2d Division.
While acting as battalion runner, Private Massicotte, regardless of personal danger, repeatedly volunteered and carried important messages under intense shell fire, and greatly assisted in maintaining liaison with other units.

MASSON, JACOB H. (261778)_____
  At Mont St. Martin, south of Fismes, France, Aug. 5, 1918.
  R—Newport, Mich.
  B—Michigan.
  G. O. No. 117, W. D., 1918.

Private, Company C, 125th Infantry, 32d Division.
During the attack on Mont St. Martin he was severely wounded in the neck and shoulder. He refused first aid and was determined to keep up with the attacking wave, which he did until he became too weak from loss of blood. By this remarkable display of courage he conveyed to his comrades the spirit of fearlessness.

MASURY, GEORGE T. (317908)_____
  At Vladivostok, Siberia, Nov. 17–18, 1919.
  R—San Francisco, Calif.
  B—Santa Cruz, Calif.
  G. O. No. 14, W. D., 1923.

Corporal, Headquarters Company, 31st Infantry.
In answer to a call to save women and children who had been unavoidably entrapped in the railroad station by the fire from contending forces of Russian Government and insurgent troops, he boldly entered the zone of fire and rushed to the station which was being fired upon from 3 sides by machine guns, rifles, and field pieces, and assisted in bringing back through the fire-swept zone a number of women noncombatants.

MATES, HARRY (246456)_____
  Near Blanc Mont Ridge, France, Oct. 3, 1918.
  R—Pittsburgh, Pa.
  B—Russia.
  G. O. No. 21, W. D., 1919.

Private, Company H, 9th Infantry, 2d Division.
While acting as company runner, he carried messages under heavy shellfire and machine-gun fire. When a machine-gun nest caused a temporary halt in the advance of his company, he attacked the nest, capturing 2 prisoners. He assisted wounded men, applied first aid, and removed them through heavy shellfire to the dressing station.

MATHEWS, HOWARD A_____
  At Marcheville, France, Sept. 26, 1918.
  R—Kansas City, Mo.
  B—Jacksonville, Ill.
  G. O. No. 139, W. D., 1918.

First lieutenant, 102d Infantry, 26th Division.
He displayed unusual courage and devotion to duty under a violent enemy bombardment by continuing in command of 2 platoons after he was severely wounded.

MATHEWS, ROY E. (2263417)_____
In the Bois-du-Fays, France, Oct. 5, 1918.
R—Seattle, Wash.
B—Fairhaven, Wash.
G. O. No. 46, W. D., 1919.

Private, Company E, 58th Infantry, 4th Division.
Acting without orders, he went through heavy artillery fire to notify his regimental commander that our own barrage was falling short, his bravery and presence of mind thus saving the lives of many American soldiers.

MATHEY, MAURICE (2306637)_____
Near Juvigny, north of Soissons, France, Sept. 1, 1918.
R—Monroe, Wis.
B—Milwaukee, Wis.
G. O. No. 15, W. D., 1919.

Private, Company F, 128th Infantry, 32d Division.
In an attack against a strong enemy position, supported by many machine guns, he, a runner, worked unceasingly in the maintenance of liaison and carried messages through the most severe machine-gun barrage. On his last trip from the regimental post of command he was severely shell shocked, but continued through the barrage to the battalion headquarters and delivered his message.

MATHIAS, JEAN_____
In the Bois-de-Belleau, France, June 11, 1918.
R—Brooklyn, N. Y.
B—New York, N. Y.
G. O. No. 89, W. D., 1919.

Private, 43d Company, 5th Regiment, U. S. Marine Corps, 2d Division.
After all the other members of his group had been killed or wounded by fire from an enemy machine gun, Private Mathias charged the gun position alone, killing 3 of the crew and capturing the gun.

*MATHIS, JOHN D_____
In the Chateau-Thierry, sector, France, June 6, 1918.
R—Americus, Ga.
B—Buena Vista, Ga.
G. O. No. 102, W. D., 1918.

First lieutenant, 23d Infantry, 2d Division.
As a leader of a platoon on the first day of the Chateau-Thierry battle he demonstrated conspicuous courage and ability, fearlessly going forward at the head of his command through hostile machine-gun fire. Killed while leading a gallant charge, his daring inspired his men to successful assault.
Posthumously awarded. Medal presented to father, Evan T. Mathis.

MATSON, LEON R. (1224735)_____
Near Ronssoy, France, Sept. 27 and 29, 1918.
R—New York, N. Y.
B—Woodhull, N. Y.
G. O. No. 32, W. D., 1919.

Sergeant, Company M, 105th Infantry, 27th Division.
On the morning of Sept. 27, after all the officers and most of the sergeants of his company had been killed, Sergeant Matson took command and led the company into effective combat, making repeated reconnaissances in front of the line under severe machine-gun fire. On Sept. 29 he led his men forward, capturing an important knoll, and held it with a small number of men. Finding the ammunition and food depleted, he led a detail through the heavy machine-gun fire, bringing back both food and ammunition.

MATSON, RAYMOND O. (2062426)_____
At Marcheville, France, Nov. 10, 1918.
R—Chicago, Ill.
B—Chicago, Ill.
G. O. No. 14, W. D., 1923.

Sergeant, Company C, 123d Machine Gun Battalion, 33d Division.
Voluntarily leaving the shelter of the trenches and exposed to terrific enemy machine-gun and artillery fire and under direct observation of the enemy, he rescued 3 wounded men, assisting them to a place of comparative safety. The outstanding bravery and soldierly devotion to duty displayed greatly encouraged the men of his company.

MATTER, PETER (1680849)_____
Near Fismes, France, Aug. 27, 1918.
R—Niagara Falls, N. Y.
B—Syria.
G. O. No. 128, W. D., 1918.

Private, Company E, 307th Infantry, 77th Division.
After having been wounded and severely burned and gassed by an explosion of mustard-gas shell, he nevertheless continued at his work as stretcher bearer, evacuating wounded until he was actually unable to see and was ordered to be evacuated himself.

MATTFELDT, CYLBURN O_____
Near Jaulny, France, Sept. 13, 1918.
R—Baltimore, Md.
B—Baltimore, Md.
G. O. No. 46, W. D., 1919.

First lieutenant, 9th Infantry, 2d Division.
In plain view of the enemy he rode across a field to a friendly battery whose barrage was falling on the American trenches and stopped its fire, thereby permitting a reestablishment of the front line and saving many lives.

MATTHEWS, GEORGE, Jr_____
Near Arbre Guernon, France, Oct. 18, 1918.
R—New York, N. Y.
B—New York, N. Y.
G. O. No. 28, W. D., 1921.

Second lieutenant, 105th Machine Gun Battalion, 27th Division.
After having been wounded in the head and suffering great pain, he coolly and efficiently made dispositions for the security of his guns and safety of his men and the evacuation of the wounded. Later at a dressing station, though bleeding profusely, he refused surgical attention until other wounded men of his platoon had been cared for.

MAUGHAN, RUSSELL L_____
Near Sommerance, France, Oct. 27, 1918.
R—Logan, Utah.
B—Logan, Utah.
G. O. No. 46, W. D., 1919.

First lieutenant, 139th Aero Squadron, 2d Pursuit Group, Air Service.
Accompanied by 2 other planes, Lieutenant Maughan was patrolling our lines, when he saw slightly below him an enemy plane (Fokker type). When he started an attack upon it he was attacked from behind by 4 more of the enemy. By several well-directed shots he sent 1 of his opponents to the earth, and, although the forces of the enemy were again increased by 7 planes, he so skillfully maneuvered that he was able to escape toward his lines. While returning he attacked and brought down an enemy plane which was diving on our trenches.

MAURER, PHILIP (1243833)_____
At Fismette, France, Aug. 10, 1918.
R—Dravosburg, Pa.
B—Austria.
G. O. No. 99, W. D., 1918.

Private, Company C, 111th Infantry, 28th Division.
Having heard that 2 wounded comrades were lying in advance of his company's line immediately north of Fismette, Private Maurer and 2 other members of his company volunteered to go through the machine-gun and rifle fire to bring them in. On their first attempt all were wounded and driven back, but in spite of their injuries they advanced a second time and reached the wounded men. Their courageous effort, however, was unfortunately in vain, as their comrades had been killed.

MAURY, ALFRED B_____
Near l'Arbre-de-Guise, France, Oct. 17, 1918.
R—Morristown, N. J.
B—Morristown, N. J.
G. O. No. 23, W. D., 1919.

Second lieutenant, 301st Battalion, Tank Corps.
Although his motor was running poorly and his tank crew badly gassed, Lieutenant Maury captured a German battery with his gun crews and turned it over to the Infantry. A little later his motor stopped completely, but he soon located another tank, whose crew was badly gassed. He transferred his ammunition and crew to the new tank and continued in the advance of our Infantry to the objective.

°MAXEY, ROBERT J.
At Cantigny, France, May 28, 1918.
R—Hot Springs, Ark.
B—Brandon, Miss.
G. O. No. 99, W. D., 1918.

Lieutenant colonel, 18th Infantry, 1st Division.
He advanced with first wave and, in the face of heavy shell and machine-gun fire, located the objective of his battalion. He was a cool, dependable, and heroic leader. Although fatally wounded, he gave detailed instructions to his second in command and caused himself to be carried to his regimental commander and delivered important information before he died.
Posthumously awarded. Medal presented to widow, Mrs. Lu Knowles Maxey.

MAXIE, RAYMOND E. (1287300).
Near Brabant, France, Oct. 3, 1918.
R—Richmond, Va.
B—Richmond, Va.
G. O. No. 37, W. D., 1919.

Corporal, Company B, 116th Infantry, 29th Division.
Corporal Maxie, in company with 4 other soldiers, attacked without support 3 machine guns, and, overcoming the desperate resistance of the enemy, captured both guns and crews.

MAY, GEORGE J. (1387659).
Near Chipilly Ridge, France, Aug. 9, 1918.
R—Chicago, Ill.
B—Chicago, Ill.
G. O. No. 71, W. D., 1919.

Sergeant, Company G, 131st Infantry, 33d Division.
On his own initiative he rushed an enemy machine-gun nest, capturing the gun and 2 prisoners. He displayed marked personal courage under heavy artillery and machine-gun fire, passing through the enemy barrage to get water and rations to the men in the front line.

MAY, JOE C. (2239686).
Near Montigny-devant-Sassey, France, Nov. 5, 1918.
R—Tahoka, Tex.
B—Brookhaven, Miss.
G. O. No. 37, W. D., 1919.

Corporal, Company B, 315th Train Headquarters and Military Police, 90th Division.
During a very heavy attack in the vicinity of his post, where artillery fire and aircraft machine-gun fire had created a most confusing situation, he calmly directed traffic, aided wounded, and removed obstructions, thereby preventing wild disorder. He also assisted the drivers of ammunition trucks in getting their machines to a place of safety.

MAY, LESTER T. (2337000).
Near Grand Ballois Farm, France, July 14–15, 1918.
R—Wyalusing, Pa.
B—Wyalusing, Pa.
G. O. No. 32, W. D., 1919.

Private, first class, Company F, 4th Infantry, 3d Division.
During a heavy shelling he volunteered and delivered messages over routes where all other runners had been either killed or wounded.

MAY, OSCAR P.
Near Bois-de-Mort Mare, France, Sept. 12, 1918.
R—Williamstown, Kans.
B—Williamstown, Kans.
G. O. No. 37, W. D., 1919.

Second lieutenant, 356th Infantry, 89th Division.
Without assistance, he very courageously attacked and captured a machine gun which threatened to wipe out his platoon.

MAYES, JOHN B., Jr.
Near Bellicourt, France, Sept. 29, 1918.
R—Stem, N. C.
B—Stem, N. C.
G. O. No. 37, W. D., 1919.

Captain, 120th Infantry, 30th Division.
Captain Mayes with 8 other soldiers, comprising his company headquarters detachment, cleaned out enemy dugouts along the banks of a canal, capturing 242 prisoners.

MAYGER, ARTHUR G.
Near Exermont, France, Oct. 6, 1918.
R—Chicago, Ill.
B—Deer Lodge, Mont.
G. O. No. 98, W. D., 1919.

First lieutenant, 28th Infantry, 1st Division.
After his company had been forced to fall back because of heavy losses and his company commander had been seriously wounded, Lieutenant Mayger reorganized the remainder of the company and, under intense shell and machine-gun fire, led it in a successful attack on a machine-gun position which had been causing many casualties in the battalion.

°MAYNE, JOHN (2800).
In the Trugny Woods, near Chateau-Thierry, France, July 23, 1918.
R—Boston, Mass.
B—Ireland
G. O. No. 116, W. D., 1918.

Private, medical detachment, 101st Infantry, 26th Division.
Private Mayne, although in an exposed position in the Trugny Woods under fire of rifles and machine-guns, courageously treated the wounded, inspiring the combat troops by his example, until shot through the head and killed.
Posthumously awarded. Medal presented to widow, Mrs. Mary Mayne.

MAYS, DOALEY (555062).
Near Connigis, France, July 15, 1918.
R—Place, Ky.
B—Knox County, Ky.
G. O. No. 22, W. D., 1920.

Sergeant, Company C, 9th Machine Gun Battalion, 3d Division.
Sergeant Mays, single-handed, operated a machine gun after the remainder of his squad had become casualties from enemy machine-gun fire. Although exposed to heavy artillery and machine-gun fire, he continued to fire his gun until assistance arrived.

MAYS, HERBERT L. (1320812).
Near Bellicourt, France, Sept. 29, 1918.
R—Taylorsville, N. C.
B—La Fayette, Ind.
G. O. No. 37, W. D., 1919.

Sergeant, Company G, 120th Infantry, 30th Division.
Sergeant Mays, with 1 other soldier, attacked a machine-gun post which was causing much damage. They captured the post, taking prisoner 1 officer and 8 men, and put the gun out of action.

°MAZKWAZ, LOUIS (107707).
Near Medeah Ferme, France, Oct. 4, 1918.
R—Philadelphia, Pa.
B—Russia.
G. O. No. 64, W. D., 1919.

Corporal, Company C, 5th Machine Gun Battalion, 2d Division.
He left the shelter of his trench to rescue soldiers who had been buried by the explosion of a shell. Shortly after, while conducting his men to cover, a shell exploded near by, severly wounding a member of his squad. Directing the remainder of the squad to take cover, he went to the assistance of the wounded man, and while rendering first aid was mortally wounded.
Posthumously awarded. Medal presented to mother, Mrs. Antonio Mazkwaz,

MAZUR, JACK (44364)_____
Near Fleville, France, Oct. 4, 1918.
R—Chicago, Ill.
B—Austria.
G. O. No. 35, W. D., 1920.

Corporal, Company M, 16th Infantry, 1st Division.
Captain Mazur led a squad of auto riflemen through intense enemy fire in an attack against an enemy machine-gun position. He rushed the position and killed or captured the gun crew. His action permitted the further advance of units of the battalion held up by the enemy fire.

MAZURKEVCZK, STANLEY (293886)_____
Near Janlay, France, Sept. 12, 1918.
R—East Walpole, Mass.
B—Russia.
G. O. No. 46, W. D., 1919.

Private, Company M, 9th Infantry, 2d Division.
With 3 other soldiers, he volunteered to carry wounded men of other companies from in front of our advanced positions and carried this work on under violent machine-gun fire while a counterattack was developing.

MAZZONI, LOUIS (49252)_____
Near La Forge Farm, France, Nov. 3–4, 1918.
R—Saugus, Mass.
B—Italy.
G. O. No. 44, W. D., 1919.

Private, first class, Machine Gun Company, 23d Infantry, 2d Division.
During the offensive operations west of the Meuse, Private Mazzoni, single handed, attacked 5 of the enemy who were firing on our column. He crawled through a woods and attacked them from the rear, killing 1 and taking the rest as prisoners.

MEADOR, ERNEST (1392371)_____
Near Bois-de-Chaume, France, Oct. 10, 1918.
R—Delavan, Ill.
B—Pineville, Mo.
G. O. No. 44, W. D., 1919.

Private, Company B, 122d Machine Gun Battalion, 33d Division.
Having induced a stretcher bearer to accompany him, he made his way through heavy shell and machine-gun fire to the front of the lines to aid a wounded comrade. His stretcher bearer was killed in the exploit, but Private Meador placed the man on the stretcher and dragged him back to safety.

MEANS, RICE W._____
At Manila, P. I., Aug. 9 and 10, 1898.
R—Denver, Colo.
B—St. Joseph, Mo.
G. O. No. 3, W. D., 1925.

Second lieutenant, 1st Colorado Volunteer Infantry.
Lieutenant Means conducted a bold and fearless reconnaissance, during which, regardless of his personal safety, he went beyond the American lines and close to those of the enemy and assisted in examining the ground between the American trenches and Fort San Antonio de Abad held by Spanish forces, thus securing information of the utmost importance in planning the successful attack of Aug. 13, 1898, on Manila, P. I.

MEBRESKI, MICHELL (40185)_____
Near Thiaucourt, France, Sept. 12, 1918.
R—Beacon, N. Y.
B—Russia.
G. O. No. 46, W. D., 1919.

Corporal, Company I, 9th Infantry, 2d Division.
Corporal Mebreski, with about 12 men assisted in flanking a machine-gun nest and then captured a German ammunition dump with about 65 prisoners.

MECOM, JOHN H. (251505)_____
Near Cierges, France, July 31, 1918.
R—Eufaula, Okla.
B—Williams, Okla.
G. O. No. 20, W. D., 1919.

Private, Company E, 125th Infantry, 32d Division.
Though severely wounded while advancing with his platoon in the face of heavy fire from enemy machine guns, Private Mecom refused to return to the rear for first aid, and he not only continued in the advance, but with another soldier successfully attacked a machine-gun nest.

*MEDEIROS, FRANK L. (58744)_____
During the counterattack by the enemy near Catigny, France, May 28, 1918.
R—South Boston, Mass.
R—East Cambridge, Mass.
G. O. No. 99, W. D., 1918.

Sergeant, Company M, 28th Infantry, 1st Division.
By courageous devotion to duty and presence of mind under fire he prevented the advance against the left flank of his command, which threatened the success of the battle. While exposing himself fearlessly to distribute ammunition and to counsel his men he was killed.
Posthumously awarded. Medal presented to mother, Mrs. Delfina L. Medeiros.

MEEHAN, EDWARD J._____
Near Monthurel, France, July 17–18, 1918.
R—Philadelphia, Pa.
B—Philadelphia, Pa.
G. O. No. 109, W. D., 1918.

Captain, 109th Infantry, 28th Division.
Early in the morning of July 17, 1918, near Monthurel, France, while in an advanced position in the fight he was severely wounded, but refused to leave his command, and continued to direct its operations until it was relieved the night of July 18, 1918.

*MEEHAN, GEORGE R._____
Near Exermont, France, Oct. 5, 1918.
R—New York, N. Y.
B—Charlestown, Mass.
G. O. No. 60, W. D., 1920.

Second lieutenant, 18th Infantry, 1st Division.
Lieutenant Meehan led his platoon through heavy artillery and machine-gun fire and urged his men forward in the attack by advancing ahead of the line. He was mortally wounded while in advance of his platoon. His example was such that inspired his men to continue in the advance and take their objective.
Posthumously awarded. Medal presented to widow, Mrs. Gertrude F. Meehan.

*MEEK, FIELDING V. (794639)_____
Near Liny, France, Oct. 5, 1918.
R—Smithfield, Ky.
B—New Castle, Ky.
G. O. No. 89, W. D., 1919.

Private, medical detachment, 11th Infantry, 5th Division.
Private Meek distinguished himself by his untiring efforts in administering first aid to the wounded, never hesitating to expose himself to danger in searching for wounded on the field. While making his way through unusually heavy machine-gun fire to a wounded soldier whom he had seen fall Private Meek was mortally wounded.
Posthumously awarded. Medal presented to father, John S. Meek.

MEEKS, CORBETT (737376)_____
Near Cunel, France, Oct. 21, 1918.
R—Neola, Ky.
B—Wolfe County, Ky.
G. O. No. 37, W. D., 1919.

First sergeant, Company H, 11th Infantry, 5th Division.
During a counterattack, he advanced alone over open country under heavy machine-gun fire to a sniping post, and by his efficient resistance greatly aided in the breaking up of the counterattack.

MEFFIN, JAMES D. (71288)_____
Near Verdun, France, Oct. 15, 1918.
R—Springfield, Mass.
B—Springfield, Mass.
G. O. No. 23, W. D., 1919.

Corporal, Company B, 104th Infantry, 26th Division.
Corporal Meffin showed extraordinary courage and bravery in going beyond our front line, under heavy machine-gun fire, and bringing back two wounded comrades.

MEISSNER, JAMES A.
  In the Toul sector, France, May 2, 1918.
  R.—Brooklyn, N. Y.
  B—Canada.
  G. O. No. 121, W. D., 1918.

First lieutenant, 94th Aero Squadron, Air Service.
He attacked 3 enemy planes at an altitude of 4,800 meters over the Foret de la Rappe, France. After a short fight he brought down 1 of the machines in flames. During the combat the entering wedge and the covering of the upper wings of his plane were torn away, and after the battle he was subjected to heavy fire from antiaircraft batteries, but by skillful operation and cool judgment he succeeded in making a landing within the American lines.
Oak-leaf cluster.
A bronze oak leaf is awarded Lieutenant Meissner for the following act of extraordinary heroism in action: On May 30, 1918, he attacked 2 enemy planes at an altitude of 4,500 meters above Jauiny, France, and after a sharp engagement shot 1 down in flames and forced the other back into its own territory.

MELCHER, EDWARD J.
  Near Chateau-Thierry, France, June 25, 1918.
  R—Louisville, Ky.
  B—Louisville, Ky.
  G. O. No. 44, W. D., 1919.

Corporal, 47th Company, 5th Regiment, U. S. Marine Corps, 2d Division.
Wounded in the head and thigh, Corporal Melcher nevertheless continued valiantly to lead his group through machine-gun and rifle fire to their objective.

MELFI, JERRY (3180086).
  Near Bois-de-Bantheville, France, Oct. 14, 1918.
  R—Swissvale, Pa.
  B—Italy.
  G. O. No. 49, W. D., 1922.

Private, first class, Company F, 126th Infantry, 32d Division.
After 2 runners had been wounded attempting to get liaison with the troops on the left of his regiment, Private Melfi, though sick from exposure, volunteered and successfully crossed an open field which was constantly swept by enemy machine-gun fire. Having accomplished his mission, he returned through the same machine-gun fire and delivered to his company commander the information he had gained. Private Melfi's devotion to duty and fearlessness were an inspiration to his comrades.

MELL, PATRICK H.
  In the region of Bantheville, France, Oct. 28, 1918.
  R—Augusta, Ga.
  B—Athens, Ga.
  G. O. No. 14, W. D., 1923.

First lieutenant, 213th Aero Squadron, Air Service.
While a voluntary member of a patrol of 6 airplanes, Lieutenant Mell attacked 4 enemy airplanes at an altitude of 2,500 meters. Nine additional enemy airplanes almost immediately joined in the fight. Despite the overwhelming number of enemy airplanes, Lieutenant Mell by great skill in maneuvering and with great bravery succeeded in bringing down out of control 1 of the enemy airplanes. On Nov. 6 Lieutenant Mell with 1 other pilot attacked 3 enemy airplanes at an altitude of 3,000 meters, 15 kilometers within the enemy lines, destroying 1 and combating another until within 50 meters of the ground, where they became separated in the fog. The outstanding bravery and superb devotion to duty displayed by Lieutenant Mell greatly inspired the members of his squadron.

MELLEN, CLIFFORD B. (71186).
  Near Verdun, France, Oct. 16, 1918.
  R—Worcester, Mass.
  B—Baldwinsville, Mass.
  G. O. No. 21, W. D., 1919.

Private, Company A, 104th Infantry, 26th Division.
When he was in a shell hole with an officer and 8 men, the enemy threw some hand grenades, 1 landing among the men. Private Mellen seized it and attempted to throw it out, when it exploded. His action saved the lives of his comrades, but resulted in a severe injury to himself.

MELROSE, ANDREW R. (1210143).
  Near Vendhuile, France, Sept. 28, 1918.
  R—Marcus, Iowa.
  B—China.
  G. O. No. 87, W. D., 1919.

Corporal, Company D, 107th Infantry, 27th Division.
Leaving the protection of a trench, he crawled out under heavy machine-gun and snipers' fire and rescued a British officer who had fallen in an exposed position. His example was an inspiration to the men serving with him.

MENARD, ALEXANDER (1211629).
  East of Ronssoy, France, Sept. 29, 1918.
  R—Malone, N. Y.
  B—Malone, N. Y.
  G. O. No. 20, W. D., 1919.

Corporal, Company K, 107th Infantry, 27th Division.
Corporal Menard, with 3 other soldiers, went out into an open field under heavy shell and machine-gun fire and succeeded in carrying back to our lines 4 seriously wounded men.
Posthumously awarded. Medal presented to father, Henry Menard.

MENDELSON, JOSEPH A.
  Near Ville-Savoye, France, Aug. 15–16, 1918.
  R—Washington, D. C.
  B—New York, N. Y.
  G. O. No. 95, W. D., 1919.

First lieutenant, Medical Corps, attached to 305th Infantry, 77th Division.
During a heavy enemy bombardment with gas and high-explosive shells, Lieutenant Mendelson worked for more than 3 hours picking up wounded and gassed men and securing their evacuation, being forced to remove their gas masks in order to accomplish this work. Though he was almost exhausted from fatigue, he then proceeded to the aid station of another battalion and assisted in treating hundreds of men. Though he was himself suffering from the effects of gas, he refused to go to the hospital upon the completion of this work, as all the other medical officers had been evacuated.

MENDENHALL, FRED D.
  Near Cunel, France, Oct. 20, 1918.
  R—Denver, Colo.
  B—Gulf Hammock, Fla.
  G. O. No. 37, W. D., 1919.

First lieutenant, 7th Engineers, 5th Division.
Although his platoon was constantly under heavy shell and machine-gun fire, he courageously directed the wiring of an extreme northern outpost line of Infantry. On the night of Oct. 10, he skillfully directed the construction of a pontoon bridge over the Loison River. So close to the enemy was his platoon that it was necessary to lash the bridge together, because the hammering of nails would have drawn instant machine-gun fire from the enemy.

MENEFEE, MARVIN JAMES.
  At Molleville Farm, France, Oct. 12, 1918.
  R—Luray, Va.
  B—Covington, Va.
  G. O. No. 44, W. D., 1919.

First lieutenant, 116th Infantry, 29th Division.
While in charge of a 37-millimeter-gun section in advance of the assaulting troops Lieutenant Menefee displayed unusual courage by operating the gun himself after his gunners had been killed, thereby reducing a machine-gun nest which had been holding up the line.

MENGE, WILLIAM M. (5839)_____
Near St. Etienne-a-Arnes, France,
Oct. 3-9, 1918.
R—Elizabeth, N. J.
B—Elizabeth, N. J.
G. O. No. 35, W. D., 1919.

Private, first class, medical detachment, 23d Infantry, 2d Division.
Throughout the engagement he tended the wounded under shell and machine-gun fire, continuing with his work after 2 of his assistants had been killed and 1 wounded.

MENGES, BEN H. (41132)_____
Near Blanc Mont, France, Oct. 3,
1918.
R—Athens, Me.
B—Bristol, Ind.
G. O. No. 64, W. D., 1919.

Private, Headquarters Company, 9th Infantry, 2d Division.
By crawling forward alone across a clearing swept by German machine-gun fire and armed only with his rifle and bayonet, he killed 4 of the enemy who resisted him, and after clearing out several dug-outs in the woods returned with 8 prisoners and valuable information. His act of valor was instrumental in warding off a strong enemy counterattack.

MENTER, LINUS H. (5865)_____
Near St. Etienne-a-Arnes, France,
Oct. 6, 1918.
R—Parish, N. Y.
B—Parish, N. Y.
G. O. No. 35, W. D., 1919.

Private, medical detachment, 23d Infantry, 2d Division.
During the day and night of Oct. 6 he constantly exposed himself under heavy fire, giving first aid to the wounded and assisting in their evacuation.

MERCER, HOWARD F. (1708040)_____
Near Stonne, France, Nov. 6, 1918.
R—New York, N. Y.
B—New York, N. Y.
G. O. No. 35, W. D., 1919.

First sergeant, Company A, 308th Infantry, 77th Division.
Voluntarily leading a patrol for a flank attack on the town of Stonne, through unusual artillery fire and exacting machine-gun fire, Sergeant Mercer, leaving his patrol, went forward alone to draw fire from the nests in order to divert the enemy's attention from the attacking patrol.

MERKEL, EDMO E_____

Near Blanc Mont, France, Oct. 3-4,
1918.
R—Hattiesburg, Miss.
B—Hattiesburg, Miss.
G. O. No. 87, W. D., 1919.

Pharmacist's mate, second class, U. S. Navy, attached to 43d Company, 5th Regiment, U. S. Marine Corps, 2d Division.
He accompanied a company of marines during an advance under violent fire, going to all parts of the line, giving first aid to wounded and directing their evacuation. Although wounded, he remained on duty until forced to go to the rear.

MERLE-SMITH, VAN SANTVOORD_____
At the crossing of the River Ourcq,
near Villers-sur-Fere, France, July
28, 1918.
R—New York, N. Y.
B—Seabright, N. J.
G. O. No. 99, W. D., 1918.

Captain, 165th Infantry, 42d Division.
Despite the loss of all other officers in his company, and although wounded himself, he continued to direct his men effectively against the enemy. When his major was killed he succeeded to the command of the battalion and led it forward throughout the day with courage and gallantry.

MERRICK, ROBERT G_____
At Courban, France, July 14-15,
1918.
R—Baltimore, Md.
B—Baltimore, Md.
G. O. No. 44, W. D., 1919.

First lieutenant, 10th Field Artillery, 3d Division.
After the members of his telephone detail had been pressed into service as runners under a hostile bombardment so severe that telephone communication could not be maintained, he volunteered to drive an ambulance. He made three trips under terrific shellfire to evacuate wounded from Greves Farm.

MERRIFIELD, EDWARD L. (2817823)_____
Near Lesseau, France, Sept. 4, 1918.
R—Greenville, Ill.
B—Greenville, Ill.
G. O. No. 15, W. D., 1919.

Private, Company E, 366th Infantry, 92d Division.
Although he was severely wounded, he remained at his post and continued to fight a superior enemy force which had attempted to enter our lines, thereby preventing the success of an enemy raid in force.

MERRILL, JESSE HERBERT (2417792)_____
In the Bois-de-Ronvaux, France,
Sept. 16-17, 1918.
R—Elizabethtown, N. Y.
B—Ticonderoga, N. Y.
G. O. No. 35, W. D., 1919.

Private, Company E, 312th Infantry, 78th Division.
With the remark, "I can get through and find him," Private Merrill volunteered and carried a message from his regimental commander to the commander of an advance battalion through a fire that seemd impassable. He returned with amazing promptness with an answer to the message. This soldier made several other trips on the same night, finding his way through a dark forest, actually walking on bodies of men who had fallen in the only path that could be used.

MERRIMON, CLIFTON (2336957)_____
Near Bussy Farm, France, Sept. 27,
1918.
R—Cambridge, Mass.
B—Cambridge, Mass.
G. O. No. 13, W. D., 1919.

Corporal, Company L, 372d Infantry, 93d Division.
He attacked with hand grenades an enemy machine gun which was causing heavy losses to his platoon and succeeded in killing the gunner and putting the gun out of action. He then organized the remainder of the platoon and led them to their positions in the trenches south of Bussy Farm.

MERRITT, CHARLES B. (170273)_____
Near Cheppy, France, Sept. 26,
1918.
R—Philadelphia, Pa.
B—Girdletree, Md.
G. O. No. 46, W. D., 1919.

Private, Company C, 345th Battalion, Tank Corps.
While he was directing a column of tanks through a mine field, assisted by another soldier, his companion was wounded, but he continued with his work until all tanks had safely passed through. Returning, he assisted his wounded comrade to safety, after which he carried messages through an intense bombardment of artillery and machine guns.

MERRITT, HARRY P. (1752148)_____
Near Grand Pre, France, Nov. 1,
1918.
R—Montclair, N. J.
B—Marlboro, N. Y.
G. O. No. 44, W. D., 1919.

Sergeant, Company I, 312th Infantry, 78th Division.
After having established an observation post without aid in broad daylight and under perilous shellfire of the enemy, he volunteered and carried rations to the post under cover of darkness. Though knocked down by shell concussion, he reached every man who was unable to leave his post, and his entire exploit was carried out under most harassing machine-gun and artillery fire.

MERRITT, HENRY C.
Near Gland, France, June 18–19, 1918.
R—Tuckahoe, N. Y.
B—Tuckahoe, N. Y.
G. O. No. 81, W. D. 1919.

Second lieutenant, 38th Infantry, 3d Division.
After successfully crossing the Marne with a night patrol, Lieutenant Merritt captured 3 prisoners. The patrol was then fired on by a detachment of the enemy, and, in the hand-to-hand conflict which followed all of the enemy were killed. The patrol returned to our lines with 1 prisoner and only 2 of the members wounded.

MERSHON, VANCE.
Near Exermont, France, Oct. 4–11, 1918.
R—Buckner, Mo.
B—Salem, Oreg.
G. O. No. 44, W. D., 1919.

First lieutenant, 28th Infantry, 1st Division.
After the battalion commander and all the senior officers had been killed or wounded, Lieutenant Mershon took command of the battalion and led it successfully to its objective, remaining with it after being painfully wounded, until properly relieved and the new commanding officer thoroughly acquainted with the situation. He then had his wound dressed and returned to his company, actively supervising the laying of barrages by indirect fire. On the third day he was ordered to the rear, there had his wounds dressed, again returning to his company, remaining with it until it was relieved and reorganized. His courage, self-sacrifice, and utter disregard for his own personal danger was a material inspiration to his men while under the terrific bombardment of enemy artillery.

MERZ, HARRY.
In the Dickebusch sector, Belgium, Aug. 27, 1918.
R—New York, N. Y.
B—New York, N. Y.
G. O. No. 56, W. D., 1922.

First lieutenant, 105th Infantry, 27th Division.
When his company was occupying a front-line position and suffering heavy losses from a near-by enemy sniper, Lieutenant Merz, locating the sniper, left his shelter and at great personal danger courageously advanced and succeeded in destroying the sniper and his nest with hand grenades.

MESSANELLI, RAY A.
Near St. Etienne, France, Oct. 4–5, 1918.
R—Utica, N. Y.
B—Clayville, N. Y.
G. O. No. 35, W. D., 1919.

Pharmacist's mate, second class, U. S. Navy, attached to 5th Regiment, U. S. Marine Corps, 2d Division.
Regardless of his personal danger, he repeatedly exposed himself to machine-gun and shell fire to render first aid to the wounded.

MESSINA, JOHN (1969496).
At the attack on Hill 223, Chatel-Chehery, Argonne Forest, France, Oct. 7, 1918.
R—Boston, Mass.
B—Italy.
G. O. No. 3, W. D., 1921.

Corporal, Company B, 328th Infantry, 82d Division.
Corporal Messina, under heavy shell and machine-gun fire at the risk of his life carried a wounded officer from a shell hole to the regimental first-aid station.

MESSINGER, ELIAS J.
Near Vierzy, France, July 19, 1918.
R—Boise, Idaho.
B—Waterloo, Iowa.
G. O. No. 126, W. D., 1919.

Private, 55th Company, 5th Regiment, U. S. Marine Corps, 2d Division.
Corporal Montag and Privates McIntyre, Messinger, and Wood captured a machine gun which was holding up the 55th Company of Marines, killing the entire crew. To accomplish this hazardous and daring work it was necessary for them to expose themselves to the fire of this gun. Even though Corporal Montag and Privates McIntyre and Messinger were wounded during the advance, the party continued forward and succeeded.

METER, ALBERT.
Near Thiaucourt, France, Sept. 15, 1918.
R—New York, N. Y.
B—New York, N. Y.
G. O. No. 37, W. D., 1919.

Private, 79th Company, 6th Regiment, U. S. Marine Corps, 2d Division.
While on duty as stretcher bearer for his company, he rushed into the open to rescue another soldier threatened with capture, in the face of a large force of advancing Germans. He killed 2 of the enemy and brought in the soldier to a place of safety.

*MEYER, ALBERT C. (1902039).
Near St. Juvin, France, Oct. 11, 1918.
R—Jeannette, Pa.
B—Pittsburgh, Pa.
G. O. No. 78, W. D., 1919.

Sergeant, Company G, 326th Infantry, 82d Division.
After his platoon had suffered heavy casualties through the devastating fire, Sergeant Meyer, although suffering from a wound, reorganized the remnants of the platoon and continued the advance. He was again wounded but refused evacuation. He insisted on remaining with and cheering the men until struck the third time, when he fell mortally wounded.
Posthumously awarded. Medal presented to father, Otto A. Meyer.

*MEYER, FRANK E. (551410).
Near Romagne, France, Oct. 8, 1918.
R—Lanesville, Ind.
B—Lanesville, Ind.
G. O. No. 37, W. D., 1919.

Sergeant, Company H, 38th Infantry, 3d Division.
He courageously led his platoon through a terrific barrage and silenced a machine-gun position which was enfilading the attacking line. He was killed later in this action.
Posthumously awarded. Medal presented to father, George Meyer.

†MEYER, FRED H. (52561).
Near Cantigny, France, May 27, 1918.
R—New York, N. Y.
B—New York, N. Y.
G. O. No. 74, W. D., 1919.

Private, Company A, 26th Infantry, 1st Division.
While acting as helper on an automatic-rifle team which was under heavy machine-gun fire, he placed himself so as to shield the gunner from the hostile fire, enabling him to operate his gun so as to neutralize the fire of two enemy machine guns. Private Meyer was mortally wounded, his heroic action costing him his life.
Posthumously awarded. Medal presented to father, Henry H. Meyer.

MEYER, GEORGE F. (91197).
At Landres-et-St. Georges, France (Cote-de-Chatillon), Nov. 1, 1918.
R—New York, N. Y.
B—Hoboken, N. J.
G. O. No. 9, W. D., 1923.

Sergeant, Company K, 165th Infantry, 42d Division.
Although suffering severely from an attack of gas received on the previous night while leading a patrol in front of Landres-et-St. Georges, during which he secured information of the utmost importance, Sergeant Meyer, in the absence of a commissioned officer, assumed command of his company, which held the most advanced post of his battalion. A high-explosive shell burying 4 men of his company, he immediately and with complete disregard for his own safety, hurried to their rescue and under a heavy barrage of enemy machine-gun and artillery fire, dug with his hands and helmet until he uncovered and rescued the first man; with the assistance of a private of his company he then rescued 2 more men and gave directions for their evacuation. He then, unassisted and under a withering enemy fire, uncovered the body of the fourth man of his company, the latter having been killed instantly.

MEYERING, WILLIAM D.
    Near Riga, France, Apr. 6, 1918.
    R—Chicago, Ill.
    B—Chicago, Ill.
    G. O. No. 59, W. D., 1918.

First lieutenant, 23d Infantry, 2d Division.
While commanding a platoon of Infantry it was attacked by the enemy on the morning of Apr. 6, 1918. He took effective measures before and during the attack to defeat the enemy and handled his men well, under fire, until he was seriously wounded. Forced to attend to his wound, he refused assistance and walked through the enemy's barrage to a dressing station. He objected to being taken to the rear till he knew the outcome of the attack. His brave example inspired his men to drive off the enemy, who did not reach our trenches. He lost his right hand by amputation as the result of the wound.

MEYERS, JAMES P. (42309)
    Service rendered under the name, James E. Porter.

Sergeant, Company C, 16th Infantry, 1st Division.

MEZOFF, JOHN J. (1288945)
    Near Samogneux, France, Oct. 15, 1918.
    R—Waverly, Va.
    B—Waverly, Va.
    G. O. No. 37, W. D., 1919.

Corporal, Company E, 116th Infantry, 29th Division.
When his company was subjected to severe machine-gun fire, Corporal Mezoff, with two other soldiers, attacked a nest of 4 machine guns, killing 8 of the enemy and capturing 27.

MIANOVICH, STANKO (1999234)
    Near Bellicourt, France, Sept. 29, 1918.
    R—Zeigler, Ill.
    B—Montenegro.
    G. O. No. 81, W. D., 1919.

Corporal, Company L, 119th Infantry, 30th Division .
Separated from his platoon, he encountered a patrol of 18 Germans, attacked them, and killed 3 and captured 15.

MICHAEL, WILLIAM H.

    Near the Bois-de-Belleau, France, June 6, 1918.
    R—Perryman, Md.
    B—Baltimore, Md.
    G. O. No. 147, W. D., 1918.

Lieutenant commander, Medical Corps, U. S. Navy, attached to 6th Regiment, U. S. Marine Corps, 2d Division.
He displayed unusual courage on the morning of June 6, 1918, near the Bois-de-Belleau, when he established a dressing station in the open, exposed to both shell and machine-gun fire, in order to be near the wounded. Under these conditions he worked for several hours.

MICHAELIS, CHARLES W. (1387421)
    At Chipilly Ridge, France, Aug. 10, 1918.
    R—East St. Louis, Ill.
    B—St. Louis, Mo.
    G. O. No. 128, W. D., 1918.

Private, Company E, 131st Infantry, 33d Division.
This soldier showed gallantry in attacking an enemy machine-gun nest with his platoon sergeant, killing the crew and capturing the gun, which he used later effectively against the enemy.

MICHAELS, EMMETT C.
    Near Varennes, France, Oct. 3, 1918.
    R—St. Joseph, Mo.
    B—Yarmouth, Iowa.
    G. O. No. 32, W. D., 1919.

Second lieutenant, 9th Infantry, 2d Division.
After 5 members of an automatic rifle squad had been killed by sniper fire and the others were unable to take a machine-gun nest which was holding up the advance of the company, Lieutenant Michaels led the remaining members of the squad against the nest, capturing 15 of the gunners and killing the others. Later he was severely wounded while carrying a wounded member of his platoon to safety, but refused medical attention until the soldier had been cared for.

MICHALKA, GUSTAVE A. (262209)
    Near Cierges, northeast of Chateau-Thierry, France, July 31, 1918.
    R—Flint, Mich.
    B—Cheboygan, Mich.
    G. O. No. 117, W. D., 1918.

Corporal, Company E, 125th Infantry, 32d Division.
When his platoon advanced up the slope in front of the Bois les Jomblets it became necessary to put out of action an enemy machine gun that was cutting up the platoon. Corporal Michalka grasped the situation and at the risk of his own life advanced upon the nest with 2 of his men, killed the operators and captured the gun.

MICHENER, JOHN H.
    Near Varennes, France, Oct. 4, 1918.
    R—Erie, Pa.
    B—Chicago, Ill.
    G. O. No. 20, W. D., 1919.

First lieutenant, pilot, 1st Aero Squadron, Air Service.
He was assigned the mission of locating the front lines of our troops at a time when dense mist and low clouds compelled him to fly at an altitude of only 100 meters. His observer's signal rockets drew fire from an advanced hostile machine-gun battery and Lieutenant Michener was wounded in the leg. Despite his wound, he continued the mission until the position of our troops was ascertained. He was then compelled to land on shell-torn ground behind the lines, the plane being completely wrecked.

MICKLISH, FRED (2395893)
    Near Fossoy, France, July 15, 1918.
    R—Jonesboro, Ark.
    B—Jonesboro, Ark.
    G. O. No. 44, W. D., 1919.

Private, Company L, 7th Infantry, 3d Division.
Acting as runner, he made repeated trips through the heavy enemy bombardment until wounded while carrying a message. Though suffering great pain, he completed his mission before reporting for treatment.

MIDKIFF, HOLLY (1388442)
    At Bois-d'Harville, France, Nov. 10, 1918.
    R—Chicago, Ill.
    B—Great Bend, Ind.
    G. O. No. 87, W. D., 1919.

Sergeant, Company L, 131st Infantry, 33d Division.
Preceding his platoon in advance by 15 yards, he discovered a machine-gun nest, and crawling forward alone under heavy fire captured the enemy position, taking as prisoners 12 Germans who had manned 2 machine guns. His bravery inspired the men of his platoon.

MIKOS, JOHN J (51295)
    Near St. Etienne-a-Arnes, France, Oct. 3-9, 1918.
    R—Chicago, Ill.
    B—Chicago, Ill.
    G. O. No. 35, W. D., 1919.

Corporal, Company H, 23d Infantry, 2d Division.
Although severely wounded early in the engagement, he refused first aid, and continued to lead his section under heavy machine-gun fire until again seriously wounded.

MILES, HARRY B. (1820826)
Near Bois-des-Ogons, France, Oct. 4–5, 1918.
R—Richmond, Va.
B—Richmond, Va.
G. O. No. 37, W. D., 1919.

Private, Company B, 318th Infantry, 80th Division.
Always a volunteer for the most dangerous service, Private Miles volunteered to carry a message through a heavy barrage and was killed in the execution of his mission.
Posthumously awarded. Medal presented to father, William E. Miles.

MILES, JOHN (1387423)
Near Wadonville, France, Nov. 9, 1918.
R—Chicago, Ill.
B—Chicago, Ill.
G. O. No. 81, W. D., 1919.

Corporal, Company E, 131st Infantry, 33d Division.
Although suffering severely from the shock of a shell concussion, Corporal Miles volunteered and went 400 yards in advance of our lines in order to draw the fire of any enemy machine guns, so that fire could be directed upon them. He returned to our lines after accomplishing the mission, but in such an exhausted condition that he had to be carried to the aid station.

MILES, PERRY L.
Near Manila, Philippine Islands, Feb. 5, 1899.
R—Columbus, Ohio.
B—Westerville, Ohio.
G. O. No. 10, W. D., 1921.
Distinguished-service medal also awarded.

First lieutenant, 14th Infantry, U. S. Army.
During the attack by two companies of the 14th Infantry on blockhouse No. 14 and adjacent trenches strongly held by insurgent forces, when the commanding officer was mortally wounded, the advance was checked and the troops were partially demoralized in the face of a heavy concentrated fire from the front and both flanks, Lieutenant Miles assumed command, ordered the advance to continue, and went along the line with utter disregard of the hostile fire and urged his men forward. Then, with exceptional gallantry and the highest qualities of leadership, he dashed forward, many yards ahead of his men, calling on them to follow, and drove the enemy from their position. His splendid example of personal heroism, courage, and coolness furnished the needed inspiration to the wavering command and resulted in the successful accomplishment of a seemingly impossible attack.

MILES, THOMAS H., Jr.
At Chateau-Thierry, France, June 6, 1918.
R—Germantown, Pa.
B—Germantown, Pa.
G. O. No. 119, W. D., 1918.

Second lieutenant, 5th Regiment, U. S. Marine Corps, 2d Division.
Killed in action at Chateau-Thierry, France, June 6, 1918, he gave the supreme proof of that extraordinary heroism which will serve as an example to hitherto untried troops.
Posthumously awarded. Medal presented to widow, Mrs. Anna H. Miles

MILESKI, BEN (1526562)
Near Cierges, France, Sept. 28, 1918.
R—Toledo, Ohio.
B—Russia.
G. O. No. 19, W. D., 1920.

Private, Company I, 145th Infantry, 37th Division.
When a platoon of Company I, 147th Infantry, was held up by machine-gun fire from the left flank, Private Mileski, without orders, rushed forward through heavy machine-gun fire, killed the machine gunner, and caused a number of the enemy to surrender.

MILGRAM, JOSEPH J. (2716667)
Near Grand Monghene, France, Nov. 8, 1918.
R—Philadelphia, Pa.
B—Russia.
G. O. No. 64, W. D., 1919.

Private, first-class, Company A, 312th Machine Gun Battalion, 79th Division.
Private Milgram, on duty as a battalion runner, displayed remarkable daring in frequently going through heavy fire in order that communication might be maintained. Upon being sent out to locate the advance units he was repeatedly fired upon by snipers and attacked with hand grenades, but succeeded in returning with valuable information, although 2 other runners accompanying him became casualties.

MILLER, ARTHUR M. (557794)
Near Sergy, France, Aug. 1, 1918.
R—Websterville, Vt.
B—Canada.
G. O. No. 37, W. D., 1919.

Private, first class, Company B, 47th Infantry, 4th Division.
Private Miller was killed while returning with an answer to a very important message which he had voluntarily delivered at a very critical stage of the attack. His mission was one of extreme danger, taking him to the most advanced position through a sweeping fire of artillery and machine guns.
Posthumously awarded. Medal presented to mother, Mrs. Mary Miller.

MILLER, BRYAN (2029593)
Near Soissons, France, July 20, 1918.
R—Detroit, Mich.
B—Philadelphia, Pa.
G. O. No. 124, W. D., 1918.

Private, Company C, 1st Engineers, 1st Division.
When volunteers were called for by his company commander, Private Miller volunteered and rescued wounded comrades from a barrage. Disregarding danger to himself, he continued the performance of these heroic deeds until killed.
Posthumously awarded. Medal presented to mother, Mrs. Nellie Miller.

MILLER, CHARLES (59796)
In the Trugny Woods, near Chateau-Thierry, France, July 23, 1918.
R—Boston, Mass.
B—Boston, Mass.
G. O. No. 125, W. D., 1918.

Private, Company A, 101st Infantry, 26th Division.
He, with 2 other men in an advanced position ahead of the battalion, charged a machine gun, killed 4 of the enemy, and drove off the rest, thereby making possible the advance of their comrades.

MILLER, CLAUDE H.
At Nauguilian, Luzon, Philippine Islands, Dec. 7, 1899.
R—Lynchburg, Va.
B—Lynchburg, Va.
G. O. No. 3, W. D., 1925.

First lieutenant, 24th Infantry, U. S. Army.
When the command of which he was a member was held up in the crossing of the Rio Grande de Cagayan by rifle fire of a well-intrenched enemy, and being without boats or rafts with which to cross, he hastily constructed a raft made of bamboo, strips of shelter tents, and personal equipment, and on this flimsy affair with 2 soldiers volunteered to cross the river. Displaying great gallantry and with utter disregard for his own safety Lieutenant Miller crossed the river in the face of heavy rifle fire and took part in an attack which drove the enemy from the trenches and the town occupied by them, thereby making possible the further advance of the command.

MILLER, EDWIN C. (554702)
At Cierges, France, Oct. 9, 1918.
R—Philipsburg, Pa.
B—Philipsburg, Pa.
G. O. No. 2, W. D., 1920.

Corporal, Company A, 9th Machine Gun Battalion, 3d Division.
Corporal Miller exposed himself to artillery and direct machine-gun fire while going 300 yards in front of our lines to assist a comrade in carrying a wounded soldier to shelter.

MILLER, FRANK D. (2706)_____
Near Exermont, France, Oct. 1-12, 1918.
R—Great Falls, Mont.
B—Great Falls, Mont.
G. O. No. 70, W. D., 1919.

Private, medical detachment, 28th Infantry, 1st Division.
His detachment having been reduced to but 3 men, Private Miller displayed conspicuous courage and devotion to duty in caring for and evacuating wounded across an area swept by shell and machine-gun fire to the regimental aid station and returning with badly needed medical supplies to the forward aid station. His conduct was an inspiration to his associates, their commanding officer being absent and the sergeant in charge having been killed.

*MILLER, FRED C. (47549)_____
Near Exermont, France, Oct. 4 and 5, 1918.
R—Bellaire, Ohio.
B—Glenville, W. Va.
G. O. No. 35, W. D., 1920.

Corporal, Company H, 18th Infantry, 1st Division.
On Oct. 4 Corporal Miller, with a small group from his platoon, advanced through heavy fire and captured an enemy machine-gun nest with its crew of 6 men. The following day he advanced in front of our lines through heavy fire and assisted several wounded comrades to safety. Corporal Miller was killed later during an engagement in the Argonne.
Posthumously awarded. Medal presented to father, M. C. Miller.

*MILLER, GEORGE F. (1316503)_____
Near Bellicourt, France, Sept. 29, 1918.
R—Dyersburg, Tenn.
B—Knoxville, Tenn.
G. O. No. 87, W. D., 1919.

Sergeant, Company K, 119th Infantry, 30th Division.
When a portion of his company was threatened by a counterattack and 2 runners had been killed in an attempt to reach the detachment with orders to withdraw, he volunteered for the dangerous mission, and attempting to cross an exposed field to carry the orders was mortally wounded.
Posthumously awarded. Medal presented to father, Luther G. Miller.

MILLER, HARRY W. (1213273)_____
East of Ronssoy, France, Sept. 29, 1918.
R—Buffalo, N. Y.
B—Peoria, Ill.
G. O. No. 20, W. D., 1919.

Sergeant, Machine Gun Company, 108th Infantry, 27th Division.
During the operations against the Hindenburg line he concealed the fact that he was wounded from his officers and continued to advance with his company during the entire day. He displayed exceptional bravery and gallantry, setting a fine example to all.

*MILLER, HENRY (1708665)_____
Near Binarville, France, Oct. 3, 1918.
R—Brooklyn, N. Y.
B—Brooklyn, N. Y.
G. O. No. 21, W. D., 1919.

Private, Company E, 308th Infantry, 77th Division.
When his company had been cut off from communication and exposed to intense shell and machine-gun fire, Private Miller observed and attacked an enemy sniper, silencing further fire from that source. While attempting to return he was killed by machine-gun fire.
Posthumously awarded. Medal presented to father, Henry Miller.

MILLER, HERBERT H. (2201262)_____
In the Bois-de-Barricourt, France, Nov. 2, 1918.
R—Ionia, Kans.
B—Ionia, Kans.
G. O. No. 66, W. D., 1919.

Sergeant, Company G, 353d Infantry, 89th Division.
Continuing forward alone, after all the other members of his combat group had been killed or wounded, Sergeant Miller penetrated the enemy's lines, despite machine-gun and rifle fire, located a machine-gun nest which was holding up the advance of his platoon, and put it out of action by effective rifle fire.

MILLER, HOBERT (1315912)_____
Near Bellicourt, France, Sept. 29, 1918.
R—Jacksboro, Tenn.
B—Briceville, Tenn.
G. O. No. 50, W. D., 1919.

Private, first class, Company G, 119th Infantry, 30th Division.
When his section of the line was held up by extremely heavy machine-gun fire, Private Miller voluntarily went forward and, unaided, routed out the crews of the 2 nests, killing 3 of the enemy and returning with 3 prisoners.

MILLER, HUGH S._____
In the Bois-de-Belleau, France, June 6, 1918.
R—St. Louis, Mo.
B—St. Louis, Mo.
G. O. No. 110, W. D., 1918.

Private, 83d Company, 6th Regiment, U. S. Marine Corps, 2d Division.
Although ordered to the rear twice because of illness, he returned to his command voluntarily and continued to fight with it vigorously throughout the advance.

MILLER, JAMES R. (1291111)_____
Near the Cote-de-Roches, France, Oct. 8, 1918.
R—Cambridge, Md.
B—Hartford, Conn.
G. O. No. 15, W. D., 1919.

Private, Company C, 112th Machine Gun Battalion, 29th Division.
When the advance of the battalion to which his company was attached was halted by heavy machine-gun fire, this soldier boldly leaped to the top of his machine-gun emplacement to draw the enemy fire and thus enable his crew to locate the enemy's emplacement. Later in the action he fearlessly left his trench in search of an enemy sniper, who was causing many casualties among our troops and killed him with a captured German rifle.

MILLER, JOHN C., Jr._____
Near Lucy-le-Bocage, France, June 19-20, 1918.
R—Huntington, W. Va.
B—Dunlow, W. Va.
G. O. No. 99, W. D., 1918.

Second lieutenant, 2d Engineers, 2d Division.
With a few volunteers, entered a woods heavily shelled and gassed, and recovered 2 wounded members of his platoon.

MILLER, JOSEPH P. (540426)_____
Near Cunel, France, Oct. 12, 1918.
R—Baton Rouge, La.
B—Ascension Parish, La.
G. O. No. 98, W. D., 1919.

Sergeant, Machine Gun Company, 7th Infantry, 3d Division.
After making a reconnaissance of the enemy positions with 2 other soldiers and finding the enemy forming for a counterattack, Sergeant Miller, upon his own initiative, extended his positions and filled a gap of 400 meters in the line, his platoon commander having been wounded, and successfully repelled the hostile counterattack. Following up the enemy's withdrawal, he placed his guns in an advanced position in a railroad cut and remained there for 4 days without communication with the rear except at night. During this period he repelled 3 enemy counterattacks and half his command became casualties, his courage under these trying conditions being an inspiration to his men.

MILLER, LAWRENCE G._____
At Belleau, France, July 18, 1918.
R—Webster Groves, Mo.
B—Asheville, N. C.
G. O. No. 11, W. D., 1921.

Second lieutenant, 103d Machine Gun Battalion, 26th Division.
Lieutenant Miller was struck by a piece of shrapnel and knocked unconscious. Upon regaining consciousness and with great difficulty he rejoined and remained with his platoon during the remainder of the attack until his battalion was relieved.

MILLER, LESTER A. (2157639)_____
Near Vaux-Andigny, France, Oct.
11, 1918.
R—Lansing, Iowa.
B—Walnut, Ill.
G. O. No. 43, W. D., 1922.

Private, Company M, 118th Infantry, 30th Division.
While his company was engaged with the enemy in the attack on this town, Private Miller, accompanied by 2 other men and with a Lewis machine gun, by display of excellent initiative and great courage, voluntarily crawled around the town through a heavy hostile machine-gun fire and with the machine gun in a near-by house silenced the enemy machine gun and drove the crew from the harassing post which was holding up the company's advance. During this action Private Miller was seriously wounded.

MILLER, PHILLIP (1314485)_____
Near Vaux-Andigny, France, Oct.
13, 1918.
R—Golddust, Tenn.
B—Jackson County, Ill.
G. O. No. 37, W. D., 1919.

Private, Company D, 118th Infantry, 30th Division.
Seeking shelter with his company from the severe artillery and machine-gun fire of the enemy, he entered a shell hole. Here he found a severely wounded soldier, and without hesitation carried the man 50 yards under plain view of the enemy and exposed to terrific fire to a place of safety with some of his company.

MILLER, THOMAS A. O._____
Near Blanc Mont Ridge, France,
Oct. 4, 1918.
R—Eau Claire, Pa.
B—Eau Claire, Pa.
G. O. No. 35, W. D., 1919.

Private, 45th Company, 5th Regiment, U. S. Marine Corps, 2d Division.
He volunteered and carried a message through terrific shell and machine-gun fire. In the performance of this mission he suffered the loss of a leg from an exploding shell.

MILLER, WILLIS C. (1467810)_____
Near Cheppy, Meuse, France, Sept.
27, 1918.
R—Kansas City, Kans.
B—McLouth, Kans.
G. O. No. 81, W. D., 1919.

Cook, Company B, 110th Engineers, 35th Division.
When his platoon had been fired upon at short range by a hostile machine gun, he advanced alone, armed merely with a pistol, and, although knocked down by an aerial bomb, went forward to the emplacement, killed the 2 gunners, captured the gun, and made prisoners of the reserve crew of 2 men who were in a neighboring emplacement.

MILLIS, JOHN M._____
In Bantheville Woods, France, Oct.
30, 1918.
R—Catlettsburg, Ky.
B—Catlettsburg, Ky.
G. O. No. 74, W. D., 1919, and G. O.
No. 13, W. D., 1923.

Second lieutenant, 354th Infantry, 89th Division.
Wounded in both legs, one of them being broken when his daylight patrol was caught in heavy machine-gun fire 500 yards in advance of our lines, he ordered his men to return without him. He was later rescued by 2 soldiers.
Oak-leaf cluster.
A bronze oak-leaf cluster is awarded Lieutenant Millis for extraordinary heroism in action near Charey, France, Sept. 30, 1918. On the morning of that day an officer returning from patrol reported to Lieutenant Millis that he had encountered enemy machine guns and that several of his patrol had been wounded. Lieutenant Millis, with great courage and without regard to his own safety, proceeded over rough and broken ground to a point 400 yards beyond his outposts, where he exchanged several shots with the enemy. Meeting another member of the patrol he was informed of a wounded man about 100 yards beyond his position at that time. Under a withering machine-gun fire Lieutenant Millis with splendid courage and devotion to duty rushed into the open, located and examined the man, whom he found to be dead. Making his way to his outpost he continued to exchange shots with the enemy, one of whom he killed.

*MILLOY, JACK L. (1391526)_____
Near Brieulles, France, Oct. 10, 1918.
R—Chicago, Ill.
B—England.
G. O. No. 38, W. D., 1922.

Sergeant, Machine Gun Company, 132d Infantry, 33d Division.
On hearing the call for aid by several men who had been wounded, Sergeant Milloy, without regard for his own life, rushed out under heavy machine-gun and shellfire, administered first aid to these men, and, while in the act of placing them on stretchers, was mortally wounded.
Posthumously awarded. Medal presented to widow, Mrs. Jack L. Milloy.

MILLS, BRUCE H._____
At Blanc Mont, France, Oct. 3, 1918.
R—Los Angeles, Calif.
B—Mason, Mich.
G. O. No. 37, W. D., 1919.

Private, 78th Company, 6th Regiment, U. S. Marine Corps, 2d Division.
When the advance of their company was held up by enfilading fire from a hostile machine-gun nest, Private Mills, with 3 other soldiers, volunteered and made a flank attack on the nest with bombs and rifles, killing 3 members of the crews and capturing 25 others, together with 3 machine guns.

MILLS, EDWIN S. (1752701)_____
Near Grand Pre, France, Oct. 23, 1918.
R—Camden, N. J.
B—Camden, N. J.
G. O. No. 46, W. D., 1919.

Private, Company D, 312th Infantry, 78th Division.
While acting as a runner he volunteered and carried messages through several heavy barrages and under direct enemy fire, always accomplishing his mission,

MILLS, EMERY W._____
Near Grand Pre, France, Oct. 25, 1918.
R—Boardman, N. C.
B—Boardman, N. C.
G. O. No. 37, W. D., 1919.

Second lieutenant, 311th Infantry, 78th Division.
Lieutenant Mills asked permission to lead a platoon against strong enemy machine-gun nests which were blocking the advance of the battalion. He not only led his platoon in a daring and extraordinary successful attack, but personally advanced ahead of his platoon and captured 2 machine guns. During the consolidation of the line he fearlessly walked up and down the line under intense machine-gun and artillery fire, establishing strong points and encouraging his men.

+MILLSAP, EARL (2788816)_____
Near Grand Pre, France, Oct. 15, 1918.
R—Asotin, Wash.
B—Asotin, Wash.
G. O. No. 21, W. D., 1919.

Private, Company B, 307th Infantry, 77th Division.
Knowing that he faced certain death, Private Millsap displayed the highest gallantry and devotion to duty by 4 times carrying messages across a field swept by machine-gun fire. He was killed while performing this hazardous service.
Posthumously awarded. Medal presented to father, Joe Millsap.

MILNE, WILLIAM L. (1489917)_____
Near St. Etienne, France, Oct. 8, 1918.
R—Muskogee, Okla.
B—Parsons, Kans.
G. O. No. 81, W. D., 1919.

Private, Company D, 142d Infantry, 36th Division.
With the aid of another soldier Private Milne attacked and captured a machine-gun nest, at the same time killing and capturing several of the enemy. He later organized a support line by assembling the scattered members of 1 of the companies of his regiment and a platoon of the machine-gun company, forming them into a combatant force.

MILNER, JACK W. (97365).............
Near Landres-et-St. Georges, France,
Oct. 15, 1918.
R—Alexander City, Ala.
B—Alexander City, Ala.
G. O. No. 131, W. D., 1918.

Sergeant, Company H, 167th Infantry, 42d Division.
After his company had sustained heavy losses in a severe engagement with the enemy, and he himself had been seriously wounded, he, realizing that he was the only sergeant left in the company, refused to be evacuated and remained on duty for 12 hours, reorganizing his company under heavy enemy artillery and machine-gun fire, thereby showing entire disregard for danger and setting an excellent example of courage and heroism under fire to his men.

MINALGA, FRANK (41138)..............
Near Medeah Ferme, France, Oct. 8, 1918.
R—New Haven, Conn.
B—Russia.
G. O. No. 21, W. D., 1919.

Private, Company E, 9th Infantry, 2d Division.
When his company was held up by an enemy machine-gun nest, Private Minalga advanced on the nest from the flank and captured it singlehanded.

MINARDI, GUISIPPE (545342)...........
Near Crezancy, France, July 15, 1918.
R—South Amboy, N. J.
B—Italy.
G. O. No. 32, W. D., 1919.

Private, Company A, 30th Infantry, 3d Division.
During the engagement he set an example to the other members of his company by his gallant conduct. After the company was ordered to withdraw, he voluntarily returned to the position his company had held and throughout the night assisted in evacuating the wounded.

°MINCEY, GEORGE A...................
At Chateau-Thierry, France, June 6, 1918.
R—Ogeechee, Ga.
B—Ogeechee, Ga.
G. O. No. 110, W. D., 1918.

Corporal, 55th Company, 5th Regiment, U. S. Marine Corps, 2d Division.
Killed in action at Chateau-Thierry, France, June 6, 1918, he gave the supreme proof of that extraordinary heroism which will serve as an example to hitherto untried troops.
Posthumously awarded. Medal presented to father, George Mincey.

MINER, ASHER.......................
At Apremont, France, Oct. 4, 1918.
R—Wilkes-Barre, Pa.
B—Wilkes-Barre, Pa.
G. O. No. 140, W. D., 1918.
Distinguished-service medal also awarded.

Colonel, 109th Field Artillery, 28th Division.
One of the batteries of the regiment commanded by this officer assigned to an advanced position in direct support of an infantry attack was heavily shelled by the enemy while it was going into action. It being necessary therefore to take another position, he went forward under heavy shellfire and personally supervised the placing of the guns in the new position. He continued his efforts until he received a severe wound that later necessitated the amputation of his leg.

MINER, DONALD.....................
At Ormont Farm, France, Oct. 10, 1918.
R—Jersey City, N. J.
B—New York, N. Y.
G. O. No. 15, W. D., 1919.

Major, Medical Corps, attached to 115th Infantry, 29th Division.
He voluntarily proceeded under heavy shellfire to an advanced aid station. For 4 hours he worked unceasingly caring for the wounded and evacuating them. Finding that he could work more effectively without his gas mask, he discarded it so that it would not hinder him in attending wounded men.

²MINGLE, CLAUDE L. (1306628)........
Near Bellicourt, France, Sept 29, 1928.
R—Knoxville, Tenn.
B—Blount County, Tenn.
G. O. No. 37, W. D., 1919.

Private, Machine Gun Company, 117th Infantry, 30th Division.
When enemy machine guns suddenly opened fire on both flanks of his platoon, he bravely refused to take cover, but delivered effective rifle fire on the enemy, putting out of action 1 of the machine guns before he was mortally wounded.
Posthumously awarded. Medal presented to father, Jake Mingle.

MINNIGERODE, FITZHUGH L..........
Near Verdun, France, Oct. 23-24, 1918.
R—Washington, D. C.
B—Oatlands, Va.
G. O. No. 44, W. D., 1919.
Distinguished-service medal also awarded.

Lieutenant colonel, 114th Infantry, 29th Division.
When his battalion commanders, who had gone forward on a reconnaissance preparatory to an attack, were prevented from returning by heavy shell and machine-gun fire, Lieutenant Colonel Minnigerode personally led his regiment into position under cover. With a soldier he then went forward for a distance of 2 kilometers under artillery and machine-gun fire, found the battalion commanders, and guided them back to their comrades.

MINNIS, JOHN A...................
Near Mezy, France, July 15, 1918.
R—Montgomery, Ala.
B—Montgomery, Ala.
G. O. No. 22, W. D., 1920.

Captain, U. S. Marine Corps, attached to 38th Infantry, 3d Division.
During the enemy drive he fearlessly reorganized a unit that had lost its officers and held his position against the enemy assault. A short time later he gallantly led 15 men in a counterattack under heavy enemy machine-gun fire. He repulsed the enemy and captured 24 prisoners.

³MINTER, PAUL B. (113934).........
Near Sergy, France, July 26, 1918.
R—Monticello, Ga.
B—Monticello, Ga.
G. O. No. 99, W. D., 1918.

Sergeant, Company B, 151st Machine Gun Battalion, 42d Division.
He led his section forward, secured his objective, supervised the consolidation of his position, laid his guns personally with calmness, accuracy, and decision, and continued to direct the operations of his men, all in the face of severe enemy machine-gun and shell fire, until he was killed.
Posthumously awarded. Medal presented to father, O. J. Minter.

°MINTON, CHARLES ARMAND.........
In the Bois-de-Naza, near Binarville, France, Oct. 5, 1918.
R—New York, N. Y.
B—Flushing, N. Y.
G. O. No. 43, W. D., 1922.

First lieutenant, 305th Infantry, 77th Division.
Although suffering with a mortal illness, Lieutenant Minton retained command of his company, and when his line had been temporarily beaten back he personally returned across an area swept by hostile machine-gun fire and rescued a wounded soldier who had fallen within 25 yards of an enemy machine gun.
Posthumously awarded. Medal presented to father, J. McKim Minton.

MINTZ, FORNEY B. (1707533).......
Near Binarville, France, Sept. 28, 1918.
R—Mill Branch, N. C.
B—Mill Branch, N. C.
G. O. No. 35, W. D., 1919.

Sergeant, Company A, 308th Infantry, 77th Division.
Sergeant Mintz, in command of a platoon, worked his way through the enemy rear guard and captured 5 machine guns and an ammunition-carrying party. Although badly wounded when an organized position of the enemy was encountered, he made his way back to request reinforcements and brought with him two German prisoners, from whom valuable information was obtained.

*MITCHELL, ARTHUR (544911)_____
Near Mezy, France, July 15, 1918.
R—Bessemer, Ala.
B—Rockford, Ala.
G. O. No. 32, W. D., 1919.

Sergeant, Machine Gun Company, 30th Infantry, 3d Division.
After his gun had been destroyed by shell fire Sergeant Mitchell led his men through the enemy lines to our own. He braved the extreme shelling and machine-gun fire by going to the aid of the wounded, and while withdrawing covered the retreat, effectively holding off the enemy. He was killed in action shortly afterwards.
Posthumously awarded. Medal presented to widow, Mrs. A. L. Mitchell.

MITCHELL, CLARENCE_____
In the Bois-Hazois, France, Nov. 1, 1918.
R—Cloverport, Ky.
B—Cloverport, Ky.
G. O. No. 98, W. D., 1919

Captain, 23d Infantry, 2d Division.
Though Captain Mitchell was wounded in the leg by shell fragments just before the opening of the attack, he continued to lead his company, advancing 3 kilometers, to the first objective with the aid of a cane and assisted by runners. His conspicuous courage and fortitude inspired his men to a successful assault against a strongly entrenched position desperately defended by the enemy.

MITCHELL, EDWARD J_____
During the Meuse-Argonne offensive, France, Sept. 26, 1918.
R—Prescott, Ariz.
B—Rochester, N. Y.
G. O. No. 37, W. D., 1919.

Captain, 363d Infantry, 91st Division.
Leading a platoon in advance of other troops, he encountered and captured 3 German 155's, which were in operation, also taking 6 officers and about 425 men. During the night he organized troops from his own and other divisions and established a formidable piece of front line.

MITCHELL, GEORGE R. (42878)_____
Near Soissons, France, July 22, 1918.
R—Holdrege, Nebr.
B—York, Pa.
G. O. No. 15, W. D., 1919.

Corporal, Company F, 16th Infantry, 1st Division.
Although wounded, he promptly took command of his company after all of its officers had been killed and courageously and successfully led it forward in the advance.

MITCHELL, JOHN_____
Near Beaumont, France, May 27, 1918.
R—Miami, Fla.
B—Cincinnati, Ohio.
G. O. No. 37, W. D., 1919.

Captain, 95th Aero Squadron, Air Service.
Seeing 3 enemy planes flying east over Apremont at 2,500 meters, he unhesitatingly attacked the 3 machines, which were in close formation, despite the fact that a fourth, hovering above, threatened to close in and join the enemy formation. He succeeded in shooting down the enemy machine, which proved to be a biplane returning from an important mission.

MITCHELL, JOHN A_____
Near Bellicourt, France, Sept. 29, 1918.
R—Livingston, Tenn.
B—Livingston, Tenn.
G. O. No. 44, W. D., 1919.

First lieutenant, 119th Infantry, 30th Division.
Hearing cries of distress from a disabled tank, he, assisted by a soldier, advanced in the face of terrific machine-gun and shell fire to that point. Notwithstanding the fact that the tank was subjected to point-blank fire of artillery, he succeeded in rescuing the badly wounded tank commander and removing him to a place of safety.

MITCHELL, JOHN B. (2383907)_____
Near St. Mihiel, France, Sept. 16, 1918.
R—Gary, Ind.
B—Natick, Mass.
G. O. No. 37, W. D., 1919.

Private, Company G, 60th Infantry, 5th Division.
Although severely wounded while carrying a message from his battalion headquarters through an extremely heavy machine-gun and artillery fire, Private Mitchell persevered and successfully delivered the message.

'MITCHELL, JOHN E_____
Near St. Etienne-a-Arnes, France, Oct. 3, 1918.
R—Cedarhurst, Long Island, N. Y.
B—Cedarhurst, Long Island, N. Y.
G. O. No. 20, W. D., 1919.

Second lieutenant, 23d Infantry, 2d Division.
Displaying utter disregard for his personal safety, he led his platoon through an extremely heavy machine-gun and artillery barrage and destroyed several enemy machine-gun nests. He was later killed while making a reconnaissance.
Posthumously awarded. Medal presented to mother, Mrs. John Mitchell.

MITCHELL, MANTON C_____
Near St. Thibault, France, Aug. 5, 1918.
R—Providence, R. I.
B—Providence, R. I.
G. O. No. 60, W. D., 1920.

Major, 39th Infantry, 4th Division.
The attack battalion having been held up by heavy machine-gun fire while attempting to cross the Vesle River, Major Mitchell, who was in command of the support battalion, went forward through heavy machine-gun fire and encouraged and assisted the advanced troops to cross the river. He was severely wounded in the leg while directing these movements, but he refused to be evacuated and continued in the attack, remaining with the attack battalion until the evening of Aug. 5.

MITCHELL, WILLIAM_____
At Noyon, France, Mar. 26, 1918; near the Marne River, France, during July, 1918; and in the St. Mihiel salient, France, Sept. 12–16, 1918.
R—Milwaukee, Wis.
B—France.
G. O. No. 120, W. D., 1918.
Distinguished-service medal also awarded.

Brigadier general, Chief of Air Service, 1st Army.
For displaying bravery far beyond that required by his position as Chief of Air Service, 1st Army, American Expeditionary Forces, setting a personal example to the United States aviation by piloting his airplane over the battle lines since the entry of the United States into the war, some instances being a flight in a monoplane over the battle of Noyon on Mar. 26, 1918, and the back areas, seeing and reporting upon the action of both air and ground troops, which led to a change in our aviation's tactical methods; a flight in a mono plane over the bridges which the Germans had laid across the Marne during July, 1918, which led to the first definite reports of the location of these bridges and the subsequent attack upon the German troops by our air forces; daily reconnaissances over the lines during the battle of St. Mihiel salient, Sept. 12 to 16, securing valuable information of the enemy troops in the air and on the ground, which led to the excellent combined action by the allied air services and ground troops particularly in this battle.

MIX, RALPH B. (1747423)_____
At Grand Pre, France, Oct. 17, 1918.
R—Hudson Falls, N. Y.
B—Marinette, Wis.
G. O. No. 35, W. D., 1919.

Private, Company F, 312th Infantry, 78th Division.
While carrying a message through heavy shell and machine-gun fire to an advanced platoon he found 2 wounded men, whose injuries he dressed. Upon returning from his mission he asked for and received permission to take food to the wounded men. He subsequently made 2 trips through intense fire, carrying them back to shelter.

MOAN, RALPH T. (69071).
Near Riaville, France, Sept. 26, 1918.
R—East Machias, Me.
B—East Machias, Me.
G. O. No. 21, W. D., 1919.

Mechanic, Company K, 103d Infantry, 26th Division.
Mechanic Moan, who was detailed as a runner, made several trips carrying important messages across terrain swept by constant fire from machine guns, snipers, trench mortars, and artillery. His disregard for personal safety and devotion to duty in the prompt delivery of messages contributed greatly to the success of the action.

MOBLEY, CHARLES R. (1315646).
Near Ypres, Belgium, Aug. 25, 1918.
R—Williamston, N. C.
B—Martin County, N. C.
G. O. No. 32, W. D., 1919.

Sergeant, Company F, 119th Infantry, 30th Division.
At imminent peril to his own life Sergeant Mobley and two companions extinguished a fire in an ammunition dump caused by a bursting shell, thereby preventing the explosion of the dump and saving the lives of a large number of men who were in the vicinity.

*MOBLEY, LOTUS N. (240329).
Near Trugny, France, July 23, 1918.
R—Cedar Rapids, Iowa.
B—Hartwell, Ind.
G. O. No. 74, W. D., 1919.

Sergeant, Company L, 102d Infantry, 26th Division.
Sergeant Mobley displayed exceptional courage in dashing into field under heavy shell and machine-gun fire and carrying to safety a wounded man. Posthumously awarded. Medal presented to friend, Mrs. R. A. Page.

MODROW, PERRY F. (1379458).
Near Romagne, France, Nov. 1, 1918.
R—East St. Louis, Ill.
B—East St. Louis, Ill.
G. O. No. 37, W. D., 1919.

Private, medical detachment, 124th Field Artillery, 33d Division.
He was wounded while serving at the battalion aid station under heavy shell-fire, but he insisted on continuing at work until compelled to go to the rear against his will.

MOEHLER, FRANK W. (152347).
Near Suippes, France, July 14-15, 1918.
R—Newark, N. J.
B—Chicago, Ill.
G. O. No. 142, W. D., 1918.

Sergeant, Battery E, 42d Artillery, Coast Artillery Corps.
Having been severely wounded early in the German attack on Suippes, he continued to direct the firing of his gun crew for 8 hours under intense shell fire, remaining on duty until all the ammunition had been expended and orders to withdraw had been received.

MOHRMAN, WILLIAM (1703888).
Near St. Pierremont, France, Nov. 4, 1918.
R—New York, N. Y.
B—Brooklyn, N. Y.
G. O. No. 37, W. D., 1919.

Sergeant, Headquarters Company, 307th Infantry, 77th Division.
After passing through a heavily bombarded area, he learned that a soldier of his platoon had been wounded and had fallen in the shelled area. He at once volunteered and went back for him, assisted in bringing him to a place of safety, and later helped to carry him through another shelled area to the first-aid station.

MOLIK, JOSEPH (41870).
Near Soissons, France, July 18, 1918.
R—Buffalo, N. Y.
B—Buffalo, N. Y.
G. O. No. 35, W. D., 1920.

Sergeant, Company B, 16th Infantry, 1st Division.
Although severely wounded in an attack on an enemy machine-gun position Sergeant Molik continued to direct his platoon in the advance until late in the afternoon, when he was ordered to be evacuated for his wounds.

MOLLER, WILLIAM G.
At Riaville, France, Sept. 26, 1918.
R—Champaign, Ill.
B—Dewey, Ill.
G. O. No. 133, W. D., 1918.

Second Lieutenant, 102d Infantry, 26th Division.
He displayed remarkable courage and judgment by organizing a platoon of men who had become detached from their various units. With this detachment he wiped out a machine-gun nest, opening the way for further advance into the town of Riaville. He maintained his position in the front line throughout the action, although subjected to heavy fire from all arms.

MOLLOY, JOSEPH A.
Near Vierzy, France, July 19, 1918.
R—Lowell, Mass.
B—Charlestown, Mass.
G. O. No. 46, W. D., 1919.

First Lieutenant, 23d Infantry, 2d Division.
Lieutenant Molloy was leading his platoon through a heavily gassed area when a large shell struck in the middle of his column, killing 15 men and wounding a like number. He himself was badly shocked, but immediately set to work administering to the wounded despite the darkness, terrific shelling, and the necessity of wearing a gas mask. He aided practically all of the wounded, single-handed, and secured their transportation to the rear.

MOLSBERRY, HOWARD C.
In the vicinity of le Thiolet, France, June 6-7, 1918.
R—Ambridge, Pa.
B—Plymouth, Iowa.
G. O. No. 99, W. D., 1918.

First lieutenant, 2d Engineers, 2d Division.
He courageously took command of and efficiently directed the advance of an infantry unit when all its officers had been killed or wounded.

MOLTER, HENRY C. (244901).
Near Montfaucon, France, Sept. 28, 1918.
R—Pittsburgh, Pa.
B—Brooklyn, N. Y.
G. O. No. 81, W. D., 1919.

Sergeant, first class, Company D, 1st Gas Regiment.
Sergeant Molter volunteered and led a detachment to recover ammunition from a dump which was under fire and liable to explode at any minute. Working under a heavy gas attack, he succeeded in removing the dump to a place of safety.

MONAHAN, EDWARD V. (1782513).
Near Moirey, France, Nov. 10, 1918.
R—Centralia, Pa.
B—Centralia, Pa.
G. O. No. 37, W. D., 1919.

Sergeant, Company E, 314th Infantry, 79th Division.
Wounded in the face by a machine-gun bullet he refused to be evacuated, but continued to lead his platoon in a successful attack with the bullet still in his flesh. After securing first-aid treatment next morning he learned that his company was advancing again, whereupon he returned and led his platoon in an assault on the Cote Romagne.

MONAHAN, FRANCIS J. (2004824).
Near Samogneux, France, Oct. 12, 1918.
R—Indianapolis, Ind.
B—Indianapolis, Ind.
G. O. No. 46, W. D., 1919.

Private, Company M, 116th Infantry, 29th Division.
He left shelter and exposed himself to direct enemy machine-gun fire to aid a wounded man, and while so engaged was himself seriously wounded.

*MONAHAN, PETER T. (1272997)_____
Near Verdun, France, Oct. 11, 1918.
R—Jersey City, N. J.
B—Jersey City, N. J.
G. O. No. 32, W. D., 1919.

Private, first class, Company D, 111th Machine Gun Battalion, 29th Division.
During an intense bombardment he volunteered to leave cover and assist in carrying a litter supporting a wounded officer. He was himself killed while engaged on this self-sacrificing mission.
Posthumously awarded. Medal presented to father, Philip Monahan.

MONGEAU, HENRY J. (58464)_____
At Seicheprey, France, Mar. 28, 1918.
R—Cherry Valley, Mass.
B—Canada.
G. O. No. 3, W. D., 1924.

Corporal, Company L, 28th Infantry, 1st Division.
This soldier was a member of a patrol consisting of an officer and 4 men, who, with great daring, entered a dangerous portion of the enemy trenches, where they surrounded a party nearly double their own strength, drove off an enemy rescuing party, and made their way back to our lines with 4 prisoners, from whom valuable information was taken.

*MONK, FRANCIS L. (3138521)_____
Near Grand Pre, France, Oct. 15, 1918.
R—Benson, Utah.
B—Murray, Utah.
G. O. No. 145, W. D., 1918.

Private, Company A, 307th Infantry, 77th Division.
He crawled out into an open field where another soldier lay severely wounded, under fire from machine guns and snipers, and dragged him to the shelter of a wall, where he dressed his wounds. In so doing he was himself wounded.
Posthumously awarded. Medal presented to half brother, Chris W. Anderson.

MONROE, CHARLIE T. (2463911)_____
At Mont de Sanges, France, Sept. 24, 1918.
R—Meyersville, Va.
B—Meyersville, Va.
G. O. No. 46, W. D., 1919.

Private, Headquarters Company, 370th Infantry, 93d Division.
Private Monroe, in the absence of his platoon commander, took charge of a platoon of Stokes mortars, directing the work of the men under heavy shell fire. Although the shelling was so intense that guns were at times buried, Private Monroe and his men worked unceasingly in placing them back into action. He himself was buried by the explosion of a shell, but on being dug out continued to direct the work of the men and encourage them by his fearless example.

*MONROE, DAVID E_____
South of Soissons, France, July 18, 1918.
R—Marion, S. C.
B—Marion, S. C.
G. O. No. 35, W. D., 1920.

Second lieutenant, 16th Infantry, 1st Division.
His platoon having been halted by machine-gun fire, he advanced alone against the nest and captured the gun and crew. Although wounded in this encounter, he returned to his platoon and led them on to its objective. His gallant conduct had a marked effect upon his men.
Posthumously awarded. Medal presented to father, T. J. Monroe.

MONROE, EDWARD M_____
At Naguilian, Luzon, Philippine Islands, Dec. 7, 1899.
R—Philadelphia, Pa.
B—Philadelphia, Pa.
G. O. No. 3, W. D., 1925.

Private, Company A, 24th Infantry, U. S. Army.
When the command of which he was a member was held up in the crossing of the Rio Grande de Cagayan by rifle fire from a well-intrenched enemy, and being without boats or rafts with which to cross, Private Monroe with 5 other members of his company volunteered to swim the river. Displaying great gallantry and with utter disregard for his life, he swam the river in the face of heavy rifle fire, returned on a raft, secured arms and ammunition, crossed a second time, and took part in an attack which drove a superior force of the enemy from their trenches and the town occupied by them, thereby making possible the further advance of his company.

MONSON, JOHN J. (1707736)_____
Near Binarville, France, Sept. 29, 1918.
R—New York, N. Y.
B—New York, N. Y.
G. O. No. 13, W. D., 1919.

Private, first class, Company A, 308th Infantry, 77th Division.
In order to obtain ammunition and rations, Private Monson, with another soldier, accompanied an officer in an effort to reestablish communication between battalion and regimental headquarters. They were attacked by a small party of Germans, but drove them off, killing 1. When night came, they crawled unknowingly into the center of a German camp, where they lay for 3 hours, undetected. Finally discovered, they made a dash to escape. In order to protect the officer, Private Monson deliberately drew the enemy fire to himself, allowing the officer to escape. Private Monson succeeded in getting through and delivering his message the next morning.

MONTAG, BERNARD WILLIAM_____
Near Vierzy, France, July 19, 1918.
R—Toledo, Ohio.
B—Oshkosh, Wis.
G. O. No. 117, W. D., 1919.

Corporal, 55th Company, 5th Regiment, U. S. Marine Corps, 2d Division.
Corporal Montag and 3 comrades, Privates McIntyre, Messinger, and Wood, captured a machine gun which was holding up the company of marines, killing the entire crew. To accomplish this hazardous and daring work it was necessary for them to expose themselves to the fire of this gun. Even though Corporal Montag and Privates McIntyre and Messinger were wounded during the advance, the party continued forward and succeeded.

MONTAGUE, ROBERT L_____
Near Landreville, France, Nov. 1, 1918.
R—Richmond, Va.
B—Danville, Va.
G. O. No. 49, W. D., 1922.

First lieutenant, 5th Regiment, U. S. Marine Corps, 2d Division.
When the advance of his company was held up, Lieutenant Montague voluntarily led a group of men in a flanking movement against a withering machine-gun fire, and under a heavy artillery bombardment entered and took the town of Landreville, capturing about 150 prisoners.

MONTEE, JESSE A_____
Near Cunel, France, Oct. 11–Nov. 11, 1918.
R—Superior, Wis.
B—Rock Valley, Iowa.
G. O. No. 98, W. D., 1919.

Second lieutenant, 61st Infantry, 5th Division.
Lieutenant Montee displayed high qualities of leadership, repeatedly reconnoitering advanced positions under terrific artillery and machine-gun fire, and leading patrols into enemy territory, thereby facilitating the advance of his battalion. On one occasion, when the battalion had been nearly surrounded by hostile machine guns, he went to the rear and successfully brought up reinforcements. Later, while acting as battalion adjutant, he accompanied the assaulting waves in all attacks, and by his utter disregard for danger assisted in maintaining order among the attacking troops and establishing lines.

*MONTGOMERY, CHARLES G. (1490849)__
Near Attigny, France, Oct. 27, 1918.
R—Goodlett, Tex.
B—Texas.
G. O. No. 27, W. D., 1919.

Private, Company I, 142d Infantry, 36th Division.
Private Montgomery volunteered and carried a message from battalion headquarters through the enemy's fire to our support line and guided a combat group into position, in absolute disregard of his personal safety. He was caught in a heavy barrage and was killed.
Posthumously awarded. Medal presented to father, H. I. Montgomery.

MONULA, NICK (2337984) .... | Private, Company A, 4th Infantry, 3d Division.
Near Grand Ballois Farm, France, July 14–15, 1918. | During a heavy gas and shell bombardment he repeatedly volunteered and delivered messages over routes other than his own when the runners assigned to those routes had been killed or wounded.
R—Pittsburgh, Pa.
B—Serbia.
G. O. No. 32, W. D., 1919.

°MOOD, JULIUS A .... | Captain, 26th Infantry, 1st Division.
Near Soissons, France, July 19–21, 1918. | During the fighting of July 19–21, 1918, near Soissons, France, he voluntarily exposed himself to fire repeatedly in order to get information and direct operations, and was killed while leading a battalion to the attack.
R—Summerton, S. C. | Posthumously awarded. Medal presented to mother, Mrs. W. R. Mood.
B—Ridgeway, S. C.
G. O. No. 132, W. D., 1918.

°MOODY, ROLF .... | Captain, 117th Infantry, 30th Division.
Near Beaurevoir, France, Oct. 7, 1918. | Captain Moody was in command of his company on the left flank of the assaulting battalion, when withering machine-gun fire from an old factory building held up the advance of the entire left flank. Realizing the gravity of the situation, he took 2 squads and led them in an attack on the machine-gun positions over ground swept by machine-gun and shell fire. After a personal encounter, in which he used his pistol and hand grenades, the machine guns were silenced. From there he started toward another machine-gun post, but was mortally wounded before reaching it, dying on the field.
R—Knoxville, Tenn. | Posthumously awarded. Medal presented to father, Henry G. Moody.
B—Joppa, Tenn.
G. O. No. 87, W. D., 1919.

MOONEY, ROBERT A. (3212609) .... | Private, Company F, 322d Infantry, 81st Division.
Near Grimaucourt, France, Nov. 10, 1918. | He voluntarily returned through heavy artillery fire to a position formerly held by his company and rescued a wounded man.
R—Rockford, Ala.
B—Rockford, Ala.
G. O. No. 32, W. D., 1919.

°MOORE, CHARLES J .... | Captain, 7th Engineers, 5th Division.
On the Andon River, France, Oct. 14, 1918. | Captain Moore went forward under heavy artillery fire, reorganized his men, who were in scattered units, after they had laid bridges across the Andon River. After being severely wounded, he gave instructions to his subordinates for carrying on the work.
R—Lampasas, Tex. | Posthumously awarded. Medal presented to widow, Mrs. Charles J. Moore.
B—Lampasas, Tex.
G. O. No. 20, W., D. 1919.

MOORE, CLAYTON H. (1450720) .... | Corporal, Headquarters Company, 138th Infantry, 35th Division.
Attack on Hilsenfirst, France, July 6, 1918. | During the attack on Hilsenfirst, France, July 6, 1918, while carrying a wounded soldier through machine-gun fire to shelter he was wounded, but by unusual pluck he brought his comrade to safety, and, realizing the scarcity of stretchers, insisted on others being carried to the rear and himself walking.
R—St. Louis, Mo.
B—Denver, Colo.
G. O. No. 99, W. D., 1918.

°MOORE, DAVID M. (2176366) .... | Supply sergeant, Company F, 353d Infantry, 89th Division.
In the Mort-Mare Woods, France, Sept. 12, 1918. | During the advance through the woods, Sergeant Moore fearlessly exposed himself while directing his men in the capture of machine-gun nests. In one instance he alone captured a machine gun with a well-thrown grenade and then, putting this gun into action, inflicted further losses on the fleeing enemy. His utter disregard of personal danger and his high qualities of leadership were an inspiration to his men and contributed greatly to the success of his platoon in this operation.
R—Sedan, Kans. | Posthumously awarded. Medal presented to widow, Mrs. Ruth Moore.
B—Sedan, Kans.
G. O. No. 31, W. D., 1922.

MOORE, EDWARD RUSSELL .... | First lieutenant, pilot, 8th Aero Squadron, Air Service.
Near Thiaucourt, France, Oct. 9, 1918. | He, with First Lieut. Gardner Philip Allen, observer, took advantage of a short period of fair weather during generally unfavorable atmospheric conditions to undertake a photographic mission behind the German lines. Accompanied by two protecting planes, they had just commenced their mission when they were attacked by 8 enemy planes, which followed them throughout their course, firing at the photographing plane. Lieutenant Moore, pilot, with both flying wires cut by bullets, a landing wire shot away, his elevators riddled with bullets, and both wings punctured, continued on the prescribed course, although it made him an easy target. Lieutenant Allen was thus enabled in the midst of the attack to take pictures of the exact territory assigned, and he made no attempt to protect the plane with his machine guns. Displaying entire disregard for their personal danger and steadfast devotion to duty, the 2 officers successfully accomplished their mission.
R—Columbia, Mo.
B—Audrain County, Mo.
G. O. No. 145, W. D., 1918.

MOORE, ELGIN J. (2193755) .... | Sergeant, first class, Company C, 314th Field Signal Battalion, 89th Division.
Near Beauclair, France, Nov. 4–11, 1918. | From the 4th to the 11th of November, while continually under heavy shell fire, Sergeant Moore laid and maintained lines of communication within his area with utter disregard for his personal safety.
R—Winslow, Ariz.
B—Las Lunas, N. Mex.
G. O. No. 37, W. D., 1919.

MOORE, FRED F .... | Captain, 355th Infantry, 89th Division.
North of Flirey, France, Sept. 12, 1918. | Wounded in the left shoulder early in the morning while in command of his company, he continued to lead and handle it during the entire day in an efficient and gallant manner under fire. He refused to take time to have his wound attended to until late that night and after his command had intrenched under fire and was safe.
R—Stewart, Minn.
B—Stewart, Minn.
G. O. No. 128, W. D., 1918.

°MOORE, FREDERICK P., Jr .... | Captain, 30th Infantry, 3d Division.
Near Crezancy, France, July 15, 1918. | During an intense bombardment he left shelter and exposed himself constantly in a wood swept by shell fire while encouraging and directing the movement of his company. He was killed by shell fire while on a personal reconnaissance.
R—Bellevue, Pa. | Posthumously awarded. Medal presented to widow, Mrs. Frederick P. Moore, jr.
B—Rochester, N. Y.
G. O. No. 32, W. D., 1919.

MOORE, HAROLD C. (2262347)_____
  Near Mount Des Ailleux, France,
    Sept. 26, 1918.
  R—Upland, Calif.
  B—Morrill, Kans.
  G. O. No. 37, W. D., 1919.

Sergeant, Company C, 348th Machine Gun Battalion, 91st Division.
Although wounded by the same shell which mortally wounded his platoon commander, he went to his company for assistance, returned through a heavy shell fire, and helped to carry his commander to a place of safety. He did not report his own wound, or receive medical attention until the officer had been cared for.

MOORE, JAMES D. (1244663)_____
  At Fismette, France, Aug. 12, 1918.
  R—Rochester, Pa.
  B—Warren, Pa.
  G. O. No. 100, W. D., 1918.

Corporal, Company G, 111th Infantry, 28th Division.
With an automatic rifle team, he occupied a house in an advanced position west of Fismette on the night of Aug. 12, the loss of which would have jeopardized his company's position and hindered the military operations then taking place. The enemy shot a flare into the house, setting fire to it, but Corporal Moore and a companion, under machine-gun and sniper fire in a brilliantly lighted room, extinguished the flames.

MOORE, JAMES EDWARD_____
  South of Cunel, France, Oct. 9, 1918.
  R—Kenova, W. Va.
  B—Norfolk, Va.
  G. O. No. 22, W. D., 1920.

Second lieutenant, 38th Infantry, 3d Division.
Although severely wounded in the head by a machine-gun bullet, Second Lieutenant Moore continued in command of his platoon and by his courageous conduct repulsed a strong enemy counterattack against the line held.

MOORE, JAMES H., Jr. (3137555)_____
  In the Argonne Forest, France, Oct.
    2, 1918.
  R—Ridgway, Mont.
  B—Spanish Peaks, Colo.
  G. O. No. 64, W. D., 1919.

Corporal, Company E, 307th Infantry, 77th Division.
During an attack, when his platoon encountered enemy wire, Corporal Moore calmly went forward and alone proceeded to cut a passage through the wire. While performing this work, he was subjected to the fiercest fire of enemy machine guns and grenades, which wounded over half the platoon. He continued in this work until he accomplished his purpose.

MOORE, JOHN CARROLL_____
  Near Montfaucon, France, Sept. 27,
    1918.
  R—Baltimore, Md.
  B—Baltimore, Md.
  G. O. No. 130, W. D., 1919.

First lieutenant, chaplain, 313th Infantry, 79th Division.
Though wounded on Sept. 26, 1918, he remained with the attacking lines of his regiment, ministering to the dying and aiding the wounded. After entering an enemy trench with a group of men, a grenade was thrown in their midst, and, in utter disregard of personal safety, he grabbed the grenade to throw it from the trench. It exploded just after leaving his hand, seriously wounding him in several places.

MOORE, JOHN D. (8599)_____
  Near Somme-Py, France, Oct. 2–9,
    1918.
  R—Haddonfield, N. J.
  B—Haddonfield, N. J.
  G. O. No. 37, W. D., 1919.

Private, first class, section No. 554, Ambulance Service.
During this period Private Moore evacuated the wounded in an advance post under shell and sniper fire. On the nights of Oct. 6 and 7 he drove an ambulance to points beyond the advanced posts to carry in the wounded under intense shell and machine-gun fire. He also assisted the litter bearers in exposed positions in carrying the wounded from the lines to the dressing station.

*MOORE, JOHN H_____
  Near Cutry, France, July 18, 1918.
  R—De Kalb, Tex.
  B—De Kalb, Tex.
  G. O. No. 126, W. D., 1918.

Second lieutenant, 3d Machine Gun Battalion, 1st Division.
While courageously leading his section in the face of intense fire near Cutry, France, on July 18, 1918, he was knocked down by a shell explosion, but continued his leadership as soon as he regained consciousness and personally reconnoitered the area in advance to find a less dangerous route. He succeeded and thereby made it possible for his men to go forward, but he himself was killed in the undertaking.
Posthumously awarded. Medal presented to mother, Mrs. S. L. Moore.

MOORE, RAYMOND N. (1391910)_____
  Near Consenvoye, France, Oct. 10,
    1918.
  R—Canton, Ill.
  B—Canton, Ill.
  G. O. No. 37, W. D., 1919.

Sergeant, Company B, 124th Machine Gun Battalion, 31st Division.
He led his section of 2 guns to the aid of an Infantry company. Failing in his attempt to establish an advantageous position, he alone took his gun 100 yards in advance of the line, exposed to violent machine-gun and artillery fire, and, setting it up in an open field, silenced the fire of enemy machine-gun snipers who had been inflicting heavy losses on our troops.

*MOORE, RICHARD W. (1248409)_____
  At Fismette, France, Aug. 27, 1918.
  R—Ridgway, Pa.
  B—Warren, Pa.
  G. O. No. 9, W. D., 1923.

Sergeant, Company H, 112th Infantry, 28th Division.
As 1 of a group which was attacked by an overwhelming force of enemy, Sergeant Moore assisted in the defense of a crossing over the Vesle River and made possible the retirement of about 20 of his comrades. He continued against great odds to hold his position and inflicted heavy casualties upon the enemy, finally retiring under orders, fighting his way through heavy machine-gun, rifle, and artillery fire. His exceptional disregard for personal safety, resourcefulness, and bravery were an inspiration to all his comrades. Sergeant Moore was severely gassed in this fight and died shortly thereafter.
Posthumously awarded. Medal presented to father, William T. Moore.

MOORE, WALLIS J_____
  Near St. Etienne, France, Oct. 8, 1918.
  R—Austin, Tex.
  B—San Marcus, Tex.
  G. O. No. 126, W. D., 1919.

Captain, 132d Machine Gun Battalion, 36th Division.
Captain Moore, although wounded by shrapnel, refused to go to the rear, and proceeded to reorganize portions of 3 Infantry platoons whose officers had become casualties, thus protecting the right flank of the 141st Infantry. He went to the rear only after he had been severely gassed.

MOORE, WALTER (1850005)_____
  Near Brancourt, France, Oct. 8, and
    near Vaux-Andigny, France, Oct.
    9, 1918.
  R—Alexandria, Va.
  B—Prince William County, Va.
  G. O. No. 59, W. D., 1919.

Private, Company I, 118th Infantry, 30th Division.
During the action of his company at Brancourt, he went out alone and attacked an outpost containing 1 officer and 8 men, capturing the entire party and turning them over as prisoners. On the following day he advanced alone 50 yards in front of his company to attack an enemy sniper who was placing an effective fire on our lines.

MOORE, WILLIAM B_____
  At Bouresches, France, June 6, 1918.
  R—New York, N. Y.
  B—Waco, Tex.
  G. O. No. 119, W. D., 1918.

Second lieutenant, 6th Regiment, U. S. Marine Corps, 2d Division.
On June 6, 1918, he volunteered and took a truck load of ammunition and material into Bouresches, France, over a road swept by artillery and machine-gun fire, thereby relieving a critical situation.

MOORE, WILLIAM E. (1517995)_____
  Near Hearne, Belgium, Nov. 4, 1918.
  R—New Boston, Ohio.
  B—Greenup County, Ky.
  G. O. No. 59, W. D., 1919.

Private, Company D, 145th Infantry, 37th Division.
He displayed exceptional personal bravery when, with 1 other soldier, he went to the assistance of a comrade who had been attacked and wounded by a patrol of 8 Germans, rescuing the wounded man and putting the enemy patrol to flight.

MOOREFIELD, DICK (099648)_____
  Near Jaulny, France, Nov. 4, 1918.
  R—Hopkinsville, Ky.
  B—Hopkinsville, Ky.
  G. O. No. 35, W. D., 1919.

Sergeant, Company B, 55th Infantry, 7th Division.
While leading a patrol in front of our lines, Sergeant Moorefield and his patrol came under machine-gun and rifle fire, and 1 of his men was severely wounded. He crawled forward with his patrol until within 20 paces of an enemy gun, when, upon raising his head slightly, he saw the German who was feeding the ammunition to the gun. He shot the man through the head with his rifle and wounded another. When the German gun nearest him jammed he captured it and another near by, together with 3 prisoners. He jumped into the emplacement, cleared the jam in the gun, and turned it on the enemy guns on the right, silencing them. He then sent his patrol and the prisoners back into our lines, covering their retreat with 1 of the captured guns.

MOORELAND, THOMAS ARCHIE (1312068)_
  Near St. Martin-Riviere, France, Oct. 17, 1918.
  R—Concord, N. C.
  B—Henrietta, N. C.
  G. O. No. 81, W. D., 1919.

Private, first class, Company K, 118th Infantry, 30th Division.
He volunteered to go forward with another soldier to attack a machine-gun emplacement which was holding up a part of our line. Advancing over open ground under heavy fire, these 2 men destroyed the enemy position, capturing 3 prisoners and allowing a resumption of the general advance.

MOORHEAD, REYNOLDS C_____
  In the Bois-de-Manheuelles, France, Nov. 9, 1918.
  R—Philadelphia, Pa.
  B—Philadelphia, Pa.
  G. O. No. 10, W. D., 1920.

First lieutenant, 324th Infantry, 81st Division.
Lieutenant Moorhead exposed himself to terrific fire from 3 directions while assisting Pvt. Thomas M. Moss carry a wounded officer to a place of shelter.

MOORMAN, HUGH B. (54619)_____
  Near Cantigny, France, May 27, 1918
  R—Sparta, Tenn.
  B—Sparta, Tenn.
  G. O. No. 16, W. D., 1920.

Sergeant, Company K, 26th Infantry, 1st Division.
During an enemy raid on his trench position Sergeant Moorman was wounded in the arm early in the engagement. He refused to be evacuated and personally directed the defense of the trench. He exposed himself to artillery and rifle fire and killed 3 Germans with his rifle.

MORAN, PATRICK J_____

  Near Thiaucourt, France, Sept. 15, 1918.
  R—Nashville, Tenn.
  B—Louisville, Ky.
  G. O. No. 44, W. D., 1919.

Private, 81st Company, 6th Machine Gun Battalion, U. S. Marine Corps, 2d Division.
Passing from 1 gun to another, at all times exposing himself to great danger, carrying ammunition and encouraging his comrades, he showed great devotion to duty. When his company commander had become seriously wounded he left his place of shelter and carried him to a first-aid station.

MORAN, RUSSELL (127276)_____
  Near Somme-Py, France, Oct. 4–5, 1918.
  R—Utica, N. Y.
  B—Lowville, N. Y.
  G. O. No. 37, W. D., 1919.

Private, Battery E, 12th Field Artillery, 2d Division.
During a violent enemy counterbarrage Private Moran, with Pvt. Harley S. Edwards, remained on duty for 14 hours repairing the telephone line from their battery position to the battalion post of command, 250 meters away. Within this period the wires were cut by shellfire more than 20 times, but these 2 soldiers, displaying remarkable coolness and disregard of danger, promptly mended all breaks and maintained constant communication between the battalion and the battery commanders.

MOREHEAD, HERBERT (106404)_____
  Near Very, France, Oct. 9, 1918.
  R—Detroit, Mich.
  B—Van Wert, Ohio.
  G. O. No. 32, W. D., 1919.

Sergeant, Company D, 3d Machine Gun Battalion, 1st Division.
He led forward 2 reorganized squads of machine gunners, during an intense shelling, in order to protect an open flank on which a counterattack was imminent. Entirely exposed, he placed his guns in a most effective position, resisting until the enemy had been thrown back, although severely wounded in the combat.

MORELAND, OSCAR E_____
  Near Blanc Mont, France, Oct. 3–5, 1918.
  R—Indianola, Ill.
  B—Indianola, Ill.
  G. O. No. 26, W. D., 1919.

Corporal, 96th Company, 6th Regiment, U. S. Marine Corps, 2d Division.
Although he was wounded, he refused to go to the rear, but remained on duty throughout the 2 days' action, during which time he distinguished himself in grenade fighting at close range, organizing the flank of his company and holding it against 3 counterattacks, and killing or capturing all the members of a hostile patrol.

*MOREY, FRANK C. (2805914)_____
  Near Foret Vaucheres, France, Sept. 13, 1918.
  R—May, Okla.
  B—Tucson, Ariz.
  G. O. No. 89, W. D., 1919.

Private, Company M, 357th Infantry, 90th Division.
As a battalion runner, he constantly exposed himself during a six-hour bombardment to maintain liaison. After passing repeatedly through the enemy barrage to deliver important messages, he volunteered to accompany or lead a reconnaissance patrol to investigate enemy activities which he had noticed while on his missions of liaison and obtained valuable information for his commander.
Posthumously awarded. Medal presented to mother, Mrs. Steven E. Morey.

MORGAN, DAVID R_____
  At Chaudun, France, July 19, 1918.
  R—Philadelphia, Pa.
  B—Kingston, Pa.
  G. O. No. 15, W. D. 1923.

First lieutenant, Medical Corps, attached to 18th Infantry, 1st Division.
While still suffering from a former attack of gas, he was again overcome by gas fumes after 36 hours of work among the wounded men in the front lines and was sent to the dressing station. Refusing to remain away from the front line, he again made his way to the elements in the advanced positions and under intense enemy fire searched for wounded men, applied first aid, and directed their removal to places of shelter. This work he continued until severely wounded and carried from the field.

MORGAN, ERNEST (1312330)_____
Near Vaux-Andigny, France, Oct.
12, 1918.
R—High Point, N. C.
B—Moore County, N. C.
G. O. No. 133, W. D. 1918.

Private, Company L, 118th Infantry, 30th Division.
While his company was consolidating its position, he crept out in full view of the enemy and took up a position in a shell hole 50 yards from the enemy's lines. He remained there throughout the day without food or water and sniped at and killed 10 of the enemy. His deadly aim kept down the observation from the German lines and enabled his company to carry on the work of consolidation.

MORGAN, FRANCIS M_____
During the Meuse-Argonne offensive, Nov. 1–11, 1918.
R—Ravenswood, W. Va.
B—New Martinsville, W. Va.
G. O. No. 37, W. D. 1919.

First lieutenant, 353d Infantry, 89th Division.
Although severely wounded, he maintained command of 2 platoons throughout the offensive, personally leading patrols through occupied enemy territory and breaking up enemy resistance on the flanks, which were holding up the advance of his neighboring units.

MORGAN, GEORGE H. (145668)_____
Near Suippes, France, July 15, 1918.
R—St. Paul, Minn.
B—Sioux City, Iowa.
G. O. No. 37, W. D., 1919.

Private, Battery C, 151st Field Artillery, 42d Division.
While on duty as a runner, carrying a message to his battery, he fell wounded before reaching his destination, but in spite of suffering severe pain he crawled the remainder of the distance on his hands and knees and delivered the message.

MORGAN, HANS E. (2023257)_____
Near Sergy, France, Aug. 1, 1918.
R—Cherry Grove, Mich.
B—Cherry Grove, Mich.
G. O. No. 37, W. D., 1919.

Private, Company B, 47th Infantry, 4th Division.
After all the other members of his automatic-rifle squad had been wounded and evacuated and he himself wounded 3 times, Private Morgan remained at his post, operating his automatic rifle against a machine-gun nest until his supply of ammunition was exhausted. He then turned his rifle over to another squad before being evacuated.

MORGAN, JOHN H. (107912)_____
Near Medeah Ferme, France, Oct. 3, 1918.
R—Cincinnati, Ohio.
B—Vanceburg, Ky.
G. O. No. 21, W. D., 1919.

Corporal, Company D, 9th Infantry, 2d Division.
All of his superiors having been killed by a nest of machine guns, Corporal Morgan took command of his platoon, and in an extremely difficult attack, wiped out a nest of 5 machine guns.
The true name of this soldier is Harry J. Morgan.

MORGAN, VERN A_____
Near Beaufort, France, Nov. 4, 1918.
R—Council Bluffs, Iowa.
B—Council Bluffs, Iowa.
G. O. No. 46, W. D., 1919.

First lieutenant, 355th Infantry, 89th Division.
Although he was wounded early in the engagement by shrapnel, Lieutenant Morgan, after receiving first-aid treatment, immediately returned to his company and led it throughout the day. After taking the town of Beaufort he pushed on with his command to its objective through heavy artillery and machine-gun fire, 40 per cent of his company becoming casualties.

MORISON, JAMES H. S_____
Near Bellicourt, France, Sept. 29, 1918.
R—Cumberland Gap, Tenn.
B—Ewing, Va.
G. O. No. 44, W. D., 1919.

First lieutenant, Medical Corps, attached to 117th Infantry, 30th Division.
After being knocked unconscious into a shell hole and, although suffering acutely from the shock, he rejoined his company and continued to care for the wounded in the open and under intense shellfire. His respirator having been blown away by the exploding shell, this mission was rendered much more precarious by enemy gas shells. He remained at his first-aid station through an intense barrage, which killed several of the stretcher bearers and helpers at this point, evacuating the wounded with great rapidity until he was severely wounded and forced to be evacuated.

MORITZ, MAX F. (2337855)_____
Near Mont St. Pere, France, July 22, 1918.
R—Camden, N. J.
B—Philadelphia, Pa.
G. O. No. 32, W. D., 1919.

Sergeant, Company A, 4th Infantry, 3d Division.
After his platoon had captured a German fieldpiece in the woods near Mont St. Pere and returned to the town, Sergeant Moritz voluntarily remained behind with a wounded comrade in a woods infested by enemy snipers, and after nightfall brought the wounded man to a place of safety.

MORNINGSTAR, LEROY (5830)_____
Near Vaux, France, July 1, 1918.
R—St. Petersburg, Fla.
B—Huntingdon, Pa.
G. O. No. 99, W. D., 1918.

Sergeant, medical detachment, 23d Infantry, 2d Division.
Sick, gassed, and stunned by shells, he remained at his post on duty under heavy fire and bravely assisted in the succoring of soldiers who had been injured, near Vaux, France, July 1, 1918.

MORPHEW, JOHN E. (2216646)_____
In the offensive against the St. Mihiel salient, France, Sept. 12, 1918.
R—Trousdale, Okla.
B—Giltham, Ark.
G. O. No. 128, W. D., 1918.

Sergeant, Company C, 357th Infantry, 90th Division.
This soldier showed utter fearlessness and bravery of a high order throughout the drive. He took 2 machine-gun nests single handed, in both cases killing the gunners and taking the other members of the crews prisoners. He took 35 prisoners during the first day, entering dugouts alone and disarming the occupants.

MORRIS, CHALMER R. (549979)_____
Near Gland, France, June 17, 1918, and near Jaulgonne, France, July 22, 1918.
R—Washington, D. C.
B—Greenville, Ind.
G. O. No. 39, W. D., 1920.

First sergeant, Company C, 38th Infantry, 3d Division.
On June 17, after an attempt to cross the Marne in a boat failed, due to its sinking, Sergeant Morris, with 2 others, swam the Marne River, penetrated the German line, and returned with valuable information. On July 22 he advanced in front of the lines through a wood, exposing himself to heavy enemy fire in order to maintain communication with the organization on his left.

MORRIS, EDWARD M_____
Near Landres-et-St. Georges, France, Oct. 30, 1918.
R—New York, N. Y.
B—Marinette, Wis.
G. O. No. 35, W. D., 1919.

Second lieutenant, pilot, 104th Aero Squadron, Air Service.
Unable to complete a photographic mission, owing to motor trouble, Lieutenant Morris, with his observer, made a reconnaissance behind the German lines. They dispersed a battalion of enemy troops, and, although twice attacked by enemy patrols, drove them off and in each case brought down 1 enemy plane. They remained in the air until their motor failed completely.

MORRIS, EFFINGHAM B., Jr_____
Near Montfaucon, France, Sept. 27, 1918.
R—Philadelphia, Pa.
B—Ardmore, Pa.
G. O. No. 37, W. D., 1919.

Captain, 313th Infantry, 79th Division.
Leading his battalion in attack, Captain Morris was painfully wounded in the leg, but continued in command during the four days' action that followed. By his persistence in remaining, despite his severe wound, he set an example which contributed largely to the success of the operations.

MORRIS, HARLAN D. (2383930)_____
Near Cunel, France, Oct. 15, 1918.
R—Sulphur Springs, Ind.
B—Snyder, Ind.
G. O. No. 98, W. D., 1919.

Sergeant, Company H, 60th Infantry, 5th Division.
Advancing under intense artillery and machine-gun fire, Sergeant Morris displayed marked coolness and disregard of danger in personally clearing the right flank of his company of dangerous snipers. In so doing he was severely wounded, but he nevertheless reorganized his detachment and held the position.

MORRIS, HUBERT C. (2387689)_____
Near Dun-sur-Meuse, France, Nov. 5, 1918.
R—Humboldt, Tenn.
B—Bowling Green, Ky.
G. O. No. 37, W. D., 1919.

Sergeant, Company H, 61st Infantry, 5th Division.
Advancing alone, Sergeant Morris attacked a machine-gun nest, capturing the entire crew, and preventing surprise fire on an exposed flank of his company.

MORRIS, JOHN P. (1243170)_____
At Les Grands Bois-Chateau-de-
Diable, France, Aug. 10, 1918.
R—Philadelphia, Pa.
B—Philadelphia, Pa.
G. O. No. 130, W. D., 1918.

Corporal, Company H, 111th Infantry, 28th Division.
After his organization had been compelled to retire in the face of a strong enemy attack, he made a reconnaissance of the bed of the Vesle River, and, wading through water shoulder deep, under heavy machine-gun fire, made five trips, carrying wounded from the north bank to a dressing station south of the river.

MORRIS, THOMAS H. (3111287)_____
Near Montfaucon, France, Sept. 30, 1918.
R—Philadelphia, Pa.
B—Scranton, Pa.
G. O. No. 44, W. D., 1919.

Private, first class, Company I, 316th Infantry, 79th Division.
Although severely wounded in the thigh, he continued to carry messages from the line to battalion headquarters, exposed at all times to terrific machine-gun fire.

MORRIS, WILLIAM H. H., Jr_____
Near Villers-devant-Dun, France, Nov. 1, 1918.
R—Ocean Grove, N. J.
B—Ocean Grove, N. J.
G. O. No. 87, W. D., 1919.

Major, 360th Infantry. 90th Division.
During darkness he led his battalion in an attack under heavy artillery and machine-gun fire. Upon reaching a hill he exposed himself to heavy fire to reconnoiter personally the enemy position, and then, although wounded by a machine-gun bullet, heroically led his battalion in their advance, refusing to be evacuated, inspiring his men by his personal courage.

MORRISON, CHARLES S. (2179380)_____
At Charey, France, Sept. 30, 1918.
R—St. Louis, Mo.
B—St. Louis, Mo.
G. O. No. 9, W. D., 1923.

Corporal, Company L, 354th Infantry, 89th Division.
On the morning of Sept. 30, 1918, when he volunteered to accompany an officer of his company in the rescue of a wounded man of his battalion who had been on patrol. Proceeding under intense and accurate fire to a point 400 yards beyond his outpost, he met a member of the patrol who stated that the wounded man was about 100 yards farther out and close to the hostile lines. With splendid courage, he proceeded to the point indicated, in company with his officer, being constantly subjected to the heaviest fire, and found the soldier, who had died from his wounds. Returning in safety to his own lines, he exchanged shots with the enemy, 1 of whom he killed.

MORRISON, HUGH J. V. (1211782)_____
East of Ronssoy, France, Sept. 29, 1918.
R—Poughkeepsie, N. Y.
B—Flatbush, N. Y.
G. O. No. 20, W. D., 1919.

Private, Company K, 107th Infantry, 27th Division.
Private Morrison, with 3 other soldiers, went out into an open field under heavy shell and machine-gun fire and succeeded in carrying back to our lines 4 seriously wounded men.

MORRISON, JESSE S. (1490855)_____
Near Attigny, France, Oct. 14, 1918.
R—Odell, Tex.
B—Moody, Tex.
G. O. No. 50, W. D., 1919.

Sergeant, Headquarters Company, 142d Infantry, 36th Division.
Sergeant Morrison drove a motor cycle through intense artillery fire and assisted in the rescue of a wounded soldier under machine-gun fire, driving back with him under intense bombardment to the dressing station.

*MORRISON, JOHN_____
Near Molleville Farm, France, Oct. 14–15, 1918.
R—Cincinnati, Ohio.
B—Cincinnati, Ohio.
G. O. No. 130, W. D., 1918.

Second lieutenant, 322d Field Artillery, 83d Division.
As liaison officer between the Infantry and Artillery he exemplified in the highest degree the spirit of bravery, devotion to duty, and self-sacrifice. He crawled beyond the front line in the face of intense machine-gun and artillery fire, with a telephone strapped on his back, in order to direct the preparatory fire of the Artillery. On the following day he accompanied the advance Infantry battalion in the attack, and under the most difficult circumstances established and maintained liaison with the Artillery. In the faithful performance of these duties this gallant officer lost his life.
Posthumously awarded. Medal presented to father, Harley J. Morrison.

MORRISON, JULIAN K_____
In the Bois-Quart-de-Reserve, France, Sept. 12, 1918.
R—Statesville, N. C.
B—Statesville, N. C.
G. O. No. 46, W. D., 1919.

Second lieutenant, 326th Battalion, Tank Corps.
Preceding his tanks on foot, Lieutenant Morrison captured a machine-gun nest. Though he was twice wounded, he continued in action for 2 days thereafter.
Oak-leaf cluster.
For the following act of extraordinary heroism in action near Very, France, Sept. 28, 1918, Lieutenant Morrison is awarded an oak-leaf cluster, to be worn with the distinguished-service cross: During the attack on Charpentry and the Bois-de-Montrebeau, he led a platoon of 5 tanks, directing his tanks on foot, 400 yards in advance of the Infantry, under intense fire. Three of his tanks were put out of action by artillery fire, but he continued in action with the remaining 2 until dark, when he directed the work of rescuing the crews.

MORRISON, LYMAN N. (263470)_____
   Near Juvigny, France, Aug. 3, 1918.
   R—Kalamazoo, Mich.
   B—Comstock, Mich.
   G. O. No. 71, W. D., 1919.

Private, Company A, 125th Infantry, 32d Division.
He displayed marked bravery in repeatedly carrying messages to the front lines over terrain swept by intense artillery and machine-gun fire. Still under heavy fire, he assisted wounded soldiers he found in exposed positions. His heroism was an inspiration to those near him.

MORRISON, OTHO K._____
   Near Cunel, France, Oct. 15, 1918.
   R—Gatesville, Tex.
   B—Gatesville, Tex.
   G. O. No. 98, W. D., 1919.

First lieutenant, 60th Infantry, 5th Division.
Lieutenant Morrison displayed exceptional courage and leadership when, being cut off by the enemy with his battalion command and a small detachment from battalion headquarters, he led patrols and drove off enemy machine gunners who had infiltrated to within striking distance of the group. On Nov. 10 he led his company against the fortified heights of Juvigny, driving back the enemy for more than a kilometer.

MORRISON, WILLIAM L._____
   In the Champagne-Marne offensive, July 16, 1918, and the Meuse-Argonne offensive, Oct. 9, 1918.
   R—Boulder, Colo.
   B—Boulder, Colo.
   G. O. No. 19, W. D., 1920.

Captain, 38th Infantry, 3d Division.
On July 16, when the members of his patrol acted as a covering detachment, he entered an enemy dugout and captured a prisoner, thus securing valuable information. On Oct. 9 he led a combat patrol into the enemy lines and succeeded, under heavy fire, in putting 2 machine guns out of action, thus enabling his own company and one of another regiment to advance.

MORRISSEY, EDWARD P. (1716036)_____
   Near Bazoches, France, Aug. 25–26, 1918.
   R—Buffalo, N. Y.
   B—Buffalo, N. Y.
   G. O. No. 37, W. D., 1919.

Private, Company C, 302d Engineers, 77th Division.
He and another soldier had become separated from their detachment and were forced to take shelter for 5½ days. He rescued a wounded soldier from exposure to machine-gun and shell fire, and later attacked a machine-gun nest in his direct front. In the attack he killed 2 of the enemy with hand grenades and subsequently returned to our lines, assisting the wounded comrade to safety.

*MORROW, HOWARD H. (1285169)_____
   Near Bois-de-Consenvoye, France, Oct. 8, 1918.
   R—Washington, D. C.
   B—Baltimore, Md.
   G. O. No. 44, W. D., 1919.

Private, first class, Company F, 115th Infantry, 29th Division.
Going forward from his own lines through terrific machine-gun and artillery fire, Private Morrow rescued and brought to safety a wounded comrade. In the action of the next few days he was so severely wounded that he died shortly afterwards.
Posthumously awarded. Medal presented to mother, Mrs. Grace A. Payne.

MORROW, WILLIAM M._____
   At Claire-Chenes, north of Montfaucon, France, Oct. 20–21, 1918.
   R—Algonac, Mich.
   B—Niles, Mich.
   G. O. No. 102, W. D., 1918.
   Distinguished-service medal also awarded.

Colonel, 7th Infantry, 3d Division.
On Oct. 20, 1918, when the Claire-Chenes had been taken by the troops of his command and a hostile counterattack had forced them back over the ground gained in the morning's fighting Colonel Morrow at once took personal command of the battalion engaged in the operations, reorganized it, and with distinguished gallantry and inspiring example, led his men to a victorious counter attack, drove the enemy from the woods, secured its possession, and consolidated it. On Oct. 21 he again displayed the same qualities of leadership and personal gallantry in the successful assault on Hill 299.

MORSE, DANIEL A. (2369942)_____
   Service rendered under the name, Daniel Moskowitz.

Private, Company F, 108th Infantry, 27th Division.

*MORSE, GUY E._____
   Near Vilcey-sur-Trey, France, Sept. 12, 1918.
   R—Kansas City, Mo.
   B—Canada.
   G. O. No. 133, W. D., 1918.

Second lieutenant, observer, 135th Aero Squadron, Air Service.
He, with First Lieut. Wilbur C. Suiter, pilot, fearlessly volunteered for the perilous mission of locating the enemy's advance unit in the rear of the Hindenburg line. Disregarding the hail of machine-gun fire and bursting anti-aircraft shell, they invaded the enemy's territory at low altitude and accomplished their mission, securing for our staff information of the greatest importance. These 2 gallant officers at once returned to the lines and undertook another reconnaissance mission, from which they failed to return. Lieutenant Morse's body was found and buried by an Artillery unit.
Posthumously awarded. Medal presented to father, Ernest Morse.

MORSE, WARREN B. (72199)_____
   At Epieds, France, July 22–23, 1918.
   R—Somerville, Mass.
   B—Somerville, Mass.
   G. O. No. 9, W. D., 1923

Private, Company E, 104th Infantry, 26th Division.
When his organization under heavy fire was forced to retire to rectify the line, it was discovered that several severely wounded men could not be moved, and volunteers were called for to remain with the wounded until reinforcements arrived. The duty involved was deemed almost certain death. With great courage and devotion to his comrades he elected to remain and for several hours cared for the wounded under intense fire; he continued his care, after being badly wounded until evacuated, his splendid courage proving an inspiration to his comrades.

MORTON, LAWRENCE A. (2339313)_____
   Near Les Evaux, France, July 10, 1918.
   R—Jeannette, Pa.
   B—Jeannette, Pa.
   G. O. No. 32, W. D., 1919.

Private, first class, Company H, 4th Infantry, 3d Division.
After being badly wounded, he continued to perform his duties as runner at a relay post on the front line under heavy machine-gun fire.

MOSCOW, LONNIE J. (1210233)_____
   Near Ronssoy, France, Sept. 29, 1918.
   R—Watertown, N. Y.
   B—Ogdensburg, N. Y.
   G. O. No. 19, W. D., 1920.

Corporal, Company C, 107th Infantry, 27th Division.
In the attack on the Hindenburg line, Corporal Moscow was an advanced scout for his platoon. The platoon was temporarily halted by machine-gun fire from a section of the enemy trench in their immediate front. Corporal Moscow rushed through the heavy enemy fire to the trench and at the point of his rifle compelled 12 of the enemy to surrender. He then signaled for the platoon to advance.

MOSELEY, GAINES.
Near St. Etienne, France, Oct. 4, 1918.
R—Aiken, S. C.
B—Aiken, S. C.
G. O. No. 37, W. D., 1919.

Captain, 5th Regiment, U. S. Marine Corps, 2d Division.
As commander of an assault company, Captain Moseley displayed exceptional courage in carrying his line forward during a heavy artillery and machine-gun barrage.

*MOSELEY, JAMES A.
Near Suippes, France, July 15, 1918.
R—Glen Ridge, N. J.
B—Raleigh, N. C.
G. O. No. 20, W. D., 1919.

First lieutenant, 166th Infantry, 42d Division.
When 2 others had failed, 1 killed and the other wounded, Lieutenant Moseley left his shelter during a most intense enemy artillery bombardment, searched for and located a wounded corporal of his platoon, bringing him a distance of more than 400 yards to safety.
Posthumously awarded. Medal presented to mother, Mrs. Anna Moseley.

MOSES, ELLISON (1871575).
Near Ardeuil, France, Sept. 30, 1918.
R—Mayesville, S. C.
B—Mayesville, S. C.
G. O. No. 46, W. D., 1919.

Private, Company G, 371st Infantry, 93d Division.
After his company had been forced to withdraw from an advanced position under severe machine-gun and artillery fire he went forward and rescued wounded soldiers, working persistently until all of them had been carried to shelter.

*MOSHER, HENRY E.
Near Cantigny, France, May 28, 1918.
R—Falconer, N. Y.
B—Falconer, N. Y.
G. O. No. 93, W. D., 1918.

Captain, 28th Infantry, 1st Division.
He displayed heroic conduct and utter disregard of his own safety while successfully directing the consolidation and defense of the position taken by his command. After succeeding in the accomplishment of his task he was struck by enemy fire and killed.
Posthumously awarded. Medal presented to father, Stiles Burt Mosher.

MOSKOWITZ, DANIEL (2669942).
Near Roussoy, France, Sept. 28, 1918.
R—New York, N. Y.
B—New York, N. Y.
G. O. No. 139, W. D., 1918.

Private, Company F, 108th Infantry, 27th Division.
He exhibited exceptional bravery by leaving shelter and going out into an open field under heavy machine-gun and shell fire to rescue wounded soldiers.

MOSKOWITZ, HERMAN (2414791).
Near Talma Hill, France, Oct. 17-21, 1918.
R—Passaic, N. J.
B—Long Island, N. Y.
G. O. No. 133, W. D., 1919.

Private, first class, Company C, 312th Infantry, 78th Division.
As a runner he displayed exceptional courage and devotion to duty in frequently volunteering and carrying messages through dangerous zones in addition to his regular duties. Though lame as the result of an accident, he carried a number of messages through a heavy barrage until he was severely wounded by a bursting shell.

MOSS, THOMAS M. (1859049).
In the Bois-de-Manhuelles, France, Nov. 9, 1918.
R—Macon County, N. C.
B—Cullasaja, N. C.
G. O. No. 32, W. D., 1919.

Private, Company I, 324th Infantry, 81st Division.
With utter disregard for personal safety, he went forward under intense machine-gun fire to rescue an officer who had been mortally wounded.

MOTLEY, FRANK L. (1458197).
Near Apremont, France, Sept. 29, 1918.
R—St. Joseph, Mo.
B—Sharps, Va.
G. O. No. 59, W. D., 1919.

Corporal, Company M, 139th Infantry, 35th Division.
When the enemy was counterattacking, having succeeded in planting machine guns behind a smoke screen, he advanced with utter disregard of personal danger and jumped into an enemy machine-gun nest where there were about 15 Germans. Single handed he killed the gunner and loader and engaged the remainder of the Germans until he received help from his platoon.

MOTLEY, ROBERT E.
Near Chateau-Thierry, France, July 31-Aug. 7, and near Verdun, France, Oct. 14-16, 1918.
R—Virden, Ill.
B—Pittsfield, Ill.
G. O. No. 59, W. D. 1919.

First lieutenant, 125th Infantry, 32d Division.
Realizing the need of medical attention at the front, he went beyond the scope of his duties as dentist by advancing with the Infantry and establishing and maintaining a dressing station with the leading elements of his command. For 7 days, from July 31, to Aug. 7, he safely evacuated many patients by his prompt and fearless action. He again volunteered and went forward in the attack on Oct. 14-16, and on the latter date carried a message back to the supply officer, requesting food for the men. Although wounded and badly gassed, he accomplished his mission, refusing evacuation until the food was started for the lines.

*MOTTERN, VIRGIL C. (1336229).
Near Mazinghiem, France, Oct. 19, 1918.
R—Jonesboro, Tenn.
B—Jonesboro, Tenn.
G. O. No. 35, W. D., 1919.

Sergeant, first class, Company C, 105th Field Signal Battalion, 30th Division.
He lost his life while personally laying a telephone line over exceedingly dangerous ground, under continuous artillery fire which had caused a great loss among the runners. He attempted the laying of this line in order to give his men a rest in a place of safety.
Posthumously awarded. Medal presented to father, George P. Mottern.

MOUNTS, WAYNE D. (3167397).
Near Brieulles, France, Nov. 4, 1918.
R—Williamson, W. Va.
B—Lindsey, W. Va.
G. O. No. 35, W. D., 1919.

Private, Company D, 15th Machine Gun Battalion, 5th Division.
Although suffering painfully from a severe shoulder wound, he refused to reveal his condition, but courageously remained on duty until the termination of hostilities, 7 days later.

MOYER, RALPH (1635729).
Near Bois-de-St. Remy, France, Sept. 12, 1918.
R—Willard, Kans.
B—Argentine, Kans.
G. O. No. 46, W. D., 1919.

Private, Company F, 103d Infantry, 26th Division.
Although painfully wounded while cutting wires under terrific shellfire, he refused to be evacuated and continued at his work until a lane had been opened and his platoon had passed through. He then joined his platoon and engaged in the battle until he became so weak from his wounds that he had to be sent to the rear.

MOYNAHAN, TIMOTHY J.
Near Cierges, France, Sept. 28-30, 1918.
R—Brooklyn, N. Y.
B—Ireland.
G. O. No. 21, W. D., 1919.

Lieutenant colonel, 146th Infantry, 37th Division.
Displaying remarkable personal courage and leadership, he personally led his battalion, without support on either flank, through terrific artillery bombardment, in the face of direct machine-gun fire and enfilading fire from 1-pounder guns on the right, capturing his objective on the ridge east of Cierges and repelling 4 hostile counterattacks.

MOYSE, HERMAN_____.
    Near Cierges, northeast of Chateau-
    Thierry, France, July 31, 1918.
    R—Baton Rouge, La.
    B—St. Gabriel, La.
    G. O. No. 117, W. D., 1918.

First lieutenant, 125th Infantry, 32d Division.
After advancing through 5 stages of artillery barrage and machine-gun fire, he led a patrol of 5 men forward to capture 2 machine guns which were endangering the success of the operation. Although seriously wounded in the chest and foot by machine-gun fire, he would not consent to being taken to the rear until the guns had been captured.

MUDGE, JOSIAH B_____.
    At Frapelle, France, Aug. 17, 1918.
    R—Lawrence, Kans.
    B—Manhattan, Kans.
    G. O. No. 15, W. D., 1919.

First lieutenant, 6th Infantry, 5th Division.
He displayed notable courage and determination by leading his company to its objective through a heavy enemy barrage of high-explosive gas shells. Although gassed and wounded in the leg by a shell fragment, he remained in command of his company until it was relieved.

°MUDGETT, BRYAN_____.
    Near St. Mihiel, France, Sept. 12-
    13, 1918.
    R—Carlsbad, N. Mex.
    B—Odessa, Tex.
    G. O. No. 46, W. D., 1919.

Second lieutenant, 357th Infantry, 90th Division.
On several occasions, during the advance of Sept. 12, he outmaneuvered enemy machine guns, capturing both guns and crew. On the night of Sept. 12–13 he led a patrol of 2 squads through the German lines, advancing over 1,000 yards to the front of the line of resistance, capturing a German battery, 1 noncommissioned officer, and 7 men. He then fought his way back through the enemy's lines, losing but 1 prisoner before meeting the advancing American troops.
Posthumously awarded. Medal presented to widow, Mrs. Zetha De Berry Mudgett.

MUELLER, JOSEPH, Jr. (274236)_____.
    Near Jametz, France, Nov. 10-11,
    1918.
    R—Milwaukee, Wis.
    B—Austria-Hungary.
    G. O. No. 66, W. D., 1919.

Private, first class, Company F, 127th Infantry, 32d Division.
Private Mueller, a runner, successfully maintained liaison between his company, which formed the liaison group with another division, and regimental headquarters, promptly carrying numerous messages across an area under heavy fire.

MUHLENBERG, FREDERICK A_____.
    Near Nantillois, France, Sept. 26-30,
    1918.
    R—Reading, Pa.
    B—Reading, Pa.
    G. O. No. 37, W. D., 1919.

Captain, 314th Infantry, 79th Division.
As regimental adjutant he displayed the utmost disregard for personal danger in assisting his regimental commander in maintaining liaison with the front lines. After being painfully wounded and gassed by a bursting gas shell, this officer refused to be evacuated, but remained on duty, carrying orders to the front line and bringing back valuable information, until he was ordered to the rear.

MUIR, CHARLES H_____.
    At Santiago, Cuba, July 2, 1898.
    R—Erie, Mich.
    B—Erie, Mich.
    G. O. No. 10, W. D., 1924.
    Distinguished-service medal also
    awarded.

First lieutenant, 2d Infantry, U. S. Army.
At the risk of his life, he voluntarily exposed himself to a heavy hostile artillery and infantry fire in a successful attempt as a sharpshooter to silence a piece of Spanish artillery at the battle of Santiago.

MULHALL, HENRY L. (50894)_____.
    Near St. Etienne-a-Arnes, France,
    Oct. 3-9, 1918.
    R—West Hazleton, Pa.
    B—Lattimer, Pa.
    G. O. No. 35, W. D., 1919.

Sergeant, Company G, 23d Infantry, 2d Division.
He led his platoon against a machine-gun nest and continued to his objective after being wounded. He was instrumental in capturing 3 prisoners and 1 machine gun.

MULHOLLAND, EMMETT PAUL_____.
    Near St. Juvin, France, Oct. 15, 1918.
    R—Fort Dodge, Iowa.
    B—Gilmore City, Iowa.
    G. O. No. 72, W. D., 1920.

Second lieutenant, 326th Infantry, 82d Division.
After having been severely wounded in the leg he continued to direct his men under terrific artillery fire in the attack against 2 enemy machine guns. Due to his initiative and gallantry the enemy position was captured.

MULLEN, ROGER H_____.
    Near Romagne, France, Oct. 14,
    1918.
    R—Chicago, Ill.
    B—Chicago, Ill.
    G. O. No. 37, W. D., 1919.

First lieutenant, 6th Infantry, 5th Division.
Lieutenant Mullen on Oct. 14, under heavy machine-gun and artillery fire personally led an attack on enemy machine-gun nests, capturing 3 machine guns and numerous prisoners. On Nov. 7 he attacked and captured an enemy machine-gun nest which was holding up the advance of his company, taking machine guns and 16 prisoners.

MULLIGAN, JAMES J. (368584)_____.
    Near Bois-de-Ormont, France, Oct.
    12, 1918.
    R—Mount Vernon, N. Y.
    B—New York, N. Y.
    G. O. No. 32, W. D., 1919.

Private, Company I, 114th Infantry, 29th Division.
Private Mulligan volunteered to carry a message from the firing line to the rear over a route commonly known as the "Valley of Death" under heavy machine-gun and shell fire. During the journey he was severely wounded in the thigh and leg, but delivered his message. Instead of waiting for treatment, Private Mulligan hopped and crawled back to the firing line with his answer.

MULLINS, RAY H. (514924)_____.
    Near Crezancy, France, July 15,
    1918.
    R—Peoples, Ky.
    B—Egypt, Ky.
    G. O. No. 32, W. D., 1919.

Sergeant, Machine Gun Company, 30th Infantry, 3d Division.
After his gun crew had been bombed out of the emplacement by the enemy coming from the rear, Sergeant Mullins continued, with the aid of one man, to fire his gun, even after his hand had been wholly shot off.

°MULLINS, SAM (1905623)_____.
    Near Bellicourt, France, Sept. 29,
    1918.
    R—East Alton, Ill.
    B—Columbus, Miss.
    G. O. No. 87, W. D., 1919.

Private, Company H, 119th Infantry, 30th Division.
When certain units of his company were halted by heavy enemy fire, he was sent to them successively, and displaying marked personal bravery and leadership carried them forward under heavy fire. He led 2 squads forward under heavy fire and flanked a machine-gun emplacement which had blocked his company's advance. In this undertaking he was mortally wounded.
Posthumously awarded. Medal presented to father, Jim W. Mullins.

*MULRAIN, CARL (1676239) Near Ville-Savoye, France, Aug. 23, 1918. R—Uxbridge, Mass. B—Whitinsville, Mass. G. O. No. 9, W. D., 1923.	Private, Company D, 308th Infantry, 77th Division. While the 1st Battalion of his regiment was making an attack to regain ground from the enemy in the outpost zone along the Vesle River, Private Mulrain continued to advance when he discovered that 3 enemy machine guns occupied the high ground in front of him. With great courage and utter disregard for his own safety he continued to go forward in the face of concentrated enemy machine-gun fire, thus helping materially to force the enemy to evacuate his machine-gun emplacement, though himself killed by a machine-gun bullet. Posthumously awarded. Medal presented to father, Bernard Mulrain.
MULTER, WALTON L. Near St. Etienne, France, Oct. 5, 1918. R—Kingston, Pa. B—Washington, D. C. G. O. No. 37, W. D., 1919.	Private, 75th Company, 6th Regiment, U. S. Marine Corps. 2d Division. He voluntarily went forward for a distance of 800 meters under heavy shellfire and rescued a wounded soldier who had been left there the night before when the advance patrols had been withdrawn.
MUNCASTER, JOHN H. Near Cunel, France, Oct. 14, 1918. R—Charleston, S. C. B—Canada. G. O. No. 37, W. D., 1919.	Major, 11th Infantry, 5th Division. After the loss of all his company commanders, Major Muncaster advanced at the head of his battalion, leading the men from a very disadvantageous position to the capture of a near-by hill held by the enemy. In the counter-attack which followed he not only commanded the men of his battalion personally, but assisted in the defense of the position.
*MUNRO, GEORGE N. Near Cunel, France, Oct. 15, 1918. R—Buena Vista, Ga. B—Buena Vista, Ga. G. O. No. 89, W. D., 1919.	Captain, 5th Train Headquarters and Military Police, 5th Division. Organizing a company composed of men who had become separated from their own organizations, Captain Munro led them with exceptional skill and bravery in an attack, materially aiding in the advance. In the course of the assault this officer was killed by machine-gun fire. Posthumously awarded. Medal presented to mother, Mrs. George P. Munro.
*MUNROE, GEORGE (73444) Near Chateau-Thierry, France, July 20-23, 1918. R—Easthampton, Mass. B—Boston, Mass. G. O. No. 102, W. D., 1918.	Private, Company K, 104th Infantry, 26th Division. Private Munroe, acting as a runner, carried messages through heavy artillery fire with absolute fearlessness until killed. Posthumously awarded. Medal presented to mother, Mrs. Dolina F. Munroe.
MUNROE, WILLIAM A. (1303) Near Sergy, France, July 28 to Aug. 2, 1918. R—Detroit, Mich. B—Saginaw, Mich. G. O. No. 64, W. D., 1919.	Sergeant, medical detachment. 125th Infantry, 32d Division. He voluntarily left his aid station and went to the field of action to deliver first aid to men in the most advanced positions. He tended the wounded under the most intense machine-gun and shell fire, and successfully carried a large number to places of comparative safety. His courage and cheerfulness under such hazardous circumstances did much to keep up the spirits of both the wounded whom he served and the men fighting in that vicinity.
*MURDOCH, ROBERT H. At Sergy, France, July 29-31, 1918, and at St. Thibaut, France, Aug. 6-12, 1918. R—Wilkes-Barre, Pa. B—Wilkes-Barre, Pa. G. O. No. 133, W. D., 1918.	First lieutenant, Medical Corps, attached to 47th Infantry, 4th Division. Accompanying his battalion in the attack on Sergy, he advanced for more than a mile under heavy shellfire, and as soon as the southern half of the town had been taken he established his dressing station, maintaining it during the 3 days of fighting under constant and severe bombardment. When his battalion went into action at St. Thibaut this faithful officer again displayed heroic devotion to duty by working in his dressing station under the most trying conditions for 6 days while the town was bombarded with gas and high-explosive shells. Posthumously awarded. Medal presented to mother, Mrs. N. Ophelia Murdoch.
MURNANE, STANLEY T. (914902) Near Brieulles, France, Nov. 4-5, 1918. R—St. Paul, Minn. B—St. Paul, Minn. G. O. No. 37, W. D., 1919.	Private, Company D, 7th Engineers, 5th Division. When 3 of the boats supporting a pontoon bridge across the Meuse River were destroyed by artillery fire, he voluntarily waded into the stream under heavy artillery and machine-gun fire and held up the deck of the bridge until new boats were launched and placed into position.
MURPHY, ALBERT R. (1243115) At Fismes and Fismette, France, Aug. 10-13, 1918. R—Philadelphia, Pa. B—Philadelphia, Pa. G. O. No. 99, W. D., 1918.	Private, medical detachment, 111th Infantry, 28th Division. He volunteered to rescue 5 wounded men who had become detached from their company and were unable to rejoin it because of their injuries. By fearlessly passing back and forth through enemy fire he succeeded in this undertaking.
MURPHY, EDWARD (552340) North of Mezy, France, July 22 1918. R—New York, N. Y. B—Ireland. G. O. No. 22, W. D., 1920.	Private, Company L, 38th Infantry, 3d Division. Private Murphy advanced ahead of his platoon exposed to heavy machine-gun and trench-mortar fire and attacked 2 enemy gunners who were operating a trench mortar. He killed 1 and forced the other to flee. His action enabled his platoon to continue its advance with slight loss.
MURPHY, EDWARD F. (70786) Near Verdun, France, Oct. 16, 1918. R—Fitchburg, Mass. B—Quincy, Mass. G. O. No. 21, W. D., 1919.	Corporal, Company D, 104th Infantry, 26th Division. When his platoon was nearly surrounded by a superior force of the enemy, he held off the enemy by his rifle fire until his comrades could withdraw, he himself being severely wounded while covering their retreat.

MURPHY, FRANK P. (554707)_____
West of Jaulgonne, France, July 25, 1918.
R—East Irvington, N. Y.
B—Hastings, N. Y.
G. O. No. 22, W. D., 1920.

Corporal, Company A, 9th Machine Gun Battalion, 3d Division.
Although wounded in the shoulder by a shell fragment on the morning of July 23, he continued to care for the wounded of his company. When sent to the rear for treatment, he refused to be evacuated but returned to his company. This exertion caused his collapse.

MURPHY, JAMES A. (52663)_____
In front of Mount Sec, northwest of Toul, France, Mar. 19, 1918.
R—Sault Ste. Marie, Mich.
B—Detroit, Mich.
G. O. No. 129, W. D., 1918.

Sergeant, Company B, 26th Infantry, 1st Division.
With his patrol leader he cut and crawled through 12 strands of wire in front of an enemy listening post, and with coolness and nerve killed 1 of the sentinels who was firing at the patrol leader.

MURPHY, JAMES J. (1706791)_____
In the Forest of Argonne, France, Oct. 4, 1918.
R—Brooklyn, N. Y.
B—Brooklyn, N. Y.
G. O. No. 44, W. D., 1919.

Corporal, Company K, 307th Infantry, 77th Division.
While his company, with 2 battalions of the 308th Infantry, were surrounded by the enemy in the Forest of Argonne, Corporal Murphy rushed through a severe machine-gun and shellfire for a distance of 75 yards and carried a severely wounded comrade to a place of safety.

MURPHY, JOHN D_____
Near Epieds, France, July 22, 1918.
R—Natick, Mass.
B—Natick, Mass.
G. O. No. 7, W. D., 1925.

Major, 102d Machine Gun Battalion, 26th Division.
Major Murphy led a small daylight patrol to reconnoiter the enemy's outpost line. Encountering machine-gun fire, he sheltered his patrol and alone went forward in the face of continuous fire to reconnoiter the town of Epieds. Assuring himself of the strength of the enemy, he returned over the same route, reporting to his brigade headquarters information of the greatest value.

MURPHY, JOHN H_____
Near Pouilly, France, Nov. 10–11, 1918.
R—Detroit, Mich.
B—Detroit, Mich.
G. O. No. 37, W. D., 1919.

First lieutenant, 356th Infantry, 89th Division.
Lieutenant Murphy and 4 soldiers flanked a machine-gun nest of 3 guns, only to be fired on directly at 30 yards. Charging the guns, they met hand to-hand resistance, but repulsed the enemy, capturing the guns. Lieutenant Murphy was wounded twice, and 3 of his men were killed.

MURPHY, JOHN J. (141498)_____
Near Nantillois, France, Oct. 31, 1918.
R—Butte, Mont.
B—England.
G. O. No. 44, W. D., 1919.

Private, first class, Battery F, 148th Field Artillery.
Private Murphy displayed a remarkable example of heroism by carrying 2 wounded men from the gun pit after being seriously wounded himself, when a German shell exploded within a few feet of the piece which was being loaded, setting fire to several boxes of powder and to the camouflage covering of the pit. After carrying the wounded men to safety, he returned to the pit, closed the breech of the piece, verified its laying, and fired it, preventing what probably would have been a very serious explosion. He was quickly carried to the aid station, where it was found that he had suffered serious burns from the terrific heat, besides being wounded in several places by shell fragments.

MURPHY, JOHN P. (1210234)_____
Near Ronssoy, France, Sept. 29, 1918.
R—New York, N. Y.
B—New York, N. Y.
G. O. No. 24, W. D., 1920.

Corporal, Company C, 107th Infantry, 27th Division.
Corporal Murphy exposed himself to heavy machine-gun and rifle fire to rescue a wounded man who lay in front of our lines. By crawling from shell hole to shell hole he was able to accomplish the rescue in spite of the heavy enemy fire.

MURPHY, MICHAEL S. (1203546)_____
East of Ronssoy, France, Sept. 29, 1918.
R—Cohoes, N. Y.
B—Cohoes, N. Y.
G. O. No. 20, W. D., 1919.

Private, Company B, 105th Infantry, 27th Division.
During the operations against the Hindenburg line he left shelter, went forward under heavy shell and machine-gun fire, and succeeded in rescuing a wounded soldier, thereby exhibiting great bravery and gallantry. In performing this act he was wounded.

MURPHY, RAY E. (143488)_____
Near Sommerance, France, Nov. 1, 1918.
R—Bedford, Ind.
B—Indiana.
G. O. No. 32, W. D., 1919.

Sergeant, Battery A, 150th Field Artillery, 42d Division.
When the powder dump near his gun was blown up by enemy fire and the fuse boxes were on fire, Sergeant Murphy, regardless of personal danger from an explosion, went into the fire, extinguishing it. He thereby saved his gun from becoming unserviceable and kept it in action.

MURPHY, THOMAS W. (1033794)_____
Near Rembercourt, France, Nov. 1, 1918.
R—New Britain, Conn.
B—New Britain, Conn.
G. O. No. 95, W. D., 1919.

Sergeant, Company I, 64th Infantry, 7th Division.
Sergeant Murphy went forward to the aid of a wounded comrade who was lying about 125 feet in front of the enemy's firing line. Under direct fire of rifles, machine guns, and artillery he applied first aid and took the man back into our trenches to a place of safety.

MURPHY, WILLIAM (88867)_____
Near Villers-sur-Fere, France, July 29, 1918.
R—New York, N. Y.
B—New York, N. Y.
G. O. No. 59, W. D., 1919.

Private, first class, Machine Gun Company, 165th Infantry, 42d Division.
As a company runner he repeatedly crossed open ground swept by rifle and machine-gun fire. He volunteered to carry messages out of turn, and though longer and safer routes were often available, he chose the shortest, exposing himself continually to expedite the delivery of important messages. He displayed equal bravery in subsequent operations. When his platoon leader was mortally wounded he organized a carrying party to take the officer back to the dressing station through heavy shellfire.

*MURPHY, WILLIAM M. (1285536)_____
Near Verdun, France, Oct. 8, 1918.
R—Baltimore, Md.
B—Baltimore, Md.
G. O. No. 3, W. D., 1919.

Private, Company H, 115th Infantry, 29th Division.
In the Bois Consenvoye, east of the Meuse River, when his platoon was stopped he voluntarily advanced in the face of direct machine-gun fire and was killed. His gallant conduct was a great inspiration to his comrades, who, following his example, captured the machine-gun nest, approximately 100 prisoners, and several machine guns.
Posthumously awarded. Medal presented to sister, Mrs. Estelle Schmeiger.

MURRAY, CHARLES I_____
During the advance upon Bouresches, France, June 6, 1918.
R—Sewickley, Pa.
B—Sewickley, Pa.
G. O. No. 126, W. D., 1918.

First lieutenant, 6th Regiment, U. S. Marine Corps, 2d Division.
He displayed conspicuous bravery and efficiency during the advance upon Bouresches, France, on the night of June 6, 1918. Having been shot through both arms by machine-gun fire and being no longer able to advance, he refused assistance and walked to the rear alone.

MURRAY, CROMWELL E_____
Near Soissons, France, July 18–22, 1918.
R—Columbia, S. C.
B—Saint George, S. C.
G. O. No. 108, W. D., 1918.

First lieutenant, 3d Machine Gun Battalion, 1st Division.
Throughout the 5 days of battle, near Soissons, France, July 18–22, 1918, his conduct was marked by exceptional initiative and bravery. He organized Infantry and machine-gun units and voluntarily led them in successful attacks against enemy machine-gun nests.

MURRAY, JAMES A (89435)_____
Near Meurcy Ferme, France, July 30, 1918.
R—New York, N. Y.
B—New York, N. Y.
G. O. No. 15, W. D., 1923.

Private, Company B, 165th Infantry, 42d Division.
While on duty as a runner, he repeatedly crossed a field swept by heavy enemy machine-gun and artillery fire. Although wounded by enemy fire he continued on his hazardous duty. Discovering 5 severely wounded men lying exposed to terrific fire he bravely made his way to them, dragged 2 of the men to safety and assisted others in rescuing the 3 remaining men. He was again wounded while so engaged and was carried from the field. The bravery and devotion to duty displayed by Private Murray greatly strengthened the morale of the men of his battalion.

ᵉMURRAY, KENNETH P_____
Near Mezy, France, July 15, 1918.
R—Mt. Vernon, N. Y.
B—Mt. Vernon, N. Y.
G. O. No 27, W. D., 1920.

First lieutenant, 38th Infantry, 3d Division.
Lieutenant Murray led his platoon in flank attack against a superior attacking force of the enemy. His fearlessness when exposed to great danger sustained the morale of his men. He continued in this attack until all but 3 of his men were killed or wounded. His conduct was an important contributing item to holding of the position against the repeated onslaughts of the enemy.
Posthumously awarded. Medal presented to mother, Mrs. P. J. Murray.

MURRAY, ROBINSON_____
Near Mezy, France, July 15, 1918.
R—Cambridge, Mass.
B—Boston, Mass.
G. O. No. 23, W. D., 1919.

First lieutenant, 38th Infantry, 3d Division.
Lieutenant Murray alone attacked an enemy observation post held by 10 of the enemy. He later organized a detachment of scattered men and filled a gap in our lines.

MURRAY, WILFRED L. (125709)_____
Near Fleville, France, Oct. 6, 1918.
R—Warren, Ill.
B—Winslow, Ill.
G. O. No. 44, W. D., 1919.

Corporal, Headquarters Company, 6th Field Artillery, 1st Division.
Corporal Murray voluntarily went forward and made his way to the enemy front lines to locate hostile artillery firing at short range on our batteries. While on this mission he rescued a wounded comrade and carried him to safety.

MURRIAN, JOHN H_____
Near Bellicourt, France, Sept. 29, 1918.
R—Knoxville, Tenn.
B—Ogden, Utah.
G. O. No. 21, W. D., 1919.

First lieutenant, 117th Infantry, 30th Division.
Lieutenant Murrian, acting as regimental intelligence officer, went out with another officer and 9 soldiers to establish an advance outpost. Near the front line they were caught in a German barrage; both officers were wounded, 2 soldiers killed and 4 wounded. As soon as he regained consciousness, he gave first aid to the other wounded, and then proceeded with a sergeant to establish the advance post and communication by telephone with the regimental post of command.

MUSE, EZRA M. (553231)_____
At Chateau-Thierry, France, May 31, to June 4, 1918.
R—New Brookland, S. C.
B—Nelson, S. C.
G. O. No. 132, W. D., 1918.

Sergeant, Company B, 7th Machine Gun Battalion, 3d Division.
While commanding a machine gun in a building which had been struck 3 times, he remained at his post, though told he might leave, because he had a better field of fire from this building than could be obtained elsewhere.

*MUTIC, ELI (94961)_____
Near Sedan, France, Nov. 7, 1918.
R—Cleveland, Ohio.
B—Hungary.
G. O. No. 44, W. D., 1919.

Private, Company K, 166th Infantry, 42d Division.
Private Mutic was a member of a patrol sent out to silence machine-gun nests which were holding up the battalion's advance. When the officer leading the patrol fell, mortally wounded, he attempted to go to the officer's assistance, despite heavy fire from machine guns only 100 yeards away, and was himself killed.
Posthumously awarded. Medal presented to cousin, Mrs. Stella Walker.

MYERS, CHARLES W_____
At Vaux, France, July 1, 1918.
R—Coketon, W. Va.
B—Marysville, Pa.
G. O. No. 99, W. D., 1918.

First lieutenant, Medical Corps, attached to 9th Infantry, 2d Division.
At Vaux, July 1, 1918, established under heavy shell fire an advance dressing station for the treatment and evacuation of men wounded in the first waves of the assault.

MYERS, CLAUDE B. (46730)_____
Near Exermont, France, Oct. 4 to 11, 1918, and south of Sedan, Nov. 5 to 7, 1918.
R—Fargo, N. Dak.
B—Fargo, N. Dak.
G. O. No. 39, W. D., 1920.

Sergeant, Company D, 18th Infantry, 1st Division.
During the operations of Oct. 4 to 11, 1918, Sergeant Myers carried messages through heavy artillery and machine-gun fire. During the operations of Nov. 5 to 7, 1918, he was in command of the platoon which maintained liaison between the First and Second Brigades. While performing this duty he single-handed captured 2 enemy prisoners.

MYERS, DEMARR E_____
Near Bayonville, France, Nov. 1, 1918.
R—Steubenville, Ohio.
B—Steubenville, Ohio.
G. O. No. 35, W. D., 1919.

Private, 82d Company, 6th Regiment, U. S. Marine Corps, 2d Division.
Exposing himself to enemy fire, Private Myers, with another soldier, courageously advanced ahead of their platoon and captured 5 machine guns and 14 prisoners.

MYERS, GEORGE F. (42563)
Near Hill 272, Argonne Forest, France, Oct. 9, 1918.
R—Northampton, Mass.
B—Northampton, Mass.
G. O. No. 44, W. D., 1919.

Private, first class, Company D, 18th Infantry, 1st Division.
When the advance of his company had been stopped by machine-gun fire, Private Myers, alone and on his own initiative, advanced into the fog under intense fire, and, with a total disregard for personal safety, captured the gun and its entire crew.

MYERS, IRWIN (1375647)
Near Romange, France, Oct. 30, 1918.
R—Chicago, Ill.
B—Humphrey, Nebr.
G. O. No. 71, W. D., 1919.

Corporal, Headquarters Company, 122d Field Artillery, 33d Division.
Facing heavy machine-gun and artillery fire, he crawled beyond the Infantry front lines to a crest overlooking the enemy position. Working under continuous fire, he made a panoramic sketch of hostile positions, which proved of great value in directing our Artillery fire.

MYERS, LOUIS W. (554137)
Near Le Rocq, France, July 14–15, 1918.
R—Fort Worth, Tex.
B—Waco, Tex.
G. O. No. 98, W. D., 1919.

Corporal, Company E, 8th Machine Gun Battalion, 3d Division.
He repeatedly exposed himself to the terrific enemy shell fire in carrying messages and rendering first aid to the wounded.

MYERS, OSCAR B
Near Cierges, France, Sept. 28, 1918.
R—Mount Vernon, N. Y.
B—Mount Vernon, N. Y.
G. O. No. 1, W. D., 1919.

First Lieutenant, 147th Aero Squadron, Air Service.
Sent on a particularly hazardous mission, he harassed and routed enemy troops. He then climbed higher to look for German planes. With two other officers, he encountered 9 Fokkers, protecting a reconnaissance machine, flying in one of the most effective formations used by the enemy. Outmaneuvering the hostile planes, the 5 officers succeeded in routing them. After a quick turn, he dived at the reconnaissance machine and crashed it to the ground in flames.

MYERS, WILLIAM R. (1449934)
Near Baulny, France, Sept. 28, 1918.
R—Lawrence, Kans.
B—Carbondale, Kans.
G. O. No. 55, W. D., 1920.

Sergeant, medical detachment, 137th Infantry, 35th Division.
Early on the morning of September 28, Sergeant Myers was wounded in the shoulder by a machine-gun bullet. In spite of his wound he continued, under heavy shell and machine-gun fire, to render first aid to the wounded until wounded a second time by a shell fragment.

MYHRMAN, ROBERT E
Near Very, France, Sept. 26, 1918.
R—Chicago, Ill.
B—Sweden.
G. O. No. 32, W. D., 1919.

Captain, 122d Field Artillery, 33d Division.
While his battery position was being heavily shelled by the enemy artillery, Captain Myhrman remained constantly with his men, ordering them to safety and caring for a wounded man. After his battery had been placed in position he conducted his own reconnaissance and prepared his own firing data, with no regard for the danger to which he was exposed from heavy enemy shell fire.

MYHRUM, MELVIN (2152651)
Near Brieulles, France, Oct. 7, 1918.
R—Fosston, Minn.
B—Fosston, Minn.
G. O. No. 32, W. D., 1919.

Private, Company K, 132d Infantry, 33d Division.
The patrol of which Private Myhrum was a member was under constant and exacting machine-gun and rifle fire. After the officer in charge had been wounded and the patrol scattered he returned to his company and voluntarily acted as guide for stretcher bearers to bring in the wounded officer. Being unable to locate him, Private Myhrum remained and searched, during which time he was twice wounded. He led a second group of stretcher bearers to the spot where the officer was finally located, and then assisted in carrying him to the rear before reporting for treatment.

*MZIK, CHARLES (1519493)
Near Eyne, Belgium, Nov. 2, 1918.
R—Cleveland, Ohio.
B—Cleveland, Ohio.
G. O. No. 72, W. D., 1920.

Corporal, Company K, 145th Infantry, 37th Division.
In full view of the enemy and under heavy artillery and machine-gun fire Corporal Mzik, with 2 other men, swam the Escaut River and assisted in the construction of a footbridge. The construction of this bridge aided materially in the later successful operations of American troops in this vicinity. Corporal Mzik was killed in the performance of this act.
Posthumously awarded. Medal presented to sister, Mrs. Nettie Bagavia.

NACHTMANN, LUDWIG J. (1784830)
Near Damvillers, France, Nov. 9, 1918.
R—Bustleton, Pa.
B—Philadelphia, Pa.
G. O. No. 37, W. D., 1919.

Sergeant, Machine Gun Company, 315th Infantry, 79th Division.
Although seriously wounded, he remained at his post and continued to direct the fire from his section under heavy shellfire until carried away by first-aid men.

NADEAU, DAVID (550836)
Near Mezy, France, July 15–19, 1918.
R—Woonsocket, R. I.
B—North Adams, Mass.
G. O. No. 44, W. D., 1919.

First sergeant, Company F, 38th Infantry, 3d Division.
Sergeant Nadeau remained on duty throughout the battle of the Marne, July 15–19, although seriously wounded, and rendered valuable assistance in sustaining the morale and managing the troops in the line.

NAEGLE, HANS MAURICE
Near Villemontry, France, Nov. 10, 1918.
R—Toquerville, Utah.
B—Moroni, Utah.
G. O. No. 32, W. D., 1919.

Private, 17th Company, 5th Regiment, U. S. Marine Corps, 2d Division.
Private Naegle and a companion went out ahead of the line and silenced a machine gun which threatened to hold up the advance of his company.

NAGAZYNA, JOHN JAMES
In the attack on Tigny, France, July 19, 1918.
R—Brooklyn, N. Y.
B—Cohoes, N. Y.
G. O. No. 117, W. D., 1918.

Gunnery sergeant, 96th Company, 6th Regiment, U. S. Marine Corps, 2d Division.
During a critical time in the assault against Tigny, when his company had suffered heavy losses, he set such an example of personal bravery and determination as to inspire his men to success. At a time when it seemed impossible to advance any farther his fearlessness in moving up and down his lines to steady his men encouraged them to go forward against heavy odds and take and hold their objective.

NAGOWSKI, ALOIZY (1215123)............
Near Ronssoy, France, Sept. 29, 1918.
R—Buffalo, N. Y.
B—Buffalo, N. Y.
G. O. No. 139, W. D., 1918.

Corporal, Company H, 108th Infantry, 27th Division.
He left shelter, went forward under intense machine-gun fire, and carried a wounded officer to a place of safety. In accomplishing this mission he was severely wounded.

NAIL, E. KELLEY (1491216)............
Near St. Etienne, France, Oct. 8, 1918.
R—Cleburne, Tex.
B—Cleburne, Tex.
G. O. No. 37, W. D., 1919.

Sergeant, Company L, 142d Infantry, 36th Division.
Sergeant Nail, in charge of a patrol, successfully flanked a machine-gun nest of several guns which was holding up the advance of his company and captured four German officers and 108 men. The success of the assault was largely due to the energy and good judgment of Sergeant Nail.

NAIMAN, HERMAN A. (2716)............
Near Soissons, France, July 18-20, 1918.
R—Gilead, Nebr.
B—Gilead, Nebr.
G. O. No. 15, W. D., 1919.

Private, medical detachment, 28th Infantry, 1st Division.
He displayed unusual courage and devotion to duty by remaining with the first wave of the attack during the 3 days of severe fighting and continuing under constant and heavy fire to give first aid to the wounded and assisting in the rescue of injured men.

*NALLE, JAMES B............
At La Tuilerie Farm, France, July 22-23, 1918.
R—Washington, D. C.
B—Albemarle County, Va.
G. O. No. 32, W. D., 1919.

Major, 4th Infantry, 3d Division.
While making an inspection of the 2 leading battalions of the regiment on the night of July 22, 1918, Major Nalle entered La Tuilerie Farm and found it occupied. As his party was leaving the farm persons were heard approaching, and, fearing that his men might fire upon friendly troops, he stepped from behind the wall and challenged the party. He was immediately fired upon and killed.
Posthumously awarded. Medal presented to widow, Mrs. Gladys F. Nalle.

NAREWOUCHEK, TROFEM (2338824)......
Near Mont St. Pere, France, July 22, 1918.
R—Philadelphia, Pa.
B—Russia.
G. O. No. 32, W. D., 1919.

Sergeant, Company F, 4th Infantry, 3d Division.
Leaving his place of safety, he made his way across an open field exposed to machine-gun fire and assisted in the rescue of 3 wounded comrades.

NARVESON, PALMER O. (2157848)........
Near Bellicourt, France, Sept. 29, 1918.
R—Twin Lakes, Minn.
B—Twin Lakes, Minn.
G. O. No. 81, W. D., 1919.

Sergeant, Company H, 119th Infantry, 30th Division.
When he and 2 soldiers, separated from the rest of the company, were fired upon from 3 directions he attacked and demolished a machine-gun nest by himself and then reduced a second hostile position. Though wounded and slightly gassed, he refused to be evacuated and continued the advance.

NASH, ARCHIE C. (198955)............
At Marcheville, France, Sept. 26, 1918.
R—Cambridge, Mass.
B—Boston, Mass.
G. O. No. 143, W. D., 1918.

Private, first class, Headquarters Company, 102d Infantry, 26th Division.
He displayed remarkable coolness and courage under violent bombardment when he voluntarily repaired telephone lines and rendered great assistance in maintaining communication. Although wounded, he continued his work until ordered evacuated by his commanding officer.

NASH, JAMES F. (1709870)............
Near Ville-Savoye, France, Aug. 22, 1918.
R—Brooklyn, N. Y.
B—New York, N. Y.
G. O. No. 32, W. D., 1919.

Private, Company K, 308th Infantry, 77th Division.
While his company was attacked by greatly superior numbers of the enemy Private Nash continued to operate his automatic rifle, even after having been wounded 2 times in the chest. After the attacking force had been driven off he refused the use of a litter in favor of a comrade whom he thought more seriously wounded than himself.

NATION, JAMES (1629729)............
On Hill 272, near Fleville, France, Oct. 9, 1918.
R—Duncan, Ariz.
B—Thatcher, Ariz.
G. O. No. 35, W. D., 1920.

Private, Company C, 16th Infantry, 1st Division.
The squad of which Private Nation was a member was directed to attack an enemy machine-gun position. During the attack all the other members of the squad were killed or wounded. He alone rushed the position, captured the gun, and killed 4 of the crew.

*NAUGHTON, FRANCIS X. (1246140)......
In the Argonne sector, France, Oct. 1, 1918.
R—Harrisburg, Pa.
B—Harrisburg, Pa.
G. O. No. 95, W. D., 1919.

Private, sanitary detachment, 112th Infantry, 28th Division.
While mess was being served a shell exploded, killing 9 men and wounding 20. Private Naughton, although severely wounded in the chest with shell splinters, 1 leg blown nearly off, and bleeding badly, refused help until the others had been attended to. Skilled in first aid, he instructed others how to adjust a tourniquet and rendered other assistance to the wounded, finally permitting his own wounds, which subsequently caused his death, to be attended after all others were cared for.
Posthumously awarded. Medal presented to father, Timothy Naughton.

NAY, ORIN E. (181321)............
Near Bethincourt, France, Sept. 26, 1918.
R—Kansas City, Mo.
B—Wheeling, Mo.
G. O. No. 37, W. D., 1919.

Corporal, Company A, 1st Gas Regiment.
Voluntarily leaving shelter, Corporal Nay and another soldier made their way, through terrific enemy barrage of artillery and machine-gun fire, to the aid of wounded comrades, carrying them to first-aid stations and administering treatment.

NEAL, GEORGE W. (181165)............
Near Bethincourt, France, Sept. 26, 1918.
R—Bulger, Pa.
B—Bulger, Pa.
G. O. No. 37, W. D., 1919.

Sergeant, first class, Company A, 1st Gas Regiment.
Voluntarily leaving shelter, Sergeant Neal and another soldier made their way through a terrific enemy barrage of artillery and machine-gun fire to the aid of wounded comrades, carrying them to first-aid stations and administering treatment.

NEALIS, JOHN J. (1200903)_____
Near Ronssoy, France, Sept. 29, 1918.
R—New York, N. Y.
B—Avoca, Pa.
G. O. No. 20, W. D., 1919.

Sergeant, Company C, 102d Field Signal Battalion, 27th Division.
During the operations against the Hindenburg line he, while in charge of telephone communication between battalion headquarters and forward positions, accompanied the advancing Infantry forward, established his advance post, where 1 of his assistants was killed by shellfire and he himself wounded, and under constant bombardment kept the telephone lines in operation, remaining at his post for 9 hours, until wounded a second time. When completely exhausted he turned over his apparatus to the man sent to relieve him. His extreme gallantry, courage, and bravery afforded a magnificent example to the combat troops who witnessed it.

NEEL, ROLAND H_____

East of Saint-Die, France, Aug. 17, 1918.
R—Macon, Ga.
B—Macon, Ga.
G. O. No. 81, W. D., 1919.

Second lieutenant, Coast Artillery Corps, observer, 99th Aero Squadron, Air Service.
Lieutenant Llewellyn acting as pilot, and Lieutenant Neel acting as observer, carried on successful liaison with the Infantry during the attack on Frapelle. They flew over the enemy lines at an altitude of only 400 meters, firing on and disconcerting the enemy, and thereby giving courage and confidence to the American forces. Despite heavy fire from 15 antiaircraft machine guns and several batteries of antiaircraft artillery they performed their work efficiently. Their airplane was struck by a number of machine-gun bullets, 1 of which cut the rudder and elevator control wires and caused the rudder to jam. The broken control wire was held and operated by Lieutenant Neel, under direction of Lieutenant Llewellyn. Running the machine together in this manner they continued their liaison work until the plane began to become unmanageable, when, in spite of its damaged condition, they brought it back to their airdrome.

NEELON, RAYMOND V. (1658251)_____
Near Imecourt, France, Nov. 1, 1918.
R—Medway, Mass.
B—Medway, Mass.
G. O. No. 44, W. D., 1919.

Sergeant, Company F, 319th Infantry, 80th Division.
Taking command of 2 platoons after their commanders had become casualties, he attacked a machine-gun nest, taking 2 guns and 146 prisoners. Later, after repulsing 2 strong counterattacks, he alone crawled out and captured a prisoner with a machine gun, which he at once set up to strengthen his position.

NEELY, JAMES (2705815)_____
Near Cierges, France, Sept. 28, 1918.
R—Philadelphia, Pa.
B—Philadelphia, Pa.
G. O. No. 44, W. D., 1919.

Private, Company F, 146th Infantry, 37th Division.
After his platoon had withdrawn about 50 yards to an established line, a wounded comrade was seen lying ahead in the position which they formerly occupied. The enemy had just launched a strong counterattack, but Private Neely, with another soldier, volunteered to go to the assistance of the wounded man. In the face of terrific fire of enemy artillery and machine guns and the fire of their own comrades, who were resisting the attack, Private Neely succeeded in bringing his man to a place of safety.

NEESE, HARRY L. (1305811)_____
Near Premont, France, Oct. 8, 1918.
R—Swansea, S. C.
B—Swansea, S. C.
G. O. No. 81, W. D., 1919.

Private, Company C, 117th Infantry, 30th Division.
When the advance of his company was held up by a machine-gun emplacement, he went forward with 2 other soldiers and attacked the enemy position. He shot both of the enemy gunners, showing marked personal bravery under heavy fire.

NEIBLING, HARLOU P_____
At Brouville, France, Sept. 2, 1918, and near Fort du Marr, France, Sept. 26, 1918.
R—Minneapolis, Minn.
B—Huron, S. Dak.
G. O. No. 46, W. D., 1919.

First lieutenant, Field Artillery, attached to 2d Balloon Squadron, Air Service.
While Lieutenant Neibling was making an aerial reconnaissance from a balloon he was repeatedly attacked by enemy planes, 2 of which dived at the balloon and opened fire with incendiary bullets. With great coolness he fired at 1 of them with his pistol and took a picture of the plane with his camera. When the balloon took fire, he was forced to jump, but he took 2 more pictures on the way down in spite of being fired upon. He reascended as soon as a new balloon could be inflated. On September 26 this officer was again attacked while conducting a reglage, but, hanging from the basket with one arm, he fired his pistol at one of the enemy planes, and jumped only when his balloon burst into flames. He immediately continued his mission in another balloon.

NEIGGEMANN, HENRY J. (2852568)_____
During the St. Mihiel offensive, Sept. 14, 1918.
R—Streator, Ill.
B—Streator, Ill.
G. O. No. 72, W. D., 1920.

Corporal, Company D, 358th Infantry, 90th Division.
Corporal Neiggemann, with 4 other men, volunteered to cross a valley to the woods opposite and silence machine guns which had held up the advance of his company. In the face of heavy enemy fire this small group accomplished its mission, thus enabling the company to cross the valley without further loss. Corporal Neiggemann was severely wounded in the performance of this act.

NEIL, ALBERT F. (1639027)_____
Near Ronssoy, France, Sept. 29, 1918.
R—Santa Barbara, Calif.
B—Holden, Minn.
G. O. No. 32, W. D., 1919.

Corporal, Company A, 301st Battalion, Tank Corps.
After aiding in rescuing the wounded from his tank, which had been struck by a shell, Corporal Neil, with Pvt. Robert F. Wisher, dismounted machine guns from the tank and operated them against the enemy until these were put out of action. They then secured rifles and hand grenades and organized an attack on the enemy trenches, which they captured and held until depletion of their numbers forced them to fall back. Later they joined Australian troops and fought with them throughout the remainder of the day.

NEILL, HENRY WHEATON_____
At Frapelle, France, Aug. 19, 1918.
R—White Springs, Fla.
B—Greenville, S. C.
G. O. No. 15, W. D., 1919.

Second lieutenant, 15th Machine Gun Battalion, 5th Division.
He displayed great courage, tenacity, and devotion to duty when, although severely wounded early in the attack and suffering great pain, he retained command of his platoon and directed its movements until its objective was attained.

NEITZEIT, ISAAC (1699169)_____
Near Bois-de-la-Naza, France, Oct. 5, 1918.
R—New York, N. Y.
B—Russia.
G. O. No. 59, W. D., 1919.

Corporal, Company L, 305th Infantry, 77th Division.
In the face of heavy machine-gun and grenade fire he went forward, with 3 other soldiers, and brought back 5 seriously wounded men to a point where they could be given first-aid treatment. He showed bravery and coolness in effecting the rescue, in which he was himself wounded.

NEITZEL, ALBERT R. (278702)_____
  Near Romagne, France, Oct. 14,
    1918.
  R—Wheeler, Kans.
  B—St. Francis, Kans.
  G. O. No. 21, W. D. 1919.

Private, first class, Company M, 126th Infantry, 32d Division.
In an attack on Cote Dame Marie the 126th Infantry was held up, owing to intense enemy machine-gun fire. Private Neitzel volunteered as a member of a combat patrol which cut through the enemy lines, captured 10 machine guns, killed and captured 15 of the enemy, and forced a large number to surrender, clearing that part of the Cote Dame Marie of the enemy, thus enabling the regiment to continue their advance.

NELSEN, ADOLPH (2152402)_____
  Near Chipilly Ridge, France, Aug.
    9, 1918.
  R—Soldier, Iowa.
  B—Soldier, Iowa.
  G. O. No. 81, W. D., 1919.

Private, Company H, 131st Infantry, 33d Division.
Although severely wounded, he, on his own initiative, went out in advance of his lines, armed with an automatic rifle, and mopped up a machine-gun nest in which there were 3 guns. He killed 4 of the enemy crew and brought the other 2 Germans back as prisoners. He set an example of heroism and devotion to duty, performing this service under heavy artillery and machine-gun fire.

*NELSON, ARTHUR E. (198200)_____
  Near Vaux, France, July 1, 1918.
  R—Albert Lea, Minn.
  B—Kenosha, Wis.
  G. O. No. 99, W. D., 1918.

Sergeant, first class, Company C, 1st Field Signal Battalion, 2d Division.
Under heavy bombardment, while sick, he went to maintain communication with an attacking battalion of the Infantry and was killed in this heroic action.
Posthumously awarded. Medal presented to father, James C. Nelson.

NELSON, BERNARD (100595)_____
  Near Landres-et-St. Georges, France,
    Oct. 14, 1918.
  R—Centerville, Iowa.
  B—Centerville, Iowa.
  G. O. No. 44, W. D., 1919.

Sergeant, Company D, 168th Infantry, 42d Division.
During the attack on Hill 288, when the assault wave was held up by machine-gun fire, Sergeant Nelson volunteered and led 2 squads to silence these guns. He cut his way through strong barbed-wire entanglements, advanced up a very steep slope in the face of direct machine-gun fire, entered the trench, and killed or wounded the entire crews of the 2 guns, making it possible for the battalion to advance.

NELSON, CHARLES E._____
  Near Trugny, France, July 22, 1918.
  R—Defiance, Ohio.
  B—Defiance, Ohio.
  G. O. No. 37, W. D., 1919.

Second lieutenant, 104th Infantry, 26th Division.
Lieutenant Nelson led 8 soldiers in an attack on a machine gun that was inflicting severe losses in his company. Two of his detachment were killed and 2 wounded before he reached the machine-gun nest; with the remaining 4 he attacked, captured the gun, killed 5 Germans, including 1 officer, and took 11 prisoners.

NELSON, CHRISTIAN F. M. (524321)____
  Near the Bois-de-Brieulles, France,
    Oct. 9, 1918.
  R—Flat City, Alaska.
  B—Denmark.
  G. O. No. 87, W. D., 1919.

Private, Company F, 1st Gas Regiment.
Displaying remarkable perseverance and daring, Private Nelson, a runner, made his way 300 yards through a heavy barrage with a message for the commander of a Stokes mortar platoon. Later he volunteered to lead 4 wounded men back through the barrage to an aid station. On the way he met 3 other wounded soldiers, 1 of whom had been severely gassed and was unable to walk. Private Nelson carried this man to the dressing station, knowing that his clothes were saturated with mustard gas.

NELSON, GEORGE (45882)_____
  North of Exermont, France, Oct.
    10, 1918.
  R—Greenriver, Utah.
  B—Thistle, Utah.
  G. O. No. 39, W. D., 1920.

Corporal, Company A, 18th Infantry, 1st Division.
When his platoon was held up by machine-gun fire from the front Corporal Nelson, with 1 other, advanced in front of the line and captured the gun with its crew. Their action enabled his platoon to continue the advance.

NELSON, GUY A. (2157873)_____
  In the Bois-de-Jure, near Gercourt,
    France, Sept. 26, 1918.
  R—Albert Lea, Minn.
  B—Albert Lea, Minn.
  G. O. No. 27, W. D., 1919.

Private, Company F, 1st Gas Regiment.
He volunteered, with another soldier, to attack a machine-gun nest which was holding up the advance. They advanced against very heavy machine-gun fire and captured the position, killing 2 Germans and routing the remainder of the gun crew.

NELSON, HERBERT W. (1896562)_____
  Near Xou Hill, France, Sept. 13,
    1918.
  R—Alton, R. I.
  B—Providence, R. I.
  G. O. No. 71, W. D., 1919.

Private, first class, Company C, 320th Machine Gun Battalion, 82d Division.
In the face of heavy machine-gun and shellfire he went ahead of his own lines and, with the aid of another soldier, carried back a wounded infantryman, who had fallen far in advance of our lines while on a patrol. Private Nelson displayed marked personal bravery and coolness under fire.

NELSON, MARTIN (2106070)_____
  Near Chezy, France, July 18 and 19,
    1918.
  R—Milton, N. Dak.
  B—Roseau, Minn.
  G. O. No. 16, W. D., 1920.

Corporal, Company H, 58th Infantry, 4th Division.
On the morning of the 18th, Corporal Nelson was wounded in the hip by a piece of shrapnel. A few hours later he was wounded in the arm by a bullet. He refused to be evacuated, but continued forward in the attack. On the 19th he was wounded in the left knee. In spite of his wounds this noncommissioned officer continued with his organization throughout the campaign.

*NELSON, OSCAR B._____
  At La Tuilerie Farm, France, Oct.
    16, 1918.
  R—Ottumwa, Iowa.
  B—Sweden.
  G. O. No. 13, W. D., 1919.

First lieutenant, 168th Infantry, 42d Division.
Lieutenant Nelson alone attacked 2 enemy machine guns, killing 2 of the enemy and capturing 19.
Oak-leaf cluster.
Lieutenant Nelson is awarded a bronze oak leaf for the following act of extraordinary heroism in action at La Tuilerie Farm, France, Oct. 16, 1918: Accompanied by 6 soldiers, this officer advanced 600 yards beyond his own lines through heavy fire from enemy artillery, machine guns, and rifles, and captured 2 more machine guns, killing, capturing, or dispersing their crews. Still later in the day he led his company in an attack on Chatillon Hill and took his objective, but in so doing received wounds which caused his death. His coolness, courage, and utter disregard for his own safety were a source of great inspiration to his men.
Posthumously awarded. Medal presented to father, Jacob Nelson.

*NELSON, SEVERT J. (2384980)_____
   Near Cunel, France, Oct. 12, 1918.
   R—Ellsworth, Iowa.
   B—Morris, Ill.
   G. O. No. 78, W. D., 1919.

Sergeant, Company M, 60th Infantry, 5th Division.
Leading his platoon in the face of murderous machine-gun fire from his front and flanks, Sergeant Nelson reached his objective after taking 4 enemy machine-gun nests and killing and capturing many prisoners. He then continued ahead of his men, and alone cleaned out some houses in the woods, which were occupied by the enemy.
Posthumously awarded. Medal presented to brother, Andrew M. Nelson.

*NELSON, THEODORE VERNON_____
   At Bois-de-Chaume, near Consen-voye, France, Oct. 9, 1918
   R—Chicago, Ill.
   B—Chicago, Ill.
   G. O. No. 19, W. D., 1920.

Second lieutenant, 132d Infantry, 33d Division.
When the right platoon of his company was held up by machine-gun fire, Lieutenant Nelson, alone and in the face of direct fire, attacked the gun crew, killing the gunner and capturing 2 prisoners. After reaching his objective he was wounded but refused to be evacuated and continued to direct the operations of his company. When an enemy counterattack forced a withdrawal of his company, he ordered the men who were assisting him to the rear to leave him. He later died of wounds.
Posthumously awarded. Medal presented to mother, Mrs. Theodore Nelson.

*NESSELSON, NATE T. (1247301)_____
   Near Fismes, France, Aug. 9, 1918.
   R—Bradford, Pa.
   B—Bradford, Pa.
   G. O. No. 11, W. D., 1921.

Private, first class, Company C, 112th Infantry, 28th Division.
Private Nesselson repeatedly exposed himself to heavy enemy fire in order to deliver messages from his company to the battalion commander. In the performance of this mission it as necessary for him to cross the Vesle River, which was constantly swept by enemy machine-gun fire. He volunteered to carry a message after others had been killed in the attempt and continued to perform this perilous duty until he was mortally wounded.
Posthumously awarded. Medal presented to father, M. Nesselson.

NETTE, WILLIAM J. (1217198)_____
   Near Montzeville, France, Sept. 14, 1918.
   R—New York, N. Y.
   B—New York, N. Y.
   G. O. No. 32, W. D., 1919.

Private, first class, Battery B, 104th Field Artillery, 27th Division.
When a continuous bombardment had set fire to the camouflage covering of a large ammunition dump of 75-millimeter shells and exploded 9 of the shells, he, utterly disregarding his personal safety, left a sheltered position and ran to the dump and, with the aid of 3 other men, extinguished the fire, not only saving the ammunition but also preventing the exact locating of the dump by the enemy.

NEUBERGER, HARRY H_____
   At Courbon, France, July 14–15, 1918.
   R—New York, N. Y.
   B—Far Rockaway, N. Y.
   G. O. No. 44, W. D., 1919.

First lieutenant, 10th Field Artillery, 3d Division.
He volunteered and assisted another officer in driving an ambulance, making 3 trips to Greves Farm under the most intense shell fire. He continued to assist in the evacuation of the wounded even after being gassed.

NEVINS, CHESTER D. (96887)_____
   Near Sergy, France, July 28, 1918.
   R—Moundville, Ala.
   B—Moundville, Ala.
   G. O. No. 71, W. D., 1919.

Sergeant, Company F, 167th Infantry, 42d Division.
Crossing 500 yards of ground swept by intense machine-gun and artillery fire, he went to the rescue of a wounded soldier. He carried the wounded man back to our lines, inspiring by his example all who saw his heroic act.

*NEWBOLD, CLINTON V. P_____
   Near Soissons, France, July 19, 1918.
   R—Akron, Ohio.
   B—Norwood, Pa.
   G. O. No. 59, W. D., 1919.

First lieutenant, 26th Infantry, 1st Division.
After the loss of many of his men and in the face of machine-gun fire, near Soissons, France, July 19, 1918, he led and directed his command to successful attack, although he himself was mortally wounded.
Posthumously awarded. Medal presented to widow, Mrs. C. V. P. Newbold.

NEWCOMER, FRANCIS K_____
   Near Fismes, France, Aug. 5, 1918.
   R—Pittsburgh, Pa.
   B—Byron, Ill.
   G. O. No. 143, W. D., 1918.

Lieutenant colonel, 4th Engineers, 4th Division.
He made a reconnaissance along the south bank of the Vesle River in advance of the front lines for the purpose of selecting a bridge site. He then led a small party of Engineers, assisting in the work of removing the German entanglements and constructing a footbridge across the Vesle River, completing this work in the face of fire of great intensity. His coolness and personal bravery afforded an inspiring example to the men of his command.

NEWELL, ALEXANDER (551266)_____
   Near Chateau-Thierry, France, July 15, 1918.
   R—Chicago, Ill.
   B—Ireland.
   G. O. No. 99, W. D., 1918.

Private, Company G, 38th Infantry, 3d Division.
Leading a squad of 9 men, he fearlessly passed through an enemy barrage, captured 5 machine guns and 33 prisoners, and recovered a sergeant of his company who was helpless from wounds, all under violent artillery fire, near Chateau-Thierry, France, July 15, 1918.

NEWHALL, STEPHEN K. (378126)_____
   Near Bois-de-Etrayes, France, Oct. 23, 1918.
   R—Brooklyn, N. Y.
   B—Billerica, Mass.
   G. O. No. 35, W. D., 1919.

Private, first class, Machine Gun Company, 113th Infantry, 29th Division.
After being seriously wounded, Private Newhall volunteered and carried a message through a heavy barrage to his company commander. He refused to be evacuated until his message had been delivered to the proper officer.

NEWLIN, ELMER L. (1243678)_____
   At Fismette, France, Aug. 10, 1918.
   R—Chester, Pa.
   B—Trainer, Pa.
   G. O. No. 116, W. D., 1918.

Private, Company C, 111th Infantry, 28th Division.
Having heard that 2 wounded comrades were lying in advance of the line immediately north of Fismette, Private Newlin and 2 other members of his company volunteered to go through machine-gun and rifle fire to bring them in. In their first attempt all were wounded and driven back, but in spite of their injuries they advanced a second time and reached the wounded men. Their courageous effort, however, was unfortunately in vain, as their comrades had been killed.

NEWTON, HARRY LEE (2662674)_____
   Near Vieville-en-Haye, France, Nov. 1, 1918.
   R—White Hall, Ill.
   B—Roodhouse, Ill.
   G. O. No. 37, W. D., 1919.

Private, Company C, 21st Machine Gun Battalion, 7th Division.
He acted as runner during offensive operations and under severest enemy barrages, and carried important messages to and from the front lines.

NEWTON, ISAAC M. (1316085)_____
  Near Bellicourt, France, Sept. 29,
    1918.
  R—Kerr, N. C.
  B—Kerr, N. C.
  G. O. No. 81, W. D., 1919.

Corporal, Company H, 119th Infantry, 30th Division.
With another soldier he attacked and destroyed 2 enemy machine-gun posts
200 yards in advance of our lines. While the other soldier stood guard at the
entrance of a dugout, he entered it and brought out 75 German soldiers and 3
officers, who were taken back to our lines as prisoners.

NICHOLS, HARLEY N. (241533)_____
  Near Charpentry, France, Oct. 4,
    1918.
  R—St. Louis Park, Minn.
  B—Kingman, Kans.
  G. O. No. 46, W. D., 1919.

Sergeant, Company C, 345th Battalion, Tank Corps.
While making an attack with 4 other tanks, Sergeant Nichols's tank was
struck by an enemy shell, which put it out of action. He continued to fire on
a machine-gun nest until it was apparently destroyed, when he, with his
driver, dismounted and started to the nest, wherefrom they were fired on by
the German gunners. They killed 2 gunners and disabled the guns,
and then drove the gunners from another gun. Under the protection of another tank, they started to our own lines, 1,500 meters away. On the way
back they encountered 2 Germans with antitank rifles and captured the
rifles. Sergeant Nichols and his driver were under heavy machine-gun and
artillery fire throughout the operation.

NICHOLLS, HAROLD O. (36238)_____
  Near Griscourt, France, Aug. 11,
    1918, and near Avocourt, France,
    Oct. 1 and 9, 1918.
  R—El Paso, Tex.
  B—Galveston, Tex.
  G. O. No. 26, W. D., 1919.

Sergeant, first class, 7th Balloon Company, Balloon Service, 1st Army.
On Aug. 11 Sergeant Nicholls volunteered and ascended for the purpose of
making an observation. He continued with his work until the balloon was
set on fire by attacking enemy planes. On Oct. 1 he remained on duty
until his balloon was fired by incendiary bullets, and again on Oct. 9,
while on duty with another observer, he remained with his balloon under
attack until it was set on fire by enemy planes and he then refused to jump
until his companion had escaped.

NICHOLSON, WILLIAM J._____
  Near the Bois-de-Beuge, Mont-
    faucon, France, Sept. 26, 1918.
  R—Washington, D. C.
  B—Washington, D. C.
  G. O. No. 15, W. D., 1923.
  Distinguished-service medal also
    awarded.

Brigadier general, 157th Infantry Brigade, 79th Division.
He established and maintained his brigade post of command on an exposed
elevation near the Bois-de-Beuge, in order that he might effectively direct
the attack of his brigade upon the Madeleine Farm and its surrounding woods.
Realizing the importance of increased artillery support, he personally visited
the division post of command behind Montfaucon to seek such support.
In his absence the brigade post of command open to enemy observation was
swept by a concentration of enemy machine-gun fire and artillery fire. In the
face of this terrific fire General Nicholson, with great coolness and with complete disregard for his own safety, rode forward on horseback to his brigade
post of command to issue orders for the renewed attack upon the Madeleine
Farm, supervising the formation for attack, and by his brave and gallant
example inspired the men of his command with renewed courage and determination, which enabled them to reach their objective and hold it against
repeated enemy counterattacks.

NICKELS, CHARLES E. (156943)_____
  Near St. Etienne-a-Arnes, France,
    Oct. 5–7, 1918.
  R—Boerne, Tex.
  B—Boerne, Tex.
  G. O. No. 23, W. D., 1919.

Sergeant, Company D, 2d Engineers, 2d Division.
Advancing ahead of the Infantry, he made several reconnaissances of the town
of St. Etienne-a-Arnes, France; and, in spite of the danger, exposed to machine-gun and artillery fire of our own and enemy guns, he procured and returned with valuable information.

*NICKERSON, SIMEON L. (60347)_____
  Near Epieds, France, July 23, 1918.
  R—Middleboro, Mass.
  B—Middleboro, Mass.
  G. O. No. 116, W. D., 1918.

Sergeant, Company D, 101st Infantry, 26th Division.
Sergeant Nickerson, Corpl. M. J. O'Connell, and Pvt. Thomas Ryan volunteered to cross an open field in front of their company, in order to ascertain
the location of enemy machine guns. While engaged in this courageous enterprise they were shot and killed. The heroic self-sacrifice of these 3 men
saved the lives of many of their comrades who would have been killed had
the company attempted to make the advance as a whole.
Posthumously awarded. Medal presented to brother, Horace E. Nickerson.

*NICKLES, EDWARD E. (199025)_____
  Near Verdun, France, Oct. 24, 1918.
  R—Cambridge, Mass.
  B—Charlestown, Mass.
  G. O. No. 21, W. D., 1919.

Sergeant, Company B, 101st Field Signal Battalion, 26th Division.
At a time when the telephone lines were badly needed he remained without
shelter for several hours testing out the lines until a shell burst in his vicinity,
wounding him. He died from the effects of the wound in a few hours.
Posthumously awarded. Medal presented to widow, Mrs. Elizabeth M.
Nickles.

NICKOVICH, ROBERT (324403)_____
  At Vladivostok, Siberia, Nov. 17
    and 18, 1919.
  R—Acme, Wyo.
  B—Montenegro.
  G. O. No. 15, W. D., 1923.

Private, first class, Company B, Replacement Battalion, American Expeditionary Forces, Siberia.
For extraordinary heroism in action at Vladivostok, Siberia, Nov. 17 and 18,
1919. In answer to a call to save women and children who had been unavoidably entrapped in the railroad station by the fire from contending forces of
Russian Government and insurgent troops, he boldly entered the zone of fire
and rushed to the station which was being fired upon from 3 sides by
machine guns, rifles, and field pieces, and assisted in bringing back through
the fire-swept zone a number of women noncombatants.

NICOL, ALEXANDER L._____
  Near Juvigny, north of Soissons,
    France, Aug. 30, 1918.
  R—Sparta, Wis.
  B—Sparta, Wis.
  G. O. No. 116, W. D., 1919.

First lieutenant, 128th Infantry, 32d Division.
After being severely wounded, Lieutenant Nicol directed the orderly retirement
of his company and organized it under heavy fire of artillery and machine guns.
At great personal risk he made several trips forward to bring in wounded men.
Throughout the entire action he fearlessly exposed himself to fire in order
to encourage and cheer his men. His energetic and faithful work furnished
an example of calmness and courage to the men under his command.

NICOL, WILLIAM O. (1276736)_____
  Near Verdun, France, Oct. 12, 1918.
  R—Jersey City, N. J.
  B—Jersey City, N. J.
  G. O. No. 35, W. D., 1920.

Sergeant, Company A, 111th Machine Gun Battalion, 29th Division.
Sergeant Nicol displayed conspicuous courage and leadership in keeping the
guns of his section in action under heavy shell fire, covering the advance of
the Infantry. His section was caught in a hostile barrage, by which 2 of
his men were killed and 5 wounded. Sergeant Nicol led the rest of the section to shelter and then returned under shell fire and rescued the wounded
and dead bodies.

NIELSEN, JULIUS (446592)............
Near Landersbach, Alsace, Oct. 4, 1918.
R—Lake Benton, Minn.
B—Hull, Iowa.
G. O. No. 27, W. D., 1919.

Corporal, Company H, 53d Infantry, 6th Division.
Corporal Nielsen was in a detachment of 50 soldiers who were attacked by a hostile raiding party composed of 300 storm troops. Although wounded, he maintained his position under the heaviest bombardment and refused to leave his post until the enemy was repulsed.

*NIGHTINGALE, HARRY M. (08128)......
Near Bois-de-St. Remy, France, Sept. 12, 1918.
R—Auburn, Me.
B—New Brunswick.
G. O. No. 34, W. D., 1919.

Corporal, Company F, 103d Infantry, 26th Division.
Leading his squad forward to attack an almost impregnable machine-gun nest, he continued to press on when only 2 of his squad remained until he himself fell mortally wounded.
Posthumously awarded. Medal presented to mother, Mrs. Martha Nightingale.

*NILES, JULIUS.....................
Near St. Mihiel, France, Sept. 12, 1918.
R—St. Louis, Mo.
B—Denver, Colo.
G. O. No. 20, W. D., 1919.

First lieutenant, 6th Infantry, 5th Division.
While leading his platoon across an open space in front of a wood he was confronted by a sudden and terrific fire from German machine guns, which killed several of his men. Wishing to make a flank attack, and finding it difficult to pass the orders along, he rose up and started to the front wave of his platoon to give the necessary orders, but was killed before he could get the flank attack started.
Posthumously awarded. Medal presented to father, Jules Niles.

NIMMO, WILLIAM T. (60828)..........
Near Bois-de-St. Remy, France, Sept. 12, 1918.
R—Waltham, Mass.
B—St. Albans, Vt.
G. O. No. 46, W. D., 1919.

Sergeant, Company F, 101st Infantry, 26th Division.
During the drive across the St. Mihiel salient, he led a group of 25 men through a severe machine-gun fire and into the woods occupied by the enemy. There he charged a machine-gun nest single handed and captured the gun. The gun crew attempted to escape by entering a near-by dugout, but Sergeant Nimmo followed them into the dugout alone and captured the entire crew.

NIMS, WILLIE HARRISON (1311262)......
Near Montbrehain, France, Oct. 8, 1918.
R—Fort Mill, S. C.
B—Fort Mill, S. C.
G. O. No. 46, W. D., 1919.

First sergeant, Company G, 118th Infantry, 30th Division.
After all his company officers had been wounded, Sergeant Nims, though himself wounded in the leg by shell fire, assumed command, and led his company with remarkable dash through heavy machine-gun fire. Using a stick as a crutch, he continued forward until the objective was reached and the position consolidated, when he consented to go to the rear for treatment.

NIXON, GEORGE R...................
Near Domevre-en-Haye, France, Aug. 28, 1918, and near Malancourt, France, Sept. 28, 1918.
R—Indianapolis, Ind.
B—Dayton, Ohio.
G. O. No. 46, W. D., 1919.

First lieutenant, Field Artillery, attached to 3d Balloon Squadron, Air Service.
On August 28, Lieutenant Nixon was locating active enemy batteries from his balloon and was attacked several times by enemy planes, but refused to descend until one had set fire to the balloon. On Sept. 28, while he was on a reglage mission, 5 enemy planes fired at him. He remained in the basket until the balloon was a mass of flames, and 1 of the enemy aviators followed him to the ground, firing at him. Despite his narrow escape he immediately reascended.

NIXON, LONNIE H...................
Near Forsoy, France, July 15, 1918.
R—Eugene, Oreg.
B—Anson, Tex.
G. O. No. 44, W. D., 1919.

Captain, 7th Infantry, 3d Division.
He fearlessly led a counterattack through an intense barrage, inspiring his men to success by his personal example.

NIXON, WILLIAM J. (1243113)..........
Near Fismette, France, Aug. 10–13, 1918.
R—Philadelphia, Pa.
B—Philadelphia, Pa.
G. O. No. 50, W. D., 1919.

Private, sanitary detachment, 111th Infantry, 28th Division.
Seeing 5 of our men lying wounded on the enemy side of the street in the town of Fismette, Private Nixon voluntarily attempted the rescue of them, despite the fact that the enemy was pouring a deluge of machine-gun fire on the location. He organized a counterattacking force and repeatedly exposed himself in attacking the nest, finally succeeding in his mission of rescuing the wounded.

NOBLE, ALFRED HOUSTON...........
In the Bois-de-Belleau, France, June 6–8, 1918.
R—Federalsburg, Md.
B—Federalsburg, Md.
G. O. No. 110, W. D., 1918.

First lieutenant, 6th Regiment, U. S. Marine Corps, 2d Division.
He was conspicuous for his judgment and personal courage in handling his company in attacks against superior numbers in strongly fortified machine-gun positions. His fortitude and initiative enabled his command each time to achieve success.

*NOBLE, CLARENCE G...............
Near St. Gilles, France, Aug. 3, 1918.
R—Soperton, Wis.
B—Reedsville, Wis.
G. O. No. 37, W. D., 1919.

First lieutenant, 128th Infantry, 32d Division.
He voluntarily exposed himself to heavy shell fire in placing his men under cover during a heavy bombardment. While assisting a wounded soldier he was struck by a shell and killed.
Posthumously awarded. Medal presented to father, John H. Noble.

*NOBLE, EARL S. (42413)............
Near the forest of Argonne, France, Oct. 9, 1918.
R—Bird City, Kans.
B—Bird City, Kans.
G. O. No. 46, W. D., 1919.

Corporal, Company D, 16th Infantry, 1st Division.
After 4 members of his automatic-rifle squad had become casualties in an effort to get their automatic rifle into action against a machine-gun nest, Corporal Noble fearlessly exposed himself, set up the rifle, and silenced the machine gun just as another enemy machine gun on the flank opened fire and killed him.
Posthumously awarded. Medal presented to mother, Mrs. Janie Berry.

*NOBLE, ELMER J..................
Near Bois-de-Cheppy, France, Sept. 26, 1918.
R—Seattle, Wash.
B—Snohomish, Wash.
G. O. No. 39, W. D., 1920.

First lieutenant, 364th Infantry, 91st Division.
Lieutenant Noble gallantly led his men under heavy fire in an attack through barbed-wire entanglements on the enemy positions before Bois-de-Cheppy. His conduct had a marked moral effect upon his men and he continued leading the attack until killed by enemy fire.
Posthumously awarded. Medal presented to widow, Mrs. Elmer J. Noble.

NOBLE, GEORGE B.
East of Sergy, northeast of Chateau-
Thierry, France, July 28, 1918.
R—Portland, Oreg.
B—Leesburg, Fla.
G. O. No. 99, W. D., 1918.

First lieutenant, 168th Infantry, 42d Division.
He gave proof of unhesitating devotion and energy during the offensive operations of Sergy, brilliantly leading his platoon to the assault in disregard of all
danger. While charged with the support and protection of a reconnaissance
in no man's land he gave the best example of calmness, decision, and courage
under intense machine-gun fire. Wounded in this action, he refused to be
evacuated and remained in command of his platoon until ordered off the
field by his major.

NOEL, HENRY M.
East of Belleau Wood, France,
July 20, 1918.
R—St. Louis, Mo.
B—Webster Groves, Mo.
G. O. No. 125, W. D., 1918.

Second lieutenant, 103d Infantry, 26th Division.
Discovering a German machine-gun nest which was inflicting severe damage
upon his battalion, he led 12 men to the right flank of the nest and charged it
up a steep hill under fire from other guns. He and his men wiped out this
center of resistance and made possible the advance of his company. Although
wounded himself, he personally took command of large numbers of men of
the company, after his captain and other platoon commanders had been
killed or wounded, and advanced with them to the company's objective and
held it.

NOLAN, DENNIS E.
Near Apremont, France, Oct. 1,
1918.
R—Akron, N. Y.
B—Akron, N. Y.
G. O. No. 50, W. D., 1919.
Distinguished-service medal also
awarded.

Brigadier general, 55th Infantry Brigade, 28th Division.
While the enemy was preparing a counterattack, which they preceded by a
terrific barrage, General Nolan made his way into the town of Apremont
and personally directed the movements of his tanks under a most harassing
fire of enemy machine guns, rifles, and artillery. His indomitable courage
and coolness so inspired his forces that about 400 of our troops repulsed an
enemy attack of two German regiments.

NOLAN, VINCENT ALBERT.

Near St. Etienne, France, Oct. 5-9,
1918.
R—Livingston, Mont.
B—Livingston, Mont.
G. O. No. 98, W. D., 1919.

Pharmacist's mate, third class, U. S. Navy, attached to 18th Company, 5th
Regiment, U. S. Marine Corps, 2d Division.
During the operations at Blanc Mont Ridge he repeatedly went through intense
machine-gun and shellfire to administer first aid to officers and soldiers who
were wounded, and lying in exposed positions.

NOLTE, WILLIAM VERMONT.

Near Blanc Mont, France, Oct. 4,
1918.
R—St. Louis, Mo.
B—Hartford City, Ind.
G. O. No. 35, W. D., 1919.

Hospital apprentice, first class, U. S. Navy, attached to 5th Regiment, U. S.
Marine Corps, 2d Division.
He rendered exceptional assistance to his wounded comrades by continually
giving first aid to them under machine-gun fire.

NORMAN, CHARLES J. (1386686).
During the Meuse-Argonne offen-
sive, France, Sept. 26, 1918.
R—Champaign, Ill.
B—East Bend, N. C.
G. O. No. 5, W. D., 1920.

Private, Company B, 131st Infantry, 33d Division.
With three other soldiers he charged and captured a battery of three .77 field
pieces which, protected by machine guns, were firing point blank on the
position held by his company. This deed enabled his company to continue
the advance.

NORRIS, ELMER C. (1251007).
Near Fismes, France, Aug. 28, 1918.
R—Pittsburgh, Pa.
B—Woodsfield, Ohio.
G. O. No. 64, W. D., 1919.

Private, Battery E, 107th Field Artillery, 28th Division.
Although severely burned by gas, he refused to be evacuated, voluntarily
remaining on duty for 2 days, repairing telephone connections day and
night over shell-swept areas from the battery position to the observation
post. He worked faithfully, maintaining adequate telephone service, until
ordered to the rear.

NORRIS, ELMER L. (57052).
Near Exermont, France, Oct. 5,
1918.
R—Bigprairie, Ohio.
B—Bigprairie, Ohio.
G. O. No. 72, W. D., 1920.

Corporal, Company C, 28th Infantry, 1st Division.
Corporal Norris, an automatic rifleman, crept about 600 yards ahead of the
line held by our troops and after locating 2 enemy machine guns, he opened
fire and succeeded in destroying 1 of the enemy guns and forcing 2 of the
enemy to surrender. He then turned the captured gun on the enemy to
such effect that his company was able to advance with slight loss.

NORRIS, RAVEE.
Near Landres-et-St. Georges, France,
Oct. 14-16, 1918.
R—Birmingham, Ala.
B—Luthersville, Ga.
G. O. No. 64, W. D., 1919.

Major, 167th Infantry, 42d Division.
During the attack on the Cote-de-Chatillon, Major Norris personally led his
battalion through the intense artillery and machine-gun fire. Although the
attack led through a dense forest, he maneuvered his battalion with such
success that liaison was maintained at all times between the units of his
command and with the units on the right and left flank, and a successful
attack made upon a position considered almost impregnable. He was
wounded during this attack.

NORRIS, SIGBERT A. G.
Near Dun-sur-Meuse, France, Sept.
26, 1918.
R—New York, N. Y.
B—England.
G. O. No. 46, W. D., 1919.

Second lieutenant, observer, 11th Aero Squadron, Air Service.
Deeming it impossible to catch their own formation, Lieutenant Norris, with
Lieut. William Waring, pilot, attached themselves to a formation from the
20th Squadron and engaged in a 35-minute fight with 30 enemy aircraft. Five
of the 20th Squadron were lost and the observer of 1 of the remaining planes
seriously wounded. The wounded man had fallen in a position which had
made the control of the machine difficult. Lieutenant Norris immediately
motioned for his pilot to take a position between the enemy formation and the
crippled companion in order to protect it, and continued to fight off the enemy
planes until our lines were crossed.

NORRIS, STEVE G (2267070).
Near Very, France. Sept. 26, 1918.
R—Colton, Calif.
B—Greece.
G. O. No. 44, W. D 1919

Corporal, Company G, 364th Infantry, 91st Division.
He voluntarily preceded his company in searching woods for the purpose of
locating enemy snipers, who were causing numerous casualties. He also made
numerous trips under artillery and machine-gun fire to maintain liaison with
adjacent units. When his company was making its first advance under heavy
shellfire, this soldier took the place of a rifle sergeant's section and kept this
unit intact during the advance.

NORSTRAND, CARL JOHANNES_____
    At Chateau-Thierry, France, June 6, 1918.
    R—Roscommon, Mich.
    B—Norway.
    G. O. No. 110, W. D., 1918.

Sergeant major, 1st Battalion, 5th Regiment, U. S. Marine Corps, 2d Division. When his presence was not demanded in the performance of the normal duties of his office, he volunteered to rescue wounded men from a field swept by machine gun fire, and continued this heroic work with the aid of other volunteers until all had been recovered.

NORTON, EARL D. (293017)_____
    Near Vaux, France, July 1, 1918.
    R—Guilford, Conn.
    B—Guilford, Conn.
    G. O. No. 66, W. D., 1919.

Private, Company H, 9th Infantry, 2d Division. While Private Norton, an automatic-rifle gunner, was advancing during the attack a shell fragment struck a bag of hand grenades which he was carrying on his hip, resulting in an explosion which shattered his left leg. Despite the severe wound, he crawled forward toward a shell hole, where the remainder of his squad had taken refuge, and, with his remaining strength, threw his automatic rifle to the men in the shell hole. His injuries necessitated the amputation of his leg.

NORTON, EVERETTE C. (1455266)_____
    Near Cheppy, France, Sept. 27, 1918.
    R—Moran, Kans.
    B—Moran, Kans.
    G. O. No. 59, W. D., 1919.

Private, first class, Machine Gun Company, 139th Infantry, 35th Division. After being wounded by a machine-gun bullet, he refused to be evacuated to the rear, but continued in his duties for 2 days thereafter, and only stopped when weakened by the loss of blood.

NORTON, FRANK B. (1911299)_____
    At Cornay, France, Oct. 9–10, 1918.
    R—Altoona, Pa.
    B—Philadelphia, Pa.
    G. O. No. 37, W. D., 1919.

Sergeant, Company M, 328th Infantry, 82d Division. After fighting for 6 hours, he volunteered to accompany 15 other soldiers and an officer on a night patrol of Cornay, which was held by many enemy machine-gun posts. The party worked from 11 o'clock at night till next morning, clearing buildings and dugouts of the enemy, capturing 65 prisoners and 2 machine guns. With 6 others Sergeant Norton volunteered and entered a dugout where 23 prisoners were captured. He was wounded while leaving the town, but he refused to go to the aid station until the prisoners had been delivered at brigade headquarters.

*NORTON, FRED W_____
    In the Toul Sector, France, July 2, 1918.
    R—Columbus, Ohio.
    B—Marblehead, Ohio.
    G. O. No. 123, W. D., 1918.

First lieutenant, 27th Aero Squadron, Air Service. Lieutenant Norton, as flight commander, led a patrol of 8 machines, the first large American formation to encounter a large German patrol. His command gave battle to 9 enemy battle planes driven by some of the leading aces of the German Army. Although both of his guns jammed at the beginning of the fight and were therefore useless, he stayed with the formation, skillfully maneuvering his machine to the best advantage. He was attacked by enemy planes at 4 different times, but skillfully avoided them or dived at them. His continued presence was a great moral help to his comrades, who destroyed 2 of the enemy planes. On July 23, 1918, this officer died of wounds received in action July 20, 1918.
Posthumously awarded. Medal presented to father, Frank Norton.

NORTON, HENRY M. (1215265)_____
    Near Ronssoy, France, Sept. 29, 1918.
    R—Olean, N. Y.
    B—Belmont, N. Y.
    G. O. No. 23, W. D., 1919.

Private, medical detachment, 108th Infantry, 27th Division. Private Norton, on his own initiative, went forward twice in advance of the front line, bringing in wounded under heavy shell and machine-gun fire.

NORTON, JAMES A. (112415)_____
    Near Juvigny, France, north of Soissons, Sept. 4, 1918.
    R—Columbia, S. Dak.
    B—Houghton, S. Dak.
    G. O. No. 15, W. D., 1919.

Wagoner, Company A, 107th Ammunition Train, 32d Division. During a heavy enemy bombardment a shell burst near 2 ammunition trucks that were being unloaded at a dump, blowing up 1 truck and setting fire to the other. Disregarding the warning of bystanders he rushed forward, threw off the burning cushions and cover on the truck, and backed it to a place of safety. His conspicuous bravery was the means of saving a large quantity of ammunition.

*NORTON, JOHN H_____
    At Sergy, France, July 29–30, 1918.
    B—Springfield, Mass.
    B—West Springfield, Mass.
    G. O. No. 70, W. D., 1919.

Captain, 47th Infantry, 4th Division. When the company on the left of his own had fallen back, leaving a gap through which the enemy was approaching for a counterattack, Captain Norton, with the remnants of 2 squads, formed an automatic-rifle post and successfully covered the withdrawal of the remainder of his command to a stronger line of resistance. Though his small group was almost annihilated by hostile fire, he held this position until the arrival of reinforcements, inflicting heavy losses on the enemy.
Posthumously awarded. Medal presented to mother, Mrs. Mabel C. Norton.

NORTON, JOHN W. (557080)_____
    Near St. Thibault, France, Aug. 6, 1918.
    R—Central Falls, R. I.
    B—Central Falls, R. I.
    G. O. No. 50, W. D., 1919.

Sergeant, Company I, 39th Infantry, 4th Division. While leading his platoon toward the Vesle River, Sergeant Norton encountered extreme machine-gun fire. Exposing himself to determine the exact location from which this fire was being made, he was seriously wounded, but he continued the fire of his men, even after he was no longer able to move with them. His action greatly aided his platoon to advance and join the remainder of the company.

NORTON, ROBERT WILLIAM_____
    Near Cunel, France, Oct. 11, 1918.
    R—East Bloomfield, N. Y.
    B—Newark, N. J.
    G. O. No. 46, W. D., 1919.

Captain, 39th Infantry, 4th Division. During the action in the Bois-de-Foret, France, Captain Norton, with another officer, braved the hazardous fire by going out into no man's land and capturing 20 Germans at the point of his pistol. Although he lost 2 of the enemy during the encounter, he personally conducted the remaining back to the lines.

NORTON, WILLIAM M. (47708)_____
    Near Seicheprey, Ansauville sector, France, Mar. 1, 1918.
    R—Hardy Ark.
    B—Hardeman County, Tenn.
    G. O. No. 133, W. D., 1918.

Sergeant, Company I, 18th Infantry, 1st Division. Finding himself in a dugout surrounded by Germans, and in which a hand grenade had been thrown, he refused to surrender, made a bold dash outside, killed 1 of his assailants, put the others to flight, and resumed his duty with his company.

NORWAT, ARTHUR (1710316)_____
Near Revillon, France, Sept. 14-15, 1918.
R—Brooklyn, N. Y.
B—Plymouth, Pa.
G. O. No. 1, W. D., 1926.

Sergeant, Company M, 308th Infantry, 77th Division.
On Sept. 14 he advanced ahead of his company and with an automatic rifle single handed silenced an enemy machine-gun nest, capturing the gunner. On the following day, after having assumed command because of the fact that all officers had become casualties, he assembled 13 men and led them in a charge against superior forces of the enemy, recapturing a trench which shortly before had been taken by the enemy.
Oak-leaf cluster.
For the following act of extraordinary heroism in action near Moulin de l'Homme Mort, France, Oct. 4, 1918, Sergeant Norwat is awarded an oak-leaf cluster: When enemy machine-gun fire had checked his attempt to reach companies which had been surrounded by the enemy, Sergeant Norwat sprang upon the parapet, in full view of the enemy, and opened fire with an automatic rifle. He continued with this heroic work until he fell mortally wounded.
Posthumously awarded. Medal presented to brother, Alfred Norwat.

NOTTINGHAM, MARSH W. 1550025)____
Near Roncheres, France, July 31, 1918.
R—Indianapolis, Ind.
B—Coldwater, Mich.
G. O. No. 44, W. D., 1919.

Corporal, Headquarters Company, 76th Field Artillery, 3d Division.
He volunteered and carried messages through the intense shelling before telephone communications were established. While leading a party to an observation post Corporal Nottingham was killed by shellfire.
Posthumously awarded. Medal presented to father, Otis W. Nottingham.

NOURSE, WILLIAM H. (67308)_____
Near Bouresches, France, July 20, 1918.
R—Hyde Park, Mass.
B—Bangor, Me.
G. O. No. 44, W. D., 1919.

Sergeant, Company C, 103d Infantry, 26th Division.
Being on special duty, Sergeant Nourse followed his company in the attack. Upon discovering a strong machine-gun nest pouring a destructive fire into the second wave of his battalion, Sergeant Nourse, sending 2 men to the flanks, advanced alone, rushed and cleaned out the nests with hand grenades and bayonet.

NOWAK, JOHN M. (274797)_____
Near Fismes, France, Aug. 4, 1918.
R—Milwaukee, Wis.
B—Milwaukee, Wis.
G. O. No. 95, W. D., 1919.

Sergeant, Company K, 127th Infantry, 32d Division.
After being severely wounded, Sergeant Nowak refused to go to the rear for first aid, but bravely continued in the advance. While he was helping to reform his company under a heavy machine-gun barrage shortly afterwards he was killed.
Posthumously awarded. Medal presented to widow, Mrs. John M. Nowak.

NOWLIN, GEORGE A. (152355)_____
Near Suippes, France, July 15, 1918.
R—Chicago, Ill.
B—Jackson County, Mich.
G. O. No. 3, W. D., 1919.

First sergeant, Battery E, 42d Artillery, Coast Artillery Corps.
This soldier displayed great bravery and devotion to duty during the action of his battery at the opening of the German attack at Suippes. The firing position was under heavy fire, and the members of the battery widely separated when orders to fire were received by his battery. He personally succeeded in quickly organizing the firing sections at their posts. On several occasions he carried severely wounded infantrymen long distances to dressing stations. He assisted in repairing damage to exposed telephone lines under heavy shrapnel fire, and with his battery was the last man to leave the firing position.

NOYES, STEPHEN H._____
Near Chatel-Chehery, France, Oct. 15, 1918.
R—Newport, R. I.
B—Newport, R. I.
G. O. No. 143, W. D., 1918.

Captain, pilot, 12th Aero Squadron, Air Service.
He volunteered under the most adverse weather conditions to stake the advance lines of the 82d Division. Disregarding the fact that darkness would set in before he and his observer could complete their mission, and at the extremely low altitude of 150 feet, he proceeded amid heavy antiaircraft and ground machine-gun fire until the necessary information was secured. On the return, due to darkness, he was forced to land on a shell-torn field and proceeded on foot to headquarters with valuable information.

NUBEL, HERMAN (1707128)_____
Near St. Pierremont, France, Nov. 4, 1918.
R—Elmhurst, N. Y.
B—New York, N. Y.
G. O. No. 37, W. D., 1919.

Corporal, Company L, 307th Infantry, 77th Division.
He advanced under heavy machine-gun fire to a position on the flank of his company's sector, located a machine-gun nest, and opened fire on it. He remained in this position in the face of fire from a 1-pounder gun.

NUNLEY, LEWIS (551694)_____
Near Mezy, France, July 15, 1918.
R—Hulette, Ky.
B—Lawrence County, Ky.
G. O. No. 27, W. D., 1920.

Corporal, Company G, 38th Infantry, 3d Division.
Corporal Nunley advanced in the face of intense machine-gun fire and single handed attacked an enemy machine-gun position, killing the crew and returning with the gun to our lines. Later he led 9 men in a flank attack on the enemy and captured 25 men.

NUTT, ALAN_____
Near Forges, France, Sept. 26, 1918.
R—Cliffside, N. J.
B—Cliffside, N. J.
G. O. No. 140, W. D., 1918.

First lieutenant, pilot, 94th Aero Squadron, Air Service.
While on a patrol he encountered and unhesitatingly attacked 8 Fokker planes. After a few minutes of severe fighting, during which he displayed indomitable courage and determination, this officer shot down 1 of the enemy planes. Totally surrounded, outnumbered, and without a thought of escape, he continued the attack until he was shot down in flames near Drillancourt.
Posthumously awarded. Medal presented to father, Robert H. Nutt.

NUTTING, LESTER HERBERT_____
Near Thiaucourt, France, Sept. 15, 1918.
R—Washington, D. C.
B—Goodman, Mo.
G. O. No. 20, W. D., 1919.

Private, 96th Company, 6th Regiment, U. S. Marine Corps, 2d Division.
He voluntarily advanced 200 yards beyond the front lines to locate enemy machine-gun nests, signaling back their positions to the other members of his detachment, who immediately destroyed them. He was killed in returning to our lines.
Posthumously awarded. Medal presented to father, Robert H. Nutting.

NYB, WILL P._____ First lieutenant, 116th Infantry, 29th Division.
  Near Samogneux, France, Oct. 15,   He advanced without assistance upon a machine-gun nest and drove the crew
    1918.   from the gun.
  R—Radford, Va.
  B—Wythe County, Va.
  G. O. No. 37, W. D., 1919.

OBENOUR, GEORGE G. (545349)_____ Private, first class, Company A, 30th Infantry, 3d Division.
  Near Crezancy, France, July 15,   Three times, under terrific enemy fire, he carried messages to battalion and
    1918.   regimental headquarters. After the company had withdrawn he voluntarily
  R—Williamsburg, Pa.   returned to the position his company had held and throughout the night
  B—Drab, Pa.   assisted in evacuating the wounded.
  G. O. No. 32, W. D., 1919.

OBERMEYER, HERMAN (551706)_____ Corporal, Company I, 38th Infantry, 3d Division.
  Near Charteves, France, July 22,   His company being temporarily halted by machine-gun fire from the front,
    1918.   Corporal Obermeyer, with 1 man, advanced ahead of our lines through heavy
  R—Washington, D. C.   machine-gun fire, attacked an enemy post and captured 8 prisoners. Due to
  B—Washington, D. C.   his act of daring, the company was able to continue its advance.
  G. O. No. 22, W. D., 1920.

*O'BRIEN, CHARLES_____ First lieutenant, 306th Infantry, 77th Division.
  Near La Cendriere Woods, near the Lieutenant O'Brien led his platoon forward toward the La Cendriere Woods
    Aisne Canal, France, Sept. 6, 1918. under heavy shellfire. When wounded in the left leg, one of his men urged
  R—Nanticoke, Pa.   him to stop and have the wound dressed. He answered, "Never mind that;
  B—Nanticoke, Pa.   they can't stop us," and led his platoon through the woods to the bank of the
  G. O. No. 99, W. D., 1918.   Aisne Canal, where, while placing his men in position, he was struck again
                                       and killed. His dauntless courage presented an inspiring example to the men
                                       of his platoon.
                                     Posthumously awarded. Medal presented to mother, Mrs. M. O'Brien.

*O'BRIEN, CORNELIUS J. (568759)_____ Private, Company D, 4th Engineers, 4th Division.
  Near Ville-Savoye, France, Aug. 11, While engaged on the construction of a bridge over the Vesle River he volun-
    1918.   tarily left shelter during intense fire and carried one of his wounded officers
  R—Butte, Mont.   through a heavy machine-gun and artillery barrage to a dressing station.
  B—Ireland.   Posthumously awarded. Medal presented to mother, Mrs. Kate Mullins
  G. O. No. 16, W. D., 1923.   O'Brien.

O'BRIEN, JOHN F._____ Private, 67th Company, 5th Regiment, U. S. Marine Corps, 2d Division.
  Near Vierzy, France, July 18, 1918. His company being held up and subjected to a severe machine-gun fire, Private
  R—Brooklyn, N. Y.   O'Brien, having ascertained the location of the machine gun, alone and single
  B—Morris, Minn.   handed crawled into the enemy's lines, came upon the machine-gun crew
  G. O. No. 49, W. D., 1922.   from the rear, surprised them, and compelled their surrender. His gallant
                                       and courageous action enabled his company to advance.

O'BRIEN, JOHN J. (156804)_____ Sergeant, Company D, 2d Engineers, 2d Division.
  Near St. Etienne-a-Arnes, France, Advancing ahead of the Infantry, he made several reconnaissances of the town
    Oct. 5–7, 1918.   of St. Etienne-a-Arnes, and, in spite of the danger, exposed to machine-gun
  R—El Paso, Tex.   and artillery fire of our own and enemy guns, he procured and returned with
  B—Sault Ste. Marie, Mich.   valuable information.
  G. O. No. 23, W. D., 1919.

O'BRIEN, JOSEPH P. (110117)_____ Private, Company A, 103d Machine Gun Battalion, 26th Division.
  Near Bouresches, France, July 18, Although severely wounded, Private O'Brien advanced in front of the line
    1918.   under terrific machine-gun fire and succeeded in rescuing a wounded com-
  R—Providence, R. I.   rade and carried him back to a place of safety. After receiving first-aid
  B—Providence, R. I.   treatment, he again went forward and returned with the body of a company
  G. O. No. 31, W. D., 1922.   officer.

O'BRIEN, THOMAS A. (1897357)_____ Sergeant, Company A, 325th Infantry, 82d Division.
  East of St. Juvin, France, Oct. 16, He assumed command of his platoon and successfully led it forward until it
    1918.   was held up by severe enemy machine-gun fire. He then went forward
  R—Philadelphia, Pa.   himself and silenced the hostile machine gun with hand grenades, being
  B—Philadelphia, Pa.   wounded in the performance of this gallant act.
  G. O. No. 46, W. D., 1919.

O'BRIEN, WILLIAM H. J._____ First lieutenant, Medical Corps, attached to 76th Field Artillery, 3d Division.
  Near La Trinity Ferme, France,   During the entire night of July 14–15 and throughout the following day he
    July 14–15, 1918.   was continually exposed to high explosives and gas shells in caring for the
  R—New Haven, Conn.   wounded, even after he had been painfully wounded, by the fragment of
  B—New Haven, Conn.   a shell.
  G. O. No. 44, W. D., 1919.

*O'CONNELL, ALBERT L. (633407)_____ Private, Battery C, 60th Artillery, Coast Artillery Corps.
  Near Montblainville, France, Oct. In an effort to rescue a comrade who had been severely wounded, he ran with
    4, 1918.   a litter into an area under heavy shell fire. He succeeded in getting the
  R—Battle Creek, Mich.   wounded soldier on the litter, but before he could carry him out of danger
  B—Port Huron, Mich.   another shell burst directly under the litter, killed the wounded soldier and
  G. O. No. 15, W. D., 1919.   severely wounding Private O'Connell.
                                     Posthumously awarded. Medal presented to mother, Mrs. Mary O'Connell.

*O'CONNELL, MICHAEL (60376)_____ Corporal, Company D, 101st Infantry, 26th Division.
  Near Epieds, France, July 23, 1918. Corporal O'Connell, Sergt. Simeon L. Nickerson, and Pvt. Thomas Ryan
  R—Boston, Mass.   volunteered to cross an open field in front of their company in order to ascer-
  B—Ireland.   tain the location of enemy machine guns. While engaged in this courageous
  G. O. No. 116, W. D., 1918.   enterprise they were shot and killed. The heroic self-sacrifice of these 3
                                       men saved the lives of many of their comrades, who would have been killed
                                       had the company attempted to make the advance as a whole.
                                     Posthumously awarded. Medal presented to mother, Mrs. Patrick O'Connell.

O'CONNOR, DANIEL (60196)_____
North of Verdun, France, Oct. 27, 1918.
R—Dorchester, Mass.
B—Dorchester, Mass.
G. O. No. 21, W. D., 1919.

Sergeant, Company C, 101st Infantry, 26th Division.
Encountering strong machine-gun nests while leading his platoon forward, he ordered his men to take cover while he advanced alone, flanked the nest and killed 2 of the enemy gunners, thereby enabling his platoon to resume the advance.

O'CONNOR, HARRY GROVER (2414680)__
Near St. Juvin, France, Oct. 15, 1918.
R—Wayland, N. Y.
B—Hornell, N. Y.
G. O. No. 37, W. D., 1919.

Sergeant, headquarters detachment, 78th Division.
He volunteered and went from his division headquarters to one of the regiments in line to obtain much-needed information and to arrange for liaison. He covered the distance of 6 kilometers through shell fire three times before the unit was found, and, while returning with the information, was seriously wounded by a shell, which mortally wounded 2 companions. He gave water and first aid to the 2 men and dragged himself down the road until he met a soldier, whom he sent for an ambulance. After reaching the hospital and having his wounds dressed, he left the hospital without the knowledge of the attendants and delivered the information to his division headquarters in person.

O'CONNOR, JAMES (91120)_____
Near Landres, France, Oct. 14, 1918.
R—New York, N. Y.
B—Ireland.
G. O. No. 35, W. D., 1919.

Corporal, Company I, 165th Infantry, 42d Division.
After his entire squad had been either killed or wounded while attacking an enemy machine-gun nest, Corporal O'Connor continued the combat single handed, and, having killed 3 of the enemy, silenced their machine gun, which was enfilading his battalion. He then carried 3 of his comrades from their exposed position to safety.

O'CONNOR, JOHN HENRY_____
In the attack on Montrebeau Woods, Sept. 28, 1918.
R—Winfield, Kans.
B—Franklin County, Ky.
G. O. No. 28, W. D., 1921.

Major, 137th Infantry, 35th Division.
By his personal example, leadership, and courage under most severe fire, he was a source of inspiration to his command and a very great factor in the successful attack.

*O'CONNOR, THOMAS P. (89678)_____
Near Villers-sur-Fere, France, July 31, 1918.
R—New York, N. Y.
B—New York, N. Y.
G. O. No. 117, W. D., 1918.

Private, Company C, 165th Infantry, 42d Division.
After his platoon had moved from an open field to the cover of the wood he returned to the field under heavy machine-gun fire to rescue a wounded comrade and was killed at the side of the man he tried to save.
Posthumously awarded. Medal presented to mother, Mrs. J. Fitzpatrick.

O'DANIEL, JOHN W._____
Near Bois-St. Claude, in the St. Mihiel salient, Sept. 12, 1918.
R—Newark, Del.
B—Newark, Del.
G. O. No. 128, W. D., 1918.

Second lieutenant, 11th Infantry, 5th Division.
After being severely wounded in the head early in the action he continued in command of his platoon, leading his men for several hours until forced to give in to complete physical exhaustion, thus displaying most exceptional courage, determination, and devotion to duty.

O'DELL, EDGAR H. (2147883)_____
At Varennes, France, Sept. 26, 1918.
R—York, N. Dak.
B—Houlton, Me.
G. O. No. 59, W. D., 1919.

Private, Company K, 137th Infantry, 35th Division.
In the face of heavy machine-gun fire he entered a building alone and captured 16 prisoners and 4 machine guns. He used an automatic rifle and hand grenades during his advance to the building, fire from which was checking our advance.

ODELL, JULIUS DONOVAN (2258469)_____
Service rendered under the name, Julius O. Yuill.

Sergeant, Company M, 361st Infantry, 91st Division.

ODENWALD, WILLIAM (1911455)_____
At Cornay, France, Oct. 9–10, 1918.
R—New York, N. Y.
B—New York, N. Y.
G. O. No. 37, W. D., 1919.

Private, first class, Company M, 328th Infantry, 82d Division.
After fighting for 6 hours Private Odenwald volunteered to accompany 15 other soldiers and an officer on a night patrol of Cornay, which was held by many enemy machine-gun posts. The party worked from 11 o'clock at night until the next morning at clearing buildings and dugouts of the enemy, capturing 65 prisoners and 2 machine guns. With 6 others, Private Odenwald volunteered and entered a dugout, where 23 prisoners were captured. He was wounded while leaving the town, but he refused to go to the aid station until the prisoners had been delivered at brigade headquarters.

*O'DONNELL, PAUL J._____
Near Dun-sur-Meuse, France, Sept. 26, 1918.
R—Wilmington, Del.
B—Philadelphia, Pa.
G. O. No. 123, W. D., 1918.

Second lieutenant, observer, 96th Aero Squadron, Air Service.
His formation was attacked while flying to bomb Dun-sur-Meuse by 7 enemy planes. With the first spurt of enemy fire he was fatally wounded. With his last strength he opened a deliberate and destructive fire on 1 of the enemy planes, driving it down out of control. He died before his antagonist struck the ground.
Posthumously awarded. Medal presented to mother, Mrs. Adie O'Donnell.

OFFINGER, EARL C._____
Near Nantillois, France, Sept. 29, 1918.
R—Springfield, Mass.
B—Greenfield, Mass.
G. O. No. 35, W. D., 1919.

Captain, Company G, 315th Infantry, 79th Division.
While leading his company under heavy shell and machine-gun fire he received severe arm wounds. He was obliged to return for treatment to a first-aid post and advised to go to the rear. Refusing, he returned to his lines, reorganized his company, and formed their position before evacuated.

*O'FLAHERTY, COLMAN E._____
Near Very, France, Oct. 3, 1918.
R—Mitchell, S. Dak.
B—Ireland.
G. O. No. 28, W. D., 1921.

First lieutenant, chaplain, 28th Infantry, 1st Division.
Chaplain O'Flaherty displayed conspicuous gallantry in administering to the wounded under terrific fire, exposing himself at all times to reach their side and give them aid. In the performance of this heroic work he was killed.
Posthumously awarded. Medal presented to cousin, Miss Mary O'Flaherty.

OGDEN, GEORGE (1241102)................
 Near Montblainville, France, Sept.
  27, and near Baslieux, France,
  Nov. 2–9, 1918.
 R—Philadelphia, Pa.
 B—Wilmington, Del.
 G O No. 93, W. D., 1919.

Corporal, Company H, 110th Infantry, 28th Division.
While acting as battalion scout, Corporal Ogden succeeded in driving away
the crews of 2 enemy machine guns by sniping. Operating 1 of these guns
himself and a sergeant the other, they materially assisted in repulsing an
enemy counterattack. On another occasion, while leading a patrol of 10
men on the Vesle River, Corporal Ogden succeeded in getting on the flank
of the enemy and by rifle fire forced about 100 to retreat from a trench in disorder, inflicting many casualties. Later he succeeded in getting in the rear of
the enemy positions, remained in hiding until night, and then returned with
valuable information relative to the enemy positions.

*OGDEN, IRA C...................
 Near St. Etienne, France, Oct. 9–10,
  1918.
 R—San Antonio, Tex.
 B—San Antonio, Tex.
 G. O. No. 20, W. D., 1919.

Captain, 141st Infantry, 36th Division.
Due to casualties among field officers, Captain Ogden was placed in command
of the support line. On the afternoon of Oct. 10 an advance was ordered,
and he requested permission to accompany the front line. He was placed
in command of the front line of the regiment and advanced with it at 4.30
p. m. Regardless of personal danger, he crossed areas swept by machine-gun
fire, and was killed in action shortly after he had reported as having reached
the objective.
Posthumously awarded. Medal presented to widow, Mrs. Ira C. Ogden.

OGLE, EDWARD W. Jr. (1763065).......
 Near Grand Pre, France, Oct. 18,
  1918.
 R—East Orange, N. J.
 B—New York, N. Y.
 G. O. No. 35, W. D., 1919.

Private, Company H, 312th Infantry, 78th Division.
Acting upon his own initiative, Private Ogle, on 3 different occasions, risked
his life by going in front of his lines and assisting the wounded comrades to
a place of safety.

O'HAGAN, THOMAS P. (89495).........
 Near Landres-et-St. Georges, France,
  Oct. 14–15, 1918.
 R—New York, N. Y.
 B—Ireland.
 G. O. No. 59, W. D., 1919.

First sergeant Company C, 165th Infantry, 42d Division.
After successfully conducting a raiding patrol and returning safely to his lines,
he immediately went back to the scene and, exposed to unusual machine-gun
and artillery fire, searched for and carried to safety a severely wounded comrade who was lying 100 yards from the enemy's wire. On the following day
he carried a severely wounded comrade to safety across an open field in spite
of the terrific and especially directed machine-gun fire.

OILER, GEORGE (39327)..................
 Near Vaux, France, July 1, 1918.
 R—Fort Thomas, Ky.
 B—Judyton, W. Va.
 G. O. No. 99, W. D., 1918.

Sergeant, Company F, 9th Infantry, 2d Division.
He volunteered and led a liaison patrol through thick woods known to be
strongly held by enemy machine guns. After being severely wounded, he
brought back 4 prisoners and valuable information.

OKE, RUSSELL (2388587)...............
 Near Chateau Charmois, France,
  Nov. 9, 1918.
 R—Canada.
 B—Canada.
 G. O. No. 37, W. D., 1919.

Sergeant, Company L, 61st Infantry, 5th Division.
He volunteered and led a liaison patrol through thick woods known to be
strongly held by enemy machine guns. After being severely wounded, he
brought back 4 prisoners and valuable information.

O'KEEFE, ARTHUR J...............
 Near Soissons, France, July 21, 1918.
 R—Fort Leavenworth, Kans.
 B—Leavenworth, Kans.
 G. O. No. 72, W. D., 1920.

Captain, 18th Infantry, 1st Division.
Captain O'Keefe personally led a group of automatic riflemen in an attack
against a number of the enemy who were attempting to cut off the assaulting wave. By his skillful leadership and gallant conduct he succeeded in
defeating the enemy party. In the performance of this duty Captain O'Keefe
was wounded in the leg by a machine-gun bullet. In spite of his wound this
officer continued to expose himself to heavy machine-gun fire while directing
the movements of his company until wounded a second time by a high-
explosive shell. Twice wounded, he refused to be evacuated until ordered
to the rear.

O'KEEFE, DANIEL J. (2290557)........
 Near Epinonville, France, Sept. 27,
  1918.
 R—San Francisco, Calif.
 B—San Francisco, Calif.
 G. O. No. 15, W. D., 1919.

Corporal, Company B, 361st Infantry, 91st Division.
When half of his platoon were on a hillside under heavy machine-gun and
sniper's fire he effectively covered the withdrawal of his detachment with
his automatic rifle.

O'KEEFE, JOHN J. (2058722)...........
 Near the Bois-d'-Harville, France,
  Nov. 10, 1918.
 R—Chicago, Ill.
 B—Chicago, Ill.
 G. O. No. 46, W. D., 1919.

First sergeant, Company M, 131st Infantry, 33d Division.
After all the officers of 2 of the companies of his battalion had become casualties Sergeant O'Keefe rallied the men, who had become disorganized
under the machine-gun fire, and led them forward toward the objective,
displaying marked courage and leadership.

*O'KEEFE, THOMAS J. (54159)...........
 Near Verdun, France, Oct. 4, 1918.
 R—Chicago, Ill.
 B—Pittsburgh, Pa.
 G. O. No. 59, W. D., 1919.

Corporal, Company H, 26th Infantry, 1st Division.
He voluntarily advanced alone into the woods to destroy an enemy machine-
gun nest, and in a single-handed pistol fight with 8 of the enemy succeeded in killing 4 Germans before he himself was killed in this unequal
combat.
Posthumously awarded. Medal presented to sister, Mrs. Margaret G. Beutel.

O'KELLEY, GROVER CLEVELAND.......
 In the Bois-de-Belleau, France, June
  6–8, 1918.
 R—Blountsville, Ga.
 B—Planter, Ga.
 G. O. No. 46, W. D., 1920.

Sergeant, 80th Company, 6th Regiment, U. S. Marine Corps, 2d Division.
In the attack upon the enemy machine-gun positions in the woods, Sergeant
O'Kelley led his platoon with great courage. During the advance the
assault section came under cross fire of enemy machine guns and all were
killed or wounded.

O'KELLY, THOMAS (89584)............
  Near Landres-et-St. Georges, France,
  Oct. 15, 1918.
  R—New York, N. Y.
  B—Ireland.
  G. O. No. 59, W. D., 1919.

Corporal, Company C, 165th Infantry, 42d Division.
Even after being warned of the danger of attempting to get through the murderous fire and after he had seen all the other battalion runners killed or wounded on the same mission, Corporal O'Kelly willingly volunteered and started with a message to regimental headquarters. When he had gone but a short distance he managed to reach his destination, after which he was taken to a hospital.

OLANSON, ARTHUR W. (1785546)...........
  Near Nantillois, France, Sept. 29,
  1918.
  R—Philadelphia, Pa.
  B—Sweden.
  G. O. No. 81, W. D., 1919.

Sergeant, Company E, 315th Infantry, 79th Division.
Sergeant Olanson, with his company commander, outflanked a machine-gun nest which was holding up their advance, shot 1 German noncommissioned officer who tried to escape, and captured 2 prisoners, the other occupants fleeing. The reduction of this machine-gun nest made it possible for the flank of the battalion to advance.

OLDFIELD, WILLIE A. (1457429)..........
  Near Charpentry, France, Sept. 26-
  27, 1918.
  {R—Canton, Kans.
  {B—Canton, Kans.
  †G. O. No. 20, W. D., 1919.

Sergeant, Company I, 139th Infantry, 35th Division.
Sergeant Oldfield was in charge of his platoon when it was subjected to heavy enemy machine-gun fire, causing numerous casualties. By his word and example he held his men in line as a unit until nightfall, when they intrenched. Next day when the tanks appeared he led the charge upon the machine guns which were holding up the advance. Inspired by his gallantry, his men went forward and cleared out the guns and assisted in the capture of Charpentry, with many prisoners.

OLDS, ARTHUR............
  Near St. Etienne, France, Oct. 2-3,
  1918.
  R—Millington, Mich.
  B—Millington, Mich.
  G. O. No. 46, W. D., 1919.

Sergeant, 18th Company, 5th Regiment, U. S. Marine Corps, 2d Division.
After all the runners had been wounded, Sergeant Olds volunteered to act as runner and made several trips through machine-gun and artillery fire.

OLDYNSKI, CHARLES (51395).............
  Near St. Etienne-a-Arnes, France,
  Oct. 3, 1918.
  R—Shamokin, Pa.
  B—Russia.
  G. O. No. 22, W. D., 1920.

Private, first class, Company H, 23d Infantry, 2d Division.
Private (First Class) Oldynski advanced ahead of his squad through heavy machine-gun fire and engaged in a hand-to-hand fight with an enemy sergeant, whom he killed. He killed 2 others of the enemy before the other members of his squad arrived and assisted in the capture of remaining enemy soldiers with their guns.

*O'LEARY, JOSEPH A. (1379337)........
  Near Epinonville, France, Oct. 7,
  1918.
  R—East St. Louis, Ill.
  B—East St. Louis, Ill.
  G. O. No. 74, W. D., 1919.

Corporal, Battery F, 124th Field Artillery, 33d Division.
He left shelter and volunteered as a stretcher bearer, making frequent trips to and from gun positions under heavy fire until he was killed by an enemy shell.
Posthumously awarded. Medal presented to brother, Robert O'Leary.

OLEJNIK, FRANK (2338100)............
  Near Gland, France, July 21, 1918.
  R—Indiana Harbor, Ind.
  B—Chicago, Ill.
  G. O. No. 32, W. D., 1919.

Corporal, Company B, 4th Infantry, 3d Division.
Assisting his platoon commander and one other comrade, Corporal Olejnik went forward, attacking and capturing an enemy machine gun and 8 prisoners.

OLIPHANT, DAVID A. (914141)...........
  At Romagne, France, Oct. 14, 1918.
  R—Ashland, Ky.
  B—Scotland.
  G. O. No. 37, W. D., 1919.

Sergeant, Company A, 7th Engineers, 5th Division.
By his energy, initiative, and courage he located several machine-gun nests and captured 9 prisoners and an anti-tank gun when sent out with 3 other soldiers to locate the enemy positions.

OLIVER, MACK O. (5789)............
  West of the Meuse, France, Oct. 11,
  1918.
  R—Winston-Salem, N. C.
  B—Winston-Salem, N. C.
  G. O. No. 44, W. D., 1919.

Sergeant, Company H, 28th Infantry, 1st Division.
After having been severely wounded by shrapnel, he refused to leave the lines, realizing the urgent need of men. After being relieved, he walked to the dressing station, despite his weakness from loss of blood and his painful suffering from the wound.

*OLLRICH, HARRY J. (262315)............
  During the advance across the River
  Ourcq and from Cierges to Fismes,
  France, July 31 to Aug. 4, 1918.
  R—Mount Clemens, Mich.
  B—Detroit, Mich.
  G. O. No. 116, W. D., 1918.

Private, Company E, 125 Infantry, 32d Division.
Many times daily during this advance Private Ollrich displayed an entire disregard of personal danger in the carrying of messages through enemy barrages. Later, in the fighting near Fismes, when it became necessary to send an important message to the commanding officer of the battalion, although Private Ollrich had been on constant duty night and day for 4 days, he attempted to deliver the message but was killed while crossing a shell-swept zone.
Posthumously awarded. Medal presented to father, Henry Ollrich.

OLSEN, ERIC S. (64751)............
  At Chavignon, France, Feb. 28, 1918.
  R—New Britain, Conn.
  B—Sweden.
  G. O. No. 129, W. D., 1918

Sergeant, Company F, 102d Infantry, 26th Division.
He was a member of a working party on the night of Feb. 28, 1918, well out in front of the advance post. His party encountered a violent barrage of the enemy which protected enemy assault troops. He helped to fight off the German troops, and walked back and forth twice under the enemy's and our own barrage to collect his men. When he heard that his lieutenant was in trouble, he walked back again to his rescue, where the barrage had at first overtaken him.

OLSEN, FRED (2266496)............
  Near Eclisfontaine France, Sept. 27
  1918.
  R—El Centro, Calif.
  B—Santa Barbara, Calif.
  G. O. No. 46, W. D., 1919.

Sergeant, Machine Gun Company, 364th Infantry, 91st Division.
With 2 other soldiers, Sergeant Olsen volunteered and went 300 yards beyond our outpost lines, through heavy shell fire, to bring in a wounded private of his regiment. The mission was promptly and successfully accomplished.

OLSEN, HAROLD (1907902)_____
   In the Meuse-Argonne offensive, France, Oct. 9, 1918.
   R—Attleboro, Mass.
   B—Norway.
   G. O. No. 50, W. D., 1919.

Corporal, Company K, 327th Infantry, 82d Division.
Assisted by another soldier, Corporal Olsen crawled far in advance of our lines under terrific machine-gun and shell fire and brought back a severely wounded comrade.

OLSEN, JOSEPH ENOCH_____
   Near St. Etienne, France, Oct. 3, 1918.
   R—Heyburn, Idaho.
   B—New York Mills, Minn.
   G. O. No. 23, W. D., 1919.

Private, 77th Company, 6th Machine Gun Battalion, U. S. Marine Corps, 2d Division.
While his platoon was following the advance of an Infantry platoon which had become separated, Private Olsen was seriously wounded in the foot by machine-gun fire. At the edge of heavy brushwood a company of German Infantry was encountered, and Private Olsen, who had been forced to fall some distance behind rushed forward as best he could and set up his tripod, acting as loader until the enemy was repulsed.

OLSEN, OLAF S. (274572)_____
   Near Juvigny, France, Aug. 31, 1918.
   R—Superior, Wis.
   B—Swift County, Minn.
   G. O. No. 124, W. D., 1918.

Private, Company I, 127th Infantry, 32d Division.
He was a squad leader in the second platoon. After reaching his objective he displayed extraordinary qualities of leadership in organizing scattering squads of Company I and placing them in advantageous positions in spite of severe machine-gun fire and artillery bombardment. His disregard of danger and fine leadership were an inspiration to his comrades.

OLSON, JOHN O. (44279)_____
   Near Bois-de-Fontaine, France, May 11, 1918.
   R—Valparaiso, Nebr.
   B—Valparaiso, Nebr.
   G. O. No. 50, W. D., 1919.

Private, first class, Company L, 16th Infantry, 1st Division.
He displayed conspicuous bravery by going from the front line to an advanced post and rescuing, unaided, a wounded comrade in the face of heavy machine-gun fire.

OLSON, MANDEL (1197267)_____
   Near Cunel, France, Oct. 14, 1918.
   R—Grand Forks, N. Dak.
   B—Fesston, Minn.
   G. O. No. 98, W. D., 1919.

Private, first class, Company A, 13th Machine Gun Battalion, 5th Division.
Accompanying another soldier, Private Olson left shelter and went forward 100 meters over territory swept by shells and machine-gun fire, and carried a wounded man to safety.

O'MALLEY, GEORGE P._____
   Near Marnetz, Belgium, Aug. 26, 1918.
   R—Cleveland, Ohio.
   B—Cleveland, Ohio.
   G. O. No. 15, W. D., 1923.

Captain, Medical Corps, attached to 7th Sussex Regiment, British Expeditionary Forces.
Under intense enemy machine-gun and rifle fire he went to the rescue of wounded British soldiers, dressed their wounds, and assisted in carrying them to places of comparative safety. While thus engaged one of the stretcher bearers in the rescue party was killed and another wounded. These men Captain O'Malley also assisted in carrying to the British dressing station.

O'NEAL, JAMES LEE (542093)_____
   Near Cierges, France, Oct. 5, 1918.
   R—California, Mo.
   B—Jefferson City, Mo.
   G. O. No. 103, W. D., 1919.

Private, Company L, 7th Infantry, 3d Division.
When machine-gun and rifle fire was so heavy that his company commander would not order a runner out, Private O'Neal volunteered and carried messages from the regimental post of command to his company, making numerous trips under heavy machine-gun and shell fire during the day and night.

O'NEAL, MIKE (554676)_____
   Near Cierges, France, Oct. 4, 1918.
   R—Jeffersonville, Ind.
   B—Jeffersonville, Ind.
   G. O. No. 98, W. D., 1919.

Sergeant, Company C, 8th Machine Gun Battalion, 3d Division.
Having been seriously wounded during an attack, Sergeant O'Neal refused to leave the field until he had led his men to cover and reestablished his guns for action, inspiring the other members of his platoon by his courage and fortitude.

O'NEIL, FRANK P._____
   Near Bois-de-St. Remy, France, Sept. 12, 1918.
   R—Dorchester, Mass.
   B—Dorchester, Mass.
   G. O. No. 37, W. D., 1919.

Captain, 101st Infantry, 26th Division.
When the advance of his battalion was held up for nearly 2 hours, Lieutenant O'Neill, with 4 other soldiers, made an attack on the enemy, and although subjected to direct fire, he succeeded in silencing 2 of their machine guns and enabling his battalion to proceed farther.

O'NEILL, JAMES T. (1261315)_____
   At Fismes, France, Aug. 10, 1918.
   R—Aldan, Pa.
   B—Phoenixville, Pa.
   G. O. No. 37, W. D., 1919.

Private, 110th Ambulance Company, 103d Sanitary Train, 28th Division.
Under heavy shell and machine-gun fire, he voluntarily made 5 trips to ascertain the condition of a bridge over the Vesle River, to make sure it was safe for the passage of ambulances. Later, when the bridge became impassable for vehicles, he crossed the bridge on foot and brought back food and medical supplies.

O'NEILL, RALPH A._____
   Near Chateau-Thierry, France, July 2, 1918.
   R—Bethlehem, Pa.
   B—Mexico.
   G. O. No. 116, W. D., 1919.

Second lieutenant, 147th Aero Squadron, Air Service.
Lieutenant O'Neill and four other pilots attacked 12 enemy battle planes. In a violent battle within the enemy's lines they brought down 3 German planes, 1 of which was credited to Lieutenant O'Neill.
Oak-leaf clusters (2).
A bronze oak-leaf cluster is awarded to Lieutenant O'Neill for the following act of extraordinary heroism in action. On July 5, 1918, he led 3 other pilots in battle against 8 German pursuit planes near Chateau-Thierry, France. He attacked the leader, opening fire at about 150 yards and closing up to 30 yards range. After a quick and decisive fight the enemy aircraft fell in flames. He then turned on 3 other machines that were attacking him from the rear and brought 1 of them down. The other 5 enemy planes were driven away.
A bronze oak-leaf cluster is also awarded to Lieutenant O'Neill for the following act of extraordinary heroism in action near Fresnes, France, July 24, 1918: Lieutenant O'Neill, with 4 other pilots, engaged 12 enemy planes discovered hiding in the sun. Leading the way to an advantageous position by a series of bold and skillful maneuvers, Lieutenant O'Neill shot down the leader of the hostile formation. The other German planes then closed in on him, but he climbed to a position of vantage above them and returned to the fight and drove down another plane. In this encounter he not only defeated his opponents in spite of overwhelming odds against him but also enabled the reconnaissance plane to carry on its work unmolested.

O'NEILL, WILLIAM (90713).
    In the Valley of Suippes, France,
        July 14-16, 1918.
    R—New York, N. Y.
    B—Ireland.
    G. O. No. 71, W. D., 1919.

Sergeant, Company H, 165th Infantry, 42d Division.
Wounded by a shell fragment, he returned to his platoon and engaged in hand-to-hand fighting, after having his wound dressed. Wounded 2 days later by a machine-gun bullet, he again returned to the firing line, after receiving first aid, and led a successful charge against an enemy position. His personal heroism was an inspiration to his men. He was killed while consolidating a position his platoon had just taken.
Posthumously awarded. Medal presented to father, John O'Neill.

OOSTERBAAN, DICK (2054915).
    In the Bois-des-Rappes, France,
        Oct. 21, 1918.
    R—Zeeland, Mich.
    B—Zeeland, Mich.
    G. O. No. 35, W. D., 1919.

Private, Company C, 15th Machine Gun Battalion, 5th Division.
While repulsing an enemy counterattack in the Bois-des-Rappes, Private Oosterbaan, a gunner, was so badly wounded that he could no longer fire, but lay beside his gun for 18 hours without medical attention, under heavy machine-gun fire, encouraging the remainder of his squad until the attack was over.

OPIE, HIEROME L.
    Near Samogneux, France, Oct. 15,
        1918.
    R—Staunton, Va.
    B—Staunton, Va.
    G. O. No. 37, W. D., 1919.

Major, 116th Infantry, 29th Division.
Although painfully wounded, Major Opie continued in command of his battalion, successfully leading it to its objective. During the action Major Opie displayed rare courage and valor, refusing relief until the new objective was consolidated.

OPPENHEIM, ARLIE C. (58091).
    Near Soissons, France, July 18, 1918.
    R—Lake Park, Ga.
    B—Statenville, Ga.
    G. O. No. 72, W. D., 1920.

Corporal, Company I, 28th Infantry, 1st Division.
During the attack on this date, Corporal Oppenheim assisted his platoon commander in pushing forward the attack against strong enemy resistance. Later, when the advance of his platoon was halted by machine-gun fire, he, with 3 others, advanced ahead of our lines and silenced the enemy machine gun. Due to his gallantry, his organization was enabled to continue the advance. Later in the engagement, he was seriously wounded and evacuated to the hospital.

ORCUTT, IVER (2290586).
    Near Preny, France, Nov. 1, 1918.
    R—Seattle, Wash.
    B—Mazomanie, Wis.
    G. O. No. 44, W. D., 1919.

Private, Company E, 56th Infantry, 7th Division.
When it was rumored that several of the front-line companies had been forced to withdraw, he volunteered and went over the entire front during the night, reporting back the exact location of each unit. His mission was accomplished under shrapnel fire and gas, but he succeeded where many other runners had failed.

ORD, RALPH EDWARD (1243700).
    West of Fismette, France, Aug. 10,
        1918.
    R—Dravosburg, Pa.
    B—Dravosburg, Pa.
    G. O. No. 128, W. D., 1918.

Sergeant, Company C, 111th Infantry, 28th Division.
Sergeant Ord, with another soldier, voluntarily left a place of safety and crawled through heavy machine-gun and shellfire to the aid of a comrade who had fallen wounded during the withdrawal of their company from an exposed position, carrying him 75 yards across an open area to shelter.

O'ROURKE, CORNELIUS (749542).
    Near Vieville-en-Haye, France,
        Sept. 14, 1918.
    R—New Haven, Conn.
    B—Ireland.
    G. O. No. 145, W. D., 1918.

Sergeant, Company A, 15th Machine Gun Battalion, 5th Division.
Seeing 2 wounded comrades lying exposed to heavy enemy machine-gun and artillery fire, he went out in disregard of all danger to himself and brought them in, 1 at a time, to a shelter place, thus inspiring his men by his great dash and courage.

O'ROURKE, JAMES H. (1746552).
    Near Vieville-en-Haye, France,
        Sept. 26, 1918.
    R—Lakewood, N. J.
    B—Camden, N. J.
    G. O. No. 26, W. D., 1919.

Private, Company D, 311th Infantry, 78th Division.
After being twice wounded he captured 2 prisoners and took them to the battalion headquarters.

O'ROURKE, JOHN P. (976787).
    Near Blanc Mont Ridge, France,
        Oct. 3-9, 1918.
    R—Elk Point, S. Dak.
    B—Cleveland, Ohio.
    G. O. No. 44, W. D., 1919.

Private, Medical Detachment, 9th Infantry, 2d Division.
With utter disregard for his personal safety, he worked untiringly under heavy shellfire and gave aid to the wounded. He lost his life while advancing in front of his company to give aid to some wounded.
Posthumously awarded. Medal presented to widow, Mrs. Harry T. McNeil.

ORR, EDWARD.
    In the Toul sector, France, Aug. 28,
        1918.
    R—Chicago, Ill.
    B—Chicago, Ill.
    G. O. No. 20, W. D., 1919.

First lieutenant, pilot, 12th Aero Squadron, Air Service.
Lieutenant Orr, flying with Lieut. Phil A. Henderson, Infantry, observer, on an unprotected reconnaissance mission, encountered a patrol of 8 enemy pursuit planes near the American balloon line. The patrol was sighted just as one of them dived on the balloon with the intention of destroying it. Without hesitation, Lieutenant Orr attacked this plane and followed it to within 50 meters, firing his single front gun against the double guns with which the German plane was equipped. In the meantime Lieutenant Henderson engaged the other 8 planes, which attacked from the rear. After a violent combat, all of the enemy planes were driven off. On Sept. 14, 1918, Lieutenant Orr was accidentally killed.
Posthumously awarded. Medal presented to father, E. K. Orr.

ORTIZ, CONCEPTION (2229339).
    Near Romagne, France, Oct. 11, 1918.
    R—Eagle Pass, Tex.
    B—Eagle Pass, Tex.
    G. O. No. 64, W. D., 1919.

Private, Company I, 125th Infantry, 32d Division.
On the morning of Oct. 11 he made numerous trips across a valley which was swept by continuous and terrific machine-gun fire, carrying messages of great importance from his company in the front line to his battalion post of command. After having successfully and fearlessly carried many messages he lost his life while performing the hazardous duty.
Posthumously awarded. Medal presented to father, Linardo Ortiz.

*ORTT, HORACE F. (1377601)_____
In the Very-Epinonville Valley, northwest of Verdun, France, Oct. 4, 1918.
R—Dixon, Ill.
B—Bridgeport, Pa.
G. O. No. 5, W. D., 1920.

Private, Battery C, 123d Field Artillery, 33d Division.
He volunteered to carry an important message to a position under heavy enemy fire, realizing in advance that he was exposing himself to unusual dangers in this undertaking. He continued in his task until killed by an emeny shell. Posthumously awarded. Medal presented to father, R. K. Ortt.

OSBORN, MORTON (558182)_____
Southeast of Bazoches, France, Aug. 7–9, 1918.
R—Load, Ky.
B—Greenup County, Ky.
G. O. No. 137, W. D., 1918.

Sergeant, Company H, 47th Infantry, 4th Division.
Wounded in the head and shoulder, he rejoined his platoon as soon as his wounds had been dressed and remained with it until the command was relieved, displaying rare qualities of leadership and judgment under heavy machine-gun and rifle fire.

OSBORNE, HARRY (39424)_____
Near Medeah Ferme France, Oct. 3, 1918.
R—Elizabeth, N. J.
B—Elizabeth, N. J.
G. O. No. 21, W. D., 1919.

Corporal, Company F, 9th Infantry, 2d Division.
Corporal Osborne, together with 4 other men, charged a machine-gun nest containing 3 heavy machine guns, capturing the 3 guns and 20 prisoners.

*OSBORNE, WEEDON EDWARD_____

During the advance on Bouresches, France, June 6, 1918.
R—Chicago, Ill.
B—Chicago, Ill.
G. O. No. 126, W. D., 1918.

Lieutenant, dental surgeon, U. S. Navy, attached to 6th Regiment, U. S. Marine Corps, 2d Division.
He voluntarily risked his life during the advance on Bouresches, France, on June 6, 1918, by helping to carry the wounded to places of safety, and while engaged in this difficult duty was struck by a shell and killed.
Posthumously awarded. Medal presented to sister, Mrs. Elizabeth Osborne Fisher.

OSMOND, FRANK W. (1782740)_____
Near Bellicourt, France, Sept. 29, 1918.
R—Philadelphia, Pa.
B—Parkesburg, Pa.
G. O. No. 32, W. D., 1919.

Corporal, Company A, 301st Battalion, Tank Corps.
Corporal Osmond was on duty as gunner in a tank whose track was broken by a direct hit from an enemy shell. Because of the heavy machine-gun fire it was impossible to repair the track, but Corporal Osmond, accompanied by another soldier, left the tank, picked up some rifles, and, crawling through the trenches and brush to the rear of the machine-gun position, killed 4 of the enemy crew. They then returned to the tank and assisted in repairing the track under heavy shellfire.

OTTE, FRED (1243371)_____
At Fismes and Fismette, France, Aug. 9–13, 1918.
R—Fairmount City, Pa.
B—Germany.
G. O. No. 99, W. D., 1918.

Private, Company A, 111th Infantry, 28th Division.
For 4 days, during the most intense fighting, he acted as runner between his battalion headquarters at Fismes and troops in Fismette. He made many trips across the Vesle River under heavy shell and machine-gun fire, and when the bridge had been destroyed he continued his trips by swimming the river, which contained wire entanglements.

OTTO, ANDREW C., Jr._____
South of Soissons, France, July 18, 1918.
R—New York, N. Y.
B—New York, N. Y.
G. O. No. 24, W. D., 1920.

Second lieutenant, 23d Infantry, 2d Division.
Lieutenant Otto led a force of 30 men, stormed the strong enemy position in the town of Vauxcastille, capturing 230 prisoners and 15 machine guns. He personally attacked and killed 2 of the enemy machine gunners. Upon two different occasions he attacked machine-gun nests single handed, killing or capturing the crew. His wonderful courage inspired his men to this remarkable victory against overwhelming odds.

*OTTO, WILLIAM HERMAN_____
At Chateau-Thierry, France, June 6, 1918.
R—Chicago, Ill.
B—Joliet, Ill.
G. O. No. 110, W. D., 1918.

Corporal, 45th Company, 5th Regiment, U. S. Marine Corps, 2d Division.
Killed in action at Chateau-Thierry, France, June 6, 1918, he gave the supreme proof of that extraordinary heroism which will serve as an example to hitherto untried troops.
Posthumously awarded. Medal presented to mother, Mrs. Martha Otto.

OVERMEYER, GEORGE J. (129725)_____
Near Chateau-Thierry, France, June 20, 1918, and near Vierzy, France, July 18, 1918.
R—Hartford City, Ind.
B—Upland, Ind.
G. O. No. 120, W. D., 1918.

Corporal, Headquarters Company, 15th Field Artillery, 2d Division.
On June 20 he was in charge of an observation post which was bombarded by gas shells. In spite of the fact that the other observers were overcome by the gas, he remained at his post and continued to transmit observation to the Artillery battalion commander. On July 18 this soldier was on liaison duty in the Infantry and advanced with the third wave of the attack. He successfully carried a message through 2 enemy barrages to the Artillery commander, thus giving proof of his courageous devotion to duty.

*OVERTON, JOHN WILLIAMS_____
Near Vierzy, France, July 19, 1918.
R—Nashville, Tenn.
B—Nashville, Tenn.
G. O. No. 22, W. D., 1920

Second lieutenant, 6th Regiment, U. S. Marine Corps, 2d Division.
While valiantly leading his platoon in an attack against the enemy, under severe machine-gun and artillery fire, he was mortally wounded. His courageous conduct had a great moral effect upon his men and helped to insure the success of the attack.
Posthumously awarded. Medal presented to father, Jesse M. Overton.

*OVERTON, MACON CALDWELL_____
Near Blanc Mont, France, Oct. 2–10, 1918.
R—Union Point, Ga.
B—Union Point, Ga.
G. O. No. 34, W. D., 1919.

Captain, 6th Regiment, U. S. Marine Corps, 2d Division.
For repeated acts of extraordinary heroism in action near Blanc Mont, France, Oct. 2–10, 1918. When his battalion was halted by severe fire, Captain Overton attacked and reduced 1 strong enemy machine-gun nest, and moving forward, captured 1 field piece which was firing point-blank at his company. He was wounded the next day, but he refused to be evacuated, and continued to lead his command with skill and courage throughout the engagement. On Oct. 8, after occupying St. Etienne without casualties in his company, Captain Overton went through heavy artillery and machine-gun fire to establish liaison with another company, his conspicuous gallantry inspiring his men to repel 2 strong counterattacks.
Oak-leaf cluster.
For the following act of extraordinary heroism in action near St. Georges, France, Nov. 1, 1918, Captain Overton is awarded an oak-leaf cluster, to be worn with the distinguished-service cross: He displayed remarkable courage in leading his company under heavy artillery fire and silencing 5 machine-gun nests. He then personally undertook to guide a tank forward against machine-gun positions and while so doing was seriously wounded by a German antitank sniper.
Posthumously awarded. Medal presented to mother, Mrs. Margaret Overton.

*OWENS, DEWEY (553718)_____
Near Chateau-Thierry, France, July 14–15, 1918.
R—Point, La.
B—Point, La.
G. O. No. 44, W. D., 1919.

Corporal, Company B, 8th Machine Gun Battalion, 3d Division.
While the enemy was attempting a crossing of the River Marne, he set his gun in position under heavy fire, losing some of his men in the exploit. He assisted the wounded to safety, after which he returned to his gun. When a shell struck his gun and disabled it, he secured another gun and placed it in position. This time all his men had become casualties, and, aided by a runner, he manned the gun. In attempting to repair it after another hit he was killed.
Posthumously awarded. Medal presented to father, S. E. Owens.

OWENS, FRANK A._____
Near Les Evaux, France, July 13, 1918.
R—Charlotte, N. C.
B—Charlotte, N. C.
G. O. No. 32, W. D., 1919.

First lieutenant, 4th Infantry, 3d Division.
After several unsuccessful attempts had been made to get patrols across the Marne River at night, Lieutenant Owens, with 2 soldiers, crossed in daylight and remained on the enemy side throughout the day.

OWENS, GILBERT (40760)_____
Near Medeah Ferme, France, Oct. 3–5, 1918.
R—Tyrone, Pa.
B—Tyrone, Pa.
G. O. No. 81, W. D., 1919.

Sergeant, Company M, 9th Infantry, 2d Division.
Suffering from 3 severe scalp wounds, Sergeant Owens remained with his company and for 2 days performed his duties under intense artillery and machine-gun fire, until sent to the hospital completely exhausted.

OWENS, JOHN J._____
At Fismette, France, Aug. 12–19, 1918.
R—Philadelphia, Pa.
B—Philadelphia, Pa.
G. O. No. 49, W. D., 1922.

First lieutenant, 109th Infantry, 28th Division.
While occupying the town with his company, and although it was continuously under shellfire, machine-gun fire, and deluged with gas, he displayed the highest order of courage, resolution, and leadership in holding the town against repeated attacks for 6 days, and though badly gassed remained with it until the company was relieved.

*OWENS, JOHN T._____
Near Verdun, France, November 4, 1918.
R—Hartford, Conn.
B—Hartford, Conn.
G. O. No. 37, W. D., 1919.

Second lieutenant, 315th Infantry, 79th Division.
After locating 3 machine-gun positions, Lieutenant Owens put 1 of them out of action with an automatic rifle by killing the gunners and forcing the carriers to abandon the gun. He was killed by machine-gun fire while reorganizing his company after a local counterattack.
Posthumously awarded. Medal presented to mother, Mrs. Julia A. Owens.

*OWENS, TEDDY (2220281)_____
During the St. Mihiel offensive, France, Sept. 14, 1918.
R—Fort Towson, Okla.
B—Jane, Va.
G. O. No. 72, W. D., 1920.

Bugler, Company D, 358th Infantry, 90th Division.
Bugler Owens, with 4 other men, volunteered to cross a valley to the woods opposite and silence machine guns which had held up the advance of his company. In the face of heavy enemy fire this small group accomplished its mission, thus enabling the company to cross the valley without further loss. He was killed in the performance of this act.
Posthumously awarded. Medal presented to mother, Mrs. Clementine Owens.

*PACCHIASOTTI, AMEDEA (543239)_____
Near Cierges, France, Oct. 5, 1918.
R—McKees Rocks, Pa.
B—Italy.
G. O. No. 60, W. D., 1920.

Private, Company M, 7th Infantry, 3d Division.
When his company was halted by heavy machine-gun fire from the front, Private Pacchiasotti exposed himself to heavy fire and advanced to attack the enemy position. He was killed in the attack, but his example so inspired his comrades that they continued on and captured the enemy machine-gun position.
Posthumously awarded. Medal presented to mother, Mrs. Orsela Pacchiasotti.

*PACKARD, WILLIAM L. (1769)_____
Near Exermont, France, Oct. 5, 1918.
R—Paris, Tex.
B—New York, N. Y.
G. O. No. 37, W. D., 1919.

Private, medical detachment, 7th Field Artillery, 1st Division.
In going to the aid of a wounded comrade Private Packard was himself mortally wounded, but continued on his self-sacrificing mission till he was too weak from loss of blood to continue his work, dying shortly afterwards.
Posthumously awarded. Medal presented to sister, Mrs. C. J. Altere.

PACKETT, JOHN W. (1309169)_____
Near Ponchaux, France, Oct. 7, 1918.
R—Lenoir City, Tenn.
B—Loudon County, Tenn.
G. O. No. 37, W. D., 1919.

Corporal, Company L, 117th Infantry, 30th Division.
He volunteered and carried a message to battalion headquarters under heavy artillery and machine-gun fire, although he had seen many of his comrades fall in attempting the same mission.

PADGETT, ANDREW J. (1309841)_____
Near Montbrehain, France, Oct. 7, 1918.
R—Knoxville, Tenn.
B—Spartanburg, S. C.
G. O. No. 133, W. D., 1918.

Sergeant, Company M, 117th Infantry, 30th Division.
Taking command of his platoon after its commander had been seriously wounded, Sergeant Padgett led it with remarkable daring through heavy machine-gun fire and captured 6 machine-gun nests. Wounded by a machine-gun bullet, he continued on to the objective, using his rifle as a crutch, and directed the consolidation of the new position.

PAGE, ALFRED W. (42541)_____
Near Soissons, France, July 18–23, 1918.
R—Easthampton, Mass.
B—Wilkinsonville, Mass.
G. O. No. 15, W. D., 1919.

Private, Company D, 16th Infantry, 1st Division.
During the entire 5 days of the advance he fulfilled with exceptional efficiency the difficult and hazardous duties of liaison agent between the infantry and the tanks. Subjected throughout the action to the direct fire of the enemy machine guns and antitank artillery, he demonstrated the highest type of courage and devotion to duty.

*PAGE, CHARLES C. (1210458)_____
Near Ronssoy, France, Sept. 29, 1918.
R—New York, N. Y.
B—New York, N. Y.
G. O. No. 46, W. D., 1919.

Private, Company D, 107th Infantry, 27th Division.
During operations against the Hindenburg line he crawled out through the murderous fire and rescued a wounded comrade, carrying him to the nearest dressing station.
Oak-leaf cluster.
For the following act of extraordinary heroism in action near St. Souplet, France, October 15, 1918, Private Page is awarded an oak-leaf cluster, to be worn with the distinguished-service cross: He was sent ahead with a scout patrol of 6 men, when they were suddenly fired upon. One of their number was killed and 4 others, including Private Page, were wounded. With one leg blown off and the other hanging by a fragment of flesh, he stimulated the greatest confidence in his companions by his words of encouragement while being placed on a stretcher. He died from the effects of his wounds shortly after reaching the hospital.
Posthumously awarded. Medal presented to mother, Mrs. Anna L. Page.

PAGE, KENNETH B. (2939)_____
At Bois Brule, near Apremont, France, Apr. 10, 1918.
R—Springfield, Mass.
B—Springfield, Mass.
G. O. No. 99, W. D., 1918.

Private, first class, medical detachment, 104th Infantry, 26th Division.
He displayed conspicuous gallantry during the action of April 10, 1918, in running through heavily shelled area to rescue an officer who had fallen mortally wounded, and at great personal risk carrying him to dressing station.

PAGE, RICHARD C. M._____
Near Fismes, France, Aug. 9, 1918.
R—Fort Myers, Fla.
B—Philadelphia, Pa.
G. O. No. 121, W. D., 1918.

First lieutenant, pilot, 88th Aero Squadron, Air Service.
Richard C. M. Page, first lieutenant, pilot, Air Service; John I. Rancourt, first lieutenant, observer, 88th Aero Squadron, 103d Field Artillery. For extraordinary heroism in action near Fismes, France, Aug. 9, 1918. These officers were detailed to fly without escort on a visual reconnaissance over the enemy's lines. They were attacked by 6 enemy battle planes 1,800 meters over Fismes. The Americans unhesitatingly fought this superior number of the enemy. Lieutenant Rancourt was 3 times seriously wounded in the legs above the knees, yet he continued to operate his machine gun and shot down 1 of the enemy planes. In spite of the fact that his elevator controls on 1 side had been shot away, Lieutenant Page skillfully maneuvered the plane throughout the combat and piloted it safely back to his airdrome.

PAINSIPP, ALBERT C. (1390422)_____
At Hamel, France, July 4, 1918.
R—Batavia, Ill.
B—Austria.
G. O. No. 99, W. D., 1918.

Corporal, Company A, 132d Infantry, 33d Division.
Single handed he attacked a German machine-gun emplacement. Although wounded in the leg, when a machine gun was trained upon him, he boldly attacked it with hand grenades and drove off the crew.

*PAISLEY, JOHN C._____
In the Belleau Wood, France, June 21, 1918.
R—Gibsonville, N. C.
B—Guilford County, N. C.
G. O. No. 60, W. D., 1920.

First lieutenant, 7th Infantry, 3d Division.
While leading his platoon against a machine-gun nest, Lieutenant Paisley encountered several enemy spies, who attempted to give him orders and confuse his men. He killed the officer in charge and several of the men and then continued in the attack. Later in the day he was hit by a 37-millimeter shell and instantly killed.
Posthumously awarded. Medal presented to father, John W. Paisley.

PALARDY, CHARLES W. (1813807)_____
Near Damvillers, France, Nov. 10, 1918.
R—Philadelphia, Pa.
B—Canada.
G. O. No. 35, W. D., 1919.

Corporal, Company F, 315th Infantry, 79th Division.
He went to the aid of a wounded comrade about 200 yards in advance of our lines through sniper and machine-gun fire and brought him safely to cover. He was wounded while returning on his self-appointed mission.

PALMER, ALVA W. (1457832)_____
South of Exermont, France, Sept. 28, 1918.
R—Platte City, Mo.
B—Putnam County, Mo.
G. O. No. 59, W. D., 1919.

Private, first class, Company K, 139th Infantry, 35th Division.
When his battalion commander asked for a volunteer to carry a message to the battalion commander on the right, Private Palmer volunteered and carried the message through an area exposed to intense machine-gun and artillery fire. By his utter disregard for his own personal safety and example of bravery he inspired all those near him.

PALMER, DONALD D. (252135)_____
Near Cierges, France, Aug. 2, 1918.
R—West Allis, Wis.
B—Green Bay, Wis.
G. O. No. 20, W. D., 1919.

Corporal, Company C, 107th Field Signal Battalion, 32d Division.
He was a member of a detachment stringing telephone wire far in advance of the front lines through the heaviest artillery fire. When connection was established at a point within 100 yards of the German line and before the American advance was begun, he volunteered to remain there until our troops had advanced far enough to establish the advance regimental post of command at that place. His devotion to duty under conditions of great danger assisted immeasurably in maintaining unity of action between the front lines and regimental post of command, and his utter indifference to his own safety made easier the capture of a strong enemy position.

PALMER, HARRY H. (2267656)_____
Near Eclisfontaine, France, Sept. 28, 1918.
R—Los Angeles, Calif.
B—Garden Plain, Kans.
G. O. No. 44, W. D., 1919.

Private, first class, Company K, 364th Infantry, 91st Division.
Responding to a call for volunteers, Private Palmer, with 5 others advanced 400 yards beyond their front to bring in wounded comrades. They succeeded in rescuing 7 of their men and also in bringing in the dead body of a lieutenant while exposed to terrific machine-gun fire.

PALMER, JOSEPH A_____
Near Fismes, France, Aug. 11, 1918.
R—Zanesville, Ohio.
B—Zanesville, Ohio.
G. O. No. 44, W. D., 1919.

Second lieutenant, 15th Field Artillery, observer, attached to 88th Aero Squadron, Air Service.
Louis G. Bernheimer, first lieutenant, pilot; John W. Jordan, second lieutenant, 7th Field Artillery, observer; Roger W. Hitchcock, second lieutenant, pilot; James S. D. Burns, deceased, second lieutenant, 165th Infantry, observer; Joel H. McClendon, deceased, first lieutenant, pilot; Charles W. Plummer, deceased, second lieutenant, 101st Field Artillery, observer; Philip R. Babcock, first lieutenant, pilot; and Joseph A. Palmer, second lieutenant, 15th Field Artillery, observer. All of these men were attached to the 88th Aero Squadron, Air Service.
For extraordinary heroism in action near Fismes, France, Aug. 11, 1918. Under the protection of 3 pursuit planes, each carrying a pilot and an observer, Lieutenants Bernheimer and Jordan, in charge of a photo plane, carried out successfully a hazardous photographic mission over the enemy's lines to the River Aisne. The 4 American ships were attacked by 12 enemy battle planes. Lieutenant Bernheimer, by coolly and skillfully maneuvering his ship, and Lieutenant Jordan, by accurate operation of his machine gun, in spite of wounds in the shoulder and leg, aided materially in the victory which came to the American ships, and returned safely with 36 valuable photographs. The pursuit plane operated by Lieutenants Hitchcock and Burns was disabled while these 2 officers were fighting effectively. Lieutenant Burns was mortally wounded and his body jammed the controls. After a headlong fall of 2,500 meters, Lieutenant Hitchcock succeeded in regaining control of this plane and piloted it back to his airdrome. Lieutenants McClendon and Plummer were shot down and killed after a vigorous combat with 5 of the enemy's planes. Lieutenants Babcock and Palmer, by gallant and skillful fighting, aided in driving off the German planes and were materially responsible for the successful execution of the photographic mission.

PALMER, LESTER E. (68130)_____
Near Bois-de-St. Remy, France, Sept. 12, 1918.
R—Dover, Me.
B—Canada.
G. O. No. 26, W. D., 1919.

Private, Company F, 103d Infantry, 26th Division.
After 3 of his platoon had been killed and 6 had been wounded, Private Palmer crawled forward to a shell hole and killed 1 gunner in the nest. Subjected to a hand-grenade bombing, he made his way to another shell hole and from there shot another of the enemy crew, after which he rushed the nest and captured the remaining gunner and machine gun.

PALMER, SIDNEY H. (1698066)_____
In the Bois-de-la-Naza, France, Oct. 5, 1918.
R—Brooklyn, N. Y.
B—Brooklyn, N. Y.
G. O. No. 95, W. D., 1919.

Sergeant, Company H, 305th Infantry, 77th Division.
With 2 other soldiers Sergeant Palmer volunteered to crawl out under enemy machine-gun fire in an effort to locate 3 members of the platoon who were missing after an unsuccessful attack on enemy machine-gun nests. Finding the body of 1, who lay helplessly wounded, by calling out his name. As a result, they drew increased fire from the enemy, but they courageously crawled 25 yards farther toward the hostile positions and succeeded in carrying back the wounded man through the machine-gun fire to our lines.

PALMER, WILLIAM W_____
In the region of Douleon, France, Oct. 3, 1918.
R—Bennettsville, S. C.
B—Warrenton, N. C.
G. O. No. 143, W. D., 1918.

First lieutenant, pilot, 94th Aero Squadron, Air Service.
He encountered 3 enemy planes (Fokker type). Despite their numerical superiority, he attacked and in a decisive combat sent 1 down in flames and forced the others to retire.

PALUBIAK, GUS W. (1387944)_____
Near Forges Woods, France, Sept. 26, 1918.
R—Chicago, Ill.
B—St. Louis, Mo.
G. O. No. 71, W. D., 1919.

Corporal, Company H, 131st Infantry, 33d Division.
He advanced alone on his own initiative, in the face of heavy machine-gun fire, and destroyed a nest of German machine guns. His brave action allowed his company to resume the advance.

PAMARANSKI, JOHN (1828967)_____
Near Bois-des-Ogons, France, Oct. 10, 1918.
R—Pittsburgh, Pa.
B—Poland.
G. O. No. 26, W. D., 1919.

Corporal, Company B, 320th Infantry, 80th Division.
When his platoon was held up by an enemy machine gun, which had caused many casualties in the platoon, Corporal Pamaranski advanced to within bombing distance of the gun, killed 1 and captured 2 of the enemy, together with the machine gun.

PAPADAKIS, CHRIST (733237)_____
At Romagne - sous - Montfaucon, France, Oct. 14-15, 1918.
R—Mulkeytown, Ill.
B—Greece.
G. O. No. 37, W. D., 1919.

Private, Company H, 6th Infantry, 5th Division.
For more than 48 hours he carried litters with wounded or administered, alone, first aid to wounded while under continuous artillery and machine-gun fire. He continued his work until forced to stop from exhaustion.

PAPPAS, PAUL J. (2659513)_____
Near Argonne Forest, France, Oct. 12, 1918.
R—Niles, Ohio.
B—Turkey.
G. O. No. 44, W. D., 1919.

Private, Company M, 39th Infantry, 4th Division.
When his company withdrew from their position, Private Pappas, with 1 other soldier, saw the enemy forming for a counterattack, and, without thought of their danger, refused to withdraw, but held this part of the line for several hours by the efficient use of an automatic rifle, subjected to withering machine-gun fire during the entire time.

PARADIS, ALBERT D. (551481)_____
  Near Mezy, France, July 15, 1918.
  R—New Bedford, Mass.
  B—New Bedford, Mass.
  G. O. No. 27, W. D., 1920.

Private, first class, Company H, 38th Infantry, 3d Division.
Private Paradis carried numerous messages for his company commander across an area swept by heavy machine-gun fire. His courage when exposed to unusual danger enabled his battalion commander to obtain accurate information of the action.

PARADIS, ARTHUR (68972)_____
  Near the Belleau Wood, France, July 18–24, 1918.
  R—Nashua, N. H.
  B—Lowell, Mass.
  G. O. No. 81, W. D., 1919.

Private, Company B, 103d Infantry, 26th Division.
Private Paradis volunteered as a runner and carried messages through heavy concentrations of machine-gun fire. Single handed he penetrated an enemy outpost and killed all of the Germans who were on guard there.

PARADISE, ROBERT C._____
  In the vicinity of Boureuilles, France, Sept. 26, 1918.
  R—England.
  B—New Orleans, La.
  G. O. No. 15, W. D., 1923.

First lieutenant, 12th Aero Squadron, Air Service.
As pilot, 12th Squadron, he was assigned the duty of locating the American front lines during the first two hours of the Argonne offensive. Unable to locate the line at the usual altitude maintained at such a time, he flew down to the dangerous altitude of 50 meters, secured the important information sought and discovered our lines held up by a strongly held nest of enemy machine guns. Noting the exact location of the nest upon his map he flew back to division headquarters and reported the exact location of our lines, as well as that of the enemy machine-gun nest. With his plane riddled by enemy bullets, one control shot away, he returned to the lines, discovered the enemy nests had not been destroyed and that they were inflicting heavy casualties upon our troops. In the face of concentrated enemy fire and attacked by 4 enemy planes he went down, his plane barely skimming the tree tops, and deliberately fired over 400 rounds into the enemy nests, thus causing the enemy gunners to abandon their guns and positions and enabling the troops of his division to resume their advance. Again gaining altitude he discovered and destroyed by his fire an enemy signal station, signaling unmolested, 2 kilometers north of the lines. This act was performed at an altitude of 50 meters amid a storm of protection fire from enemy antiaircraft guns.

*PARADISO, TONY (64349)_____
  At Epieds, France, July 23, 1918.
  R—Norwalk, Conn.
  B—Italy.
  G. O. No. 46, W. D., 1919.

Private, Company D, 102d Infantry, 26th Division.
Fighting with rare courage at Epieds, he bayoneted several Germans, and then, discovering 2 machine gunners in a tree, he crept through the wheat fields alone and killed them. Later he made several trips from Epieds to a dressing station in the woods, traversing a road under constant shell and machine-gun fire.
Posthumously awarded. Medal presented to father, Louis Paradiso.

PARCELL, CHARLEY N. (1817718)_____
  Near Nantillois, France, Oct. 5, 1918.
  R—Rockymount, Va.
  B—Rockymount, Va.
  G. O. No. 46, W. D., 1919.

Private, Company D, 317th Infantry, 80th Division.
Carrying messages for the platoon commander to squad leaders, under heavy enemy fire, Private Parcell greatly aided the advance of his platoon. Although twice wounded in the face by shrapnel, he continued his duties until ordered to the dressing station.

PARENT, EDDIE J. (2722420)_____
  Near Verdun, France, Oct. 10, 1918.
  R—Brunswick, Me.
  B—Brunswick, Me.
  G. O. No. 44, W. D., 1919.

Private, Company G, 26th Infantry, 1st Division.
While his company was suffering severe losses from an enemy machine gun, Private Parent, unaided, crawled forward and silenced the gun.

PARENT, JOSEPH C. (134757)_____
  At Seicheprey, France, Apr. 20, 1918.
  R—Webster, Mass.
  B—Webster, Mass.
  G. O. No. 99, W. D., 1918.

Private, first class, Battery B, 102d Field Artillery, 26th Division.
For faithfulness and great coolness in the execution of his duty on Apr. 20, 1918, when, although severely wounded in the head and left leg, he continued, under heavy shellfire, to repair the telephone lines and succeeded in reestablishing communication.

*PARISER, HARRY (1707038)_____
  Southwest of Binarville, France, Oct. 3, 1918.
  R—New York, N. Y.
  B—New York, N. Y.
  G. O. No. 5, W. D., 1920.

Corporal, Company L, 307th Infantry, 77th Division.
After being wounded in the chin by a shell fragment, he refused to go to the rear for treatment, but continued leading his men in action and exhibiting a brilliant spirit of courage and devotion to duty. He was killed Oct. 4, 1918, while again leading his squad in the attack.
Posthumously awarded. Medal presented to brother, Abraham M. Pariser.

PARISSI, GUISEPPE (2337996)_____
  Near Crezancy, France, July 14–15, 1918.
  R—Olean, N. Y.
  B—Italy.
  G. O. No. 126, W. D., 1919.

Private, first class, Company A, 4th Infantry, 3d Division.
Throughout the night of July 14–15, during the height of the offensive German bombardment, Private Parissi carried messages through woods made almost impassable by fallen trees.

PARKE, IRA S. (405912)_____
  Near Bony, France, Sept. 29–30, 1918.
  R—New York, N. Y.
  B—Detroit, Mich.
  G. O. No. 53, W. D., 1920.

Private, Machine Gun Company, 107th Infantry, 27th Division.
Private Parke, a machine gunner, although wounded 3 times during the attack on the Hindenburg line, refused to leave the field, and set a splendid example for his comrades.

PARKER, CHARLES W._____
  Near Ardeuil, France, Sept. 29–Oct. 1, 1918.
  R—Woodland, N. C.
  B—Menola, N. C.
  G. O. No. 21, W. D., 1919.

Second lieutenant, 371st Infantry, 93d Division.
Severely wounded in the foot Sept. 29, Lieutenant Parker remained on duty and ably commanded his platoon until Oct. 1, 1918.

*PARKER, DONALD M_____
Near Thiaucourt, France, Sept. 15, 1918.
R—Detroit, Mich.
B—Leominster, Mass.
G. O. No. 20, W. D., 1919.

Corporal, 80th Company, 6th Regiment, U. S. Marine Corps, 2d Division.
He voluntarily joined an officer and with him attacked and silenced a strong machine-gun nest menacing the left flank of the line. He held the position in the face of strong opposition until he was fatally wounded by a sniper. Posthumously awarded. Medal presented to mother, Mrs. Esther J. Parker.

PARKER, GEORGE E., Jr_____
Near Medeah Ferme, France, Oct. 8, 1918.
R—Baltimore, Md.
B—Baltimore, Md.
G. O. No. 21, W. D., 1919.

First lieutenant, 9th Infantry, 2d Division.
Gassed several times and his gas mask and pistol clip shot from his belt while going through a barrage, he continued to lead his company forward to its objective. He continually took and held first-line positions and repulsed several counterattacks. When the commanding officer of his battalion was cut off by the enemy he organized the battalion and held off repeated counterattacks, the while greatly outnumbered and fighting on 3 sides.

PARKER, HUGH C_____
Near Bois-des-Ogons, France, Oct. 10, 1918.
R—Mount Landing, Va.
B—Mount Landing, Va.
G. O. No. 23, W. D., 1919.

First lieutenant, 320th Infantry, 80th Division.
While his platoon was being held up by machine-gun fire and the casualties were becoming very heavy, Lieutenant Parker crawled forward to within bombing distance of the enemy, and by killing 1 and capturing 2 of the enemy with their machine guns, he enabled his platoon to continue its advance.

PARKER, JOHN A. (1497348)_____
Near Medeah Ferme, France, Oct. 8, 1918.
R—Greenville, Tex.
B—Greenville, Tex.
G. O. No. 21, W. D., 1919.

Private, Company G, 9th Infantry, 2d Division.
In addition to his duties as runner, he volunteered and assisted in cleaning out many dugouts. At one dugout he was attacked by a number of Germans; he counterattacked with grenades, capturing 10 prisoners and 2 light machine guns.

PARKER, JOHN H_____
At Seicheprey, France, Apr. 20, 1918.
R—Green Ridge, Mo.
B—Tipton, Mo.
G. O. No. 127, W. D., 1918, and G. O. No. 56, W. D., 1922.
Distinguished-service medal also awarded.

Colonel, 102d Infantry, 26th Division.
During the engagement at Seicheprey he went out in a withering hostile barrage to inspect his lines. Repeatedly he climbed upon the firing step of the trench, and, standing there with his back toward the enemy and with shell splinters falling about him, he talked to his men in such cool, calm terms as to reassure them and brace them up so that when he left they were in a cheerful state of mind and in better condition to ward against attack. Oak-leaf clusters (3).
A bronze oak leaf is awarded to Colonel Parker for the following act of extraordinary heroism: On July 21, 1918, near Trugny, France, he made a personal reconnaissance over a front of about 2 kilometers on horseback in the face of enemy fire and determined the strength of the German forces to insure the most advantageous approach for his troops to attack. Several times he was an inspiring figure to his men under a heavy artillery barrage and concentration of machine-gun fire.
A bronze oak leaf is also awarded to Colonel Parker for the following act of extraordinary heroism: On July 25, 1918, on the road through La Fere Wood, between Beuvardes and Le Charmel, France, a battalion just coming into the line was halted, awaiting orders. Subjected suddenly to an intense artillery concentration, the men, who had only such cover as was afforded by the shallow ditches along the road, were thrown into some confusion. At that moment Colonel Parker came down the road on horseback. Immediately appreciating the situation, he twice rode down the line and back again at a slow walk, stopping to talk with the men; and thus by his fearless personal exposure to and disregard of danger he promptly steadied the troops and prevented probable disorder at an important juncture.
A bronze oak leaf is also awarded to Colonel Parker for extraordinary heroism in action near Gesnes, France, Sept. 29, 1918. During the attack on the village of Gesnes he displayed great gallantry and fearlessness in leading and directing his front line with utter disregard for personal safety and urged his men forward by his personal example, all under heavy machine-gun, high-explosive gas-shell, and shrapnel fire. He was abreast of his front line until he fell, twice wounded, but thereafter remained in active command for a period of five hours, when he was relieved by the lieutenant colonel of his regiment.

PARKER, SAMUEL I_____
Near Exermont, France, Oct. 5, 1918.
R—Monroe, N. C.
B—Monroe, N. C.
G. O. No. 44, W. D., 1919.

Second lieutenant, 28th Infantry, 1st Division.
With total disregard for his own personal danger, he advanced directly on a machine gun 150 yards away while the enemy were firing directly at him and killed the gunner with his pistol. In the town of Exermont his platoon was almost surrounded, after having taken several prisoners and inflicted heavy losses on the enemy, but, despite the fact that only a few men of the platoon were left, continued to fight until other troops came to their aid.

PARKER, WILLIAM E. (2993207)_____
At Bois-de-Manheulles, France, Nov. 9-11, 1918.
R—Verona, N. C.
B—Verona, N. C.
G. O. No. 81, W. D., 1919.

Private, Company E, 323d Infantry, 81st Division.
Private Parker gave proof of unhesitating devotion to duty and disregard for personal safety by continually volunteering and carrying messages to various units, crossing zones swept by machine-gun and heavy artillery fire.

PARKER, WILLIAM J. (1319291)_____
Near Bellicourt, France, Sept. 29, 1918.
R—Lexington, N. C.
B—Davidson County, N. C.
G. O. No. 81, W. D., 1919.

Sergeant, Company A, 120th Infantry, 30th Division.
Severely wounded in the abdomen while in charge of a detail carrying up trench-mortar ammunition, he refused to be evacuated, advancing 500 yards until his left arm was blown off by shellfire. Refusing to be carried in a stretcher, which he said was needed for more severely wounded men, he walked 2 kilometers to the first-aid station.

PARKHILL, OAKLEY L.
In the Bois-du-Fays, France, Oct. 13–Nov. 11, 1918.
R—Abbotsford, Wis.
B—Thorp, Wis.
G. O. No. 37, W. D., 1919.

Second lieutenant, 61st Infantry, 5th Division.
On Oct. 13, in company with another lieutenant, he was wounded by the explosion of a shell. Regardless of his own wound he administered first aid to his companion and carried him to a first-aid station. After he was evacuated to the field hospital, Lieutenant Parkhill refused to be evacuated to the service of supplies hospital, and on Oct. 18 returned to duty with his company. During the period Oct. 25 to Nov. 11 he repeatedly exposed himself to enemy fire while reconnoitering and leading his company.

PARKIN, HARRY D.
At Hill 378, the Borne du Cornouiller, France, Nov. 4, 1918.
R—Pittsburgh, Pa.
B—Pittsburgh, Pa.
G. O. No. 15, W. D., 1923.

Major, 316th Infantry, 79th Division.
For extraordinary heroism in action at Hill 378, the Borne du Cornouiller, France, Nov. 4, 1918, while in command of one of the assaulting battalions of the 316th Infantry. Leading the attack, Major Parkin received 4 wounds from enemy machine-gun fire, but declined to be evacuated, remaining with his command in the position he had captured, temporarily assigning active command of his battalion to his senior captain. Later, learning that this officer had been killed, Major Parkin, despite intense suffering from his wounds, again assumed active command, and under a terrific enemy concentration of artillery and machine-gun fire, defended the position with great bravery and gallantry against counterattacks by vastly superior numbers of the enemy forces. His undaunted courage greatly inspired the men of his command, raising their morale to a great pitch.

*PARMLEY, WILLIAM BRACKSON
At Chateau-Thierry, France, June 6, 1918.
R—Somerset, Ky.
B—Pulaski County, Ky.
G. O. No. 110, W. D., 1918.

Sergeant, 18th Company, 5th Regiment, U. S. Marine Corps, 2d Division.
Killed in action at Chateau-Thierry, France, June 6, 1918, he gave the supreme proof of that extraordinary heroism which will serve as an example to hitherto untried troops.
Posthumously awarded. Medal presented to father, N. R. Parmley.

PARRIS, WORDEN W.
At Berzy-le-Sec, France, July 21, 1918.
R—Washington, D. C.
B—Washington, D. C.
G. O. No. 132, W. D., 1918.

First lieutenant, 2d Infantry Brigade, 1st Division.
While serving as an aide on the brigade staff he went through machine-gun fire and artillery bombardment with heroic fearlessness to obtain vital information from the front lines for the division commander.

PARRISH, GRADY (97137)
Near Cote-de-Chatillon, France, Oct. 16, 1918.
R—Daleville, Ala.
B—Daleville, Ala.
G. O. No. 20, W. D., 1919.

Sergeant, Company G, 167th Infantry, 42d Division.
After his platoon commander had been severely wounded and his platoon had suffered heavy casualties he quickly reorganized the remainder of the platoon and personally led it in the attack on Cote-de-Chatillon. By his daring acts, coolness, and good judgment he broke up a heavy enemy counterattack on his front, thereby saving his men and being an example of exceptional heroism and devotion to duty.

PARROTT, ROGER S.
On Patian Island, P. I., July 4, 1909.
R—Dayton, Ohio.
B—Dayton, Ohio.
G. O. No. 13, W. D., 1924.

Second lieutenant, 2d Field Artillery, U. S. Army.
During the attack on the Moro stronghold he commanded with great gallantry and coolness a mountain-gun detachment. In the face of enemy fire the mountain gun was dragged to and held by block and tackle within a few yards of the hostile position, from which place he directed a heavy fire on the enemy, and replaced his gunner when the latter was severely wounded. When the assault on the enemy's position took place and the gun commanded by Lieutenant Parrott could no longer be fired, he took command of the men in his immediate vicinity, gallantly leading them forward, and engaging the charging enemy in a hand-to-hand combat.

PARSONS, JAMES K.
Near Cuisy, France, Sept. 27–Oct. 11, 1918.
R—Birmingham, Ala.
B—Birmingham, Ala.
G. O. No. 98, W. D., 1919.
Distinguished-service medal also awarded.

Colonel, 39th Infantry, 4th Division.
Having volunteered to take command of a battalion whose commander had been wounded, Colonel Parsons was knocked down by hostile shellfire, but he succeeded in rallying his men and kept them well organized, so as to withstand the heavy fire of the enemy. On the following day he assumed command of the regiment and commanded it in successful attacks, refusing to be evacuated after being so severely gassed that he was unable to see.

PASCHAL, PAUL C.
In the Bois-d'Aigremont, France, July 15, 1918.
R—Goldston, N. C.
B—Siler City, N. C.
G. O. No. 32, W. D., 1919.

Major, 30th Infantry, 3d Division.
During the intense artillery bombardment preceding the German drive of July 15, when the wounded were so numerous that it was impossible to care for them in the dressing stations, Major Paschal voluntarily gave up his dugout for the use of the wounded and exposed himself to the heavy fire for 16 hours. After crossing the Marne this officer placed himself in the front line, in spite of the severe artillery barrage, in order to direct the attack, capturing 2 strongly fortified farmhouses and advancing his line for a distance of 4 kilometers. After gaining the position he remained on duty for 2 days without food, despite the fact that he had been wounded and gassed.

PASSAFIUME, JOSEPH (1706135)
In the Argonne Forest, France, Sept. 29–30, 1918.
R—Buffalo, N. Y.
B—Italy.
G. O. No. 46, W. D., 1919.

Private, Company G, 307th Infantry, 77th Division.
He was detailed as a member of a team of runners, 4 teams having been sent to the battalion companies with a message regarding the attack of Sept. 30. On account of the extreme darkness and the fact that the companies had changed positions, all the runners except Private Passafiume reported back, being unsuccessful in the mission. He continued on, however, reaching the company to which he had been sent, and thinking that the others may have been unsuccessful, found all the other companies, obtaining a signed receipt of the message. He then found his way back and reported to his battalion commander.

PATARCITY, ADAM (1280804)............
Near Verdun, France, Oct. 12, 1918.
R—Trenton, N. J.
B—Austria-Hungary.
G. O. No. 130, W. D., 1918.

Bugler, Company F, 114th Infantry, 29th Division.
He held his position in the face of an enemy counterattack, silenced with his pistol 1 machine-gun nest, and, unaided, brought in 3 prisoners from another.

PATCH, JOSEPH D............
Near Chaudun, France, July 18, 1918.
R—Lebanon, Pa.
B—Fort Huachuca, Ariz.
G. O. No. 56, W. D., 1922.

Major, 18th Infantry, 1st Division.
The leading battalion having encountered heavy resistance and his battalion having been ordered to pass through the leading battalion, Major Patch, because so many of his officers and men had been killed or wounded, with the greatest courage, coolness, and efficiency, personally led the assault of his battalion on the final objective. As a result of his fearlessness and leadership the objective was carried and he was severely wounded.

PATE, JOSEPH B............
In the Meuse-Argonne offensive, France, Sept. 28, 1918.
R—Chattanooga, Tenn.
B—Maryville, Tenn.
G. O. No. 19, W. D., 1922.

Major, 371st Infantry, 93d Division.
Major Pate having been ordered to place his battalion in position to cover a gap in the line, preparatory to leading the assault that day, and having first sent out 2 patrols which failed to accomplish their mission on account of darkness and heavy enemy fire, did make with only 1 French interpreter, a personal reconnaissance under heavy machine-gun fire, traversing the whole front of the gap, locating it accurately, and returning with the information necessary for the intelligent issue of orders for the assault.

PATON, NOEL E. (8669)............
Near Woel, France, Sept. 14, 1918.
R—Fayetteville, N. C.
B—Bethel, N. C.
G. O. No. 46, W. D., 1919.

Sergeant, Company A, 344th Battalion, Tank Corps.
While on a reconnaissance patrol under heavy machine-gun fire he was seriously wounded and ordered to the rear. Refusing to seek safety, he crawled to the assistance of 2 comrades whom he had seen disappear under a burst of shrapnel, and with 1 arm useless attempted to render aid while he was himself suffering from loss of blood.

PATRICK, WILLIAM E............
On the Meuse River, France, Nov. 1-11, 1918.
R—Boston, Mass.
B—Cambridge, Mass.
G. O. No. 46, W. D., 1919.

First lieutenant, chaplain, 23d Infantry, 2d Division.
During this period Chaplain Patrick constantly exposed himself to the enemy fire while giving first aid to the wounded and assisting in their evacuation.

*PATTEN, JAMES H. (1235825)............
Near Conde-en-Brie, France, July 17, 1918.
R—Philadelphia, Pa.
B—Locust Grove, Md.
G. O. No. 74, W. D., 1919.

Corporal, Company D, 109th Infantry, 28th Division.
He was an example to the men of his platoon when they were under fire for the first time, near Conde-en-Brie, France, July 17, 1918. He continually circulated among his men, encouraging and cautioning them. Mortally wounded by shrapnel, he refused to be evacuated, but stayed with his platoon until he died, with a last word of encouragement on his lips.
Posthumously awarded. Medal presented to uncle, Simon P. Moffett.

PATTEN, LOUIS P............
Near the Forest of Argonne, France, Sept. 28, 1918.
R—Toledo, Ohio.
B—Ohiowa, Nebr.
G. O. No. 126, W. D., 1919.

Captain, 147th Infantry, 37th Division.
Captain Patten was seriously wounded in the shoulder while leading his company, but after being tagged for evacuation at the dressing station his insistent request for permission to return to his command was granted, and he continued to lead his company until the division was relieved.

*PATTERSON, ALFRED B., Jr............
In the region of Moiry, France, Sept. 29, 1918.
R—Wilkinsburg, Pa.
B—Pittsburgh, Pa.
G. O. No. 133, W. D., 1918.

First lieutenant, pilot, 93d Aero Squadron, Air Service.
While on a patrol with 2 other machines, he attacked an enemy formation of 7 planes (Fokker type) that were protecting a biplane plane. They destroyed the biplane and 4 of the Fokkers, forcing the remaining 3 to retire.
Oak-leaf cluster.
A bronze oak leaf, for extraordinary heroism in action in the region of Moiry, France, Oct. 23, 1918. He led a formation for the purpose of protecting our bombing planes, the accompanying planes being obliged to return, due to engine trouble. Despite this fact, Lieutenant Patterson proceeded on the mission alone. He sighted an enemy patrol of 9 machines (Fokker type) and attacked them, driving 1 down.
Posthumously awarded. Medal presented to father, A. B. Patterson.

PATTERSON, EARL H. (1253777)............
Near Apremont, France, Oct. 4, 1918.
R—Wilkes-Barre, Pa.
B—Wilkes-Barre, Pa.
G. O. No. 13, W. D., 1919.

Corporal, Battery D, 109th Field Artillery, 28th Division.
While acting as a runner for the battalion, he constantly exposed himself to shellfire. While taking a message to the battalion commander at Apremont he was wounded, but, regardless of his own suffering and danger, endeavored to carry a comrade, who was mortally wounded, to a place of safety. He then delivered the message before he would allow his wounds to be dressed.

PATTERSON, FREDERICK W. McL.............
Near Nantillois, France, Oct. 28-29, 1918.
R—Pittsburgh, Pa.
B—England.
G. O. No. 15, W. D., 1921.

Major, 315th Infantry, 79th Division.
After being severely wounded in the left leg, he continued throughout the night to exercise command of his battalion at a critical time. He refused medical aid until the morning of the 29th and was evacuated by order of the regimental commander.

PATTERSON, ROBERT P............
Near Bazoches, France, Aug. 14, 1918.
R—Glens Falls, N. Y.
B—Glens Falls, N. Y.
G. O. No. 35, W. D., 1920.

Captain, 306th Infantry, 77th Division.
Captain Patterson, accompanied by 2 noncommissioned officers, made a daring daylight reconnaissance into the enemy lines. He surprised an enemy outpost of superior numbers and personally destroyed the outpost. Later he again had an encounter with another outpost, during which several of the enemy were killed or wounded and 1 member of his patrol wounded. The enemy advanced their outposts, and Captain Patterson covered the retreat of his patrol, during which he dropped into a depression and feigned being killed in order to escape capture. Here he lay until he was able to escape to his lines under cover of darkness.

PATTILLO, FRANK A.—————————
 North of Montfaucon, France, Oct. 11, 1918.
 R—Georgia.
 B—Forsyth, Ga.
 G. O. No. 27, W. D., 1920.

Captain, 38th Infantry, 3d Division.
Captain Pattillo personally led his company in attack upon the enemy position, exposed himself to heavy machine-gun fire in order to advance his forward units. Although twice wounded, he refused to be evacuated, but remained where he fell, continuing to urge his men forward.

PATTON, GEORGE S., Jr.—————————
 Near Cheppy, France, Sept. 26, 1918.
 R—San Gabriel, Calif.
 B—San Gabriel, Calif.
 G. O. No. 133, W. D., 1918.
 Distinguished-service medal also awarded.

Lieutenant colonel, Tank Corps.
He displayed conspicuous courage, coolness, energy, and intelligence in directing the advance of his brigade down the valley of the Aire. Later he rallied a force of disorganized Infantry and led it forward behind the tanks under heavy machine-gun and artillery fire until he was wounded. Unable to advance farther, he continued to direct the operations of his unit until all arrangements for turning over the command were completed.

*PAUL, EDWIN (1212002)—————————
 Near Ronssoy, France, Sept. 29, 1918.
 R—White Plains, N. Y.
 B—England.
 G. O. No. 24, W. D., 1920.

Private, Company L, 107th Infantry, 27th Division.
Private Paul fearlessly ran in front of a tank under heavy machine-gun fire in order to drag a wounded officer out of the path of its advance. Shortly after, although himself severely wounded, he again exposed himself to heavy fire in order to render aid to a wounded corporal. While bandaging the corporal's leg he was hit by a shell fragment and killed.
Posthumously awarded. Medal presented to father, Eugene Paul.

PAUL, HUBERT C. (2019049)—————————
 Near Kadish, Russia, Sept. 27–28, 1918.
 R—Detroit, Mich.
 B—Alton, Ill.
 G. O. No. 78, W. D., 1919.

Private, 337th Ambulance Company, 310th Sanitary Train, attached to the 339th Infantry, 85th Division (detachment in North Russia).
Acting as stretcher bearer to 2 companies of infantry in action against the Bolsheviks, Private Paul for 2 days and nights made his way through swamps and forests to administer first aid and carry wounded to the dressing station. His work at all times was accomplished under sweeping machine-gun and intense artillery fire, making it necessary for him to crawl on his hands and knees for long distances.

*PAUL, JOHN (8165)—————————
 Near Fismes, France, Aug. 9, 1918.
 R—Paterson, N. J.
 B—Paterson, N. J.
 G. O. No. 27, W. D., 1919.

Private, first class, section No. 524, Ambulance Service.
After driving his ambulance continuously for a period of 15 hours Private Paul voluntarily left his post and went 4 kilometers in advance for wounded, traveling a road subjected to heavy machine-gun and shell fire. He was instantly killed by a shell after returning with these wounded men and carrying them to a dugout.
Posthumously awarded. Medal presented to father, John S. Paul.

PAULEY, WILLARD E.—————————
 At Bois-de-Belleau, France, June 2, 1918.
 R—St. Albans, W. Va.
 B—St. Albans, W. Va.
 G. O. No. 107, W. D., 1918.

Private, 15th Company, 6th Machine Gun Battalion, U. S. Marine Corps, 2d Division.
Showed the greatest determination and courage at Bois-de-Belleau, on June 2, 1918, when he maintained communication between the firing line and his headquarters by visual signaling. Knocked down twice, he remained at his post in the open for several hours under heavy shellfire.

*PAULSON, ARTHUR (560703)—————————
 Near Brieulles, France, Sept. 29, 1918.
 R—Cadillac, Mich.
 B—Cadillac, Mich.
 G. O. No. 89, W. D., 1919.

Sergeant, Company A, 59th Infantry, 4th Division.
While fearlessly exposing himself by walking along the front line, in order to convey orders to his platoon, Sergeant Paulson was shot 3 times through the stomach. He nevertheless refused to go to the rear until he had conducted the platoon to its new position, and then declined assistance, walking 500 yards under fire to the dressing station. Upon arriving there, he insisted on sitting up, saying that the stretchers were needed for others. He died shortly afterwards, having exhibited exceptional qualities of leadership, courage, and devotion to duty.
Posthumously awarded. Medal presented to mother, Mrs. Anna Paulson.

PAUSTIAN, HERMAN G. (3085394)—————————
 Near Verdun, France, Nov. 7, 1918.
 R—Kansas City, Mo.
 B—Scott County, Iowa.
 G. O. No. 37, W. D., 1919.

Private, Company D, 316th Infantry, 79th Division.
He advanced ahead of his battalion during a heavy barrage, trying to locate a small group of Americans who had become lost. For 2 days and nights he carried messages from 1 shell hole to another, having no food or water during that period. His work was carried on under intense bombardment at all times, but with great courage he remained at his task, killing at least 2 enemy snipers.

PAWEL, VINTON (58760)—————————
 Near Soissons, France, July 18–22, 1918.
 R—Milwaukee, Wis.
 B—Russia.
 G. O. No. 9, W. D., 1923.

Supply sergeant, Company M, 28th Infantry, 1st Division.
As supply sergeant of his company, he voluntarily and fearlessly led a platoon of his company in an attack on St. Amand Farm at the Soissons-Paris Road, spurring his men on to their objectives under intense machine-gun fire. His company officers having been killed, wounded, or called to higher units, he assumed command of his company, frequently exposing himself to heavy enemy fire during its reorganization and preparation for a renewal of the advance. On July 20–21 he again led his company in the assault, inspiring his men by his utter disregard of personal safety, maintaining constant liaison with his flanks and with his battalion commander, refusing repeatedly to be evacuated, although severely and painfully wounded by shrapnel, until relieved on July 22 by a commissioned officer.

*PAYNE, EARL C. (5789)—————————
 Near Blanc Mont Ridge, France, Oct. 7, 1918.
 R—St. Joseph, Mo.
 B—Fleming, Ky.
 G. O. No. 32, W. D., 1919.

Private, first class, medical detachment, 9th Infantry, 2d Division.
He displayed exceptional valor and devotion to duty by constantly attending the wounded under machine-gun and artillery fire. He continued his task until mortally wounded while rendering first aid to a wounded soldier under the direct observation of an enemy machine gunner.
Posthumously awarded. Medal presented to father, James R. Payne.

*PAYNE, FRANCIS W.

Near Soissons, France, July 19, 1918.
R—Charleston, W. Va.
B—Charleston, W. Va
G. O. No. 132, W. D., 1918.

Second lieutenant, 26th Infantry, 1st Division.
While in charge of an ammunition-carrying party near Soissons, France, July 19, 1918, he showed the highest degree of courage in taking ammunition to the front lines through artillery and machine-gun fire, and was killed while engaged in this duty.
Posthumously awarded. Medal presented to father, James M. Payne.

PAYNE, IRA M. (2335261)

Near Sechault, France, Sept. 29, 1918.
R—Washington, D. C.
B—Washington, D. C.
G. O. No. 13, W. D., 1919.

Sergeant, Company A, 372d Infantry, 93d Division.
Having found a machine gun hidden in a brush which was causing serious casualties to his company, he crept up, killed the gunners with his rifle, and captured the gun.

PAYNE, KARL O.

Near Longuyon, France, Sept. 16, 1918.
R—Belmont, Mass.
B—Cambridge, Mass.
G. O. No. 123, W. D., 1918.

First lieutenant, observer, 20th Aero Squadron, Air Service.
Starting on a very important daylight bombing mission with 5 other planes, as observer he went on alone when the other 5 planes were forced to turn back. On crossing the German line, he was attacked by 3 enemy planes. Using his guns to keep the enemy at bay, he went on, reached his objective, and dropped his bombs on the railroad junction, cutting the line. On the way back 4 more planes joined in the attack, but, keeping them at bay with his guns, he reached the allied lines.

*PAYNE, WORTHAM J. (106420)

Near Very, France, Oct. 9, 1918.
R—Cheneyville, La.
B—Montgomery, La.
G. O. No. 32, W. D., 1919.

Sergeant, Company D, 3d Machine Gun Battalion, 1st Division.
During a heavy bombardment he located a position in which his platoon would be less exposed to the intense shelling, and, returning, he collected his men and led them to this new location without a casualty. In the course of this exploit he was severely wounded, but directed his platoon to the place of protection after falling from exhaustion. He died on his way to the hospital.
Posthumously awarded. Medal presented to father, J. J. Payne.

*PAYSON, CARL F. (261851)

Near Cierges, northeast of Chateau-Thierry, France, Aug. 1, 1918.
R—Monroe, Mich.
B—Paulding, Ohio.
G. O. No. 116, W. D., 1918.

Sergeant, Company C, 125th Infantry, 32d Division.
During the attack made by the company on the village of Cierges, Sergeant Payson was mortally wounded in the head by a machine-gun bullet. He succeeded in keeping on his feet, however, and with the attacking wave, encouraging them, and by his strong will power he instilled in them all the spirit of fearlessness.
Posthumously awarded. Medal presented to mother, Mrs. Minnie Timberman.

*PEABODY, MARSHALL G.

Near Moulin-de-Charlevaux, in the Forest d'Argonne, France, Oct. 4-5, 1918.
R—New York, N. Y.
B—Brooklyn, N. Y.
G. O. No. 56, W. D., 1922.

Second lieutenant, 306th Machine Gun Battalion, 77th Division.
While commanding a detachment of his battalion operating with a battalion of the 308th Infantry, Lieutenant Peabody, although badly wounded, continued to personally direct the fire and operation of his machine guns, which were continuously meeting and shattering the repeated hostile attacks and defending the entire Infantry detachment in its exposed and precarious position. While crawling in a severely wounded condition to a machine gun in a most exposed position, he was killed by enemy machine-gun fire.
Posthumously awarded. Medal presented to father, Alexander M. Peabody.

PEACOCK, JACK (1488214)

Near St. Etienne, France, Oct. 8, 1918.
R—Waco, Tex.
B—Gatesville, Tex.
G. O. No 37, W. D., 1919.

Sergeant, Company K, 141st Infantry, 36th Division.
All the officers of his company being killed, he took command and led the company into action, capturing 62 German prisoners who were occupying and directing a fire against our troops from 8 machine-gun nests.

PEACOCK, RAYMOND F. (1244410)

Near Fismette, France, Aug. 10, 1918.
R—Norristown, Pa.
B—Bryn Mawr, Pa.
G. O. No. 99, W. D., 1918.

Corporal, Company F, 111th Infantry, 28th Division.
Being the only member of his detachment who knew how to operate an enemy machine gun, he volunteered to go forward in the attack near Fismette, in spite of just having been so badly wounded in his left shoulder that his left arm was partially useless. He participated in the assault and with one arm operated a captured German machine gun against the enemy until he was again wounded.

PEADEN, ALBERT JOHN (1878105)

Near Vaux-Andigny, France, Oct. 11, 1918.
R—Farmville, N. C.
B—Pitt County, N. C.
G. O. No. 50, W. D., 1919.

Private, Company M, 118th Infantry, 30th Division.
While delivering a message, Private Peaden was seriously wounded by a bullet which entered his cheek and passed through his lower right jaw, but he refused to be evacuated, and continued on duty until the following day. Upon reporting to the aid station he was evacuated to the hospital, where the wound was found to be so serious that he was compelled to remain there for several weeks.

PEARCE, PERCY R.

At Berzy-le-Sec, France, July 21, 1918.
R—Newark, N. J.
B—Newark, N. J.
G. O. No. 132, W. D., 1918.

First lieutenant, 2d Infantry Brigade, 1st Division.
During a violent attack from artillery and machine guns at Berzy-le-Sec, France, July 21, 1918, while serving as liaison officer he fearlessly exposed himself, exceeding the demands of duty to assist in reforming units that had been disseminated in battle and directing them to effective positions.

*PEARCE, ZENO W. (155524)

Near Soissons, France, July 20, 1918.
R—Oakland, Calif.
B—Olympia, Wash.
G. O. No. 124, W. D., 1918.

Private, Company C, 1st Engineers, 1st Division.
When volunteers were called for by his company commander, Private Pearce volunteered and rescued wounded comrades from a barrage. Disregarding danger to himself, he continued the performance of these heroic deeds until killed.
Posthumously awarded. Medal presented to father, Joseph G. Pearce.

PEARSON, HARRY L (1403333)

Near Ferme La Folie, France, Sept. 30, 1918.
R—Decatur, Ill.
B—Tupelo, Miss.
G. O. No. 46, W. D., 1919.

Private, Machine Gun Company, 370th Infantry, 93d Division.
While under heavy fire Private Pearson volunteered and, accompanied by another soldier, rescued a wounded comrade from an exposed position, carrying him to the first-aid station.

⁴ PEARSON, VARLOURD (1449077).........
Near Baulny, France, Sept. 28, 1918.
R—Manhattan, Kans.
B—Dadeville, Ala.
G. O. No. 95, W. D., 1919.

Sergeant, Company I, 137th Infantry, 35th Division.
Though wounded 3 times by shrapnel and machine-gun bullets, he refused to be evacuated and continued to lead the advance of his platoon, remaining in command for several hours, until he received a fourth wound, which proved fatal.
Posthumously awarded. Medal presented to father, C. I. Pearson.

PEASE, LIBERTY (101022)..............
In the Foret-de-Fere, near Nesles, northeast of Chateau-Thierry, France, July 26–Aug. 2, 1918.
R—Farragut, Iowa.
B—Farragut, Iowa.
G. O. No. 102, W. D., 1918.

Private, Company E, 168th Infantry, 42d Division.
During the advance of his regiment in the Foret-de-Fere, by his voluntary, authorized, and untiring efforts in carrying wounded, both by day and by night, under the most severe and dangerous circum tances, and especially when the town of Sergy was under bombardment, July 31, 1918.

PEATROSS, JAMES L..................
Near Bantheville, France, Nov. 2, 1918.
R—Rome, Mo.
B—Stokes County, N. C.
G. O. No. 46, W. D., 1919.

Major, 353d Infantry, 89th Division.
Though he had been wounded the day before and was so weak from exposure that he could hardly talk, Major Peatross remained with his battalion and led it in assault on enemy machine-gun nests north of the Bois-de-Barricourt. Under his personal direction, without artillery support, the machine-gun nests were flanked and the day's objective reached in spite of the most determined resistance.

PEAVY, JOE B.....................
Near Cornay, France, Oct. 9, 1918.
R—Hamilton, Ga.
B—Greenville, Ga.
G. O. No. 81, W. D., 1919.

Second lieutenant, 327th Infantry, 82d Division.
Although his command was nearly surrounded and enemy machine guns were pouring an incessant fire on them, Lieutenant Peavy directed the fire of his men, after having been seriously wounded. He remained during the advance, refusing first aid until those about him needing attention were properly cared for.

*PECK, MYRON H..................
At St. Etienne, France, Oct. 9, 1918.
R—Montclair, N. J.
B—Racine, Wis.
G. O. No. 44, W. D., 1919.

Captain, 2d Engineers, 2d Division.
While in command of this battalion, holding part of the line in St. Etienne, Captain Peck personally conducted a reconnaissance, after patrols had previously failed, in order to establish liaison with the troops on his right. He lost his life during this reconnaissance.
Posthumously awarded. Medal presented to widow, Mrs. M. H. Peck.

PECK, ROBERT H..................
Near Liny-devant-Dun, Fontaine, Murvaux and Brandeville, France, Nov. 6–8, 1918.
R—San Diego, Calif.
B—San Francisco, Calif.
G. O. No. 143, W. D., 1918.
Distinguished-service medal also awarded.

Colonel, 11th Infantry, 5th Division.
Throughout the successive attacks on Liny-devant-Dun, Cote 292, Bois-du-Chenois, Fontaine, Murvaux, Bois-du-Corrai, and Bois-de-Brandeville, he exhibited conspicuous gallantry, stimulating his command to a high state of enthusiasm and creating a superb morale. Placing himself in front of the leading waves, h personally led his men to the a sault. Accompanied by 9 men, he attacked a battery of enemy artillery near La Maisonette Farm, forcing the abandonment and subsequent capture of the battery. Under his skillful leadership his regiment captured numerous prisoners, 6 pieces of artillery, 3 antiaircraft guns, 150 machine guns, and vast quantities of ammunition and supplies.

PEDERSEN, INGVALD O. (540099).......
Near Fossoy, France, July 14–15, 1918.
R—Pittsburgh, Pa.
B—Portland, Oreg.
G. O. No. 44, W. D., 1919.

Private, Headquarters Company, 7th Infantry, 3d Division.
He volunteered and carried a message over a heavily shelled route, and although wounded in the execution of this task he accomplished the mission.

PEDERSON, WILLIAM J. (3125361).......
Near Oches, France, Nov. 4, 1918.
R—Becker, Minn.
B—Santiago Minn.
G. O. No. 46, W. D., 1919.

Private, Company H, 307th Infantry, 77th Division.
Exposing himself to heavy machine-gun fire, Private Pederson crossed an open field 300 yards wide and rescued a severely wounded comrade.

PEDRO, JEROME C. (41168)...........
Near Medeah Ferme, France, Oct. 3–9, 1918.
R—New Bedford, Mass.
B—New Bedford, Mass.
G. O. No. 98, W. D., 1919.

Private, Headquarters Company, 9th Infantry, 2d Division.
As a runner Private Pedro displayed the utmost disregard for personal danger in carrying messages from his regimental commander to all parts of the line under heavy shell fire, setting a splendid example of courage and devotion to duty, until he was seriously wounded.

*PEGG, DONALD A. (11076)...........
Near the Bois-des-Ogons France, Sept. 30, 1918.
R—Arlington, N. J.
B—Indianapolis, Ind.
G. O. No. 44, W. D., 1919.

Private, medical detachment, 12th Machine Gun Battalion, 4th Division.
While engaged in administering first aid under terrific machine-gun fire, Private Pegg voluntarily went to an especially dangerous position to care for a wounded soldier, and in so doing was himself killed.
Posthumously awarded. Medal presented to father, George A. Pegg.

PEGUES, JOSIAH J.................
Near Dun-sur-Meuse, France, Nov. 5, 1918.
R—Chicago, Ill.
B—Quincy, Ill.
G. O. No. 46, W. D., 1919.

First lieutenant, 95th Aero Squadron, Air Service.
On account of heavy clouds and mist, Lieutenant Pegues became detached from his formation. While endeavoring to find it, he came upon 8 hostile planes which were maneuvering to attack 4 of our planes. With great courage and skill he passed through the formation and attacked its leader, dispersing the formation and preventing further attack.

PELKEY, EDWIN (2278753)...........
Near Cierges, France, Aug. 2, 1918.
R—Reno, Nev.
B—Canada.
G. O. No. 20, W. D., 1919,

Private, Company C, 107th Field Signal Battalion, 32d Division.
He was a member of a detachment stringing telephone wire in advance of the front lines through the heaviest artillery fire. When connection was established at a point within 100 yards of the German line and before the American advance was begun, Private Pelkey volunteered to remain there until our troops had advanced far enough to establish the advance regimental post of command at that place. His devotion to duty under conditions of great danger assisted immeasurably in maintaining unity of action between the front lines and regimental post of command, and his utter indifference to his own safety made easier the capture of a strong enemy position.

PELLEGROM, HOWARD H _ _ _ _ _ _ _ _ _ _ _ _
  At Bolshieozerka, Russia, Apr. 2,
    1919.
  R—Grand Haven, Mich.
  B—Grand Haven, Mich.
  G. O. No. 19, W. D., 1920.

Second lieutenant, 339th Infantry, 85th Division (detachment in North Russia).
Lieutenant Pellegrom exposed himself to direct enemy observation and fire to go forward 200 yards in advance of our lines and drag a wounded medical attendant to a place of safety.

PENDELL, ELMER _ _ _ _ _ _ _ _ _ _ _ _ _ _ _
  Near Rembercourt and Charey,
    France, Nov. 4, 1918.
  R—Waverly, N. Y.
  B—Waverly, N. Y.
  G. O. No, 1, W. D., 1919.

First lieutenant, observer, 120th Infantry, 168th Aero Squadron, Air Service.
As an observer in a De Haviland-4 plane, he flew an Infantry contact mission over the line of the 7th Division. Because of exceedingly adverse weather conditions, he disregarded the danger of fire from the ground and crossed the lines at 1,000 feet altitude. While thus flying he was wounded in the shoulder by an explosive bullet fired from the ground. Disregarding his wound, he came down to an altitude as low as 500 feet. After securing the desired information, he wrote out his message with great effort and dropped it to the division.

PENNINGTON, EDGAR (2218090) _ _ _ _ _ _ _ _ _
  Near Bantheville, France, Oct.
    23-24, 1918.
  R—Mangum, Okla.
  B—Sunset, Tex.
  G. O. No. 46, W. D., 1919.

Private, first class, medical detachment, 357th Infantry, 90th Division.
Private Pennington demonstrated the highest bravery and devotion to duty in giving first-aid treatment to wounded men under terrific bombardment of gas and high-explosive shells. Going over the top with his company, he attended wounded men and directed their evacuation until he became exhausted and had to be sent to a hospital.

PERCY, WILLIAM J. (1284290) _ _ _ _ _ _ _ _ _
  Near Ronssoy, France, Sept. 29, 1918.
  R—Niagara Falls, N. Y.
  B—Niagara Falls, N. Y.
  G. O. No. 37, W. D., 1919.

Sergeant, Company E, 108th Infantry, 27th Division.
After having been wounded in the face and legs, he led a patrol, under heavy shell and machine-gun fire, against an enemy machine-gun nest and succeeded in capturing 1 gun and 15 prisoners.

PERDEW, EARNEST E. (2267260) _ _ _ _ _ _ _ _
  Near Eclisfontaine, France, Sept.
    28, 1918.
  R—Etiwanda, Calif.
  B—Etiwanda, Calif.
  G. O. No. 37, W. D., 1919.

Sergeant, Company H, 364th Infantry, 91st Division.
Assisted by another sergeant, and leading a combat group across an open valley under constant hostile fire, he completed the capture of 4 machine-gun nests and 3 prisoners.

PERKAUS, FRANK (40295) _ _ _ _ _ _ _ _ _ _ _ _
  Near Soissons, France, July 18, 1918.
  R—Chicago, Ill.
  B—Chicago, Ill.
  G. O. No. 46, W. D., 1919.

Supply sergeant, Company K, 9th Infantry, 2d Division.
Sergeant Perkaus volunteered to go about 350 yards in advance of our lines to locate the enemy and secure other information. He made the trip through heavy machine-gun and artillery fire and secured the information, but was wounded while returning to our line. When ordered to the dressing station by his commanding officer, he helped others who were more seriously wounded than himself to reach the station.

*PERKINS, BYRON R. (155533) _ _ _ _ _ _ _ _ _
  Near Soissons, France, July 20, 1918.
  R—Springfield, Mass.
  B—New York, N. Y.
  G. O. No. 124, W. D., 1918.

Private, Headquarters Company, 1st Engineers, 1st Division.
When volunteers were called for by his company commander, Private Perkins volunteered and rescued wounded comrades from a barrage. Disregarding danger to himself, he continued the performance of these heroic deeds until killed.
Posthumously awarded. Medal presented to sister, Mrs. Eula B. Chamberlain.

PERKINS, EARL H. (1388545) _ _ _ _ _ _ _ _ _
  In the Meuse-Argonne offensive,
    France, Sept. 26, 1918.
  R—Chicago, Ill.
  B—Chicago, Ill.
  G. O. No. 87, W. D., 1919.

Sergeant, Company M, 131st Infantry, 33d Division.
With 3 other soldiers he, on his own initiative, crawled across an open field, subjected to intense artillery and machine-gun fire, flanking 3 machine-gun positions which were holding up our advance. This mission was successful, 7 Germans being killed by the patrol and 23 captured.

PERONACE, ANTHONY (2338196) _ _ _ _ _ _ _ _
  Near Gland, France, July 21, 1918.
  R—New York, N. Y.
  B—Italy.
  G. O. No. 32, W. D., 1919.

Private, Company B, 4th Infantry, 3d Division.
With his platoon leader and 1 other soldier, he captured an enemy machine gun and 8 prisoners.

*PERRY, SETH E. (1316548) _ _ _ _ _ _ _ _ _ _ _
  Near Bellicourt, France, Sept. 29,
    1918.
  R—Okisco, N. C.
  B—Pasquotank County., N. C.
  G. O. No. 87, W. D., 1919.

Corporal, Company K, 119th Infantry, 30th Division.
When a portion of his company was threatened with a counterattack and he had seen 1 runner killed in an attempt to reach them from company headquarters with orders to fall back, he volunteered for the dangerous mission. While crossing an open field under heavy fire, he was mortally wounded.
Posthumously awarded. Medal presented to mother, Mrs. Mary E. Perry.

PERSONETT, JOHN E. (42098) _ _ _ _ _ _ _ _ _
  At Hill 272, near Fleville, France,
    Oct. 8, 1918.
  R—Lenora, Kans.
  B—Lenora, Kans.
  G. O. No. 35, W. D., 1920.

Sergeant, Company C, 16th Infantry, 1st Division.
Although severely wounded at the beginning of the attack, he refused to be evacuated, but continued to direct his platoon in the advance. In an attack on an enemy machine-gun position, after all other members of his group were killed or wounded in the attack, he alone rushed the position and captured the gun and crew. This permitted the further advance of the company.

PERSONS, JOHN C. _ _ _ _ _ _ _ _ _ _ _ _ _ _ _
  At St. Thibaut, France, Aug. 8,
    1918.
  R—Tuscaloosa, Ala.
  B—Atlanta, Ga.
  G. O. No. 9, W. D., 1923.

Captain, 47th Infantry, 4th Division.
While serving as adjutant, 47th Infantry, he was instructed by his regimental commander to deliver a message to the brigade commander. The telephone lines to the rear having been destroyed, he proceeded under intense enemy fire through a narrow pass, accompanied by a corporal and private of his regiment. Exposed to constant enemy fire, he had reached a place of safety when he learned that the corporal had been hit by enemy fire. Immediately returning, he carried the corporal to a dressing station in a storm of machine-gun and rifle fire from the enemy lines, thus saving the soldier's life and in utter disregard for his own safety.

*PETERS, HERBERT N_____
    Near Les Huit Chemins, France,
        Sept. 12–14, 1918.
    R—Sabinal, Tex.
    B—Sabinal, Tex.
    G. O. No. 140, W. D., 1918.

Captain, 358th Infantry, 90th Division.
As commander of the support company of his battalion, Captain Peters displayed courage and leadership by rushing 2 platoons into position to protect the right flank of the battalion which had suddenly become exposed. Under his personal leadership, in the face of intense machine-gun and shell fire, a number of enemy machine-gun nests were stormed and enemy combat groups dispersed. Serious danger to the advancing line was thereby averted. This gallant officer was killed shortly afterwards in a raid on the enemy.
Posthumously awarded. Medal presented to mother, Mrs. Nettie Peters.

PETERS, WILLIAM H. (1388055)_____
    At Bois-de-Chaume France, Oct.
        10, 1918.
    R—Chicago, Ill.
    B—Chicago, Ill.
    G. O. No. 35, W. D., 1919.

Private, Company I, 131st Infantry, 33d Division.
When the advance of his platoon was held up by an enemy machine gun, Private Peters, on his own initiative, flanked the position, killed the gunner, and captured the rest of the crew, thereby allowing the platoon to advance.

*PETERSEN, LEONARD (1472728)_____
    Near Baulny, France, Sept. 29, 1918.
    R—Kansas City, Kans.
    B—Kansas City, Kans.
    G. O. No. 59, W. D., 1919.

Private, 140th Ambulance Company, 110th Sanitary Train, 35th Division.
Serving as a litter-bearer, he voluntarily left cover and exposed himself to intense artillery fire to rescue wounded men lying in the open. Mortally wounded, he continued to assist in the rescue of wounded comrades till he fell exhausted.
Posthumously awarded. Medal presented to mother, Mrs. Helen Petersen.

*PETERSEN, THEODOR (1232)_____
    Near Pexonne, France, Mar. 5, 1918.
    R—Minneapolis, Minn.
    B—Denmark.
    G. O. No. 17, W. D., 1924.

Sergeant, medical detachment, 151st Field Artillery, 42d Division.
After being mortally wounded he gave detailed instructions to other wounded soldiers and gave first gas tests in order to save the lives of the men about him. He died the same night.
Posthumously awarded. Medal presented to mother, Mrs. J. A. Petersen.

PETERSEN, VICTOR (1389526)_____
    Near Forges, France, Sept. 26, 1918.
    R—Chicago, Ill.
    B—Chicago, Ill.
    G. O. No. 37, W. D., 1919.

Corporal, Company H, 132d Infantry, 33d Division.
When his platoon was held up by a heavy flanking machine-gun fire, Corporal Petersen advanced alone ahead of the platoon, on his own initiative, and successfully cleaned up the machine-gun nest with hand grenades and captured the machine gun.

*PETERSON, ALBERT C. (278805)_____
    Near Grand Pre, France, Oct. 15,
        1918.
    R—Stacy, Minn.
    B—Burt, Iowa.
    G. O. No. 37, W. D., 1919.

Private, Company B, 307th Infantry, 77th Division.
He sacrificed his life in fearlessly going out in the face of machine-gun fire and attempting to rescue another soldier who had been mortally wounded.
Posthumously awarded. Medal presented to father, Louis C. Peterson.

PETERSON, DAVID McK_____
    Near Luneville, France, May 3, 1918.
    R—Honesdale, Pa.
    B—Honesdale, Pa.
    G. O. No. 121, W. D., 1918.

Captain, 94th Aero Squadron, Air Service.
Leading a patrol of 3, he encountered 5 enemy planes at an altitude of 3,500 meters and immediately gave battle. Notwithstanding the fact he was attacked from all sides, this officer, by skillful maneuvering, succeeded in shooting down 1 of the enemy planes and dispersing the remaining 4.
Oak-leaf cluster.
A bronze oak leaf is awarded to Captain Peterson for extraordinary heroism in action near Thiaucourt, France, on May 15, 1918. While on a patrol alone he encountered 2 enemy planes at an altitude of 52 meters. He promptly attacked, despite the odds, and shot down 1 of the enemy planes in flames. While thus engaged he was attacked from above by the second enemy plane, but by skillful maneuvering he succeeded in shooting it down also.

PETERSON, GEORGE I._____
    Near St. Etienne, France, Oct. 3–7,
        1918.
    R—Paxton, Ill.
    B—Chicago, Ill.
    G. O. No. 37, W. D., 1919.

Pharmacist's mate, third class, U. S. Navy, attached to 5th Regiment, U. S. Marine Corps, 2d Division.
He was directly responsible for the saving of several lives while obliged to care for the company's wounded alone. On succeeding days he traveled from one side of the company sector to the other, through artillery and machine-gun barrage, hunting and caring for the wounded.

*PETERSON, HELMER (2101106)_____
    Near Tuilerie Ferme, France, Nov.
        4, 1918.
    R—Decorah, Iowa.
    B—Decorah, Iowa.
    G. O. No. 44, W. D., 1919.

Private, Company E, 9th Infantry, 2d Division.
He displayed exceptional bravery in carrying important messages to the rear through heavy enemy artillery and machine-gun barrages, keeping his commanding officer informed as to the situation at all times. Tireless in his efforts, he was instrumental in the success of the operation.
Posthumously awarded. Medal presented to sister, Mrs. Jonas J. Akre.

*PETERSON, HOLGAR (1709115)_____
    Near Charlevaux, France, Oct. 3–7,
        1918.
    R—New York, N. Y.
    B—Spencer, Iowa.
    G. O. No. 87, W. D., 1919.

Corporal, Company G, 308th Infantry, 77th Division.
While leading a scouting party, Corporal Peterson encountered an enemy patrol and displayed exceptional courage and leadership in killing the officer and two soldiers who composed it. He repeatedly volunteered for dangerous patrol work with great bravery and aggressiveness until he was killed.
Posthumously awarded. Medal presented to widow, Mrs. Catherine Peterson.

PETERSON, OSCAR W. (560685)_____
    Near Courchamps, France, July 19,
        1918.
    R—Jamestown, N. Dak.
    B—Jamestown, N. Dak.
    G. O. No. 95, W. D., 1919.

Sergeant, Company A, 59th Infantry, 4th Division.
Discovering the enemy making a counterattack to the left flank of his platoon Sergeant Peterson immediately organized a combat group of 25 men, and though greatly outnumbered by the Germans he succeeded in routing them, inspiring his men by his disregard of personal danger. He was severely wounded later in the day, but he refused to go to the rear until he had reorganized his platoon an hour and a half later.

PETERSON, ROY W. (2180615)_____
    Near Bellicourt, France, Sept. 29,
        1918.
    R—Center, Nebr.
    B—Fremont, Nebr.
    G. O. No. 32, W. D., 1919.

Private, first class, Company D, 114th Machine Gun Battalion, 30th Division.
Severely wounded while operating a machine gun under shell fire, he refused to leave his post until he was removed by his comrades against his protests.

PETERSON, SOLOMON (2261095)_____
  During the Argonne offensive, France, Sept. 26-29, 1918.
  R—Mosby, Mont.
  B—Atlanta, Kans.
  G. O. No. 15, W. D., 1919.

Sergeant, Company I, 362d Infantry, 91st Division.
He repeatedly led patrols in successful attacks on enemy machine-gun emplacements, displaying calmness and keen judgment. After being wounded he insisted on remaining in command of his platoon.

PETERSON, VAN WALKER (1386508)_____
  Near Bois-de-Chaume, France, Oct. 10, 1918.
  R—Chicago, Ill.
  B—River Junction, Iowa.
  G. O. No. 37, W. D., 1919.

Sergeant, Company B, 131st Infantry, 33d Division.
When the company guarding the flank was on the verge of retreating in disorder, Sergeant Peterson avoided the perilous situation by jumping to the front and holding the badly shaken troops in their positions on the line. His quick action during the terrific fire was responsible for the safety of the entire line.

PETERSON, WALTER O. L. (284497)_____
  Near Romagne, France, Oct. 5, 1918.
  R—Milwaukee, Wis.
  B—Menominee, Mich.
  G. O. No. 44, W. D., 1919.

First Sergeant, Company H, 128th Infantry, 32d Division.
Because of casualties among officers, Sergeant Peterson was placed in command of the second wave, which he led with exceptional bravery and leadership. When it became isolated in a fog he crawled forward by himself to ascertain the character of troops which were seen a kilometer to the front, and upon finding that they were hostile immediately established liaison with adjacent units and straightened out his line, after breaking up several enemy machine-gun nests.

*PETERSON, WILLIAM C._____
  At Chateau-Thierry, France, June 6, 1918.
  R—Crystal Lake, Ill.
  B—Crystal Lake, Ill.
  G. O. No. 119, W. D., 1918.

Second lieutenant, Infantry, attached to 5th Regiment, U. S. Marine Corps, 2d Division.
Killed in action at Chateau-Thierry, France, June 6, 1918, he gave the supreme proof of that extraordinary heroism which will serve as an example to hitherto untried troops.
Posthumously awarded. Medal presented to father, Fred Peterson.

PETIT, CHARLES L._____
  Near Verdun, France, Oct. 27, 1918.
  R—Clare, Mich.
  B—Hemlock, Mich.
  G. O. No. 98, W. D., 1919.

Second lieutenant, 102d Infantry, 26th Division.
After being seriously wounded in the assault against strong enemy positions in the Bois-de-la-Reine, Lieutenant Petit stumbled on at the head of his command until compelled to turn over the command to a sergeant on account of his condition. After this he remained in a shell hole under a terrific concentration of machine-gun and artillery fire, encouraging his men as best he could, and aiding to the utmost of his ability by sniping the enemy with the rifle of a soldier wounded more severely than himself.

PETRACH, EMIL H. (1098720)_____
  Near Preny Ridge, France, Nov. 1, 1918.
  R—Youngstown, Ohio.
  B—Austria.
  G. O. No. 32, W. D., 1919.

Bugler, Company G, 56th Infantry, 7th Division.
Under heavy and deadly machine-gun fire he carried messages to and from his company. After all other means of communication had been cut off he volunteered and carried many important messages through artillery fire, thus establishing liaison with the rear. He was later wounded while passing through a barrage.

*PETREE, HARRIS E._____
  Near Marville, France, Sept. 26, 1918.
  R—Washington, D. C.
  B—Lincoln, Kans.
  G. O. No. 19, W. D., 1920.

First lieutenant, pilot, 139th Aero Squadron, Air Service.
After having become separated from his patrol, Lieutenant Petree encountered 7 enemy planes. He alone attacked this enemy group and continued in combat against these great odds for over 1 hour, when he was killed.
Posthumously awarded. Medal presented to father, Frank Petree.

*PETRIMEAN, GEORGE (45032)_____
  Northeast of Exermont, France, Oct. 9, 1918.
  R—Minneapolis, Minn.
  B—Greece.
  G. O. No. 35, W. D., 1920.

Sergeant, Machine Gun Company, 16th Infantry, 1st Division.
Sergeant Petrimean led his machine-gun section through heavy machine-gun and artillery fire to an exposed position in order to engage an enemy machine gun which was causing heavy casualties to his battalion. He was killed during the attack. Sergeant Petrimean had previously performed gallant service while keeping his machine-gun section in action during a heavy enemy bombardment at Cantigny, France, May 28, 1918.
Posthumously awarded. Medal presented to brother, Cleanthis Petrimean.

PETROVIC, JOSEPH F. (126059)_____
  Near Fleville, France, Oct. 5-7, 1918.
  R—Joliet, Ill.
  B—Joliet, Ill.
  G. O. No. 98, W. D., 1919.

Corporal, Battery D, 6th Field Artillery, 1st Division.
When a shell burst near his gun, throwing him across the train of the piece and killing or wounding all of the section but himself and 1 other cannoneer, Corporal Petrovic and the latter succeeded in repairing the piece and continued the rolling barrage until it became impossible to fire the gun again. Two days later, under almost identical circumstances, he and one other soldier continued at their post, after the other members of the crew had been wounded, and continued to serve the piece until the completion of the barrage.

PETTY, ORLANDO H._____
  At Lucy-le-Bocage, during the attack on the Bois-de-Belleau, France, June 11, 1918.
  R—Roxborough, Philadelphia, Pa.
  B—Harrison, Ohio.
  G. O. No. 3, W. D., 1925.

Passed Assistant Surgeon, U. S. Navy, attached to 5th Regiment, U. S. Marine Corps, 2d Division.
While he was treating wounded under bombardment of gas and high-explosive shells, he was knocked down and his gas mask torn by a bursting gas shell, but he discarded his gas mask and continued his work. Later, when his dressing station was demolished by another shell, he helped carry a wounded officer through the shellfire to a place of safety.

*PETTY, WILLARD D. (1386691)_____
  Near Consenvoye, France, Oct. 10, 1918.
  R—Joliet, Ill.
  B—Pearl, Ill.
  G. O. No. 64, W. D., 1919.

Private, Company B, 131st Infantry, 33d Division.
Showing utter disregard of personal danger, he went to the rescue of his wounded platoon leader, who lay in a zone covered by heavy enemy machine-gun fire, being himself mortally wounded in the attempt. Private Petty volunteered to attempt the rescue, which cost him his life, after two of his comrades had been killed and another wounded in similar trials.
Posthumously awarded. Medal presented to son, Howard Wayne Petty.

PEURIFOY, JOHN M. (1902869)_____
Near Pylone, France, Oct. 9, 1918.
R—Griffin, Ga.
B—Milner, Ga.
G. O. No. 81, W. D., 1919.

First sergeant, Company L, 326th Infantry, 82d Division.
After he had seen an officer and 13 men of his company fall from the fire of enemy machine guns and snipers, Sergeant Peurifoy advanced alone, and, after crawling about 50 yards in advance of his company, he shot a sniper from a tree and drove off the crews of 2 machine guns located near by. His action permitted the further advance of his company.

PEYTON, BYRON W. (92472)_____
Northeast of Chateau-Thierry, France, July 29, 1918.
R—Columbus, Ohio.
B—Columbus, Ohio.
G. O. No. 99, W. D., 1918.

Supply sergeant, Supply Company, 166th Infantry, 42d Division.
In response to a call from the attacking battalion for ammunition he drove a combat wagon in broad daylight into the front-line positions near Fere-en-Tardenois, and delivered the ammunition required by his comrades on the front.

PFEIL, CLARENCE W. (2127055)_____
In the Meuse-Argonne offensive, France, Oct. 9, 1918.
R—Sandusky, Ohio.
B—Sandusky, Ohio.
G. O. No. 50, W. D., 1919.

Private, Company K, 327th Infantry, 82d Division.
Assisted by another soldier, Private Pfeil crawled far in advance of our lines under terrific machine-gun and shell fire and brought back a severely wounded comrade.

PHELAN, EDWARD F. (50466)_____
Near Vierzy, France, July 18, 1918.
R—Braintree, Mass.
B—Calais, Me.
G. O. No. 81, W. D., 1919.

Corporal, Company E, 23d Infantry, 2d Division.
Corporal Phelan voluntarily left the assaulting wave of his company, and, single handed, captured or killed the entire crew of a concealed machine-gun position, which was delivering a terrific and accurate fire upon his comrades from the right flank. His timely and gallant act drew the fire of the machine gun from his comrades until they were able to find shelter and saved the lives of many of the assaulting wave.

PHELAN, JEREMIAH A. (1911315)_____
At Cornay, France, Oct. 9–10, 1918.
R—New York, N. Y.
B—New York, N. Y.
G. O. No. 44, W. D., 1919.

Corporal, Company M, 328th Infantry, 82d Division.
After fighting for 6 hours Corporal Phelan volunteered to accompany 15 other soldiers and an officer on night patrol of the town of Cornay, which was held by many machine-gun posts. The party worked from 11 o'clock at night till next morning in clearing buildings and dugouts of the enemy, capturing 65 prisoners and 2 machine guns. With 6 others, Corporal Phelan volunteered and entered a dugout, where 23 prisoners were captured. He was wounded while leaving the town, but he refused to go to the aid station until the prisoners had been delivered at brigade headquarters.

PHELPS, GLENN_____
Near Villers-sur-Marne, France, July 15, and Aug. 7, 1918, and near Chatel-Chehery, France, Oct. 27–30, 1918.
R—St. Louis, Mo.
B—Lutesville, Mo.
G. O. No. 46, W. D., 1919.

First lieutenant, observer, 5th Balloon Company, Air Service.
While regulating artillery fire from his balloon, Lieutenant Phelps, with another observer, was attacked by 3 enemy planes and forced to jump after his balloon had been set on fire. On 4 other occasions his balloon was sent down in flames, after being attacked by superior numbers of the enemy, but on each occasion he resumed his work just as soon as another balloon could be obtained.

*PHILBLAD, HARRY W_____
At Blanc Mont, France, Oct. 3, 1918.
R—Knoxville, Ill.
B—Knoxville, Ill.
G. O. No. 32, W. D., 1919.

Corporal, 78th Company, 6th Regiment, U. S. Marine Corps, 2d Division.
He advanced alone on 2 machine-gun nests, which he captured, killing several of the crew with his pistol. Two hours later he again went forward with 2 other soldiers, and while attacking another machine-gun nest, he was killed by shrapnel.
Posthumously awarded. Medal presented to mother, Mrs. Emma Philblad.

PHILLIPS, CHARLES_____
Near Vierzy, France, July 18–19, 1918.
R—Martins Ferry, Ohio.
B—Helena, Mont.
G. O. No. 15, W. D., 1919.

Private, Company A, 4th Machine Gun Battalion, 2d Division.
On July 18, 1918, he twice drove a light truck loaded with ammunition through Vierzy and up the road directly in the rear of the position occupied by his company, this road being under heavy shell fire at all times. On the morning of the 19th of July he returned over the same route with rations and ran his truck under machine-gun fire to within 50 yards of the trenches. On the return trip his truck was hit by a shell and destroyed, whereupon he borrowed another truck and returned.

*PHILLIPS, CLIFFORD F_____
Near Bolshieozerke, Russia, Apr. 2, 1919.
R—Falls City, Nebr.
B—Gage County, Nebr.
G. O. No. 95, W. D., 1919.

First lieutenant, 339th Infantry, 85th Division (detachment in north Russia).
With a few men and 2 Lewis guns he held the enemy counterattack for an hour, until reinforcements arrived. He constantly encouraged and inspired his men by the example of heroism he set, refusing all aid when seriously wounded, to avoid weakening his small effective force.
Posthumously awarded. Medal presented to widow, Mrs. Ann Kathryn Phillips.

*PHILLIPS, DEWEY (263473)_____
Near Sergy, France, July 31, 1918.
R—Saginaw, Mich.
B—Saginaw, Mich.
G. O. No. 64, W. D., 1919.

Private, first class, Company K, 125th Infantry, 32d Division.
While his company was waiting orders, after having reached the crest of Hill 212, he voluntarily left his place and went for a message when an approaching runner was seen to fall, too badly wounded to reach the company. This act was done voluntarily and under the direct fire of the enemy's machine guns, as well as the terrific bombardment to which the hill was then subjected.
Posthumously awarded. Medal presented to father, John Phillips.

PHILLIPS, ELMER A. (1348291)_____
Near Sedan, France, Nov. 7, 1918.
R—Jasper, Ala.
B—Mary Lee, Ala.
G. O. No. 44, W. D., 1919.

Private, Company K, 166th Infantry, 42d Division.
Private Phillips was a member of a patrol sent out to silence machine-gun nests which were holding up the battalion's advance. When the officer leading the patrol fell, mortally wounded, he went to his assistance in the face of heavy fire from machine guns only 100 yards away, remaining in this position until nightfall, though himself seriously wounded.

PHILLIPS, GEORGE R.
Near Beffu et la Morthomme, France, Oct. 23, 1918.
R—Lewistown, Pa.
B—Burnham, Pa.
G. O. No. 20, W. D., 1919.

First lieutenant, pilot, 50th Aero Squadron, Air Service.
Lieutenant Phillips, pilot, accompanied by Lieut. Mitchell H. Brown, observer, while on a reconnaissance for the 78th Division, attacked an enemy balloon and forced it to descend, and was in turn attacked by 3 enemy planes (Fokker type). The incendiary bullets from the enemy's machines set the signal rockets in the observer's cockpit afire. Disregarding the possibility of going down in flames, Lieutenant Phillips maneuvered his plane so that his observer was able to fire on and destroy 1 enemy plane and drive the others away. He then handed his fire extinguisher to Lieutenant Brown, who extinguished the flames. They completed their mission and secured valuable information.

PHILLIPS, OCEA V. (145606).
Near Suippes, France, July 15, 1918.
R—Duluth, Minn.
B—Augusta, Minn.
G. O. No. 35, W. D., 1919.

Sergeant, Battery C, 151st Field Artillery, 42d Division.
After all telephone communication had been severed on account of heavy enemy shellfire, Sergeant Phillips volunteered and carried a very important message through an extremely heavy bombardment. Although wounded while carrying out his mission, he refused medical attention until the message had been delivered.

PHILLIPS, RUFUS R. (1311239).
Near Brancourt, France, Oct. 9, 1918.
R—Gaffney, S. C.
B—Cherokee County, S. C.
G. O. No. 133, W. D. 1918.

Private, Company F, 118th Infantry, 30th Division.
When his company was about to reach its objective, a sunken road, the company was swept by enfilading fire from several hostile machine guns. Upon his own initiative this soldier jumped down the bank, mounted his automatic rifle in the center of the road in the face of the enemy's fire, and opened fire, sweeping the parapets of the hostile positions with well-directed fire. His act resulted in the capture of the 30 Germans occupying the post.

PHILLIPS, SAMUEL E. (1787095).
Near Montfaucon, France, Sept. 26–30, 1918.
R—Greencastle, Pa.
B—Charleston, W. Va.
G. O. No. 44, W. D., 1919.

First sergeant, Company B, 316th Infantry, 79th Division.
During the entire 4 days of action he exposed himself to the dangers of artillery and machine-gun fire, assisting in every way possible to insure the success of the advance. He made repeated trips to the rear, and either urged his comrades forward or led them up to their positions. So strenuously did he labor during the entire action that at the end of the fourth day he was so exhausted from strain and shell shock that he was taken from the field.

*PHILLIPS, SYLVESTER (1038936).
Near Greves Farm, France, July 14, 1918.
R—Des Moines, Iowa.
B—Chisholm, Iowa.
G. O. No. 44, W. D., 1919.

Private, Battery E, 10th Field Artillery, 3d Division.
He was acting as telephone operator at a gun in a detached position when all the crew became casualties. With another soldier he courageously continued to fire the piece under the heaviest bombardment until it was struck by a shell and he was killed.
Posthumously awarded. Medal presented to mother, Mrs. Mary Phillips.

PHILLIS, OLEX (737035).
Near Brandeville, France, Nov. 5–10, 1918.
R—Mobridge, S. Dak.
B—Greece.
G. O. No. 37, W. D., 1919.

Sergeant, Company L, 11th Infantry, 5th Division.
He rendered excellent service on patrols, and, by volunteering his services for an expedition into the German lines, he captured several machine guns and rendered great assistance to the wounded by giving first aid in the absence of medical personnel.

PHIPPS, GEORGE T.
Near Soissons, France, July 18, 1918.
R—Evansville, Ind.
B—Waverly, Ky.
G. O. No. 71, W. D. 1919.

First lieutenant, 16th Infantry, 1st Division.
Remaining on duty after his right elbow had been shattered by a machine-gun bullet, he personally led a company to the left and front, covering a flank that was entirely exposed. Being then placed in command of a battalion, he led a successful attack upon a strong enemy position, showing marked ability as a leader and inspiring his men by his bravery.

PHIPPS, WALTER B. (2471739).
Near Vilosnes, France, Sept. 27–28, 1918.
R—Clintwood, Va.
B—Clintwood, Va.
G. O. No. 7, W. D., 1919.

Private, Headquarters Company, 319th Infantry, 80th Division.
For 2 days and 2 nights he repeatedly exposed himself to heavy shellfire in directing and maintaining the battalion relay runner service. He rendered valuable service in carrying messages over fire-swept areas, directing wounded soldiers to the first-aid station, and locating a new aid station when severe bombardment necessitated its removal.

PIAZZA, JOHN L. (2450145).
Near Talma Farm, France, Oct. 17, 1918.
R—New York, N. Y.
B—Italy.
G. O. No. 35, W. D., 1919.

Sergeant, Company C, 312th Infantry, 78th Division.
During the operations of Oct. 17 he went through a heavy machine-gun and artillery fire to carry a wounded private of his platoon from an exposed position to a place of safety. A short time after he carried a wounded officer of his company across an open space of 80 yards and a stream, subjected to machine-gun and sniper's fire, to a place where he could be removed by stretcher bearers.

PIAZZANI, JULIUS A. (2411286).
Near Bois-des-Loges, France, Oct. 19, 1918.
R—North Bergen, N. J.
B—New York, N. Y.
G. O. No. 44, W. D., 1919.

Corporal, Company F, 310th Infantry, 78th Division.
When machine-gun fire had held up the advance of his company, and all had entrenched themselves in places of safety, Corporal Piazzani noticed a wounded man about 250 yards in front of the line. Voluntarily and without hesitation he went to the side of the wounded man and rendered first aid. Upon his return he advised the stretcher bearers, enabling them to safely bring this man in, thereby saving his life.

PICKERING, WOODELL A.
In the Champagne sector, France, Sept. 26–Oct. 1, 1918.
R—New York, N. Y.
B—Walla Walla, Wash.
G. O. No. 37, W. D., 1919.

Lieutenant colonel, 369th Infantry, 93d Division.
He repeatedly exposed himself to intense shell and machine-gun fire, establishing observation stations and giving able counsel to subordinate officers. On 2 occasions he advanced under heavy fire beyond the assault lines to make personal reconnaissance and establish advance posts.

PIERCE, CHESTER O. (550846).
Near Moulins, France, July 15, 1918.
R—Carthage, N. Y.
B—Gardiner, Me.
G. O. No. 24, W. D., 1920.

First sergeant, Company F, 38th Infantry, 3d Division.
Sergeant Pierce led a counterattack against the enemy, who were deployed to attack. He fearlessly exposed himself to heavy enemy fire and by effective fire of his platoon forced the enemy to retire in disorder. He later took up a position which he held successfully against repeated attacks by the enemy.

*PIERCE, EDWARD A. (1210150)_____
During the attack on the Hindenburg line, in France, Sept. 29, 1918.
R—New York, N. Y.
B—Springfield, Mass.
G. O. No. 68, W. D., 1920.

Private, first class, Company C, 107th Infantry, 27th Division.
While a member of a Lewis gun squad which attacked a superior force of the enemy, Private Pierce was severely wounded in the ankle. He refused to be evacuated, and although wounded so that he was hardly able to walk, he continued on in the advance, inflicting severe casualties on the enemy, until he was killed by enemy fire.
Posthumously awarded. Medal presented to father, Edward F. Pierce.

*PIERCE, EDWARD P. (1214168)_____
Near Ronssoy, France, Sept. 29, 1918.
R—Buffalo, N. Y.
B—Derby, N. Y.
G. O. No. 20, W. D., 1919.

Private, first class, Company D, 108th Infantry, 27th Division.
He left shelter, went into an open field under heavy machine-gun and shell fire, and dragged a wounded soldier to safety. This courageous soldier was killed while advancing with his company later in the action.
Posthumously awarded. Medal presented to father, George Pierce.

PIERCE, ROBERT S. (1336625)_____
Near Bellicourt, France, Sept. 27, 1918.
R—Holdenville, Okla.
B—Watchwax, Tex.
G. O. No. 46, W. D., 1919.

Private, first class, Company C, 105th Field Signal Battalion, 30th Division.
After the signal detachment of the 118th Infantry had suffered severe casualties and were no longer able to aid in maintaining lines between the 118th and 117th Regiments, Private Pierce rendered valuable services by keeping up the entire line of communication, working day and night under constant and sweeping artillery fire. Almost uninterrupted service was maintained between the regiments, owing in great part to his untiring energy.

PIERCE, THOMAS L_____
Near St. Juvin, France, Oct. 11–14, 1918.
R—West Baldwin, Me.
B—Portland, Me.
G. O. No. 46, W. D., 1919.

Major, 325th Infantry, 82d Division.
Although suffering from a machine-gun bullet wound, he refused to go to an aid station, but remained in personal command of his battalion during the action. Upon receiving 2 other wounds 3 days later this officer again refused assistance, and remained with his command until the afternoon, when he was again severely wounded. He permitted himself to be evacuated only after he had given his successor detailed instructions and information.

*PIERCE, WILLIAM O_____
Near Very and Eclisfontaine, France, Sept. 26–28, 1918.
R—Malta, Idaho.
B—Clear Creek, Idaho.
G. O. No. 39, W. D., 1920.

First lieutenant, 364th Infantry, 91st Division.
Lieutenant Pierce valiantly led his platoon in the attack on enemy positions before Very and Eclisfontaine. The following day he again led his platoon through ravines and approaches covered by enemy machine-gun fire, in the attack on the Bois-de-Baulny. He was killed by enemy fire while leading his men in this attack.
Posthumously awarded. Medal presented to mother, Mrs. Fred Linden.

PIERSON, ELVIN L. (1951636)_____
Near Bois-de-Grande-Montagne, France, Oct. 16, 1918.
R—Dayton, Ohio.
B—Ulysses, N. Y.
G. O. No. 37, W. D., 1919.

Corporal, Battery A, 322d Field Artillery, 83d Division.
When the telephone communications had been cut off he made 4 trips as a runner through severe artillery barrage and machine-gun fire, maintaining liaison between Artillery and Infantry.

*PIERSON, WARD W_____
Near Etraye, France, Nov. 8–9, 1918.
R—Somerton, Pa.
B—Radcliffe, Iowa.
G. O. No. 35, W. D., 1919.

Major, 315th Infantry, 79th Division.
He displayed the highest courage and leadership reconnoitering the enemy's position under heavy shell and machine-gun fire. He was killed while in the performance of this act.
Posthumously awarded. Medal presented to widow, Mrs. Harriet A. Pierson.

PIGMAN, VAN BUREN (2387456)_____
North of Vieville-en-Haye, France, Sept. 16, 1918.
R—Hindman, Ky.
B—Ivis, Ky.
G. O. No. 53, W. D., 1920.

Sergeant, Company G, 61st Infantry, 5th Division.
Sergeant Pigman, with 2 others, rushed an enemy machine-gun nest, captured the gun, and killed the crew. He later exposed himself to heavy fire while leading a patrol which captured 3 other machine guns and forced 9 of the enemy to surrender.

*PILCHER, LUTHER W_____
At Chateau-Thierry, France, June 6, 1918.
R—Chipley, Fla.
B—Dothan, Ala.
G. O. No. 110, W. D., 1918.

Sergeant, 20th Company, 5th Regiment, U. S. Marine Corps, 2d Division.
Killed in action at Chateau-Thierry, France, June 6, 1918, he gave the supreme proof of that extraordinary heroism which will serve as an example to hitherto untried troops.
Posthumously awarded. Medal presented to brother, W. D. Pilcher.

PILKERTON, ALVIN W_____
Near Thiaucourt, France, Sept. 15, 1918.
R—Greensboro, Ala.
B—Greensboro, Ala.
G. O. No. 37, W. D., 1919.

Pharmacist's mate, third class, U. S. Navy, attached to 6th Regiment, U. S Marine Corps, 2d Division.
While he was dressing the injuries of a wounded soldier under heavy shellfire he was himself severely wounded in 2 places, but he refused to treat his own wounds until he had taken care of his patient.

PINCOFFS, MAURICE C_____
Near Blanc Mont, France, Oct. 9, 1918.
R—Baltimore, Md.
B—Chicago, Ill.
G. O. No. 37, W. D., 1919.

Captain, Medical Corps, 2d Sanitary Train, 2d Division.
After a withdrawal of the line had been made, he voluntarily crossed an open field under heavy fire to a small wood, where he located a number of wounded men, whose injuries he dressed, and directed their evacuation without further casualties.

PINE, HARRY W_____
Near Haumont, France, Oct. 4, 1918.
R—Jackson, Miss.
B—Chicago, Ill.
G. O. No. 34, W. D., 1919.

Second lieutenant, 353d Infantry, 89th Division.
Lieutenant Pine led a patrol of 8 men in an attempt to gain entrance into the town of Haumont, which was stubbornly held by the enemy. Working his way through the rear outposts, he encountered an enemy patrol near the entrance, attacking and defeating it. He then proceeded through the entire town, capturing and returning with 2 prisoners, as well as obtaining valuable information.

PIOVANO, JOSEPH (542785)_____
Near Cierges, France, Oct. 7, 1918.
R—Coral, Pa.
B—Italy.
G. O. No. 60, W. D., 1920.

Sergeant, Company K, 7th Infantry, 3d Division.
When the advance of his company was halted by machine-gun fire Sergeant Piovano crawled to an exposed position and opened fire on the enemy, although a target for the fire of several enemy machine guns. He killed several of the enemy with rifle fire and so disorganized the enemy that his organization was able to advance with slight loss.

PIRINOLI, MIKE (1645874)_____
Near St. Juvin, France, Nov. 1, 1918.
R—Sebastopol, Calif.
B—Italy.
G. O. No. 71, W. D., 1919.

Private, Company E, 305th Infantry, 77th Division.
When his company had suffered heavy casualties and been withdrawn he went out with another soldier in advance of our lines and under machine-gun fire to rescue the wounded. His companion shot down, he continued his work, though constantly exposed to heavy fire, carrying back 2 wounded men and the soldier who had started out with him.

PIRTLE, JAMES J_____
In the Bois-du-Fays, France, Oct. 4–5, 1918.
R—Carlisle, Ind.
B—Carlisle, Ind.
G. O. No. 126, W. D., 1919.

First lieutenant, 59th Infantry, 4th Division.
Throughout the engagement in the Bois-du-Fays Lieutenant Pirtle led his men with absolute disregard for his personal safety. He walked up and down the lines under intense enemy machine-gun and artillery fire, encouraging his men and consolidating his position. His courageous example contributed greatly to the success of the operation in which his organization was engaged. He continued in action until severely wounded in the knee and was carried from the field.

PISTIKOUDIS, THEODORE (107720)_____
Near Chateau-Thierry, France, June 6, 1918.
R—Philadelphia, Pa.
B—Turkey.
G. O. No. 99, W. D., 1918.

Private, Company C, 5th Machine Gun Battalion, 2d Division.
When 3 infantrymen were buried by a shell explosion near Chateau-Thierry June 6, 1918, he fearlessly left shelter in face of heavy shelling and rescued them.

PITTS, WILLIAM A. (98433)_____
Near Sergy, France, July 31, 1918.
R—Anniston, Ala.
B—Carrollton, Ga.
G. O. No. 23, W. D., 1919.

Private, first class, Company M, 167th Infantry, 42d Division.
Being informed that a wounded man was lying in no man's land, Private Pitts immediately volunteered, and, with Sergeant Collins, went to his aid. The intense fire of the enemy necessitated crawling the entire distance. While on the return trip the wounded man was hit by a machine-gun bullet and instantly killed, but these 2 men brought in the dead body, crawling with great difficulty over the shell-torn ground.

*PLASSMEYER, ALBERT J. (1952505)_____
Near Brabant-sur-Meuse, France, Oct. 22–23, 1918.
R—Zelienople, Pa.
B—Pittsburgh, Pa.
G. O. No 71, W. D., 1919.

Private, first class, Battery E, 322d Field Artillery, 83d Division.
Though mortally wounded and gassed, he continued his work as telephone operator and linesman, repairing telephone lines and remaining on duty until ordered to be evacuated. His example of heroism and fortitude inspired those working with him.
Posthumously awarded. Medal presented to father, Albert J. Plassmeyer.

*PLATNER, AARON A_____
Near Medeah Ferme, France, Oct. 3–9, 1918.
R—Ellis, Kans.
B—Ellsworth, Kans.
G. O. No. 126, W. D., 1919.

Captain, 9th Infantry, 2d Division.
He repeatedly led his battalion against machine-gun nests, through terrific enemy bombardment, until his objectives were attained. During the process of the attack, when his men were being cut down by hidden fire, he personally located the gun and shot the gunner. His gallant example to his troops was an important factor in the success of the attack.
Posthumously awarded. Medal presented to mother, Mrs. Andrew Platner.

PLATT, ABNER H_____
Near St. Souplet, France, Oct. 17, 1918.
R—West New Brighton, N. Y.
B—San Francisco, Calif.
G. O. No. 10, W. D., 1920

Captain, 106th Machine Gun Battalion, 27th Division.
Captain Platt, while moving forward in the attack with machine-gun units, encountered a number of stragglers in a sunken road. He organized these men into a company and led them in attacking waves. Later he personally led a patrol which located and silenced enemy machine guns which were firing from a flank position.

*PLATT, CHESTER ERASTUS_____
At Chateau-Thierry, France, June 6, 1918.
R—La Fayette, Ind.
B—Stockwell, Ind.
G. O. No. 110, W. D., 1918.

Corporal, 45th Company, 5th Regiment, U. S. Marine Corps, 2d Division.
Killed in action at Chateau-Thierry, France, June 6, 1918, he gave the supreme proof of that extraordinary heroism which will serve as an example to hitherto untried troops.
Posthumously awarded. Medal presented to mother, Mrs. Lettie C. Platt.

PLATT, JONAS HENRY_____
Near the Bois-de-Belleau, Chateau-Thierry, France, June 6, 1918.
R—Brooklyn, N. Y.
B—Brooklyn, N. Y.
G. O. No. 15, W. D. 1919.

First lieutenant, 5th Regiment, U. S. Marine Corps, 2d Division.
Seriously wounded in the leg early in the engagement, he continued to direct the operations not only of his platoon but of another. He charged and drove off the crew of an enemy machine gun, supervised the disposition and digging in of a large part of his company, and yielded command only when exhausted from pain and loss of blood.

PLATTEN, MICHAEL A. (2304032)_____
Near Chery-Chartreuve, France, Aug. 14, 1918.
R—Green Bay, Wis.
B—Green Bay, Wis.
G. O. No. 21, W. D., 1919.

Cook, Battery B, 121st Field Artillery, 32d Division.
When the gun crews of the platoons in the woods were forced to withdraw on account of the intense enemy shelling, a wounded man was left behind. Noticing this, Cook Platten rushed into the woods, despite the continual shelling, and brought the man to safety.

PLAUMAN, HERMAN (261709)_____
Near St. Gilles, France, Aug. 3, 1918.
R—Detroit, Mich.
B—Romeo, Mich.
G. O. No. 44, W. D., 1919.

Private, first class, Company H, 128th Infantry, 32d Division.
After having just returned from a hazardous trip through heavy shellfire, he volunteered and carried a message to his company commander. As he was about to complete his mission the company commander was mortally wounded by a shell. After administering first aid, he reported back to his battalion commander, who, upon noticing that 1 of the runner's fingers was missing, ordered him to the rear for treatment.

PLEMONS, RUSSELL L. (1309472)......
Near Ponchaux, France, Oct. 7, 1918.
R—Lenoir City, Tenn.
B—Roane County, Tenn.
G. O. No. 37, W. D., 1919.

Corporal, Company L, 117th Infantry, 30th Division.
After seeing 2 other soldiers killed while attempting to carry automatic-rifle ammunition through a heavy artillery and machine-gun barrage, Corporal Plemons volunteered for this dangerous mission and successfully accomplished it.

*PLIMPTON, CHESTER H......
At Thiaucourt, France, Sept. 27, 1918.
R—Buffalo, N. Y.
B—Buffalo, N. Y.
G. O. No. 53, W. D., 1920.

First lieutenant, Company F, 21st Engineers.
Lieutenant Plimpton exposed himself to heavy artillery fire in order to direct the repair of a railroad track over which ammunition was delivered to the batteries. The enemy made a determined effort to destroy the line of communication and subjected the locality to intense and accurate bombardment for a number of hours. The gallantry displayed by this officer was an important factor in the successful completion of the mission assigned to him. He was killed by concussion of a large-caliber shell as his work was nearing completion.
Posthumously awarded. Medal presented to mother, Mrs. George A. Plimpton.

PLUMLEY, RICHARD G......
Near the Bois-de-Septsarges, France, Sept. 27–Oct. 10, 1918.
R—Hartford, Conn.
B—Hammonton, N. J.
G. O. No. 98, W. D., 1919.

Captain, 39th Infantry, 4th Division.
On duty as regimental adjutant, Captain Plumley left a place of safety, and going forward under heavy fire assisted in reforming the assault battalion, which had lost most of its officers and was becoming disorganized. During the following days he repeatedly crossed areas which had been subjected to heavy gas bombardments, and as a result became almost blind and greatly weakened by gas poisoning. He refused to be evacuated, however, and remained on duty throughout the night, rendering valuable assistance to the regimental commander, who had just taken command.

*PLUMMER, CHARLES W......
Near Fismes, France, Aug. 11, 1918.
R—South Dartmouth, Mass.
B—New Bedford, Mass.
G. O. No. 44, W. D., 1919.

Second lieutenant, 101st Field Artillery, observer, 88th Aero Squadron, Air Service.
Louis G. Bernheimer, first lieutenant, pilot; John W. Jordan, second lieutenant, 7th Field Artillery, observer; Roger W. Hitchcock, second lieutenant, pilot; James S. D. Burns, deceased, second lieutenant, 165th Infantry, observer; Joel H. McClendon, deceased, first lieutenant, pilot; Charles W. Plummer, deceased, second lieutenant, 101st Field Artillery, observer; Philip R. Babcock, first lieutenant, pilot; and Joseph A. Palmer, second lieutenant, 15th Field Artillery, observer. All of these men were attached to the 88th Aero Squadron, Air Service.
For extraordinary heroism in action near Fismes, France, August 11, 1918. Under the protection of 3 pursuit planes, each carrying a pilot and an observer, Lieutenants Bernheimer and Jordan, in charge of a photo plane, carried out successfully a hazardous photographic mission over the enemy's lines to the River Aisne. The 4 American ships were attacked by 12 enemy battle planes. Lieutenant Bernheimer, by coolly and skillfully maneuvering his ship, and Lieutenant Jordan, by accurate operation of his machine gun, in spite of wounds in the shoulder and leg, aided materially in the victory which came to the American ships, and returned safely with 36 valuable photographs. The pursuit plane operated by Lieutenants Hitchcock and Burns was disabled while these 2 officers were fighting effectively. Lieutenant Burns was mortally wounded and his body jammed the controls. After a headlong fall of 2,500 meters, Lieutenant Hitchcock succeeded in regaining control of this plane and piloted it back to his airdrome. Lieutenants McClendon and Plummer were shot down and killed after a vigorous combat with 5 of the enemy's planes. Lieutenants Babcock and Palmer, by gallant and skillful fighting, aided in driving off the German planes and were materially responsible for the successful execution of the photographic mission.
Posthumously awarded. Medal presented to father, Henry W. Plummer.

*PLUMMER, GEORGE, Jr. (1286157)......
South of Soissons, France, July 18–25, 1918.
R—Bethesda, Md.
B—Mount Ephrim, Md.
G. O. No. 117, W. D., 1918.

Private, Company L, 9th Infantry, 2d Division.
He distinguished himself by volunteering 3 times to carry messages through heavy shellfire after all runners had been killed or wounded.
Posthumously awarded. Medal presented to mother, Mrs. Kate Plummer.

PLUSH, LEWIS C......
Near Romagne, France, Oct. 4, 1918.
R—Pomona, Calif.
B—Sumner, Mo.
G. O. No. 15, W. D., 1919.

First lieutenant, 49th Aero Squadron, Air Service.
He was a member of a patrol of 7 machines which attacked 17 enemy Fokkers. After shooting down 1 of the enemy, this officer returned to the fight and shot down another.

POE, NEILSON......
Near Soissons, France, July 19, 1918.
R—Baltimore, Md.
B—Baltimore, Md.
G. O. No. 72, W. D., 1920.

Second lieutenant, 28th Infantry, 1st Division.
During the attack of his organization, after he had been severely wounded, Lieutenant Poe refused to be evacuated, but rejoined his organization and went forward to the attack through heavy enemy fire. The company commander being killed, Lieutenant Poe took command of the company and continued to direct it until the day's objective was reached. Although suffering great pain, he remained with his organization for more than 24 hours after he was wounded.

POHL, GEORGE H. (302697)......
Near Juvigny, north of Soissons, France, Aug. 28, 1918.
R—Mount Clemens, Mich.
B—Mount Clemens, Mich.
G. O. No. 1, W. D., 1926.

Corporal, Company G, 126th Infantry, 32d Division.
He was severely wounded during the first stage of the advance, but he remained with his company until the objective was gained. He then went to the rear, where his wound was dressed, and he was ordered to a hospital. When his evacuation was delayed by lack of ambulances, this soldier decided to return to the front line, where he reported to his commanding officer and volunteered to assist in carrying wounded to a dressing station. During the remainder of the operation, in spite of his wounds, he continued to assist in bringing the wounded in from a field constantly swept by machine-gun fire.

POILLON, JOHN J. (1783996)_____
Near Montfaucon, France, Sept. 29, 1918.
R—Milford, Pa.
B—Milford, Pa.
G. O. No. 46, W. D., 1919.

Corporal, Company C, 311th Machine Gun Battalion, 79th Division.
Although engaged as company clerk, when the platoon to which he was attached was required to advance over a heavily shelled territory, Corporal Poillon volunteered and carried messages from his company commander to the platoon, the journey being made under the most hazardous conditions and under severest shellfire.

POKORNY, ROBERT (2337477)_____
Near Roncheres, France, July 28, 1918.
R—Chicago, Ill.
B—Chicago, Ill.
G. O. No. 32, W. D., 1919.

Private, first class, Headquarters Company, 4th Infantry, 3d Division.
He volunteered and went forward to recover the trail of a 37-millimeter gun, despite the great danger of heavy machine-gun fire; he succeeded in recovering and dragging it back to our lines.

POLASKA, EDWARD P. (322951)_____
In Shkotovo and Suchan campaign, Siberia, May 20, 1919, to July 20, 1919.
R—South Bend, Ind.
B—Saginaw, Mich.
G. O. No. 133, W. D., 1919.

Sergeant, Company D, 31st Infantry.
He commanded a platoon of his company during the entire period and led it in 5 different engagements. Hardships were undergone which tested discipline severely, through all of which Sergeant Polaska maintained perfect control, held his platoon at a high state of efficiency, and displayed extraordinary skill in handling it while actually in action, notably at Sitsa, Siberia, June 26, 1919, and at Kazanka, Siberia, July 3, 1919.

POLITTE, MELVIN J. (2337478)_____
Near Grand Ballois Farm, France, July 15, 1918.
R—Old Mines, Mo.
B—Old Mines, Mo.
G. O. No. 32, W. D., 1919.

Corporal, Headquarters Company, 4th Infantry, 3d Division.
After seeing many of his comrades killed or wounded in attempting the same mission, Corporal Politte went forward under heavy shell and gas bombardment and repaired telephone lines.

POLLARD, RUSSELL (1967745)_____
At Bois-Frehaut, France, Nov. 10, 1918.
R—Anadarko, Okla.
B—Annona, Tex.
G. O. No. 44, W. D., 1919.

Corporal, Company H, 365th Infantry, 92d Division.
During the assault at Bois-Frehaut, Corporal Pollard, a rifle grenadier, conducted his squad skillfully in firing on hostile machine guns until his rifle was broken. He then used his wire cutters with speed and skill under heavy shell and machine-gun fire. Although wounded in his right arm, he continued to cut the wire with his left hand, and assisted his men in getting through it until ordered to the dressing station a second time by his company commander.

POLLEY, BRITTON_____
Near Romagne, France, Oct. 9, 1918.
R—New York, N. Y.
B—Circleville, N. Y.
G. O. No. 35, W. D., 1919.

First lieutenant, Field Artillery, observer, 99th Aero Squadron, Air Service.
Lieutenant Polley was assigned to a mission to find line troops of the division to which his squadron was attached. Weather conditions made flying almost impossible, a second plane assigned to the mission returning on that account. Flying at an altitude of 25 meters over enemy lines, he encountered and defeated 3 enemy patrols, gathering and delivering to his division headquarters most valuable information.

POLLEY, JOHN R. (1467612)_____
Near Cheppy, France, Sept. 26, 1918.
R—Hutchinson, Kans.
B—Louisville, Ky.
G. O. No. 59, W. D., 1919.

Private, Company A, 110th Engineer, 35th Division.
When the Infantry attack was held up by direct fire from an enemy concrete machine-gun emplacement, Private Polley, who was a member of a wire-cutting detail, charged the emplacement with 1 Infantry soldier, capturing 2 guns and 2 prisoners and enabling the advance to continue.

POLLINGER, FRANK J. (1683693)_____
Near Charlevaux, France, Oct. 3–7, 1918.
R—Worcester, Mass.
B—Worcester, Mass.
G. O. No. 35, W. D., 1919.

Private, Company G, 308th Infantry, 77th Division.
During the period of 4 days, when his battalion was surrounded by the enemy and after his squad leader had been wounded, Private Pollinger took command of the squad, although he himself was suffering from a wound received 4 days previous. His indomitable courage and perseverance upheld the spirit and morale of his men under such trying circumstances and he continued to direct their movements until forced out of action by a second wound.

PONDER, WILLIAM THOMAS_____
Near Fontaines, France, Oct. 23, 1918.
R—Mangum, Okla.
B—Llano, Tex.
G. O. No. 46, W. D., 1919.

First lieutenant, 103d Aero Squadron, Air Service.
Having been separated from his patrol, he observed and went to the assistance of an allied plane which was being attacked by 13 of the enemy. Against great odds Lieutenant Ponder destroyed 1 enemy plane and so demoralized the remaining that both he and his comrade were able to return to their lines.

POORE, BENJAMIN A._____
At Bois-des-Septsarges, France, Sept. 27, 1918, and at Bois-du-Fays, France, Oct. 11, 1918.
R—Fitchburg, Mass.
B—Center, Ala.
G. O. No. 44, W. D., 1919.
Distinguished-service medal also awarded.

Brigadier general, 7th Infantry Brigade, 4th Division.
At Bois-des-Septsarges on Sept. 27, General Poore personally re-formed his disorganized troops, who were falling back through lack of command and because of severe casualties. Under heavy fire, he led them to the lines, and presented an unbroken front to the enemy. Again on Oct. 11, in the region of Bois-du-Fays, he gathered together troops who were taking refuge from hostile fire, and turned them over to the support commander.

POPE, OLLIE (1319961)_____
Between St. Quentin and Cambrai, France, Oct. 9, 1918.
R—Durham, N. C.
B—Sampson County, N. C.
G. O. No. 37, W. D., 1919.

Private, Company C, 120th Infantry, 30th Division.
He was wounded in action between St. Quentin and Cambrai, France, and after having his wounds dressed, he was unable to locate his company. He returned, however, to the front line, and fought throughout the day, locating and returning to his own organization after dark.

*POPLIN, DANIEL C. (2339323)_____
Near Roncheres, France, July 29, 1918.
R—Charlotte, N. C.
B—Charlotte, N. C.
G. O. No. 32, W. D., 1919.

Private, Company H, 4th Infantry, 3d Division.
He repeatedly carried messages between his own and another company across an open field swept by heavy machine-gun and sniper fire and was killed while on 1 of these missions.
Posthumously awarded. Medal presented to mother, Mrs. Mary Poplin.

PORTER, CHARLES PULLMAN_____
In the region of Epieds, France, July 16, 1918.
R—New Rochelle, N. Y.
B—Brooklyn, N. Y.
G. O. No. 145, W. D., 1918.

Second lieutenant, pilot, 147th Aero Squadron, Air Service.
While on patrol he observed 2 enemy planes (Fokker type) about 1,000 meters above him. He immediately maneuvered to obtain height and a position for attack. The enemy turned and Lieutenant Porter gave chase and attacked from below, destroying 1 and forcing the other to retire.
Oak-leaf cluster.
A bronze oak leaf, for extraordinary heroism in action in the region of Foret-de-Fere, France, July 24, 1918. While leading a patrol he attacked an enemy formation of 12 planes (Fokker type). He engaged 1 enemy and sent it down out of control. One of his guns jammed, and while he was repairing the gun 2 of the enemy planes got behind him. Unable to repair the gun and only to fire a single shot, he turned to attack, destroying second plane, and remained in the fight until the enemy retired.

PORTER, CHAUNCEY W. (2202524)_____
North of Flirey, France, Sept. 12, 1918.
R—Chambers, Nebr.
B—Brainard, Nebr.
G. O. No. 129, W. D., 1918.

Bugler, Company B, 355th Infantry, 89th Division.
He charged a machine gun alone with an automatic pistol, killed 1 man, captured another, and drove the remainder of the enemy platoon back along their trench, thereby enabling his platoon to advance.

PORTER, CLARENCE R. (1867829)_____
Near Bellicourt, France, Sept. 29, 1918.
R—Pickens, S. C.
B—Pickens, S. C.
G. O. No. 37, W. D., 1919.

Private, Company D, 119th Infantry, 30th Division.
While his company was making an attack on the Hindenburg line, he continued a covering fire with his Lewis gun. In spite of 2 wounds from which he was suffering, he remained with his gun until his comrades had succeeded in crossing the line.

PORTER, EARL W._____
Near Lassigny, France, Aug. 9, 1918.
R—Atlantic, Iowa.
B—Atlantic, Iowa.
G. O. No. 124, W. D., 1918.

Second lieutenant, Air Service, U. S. Army, attached to French Army.
He, with First Lieut. Charles Raymond Blake, pilot, while on a reconnaissance expedition at a low altitude and beyond the enemy lines, was attacked by 5 German battle planes. Although wounded at the beginning of the combat, he shot down 1 of the enemy machines and by cool and courageous operation of his gun, while his pilot skillfully maneuvered the plane, fought off the others and made possible a safe return to friendly territory.

*PORTER, ERNEST WASHINGTON (103602)_
Near Thiaucourt, France, Sept. 15, 1918.
R—Newark, N. J.
B—Rutherford, N. J.
G. O. No. 37, W. D., 1919.

Private, 23d Company, 6th Machine Gun Battalion, U. S. Marine Corps, 2d Division.
While taking cover with the remainder of his gun crew from a heavy artillery barrage, Private Porter answered a call for volunteers to combat an enemy aeroplane. Upon reaching his gun he was instantly killed by an aerial bomb.
Posthumously awarded. Medal presented to father, Ernest W. Porter.

PORTER, JAMES E. (42309)_____
Near Sommerance, France, Oct. 9, 1918.
R—Detroit, Mich.
B—Philadelphia, Pa.
G. O. No. 72, W. D., 1920.

Sergeant, Company C, 16th Infantry, 1st Division.
During the attack on Hill 272, Sergeant Porter, with 4 men, exposed himself to heavy machine-gun fire in order to attack an enemy machine gun which was causing heavy casualties among his company. Although all his men were either killed or wounded, he succeeded in capturing the gun. Due to his gallantry, his company was able to continue the advance.

PORTER, KENNETH L._____
Near Chateau-Thierry, France, July 2, 1918.
R—Dowagiac, Mich.
B—Dowagiac, Mich.
G. O. No. 1, W. D., 1919.

Second lieutenant, 147th Aero Squadron, Air Service.
He, with 4 other pilots, attacked 12 enemy aircraft (type, Pfalz) flying in 2 groups well within the enemy lines. As soon as the enemy planes were sighted, he maneuvered to get between them and the sun, and with great difficulty gained the advantage. While 3 of the other American officers dived on the lower formation, he and Second Lieut. John H. Stevens engaged the upper formation in a bold and brilliant combat, 2 planes of which they crashed to the earth.

PORTER, RAY E._____
Near Rembercourt, France, Nov. 1–2, 1918.
R—Fordyce, Ark.
B—Fordyce, Ark.
G. O. No. 35, W. D., 1919.

First lieutenant, 34th Infantry, 7th Division.
He led his company in a successful assault on a ridge of high ground, taking several strong points and machine gun nests and numerous prisoners. He held this position for 30 hours without food or water against 2 enemy counterattacks until relieved.

PORTER, VINCENT C. (1520649)_____
Near Montfaucon, France, Sept. 26–Oct. 1, 1918.
R—Canton, Ohio.
B—Elwood City, Pa.
G. O. No. 145, W. D., 1918.

Corporal, Company C, 146th Infantry, 37th Division.
Though he was acting as company clerk, throughout the drive west of the Meuse River he volunteered for service as a runner and also took charge of the delivery of rations under constant shellfire in a highly exposed position. He performed valuable service in giving first aid to wounded, and at one time carried a wounded soldier much heavier than himself up a hill through shell and machine-gun fire.

POSSER, FREDERICK (2669271)_____
Near Ronssoy, France, Sept. 29, 1918.
R—New York, N. Y.
B—New York, N. Y.
G. O. No. 35, W. D., 1919.

Corporal, Machine Gun Company, 107th Infantry, 27th Division.
During the thick of the fighting against the Hindenburg line, Corporal Posser voluntarily went forward to locate friendly troops, and in doing so he was obliged to pass between two strongly fortified enemy nests, from which a deadly fire was pouring. Despite this obstacle, he communicated with the infantry and returned to his position.

POSTMOY, ALEXANDER (541717)..........
Near Fossoy, France, July 15, 1918.
R—Detroit, Mich.
B—Russia.
G. O. No. 44, W. D., 1919.

Corporal, Company F, 7th Infantry, 3d Division.
After being seriously wounded by machine-gun fire, he remained on duty at his observation post at the river edge, so that the enemy could not cross unobserved.

POSTULA, JOHN I. (1389506)...........
Near Bois-des-Forges, France, Sept. 26, 1918.
R—Chicago, Ill.
B—Chicago, Ill.
G. O. No. 71, W. D., 1919.

Sergeant, Company H, 132d Infantry, 33d Division.
When the advance of his platoon was held up by enemy fire, he advanced alone against a machine-gun nest and killed the crew. He brought back the enemy machine gun and the platoon was able to renew the advance. He showed marked personal bravery under heavy fire.

POTTER, ERNEST R. (556150)..........
Near St. Thibaut, France, Aug. 7, 1918.
R—Pittsburgh, Pa.
B—Clarion, Pa.
G. O. No. 98, W. D., 1919.

First sergeant, Company D, 39th Infantry, 4th Division.
When all the officers of his company had become casualties and the morale of the men was sinking, Sergeant Potter assumed command, and after reorganizing the company successfully led it in repelling several vicious hostile counterattacks. During the action he was wounded in the shoulder, but he refused to go to the rear until he was ordered to do so by the officer sent to relieve him.

POTTER, FRANK R..................
At the town of Arbre Guernon, France, Oct. 17-18, 1918.
R—New York, N. Y.
B—Waterville, N. Y.
G. O. No. 9, W. D., 1923.

Captain, 105th Infantry, 27th Division.
Organizing his units just before entering the town, he advanced under heavy machine-gun and rifle fire, captured the town with numerous prisoners, and from his new position silenced several enemy machine guns. On the morning of Oct. 18, while he was in command, the 105th Infantry Battalion advanced about 300 yards and was stopped by intense machine-gun fire. His coolness and complete indifference to danger inspired his men to resume the advance despite the heavy machine-gun fire from the strongly held position in a sunken road immediately in the battalion's front, which position was promptly captured, together with numerous prisoners and machine guns.

POTTER, WALTER (1977989)..........
At Bois-d'Harville, France, Nov. 10, 1918.
R—Jenkins, Ky.
B—Dayton, Tenn.
G. O. No. 87, W. D., 1919.

Private, Company L, 131st Infantry, 33d Division.
He volunteered and crawled out in the face of heavy enemy fire to attack a machine-gun nest. He killed the 4 members of the enemy crew, inspiring the men serving with him by his example of heroism.

POTTER, WILLIAM CLARKSON..........
Near Dun-sur-Meuse, France, Sept. 26, 1918.
R—Riverdale-on-Hudson, N. Y.
B—France.
G. O. No. 107, W. D., 1918

First lieutenant, 20th Aero Squadron, Air Service.
A formation of 8 Liberty bombing planes, while on a daylight bombing mission on Dun-sur-Meuse, was attacked by a force of enemy planes three times its number. Lieutenant Potter saw that the observer's guns of the leading machine were inactive, while its pilot exerted great effort to control his machine. Under conditions demanding greatest courage and determination Lieutenant Potter flew in close to the leader so as to protect him from the rear. This position he held under ever-increasing enemy attack, and in face of the fact that his leader continued on into Germany. The conditions became more desperate, still Lieutenant Potter hung on, until his leader was finally able to make a turn about to the allied lines. On landing it was found that the observer of the leading machine had been killed and had fallen and jammed the controls, making a turn impossible. Lieutenant Potter, by his courage and disregard of danger, saved the life of his leader and brought his machine safely back to our lines.

POTTER, WILLIAM J..................
Near Eclisfontaine, France, Sept. 28, 1918.
R—Scranton, Pa.
B—Scranton, Pa.
G. O. No. 10, W. D., 1920.

Captain, 361st Infantry, 91st Division.
After being painfully wounded by a shell fragment during the night, Captain Potter refused to go to the rear and organized his company for an attack and led it in the advance under heavy machine-gun and artillery fire, freely exposing himself and cheering his men by his presence until he was a second time wounded through the lungs, even then refusing to be evacuated until the company was organized and properly turned over to his successor for another attack which was then impending.

POWELL, GEORGE W. (1282737)........
Near Verdun, France, Oct. 12, 1918.
R—Camden, N. J.
B—Philadelphia, Pa.
G. O. No. 37, W. D., 1919.

Sergeant, Company B, 111th Machine Gun Battalion, 29th Division.
After all the men of his section except himself and 2 other soldiers had been killed or wounded, he took charge of a machine gun and remained in an isolated position for 4 days, keeping his men constantly in action and inflicting many casualties on the enemy.

POWELL, JAMES T. (1247276).........
Near Fismette, France, Aug. 9, 1918.
R—Bradford, Pa.
B—Bradford, Pa.
G. O. No. 145, W. D., 1918.

Corporal, Company C, 112th Infantry, 28th Division.
When a platoon of his company was held up by sniper fire, he, undaunted, voluntarily crawled through holes in walls and over roofs, located the enemy sniper, and killed him, enabling the platoon to proceed without further loss. Later in the engagement, when reinforcements and ammunition were needed, he volunteered and swam the Vesle River under machine-gun fire.

POWELL, ROBERT E. (1246331)........
Near Fismette, France, Aug. 9, 1918.
R—Philadelphia, Pa.
B—Philadelphia, Pa.
G. O. No. 9, W. D., 1923.

Regimental supply sergeant, Supply Company, 112th Infantry, 28th Division.
While regimental supply sergeant, Supply Company, 112th Infantry, he voluntarily participated in the attack and displayed high courage and determination during the battle, assisting after the village had been captured in removing the seriously wounded across a narrow bridge swept by heavy shell and machine-gun fire. On the evening of Aug. 9, he voluntarily entered the most dangerous parts of the company's position and by his coolness and indifference to danger inspired the men of the command with new courage when they were sorely beset by the enemy, who had gained access to the village. Learning that an enemy patrol had entered a near-by building from the rear, he plunged into the building in complete darkness, dispatched 4 of the enemy, thus retaining possession of the building the loss of which would have placed the command in a precarious position.

*POWELL, TOM (2649392)\
Near Beaume, France, Nov. 8, 1918.\
R—Hawkinsville Ga.\
B—Hawkinsville, Ga.\
G. O. No. 37, W. D., 1919.

Private, Company H, 370th Infantry, 93d Division.\
Private Powell repeatedly carried messages under severe enemy fire to the various units in the vicinity of his company until he was killed while in the performance of his duty.\
Posthumously awarded. Medal presented to mother, Mrs. Eliza Fountain Powell.

POWELL, WILLIAM H. (1319097)\
Near Bellicourt, France, Sept. 29, 1918.\
R—Oxford, N. C.\
B—Oxford, N. C.\
G. O. No. 44, W. D., 1919.

Private, Machine Gun Company, 120th Infantry, 30th Division.\
Private Powell took charge of 4 other soldiers who had become separated from their platoon and led them forward toward the objective. Attacking a machine-gun nest, they captured 7 prisoners and a Maxim gun, which they immediately put into action and fired 2,000 rounds at the enemy. They then continued to advance under heavy artillery and machine-gun fire.

POWER, JAMES B. (1388463)\
Near Albert, France, Aug. 4, 1918.\
R—Chicago, Ill.\
B—Michigamme, Mich.\
G. O. No. 71, W. D., 1919.

Sergeant, Company L, 131st Infantry, 33d Division.\
While his company was occupying trenches on the outskirts of Albert, France, he, on his own initiative, left shelter and, creeping forward, worked his way through the town, though subjected to snipers who had been firing on our troops.

POWER, LLEWELLYN (1210299)\
Near Ronssoy, France, Sept. 29, 1918.\
R—New York, N. Y.\
B—Orange, N. J.\
G. O. No. 20, W. D., 1919.

Corporal, Company D, 107th Infantry, 27th Division.\
During the operations against the Hindenburg line Corporal Power, with 4 other soldiers, left shelter and went forward into an open field under shell and machine-gun fire and succeeded in bandaging and carrying to our lines 2 wounded men.

POWERS, EDWARD J. (1391424)\
Near Bois-du-Fays, France, Oct. 9, 1918.\
R—Chicago, Ill.\
B—Chicago, Ill.\
G. O. No. 32, W. D., 1919.

Private, Machine Gun Company, 132d Infantry, 33d Division.\
After being wounded he received treatment at a first-aid station, from where he was consigned to the hospital. Throwing away his evacuation ticket, he returned to the front line, where he acted as runner until the company was relieved, when he was removed to a hospital.

POWERS, JOSEPH J. (1708565)\
Near St. Juvin, France, Oct. 15, 1918.\
R—Brooklyn, N. Y.\
B—Brooklyn, N. Y.\
G. O. No. 35, W. D., 1919.

Sergeant, Company E, 308th Infantry, 77th Division.\
After 4 men had been killed or wounded, while attempting to deliver a message from the company commander to the rear, Sergeant Powers volunteered and carried the message through area swept by machine-gun fire with no regard for his personal safety.

*POWERS, RALPH E\
\
At Ust Padenga Russia, Jan. 20–23, 1919.\
R—Amherst, Ohio,\
B—Mogadore, Ohio.\
G. O. No. 89, W. D., 1919.

First lieutenant, Medical Corps, 310th Sanitary Train, 85th Division (detachment in North Russia).\
While his dressing station was burning as a result of having been struck by a shell, Lieutenant Powers successfully evacuated all his patients, numbering 40. He then moved to a new location and continued to work for two days under shellfire until this dressing station, too, was struck and he himself mortally wounded, whereupon he gave orders that the other wounded should be removed first and that he be left until the last.\
Posthumously awarded. Medal presented to mother, Mrs. H. W. Powers.

*POWLESS, JOSIAH ALVIN\
Near Chevieres, France, Oct. 14, 1918.\
R—West Depere, Wis.\
B—Oneida Reservation, Wis.\
G. O. No. 46, W. D., 1920.

First lieutenant, medical detachment, attached to 308th Infantry, 77th Division.\
When notified that his colleague, Capt. James M. McKibban, had been wounded, Lieutenant Powless immediately went forward to his assistance. He crossed an area subjected to intense machine-gun and constant artillery fire, reached his colleague, whose wound proved to be fatal, and after dressing his wounds had him carried to the rear. Lieutenant Powless was seriously wounded while performing this service.\
Posthumously awarded. Medal presented to widow, Mrs. J. A. Powless.

POZZI, WILLIAM (72534)\
Near Belleau, France, July 21, 1918.\
R—Springfield, Mass.\
B—Italy.\
G. O. No. 37, W. D., 1919.

Corporal, Company G, 104th Infantry, 26th Division.\
He voluntarily left his trench and ran through heavy shellfire to the aid of a wounded soldier and carried him to safety.

PRAGER, BENJAMIN (1244151)\
Near Fismes, France, Aug. 11, 1918.\
R—Pittsburgh, Pa.\
B—Pittsburgh, Pa.\
G. O. No. 71, W. D., 1919.

Sergeant, Company E, 111th Infantry, 28th Division.\
On his own initiative and under heavy fire he led an automatic-rifle squad to a house far in advance of our lines and by purposely exposing himself at a window drew fire from an enemy machine gun, thus disclosing its position and enabling his squad to destroy it. After being wounded he refused to be evacuated until he had visited another portion of the line and assured himself that the position was well consolidated.

PRATT, JESSE W. (541923)\
Near Cierges, France, Oct. 11 and 24, 1918.\
R—Vanderbilt, Pa.\
B—Vanderbilt, Pa.\
G. O. No. 98, W. D., 1919.

Sergeant, Company G, 7th Infantry, 3d Division.\
Though his platoon had been reduced by casualties to only 3 squads, Sergeant Pratt led it to the objective, and despite an intense artillery and machine-gun barrage succeeded in capturing a machine-gun nest which was enfilading his flank, killing the entire crew. On October 24, after all the officers of his battalion had become casualties, Sergeant Pratt assumed command and by daring leadership pushed forward the attack, successfully consolidating the captured position and holding it against repeated hostile counterattacks.

PRATT, JOHN (95285)\
Near Chevenges, France, Nov. 7, 1918.\
R—Toledo, Ohio.\
B—Toledo, Ohio.\
G. O. No. 50, W. D., 1919.

Corporal, Company L, 166th Infantry, 42d Division.\
While engaged as a runner Corporal Pratt saw 2 comrades lying wounded several hundred yards away. Despite the fact that the vicinity was being heavily shelled, he crawled out across the open field and administered first aid to both men, after which he carried them 1 at a time back to a dressing station.

PRATT, JOHN H., Jr.
Near Bazoches, France, Aug. 7-9, 1918.
R—New York, N. Y.
B—New York, N. Y.
G. O. No. 138, W. D., 1918.

Second lieutenant, 47th Infantry, 4th Division.
He was untiring and fearless at all times in the performance of his duties as liaison officer. Under heavy fire he made 3 exceptionally hazardous trips with messages of vital importance when other means of communication had failed, volunteering for this service.

PRATT, LESTER L.
In the Bois-de-Belleau, France, June 11, 1918.
R—Bellefontaine, Ohio.
B—Bellefontaine, Ohio.
G. O. No. 44, W. D., 1919.

Assistant surgeon, U. S. Navy, attached to 5th Regiment, U. S. Marine Corps, 2d Division.
Although he had been wounded under the left eye, almost blinded by gas fumes, and his dressing station wrecked by shellfire, he remained at his post, working under the most trying conditions until all the wounded had been safely evacuated.

PRATT, ROBERT M. (2021896)
Near Enitsa, Russia, Oct. 17, 1918.
R—Detroit, Mich.
B—Ashton, Mich.
G. O. No. 16, W. D., 1920.

Corporal, Company M, 339th Infantry, 85th Division (detachment in north Russia).
In an attack on an enemy strong point Corporal Pratt led his Lewis gun crew in a gallant dash in the face of enemy fire, 80 yards ahead of the other members of his platoon. His section delivered an accurate and enfilading fire, which forced the enemy to retire. This act enabled his company to capture the entire enemy position.

PRAUSE, CARL W. T.
Near Vaux-Andigny, France, Oct. 11, 1918.
R—Charleston, S. C.
B—Charleston, S. C.
G. O. No. 32, W. D., 1919.

Second lieutenant, 118th Infantry, 30th Division.
While leading his company in attack Lieutenant Prause was wounded by shellfire, but he remained for 3 days thereafter, without medical aid, directing the steady progress of his command, in the face of the enemy's determined resistance.

PRESCOTT, LEE O. (2302171)
At St. Gilles, near Fismes, France, Aug. 4, 1918.
R—Lansing, Mich.
B—Spicerville, Mich.
G. O. No. 139, W. D., 1918.

Private, Headquarters Company, 120th Field Artillery, 32d Division.
He with other soldiers made frequent trips to maintain telephone communication between battalion and regimental headquarters during a destructive enemy bombardment. All other lines had been destroyed. As this line was used by both Infantry and Artillery for a communication with the rear, it was of the utmost importance that it be maintained.

PRESCOTT, FLOYD W. (2302159)
At St. Gilles, near Fismes, France, Aug. 4, 1918.
R—Lansing, Mich.
B—Leslie, Mich.
G. O. No. 142, W. D., 1918.

Corporal, Headquarters Company, 120th Field Artillery, 32d Division.
He with other soldiers made frequent trips to maintain telephone communication between battalion and regimental headquarters during a destructive enemy bombardment. All other lines had been destroyed, and as this line was used by both Infantry and Artillery for communication with the rear it was of the utmost importance that it be maintained.

PRESLEY, ALBERT C. (2285324)
Near Eclisfontaine, France, Sept. 27, 1918.
R—Grants Pass, Oreg.
B—Rogue River, Oreg.
G. O. No. 37, W. D., 1919.

Private, Headquarters Company, 363d Infantry, 91st Division.
Sergeant Presley, with a patrol of 4 other men, went out to reduce what was thought to be a sniping post; they discovered upon arriving nearer that it was a machine-gun nest and attacked it by a series of short rushes. The attack resulted in the capture of 25 prisoners and 2 machine guns.

PRESLEY, EARL C. (2782927)
Near Eclisfontaine, France, Sept. 29, 1918.
R—Hurricane, Utah.
B—Albion, N. Y.
G. O. No. 37, W. D., 1919.

Private, Company K, 364th Infantry, 91st Division.
He volunteered and went to the assistance of a wounded comrade who was seen making his way to our lines, exposed to machine-gun and sniper fire which infested the woods at our direct front.

PRESTON, GLEN A.
Near Andevanne, France, Oct. 29, 1918.
R—Howe, Ind.
B—Ontario, Ind.
G. O. No. 64, W. D., 1919.

Second lieutenant, Field Artillery, observer, 99th Aero Squadron, Air Service.
Becoming separated from his protecting planes while on a photographic mission, Lieutenant Preston continued alone, and, although he was attacked by 7 enemy planes (type Fokker), he drove them off and secured numerous photographs.
Oak-leaf clusters (2).
For the following act of extraordinary heroism in action near Remonville, France, Oct. 30, 1918, Lieutenant Preston is awarded one oak-leaf cluster, to be worn with his distinguished-service cross: He successfully accomplished his mission in spite of encounters with 4 separate enemy formations, one of 38 machines, another of 6 (type Pfalz), another of 7 (type Fokker), and a formation of biplanes. He shot down one of the enemy and returned with valuable information.
For the following act of extraordinary heroism in action near Cunel, France, Oct. 5, 1918, Lieutenant Preston is awarded a second oak-leaf cluster: While on a photographic mission Lieutenant Preston and his pilot were attacked by 7 enemy planes (type Fokker) and driven back to our own lines. They almost immediately returned to the same locality without the protection of battleplanes and continued to take photographs until attacked by 5 machines (Pfalz type). They opened fire on this formation and brought down 2 of them and drove the others away and then returned with photographs of great importance.

PRESTON, JOHN T., Jr. (198648)
During the operations of Berzy-le-Sec, France, July 18-21, 1918.
R—Manville, R. I.
B—Manville, R. I.
G. O. No. 132, W. D., 1918.

Sergeant, 2d Field Signal Battalion, 1st Division.
Attached to headquarters of the 1st Division as dispatch rider during the operations of Berzy-le-Sec, France, July 18-21, 1918, he courageously and unhesitatingly passed through areas under steady artillery bombardment to carry messages whose delivery was of vital necessity to the success of the attack.

PRETE, FRANK P. (1375582)_____
   Near Bantheville, France, Nov. 1,
    1918.
   R—Chicago, Ill.
   B—Italy.
   G. O. No. 87, W. D., 1919.

Sergeant, Battery B, 122d Field Artillery, 33d Division.
Sergeant Prete 3 times passed through a heavy enemy barrage and machine-gun fire while guiding a combat train forward to an advanced artillery platoon.

PRETTY, JAMES LEWIS_____
   In the Bois-de-Belleau, France,
    June 17, 1918.
   R—Salt Lake City, Utah.
   B—Marshfield, Mo.
   G. O. No. 101, W. D., 1918.

Private, 23d Company, 6th Machine Gun Battalion, U. S. Marine Corps, 2d Division.
In the Bois-de-Belleau, France, on June 17, 1918, he and a comrade left shelter and went 200 yards in the open under fire of the enemy and carried a wounded Infantry soldier back to his lines, thereby demonstrating heroic and voluntary disregard of self to save one who could not help himself.

*PRETTY, SAUNDERS P. (549279)_____
   At Launay, near Chateau-Thierry,
    France, July 15, 1918.
   R—Cincinnati, Ohio.
   B—Louisville, Ky.
   G. O. No. 24, W. D., 1920.

Private, first class, Machine Gun Company, 38th Infantry, 3d Division.
Private (First Class) Pretty fearlessly exposed himself to machine-gun and artillery fire in order to keep up communication between exposed machine-gun positions. His work enabled the quick and accurate changing of our fire during a critical period of the action.
Posthumously awarded. Medal presented to mother, Mrs. Lena Pretty.

PREVOST, PHILIP W. (2284906)_____
   Near Eclisfontaine, France, Sept.
    23, 1918.
   R—Geyser, Mont.
   B—Bay City, Mich.
   G. O. No. 64, W. D., 1919.

Private, first class, Company D, 364th Infantry, 91st Division.
A combat group had worked its way far ahead when the remainder of the line was held up by heavy bursts of machine-gun fire and the order to dig in and hold the position was given. Private Prevost volunteered to carry the message through heavy machine-gun fire to the combat group, which was still advancing. He delivered the order and returned with information which enabled the battalion to make dispositions for the capture of the line of enemy machine-gun nests and the saving of the combat group.

PRICE, EDWARD H_____
   Near Cheppy, France, Sept. 26, 1918.
   R—St. Louis, Mo.
   B—Louisville, Ky.
   G. O. No. 46, W. D., 1919.

First lieutenant, 138th Infantry, 35th Division.
At the head of his command he charged and captured a machine-gun nest defending the southeastern approach to Cheppy and was the first to enter the town. With his company of about 40 men he captured 4 guns and 154 prisoners. On the following day he led his men through direct artillery fire with only 2 casualties, due to his dexterity of command. Later he was seriously wounded, but continued with his company until physical exhaustion prevented his going farther.

PRICE, EDWARD J_____
   Near the Bois-de-Chaume, France,
    Oct. 11, 1918.
   R—Minneapolis, Minn.
   B—Duluth, Minn.
   G. O. No. 81, W. D., 1919.

First lieutenant, 124th Machine Gun Battalion, 33d Division.
Upon learning that a counterattack had been launched against the battalion on his right flank, Lieutenant Price took his platoon into action in advance of the infantry and broke up the counterattack. There being no officers present with the infantry unit to which he was attached, he assumed command, reorganized it, and led it forward, designating targets and ranges and going up and down the line to direct the operation.

PRICE, OTTO D. (1039937)_____
   Near Greves Farm, France, July 15,
    1918.
   R—Ada, Kans.
   B—Ada, Kans.
   G. O. No. 98, W. D., 1919.

Private, Battery F, 10th Field Artillery, 3d Division.
Responding to a call for volunteers, Private Price, with 8 other soldiers, manned 2 guns of a French battery which had been deserted by the French during the unprecedented fire after many casualties had been inflicted on their forces. For 2 hours he remained at his post and poured an effective fire into the ranks of the enemy.

PRICE, THOMAS F. (1560327)_____
   Near Hill 272, France, Oct. 11, 1918.
   R—Abingdon, Va.
   B—Asheville, N. C.
   G. O. No. 23, W. D., 1919.

Private, Company B, 16th Infantry, 1st Division.
He volunteered and led a patrol of 4 men against an enemy machine gun which was inflicting severe losses on his ranks. He successfully accomplished the silencing of the gun.

*PRIDDY, WELLBORN S_____
   Near Badonviller, France, May 26,
    1918.
   R—Chicago, Ill.
   B—Findlay, Ohio.
   G. O. No. 88, W. D., 1918.

Second lieutenant, 168th Infantry, 42d Division.
While in command of an important post near Badonviller, France, on May 26, 1918, he displayed courage, judgment, and devotion to duty in heroically defending his position against a large force of the enemy, continuing to perform his duty after having been badly gassed. Died May 29, 1918, as a result of the gas poisoning.
Posthumously awarded. Medal presented to mother, Mrs. Emerson Priddy.

PRIDE, HENRY N_____
   Near Bois-de-Chaume, France, Oct.
    10–12, 1918.
   R—Chicago, Ill.
   B—Blue Island, Ill.
   G. O. No. 59, W. D., 1919.

First lieutenant, 131st Infantry, 33d Division.
Acting on his own initiative, he led a patrol of 3 which penetrated the enemy line and after killing 3 Germans returned with 3 prisoners, 1 machine gun, and 1 automatic rifle. When the commander of the company on his left was killed, Lieutenant Pride assumed command and consolidated the position, repulsing 2 counterattacks in which the enemy lost 75 dead and wounded and 10 prisoners.

*PRIEST, CHARLES D_____
   Near Les Huit Chemins, France,
    Sept. 29, 1918.
   R—Estherville, Iowa.
   B—Weldon, Iowa.
   G. O. No. 37, W. D., 1919.

First lieutenant, chaplain, 353th Infantry, 90th Division.
He disregarded personal danger by going 600 yards beyond the front lines and with the aid of a soldier carried back a wounded man to shelter.
Posthumously awarded. Medal presented to widow, Mrs. Wilma Priest.

PRITCHARD, JAY C. (306803)............
  Near Thiaucourt, France, Sept. 12, 1918.
  R—Athens, Pa.
  B—Lawrenceville, Pa.
  G. O. No. 98, W. D., 1919.

Sergeant, Company D, 14th Machine Gun Battalion, 5th Division.
Though he was greatly fatigued from walking 30 kilometers in an effort to locate his company, from which he had become separated, Sergeant Pritchard organized a squad of men, who had also become separated from their organizations, and advancing with them for more than 9 kilometers, took up a position in advance of the infantry and directed effective machine-gun fire against the enemy, breaking up a hostile counterattack, until an enemy shell demolished the gun. Though he was severely wounded and suffering much pain, he conducted his squad back to Thiaucourt before permitting himself to be evacuated.

PROCTOR, HAROLD F. (210586)..........
  Near Bois-de-St. Remy, France, Sept. 12, 1918.
  R—Amesbury, Mass.
  B—Westfield, Mass.
  G. O. No. 26, W. D., 1919.

Private, Headquarters Troop, 26th Division.
Accompanied by another soldier, Private Proctor made his way far into the enemy lines to determine the location of an enemy emplacement which was holding up our advance. Having cut enemy telephone cables, he approached the nest from the rear and captured the entire personnel of the stronghold, consisting of 1 officer and 39 men.

PROUT, WILLIAM L. (55375)............
  Near Soissons, France, July 18–22, 1918.
  R—Catlettsburg, Ky.
  B—Montgomery, W. Va.
  G. O. No. 132, W. D., 1918.

Private, Machine Gun Company, 26th Infantry, 1st Division.
Throughout the 5 days of the attack near Soissons, France, July 18–22, 1918, he carried messages through artillery and machine-gun fire and took water and ammunition to his platoon over ground then under heavy bombardment.

PRUETTE, JOSEPH (100536)............
  East of Grand-Pre, France, Oct. 16, 1918.
  R—Iola. Ill.
  B—Iola, Ill.
  G. O. No. 13, W. D., 1919.

Corporal, Company C, 168th Infantry, 42d Division.
After a daring dash with his platoon across open ground swept by machine-gun fire, he saw an enemy machine-gun crew preparing to open fire upon the flank and rear of his position. Single handed he attacked, using enemy grenades, and drove the crew into a dugout. Bombing the entrance of the dugout, he effected the capture of 4 German officers, 64 men, and 4 heavy machine guns. With remarkable gallantry this soldier removed an obstacle that critically threatened a success already gained.

PRUITT, FRED C. (1317351)............
  Near Ypres, Belgium, Aug. 25, 1918.
  R—Mount Airy, N. C.
  B—Mount Airy, N. C.
  G. O. No. 32, W. D., 1919.

Sergeant, Headquarters Company, 119th Infantry, 30th Division.
At imminent peril to his own life, Sergeant Pruitt and two companions extinguished a fire in an ammunition dump, caused by a bursting shell, thereby preventing the explosion of the dump and saving the lives of a large number of men who were in the vicinity.

PRUITT, PINK S. (1309035)............
  Near Vaux-Andigny, Molain Ribeauville, France, Oct. 17, 1918.
  R—Rossville, Tenn.
  B—Rossville, Tenn.
  G. O. No. 46, W. D., 1919.

Private, first class, Company I, 117th Infantry, 30th Division.
At the starting of the attack Private Pruitt was painfully wounded in the arm. Disregarding his sufferings, he continued with the advance of his company, remaining with them throughout the day, until the objective was reached.

*PRYOR, JOHN P........................
  During the Meuse-Argonne offensive, near Exermont, France, Oct. 4–6, 1918.
  R—El Paso, Tex.
  B—Alexandria, Va.
  G. O. No. 15, W. D., 1923.

Captain, 2d Machine Gun Battalion, 1st Division.
Attacked by Spanish influenza he refused to leave his command for medical treatment because of the heavy casualties among the officers of the battalion to which his company was attached. Captain Pryor remained with his command, participating in the heavy fighting around Exermont and Hill 240, inspiring his men through his conspicuous bravery and indomitable will until he collapsed upon the field of battle, and was carried from the field, dying while en route to the hospital.
Posthumously awarded. Medal presented to aunt, Mrs. A. P. Krause.

PUCHAJDA, EDWARD (1245040)..........
  North of Coulandon, France, Sept. 6–7, 1918.
  R—Carnegie, Pa.
  B—Carnegie, Pa.
  G. O. No. 60, W. D., 1920.

Private, Company H, 111th Infantry, 28th Division.
Private Puchajda repeatedly exposed himself to heavy machine-gun fire in order to carry messages to his company commander across an area raked continuously by enemy fire. He accomplished his mission and maintained communication after several other runners had been killed in the attempt.

PULKER, HOWARD C. (254037)..........
  Near Suippes, France, July 14–15, 1918.
  R—Sharon, Pa.
  B—Pittsburgh, Pa.
  G. O. No. 143, W. D., 1918.

Private, Battery C, 42d Artillery, Coast Artillery Corps.
He, a chauffeur to whom no regular duty during the engagement had been assigned, voluntarily assisted in carrying wounded French and American soldiers to safety under severe bombardment. At one time he gave aid to a severely wounded soldier who was carrying a message to the battery commander, assisting him in performing his mission. When orders to withdraw were received he continued valiant services.

PULLEN, DANIEL D....................
  In the Bois-de-Cuisy, France, Sept. 26, 1918.
  R—Skagway, Alaska.
  B—La Push, Wash.
  G. O. No. 133, W. D., 1918.

Lieutenant colonel, Tank Corps.
Colonel Pullen displayed conspicuous gallantry and leadership in directing a tank attack on the Bois-de-Cuisy, after which he rallied a force of disorganized infantry, leading it forward in the face of violent machine-gun fire and occupying the ground which had been taken by the tanks.

PULLONO, CLEMENTE R. (2469867)......
  Near Cunel, France, Oct. 11, 1918.
  R—Pittsburgh, Pa.
  B—Italy.
  G. O. No. 81, W. D., 1919.

Private, Company C, 319th Infantry, 80th Division.
Seeing the enemy mounting a gun which when operated would sweep his platoon at close range, Private Pullono shot the gunner just as he was about to open fire. He also shot another German who attempted to fire the gun, after which he charged the position and captured the remainder of the crew.

PURCELL, HOMER (41882)...............
  Near Hill 272, north of Fleville, France, Oct. 11, 1918.
  R—Oklahoma City, Okla.
  B—Dallas, Tex.
  G. O. No. 35, W. D., 1920.

Sergeant, Company B, 16th Infantry, 1st Division.
After his platoon had suffered heavy casualties, Sergeant Purcell led the remnants up Hill 272 under heavy machine-gun fire and surrounded a heavy trench mortar which was defended by 2 machine guns. In the encounter he personally shot down the noncommissioned officer in charge of the enemy gun. 7 of the enemy crew were captured and 4 were killed.

*PURCELL, WARREN B. (1098636)_____
  Near Preny, France, Nov. 1, 1918.
  R—Weikert, Pa.
  B—Weikert, Pa.
  G. O. No. 87, W. D., 1919.

Sergeant, Company F, 56th Infantry, 7th Division.
Sergeant Purcell courageously led his half platoon against an enemy machine-gun nest, capturing it, killing several of the enemy, and taking 6 prisoners. He was later killed in action.
Posthumously awarded. Medal presented to mother, Mrs. Minnie Purcell.

PURDOM, THOMAS M. (2297238)_____
  Near Romagne, France, Oct. 9-11, 1918.
  R—Sparks, Ga.
  B—Sparks, Ga.
  G. O. No. 64, W. D., 1919.

Corporal, Company I, 125th Infantry, 32d Division.
Voluntarily assuming the duties of runner, after he had seen many others fail in attempting to get through the unusually heavy fire, Corporal Purdom repeatedly passed through the fire and aided materially in the success of the entire operations. When the supply of first-aid material had become exhausted he again went through, returning with sufficient bandages to care for the wounded, who could not, at that time, be removed.

*PURDY, ROBELL (1348297)_____
  Near Sedan, France, Nov. 7, 1918.
  R—Hanceville, Ala.
  B—Blount County, Ala.
  G. O. No. 44, W. D., 1919.

Private, Company K, 166th Infantry, 42d Division.
He was a member of a patrol sent out to silence machine-gun nests which were holding up the battalion's advance. When the officer leading the patrol fell mortally wounded, he attempted to go to the officer's assistance, despite heavy fire from machine guns only 100 yards away, and was himself killed.
Posthumously awarded. Medal presented to father, William T. Purdy.

*PURDY, WILLARD D. (273297)_____
  Near Hagenbach, Alsace, July 4, 1918.
  R—Marshfield, Wis.
  B—Shawano, Wis.
  G. O. No. 66, W. D., 1919.

Sergeant, Company A, 127th Infantry, 32d Division.
Upon returning with his patrol after a reconnaissance of the enemy's line, Sergeant Purdy was calling the roll of his men and collecting their hand grenades when the pin of one of the grenades became disengaged. Seeing that the grenade could not be thrown away without injuring some of the men, Sergeant Purdy called on them all to run, while he picked up 3 of the grenades, and bending over held them against his stomach. The grenade exploded, killing Sergeant Purdy instantly, but his presence of mind and self-sacrificing act saved the lives of his comrades.
Posthumously awarded. Medal presented to mother, Mrs. Esther Purdy.

PURRINGTON, ALDEN CLIFFORD_____
  Near Bois-d'Aigremont, France, July 15, 1918.
  R—Haydenville, Mass.
  B—Haydenville, Mass.
  G. O. No. 66, W. D., 1919.

Second lieutenant, 30th Infantry, 3d Division.
When the German barrage preceding their drive of July 15 was at its worst, he volunteered to go through the barrage in Bois-d'Aigremont to secure hand grenades for the defense of a wooded ravine after the forward grenade dump was blown up. This was at a time when it seemed impossible for any human being to get through the barrage. Throughout the entire engagement he volunteered to lead a number of patrols, both to the front and flanks.

PURSLEY, EARL (2170837)_____
  Near Lesseux, France, Sept. 4, 1918.
  R—Minneapolis, Minn.
  B—Hickman, Ky.
  G. O. No. 37, W. D., 1919.

Private, first class, medical detachment, 360th Infantry, 90th Division.
He voluntarily carried a wounded soldier from an exposed position under intense enemy shellfire for a distance of 400 yards to the dressing station. He then immediately returned to the position and helped to dig out men who had been buried by the explosion of a shell.

PUTMAN, HARRY P. (1204804)_____
  Near Ronssoy, France, Sept. 30, 1918.
  R—Fort Johnson, N. Y.
  B—Fort Johnson, N. Y.
  G. O. No. 143, W. D., 1918.

Private, Company H, 105th Infantry, 27th Division.
He exhibited exceptional bravery in voluntarily leaving shelter, going forward under heavy shell and machine-gun fire and bringing back to our lines several wounded comrades.

*PUTNAM, DAVID E_____
  Near Lachaussee, France, Sept. 12, 1918.
  R—Boston, Mass.
  B—Boston, Mass.
  G. O. No. 71, W. D., 1919.

First lieutenant, 139th Aero Squadron, Air Service.
After destroying 1 of the 8 German planes which had attacked him, he was turning to our lines when he saw 7 Fokkers attack an allied biplane. He attacked the Germans and saved the biplane, but was himself driven down, shot through the heart.
Posthumously awarded. Medal presented to mother, Mrs. F. H. Putnam.

*PYLES, ADAM H. (95180)_____
  Near St. Georges, France, Oct. 15, 1918.
  R—Lancaster, Ohio.
  B—Hocking County, Ohio.
  G. O. No. 27, W. D., 1919.

Private, first class, Company L, 166th Infantry, 42d Division.
Seeing his comrades either killed or wounded, immediately after seeking shelter, Private Pyles, undeterred, continually volunteered and carried messages over territory covered by violent artillery fire, incessant machine-gun fire, and accurate sniping until he was killed by this heavy fire.
Posthumously awarded. Medal presented to mother, Mrs. Elizabeth Pyles.

PYNE, PERCY RIVINGTON_____
  Near Dun-sur-Meuse, France, Oct. 23, 1918.
  R—New York, N. Y.
  B—Tuxedo Park, N. Y.
  G. O. No. 46, W. D., 1919.

First lieutenant, 103d Aero Squadron, Air Service.
While protecting 3 planes on a photographic mission, he attacked and drove off 5 enemy machines (type Fokker). Later another German formation of 7 (type Fokker) was encountered, but despite the odds Lieutenant Pyne swung up into the midst of the enemy and scattered them, diving on 1 of the Fokkers and sending it crashing to the ground.

PYRAH, GEORGE W. (660663)_____
  Near Sommerance, France, Oct. 15, 1918.
  R—Philadelphia, Pa.
  B—Philadelphia, Pa.
  G. O. No. 37, W. D., 1919.

Private, Company F, 117th Engineers, 42d Division.
Private Pyrah and 3 other soldiers were detailed to cut wire in advance of the Infantry during an attack on the enemy lines. While carrying out the mission they were fired upon by a machine gun at close range, which killed 1 and wounded another. He fired into the enemy nest, wounded the gunner, and disabled the gun; then charged the group with his bayonet and captured 3 prisoners.

QUICK, ARTHUR H. (125229)_____
  At Villers-Tournelle, Cantigny Sector, France, May 1, 1918.
  R—Kansas City, Kans.
  B—Kansas City, Kans.
  G. O. No. 100, W. D., 1918.

Corporal, Headquarters Company, 6th Field Artillery, 1st Division.
He displayed distinguished bravery in leaving his shelter during a heavy bombardment and going to the assistance of a wounded man who was lying exposed in the open.

QUICK, JOHN HENRY..............
At Bouresches, France, June 6, 1918.
R—Charles Town, W. Va.
B—Charles Town, W. Va.
G. O. No. 119, W. D., 1918.

Sergeant major, Headquarters Company, 6th Regiment, U. S. Marine Corps, 2d Division.
He volunteered and assisted in taking a truck load of ammunition and material into Bouresches, France, over a road swept by artillery and machine-gun fire, thereby relieving a critical situation.

QUINN, HENRY E. (57664)..............
At Cantigny, France, May 29, 1918.
R—Swartz, La.
B—Anniston, Ala.
G. O. No. 99, W. D., 1918.

Private, Company F, 28th Infantry, 1st Division.
In response to a call for volunteers to penetrate a heavy enemy barrage and obtain definite information concerning tanks and conditions of enemy front line, he accomplished his mission to the imminent peril of his life.

QUINN, JAMES H. (1709548)..............
Near Revillon, France, Sept. 10, 1918.
R—New York, N. Y.
B—Tarrytown, N. Y.
G. O. No. 32, W. D., 1919.

Sergeant, Company I, 308th Infantry, 77th Division.
He volunteered and accompanied a patrol which was sent out for the purpose of capturing prisoners. Crawling through no man's land, he came upon 2 Germans occupying an outpost. In the struggle that ensued the enemy was overpowered, but the exploit brought forth a destructive fire of rifles and rifle grenades from the enemy, through which Sergeant Quinn successfully maneuvered back to his lines, bringing his captives with him.

QUINN, JIM..............
Near Soissons, France, July 18, 1918.
R—Memphis, Tenn.
B—Mayfield, Ky.
G. O. No. 100, W. D., 1918.

Second lieutenant, 28th Infantry, 1st Division.
With a small platoon he attacked and captured a fortified French farmhouse in an open field. He so courageously and skillfully handled his men that this German strong point, held by 100 men and 5 machine guns, was promptly captured.

QUINN, JOHN (1039034)..............
Near Greves Farm, France, July 15, 1918.
R—Chicago, Ill.
B—Chicago, Ill.
G. O. No. 98, W. D., 1919.

Sergeant, Battery F, 10th Field Artillery, 3d Division.
Responding to a call for volunteers, Sergeant Quinn, with 8 other soldiers, manned 2 guns of a French battery which had been deserted by the French during the unprecedented fire after many casualties had been inflicted on their forces. For 2 hours he remained at his post and poured an effective fire into the ranks of the enemy.

QUINN, JOHN J..............
Near Bantheville, France, Oct. 23, 1918.
R—Annapolis, Md.
B—Baltimore, Md.
G. O. No. 46, W. D., 1919.

First lieutenant, 139th Aero Squadron, Air Service.
While patrolling the lines Lieutenant Quinn, with 1 other pilot, sighted and attacked 4 enemy machines (type Fokker). Several additional enemy planes joined the first 4 and, notwithstanding his great odds, he sent 1 machine crashing to the earth. Motor trouble forced him to drive straight through the enemy formation, and although followed and his machine badly damaged, he was able to outmaneuver and escape from his adversaries.

*QUIRI, ROBERT (1762908)..............
Near Thiaucourt, France, Sept. 18, 1918.
R—Syracuse, N. Y.
B—Amsterdam, N. Y.
G. O. No. 37, W. D., 1919.

Sergeant, Company F, 310th Infantry, 78th Division.
Leading a patrol under heavy fire, he was able to protect an unprotected portion of his line until the next unit could take it over. He was at all times an inspiration to his men, continually exposing himself to danger, while assuring their safety and comfort. After having his legs blown off and receiving other wounds, he gave all necessary information to his successor before allowing himself to be carried away. He died soon after reaching the dressing station. Posthumously awarded. Medal presented to father, Charles A. Quiri.

RABINOWITZ, ISAAC (2671573)..............
Near St. Souplet, France, Oct. 18, 1918.
R—New York, N. Y.
B—Russia.
G. O. No. 98, W. D., 1919.

Private, first class, Company A, 107th Infantry, 27th Division.
When the advance of his battalion was checked by heavy machine-gun fire Private Rabinowitz, with 2 other soldiers, went forward under heavy fire to reconnoiter the enemy positions. By effective rifle fire they drove the gunners from 2 machine-gun nests into a dugout near by, which they captured, together with 35 prisoners, including 3 officers.

RABORN, JOHN (540752)..............
Near Bois-de-Belleau, France, June 21, 1918.
R—Augusta, Ga.
B—Spread, Ga.
G. O. No. 81, W. D., 1919.

Private, Company A, 7th Infantry, 3d Division.
While making his way to a first-aid station, after being wounded in 5 places by machine-gun bullets and shrapnel, he encountered 2 Germans. His own rifle having been shot away, Private Raborn picked up a rifle lying near by and shot 1 of the Germans, and in a bayonet duel with the other German killed him also, after he himself had been wounded again during the encounter.

RACHEK, JOHN..............
At Trugny, France, July 22, 1918.
R—Brooklyn, N. Y.
B—Austria.
G. O. No. 130 W. D., 1918.

First lieutenant, 104th Infantry, 26th Division.
Although he had been twice wounded, he refused to be evacuated, and continued on duty with his company during the attack and capture of Trugny under heavy fire until he was incapacitated by a third wound.

*RADEVICK, RADOVAN (2338005)..............
Near Grand Ballois Farm, France, July 14-15, 1918.
R—Cleveland, Ohio.
B—Montenegro.
G. O. No. 32, W. D., 1919.

Private, Company A, 4th Infantry, 3d Division.
During an intense shell and gas bombardment, he repeatedly volunteered and carried messages and assisted in caring for the wounded at great personal risk. Posthumously awarded. Medal presented to father, Gavio Radevick.

RAFALSKY, NISEL (1286789)..............
Near Verdun, France, Oct. 9-12, 1918.
R—Baltimore, Md.
B—Russia.
G. O. No. 2, W. D., 1919.

Sergeant, sanitary detachment, 115th Infantry, 29th Division.
For extraordinary heroism in action near Verdun during the drive in which this regiment took part in the vicinity of the Meuse. He displayed great courage and presence of mind in attending to the wounded, not only of the organization to which he was attached, but also of those in adjoining organizations. The exceptionally valuable service performed by this soldier was done under heavy shell and machine-gun fire.

RAFFINGTON, CHARLES S. (5385)..............
At Lucy, France, June 2-3, 1918.
R—Hutchinson, Kans.
B—Phillipsburg, Kans.
G. O. No. 99, W. D., 1918.

Private, medical detachment, 2d Engineers, 2d Division.
During the day and night of June 2-3, 1918, he exposed himself to severe and continuous fire beyond the call of duty in order to bring aid to wounded engineers and marines.

RAFTER, EDWIN J. (1205410)............
   East of Ronssoy, France, Sept. 27, 1918.
   R—New York, N. Y.
   B—New York, N. Y.
   G. O. No. 13, W. D., 1923.

Corporal, Company L, 105th Infantry, 27th Division.
This soldier while in command of a flank detachment of which all the members except himself were killed or wounded remained alone at his post during the entire night, and by his constant rifle fire defeated several attempts of enemy groups to develop an attack on the left of his organization, thus enabling his company to reorganize and consolidate their position. His indomitable courage and devotion to duty set an inspiring example to the men of his company and battalion.

*RAGSDALE, IRVING LE NOIS............
   Near Beaufort, France, Nov. 4, 1918, and near Laneuville, France, Nov. 6, 1918.
   R—Portland, Oreg.
   B—Carthage, Mo.
   G. O. No. 35, W. D., 1919.

Second lieutenant, 356th Infantry, 89th Division.
Advancing across open ground under intense machine-gun and artillery fire on Nov. 4, 1918, Lieutenant Ragsdale killed an enemy machine gunner with his automatic pistol. Later he again crossed an open field under terrific machine-gun fire, killing 2 enemy gunners who were retarding the advance. On Nov. 6, during an intense shell and gas bombardment of Laneuville, he repeatedly exposed himself while assisting the wounded.
Posthumously awarded. Medal presented to widow, Mrs. Reina V. Ragsdale.

RAIBLE, JOSEPH C., Jr.............
   Near Chateau-Thierry, France, July 5, 1918.
   R—Hannibal, Mo.
   B—Hannibal, Mo.
   G. O. No. 121, W. D., 1918.

First lieutenant, 147th Aero Squadron, Air Service.
Lieutenant Raible and 3 other pilots, at an altitude of 4,700 meters, attacked an enemy formation of 8 battle planes flying at an altitude of 5,000 meters. The German machines dived on them and Lieutenant Raible engaged 2 in combat. In a hard fight, lasting 5 minutes and finishing at an altitude of 3,000 meters, he shot down 1 of the attacking party and drove off the other.

RAIKOVICH, MATO (45749)............
   South of Soissons, France, July 21, 1918.
   R—Woodlawn, Pa.
   B—Austria.
   G. O. No. 3, W. D., 1921.

Private, Machine Gun Company, 18th Infantry, 1st Division.
Private Raikovich assisted in carrying a machine gun forward through intense enemy machine-gun and artillery fire. Placing his machine gun in action, he delivered an effective fire which caused the enemy to abandon the counterattack.

RAINES, LESTER (2430050)............
   Near Grand Pre, France, Oct. 29, Nov. 2, 1918.
   R—Akron, Ohio.
   B—Ryan, W. Va.
   G. O. No. 35, W. D., 1919.

Private, Company C, 309th Machine Gun Battalion, 78th Division.
During the 5 days of operations he worked without hesitation carrying messages through constant shellfire and acting as guide for ration parties, his information at all times proving most valuable and accurate.

RALEIGH, WALTER J. (258303)..........
   Near Juvigny, north of Soissons, France, Sept. 4, 1918.
   R—Menasha, Wis.
   B—Menasha, Wis.
   G. O. No. 137, W. D., 1918.

Private, first class, Company A, 107th Ammunition Train.
During a heavy enemy bombardment a shell burst near 2 ammunition trucks that were being unloaded at a dump, blowing up 1 truck and setting fire to the other. Disregarding the warnings of bystanders, Private Raleigh rushed forward, cranked the engine of the burning truck and assisted in backing it to a place of safety while others extinguished the fire. His conspicuous bravery was the means of saving a large quantity of ammunition.

RALSTON, ORVILLE A................
   Over Bourlon Wood, Sept. 26, 1918.
   R—Avoca, Nebr.
   B—Weeping Water, Nebr.
   G. O. No. 38, W. D., 1921.

First lieutenant, 148th Aero Squadron, Air Service.
Having engine trouble, he signaled his flight commander, left his formation, and started for the lines. Shortly afterwards his engine picked up and he decided to rejoin his formation. He found 3 of them engaged with 7 Fokker biplanes over Bourlon Wood. Seeing that 1 of our machines was hard pressed and in distress, Lieutenant Ralston instantly went to its assistance and drove 1 Fokker down into the clouds below. He followed directly behind the enemy machine and, as they came out of the clouds at a height of 3,000 feet, opened fire again on this Fokker at 15 yards range. The enemy machine made 1 complete spiral and crashed northeast of Bourlon Wood. Four more Fokkers now attacked Lieutenant Ralston, but he managed to get back in the clouds and return safely to our lines, as did the rest of his flight.

RALSTON, SAMUEL J. (57507)..........
   At Cantigny, France, May 28, 1918.
   R—Philadelphia, Pa.
   B—Philadelphia, Pa.
   G. O. No. 13, W. D., 1923.

First sergeant, Company F, 28th Infantry, 1st Division.
He assumed command of a platoon of his company, efficiently leading it in the initial attack and inspiring his men to reach their objective, where they dug in. 2 runners having been wounded in an attempt to carry back a message calling for the lengthening of our artillery fire, he obtained permission to leave his platoon, volunteered to carry the message, and successfully delivered it to his regimental commander through a heavy German counter-barrage, the passage of the American barrage, and continuous and intense machine-gun fire, by which he was wounded. He repeatedly refused to be evacuated to the rear, though a serious counterattack was in progress, remaining at regimental headquarters until his recovery.

RAMPSCH, JOHN (43095)............
   Near Soissons, France, July 19, 1918.
   R—Chicago, Ill.
   B—Russia.
   G. O. No. 44, W. D., 1919.

Sergeant, Company G, 16th Infantry, 1st Division.
During offensive operations near Soissons he was severely wounded in the neck, but continued to lead his platoon until forced to retire from loss of blood.

RAMSDELL, RALPH L. (67635)..........
   At Marcheville, France, Sept. 26, 1918.
   R—Waterboro, Me.
   B—Hiram, Me.
   G. O. No. 1, W. D., 1926.

Corporal, Company D, 103d Machine Gun Battalion, 26th Division.
Under terrific artillery and machine-gun fire he displayed exceptional bravery in locating machine-gun nests.

RAMSEY, EARL E. (2176548)
Near the Barricourt Woods, France, Nov. 2, 1918.
R—Cedar Vale, Kans.
B—Cedar Vale, Kans.
G. O. No. 98, W. D., 1919.

Sergeant, Company G, 353d Infantry, 89th Division.
Sergeant Ramsey was leading the first section of a combat platoon when it encountered machine-gun fire of such intensity that the entire advance was threatened. Realizing the gravity of the situation, he stepped out into the open ahead of his men in order to direct them more effectively, inspiring them by his bravery. He was seriously wounded a few minutes later.
Posthumously awarded. Medal presented to father, Tillman Howard Ramsey.

RAMSEY, HENRY (2260593)
During the Meuse-Argonne offensive, France, Sept. 26-29, 1918.
R—New Bedford, Mass.
B—Virgie, Ky.
G. O. Nos. 2 and 46, W. D., 1919.

First sergeant, Company F, 362d Infantry, 91st Division.
For 3 days he kept his men well organized, and when he was gassed and severely wounded in the chest, insisted that other men more seriously wounded than he be removed from the field before he would permit anyone to assist him to the dressing station.
Oak-leaf cluster.
For extraordinary heorism in action near the Scheldt River, Belgium, Oct. 31, 1918, Sergeant Ramsey is awarded an oak-leaf cluster, to be worn with the distinguished-service cross previously awarded, which award is published in General Orders, No. 2, War Department, 1919: When the advance of the front line was held up by fire from a machine-gun nest 300 yards to the front, Sergeant Ramsey, with 2 others, crossed the open field in the face of fire from enemy artillery, machine guns, and snipers. Charging the nest, they killed 2 of the crew, wounded 2 others, and captured 5, together with the gun.

RAMSEY, JAMES R. (1249422)
Near Apremont, France, Oct. 2, 1918.
R—Grove City, Pa.
B—Butler Co., Pa.
G. O. No. 11, W. D., 1921.

Corporal, Company M, 112th Infantry, 28th Division.
During the attack east of Chene Tondu, when his organization was held up by a strong counterattack, Corporal Ramsey, in command of an automatic rifle squad, kept up harassing fire on the enemy from a position which was under heavy enemy fire. 4 members of his squad were disabled and 1 gun put out of action, but he, single handed, kept up a constant fire with the 1 remaining gun. His heroic conduct was a material factor in the successful repulse of the enemy.

RANCOURT, JOHN I.
Near Fismes, France, Aug. 9, 1918.
R—Providence, R. I.
B—Port Henry, N. Y.
G. O. No. 121, W. D., 1918.

First lieutenant, 103d Field Artillery, observer, 88th Aero Squadron, Air Service.
Richard C. M. Page, first lieutenant, pilot, Air Service. John I. Rancourt, first lieutenant, observer, 88th Aero Squadron, 103d Field Artillery. For extraordinary heorism in action near Fismes, France, Aug. 9, 1918. These officers were detailed to fly without escort on a visual reconnaissance over the enemy's lines. They were attacked by 6 enemy battle planes 1,800 meters over Fismes. The Americans unhesitatingly fought this superior number of the enemy. Lieutenant Rancourt was 3 times seriously wounded in the legs above the knees, yet he continued to operate his machine gun and shot down 1 of the enemy planes. In spite of the fact that his elevator controls on one side had been shot away, Lieutenant Page skillfully maneuvered the plane throughout the combat and piloted it safely back to his airdrome.

RAND, GEORGE E.
In Bois-de-Barricourt, France, Nov. 2, 1918.
R—Vasselboro, Me.
B—Detroit, Me.
G. O. No. 37, W. D., 1919.

First lieutenant, 353d Infantry, 89th Division.
He led his platoon in the face of terrific machine-gun fire, capturing 4 machine guns and their crews, thereby facilitating the advance of the command. Over half of his men were either killed or wounded before reaching the objective.

RANDALL, MOOD A. (1319329)
Near Catillon, France, Oct. 19, 1918.
R—Memphis, Tenn.
B—Vaiden, Miss.
G. O. No. 98, W. D., 1919.

Sergeant, Company A, 120th Infantry, 30th Division.
When the advance was held up by enemy machine-gun fire, Sergeant Randall volunteered and led his platoon, which consisted of only 8 men, under heavy enemy fire, in a successful attack on a machine-gun nest, outflanking and putting same out of action, thereby enabling his company to continue advancing.

RANDALL, SAMUEL J. (1215857)
Near St. Souplet, France, Oct. 15, 1918.
R—Penn Yan, N. Y.
B—Penn Yan, N. Y.
G. O. No. 37, W. D., 1919.

Private, Company L, 108th Infantry, 27th Division.
Accompanied by an officer and 3 other soldiers, he made a reconnaissance of the River La Selle, the journey being made under constant heavy machine-gun fire. To secure the desired information it was necessary to wade the stream for the entire distance.

RANDLES, HAROLD J.
In the Bois-de-Belleau, France, June 6, 1918.
R—Rochester, N. Y.
B—Rochester, N. Y.
G. O. No. 119, W. D., 1918.

Corporal, 80th Company, 6th Regiment, U. S. Marine Corps, 2d Division.
In delivering messages, he voluntarily chose the most direct route, although it was through a machine-gun barrage, to deliver information which prevented the bombardment of positions that had just been occupied. He took the path of danger to save his comrades.

RANSOM, WARREN A.
Near Missy-aux-Bois, Chaudon, France, July 18, 1918.
R—New York, N. Y.
B—New York, N. Y.
G. O. No. 14, W. D., 1923.

Second lieutenant, 6th Field Artillery, 1st Division.
He accompanied Maj. John A. Crane, Field Artillery, on a reconnaissance of the enemy's position under intense enemy rifle, machine-gun, and artillery fire; the mission accomplished, Major Crane was seriously wounded while some distance in rear of Lieutenant Ransom and in plain view of the enemy gunners. Lieutenant Ransom returned at once to Major Crane, carried him to a place of comparative shelter, and, despite the heavy enemy fire, sought and found a medical officer, whom he led to the wounded officer. Again leaving shelter he secured a litter and with the assistance of two French soldiers carried the wounded officer to a place of safety.

RANSON, JOHN O.
At Ardeuil, France, Sept. 30, 1918.
R—Huntersville, N. C.
B—Huntersville, N. C.
G. O. No. 21, W. D., 1925.

First lieutenant, 371st Infantry, 93d Division.
When his company was held up by an enemy machine-gun nest Lieutenant Ranson volunteered and led his platoon in an attack on the position and while attempting to carry out his mission was killed.
Posthumously awarded. Medal presented to widow, Mrs. John O. Ranson.

*RAPP, FRED N. (572451) _____

In the Bois-du-Fays, France, Oct. 6, 1918.

R—Alexandria, S. Dak.

B—Clarke County, Iowa.

G. O. No. 44, W. D., 1919.

Corporal, Machine Gun Company, 59th Infantry, 4th Division.

While exposed to an exceptionally heavy barrage in the Bois-du-Fays, Corporal Rapp left his shelter and went to the aid of a seriously wounded comrade. He was killed by a fragment from a high-explosive shell while in the performance of this gallant mission.

Posthumously awarded. Medal presented to mother, Mrs. Sarah C. Rapp.

RAPPORT, GEORGE D. (1204528) _____

Near St. Souplet, France, Oct. 17, 1918.

R—Gloversville, N. Y.

B—Gloversville, N. Y.

G. O. No. 16, W. D., 1923.

Sergeant, Company G, 105th Infantry, 27th Division.

With complete disregard for his own safety he attacked single handed 2 enemy machine-gun nests, killing or capturing the crews. The gallantry and devotion to duty thus displayed greatly inspired the men of his regiment.

RASCOE, ROBERT R. (1322305) _____

Near Becquigny, France, Oct. 10, 1918.

R—Reidsville, N. C.

B—Rockingham County, N. C.

G. O. No. 26, W. D., 1919.

Sergeant, sanitary detachment, 120th Infantry, 30th Division.

Going forward to establish an aid post, Sergeant Rascoe, finding that the advance had already started, took his position in the front line, and, exposed to terrific fire, cared for the wounded until the medical department was brought up. Later, while bringing up rations, he encountered shellfire and, although wounded and knocked down, he quickly regained his feet and completed his mission.

*RASMUSSEN, AXEL _____

At Rocquencourt, France, May 4, 1918.

R—Sherwood, Oreg.

B—Denmark.

G. O. No. 83, W. D., 1918.

Major, 28th Infantry, 1st Division.

He proceeded to his post of command in spite of heavy bombardment in order to save important papers, and while thus engaged was killed by shellfire May 4, 1918.

Posthumously awarded. Medal presented to father, Severin Rasmussen.

RASSMUSSEN, CARL (2250434) _____

Near the Bois-de-Brieulles, France, Sept. 27, 1918.

R—Edinburg, Tex.

B—Louisa, Iowa.

G. O. No. 98, W. D., 1919.

Private, Company B, 39th Infantry, 4th Division.

Private Rassmussen, a company runner, volunteered and made 2 trips from the post of command of his own regiment to that of the regiment adjoining his own, passing each time more than a thousand yards under intense enemy machine-gun fire.

RATENBURG, HERBERT (2980) _____

North of Chateau-Thierry, France, July 22, 1918.

R—Manchester, Conn.

B—Manchester, Conn.

G. O. No. 125, W. D., 1918.

Sergeant, sanitary detachment, 101st Machine Gun Battalion, 26th Division.

Although wounded in 3 places by machine-gun bullets, he followed the attack and continued his duty, thereby inspiring his comrades.

RATH, HOWARD G _____

Between Chambley and Xammes, France, Sept. 13, 1918.

R—Los Angeles, Calif.

B—Ackley, Iowa.

G. O. No. 123, W. D., 1918.

Second lieutenant, observer, 96th Aero Squadron, Air Service.

While acting as leading observer of a flight of 3 planes he was attacked by 15 enemy planes. In spite of the fact that his formation was surrounded by an enemy five times as large, he carried out successfully his mission and bombed his objective. In the return running fight he and his pilot continued the unequal fight and succeeded in returning to their airdrome with valuable information.

RATKOVICH, PETER (1653640) _____

Near Varennes, France, Sept. 26, 1918.

R—Oakland, Calif.

B—Austria.

G. O. No. 71, W. D., 1919.

Private, Company C, 110th Infantry, 28th Division.

With 2 other soldiers, he, on his own initiative, led a charge on an enemy machine-gun nest, and, although severely wounded, pressed forward, wounding 3 and capturing 3 of the enemy.

RAWLINSON, JOHN W. (2246030) _____

At Quinnemout, France, near Ronssoy, France, Sept. 28, 1918.

R—Kingsville, Tex.

B—Corpus Christi, Tex.

G. O. No. 68, W. D., 1920.

Private, Company K, 106th Infantry, 27th Division.

After assisting in repulsing a strong enemy counterattack, Private Rawlinson, with 2 other soldiers, became separated from his company, due to the heavy fog. Seeing a superior force of the enemy in a trench, they unhesitatingly attacked, and after killing and wounding several of the enemy, they captured numerous prisoners and brought them back to our lines.

*RAY, JOHN _____

Near Bellicourt, France, Sept. 29, 1918.

R—Raleigh, N. C.

B—Hendersonville, N. C.

G. O. No. 37, W. D., 1919.

Captain, 119th Infantry, 30th Division.

Establishing his first-aid station in the front line, he advanced with the Infantry. He continued on with the troops, caring for the wounded until he himself was so badly wounded that he was evacuated. He died from his wounds a few days later.

Posthumously awarded. Medal presented to mother, Mrs. John E. Ray.

RAY, LEE M. (556193) _____

Near St. Thibaut, France, Aug. 5, 1918.

R—Philadelphia, Pa.

B—Philadelphia, Pa.

G. O. No. 44, W. D., 1919.

Corporal, 39th Infantry, 4th Division.

Corporal Ray, clerk of headquarters, volunteered and delivered important operations messages to the French regiments attacking on the left flank of the 39th Infantry. He made his way for about 1½ miles through heavy artillery, machine-gun, and sniping fire parallel to the enemy's line, located the French headquarters, and delivered the message in time to stop flanking attacks by the enemy.

RAYKMAN, ROY (145213) _____

Near the Cote-de-Chatillon, France, Oct. 26, 1918.

R—Clam Falls, Wis.

B—South Chicago, Ill.

G. O. No. 37, W. D., 1919.

Sergeant, Battery A, 151st Field Artillery, 42d Division.

During a heavy bombardment of his ammunition train, when 1 man and 12 horses were killed and several men wounded, he displayed coolness and quick judgment in cutting loose the dead horses and straightening out the train. He then returned through the shell fire and searched in the darkness until he found all the wounded.

RAYMOND, FRANK H. (281485)..........
Near Romagne, France, Oct. 14, 1918.
R—Freemont, Mich.
B—Kent City, Mich.
G. O. No. 21, W. D., 1919.

Corporal, Company M, 126th Infantry, 32d Division.
In an attack on Cote Dame Marie, the 126th Infantry was held up, owing to intense machine-gun fire and grenades. Corporal Raymond volunteered as a member of a combat patrol which cut through the enemy lines, captured 10 machine guns, killed and captured 15 of the enemy, and forced others to surrender. They cleared that part of the Cote Dame Marie of the enemy, enabling the regiment to continue their advance.

RAYMOND, ROBERT FULTON..........
Near Chateau-Thierry, France, June 24, 1918.
R—Newton Center, Mass.
B—New Bedford, Mass.
G. O. No. 121, W. D., 1918.

First lieutenant, 27th Aero Squadron, Air Service.
He piloted 1 machine in a formation of 3 which was escorting 3 reconnaissance planes over enemy territory. On account of motor trouble, he was unable to keep up with his companions, and while thus detached was attacked by an enemy machine. In spite of the condition of his engine and his presence far within the German lines, he vigorously attacked the German plane and destroyed it, after which he succeeded in rejoining his patrol.

RAYNER, IRA C. (2131428)..........
Near Nonsard, France, Sept. 12, 1918.
R—Hollandale, Miss.
B—Durant, Miss.
G. O. No. 37, W. D., 1919.

Sergeant, Company G, 28th Infantry, 1st Division.
Although under fire for the first time and very ill with a high fever, Sergeant Rayner took charge of the platoon, after the platoon leader had been killed, and effectively directed it for 2 days, in spite of his physical condition, refusing to be evacuated until in a state of collapse.

REA, LEONARD E..........
Near Blanc Mont, France, Oct. 4, 1918.
R—Auburn, N. Y.
B—Auburn, N. Y.
G. O. No. 35, W. D., 1919.

Second lieutenant, 5th Regiment, U. S. Marine Corps, 2d Division.
He retained command of his platoon after receiving a severe wound, which rendered him unable to move without assistance, and would not leave the line until ordered by his commanding officer.

REACH, HARRY B. (1241545)..........
Near Varennes, France, Sept. 27, 1918.
R—Penns Grove, N. J.
B—Philadelphia, Pa.
G. O. No. 98, W. D., 1919.

Private, Company K, 110th Infantry, 28th Division.
Acting as a company runner, Private Reach voluntarily carried numerous messages under heavy machine-gun fire, displaying marked courage and devotion to duty.

*READ, JOHN J. (1784715)..........
Near Molleville Farm, north of Verdun, France, Nov. 5, 1918.
R—Philadelphia, Pa.
B—Philadelphia, Pa.
G. O. No. 72, W. D., 1920.

Sergeant, Company C, 315th Infantry, 79th Division.
Sergeant Read led the advance of his platoon through heavy enemy fire in the attack against a strongly held enemy position. Although cut off from his company he continued to advance until mortally wounded. Prior to the performance of the above act he distinguished himself by volunteering to conduct and conducting ration details over routes exposed to heavy artillery fire.
Posthumously awarded. Medal presented to mother, Mrs. Emma Mary Read.

READ, ROSS E. (155611)..........
Near Cantigny, France, May 28, 1918.
R—Portland, Oreg.
B—Turner, Oreg.
G. O. No. 32, W. D., 1919.

Private, Company D, 1st Engineers, 1st Division.
He voluntarily went forward over an area swept by machine-gun fire to the aid of a wounded comrade who was entangled in barbed wire. He worked in a perilously exposed position until he extricated his companion and carried him to safety.

REAM, BERTRAM LEE (108607)..........
In the Bois-de-Belleau, France, June 17, 1918.
R—Elizabethtown, Pa.
B—Elizabethtown, Pa.
G. O. No. 101, W. D., 1918.

Private, Company B, 6th Machine Gun Battalion, U. S. Marine Corps, 2d Division.
He and a comrade left shelter and went 200 yards in the open under fire of the enemy and carried a wounded Infantry soldier back to his lines, thereby demonstrating heroic and voluntary disregard of self to save one who could not help himself.

*REATH, THOMAS R..........
Near Belleau Wood, France, June 11, 1918.
R—Philadelphia, Pa.
B—Riverton, N. J.
G. O. No. 16, W. D., 1923.

Sergeant, 43d Company, 5th Regiment, U. S. Marine Corps, 2d Division.
During the advance of the 43d Company of Marines Sergeant Reath, with great coolness and devotion to duty, attacked an enemy machine-gun nest killing 3 of the enemy and capturing the 2 remaining members of the gun crew, thus enabling his company to continue the advance. This heroic deed was performed by Sergeant Reath under intense enemy machine-gun fire, and greatly inspired the members of his company.
Posthumously awarded. Medal presented to father, Theodore W. Reath.

RECKTENWALD, JACOB (1784788)..........
Near Gibercy, France, Nov. 7-10, 1918.
R—Philadelphia, Pa.
B—Philadelphia, Pa.
G. O. No. 37, W. D., 1919.

Sergeant, Company C, 315th Infantry, 79th Division.
On the night of Nov. 7 he risked his own life in heavy artillery fire, going from shell hole to shell hole helping his wounded comrades. On Nov. 10 he distributed rations to the men of his company under shell and machine-gun fire.

RED, HAROLD D. (1375415)..........
Near the Bois-de-Bantheville, France, Oct. 30, 1918.
R—Chicago, Ill.
B—Abilene, Tex.
G. O. No. 44, W. D., 1919.

Corporal, Battery A, 122d Field Artillery, 33d Division.
Under heavy shellfire he crawled 200 meters to a shell hole in order to draw a sketch of the enemy's position.

REDEKER, PAUL W. (263946)..........
Near Sergy, northeast of Chateau-Thierry, France, July 31, 1918.
R—Manistique, Mich.
B—Manistique, Mich.
G. O. No. 117, W. D., 1918.

Corporal, Company M, 125th Infantry, 32d Division.
He twice volunteered to carry messages from company headquarters to the battalion post of command through heavy machine-gun fire and artillery barrage. He assisted in gathering the elements of the company together after the assault. He volunteered for every dangerous duty and in broad daylight, in full sight of the enemy, dragged wounded to places of shelter.

REDICK, FRED C.
  Near Montfaucon, France, Sept. 26, 1918.
  R—Wooster, Ohio.
  B—Wooster, Ohio.
  G. O. No. 59, W. D. 1919.

Captain, 146th Infantry, 37th Division.
Severely wounded in the head and leg while leading his company, he refused to go to the rear, though he was ordered to do so by the battalion commander and attending surgeon, continuing in the attack and inspiring his men by his conspicuous bravery.

REDWOOD, GEORGE B.
  At Seicheprey, France, Mar. 28, 1918, and at Cantigny, France, May 29, 1918.
  R—Baltimore, Md.
  B—Baltimore, Md.
  G. O. No. 27, W. D., 1919.

First lieutenant, 28th Infantry, 1st Division.
With great daring he led a patrol of our men into a dangerous portion of the enemy trenches, where the patrol surrounded a party nearly double their own strength, captured a greater number than themselves, drove off an enemy rescuing party, and made their way back to our lines with 4 prisoners, from whom valuable information was taken.
Oak-leaf cluster.
He is also awarded an oak-leaf cluster, to be worn with the distinguished-service cross, for the following act of extraordinary heroism: At Cantigny, France, May 29, 1918, he conducted himself fearlessly to obtain information of the enemy's action. Although wounded, he volunteered to reconnoiter the enemy's line, which was reported to be under consolidation. While making a sketch of the German position on this mission he was under heavy fire, and continued his work after being fatally wounded until it was completed. The injuries sustained at this time caused his death.
Posthumously awarded. Medal presented to mother, Mrs. Francis T. Redwood.

REECE, BRAZILLA CARROLL
  In the Bois-d'Ormont, France, Oct. 23–28, 1918.
  R—Butler, Tenn.
  B—Butler, Tenn.
  G. O. No. 46, W. D., 1919.
  Distinguished-service medal also awarded.

First lieutenant, 102d Infantry, 26th Division.
In leading his company through 4 successful actions he was twice thrown violently to the ground and rendered unconscious by bursting shells, but upon recovering consciousness he immediately reorganized his scattered command and consolidated his position. On several occasions, under heavy enemy machine-gun fire, he crawled far in advance of his front line and rescued wounded men who had taken refuge in shell holes.

REED, ALBERT J. (140281)
  Near Juvigny, north of Soissons, France, Aug. 30, 1918.
  R—Davis, Calif.
  B—San Francisco, Calif.
  G. O. No. 20, W. D., 1919.

Corporal, Headquarters Company, 147th Field Artillery, 32d Division.
While stationed in an observation post which was heavily bombarded with gas and high-explosive shells he assisted in carrying to the rear through this heavy fire another member of the party who was seriously wounded, it being possible to proceed only by going from one shell hole to another. After accomplishing this mission he returned to his post of duty under the same severe fire.

REED, CECIL E. (2261115)
  Near Barricourt, France, Nov. 2, 1918.
  R—Stratton, Nebr.
  B—Waverly, Nebr.
  G. O. No. 66, W. D., 1919.

Private, first class, Company E, 353d Infantry, 89th Division.
When the advance of his platoon was held up by severe machine-gun fire, Private Reed left cover, advanced across open ground, and opened fire on the enemy nest with rifle grenades. After twice returning to obtain more grenades, he succeeded with a well-directed shot in driving the enemy crew from the nest, whereupon they were killed by other members of his company.

REED, EDGAR F. (2386954)
  At Bois-des-Rappes, France, Oct. 14–16, 1918, and at Aincreville, France, Oct. 31 and Nov. 2, 1918.
  R—Kokomo, Ind.
  B—Nevada, Ind.
  G. O. No. 20, W. D., 1919.

Sergeant, Company E, 61st Infantry, 5th Division.
On Oct. 16 he made his way through the German line, carrying a wounded companion. Although wounded and burned by mustard gas, he remained on duty, refusing to be evacuated. On Oct. 31 he volunteered and materially assisted in the holding of Aincreville while the defense was organized. On Nov. 2 he received a wound from which he later died, but he remained at his post acting as sniper for 4 hours.
Posthumously awarded. Medal presented to mother, Mrs. Jane Reed.

REED, EUGENE R.
  Near St. Etienne, France, Oct. 4, 1918.
  R—Danbury, Conn.
  B—South Norwalk, Conn.
  G. O. No. 35, W. D., 1919.

Pharmacist's mate, second class, 5th Regiment, U. S. Marine Corps, 2d Division.
During a bombardment he four times crossed an area heavily shelled and subjected to machine-gun fire to render assistance to his comrades.

REED, GEORGE (1518616)
  At Eyne, Belgium, Nov. 1, 1918.
  R—Norwalk, Ohio.
  B—Chicago Junction, Ohio.
  G. O. No. 50, W. D., 1919.

Cook, Company G, 145th Infantry, 37th Division.
After the remainder of his company had withdrawn, he crossed the Scheldt River alone, under terrific machine-gun and artillery fire, and rescued a wounded comrade.

REED, GLENN M. (2181356)
  Near Boney, France, Sept. 13, 1918.
  R—Grant, Nebr.
  B—Shenandoah, Iowa.
  G. O. No. 129, W. D., 1918.

Sergeant, Company B, 355th Infantry, 89th Division.
He voluntarily left shelter and passed through a heavy barrage to assist a wounded comrade who was unable to reach shelter by himself. As a result of this heroic action, he was killed.
Posthumously awarded. Medal presented to father, Milton H. Reed.

REED, RAYMOND E. (1214619)
  Near Ronssoy, France, Sept. 29, 1918.
  R—Medina, N. Y.
  B—Ridgeway, N. Y.
  G. O. No. 143, W. D., 1918.

Private, first class, Company F, 108th Infantry, 27th Division.
With great courage he went through heavy machine-gun and shellfire to the rescue of 2 wounded soldiers, whom he carried to our lines after dressing their injuries.

REED, WASHINGTON
  Near Pont-a-Mousson, France, Sept. 25, 1918.
  R—Wayne, Pa.
  B—Smithfield, Pa.
  G. O. No. 44, W. D., 1919.

Second lieutenant, 60th Infantry, 5th Division.
Wounded severely in the knee while leading his company in action, he refused first aid and continued to his objective 500 yards away. Here he organized the position under intense shellfire and flank infiltration by the enemy. When ordered to withdraw he used the stretcher which had been sent for him to carry back a dead soldier of his company.

*REES, JOHN (2258473)
Near Gesnes, France, Sept. 29, 1918.
R—Ellensburg, Wash.
B—Denmark.
G. O. No. 20, W. D., 1919.

Sergeant, Company M, 361st Infantry, 91st Division.
He fearlessly led his platoon in the face of a murderous fire in an attack on a machine-gun nest, and by his personal example contributed largely to the success of the attack by his platoon.
Posthumously awarded. Medal presented to father, Paul K. Johansen.

REESE, HAROLD L.
Near Mezy, France, July 15, 1918.
R—Mahanoy City, Pa.
B—Centralia, Pa.
G. O. No. 37, W. D., 1919.

First lieutenant, 30th Infantry, 3d Division.
During the unprecedented artillery bombardment preparatory to the great German offensive of July 15 Lieutenant Reese maintained liaison between different signal units by visiting the positions during the bombardment. It seemed utterly impossible for runners to venture through this fire, yet Lieutenant Reese voluntarily led a detail through the barrage, and thus established communication, also encouraging his men to greatest efforts.

REESE, JOHN D. (1481842)
At Roche, France, Oct. 27, 1918.
R—Farwell, Tex.
B—Josephine, Tex.
G. O. No. 44, W. D., 1919.

Private, Company L, 142d Infantry, 36th Division.
Though he was suffering from illness and had been told to go to the rear, he for 3 days remained on duty as a runner, and, when almost exhausted, went forward with his company in attack, voluntarily accompanying a liaison patrol on a dangerous mission.

REESE, JOHN E. (2273820)
At Audenarde, Belgium, Nov. 1, 1918.
R—Butte, Mont.
B—Hutchinson, Minn.
G. O. No. 15, W. D., 1919.

Sergeant, Company F, 316th Engineers, 91st Division.
He volunteered to accompany an officer and 3 other soldiers on a reconnaissance patrol of the city of Audenarde. Entering under heavy shellfire, the party reconnoitered the city for seven hours, while it was still being patrolled by the enemy, advancing 2 kilometers in front of our own outposts and beyond those of the enemy.

REESE, WILLIAM M. (2242497)
Near Romagne, France, Oct. 11, 1918.
R—San Saba, Tex.
B—Hillsboro, Tex.
G. O. No. 64, W. D., 1919.

Private, first class, Company I, 125th Infantry, 32d Division.
He displayed exceptional courage in repeatedly crossing an area swept by machine-gun and shellfire to carry messages to battalion headquarters after other soldiers had been killed or wounded in attempting the same mission.

*REEVE, CHARLES B.
Near St. Etienne, France, Oct. 3–9, 1918.
R—Plymouth, Ind.
B—Plymouth, Ind.
G. O. No. 37, W. D., 1919.

First lieutenant, 23d Infantry, 2d Division.
After his battalion commander had become a casualty, he assumed command and showed exceptional dash and skill in attack. When his battalion had been halted by heavy machine-gun fire, he commanded and led a charge through an open field, gaining his objective. It was during this charge that he was killed.
Posthumously awarded. Medal presented to father, C. A. Reeve.

REEVES, DACHE M.
North of Avocourt (Meuse), France, Oct. 9, 1918.
R—Atlanta, Ga.
B—Bloomingdale, Ga.
G. O. No. 14, W. D., 1923.

First lieutenant, 9th Aero Squadron, Air Service.
While performing an important aerial mission in his balloon, he was attacked by enemy airplanes. He hung from his basket under fire from enemy machine guns until the balloon burst into flames, when he jumped. He reascended as soon as another balloon could be inflated, although the air was strongly patrolled by the enemy. On Oct. 23, near Gesnes (Meuse), he was in the basket with another observer when a circus of 15 enemy airplanes made an attack from above. He remained in the basket until forced to jump. This officer showed extraordinary heroism by reascending as soon as another balloon could be made ready. Two hours later, while engaged in locating enemy batteries from his balloon, he was again attacked and the balloon burst into flames, forcing him to jump once more. In spite of these experiences this officer continued his mission in another balloon.

REEVES, JAMES H.
Near St. Mihiel, France, Sept. 12–13, 1918.
R—Centre, Ala.
B—Centre, Ala.
G. O. No. 87, W. D., 1919.
Distinguished-service medal also awarded.

Colonel, 353d Infantry, 89th Division.
On the opening day of the St. Mihiel offensive Colonel Reeves placed himself at the head of the assaulting battalion and personally led the advance from the inception of the attack until the fourth objective was reached. He was constantly exposed to artillery, machine-gun, and rifle fire, and, by his total disregard for personal danger, furnished an inspiring example to his men. On the following day he rallied a battalion of another regiment which had become disorganized and was retreating. Under heavy artillery fire, he reorganized it and sent it forward again at a critical juncture in the attack.

REEVES, ROY W.
Near Blanc Mont, France, Oct. 3, 1918.
R—San Diego, Calif.
B—Ravana, Miss.
G. O. No. 26, W. D., 1919.

Corporal, 96th Company, 6th Regiment, U. S. Marine Corps, 2d Division.
When a hand grenade was hurled into a group composed of himself and 5 other soldiers Corporal Reeves risked his life to save his comrades by picking up the grenade and throwing it out of the trench. It exploded a few yards from his hand, seriously wounding him in the face and head.

*REGAN, GERALD V.
Near St. Etienne, France, Oct. 4, 1918.
R—Duryea, Pa.
B—Duryea, Pa.
G. O. No. 37, W. D., 1919.

Corporal, 16th Company, 5th Regiment, U. S. Marine Corps, 2d Division.
Acting in the capacity of section leader, he rendered great assistance to his platoon and company commanders during an attack, and led his section in advance until he fell mortally wounded.
Posthumously awarded. Medal presented to father, Frederick J. Regan.

*REGAN, JOHN M.
Near Cierges, France, Aug. 1, 1918.
R—Boise, Idaho.
B—Silver City, Idaho.
G. O. No. 74, W. D., 1919.

Second lieutenant, 128th Infantry, 32d Division.
Mortally wounded by enemy fire while leading his platoon, he remained at the head of his men until he collapsed. He set an example of coolness and fortitude to his command, encouraging them by word and action.
Posthumously awarded. Medal presented to father, Timothy Regan.

REGGIARDO, ANTONIO (2265023).
Near Waereghem, Belgium, Oct. 31, 1918.
R—Martinez, Calif.
B—Martinez, Calif.
G. O. No. 37, W. D., 1919.

Corporal, Company K, 363d Infantry, 91st Division.
Corporal Reggiardo voluntarily went forward with a squad of men to combat hostile machine-gun nests which had held up the advance of our companies. After the squad leader had been killed, he took command, and directing a heavy fire of automatic rifles and rifle grenades, he drove back the advance German posts and cleared the way for the further progress of his company.

REICH, LOUIS B.
Near Cierges, France, Aug. 1, 1918.
R—Fort Atkinson, Wis.
B—Rome, Wis.
G. O. No. 93, W. D., 1919.

First lieutenant, 128th Infantry, 32d Division.
Lieutenant Reich voluntarily went forward and exposed himself in order to draw the enemy machine-gun fire, so as to locate their position. He succeeded in ascertaining their positions, and while returning to his front line he was hit and severely wounded by a shell fragment; but, refusing to be evacuated, he returned to his organization and remained in action throughout the engagement.

REID, ALLISON W. (156285).
Near Medeah Ferme, France, Oct. 8–9, 1918.
R—San Leandro, Calif.
B—San Leandro, Calif.
G. O. No. 44, W. D., 1919.

Private, Company A, 2d Engineers, 2d Division.
Engaged as runner, he constantly carried messages through a sector which was under intense shell and machine-gun fire and infested with sniper fire.

REID, GEORGE B. (145284).
Near Suippes, France, July 15, 1918.
R—Minneapolis, Minn.
B—Minneapolis, Minn.
G. O. No. 37, W. D., 1919.

Corporal, Battery A, 151st Field Artillery, 42d Division.
While acting as gunner during the firing of a barrage, he was shot through the arm by an enemy machine gun from an airplane, but continued to fire his gun throughout the barrage. He was then evacuated in an exhausted condition.

REID, JOSEPH W. (1288979).
In the Bois-de-la-Grande-Montagne, France, Oct. 15, 1918.
R—Winchester, Va.
B—Rockingham County, Va.
G. O. No. 35, W. D., 1919.

Corporal, Company I, 116th Infantry, 29th Division.
When his platoon was held up by machine-gun fire, Corporal Reid fearlessly led them forward and captured machine gun and prisoners. He later organized and consolidated the position won.

REID, THOMAS C.
Near Moulin, France, July 15, 1918.
R—Demopolis, Ala.
B—Hatfield, Mo.
G. O. No. 24, W. D., 1920.

Captain, 38th Infantry, 3d Division.
During a strong enemy attack on his company sector, Captain Reid fearlessly exposed himself to heavy fire in order to direct the fire of his men. He personally led two counterattacks upon the enemy, breaking up their attack and forcing them to retire. Due to this gallantry, his company, notwithstanding its heavy losses, decisively defeated an enemy grenadier regiment and forced it to retire across the Marne.

*REID, WILLIAM R.
Near Chateau-de-Diable, France, Aug. 27, 1918.
R—New York, N. Y.
B—New York, N. Y.
G. O. No. 32, W. D., 1919.

First lieutenant, 307th Infantry, 77th Division.
Lieutenant Reid, while on duty as battalion adjutant, voluntarily led a small patrol into woods held by the enemy to ascertain the source of heavy machine-gun fire which stopped the advance of his battalion. In the performance of this courageous act he was killed by enemy machine-gun fire.
Posthumously awarded. Medal presented to father, William J. Reid.

*REIFIN, ABE (58040).
Near the Meuse River, France, Oct. 14, 1918.
R—Cincinnati, Ohio.
B—Russia.
G. O. No. 89, W. D., 1919.

Private, Company H, 28th Infantry, 1st Division.
Private Reifin displayed exceptional courage in volunteering and going over open ground through direct artillery and machine-gun fire. Upon returning he again volunteered to pass through the same heavy fire in order to establish liaison between his platoon and company. He lost his life in attempting this hazardous mission.
Posthumously awarded. Medal presented to mother, Mrs. Fannie Reifin.

REILLEY, CHARLES R. (2273734).
At Audenarde, Belgium, Nov. 1, and Nov. 10, 1918.
R—Jumbo Town, Colo.
B—Victor, Colo.
G. O. No. 44, W. D., 1919.

Sergeant, first class, Company F, 316th Engineers, 91st Division.
On Nov. 1 Sergeant Reilley voluntarily accompanied a patrol into the city of Audenarde, when it was still occupied by the enemy, obtaining important data on destroyed bridges and attacking an enemy patrol three times their number. He also captured a German spy while the latter was attempting to escape. On this same day he forced a sniper to cover, thus saving the life of his captain, who was about to be fired upon. On Nov. 10 he swam the Escaut River, braving the fire of enemy snipers on the opposite bank, and tied a rope to an enemy barrel bridge, thereby making a crossover for the infantry.

REILLEY, THOMAS T.
Near Villers-sur-Fere, France, July 27, 1918, to Aug. 1, 1918.
R—New York, N. Y.
B—New York, N. Y.
G. O. No. 99, W. D., 1918.

Captain, 165th Infantry, 42d Division.
Wounded and ordered to the rear, he nevertheless remained with his men in an exposed and dangerous position, which it was necessary to hold, near Villers-sur-Fere, France, on July 27 to Aug. 1, 1918. His presence and example held his company fast against continuous fire.

REILLY, ARCHIBALD F. (89499).
Near Landres-et-St. Georges, France, Oct. 15, 1918.
R—Richmond Hill, Long Island, N. Y.
B—Morris Park, Long Island, N. Y.
G. O. No. 37, W. D., 1919.

Private, Company C, 165th Infantry, 42d Division.
Private Reilly, with 1 other soldier, went to the aid of a wounded comrade who was lying about 50 yards in advance of our lines, in plain view of enemy gunners and snipers, and carried him through machine-gun and shellfire to a place of safety.

REILLY, MICHAEL (43131).
Near Soissons, France, July 18, 1918.
R—San Francisco, Calif.
B—Ireland.
G. O. No. 98, W. D., 1919.

Corporal, Company G, 16th Infantry, 1st Division.
After being wounded Corporal Reilly remained with his squad and continued to lead it in action until he was again seriously wounded 2 days later.

REILLY, THOMAS L. (793032)_____
Near Vieville, France, Nov. 1, 1918.
R—New York, N. Y.
B—New York, N. Y.
G. O. No. 46, W. D., 1919.

Corporal, Company D, 21st Machine Gun Battalion, 7th Division.
During a heavy barrage of high-explosive and gas shells Corporal Reilly assisted an officer to give first aid to a wounded officer and 2 soldiers after the platoon had withdrawn from that position. He went to the rear for medical aid and passed through the barrage the second time as he returned.

REINHARD, FRED W. (547600)_____
In the Bois-d'Aigremont, France, July 15, 1918.
R—Spring City, Pa.
B—Knoxen, Pa.
G. O. No. 32, W. D., 1919.

Private, Company I, 30th Infantry, 3d Division.
He carried messages during a heavy German barrage until he was seriously wounded, when he showed great fortitude, his one thought being that his message must be delivered.

REINHOLDT, ROLAND R_____
Near Cheppy, France, Sept. 26, 1918.
R—St. Louis, Mo.
B—St. Louis, Mo.
G. O. No. 66, W. D., 1919.

Captain, 138th Infantry, 35th Division.
Sent forward with 2 platoons on a reconnaissance mission, he encountered the enemy in force, but effected the capture of 13 Germans, including an officer. He then held the position he had seized for 3 hours, although subjected to fire that rendered three-fourths of his men casualties, until French tanks arrived, when, with the remnants of his platoons, he attacked and captured machine-gun nests inaccessible to the tanks. The stubborn resistance of enemy attacks which he maintained was one of the deciding factors in the fight.

REITER, CHARLES (1245045)_____
Near Apremont, France, Oct. 1, 1918.
R—Pittsburgh, Pa.
B—Duluth, Minn.
G. O. No. 53, W. D., 1920.

Sergeant, Company H, 111th Infantry, 28th Division.
Sergeant Reiter and a companion exposed themselves to heavy machine-gun fire and advanced in front of our lines to assist a wounded soldier to a place of safety. In the attack on Hill 244, on Oct. 8, 1918, after the officers had become casualties he assumed command of a unit and displayed unusual ability and leadership, until severely wounded.

REITERMAN, FRANK (53557)_____
Near Soissons, France, July 18, 1918.
R—Louisville, Ky.
B—Hungary.
G. O. No. 132, W. D., 1918.

Sergeant, Company E, 26th Infantry, 1st Division.
As leader of a platoon he attacked a machine-gun nest, captured several guns, and held his position against vigorous counterattacks.

REMINGTON, PHILIP_____
At Malala River, Mindanao, P. I., Oct. 22, 1905.
R—Windsor, Conn.
B—Indian Territory.
G. O. No. 4, W. D., 1923.

Second lieutenant, 22d Infantry, U. S. Army.
While commanding the leading troop of the advance guard of an expedition against Dato Ali, upon locating Ali's cotta, which was occupied by hostile Moros, Lieutenant Remington courageously dashed forward, leading his men, and engaged the enemy at point-blank range. In a pistol and rifle duel he succeeded in killing Ali with his pistol, but not until 1 of Lieutenant Remington's own men had been killed by rifle fire from Ali. He then disposed his command, which in a brisk fire fight killed the remaining members of the band of Moros.

RENICK, FRED A. (9412)_____
Near Beauvois, France, Apr. 4, 1918.
R—St. Louis, Mo.
B—Sullivan, Mo.
G. O. No. 129, W. D., 1918.

Private, first class, section No. 598, Ambulance Service.
On Apr. 4, 1918, he was ordered to drive his ambulance to a dressing station. The road over which it was necessary to pass was under continuous shell-fire. On his way to the dressing station he received a slight wound. In spite of the wound, which was dressed at a dressing station, he resumed his post, and on the return trip a shell struck his car seriously wounding him and killing his passenger.

RENSHAW, LEONARD A. (1285855)_____
At Le-Bois-Plat-Chene, France, Oct. 10-29, 1918.
R—Princess Anne, Md.
B—Princess Anne, Md.
G. O. No. 37, W. D., 1919.

Corporal, Company I, 115th Infantry, 29th Division.
During the advance of the 3d Battalion Corporal Renshaw was placed in charge of the battalion liaison group. He not only managed the group with skill, but repeatedly carried messages through shell and machine-gun fire. Although several times gassed during the succeeding operations, he continued his duties, showing utter disregard for his personal safety.

RENTFRO, CHARLES C_____
Before St. Agnan, France, July 15-18, 1918.
R—Chicago, Ill.
B—Sigourney, Iowa.
G. O. No. 99, W. D., 1918.

First lieutenant, Medical Corps, attached to 109th Infantry, 28th Division.
For 3 days, July 15-18, 1918, before St. Agnan, France, he went without sleep in order to care for the wounded and performed his work fearlessly without shelter under continuous bombardment.

RESSEGUIE, HAROLD D_____
Near Grand Pre, France, Nov. 1, 1918.
R—Watertown, N. Y.
B—Watertown, N. Y.
G. O. No. 44, W. D., 1919.

Captain, 311th Infantry, 78th Division.
Although his wrist was shattered and he was suffering from severe machine-gun wounds, he continued to direct operations for several hours, refusing to be evacuated until all the other wounded had received attention. After reaching the dressing station, although suffering intense pain, he gave full tactical information to his successor.

RETTMAN, LOUIE (545460)_____
Near Crezancy, France, July 15, 1918.
R—Hutchinson, Minn.
B—Hutchinson, Minn.
G. O. No. 24, W. D., 1920.

First sergeant, Company B, 30th Infantry, 3d Division.
After company officers had become casualties, Sergeant Rettman reorganized the remnants of the company and took and held an important position against strong enemy attacks. His company commander having been left wounded in advance of the new line, Sergeant Rettman exposed himself to heavy fire in order to carry him to shelter.

REVELS, JAMES F. (2021610)_____
Near Oborzerskaya, Russia, Sept. 16, 1918.
R—Detroit, Mich.
B—Mt. Carmel, Pa.
G. O. No. 16, W. D., 1920.

Bugler, Company I, 339th Infantry, 85th Division (detachment in North Russia).
Bugler Revels succeeded, after 3 others had failed in delivering a message. In the performance of this duty he passed through dense woods, exposed to artillery and rifle fire. Throughout the action he carried numerous messages to the flanks, always selecting the shortest route regardless of the dangers involved.

REX, NEWTON (1521455)..............
Near Montfaucon, France, Sept. 27, 1918.
R—Bowling Green, Ohio.
B—Wells County, Ind.
G. O. No. 59, W. D., 1919.

Corporal, Company F, 146th Infantry, 37th Division.
Leading a patrol of 12 men from his own and another company, he encountered 35 of the enemy in a ravine. Under a terrific enfilading fire from 7 machine guns, he led an attack on the enemy in which 5 of the latter were killed and 15 captured, together with the 7 machine guns.

REXROTH, HARRY J. (2276437)..........
Near Audenarde, Belgium, Nov. 1-4, 1918.
R—National, Wash.
B—Philadelphia, Pa.
G. O. No. 15, W. D., 1919.

Private, first class, 364th Ambulance Company, 316th Sanitary Train, 91st Division.
He repeatedly showed utter disregard for his safety, in establishing and maintaining liaison between advanced dressing stations and battalion aid stations, and in searching the battle fields for wounded, passing over areas under heavy fire from enemy artillery, machine guns, and snipers. On Nov. 4, he entered the town of Audenarde while it was under terrific bombardment, made a thorough search for wounded, and later accompanied ambulances back into the town to evacuate the wounded.

REYNOLDS, CLEARTON H..............
Near Romagne, France, Oct. 9, 1918.
R—Garden City, Long Island, N. Y.
B—Provincetown, Mass.
G. O. No. 35, W. D., 1919.

Captain, pilot, 104th Aero Squadron, Air Service.
Although weather conditions made flying exceedingly dangerous, Captain Reynolds, with his observer, started on a mission to determine the position of the front-line troops of the division to which his squadron was attached. Flying at an altitude of 25 meters, they encountered and defeated 3 enemy patrols and gathered and delivered to division headquarters very valuable information.

REYNOLDS, EUGENE C. (2669321)......
At Quinnemont Farm, near Ronssoy, France, Sept. 28, 1918.
R—Brooklyn, N. Y.
B—Brooklyn, N. Y.
G. O. No. 68, W. D., 1920.

Private, Company K, 106th Infantry, 27th Division.
After assisting in repulsing a strong enemy counterattack, Private Reynolds, with 2 other soldiers, became separated from his company, due to the heavy fog. Seeing a superior force of the enemy in a trench, they unhesitatingly attacked, and after killing and wounding several of the enemy, they captured numerous prisoners and brought them back to our lines.

REYNOLDS, FRANK J. (2686)..........
At Cantigny, France, May 28-30, 1918.
R—Lee, Mass.
B—Lee, Mass.
G. O. No. 99, W. D., 1918.

Private, first class, medical detachment, 28th Infantry, 1st Division.
During the fight at Cantigny, France, on May 28-30, 1918, while acting as a stretcher bearer, he constantly and fearlessly exposed himself to artillery and machine-gun fire to succor the wounded, frequently on his own initiative, when he might have remained in security himself.

REYNOLDS, JOHN N..............
In the region of Verdun, France, Oct. 10, 1918.
R—Washington, D. C.
B—Washington, D. C.
G. O. No. 143, W. D., 1918.

Major, 1st Army Observation Group, Air Service.
He proceeded over the enemy lines without benefit of protection planes on a mission of great urgency. He flew about 12 kilometers over the lines, when he was suddenly set upon by 14 hostile planes. He fought them off and succeeded in downing 1 of the enemy. He continued his flight with his badly damaged plane and concluded his mission.
Oak-leaf cluster.
A bronze oak leaf for extraordinary heroism in action in the region of Grand Pre, France, Oct. 29, 1918. While on a mission he was suddenly set upon by 6 enemy aircraft. Although in the German territory, without protection and in danger of being cut off in the rear, he entered into combat with the hostile aircraft. He succeeded in shooting down 2 of the enemy and dispersing the rest of the formation. With his machine severely damaged he continued until he had completed his mission.

*REYNOLDS, PATRICK (90891)..........
Near Villers-sur-Fere, France, July 30, 1918.
R—New York, N. Y.
B—Ireland.
G. O. No. 88, W. D., 1918.

Private, first class, Company H, 165th Infantry, 42d Division.
He was killed near Villers-sur-Fere, France, on July 30, 1918, when he went out alone in the face of enemy machine-gun fire in a heroic effort to capture an enemy machine-gun nest.
Posthumously awarded. Medal presented to sister, Mrs. Mary O'Donnell.

REYNOLDS, WILLIAM G..............
Near St. Etienne, France, Oct. 4, 1918.
R—Berryville, Va.
B—Kingston, Pa.
G. O. No. 46, W. D., 1919.

Captain, 23d Infantry, 2d Division.
After Captain Reynolds had been severely wounded by a shell, he managed by a supreme effort to regain sufficient consciousness to acquaint his successor with the necessary information for the continuance of the struggle. His courage, under such great agony, set a most wonderful example for his men.

RHEA, JAMES C..............
Near St. Etienne, France, Oct. 9, 1918.
R—Strawn, Tex.
B—Hamburg, Iowa.
G. O. No. 120, W. D., 1918.
Distinguished-service medal also awarded.

Colonel (Cavalry), Chief of Staff, 2d Division.
Colonel Rhea, with Lieutenant Le Pelletier de Woillemont, French Army, voluntarily undertook an important reconnaissance under hazardous circumstances during the Masif Blanc Mont operations at a time when accurate information concerning our advanced positions was greatly needed and could not be obtained from other sources. In an automobile, whose conspicuous appearance drew the concentrated fire of enemy artillery and machine guns, they proceeded 1 mile across open ground to the town of St. Etienne, where our troops were in contact with the enemy. Under fire these 2 officers reconnoitered the front lines, locating the position of the enemy, as well as that of the French units on the flank, and returned across the open with complete, reliable, and timely information of the highest military value in subsequent operations.

RHODES, CHARLES D..............
Near the barrio of San Nicolas, Pueblo of Bacoor, Cavite, Luzon, P. I., Dec. 31, 1901.
R—Delaware, Ohio.
B—Delaware, Ohio.
G. O. No. 126, W. D., 1919.
Distinguished-service medal also awarded.

Captain, 6th Cavalry, U. S. Army.
He gallantly and fearlessly led an attack on a superior body of insurgents with 2 men of his troop, killing 2 of the enemy and wounding 2, including their leader, and dispersing the remainder.

RHODES, NELLUS A
At Fismette, France, Aug. 9, 1918.
R—Meadville, Pa.
B—Meadville, Pa.
G. O. No. 9, W. D., 1923.

Second lieutenant, 112th Infantry, 28th Division.
When the town had been taken by his battalion in the morning and was being held at night against terrific counterattacks in which our troops were engaged most of the time in desperate hand-to-hand combat, Lieutenant Rhodes, although a member of the battalion intelligence section, went into the most dangerous places and by his splendid courage bolstered the morale of the other members of the command, who were being hard pressed by enemy troops who had gained entrance to the town. Learning that the enemy was coming through the back of a building, he bravely entered it, killing 4 of the enemy, and retained possession of the building. With 5 or 6 stragglers, he crossed and recrossed a street swept by hostile machine-gun fire and prevented the enemy from filtering through a hole which they had blown in the wall of the building. His courageous actions were an inspiration to his comrades in the desperate fighting.

RHODES, ROBERT A. (254216)
Near Chateau-Thierry, France, July 20 and 22, 1918.
R—St. Joseph, Mo.
B—Aurora, Ill.
G. O. No. 125, W. D., 1918.

Private, Company M, 103d Infantry, 26th Division.
East of Belleau Wood he continually carried messages under heavy machine-gun fire while acting as a runner. On July 22, near Epieds, he crossed an open gap swept by machine-gun fire in order to deliver an important message, and later was a voluntary member of a patrol which rescued wounded under fire from advanced positions.

RICE, CARL C
Near Chateau-Thierry, France, June 6, 1918.
R—Rolla, Mo.
B—Danville, Ill.
G. O. No. 99, W. D., 1918.

Second lieutenant, 5th Machine Gun Battalion, 2d Division.
He was wounded soon after the advance began, but refused to have his wound dressed for fear it would delay the movement. He bravely continued to lead the section until he fell from exhaustion.

RICE, ELMER V. (548068)
In the Bois-d'Aigremont, France, July 15, 1918.
R—Midland, Mich.
B—Kinde, Mich.
G. O. No. 46, W. D., 1919.

Private, Company L, 30th Infantry, 3d Division.
During the intense artillery fire preceding the German attack of July 15, after another runner had been sent with a message from the battalion post of command and had been unable to get through the wood, which was being heavily bombarded, Private Rice volunteered for this seemingly impossible mission and successfully accomplished it. Throughout the night he declined to take cover, but continued to search for wounded men, exposing himself to the heaviest fire.

RICE, GEORGE D
At Bayan, Lake Lanao, Mindanao, P. I., May 2, 1902.
R—Medford, Mass.
B—Malden, Mass.
G. O. No. 14, W. D., 1925.

Chaplain, 27th Infantry, U. S. Army.
With utter disregard for his personal safety he administered to the wounded under heavy fire of the enemy.

RICE, JAMES T. (570794)
Near the Bois-du-Fays, France, Sept. 29, 1918, and in the Bois-de-Malaumont, France, Oct. 11–13, 1918.
R—Seattle, Wash.
B—Carter County, Ky.
G. O. No. 89, W. D., 1919.

Private, first class, Company C, 8th Field Signal Battalion, 4th Division.
While at work with a group of men maintaining telephone communication, Private Rice went out under heavy fire and carried to shelter a comrade who had been wounded by a bursting shell, returning immediately and repairing breaks in the line. During the action in the Bois-de-Malaumont, he repeatedly exposed himself to heavy artillery and machine-gun fire in order to maintain telephone lines for the Infantry, displaying remarkable courage.

RICE, MALCOLM
Near Exermont, France, Oct. 1–3, 1918.
R—Paintsville, Ky.
B—Paintsville, Ky.
G. O. No. 46, W. D., 1919.

Captain, 18th Infantry, 1st Division.
During the advance Oct. 1–3 Captain Rice was severely gassed, and, although suffering greatly from the effects, he remained with his company for 4 days, after which he was forced to evacuate on account of temporary blindness.

RICE, MATTHEW GEORGE (89240)
At Landres-et-St. Georges, France, Oct. 15, 1918.
R—New York, N. Y.
B—New York, N. Y.
G. O. No. 15, W. D., 1923.

Corporal, Company A, 165th Infantry, 42d Division.
While acting as runner he delivered a message from the regimental commander to the commander of the assault battalion, crossing a level field swept by intense machine-gun and artillery fire, and though severely wounded returned with a message to the regimental commander who stated that all the runners had been killed or wounded and that he desired to send another message to the battalion commander. Private Rice promptly volunteered to carry the message and in accomplishing his mission was again wounded; notwithstanding which fact he again crossed the fire-swept zone and delivered a return message to the regimental commander. Private Rice was then carried from the field.

RICE, WILLIAM M. (1284782)
In Consenvoye Woods, France, Oct. 9, 1918.
R—North East, Md.
B—Principio Furnace, Md.
G. O. No. 37, W. D., 1919.

Corporal, Company E, 115th Infantry, 29th Division.
Seeing a good position in advance of the lines, Corporal Rice took his automatic rifle and crawled through machine-gun fire to this place, where he established an automatic-rifle post and called on his squad to follow him. An enemy counterattack was eventually broken up at this point and the line was thereby advanced to a more advantageous position.

RICHARDS, CALVIN D
Near Verdun, France, Oct. 9, 1918.
R—Morganfield, Ky.
B—Morganfield, Ky.
G. O. No. 21, W. D., 1919.

Second lieutenant, 26th Infantry, 1st Division.
While defending a hill, Lieutenant Richards, with 7 machine gunners, beat off an enemy attack of greatly superior numbers, after a sharp hand-to-hand encounter with pistols and grenades. Although his small force suffered 4 casualties, he still continued to defend the hill, an important tactical point for his division.

RICHARDS, ELMER PRESTON (234865)....
Near Xammes, France, Sept. 18, 1918.
R—Moundville, Mo.
B—Windsor, Ill.
G. O. No. 37, W. D., 1919.

Private, Company D, 354th Infantry, 89th Division.
Knowing that on account of the intense shelling it would be impossible to supply the men in the front line with rations, Private Richards, in a wounded condition, procured a quantity of rations and carried them to the line through heavy shell fire, and personally distributed to each man a portion.

*RICHARDS, JAMES N. C.............
Near Soissons, France, July 18, 1918.
R—Riverton, Va.
B—Petersburg, Va.
G. O. No. 132, W. D., 1918.

Captain, 26th Infantry, 1st Division.
Displaying valorous leadership throughout the attack on July 18, 1918, near Soissons, France, he was killed while charging enemy machine guns at the head of his command.
Posthumously awarded. Medal presented to widow, Mrs. James N. C. Richards.

*RICHARDS, SAMPSON (97564).........
Near Landres-et-St. Georges, France, Oct. 14, 1918.
R—Sanger, Calif.
B—England.
G. O. No. 131, W. D., 1918.

Corporal, Company H, 167th Infantry, 42d Division.
When his platoon had become scattered during an attack and his platoon commander had been killed, Corporal Richards, although himself seriously wounded, reorganized the platoon under heavy shell and machine-gun fire and turned the platoon over to the next in command ready for the assault before he permitted himself to be evacuated, thereby setting to his associates an example of utter disregard for danger and remarkable coolness and courage in the face of the enemy.
Posthumously awarded. Medal presented to brother, William Richards.

RICHARDS, THADDIS R. (1307008).......
Near Bellicourt, France, Sept. 29, 1918.
R—Maryville, Tenn.
B—Pickens County, Ga.
G. O. No. 44, W. D., 1919.

Bugler, Company B, 117th Infantry, 30th Division.
During the very thickest of the fighting Bugler Richards, assisted by another soldier, attacked 2 enemy machine-gun positions, killing the gunners and capturing the guns, thus allowing the further advance of his company.

RICHARDS, WALTER A..................
Near St. Juvin, France, Oct. 11, 1918.
R—Clifton Station, Va.
B—Washington, D. C.
G. O. No. 46, W. D., 1919.

First lieutenant, 326th Infantry, 82d Division.
Leading his platoon in attack, Lieutenant Richards was subjected to fierce and devastating fire of enemy artillery and machine guns. Although he himself was wounded and 90 per cent of his platoon made casualties, he continued to press forward until he was felled by machine-gun fire after reaching the foremost position of the entire action.

RICHARDSON, CHARLES E. (71000).......
East of Belleau, France, July 21, 1918.
R—Wakefield, Mass.
B—Medford, Mass.
G. O. No. 125, W. D., 1918.

Private, Company I, 104th Infantry, 26th Division.
Under the leadership of an officer, he and Private Joseph R. Huntley charged a machine-gun nest, captured 2 guns and killed or captured 12 men.

RICHARDSON, CHARLES M. (1247262)....
At Fismette, France, Aug. 9, 1918.
R—Bradford, Pa.
B—State Line, Pa.
G. O. No. 98, W. D., 1919.

Sergeant, Company C, 112th Infantry, 28th Division.
Sergeant Richardson volunteered to go out in the open with a comrade, under hostile machine-gun fire, to rescue a wounded soldier. As they were carrying the latter to shelter he was again struck by a machine-gun bullet and killed, and the companion, also being wounded, was dragged to safety by Sergeant Richardson.

RICHARDSON, JAMES M..................
Near Grand Pre, France, Oct 6, 1918.
R—Nashville, Tenn.
B—Memphis, Tenn.
G. O. No. 37, W. D., 1919.

Second lieutenant, pilot, 1st Aero Squadron, Air Service.
He undertook an infantry contact patrol mission under weather conditions which necessitated flying at an altitude of only 100 meters. Near the front lines machine guns opened an effective fire on his plane, and he was wounded in the foot, but he continued on the mission until the front lines of the American troops were located and his observer had written out a report for the division commander.

RICHARDSON, JOHN B..................
Near Ville-Savoye, France, Aug. 21, 1918.
R—Woodville, Miss.
B—Woodville, Miss.
G. O. No. 56, W. D., 1922.

Major, 306th Machine Gun Battalion, 77th Division.
When the advance on the Taunerio by a company of the 308th Infantry was being held up by a heavy hostile fire, Major Richardson, then commanding the 306th Machine Gun Battalion, which had a platoon of machine guns supporting the attack, seeing that the attacking troops were wavering on account of an inexperienced leader and under a heavy hostile fire, with great gallantry and the utmost disregard of personal danger, took command of the company and led it through heavy artillery and machine-gun fire to its objective, which was captured and later consolidated by a skillful disposition of machine guns under his direction.

RICHEY, WILLIAM R., Jr..................
Near Ardeuil, France, Sept. 29–30, 1918.
R—Laurens, S. C.
B—Hodges, S. C.
G. O. No. 49, W. D., 1922.

Captain, 371st Infantry, 93d Division.
Although badly gassed during the night of Sept. 28, he nevertheless remained in command of his company and with utter disregard of personal danger twice led it in the attack on successive days and was not evacuated until completely exhausted on Sept. 30, 1918.

RICHFORD, ALBERT F. (88791)...........
Near Nonsard, France, Sept. 30, 1918.
R—Brooklyn, N. Y.
B—Brooklyn, N. Y.
G. O. No. 35, W. D., 1919.

Wagoner, Supply Company, 165th Infantry, 42d Division.
Wagoner Richford was severely wounded by a shell fragment while driving a wagon containing rations and other supplies for his regiment over a heavily shelled road, but, disregarding his wounds, remained in charge of his wagon until the mission had been accomplished.

*RICHMAN, HENRY C. (52301)...........
Near Vaux, France, July 1–2, 1918.
R—Laporte, Ind.
B—Daleville, Ind.
G. O. No. 99, W. D., 1918.

Private, Company M, 23d Infantry, 2d Division.
He moved through heavy woods alone under heavy machine-gun fire, flanking dugouts, from which 12 German prisoners were taken. In the action of July 16–19, 1918, near Soissons, France, showing the same fearless qualities, he was killed.
Posthumously awarded. Medal presented to sister, Mrs. Clara Boram.

RICHMOND, CHARLES H.
Near Blanc Mont, France, Oct. 5, 1918.
R—Boston, Mass.
B—Charles County, Md.
G. O. No. 81, W. D., 1919.

Corporal, 55th Company, 5th Regiment, U. S. Marine Corps, 2d Division.
With the aid of one other soldier, Corporal Richmond located and captured a machine-gun nest of 4 guns.

RICHMOND, CLARENCE L.
Near Blanc Mont, France, Oct. 3-5, 1918.
R—Cleveland, Tenn.
B—Cleveland, Tenn.
G. O. No. 46, W. D., 1919.

Private, 43d Company, 5th Regiment, U. S. Marine Corps, 2d Division.
He unhesitatingly went through the heaviest machine-gun and artillery fire, dressing and carrying wounded. Disregarding his own safety, he refused to take rest or food while there were wounded needing attention.

RICHMOND, LLOYD (1472600).
At Chaudron Farm, France, Sept. 29, 1918.
R—Kansas City, Mo.
B—Kansas City, Mo.
G. O. No. 70, W. D., 1919.

Private, 139th Ambulance Company, 110th Sanitary Train, 35th Division.
After the infantry had been withdrawn he voluntarily remained in an advance dressing station with wounded men, whose condition made it impossible to remove them, and worked alone for several hours caring for these men under heavy shell and machine-gun fire until he himself was wounded by a bursting shell which killed 2 of his patients.

RICKENBACKER, EDWARD V.
Near Montsec, France, Apr. 29, 1918.
R—Columbus, Ohio.
B—Columbus, Ohio.
G. O. No. 121, W. D., 1918, and
G. O. No. 32, W. D., 1919.

First lieutenant, 94th Aero Squadron, Air Service.
He attacked an enemy Albatross monoplane, and after a vigorous fight, in which he followed his foe into German territory, he succeeded in shooting it down near Vigneulles-les-Hatton Chatel.
Oak-leaf clusters (7).
One bronze oak leaf is awarded Lieutenant Rickenbacker for each of the following acts of extraordinary heroism in action: On May 17, 1918, he attacked 3 Albatross enemy planes, shooting 1 down in the vicinity of Richecourt, France, and forcing the others to retreat over their own lines. On May 22, 1918, he attacked 3 Albatross monoplanes 4,000 meters over St. Mihiel, France. He drove them back into German territory, separated 1 from the group, and shot it down near Flirey. On May 28, 1918, he sighted a group of 2 battle planes and 4 monoplanes, German planes, which he at once attacked vigorously, shooting down 1 and dispersing the others. On May 30, 1918, 4,000 meters over Jaulny, France, he attacked a group of 5 enemy planes. After a violent battle, he shot down 1 plane and drove the others away. On Sept. 14, 1918, in the region of Villecy, he attacked 4 Fokker enemy planes at an altitude of 3,000 meters. After a sharp and hot action, he succeeded in shooting 1 down in flames and dispersing the other 3. On Sept. 15, 1918, in the region of Bois-de-Wavrille, he encountered 6 enemy planes, who were in the act of attacking 4 Spads, which were below them. Undeterred by their superior numbers, he unhesitatingly attacked them and succeeded in shooting 1 down in flames and completely breaking the formation of the others. On Sept. 25, 1918, near Billy, France, while on voluntary patrol over the lines he attacked 7 enemy planes (5 type Fokker, protecting 2 type Halberstadt). Disregarding the odds against him, he dived on them and shot down 1 of the Fokkers out of control. He then attacked 1 of the Halberstadts and sent it down also.

RICKER, MAURICE STANLEY.
Near Brieulles, France, Nov. 4, 1918.
R—Brookline, Mass.
B—Bangor, Me.
G. O. No. 46 W. D., 1919.

First lieutenant, 6th Infantry, 5th Division.
In covering the right flank of his company he led his platoon across a pontoon bridge which was broken by artillery fire before the entire command had crossed. Without hesitation he proceeded to lead his men under direct machine-gun and minenwerfer fire, routing a large detachment of the enemy, capturing 8 Germans, 5 machine guns, and 2 minenwerfers, and successfully covering the crossing of the remainder of the company.

RICKET, HARRY C. (95597).
At Chateau-de-la-Foret, near Villers-sur-Fere, France, July 28-29, 1918.
R—Columbus, Ohio.
B—Spring Hill, Kans.
G. O. No. 99, W. D., 1918.

Cook, Headquarters Company, 166th Infantry, 42d Division.
During a bombardment so intense as to drive all other kitchens out of the village. When his stove had to be taken to the rear, he improvised a fire in the ground and continued his work until ordered to leave. He carried water from a spring, which was repeatedly shelled, when others would not approach it. Unaided, of his own volition, he conducted a first-aid station for wounded and exhausted men at his kitchen. Constantly in extreme personal danger from machine-gun fire from low-flying airplanes and bombardment by high-explosive shells, he devoted himself entirely to the needs of others and made possible the care of several hundred wounded, exhausted, and hungry men.

RIDDICK, ARCHIE (1320610).
Near Vaux-Andigny, France, Oct. 19, 1918.
R—Belvidere, N. C.
B—Gates County, N. C.
G. O. No. 44, W. D., 1919.

Private, Company F, 120th Infantry, 30th Division.
When the position of his company had become untenable, because of enemy machine-gun and artillery fire Private Riddick, with another soldier, the sole survivors of a Lewis machine-gun team, covered the retreat of their company. Clinging to their advanced post throughout the day, they took up the advance with the company at dusk that evening.

RIDDLE, LAWRENCE SCOTT (2661994).
At Bois-de-Chaume, France, Oct. 11, 1918.
R—Mattoon, Ill.
B—Mattoon, Ill.
G. O. No. 35, W. D., 1919.

First sergeant, Company I, 131st Infantry, 33d Division.
Sergeant Riddle, with 4 other soldiers, flanked an enemy machine-gun position, killed 3 of the crew and captured 1, together with the guns. He was subsequently killed while leading a small group of men in an attack on an enemy machine-gun nest.
Posthumously awarded. Medal presented to mother, Mrs. Lillie L. Riddle.

RIDEOUT, PERCY A.
At Cierges, France, Oct. 4, 1918.
R—Concord Junction, Mass.
B—Ashburnham, Mass.
G. O. No. 142, W. D., 1918.

First lieutenant, Company D, 1st Gas Regiment.
He made an extended reconnaissance in advance of the outposts, fearlessly exposed himself to enemy machine-gun fire, and was several times knocked down by exploding shells. The information he secured was valuable to the infantry, giving them knowledge of the exact location of machine-gun nests. During the action this officer directed the laying of the smoke barrage from an exposed position, remaining at his station throughout the operation, in spite of severe shell and machine-gun fire, and continuing to display the highest courage until he was killed by shellfire.
Posthumously awarded. Medal presented to widow, Mrs. Helen P. Rideout.

RIDGELY, CHARLES_____
Near Berzy-le-Sec, France, July 21, 1918.
R—New York, N. Y.
B—Springfield, Ill.
G. O. No. 44, W. D., 1919.

First lieutenant, 26th Infantry, 1st Division.
During the final attack on Soissons, Lieutenant Ridgely, advancing in front of his assault waves, alone charged a machine-gun position which was delivering a punishing fire on our troops. He killed both gunners and captured the gun.

RIDLEY, JAMES A_____
Near Bellicourt, France, Sept. 29–30, 1918, and near Mazinghien, France, Oct. 19, 1918.
R—Murfreesboro, Tenn.
B—Murfreesboro, Tenn.
G. O. No. 98, W. D., 1919.

First lieutenant, 113th Machine Gun Battalion, 30th Division.
Taking command of his company, after the company commander and second in command had been wounded, Lieutenant Ridley led his men through an intense artillery barrage and assisted in reducing 11 enemy machine-gun nests, capturing 150 prisoners and several machine guns, which were successfully put in operation against the enemy. Near Mazinghien, France, on Oct. 19, he led his company forward to advanced positions under terrific artillery fire, inspiring his men by his coolness and bravery.

RIECK, JAMES G. (1040)_____
Near Villers-sur-Fere, France, July 27, 1918.
R—Delaware, Ohio.
B—Bowdle, S. Dak.
G. O. No. 44, W. D., 1919.

Private, medical detachment, 166th Infantry, 42d Division.
Severely wounded while doing first-aid work, he declined to go to the rear, but dressed his own wound and continued to advance with his battalion, treating the wounded and assisting in their evacuation until he was sent to the hospital 12 hours later.

RIECKE, HENRY A_____
Near Bouresches, France, July 20, 1918.
R—Meriden, Conn.
B—Brooklyn, N. Y.
G. O. No. 125, W. D., 1918.

First lieutenant, 102d Infantry, 26th Division.
When the advance of his company was temporarily held up by machine-gun fire in front of Bouresches he went ahead alone and, although hit 3 times by machine-gun bullets, he continued to urge his men forward, and by his example of fearlessness and grit inspired them to successful attack.

RIEGER, JAMES E_____
Near Charpentry, France, Sept. 27, 1918.
R—Kirksville, Mo.
B—Peoria, Ill.
G. O. No. 59, W. D., 1919.

Lieutenant colonel, 138th Infantry, 35th Division.
He commanded the battalion which had, with conspicuous gallantry, captured Vauquois Hill and the Bois-de-Rosignol, and which was later held up for some hours in front of Charpentry by severe artillery and machine-gun fire. He placed himself in front of all his men, and thus starting them forward led them to the attack with such speed and dash that a large number of the enemy were cut off and captured.

RIGGIO, STEPHANO (1698585)_____
Near Septsarges, France, Sept. 28, 1918.
R—Rockland, Me.
B—Italy.
G. O. No. 46, W. D., 1919.

Private, Company K, 39th Infantry, 4th Division.
While his company was halted by machine-gun and sniper fire from the front and both flanks, Private Riggio moved forward to outflank the enemy sniping posts. He was wounded in the execution of his mission, but he managed to make his way back and reported the information he had obtained.

RIGGLE, GEORGE (1979810)_____
Near Bellicourt, France, Sept. 29, 1918.
R—Sellersburg, Ind.
B—Sellersburg, Ind.
G. O. No. 37, W. D., 1919.

Private, Company D, 120th Infantry, 30th Division.
With 8 other soldiers, comprising the company headquarters detachment, he assisted his company commander in cleaning out enemy dugouts along a canal and capturing 242 prisoners.

RIGGSBY, ROBERT (91713)_____
Near Landres-et-St. Georges, France, Oct. 14, 1918.
R—New York, N. Y.
B—Colson, Ky.
G. O. No. 35, W. D., 1919.

Private, first class, Company M, 165th Infantry, 42d Division.
When the advance of his platoon was held up by machine-gun fire he went forward alone, killed 1 and captured 5 of the enemy machine-gun crew, and succeeded in silencing 2 machine guns, thus permitting his platoon to continue their advance.

RIGO, ALPHONSE M. (1902308)_____
Near St. Juvin, France, Oct. 11, 1918.
R—New York, N. Y.
B—New York, N. Y.
G. O. No. 68, W. D., 1920.

Private, Company H, 326th Infantry, 82d Division.
Private Rigo exposed himself to heavy artillery and machine-gun fire in crossing the Aire River on several occasions to deliver important messages to the commander of the attacking force. He then carried wounded from exposed positions to shelter across the Aire River.

RILEY, CHARLES R. (1709722)_____
Near Grand Pre, France, Oct. 14, 1918.
R—Binghamton, N. Y.
B—Oswego, N. Y.
G. O. No. 35, W. D., 1919.

Sergeant, Company I, 308th Infantry, 77th Division.
When his company was halted by machine-gun fire which threatened to wipe out his entire number, Sergeant Riley led a patrol and charged the nest, and was successful not only in cleaning out the stronghold but in enabling his company to command a more favorable position.

*RILEY, LOWELL H_____
At Ville-Savoye, northeast of Chateau-Thierry, France, Aug. 7, 1918.
R—Orange, N. J.
B—Orange, N. J.
G. O. No. 116, W. D., 1918.

Second lieutenant, 58th Infantry, 4th Division.
Lieutenant Riley maintained an observing station for his battalion commander for 2 days, although subjected during the whole of this time to intense artillery bombardment. He obtained valuable information as to the movements of the enemy, which was used in directing artillery fire. While engaged in this very important and hazardous work he was killed by shellfire. Posthumously awarded. Medal presented to father, Abram M. Riley.

RILEY, RAYMOND W. (57550)_____
Near Soissons, France, July 19, 1918.
R—Baltimore, Md.
B—Baltimore, Md.
G. O. No. 117, W. D., 1918.

Private, Company F, 28th Infantry, 1st Division.
He showed absolute disregard for the safety of his own life by advancing upon a machine gun which was holding up his platoon, and finally putting it out of action after being wounded himself.

RILEY, ROBERT R. (1243689) _____
At Fismette, France, Aug. 10, 1918.
R—Chester, Pa.
B—Chester, Pa.
G. O. No. 99, W. D., 1918.

Corporal, Company C, 111th Infantry, 28th Division.
Having heard that 2 wounded comrades were lying in advance of the line immediately north of Fismette, Corporal Riley and 2 other members of his company volunteered to go through machine-gun and rifle fire to bring them in. On their first attempt all were wounded and driven back, but in spite of their injuries they advanced a second time and reached the wounded men. Their courageous effort, however, was unfortunately in vain, as their comrades had been killed.

*RINDEAU, ARTHUR J. _____
At Chateau-Thierry, France, June 6, 1918.
R—Southbridge, Mass.
B—Saratoga, N. Y.
G. O. No. 110, W. D., 1918.

Gunnery sergeant, 47th Company, 5th Regiment, U. S. Marine Corps, 2d Division.
Killed in action at Chateau-Thierry, France, June 6, 1918, he gave the supreme proof of that extraordinary heroism which will serve as an example to hitherto untried troops.
Posthumously awarded. Medal presented to sister, Mrs. Alma R. Bernier.

RINEBOLD, WILLIAM J. (8168) _____
Near Fismes, France, Aug. 8-9, 1918.
R—Athens, Pa.
B—Superior, Wis.
G. O. No. 26, W. D., 1919.

Private, first class, section No. 524, Ambulance Service.
He volunteered and acted as guide for ambulances going to the most advanced points for the wounded. He made 9 trips over a road subjected to heavy shell and machine-gun fire and was severely wounded by a shell fragment on his last trip.

*RINEHART, ERNEST C. (1311529) _____
Near St. Martin-Riviere, France, Oct. 11, 1918.
R—Leesville, S. C.
B—Saluda County, S. C.
G. O. No. 46, W. D., 1919.

Corporal, Company H, 118th Infantry, 30th Division.
Seeing a wounded comrade lying helpless in a most exposed position in front of our lines, he unhesitatingly braved the murderous fire of machine guns and snipers by going forward to his rescue. He succeeded in bringing in the wounded man after he had seen a stretcher bearer instantly killed in attempting the same mission.
Posthumously awarded. Medal presented to widow, Mrs. Ernest C. Rinehart.

*RINGER, HARVEY C. _____
South of Soissons, France, July 18, 1918.
R—Fulton, Kans.
B—Paola, Kans.
G. O. No. 39, W. D., 1920.

First lieutenant, 18th Infantry, 1st Division.
Lieutenant Ringer personally led his company through heavy artillery and machine-gun fire in an attack upon a strongly fortified position until he fell mortally wounded. His gallantry and personal leadership were material factors in the successful attack.
Posthumously awarded. Medal presented to widow, Mrs. Ida Ringer.

RIPPETOE, GROVER C. _____
Near Soissons, France, July 18-22, 1918.
R—Charleston, W. Va.
B—Clay County, W. Va.
G. O. No. 15, W. D., 1919.

First lieutenant, 26th Infantry, 1st Division.
After all the other officers of his company had been killed near Soissons, France, July 18-22, 1918, he took command, attacked a machine-gun nest, and captured the gun with its crew. Gassed and suffering from shell shock, he refused to quit his post until the company was relieved.

RISCHMANN, EDWARD (2412427) _____
Near Grand Pre, France, Oct. 23, 1918.
R—Newark, N. J.
B—Newark, N. J.
G. O. No. 37, W. D., 1919.

Private, Company I, 312th Infantry, 78th Division.
He was a member of an assaulting party which stormed and captured the citadel at Grand Pre. He scaled the wall and alone entered a dugout, from which he captured 45 Germans, guarding them until assistance arrived.

*RISMILLER, CHARLES C. (571132) _____
Near St. Thibaut, France, Aug. 5, 1918.
R—Leesport, Pa.
B—Leesport, Pa.
G. O. No. 5, W. D., 1920.

Private, medical detachment, 4th Engineers, 4th Division.
He went forward exposed to intense rifle, machine-gun, and artillery fire and assisted a seriously wounded comrade to a place of safety, thus saving his life. In the performance of this gallant act Private Rismiller was mortally wounded.
Posthumously awarded. Medal presented to mother, Mrs. Maggie Yeager.

RITCHIE, EDWARD D. (562140) _____
Near St. Thibaut, France, Aug. 10, 1918.
R—Stratford, Tex.
B—Dallas, Tex.
G. O. No. 46, W. D., 1919.

Private, Company M, 47th Infantry, 4th Division.
While on an outpost near the Vesle River, he volunteered to accompany Corpl. John S. Weimer in rescuing a wounded soldier who had been left by members of a patrol in a shell hole some distance to the front. Under fire from machine guns and snipers, Private Ritchie and Corporal Weimer proceeded to the shell hole and found the wounded man, who was unable to walk. Suggesting that the 3 of them in a group would make a more conspicuous target for the enemy, Private Ritchie offered to run ahead to draw the enemy fire while his comrade assisted the wounded man. He made his way back to shelter under continuous machine-gun and sniper fire, while Corporal Weimer carried the wounded soldier to safety.

*RITZERT, CHARLES T. (1953948) _____
Near Courmont and St. Martin, France, July 31-Aug. 4, 1918.
R—Chicora, Pa.
B—St. Joseph, Pa.
G. O. No. 20, W. D., 1919.

Private, Company G, 125th Infantry, 32d Division.
Throughout the battle to force passage of the Ourcq River and capture the heights beyond, Private Ritzert, a stretcher bearer, worked day and night evacuating wounded under heavy artillery and machine-gun fire. On Aug. 4, under violent shellfire opposite Mont St. Martin, he made repeated trips between the firing line and dressing station until he was killed by a shell.
Posthumously awarded. Medal presented to father, Adam Francis Ritzert.

RIVEL, THOMAS M. (1778977) _____
Near Montfaucon, France, Sept. 28-30, 1918.
R—Philadelphia, Pa.
B—Philadelphia, Pa.
G. O. No. 37, W. D., 1919.

Sergeant, headquarters detachment, 79th Division.
While acting as a division observer Sergeant Rivel remained constantly on duty for several days in a building on the heights of Montfaucon. While in this building portions of it were destroyed by direct artillery hits, and hospital shelling was such that other observers located in the same building withdrew to a safer place. He, however, remained constantly at his post of duty and obtained important information.

RIVERS, TOM (2169507)............
    Near the Bois-de-la-Voivrotte, France, Nov. 11, 1918.
    R—New Castle, Ala.
    B—Opelika, Ala.
    G. O. No. 44, W. D., 1919.

Private, Company G, 366th Infantry, 92d Division.
Although gassed, he volunteered and carried important messages through heavy barrages to the support companies. He refused first aid until his company was relieved.

*RIVES, JOHN S. (1315603)............
    Near Bellicourt, France, Sept. 29, 1918.
    R—Lincoln, Tenn.
    B—Lincoln County, Tenn.
    G. O. No. 87, W. D., 1919.

Private, Company E, 119th Infantry, 30th Division.
Showing marked personal bravery, he repeatedly crossed shell-swept areas, subjected to heavy machine-gun fire, to deliver important messages. Wounded in the head by shrapnel, he bound up the wound and continued his work of maintaining liaison until he was killed by machine-gun fire.
Posthumously awarded. Medal presented to sister, Miss Florence Rives.

ROACH, HARRY E. (1261251)............
    Near Fismes, France, Aug. 10–11, 1918.
    R—Philadelphia, Pa.
    B—Fort Washington, Pa.
    G. O. No. 15, W. D., 1919.

Wagoner, 110th Ambulance Company, 103d Sanitary Train, 28th Division.
Because of the destruction from shellfire of 10 of the 13 ambulances of his company, he worked for 48 hours driving through a shell-swept and gas-infested area, thereby making possible the evacuation of the wounded.

ROACH, JAMES J............
    Near Cunel, France, Oct. 4, 1918.
    R—Boston, Mass.
    B—Boston, Mass.
    G. O. No. 95, W. D., 1919.

First lieutenant, 8th Machine Gun Battalion, 3d Division.
As Lieutenant Roach and an infantry captain were making a reconnaissance under enemy fire both of them were wounded by machine-gun bullets. Disregarding his own wound, Lieutenant Roach secured assistance for his wounded companion and then organized his platoon and the infantrymen near by for an expected hostile counterattack. After seeing that all his guns were in position and his men under cover from the increasing enemy fire, though he was weak from loss of blood, he assisted in carrying the wounded captain to the aid station.

ROBART, RALPH W............
    Near Belleau Wood, France, July 20, 1918.
    R—Arlington, Mass.
    B—Wakefield, Mass.
    G. O. No. 64, W. D., 1919.

Second lieutenant, 104th Infantry, 26th Division.
After being painfully wounded by machine-gun fire, the leader lost, this officer assumed command of the company and gallantly led it to its objective. He immediately reorganized his command and while energetically engaged in this work he was sent to the rear for treatment.

ROBB, WINFRED E............
    Throughout the advance across the River Ourcq, northeast of Chateau-Thierry, France, July 26–Aug. 2, 1918.
    R—Des Moines, Iowa.
    B—Nebraska.
    G. O. No. 99, W. D., 1918.

First lieutenant, chaplain, 168th Infantry, 42d Division.
During the pursuit of the enemy by the 168th Infantry across the River Ourcq, he distinguished himself by his bravery under fire. During all of this time, and particularly during the operations near Sergy, he showed the greatest coolness under severe artillery fire in attending and carrying the wounded and dying, and in every way ministering to the needs of the men of his regiment.

*ROBBINS, CARL (324314)............
    At Posolskaya, Siberia, Jan. 10, 1920.
    R—Concord, Tenn.
    B—Blount County, Tenn.
    G. O. No. 35, W. D., 1920.

Sergeant, Company M, 27th Infantry.
When his platoon was attacked in the night by an armored train, he climbed up on the engine of the armored train in the face of pistol and machine-gun fire and hurled a grenade into the cab, which rendered the engine incapable of further operation, losing his life by his gallant conduct.
Posthumously awarded. Medal presented to mother, Mrs. Alice Robbins.

ROBBINS, CHARLES A. (1746061)............
    Near Bois-de-Grand Fontaine, France, Sept. 26, 1918.
    R—South Manchester, Conn.
    B—South Manchester, Conn.
    G. O. No. 37, W. D., 1919.

First sergeant, Company D, 311th Infantry, 78th Division.
Although severely wounded, he continued to advance with the company until the objective was reached, and then returned to the rear only when ordered to do so by his commanding officer. He then assisted in the removal of the wounded to a dressing station.

*ROBBINS, WILLIAM E. (1314788)............
    Near Bellicourt, France, Sept. 29, 1918.
    R—Wilson, N. C.
    B—Wilson, N. C.
    G. O. No. 21, W. D., 1919.

Private, Company A, 119th Infantry, 30th Division.
During an attack by his regiment, Private Robbins was wounded in the leg. Having dressed his own wound, he continued to advance with his Lewis gun and ammunition until he was killed by shellfire.
Posthumously awarded. Medal presented to father, Tom Robbins.

*ROBERGE, JOSEPH H. (1743922)............
    Near St. Juvin, France, Oct. 18, 1918.
    R—Manchester, N. H.
    B—Canada.
    G. O. No. 44, W. D., 1919.

Private, Headquarters Company, 309th Infantry, 78th Division.
He displayed remarkable courage in laying and repairing a telephone line under such intense artillery and machine-gun fire that the line was ultimately abandoned. He was killed in action Oct. 28, 1918.
Posthumously awarded. Medal presented to widow, Mrs. Louise Roberge.

ROBERGE, PHILIP (110649)............
    Near Belleau, France, July 18, 1918.
    R—Danielson, Conn.
    B—Oscoda, Mich.
    G. O. No. 125, W. D., 1918.

Private, Company D, 103d Machine Gun Battalion, 26th Division.
He showed absolute disregard of personal danger while acting as litter bearer, bringing in wounded from his own and other companies under heavy machine-gun and artillery fire. While carrying a stretcher, which bore a wounded soldier, he received a wound which put out 1 of his eyes, yet he continued until the wounded man had been taken to the dressing station.

ROBERSON, JOSEPH N. (1320073)............
    Near Bellicourt, France, Sept. 29, 1918.
    R—Saxapahaw, N. C.
    B—Alamance County, N. C.
    G. O. No. 37, W. D., 1919.

First sergeant, Company D, 120th Infantry, 30th Division.
With 8 other soldiers, comprising the company headquarters detachment, he assisted his company commander in cleaning out enemy dugouts along a canal and capturing 242 prisoners.

ROBERTS, ARTHUR S. (1785375)_____
Near Montfaucon, France, Sept. 28-30, 1918.
R—Philadelphia, Pa.
B—Fredericksburg, Va.
G. O. No. 37, W. D., 1919.

Private, Headquarters Company, 315th Infantry, 79th Division.
While acting as a division observer, Private Roberts remained constantly on duty for several days in a building on the heights of Montfaucon. While in this building portions of it were destroyed by direct artillery hits and hospital shelling was such that other observers located in the same building withdrew to a safer place. Private Roberts, however, remained constantly at his post of duty and obtained important information.

ROBERTS, CHARLES DEWAYNE_____
In the Bois-de-Belleau, France, June 6 and 8, 1918.
R—Cleveland, Ohio.
B—Kansas City, Kans.
G. O. No. 110, W. D., 1918.

First lieutenant, 6th Regiment, U. S. Marine Corps, 2d Division.
He showed rare courage in repeatedly leading his platoon to an attack against an impregnable machine-gun position. Severely wounded and having lost the greater part of his men, he remained in action and persisted in requesting reinforcements with which to renew the attack.

ROBERTS, CHESTER A. (1491227)_____
Near St. Etienne, France, Oct. 8, 1918.
R—Cleburne, Tex.
B—Cleburne, Tex.
G. O. No. 66, W. D., 1919.

Sergeant, Company L, 142d Infantry, 36th Division.
He led an automatic-rifle team of 7 men in an attack on an enemy machine-gun nest, advancing 150 yards under heavy machine-gun fire to within 50 yards of the enemy position, from which point he directed the fire of his team with such skill that the enemy surrendered, resulting in the capture of 4 officers, 112 men, and 17 machine guns.

ROBERTS, CLAIR C._____
Near Landres-et-St. Georges, France, Oct. 25, 1918.
R—Altoona, Pa.
B—Huntington, Pa.
G. O. No. 15, W. D., 1919.

Second lieutenant, 167th Infantry, 42d Division.
His platoon suffered heavy casualties and he himself was gassed in the advance on Hill 260. Being the first to reach this hill, he observed that the enemy were forming for a counterattack. Displaying coolness and quick judgment he organized all the available men in his vicinity and launched a vigorous attack upon the enemy, who were routed. The daring and leadership of this officer enabled the support to reach Hill 260 without further fighting.

ROBERTS, CLAUDE R. (284299)_____
Near Terny-Sorny, France, Sept. 1, 1918.
R—Wausau, Wis.
B—Racine, Wis.
G. O. No. 44, W. D., 1919.

Corporal, Company G, 128th Infantry, 32d Division.
Although wounded in the left hand and forearm, he remained with his platoon throughout the attack and rendered valuable assistance in silencing enemy snipers by his effective rifle fire.

ROBERTS, GARY A. (95870)_____
Northeast of Chateau-Thierry, France, July 26-27, 1918.
R—Bay Minette, Ala.
B—Farmersville, Ala.
G. O. No. 108, W. D., 1918.

Corporal, Company B, 167th Infantry, 42d Division.
Three times wounded in action, he nevertheless continued in the attack under heavy enemy fire from artillery and machine guns, thereby setting the men of his command an example of exceptional bravery and devotion to duty.

*ROBERTS, JAMES H. (557227)_____
Near Montfaucon, France, Sept. 26-28, 1918.
R—South Manchester, Conn.
B—Rouses Point, N. Y.
G. O. No. 46, W. D., 1920.

Sergeant, Company K, 39th Infantry, 4th Division.
Sergeant Roberts displayed marked courage and self-sacrifice, when, after being wounded in the arm, he refused to leave the battle field and continued to perform his duties as platoon sergeant until he was wounded in the knee 2 days later and had to be carried from the field.
Posthumously awarded. Medal presented to father, Peter Roberts.

*ROBERTS, JAMES H._____
South of Soissons, France, July 18, 1918.
R—Baltimore, Md.
B—Baltimore, Md.
G. O. No. 55, W. D., 1920.

Private, 51st Company, 5th Regiment, U. S. Marine Corps, 2d Division.
Private Roberts, armed with an automatic rifle, crawled through a wire entanglement and disabled 1 machine gun with a hand grenade and forced the crew of a second gun to surrender. His action enabled his company to pass through the entanglement without serious loss.
Posthumously awarded. Medal presented to mother, Mrs. Elizabeth Goetz.

ROBERTS, LEO D. (553000)_____
Near Nantillois, France, Oct. 12, 1918.
R—Bellefontaine, Ohio.
B—Harper, Ohio.
G. O. No. 87, W. D., 1919.

Sergeant, Company A, 11th Machine Gun Battalion, 4th Division.
After the infantry had fallen back 200 meters under heavy fire, Sergeant Roberts stayed at his 1 remaining machine gun and operated it until the Infantry had reestablished its position, capturing a German machine gun and 3 prisoners.

ROBERTS, SEWELL K. (1317576)_____
Near Bellicourt, France, Sept. 29, 1918.
R—Rockwood, Tenn.
B—Rockwood, Tenn.
G. O. No. 81, W. D., 1919.

Private, Company H, 119th Infantry, 30th Division.
Advancing alone against 2 enemy positions in succession, he killed the machine gunners with hand grenades, allowing our advance to continue. Severely wounded, he refused to be evacuated until ordered to the rear by his company commander.

ROBERTSON, ANGUS (1204123)_____
Near Renssoy, France, Sept. 25, 1918.
R—Yonkers, N. Y.
B—Brooklyn, N. Y.
G. O. No. 37, W. D., 1919.

Sergeant, Company E, 105th Infantry, 27th Division.
Although suffering intense agony from the effects of a severe gassing, he continued in command of his platoon during a most terrific shelling. By administering first aid to a wounded comrade he was instrumental in saving his life, although risking his own by removing his gas mask to render more valuable treatment. He continued to assist the wounded until he collapsed.

ROBERTSON, ARCHIBALD G._____
Near Thiaucourt, France, Sept. 12, 1918.
R—Staunton, Va.
B—Staunton, Va.
G. O. No. 46, W. D., 1919.

Second lieutenant, 9th Infantry, 2d Division.
Although wounded by shell fire early in the attack, he refused to go to the rear but continued to lead his platoon to the objective, where, under heavy machine-gun fire, he prepared his position for the enemy counterattack and held it throughout the night, remaining with his platoon until it was relieved the following day.

ROBERTSON, JAMES FERGUSON_____
In the capture of Bouresches, France, June 6, 1918.
R—Chicago, Ill.
B—New Zealand.
G. O. No. 119, W. D., 1918.

First lieutenant, 96th Company, 6th Regiment, U. S. Marine Corps, 2d Division.
He displayed marked courage and resourcefulness in the capture of Bouresches, France. With 1 platoon of his company on the night of June 6, 1918, in the face of heavy machine-gun barrage, he entered the town and heroically withstood vigorous attempts of superior forces to dislodge him.

*ROBERTSON, MALCOLM T. (88640)_____
At Bois Coles, north of the River Ourcq, near Villers-sur-Fere, France, July 30, 1918.
R—Brooklyn, N. Y.
B—Brooklyn, N. Y.
G. O. No. 13, W. D., 1923.

Private, Headquarters Company, 165th Infantry, 42d Division.
In the absence of his platoon commander, who had gone to an advanced position for observation, and after his section sergeant had been wounded and evacuated, he assumed leadership of his Stokes mortar crew; when called upon by his platoon commander for Stokes mortar fire to repel an assault by the enemy, who were advancing 100 yards away, he with his crew responded so effectively as to repulse the enemy with heavy losses. He was killed by an enemy shell while consolidating a position and rallying his men to repel the assault.
Posthumously awarded. Medal presented to father, Dr. Victor A. Robertson.

ROBERTSON, RAYMOND D. (560255)_____
West of Fismes, France, Aug. 5, 1918.
R—Berkeley, Calif.
B—Valley Ford, Calif.
G. O. No. 15, W. D., 1919.

Sergeant, Company F, 4th Engineers, 4th Division.
Sergeant Robertson was a member of a small detachment of engineers which went out in advance of the front line of the Infantry through an enemy barrage from 77-millimeter and 1-pounder guns to construct a footbridge over the River Vesle. As soon as their operations were discovered machine-gun fire was opened up on them, but, undaunted, the party continued work, removing the German wire entanglements and successfully completing a bridge which was of great value in subsequent operations.

ROBINS, EMMETT W. (486867)_____
Near Vieville-en-Haye, France, Oct. 20, 1918.
R—Wichita, Kans.
B—Rich Hill, Mo.
G. O. No. 35, W. D., 1919.

Private, Company F, 64th Infantry, 7th Division.
Private Robins, with 4 other soldiers, was on duty in an observation post when a German patrol 15 or 20 strong attacked with a machine gun. His 4 companions being immediately killed or wounded, Private Robins, with great coolness and courage, remained at his post and returned the fire with such good effect that the enemy broke and fled, leaving 2 dead and the machine gun behind.

ROBINS, JOSEPH (1212534)_____
Near Ronssoy, France, Sept. 29, 1918.
R—New York, N. Y.
B—New York, N. Y.
G. O. No. 35, W. D., 1919.

Sergeant, Machine Gun Company, 107th Infantry, 27th Division.
During the thick of the fighting against the Hindenburg line Sergeant Robins voluntarily went forward to locate friendly troops, and in doing so he was obliged to pass between two strongly fortified enemy nests, from which a deadly fire was pouring. Despite the fact that he was badly wounded, he communicated with the Infantry and returned to his position.

*ROBINSON, ARTHUR A_____
Near Chery-Chartreuve, France, Aug. 16, 1918.
R—Flushing, N. Y.
B—Brooklyn, N. Y.
G. O. No. 37, W. D., 1919.

Second lieutenant, 305th Field Artillery, 77th Division.
Lieutenant Robinson, assisted by Corporal Johnson, of the same battery, rescued the body of an officer from a flaming ammunition dump which was under fire.
Posthumously awarded. Medal presented to widow, Mrs. Florence E. Robinson.

ROBINSON, ARTHUR HARRISON_____
Near Ville-Savoye, France, Aug. 22, 1918.
R—Madison, Wis.
B—Baraboo, Wis.
G. O. No. 35, W. D., 1919.

First lieutenant, 308th Infantry, 77th Division.
Under a screen of dense fog and the smoke of a heavy barrage, the Germans set up a machine gun within 30 yards of the flank of Lieutenant Robinson's company. The Germans opened up a deadly fire as the fog lifted, but Lieutenant Robinson attacked the position with grenades and drove off the enemy. He then turned the gun on the advancing Germans, completely breaking up their counterattack.

*ROBINSON, CALDWELL C_____
At Chateau-Thierry, France, June 6, 1918.
R—Hartford, Conn.
B—Hartford, Conn.
G. O. No. 119, W. D., 1918.

Second lieutenant, 6th Regiment, U. S. Marine Corps, 2d Division.
Killed in action at Chateau-Thierry, France, June 6, 1918, he gave the supreme proof of that extraordinary heroism which will serve as an example to hitherto untried troops.
Posthumously awarded. Medal presented to mother, Mrs. C. L. F. Robinson.

ROBINSON, FRANK N. (549412)_____
Near Mezy, France, July 15, 1918.
R—Newton Center, Mass.
B—Boston, Mass.
G. O. No. 23, W. D., 1919.

Sergeant, Company A, 38th Infantry, 3d Division.
During the heavy enemy artillery bombardment which preceded the German offensive of July 15, 1918, Sergeant Robinson showed great bravery by voluntarily rescuing wounded men under severe fire.

ROBINSON, HENRY (2856888)_____
Near Fey-en-Haye, France, Sept. 12, 1918.
R—Valley City, N. Dak.
B—Bloomington, Ill.
G. O. No. 16, W. D., 1920.

Private, Company A, 357th Infantry, 90th Division.
Private Robinson, with Corporal Kach, successfully attacked a German machine-gun nest from the flank, killing 7 Germans, wounding 1, and capturing 3 machine guns.

ROBINSON, JOHN J. (540549)_____
At Cunel Heights, France, Oct. 12, 1918.
R—Manchester, N. Y.
B—Black Rock, N. Y.
G. O. No. 98, W. D., 1919.

Corporal, Machine Gun Company, 7th Infantry, 3d Division.
After 3 other runners had been wounded in attempting to carry a message to a platoon in advance of the front line, Corporal Robinson volunteered for this hazardous mission, and, passing 400 meters under direct machine-gun fire of the enemy, succeeded in reaching the platoon and notifying it to withdraw before our artillery barrage began to fall.

ROBINSON, JOHN M. (551291)_____
Near Mezy, France, July 15, 1918, and south of Cunel, France, Oct. 11, 1918.
R—Philadelphia, Pa.
B—Philadelphia, Pa.
G. O. No. 3, W. D., 1921.

Sergeant, Company G, 38th Infantry, 3d Division.
During the enemy offensive Sergeant Robinson held an advanced lookout post exposed to heavy artillery fire in order to warn his platoon of the approach of the enemy. In the Argonne, on Oct. 11, he led several members of his company in advance of our lines and rescued several wounded comrades.

ROBINSON, OLIVER THOMAS (1319236)___
  Near Vaux-Andigny, France, Oct.
  10, 1918.
  R—Waterford, Miss.
  B—Waterford, Miss.
  G. O. No. 35, W. D., 1919.

Sergeant, Company A, 120th Infantry, 30th Division.
Although severely wounded, he displayed remarkable coolness in extricating his platoon from an extremely dangerous position under terrific shell and machine-gun fire, thereby saving it from almost certain annihilation. Being wounded a second time, he refused to go to the rear until ordered to so so by his company commander.

ROBINSON, PHILIP K._____
  Near Mont-Notre-Dame, France,
  Sept. 10, 1918.
  R—Green Bay, Wis.
  B—Green Bay, Wis.
  G. O. No. 14, W. D., 1923.

First lieutenant, 306th Infantry, 77th Division.
The 3d Battalion, 306th Infantry, having relieved another regiment in a position in front of Mont-Notre-Dame, it was reported that wounded members of the organization relieved were in the Bois-de-Chandriere in front of the position occupied by the 3d Battalion. Lieutenant Robinson, together with a non-commissioned officer and a private of his battalion, volunteered to search the wood and proceeded on his hazardous mission, crossing an open field, a distance of 600 yards, under direct observation of the enemy and under a concentration of heavy enemy machine-gun and artillery fire. Finding several wounded men he assisted them to return to their own lines, undoubtedly saving their lives. The extraordinary heroism and soldierly devotion to duty displayed by Lieutenant Robinson greatly inspired the men of his battalion.

ROBINSON, WILLIAM F. (42872)_____
  Near Soissons, France, July 19, 1918.
  R—Staples, Minn.
  B—Staples, Minn.
  G. O. No. 15, W. D., 1919.

Corporal, Company F, 16th Infantry, 1st Division.
He displayed the highest type of bravery in rescuing 3 wounded comrades from the hands of the enemy under violent artillery and machine-gun fire.

ROBINTON, CHARLES H. (1203329)_____
  Near Roussoy, France, Sept. 29, 1918.
  R—Ilion, N. Y.
  B—Lawrence, Mass.
  G. O. No. 32, W. D., 1919.

Private, Company A, 105th Infantry, 27th Division.
During operations against the Hindenburg line, he went forth in the face of unusually heavy machine-gun fire to aid a wounded comrade. He administered first aid, and, while shielding the man from the enemy fire, he received a severe wound in the back. Despite this wound, he struggled back to safety, bringing his comrade with him.

ROBISON, EDWARD M._____
  Near Monthois, France, Oct. 1–3,
  1918.
  R—Flagstaff, Ariz.
  D—Beallesville, Pa.
  G. O. No. 13, W. D., 1919.

Captain, 372d Infantry, 93d Division.
Although he was severely wounded, he remained with his battalion for 2 days, continuously under heavy shell and machine-gun fire, encouraging his men and inspiring them by his example. He led them to the attack until he collapsed from the effects of his wound.

ROCHFORD, PATRICK (1710149)_____
  West of St. Juvin, France, Oct. 16,
  1918.
  R—New York, N. Y.
  B—Ireland.
  G. O. No. 20, W. D., 1919.

Private, Company L, 308th Infantry, 77th Division.
Private Rochford, with another soldier, volunteered to cross a level open space for 600 yards swept by converging machine-gun fire to deliver a message to the front line, undeterred by the knowledge that 6 other soldiers had been wounded in a similar attempt. Crawling from one shell hole to another, he succeeded in reaching the front line and delivering the message.

ROCHFORT, JAMES J. (1386502)_____
  Near St. Hilaire Woods, France,
  Nov. 9, 1918.
  R—Chicago, Ill.
  B—Chicago, Ill.
  G. O. No. 71, W. D., 1919.

Sergeant, Company B, 131st Infantry, 33d Division.
With utter disregard of personal danger, he advanced alone and attacked a machine-gun nest which was inflicting heavy casualties on his company. Exposing himself to heavy fire, he killed 2 of the gun crew and routed the others, allowing his company to resume the advance.

ROCHKIND, WILLIAM (1699283)_____
  Near St. Juvin, France, Oct. 16, 1918.
  R—New York, N. Y.
  B—Russia.
  G. O. No. 44, W. D., 1919.

Corporal, Company I, 305th Infantry, 77th Division.
While leading a reconnaissance patrol of 8 men, Corporal Rochkind encountered severe machine-gun fire which forced him to take cover. While attempting to move his patrol to another place during a lull in the firing 1 man was killed and another wounded. Placing his men under cover, he alone ventured forth to the aid of the wounded man, who was lying exposed to the enemy fire, placing him on his back and carrying him to safety.

*ROCK, WILLIAM C._____
  Near Molain, France, Oct. 17, 1918.
  R—Philadelphia, Pa.
  B—Pittsburgh, Pa.
  G. O. No. 32, W. D., 1919.

Second lieutenant, 301st Battalion, Tank Corps.
Lieutenant Rock was in charge of a tank when it was struck by 3 shells from a German trench mortar, which set fire to the tank and knocked the track off. He assisted in extricating the wounded men from the tank and carrying them to the only available cover. In attempting to put an enemy machine gun out of action with his pistol he was killed.
Posthumously awarded. Medal presented to father, William D. Rock.

ROCKEY, KELLER EMRICK_____
  At Chateau-Thierry, France, June 6,
  1918.
  R—Stone Harbor, N. J.
  B—Columbia City, Ind.
  G. O. No. 110, W. D., 1918.

Captain, 5th Regiment, U. S. Marine Corps, 2d Division.
He performed distinguished service by bringing up supports and placing them in the front lines at great personal exposure, showing exceptional ability and extraordinary heroism. He was indefatigable and invaluable in carrying forward the attack and organizing and holding the position.

ROCKWELL, JOHN C. (39672)_____
  South of Soissons, France, July 18,
  1918.
  R—Rockwell Springs, N. Y.
  B—Rockwell Springs, N. Y.
  G. O. No. 20, W. D., 1919.

Private, Company G, 9th Infantry, 2d Division.
Jerome Buschman, sergeant; John Rockwell, private; William F. Rockwell, private; Alfred Shimanoski, private; and Watzlaw Viniarsky, private; all of Company G, 9th Infantry. For extraordinary heroism in action south of Soissons, France, July 18, 1918. They conspicuously distinguished themselves by attacking a party of more than 60 Germans and, in an intense and desperate hand-to-hand fight, succeeded in killing 22 men and capturing 40 men and 5 machine guns.

*ROCKWELL, MEARL COLIN.
Near Torcy, France, June 4, 1918.
R—Holly, Colo.
B—Coolidge, Kans.
G. O. No. 119, W. D., 1918.

Private, Headquarters Company, 6th Regiment, U. S. Marine Corps, 2d Division.
As a member of a raiding patrol, he displayed great courage and devotion by fearlessly entering extremely dangerous areas and obtaining information imperatively necessary to the success of subsequent operations.
Posthumously awarded. Medal presented to mother, Mrs. Katie M. Rockwell.

*ROCKWELL, WILLIAM F. (39674).
South of Soissons, France, July 18, 1918.
R—Rockwell Springs, N. Y.
B—Rockwell Springs, N. Y.
G. O. No. 20, W. D., 1919.

Private, Company G, 9th Infantry, 2d Division.
Jerome Buschman, sergeant; John Rockwell, private; William F. Rockwell, private; Alfred Shimanoski, private; and Watzlaw Viniarsky, private; all of Company G, 9th Infantry. For extraordinary heroism in action south of Soissons, France, July 18, 1918. They conspicuously distinguished themselves by attacking a party of more than 60 Germans and, in an intense and desperate hand-to-hand fight, succeeded in killing 22 men and capturing 40 men and 5 machine guns.
Posthumously awarded. Medal presented to mother, Mrs. Elizabeth Rockwell.

*ROCKWOOD, RICHARD B.
Near Thiaucourt, France, Sept. 26, 1918.
R—Wurtsboro, N. Y.
B—Wurtsboro, N. Y.
G. O. No. 74, W. D., 1919.

Second lieutenant, 310th Infantry, 78th Division.
Intrusted with an important message from the brigade commander to an infantry unit, he fearlessly crossed a shell-swept area, delivered the message, and, while returning with reply, was mortally wounded by a shell fragment. With great effort, notwithstanding his wound, he delivered the reply to the message and fell unconscious, dying shortly after.
Posthumously awarded. Medal presented to mother, Mrs. W. E. Rockwood.

RODA, MARVIN (2366608).
At Vladivostok, Siberia, Nov. 17 and 18, 1919.
R—San Francisco, Calif.
B—Quigley, Mont.
G. O. No. 16, W. D., 1923.

Sergeant, Company B, Replacement Battalion, American Expeditionary Forces, Siberia.
In answer to a call to save women and children who had been unavoidably entrapped in the railroad station by the fire from contending forces of Russian Government and insurgent troops, he boldly entered the zone of fire and rushed to the station which was being fired upon from 3 sides by machine guns, rifles, and fieldpieces and assisted in bringing back through the fire-swept zone a number of women noncombatants.

RODEN, THOMAS (551174).
Near Romagne, France, Oct. 20, 1918.
R—New York, N. Y.
B—Ireland.
G. O. No. 27, W. D., 1920.

Sergeant, Company G, 38th Infantry, 3d Division.
After his company had suffered heavy casualties and all officers either killed or wounded during the attack on Hill 299, Sergeant Roden reorganized his company and took a position which he held against repeated attack by a superior enemy force. He fearlessly exposed himself to the fire of the attacking party in order to encourage his men and better direct their fire on the enemy, who were repulsed, several of the enemy being taken prisoners. On the following night he voluntarily led a patrol into the enemy lines, captured 4 of the enemy, and returned with valuable information.

*RODGERS, ALEXANDER, Jr.
Near Gercourt, France, Sept. 26 and 27, 1918.
R—Washington, D. C.
B—Washington, D. C.
G. O. No. 9, W. D., 1923.

First lieutenant, 319th Infantry, 80th Division.
On the night of Sept. 26 and during the following day he repeatedly led his detachment in the face of heavy artillery and machine-gun fire in repairing telephone lines between regimental headquarters and front-line battalions. After reestablishing these lines innumerable times, and after they were broken beyond repair, with materials at hand, he, with one of his runners, voluntarily and under intense machine-gun and shell fire carried messages to the front-line elements of his regiment. Later, in action near Cunel, in the Bois-des-Ogons, Oct. 4–12, 1918, severely gassed, he refused to be evacuated, working continuously without sleep for 4 days, keeping lines of communication open to the front, and contracted pneumonia which caused his death on Oct. 23, 1918.
Posthumously awarded. Medal presented to father, Col. Alexander Rodgers.

RODGERS, JAMES F. (3955453).
Near Waereghem, Belgium, Oct. 31, 1918.
R—Bassett, Nebr.
B—Spokane, Wash.
G. O. No. 37, W. D., 1919.

Private, Company L, 363d Infantry, 91st Division.
When intense fire of enemy machine guns had held up the advance of 2 of our companies and the enemy threatened to close in on our troops, he accompanied a squad of men forward to a point where he could most effectively use his automatic rifle. After the squad leader and 2 men had been killed, he remained at his post and forced the advance German posts to retire, making possible the continuance of the progress of our companies.

*RODGERS, JOHN WILEY.
At Chateau-Thierry, France, June 6, 1918.
R—San Diego, Calif.
B—Equality, Ill.
G. O. No. 110, W. D., 1918.

Sergeant, 43d Company, 5th Regiment, U. S. Marine Corps, 2d Division.
Killed in action at Chateau-Thierry, France, June 6, 1918, he gave the supreme proof of that extraordinary heroism which will serve as an example to hitherto untried troops.
Posthumously awarded. Medal presented to sister, Mrs. Elsie Moore.

RODGERS, MARTIS SANDERS.
Near the Champagne sector, France, Oct. 4, 1918.
R—Gordo, Ala.
B—Pleasant Grove, Ala.
G. O. No. 50, W. D., 1919.

Gunnery sergeant, 47th Company, 5th Regiment, U. S. Marine Corps, 2d Division.
Even after being wounded, he volunteered and led a patrol into no man's land and succeeded in returning with 4 wounded comrades. After his platoon commander had been wounded, he assumed command and led the men through most trying and difficult conditions until a second wound forced his removal to the rear.

RODGERS, WILLES (1385026).
Near Consenvoye, France, Oct. 10, 1918.
R—Argenta, Ill.
B—Cedar Rapids, Nebr.
G. O. No. 44, W. D., 1919.

Private, Company M, 130th Infantry, 33d Division.
Although suffering painfully from an infected hand, he acted as stretcher bearer; while his company was in action he made five trips to the dressing stations, a total distance of about 25 miles, and was under shellfire at all stages of his journey.

ROGERS, ALAN.
Near La Palletta Pavillon, France,
Oct. 4, 1918.
R—New York, N. Y.
B—New York, N. Y.
G. O. No. 81, W. D., 1919.

Second lieutenant, 307th Infantry, 77th Division.
Having taken command of his company after the company commander and second in command had been wounded, Lieutenant Rogers personally undertook a reconnaissance of the front line. Crawling forward alone under intense rifle and machine-gun fire for 200 yards to within 30 yards of an enemy machine-gun nest, he was seriously wounded in the knee, but applying a tourniquet to his leg he succeeded in crawling back to his company. Here he resumed command, and though suffering intense pain, gave instructions for repelling an expected counterattack, directing that no man be taken from the firing line to carry him to the rear. For 7 hours after being wounded he remained with his command, inspiring his men by his fortitude and courage.

ROGERS, BENJAMIN F.
Near St. Etienne, France, Oct. 4, 1918.
R—Gresham, Ore.
B—Big Prairie, Mich.
G. O. No. 35, W. D., 1919.

Pharmacist's mate, second class, U. S. Navy, attached to 5th Regiment, U. S. Marine Corps, 2d Division.
He left his shelter and went beyond our most advanced positions, giving first aid to the wounded under machine-gun and shell fire until all had been cared for and evacuated.

ROGERS, FRED (96361).
Northeast of Chateau-Thierry, at Croix Rouge Farm, France, July 26, 1918.
R—Ensley, Ala.
B—Bessemer, Ala.
G. O. No. 132, W. D., 1918.

Sergeant, Company D, 167th Infantry, 42d Division.
After being wounded in the head he continued his advance on the enemy with his platoon until more severely wounded. He died as a result of his injuries. Posthumously awarded. Medal presented to mother, Mrs. J. T. Rogers.

ROGERS, HARRY.
Near Binarville, France, Oct. 2–6, 1918.
R—Liberty, Mo.
B—Carthage, Mo.
G. O. No. 34, W. D., 1924.

Second lieutenant, 308th Infantry, 77th Division.
He was in command of a detachment comprising part of 2 battalions which were cut off and surrounded by the enemy in the Argonne Forest, France. During the days of the isolation from friendly troops, he was on the exposed flank without food. Although under a heavy concentration of fire from enemy machine guns and snipers, by his personal example of calmness he kept his men in order and helped repel counterattacks. This intrepid officer was killed in action Oct. 6, 1918.
Posthumously awarded. Medal presented to uncle, S. D. Rogers.

ROGERS, HORATIO R. (291666).
Near Exermont, France, Oct. 4, 1918.
R—Evanston, Ill.
B—Newport, R. I.
G. O. No. 15, W. D., 1923.

Private, Company C, 344th Battalion, Tank Corps.
Acting as a runner, Private Rogers, upon learning that there was a scarcity of tank drivers, begged permission to drive a tank. Permission being granted, he drove his tank well in advance of the Infantry until the officer in command of his tank became wounded by enemy fire. Private Rogers left the shelter of his tank and crawled to other tanks of his company, carrying messages from his wounded officer. This duty was performed in the face of heavy artillery, machine-gun, and rifle fire, and was carried on until Private Rogers was severely wounded. The coolness, devotion to duty, and fearlessness displayed inspired the men of his company to still greater endeavors.

ROGERS, VERNE E.
Near Avillers, France, Sept. 13, 1918.
R—New Athens, Ohio.
B—Blissfield, Mich.
G. O. No. 37, W. D., 1919.

Second lieutenant, 104th Infantry, 26th Division.
When a section of his platoon was being held up by machine-gun fire, he directed an attack on the nest, while he and a corporal attacked from the opposite side, driving out and capturing 3 of the enemy.

ROGERS, WILL (40462).
Near Soissons, France, July 18, 1918.
R—Limon, Colo.
B—Springfield, Mo.
G. O. No. 37, W. D., 1919.

Private, Company K, 9th Infantry, 2d Division.
After a machine gun had caused great losses in his company, Private Rogers, with 3 other soldiers, made an attack on the gun, which was lodged in a deep ravine. After his companions had been killed or wounded, he continued with the attack and succeeded in silencing the gun.

ROHAN, EDGAR A. (71217).
Near Belleau and Epieds, France, July 20–23, 1918.
R—Worcester, Mass.
B—New Durham, N. H.
G. O. No. 37, W. D., 1919.

Private, Company A, 104th Infantry, 26th Division.
After receiving several wounds he took charge of an automatic rifle and continued in action for 3 days.

ROLAIN, RAY C. (275031).
At Juvigny, France, Aug. 31, 1918.
R—Rhinelander, Wis.
B—Rhinelander, Wis.
G. O. No. 20, W. D., 1919.

Sergeant, Company L, 127th Infantry, 32d Division.
He attacked a machine-gun nest single handed after the 4 soldiers accompanying him had been wounded. He killed the operator of 1 gun and captured the remainder of 2 gun crews and both guns.

ROLFE, ONSLOW S.
Near Fossoy, France, July 14–15, 1918.
R—Concord, N. H.
B—Concord, N. H.
G. O. No. 116, W. D., 1919.

Captain, 7th Infantry, 3d Division.
During the heavy enemy bombardment preceding the second battle of the Marne, Captain Rolfe, regimental intelligence officer, voluntarily carried an important message, in full view of the enemy, across an open field to the support and reserve battalions for the purpose of bringing up reinforcements.

RONERI, VINCENZO (1247694).
Near Chatel-Chehery, France. Oct. 7, 1918.
R—Big Mine Run, Pa.
B—Italy.
G. O. No. 11, W. D., 1921.

Private, Company D, 112th Infantry, 28th Division.
Responding to a call from a noncommissioned officer, Private Roneri exposed himself to heavy machine-gun fire to go 200 yards in advance of our lines and assist in carrying a wounded man to a place of safety. While in the performance of this act, he was exposed to machine-gun fire from 2 directions.

ROONEY, EDWARD K. (91386)----------
    Near Forrest of Parroy, France,
        Mar. 20, 1918.
    R—New York, N. Y.
    B—Ireland.
    G. O. No. 35, W. D., 1919.

Sergeant, Company K, 165th Infantry, 42d Division.
After having successfully passed through an extraordinary heavy barrage of gas and high-explosive shells, he volunteered and carried a message to the front-line detachment, which up to that time had been cut off from all communication with the company.

ROONEY, PAUL N. A----------
    Near Ansauville and Germonville,
        France, July 22–Sept. 26, 1918.
    R—Boston, Mass.
    B—Boston, Mass.
    G. O. No. 15, W. D., 1919.

First lieutenant, Air Service, balloon section, 1st Army.
On July 22, near Ansauville, with Lieutenant Ferrenbach, he was conducting an important observation. At an altitude of 800 meters he was several times attacked by enemy planes, but refused to leave his post until his balloon was set afire, and only then after he had seen that his companion had safely jumped. While descending, his parachute was almost hit by the falling balloon. He insisted upon returning to his post and was in the air again as soon as another balloon could be inflated. On Sept. 26, while adjusting artillery fire, his balloon was attacked by 3 enemy planes (Fokker type). At imminent peril of his life he stuck to his post until 1 plane dived directly at his balloon. He would not leave the basket until his companion, Lieutenant Montgomery, had jumped to safety.

*ROOS, JAMES J----------
    Near St. Souplet, France, Oct. 17,
        1918.
    R—Buffalo, N. Y.
    B—Buffalo, N. Y.
    G. O. No. 37, W. D., 1919.

First lieutenant, 108th Infantry, 27th Division.
During the forcing of La Selle River he made personal reconnaissances of the territory, under terrific machine-gun fire, before leading his men in attack. Advancing to a farm which was strongly fortified by the enemy, he scattered his men about the buildings, from which a deadly fire was pouring, and, advancing alone into the building, captured nearly 200 Germans at the point of his pistol. He was killed in attack the following morning.
Posthumously awarded. Medal presented to widow, Mrs. James S. Roos.

ROOSE, THOMAS W. W----------

    Near Torcy, France, July 18 to 20,
        1918.
    R—Charlestown, Mass.
    B—England.
    G. O. No. 39, W. D., 1920.

Regimental sergeant major, Headquarters, 52d Infantry Brigade, attached to 3d Battalion, 103d Infantry, 26th Division.
On July 20 he crossed an open space, under direct observation and fire of the enemy, to assist a wounded comrade to shelter. Two days previous he made a reconnaissance along the advancing front line, exposed to rifle, machine-gun, and artillery fire, to make sure that orders had been carried out.

ROOSEVELT, THEODORE, Jr----------
    Near Cantigny, France, May 28,
        1918.
    R—New York, N. Y.
    B—Oyster Bay, L. I., N. Y.
    G. O. No. 10, W. D., 1920.
    Distinguished-service medal also
        awarded.

Major, 26th Infantry, 1st Division.
After the completion of a raid Major Roosevelt exposed himself to intense machine-gun, rifle, and grenade fire while he went forward and assisted in rescuing a wounded member of the raiding party. At Soissons, France, July 19, 1918, he personally led the assault companies of his battalion, and although wounded in the knee he refused to be evacuated until carried off the field.

RORISON, HARMON C----------
    Near Beaumont, France, Nov. 3,
        1918.
    R—Wilmington, N. C.
    B—Wilmington, N. C.
    G. O. No. 46, W. D., 1919.

First lieutenant, 22d Aero Squadron, Air Service.
While on a bombing mission his patrol was attacked by 18 enemy planes (type Fokker). Three of his comrades were immediately shot down, but he continued in the fight for 30 minutes and destroyed 2 Fokkers which were attacking the other 2 members of his patrol. With his plane badly damaged and himself wounded, he succeeded in shooting down another Fokker just before 1 of his guns was put out of action. By skillful maneuvering he shook off the rest of the Fokkers and reached his lines, 15 miles away, in safety.

RORTY, JAMES H. (10398)----------
    Near Sommerance, France, Oct. 11,
        1918.
    R—New York, N. Y.
    B—Middletown, N. J.
    G. O. No. 15, W. D., 1919.

Private, section No. 647, Ambulance Service, with French Army.
He was relieved from duty as a mechanic in order that he might serve as aid on cars during the Argonne offensive. While engaged in evacuating wounded from the culvert not far from enemy outposts, fragments of a shell pierced his clothing, and although he was suffering from shock he repeatedly ran ahead in the dark to guide the car over a road partly destroyed by shells and still under enemy machine-gun fire. Returning with relief cars, he again served as guide and as stretcher bearer until the evacuation was completed.

ROSE, DECATUR F. (1316532)----------
    Near St. Souplet, France, Oct. 11,
        1918.
    R—Unaka, N. C.
    B—Unaka, N. C.
    G. O. No. 37, W. D., 1919.

Private, Company K, 119th Infantry, 30th Division.
During an attack by his regiment he was carrying a message from his platoon commander to company headquarters. On the way he met an enemy patrol and, although alone, immediately opened fire upon them, continuing to fire, after being wounded in both legs, until the enemy had been completely routed.

ROSE, HAROLD W. (2276440)----------

    Near Very, France, Sept. 28–Oct. 4,
        1918 and at Audenarde, Belgium,
        Nov. 4, 1918.
    R—Oakland, Calif.
    B—Detroit, Mich.
    G. O. No. 3, W. D., 1919.

Private, first class, 364th Ambulance Company, 316th Sanitary Train, 91st Division.
During the offensive in the Forest of Argonne this soldier displayed unusual courage and devotion to duty in driving a motor cycle for his commanding officer and also in performing liaison service. He repeatedly showed utter disregard for his own life by riding through areas and over roads that were being heavily shelled by the enemy. He was for 3 days and nights without rest and with very little food. When his motor cycle was disabled by shellfire, he continued on foot and delivered a message as he collapsed from exhaustion. On Nov. 4 he drove a motor cycle with his commanding officer into the town of Audenarde to search for wounded, faithfully performing his duty where the streets had been blown up and timbers from bombarded buildings were falling around him.

ᵃROSELL, WILLIAM E. (2061784)_____
　　During the Meuse-Argonne offen-
　　sive, France, Sept. 26, 1918.
　　R—Chicago, Ill.
　　B—Sweden.
　　G. O. No. 46, W. D., 1920.

Private, Company B, 131st Infantry, 33d Division.
With 3 other soldiers he charged and captured a battery of three .77 fieldpieces which, protected by machine guns, were firing point blank on the
position held by his company. This deed enabled his company to continue
the advance.
Posthumously awarded. Medal presented to mother, Mrs. Petronella Rosell.

ROSEN, HARRY (43851)_____
　　Near Fleville, France, Oct. 4, 1918.
　　R—San Francisco, Calif.
　　B—Brooklyn, N. Y.
　　G. O. No. 35, W. D., 1920.

Sergeant, Company K, 16th Infantry, 1st Division.
After all officers of his company had been killed or wounded, Sergeant Rosen
took command and reorganized the company under fire. He then led a
patrol of 6 men through heavy fire in an attack on an enemy machine-gun
nest. With the assistance of other members of his patrol he forced 18 of the
enemy to surrender.

ROSEN, THEODORE_____
　　In the Grande Montagne sector,
　　north of Verdun, Nov. 4, 1918.
　　R—Philadelphia, Pa.
　　B—Carmel, N. J.
　　G. O. No. 19, W. D., 1920.

First lieutenant, 315th Infantry, 79th Division.
While on a reconnaissance with 2 other officers, Lieutenant Rosen drew fire
from a machine-gun nest in order to allow the other 2 officers to escape. A
few minutes later he and 2 runners were sent into the Bois d'Etraye in
order to locate the left flank. Lieutenant Rosen again came under close-range
fire of the enemy. The runner, who was some yards in rear, escaped, but
Lieutenant Rosen, who had been terribly wounded by a hand grenade,
unable to move or resist by further fighting, was taken prisoner.

ROSENBERGER, GEORGE V. (2411026)___
　　Near Vieville-en-Haye, France,
　　Sept. 25-26, 1918.
　　R—Bloomsbury, N. J.
　　B—Bloomsbury, N. J.
　　G. O. No. 44, W. D., 1919.

Private, first class, medical detachment, 311th Infantry, 78th Division.
Private Rosenberger, with another soldier, had advanced to an exposed position,
and then were administering first aid to a wounded man, when suddenly
surrounded and captured by a party of the enemy. While being taken
toward the German lines, Private Rosenberger and his companion attacked
their captors and succeeded in freeing themselves, at the same time capturing
2 Germans, whom they brought to battalion headquarters, together with
the wounded man.

ᵃROSENFELD, MERRILL_____
　　Near Verdun, France, Oct. 15, 1918.
　　R—Baltimore, Md.
　　B—Baltimore, Md.
　　G. O. No. 2, W. D., 1919.

First lieutenant, 115th Infantry, 29th Division.
During the various offensives of this regiment in the vicinity of the Meuse
River, he displayed the greatest of bravery and coolness. He met his death
while leading a group that silenced an enemy machine gun menacing his
right flank.
Posthumously awarded. Medal presented to father, Israel Rosenfeld.

ᵃROSENWALD, JOHN P._____
　　During the action at Pexonne,
　　France, Mar. 5, 1918.
　　R—Minneapolis, Minn.
　　B—Yellow Bank, Minn.
　　G. O. No. 88, W. D., 1918.

First lieutenant, medical detachment, 151st Field Artillery, 42d Division.
He twice entered the quarry of Battery C, 151st Field Artillery, under heavy
shellfire, during the action at Pexonne, France, on Mar. 5, 1918, in order to
care for the wounded. Died May 6, 1918, of wounds received in action.
Posthumously awarded. Medal presented to widow, Mrs. J. P. Rosenwald.

ROSIO, WILLIAM (555288)_____
　　Near Mezy, France, July 15, 1918.
　　R—East Mauch Chunk, Pa.
　　B—Keeseville, N. Y.
　　G. O. No. 27, W. D., 1920.

Corporal, Company A, 9th Machine Gun Battalion, 3d Division.
Corporal Rosio commanded 1 of the 2 machine guns at the railroad bridge.
He exposed himself to heavy machine-gun and artillery fire in order to effectively direct the fire of his guns against the enemy infantry crossing the Marne
River. This was done with such effectiveness that many of the enemy
boats were sunk and assault halted at the Marne River bank. Later, after
his gun was disabled, he made a reconnaissance of the enemy lines and
returned with valuable information.

ROSKOSKI, FRANK J. (1717519)_____
　　Near Ville-Savoye, France, Aug. 18,
　　1918.
　　R—Bronx, N. Y.
　　B—New York, N. Y.
　　G. O. No. 44, W. D., 1919.

Sergeant, Company F, 302d Engineers, 77th Division.
He voluntarily plunged into the Vesle River to rescue some soldiers who had
fallen into the water with full packs while crossing a footbridge and were in
danger of drowning. In order to see he removed his gas mask, and as a result
was severely gassed.

ᵃROSKOWSKI, JOHN (1554680)_____
　　Near Soissons, France, July 20, 1918.
　　R—Chicago, Ill.
　　B—Austria.
　　G. O. No. 124, W. D., 1918.

Private, Company C, 1st Engineers, 1st Division.
When volunteers were called for by his company commander, Private Roskow-
ski volunteered and rescued wounded comrades from a barrage. Although
wounded in the performance of these heroic deeds, he continued until killed
by shellfire.
Posthumously awarded. Medal presented to sister, Mrs. Julia Provok.

ROSS, CARL G. R._____
　　East of Ronssoy, France, Sept. 29,
　　1918.
　　R—New York, N. Y.
　　B—New York, N. Y.
　　G. O. No. 20, W. D., 1919.

First lieutenant, 105th Infantry, 27th Division.
When his company was held up by an enemy machine-gun post, he advanced
alone against it and succeeded in putting it out of action, exhibiting great
bravery and gallantry, which was a splendid example to all ranks.

ᵃROSS, CLEO JEPSON_____
　　Near Brabant, France, Sept. 26,
　　1918.
　　R—Titusville, Pa.
　　B—Titusville, Pa.
　　G. O. No. 15, W. D., 1919.

First lieutenant, 8th Balloon Squadron, Air Service.
He was engaged in an important observation, regulating artillery fire, when his
balloon was attacked by enemy planes. One of the planes dived from a
cloud and fired at the balloon, setting fire to it, and although he could have
jumped from the basket at once he refused to leave until his companion, a
student observer, had jumped. He then leaped, but it was too late for the
burning balloon dropped on his parachute. He was dashed to the ground
from a height of 300 meters and killed instantly.
Posthumously awarded. Medal presented to father, E. M. Ross.

Ross, Douglas R. (61925)_____ | Private, first class, Company K, 101st Infantry, 26th Division.
Near Vaux, France, July 16, 1918. | He killed 2 of the enemy who were attempting to establish a machine-gun
R—Hull, Mass. | position in a railroad station and captured their machine guns. Later, the
B—Hull, Mass. | same day, when snipers were working in a ravine near the American line,
G. O. No. 125, W. D., 1918. | he, with 2 other soldiers, went forward to drive them back. One of his comrades was killed, but he, with great daring, attacked the Germans, killing 1 and wounding another.

Ross, Earl (1408317)_____ | Corporal, Company B, 5th Ammunition Train, 5th Division.
Near Septsarges, France, Oct. 24, 1918. | When an enemy shell struck some pyrotechnics stored in the ammunition
R—Savanna, Ill. | dump of his organization, he directed and assisted in the removal of inflammable material and placing the fire under control. Through his coolness
B—Savanna, Ill. | and courage the destruction of a large quantity of near-by ammunition was
G. O. No. 37, W. D., 1919. | avoided.

Ross, Hiram E._____ | Major, Medical Corps, attached to 18th Infantry, 1st Division.
Near Villers-Tournelle, France, May 3–4, 1918. | While under heavy bombardment and working in an area saturated with gas,
R—Danville, Ill. | Major Ross showed extreme gallantry and efficiency in caring for the more
B—Danville, Ill. | dangerously wounded, removing his mask at times during the attack to
G. O. No. 46, W. D., 1919. | better ascertain the extent of the wound. It was while thus acting that he was severely gassed.

*Ross, Karl E. (2262839)_____ | Sergeant, Machine Gun Company, 363d Infantry, 91st Division.
Near Waereghem, Belgium, Oct. 31, 1918. | At a distance of less than 200 meters from the enemy he set up and directed the
R—Stockton, Calif. | fire of his guns, exposed during the whole operation to direct enemy fire.
B—Petaluma, Calif. | He killed 1 gunner and, while searching for the gun on his flank, was
G. O. No. 37, W. D., 1919. | himself killed.
 | Posthumously awarded. Medal presented to mother, Mrs. Carrie W. Ross.

Ross, Leo L. (2257223)_____ | Corporal, Company D, 361st Infantry, 91st Division.
Near Gesnes, France, Sept. 26, 1918. | When the advance of his battalion was held up by an enemy machine-gun
R—Santa Barbara, Calif. | nest, Corporal Ross, in company with an officer, crossed the enemy wire,
B—Arkansas City, Kans. | took the position, and captured 4 prisoners and 2 machine guns. On the
G. O. No. 32, W. D., 1919. | same day, accompanied by 2 other soldiers, he captured 3 machine guns and 26 prisoners.

Ross, Lloyd D._____ | Major, 168th Infantry, 42d Division.
In the salient du-Feys, France, Mar. 9, 1918. | He displayed notable gallantry on Mar. 9, 1918, in leading a command of untried
R—Red Oak, Iowa. | men in company with French troops in a successful raid on enemy trenches
B—Adair County, Iowa. | in the salient du-Feys, France. By his heroic conduct he inspired both his
G. O. No. 27, W. D., 1919. | own men and the men of our ally participating in the operation.
 | Oak-leaf cluster.
 | An oak-leaf cluster is awarded Major Ross for the following acts of distinguished service: The courage, resolution, and resource of Major Ross as battalion commander made possible the successful capture of Hills 288, 242, and Cote-de-Chatillon, France, Oct. 14, 15, and 16, 1918, which was accomplished only after the most desperate fighting through wire and trenches against a resolute and determined defense involving frequent and bitter counterattacks. His brilliant and determined leadership was an example and inspiration to the entire command.

Rossire, Charles O., Jr._____ | Captain, 319th Infantry, 80th Division.
Near Imecourt, France, Nov. 1, 1918. | His company being halted by enemy machine-gun fire, he secured 2 hand
R—Washington, D. C. | grenades and rushed alone for an enemy machine-gun nest, throwing his
B—Philadelphia, Pa. | grenades and compelling the surrender of the gun crews. Returning to his
G. O. No. 15, W. D., 1923. | company, he directed the fire upon the remaining machine-gun crews, which were causing heavy casualties in his company, silenced the enemy guns, and resumed the advance with minimum losses. The soldierly courage displayed by Captain Rossire greatly inspired the members of his company.

Rossum, Haakon A. (1709067)__ | Corporal, Company G, 308th Infantry, 77th Division.
Near Charlevaux, France, Oct. 3–7, 1918. | During the 5 days that his battalion was cut off and surrounded by the
R—Brooklyn, N. Y. | enemy and throughout these 5 days of hunger, suffering, and enemy attacks
B—Norway. | Corporal Rossum commanded an advanced outpost in a position exposed
G. O. No. 32, W. D., 1919. | to each hostile onslaught. He was subjected constantly to fire from snipers, machine guns, trench mortars, and hand grenades. By his high courage, personal example, and inspiring leadership he defeated all attempts of the enemy to force his post back, and by so doing aided materially in the defense of his section of the line.

Rote, Tobin C._____ | First lieutenant, 357th Infantry, 90th Division.
Near Fey-en-Haye, France, Sept. 12, 1918. | Lieutenant Rote displayed daring in rushing machine-gun emplacements in
R—San Antonio, Tex. | the path of his platoon. Single handed he captured crews and emplacements
B—San Antonio, Tex. | of enemy machine guns.
G. O. No. 98, W. D., 1919. |

Roth, Paul Krusa_____ | First lieutenant, 306th Infantry, 77th Division.
Near St. Thibaut, Bazoches, France, Aug. 10, 1918. | Constantly exposed to a terrific concentration of enemy machine-gun and shell
R—Brooklyn, N. Y. | fire, without regard to his own safety, he voluntarily and unassisted rescued
B—Brooklyn, N. Y. | 5 badly wounded men of his regiment, carrying each man upon his back for a
G. O. No. 9, W. D., 1923. | distance of 150 yards, at all times under observation of the enemy occupying the town of Bazoches. His heroic conduct served as an impressive example to every man of his regiment.

ROTH, WILLIAM (540626)............
Near Mont-St. Pere, Marne, France,
July 21, 1918.
R—New York, N. Y.
B—New York N. Y.
G. O. No. 27, W. D., 1920.

Corporal, Company A, 7th Infantry, 3d Division.
After his company had been halted by machine-gun fire from the front, Corporal Roth advanced ahead of his organization to a position exposed to heavy enemy fire, and opened fire with an automatic rifle on the enemy nest. He continued this fire until killed by the heavy fire directed on him from the enemy machine-gun nest.
Posthumously awarded. Medal presented to father, Max Roth.

ROTHWELL, ROBERT (1039047)........
Near Greves Farm, France, July 15, 1918.
R—New York, N. Y.
B—England.
G. O. No. 98, W. D., 1919.

Corporal, Battery F, 10th Field Artillery, 3d Division.
Responding to a call for volunteers, Corporal Rothwell, with 3 other soldiers, manned 2 guns of a French battery which had been deserted by the French during the unprecedented fire after many casualties had been inflicted on their forces. For 2 hours he remained at his post and poured an effective fire into the ranks of the enemy.
Posthumously awarded. Medal presented to brother, Thomas Rothwell.

ROUNDS, CHARLES D (2385164)........
Near Ban-de-Laveline and Clery-le-Grande, France, June 29 and Oct. 28, 1918.
R—Witherbee, N. Y.
B—Chateaugay, N. Y.
G. O. No. 44, W. D., 1919.

Corporal, Company M, 60th Infantry, 5th Division.
On June 29 he was driven out of his post by hand grenades and machine-gun fire. After the grenades had exploded he reentered his post under machine-gun fire and drove the German gunners away with his automatic rifle. On Oct. 28 he located several machine-gun nests and a battery of field artillery and returned with information which made it possible to destroy them.

ROUNDS, HAROLD E. (62269)........
Near Vaux, France, July 20, 1918.
R—Malden, Mass.
B—Hyannis, Mass.
G. O. No. 125, W. D., 1918.

Private, Company L, 101st Infantry, 26th Division.
Private Rounds, Corpl. Austin J. Kelley, and Pvt. John J. Grady penetrated the enemy's lines in the face of machine-gun fire, captured a machine gun and its crew, and returned with valuable information concerning the enemy's positions.

ROUSH, JOE R. (935306)........
North of Chateau-Thierry, France, July 18, 1918.
R—Draper, S. Dak.
B—Monroe, Iowa.
G. O. No. 125, W. D., 1918.

Private, 103d Ambulance Company, 101st Sanitary Train, 26th Division.
While assisting a wounded Infantry soldier under heavy shellfire he was severely wounded in the face. Although his injury was more serious than that of the man whom he was attending, he carried the latter to an aid station, after which he voluntarily returned and assisted in first-aid work, inspiring his comrades by his example.

ROWAN, ANDREW S........
In connection with the operations in Cuba in May, 1898.
R—Union, W. Va.
B—Virginia.
G. O. No. 38, W. D., 1922.

Captain, 19th Infantry, U. S. Army.
At the outbreak of the Spanish-American campaign Lieutenant Rowan, under disguise, entered the enemy lines in Oriente, crossed the island of Cuba, and not only succeeded in delivering a message to General Garcia, but secured secret information relative to existing military conditions in that region of such great value that it had an important bearing on the quick ending of the struggle and the complete success of the U. S. Army.

ROWAN, CHARLES R........
Near Apremont, France, Sept. 29, 1918.
R—Altoona, Pa.
B—Altoona, Pa.
G. O. No. 139, W. D., 1918.

First lieutenant, 110th Infantry, 28th Division.
Being familiar with the ground over which an attack was to be made, he volunteered to leave his own company in the reserve and lead another company which was without officers. The enemy attacked before our own operations were begun, and he was wounded by a machine-gun bullet. Exemplifying in the highest degree the spirit of self-sacrifice and devotion to duty he remained with his command for an hour and a half until the hostile attack was repulsed. He has since died from the wounds received in this engagement.
Posthumously awarded. Medal presented to father, R. M. Rowan.

ROWAN, PAUL C. (2386685)........
At Cunel, France, Oct. 14, 1918.
R—Rosiclare, Ill.
B—Rosiclare, Ill.
G. O. No. 20, W. D., 1919.

First sergeant, Company D, 61st Infantry, 5th Division.
In the absence of a commissioned officer and under heavy shell and machine-gun fire, he succeeded in reorganizing the units of his command, and by his fearless example greatly increased the morale of his company. He successfully led them to the attack until he was killed by shellfire.
Posthumously awarded. Medal presented to mother, Mrs. Clara Oxford.

ROWAN, ROBERT P. (1286988)........
Near Samogneux, France, Oct. 9, 1918.
R—Greenville, Va.
B—Russellville, Ark.
G. O. No. 37, W. D., 1919.

Corporal, Company A, 116th Infantry, 29th Division.
After being painfully wounded, this soldier continued to lead his squad against a machine gun and silenced it, thereby saving his company many casualties.

ROWBOTTOM, RAYMOND G. (1244187)....
Near Fismette, France, Aug. 12, 1918.
R—Pittsburgh, Pa.
B—Allegheny County, Pa.
G. O. No. 160, W. D., 1918.

Corporal, Company E, 111th Infantry, 28th Division.
With an automatic-rifle team, he occupied a house in an advanced position west of Fismette on the night of Aug. 12, the loss of which would have jeopardized his company's position and hindered the military operations then taking place. The enemy shot a flare into the house, setting fire to it, but Corporal Rowbottom and a companion, under machine-gun and sniper fire in a brilliantly lighted room, extinguished the flames.

ROWE, GEORGE (1211444)........
Near Ronssoy, France, Sept. 29, 1918.
R—Ossining, N. Y.
B—Ossining, N. Y.
G. O. No. 32, W. D., 1919.

Sergeant, Company I, 107th Infantry, 27th Division.
Although seriously wounded, he continued to lead his platoon in operations against the Hindenburg line, refusing to be evacuated until ordered to the rear by his commanding officer.

ROWE, GUY I........
East of Chateau-Thierry, France, July 15, 1918.
R—Danville, Vt.
B—Peacham, Vt.
G. O. No. 99, W. D., 1918.

Major, 38th Infantry, 3d Division.
For 14½ hours on July 15, 1918, he held his battalion in an advanced and exposed position on the Marne, east of Chateau-Thierry, France, although violently and persistently attacked on his front and on both flanks by greatly superior enemy forces.

*ROWLEY, JOSEPH C. (1746465)_____
Near Grand Pre, France, Oct. 25, 1918.
R—Port Norris, N. J.
B—Port Norris, N. J.
G. O. No. 37, W. D., 1919.

Sergeant, Company M, 311th Infantry, 78th Division.
Assisting his company commander in organizing positions and liaison, he showed great bravery and devotion to duty. Although wounded in the first combat with the enemy, he continued with his work, declining to be evacuated. While patrolling between outposts he was killed.
Posthumously awarded. Medal presented to father, Stultz Rowley.

ROY, JOHN W. (72797)_____
In the Belleau Wood, France, July 18 and 19, 1918.
R—Worcester, Mass.
B—Worcester, Mass.
G. O. No. 125, W. D., 1918.

Bugler, Company H, 104th Infantry, 26th Division.
He displayed notable bravery in delivering messages through violent shellfire. At one time, after 3 other runners had been killed and a fourth wounded, he passed over the same route, undaunted, to carry a message vitally necessary to the successful operation of his company.

ROYSTER, THOMAS H_____
Near Crezancy, France, July 15, 1918.
R—Tarboro, N. C.
B—Granville County, N. C.
G. O. No. 32, W. D., 1919.

First lieutenant, Medical Corps, attached to 30th Infantry, 3d Division.
When casualties during the offensive of July 15, 1918, had become so great that it was necessary to work in the open, Lieutenant Royster exposed himself to the severe fire for 10 hours, dressing and caring for the wounded.

ROZELLE, GEORGE F., Jr_____
Near Cantigny, France, May 28–30, 1918.
R—Rogers, Ark.
B—Little Rock, Ark.
G. O. No. 15, W. D., 1919.

Major, 28th Infantry, 1st Division.
For three days near Cantigny, France, May 28–30, 1918, he withstood German assaults under intense bombardment, heroically exposing himself to fire constantly in order to command his battalion effectively, and, although his command lost half its officers and 30 per cent of its men, he held his position and prevented a break in the line at that point.

*RUANE, EDWARD T. (1203458)_____
East of Ronssoy, France, Sept. 29, 1918.
R—Cohoes, N. Y.
B—Cohoes, N. Y.
G. O. No. 20, W. D., 1919.

First sergeant, Company B, 105th Infantry, 27th Division.
During the operations against the Hindenburg line Sergeant Ruane, with an officer and 2 other sergeants, occupied an outpost position in advance of the line, which was attacked by a superior force of the enemy. Sergeant Ruane assisted in repulsing this attack and in killing 10 Germans, capturing 5, and driving off the others. The bravery and determination displayed by this group was an inspiration to all who witnessed it.
Posthumously awarded. Medal presented to father, John Ruane.

RUANE, MICHAEL (89694)_____
Near Villers-sur-Fere, Aisne, France, July 31, 1918.
R—Brooklyn, N. Y.
B—Ireland.
G. O. No. 44, W. D., 1919.

Private, Company C, 165th Infantry, 42d Division.
He went to the rescue of 2 wounded men over ground so swept by machine-gun fire that 2 men had been killed and 1 wounded previously in the attempt, and succeeded in carrying the 2 wounded men safely to shelter.

RUBEL, ALBERT C_____
Near Montfaucon, France, Sept. 26, 1918.
R—Indianapolis, Ind.
B—Louisville, Ky.
G. O. No. 37, W. D., 1919.

First lieutenant, 304th Engineers, 79th Division.
While reconnoitering to locate the Avocourt-Malancourt Road he was held up by an enemy machine gun on the parapet of a trench running parallel to the road. He proceeded ahead of two men of his platoon and personally disposed of the 2 German gunners.

RUCKER, EDWARD W_____
Near Luneville, France, June 13, 1918.
R—Lebanon, Mo.
B—Bosworth, Mo.
G. O. No. 161, W. D., 1918.

First lieutenant, 27th Aero Squadron, Air Service.
Outnumbered and handicapped by his presence far behind the German lines, he and 3 flying companions fought brilliantly a large group of enemy planes, bringing down or putting to flight all in the attacking party, while performing an important mission near Luneville, France, on June 13, 1918.

*RUDDOCK, ALEXANDER L. (2384986)____
Near Bois-de-Juvigny, France, Oct. 12, 1918.
R—Chester, Pa.
B—Beech Tree, Pa.
G. O. No. 78, W. D., 1919.

Supply sergeant, Company M, 60th Infantry, 5th Division.
Having been left behind to care for the equipment of the company and, seeing the difficult position in which the company had been placed, and noticing that a large number of his comrades were being evacuated because of wounds, Sergeant Ruddock joined the company and assisted in pushing the advance to its objective. He remained until the remnants of the command had been ordered back out of range of friendly artillery. He was mortally wounded on Nov. 10, after routing out 2 machine-gun nests.
Posthumously awarded. Medal presented to father, Alexander M. Ruddock.

RUDOLPH, EDWARD W. (2305940)_____
Near Mezy, France, July 15, 1918.
R—Joplin, Mo.
B—Lyndon, Kans.
G. O. No. 32, W. D., 1919.

Private, Headquarters Company, 30th Infantry, 3d Division.
He successfully carried messages through terrific artillery and machine-gun fire and was twice wounded while performing the mission.

RUDOLPH, ERNEST E. (1630524)_____
Near Varennes, France, Sept. 26, 1918.
R—Denver, Colo.
B—Keyesport, Ill.
G. O. No. 126, W. D., 1919.

Corporal, Company C, 110th Infantry, 28th Division.
Acting voluntarily, Corporal Rudolph and 2 other soldiers went out under heavy artillery and machine-gun fire and attacked an enemy machine-gun nest, killing 4 of the crew and capturing 11 prisoners, together with the machine gun.

RUDOLPH, MARTIN C_____
At Vieville-en-Haye, France, Sept. 12, 1918, and near Cunel, France, Oct. 21, 1918.
R—Moultrieville, S. C.
B—Carlstadt, N. J.
G. O. No. 95, W. D., 1919.

Captain, 11th Infantry, 5th Division.
When an enemy machine gun suddenly opened fire on his company Captain Rudolph signaled the platoon on his right to execute a flanking movement, while he advanced alone toward the gun. He killed the enemy gunner with his pistol and captured the remainder of the crew. He then ordered the captured gun carried along in the advance, and 200 yards farther used it successfully in silencing another enemy machine gun which was holding up his company. Captain Rudolph was severely wounded by a hand grenade on Oct. 21, but refused to go to the rear, and remained with his company for 12 hours, inspiring his men to hold an important position against a superior force of the enemy.

RUE, LAWRENCE E. (1388360) Near Consenvoye, France, Oct. 9, 1918. R—Chicago, Ill. B—Mills Springs, Mo. G. O. No. 59, W. D., 1919.	Sergeant, Company E, 132d Infantry, 33d Division. He had led his platoon to its objective when orders were recieved to shift the line in preparation for a hostile counterattack. He thereupon opened fire with an automatic rifle and remained behind, under heavy artillery and machine-gun fire, until the last man of his platoon had reached the newly designated line.
RUFUS, RAY (95015) Service rendered under the name, Frank Smith.	Corporal, Company K, 166th Infantry, 42d Division.
RUGE, EDWIN Near Flirey, France, Aug. 4, 1918. R—Atlanta, Ga. B—Apalachicola, Fla. G. O. No. 35, W. D., 1920.	First lieutenant, 326th Infantry, 82d Division. In an early morning raid Lieutenant Ruge charged up an exposed hillside and single-handed captured an enemy machine-gun position. This officer further distinguished himself near Champigneulle, France, Oct. 16, 1918, while reorganizing the 1st Battalion, 326th Infantry, under heavy fire and leading it to the attack.
RUGGERO, PETRO (2429475) Near Hill 272, near Fleville, France, Oct. 10, 1918. R—Cleveland, Ohio. B—Italy. G. O. No. 35, W. D., 1920.	Private, Company B, 16th Infantry, 1st Division. Private Ruggero, single handed, charged an enemy machine gun, killing the operator and capturing the gun.
RUHL, GEORGE E. (1272880) Near Bois-de-Consenvoye, France, Oct. 23, 1918. R—Baltimore, Md. B—Cumberland, Md. G. O. No. 37, W. D., 1919.	Private, Company A, 110th Machine Gun Battalion, 29th Division. Under a heavy artillery barrage, Private Ruhl displayed great bravery in rescuing and bringing to shelter a wounded comrade.
RUHL, LUTHER (43057) Near Fleville, France, Oct. 4, 1918. R—Hugo, Okla. B—Hugo, Okla. G. O. No. 35, W. D., 1920.	Sergeant, Company F, 16th Infantry, 1st Division. After his platoon commander had become a casualty, Sergeant Ruhl reorganized his platoon under heavy fire and led it forward in a successful attack against a machine-gun nest, killing or capturing the crew and taking the gun. Upon reaching his objective, he organized his platoon for defense and held his position against counterattacks.
RULAND, HENRY F. (2240721) Near Vilcey-sur-Trey, France, Sept. 12, 1918. R—Brenham, Tex. B—Brenham, Tex. G. O. No. 98, W. D., 1919.	Private, 357th Ambulance Company, 315th Sanitary Train, 90th Division. With another soldier Private Ruland left the shelter of a wood and went forward to rescue a soldier who had fallen wounded on a hill under constant machine-gun and shellfire. While they were carrying him back on a litter he was again wounded and the litter was struck twice by machine-gun bullets, but they succeeded in carrying him back to safety, thereby saving his life.
RULE, EDGAR J. (170053) Near Courbon, France, July 14-15, 1918. R—Boone, Iowa. B—Boone, Iowa. G. O. No. 44, W. D., 1919.	Sergeant, Headquarters Company, 10th Field Artillery, 3d Division. Sergeant Rule, who was in charge of a telephone detail, fearlessly repaired lines under heavy fire of gas and high-explosive shells until the lines were cut beyond repair, when he volunteered and carried messages through the bombardment.
RUMBAUGH, ERNEST R. (1521961) Near Montfaucon, France, Sept. 27, 1918. R—Holmesville, Ohio. B—Holmes County, Ohio. G. O. No. 16, W. D., 1920.	Corporal, Company H, 146th Infantry, 37th Division. Corporal Rumbaugh, with an officer and noncommissioned officer, advanced 200 yards beyond the objective of the patrol in the face of heavy machine-gun fire and captured three 77-millimeter field-pieces and 2 light machine guns.
RUMBERGER, HAROLD P. (1787239) Near Bois-de-Mont, France, Sept. 26, 1918. R—Waynesboro, Pa. B—Carlisle, Pa. G. O. No. 37, W. D., 1919.	Private, Company B, 316th Infantry, 79th Division. Failing to reduce a machine-gun nest with his rifle, he returned, procured an automatic rifle, attacked the nest the second time, and successfully reduced it.
RUMMELL, LESLIE J. In the region of Moirey, France, Sept. 29, 1918. R—Newark, N. J. B—Newark, N. J. G. O. No. 126, W. D., 1919.	First lieutenant, 93d Aero Squadron, Air Service. Lieutenant Rummell, leading a patrol of 3 planes, sighted an enemy biplane which was protected by 7 machines (Fokker type). Despite the tremendous odds, he led his patrol to the attack and destroyed the Di Nash plane. By his superior maneuvering and leadership 4 more of the enemy planes were destroyed, and the remaining 3 retired.
RUNDQUIST, OSCAR A. (280648) Near Romagne, France, Oct. 9, 1918. R—Big Rapids, Mich. B—Big Rapids, Mich. G. O. No. 71, W. D., 1919.	Sergeant, Company I, 126th Infantry, 32d Division. Though mortally wounded early in the assault, he continued to lead his men in an attack on German machine-gun nests that were holding up the advance of the American troops. With marked coolness and bravery he encouraged his men until the objective had been reached, shortly after which he fell dead from loss of blood. Posthumously awarded. Medal presented to father, Olaf Rundquist.

RUNNELLS, ERNEST P. (3273)
At Wadonville, France, Sept. 25, 1918.
R—Concord, N. H.
B—Concord, N. H.
G. O. No. 137, W. D., 1918.

Private, 101st Ambulance Company, 101st Sanitary Train, 26th Division.
He assisted in establishing a dressing station in a dugout in an advanced position. When it was destroyed by shell he worked unceasingly in the open under fire from enemy machine guns and snipers, caring for the wounded. He remained at his post for several hours after his station had been ordered closed, permitting neither his own exhaustion nor the enemy fire to deter him from aiding the wounded.

*RUNNING, TILMER A
Near Verdun, France, Oct. 12, 1918.
R—Viroqua, Wis.
B—Cashton, Wis.
G. O. No. 44, W. D., 1919.

Second lieutenant, 114th Infantry, 29th Division.
While advancing on an enemy position under direct machine-gun fire, he was seriously wounded but remained with his platoon until he died.
Posthumously awarded. Medal presented to father, Henry Running.

RUNYAN, EDGAR A. (45988)
North of Exermont, France, Oct. 9, 1918.
R—Black Rock, Utah.
B—Laurens, Iowa.
G. O. No. 53, W. D., 1920.

Sergeant, Company A, 18th Infantry, 1st Division.
Sergeant Runyan assumed command of a platoon after all other sergeants of the platoon had become casualties and led it through several successive attacks. Although wounded in the head by a machine-gun bullet he refused to go to the rear for medical attention but remained in command of the platoon until his company was relieved.

*RUPHOLDT, LOUIS C. (551128)
Near Mezy, France, July 15, 1918.
R—Goshen, Ind.
B—Goshen, Ind.
G. O. No. 24, W. D., 1920.

Sergeant, Company G, 38th Infantry, 3d Division.
Sergeant Rupholdt held his post on the bank of the Marne until nearly his entire platoon had been annihilated and he himself wounded. After being carried a short distance to the rear he continued to direct the defense of the position until killed.
Posthumously awarded. Medal presented to father, August Rupholdt.

RUPPEL, WILLIAM (732099)
Near Fontaine, France, Nov. 8, 1918.
R—Cincinnati, Ohio.
B—Caldwell, Ohio.
G. O. No. 81, W. D., 1919.

Sergeant, Company D, 6th Infantry, 5th Division.
While in command of a flank platoon of the battalion, Sergeant Ruppel overcame 3 enemy machine-gun groups and personally led the flanking patrol when his platoon was held up by enemy machine-gun fire.

RUSCH, ERNEST J. G.
In the Meuse-Argonne offensive, France, Oct. 9, 1918.
R—Neenah, Wis.
B—Germany.
G. O. No. 38, W. D., 1922.

First lieutenant, 18th Infantry, 1st Division.
Although previously wounded several times Lieutenant Rusch steadfastly remained in command of his platoon of machine guns and gallantly led it forward into position through heavy artillery and machine-gun fire to support his Infantry, which had been held up by enemy fire. By his timely action, his skill in the location of his guns and the direction of their fire, and his absolute disregard of personal danger, he quickly caused the withdrawal of the enemy machine guns, thus enabling his Infantry to advance from its precarious position where it was suffering heavy losses from direct fire of enemy machine guns and artillery.

RUSSELL, THOMAS N.
Near Blanc Mont Ridge, France, Oct. 4, 1918.
R—Vallejo, Calif.
B—Denver, Colo.
G. O. No. 46, W. D., 1919.

Pharmacist's mate, second class, U. S. Navy, attached to 2d Battalion, 5th Regiment, U. S. Marine Corps, 2d Division.
He gave proof of remarkable courage and disregard for personal safety by remaining at his post for 3 days under heavy shellfire and rendering first aid to the wounded, even when his dressing station was hit by a large shell.

*RUSSELL, WILLIAM H. (2021902)
Near Bolsnieozerki, Russia, Apr. 1, 1919.
R—Detroit, Mich.
B—Canada.
G. O. No. 16, W. D., 1920.

Corporal, Company M, 339th Infantry, 85th Division (detachment in North Russia).
Corporal Russell assumed command of a Russian machine-gun crew, in addition to his own, and held his position for 10 hours against determined attacks of a greatly superior force. When the enemy fire was so severe as to cause the Russians to temporarily leave their gun, he himself took charge of the gun and continued the fire.
Posthumously awarded. Medal presented to father, William Russell.

RUST, ALBERT L. (1328259)
At Bellicourt, France, Sept. 29, 1918.
R—Morgantown, N. C.
B—Bridgewater, N. C.
G. O. No. 145, W. D., 1918.

Master engineer, Company D, 105th Engineers, 30th division.
He commanded a platoon of engineers, following the first wave of the Infantry for the purpose of clearing a road for the artillery. Under heavy shell and machine-gun fire, he directed the work with exceptional ability, at one time leading his platoon in advance of the Infantry. By organizing covering parties and utilizing 2 automatic riflemen, who had become separated from their own unit, he kept his platoon intact, capturing 35 prisoners and cleaning out 3 machine-gun nests in the course of his operations. While making a reconnaissance ahead of his platoon he personally took 9 Germans, after wounding their officer. As a result of his skillful leadership and gallant conduct his mission was successfully carried out.

RUTHERFORD, JAMES E. (1897287)
Near St. Juvin, France, Oct. 12, 1918.
R—Somerville, Mass.
B—Somerville, Mass.
G. O. No. 98, W. D., 1919.

Private, first class, Machine Gun Company, 325th Infantry, 82d Division.
At a critical moment during a hostile counterattack, Private Rutherford singlehanded mounted a machine gun, under heavy enemy fire, and operated it at close range against the advancing enemy, repulsing the counterattack by his bravery.

RUUSULEHTO, VAINO (548069)
At Madeleine Farm, France, Oct. 9, 1918.
R—New York, N. Y.
B—Finland.
G. O. No. 27, W. D., 1920.

Corporal, Company L, 26th Infantry, 3d Division.
Accompanied by a comrade, Corporal Ruusulehto volunteered to attack an enemy machine-gun position, the fire from which was enfilading the position held by his platoon. During the attack his comrade was killed, but Corporal Ruusulehto, by a flank movement, reached the position, killed 2 members of the crew, and forced 2 others to surrender.

RYAN, C. WILLIAM
Near Romagne, France, Oct. 9, 1918.
R—Wathena, Kans.
B—Severance, Kans.
G. O. No. 37, W. D., 1919.

First lieutenant, 38th Infantry, 3d Division.
Being severely wounded, Lieutenant Ryan led his platoon through heavy shell and machine-gun fire, holding his command intact and capturing or destroying several machine guns.

RYAN, FRANK W. (1375348)
Near Nouart, France, Nov. 1–9, 1918.
R—Chicago, Ill.
B—Chicago, Ill.
G. O. No. 44, W. D., 1919.

Private, Headquarters Company, 122d Field Artillery, 33d Division.
Maintaining a telephone line 3 kilometers long over a period of 8 days, he was under a terrific bombardment during the whole period, keeping communication under circumstances which called for the greatest courage and determination. He had no relief and was at one time without rations for 48 hours.

RYAN, JOHN EDWARD
Near St. Étienne, France, Oct. 4, 1918.
R—Galveston, Tex.
B—Galveston, Tex.
G. O. No. 37, W. D., 1919.

Corporal, 18th Company, 5th Regiment, U. S. Marine Corps, 2d Division.
He requested and obtained permission to lead his company's advance. In performing this task and providing for the safety of his men he fell, wounded through the leg.

RYAN, OSCAR H. (1496095)
Near Faulburg, France, Nov. 8, 1918, and at Villemontry, France, Nov. 10, 1918.
R—Cuero, Tex.
B—Cuero, Tex.
G. O. No. 46, W. D., 1919.

Corporal, Company K, 9th Infantry, 2d Division.
On Nov. 8 Corporal Ryan went on a patrol through heavy machine-gun and artillery fire and returned with valuable information of the enemy. On Nov. 10 he remained on post in a building after half of it had been demolished by shellfire, and after being relieved of this duty helped to dig men from the débris of other houses which had been destroyed by the shells.

RYAN, RICHARD J.
Across the River Ourcq, near Villers-sur-Fere, France, July 28, 1918.
R—Watertown, N. Y.
B—Canada.
G. O. No. 99, W. D., 1918.

Captain, 165th Infantry, 42d Division.
Three times wounded on July 28, 1918, in the attack across the River Ourcq, near Villers-sur-Fere, France, and up the heights beyond, in which he led his company forward in the face of extremely heavy fire from machine guns and artillery, he refused to be evacuated and remained with his company until it was withdrawn. They reached their objectives and made their stand because of his fine spirit and unflinching determination.

*RYAN, THOMAS A. (60535)
Near Épieds, France, July 23, 1918.
R—Boston, Mass.
B—Boston, Mass.
G. O. No. 116, W. D., 1918.

Private, Company D, 101st Infantry, 26th Division.
Private Ryan, Sergt. Simeon L. Nickerson, and Corpl. M. J. O'Connell volunteered to cross an open field in front of their company in order to ascertain the location of enemy machine guns. While engaged in this courageous enterprise they were shot and killed. The heroic self-sacrifice of these 3 men saved the lives of many of their comrades, who would have been killed had the company attempted to make the advance as a whole.
Posthumously awarded. Medal presented to mother, Mrs. Mary Ryan

*RYANS, ROBERT M. (65325)
At Marcheville, France, Sept. 26, 1918.
R—Hartford, Conn.
B—Boston, Mass.
G. O. No. 15, W. D., 1919.

Sergeant, Company A, 102d Infantry, 26th Division.
He was in command of a platoon advancing under heavy artillery bombardment, machine-gun and rifle fire. Though severely wounded, he continued to lead his platoon, pushing on with his men until he was killed.
Posthumously awarded. Medal presented to mother, Mrs. Rebecca Wand.

RYDER, CHARLES W.
Near Soissons, France, July 21, 1918, and near Hill 272, Oct. 9, 1918.
R—Topeka, Kans.
B—Topeka, Kans.
G. O. No. 39, W. D., 1920.

Major, 16th Infantry, 1st Division.
Major Ryder took command of the front-line units and reorganized them under heavy artillery and machine-gun fire. Although wounded in the early operations, he remained in command and directed the attack until all objectives had been taken.
Oak-leaf cluster.
A bronze oak-leaf cluster is awarded Major Ryder for the following act of extraordinary heroism in action near Fleville, France, Oct. 9, 1918: In the attack on Hill 272, after all his runners had been killed or wounded while trying to establish liaison with the front-line companies, he advanced alone and personally directed the action of his command although under direct fire from 2 enemy machine guns. He later personally led the final assault on Hill 272, thereby making possible the success of the entire attack.

*RYKUS, WILLIAM (549282)
At Launay, near Chateau-Thierry, France, July 15, 1918.
R—Brooklyn, N. Y.
B—Brooklyn, N. Y.
G. O. No. 60, W. D., 1920.

Private, first class, Machine Gun Company, 38th Infantry, 3d Division.
Private Rykus repeatedly ran through enemy machine-gun and rifle fire with important messages. Being twice wounded in these missions, he died from the effects thereof.
Posthumously awarded. Medal presented to mother, Mrs. Mollie Rykus.

RYLEY, NORRIS W. (136426)
Near Seicheprey, France, Apr. 20, and 21, 1918.
R—Norwich, Conn.
B—Mystic, Conn.
G. O. No. 9, W. D., 1923.

Private, first class, Battery B, 103d Field Artillery, 26th Division.
All wire communications having been destroyed, 2 of the 3 guns of his battery also destroyed, and 50 per cent of the men of the battery killed or wounded, and several runners having been killed or wounded in attempting to carry messages to the battalion post of command, Private Ryley voluntarily and repeatedly crossed an exposed field a distance of 700 yards under terrific machine-gun and artillery fire, and again repeated his hazardous task on Apr. 21, 1918, and on each occasion accomplishing his mission. His heroic conduct was an inspiration to every man of his regiment.

*RYMAN, HERBERT D.
Near St. Gilles, France, Aug. 17, 1918.
R—Mount Pulaski, Ill.
B—Vernon, Ill.
G. O. No. 64, W. D., 1919.

Captain, Medical Corps, attached to 107th Field Artillery, 28th Division.
While administering first-aid to a wounded soldier he was himself mortally wounded. Refusing aid, he assisted in rendering first aid and directing the treatment of 3 other soldiers. Though weakened by loss of blood, he showed utter disregard for his personal danger, refusing to accept treatment until the other wounded had been cared for.
Posthumously awarded. Medal presented to widow, Mrs. Cora Belle Ryman.

RYMER, CHARLES B. (1452467)_____
    Near Very, France, Sept. 26, 1918.
    R—St. Louis, Mo.
    B—Chattanooga, Tenn.
    G. O. No. 46, W. D., 1919.

Bugler, Company F, 138th Infantry, 35th Division.
He was a member of a liaison group who worked their way 1,000 yards in advance of their first wave. Surrounded by machine-gun fire, they were forced to take refuge in a trench. Bugler Rymer worked his way along the trench to a clump of woods, where he captured a German officer alone. Assisted by 2 other soldiers, he later killed an enemy machine gunner and took 23 prisoners.

RYPKEMA, HANNES (2147198)_____
    Near St. Juvin, France, Oct. 14, 1918.
    R—Hinckley, Minn.
    B—Renville, Minn.
    G. O. No. 71, W. D., 1919.

Sergeant, Company A, 326th Machine Gun Battalion, 82d Division.
He went in advance of his section and unassisted captured and sent to the rear 13 German prisoners. With the aid of 2 infantrymen he later captured 28 more prisoners, under heavy artillery and machine-gun fire. When his platoon commander had been killed and more than three-fourths of his platoon had become casualties he reorganized the remainder of the platoon and continued the attack.

SACK, WILLIAM (280641)_____
    Near Romagne, France, Oct. 9, 1918.
    R—Big Rapids, Mich.
    B—Keno, Mich.
    G. O. No. 126, W. D., 1919.

First sergeant, Company I, 126th Infantry, 32d Division.
Assuming command of his company after all the officers had become casualties, Sergeant Sack skillfully maneuvered it in an attack on enemy machine-gun nests which were hindering the advance. He was severely wounded early in the engagement, but he continued to lead and encourage his men until he received a second wound, which rendered him a cripple for life.

SACKETT, DAYTON_____
    Near Soissons, France, July 22, 1918;
    near St. Mihiel, France, Sept. 12–
    15, 1918; near Hill 212, France, Oct.
    9, 1918.
    R—Greenfield, Tenn.
    B—Clarksburg, Tenn.
    G. O. No. 44, W. D., 1919.

First lieutenant, 26th Infantry, 1st Division.
On July 22 Lieutenant Sackett continued in the fight until he had taken and consolidated the objective assigned him, despite the fact that he was wounded early in the action. On Sept. 12 he was wounded in the first day's fighting, but continued to command his company, refusing to be evacuated until the corps' objective was reached. On Oct. 9 he fearlessly walked across an open space before allowing his troops to cross to ascertain whether or not it was subjected to enemy fire. He was seriously wounded while on this mission.

SADKOWSKI, FRANK (1898379)_____
    Near Eply, France, Sept. 4, 1918.
    R—Keyport, N. J.
    B—Russia.
    G. O. No. 20, W. D., 1919.

Corporal, Company E, 325th Infantry, 82d Division.
Under heavy fire from machine guns, and although seriously wounded, he continued to advance within the enemy's lines. By words of encouragement he urged his men to follow. By his brave leadership an enemy outpost defended by 2 machine guns and 6 riflemen was captured.

SADLER, GEORGE W._____
    Near Cunel, France, Oct. 9–16, 1918.
    R—Laneview, Va.
    B—Laneview, Va.
    G. O. No. 32, W. D., 1919.

First lieutenant, chaplain, 30th Infantry, 3d Division.
Throughout this period Chaplain Sadler, regardless of his personal safety, gave first aid and assisted in the evacuation of the wounded from the field under heavy machine-gun, shell fire, and gas.

*SAGER, GAIL H. (1214099)_____
    Near Ronssoy, France, Sept. 29,
    1918.
    R—Buffalo, N. Y.
    B—Clarington, Pa.
    G. O. No. 37, W. D., 1919.

Corporal, Company D, 108th Infantry, 27th Division.
Upon being wounded in the hand, he bandaged the wound himself and immediately returned to the firing line. He then picked up an automatic rifle, and, advancing alone toward machine-gun nests which were holding up his company, was killed after proceeding only a short distance.
Posthumously awarded. Medal presented to widow, Mrs. Gail H. Sager.

ST. GEORGE, EMERY_____
    Near Crezancy, France, July 15, 1918.
    R—Plymouth, Mass.
    B—Plymouth, Mass.
    G. O. No. 46, W. D., 1919.

First lieutenant, 30th Infantry, 3d Division.
On duty as assistant regimental signal officer, he continually exposed himself during the terrific enemy bombardment preceding the attack of July 15, repairing wires and endeavoring to keep the lines in operation.

*ST. GEORGE, RAYMOND (61320)_____
    Near Bois-de-Wavrille, France, Oct.
    2, 1918.
    R—Worcester, Mass.
    B—Worcester, Mass.
    G. O. No. 89, W. D., 1919.

Private, Company G, 101st Infantry, 26th Division.
In the performance of his duties as scout, Private St. George displayed unusual courage in locating and charging machine-gun nests. After being mortally wounded, he insisted that those who stopped to aid him continue their advance to capture nests.
Posthumously awarded. Medal presented to mother, Mrs. Mary St. George.

ST. JAMES, LEONARD (263298)_____
    Near Romagne, France, Oct. 9, 1918.
    R—Bay City, Mich.
    B—Whittemore, Mich.
    G. O. No. 64, W. D., 1919.

Private, first class, Company I, 125th Infantry, 32d Division.
He repeatedly crossed an open area, 500 meters wide, under intense machine-gun fire in carrying messages to battalion headquarters. On one of his trips he came upon a wounded soldier, whom he took to the aid station, after administering first aid to him.

SAKRISON, ROY H._____
    Near Nantillois, France, Oct. 4,
    1918.
    R—Deer Park, Wis.
    B—Deer Park, Wis.
    G. O. No. 7, W. D., 1919.

First lieutenant, Infantry Headquarters, 80th Division.
Lieutenant Sakrison, with a group of observers and signalmen, was in charge of the forward observation post. When the Infantry advanced, he followed closely with his telephone lines and established another post on Hill 274. Though he was several times buried by bursting shells, he continued to make reports over the telephone until he was seriously wounded. After walking to a dressing station and securing first aid he returned to his post through heavy shellfire and continued to transmit important information for 3 hours until relieved.

SALE, LARRY (1596196)_____
    In the Argonne Forest, France,
    Oct. 9, 1918.
    R—Homer, La.
    B—Claiborne Parish, La.
    G. O. No. 44, W. D., 1919.

Private, Company D, 16th Infantry, 1st Division.
After all the platoon runners had been killed or wounded, he volunteered and carried an important message through heavy machine-gun and artillery barrage.

SALIK, ALEXANDER (281089)_____
    Near Juvigny, north of Soissons, France, Aug. 28, 1918.
    R—Grand Rapids, Mich.
    B—Russia.
    G. O. No. 21, W. D., 1919.

Sergeant, Company K, 126th Infantry, 32d Division.
Sergeant Salik, regardless of wounds and of mustard-gas burns previously received, rejoined and advanced with his company in the attack and assisted in re-forming a platoon after it had suffered severe casualties.

*SANBORN, EASTMAN M_____
    Near the town of Montfaucon, France, Sept. 29, 1918.
    R—Cleveland, Ohio.
    B—Downers Grove, Ill.
    G. O. No. 9, W. D., 1923.

First lieutenant, 316th Infantry, 79th Division.
For extraordinary heroism in action while leading his company in an attack upon the enemy under heavy machine-gun and shell fire. Although badly wounded during the advance he proceeded with his men, his command suffering heavy losses; being again wounded he retained command until exhausted by loss of blood and fully incapacitated. While making his way to the dressing station he was again wounded, and upon arrival at the dressing station yet again. His courage, coolness, and great fortitude and devotion to duty inspired his men to heroic efforts.
Posthumously awarded. Medal presented to father, William J. Sanborn.

SANBORN, JOSEPH BROWN_____
    Near Gressaire Wood, France, Aug. 9, 1918.
    R—Chicago, Ill.
    B—Manchester, N. H.
    G. O. No. 46, W. D., 1919.
    Distinguished-service medal also awarded.

Colonel, 131st Infantry, 33d Division.
Immediately after a forced march of 25 miles Colonel Sanborn's regiment was ordered into a critical engagement. Hurrying to the front, he personally led his forces through a heavy and concentrated shellfire and started the attack at the exact allotted time. After launching his attack he established his post of command in a shell hole and directed the battle to a successful termination. The courage and fearlessness of Colonel Sanborn, despite his advanced age of 62 years, were remarkable to all under his command.

SANDBURG, CHARLES A_____
    Near St. Souplet, France, Oct. 17, 1918.
    R—Jamestown, N. Y.
    B—Titusville, Pa.
    G. O. No. 37, W. D., 1919.

Captain, 108th Infantry, 27th Division.
After having been severely wounded he continued to advance with his command until ordered to leave the field by his regimental commander.

SANDEFORD, ALVAN C_____
    Near Chery-Chartreuve, France, Aug. 8-17, 1918.
    R—Midville, Ga.
    B—Midville, Ga.
    G. O. No. 47, W. D., 1921.

Major, 13th Field Artillery, 4th Division.
Twice gassed, he declined to be evacuated and continued in active command of his battalion. Having been advised and knowing that failure to be evacuated would probably result in his death, he nevertheless continued until he fell from his saddle in a state of total collapse. His fortitude and spirit of self-sacrifice were conspicuous.

SANDERFER, PAUL C. (1316865)_____
    Near St. Souplet, France, Oct. 10, 1918.
    R—Trenton, Tenn.
    B—Union City, Tenn.
    G. O. No. 37, W. D. 1919.

Sergeant, Company M, 119th Infantry, 30th Division.
Although wounded by enemy machine-gun fire, he continued to lead his platoon forward until he fell from weakness caused by loss of blood. He even then continued to advance by crawling until his strength entirely failed him.

SANDERS, JOSEPH D. (156868)_____
    Near Chateau-Thierry, France, June 3 and 13, 1918.
    R—Hoisington, Kans.
    B—Van Buren, Ark.
    G. O. No. 23, W. D. 1919.

Corporal, Company D, 2d Engineers, 2d Division.
In command of an important outpost, Corporal Sanders exposed himself to rifle and shell fire to better observe the movements of the enemy. He was knocked unconscious by a shell burst, but returned to his post immediately upon regaining consciousness. On June 13 he carried a wounded officer through an intense barrage to a dressing station.

SANDERS, NATHAN P. (40598)_____
    Near Seissons, France, July 18-25, 1918.
    R—Weatherford, Okla.
    B—Barnett, Okla.
    G. O. No. 117, W. D., 1918.

Corporal, Company L, 9th Infantry, 2d Division.
He gave proof of utter fearlessness and courage during the whole 7-days' offensive. On July 18 his company was stopped by an exceptionally well-located machine gun, making advance impossible until it was silenced. He on his own initiative crept upon the gun, killed the gunner, and captured 4 others of the crew. He then turned the gun and operated it on the retreating Germans.

SANDERS, PLEAS (546605)_____
    Near Cunel, France, Oct. 10, 1918.
    R—Brownsville, Ky.
    B—Brownsville, Ky.
    G. O. No. 32, W. D., 1919.

Sergeant, Company F, 30th Infantry, 3d Division.
He attacked an enemy strong point covered by machine-gun fire. Although severely wounded, he continued to direct his platoon from a shell hole until the objective was reached.

SANDERSON, EARL H. (64777)_____
    At Chavignon, Chemin-des-Dames sector, France, Feb. 28, 1918.
    R—Worcester, Mass.
    B—Springfield, Mass.
    G. O. No. 126, W. D., 1918.

Corporal, Company F, 102d Infantry, 26th Division.
He was a member of a working party on the night of Feb. 28, 1918, well out in front of the advance post. The party encountered a violent barrage of the enemy, which protected enemy assault troops. He helped to fight off the German troops and walked back and forth under the enemy's and our own barrage to collect his men. When he heard his lieutenant was in trouble he walked again to his rescue where the barrage had at first overtaken him.

SANDFORD, WILLIAM (104280)_____
    Near the Ripont River, Champagne, France, Sept. 26, 1918.
    R—New York, N. Y.
    B—New York, N. Y.
    G. O. No. 47, W. D., 1921.

Private, medical detachment, 369th Infantry, 93d Division.
Under direct and close-range fire of several enemy machine guns he crawled to exposed positions to dress the wounds of officers and men.

*SANDH, ROBIN (1633477)_____
    Near Montfaucon, France, Oct. 8, 1918.
    R—Gill, Colo.
    B—Arcadia, Nebr.
    G. O. No. 24, W. D., 1920.

Private, Machine Gun Company, 7th Infantry, 3d Division.
Private Sandh repeatedly carried messages to front-line platoons through heavy artillery and machine-gun fire. While en route to an exposed machine-gun position, under heavy fire of the enemy, he was killed by a shell fragment.
Posthumously awarded. Medal presented to mother, Mrs. Sarah Sandh.

°SANDMAN, LEO L. (2212260) _____
In the Bois-de-Barricourt, France, Nov. 2, 1918.
R—Barrington, Ill.
B—Barrington, Ill.
G. O. No. 89, W. D., 1919.

Private, Company F, 353d Infantry, 89th Division.
Private Sandman, with a comrade, advanced as a scout across an open space which was covered by heavy machine-gun fire from the enemy. This advance was made in order to locate the position of the enemy machine guns and draw their attention while the rest of the platoon advanced on the flanks. Private Sandman signaled the location of the German guns to his comrades, and he had advanced more than 200 yards through the murderous fire when he was killed.
Posthumously awarded. Medal presented to mother, Mrs. Mina Sandman.

°SANFORD, LUMAN K. (1551372) _____
Near Chateau-Thierry, France, July 14–15, 1918.
R—Endicott, N. Y.
B—Lake Arier, Pa.
G. O. No. 44, W. D., 1919.

Private, Battery F, 76th Field Artillery, 3d Division.
During a heavy shelling he continued to repair broken telephone lines, which were constantly being severed by shell fire, until he was killed by an enemy shell.
Posthumously awarded. Medal presented to father, Rev. Luman E. Sanford.

SANTARSIERO, GUISEPPE (2411222) _____
Near Grand Pre, France, Oct. 26, 1918.
R—Trenton, N. J.
B—Italy.
G. O. No. 35, W. D., 1919.

Private, Company M, 311th Infantry, 78th Division.
He rushed ahead of his company and, single-handed, flanked a machine-gun nest, which was causing losses to his company, killing the gunners.

SAPLIO, SAM (1244810) _____
Near Fismette, France, Aug. 10–12, 1918.
R—Marion, W. Va.
B—Italy.
G. O. No. 123, W. D., 1918.

Private, Company C, 111th Infantry, 28th Division.
Without fear or thought for his personal safety, he sought out enemy snipers posted in trees and killed a number of them. Later, with Sergt. John W. Thompson, he attacked an enemy machine-gun nest, killed the crew, and turned the gun on the enemy, operating it with deadly effect on the Infantry and machine-gun positions, killing the gun crews and capturing 10 machine guns.

°SAPP, AMBERS (736957) _____
Near Frapelle, France, Aug. 17, 1918.
R—Nepton, Ky.
B—Fleming County, Ky.
G. O. No. 15, W. D., 1919.

Private, Headquarters Company, 6th Infantry, 5th Division.
He displayed great coolness and courage under a heavy enemy barrage when he unhesitatingly went forward to destroy enemy wire entanglements and continued this extremely hazardous work until killed.
Posthumously awarded. Medal presented to father, Rufe Sapp.

°SARGENT, BRADLEY V., Jr _____
At Romagne, France, Oct. 27, 1918.
R—San Francisco, Calif.
B—Monterey, Calif.
G. O. No. 37, W. D., 1919.

Second lieutenant, 11th Field Artillery, 6th Division.
Under heavy shellfire, disregarding his own safety, he remained on duty, superintending the unloading of ammunition until he was mortally wounded.
Posthumously awarded. Medal presented to widow, Mrs. Bradley V. Sargent, jr.

SARGENT, JESSIE W. (44417) _____
Near Fleville, France, Oct. 4, 1918.
R—Crown, W. Va.
B—Frenchburg, Ky.
G. O. No. 35, W. D., 1920.

Corporal, Company M, 16th Infantry, 1st Division.
When enemy machine-gun fire was causing losses to our line, Corporal Sargent went forward in broad daylight under fire and silenced the machine gun by killing the gunner. His action enabled his unit to continue the advance.

SARTAIN, GEORGE W. (547762) _____
Near Jaulgonne, France, July 26, 1918.
R—Washington, D. C.
B—Washington, D. C.
G. O. No. 22, W. D., 1919.

Corporal, Company K, 30th Infantry, 3d Division.
He gave aid to 3 wounded comrades during a heavy bombardment and, after 4 unsuccessful attempts, finally succeeded in carrying them to a dressing station.

SARTAIN, JAKE C. (2236537) _____
Near St. Marie Farm, France, Sept. 18, 1918.
R—Atlanta, Ga.
B—Greenville, Ga.
G. O. No. 87, W. D., 1919.

Sergeant, first class, Company A, 315th Engineers, 90th Division.
As Sergeant Sartain was successfully directing the erection of barbed-wire entanglements under heavy shellfire he heard cries for help from the direction of the enemy's lines. He immediately went out to investigate, and upon finding a soldier of another organization lying wounded carried him to our lines, twice passing through enemy machine-gun fire in accomplishing this heroic act.

SARTI, WILLIAM (156082) _____
Near Medeah Ferme, France, Oct. 8–9, 1918.
R—Garfield, N. J.
B—New York, N. Y.
G. O. No. 37, W. D., 1919.

Sergeant, first class, Company A, 2d Engineers, 2d Division.
His platoon commander and only other sergeant being wounded, Sergeant Sarti assumed command of his platoon, although himself wounded. He made a reconnaissance of the position his platoon was to occupy under heavy shell fire, and, returning, conducted it to the new position without the loss of a single man, and remained with it for 48 hours.

SATTLER, WILLIAM N. (1389862) _____
At Bois-de-Foret, France, Oct. 6–13, 1918.
R—Chicago, Ill.
B—Chicago, Ill.
G. O. No. 81, W. D., 1919.

Corporal, Headquarters Company, 132d Infantry, 33d Division.
Corporal Sattler was in charge of all runners at advance post of command of the regiment. The area was heavily gassed. Although so badly gassed that his eyes were swollen shut and his voice affected, he refused to be evacuated but continued on duty. On Oct. 10, when all runners were killed, wounded, or gassed, he repeatedly carried many important messages in order to maintain communication.

*SAUER, JOSEPH (1708815) _____
Near Binarville, France, Oct. 2, 1918.
R—New York, N. Y.
B—Pittsburgh, Pa.
G. O. No. 35, W. D., 1919.

Corporal, Company F, 308th Infantry, 77th Division.
He volunteered in the face of heavy enemy machine-gun fire to deliver a message to a platoon sergeant who was leading an attack on enemy machine-gun nests. He was wounded in one leg just as he started and was wounded in the other leg before reaching the sergeant, but did, by calling aloud, deliver the message verbally and accurately.
Posthumously awarded. Medal presented to father, Joseph Sauer.

SAUERS, ROY M. (2193727).
Near Tailly, France, Nov. 4, 1918.
R—Stuart, Nebr.
B—Hooper, Nebr.
G. O. No. 37, W. D., 1919.

Sergeant, first class, Company B, 314th Field Signal Battalion, 89th Division.
He was in charge of a wire-laying detail between Tailly and Beauclair. Over a
road swept by heavy shellfire he carried the line forward and in constant
repair to the support of a battalion of the 355th Infantry.

SAUL, TOM W.
Near the Bois-de-Remieres, France,
Sept. 12, 1918.
R—Portland, Oreg.
B—Wyoming, Ohio.
G. O. No. 46, W. D., 1919.

First lieutenant, 327th Battalion, Tank Corps.
He coolly exposed himself to enemy fire by standing on the parapet of a trench
and directing his men in the work of getting the tanks forward.

SAUNDERS, EUGENE F. (1285175).
Near Bois-de-Consenvoye, France,
Oct. 8, 1918.
R—Washington, D. C.
B—Washington, D. C.
G. O. No. 44, W. D., 1919.

Private, first class, Company F, 115th Infantry, 29th Division.
He carried a wounded comrade through a terrific machine-gun and artillery
barrage to a place of safety, and thereby saved his life, although risking his
own in the exploit.

SAUNDERS, THOMAS D. (156126).
At Jaulny, France, Sept. 12, 1918.
R—Cheyenne, Wyo.
B—Medicine Bow, Wyo.
G. O. No. 142, W. D., 1918.

Corporal, Company A, 2d Engineers, 2d Division.
He and another soldier, who were acting as wire cutters with the first line of
Infantry, fought their way forward in advance of their units and were the
first men to enter Jaulny while it was swept by machine-gun fire, infested with
snipers, and still occupied by rear-guard detachments of the enemy. After
capturing 8 Germans in a dugout they searched the caves in the town and took
55 additional prisoners.

*SAUNDERS, WILLIAM H.
In the Toul sector, France, May 25,
1918.
R—Dalzell, S. C.
B—Claremont, S. C.
G. O. No. 15, W. D., 1923.

Captain, 12th Aero Squadron, Air Service.
The artillery of the 26th Division desiring its batteries to be adjusted upon
objectives in front of the division, Captain Saunders volunteered to make the
attempt, although the weather was most unfavorable to flying. After flying
two hours amid heavy antiaircraft fire and having adjusted the fire of 3
batteries his plane was hit and disabled. Returning to his airdrome he
secured another plane and returned to the enemy line to complete his mission.
After another hour in the air he was again forced on account of motor trouble
to return to the airdrome. Obtaining a third plane he again returned to the
lines, the weather conditions forcing him to proceed for a considerable distance
behind the enemy lines and at low altitude. Flying thus for an hour his plane
was hit by antiaircraft fire and badly damaged. With one control shot away,
and his propeller likewise injured, he still continued to adjust, always at low
altitude and under constant enemy fire, until his mission was successfully
accomplished. The heroic conduct of Captain Saunders served as a splendid
example of soldierly devotion to duty to the men of his squadron.
Posthumously awarded. Medal presented to father, W. L. Saunders.

SAURMAN, HAROLD P. (2991).
Near Trugny, France, July 22–23,
1918.
R—Boston, Mass.
B—Medford, Mass.
G. O. No. 37, W. D., 1919.

Private, sanitary detachment, 102d Machine Gun Battalion, 26th Division.
He displayed remarkable courage in going out under heavy machine-gun and
artillery fire, giving aid to wounded soldiers and carrying them back to the
dressing stations.

*SAVAGE, ARTHUR V.
Near Mezy, France, July 15, 1918.
R—Philadelphia, Pa.
B—Philadelphia, Pa.
G. O. No. 32, W. D., 1919.

First lieutenant, 30th Infantry, 3d Division.
In the darkness he charged a machine gun which had been brought across the
Marne by the Germans to cover their crossing during the night, but was com-
pletely surrounded and killed just as he reached the gun.
Posthumously awarded. Medal presented to father, Charles S. Savage.

SAVAGE, ERNEST S.
Near Grimaucourt, France, Nov.
11, 1918.
R—Council, N. C.
B—Council, N. C.
G. O. No. 32, W. D., 1919.

First lieutenant, 316th Machine Gun Battalion, 81st Division.
Although so sick from gas that he could hardly move, and vomiting heavily
into his gas mask, he successfully conducted the fire of his machine-gun
platoon in the face of heavy shrapnel, gas, and machine-gun fire. He received
no medical attention until late in the afternoon after the attack was over.

SAVITSKY, ANTHONY (155709).
Near Verdun, France, Oct. 9, 1918.
R—Colonie, N. Y.
B—Russia.
G. O. No. 98, W. D., 1919.

Corporal, Company A, 1st Engineers, 1st Division.
Upon his own initiative, Corporal Savitsky led his squad in the face of heavy
machine-gun fire in order to silence enemy machine guns which were endangering our position. Through his skill and courage several machine-gun
nests were surrounded and silenced and 20 prisoners captured.

SAWEUK, ILLIAN (551289).
Near Mezy, France, July 15, 1918.
R—New York, N. Y.
B—Russia.
G. O. No. 98, W. D., 1919.

Private, Company G, 38th Infantry, 3d Division.
On two separate occasions during the battle of the Marne, Private Saweuk went
over the top alone, returning with 12 prisoners the first time and 8 the second.

*SAXON, JOHN W. (1286013).
Near Verdun, France, Oct. 10, 1918.
R—Kensington, Md.
B—Albion, Tex.
G. O. No. 27, W. D., 1919.

Sergeant, Company K, 115th Infantry, 29th Division.
In the advance on Rechene Hill he showed great courage and judgment in leading his platoon and wiping out several machine guns that were holding up
the advance. He was killed while gallantly leading his platoon against the
last of these.
Posthumously awarded. Medal presented to father, Jesse W. Saxon.

SCANDEL, ALEXANDER (2383284).
Near Bois-de-Pultiere, France, Oct.
15, 1918.
R—Shenandoah, Pa.
B—Poland.
G. O. No. 59, W. D., 1919.

Private, Company E, 60th Infantry, 5th Division.
When his platoon was forced to take shelter from the intense fire, Private
Scandel, with another soldier, having located the source of the fire, crawled
forward 200 yards, capturing a nest containing 2 guns and 5 of the enemy.
When his companion had started to the rear with the prisoners, Private
Scandel continued forward, rushing from shell hole to shell hole and outflank-
ing a second nest, taking 1 gun and 3 prisoners.

SCANLAN, ANTHONY (42842)............
  Near Soissons, France, July 21, 1918.
  R—Lost Creek, Pa.
  B—Ireland.
  G. O. No. 15, W. D., 1919.

Sergeant, Company F, 16th Infantry, 1st Division.
Although severely wounded, he displayed exceptional courage and leadership by reorganizing his battalion under fire when all of its officers had been killed or incapacitated by injuries.

*SCANLON, HORACE E............
  Near Ronssoy, France, Sept. 27, 1918.
  R—Brooklyn, N. Y.
  B—Brooklyn, N. Y.
  G. O. No. 44, W. D., 1919.

Second lieutenant, 106th Infantry, 27th Division.
Gathering about 40 men from various units in a forward trench, he organized them into an attacking party and led them forward under heavy machine-gun fire, repulsing an enemy counterattack. While in the performance of this exploit he was mortally wounded, but, attempting to push forward with his men, he called out: "Go on fighting! Never mind what happened to me."
Posthumously awarded. Medal presented to mother, Mrs. John L. Scanlon.

SCANLON, JAMES E............
  Near Limey, France, Sept. 12, 1918.
  R—Boswell, Ind.
  B—Boswell, Ind.
  G. O. No. 98, W. D., 1919.

First lieutenant, 353d Infantry, 89th Division.
Although wounded himself, he went to the rescue of another officer who had fallen in an exposed position. With marked bravery he passed through heavy enemy fire, and reaching the other officer, carried him to safety.

SCANLON, RAYMOND (60856)............
  In the Bois-de-Belleu north of Verdun, France, Oct. 25, 1918.
  R—Waltham, Mass.
  B—Waltham, Mass.
  G. O. No. 60, W. D., 1920.

Sergeant, Company F, 101st Infantry, 26th Division.
Sergeant Scanlon, while advancing in command of a liaison group, was halted by machine-gun fire from the front. Alone, he dashed ahead of his men toward the enemy position. He wounded 2 of the enemy and captured the gun. Upon being rejoined by his men he continued the advance.

SCHABINGER, ANDREW C. (1387453)............
  At Hamel, France, July 4, 1918.
  R—Washington, Ill.
  B—Washington, Ill.
  G. O. No. 99, W. D., 1918.

Corporal, Company E, 131st Infantry, 33d Division.
Although severely wounded in the arm at the beginning of the engagement, he continued forward as squad leader, exhibiting great gallantry and setting an inspiring example to his men.

SCHAD, ALBERT P. (1245668)............
  Near Montblainville, France, Sept. 30, 1918.
  R—Philadelphia, Pa.
  B—Philadelphia, Pa.
  G. O. No. 50, W. D., 1919.

First sergeant, Company L, 111th Infantry, 28th Division.
With a detail of 16 men, Sergeant Schad attempted the mission of cleaning up a machine-gun nest which was enfilading the Montblainville Road. Leaving his men in a sheltered position where a counter fire could be directed upon the enemy gun, he advanced alone, located the nest, killed 2 of the crew with hand grenades, put the others to flight, and destroyed the gun.
Oak-leaf cluster.
For the following act of extraordinary heroism in action near Le Chene Tondu, France, Oct. 4, 1918, First Sergeant Schad is awarded an oak-leaf cluster, to be worn with the distinguished-service cross: On his own initiative he took forward a platoon to fill a gap left in the attacking platoons on account of casualties. Leaving his platoon to cover the road, he advanced alone upon a machine-gun nest, killed the entire crew with hand grenades, and captured the gun. He then held the position until the advance was taken up the following morning.

*SCHAFFNER, FRED C............
  In a shell bombardment, Apr. 13, 1918.
  R—Rock Island, Ill.
  B—Kewanee, Ill.
  G. O. No. 88, W. D., 1918.

Pharmacist's mate, third class, U. S. Navy, attached to 6th Regiment, U. S. Marine Corps, 2d Division.
After having been gassed himself in the gas-shell bombardment of Apr. 13, 1918, he courageously helped in the treatment of more than 100 cases of gas casualties, disregarding his own condition until overcome. Died Apr. 18, 1918.
Posthumously awarded. Medal presented to mother, Mrs. Emma Schaffner.

*SCHAIRER, JAMES V. (1543536)............
  Near Montfaucon, France, Sept. 26, 1918.
  R—Toledo, Ohio.
  B—Benton Harbor, Mich.
  G. O. No. 20, W. D., 1919.

Private, medical detachment, 147th Infantry, 37th Division.
Seeing 2 men fall wounded, Private Schairer immediately went to their assistance, unmindful of the extreme danger that he was exposed to, and after dragging the men to a shell hole administered effective first aid. A few days later he was killed in the performance of his duties.
Posthumously awarded. Medal presented to father, John Schairer.

SCHALLERT, EDWARD I. (1633084)............
  At Madeleine Farm, France, Oct. 10–13, 1918.
  R—St. Louis, Mo.
  B—St. Louis, Mo.
  G. O. No. 24, W. D., 1920.

Private, medical detachment, 30th Infantry, 3d Division.
Although wounded in the thigh on Oct. 10, he continued to render first aid to the wounded under enemy fire until wounded a second time by a machine-gun bullet in the arm.

SCHENCK, ALEXANDER P............
  In the region of Doulcon, France, Oct. 4, 1918.
  R—Plainfield, N. J.
  B—Greensboro, N. C.
  G. O. No. 138, W. D., 1918.

First lieutenant, pilot, 49th Aero Squadron, Air Service.
He was one of an offensive patrol of 6 planes that attacked and engaged in combat 17 enemy machines (Fokker type). While he was engaging 1 of the enemy he observed a comrade about to be sent down by an enemy plane that had maneuvered to an advantageous position. He immediately left off the combat he was engaged in and shot down the plane, thereby saving the life of his comrade.

*SCHENCK, GORDON L............
  In the Argonne Forest, near Binarville, France, Oct. 3 to 7, 1918.
  R—Brooklyn, N. Y.
  B—Brooklyn, N. Y.
  G. O. No. 19, W. D., 1920.

Second lieutenant, 308th Infantry, 77th Division.
While his battalion was surrounded by the enemy, Lieutenant Schenck, by his heroic conduct, while repulsing frequent enemy attacks, inspired his command. Fearlessly exposing himself to fire, he seized his rifle and ran to the top of a bank in front of his company's position where he was able to throw hand grenades at the enemy until killed by an enemy shell.
Posthumously awarded. Medal presented to father, Charles N. Schenck.

SCHERMERHORN, CHARLES EARL_____
Near Cornay, France, Oct. 9-10, 1918.
R—Troy, N. Y.
B—Troy, N. Y.
G. O. No. 37, W. D., 1919.

Second lieutenant, 328th Infantry, 82d Division.
After successfully driving off the enemy his attacking force was counterattacked and surrounded. The officers in charge decided to surrender to the greatly superior numbers, but he, refusing to do so, made his way to our lines through deadly enemy fire, although severely wounded while doing so.

SCHIANI, ALFRED_____
In the Bois-de-Belleau, France, June 13, 1918.
R—Newark, N. J.
B—Brooklyn, N. Y.
G. O. No. 89, W. D., 1919.

Private, 18th Company, 5th Regiment, U. S. Marine Corps, 2d Division.
Severely wounded at the beginning of the attack, Private Schiani, an automatic rifle carrier, continued to advance, carrying two pouches of ammunition until he fell unconscious.

SCHICK, FRED (1697987)_____
In the Bois-de-la-Naza, France, Oct. 5, 1918.
R—Rosebank, N. Y.
B—Stapleton, Long Island, N. Y.
G. O. No. 95, W. D., 1919.

Corporal, Company H, 305th Infantry, 77th Division.
With 2 other soldiers, Corporal Schick volunteered to crawl out under enemy machine-gun fire in an effort to locate three members of the platoon who were missing after an unsuccessful attack on enemy machine-gun nests. Finding the body of 1, they located another, who lay helplessly wounded, by calling out his name. As a result they drew increased fire from the enemy, but they courageously crawled 25 yards farther toward the hostile positions and succeeded in carrying back the wounded man through the machine-gun fire to our lines.

SCHIDE, CLARENCE C._____
Near Bois-d'Ormont, France, Oct. 12, 1918.
R—Mason City, Iowa.
B—Charles City, Iowa.
G. O. No. 26, W. D., 1919.

Second lieutenant, 114th Infantry, 29th Division.
Although severely wounded, Lieutenant Schide continued to lead his platoon over open ground and subjected to heavy artillery and machine-gun fire until he received a second wound, which necessitated his removal from the field in a critical condition.

SCHKODA, THOMAS (41432)_____
Near Medeah Ferme, France, Oct. 7, 1918.
R—New York, N. Y.
B—Russia.
G. O. No. 21, W. D., 1919.

Bugler, Machine Gun Company, 9th Infantry, 2d Division.
After having received a wound in his knee by a machine-gun bullet, and a shell wound in the face, he continued his duties as runner until ordered evacuated by his company commander.

SCHLESINGER, ALBERT (1526010)_____
At Bois-Dommartin, near Beney, France, Oct. 11, 1918.
R—Cincinnati, Ohio.
B—Rumania.
G. O. No. 9, W. D., 1923.

Sergeant, Company G, 147th Infantry, 37th Division.
On the night of Oct. 11, 1918, at Bois-Dommartin, near Beney, France, he volunteered to recover the body of an American officer who had been killed while leading a raiding party, the body being left about 50 paces in front of the enemy positions. With a patrol of 6 men he proceeded on his mission, meeting heavy rifle and machine-gun fire. Ordering his men to retire, he, with 1 man of his patrol, covered the retirement, which was successfully accomplished. Although severely wounded, he continued his covering fire with automatic rifle and grenades, unassisted, and eventually reached his own lines.

*SCHMELZ, FREDERICK (1281962)_____
North of Verdun, France, Oct. 27, 1918.
R—Jersey City, N. J.
B—New York, N. Y.
G. O. No. 1, W. D., 1926.

Cook, Company K, 114th Infantry, 29th Division.
He volunteered to take hot food to the front-line troops, who had not received hot food for 3 days. After traveling 4 kilometers, he was fatally wounded by a bursting shell.
Posthumously awarded. Medal presented to father, William Schmelz.

SCHMIDT, FERDINAND A. (2407360)_____
Near Talma Farm, France, Oct. 22, 1918.
R—Newark, N. J.
B—Newark, N. J.
G. O. No. 35, W. D., 1919.

Private, Company B, 312th Infantry, 78th Division.
When his company's advance had been held up by intense machine-gun fire, he crawled through a barrage of hand grenades and at the point of his bayonet held 19 of the enemy in a dugout until assistance arrived. Before performing this courageous act he was slightly wounded.

SCHMIDT, RUSSELL A._____
Near Cumieres, France, Oct. 8, 1918.
R—Council Bluffs, Iowa.
B—Council Bluffs, Iowa.
G. O. No. 66, W. D., 1919.

Captain, 108th Field Signal Battalion, 33d Division.
With a detail of 5 men, Captain Schmidt was engaged in attempting to lay a telephone line across the Meuse River, when they were discovered and attacked by a superior force of the enemy. Even after being wounded 3 times, Captain Schmidt continued the unequal struggle, killing 1 and wounding 3 of the enemy, until all his ammunition was exhausted and all of his men severely wounded. Believing himself to be mortally wounded, he advanced into the enemy's lines and gave himself up in order to save the lives of his men. He was recaptured by our forces later in the day.

*SCHMITT, EDWARD F. (1703461)_____
Near Mont Notre Dame, France, Sept. 10, 1918.
R—Buffalo, N. Y.
B—Buffalo, N. Y.
G. O. No. 16, W. D., 1923.

Private, first class, Company L, 306th Infantry, 77th Division.
Voluntarily accompanying an officer and noncommissioned officer of his battalion in a search of the Bois-de-Chandriere for the survivors of a battalion which had been relieved from its position in the line, he crossed an open field under terrific enemy fire, a distance of 600 yards, under constant observation of the enemy, sought and found several survivors, and led them back to his own lines. The heroic and soldierly conduct of Private Schmitt and his devotion to his comrades greatly inspired the men of his battalion. Private Schmitt was later killed in action while gallantly fighting with his battalion in the Argonne Forest.
Posthumously awarded. Medal presented to father, William Schmitt.

SCHMITZ, CHARLES (40214)_____
Near Tilly, Marie Louise sector, France, Apr. 14, 1918.
R—Saginaw, Mich.
B—Trenton, Pa.
G. O. No. 126, W. D., 1919.

Private, Company I, 9th Infantry, 2d Division.
On Apr. 14, 1918, during an attack on his company by superior forces, he advanced single handed against 5 Germans who had taken cover in a shell hole and killed or wounded all of them with an automatic rifle.

*SCHNEIDER, JOHN G., Jr.
Near the Forest of Argonne, France, Nov. 1, 1918.
R—St. Joseph, Mo.
B—St. Joseph, Mo.
G. O. No. 21, W. D., 1919.

First lieutenant, 6th Regiment, U. S. Marine Corps, 2d Division.
Although he was painfully wounded, he continued to advance with his command until he was wounded a second time.
Posthumously awarded. Medal presented to father, John G. Schneider.

*SCHOBERTH, RAYMOND A.
Near Bantheville, France, Nov. 1, 1918.
R—Versailles, Ky.
B—Versailles, Ky.
G. O. No. 37, W. D., 1919.

First lieutenant, 359th Infantry, 90th Division.
Lieutenant Schoberth continued to lead his platoon after being wounded in the arm by a machine-gun bullet. He set an excellent example for his platoon by his courage and disregard for personal danger. This gallant officer was later killed by a shell fragment.
Posthumously awarded. Medal presented to father, Anthony Schoberth.

*SCHOEN, KARL J.
Near Ancorville, France, Oct. 10, 1918.
R—Indianapolis, Ind.
B—Indianapolis, Ind.
G. O. No. 37, W. D., 1919.

First lieutenant, 139th Aero Squadron, Air Service.
While leading a patrol of 3 machines he sighted 9 enemy planes (Fokker type) and immediately attacked them. Although greatly outnumbered, he destroyed 1 of the planes and put the others to flight. He was killed in action Oct. 29, 1918, and has been officially credited with destroying 7 enemy aircraft.
Posthumously awarded. Medal presented to widow, Mrs. Maurine Estelle Schoen.

SCHOLES, WILLIAM (1386755).
At Chipilly Ridge, France, Aug. 10, 1918.
R—Chicago, Ill.
B—Chicago, Ill.
G. O. No. 140, W. D., 1919.

Sergeant, Company C, 131st Infantry, 33d Division.
When the advance of his platoon was suddenly halted by intense machine-gun fire at close range, wounding his platoon commander and other platoon sergeants, Sergeant Scholes showed splendid devotion to duty by personally manning a machine gun in the advance position and maintaining fire until the rest of the platoon had reached shelter.

SCHOOLEY, HARRY T. (1736088).
Near Grand Pre, France, Oct. 25, 1918.
R—Laurel, Md.
B—Laurel, Md.
G. O. No. 35, W. D., 1919.

Corporal, Company L, 311th Infantry, 78th Division.
After his platoon had fallen back under heavy shellfire, Corporal Schooley made a personal reconnaissance of an enemy machine-gun nest through a heavy barrage, killed 1 of the enemy gunners, and drove off the others with hand grenades, thereby enabling his platoon to resume its position.

SCHRADER, EDGAR A. (198706).
At Berzy-le-Sec, France, July 18–21, 1918.
R—Vacaville, Calif.
B—Oldshasta, Calif.
G. O. No. 126, W. D., 1919.

Private, 2d Field Signal Battalion, 1st Division.
Attached to headquarters of the 1st Division as dispatch rider during the operations at Berzy-le-Sec, France, July 18–21, 1918, he courageously and unhesitatingly passed through areas under steady artillery bombardment to carry messages whose delivery was of vital necessity to the success of the attack.

SCHREECH, GEORGE WALTER.
Near Bayonville, France, Nov. 1, 1918.
R—Indianapolis, Ind.
B—Kansas, Ill.
G. O. No. 37, W. D., 1919.

Corporal, 32d Company, 6th Regiment, U. S. Marine Corps, 2d Division.
He volunteered and went forward to reconnoiter a ravine infested with hostile machine-gun and artillery positions, returning with several prisoners.

SCHROEDEL, JOHN C. (2159955).
Near Bellicourt, France, Sept. 29, 1918.
R—Sherburn, Minn.
B—Hinckley, Ill.
G. O. No. 37, W. D., 1919.

Private, Company B, 119th Infantry, 30th Division.
During operations in the region of Bellicourt Private Schroedel, unassisted, attacked an enemy stronghold and captured 2 machines and 5 prisoners.

SCHROTH, RAYMOND A. (1746136).
Near Grand Pre, France, Nov. 1, 1918.
R—Trenton, N. J.
B—Trenton, N. J.
G. O. No. 37, W. D., 1919.

First sergeant, Company E, 311th Infantry, 78th Division.
He was directed to lead an attack against an enemy machine-gun nest which was impeding the progress of his company. Reaching a point within 50 feet of the stronghold, he ordered an attack, and 6 Germans near the gun sought to surrender until they saw the strength of his force, now reduced through casualties to 2 or 3 men. The enemy was then reinforced by 20 men and launched a severe counterattack, which forced Sergeant Schroth to abandon his attack. After ordering his men to safety, he remained at his post alone, fighting against the superior forces until he drove a prisoner back to our lines at the point of his empty pistol. Despite his desire to provide safety for his patrol, he was the only survivor to return from the mission.

SCHUCHART, FRANK (2057101).
Near Juvigny, France, Aug. 30, 1918, and in the Argonne Forest, France, Nov. 8, 1918.
R—Belmont, Wis.
B—Liberty, Wis.
G. O. No. 95, W. D., 1919.

Private, Company L, 128th Infantry, 32d Division.
During the attack near Juvigny Private Schuchart, while acting as runner, repeatedly exposed himself to severe enemy artillery and machine-gun fire. After the attack he voluntarily joined in the work of clearing the field of wounded under a heavy enemy fire. On Nov. 8, while carrying an important message, he was severely wounded in the left leg. He stopped in a shell hole, dressed the wound himself, proceeded to deliver the message, and did not report for treatment.

SCHUEREN, DAN E., Jr.
Near Barricourt, France, Nov. 1, 1918.
R—Chicago, Ill.
B—Chicago, Ill.
G. O. No. 71, W. D., 1919.

Sergeant, 122d Field Artillery, 33d Division.
Sergeant Schueren, acting as liaison agent with an assaulting infantry battalion, on his own initiative took command of a platoon of infantry when its leader was wounded. He ordered the advance resumed and under his leadership machine-gun nests that threatened to hold up the advance of the entire battalion were flanked and silenced.

SCHULTHEIS, BERNARD (264586)..........
Near Terny-Sorny, north of Soissons,
France, Sept. 1, 1918.
R—Flint, Mich.
B—St. Louis, Mich.
G. O. No. 124, W. D., 1918.

Private, Machine Gun Company, 125th Infantry, 32d Division.
When the infantry was advancing in a position exposed to cross fire, he volunteered and carried a message to the advancing troops, informing them that a machine-gun barrage laid down on the enemy emplacements was friendly fire from a unit not in their support and acting without orders to cover their advance. He delivered the message, returned across an open field swept by enemy machine guns, and thereby made it possible for the infantry unit to advance 400 meters and gain its objective.

SCHULTZ, ARTHUR (284565)..........
Near Juvigny, north of Soissons,
France, Aug. 30, 1918.
R—Neenah, Wis.
B—Menasha, Wis.
G. O. No. 20, W. D., 1919.

Sergeant, Company I, 128th Infantry, 32d Division.
He displayed great courage and coolness during an attack in going forward under heavy fire and firing upon machine-gun emplacements. Later, when a retirement was ordered, he remained in advance of the line to carry back the wounded, in spite of heavy fire from artillery and machine guns.

*SCHULTZ, CHARLES (1389603)..........
Near Forges, France, Sept. 26, 1918.
R—Chicago, Ill.
B—Chicago, Ill.
G. O. No. 37, W. D., 1919.

Private, Company H, 132d Infantry, 33d Division.
While his platoon was being held up by machine-gun fire, he braved the hazardous fire by going forward and driving out the crew, after which he captured the gun. He died from wounds received in the exploit.
Posthumously awarded. Medal presented to mother, Mrs. Charles Schultz.

SCHULTZ, FRED M. (44376)..........
Near Fleville, France, Oct. 4, 1918.
R—Mount Clemens, Mich.
B—Mount Clemens, Mich.
G. O. No. 23, W. D., 1919.

Corporal, Company M, 16th Infantry, 1st Division.
Leading his squad through a heavy barrage and against violent machine-gun fire, he attacked an enemy field gun which had been holding up the progress of our tanks. He disabled the gun crew and took 15 prisoners, after which he personally captured a machine gun and killed its operator. Although wounded himself, he assisted 2 wounded members of his squad to the first-aid station.

SCHULTZ, GEORGE F. (3535520)..........
Near Sedan, France, Nov. 7, 1918.
R—New Buffalo, Mich.
B—Michigan City, Ind.
G. O. No. 59, W. D., 1919.

Private, Company E, 16th Infantry, 1st Division.
While accompanying his company as liaison agent in the advance, he attacked single handed a machine-gun nest which was delivering a heavy fire, killed the gunner, and caused many other casualties among the enemy. His act also caused other enemy machine gunners to withdraw and saved his company from a very dangerous flanking fire.

SCHULZ, FRANK (2443933)..........
Near Bazoches, France, Aug. 25–26,
1918.
R—Tompkinsville, N. Y.
B—Germany.
G. O. No. 21, W. D., 1919.

Private, Company C, 302d Engineers, 77th Division.
He and another soldier had become separated from their detachment and were forced to take shelter for 5½ days. He rescued a wounded comrade from exposure to machine-gun and shellfire and later attacked a machine-gun nest in his direct front. In the attack he killed 2 of the enemy with hand grenades and subsequently returned to our lines, assisting the wounded comrade to safety.

SCHULZE, RAYMOND J. (9060)..........
Near Orvillers-Sorel (Oise) France,
Aug. 16, 1918.
R—Cedar Rapids, Iowa.
B—Cedar Rapids, Iowa.
G. O. No. 37, W. D., 1919.

Private, first class, section No. 583, Ambulance Service.
When many French and American drivers had been killed or wounded during an intense bombardment on a dressing station, he immediately went to their assistance, but received wounds himself which will make him a cripple for life.

SCHUMACHER, MAX (1136293)..........
At Clery-le-Petit, France, Nov. 5,
1918.
R—Brenham, Tex.
B—Brenham, Tex.
G. O. No. 98, W. D., 1919.

Private, Company I, 60th Infantry, 5th Division.
When the advance of his company was held up by enemy machine-gun fire, Private Schumacher, with his platoon commander and another soldier, advanced in front of the line and attacked a machine-gun nest, killing 2 gunners and taking 6 prisoners, thereby enabling the company to advance and establish a brigade bridgehead.

SCHUMAKER, FRANCIS X..........
Near Heurne, Belgium, Nov. 3,
1918.
R—Dayton, Ohio.
B—Dayton, Ohio.
G. O. No. 37, W. D., 1919.

First lieutenant, 148th Infantry, 37th Division.
In the face of terrific machine-gun and artillery fire, he gave valuable assistance in the construction of a log bridge over the Scheldt River, which enabled his battalion to cross and establish itself in its objective. He remained with his company after being wounded until he was forced to be evacuated.

*SCHURTER, ALPHIA (489438)..........
In the lower Suchan Valley, Siberia,
July 5, 1919.
R—Hilltop, Kans.
B—Hilltop, Kans.
G. O. No. 133, W. D., 1919.

Private, first class, Company D, 31st Infantry.
Although mortally wounded he continued to advance with his platoon and assisted them by fire action until ordered to stop by his automatic rifle sergeant.
Posthumously awarded. Medal presented to widow, Mrs. Murial Schurter.

*SCHWAB, VINCENT M..........
At Chateau-Thierry, France, June
6, 1918.
R—St. Louis, Mo.
B—Germany.
G. O. No. 110, W. D., 1918.

Sergeant, 8th Company, 5th Regiment, U. S. Marine Corps, 2d Division.
Killed in action at Chateau-Thierry, France, June 6, 1918, he gave the supreme proof of that extraordinary heroism which will serve as an example to bitherto untried troops.
Posthumously awarded. Medal presented to sister, Miss Tillie Schwab.

SCHWANKE, OTTO A. (2024343)..........
At Sergy, France, Aug. 1, 1918.
R—Potter, Wis.
B—Rockland, Wis.
G. O. No. 21, W. D., 1919.

Private, first class, Company B, 47th Infantry, 4th Division.
He displayed the greatest devotion to duty, loyalty, and courage by repeatedly volunteering, night and day, to carry messages under the heaviest machine-gun and shell fire from his battalion commander to the company commanders, thereby maintaining efficient liaison at all times.

SCHWARTZ, ADOLPH (545204)..........
Service rendered under the name,
Frank Brown.

Corporal, Company A, 30th Infantry, 3d Division.

SCHWARTZ, BENJAMIN (551869) _____
During the attack on Hill 253, north of Clerges, France, Oct. 8–9, 1918.
R—Brooklyn, N. Y.
B—New York, N. Y.
G. O. No. 60, W. D., 1920.

Private, Company I, 38th Infantry, 3d Division.
Private Schwartz, carried numerous messages over routes swept by heavy machine-gun and artillery fire. Due to his personal heroism when exposed to heavy fire, his company commander was able to maintain communication with the battalion at all times during the attack.

SCHWARZWAELDER, CHRISTIAN ALLEN __
Near Vicville, France, Nov. 1, 1918.
R—Lake Mahopac, N. Y.
B—Brooklyn, N. Y.
G. O. No. 46, W. D., 1919.

First lieutenant, 21st Machine Gun Battalion, 7th Division.
Lieutenant Schwarzwaelder remained in an exposed place administering first aid to a wounded officer and 2 soldiers throughout a heavy barrage of gas shells and high explosives after his platoon had withdrawn from the position. He had the wounded men removed to a less exposed place and remained with them until they were evacuated.

SCHWEGLER, JOHN W. (1210086) _____
Near Ronssoy, France, Sept. 28, 1918.
R—New York, N. Y.
B—New York, N. Y.
G. O. No. 81, W. D., 1919.

Sergeant, Company C, 107th Infantry, 27th Division.
Sergeant Schwegler went forward from a front-line trench in daylight for a distance of 60 yards through enemy machine-gun and snipers' fire and brought back a wounded soldier to shelter.

SCHWER, HENRY G. (2158708) _____
Near Bellicourt, France, Sept. 29–30, 1918.
R—Fairfax, Iowa.
B—Falmouth, Ky.
G. O. No. 37, W. D., 1919.

Private, Company B, 119th Infantry, 30th Division.
During an attack by his regiment he was wounded, but continued his work as stretcher bearer throughout the night, refusing to be evacuated while able to render assistance to his comrades.

*SCHWING, FRED (1248737)_____
Near Montblainville, France, Sept. 28, 1918.
R—Warren, Pa.
B—Warren, Pa.
G. O. No. 16, W. D., 1923.

Private, first class, Company I, 112th Infantry, 28th Division.
Despite a concentration of enemy machine-gun and rifle fire, Private Schwing, together with Sergeant Small of his company, left the protection of the trenches, and in full view of the enemy advanced across an open space for a distance of 75 yards, rescued a wounded soldier and carried him to shelter. The bravery and devotion to duty thus displayed greatly inspired and encouraged the members of their command, inciting them to still greater endeavors.
Posthumously awarded. Medal presented to father, Erhart Schwing.

SCHWING, JAMES ALBERT_____
Near Montbrehain, France, Oct. 8, 1918.
R—Spartanburg, S. C.
B—Augusta, Ga.
G. O. No. 35, W. D., 1919.

First lieutenant, Company F, 118th Infantry, 30th Division.
With 2 soldiers he attacked a machine-gun nest of 4 guns and about 40 Germans. By the efficient use of grenades and automatic rifles the Germans were forced to surrender, thereby allowing the company to continue the advance.

SCIALABBA, IGNAZIO (1827847)_____
Near Imecourt, France, Nov. 1, 1918.
R—Mount Oliver, Pa.
B—Italy.
G. O. No. 37, W. D., 1919.

Corporal, Company K, 319th Infantry, 80th Division.
He crawled 300 yards alone, outflanked a machine gun, killed 4, and captured 3 of the crew. Although wounded by a shell fragment, he refused to go to the rear.

*SCIALABBA, JOSEPH (2338403)_____
Near Cunel, France, Oct. 13, 1918.
R—Butler, Pa.
B—Italy.
G. O. No. 89, W. D. 1919.

Private, Company C, 4th Infantry, 3d Division.
From an exposed outpost position, during a strong enemy counterattack, Private Scialabba opened fire on the advancing enemy with a captured machine gun. When the machine gun jammed, he picked up a light Browning gun and used it with the same deadly effect until the magazines were exhausted, whereupon he resorted to rifle fire. This gallant soldier was killed in action 2 days later.
Posthumously awarded. Medal presented to brother, Peter Scialabba.

SCIONTI, LOUIS (558045)_____
Near Bazoches, France, Aug. 9, 1918.
R—Boston, Mass.
B—Italy.
G. O. No. 46, W. D., 1919.

Sergeant, Company F, 47th Infantry, 4th Division.
Responding to a call for volunteers to destroy a hostile machine gun, Sergeant Scionti, with 2 other soldiers, boldly went forward through machine-gun fire and accomplished this mission.

SCLAFONI, ANTHONY (1203426) _____
Near Ronssoy, France, Sept. 29, 1918.
R—New York, N. Y.
B—New York, N. Y.
G. O. No. 32, W. D., 1919.

Private, Company A, 105th Infantry, 27th Division.
While the advance against the Hindenburg Line was at its height Private Sclafoni, seeing a Lewis gunner exposed to the enemy, ran to his assistance. On the way he was seriously wounded, but continued on, reaching the position and using his body to shield the gunner while the latter poured a fire into the enemy. He was wounded 3 times, finally losing consciousness, but, after his wounds were dressed, he insisted on leaving the field unaided.

SCOBY, OTIS C. (915416)_____
Near Brieulles, France, Nov. 2–4, 1918.
R—St. Francis, Kans.
B—St. Francis, Kans.
G. O. No. 37, W. D., 1919.

Sergeant, Company F, 7th Engineers, 5th Division.
While making a daylight reconnaissance of the Canal Est, he was at all times in full view of the enemy's snipers and machine guns. Unmindful of the danger, he continued on to the successful accomplishment of his mission. On the morning of Nov. 4 he aided materially in the construction of a pontoon bridge across the Canal Est, under heavy shell fire, thus enabling the Infantry to cross and capture commanding heights on the east bank of the Meuse.

*SCOTT, ALBERT E. (61565)_____
In Trugny Woods, northwest of Chateau-Thierry, France, July 23, 1918, during the Aisne-Marne offensive.
R—Brookline, Mass.
B—Boston, Mass.
G. O. No. 12, W. D., 1920.

Private, Company H, 101st Infantry, 26th Division.
Private Scott, an automatic rifleman, voluntarily posted himself on an exposed flank to cover a means of approach of an enemy attacking party. Absolutely alone, he opened fire on the enemy, killing and wounding many and fully stopping the flank attack before he himself was killed by a sniper's bullet. By his heroic act he saved the company a great many casualties and assured the maintenance of the perilous position.
Posthumously awarded. Medal presented to father, Stewart C. Scott.

°SCOTT, EDWARD W. (1211829)_____
Near Ronssoy, France, Sept. 29, 1918.
R—Westchester, N. Y.
B—Elberon, N. J.
G. O. No. 20, W. D., 1919.

First sergeant, Company L, 107th Infantry, 27th Division.
He assumed command of his company after all the officers had become casualties, though he himself had been shot through the arm, and led it into effective combat. After being wounded a second time he refused to go to the rear, but continued to advance until he was killed.
Posthumously awarded. Medal presented to mother, Mrs. Emily A. Scott.

SCOTT, JOHN S. (52429)_____
Near Soissons, France, July 20, 1918.
R—New Eagle, Pa.
B—Glassport, Pa.
G. O. No. 132, W. D., 1918.

Private, Company A, 26th Infantry, 1st Division.
When that portion of the line of which he was a part was violently attacked on July 20, 1918, near Soissons, France, he held his post and repulsed the enemy.

SCOTT, MILTON R_____
Near St. Etienne, France, Oct. 4, 1918.
R—La Monte, Mo.
B—La Monte, Mo.
G. O. No. 35, W. D., 1919.

Gunnery sergeant, 17th Company, 5th Regiment, U. S. Marine Corps, 2d Division.
After being severely wounded he continued to assist in consolidating the position of his platoon, later placing himself in an exposed position in order to gain good observation for sniping enemy machine-gun positions.

SCOTT, REGNOLL C. (3134334)_____
In the Argonne Forest, France, Oct. 3, 1918.
R—Ione, Wash.
B—Denver, Colo.
G. O. No. 44, W. D., 1919.

Private, Company L, 305th Infantry, 77th Division.
Although himself severely wounded, he assisted in caring for wounded comrades, refusing aid until all others had received treatment. In the performance of his duties he carried a message through an area which was under heavy machine-gun fire and constant hand-grenade bombing.

SCUDELLARI, PIETRO (43314)_____
Near Soissons, France, July 18, 1918.
R—Springfield, Mass.
B—Italy.
G. O. No. 35, W. D., 1920.

Private, Company G, 16th Infantry, 1st Division.
Knowing that the enemy had captured a wounded member of his company, Private Scudellari with 2 others advanced across dangerous ground to a barn, where they routed the enemy captors and carried back their comrade to safety.

SCULLY, GEORGE F. (2332019)_____
Near Grand Ballois Farm, France, July 14-15, 1918.
R—Philadelphia, Pa.
B—Philadelphia, Pa.
G. O. No. 32, W. D., 1919.

Private, Company A, 4th Infantry, 3d Division.
After being badly gassed, he continued to carry messages through heavy gas and high-explosive shell bombardment to the front line.

SEAGRAVES, CHARLES (541692)_____
At Fossoy, France, July 14-15, 1918.
R—Nashville, Tenn.
B—Akersville, Ky.
G. O. No. 98, W. D., 1919.

Sergeant, Company F, 7th Infantry, 3d Division.
During the intense artillery bombardment preceding the second battle of the Marne, Sergeant Seagraves volunteered to reestablish broken liaison with his company post of command. While carrying messages he was twice captured by groups of the enemy, but each time he escaped, killing 5 of his captors. On returning to his platoon's position and finding that every member of it had been killed or captured, he organized a group of 100 men from his own and other companies, and closed the breach of 500 meters in the line. Shortly afterwards he went out alone and, locating an enemy machine gun, captured the entire crew single handed.

°SEAGRAVES, VICTOR L. (1455903)_____
Near Baulny, France, Sept. 28, 1918.
R—Oskaloosa, Kans.
B—Jefferson County, Kans.
G. O. No. 81, W. D., 1919.

Sergeant, Company B, 1st Battalion, scout platoon, 139th Infantry, 35th Division.
Sergeant Seagraves voluntarily formed and led a patrol against an enemy machine-gun nest which was causing many casualties in his battalion and captured 1 of the guns. With utter disregard for his personal safety, he advanced alone on another gun of the nest, but was severely wounded by the intense fire in the performance of this heroic act.
Posthumously awarded. Medal presented to father, Patrick Seagraves.

SEALIE, MITCHELL J. (98756)_____
Northeast of Chateau-Thierry, France, July 26-27, 1918.
R—Birmingham, Ala.
B—Carbon Hill, Ala.
G. O. No. 126, W. D., 1919.

Sergeant, Company K, 167th Infantry, 42d Division.
Although seriously wounded during the advance near Croix Rouge Farm, he nevertheless continued in the attack under heavy enemy fire from artillery and machine guns, thereby setting to the men of his command an example of exceptional bravery and devotion to duty.

SEAMON, ALEXANDER R_____
Near Charpentry, France, Sept. 26, 1918.
R—Deming, N. Mex.
B—Rolla, Mo.
G. O. No. 13, W. D., 1923.

First lieutenant, 138th Infantry, 35th Division.
With a combat patrol, Lieutenant Seamon passed through our own weakened barrage and through a heavy enemy barrage, penetrating the enemy line to a depth of about 2 kilometers, entering the environs of Charpentry and capturing a German headquarters detachment of a dozen officers and men, together with valuable artillery maps showing the location of enemy batteries. Returning to his command, he organized his men and advanced again against a heavily manned and fortified machine-gun nest near Charpentry-Eclisfontaine Road, meeting his death in the advance.
Posthumously awarded. Medal presented to father, W. H. Seamon.

SEASTRAND, EINAR W. (938860)_____
Near Medeah Ferme, France, Oct. 5, 1918.
R—Greeley, Colo.
B—Cheyenne, N. Dak.
G. O. No. 21, W. D., 1919.

Private, medical detachment, Company G, 9th Infantry, 2d Division.
He displayed exceptional courage and devotion to duty by rendering first-aid to wounded soldiers under the most hazardous circumstances, many times braving machine-gun-swept fields in the performance of his duty.

SEAVER, ARTHUR F.
In the region of Etain, France, Sept. 16, 1918.
R—Brooklyn, N. Y.
B—Brooklyn, N. Y.
G. O. No. 143, W. D., 1918.

First lieutenant, pilot, 20th Aero Squadron, Air Service.
With his squadron he started on a bombing raid. The formation was broken up because of various troubles to the machines. He and his observer, Lieutenant Stokes, continued on and joined a formation of another bombing squadron. After crossing the lines their plane was struck by an antiaircraft explosive shell, throwing the machine out of control. When Lieutenant Seaver gained control of the machine it had fallen away from the protection of the other planes. With their crippled plane and missing motor they continued until they reached their objective, when their motor died completely. An enemy plane attacked, but Lieutenant Stokes kept him off until his machine coasted to their own lines.

SEAY, MILO B. (547509).
Near Crezancy, France, July 15, 1918.
R—New York, N. Y.
B—Columbia, S. C.
G. O. No. 44, W. D., 1919.

Corporal, Company I, 30th Infantry, 3d Division.
Although wounded, he continued to perform his duties as runner, and after having his wounds dressed immediately returned to duty.

*SECOR, JOHN H. (1764412).
Near St. Juvin, France, Oct. 16, 1918.
R—Pearl River, N. Y.
B—Newark, N. J.
G. O. No. 145, W. D., 1918.

Sergeant, Company M, 310th Infantry, 78th Division.
Having been painfully wounded in the foot, he remained with his platoon and went over the top with it in the advance near St. Juvin. He gallantly assisted his platoon commander in the attack until he was again wounded and rendered unable to advance farther. His example of bravery and devotion to duty furnished an inspiring example to the other members of the platoon, many of whom were under fire for the first time.
Posthumously awarded. Medal presented to father, John J. Secor.

SEDUSKY, ROBERT (15010).
South of Soissons, France, July 21, 1918.
R—Stamford, Conn.
B—Greenwich, Conn.
G. O. No. 39, W. D., 1920.

Sergeant, Machine Gun Company, 16th Infantry, 1st Division.
After reaching his objective, Sergeant Sedusky took command of scattered groups, reorganized them, and prepared the position for defense. Although under fire from 3 directions, he fearlessly exposed himself to this fire in order to direct the defense of the position. On Oct. 9, north of Exermont, this noncommissioned officer fearlessly led his section through machine-gun fire to position on Hill 272.

SEELER, WILFRED (1321).
Southwest of Fismes, France, Aug. 5, 1918.
R—Detroit, Mich.
B—Canada.
G. O. No. 117, W. D., 1918.

Private, first class, medical detachment, 125th Infantry, 32d Division.
During the forward movement of the 1st Battalion, 125th Infantry, a large number of the company to which he was attached were wounded while crossing an open field. At this point the artillery fire was accurate and intense, but he disregarded all possibilities of personal injuries and remained upon the field until he had administered first aid to all his fallen comrades.

SEELINGER, HARRY R.
At Nantillois, France, Oct. 5, 1918.
R—Norfolk, Va.
B—Erie, Pa.
G. O. No. 9, W. D., 1923.

First lieutenant, 317th Infantry, 80th Division.
During an attack made by 2 companies of the 3d Battalion, 320th Infantry, from the Bois-du-Fays, Lieutenant Seelinger with his medical detachment accompanied the troops and opened a first-aid station in an old cellar, with no overhead cover, remaining there under an intense barrage of enemy high-explosive shell and shrapnel fire until ordered to retire, working continuously from 9 a. m. until 6 p. m. caring for the wounded with great devotion and rare bravery. His coolness and utter disregard for his own safety under terrific enemy fire encouraged the wounded and raised the morale of his men to a high pitch.

SEIBEL, ALBERT (488697).
Near Jaulny, France, Nov. 8, 1918.
R—Clayton, Ill.
B—Camp Point, Ill.
G. O. No. 32, W. D., 1919.

Private, medical detachmet, 55th Infantry, 7th Division.
When an officer and part of 1 platoon had been cut off from the company, Private Seibel made 2 attempts to find them, going through a heavy barrage. He was successful the second time and administered first aid to the wounded officer while under a heavy shell and machine-gun fire.

SEIBEL, HERMAN F. (1979778).
Near Bellicourt, France, Sept. 29, 1918.
R—Sellersburg, Ind.
B—Sellersburg, Ind.
G. O. No. 37, W. D., 1919.

Private, Company D, 120th Infantry, 30th Division.
With 8 other soldiers, comprising the company headquarters detachment, he assisted his company commander in cleaning out the enemy dugouts along a canal and capturing 242 prisoners.

SEIBERLING, PAUL A.
Near Madeleine Farm, France, Oct. 16, 1918.
R—Jonesboro, Ind.
B—Jonesboro, Ind.
G. O. No. 53, W. D., 1920.

Second lieutenant, 9th Machine Gun Battalion, 3d Division.
Lieutenant Seiberling, with a comrade, exposed himself to heavy enemy machine-gun fire in crawling forward in advance of our lines to rescue a wounded soldier. The wounded man was lying in a shell hole about 100 yards in advance of our lines. The rescuers were subject to enemy machine-gun fire from the time they left our lines until they returned.

SEIDEL, THOMAS (284662).
Near Soissons, France, July 19, 1918.
R—Neenah, Wis.
B—Greenwood, Wis.
G. O. No. 132, W. D., 1918.

Private, Company G, 26th Infantry, 1st Division.
With 2 other soldiers he rushed a machine-gun position near Soissons, France, July 19, 1918, killed the crew, and captured the gun in order to make the advance of his platoon possible.

SEIDERS, CLIFFORD M. (1761957).
Near Malancourt, France, Sept. 26, 1918.
R—Philadelphia, Pa.
B—Easton, Pa.
G. O. No. 46, W. D., 1919.

Private, first class, Machine Gun Company, 314th Infantry, 79th Division.
Advancing ahead of his platoon in the face of heavy machine-gun fire, Private Seiders entered alone a ruined building and discovered 13 of the enemy. He shot 1 who resisted capture and made prisoners of the remaining 12, bringing in with him 3 light machine guns. Later in the same day he captured 10 of the enemy and 5 machine guns.

SEIDERS, WALTER H. (42083).
South of Soissons, France, July 20, 1918.
R—Philadelphia, Pa.
B—Easton, Pa.
G. O. No. 35, W. D., 1920.

Private, first class, Company B, 16th Infantry, 1st Division.
After 3 others had been killed in the attempt, Private Seiders voluntarily carried a very important message to advance line positions. He passed over terrain exposed to heavy artillery and machine-gun fire, and delivered his message. The delivering of the message was vitally important to the success of the operation.

SEIGLER, WILLIAM (2667022)............
Near Ivoiry, France, Sept. 28, 1918.
R—Philadelphia, Pa.
B—Philadelphia, Pa.
G. O. No. 59, W. D., 1919.

Private, Company A, 146th Infantry, 37th Division.
He repeatedly volunteered and carried messages under heavy enemy bombardment until he was severely wounded.

SEITZ, LESTER EARL...................
Near Blanc Mont, France, Oct. 3-5, 1918.
R—McArthur, Ohio.
B—Fincastle, Ohio.
G. O. No. 46, W. D., 1919.

Private, 43d Company, 5th Regiment, U. S. Marine Corps, 2d Division.
After being struck in the leg by shrapnel, he continued to act as stretcher-bearer for 2 days and nights under heavy artillery and machine-gun fire, carrying wounded comrades to the first-aid station.

SELBY, HARRY J...................
Near Exermont, France, Oct. 4, 1918.
R—Ivory, Md.
B—Ivory, Md.
G. O. No. 39, W. D., 1920.

Captain, 18th Infantry, 1st Division.
Captain Selby led his battalion in the attack through heavy artillery and machine-gun fire until killed by a machine-gun bullet. On one occasion he opened fire upon an enemy machine-gun nest with his pistol, thus drawing its fire while others made a successful flank attack.
Posthumously awarded. Medal presented to father, John W. Selby.

SELFE, CARTER C. (54750)............
Near Cantigny, France, May 27, 1918.
R—Castlewood, Va.
B—Castlewood, Va.
G. O. No. 15, W. D., 1921.

Corporal, Company K, 26th Infantry, 1st Division.
During an enemy attack on his position Corporal Selfe, although subjected to most terrific artillery bombardment and heavy machine-gun fire, held the position and conducted the fire of his squad until all the members had become casualties. Although he was severely wounded, he took the 1 remaining automatic rifle and rushed to the assistance of a near-by automatic rifle post where the enemy was about to penetrate our lines. Although again wounded, he refused to be evacuated until after the enemy had been repulsed. His gallant conduct was a material factor in the successful defense of the position.

SELL, HERMAN M. (1700766)............
In La Cendriere Woods, near Vauxcere, between the Vesle and Aisne, France, Sept. 6, 1918
R—Seaford, N. Y.
B—Brooklyn, N. Y.
G. O. No. 99, W. D., 1918.

First sergeant, Company A, 306th Infantry, 77th Division.
First sergeant Sell volunteered to deliver a message of great importance to his battalion commander after 6 runners, who had been sent with the same message, failed to return. He voluntarily crossed 600 yards of open field swept by shell and machine-gun fire, reached his destination, accomplished his mission, and returned to his company with information of vital importance.

SELLERS, CECIL G.................
Near Longuyon, France, Sept. 16, 1918.
R—Memphis, Tenn.
B—Dyersburg, Tenn.
G. O. No. 123, W. D., 1918

First lieutenant, pilot, 20th Aero Squadron, Air Service.
Starting on a very important bombing mission with 5 other planes, as pilot he went on alone when the other 5 machines were forced to turn back. On crossing the enemy lines he was attacked by 3 enemy planes, but continued toward his objectives, while his observer kept them at bay. In the face of this hostile opposition the objective was reached and their bombs dropped. On the way back 4 more planes joined in the attack, but fighting them off they reached our lines with valuable information after a fight lasting 38 minutes.

SELLERS, GUY E. (2397944)............
Near Moulin, France, July 15, 1918.
R—Sparta, Mich.
B—Jackson, Ohio.
G. O. No. 24, W. D., 1920.

Private, first class, Company E, 38th Infantry, 3d Division.
He carried numerous messages over routes swept by enemy rifle and machine-gun fire. Due to his individual heroism when exposed to heavy fire, his platoon commander was able to learn of the plan of action at a very important time during the powerful enemy offensive across the Marne.

SELLERS, JAMES McB.................
At Bouresches, France, June 6, 1918.
R—Lexington, Mo.
B—Lexington, Mo.
G. O. No. 126, W D, 1918.

First lieutenant, 6th Regiment, U. S. Marine Corps, 2d Division.
On June 6, 1918, at Bouresches, France, at a critical period of the attack he was selected to transport a message, of the extreme importance of which he was cognizant. In order to execute this mission he had to pass through a heavy artillery bombardment of high-explosive and gas shells. Although seriously wounded while making the trip, he successfully executed his mission.

SELLERS, WILLIAM EDMOND (1311712)...
Near Brancourt, France, Oct. 8, 1918.
R—Chesterfield, S. C.
B—Chesterfield, S. C.
G. O. No. 81, W. D., 1919.

Sergeant, Company I, 118th Infantry, 30th Division.
While his platoon was advancing, he on his own initiative rushed ahead of the line and, flanking an enemy machine-gun post, shot 1 of the crew and bayoneted the other. His action saved his platoon from heavy casualties.

SELTZER, MAX (44355)............
Near Fleville, France, Oct. 4, 1918.
R—Brooklyn, N. Y.
B—Russia.
G. O. No. 35, W. D., 1920.

Corporal, Company M, 16th Infantry, 1st Division.
In the attack launched along the Aire River, Corporal Seltzer, single handed, silenced an enemy machine gun that was causing casualties in his company. He was severely wounded on Oct. 9 while leading his squad in an attack on an enemy strong point.

SEMBERTRANT, FRANK (2262922)........
Near Waereghem, Belgium, Oct. 31, 1918.
R—San Francisco, Calif.
B—San Francisco, Calif.
G. O. No. 44, W. D., 1919.

Private, first class, Machine Gun Company, 363d Infantry, 91st Division.
Having set up his gun in the open, near the enemy wire, at a range of less than 200 meters from the enemy, and in the face of direct machine-gun fire, Private Sembertrant offered a most stubborn resistance to the enemy, despite the fact that his sergeant had been killed and his gun damaged by the heavy fire. He continued until 1 gunner had been killed and the fire from another gun silenced.

SEMMES, HARRY HODGES_____
    Near Xivray, France, Sept. 12, 1918.
    R—Washington, D. C.
    B—Washington, D. C.
    G. O. No. 35, W. D., 1919.

Captain, Tank Corps.
During the operations along the Rupt de Mad, Captain Semmes's tank fell into the water and was completely submerged. Upon escaping through the turret door and finding that his driver was still in the tank, he returned and rescued the driver under machine-gun fire.
Oak-leaf cluster.
For the following act of extraordinary heroism in action near Vauquois, France, Sept. 26, 1918, Captain Semmes is awarded an oak-leaf cluster to be worn with the distinguished-service cross: He left his tank under severe rifle fire and personally reconnoitered a passage for his tank across the German trenches, remaining dismounted until the last tank had passed. While so engaged, he was severely wounded.

SEMPLE, FRANK J. (2941349)_____
    Near Grand Pre, France, Nov. 1, 1918.
    R—Rochester, N. Y.
    B—Rochester, N. Y.
    G. O. No. 37, W. D., 1919.

Private, Company I, 310th Infantry, 78th Division.
Under heavy machine-gun fire, Private Semple, a battalion runner, volunteered to carry an important message to a detachment on the extreme flank after seeing 2 runners killed by machine-gun fire while endeavoring to deliver the same message. He was successful in his mission.

SENAY, CHARLES T_____
    Near Ploisy, south of Soissons, France, July 19, 1918.
    R—New London, Conn.
    B—Norwich, Conn.
    G. O. No. 117, W. D., 1918.

Captain, 28th Infantry, 1st Division.
He displayed inspiring courage and leadership under heavy fire during the capture of Ploisy and while reorganizing units and repelling a counterattack.

SERNA, MARCELINO (2195593)_____
    Near Flirey, France, Sept. 12, 1918.
    R—Fort Morgan, Colo.
    B—Mexico.
    G. O. No. 27, W. D., 1919.

Private, Company B, 355th Infantry, 89th Division.
He displayed exceptional coolness and courage in single-handed charging and capturing 24 Germans.

SESSIONS, HARRY O_____
    Near Bussy Farm, France, Sept. 28–29, 1918.
    R—Oakland, Calif.
    B—Oakland, Calif.
    G. O. No. 13, W. D., 1919.

Second lieutenant, 372d Infantry, 93d Division.
Although he was on duty in the rear, he joined his battalion and was directed by his battalion commander to locate openings through the enemy's wire and attack enemy positions. He hastened to the front and cut a large opening through the wire in the face of terrific machine-gun fire. Just as his task was completed he was so severely wounded that he had to be carried from the field. His gallant act cleared the way for the rush that captured the enemy positions.

SETTLE, FRANK J. (129306)_____
    Near Thiaucourt, France, Sept. 12, 1918.
    R—Blue Creek, W. Va.
    B—Blue Creek, W. Va.
    G. O. No. 37, W. D., 1919.

Private, first class, Battery E, 12th Field Artillery, 2d Division.
While acting as No. 1 of the fourth piece, Private Settle continued in the service of his piece under a heavy and well-directed enfilading fire. When a shell wiped out the entire gun crew of the first section, at a word from his executive officer he sprang to assume the duties of gunner of the first piece. He assisted in carrying the dead and wounded and acted in his new capacity until the Infantry attained their objective.

SEVALIA, WALTER S. (915500)_____
    Near Brieulles, France, Nov. 3, 1918.
    R—Brule, Wis.
    B—Ashland, Wis.
    G. O. No. 37, W. D., 1919.

Corporal, Company F, 7th Engineers, 5th Division.
He swam the Meuse River with a cable for a pontoon bridge under direct machine-gun fire. Later he carried a cable for another bridge over the Est Canal across an open field covered by enemy machine guns. Here he was wounded by a machine-gun bullet, but returned carrying a message of great importance.

SEWALL, SUMNER_____
    Near Menil-la-Tour, France, June 3, 1918 and near Landres-et-St. Georges, France, Oct. 18, 1918.
    R—Bath, Me.
    B—Bath, Me.
    G. O. No. 32, W. D., 1919.

First lieutenant, 95th Aero Squadron, Air Service.
On June 3 Lieutenant Sewall, with 2 other pilots, attacked a formation of 6 hostile planes. Though his companions were forced to withdraw because of jammed guns, he continued in the fight for 15 minutes and succeeded in sending 1 of his adversaries down in flames. On Oct. 13 while on a voluntary patrol he saw an American observation plane being attacked by a German machine (type Fokker), accompanied by 8 other hostile planes. He immediately attacked and destroyed the Fokker and was in turn attacked by the other 8 planes. By skillful maneuvering he evaded them and escorted the observation plane back to our lines.
Oak-leaf cluster.
Lieutenant Sewall is also awarded an oak-leaf cluster for the following act of extraordinary heroism in action near Rocourt, France, July 7, 1918: He fearlessly attacked a formation of 5 enemy planes (type Fokker), and separating 1 from the group, pursued it far behind the enemy's lines, and sent it down in a crash, following it to within 30 meters of the ground in spite of severe fire from a machine gun, rifles, and antiaircraft guns, bullets from which passed through his clothing.

*SEXTON, FRED (54830)_____
    South of Soissons, France, July 18, 1918.
    R—Oneida, Tenn.
    B—Point Rock, Tenn.
    G. O. No. 24, W. D., 1920.

Sergeant, Company L, 26th Infantry, 1st Division.
Sergeant Sexton exposed himself to heavy machine-gun fire while leading his platoon in an attack on a machine-gun position near Missy-aux-Bois. Although wounded in the leg by a machine-gun bullet, he continued to command the platoon until again wounded on the following day.
Posthumously awarded. Medal presented to mother, Mrs. Eldora Sexton.

*SEXTON, FRED H_____
    Near Molleville Farm, France, Oct. 17, 1918.
    R—Union, S. C.
    B—Union, S. C.
    G. O. No. 32, W. D., 1919.

Second lieutenant, 113th Infantry, 29th Division.
During the thickest of the fight in the attack on Molleville Farm, Lieutenant Sexton alone set out to locate enemy machine-gun positions. While on this mission he was killed.
Posthumously awarded. Medal presented to father, J. T. Sexton.

SEXTON, FRED LEO.............
Near Bayonville, France, Nov. 1, 1918.
R—Osage, Iowa.
B—Osage, Iowa.
G. O. No. 35, W. D., 1919.

Private, 82d Company, 6th Regiment, U. S. Marine Corps, 2d Division.
Exposing himself to enemy fire, Private Sexton, with another soldier, courageously advanced ahead of their platoon and captured 5 machine guns and 14 prisoners.

SEYMOUR, QUINCY R. (2186863)...........
In the Bois-de-Barricourt, France, Nov. 2, 1918.
R—Rantoul, Kans.
B—Rantoul, Kans.
G. O. No. 89, W. D., 1919.

Private, Company F, 353d Infantry, 89th Division.
With another soldier, Private Seymour advanced more than 150 yards over an open space swept by fire from 30 enemy machine guns, for the purpose of drawing the fire of these guns, while the remainder of his company attacked them from the flanks. His self-sacrificing act cost him his life, but enabled his comrades to capture the hostile position.
Posthumously awarded. Medal presented to father, James O. Seymour.

SHADRICK, BART L. (2218529)...........
Near Pey-en-Haye, France, Sept. 12, 1918.
R—Sapulpa, Okla.
B—Columbia, Mo.
G. O. No. 128, W. D., 1918.

Private, Company E, 357th Infantry, 90th Division.
When a part of his company was held up by machine-gun fire this soldier, with the aid of 2 others, flanked 2 machine guns, killed the gunners, and captured the guns, thereby allowing the company to advance without delay or losses. Later on in the same advance he crept up to a German machine-gun emplacement, rolled over the parapet onto the gun crew, putting gun and gunners out of action.

SHAHAN, WINFIELD F. (1458127).........
In the Meuse-Argonne offensive, Sept. 26-28, 1918, and near Exermont, France, Sept. 29, 1918.
R—Marion, Kans.
B—Marion, Kans.
G. O. No. 59, W. D., 1919.

Corporal, Company M, 139th Infantry, 35th Division.
Corporal Shahan, regimental liaison noncommissioned officer, with great courage constantly exposed himself to heavy enemy machine-gun and shell-fire for 3 days in maintaining liaison between his regimental headquarters and the companies in the front line. On Sept. 29 he attacked, single handed, an enemy machine-gun nest killing several of the enemy and taking 1 prisoner. While making his way back to his lines he was fired upon by another German machine gun, which wounded him in the right arm, and was at the same time attacked by his prisoner. In spite of his wound, he killed the German with his pistol and reached his lines in safety.

SHAHWOOD, SOLOMON (2444687)..........
Near Carrefour-de-Meurrussons, France, Sept. 27, 1918.
R—Buffalo, N. Y.
B—Syria.
G. O. No. 44, W. D., 1919.

Private, Company A, 305th Infantry, 77th Division.
After his company had taken shelter from the enfilading machine-gun and trench-mortar fire of the enemy, Private Shahwood, with 2 other soldiers, crawled to the aid of wounded comrades, thus saving the lives of at least 2, while exposed to terrific fire of the enemy.

SHALLENBERGER, HUGH D. Jr...........
Near Preny Ridge, France, Nov. 1, 1918.
R—Vanderbilt, Pa.
B—Vanderbilt, Pa.
G. O. No. 32, W. D., 1919.

Second lieutenant, 56th Infantry, 7th Division.
Although twice wounded by machine-gun fire while leading his men in an attack on Preny Ridge, under heavy machine-gun fire, Lieutenant Shallenberger continued in the advance until he was killed by the explosion of a shell.
Posthumously awarded. Medal presented to father, H. D. Shallenberger.

SHAMANSKI, WALTER A. (41222)......
At Vaux, France, July 1, 1918.
R—Mount Carmel, Pa.
B—Mount Carmel, Pa.
G. O. No. 69, W. D., 1918.

Private, first class, Headquarters Company, 9th Infantry, 2d Division.
At Vaux, July 1, 1918, having entered a cellar to install his telephone, he was attacked by 11 of the enemy, of whom he killed 2 and took 9 prisoners, single handed.
Posthumously awarded. Medal presented to father, Joseph Shamanski.

SHANAHAN, EDWARD T. (91142)..........
Near Sergy, and Seringes, France, July 28, 1918.
R—Manville, N. J
B—Liberty, Wis.
G. O. No. 108, W. D., 1918.

Sergeant, Company I, 165th Infantry, 42d Division.
In the face of violent artillery and machine-gun fire near Sergy and Seringes, France, on July 28, 1918, he selected a squad of men and rushed a machine gun that had been harassing his company with its fire. He reached and captured the gun and killed the crew.

SHANE, WILLIAM M. (1245145)..........
Near Le Chene Tondu, France, Oct. 1, 1918.
R—Pittsburgh, Pa.
B—Pittsburgh, Pa.
G. O. No. 50, W. D., 1919.

Corporal, Company I, 111th Infantry, 28th Division.
Seeing the commanding officer of an adjacent unit fall from a wound, Corporal Shane left his place of safety and made his way through the continuous rain of machine-gun bullets to the side of the wounded officer. After a severe struggle he managed to drag him to a place of safety.

SHANKLE, VANCE C. (1312113)..........
Near St. Martin Riviere, France, Oct. 17, 1918.
R—Concord, N. C.
B—Albemarle, N. C.
G. O. No. 87, W. D., 1919.

Corporal, Company K, 118th Infantry, 30th Division.
When the advance of his company was held up, he volunteered to go forward with another soldier, to reduce a machine-gun emplacement. Advancing in front of our lines, these 2 soldiers attacked the enemy position, destroyed it, and captured 3 prisoners. Corporal Shankle was killed in action shortly afterwards.
Posthumously awarded. Medal presented to brother, Brooks B. Shankle.

SHANKLIN, ALMERON W..............
Near Cunel, France, Oct. 14, 1918.
R—Rome, Ga.
B—Rome, Ga.
G. O. No. 20, W. D., 1919.

First lieutenant, 11th Infantry, 5th Division.
Forbidding his men to leave their place of safety, Lieutenant Shanklin went forth in the face of heavy machine-gun fire, located and sighted his 37-millimeter gun, receiving wounds which proved fatal.
Posthumously awarded. Medal presented to widow, Mrs. Walton Shanklin.

SHANNON, FRED B. (1309123) ....
Near Geneve and Promont, France, Oct. 7-20, 1918.
R—Signal Mountain, Tenn.
B—Fall River, Tenn.
G. O. No. 44, W. D., 1919.

Sergeant, Company K, 117th Infantry, 30th Division.
Throughout the engagement he led his platoon with great bravery and distinction, participating constantly in the severe fighting of that period, despite a painful wound in the hand and another in the arm, received the first day of the engagement. He gave unsparingly of his strength while helping others in addition to his own work until his company had been relieved. He dropped unconscious from exhaustion and the effects of his wound soon after turning over his platoon to the second in command.

*SHANNON, JAMES A_____
Near Chatel-Chehery, France, Oct. 5–6, 1918.
R—Duluth, Minn.
B—Granite Falls, Minn.
G. O. No. 130, W. D., 1918.

Lieutenant colonel, 112th Infantry, 28th Division.
He voluntarily led an officers' patrol to a depth of 3 kilometers within the enemy lines. As a result of his exceptional bravery and skill in leading this patrol in its contact with the enemy, vital information was obtained at a critical period of the battle, to which much of the success of the next few days was due. The information thus secured was followed up by an attack the next morning, which this officer personally led and wherein he was fatally wounded. His superb leadership and personal courage furnished the necessary inspiration to an exhausted command.
Posthumously awarded. Medal presented to widow, Mrs. James A. Shannon.

*SHANNON, JOHN (554630)_____
Near Jaulgonne, France, July 23, 1918.
R—Newport, Ky.
B—Newport, Ky.
G. O. No. 24, W. D., 1920.

Private, Company A, 9th Machine Gun Battalion, 3d Division.
Private Shannon heroically worked alone, exposed to heavy enemy fire, in helping wounded to shelter. He carried water for wounded exposed to sniping fire. When himself severely wounded he refused attention, directing that other wounded be removed.
Posthumously awarded. Medal presented to father, John T. Sullivan.

SHANTZ, JOSEPH E_____
Near Consenvoye, France, Oct. 13, 1918.
R—Wilmette, Ill.
B—Philadelphia, Pa.
G. O. No. 37, W. D., 1919.

First lieutenant, 131st Infantry, 33d Division.
Although seriously wounded in the head by shrapnel, he went forward to rectify the position of our troops, who were occupying the ground on which our barrage was scheduled to fall. Through a perilous fire he brought the line back to a new position.

SHAPIRO, ELI R. (1390878)_____
Near Forges, France, Sept. 26, 1918.
R—Chicago, Ill.
B—Russia.
G. O. No. 44, W. D., 1919.

Corporal, Company D, 132d Infantry, 33d Division.
After having been severely wounded, he continued to lead his squad during the entire attack, which lasted several hours, and he remained until his objective had been reached and his squad sheltered.

SHARKEY, CHARLES WESLEY (794962)__
During the St. Mihiel offensive, Sept. 12, 1918.
R—Cleveland, Ohio.
B—Maysville, Ky.
G. O. No. 35, W. D., 1919.

Private, first class, Company C, 15th Machine Gun Battalion, 5th Division.
After being shot in the right arm, he continued to advance and, by the effective use of his pistol with his left hand, alone captured 20 Germans and 2 machine guns.

*SHARP, DON E. (17106)_____
During the Aisne-Marne offensive, July 31, 1918.
R—Saginaw, Mich.
B—Gladwin, Mich.
G. O. No. 14, W. D., 1923.

Private, medical detachment, 125th Infantry, 32d Division.
Though he had been twice wounded he continued to render first aid to the wounded men of his command until killed by enemy fire. His splendid example of high courage and coolness under intense enemy machine-gun and artillery fire and his unselfish devotion to duty inspired his comrades to great endeavors.
Posthumously awarded. Medal presented to father, Samuel Sharp.

SHARP, JAMES H_____
Near St. Etienne-a-Arnes, France, Oct. 3–9, 1918.
R—Moorhead, Minn.
B—Moorhead, Minn.
G. O. No. 35, W. D., 1919.

First lieutenant, 23d Infantry, 2d Division.
He volunteered and made several reconnaissances through heavy machine-gun and artillery fire. When the flank of his organization was dangerously exposed, he volunteered to get assistance. He brought up several companies, thus saving the flank from annihilation. He carried several important messages through extremely heavy machine-gun and artillery fire.

SHARP, ROBERT E_____
Near Estrees, France, Oct. 6–7, 1918.
R—Chattanooga, Tenn.
B—Bloomingport, Ind.
G. O. No. 46, W. D., 1919.

Second lieutenant, 117th Infantry, 30th Division.
Acting as battalion gas officer, Lieutenant Sharp volunteered to carry an important message to 1 of the companies after 3 runners had been killed in attempting to do so. He succeeded in making the trip through heavy shell and machine-gun fire and returning safely. Next morning, upon his own request, he was permitted to join one of the attacking companies and was severely wounded while leading a platoon to its objective.

*SHARP, THOMAS V. (732603)_____
Near Regnieville, France, Sept. 12, 1918.
R—Wichita, Kans.
B—Osawatomie, Kans.
G. O. No. 95, W. D., 1919.

Private, Company F, 6th Infantry, 5th Division.
Having located an enemy machine-gun nest, Private Sharp, accompanied by another soldier, was advancing on the nest under fire when the German gunners threw up their hands and yelled, "Kamerad." They continued toward the nest and when they were within 15 yards of the position the enemy again opened fire, killing Private Sharp's companion, thereupon Private Sharp dashed straight at the enemy emplacement, shooting 1 gunner, bayoneting 2 others, and capturing 4, together with 3 machine guns.
Posthumously awarded. Medal presented to mother, Mrs. Belle C. Sharp.

SHARRAR, OLIVER (1911362)_____
At Cornay, France, Oct. 9–10, 1918.
R—Fertigs, Pa.
B—Venango County, Pa.
G. O. No. 37, W. D., 1919.

Corporal, Company M, 328th Infantry, 82d Division.
After fighting for 6 hours, Corporal Sharrar volunteered to accompany 15 other soldiers and an officer on a night patrol of Cornay, which was held by many enemy machine-gun posts. The party worked from 11 o'clock at night until next morning at clearing buildings and dugouts of the enemy, capturing 65 prisoners and 2 machine guns. With 6 others, Corporal Sharrar volunteered and entered a dugout where 23 prisoners were captured. He was wounded while leaving the town, but he refused to go to the aid station until the prisoners had been delivered at brigade headquarters.

SHARTLE, ALBERT J_____
Near Bethincourt, France, Sept. 26, 1918.
R—Philadelphia, Pa.
B—Philadelphia, Pa.
G. O. No. 16, W. D., 1919.

First lieutenant, 315th Machine Gun Battalion, 80th Division.
He gave proof of courage and unhesitating devotion to duty when he rallied a platoon of Infantry held up by intense fire from a machine gun directly to the front. This officer led the platoon against the hostile strong point, captured it, and fell severely wounded.

SHASKAN, SAMUEL (282956)_____
Near Juvigny, France, Aug. 28, 1918.
R—Chicago, Ill.
B—Russia.
G. O. No. 66, W. D., 1919.

Private, Headquarters Company, 128th Infantry, 32d Division.
Going out from the front line through barbed-wire entanglements, under heavy artillery and machine-gun fire, Private Shaskan brought back to safety a wounded soldier.

SHAW, ANDREW A. (73108)_____
At Givry, France, July 20, 1918.
R—Cummington, Mass.
B—Cummington, Mass.
G. O. No. 9, W. D., 1923.

Private, Company I, 104th Infantry, 26th Division.
When he on his own initiative led a group of 7 men under a severe rifle and machine-gun fire through a gap in the enemy's wire in an effort to silence a hostile machine gun which had been taking heavy toll of his comrades. Six of the 7 men in his party having been killed or wounded, he continued to advance, killing or wounding the crew and putting the gun out of action. His splendid courage and leadership were an inspiration to his comrades.

°SHAW, CHARLES A_____
During the St. Mihiel offensive, France, Sept. 12–13, 1918.
R—Pattonsburg, Mo.
B—Pattonsburg, Mo.
G. O. No. 125, W. D., 1918.

First lieutenant, 353d Infantry, 89th Division.
He personally led his platoon under heavy machine-gun fire into the undamaged enemy wire, so inspiring his platoon that, regardless of heavy losses, the machine-gun nest was neutralized. He was killed 1 minute after his platoon had accomplished its mission.
Posthumously awarded. Medal presented to father, Philip Shaw.

SHEA, RICHARD O'B_____
At Chateau-Thierry, France, June 6, 1918.
R—Westerly, R. I.
B—Norwich, Conn.
G. O. No. 109, W. D., 1918.

Passed Assistant Surgeon, U. S. Navy, attached to 5th Regiment, U. S. Marine Corps, 2d Division.
At Chateau-Thierry, France, on June 6, 1918, he displayed extraordinary heroism treating the wounded while under heavy bombardment. He showed utter disregard of his personal safety in order to succor others.

SHEA, WILLIAM A. (556292)_____
Near Cuisy, France, Sept. 26, 1918.
R—Niagara Falls, N. Y.
B—Holyoke, Mass.
G. O. No. 46, W. D., 1919.

Sergeant, Machine Gun Company, 39th Infantry, 4th Division.
Although painfully wounded by machine-gun fire, he placed himself in an exposed position between 2 machine guns, and by the use of his glasses directed the fire of a heavy machine-gun barrage on the enemy. He remained in this exposed position for 2 hours, and his were the only guns which remained in action under the sweeping fire of the enemy.

SHEAFF, DONALD RAMSAY_____
In the Bois-de-Belleau, France, June 6, 1918.
R—Colorado, Tex,
B—Colorado, Tex.
G. O. No. 119, W. D., 1918.

Corporal, 80th Company, 6th Regiment, U. S. Marine Corps, 2d Division.
In the Bois-de-Belleau, France, on June 6, 1918, in delivering messages he voluntarily chose the most direct route, although it was through a machine-gun barrage, to deliver information which prevented the bombardment of positions that had just been occupied. He took the path of danger to save his comrades.

SHEARER, MAURICE E_____
In the Bois-de-Belleau, France, June 25, 1918.
R—Indianapolis, Ind,
B—Marion County, Ind.
G. O. No. 71, W. D., 1919.

Major, 5th Regiment, U. S. Marine Corps, 2d Division.
He displayed conspicuous courage, going forward at the head of his command during the attack. Personally going along the front line after the objective had been reached, he encouraged his men and directed the repulse of a counterattack by the enemy. During the encounter his battalion took over 200 prisoners and 19 machine guns.

SHECKART, GROVER C. (1787320)_____
Near Montfaucon, France, Sept. 29, 1918.
R—Hershey, Pa.
B—York County, Pa.
G. O. No. 46, W. D., 1919.

Sergeant, Company C, 316th Infantry, 79th Division.
After his commanding officer had been wounded and taken from the field, Sergeant Sheckart reorganized 2 platoons of his company and led them into a thick woods against strong machine-gun nests. He advanced alone against a machine-gun crew, killed the officer in charge, and took 4 prisoners. He continued to lead his men during the advance of that day, in spite of a wound in the foot, which caused his evacuation in the evening.

SHEDLEWSKI, JOHN F. (253348)_____
Near Juvigny, north of Soissons, France, Sept. 4, 1918.
R—Menasha, Wis.
B—Menasha, Wis.
G. O. No. 21, W. D., 1919.

Private, first class, Company A, 107th Ammunition Train, 32d Division.
During a heavy enemy bombardment a shell burst near 2 ammunition trucks that were being unloaded at a dump, blowing up 1 truck and setting fire to the other. Disregarding the warnings of bystanders, Private Shedlewski rushed forward and assisted in throwing the burning cushions and cover off the truck and backing it to a place of safety. His conspicuous bravery was the means of saving a large quantity of ammunition.

SHEDLOCK, ANTHONY F. (559887)_____
Near Ville-Savoye, France, Aug. 6, 1918.
R—Utahville, Pa.
B—Chestenfield, Pa.
G. O. No. 27, W. D., 1920.

Sergeant, Company H, 58th Infantry, 4th Division.
Sergeant Shedlock, when the officers of the company became casualties, took command, reorganized the scattered groups into a platoon, and personally led them across the Vesle River, in the face of heavy machine-gun fire, and drove the enemy from their position on the railroad embankment 500 yards beyond the river. He defended his position under the heavy fire and attacks of the enemy.

SHEEN, HENRY H_____
Near Dasmarinas, P. I., Aug. 19, 1900.
R—Norfolk, Va.
B—Quincy, Mass.
G. O. No. 118, W. D., 1919.
Distinguished-service medal also awarded.

Captain, 46th Infantry, U. S. Volunteers.
For extraordinary heroism in connection with operations against an armed enemy near Dasmarinas, Philippine Islands, Aug. 19, 1900. Captain Sheen at the imminent risk of his own life succeeded in saving the life of a brother officer from drowning while the detachment under his command was attempting to cross the Imus River and effect a surprise upon a rendezvous of a ladrone band in the vicinity.

SHEEHAN, JAMES J_____
Near Chateau-Thierry, France, June 6, 1918.
R—Chicago, Ill.
B—Chicago, Ill.
G. O. No. 99, W. D., 1918.

First lieutenant, 23d Infantry, 2d Division.
After being severely wounded, near Chateau-Thierry, France, June 6, 1918, he displayed remarkable fortitude and exemplary poise by continuing to direct the operation of his platoon under violent machine-gun fire.

*SHEFFER, ERVIN C. (551874)_____
North of Cierges, France, Oct. 9, 1918.
R—York, Pa.
B—York, Pa.
G. O. No. 27, W. D., 1920.

Corporal, Company I, 38th Infantry, 3d Division.
In the attack on Hill 253, Corporal Sheffer was in command of an automatic-rifle squad which was making an attack under enemy rifle and machine-gun fire on an enemy machine-gun nest. After the gunner had been killed, he seized the automatic rifle of the dead gunner and while rushing forward toward the nest was killed by machine-gun fire.
Posthumously awarded. Medal presented to mother, Mrs. Lillian Deckman.

*SHEFRIN, WILLIAM (1701301)_____
In the Ravine de L'Homme mort, near Vauxcere, between the Vesle and Aisne Rivers, Sept. 5, 1918.
R—Brooklyn, N. Y.
B—New York, N. Y.
G. O. No. 99, W. D., 1918.

Cook, Company C, 306th Infantry, 77th Division.
After both of his feet had been blown off by a bursting shell Cook Shefrin, although mortally wounded, cooly directed the work of rescuing and caring for other wounded men of the kitchen detachment who had been wounded when his transport was struck.
Posthumously awarded. Medal presented to father, Nathan Shefrin.

SHELBY, RICHARD D_____
Near Verdun, France, Oct. 10, 1918.
R—Rosedale, Miss.
B—Rosedale, Miss.
G. O. No. 35, W. D., 1919.

First lieutenant, 139th Aero Squadron, Air Service.
He encountered 6 enemy planes at a very low altitude strafing our trenches. He immediately attacked and dispersed the enemy planes, and by skillful maneuvering brought one of the planes down just behind his own lines.

SHELDON, RAYMOND_____
Near Grand Pre, France, Oct. 15, 1918, and near Oches and Raucourt, France, Nov. 4–6, 1918.
R—Orange, N. J.
B—Princeton, N. J.
G. O. No. 140, W. D., 1918.

Colonel, 307th Infantry, 77th Division.
In the attack on Grand Pre, Oct. 15, he displayed gallant conduct in going forward under heavy artillery and machine-gun fire and taking personal command of the leading battalion of his regiment, by his presence inspiring his men and facilitating the capture of this town. During the advance on Oches, Nov. 4, when his leading units were held up by machine-gun fire, he went forward to the skirmish line in order to estimate the strength of the enemy's position. The location by him of certain enemy machine guns resulted in their destruction by our artillery. During the advance on the Meuse, Nov. 4–6, he was constantly with the advanced elements of his regiment.

SHELLY, HARRY (1390351)_____
Near Hamel, France, July 4, 1918.
R—Chicago, Ill.
B—Chicago, Ill.
G. O. No. 44, W. D., 1919.

Private, Company A, 132d Infantry, 33d Division.
With an Australian soldier, Private Shelly went out and silenced an enemy sniping post and brought back 8 prisoners.

SHELOR, CHARLES A_____
In the Bois-de-Bantheville, France, Oct. 15, 1918.
R—Richmond, Va.
B—Roanoke, Va.
G. O. No. 93, W. D., 1919.

Second lieutenant, 127th Infantry, 32d Division.
Under heavy fire, Lieutenant Shelor made a reconnaissance of woods infested by enemy machine guns and snipers, locating another battalion of his regiment from which his own had become separated, and securing information which made it possible to continue the attack next day.

SHELTON, CLYDE (1321818)_____
Near Mazinghien, France, Oct. 19, 1918.
R—Mount Airy, N. C.
B—Surry County, N. C.
G. O. No. 44, W. D., 1919.

Sergeant, Company L, 120th Infantry, 30th Division.
Sergeant Shelton, who was in command of a platoon, was ordered to post an automatic rifle so as to protect the right flank of his battalion, and in order to do this it was necessary to advance his line beyond a hedge and wire fence. Halting his platoon, he went forward himself, and under heavy fire, in clear view of the enemy, he cut an opening in the barrier. His courageous act permitted a patrol to pass through, and the line was subsequently established with a minimum of casualties.

SHELTON, FRANCIS R_____
Near Fossoy, France, July 15, 1918.
R—Grayville, Ill.
B—Grayville, Ill.
G. O. No. 44, W. D., 1919.

First lieutenant, 7th Infantry, 3d Division.
Although wounded in the side by shrapnel and suffering great pain, he remained in command of his company for 48 hours, successfully repelling the offensive launched by the enemy.

SHEMIN, WILLIAM (558173)_____
On the Vesle River, near Bazoches, France, Aug. 7, 8, and 9, 1918.
R—Bayonne, N. J.
B—New York, N. Y.
G. O. No. 5, W. D., 1920.

Sergeant, Company G, 47th Infantry, 4th Division.
Sergeant Shemin upon 3 different occasions left cover and crossed an open space 150 yards, exposed to heavy machine-gun and rifle fire, to rescue wounded. After officers and senior noncommissioned officers had become casualties, Sergeant Shemin took command of the platoon and displayed great initiative under fire until wounded on Aug. 9.

SHENKEL, JOHN H_____
Near Chateau-Thierry, France, July 1, 1918.
R—Pittsburgh, Pa.
B—Pittsburgh, Pa.
G. O. No. 126, W. D., 1919.

First lieutenant, 111th Infantry, 28th Division.
Lieutenant Shenkel displayed marked bravery; with a number of others he volunteered to assist the French in retaking Hill 204. Finding himself with but 7 men, completely surrounded by the enemy, he led his detachment in fighting their way out with rifle butts and bayonets, himself killing a German officer with his pistol.

SHEPARD, ERWIN E. (2309679)_____
Near Medeah Ferme, France, Oct. 9, 1918.
R—Waterbury, Conn.
B—Meriden, Conn.
G. O. No. 23, W. D., 1919.

Private, first class, Company C, 2d Engineers, 2d Division.
Crawling forward under heavy machine-gun fire, he assisted in bringing a wounded comrade to safety.

SHEPHERD, GRANT_____
At Soissons and Chateau-Thierry, France, June and July, 1918.
R—Washington, D. C.
B—Washington, D. C.
G. O. No. 89, W. D., 1919.

Captain, 23d Infantry, 2d Division.
After being so seriously gassed as to be rendered temporarily so blind that he had to be led by hand through his trenches, he refused to be evacuated, nevertheless visiting all portions of his trenches to encourage his troops to hold at a most critical stage in the operations. Commanding his company in the Soissons-Reims offensive, he advanced over the top in front of his company, personally engaging machine-gun nests with his men until he was so severely wounded by the explosion of a shell as to render him a cripple for the rest of his life.

SHEPHERD, LEMUEL C_____
Near Lucy-Torcy Roads, France,
June 3, 1918.
R—Norfolk, Va.
B—Norfolk, Va.
G. O. No. 101, W. D., 1918.

First lieutenant, 5th Regiment, U. S. Marine Corps, 2d Division.
On June 3, 1918, near the Lucy-Torcy Roads, he declined medical treatment after being wounded and continued courageously to lead his men.

SHEPHERD, MARION F_____
Near Ripont, France, Sept. 29–30,
1918.
R—Parrot, Ky.
B—Delphia, Ky.
G. O. No. 46, W. D., 1919.

Captain, 371st Infantry, 93d Division.
Captain Shepherd was wounded twice, but each time he refused to be evacuated after receiving first-aid treatment, holding his shattered command in position under heavy shell fire for 6 days.

SHEPHERD, ROYAL HAMILTON CLATER__
Near Tigny, France, July 19, 1918.
R—Houston, Tex.
B—Oakhawn, Ill.
G. O. No. 117, W. D., 1918.

Private, 95th Company, 6th Regiment, U. S. Marine Corps, 2d Division.
He entered the action with a badly burned foot, which fact he concealed from his officers. Shot through the shoulder early in the advance and unable to hold his position in the firing line, he carried wounded men to shelter for 6 hours, all of the time under heavy fire of the enemy, and yielded to treatment himself only when he had become exhausted from the effects of his injury.

*SHERET, JAMES A. (1214550)_____
Near Ronssoy, France, Sept. 29, 1918.
R—Albion, N. Y.
B—Scranton, Pa.
G. O. No. 32, W. D., 1919.

Sergeant, Company F, 108th Infantry, 27th Division.
During the operations against the Hindenburg line on Sept. 29, 1918, this soldier displayed exceptional bravery in several single-handed attacks on enemy positions. After rushing 2 hostile posts and killing the occupants with his revolver, he attacked 4 of the enemy in a machine-gun position, killing 1 of them before he was himself surrounded and killed by the other 3.
Posthumously awarded. Medal presented to father, John Sheret.

SHERIDAN, CHARLES L_____
On Hill 230, near Cierges, France,
July 31 and Aug. 1, 1918.
R—Bozeman, Mont.
B—Marshalltown, Iowa.
G. O. No. 124, W. D., 1918.

Captain, 128th Infantry, 32d Division.
He demonstrated notable courage and leadership by taking command of the remnants of 2 companies and leading them up the hill and into the woods against violent fire from the enemy. His grit and leadership inspired his men to force the enemy back. He personally shot and killed 3 of the enemy, and under his direction six machines were put out of action and the hill captured.

*SHERIDAN, RICHARD B_____
Near Ville-Savoye, France, Aug. 23,
1918.
R—Brooklyn, N. Y.
B—Brooklyn, N. Y.
G. O. No. 32, W. D., 1919.

First lieutenant, 308th Infantry, 77th Division.
While leading his platoon in attack Lieutenant Sheridan had one of his legs badly shattered by shellfire. Refusing evacuation, he remained to direct the movements of his men until he died.
Posthumously awarded. Medal presented to mother, Mrs. Isabella Sheridan.

*SHERMAN, STEPHEN GEORGE_____
At Chateau-Thierry, France, June
6, 1918.
R—Minneapolis, Minn.
B—Minneapolis, Minn.
G. O. No. 110, W. D., 1918.

Sergeant, 20th Company, 5th Regiment, U. S. Marine Corps, 2d Division.
Killed in action at Chateau-Thierry, France, June 6, 1918, he gave the supreme proof of that extraordinary heroism which will serve as an example to hitherto untried troops.
Posthumously awarded. Medal presented to father, George C. Sherman.

SHETHAR, SAMUEL_____
In the Champagne sector, France,
Sept. 26–Oct. 1, 1918.
R—New York, N. Y.
B—New York, N. Y.
G. O. No. 37, W. D., 1919.

Captain, 369th Infantry, 93d Division.
Acting as operations officer, Captain Shethar on several occasions voluntarily collected small units which had become separated from their organizations, organized them, and led them to their positions through intense machine-gun and shell fire. At another time he spent several hours searching for a wounded battalion commander until he found him and carried him through heavy fire to the rear.

*SHIMANOSKI, ALFRED (30681)_____
South of Soissons, France, July 18,
1918.
R—Brooklyn, N. Y.
B—Russia.
G. O. No. 20, W. D., 1919.

Private, Company G, 9th Infantry, 2d Division.
Jerome Buschman, sergeant; John Rockwell, private; William F. Rockwell, private; Alfred Shimanoski, private; and Watzlaw Viniarsky, private, all of Company G, 9th Infantry. For extraordinary heroism in action south of Soissons, France, July 18, 1918. They conspicuously distinguished themselves by attacking a party of more than 60 Germans and, in an intense and desperate hand-to-hand fight, succeeded in killing 22 men and capturing 40 men and 5 machine guns.
Posthumously awarded. Medal presented to sister, Mrs. Sophie Shulske.

SHIMANOWICH, ALEX (52065)_____
At Vaux, France, July 1, 1918.
R—Newark, N. J.
B—Russia.
G. O. No. 132, W. D., 1918.

Private, Company L, 23d Infantry, 2d Division.
He displayed daring bravery by creeping forward alone, attacking without assistance and putting out of operation an enemy machine-gun detachment which was holding up the advance of his platoon. The machine gun was captured and its crew killed or made prisoners as a result of his heroic and successful attack.

SHIMEALL, RALPH M. (2213015)_____
Near Bantheville, France, Nov. 1–2,
1918.
R—Norton, Kans.
B—Norton, Kans.
G. O. No. 37, W. D., 1919.

Sergeant, Company M, 353d Infantry, 89th Division.
Sergeant Shimeall, although wounded twice, continued in action for 2 days without reporting for medical aid. He established and maintained liaison during these 2 days in a very efficient manner.

SHIMEL, FIRM F. (2273999)_____
Near Epinonville, France, Oct. 1–2,
1918.
R—Lodi, Calif.
B—Collins Hollow, Pa.
G. O. No. 87, W. D., 1919.

Sergeant, first class, Company B, 316th Field Signal Battalion, 91st Division.
Sergeant Shimel was in charge of a party of men stringing wire when an enemy sniper was firing at them. Sending his men to cover, he advanced alone, located the sniper, and killed him. Next day while he and his party were repairing breaks in the line under shellfire, a shell burst a few feet away. His coolness and courage under fire inspired his men to continue their work and prevented communication being interrupted.

SHINGLE, JOHN BENJAMIN (1249376).
Near Fismes, France, Aug. 6, 1918.
R—McVeytown, Pa.
B—McVeytown, Pa.
G. O. No. 11, W. D., 1921.

Mechanic, Company M, 112th Infantry, 28th Division.
Prior to the attack of his battalion, Mechanic Shingle exposed himself to heavy enemy fire while making a reconnaissance of the Vesle River, which was some 300 yards in advance of the line held by his company. He later guided his company to a suitable position from which a crossing was made without delay and without many casualties.

SHINN, LEON P.
Near Flirey, France, Sept. 12, 1918.
R—Newark, Ohio.
B—Huron, Ohio.
G. O. No. 37, W. D., 1919.

First lieutenant, 356th Infantry, 89th Division.
He continued to lead his platoon until the third objective had been reached, after being wounded in the leg during the first 20 minutes of the advance.

SHIPLEY, GEORGE A.
Near Landres-et-St. Georges, France, Nov. 1, 1918.
R—Platteville, Wis.
B—Montport, Wis.
G. O. No. 35, W. D., 1920.

Captain, 23d Infantry, 2d Division.
During the attack his organization was held up by machine-gun fire from the front. Captain Shipley exposed himself to heavy fire in order to make a flank attack on the enemy. Armed with a rifle, he courageously attacked a machine-gun position, which resulted in the capture of the gun and 28 prisoners. During the period from Nov. 1 to 7 he led a battalion in its attack on the Bois Hazois and Bois L'Epasse, and led a detachment which captured the strongly defended town of L'Etanne. His valiant conduct had a marked moral effect upon his men.

SHIPMAN, HAROLD L. (1213824)
East of Ronssoy, France, Sept. 29, 1918.
R—Buffalo, N. Y.
B—Buffalo, N. Y.
G. O. No. 20, W. D., 1919.

Private, Company B, 108th Infantry, 27th Division.
During the operations against the Hindenburg line Private Shipman, a Lewis gunner, exhibited great courage and dash when a party of about 40 German prisoners, seeing their guards killed by German snipers while going to the rear, seized rifles and opened fire on the Americans. Private Shipman rushed forward with his Lewis gun and put the entire group out of action. During the engagement he also silenced 3 enemy machine-gun positions.

SHIPMAN, STEPHEN V. (278336)
In the woods north of Cierges, northeast of Chateau-Thierry, France, Aug. 1, 1918.
R—Bangor, Mich.
B—Benton Harbor, Mich.
G. O. No. 117, W. D., 1918.

Private, Company C, 126th Infantry, 32d Division.
After his company had entered the woods north of Cierges he and another soldier maneuvered around a machine-gun which was causing many casualties in the company and reached a shell hole after crossing an open space that was swept by hostile fire. From here they killed the crew of the machine-gun, captured the gun, and turned it on the enemy.

SHIPP, BEVERLY A.
Near Cornay, France, Oct. 9–10, 1918.
R—Cordele, Ga.
B—Columbus, Ga.
G. O. No. 37, W. D., 1919.

First lieutenant, 328th Infantry, 82d Division.
After successfully driving off the enemy, his attacking force was counterattacked and surrounded. The officers in charge decided to surrender to the greatly superior numbers, but lieutenant Shipp, refusing to do so, made his way to our lines through deadly enemy fire, although severely wounded while doing so.

SHIRLEY, WALTER L.
Near Bois-de-Bantheville, France, Oct. 18, 1918.
R—Jackson, Mich.
B—New Carlisle, Ohio.
G. O. No. 81, W. D., 1919.

First lieutenant, 126th Infantry, 32d Division.
Going forward to the outpost line on a reconnaissance mission, he was wounded, but upon receiving first-aid treatment, returned to his position within 30 yards of the enemy, and although under heavy fire, continued his observations until he had obtained the desired information.

SHIVELY, GEORGE J. (9141)
Near Soissons, France, July 21, 1918.
R—Brookville, Pa.
B—Brookville, Pa.
G. O. No. 109, W. D., 1918.

Private, Section No. 585, Ambulance Service.
During the fighting near Soissons, France, July 21, 1918, he drove his ambulance through shellfire and continued on after his car was badly shattered until he had delivered his patients to a dressing station, when he fainted from serious wounds in his left arm and both legs, the existence of which he had concealed when the ambulance was hit.

SHIVELY, HARVEY H. (1320864)
Near Bellicourt, France, Sept. 29, 1918, and near Becquigny, France, Oct. 9, 1918.
R—Spray, N. C.
B—Floyd County, Va.
G. O. No. 37, W. D., 1919.

Private, 2d Battalion, Intelligence Section, 120th Infantry, 30th Division.
Near Bellicourt, Private Shively, with an Australian soldier, captured 42 of the enemy, including 2 officers. On Oct. 9, near Becquigny, he accompanied another soldier in penetrating the enemy's outpost line and captured 2 enemy machine gunners, putting the gun out of action.

*SHOEMAKER, LONNIE O. (1491231)
Near St. Etienne, France, Oct. 8, 1918.
R—Childress, Tex.
B—Hillsboro, Tex.
G. O. No. 50, W. D., 1919.

Corporal, Company L, 142d Infantry, 36th Division.
Although he was severely gassed, he continued in the attack until his company had reached its objective and organized the new position, when he was ordered to the rear. The exposure to which he voluntarily submitted resulted in his death.
Posthumously awarded. Medal presented to brother, E. J. Shoemaker.

SHOENER, WILLIAM M. (1551350)
Near Chateau-Thierry, France, July 14–15, 1918.
R—Philadelphia, Pa.
B—Pringsburg, Pa.
G. O. No. 44, W. D., 1919.

Cook, Battery F, 76th Field Artillery, 3d Division.
Leaving his own work, he went to the assistance of the wounded, remaining in the field throughout the entire night, giving first aid and carrying wounded comrades to places of safety.

SHOLETTE, EDGAR M. (1210275)
East of Ronssoy, France, Sept. 29, 1918.
R—Ogdensburg, N. Y.
B—Ogdensburg, N. Y.
G. O. No. 20, W. D., 1919.

Sergeant, Company D, 107th Infantry, 27th Division.
He went out into the open field under heavy shell and machine-gun fire and succeeded in carrying back to our lines a wounded soldier.

SHOMAN, MAURICE (60129)_____
Near Verdun, France, Oct. 27, 1918.
R—Plymouth, Mass.
B—Russia.
G. O. No. 46, W. D., 1919.

Private, Company D, 101st Infantry, 26th Division.
After killing many of the enemy, he was left alone in a shell hole with no more ammunition. Finding himself surrounded by a sudden counterattack of the enemy, he grabbed a light machine-gun and held off the enemy until he was rescued by his comrades. The fire from his gun was decidedly instrumental in overcoming the counterattack.

SHORE, LAUREL (280799)_____
Near Ivoiry, France, Oct. 4, 1918.
R—Evart, Mich.
B—Osceola County, Mich.
G. O. No. 126, W. D., 1919.

Corporal, Company I, 126th Infantry, 32d Division.
He displayed utter disregard for personal danger in repeatedly carrying messages across an area 1,000 meters wide which was being subjected to heavy artillery and machine-gun fire. In carrying messages between company and battalion headquarters he repeatedly passed through German barrage.

*SHORT, ABE (551605)_____
Near Romagne, France, Oct. 8, 1918.
R—Aurora, Ark.
B—Houston, Okla.
G. O. No. 64, W. D., 1919.

Sergeant, Company H, 38th Infantry, 3d Division.
Sergeant Short courageously led his platoon through a terrific barrage and silenced a machine-gun position which was enfilading the attacking lines. He was killed later in this action.
Oak-leaf cluster.
For the following act of extraordinary heroism in action near Mezy, France, July 15, 1918, Sergeant Short is awarded an oak-leaf cluster: Although seriously wounded, he continued in command of his group during the battle of the Marne and succeeded in destroying 3 boats loaded with Germans.
Posthumously awarded. Medal presented to mother, Mrs. Mary J. Short.

SHORT, GILBERT D. (1314510)_____
Near Vaux-Andigny, France, Oct. 19, 1918.
R—Henderson, Tenn.
B—Hardin County, Tenn.
G. O. No. 44, W. D., 1919.

Private, Company F, 120th Infantry, 30th Division.
When the position of his company had become untenable because of enemy machine-gun and artillery fire, Private Short, with another soldier, the sole survivors of a Lewis machine-gun team, covered the retreat of their company. Clinging to their advanced post throughout the day, they took up the advance with the company at dusk that evening.

SHOULTS, EDGAR (2183705)_____
Near Remonville, France, Nov. 1, 1918.
R—Perryville, Mo.
B—Perry County, Mo.
G. O. No. 44, W. D., 1919.

Corporal, Company B, 354th Infantry, 89th Division.
In command of a combat group, he led his men in a bayonet charge on an enemy stronghold, capturing many machine guns and killing or capturing the entire crews of the guns.

SHOWERS, WILLIAM LESTER (4606952)___
Near St. Etienne, France, Oct. 4–6, 1918.
R—Fort Branch, Ind.
B—Fort Branch, Ind.
G. O. No. 37, W. D., 1919.

Private, 47th Company, 5th Regiment, U. S. Marine Corps, 2d Division.
Private Showers, a runner, displayed exceptional courage in carrying messages for 3 days under shell and machine-gun fire.

SHROY, DANIEL D. (1193798)_____
Near Courbon, France, July 15, 1918.
R—Middletown, Pa.
B—Middletown, Pa.
G. O. No. 44, W. D., 1919.

Private, Headquarters Company, 10th Field Artillery, 3d Division.
He repeatedly volunteered and carried messages over areas heavily bombarded with gas and high-explosive shells until he was gassed and forced to go to an aid station.

SHRUM, JOHN E. (48717)_____
Near Soissons, France, July 19, 1918.
R—Greenwald, Pa.
B—Derry, Pa.
G. O. No. 3, W. D., 1921.

Private, Company D, 18th Infantry, 1st Division.
Private Shrum, although wounded, delivered an important message for his platoon commander. In order to accomplish this mission, it was necessary for him to cross an area swept by enemy machine-gun fire.

SHUEY, PERRY R. (105433)_____
Near Fleville, France, Oct. 5, 1918.
R—Lebanon, Pa.
B—Lebanon, Pa.
G. O. No. 87, W. D., 1919.

Sergeant, Company B, 2d Machine Gun Battalion, 1st Division.
After his platoon commander had been killed and the organization had suffered 50 per cent casualties, he reorganized the platoon by gathering stray squads from both flanks and the front. This work completed, he led the platoon forward, under intense artillery and machine-gun fire, to positions in advance of the Infantry to withstand a counterattack. He displayed absolute fearlessness under heavy fire, inspiring the men with him by his example of heroism.

SHUGG, WILLIAM R. (407355)_____
East of Ronssoy, France, Sept. 29, 1918.
R—Rutherford, N. J.
B—Rutherford, N. J.
G. O. No. 145, W. D., 1918.

Private, Company C, 102d Field Signal Battalion, 27th Division.
After the commander of the Infantry platoon to which he was attached as a visual-signal man had been killed he took command of the platoon and exhibited remarkable gallantry and leadership in leading it into effective combat.

*SHULL, LAURENS C_____
Near Soissons, France, July 19, 1918.
R—Sioux City, Iowa.
B—Sioux City, Iowa.
G. O. No. 100, W. D., 1918.

Second lieutenant, 26th Infantry, 1st Division.
He led his platoon with brilliant courage in 2 attacks and was badly wounded in the third, when, with equal vigor, he advanced against a machine-gun nest.
Posthumously awarded. Medal presented to father, D. C. Shull.

SHUMAN, GEORGE A_____
Near Fey-en-Haye, France, Sept. 15, 1918.
R—Minneapolis, Minn.
B—Mifflintown, Pa.
G. O. No. 123, W. D., 1918.

Second lieutenant, 360th Infantry, 90th Division.
This officer saved the lives of wounded men in his command by going into no man's land under severe shellfire in plain view of the enemy, giving them first-aid treatment and assisting them back to shelter

SHUMATE, CARSON L. (58128)_____
At Seicheprey, France, Mar. 28 and 29, 1918.
R—Bluefield, W. Va.
B—Bluefield, W. Va.
G. O. No. 129, W. D., 1918.

Private, Company I, 28th Infantry, 1st Division.
He was a member of a patrol consisting of an officer and 4 men who, with great daring, entered a dangerous portion of the enemy trenches where they surrounded a party of nearly double their own strength, captured a greater number than themselves, drove off an enemy rescuing party, and made their way back to our lines with 4 prisoners, from whom valuable information was taken.

*SHUMATE, JOHN W. (52189)_____
Near Chateau-Thierry, France, June 6, 1918.
R—Charleston, W. Va.
B—Montgomery County, Va.
G. O. No. 88, W. D., 1918.

Private, first class, Company M, 23d Infantry, 2d Division.
After his platoon was practically wiped out and had been withdrawn near Chateau-Thierry, France, on June 6, 1918, he continued forward to his objective and remained throughout the night under heavy fire in hope of keeping the ground gained until reinforcements came up, and was later killed in action on June 14.
Posthumously awarded. Medal presented to father, John W. Shumate.

SHUPP, ROY F._____
Near Gland, France, July 21, 1918.
R—New Bern, N. C.
B—Kresgeville, Pa.
G. O. No. 35, W. D., 1919.

First lieutenant, 4th Infantry, 3d Division.
After crossing the Marne, with the leading platoon of his company, Lieutenant Shupp, with 2 companions, made a surprise attack on an enemy machine-gun emplacement and succeeded in taking 1 gun and 8 prisoners.

SIADE, JOSEPH (1290177)_____
At Molleville Farm, France, Oct. 15, 1918.
R—Norfolk, Va.
B—Syria.
G. O. No. 28, W. D., 1921.

Private, first class, Headquarters Company, 116th Infantry, 29th Division.
Private Siade remained in the face of heavy enemy machine-gun fire to administer first aid to a wounded officer, showing utter disregard for his personal safety. He later carried the officer through a heavily fire-swept zone to a place of shelter.

SIBOLD, GEORGE G. (1290307)_____
Near Bois-de-Consenvoye, France, Oct. 10, 1918.
R—Roanoke, Va.
B—Blacksburg, Va.
G. O. No. 44, W. D., 1919.

Sergeant, Machine Gun Company, 116th Infantry, 29th Division.
After his platoon commander had become a casualty, and while he was suffering from gas poisoning, he led his platoon forward and reported to his company commander, after which he fell from exhaustion.

SIEBERT, ERNEST T. (3400)_____
Near Trugny, France, July 23, 1918.
R—Newton, Mass.
B—Newton Center, Mass.
G. O. No. 125, W. D., 1918.

Corporal, 103d Ambulance Company, 101st Sanitary Train, 26th Division.
He voluntarily rescued a wounded soldier who was lying on a shell-swept road by carrying him 300 yards on his back. Although wounded in the shoulder by a shell fragment, he courageously stuck to his task until it was successfully accomplished. His courageous act was an inspiration to his men.

SIEBERT, WALTER (274261)_____
Near Gesnes, France, Oct. 16, 1918.
R—Shepley, Wis.
B—Gresham, Wis.
G. O. No. 66, W. D., 1919.

Sergeant, Company F, 127th Infantry, 32d Division.
Locating an enemy machine-gun nest, Private Siebert advanced on it alone, and by accurate fire from his automatic rifle killed or wounded the members of the crew, thereby saving his company heavy casualties.

SIEG, ROBERT E._____
Near Blanc Mont, France, Oct. 3–5, 1918.
R—Hooper, Nebr.
B—Lundy, Mo.
G. O. No. 98, W. D., 1919.

Private, 43d Company, 5th Regiment, U. S. Marine Corps, 2d Division.
He unhesitatingly went through the heaviest machine-gun and artillery fire, dressing and carrying wounded. Disregarding his own safety he refused to take rest or food while there were wounded needing attention.

SIELOFF, THEODORE H. (2021616)_____
At Verst 444, near Emtsa, Russia, Nov. 4, 1918.
R—Detroit, Mich.
B—Detroit, Mich.
G. O. No. 24, W. D., 1920.

Corporal, Company I, 339th Infantry, 85th Division (detachment in North Russia).
During an enemy attack, when he was fired on from the rear, Corporal Sieloff moved his gun to an exposed position and delivered effective fire on the enemy. Although exposed to enemy fire from two directions, he held the enemy in check until his gun had become disabled, when he dismantled it, replaced the broken parts, and renewed his fire, repulsing a strong force of the enemy.

SIELSKY, LOUIS (1901027)_____
Near St. Juvin, France, Oct. 17–21, 1918.
R—Chicago, Ill.
B—Poland.
G. O. No. 98, W. D., 1919.

Corporal, Company C, 326th Infantry, 82d Division.
Leading a daylight patrol across an exposed hillside through terrific artillery and machine-gun fire to locate enemy machine guns, Corporal Sielsky secured valuable information and carried it back to the battalion commander. While leading a similar daylight patrol 4 days later, he was seriously wounded in 2 places by machine-gun fire after displaying inspiring bravery and devotion to duty.

SIEMERING, WILLIAM H. (1106054)_____
Near St. Etienne, France, Oct. 8, 1918.
R—Le Sueur, Minn.
B—Le Sueur, Minn.
G. O. No. 66, W. D., 1919.

Private, first class, Company G, 142d Infantry, 36th Division.
Although one of his hands was disabled, he left a sheltered position against the advice of his companions and went through heavy shell and machine-gun fire to the aid of a wounded comrade, bringing the latter to a place of safety.

SIERS, FRANK (52316)_____
Near Chateau-Thierry, France, June 6, 1918.
R—Nebo, W. Va.
B—Nebo, W. Va.
G. O. No. 109, W. D., 1918.

Private, Company M, 23d Infantry, 2d Division.
After being wounded in the arm and back, he continued his duties of bearing messages and collecting information, and was severely wounded while attempting to establish liaison with a neighboring company.

SIGG, CHARLES F. (204355)_____
Near Montblanc, France, Oct. 8, 1918.
R—West Park, Ohio.
B—Germany.
G. O. No. 44, W. D., 1919.

First sergeant, Company A, 2d Ammunition Train, 2d Division.
He was in charge of a convoy of trucks loaded with artillery ammunition, of which our batteries were in urgent need. The site selected for the dump was under very heavy shellfire, but Sergeant Sigg, drawing his convoy up in sections, directed the unloading throughout the rain of shells, which were exploding on all sides of his machines.

SIGNOR, HENRY L. (10401)............
Near Sommerance, France, Oct. 11, 1918.
R—Worcester, Mass.
B—Clinton, Mass.
G. O. No. 15, W. D., 1919.

Private, first class, section 647, Ambulance Service.
Following the advance of the Infantry, he caused his car to be lifted across a mine crater by some infantrymen, and proceeding for 3 kilometers down a road heavily bombarded with gas and high-explosive shells, he evacuated wounded from a culvert only 400 yards from enemy outposts. On the return trip his car was struck by splinters from an exploding shell which pierced the clothing of his aid and caused fresh wounds to one of his patients. After transferring his wounded across the crater to another car, he succeeded in driving it over a road almost destroyed by shellfire to a newly established dressing station in Sommerance. He continued to operate his car for 12 hours until he was relieved, having at all times displayed unhesitating courage and devotion to duty.

SIKIVICA, PIT (107661).............
Near Medeah Ferme, France, Oct. 3, 1918.
R—Johnstown, Pa.
B—Austria.
G. O. No. 21, W. D., 1919.

Private, Company D, 9th Infantry, 2d Division.
During an attack made by his platoon on an enemy machine-gun nest, 2 automatic rifles were destroyed in his hands. Private Sikivica fell back to the supporting company, borrowed an automatic rifle, and killed 2 of the enemy machine gunners.

SILI, FREDERICK D............
Near Ponchaux and Geneve, France, Oct. 8, 1918.
R—Albany, N. Y.
B—Cohoes, N. Y.
G. O. No. 35, W. D., 1919.

First lieutenant, 105th Engineers, 30th Division.
He performed the difficult task of laying the tape for the jumping-off line and also for the support line, on the night preceding the attack of Oct. 8. Despite the fact that 1 battalion had changed its line on Oct. 7 and that he had to face a continuous fire of artillery, trench mortars, and machine guns, he performed a mission which would have been extremely difficult even under normal conditions. While returning to headquarters, he carried his wounded orderly through a heavy barrage of machine-gun and artillery fire until assistance could be procured.

SILLMAN, ROBERT H............
Before Manila, P. I., Aug. 13, 1898.
R—New York, N. Y.
B—New York, N. Y.
G. O. No. 116, W. D., 1919.

Sergeant, Astor Battery, U. S. Army.
He gallantly took part, until disabled by a Spanish bullet, in a charge on the enemy's breastworks, volunteers having been called for by the brigadier general commanding.

SILLOWAY, RALPH (1378997)...........
Near Romagne, France, Nov. 1-3, 1918.
R—Peoria, Ill.
B—Roodhouse, Ill.
G. O. No. 37, W. D., 1919.

Private, first class, Battery C, 124th Field Artillery, 33d Division.
During heavy enemy shellfire, when the other members of his section were all wounded or engaged in first-aid work, he alone served his piece and kept it firing. Two days later, when the chief of his section was wounded, he took command of the section and followed the barrage.

SILVA, LOUIS J. (2263395)...........
Near Very, France, Sept. 26, 1918.
R—Hayward, Calif.
B—Alameda County, Calif.
G. O. No. 72, W. D., 1920.

Private, Company B, 363d Infantry, 91st Division.
After the advance of his platoon had been held up by machine-gun fire, Private Silva, with a noncommissioned officer, attacked 1 machine gun and put it out of action. This act resulted in the enemy abandoning 2 other machine guns and permitted the advance of his platoon.

SILVER, HENRY SPRAGUE............
Near Exermont, France, Oct. 11, 1918.
R—Charlotte, N. C.
B—Morganton, N. C.
G. O. No. 44, W. D., 1919.

First lieutenant, 28th Infantry, 1st Division.
He led a patrol into the woods under a severe artillery and machine-gun fire to establish liaison with the units on the left flank. He continued on his mission after three-fourths of his patrol had been killed or wounded, and succeeded in bringing valuable information to his battalion commander.

SILVER, TOM (736111)............
Near Fontaines, France, Nov. 6, 1918.
R—Comer, Ga.
B—Athens, Ga.
G. O. No. 37, W. D., 1919.

Corporal, Company H, 11th Infantry, 5th Division.
Corporal Silver, single handed, captured and destroyed a machine gun which was operating on the flank of his company, making progress impossible.

SILVERBERG, MORRIS (2671459)...........
Near Ronssoy, France, Sept. 29, 1918.
R—New York, N. Y.
B—Russia.
G. O. No. 21, W. D., 1919.

Private, Company G, 108th Infantry, 27th Division.
Private Silverberg, a stretcher bearer, displayed extreme courage by repeatedly leaving shelter and advancing over an area swept by machine-gun and shell fire to rescue wounded comrades. Hearing that his company commander had been wounded, he voluntarily went forward alone and, upon finding that his officer had been killed, brought back his body.

SILVERMAN, HYMAN (2383446).........
Near Verdun, France, Oct. 27, 1918.
R—Chelsea, Mass.
B—Boston, Mass.
G. O. No. 37, W. D., 1919.

Private, first class, Company E, 60th Infantry, 5th Division.
When enemy shellfire had ignited an ammunition dump, Private Silverman assisted in removing the ammunition from the blazing dump. Several of his comrades were seriously wounded by exploding shells, and he himself was hit in many places by hand-grenade explosions, but he continued until the greater part of the explosives were moved to safety. He then assisted in removing his wounded comrades before submitting to treatment for his wounds.

SILVERTHORN, MERWIN H............
Near St. Etienne, France, Oct. 4, 1918.
R—Minneapolis, Minn.
B—Minneapolis, Minn.
G. O. No. 35, W. D., 1919.

Second lieutenant, 5th Regiment, U. S. Marine Corps, 2d Division.
Lieutenant Silverthorn carried an important message to his battalion commander and returned with instructions at a critical time through heavy machine-gun and shell fire.

SILVESTER, LINDSAY McD............
In the Bois-d'Aigremont, France, July 15, 1918.
R—Norfolk, Va.
B—Norfolk County, Va.
G. O. No. 64, W. D., 1919.

Major, 30th Infantry, 3d Division.
During the intense bombardment preceding the German drive of July 15, when the wounded were so numerous that it was impossible to care for them in the dressing station, Major Silvester voluntarily gave up his dugout for the use of the wounded and exposed himself to heavy fire during the 10 hours' terrific bombardment. After leading his command across the Marne this officer directed the reduction of a number of machine-gun nests and advanced his lines 4 kilometers despite the determined resistance.

SIMAS, MANUEL (2266516) _____
Near Eclisfontaine, France, Sept. 28, 1918.
R—San Jose, Calif.
B—Portugal.
G. O. No. 46, W. D., 1919.

Private, Company C, 364th Inf antry, 91st Division.
After being wounded by a machine gun bullet early in the afternoon he remained in action and, without making his wound known, willingly offered and held a very dangerous outpost until late the next morning, when he was ordered to the hospital.

SIMMERS, LEROY E. (1274824) _____
Near Haumont, France, Oct. 11, 1918.
R—Wilmington, Del.
B—Port Deposit, Md.
G. O. No. 27, W. D., 1919.

Private, 116th Ambulance Company, 104th Sanitary Train, 29th Division.
As a stretcher bearer he gave proof of great courage and unhesitating devotion to duty under heavy shellfire by assisting 3 wounded soldiers to a place of safety, he himself being wounded while so doing. After receiving first aid he returned to the shell-swept area and continued in the work of rescuing the wounded.

SIMMONS, SAMUEL S. _____
At Blanc Mont, France, Oct. 3, 1918.
R—Lancaster, Pa.
B—Lancaster, Pa.
G. O. No. 37, W. D., 1919.

Private, 78th Company, 6th Regiment, U. S. Marine Corps, 2d Division.
With 2 other soldiers, Private Simmons volunteered and attacked a machine gun next in advance of his front line, killing the entire crew. Later, with another soldier, he went into an enemy dugout and captured 40 prisoners. He also carried 3 messages through the enemy barrage.

*SIMON, FRANK J. _____
Near St. Georges, France, Nov. 1, 1918.
R—Cleveland, Ohio.
B—Chicago, Ill.
G. O. No. 46, W D , 1919

Sergeant, 76th Company, 6th Regiment, U. S. Marine Corps, 2d Division.
Advancing with 2 other men alongside of a tank, in front of his company, Sergeant Simon encountered terrific enemy fire. After cutting their way through the wire the men in the tank and the men following Sergeant Simon were killed and he was wounded. Undaunted by his wound, he continued on alone, encountering and capturing 6 of the crew of the enemy machine gun in a dugout.
Posthumously awarded   Medal presented to mother, Mrs. Margaret Simon.

SIMON, LOUIS C., Jr. _____
In the region of Hadonville-les-Lachaussee, France, Sept 16, 1918
R—Columbus, Ohio.
B—Columbus, Ohio.
G O No. 44, W. D., 1919.

First lieutenant, 147th Aero Squadron, Air Service.
While on a protection patrol for American observation planes from the 99th Aero Squadron, Lieutenant Simon was fired upon by 3 Halberstadt biplane fighters. Regardless of his personal danger, he immediately engaged the enemy, although alone, drawing them down and away from the observation planes, which continued their important work unmolested. Lieutenant Simon continued fighting the 3 Halberstadts fiercely in spite of the odds against him. He finally succeeded in getting on the tail of one, and, after firing a short burst at close range, the enemy plane fell out of control. The remaining 2 planes quickly broke off the combat and headed east with motors full on.
Oak-leaf cluster.
Lieutenant Simon is awarded an oak-leaf cluster, to be worn with the distinguished-service cross awarded him Oct. 23, 1918: Lieutenant Simon and 2 other pilots encountered 9 (Fokker type) enemy planes, which were protecting an observation plane (Rumpler type). He attacked the lower formation of 4 planes alone and drove them off. He next dived at the observation plane and sent it crashing to the ground in flames.

SIMONI, ARISTEO V. _____
Near La Chene Tondu, France, Oct. 3, 1918.
R—Chicago, Ill.
B—Italy.
G. O. No. 44, W. D., 1919.

First lieutenant, chaplain, 111th Infantry, 28th Division.
Upon learning that there were 6 wounded men in front of our lines Chaplain Simoni asked for 2 volunteers, and, with the aid of these men, successfully brought the wounded men to our own lines through a terrific machine-gun and grenade fire.

*SIMPSON, ALBERT B. _____
Near Nantillois, France, Sept. 27–28, 1918.
R—Atlanta, Ga.
B—Eelbeck, Ga.
G. O. No. 27, W. D., 1919.

First lieutenant, 11th Machine Gun Battalion, 4th Division.
Though he was wounded, he remained with his company and by skillful arrangement of his machine gun covered a retirement of the infantry. Next day he was again wounded, and although urged by the surgeon to go to the rear, this gallant officer replied that there was too much work yet to be done at the front. He left to rejoin his command, and had gone about half the distance when he was killed by a high-explosive shell.
Posthumously awarded. Medal presented to father, Robert N. Simpson.

SIMPSON, ALFRED R. (2230068) _____
Near Scammerance, France, Oct. 12, 1918.
R—Bee Cave, Tex
B—Velasco, Tex.
G. O. No. 27, W. D., 1919.

Private, Company B, 321st Machine Gun Battalion, 82d Division.
While his company was covering with machine-gun fire a temporary withdrawal of the infantry, before a hostile counterattack, he secured an abandoned German machine gun and operated it until his own company, as well as the infantry, had returned safely. He remained at his post until his ammunition was exhausted and was the last one to leave the position. Through his bravery and skill the advance of the enemy was checked, and our own forces were able to organize a fresh counter dash attack.

SIMPSON, CHARLES E. (1811496) _____
Near Verdun, France, Nov. 5, 1918.
R—Great Bend, Pa.
B—Great Bend, Pa.
G. O. No. 44, W. D., 1919.

Private, Company A, 316th Machine Gun Battalion, 79th Division.
With 2 other soldiers, he voluntarily left a place of safety, went forward 40 meters under machine-gun fire in plain view of the enemy, and rescued another soldier who had been blinded by a machine-gun bullet and was helplessly staggering about.

SIMPSON, HARRY P. (1550407) _____
Near Le Charmel, France, July 28, 1918.
R—Covington, Ga.
B—Covington, Ga.
G. O. No. 44, W. D., 1919.

Sergeant, Battery B, 76th Field Artillery, 3d Division.
After his commanding officer and 32 members of his battery had been wounded by a bomb from an enemy plane, Sergeant Simpson, himself wounded, assisted in the evacuation of the wounded, after which he remained until his piece was placed in a new position before he retired for treatment.

*SIMPSON, JOHN S. (40012)............
In the Soissons Sector, France, July 18, 1918.
R—Ready, Ky.
B—Ready, Ky.
G. O. No. 116, W. D., 1918.

Sergeant, Company I, 9th Infantry, 2d Division.
During the assault near Soissons, Sergeant Simpson, although severely wounded continued in action for several hours, leading a group of men beyond and back of an enemy machine-gun emplacement in order to flank it and make the Infantry advance at this point possible. He succeeded, but while engaged in this courageous duty he was killed.
Posthumously awarded. Medal presented to mother, Mrs. Mary Simpson.

SIMPSON, JOSEPH M................
Near Foret Vencheres, France, Sept. 14, 1918.
R—San Antonio, Tex.
B—Ireland.
G. O. No. 87, W. D., 1919.

Captain, 357th Infantry, 90th Division.
He took command of a platoon that had become separated from its command, reorganized it, and showed marked personal courage in leading it forward under heavy fire. He called for volunteers and then led them in an attack upon a machine-gun nest that had been holding up our advance. With 2 men he charged the nest, captured the gun, and killed the crew.

SIMPSON, RICHARD LYLE............
Near Preny, France, Oct. 29, 1918.
R—Louisville, Ky.
B—Ekron, Ky.
G. O. No. 87, W. D., 1919.

Second lieutenant, 56th Infantry, 7th Division.
In charge of a patrol, Lieutenant Simpson had located the exact position of the enemy and was withdrawing, when he discovered that 1 man was missing. Although wounded himself, he went back, finding that the man had been killed and was entangled in the enemy's wire. Unable to recover the body and being wounded the second time, he covered the withdrawal of his patrol and returned with his report.

SIMPSON, ROBERT A. (41804)........
Near Soissons, France, July 22, 1918.
R—Shelby, Mont.
B—Scotland.
G. O. No. 98, W. D., 1919.

Private, Company A, 16th Infantry, 1st Division.
After being wounded Private Simpson returned to the line and continued to carry messages with absolute disregard for his own safety until he was wounded the second time.

SIMPSON, ROY HOBSON................
In the attack on Bois-de-Belleau, France, June 12, 1918.
R—Philadelphia, Pa.
B—Philadelphia, Pa.
G. O. No. 53, W. D., 1920.

Private, 47th Company, 5th Regiment, U. S. Marine Corps, 2d Division.
He carried a message from battalion to company headquarters directly across the face of enemy fire. Shot through the chest he continued running and called out, "I must deliver this message," struggling forward for 50 feet more in his heroic effort to carry out his mission before falling.

SIMPSON, THOMAS G................
Near Ronssoy, France, Sept. 29, 1918.
R—New York, N. Y.
B—New York, N. Y.
G. O. No. 20, W. D., 1919.

Second lieutenant, 107th Infantry, 27th Division.
He went out into the open under heavy machine-gun fire and succeeded in carrying back for a distance of about 25 yards a wounded officer and a wounded soldier.

*SIMS, GEORGE D. (1312562)..........
Near Montbrehain, France, Oct. 8, 1918.
R—Sumter, S. C.
B—Sumter County, S. C.
G. O. No. 64, W. D., 1919.

Private, Company M, 118th Infantry, 30th Division.
While assisting his automatic-rifle squad in a most advanced position Private Sims and those about him were seriously wounded by shrapnel. Realizing that his wounds were fatal and that his comrades might be saved, he insisted that the stretcher bearers attend to the others. His unusual heroism was instrumental in saving the lives of his fellow soldiers even at the cost of his own.
Posthumously awarded. Medal presented to father, Willie C. Sims.

SIMS, WILLIAM L. (1484963)..........
At Frapelle, France, Aug. 16-18, 1918.
R—McLean, Tex.
B—Waxahachie, Tex.
G. O. No. 99, W. D., 1918.

Private, Company A, 13th Machine Gun Battalion, 5th Division.
While acting as a runner he showed exceptional bravery in carrying messages through a heavily shelled and gassed area. After being wounded in the hand he made 12 trips from Frapelle to his company headquarters at Chapelle St. Clair.

SINATRA, MARION (51803)..........
Near Chateau-Thierry, France, June 6, 1918.
R—Boston, Mass.
B—Italy.
G. O. No. 89, W. D., 1919.

Private, Company K, 23d Infantry, 2d Division.
Rushing through rifle and machine-gun fire for a distance of 75 meters, Private Sinatra rescued a wounded comrade; and, while carrying him to safety, was himself seriously wounded. He continued on with his comrade, however, until he reached cover, administering first aid to the other man before attending to his own wound.

SINCLAIR, PAUL K. (1312393)..........
At Vaux-Andigny, France, Oct. 11, 1918.
R—Camden, S. C.
B—Camden, S. C.
G. O. No. 44, W. D., 1919.

Corporal, Company M, 118th Infantry, 30th Division.
When the advance was checked by fire from enemy machine guns and snipers in a sunken trench, Corporal Sinclair, crawling and jumping from 1 shell hole to another, under heavy machine-gun and artillery fire, opened fire with his automatic rifle and silenced both the machine-gun post and the snipers.

SINCLAIR, WILBERT W................
Near Beaumont, France, Nov. 10, 1918.
R—Roxbury, Mass.
B—Tyngsboro, Mass.
G. O. No. 37, W. D., 1919.

Private, 55th Company, 5th Regiment, U. S. Marine Corps, 2d Division.
He, alone, reconnoitered the position of enemy machine guns which were holding up the advance of his company across the Meuse. After he had located them he silenced the fire of 2 guns, thus making possible the continuance of his company's advance.

*SINER, EARL R. (1756625)............
Near Thiaucourt, France, Sept. 26, 1918.
R—Pawtucket, R. I.
B—Pawtucket, R. I.
G. O. No. 89, W. D., 1919.

Private, first class, Company G, 310th Infantry, 78th Division.
While his company was on outpost duty Private Siner crawled out from a trench under heavy enemy fire, to rescue a wounded comrade and was instantly killed by a bursting shell just as he had reached the wounded man.
Posthumously awarded. Medal presented to mother, Mrs. Lillie C. Siner.

*SINGLETON, LOWA L. (1496485)........
In the Meuse-Argonne offensive, France, Oct. 15 and 18, 1918.
R—Alvin, Tex.
B—Stanford, Ky.
G. O. No. 130, W. D., 1919.

Private, Machine Gun Company, 30th Infantry, 3d Division.
On Oct. 15, though severely wounded, he refused to be evacuated and continued to perform his duties as platoon runner under difficult and dangerous conditions. On Oct. 18 he exposed himself to heavy machine-gun and artillery fire in order to carry water to other members of his platoon, in the performance of which task he was mortally wounded.
Posthumously awarded. Medal presented to father, J. D. Singleton.

SIRMON, WILLIAM A_____
Near Clemery, France, Aug. 16, 1918.
R—Crichton, Ala.
B—Bluffsprings, Fla.
G. O. No. 37, W. D., 1919.

Captain, 325th Infantry, 82d Division.
At an imminent risk of his own life he rescued another officer by carrying him at night through enemy fire and under heavy machine-gun fire for 300 yards to a place of safety, where he dressed the wounds of the disabled officer.

SIROTA, IRVING (1711214)_____
Near Binarville, France, Oct. 2–7, 1918.
R—New York, N. Y.
B—Russia.
G. O. No. 21, W. D., 1919.

Private, first class, medical detachment, 308th Infantry, 77th Division.
He was on duty with a detachment of his regiment which was cut off and surrounded by the enemy in the forest of Argonne. During this period he was without food, but he continued to assist and give first aid to the wounded, exposing himself to heavy shell and machine-gun fire at the risk of his life, until he was completely exhausted.

SISSON, CHARLES N_____
Near Cornay, France, Oct 9, 1918.
R—Jacksonville, Ala.
B—Jacksonville, Ala.
G. O. No. 15, W. D., 1919.

Captain, 328th Infantry, 82d Division.
When the advance was checked on the outskirts of Cornay because of the exhaustion of the troops and machine-gun fire from the town, Captain Sisson, who had been in action several hours, took charge without orders, and started 2 patrols into the town. One was driven back by the machine-gun fire, but this gallant officer personally led the other and succeeded in capturing 2 machine guns and their crews, and 112 prisoners, completely cleaning out the town. Throughout this operation he displayed great bravery and coolness under the most trying circumstances.

SITTLER, EDWARD (2855903)_____
In the Bois-de-Bantheville, France, Oct. 24, 1918.
R—Merna, Nebr.
B—Wheaton, Ill.
G. O. No. 44, W. D., 1919.

Private, Company C, 341st Machine Gun Battalion, 89th Division.
Although himself severely wounded in the leg by shellfire, he did not report for medical attention until he had given first aid and assisted in carrying other wounded soldiers to the first-aid station, a distance of 6 kilometers. After all his comrades had received attention he had his own wounds dressed and was evacuated to the hospital.

*SKIFF, CLAYTON B. (1247001)_____
At Chatel-Chehery, France, Oct. 8, 1918.
R—Spartansburg, Pa.
B—East Branch, Pa.
G. O. No. 95, W. D., 1919.

Private, Company A, 112th Infantry, 28th Division.
When his company was stopped by enemy machine-gun fire Private Skiff crawled forward alone, climbed a steep hill under intense fire, and put a hostile machine-gun nest out of action. In the performance of this gallant exploit Private Skiff was mortally wounded and died on the way to the hospital. Posthumously awarded. Medal presented to mother, Mrs. Maude Skiff.

SKOGSBURG, VIVIAN (2001931)_____
Near Forges Woods, France, Sept. 26, 1918.
R—Chicago, Ill.
B—Highland, Iowa.
G. O. No. 71, W. D., 1919.

Sergeant, Company L, 131st Infantry, 33d Division.
Although seriously burned by a phosphorus shell, he continued in command of his platoon, leading it forward 8 kilometers to its objective, directing the mopping up of the territory and the consolidation of the new position. His example was an inspiration to his men. When, on orders from his company commander, he started to walk to the rear to receive medical attention, he fell unconscious.

SKRYPECK, ANDY (1340)_____
Southwest of Fismes, France, Aug. 5, 1918.
R—Detroit, Mich.
B—Austria.
G. O. No. 117, W. D., 1918.

Private, medical detachment, 125th Infantry, 32d Division.
During the forward movement of the 1st Battalion, 125th Infantry, a large number of the company to which he was attached were wounded while crossing an open field. At this point the artillery fire was very accurate and intense, but he disregarded all possibilities of personal injury and remained upon the field until he had administered first aid to all his fallen comrades.

SLADEN, FRED W_____
Near Ferme de la Madelaine, France, Oct. 14, 1918.
R—Omaha, Nebr.
B—Lowell, Mass.
G. O. No. 81, W. D., 1919.
Distinguished-service medal also awarded.

Brigadier general, headquarters 5th Infantry Brigade, 3d Division.
Although almost exhausted from 48 hours of continuous duty without rest of any kind, General Sladen, upon learning that the front line was held up by enemy machine-gun fire, proceeded to the advanced position through 3 kilometers of severe artillery fire. Upon arrival he found that the battalion commander had been killed and the unit badly disorganized and intermingled. He personally reorganized the troops under the terrific machine-gun and shell fire, reconnoitered the enemy's positions, and launched the advance anew. While engaged in this perilous mission he fainted from exhaustion, but upon being revived refused to be evacuated, and continued in the work of reorganizing and stabilizing the line at this critical period. Due to his efforts the action was carried to a successful conclusion in the face of apparently insurmountable difficulties.

SLAGSVOL, OSCAR T_____
Near St. Gilles, France, Aug. 3, 1918.
R—Eau Claire, Wis.
B—Eau Claire, Wis.
G. O. No. 44, W. D., 1919.

Second lieutenant, 128th Infantry, 32d Division.
Commanding the battalion patrols, Lieutenant Slagsvol was engaged continuously throughout the day in making reconnaissances under heavy fire. Although wounded, he preceded the battalion into the enemy's position and continued to perform his duties until he was overcome by exhaustion.

SLATE, JOSEPH W. (198864)_____
Near Exermont, France, Oct. 2–11, 1918.
R—Crockett, Calif.
B—San Francisco, Calif.
G. O. No. 44, W. D., 1919

Corporal, Company C, 2d Field Signal Battalion, 1st Division.
He volunteered and maintained a telephone line which ran through thick undergrowth and barbed-wire entanglements to an advanced observation post. Despite heavy artillery and direct machine-gun fire, Corporal Slate kept the line in operation for many hours without relief.

SLATE, RALPH_____
Near Bois-des-Septsarges, France, Sept. 27, 1918.
R—Cadillac, Mich.
B—Grand Rapids, Mich.
G. O. No. 81, W. D., 1919.

Captain, 39th Infantry, 4th Division.
After being wounded in a previous action, Captain Slate led his command in the face of unusual machine-gun fire, repeatedly exposing himself to prevent his units from becoming scattered and strengthening and holding his line until again severely wounded.

*SLATER, NORMAN C. (2942033)_____
Near Grand Pre, France, Oct. 17, 1918.
R—Bainbridge, N. Y.
B—Novemburg, N. Y.
G. O. No. 70, W. D., 1919.

Private, Company H, 312th Infantry, 78th Division.
Although wounded by a shell fragment, he refused treatment, volunteering his services as a litter bearer for other wounded. Next day he accompanied his platoon in the attack. After heavy fighting for some hours under terrific machine-gun fire, a withdrawal was ordered, Private Slater remaining in advance with an automatic rifle squad to cover the withdrawal. Being one of the last to go back, he was killed by shellfire on his way to the new position. Posthumously awarded. Medal presented to mother, Mrs. Effie Hinman.

SLAY, JOHN R. (2179328)_____
Near Barricourt, France, Nov. 2, 1918.
R—St. Louis, Mo.
B—New York, N. Y.
G. O. No. 44, W. D., 1919.

Sergeant, Company G, 354th Infantry, 89th Division.
After having been severely wounded by a machine-gun bullet, he continued in the attack with his platoon until ordered to the rear.

SLICKLEN, ARTHUR C. (89592)_____
Near Villers-sur-Fere, France, July 28-29, 1918.
R—New York, N. Y.
B—New York, N. Y.
G. O. No. 9, W. D., 1923.

Private, first class, Company C, 165th Infantry, 42d Division.
Under a severe bombardment of artillery and heavy machine-gun fire, Private Slicklen voluntarily went forward three times and brought wounded comrades back to safety, after helping to disperse enemy snipers and grenadiers that had been threatening the left flank of his company. On the following day with the troops leading the attack on Meurcy Farm, under a deadly machine-gun fire from front and flank, he displayed the greatest coolness and courage until wounded four times and carried from the field.

SLINGO, HERBERT J._____
In the Meuse-Argonne offensive, France, Nov. 6, 1918.
R—New York, N. Y.
B—Chicago, Ill.
G. O. No. 49, W. D., 1922

First lieutenant, Signal Corps, Signal Officer, 1st Infantry Brigade, 1st Division.
He displayed the highest qualities—courage, fearlessness, and leadership—in the handling of his section in the Meuse-Argonne offensive. On Nov. 6, 1918, with utter disregard for his own personal safety, he successfully ran communication lines from headquarters, 1st Infantry Brigade, to the headquarters of the 16th and 18th Infantry Regiments, keeping the brigade commander in constant touch with the commanding officers of these regiments. Arriving at the headquarters of the 18th Infantry and finding a shortage of equipment and men, he personally directed the running of communication lines to the front-line battalion at the imminent risk of his life through heavy shell and machine-gun fire. The lines were successfully laid under most hazardous circumstances, Lieutenant Slingo working all the while under direct observation and shellfire of the enemy. His courageous action at a most trying moment was an inspiration to all and assisted materially in the success of this operation.

SLOAN, OZRO L. (38793)_____
Near Thiaucourt, France, Sept. 12, 1918.
R—Cartwright, Tex.
B—Cartwright, Tex.
G. O. No. 46, W. D., 1919.

Private, Company C, 9th Infantry, 2d Division.
When his platoon was under heavy shellfire and threatened by an enemy counter attack, Private Sloan collected in the vicinity a number of soldiers separated from their organizations and led them to reinforce his platoon in the first line. While so engaged he was seriously wounded, but continued on duty with his platoon until completely exhausted.

*SLOAN, WILLIAM E. (1449259)_____
Near Varennes, France, Sept. 26, 1918.
R—Wichita, Kans.
B—Rush County, Kans.
G. O. No. 59, W. D., 1919.

Mechanic, Company I, 137th Infantry, 35th Division.
He continued to advance with his platoon after having been severely wounded and personally guided a tank to an enemy machine-gun nest whose location he had learned. In the course of this extraordinary duty he was killed. Posthumously awarded. Medal presented to mother, Mrs. Cora D. Sloan.

*SLOVER, LUKE E., Jr. (2411118)_____
Near Vieville-en-Haye, France, Sept. 24-25, 1918.
R—Keyport, N. J.
B—Old Bridge, N. J.
G. O. No. 27, W. D., 1919.

Private, first class, Company B, 311th Infantry, 78th Division.
On the night of Sept. 24, Private Slover repeatedly carried messages between his company and battalion headquarters through a heavy barrage. He also took the place of a wounded litter bearer and assisted in bringing in wounded under shellfire. He was later killed in action. Posthumously awarded. Medal presented to father, Luke Slover.

SLOVER, ROBERT._____
Near St. Etienne, France, Oct. 4, 1918.
R—Coal Creek, Tenn.
B—Coal Creek, Tenn.
G. O. No. 37, W. D., 1919.

Corporal, 49th Company, 5th Regiment, U. S. Marine Corps, 2d Division.
Corporal Slover assisted in preparing an emergency force of about 30 men, leading them in attack against greatly superior numbers of the enemy, who were preparing a surprise attack against an unprotected portion of our lines. His leadership and daring resulted in the complete success of the exploit.

SLUSHER, ERNEST W._____
Near Charpentry, France, Sept. 29-30, 1918.
R—Kansas City, Mo.
B—Dover, Mo.
G. O. No. 31, W. D., 1919.

Major, Medical Corps, attached to 140th Infantry, 35th Division.
Although severely gassed, he continued on duty until he collapsed twice, and was carried each time to a dressing station. Advised to go to the field hospital for treatment, he waited until he had partially recovered and then returned to duty in the field, working continually among the wounded and exposing himself to hostile fire.

*SLYKE, ALFRED G. (108185)_____
Near Mont Blanc, France, Oct. 4, 1918.
R—Amsterdam, N. Y.
B—Amsterdam, N. Y.
G. O. No. 46, W. D., 1920.

Sergeant, 77th Company, 6th Machine Gun Battalion, U. S. Marine Corps, 2d Division.
When the Germans attacked his machine-gun detachment at a close range, ammunition was dropped between the gun and the enemy. Although the enemy was bombing the gun position with hand grenades, he went forward and secured the ammunition and then opened fire and routed the enemy, who had already injured 3 of his crew with grenades. Posthumously awarded. Medal presented to father, William Slyke.

SMALL, EARL R. (1248735)_____
Near Montblainville, France, Sept. 28, 1918.
R—Sheffield, Pa.
B—Forest County, Pa.
G. O. No. 16, W. D., 1923.

Sergeant, Company I, 112th Infantry, 28th Division.
In the face of a concentration of enemy machine-gun and rifle fire, Sergeant Small, together with Private Schwing, of his company, voluntarily left the shelter of the trenches constantly under the observation of the enemy, advanced across open ground a distance of 75 yards, rescued a severely wounded soldier, and carried him to shelter. The bravery and devotion to duty thus displayed inspired and encouraged the members of their command, inciting them to still greater endeavors.

SMALL, LYLE H. (2853942)............
Near Vilcey, France, Sept. 12, 1918.
R—Mazon, Ill.
B—Mazon, Ill.
G. O. No. 81, W. D., 1919.

Private, Company M, 358th Infantry, 90th Division.
Private Small, with another soldier, volunteered to outflank an enemy machine-gun nest, and under most harassing fire captured the gun and 11 prisoners.

*SMALLEY, JOHN W. (2222182)...........
Near Vilcey, France, Sept. 12, 1918.
R—Drumright, Okla.
B—Summitville, Ind.
G. O. No. 129, W. D., 1918.

Sergeant, Company M, 358th Infantry, 90th Division.
He displayed great heroism and disregard of personal danger in attacking an enemy machine-gun nest which was holding up the advance of his group. Assisted by another soldier, he flanked the gun, shot 1 of the crew, and drove off the others, but was himself killed in the performance of this courageous act.
Posthumously awarded. Medal presented to father, A. J. Smalley.

SMALLYON, EDWARD H. (362648).......
Near Mezy, France, July 15, 1918.
R—Hartford, Conn.
B—Hartford, Conn.
G. O. No. 32, W. D., 1919.

Private, Machine Gun Company, 30th Infantry, 3d Division.
Given a message to send by buzzer, he found that all his wires had been destroyed. He immediately started through the bombardment and safely delivered the message, although nearly surrounded by the enemy.

SMART, PAUL H.............................
During the attack on Marcheville-en-Woevre, France, Sept. 26, 1918.
R—Newton Highlands, Mass.
B—Nova Scotia.
G. O. No. 13, W. D., 1923.

Second lieutenant, 101st Field Artillery, 26th Division.
During the attack on Marcheville-en-Woevre. Lieutenant Smart volunteered to run back with a message to the rear through a dense enemy concentration of high-explosive shell and gas, after all of the Infantry and Artillery runners had been either killed or wounded in attempting this same mission. Lieutenant Smart was acting as Artillery liaison officer with the attacking units of the 102d Infantry. When the Infantry still met enemy resistance at Marcheville and all communications to the rear had been cut by the heavy enemy barrage laid down in rear of the attacking force, Lieutenant Smart, at the risk of his own life, ran through the barrage to a forward telephone station, communicated the situation to the Artillery commander and then ran back through the same barrage and rejoined and remained with the Infantry commander.

SMEAD, BURTON A.....................
In Argonne-Meuse offensive, France Nov. 1–6, 1918.
R—Denver, Colo.
B—Chicago, Ill.
G. O. No. 37, W. D., 1919.

Major, division adjutant, 39th Division.
Upon his own request, Major Smead was assigned to the hazardous duty of conducting the divisional advance message center. He was constantly exposed to fire, going to the front line for information when it could not be otherwise obtained.

SMECK, JAMES (112905).................
Near Sergy, France, July 31, 1918.
R—Reading, Pa.
B—Reading, Pa.
G. O. No. 71, W. D., 1919.

Private, first class, Company B, 149th Machine Gun Battalion, 42d Division.
He displayed extraordinary heroism in the performance of his duties as runner, and when another runner had been mortally wounded trying to reach company headquarters he volunteered for and accomplished this mission, passing through intense artillery and machine-gun fire.

SMIDT, WILLIAM F. (1211198)...........
Near Ronssoy, France, Sept. 29, 1918.
R—New York, N. Y.
B—New York, N. Y.
G. O. No. 16, W. D., 1923.

Sergeant, Company H, 107th Infantry, 27th Division.
Although suffering from a most painful wound, Sergeant Smidt refused to be evacuated, but continued to lead his platoon in their assault against the enemy. The advance being halted by intense enemy machine-gun fire, he personally rushed upon an enemy machine-gun nest and with bombs put the machine gun out of action, thus enabling his platoon to resume their advance. Again stopped by enemy machine-gun fire, he again attacked an enemy nest, but was seriously wounded in the attempt and evacuated to the hospital. His indomitable bravery and utter disregard for his own safety greatly inspired the men of his company.

*SMILEY, DEAN F......................
Near St. Etienne, France, Oct. 9, 1918.
R—Goshen, Ind.
B—Goshen, Ind.
G. O. No. 37, W. D., 1919.

Private, 75th Company, 6th Regiment, U. S. Marine Corps, 2d Division.
He rushed a hostile machine-gun nest single handed, killing 3 of the crew and capturing the remainder. While taking his prisoners to the rear, this gallant soldier was killed by enemy artillery fire.
Posthumously awarded. Medal presented to mother, Mrs. Jennie Smiley.

SMITH, ALBERT L. (2265542)...........
Near Eclisfontaine, France, Sept. 27, 1918.
R—Fillmore, Calif.
B—Fillmore, Calif.
G. O. No. 46, W. D., 1919.

Private, Machine Gun Company, 364th Infantry, 91st Division.
With 2 other soldiers Private Smith volunteered and went 300 yards beyond our outpost lines through heavy shellfire to bring in a wounded private of his regiment. The mission was promptly and successfully accomplished.

SMITH, ALBERT M......................
At Selzo, Russia, Sept. 21, 1918.
R—Kalamazoo, Mich.
B—Peoria, Ill.
G. O. No. 14, W. D., 1920.

First lieutenant, 339th Infantry, 85th Division (detachment in North Russia).
While leading his platoon in an attack on a strong enemy position, he was severely wounded in the side. Concealing his condition from the men, he continued to direct his platoon, fearlessly exposing himself to fire throughout the action which followed an unsuccessful attack.

SMITH, ANSLEY (42109).................
Near Soissons, France, July 18, 1918.
R—Danville, Ala.
B—Massie, Ala.
G. O. No. 15, W. D., 1919.

Sergeant, Company C, 16th Infantry, 1st Division.
Severely wounded early in the morning, he refused to relinquish command of his platoon, but led its attack to its final objective, remaining in command until after nightfall, when he was ordered to an aid station.

*SMITH, BENJAMIN B. (1314600)........
Near Bellicourt, France, Sept. 29, 1918.
R—Ash, N. C.
B—Ash, N. C.
G. O. No. 21, W. D., 1919.

Private, Company A, 119th Infantry, 30th Division.
After being wounded twice in making attacks with his own organization, he joined Australian troops and attacked with them, being wounded a third time before he consented to be evacuated.
Posthumously awarded. Medal presented to father, William M. Smith.

SMITH, CALLIE A. (1311386)_____
Near Montbrehain, France, Oct. 8, 1918.
R—Rock Hill, S. C.
B—York County, S. C.
G. O. No. 133, W. D., 1918.

Private, first class, Company G, 118th Infantry, 30th Division.
When his company was held up by heavy machine-gun fire, he voluntarily accompanied an officer and assisted him in flanking a machine-gun post and driving out the gunners with grenades and pistol.

SMITH, CHARLES M. (1521380)_____
Near Montfaucon, France, Sept. 27, 1918.
R—Akron, Ohio.
B—Martinsville, W. Va.
G. O. No. 59, W. D., 1919.

Sergeant, Company F, 146th Infantry, 37th Division.
While leading a reconnaissance patrol sent out to locate enemy machine-gun nests he was severely wounded. Lying helpless where he fell, he disregarded his own wounds and continued to direct his men. Through his courage and fortitude many enemy machine guns were located and subsequently destroyed.

SMITH, CHARLIE E. (1309595)_____
Near Ponchaux, France, Oct. 7, 1918.
R—Copperhill, Tenn.
B—Oak Park, N. C.
G. O. No. 37, W. D., 1919.

Private, first class, Company L, 117th Infantry, 30th Division.
Although severely wounded in the leg by machine-gun fire, he continued to advance with his platoon, securing the rifle of a dead soldier when his own was struck by shrapnel and rendered unserviceable.

SMITH, CLARENCE W. (262188)_____
Near Cierges, northeast of Chateau-Thierry, France, July 31, 1918.
R—Flint, Mich.
B—McArthur, Ohio.
G. O. No. 117, W. D., 1918.

Corporal, Company E, 125th Infantry, 32d Division.
He was seriously wounded in the arm by machine-gun fire during the advance on the heights north of the River Ourcq. After receiving first aid he crawled slowly forward in the face of hostile fire and assisted in giving first aid to a lieutenant who had been severely wounded, and then, with his one available arm, assisted in carrying the officer to the rear.

SMITH, DALTON (1319720)_____
Near Mazinghien, France, Oct. 19, 1918.
R—Macon, N. C.
B—Macon, N. C.
G. O. No. 98, W. D., 1919.

Private, Company B, 120th Infantry, 30th Division.
Acting as a scout, Private Smith fearlessly advanced ahead of his company under heavy fire and sent back all obtainable information to the company commander. While standing erect in the open and directing effective rifle fire at the retreating enemy he was seriously wounded.

SMITH, DANIEL R. (43340)_____
Near Soissons, France, July 20, 1918.
R—Reading, Pa.
B—Germany.
G. O. No. 35, W. D., 1920.

Sergeant, Company H, 16th Infantry, 1st Division.
Sergeant Smith led 3 squads against an artillery position which was holding up the advance of his company by direct fire. After suffering heavy casualties, he with 2 others charged the enemy position. Due to his courage and leadership four 77-millimeter guns were captured and 50 of the enemy forced to surrender.

SMITH, DWIGHT F_____
In the Bois-de-Belleau, France, June 8, 1918.
R—Stowe, Vt.
B—Stowe, Vt.
G. O. No. 110, W. D., 1918.

Captain, 6th Regiment, U. S. Marine Corps, 2d Division.
He was conspicuous for his gallantry and energy in conducting attacks against superior forces in strongly fortified machine-gun positions. Under heavy machine-gun fire he fought until incapacitated by wounds.

*SMITH, EBEN A. (547459)_____
Near Crezancy, France, July 16, 1918.
R—Waterloo, Iowa.
B—Tyrone, Iowa.
G. O. No. 89, W. D., 1919.

Sergeant, Company I, 30th Infantry, 3d Division.
Although knocked down by the explosion of a shell, Sergeant Smith immediately got up and rendered valuable assistance to his platoon leader in conducting the movement of the platoon through the most intense shellfire. He was subsequently killed in action.
Posthumously awarded. Medal presented to mother, Mrs. Kate Bronson.

SMITH, EMERSON R. (2339975)_____
Near Le Charmel, France, July 26, 1918.
R—Dayton, Ohio.
B—Dayton, Ohio.
G. O. No. 28, W. D., 1921.

Private, first class, Company M, 4th Infantry, 3d Division.
Engaged as runner, Private Smith carried an important message through heavy shell and machine-gun fire, completing his mission, although so badly gassed that immediate removal to a hospital was necessary.

SMITH, FORD D. (568844)_____
Near Ville-Savoye, France, Aug. 11, 1918.
R—Antioch, Calif.
B—Wyandotte, Mich.
G. O. No. 71, W. D., 1919.

Corporal, Company D, 4th Engineers, 4th Division.
Leaving a sheltered position, he exposed himself to an intense artillery barrage to rescue a wounded officer. He carried him across the Vesle River to where he could obtain aid in taking him to a dressing station. He displayed utter disregard of personal danger while under heavy fire.

SMITH, FRANK (95015)_____
Near St. Baussant, northeast of St. Mihiel, France, Sept. 12, 1918.
R—Newport, Tenn.
B—Gate City, Va.
G. O. No. 125, W. D., 1918.

Corporal, Company K, 166th Infantry, 42d Division.
While advancing in the assault line, he spied a German about to open fire with a machine gun, which would have taken in enfilade his entire platoon. He killed the German with a single rifle shot. The other 3 of the machine-gun crew fled, but he pursued them alone, cut them off from the rear and captured, single handed, 16 of the enemy in one group. His quick decision, excellent marksmanship, and absolute fearlessness were of the greatest value in overcoming the enemy's resistance.

SMITH, FRED (737107)_____
Near Vieville, France, Sept. 12, 1918.
R—Dayton, Ohio.
B—Spring Valley, Ohio.
G. O. No. 37, W. D., 1919.

Sergeant, Company M, 11th Infantry, 5th Division.
After being gassed and shot through the shoulder early in the morning, he continued to lead his platoon throughout the day, refusing to return to the first-aid station for treatment.

SMITH, FRED E. (41183)_____
Near Faubarg, France, Nov. 8, 1918.
R—Syracuse, N. Y.
B—Baldwinsville, N. Y.
G. O. No. 46, W. D., 1919.

Private, Company K, 9th Infantry, 2d Division.
In the absence of officers Private Smith took command of 2 platoons and led them with great fortitude and bravery. When 1 of his men was wounded, he made his way alone through heavy shell and machine-gun fire, brought the wounded man to our line, and applied first aid

102444°—27——37

*SMITH, FRED SHERRY (2212976)........
Near Remonville, France, Nov. 1, 1918.
R—Denver, Colo.
B—Boulder, Colo.
G. O. No. 89, W. D., 1919.

Private, Machine Gun Company, 354th Infantry, 89th Division.
Private Smith was a member of a machine-gun crew firing at close range from a shell hole in an open field when their gun became disabled. Thereupon he and 2 other soldiers advanced with pistols upon the enemy machine-gun nest at which they had been firing and captured it, with 3 guns and 9 prisoners. Putting 1 of the captured guns into immediate action against the enemy, they enabled the infantry to advance with a minimum of casualties. This soldier was killed next day when he went out from cover to warn some comrades that they were in the line of fire from his gun.
Posthumously awarded. Medal presented to mother, Mrs. Jessie Smith.

*SMITH, HAMILTON A.................
Near Soissons, France, July 19–22, 1918.
R—Millen, Ga.
B—Greenwood, Fla.
G. O. No. 132, W. D., 1918.

Colonel, 26th Infantry, 1st Division.
He spent the greater part of his time in the front lines to encourage and direct his command, without sign of fear for his personal safety, and by his courageous leadership inspired his officers and men to effective combat. He was killed while directing an attack on a machine-gun emplacement.
Posthumously awarded. Medal presented to widow, Mrs. Hamilton A. Smith.

SMITH, HARFORD D. (1284328)........
Near Sivry, France, Oct. 18, 1918.
R—Cambridge, Md.
B—Cambridge, Md.
G. O. No. 37, W. D., 1919.

Corporal, Company C, 115th Infantry, 29th Division.
He volunteered and led an automatic-rifle crew forward, silencing a machine-gun nest which was holding up the advance of his company. He worked his way forward through a barrage from 4 machine-gun nests, killing all the occupants of 2 nests and forcing the others to withdraw, thus permitting his company to advance.

SMITH, HARRY L................
Near Cunel, France, Oct. 14, 1918.
R—Lucedale, Miss.
B—Mobile, Ala.
G. O. No. 35, W. D., 1919.

First lieutenant, 13th Machine Gun Battalion, 5th Division.
Leaving his shelter in a shallow machine-gun emplacement, accompanied by 1 soldier, Lieutenant Smith ventured forth through a most intense fire to the aid of a wounded officer, and assisted in carrying him to a distance of 170 yards to safety.

SMITH, HARRY S................
Near Tuilerie Farm, France, Nov. 3, 1918.
R—Waynesburg, Pa.
B—Waynesburg, Pa.
G. O. No. 46, W. D., 1919.

Second lieutenant, 9th Infantry, 2d Division.
Lieutenant Smith led the advance elements of his regiment during an advance of 8 kilometers through the German lines, and, with extraordinary skill and courage, reduced several enemy strong points. In addition he captured 50 prisoners and a large amount of material.

*SMITH, HEARL (1460802)................
Near Cheppy, France, Sept. 27, 1918.
R—Willow Springs, Mo.
B—Newport, Tenn.
G. O. No. 89, W. D., 1919.

Sergeant, Company F, 140th Infantry, 35th Division.
Having been mortally wounded while rushing a machine-gun nest, Sergeant Smith continued faithfully to perform his duties and calmly directed the movements of his half platoon until he died on the field.
Posthumously awarded. Medal presented to father, Monroe Smith.

SMITH, HENRY M................
Near Malancourt, France, Sept. 26, 1918.
R—Greensburg, Pa.
B—Jeannette, Pa.
G. O. No. 46, W. D., 1919.

Captain, 314th Infantry, 79th Division.
Although painfully wounded while leading a platoon of his company against strong machine-gun nests, Captain Smith continued the advance until all the machine guns in his immediate front were silenced and the crews killed or taken prisoners. He continued on duty until ordered to the rear by his regimental commander.

SMITH, HERBERT REEVES (6502910)....
At Ojo de Agua, Tex., Oct. 21, 1915.
R—Tampico, Ill.
B—Tampico, Ill.
G. O. No. 10, W. D., 1921.

Sergeant, first class, Signal Corps, U. S. Army.
When the detachment of which he was a member was treacherously attacked at night by a hostile Mexican force of approximately 5 times its strength and the senior line noncommissioned officer of the detachment and 2 other men killed, Sergeant Smith, although suffering from 3 wounds, took command, and by his coolness, personal bravery, and qualities of leadership so encouraged the detachment and organized its defense that the greatly superior force of Mexicans was driven off and the detachment saved from probable annihilation.

SMITH, HORACE L., Jr................
Near Charpentry, France, Oct. 4, 1918.
R—Petersburg, Va.
B—Richmond, Va.
G. O. No. 44, W. D., 1919.

Captain, 1st Engineers, 1st Division.
While repairing roads, a large ammunition dump was set on fire by an enemy shell. Captain Smith, with a party of his men, extinguished the flames and rescued a large quantity of ammunition and supplies, despite the threatened explosion, which would have destroyed the entire dump and blocked traffic at an important crossroad for hours.

SMITH, HOWARD G................
In Bois-de-Romagne, France, Oct. 15, 1918.
R—East Lansing, Mich.
B—Cleveland, Ohio.
G. O. No. 15, W. D., 1919.

First lieutenant, 168th Infantry, 42d Division.
He was wounded early in the engagement, but he declined to be evacuated, although he was suffering much pain. He brilliantly led his platoon in a charge on 4 machine-guns, which he captured, together with many prisoners, and was instrumental in clearing the Bois-de-Romagne of the enemy under terrific machine-gun fire. Throughout the action his leadership, courage, and determination inspired the greatest confidence. When he was partly overcome by the loss of blood, he volunteered to guide 60 prisoners back over a shell-swept area, and refused medical treatment until the prisoners were delivered at battalion headquarters.

SMITH, IVAN H. (262172)................
Near Cierges, northeast of Chateau-Thierry, France, July 31, 1918.
R—Flint, Mich.
B—Bay Port, Mich.
G. O. No. 132, W. D., 1918.

Sergeant, Company E, 125th Infantry, 32d Division.
For extraordinary heroism in action during the forcing of a passage of the River Ourcq and the capture of the heights beyond, near Cierges, northeast of Chateau-Thierry, France, July 31, 1918. Sergeant Smith captured a machine gun single-handed and after being wounded while so doing reorganized his platoon before being taken back for first aid.

SMITH, JACOB C.
At San Juan, Cuba, July 1, 1898.
R—Rushville, Ind.
B—Taylorsville, Ky.
G. O. No. 14, W. D., 1925.

Saddler sergeant, 10th Cavalry, U. S. Army.
Sergeant Smith, with utter disregard for his personal safety and while exposed to a heavy fire of shell and small arms from the enemy, deliberately cut the fence or obstruction consisting of 4 or 5 strands of barbed wire on top of an almost perpendicular bank, thus enabling Troop A, 10th Cavalry, U. S. Army, to advance and take the position to which ordered.

SMITH, JOE (556655).
Near Bois-du-Fays, France, Oct. 10-12, 1918.
R—Center Point, Ark.
B—Montgomery County, Ark.
G. O. No. 64, W. D., 1919.

Private, Company C, 39th Infantry, 4th Division.
Acting as battalion runner, Private Smith repeatedly carried messages over a route swept by machine-gun and artillery fire. It was necessary to send runners night and day in order to maintain communication with the front lines. He volunteered out of his turn for this dangerous but all-important work.

SMITH, JOHN E. (53922).
Near Soissons, France, July 19, 1918.
R—Norman Park, Ga.
B—Hoschton, Ga.
G. O. No. 132, W. D., 1918.

Sergeant, Company G, 26th Infantry, 1st Division.
On his own initiative he took command of his company near Soissons, France, July 19, 1918, when all its officers and its first sergeant had been killed or wounded, and carried forward successfully its part in the day's attack.

ʻSMITH, JOHN F.
Near St. Etienne, France, Oct. 4, 1918.
R—Cincinnati, Ohio.
B—Youngstown, Ohio.
G. O. No. 37, W. D., 1919.

Private, Machine Gun Company, 5th Regiment, U. S. Marine Corps, 2d Division.
During a heavy enemy counterbarrage Private Smith was engaged as a runner. In the execution of his duty he displayed exceptional bravery, carrying messages through intense shellfire, falling severely wounded after his fourth journey.
Posthumously awarded. Medal presented to mother, Mrs. George T. Knox.

SMITH, JOSEPH W.
Near St. Baussant, northeast of St. Mihiel, France, Sept. 12, 1918.
R—Austin, Tex.
B—Meridian, Tex.
G. O. No. 99, W. D., 1918.

First lieutenant, 166th Infantry, 42d Division.
Finding that his platoon would be under heavy fire of enemy machine guns while crossing the Rupt de Mad, Lieutenant Smith, rather than permit the advance to be delayed, unhesitatingly plunged into the stream, crossed it under heavy fire, ascertained the exact location of the enemy, brought his platoon through the river by a protected route, and with it flanked and captured 6 machine guns and 19 prisoners.

SMITH, LEROY W. (1311153).
Near St. Martin-Riviere, France, Oct. 17, 1918.
R—Cades, S. C.
B—Williamsburg County, S. C.
G. O. No. 46, W. D., 1919.

Sergeant, Company F, 118th Infantry, 30th Division.
Immediately after the starting of the attack Sergeant Smith collapsed from gas but realizing his extreme need of a compass in the dense fog and having the only 1 of the company he struggled along by his company commander, indicating the proper direction with his hands, being unable to talk. He refused evacuation, and voluntarily led a patrol to establish liaison with his right flank, being subjected to annihilating machine-gun fire during the entire exploit.

SMITH, LOUIS S. (1783200).
Near Montfaucon, France, Sept. 27, 1918.
R—Dalton, Pa.
B—Shultzville, Pa.
G. O. No. 37, W. D., 1919.

Sergeant, Company I, 314th Infantry, 79th Division.
Sergeant Smith advanced alone and silenced a machine-gun nest which was holding up the advance of his section.

SMITH, MARTIN E. (1390962).
Near Consenvoye, France, Oct. 9, 1918.
R—Chicago, Ill.
B—Washington, Mich.
G. O. No. 64, W. D., 1919.

First sergeant, Company G, 132d Infantry, 33d Division.
When his company was held up by heavy machine-gun fire, he showed marked personal bravery in working his way to the rear of the enemy emplacement. He opened fire upon the enemy from the rear, who then surrendered to him. He returned to his own lines with 2 officers and 15 men as prisoners.

SMITH, MARTIN M. (1214761).
Near Ronssoy, France, Sept. 29-30, 1918.
R—Batavia, N. Y.
B—Batavia, N. Y.
G. O. No. 21, W. D., 1919.

Sergeant, Company G, 108th Infantry, 27th Division.
He exhibited exceptional gallantry and ability in leadership when, after being severely shell shocked, he continued to direct the steady advance of his platoon under intense machine-gun and shellfire, with utter disregard for his personal safety. He continued with his platoon until the morning of Sept. 30, when he collapsed as the result of shell shock and was evacuated to the rear.

SMITH, MAXWELL E. (279478).
Near Gesnes, northwest of Verdun, France, Oct. 3, 1918.
R—Ionia, Mich.
B—Ionia, Mich.
G. O. No. 37, W. D., 1919.

Sergeant, Company D, 126th Infantry, 32d Division.
In an attack on German strong points, Sergeant Smith was wounded early in the action by a shell fragment, but after dressing his wound himself he immediately rejoined his organization. Upon hearing that his company commander had been killed, he reported this fact to the regimental post of command, and although he was urged to go to the rear for medical aid, he again went forward and assumed command of the company, remaining in charge until the following morning. His courage and disregard for danger were an inspiration to his men and a thorough demonstration of his loyalty and devotion to duty.

SMITH, MILLARD (40475).
Near Beaumont, France, Nov. 8, 1918.
R—Vox, Ky.
B—Whitley County, Ky.
G. O. No. 21, W. D., 1919.

Sergeant, Company K, 9th Infantry, 2d Division.
During an advance of his company, terrific machine-gun fire was encountered, the enemy being strongly entrenched in a ravine. With a patrol of 10 men, Sergeant Smith attacked the position, but lost several of his men in the first encounter. With the remaining few he continued and silenced the fire of the enemy.

SMITH, NAT R. (2258138).
Near Gesnes, France, Sept. 28, 1918.
R—Ostrander, Wash.
B—Kelso, Wash.
G. O. No. 20, W. D., 1919.

Sergeant, Company K, 361st Infantry, 91st Division.
He successfully led his patrol, in the face of heavy machine-gun fire, being a point direct, in order to make better reconnaissance, and, although severely wounded, continued to lead his patrol.

SMITH, NICKOLAS (280883)............
Near St. Georges, France, Oct. 14, 1918.
R—Grand Rapids, Mich.
B—Grand Rapids, Mich.
G. O. No. 89, W. D., 1919.

Sergeant, Company K, 126th Infantry, 32d Division.
After his platoon had been stopped and disorganized by machine-gun fire, Sergeant Smith, with another soldier, reorganized the platoon and led it in a charge on an enemy machine-gun nest, capturing it, together with 1 officer and 20 men.

SMITH, PHILIP F. (1284100)............
Near Verdun, France Oct. 13, 1918.
R—Baltimore, Md.
B—Baltimore, Md.
G. O. No. 16, W. D., 1919.

Private, first class, Company B, 115th Infantry, 29th Division.
During a heavy artillery barrage on the night of Oct. 13, 2 men having been killed next to him and 1 severely wounded, he, disregarding his personal safety, carried the wounded man through the barrage to the company headquarters; knocked down by a shell splinter and severely bruised, he continued with the wounded man to a first-aid station.

SMITH, RAYMOND R. (559162)..........
Northeast of Cunel, France, Oct. 12, 1918.
R—Burlington, Iowa.
B—Burlington, Iowa.
G. O. No. 21, W. D., 1919.

Corporal, Company C, 11th Machine Gun Battalion, 4th Division.
During a heavy bombardment after a shell had struck his machine gun, knocking it and his squad completely out of action, Corporal Smith assembled 3 men from another squad and obtaining another gun again took up position on the line and remained throughout the action, as the front was at that time thinly held and in constant danger of counterattack. The prompt initiative and splendid courage on the part of this soldier not only inspired and encouraged his men, but aided materially in the success of the action.

SMITH, RICHARD THOMPSON..........
In the vicinity of Fort de Manonviller, France, Mar. 17, 1918.
R—Kansas City, Mo.
B—Boonville, Mo.
G. O. No. 126, W. D., 1919.

Captain, 117th Field Signal Battalion, 42d Division.
While under heavy shellfire on Mar. 17, 1918, in the vicinity of Fort de Manonviller, France, he showed prompt initiative and courage in conducting to shelter a party of 65 men who were constructing and repairing communicating lines. After making them secure he went out under fire and brought in a wounded soldier who was lying exposed in the open.

SMITH, ROBERT O....................
Near the Meuse River, France, Nov. 6–11, 1918.
R—Denver, Colo.
B—De Beque, Colo.
G. O. No. 32, W. D., 1919.

First lieutenant, Dental Corps, attached to 356th Infantry, 89th Division.
After all the medical officers of the battalion had been wounded, Lieutenant Smith for 6 days efficiently performed the duties of a medical officer, repeatedly moving his first-aid station forward and administering to the wounded under perilous shellfire. After caring for the wounded, he personally searched the field of action for further casualties.

SMITH, ROYAL H. G..................
Near Remonville, France, Nov. 1, 1918.
R—Gorham, Me.
B—Dayton, Me.
G. O. No. 95. W. D., 1919.

First lieutenant, 353d Infantry, 89th Division.
Lieutenant Smith was a member of a group of several officers and soldiers, who, armed only with pistols, were cut off from the battalion headquarters by the fire of 3 enemy machine guns. Being unable to reach the enemy with pistol fire they were in danger of annihilation when Lieutenant Smith with great daring, dashed from cover through the machine-gun fire and returned with a platoon of Infantry, with which he successfully attacked the enemy machine guns and thereby saved the lives of the party. In accomplishing this feat Lieutenant Smith was wounded in the leg by a machine-gun bullet, but he continued in action, refusing to be evacuated.

SMITH, RUSSELL C. (1284053)..........
Near Verdun, France, Oct. 15, 1918.
R—Hagerstown, Md.
B—Hagerstown, Md.
G. O. No. 15, W. D., 1919.

Mechanic, Company B, 115th Infantry, 29th Division.
While carrying a message to the battalion commander in the Bois-de-Consenvoye, he was caught in an artillery barrage and severely wounded. Greatly exhausted, he refused medical attention and continued with the message until he reached the battalion headquarters.

SMITH, SAMUEL T...................
Near Fleville, France, Oct. 4, 1918.
R—Conway, Ark.
B—Conway, Ark.
G. O. No. 37, W. D., 1919.

First lieutenant, 6th Field Artillery, 1st Division.
When his platoon had been caught in an enemy barrage and all the cannoneers except 2 had been killed, he made repeated trips into the shelled area to remove the wounded to a place of safety. He refused any treatment for 4 wounds which he had received until all of his men had received medical treatment.

SMITH, SIDNEY (3129935)............
Near Binarville, France, Oct. 2–8, 1918.
R—Blaine, Mont.
B—Pearl, Ill.
G. O. No. 37, W. D., 1919.

Private, Company H, 308th Infantry, 77th Division.
When his company had been cut off from communication he, though seriously wounded, refused to seek shelter. He participated in several attacks with courage and aggressiveness, using his rifle very effectively and encouraging his comrades. When relief came he walked back to the dressing station, so that medical attention could first be given to the more seriously wounded.

SMITH, THOMAS J. (39947)..........
Near Chateau-Thierry, France, June 6, 1918.
R—New Britain, Conn.
B—New Britain, Conn.
G. O. No. 107, W. D., 1918.

Corporal, Company H, 9th Infantry, 2d Division.
After having been severely wounded, he remained with his platoon, encouraging and urging on men in the absence of their platoon sergeant, who had been killed.

SMITH, TOM H. (2221079)............
In the Bois-de-Consenvoye, France, Oct. 18, 1918.
R—Shamrock, Okla.
B—Weatherford, Tex.
G. O. No. 37, W. D., 1919.

Private, Company C, 115th Infantry, 29th Division.
He was with 20 men in the front line, and for 40 hours they had been without food, the heavy bombardment preventing ration details from reaching them. As the men were losing their strength and morale, this soldier voluntarily went through heavy shellfire to procure food, making repeated trips till all the men were supplied.

SMITH, WALLACE W. (2294204)..........
Near Gesnes, France, Sept. 28 to Oct. 1, 1918.
R—Corvallis, Oreg.
B—Turkey.
G. O. No. 20, W. D., 1919.

Private, Company I, 361st Infantry, 91st Division.
Although twice wounded, he stayed out in front under heavy machine-gun and artillery fire and helped to take back within our lines wounded comrades who otherwise would have fallen into the hands of the enemy.

SMITH, WARDE B._____
Near Cierges, France, July 21 to
Aug. 7, 1918.
R—Frankfort, Ohio.
B—Frankfort, Ohio.
G. O. No. 9, W. D., 1923.

First lieutenant, Medical Corps, attached to 125th Infantry, 32d Division.
Maintaining dressing stations close to the advanced lines, under heavy enemy machine-gun and artillery fire he continued day and night to render first aid and to evacuate the wounded. On Aug. 2 his aid station, due to withdrawals of the Infantry to take a more advantageous position, was left in advance of the front line. With a score of severely wounded men to evacuate, Lieutenant Smith remained at the advanced position for 6 hours under intense enemy fire until all patients had been given every possible care and carried to a place of safety. His coolness, courage, and devotion to duty saved the lives of many men.

*SMITH, WILLARD L._____
Near Remenauville, France, Sept.
12, 1918.
R—Worcester, Mass.
B—Worcester, Mass.
G. O. No. 37, W. D., 1919.

First lieutenant, 9th Infantry, 2d Division.
He was killed while gallantly assisting in maintaining liaison between the troops advancing on the open ground to the west of Bois-du-Four and those in the woods. It was due to his fearless example while leading his men that the line was held intact at this point.
Posthumously awarded. Medal presented to father, F. B. Smith.

*SMITH, WILLIAM F. (1243621)_____
Near Fismette, France, Aug. 10,
1918.
R—Chester, Pa.
B—Essington, Pa.
G. O. No. 49, W. D., 1922.

Private, Company B, 111th Infantry, 28th Division.
When the attack of his company was held up by fire from a hostile strong point, Private Smith, with 2 other men, voluntarily cut their way through enemy wire entanglements under heavy fire, reached their objective, and engaged the enemy in hand-to-hand combat. During the latter action 6 of the enemy were killed and the attacking line was enabled to advance to the new position. Private Smith was mortally wounded, his heroic action costing him his life.
Posthumously awarded. Medal presented to mother, Mrs. Minnie Smith.

SMITH, WILLIAM K. (105775)_____
Near Croix Le Perre, France, July
18, 1918.
R—Hubbard, Tex.
B—Hubbard, Tex.
G. O. No. 98, W. D., 1919.

Sergeant, Company D, 2d Machine Gun Battalion, 1st Division.
After his platoon commander had become a casualty and all the machine guns of his section had been put out of action by shellfire, Sergeant Smith collected the few remaining men of his section, armed them with rifles and pistols, and voluntarily led them against an enemy machine-gun nest which was checking the advance of the battalion. Despite intense machine gun and shell fire, he made a flank attack on the nest and reduced it, killing or capturing 12 of the enemy. His marked courage and quick initiative enabled the battalion to resume its advance.

SMITH, WILLIAM OLIVER_____
North of Haudiomont, France, Nov.
9-10, 1918.
R—Raleigh, N. C.
B—Liberty, Mo.
G. O. No. 60, W. D., 1920.

First lieutenant, 318th Machine Gun Battalion, 81st Division.
Lieutenant Smith courageously led his machine-gun platoon in an attack on the afternoon of Nov. 9, and later assisted in organizing a position for defense. On Nov. 10, the enemy launched a strong counterattack and the Infantry withdrew under cover of the machine-gun fire. Later, when attacked by greatly superior numbers, Lieutenant Smith defended his position an hour. Although wounded 3 times, he persisted in his resistance, holding his position until his ammunition was exhausted, when he was taken prisoner by the enemy.

SMITH, WILLIAMSON ALFRED (1311036)__
East of the La Selle River, France,
Oct. 17, 1918.
R—Stanfield, N. C.
B—Stanley County, N. C.
G. O. No. 98, W. D., 1919.

Private, Company E, 118th Infantry, 30th Division.
Having become separated from his company in a fog, Private Smith, an automatic rifle gunner, attached himself to a company in the attacking wave and continued in the advance. Working his way through heavy machine-gun and shellfire he put his automatic rifle into action, poured an enfilading fire on the enemy and aided materially in breaking the hostile resistance at a critical time.

SMITHHISLER, PAUL A. (1538321)_____
Near Heuvel, Belgium, Nov. 2, 1918.
R—Cleveland, Ohio.
B—Mount Vernon, Ohio.
G. O. No. 37, W. D., 1919.

Sergeant, first class, headquarters detachment, 112th Engineers, 37th Division.
Under cover of darkness he swam the Scheldt River at a point where it was covered by hostile machine guns and reconnoitered a road for a distance of 500 meters, returning with valuable information.

SMOTHERMAN, HORACE (1387630)_____
Near Gercourt, France, Sept. 26,
1918.
R—Chicago, Ill.
B—Paris, Tex.
G. O. No. 71, W. D., 1919.

Private, Company F, 131st Infantry, 33d Division.
Advancing, on his own initiative, under heavy artillery, snipers', and machine-gun fire, he threw a grenade into a sniper's post, fire from which had held up our advance. His grenade killed the snipers and allowed a renewal of the advance.

*SMYTH, ROY M._____
Near Les Franquettes Farm, France,
July 23, 1918.
R—Reno, Nev.
B—Carters, Calif.
G. O. No. 32, W. D., 1919.

Major, 4th Infantry, 3d Division.
After part of his battalion had already entered an open field before Les Franquettes Farm, enemy machine guns suddenly opened fire from several points, and at the same time hidden mines in the field were exploded. Major Smyth constantly exposed himself to great personal danger while getting his men into the best available cover and reorganizing his position.
Posthumously awarded. Medal presented to father, Hugh N. Smyth.

SNEEDEN, SILAS V. (1323126)_____
Near Ypres, Belgium, Aug. 23, 1918.
R—Sea Gate, N. C.
B—Sea Gate, N. C.
G. O. No. 35, W. D., 1919.

Private, Company C, 115th Machine Gun Battalion, 30th Division.
Upon learning that his platoon commander and several comrades had been wounded by heavy shellfire, he voluntarily left his dugout and went to their assistance, helping to carry them 500 yards to the dressing station across an open field heavily bombarded with gas and high-explosive shells.

SNIDER, JAMES J. (1391401)_____
In the Bois-de-Foret, France, Oct.
10, 1918.
R—Chicago, Ill.
B—Chicago, Ill.
G. O. No. 44, W. D., 1919.

Private, Machine Gun Company, 132d Infantry, 33d Division.
After 6 runners had been killed or wounded in attempts to get through heavy shellfire with an important message from the regimental commander of the 39th Infantry to the regimental commander of the 59th Infantry, Private Snider, with Private Sidney Holzeman, responded to a call for volunteers and succeeded in delivering the message. While engaged on this mission Private Snider was badly gassed.

SNOW, WILLIAM A._____
In the Belleau Wood, France, June 12–15, 1918.
R—Washington, D. C.
B—Fort Hamilton, N. Y.
G. O. No. 87, W. D., 1919.

Major, 2d Engineers, 2d Division.
In order to consolidate the position of his brigade Major Snow personally led 1 company of his battalion through a heavy barrage. After passing through the barrage he discovered that part of his company had become separated because of the violent fire. He returned through the barrage, and, in so doing, was wounded in the neck. After having his wound dressed at the aid station, he refused to go to the rear, but went back and conducted the remainder of the men through the barrage. Despite his wound, he remained on duty for 16 hours until ordered to the rear.

SNOWDEN, SAMUEL (262922)_____
Near Fismes, France, Aug. 8, 1918.
R—Wyandotte, Mich.
B—Cheboygan, Mich.
G. O. No. 99, W. D., 1919.

Sergeant, Company H, 125th Infantry, 32d Division.
Exposed to intensive artillery and machine-gun fire, Sergeant Snowden crawled to the crest of a hill and administered to a wounded man, thus saving his life. While so doing he was wounded in the right leg above the knee, but, undaunted, he picked up another more seriously wounded than himself and brought him to a dressing station.

*SNYDER, ABOIL E. (2339528)_____
Near Le Charmel, France, July 26, 1918.
R—Alburtis, Pa.
B—Alburtis, Pa.
G. O. No. 32, W. D., 1919.

Private, Company I, 4th Infantry, 3d Division.
Although shot through the stomach with a machine-gun bullet, he continued to the next relay station and arranged for the safe delivery of his message. He died a few minutes later.
Posthumously awarded. Medal presented to grandfather, Aboil K. Snyder.

SNYDER, CLAYTON EVANS_____
Near Cunel, France, Oct. 13, 1918.
R—Malta, Mont.
B—Columbus, Ohio.
G. O. No. 26, W. D., 1919.

Second lieutenant, 9th Machine Gun Battalion, 3d Division.
Although wounded by machine-gun fire, he refused to be evacuated, and going out into no man's land located several enemy machine guns which were endangering his platoon, and directed the fire of his men with such accuracy that the guns were silenced.

SNYDER, JOHN H._____
Near St. Mihiel, Briey and Thiouville, France, Sept. 12, 1918.
R—Reading, Pa.
B—Reading, Pa.
G. O. No. 1, W. D., 1919.

First lieutenant, observer, 1st Army observation group, Air Service.
While on a special mission to determine the probable enemy concentration in the back areas he, with his pilot, in spite of almost impossible flying conditions, flew 60 kilometers over the enemy lines at a very low altitude. The unfavorable weather alone would have warranted them in turning back, but they continued on regardless of very active and accurate machine-gun and antiaircraft fire. They returned to our lines only when their mission was successfully completed.

SNYDER, WILLIS P. (113567)_____
East of Reims, France, July 15, 1918.
R—Reading, Pa.
B—Reading, Pa.
G. O. No. 128, W. D., 1918.

Private, Company D, 150th Machine Gun Battalion, 42d Division.
While manning a machine gun against the enemy, and after all his comrades had either been killed or wounded, he remained at his post and in the hand-to-hand fight which ensued forced the enemy to retire; and, although wounded, he attempted to carry back his wounded comrades. His supreme courage and devotion to duty were an inspiration to all associated with him.

*SOCHA, RUDOLPH (297587)_____
Near Montfaucon, France, Sept. 29, 1918.
R—Detroit, Mich.
B—Talbot, Mich.
G. O. No. 86, W. D., 1919.

Private, Battery D, 119th Field Artillery, 32d Division.
Having been severely wounded in the knee, Private Socha refused to be evacuated, but immediately returned from the dressing station to his battery position, where he assisted in caring for other wounded men until he was again wounded, inspiring his comrades by his undiminishing courage and cheerfulness in the face of danger.
Posthumously awarded. Medal presented to sister, Mrs. Agnes Haumschild.

SOLINSKI, WACTAW (54813)_____
Near Soissons, France, July 18, 1918.
R—New York, N. Y.
B—Poland.
G. O. No. 3, W. D., 1921.

Corporal, Company K, 26th Infantry, 1st Division.
Sergeant Solinski, although seriously wounded in the thigh by an aerial bomb, took command of his platoon after the death of his commander and gallantly led it forward until he fell unconscious.

SOLOMON, ISADORE (1817476)_____
Near Sommauthe, France, Nov. 4, 1918.
R—Chicago, Ill.
B—Chicago, Ill.
G. O. No. 11, W. D., 1921.

Sergeant, Company C, 317th Infantry, 80th Division.
When a friendly airplane had dropped a message in front of our line Sergeant Solomon, disregarding personal safety, exposed himself to heavy machine-gun fire in order to go out and recover the message. The information thus obtained was of vital importance for the successful continuance of the advance.

*SOMERS, VERNON L._____
At Chateau-Thierry, France, June 6, 1918.
R—Bloxom, Va.
B—Bloxom, Va.
G. O. No. 119, W. D., 1918.

Second lieutenant, 5th Regiment, U. S. Marine Corps, 2d Division.
Killed in action at Chateau-Thierry, France, June 6, 1918, he gave the supreme proof of that extraordinary heroism which will serve as an example to hitherto untried troops.
Posthumously awarded. Medal presented to mother, Mrs. Maggie A. Somers.

SOMERVELL, BREHON B._____
Near Pouilly, France, Nov. 5–6, 1918.
R—Little Rock, Ark.
B—Little Rock, Ark.
G. O. No. 37, W. D., 1919.
Distinguished-service medal also awarded.

Lieutenant colonel, Corps of Engineers.
Voluntarily serving on the staff of the 89th Division, he conducted the first engineering reconnaissance of the damaged bridges at Pouilly, where with 2 scouts he advanced more than 500 meters beyond the American outposts, crossing 3 branches of the Meuse River and successfully encountering the enemy.

SOMES, RUSSELL V. (263260)_____
Near Cierges, France, July 31, 1918.
R—Sault Ste. Marie, Mich.
B—Sault Ste. Marie, Mich.
G. O. No. 124, W. D., 1918.

Sergeant, Company I, 125th Infantry, 32d Division.
He advanced in front of his lines on the right of Hill 212, under heavy machine-gun fire, and rescued 3 wounded soldiers. Later he went out into an advanced machine-gun position where 3 men had already been killed and rescued the only survivor, who had been blinded by shellfire and could not help himself.

SOMNITZ, CARL G. (1387579) _____
   At Chipilly Ridge, France, Aug. 9,
      1918.
   R—Chicago, Ill.
   B—Chicago, Ill.
   G. O. No. 128, W. D., 1918.

Corporal, Company F, 131st Infantry, 33d Division.
When all the runners of his platoon had failed to establish liaison with the platoon on the left he succeeded in getting through with a message. On his return trip he was twice wounded, but dragged himself along the ground and delivered his message before lapsing into unconsciousness.

SONSTELIE, CARL J _____
   Near Montfaucon, France, Sept. 26,
      1918.
   R—Kalispell, Mont.
   B—Vesta, N. Dak.
   G. O. No. 128, W. D., 1918.

First lieutenant, 3d Brigade, Tank Corps.
He displayed bravery and leadership of a high order in the advance toward Montfaucon by going out ahead of the engineers, reconnoitering a tank route under fire, and urging the tanks forward. He located the resistance in the Bois-de-Cuisy in advance, later rallying disorganized soldiers and enabling them to hold the Bois-de-Cuisy.

SORENSEN, SOREN C _____
   At Cantigny, France, May 28, 1918.
   R—Grand Island, Nebr.
   B—Denmark.
   G. O No. 99, W. D., 1918

First lieutenant, 28th Infantry, 2d Division.
When the officers of his unit were killed or wounded at Cantigny, France, May 28, 1918, and although he himself had been wounded early in the attack and suffered intensely, he took command, refused to leave his post, and, by heroic courage and resolution in resisting counterattacks, contributed in great measure to the successful defense of his sector.

SORENSON, JOHN H. (2289565) _____
   Near Eclisfontaine, France, Oct. 4,
      1918.
   R—Minot, N. Dak.
   B—Denmark.
   G. O. No. 46, W. D., 1919.

Private, Machine Gun Company, 364th Infantry, 91st Division.
He volunteered and remained with a wounded comrade in a gun position when his division was relieved. He gave all the aid possible and then went some 500 meters through heavy shell and machine-gun fire for further medical assistance and returned with it to his companion.

SORROW, LOUIS (1919060) _____
   Near Fleville, France, Oct. 13–21,
      1918.
   R—Bronx, N. Y.
   B—New York, N. Y.
   G. O. No. 147, W. D., 1918.

Corporal, Company B, 307th Field Signal Battalion, 82d Division.
After being on duty continuously for 36 hours, on Oct. 13, 1918, he volunteered to repair telephone lines which had been cut by shellfire. Under extremely heavy bombardment he worked all night repairing breaks in lines and thereby making possible constant communication with 1 of the advanced regiments. On Oct. 21, 1918, after 1 of his helpers had been killed and the other wounded by heavy shellfire, he continued on alone and repaired the telephone lines, displaying unusual bravery and devotion to duty.

SOUCY, FRED G. (42353) _____
   South of Soissons, France, July 18,
      1918.
   R—Lewiston, Me.
   B—Lewiston, Me.
   G. O. No. 15, W. D., 1919.

Private, Company E, 16th Infantry, 1st Division.
When his platoon was held up by a machine gun he, with 2 other privates, who were killed before reaching the emplacement, charged the gun, killed the crew of 5 Germans, and captured their gun.

SOULES, JAMES A _____
   Near Sedan, France, Nov. 6–7, 1918.
   R—Dickinson, N. Dak.
   B—Terre Haute, Ind.
   G. O. No. 44, W. D., 1919.

Second lieutenant, 16th Infantry, 1st Division.
Accompanied by another soldier in his platoon, Lieutenant Soules entered the town of Noyers-Pout-Maugis, which was held by the enemy, against murderous machine-gun fire. He routed the gunners, killing 1, thereby saving his company from a harassing flanking fire.

SOUTHARD, WILLIAM E _____
   Near Torcy, France, July 18, 1918.
   R—Bangor, Me.
   B—Garland, Me.
   G. O. No. 44, W. D., 1919.

Major, 103d Infantry, 26th Division.
Immediately after an enemy barrage was laid down on his assaulting line, Major Southard pushed forward through the halting ranks and, calling on his men to follow, advanced at double time to the storming of Torcy, attaining his objective. He then organized the defense of the town, supervising the work under hazardous artillery, machine-gun, and sniper fire. During the attack Major Southard's forces suffered heavily from casualties, yet he resolutely held his position for 2 days, after which he was wounded while leading an assault upon the heights beyond Belleau.

SPADAFORA, GUISEPPE (1784205) _____
   Near Montfaucon, France, Sept. 29,
      1918.
   R—Philadelphia, Pa.
   B—Italy.
   G. O. No. 37, W. D., 1919.

Private, Headquarters Company, 315th Infantry, 79th Division.
He was helping to remove a great many wounded men from a dressing station to a place of comparative safety when a heavy enemy bombardment began. He forced 4 German prisoners to assist him and repeatedly entered the heavily shelled area, bringing out wounded men.

*SPAFFORD, JAMES H _____
   At St. Etienne, France, Oct. 9, 1918.
   R—Baltimore, Md.
   B—Baltimore, Md.
   G. O. No. 20, W. D., 1919.

First lieutenant, 2d Engineers, 2d Division.
Seeing a combat patrol suddenly fired upon by an enemy machine-gun nest and hard pressed, Lieutenant Spafford went to its relief, courageously leading an attack on the nest. Although wounded in the arm during the attack, he continued in the action of the attack until he received a second wound, which caused his death.
Posthumously awarded. Medal presented to father, James Spafford.

SPAIN, GARLAND (1864750) _____
   Near Moranville, France, Nov. 9,
      1918.
   R—Rocky Mount, N. C.
   B—Greensville County, Va.
   G. O. No. 32, W. D., 1919.

Corporal, Company E, 322d Infantry, 81st Division.
Leading his squad against 6 enemy machine guns, during which time he was hit twice by the exacting fire therefrom, he drove the enemy from the stronghold, making possible the further advance of his company.

SPAMPANATO, ANIELLO (2854210) _____
   Near Montfaucon, France, Oct. 25,
      1918.
   R—Marseilles, Ill.
   B—Italy.
   G. O. No. 32, W. D., 1919.

Private, Company L, 357th Infantry, 90th Division.
He was on a patrol with 3 other soldiers when they were fired upon by a hostile machine gun 50 yards in advance of the line. After several hand grenades had been thrown at the machine-gun nest one of the crew was seen crawling away. Private Spamanato killed this man with his rifle and then rushed the nest alone, capturing the gun and 3 surviving members of the crew, 2 others having been killed by hand grenades.

*SPANGLER, LEWIS G. (239734) _____
    Near Fossoy, France, July 15, 1918.
    R—Lometa, Tex.
    B—Sealy, Tex.
    G. O. No. 44, W. D., 1919.

Private, Company K, 7th Infantry, 3d Division.
During the intense shelling by the enemy just prior to their offensive of July 15 he volunteered and carried a message through the heavy fire and returned with an answer.
Posthumously awarded. Medal presented to mother, Mrs. J. H. Spangler.

SPATARO, DOMINICO (1736380) _____
    Near Grand Pre, France, Oct. 25–26, 1918.
    R—Oswego, N. Y.
    B—Italy.
    G. O. No. 37, W. D., 1919.

Private, Company K, 311th Infantry, 78th Division.
Private Spataro, with hand grenades, broke up an enemy machine-gun nest and took 4 prisoners without assistance. He voluntarily acted as stretcher bearer for a period of 26 hours, performing valiant services until severely wounded.

SPATZ, CARL. _____
    During the St. Mihiel offensive, France, Sept. 26, 1918.
    R—Boyertown, Pa.
    B—Boyertown, Pa.
    G. O. No. 123, W. D., 1918.

Major, pilot, 3d Aero Squadron, Air Service.
Although he had received orders to go to the United States, he begged for and received permission to serve with a pursuit squadron at the front. Subordinating himself to men of lower rank, he was attached to a squadron as a pilot and saw conditions and arduous service through the offensive. As a result of his efficient work he was promoted to the position of flight commander. Knowing that another attack was to take place in the vicinity of Verdun, he remained on duty in order to take part. On the day of the attack west of the Meuse, while with his patrol over enemy lines, a number of enemy aircraft were encountered. In the combat that followed he succeeded in bringing down 2 enemy planes. In his ardor and enthusiasm he became separated from his patrol while following another enemy far beyond the lines. His gasoline giving out, he was forced to land and managed to land within friendly territory. Through these acts he became an inspiration and example to all men with whom he was associated.

SPAULDING, DAVID L. _____
    In the advance on Bouresches, France, June 6, 1918.
    R—Hood River, Oreg.
    B—Hood River, Oreg.
    G. O. No. 100, W. D., 1918.

Corporal, 79th Company, 6th Regiment, U. S. Marine Corps, 2d Division.
He returned to the front lines encouraging his men after being sent to the rear with a severe wound in the advance on Bouresches, France, on June 6, 1918.

*SPAUTZ, MATTHEW (90868) _____
    Near the River Ourcq, northeast of Chateau-Thierry, France, July 30, 1918.
    R—Dubuque, Iowa.
    B—Dubuque, Iowa.
    G. O. No. 102, W. D., 1918.

Sergeant, Company A, 168th Infantry, 42d Division.
During the advance of July 30, 1918, while in command of his platoon, Sergeant Spautz showed extraordinary heroism, leading his men on in the advance, having 3 times been knocked down by enemy shells. After having been wounded by machine-gun fire, he still continued to advance. He was finally killed while doing his utmost to advance.
Posthumously awarded. Medal presented to father, Michael Spautz.

SPEARS, GEORGE W. (1309476) _____
    Near Pouchaux, France, Oct. 7, 1918.
    R—Lenoir City, Tenn.
    B—London, Tenn.
    G. O. No. 37, W. D., 1919.

Corporal, Company L, 117th Infantry, 30th Division.
When part of the line had been halted by heavy fire from 3 machine-gun nests, Corporal Spears and Pvt. Thomas G. Cagle, armed only with rifles and bayonets, rushed the nearest hostile position, and, of the crew of 6, killed 3 and put the remainder to flight. Being unable to advance on 2 other guns because of their heavy fire, these 2 soldiers then opened fire with their rifles and forced the remainder of the crew of approximately 12 to abandon the position after 2 of their number had been killed and 2 wounded.

SPEER, CHARLES EDWARD _____
    Near Vierzy, France, July 18, 1918.
    R—Baltimore, Md.
    B—Pittsburgh, Pa.
    G. O. No. 9, W. D., 1923.

Captain, 9th Infantry, 2d Division.
While commanding the 1st Battalion, 9th Infantry, which was held up in crossing a deep ravine by a heavy rifle and machine-gun fire, Captain Speer, with utter disregard for his own safety, although previously wounded while capturing an enemy battery, made a reconnaissance along the front line under heavy rifle, machine-gun, and shellfire to the left flank, where he led his men to the attack, gained the ridge across the ravine, and made possible the advance of the entire line by enfilading the enemy's position. After gaining the assigned objective, he encouraged and reorganized his command, which had heavy losses, and made a personal reconnaissance under heavy fire in preparation for a further advance. While leading his battalion in the second attack he was severely wounded. Being unable to walk as a result of his wound, he ordered his men to push forward and remained alone as his men, inspired by his example, drove the enemy from their positions and continued to advance.

SPEERS, THOMAS G _____
    At Marcheville, France, Sept. 26, 1918.
    R—Montclair, N. J.
    B—Atlantic Highlands, N. J.
    G. O. No. 138, W. D., 1918.

First lieutenant, chaplain, 102d Infantry, 26th Division.
He accompanied the advance elements, which were constantly under terrific artillery and machine-gun fire during the action. He was continually aiding and cheering the wounded, and particularly distinguished himself by carrying a wounded officer to a dressing station through heavy artillery and machine-gun barrage.

SPENCER, EDWARD L _____
    North of Ardeuil, France, Sept. 30, 1918.
    R—Lenoir, N. C.
    B—Lenoir, N. C.
    G. O. No. 46, W. D., 1919.

Second lieutenant, 371st Infantry, 93d Division.
Having been wounded in the leg by machine-gun fire, he nevertheless continued to remain with his platoon, leading it successfully through an intense barrage of machine-gun and artillery fire to its position. He remained on duty with his command until 2 days later, when his regimental commander ordered him to the rear.

SPENCER, ERIC W. (1206280) _____
    Near St. Souplet, France, Oct. 17, 1918.
    R—New York, N. Y.
    B—Newfoundland.
    G. O. No. 37, W. D., 1919.

Sergeant, Machine Gun Company, 106th Infantry, 27th Division.
During the fording of the La Selle River and the heights beyond, he advanced against a nest of enemy snipers under heavy machine-gun and shellfire and by his courage and bravery succeeded in killing 4 of the enemy.

SPENCER, ERNEST_____
Near Thiaucourt, France, Sept. 12-15, 1918.
R—Toppenish, Wash.
B—Toppenish, Wash.
G. O. No. 46, W. D., 1919.

Private, 81st Company, 6th Machine Gun Battalion, U. S. Marine Corps, 2d Division.
Private Spencer repeatedly volunteered and carried messages through intense machine-gun and artillery fire, obtaining valuable information at critical moments.

SPENCER, GILBERT A. (40257)_____
Near Soissons, France, July 18, 1918.
R—Ionia, Mich.
B—Sailors Encampment, Mich.
G. O. No. 46, W. D., 1919.

First sergeant, Company K, 9th Infantry, 2d Division.
After being severely wounded and ordered to the rear by his commanding officer, Sergeant Spencer gathered together about 15 men who were retreating, took them back to the line, and turned them over to the commanding officer of his company.

SPENCER, JOHN D._____
At Fismes, France, Aug. 4, 1918.
R—Oshkosh, Wis.
B—Oshkosh, Wis.
G. O. No. 124, W. D., 1918.

First lieutenant, 127th Infantry, 32d Division.
While leading his company in the attack against Fismes he was knocked down and severely wounded by machine-gun fire. Without regard to his wounds he regained his feet and continued to lead his command until again severely wounded.

SPENCER, LORILLARD_____
In the Champagne sector, France, Sept. 26, 1918.
R—New York, N. Y.
B—New York, N. Y.
G. O. No. 37, W. D., 1919.

Major, 369th Infantry, 93d Division.
Commanding a battalion which was in action for the first time, Major Spencer inspired his men by his own coolness and courage under intense machine-gun fire. He continually exposed himself without regard for personal safety until he was wounded 6 times.

SPENCER, WILLIAM M._____
Near Villers-sur-Fere, France, July 28, 1918.
R—Erie, Pa.
B—Erie, Pa.
G. O. No. 99, W. D., 1918.

Second lieutenant, 165th Infantry, 42d Division.
He led his platoon in an attack which stormed and took the strongly prepared enemy positions on the heights north of the River Ourcq, near Villers-sur-Fere, France, on July 28, 1918. He maintained the position thus gained under a fire that lasted for 7 hours. During this entire time he continually circulated among his men, cheering them, and giving the wounded first aid. In order to reach and administer aid to his wounded captain, he passed without cover into an area which was under extremely heavy machine-gun fire and was himself wounded.

SPENCLEY, GEORGE H. (553657)_____
Near Cunel, France, Oct. 15, 1918.
R—Lovering, Mich.
B—Standish, Mich.
G. O. No. 27, W. D., 1920.

Corporal, Company A, 8th Machine Gun Battalion, 3d Division.
Although painfully wounded in the back by a piece of enemy shell and tagged for evacuation to the hospital, Corporal Spencley refused to be evacuated, but returned to his gun during 2 enemy counterattacks, during which he stopped by his fire an enemy attack which had reached within 40 yards of his position.

SPESSARD, RUTHERFORD H._____
Near Ville-Savoye, France, Aug. 6, 1918, and near Bois-du-Fays, France, Oct. 2, 1918.
R—Newcastle, Va.
B—Newcastle, Va.
G. O. No. 98, W. D., 1919.

Major, 58th Infantry, 4th Division.
During the crossing of the Vesle River Maj. Rutherford H. Spessard, when his battalion commander was killed, immediately assumed command of the battalion without orders and led them across the Vesle River against strongly fortified enemy positions, displaying absolute disregard for his personal danger. On Oct. 2, in the vicinity of the Bois-du-Fays, Major Spessard exposed himself to intense enemy artillery and machine-gun fire while making observations and directing the movement of his men. He established his battalion headquarters a short distance to the rear of his lines in a position continually subjected to severe enemy artillery fire.

*SPICKERMAN, RAYMOND H. (1213385)__
Near Ronssoy, France, Sept. 29, 1918.
R—Bloomville, N. Y.
B—Bloomville, N. Y.
G. O. No. 50, W. D., 1919.

Corporal, Machine Gun Company, 107th Infantry, 27th Division.
He and his machine gunner pushed forward to a blind trench, which was partially surrounded by machine gunners and snipers, under terrific machine-gun and trench-mortar fire and through a heavy smoke screen. He barricaded a sap at the most dangerous position only a few yards from the enemy machine guns, and after killing 4 of the enemy with a rifle was mortally wounded, but continued to hold his position until he died.
Posthumously awarded. Medal presented to father, Herman Spickerman.

*SPINNEY, GEORGE F. (60208)_____
North of Verdun, France, Oct. 27, 1918.
R—Faneuil, Mass.
B—Brighton, Mass.
G. O. No. 21, W. D., 1919.

Corporal, Company C, 101st Infantry, 26th Division.
While advancing with the first wave, Corporal Spinney, with another soldier, attacked a machine-gun nest and killed 2 of the crew. While attempting to capture the remainder of the crew this gallant soldier was himself killed.
Posthumously awarded. Medal presented to father, Freeman Spinney.

SPITZNAGEL, CHARLES (794968)_____
In the Bois-des-Rappes, France, Oct. 21, 1918.
R—Cincinnati, Ohio.
B—Cincinnati, Ohio.
G. O. No. 35, W. D., 1919.

Corporal, Company C, 15th Machine Gun Battalion, 5th Division.
He displayed utter disregard for his personal safety in the attack on the Rappes, when his gunner was severely wounded and his leader killed. He then fired the gun himself until he was seriously wounded, when he refused to be evacuated, but remained with his crew, encouraging them and directing their fire until relieved 2 hours later.

SPIVEY, FRED F. (731815)_____
Near Romagne, France, Oct. 14-18, 1918.
R—Lexington, Ky.
B—Booneville, Ky.
G. O. No. 37, W. D., 1919.

Sergeant, Company B, 6th Infantry, 5th Division.
He set a splendid example to his men while in command of a platoon under severe machine-gun fire, personally capturing 2 machine guns.

SPRAGUE, ALMON E. (2192613)_____
Near Tailly, France, Nov. 4, 1918.
R—Platte, S. Dak.
B—Fayette, Iowa.
G. O. No. 87, W. D., 1919.

Private, medical detachment, 355th Infantry, 89th Division.
Under heavy artillery and machine-gun fire he exposed himself fearlessly on the battle field to give first aid to the wounded, showing marked personal valor. When his bandages were expended, he obtained a fresh supply and, under the continuous fire of a sniper, went to the assistance of 20 wounded men, bound up their wounds, and saw that the more serious cases were first carried from the field.

SPRAGUE, CHANDLER. | First lieutenant, 115th Infantry, 29th Division.
Near Balschwiller, Alsace, Aug. 31, 1918.
R—Baltimore, Md.
B—Haverhill, Mass.
G. O. No. 100, W. D., 1918. | Upon returning from a raid which he led against enemy trenches, Lieutenant Sprague found 1 of his men was missing. Accompanied by 1 man, he promptly and voluntarily returned through artillery, machine-gun, and rifle fire, found the missing man, who had been wounded, and carried him back to the American lines.

*SPRINGER, FRANK (653919). | Private, Company C, 1st Engineers, 1st Division.
Near Soissons, France, July 20, 1918.
R—Aurora, Ill.
B—Sheboygan Falls, Wis.
G. O. No. 124, W. D., 1918. | When volunteers were called for by his company commander, Private Springer volunteered and rescued wounded comrades from a barrage. Disregarding danger to himself, he continued the performance of these heroic deeds until killed.
Posthumously awarded. Medal presented to mother, Mrs. Margaret Springer.

SPRINGS, ELLIOTT WHITE. | First lieutenant, 148th Aero Squadron, Air Service.
Near Bapaume, France, Aug. 22, 1918.
R—Lancaster, S. C.
B—Lancaster, S. C.
G. O. No. 23, W. D., 1919. | Attacking 3 enemy planes (type Fokker) who were driving on 1 of our planes, Lieutenant Springs, after a short and skillful flight, drove off 2 of the enemy and shot down the third. On the same day he attacked a formation of 5 enemy planes (type Fokker) and, after shooting down 1 plane, was forced to retire because of lack of ammunition.

SPRINGS, WILLIAM H. (42618). | Sergeant, Company E, 16th Infantry, 1st Division.
South of Soissons, France, July 18, 1918.
R—Madison County, N. C.
B—Greenville, S. C.
G. O. No. 35, W. D., 1920. | After his platoon commander had been wounded Sergeant Springs took command, reorganized, and led the platoon forward through heavy fire to all its objectives, in which attack he was severely wounded. He also rendered gallant service before Montdidier and St. Mihiel, during both of which operations he was wounded.

SPROUSE, ROBERT (545179). | First sergeant, Company A, 30th Infantry, 3d Division.
Near Crezancy, France, July 15–16, 1918.
R—Clinton, Mo.
B—Alexander, N. C.
G. O. No. 32, W. D., 1919. | Throughout the engagement he encouraged his men by his gallant conduct. After the company was ordered to withdraw, he voluntarily returned to the position his company had held and, throughout the night of July 15–16, assisted in evacuating the wounded.

SRYGLEY, ELAM F. | First lieutenant, Medical Corps, attached to 4th Machine Gun Battalion, 2d Division.
Near Medeah Ferme, France, Oct. 8–9, 1918.
R—Nashville, Tenn.
B—Lebanon, Tenn.
G. O. No. 21, W. D., 1919. | When a platoon was being heavily gassed and under intense artillery and machine-gun fire, he voluntarily left the shelter of his dressing station, proceeded to the line, and rendered invaluable aid to the wounded. On Oct. 9, he again left the shelter of his dressing station, and under intense fire voluntarily went to the assistance of the wounded of the 141st Infantry.

STACKPOLE, EDWARD J., Jr. | Captain, 110th Infantry, 28th Division.
Near Baslieux, France, Aug. 24, 1918.
R—Harrisburg, Pa.
B—Harrisburg, Pa.
G. O. No. 71, W. D., 1919. | Directed to advance to a new position, he led his men forward with great gallantry. Although painfully wounded in the back and leg by shell fragments, he remained on duty with his men, inspiring them by his courage and coolness to hold a difficult position against repeated attacks by the enemy in force for a period of 24 hours.

STADIE, HERMAN EDWARD. | Captain, 306th Infantry, 77th Division.
At Ferme des Dames, west of Fismes, Aug. 20, 1918.
R—New York, N. Y.
B—Germany.
G. O. No. 99, W. D., 1918. | While the vicinity of the regimental command post where he was stationed was under heavy bombardment, Captain Stadie, without thought of personal danger, voluntarily ran outside, through shrapnel and high-explosive shells, and rescued a wounded runner.

*STAEHELI, OTTO. | First lieutenant, 7th Infantry, 3d Division.
North of Cunel, France, Oct. 12, 1918.
R—Chicago, Ill.
B—Chicago, Ill.
G. O. No. 24, W. D., 1920. | Lieutenant Staeheli personally led a platoon in the attack of Hill 258. He rushed 25 yards ahead of his platoon and single handed captured 3 of the enemy. His platoon, inspired by his deeds, succeeded in forcing 63 others to surrender.
Posthumously awarded. Medal presented to mother, Mrs. Emma Staeheli.

STAFFORD, CHARLES (93392). | Sergeant, Company D, 166th Infantry, 42d Division.
In the St. Mihiel offensive, France, Sept. 12, 1918.
R—Marion, Ohio.
B—Marion, Ohio.
G. O. No. 21, W. D., 1919. | Personally reconnoitering an enemy position, Sergeant Stafford encountered and captured single handed 6 of the enemy.

STAFFORD, THOMAS J. (275221). | Private, first class, Company L, 127th Infantry, 32d Division.
Near Juvigny, France, Sept. 1, 1918.
R—Rhinelander, Wis.
B—Canada.
G. O. No. 98, W. D., 1919. | Locating an enemy machine-gun nest, Private Stafford, upon his own initiative, organized a patrol and led it in an attack on the hostile position, silencing the gun and capturing 18 prisoners, thereby facilitating the advance of his company.

STAINS, TRACY B. | Second lieutenant, 3d Machine Gun Battalion, 1st Division.
Near Berzy-le-Sec, France, July 18, 1918.
R—Chicago, Ill.
B—Falls City, Nebr.
G. O. No. 99, W. D., 1918. | After being severely wounded at the beginning of the engagement near Berzy-le-Sec, France, July 18, 1918, he continued to lead his command forward until he had taken positions assigned to him. He declined medical assistance and did not retire to an aid station until he had seen to the disposition and security of his men.

*STAINTON, MARVIN E. | Second lieutenant, 28th Infantry, 1st Division.
Near Verdun, France, Oct. 9, 1918.
R—Laurel, Miss.
B—Laurel, Miss.
G. O. No. 37, W. D., 1919 | While his battalion was being held up by heavy machine-gun fire he voluntarily led a small detachment of his platoon forward. He advanced far into enemy territory and succeeded in capturing 7 machine-gun nests and 47 prisoners, continuing with his mission until he was killed.
Posthumously awarded. Medal presented to mother, Mrs. L. Stainton.

STAIR, WILLETT A._____
Near Torcy, France, June 4, 1918.
R—Minneapolis, Minn.
B—Bristol, S. Dak.
G. O. No. 119, W. D., 1918.

Private, Headquarters Company, 6th Regiment, U. S. Marine Corps, 2d Division.
Near Torcy, France, on the night of June 4, 1918, as a member of a raiding patrol, he displayed great courage and devotion by fearlessly entering extremely dangerous areas and obtaining information imperatively necessary to the success of subsequent operations.

STALCUP, JAMES (1314522)_____
Near La Haie, France, Oct. 17, 1918.
R—Hartsville, Tenn.
B—Hartsville, Tenn.
G. O. No. 39, W. D., 1920.

Private, Company C, 114th Machine Gun Battalion, 30th Division.
During the attack of the enemy position, Private Stalcup, although wounded in the shoulder by a shell fragment, continued to go forward with his section for 7 hours until severely wounded by a trench-mortar shell. Due to his second wound, he lost his left arm. The courage and fortitude displayed by Private Stalcup enabled his section to reach its objective with all its guns.

STAMBAUGH, ISABELLE_____
In front of Amiens, France, Mar. 21, 1918.
R—Philadelphia, Pa.
B—Mifflintown, Pa.
G. O. No. 70, W. D., 1919.

Reserve nurse, Base Hospital No. 10, Army Nurse Corps.
While with a surgical team at a British casualty clearing station during the big German drive of March 21, 1918, in front of Amiens, France, she was seriously wounded by shellfire from German aeroplanes.

STAMPS, BERNICE B_____
Near Jaulny, France, Sept. 13-15, 1918.
R—New Hebron, Miss.
B—New Hebron, Miss.
G. O. No. 46, W. D., 1919.

Chief pharmacist's mate, U. S. Navy, attached to 6th Machine Gun Battalion, U. S. Marine Corps, 2d Division.
Working continually without rest or food, he cared for the wounded under most hazardous conditions. When a counterattack by the enemy seemed imminent, the medical detachment was ordered to the rear, but he willingly stayed with the wounded and assisted greatly in their evacuation.

STANFIELD, LAWRENCE (1318597)_____
Near Bellicourt, France, Sept. 28, 1918.
R—Durham, N. C.
B—Alamance County, N. C.
G. O. No. 133, W. D., 1918.

Color sergeant, Headquarters Company, 120th Infantry, 30th Division.
While attached to the regimental intelligence service he was severely gassed, but after receiving first-aid treatment, he insisted on returning to duty. Gassed a second time and relieved for a short period, he personally made a search for wounded men, and, finding a large number, went to the aid station and brought stretcher bearers. He continued this work until he was blinded by the effects of the gas.

STANKUNOS, BENJAMIN G. (2301304)____
Near Verdun, France, Nov. 5, 1918.
R—Shamokin, Pa.
B—Shamokin, Pa.
G. O. No. 44, W. D., 1919.

Private, Company B, 310th Machine Gun Battalion, 79th Division.
With 2 other soldiers, Private Stankunos voluntarily left a place of safety, went forward 40 meters under machine-gun fire in plain view of the enemy, and rescued another soldier who had been blinded by a machine-gun bullet and was helplessly staggering about.

STANTON, CHARLES, Jr. (1215748)_____
Near St. Souplet, France, Oct. 15, 1918.
R—Elmira, N. Y.
B—Troy, Pa.
G. O. No. 37, W. D., 1919.

Corporal, Company L, 108th Infantry, 27th Division.
Accompanied by an officer and 3 other soldiers, he made a reconnaissance of the River La Selle, the journey being made under constant and heavy machine-gun fire. To secure the desired information it was necessary to wade the stream for the entire distance.

STAPLETON, GORDON_____
Near Vilosnes, France, Nov. 6, 1918.
R—Ennis, Tex.
B—Prairie Hill, Mo.
G. O. No. 37, W. D., 1919.

Second lieutenant, 6th Infantry, 5th Division.
Entering Vilosnes with a patrol, he encountered heavy enemy machine-gun fire. He attacked and killed 4 Germans, took 4 prisoners, and forced those remaining to take shelter in a trench. Although greatly outnumbered, he held his position while the French crossed the River Meuse and took many prisoners.

STAPLETON, WILLIAM A. (581321)_____
Near Soissons, France, July 19, 1918.
R—Rush, Ky.
B—Rush, Ky.
G. O. No. 117, W. D., 1918.

Corporal, Company I, 28th Infantry, 1st Division.
With the aid of 2 men, he charged and captured a German machine gun.

STARK, ALEXANDER N., Jr_____
Near Cote St. Germain, France, Nov. 7, 1918.
R—Fortress Monroe, Va.
B—Fort Sam Houston, Tex.
G. O. No. 46, W. D., 1919.

Major, 61st Infantry, 5th Division.
His battalion being stopped by machine-gun fire in the attack of Nov. 7, 1918, Major Stark personally led it in a renewed attack, and thus succeeded in gaining the Cote St. Germain. Major Stark personally captured a machine gun and 13 prisoners, his personal example of fearlessness encouraging his men to advance against odds.

STARKEY, JOSEPH W_____
Near Medeah Ferme, France, Oct. 8, 1918.
R—Chattanooga, Tenn.
B—Tuscumbia, Ala.
G. O. No. 21, W. D., 1919.

First lieutenant, 9th Infantry, 2d Division.
Wounded, but regardless of danger to himself, he led his men through heavy machine-gun and artillery fire in an attack overwhelmingly successful, in which he received a second wound.

STARLINGS, PAUL N_____
At Berzy-le-Sec, France, July 21, 1918.
R—Annapolis, Md.
B—Leitchs, Md.
G. O. No. 132, W. D., 1918.

Captain, 26th Infantry, 1st Division.
In spite of the fact that returning wounded men informed that it was impossible to take Berzy-le-Sec, France, July 21, 1918, he led his company forward with courage and determination under steady fire, and thereby gave invaluable aid in the assault in which he knew his command was the last reserve.

STARR, CHARLIE L. (551885)_____
Near Romagne, France, Oct. 9, 1918.
R—Lamoille, Ill.
B—Lamoille, Ill.
G. O. No. 35, W. D., 1920.

Sergeant, Company I, 38th Infantry, 3d Division.
During the attack on Hill 253 Sergeant Starr was wounded in the arm by a machine-gun bullet. Disregarding his wound, he continued to lead his platoon through heavy machine-gun and artillery fire, until he was wounded a second time.

*STATHAM, GEORGE B. (41442)_____
  Near Tuilerie Ferme, France, Nov. 4, 1918.
  R—Cordele, Ga.
  B—Selma, Ala.
  G. O. No. 37, W. D., 1919.

Private, first class, Machine Gun Battalion, 9th Infantry, 2d Division.
Although he was the only remaining member of his gun crew, he courageously operated his gun until he had put an enemy machine-gun nest out of action. He continued with his heroic work until he was killed.
Posthumously awarded. Medal presented to father, Thomas S. Statham.

STAVROULAKIS, EMMANUEL (42329)_____
  South of Soissons, France, July 18, 1918.
  R—Estherville, Iowa.
  B—Greece.
  G. O. No. 35, W. D., 1920.

Corporal, Company C, 16th Infantry, 1st Division.
After its leader had been killed, Corporal Stavroulakis took command of a patrol and led it in an attack on a machine-gun position and captured the crew.
Oak-leaf cluster.
A bronze oak-leaf cluster is awarded to Corporal Stavroulakis for the following act of extraordinary heroism in action south of Soissons, France, July 19, 1918: After 2 others had been killed in the attempt, he carried an important message through heavy fire to battalion headquarters. Upon his return he led a patrol in attack on an enemy machine-gun position. Although wounded in both legs, he continued in the attack until the enemy machine gun was captured.

STAVRUM, EDWIN R._____
  West of Chateau-Thierry, France, June 6, 1918.
  R—La Crosse, Wis.
  B—La Crosse, Wis.
  G. O. No. 27, W. D., 1920.

First lieutenant, 23d Infantry, 2d Division.
Lieutenant Stavrum was severely wounded in the left shoulder during the first phase of the attack. In spite of his wound he conducted his platoon to its objective and exposed himself to heavy fire in order to organize his position for defense.

STEARNS, DAVE W. (563697)_____
  Near St. Thibaut, France, Aug. 6, 1918.
  R—Portland, Oreg.
  B—Waldport, Oreg.
  G. O. No. 46, W. D., 1919.

Corporal, Company C, 4th Engineers, 4th Division.
He was a member of a platoon ordered to precede the Infantry to construct footbridges across the Vesle River. Enemy sniper, machine-gun, and artillery fire was so intense that 4 attempts of his platoon failed. Acting upon his own initiative, he made his way along the river in the face of the deadly fire, and for 1 hour reconnoitered the enemy's positions, reporting back to his commanding officer with information of the greatest value.

STEEDE, WALTER J._____
  In the Forest de Fere, near Nesles, northeast of Chateau-Thierry, France, July 26 to Aug. 2, 1918.
  R—Grand Rapids, Mich.
  B—Grand Rapids, Mich.
  G. O. No. 102, W. D., 1918.

Private, Company E, 168th Infantry, 42d Division.
For extraordinary heroism in action in the Forest de Fere, near Nesles, northeast of Chateau-Thierry, France, July 26 to Aug. 2, 1918, during the advance of his regiment in the Forest de Fere, by his voluntary, authorized, and untiring efforts in carrying in the wounded, both by day and by night, under the most severe and dangerous circumstances, and especially when the town of Sergy was under heavy bombardment, July 29–31, 1918.

STEELE, FRANK S. (96135)_____
  Near Beuvardes, France, July 26, 1918.
  R—Abernant, Ala.
  B—Rising Fawn, Ga.
  G. O. No. 53, W. D., 1920.

Corporal, Company C, 167th Infantry, 42d Division.
Corporal Steele, although severely wounded in the right shoulder, continued to lead his squad forward through heavy machine-gun fire. Later with a few others took up a position in a sunken road and repulsed an enemy counterattack. His devotion to duty was an excellent example to the entire command.

STEELE, RICHARD WILSON_____
  Near Bois-de-Barricourt, France, Oct. 23, 1918.
  R—Oak Park, Ill.
  B—Omaha, Nebr.
  G. O. No. 7, W. D., 1919.

Second lieutenant, observer, 166th Aero Squadron, Air Service.
While on a bombing raid back of the German line he, accompanied by his pilot, was attacked by 6 German pursuit planes. They were forced to leave the formation in which they were traveling, owing to engine trouble; the enemy began riddling their plane with machine-gun fire. Lieutenant Steele fought them on all sides, and is credited by members of the 11th Aero Squadron, who were flying over him several thousand feet, with having brought down one of his opponents. He was wounded twice in the leg and twice in the arm and continued fighting, although each time he was hit he was knocked down into the observer's cockpit. At last, however, only his tail gun was in working condition, the other 2 having been disabled by bullets, and he sank unconscious into the cockpit.

STEELE, WALTER P. (43100)_____
  Near Soissons, France, July 18, 1918.
  R—Rosiclare, Ill.
  B—Paducah, Ky.
  G. O. No. 15, W. D., 1919.

Corporal, Company G, 16th Infantry, 1st Division.
He advanced alone upon an enemy machine-gun nest, shot 3 of the crew and continued his efforts to silence the guns until he was wounded.

*STEGAR, BERNARD A. (2219319)_____
  Near St. Souplet, France, Oct. 18, 1918.
  R—Marlin, Tex.
  B—Marlin, Tex.
  G. O. No. 71, W. D., 1919.

Private, Company F, 107th Infantry, 27th Division.
When a comrade was severely wounded by machine-gun fire, he went to his rescue, crossing open spaces subjected to intense fire by the enemy. He was wounded as he advanced, but he continued forward and reached the side of his wounded comrade, when he was again hit by a machine-gun bullet and instantly killed.
Posthumously awarded. Medal presented to father, Mike Stegar.

STEIMEL, WILLIAM J. (156875)_____
  Near Bois-de-Belleau, France, June 12, 1918.
  R—Debow, Ark.
  B—Tipton, Mo.
  G. O. No. 37, W. D., 1919.

Private, Company D, 2d Engineers, 2d Division.
Although wounded in several places by an enemy hand grenade, he refused to go to the rear until his mission was completed. After receiving first aid he again returned to the front line, although the entire line was at that time being subjected to a severe shelling.

STEIN, FRED C. (262506)_____
  Near Romagne, France, Oct. 9, 1918.
  R—Atlanta, Mich.
  B—Coleman, Mich.
  G. O. No. 44, W. D., 1919.

Corporal, Company F, 125th Infantry, 32d Division.
Corporal Stein charged and captured a strong enemy machine-gun nest and immediately turned the gun on the enemy. He was twice wounded while changing the position of the gun, but continued to operate it under heavy machine-gun fire until he received a third wound in the arm, which made it impossible for him further to operate the gun.

STEINER, GEORGE C. (2412982)_____
    Near Blanc Mont Ridge, France,
    Oct. 3, 1918.
    R—Cleveland, Ohio.
    B—Long Island City, N. Y.
    G. O. No. 21, W. D., 1919.

Sergeant, Company C, 9th Infantry, 2d Division.
Sergeant Steiner, severely wounded on the battle line with his company, remained on duty until the objective was gained and the position consolidated.

STEINER, JOHN JEFFERSON FLOWERS___
    Near Medeah Ferme, France, Oct. 9, 1918.
    R—Montgomery, Ala.
    B—Montgomery, Ala.
    G. O. No. 37, W. D., 1919.

Major, 2d Engineers, 2d Division.
In command of his battalion in the front line, Major Steiner personally conducted a reconnaissance. Exposed to enemy fire, he obtained valuable information after other patrols had failed. An attack was then organized in which 2 machine guns were captured and a dangerous salient eliminated.

STEINHILBER, CLOYD W._____
    Near Barricourt, France, Nov. 1-2, 1918.
    R—Highland Park, Mich.
    B—Seward, Nebr.
    G. O. No. 37, W. D., 1919.

First lieutenant, 354th Infantry, 89th Division.
When his company commander was seriously wounded he took an automatic rifle from a dead soldier and held off the enemy for 2 hours, defending the wounded officer until assistance could be obtained. Next day while in command of his company he was himself wounded, but continued to urge his men forward, inspiring them by his example under heavy machine-gun fire.

STEININGER, ROY H. (5815)_____
    Near Medeah Ferme, France, Oct. 4, 1918.
    R—Chester, Pa.
    B—Mifflintown, Pa.
    G. O. No. 37, W. D., 1919.

Private, Company C, 9th Infantry, 2d Division.
After several men of his company had been wounded he repeatedly left cover, exposing himself in an open field to enemy machine-gun fire in order to bring in wounded and administer first aid.

STEINKRAUS, HERMAN W._____
    Near Bois-de-Bantheville, France, Oct. 15, 1918.
    R—Cleveland, Ohio.
    B—Cleveland, Ohio.
    G. O. No. 66, W. D., 1919.

First lieutenant, 127th Infantry, 32d Division.
Continuing in command of his company, after he had been instructed to go to the rear for treatment for an infected leg, Lieutenant Steinkraus skillfully extricated his company with few casualties, when it became suddenly exposed to intense machine-gun fire from both flanks. Reinforcing his command with stragglers, he organized a strong right flank guard by utilizing captured German machine guns and succeeded in maintaining his position.

STEMBRIDGE, ROGER W._____
    Near Vieville-en-Haye, France, Oct. 31, 1918.
    R—Milledgeville, Ga.
    B—Baldwin County, Ga.
    G. O. No. 35, W. D., 1919.

First lieutenant, 21st Machine Gun Battalion, 7th Division.
Although wounded by a shell fragment and suffering from the effects of an antitetanic serum, Lieutenant Stembridge continued to lead his platoon through the night of Oct. 31 and the offensive operation of Nov. 1 under heavy enemy shellfire, encouraging his men by his gallant conduct.

STENSETH, MARTINUS_____
    Over the Argonne Forest, France, Oct. 22, 1918.
    R—Minneapolis, Minn.
    B—Minneapolis, Minn.
    G. O. No. 9, W. D., 1923.

First lieutenant, 28th Aero Squadron, Air Service.
For extraordinary heroism in action Oct. 22, 1918, over the Argonne Forest when he went to the rescue of a French plane attacked by 6 enemy Fokker planes with 12 additional enemy planes hovering in reserve. Attacking the enemy with vigor, single handed, he drove down and destroyed 1 enemy plane and put to flight the remainder. His gallant act in the face of overwhelming odds proved an inspiration to the men of his squadron.

*STENSSON, CARL H._____
    Near St. Etienne, France, Oct. 3, 1918.
    R—Framingham, Mass.
    B—Framingham, Mass.
    G. O. No. 44, W. D., 1919.

Private, 18th Company, 5th Regiment, U. S. Marine Corps, 2d Division.
He displayed great courage in serving as a stretcher bearer during the operations at Blanc Mont Ridge. When his helper was wounded he went into an open road swept by machine-gun fire to rescue him and was killed in his self-sacrificing attempt.
Posthumously awarded. Medal presented to mother, Mrs. Christine E. Stensson.

STEPHENS, JOSEPH W. G._____
    Near Soissons, France, July 19, 1918.
    R—Wicomico Church, Va.
    B—Wicomico Church, Va.
    G. O. No. 132, W. D., 1918.

Captain, 26th Infantry, 1st Division.
When necessity arose for a company to advance to an important position in the fighting near Soissons, France, July 19, 1918, he led his command through a heavily shelled area with conspicuous bravery, reached his objective, and directed his men to a successful attack, until so seriously wounded as to necessitate his evacuation.

STEPHENSON, CARLTON (1319639)_____
    Near Catillon, France, Oct. 18, 1918.
    R—Clayton, N. C.
    B—Johnston County, N. C.
    G. O. No. 35, W. D., 1919.

Corporal, Company B, 120th Infantry, 30th Division.
Severely wounded, he remained with his automatic rifle section in an exposed position, covering the withdrawal of his company. Although almost surrounded, he inflicted severe losses on the enemy and held his position throughout the day.

STEPHENSON, CHARLES F. (1329349)_____
    At Bellicourt, France, Sept. 29, 1918.
    R—Rocky Mount, N. C.
    B—Johnston County, N. C.
    G. O. No. 98, W. D., 1919.

Corporal, Company D, 105th Engineers, 30th Division.
As Corporal Stephenson and his squad were engaged in planking over a shell hole, they were fired on from the side. Locating the course of the fire by a flash, he attacked the enemy position with his rifle, killing 1 German, taking 2 prisoners, and clearing the adjacent shell holes. His quick initiative and bravery saved the lives of his men and prevented an interruption of their work.

STERN, HENRY R._____
    Near Vieville-en-Haye, France, Sept. 20, 1918.
    R—New York, N. Y.
    B—New York, N. Y.
    G. O. No. 2, W. D., 1920.

First lieutenant, 311th Infantry, 78th Division.
During an enemy attack on the position held by his platoon, after being severely wounded in the leg by a machine-gun bullet, he remained in a position subjected to heavy machine-gun fire, and refused to accept aid until after the attack had been repulsed. His deed greatly encouraged his men.

*STEVENS, HARRY A_____
Near Somme-Py, France, Oct. 3, 1918.
R—Brooklyn, N. Y.
B—Rutherford, N. J.
G. O. No. 37, W. D., 1919.

Second lieutenant, 5th Machine Gun Battalion, 2d Division.
While leading his platoon in attack, Lieutenant Stevens fell mortally wounded, but refused to be taken to the dressing station until he had directed the advance of his platoon and assured himself that it would not be checked. Posthumously awarded. Medal presented to mother, Mrs. Phoebe J. Stevens.

*STEVENS, JOHN H_____
Near Chateau-Thierry, France, July 2, 1918.
R—Albion, N. Y.
B—Lynchburg, Va.
G. O. No. 1, W. D., 1919.

Second lieutenant, 147th Aero Squadron, Air Service.
He, with 4 other pilots, attacked 12 enemy aircraft (type Pfalz), flying in 2 groups well within the enemy lines. As soon as the enemy planes were sighted he maneuvered to get between them and the sun, and with great difficulty gained the advantage. While 3 of the other American officers dived on the lower formation, he and Second Lieut. Kenneth L. Porter engaged the upper formation in a bold and brilliant combat, 2 planes of which they crashed to the earth.
Posthumously awarded. Medal presented to mother, Mrs. Effie Stevens.

STEVENS, LEVI_____
Near Romagne, France, Oct. 9, 1918.
R—Alpena, Mich.
B—Trenton, N. J.
G. O. No. 37, W. D., 1919.

First lieutenant, 125th Infantry, 32d Division.
Commanding a small detachment, he charged and captured a strong enemy machine-gun nest, his personal activity and courage aiding greatly in the success of the exploit. Although wounded and under heavy fire, he organized a position from which his detachment could effectively turn the captured gun on the enemy.

STEVENS, MATT (263344)_____
Throughout the advance across the River Ourcq and to Fismes, France, on the south bank of the Vesle River, July 31 to Aug. 8, 1918.
R—St. Johns, Mich.
B—St. Johns, Mich.
G. O. No. 117, W. D., 1918.

Private, Company E, 125th Infantry, 32d Division.
Private Stevens was a runner for his company and was engaged day and night in carrying messages throughout machine-gun and artillery fire. He did his work without fear or hesitation, thereby keeping constant liaison with higher authority. During times not so occupied he administered aid to the wounded, crawling to stricken comrades at imminent risk of his own life, through areas swept by machine-gun fire. Through disregard of danger he was the means of saving many wounded men.

*STEVENSON, ALFRED (1243679)_____
Near Fismette, France, Aug. 10, 1918.
R—Chester, Pa.
B—Chester, Pa.
G. O. No. 37, W. D., 1919.

Sergeant, Company C, 111th Infantry, 28th Division.
Sergeant Stevenson and another soldier voluntarily went through heavy machine-gun fire to carry an important message to an advanced unit. Attracted by the cries of a wounded soldier while they were returning they went to his assistance, and, in doing so, Sergeant Stevenson was mortally wounded.
Posthumously awarded. Medal presented to widow, Mrs. Doris Stevenson.

*STEVENSON, JENS L. (1926693)_____
At Bois-des-Ogons, France, Oct. 6, 1928.
R—Pittsburgh, Pa.
B—Ephraim, Utah.
G. O. No. 2, W. D., 1926.

Corporal, Company F, 319th Infantry, 80th Division.
He voluntarily left shelter and crawled in the open under heavy machine-gun fire to the aid of a wounded soldier. While trying to dress the latter's injuries he was killed by a machine-gun bullet.
Posthumously awarded. Medal presented to father, John G. Stevenson.

STEVENSON, MAURICE S._____
Near Exermont, France, Oct. 9, 1918.
R—Kansas City, Mo.
B—Milwaukee, Wis.
G. O. No. 128, W. D., 1918.

Second lieutenant, 16th Infantry, 1st Division.
He displayed splendid devotion to duty by twice passing through a terrific artillery and machine-gun barrage in order to transmit important orders from his brigade commander to the assaulting battalion, and while in the performance of such duty was seriously wounded, but refused to be evacuated before he had made his report.

*STEWART, ALPHEUS E. (2241347)_____
East of Ronssoy, France, Sept. 29, 1918.
R—San Antonio, Tex.
B—Fairview, Tex.
G. O. No. 16, W. D., 1919.

Private, Company G, 107th Infantry, 27th Division.
Private Stewart, having been wounded in the head, advanced with fearless disregard for his own personal safety against an enemy machine-gun nest and succeeded in putting it out of action by bombing the gunners. He was killed immediately thereafter by an enemy machine-gun fire.
Posthumously awarded. Medal presented to father, John H. Stewart.

STEWART, BERT L. (40851)_____
Near Medeah Ferme, France, Oct. 3, 1918.
R—Idaville, Ind.
B—Osgood, Ind.
G. O. No. 21, W. D., 1919.

Corporal, Company M, 9th Infantry, 2d Division.
After his officer had been wounded by shellfire, Corporal Stewart, suffering from seven wounds about the knee, from the same shell, took command of his platoon, led it in the assault to the objective, and established it in line.

STEWART, CLARENCE L. (553370)_____
At Chateau-Thierry, France, May 31 to June 4, 1918.
R—Verona, Pa.
B—Wilkinsburg, Pa.
G. O. No. 15, W. D., 1919.

Private, Company B, 7th Machine Gun Battalion, 3d Division.
As a motorcycle rider he worked steadily for 24 hours without rest. He was struck by shrapnel, which wounded him in the neck, back, and in both legs, but he continued on duty in spite of these injuries.

STEWART, DAVID B., Jr. (57297)_____
On Hill 240, near Exermont, France, Oct. 5, 1918.
R—Rochester, N. Y.
B—Gloucester, Mass.
G. O. No. 66, W. D., 1920.

Corporal, Company D, 28th Infantry, 1st Division.
Corporal Stewart exposed himself to heavy fire to advance 100 yards in advance of his company in order to locate enemy machine-gun positions. Coming suddenly upon an enemy machine-gun nest, he forced 12 of the enemy to surrender and then continued his reconnaissance.

*STEWART, GEORGE L. (2339380)_____
Near Nesles, France, July 14–15, 1918.
R—Bluefield, W. Va.
B—Otey, Va.
G. O. No. 32, W. D., 1919.

Sergeant, Company I, 4th Infantry, 3d Division.
After he had seen several of his comrades fall in the attempt to accomplish the mission, Sergeant Stewart volunteered to carry an important message through a heavy shell and gas bombardment.
Posthumously awarded. Medal presented to father, John Stewart.

*STEWART, KIRBY P._____
  Near Chatel-Chehery, France, Oct.
    8, 1918.
  R—Bradentown, Fla.
  B—Lake City, Fla.
  G. O. No. 71, W. D., 1919.

Second lieutenant, 328th Infantry, 82d Division.
Leading his platoon in an attack through an open valley which was swept by enemy machine-gun fire from both flanks, he displayed marked heroism in continuing in command of his men after being himself severely wounded, inspiring them by his courage till he fell mortally wounded by a second machine-gun bullet.
Posthumously awarded. Medal presented to mother, Mrs. J. M. Stewart.

STEWART, MALLEY (1570692)_____
  Near Bussy Farm, France, Sept. 29,
    1918.
  R—Columbia, S. C.
  B—Fort Motte, S. C.
  G. O. No. 37, W. D., 1919.

Private, Headquarters Company, 371st Infantry, 93d Division.
Although severely wounded, he continued to carry telephone material forward through a heavy barrage for several hours until overcome by loss of blood and weakness.

STEWART, WARREN C. (1236772)_____
  Near Verdun, France, Oct. 10, 1918.
  R—Baltimore, Md.
  B—Washington, D. C.
  G. O. No. 27, W. D., 1919.

Private, sanitary detachment, 115th Infantry, 29th Division.
In the Bois-de-Montagne, east of the Meuse, he voluntarily, and at the risk of his life, walked through an opening under direct machine-gun fire to administer first aid to the wounded in an advanced post. During the entire offensive his conduct was instrumental in maintaining the morale of the troops to which he was attached.

STICKLES, HARVEY (92284)_____
  Northeast of Chateau-Thierry,
    France, July 29, 1918.
  R—Akron, Ohio.
  B—Waynesburg, Pa.
  G. O. No. 108, W. D., 1918.

Private, Headquarters Company, 166th Infantry, 42d Division.
Several times during the night of July 29, 1918, and 3 times during the afternoon of July 30, 1918, he left the shelter of battalion headquarters and went out into an intense bombardment to repair telephone lines connecting battalion and regimental command posts. He had several narrow escapes and was once thrown to the ground by the burst of an exploding shell, yet he continued to perform his important work after others had failed.

*STIER, VICTOR (2041538)_____
  At Nijnigora, Russia, Jan. 19, 1919.
  R—Detroit, Mich.
  B—Cincinnati, Ohio.
  G. O. No. 16, W. D., 1920.

Private, Company A, 339th Infantry, 85th Division (detachment in North Russia).
When his detachment was almost surrounded, Private Stier went forward to a machine-gun position and opened fire on the enemy, checking their advance and allowing his comrades to withdraw. He was wounded in the jaw with a rifle bullet. He continued to fire at the enemy until ordered to the rear. He then, while under direct fire, dismantled his gun, and while going to the secondary position he received a second wound, which caused his death.
Posthumously awarded. Medal presented to father, John N. Stier.

*STIFENELL, LUCH (1250625)_____
  Near Peterghem, Belgium, Oct. 31,
    1918.
  R—Norristown, Pa.
  B—Norristown, Pa.
  G. O. No. 74, W. D., 1919.

Private, Battery C, 107th Field Artillery, 28th Division.
Mortally wounded, yet realizing the need of every effective at the piece to continue its operation, he refused help from his comrades, and while lying on the ground cheered the members of the gun crew and urged them to maintain their fire, until he was removed to a dressing station by a stretcher bearer.
Posthumously awarded. Medal presented to mother, Mrs. Pauline Stifenell.

STIFF, WILLIAM C._____
  Near Fossoy, France, July 14-15,
    1918.
  R—Plymouth, Pa.
  B—Bloomsburg, Pa.
  G. O. No. 15, W. D., 1923.

Captain, Medical Corps, attached to 7th Infantry, 3d Division.
The observation post of his regiment having been destroyed by enemy shell-fire and a number of men wounded, Captain Stiff volunteered to go to the aid of the wounded men if a guide were furnished. The offer was declined as it was thought to be too hazardous an undertaking, the entire zone being swept by intense enemy machine-gun and artillery fire. Captain Stiff insisted upon making the attempt, and in company with Corporal Blankenship of the Headquarters Company, started on his mission despite the protests of officers and men acquainted with the terrain. Under terrific enemy fire they made their way to the outpost position, found the wounded men whose wounds they dressed, and led the disabled men to places of shelter.

STILLWELL, FRANK (125397)_____
  Near Fleville, France, Oct. 4, 1918.
  R—Kansas City, Kans.
  B—Kansas City, Kans.
  G. O. No. 37, W. D., 1919.

Sergeant, Battery A, 6th Field Artillery, 1st Division.
When his section of a platoon had been caught in an enemy barrage and all cannoneers of the platoon had been either killed or wounded, Sergeant Stillwell made repeated trips into the shelled area to remove the wounded. He assumed command of the platoon after the commanding officer had been evacuated and skillfully performed the duties involved therein.

*STINE, RALPH W._____
  Near Forges, France, Sept. 26, 1918.
  R—Paxton, Ill.
  B—Paxton, Ill.
  G. O. No. 44, W. D., 1919.

First lieutenant, 132d Infantry, 33d Division.
During the progress of the attack, Lieutenant Stine led a squad which wiped out 6 machine-gun nests and put the crews of 5 others to flight. At the last nest he met stubborn resistance and was instantly killed by a sniper while advancing upon it at close range.
Posthumously awarded. Medal presented to mother, Mrs. Minnie Stine.

STINER, WILLIAM J. (550606)_____
  Near Moulins, France, July 15, 1918,
    and near Romagne, France, Oct.
    9, 1918.
  R—New York, N. Y.
  B—New York, N. Y.
  G. O. No. 19, W. D., 1920.

Private, first class, Company F, 38th Infantry, 3d Division.
During the enemy offensive of July 15, Private Stiner exposed himself to intense artillery and machine-gun fire while guiding the support platoons of his company to the front line. On Oct. 9, during the Meuse-Argonne offensive, with 4 others, he led an attack on an enemy trench and succeeded in capturing 39 prisoners and 8 machine guns.

STINSON, JAMES K. (199291)_____
  At Marcheville, France, Sept. 26,
    1918.
  R—Newark, N. J.
  B—Canada.
  G. O. No. 21, W. D., 1919.

Private, first class, Company C, 101st Field Signal Battalion, 26th Division.
He showed exceptional coolness and courage in voluntarily laying and repairing telephone lines under a violent bombardment. Later, when all other wires had been cut, he succeeded in tapping in on lines and putting through a call for a barrage.

STIRLING, THOMAS (2183329)_____
  Near Remonville, France, Nov. 1, 1918.
  R—Denver, Colo.
  B—Scotland.
  G. O. No. 95, W. D., 1919.

Corporal, Machine Gun Company, 354th Infantry, 89th Division.
Corporal Stirling was directing the fire of his machine-gun crew at close range from a shell hole in an open field when their gun became disabled. Thereupon he and 2 other soldiers advanced with pistols upon the enemy machine-gun nest at which they had been firing and captured it, with 3 guns and 9 prisoners. Putting 1 of the captured guns into immediate action against the enemy, they enabled the Infantry to advance with a minimum of casualties.

*STOCKTON, FRANK R. (96634)_____
  At La Musard Farm, near Landres-et-St. Georges, France, Oct. 14, 1918.
  R—New Decatur, Ala.
  B—New Decatur, Ala.
  G. O. No. 130, W. D., 1918.

Private, Company E, 167th Infantry, 42d Division.
This soldier, while acting as litter bearer, went through deadly artillery and machine-gun fire to the aid of some wounded soldiers, disregarding warnings as to the danger in so doing. After administering first aid to one of the wounded, he carried him toward our lines, and had almost reached a place of safety when he was killed by machine-gun fire, having given proof of the highest devotion to duty, courage, and self-sacrifice.
Posthumously awarded. Medal presented to father, Rev. J. I. Stockton.

STOCKTON, JAMES R_____
  Near Blanc Mont Ridge, France, Oct. 3, 1918.
  R—Jacksonville, Fla.
  B—Duval County, Fla.
  G. O. No. 13, W. D., 1924.

Captain, 5th Regiment, U. S. Marine Corps, 2d Division.
While directing his platoon in the attack under heavy artillery and machine-gun fire, Captain Stockton, although severely wounded, displayed exceptional bravery and coolness in remaining in command and leading his men, refusing to be evacuated until forced to do so through loss of blood.

STOKER, ALEXANDER (731362)_____
  Near Bois-de-Bantillon, France, Nov. 5, 1918.
  R—LaFayette, Ga.
  B—LaFayette, Ga.
  G. O. No. 44, W. D., 1919.

Sergeant, Company A, 6th Infantry, 5th Division.
Accompanied by another soldier, Sergeant Stoker advanced against an enemy machine-gun unit, which was holding up the advance of his platoon and the company on his left. He completed the capture of the position, killing 3 of the enemy and routing the remaining part of the unit, thereby enabling his platoon to advance to their objective.

STOKES, JOHN Y., Jr_____
  Near Etain, France, Sept. 16, 1918.
  R—Reidsville, N. C.
  B—Reidsville, N. C.
  G. O. No. 37, W. D., 1919.

First lieutenant, 20th Aero Squadron, Air Service.
After their own formation had been broken up, Lieutenant Stokes and his pilot voluntarily continued on their bombing mission with planes from another squadron. Although their plane was thrown out of control by antiaircraft fire, they proceeded to their objective and dropped their bombs. Their motor then died completely, and they were attacked by an enemy combat plane, but they fought off the attacking machine and reached the Allied lines, where their plane crashed in a forest.

STOLL, CHARLES T. (1210158)_____
  Near St. Souplet, France, Oct. 18, 1918.
  R—New York, N. Y.
  B—New York, N. Y.
  G. O. No. 46, W. D., 1919.

Corporal, Company C, 107th Infantry, 27th Division.
After the advance of his company had been stopped by strong hostile machine-gun fire, Corporal Stoll, with 3 companions, advanced far ahead of the front line to attack an enemy position located in a large farmhouse. By skillful maneuvering in the broad daylight they covered all entrances to the house and forced the surrender of the entire force of the enemy, numbering 36 men and 2 officers. During the exploit they killed 2 Germans, who attempted to take cover in the cellar.

STONE, ALEXANDER H_____
  Near Brabrant, France, Oct. 8, 1918.
  R—Fredericksburg, Va.
  B—Fredericksburg, Va.
  G. O. No. 37, W. D., 1919.

First lieutenant, 116th Infantry, 29th Division.
He personally led his men in attack on machine-gun nests, destroying several of them and capturing many prisoners and several guns. He repeatedly refused to go to the rear with a badly sprained ankle, though ordered to do so by a surgeon.

STONE, BERNARD (1451493)_____
  At Cheppy, France, Sept. 26, 1918.
  R—St. Louis, Mo.
  B—Cleveland, Ohio.
  G. O. No. 37, W. D., 1919.

Private, first class, Company B, 138th Infantry, 35th Division.
In the face of machine-gun fire Private Stone entered an enemy dugout alone, killed 1 German, and captured 6 prisoners and 2 machine guns. Though he was twice wounded, he remained on duty until the last day of the drive. After his wounds were dressed he left the hospital to join his company.

STONE, EDWARD R_____
  Near Medeah Ferme, France, Oct. 3-10, 1918.
  R—Spencer, Mass.
  B—Spencer, Mass.
  G. O. No. 44, W. D., 1919.
  Distinguished-service medal also awarded.

Colonel, 23d Infantry, 2d Division.
Throughout the heavy fighting near Medeah Ferme he was at all times in the most exposed position, going over the top with his regiment 4 times in 7 days, after all his battalion commanders had been killed. He repeatedly refused to go to a place of safety during the most severe bombardments, even after being seriously gassed.

STONE, ELLSWORTH A_____
  Near St. Juvin, France, Oct. 11, 1918.
  R—Woodhaven, Long Island, N. Y.
  B—Lynn, Mass.
  G. O. No. 81, W. D., 1919.

Second lieutenant, 326th Infantry, 82d Division.
Leading his platoon under withering machine-gun and artillery fire, Lieutenant Stone, although wounded, admirably led his men on until ordered to withdraw. He personally supervised the evacuation of the wounded, and, in taking a new position, he was again wounded. Scarcely able to stand, he remained in the action until ordered to the rear by his company commander.

STONE, JAMES E. (1789441)_____
  Near Nantillois, France, Sept. 29-Oct. 1, 1918.
  R—Owensboro, Ky.
  B—Madisonville, Ky.
  G. O. No. 37, W. D., 1919.

Sergeant, 314th Ambulance Company, 304th Sanitary Train, 79th Division.
While on duty at the ambulance dressing station Sergeant Stone heard the report that there were several wounded men on a wooded hill exposed to enemy machine-gun fire. He volunteered and brought the wounded men in, which necessitated several trips under heavy fire. On many more occasions during the fighting around Montfaucon he exposed himself to the enemy fire in rescuing the wounded.

STONE, OTIS L. (73271)_____
  Near Epieds, France, July 23, 1918.
  R—Stoneham, Mass.
  B—Medford, Mass.
  G. O. No. 81, W. D., 1919.

Corporal, Company K, 104th Infantry, 26th Division.
Assisted by 2 comrades, Corporal Stone rushed out in the face of direct and annihilating machine-gun fire and brought in a wounded comrade, who was lying 25 yards in front of his lines.

*STONECIPHER, MANIPHE (58792)_____
    Near Cantigny, France, May 28–30, 1918.
    R—Iuka, Ill.
    B—Glenmary, Tenn.
    G. O. No. 99, W. D., 1918.

Sergeant, Company M, 28th Infantry, 1st Division.
He showed remarkable coolness and disregard of danger under heavy bombardment near Cantigny, France, May 28–30, 1918. While directing the consolidation of a new position, a driving fire caused many casualties in his command, but by his example of fortitude he inspired confidence in his men and refused to withdraw to the second line, even when his ammunition was nearly exhausted.
Posthumously awarded. Medal presented to mother, Mrs. Hattie Stonecipher.

STONEY, BRUCE (1870522)_____
    Near Ardeuil, France, Sept. 29, 1918.
    R—Denmark, S. C.
    B—Allendale, S. C.
    G. O. No. 46, W. D., 1919.

Private, medical detachment, 371st Infantry, 93d Division.
With 3 other soldiers he crawled 200 yards ahead of our lines under violent machine-gun fire and rescued an officer who was lying mortally wounded in a shell hole.

*STORM, GEORGE P. (44579)_____
    Near Baulay, France, Oct. 4, 1918.
    R—Catasauqua, Pa.
    B—Catasauqua, Pa.
    G. O. No. 32, W. D., 1919.

Battalion sergeant major, Headquarters Company, 16th Infantry, 1st Division.
He volunteered and carried a message through violent artillery and machine-gun fire. He then assisted in caring for the wounded and sending them to the rear with prisoners. Later, when his battalion had made an advance, Sergeant Major Storm, after finishing his duties at the old post of command, advanced through violent fire to his new station, where he was killed by shellfire while assisting in the consolidation.
Posthumously awarded. Medal presented to brother, Edward Storm.

STORRIE, ROBERT S. (1898430)_____
    Near Eply, France, Sept. 4, 1918.
    R—Brooklyn, N. Y.
    B—New York, N. Y.
    G. O. No. 20, W. D., 1919.

Corporal, Company E, 325th Infantry, 82d Division.
Under heavy fire from machine guns, and although seriously wounded, he continued to advance within the enemy's lines. By words of encouragement he urged his men to follow. By his brave leadership an enemy outpost defended by 2 machine guns and 6 riflemen was captured.

STOTHER, GREEN W. (735870)_____
    Near Vieville, France, Sept. 12, 1918.
    R—Mitchell, La.
    B—Gum, La.
    G. O. No. 37, W. D., 1919.

Corporal, Company G, 11th Infantry, 5th Division.
Although on duty with the regimental chaplain, he requested and was granted permission to accompany the first wave. Aided by a fellow soldier, he successfully accomplished the capture of 14 prisoners and their machine guns.

STOUT, ALBERT H._____
    In Bois-de-Foret, France, Oct. 12, 1918.
    R—Cairo, Ill.
    B—Cairo, Ill.
    G. O. No. 59, W. D., 1919

Second lieutenant, 132d Infantry, 33d Division.
After the battalion objective had been reached at the north edge of Bois-de-Foret, Lieutenant Stout's platoon, which was in the front wave, was attacked from the rear by the enemy, who had penetrated the line to the left. Lieutenant Stout quickly changed his position and led his men in a hand-to-hand fight. The hostile force, consisting of 40 men armed with 6 machine guns, was killed or captured, Lieutenant Stout himself killing 3 Germans and capturing 1 machine gun.

STOUT, LOUIS A._____
    Near Noyers, France, Nov. 7, 1918.
    R—Kansas City, Mo.
    B—Stoutsville, Ohio.
    G. O. No. 37, W. D., 1919.

Captain, 165th Infantry, 42d Division.
After the ammunition supply had been exhausted, he displayed exceptional gallantry and leadership in leading this company and 1 platoon of another in a bayonet charge up Hill 346, capturing this strong point, together with 6 machine guns and 23 prisoners.

STOUT, PENROSE V._____
    Near Charnay, France, Sept. 28, 1918.
    R—Bronxville, N. Y.
    B—Montgomery, Ala.
    G. O. No. 46, W. D., 1919.

First lieutenant, 27th Aero Squadron, Air Service.
While engaged in a solitary patrol of the enemy lines, Lieutenant Stout attacked an artillery regulating machine. He was almost immediately attacked by 5 enemy planes and subjected to infantry and antiaircraft fire, but fearlessly continued the unequal fight until his machine guns were broken and he was shot through the shoulder and lung.

STOVALL, WILLIAM H._____
    In the region of Etain, France, Sept. 26, 1918.
    R—Stovall, Miss.
    B—Stovall, Miss.
    G. O. No. 145, W. D., 1918.

First lieutenant, pilot, 13th Aero Squadron, Air Service.
While leading a protection patrol over a day bombing formation his patrol became reduced through motor trouble to himself and 1 other pilot. When the bombing patrol was attacked by 7 enemy planes he in turn attacked the enemy and destroyed 1 plane.

*STOWELL, EARLE B. (71524)_____
    Near St. Remy, France, Sept. 12, 1918.
    R—Worcester, Mass.
    B—East Hartford, Conn.
    G. O. No. 27, W. D., 1919.

Corporal, Company C, 104th Infantry, 26th Division.
When his platoon was held up by machine guns, Corporal Stowell volunteered with others and charged an enemy machine-gun nest, capturing 2 guns, 1 trench mortar, and 12 prisoners.
Posthumously awarded. Medal presented to mother, Mrs. Clara J. Kempton.

STRAABE, GILBERT (2257230)_____
    Near Gesnes, France, Oct. 3, 1918.
    R—Geraldine, Mont.
    B—Norway.
    G. O. No. 20, W. D., 1919.

Private, Company D, 361st Infantry, 91st Division.
He voluntarily and unhesitatingly left shelter under heavy shellfire and without thought of personal danger rendered first aid and carried a wounded comrade to a place of safety.

STRAHM, VICTOR H._____
    Near Metz, France, Sept. 13, 1918.
    R—Bowling Green, Ky.
    B—Nashville, Tenn.
    G. O. No. 1, W. D., 1919.

Captain, 91st Aero Squadron, Air Service.
He displayed remarkable courage and skill in penetrating the enemy territory for a distance of 25 kilometers, flying at an altitude of less than 300 meters. His plane was subjected to intense fire from antiaircraft guns in the region of Metz, and he was attacked by a superior number of German planes, 1 of which he destroyed. He completed his mission and returned with information of great military value.

STRAIN, BENJAMIN T.
  At Chateau-Thierry, France, June 6, 1918.
  R–Greensburg, Ind.
  B–Newton, Kans.
  G. O. No. 64, W. D., 1919.

Corporal, 45th Company, 5th Regiment, U. S. Marine Corps, 2d Division.
Killed in action at Chateau-Thierry, France, June 6, 1918. He gave the supreme proof of that extraordinary heroism which will serve as an example to hitherto untried troops.
Posthumously awarded. Medal presented to mother, Mrs. Anna T. Strain.

STRAIN, JAMES F.
  Near Very, France, Sept. 27, 1918.
  R–Pittsburg, Kans.
  B–Atchison, Kans.
  G. O. No. 37, W. D., 1919.

First lieutenant, 363d Infantry, 91st Division.
Although severely wounded himself, he crawled through heavy fire to the side of his commanding officer, and, taking the latter on his back, brought him to safety. He then took command of the company, and, except for a visit to the first-aid station, remained throughout the entire operation.

STRAKEY, GEORGE (527624).
  Near Romanovka, Siberia, June 25, 1919.
  R–Castlegate, Utah.
  B–Rockville, Colo.
  G. O. No. 133, W. D., 1919.

Private, first class, Company A, 31st Infantry.
After having been wounded he continued to use his rifle throughout the action.

STRANGE, GEORGE F. (1309519).
  Near Ponchaux, France, Oct. 7, 1918.
  R–Adams, Tenn.
  B–Robertson County, Tenn.
  G. O. No. 81, W. D., 1919.

Private, first class, Company L, 117th Infantry, 30th Division.
Private Strange and 20 other soldiers, the remnants of 2 platoons, were isolated in a railroad cut under heavy enemy fire, when 75 of the enemy started a counterattack on their position. Possessing the only automatic rifle in the group, Private Strange fearlessly opened fire on the enemy from an exposed position until his automatic rifle jammed and his left arm was paralyzed by a wound. He succeeded in clearing the jam, however, and immediately resumed firing despite his left arm being disabled, driving off the enemy, and breaking up the counterattack through his exceptional fortitude and determination. Shortly afterwards he was again wounded by a bursting shell.

STRAUB, ROBERT A. (1702047).
  At Bazoches, France, Aug. 14, 1918.
  R–New York, N. Y.
  B–New York, N. Y.
  G. O. No. 16, W. D., 1923.

Corporal, Company F, 306th Infantry, 77th Division.
Voluntarily joining a daylight patrol seeking information as to the strength and positions of the enemy which was attacked about 100 yards beyond its own lines by an enemy hostile post of 7 men. The enemy was immediately attacked from the rear, several of the men killed and the survivors scattered. A moment later another enemy post was attacked and in hand to hand fighting Corporal Straub killed 1 of the enemy and was himself badly wounded. Although unable to walk and under heavy fire from near-by enemy posts, Corporal Straub dragged himself to our lines and gave valuable information as to the disposition of the enemy forces.

STRAWBRIDGE, GEORGE (105576).
  Near Fleville, France, Oct. 5, 1918.
  R–Flaxton, N. Dak.
  B–Northwood, Iowa.
  G. O. No. 23, W. D., 1919.

Private, Company B, 2d Machine Gun Battalion, 1st Division.
He administered first aid to a wounded comrade under heavy maching-gun and artillery fire, and, although wounded himself, he refused evacuation, remaining on duty with the company during the entire action.

STREB, THOMAS F. (1285699).
  Near Verdun, France, Oct. 17, 1918.
  R–Baltimore, Md.
  B–Baltimore, Md.
  G. O. No. 27, W. D., 1919.

Private, first class, Company H, 115th Infantry, 29th Division.
In the Bois-de-Consenvoye, east of the Meuse, he operated his automatic rifle on a post enfiladed by direct maching-gun fire during a desperate counterattack by the enemy until the rifle was damaged by the enemy's fire and he himself was wounded. He remained on post continuing to defend same with an ordinary rifle. He was later gassed, and refused to go to the hospital until ordered to do so by his company commander.

STRICKLAND, ALBERT B. (51121).
  Near Vierzy, France, July 18, 1918.
  R–Loper, Ala.
  B–New Augusta, Miss.
  G. O. No. 24, W. D., 1920.

Sergeant, Company H, 23d Infantry, 2d Division.
While leading his platoon in attack, Sergeant Strickland was painfully wounded in the leg; disregarding his wound, he continued to lead his platoon forward until again very severely wounded by a shell fragment.

STRICKLAND, CURTIS MIMS (1312553).
  Near Brancourt, France, Oct. 8, 1918.
  R–Colleton, S. C.
  B–Colleton, S. C.
  G. O. No. 50, W. D., 1919.

Corporal, Company M, 118th Infantry, 30th Division.
Crawling several hundred feet under deadly rifle and machine-gun fire, Corporal Strickland, with another soldier, flanked a shell hole wherein a number of the enemy were hiding. In this heroic exploit he either killed or captured 14 of the enemy, as well as taking a machine gun.

STRIPLING, WALTER B. (57859).
  Near Nonsard, France, Sept. 12, 1918.
  R–Oliver Springs, Tenn.
  B–Polk County, Ark.
  G. O. No. 37, W. D., 1919.

Corporal, Company G, 28th Infantry, 1st Division.
The whole line being held up by heavy fire from an enemy strong point, Corporal Stripling, with marked initiative, led his squad and, despite strong resistance, skillfully flanked and captured his objective, together with 30 prisoners, without the loss of a man.

STROBEL, HENRY A. (1979816).
  Near Bellicourt, France, Sept. 29, 1918.
  R–Tell City, Ind.
  B–St. Meinrad, Ind.
  G. O. No. 32, W. D., 1919.

Private, Company D, 120th Infantry, 30th Division.
With 8 other soldiers, comprising the company headquarters detachment, Private Strobel assisted his company commander in cleaning out enemy dugouts along the canal and capturing 242 prisoners.

STROMAN, HENRY H. (2285158).
  At Eclisfontaine, France, near Bois-de-Bauhry, France, Sept. 28, 1918.
  R–Tallahassee, Fla.
  B–Tallahassee, Fla.
  G. O. No. 37, W. D., 1919.

Sergeant, Company K, 364th Infantry, 91st Division.
Responding to a call for volunteers, Sergeant Stroman with 5 others advanced 400 yards beyond their front to bring in wounded comrades. They succeeded in rescuing 7 of their men, also in bringing in the dead body of a lieutenant, while exposed to terrific machine-gun fire.

STROTHER, HAROLD C. (2261605) _____
Near Steenbrugge, Belgium, Oct. 31, 1918.
R—Ripon, Calif.
B—Ripon, Calif.
G. O. No. 37, W. D., 1919.

Corporal, Company L, 362d Infantry, 91st Division.
Advancing under heavy machine-gun fire, with the aid of 2 other soldiers he silenced the fire of a strongly fortified machine-gun position which was causing severe losses in his ranks. His action made possible the further advance not only of his own platoon but also the company on his left.

*STRUCEL, PETER (1783307) _____
Near Montfaucon, France, Sept. 26, 1918.
R—Calumet, Mich.
B—Manistique, Mich.
G. O. No. 27, W. D., 1919.

Sergeant, Company L, 314th Infantry, 79th Division.
During an attack and under heavy machine-gun fire, Sergeant Strucel showed exceptional courage and devotion to duty by constantly walking up and down the line cheering and encouraging his men. In the performance of this task he was killed.
Posthumously awarded. Medal presented to father, George Strucel.

*STUART, ARTHUR J. (363941) _____
North of Jaulgonne, near Sergy, France, July 31, 1918.
R—Detroit, Mich.
B—Canada.
G. O. No. 116, W. D., 1918.

Corporal, Company M, 125th Infantry, 32d Division.
Although exposed to artillery, machine-gun, and rifle fire, Corporal Stuart attempted to carry a wounded man from within 100 feet of the German line. He was killed while crawling toward his own lines with his wounded comrade on his back.
Posthumously awarded. Medal presented to father, Sumner B. Stuart.

STUART, GEORGE (1276934) _____
Near Ravine-de-la-Reine, north of Samogneux, France, Oct. 10, 1918.
R—Jersey City, N. J.
B—Jersey City, N. J.
G. O. No. 130, W. D., 1918.

Corporal, Company K, 113th Infantry, 29th Division.
Under difficult circumstances he led his squad to its objective, although they were greatly outnumbered by the enemy. Single handed, he afterwards killed 6 of the enemy and captured 2 machine guns.

STUBBS, EDWIN J. (89224) _____
At Meurcy Farm, near Villers-sur-Fere, France, Aug. 1, 1918.
R—New York, N. Y.
B—New York, N. Y.
G. O. No. 4, W. D., 1923.

Private, Company A, 165th Infantry, 42d Division.
After volunteering to proceed in the advancing line as a sniper, by his skill and courage he disposed of 2 enemy machine gunners who were causing heavy losses to the assaulting battalion. While in this position of extreme danger, he was wounded by an enemy sniper, but held his post in spite of great pain and suffering until relieved. Later, as a result of his wound, his arm was amputated.

STUCKRAD, ARTHUR L. (274269) _____
Near Gesnes, France, Oct. 5, 1918.
R—Milwaukee, Wis.
B—Charles City, Iowa.
G. O. No. 66, W. D., 1919.

Corporal, Company F, 127th Infantry, 32d Division.
With 2 other soldiers Corporal Stuckrad advanced ahead of their company and rushed an enemy machine-gun nest from the flank, capturing 14 members of the crew and 2 machine guns, and thereby enabling the advance to continue. Carrying the captured guns with them to the objective, they later used them successfully in repelling a hostile counterattack.

STUDY, MARION FRANCIS (2154036) _____
Near Molain, France, Oct. 17, 1918.
R—Thurman, Iowa.
B—Fremont, Iowa.
G. O. No. 81, W. D., 1919.

Private, first class, Company L, 117th Infantry, 30th Division.
With another soldier he volunteered to go out across an open space swept by heavy machine-gun fire, about 150 yards to the front, to rescue 2 wounded soldiers, which he helped to bring back to the line.

STURTEVANT, WALLIS H. (1684164) _____
Near Chery-Chartreuve, France, Aug. 10, 1918.
R—Fitchburg, Mass.
B—Greenfield, Mass.
G. O. No. 59, W. D., 1919.

Corporal, Company D, 12th Machine Gun Battalion, 4th Division.
He voluntarily ran through a terrific shellfire into a burning ammunition dump and rescued a badly wounded and burned comrade. The ammunition was exploded a few seconds after this heroic act was performed.

*SUITER, WILBUR C. _____
Near Vilcey-sur-Trey, France, Sept. 12, 1918.
R—York, Pa.
B—Lockhaven, Pa.
G. O. No. 133, W. D., 1918.

First lieutenant, pilot, 135th Aero Squadron, Air Service.
He, with Second Lieut. Guy E. Morse, observer, fearlessly volunteered for the perilous mission of locating the enemy's advance unit in the rear of the Hindenburg line. Disregarding the hail of machine-gun fire and bursting antiaircraft shell, they invaded the enemy territory at a low altitude and accomplished their mission, securing for our staff information of the greatest importance. These 2 gallant officers at once returned to the lines and undertook another reconnaissance mission, from which they failed to return.
Posthumously awarded. Medal presented to father, S. F. Suiter.

SULLIVAN, DAN W. (49375) _____
Near Chateau-Thierry, France, June 6, 1918.
R—New Orleans, La.
B—New Orleans, La.
G. O. No. 99, W. D., 1918.

Private, first class, Machine Gun Company, 23d Infantry, 2d Division.
During a period of 5 hours and 30 minutes on June 6, 1918, near Chateau-Thierry, France, he carried messages between the commanding officer and platoon leader through constant machine-gun fire, thereby maintaining communication successfully at the imminent risk of his life.

SULLIVAN, EDWARD J. _____
Near Les Eparges, France, Sept. 12, 1918.
R—New York, N. Y.
B—New York, N. Y.
G. O. No. 37, W. D., 1919.

Second lieutenant, 104th Infantry, 26th Division.
While leading his platoon forward and finding no gap in the enemy's wire, Lieutenant Sullivan, although severely wounded, directed his men in cutting the wire and encouraged their advance until he fell exhausted from his wounds.

SULLIVAN, GROVER C. (1683735) _____
At La Besace, France, Nov. 5, 1918.
R—Norwood, N. Y.
B—Norwood, N. Y.
G. O. No. 21, W. D., 1919.

Private, Company L, 306th Infantry, 77th Division.
He displayed rare bravery and devotion to duty by remaining on duty after being seriously wounded during a heavy artillery bombardment and giving first-aid treatment to 5 severely wounded comrades.

SULLIVAN, JAMES (1702578) _____
At St. Juvin, France, Oct. 15, 1918.
R—New York, N. Y.
B—Newport, R. I.
G. O. No. 20, W. D., 1919.

Sergeant, Company H, 306th Infantry, 77th Division.
This soldier fearlessly entered a dugout in which he knew there were Germans hiding and single handed captured 20 prisoners.

*SULLIVAN, JERRY (42854)_____
South of Soissons, France, July 18, 1918.
R—Barry, Vt.
B—Ireland.
G. O. No. 15, W. D., 1919.

Sergeant, Company F, 16th Infantry, 1st Division.
He displayed exceptional courage and initiative by leading his platoon to the attack and capture of a battery of 77-millimeter guns. After the successful accomplishment of this unusual and heroic duty he was killed in action. Posthumously awarded. Medal presented to mother, Mrs. Eugene Sullivan.

SULLIVAN, JOHN L. B. (88992)_____
North of the River Ourcq, July 29, 1918.
R—New York, N. Y.
B—New York, N. Y.
G. O. No. 9, W. D., 1923.

Private, first class, Machine Gun Company, 165th Infantry, 42d Division.
For extraordinary heroism in action north of the River Ourcq, July 29, 1918, while advancing against the enemy. No messengers being available, he volunteered to carry messages from the machine-gun company commander to the battalion commander, crossing an open valley swept by machine-gun and shellfire, part of the distance of 200 meters being in water waist deep. This journey accomplished twice within an hour made possible proper liaison between the company and battalion commanders.

SULLIVAN, JOHN M. (3484535)_____
Near the Meuse River, France, Oct. 5, 1918.
R—Akron, Ohio.
B—Louisville, Ky.
G. O. No. 44, W. D., 1919.

Private, Company H, 28th Infantry, 1st Division.
On Oct. 5 he twice left his place of shelter and advanced under intense machine-gun fire to the rescue of wounded comrades, bringing them back to a place of safety where they could be cared for.

SULLIVAN, JOHN P. (554630)_____
Service rendered under the name, John Shannon.

Private, Company A, 9th Machine Gun Battalion, 3d Division.

SULLIVAN, JOSEPH J. (2261691)_____
Near Gesnes, France, Sept. 29, 1918.
R—Jordan, Mont.
B—Chicago, Ill.
G. O. No. 15, W. D., 1919.

Corporal, Company M, 362d Infantry, 91st Division.
Observing that the left flank of the regimental line was unprotected, he voluntarily took out a combat patrol and while so doing encountered 3 machine guns, which were employing effective enfilade fire. Boldly advancing on this position, he silenced the guns.

SULLIVAN, RALPH B. (1217078)_____
Near Montzeville, France, Sept. 14, 1918.
R—New York, N. Y.
B—New York, N. Y.
G. O. No. 32, W. D., 1919.

Private, Battery B, 104th Field Artillery, 27th Division.
When a continuous bombardment had set fire to the camouflage covering of a large ammunition dump of 75-millimeter shells and exploded 9 of the shells, he, utterly disregarding his personal safety, left a sheltered position and ran to the dump and, with the aid of 3 other men, extinguished the fire, not only saving the ammunition but also preventing the exact locating of the dump by the enemy.

SULLIVAN, WILLIAM Q_____
Near Chevieres, France, Oct. 14, 1918.
R—Norwood, Colo.
B—Pueblo, Colo.
G. O. No. 35, W. D., 1919.

First lieutenant, 308th Infantry, 77th Division.
After his company commander had been seriously wounded and he himself wounded in the head by a machine-gun bullet, Lieutenant Sullivan continued to lead and encourage his men until wounded the second time. He then continued in command of the company until ordered to be evacuated by his battalion commander.

SUMMERALL, CHARLES P_____
Before Berzy-le-Sec, near Soissons, France, during the Aisne-Marne offensive, July 19, 1918.
R—Astatula, Fla.
B—Lake City, Fla.
G. O. No. 9, W. D., 1923.
Distinguished-service medal also awarded.

Major general, U. S. Army.
General Summerall, commanding the 1st Division, visited, with great gallantry and with utter disregard for his own safety, the extreme front lines of his division and personally made a reconnaissance of the position in the face of heavy hostile machine-gun and artillery fire, exhorting his men to renew the attack on Berzy-le-Sec, promising them a powerful artillery support, and so encouraging them by his presence and example that they declared their readiness to take the town for him. Due to his great courage and utter disregard for his own safety, the men of his division were inspired to enormous and heroic efforts, capturing Berzy-le-Sec the next morning under terrific enemy fire, and later in the day the division reached all its objectives.

SUMMERS, ALBERT E. (1679686)_____
In the Argonne Forest, France, Oct. 6, 1918.
R—Auburn, N. Y.
B—England.
G. O. No. 71, W. D., 1919.

Private, Company H, 308th Infantry, 77th Division.
In the face of direct machine-gun fire he left cover and went out 100 yards to rescue a wounded soldier. Dragging the wounded man back to his funk hole, he gave him first aid, and then again exposing himself to enemy fire obtained water for him. He showed utter disregard for personal danger in aiding other wounded men in addition to performing his duties as scout.

SUMMERTON, RALPH N. (1248643)_____
Near Chatel-Chehery, France, Oct. 6, 1918.
R—Tidioute, Pa.
B—Tidioute, Pa.
G. O. No. 130, W. D., 1918.

Sergeant, Company I, 112th Infantry, 28th Division.
Sergeant Summerton, having on his body several aggravated wounds from an enemy grenade and being tagged for evacuation for these, as well as for grippe, when assured that his company was about to attack Chatel-Chehery and that it had lost all its officers, went back to his company and courageously and skillfully led it as the first wave, and while so doing was again wounded.

SUMNER, CHARLES S_____
At Bussy Farm and Sechault, France, Sept. 28–29, 1918.
R—St. Albans, Vt.
B—St. Albans, Vt.
G. O. No. 13, W. D., 1919.

Captain, 372d Infantry, 93d Division.
During the attack on Bussy Farm and Sechault he courageously led his command under the most intense artillery fire and in the face of a fusillade of machine-gun bullets. Although he was suffering from the effects of gas and had been twice knocked down by explosion of shells, he remained on duty, and, inspired by his example, his men overcame the strong enemy resistance.

SUNDIN, MILTON C. (2202754)_____
In the St. Mihiel salient, France, Sept. 12–13, 1918.
R—Denver, Colo.
B—Denver, Colo.
G. O. No. 20, W. D., 1919.

Private, Company L, 353d Infantry, 89th Division.
Private Sundin, while advancing through wooded territory with 4 other men, was surprised by the fire of 6 machine guns. Though 2 of the party were wounded, Private Sundin, with great daring, worked around the flank of the position and succeeded in routing the enemy machine gunners in time to permit the advance without casualties of 2 platoons operating near by.

*SUPLEE, HOWARD R. (3111260)_____
    North of Verdun, France, at Hill 378,
    Grande Montagne sector, Nov. 6,
    1918.
    R—Philadelphia, Pa.
    B—Philadelphia, Pa.
    G. O. No. 14, W. D., 1923.

Private, Company I, 316th Infantry, 79th Division.
The 316th Infantry, depleted in strength and numbers, attacked from the crest of Hill 378, advancing over the exposed northern slope in the face of terrific machine-gun and artillery fire. Halfway down the slope the thin line was held up by a rain of machine-gun fire from the road leading eastwardly through the Bois-de-la-Grande-Montagne to Reville. Private Suplee offered to subdue the fire, and while advancing single handed to the accomplishment of his mission received a wound from which he died a short while later. Posthumously awarded. Medal presented to father, Howard R. Suplee.

SUPLER, JOHN M. (2267899)_____
    Near Waereghem, Belgium, Oct. 30,
    1918.
    R—Brawley, Calif.
    B—Greene County, Pa.
    G. O. No. 37, W. D., 1919.

Private, Company L, 364th Infantry, 91st Division.
He received a severe shoulder wound, the same shell blowing the rifle to pieces in his hand. After receiving treatment he continued with the company, working under terrific pain, until ordered by his sergeant to report to the dressing station.

SURDEZ, LOUIS (2383729)_____
    Near St. Mihiel, France, Sept. 16,
    1918.
    R—St. George, Staten Island, N. Y.
    B—New York, N. Y.
    G. O. No. 37, W. D., 1919.

Sergeant, Company G, 60th Infantry, 5th Division.
Although wounded by shellfire a few minutes before his company took up the advance, Sergeant Surdez led his platoon through 17 hours of shellfire and by his exceptional example during the advance and consolidation of the new positions encouraged his men to their full duty in action.

SUSTICK, EMANUEL (1698549)_____
    In Boise-de-la-Naza, France, Oct.
    4, 1918.
    R—Brooklyn, N. Y.
    B—Farmingdale, N. J.
    G. O. No. 59, W. D., 1919.

Sergeant, Company L, 305th Infantry, 77th Division.
He volunteered to advance through thick brush subjected to a heavy machine-gun fire to a point within a few yards of enemy emplacements in order to observe the effects of our trench mortars on machine-gun nests. He made his observations successfully, though exposed alike to enemy fire and our own barrage.

SUTHERLAND, FRANCIS S. (3206525)____
    In the St. Die sector, France, Oct. 9,
    1918.
    R—Ensley, Ala.
    B—Canada.
    G. O. No. 20, W. D., 1919.

Corporal, Company I, 321st Infantry, 81st Division.
During a heavy bombardment he maintained liaison between his combat group and his company commander, crossing completely unprotected ground under terrific barrage and supplying his group at the same time with much-needed ammunition.

*SUTHERLAND, JAMES (2448847)_____
    In the Forest of Argonne, France,
    Oct. 3, 1918.
    R—Chicago, Ill.
    B—Scotland.
    G. O. No. 87, W. D., 1919.

Sergeant, Company E, 305th Infantry, 77th Division.
Displaying exceptional devotion to duty and conspicuous courage, Sergeant Sutherland led his platoon up the steep slope of a ravine under murderous machine-gun fire in an attack on a series of strong enemy machine-gun nests and in so doing was seriously wounded. Posthumously awarded. Medal presented to uncle, John Simpson.

SWAAB, JACQUES M._____
    Near Montfaucon, France, Sept. 28,
    1918, and in the region of Campig-
    neulles, Oct. 27, 1918.
    R—Philadelphia, Pa.
    B—Philadelphia, Pa.
    G. O. No. 53, W. D., 1920.

First lieutenant, 22d Aero Squadron, Air Service.
On Sept. 28 Lieutenant Swaab, although himself pursued by 2 enemy planes, perceiving 1 of his comrades in distress and in danger of being shot down, dived upon the enemy plane which was directly behind that of his comrade and shot the enemy plane out of control, forcing it to withdraw. His prompt act in going to the assistance of his comrade enabled the latter to escape. On Oct. 27 Lieutenant Swaab and another member of his group engaged in combat with 7 enemy planes. In this encounter, although outnumbered, Lieutenant Swaab continued in his attack and succeeded in shooting down an enemy D. F. W. observation plane.

SWAGGERTY, ALLIE (2381943)_____
    At Madeleine Farm, France, Oct.
    12, and near Clery-le-Petit, France,
    Nov. 2, 1918.
    R—Byington, Tenn.
    B—Knox County, Tenn.
    G. O. No. 46, W. D., 1919.

Sergeant, Headquarters Company, 60th Infantry, 5th Division.
Being on duty with a platoon which was not to take part in the attack, he asked for and received permission to go over the top with the attacking companies. When the left flank was held up by several machine-gun nests, he alone cleaned out one of the nests, in addition to cleaning out many German machine gunners and snipers. Near Clery-le-Petit, on Nov. 2, this soldier again voluntarily accompanied attacking troops, crawling 300 yards under heavy fire and bringing down three machine gunners out of trees.

SWAIN, JACK R. (10391)_____
    Near Beaumont, France, June 19,
    1918.
    R—Dallas, Tex.
    B—Chillicothe, Mo.
    G. O. No. 15, W. D., 1919.

Private, first class, Section No. 647, Ambulance Service.
He went to the rescue of wounded men who were exposed to shellfire as a result of an accident to their ambulance. Being able to approach only to within 300 yards of the wrecked car on the road, he took a stretcher and crawled along a ditch to reach them. He then returned and recovered the body of a third man who had been killed in the accident.

SWAN, THOMAS E._____
    Between the Marne and Vesle
    Rivers, France, July 31 to Aug.
    6, 1918.
    R—Saginaw, Mich.
    B—England.
    G. O. No. 124, W. D., 1918.

Captain, chaplain, 125th Infantry, 32d Division.
During the heavy fighting near the Ourcq River this officer was in the front lines at all times, under heavy machine-gun and artillery fire throughout the day and night, comforting and aiding the wounded. On one occasion he crossed a field 200 yards wide under violent shellfire to minister to 2 soldiers who had been mortally wounded. In the operations near Mont St. Martin he continually went back and forth over the crest of a hill during heavy artillery fire to care for the wounded.

SWAN, WYMAN R._____
    Near Brieulles, France, Nov. 4-6,
    1918.
    R—Rockport, Ind.
    B—Rockport, Ind.
    G. O. No. 37, W. D., 1919.

Major, 7th Engineers, 5th Division.
He demonstrated commendable judgment in locating the site of a pontoon bridge and personally supervising the construction of the bridge; although under constant shellfire, he remained in charge for 36 hours, insuring the complete success of the exploit and the crossing of the division east of the Meuse.

*SWANGER, IRA V. (2151933)----------
At Marcheville, France, Nov. 10, 1918.
R—Persia, Iowa.
B—Persia, Iowa.
G. O. No. 44, W. D., 1919.

Corporal, Company F, 130th Infantry, 33d Division.
After showing exceptional bravery and judgment in leading his squad against enemy machine-gun positions, he was mortally wounded. Realizing that he had no chance of recovery, he refused to permit stretcher bearers to take him to the rear, urging them to care for others whose condition was less serious.
Posthumously awarded. Medal presented to mother, Mrs. Anna Rishel.

SWANSON, ADOLPH (2850023)----------
Near Les Huit Chemins, France, Sept. 29, 1918.
R—Grant, Iowa.
B—Sweden.
G. O. No. 46, W. D., 1919.

Private, Company I, 357th Infantry, 90th Division.
He volunteered and accompanied Chaplain D. Priest in going 600 yards beyond the front line and assisted him in carrying to safety a wounded man.

*SWANSON, CARL E. (2162063)----------
Near Brieulles, France, Oct. 9–12, 1918.
R—Grove City, Minn.
B—Sweden.
G. O. No. 32, W. D., 1919.

Private, Company K, 132d Infantry, 33d Division.
While attempting to rescue a wounded officer who was lying exposed to terrific machine-gun fire, Private Swanson was killed. For 4 days before his death, in the performance of his duties as stretcher bearer, he rendered invaluable service in administering first aid to the wounded and carrying them to places of safety, working at all times under most perilous fire of artillery and machine guns.
Posthumously awarded. Medal presented to mother, Mrs. Maria Swanson.

SWANSON, CLAYTON E. (107395)----------
Near Mont Blanc, France, Oct. 4, 1918.
R—Jamestown, N. Y.
B—Jamestown, N. Y.
G. O. No. 37, W. D., 1918.

Corporal, Company A, 5th Machine Gun Battalion, 2d Division.
On learning that a member of his squad was in front of the lines in a heavily shelled position, Corporal Swanson obtained permission to make a search, to find that the man was dead.

SWARTS, RALPH E.----------
Near St. Etienne-a-Arnes, France, Oct. 3–9, 1918.
R—Arkansas City, Kans.
B—Arkansas City, Kans.
G. O. No. 35, W. D., 1919.

First lieutenant, Medical Corps, attached to 23d Infantry, 2d Division.
During the offensive operations of Oct. 3–9 he worked unceasingly in the most advanced stations in the divisional sector, dressing the wounded in the open under terrific machine-gun and shellfire. He took cover only when all wounded had been dressed and evacuated.

SWEARINGEN, WILLIAM H. (3027210)----------
Near Cote-de-Morimont, France, Oct. 26–Nov. 10, 1918.
R—Elida, N. Mex.
B—Jasper County, Mo.
G. O. No. 46, W. D., 1919.

Private, medical detachment, 315th Infantry, 79th Division.
Hearing a call for help, he went from cover to a position 360 yards distant, and in the face of incessant machine-gun and sniper fire gave first aid to a wounded comrade. He then provided some shelter for the wounded man and himself and remained until dark, at which time he returned to safety, carrying the man with him.

SWEENEY, BERNARD F., Jr. (1784950)----------
Near Brabant, France, Oct. 31, 1918.
R—Philadelphia, Pa.
B—Philadelphia, Pa.
G. O. No. 37, W. D., 1919.

Sergeant, Headquarters Company, 315th Infantry, 79th Division.
On the night of Oct. 31 he made a dozen trips to repair telephone wire broken by the continuous shelling of the area. Early the next morning he was wounded while still in the performance of his duty.

*SWEENEY, PATRICK (3282912)----------
Near Abaucourt, France, Nov. 9, 1918.
R—Chicago, Ill.
B—Ireland.
G. O. No. 32, W. D., 1919.

Private, Company D, 322d Infantry, 81st Division.
He voluntarily advanced through intense artillery and machine-gun fire into the ruins of Abaucourt to locate an enemy machine-gun nest. He was killed by shellfire after having reached a position in the enemy's trenches.
Posthumously awarded. Medal presented to father, John Sweeney.

SWEENEY, THOMAS JOSEPH (88001)----------
Near Landres-et-St. Georges, France, Oct. 15, 1918.
R—New York, N. Y.
B—New York, N. Y.
G. O. No. 37, W. D., 1919.

First sergeant, Company A, 165th Infantry, 42d Division.
He courageously supervised the carrying of the wounded, his duties exposing him at all times to the continuous fire of the enemy. By his valor and strict devotion to duty all the wounded were safely evacuated.

SWEET, WALTER----------
Near Chateau-Thierry, France, June 25, 1918.
R—Lowell, Mass.
B—Lowell, Mass.
G. O. No. 44, W. D., 1919.

Gunnery sergeant, 16th Company, 5th Regiment, U. S. Marine Corps, 2d Division.
In the attack of June 25, after his company commander had been removed because of wounds, Sergeant Sweet reorganized the platoon and, leading them forward, rushed a strong enemy emplacement, capturing 2 guns and their crews. After having consolidated his position and established liaison, the enemy opened a harassing trench-mortar fire, during which his platoon suffered heavy casualties. He attacked the nest with the aid of hand grenades and put the gun out of action, killing 2 of the enemy and taking 5 prisoners, with whom he returned to his position.

SWENSON, EARL J.----------
Near Very, France, Sept. 26–Oct. 4, 1918, and near Audenarde, Belgium, Oct. 30–Nov. 3, 1918.
R—Portland, Oreg.
B—Assaria, Kans.
G. O. No. 3, W. D., 1919.

Captain, Medical Corps, 316th Sanitary Train, 91st Division.
During the drive in the forest of Argonne he established and maintained a dressing station at Very under almost constant aerial raids and severe shellfire. During the operations between the Lys and Scheldt Rivers this officer repeatedly showed utter disregard for his own life, maintaining liaison between his own advanced dressing station and the battalion aid stations and searching for wounded on the battlefield while he was exposed to heavy fire from artillery, machine guns, and snipers.

*SWEZEY, LOUIS H. (1697965)----------
In the Bois-de-la-Naza, France, Oct. 3, 1918.
R—Patchogue, N. Y.
B—Patchogue, N. Y.
G. O. No. 89, W. D., 1919.

Private, first class, Company G, 305th Infantry, 77th Division.
After his company's line had been almost wiped out by enemy machine-gun fire Private Swezey displayed the highest courage and initiative in re-forming a defensive position and reorganizing the scattered groups of men who remained. In performing this important service he moved up and down the line, under heavy fire from enemy machine guns and trench mortars, in entire disregard for his own safety. This gallant soldier was killed while on a patrol next day.
Posthumously awarded. Medal presented to widow, Mrs. Louis H. Swezey.

SWIFT, HARRY (280257)............
Near Juvigny, France, Aug. 28, 1918.
R—Detroit, Mich.
B—Detroit, Mich.
G. O. No. 71, W. D., 1919.

Private, Company G, 126th Infantry, 32d Division.
He twice volunteered and carried messages across open fields swept by machine-gun fire after other runners had been killed on similar missions. Returning from his second mission, he saw his company commander fall wounded. Passing through heavy machine-gun fire, he went to his rescue and administered first aid, being himself severely wounded.

*SWIFT, JOSEPH............
Near Epinonville, France, Sept. 27, 1918, and near Gesnes, France, Sept. 29, 1918.
R—Safford, Ariz.
B—State Center, Iowa.
G. O. No. 59, W. D., 1919.

First lieutenant, 362d Infantry, 91st Division.
After a machine gun company had in vain attempted for an hour to silence a machine gun which was causing heavy losses to his regiment, Lieutenant Swift, armed only with a pistol, advanced alone upon the enemy position. He killed the crew of 4 men, saving the lives of many of our men and rendering a more rapid advance possible. In the attack on Gesnes, while making a daring attempt to perform a similar act, Lieutenant Swift was killed by enemy fire.
Posthumously awarded. Medal presented to father, P. P. Swift.

SWIFT, WALTER E. (2390317)............
Near Cunel, France, Oct. 13, 1918.
R—Brentwood, Calif.
B—Brentwood, Calif.
G. O. No. 27, W. D., 1920.

Private, first class, Company B, 14th Machine Gun Battalion, 5th Division.
Private Swift, with another runner, received messages to be delivered at 2 different points near Nantillois. En route his companion was killed and Private Swift was severely wounded by enemy shellfire. In spite of his wound, he delivered both messages before submitting to evacuation for his wounds.

*SWINGLE, CRAY (158265)............
Near Bois-des-Tailloux, France, Mar. 28, 1918.
R—Springfield, Ohio.
B—Hicksville, Ohio.
G. O. No. 88, W. D., 1918.

Sergeant, Company D, 6th Engineers, 3d Division.
The patrol came under hostile machine-gun fire and Sergeant Swingle was mortally wounded. He gave instructions to the patrol to return to their company commander and ordered them to leave him, as the patrol was under fire and would in all probability be wiped out. Died Mar. 28, 1918.
Posthumously awarded. Medal presented to father, Burt S. Swingle.

SYBERT, CLARENCE L. (77372)............
Near Landres-et-St.Georges, France, Nov. 2, 1918.
R—Centralia, Wash.
B—Reynoldsville, Pa.
G. O. No. 44, W. D., 1919.

Private, Company M, 23d Infantry, 2d Division.
Although severely wounded, he remained in the action during the offensive operations west of the Meuse, and after the sergeant in charge had been removed he took over his duties as commander of the battalion runners, faithfully performing the task despite his pain from his wounds.

*SYNNOTT, JOSEPH A............
At Chateau-Thierry, France, June 6, 1918.
R—Passaic, N. J.
B—Passaic, N. J.
G. O. No. 119, W. D., 1918.

Second lieutenant, 5th Regiment, U. S. Marine Corps, 2d Division.
Killed in action at Chateau-Thierry, France, June 6, 1918, he gave the supreme proof of that extraordinary heroism which will serve as an example to hitherto untried troops.
Posthumously awarded. Medal presented to sister, Mrs. Mollie S. Reiley.

SYNOTT, PATRICK (1214545)............
Near Ronssoy, France, Sept. 28, 1918.
R—New York, N. Y.
B—Ireland.
G. O. No. 37, W. D., 1919.

Corporal, Company F, 108th Infantry, 27th Division.
He displayed exceptional bravery in leaving shelter and going forward under heavy machine-gun fire and bringing back several wounded soldiers.

SYVERSON, GRANNIS I............
Near St. Etienne, France, Oct. 3, 1918.
R—Minneapolis, Minn.
B—White Rock, S. Dak.
G. O. No. 23, W. D., 1919.

Private, 66th Company, 6th Machine Gun Battalion, U. S. Marine Corps, 2d Division.
When our advance Infantry was forced to withdraw, Private Syverson's machine-gun crew refused to withdraw, but calmly set up their machine gun. The gun was upset by a bursting hand grenade, which also injured 2 members of the squad. Despite these injuries, they immediately reset the gun and opened fire on the advancing Germans when 20 feet distant, causing the Germans to break and retreat in disorder.

SZCZEPANIK, JOSEPH A. (2209637)............
Near Limey, France, Sept. 12–13, 1918.
R—New York Mills, N. Y.
B—Three Rivers, Mass.
G. O. No. 98, W. D., 1919.

Private, Company M, 353d Infantry, 89th Division.
Private Szczepanik displayed remarkable daring in going out alone and locating the hiding places into which enemy soldiers had been driven by our barrage. Through his efforts about 150 Germans were captured before they had a chance to come out from cover and man their machine guns. He was wounded while attempting to enter barracks in which several of the enemy had taken refuge.

*TABARA, WLADYSLAW (1716369)............
Near Revillon, France, Sept. 13, 1918.
R—Sag Harbor, N. Y.
B—Russia.
G. O. No. 64, W. D., 1919.

Private, Company M, 308th Infantry, 77th Division.
With a companion he determined the location of a machine gun which had checked the advance of his company, and, advancing ahead of the company, made a sudden rush from the flank, killed, wounded, or captured the entire crew and captured 4 machine guns.

*TABOR, RALPH E. (1205416)............
East of Ronssoy, France, Sept. 29, 1918.
R—Mechanicville, N. Y.
B—Stillwater, N. Y.
G. O. No. 16, W. D., 1919.

Corporal, Company L, 105th Infantry, 27th Division.
During the operations against the Hindenburg line, Corporal Tabor left shelter, went forward under heavy shell and machine-gun fire, and succeeded in bringing back to our lines a wounded soldier. His splendid courage and gallant conduct was a fine example to his comrades.
Posthumously awarded. Medal presented to father, Nathaniel Tabor.

TACK, ABRAHAM T. (540627)............
Near Hill 299, France, Oct. 16, 1918.
R—Sodus, N. Y.
B—Netherlands.
G. O. No. 27, W. D., 1920.

Private, first class, Company A, 7th Infantry, 3d Division.
Private Tack assumed command of a platoon after its officers had been wounded and led it to its objective. He advanced through heavy machine-gun and artillery fire for a distance of 800 meters and engaged in a hand-to-hand fight that resulted in the defeat of the enemy and the capture of 5 machine guns and 30 prisoners.

TALBOT, ARTHUR..........
  Near La Roux Farm, France, Oct. 18, 1913.
  R—New York, N. Y.
  B—Lyme, Conn.
  G. O. No. 24, W. D., 1920.

First lieutenant, 107th Infantry, 27th Division.
Being unable to find a suitable target for the 37-millimeter gun of which he was in command, Lieutenant Talbot armed the men of his section with enemy rifles and led a daylight patrol in advance of the lines. He reconnoitered La Roux Farm, exposed to heavy machine-gun fire, and put 2 enemy machine guns out of action, thus enabling our line to advance 1,000 yards without serious losses.

TAMME, NICHOLAS L. (1527377)..........
  South of Cierges, France, Sept. 29, 1918.
  R—Cincinnati, Ohio.
  B—Ripley, Ohio.
  G. O. No. 9, W. D., 1923.

Private, Headquarters Company, 147th Infantry, 37th Division.
Volunteering to attempt the recovery of the barrel of a 37-millimeter gun abandoned the previous day when the gun crew was gassed and when his battalion had retired to a more advantageous position, Private Tamme advanced alone in broad daylight and under observation of the enemy 200 yards in advance of his own lines under intense machine-gun fire, recovered the missing part and returned in safety to his own lines. His conduct was a splendid example of devotion to duty and proved an inspiration to the men of his battalion.

*TAPPEN, JAMES J. (1708342)..........
  Near Binarville, France, Sept. 28, 1918.
  R—Stapleton, N. Y.
  B—Stapleton, N. Y.
  G. O. No. 32, W. D., 1919.

Private, first class, Company D, 308th Infantry, 77th Division.
He pushed forward alone against several enemy snipers who were causing many casualties among his comrades. He killed 2 of the snipers, but was killed while attempting to capture the third sniper.
Posthumously awarded. Medal presented to father, James Tappen.

TARTER, CHARLES M. (51354)..........
  Near Vaux, France, July 1, 1918.
  R—Chapman, Kans.
  B—Columbia, Ky.
  G. O. No. 59, W. D., 1919.

Sergeant, Company I, 23d Infantry, 2d Division.
When his captain was wounded he went out under violent machine-gun fire to bring the officer to shelter and was severely wounded himself while performing this heroic act.

TAUBERT, ALBERT A..........
  In the Villers-Cotterets Forest, south of Soissons, France, July 18, 1918.
  R—Madison, Wis.
  B—Madison, Wis.
  G. O. No. 117, W. D., 1918.

Private, 66th Company, 5th Regiment, U. S. Marine Corps, 2d Division.
He went out in advance of the line of his company into the fire of a machine gun that was shooting at him and captured the gun and its crew.

TAUGHER, CLAUDE BUCKLEY..........
  At Bayonville, France, Nov. 2, 1918.
  R—Wausau, Wis.
  B—Wausau, Wis.
  G. O. No. 35, W. D., 1919.

Second lieutenant, 6th Regiment, U. S. Marine Corps, 2d Division.
Lieutenant Taugher with great dash led his platoon in surrounding enemy dugouts in the village of Bayonville before the occupants had time to escape or organize effective resistance, capturing 61 of the enemy; although wounded in the ankle, he refused to be evacuated.

TAVANO, ANTONIO J. (1273012)..........
  Near Verdun, France, Oct. 11, 1918.
  R—Dundee Lake, N. J.
  B—Croton-on-Hudson, N. J.
  G. O. No. 37, W. D., 1919.

Sergeant, Company D, 111th Machine Gun Battalion, 29th Division.
He voluntarily left his cover during a heavy bombardment and brought a wounded officer to a place of safety after the litter bearers were killed. He gave all the assistance possible to these wounded men before they died.

TAVENNER, ROBERT L..........
  Near Cierges, France, Sept. 29, 1918, and near Olsene, Belgium, Oct. 31, 1918.
  R—Mount Vernon, Ohio.
  B—Springfield, Ohio.
  G. O. No. 98, W. D., 1919.

Captain, 148th Infantry, 37th Division.
Without regard for his own safety Captain Tavenner personally conducted a tank in an attack on a machine-gun nest. After several of the tanks had been put out of action and the others had withdrawn, he walked up and down the firing line under heavy machine-gun fire cheering his men, who despite severe losses, fought till all of their ammunition was exhausted. On Oct. 31, he was severely wounded while making a personal reconnaissance of the enemy's position.

TAWATER, CARL (41245)..........
  Near Landres-et-St. Georges, France, Nov. 1, 1918.
  R—Roseland, Tex.
  B—Winchester, Tenn.
  G. O. No. 46, W. D., 1919.

Sergeant, Headquarters Company, 9th Infantry, 2d Division.
Just as his platoon went over the top Sergeant Tawater and several other soldiers were wounded by a shell which exploded near them. After seeing that the other wounded men were properly cared for, he organized the rest of his platoon and rejoined his unit, remaining on duty all day with his Stokes mortar section in spite of a painful wound in the foot.

*TAYLOR, DOUGLAS A..........
  Near Juvigny, France, Aug. 28-30, 1918.
  R—Rhinelander, Wis.
  B—Rhinelander, Wis.
  G. O. No. 74, W. D., 1919.

Second lieutenant, 127th Infantry, 32d Division.
He displayed marked heroism during the attack on Juvigny and when mortally wounded refused to be evacuated, but continued to advance and gave orders to continue the attack.
Posthumously awarded. Medal presented to father, Arthur Taylor.

TAYLOR, EWING M..........
  Near Exermont, France, Oct. 5, 1918.
  R—New York, N. Y.
  B—Poughkeepsie, N. Y.
  G. O. No. 126, W. D., 1919.

Major, 18th Infantry, 1st Division.
He displayed marked personal bravery in engagements with the enemy at Cantigny and Soissons, in each of which he was wounded, and later, near Exermont, exhibited heroism and able leadership in advancing his machine guns under heavy fire, aiding the advance of the entire battalion, until he was himself severely wounded.

TAYLOR, HERBERT S. (1459496)..........
  At Cheppy, France, Sept. 26, 1918.
  R—St. Louis, Mo.
  B—England.
  G. O. No. 13, W. D., 1919.

Mess sergeant, Company B, 138th Infantry, 35th Division.
Sergeant Taylor, with a small detachment, volunteered to go to the assistance of several men of another company who, cut off from support, were being annihilated by enemy machine gunners and snipers. Pushing forward under fire, Sergeant Taylor alone crawled around to the rear of a building from which an intense fire was coming and returned with 18 prisoners.

*TAYLOR, JOHN L..........
  Near Soissons, France, July 18, 1918.
  R—Hustonville, Ky.
  B—Casey County, Ky.
  G. O. No. 132, W. D., 1918.

Captain, 9th Infantry, 2d Division.
He assumed command of his battalion upon the death of his major and continued to lead the advance under heavy artillery and machine-gun fire, refusing to leave until he had been wounded 5 times. His example was an inspiration to all near him and an important factor in the successful attack made by his regiment.
Posthumously awarded. Medal presented to widow, Mrs. Caroline Taylor.

TAYLOR, LOUIS H. (1903594) _____
Near St. Juvin, France, Oct. 11, 1918.
R—Westfield, Mass.
B—Southwick, Mass.
G. O. No. 50, W. D., 1919.

Private, first class, medical detachment, 326th Infantry, 82d Division.
He repeatedly exposed himself to concentrated machine-gun and artillery fire, crossing the Aire River several times, and administered first aid to wounded men with complete disregard for his own safety.

TAYLOR, ORVILLE R. (253893) _____
Near Suippes, France, July 14-15, 1918.
R—Upland, Ind.
B—Eaton, Ind.
G. O No. 21, W. D., 1919.

Private, first class, Battery E, 42d Artillery, Coast Artillery Corps.
No other duties having been assigned to him, Private Taylor volunteered for service as a stretcher bearer, and, working all night under the heaviest shellfire, he carried wounded American and French soldiers to safety. While taking a severely wounded soldier by automobile to a hospital, a shell burst near him, wounding him, but he continued on his mission and delivered the wounded man to the aid station.

TAYLOR, OSCAR O. (1875665) _____
Near Ville-en-Woevre, France, Nov. 9, 1918.
R—Jonesboro, Tenn.
B—Jonesboro, Tenn.
G. O. No. 81, W. D., 1919.

Sergeant, Company D, 318th Machine Gun Battalion, 81st Division.
Displaying inspiring courage, Sergeant Taylor led his section through 3 heavy artillery barrages and directed the mounting of his guns on positions which he had personally reconnoitered under heavy machine-gun fire. At a critical juncture, when the Infantry was held up by enemy fire, he successfully led the nearest Infantry combat group forward in the assault, under cover of fire from his own guns.

TAYLOR, THOMAS J._____
Near the Cote-de-Chatillon, France, Nov. 1, 1918.
R—Brooklyn, N. Y.
B—New York, N. Y.
G. O. No. 46, W. D., 1919.

Second lieutenant, 23d Infantry, 2d Division.
When all the other officers of his company had been incapacitated, Lieutenant Taylor took command and successfully led his men throughout the 5 days' operations, capturing a strongly held position with more than 100 prisoners.

TAYLOR, WILLIAM C. (1308581) _____
In the Butry Woods, France, Oct. 9, 1918.
R—Emmett, Tenn.
B—Blountville, Tenn.
G. O. No. 87, W. D., 1919.

Sergeant, Company H, 117th Infantry, 30th Division.
Upon learning that an advanced platoon was under heavy enemy machine-gun fire from the front and flanks, Sergeant Taylor on his own initiative took a squad of men and an automatic rifle and went to the assistance of the platoon. Despite the heavy fire, he succeeded in compelling the enemy to withdraw from their positions and thereby enabled the battalion to resume its advance. He was later wounded in the leg and arms in attempting to bring up reinforcements across an open space swept by machine-gun fire.

*TAYLOR, WILLIAM H._____
Near Pont-a-Mousson, France, May 28, 1918.
R—New York, N. Y.
B—Scranton, Pa.
G. O. No. 15, W. D., 1923.

First lieutenant, 95th Aero Squadron, Air Service.
Accompanied by Lieutenant Hambleton, he answered an alert to Lironville and encountered 5 enemy planes in the vicinity of St. Mihiel. As they approached, the enemy turned away. Lieutenant Taylor and Lieutenant Hambleton followed and at Pont-a-Mousson again came up with them, flying, 1 at 1,500 meters, 2 at 2,000 meters, and the remaining 2 above at 2,500 meters. Lieutenant Hambleton attacked the lowest one, firing 20 rounds and forcing it from the formation, while Lieutenant Taylor remained above to protect him and to keep off the other enemy planes. An enemy bullet having shot the cross-section wires of Lieutenant Hambleton's plane away, the splinters from same cutting his cheek and right shoulder, he turned from combat to ascertain damage to his plane. As the enemy plane was falling Lieutenant Taylor opened fire and immediately brought the German down.
Posthumously awarded. Medal presented to father, William H. Taylor.

TAYLOR, WILLIAM J. R._____
Near Malancourt and Montfaucon, France, Sept. 26 Oct. 10, 1918.
R—Rochester, N. Y.
B—Rochester, N. Y.
G. O. No. 3, W. D., 1919.

First lieutenant, 3d Balloon Squadron, Air Service.
On Sept. 26, while conducting an important observation, he was twice attacked by enemy planes. He would not jump from his balloon because of the valuable work he was doing for the Infantry, although he was at all times in danger of losing his life from incendiary bullets. On Oct. 3, near Montfaucon, he was attacked but refused to leave until his balloon caught fire. Again on Oct. 6, he was attacked and forced down in his parachute. On Oct. 10, while he was conducting an important observation, an enemy hovered over his balloon; he refused to jump until attacked at close quarters. His heroic devotion to duty was an inspiration to the officers and men of his company.

TAYNTOR, CLARK O._____
At Sergy, France, July 29-30, 1918.
R—Erie, Pa.
B—Barre, Vt.
G. O. No. 66, W. D., 1919.

First lieutenant, 47th Infantry, 4th Division.
Disregarding 2 wounds from shellfire, which he had suffered, Lieutenant Tayntor continued in the advance with his platoon, keeping his men well organized, directing the consolidation of the line throughout the night, and refusing medical attention until all the wounded men in his platoon had received treatment.

*TEACHEY, ROBERT MARSHALL (1319726)
Near Ypres, Belgium, Aug. 2, 1918.
R—Raleigh, N. C.
B—Raleigh, N. C.
G. O. No. 142, W. D., 1918.

Private, Company B, 120th Infantry, 30th Division.
He volunteered to accompany an officer on a daylight patrol to destroy an enemy pill box. With great courage under heavy shell and machine-gun fire, they rushed the pill box, killed or wounded the occupants, and accomplished their mission.
Posthumously awarded. Medal presented to father, J. H. Teachey.

TECHEL, EDWARD W. (2039644) _____
In the Bois-Brabant-sur-Meuse, France, Oct. 8, 1918.
R—Milwaukee, Wis.
B—Milwaukee, Wis.
G. O. No. 27, W. D., 1919.

Private, Company B, 116th Infantry, 29th Division.
With 4 other soldiers he attacked 8 German machine guns, capturing them and their crews in spite of determined resistance by the enemy.

TEER, HUBERT O._____
At Ardeuil, France, Sept. 29, 1918.
R—Durham, N. C.
B—Durham, N. C.
G. O. No. 21, W. D., 1919.

First lieutenant, 371st Infantry, 93d Division.
Severely wounded in the back about 11 a. m., Lieutenant Teer continued to command his platoon until 4 p. m., when he was forced to withdraw from action on account of complete exhaustion.

TEEVAN, JOHN (89705)_____
Near Villers-sur-Fere, France, July 31, 1918.
R—New York, N. Y.
B—New York, N. Y.
G. O. No. 32, W. D., 1919.

Private, Company C, 165th Infantry, 42d Division.
After his platoon had withdrawn from their position he volunteered and returned to the position formerly occupied in an attempt to rescue a wounded comrade. He crossed a field swept by unusually intensive machine-gun fire, continuing in his attempted rescue until himself wounded.

TEICHLER, JOHN (263816)_____
At Hill 212, near Cierges, northeast of Chateau-Thierry, France, July 31, 1918.
R—Menominee, Mich.
B—Menominee, Mich.
G. O. No. 132, W. D., 1918.

Sergeant, Company L, 125th Infantry, 32d Division.
Although he was himself severely wounded in the attack on the Bois-les-Jamblets, yet he attempted to carry in another wounded man, passing through severe machine-gun fire from the front and from the flanks. While doing so he received a second wound, which caused his death.
Posthumously awarded. Medal presented to brother, Edward A. Teichler.

TEISETH, JACOB B. (2291197)_____
At Claire-Chenes Woods, France, Oct. 20, 1918.
R—Stanwood, Wash.
B—Norway.
G. O. No. 20, W. D., 1919.

Private, medical detachment, 6th Engineers, 3d Division.
He advanced in the attack with the company to which he was attached and worked constantly under heavy machine-gun and rifle fire searching for wounded and superintending their evacuation. While engaged in this work in the open under machine-gun fire, he was killed.
Posthumously awarded. Medal presented to mother, Mrs. Anna Teiseth.

TELFAIR, SAMUEL F_____
At Brieulles, France, Nov. 4, 1918.
R—Raleigh, N. C.
B—Raleigh, N. C.
G. O. No. 37, W. D., 1919.

Second lieutenant, 2d Antiaircraft Machine Gun Battalion.
He was leading a patrol to reconnoiter a position for antiaircraft machine guns when his group became scattered by intense shellfire. Upon returning to the shell-swept area to look for his patrol he found 1 of the men severely wounded. Making 2 trips through the heavy shellfire he secured the assistance of Private Laurel B. Heath and carried the wounded soldier to safety.

TEMPLE, JOHN H. (1736172)_____
Near Grand Pre, France, Oct. 23, 1918.
R—Marshallton, Del.
B—Newark, Del.
G. O. No. 64, W. D., 1919.

Private, Company I, 312th Infantry, 78th Division.
After his platoon had reached its objective and was forced to retire under perilous machine-gun fire, Private Temple and 2 companions were surrounded by the enemy. His companions were wounded, but he bravely held off the enemy, after which he assisted both his companions to a first-aid station.

TEMPLETON, CHARLES K_____
Near Nouart, France, Nov. 5, 1918.
R—Chicago, Ill.
B—Superior, Nebr.
G. O. No. 37, W. D., 1919.

Second lieutenant, 122d Field Artillery, 33d Division.
After telephone communications had been destroyed and his runners scattered on other missions, Lieutenant Templeton started on a mission of extreme importance from the Infantry to the Artillery. His path lay through a heavy machine-gun and shell fire, and before he reached his destination he was seriously wounded. He succeeded, however, in relaying his message to its destination.

TEN EYCK, WALTON D., Jr_____
Near Briquenay, France, Oct. 27, 1918.
R—Brooklyn, N. Y.
B—Brooklyn, N. Y.
G. O. No. 15, W. D., 1919.

Second lieutenant, pilot, 96th Aero Squadron, Air Service.
While engaged on a voluntary bombing mission he was attacked by 7 enemy planes (Fokker type). Although seriously wounded, he maneuvered his plane so skillfully that his observer was able to drive off the enemy planes. In the combat his plane was struck by 25 enemy bullets, some of which exploded the magazines of the observer's guns. In spite of his wounds and the damage to his machine he succeeded in landing safely on a strange field.

TENLEY, EUGENE H_____
Near St. Etienne, France, Oct. 4, 1918.
R—Willcox, Ariz.
B—Quanah, Tex.
G. O. No. 32, W. D., 1919.

Hospital apprentice, first class, U. S. Navy, attached to 49th Company, 5th Regiment, U. S. Marine Corps, 2d Division.
Disregarding his own safety, he voluntarily accompanied a small force into action, rendering most valuable treatment to the wounded until killed by a fragment of a shell.
Posthumously awarded. Medal presented to father, Samuel W. Tenley.

TENNYSON, JOSEPH E. (1264045)_____
Near Verdun, France, Oct. 8–24, 1918.
R—Baltimore, Md.
B—Baltimore, Md.
G. O. No. 126, W. D., 1919.

Corporal, Company B, 115th Infantry, 29th Division.
In several advances during this period Corporal Tennyson led his squad in attacks on machine-gun nests with conspicuous gallantry, always disregarding his own safety and encouraging his men both by words and actions. On Oct. 24, while leading his squad in an attack on a machine-gun nest, he was instantly killed.
Posthumously awarded. Medal presented to widow, Mrs. Gertrude Tennyson.

TERNIG, JACOB B. (1389369)_____
Near Bois-des-Forges, France, Sept. 26, 1918.
R—Chicago, Ill.
B—Luxembourg.
G. O. No. 64, W. D., 1919.

Sergeant, Company C, 132d Infantry, 33d Division.
He had just captured and was taking to his platoon commander a German captain, when fire was opened on his platoon from 3 concealed machine guns. Showing great bravery and presence of mind, Sergeant Ternig, who speaks German, ran toward the enemy emplacements, taking his prisoner with him, and called upon the crews to cease firing. Firing stopped and his platoon was enabled to take the enemy position and 30 prisoners without loss.

TERRELL, ALEXANDER W_____
Near Pexonne, France, Mar. 5, 1918.
R—Fort Worth, Tex.
B—Booneville, Mo.
G. O. No. 139, W. D., 1918.

Second lieutenant, 151st Field Artillery, 42d Division.
He showed unusual courage in assisting to direct the operations of Battery C, 151st Field Artillery, near Pexonne, France, on Mar. 5, 1918, when that organization was under particularly accurate artillery bombardment. Although wounded himself, he refused first aid and continued on duty until all of the wounded soldiers of the command had been treated.

TERRELL, HUBERT P. (1311740)_____
Near Vaux-Andigny, France, Oct. 12, 1918.
R—Cheraw, S. C.
B—Chesterfield County, S. C.
G. O. No. 133, W. D., 1918.

Corporal, Company I, 118th Infantry, 30th Division.
During an advance, when his company came under an enfilading fire from an enemy machine gun, he asked permission from his platoon commander to attempt the taking of the position. Although under heavy fire from this post and from trench-mortar shells, he, with exceptional dash and bravery, attacked the position alone, putting it out of action, killing 2 of the enemy and wounding a third. This soldier was killed the same day while reorganizing and advancing the weakened platoon of which he was then in charge.
Posthumously awarded. Medal presented to brother, Clarence Terrell.

TERRILL, ELSWORTH O. (1277360)_____
In the vicinity of Hagenbach, Alsace, east of Belfort, France, Aug. 21, 1918.
R—Rahway, N. J.
B—Rahway, N. J.
G. O. No. 99, W. D., 1918.

Corporal, Company H, 113th Infantry, 29th Division.
When his right hand and arm were badly mangled by the explosion of a grenade during an enemy raid into our lines, he placed his injured hand in his trousers pocket to support it, went over the top with his comrades, and joined in the pursuit of the defeated and retreating Germans, throwing hand grenades with his left hand as he followed them back to their own lines.

TERRY, MILO E_____
Near Montfaucon, France, Sept. 26 to 30, 1918.
R—Van Wert, Ohio.
B—Van Wert, Ohio.
G. O. No. 9, W. D., 1923.

Captain, 145th Infantry, 37th Division.
Although severely wounded while leading his company in the assault, he refused to be evacuated and courageously continued in command of his company for 4 days in action under heavy fire of all arms and constantly in contact with the enemy. On Sept. 30 he was again severely wounded, but remained with his company until evacuated in a delirious condition after the company's relief had been completed. By his intrepid conduct and disregard of personal danger he inspired the men of his company and contributed greatly to the success of the operation.

TESKE, AMOS (96367)_____
Near Ancerviller, France, Mar. 4, 1918.
R—Coal Valley, Ala.
B—Cardiff, Ala.
G. O. No. 126, W. D., 1919.

Corporal, Company D, 167th Infantry, 42d Division.
He was a member of a patrol of 5 men on Mar. 4, 1918, near Ancerviller, France, and took a conspicuous part when it encountered an enemy patrol of 11 men, which it attacked and routed, taking 2 prisoners.

THACHER, ARCHIBALD G_____
At St. Juvin, France, Oct. 14, 1918.
R—New York, N. Y.
B—Boston, Mass.
G. O. No. 43, W. D., 1922.

Major, 306th Infantry, 77th Division.
While commanding the 2d Battalion in a flank march across the Aire River, Major Thacher, acting with the greatest gallantry and with utter disregard for his own safety, personally made a reconnaissance in the face of heavy hostile machine-gun and shell fire, well in advance of his battalion, thereby saving his command from heavy losses. It was due to his thorough reconnaissance that his subsequent successful attack on this strong hostile position was consummated.

THACKER, EDGAR (57952)_____
Near Cantigny, France, May 28-30, 1918.
R—Vanceburg, Ky.
B—Vanceburg, Ky.
G. O. No. 109, W. D., 1918.

Private, first class, Company H, 28th Infantry, 1st Division.
He displayed distinguished conduct as a company runner, passing frequently through Cantigny when it was being heavily shelled and also running through German barrages to deliver messages.

THALKE, MAX P. (284575)_____
Near Juvigny, north of Soissons, France, Aug. 30, 1918.
R—Menasha, Wis.
B—Aurora, Ill.
G. O. No. 20, W. D., 1919.

Sergeant, Company I, 128th Infantry, 32d Division.
He displayed unusual courage and gallantry in leading his platoon forward under heavy fire from artillery and machine guns. He also gave first aid to the wounded while under fire, and when a retirement was ordered he remained behind to carry back the wounded.

THARAU, HERMAN_____
Near Vierzy, France, July 18, 1918.
R—Buffalo, N. Y.
B—Germany.
G. O. No. 132, W. D., 1918.

Gunnery sergeant, 55th Company, 5th Regiment, U. S. Marine Corps, 2d Division.
While out with a reconnoitering party to establish liaison with the company on his right Sergeant Tharau captured a machine gun and killed the crew.
Posthumously awarded. Medal presented to mother, Mrs. Annie Tharau.

THARP, LEWIS M_____
North of Charpentry, France, Sept. 27 and 28, 1918.
R—Winfield, Kans.
B—Melrose, Kans.
G. O. No. 11, W. D., 1921.

First lieutenant, 140th Infantry, 35th Division.
Lieutenant Tharp repeatedly exposed himself to heavy enemy artillery and machine-gun fire in order to maintain communication between company and battalion headquarters.
Posthumously awarded. Medal presented to father, Walter P. Tharp.

THAW, WILLIAM_____
Near Reims, France, Mar. 26, 1918.
R—Pittsburgh, Pa.
B—Pittsburgh, Pa.
G. O. No. 121, W. D., 1918.

Major, 103d Aero Squadron, Air Service.
He was the leader of a patrol of 3 planes which attacked 5 enemy monoplanes and 3 battle planes. He and another member of the patrol brought down 1 enemy plane and the 3 drove down, out of control, 2 others, and dispersed the remainder.
Oak-leaf cluster.
A bronze oak leaf is awarded Major Thaw for extraordinary heroism in action near Montaigne, France, Apr. 20, 1918. In the region of Montaigne he attacked and brought down, burning, an enemy balloon. While returning to his own lines the same day, he attacked 2 enemy monoplanes, 1 of which he shot down in flames.

THAYER, SIDNEY, Jr_____
Near Beaumont, France, Nov. 11, 1918.
R—Haverford, Pa.
B—Marion, Pa.
G. O. No. 46, W. D., 1919.

First lieutenant, 5th Regiment, U. S. Marine Corps, 2d Division.
After having been wounded he remained with his company until its objective had been reached, refusing evacuation until rendered unconscious by loss of blood.

THEBAUD, DELPHIN E_____
Near Romagne, France, Oct. 9, 1918.
R—Philippine Islands.
B—Alameda, Calif.
G. O. No. 24, W. D., 1920.

Captain, 38th Infantry, 3d Division.
After the successful attack on Hill 253, Captain Thebaud was directed to establish liaison with the 30th Infantry. Three runners were sent out, but each returned, stating that it was impossible to reach the 30th Infantry, due to the intensity of enemy fire. Captain Thebaud turned over the command of his company and fearlessly exposed himself to heavy machine-gun fire to accomplish the mission. After having proceeded about 300 yards he fell, severely wounded by machine-gun fire. His conduct had a marked effect upon the morale of his men.

THEBERT, WILLIAM F. (51702)_____
Near St. Etienne-a-Arnes, France, Oct. 3–9, 1918.
R—Fort Covington, N. Y.
B—Fort Covington, N. Y.
G. O. No. 46, W. D., 1919.

Private, first class, Company K, 23d Infantry, 2d Division.
Acting as battalion runner, Private Thebert carried messages through intense artillery and machine-gun fire. This soldier had been on duty as a runner since June 6, being entrusted with especially important messages because of his fearlessness and reliability.

THEDINGER, LOUIS C. (1490035)_____
Near St. Etienne, France, Oct. 8, 1918.
R—Perry, Okla.
B—St. Joseph, Md.
G. O. No. 66, W. D., 1919.

Sergeant, Company E, 142d Infantry, 36th Division.
When his company had been stopped by heavy enemy machine-gun fire, Sergeant Thedinger left shelter for the purpose of interviewing 2 German prisoners. Learning from them the location and strength of the enemy's position, he obtained permission to attack it, and, with a party of 10 volunteers, went forward in the face of heavy machine-gun fire, flanking the machine-gun nest, and capturing 40 prisoners and 3 machine guns.

THEOBALD, CARL G. (2257163)_____
Near Gesnes, France, Oct. 10, 1918.
R—Hinckley, Utah.
B—Hinckley, Utah.
G. O. No. 37, W. D., 1919.

Corporal, 1st Battalion, Intelligence Section, 361st Infantry, 91st Division.
While on a liaison patrol, Corporal Theobald and Pvt. Ivan Y. Bailey attacked and captured a hostile machine-gun nest and its entire crew.

THIBODEAU, JOSEPH A. (69316)_____
Near Belleau Wood, France, July 18–23, 1918.
R—Lawrence, Mass.
B—Canada.
G. O. No. 125, W. D., 1918.

Mechanic, Company L, 103d Infantry, 26th Division.
During the early part of the action he assisted in the evacuation of wounded under severe artillery and machine-gun fire. When a wound in the arm made it impossible for him to carry stretchers, he refused to be evacuated, but rejoined his company, went over the top with his comrades, and continued in action with them until wounded in the leg.

*THOETE, CARL G. (155883)_____
At Cantigny, France, May 28, 1918.
R—Santa Barbara, Calif.
B—Lockland, Ohio.
G. O. No. 99, W. D., 1918.

Sergeant, first class, Company D, 1st Engineers, 1st Division.
Although twice wounded early in the attack at Cantigny, France, May 28, 1918, he went over the top with his section and courageously directed its operations for 5 hours under steady fire, refused medical treatment, and led a second advance until killed by a machine-gun bullet.
Posthumously awarded. Medal presented to sister, Mrs. Leona Thoete Ott.

THOMAS, CARR M. (128384)_____
Near Chateau-Thierry, France, July 21–23, 1918.
R—New Rochelle, N. Y.
B—Chicago, Ill.
G. O. No. 44, W. D., 1919.

Sergeant, Battery A, 12th Field Artillery, 2d Division.
With another soldier, Sergeant Thomas voluntarily crossed an area swept by shell and machine-gun fire to establish liaison with the infantry, obtaining valuable information for the battery commander. Two days later, after working in an observatory under constant shellfire, this soldier was wounded, but he refused first aid until other men had been cared for, and went to the rear only upon being ordered to do so.

THOMAS, CHARLES I. (1813890)_____
Near Nantillois, France, Sept. 29, 1918.
R—Pittsburgh, Pa.
B—Latimore, Pa.
G. O. No. 37, W. D., 1919.

Private, Company D, 311th Machine Gun Battalion, 79th Division.
He was detailed as a runner between the battalion commander and his company. While delivering messages he was severely wounded, but continued in the performance of his duty, refusing aid, until ordered to the rear by his battalion commander.

THOMAS, DAVID (155864)_____
Northwest of Verdun, France, Oct. 9, 1918.
R—Avoca, Pa.
B—Taylor, Pa.
G. O. No. 35, W. D., 1919.

Corporal, Company A, 1st Engineers, 1st Division.
Upon 2 occasions Corporal Thomas, upon his own initiative, went out in advance of his platoon, armed only with a rifle, and attacked machine guns which were endangering his company by enfilading fire. In the face of fire from these guns he continued to fire on them until he had killed the gunners.

*THOMAS, EVERETT (51401)_____
Near Vaux, France, July 1, 1918.
R—Paris, Ill.
B—Edgar County, Ill.
G. O. No. 102, W. D., 1918.

Bugler, Company I, 23d Infantry, 2d Division.
Attempting to shelter his captain, who was lying wounded and exposed to fire, near Vaux, France, July 1, 1918, he was himself killed, thereby sacrificing his life in an effort to rescue his commanding officer.
Posthumously awarded. Medal presented to father, Samuel B. Thomas.

THOMAS, FRANK B. (1201207)_____
Near Ronssoy, France, Sept. 27, 1918.
R—New York, N. Y.
B—New York, N. Y.
G. O. No. 44, W. D., 1919.

Private, first class, Company C, 102d Field Signal Battalion, 27th Division.
When the telephone lines had been destroyed by the advancing tanks, and the enemy had started a counterattack from 3 sides before new ones could be laid, Private Thomas volunteered to carry a message from the Infantry battalion to which he was attached and succeeded in going through intense artillery, machine-gun, and sniper fire to regimental headquarters, delivering the message in time to enable reinforcements to be brought up.

THOMAS, FRED_____
Near St. Etienne, France, Oct. 4, 1918.
R—Hundred, W. Va.
B—Hundred, W. Va.
G. O. No. 44, W. D., 1919

Second lieutenant, 5th Regiment, U. S. Marine Corps, 2d Division.
While endeavoring to reestablish a large company front which had become disconnected, Lieutenant Thomas encountered a large number of the enemy filtering through our lines. By strategic maneuvers he formed a strong resistance, causing heavy casualties on the enemy and forcing their retreat after he himself had been seriously wounded.
Oak-leaf cluster.
Lieutenant Thomas is also awarded an oak-leaf cluster for the following act of extraordinary heroism in action near Chateau-Thierry, France, June 25, 1918: He commanded the left flank platoon of his company, which was subjected to heavy fire from enemy machine guns and trench mortars. When further advance in the face of the fire became impossible, he went forward alone, located the machine-gun positions, and then organized a flank attack on the emplacements, putting out of action 4 guns, 1 of which he himself captured. In this exploit his command suffered 40 per cent casualties and captured 21 prisoners. Reforming the remnants of his platoon, he moved forward through the enemy's barrage and to his objective, which he consolidated and held in the face of 3 counterattacks in 5 hours.

*THOMAS, GERALD PROVOST_____ Near Cambrai, France, Sept. 22, 1918. R—Flushing, N. Y. B—Flushing, N. Y. G. O. No. 35, W. D., 1920.	Second lieutenant, 17th Aero Squadron, Air Service. When the 10 planes of his group were attacked by a superior number of the enemy, Lieutenant Thomas refused to seek safety in flight, but attacked a superior number of the enemy in order to assist another member of his squadron to escape. In the performance of this act he was shot down and killed by the enemy. Posthumously awarded. Medal presented to father, Rupert B. Thomas
THOMAS, HASTINGS (3086695)_____ Near Vieville-en-Haye, France, Nov. 1, 1918. R—Cosby, Mo. B—Cosby, Mo. G. O. No. 37, W. D., 1919.	Private, Company C, 21st Machine Gun Battalion, 7th Division. Private Thomas acted as a runner during offensive operations and under severest enemy barrages, carrying messages to and from the front lines.
THOMAS, JOHN (3105811)_____ Near Nantillois, France, Sept. 28, 1918. R—Philadelphia, Pa. B—Greece. G. O. No. 15, W. D., 1923.	Private, Company F, 316th Infantry, 79th Division. Under terrific enemy fire he advanced alone to the enemy lines and silenced an enemy machine gun, driving the enemy crew to flight. His bravery and devotion to duty enabled the Infantry to resume the advance against the enemy forces with a minimum of losses. His conduct greatly inspired the men with whom he served.
THOMAS, ROLAND CALVIN (1312437)_____ At Vaux-Andigny, France, Oct. 15, 1918. R—Kershaw, S. C. B—Union County, N. C. G. O. No. 37, W. D., 1919.	Corporal, Company M, 118th Infantry, 30th Division. After being twice wounded he continued to advance with his automatic-rifle squad, leading his men 100 yards under extremely heavy fire.
THOMAS, SPIROS (89250)_____ Near Landres-et-St. Georges, France, Oct. 15, 1918. R—New York, N. Y. B—Greece. G. O. No. 37, W. D., 1919.	Sergeant, Company B, 165th Infantry, 42d Division. Sergeant Thomas, after all his officers and first sergeant had become casualties, took command of his company, led them forward under heavy artillery and machine-gun fire, and retained complete control of the company, although suffering heavy casualties and under trying conditions, until relieved at the close of the day.
*THOMAS, WILLIAM (1717272)_____ Near Chevieres, France, Oct. 13, 1918. R—Yonkers, N. Y. B—South Wales. G. O. No. 21, W. D., 1919.	Sergeant, Company D, 302d Engineers, 77th Division. He accompanied an officer on a reconnaissance, searching for possible locations for crossing the Aire River. They crossed open ground subject to shellfire and under direct observation of the enemy. On reaching the river they were exposed to machine-gun and sniper's fire. Both he and the officer failed to return, and their bodies were afterwards discovered in the Aire River, where they had fallen after being killed or wounded by enemy fire. Posthumously awarded. Medal presented to uncle, Thomas Price.
*THOMPSON, CECIL E. (2267689)_____ Near Eclisfontaine, France, Sept. 28, 1918. R—Fellows, Calif. B—Sacramento, Calif. G. O. No. 21, W. D., 1919.	Private, Company K, 364th Infantry, 91st Division. Engaged on scouting duty, Private Thompson went forward on his own initiative, and located the position of the enemy on our front. After an all-night exploit, he returned with his information. Without rest, he went out in the morning and localted a troublesome machine-gun nest. Posthumously awarded. Medal presented to father, Charles Thompson.
THOMPSON, CHARLES W. (1979177)_____ Near Vaux-Andigny, France, Oct. 11, 1918. R—Lynnville, Ind. B—Warrick County, Ind. G. O. No. 44, W. D., 1919.	Sergeant, Machine Gun Company, 120th Infantry, 30th Division. When his machine-gun position on the flank of the line became untenable, he crawled 20 yards in front of the position and opened fire with his rifle, covering the withdrawal of the crew and thereby saving both gun and crew from capture.
THOMPSON, CLARENCE W. (1039036)____ Near Greves Farm, France, July 15, 1918. R—Van Norman, Mont. B—Belmont, Ohio. G. O. No. 98, W. D., 1919.	Sergeant, Battery F, 10th Field Artillery, 3d Division. Responding to a call for volunteers, Sergeant Thompson, with 8 other soldiers, manned 2 guns of a French battery which had been deserted by the French during the unprecedented fire, after many casualties had been inflicted on their forces. For 2 hours he remained at his post and poured an effective fire into the ranks of the enemy.
THOMPSON, CLIFFORD (93311)_____ Near Sommerance, France, Oct. 23, 1918. R—Troy, Ohio. B—Troy, Ohio. G. O. No. 26, W. D., 1919.	Sergeant, Company C, 166th Infantry, 42d Division. Seeing an ignited hand grenade in the midst of his platoon, Sergeant Thompson without hesitation seized the grenade and attempted to throw it from the ditch. When leaving his hand the grenade exploded, seriously wounding him, but his act saved the lives of many of his men.
THOMPSON, EDWARD N. (1204944)_____ Near Mount Kemmel, Belgium, Aug. 31, 1918. R—New York, N. Y. B—New York, N. Y. G. O. No. 23, W. D., 1919.	First sergeant, Company I, 105th Infantry, 27th Division. When the 2 platoons commanded by him met with heavy machine-gun fire, Sergeant Thompson placed his men under cover, and, single-handed, went forward to reconnoiter his objective in the face of heavy shell and machine-gun fire.
THOMPSON, EMMETT (1403376)_____ At Mont-de-Sanges, France, Sept. 20 to Oct. 1, 1918. R—Quincy, Ill. B—La Belle, Mo. G. O. No. 46, W. D., 1919.	Corporal, Company L, 370th Infantry, 93d Division. After others had failed, Corporal Thompson volunteered and took charge of a detail to secure rations. He succeeded in this mission under very dangerous and trying conditions, and, notwithstanding the fact that his detachment suffered numerous casualties, he remained on this duty and continued to supply the company with rations until completely exhausted.

THOMPSON, GEORGE M. (1249878) _____
Near Montblainville, France, Sept. 26, 1918.
R—Springdale, Pa.
B—Springdale, Pa.
G. O. No. 74, W. D., 1919.

Private, headquarters 2d Battalion, 107th Field Artillery, 28th Division.
For 5 hours after he had been severely wounded in the arm Private Thompson maintained liaison between the Infantry and supporting Artillery, repeatedly carrying messages through the terrific fire and, being the only means of communication, greatly aided in the success of the attack. He went to the rear only when ordered to do so by his commanding officer.

THOMPSON, GEORGE RICHARD _____
Near Bellicourt, France, Sept. 29, 1918.
R—Forest Glen, Md.
B—Washington, D. C.
G. O. No. 37, W. D., 1919.

First lieutenant, 105th Field Signal Battalion, attached to 117th Infantry, 30th Division.
With another officer and 24 soldiers, he was proceeding to the front line to establish an advance message center, when the detachment was caught in the enemy's counterbarrage. Although seriously wounded himself, he assisted in dressing the wounds of his men and then continued the work of establishing communication until he was forced to be evacuated 2 hours later.

*THOMPSON, HENRY L _____
Near Vaux-en-Dieulet, France, Nov. 3, 1918.
R—Columbia, S. C.
B—Athol, Mass.
G. O. No. 21, W. D., 1919.

Captain, 23d Infantry, 2d Division.
Although painfully wounded, Captain Thompson led his battalion to the outskirts of Vaux-en-Dieulet, the advance being without artillery support and accomplished only by effective rifle fire. This officer himself set an example for his men by killing with a rifle 2 German machine gunners at a distance of 500 yards. He was again seriously wounded after reaching the objective while making dispositions for defense against counterattacks.
Posthumously awarded. Medal presented to widow, Mrs. Eleanor Thompson.

THOMPSON, JOHN W. (1244631) _____
Near Le Grande Savart, west of Fismette, France, Aug. 10, 1918.
R—Pittsburgh, Pa.
B—Pittsburgh, Pa.
G. O. No. 128, W. D., 1918.

First sergeant, Company G, 111th Infantry, 28th Division.
He showed remarkable bravery and disregard of personal danger when with 2 other soldiers he attacked a German machine gun, killed the crew, and then with deadly effect turned the gun upon other machine guns and hostile infantry which were in position near by. The crews of all the other German machine guns were killed, 10 machine guns were captured, and the way cleared for the further advance of the American forces.

THOMPSON, JOHN WILLIAM _____
Near Blanc Mont Ridge, France, Oct. 4, 1918.
R—Middlebury, Vt.
B—Middlebury, Vt.
G. O. No. 32, W. D., 1919.

Private, 55th Company, 5th Regiment, U. S. Marine Corps, 2d Division.
After locating a machine-gun nest, he destroyed 1 of the guns and returned to our lines with valuable information concerning the location of the nest.

*THOMPSON, JOHN W., Jr. (1315445) ____
Near Bellicourt, France, Sept. 29, 1918.
R—Mount Pleasant, Tenn.
B—Mount Pleasant, Tenn.
G. O. No. 32, W. D., 1919.

Corporal, Company E, 119th Infantry, 30th Division.
With another soldier, Corporal Thompson rushed a hostile machine gun which was firing on his company and killed 2 of the enemy. He then continued to advance close behind the barrage and displayed great bravery in the attack. Shortly before the company's objective was reached he was severely wounded and has since died of his wounds.
Posthumously awarded. Medal presented to father, J. W. Thompson.

*THOMPSON, LAWRENCE E. (42873) _____
Near Soissons, France, July 19, 1918.
R—Minden, W. Va.
B—McDowell County, W. Va.
G. O. No. 15, W. D., 1919.

Corporal, Company F, 16th Infantry, 1st Division.
In order to ascertain the location of a machine-gun which was inflicting heavy losses upon his platoon, he unhesitatingly went forward and was killed in the performance of this courageous duty.
Posthumously awarded. Medal presented to father, P. H. Thompson.

THOMPSON, ORLEN NELSON _____
Near the Argonne Forest, France, Sept. 26, 1918.
R—Detroit, Mich.
B—Cleveland, Ohio.
G. O. No. 21, W. D., 1919.

First lieutenant, 305th Infantry, 77th Division.
In the course of a successful advance in which 10 of the enemy had been captured, Lieutenant Thompson was severely wounded in the head by a shell fragment, but after regaining consciousness he refused assistance and carefully transmitted all orders and information to the second in command. Though he was weak from loss of blood, he went to the rear unaided, taking with him 10 prisoners.

*THOMPSON, ORRIE (279707) _____
In the woods north of Cierges, northeast of Chateau-Thierry, France, Aug. 1, 1918.
R—Kalamazoo, Mich.
B—Bangor, Mich.
G. O. No. 117, W. D., 1918.

Sergeant, Company C, 126th Infantry, 32d Division.
After his company had entered the woods north of Cierges he and another soldier maneuvered around a machine gun, which was causing many casualties in the company, and reached a shell hole after crossing an open space that was swept by hostile fire. From here they killed the crew of the machine gun, captured the gun, and turned it on the enemy.
Posthumously awarded. Medal presented to mother, Mrs. Eliza Thompson.

*THOMPSON, ROBERT E. _____
Between Chambley and Xammes, France, Sept. 13, 1918.
R—Temple, Tex.
B—Oenaville, Tex.
G. O. No. 15, W. D., 1923.

Second lieutenant, 96th Aero Squadron, Air Service.
While acting as observer of a flight of 3 airplanes they were attacked by a flight of 15 enemy airplanes. Despite the fact that his formation was surrounded by overwhelming numbers of the enemy, he continued his mission and bombed his objective. In the fight which followed Lieutenant Thompson and his pilot fought gallantly, thus enabling another airplane of the flight to return with valuable information of the enemy. In this fight Lieutenant Thompson's airplane was shot down and both he and his pilot were killed when their airplane crashed to the ground.
Posthumously awarded. Medal presented to father, Charles E. Thompson.

THOMPSON, SIMON M. (1439030) _____
Near Medeah Ferme, France, Oct. 8, 1918.
R—Libby, Minn.
B—Carlton, Minn.
G. O. No. 98, W. D., 1919.

Private, Company F, 9th Infantry, 2d Division.
Under intense machine-gun fire during a counterattack following a heavy artillery barrage, Private Thompson, with another soldier, checked the attack for a considerable distance, killing 10 of the enemy, including 2 officers, and keeping off the hostile party with his pistol while loading his rifle with his other hand.

THOMPSON, WALDO (193774)_____
Near Exermont, France, Oct. 5, 1918.
R—Opportunity, Mont.
B—Denmark.
G. O. No. 44, W. D., 1919.

Corporal, Company C, 2d Field Signal Battalion, 1st Division.
He voluntarily went forward in the face of a most destructive bombardment and kept in repair the telephone line connecting the Infantry and Artillery, thereby assuring the close cooperation between these 2 elements.

THOMPSON, WILLIAM DARIUS_____
Near Fleville, France, Oct. 4, 1918.
R—Port Huron, Mich.
B—Midland, Mich.
G. O. No. 68, W. D., 1920.

Captain, 2d Machine Gun Battalion, 1st Division.
By skillfully employing his machine guns he silenced the fire of hostile guns which were holding up the progress of the Infantry. He also led an attack on several nests, with the aid of tanks, and when his objective was reached and his guns placed he returned through the heavy barrage and brought up a platoon of Infantry which had been lost.

*THOMPSON, WILLIAM J. (89225)_____
Near Villers-sur-Fere, France, July 28, 1918.
R—New York, N. Y.
B—New York, N. Y.
G. O. No. 14, W. D., 1923.

Private, Company A, 165th Infantry, 42d Division.
On duty as sniper of the assaulting battalion, he assisted a runner who had been mortally wounded, relieved him of his message, and although mortally wounded himself delivered the dispatch to his company commander, dying shortly thereafter. His heroic conduct was an inspiration to his regiment.
Posthumously awarded. Medal presented to mother, Mrs. Annie Thompson.

THOMSON, HAROLD (97836)_____
Near Landres-et-St. Georges, France, Oct. 16, 1918.
R—St. Anthony, Idaho.
B—Salt Lake City, Utah.
G. O. No. 87, W. D., 1919.

Corporal, Company I, 167th Infantry, 42d Division.
With 4 other soldiers, Corporal Thomson pushed out on the right flank of his company and by well-directed fire gained fire superiority for our forces, captured 8 of the enemy, including an officer, and drove off a large number of others. His skillful leadership was of material assistance in facilitating the advance of the platoon.

THOMSON, JAMES C. (71608)_____
Near Bouresches, France, July 20, 1918.
R—Cambridge, Mass.
B—Scotland.
G. O. No. 125, W. D., 1918.

Private, Company C, 104th Infantry, 26th Division.
When wounded in the right arm he refused to be relieved from duty and continued the operation of his automatic rifle with his left hand. Later he volunteered to act as runner, and continued this duty until he fell exhausted.

THORP, ABRAHAM (2256596)_____
Near Gesnes, France, Sept. 28, 1918.
R—Rexburg, Idaho.
B—Russia.
G. O. No. 21, W. D., 1919.

Supply sergeant, Company B, 361st Infantry, 91st Division.
Although badly wounded, he crawled 500 meters under heavy shellfire to deliver important papers to his company commander.

THORNBURG, ZEBULON B._____
Near Montbrehain, France, Oct. 8-16, 1918.
R—Concord, N. C.
B—Cabarrus County, N. C.
G. O. No. 37, W. D., 1919.

First lieutenant, 118th Infantry, 30th Division.
Although he was severely wounded on Oct. 8 to such an extent that eating was impossible, he remained as second in command until the night of Oct. 16, when he was again wounded during an advance by his company.

*THORNE, CHARLES E. (139552)_____
Near Nantillois, France, Sept. 29, 1918.
R—Pierre, S. Dak.
B—Omaha, Nebr.
G. O. No. 21, W. D., 1919.

Private, first class, Battery C, 147th Field Artillery.
While on duty with his battery as a lineman, Private Thorne saw a soldier fall wounded by shell fragments. Leaving his shelter, he went through concentrated shellfire to the assistance of the wounded soldier, and in endeavoring to rescue him was himself killed by the explosion of a shell.
Posthumously awarded. Medal presented to father, Al Thorne.

THORNGATE, GEORGE_____
Near Romagne, France, Oct. 14-15, 1918.
R—Milton, Wis.
B—North Loup, Nebr.
G. O. No. 37, W. D., 1919.

First lieutenant, 6th Infantry, 2d Division.
Being seriously wounded, Lieutenant Thorngate displayed marked devotion to duty by refusing to go to the rear and remaining in command of his company until next day.

THORNHILL, WALTER P. (68634)_____
Near Bois-de-St. Remy, France, Sept. 12, 1918.
R—Fairfield, Me.
B—Newfoundland.
G. O. No. 26, W. D., 1919.

Corporal, Company H, 103d Infantry, 26th Division.
Advancing alone and under fire, he captured a machine gun and 8 prisoners.

THORNLEY, JAMES R. (64014)_____
In the Seicheprey engagement, France, Apr. 20, 1918.
R—Fall River, Mass.
B—Fall River, Mass.
G. O. No. 99, W. D., 1918.

Corporal, Company C, 102d Infantry, 26th Division.
Wounded early in the Seicheprey engagement on the morning of Apr. 20, 1918, he displayed great gallantry and devotion to duty in continuing to urge his men to defend their positions, aiding greatly in the defense of same by climbing a tree and from there shouting out directions as to the enemy's location.

THORNTON, JOSEPH (737074)_____
Near Thiaucourt, France, Sept. 12, 1918.
R—Glencoe, Ohio.
B—Brookside, Ohio.
G. O. No. 37, W. D., 1919.

Private, Company L, 11th Infantry, 5th Division.
By the effective use of an automatic rifle, he advanced without aid on the intrenched position of the enemy, forcing their surrender, despite severe wounds. He continued to advance and assisted in locating and routing other machine-gun nests.

THORNTON, ROBERT M. (97692)_____
Northeast of Chateau-Thierry, France, July 26, 1918.
R—Central, Ala.
B—Elmore County, Ala.
G. O. No. 108, W. D., 1918.

Corporal, Company I, 167th Infantry, 42d Division.
After being wounded, he remained in command of his squad and continued to direct the fire of their 2 automatic rifles for more than 2 hours and until he had been wounded 3 more times.

*THORSEN, EDWIN B_____
    Near Roncheres, France, July 3, 1918.
    R—Ashland, Wis.
    B—Ashland, Wis.
    G. O. No. 74, W. D., 1919.

Second lieutenant, 127th Infantry, 32d Division.
Wounded in the abdomen by a machine-gun bullet, he continued to lead his men in the advance for 100 yards, until he fell from loss of blood, dying later from his wound. His example was an inspiration to his men.
Posthumously awarded. Medal presented to mother, Mrs. Bertha Thorsen.

THRALL, HARRY (262324)_____
    Near Fismes, France, Aug. 8, 1918.
    R—Dunningville, Mich.
    B—Dunningville, Mich.
    G. O. No. 98, W. D., 1919.

Private, Company E, 125th Infantry, 32d Division.
Private Thrall was called upon to carry a message to battalion headquarters. This necessitated his passing through an intense artillery barrage. He was severely wounded in the leg by shrapnel. In spite of his wound he struggled on, refusing first aid until the message was delivered.

*THRASHER, DANA BRISTOL_____
    Near Vierzy, France, July 19, 1918.
    R—Chicago, Ill.
    B—New Haven, Conn.
    G. O. No. 22, W. D., 1920.

Private, Headquarters Company, 6th Regiment, U. S. Marine Corps, 2d Division.
Private Thrasher carried numerous messages to front-line platoons, crossing and recrossing areas swept by severe machine-gun and artillery fire. When wounded, he refused medical attention, directing those who came to his assistance to leave him and go forward.
Posthumously awarded. Medal presented to father, Samuel P. Thrasher.

THURMAN, LITTON T. (1164983)_____
    Near Bellicourt, France, Sept. 29, 1918.
    R—Crossville, Tenn.
    B—Evansville, Tenn.
    G. O. No. 81, W. D., 1919.

Sergeant, Company H, 119th Infantry, 30th Division.
When all the officers of his company had become casualties, he reorganized the company, extricated it from a dangerous position, and with coolness and courage led the command forward throughout the day. With a few other men he faced heavy machine gun and grenade fire to charge an enemy emplacement.

TIBBETTS, LOYD J. (1634401)_____
    Near Bois-de-St. Remy, France, Sept. 12, 1918.
    R—East Highland, Calif.
    B—Lincoln, Nebr.
    G. O. No. 26, W. D., 1919.

Private, Company F, 103d Infantry, 26th Division.
Although twice wounded, he continued to direct the fire of an automatic rifle squad, after which he led them forward, engaging in a hand-to-hand encounter with the enemy until he dropped from loss of blood.

TICKNER, ARTHUR J. (40226)_____
    Near Soissons, France, July 18, 1918.
    R—Syracuse, N. Y.
    B—Syracuse, N. Y.
    G. O. No. 21, W. D., 1919.

Corporal, Company I, 9th Infantry, 2d Division.
After his company commander had been wounded and he himself had been shot through the wrist, he assisted the captain to walk forward in the attack. During the advance a shell burst near by and took off the leg of his company commander and again wounded the corporal, who, in spite of his injuries, forced 5 Germans to carry his captain more than 4 kilometers to an aid station, thereby saving his live.

*TIEMAN, FREDERICK A. (1211399)_____
    Near Bony, France, Sept. 29, 1918.
    R—Jersey City, N. J.
    B—New York, N. Y.
    G. O. No. 56, W. D., 1922.

Private, Company H, 107th Infantry, 27th Division.
During the attack against the Hindenburg line, when he became separated from from his squad, with utter disregard for his personal safety Private Tieman fearlessly operated a Lewis machine gun, inflicting heavy losses upon the enemy. Later, when it was possible for him to rejoin his company, he assisted in rallying the men to further efforts and aided materially in the victory that was ultimately won. His splendid courage and gallant conduct were an inspiration to all his comrades.
Posthumously awarded. Medal presented to mother, Mrs. Esther C. Merritt.

TIERCE, WILLIAM A. (1348045)_____
    Near Thiaucourt, France, Sept. 12, 1918.
    R—Corona, Ala.
    B—Alabama.
    G. O. No. 46, W. D., 1919.

Private, Company D, 9th Infantry, 2d Division.
He showed exceptional courage when he left his trench in order to obtain a better field of fire for his automatic rifle. He operated the rifle to good advantage until both hands were broken and the gun disabled by a shell fragment. He continued with his company for 2 days after receiving this wound.

TIGNOR, WILLIAM P. (1821167)_____
    Near Sommauthe, France, Nov. 4-7, 1918.
    R—Old Church, Va.
    B—Old Church, Va.
    G. O. No. 37, W. D., 1919.

Private, Company D, 318th Infantry, 80th Division.
Acting as a scout he repeatedly went forward and by calling and making noises drew machine-gun fire upon himself in order to locate machine-gun nests, which were subsequently put out of action.

TILGHMAN, ALLEN BENJAMIN_____
    In the Bois-de-Belleau, France, June 6 and 8, 1918.
    R—St. Louis, Mo.
    B—Kenton, Tenn.
    G. O. No. 70, W. D., 1919.

Corporal, 83d Company, 6th Regiment, U. S. Marine Corps, 2d Division.
Howard Childs, Joseph A. Dargis, and Allen Benjamin Tilghman, corporals, and Herman L. McLeod, private, Company K, 6th Regiment, U. S. Marine Corps. These 4 men were prominent in the attack on enemy machine-gun positions in the Bois-de-Belleau on June 6 and 8, 1918, were foremost in their company at all times, and acquitted themselves with such distinction that they were an example for the remainder of their command.

TILGHMAN, CHARLES H._____
    Near Nantillois, France, Sept. 28, 1918.
    R—Easton, Md.
    B—Baltimore, Md.
    G. O. No. 81, W. D., 1919.

Captain, 315th Infantry, 79th Division.
After having been wounded in the head by a piece of high-explosive shell, which slightly fractured his skull and rendered one eye useless, Captain Tilghman insisted on remaining with his command. Throughout the night of constant rain and continual gas attacks, he encouraged his demoralized troops, remaining with them until evacuated on the following morning.

*TILLERY, JAMES M. (1306513)_____
    Near Montbrehain, France, Oct. 8, 1918.
    R—Knoxville, Tenn.
    B—Inskip, Tenn.
    G. O. No. 37, W. D., 1919.

Sergeant, Machine Gun Company, 117th Infantry, 30th Division.
On Oct. 8, near Montbrehain, although wounded, he took charge of a platoon whose leader had become a casualty, leading it in the advance until he received another wound, which later proved fatal.
Posthumously awarded. Medal presented to father, R. M. Tillery.

TILLMAN, FRED A_____
West of Reims, France, June 26, 1918.
R—Fayetteville, Ark.
B—Fayetteville, Ark.
G. O. No. 99, W. D., 1918.

Second lieutenant, Field Artillery, attached to 260th Squadron, French Air Service.
He aroused the admiration of all the French first-line infantrymen when making an Infantry liaison west of Reims, France, June 26, 1918. He flew over the enemy lines at an altitude of only 150 meters, in spite of violent machine-gun and antiaircraft fire. Shot down between the lines, with his plane riddled with bullets and his pilot severely wounded, he picked up his pilot in his arms and carried him through heavy fire more than 200 meters to the French first lines after he himself was wounded in the neck.

TIMBLIN, ARCHIE (2163514)_____
Near Butgneville, France, Nov. 11, 1918.
R—Blackduck, Minn.
B—Barren, Minn.
G. O. No. 71, W. D., 1919.

Private, first class, Company F, 131st Infantry, 33d Division.
Voluntarily facing heavy machine-gun fire, he advanced 100 yards ahead of our lines to bring back wounded soldiers. He showed absolute disregard for personal danger in his work of rescue.

*TIMM, CHARLES L. (65097)_____
Near Crepion, France, Oct. 23, 1918.
R—Ansonia, Conn.
B—Hamburg, N. Y.
G. O. No. 32, W. D., 1919.

Corporal, Company G, 102d Infantry, 26th Division.
Corporal Timm, with the assistance of 1 other soldier, captured 3 enemy machine guns and their crews which had held up the advance of his company. He was later killed in action in the Bois-de-la-Reine.
Posthumously awarded. Medal presented to sister, Miss Charlotte Timm.

TIMMERMAN, LOUIS F_____
In the Bois-de-Belleau, France, June 6, 1918.
R—Leonia, N. J.
B—New York, N. Y.
G. O. No. 110, W. D., 1918.

Second lieutenant, 6th Regiment, U. S. Marine Corps, 2d Division.
In the Bois-de-Belleau, on June 6, 1918, he led his men in a bayonet charge against superior numbers of the enemy, capturing 2 machine guns and 17 prisoners. Wounded in the face by shrapnel, he continued to heroically perform his duties until relieved.

*TIMOTHY, JAMES S_____
Near Chateau-Thierry, France, June 1-15, 1918.
R—Highland Falls, N. Y.
B—Nashville, Tenn.
G. O. No. 99, W. D., 1918.

Second lieutenant, Infantry, attached to 6th Regiment, U. S. Marine Corps, 2d Division.
Although weakened by gas poisoning, inflicted while serving with the French in Verdun sector, he declined medical assistance and served with heroic fortitude with the marines. In the operations of June 1-15, 1918, near Chateau-Thierry, he inspired the officers and men with whom he was in action by his fearlessness and fortitude until instantly killed by a high-explosive shell.
Posthumously awarded. Medal presented to father, P. H. Timothy.

TINDALL, PHILIP_____
Near Gesnes, northwest of Verdun, France, Oct. 2, 1918.
R—Seattle, Wash.
B—Washington, D. C.
G. O. No. 20, W. D., 1919.

First lieutenant, 126th Infantry, 32d Division.
He was severely wounded in the shoulder by a shell fragment at the beginning of the advance on Gesnes, but in spite of his wound he continued to lead his company throughout the advance. He helped to organize the ground against counterattack and remained on duty with his command until the next morning, when he went to the rear only under vigorous protest. Throughout the entire engagement this officer displayed the utmost coolness and devotion to duty under the heaviest fire.

TINSLEY, WILLIAM S. (2216380)____
At Apremont, France, Sept. 29, 1918.
R—Britton, Okla.
B—Saginaw, Tex.
G. O. No. 98, W. D., 1919.

Private, Company I, 110th Infantry, 28th Division.
During a determined enemy counterattack Private Tinsley volunteered with another soldier to locate the source of enemy fire which was causing heavy casualties in his company. Going forward under fire and discovering 7 of the enemy, they opened fire, killing 2, wounding 2, and capturing a machine gun. Their courageous feat materially aided his unit to continue the advance.

TITTMAN, HAROLD H_____
Near Bouresches, France, July 1, 1918.
R—St. Louis, Mo.
B—St. Louis, Mo.
G. O. No. 143, W. D., 1918.

First lieutenant, 94th Aero Squadron, 1st Pursuit Group, Air Service.
While on patrol he encountered 7 machines. Despite numerical superiority and the enemy advantage of position, he immediately attacked. After firing a few rounds his guns became jammed. In the midst of a veritable hail of machine-gun fire he repaired the jam and resumed the attack. Although he was severely wounded, he continued until the enemy was forced to retire behind their own lines.

TOBIN, EDGAR G_____
In the region of Vieville, France, July 16, 1918.
R—San Antonio, Tex.
B—San Antonio, Tex.
G. O. No. 99, W. D., 1918.

First lieutenant, 103d Aero Squadron, Air Service.
While leading a patrol of 3 machines in the region of Vieville, France, July 16, 1918, he attacked an enemy formation of 6 single seaters. He destroyed 2 himself and forced down a third out of control.

TOBIN, RICHARD J. (915468)_____
At Clery-le-Grand, France, Nov. 1, 2, and 10, 1918.
R—Baltimore, Md.
B—Leavenworth, Kans.
G. O. No. 81, W. D., 1919.

Master engineer, junior grade, Company C, 7th Engineers, 5th Division.
On Nov. 1 Master Engineer Tobin reconnoitered a destroyed bridge in advance of our outposts and on Nov. 2 constructed a bridge at the same point while under shellfire. On Nov. 10 he gave valuable aid to wounded men in the face of continuous fire.

TOBLINI, ANDY (242302)_____
Near Medeah Ferme, France, Oct. 3, 1918.
R—Colver, Pa.
B—Italy.
G. O. No. 21, W. D., 1919.

Private, Company F, 9th Infantry, 2d Division.
Private Toblini, together with 4 other men, charged a machine-gun nest containing 3 heavy machine guns and captured the 3 guns and 20 prisoners.

TODD, ELMER (105438)_____
South of Soissons, France, July 20, 1918.
R—Tulsa, Okla.
B—Tulsa, Okla.
G. O. No. 24, W. D., 1920.

Corporal, Company B, 2d Machine Gun Battalion, 1st Division.
Corporal Todd led his squad by crawling from shell hole to shell hole through heavy machine-gun fire for a distance of about 300 yards in advance of the Infantry. Selecting a position, he placed his machine gun so as to bring an enfilade fire on a section of enemy trench that was holding up the advance of the Infantry. The fire of his squad was so effective as to cause the enemy to surrender, thereby facilitating the further advance of the attacking battalion.

*TODD, HAROLD.
At Chateau-Thierry, France, June 6, 1918.
R—Detroit, Mich.
B—Detroit, Mich.
G. O. No. 110, W. D., 1918.

Gunnery sergeant, 45th Company, 5th Regiment, U. S. Marine Corps, 2d Division.
Killed in action at Chateau-Thierry, France, June 6, 1918, he gave the supreme proof of that extraordinary heroism which will serve as an example to hitherto untried troops.
Posthumously awarded. Medal presented to mother, Mrs. Catherine Todd Blaney.

TODOR, SAM (47979).
At Cantigny, France, May 28, 1918.
R—Harvey, Ill.
B—Austria.
G. O. No. 39, W. D., 1920.

Private, Company K, 18th Infantry, 1st Division.
After 10 others had been killed or wounded in an attempt to deliver a message, Private Todor carried an important message through heavy artillery and machine-gun fire and returned with a receipt showing the message had been delivered.

TOELKEN, JULIUS W.
Near Bouresches, France, July 20, 1918.
R—Springfield, Mass.
B—Suffield, Conn.
G. O. No. 125, W. D., 1918.

Second lieutenant, 104th Infantry, 26th Division.
When the advance of his platoon was checked by enemy machine-gun fire he crawled forward alone to a position from which he could fire and killed 3 of the machine-gun crew, after which, with his platoon, he captured the gun and turned it on the foe.

TOMANEK, FRANK F. (2176998).
In the Bois-de-Bantheville, France, Nov. 1, 1918.
R—Quinter, Kans.
B—Collyer, Kans.
G. O. No. 37, W. D., 1919.

Private, Company I, 353d Infantry, 89th Division.
When volunteers were called for to maintain liaison with the assault battalion, during heavy counterbarrage by the enemy, he volunteered and within 2 hours successfully carried out 4 such missions.

*TOMLIN, GILMORE (732896).
Near Reigneville, France, Sept. 12, 1918.
R—Lynchburg, Va.
B—New Glasgow, Va.
G. O. No. 37, W. D., 1919.

Sergeant, Company G, 6th Infantry, 2d Division.
Facing heavy machine-gun fire, he alone charged a machine gun which was causing his company many casualties, killed the gunner, and captured the gun.
Posthumously awarded. Medal presented to mother, Mrs. Daisy Tomlin.

TOMLINSON, RAYMOND W. (398965).
Near Vaux, France, July 1 to 10, 1918.
R—Baltimore, Md.
B—Baltimore, Md.
G. O. No. 24, W. D., 1920.

Mechanic, Company H, 9th Infantry, 2d Division.
During the attack on Vaux, Mechanic Tomlinson received a rifle-ball wound in the right knee. Although suffering great pain, he made no mention of his wound. Later during the attack he assisted in the capture of 2 officers and 5 men. During the 9 days that his company continued on duty in the front line Mechanic Tomlinson carried numerous messages to front-line platoons while exposed to heavy artillery and machine-gun fire.

TOMMIE, HOMER D. (6318742).
At Posolskaya, Siberia, Jan. 10, 1920.
R—Anniston, Ala.
B—Anniston, Ala.
G. O. No. 35, W. D., 1920.

Private, first class, Company M, 27th Infantry.
When his platoon was attacked in the night by an armored train, he, without thought of his own danger, attempted to board the armored train and was fired upon, wounded, and fell under the wheels of the car, which severed his leg from his body.

*TOMPKINS, FRED W. (543465).
Near Jaulgonne, France, July 22, 1918.
R—Owosso, Mich.
B—Canada.
G. O. No. 24, W. D., 1920.

Sergeant, Machine Gun Company, 38th Infantry, 3d Division.
Sergeant Tompkins, while leading his machine-gun section in an attack, exposed himself to heavy artillery and machine-gun fire and alone advanced in front of our lines in order to select machine-gun positions. Later he was mortally wounded while exposing himself in order to direct his men to cover.
Posthumously awarded. Medal presented to father, E. G. Tompkins.

TOMPKINS, HARRISON (1698859).
Near Bois-de-la-Naza, France, Oct. 5, 1918.
R—Yonkers, N. Y.
B—Tompkins Corners, N. Y.
G. O. No. 59, W. D., 1919.

Sergeant, Company L, 305th Infantry, 77th Division.
In the face of heavy machine-gun and grenade fire he went forward, with 3 other soldiers, and brought back 5 seriously wounded men to a point where they could be given first-aid treatment. With utter disregard for his personal safety, he displayed courage, coolness, and good judgment in effecting the rescue.

*TONKS, MARK (543116).
Near Fossoy, France, July 15, 1918.
R—Witt, Ill.
B—Birmingham, Pa.
G. O. No. 126, W. D., 1919.

Private, Company M, 7th Infantry, 3d Division.
After having been painfully wounded by shrapnel he refused to go the rear for treatment, but remained at his post until fatally wounded by another shell.
Posthumously awarded. Medal presented to father, Mathew Tonks.

TOPIC, FRANK J. (2850600).
Near Bantheville, France, Oct. 23, 1918.
R—St. Paul, Minn.
B—St. Paul, Minn.
G. O. No. 98, W. D., 1919.

Private, Company K, 357th Infantry, 90th Division.
During the night he made frequent trips with wounded back through a heavily shelled area, setting an example of fearlessness to his comrades. The next day in an advance he carried a stretcher with the front wave, and when the objective was reached went from shell hole to shell hole, under intense enemy fire, giving first aid to the wounded and carrying them to the rear.

TORREY, NORMAN L. (133716).
Near Verdun, France, Oct. 23–27, 1918.
R—Rowley, Mass.
B—Newbury, Mass.
G. O. No. 21, W. D., 1919.

Private, Battery C, 101st Field Artillery, 26th Division.
He acted as a runner for the artillery liaison officer and after this officer was returned, wounded, voluntarily remained for 24 hours, acting as a runner for the Infantry, constantly passing through the most intense artillery and machine-gun fire.

TOUSIG, FRANK.
Near Suippes, France, Oct. 3–7, 1918.
R—New York, N. Y.
B—New York, N. Y.
G. O. No. 37, W. D., 1919.

Chief pharmacist's mate, U. S. Navy, attached to the 5th Regiment, U. S. Marine Corps, 2d Division.
With no regard for his own safety, he labored unceasingly in caring for and evacuating the wounded under constant shellfire. His great activity and courage saved the lives of many of his comrades.

TOWELL, JAMES J. (189052)
Near St. Juvin, France, Oct. 15, 1918.
R—Belfast, N. Y.
B—Belfast, N. Y.
G. O. No. 46, W. D., 1919.

Private, first class, Company C, 325th Infantry, 82d Division.
With remarkable bravery he carried an important message through heavy enemy artillery and machine-gun fire. Later, in the same day, he again demonstrated a spirit of self-sacrifice by going out under heavy fire and bringing in a wounded comrade. In so doing he was twice wounded by a sniper.

TOWNE, EUGENE W. (1205170)
Near St. Souplet, France, Oct. 18, 1918.
R—Thomson, N. Y.
B—New Hampton, N. H.
G. O. No. 23, W. D., 1919.

Sergeant, Company K, 105th Infantry, 27th Division.
With 2 other soldiers, Sergeant Towne rushed forward into some hedges and silenced 3 light machine guns which were hindering the advance by flanking fire.

TOWNSEND, JAMES B. (1932946)
Near Vandieres, France, Oct. 15, 1918.
R—Troy, Ala.
B—Troy, Ala.
G. O. No. 27, W. D., 1919.

Private, sanitary detachment, 328th Infantry, 82d Division.
Private Townsend left Vandieres, went out on the field, which was continually under shellfire, and collected and brought into Vandieres 16 wounded. He remained with them 32 hours, bringing them food and water, adjusting gas masks, and making the wounded comfortable until all had been evacuated.

TOWNSEND, RICHARD L. (8602)
Near Somme-Py, France, Oct. 2–9, 1918.
R—Bryn Mawr, Pa.
B—Bryn Mawr, Pa.
G. O. No. 37, W. D., 1919.

Private, first class, Section No. 554, Ambulance Service.
He drove an ambulance night and day to an advanced dressing station under heavy shellfire. On the night of Oct. 5 he voluntarily drove a machine to a place near the lines, over a road raked by machine-gun and shellfire, to evacuate a number of wounded whom the litter bearers were unable to bring in immediately.

TOY, CHARLES S. (109319)
At Marcheville, France, Sept. 26, 1918.
R—West Roxbury, Mass.
B—Jamaica, Long Island, N. Y.
G. O. No. 15, W. D., 1919.

Private, first class, Company A, 102d Machine Gun Battalion, 26th Division.
He remained with the wounded under a sudden counterattack by the enemy and with the fire of his rifle prevented their falling into the hands of the enemy.
Posthumously awarded. Medal presented to father, Samuel Toy.

TRACY, JAMES F. (155901)
Near Exermont, France, Oct. 9, 1918.
R—Baltimore, Md.
B—Baltimore, Md.
G. O. No. 32, W. D., 1919.

Sergeant, Company B, 1st Engineers, 1st Division.
Without waiting for orders he voluntarily led a patrol against an enemy machine gun which was threatening the advance of our troops and delivering a severe fire into our ranks. By his skillful and courageous efforts the machine gun was captured. He continued to lead his men forward under the severe artillery and machine-gun fire until he was killed.
Posthumously awarded. Medal presented to father, James Tracy.

TRAGER, JOHN W.
Near Consenvoye, France, Oct. 8, 1918.
R—Peoria, Ill.
B—Peoria, Ill.
G. O. No. 46, W. D., 1919.

Second lieutenant, 124th Machine Gun Battalion, 33d Division.
Upon their own initiative, Lieutenant Trager, who was on duty as transportation officer, and another officer crossed the Meuse River to reconnoiter a supply route. They were fired on by 2 enemy machine guns, but, disregarding the heavy machine-gun and shell fire, they advanced on the enemy positions and captured the 2 machine guns, together with 31 Austrian prisoners.

TRAHERN, ROY D. (2217061)
Near Chatel-Chehery, France, Oct. 7, 1918.
R—Wilson, Okla.
B—Spiro, Okla.
G. O. No. 11, W. D., 1921.

Sergeant, Company D, 112th Infantry, 28th Division.
After his platoon had been forced to retire from the crest of the hill beyond Chatel-Chehery, Sergeant Trahern exposed himself to heavy machine-gun fire and went back 200 yards toward the enemy to the aid of a severely wounded comrade. Being unable to move the man alone, he called for assistance and later, with the help of another, he carried the wounded man to a place of safety.

TRAVERS, HUGH P., Jr. (2395893)
Near Mezy, France, July 15, 1918.
R—New York, N. Y.
B—New York, N. Y.
G. O. No. 32, W. D., 1919.

Private, Company E, 38th Infantry, 3d Division.
Early on the morning of July 15 he was wounded while the Germans were attempting to force a passage of the Marne, but remained at his post. Later in the day he was again wounded, but refused to leave, and continued on duty until the engagement was over.

TRAVERS, PATRICK (90923)
Near Sedan, France, Nov. 6, 1918.
R—Brooklyn, N. Y.
B—Ireland.
G. O. No. 37, W. D., 1919.

Corporal, Company H, 165th Infantry, 42d Division.
Without assistance, he advanced on an enemy's sniper post and successfully made prisoners of the entire crew, which included 1 officer, only being able to take the latter after a brief struggle. His action prevented a flanking fire on his platoon and aided greatly in their rapid advance.

TREADWAY, WOLCOTT W.
Near Soissons, France, July 19, 1918.
R—Ludlow, Mass.
B—Noroton, Conn.
G. O. No. 100, W. D., 1918.

Second lieutenant, 26th Infantry, 1st Division.
By exceptional bravery near Soissons, France, July 19, 1918, he aroused the admiration of his command, inspired his men by his example, and carried them forward in the face of heavy fire to their objective before he fell, mortally wounded.
Posthumously awarded. Medal presented to widow, Mrs. W. W. Treadway.

TREADWELL, ALVIN H.
In the region of St. Juvin, France, Oct. 10, 1918.
R—Poughkeepsie, N. Y.
B—Oxford, Ohio.
G. O. No. 15, W. D., 1923.

First lieutenant, 213th Aero Squadron, Air Service.
While leading a patrol of 4 machines at an altitude of 3,000 meters Lieutenant Treadwell observed 2 American observation airplanes hard pressed by 9 of the enemy. Disregarding the enemy's advantage in number and position he promptly attacked, whereupon the enemy immediately retired. On Oct. 29, 1918, in the region of Bayonville, France, at an altitude of 3,000 meters, Lieutenant Treadwell attacked an enemy biplane, killing the observer and following the machine down to within 50 meters of the ground, well within the enemy's territory. The gallantry and devotion to duty displayed by Lieutenant Treadwell greatly inspired the members of his squadron.
Posthumously awarded. Medal presented to father, Aaron L. Treadwell.

TREKAUSKAS, TONY A. (4259)_____
Near Soissons, France, July 18, 1918.
R—Cincinnati, Ohio.
B—Cincinnati, Ohio.
G. O. No. 15, W. D., 1919.

Sergeant, Company E, 16th Infantry, 1st Division.
Voluntarily and single handed he captured a machine gun and killed the crew.

TRERISE, BENJAMIN E. (1708988)_____
Near Binarville, France, Oct. 4, 1918, and near St. Juvin, France, Oct. 15, 1918.
R—New York, N. Y.
B—Silver City, N. Mex.
G. O. No. 27, W. D., 1920.

First sergeant, Company F, 308th Infantry, 77th Division.
During an attack in the Argonne Forest, Oct. 4, 1918, Sergeant Trerise was wounded in 5 places by shrapnel. Although in need of medical attention, he refused to be evacuated but remained, steadying his men and holding his unit intact. On Oct. 15, after 2 attempts at rescue of a wounded man had failed, he advanced through heavy enemy fire and brought the wounded man to shelter.

°TRESTRAIL, FREDERICK J._____
North of Verdun, France, Oct. 11, 1918.
R—Jersey City, N. J.
B—Jersey City, N. J.
G. O. No. 1, W. D., 1919.

First lieutenant, 113th Infantry, 29th Division.
When the advance of his company was checked by terrific enfilading fire from machine guns, he halted his men, and with great coolness ascended a hill to ascertain the location of the enemy machine-gun nests. He had barely reached the top of the hill when he was killed by an exploding shell.
Posthumously awarded. Medal presented to mother, Mrs. Harry J. Ralph.

TREW, RALPH T. (1519517)_____
Near Heurne, Belgium, Oct. 4, 1918.
R—Cleveland, Ohio.
B—McComb, Ill.
G. O. No. 59, W. D., 1919.

Sergeant, Company K, 145th Infantry, 37th Division.
Volunteering to construct a footbridge across the Scheldt River, Sergeant Trew crossed the stream in plain view of the enemy under violent machine-gun fire and, after the bridge had been completed, returned and led the first detachment of his regiment across.

TRIMBLE, DANA N. (155906)_____
Near Soissons, France, July 20, 1918.
R—Ipswich, Mass.
B—Canada.
G. O. No. 15, W. D., 1919.

Sergeant, Company B, 1st Engineers, 1st Division.
He volunteered and obtained the consent of his company commander to recover wounded men from an exposed area in front of the line. He went through a violent bombardment in the performance of this duty 3 times and stopped only when he had been severely wounded.

TRIMMER, LEE (44845)_____
Near Villers - devant - Mouzon, France, Nov. 7, 1918.
R—Lawrenceburg, Tenn.
B—Lawrenceburg, Tenn.
G. O. No. 32, W. D., 1919.

Private, Headquarters Company, 16th Infantry, 1st Division.
After being relieved from duty as a cart driver at his own request, he volunteered and made individual patrols. He located 2 machine guns by exposing himself and drawing their fire. His exceptional bravery resulted in the destruction of a machine gun and the dispersal of its crew.

TRIPLETT, NATHANIEL C. (1521412)____
Near Montfaucon, France, Sept. 27–28, 1918.
R—Akron, Ohio.
B—Bandana, Ky.
G. O. No. 44, W. D., 1919.

Mechanic, Company F, 146th Infantry, 37th Division.
He was a member of a patrol which encountered severe hostile machine-gun fire. He assisted in getting several wounded men to cover and administered first aid until his supply of bandages was exhausted. Returning to company headquarters across a field swept by artillery fire, he secured more bandages, came back with them to his comrades, and resumed his first-aid work. On the following day he again displayed exceptional courage under machine-gun and shell fire by carrying a wounded officer to safety.

TROSKA, CHARLES F. (44845)_____
South of Soissons, France, July 21, 1918.
R—Genou, Mont.
B—Wells, Minn.
G. O. No. 35, W. D., 1920.

Private, Company B, 16th Infantry, 1st Division.
Private Troska carried a message through a wood occupied by the enemy to a company which had become separated from its battalion. His deed permitted the coordinating of the attacking units.
Oak-leaf cluster.
Corporal Charles Troska was awarded an oak-leaf cluster for the following act: For extraordinary heroism in action near Fleville, France, Oct. 11, 1918. In the attack on Hill 272, Oct. 11, 1918, after other runners had been killed in the attempt to deliver a message to 2 attacking companies, he voluntarily carried and delivered the message, thus again effecting the proper coordination of the attack units of the battalion.

°TROTTER, AUGUSTUS M._____
At Belleau Woods, France, June 21, 1918.
R—Camden, S. C.
B—Camden, S. C.
G. O. No. 60, W. D., 1920.

First lieutenant, 7th Infantry, 3d Division.
Lieutenant Trotter gallantly led his platoon through heavy machine-gun fire in an attack on a strongly fortified enemy position. His platoon suffered heavy casualties, but he pushed forward until shot down near the enemy position.
Posthumously awarded. Medal presented to brother, T. K. Trotter.

TROUP, CLARENCE DAVID_____
Near Bayonville, France, Nov. 1, 1918.
R—Chicago, Ill.
B—Chicago, Ill.
G. O. No. 35, W. D., 1919.

Private, 82d Company, 6th Regiment, U. S. Marine Corps, 2d Division.
He volunteered and went forward to reconnoiter a ravine which was infested with hostile machine-gun and artillery positions, returning with several prisoners.

TROWER, STALLARD (43586)_____
Near Soissons, France, July 21, 1918.
R—Harrodsburg, Ky.
B—Mercer County, Ky.
G. O. No. 125, W. D., 1918.

Sergeant, Company I, 16th Infantry, 1st Division.
After all of his officers had been killed or wounded he assumed command of the company, and with exceptional bravery and courage kept continually pressing on and engaging the enemy.

TRUTKO, ALEXANDER (48630)_____
Near Soissons, France, July 18–22, 1918.
R—Chicago, Ill.
B—Russia.
G. O. No. 72, W. D., 1920.

Private, Company M, 18th Infantry, 1st Division.
Private Trutko repeatedly carried messages across areas swept by artillery and machine-gun fire. Due in part to his individual gallantry, his company commander was able to maintain communication at all times during this operation.

TRYON, JEREMIAH (64107)............
During the Seicheprey engagement,
France, Apr. 20, 1918.
R—Saybrook, Conn.
B—Saybrook, Conn.
G. O. No. 99, W. D., 1918.

Private, Company C, 102d Infantry, 26th Division.
He displayed unusual daring and courage during the Seicheprey engagement on the morning of Apr. 20, 1918, when, under heavy artillery fire, he climbed out of his trench in the front line onto the top and killed a sniper who was pouring a destructive fire into our trenches.

*TUBBS, BENJAMIN T. (1598382)........
Near Pouilly, France, Nov. 10–11, 1918.
R—Farmerville, La.
B—Farmerville, La.
G. O. No. 44, W. D., 1919.

Private, Company I, 356th Infantry, 89th Division.
Private Tubbs accompanied Lieutenant Murphy and 3 other soldiers in a flank attack on 3 heavy machine guns. Fired on directly at 30 yards, they charged the guns, and in the hand-to-hand fight which followed this soldier and 2 of his comrades were killed.
Posthumously awarded. Medal presented to father, Howard Homer Tubbs.

*TUCKER, LOUIS J. (3490947)...........
In the Bois-d'Ormont, France, Oct. 23–27, 1918.
R—Memphis, Tenn.
B—Boonville, Miss.
G. O. No. 87, W. D., 1919.

Private, Company K, 102d Infantry, 26th Division.
Private Tucker bravely volunteered to carry an important message to the battalion commander at a critical juncture, and in attempting to pass through a terrific enemy barrage was instantly killed by shellfire.
Posthumously awarded. Medal presented to mother, Mrs. Sallie N. Tucker.

TUCKER, MARION C. (1911289)..........
Near Sommerance, France, Oct. 12, 1918.
R—Moultrie, Ga.
B—Moultrie, Ga.
G. O. No. 46, W. D., 1919.

Sergeant, Company L, 328th Infantry, 82d Division.
While being carried unconscious to a dressing station after having been wounded, he regained consciousness, arose from the stretcher, and rejoined his company. Shortly after, a gas shell exploded in a dugout where he and other men were taking cover. Although badly gassed, he again refused evacuation, remaining on duty with the company until it was relieved 16 days later.

TUCKER, ROYAL K...................
East of Ronssoy, France, Sept. 29, 1918.
R—Mobile, Ala.
B—Upper Alton, Ill.
G. O. No. 143, W. D., 1918.

First lieutenant, chaplain, 105th Infantry, 27th Division.
During the operations against the Hindenburg line he displayed remarkable devotion to duty and courage in caring for the wounded under heavy shell and machine-gun fire. The splendid example set by this officer was an inspiration to the combat troops.

TUDURY, HENRY J. (562712)...........
Near Courchamps, France, July 18–20, 1918.
R—Bay St. Louis, Miss.
B—Bay St. Louis, Miss.
G. O. No. 46, W. D., 1919.

Private, Company C, 12th Machine Gun Battalion, 4th Division.
Engaged as runner, he made repeated trips through intense shelling and machine-gun fire. On July 18 he was gassed, but bravely continued with his heroic work until he fell exhausted on the 20th.

TUFTIN, CARL (3752519)..............
Near Sommauthe, France, Nov. 4–5, 1918.
R—Clayton, Wis.
B—Norway.
G. O. No. 44, W. D., 1919.

Private, Company D, 318th Infantry, 80th Division.
Private Tuftin volunteered for dangerous outpost service within the enemy line. He worked his way behind enemy machine guns; was wounded, but refused to leave his post; and by the skillful use of his automatic rifle assisted in driving the enemy from their entrenched positions. He refused to leave his post until the battalion was relieved.

TUKEY, ALLAN A..................
Near Soissons, France, July 18–19, 1918.
R—Omaha, Nebr.
B—Omaha, Nebr.
G. O. No. 99, W. D., 1918.

Second lieutenant, 3d Machine Gun Battalion, 1st Division.
During the engagement near Soissons, France, July 18 and 19, 1918, the leadership of his platoon was exceptionally distinctive by reason of his courage, initiative, and presence of mind. While advancing on the second day of the attack he was wounded, but continued with his command until he had given complete instructions to his platoon sergeant and notified his company commander of the disposition of his guns, after which he was compelled to yield command because of weakness from his injuries.

TULEY, HOMER A. (2019076)..........
Near Bolshieozerke, Russia, Apr. 2, 1919.
R—Detroit, Mich.
B—Dallas, Tex.
G. O. No. 19, W. D., 1920.

Private, first class, 337th Ambulance Company, 310th Sanitary Train, 85th Division (detachment in North Russia).
Private, first class, Tuley went forward some 200 yards in front of our lines into an area swept by artillery and machine-gun fire to assist in bringing a wounded man to a place of safety.

TUNE, HORACE R..................
Near Clery-le-Grand, France, Nov. 1, 6, and 10, 1918.
R—Shelbyville, Tenn.
B—Shelbyville, Tenn.
G. O. No. 64, W. D., 1919.

First lieutenant, 60th Infantry, 5th Division.
On the morning of Nov. 1 Lieutenant Tune personally led an attack against an enemy machine-gun nest, killing 1 of the enemy, wounding 1, and taking 4 prisoners. On Nov. 6 he went far in advance of his company and with 2 other men killed a sniper in a building in the village of Marvaux. On Nov. 10 he led the remnants of his company in clearing the Bois-de-Juvigny of enemy detachments. In this action Lieutenant Tune killed 1 of the enemy with a rifle and assisted in the capture of 5 prisoners.

*TURANO, JOHN (51578)..............
Near Vaux, France, July 1, 1918.
R—Worcester, Mass.
B—Westerly, R. I.
G. O. No. 99, W. D., 1918.

Private, Company I, 23d Infantry, 2d Division.
Attempting to bring to shelter his captain, who was lying wounded and exposed to fire near Vaux, France, July 1, 1918, he was himself killed, thereby sacrificing his life in an effort to rescue his commanding officer.
Posthumously awarded. Medal presented to father, Santo Turano.

TURBEVILLE, WILLIAM JAMES (1311040).
Near Bellicourt, France, Sept. 30, 1918.
R—Lexington, S. C.
B—Clarendon County, S. C.
G. O. No. 37, W. D., 1919.

Private, first class, Company E, 118th Infantry, 30th Division.
Private Turbeville, a battalion runner, displayed exceptional courage and disregard for personal danger in making 3 trips with important messages through heavy enemy machine-gun and shell fire.

TURKOPP, CARL F. (1956265)_____
Near Brabant-sur-Meuse, France, Oct. 23, 1918.
R—Columbus, Ohio.
B—Toledo, Ohio.
G. O. No. 37, W. D., 1919.

Corporal, 308th Trench Mortar Battery, 158th Field Artillery Brigade, 83d Division.
During the offensive operations in the Bossois Bois, Corporal Turkopp, although wounded by shellfire and knocked down by concussion, returned to his gun and continued to fire. So great was his exhaustion that it was necessary for him to be supported while doing his work.

*TURLEY, CLARENCE L. (2258561) _____
Near Juvigny, north of Soissons, France, Aug. 30, 1918.
R—Pasco, Wash.
B—Marion, Ky.
G. O. No. 20, W. D., 1919.

Corporal, Company I, 128th Infantry, 32d Division.
In an attack by his company Corporal Turley gave proof of unusual gallantry and courage by fearlessly going out under heavy machine-gun and artillery fire to give aid to and carry back the wounded. He was himself seriously wounded while engaged in this work.
Posthumously awarded. Medal presented to stepmother, Mrs. Myrtle Turley.

TURNER, BEN E._____
At Fismette, France, Aug. 27, 1918.
R—Kahoka, Mo.
B—Kahoka, Mo.
G. O. No. 98, W. D., 1919.

First lieutenant, 112th Infantry, 28th Division.
When the enemy attack, preceded by a very heavy barrage, had broken through and forced a retirement over the Vesle, Lieutenant Turner, himself wounded and under enemy fire from front and flanks, directed the retirement of his men, while he alone covered their withdrawal over the river with an automatic rifle, crossing after the last man was safely over.

*TURNER, CHARLES W._____
Near Binarville, France, Oct. 6, 1918.
R—Brooklyn, N. Y.
B—Brooklyn, N. Y.
G. O. No. 37, W. D., 1919.

First lieutenant, 308th Infantry, 77th Division.
Surrounded by enemy machine guns and snipers and under heavy shellfire, he refused to surrender, but held his position with extraordinary heroism and total disregard for his own life until he and all his detachment were killed.
Posthumously awarded. Medal presented to widow, Mrs. Josephine Turner.

TURNER, DENNIS C._____
Near Mezy, France, July 15, 1918.
R—Charlotte, N. C.
B—Shelby, N. C.
G. O. No. 32, W. D., 1919.

First lieutenant, 30th Infantry, 4th Division.
Although completely surrounded and his ammunition exhausted, Lieutenant Turner refused to surrender. Assembling his platoon of about 18 men, he made a dash for our lines through the enemy's machine-gun and rifle fire, and, by taking advantage of all available cover and using grenades and ammunition found on the way, succeeded in joining our troops.

TURNER, HENRY D. (2101130)_____
Near Le Vallee, France, July 23, 1918; near St. Thibault, France, Aug. 9, 1918; near the Bois-de-Septsarges, France, Sept. 29, 1918; and near the Bois-du-Fays, France, Oct. 6, 1918.
R—Fairfield, Ill.
B—Burnt Prairie, Ill.
G. O. No. 66, W. D., 1919.

Sergeant, Company B, 10th Machine Gun Battalion, 4th Division.
Sergeant Turner, a runner, repeatedly went out under shell and machine-gun fire to maintain liaison between units, frequently volunteering for especially hazardous missions. After other runners had been killed he rendered valuable service by repeatedly crossing dangerous areas in order to maintain communications.

*TURNER, JAMES A._____
Near Buzancy, France, Nov. 2–3, 1918.
R—Chicago, Ill.
B—Ludlow, Ky.
G. O. No. 59, W. D., 1919.

First lieutenant, 318th Infantry, 80th Division.
After having been severely wounded during the night of Nov. 2, 1918, he continued in command of his company. Despite his wound, he led his company in the attack the following day, when he was killed by an enemy shell. He set an example of fearlessness and bravery to his men.
Posthumously awarded. Medal presented to widow, Mrs. James A. Turner.

TURNER, JOSEPH W. (1310695)_____
Near Vaux-Andigny, France, Oct. 11, 1918.
R—Enoree, S. C.
B—Walnut Grove, S. C.
G. O. No. 50, W. D., 1919.

Sergeant, Company D, 118th Infantry, 30th Division.
Sergeant Turner volunteered and carried an automatic rifle to an advantageous position far in advance of his own line and maintained an effective fire on the enemy until his gun was put out of action and he was wounded in both hands and forced to retire. Before going to the rear he gave full and valuable information regarding the enemy's position to his officers.

TURNER, OTIS E. (2154626)_____
Near Busigny, France, Oct. 18, 1918.
R—Belville Island, Iowa.
B—Jasper County, Iowa.
G. O. No. 133, W. D., 1918.

Private, Company M, 117th Infantry, 30th Division.
When his platoon was held up by an enemy machine-gun post, Private Turner with another soldier took their automatic rifle, rushed 50 yards through intense fire, skillfully placed the rifle in position, and opened an effective fire.

TURNER, RAY C. (935323)_____
Near Saulx, France, Sept. 26, 1918.
R—Danville, Ill.
B—Danville, Ill.
G. O. No. 21, W. D., 1919.

Private, 101st Ambulance Company, 101st Sanitary Train, 26th Division.
Under intense bombardment, he volunteered to go forward with a sergeant to rescue a number of seriously wounded soldiers. While they were engaged in this heroic work he was wounded and his companion killed by an exploding shell. After receiving first aid he immediately resumed his duties and remained at the front until the advanced station was closed.

*TURRENTINE, HERBERT S. (1315188)___
Near Ypres, Belgium, Aug. 31, 1918.
R—Winston-Salem, N. C.
B—Mocksville, N. C.
G. O. No. 44, W. D., 1919.

Private, Company C, 119th Infantry, 30th Division.
After his platoon sergeant and a corporal had been shot while firing an automatic rifle, he ran forward across an open space and picked up the gun, but was instantly killed by sniper fire while attempting to get the automatic gun back into action.
Posthumously awarded. Medal presented to sister, Mrs. J. P. Shaw.

TURRILL, JULIUS SPEAR_____
In the Bois-de-Belleau, France, June 6, 1918.
R—Burlington, Vt.
B—Shelburne, Vt.
G. O. No. 98, W. D., 1919.

Major, 5th Regiment, U. S. Marine Corps, 2d Division.
He displayed extraordinary heroism and set a splendid example in fearlessly leading his command under heavy fire against superior odds. Because of his bravery and initiative every possible advantage in the attack was obtained.

TUTTLE, MELLEN F. (67163)
On Hill 190, near Chateau-Thierry, France, July 20, 1918.
R—New Gloucester, Me.
B—Freeport, Me.
G. O. No. 125, W. D., 1918.

Private, first class, Company B, 103d Infantry, 26th Division.
When all of the men of the automatic-rifle team of which he was a member had been wounded he voluntarily advanced alone, attacked a number of enemy machine-gun nests which were holding up the advance, and forced the enemy to retreat, thereby making possible the continued forward movement of his detachment.

TVETEN, HANS L. (2261485)
At Gesnes, France, Sept. 29, 1918.
R—Sand Creek, Mont.
B—Norway.
G. O. No. 1, W. D., 1919.

Private, Company K, 362d Infantry, 91st Division.
When his company was under fire from 2 German machine guns, he crept forward alone and put the guns out of action with rifle grenades, capturing single handed 4 Germans and both machine guns.

TWIFORD, ELWOOD (1314770)
Near Bellicourt, France, Sept. 29, 1918.
R—Dare County, N. C.
B—East Lake, N. C.
G. O. No. 87, W. D., 1919.

Private, Company A, 119th Infantry, 30th Division.
Having become separated from the remainder of his squad in a heavy fog and being surrounded by several enemy machine gunners, Private Twiford set up his automatic rifle and within a few minutes killed or captured all of the enemy near him.

TWISS, JULIUS I. (63386)
At Marcheville, France, Sept. 26, 1918.
R—Hartford, Conn.
B—Niantic, Conn.
G. O. No. 15, W. D., 1919.

Sergeant, Headquarters Company, 102d Infantry, 26th Division.
During the counterattack on Marcheville he became separated from his command. Under terrific artillery and machine-gun fire he voluntarily gathered together a few scattered men and organized a point of defense, showing coolness, bravery, and judgment which materially assisted in the success of the counterattack.

UCAC, JOSEPH (1707692)
Near Binarville, France, Sept. 27, 1918.
R—Brooklyn, N. Y.
B—Russia.
G. O. No. 35, W. D., 1919.

Private, first class, Company A, 308th Infantry, 77th Division.
Returning to the line after being wounded by a hand grenade the previous day, Private Ucac persistently requested to be allowed to assist stretcher bearers in the removal of the wounded. While performing this heroic mission, constantly subjected to treacherous machine-gun and artillery fire, he was again wounded.

ULRICH, WILLIAM
Near Thiaucourt, France, Sept. 15, 1918.
R—New York, N. Y.
B—Germany.
G. O. No. 37, W. D., 1919.

Sergeant major, 2d Battalion, 6th Regiment, U. S. Marine Corps, 2d Division.
Accompanied by 3 other soldiers he fearlessly charged into the enemy's lines through intense machine-gun fire in pursuit of a party of Germans and returned with 51 prisoners.

UNDERWOOD, DAVID H. (279060)
Near Fismes, France, Aug. 6, 1918.
R—Tecumseh, Mich.
B—Tecumseh, Mich.
G. O. No. 66, W. D., 1919.

Private, first class, Company B, 125th Infantry, 32d Division.
With exceptional courage, Private Underwood went out under heavy fire from enemy machine guns and snipers and rescued a wounded runner who had strayed into the enemy's lines by mistake.

UPTON, LA ROY S.
Near Soissons, France, July 18-19, 1918.
R—Big Rapids, Mich.
B—Decatur, Mich.
G. O. No. 132, W. D., 1918.
Distinguished-service medal also awarded.

Colonel, 9th Infantry, 2d Division.
His regiment having suffered heavy casualties in its first attack on July 18, 1918, and he having received orders to attack a second time, Colonel Upton re-formed his command and conducted the second attack in person until stopped by darkness. His line being broken by a gap in its center, all of his battalion commanders being killed or wounded, and all of his reserves being in the thinly held line, he established his command post on the extreme front at the right of the gap and remained there for 24 hours under steady and intense artillery bombardment and machine-gun fire, holding his position until his regiment was relieved. His presence and his example of fearlessness inspired his weakened line thus to guard the unprotected flank of the whole advance and beat off a violent counterattack.

*UPTON, THOMAS A. (71276)
Near Belleau, France, July 21, 1918.
R—Salem, Mass.
B—Wales.
G. O. No. 32, W. D., 1919.

Corporal, Company B, 104th Infantry, 26th Division.
He voluntarily crossed a zone swept by machine-gun and shell fire to aid wounded soldiers and was killed.
Posthumously awarded. Medal presented to brother, Samuel F. Upton.

URSPRUNG, RUDOLPH S.
Near Eyne, Belgium, Nov. 1, 1918.
R—Berea, Ohio.
B—Cleveland, Ohio.
G. O. No. 50, W. D., 1919.

First lieutenant, 145th Infantry, 37th Division.
Seeing a wounded soldier lying 150 yards in front of the line, after his company had withdrawn to a more secure position, Lieutenant Ursprung crawled through heavy fire and administered first aid to him. He then picked up the wounded man, carried him across the open, wading a canal through water waist deep, and succeeded in taking him to a place of safety.

VAIL, ROBERT M.
Near Villette, France, Sept. 5, 1918.
R—Scranton, Pa.
B—Scranton, Pa.
G. O. No. 44, W. D., 1919.

Major, 108th Machine Gun Battalion, 28th Division.
During the crossing of the Vesle River, Major Vail expedited the construction of bridges by his advice and assistance. He personally cut a passageway through enemy barbed wire along the river, and then led his troops through this opening. When the officers of a supporting Infantry company had been killed and the men were falling back in confusion, Major Vail kept his forces intact 1,200 yards in front of any Infantry support, holding the position until reinforced by an Infantry unit.

VAIL, WILLIAM H.
At Stenay, France, Nov. 6, 1918.
R—Chicago, Ill.
B—Chicago, Ill.
G. O. No. 37, W. D., 1919.

First lieutenant, pilot, 95th Aero Squadron, Air Service.
Lieutenant Vail while on patrol engaged 4 hostile pursuit planes which were about to attack an accompanying plane. Almost immediately he was attacked by 5 more enemy planes, all of which he continued to fight until he was severely wounded and his plane disabled. He glided to the ground, abandoning the fight only when his machine fell to pieces near the ground.

VALENTINE, AARON F. (2249811)........
   Near St. Marie Valley, France,
    Sept. 15, 1918.
   R—Madison, Okla.
   B—Kingfisher, Okla.
   G. O. No. 50, W. D., 1919.

Private, Company A, 344th Machine Gun Battalion, 90th Division.
When the advance of an Infantry company was held up by an enemy machine gun located in a tree, he rushed into the open and, mounting his machine gun, killed the gunner and knocked the nest from its position, his exploit being stiffly resisted by direct firing.

VALLANCE, SAMUEL HYATT...........
   Service rendered under the name,
    Samuel V. H. Danzig.

First lieutenant, 8th Machine Gun Battalion, 3d Division.

*VALLELY, FRANCIS P. (126276).. ......
   At Coullemelle France, Apr. 27, 1918.
   R—Pratt City, Ala.
   B—Pratt City, Ala.
   G. O. No. 100, W. D., 1918.

Private, first class, Battery E, 6th Field Artillery, 1st Division.
Under a heavy bombardment, he voluntarily went to the assistance of other soldiers who had been buried in a dugout by enemy shellfire, and was killed while engaged in this heroic action.
Posthumously awarded. Medal presented to father, P. T. Vallely.

VALLEY, ISAAC (1403725)...............
   At Vraincourt, France, July 22, 1918.
   R—Girard, Kans.
   B—Girard, Kans.
   G. O. No. 101, W. D., 1918.

Corporal, Company M, 370th Infantry, 93d Division.
When, on July 22, 1918, a hand grenade was dropped among a group of soldiers in a trench, and when he might have saved himself by flight he attempted to cover it with his foot and thereby protect his comrades. In the performance of this brave act he was severely wounded.

VAN ALLEN, CLARENCE R. (2337012)...
   Near Bussy Farm, France, Sept. 28,
    1918.
   R—Boston, Mass.
   B—West Newton, Mass.
   G. O. No. 13, W. D., 1919.

Private, Company L, 372d Infantry, 93d Division.
This soldier, unassisted, rushed an enemy machine gun, putting it out of action, and capturing 3 prisoners.

VAN AMBURGH, HUGH C.............

   Near Vierzy, France, July 19, 1918.
   R—Tacoma, Wash.
   B—Oakesdale, Wash.
   G. O. No. 117, W. D., 1918.

Corporal, Headquarters Company, 4th Brigade, U. S. Marine Corps, 2d Division.
As a motor-cycle dispatch rider he made repeated trips along shell-swept roads and in a gassed area before and during the capture of Vierzy. When Vierzy was still in German hands he dismounted from his motor cycle in front of the town and with great coolness and disregard of personal safety crawled into it and brought back information of great value to his brigade commander.

VAN BUREN, GEORGE (43817)..........
   Near Fleville, France, Oct. 4, 1918.
   R—San Francisco, Calif.
   B—Hudson, N. Y.
   G. O. No. 32, W. D., 1919.

Private, Company I, 16th Infantry, 1st Division.
He continued to advance after being seriously wounded until he fainted from exhaustion. After recovering consciousness he again joined his company and assisted in repelling a counterattack in which he received a fourth wound, remaining on duty until he was ordered evacuated.

VAN DE GRAFF, COLEMAN H..........
   Near Villemontoire, France, July 21,
    1918.
   R—Tuscaloosa, Ala.
   B—Tuscaloosa, Ala.
   G. O. No. 96, W. D., 1919.

Second lieutenant, 15th Field Artillery, 2d Division.
On duty with the Infantry as liaison officer, Lieutenant Van de Graff displayed marked courage in passing through an enemy artillery barrage several times, in carrying information to his battalion commander, and administering aid to wounded men under heavy shellfire.

VANDER VEEN, THOMAS (573830)......
   Near Nantillois, France, Oct. 10–13,
    1918.
   R—San Fernando, Calif.
   B—Holland.
   G. O. No. 87, W. D., 1919.

Private, first class, Company C, 11th Machine Gun Battalion, 4th Division.
As company liaison agent he maintained continual contact between his company commander and the battalion post of command, repeatedly exposing himself to artillery, machine-gun, and snipers' fire to deliver important messages. On one occasion it was necessary for him to pass through the German and our own barrages, but he accomplished this mission fearlessly, showing marked personal bravery.

*VAN DUESEN, ROBERT R............
   Near St. Etienne, France, Oct. 4,
    1918.
   R—Vineland, N. J.
   B—Vineland, N. J.
   G. O. No. 37, W. D., 1919.

Sergeant, 17th Company, 5th Regiment, U. S. Marine Corps, 2d Division.
With his platoon in a very dangerous position, he volunteered to carry a message from his platoon commander across a machine-gun-swept field. Having successfully accomplished his mission, he returned, and while directing his men to shelter he was severely wounded by a machine-gun bullet.
Posthumously awarded. Medal presented to mother, Mrs. R. B. Van Duesen.

VAN DUZER, EDWIN T. (1709923)......
   Near Ville-Savoye, France, Aug. 22,
    1918.
   R—Brooklyn, N. Y.
   B—Brooklyn, N. Y.
   G. O. No. 35, W. D., 1919.

Private, first class, Company K, 308th Infantry, 77th Division.
He was a member of a combat liaison group which was attacked by liquid fire. Although severely burned, he alone charged the flame thrower and put him out of action, after which he reassembled his men and continued on duty until relieved.

VAN DYNE, JOHN A. (1213836)..........
   Near Ronssoy, France, Sept. 29, 1918.
   R—Geneva, N. Y
   B—Seneca Falls, N. Y.
   G. O. No. 98, W. D., 1919.

Private, Company B, 108th Infantry, 27th Division.
Private Van Dyne, a runner, displayed notable courage in carrying messages through heavy artillery and machine-gun fire.

VAN GUNDAY, BEN (2219587)..........
   Near Bois-de-Consenvoye, France,
    Oct. 8, and Bois-de-Grande-Mon-
    tagne, France, Oct. 16, 1918.
   R—Wyandotte, Okla.
   B—Neosho, Mo.
   G. O. No. 44, W. D., 1919.

Private, Company F, 115th Infantry, 29th Division.
During the entire action he operated his automatic rifle with great effect against extremely heavy odds, aiding the advance of his platoon, killing many of the enemy, and assisting in the capture of many more. He repeatedly exposed himself to draw the enemy fire so that he could better operate his gun.

VAN HART, JOHN A. (1277637) _____ At Molleville Farm, France, Oct. 10–25, 1918. R—Elizabeth, N. J. B—Elizabeth, N. J. G. O. No. 37, W. D., 1919.	Sergeant, Company G, 113th Infantry, 29th Division. Attached to the battalion scouts, he repeatedly went out on dangerous patrols, secured valuable information, assisted in first-aid work, carried litters, rations, and water, and voluntarily made numerous trips through artillery and machine-gun fire with important messages.
VAN HOY, JAMES L. (1310696) _____ Near Vaux-Andigny, France, Oct. 11, 1918. R—Laurens, S. C. B—Elkin, N. C. G. O. No. 87, W. D., 1919.	Private, Company D, 118th Infantry, 30th Division. When his company was caught in a barrage, Private Van Hoy volunteered and carried a message to battalion headquarters under direct observation by the enemy through gas and terrific machine-gun and shellfire.
VANN, JOHN C. _____ Near Bazoches, France, Aug. 7, 1918. R—Columbus, Ga. B—Valdosta, Ga. G. O. No. 46, W. D., 1919.	Second lieutenant, 47th Infantry, 4th Division. Lieutenant Vann concealed the fact that he was wounded and led the advance platoon of his company to their objectives despite heavy losses. He remained with his command, displaying the highest leadership and courage, until he was wounded a second time.
VAN OOSTENBRUGGE, HORACE B. _____ Near Villette, France, Sept. 6, 1918. R—Schenectady, N. Y. B—Troy, N. Y. G. O. No. 71, W. D., 1919.	Second lieutenant, 109th Infantry, 28th Division. Being the only officer with his company, he continued in command for 36 hours after being severely wounded, constantly encouraging his men and setting an example of coolness and heroism. It was only after being wounded a second time that he consented to be evacuated.
*VAN'T HOF, BERNARD _____ Northeast of Chateau-Thierry, France, July 28, 1918. R—Grand Rapids, Mich. B—Grand Rapids, Mich. G. O. No. 59, W. D., 1919.	First lieutenant, 168th Infantry, 42d Division. He directed his platoon so skillfully in attack near Sergy and conducted himself with such bravery and fearlessness that his men captured 6 machine guns from the Prussian Guards and took 25 prisoners, which guns were then used with effect in driving the enemy from their positions. Posthumously awarded. Medal presented to father, Kryn Van't Hof.
*VAN VORIS, HOWARD H. _____ Near Waereghem, Belgium, Oct. 30–31, 1918. R—Stites, Idaho. B—Asotin, Wash. G. O. No. 15, W. D., 1919.	Second lieutenant, 364th Infantry, 91st Division. As battalion intelligence officer on the night before the engagement he was tireless in his efforts to maintain liaison on the flanks of his battalion. Next day, against the advice of senior officers, he made repeated reconnaissances of the front lines in the face of heavy shell and machine-gun fire. Penetrating beyond the Infantry lines on one of these patrols, this gallant officer was killed by machine-gun fire. Posthumously awarded. Medal presented to widow, Mrs. Gladys Van Voris.
VAN WAY, CHARLES W. _____ On the Abra River, Abra Province, Luzon, Philippine Islands, Nov. 2, 1900. R—Winfield, Kans. B—Shelbyville, Ind. G. O. No. 44, W. D., 1919.	Captain, 33d Infantry, U. S. Volunteers. In action against insurgents.
VAN YORX, VICTOR (91288) _____ Near Villers-sur-Fere, France, July 28, 1918. R—Mount Vernon, N. Y. B—New York, N. Y. G. O. No. 99, W. D., 1918.	Private, Company K, 165th Infantry, 42d Division. Though severely wounded in the ankle, he refused to leave the field of battle, and next day made the attack on the heights north of the River Ourcq, remaining with his company until again wounded.
VARNER, ANDREW H. (1315404) _____ Near Bellicourt, France, Sept. 29, 1918. R—Thomasville, N. C. B—Randolph County, N. C. G. O. No. 81, W. D., 1919.	Private, first class, Company D, 119th Infantry, 30th Division. Seeing that a wounded companion had been abandoned by stretcher bearers because of intense enemy shelling, he took 2 enemy prisoners and going out with them for 75 yards through heavy fire rescued the wounded soldier.
*VARNEY, KIT R. _____ Near Ronssoy, France, Sept. 29, 1918. R—San Francisco, Calif. B—Virginia City, Nev. G. O. No. 32, W. D., 1919.	Captain, 301st Battalion, Tank Corps. When a dense fog and a smoke barrage had made visibility so poor that it was difficult to get his tanks into action, Captain Varney personally led his machines on foot several hundred yards in advance of the first wave of infantry in the face of deadly artillery and machine-gun fire. He thus enabled his tanks to maintain their direction and cleared a path for the Infantry, but in accomplishing this heroic task he was killed. Posthumously awarded. Medal presented to mother, Mrs. Calista R. Varney
VAUGHAN, JOE H. (1315516) _____ Near Mazinghem, France, Oct. 18, 1918. R—Bon Aqua, Tenn. B—Hickman County, Tenn. G. O. No. 81, W. D., 1919.	Sergeant, Company E, 119th Infantry, 30th Division. After several others had failed in the attempt, he crossed ground swept by heavy enemy fire and established liaison with the British unit operating on the flank, volunteering for the mission.
*VAUGHAN, RICHARD H. _____ At Fismette, France, Aug. 9–13, 1918. R—Royersford, Pa. B—Royersford, Pa. G. O. No. 16, W. D., 1919.	Sergeant, Company A, 111th Infantry, 28th Division. Although he had been severely gassed and had received a scalp wound from shrapnel on Aug. 9, 1918, he refused to be evacuated, and after having his wound dressed continued to command his platoon for 4 days until relieved. By his bravery and encouragement to his men he exemplified the highest qualities of leadership. Posthumously awarded. Medal presented to father, Dr. E. M. Vaughan.

VAUGHN, GEORGE AUGUSTUS_____
    Near Cambrai, France, Sept. 22, 1918.
    R—Brooklyn, N. Y.
    B—Brooklyn, N. Y.
    G. O. No. 60, W. D., 1920.

First lieutenant, 17th Aero Squadron, Air Service.
Lieutenant Vaughn while leading an offensive flight patrol sighted 18 enemy Fokkers about to attack a group of 5 allied planes flying at a low level. Although outnumbered nearly 5 to 1, he attacked the enemy group, personally shot down 2 enemy planes, the remaining 3 planes of his group shooting down 2 more. His courage and daring enabled the group of allied planes to escape. Again on Sept. 28, 1918, he alone attacked an enemy advance plane which was supported by 7 Fokkers and shot the advance plane down in flames.

VAUGHT, GLENN (1446055)_____
    Near Varennes, France, Sept. 26, 1918.
    R—Monett, Mo.
    B—Cassville, Mo.
    G. O. No. 98, W. D., 1919.

Corporal, Company A, 129th Machine Gun Battalion, 35th Division.
Accompanied by another soldier, Corporal Vaught went through heavy fire, and by skillful maneuvering located and captured an enemy 77-mm. gun, with part of its crew, which had been checking the advance of our troops.

*VEDILAGO, JOSEPH (1707556)_____
    Near Binarville, France, Sept. 28, 1918.
    R—Jamaica, N. Y.
    B—Italy.
    G. O. No. 32, W. D., 1919.

Corporal, Company A, 308th Infantry, 77th Division.
He crawled from his shelter to get an automatic rifle after the members of the rifle team had been killed or wounded, and with this weapon continued in the advance until he was killed by shell fragments.
Posthumously awarded. Medal presented to mother, Mrs. Grace Vedilago.

VEDRAL, ANTHONY (57142)_____
    Near Verdun, France, Oct. 9, 1918.
    R—Detroit, Mich.
    B—Bohemia.
    G. O. No. 37, W. D., 1919.

Sergeant, Company D, 28th Infantry, 1st Division.
Assuming command of the platoon, Sergeant Vedral led it forward to its objective. When he arrived at the line, he found that his company had been cut off by the enemy barrage. He therefore consolidated his position with the few men he commanded and held it for 18 hours under most terrific bombardment.

VERBEKE, REMI (2150438)_____
    Near Bellicourt, France, Sept. 26, 1918.
    R—Climax, Minn.
    B—Belgium.
    G. O. No. 37, W. D., 1919.

Private, Company D, 118th Infantry, 30th Division.
After receiving a painful wound in the shoulder he delivered a message to his company headquarters. He then had his wound dressed, returned with another message, through shellfire and under direct observation of the enemy and reported for continuous duty. He later voluntarily assisted in driving back a strong enemy patrol and was severely wounded in the encounter.

VERCOE, STANLEY (807295)_____
    Near Thiaucourt, France, Sept. 13, 1918.
    R—Grastra, Mich.
    B—England.
    G. O. No. 35, W. D., 1919.

Private, medical detachment, 5th Machine Gun Battalion, 2d Division.
In a territory swept by the direct fire of 2 German batteries, he displayed fearlessness and devotion to duty in giving first aid to the wounded and carrying them to a place of safety.

VERCRUYSSE, JOSEPH C. (1646636)_____
    At St. Juvin, France, Oct. 15, 1918.
    R—Oakland, Calif.
    B—Belgium.
    G. O. No. 20, W. D., 1919.

Private, Company H, 306th Infantry, 77th Division.
He volunteered and carried a message to supporting troops through an intense artillery barrage, displaying courage and persistent devotion to duty. This message was of vital importance in connection with the capture of St. Juvin.

VERDIER, WILLIAM (1260361)_____
    At Fismes, France, Aug. 12, 1918.
    R—Bausman, Pa.
    B—Wrightsville, Pa.
    G. O. No. 130, W. D., 1918.

Private, 109th Machine Gun Battalion, 28th Division.
Seeing a wounded soldier in an ambulance which had broken down while crossing a small bridge, in plain view of the enemy and under heavy artillery fire, he went to his rescue and, assisted by another soldier, carried out the wounded man under intense shellfire 1½ miles to a dressing station.

VER MEHREN, HUBERT (915581)_____
    Near Brandeville, France, Nov. 8–10, 1918.
    R—Omaha, Nebr.
    B—Arcadia, Iowa.
    G. O. No. 44, W. D., 1919.

Sergeant, first class, medical detachment, 7th Engineers, 5th Division.
He showed utter disregard for his own personal danger in giving first aid to the wounded and carrying them to a place of safety under intense machine-gun and shellfire

*VERNAM, REMINGTON DE B._____
    Near Buzancy, France, Oct. 10, 1918.
    R—New York, N. Y.
    B—Rutherford, N. Y.
    G. O. No. 46, W. D., 1919.

First lieutenant, pilot, 22d Aero Squadron, Air Service.
Successively attacking 2 enemy balloons which were moored to their nests, Lieutenant Vernam displayed the highest degree of daring. He executed his task despite the fact that several enemy planes were above him, descending to an altitude of less than 10 meters when 5 miles within the enemy lines. His well-directed fire caused both balloons to burst into flames.
Posthumously awarded. Medal presented to mother, Mrs. Philip J. Ross.

VIAL, FRANK A_____
    In the Bois-de-Belleau, France, June 8, 1918.
    R—Richmond, Va.
    B—Hanover, Va.
    G. O. No. 119, W. D., 1918.

Corporal, 83d Company, 6th Regiment, U. S. Marine Corps, 2d Division.
Although exposed to fire constantly from machine guns in the Bois-de-Belleau, France, on June 8, 1918, he repeatedly carried messages from one post to another. He particularly distinguished himself for bravery by voluntarily passing through a machine-gun barrage to guide a detachment to its position.

*VIBBERT, EDWARD T. (242283)_____
    Near Sergy, northeast of Chateau-Thierry, France, July 31, 1918.
    R—Corunna, Mich.
    B—Brockport, Pa.
    G. O. No. 117, W. D., 1918.

Private, Company M, 125th Infantry, 32d Division.
When his company was obliged to dig in under heavy fire from all arms within 150 yards of the main German line, he was sent with a message from the company commander to the chief of a front-line platoon. In endeavoring to accomplish his mission this soldier was mortally wounded. Lying on the ground he yelled "Message," attracted the attention of the platoon leader, and with his dying breath delivered the message he bore.
Posthumously awarded. Medal presented to mother, Mrs. Unice H. Vibbert.

VIDA, FRANK J. (1225539)...............
Near Ronssoy, France, Sept. 29-30,
1918.
R—New York, N. Y.
B—Hungary.
G. O. No. 44, W. D., 1919.

First sergeant, Company G, 108th Infantry, 27th Division.
After all his company officers had been killed or wounded he took command,
despite the fact that he, too, had been wounded. He succeeded in capturing
part of the Hindenburg line and holding it against several strong counterattacks, remaining with his company and refusing to go to the rear for medical
treatment until it was relieved.

VIDMER, GEORGE...............
Near Zube, France, Sept. 27, and St.
Juvin, France, Oct. 14, 1918.
R—Mobile, Ala.
B—Mobile, Ala.
G. O. No. 64, W. D., 1919.
Distinguished-service medal also
awarded.

Colonel, 306th Infantry, 77th Division.
By his personal presence, example, and determination he repulsed strong counterattacks and drove the enemy from important positions north of Zube. On
Oct. 14 he personally directed the attack of his unit and carried it forward
to a successful conclusion under heavy machine-gun, rifle, and artillery fire.

VIEIRA, HENRY (48856)...............
South of Soissons, France, July 18,
1918.
R—New Bedford, Mass.
B—Bedford, Mass.
G. O. No. 117, W. D., 1918.

Private, first class, Company M, 9th Infantry, 2d Division.
During the attack and advance on July 18, 1918, and after all the runners had
been killed or wounded, Private Vieira volunteered to carry messages through
heavy machine-gun and shellfire. He succeeded in this undertaking, thereby
keeping up vitally important communication with regimental headquarters.

VIERA, JOE NICHOLS...............
Near Blanc Mont, France, Oct. 3,
1918.
R—Fallon, Nev.
B—Providence, R. I.
G. O. No. 37, W. D., 1919.

Private, 78th Company, 6th Regiment, U. S. Marine Corps, 2d Division.
After assisting in the capture of 3 machine-gun nests, Private Viera with
another soldier, went into a dugout when the occupants refused to come out
and captured 40 of the enemy.

VIERBUCHEN, WILLIAM J...............
In the Bois-de-Belleau, France, June
11, 1918.
R—Newark, N. J.
B—Washington, D. C.
G. O. No. 98, W. D., 1919.

Sergeant, 55th Company, 5th Regiment, U. S. Marine Corps, 2d Division.
Though he had been wounded by fire from an enemy machine-gun nest, Sergeant
Vierbuchen made a reconnaissance, securing information which was largely
instrumental in the successful capture of this nest a few hours later.

*VIGILETTRE, MICHAEL (1214947)........
Near Ronssoy, France, Sept. 29, 1918.
R—Rochester, N. Y.
B—Italy.
G. O. No. 21, W. D., 1919.

Private, Company G, 108th Infantry, 27th Division.
He voluntarily exposed himself to bring in wounded soldiers belonging to
another organization. Throughout the engagement under constant rifle and
machine-gun fire he courageously treated the wounded, inspiring the combat
troops by his example until killed by a bursting shell.
Posthumously awarded. Medal presented to father, Andrea Vigilettre.

VINALL, EARL R. (110333)...............
At Belleau, France, July 18-24, 1918.
R—Meredith, N. H.
B—Peterboro, N. H.
G. O. No. 126, W. D., 1919.

Private, Company B, 103d Machine Gun Battalion, 26th Division.
On duty as a runner, Private Vinall displayed marked courage in repeatedly
passing through heavy machine-gun and artillery barrages throughout 5
days of action.

VINCENT, JAMES A...............
Near Echisfontaine, France, Sept. 27,
1918.
R—Berkeley, Calif.
B—Davenport, Iowa.
G. O. No. 37, W. D., 1919.

First lieutenant, 363d Infantry, 91st Division.
Returning to the company after being treated for a very severe wound in the
neck, he commanded his platoon, which had been ordered to fall back because
of a violent barrage. He volunteered and went forward to the aid of 2
enlisted men of his platoon who had been seriously wounded. While performing this duty he was again wounded in the knee, but worked his way
back to the dressing station, and from there walked a distance of 4 kilometers to the field hospital.

VINIARSKY, WATZLAW (39711) ....
South of Soissons, France, July 18,
1918.
R—Jersey City, N. J.
B—Russia.
G. O. No. 20, W. D., 1919.

Private, Company G, 9th Infantry, 2d Division.
Jerome Buschman, sergeant; John Rockwell, private; William F. Rockwell,
private; Alfred Shimanoski, private; and Watzlaw Viniarsky, private, all
of Company G, 9th Infantry. For extraordinary heroism in action south of
Soissons, France, July 18, 1918. They conspicuously distinguished themselves by attacking a party of more than 60 Germans and in an intense and
desperate hand-to-hand fight succeeded in killing 22 men and capturing 40
men and 5 machine guns.

VINTON, THOMAS W...............
Near Ardeuil, France, Sept. 29-Oct.
1, 1918.
R—Memphis, Tenn.
B—Memphis, Tenn.
G. O. No. 44, W. D., 1919.

Second lieutenant, 371st Infantry, 93d Division.
Painfully wounded in the hand at 11 a. m. and slightly wounded again in the
hip at 4 p. m. Sept. 29, Lieutenant Vinton continued on duty and ably commanded his platoon until evacuated on Oct. 1, 1918.

VIZENOR, LAWRENCE A. (2152197)........
In the Bois-du-Fays, France, Oct. 8,
1918.
R—Richwood, Minn.
B—Richwood, Minn.
G. O. No. 98, W. D., 1919.

Private, Company I, 132d Infantry, 33d Division.
Private Vizenor was a member of a reconnaissance patrol which encountered
such intense fire from an enemy machine-gun nest that part of the patrol was
driven back. Despite the heavy fire, he and another soldier, with an officer,
continued forward and secured the information for which they were sent.
The officer was mortally wounded, but Private Vizenor and his comrade
silenced the machine-gun nest by effective rifle fire, carried the wounded
officer to the rear, and reported their valuable information concerning the
enemy's position.

*VOGEL ANDREW F. (1829263) _____ | Sergeant, Company C, 320th Infantry, 80th Division.
Near Bois-des-Ogons, France, Oct. 10, 1918. | Crawling on his hands and knees from a place of safety, Sergeant Vogel went to the aid of a wounded comrade, exposed to intense machine-gun fire during the entire exploit. He successfully accomplished his task, but during a later artillery attack he was instantly killed.
R—Pittsburgh, Pa.
B—Pittsburgh, Pa.
G. O. No. 44, W. D., 1919. | Posthumously awarded. Medal presented to sister, Mrs. Joseph B. Kenny.

VOGEL, ARTHUR H. (2368061) _____ | Corporal, Company D, 31st Infantry.
Near Sitsa, Siberia, June 26, 1919. | He voluntarily ran a railway locomotive past a cliff three different times to draw the fire of a large force of bandits in order to locate their exact position, so that the bandits could be driven off by machine-gun fire in order to clear the way for a wagon train. On each trip he was subjected to heavy rifle fire, the cab of the engine being punctured in several places, and his Russian assistant slightly wounded.
R—Heber, Calif.
B—San Diego, Calif.
G. O. No. 133, W. D., 1919.

VOIGT, RALPH L. (1775672) _____ | Private, Company A, 301st Battalion, Tank Corps.
Near Ribeauville, France, Oct. 17, 1918. | Although severely wounded by a shell splinter during the attack, Private Voigt continued on duty without revealing his wound. During a halt he assisted in mopping up with the Infantry, and again took up the advance, refusing to go to the rear for treatment. When he returned to the rallying point with the crew he was sent to the hospital.
R—Kingston, N. Y.
B—Colony, N. Y.
G. O. No. 32, W. D., 1919.

VOLIVA, JAMES B. (75091) _____ | Sergeant, Company F, 128th Infantry, 32d Division.
Near Gesnes, France, Oct. 7, 1918. | One of 4 survivors of a platoon of 41 who attacked Hill 269, he, with the 3 others continued on their mission and held the hill for some time without hope o reinforcements.
R—Pomona, Wash.
B—Pomona, Wash.
G. O. No. 47, W. D., 1921.

VOLK, JOSEPH W. (1750463) _____ | Private, Company B, 312th Infantry, 78th Division.
Near Grand Pre, France, Oct. 22, 1918. | While engaged as runner, he carried messages to and from advanced positions and, although wounded, continued through the whole afternoon before his wound was discovered. Even then he volunteered to carry an important message to the front lines. While directing his company commander and a number of wounded men through an intense barrage he successfully forded a river four times, showing during the entire operations an utter disregard of personal safety.
R—Kingston, N. Y.
B—Kingston, N. Y.
G. O. No. 35, W. D., 1919.

VOLLMER, FRANK D _____ | Private, Headquarters Company, 5th Regiment, U. S. Marine Corps, 2d Division.
Near St. Etienne, France, Oct. 4–6, 1918. | As a runner he displayed exceptional courage in volunteering to carry important messages over terrain constantly swept by machine-gun and shell fire.
R—Cincinnati, Ohio.
B—Cincinnati, Ohio.
G. O. No. 46, W. D., 1919.

*VON KREBS, PAUL (558501) _____ | First sergeant, Company M, 47th Infantry, 4th Division.
At Sergy, France, July 29–30, 1918. | Sergeant Von Krebs displayed exceptional bravery in voluntarily carrying wounded men to safety across shell-swept areas. Later he took charge of 2 platoons, whose officers had become casualties, and reorganized them. Strengthening these with stragglers from other organizations, he led them all into the attack at a critical moment.
R—Franklin Park, N. J.
B—Germany.
G. O. No. 55, W. D., 1920. | Posthumously awarded. Medal presented to stepmother, Mrs. Leonie Von Krebs.

VOORHEES, GEORGE C _____ | Private, 81st Company, 6th Machine Gun Battalion, U. S. Marine Corps, 2d Division.
Near Blanc Mont and St. Etienne, France, Oct. 3–10, 1918. | He displayed remarkable devotion to duty in repeatedly carrying important messages through the most violent artillery and machine-gun barrages with utter disregard for his own safety.
R—Lansing, Mich.
B—Brooklyn, N. Y.
G. O. No. 15, W. D., 1919.

VOSBURGH, FRED _____ | First lieutenant, Medical Corps, attached to 116th Infantry, 29th Division.
Near Samogneux, France, Oct. 15, 1918. | In the attack on the Bois-de-la-Grande Montagne, First Lieutenant Vosburgh accompanied the attacking Infantry, exposing himself to heavy machine-gun fire in order to direct the prompt evacuation of the wounded. After being severely wounded he continued in his efforts until his evacuation was forced by loss of blood.
R—Standish, N. Y.
B—Cobleskill, N. Y.
G. O. No. 24, W. D., 1920.

VOSBURGH, PHILIP DE M. (1211621) _____ | Sergeant, Company K, 107th Infantry, 27th Division.
In the vicinity of Bony, France, Sept. 28, 1918. | Sergeant Vosburch exposed himself to direct observation and fire from the enemy in leaving a place of shelter to go to the assistance of a wounded officer. Upon reaching the officer he was hit in the knee by a machine-gun bullet, causing him to fall back into the trench, dragging the officer with him.
R—New Brighton, N. Y.
B—Buffalo, N. Y.
G. O. No. 19, W. D., 1920.

VOSSELER, EDWARD A _____ | First lieutenant, 60th Infantry, 5th Division.
North of Cunel, France, Oct. 15, 1918. | His company being held up by machine-gun fire from the front, Lieutenant Vosseler alone rushed 100 yards ahead of his company, exposed to heavy machine-gun fire, and silenced a machine gun. Although knocked down and slightly wounded by shellfire, he went forward and silenced a second gun, thus enabling his unit to continue the advance.
R—Brooklyn, N. Y.
B—Brooklyn, N. Y.
G. O. No 27, W. D., 1920.

VOTAW, LOUIS H. (2228329) _____ | Corporal, Company B, 360th Infantry, 90th Division.
In the Bois-le-Pretre, France, Sept. 12, 1918. | Although he was wounded in the body early in the action and later received another wound in the head, he continued at his post as squad leader until the action was over and then went to the rear only upon orders from his platoon commander.
R—Beaumont, Tex.
B—Colmesneil, Tex.
G. O. No. 128, W. D., 1918.

VROOMAN, VERNON A
Near Grand Pre, France, Oct. 26, 1918.
R—Albany, N. Y.
B—Middleburgh, N. Y.
G. O. No. 56, W. D., 1922.

Captain, 311th Infantry, 78th Division.
As adjutant of the 3d Battalion he volunteered and personally reconnoitered under fire the temporary front-line positions of his battalion for location and ammunition supply, when he returned to the command post and collected materials and all available men to carry them forward preparatory to the jump off early the next morning. Finding that there were not enough men to carry the needed supplies, Captain Vrooman loaded a wheelbarrow with ammunition and personally wheeled it 3 kilometers under fire to the firing line, where he supervised its distribution. At "H" hour, observing a company in need of assistance he attached himself to it and went over the top to the objective. After the position was consolidated, and knowing the importance of executing liaison between the attacking troops and those in Grand Pre, he went to that town and brought to the battalion command post definite information that our lines had been joined up.

WADDILL, EDMUND C
Near Chateau-Thierry, France, June 6, 7, and 25, 1918, and near Soissons, France, July 18, 1918.
R—Richmond, Va.
B—Richmond, Va.
G. O. No. 98, W. D., 1919.

Major, 23d Infantry, 2d Division.
During the attack by his battalion near Chateau-Thierry Major Waddill displayed exceptional bravery by advancing in the open under intense shell and machine-gun fire, reorganizing his leading echelons and pressing the attack with the utmost disregard for personal danger. On June 25 he went among his troops during a heavy gas attack, disregarding his own danger in order to protect his men, remaining in the sector and refusing to be evacuated until he had been so badly burned by gas that his face was black. In the Soissons-Reims attack he again displayed marked courage and leadership in personally taking the lead with his battalion and pushing forward the attack until further advance was stopped by darkness.

*WADSWORTH, LEE A. (97451)
Near Landres-et-St. Georges, France, Oct. 15, 1918.
R—Mulberry, Ala.
B—Mulberry, Ala.
G. O. No. 131, W. D., 1918.

Sergeant, Company H, 167th Infantry, 42d Division.
He was severely wounded in the attack on the Cote-de-Chatillon, but he refused to be evacuated and remained with his platoon under heavy fire, reorganizing it for the counterattack for which the enemy were forming, thereby setting to his men an inspiring example of utter disregard for danger and heroism in the face of the enemy.
Posthumously awarded. Medal presented to father, Thomas M. Wadsworth.

WAGNER, DONALD L. (1789432)
Near Montfaucon, France, Sept. 29, 1918.
R—Winston-Salem, N. C.
B—Winston-Salem, N. C.
G. O. No. 37, W. D., 1919.

Sergeant, 314th Ambulance Company, 304th Sanitary Train, 79th Division.
He heard a cry for help while in a dugout having his own wounds dressed. Although it was during particularly heavy shellfire, he immediately went outside and carried the wounded man to shelter. Later that day, when the dressing station caught fire, he made his way into the burning dressing station under heavy shellfire and secured surgical equipment necessary to save a patient's life.

WAGNER, FRANCIS W. (1708042)
Near Badonviller, France, June 24, 1918.
R—New York, N. Y.
B—New York, N. Y.
G. O. No. 37, W. D., 1919.

Sergeant, Company C, 308th Infantry, 77th Division.
He was found badly wounded in the neck and legs, crawling back to bring up support to his position.

WAGNER, JEROME E. (94804)
Near Seicheprey, France, Sept. 12, 1918.
R—Osgood, Ind.
B—Osgood, Ind.
G. O. No. 44, W. D., 1919.

Corporal, Company I, 166th Infantry, 42d Division.
When concentrated machine-gun fire was encountered, he maneuvered his squad to a point near the nests, and, although severely wounded, continued to direct his men in silencing the guns.

WAGNER, TONY (40969)
Near Jaulny, France, Sept. 12, 1918.
R—New York, N. Y.
B—Poland.
G. O. No. 46, W. D., 1919.

Private, Company M, 9th Infantry, 2d Division.
Private Wagner, with 3 other soldiers, volunteered to carry wounded men of other companies from in front of our advanced positions, and carried this work on under violent machine-gun fire while a counterattack was developing.

WAHLER, RICHARD (2176034)
Near Bois-de-Bantheville, France, Oct. 21, 1918.
R—Leavenworth, Kans.
B—Leavenworth, Kans.
G. O. No. 44, W. D., 1919.

Private, Company C, 353d Infantry, 89th Division.
After being severely wounded in the hand and face, he continued his duties as stretcher bearer, carrying wounded from the field during the entire afternoon and evening. For 7 days thereafter he remained constantly on duty and during the entire period he worked under an almost incessant rain of shells.

*WAITE, GEORGE T. (2337545)
Near Roncheres, France, July 28, 1918.
R—Bruce, S. Dak.
B—Pipestone, Minn.
G. O. No. 32, W. D., 1919.

Private, first class, Headquarters Company, 4th Infantry, 3d Division.
Exposing himself to heavy shellfire, he kept in repair the telephone wire to the front line, succeeding in his mission despite the fact that shellfire was causing repeated breaks while he was thus engaged.
Posthumously awarded. Medal presented to father, G. A. Waite.

WAITE, HOWARD E. (180832)
Near Juvigny, France, Aug. 31, 1918.
R—Highland Park, Mich.
B—Canada.
G. O. No. 71, W. D., 1919.

Private, first class, sanitary detachment, 126th Infantry, 32d Division.
Facing heavy machine-gun fire, he volunteered to go out and administer first aid to wounded soldiers lying on an open field, saving the lives of 5 men and alleviating the suffering of many others. After dark he continued his work until all the wounded had been removed to the rear and their wounds dressed.

WAITE, ROBERT SNELLEY
Near Cunel Woods, France, Oct. 11, 1918.
R—Birmingham, Ala.
B—Easonville, Ala.
G. O. No. 44, W. D., 1919.

First lieutenant, 7th Infantry, 3d Division.
Although shot through the arm, Lieutenant Waite with his company charged a machine-gun nest. His attempt being unsuccessful, he reformed his company and again attacked, this time silencing the nest and capturing it.

WALDEN, FITZGERALD (1490019)..........
Near St. Etienne, France, Oct. 8, 1918.
R—Durant, Okla.
B—Hood County, Tex.
G. O. No. 66, W. D., 1919.

Sergeant, Company E, 142d Infantry, 36th Division.
After his company commander had been killed, Sergeant Walden took command of that portion of the company near him and, reorganizing the line under heavy fire, continued forward with his men and repulsed an enemy counterattack. He refused to be evacuated until nightfall, when he was no longer able to stand.

WALDO, ANTONIO (1762180)...........
Near the Bois-des-Loges, France, Nov. 1–4, 1918.
R—Canastota, N. Y.
B—Camden, N. Y.
G. O. No. 87, W. D., 1919.

First sergeant, Company C, 310th Infantry, 78th Division.
Taking command of his company after all the officers had been wounded, Sergeant Waldo displayed exceptional courage and leadership in holding an advanced position throughout the night and leading his men in a successful advance next morning. On the 3 following days he directed the operations of his company against enemy machine guns and by skillful maneuvering of patrols succeeded in reducing the hostile resistance.

WALDRON, JOSEPH F. (2387671)..........
In the Bois-de-la-Grande Fontaine, France, Sept. 16, 1918.
R—New Bedford, Mass.
B—England.
G. O. No. 9, W. D., 1923.

Private, first class, Company G, 61st Infantry, 5th Division.
While serving as a company mail carrier, he repeatedly volunteered and carried important messages through machine-gun and artillery fire. Although severely wounded in the head and neck, he continued to accomplish his dangerous missions, refusing medical treatment until ordered to the rear by his company commander. His fearless conduct and devotion to duty inspired and steadied the men of his organization.

WALDRON, WILLIAM H..............
At Tientsin, China, July 13, 1900.
R—Welch, W. Va.
B—Huntington, W. Va.
G. O. No. 43, W. D., 1922.
Distinguished-service medal also awarded.

Second lieutenant, 9th Infantry, U. S. Army.
For conspicuous gallantry in action at Tientsin, China, July 13, 1900, in rescuing while under a heavy fire 3 of his men from drowning, all of whom, fully accoutered, had fallen into a deep ditch, when Lieutenant Waldron, also fully accoutered, jumped in and saved them.

*WALDROOP, WALTER (55383)...........
Near Verdun, France, Oct. 9, 1918.
R—Sylva, N. C.
B—Macon County, N. C.
G. O. No. 89, W. D., 1919.

Private, first class, Machine Gun Company, 26th Infantry, 1st Division.
Private Waldroop, with an officer and 6 other soldiers, drove off a violent assault of 50 of the enemy after a terrific pistol and grenade fight, thereby holding Hill 269, which was of the utmost tactical importance. During the fighting Private Waldroop was killed.
Posthumously awarded. Medal presented to mother, Mrs. E. H. Waldroop.

*WALDROP, BERGEN X. (1499676).......
Near St. Etienne, France, Oct. 8, 1918.
R—Clarendon, Tex.
B—Falkville, Okla.
G. O. No. 20, W. D., 1919.

Sergeant, Company H, 142d Infantry, 36th Division.
While leading his platoon in an advance in the face of heavy machine-gun and shell fire, he was wounded, but refused to go to the rear, continuing his advance until the objectives had been attained.
Posthumously awarded. Medal presented to father, W. W. Waldrop.

WALES, WADE C. (1197349)...........
Near Cunel, France, Oct. 14, 1918.
R—Weston, W. Va.
B—Upshur County, W. Va.
G. O. No. 44, W. D., 1919.

Private, first class, Company A, 13th Machine Gun Battalion, 5th Division.
Accompanying another soldier, Private Wales left shelter and went forward 100 meters over territory swept by shells and machine-gun fire and carried a wounded man to safety.

WALKER, CAROL (2264876)...........
Near Waereghem, Belgium, Oct. 31, 1918.
R—Healdsburg, Calif.
B—Healdsburg, Calif.
G. O. No. 44, W. D., 1919.

Sergeant, Company K, 363d Infantry, 91st Division.
Without assistance and in the face of heavy fire, he killed an enemy machine gunner and captured his machine gun, which had been pouring a destructive fire on our forces.

WALKER, EUGENE P. (914715)..........
Near Verdun, France, Nov. 4, 1918.
R—Reidsville, N. N.
B—Reidsville, N. C.
G. O. No. 37, W. D., 1919.

Sergeant, Company D, 7th Engineers, 5th Division.
When 3 boats in a pontoon bridge across the Meuse River were destroyed by artillery fire, he volunteered and waded into the river under heavy shellfire and, by holding up the deck until new boats were launched and placed in position, although under great physical strain, permitted the uninterrupted crossing of the Infantry.

WALKER, FRED L...............
Near the Marne River, France, July 15, 1918.
R—Kierkesville, Ohio.
B—Fairfield County, Ohio.
G. O. No. 80, W. D., 1919.

Major, 30th Infantry, 3d Division.
Holding a front of more than 4½ kilometers along the Marne River, Major Walker commanded a front-line battalion, which received the principal shock of the German attack on the French Army Corps front, but inflicted great losses on the enemy as the latter crossed the river. Those who succeeded in crossing were thrown into such confusion that they were unable to follow the barrage; and, through the effective leadership of this officer, no Germans remained in his sector south of the river at the end of the day's action. When 1 platoon had been cut off by an entire enemy battalion near the river, he sent other units to its relief and captured the entire German battalion, numbering 200 soldiers and 5 officers, including the battalion commander.

WALKER, HUBBARD J. (1307335).......
Near Ypres, Belgium, July 24, 1918.
R—Fruitland, Tenn.
B—Fruitland, Tenn.
G. O. No. 21, W. D., 1919.

Private, first class, Company C, 117th Infantry, 30th Division.
He was on duty at a detached automatic-rifle post heavily shelled by the enemy. 2 soldiers were killed and 3, including himself, seriously wounded. Though this was his first experience under fire, he displayed unhesitating devotion by remaining at his post, while, because of his wound, he could use but 1 hand in handling his rifle.

WALKER, JAMES M. (2156266)........
Near Norroy, France, Sept. 15, 1918.
R—Tipton, Iowa.
B—Geneva, Nebr.
G. O. No. 37, W. D., 1919.

Private, first class, Company K, 328th Infantry, 82d Division.
When his platoon had successfully reached its objective, he was dispatched with a message to battalion headquarters, the journey being under intense fire for the whole distance. He not only delivered the message, but while returning assisted many other carriers by directing them to their proper destinations.

WALKER, JOSEPH.
Near Blosmes Village, France, July
R—New York, N. Y.
B—New York, N. Y.
G. O. No. 44, W. D., 1919.

Captain, 76th Field Artillery, 3d Division.
Making his way from the forward observation post through an extremely heavy shelling, he delivered very important information after telephone communications had been cut and after he had been severely wounded.

*WALKER, WILLIS J. (181766).
Near Bezu-St. Germain, France,
Sept. 7, 1918.
R—Brady, Tex.
B—Comanche County, Tex.
G. O. No. 142, W. D., 1918.

Private, first class, Salvage Squad No. 1, Quartermaster Corps.
When fire broke out in a wood where a salvage detachment was encamped, seriously endangering the lives of 200 men because of its proximity to a pile of salvaged German high-explosive 155-millimeter shells, he and Sergt. Afton E. Wheeler voluntarily ran to the scene of the fire and attempted to extinguish the flames, fully aware of the grave danger to themselves. They fought the fire with blankets and sticks, but the fire quickly spread to the shells. Both men were killed by the explosion which followed.
Posthumously awarded. Medal presented to widow, Mrs. Stella Walker.

*WALL, EARL L.
In the Bois-de-Maulamont, France,
Oct. 8, 1918.
R—Chicago, Ill.
B—Marshalltown, Iowa.
G. O. No. 87, W. D., 1919.

Second lieutenant, 132d Infantry, 33d Division.
As battalion scout officer, Lieutenant Wall led a patrol into the wood for the purpose of securing information of enemy units in preparation for an attack. Severe machine-gun fire was encountered, and this officer was wounded, but with 2 soldiers he continued on until he was wounded the second time, securing the desired information.
Posthumously awarded. Medal presented to father, Jesse J. Wall.

WALL, WALTER W. (76005).
Near Soissons, France, July 8, 1918.
R—Winlock, Wash.
B—Winlock, Wash.
G. O. No. 132, W. D., 1918.

Private, Company E, 9th Infantry, 2d Division.
After 12 hours of hard fighting, when Private Wall's platoon had gained its objective, the water taken forward in canteens had become exhausted, and the men were suffering from thirst. Knowing that the chances were against anyone being able to cross the shell-swept territory for water, the platoon commander called for volunteers. Private Wall responded, and, collecting the canteens of his comrades, departed on his precarious mission. Several hours later he returned, utterly exhausted, but bearing with him the canteens filled with precious water. Other men attempting to make similar trips in the same vicinity were either killed or wounded.

WALLACE, ANTHONY M. (2383465).
Near Cunel, France, Oct. 14, 1918.
R—Bridgeport, Conn.
B—Russia.
G. O. No. 59, W. D., 1919.

Private, Company E, 60th Infantry, 5th Division.
After his company had been held up by terrific machine-gun fire, while advancing on Cunel, Private Wallace, with another soldier, went forward in face of the annihilating fire and by flanking the strong point succeeded in capturing 3 prisoners and 2 guns.

WALLACE, FRED E. (1995437).
Near St. Souplet, France, Oct. 17-19,
1918.
R—Roodhouse, Ill.
B—Roodhouse, Ill.
G. O. No. 37, W. D., 1919.

Private, first class, Company F, 119th Infantry, 30th Division.
He volunteered and located the right flank of Company G and the left flank of Company H, 119th Regiment, under heavy machine-gun fire. He was wounded while on this mission, but returned with the desired information.

WALLACE, HERBERT E.
Near Maribois Farm, north of
Beney, France, Sept. 16-22, 1918.
R—Hartsville, S. C.
B—Darlington, S. C.
G. O. No. 102, W. D., 1918.

Second lieutenant, 168th Infantry, 42d Division.
On Sept. 16, 1918, under heavy artillery and machine-gun fire, without regard to his personal safety he led a raiding party from our lines and attacked the Germans at Maribois Farm, and in severe hand-to-hand fighting inflicted severe loss upon the enemy, captured numerous prisoners, and obtained the information for which he was sent. On Sept. 22, 1918, he voluntarily led a second raiding party into Maribois Farm, inflicted great loss upon the enemy in hand-to-hand fighting, captured many prisoners, and obtained the desired information.

WALLACE, JOSEPH A. (2257240).
Near Gesnes, France, Oct. 3, 1918.
R—Battle Ground, Wash.
B—South Harbor, Minn.
G. O. No. 20, W. D., 1919.

Corporal, Company B, 361st Infantry, 91st Division.
While his company was under heavy shellfire he voluntarily, unhesitatingly, and repeatedly left his shelter under heavy shellfire, without thought of personal danger, rendered first aid, and carried wounded comrades to a place of safety.

WALLACE, WILLIAM M. (1320330).
Near Mazinghien, France, Oct. 19,
1918.
R—Othello, N. C.
B—Ashe County, N. C.
G. O. No. 50, W. D., 1919.

Private, first class, Company E, 120th Infantry, 30th Division.
With another soldier Private Wallace volunteered and rescued a wounded comrade from an exposed position in front of the line, after 2 other men had lost their lives in attempting to do so.

*WALLACE, WILLIAM NOBLE.
Near St. Etienne, France, Oct. 8,
1918.
R—Indianapolis, Ind.
B—Indianapolis, Ind.
G. O. No. 15, W. D., 1921.

First lieutenant, 6th Regiment, U. S. Marine Corps, 2d Division.
Lieutenant Wallace, with 1 comrade, exposed frequently to direct hostile observation and heavy fire, accomplished an exceedingly hazardous reconnaissance of the front lines of his regiment. After having made his reconnaissance, located enemy strong points, and obtained vitally important information, Lieutenant Wallace was killed by hostile fire, but by the delivery of his careful notes and sketches to his commander, his mission was fulfilled.
Posthumously awarded. Medal presented to father, H. L. Wallace.

WALLACE, WILLIAM R. (1236515).
Near Baslieux, Marne, France, Sept.
6, 1918.
R—Delaware Water Gap, Pa.
B—Delaware Water Gap, Pa.
G. O. No. 35, W. D., 1920.

Sergeant, Company G, 109th Infantry, 28th Division.
Although twice wounded by enemy machine-gun bullets, Sergeant Wallace continued to lead his platoon forward through artillery and machine-gun fire. In the hand-to-hand conflict which followed 11 of the enemy were killed or wounded. His devotion to duty was an excellent example to his command.

WALLENMAIER, HERMAN (42580)........
Near Argonne Forest, France, Oct. 9, 1918.
R—Valley Town, Mont.
B—Salem, Mich.
G. O. No. 44, W. D., 1919.

Private, Company D, 16th Infantry, 1st Division.
Although suffering painfully from wounds, he remained with his company during the entire action, and then was evacuated only when ordered to leave by his commanding officer, being unable to proceed farther because of the loss of blood.

WALLER, LUTHER HILL................
North of Vandieres, France, Sept. 15, 1918.
R—Montgomery, Ala.
B—Montgomery, Ala.
G. O. No. 53, W. D., 1920.

First lieutenant, 328th Infantry, 82d Division.
After his platoon had suffered severe losses by machine-gun and artillery fire during the advance, he reorganized it and led it in the attack through heavy fire to its objective. He then exposed himself to heavy artillery fire in order to bring back wounded men who lay in advance of the line, having gone out on a reconnaissance to locate flanking machine-gun positions. He made 4 trips across the fire-swept area to assist the wounded men to shelter.

WALLERIUS, JAMES J. (553715)........
Near Montfaucon, France, Oct. 8, 1918.
R—Utica, N. Y.
B—New York, N. Y.
G. O. No. 89, W. D., 1919.

Sergeant, Company B, 8th Machine Gun Battalion, 3d Division.
When an enemy shell struck a dugout, burying 2 soldiers, Sergeant Wallerius, in plain view of the enemy, directed the work of getting the men out under heavy shell and direct machine-gun fire, displaying the utmost fearlessness. This soldier was in command of his platoon for 11 days of action, leading his men with marked ability and conspicuous bravery.

WALLIS, JAMES E., Jr................
In the region of Meiz, France, Sept. 13, 1918.
R—Cambridge, Mass.
B—East Aurora, N. Y.
G. O. No. 145, W. D., 1918.

Captain, observer, 91st Aero Squadron, Air Service.
While on a reconnoissance under the most adverse weather conditions, which necessitated flying at an extremely low altitude, he with his pilot penetrated the enemy's territory to a depth of 25 kilometers. Attacked by 5 enemy planes, they destroyed 1 and forced the others to retire. In heavy fire from the ground they continued on their mission until it was completed.

WALLS, OKLA M. (504105)...........
At Marcheville, France, Sept. 26, 1918.
R—Ringwood, Okla.
B—Ringwood, Okla.
G. O. No. 139, W. D., 1918.

Private, Company C, 101st Field Signal Battalion, 26th Division.
Preparatory to establishing telephone communication from the leading elements to the rear, he voluntarily reconnoitered an area swept by heavy artillery and machine-gun fire, locating forward positions in which wires could be strung.

WALSH, CHARLES H. (1211600).........
Near Bony, France, Sept. 29, 1918.
R—New York, N. Y.
B—Philadelphia, Pa.
G. O. No. 15, W. D., 1923

Private, first class, Company I, 107th Infantry, 27th Division.
As a member of a Lewis gun squad he continued to advance with his section although severely wounded by enemy fire. Upon reaching the enemy trenches an intense enemy machine-gun fire killed or wounded every man of his squad. Although Private Walsh was again severely wounded, he continued to operate his Lewis gun until he collapsed and was carried from the field.

WALSH, FRANK (2214224).............
Near Mont St. Pere, France, July 22, 1918.
R—Benkelman, Nebr.
B—Benkelman, Nebr.
G. O. No. 32, W. D., 1919.

Private, first class, Company C, 4th Infantry, 3d Division.
After being severely wounded, he continued to operate his automatic rifle throughout the night.

WALSH, HERBERT E. (1208032).........
Near Ronssoy, France, Sept. 27–29, 1918.
R—Brooklyn, N. Y.
B—Long Branch, N. J.
G. O. No. 60, W. D., 1920.

Sergeant, Company H, 106th Infantry, 27th Division.
On Sept. 27 Sergeant Walsh assumed command of his company after other officers and noncommissioned officers had been killed or wounded and led it forward through heavy fire to its objective. He then organized his position for defense. He later made a personal reconnaissance in advance of our lines and returned with valuable information. The courageous conduct of this noncommissioned officer was an important factor in the success of the operations of the company.

WALSH, JAMES (63580)...............
Near Marcheville, France, Sept. 26, 1918.
R—Hartford, Conn.
B—Rutland, Vt.
G. O. No. 15, W. D., 1919.

Sergeant, Company A, 102d Infantry, 26th Division.
He displayed remarkable coolness, courage, and devotion to duty under terrific shell and machine-gun fire. When surrounded by the enemy, he organized men near him, collected the wounded, and brought them to safety. He was himself wounded, but remained in action until his company was relieved, several hours later.

WALSH, JAMES G. (185183)...........
In the Belleau Woods, France, July 16–18, 1918.
R—Boston, Mass.
B—Forest Hills, Mass.
G. O. No. 37, W. D., 1919.

Private, Company D, 101st Engineers, 26th Division.
While out on a working party fired on by the enemy he cared for the wounded. Two days later, when his platoon was ordered to attack, he was the first man over the top. A machine-gun nest delivering a violent enfilading fire from the opposite side of a railroad cut, Private Walsh ran across the track alone to put it out of action and fell wounded before the gun.

*WALSH, JOHN A...................
Near Chevieres, France, Oct. 13, 1918.
R—New York, N. Y.
B—New York, N. Y.
G. O. No. 37, W. D., 1919.

First lieutenant, Company F, 302d Engineers, 77th Division.
Accompanied by a sergeant, he went in advance of our lines to reconnoiter for locations for crossing the River Aire. After being constantly exposed to heavy shellfire, they reached the bank of the river, where Lieutenant Walsh pushed farther on and was killed by machine-gun and sniper fire.
Posthumously awarded. Medal presented to mother, Mrs. Mary Walsh.

WALSH, JOHN R. (63005).............
In the Belleu Bois, France, Oct. 23, 1918.
R—Woburn, Mass.
B—Woburn, Mass.
G. O. No. 59, W. D., 1921.

Private, Machine Gun Company, 101st Infantry, 26th Division.
Private Walsh with another soldier advanced carrying their machine gun to an enemy pill box and outflanked the enemy. For 2 days and nights without food or water he remained in the pill box under heavy artillery and machine-gun fire and rendered invaluable assistance to the Infantry.

WALSH, MICHAEL J._____
Near the Meuse River, France,
Oct. 14, 1918.
R—New York, N. Y.
B—Scranton, Pa.
G. O. No. 87, W. D., 1919.

Captain, 165th Infantry, 42d Division.
After being wounded in the arm by an enemy sniper, Captain Walsh refused
to go to the rear, but continued with his company, encouraging his men by
his coolness and courage. He was killed soon afterwards in dislodging a
sniper who had been inflicting many casualties among his men.
Posthumously awarded. Medal presented to widow, Mrs. Michael J. Walsh.

WALSH, PATRICK (47712)_____
Near Seicheprey, Ansauville sector,
France, Mar. 1, 1918.
R—Detroit, Mich.
B—Ireland.
G. O. No. 126, W. D., 1918.

Sergeant, Company I, 18th Infantry, 1st Division.
He voluntarily followed his company commander to the first line through a
severe barrage, and when the captain was killed he assumed command of a
group on his own initiative, attacked a superior force of the enemy, and
inflicted heavy losses upon them.

WALSH, PRESTON F._____
Near Monthois, France, Sept. 27 to
Oct. 7, 1918.
R—New York, N. Y.
B—New York, N. Y.
G. O. No. 13, W. D., 1919.

First lieutenant, 372d Infantry, 93d Division.
As regimental intelligence officer, he went each day during the attack on Monthois to reconnoiter and secure direct information. On Sept. 29 he penetrated the enemy lines east of Ardeuil, discovered the location of a machine-
gun nest which was holding up the advance, and was most daring in accomplishing his mission. Though he was wounded by a machine-gun bullet,
he remained on duty.

WALSH, RICHARD J._____
Near Marg, France, Oct. 18, 1918.
R—Philadelphia, Pa.
B—New York, N. Y.
G. O. No. 44, W. D., 1919.

First lieutenant, Dental Corps, attached to 303d Engineers, 78th Division.
Voluntarily acting as battalion medical officer, Lieutenant Walsh, although
severely gassed, administered first aid to injured men under heavy shellfire.
He worked constantly until all the wounded were removed to places of safety.

WALSH, THOMAS F. (1903338)_____
Near St. Juvin, France, Oct. 16,
1918.
R—Brooklyn, N. Y.
B—Long Island, City N. Y.
G. O. No. 59, W. D., 1919.

Corporal, Company M, 326th Infantry, 82d Division.
Advancing under heavy artillery and incessant machine-gun fire for a distance
of 200 yards, Corporal Walsh rescued a wounded comrade and brought him
safely back to our lines

WALSH, THOMAS J. (1387002)_____
At Bray-sur-Somme, France, Aug.
17, 1918.
R—Chicago, Ill.
B—Chicago, Ill.
G. O. No. 70, W. D., 1919.

Sergeant, Company D, 131st Infantry, 33d Division.
He volunteered to lead a daylight raid on enemy trenches and was successful
in reaching the objective, capturing machine-gun positions and prisoners.
Although he was seriously wounded, he carried a wounded comrade to safety
through heavy shell fire and immediately returned to direct further attacks
on enemy positions, refusing first aid until he was ordered back by his commanding officer.

WALSH, WILLIAM J. (1782946)_____
Near Montfaucon, France, Sept. 27,
1918.
R—Lackawanna, N. Y.
B—Minooka, Pa.
G. O. No. 37, W. D., 1919.

Corporal, Company H, 314th Infantry, 79th Division.
While leading a scouting patrol 300 meters in advance of his company, he was
fired upon from enemy machine-gun points. Several of his patrol were
wounded, but after carrying 1 man to shelter and assisting the others he
continued under heavy fire, locating 6 machine-gun nests and shooting the
entire crew of 1 of them.

WALSHE, ROBERT J. (1286089)_____
Near Molleville Farm, France, Oct.
23, 1918.
R—Baltimore, Md.
B—Baltimore, Md.
G. O. No. 37, W. D., 1919.

Corporal, Company A, 110th Machine Gun Battalion, 29th Division.
He remained with a wounded comrade and gave him all possible aid under a
severe bombardment of high-explosive and gas shells. He later secured
assistance and carried the wounded soldier to a first-aid station.

WALSTON, RAY E. (2258495)_____
Near Gesnes, France, Sept. 29–Oct.
3, 1918.
R—Colville, Wash.
B—Bradgate, Iowa.
G. O. No. 20, W. D., 1919.

Bugler, Company M, 361st Infantry, 91st Division.
Without any thought of personal danger, he repeatedly carried messages over
ground swept by shell and machine-gun fire, delivering his messages with the
utmost promptness.

WALTER, HIRAM F. (1491368)_____
Near St. Étienne, France, Oct. 8,
1918.
R—Okemah, Okla.
B—Cedar Creek, Mo.
G. O. No. 66, W. D., 1919.

Corporal, Company E, 142d Infantry, 36th Division.
Corporal Walter volunteered to lead a patrol for the purpose of locating and
silencing an enemy machine-gun nest which was holding up the advance.
Before reaching the nest all his men had been killed or wounded, but he continued on alone to within a short distance of the nest, ascertained its position,
and reported its location so accurately that it was soon silenced.

*WALTER, JOHN (1286370)_____
Near Verdun, France, Oct. 10, 1918.
R—Germantown, Md.
B—Washington, D. C.
G. O. No. 2, W. D., 1919.

Private, first class, Company K, 115th Infantry 29th Division.
During an advance on Rechene Hill, after being shot twice in the abdomen,
he captured a machine gun by killing 3 of the enemy.
Posthumously awarded. Medal presented to father, John Walter.

WALTERS, ARTHUR L. (204377)_____
Near Beaumont, France, Nov. 9,
1918.
R—Wadena, Iowa.
B—Wadena, Iowa.
G. O. No. 81, W. D., 1919.

Sergeant, Company B, 2d Ammunition Train, 2d Division.
Sergeant Walters was in charge of a convoy of ammunition trucks which was
halted in the town. An enemy shell struck the train and set one of the trucks
on fire. Although knocked down by the explosion, Sergeant Walters quickly
recovered himself and moved his convoy to safety, after which he returned
and jumping to the wheel of the blazing truck, drove to a place where it no
longer endangered the lives of others, and extinguished the fire, saving both
trucks and ammunition.

*WALTERS, JOHN B. F. (96895)_____

Vicinity of Ancerviller, France, May 3-4, 1918.
R—Gadsden, Ala.
B—Gadsden, Ala.
G. O. No. 100, W. D., 1918.

Private, first class, Company F, 167th Infantry, 42d Division.
While a member of a patrol in no man's land, in the vicinity of Ancerviller on the night of May 3-4, 1918, he displayed great self-sacrifice in refusing aid and continuing to do his duty after being mortally wounded.
Posthumously awarded. Medal presented to father, Sam Walters.

WALTMAN, EMMETT W. (569248)_____

West of Fismes, France, Aug. 5, 1918.
R—Kellogg, Idaho.
B—Rockford, Wash.
G. O. No. 145, W. D., 1918.

Corporal, Company F, 4th Engineers, 4th Division.
He was a member of a small detachment of Engineers which went out in advance of the front line of the Infantry through an enemy barrage from 77-millimeter and 1-pounder guns to construct a footbridge over the River Vesle. As soon as their operations were discovered machine-gun fire was opened up on them, but, undaunted, the party continued at work, removing the German wire entanglements and completing a bridge which was of great value in subsequent operations.

WALTON, ALONZO (1401373)_____

At Rue Larcher and Pont-d'Any, France, Nov. 7-9, 1918.
R—Bloomington, Ill.
B—Normal, Ill.
G. O. No. 64, W. D., 1919.

Private, Machine Gun Company, 370th Infantry, 93d Division.
When his company had been separated from their food supply for 2 days, Private Walton twice volunteered, taking a machine-gun cart, and under heavy fire located the kitchen and brought back much-needed food.

WALTON, CHARLES WAYNE (9976)_____

Near Woel, France, Oct. 6, 1918.
R—Woodbury, N. J.
B—Woodbury, N. J.
G. O. No. 127, W. D., 1918.

Private, first class, Section No. 635, Ambulance Service.
He proceeded to a point within 15 meters of the German line to rescue the surviving member of a small French patrol. He placed the man in his car and was proceeding under fire when his car became disabled. He removed the wounded man under a severe fire to a place of safety. On the same day, in order to quickly evacuate 2 severely wounded men whose only chance of recovery lay in being promptly removed to a hospital, he went fearlessly through barrage on the only road over which he could travel, bringing the wounded men to a hospital alive.

WALTON, EDWARD A_____

Near Ripont, Marne, France, Sept. 16, 1918.
R—New York, N. Y.
B—Ridgewood, N. J.
G. O. No. 14, W. D., 1923.

First lieutenant, 369th Infantry, 93d Division.
While acting as adjutant of the 3d Battalion of his regiment he accompanied the battalion commander on a personal reconnaissance, advancing 100 meters in advance of the assaulting lines, where they were met by heavy enemy machine-gun fire. The battalion commander received 6 severe leg wounds; he was carried and dragged under intense fire to a place of comparative shelter by Lieutenant Walton, who assisted a member of the Medical Corps to apply first aid. While so engaged he himself was wounded. The splendid example of courage and devotion to duty greatly encouraged and inspired the men of the regiment.

WALTON, ELMER A. (2339335)_____

Near Mezy, France, July 8, 1918.
R—Martins Ferry, Ohio.
B—Martins Ferry, Ohio.
G. O. No. 32, W. D., 1919.

Mechanic, Company H, 4th Infantry, 3d Division.
Mechanic Walton volunteered and carried a message over territory generally thought impassable during daylight. He accomplished his mission in spite of having been wounded and nearly buried by a shell explosion.

WALTON, ROBERT, Jr_____

At Cornay, France, Oct. 9-10, 1918.
R—Augusta, Ga.
B—Augusta, Ga.
G. O. No. 44, W. D., 1919.

First lieutenant, 328th Infantry, 82d Division.
After fighting for 6 hours, he volunteered to lead 16 men in a night patrol of the town of Cornay, which was held by many enemy machine-gun posts. The party worked at clearing the town of the enemy from 11 o'clock at night until next morning, capturing 35 prisoners and 2 machine guns. With 3 soldiers, he entered an enemy dugout and captured 23 prisoners.

WARD, FRANK B. (2805234)_____

Near the Meuse River, France, Nov. 6, 1918.
R—Hardy, Okla.
B—Winfield, Kans.
G. G. No. 98, W. D., 1919.

Private, Company K, 357th Infantry, 90th Division.
When the patrol of which he was a member had sustained severe casualties, he took command, extricated the patrol from ambush, and, exposing himself to intense enemy fire, made 3 trips back and forward to recover the dead and wounded.

WARD, FRANK G. (155956)_____

At Cantigny, France, May 28, 1918.
R—Washington, D. C.
B—Washington, D. C.
G. O. No. 99, W. D., 1918.

Private, Company D, 1st Engineers, 1st Division.
Even though his normal duties were as orderly for 2 officers, he volunteered for action at Cantigny, France, May 28, 1918, successfully went into no man's land and killed a sniper who was inflicting losses on his detachment, carried messages through machine-gun and artillery fire, and, although twice buried in shell craters, he displayed heroic bravery, coolness, and fearless devotion throughout.

*WARD, GALBRAITH (1703569)_____

Near Mont-Notre-Dame, France, Sept. 10, 1918.
R—New York, N. Y.
B—Newport, R. I.
G. O. No. 16, W. D., 1923.

Sergeant, Company M, 306th Infantry, 77th Division.
Voluntarily accompanying an officer and enlisted man of his battalion in a search of the Bois-de-Chandriere for the survivors of a battalion which had been relieved from its position in the line, he crossed an open field under terrific enemy fire a distance of 600 yards under constant observation of the enemy, sought and found several survivors, and led them back to his own lines. Sergeant Ward was severely wounded in the performance of this hazardous duty and died of pneumonia shortly thereafter.
Posthumously awarded. Medal presented to father, H. G. Ward.

WARD, GEORGE B. (1317764)_____

Near Bellicourt, France, Sept. 29, 1918.
R—Fayetteville, N. C.
B—Fayetteville, N. C.
G. O. No. 44, W. D., 1919.

Private, Company D, 119th Infantry, 30th Division.
When his company was halted by enemy machine-gun fire, Private Ward rushed the hostile position and killed 1 gunner with his bayonet. Later in the engagement he came upon 20 of the enemy in a trench. He bayoneted 3 of these and took the others prisoners. He was severely wounded in the action.

WARD, GEORGE BLAIN (1309976)_____
Near Brancourt, France, Oct. 8, 1918.
R—Easley, S. C.
B—Brickton, N. C.
G. O. No. 35, W. D., 1919.

Sergeant, Company A, 118th Infantry, 30th Division.
Taking command of the company after all officers had become casualties, he reorganized it and led it under hostile shelling and withering machine-gun fire to its objective. He remained in command until painfully wounded on the following day.

WARD, HARRY M. (2178243)_____
Near Barricourt, France, Nov. 1, 1918.
R—Gregory Landing, Mo.
B—Plymouth, Ill.
G. O. No. 87, W. D., 1919.

Corporal, Company A, 354th Infantry, 89th Division.
After his company had reached its objective and was being subjected to severe fire from an enemy machine-gun nest, he led his combat group of 3 men around the right flank of the company and, under heavy fire, charged the enemy position, capturing 3 guns and 13 prisoners.

WARD, HERBERT (57104)_____
Near Berzy-le-Sec, France, July 19, 1918.
R—Paintsville, Ky.
B—Offutt, Ky.
G. O. No. 39, W. D., 1920.

Sergeant, Company C, 28th Infantry, 1st Division.
After his platoon leader had been wounded, Sergeant Ward reorganized the platoon under heavy fire and led it to its objective. He then led forward a patrol to locate the enemy positions, during which reconnaissance he was wounded by a machine-gun bullet. The ball lodged above the right eye. After receiving first aid he returned to his unit for duty. During a subsequent attack in the Argonne he was again wounded.

WARD, JOHN C._____
East of Ronssoy, France, Sept. 29, 1918.
R—Buffalo, N. Y.
B—Elmira, N. Y.
G. O. No. 20, W. D., 1919.

First lieutenant, chaplain, 108th Infantry, 27th Division.
During the operations against the Hindenburg line he voluntarily and at great risk to himself went forward under heavy shell and machine-gun fire to care for the wounded and to search for the dead. Twice he was ordered off the field of battle by officers, being told each time that it was sure death to remain. During the entire time his regiment was engaged he remained on the field under fire, displaying a fine example of bravery and courage which was an inspiration to all.

WARD, JOHN M. (3668773)_____
Near Falbas, France, Nov. 10, 1918.
R—Hoboken, N. J.
B—New York, N. Y.
G. O. No. 89, W. D., 1919.

Private, medical detachment, 314th Infantry, 79th Division.
Under heavy machine-gun and artillery fire, Private Ward waded through a swamp, administered first aid to a wounded soldier, and then carried the latter to safety.

WARD, MAHLON C. (1214562)_____
Near Ronssoy, France, Sept. 28, 1918.
R—Medina, N. Y.
B—Ridgeway, N. Y.
G. O. No. 159, W. D., 1918.

Private, Company F, 108th Infantry, 27th Division.
During the operations against the enemy lines east of Ronssoy he went out under heavy shell and machine-gun fire and succeeded in bandaging and bringing back to our line wounded soldiers.

WARD, THOMAS F., Jr._____
Near Bony, France, Sept. 27, 1918.
R—Brooklyn, N. Y.
B—Brooklyn, N. Y.
G. O. No. 14, W. D., 1923.

First lieutenant, 106th Infantry, 27th Division.
Severely wounded by enemy fire, he learned that an officer of his company, Lieutenant Boullee, was seriously wounded and lying in a shell hole some distance away. Dragging himself to the side of Lieutenant Boullee, he laboriously rendered first aid to the latter, who could not be moved on account of his wounded condition. Refusing to be evacuated to hospital, Lieutenant Ward chose to remain with his brother officer. The enemy having recaptured the territory in which the wounded officers were lying, the American officers remained concealed in their place of comparative shelter for 48 hours, when the enemy was repulsed and the ground regained, when the wounded men were sent to the rear.

*WARD, WILLIAM H., Jr. (1215870)_____
Near Ronssoy, France, Sept. 29, 1918.
R—Auburn, N. Y.
B—Auburn, N. Y.
G. O. No. 21, W. D., 1919

First sergeant, Company M, 108th Infantry, 27th Division.
Although severely wounded, Sergeant Ward assumed command of his company after the company commander had become a casualty, displaying great gallantry and bravery in leading them into action. While endeavoring to locate enemy machine-gun nests, he was killed.
Posthumously awarded. Medal presented to father, William H. Ward, sr.

*WARE, ARTHUR F._____
In the vicinity of Chateau-Thierry, France, June 6, 1918.
R—Kansas City, Mo.
B—Des Moines, Iowa.
G. O. No. 101, W. D., 1918.

Sergeant, 49th Company, 5th Regiment, U. S. Marine Corps, 2d Division.
Under heavy machine-gun fire, he attempted to establish liaison with an adjoining French unit, during which he was killed.
Posthumously awarded. Medal presented to aunt, Mrs. Libby Riley.

WARE, JAMES V._____
Near Exermont, France, Oct. 5, 1918.
R—Norfolk, Va.
B—Baltimore, Md.
G. O. No. 27, W. D., 1920.

Captain, 28th Infantry, 1st Division.
His company having been repulsed in an attack on a strongly organized position, Captain Ware reformed his company and personally led the 63 remaining members of company in a second attack and in the taking of the enemy position at the point of the bayonet. He then, under heavy fire, prepared the position for defense against enemy assault.

WARFIELD, ARTHUR H. (557740)_____
At Sergy, France, Aug. 1, 1918.
R—West Brookfield, Mass.
B—Conway, Mass.
G. O. No. 145, W. D., 1918.

Sergeant, Company B, 47th Infantry, 4th Division.
He displayed exceptional courage and loyalty by remaining in active command of his section after being wounded twice.

WARFIELD, WILLIAM J._____
Near Ferme-de-la-Riviere, France, Sept. 28, 1918.
R—Chicago, Ill.
B—Chicago, Ill.
G. O. No. 37, W. D., 1919.

First lieutenant, 370th Infantry, 93d Division.
Although separated with his platoon from the company, he continued to lead a stubborn resistance against enemy machine-gun nests, successfully capturing a gun and killing the crew. After having been severely wounded, he still continued in command, refusing relief until his objective was reached.

*WARING, WILLIAM W._____
  Near Dur-sur-Meuse, France, Sept. 26, 1918.
  R—Buffalo, N. Y.
  B—Franklinville, N. Y.
  G. O. No. 37, W. D., 1919.

First lieutenant, pilot, 11th Aero Squadron, Air Service.
Deeming it impossible to catch their own formation, Lieutenant Waring, with Lieut. Sigbert Norris, observer, attached themselves to a formation from the 20th Squadron and engaged in a 35-minute fight with 20 enemy aircraft. Five of this squadron were lost and the observer of one of the three remaining planes seriously wounded. The wounded man had fallen in a position which made the control of the machine difficult. Lieutenant Waring immediately placed his machine between the enemy formation and the crippled companion in order to protect it and continued to fly in this place until our lines were crossed and the enemy scouts driven off.
Posthumously awarded. Medal presented to father, W. W. Waring.

WARMAN, JOHN W. (1516244)_____
  Near Eyne, Belgium, Nov. 2, 1918.
  R—Youngstown, Ohio.
  B—Oliphant, Pa.
  G. O. No. 37, W. D., 1919.

Private, Company B, 135th Machine Gun Battalion, 37th Division.
In the face of intense machine-gun fire he voluntarily swam the Scheldt River to obtain information regarding the enemy. His successful return with the desired information enabled his company commander to so place his guns that they could be fired with great advantage.

WARNER, DONALD D._____
  From Friauville to Lamorville, France, Sept. 4, 1918.
  R—Swampscott, Mass.
  B—Rochester, N. Y.
  G. O. No. 121, W. D., 1918.

First lieutenant, 96th Aero Squadron, Air Service.
While on a bombing expedition with other planes from his squadron he engaged in a running fight over hostile territory with a superior number of enemy battle planes from Friauville to Lamorville, France. During the combat he was severely wounded, his right thigh being shattered. In spite of his injuries he continued to operate his machine guns until the hostile formation had been driven off and one plane shot down burning.

WARNER, LEO V._____
  Near Cunel, France, Oct. 14, 1918.
  R—Loda, Ill.
  B—Loda, Ill.
  G. O. No. 89, W. D., 1919.

Captain, 8th Machine Gun Battalion, 3d Division.
Captain Warner, accompanied by a runner, deliberately crossed an open space in order to draw enemy fire and thereby locate hostile positions, securing information which enabled him to maneuver his men into position with a minimum of casualties. He repeatedly visited all parts of his company's position under fire and in so doing was wounded.

WARREN, CHARLES F._____
  Near Exermont, France, Oct. 5, 1918.
  R—Hewitt, Tex.
  B—Hewitt, Tex.
  G. O. No. 39, W. D., 1920.

First lieutenant, 18th Infantry, 1st Division.
Lieutenant Warren led his platoon forward through artillery and machine-gun fire to rescue 6 men who had been cut off from our lines by the enemy. While crossing an open space his platoon was fired upon by enemy machine guns. Lieutenant Warren advanced ahead of his platoon, calling to his men "Follow me," until he fell wounded by a machine-gun bullet.

WARREN, EDWARD R._____
  Near Fey-en-Haye, France, Sept. 12, 1918.
  R—El Paso, Tex.
  B—San Antonio, Tex.
  G. O. No. 123, W. D., 1918.

First lieutenant, 315th Engineers, 90th Division.
He was in command of a platoon of Engineers and went over the top with the second wave of Infantry. When the first wave was halted by severe machine-gun and shell fire early in the action and all its officers killed or disabled he led his men up to the first wave, reorganized the remaining effectives and led them across a valley and up a hill through severe fire from German machine guns. He was knocked down by the explosion of a shell, but undaunted by murderous fire from the front and both flanks, he continued to lead his men on toward their objectives until he was shot down by a machine gun.

*WARREN, ROBERT F. (2383468)_____
  Near Clery-le-Petit, France, Nov. 4, 1918.
  R—Syracuse, N. Y.
  B—Solvay, N. Y.
  G. O. No. 20, W. D., 1919.

Corporal, Company E, 60th Infantry, 5th Division.
He voluntarily left a place of comparative safety and went over open and bullet-swept ground to the assistance of a comrade who had been wounded in the advance. While administering first aid he was a continual prey for enemy snipers, but he bravely continued with his mission until killed.
Posthumously awarded. Medal presented to mother, Mrs. Rose Warren.

WARREN, RUFUS (2244006)_____
  Near Bautheville, France, Nov. 1, 1918.
  R—Ratcliff, Tex.
  B—Nacogdoches, Tex.
  G. O. No. 44, W. D., 1919.

Private, Headquarters Company, 360th Infantry, 90th Division.
Although wounded in the leg and hand, he insisted on advancing with his unit. In addition to his equipment he carried a trench-mortar barrel, the extra weight proving a severe strain on account of his wounds. After receiving 2 more wounds from shrapnel he crawled alone to the dressing station, refusing proffered help.

WARTHEN, BRUCE (306756)_____
  Near Nervius Ferme, France, July 15, 1918.
  R—St. Paul, Minn.
  B—Omaha, Nebr.
  G. O. 126, W. D., 1919.

Mechanic, Battery E, 76th Field Artillery, 3d Division.
During a severe gas shelling Mechanic Warthen aided a wounded comrade in adjusting his mask before he had placed his own, resulting in his being seriously gassed.

WASCHER, HAROLD A._____
  Near Nouart, France, Nov. 5, 1918.
  R—St. Cloud, Minn.
  B—Champaign, Ill.
  G. O. No. 37, W. D., 1919.

Second lieutenant, 122d Field Artillery, 33d Division.
While commanding an observation party he established a post well in advance of the Infantry, and despite the severe fire to which he was subjected he set up and maintained telephone communications. While thus engaged he was severely wounded by machine-gun fire.

WASHA, JAMES J. (1387491)_____
  Near Chipilly Ridge, France, Aug. 9, 1918.
  R—Chicago, Ill.
  B—Chicago, Ill.
  G. O. No. 71, W. D., 1919.

Sergeant, Company F, 131st Infantry, 33d Division.
Exposing himself to heavy artillery and machine-gun fire, he single handed silenced 2 enemy machine-gun nests which had been holding up his platoon. On his own initiative he advanced against the first of the enemy posts and killed its crew. He then attacked the second position and took the enemy crew prisoner.

WASILEWSKI, JOSEPH (1279636)_____
  North of Verdun, France, Oct. 12, 1918.
  R—Passaic, N. J.
  B—Russia.
  G. O. No. 35, W. D., 1919.

Private, first class, Company A, 114th Infantry, 29th Division.
Throughout the entire day he passed from the rear to the front line carrying food, water, and supplies to the front. During his return trips he assisted the wounded and once carried a comrade on his back. On every journey he was compelled to pass through terrific shellfire.

WASKIEWIC, JOSEPH (559066)............
Near Bois-de-Brieulles, France, Oct. 9-13, 1918.
R—New Bedford, Mass.
B—Thorndike, Mass.
G. O. No. 87, W. D., 1919.

Private, Company A, 11th Machine Gun Battalion, 4th Division.
As a runner between company and battalion headquarters he crossed heavily shelled areas to deliver important messages. Wounded when crossing an open space, subjected to artillery and machine-gun fire, he refused to be evacuated, but continued the performance of his duties.

*WASS, LESTER S...........
In Bois-de-Belleau, France, June 11, 1918, and near Vierzy, France, July 18, 1918.
R—Gloucester, Mass.
B—Gloucester, Mass.
G. O. No. 71, W. D., 1919.

Captain, 5th Regiment, U. S. Marine Corps, 2d Division.
In the Bois-de-Belleau when all the officers of his company had become casualties he displayed marked heroism in leading his men forward in the face of heavy machine-gun fire, assisting in the capture of many machine guns. Near Vierzy he fearlessly exposed himself to enemy machine-gun and artillery fire, directing personally the reduction of strong points. He was killed at the head of his men while leading an advance.
Posthumously awarded. Medal presented to father, L. A. Wass.

WATERHOUSE, JOHN R. (1389541)........
Near Bois-de-Chaume, France, Oct. 8, 1918.
R—Chicago, Ill.
B—Fort Wayne, Ind.
G. O. No. 71, W. D., 1919.

Private, first class, Company H, 132d Infantry, 33d Division.
Showing utter disregard for personal danger, he advanced under heavy fire 200 yards farther into the woods than the rest of his platoon, captured 26 prisoners, and brought them back to our lines.

*WATERS, FLOYD E. (41280)............
Near Villemontry, France, Nov. 10, 1918.
R—Susquehanna, Pa.
B—Susquehanna, Pa.
G. O. No. 37, W. D., 1919

Corporal, Headquarters Company, 9th Infantry, 2d Division.
After participating in the action throughout the day and after seeing that his men had shelter for the night, he voluntarily exposed himself to care for the wounded who were lying out in the open.
Posthumously awarded. Medal presented to father, John Waters.

WATERS, JAMES L. (1386776)............
Near Gressaire Woods, France, Aug. 9, 1918.
R—Antioch, Ill.
B—Waukegan, Ill.
G. O. No. 71, W. D., 1919.

Corporal, Company C, 131st Infantry, 33d Division.
Showing utter disregard for personal danger he advanced alone in the face of heavy fire for 100 yards in advance of our lines to attack a machine-gun emplacement, the fire from which was causing heavy casualties. He killed the 2 men at the enemy gun, permitting a renewal of the advance of his company.

WATERS, TALIESIN.................
Near Baslieux, France, Sept. 6, 1918.
R—Nanticoke, Pa.
B—Nanticoke, Pa.
G. O. No. 7, W. D., 1919.

Second lieutenant, 107th Field Artillery, 28th Division.
He voluntarily went to the assistance of a large number of wounded soldiers who were in an exposed position waiting aid and continued for several hours to dress their wounds throughout a severe bombardment of gas and high-explosive shells, while hostile airplanes flew low and swept with machine-gun fire the line of litters bearing the wounded. After administering aid to 36 wounded men he helped carry them to a place of safety.

*WATKINS, EUGENE G. (3114715)........
Near Verdun, France, Nov. 1, 1918.
R—Bristol, Pa.
B—New York, N. Y.
G. O. No. 37, W. D., 1919.

Private, Company K, 315th Infantry, 79th Division.
While acting as runner between battalion and regimental headquarters he received severe wounds, but continued on with his mission to his destination, which was reached just before he died. After being wounded he covered a distance of approximately 300 meters to deliver his message.
Posthumously awarded. Medal presented to sister, Miss Harriet Jane Watkins.

WATKINS, GEORGE (1515210)...........
Near Cierges, France, Sept. 27-28, 1918.
R—East Liverpool, Ohio.
B—East Liverpool, Ohio.
G. O. No. 89, W. D., 1919.

Sergeant, Company D, 135th Machine Gun Battalion, 37th Division.
After being twice wounded, Sergeant Watkins continued to lead his section in action against the enemy under severe machine-gun and direct artillery fire. Though he had been ordered to the rear by his platoon commander, he returned to his section as soon as his wounds had been dressed.

*WATKINS, GEORGE F.................
North of the River Ourcq, near Villers-sur-Fere, France, July 23, 1918.
R—Springfield, Mass.
B—Boston, Mass.
G. O. No. 132, W. D., 1918.

Second lieutenant, 165th Infantry, 42d Division.
During the storming of the heights north of the River Ourcq, near Villers-sur-Fere, France, July 23, 1918, he was an example of courage and soldiery fortitude. He was continually with the foremost elements of his platoon in the most dangerous areas it had to occupy, both during the advance and during the maintenance of the position gained. His platoon was almost annihilated and he himself was killed.
Posthumously awarded. Medal presented to mother, Mrs. Francis Watkins.

WATKINS, HOMER.................
Near St. Juvin, France, Oct. 11-17, 1918.
R—Atlanta, Ga.
B—Carroll County, Ga.
G. O. No. 89, W. D., 1919.

Major, 326th Infantry, 82d Division.
On the night of Oct. 11 Major Watkins led his battalion under heavy shellfire, forded the Aire River, and took up an important position protecting a flank. On Oct. 14 and 15 he advanced against strongly held enemy positions, penetrating and capturing numerous prisoners and machine guns. He was wounded on both days, but he continued on duty until Oct. 17, suffering intense pain, until he was ordered to the rear. His fortitude and bravery furnished an inspiring example to his men.

WATKINS, LEWIS (2816131)...........
Near Eply, France, Nov. 4, 1918.
R—Ullin, Ill.
B—Illinois.
G. O. No. 139, W. D., 1918.

Private, first class, Company A, 350th Machine Gun Battalion, 92d Division.
He accompanied an Infantry patrol, acting as gunner with a heavy machine gun. When a large party of the enemy had worked around the flank of the patrol and was advancing across a road along which the patrol was withdrawing, he went into action with his gun at a range of less than 100 yards, although the order to withdraw had been given. Displaying exceptional coolness and bravery under heavy rifle and machine-gun fire, he succeeded in dispersing the enemy. He was the last of the patrol to retire.

WATRES, LAURENCE H.
Near Baslieux, France, Sept. 5, 1918.
R—Scranton, Pa.
B—Scranton, Pa.
G. O. No. 130, W. D., 1919.

Captain, 108th Machine Gun Battalion, 28th Division.
When, under heavy enemy machine-gun fire, he took command of Company D, 109th Infantry, which was without officers and was greatly disorganized in a position to his rear. He led the company, together with some of his own men, to the attack, killing a number of the enemy, taking others prisoners, and capturing several machine-gun nests.

WATSON, RAY E.
Near Nantillois, France, Oct. 5, 1918.
R—Joplin, Mo.
B—Webb City, Mo.
G. O. No. 37, W. D., 1919.

Second lieutenant, 317th Infantry, 80th Division.
Although severely wounded, Lieutenant Watson continued to lead his platoon of the machine gun company with great coolness and disregard of personal danger. When the attacking infantry dropped back in the face of heavy machine-gun fire, he held his position in front of them until they returned to the attack.

WATTS, KENNETH (2857069).
At Andevanne, France, Nov. 2, 1918.
R—Barnes City, Iowa.
B—Barnes City, Iowa.
G. O. No. 98, W. D., 1919.

Private, Company B, 369th Infantry, 90th Division.
Sent with another runner from battalion headquarters to deliver a message to a front-line company, he made his way through the enemy fire, and when his comrade was wounded delivered the message; then he rescued his wounded companion, carrying him under heavy fire to a dressing station. He then returned to duty.

*WAY, PENNINGTON H.
Near Buxieres, France, Sept. 12, 1918.
R—St. Davids, Pa.
B—Philadelphia, Pa.
G. O. No. 37, W. D., 1919.

Second lieutenant, observer, 96th Aero Squadron, Air Service.
Lieutenant Way, with First Lieutenant Gundelach, pilot, volunteered for a hazardous mission to bomb concentrations of enemy troops. They successfully bombed their objective, but while returning were attacked by 8 enemy planes. Their plane was brought down in flames and both officers killed.
Posthumously awarded. Medal presented to widow, Mrs. Eleanor H. Way.

WEAR, EUGENE W.
In the vicinity of Chateau-Thierry, France, June 6, 1918.
R—Hazelton, Pa.
B—Beaver Meadows, Pa.
G. O. No. 101, W. D., 1918.

Corporal, 49th Company, 5th Regiment, U. S. Marine Corps, 2d Division.
On June 6, 1918, in the vicinity of Chateau-Thierry with a private he went out into an open field under heavy shell and machine-gun fire and succeeded in bandaging and carrying back to our lines a wounded comrade.

*WEATHERMAN, HUGH (124281).
At Mandres, France, Mar. 1, 1918.
R—Beaman, Iowa.
B—Winston-Salem, N. C.
G. O. No. 74, W. D., 1919.

Private, Battery C, 5th Field Artillery, 1st Division.
During a heavy enemy bombardment of gas and high-explosive shells, Private Weatherman left shelter for the purpose of putting gas masks on his horses, and while so doing was mortally wounded by a shell fragment. Realizing the character of his wound, he refused medical attention, urging the Medical Corps men to assist other wounded men who could be saved.
Posthumously awarded. Medal presented to mother, Mrs. Ellen Weatherman.

WEAVER, CHARLES H.
Near Soissons, France, July 19, 1918.
R—Sebring, Ohio.
B—East Liverpool, Ohio.
G. O. No. 132, W. D., 1918.

Second lieutenant, Company C, 26th Infantry, 1st Division.
On July 19, 1918, near Soissons, France, when severely wounded he refused to leave his command, but led it forward under heavy fire until his objective was reached.

WEAVER, JESSE FRANK (1307339).
Near Geneve, France, Oct. 8, 1918.
R—Finger, Tenn.
B—Finger, Tenn.
G. O. No. 50, W. D., 1919.

Private, Company C, 117th Infantry, 30th Division.
At the starting of the attack Private Weaver was painfully wounded by machine-gun fire. Disregarding his wound, he continued on, and when the company was held up by the extreme fire he voluntarily flanked the enemy position and enabled his comrades to capture the gun. He was evacuated when the objective was reached.

WEAVER, ROSS E.
At Marcheville, France, Sept. 26, 1918.
R—Concordia, Kans.
B—St. Joseph, Mo.
G. O. No. 15, W. D., 1919.

First lieutenant, Medical Corps, attached to 102d Infantry, 26th Division.
He showed complete disregard of personal safety by remaining with the foremost elements and administering aid to the wounded throughout the day under constant artillery bombardment and direct machine-gun and rifle fire from the enemy.

WEAVER, WILLIAM D. (546302).
Near Jaulgonne, France, July 15–21 and 24–27, 1918.
R—Charleston, W. Va.
B—Charleston, W. Va.
G. O. No. 32, W. D., 1919.

Private, medical detachment, 30th Infantry, 3d Division.
From July 15 to 21 he worked continuously among the wounded of his regiment, never hesitating for the heaviest fire. He volunteered and remained with the unit which relieved his regiment and continued his work with the new unit from July 24–27.

WEAVER, WILLIAM G.
Near Cierges, France, Oct. 4, 1918.
R—Louisville, Ky.
B—Louisville, Ky.
G. O. No. 35, W. D., 1920.

Major, 8th Machine Gun Battalion, 3d Division.
The assault battalion, 7th Infantry, having been halted by heavy machine-gun and artillery fire, Major Weaver personally placed 4 machine guns and the 1-pounder in position and directed their fire in close cooperation with the Infantry. In performing this act Major Weaver was forced to cross 3 times an area exposed to heavy machine-gun fire, in which no less than 50 men had been previously killed or wounded. His heroic efforts enabled the Infantry to advance and insured the success of this operation.

WEAVERLING, HAROLD (57324).
South of Soissons, France, July 18–21, 1918.
R—Kearney, Pa.
B—Tatesville, Pa.
G. O. No. 39, W. D., 1920.

Sergeant, Company E, 28th Infantry, 1st Division.
On July 18 Sergeant Weaverling assumed command of his company after all officers had been killed or wounded and led it to the day's objective and consolidated the position taken. On July 19 he was wounded and rendered unconscious. After regaining consciousness at the aid station he returned to his company and again assumed command of it and later of the 2d battalion in the final operations near Berzy-le-Sec. The success of his organization was in a measure due to the devotion to duty and brilliant leadership of this noncommissioned officer.

*WEBB, HARRY L.
Near Verdun, France, Oct. 8–25, 1918.
R—Bel Air, Md.
B—Baltimore, Md.
G. O. No. 2, W. D., 1919.

First lieutenant, 115th Infantry, 29th Division.
In several advances during this period he led his men, regardless of personal danger, capturing a number of machine guns and prisoners. On Oct. 11 he was wounded, but refused to go to the rear. During the advance on Oct. 24 in the Bois-de-la-Grande Montagne the right combat group of his platoon being disorganized by artillery fire and several men killed and wounded, he displayed exceptional gallantry in reorganizing the remainder of his platoon and in reestablishing liaison with the units on his right, thus relieving a dangerous situation. He was killed on Oct. 25 while leading an attack on a machine-gun nest.
Posthumously awarded. Medal presented to widow, Mrs. Harry L. Webb.

WEBB, JOHN R.
Near Bellicourt, France, Sept. 29, 1918.
R—Tulsa, Okla.
B—Santa Barbara, Calif.
G. O. No. 32, W. D., 1919.

Second lieutenant, 301st Battalion, Tank Corps.
While his crew was engaged in digging out the tank which had become ditched in a shell hole in front of the main Hindenburg line, an enemy machine gun opened fire at a distance of 30 yards. Being unable to use his guns on account of his position, Lieutenant Webb crawled forward to the machine gun and killed the enemy gunners with his pistol. His act enabled the men to free the tank, which subsequently aided the advancing Infantry.

WEBB, MILTON C. (1379461)
Near Romagne, France, Nov. 1, 1918.
R—East St. Louis, Ill.
B—Birmingham, Ala.
G. O. No. 37, W. D., 1919.

Private, first class, medical detachment, 124th Field Artillery, 33d Division.
Wounded while administering aid to other men during shellfire, he remained on duty in disregard of his own injury.

WEBBER, GEORGE B. (1391383)
Near Brieulles, France, Oct. 8, 1918.
R—Chicago, Ill.
B—Pittsburgh, Pa.
G. O. No. 32, W. D., 1919.

First sergeant, Machine Gun Company, 132d Infantry, 33d Division.
When it appeared evident that his forces would give way under the pressure of the unusual enemy fire, Sergeant Webber jumped forward and, taking command of a machine-gun crew, led them into the front line, where he remained for 2 days. He refused evacuation while suffering from a severe gassing, until he finally collapsed under the strain.

WEBER, BENJAMIN S. (1705231)
At Fond-de-Vas, France, Sept. 14, 1918.
R—New York, N. Y.
B—New York, N. Y.
G. O. No. 46, W. D., 1919.

Sergeant, Company D, 307th Infantry, 77th Division.
Although severely wounded, he continued to lead his platoon in an attack on enemy machine-gun nests through a sweeping artillery and machine-gun fire until he fell, completely exhausted.

WEBER, JOHN F.
Near Vieville-en-Haye, France, Sept. 29, 1918; Chevieres, France, Oct. 15, 1918; and Grand Pre, France, Oct. 25, 1918.
R—South Amboy, N. J.
B—Florence, N. J.
G. O. No. 37, W. D., 1919.

First lieutenant, Medical Corps, attached to 311th Infantry, 78th Division.
On September 29 and the following days Lieutenant Weber remained at his aid station under shell and machine-gun fire, giving medical aid and directing the evacuation of wounded. On Oct. 15 he established an unprotected aid station, and, though slightly gassed, he continued to give first aid to the wounded and direct their evacuation. On Oct. 25 he left his battalion in support and continued to the town of Grand Pre, where he established an aid station, keeping on with his work through heavy bombardment of the town.

WEBER, NICKLOUS (2153453)
Near St. Juvin, France, Oct. 16, 1918.
R—Waterloo, Iowa.
B—Waterloo, Iowa.
G. O. No. 46, W. D., 1919.

Private, Company K, 325th Infantry, 82d Division.
After 3 stretcher bearers had been shot down while trying to bring in a wounded soldier, he advanced in the face of the terrific machine-gun and artillery fire and rescued the wounded man. He then returned to the field and successfully brought the 3 stretcher bearers to our lines.

*WEBSTER, HARRISON B.
Near Bois-de-Brieulles, France, Sept. 26 to Oct. 12, 1918.
R—Castine, Me.
B—Castine, Me.
G. O. No. 74, W. D., 1919.

Major, Medical Corps, attached to 47th Infantry, 4th Division.
After seeing that his personnel was functioning properly, he went fearlessly to positions in the front lines. When stretcher bearers were unable to handle the large number of casualties, he personally took a light German wagon to the front lines and gathered the wounded. His personal bravery was an inspiration to his men throughout his service. He was killed by shellfire on Oct. 12, 1918.
Posthumously awarded. Medal presented to widow, Mrs. Harrison B. Webster.

WEBSTER, TILLMAN (2118125)
Near Ardeuil, France, Sept. 29, 1918.
R—Alexandria, La.
B—Alexandria, La.
G. O. No. 46, W. D., 1919.

Private, Machine Gun Company, 371st Infantry, 93d Division.
With 3 other soldiers, Private Webster crawled 200 yards ahead of our lines, under violent machine-gun fire, and rescued an officer who was lying mortally wounded in a shell hole.

*WEBSTER, WILLARD M.
Near Ronssoy, France, Sept. 29, 1918.
R—New York, N. Y.
B—Houston, Tex.
G. O. No. 70, W. D., 1919.

First lieutenant, 106th Infantry, 27th Division.
He received a painful wound in the face shortly after leading his company to the attack, but he refused to be evacuated until he suffered additional wounds, which eventually caused his death.
Posthumously awarded. Medal presented to mother, Mrs. Beulah Webster.

WEED, EARL H.
Near Soissons, France, July 19–22, 1918.
R—Berkeley, Calif.
B—Wilton, Iowa.
G. O. No. 125, W. D., 1918.

First lieutenant, chaplain, 16th Infantry, 1st Division.
He displayed exceptional bravery in passing through open fields under heavy fire to the front lines to render first aid and to cheer the wounded.

WEED, NEWELL P.
Near Foret-d'Argonne, France, Sept. 26, 1918.
R—Montclair, N. J.
B—Brooklyn, N. Y.
G. O. No. 46, W. D., 1919.

Captain, Tank Corps.
During the operations on the edge of Foret-d'Argonne Captain Weed advanced alone some 300 yards ahead of the tanks and Infantry through heavy machine-gun fire in order to reconnoiter a passage for his command. While examining German trenches he was surprised by German infantrymen and was being conducted to the rear when he heard one of his tanks. In spite of the fact that he was unarmed and the Germans threatened his life if he moved, he signaled the tank and made his escape.

WEEKS, MODY A. (1342711)
Near Brieulles, France, Nov. 3, 1918.
R—Crews Depot, Ala.
B—Pharas, Ala.
G. O. No. 37, W. D., 1919.

Private, Company F, 7th Engineers, 5th Division.
He showed extraordinary daring and nerve in helping place cables across the River Meuse for a pontoon bridge and later in placing cables across the Est Canal for the same purpose. The position was under direct observation of German machine gunners and snipers.

*WEEKS, YOUMAN Z. (1311088)
Near Bellicourt, France, Sept. 30 and Oct. 8, 1918.
R—Colleton, S. C.
B—Moorehead City, N. C.
G. O. No. 133, W. D., 1918.

Corporal, Company F, 118th Infantry, 30th Division.
Corporal Weeks on the morning of Sept. 30, when 2 enemy machine guns were making a part of the line untenable, advanced across open ground upon 1 of the guns, rushed the position alone, captured the gun and 5 of the enemy, and shot down the sixth, who endeavored to escape. By this gallant act he prevented the enemy from enfilading our position and thereby saved the lives of many of his comrades. In a later advance, while leading his men in an attack upon an enemy machine-gun nest, he was killed.
Posthumously awarded. Medal presented to mother, Mrs. Mary Weeks.

WEEMS, JAMES F. (1307985)
Near Molain, France, Oct. 17, 1918.
R—Greenville, Tenn.
B—Greene County, Tenn.
G. O. No. 50, W. D., 1919.

Private, Company E, 117th Infantry, 30th Division.
Having volunteered to carry a message to an automatic-rifle post 100 yards in advance of the line, across a field swept by machine-gun fire, Private Weems continued on his mission, even after being seriously wounded, and delivered the message, thereby facilitating the destruction of machine-gun nests which were hindering the advance. After returning with the answer he insisted upon walking to the dressing station.

*WEHNER, JOSEPH F.
Near Rouvres, France, Sept. 15, 1918.
R—Everett, Mass.
B—Boston, Mass.
G. O. No. 138, W. D., 1918.

First lieutenant, pilot, 27th Aero Squadron, Air Service.
While on a mission he found an enemy patrol of 8 machines attacking a single American observation machine. He immediately attacked, destroying 1 and forcing another down out of control, his own plane being badly damaged by enemy machine-gun fire. He managed to convoy the American plane to safety.
Oak-leaf cluster.
A bronze oak leaf is awarded him for the following act of extraordinary heroism in action near Mangiennes and Reville, France, Sept. 16, 1918: Amid terrific antiaircraft and ground machine-gun fire he descended, attacked, and destroyed 2 enemy balloons. One of these balloons was destroyed in flames after it had been hauled to the ground and was resting in its bed.
Posthumously awarded. Medal presented to father, Frank W. Wehner.

*WEIGEL, ROY (42617)
Near St. Mihiel, France, Sept. 12, 1918.
R—St. Louis, Mo.
B—Calhoun County, Ill.
G. O. No. 129, W. D., 1918.

Sergeant, Company E, 16th Infantry, 1st Division.
He showed entire disregard for his own safety in making several attempts to locate the positions of machine guns whose heavy fire was hindering the advance of his battalion. He was killed while leading a rush upon one of the guns which he had located.
Posthumously awarded. Medal presented to sister, Mrs. Emma Kemper.

WEIK, IRVING C. (2264670)
Near Waereghem, Belgium, Oct. 31, 1918.
R—Oakland, Calif.
B—San Francisco, Calif.
G. O. No. 37, W. D., 1919.

First sergeant, Company I, 363d Infantry, 91st Division.
On 2 occasions he passed through an uncut wire entanglement, enfiladed by enemy machine-gun fire, to obtain the assistance of our machine guns to aid in the advance of his company. His entire exploit was under terrific fire of the enemy, but he succeeded in enabling his company to go forward and clean out the opposing machine-gun nest.

WEIMER, HERMAN H.
Near the Bois-de-Chaume, France, Oct. 9, 1918.
R—Chicago, Ill.
B—Galena, Ill.
G. O. No. 46, W. D., 1919.

First lieutenant, 131st Infantry, 33d Division.
He had been wounded in the shoulder and a machine-gun bullet had penetrated his steel helmet, but he nevertheless continued to lead his company, creating confidence in his men at a critical moment. Upon being ordered to the rear by his battalion commander, he returned to his company as soon as his wounds had been dressed.

WEIMER, JOHN SAMUEL (2225018)
Near the Vesle River, France, Aug. 10, 1918.
R—Mount Pleasant, Tex.
B—Waxahachie, Tex.
G. O. No. 81, W. D., 1919.

Private, Company M, 47th Infantry, 4th Division.
While on outpost duty Private Weimer learned that a soldier from another organization was lying wounded in a shell hole 200 yards away. With another member of his squad Private Weimer voluntarily went through machine-gun and sniper fire and carried the wounded man to shelter.

WEINE, WILLIAM F.
Near the Ourcq River, France, Aug. 8, 1918.
R—Alpena, Mich.
B—Alpena, Mich.
G. O. No. 39, W. D., 1919.

First lieutenant, 125th Infantry, 32d Division.
After being seriously wounded in the abdomen Lieutenant Weine displayed remarkable fortitude in organizing his command for the continuation of the attack before consenting to his removal to the rear.

WEINER, DANIEL J. (551322)
Near Mezy, France, July 15–19, 1918.
R—Brooklyn, N. Y.
B—New York, N. Y.
G. O. No. 95, W. D., 1919.

Private, Company G, 38th Infantry, 3d Division.
Private Weiner displayed the utmost devotion to duty and disregard for personal safety in carrying messages through heavy artillery barrages, thereby enabling his company commander to maintain liaison with units in the rear.

WEINMAN, GLEN G. (43179)
South of Soissons, France, July 20, 1918.
R—Columbus, Ohio.
B—Columbus, Ohio.
G. O. No. 35, W. D., 1920.

Corporal, Company M, 16th Infantry, 1st Division.
Corporal Weinman carried an important message through heavy artillery and machine-gun fire calling for reinforcements and ammunition. Notwithstanding the fact that 2 previous runners had been killed, he carried out his mission. His individual gallantry contributed materially to the success of the operation.

WEIS, ANTHONY J. (1569828)_____
At Marcheville, France, Sept. 26, 1918.
R—Hammond, Ind.
B—Hammond, Ind.
G. O. No. 50, W. D., 1919.

Private, medical detachment, 103d Infantry, 26th Division.
He displayed exceptional courage under violent machine-gun and rifle fire by standing up in the open for the purpose of locating machine-gun nests.

WEISS, FRED R. (1097204)_____
Near Montauville, France, Oct. 24, 1918.
R—Chicago, Ill.
B—Russia.
G. O. No. 21, W. D., 1919.

Private, first class, Battery F, 21st Field Artillery, 5th Division.
When shellfire had ignited the powder store of his battery, Private Weiss, in his stocking feet, was the first to enter the dump and single handed pulled numerous boxes of ammunition to safety, despite the danger from explosion and increased enemy shellfire.

WEITZENBERG, GEORGE (1228915)_____
At Neuvilly, France, Sept. 25, 1918.
R—Brooklyn, N. Y.
B—Germany.
G. O. No. 128, W. D., 1918.

Sergeant, Company A, 2d Antiaircraft, Machine Gun Battalion.
He voluntarily ran through violent enemy shellfire to the aid of 2 soldiers of another organization who had been struck by an exploding shell. Finding 1 dead and the other severely wounded, he administered first aid to the wounded soldier and remained with him until an ambulance could be brought up.

*WELKER, THOMAS B. (1520540)_____
Near Cierges, France, Sept. 28, 1918.
R—Akron, Ohio.
B—Danville, Ohio.
G. O. No. 50, W. D., 1919.

Private, Company B, 146th Infantry, 37th Division.
When his company had become disorganized under intense machine-gun fire, Private Welker assumed leadership of a group of men and courageously charged a machine-gun nest in plain view of the enemy, losing his life in this heroic attempt.
Posthumously awarded. Medal presented to father, Norman H. Welker.

*WELLES, HALLACK, JR._____
Near Bouresches, France, July 20, 1918.
R—Brookline, Mass.
B—New York, N. Y.
G. O. No. 89, W. D., 1919.

First lieutenant, 104th Infantry, 26th Division.
Lieutenant Welles was seriously wounded in exposing himself to enemy machine-gun fire in order to locate the source. He nevertheless refused to be evacuated but continued in the advance with his men, inspiring them by his bravery.
Posthumously awarded. Medal presented to brother, M. C. Welles.

WELLING, HANK_____
Near Montfaucon, France, Sept. 27, 1918.
R—Trenton, N. J.
B—Trenton, N. J.
G. O. No. 93, W. D., 1919.

First lieutenant, 316th Infantry, 79th Division.
After being severely wounded in the side Lieutenant Welling refused to be evacuated but continued to lead his platoon in the attack. Throughout the afternoon and evening he remained with his men, inspiring them by his courage and fortitude in spite of intense pain, it being necessary to carry him when a temporary withdrawal of the line was made.

*WELLS, EDWARD L._____
Near Exermont, France, Oct. 4, 1918.
R—Charleston, S. C.
B—Charleston, S. C.
G. O. No. 53, W. D., 1920.

Second lieutenant, 2d Machine Gun Battalion, 1st Division.
When the attack was held up by heavy machine-gun fire, he volunteered for the mission and led a platoon of Infantry, reinforced by 4 machine guns, into Exermont. In spite of desperate resistance, he led the attack through the streets, capturing many prisoners and learning from 1 of these the approximate location of machine guns on heights to the north led the 3 remaining members of the command against these. Within 50 yards of the enemy emplacements 1 of his men was killed and Lieutenant Wells was mortally wounded, but he had succeeded in indicating to those in the rear the location of the hostile positions.
Oak-leaf cluster.
Lieutenant Wells was awarded an oak-leaf cluster for the following act of extraordinary heroism near Buzancy, France, July 21, 1918: When the advancing lines were checked by the fire of numerous enemy machine guns, Lieutenant Wells skillfully directed the placing of one of his machine guns and silenced the hostile guns. While the line was being consolidated this officer, with another, reconnoitered beyond the left flank, which was being swept by enfilading fire. Locating a German machine gun, he put it out of action by well-aimed shots from a rifle which he was then carrying. He then pushed on farther, accompanied by a soldier, captured a prisoner and discovered a nest of 8 enemy machine guns located in a trench, whereupon with great daring he brought up a section of guns and opened up an annihilating fire which dispersed the enemy with many casualties.
Posthumously awarded. Medal presented to mother, Mrs. Edward L. Wells.

WELLS, FLOYD H. (2158832)_____
Near St. Juvin, France, Oct. 16, 1918.
R—Chester, Iowa.
B—Jolley, Iowa.
G. O. No. 50, W. D., 1919.

Corporal, Company M, 326th Infantry, 82d Division.
With another soldier Corporal Wells advanced several hundred yards ahead of the front line, under heavy artillery and machine-gun fire, and rescued a wounded comrade.

WELLS, JOHN T. (1315439)_____
Near Bellicourt, France, Sept. 29, 1918.
R—Watha, N. C.
B—Watha, N. C.
G. O. No. 81, W. D., 1919.

Sergeant, Company E, 119th Infantry, 30th Division.
Wounded at the start of an advance, he continued in command of his platoon and, engaging in hand-to-hand fighting, bayoneted 3 Germans and captured several others. He displayed marked personal bravery, leading his platoon ably until forced to retire because of loss of blood from his wound.

WELSCH, THEODORE P. (1038068)_____
Near St. Eugene, France, July 15, 1918.
R—Newark, Ohio.
B—Pittsburgh, Pa.
G. O. No. 44, W. D., 1919.

Private, Battery A, 10th Field Artillery, 3d Division.
Engaged in maintaining liaison between the Artillery and the Infantry, he was wounded in the arm while carrying an important message. He completed his mission, however, before securing first-aid treatment and immediately afterwards returned to duty.

WELSH, EDWARD J. (2409727) _____
Near Grand Pre, France, Oct. 19, 1918.
R—Freehold, N. J.
B—Philadelphia, Pa.
G. O. No. 37, W. D., 1919.

Sergeant, Company B, 311th Infantry, 78th Division.
After having received 7 machine-gun wounds, he refused to go to the dressing station, remaining with his company and rendering valuable aid to both his platoon and company commander.

WENDELL, WILLIAM (2055680) _____
Against hostile Moros on Patian Island, Philippine Islands, July 4, 1909.
R—Cleveland, Ohio.
B—Chicago, Ill.
G. O. No. 14, W. D., 1925.

Sergeant, Troop C, 6th Cavalry, United States Army.
Sergeant Wendell with other men entered a cave occupied by a desperate enemy and in the face of a heavy fire, with utter disregard for his personal safety, aided in forcing the outlaws to abandon their stronghold, which resulted in their destruction by our forces.

WENDELS, ANTHONIE (40363) _____
Near Soissons, France, July 18, 1918.
R—Ridgewood, N. J.
B—Holland.
G. O. No. 46, W. D., 1919.

Private, Company K, 9th Infantry, 2d Division.
He went forward ahead of his company against a machine gun that was checking the advance, killed the crew, and captured the gun.

WENELL, CARL O. (1240) _____
At Pannes, France, Sept. 21, 1918.
R—Minneapolis, Minn.
B—Minneapolis, Minn.
G. O. No. 128, W. D., 1918.

Private, medical detachment, 151st Field Artillery, 42d Division.
He displayed the highest bravery and self-sacrificing spirit by voluntarily leaving shelter during a heavy hostile bombardment and going to the assistance of several wounded men of another regiment whose position adjoined that of his own organization. Under terrific shellfire he skillfully dressed their wounds and then removed them to a place of safety, thereby saving their lives. Again returning to the shell-swept street, he made a careful reconnaissance for any other casualties which might have been overlooked.

*WERNER, BERNARD _____
At Chateau-Thierry, France, June 6, 1918.
R—Detroit, Mich.
B—Switzerland.
G. O. No. 110, W. D., 1918.

Sergeant, 43d Company, 5th Regiment, U. S. Marine Corps, 2d Division.
Killed in action at Chateau-Thierry, France, June 6, 1918, he gave the supreme proof of that extraordinary heroism which will serve as an example to hitherto untried troops.
Posthumously awarded. Medal presented to father, Amil Werner.

*WERNER, GEORGE (2444680) _____
In the Bois-de-la-Naza, France, Oct. 5, 1918.
R—Albany, N. Y.
B—Albany, N. Y.
G. O. No. 89, W. D., 1919.

Corporal, Company H, 305th Infantry, 77th Division.
With 2 other soldiers Corporal Werner volunteered to crawl out under enemy machine-gun fire in an effort to locate 3 members of the platoon who were missing after an unsuccessful attack on enemy machine-gun nests. Finding the body of 1, they located another, who lay helplessly wounded, by calling out his name. As a result they drew increased fire from the enemy but they courageously crawled 25 yards farther toward the hostile positions and succeeded in carrying back the wounded man through the machine-gun fire to our lines.
Posthumously awarded. Medal presented to mother, Mrs. Susanna Werner.

WESCOTT, ALLEN P. (679507) _____
Near Chevieres, France, Oct. 21, 1918.
R—North Castine, Me.
B—Portland, Me.
G. O. No. 15, W. D., 1919.

Private, Troop C, 2d Cavalry, attached to Company G, 311th Infantry, 78th Division.
Becoming separated from his own organization he attached himself to an Infantry company. While on a patrol he was wounded 3 times. After the party had been surrounded by German machine guns, he volunteered to carry a message to the company commander, wading across the Aire River in so doing. After guiding a platoon to the relief of the patrol he again made several trips to and from the company post of command, crossing through the river waist deep 5 times after being wounded. He was sent to the rear against his vigorous protests, and after being tagged for evacuation he gave further proof of his devotion to duty and unselfishness by helping carry another wounded soldier 3 miles on a stretcher.

*WESCOTT, IRA L. (281389) _____
Near Juvigny, France, Aug. 28-30, 1918.
R—Grandville, Mich.
B—Byron Center, Mich.
G. O. No. 66, W. D., 1919.

Sergeant, Company M, 126th Infantry, 32d Division.
Sergeant Wescott, as second in command, accompanied a small combat patrol, which successfully attacked an enemy trench held up by 30 of the enemy with machine guns, driving the enemy from the trench and inflicting many casualties. When the patrol leader was severely wounded Sergeant Wescott took command and immediately reorganized the patrol, holding the captured position, despite a severe harassing fire from the enemy until relief came.
Posthumously awarded. Medal presented to father, Leslie D. Wescott.

WESSEL, LEONARD, H. F. (366819) _____
Near Verdun, France, Oct. 12-14, 1918.
R—West Hartford, Conn.
B—Boston, Mass.
G. O. No. 130, W. D., 1918.

Private, Company K, 114th Infantry, 29th Division.
He performed his duties as a runner under heavy shellfire for 3 days and nights without rest, and when the relay stations between the battalion and regimental posts of command had been wiped out he continued to carry messages the entire distances between the two posts.

WESSELHOEFT, CONRAD _____
Near Verdun, France, Nov. 8, 1918.
R—Boston, Mass.
B—Wilmington, N. C.
G. O. No. 23, W. D., 1919.

Captain, Medical Corps, attached to 102d Infantry, 26th Division.
Captain Wesselhoeft went forward under heavy machine-gun fire to the aid of a wounded soldier. The fire was so heavy that they were compelled to remain in the shell hole until nightfall, when he brought the wounded man to our lines.

WEST, BRODIE (1314708) _____
Near Bellicourt, France, Sept. 29, 1918.
R—Pikeville, N. C.
B—Wayne County, N. C.
G. O. No. 50, W. D., 1919.

Corporal, Company A, 119th Infantry, 30th Division.
When his automatic squad had become lost from the platoon in a heavy smoke barrage, Corporal West advanced alone upon a machine-gun nest which was firing directly from the front, silenced the gun, and returned to our lines with 37 prisoners.

*WEST, CARROLL B. (573878)................
  Near the Bois-des-Ogons, France,
  Sept. 30 and Oct. 2, 1918.
  R—Milton Junction, Wis.
  B—Lakemills, Wis.
  G. O. No. 66, W. D., 1919.

Sergeant, Company B, 12th Machine Gun Battalion, 4th Division.
He displayed exceptional courage and leadership in leading the section forward and maintaining fire on the enemy from an advanced position in the wood, successfully covering the withdrawal of the Infantry to a more secure position. This gallant soldier was killed 2 days later while he was successfully directing his section in breaking up an enemy counterattack.
Posthumously awarded. Medal presented to father, Allen B. West.

WEST, HEDFORD (1449786)................
  Near Montrebeau Woods, France,
  Sept. 29, 1918.
  R—Salina, Kans.
  B—Ogallah, Kans.
  G. O. No. 71, W. D., 1919.

Mechanic, Company M, 137th Infantry, 35th Division.
Seeing a comrade lying wounded in advance of our lines, he left a shell hole and, exposing himself to heavy machine-gun fire, went into the open and, assisted by another soldier, carried the wounded man back to safety.

*WEST, HENRY................
  Near Chateau-Thierry, France, June
  25, 1918.
  R—Watertown, Mass.
  B—Ware, Mass.
  G. O. No. 37, W. D., 1919.

Sergeant, 47th Company, 5th Regiment, U. S. Marine Corps, 2d Division.
He unselfishly exposed himself in an effort to bring down an enemy sniper who had wounded several members of his group and was himself killed while in the performance of this self-sacrificing act.
Posthumously awarded. Medal presented to mother, Mrs. Henry West.

WEST, HENRY ARTHUR (2154039)........
  Near Molain, France, Oct. 17, 1918.
  R—Riverton, Iowa.
  B—Riverton, Iowa.
  G. O. No. 81, W. D., 1919.

Private, first class, Company L, 117th Infantry, 30th Division.
With another soldier he volunteered to go out across an open space swept by heavy machine-gun fire for 150 yards to rescue 2 wounded soldiers. The mission was accomplished successfully.

WEST, JAMES H. (96865)................
  Near Ancerviller, France, Mar. 4,
  1918.
  R—Hokes Bluff, Ala.
  B—Cherokee County, Ala.
  G. O. No. 126, W. D., 1919.

Sergeant, Company F, 167th Infantry, 42d Division.
He was a member of a patrol of 5 men which, on March 4, 1918, near Ancerviller, France, encountered an enemy patrol of 11 men, which it attacked and routed, taking 2 prisoners

WEST, JOHN ALBERT................
  Near Blanc Mont Ridge, France,
  Oct. 2-5, 1918.
  R—Cincinnati, Ohio.
  B—Cincinnati, Ohio.
  G. O. No. 37, W. D., 1919.

Second lieutenant, 6th Regiment, U. S. Marine Corps, 2d Division.
He voluntarily led a reconnaissance patrol under difficult conditions and secured information necessary to an attack. Three days later, after the command of his company had devolved on him, he continued in action in spite of being severely wounded, leading his men to the objective and refusing to be evacuated until proper disposition had been made.

WEST, JOHN E. (1285024)................
  Near Verdun, France, Oct. 8-16,
  1918.
  R—Baltimore, Md.
  B—Belle Haven, Va.
  G. O. No. 2, W. D., 1919.

Sergeant, Company F, 115th Infantry, 29th Division.
In the Bois-de-Consenvoye on Oct. 8, and in the Bois-de-la-Grande Montagne on Oct. 16, 1918, he ably led his platoon with extreme courage and gallantry and by his conduct inspired the men of his platoon to greater effort. This soldier led a detachment against a strongly protected enemy machine-gun nest and successfully took the position under heavy fire.

WEST, ROBERT JOHN................
  At Liny-devant-Dun and Fontaines,
  France, Nov. 6-7, 1918.
  R—Leavenworth, Kans.
  B—Leavenworth, Kans.
  G. O. No. 143, W. D., 1918

Lieutenant colonel, 11th Infantry, 5th Division.
When his command was halted by heavy fire from an enemy position, strongly entrenched and supported by a large number of machine guns, he placed himself in front of his men and gallantly led them in person to a successful attack, thereby securing a foothold on the east bank of the Meuse and insuring the safe passage of additional troops. By his disregard for personal danger he set an inspiring example to his men and played a conspicuous rôle in gaining the heights on the east bank of the Meuse held by the enemy since 1914.

WESTERGREN, HARRY ORMAN........

  Near St. Etienne, France, Oct. 4-6,
  1918.
  R—Emporia, Kans.
  B—Clearfield, Kans.
  G. O. No. 37, W. D., 1919.

Private, Headquarters Company, 5th Regiment, U. S. Marine Corps, 2d Division.
As a runner he displayed exceptional courage in volunteering to carry important messages over terrain constantly swept by machine-gun and shell fire.

WESTERVELT, EDGAR C................
  Near Soissons, France, July 18-21,
  1918.
  R—Lincoln, Nebr.
  B—Grand Island, Nebr.
  G. O. No. 132, W. D., 1918.

Second lieutenant, 26th Infantry, 1st Division.
Unaided he reconnoitered enemy machine-gun positions near Soissons, France, July 18-21, 1918. On July 19, 1918, with a small party he crossed an area swept by fire and demolished several machine-gun nests that were holding up the advance.

WESTFALL, ALBERT C. (1211181)........
  Near St. Souplet, France, Oct. 18,
  1918.
  R—Oneonta, N. Y.
  B—Scranton, Pa.
  G. O. No. 44, W. D., 1919.

Corporal, Company G, 107th Infantry, 27th Division.
Undaunted by terrific machine-gun fire, Corporal Westfall went out into the open and rescued a British officer who had fallen wounded. Later, after 2 runners had been killed in trying to locate missing elements of his battalion, Corporal Westfall assumed this task and in performing it 4 times crossed a sunken road, which was continuously raked by enemy machine-gun fire.

WESTON, STEPHEN J. (558269)........
  Near Bois-de-Brieulles, France,
  Sept. 28, 1918.
  R—Waterbury, Conn.
  B—Waterbury, Conn.
  G. O. No. 46, W. D., 1919.

Sergeant, Company I, 47th Infantry, 4th Division.
Sergeant Weston charged an enemy machine gun which was inflicting heavy losses upon our troops and delaying the advance. He wounded the gunner and captured the gun, thereby enabling our advance to continue.

WESTPHAL, ARTHUR E_____
Near Fossoy, France, July 15, 1918.
R—West Newton, Mass.
B—England.
G. O. No. 95, W. D., 1919.

First lieutenant, 7th Infantry, 3d Division.
In command of a Stokes mortar detachment Lieutenant Westphal displayed marked coolness and leadership under intense enemy shellfire in so separating his guns as to stop the advance of the Germans and prevent their crossing the Marne.

WETZEL, HENRY W. (274280)_____
Near Gesnes, France, Oct. 4–20, 1918.
R—Shawano, Wis.
B—Shawano, Wis.
G. O. No. 66, W. D., 1919.

Private, first class, Company F, 127th Infantry, 32d Division.
Private Wetzel, a company runner, repeatedly volunteered for missions so hazardous that no others would attempt them. At all hours of the day and night, over unknown ground, he carried numerous messages in the face of heavy machine-gun and shellfire.

WHALEN, JAMES (57193)_____
Near Cantigny, France, May 28, 1918.
R—Reading, Pa.
B—Reading, Pa.
G. O. No. 39, W. D., 1920.

First sergeant, Company D, 28th Infantry, 1st Division.
During an enemy counterattack Sergeant Whalen led a small group forward to a shell hole about 50 yards in front of his company position. Although wounded, he kept an automatic rifle in action. At this new position he assisted in breaking up the enemy attack.

*WHALEY, WELLMON P. (2338817)_____
Near Mont St. Pere, France, July 22, 1918.
R—White Castle, La.
B—White Castle, La.
G. O. No. 32, W. D., 1919.

Sergeant, Company F, 4th Infantry, 3d Division.
Advancing far ahead of his patrol, he encountered an enemy patrol which opened fire on him. Despite the fact that he had been severely wounded, he continued to combat the enemy, killing 1 and dispersing the others.
Posthumously awarded. Medal presented to father, Wellmon P. Whaley.

WHEAT, HARRY R._____
From Sergy to Mont-St. Martin, France, between the Ourcq and Vesle Rivers, Aug. 1–6, 1918.
R—Springfield, Mass.
B—New Hampshire.
G. O. No. 99, W. D., 1918.

First lieutenant, Medical Corps, attached to 125th Infantry, 32d Division.
During the attack on Aug. 1, 1918, he went forward with the first wave and established a dressing station in an advanced position to render immediate aid to the wounded. On Aug. 5, 1918, at St. Martin, after having been knocked down by the explosion of a shell and while under severe fire and machine-gun fire, he displayed exceptional coolness and devotion to duty in rendering surgical attention to others who had been wounded by the same shell.

*WHEATON, HOMER J. (61304)_____
At Chavignon, Chemin des Dames Sector, France, Feb. 27, 1918.
R—Worcester, Mass.
B—Pompey, N. Y.
G. O. No. 88, W. D., 1918.

Corporal, Company G, 101st Infantry, 26th Division.
During a heavy bombardment on the morning of Feb. 27, 1918, 1 of the hand grenades which were being distributed to the men of his company was dropped by accident. Corporal Wheaton, with extreme courage and self-sacrifice, ran to and picked up the grenade in an effort to cast it out of the danger area before it exploded. It exploded, however, before he could throw it away, and he was fatally wounded, dying shortly thereafter.
Posthumously awarded. Medal presented to aunt, Mrs. Ada S. Pen Eycke.

WHEDON, HERBERT S. (64763)_____
At Marcheville, France, Sept. 26, 1918.
R—Madison, Conn.
B—Madison, Conn.
G. O. No. 145, W. D., 1918.

Sergeant, Company B, 102d Infantry, 26th Division.
During an intermittent barrage, lasting for 2 hours, he placed his men in the best shelter available, but himself remained in observation, refusing to take cover from terrific artillery fire. He was twice buried by exploding shells while succoring wounded.

*WHEELER, AFTON E. (207762)_____
Near Bezu-St. Germain, France, Sept. 7, 1918.
R—Cambridge, Mass.
B—Somerville, Mass.
G. O. No. 142, W. D., 1918.

Ordnance sergeant, Salvage Squadron No. 1, Quartermaster Corps, American Expeditionary Forces.
For extraordinary heroism in connection with military operations against an armed enemy near Bezu-St. Germain, France, Sept. 7, 1918. When fire broke out in a wood where a salvage detachment was encamped, seriously endangering the lives of 200 men because of their proximity to a pile of salvaged German high-explosive 155-millimeter shells, he and Private Willie J. Walker voluntarily ran to the scene of the fire and attempted to extinguish the flames, fully aware of the grave danger to themselves. They fought the fire with blankets and sticks, but the fire quickly spread to the shells. Both men were killed by the explosion which followed.
Posthumously awarded. Medal presented to father, Clarence A. Wheeler.

WHEELER, FREDERIC COLLINS_____
Near Bouresches, France, June 5, 1918.
R—Philadelphia, Pa.
B—Philadelphia, Pa.
G. O. No. 119, W. D., 1918.

First lieutenant, 6th Regiment, U. S. Marine Corps, 2d Division.
He was conspicuous for his bravery in remaining in action, although twice wounded, refusing to be evacuated until wounded a third time and then endeavoring to return to his command.

WHEELOCK, FRANK R._____
Near Malancourt, France, Sept. 26–30, 1918.
R—Scranton, Pa.
B—Boston, Mass.
G. O. No. 37, W. D., 1919.

Captain, Medical Corps, attached to 313th Infantry, 79th Division.
Working in areas that were continually being swept by machine-gun, rifle, and shell fire, Captain Wheelock worked voluntarily and unceasingly, giving aid, food, and water to the wounded. Throughout the entire operations he showed utter disregard for his own safety, being knocked down many times by shell explosions. For 2 nights he worked as a stretcher bearer, carrying patients to places of safety after giving them medical attention during the day.

WHIPPLE, COLUMBUS (1630549)_____
Near Bazoches, France, Aug. 7, 1918.
R—Snowflake, Ariz.
B—Adair, Ariz.
G. O. No. 147, W. D., 1913.

Private, Company H, 47th Infantry, 4th Division.
He crossed the Vesle River in the face of enemy fire and rescued a drowning comrade in the deep, swift current of the stream.

WHISENANT, HERBERT W._____
Near Soissons, France, July 18, 1918.
R—Austin, Tex.
B—Kyle, Tex.
G. O. No. 44, W. D., 1919.

Second lieutenant, 16th Infantry, 1st Division.
While advancing with this platoon, Lieutenant Whisenant, after he was so severely wounded that he was unable to continue, so encouraged and inspired his men that they won a decided victory and captured many men and guns. His wound resulted in the loss of a leg.

WHITAKER, DANIEL J. (732334)_____
  Near Fontaine, France, Nov. 8, 1918.
  R—Pageland, S. C.
  B—Lancaster, S. C.
  G. O. No. 37, W. D., 1919.

Private, Company D, 6th Infantry, 5th Division.
While engaged as company runner he displayed rare devotion to duty by carrying messages through heavy machine-gun fire, continuing his work after being severely wounded.

WHITAKER, DEWEY A. (58256)_____
  At Cantigny, France, May 28, 1918.
  R—Greenville, S. C.
  B—Spartanburg, S. C.
  G. O. No. 39, W. D., 1920.

Private, first class, Company I, 28th Infantry, 1st Division.
Private Whitaker exposed himself to direct fire of the enemy machine guns while going in advance of the lines to assist a wounded comrade to shelter.

WHITAKER, JESSE L. (240731)_____
  Near St. Etienne-a-Arnes, France, Oct. 3-9, 1918.
  R—Paris, Mo.
  B—Paris, Mo.
  G. O. No. 46, W. D., 1919.

Corporal, Company L, 23d Infantry, 2d Division.
A few minutes before the attack, in an offensive operation, he was wounded by a shell fragment but remained with his platoon and led his squad with great courage and initiative during the attack.

WHITCOMB, CECIL B. (1516484)_____
  Near Montfaucon, France, Sept. 26-28, 1918.
  R—East Cleveland, Ohio.
  B—Somerset, N. Y.
  G. O. No. 16, W. D., 1923.

Sergeant, Headquarters Company, 145th Infantry, 37th Division.
Attached to the regimental intelligence section of the 145th Infantry, he, with several men of his section, accompanied the first attacking wave of the regiment on Sept. 26. Losing contact temporarily with the assaulting wave on account of a smoke barrage, he halted momentarily and upon resuming the advance encountered enemy machine-gun and sniper fire near a swale in the Bois-de-Montfaucon. Leaving his men in a place of safety, Sergeant Whitcomb discovered an enemy machine-gun nest which covered a bridge across the swale. He captured several unarmed enemy soldiers and an officer; the latter he forced to return to the machine-gun nest and to deliver to him the gun crews, a number of men, as well as several enemy snipers who had been inflicting heavy casualties upon his men; this action permitted the Infantry to advance without further heavy losses.

WHITCOMB, GEORGE (34541)_____
  Near Cunel, France, Oct. 12, 1918.
  R—Helena, Mont.
  B—Malta, Mont.
  G. O. No. 26, W. D., 1919.

Private, Company B, 9th Machine Gun Battalion, 3d Division.
Although seriously wounded, he refused to be evacuated until he had gone under heavy artillery and machine-gun fire to four other gun crews, requesting that men be sent to his gun, thereby enabling an important gun to remain in action.

WHITE, AMBROSE F._____
  At La Franquette Farm, July 22, 1918.
  R—Bluefield, W. Va.
  B—Virginia City, Va.
  G. O. No. 22, W. D., 1920.

First lieutenant, 38th Infantry, 3d Division.
While Lieutenant White's unit was in close contact with the enemy, a spy circulated a report that the company had been ordered to withdraw. Those who thus retired were shot down by the enemy. During the disorder that followed, Lieutenant White fearlessly reorganized the platoon under close-range enemy fire. Due to his heroic example, the men of his platoon held their ground and repulsed the strong enemy attack which followed.

*WHITE, DONALD W._____
  Near Landres-et-St. Georges, France, Nov. 1, 1918.
  R—Manitowoc, Wis.
  B—Antigo, Wis.
  G. O. No. 37, W. D., 1919.

Second lieutenant, 23d Infantry, 2d Division.
When the advance of his battlion was hindered by a strong enemy machine-gun nest, he led his platoon forward in an attack on the hostile position and was killed at the head of his platoon just before the last machine gun was put out of action.
Posthumously awarded. Medal presented to mother, Mrs. Emma E. White.

WHITE, EDWARD R. (1285786)_____
  Near Consenvoye Wood, France, Oct. 10, 1918.
  R—Salisbury, Md.
  B—Salisbury, Md.
  G. O. No. 37, W. D., 1919.

Sergeant, Company I, 115th Infantry, 29th Division.
After his platoon leader had been killed he took command. The advance of the company had been held up by a machine-gun nest until Sergeant White, with 2 other soldiers, cleaned out the nest, killing 4 and capturing 6 of the enemy.

WHITE, JESS (2471590)_____
  Near Nantillois, France, Oct. 5, 1918.
  R—Chelyan, W. Va.
  B—Charleston, W. Va.
  G. O. No. 37, W. D., 1918.

Corporal, Company D, 317th Infantry, 80th Division.
He led his squad across an area swept by machine-gun and shell fire with utter disregard of his personal danger. Although he was severely wounded by a machine-gun bullet, he continued to direct his squad until completely exhausted from loss of blood.

WHITE, JOHN B. (2691)_____
  At Cantigny, France, May 28-31, 1918.
  R—Milligan College, Tenn.
  B—Washington County, Tenn.
  G. O. No. 109, W. D., 1918.

Private, medical detachment, 28th Infantry, 1st Division.
For 3 nights at Cantigny, France, on May 28-31, 1918, he worked unceasingly under fire, bringing the wounded to safety and ministering to them on his own initiative. He repeatedly left shelter to help wounded men.

WHITE, LOUIS D. (1448191)_____
  Near Baulny, France, Sept. 28, 1918.
  R—Hutchinson, Kans.
  B—Hutchinson, Kans.
  G. O. No. 59, W. D., 1919.

First sergeant, Company E, 137th Infantry, 35th Division.
He volunteered to carry a message to the rear through heavy artillery fire to obtain ammunition and reinforcements. That mission accomplished, he learned that his captain, the only officer left with the company, had been wounded. Though himself wounded and suffering from gas, he returned to the front lines, reorganized the company, and held his section of the front line until the division was relieved.

WHITE, LYMAN (1316071)_____
  Near Bellicourt, France, Sept. 29, 1918.
  R—Salemburg, N. C.
  B—Fayetteville, N. C.
  G. O. No. 81, W. D., 1919.

Sergeant, Company H, 119th Infantry, 30th Division.
When, with 3 other men, he encountered a German patrol, which outnumbered them five to one, he ordered his companions to keep the enemy down with fire from their Lewis gun. He then crept to the rear of the hostile patrol and attacked the Germans with bombs. At the same time his companions attacked from the front, killing several of the Germans and capturing 9.

*WHITE, NATHANIEL C. (1402540) _____
  At Vauxaillon, France, Sept. 19, 1918.
  R—Chicago, Ill.
  B—Tallulah, La.
  G. O. No. 44, W. D., 1919.

Private, first class, Company F, 370th Infantry, 93d Division.
Private White, while acting as company runner, exposed himself constantly to intense enemy machine-gun and artillery fire and was killed while in the performance of his duty.
Posthumously awarded. Medal presented to mother, Mrs. Julia White.

WHITE, RICHARD G _____
  Near Soissons, France, July 18, 1918.
  R—Charleston, S. C.
  B—Marion, S. C.
  G. O. No. 15, W. D., 1919.

First lieutenant, 16th Infantry, 1st Division.
He led his platoon through intense machine-gun and artillery fire, destroying machine guns that were causing heavy losses on an exposed flank and remaining in command of his platoon until twice severely wounded.

WHITE, RICHARD J _____
  Near Ravine-de-Molieville, north of Samogneux, France, Oct. 15, 1918.
  R—Creston, Iowa.
  B—Creston, Iowa.
  G. O. No. 130, W. D., 1918.

First lieutenant, 113th Infantry, 29th Division.
He was a member of a small party which was suddenly fired upon by 3 German machine guns, 1 soldier being killed and an officer severely wounded. Himself unarmed, Lieutenant White returned with another soldier and, in the face of machine-gun fire, approached within 50 yards of the machine-gun nests and carried the wounded officer to shelter.

WHITE, THOMAS M. (1711891) _____
  Near Toter Manns Valley, France, Oct. 4, 1918.
  R—New York, N. Y.
  B—Stepney, Conn.
  G. O. No. 81, W. D., 1919.

Sergeant, Company D, 306th Machine Gun Battalion, 77th Division.
While in command of his platoon, Sergeant White went with 2 other soldiers to the rescue of 3 members of a gun crew among whom a German hand grenade had burst. Finding 1 of the men dead and another so severely wounded that he could not be moved, Sergeant White carried the third man to shelter, in plain view of the enemy under continuous shell and machine-gun fire. His 2 comrades also having been wounded, he also succeeded in getting them back to safety, and thereafter twice returned to the gun position to administer first aid and carry water to the wounded soldier who could not be moved.

WHITE, TRACY S. (2410793) _____
  Near Ferme-des-Loges, France, Oct. 19, 1918.
  R—Ocean Grove, N. J.
  B—Ocean Grove, N. J.
  G. O. No. 81, W. D., 1919.

First sergeant, Company B, 311th Infantry, 78th Division.
When the position his company held was enfiladed and communication to the rear cut off, he volunteered to carry a message to the battalion commander after several runners had been killed in the attempt. Crossing ground swept by intense machine-gun and artillery fire, he delivered the message and returned with orders as to the disposition of the company.

*WHITE, WALTER D. (1750790) _____
  Near the Bois-des-Loges, France, Nov. 1, 1918.
  R—Byron, N. Y.
  B—Byron, N. Y.
  G. O. No. 78, W. D., 1919.

Private, Company B, 309th Infantry, 78th Division.
While acting as runner, Private White volunteered to carry a message across a long stretch of open country which was subjected to heavy machine-gun and artillery fire. He successfully crossed the space and delivered his message, but in an attempt to return he was killed by a rain of machine-gun bullets. His conduct served as an inspiration to other runners.
Posthumously awarded. Medal presented to mother, Mrs. Carrie L. White.

*WHITE, WILBERT W _____
  In the region of Etain and Chambley, France, Sept. 14, 1918.
  R—New York, N. Y.
  B—New Haven, Conn.
  G. O. No. 71, W. D., 1919.

Second lieutenant, pilot, 147th Aero Squadron, Air Service.
While protecting 3 allied observation planes in the region of Etain, Lieutenant White was attacked by 3 Halberstadt fighters. He engaged them immediately, successfully fighting them off and leading them all away from the observation planes, which were thus permitted to carry on their work unmolested. While returning home he dived through a cloud to attack an enemy balloon near Chambley, bringing it down in flames. Two Fokker scouts then attacked him, and although he was alone, with intrepid courage he attacked the first Fokker head-on, shooting until it went down in a vertical dive out of control. Pulling sharply, he fired a long burst at the second Fokker as it went over him, putting it to immediate flight.
Oak-leaf cluster.
For the following act of extraordinary heroism in action near Toul, France, Oct. 10, 1918, Lieutenant White is awarded an oak-leaf cluster to be worn with his distinguished-service cross: In command of a patrol of 4 planes which was attacked by 5 German Fokkers, he attacked the enemy plane which was hard pressing a new pilot. The German Fokker had gotten at the tail of the American plane and was overtaking it. Lieutenant White's gun having jammed, he drove his plane head-on into the German Fokker, both crashing to earth, 500 meters below.
Posthumously awarded. Medal presented to father, Dr. W. W. White.

WHITE, WILLIAM P. (89893) _____
  In the Argonne, France, Oct. 14–29, 1918.
  R—New York, N. Y.
  B—New York, N. Y.
  G. O. No. 59, W. D., 1919.

Private, Company D, 165th Infantry, 42d Division.
Attached to the regimental liaison group, he time and again traversed 3 kilometers to the front lines, proving the swiftest and surest runner. For three days and nights he worked unceasingly under terrific artillery and machine-gun fire, accomplishing his mission when other runners had failed.

WHITED, HOMER (96385) _____
  Near Ancerviller, France, Mar. 4, 1918.
  R—Bessemer, Ala.
  B—Cedar Bend, Ala.
  G. O. No. 126, W. D., 1919.

Corporal, Company D, 167th Infantry, 42d Division.
He was a member of a patrol of five men which on Mar. 4, 1918, near Ancerviller, France, encountered an enemy patrol of 11 men, which it attacked and routed, taking 2 prisoners.

WHITEHEAD, FRANK _____
  Near St. Etienne, France, Oct. 4, 1918.
  R—Chelsea, Mass.
  B—Camden, N. J.
  G. O. No. 35, W. D., 1919.

Captain, 5th Regiment, U. S. Marine Corps, 2d Division.
Although severely wounded, he showed exceptional coolness and bravery in his selection of machine-gun sites and in routing the enemy while under heavy machine-gun fire.

WHITEHEAD, LEWIS E. (1764979)........
At St. Juvin, France, Oct. 16, 1918.
R—Elmira, N. Y.
B—Elmira, N. Y.
G. O. No. 25, W. D., 1919.

Corporal, Machine Gun Company, 310th Infantry, 78th Division.
Corporal Whitehead, after giving first aid to his platoon leader who had been wounded, took command of the platoon and led it in an attack in the face of concentrated enemy artillery and machine-gun fire, reaching the objective and effectively protecting the exposed flank of the assaulting battalion with his 2 guns.

WHITEMAN, RALPH A. (1773801)........
At Grand Pre, France, Oct. 18, 1918.
R—Clearfield, Pa.
B—Williamsport, Pa.
G. O. No. 98, W. D., 1919.

Sergeant, Company D, 312th Infantry, 78th Division.
When his detachment had become disorganized by sniper and machine-gun fire, Sergeant Whiteman, without regard for personal danger, reorganized his command and by his gallant example led his men against the enemy machine-gun position, capturing it and bringing the gun back to our lines.

*WHITING, CHARLES W. (1681631)......
Near Barbonval, France, Sept. 10, 1918.
R—Avon, Mass.
B—Avon, Mass.
G. O. No. 32, W. D., 1919.

Private, Headquarters Company, 308th Infantry, 77th Division.
He had charge of maintaining a telephone line from Barbonval to Blanzy. The line was under direct observation of the enemy, and the appearance of a lineman was the immediate occasion for shelling by the enemy with field artillery and 1-pounders. He stuck to his work, repairing break after break, until he was mortally wounded by the enemy shellfire.
Posthumously awarded. Medal presented to mother, Mrs. Annie Battles.

*WHITING, CLINTON L...............
Near La Harazee, France, Sept. 26-28, 1918.
R—Brooklyn, N. Y.
B—Elizabeth, N. J.
G. O. No. 87, W. D., 1919.

First lieutenant, 308th Infantry, 77th Division.
During the advance in the Argonne Forest, Lieutenant Whiting exposed himself fearlessly to enemy machine-gun and sniper fire while leading his men and consolidating his position, which was in a marsh covered with wire grass and stunted brush. He continued to lead his men with utter disregard for personal danger until he fell seriously wounded by a machine-gun bullet on the afternoon of Sept. 28 near Binarville.
Posthumously awarded. Medal presented to father, D. Clinton Whiting.

WHITMAN, GUY (274283)...............
Near Gesnes, France, Oct. 4-20, 1918.
R—Shawano, Wis.
B—Shawano, Wis.
G. O. No. 87, W. D., 1919.

Private, first class, Company F, 127th Infantry, 32d Division.
During this period Private Whitman, although many runners had been shot down, repeatedly volunteered and carried messages through heavy enemy barrages, successfully accomplishing his work, and thereby saving the lives of many of his comrades.

WHITMAN, WALTER MONTIETH.......
Near Fleville and St. Juvin, France, Oct. 11-12, 1918.
R—New York, N. Y.
B—New York, N. Y.
G. O. No. 126, W. D., 1919.
Distinguished-service medal also awarded.

Colonel, 325th Infantry, 82d Division.
When his regiment was attacked in column before reaching the line which it was to hold, he took command and personally led his men into action. Always on the firing line, he led 4 attacks under heavy fire from artillery, machine guns, and snipers on the hill east of St. Juvin, the fourth of which was successful. He maintained his post of command on or near the front line throughout the engagement and by his personal example of courage inspired his men to valiant and successful combat.

WHITNEY, HENRY H...............
On the island of Porto Rico, in May, 1898.
R—Philipsburg, Pa.
B—Hopewell, Pa.
G. O. No. 19, W. D., 1922.

Captain and assistant adjutant general, U. S. Volunteers.
For extraordinary heroism in connection with the operations on the island of Porto Rico in May, 1898, under disguise and in the midst of an enemy.

WHITNEY, LEROY F. (1216002).......
Near St. Souplet, France, Oct. 17, 1918.
R—Auburn, N. Y.
B—Auburn, N. Y.
G. O. No. 37, W. D., 1919.

Corporal, Company M, 108th Infantry, 27th Division.
Voluntarily carrying messages under heavy shell and machine-gun fire, he displayed great bravery and gallantry. In 1 instance he completed the mission of a runner who had been wounded and returned with very important information as to where the barrage would fall.

*WHITNEY, RALPH L. (2036685).......
Near Montagne, France, Oct. 15, 1918.
R—Ann Arbor, Mich.
B—Detroit, Mich.
G. O. No. 32, W. D., 1919.

Private, Company C, 112th Machine Gun Battalion, 29th Division.
During an attack he was tireless in his efforts to bring food and water to his comrades. On the same day he captured 13 Germans without assistance and without regard to his personal safety. Later, while aiding a wounded comrade, he was severely wounded.
Posthumously awarded. Medal presented to father, E. E. Whitney.

*WHITSON, LESTER C. (1387295).......
At Hamel, France, July 4, 1918.
R—Chicago, Ill.
B—Chicago, Ill.
G. O. No. 14, W. D., 1920.

Corporal, Company E, 131st Infantry, 33d Division.
Although severely wounded in the shoulder at the beginning of the engagement, he continued forward as squad leader, exhibiting great gallantry and setting an inspiring example to his men.
Posthumously awarded. Medal presented to mother, Mrs. Emma Whitson.

WHITSON, ROBERT KENNETH...........
Near Soissons, France, July 19, 1918.
R—Union City, Tenn.
B—Union City, Tenn.
G. O. No. 15, W. D., 1919.

Captain, 26th Infantry, 1st Division.
When his major was killed near Soissons, France, July 19, 1918, he took command of his battalion and, although wounded, led it forward for the succeeding 3 days to its final objective, and, although wounded again, refused to be evacuated until he had directed the consolidation of his position.

WHITTHORNE, HARRY S...........
Near Exermont, France, Sept. 23-Oct. 1, 1918.
R—San Francisco, Calif.
B—Vallejo, Calif.
G. O. No. 59, W. D., 1919.

First lieutenant, 140th Infantry, 35th Division.
He organized a detachment to go 1,200 yards in front of our lines to rescue the wounded in a wood previously occupied. He brought back over 20 of the wounded, who would otherwise have been captured or died from exposure, the rescue being effected under heavy machine-gun and artillery fire. Later, when he was the only officer with the battalion, he refused to be evacuated, though wounded and burned by mustard gas, remaining in command until the battalion was relieved.

WHITTINGTON, CHARLES E. (58436).
At Cantigny, France, May 28–30, 1918.
R—Sumter, S. C.
B—Martinsburg, W. Va.
G. O. No. 99, W. D., 1918.

Private, Company K, 28th Infantry, 1st Division.
For 3 days at Cantigny, France, on May 28–30, 1918, he performed with great bravery the duties of battalion gunner without rest. Although wounded, he remained on duty under fire until his battalion was relieved.

WIBERG, ALBIN (1386855).
During the Somme offensive, France, Aug. 15, 1918.
R—Chicago, Ill.
B—Sweden.
G. O. No. 71, W. D., 1919.

Sergeant, Company C, 131st Infantry, 33d Division.
Blown over the parapet of an outpost when an enemy shell made a direct hit, he upon regaining consciousness carried the wounded members of his squad through heavy shell and machine-gun fire to a dressing station. He then drew a new automatic rifle from a near-by dump, and making his way through heavy fire established a new outpost, holding it alone against the enemy for 14 hours until relieved.

WICKHAM, GORDON (1385825).
At Chipilly Ridge, France, Aug. 11, 1918.
R—Chicago, Ill.
B—Chicago, Ill.
G. O. No. 128, W. D., 1918.

Private Headquarters Company, 131st Infantry, 33d Division.
He was on duty with a carrying party, which was severely shelled and gassed while passing through Grassier Wood. In utter disregard of his own personal safety, this courageous soldier made repeated trips into the woods under heavy shellfire and rescued wounded soldiers.

WICKLIFFE, ROBERT E. (2214773).
Near Grand Ballois Farm, France, July 14–15, 1918.
R—Spencer, Iowa.
B—Warsaw, Mo.
G. O. No. 32, W. D., 1919.

Private, Company A, 4th Infantry, 3d Division.
After being severely wounded he remained at his post, performing his duties as a relay runner until relieved.

WIDDIFIELD, CECIL J.
Near St. Etienne, France, Oct. 5, 1918.
R—Troy, Mont.
B—East Tawas, Mich.
G. O. No. 44, W. D., 1919.

Second lieutenant, 6th Regiment, U. S. Marine Corps, 2d Division.
He voluntarily went forward for a distance of 800 meters under heavy shellfire and rescued a wounded soldier who had been left there the night before when the advance patrols had been withdrawn.

WIECHMANN, WALTER H.
At Marcheville, France, Sept. 26, 1918.
R—Detroit, Mich.
B—Detroit, Mich.
G. O. No. 126, W. D., 1919.

Corporal, Company D, 103d Infantry, 26th Division.
Corporal Wiechman climbed out from the top of a trench under machine-gun fire from all directions to take prisoners in another trench who had thrown up their hands and shouted that they had surrendered. Approaching, he was met with a shower of hand grenades, but he nevertheless stood his ground and opened fire.

WIEDMAIER, BENJAMIN P. (2214774).
Near Grand Ballois Farm, France, July 14–15, 1918.
R—Clarksdale, Mo.
B—Clarksdale, Mo.
G. O. No. 32, W. D., 1919.

Private, Company A, 4th Infantry, 3d Division.
During a heavy gas and shell bombardment he repeatedly volunteered and delivered messages over routes other than his own when the runners assigned to those routes had been killed or wounded.

WIESE, EDWIN (2207673).
At Essey, France, Sept. 12, 1918.
R—St. Louis, Mo.
B—St. Louis, Mo.
G. O. No. 128, W. D., 1918.

Private, Company C, 355th Infantry, 89th Division.
He displayed conspicuous gallantry by creeping forward alone under machine-gun fire and capturing 2 enemy machine guns which were holding up the advance of his organization.

*WIGGINS, EDWIN W. (1445247).
Near Baulny, France, Sept. 29, 1918.
R—Carthage, Mo.
B—Carthage, Mo.
G. O. No. 89, W. D., 1919.

Sergeant, Company A, 128th Machine Gun Battalion, 35th Division.
Sergeant Wiggins led a machine-gun platoon to a threatened portion of the line under a heavy enemy barrage, walking back and forth along the front under heavy enemy fire, encouraging his men, and directing the construction of emplacements. He also organized a group of infantrymen who had become separated from their organizations and put them in the line, supervising their intrenchments. This gallant soldier was killed just as this work was completed.
Posthumously awarded. Medal presented to father, T. S. Wiggins.

WIGGLESWORTH, ROBERT.
Near Consenvoye, France, Oct. 9, 1918.
R—Chicago, Ill.
B—Chicago, Ill.
G. O. No. 44, W. D., 1919.

Captain, 132d Infantry, 33d Division.
When the 2 platoons he was leading in attack were held up by terrific fire from 2 machine guns, he ordered his men to lie down, and he single handed rushed 1 nest, killing the gunner and capturing the crew. He then forced the surrender of the second gun crew.

WIGHT, HOWARD M. (2294304).
Near Gesnes, France, Sept. 28, 1918.
R—Corvallis, Oreg.
B—Harrison, Me.
G. O. No. 15, W. D., 1919.

Private, Company I, 361st Infantry, 91st Division.
When his battalion withdrew after attacking a hostile position under heavy fire, Private Wight, instead of falling back, organized a party and in the face of intense machine-gun fire rescued 15 wounded soldiers who would otherwise have fallen into the hands of the enemy. He placed the wounded men in a gravel pit and remained the entire night, administering first aid, despite the fact that he himself was nearly exhausted after 3 days of fighting.

WILBUR, THOMAS WHITESIDE.

Near Jaulny, France, Sept. 13–15, 1918.
R—Larchmont, N. Y.
B—New Britain, Conn.
G. O. No. 46, W. D., 1919.

Secretary, Y. M. C. A., attached to 6th Machine Gun Battalion, U. S. Marine Corps, 2d Division.
Declining to remain in the rear, Mr. Wilbur attached himself to the Medical Department, rendering first aid and bringing in wounded, serving at all times in a most valuable manner. He disregarded an order to return to the rear when it seemed that the enemy would launch a counterattack, but remained with the wounded until all were safely evacuated.

WILCOX, GILBERT W (569496)_____
On the Vesle River, near Ville-Savoye, France, Aug. 11, 1918.
R—Linton, Oreg.
B—Nova Scotia.
G. O. No. 128, W. D., 1918.

Private, first class, Company D, 4th Engineers, 4th Division.
He volunteered to go into Ville-Savoye at a time when it was under a heavy bombardment to rescue a wounded officer.

*WILCOX, GLENN E_____
Near Jaulgonne, France, July 23, 1918.
R—Detroit, Mich.
B—Cadillac, Mich.
G. O. No. 44, W. D., 1919.

Second lieutenant, 30th Infantry, 3d Division.
When his company had reached its objective and was suffering heavy casualties from shellfire, he rendered valuable assistance in reorganizing the company and caring for the wounded. He remained on duty even though suffering from severe mustard-gas burns.
Posthumously awarded. Medal presented to mother, Mrs. Louise M. Wilcox.

WILCOX, RALPH M_____
Near Letanne, France, Nov. 10-11, 1918.
R—Portland, Oreg.
B—Ogden, Utah.
G. O. No. 32, W. D., 1919.

First lieutenant, 5th Regiment, U. S. Marine Corps, 2d Division.
He volunteered for a liaison mission and successfully accomplished it, displaying marked bravery. Passing through heavy artillery and machine-gun barrage, he pushed through the enemy outpost line, routed 1 of the outposts, and succeeded in establishing liaison between 2 battalions at a critical moment.

WILCOXSON, ORVAL (2216138)_____
In the Argonne, France, Oct. 23, 1918.
R—Bomar, Okla.
B—Leon, Okla.
G. O. No. 87, W. D., 1919.

Private, first class, Machine Gun Company, 357th Infantry, 90th Division.
He was always the first to volunteer as a company runner for dangerous missions, and repeatedly passed through heavy fire to deliver important messages, showing marked personal heroism.

WILDER, MARSHALL P_____
Near Xammes, France, Sept. 26, 1918.
R—Manhattan, Kans.
B—Clinton, Mo.
G. O. No. 95, W. D., 1919.

Captain, 354th Infantry, 89th Division.
Captain Wilder was in command of a raiding party which was caught under such heavy machine-gun fire that the success of the raid was threatened. Taking charge of a combat group, whose leader had become a casualty, this officer charged the hostile strong point and succeeded in clearing it with heavy casualties to the enemy, the remainder of whom retreated.

*WILDER, THOMAS E (1448605)_____
Near Baulny, France, Sept. 28, 1918.
R—Macksville, Kans.
B—Cassville, Mo.
G. O. No. 87, W. D., 1919.

Corporal, Company F, 137th Infantry, 35th Division.
After all of his squad but himself had been killed by the explosion of a shell, and after half of 1 hand had been carried away by a piece of shell, Corporal Wilder valiantly continued the combat until he himself was killed.
Posthumously awarded. Medal presented to father, George M. Wilder.

WILES, GEORGE L (1219296)_____
Near Bellicourt, France, Sept. 29, 1918.
R—New Market, Tenn.
B—Shady Grove, Tenn.
G. O. No. 35, W. D., 1919.

Private, Machine Gun Company, 120th Infantry, 30th Division.
After his own gun had been knocked out, he assisted another soldier in breaking up an enemy machine-gun nest and turning the captured gun on the enemy, firing about 1,000 rounds. When this gun jammed, he procured grenades and the rifle of a dead soldier and continued on to the objective.

WILEY, JAMES E (2338233)_____
Near Les Evaux, France, July 14-15, 1918.
R—Colorado Springs, Colo.
B—Colorado Springs, Colo.
G. O. No. 32, W. D., 1919.

Private, Company B, 4th Infantry, 3d Division.
After being badly gassed he continued with his duties as runner, carrying messages through a heavy bombardment to and from the front line.

WILKEN, ALT C (102572)_____
Near Sergy, France, July 28, 1918.
R—Atlantic, Iowa.
B—Atlantic, Iowa.
G. O. No. 99, W. D., 1918.

Private, Company M, 168th Infantry, 42d Division.
He added materially in the advance against the Prussian Guards near Sergy, France, July 28, 1918. Despite 3 wounds he continued firing with his automatic rifle until his right hand was shattered.

WILKERSON, ALFRED (2105602)_____
At Jaulny, France, Sept. 12, 1918.
R—Youngstown, Ohio.
B—Cameron, W. Va.
G. O. No. 3, W. D., 1919.

Private, Company B, 2d Engineers, 2d Division.
He and another soldier, who were acting as wire cutters with the first line of Infantry, fought their way forward in advance of their unit and were the first men to enter Jaulny while it was swept by machine-gun fire, infested with snipers, and still occupied by rear-guard detachments of the enemy. After capturing 8 Germans in a dugout, they courageously searched the caves in the town and took 55 prisoners.

WILKINS, FRED R (2078217)_____
At Hamel, France, July 4, 1918.
R—Casper, Wyo.
B—Freeport, Ill.
G. O. No. 14, W. D., 1920.

Private, Company A, 132d Infantry, 33d Division.
Unaided he attacked a machine-gun position with hand grenades, drove off the gun crew, and captured the gun.

WILKINS, JOHN (1787633)_____
Near Verdun, France, Nov. 3-4, 1918.
R—Philadelphia, Pa.
B—Philadelphia, Pa.
G. O. No. 37, W. D., 1919.

Private, first class, Machine Gun Company, 316th Infantry, 79th Division.
Repeatedly volunteering to act as runner, Private Wilkins made several hazardous journeys from headquarters to the machine-gun positions, a distance of 500 meters, at all times subjected to intense artillery bombardment.

*WILKINSON, GEORGE A (99360)_____
Near Cote-de-Chatillon, France, Oct. 14, 1918.
R—Winterset, Iowa.
B—Madison County, Iowa.
G. O. No. 9, W. D., 1923.

Sergeant, Company A, 168th Infantry, 42d Division.
While leading his platoon up the steep and strongly fortified slopes of Hill 288 under terrific machine-gun fire, with great dash and courage he charged and captured 3 machine-gun nests which had caused severe losses to his platoon. On 2 occasions he reorganized his platoon under heavy machine-gun fire, and while unfalteringly pressing his advantage in personally charging and putting out of action a fourth machine-gun implacement he received the wound which caused his instant death.
Posthumously awarded. Medal presented to father, Henry Wilkinson.

WILKINSON, HAROLD (441859) _____
South of Soissons, France, July 20, 1918.
R—Williams, Iowa.
B—Williams, Iowa.
G. O. No. 117, W. D., 1918.

Private, Company B, 16th Infantry, 1st Division.
When the enemy was forming for a counterattack, he carried messages 3 times through their heavy barrage, and although in an exhausted condition from that work he remained with his company throughout the entire operations.

WILKINSON, JACK H. (1210468) _____
Near Ronssoy, France, Sept. 29, 1918.
R—New York, N. Y.
B—England.
G. O. No. 20, W. D., 1919.

Private, Company D, 107th Infantry, 27th Division.
During the operations against the Hindenburg line Private Wilkinson left shelter and went forward, crawling on his hands and knees, under heavy machine-gun fire, to the aid of a wounded officer and a wounded soldier. With the assistance of another soldier he succeeded in dragging and carrying them back to the shelter of a trench.

WILKINSON, JOHN L. (1181043) _____
Near Exermont, France, Sept. 28, 1918.
R—Richmond, Mo.
B—Richmond, Mo.
G. O. No. 95, W. D., 1919.

First sergeant, Company C, 140th Infantry, 35th Division.
After being seriously gassed, Sergeant Wilkinson refused to be evacuated, but remained on duty for 3 days, assisting the officers in maintaining organization with utter disregard for his own safety until the company was relieved, becoming so exhausted that he was unable to walk.

WILKINSON, THADDEUS (2498) _____
At Verdun, France, Oct. 4–9, 1918.
R—Capleville, Tenn.
B—Union County, Miss.
G. O. No. 98, W. D., 1919.

Sergeant, medical detachment, 26th Infantry, 1st Division.
Throughout this period Sergeant Wilkinson, with utter disregard for personal danger, rendered first aid to the wounded under heavy artillery and machine-gun fire. On Oct. 9 he was severely wounded while endeavoring to reach a wounded comrade, passing through the direct fire of the enemy in his gallant attempt.

WILLARD, HENRY W. (540990) _____
In the Bois-de-Belleau, France, June 20, 1918.
R—Slaterville Springs, N. Y.
B—Slaterville Springs, N. Y.
G. O. No. 99, W. D., 1918.

Corporal, Company B, 7th Infantry, 3d Division.
In the Bois-de-Belleau on June 20, 1918, he went out of his position for a distance of 75 yards under heavy machine-gun fire and by rifle fire took possession of 1 of the guns, brought it back, and proceeded to put it in operation against the Germans.

WILLARD, ROSCOE A. (42616) _____
South of Soissons, France, July 18, 1918.
R—Mount Carmel, Ill.
B—Greenbrier, Tenn.
G. O. No. 35, W. D., 1920.

Sergeant, Company E, 16th Infantry, 1st Division.
After the platoon commander of the 1st Platoon had been killed and the commander of the 2d Platoon wounded Sergeant Willard reorganized the platoons and assumed command of both. He then, under heavy fire, led them forward in the attack on the enemy position. He continued to lead his men forward until the objective was reached. He was severely wounded during this attack.

*WILLIAMS, BERTRAM _____
Between Chambley and Xammes, France, Sept. 13, 1918.
R—Cambridge, Mass.
B—Cambridge, Mass.
G. O. No. 15, W. D., 1923.

First lieutenant, 96th Aero Squadron, Air Service.
As observer, he accompanied Lieutenant Hopkins, pilot, on a flight of 3 planes which were attacked by 15 enemy planes. Despite the overwhelming number of the enemy by which they were surrounded, the American planes proceeded on their mission and bombed the objective. In the action which followed he and his pilot continued the unequal fight until they were shot down and killed. The heroic conduct displayed greatly inspired the members of the squadron and enabled 1 of the American planes to return to its airdrome with valuable information of the enemy.
Posthumously awarded. Medal presented to mother, Mrs. Olive Swan Williams.

WILLIAMS, CHARLES F. _____
Near Romagne, France, Oct. 11, 1918.
R—Pittsburgh, Pa.
B—Philadelphia, Pa.
G. O. No. 26, W. D., 1919.

Second lieutenant, 9th Machine Gun Battalion, 3d Division.
Although wounded by a high-explosive shell, Lieutenant Williams refused to be evacuated, but continued leading his platoon in the attack and successfully defended his positions from counterattack until completely exhausted.

WILLIAMS, CHARLES V. (10358) _____
Near Baulny, Meuse, France, Sept. 28, 1918.
R—Philadelphia, Pa.
B—Philadelphia, Pa.
G. O. No. 44, W. D., 1919.

Corporal, Company B, 345th Battalion, Tank Corps.
During an attack on a hedge south of the Montrebeau Woods, Corporal Williams left his tank, which was out of action, and went through the severe rifle, machine-gun, and artillery fire to give first aid to his wounded lieutenant. He then took the wounded officer's place, leading the platoon of tanks to the objective through the intense enemy fire.

WILLIAMS, CLARENCE M. _____
On the Ourcq River, July 31 and Aug. 1, 1918; northwest of Coulonges, France, Aug. 2, 1918; and on the heights overlooking the Vesle River, France, Aug. 3–7, 1918.
R—Alpena, Mich.
B—Mumby, Mich.
G. O. No. 99, W. D., 1918.

Captain, Medical Corps, attached to 125th Infantry, 32d Division.
During these 3 periods of severe fighting he maintained a dressing station close to the advanced lines and worked continuously night and day under heavy artillery and machine-gun fire.

WILLIAMS, FRANK (46527) _____
Near the Argonne Forest, France, Oct. 9, 1918.
R—Wellston, Ohio.
B—Wellston, Ohio.
G. O. No. 98, W. D., 1919.

Sergeant, Company M, 28th Infantry, 1st Division.
In the absence of the platoon commander, Sergeant Williams displayed great courage and good judgment in leading his platoon in the attack on Hill 263. Although wounded in the advance, he refused to be evacuated, but continued to lead his men under heavy artillery and machine-gun fire. He remained with his command until it was relieved.

WILLIAMS, FRANK G. (41270)............
Near Medeah Ferme, France, Oct. 3, 1918.
R—Furnessville, Ind.
B—Furnessville, Ind.
G. O. No. 21, W. D., 1919.

Private, first class, Headquarters Company 9th Infantry, 2d Division.
While acting as runner, Private Williams, badly wounded, concealed the fact when he realized that his services were badly needed. He ran and delivered messages throughout the attack, and not until the objectives were attained and consolidated would he allow his wound to be dressed.

WILLIAMS, FRANK J., Jr. (642401).......
Near Ronssoy, France, Sept. 29, 1918.
R—Buffalo, N. Y.
B—Canada.
G. O. No. 4, W. D., 1923.

Sergeant, Company C, 301st Battalion, Tank Corps.
While operating against the enemy, his tank received a direct hit, killing or wounding the entire crew. Although severely wounded, Sergeant Williams assisted his tank commander, who was severely wounded and temporarily blinded, to a position of shelter in the sap of a near-by trench. He then returned to his tank and under heavy fire continued to operate a 6-pounder against the enemy until driven out by armor-piercing shells. He then assisted in the operation of a machine gun against heavy enemy fire from a trench that lay between them and our first line. When it became sufficiently dark, he aided his tank commander to a first-aid station. His courage and heroic actions throughout the day were largely responsible for saving the life of the officer commanding his tank.

WILLIAMS, FRANK L.................
In Champagne, east of Reims, July 15, 1918, and near the River Ourcq, northeast of Chateau - Thierry, France, July 30, 1918.
R—Des Moines, Iowa.
B—Ellis, Kans.
G. O. No. 117, W. D., 1918.

First lieutenant, Medical Corps, attached to 168th Infantry, 42d Division.
He voluntarily left a dugout on the Champagne front and for more than 2 hours, all the time under shellfire, administered to the needs of wounded men who were lying in the open. During the advance across the River Ourcq he voluntarily remained in exposed positions under heavy shellfire, caring for and dressing the wounded, until he was severely injured.

WILLIAMS, FRANK M.................
Near St. Juvin, France, Oct. 12 and 16, 1918.
R—Tampa, Fla.
B—Storm Lake, Iowa.
G. O. No. 139, W. D., 1918.

Captain, 325th Infantry, 82d Division.
During the operations in the vicinity of St. Juvin this officer demonstrated the highest personal bravery and leadership. On Oct. 12, although he was wounded, he organized a provisional combat group and led it to a ridge, repulsing an enemy counterattack which threatened our left flank. On Oct. 16, while he was reconnoitering a position for machine guns, he rescued an American soldier from 5 armed Germans, 4 of whom he killed with his pistol. Later on the same day he saw a hostile skirmish line advancing toward Hill 182. He rushed a machine gun forward, with which the attack was broken.

WILLIAMS, GUS J. (1024033)............
At Naguilian, Luzon, P. I., Dec. 7, 1899.
R—Jacksonville, Ala.
B—Anniston, Ala.
G. O. No. 3, W. D., 1925.

Private, Company A, 24th Infantry, U. S. Army.
When the command of which he was a member was held up in the crossing of the Rio Grande de Cagayan by rifle fire from a well-intrenched enemy, and being without boats or rafts with which to cross, Private Williams with 5 other members of his company volunteered to swim the river. Displaying great gallantry and with utter disregard for his life, he swam the river in the face of heavy rifle fire, returned on a raft, secured arms and ammunition, crossed a second time, and took part in an attack which drove a superior force of the enemy from their trenches and the town occupied by them, thereby making possible the further advance of his company.

WILLIAMS, HENRY M.................
Near Dun-sur-Meuse, France, Oct. 31–Nov. 11, 1918.
R—Eureka, Kans.
B—Kansas City, Kans.
G. O. No. 93, W. D., 1919.

First lieutenant, 76th Field Artillery, 3d Division.
In command of a detached piece operating with Infantry of the 5th Division Lieutenant Williams kept his gun close behind the attacking waves and skilfully accomplished the missions assigned to him. This was the first fieldpiece to cross the Meuse. Putting out of action a battery of German 77-millimeter guns by direct fire, he later turned them on the retreating enemy after his own ammunition had been exhausted.

WILLIAMS, ISHAM R.................
Near Fossoy, France, July 21, 1918.
R—Faison, N. C.
B—Spout Springs, N. C.
G. O. No. 44, W. D., 1919.

Second lieutenant, 7th Infantry, 3d Division.
He led a patrol across the Marne River under intense machine-gun fire, and when his boat was sunk twice swam the river to correct the fire of his covering detachment and to bring his patrol to safety after their mission had been accomplished.

WILLIAMS, JAMES R. (2002920)............
Near Mazinghien, France, Oct. 19, 1918.
R—Peytonsburg, Ky.
B—Peytonsburg, Ky.
G. O. No. 50, W. D., 1919.

Private, Company E, 120th Infantry, 30th Division.
With another soldier Private Williams volunteered and rescued a wounded comrade from an exposed position in front of the line after 2 other men had lost their lives in attempting to do so.

WILLIAMS, JESSIE V. (96969)............
Near Haumont, France, Sept. 15, 1918.
R—Lanett, Ala.
B—Dadeville, Ala.
G. O. No. 71, W. D., 1919.

Sergeant, Company F, 167th Infantry, 42d Division.
When the platoon he commanded was enfiladed successively by 2 machine-gun nests, he disposed his men so that both nests were captured without casualties. He displayed marked personal heroism in the advance, killing 1 enemy gunner with the butt of his rifle and bayoneting a second. His example of fearlessness was an example that inspired his men.

WILLIAMS, JOE (2109035)............
Near Lesseau, France, Sept. 4, 1918.
R—Acton, Ala.
B—Christian Place, Ala.
G. O. No. 143, W. D., 1918.

Private, Company E, 366th Infantry, 92d Division.
He was a member of a combat group which was attacked by 20 of an enemy raiding party advancing under a heavy barrage and using liquid. The sergeant in charge of the group was killed and several others, including Private Williams, were wounded. Nevertheless, this soldier with 3 others fearlessly resisted the enemy until they were driven off.

WILLIAMS, JOHN F., Jr_____
Near Ypres, Belgium, Aug. 2, 1918.
R—Charlotte, N. C.
B—Charlotte, N. C.
G. O. No. 143, W. D., 1918.

First lieutenant, 120th Infantry, 30th Division.
He volunteered to destroy an enemy pill box which had caused many casualties in his battalion. With much skill and daring he led a daylight patrol under heavy shell and machine-gun fire, rushed the pill box, killed or wounded the occupants, and accomplished his mission.

WILLIAMS, JOHN J_____
Near Villers-sur-Fere, France, July 28, 1918.
R—Berlin, Wis.
B—Lapeer, Mich.
G. O. No. 99, W. D., 1918.

Second lieutenant, 165th Infantry, 42d Division.
On July 28, 1918, near Villers-sur-Fere, France, when all the other officers of his company had been killed or wounded, he promptly took command, led his men through artillery and machine-gun fire, rushed a machine gun which was blocking his advance, personally killed 4 members of its crew, gained his objective and held it.

WILLIAMS, JULIUS DE WITT (1211928)____
Near Ronssoy, France, Sept. 29, 1918.
R—Brightwater, N. Y.
B—Brooklyn, N. Y.
G. O. No. 35, W. D., 1920.

Corporal, Company L, 107th Infantry, 27th Division.
Corporal Williams displayed great courage and was an excellent example for the men whom he led in the attack on the Hindenburg line. Although wounded in the right hip by a machine-gun bullet, he remained in command of his unit and assisted in the organization of a position for defense. He did not submit to evacuation until the position was secure against counterattack.

WILLIAMS, LOCKWOOD (2384231)_____
Near Clery-le-Petit, France, Nov. 5, 1918.
R—Asheville, N. C.
B—Emma, N. C.
G. O. No. 71, W. D., 1919.

Sergeant, Company I, 60th Infantry, 5th Division.
When his company was held up by a machine-gun nest, Sergeant Williams with 2 soldiers attacked the nest, killed 2 gunners, and captured 8 prisoners. This act made it possible for his company to advance and clear the bridgehead for the crossing of the brigade.

WILLIAMS, MACK H. (40859)_____
Near Medeah Ferme, France, Oct. 3, 1918.
R—Honey, Miss.
B—Greene County, Miss.
G. O. No. 21, W. D., 1919.

Private, first class, Company M, 9th Infantry, 2d Division.
Wounded in the hip by shellfire, while acting as stretcher bearer, Private Williams remained on duty until his company was relieved 7 days later.

WILLIAMS, PONTIAC J., Jr. (263592)_____
At Hill 212, near Sergy, northeast of Chateau-Thierry, France, July 31, 1918.
R—Bay Shore, Mich.
B—Hart, Mich.
G. O. No. 117, W. D., 1918.

Private, Company K, 126th Infantry, 32d Division.
He volunteered to go out in front of our lines and bring in a wounded runner. Although he was shot in the face before he reached the runner, he accomplished his mission.

WILLIAMS, RAY (1319337)_____
Near Vaux-Andigny, France, Oct. 8–12, 1918.
R—Lexington, N. C.
B—Iredell County, N. C.
G. O. No. 87, W. D., 1919.

Bugler, Company A, 120th Infantry, 30th Division.
Throughout this period Bugler Williams, acting as company runner, showed utter disregard for personal safety in carrying messages under fire. On Oct. 10, when the advance of his company was checked by enemy machine-gun and direct artillery fire, he carried a message of great importance to battalion headquarters and returned with an answer through a hail of bullets and shells. He continued to carry messages until he dropped from sheer exhaustion and even then begged to be permitted to resume his duties.

WILLIAMS, RAY T. (2214784)_____
Near Grand Ballois Farm, France, July 15, 1918.
R—Humphreys, Mo.
B—Humphreys, Mo.
G. O. No. 32, W. D., 1919.

Private, Headquarters Company, 4th Infantry, 3d Division.
After several of his comrades had been killed or wounded while attempting to repair telephone lines, Private Williams went out and performed the mission under heavy gas and shell bombardment.
Posthumously awarded. Medal presented to widow, Mrs. Elfie C. Williams.

WILLIAMS, WALTER (1096367)_____
Near Le Donjon Farm, France, July 15–17, 1918.
R—Memphis, Tenn.
B—Memphis, Tenn.
G. O. No. 44, W. D., 1919.

Corporal, Battery B, 18th Field Artillery, 3d Division.
Despite severe injuries, he remained at his post through a terrific shelling, firing his piece and directing his squad, for 2 days before allowing himself to be evacuated.

WILLIAMS, WILFRED (1507952)_____
Near Montblainville, France, Sept. 28, 1918.
R—Bayou Perout, La.
B—Bayou Perout, La.
G. O. No. 74, W. D., 1919.

Private, Company K, 109th Infantry, 28th Division.
As point of a patrol sent out to locate and destroy an enemy machine-gun nest he exposed himself fearlessly to draw the enemy fire, which instantly killed him. Through his sacrifice and inspired by his bravery his companions rushed and captured the enemy emplacement.
Posthumously awarded. Medal presented to father, Wilfred Williams, sr.

WILLIAMS, WILL J. (1357481)_____
Near Chipilly Ridge, France, Aug. 8, 1918.
R—Urbana, Ill.
B—Urbana, Ill.
G. O. No. 71, W. D., 1919.

Private, Company E, 131st Infantry, 33d Division.
Although seriously wounded, he remained on duty carrying messages across zones swept by heavy fire. He showed marked heroism, his example being an inspiration to those serving with him.

WILLIAMS, WILLIAM C_____
Near Monthurel, France, July 15, 1918.
R—Philadelphia, Pa.
B—Jonesville, Va.
G. O. No. 109, W. D., 1918.

Major, 109th Infantry, 28th Division.
Early in the fighting near Monthurel, France, July 15, 1918, he was wounded 3 times, but in spite of suffering and loss of blood he refused to leave his men until his battalion was relieved 3 days later.

WILLIAMSON, ALFRED (1403753)_____
Near Beaume, France, Nov. 8, 1918.
R—Chicago, Ill.
B—Hot Springs, Ark.
G. O. No. 46, W. D., 1919.

Private, first class, medical detachment, 370th Infantry, 93d Division.
Private Williamson was assigned to duty at the first-aid station, but volunteered to accompany the attacking lines to more expeditiously attend to the wounded. During the advance he constantly exposed himself in view of the enemy and under heavy fire to render first aid.

WILLIAMSON, HARRY A. (731861)........
At Romagne, France, Oct. 14, 1918.
R—Vandevoort, Ark.
B—Kell, Ill.
G. O. No. 37, W. D., 1919.

Sergeant, Company C, 6th Infantry, 5th Division.
While advancing with his platoon he located the position of an enemy machine gun. Without assistance he attacked the gun and successfully accomplished the capture of both gun and crew.

WILLIAMSON, PHILIP H........
Near Thiaucourt, France, Sept. 10–26, 1918.
R—Baltimore, Md.
B—Baltimore, Md.
G. O. No. 140, W. D., 1918.

First lieutenant, 1st Antiaircraft Machine Gun Battalion.
He displayed extreme coolness and courage while conducting the advance of company in the sector near Thiaucourt. He visited daily under heavy shellfire his gun positions and made daily reconnaissances of the lines. When wounded, he refused to be taken to the hospital until he had superintended the removal of his men to a place of safety.

WILLIAMSON, WILLIAM H. (1215898)....
Near Ronssoy, France, Sept. 29, 1918.
R—Auburn, N. Y.
B—Auburn, N. Y.
G. O. No. 44, W. D., 1919.

Sergeant, Company M, 108th Infantry, 27th Division.
Sergeant Williamson, in charge of a combat patrol, successfully accomplished his mission under heavy shell and machine-gun fire after three-fourths of his patrol had been killed or wounded. In the same engagement he successfully reorganized his company after all the officers were killed or wounded and led it in effective combat.

WILLIS, EDWARD........
Near Ronssoy, France, Sept. 29, 1918.
R—Summit, N. J.
B—Middle Granville, N. Y.
G. O. No. 37, W. D., 1919.

First lieutenant, 107th Infantry, 27th Division.
He displayed remarkable gallantry in leading his platoon of machine guns for more than 2,000 yards under terrific machine-gun fire. Even after being mortally wounded and unable to advance farther, he continued to urge his men on.
Posthumously awarded. Medal presented to widow, Mrs. Edward Willis.

WILLIS, EDWARD S. (749550)..........
Near Vieville-en-Haye, in the St. Mihiel salient, France, Sept. 12, 1918.
R—Durango, Colo.
B—Espanola, N. Mex.
G. O. No. 128, W. D., 1918.

Sergeant, Company A, 15th Machine Gun Battalion, 5th Division.
He displayed great courage, determination, and devotion to duty leading his section forward to its objective, inspiring his men by his coolness under fire. While consolidating his objective he was severely wounded by shrapnel, but remained in command of his section until overcome by weakness and sent to the rear. His example was an inspiration to every soldier of his command.

WILLIS, PAUL (1488233)........
Near St. Etienne, France, Oct. 8, 1918.
R—China Springs, Tex.
B—Tunnel Hill, Ga.
G. O. No. 20, W. D., 1919.

Sergeant, Company K, 141st Infantry, 36th Division.
Upon the death of his platoon commander, Sergeant Willis took command of the platoon and led his men in an attack against the enemy. Although wounded several times, he continued to lead his men until killed by machine-gun fire.
Posthumously awarded. Medal presented to father, Daniel Willis.

WILLMOT, WILLIAM H........
Near Barricourt, France, Nov. 1–2, 1918.
R—Ypsilanti, Mich.
B—Milan, Mich.
G. O. No. 32, W. D., 1919.

Gunnery sergeant, 47th Company, 5th Regiment, U. S. Marine Corps, 2d Division.
He established an outpost under heavy enemy fire and by cool leadership and unusual daring effected the capture of a hostile machine gun and 5 prisoners.

WILLOUGHBY, JESSE C. (47994)........
Near Exermont, France, Oct. 4, 1918.
R—Lafollette, Tenn.
B—Claiborne County, Tenn.
G. O. No. 53, W. D., 1920.

Sergeant, Company K, 18th Infantry, 1st Division.
His company commander being incapacitated on the morning of Oct. 4, 1918, Sergeant Willoughby assumed command of the company, reorganized it under heavy fire, and led it forward in the attack. The company participated in attack and capture of Hill 240 on that date. This noncommissioned officer held the command in the position gained until he was relieved on Oct. 11, 1918.

WILLS, JOHN H........
At Berzy-le-Sec, France, July 21, 1918.
P—Auburn, Ala.
B—Auburn, Ala.
G. O. No. 56, W. D., 1919.

Major, 1st Engineers, 1st Division.
Adjutant of his brigade at Berzy-le-Sec, France, July 21, 1918, he repeatedly displayed great bravery, making trips among troops under violent fire, and by his courage and initiative contributed materially to the success of the engagement.
Posthumously awarded. Medal presented to foster mother, Mrs. James T. Anderson.

WILSON, CARLISLE R........
Near Montblainville, France, Sept. 27, 1918.
R—Bethany, Mo.
B—Bethany, Mo.
G. O. No. 21, W. D., 1919.

First lieutenant, 139th Infantry, 35th Division.
In order to establish and maintain liaison with the adjacent division, Lieutenant Wilson, although wounded, led his men along the valley of the Aire River and across a bridge through the heaviest kind of artillery and machine-gun fire. He died soon after this exploit from the wounds received.
Posthumously awarded. Medal presented to father, J. C. Wilson.

WILSON, CASEL (517874)........
Near Soissons, France, July 18, 1918.
R—Chillicothe, Ohio.
B—Harrisville, W. Va.
G. O. No. 145, W. D., 1918.

Private, Company G, 28th Infantry, 1st Division.
In order to stop artillery fire which was causing heavy losses in our ranks, he, with another soldier, rushed 300 yards to the front, attacked a machine-gun strong point and a 77-millimeter artillery gun, captured the position and the gun, killed 2, and captured 13 of the enemy.

WILSON, EARLE W. (102466)..........
Near Sergy, France, July 28, 1918.
R—Red Oak, Iowa.
B—Red Oak, Iowa.
G. O. No. 99, W. D., 1918.

Sergeant, Company M, 168th Infantry, 42d Division.
Showing great personal bravery and contempt of danger at all times during the attack on Hill 212, he maneuvered his platoon so skillfully as to capture a machine-gun position with 4 of its occupants, after which he operated the 2 enemy guns thus taken against the retreating Germans.

WILSON, FRED T........
Near Soissons, France, July 18, 1918.
R—Mamaroneck, N. Y.
B—Mamaroneck, N. Y.
G. O. No. 32, W. D., 1919.

Second lieutenant, 16th Infantry, 1st Division.
Wounded early in the engagement, he refused to be evacuated and remained with his platoon throughout the day's fighting until the objective was reached. Although he was suffering acute pain from his wound, he personally attacked several machine-gun nests and aided other wounded men.

WILSON, GUY M.
From Couiment to Mont St. Martin, France, July 31–Aug. 5, 1918.
R—Detroit, Mich.
B—Genesee County, Mich.
G. O. No. 117, W. D., 1918.

Major, 125th Infantry, 32d Division.
For extraordinary heroism in action while commanding the leading battalion in the successful advance from Couiment to Mont St. Martin, France, from July 31 to Aug. 5, 1918, including the forced crossing of the Ourcq River and several engagements, and especially at Les Jamblet, where he personally led the successful charge of his battalion.

WILSON, HARVEY W.
Near Bussy Farm, France, Sept. 28–29, 1918.
R—Boston, Mass.
B—Boston, Mass.
G. O. No. 13, W. D., 1919.

Second lieutenant, 372d Infantry, 93d Division.
After being hit by a shell splinter he continued to lead his platoon against the enemy position until he was again hit by another shell fragment and had to be carried from the field. His example of devotion to duty and his courage inspired the men of the platoon to continue the attack successfully.

WILSON, HAZEN (261723).
Near Gesnes, France, Oct. 9, 1918.
R—Detroit, Mich.
B—Monroe County, Mich.
G. O. No. 64, W. D., 1919.

Private, Company C, 125th Infantry, 32d Division.
He displayed exceptional courage and bravery while carrying messages from his company in the line to his regimental post of command. While carrying messages he passed through areas swept by a terrific enemy fire. He was repeatedly knocked down by the shell explosions and had his rifle broken and his pack torn from his back by the force of the exploding shells, but successfully completed his mission, delivering messages which were of great value in the success of the operation.

WILSON, JAMES M.
Near Ammertzwiller, Alsace, July 8, 1918.
R—Kalamazoo, Mich.
B—Bloomingdale, Mich.
G. O. No. 101, W. D., 1918.

Second lieutenant, 126th Infantry, 32d Division.
He returned under fire into enemy barbed wire near Ammertzwiller, Alsace, the night of July 8, 1918, to recover 2 of his patrol who were missing after a raid, and, although painfully wounded himself, brought them safely to the American trenches, concealing the fact of his injury until he had succeeded in his undertaking and fainted from exhaustion.

² WILSON, JOSEPHUS B.
Near Cunel, France, Oct. 12, 1918.
R—Athens, Tenn.
B—Denver, Colo.
G. O. No. 32, W. D., 1919.

First lieutenant, 15th Machine Gun Battalion, 5th Division.
He skillfully led a portion of his company through a terrific hostile barrage, establishing them in shell holes where the guns were set up. He then returned through the same barrage and, assembling his reserve platoon, started in the direction of the enemy, but while leading his men to the attack he was mortally wounded and died upon the field.
Posthumously awarded. Medal presented to mother, Mrs. Ellsworth Wilson.

WILSON, MERRITT B.
Near Reddy Farm, France, Aug. 2, 1918.
R—Menominee, Mich.
B—Menominee, Mich.
G. O. No. 64, W. D., 1919.

First lieutenant, 125th Infantry, 32d Division.
With a party of 30 men he led the advance on the Bois Chenet, where a full company of Germans, supported by machine guns, were encountered. Due to his splendid leadership and example this resistance was overcome and the woods were taken. Although suffering great pain from a broken ear drum, caused by the explosion of a shell, Lieutenant Wilson immediately led his party to the flank of the battalion, where numerous attempts of the enemy to retake the woods were repulsed. He refused to leave his company for first aid until darkness had brought an end to the advance.

² WILSON, ROBERT M. (1311934).
Near Brancourt, France Oct. 8, 1918.
R—Great Falls, S. C.
B—Fairfield County, S. C.
G. O. No. 64, W. D., 1919.

Private, Company I, 118th Infantry, 30th Division.
Private Wilson, who was a Lewis gunner, encountered an enemy machine-gun nest containing 4 Germans, who were inflicting heavy casualties on the right platoon of the company. He opened fire with his Lewis gun and then charged the nest, firing as he advanced, and killing all the occupants of the post. On Oct. 17, 1918, Private Wilson was killed while on duty with his company.
Posthumously awarded. Medal presented to father, David Y. Wilson.

WILSON, ROGERS M.
Near Soissons, France, July 18–22, 1918.
R—Savannah, Ga.
B—Savannah, Ga.
G. O. No. 98, W. D., 1919.

Captain, 18th Infantry, 1st Division.
Throughout this period Captain Wilson led his company against the enemy in a masterly manner, displaying exceptional judgment, energy, and conspicuous gallantry until he received a severe wound which permanently disabled his right arm.

WILSON, SHUG (1329370).
Near Bellicourt, France, Sept. 29, 1918.
R—Wilder, Tenn.
B—Wilder, Tenn.
G. O. No. 37, W. D., 1919.

Private, Company D, 105th Engineers, 30th Division.
After his company had taken shelter from a terrific bombardment of shell and machine-gun fire, he volunteered and went to the aid of a wounded comrade who was lying 100 yards out on a shell-swept area. He gave first-aid treatment, after which he carried him back to the dressing station.

WILSON, THOMAS J. (1308186).
Near Premont, France, Oct. 7, 1918.
R—Chattanooga, Tenn.
B—Bridgeport, Ala.
G. O. No. 91, W. D., 1919.

Private, Company K, 117th Infantry, 30th Division.
After his platoon had reached its objective in an exhausted condition and without food or water he voluntarily exposed himself to heavy fire to get rations and canteens from dead soldiers who had fallen in exposed positions and distributed these among the men of his platoon. Later he carried an important message over ground subjected to intense artillery fire, and with 3 other soldiers carried a wounded officer to a dressing station over ground commanded by the enemy positions.

WILT, PERRY W. (3112819).
Near Montfaucon, France, Sept. 29, 1918.
R—Swanton, Md.
B—Swanton, Md.
G. O. No. 46, W. D., 1919.

Private, Company C, 311th Machine Gun Battalion, 79th Division.
While performing his duties as company runner he passed through an intensely shelled area on 6 different occasions and expressed a willingness to make several more trips as the situation required. While making his last journey across the area he was severely wounded.

WINANT, FREDERICK, Jr.............
At Mezy, France, July 15, 1918.
R—New York, N. Y.
B—Monmouth Beach, N. J.
G. O. No. 46, W. D., 1919.

First lieutenant, 30th Infantry, 3d Division.
He was in command of the Stokes mortar platoon of his regiment at the beginning of the German attack of July 15, when all but 2 of his guns were blown out of their pits by enemy fire. Changing the location of his 2 remaining guns, he continued to fire on the Germans as they crossed the Marne, and when he was no longer able to do so he withdrew his men, numbering about 20, and assisted in holding back the enemy approaching from 3 sides. Regardless of personal danger, he remained on duty throughout the action, refusing to accept first aid, though he had been twice wounded.

WINCENCIAK, WILLIAM..............
Near Blanc Mont, France, Oct. 4, 1918.
R—Dunkirk, N. Y.
B—Buffalo, N. Y.
G. O. No. 21, W. D., 1919.

Sergeant, 23d Company, 6th Machine Gun Battalion, U. S. Marine Corps, 2d Division.
When his platoon commander was killed, he took charge of the platoon under heavy shellfire, but was immediately seriously wounded. He then turned over his orders to the next in command, ordered stretcher bearers to carry another man away first, and waited until they had returned.

WINCHENBAUGH, WOLCOTT..........
In the Toulon Sector, France, Apr. 22, 1918.
R—Hyde Park, Mass.
B—Hyde Park, Mass.
G. O. No. 129, W. D., 1918.

Corporal, 18th Company, 5th Regiment, U. S. Marine Corps, 2d Division.
On Apr. 22, 1918, when the patrol of which he was a member was rushed by superior numbers near the enemy's trenches, he displayed exceptional coolness and courage before and after the wounding of his leader, Second Lieut. A. L. Sundval, whom he rescued from the hands of the enemy and half dragged and half carried back to his own lines.

*WINCHESTER, ERNEST E. (3484939)....
Near Le Cheue Tondu, France, Oct. 4–5, 1918.
R—Iola, Ill.
B—Iola, Ill.
G. O. No. 46, W. D., 1919.

Sergeant, Company M, 111th Infantry, 28th Division.
After 3 attempts had failed to bomb out enemy machine-gun nests which were holding up the advance of his company, Sergeant Winchester voluntarily led a fourth patrol. He made his way to a point close to the nests and pressed on, even after all the members of his patrol had been killed or wounded. Though finally killed, his efforts were instrumental in uncovering the nests, which were soon after destroyed.
Posthumously awarded. Medal presented to mother, Mrs. Mary Powell Winchester.

WINES, PEARL J. (2220447)...........
North of Pey-en-Haye, France, Sept. 12, 1918.
R—Bartlesville, Okla.
B—West Plains, Mo.
G. O. No. 81, W. D., 1919.

Sergeant, Company E, 353th Infantry, 89th Division.
Upon encountering a party of 5 Germans, one of whom wounded him in the side, Sergeant Wines, unaided, engaged the entire number, killing 3 and capturing the other 2.

WINESTOCK, JAMES E. (91775)........
Near Landres-et-St. Georges, France, Oct. 14, 1918.
R—New York, N. Y.
B—New York, N. Y.
G. O. No. 35, W. D., 1919.

Private, first class, Company M, 165th Infantry, 42d Division.
He showed an utter disregard of personal danger by repeatedly carrying messages from his company commander to the platoon commanders through an area swept by heavy shell, machine-gun, and rifle fire in full view of enemy snipers who were firing upon him. On 1 trip he found 11 men who were without a leader, and he personally led them in combat against the enemy.

WININGER, LAWRENCE (633222).......
Near Montblainville, France, Oct. 4, 1918.
R—French Lick, Ind.
B—Martin County, Ind.
G. O. No. 13, W. D., 1919.

Sergeant, Battery C, 60th Artillery, Coast Artillery Corps.
He ran with a litter into an area under heavy shellfire in an effort to save a wounded comrade. He succeeded in getting the soldier on the litter, but before he was able to carry him to a place of safety a shell struck almost directly beneath the litter, killing the wounded man and wounding Sergeant Wininger severely.

WINSHIP, BLANTON................
Near Lachaussee, France, Nov. 9, 1918.
R—Macon, Ga.
B—Macon, Ga.
G. O. No. 9, W. D., 1923.
Distinguished-service medal also awarded.

Colonel, 110th Infantry, 28th Division.
While commanding his regiment and observing from his outpost line the progress of a daylight raid on the enemy by a detachment of his officers and men, he discovered the enemy enveloping the right flank of the raiding party. Hastily collecting and organizing a small party from the few available men, he, regardless of his own safety, personally led them forward under heavy rifle, machine-gun, and shell fire, and covered the exposed flank, advancing over a deep tank obstruction and through enemy wire to their second line, destroying several machine guns and killing many of the enemy. His prompt and fearless action enabled the main raiding party to accomplish its mission, and his personal conduct was a great inspiration to his officers and men and contributed largely to the success of the raid.

WINSLOW, ALAN F................
In the Toul sector, France, June 6, 1918.
R—River Forest, Ill.
B—River Forest, Ill.
G. O. No. 121, W. D., 1918.

Second lieutenant, 94th Aero Squadron, Air Service.
While on a patrol, consisting of himself and 2 other pilots, he encountered an enemy biplane at an altitude of 4,000 meters near St. Mihiel, France. He promptly and vigorously attacked, and after a running fight extending far beyond the German lines shot his foe down in flames near Thiacourt.

WINSLOW, ARTHUR J. (68768).......
Near Bois-de-St. Remy, France, Sept. 12, 1918.
R—Harrisville, N. H.
B—Claremont, N. H.
G. O. No. 26, W. D., 1919.

Corporal, Company H, 103d Infantry, 26th Division.
Rushing from shell hole to shell hole, he reached an enemy trench, and, having flanked a machine gun, killed the gunner and took the remaining members of the crew as prisoners.

WINSOR, MERLE R. (567353)........
Near Hauteveunes, France, July 19, 1918.
R—Campello, Mass.
B—West Falmouth, Mass.
G. O. No. 50, W. D., 1919.

Corporal, Company D, 12th Machine Gun Battalion, 4th Division.
Although severely wounded by a flanking machine-gun fire, he remained with his gun crew in an exposed position and under a sweeping artillery and machine-gun fire. He received aid from members of his company and remained on duty with the platoon until the company had withdrawn and he had been ordered to the aid station.

*WINSTEAD, GUY J_____
Near Chateau-Thierry, France, during June and July, 1918.
R—Roxboro, N. C.
B—Roxboro, N. C.
G. O. No. 27, W. D., 1920.

First lieutenant, 38th Infantry, 3d Division.
Lieutenant Winstead led 4 patrols across the Marne River while exposed to heavy enemy machine-gun fire. On the second of these patrols the boat was sunk and it was necessary to swim the river. While within the enemy lines he and 5 others raided a German outpost, killing 5 of the enemy, and, in spite of heavy enemy fire, returned with a prisoner. On July 15, 1918, shortly after leading his platoon under gas and shell fire to a position on a hill, he was killed by enemy fire.
Posthumously awarded. Medal presented to father, Charles M. Winstead.

WINTERS, RAY (1467624)_____
Near Beulny, France, Sept. 30, 1918.
R—Kansas City, Kans.
B—Bethel, Kans.
G. O. No. 59, W. D., 1919.

Sergeant, Company B, 110th Engineers, 25th Division.
In the face of an enemy counterattack and while exposing himself to intense machine-gun and artillery fire, he advanced 300 yards beyond our lines to rescue a wounded soldier. He administered first aid to the wounded man and then carried him back to our lines.

WINTERS, RAYMOND C_____
In the Meuse-Argonne offensive, France, Sept. 26, 1918.
R—Whitestone, N. Y.
B—New York, N. Y.
G. O. No. 98, W. D. 1919.

First lieutenant, 28th Infantry, 1st Division.
When the advance of his company was held up by machine-gun fire from a strong enemy position, Lieutenant Winters led his platoon forward with utter disregard for his personal safety. He succeeded in flanking the enemy position with a squad of riflemen and an automatic rifle team, and by personally directing their fire silenced the enemy machine guns, and thereby assisted greatly in the capture of the position, together with 80 prisoners.

WINTHROP, DUDLEY M. (90034)_____
At sector Auberive, France, July 16, 1918.
R—Brooklyn, N. Y.
B—New York, N. Y.
G. O. No. 71, W. D., 1919.

Sergeant, Company H, 165th Infantry, 42d Division.
After repeated attempts to rescue a wounded soldier in a communication trench held by the enemy had failed, he went out in the face of heavy machine-gun fire, rescued the wounded man, and dressed his wounds.

WINTON, DAVID J. (9604)_____
Near Evermont, France, Oct. 4, 1918.
R—Minneapolis, Minn.
B—Warsaw, Wis.
G. O. No. 59, W. D., 1919.

Sergeant, Company C, 345th Battalion, Tank Corps.
Sergeant Winton ran his tank into the wood to reduce a machine-gun nest, but it was hit and set on fire. He and the driver were wounded as they left the tank, but advanced on the nest and were both wounded the second time. While attempting to reach his companion, who had been hit the third time, Sergeant Winton was again wounded, but reached the driver. They then took cover and remained until darkness, when Sergeant Winton made his way back to our lines, being hit 3 more times while returning.

WINTRODE, JOHN H. (93860)_____
Near the River Ourcq, northeast of Chateau-Thierry, France, July 30, 1918.
R—Winterset, Iowa.
B—Winterset, Iowa.
G. O. No. 116, W. D., 1918.

Sergeant, Company A, 168th Infantry, 42d Division.
He took command of his company when all his officers were killed or wounded and handled it with extreme courage, coolness, and skill under an intense artillery bombardment and machine-gun fire during an exceptionally difficult attack.

WIRTH, THOMAS F_____
Near the Bois-de-Mort Mare, France, Sept. 12, 1918.
R—Louisville, Ky.
B—Hodgenville, Ky.
G. O. No. 44, W. D., 1919.

Major, 355th Infantry, 89th Division.
When his battalion was held up by numerous German machine-gun nests he walked out in front in the face of violent fire and led his battalion in capturing 8 machine guns and 12 prisoners.

WISCHMEIER, OTTO T. (2154170)____
Near Busigny, France, Oct. 9, 1918.
R—West Burlington, Iowa.
B—West Burlington, Iowa.
G. O. No. 37, W. D., 1919.

Private, Company L, 117th Infantry, 30th Division.
Voluntarily accompanying a party sent out to attack machine-gun posts, Private Wischmeier, armed only with a rifle and bayonet, entered an enemy dugout alone and captured a number of Germans.

WISE, CHARLES E_____
Near Bois-de-Foret, France, Oct. 12, 1918.
R—Mankato, Minn.
B—Mankato, Minn.
G. O. No. 59, W. D., 1919.

Captain, 132d Infantry, 33d Division.
While leading his company in an advance from the Bois-du-Fays to the Bois-de-Foret he was severely wounded, but continued to lead his men until he became so weak that he was unable to advance farther. He then directed the advance of his company from the shelter of a shell hole until the command could be turned over to the first sergeant, all of the other officers having become casualties.

WISE, JENNINGS C_____
During the Meuse-Argonne offensive, near Nantillois, France, Oct. 4, 1918.
R—Richmond, Va.
B—Richmond, Va.
G. O. No. 72, W. D., 1920.

Major, 318th Infantry, 80th Division.
Major Wise, while gallantly leading his battalion in the attack was painfully wounded by a shell fragment. He refused to be evacuated but continued to successfully command his battalion in an advance against strong enemy resistance until his battalion was relieved on Oct. 7.

WISEMAN, RUFUS E. (92432)_____
Northeast of Chateau-Thierry, France, July 29–Aug. 2, 1918.
R—Yellowbird, Ohio.
B—Franklin County, Ohio.
G. O. No. 99, W. D., 1918.

Corporal, Company H, 166th Infantry, 42d Division.
He was in charge of a detail for carrying ammunition to a machine-gun section. He had performed his duties and had been given permission to withdraw to the rear, but he remained with his detail on the firing line under a heavy bombardment and machine-gun fire, assisting the machine-gun crew. During these 4 days he was suffering from the effects of gas but refused to be evacuated.

WISHER, ROBERT F. (1782660) ....
Near Ronssoy, France, Sept. 29, 1918.
R—Philadelphia, Pa.
B—Atglen, Pa.
G. O. No. 32, W. D., 1919.

Private, Company A, 301st Battalion, Tank Corps.
After aiding in rescuing the wounded from his tank, which had been struck by a shell, Private Wisher, with Corpl. Albert F. Neil, dismounted machine guns from the tank and operated them against the enemy until these were put out of action. They then secured rifles and hand grenades and organized an attack on the enemy trenches, which they captured and held until depletion of their numbers forced them to fall back. Later they joined Australian troops and fought with them throughout the remainder of the day.

WITHERELL, WILLIAM R. ....
Near Cote-de-Chatillon, France, Oct. 15-16, 1918.
R—North Adams, Mass.
B—North Adams, Mass.
G. O. No. 13, W. D., 1919.

First lieutenant, 168th Infantry, 42d Division.
While in command of an assaulting company, which was without other officers, he displayed unhesitating devotion to duty and courage during the offensive operations at Cote-de-Chatillon. Brilliantly leading his company in an attack over open ground swept by violent machine-gun fire, he captured 63 prisoners and 4 officers and directed the organization of the captured positions in disregard of all danger. On the next day, after severe hand-to-hand fighting, he drove off and completely broke up a pending counterattack, furnishing a splendid example of calmness, decision, and courage at a very critical time.

WITKOSKI, MICHAEL (1008766) ....
Near Vilcey-sur-Trey, France, Nov. 2, 1918.
R—Kinderhook, N. Y.
B—Stuyvesant Falls, N. Y.
G. O. No. 44, W. D., 1919.

Corporal, Company G, 56th Infantry, 7th Division.
After Corporal Witkoski and 2 wounded comrades had become separated from their platoon and were almost entirely surrounded by the enemy, he directed the fire of his men, killing and wounding several of the enemy and holding the position until assistance arrived from his platoon.

WITMER, GEORGE (57875) ....
Near Nonsard, France, Sept. 12, 1918.
R—Manchester, Pa.
B—Manchester, Pa.
G. O. No. 37, W. D., 1919.

Corporal, Company G, 28th Infantry, 1st Division.
Accompanied by another soldier, Corporal Witmer attacked and destroyed an enemy machine-gun nest, using only his rifle and bayonet.

*WITT, GEORGE DOUGLAS ....
Near St. Etienne-a-Arnes, France, Oct. 6, 1918.
R—Harrington, Wash.
B—Harrington, Wash.
G. O. No. 15, W. D., 1919.

Pharmacist's mate, third class, U. S. Navy, attached to 6th Machine Gun Battalion, U. S. Marine Corps, 2d Division.
He displayed remarkable bravery and coolness in giving medical aid to wounded marines while going forward with the assault waves during the attack north of Blanc Mont Ridge, near St. Etienne-a-Arnes, France, on Oct. 6, 1918. Late in the afternoon on the same date while giving first aid to a wounded marine in an advance machine-gun post he was shot and seriously wounded by an enemy sniper.
Posthumously awarded. Medal presented to father, George M. Witt.

WITT, WALTER S. (2176194) ....
In the Bois-de-Bantheville, France, Oct. 22, 1918.
R—Sugar City, Colo.
B—Parkville, Mo.
G. O. No. 37, W. D., 1919.

Sergeant, Company D, 353d Infantry, 89th Division.
Although wounded in the face by machine-gun fire, he refused to go to the first-aid station for treatment, remaining with his platoon throughout the engagement until the objective was reached and the position consolidated.

WITTE, LOUIS (274172) ....
Near Roncheres, France, July 30, 1918.
R—Shepley, Wis.
B—Shawano County, Wis.
G. O. No. 66, W. D., 1919.

Private, first class, Company F, 127th Infantry, 32d Division.
When all the other members of his automatic-rifle squad had been killed or wounded, Private Witte took the dead gunner's automatic rifle and kept it in action against the enemy. Considerable enemy machine-gun fire was thereby drawn upon him, and he was wounded, but he nevertheless continued to maintain an effective fire while the remainder of his company was withdrawing. He remained at his post firing until the entire company had withdrawn.

WITTEN, CLARENCE (52613) ....
Near Soissons, France, July 20, 1918.
R—Harold, Ky.
B—Johnson County, Ky.
G. O. No. 132, W. D., 1918.

Private, first class, Company A, 26th Infantry, 1st Division.
He fearlessly exposed himself to fire near Soissons, France, July 20, 1918, in order to obtain effective positions from which to fire upon enemy machine-gun nests, continuing this valuable work until seriously wounded.

WOERMAN, AUGUST (2146695) ....
Near Sommerance, France, Oct. 15, 1918.
R—Quincy, Ill.
B—Quincy, Ill.
G. O. No. 46, W. D., 1919.

Private, Company A, 321st Machine Gun Battalion, 82d Division.
Although wounded by shrapnel, he remained on duty throughout the entire night and after having his wounds dressed insisted on returning for duty. He was again wounded the following day, once more disregarding his wound and continuing on duty.

WOLL, HERMAN B. (2254426) ....
Near Vilcey-sur-Trey, France, Sept. 12, 1918.
R—Estherville, Iowa.
B—Estherville, Iowa.
G. O. No. 92, W. D., 1919.

Private, 357th Ambulance Company, 315th Sanitary Train, 90th Division.
With another soldier, Private Woll left the shelter of a wood and went forward to rescue a soldier who had fallen wounded on a hill under constant machine-gun and shell fire. While they were carrying him back on a litter he was again wounded and the litter was struck twice by machine-gun bullets, but they succeeded in carrying him back to safety, thereby saving his life.

WOLLERT, EDWARD J. ....
Near Thiaucourt, France, Sept. 15, 1918.
R—Milwaukee, Wis.
B—Milwaukee, Wis.
G. O. No. 37, W. D., 1919

Corporal, 79th Company, 6th Regiment, U. S. Marine Corps, 2d Division.
At the risk of his own life, he went to the aid of a wounded officer who was a prisoner in the hands of 6 Germans. With his pistol he shot 2 of them while the officer killed 2 others. He captured the 2 remaining Germans and forced them to carry the wounded officer back to our lines.

WOMACK, JOHN H._____ | Private, Machine Gun Company, 5th Regiment, U. S. Marine Corps, 2d Division.
Near Beaumont, France, Nov. 7, 1918. | Private Womack was wounded early in the attack, but he refused to be evacuated and continued in the advance for five days, until he was rendered helpless by a second wound.
R—Sacramento, Calif.
B—Wallowa, Oreg.
G. O. No. 93, W. D., 1919.

*WOOD, ALTON P._____ | Second lieutenant, 167th Infantry, 42d Division.
In the vicinity of Ancerviller, France, May 3-4, 1918. | While on patrol in No Man's Land in the vicinity of Ancerviller on the night of May 3-4, 1918, he displayed great courage and devotion to duty in continuing to direct his men after having been mortally wounded and refusing aid until he was assured of the safety of his men.
R—Boston, Mass.
B—Dighton, Mass.
G. O. No. 100, W. D., 1918. | Posthumously awarded. Medal presented to father, Nathan L. Wood.

WOOD, DOLPH_____ | Private, 55th Company, 5th Regiment, U. S. Marine Corps, 2d Division.
Near Vierzy, France, July 19, 1918. | Corporal Montag and Privates McIntyre, Messinger, and Wood captured a machine gun which was holding up the 55th Company of Marines, killing the entire crew. To accomplish this hazardous and daring work it was necessary for them to expose themselves to the fire of this gun. Even though Corporal Montag and Privates McIntyre and Messinger were wounded during the advance, the party continued forward and succeeded.
R—Venice, Ill.
B—Oakland City, Ind.
G. O. No. 126, W. D., 1919.

*WOOD, LAMBERT A._____ | First lieutenant, 9th Infantry, 2d Division.
At Chateau-Thierry, France, June 6-7, 1918. | With entire disregard for personal danger, Lieutenant Wood passed through heavy artillery fire with a message to stop misdirected supporting artillery fire, which fire imperiled the safety of his organization. He was killed near Soissons, France, on July 18, 1918, while leading his machine-gun platoon on a flank movement against an enemy group which was enfilading our advancing Infantry line.
R—Portland, Oreg.
B—Portland, Oreg.
G. O. No. 3, W. D., 1924. | Posthumously awarded. Medal presented to mother, Mrs. Elizabeth L. Wood.

WOOD, MEREDITH_____ | First lieutenant, 308th Infantry, 77th Division.
Near Badonviller, France, June 30, 1918, and near Chery-Chartreuve, France, Aug. 24, 1918. | On the first date, accompanied by only 1 noncommissioned officer, Lieutenant Wood, acting as signal officer, penetrated the enemy's front line and bravely patrolled their territory, following a wire which was thought to lead to a listening post. He cut the wire and returned to our lines with valuable information. On Aug. 24, when a direct hit was made on the building occupied by regimental headquarters, he was severely gassed when he removed his mask to aid a mortally wounded soldier and to search for others who might have been overcome.
R—Brooklyn, N. Y.
B—Brooklyn, N. Y.
G. O. No. 32, W. D., 1919.

WOOD, WILLIAM E. (1311052)_____ | Private, Company E, 118th Infantry, 30th Division.
Near the La Selle River, France, Oct. 17, 1918. | When his squad had become separated from the company in a dense fog Private Wood immediately attached himself and his squad to a company in the attacking wave and continued in the advance. He worked forward with a Lewis gun and so placed it that he delivered so severe a fire upon an enemy machine-gun nest that the crew deserted it. He continued firing until his gun was completely demolished by an enemy shell. Not daunted by this, Private Wood secured a rifle and continued to pour the fire of his whole squad upon the retreating enemy, killing many of them.
R—Greer, S. C.
B—Toccoa, Ga.
G. O. No. 98, W. D., 1919.

WOOD, WILLIAM J. (568737)_____ | Sergeant, Company D, 4th Engineers, 4th Division.
Near Ville-Savoye, France, Aug. 11, 1918. | Although his eyes had been burned by gas, he volunteered for duty and assisted in the construction of an artillery bridge across the Vesle River under constant machine-gun and artillery fire, setting a conspicuous example of personal bravery and devotion to duty.
R—Portland, Oreg.
B—Hesler, Ky.
G. O. No. 147, W. D., 1918.

WOODARD, WILLIAM E. (2854681)___ | Corporal, Company M, 358th Infantry, 90th Division.
Near Vilcey, France, Sept. 12, 1918. | Although wounded in the back by machine-gun fire early in the attack, Corporal Woodard refused to stop even for the application of a first-aid dressing. He continued to command his squad regardless of the pain and with utter disregard of personal danger until the objective was reached.
R—Westville, Ill.
B—Fairmount, Ill.
G. O. No. 3, W. D., 1921.

WOODARD, WILLIE L. (1307341)____ | Private, Company C, 117th Infantry, 30th Division.
Near Geneve and Ponchaux, France, Oct. 8, 1918. | Early on the morning of Oct. 8, Private Woodard was painfully wounded by severe machine-gun fire, which also riddled his gas mask. Unmindful of his suffering, and despite the fact that he had no protection from gas, he continued to press on with his company until he fell exhausted and was evacuated.
R—Arp, Tenn.
B—Arp, Tenn.
G. O. No. 50, W. D., 1919.

WOODS, HARRY MELVIN (244914)___ | Sergeant, Company D, 1st Gas Regiment.
Near Montfaucon, France, Sept. 29, 1918. | While his position was under heavy and continuous bombardment of both gas and high-explosive shells he voluntarily left his dugout and put gas masks on 900 soldiers, giving his own mask to one of them, and thus saving their lives. After being severely gassed by the explosion of a shell, one piece of which struck him, he continued to administer aid to the other wounded and quit only when his eyes were swelling shut and he was completely exhausted.
R—Elkhart, Ind.
B—Durango, Colo.
G. O. No. 145, W. D., 1918.

WOODS, HOWARD S. (1490224)_____ | Sergeant, Company F, 142d Infantry, 36th Division.
Near St. Etienne, France, Oct. 8, 1918. | After all the officers of his company had become casualties, Sergeant Woods, though himself wounded severely, remained at his post and reorganized his company under heavy machine-gun and artillery fire, thereby making possible its further advance.
R—Wewoka, Okla.
B—Tyner, Kans.
G. O. No. 50, W. D., 1919.

WOODSIDE, ROBERT G ....
At Les Franquette Farm, near Jaulgonne, France, July 22, 1918.
R—Pittsburgh, Pa.
B—Brooklyn, N. Y.
G. O. No. 27, W. D., 1920.

Captain, 38th Infantry, 3d Division.
Captain Woodside rallied the men of one platoon of his company who were falling back in disorder, reformed them under heavy enemy shell and machine-gun fire, and led them to the left front of the battalion sector and engaged the attacking enemy. His prompt action stopped an enveloping movement of the enemy which imperiled the position of the battalion.

WOODSMALL, WILLIAM M. (2869212) ....
Near Consenvoye, France, Oct. 10, 1918.
R—Little Rock, Ark.
B—Little Rock, Ark.
G. O. No. 71, W. D., 1919.

Private, Company A, 131st Infantry, 33d Division.
He left our lines on his own initiative and advancing along against a German machine-gun nest killed the crew and brought back their machine gun. He showed marked coolness and bravery, with utter disregard for the heavy fire to which he was subjected.

WOODVILLE, JOSEPH P. (1296986) ......
In the Bois-de-Consenvoye, France, Oct. 15, 1918.
R—Norfolk, Va.
B—Fincastle, Va.
G. O. No. 37, W. D., 1919.

Private, Company B, 112th Machine Gun Battalion, 29th Division.
He remained at his gun until the position was destroyed, the other gun sentry killed, and he himself wounded. He refused first aid until the bombardment had lifted and kept up the morale of the other members of his section by his heroism and cheerfulness.

*WOODWARD, DUDLEY W ....
Near Soissons, France, July 18, 1918.
R—New Amsterdam, Ind.
B—New Amsterdam, Ind.
G. O. No. 44, W. D., 1919.

Captain, 9th Infantry, 2d Division.
Advancing with his company in the face of withering machine-gun fire, he attacked a nest of 10 machine guns and a battery of field guns which were holding up the attack. In spite of the additional hazard of heavy enfilading fire from enemy artillery on the left, he succeeded with his mission. On another occasion he broke up an enemy counterattack formation by placing an automatic-rifle team in an exposed position and disorganized the enemy with a harassing fire.

WOODWARD, FREDERICK A. (152480) ...
Near Suippes, Marne, France, July 14-15, 1918.
R—Elizabeth, N. J.
B—Elizabeth, N. J.
G. O. No. 37, W. D., 1919.

Private, Battery E, 42d Artillery, Coast Artillery Corps.
Acting as a runner during an engagement, he kept up his work throughout the bombardment. On the trip he was wounded by a bursting shell, but succeeded in carrying a very important message to his battery commander.

*WOODWARD, RICHARD FULLER ......
Near Cunel, France, Oct. 9, 1918.
R—Richmond, Long Island, N. Y.
B—Suffolk, Va.
G. O. No. 32, W. D., 1919.

First lieutenant, 319th Infantry, 80th Division.
Disregarding his intense suffering from wounds, he continued to lead and encourage his men until killed by another rain of machine-gun bullets. Posthumously awarded. Medal presented to widow, Mrs. Jeanne B. Woodward.

WOOLDRIDGE, JESSE WALTON ....
East of Chateau-Thierry, France, July 15, 1918.
R—San Francisco, Calif.
B—Hopkinsville, Ky.
G. O. No. 99, W. D., 1918.
Distinguished-service medal also awarded.

Captain, 38th Infantry, 3d Division.
With rare courage and conspicuous gallantry he led a counterattack against an enemy of 5 times his own numbers on July 15, 1918, east of Chateau-Thierry, France; 189 men entered this counterattack and 51 emerged untouched. More than 1,000 of the enemy were killed, wounded, or taken prisoners.

WOOLFE, IRVING (1657465) ....
Near Revillon, France, Sept. 10, 1918.
R—Hartford, Conn.
B—Hartford, Conn.
G. O. No. 35, W. D., 1919.

Private, Company I, 308th Infantry, 77th Division.
Volunteering to serve on a patrol for the purpose of capturing prisoners, Private Woolfe crawled forward to a sentry post 25 yards from the enemy lines. Overpowering 2 sentries, he started back under a heavy barrage of rifle grenades and rifle fire, and ignoring his great danger he successfully delivered his prisoners to the battalion commander.

WOOLSHLAGER, JOHN F ....
Northwest of Grand Pre, France, Oct. 18, 1918.
R—Castorland, N. Y.
B—Beaver Falls, N. Y.
G. O. No. 16, W. D., 1920.

First lieutenant, 312th Infantry, 78th Division.
In the attack of morning of Oct. 18 Lieutenant Woolshlager was severely wounded, both legs being broken. He nevertheless retained command of his platoon and that of an adjoining platoon. Throughout the day, exposed to heavy machine-gun and artillery fire, he encouraged and directed his men. Due to his efforts the position, gained at great cost, was held against enemy attacks.

*WOOMER, ELMER E. (1246074) ....
Near Le Chene Tondu, France, Oct. 2, 1918.
R—Meyerstown, Pa.
B—Meyerstown, Pa.
G. O. No. 46, W. D., 1919.

Sergeant, Company M, 111th Infantry, 28th Division.
Leading a patrol to locate hidden machine-gun nests, Sergeant Woomer placed his men in advantageous positions and advanced alone to draw fire from the enemy strongholds. In the execution of his mission he was killed, but his heroic action saved the lives of many in the advance that followed. Posthumously awarded. Medal presented to sister, Mrs. William Line.

WORD, WILLIAM E ....
Near Pexonne, France, Mar. 5, 1918.
R—Richmond, Va.
B—Richmond, Va.
G. O. No. 126, W. D., 1918.

First lieutenant, 151st Field Artillery, 42d Division.
During the action near Pexonne, France, on Mar. 5, 1918, he displayed unusual presence of mind and initiative by the effective manner in which he assisted in directing the operations of Battery C, 151st Field Artillery, when it was under particularly heavy bombardment.

WORDEN, ROBERT L. (934200) ....
At Ville-Savoye, France, Aug. 7, 1918.
R—Wichita, Kans.
B—Wellington, Kans.
G. O. No. 99, W. D., 1918.

Wagoner, 21st Ambulance Company, 4th Sanitary Train, 4th Division.
While driving an ambulance through the town he heard cries for help. Voluntarily and under heavy shell and machine-gun fire, he climbed a tower, in which he found 2 officers and a corporal severely wounded. He rendered first aid and assisted in carrying the wounded men to a place of safety.

WORNER, ERNEST (3137861)............
Near Moulin-de-Charlevaux, France, Oct. 3, 1918.
R—Mackay, Idaho.
B—Pocatello, Idaho.
G. O. No. 71, W. D., 1919.

Private, first class, Company G, 308th Infantry, 77th Division.
Facing heavy machine-gun and rifle fire, he went out alone and rescued a soldier who had been wounded in advance of our lines while on a patrol.

WOPPELL, JOHN M............
At Bouresches, France, June 6, 1918.
R—Colorado, Tex.
B—Sweet Water, Ala.
G. O. No. 110, W. D., 1918.

Private, 97th Company, 6th Regiment, U. S. Marine Corps, 2d Division.
At Bouresches, France, on June 6, 1918, he voluntarily obtained permission to leave shelter and fearlessly went into heavy fire in order to rescue wounded from a field then under artillery and machine-gun bombardment, continuing this heroic work until he was himself wounded.

*WORSHAM, ELIJAH W............
Near Gesnes, France, Sept. 26, 1918.
R—Seattle, Wash.
B—Evansville, Ind.
G. O. No. 8, W. D., 1926.

Captain, 362d Infantry, 91st Division.
In command of the Machine Gun Company, Captain Worsham personally led his men forward, reconnoitering and establishing a line for machine-gun emplacements under terrific artillery and machine-gun fire. His fearless and aggressive leadership was of the utmost assistance in the capture of Gesnes. In exposing himself to hostile fire in order to observe the fire effect of his guns he was killed.
Posthumously awarded. Medal presented to sister, Miss Mary L. Worsham.

WORTHEN, GEORGE T............
Near Varennes, France, Sept. 26, 1918.
R—St. Joseph, Mo.
B—Ottumwa, Iowa.
G. O. No. 87, W. D., 1919.

First lieutenant, 130th Infantry, 35th Division.
Arming himself with a rifle, Lieutenant Worthen personally led an attack on a hostile machine-gun nest which was holding up the advance, capturing the position and killing or capturing the entire enemy unit, including 2 officers.

WORTHEN, WILLIAM A. (251624)............
Near Tuileria Farm, France, Oct. 3, 1918, and at Mouzon, France, Nov. 7, 1918.
R—Sunset, Tex.
B—Milan, Ala.
G. O. No. 46, W. D., 1919.

Mechanic, Company M, 9th Infantry, 2d Division.
On Oct. 3 Mechanic Worthen was assigned to duty as a runner after he had requested to go into action with his company instead of remaining in the rear. He showed exceptional bravery while carrying messages through heavy machine-gun and artillery fire, and on Nov. 7 he maintained communication with battalion headquarters when the enemy was endeavoring to cut off his company by machine-gun and shellfire.

*WORTHINGTON, HENRY H............
At Maisey, France, Apr. 12–13, 1918.
R—Trenton, N. J.
B—Philadelphia, Pa.
G. O. No. 3, W. D., 1921.

Captain, 9th Infantry, 2d Division.
When the enemy launched a powerful raid after a terrific bombardment for 5 hours, Captain Worthington gathered his men into effective combat groups and, although greatly outnumbered, drove the enemy from the trenches inflicting heavy casualties on them. Although severely wounded by a grenade early in the action, he continued to direct his company throughout the night and until the enemy was driven out and his evacuation became necessary through weakness.
Posthumously awarded. Medal presented to mother, Mrs. T. K. Worthington.

WORTHINGTON, RICHARD (1587579)......
Near Soissons, France, July 18, 1918.
R—Swifton, Ark.
B—Mammoth Spring, Ark.
G. O. No. 72, W. D., 1920.

Corporal, Company G, 16th Infantry, 1st Division.
Knowing that the enemy had captured a wounded member of his company, Corporal Worthington, with 2 others, advanced across dangerous ground to a barn, where they routed the enemy captors and carried back their comrade to safety.

WORTHY, ELMER T. (2201678)............
Near Gesnes, France, Sept. 29, 1918.
R—Santa Ana, Calif.
B—Franklin County, Ark.
G. O. No. 1, W. D., 1919.

First sergeant, Company M, 362d Infantry, 91st Division.
During the attack on Gesnes he took charge of 15 soldiers who had become separated from their organizations and organized them into a combat group. Continuing forward in the face of shell and machine-gun fire, he led his party in an attack on 3 machine guns that were holding up the advance of the American troops and effectively silenced them. The fearless leadership displayed in this act furnished an inspiration to all who witnessed it.

*WOZNIAK, ANTHONY (107548)............
Near Somme Py, France, Oct. 3, 1918.
R—Cleveland, Ohio.
B—Cleveland, Ohio.
G. O. No. 32, W. D., 1919.

Private, Company B, 5th Machine Gun Battalion, 2d Division.
While going over the top with his company, he was wounded in the foot. Immediately afterwards he rejoined his squad and remained in action for 60 hours until again severely wounded.
Posthumously awarded. Medal presented to mother, Mrs. Magdalena Wozniak.

WREN, EDWARD R............
Near Haumont, France, Sept. 22, 1918.
R—Talladega, Ala.
B—Talladega, Ala.
G. O. No. 46, W. D., 1919.

Second lieutenant, 167th Infantry, 42d Division.
After clearing the village of Haumont he learned that a soldier of his command was lying either killed or wounded in the town. Disregarding the grave danger of perilous machine-gun fire, he returned into the town, and taking the dead body of the soldier carried it several hundred yards in an endeavor to get back to our lines.

WRIGHT, CHESTER F............
Near Bellu, France, Oct. 10, 1918.
R—Brookline, Mass.
B—Readville, Mass.
G. O. No. 15, W. D., 1919.

First lieutenant, pilot, 93d Aero Squadron, Air Service.
He attacked an enemy observation balloon protected by 4 enemy planes, and despite numerical superiority he forced the planes to withdraw and destroyed the enemy balloon.
Oak-leaf cluster.
A bronze oak leaf, for extraordinary heroism in action near Bantheville, France, Oct. 23, 1918. Accompanied by 1 other machine, he attacked and sent down in flames an enemy plane (Fokker type) that was attacking an allied plane. He was in turn attacked by 3 enemy planes. His companion was forced to withdraw on account of motor trouble. He continued the combat and succeeded in bringing down 1 of the enemy planes and forced the remaining 2 into their own territory.

WRIGHT, CLARENCE L. (1379056)_____
At Romagne, France, Nov. 1, 1918.
R—Peoria, Ill.
B—Peoria, Ill.
G. O. No. 37, W. D., 1919.

Sergeant, Battery C, 124th Field Artillery, 33d Division.
After 3 members of his gun crew had been wounded during heavy enemy shellfire he alone continued to keep his piece in action for 15 minutes until assistance reached him. He was wounded in action shortly afterwards.

WRIGHT, CLARENCE S. (1309482)_____
Near Ponchaux, France, Oct. 7, 1918.
R—Knoxville, Tenn.
B—Knoxville, Tenn.
G. O. No. 37, W. D., 1919.

Corporal, Company L, 117th Infantry, 30th Division.
Accompanying an officer, he aided him in putting 2 enemy machine guns out of action. He then carried the information which they secured through a heavy machine-gun barrage to battalion headquarters. Later, when his platoon commander had been wounded, Corporal Wright took command and led the platoon until he was himself wounded.

WRIGHT, DEWEY EDWARD (43957)_____
Near Soissons, France, July 19, 1918.
R—Ewing, Nebr.
B—Ewing, Nebr.
G. O. No. 35, W. D., 1920.

Private, Company K, 16th Infantry, 1st Division.
In the advance Private Wright and 2 comrades were cut off from the rest of the company by the enemy. He resisted stoutly until he fell with a rifle-shot wound through both legs and was taken prisoner. 2 days later, during the advance of our troops, he assisted in the capture of all the enemy in a dugout by calling to the troops and disclosing his position.

WRIGHT, EARL (1457525)_____
Near Charpentry, France, Sept. 27-28, 1918.
R—Bucklin, Mo.
B—Bucklin, Mo.
G. O. No. 95, W. D., 1919.

Corporal, Company I, 139th Infantry, 35th Division.
After being wounded in the knee, Corporal Wright refused to be evacuated but remained at his post for nearly 24 hours, until his wounds became so serious that he was ordered to the dressing station. His example of bravery and fortitude was an inspiration to his comrades.

WRIGHT, EARL V. (2185246)_____
At Pouilly, France, Nov. 10-11, 1918.
R—Cross Timbers, Mo.
B—Cross Timbers, Mo.
G. O. No. 44, W. D., 1919.

Private, Company K, 356th Infantry, 89th Division.
Accompanying Second Lieut. Charles R. Hanger, Private Wright made 3 trips through heavy shellfire to locate 3 companies which had become lost in the dense fog during the crossing of the Meuse River and guided them to the river crossing.

*WRIGHT, ERNEST N. (254365)_____
Near Nonsard, France, Sept. 12, 1918, and near Varennes, France, Sept. 27, 1918.
R—Amador City, Calif.
B—Germantown, Pa.
G. O. No. 44, W. D., 1919.

Corporal, Company C, 344th Battalion, Tank Corps.
Entering the town of Nonsard during an extremely heavy barrage, Corporal Wright put out of action an enemy machine gun which was impeding the progress of our troops into the town. On Sept. 27, accompanied by another tank, Corporal Wright advanced far ahead of the infantry and put to flight several enemy machine gunners. A counterattack by the enemy damaged his tank slightly, which forced him to withdraw to a supposed place of safety for repairs. Discovering his tank, the enemy shelled it, when both he and his companion were wounded and taken for treatment. He returned to his tank voluntarily and rendered most valiant service until killed Oct. 3.
Posthumously awarded. Medal presented to mother, Mrs. Kathleen R. Wright.

WRIGHT, FABIAN W. (544384)_____
Near Crezancy, France, July 15, 1918.
R—Pittsburgh, Pa.
B—Pittsburgh, Pa.
G. O. No. 32, W. D., 1919.

Sergeant, Headquarters Company, 30th Infantry, 3d Division.
Under constant high-explosive and shrapnel fire, he remained exposed for 19 hours, observing the movements of the enemy, obtaining information of the utmost value.

WRIGHT, GEORGE L_____
Near Malancourt, France, Sept. 28, 1918.
R—Norristown, Pa.
B—Norristown, Pa.
G. O. No. 35, W. D., 1919.

Captain, 315th Infantry, 79th Division.
On Sept. 28, in order to save his men, he crossed a clearing under heavy machine gun fire to secure information of the troops on his flank. On Nov. 6 he set a splendid example to his men under heavy shellfire by going from shell hole to shell hole, encouraging them and directing them to safe places.

WRIGHT, JOE D. (1309613)_____
Near Ponchaux, France, Oct. 7, 1918.
R—Lenoir City, Tenn.
B—Jefferson City, Mo.
G. O. No. 37, W. D., 1919.

Corporal, Company L, 117th Infantry, 30th Division.
He went through heavy artillery and machine-gun fire, taking forward an automatic rifle to the front line to replace one which had become unserviceable, after seeing several other soldiers killed or wounded in a similar attempt.

WRIGHT, JOHN W. (2339546)_____
Near Nesles, France, July 15, 1918.
R—Canebrake, W. Va.
B—Roanoke County, Va.
G. O. No. 32, W. D., 1919.

Sergeant, Company I, 4th Infantry, 3d Division.
After he had been severely wounded Sergeant Wright remained on duty through a heavy shell and gas bombardment, rendering valuable assistance to his platoon commander.

WROBBEL, JOHN (92481)_____
In the Champagne sector, France, July 15-18, 1918.
R—Columbus, Ohio.
B—Germany.
G. O. No. 26, W. D., 1919.

Cook, Supply Company, 166th Infantry, 42d Division.
During the heaviest bombardment he regularly supplied hot meals to his men. On July 16 his kitchen was almost demolished by shell bursts and a large number of rations destroyed, but he remained at his post after all his assistants had sought places of safety.

WYATT, EDWARD H_____
At Wadonville, France, Sept. 25, 1918.
R—Rudyard, Mich.
B—Hull, Iowa.
G. O. No. 138, W. D., 1918.

Second lieutenant, 102d Machine Gun Battalion, 26th Division.
Wounded while conducting his platoon into position to lay a barrage for a raid, he showed complete disregard for his own safety by remaining on duty for more than an hour under heavy machine-gun and shell fire, directing the location and adjustment of his guns. After his wounds were dressed he returned to his platoon and remained with it until it was relieved.

WYATT, LINDON (107135)..............
Near Medeah Ferme, France, Oct.
8, 1918.
R—Elkatawa, Ky.
B—Elkatawa, Ky.
G. O. No. 145, W. D., 1918.

Corporal, Company B, 4th Machine Gun Battalion, 2d Division.
He remained on duty after being wounded while leading his squad into action. Under heavy artillery and machine-gun fire he directed the advance of the gun upon an enemy pill box in the open at close range, displaying notable coolness and bravery until he was again severely wounded by shellfire.
Posthumously awarded. Medal presented to mother, Mrs. Elizabeth Wyatt.

WYGAL, LAWRENCE A. (455877)........
Near Rembercourt, France, Nov. 1, 1918.
R—Honaker, Va.
B—Northfork, W. Va.
G. O. No. 37, W. D., 1919.

Private, medical detachment, 64th Infantry, 7th Division.
He went forward to the aid of a wounded comrade who was lying about 125 feet in front of the enemy's firing line. Under direct fire of rifles, machine guns, and artillery he applied first aid and took the man back into our trenches to a place of safety.

WYGAST, GREGORY (794694)..........
Near Vieville-en-Haye, France, Sept. 13, 1918.
R—Toledo, Ohio.
B—Poland.
G. O. No. 35, W. D., 1919.

Private, Company C, 15th Machine Gun Battalion, 5th Division.
Although severely wounded, he made 5 trips through an unusually heavy barrage of machine-gun and shell fire, bringing up ammunition for his squad, relieving their perilous position. Finding his squad leader killed when he returned the fifth time, he took charge of the squad and directed their fire with good effect until completely overcome.

WYKE, GODFREY N..................
At Fismes and Fismette, France, Aug. 10–12, 1918.
R—Coraopolis, Pa.
B—Sackets Harbor, N. Y.
G. O. No. 99, W. D., 1918.

First lieutenant, 111th Infantry, 28th Division.
For 3 days he voluntarily acted as runner after 3 of the 5 runners of his company had been killed and 2 had been wounded. He made numerous trips by day and night through exposed areas under fire, and thus successfully maintained liaison.

WYLDER, CECIL O. (76524)..........
At Marcheville, France, Sept. 26, 1918.
R—Spokane, Wash.
B—Spirit Lake, Iowa.
G. O. No. 15, W. D., 1919.

Private, first class, Headquarters Company, 102d Infantry, 26th Division.
He volunteered to go through a violent bombardment to repair telephone lines and thereby succeeded in establishing communication with regimental headquarters in time to call for a barrage at a critical junction.

WYLY, LAWRENCE T................
Various dates and places in France, 1918.
R—Duluth, Minn.
B—Cardington, Ohio.
G. O. No. 14, W. D., 1923.

First lieutenant, 148th Aero Squadron, Air Service.
For extraordinary heroism in action near Chaulnes, France, August 15, 1918. Acting as flight leader of 5 airplanes he observed 15 or 20 enemy planes attacking a small number of allied planes. Lieutenant Wyly rushed to the assistance of the allied airmen and repeatedly attacked superior numbers of enemy planes. His machine was riddled by enemy fire and his gas tank perforated. Despite this fact, he continued to fight until his plane was shot down close behind our lines, but before landing had succeeded in scattering and driving off the enemy planes. On Sept. 17, northwest of Cambrai, he boldly attacked 5 enemy planes, shooting down 1 of the enemy. On Oct. 21 he volunteered to attack the enemy airdrome near Famors, outside of Valenciennes, many miles behind the enemy lines, despite the fog and mist on that day. In company with another pilot of the squadron he reached his destination to find the airdrome had the night before been evacuated. They continued on to Valenciennes, discovered an enemy transport column 2 miles in length. In spite of enemy machine-gun fire and almost impossible flying conditions, the column was attacked and great damage inflicted upon it. His machine badly crippled, he returned to his airdrome with extremely valuable information.

WYNN, THOMAS (1706264)..........
In the Argonne Forest, France, Oct. 3 and 6, 1918.
R—Brooklyn, N. Y.
B—Ireland.
G. O. No. 50, W. D., 1919.

Sergeant, Company H, 307th Infantry, 77th Division.
He advanced alone to within 20 yards of the enemy lines under heavy machine-gun fire after ordering the members of his platoon to take cover, and cut openings in the enemy's barbed wire. He then led his platoon in an attack on the hostile trenches in conjunction with another company and captured 15 prisoners. 3 days later this soldier again displayed exceptional courage when attempts were being made to relieve a battalion of his regiment which had been cut off by the enemy in leading the first wave of his platoon in the attack, securing a foothold on the top of a hill and holding it all night. Next morning he renewed the attack, despite the fact that he had been wounded.

YADOVITZ, BENJAMIN (600890)........
Near St. Remy, France, Sept. 12, 1918.
R—Chelsea, Mass.
B—Russia.
G. O. No. 24, W. D., 1919.

Private, Company D, 101st Infantry, 26th Division.
While carrying a message through an advance trench he was attacked by 14 of the enemy. After receiving 2 bayonet wounds, this soldier succeeded in killing 3 of the enemy and capturing the other 11, whom he brought to the rear.

YAEGER, LOUIS (2253657)..........
Near Pont-a-Mousson, France, Sept. 12, 1918.
R—San Diego, Tex.
B—Cotulla, Tex.
G. O. No. 24, W. D., 1919.

Private, Company D, 321st Machine Gun Battalion, 82d Division.
Private Yaeger, with his brother, Corpl. Roy Yaeger, remained at an advanced position in the face of heavy machine-gun and rifle fire from the enemy and by effective use of their machine gun and pistols covered the withdrawal of the Infantry, inflicting serious losses on the enemy, and refusing to retire until they were ordered to do so.

YAEGER, ROY (2253610)..........
Near Pont-a-Mousson, France, Sept. 12, 1918.
R—Hebbronville, Tex.
B—Cotulla, Tex.
G. O. No. 24, W. D., 1919.

Corporal, Company D, 321st Machine Gun Battalion, 82d Division.
Corporal Yaeger, with his brother, Pvt. Louis Yaeger, remained at an advanced position in the face of heavy machine-gun and rifle fire from the enemy, and by effective use of their machine gun and pistols covered the withdrawal of the Infantry, inflicting serious losses on the enemy and refusing to retire until ordered to do so.

YAGLE, HARRY.
    At Hamel, France, July 4, 1918.
    R—Woodstock, Ill.
    B—Dundee, Ill.
    G. O. No. 44, W. D., 1919.

Second lieutenant, 131st Infantry, 33d Division.
While digging in at his final objective, he came under fire from a hostile machine gun in a sunken road 200 yards to the right front. With Sergeant Koljane and 2 Australian soldiers, he rushed the position and captured the gun and 8 prisoners.

*YAMIN, AARON (1701168).
    In the Forest of Argonne, France, Sept. 27, 1918.
    R—New York, N. Y.
    B—Russia.
    G. O. No. 20, W. D., 1919.

Corporal, Company B, 306th Infantry, 77th Division.
He displayed exceptional bravery by volunteering to cut a strip of enemy barbed wire to make an opening for his company, which was at that time under heavy fire from artillery and machine guns. In performing this mission this soldier received wounds from which he afterwards died.
Posthumously awarded. Medal presented to father, Louis Yamin.

YANCHULIS, MARTIN (547640).
    Near Crezancy, France, July 15, 1918.
    R—Niagara Falls, N. Y.
    B—Lithuania.
    G. O. No. 44, W. D., 1919.

Private, Company I, 30th Infantry, 3d Division.
Although severely wounded, he made his way through the terrific enemy barrage to his post in the trenches and remained on duty until ordered to the rear by his commanding officer.

*YANNANTUONO, FREDERICK (2362187).
    Near Cunel, France, Oct. 14, 1918.
    R—Brooklyn, N. Y.
    B—Brooklyn, N. Y.
    G. O. No. 95, W. D., 1919.

Private, first class, medical detachment, 13th Machine Gun Battalion, 5th Division.
Private Yannantuono voluntarily went forward and administered first aid to wounded Infantry soldiers under heavy shell and machine-gun fire in plain view of the enemy, being killed in the performance of this self-sacrificing mission.
Posthumously awarded. Medal presented to brother, Paul Yannantuono.

YANTIS, ERNEST M.
    Near Tronsol Farm, France, Sept. 30, 1918.
    R—Granbury, Tex.
    B—Brown County, Tex.
    G. O. No. 21, W. D., 1919.

First lieutenant, 363d Infantry, 91st Division.
Leading his platoon as a combat patrol 500 yards ahead of the front line under intense shell and machine-gun fire, he was wounded 3 times, but remained on duty for more than an hour until relieved. He then refused to leave until the new officer had been fully informed as to his disposition and that of the enemy.

*YARBOROUGH, GEORGE HAMPTON, Jr.
    In the Bois-de-Belleau, France, June 23, 1918.
    R—Mullins, S. C.
    B—Roxboro, N. C.
    G. O. No. 44, W. D., 1919.

First lieutenant, 5th Regiment, U. S. Marine Corps, 2d Division.
He displayed exceptional bravery when his platoon was in a support position under intense artillery fire by moving from 1 shell hole to another in the open and steadying his men. After making 1 trip over his line he was wounded by an exploding shell, but refused aid until he saw that the wounded soldiers with him had been treated and taken to shelter. He later died of his wounds.
Posthumously awarded. Medal presented to father, G. H. Yarborough, sr.

YARNIS, HYMAN (42342).
    On Hill 272, near Fleville, France, Oct. 9, 1918.
    R—New York, N. Y.
    B—Russia.
    G. O. No. 39, W. D., 1920.

Corporal, Company C, 18th Infantry, 1st Division.
Corporal Yarnis volunteered for and attacked a machine-gun position which was enfilading the lines by its fire from the left flank. He killed the enemy crew, in which encounter he was wounded by a grenade. Notwithstanding his wound, he advanced against a second machine-gun position, during which he was wounded a second time. However, he continued on and attacked the enemy crew.

YATES, FRANK ROY.
    Near St. Etienne, France, Oct. 4, 1918.
    R—San Francisco, Calif.
    B—Colusa, Calif.
    G. O. No. 15, W. D., 1919.

Pharmacist's mate, third class, U. S. Navy, attached to 6th Machine Gun Battalion, U. S. Marine Corps, 2d Division.
He attended the wounded during a heavy artillery and gas bombardment, remaining at his post even after his gas mask had been torn from his face by a shell fragment. Late in the day in a violent barrage of machine-gun fire he showed entire disregard for his own safety in ministering to wounded soldiers and in organizing 2 crews of litter bearers to carry them from the road to the dressing station.

*YEAGER, CURTIS L. (2003854).
    Near Verdun, France, Oct. 12, 1918.
    R—Atherton, Ind.
    B—Terre Haute, Ind.
    G. O. No. 27, W. D., 1919.

Private, Company L, 116th Infantry, 29th Division.
Private Yeager, in an exposed position under a heavy machine-gun and artillery barrage, kept up an effective fire from his automatic rifle until severely wounded. After waiting 7 hours with a comrade for a stretcher and only 1 arrived, although practically unconscious he insisted on his comrade leaving first.
Posthumously awarded. Medal presented to father, James Yeager.

YOCHIM, FRED J. (550251).
    Near Mezy, France, July 15, 1918.
    R—Erie, Pa.
    B—Erie, Pa.
    G. O. No. 24, W. D., 1920.

Private, Company C, 38th Infantry, 3d Division.
When other means of liaison were destroyed by heavy enemy fire, Private Yochim carried numerous messages over routes exposed to heavy artillery fire from units in line to battalion headquarters. In all subsequent engagements of this regiment he exposed himself fearlessly in order to maintain communication until severely wounded near Cunel, Oct. 21, 1918, while carrying a message over area swept by machine-gun fire.

YOCKEY, WILLIAM S. (3487215).
    Near Sedan, France, Nov. 6–7, 1918.
    R—Akron, Ohio.
    B—Woodsfield, Ohio.
    G. O. No. 74, W. D., 1919.

Private, Company E, 16th Infantry, 1st Division.
Private Yockey voluntarily led 2 other men in an attack on a machine-gun nest which was delivering a withering fire on the company and delaying its advance. By the skill and bravery of this attack the enemy gunner was killed and the advancing company saved from a dangerous flanking fire.

YOPP, SAMUEL F., Jr. (1317609).
    Near Hargicourt, France, Sept. 28, 1918.
    R—Wilmington, N. C.
    B—Wilmington, N. C.
    G. O. No. 98, W. D., 1919.

Sergeant, medical detachment, 119th Infantry, 30th Division.
While directing the evacuation of the wounded, he was severely gassed, but refused to be evacuated and continued in charge of the dressing station to which he had been assigned. He displayed marked fortitude and personal bravery, working constantly to help the wounded.

YOUELL, RICE MCNUTT_____
   Near Verdun, France, Oct. 1–12, 1918.
   R—Norton, Va.
   B—Rockbridge Baths, Va.
   G. O. No. 64, W. D., 1919.

Major, 26th Infantry, 1st Division.
Taking command of his battalion after the battalion commander had been mortally wounded, he led it with remarkable bravery throughout 9 days of the hardest fighting, though he was himself painfully wounded on the first day, when he led his command in storming the heights beyond the Eau de Gauffre. On Oct. 10, when the enemy's resistance had been broken and a rapid thrust into the disorganized defenses was necessary in order to enable a unit on the right to advance, Major Youell, with 1 company and no artillery support, pushed forward 2 kilometers under heavy fire, driving back a force of enemy Infantry superior in number to his own and capturing important artillery positions on Hill 263.

YOUNG, ARTHUR J. (795010)_____
   Near Romagne, France, Oct. 16, 1918.
   R—Erie, Pa.
   B—Concord, N. Y.
   G. O. No. 44, W. D., 1919.

Private, medical detachment, 6th Infantry, 5th Division.
Although wounded by a machine-gun bullet, he carried 4 comrades to a shell hole through terrific shell and machine-gun fire and dressed their wounds. He then carried them to a place of safety in the rear of our lines.

YOUNG, CHARLES C. (243375)_____
   Near Apremont, France, Sept. 27, 1918.
   R—Lansing, Mich.
   B—Millbrook, Mich.
   G. O. No. 46, W. D., 1919.

Sergeant, Company A, 345th Battalion, Tank Corps.
Although wounded twice by fire from antitank guns during the attack, he continued at his post, refusing to be sent to the rear.

YOUNG, CHARLES G_____
   Near Binarville, France, Sept. 27–28, 1918.
   R—Washington, D. C.
   B—Manor, Tex.
   G. O. No. 128, W. D., 1918.

First lieutenant, 368th Infantry, 92d Division.
While in command of a scout platoon, he was twice severely wounded from shellfire, but refused medical attention and remained with his men, helping to dress their wounds and to evacuate his own wounded during the entire night and holding firmly his exposed position covering the right flank of his battalion.

YOUNG, CHARLES I. Jr. (1906283)_____
   Near Cornay, France, Oct. 9, 1918.
   R—Reading, Pa.
   B—Spring City, Pa.
   G. O. No. 46, W. D., 1919.

First sergeant, Company D, 327th Infantry, 82d Division.
Assisted by 6 other soldiers, he fought his way through a greatly superior number of the enemy and rescued 13 wounded comrades, thereby saving them from being taken by the enemy. After all the officers had become casualties he assumed command of the company, reorganizing and leading it through many attacks and contributing greatly to the success of his company.

*YOUNG, EDWARD M_____
   Near Beaufort, France, Nov. 8, 1918.
   R—Hartford, Conn.
   B—Hartford, Conn.
   G. O. No. 44, W. D., 1919.

Second lieutenant, 178th Infantry Brigade, 89th Division.
While under unusually heavy shellfire, he was severely wounded and at the same time 15 of his men were also wounded. He refused to be moved until his men had received attention, and after the arrival of ambulances and litter bearers he still insisted that the men be moved to the first-aid station before allowing himself to be taken. Shortly after he arrived at the station for treatment he died.
Posthumously awarded. Medal presented to father, William E. Young.

YOUNG, GEORGE (554730)_____
   Near Moulins, France, July 14–15, 1918.
   R—Lorain, Ohio.
   B—Lorain, Ohio.
   G. O. No. 44, W. D., 1919.

Corporal, Company A, 9th Machine Gun Battalion, 3d Division.
Although seriously wounded, Corporal Young maintained a steady fire from his machine gun until forced to be carried to the rear for treatment.

YOUNG, GUY L. (263186)_____
   Near Cierges, France, July 31, 1918.
   R—Sault Ste. Marie, Mich.
   B—Dafter, Mich.
   G. O. No. 126, W. D., 1919.

Sergeant, Company I, 125th Infantry, 32d Division.
Under heavy machine-gun fire Sergeant Young went out in front of our lines and carried to shelter 2 wounded men.

YOUNG, ROBERT B. (1243468)_____
   Near Fismette, France, Aug. 10, 1918.
   R—Pittsburgh, Pa.
   B—Sewickley, Pa.
   G. O. No. 49, W. D., 1922.

Corporal, Company B, 111th Infantry, 28th Division.
When the attack of his company was held up by fire from a hostile strong point, Corporal Young, with 2 other men, voluntarily cut their way through enemy wire entanglements under heavy fire, reached their objective, and engaged the enemy in hand-to-hand combat. During this latter action 6 of the enemy were killed and the attacking line was enabled to advance to the new position.

YOUNGBAR, ANDY F. (1286094)_____
   Near Gildwiller, France, July 31, 1918.
   R—Curtis Bay, Md.
   B—Curtis Bay, Md.
   G. O. No. 109, W. D., 1918.

Private, first class, Company K, 115th Infantry, 29th Division.
During a raid against a post of his command near Gildwiller, France, July 31, 1918, he showed fine courage and endurance when attacked with hand grenades. Although seriously wounded, he joined in a counterattack against greatly superior numbers and continued to fight, even after receiving a second wound, until the enemy was repulsed.

*YOUNGDAHL, OSKAR E_____
   Near St. Etienne-a-Arnes, France, Oct. 6, 1918.
   R—Red Wing, Minn.
   B—Red Wing, Minn.
   G. O. No. 55, W. D., 1920.

Captain, 23d Infantry, 2d Division.
Armed with a rifle, Captain Youngdahl went through a heavy machine-gun fire alone to a position from which he could fire upon German machine gunners who were pouring a deadly fire into the flank of his company. He killed 1 of the gunners and captured 4 others, but was severely wounded himself. He stayed with his company until it had carried its objective, but died in the hospital of his wounds 2 days later.
Posthumously awarded. Medal presented to mother, Mrs. Olivia Youngdahl.

YOUNGER, RALEIGH............
Near Mezy, France, July 15, 1918.
R—Columbia, Tenn.
B—Santa Fe, Tenn.
G. O. No. 23, W. D., 1919.

Second lieutenant, 38th Infantry, 3d Division.
After being wounded in both hands, Lieutenant Younger took a rifle and killed an enemy machine gunner and, disregarding his wounds, remained with his platoon until it had taken up a new position several hours later.

YUILL, JULIUS O. (2258469)............
Near Epinonville, France, Sept. 26, 1918.
R—Soap Lake, Wash.
B—Howard, S. Dak.
G. O. No. 32, W. D., 1919.

Sergeant, Company M, 361st Infantry, 91st Division.
Accompanied by 1 man, Sergeant Yuill went forward to a German trench and bombed it, killing a German officer and 2 soldiers, and held the trench until reinforced by a party of 4. Fearing that the Germans in the trench would escape, he led these men 500 meters through sniper and machine-gun fire, cut off their means of escape, and captured 27 prisoners. On the same day he killed an officer who, with 2 men, was attempting to set up a machine gun to ambush the command group. He followed the 2 men into their dugout and killed 1 of them, and with the help of other members of the command thoroughly mopped up the place.

ZACHER, VERNON B............
Near Bantheville, France, Nov. 1, 1918.
R—Jamestown, N. Dak.
B—Barnesville, Minn.
G. O. No. 46, W. D., 1919.

First lieutenant, 359th Infantry, 90th Division.
Without regard for his own safety, Lieutenant Zacher led his platoon to the capture of 2 machine guns which were holding up the advance of his battalion. Although painfully wounded, he would not stop until his objective had been reached.

ZAMBRYSKI, ALEXANDER (2719080)......
Near Mousson, France, Nov. 7, 1918.
R—Worcester, Mass.
B—Poland.
G. O. No. 81, W. D., 1919.

Private, Company M, 9th Infantry, 2d Division.
While making a reconnaissance of a destroyed bridge over the River Meuse, a member of the patrol had his leg broken by machine-gun fire and fell into the river. Despite the fact that 4 enemy machine guns were firing point-blank on the spot at a distance of less than 100 yards, Private Zambryski volunteered and went to the rescue of his comrade and succeeded, single handed, in recovering his comrade while under severe fire and taking him to a place of safety.
Oak-leaf cluster.
For the following act of extraordinary heroism in action near Mousson, France, Nov. 8, 1918, Private Zambryski is awarded an oak-leaf cluster to be worn with the distinguished-service cross: Braving the murderous fire of machine guns, Private Zambryski rescued a wounded gunner who was lying on the river bank in full view of the enemy, carrying him without assistance to a place of safety.

ZANE, EDMUND L............
Near St. Etienne, France, Oct. 3–9, 1918.
R—San Francisco, Calif.
B—San Francisco, Calif.
G. O. No. 35, W. D., 1919.

Lieutenant colonel, 23d Infantry, 2d Division.
With remarkable courage and daring, Colonel Zane led his battalion through heavy machine-gun and shellfire to its objective. On several other occasions he voluntarily visited the front under most hazardous conditions, thereby acquainting his regimental commander with exceptionally important data.

ZANE, RANDOLPH TALCOTT............
While holding Bouresches, France, June 7–8, 1918.
R—Philadelphia, Pa.
B—Philadelphia, Pa.
G. O. No. 119, W. D., 1918.

Captain, 6th Regiment, U. S. Marine Corps, 2d Division.
While holding the town of Bouresches, France, on the night of June 7–8, 1918, he displayed such bravery as to inspire the garrison to resist successfully a heavy machine-gun and infantry attack by superior numbers.

*ZANOVITZ, STANLEY (545473)............
Near Mezy, France, July 15, 1918.
R—Nanticoke, Pa.
B—Nanticoke, Pa.
G. O. No. 35, W. D., 1920.

Corporal, Company B, 30th Infantry, 3d Division.
When attacked by superior numbers, Corporal Zanovitz led a squad in covering the withdrawal of his platoon. After being surrounded by the enemy he fought his way to our lines, when he took up an automatic rifle and by his fire held back superior numbers of the enemy until killed by their fire.
Posthumously awarded. Medal presented to father, Anthony Zanovitz.

ZAPPA, STEVE (1386984)............
Near Chipilly Ridge, France, Aug. 10–19, 1918.
R—Dwight, Ill.
B—Italy.
G. O. No. 71, W. D., 1919.

Private, first class, Company C, 131st Infantry, 33d Division.
He volunteered for dangerous missions, carrying messages over areas swept by heavy machine-gun and shellfire. He displayed great courage in accomplishing each task.

ZAVITZ, ARCHIE M. (280075)............
Near Romagne, France, Oct. 9, 1918.
R—Fruitport, Mich.
B—Rothbury, Mich.
G. O. No. 81, W. D., 1919.

Sergeant, Company I, 126th Infantry, 32d Division.
Facing heavy fire, he crawled 75 yards in advance of the platoon he commanded and reduced a machine-gun nest with rifle grenades.

ZAVODSKY, JOHN (1746508)............
Near Vieville-en-Haye, France, Sept. 26, 1918.
R—Perth Amboy, N. J.
B—Hungary.
G. O. No. 26, W. D., 1919.

Sergeant, Company D, 311th Infantry, 78th Division.
Although he was wounded, he remained with his company until its objective was reached before seeking first-aid treatment.

ZAX, HENRY E. (1918638)............
Near Vaux-Andigny, France, Oct. 11, 1918.
R—Louisville, Ky.
B—St. Louis, Mo.
G. O. No. 35, W. D., 1919.

Corporal, Company B, 120th Infantry, 30th Division.
Having been sent on a dangerous liaison patrol, he was severely wounded soon after he had located the unit on the right of his own. He nevertheless went forward to battalion headquarters immediately after securing first aid and made a complete report to his battalion commander before going to the rear.

ZECH, CLARENCE H. (2019084) ---------- | Private, 337th Ambulance Company, attached to 339th Infantry, 85th Division (detachment in North Russia).
Near Kadish, Russia, Sept. 27–28, 1918.
R—Detroit, Mich.
B—Detroit, Mich.
G. O. No. 78, W. D., 1919.

Acting as stretcher bearer to 2 companies of infantry in action against the Bolsheviks. Private Zech for 2 days and nights made his way through swamps and forests to administer first aid and carry wounded to the dressing station. His work at all times was accomplished under sweeping machine-gun and intense artillery fire, making it necessary for him to crawl on his hands and knees for long distances.

ZEILER, ELMER (39365) ---------------- Corporal, Company F, 9th Infantry, 2d Division.
Near Medeah Ferme, France, Oct. 3, 1918.
R—Fairhaven, Pa.
B—Fairhaven, Pa.
G. O. No. 21, W. D., 1919.

Corporal Zeiler, together with 4 other men, charged a machine-gun nest containing 3 heavy machine guns and captured the 3 guns and 20 prisoners.

ZIELINSKI, VINCENT P. (1702716) ------- Corporal, Company H, 306th Infantry, 77th Division.
At St. Juvin, France, Oct. 15, 1918.
R—Buffalo, N. Y.
B—Buffalo, N. Y.
G. O. No. 20, W. D., 1919.

He volunteered and carried a message of vital importance in connection with the capture of St. Juvin through an intense artillery barrage, displaying courage and persistent devotion to duty.

ZELDAM, JOHN J. (281431) ------------- Private, Company M, 126th Infantry, 32d Division.
North of Cierges, France, Aug. 1, 1918.
R—Grand Rapids, Mich.
B—Grand Rapids, Mich.
G. O. No. 117, W. D., 1918.

Following an assault, in which he was wounded in the leg by a machine-gun bullet and when further advance was impossible because of a barrage, he took refuge in a shell hole. From this shelter he observed a comrade who was seriously wounded and needed assistance. Despite heavy artillery and machine-gun fire, he crawled 20 yards through the open, reached the helpless man, and took him back to the shell hole. After the two had lain in the shell hole nearly the entire day, Private Zeldam, leaving his canteen with his companion, crawled across the danger zone and obtained assistance to carry the other wounded man to a dressing station.

†ZILKEY, FRANK (42419) -------------- Corporal, Company D, 16th Infantry, 1st Division.
Near the Forest of Argonne, France, Oct. 9, 1918.
R—Butte, Mont.
B—Butte, Mont.
G. O. No. 44, W. D., 1919.

After all the other members of his squad had been killed or wounded in advancing on a hostile machine gun, he pressed forward alone in the face of direct fire from the gun and by remarkable courage captured both the gun and its crew. Upon his own initiative he then started out alone to attack another gun and was killed.
Posthumously awarded. Medal presented to mother, Mrs. J. J. Carr.

ZILKEY, GUY L. (2261519) ------------ Sergeant, Company L, 362d Infantry, 91st Division.
Near Steenbrugge, Belgium, Oct. 31, 1918.
R—May, Idaho.
B—Eugene, Oreg.
G. O. No. 37, W. D., 1919.

Reorganizing badly shattered forces, he took command of the location and by proper distribution of those under his command ably protected his flanks. Assisted by 2 comrades, he attacked and drove out a machine-gun nest that was holding up his advance, reporting the situation to his company commander by establishing an efficient liaison.

*ZIMBORSKI, ALEXANDER JOHN (237697) | Corporal, Company C, 345th Battalion, Tank Corps.
Near Bois-de-Montrebeau, France, Oct. 4, 1918.
R—New York, N. Y.
B—Washingtonville, N. Y.
G. O. No. 37, W. D., 1919.

While running his tank into a woods to rout a machine-gun nest, his tank was hit and set on fire. He fought on until compelled to leave because of the excessive heat. While he and his gunner were leaving the tank, they were both wounded, and when making an advance on a machine gun Corporal Zimborski received further wounds which caused his death.
Posthumously awarded. Medal presented to widow, Mrs. Alexander Zimborski.

ZIMMER, JOHN H. (543075) ----------- Private, Company L, 7th Infantry, 3d Division.
Near Fossoy, France, July 15, 1918.
R—Providence, R. I.
B—Elkhorn, Pa.
G. O. No. 44, W. D., 1919.

Acting as runner, he made repeated trips through the heavy enemy bombardment of July 15, and, after being wounded on a mission, he accomplished his task before receiving medical attention.

ZIMMERMAN, ARTHUR P. (2257107) ----- Sergeant, Company D, 361st Infantry, 91st Division.
Near Gesnes, France, Oct. 3, 1918.
R—Bonners Ferry, Idaho.
B—Plymouth, Ind.
G. O. No. 20, W. D., 1919.

He voluntarily and unhesitatingly left shelter under heavy shellfire and, without thought of personal danger, rendered first aid and carried a wounded comrade to a place of safety.

ZIMMERMAN, RUDOLPH A. (2178114) ---- Sergeant, Machine Gun Company, 354th Infantry, 89th Division.
Near Remonville, France, Nov. 2–3, 1918.
R—New Florence, Mo.
B—New Florence, Mo.
G. O. No. 70, W. D., 1919.

Severely wounded in the cheek by a machine-gun bullet, he refused to be evacuated, but continued to lead his machine-gun section with the assault wave, displaying remarkable bravery and leadership, until his company was relieved the next morning.

ZINNER, FRED JOSEPH --------------- Second lieutenant, 5th Regiment, U. S. Marine Corps, 2d Division.
Near St. Etienne, France, Oct. 4, 1918.
R—Columbus, Ohio.
B—Columbus, Ohio.
G. O. No. 46, W. D., 1919.

While attacking a strongly held enemy position under heavy machine-gun and artillery fire, he rallied men of another company who had become separated from their organization in the support. With these reinforcements his platoon was able to relieve a very critical situation.

ZIRKLE, JAMES M. (1237160) ---------- Private, first class, Company A, 116th Infantry, 29th Division.
Near Samogneux, France, Oct. 17, 1918.
R—Lofton, Va.
B—Lofton, Va.
G. O. No. 37, W. D., 1919.

He volunteered and carried messages from battalion headquarters to the front line through artillery and machine-gun fire. He not only maintained effective liaison with his company, but also furnished an inspiring example of coolness and bravery to his comrades.

*ZITO, DOMINICK (1708792)............
Near Fismette, France, Aug. 10-11, 1918.
R—Mariners Harbor, Staten Island, N. Y.
B—Italy.
G. O. No. 46, W. D., 1919.

Private, Company M, 111th Infantry, 28th Division.
On Aug. 10 he three times volunteered and alone carried severely wounded comrades to the dressing station. Each trip was made through an intense enemy fire, but he unhesitatingly made the trip and returned to his post. On Aug. 11, while assisting 2 other men to carry a seriously wounded comrade to the aid station, 1 of the party was killed and the others driven from the road by an intense machine-gun fire, but Private Zito alone carried the wounded man to a place of shelter, from which he was evacuated that night. Private Zito was killed that afternoon in the advance line of the attack.
Posthumously awarded. Medal presented to mother, Mrs. Vita Zito.

ZLOTNIKOFF, JOHN (2364778).............
Near Dun-sur-Meuse, France, Oct. 5, 1918.
R—Vestaburg, Pa.
B—Russia.
G. O. No. 37, W. D., 1919.

Private, first class, Company L, 60th Infantry, 5th Division.
When his company was held up and unable to cross the river because of the destruction of a pontoon bridge, Private Zlotnikoff swam the river, carrying an automatic rifle, in the face of terrific machine-gun fire and direct artillery fire.

ZOBNOWSKI, WALTER F. (1239819)......
Near Apremont, France, Oct. 1, 1918.
R—Philadelphia, Pa.
B—Philadelphia, Pa.
G. O. No. 98, W. D., 1919.

Private, first class, Company M, 110th Infantry, 28th Division.
Having volunteered with 2 other soldiers to establish liaison with another unit which had been cut off by a hostile counterattack, Private Zobnowski rushed an enemy machine-gun nest, killing 4 of the enemy and capturing 4 prisoners. After taking his prisoners to the rear he volunteered and led a small force in a successful attack on the enemy.

*ZUCKERMAN, LOUIS (1697852)...........
Near St. Juvin, France, Oct. 15, 1918.
R—New York, N. Y.
B—Russia.
G. O. No. 95, W. D., 1919.

Private, Company G, 305th Infantry, 77th Division.
In order to enable his platoon to locate an enemy machine-gun nest, Private Zuckerman courageously volunteered and went out into an open field to draw the enemy fire. In the performance of this self-sacrificing mission he was killed.
Posthumously awarded. Medal presented to brother, Samuel Zuckerman.

ZYCH, JOHN (1277367)..................
In the vicinity of Hagenbach, Alsace, east of Belfort, France, Aug. 21, 1918.
R—South Plainfield, N. J.
B—Poland.
G. O. No. 102, W. D., 1918.

Private, Company H, 113th Infantry, 29th Division.
During the action in the vicinity of Hagenbach, Alsace, east of Belfort, France, Aug. 21, 1918, in an enemy raid on the position held by his regiment, he had his right eye shot out at the beginning of the action. He applied first aid himself, went back to his battle position, assisted in driving off the raiding party, operated his rifle until the end of the action, and continued to fire upon the retreating Germans as long as they could be seen.

[Awarded for extraordinary heroism in action under the provisions of the act of Congress approved July 9, 1918]

AUTY, EDWARD W. (17106)_____

Near Vierstraat, Belgium, Aug. 30 and 31, 1918.
R—England.
G. O. No. 60, W. D., 1920.

Signaler, British Army, 112th Battery, 24th Brigade, Royal Field Artillery, British Expeditionary Forces.
Signaler Auty accompanied an officer in a daring reconnaissance in advance of our lines. In an encounter with an enemy patrol near Rossignol Wood 2 of the enemy were captured. This action took place when fighting alongside the American 27th Division.

BIRKINSHAW, ROBERT S. (12077)_____

At Selency, France, Sept. 24, 1918.
R—England.
G. O. 60, W. D., 1920.

Sergeant, British Army, 1st Battalion, West Yorkshire Regiment, British Expeditionary Forces.
Sergeant Birkinshaw repeatedly led attacks against a strongly defended barricade at the entrance to Selency. The ultimate capture of this position was due to the determination and personal gallantry of this noncommissioned officer. This action took place when fighting alongside the American 27th Division.

BRADY, VINCENT J_____
East of Busigny, France, Oct. 18 to 20, 1918.
R—England.
G. O. No. 60, W. D., 1920.

Lieutenant, 13th Field Artillery Brigade, Australian Imperial Force.
Lieutenant Brady, in command of a forward section of artillery, pushed his guns in advance in close support of the Infantry. In crossing the La Selle River the fire of his guns silenced numerous machine-gun positions and aided the Infantry to advance. This action took place when fighting alongside the American 27th Division.

CADDY, THOMAS E. (453)_____
At Hamel, France, July 4, 1918.
R—England.
G. O. No. 60, W. D., 1920.

Sergeant, 43d Battalion, 11th Infantry, Australian Imperial Force.
Early in the advance his company was held up by fire from a machine-gun post. Sergeant Caddy with great bravery charged the position, bayoneted the crew, thus enabling his company to advance. This action took place when fighting alongside the American 27th Division.

CAMERON, CHARLES (548137)_____

On the St. Quentin Canal, France, Sept. 29, 1918.
R—England.
G. O. No. 63, W. D., 1920.

Second corporal (acting sergeant), 468th North Midland Field Company, Royal Engineers (Territorial Force), 46th Division, British Army.
Corporal Cameron was in charge of a party of sappers which advanced with the Infantry. He advanced through heavy fire and on several occasions showed contempt for danger in entering dugouts and tunnels which were suspected of being mined by the enemy. During the advance he attacked and captured an enemy machine gun and 6 men. This action took place in conjunction with the attack of the 30th U. S. Division.

COLLISON, HERBERT L_____
Near Obozerskaya, Russia, Oct. 16, 1918.
R—England.
G. O. No. 24, W. D., 1920.

Captain, Royal Artillery, British Army.
This officer acted as artillery observer in an advance against the enemy, which was continuous over several hours and at all times under very heavy artillery and machine-gun fire. He not only accompanied the advance troops, but at times went ahead of them, in order better to direct the fire of the American troops. At one time when, by a rapid advance of a portion of our troops, they were within the danger zone of our own fire, telephone communication having been cut, Captain Collison returned through the enemy barrage and taking a position in the open and under fire signaled a warning to our artillery. By this particularly brave and daring act the success of the American Infantry in this engagement was assured.

DODWELL, THOMAS B_____

Near Bruges, Belgium, Aug. 13, 1918.
R—England.
G. O. No. 44, W. D., 1919.

Second lieutenant, British Army, observer, Royal Air Forces, British Expeditionary Forces.
This officer and his pilot led 2 other machines on a long photographic mission over the area north of Bruges. Over Thourout they were attacked by 6 enemy planes. While heavily engaged, Lieutenant Dodwell and his pilot saw 1 of their machines in difficulty and trying to make our lines with an enemy plane close at his tail. Regardless of their own danger from the remaining planes, they dived to the assistance of the crippled plane. Taking advantage of their preoccupation, several enemy planes attacked from the rear but in spite of this rear attack they drove off the enemy plane and allowed the damaged plane to land within our lines. Half of the tail plane was shot away, but Lieutenant Dodwell climbed along the wing and lay down along the cowling in front of the pilot, enabling the pilot to regain partial control of his machine. When nearing the ground he crawled back into the cockpit to allow the nose to rise, and the pilot made a safe landing. The presence of mind and cool courage of this officer saved the machine from crashing to the ground.

ERRINGTON, ARTHUR (91)_____
Near Bellicourt, France, Sept. 29, 1918.
R—England.
G. O. No. 68, W. D., 1920.

Sergeant, 32d Battalion, 5th Division, Australian Imperial Force.
When his company was halted by machine-gun fire, Sergeant Errington with 2 comrades rushed the enemy position and captured the machine gun. Later the same day he rushed a second machine gun, capturing the gun and killing 3 of the crew. This action took place in conjunction with the attack of the 30th U. S. Division.

FOSTER, PETER (1670)_____

Near Serain, France, Oct. 11, 1918.
R—England.
G. O. No. 60, W. D., 1920.

Private, British Army, 5th Battalion, Connaught Rangers, British Expeditionary Forces.
Private Foster exposed himself to heavy fire in order to locate the positions of enemy machine guns. Later he led a liaison patrol through heavy fire in order to establish communication with the American unit on his right. This action took place when fighting alongside the American 27th Division.

HILL, JOHN H._____
Northwest of Bony, France, Sept. 29, 1918.
R—England.
G. O. No. 60, W. D., 1920.

Lieutenant, 50th Battalion, 13th Infantry, Australian Imperial Force.
While attached to a battalion of the 107th American Infantry, Lieutenant Hill exposed himself to heavy fire in order to mark out the jumping-off line. He assisted in forming the men for the attack and gallantly led them forward until he was severely wounded in the leg.

HOWARTH, JAMES (29649)_____
Near Kadish, Russia, Oct. 7, 1918.
R—England.
G. O. No. 15, W. D., 1921.

Sergeant, Liverpool Regiment, British Army.
Sergeant Howarth gallantly led his platoon in 5 successive attacks upon a strongly fortified line of blockhouses. During the withdrawal he exposed himself to heavy fire in order to direct the fire of a Lewis gun crew in order to secure the safe evacuation of his wounded.

KESSLER, ALBERT E. (15661)_____
South of Bazuel, France, Oct. 22, 1918.
R—England.
G. O. No. 60, W. D., 1920.

Private, British Army, 11th Battalion, Essex Regiment, British Expeditionary Forces.
Private Kessler delivered an important message to the units of the front lines. En route he was exposed to heavy enemy fire, but he continued until he had accomplished his mission. By his fearless conduct he enabled the companies to take up new positions without delay. This action took place when fighting alongside the American 27th Division.

LITTLE, WILLIAM B._____
During the Ypres-Lys offensive, Oct. 17, 1918.
R—England.
G. O. No. 60, W. D., 1920.

Major, British Army, East Lancashire Regiment, attached to 6th Battalion, Royal Dublin Fusiliers, 198th Infantry, British Expeditionary Forces.
During the Ypres-Lys offensive, Oct. 17, 1918, when closely affiliated with American troops. On one occasion, when Major Little was forming his battalion for the attack, the enemy put down a heavy barrage, inflicting many casualties. Regardless of the heavy enemy fire, this officer reassured his men and then gallantly led them in a successful attack. This action took place when fighting alongside the American 27th Division.

McNAMEE, JOHN T._____
Various dates and places in France.
R—England.
G. O. No. 126, W. D., 1919.
Distinguished-service medal also awarded.

Captain, Royal Field Artillery, British Army, attached to 1st Battalion, 1st Gas Regiment, American Expeditionary Forces.
Volunteering, he led a detachment of Engineers up to the front lines on July 30, 1918, for the purpose of assisting the advance of the Infantry with thermite and smoke bombs. That night he led his men through a heavy barrage, exhibiting courage and leadership. For 3 days and nights he remained with his men in the extreme front line in the Bois Colas, greatly aiding in repulsing enemy counterattacks by laying down barrages of thermite and phosphorus, cleaning out machine-gun nests in the same manner, and enabling our Infantry to attack behind smoke screens. On Aug. 5 he took another detachment into St. Thibaut and brought ammunition into the village before it was occupied by our Infantry and while the enemy patrols were still there. The advance of the Infantry across the 2 rivers, the Ourcq and the Vesle, was greatly facilitated and the lives of many of them saved by the smoke screens which Captain McNamee so successfully prepared. Throughout this entire advance across these 2 rivers he conducted himself with extraordinary heroism, setting an example to the men of the regiment to which he was attached, constantly exposing himself to danger in making reconnaissance and at the same time shielding his men.

MORISSET, VAUX L._____
At Nauroy, France, Sept. 29, 1918.
R—England.
G. O. No. 68, W. D., 1920.

Captain, 31st Battalion, 5th Division, Australian Imperial Force.
During the advance following the capture of Bellicourt, Captain Morisset, acting battalion adjutant, observed that flanking troops had failed to advance, leaving a gap in the line. On his own initiative, he gathered several detachments of troops under heavy machine-gun fire and deployed them, covering the threatened gap. His prompt and courageous conduct contributed to the success of the operations in which the 30th U. S. Division was engaged.

MUNDAY, WILLIAM T. J._____
Near Vierstraat, Belgium, Aug. 30 and 31, 1918.
R—England.
G. O. No. 60, W. D., 1920.

Lieutenant, British Army, 112th Battery, 24th Brigade, Royal Field Artillery, British Expeditionary Forces.
While in command of an accompanying gun, Lieutenant Munday advanced in close support of the attack of the 106th American Infantry. With a signaler, he made a daring reconnaissance in advance of our lines and returned with valuable information. In an encounter with an enemy patrol near Rossignol Wood 2 of the enemy were captured.

PADGETT, JOHN R. (546)_____
In the attack on the Hindenburg line, Sept. 29, 1918.
R—England.
G. O. No. 60, W. D., 1920.

Lance-corporal, 44th Battalion, 11th Infantry, Australian Imperial Force.
Corporal Padgett led his section in a successful attack on an enemy trench. When an enemy grenade fell amongst his section, he picked it up and threw it back toward the enemy. Later he exposed himself to heavy fire in order to carry an American soldier to a place of safety. This action took place when fighting alongside the American 27th Division.

PARKES, THOMAS (2487)_____
North of Bellicourt, France, Sept. 30, 1918.
R—England.
G. O. No. 60, W. D., 1920.

Private, 55th Battalion, 14th Infantry, Australian Imperial Force.
Private Parkes exposed himself to heavy machine-gun fire in going out in front of our positions and assisting 2 wounded Americans to a place of safety. This action took place when fighting alongside the American 27th Division.

RICE, WILFRID_____
During Ypres-Lys offensive, Oct. 17, 1918.
R—England.
G. O. No. 60, W. D., 1920

Lieutenant, British Army, 2d Battalion, Durham Light Infantry, British Expeditionary Forces.
Lieutenant Rice went forward during the advance under a heavy barrage of rifle and machine-gun fire and cut paths in the wire in order to further the advance of his company. Having cut the wire, with 2 men he rushed a machine-gun position, capturing the gun, and forcing 6 of the enemy to surrender. This action took place when fighting alongside the American 27th Division.

RODAKIS, NICHOLAS (5451)	Sergeant, 4th Battalion, Australian Machine Gun Corps, British Expeditionary Forces, attached to machine gun company, 105th Infantry, 27th Division, American Expeditionary Forces.
Near Ronssoy, France, Sept. 29, 1918. R—England. G. O. No. 37, W. D., 1919.	Organizing troops from different units, he exhibited great bravery and dash in leading them into effective combat, inspiring all by his courage and fearlessness.
SMYRDEN, RICHARD G	Captain, Liverpool Regiment, British Army.
Near Kadish, Russia, Feb. 7, 1919. R—England. G. O. No. 24, W. D., 1920.	While moving forward in command of a support company, Captain Smyrden encountered a disorganized Russian company in full retreat. Quickly reorganizing 2 platoons of this force, and with his own company, he launched a successful counterattack which reestablished the firing line in the original position. He then launched a new attack, which drove the enemy into their line of blockhouses. The coolness and courage of this officer was a great inspiration to his men.
TREWARN, FREDERICK (2295)	Private, 5th Machine Gun Battalion, Australian Imperial Force.
Near Bellicourt, France, Sept. 29, 1918. R—England. G. O. No. 60, W. D., 1920.	On 3 different occasions Private Trewarn exposed himself to artillery and machine-gun fire in order to assist wounded to shelter. Later on, when warning was received of an impending enemy counterattack, he displayed great coolness and good judgment under heavy fire in selecting positions for his guns. This action took place when fighting alongside the American 27th Division.
UNDERWOOD, ALBERT T. (240721)	Lance-corporal, 5th Battalion, Leicestershire Regiment, Territorial Force, 46th Division, British Army.
During the attack on Riquerval Woods, France, Oct. 11, 1918. R—England. G. O. No. 68, W. D., 1920.	While acting as a platoon scout in advance of his platoon, Corporal Underwood suddenly encountered a post of 9 Germans. He fearlessly engaged them until his platoon arrived. His courageous action not only saved his platoon from being ambushed but enabled it to reach its objective with slight loss. This action took place in conjunction with the attack of the 30th U. S. Division.
WOODCOCK, HUGH (42498)	Private, British Army, 9th Service Battalion, Norfolk Regiment, 6th Division, 9th Corps, British Expeditionary Forces.
Near Brancourt, France, Oct. 8, 1918. R—England. G. O. No. 68, W. D., 1920.	While acting as a runner to the 30th U. S. Division, Private Woodcock with a companion passed through heavy enemy machine-gun and artillery fire to deliver an important message. His companion was killed, but he continued on and completed his mission, which made possible a successful combined attack with the 30th U. S. Division on a strong enemy position.

FRENCH

[Awarded for extraordinary heroism in action under the provisions of the act of Congress approved July 9, 1918]

ANDRAL, LOUIS

Near Dun-sur-Meuse, France, Oct. 30, 1918.
R—France.
G. O. No. 50, W. D., 1919.

First lieutenant, observer, 284th Squadron, Air Service, French Army, attached to 3d Army, American Expeditionary Forces.
Distinguishing himself by his constant bravery and brilliance as an observer, Lieutenant Andral rendered valiant service, flying at times under most hazardous conditions. Many times he returned from low-flying patrols with his machine riddled with bullets. He attacked and drove from its mission an enemy observation plane, and later attacked a patrol of 4 enemy planes. In the fight that ensued he proved unequal to such an adversary and was killed.
Posthumously awarded. Medal forwarded to the French mission, General Headquarters, American Expeditionary Forces, for delivery to the next of kin.

ARMENGAUD, PAUL F. M

Near St. Mihiel, France, Sept. 12–16, 1918.
R—France.
G. O. No. 120, W. D., 1918.
Distinguished-service medal also awarded.

Major of Air Service, French Army, acting as Assistant Chief of Staff, Air Service, 1st Army, American Expeditionary Forces.
Major Armengaud, acting as Assistant Chief of Staff, Air Service, 1st Army, did display great bravery much beyond that required by his position while acting as observer in an airplane. Each day of the battle in the St. Mihiel salient he flew over the hostile lines through our own and the enemys' artillery and machine-gun fire, observing the enemy air and ground activity and the disposition of our own air forces, thereby bringing back valuable information as to the enemy's dispositions and probable intentions, which materially aided in our subsequent operations.

ARTONI, CHARLES

Near Massif Blanc Mont, France, Oct. 3–8, 1918.
R—France.
G. O. No. 126, W. D., 1919.

Machine gunner, AS. 307, 3d Battalion of Light Tanks, Assault Artillery, French Army.
During attacks by the 2d American Division he distinguished himself by volunteering twice for hazardous duty—first, to take food to the firing line, and later, to replace a wounded comrade as liaison agent. His personal courage and coolness under heavy fire were marked.

ASCHLIMAN, PAUL

Near Villers-Tournelle, France, May 3–4, 1918.
R—France.
G. O. No. 37, W. D., 1919.

Sergeant, interpreter, Battalion of 19th Train of Military Transports, French Army, attached to 1st Infantry Brigade, 1st Division, American Expeditionary Forces.
During a particularly intense bombardment of high-explosive and gas shells he went through the town notifying and warning troops of the presence of gas and directing various elements passing that point, his mission at all times exposing him to this heavy fire.

BARDOU, HENRI

Near Massif Blanc Mont, France, Oct. 8, 1918.
R—France.
G. O. No. 126, W. D., 1919.

Brigadier, 308th Company, Tank Corps, 3d Battalion of Light Tanks, French Army.
During an attack by the 2d American Division he displayed marked coolness and bravery in guiding his tank through heavy counterpreparation fire during an approach march. He never hesitated to leave the protection of the tank the better to direct its advance. Later he showed great personal courage in going through heavy shellfire to the rescue of several wounded American soldiers.

BARRE, MARCEL

Near Massif Blanc Mont, France, Oct. 3–8, 1918.
R—France.
G. O. No. 81, W. D., 1919.

Private, 308th Company, Tank Corps, 3d Battalion of Light Tanks, Assault Artillery, French Army.
During attacks by the 2d American Division he distinguished himself by coolness and courage in leading his tank in the approach march during heavy counterpreparation fire. He brought his tank into position with great skill and later displayed extraordinary personal heroism in going under heavy shellfire to the rescue of American wounded.

BAURIN, ALBERT

Near Gesnes, France, Oct. 4, 1918.
R—France.
G. O. No. 78, W. D., 1919.

Brigadier, 350th Tank Company, French Army.
During the course of an engagement in which the French tanks assisted the American Infantry to attack a difficult position, the tank in which Brigadier Baurin was advancing was put out of action and his companion wounded by heavy artillery fire. He continued to assist his comrade in firing until the ammunition was completely exhausted and then dismounted from the tank and he carried the wounded man to the American line through the intense enemy fire.

BECQUART, HENRI

Near Massif Blanc Mont, France, Oct. 3–8, 1918.
R—France.
G. O. No. 81, W. D., 1919.

Maréchal des Logis, AS. 309, Tank Corps, 3d Battalion of Light Tanks, Assault Artillery, French Army.
During successive attacks by the 2d American Division this noncommissioned officer displayed marked courage and coolness. Though wounded he refused to be evacuated, remaining on duty until the end of the fight. His bravery was an inspiration to those near him.

BERCEROT, JACQUES

Near Mortzwiller, Alsace, Sept. 1, 1918.
R—France.
G. O. No. 70, W. D., 1919.

Maréchal des Logis, 19th Battery, E. V. N., 70th Regiment, A. L. G. P., French Army.
Driving his engine through an area which was subjected to a most intense shelling, he removed to safety 9 carloads of ammunition. His exploit was accomplished in spite of damaged rails and constant shell bursts, the latter damaging his engine.

653

BERTRAND
Near Blanc Mont Ridge, France, Oct. 3, 1918.
R—France.
G. O. No. 126, W. D., 1919.

Captain, commanding 308th Company, Tank Corps, French Army.
While attached to the 2d American Division he led his company of tanks into the fight under terrific shell and machine-gun fire. He set an example of coolness and bravery to all about him, being conspicuous for his devotion to duty, in the performance of which he was killed by the enemy fire.
Posthumously awarded. Medal forwarded to the French mission, General Headquarters, American Expeditionary Forces, for delivery to next of kin.

BOGLIONE
At Tavannes, Meuse, France, Oct. 14, 1918.
R—France.
G. O. No. 64, W. D., 1919.

Maréchal des Logis, 3d Battery, 74th Regiment of Artillery, French Army, attached to railway artillery, American Expeditionary Forces.
Under heavy shelling he displayed exceptional bravery in extinguishing a fire which threatened the destruction of a powder dump and an ammunition car.

BONNARD, EMILE
Near Meuse River, France, Nov. 3–11, 1918.
R—France.
G. O. No. 62, W. D., 1919.

Major, 2d Regiment of Colonial Infantry, French Army.
Working in conjunction with an American brigade, he led his battalion against determined resistance of the enemy, working his way with great valor to take an important hostile position. He kept constantly in touch with the American commander, thus insuring perfect liasion during the entire operations.

BOURDU, JOSEPH
Near Le Ruddin, Vosges, France, June 15, 1918.
R—France.
G. O. No. 126, W. D., 1918.

Master gunner, 26th Battery, 208th Regiment of Field Artillery, French Army.
On June 15, 1918, while a battalion of the 11th Infantry was halted on the road near Le Ruddin, Vosges, it came under heavy shellfire from a German battery. First Lieut. Edison M. Boarke, badly wounded in the left arm and shoulder, attempted to reach a place of safety by rolling down an embankment near the road, when he was picked up by Master Gunner Bourdu under heavy shellfire and carried on his back to a place of safety.

BOURGON, NARCISSE
At Verst 455, North Russia, Oct. 14, 1918.
R—France.
G. O. No. 24, W. D., 1920.

Second lieutenant, 21st Battalion de Marche, Colonial Infantry, French Army.
Lieutenant Bourgon placed himself at the head of a Franco-American section and personally led it through heavy machine-gun fire against a strongly fortified position. This brilliant officer was killed at Bolshie-Ozerka, Russia, Mar. 19, 1919, while defending with a small force his position, which was repeatedly attacked by 3 battalions of the enemy.
Posthumously awarded. Medal forwarded to the military attaché, American Embassy, Paris, France, for delivery to the next of kin.

BREDIN, ANDRÉ E.
In the region of St. Mihiel, France, Sept. 12–16, 1918.
R—France.
G. O. No. 81, W. D., 1919.

Second lieutenant (pilot), 16th Combat Group, French Army.
While working in connection with the 1st American Army, Lieutenant Bredin made a series of patrols into the enemy's territory and by his valuable information and daring attacks on the enemy aided materially in preventing the enemy from penetrating into our lines.

BRELIER, FRANCOIS
Near Kemmel, Belgium, Apr. 26, 1918.
R—France.
G. O. No. 62, W. D., 1919.

Warrant officer, 153d Regiment of Infantry, French Army.
Attacking the strongly fortified position on Mount Kemmel, although wounded, he continued to struggle until his position had been established. On no less than 8 occasions he led patrols into enemy lines, and each time inspired those under his command by his bravery and intrepidity.

BRELIVET, HERVE M.
Near Sivry-sur-Meuse, France.
R—France.
G. O. No. 62, W. D., 1919.

Warrant officer, 2d Regiment of Colonial Infantry, French Army.
Bravely leading a platoon of the front line, he afforded valuable assistance to an American regiment during the course of the advance. While progressing he was severely wounded. He took a prominent part in all advances made by his organization during the entire war, and his courage and bravery was at all times an inspiration to his comrades.

BUCHET, XAVIER
In the Argonne-Meuse operations, France, Nov. 9–11, 1918.
R—France.
G. O. No. 62, W. D., 1919.

Captain, 33d Regiment of Colonial Infantry, French Army.
During 3 days of most terrific struggle and exposed to enemy artillery and machine-gun fire and the hazard of poisonous shells, Captain Buchet unceasingly urged and led his men to their objective. His example of courage, endurance, and valor served as a high standard for those under his command.

BUISSON
Near Massif Blanc Mont., France, Oct. 3–8, 1918.
R—France.
G. O. No. 126, W. D., 1919.

Lieutenant, 307th Company, Tank Corps, 3d Battalion of Light Tanks, Assault Artillery, French Army.
During attacks by the 2d American Division he continually distinguished himself by acts of courage and by his devotion to duty. When his captain was killed, he assumed command of the tank company, and with extraordinary heroism, in the face of heavy fire, accomplished the mission intrusted to him.

BUREL, MICHEL
At Chateau-Thierry, France, July 25–26, 1918, and St. Mihiel, France, Oct. 5–8, 1918.
R—France.
G. O. No. 62, W. D., 1919.

Private, first class, 156th Regiment of Infantry, French Army.
In the attack on Chateau-Thierry he was at all times in the very thickest of the struggle, exhibiting singular valor and devotion to duty. During the attack on St. Mihiel he alone took a hostile machine gun, and a few days later put to route 10 of the enemy, killing 1 during the encounter.

CARRERE, JEAN B.
Near Ammertzwiller, Alsace, July 1, 1918.
R—France.
G. O. No. 150, W. D., 1918.

Corporal, 319th Regiment of Infantry, French Army, attached to 42d Division, American Expeditionary Forces.
While serving under the command of an American divisional commander, when the advanced posts were ordered to withdraw and join the combat groups in anticipation of a raid, Corporal Carrere and the 4 men of his squad were cut off by the enemy barrage. This little group made a stand and by the use of an automatic rifle and grenades repulsed the attack made on them by 2 parties of Germans, each party estimated to be between 15 and 20 men. Corporal Carrere displayed splendid courage and initiative, especially in exploring the surrounding terrain. Although completely isolated, he and his men continued the resistance against the enemy until they finally drove them back. To him and to his 4 men, to whom he knew how to communicate his spirit, is due the credit for the failure of the raid, although it was supported by heavy artillery and executed in force.

CHANOINE, CHARLES M. M.
Near Maizerais, France. Sept. 12, 1918.
R—France.
G. O. No. 14, W. D., 1920.

Major of Cavalry, Tank Corps, French Army.
While Major Chanoine's group was advancing to attack with the 1st Brigade of American Tanks his leading tank became stalled, and this officer left his post of command and went forward to direct its repair. While he was so engaged the enemy opened fire, and 1 shell struck the disabled tank, killing and wounding 15 men and knocking Major Chanoine unconscious. Upon regaining consciousness Major Chanoine refused to be carried to the rear but continued his work under fire until it was possible for the tanks to continue their advance, when he led his group forward on foot for a distance of 3 kilometers, remaining on duty throughout the engagement.
Oak-leaf cluster.
For the following act of extraordinary heroism in action near Cheppy, France, Sept. 26, 1918, Major Chanoine was awarded an oak-leaf cluster to be worn with the distinguished-service cross: Upon learning that his brigade commander had been wounded, Major Chanoine advanced on foot under heavy fire from all arms and reconnoitered a passage for his tanks over difficult terrain 1 kilometer to the front. He then led his tanks in a successful attack on the town of Cheppy.

CHARRON, JEAN M. H.
North of Daucourt, France, Sept. 14-15, 1918.
R—France.
G. O. No. 62, W. D., 1919.

Second lieutenant, 5th Regiment of Mounted Chasseurs, French Army.
After maintaining contact with the enemy all night in an outpost position near the Bois-des-Hautes Epines, France, Lieutenant Charron, by skillful maneuvering, led his platoon in a successful attack on the wood, capturing 35 prisoners, including 3 officers, advancing the line more than a kilometer.

CHATAIGNEAU, YVES.
Near Verdun, France, Oct. 1-11, 1918.
R—France.
G. O. No. 70, W. D., 1919.

Lieutenant, 409th Regiment of Infantry, French Army.
During the 11 days of action he contributed greatly to the success of our operations by making repeated journeys to the battle line under most hazardous conditions, making tactical dispositions of the conquered territory, and aiding in the maintenance of the liaison between the regimental commander, 26th Infantry, and his assault troops.

CHEVALIER, OLIVER.
Near Champagne, France, July 15-16, 1918, and near the Ourcq River, France, July 28 to Aug. 2, 1918.
R—France.
G. O. No. 70, W. D., 1919.

Captain, 71st Regiment of Infantry, French Army, attached to 42d Division, American Expeditionary Forces.
He displayed extraordinary heroism and conspicuous gallantry in making daily reconnaissances of the front lines under unusually heavy machine-gun and artillery fire, rendering valuable aid to the division to which he was attached.

COMPANA, ANTOINE.
Near Mortzwiller, Alsace, Sept. 1, 1918.
R—France.
G. O. No. 35, W. D., 1920.

Gunner, 19th Battery, 70th Regiment, A. L. G. P., French Army.
During a terrific bombardment of the railway station, Gunner Compana, as a member of a train crew, assisted in the removal of 9 cars of ammunition to a place of safety. The mission was successfully accomplished, despite the fact that the fierce shelling had damaged the track, and trees blown across the track made it nearly impassable.

CONNELLY, JAMES A.
Near Suippes, France, Sept. 5, 1918.
R—France.
G. O. No. 50, W. D., 1919.

Sergeant-pilot, Lafayette Escadrille, French Army.
An American pilot serving with the French Army, Sergeant-pilot Connelly attacked a formation of 12 enemy planes (type Fokker), shooting down the flight commander and forcing the remainder to seek safety. He continued with the unequal combat until his ammunition was exhausted.

CORBARON, BÉNIGME V. M. J.
At the Salient-du-Feys, France, Mar. 9, 1918.
R—France.
G. O. No. 126, W. D., 1919.

Major, 13th Regiment of Infantry, French mission, attached to 42d Division, American Expeditionary Forces.
In the assault upon the German position at the Salient-du-Feys, France, Mar. 9, 1918, during which 3 hostile lines of trenches were overrun, he voluntarily joined Company E, 166th Infantry, 42d Division, while this company was undergoing a severe fire from the enemy lasting 3 hours. His coolness and conspicuous courage had a marked effect on the behavior of this organization.

COSTA DE BEAUREGARD, ROBERT.
At St. Juvin, France, Oct. 14, 1918.
R—France.
G. O. No. 27, W. D., 1920.

Captain, 23d Regiment of Infantry, Territorials, French Army, attached to 306th Infantry, 77th Division, American Expeditionary Forces.
When communication with the forward observation post was broken and no runners were available, Captain Costa De Beauregard voluntarily carried a message to the observation post through intense shellfire, displaying great bravery and coolness, and succeeded in reestablishing communication with the regimental post of command.

COURTOIS, ANDRÉ.
Near Bois-de-Belleau, France, June 11, 1918.
R—France.
G. O. No. 81, W. D., 1919.

Maréchal des Logis, 19th Battalion of the Train of Military Transports, French mission, attached to 2d Battalion, 5th Regiment, U. S. Marine Corps, 2d Division, American Expeditionary Forces.
After runners had failed to establish liaison with one of the attacking companies, M. Courtois volunteered and successfully accomplished the mission, making his way over ground subjected to heavy shell and machine-gun fire and through woods infested with enemy snipers. He returned to the battalion commander with valuable information, contributing greatly to the success of the attack.

DANO, ROBERT.
Near Somme Py, France, Oct. 3-8, 1918.
R—France.
G. O. No. 126, W. D., 1919.

Second lieutenant of Artillery, 308th Company, Tank Corps, 3d Battalion of Light Tanks, Assault Artillery, French Army.
He displayed conspicuous gallantry during operations of the 2d American Division, and when his captain was killed and the company had suffered heavy losses he promptly effected a reorganization. With great heroism and able leadership, he then pushed forward the assault in the face of heavy shellfire.

DAUNE, CHARLES E_____
 Near St. Mihiel, France, Sept. 12, 1918.
 R—France.
 G. O. No. 62, W. D., 1919.

Second lieutenant, 8th Regiment of Mounted Chasseurs, French Army.
When the advance of his battalion was held up by a hostile strong point, well fortified and protected by barbed-wire entanglements, Lieutenant Daune at the head of his platoon led his men forward, forcing an opening through the wire, and overcame the enemy's resistance, capturing 153 prisoners, including 2 officers and 15 machine guns.

DE BOULANCY D'ESCAYARAC, M_____
 Near Binarville, France, Sept. 27, 1918.
 R—France.
 G. O. No. 50, W. D., 1919.

Lieutenant of Cavalry, French Army.
During a raid on an enemy ammunition depot Lieutenant de Boulancy d'Escayarac proved of the greatest assistance and value and the success of the exploit was wholly due to his bravery and efficiency. He tirelessly explored the front on different occasions, and the information that he furnished our officers regarding hidden machine-gun nests proved to be of the utmost importance and value.

DEBRUT, F. O_____
 Near Cantigny, France, May 27–31, 1918.
 R—France.
 G. O. No. 81, W. D., 1919.

Lieutenant, 8th Regiment of Engineers, French Army.
Lieutenant Debrut displayed high qualities of efficiency and courage in maintaining the organization and upkeep of telephone communications during a violent bombardment. Although wounded, he continued to perform his duty until the end of the action.

DE FROISSARD-BROISSIS, MICHEL M. F__
 North of Verdun, France, Nov. 10, 1918.
 R—France.
 G. O. No. 62, W. D., 1919.

Lieutenant, 6th Regiment of Colonial Infantry, French Army.
During the attack on the village of Damvillers in liaison with the American troops he directed the assault at the head of his company until he was wounded.

*DE GUIROYE_____

 Near Blanc Mont Ridge, France, Oct. 3, 1918.
 R—France.
 G. O. No. 78, W. D., 1919.

Captain, commanding 307th Company, Tank Corps, French Army, attached to 2d Division, American Expeditionary Forces.
While leading his company of tanks attached to the 2d American Division into the fight under terrific shell and machine-gun fire he set an example to all about him by his coolness and bravery. He showed conspicuous devotion to duty in the performance of which he was killed by enemy fire.
Posthumously awarded. Medal forwarded to the French mission, General Headquarters, American Expeditionary Forces, for delivery to the next of kin.

DE JACQUELOT DE BOISROUVAY, ALAIN__
 At Seicheprey, France, Apr. 20–21, 1918.
 R—France.
 G. O. No. 87, W. D., 1919.

Major of Infantry, French Army, attached to 26th Division, American Expeditionary Forces.
He exposed himself to extremely heavy enemy fire with fearless disregard for personal danger in order to secure information for tactical dispositions and Artillery support, which were of great service in checking the hostile attack.

DE LESSEPS, JACQUES B. M_____
 At Conflans and Audun-le-Roman, France, Aug. 15, 1918.
 R—France.
 G. O. No. 81, W. D., 1919.

Captain, pilot, 2d Bombing Group, French Army.
Captain de Lesseps made 3 successful bombing raids in 1 night, 2 on Conflans and 1 on Audun-le-Roman, causing great damage. Despite heavy antiaircraft fire, he flew at an extremely low altitude and besides his successful raids returned with valuable information of the enemy's movements.

*DE MANDAT GRANCEY_____
 Near Fismette, France, Sept. 6, 1918.
 R—France.
 G. O. No. 89, W. D., 1919.

Lieutenant, French Army, attached to 107th Field Artillery, 28th Division, American Expeditionary Forces.
He went to an advanced observation post and when our front line temporarily withdrew stuck bravely to his position, rendering valuable service until he was killed by the intense enemy machine-gun fire. His heroism was an inspiration to all near him.
Posthumously awarded. Medal forwarded to the French mission, General Headquarters, American Expeditionary Forces, for delivery to the next of kin.

DE PAVANT, FRANCOIS_____
 Near Brieulles, France, Oct. 4, 1918.
 R—France.
 G. O. No. 62, W. D., 1919.

Lieutenant, observer, 284th Escadrille, Air Service, French Army, attached to 3d Army Corps, American Expeditionary Forces.
While engaged on an Infantry contact patrol he attacked 6 Fokker pursuit planes and valiantly drove them off, remaining in action until his plane was so badly damaged by fire that it was forced to land.

DEQUERTE, CONRAD_____
 At Verst 455, near Obozerskaya, Russia, Oct. 14, 1918.
 R—France.
 G. O. No. 24, W. D., 1920.

Second lieutenant, 21st Battalion de Marche, Colonial Regiment of Infantry, French Army.
Attached as liaison officer to a Franco-American detachment during the attack, he placed himself at the head of a section of Infantry and led it against a strongly fortified position. He was among the first to make the objective. The energy displayed by this brilliant officer while exposed to heavy fire was an important factor in the successful attack.

DIOT, LUCIEN_____
 In the region of St. Mihiel, France, Sept. 12, 1918.
 R—France.
 G. O. No. 81, W. D., 1919.

Aspirant, pilot, 219th Escadrille Air Service, French Army.
Aspirant Diot on Sept. 12, in the region of St. Mihiel, while flying at an extremely low altitude, had his wireless set destroyed by enemy fire. Rather than return to his field, and in spite of being 8 kilometers beyond the lines, he descended to less than 100 meters altitude and attacked an enemy convoy, routing it in confusion. This act was accomplished in spite of being under terrific machine-gun fire and heavy antiaircraft artillery fire from the ground, his plane being shattered with bullets.

DORMOY, GEORGES_____
 Near St. Mihiel, France, Sept. 12, 1918.
 R—France.
 G. O. No. 62, W. D., 1919.

Maréchal des Logis, 8th Regiment of Mounted Chasseurs, French Army.
Maréchal des Logis Dormoy demonstrated marked courage in leading his platoon in an attack on a strong center of resistance. Running ahead of his men, he was the first to enter the enemy trench, where he captured more than 60 prisoners.

DROUHIN, ROBERT G_____

Near Monthois, France, Sept. 27, 1918.
R—France.
G. O. No. 13, W. D., 1919.

Second lieutenant of Infantry, French Army, attached to 372d Infantry, 93d Division, American Expeditionary Forces.
During the attack on Monthois Lieutenant Drouhin voluntarily went each day to reconnoiter the first position, observe the advance, and to secure liaison with neighboring units, fearlessly exposing himself to the severest artillery and machine-gun fire. On Sept. 29 he entered the enemy positions east of Ardeuil, located the machine-gun nests which were holding up the advance, traversed an open field swept by the fire of these guns to reach the liaison officer of Artillery, and to give him the objective. Throughout the operations he rendered invaluable assistance to the regimental commander, and his energy and courage were an inspiration to the entire regiment.

DUBET, JEAN O_____
Near St. Maurice, France, Sept. 14, 1918.
R—France.
G. O. No. 62, W. D., 1919.

Lieutenant, 8th Regiment of Mounted Chasseurs, French Army.
Having been ordered to secure contact with the enemy, he led his men against a hostile strong point held by a superior force. Through his decision and personal bravery this center of resistance was overcome, 34 prisoners were captured, and valuable information secured.

DUBOIS, ALFRED_____

Near Missy-aux-Bois, France, July 18, 1918.
R—France.
G. O. No. 99, W. D., 1918.

Interpreter, 19th Train of Military Transports, 35th Company, French mission, attached to Headquarters, 6th Field Artillery, 1st Division, American Expeditionary Forces.
He voluntarily exposed himself to heavy shellfire while carrying wounded men to a place of safety.

DURAND, LÉON_____ _____
In France, Sept. 12, 1918.
R—France.
G. O. No. 81, W. D., 1919.

Private, first class, 156th Regiment of Infantry, French Army.
While his platoon was being held up by enemy grenadiers he opened fire from his machine gun and prevented an enemy counterattack. An exploding mine had damaged his gun, and when the enemy rushed forward he jumped from his position, and with the aid of hand grenades he resisted the advance and caused many casualties.

EHRHARDT, GUSTAVE_____

Near Cheppy, France, Sept. 25–26, 1918.
R—France.
G. O. No. 125, W. D., 1918.

Captain, 140th Regiment of Infantry, French Army, attached to 138th Infantry, 35th Division, American Expeditionary Forces.
As liaison officer it was no part of his duty to go into action with the forward elements of the regiment, but he insisted upon doing so. Undaunted by a wound in the left arm, caused by a bursting shell, he continued to advance in the face of very intense shellfire. He was again struck down by a shell fragment, which shattered his right arm, but he arose and followed the regimental commander into the shell-swept area. He was knocked down a third time by shell fragments, from which he received wounds in the back so severe in character that he was unable to rise. He later received additional wounds in the body.

ESCUDIER, ETIENNE_____

In the Bois-de-Brieulles, France, Sept. 29, 1918.
R—France.
G. O. No. 71, W. D., 1919.

First lieutenant, 79th Regiment of Infantry, French Army, attached to 59th Infantry, 4th Division, American Expeditionary Forces.
Though he was not required to do so by the duties of his position, Lieutenant Escudier volunteered to ascertain the source of an extremely heavy artillery fire which was being directed upon the American Infantry. In accomplishing this mission he exposed himself to heavy shell and machine-gun fire for 3 hours and secured accurate information, displaying absolute fearlessness and indifference to his own personal safety.

ETIENNE, EUGENE_____
Near Gesnes, France, Oct. 4 and 5, 1918.
R—France.
G. O. No. 78, W. D., 1919.

Lieutenant, 350th Tank Company, French Army.
Lieutenant Etienne personally located a point at which the tanks could cross the stream and then remained at this point under the heavy enemy artillery and machine-gun fire until all the tanks had struggled across. He then led the section in the successful attack on the Bois-de-la-Morine, where the tanks destroyed the enemy machine-gun nests and allowed the American Infantry to reach the objective.

FILIPPI, JEAN_____
At Magenta Farm, France, Nov. 3, 1918.
R—France.
G. O. No. 62, W. D., 1919.

Corporal, 2d Regiment of Colonial Infantry, French Army.
While engaged in maintaining liaison with American troops he repeatedly passed through the enemy's lines. At one time he was attacked and almost captured by several Germans, but, after a hand-to-hand struggle, he succeeded in freeing himself and continuing on his mission.

GAILLOT, MAURICE G_____
Near Chateau-Thierry, France, July 31 to Aug. 6, 1918, and near Soissons, France, Aug. 25 to Sept. 1, 1918.
R—France.
G. O. No. 70, W. D., 1919.

Captain, 36th Regiment of Infantry, French Army.
Throughout the 2 campaigns he accompanied every officer's patrol, fearlessly exposing himself to heavy machine-gun and artillery fire, rendering invaluable service to the regiment to which he was attached.

GAUFFENY, EMMANUEL P. F_____
In the St. Mihiel sector, France, Sept. 26, 1918.
R—France.
G. O. No. 62, W. D., 1919.

Lieutenant, 146th Regiment of Infantry, French Army.
Commanding and leading a raid against the enemy, in spite of the severe shelling, he reached his objective and took many prisoners. His great courage during most hazardous reconnaissance won the admiration of all under his command, and the information he supplied was always of the utmost value.

GAUTHIER, JEAN C_____

Near Bussy Farm, France, Sept. 29, 1918.
R—France.
G. O. No. 37, W. D., 1919.

Second lieutenant of Artillery, French Army, liaison officer with the 371st Infantry, 93d Division, American Expeditionary Forces.
While on duty as French liaison officer with the 371st Infantry, this officer was knocked down by a shell which burst near by during a severe artillery barrage. He immediately resumed the advance with the utmost coolness, affording an excellent example to the men near him. During the whole battle he was many times exposed to heavy fire in maintaining liaison and performing other voluntary service, his experience, sangfroid, and judgment assisting materially in the success of the operation.

GIET, AUGUSTE A.
In France, on July 22, 1918.
R—France.
G. O. No. 81, W. D., 1919.

Private, first class, 153d Regiment of Infantry, French Army.
While his company was surrounded by the enemy after an attack, he volunteered and established liaison with his battalion, his mission being successfully accomplished despite the severe fire from many machine guns in the vicinity.

GOYNE, ANTOINE.
Near St. Mihiel France, Sept. 13, 1918.
R—France.
G. O. No. 62, W. D., 1919.

Brigadier, 4th Squadron, 6th Regiment de Chasseurs d'Afrique, French Army.
Having been ordered to establish an advance post in the village of Deuxnouds, he entered the town ahead of the infantry and captured 6 prisoners.

GRIZEL, MARCEL.
Near Mortzwiller, Alsace, Sept. 1, 1918.
R—France.
G. O. No. 81, W. D., 1919.

Private, 19th Battery, 70th Regiment, A. L. G. P., French Army.
At the railroad garage near Mortzwiller, Private Grizel unhesitatingly entered an area under heavy bombardment by the enemy, and, as a member of the train crew, assisted in the removal of 9 cars of ammunition to a place of safety. This was accomplished in spite of the track being cut and limbs of trees being thrown across the track by bursting shells.

GROS, JOSEPH.
Near Crete-des-Eparges, France, Sept. 12, 1918.
R—France.
G. O. No. 62, W. D., 1919.

Lieutenant, 6th Regiment of Colonial Infantry, French Army.
During the action near Crete-des-Eparges, Lieutenant Gros established and maintained liaison with American troops, and, against dangerous and violent counterattacks, he organized and defended his position. During all attacks by his organization he was always conspicuous by his bravery and gallantry and his example of courage and bravery was an inspiration to his men.

GUINET, EUGÈNE.
Near Nantillois, Meuse, France, Oct. 8, 1918.
R—France.
G. O. No. 81, W. D., 1919.

Lieutenant, observer, 208th Aero Squadron, French Army, attached to 3d Army Corps, American Expeditionary Forces.
This officer displayed remarkable gallantry and devotion to duty when he engaged in combat against a superior force of enemy pursuit planes. In the course of the action his pilot was killed. Lieutenant Guinet took control of the machine and brought it back under fire to friendly territory and was seriously wounded in landing.

GUYOT, BENJAMIN H.
In the region of St. Mihiel, France, Sept. 14–15, 1918.
R—France.
G. O. No. 81, W. D., 1919.

Lieutenant, pilot, 219th Escadrille, French Army.
Lieutenant Guyot without protection made 3 separate and distinct trips well into the enemy's lines; disregarding antiaircraft and ground machine-gun fire, he accomplished artillery adjustments of the highest military value.

HALLIER, ANDRÉ.
In the Bois-de-Beuge, France, Oct. 3–4, 1918.
R—France.
G. O. No. 81, W. D., 1919.

Lieutenant, 15th Battalion of Light Tanks, French Army.
After leading units of his battalion to their jumping-off positions, he went out with 1 enlisted man and established an observation post in advance of the outpost line. He maintained this position despite intense artillery and machine-gun fire, and sent back valuable information as to the enemy, which was, in a large measure, responsible for the accuracy of our artillery fire. He displayed marked heroism and utter disregard for personal danger.

HASSELVENDER, ALFRED J. V.
At Verst 455, North Russia, Oct. 14, 1918.
R—France.
G. O. No. 24, W. D., 1920.

Captain, 21st Battalion de Marche, Colonial Infantry, French Army.
He personally led a Franco-American platoon against a strongly fortified position. The heroic conduct displayed under heavy fire by this brilliant officer was an inspiration to the entire command.

HAUMANT, MARCEL A. E.
Near Daucourt, France, Sept. 15, 1918.
R—France.
G. O. No. 62, W. D., 1919.

Maréchal des Logis, 5th Regiment of Mounted Chasseurs, French Army.
While reconnoitering near the Bois-des-Haute-Epines he led his platoon in an attack on a hostile strong point and captured 10 prisoners.

HENDRICK, PIERRE H.
At Coullemelle Ferme, and at Berzy-le-Sec, south of Soissons, France, July 20–21, 1918.
R—France.
G. O. No. 100, W. D., 1918.

Captain of Infantry, French Army, attached to 1st Infantry Brigade, 1st Division, American Expeditionary Forces.
He gallantly crossed a zone under heavy fire to verify liaison with adjoining French troops, and during the attack on Berzy-le-Sec showed extraordinary heroism by his fearless exposure under heavy machine and artillery fire.

HOFFENBACH, EDOUARD.
Near Soissons, France, July 18, 1918.
R—France.
G. O. No. 126, W. D., 1919.

Captain of Infantry, French mission, attached to 9th Infantry, 2d Division, American Expeditionary Forces.
Captain Hoffenbach volunteered and led a combat liaison group between the 6th Marines and units on the flank, continuing on his mission until liaison had been established, when he was evacuated.

HOUROUX, ETIENNE.
Near Dun-sur-Meuse, France, Oct. 30, 1918.
R—France.
G. O. No. 62, W. D., 1919.

Sergeant-pilot, 284th Escadrille, French Army, attached to 2d Army Corps, American Expeditionary Forces.
While engaged in a visual reconnaissance Sergeant Houroux accepted combat with 4 enemy planes who attacked him with the object of forcing him to abandon his mission. Although he was himself seriously wounded, he sustained the unequal fight until his observer was killed by the fire of the attacking aircraft. When no further defense was left to him, he made a successful retreat into the allied lines and landed safely. Suffering severely from his wound, and too weak to leave the pilot's seat without assistance, he insisted that his observer be cared for before permitting anyone to aid him.

HUGO, JEAN.
At Cantigny, France, May 28–30, 1918.
R—France.
G. O. No. 125, W. D., 1918.

Lieutenant, 36th Regiment of Infantry, French Army, attached to 28th Infantry, 1st Division, American Expeditionary Forces.
During the attack and defense of Cantigny he showed utter disregard for personal danger and in critical situations inspired great confidence in those about him, contributing largely to the successful defense of the sector against repeated counterattacks. He rendered valuable assistance in placing troops in their positions and inspired confidence in the men by his coolness.

JACOBSON, ALFRED_____
Near the Ourcq River, France, July 26 to Aug. 2, 1918.
R—France.
G. O. No. 81, W. D., 1919.

Captain, 15th Regiment of Artillery, French Army, attached to 42d Division, American Expeditionary Forces.
When our advance was held up by stubborn resistance of the enemy in the Bois Brule, Captain Jacobson personally took a telephone to a point less than 300 meters from the enemy's lines and so directed the fire of our artillery as to compel the evacuation of the Bois. He was under continuous shellfire while performing this mission. On Aug. 2, when the regiment had lost contact with the enemy, he went forward with a small party at great risk of his own life and developed the enemy's line.

JACQUIN, ALFRED_____
Near Dun-sur-Meuse and Brieulles, France, Nov. 1-3, 1918.
R—France.
G. O. No. 37, W. D., 1919.

Lieutenant of Engineers, French Army, attached to 7th Engineers, 5th Division, American Expeditionary Forces.
Lieutenant Jacquin voluntarily patrolled the banks of the Meuse River day and night under machine-gun and shellfire locating enemy machine guns and artillery, determining the damage to bridges, and obtaining data for the location and construction of pontoon bridges. During the construction of these bridges, by means of which the heights overlooking the Meuse were reached and stormed, this officer remained constantly at the bridges under heavy fire, directing the work with the highest courage and technical skill.

JOURDE, GEORGES A. F._____
Near St. Mihiel, France, Sept. 12, 1918.
R—France.
G. O. No. 62, W. D., 1919.

Second lieutenant, 12th Regiment of Mounted Chasseurs, French Army.
He led his platoon with conspicuous bravery in clearing a section of trenches 500 meters long containing numerous machine guns which had been inflicting many casualties. Through his skillful leadership this position was reduced and 144 prisoners taken, including several officers.

LABOUR, RENÉ_____
Near Fismes, France, Sept. 2, 1918, and near Grand Pre, France, Oct. 15-16, 1918.
R—France.
G. O. No. 46, W. D., 1919.

Captain, 415th Regiment of Infantry, French Army, attached to 307th Infantry, 77th Division, American Expeditionary Forces.
He displayed singular bravery in going forward in the face of violent machine-gun fire to inspect the technical organization of the sector prior to the attack on Fismes of Sept. 2. During the night of Oct. 15-16, under perilous artillery and trench-mortar fire, he visited the front line and obtained most valuable information, which aided materially in the capture of the town.

LAGACHE, GUSTAVE P. A._____
Near St. Mihiel, France, Sept. 24, 1918.
R—France.
G. O. No. 53, W. D., 1920.

Captain of Air Service, French Army.
Captain Lagache brilliantly commanded the 101st French Squadron from Sept. 7-15, 1918, while it was assigned to the American Army during the St. Mihiel offensive. In an encounter with a superior force of the enemy on Sept. 24 he fearlessly attacked the group, and in the unequal combat which followed he persisted in the attack and succeeded in shooting down an enemy plane within our lines. This action took place while serving under American command.

LANO, MAURICE_____
Near Mouson, France, Nov. 7, 1918.
R—France.
G. O. No. 19, W. D., 1922.

Captain of Colonial Artillery, French mission, attached to 2d Infantry Brigade, 1st Division, American Expeditionary Forces.
Captain Lano, with another officer, made a most hazardous reconnaissance of the enemy position along the Meuse River and supplied valuable information of these positions. During the entire exploit they were constantly under enemy observation and heavy fire of their guns.

LARRA, JOSEPH_____
At Cheppy, France, Sept. 26, 1918.
R—France.
G. O. No. 70, W. D., 1919.

Second lieutenant, 16th Tank Corps, 504th Heavy Artillery, French Army.
He was acting as liaison officer between French tanks and American Infantry when the latter met with severe artillery and machine-gun fire from strong enemy positions. Passing through a heavy artillery barrage, he led 2 small tanks into action in an effort to overcome the enemy's resistance and personally charged a machine-gun nest with his pistol, killing 1 of the gunners and capturing the other 2. The 2 small tanks proving to be inadequate, this officer went back through the barrage and brought up 3 large tanks, leading them on foot in the open under intense shellfire and direct machine-gun fire. He maneuvered these tanks so skillfully that the subsequent capture of the stronghold of Cheppy by the Infantry was made possible.

LEANDRI, DOMINIQUE A._____
Near St. Mihiel, France, Sept. 12, 1918.
R—France.
G. O. No. 62, W. D., 1919.

Colonel, 8th Regiment of Mounted Chasseurs, French Army.
Leading his regiment in the attack in liaison with the 26th American Division, he directed the assault with distinguished gallantry and leadership. In an advance of 6 kilometers his command captured 1,780 prisoners, including 2 regimental commanders and 37 other officers, 90 machine guns, 3 heavy howitzers, 22 minenwerfers, one 77-millimeter gun, and a large quantity of other material.

LE BELLEGE, JEAN B._____
In the Argonne Forest, France, November, 1918.
R—France.
G. O. No. 81, W. D., 1919.

Maréchal des Logis, 22d Regiment of Artillery of Colonial Campaign, French Army.
This soldier was wounded on 3 different occasions, but each time remained on duty with his battery, consenting to be treated after his mission had been completed. His coolness under gravest circumstances won the admiration of all with whom he was connected.

LEBRE, ANTON_____
Near Montfauxelles, France, Sept. 29-Oct. 4, 1918.
R—France.
G. O. No. 37, W. D., 1919.

Captain, 344th Regiment of Infantry, French Army.
While on duty as senior French officer near the colonel of the 371st U. S. Infantry, Captain Lebre volunteered and went forward to make reconnaissance, during which he was wounded in the face by a bursting shell. After being evacuated he insisted on returning to duty with the regiment, though he was still suffering from fever caused by his wound and exposure during the battle.

LE CAM, YVES M.
Near Vauxbuin, Soissons, France, July 1, 1918.
R—France.
G. O. No. 81, W. D., 1919.

Private, first class, 153d Regiment of Infantry, French Army.
While advancing with and assisting an American colonel in an attack on the enemy Private Le Cam displayed meritorious valor in combating the enemy. With fixed bayonet he attacked the hostile party and exacted no less than 4 casualties, 1 of whom was a noncommissioned officer.

LECLERC, HIPPOLYTE A. M.
In the Verdun campaign, France, September–November, 1918.
R—France.
G. O. No. 62, W. D., 1919.

Major, 41st Regiment of Artillery of the Colonial Service, French Army.
By his valuable assistance rendered to the Infantry, which he was supporting, Major Leclerc, by the effective use of his batteries inflicted heavy casualties on the enemy. To better assist the American division attacking Brancourt, he occupied an advanced position, despite the fatigued condition of his troops, who were constantly subjected to machine-gun fire and high-explosive and poisonous shells.

LE COIN, RENÉ H.
Near Cantigny, France, May 28–30, 1918.
R—France.
G. O. No. 81, W. D., 1919.

Interpreter, French mission, attached to 28th Infantry, 1st Division, American Expeditionary Forces.
During the critical operations around Cantigny, Interpreter Le Coin was constantly on duty at an observation post. Seeing a group of men retreating in disorder, he rushed to them, stopped their retreat, and returned them to their positions on the line.

LENOIR, HENRI.
Near Vaux-Audigny, France, Oct. 11, 1918.
R—France.
G. O. No. 70, W. D., 1919.

Brigadier interpreter, 19th Squadron of the Train of Military Transports, French mission, attached to 120th Infantry, 30th Division, American Expeditionary Forces.
Discovering 2 of the enemy hiding, he captured them and turned them over to the battalion commander. The information regarding the enemy's line of defense and movement obtained from these prisoners proved to be correct and of the utmost value. On another occasion he braved the dangers of terrific hostile fire by going ahead of the battalion into a village and aiding the sick and wounded among the inhabitants, disregarding the fact that he was suffering agony from the effects of gas.

LE PELLETIER DE WOILLEMONT, BERNARD C. F. M. X. E. G.
Near St. Etienne, France, Oct. 9, 1918.
R—France.
G. O. No. 120, W. D., 1918.
Distinguished-service medal also awarded.

Lieutenant of Cavalry, French Army, liaison officer with the 2d Division, American Expeditionary Forces.
Lieutenant Le Pelletier De Woillemont, with Col. James C. Rhea, U. S. Army, voluntarily undertook an important reconnaissance under hazardous circumstances during the Massif Blanc Mont operations at a time when accurate information concerning our advanced positions was greatly needed and could not be obtained from other sources. In an automobile, whose conspicuous appearance drew the concentrated fire of artillery and machine guns, they proceeded 1 mile across open ground to the town of St. Etienne, where our troops were in contact with the enemy. Under fire these 2 officers reconnoitered the position of the front lines, locating the enemy, as well as that of the French units on the flank, and returned across the open with complete, reliable, and timely information of the highest military value in subsequent operations.

LEPLUS, PAUL.
Near Gesnes, France, Oct. 4, 1918.
R—France.
G. O. No. 78, W. D., 1919.

Lieutenant, 39th Regiment of Infantry, French Army, attached to 127th Infantry, 32d Division, American Expeditionary Forces.
Realizing the necessity and importance of the tanks in the attack on the Bois-de-la-Morine, Lieutenant Leplus volunteered and led the tanks over unfamiliar and difficult terrain. He continually exposed himself to the terrific fire of the enemy in their attempt to destroy the tanks, but carried his mission to a successful conclusion.

LEROUX, MAURICE.
At Blanc Mont, France, Oct. 4, 1918.
R—France.
G. O. No. 70, W. D., 1919.

Lieutenant, French Army, attached to 5th Regiment, U. S. Marine Corps, 2d Division, American Expeditionary Forces.
At great personal risk he volunteered and crossed an area swept by heavy machine-gun and artillery fire in order to establish liaison with French troops on the flank. He succeeded in locating these units and delivered an important message, displaying exceptional courage and utter disregard for personal danger.

LESCADRON, HENRI J.
Near St. Mihiel, France, Sept. 12–14, 1918.
R—France.
G. O. No. 70, W. D., 1919.

Second lieutenant, French mission, attached to 353th Infantry, 89th Division, American Expeditionary Forces.
Upon learning that 1 of the companies of the regiment had lost all its officers except the company commander, Lieutenant Lescadron left the regimental post of command and joined this company, which was in the first wave, gallantly assisting the company commander and going through artillery and machine-gun fire to aid the platoon leaders. He continued with this company until its final objective was reached and assisted in organizing the position.

LIARAS, GAËTAN.
In the Bois-de-Guisy, France, Sept. 26, 1918.
R—France.
G. O. No. 62, W. D., 1919.

Captain, 337th Company, 505th Regiment of Assault Artillery, French Army.
Preceding his 5 tanks on foot, he personally directed the attack on enemy machine guns and snipers that had held up the advance of an entire Infantry brigade. He killed 2 of the enemy himself and assisted in the capture of the remainder.

LORANS, MARCEL.
During the battle of the Ourcq, France, July 26 to Aug. 2, 1918.
R—France.
G. O. No. 44, W. D., 1919.

Captain of Infantry, French Army, attached to 42d Division, American Expeditionary Forces.
He remained constantly in the front line, attaching himself to 1 battalion after another as they in turn came into the fight, and by his energy and gallantry under fire setting a splendid example to the officers and soldiers of the regiment.

LORBER, J. B. X..............
    At Seletskoe, Russia, Sept. 14-15, 1918, and near Obozerskaya, Russia, Nov. 4, 1918.
    R—France.
    G. O. No. 24, W. D., 1920.

Lieutenant, 21st Battalion de Marche, Colonial Infantry, French Army.
On Sept. 14-15, 1918, he displayed great courage in exposing himself to heavy enemy fire while directing the fire of his machine guns. On Nov. 4 he further distinguished himself by going forward under heavy fire and supplying needed ammunition to advanced positions.

LORIOT, JEAN J..............
    Near St. Mihiel, France, Sept. 14, 1918.
    R—France.
    G. O. No. 81, W. D., 1919.

Sergeant-pilot, 151st Escadrille Spa, Air Service, French Army.
Sergeant Loriot on patrol met and was attacked by a large number of enemy planes (Fokker type), and in the course of the combat his motor and plane were severely damaged. By clever maneuvering he managed to elude the enemy planes and land safely behind the American lines.

LUCAS, EUGÈNE L. E..............
    Near Apremont, France, Sept. 12, 1918, and near Bois-de-Haudronvilles Bas, France, Sept. 16, 1918.
    R—France.
    G. O. No. 62, W. D., 1919.

Sergeant, 156th Regiment of Infantry, French Army.
On Sept. 12, the eve of the attack on Apremont, he alone reconnoitered the enemy lines and by exposing himself to machine-gun fire was able to supply his battalion commander with accurate information as to the position of these strongholds. On Sept. 16, aiding some French and American soldiers in attack, he successfully routed superior numbers of the enemy after a lively bayonet encounter.

MACON DE LA GICLAIS, JEAN..............
    Near Pexonne, France, Mar. 5, 1918.
    R—France.
    G. O. No. 126, W. D., 1919.

First lieutenant of Cavalry, French Army, attached to 42d Division, American Expeditionary Forces.
In action of Mar. 5, 1918, near Pexonne, France, although he might have remained in a place of safety, he went to the position of Battery C, 151st Field Artillery, 42d Division, when it was under bombardment by accurately adjusted artillery and by his courage and coolness assisted the officers and men of the command.

MALBE, FERNAND..............
    Near Gesnes, France, Oct. 5, 1918.
    R—France.
    G. O. No. 81, W. D., 1919.

Lieutenant, 350th Tank Company, French Army.
After 2 of his tanks had been destroyed, Lieutenant Malbe continued to lead his tank section in a desperate attack on the machine-gun nests at the crest of Hill 255. His entire disregard of personal danger in successfully carrying out this mission under the terrific enemy fire made it possible for our troops to reach the objective and hold the position.

*MALLET, ARTHUR H. G..............
    On the Vesle River, near Bazoches, France, Aug. 7, 1918.
    R—France.
    G. O. No. 4, W. D., 1923.

Lieutenant, French Army, liaison officer with the 2d Battalion, 47th Infantry, 4th Division, American Expeditionary Forces.
While serving as liaison officer with the 2d Battalion, 47th U. S. Infantry, which led the attack against the enemy and in the face of stubborn resistance, crossed the Vesle River, seized a critical position north of that stream, and held tenaciously to it throughout the day. Lieutenant Mallet, under a heavy and continuous hostile fire, repeatedly went from 1 front-line combat group to another, assisting materially in the successful conduct of the action by his courageous actions, suggestions, and professional skill. He rendered highly important services at a critical moment when the left of the line was sorely pressed by enemy counterattack and at all times by his soldierly conduct, inspiring courage, and high qualities of leadership was an heroic example to his comrades in arms. He was killed in action late in the afternoon.
Posthumously awarded. Medal presented to father, Frederic Mallet.

MARCHAND, LÉON..............
    During the battle of the Marne, July 15, 1918.
    R—France.
    G. O. No. 44, W. D., 1919.

Lieutenant, 202d Regiment of Infantry, French Army, attached to 30th Infantry, 3d Division, American Expeditionary Forces.
Lieutenant Marchand repeatedly displayed superb courage by voluntarily proceeding from the regimental command post dugout to an observation post on the edge of the woods through intense shellfire in order to observe the progress of the action and obtain information necessary for the commanding officer. The superb courage of Lieutenant Marchand was an inspiration to the men of the regiment to which he was attached.
Oak-leaf cluster.
For the following acts of extraordinary heroism in action near the Ferme de Madeleine du Cunel, France, Oct. 14, 1918, Lieutenant Marchand is awarded an oak-leaf cluster: He twice voluntarily accompanied attacking troops through heavy enemy barrages and on 1 of these occasions was caught between friendly and hostile machine-gun fire, displaying notable coolness and gallantry. Later in the same night he volunteered and lead a patrol into the Bois-de-Pultiere to locate a dugout for the advance regimental post command, although he knew the woods were saturated with gas, by which he was overcome.

MARÉCHAL, ADRIEN..............
    At Pouilly, Meuse, France, Nov. 3, 1918.
    R—France
    G. O. 81, W. D., 1919.

Second lieutenant, observer, 214th Aero Squadron, French Army, attached to 5th Army Corps observation group, American Expeditionary Forces.
After 2 fingers of his right hand had been shot away, a hole shot through his hand, and the trigger of 1 machine gun blown off, Lieutenant Maréchal continued to fire with his left hand on 5 enemy monoplace planes which had attacked him over the hostile lines. By doing this he enabled the pilot to bring his damaged machine back to our own lines and brought valuable information of our own and enemy troops. After landing he made his report before he received medical treatment.

MARIUS, JEAN..............
    Near Vilosnes, France, Nov. 3, 1918.
    R—France.
    G. O. No. 62, W. D., 1919.

Sergeant, 5th Regiment of Colonial Infantry, French Army.
He voluntarily accompanied an American officer on an engineering reconnaissance of a dangerous character. When the detachment was almost surrounded by the enemy he succeeded in killing several Germans and thereby saved the officer's life.

MARTELLIÈRE, ANDRÉ P.
In the vicinity of Juvigny, France, Aug. 20 and Sept. 2, 1918.
R—France.
G. O. No. 53, W. D., 1920.

Captain, 8th Company, 64th Battalion, Chasseurs Alpins, French Army.
Captain Martellière, in command of the 8th Company of 64th Battalion, Chasseurs Alpins, French Army, which company operated as a liaison unit of the 32d American Division, maintained communication on an exposed flank under heavy enemy fire during the 4 days of the operation. On Sept. 2, although suffering from the effects of gas, he continued on and fearlessly led his company in the assault on Mont-de-Leuilly under heavy enemy fire and assisted in the capture of the position, together with a number of prisoners and much material.

*MAXWELL, ROGER.
At Vaulny, France, Sept. 26–30, 1918.
R—France.
G. O. No. 19, W. D., 1922.

First lieutenant of Infantry, French Army, information officer attached to 369th Infantry, 93d Division, American Expeditionary Forces.
For extraordinary heroism while serving as information officer attached to the 369th Infantry, 93d Division, U. S. Army. Lieutenant Maxwell accomplished liaison missions under very difficult circumstances and gave valuable help to the general commanding the Army corps during the battle of Sept. 26–30, 1918.
Posthumously awarded. Medal forwarded to the military attaché, American Embassy, Paris, France, for delivery to next of kin.

MENI, JEAN.
In the St. Mihiel offensive, France, Sept. 12, 1918.
R—France.
G. O. No. 62, W. D., 1919.

Captain of Air Service, 16th Pursuit Group, Air Service, French Army.
Despite the unfavorable weather conditions, Captain Meni made a reconnaissance flight over the enemy lines, returning with valuable information concerning the evacuation of the enemy and the dominant position of Montsec.

MÉREL, ROBERT A.
Near Spittaals Bosschen and Audenarde, Belgium, Oct. 31 and Nov. 1, 1918.
R—France.
G. O. No. 46, W. D., 1919.

First lieutenant of Infantry, French mission, attached to 91st Division, American Expeditionary Forces.
Armed with an automatic rifle, he went forward alone and killed the gunner of an enemy machine gun whose fire had been holding up the advancing line.

MEURISSE, JEAN L.
Near Chevillon, France, July 18, 1918.
R—France.
G. O. No. 81, W. D., 1919.

Captain, 27th Regiment of Infantry, French Army.
Acting as liaison officer with the 58th American Infantry, he showed marked personal courage under intense fire, setting an example of fearlessness to the officers and men with him. His knowledge of German artillery enabled him to advise methods of approach for our troops which were instrumental in preventing many casualties.

MICHEL, MARCEL H.
In the region of St. Mihiel, France, Sept. 12, 1918.
R—France.
G. O. No. 81, W. D., 1919.

Sergeant-pilot, 218th Escadrille, French Army.
Sergeant Michel while on a reconnaissance had his wireless outfit destroyed by enemy's fire. Realizing that he could not communicate with his lines and also being well into the enemy's territory, he descended to an extremely low altitude. Disregarding the enemy's antiaircraft and ground machine-gun fire, he attacked an enemy convoy, causing considerable damage.

*MILLERET, NORBERT.
Near Thiaucourt, France, Sept. 14–Oct. 3, 1918, and near the Bois-des-Loges, France, Oct. 15–19, 1918.
R—France.
G. O. No. 78, W. D., 1919.

First lieutenant, 49th Regiment of Infantry, French Army, liaison officer with the 155th Infantry Brigade, 78th Division, American Expeditionary Forces.
As liaison officer with the 155th Infantry Brigade, Lieutenant Milleret was untiring in his constant efforts to further the success of the operations and repeatedly ignored his personal safety in visiting observation posts and assisting in the machine-gun and intelligence work of the brigade. This gallant officer was killed by shellfire on October 19, 1918.
Posthumously awarded. Medal forwarded to widow, Madam Norbert Milleret.

MOREL, JULIEN.
Near Cheppy, France, Sept. 18, 1918.
R—France.
G. O. No. 53, W. D., 1920.

French interpreter attached to 138th Infantry, 35th Division, American Expeditionary Forces.
Attached as an interpreter to the 138th Infantry, he volunteered to go forward with the organization in an attack, and though himself wounded helped to carry a wounded officer from the battle field and then returning led the tanks to Cheppy, making possible the entrance of the Infantry to that town.

PEPIN, EDMOND.
In the region of St. Mihiel, France, Sept. 12, 1918.
R—France.
G. O. No. 81, W. D., 1919.

Second lieutenant, pilot, 47th Escadrille Sal, French Army.
Lieutenant Pepin in the most adverse weather conditions flew at an extremely low altitude for 2½ hours, thoroughly reconnoitering enemy positions and returning with information of the greatest value.

PERRIN, EDOUARD.
Near Gesnes, France, Oct. 4, 1918.
R—France.
G. O. No. 78, W. D., 1919.

Brigadier, 350th Tank Company, Frency Army.
During the first attack on Hill 255 Brigadier Perrin's tank was destroyed and he was captured. On the following night he escaped and on the following day took part in the second attack on the hill. He was wounded soon after the engagement began, but continued to operate his gun, giving an important support to our advancing Infantry. He refused to leave the field for treatment until the mission had been successfully completed.

PETIT, AUGUSTE J.
East of Deucourt, France, Sept. 27, 1918.
R—France.
G. O. No. 62, W. D., 1919.

Maréchal des Logis, 5th Company, 5th Regiment of Mounted Chasseurs, French Army.
Having been ordered to attack Bonvrot Farm, he led his platoon forward with fixed bayonets and cut off the enemy, capturing the position, which had been occupied by the enemy since dawn. His platoon, numbering but 20, took 85 prisoners and 4 machine guns.

PIVETEAU_____ Lieutenant, 3d Battalion of Light Tanks, Assault Artillery, French Army.
Near Massif Blanc Mont, France, During 2 attacks by the 2d American Division, he repeatedly distinguished
Oct. 3–8, 1918. himself by his courage and utter disregard of danger in transmitting orders
R—France. and gathering important information for his battalion commander. His zeal,
G. O. No. 126, W. D., 1919. devotion to duty, and initiative were of the highest order.

PRALY, LOUIS_____ Aspirant, 350th Company, Tank Corps, French Army.
Near Gesnes, France, Oct. 4, 1918. Commanding one of the tank sections, which was preceding the advance of the
R—France. American Infantry, Aspirant Praly displayed exceptional bravery in destroy-
G. O. No. 78, W. D., 1919. ing machine-gun nests until his tank received a direct hit from the enemy's
artillery. Notwithstanding that his tank was unable to continue forward
and that he himself was wounded, Aspirant Praly continued to operate his
guns until all his ammunition was exhausted.

PRUDHOMME, FRÉDÉRIC H_____ Second lieutenant, 12th Regiment of Mounted Chasseurs, French Army.
At Woel, France, Sept. 14, 1918. Having been ordered to occupy the village of Woel, which he found to be held
R—France. by 1 company of Germans, Lieutenant Prudhomme unhesitatingly at-
G. O. No. 62, W. D., 1919. tacked with 1 section and captured the town, taking 18 prisoners, 2 machine
guns, and killing many of the enemy. He then held the position for 24 hours
against several hostile counterattacks.

QUINTON, R._____ Lieutenant colonel, 452d Regiment of Field Artillery, French Army, attached
to 2d Division, American Expeditionary Forces.
Near Blanc Mont, France, Oct. 3–17, During the attack on Blanc Mont he, undeterred by heavy shellfire, personally
1918. made reconnaissance to the front each day and secured information of the
R—France. enemy to determine locations for his batteries.
G. O. No. 62, W. D., 1919.

RAULT, JEAN E._____ Captain, 5th Regiment of Colonial Infantry, French Army.
Near Haudiomont, France, Sept. In making a local attack in the Bois-de-Manheulles Captain Rault by skillful
26, 1918. maneuvering succeeded in cutting off the retreat of an entire company of the
R—France. enemy.
G. O. No. 62, W. D., 1919.

RAVISSE, HENRI._____ Captain, 153d Regiment of Infantry, French Army.
Near Montsec, France, Sept. 12, 1918. Riding ahead of his lines under most terrific fire, he established and maintained
R—France. liaison with the next American division. On many occasions he rendered
G. O. No. 62, W. D., 1919. most valuable assistance to the allied armies, undertaking most perilous
missions to insure communication between troops.

REDIER, MAURICE._____ Captain, 65th Regiment of Infantry, French Army, attached to 72d Infantry
Brigade, 36th Division, American Expeditionay Forces.
Near Pauvres, France, Oct. 13, 1918. Captain Redier moved forward in advance of our infantry patrols of units on
R—France. the flank. He displayed great coolness and dash under artillery and machine-
G. O. No. 126, W. D., 1919. gun fire. He also entered Vaux-Champagne in advance of our troops. His
conduct was an inspiration to all troops in the attack.

REISS, ANDRÉ._____ Interpreter, French Army, attached to 4th Machine Gun Battalion, 2d Divi-
sion, American Expeditionary Forces.
During the St. Mihiel offensive, While attached to the 4th Machine Gun Battalion, he voluntarily assumed
France, 1918, and at Mont Blanc, the duties of the battalion adjutant who had been injured. In reconnaissance,
France, Oct. 5–9, 1918. maintaining communication, and establishing liaison he displayed absolute
R—France. fearlessness and rendered valuable assistance to the battalion. At Mont
G. O. No. 49, W. D., 1922. Blanc, Oct. 5–9, 1918, he continually went back and forth through artillery
and machine-gun fire in order to maintain liaison between the 4th Machine
Gun Battalion and the French unit on the right.

RERAT, ARMAND._____ Lieutenant of Infantry, French Army, attached to 42d Division, American
Expeditionary Forces.
In Champagne, France, July 15–18, During the German attack of July 15–18 in Champagne Lieutenant Rerat vol-
1918, and on the Ourcq River, untarily joined the 2d Battalion of the 165th Infantry in the fight in the front
France, July 26 to Aug. 2, 1918. line and was conspicuous for his bravery. He again behaved himself very
R—France. gallantly during the attack on the Ourcq River July 26 to Aug. 2, 1918, where
G. O. No. 37, W. D., 1919. he was slightly wounded but refused aid until the fight was over.

RITT, MAURICE J. V._____ Lieutenant of Infantry, French mission, attached to 127th Infantry, 32d Divi-
sion, American Expeditionary Forces.
Near Juvigny, north of Soissons, Lieutenant Ritt assisted in establishing an advanced machine-gun position in
France, Aug. 31, 1918. the village of Juvigny, the fire of which forced the surrender of 32 enemy
R—France. prisoners. This gallant officer assisted in forming a new line and went from
G. O. No. 81, W. D., 1919. one end to the other picking up stragglers and getting the line organized,
exposing himself to machine-gun fire throughout the operation. Three times
he went through heavy fire for the purpose of maintaining liaison with a
French division on the right.

ROBERT._____ Second lieutenant, 21st Battalion de Marche, Colonial Infantry, French Army.
At Verst 455, north Russia, Oct. 14, Lieutenant Robert placed himself at the head of a Franco-American section and
1918. personally led it against a strongly fortified position. The courage displayed
R—France. by this brilliant officer under heavy machine-gun fire was an important factor
G. O. No. 24, W. D., 1920. in the successful attack.

ROUSSEL, JEAN._____ Private, first class, 3d Company, 12th Regiment of Mounted Chasseurs, French
Army.
Near St. Mihiel, France, Sept. 12–13, While engaged on liaison duty with his platoon in the front line he displayed
1918. remarkable daring in an encounter with 10 of the enemy. After seizing a
R—France. rifle from the hands of 1 of them he succeeded in disarming the others and
G. O. No. 62, W. D., 1919. brought them to our lines.

SANDEAU......................
At Verst 455, north Russia, Oct. 14, 1918.
R—France.
G. O. No. 24, W. D., 1920.

Aspirant, 21st Battalion de Marche, Colonial Infantry, French Army.
Aspirant Sandeau personally led a Franco-American section against a strongly fortified enemy position. The courage displayed by him while exposed to heavy machine-gun and rifle fire was an important factor in the successful attack.

SANTINI, PHILIPPE...............
In France, July 15–16, Oct. 14, 1918, and in the Argonne-Meuse, France, Nov. 9, 1918.
R—France.
G. O. No. 62, W. D., 1919.

Lieutenant, 53d Regiment of Colonial Infantry, French Army.
On July 15–16 Lieutenant Santini defended a stronghold against overwhelming forces of the enemy for a period of 34 hours until assisted by reinforcements. On Oct. 14, under cover of a fog, he carried on a raid against an enemy picket and without loss he captured 14 prisoners and 1 machine gun. On Nov. 9, after being severely wounded, he took a strong enemy position after a severe struggle and also captured 9 prisoners and 2 machine guns.

SARDIER, GILBERT J. M. L...........
Near Mesnil-St.-Firmin, France, May 15, 1918, and north of Chateau-Thierry, France, June 4, 1918.
R—France.
G. O. No. 53, W. D., 1920.

First lieutenant of Air Service, French Army.
On May 15 Lieutenant Sardier, while a member on a patrol, left his unit and alone attacked and destroyed 2 enemy planes (single seaters). On June 4, while under American command, he attacked and burned 2 enemy balloons. On Sept. 14, near St. Mihiel, he attacked 2 enemy balloons and drove off an enemy biplane. This action took place while serving under American command.

SARTORIOUS, EMILE..............
At Vadenay, north of Chalons-sur-Marne, France, July 15, 1918.
R—France.
G. O. No. 117, W. D., 1918.

Adjutant interpreter, French Army, attached to 42d Division, American Expeditionary Forces.
During the shelling of Vadenay on the morning of July 15, 1918, he voluntarily left a place of safety to conduct American troops to shelter under a heavy fire of major-caliber shells and was severely wounded.

SIMONET, ANDRÉ.................
Near Trieres Farm, France, Sept. 30, 1918.
R—France.
G. O. No. 37, W. D., 1919.

Maréchal des Logis, 19th Train of Military Transports, French Army, attached to 371st Infantry, 93d Division, American Expeditionary Forces.
While on duty with the 371st U. S. Infantry as interpreter he rendered exceptional service to our forces by assuming command until the second in command could be notified, when the adjutant and commanding officer of the battalion to which he was attached were wounded. At this time he made a voluntary trip to the regimental post of command to report conditions in the battalion.

TCHEIMESSOFF, SERGUEI...........
Near Kadish, Russia, Feb. 7, 1919.
R—France.
G. O. No. 24, W. D., 1920.

Private, French Foreign Legion.
While serving under American command, Private Tcheimessoff heroically defended his position against attack of overwhelming numbers. Although severely wounded in the head, he refused to leave his position, but continued at his post until all others had retired.

TESSIER, FERNAND...............
In the Champagne sector, France, Sept. 26–Oct. 1, 1918.
R—France.
G. O. No. 62, W. D., 1919.

First lieutenant, 14th Regiment of Chasseurs, French Army, attached to 369th Infantry, 93d Division, American Expeditionary Forces.
This officer was attached to the 369th Infantry as liaison officer and by continuous passage through zones which were under most intense fire maintained perfect liaison with brigade headquarters. In the attack on Sechault he was constantly in touch with the most advanced lines and personally carried messages to the assaulting battalions. He aided materially in holding the captured positions.

THIABAUD, CLAUDE E............
At La Ferme and Damvillers, France, Nov. 7–10, 1918.
R—France.
G. O. No. 62, W. D., 1919.

Lieutenant, 6th Regiment of Colonial Infantry, French Army.
In liaison with American troops he rendered most valuable assistance and greatly aided in driving the enemy from La Ferme and Damvillers. Throughout the war he participated in all the important attacks of his organization, and his brilliant leadership, singular courage, and devotion to duty played an important part in the success of the operations.

THIEBAULT, RENÉ...............
Near Massif Blanc Mont, France, Oct. 3, 1918.
R—France.
G. O. No. 126, W. D., 1919.

Maréchal des Logis, 308th Company, Tank Corps, 3d Battalion of Light Tanks, French Army.
Following an attack by the 2d American Division he distinguished himself by personal courage and coolness in going under heavy shellfire to the rescue of wounded American soldiers.

TRIBOT-LASPIÈRRE, JEAN R........
Near Bois-de-Belleau, France, June 6, 1918.
R—France.
G. O. No. 28, W. D., 1921.

Captain, 1st Algerian Tirailleurs (rifle regiment), French Army.
Although suffering severely from shell shock and the effects of gas, Captain Tribot-Laspièrre made his way from the 6th Regiment to the 5th Regiment, U. S. Marine Corps, through very heavy and effective fire, with information of great importance.

TRIVES, FRANCOIS M............
Near Haumont, France, Nov. 2, 1918.
R—France.
G. O. No. 126, W. D., 1919.

Captain of Artillery, French Army, liaison officer with the 164th Artillery Brigade, 89th Division, American Expeditionary Forces.
Captain Trives volunteered to accompany a raiding party of the 28th Division. On reaching the German wire the patrol became confused and disorganized. Realizing that the party was in great danger of being caught in a heavy barrage, Captain Trives quickly reorganized the patrol, working under heavy fire, and continued to lead the raiding party until he was seriously wounded.

VACARISAS, JOSEPH.............
Near Mortzwiller, Alsace, Sept. 1, 1918.
R—France.
G. O. No. 81, W. D., 1919.

Brigadier, 19th Battery, E. V. N., 70th Regiment of Artillery, French Army.
During an unusually heavy enemy bombardment of the railroad station, Brigadier Vacarisas entered the area and assisted the train crew in removing to safety 9 carloads of ammunition. The mission was accomplished despite the fact that the firing had damaged the track and had scattered branches of trees over the route.

VALLOIS, ROBERT_____
Near Thiaucourt, France, Sept. 12, 1918.
R—France.
G. O. No. 46, W. D., 1919.

Captain of Air Service. observer, 16th Pursuit Group, French Army.
Captain Vallois volunteered to fly with Maj. Lewis H. Brereton, U. S. Army, on an important reconnaissance mission. On account of poor visibility, they were forced to fly at a very low altitude and were continually harassed by antiaircraft fire. 4 enemy monoplace planes (type Fokker) attacked them, and during the combat which followed Captain Vallois's gun jammed. After withdrawing for the purpose of clearing the jam, they again returned to the fight, and despite the fact that he had been painfully wounded in the face Captain Vallois succeeded in dispersing 3 of the adversaries and fought off the other while his pilot made a landing.

"VERRY, LOUIS_____
Near Soissons, France, July 18, 1918.
R—France.
G. O. No. 35, W. D., 1920.

Maréchal des Logis, interpreter, French Army, attached to 5th Regiment, U. S. Marine Corps, 2d Division, American Expeditionary Forces.
Under heavy enemy shellfire Maréchal des Logis Verry volunteered to lead a liaison patrol for the purpose of establishing liaison with French units on the left. He and 3 American soldiers were killed by shellfire after advancing only a few yards on this perilous mission.
Posthumously awarded. Medal forwarded to Franco-American special bureau, Paris, France, for delivery to next of kin.

VIAUD, LOUIS_____
In France, on the night of Apr. 18-19, 1918.
R—France.
G. O. No. 126, W. D., 1918.

First lieutenant, 1st Company, 20th Regiment of Infantry, French Army.
Lieutenant Viaud led a French-American detachment in a raid with the greatest coolness and bravery, displaying leadership and resourcefulness. When counterattacked by a strong German force he maintained his ground and repulsed the enemy.

VIVIEN, ROBERT_____
Near Bois-les-Marettes, France, June 1, 1918, and at Villers-Cotterets Woods, France, July 18, 1918.
R—France.
G. O. No. 126, W. D., 1919.

Captain of Infantry, French Army, attached to 9th Infantry, 2d Division, American Expeditionary Forces.
Upon the arrival of the 9th Infantry in the Chateau-Thierry sector June 1, 1918, Captain Vivien assisted in placing several companies in position and established liaison with the French. At this time the position of the enemy was unknown. Captain Vivien performed this act at great risk of being captured by the enemy. In the attack south of Soissons Captain Vivien assisted in re-forming the line and reorganizing combat groups after the attack on the German positions.

WACKERNIE, GEORGES_____
At Chateau-Thierry, France, May 31 to June 4, 1918.
R—France.
G. O. No. 126, W. D., 1918.

Lieutenant, 54th Regiment of Infantry, French Army, attached to 7th Machine Gun Battalion, 3d Division, American Expeditionary Forces.
During the operations against the enemy at Chateau-Thierry, France, from May 31 to June 4, 1918, he constantly distinguished himself by his extraordinary heroism in voluntarily going through heavy machine-gun fire in order to secure and give important information which could not otherwise be communicated.

WICHART, GEORGES_____
Near Monthois, France, Sept. 27 to Oct. 7, 1918.
R—France.
G. O. No. 13, W. D., 1919.

Second lieutenant of Infantry, French Army, attached to 372d Infantry, 93d Division, American Expeditionary Forces.
During the attack on Monthois he voluntarily undertook the most hazardous missions, fearlessly traversing ground swept by machine-gun fire and severe bombardment to secure liaison between neighboring French units and to reconnoiter our first-line positions. His reports were invaluable. On the night of Oct. 2 he led a battalion to its position of attack and personally reconnoitered the line under intense machine-gun and artillery fire, furnishing a splendid example of coolness and utter disregard of danger to the men of the battalion.

The Distinguished-Service Medal

٦٨٩

ALPHABETICAL LIST OF AWARDS OF THE DISTINGUISHED-SERVICE MEDAL IN NATIONAL GROUPS

AMERICANS

[Awarded for exceptionally meritorious and distinguished services, in a position of great responsibility, under the provisions of the act of Congress approved July 9, 1918, except as otherwise indicated]

ABBOT, FREDERIC V.
  R—Willetts Point, N. Y.
  B—Massachusetts.
  G. O. No. 16, W. D., 1921.

Brigadier general, U. S. Army.
For services in the organization of engineer troops and the procurement of enlisted men for the service in the war. His zeal was untiring and the success of his effort marked.

ACHER, ALBERT H.
  R—Grove City, Pa.
  B—Greenville, Pa.
  G. O. No. 95, W. D., 1919.

Colonel, Corps of Engineers, U. S. Army.
As commanding officer of the 4th Engineers he contributed materially to the successes of the 4th Division in the Aisne-Marne offensive and in the Meuse-Argonne operations. By his skill in the construction of roads and bridges, he ably assisted in the operations of his division. His ability as a leader was shown in the efficiency of the 4th Engineers, both as a technical and as a combat unit. Later he showed the same rare qualities when he commanded the 27th Engineers.

ADAMS, EMORY S.
  R—Manhattan, Kans.
  B—Manhattan, Kans.
  G. O. No. 16, W. D., 1923.

Colonel, Infantry, U. S. Army.
As adjutant, Base Section No. 5, Brest, France, from Dec. 10, 1918, to Dec. 20, 1919, he displayed exceptional administrative and executive ability, sound judgment, uniform courtesy and unremitting devotion to duty, contributing markedly to the successful accomplishments of Base Section No. 5, upon which was placed among other duties the responsibility for the repatriation of more than a million American soldiers.

\MS, HARRY M.
  R—Omaha, Nebr.
  B—Comanche, Iowa.
  G. O. No. 35, W. D., 1919.

Director of inland traffic service, War Department.
His responsibilities have been great in supervising the utilization of railroad facilities and the immense movement of troops and supplies during the war. His excellent judgment and marked ability have contributed materially to the successful and orderly movement of troops and supplies to the ports of embarkation and for the Army overseas.

\MS, JOHN H.
  R—La Porte, Ind.
  B—Chicago, Ill.
  G. O. No. 38, W. D., 1922.

Lieutenant colonel, Quartermaster Corps, U. S. Army.
For services as assistant chief, and later as chief of the Subsistence Division, Office of the Quartermaster General. His keen foresight and able grasp of the problems at hand made him an invaluable aid during those days when the all-important work of organizing the food supply was under way. By his tireless energy and marked ability he conducted his duties in the procurement and supply of subsistence stores, as to meet satisfactorily the all-important needs of the Army. In both positions he demonstrated marked business ability, good judgment, and carried to a successful conclusion each project presented to him.

GEORGE E.
  ¬an, Tex.
  ısville, Tex.
). 56, W. D., 1922.

Captain, Adjutant General's Department, U. S. Army.
As confidential secretary of the Commander in Chief, American Expeditionary Forces, and later of the General of the Armies, he has worked untiringly and given proof of marked ability and resourcefulness. In the multifarious details connected with his duties he has at all times displayed keen judgment, tact, unfailing courtesy, and loyalty. In the office of the Commander in Chief, American Expeditionary Forces, where there devolved upon Captain Adamson a great volume of work and a mass of detail, he handled each new problem which confronted him in an able and masterful manner and rendered invaluable services to the American Expeditionary Forces in a position of great responsibility and in times and circumstances of the gravest importance.

ᴇMMETT.
  Youngstown, Ohio.
  New Haven, Conn.
  O. No. 56, W. D., 1922.

Colonel (Cavalry), General Staff Corps, U. S. Army.
He served with marked ability as Assistant Chief of Staff, G–3, 38th Division, during the early days of its organization and training. As an instructor at the Army General Staff College, American Expeditionary Forces, he displayed high professional attainments, and unfailing energy, performing service of inestimable worth in connection with the instruction and training of officers for General Staff duty. Later in the Office of the Provost Marshal General he again demonstrated those splendid characteristics which have at all times been outstanding features of his service.

BRIGHT, OWEN S.
  R—Memphis, Tenn.
  B—St. Louis, Mo.
  G. O. No. 35, W. D., 1920.

Lieutenant colonel, Signal Corps, U. S. Army.
In command of field signal battalions at the front, as an instructor at corps schools, and as division signal officer of the 2d Division, he rendered services of great value to the American Expeditionary Forces.

\LDEN, HERBERT W.
  R—Detroit, Mich.
  B—Vermont.
  G. O. No. 77, W. D., 1919.

Lieutenant colonel, Ordnance Department, U. S. Army.
For services first, as American engineering representative at the conference called to design the Anglo-American Mark VIII tank, and later as being directly responsible for the design of a new, valuable, and easily obtained implement of mechanical warfare, the fast 3-ton tank, susceptible of production in America in such quantity as to constitute a most material contribution to the effective fighting power of the United States Army.

669

ALESHIRE, JOSEPH P.
R—Washington, D. C
B—Fort Custer, Mont.
G. O. No. 56, W. D., 1922.

Major, Quartermaster Corps, U. S. Army.
As Assistant Chief of Staff, G–3, 81st Division, from October, 1918, until March, 1919, by his marked ability, high professional attainments, and loyal devotion to duty, he rendered valuable assistance in the staff work in the Vosges during the Meuse-Argonne offensive, thereby contributing materially to the success of his division in those operations.

ALEXANDER, ROGER G.
R—Paris, Mo.
B—Paris, Mo.
G. O. No. 59, W. D., 1919.

Colonel (Corps of Engineers), General Staff Corps, U. S. Army.
As chief of the topographical division of the intelligence section he organized and administered, with exceptional ability, the topographical and sound and flash ranging services of the American Expeditionary Forces. Due to his foresight and energy our armies in the field were at all times supplied abundantly with excellent maps of the theater of operations.

ALLEN, CHARLES C
R—Philadelphia, Pa.
B—Philadelphia, Pa.
G. O. No. 43, W. D., 1922.

Lieutenant colonel, Infantry, U. S. Army.
As G–2 of the 33d Division from August, 1917, to June, 1918, and from September to November, 1918, he displayed sound judgment and exceptional ability in the organization, administration, and operation of that section of the division staff. By his tireless energy, military attainments, and unceasing devotion to duty he contributed greatly to the successes of the division during the Meuse-Argonne offensive and the operations in the Woevre Valley from September to November, 1918.

ALLEN, HENRY A
R—Chicago, Ill
B—Madison, Wis.
G. O. No. 89, W. D., 1919.

Colonel, Corps of Engineers, U. S. Army.
He served with distinction as commanding officer of the 108th Engineers and as engineer officer of the 33d Division. By his technical skill and untiring energy in supervising the construction of bridges across the Meuse River he proved himself an important factor in the successes gained by our troops in their operations along the right bank of that stream during the Meuse-Argonne offensive.

ALLEN, HENRY T
R—Sharpsburg, Ky.
B—Sharpsburg, Ky.
G. O. No. 12, W. D., 1919.

Major general, U. S. Army.
In command of the 19th Division he had the important position of conducting the right flank at the St. Mihiel salient. The brilliant success there gained and later repeated in the Argonne-Meuse offensive showed him to be an officer of splendid judgment, high attainments, and excellent leadership. Later he commanded the Eighth Army Corps with skill and judgment.

ALLEN, ROBERT H
R—Buchanan, Va.
B—Buchanan, Va.
G. O. No. 87, W. D., 1919.

Colonel, Infantry, U. S. Army.
As commander of the 356th Infantry during the Argonne-Meuse offensive he proved himself a skillful tactician. Resourceful and energetic, he was at all times equal to any emergency which arose, showing qualities of rare leadership. Subsequently during the march into Germany and the occupation of the enemy territory his administrative ability was reflected in the high standard of excellence consistently maintained by his regiment, rendering services of signal worth.

ALLIN, GEORGE R
R—Iowa City, Iowa.
B—Iowa City, Iowa.
G. O. No. 31, W. D., 1922.

Brigadier general, U. S. Army.
As executive officer and director of training in the office of the Chief of Field Artillery from Mar. 21, 1918, to Sept. 1, 1918, by reason of his high professional attainments, ability, foresight, and judgment, he rendered invaluable aid in solving the many complex problems confronting his arm of the service.

ALLISON, JAMES B
R—Yorkville, S. C.
B—Yorkville, S. C.
G. O. No. 3, W. D., 1921.

Colonel, Signal Corps, U. S. Army.
For services in the organization and training of technical troops of the Signal Corps, while commanding officer of the Signal Corps Training School, Fort Leavenworth, Kans., and commanding officer of the Franklin cantonment, Camp Meade, Md.

ALLISON, NATHANIEL
R—St. Louis, Mo.
B—St. Louis, Mo.
G. O. No. 50, W. D., 1919.

Colonel, Medical Corps, U. S. Army.
As chief of the orthopedic work in the zone of the Army, he personally directed in a most efficient, conscientious, and painstaking manner splinting and orthopedic work, which resulted in the saving of many lives and greatly relieved suffering among our wounded.

ALMY, EDMUND D
R—Altamont, Ky.
B—Wellsville, N. Y.
G. O. No. 116, W. D., 1919.

Commander, U. S. Navy.
For services as force engineer officer, in which position, by his untiring energy and close cooperation with the Army authorities, he successfully equipped a large number of Army and Navy transports.

ALVORD, BENJAMIN
R—Washington, D. C.
B—Fort Vancouver, Washington Territory.
G. O. No. 87, W. D., 1919.

Colonel, Adjutant General's Department, U. S. Army.
As adjutant general of the American Expeditionary Forces during the beginning of its organization his long experience, good judgment, and breadth of vision were of value in the establishment of the innumerable activities of the adjutant general's department of the American Expeditionary Forces.

ANDERSON, ALEXANDER E
R—New York, N. Y.
B—New York, N. Y.
G. O. No. 56, W. D., 1922.
Distinguished-service cross also awarded.

Major, Infantry, U. S. Army.
He served with the 165th Infantry throughout all its operations, displaying military attainments of the highest order. By his fearless bravery and splendid leadership he at all times inspired a notable spirit among the members of his command. His unflagging energy and resourcefulness in overcoming the numerous adverse conditions which confronted his command marked him as an officer of splendid soldierly qualities. By his sound tactical judgment, keen foresight, and aggressive fighting spirit he proved himself an important factor in the successes of his regiment and division. He rendered services of conspicuous worth to the American Expeditionary Forces.

ANDERSON, ALVORD V. P.
R—Montclair, N. J.
B—New York, N. Y.
G. O. No. 49, W. D., 1922.

Colonel, Infantry, U. S. Army.
As commander of the 312th Infantry throughout its organization, training, and all its active operations he displayed marked efficiency, unflagging energy, and military attainments of the highest order. In the attack on Grand Pre, during the Meuse-Argonne offensive, by his prompt conception and brilliant execution of a skillful and successful attack on that strong and dominating position, he contributed largely to our successes in that great operation.

ANDERSON, EDWARD D.
R—Jasper, Tenn.
B—Jasper, Tenn.
G. O. No. 18, W. D., 1919.

Brigadier general, U. S. Army.
For services in initiating and executing plans for the mobilization of enlisted personnel of the Army during the war.

ANDRESS, MARY VAIL.
R—New York, N. Y.
B—New Jersey.
G. O. No. 70, W. D., 1919.

American Red Cross.
On her own initiative she organized and efficiently developed and administered the work of the American Red Cross at Toul, France. Under her wise supervision this work grew from the ministering and supplying of small comforts to soldiers passing through in hospital trains to an undertaking of extensive proportions, which has aided and cheered thousands of men in the service. In the performance of her exacting tasks, she has displayed marked foresight and sound judgment, with untiring personal devotion to the interests and comfort of those whom she served.

ANDREW, ABRAM PIATT.
R—Gloucester, Mass.
B—La Porte, Ind.
G. O. No. 59, W. D., 1919.

Lieutenant colonel, Ambulance Corps, U. S. Army.
Coming to France at the beginning of the war he showed remarkable ability in organizing the American Field Service, a volunteer service for the transportation of the wounded of the French Armies at the front. Upon the entry of the United States into the war he turned over the efficient organization he had built to the United States Army Ambulance Service, and by his sound judgment and expert advice rendered invaluable aid in the development of that organization. To him is due, in a large measure, the credit for the increasingly valuable work done by the light ambulances at the front.

ANDREWS, AVERY D.
R—Massena, N. Y.
B—Massena, N. Y.
G. O. No. 12, W. D., 1919.

Brigadier general, U. S. Army.
As assistant chief of staff, American Expeditionary Forces, he has rendered most efficient service in connection with the organization and administration of the transportation department of the American Army in France and as deputy chief of utilities in the services of supply. Later, with marked ability, he headed the important administrative section of the general staff of the American Expeditionary Forces.

ANDREWS, JAMES M.
R—Schenectady, N. Y.
B—Saratoga Springs, N. Y.
G. O. No. 31, W. D., 1922.

Colonel, Infantry, U. S. Army.
For services as commander of the 105th Infantry throughout the active operations of the 27th Division in Belgium and France, during the Ypres-Lys and Somme offensives, his energetic and zealous qualities of leadership demonstrated in battle were conspicuous.

ANDREWS, LINCOLN C.
R—Seneca Falls, N. Y.
B—Owatonna, Minn.
G. O. No. 31, W. D., 1922.

Brigadier general, U. S. Army.
He originated in military training the analysis and study of the qualities of military leadership and the psychology of military training, and by his lectures and writings did much to make possible the successful training of thousands of civilians into efficient military leaders. He served in turn as organizer of the 304th Cavalry, as commander of the 172d Infantry Brigade, as assistant to assistant chief of staff, G–5, General Headquarters, American Expeditionary Forces, and as deputy provost marshal general, in all of which capacities he held positions of great responsibility and rendered exceptionally meritorious services.

ANDREWS, SCHOFIELD.
R—Philadelphia, Pa.
B—Governors Island, N. Y.
G. O. No. 4, W. D., 1923.

Lieutenant colonel (Infantry), General Staff Corps, U. S. Army.
As assistant chief of staff, G–3, 90th Division, from July, 1918, until June, 1919, he displayed sound judgment and exceptional ability in the administration and operation of that section of the division staff. By his loyal devotion to duty, marked tactical ability, and excellent military attainments he contributed materially to the success attained by the division in the St. Mihiel and Meuse-Argonne offensives.

ANSELL, SAMUEL T.
R—Coinjock, N. C.
B—Coinjock, N. C.
G. O. No. 18, W. D., 1919.

Brigadier general, U. S. Army.
For services as Acting Judge Advocate General of the Army, whose broad and constructive interpretation of law and regulations have greatly facilitated the conduct of the war and military administration.

ARMSTRONG, FRANK S.
R—Jeffersonville, Ind.
B—Jeffersonville, Ind.
G. O. No. 59, W. D., 1919.

Colonel, Cavalry, U. S. Army.
With painstaking efforts he reorganized and placed the Remount Service upon an efficient basis, overcoming innumerable difficulties and finding ways and means of supplying combatant divisions with animals when the sources of supply were very limited. In this great task he showed qualities meriting the highest praise.

ARNOLD, LESLIE P.
R—New London, Conn.
B—New Haven, Conn.
G. O. No. 14, W. D., 1925.
Act of Congress Feb. 25, 1925.

First lieutenant, Air Service, U. S. Army.
Lieutenant Arnold as assistant pilot of airplane No. 2, the Chicago, and adjutant and finance officer of the U. S. Army Air Service around-the-world flight from Apr. 6, 1924, to Sept. 28, 1924, displayed rare organizing ability, initiative, and resourcefulness in carrying out these duties, in addition to the alternate piloting of airplane No. 2 throughout the voyage. His technical skill, broad vision, business experience, high personal courage, and untiring energy contributed in a very decided manner to the successful accomplishment of this pioneer flight of airplanes around the world. In the splendid performance of these arduous and trying duties he conspicuously contributed in an accomplishment of the first magnitude of the military forces of the United States.

ARTHUR, ROBERT_____ R—Webster, S. Dak. B—Webster, S. Dak. G. O. No. 16, W. D., 1923.	Lieutenant colonel (Field Artillery), Coast Artillery Corps, U. S. Army. He commanded the 121st Field Artillery during the Aisne-Marne, Oise-Aisne, and Meuse-Argonne offensives with distinction. In addition he served as chief of heavy artillery of the 57th Field Artillery Brigade in those offensives. His high professional skill, sound judgment, leadership, and devotion to duty were material factors in the successful operations of the artillery forces with which he served.
ASHFORD, BAILEY K_____ R—Washington, D. C. B—Washington, D. C. G. O. No. 3, W. D., 1925.	Colonel, Medical Corps, U. S. Army. As director of the Army Sanitary School, by his individual energy, ability, and vision, he placed at the disposal of the American Expeditionary Forces the experience and training facilities of the medical services of the French Armies and of the British Expeditionary Forces in France. He organized a system for the training of officers of the medical service of the Army of the United States in their duties at the front which contributed to a remarkable degree to the success attained in the treatment and evacuation of battle casualties
ATKINS, JOSEPH A_____ R—Atlanta, Ga. B—Atlanta, Ga. G. O. No. 49, W. D., 1922.	Lieutenant colonel (Infantry), General Staff Corps, U. S. Army. He served with the 3d Division as assistant chief of staff, G-3, from December, 1917, until March, 1918; acting chief of staff and G-3, from March, 1918, to May 27, 1918; G-3, from May 28 to June 11, and from Sept. 1 to 19, 1918; and as G-3, 36th Division, from September, 1918, to March, 1919. By his tireless energy, devotion to duty and high military attainments, he contributed in a large measure to the successes attained by the commands with which he served.
ATKISSON, EARL J_____ R—Fowler, Calif. B—Broken Bow, Nebr. G. O. No. 59, W. D., 1919.	Colonel, Engineers, U. S. Army. He organized and trained the 1st Gas Regiment in a type of warfare new to the American Army and directed the operations of that regiment with marked distinction during the St. Mihiel and Argonne-Meuse offensives of the First American Army.
ATTERBURY, WILLIAM W_____ R—Radnor, Pa. B—New Albany, Ind. G. O. No. 12, W. D., 1919.	Brigadier general, U. S. Army. As director general of transportation, in the face of almost insurmountable obstacles he organized and brought to a high state of efficiency the transportation service of American Expeditionary Forces. The successful operation of this most important service, upon which the movements and supply of the combat troops were dependent, was largely due to his energy, foresight, and ability.
⁺AUBERT, LILLIAN_____ R—Shreveport, La. B—West Baton Rouge, La. G. O. No. 9, W. D., 1923.	Chief nurse, Army Nurse Corps, U. S. Army. As assistant superintendent, Army Nurse Corps, in the office of the Surgeon General during the World War, she rendered services of the highest order. By her devotion to duty and great efficiency at a time when members of the Army Nurse Corps were being enrolled, equipped, and assigned to both overseas and home service she made an invaluable contribution to the work of the Medical Department in caring for the sick. She was taken ill while on duty Oct. 2, 1918, and died Oct. 6, 1918, of pneumonia, in line of duty, as a result of overwork. Posthumously awarded. Medal presented to mother, Mrs. Grace Aubert.
AULTMAN, DWIGHT E_____ R—Pittsburgh, Pa. B—Allegheny City, Pa. G. O. No. 59, W. D., 1919.	Brigadier general, U. S. Army As Chief of Artillery of the 5th Corps in the operations against the enemy in November, 1918, by his exceptional skill as an artillerist he was largely responsible for the rupture of the enemy's position and the breaking of his resistance.
AUSTIN, ELMORE F_____ R—New York, N. Y. B—Roadout, N. Y. G. O. No. 15, W. D., 1923.	Colonel, Coast Artillery Corps, U. S. Army. He served as assistant to, and at intervals as the coast defense commander, coast defenses of eastern New York from August to December, 1917; in command of the 57th Regiment, Coast Artillery Corps (155 G. P. F.), from December, 1917, until Oct. 16, 1918, taking part in the St. Mihiel and Meuse-Argonne offensives; and then in command of Replacement Battalion, 1st Army Artillery, from Oct. 17 to Nov. 17, 1918. He displayed at all times sound judgment, great energy, and a thorough knowledge of his duties, all of which were manifested in the high degree of efficiency attained by his regiment in organization, training, and active operations.
AUSTIN, FRED T_____ R—Boston, Mass. B—Hancock, Vt. G. O. No. 30, W. D., 1921.	Brigadier general, U. S. Army. For services while in command of Camp Zachary Taylor, Ky., and particularly during the period that said camp was subject to a severe epidemic of influenza.
AXTON, JOHN T_____ R—Salt Lake City, Utah. B—Salt Lake City, Utah. G. O. No. 69, W. D., 1919.	Major, chaplain, U. S. Army. In organizing and administering numerous welfare activities connected with the port of embarkation, Hoboken, N. J., and New York City, whereby provision was made for the comfort and pleasure of enlisted men.
AYRES, LEONARD P_____ R—Washington, D. C. B—Niantic, Conn. G. O. No. 87, W. D., 1919.	Colonel, General Staff, U. S. Army. His services as chief of the division of statistics, Council of National Defense, as chief of the statistics branch of the General Staff, and chief statistical officer of the American Commission to Negotiate Peace have been conspicuous. He established the statistical division at General Headquarters, American Expeditionary Forces, and the statistics branch at Headquarters, Service of Supply, American Expeditionary Forces.

BABBITT, EDWIN B.
   R—Washington Territory.
   B—Watervliet Arsenal, N. Y.
   G. O. No. 19, W. D., 1920.

Brigadier general, U. S. Army.
He commanded the 4th Field Artillery Brigade from its organization to the close of hostilities, participating with marked distinction in the actions on the Vesle River and in the St. Mihiel and the Meuse-Argonne offensives. The skillful manner in which he pushed forward the artillery units in support of the infantry was a material factor in the successes of these campaigns. In the Meuse-Argonne offensive he had under his command, in addition to the 4th Artillery Brigade, the 10th Field Artillery, the 18th Field Artillery, the 205th French R. A. C., and the 2d Battalion, 308th French R. A. C.

BABCOCK, CONRAD S.
   R—New York, N. Y.
   B—Stonington, Conn.
   G. O. No. 87, W. D., 1919.

Colonel, Cavalry, U. S. Army.
As post commandant at general headquarters, he served with distinction. Later he commanded the 354th Infantry throughout the successful operations against the St. Mihiel salient and those of the Argonne-Meuse in which his regiment participated and subsequently when it formed part of the Army of Occupation. At all times he displayed military attainments of the highest order. His unflagging energy and marked tactical ability were demonstrated in the successful accomplishment by his regiment of all missions assigned to it even under the most trying conditions. His service was of great value to the American Expeditionary Forces.

BABCOCK, WALTER C.
   R—Boston, Mass.
   B—Boston, Mass.
   G. O. No. 87, W. D., 1919.

Colonel, Infantry, U. S. Army.
In command of the 310th Infantry he displayed marked ability alike in its organization and training and in the field. In offensive operations against the enemy he led his command with exceptional judgment and tactical ability, showing himself always possessed of a full grasp of the situation and its needs, and keeping his higher commanders at all times informed of the conditions as he learned them by personal reconnaissance. He was untiring in energy and devotion to the important tasks assigned him, acting unhesitatingly and successfully in times of emergency.

BACH, CHRISTIAN A.
   R—St. Paul, Minn.
   B—St. Martin, Minn.
   G. O. No. 70, W. D., 1919.

Colonel, Cavalry, U. S. Army.
As chief of staff of the 4th Division since its organization he has performed his duties with the utmost loyalty, excellent judgment, and tireless energy, both during the training period and in actual combat. To his energy and military ability is due in no small degree the excellent record of his division in the fighting on the Vesle River and during the Meuse-Argonne offensive.

BACON, RAYMOND F.
   R—Pittsburgh, Pa.
   B—Muncie, Ind.
   G. O. No. 56, W. D., 1922.

Colonel, Chemical Warfare Service, U. S. Army.
As chief of the technical division, Chemical Warfare Service, he displayed untiring energy, marked scientific attainments, and a comprehensive technical knowledge in the organization and operation of the laboratory units and proving-ground tests, thereby aiding materially in the success of the American Expeditionary Forces.

BACON, ROBERT.
   R—New York, N. Y.
   B—Boston, Mass.
   G. O. No. 59, W. D., 1919.

Lieutenant colonel, Infantry, U. S. Army.
He served with great credit and distinction as post commandant of General Headquarters and as aid-de-camp to the Commander in Chief. By his untiring efforts as chief of the American Mission at British General Headquarters he has performed with marked ability innumerable duties requiring great tact and address.

BACON, ROBERT LOW.
   R—Westbury, Long Island, N. Y.
   B—Boston, Mass.
   G. O. No. 27, W. D., 1922.

Major, Field Artillery, U. S. Army.
As assistant to the Chief of Field Artillery from Feb. 8, 1918, to Jan. 2, 1919, he planned, instituted, and supervised the system by which an adequate number of properly qualified officers were secured for the Field Artillery.

BAER, JOSEPH A.
   R—Reading, Pa.
   B—Kutztown, Pa.
   G. O. No. 59, W. D., 1919.

Colonel, Inspector General's Department, U. S. Army.
During the active operations of the armies in the field in the St. Mihiel salient and in the Argonne offensive he revealed marked ability in the inspection of conduct and methods and showed military tactical knowledge of a high order.

BAGBY, PHILIP H.
   R—Richmond, Va.
   B—Richmond, Va.
   G. O. No. 49, W. D., 1922.

Lieutenant colonel (Infantry), General Staff Corps, U. S. Army.
He served as intelligence liaison officer between British General Headquarters and American General Headquarters from September to December, 1918; as director of Army Intelligence School, Langres, France, December, 1918, and January, 1919; as assistant, G-2, Third Army, February to July, 1919; then as Assistant Chief of Staff, G-2, American Forces in Germany. Charged at all times with duties of a most important nature, in the performance of which he manifested steadfast loyalty and military ability of a high order, rendering services of signal worth. His comprehensive grasp of all important phases of interallied relations, as well as his unusual ability in delicate and vital matters, were of the greatest value. His rare powers of discernment, his tact and sound judgment contributed materially to the success of the commands with which he served.

BAILEY, CHARLES J.
   R—Jamestown, N. Y.
   B—Tamaqua, Pa.
   G. O. No. 70, W. D., 1919.

Major general, U. S. Army.
He commanded the 81st Division with distinction throughout its operations, beginning Oct. 1, 1918. The excellent conduct of this division was due, in a large measure, to his great military knowledge, energy, and zeal. He has shown qualities of able leadership and has rendered services of great value to the American Expeditionary Forces.

BAILEY, PEARCE.
   R—New York, N. Y.
   B—New York, N. Y.
   G. O. No. 11, W. D., 1921.

Colonel, Medical Corps, U. S. Army.
As chief of the division of neuropsychiatry, Surgeon General's Office, in which capacity he displayed exceptional zeal, foresight, and good judgment in organizing, developing, and directing neuropsychiatric work in the Army on a high plane of efficiency.

BAIRD, CLAIR W.
R—Punxsutawney, Pa.
B—Burton, Ohio.
G. O. No. 16, W. D., 1923.

Colonel, Coast Artillery Corps, U. S. Army.
As assistant to the Chief of Coast Artillery during the entire period of the war and as chief of the personnel section from Sept. 24, 1918, he displayed foresight, excellent judgment, and marked ability in the preparation and execution of plans for the effective accomplishment of the duties assigned to the Coast Artillery Corps in the operations in France, thereby rendering services of great value to the Government.

BAKER, ASHER CARTER.
R—Matawan, N. J.
B—Cedar Rapids, Iowa.
G. O. No. 89, W. D., 1919.

Captain, U. S. Navy, retired.
Voluntarily returning to active service after retirement, he served with distinction as naval representative with the Transportation Department. Through his extensive naval experience, untiring zeal, and intimate knowledge of the French language and customs, he rendered services of inestimable value to the American Expeditionary Forces.

BAKER, FRANK C.
R—Washington, D. C.
B—Washington, D. C.
G. O. No. 59, W. D., 1919.

Colonel, Medical Corps, U. S. Army.
As commanding officer of Evacuation Hospital No. 6, at Chateau-Thierry, from June to August, 1918, Colonel Baker so promptly arranged his hospital under most difficult conditions and with great resourcefulness and good judgment made such use of the inadequate means at his disposal that he was able to receive and evacuate after splendid treatment and in perfect order a large number of wounded from the Marne offensive at a time when that section of France was greatly demoralized.

BAKER, WALTER C.
R—Chester, Pa.
B—Chester, Pa.
G. O. No. 16, W. D., 1923.

Colonel, Coast Artillery Corps, U. S. Army.
As assistant and executive assistant to the Chief of Transportation Service from Apr. 15, 1918, to Oct. 4, 1920. In this position he demonstrated unusual executive and administrative ability, sound judgment and untiring energy, and contributed in a marked degree to the successful operations of the Transportation Service.

BALDWIN, KARL FERGUSON.
R—East Liberty, Ohio.
B—Macksburg, Iowa.
G. O. No. 124, W. D., 1919.

Lieutenant colonel, Coast Artillery Corps, U. S. Army.
For especially meritorious and distinguished service while serving as military attaché at Tokyo, Japan.

BALL, WILLIAM G.
R—Chillicothe, Ohio.
B—Blanchester, Ohio.
G. O. No. 38, W. D., 1922.

Colonel, Quartermaster Corps, U. S. Army.
In organizing and directing the bakery service of the Quartermaster Corps, American Expeditionary Forces, including the personal supervision of mechanical bakeries at Is-sur-Tille, Bordeaux, Brest, and St. Nazaire, France. By his experience, initiative, and unremitting efforts he brought this important service to a high degree of efficiency in training of personnel, equipment of units, and the prompt supply of soft white bread to the armies of the American Expeditionary Forces in France, thereby contributing materially to the success of the American Expeditionary Forces.

BALL, WILLIAM L.
R—Woburn, Mass.
B—Woburn, Mass.
G. O. No. 16, W. D., 1923.

Major, Adjutant General's Department, U. S. Army.
As executive officer, statistical division, Adjutant General's Office, General Headquarters, American Expeditionary Forces, he displayed outstanding administrative and executive ability coordinating the work of various departments of that division and maintaining liaison with the personnel adjutants of the various headquarters of the American Expeditionary Forces and with the central records office in Bourges, France. With tireless energy and unremitting devotion to duty he met the grave responsibilities of his difficult position with signal distinction, contributing markedly to the successful operations of the Adjutant General's Department, American Expeditionary Forces.

BAMFORD, FRANK E.
R—Omaha, Nebr.
B—Milwaukee, Wis.
G. O. No. 62, W. D., 1919.

Brigadier general, U. S. Army.
As its commanding officer he organized and successfully conducted the Second Corps school. Successively in command of a battalion, regiment, brigade, and division, he participated in the operations of American troops from Cantigny to those of the Meuse-Argonne. He later commanded the Army school at Langres, at all times bringing to bear upon his duties his sound judgment, high military attainments, and untiring zeal.

BANDHOLTZ, HARRY H.
R—Constantine, Mich.
B—Constantine, Mich.
G. O. No. 59, W. D., 1919.

Brigadier general, U. S. Army.
He served in turn as chief of staff of the 27th Division, as commander of the 58th Infantry Brigade, and as provost marshal general of the American Expeditionary Forces, in all of which capacities he displayed exceptional ability. His foresight, broad experience, and sound judgment resulted in the efficient reorganization and administration of the important Provost Marshal General's Department.

BANKER, GRACE D.
R—Passaic, N. J.
B—Passaic, N. J.
G. O. No. 70, W. D., 1919.

Signal Corps, U. S. Army.
She served with exceptional ability as chief operator in the Signal Corps exchange at General Headquarters, American Expeditionary Forces, and later in a similar capacity at 1st Army headquarters. By untiring devotion to her exacting duties under trying conditions she did much to assure the success of the telephone service during the operations of the 1st Army against the St. Mihiel salient and to the north of Verdun.

BARBER, CHARLES W.
R—Woodbury, N. J.
B—Woodbury, N. J.
G. O. No. 70, W. D., 1919.

Colonel, Infantry, U. S. Army.
As assistant chief of staff, G-1, and later as Chief of Staff, Base Section No. 2, during the period of its reorganization he displayed exceptional administrative ability and was in a large measure responsible for the efficient organization created for the repatriation of troops through the port of Bordeaux rendering services of signal worth.

BARBER, JAMES FRANK R—Philadelphia, Pa. B—Haddonfield, N. J. G. O. No. 56, W. D., 1922.	Colonel, Corps of Engineers, U. S. Army. While commanding the 304th Regiment of Engineers of the 79th Division, during the Meuse-Argonne offensive, by his marked ability and tireless energy his regiment was enabled to further the combat operations of his division, frequently building roads and bridges under fire. Charged with the duty of removing enemy mines and traps in front of the right of the 1st Army, he successfully accomplished a difficult and dangerous duty immediately following the Armistice, thereby rendering services of great value to the American Expeditionary Forces.
BARE, WALTER E R—Gadsden, Ala. B—Lexington, Va. G. O. No. 53, W. D., 1921.	Lieutenant colonel, 167th Infantry, 42d Division, U. S. Army. For services while in command of the 167th Infantry, operating near Cote-de-Chatillon and Landres-et-St. Georges in the Meuse-Argonne offensive during the month of October, 1918.
BARNES, HARRY C R—Guthrie, Okla. B—Little Rock, Ark. G. O. No. 133, W. D., 1919.	Colonel, U. S. Army. As commander of the 30th Artillery Brigade he planned and directed the operations of that unit with great skill and ability during the Meuse-Argonne offensive. As chief of staff of the Railway Artillery Reserve he rendered valuable services in the organization and operations of the Railway Artillery units.
BARNES, JOHN B R—Highland, W. Va. B—Pennsboro, W. Va. G. O. No. 49, W. D., 1922.	Lieutenant colonel, Infantry, U. S. Army. While serving successively as G-3 of the 5th and 80th Divisions from June until November, 1918, and then as G-3, 9th Army Corps, he rendered services of great value. By his tireless energy, foresight, sound tactical judgment, and intelligent cooperation he contributed largely to the successes of the operations of those units.
BARNES, JOSEPH F R—Washington, D. C. B—Washington, D. C. G. O. No. 89, W. D., 1919.	Colonel, Field Artillery, U. S. Army. As corps adjutant of the 2d Army Corps, by his able management and complete knowledge of all details of the Adjutant General's Department, he established and operated with remarkable success the numerous branches of the Adjutant General's Office. Later as adjutant general, First Army, he organized with rare initiative and administered with marked ability the operations of his important office, rendering services of inestimable value.
BARNEY, JAMES P R—Townsend, Va. B—Dayton, Ohio. G. O. No. 19, W. D., 1922.	Lieutenant colonel, General Staff Corps, U. S. Army. As assistant chief of staff, G-1, of the 92d Division, he organized the entire system of supply for the division. Due to his administrative ability, exceptional foresight, and tireless energy, he handled numerous difficult problems of supply and transportation with great efficiency and success.
BARNHARDT, GEORGE C R—Norwood, N. C. B—Gold Hill, N. C. G. O. No. 9, W. D., 1923.	Brigadier General, U. S. Army. As commander of the 28th Infantry, he handled his regiment so brilliantly under severe conditions during the St. Mihiel offensive, Sept. 12 and 13, 1918, and during the battle of the Meuse-Argonne, Oct. 1 to 11, 1918, that the regiment demonstrated an unusually high degree of efficiency and morale. He repeatedly displayed superior tactical judgment, and by his exceptional ability, leadership, and devotion to duty, he effectively executed the most difficult missions assigned to his regiment. Later, in command of the 2d Infantry Brigade and then the 178th Infantry Brigade, he again displayed high efficiency and military attainments, thereby rendering with all his commands important services to the American Expeditionary Forces.
BARNUM, MALVERN-HILL R—New York, N. Y. B—Syracuse, N. Y. G. O. No. 59, W. D., 1919.	Brigadier General, U. S. Army. He commanded with marked success the 183d Infantry Brigade from its organization to the close of active operations. The conduct of his brigade in the St. Die and Marbache sectors was indicative of his good leadership. As a member of the interallied armistice board he has performed his many exacting duties with marked ability, address, and sound judgment, rendering services of the highest character to the Government.
BARRY, THOMAS H R—New York, N. Y. B—New York, N. Y. G. O. No. 73, W. D., 1919.	Major general, U. S. Army. As department commander, central department, he handled many difficult problems arising in that department during the war with rare judgment, tact, and great skill.
BARUCH, BERNARD M R—New York, N. Y. B—Camden, S. C. G. O. No. 15, W. D., 1921.	Director, War Industries Board. For services in the organization and administration of the War Industries Board and in the coordination of allied purchases in the United States. By establishing a broad and comprehensive policy for the supervision and control of the raw materials, manufacturing facilities, and distribution of the products of industry, he stimulated the production of war supplies, coordinated the needs of the military service and the civilian population, and contributed alike to the completeness and speed of the mobilization and equipment of the military forces and the continuity of their supply.
BASH, LOUIS H R—Peoria, Ill. B—Chicago, Ill. G. O. No. 59, W. D., 1919.	Colonel, Infantry, U. S. Army. He supervised with tact and sound judgment the establishment of the important base ports of St. Nazaire and Brest. Later, while he was adjutant general of the Services of Supply, his splendid knowledge of administration, his energy, and personal attention to duties were shown by the efficiency of his office, which met fully the diversified demands made upon it.

BASKETTE, ALVIN K.
R—Nashville, Tenn.
B—Nashville, Tenn.
G. O. No. 59, W. D., 1919.

Colonel, Quartermaster Corps, U. S. Army.
He organized and coordinated the several activities of the salvage depot at St. Pierre de Corps, which was the largest and most important of such depots in the American Expeditionary Forces. By his zeal, tact, and ability in solving the various labor problems that arose in connection with the employment of many French civilians, he produced a high degree of economic efficiency in the operations of the Salvage Service.

BATTLE, MARION S.
R—Tarboro, N. C.
B—Edgecombe County, N. C.
G. O. No. 60, W. D., 1920.

Colonel, Coast Artillery Corps, U. S. Army.
As artillery information officer of the First Army, he efficiently operated this important service. Later, he commanded with distinction a regiment of artillery in the Army of occupation. Subsequently, as provost marshal of Paris, he performed duties of a most difficult nature with unfailing tact, efficiency, and sound judgment. He has demonstrated organizing ability and executive capacity to a marked degree, and he has been a contributing factor toward the raising of the morale and efficiency of the American Expeditionary Forces in Paris. He has rendered services of particular merit to the American Expeditionary Forces.

BAYNE, HUGH A.
R—Bronxville, N. Y.
B—New Orleans, La.
G. O. No. 15, W. D., 1923.

Lieutenant colonel, Judge Advocate General's Department, U. S. Army.
As assistant judge advocate of the services of supply, as counsel for the United States Prisoners of War Commission, judge advocate of the 80th Division and 9th Army Corps during combat operations in France, he displayed untiring zeal, rare professional ability, and intellectual qualities of a high order. His special knowledge of the French language and the laws of France enabled him to render the Government services of immeasurable value and contributed markedly to the successes of the American Expeditionary Forces.

BEACH, WILLIAM D.
R—New York, N. Y.
B—New York, N. Y.
G. O. No. 9, W. D., 1923.

Brigadier general, U. S. Army.
As commanding officer, 176th Infantry Brigade, 88th Division, he displayed organizing and training abilities of the highest order, and by the sound judgment, constant initiative, resourcefulness, and indefatigable energy, abundant tact, and thorough understanding of men which characterized his performance of duty as brigade commander, he contributed materially to successful operations of that brigade and the 88th Division.

BEACHAM, JOSEPH W., Jr.
R—Brooklyn, N. Y.
B—Brooklyn, N. Y.
G. O. No. 9, W. D., 1923.

Colonel (Infantry), General Staff Corps, U. S. Army.
As assistant chief of staff, G-1, 42d Division, from May 10 to Aug. 25, 1918, by his extraordinary energy, initiative, exceptional executive and administrative ability, he rendered valuable services in overcoming many difficult problems of supply under most trying conditions, contributing largely to the successes of the division. Later, as chief of staff, 6th Division, from Aug. 26, 1918, until May 25, 1919, by his intimate knowledge of staff duties, his clear conception of the requirements of troops of the line and by his devotion to duty and marked ability he contributed in a large measure to the progress of the division, thereby rendering valuable services to the American Expeditionary Forces.

BEARSS, HIRAM I.
R—Peru, Ind.
B—Peru, Ind.
G. O. No. 89, W. D., 1919.
Distinguished - service cross also awarded.

Colonel, U. S. Marine Corps.
He commanded with distinction the 102d Infantry, achieving notable successes in the active operations in which that regiment was engaged. By his untiring energy and dauntless courage in overcoming the numerous difficulties confronting him he gave proof of military leadership of a high order.

BECK, ROBERT McC., Jr.
R—Wickford, R. I.
B—Westminster, Md.
G. O. No. 59, W. D., 1919.

Colonel, Infantry, U. S. Army.
He showed extraordinary efficiency in directing the staff work of the 32d Division at the Second Battle of the Marne and in the operations near Soissons and north of Verdun, France, from July to October, 1918. In the preparations for battle and in the reorganizations between battles, he ably handled the many difficult situations that presented themselves.

BECKHAM, DAVID Y.
R—Bardstown, Ky.
B—Bardstown, Ky.
G. O. No. 49, W. D., 1922.

Colonel (Coast Artillery Corps), Adjutant General's Department, U. S. Army.
As officer in charge of war risk insurance in the War Department, he organized, perfected, and directed in a highly efficient manner the system of handling insurance and other relief features of the war risk insurance act. To his tact, vision, marked ability, and loyal devotion to duty is largely due the success attained in the handling of a tremendous amount of business and the insuring of more than 90 per cent of the United States Army.

BEEBE, ROYDEN E.
R—Burlington, Vt.
B—South Burlington, Vt.
G. O. No. 59, W. D., 1921.

Lieutenant colonel, Infantry, U. S. Army.
For services as chief of staff, 82d Division; assistant chief of staff, G-3, 2d Division; and assistant chief of staff, G-3, 1st Army Corps.

BEEUWKES, HENRY.
R—New York, N. Y.
B—Jamesburg, N. J.
G. O. No. 59, W. D., 1919.

Lieutenant colonel, Medical Corps, U. S. Army.
He rendered especially valuable services as inspector of hospitalization of troops in the field. By tireless energy in the performance of his duties he assisted greatly in raising the efficiency of this service and in bettering the facilities for the care and evacuation of the wounded of our armies.

BEHN, SOSTHENES.
R—Havana, Cuba.
B—St. Thomas, A. W. I.
G. O. No. 50, W. D., 1919.

Lieutenant colonel, Signal Corps, U. S. Army.
He served in turn as liaison officer with the French Department of Posts and Telegraphs, as executive to the chief signal officer, as commander of a field signal battalion, and as assistant to the chief signal officer, First Army. In all of these capacities he demonstrated marked ability and performed exceptionally meritorious service.

BELKNAP, CHARLES, Jr. R—Concord, Mass. B—Maryland. G. O. No. 116, W. D., 1919.	Commander, U. S. Navy. For services in connection with the Naval Overseas Transportation Service. His successful organization and administration of this service contributed greatly to the successful operation of the American forces abroad.
BELL, GEORGE, Jr. R—District of Columbia. B—Maryland. G. O. No. 59, W. D., 1919.	Major general, U. S. Army. He led his command, with distinction, in the offensive operations with the British which resulted in the capture of Hamel and Hamel Woods, and in the fighting on the Meuse that gained the villages of Marchéville, St. Hilaire, and a portion of Bois d'Harville. He displayed a high order of leadership in the Argonne-Meuse offensive, when his division attacked and captured the strongly fortified Bois-de-Forges. The successful operations of the division which he trained and commanded in combat were greatly influenced by his energy and abilities as a commander.
*BELL, J. FRANKLIN. R—Shelbyville, Ky. B—Shelbyville, Ky. G. O. No. 73, W. D., 1919. Medal of honor and distinguished-service cross also awarded.	Major general, U. S. Army. For exceptionally meritorious and distinguished service during the war as division, cantonment, and department commander. Posthumously awarded. Medal presented to widow, Mrs. J. Franklin Bell.
BELLINGER, JOHN B. R—Charleston, S. C. B—Charleston, S. C. G. O. No. 56, W. D., 1921.	Colonel, Quartermaster Corps, U. S. Army. As department quartermaster, Philippine Department, a position of great responsibility, he administered the services of transportation and of the supply of the troops serving in the Philippines and China in a markedly successful manner. He originated and executed the supplying of the Siberian American Expeditionary Forces and the purchasing of foods in the Orient, and aided the Philippine government in its problems. He rendered services of much value.
BENDER, LOUIS B. R—Charleston, Wash. B—Highland, Kans. G. O. No. 56, W. D., 1922.	Lieutenant colonel, Signal Corps, U. S. Army. He served in the office of the Chief Signal Officer, American Expeditionary Forces, as assistant director of supplies from July, 1918, until December, 1918, and director of supplies from December, 1918, until September, 1919. By his sound judgment, unfailing energy, and unusual ability he rendered services of the greatest value in both capacities. He met the many military commercial problems which confronted him with a broad vision and solved them with unvarying judgment and skill, thereby contributing materially to the success of the American Expeditionary Forces in positions of great responsibility.
BENEDICT, JAY L. R—Hastings, Nebr. B—Hastings, Nebr. G. O. No. 73, W. D., 1919.	Colonel, Infantry, U. S. Army. In the organization and administration of the procurement and discharge section of the personnel branch, his energy, intelligent application, and good judgment have contributed greatly to the solution of the many difficult personnel problems pertaining to the procurement and discharge of officers, and the building up of the Officers' Reserve Corps.
BENNION, HOWARD S. R—Vernon, Utah. B—Vernon, Utah. G. O. No. 56, W. D., 1922.	Lieutenant colonel, Corps of Engineers, U. S. Army. As chief camouflage officer, American Expeditionary Forces, from October, 1917, until February, 1919, in a position of great responsibility he rendered conspicuous service in an entirely new field of endeavor. By his tireless energy, sound judgment, and marked technical ability he organized and placed the work of the camouflage section on a practical and highly efficient basis and directed its functions in a most satisfactory manner, thereby contributing materially to the success of the American Expeditionary Forces.
BENSON, WILLIAM SHEPHERD. R—Macon, Ga. B—Macon, Ga. G. O. No. 116, W. D., 1919.	Admiral, U. S. Navy. As Chief of Naval Operations, his close cooperation and assistance in that position did much toward the successful outcome of the combined operation of the Army and Navy overseas.
BENTON, GUY POTTER, Dr. R—Burlington, Vt. B—Kenton, Ohio. G. O. No. 19, W. D., 1920.	Director, Educational work in American Expeditionary Forces. As director in charge of the educational work undertaken in the Third Army of the American Expeditionary Forces, by his marked ability, untiring energy, and loyal devotion to his task, he contributed in a large measure to the successful results obtained in this vast undertaking. Through his great work among 10,000 illiterate soldiers over 8,000 of them were taught to read and write. By his efforts he has rendered services of particular worth to the American Expeditionary Forces.
BERRY, HARRY S. R—Hendersonville, Tenn. B—Nashville, Tenn. G. O. No. 4, W. D., 1923.	Colonel, 115th Field Artillery, 30th Division, U. S. Army. As commander of the 115th Field Artillery during its organization and training he displayed marked efficiency, great resourcefulness, and military attainments of a high order. He commanded a grouping of his regiment and other French and American Artillery units in the Meuse-Argonne offensive; also a grouping of his regiment and other 155-millimeter howitzers in the operation of the Second Army, and by his skilled and energetic handling of his command played an important part in the success of these operations.
BETHEL, WALTER A. R—Smyrna, Ohio. B—Smyrna, Ohio. G. O. No. 12, W. D., 1919.	Brigadier general, U. S. Army. As judge advocate of the American Expeditionary Forces he organized this important department and administered its affairs with conspicuous efficiency from the date of the arrival in France of the first American combat troops. His marked legal ability and sound judgment were important factors in the splendid work of his department, and he at all times handled with success the various military and international problems that arose as a result of the operation of our armies.

BEVANS, JAMES L.
R—Decatur, Ill.
B—Platteville, Wis.
G. O. No. 87, W. D., 1919.

Colonel, Medical Corps, U. S. Army.
He served with distinction as chief surgeon of the Third Army Corps, where he solved important problems of sanitation and evacuation with conspicuous success. He showed marked administrative ability during the final phases of the Argonne-Meuse offensive, when, through his sound judgment and efficient supervision of the medical and sanitary services under his direction, many lives were saved, thereby rendering valuable service to the American Expeditionary Forces.

BIDDLE, JOHN.
R—Grosse Ile, Mich.
B—Detroit, Mich.
G. O. No. 59, W. D., 1919.

Major general, U. S. Army.
In command of American troops in England, by his tact and diplomacy in handling intricate problems, he made possible the successful transshipment of many thousands of men to France. To his executive ability the efficient handling, control, and dispatch of casual troops through England is largely due.

*BIDDLE, NICHOLAS.
R—New York, N. Y.
B—Fort Whipple, Ariz.
G. O. No. 9, W. D., 1923.

Lieutenant colonel, General Staff Corps, U. S. Army.
For services as intelligence officer, in charge in the city of New York during the entire period of American participation in the World War. His ability as an organizer, his broad experience in large affairs, contributed largely to the failure of the enemy to thwart our military efforts in the city of New York by espionage, sabotage, and propaganda.
Posthumously awarded. Medal presented to widow, Mrs. Nicholas Biddle.

BILLINGS, FRANK.
R—Chicago, Ill.
B—Iowa County, Wis.
G. O. No. 69, W. D., 1919.

Colonel, Medical Corps, U. S. Army.
For services in the organization and administration of the division of reconstruction of the Medical Department.

BINGHAM, ERNEST G.
R—Talladega, Ala.
B—Talladega, Ala.
G. O. No. 59, W. D., 1919.

Colonel, Medical Corps, U. S. Army.
As chief surgeon of the Paris district he most efficiently directed the coordination of the work of the hospitals and hospital and ambulance trains in the region of the Paris group during the Second Battle of the Marne. By his untiring zeal and his exact understanding of conditions he most ably handled the limited hospital resources of the district of Paris, permitting the clearing of the battle field of the wounded and the proper provision for their care. In all these tasks he showed professional attainments of the highest order unflagging energy, and great devotion to duty.

BIRNIE, UPTON, Jr.
R—Philadelphia, Pa.
B—Carlisle, Pa.
G. O. No. 27, W. D., 1920.

Colonel, Field Artillery, U. S. Army.
As principal assistant in the operations section, General Headquarters, American Expeditionary Forces, he has by his thorough military knowledge, loyalty, and devotion to duty materially assisted in attaining the success of that section of the General Staff.

BISHOP, HARRY G.
R—Goshen, Ind.
B—Grand Rapids, Mich.
G. O. No. 59, W. D., 1919.

Brigadier general, U. S. Army.
While in command of the 3d Field Artillery Brigade, during the battles of the Argonne-Meuse, and in the subsequent advance to Sedan, by his skill and able leadership, he rendered exceptionally valuable services.

BISHOP, PERCY POE.
R—Powells, Tenn.
B—Powells, Tenn.
G. O. No. 18, W. D., 1919.

Brigadier general, U. S. Army.
For services as secretary of the General Staff and in the organization and coordination of matters relating to the commissioned personnel of the Army.

BISHOP, WILLIAM H.
R—New York, N. Y.
B—Jackport, N. Y.
G. O. No. 38, W. D., 1922.

Lieutenant colonel, Medical Corps, U. S. Army.
At Orleans, France. By his great ability, initiative, and tact he enlisted the sympathies of the French authorities and people, obtained buildings, organized and enlarged base hospital, and contributed materially to the care of the sick and wounded during the operations of 1918. He has rendered services of much value. This efficient hospitalization was later adopted as a model by the French medical service.

BJORNSTAD, ALFRED W.
R—St. Paul, Minn.
B—St. Paul, Minn.
G. O. No. 89, W. D., 1919.
Distinguished-service cross also awarded.

Brigadier general, U. S. Army.
As director of the Army General Staff College at Langres, he organized and conducted this institution during the first and second courses. Although he was without adequate material or personnel, by the energy and great effort he put forth he established a school which provided our armies with staff officers in a minimum of time.

BLACK, WILLIAM M.
R—Lancaster, Pa.
B—Lancaster, Pa.
G. O. No. 144, W. D., 1918.

Major general, Chief of Engineers, U. S. Army.
For services in planning and administering the engineer and military railway services during the war.

BLAKE, JOSEPH A.
R—New York, N. Y.
B—California.
G. O. No. 59, W. D., 1919.

Colonel, Medical Corps, U. S. Army.
As chief consultant for the district of Paris, and commanding officer of Red Cross Hospital No. 2, he efficiently standardized surgical procedures, especially in the recent methods of treating fractures. His remarkable talent has materially reduced the suffering and loss of life among our wounded.

BLAMER, DEWITT.
R—Independence, Iowa.
B—Independence, Iowa.
G. O. No. 116, W. D., 1919.

Captain, U. S. Navy.
For services as chief of staff of the commander, cruiser and transport fleet.

BLANCK, CARROLL T.
R—Los Angeles, Calif.
B—Greensburg, Pa.
G. O. No. 53, W. D., 1921.

Lieutenant colonel, Signal Corps, U. S. Army.
For services in connection with the control and operation of the telephone and telegraph service of the American Expeditionary Forces.

BLANDING, ALBERT H.
R—Bartow, Fla.
B—Lyons, Iowa.
G. O. No. 118, W. D., 1919.

Brigadier general, U. S. Army.
For services while commanding general of the 53d Infantry Brigade of the 27th Division throughout the entire period of active operations.

BLISS, EDWARD G.
R—Fort Totten, N. Y.
B—Rosemont, Pa.
G. O. No. 15, W. D., 1923.

Lieutenant colonel, Corps of Engineers, U. S. Army.
As executive officer in the office of the director general of transportation, American Expeditionary Forces, he was charged with the responsibility for the organization of the personnel and the administration of the Transportation Corps. His sound judgment, high administrative and executive ability, untiring energy and devotion to duty constituted highly important assets of the Transportation Corps and contributed materially to the success of that organization during its services with the American Expeditionary Forces.

BLISS, ELMER JARED.
R—Boston, Mass.
B—Wrentham, Mass.
G. O. No. 2, W. D., 1920.

The formulation and methods adopted which gave to the U. S. Army unexcelled methods of shoe procurement and distribution were brought about largely through his efforts. As a result of the operation of these methods the efficiency and comfort of the marching soldier were greatly increased.

BLISS, TASKER H.
R—Chester, Pa.
B—Lewisburg, Pa.
G. O. No. 136, W. D., 1918.

General, U. S. Army.
For his most exceptional services as Assistant Chief of Staff, acting Chief of Staff, and Chief of Staff of the U. S. Army, in which important positions his administrative ability and professional attainments were of great value to our armies. As chief of the American section of the Supreme War Council he has taken an important part in the shaping of the policies that have brought victory to our cause.

BLOOR, ALFRED W.
R—Austin, Tex.
B—Pittsburgh, Pa.
G. O. No. 14, W. D., 1923.

Colonel, Infantry, U. S. Army.
In command of the 142d Infantry, 36th Division, including the period of its reorganization and training and during its combat operations in France, he displayed untiring energy, administrative and executive ability, and sound tactical judgment, these qualities, coupled with unremitting devotion to duty and high qualities of leadership, contributing in a conspicuous way to the success of the 36th Division in its operations against the enemy.

BOAK, SEIBERT D.
R—West Virginia.
B—Virginia.
G. O. No. 15, W. D., 1923.

Colonel, Dental Corps, U. S. Army.
As director of the dental section of the Army Sanitary School at Langres, France, from January to December, 1918, he displayed organizing and training ability and accomplishments of the highest order in successfully directing the classification and training of dental officers for field service, thereby rendering services of great value to the American Expeditionary Forces.

BOLLES, FRANK C.
R—Rolla, Mo.
B—Elgin, Ill.
G. O. No. 95, W. D., 1919.
Distinguished-service cross and oak-leaf cluster also awarded.

Colonel, Infantry, U. S. Army.
He commanded, with keen tactical ability, the 39th Infantry throughout the various campaigns in which the Fourth Division participated until the early stages of the Meuse-Argonne offensive, when he was wounded. By his exceptional ability and energetic leadership he proved to be an important factor in the successes of his command during its active operations against the enemy.

*BOLLING, RAYNAL C.
R—Greenwich, Conn.
B—Hot Springs, Ark.
G. O. No. 50, W. D., 1919.

Colonel, Air Service, U. S. Army.
His service to the United States aviation was distinguished for an accurate and comprehensive grasp of aviation matters; for a sound and far-sighted conception of the measures needed to establish an efficient American air service in Europe; for initiative and resourcefulness in attacking the problems of a young air service; for brilliant capacity in arranging affairs with foreign governments; for boldness and vigor in executing determined policies. In all of these he has rendered service of great value to the Government.
Posthumously awarded. Medal presented to widow, Mrs. Anna P. Bolling.

BOOTH, ALFRED J.
R—Albany, N. Y.
B—Albany, N. Y.
G. O. No. 59, W. D., 1919.

Colonel, Adjutant General's Department, U. S. Army.
As assistant to The Adjutant General, American Expeditionary Forces, he was charged with the important duty of verifying, preparing, and distributing all orders and bulletins issued from General Headquarters, American Expeditionary Forces. To his painstaking efforts are due the accuracy with which these orders were drawn and the promptness with which they were distributed. He organized and efficiently supervised the administration of The Adjutant General's printing plant at General Headquarters, American Expeditionary Forces. To his untiring zeal is largely due the success with which it handled a tremendous volume of printed matter, rendering important service to the American Expeditionary Forces.

BOOTH, EVANGELINE C.
R—New York, N. Y.
B—England.
G. O. No. 87, W. D., 1919.

Commander of the Salvation Army in the United States.
She has been tireless in her devotion to her manifold duties. The contribution of the Salvation Army toward winning the war is conspicuous, and the results obtained were due in marked degree to the great executive ability of its commander.

BOOTH, EWING E.
R—Pueblo, Colo.
B—Bowers Mills, Mo.
G. O. No. 59, W. D., 1919.

Brigadier general, U. S. Army.
He commanded, with great ability and gallantry, the 8th Infantry Brigade in the operations which forced the reluctant enemy to evacuate Bois-du-Feys, Bois-de-Malaumont, Bois-de-Peut-de-Faux, and Bois-de-Foret in September and October, 1918. His splendid leadership was an important factor in these actions.

BOOTHE, EARLE............. | Lieutenant colonel, Adjutant General's Department, U. S. Army.
R—South Pasadena, Calif. | He reorganized and administered with marked distinction the central records
B—Derby, Conn. | office of the American Expeditionary Forces. He handled the complex
G. O. No. 59, W. D., 1919. | problems constantly arising with great discretion, displaying keen perception
| amid the maze of details involved in the reporting of casualties and changes
| of status of officers and soldiers. With unflagging energy and exceptional
| ability he performed a task of great magnitude.

BOUGHTON, EDWARD J............. | Lieutenant colonel, Judge Advocate General's Department, U. S. Army.
R—Denver, Colo. | He served with distinction as head of the international law division in the office
B—Albany, N. Y. | of the Judge Advocate, American Expeditionary Forces. Through his ex-
G. O. No. 89, W. D., 1919. | tensive knowledge of international law and diplomatic ability, he was of the
| utmost assistance in handling many delicate questions involving relations
| between the American and allied armies.

BOWDITCH, EDWARD, Jr............. | Lieutenant colonel, Infantry, U. S. Army.
R—Boston, Mass. | At the Army General Staff College, as assistant to G–3, headquarters 1st Army
B—Albany, N. Y. | Corps, he rendered service of distinction, always showing himself able in
G. O. No. 59, W. D., 1919. | time of emergency, aggressive in action, and possessed of tact and sound
| judgment. As aid-de-camp to the commander in chief, American Expedi-
| tionary Forces, he displayed unflagging energy and devotion to duties of
| great importance. His military attainments were of marked character,
| proving of utmost assistance in the handling of difficult situations.

BOWEN, WILLIAM S............. | Lieutenant colonel (Field Artillery), General Staff Corps, U. S. Army.
R—Omaha, Nebr. | From September, 1918, to June, 1919, as assistant chief of staff, G–3, 29th Divi-
B—Omaha, Nebr. | sion, he displayed the highest qualities of a staff officer and by his untiring
G. O. No. 9, W. D., 1923. | energy, good judgment, and devotion to duty he contributed in a marked
| degree to the success of his division.

BOWLEY, ALBERT J............. | Brigadier general, U. S. Army.
R—San Francisco, Calif. | He commanded the 17th Field Artillery and later the 2d Field Artillery Brigade
B—Westminster, Calif. | in the active operations from July to November, 1918. The artillery support
G. O. No. 59, W. D., 1919. | under his direction in the engagements near Chateau-Thierry, near Soissons,
| those in the St. Mihiel salient, Blanc Mont Ridge, and in the Meuse-Argonne
| region were important factors in the great successes gained.

BOWMAN, GEORGE T............. | Colonel, Cavalry, U. S. Army.
R—Buffalo, N. Y. | As chief of a subsection of G–1 of the General Staff at General Headquarters he
B—Buffalo, N. Y. | prepared the priority schedules for the movement of troops from the United
G. O. No. 35, W. D., 1920. | States to France, directed replacements during active operations, prepared
| the order of battle data, and conserved many important and confidential
| records of the personnel of the American Expeditionary Forces. By his
| marked executive ability and loyal cooperation in all details of his important
| task he has given services of noteworthy consequence to the American Ex-
| peditionary Forces.

*BOYD, CARL............. | Colonel, Cavalry, U. S. Army.
R—Adairsville, Ga. | As military attaché to the American Embassy in Paris he performed services
B—Decora, Ga. | of a most distinguished character. Later, as senior aid-de-camp to the com-
G. O. No. 59, W. D., 1919. | mander in chief, he displayed remarkable ability, sound judgment, and tact
| in the many varied negotiations with the allied commanders and other allied
| officials, rendering services of inestimable value to the American Expedi-
| tionary Forces.
| Posthumously awarded. Medal presented to widow, Mrs. Annie P. Boyd.

BRABSON, FAY W............. | Lieutenant colonel (Infantry), General Staff Corps, U. S. Army.
R—Greenville, Tenn. | As an instructor of the Army General Staff College, Langres, France, May to
B—Greenville, Tenn. | Sept. 15, 1918, he performed exceptionally meritorious services to the Gov-
G. O. No. 59, W. D., 1921. | ernment in instructing and preparing student officers to function in the im-
| portant and responsible positions as General Staff officers with troops.

BRABSON, JOE REESE............. | Lieutenant colonel (Field Artillery), General Staff Corps, U. S. Army.
R—Greenville, Tenn. | As chief of staff of the 28th Division during the Marne-Aisne offensive he ren-
B—Greenville, Tenn. | dered conspicuous service. Later as an instructor of the fourth course at the
G. O. No. 78, W. D., 1919. | Army General Staff College at Langres he ably assisted in the instruction of
| a large number of officers recommended for General Staff duty. Upon com-
| pletion of his duty at the staff school he served with marked success as G–5
| of the Second Army.

BRADLEY, ALFRED E............. | Brigadier general, U. S. Army.
R—Frewsburg, N. Y. | As chief surgeon, American Expeditionary Forces, he gave his utmost energy
B—Jamestown, N. Y. | and undivided devotion to the duty of planning and organizing the work of
G. O. No. 12, W. D., 1919. | the Medical Department in France during a period fraught with untold diffi-
| culties. To his foresight was largely due the successful operations of that
| department when it was called upon to meet the demands that were subse-
| quently made upon it.

BRADLEY, JOHN J............. | Brigadier general, U. S. Army.
R—Chicago, Ill. | For services as chief of the training and instruction branch, War Plans Divi-
B—Lake View, Ill. | sion, General Staff, in initiating and standardizing the training and instruc-
G. O. No. 47, W. D., 1919. | tion of the Army during its formative period.

BREES, HERBERT J............. | Colonel, General Staff Corps, U. S. Army.
R—Laramie, Wyo. | He served with distinction as chief of staff of the 91st Division throughout its
B—Laramie, Wyo. | training period and during the greater part of its active operations. His
G. O. No. 87, W. D., 1919. | marked administrative ability was reflected in the successes of this division
| during the first phases of the Meuse-Argonne operations. Later, as chief of
| staff of the 7th Army Corps, he rendered invaluable services in perfecting the
| necessary organization for the march into the German territory, overcoming
| grave difficulties in securing supplies and equipment.

BRENNAN, CECELIA_____ Chief nurse, Army Nurse Corps, U. S. Army.
   R—Philadelphia, Pa.
   B—Branchdale, Pa.
   G. O. No. 9, W. D., 1923.

As chief nurse of the Toul Hospital Center, France, during the World War, she contributed largely to the successful care of over 10,000 sick and wounded by her skillful, tactful, and able direction of the work of the nurses at this center.

BRENT, CHARLES H_____ Major, chaplain, U. S. Army.
   R—Buffalo, N. Y.
   B—Canada.
   G. O. No. 59, W. D., 1919.

As senior headquarters chaplain, he organized the chaplains' school and established a schematic system of religious effort, enabling all chaplains throughout France to further those excellent results which have marked their duties amongst the troops. By his loyal spirit of cooperation, his marked ability, and by his masterful attainments he has rendered services of most conspicuous merit and lasting value to the American Government.

BRETT, LLOYD M_____ Brigadier general, U. S. Army.
   R—Malden, Mass.
   B—Maine.
   G. O. No. 59, W. D., 1919.
   Medal of honor also awarded.

He commanded the 160th Infantry Brigade with particular efficiency in the markedly successful operations resulting in the occupation of the Dannevoux sector in October, 1918. In the actions near Imecourt and Buzancy in November his brigade broke the enemy's resistance. Due to his masterful ability and brilliant leadership, these operations proved a crowning success.

BRETT, SERENO E_____ Major, Tank Corps, U. S. Army.
   R—Corvallis, Oreg.
   B—Portland, Oreg.
   G. O. No. 49, W. D., 1922.
   Distinguished-service cross also awarded.

As chief instructor at tank center, American Expeditionary Forces, he organized and trained the 327th Battalion (Light) Tanks. Later, as commander of the 326th Battalion (Light) Tanks, he vigorously and skillfully led it in the St. Mihiel offensive over a terrain rendered most difficult through four years of enemy intrenching. Succeeding to the command of the 1st Brigade, Tank Corps, in the Meuse-Argonne offensive, he ably, devotedly, and courageously commanded his brigade from Sept. 26 to Nov. 10, 1918; during this period of 46 days his brigade supported eight of the divisions of the First Army in 18 separate attacks. By his brilliant professional attainments, technical ability, and unusual leadership he contributed in a marked manner to the success of the First Army and rendered most conspicuous services to the American Expeditionary Forces in a position of great responsibility.

BREWSTER, ANDRÉ W_____ Major general, U. S. Army.
   R—Philadelphia, Pa.
   B—Hoboken, N. J.
   G. O. No. 12, W. D., 1919.
   Medal of honor also awarded.

He organized and administered with marked ability the Inspector General's Department of the American Expeditionary Forces, and his soldierly characteristics and unceasing labors influenced greatly the attainment of efficiency in the American Army in France.

BRICKER, EDWIN D_____ Colonel, Ordnance Department, U. S. Army.
   R—Chambersburg, Pa.
   B—Chambersburg, Pa.
   G. O. No. 78, W. D., 1919.

As chief ordnance purchasing officer and later as ordnance representative on the general purchasing board, he conducted negotiations with marked success for material needed to supplement the supply from the United States. He worked tirelessly and with unflagging energy to the end that there would be no shortage in supplies sent to the troops at the front. At all times exercising sound judgment and discernment in times of emergency, he achieved marked successes.

BRIDGES, CHARLES H_____ Colonel, Infantry, U. S. Army.
   R—Jerseyville, Ill.
   B—White Hall, Ill.
   G. O. No. 87, W. D., 1919.

As assistant chief of staff, first section of the 2d Division, and later as assistant chief of staff, first section of the 6th Army Corps, he performed creditably duties of great importance in connection with the services of supply, communication, and the movements of troops of his units, rendering services of value to the American Expeditionary Forces.

BRIGGS, RAYMOND W_____ Brigadier general, U. S. Army.
   R—Philadelphia, Pa.
   B—Beaver, Pa.
   G. O. No. 9, W. D., 1923.

For services as chief of the remount service, American Expeditionary Forces; as colonel of the 311th and 304th Field Artillery, the training of which units he developed to a high degree, commanding the latter unit in action with distinction; as commanding general of the 18th and 8th Field Artillery Brigades, which units he developed to a high state of efficiency; and as commanding general of Camp Knox, Ky., during and after the period of demobilization.

BRISTOL, ARTHUR LEROY, Jr_____ Commander, U. S. Navy.
   R—Charleston, S. C.
   B—Charleston, S. C.
   G. O. No. 116, W. D., 1919.

For services as flag secretary to the commander, Cruiser and Transport Fleet. His close cooperation with the Army authorities in the handling of troop ships contributed greatly to the successful outcome of our oversea operations.

BROOKE, ROGER_____ Colonel, Medical Corps, U. S. Army.
   R—Sandy Spring, Md.
   B—Sandy Spring, Md.
   G. O. No. 49, W. D., 1922.

From September, 1917, until December, 1918, as senior instructor in charge of the educational training of medical officers and enlisted men of the Medical Department at the Medical Officers' Training Camp, Camp Greenleaf, Ga., he directed and coordinated the work of its several special schools with great efficiency. By his untiring efforts, devotion to his duties, and brilliant professional ability he was largely responsible for the successful training of 10,000 officers and 70,000 men, thereby rendering conspicuous service to the Government in a position of great responsibility.

BROOKINGS, ROBERT S_____ Chairman, price fixing committee, War Industries Board.
   R—St. Louis, Mo.
   B—Cecil County, Md.
   G. O. No. 15, W. D., 1923.

For services in connection with the operations of the War Industries Board during the World War. As a member of the board he rendered, through his broad vision, distinguished capacity, and business ability, services of inestimable value in marshaling the industrial forces of the Nation and mobilizing its economic resources—marked factors in assisting to make military success attainable. Through his untiring efforts and devotion to duty as commissioner of finished products and later as chairman of the price fixing committee of the board he contributed markedly to the success of the supply systems of the War Department.

BROOKS, HARLOW____
  R—New York, N. Y.
  B—Medo, Minn.
  G. O. No. 31, W. D. 1922.

Lieutenant colonel, Medical Corps, U. S. Army.
For services as medical consultant of the First Army and later as chief medical consultant of the Second Army by the application of principles of prevention and treatment which resulted in a marked reduction in the complications and mortality of influenza and other epidemic diseases and lessened the strain on the evacuation service already overtaxed by battle casualties, through early segregation and hospitalization within the front areas. After the armistice, Colonel Brooks, in addition to these duties, was placed in charge of all hospitals of the Second Army, where, by reorganizing the hospital service and establishing courses of instruction and clinical conferences for medical officers, he largely contributed to the betterment of the medical service of the Second Army with a consequent saving of the lives of many American and French soldiers.

BROWN, FRED R____
  R—Cornell, Ill.
  B—Streator, Ill.
  G. O. No. 56, W. D., 1922.

Colonel, Infantry, U. S. Army.
As lieutenant colonel, 313th Infantry, from July, 1917, until July, 1918, throughout its organization and training period he displayed marked efficiency. From August until December, 1918, he commanded the 368th Infantry with indefatigable energy, exceptional initiative, and resourcefulness in all its combat operations. Later, as commanding officer, 58th Infantry, as military commander of Coblenz, Germany, and as officer in charge of civil affairs in the area occupied by Third Army troops, he performed a difficult and responsible task with conspicuous success, maintaining at all times a high state of discipline and morale among the troops and civilian population, thereby contributing materially to the success of the American Expeditionary Forces, in positions of great responsibility.

BROWN, HOBART B____
  R—New York, N. Y.
  B—East Rutherford, N. J.
  G. O. No. 49, W. D., 1922.

Colonel, Infantry, U. S. Army.
From September, 1917, until April, 1918, by his broad experience and sound judgment he organized and commanded with exceptional ability the 104th Military Police. From April to October, 1918, as lieutenant colonel, 116th Infantry, and colonel, 114th Infantry, he showed himself to be resourceful and energetic and at all times equal to any emergency which arose. As deputy provost marshal general, American Expeditionary Forces, from October, 1918, to February, 1919, he displayed marked ability in a position of great responsibility, thereby rendering services of great value to the American Expeditionary Forces.

BROWN, KATHARINE____
  R—Philadelphia, Pa.
  B—Philadelphia, Pa.
  G. O. No. 9, W. D., 1923.

Chief nurse, Army Nurse Corps, U. S. Army.
As chief nurse of the Nantes Hospital Center, France, during the World War, she supervised and directed the nursing care of more than 4,000 patients. Her good judgment, tact, foresight, and energy resulted in most successful accomplishments on the part of the nursing force of that center.

BROWN, LYTLE____
  R—Nashville, Tenn.
  B—Nashville, Tenn.
  G. O. No. 47, W. D., 1919.

Brigadier general, U. S. Army.
For services as director of the War Plans Division, for his skill and good judgment in handling the many and varied questions of training, organization, and policy that have been acted on by the War Plans Division during the war.

BROWN, PRESTON____
  R—Lexington, Ky.
  B—Lexington, Ky.
  G. O. No. 12, W. D., 1919.

Brigadier general, U. S. Army.
As chief of staff of the 2d Division he directed the details of the battles near Chateau-Thierry, Soissons, and at the St. Mihiel salient with great credit. Later, in command of the 3d Division in the Argonne-Meuse offensive, at a most critical time, by his splendid judgment and energetic action, his division was able to carry to a successful conclusion the operations at Claire-Chênes and at Hill 294.

BROWN, WILL H____
  R—Indianapolis, Ind.
  B—Indianapolis, Ind.
  G. O. No. 15, W. D., 1923.

Lieutenant colonel, Motor Transport Corps, U. S. Army.
As chief motor transport officer, Base Section No. 1, St. Nazaire, France, he was charged with the important duty of the receipt of motor transportation arriving from overseas, the general supervision and operation of large reception parks where motor vehicles were set up and made ready for service, and for the instruction and direction of the personnel of convoys distributing vehicles to division, corps, and army troops. The grave responsibilities thus placed upon Colonel Brown were carried out by him in a signally successful way. He displayed unusual initiative, sound judgment, and high professional skill, contributing in a very material way to the successful operations of the American forces in France.

BROWN, WILLIAM C____
  R—St. Peter, Minn.
  B—St. Peter, Minn.
  G. O. No. 43, W. D., 1922.

Colonel, Cavalry, U. S. Army.
As inspector, Quartermaster Corps, American Expeditionary Forces, from November, 1917, until December, 1918, throughout the zone of operations, he displayed the greatest zeal, utmost devotion to duty, and indefatigable efforts. By his long experience, marked efficiency, and tireless energy he made highly intelligent inspections and recommendations, thereby enabling the Quartermaster Corps to improve the services of supply and the saving of a large quantity of important material.

BROWNE, BEVERLY F____
  R—Accomac, Va.
  B—Accomac, Va.
  G. O. No. 9, W. D., 1923

Brigadier general, U. S. Army.
He organized and conducted schools for artillery information service and the counter battery service in France in October and November, 1917. He participated in preparation and execution of artillery plans of First Army for the St. Mihiel offensive in September, 1918, and commanded 166th Field Artillery Brigade in October, 1918, and the corps artillery, First Army Corps, Nov. 1, to 11, 1918, during the final assault of the First Army. His high professional attainments, sound tactical judgment, and devotion to duty contributed materially to the successful operation of the American Expeditionary Forces.

BROWNING, WILLIAM S............
  R—Brooklyn, N. Y.
  B—Brooklyn, N. Y.
  G. O. No. 69, W. D., 1919.

Colonel, Field Artillery, U. S. Army.
As a member of the American section of the Supreme War Council, by his ability and his clear and sound conception of the constantly changing military situation, he has rendered invaluable aid in solving the many complex problems than have come before the Supreme War Council.

BRUFF, AUSTIN J............
  R—Detroit, Mich.
  B—New York, N. Y.
  G. O. No. 38, W. D., 1922.

Lieutenant colonel, Ordnance Department, U. S. Army.
For services in the reorganization and administration of the property and financial departments at Rock Island Arsenal, Ill., which underwent an enormous expansion in the early stages of the war. Later, he rendered unusual and highly meritorious services in developing in England the manufacture of various special and service types of small-arms ammunition, thereby securing for the American Expeditionary Forces a certain and readily available source of supply of this indispensable material.

BRYANT, MORTIMER D............
  R—Brooklyn, N. Y.
  B—Brooklyn, N. Y.
  G. O. No. 13, W. D., 1923.

Colonel, Infantry, U. S. Army.
As machine-gun officer of his division and later as colonel, 107th Infantry, he displayed rare qualities of courageous leadership, sound tactical judgment, and technical skill of a high order. His ability as an organizer and administrator and his unremitting devotion to duty enabled him to render the Government services of inestimable value in positions of responsibility.

BRYDEN, WILLIAM............
  R—Chelsea, Mass.
  B—Hartford, Conn.
  G. O. No. 19, W. D., 1921.

Brigadier general, U. S. Army.
As director of the department of field gunnery, School of Fire for Field Artillery, Fort Sill, Okla., from September, 1917, to May, 1918, and as assistant commandant of that school from May, 1918, to October, 1918, he displayed organizing ability and other professional attainments of a high order in developing and conducting a sound course of instruction in the principles of field gunnery.

BUCHAN, FRED E............
  R—Kansas City, Kans.
  B—Wyandotte, Kans.
  G. O. No. 56, W. D., 1922.

Colonel, General Staff Corps, U. S. Army.
As assistant, G-3, 2d Army Corps, and later as G-3 of that organization, he displayed military talent of a high order in the training of the organizations of the corps for battle. During operations which broke the Hindenburg line, between Cambrai and St. Quentin, he assisted in the planning and execution of operations of great moment with exceptional ability and tireless energy, and contributed to a high degree to the successes of the operations.

BUCKEY, MERVYN C............
  R—Washington, D. C.
  B—Frederick, Md.
  G. O. No. 124, W. D., 1919.

Colonel, Field Artillery, U. S. Army.
For especially meritorious and distinguished service while serving as military attaché at Rome, Italy.

BULLARD, ROBERT L............
  R—Lafayette, Ala.
  B—Youngsborough, Ala.
  G. O. No. 136, W. D., 1918.

Lieutenant general, U. S. Army.
Commander of the Second Army of the American Expeditionary Forces. In the course of this war he commanded in turn the first American division to take its place in the front lines in France, the 3d Corps, and the Second Army. He participated in operations in reduction of the Marne salient and in the Meuse-Argonne offensive. He was in command of the Second Army when the German resistance west of the Meuse was shattered.

BUNNELL, GEORGE W............
  R—Worcester, Mass.
  B—Oakland, Calif.
  G. O. No. 13, W. D., 1923.

Colonel, Corps of Engineers, U. S. Army.
As commanding officer of the 101st Engineer Regiment, from August, 1917, until April, 1919, during its organization, training, and in all its combat operations he performed all his tasks with unusual ability and in a manner that reflected credit of a high degree upon him. By his careful and thorough preparation, rare judgment, skillful and energetic leadership, and sound tactical and technical knowledge he contributed in a marked degree to the successes achieved by his regiment and the 26th Division.

BURGHER, EMIL H............
  R—St. Louis, Mo.
  B—Switzerland.
  G. O. No. 59, W. D., 1919.

Major, Medical Corps, U. S. Army.
As regimental surgeon of the 138th Infantry, he supervised the care of the wounded during the Argonne offensive. With untiring energy and ability of a high order, displaying personal courage under shell fire, personally rallying his men and directing them forward, he was an inspiration to all. His dressing station was placed to within a few hundred yards of the front lines whenever the terrain rendered the passage of ambulances impossible. His zeal, devotion to duty, and efficient services added greatly to the morale of all who served with him.

BURKE, Rev. JOHN J............
  R—New York, N. Y.
  B—New York, N. Y.
  G. O. No. 73, W. D., 1919.

For especially meritorious and conspicuous service as chairman of the committee on special war activities of the National Catholic War Council and as chairman of the committee of six, dealing with the subject of chaplains.

BURKHAM, ROBERT............
  R—St. Louis, Mo.
  B—Sioux City, Iowa.
  G. O. No. 19, W. D., 1920.

Lieutenant colonel, Judge Advocate General's Department, U. S. Army.
By his exceptional ability and energy he successfully organized and put into efficient practice the claims department of the rents, requisitions, and claims service. The successful handling of the many complex problems in respect to the adjustment of claims was due, in a large measure, to his high professional attainments and sound judgment. He has rendered services of signal worth to the American Expeditionary Forces.

BURNETT, CHARLES............
  R—Carlinville, Ill.
  B—Concord, Tenn.
  G. O. No. 56, W. D., 1921.

Colonel (Cavalry), General Staff Corps, U. S. Army.
As G-3 of the 30th Division during its operation in Belgium and northern France, subsequent to the armistice he functioned as chief of staff at base section No. 1 in a most creditable manner. He has rendered services of much value to the United States.

BURNETT, FRANK C R—Knoxville, Iowa. B—Casey, Iowa. G. O. No. 59, W. D., 1919.	Colonel, Infantry, U. S. Army. He commanded with distinction a battalion of the first American regiment to occupy trenches in France, and participated in the repulse of the first raid made by the enemy upon American troops. As deputy adjutant general, General Headquarters, American Expeditionary Forces, he has performed his manifold duties with ability and sound judgment.
BURNS, JAMES H R—Pawling, N. Y. B—Pawling, N. Y. G. O. No. 56, W. D., 1922.	Colonel, Ordnance Department, U. S. Army. First as chief of the explosives branch, design section, gun division, in which capacity he was charged with the production of sufficient explosives, propellants, and shell loading for the needs of the American Government and her allies; and later as chief of the explosives section, production division, with securing the necessary production of explosives, propellants, and assembly of ammunition to meet the needs of America in the World War.
BURNS, SOPHY M R—St. Francis, Wis. B—St. Francis, Wis. G. O. No. 9, W. D., 1923.	First lieutenant, chief nurse, Army Nurse Corps, U. S. Army. As chief nurse of Base Hospital No. 116, at Bazoilles-sur-Meuse, France, and later, as chief nurse of Mobile Hospital No. 9, American Expeditionary Forces, in the field during the World War, she rendered conspicuous service by her unusual executive ability, tact, good judgment, and faithfulness to detail in caring for the large number of sick and wounded under her charge.
BURR, GEORGE W R—Sedalia, Mo. B—Tolono, Ill. G. O. No. 77, W. D., 1919.	Major general, U. S. Army. For services as director of purchase, storage, and traffic, General Staff, he has had under his supervision during the last several months most important and complicated operations in relation to the cancellation of contracts, the adjustment of claims, the disposal of surplus supplies and the storage of materials that have accumulated during the war or that had been delivered by manufacturers since the armistice.
BURRELL, GEORGE ARTHUR R—Washington, D. C. B—Cleveland, Ohio. G. O. No. 77, W. D., 1919.	Colonel, Chemical Warfare Service, U. S. Army. For services in research work pertaining to gas warfare. Colonel Burrell was in charge of the research division, and its organization was doubtless the greatest of its kind ever formed. It accomplished remarkable results of the greatest importance to our military forces.
BURTT, WILSON B R—La Grange, Ill. B—Hinsdale, Ill. G. O. No. 59, W. D., 1919.	Brigadier general, U. S. Army. As chief of staff of the 5th Corps he displayed great tact and judgment in the organization of that command. He directed with marked ability the staff work of his corps during the St. Mihiel and Argonne-Meuse offensives and was a potent factor in insuring the successes of his organization in that campaign.
BURY, FREDERICK E R—Marion, Ind. B—Miami County, Ind. G. O. No. 53, W. D., 1921.	Lieutenant colonel, Infantry, U. S. Army. For exceptionally meritorious and distinguished services as chief of staff, American Forces in Russia.
BUTLER, SMEDLEY D R—West Chester, Pa. B—West Chester, Pa. G. O. No. 95, W. D., 1919.	Brigadier General, U. S. Marine Corps. He has commanded with ability and energy Pontanezen Camp at Brest during the time in which it has developed into the largest embarkation camp in the world. Confronted with problems of extraordinary magnitude in supervising the reception, entertainment, and departure of the large numbers of officers and soldiers passing through this camp, he has solved all with conspicuous success, performing services of the highest character for the American Expeditionary Forces.
BUTNER, HENRY W R—Stony Ridge, N. C. B—Stony Ridge, N. C. G. O. No. 19, W. D., 1920.	Brigadier general, U. S. Army. He commanded, with marked distinction, the 1st Field Artillery Brigade from Aug. 18 to Nov. 11, 1918, displaying at all times keen tactical ability, initiative, and loyal devotion to duty. By his high military attainments and sound judgment he proved to be a material factor in the successes achieved by the divisions whose advances he supported.
BYLLESBY, HENRY M R—Chicago, Ill. B—Pittsburgh, Pa. G. O. No. 38, W. D., 1922.	Lieutenant colonel, Signal Corps, U. S. Army. As general purchasing agent for the American Expeditionary Forces at Base Section No. 3, in Great Britain, from May to December, 1918, he displayed great energy, a comprehensive knowledge of large business affairs, and executive ability of the highest order. By his broad experience, foresight, and splendid ability to cooperate with representatives of our allies, he solved many difficult problems of fuel supply with conspicuous success and in a manner which insured at critical times a plentiful supply of coal, both for our transport service and our troops in France, thereby rendering services of great value to the American Expeditionary Forces.
BYRON, JOSEPH C R—Hagerstown, Md. B—Buffalo, N. Y. G. O. No. 10, W. D., 1920.	Supply service, hides and leather. For services in connection with the supply service of the Army during the World War, by his individual efforts he made possible the accomplishment of that portion of the supply program that depended for success upon the supply of hides and leather.
CABELL, DE ROSEY C R—Roseville, Ark. B—Charleston, Ark. G. O. No. 73, W. D., 1919.	Major general, U. S. Army. While in command of the Arizona district of the southern department, he handled the delicate border situation there with firmness and sound judgment.

CABELL, HENRY C.
R—Richmond, Va.
B—Richmond, Va.
G. O. No. 10, W. D., 1922.

Colonel, Adjutant General's Department, U. S. Army.
For services while in charge of war risk insurance matters and later in charge of the appointment division. By his untiring efforts in dealing with other departments and the public, he expedited the dispatch of official business and the solution of intricate problems arising under new laws. He rendered services of great merit.

CALL, LEWIS W.
R—Garrett Park, Md.
B—Upper Sandusky, Ohio.
G. O. No. 28, W. D., 1921.

Colonel, Judge Advocate General's Department, U. S. Army.
For services as chief of the contracts and claims section, Judge Advocate General's Office, from 1917 to 1920, his knowledge of the law, sound judgment, and application enabled him to deal promptly and effectively with the many perplexing questions arising with respect to contracts, requisitions, and compulsory orders which formed the basis of claims against the Government during the war period.

CALLAN, ROBERT E.
R—Knoxville, Tenn.
B—Baltimore, Md.
G. O. No. 16, W. D., 1920.

Brigadier General, U. S. Army.
As chief of staff of the Army Artillery, First Army, he exhibited ability in the organization of that unit. Later, as commanding general of the 33d Coast Artillery Brigade, he displayed high technical ability. Though confronted with innumerable difficulties, he developed the heavy artillery regiments under his command into combat units of remarkable efficiency, which units proved to be of the utmost value during the St. Mihiel and Meuse-Argonne offensives.

CAMERON, REBA G.
R—Taunton, Mass.
B—Canada.
G. O. No. 9, W. D., 1923.

First lieutenant, chief nurse, Army Nurse Corps, U. S. Army.
As chief nurse of the General Hospital at Plattsburg Barracks, N. Y., during the World War, and later as chief nurse of the general hospital at Hampton, Va., during the demobilization period, her services such as to call for special commendation. Whether by day or night, in emergency or routine work, she never failed to respond efficiently to every call. By her conscientious, unselfish devotion to duty, and cheerful, cooperative manner she was an example and an inspiration to her entire staff, and was in large measure responsible for the marked success of these special hospitals.

CAMPBELL, ROBERT M.
R—Owings Mills, Md.
B—Owings Mills, Md.
G. O. No. 124, W. D., 1919.

Lieutenant colonel, Cavalry, U. S. Army.
For especially meritorious and distinguished service while serving as military attaché at Mexico City, Mexico.

CANFIELD, EDWARD, Jr.
R—Middletown, N. Y.
B—Bath, N. Y.
G. O. No. 59, W. D., 1919.

Lieutenant colonel, Coast Artillery Corps, U. S. Army.
He served as assistant Chief of Staff, G–1, of the 4th Division, and organized the entire system of supply for the division. He trained and supervised the personnel and the operation of the administrative sections. He handled all problems connected with supply and transportation with such efficiency and success that the division was never short of either rations or ammunition. He proved himself to be an officer of the greatest administrative ability, exceptional foresight, and tireless energy.

CANNON, WALTER B.
R—Cambridge, Mass.
B—Prairie du Chien, Wis.
G. O. No. 49, W. D., 1922.

Lieutenant colonel, Medical Corps, U. S. Army.
As director of physiological research for the American Expeditionary Forces in France, his activities in connection with the development of a standard method for the resuscitation of the wounded and in organizing, instructing, and directing the work of shock teams in hospitals at the front reflected professional skill and judgment of the highest order and resulted in saving many lives.

CARLETON, GUY.
R—San Antonio, Tex.
B—Austin, Tex.
G. O. No. 47, W. D., 1919.

Major general, U. S. Army.
For exceptionally meritorious and conspicuous service as commanding general at Camp Wadsworth, S. C., in organizing and training corps and Army troops during the war.

CARSON, CLIFFORD C.
R—Muncie, Ind.
B—North Greenfield, Ohio.
G. O. No. 59, W. D., 1919.

Colonel, Coast Artillery Corps, U. S. Army.
He organized and commanded the training centers for the instruction of officers for the Tractor Artillery of the American Expeditionary Forces. In this new and important field of activity he rendered conspicuously meritorious service.

CARSON, JOHN M.
R—Philadelphia, Pa.
B—Philadelphia, Pa.
G. O. No. 53, W. D., 1921.

Brigadier general, U. S. Army.
For services as chief quartermaster, line of communications, American Expeditionary Forces, and later as deputy chief quartermaster, American Expeditionary Forces, positions of great responsibility, due to his ability and energy he perfected and directed the organization and operation of the Quartermaster Corps of the line of communications. Later he skillfully carried out the plans and projects to make the Quartermaster Corps an unfailing auxiliary to the combatant troops of the American Expeditionary Forces. He has rendered services of much value.

CARTER, ARTHUR H.
R—Leesburg, Va.
B—Hillsboro, Kans.
G. O. No. 69, W. D., 1919.

Colonel, Field Artillery, U. S. Army.
While on duty in the office of the Chief of Field Artillery, he displayed great ability in developing the organization of the Field Artillery Central Officers' Training School; he then proceeded to Camp Taylor, established this school, and administered it in an exceptionally meritorious manner during the remainder of the war.

CARTER, JESSE McI.
R—Farmington, Mo.
B—St. Francois County, Mo.
G. O. No. 25, W. D., 1919.

Major general, U. S. Army.
As chief of the Militia Bureau he conceived and directed the organization of the United States Guards and utilized these and other forces most effectively in the important work of safeguarding the utilities and industries of the Nation essential to the prosecution of the war.

CARTER, WILLIAM H. R—New York, N. Y. B—Nashville, Tenn. G. O. No. 124, W. D., 1919. Medal of honor also awarded.	Major general, U. S. Army. For services as department commander, Central Department, between Aug. 26, 1917, and Mar. 13, 1918; he handled many difficult problems arising in that department with rare judgment, tact, and great skill.
CARTY, JOHN J. R—Short Hills, N. J. B—Cambridge, Mass. G. O. No. 59, W. D., 1919.	Colonel, Signal Corps, U. S. Army. He was largely instrumental in securing from the telephone and telegraph companies of the United States the best talent available to meet the urgent requirements of the Signal Corps at the outbreak of the war. He has served with marked distinction as a member of the American Expeditionary Forces, and his brilliant professional attainments and sound judgment have rendered his services of exceptional value to the Government.
CASAD, ADAM F. R—Wichita, Kans. B—Delphi, Ind. G. O. No. 78, W. D., 1919.	Colonel, Ordnance Department, U. S. Army. As deputy chief ordnance officer at General Headquarters, American Expeditionary Forces, he exercised conspicuous initiative and sound judgment in the supervision of ordnance activities. With tireless energy he organized and administered the work of the Ordnance Department in the zone of the Armies. As representative of the Ordnance Department at Chaumont, he showed wide vision and full comprehension of conditions and needs of the service, working with exceptional devotion to prevent any stoppage in the supply of ordnance material.
CASTNER, JOSEPH COMPTON R—New Brunswick, N. J. B—New Brunswick, N. J. G. O. No. 59, W. D., 1919.	Brigadier general, U. S. Army. While in command of the 9th Infantry Brigade he displayed conspicuous tenacity of purpose and a determination to overcome all obstacles. At the Bois des Rappes, in the St. Mihiel salient and ensuing actions, his brigade effectively routed the enemy. The success of his command was in a large measure due to the splendid training and excellent leadership given it by its commander.
CATRON, THOMAS B. 2D R—Santa Fe, N. Mex. B—New Mexico. G. O. No. 103, W. D., 1919.	Major, Signal Corps, U. S. Army. As an instructor at the Army Intelligence School he performed important duties with marked zeal and ability, aiding materially in the efficient training of a large number of officers for the intelligence service of the units of our Armies in the field.
CAVANAUGH, JAMES B. R—Olympia, Wash. B—Carrollton, Ill. G. O. No. 50, W. D., 1919.	Colonel, Engineers, U. S. Army. As commander of an Engineer regiment, he rendered great assistance in the early development of the American port at Bassens. As assistant chief of staff in charge of the administrative section of the services of supply, he exhibited rare qualities and marked ability in the solution of many problems of policy. His efforts in connection with the repatriation of American troops have been of conspicuous merit.
CHAFFEE, ADNA R. R—Manila, P. I. B—Junction City, Kans. G. O. No. 62, W. D., 1919.	Colonel, Cavalry, U. S. Army. At Army General Staff College he displayed military attainments of a high order, contributing efficiently to the training of a large number of officers. He performed tasks of great difficulty with marked distinction as G-3 of the 81st Division and later of the 7th Corps. Later, as chief of the third section, general staff, 3d Corps, he acted with sound judgment and wide comprehension of existing conditions in the discharge of the grave responsibilities connected with his office during the closing days of the Meuse-Argonne offensive, handling perplexing problems with keen energy and wise discernment.
CHAMBERLAIN, JOHN L. R—South Livonia, N. Y. B—South Livonia, N. Y. G. O. No. 25, W. D., 1919.	Major general, U. S. Army. As inspector general of the Army he has, by his highly responsible services, materially contributed to the efficiency of all departments and bureaus of the Military Establishment and to the successful execution of the military program.
CHAMBERLAIN, WESTON P. R—Bristol, Me. B—Bristol, Me. G. O. No. 10, W. D., 1922.	Colonel, Medical Corps, U. S. Army. As chief sanitary inspector of the Army within the continental limits of the United States during the World War he displayed exceptional efficiency in organizing and administering a sanitary inspection service during the periods of mobilization, active operations, and demobilization. His achievements in this capacity were of great value to the Government.
CHAMBERLAINE, WILLIAM R—Norfolk, Va. B—Norfolk, Va. G. O. No. 2, W. D., 1920.	Brigadier general, U. S. Army. As commanding general of the Railway Artillery Reserve, he rendered valuable services to the American Expeditionary Forces in the operations of the Railway Artillery units during the Meuse-Argonne offensive.
CHAMBERLIN, STEPHEN J. R—Spring Hill, Kans. B—Spring Hill, Kans. G. O. No. 38, W. D., 1922.	Major, Infantry, U. S. Army. As acting dispatch officer and dispatch officer at port of embarkation, Hoboken, N. J., from Nov. 15, 1917, to Sept. 6, 1918, Major Chamberlin displayed marked ability in handling the movements of troops through the port, assigning units and detachments to camps, convoys, and ships, and by foresight, thorough organization, and hard work arranged for the smooth working of troop movements, prevented congestion at the camps and piers, thus enabling the transports to sail at the appointed time with the appropriate number of troops.
CHANDLER, CHARLES DEF. R—Cleveland, Ohio. B—Cleveland, Ohio. G. O. No. 50, W. D., 1919.	Colonel, Signal Corps, U. S. Army. As chief of the balloon section, Air Service, American Expeditionary Forces, from November, 1917, to February, 1919, he rendered notable service in the supply, administration, and operation of the balloon units that so thoroughly demonstrated their efficiency during all the major operations of the American Expeditionary Forces.

CHAPIN, LINDLEY H. F.
R—New York, N. Y.
B—New York, N. Y.
G. O. No. 59, W. D., 1919.

First lieutenant, General Staff, U. S. Army.
As the representative of G-4, of the American Expeditionary Forces at the D. G. C. R. A., he displayed marked ability and devotion to duty in a position of great responsibility, he handled with tact and sound judgment the involved and delicate questions continually arising in connection with our relationship with the allied armies, and rendered service of great value to the Government.

CHEATHAM, B. FRANK.
R—Nashville, Tenn.
B—Beech Grove, Tenn.
G. O. No. 9, W. D., 1923.

Colonel, Quartermaster Corps, U. S. Army.
As chief quartermaster, 1st Army Corps, he displayed sound judgment, great initiative, and high professional attainments. Later, as colonel, 104th Infantry, 26th Division, in operations against the enemy in the Meuse-Argonne offensive north of Verdun, Oct. 14 to Nov. 11, 1918, he rendered exceptionally valuable services, his high courage, leadership, and tactical skill proving important factors in the successful operations of the 26th Division during the second phase of the Meuse-Argonne offensive.

CHENEY, SHERWOOD A.
R—South Manchester, Conn.
B—South Manchester, Conn.
G. O. No. 108, W. D., 1919.

Brigadier general, U. S. Army.
As assistant chief engineer, General Headquarters, he rendered valuable services in the organization of the Engineer Corps and its coordination with the associated services. Later, as director of the Army Transport Service, he performed eminently valuable services, achieving remarkable results in a task of great magnitude involving the expeditious return of many thousands of soldiers from the ports of France to the United States.

CHURCH, EARL D.
R—Hartford, Conn.
B—Rockville, Conn.
G. O. No. 31, W. D., 1922.

Lieutenant colonel, Ordnance Department, U. S. Army.
As ordnance officer of the 30th Division, not only were ordnance and ammunition supplied at all times to members of that division, but also to thousands of men in other divisions at various times when their own supply failed. His organization of the ordnance supply system, as division ordnance officer, showed the results of exhaustive study and of determined and intelligent efforts to overcome adverse conditions. Later, as chief ordnance officer of the 9th Army Corps, he displayed high qualities of zeal, loyalty, and efficiency.

CHURCHILL, MARLBOROUGH.
R—Andover, Mass.
B—Andover, Mass.
G. O. No. 73, W. D., 1919.

Brigadier general, U. S. Army.
For services as chief of staff of the Army artillery of the 1st Army, American Expeditionary Forces, and for his ability, zeal, and untiring energy in building up the military intelligence division of the General Staff as director of military intelligence. He discharged these duties of great responsibility with ability, tact, and energy. He built up the intelligence service to its present high state of efficiency.

CLARK, ALBERT P.
R—Washington, D. C.
B—Washington, D. C.
G. O. No. 62, W. D., 1919.

Lieutenant colonel, Medical Corps, U. S. Army.
As medical representative on, and later as General Staff member of, the first section, General Headquarters, American Expeditionary Forces, he displayed sound judgment and wide comprehension of existing conditions in the management of ocean tonnage allotments, and devised and efficiently operated a system of supply for the Medical Department of the American Expeditionary Forces. Largely through his personal efforts, energy, and farsightedness the difficulties in the procurement and shipment of medical supplies for the sick and wounded were successfully overcome.

CLARK, FRANCIS W.
R—Chicago, Ill.
B—Wichita, Kans.
G. O. No. 38, W. D., 1922.

Lieutenant colonel (Field Artillery), General Staff Corps, U. S. Army.
As acting G-3 of the 3d Corps he displayed sound judgment and military attainments of the highest order. With the utmost clearness and skill he prepared the orders under which his corps operated both from the Vesle to the Aisne Rivers and during the Meuse-Argonne offensive until Oct. 14, 1918. By his marked tactical ability, loyal devotion to duty, and untiring zeal he contributed materially to the success of those operations.

CLARK, GRENVILLE.
R—New York, N. Y.
B—New York, N. Y.
G. O. No. 15, W. D., 1921.

Lieutenant colonel, Adjutant General's Department, U. S. Army.
For services on the committee on classification of personnel and later as a member of the committee on education and special training.

CLARK, JOSHUA REUBEN, Jr.
R—Washington, D. C.
B—Grantsville, Utah.
G. O. No. 49, W. D., 1922.

Major, Judge Advocate General's Department, U. S. Army.
As special assistant to the Attorney General of the United States from June, 1917, until September, 1918, by his zeal, great industry, and eminent legal attainments, he rendered conspicuous services in the compilation and publication of an extremely valuable and comprehensive edition of the laws and analogous legislation pertaining to the war powers of our Government since its beginning. From September to December, 1918, as executive officer of the Provost Marshal General's Office, he again rendered services of inestimable value in connection with the preparation and execution of complete regulations governing the classification and later the demobilization of several million registrants.

CLARK, PAUL H.
R—Chicago, Ill.
B—Chicago, Ill.
G. O. No. 59, W. D., 1919

Lieutenant colonel, Infantry, U. S. Army.
As chief of the American mission at French General Headquarters, he performed with marked distinction important duties requiring tact and judgment. His ceaseless efforts and untiring energy were of material benefit in securing the necessary cooperation with the French military authorities.

CLARK, STEPHEN C.
R—New York, N. Y.
B—Cooperstown, N. Y.
G. O. No. 56, W. D., 1922.

Lieutenant colonel, Adjutant General's Department, U. S. Army.
As assistant adjutant and later as adjutant of the 2d Army Corps from August, 1918, to February, 1919, he displayed administrative ability of an exceptionally high order. During active operations of the corps his tireless energy and unceasing devotion to duty assisted materially in the successes achieved by his organization.

*CLARKE, THOMAS C.
  R—New York, N. Y.
  B—Philadelphia, Pa.
  G. O. No. 56, W. D., 1921.

Colonel, 110th Engineers, 35th Division, U. S. Army.
For services at Baulny, France, Sept. 29, 1918. He admirably established and constructed a line of resistance which he held for several days, during which time two counterattacks were repelled and severe loss inflicted upon the enemy. Later he commanded and held an advance line for two days, repelling two determined counterattacks.
Posthumously awarded. Medal presented to widow, Mrs. Thomas C. Clarke.

CLEMENS, PAUL B.
  R—Superior, Wis.
  B—Superior, Wis.
  G. O. No. 9, W. D., 1923.

Lieutenant colonel (Infantry), General Staff Corps, U. S. Army.
As assistant chief of staff, G-2, 32d Division, during its operations in France he displayed unusual and masterful grasp of his duties, executive ability of a high order, and intense zeal and devotion to duty. His initiative, foresight, and good judgment were important factors in the successes of his division, and made his services of inestimable value to the Government in a position of great responsibility.

CLEVELAND, MAUDE.
  R—Berkeley, Calif.
  B—Gresham, Oreg.
  G. O. No. 133, W. D., 1919.

American Red Cross.
Chief of the home communication and casualty service of the Red Cross at Brest, France. By her unremitting efforts in caring for the sick and wounded evacuated through the port of Brest, her valuable assistance in the interment of the dead, consummated at night under the most adverse weather conditions, her careful consideration in writing the details of the death to the nearest relative, and her supreme exertion during the distressing epidemic of influenza pneumonia from September to December, 1918, she has rendered self-sacrificing services of the highest character to the American Expeditionary Forces.

CLIFFORD, EDWARD.
  R—Evanston, Ill.
  B—Virginia, Ill.
  G. O. No. 10, W. D., 1922.

Lieutenant colonel, Quartermaster Corps, U. S. Army.
For services in organizing and administering the Army allotment system, thereby enabling the War Department to make prompt payments of allotments. He rendered services of great value.

CLINNIN, JOHN V.
  R—Chicago, Ill.
  B—Huntley, Ill.
  G. O. No. 56, W. D., 1921.

Colonel, Infantry, U. S. Army.
As regimental commander of the 130th Infantry, 33d Division, by his force and energy he brought his regiment to a high state of efficiency, instilling into it an aggressive spirit which proved a valuable factor in the operations of the regiment near Breuilles, Verdun, Bois-de-Chaume, Sivry, and Bois du Platt Chene. He rendered services of great merit.

CLOMAN, SYDNEY A.
  R—Deavertown, Ohio.
  B—Deavertown, Ohio.
  G. O. No. 59, W. D., 1919.

Colonel, Infantry, U. S. Army.
As chief of staff of the 29th Division, he showed himself resourceful and equal to any emergency. His sound judgment and ability, especially during the operations north of Verdun, France, in October, 1918, were of a high order. The success of these operations was in a measure due to his energy, zeal, and rare qualities of leadership.

CLOPTON, WILLIAM H., Jr.
  R—St. Louis, Mo.
  B—St. Louis, Mo.
  G. O. No. 43, W. D., 1922.

Colonel, Tank Corps, U. S. Army.
By his special knowledge of personnel matters, untiring zeal, good judgment, and administrative ability he was largely responsible for the solution of many difficult problems in the organization and operation of the personnel branch, Office of the Quartermaster General, especially in the organization of the technical field units of the Quartermaster Corps. Later, as colonel, Tank Corps, he rendered conspicuous service in the establishment of tank schools and in the training of technical troops therein, while commanding tank training camps at Tobyhanna, Pa., and Camp Polk, N. C.

COCHEU, FRANK S.
  R—Brooklyn, N. Y.
  B—Brooklyn, N. Y.
  G. O. No. 49, W. D., 1922.

Colonel, Infantry, U. S. Army.
In command of the 319th Infantry, 80th Division, from August, 1917, to October, 1918, he displayed marked ability in its organization, training, and service in the field. In operations against the enemy in the Artois sector and Meuse-Argonne offensive he rendered conspicuous service by leading his command with exceptional judgment, unflagging energy, and tactical ability, at all times proving himself to be a skillful commander, thus enabling his regiment to always carry its tasks through to a successful end. His services were highly meritorious and rendered in a position of great responsibility.

COCHEU, GEORGE W.
  R—Brooklyn, N. Y.
  B—Brooklyn, N. Y.
  G. O. No. 10, W. D., 1922.

Colonel (Coast Artillery Corps), General Staff Corps, U. S. Army.
As a member and later as chief of the coordination section, General Staff, a position of great responsibility, he devised many methods for improving and making more efficient the administrative procedure within the War Department, thereby materially facilitating the transaction of the business of the War Department and between the War Department and the Army, thus rendering service of great value to the entire military establishment.

COE, FRANK W.
  R—Manhattan, Kans.
  B—Manhattan, Kans.
  G. O. No. 18, W. D., 1919.

Major general, U. S. Army.
For services in the reorganization of the Coast Artillery, thereby enabling it to meet the great demand for oversea artillery.

COLE, HAYDN S.
  R—Kewanee, Ill.
  B—Newark Valley, N. Y.
  G. O. No. 14, W. D., 1923.

Colonel, Quartermaster Corps, U. S. Army.
For services while assistant to the general superintendent, Army Transport Service, New York City, May 1 to Nov. 1, 1917; general manager of the Hoboken Shore R. R., July 1 to Nov. 1, 1917; and in full charge of operations at Bush Terminal System, Brooklyn, N. Y., until January, 1919. He displayed rare administrative and executive ability, sound judgment, and rendered services of immeasurable value to the Government.

COLE, WILLIAM E_____
  R—Willard, Utah.
  B—Willard, Utah.
  G. O. No. 38, W. D., 1922.

Brigadier general, U. S. Army.
As commanding officer of the 351st Field Artillery, 92d Division, from November, 1917, to August, 1918, by his rare judgment and exceptional ability he organized and trained his regiment to a high standard of efficiency under the most adverse circumstances. As brigade commander from August, 1918, until December, 1918, he again displayed resourcefulness and unusual ability in the successful organization and training of the 11th and 20th Field Artillery Brigades. As commanding general at Camp Jackson, S. C., from January, 1919, to April, 1919, he rendered most valuable and distinguished service in the demobilization of great numbers of troops and immense quantities of material.

COLEMAN, FREDERICK W_____
  R—Washington, D. C.
  B—Baltimore, Md.
  G. O. No. 49, W. D., 1922.

Colonel (Infantry), General Staff Corps, U. S. Army.
As G-1, 91st Division, from August to October, 1918, due to his unusual foresight, indefatigable zeal, exceptional executive and administrative ability, he so organized the supply and administrative services of the division as to insure complete coordination and a regular flow of supplies of all kinds, notwithstanding a shortage of transportation and despite grave and tremendous difficulties because of road congestion, thereby rendering conspicuous service in a position of great responsibility to the American Expeditionary Forces.

COLES, ROY H_____
  R—Warren, Ind.
  B—Warren, Ind.
  G. O. No. 59, W. D., 1919.

Lieutenant colonel, Signal Corps, U. S. Army.
He served as assistant to and executive officer for the chief signal officer, American Expeditionary Forces, and at all times he performed his most exacting duties in an especially meritorious manner. By his exceptional executive ability, tireless energy, and sound judgment he successfully met every demand that was made upon him.

COLLADAY, EDGAR B_____
  R—Dunn, Wis.
  B—Dunn, Wis.
  G. O. No. 27, W. D., 1922.

Lieutenant colonel (Coast Artillery Corps), General Staff Corps, U. S. Army.
As executive officer of the cable section, General Staff, during the World War, his sound judgment, his marked ability, his untiring energy, and his willingness to accept grave responsibilities, contributed greatly to the successful operation of the cable business during the World War. Later, as executive officer of the statistics branch, General Staff, he assisted materially in collecting and compiling valuable statistics pertaining to the World War.

COLLINS, CHRISTOPHER C_____
  R—Lynchburg, Va.
  B—Lynchburg, Va.
  G. O. No. 4, W. D., 1923.

Colonel, Medical Corps, U. S. Army.
As corps surgeon, 2d Army Corps, from February, 1918, to February, 1919, he displayed professional attainments of a high order in the training and the organization of the corps for subsequent operations. During active operations the efficiency of his organization and arrangements for the care of the sick and wounded and their evacuation contributed in high degree to the success of the operations of the corps.

COLLINS, EDGAR T_____
  R—Williamsport, Pa.
  B—Williamsport, Pa.
  G. O. No. 16, W. D., 1920.

Colonel (Infantry), General Staff Corps, U. S. Army.
As assistant to G-5, General Headquarters, and later as Chief of Staff of the 6th Army Corps, he demonstrated rare military attainments, performing his difficult tasks with unremitting zeal, rendering services of conspicuous worth to the American Expeditionary Forces.

COLLINS, JAMES L_____
  R—New Orleans, La.
  B—New Orleans, La.
  G. O. No. 59, W. D., 1919.

Lieutenant colonel, Field Artillery, U. S. Army.
As aide-de-camp to the commander in chief, as line officer on duty with troops, and as secretary of the General Staff of the American Expeditionary Forces, he displayed a thorough knowledge of every duty with which he was intrusted. With tireless energy, keen perception, and able execution of his manifold duties he rendered especially meritorious services to the American Expeditionary Forces.

COLLINS, OWEN G_____
  R—Chicago, Ill.
  B—Chicago, Ill.
  G. O. No. 59, W. D., 1919.

Colonel, Quartermaster Corps, U. S. Army.
Under his administration the supply of troops, care of property, and the operation of the Quartermaster Depot at Gievres, showed an excellent degree of efficiency. The great improvement and development of the organization of this important depot under his direction was largely due to his administrative ability and untiring zeal.

COLLINS, ROBERT L_____
  R—Wellesley Hills, Mass.
  B—Lancaster, Mass.
  G. O. No. 9, W. D., 1923.

Colonel, Cavalry, U. S. Army.
He served as assistant chief of staff, G-1 and G-2, 33d Division, from June 13 to Oct. 2, 1918, and assistant chief of staff, G-2, 37th Division, from Oct. 25 to Nov. 7, 1918; assistant chief of staff, G-2, 84th Division, Oct. 2-25, 1918; assistant chief of staff and chief of staff, general headquarters, Tank Corps, from Nov. 8, 1918, to June 30, 1920. He displayed sound judgment and exceptional ability in the administration and operation of all the staff sections of the units with which he served and by his initiative, tireless energy, and military attainments of a high order he contributed in a marked degree to the successes achieved by those forces against the enemy.

COMMISKEY, ARCHIBALD F_____
  R—Brooklyn, N. Y.
  B—Brooklyn, N. Y.
  G. O. No. 39, W. D., 1920.

Colonel, Field Artillery, U. S. Army
As regimental commander of the 77th Field Artillery, 4th Division, and during the Meuse-Argonne offensive from Oct. 3 to 11, 1918, as commander of an artillery grouping consisting of the 77th and 16th Field Artillery, he demonstrated marked ability and good judgment in the direction of his units in support of the 4th Division during its attack. From Nov. 1 to 11, 1918, his regiment supported, with marked success, the advance of the 5th Division. He kept his elements close to the attacking units and gave valuable assistance to them. He has rendered services of considerable value to the American Expeditionary Forces.

CONGER, ARTHUR L. R—Akron, Ohio. B—Akron, Ohio. G. O. No. 35, W. D., 1920.	Colonel, Infantry, U. S. Army. As a member of the second section, General Staff, General Headquarters, by his marked professional attainments, his zeal, and his sound judgment he contributed largely to the successful operation of this section. As chief of the second section, General Staff, of the 2d Division, during active operations, and later as commander of a brigade of the 28th Division during the Argonne-Meuse offensive, he demonstrated his great energy and his clear conception of tactics.
CONLEY, EDGAR T. R—Fairland, Md. B—Fairland, Md. G. O. No. 56, W. D., 1921.	Colonel (Infantry), Adjutant General's Department, U. S. Army. As chief of the prisoners of war division, the Provost Marshal General's Department, he had charge of and was responsible for all matters concerning the prisoners of war labor companies, escort companies, and inclosures. His sound judgment, marked ability, and devotion to duty resulted in the handling of the delicate prisoner of war questions in such a manner as to produce only commendation. His services were exceptionally valuable to the Government.
CONNELL, KARL R—New York, N. Y. B—Omaha, Nebr. G. O. No. 56, W. D., 1922.	Major, Medical Corps, U. S. Army. In the Chemical Warfare Service, practically alone and unaided and at a great personal risk of his life, he exposed himself unhesitatingly to the highest concentrations of deadly gases while working with experimental models of masks. Major Connell invented, tested out, and perfected a new type of gas mask superior to any then in existence, thereby rendering service of inestimable value to the American Expeditionary Forces.
CONNER, FOX R—Slate Spring, Miss. B—Slate Spring, Miss. G. O. No. 12, W. D., 1919.	Brigadier general, U. S. Army. As assistant chief of staff in charge of the operations section he has shown a masterful conception of all the tactical situations which have confronted the American forces in Europe. By his high professional attainments and sound military judgment he has handled with marked skill the many details of the complex problems of organization and troop movements that were necessitated by the various operations of the American Expeditionary Forces.
CONNER, LEWIS A. R—New York, N. Y. B—New Albany, Ind. G. O. No. 10, W. D., 1922.	Colonel, Medical Corps, U. S. Army. As chief of the internal medicine division of the Surgeon General's Office, a position of great responsibility, to him is due, in a large measure, the expansion and successful administration of that division. By his powerful influence he induced many eminent internists to remain in the base hospitals after the signing of the armistice until the sick from overseas had been taken care of.
CONNER, WILLIAM D. R—Clinton, Iowa. B—Rock, Wis. G. O. No. 12, W. D., 1919.	Brigadier general, U. S. Army. As assistant chief of staff and head of the coordination section of the General Staff, American Expeditionary Forces, he showed unusual ability and tireless energy. As chief of staff of the 32d Division in the trench operations in the Belfort sector and later as commander of the 63d Infantry Brigade in the advance to the Vesle he displayed particular ability as a leader of troops. He also performed valuable services as commander of a base port and as chief of staff of the services of supply.
CONRAD, CASPER H., Jr. R—Fort Randall, S. Dak. B—Columbus, Ohio. G. O. No. 2, W. D., 1920.	Colonel, Infantry, U. S. Army. As commander of advance embarkation section, S. O. S., by his marked executive ability and energetic efforts displayed in the transportation of troops from the area of the Army of Occupation to the base ports, and the management of troop trains which were comfortably equipped, safely operated, and sanitarily maintained, he has rendered services of great worth.
COOK, CHARLES F. R—Richmond Hill, N. Y. B—Knoxboro, N. Y. G. O. No. 89, W. D., 1919.	Major, Ordnance Department, U. S. Army. While serving in the Ordnance Department he was instrumental in organizing the divisions dealing with the procurement of ordnance material. While serving on the General Staff he worked out a plan for the consolidation of all articles of standard commercial circulation in single purchasing units throughout the War Department upon which basis was built the organization of the present division of purchase, storage and traffic of the General Staff. In his various assignments he gave himself whole-heartedly and self-sacrificingly to the work of the Government. Posthumously awarded. Medal presented to sister, Mrs. Blanche Seavey.
COOPER, HARRY L. R—Philadelphia, Pa. B—Philadelphia, Pa. G. O. No. 4, W. D., 1923.	Colonel, Infantry, U. S. Army. He commanded the 2d Army Corps School at Chatillon, France, from Aug. 15, 1918, to May 19, 1919. He so organized and coordinated the various activities at these schools that 1,800 to 2,500 students were constantly undergoing instruction. He was primarily responsible for the excellent system of training given, which training as received at these schools exercised a strong influence toward the efficiency of the whole body of American troops in France. By his sound administrative and superior technical ability, untiring zeal, and splendid judgment in reorganizing and expanding the schools, he produced an organization of the highest efficiency. He rendered services of signal worth to the American Expeditionary Forces in a position of great responsibility.
COOPER, WIBB E. R—Nashville, Tenn. B—Mount Pleasant, Tenn. G. O. No. 103, W. D., 1919.	Colonel, Medical Corps, U. S. Army. He commanded with notable success Base Hospital No. 8, at Savenay, which under his efficient administration became the nucleus of a large hospital center, which developed into the largest classification and evacuation hospital in France for patients returning to the United States. By his marked ability in directing the numerous activities under his control he rendered services of conspicuous worth to the American Expeditionary Forces.

COOTES, HARRY N.
R—Staunton, Va.
B—Staunton, Va.
G. O. No. 43, W. D., 1922.

Colonel (Cavalry), General Staff Corps, U. S. Army.
As chief of staff, 78th Division, during its organization and training period and later during the St. Mihiel offensive he displayed tact, sound judgment, and military attainments of the highest order. Due to his ability in coordinating all of the various staff agencies into a harmonious machine, and by his loyalty and untiring energy in carrying out the policies of the division commander, he rendered services of great value to the American Expeditionary Forces.

CORDIER, CONSTANT.
R—Flagstaff, Ariz.
B—New Orleans, La.
G. O. No. 25, W. D., 1919.

Colonel (Infantry), General Staff Corps, U. S. Army.
While on duty as liaison officer between the War Department and the foreign military missions he displayed the greatest discretion and ability and contributed materially to the successful conduct of military-diplomatic relations between the War Department and the allied military missions.

CORLETT, CHARLES H.
R—Monte Vista, Colo.
B—Burchard, Nebr.
G. O. No. 62, W. D., 1919.

Lieutenant colonel, Signal Corps, U. S. Army.
As deputy to the Chief Signal Officer of the line of communications he displayed marked ability for organization and administration in the establishment of important Signal Corps undertakings, laying the foundation of the Signal Corps work in the American Expeditionary Forces. Later, as director of supplies in the Office of the Chief Signal Officer, he performed exacting duties with unusual ability, solving with sound judgment perplexing problems, enabling a steady flow of signal supplies to be maintained to the troops in the field.

COUGHLIN, EDNA M.
R—Chicago, Ill.
B—Chicago, Ill.
G. O. No. 10, W. D., 1920.

Reserve nurse, Army Nurse Corps, U. S. Army, attached to Emergency Medical Team No. 142, American Expeditionary Forces.
As a member of an emergency medical team during an extended period of active operations she served the nontransportable wounded of six divisions during the advances at Glorieux, Fromereville, Bethincourt, Septsarges, Bantheville, and Dun-sur-Meuse. She courageously administered to the gravely wounded in the advanced area under the fire of shells and aerial bombs, rendering a service of particular value to the American Expeditionary Forces.

COWARD, JACOB M.
R—Trenton, N. J.
B—Allentown, N. J.
G. O. No. 60, W. D., 1920.

Colonel, Coast Artillery Corps, U. S. Army.
As a member of the American section of the Supreme War Council by his high professional qualifications and sound military judgment he rendered invaluable aid in solving the many complex problems that came before the Supreme War Council.

COX, ALBERT L.
R—Raleigh, N. C.
B—Raleigh, N. C.
G. O. No. 9, W. D., 1923.

Colonel, Field Artillery, U. S. Army.
As commanding officer of the 113th Field Artillery, 30th Division, during its organization, training, and active operations in the St. Mihiel and Meuse-Argonne offensives he displayed tireless energy, great resourcefulness, and military attainments of a high order. By his skillful and energetic handling of his regiment he rendered the maximum support to the Infantry to which he was attached so effectively that he aided materially in the successes achieved by our troops in those important engagements.

COX, CREED F.
R—Saddle, Va.
B—Bridle Creek, Va.
G. O. No. 15, W. D., 1923.

Colonel, Field Artillery, U. S. Army.
As commander of one or more elements of the 4th Artillery Brigade in addition to his own regiment, the 77th Field Artillery, in the Aisne Marne, St. Mihiel, and Meuse-Argonne operations he rendered conspicuously valuable services. In the Meuse-Argonne offensive in charge of the barrage groupings of the 4th Artillery Brigade his high technical skill contributed in a marked way to the successful operations of the American forces. His untiring energy and devotion to duty served as a stimulus to the officers and men under his command.

COXE, ALEXANDER B.
R—St. Paul, Minn.
B—Santa Fe, N. Mex.
G. O. No. 2, W. D., 1920.

Colonel (Cavalry), General Staff Corps, U. S. Army.
He assisted in the production of an efficient intelligence service, and in organizing and sending to France the required intelligence personnel. His excellent judgment and pronounced knowledge of intelligence principles added greatly to the efficiency of the intelligence service.

CRAIG, CHARLES F.
R—Danbury, Conn.
B—Danbury, Conn.
G. O. No. 10, W. D., 1922.

Colonel, Medical Corps, U. S. Army.
For services as the organizer and administrator of Army schools for the development of laboratory personnel, thereby contributing one of the most important measures for the prevention and control of epidemic diseases throughout the Army.

CRAIG, DANIEL F.
R—Garnett, Kans.
B—Mahaska County, Iowa.
G. O. No. 103, W. D., 1919.

Brigadier general, U. S. Army.
He served with distinction as commanding officer of the 302d Field Artillery, 76th Division, and later, upon being promoted to the grade of brigadier general, as commanding general of the 157th Field Artillery Brigade, 82d Division, his service was equally conspicuous. Due to his aggressive leadership, his batteries were at all times close behind the advancing Infantry. The accurate support which they furnished was largely due to his management and technical skill as an artillerist.

CRAIG, MALIN.
R—Philadelphia, Pa.
B—St. Joseph, Mo.
G. O. No. 12, W. D., 1919.

Brigadier general, U. S. Army.
He served in turn as chief of staff of a division, a corps, and an army, in each of which capacities he exhibited great ability. His personal influence, aggressiveness, and untiring efforts were repeatedly displayed in the operations of the 1st Corps in the vicinity of Chateau-Thierry, on the Ourcq, and the Vesle during the St. Mihiel and Argonne-Meuse offensives.

CRAIN, JAMES K.
R—Cuero, Tex.
B—Hallettsville, Tex.
G. O. No. 78, W. D., 1919.

Colonel, Coast Artillery Corps, U. S. Army.
Successively as ordnance officer of the 42d Division, 1st Army Corps and 2d Army, he displayed exceptional ability in the organization and administration of work of great magnitude. Encountering unforeseen and perplexing problems, he solved them with initiative and sound judgment, showing a full understanding of existing needs and conditions of the service. He was tireless in energy and resourceful, proving at all times devoted to his important duties.

CRAVATH, PAUL D.
R—New York, N. Y.
B—Berlin Heights, Ohio.
G. O. No. 46, W. D., 1919.

Representative of the Treasury Department.
With great ability, energy, and patience, he cooperated in international matters involving the interests of the American Expeditionary Forces. Establishing and maintaining the most cordial relations with the British authorities, he greatly contributed to the establishment of their effective cooperation with the military board of allied supply and in many other matters of extreme importance.

CRAVENS, RICHARD K.
R—Muskogee, Indian Territory.
B—Fort Smith, Ark.
G. O. No. 19, W. D., 1920.

Colonel, Coast Artillery Corps, U. S. Army.
Both as adjutant and as assistant chief of staff, G-1, of the Headquarters Army Artillery, 1st Army, American Expeditionary Forces, he exhibited military attainments and ability of a high order. By his clear conception of his function, his untiring energy, and in his proper exercise of initiative, he contributed in a large measure to the successful employment of the Army Artillery of the 1st Army during the St. Mihiel and Meuse-Argonne offensives.

CRENSHAW, RUSSELL SYDNOR.
R—Richmond, Va.
B—Richmond, Va.
G. O. No. 116, W. D., 1919.

Commander, U. S. Navy.
For services in connection with the Naval Overseas Transportation Service and convoy system for cargo of transport fleet.

CRILE, GEORGE W.
R—Cleveland, Ohio.
B—Chili, Ohio.
G. O. No. 50, W. D., 1919.

Colonel, Medical Corps, U. S. Army.
By his skill, researches, and discoveries he saved the lives of many of our wounded soldiers. His tireless efforts to devise new methods of treatment to prevent infection and surgical shock revolutionized Army surgery and met with the greatest success.

CRONKHITE, ADELBERT.
R—Arizona.
B—Litchfield Center, N. Y.
G. O. No. 12, W. D., 1919.

Major general, U. S. Army.
He commanded the 80th Division during the Argonne-Meuse offensive, where he demonstrated great ability as a leader and proved himself a commander of initiative and courage.

CROOKSTON, WILLIAM J.
R—Pittsburgh, Pa.
B—Irwin, Pa.
G. O. No. 62, W. D., 1919.

Colonel, Medical Corps, U. S. Army.
As division surgeon he displayed marked ability of organization and administration throughout the service of the 28th Division in France. With keen judgment he supervised the location of dressing stations and field hospitals and used remarkable discretion in directing the entire work of evacuation of a large number of casualties. By constant vigilance and unceasing effort he provided for the health and treatment of the troops with whom he served, displaying professional attainments of a high order.

CROWDER, ENOCH H.
R—Trenton, Mo.
B—Edinburg, Mo.
G. O. No. 144, W. D., 1918.

Major general, U. S. Army.
For services as provost marshal general in the preparation and operation of the draft laws of the Nation during the war.

CRUIKSHANK, WILLIAM M.
R—Washington, D. C.
B—Washington, D. C.
G. O. No. 59, W. D., 1919.

Brigadier general, U. S. Army.
He commanded with ability the artillery of the 3d Division on the Marne during the German attack on July 15. Subsequently, during the advance on July 18, due to his tactical knowledge and successful placing of the guns, he greatly assisted in the repulse of the enemy. Later he rendered valuable services as commander of the artillery of the 4th Army Corps.

CRUSAN, CLYDE B.
R—Pittsburgh, Pa.
B—Kellys Station, Pa.
G. O. No. 59, W. D., 1919.

Colonel, Quartermaster Corps, U. S. Army.
He was charged with the important duty of administering the supplies division of the Quartermaster Department of the American Expeditionary Forces, where he exhibited organizing ability of the highest order. His efforts and foresight had a marked influence on the successful delivery of rations, clothing, and other quartermaster supplies to combat troops.

CUBBISON, DONALD C.
R—Kansas City, Kans.
B—Harrisville, Pa.
G. O. No. 56, W. D., 1922.

Colonel, Field Artillery, U. S. Army.
Serving in turn as director, 2d Corps Artillery School, from January to April, 1918; chief of staff of the artillery of the 1st and 4th Army Corps from May to September, 1918; chief of field artillery section, Office of Chief of Artillery, American Expeditionary Forces, from September, 1918, to March, 1919, he performed his duties in a conspicuously meritorious manner at all times. By his great energy, sound judgment, marked ability, and high professional attainments, he contributed materially to the successes achieved against the enemy, rendering invaluable services to the American Expeditionary Forces.

CULBERSON, WILLIAM L.
R—Hillsboro, Tex.
B—Clifton, Tex.
G. O. No. 78, W. D., 1919.

Lieutenant colonel, Infantry, U. S. Army.
He displayed extraordinary qualities of leadership and ability for organization. While engaged upon another mission he discovered six companies of infantry, which had been ordered to relieve a front-line unit, lost and confused, due to the misdirection of the guides, who had lost their way. Coming upon these companies when the men were beginning to straggle from the ranks, he rallied them, and by his personal efforts alone succeeded in bringing them to their position just before daylight and in time to take part in the pending operation.

CULKIN, JOSEPH R.
    R—Rochester, N. Y.
    B—Oswego, N. Y.
    G. O. No. 126, W. D., 1919.

Major, Medical Corps, U. S. Army.
For services when in charge of Camp Hospital No. 1, at Camp Upton, N. Y., during the serious epidemic of influenza at this camp in September and October, 1918. Due to his great energy and good will and unwillingness to meet defeat in any form, remarkable results were obtained at this hospital.

CULVER, CLARENCE C.
    R—Milford, Nebr.
    B—Milford, Nebr.
    G. O. No. 69, W. D., 1919.

Colonel, Air Service, U. S. Army.
To Colonel Culver's untiring energy, close application, and perseverance is due the credit for having completed the coordination of the chain of events leading from the earliest conception of the radio telephone to the successful accomplishment of voice-commanded flying carried through to full fruition.

CURTIS, FRANK R.
    R—Mount Vernon, N. Y.
    B—Mount Vernon, N. Y.
    G. O. No. 3, W. D., 1921.

Colonel, Signal Corps, U. S. Army.
For services while on duty in the office of the Chief Signal Officer of the Army in the organization and training of technical troops of the Signal Corps.

CUSHING, HARVEY.
    R—Brookline, Mass.
    B—Cleveland, Ohio.
    G. O. No. 15, W. D., 1923.

Colonel, Medical Corps, U. S. Army.
As senior consultant of neurological surgery, American Expeditionary Forces and in direct charge of the treatment of gunshot wounds of the head in hospitals of the 1st Army, American Expeditionary Forces, during the Meuse-Argonne offensive, he performed conspicuous and distinguished service to the Government, and through his individual efforts in that capacity saved the lives of many severely wounded soldiers.

CUSHMAN, MRS. JAMES S.
    R—New York, N. Y.
    B—Ottawa, Ill.
    G. O. No. 73, W. D., 1919.

For services as chairman of the war work council of the Young Women's Christian Association of the United States of America.

CUTCHEON, FRANKLIN WARNER M.
    R—Locust Valley, Long Island, N. Y.
    B—Dexter, Mich.
    G. O. No. 89, W. D., 1919.

Lieutenant colonel, Infantry, U. S. Army.
As chairman of the Board of Contracts and Adjustments he supervised and conducted important negotiations with allied governments and their citizens with marked success. His complete knowledge of legal and financial matters coupled with his capacity for work were important factors in the successful management of the Army's fiscal affairs in Europe, rendering services of great value to the American Expeditionary Forces.

CUTCHINS, JOHN A.
    R—Richmond, Va.
    B—Richmond, Va.
    G. O. No. 49, W. D., 1922.

Lieutenant colonel, General Staff Corps, U. S. Army.
As G-2, 29th Division, from September, 1917, until January, 1918, and from July to December, 1918, he displayed sound judgment and exceptional ability in the organization and administration of that section of the division staff. Later as G-4, liaison officer, American section, Permanent International Armistice Commission, he rendered highly meritorious services in a position of great responsibility.

CUTHELL, CHESTER WELDE.
    R—New York, N. Y.
    B—New York, N. Y.
    G. O. No. 126, W. D., 1919.

As special representative of the Secretary of War he built up and supervised an organization to liquidate the claims of the United States against the European allied governments growing out of the purchases of war materials by these governments in the United States, and, by reducing to accurate and clear statements vast and intricate transactions, brought about agreements with the representatives of the allied governments which not only promoted and maintained harmonious feeling, but settled difficult financial relation happily, speedily, and justly, leaving on all sides a sense of appreciation of the accuracy and fair dealing of the Government of the United States in its business relations with the allied and associated governments.

CUTLER, ELLIOTT C.
    R—Brookline, Mass.
    B—Bangor, Me.
    G. O. No. 19, W. D., 1922.

Major, Medical Corps, U. S. Army.
As director of surgical teams and chief of the surgical service in hospital formations at the front during our activities on the Marne and the St. Mihiel and Meuse-Argonne offensives.

CUTLER, HARRY.
    R—Providence, R. I.
    B—Russia.
    G. O. No. 73, W. D., 1919.

Colonel, National Guard, Rhode Island (retired).
For services as chairman, executive committee of the Jewish Welfare Board.

DALEY, EDMUND L.
    R—Worcester, Mass.
    B—Worcester, Mass.
    G. O. No. 95, W. D., 1919.

Colonel, Corps of Engineers, U. S. Army.
He served with distinction as division engineer of the 3d Division and as commanding officer of the 6th Engineers. Due to his energy and resourcefulness, he accomplished arduous tasks with marked success. With remarkable skill he directed the laying out of the defense scheme of the positions taken in the Meuse-Argonne offensive, rendering services of inestimable value to the American Expeditionary Forces.

DALTON, ALBERT C.
    R—Clarkshill, Ind.
    B—Thorntown, Ind.
    G. O. No. 38, W. D., 1922.

Brigadier general, U. S. Army.
For services as general superintendent, Army transport services, and quartermaster, port of embarkation, Hoboken, N. J., from Nov. 1, 1917, to Nov. 5, 1918. Colonel Dalton displayed marked ability as an organizer and administrator, having under his supervision thousands of employees and subordinates. By his energy, capacity, and ability to get results he rendered services of great value to the Government.

DALY, CHARLES P.
    R—Junction City, Kans.
    B—St. Louis, Mo.
    G. O. No. 38, W. D., 1922.

Colonel, Quartermaster Corps, U. S. Army.
In aiding and improving the service of supply in the United States while on duty in the office of the Quartermaster General during the period when the development of supply methods in the United States was in process of initiation, his experience and knowledge of administrative, legal, and executive matters were of immense value to the Government. Later he rendered exceptionally meritorious service in the organization and operation of the general supply depot at New Orleans, La.

DALY, JOSEPH J.
R—New York, N. Y.
B—New York, N. Y.
G. O. No. 56, W. D., 1921.

Lieutenant colonel, Ordnance Department, U. S. Army.
For services as division ordnance officer, 27th Division, a position of great responsibility, involving many difficulties of supply and administration. With marked ability he adapted the American supply system to that used by the British units with which his division operated. He accomplished a great task and rendered services of eminent worth.

DANFORD, ROBERT M.
R—New Boston, Ill.
B—New Boston, Ill.
G. O. No. 47, W. D., 1919.

Brigadier general, U. S. Army.
While on duty in the office of the Chief of Field Artillery he displayed marked ability in planning the organization of field artillery replacement depots; he then proceeded to Camp Jackson, S. C., established this depot, and administered it during the remainder of the war with rare ability and judgment.

DARNALL, CARL ROGER
R—Milford, N. J.
B—Collin County, Tex.
G. O. No. 69, W. D., 1919.

Colonel, Medical Corps, U. S. Army.
He has rendered especially meritorious and distinguished service in organizing developing, and administering the supply division of the Medical Department, and it is due to his foresight, and ability that new sources of medical supplies were developed in this country so that adequate quantities of material were always available for use with the sick and wounded of the Army.

*DAVIDSON, FRED LINCOLN
R—Cincinnati, Ohio.
B—Bucksport, Me.
G. O. No. 59, W. D., 1921.

Lieutenant colonel, Infantry, U. S. Army.
For services as division machine-gun officer, 3d Division. By his supervision and dispositions, he contributed materially to the success of the 7th Machine Gun Battalion at Chateau-Thierry in May, 1918, and that of the machine-gun units of the division in the repulse of the enemy offensive across the Marne July 15 to 18, 1918, and during the Meuse-Argonne campaign. He displayed at all times energy and excellent judgment under difficult conditions, and his work deserves high commendation.
Posthumously awarded. Medal presented to widow, Mrs. Fred L. Davidson.

DAVIE, PRESTON
R—Tuxedo Park, N. Y.
B—Louisville, Ky.
G. O. No. 111, W. D., 1919.

Lieutenant colonel, Quartermaster Corps, U. S. Army.
He organized the fuel and forage division of the War Department and also assisted in reorganizing the salvage service upon a most efficient basis to meet war conditions. By his sound judgment, marked legal and administrative ability, and unselfish devotion to duty he rendered conspicuous service in reorganizing and developing the real estate service of the War Department.

DAVIS, ABEL
R—Chicago, Ill.
B—Germany.
G. O. No. 9, W. D., 1923.
Distinguished service cross also awarded.

Colonel, Infantry, U. S. Army.
As commanding officer, 132d Infantry, 33d Division, he displayed in a marked degree the many and varied qualifications of a successful commanding officer of troops. In the organization and training of his regiment he brought it to a notably high state of efficiency and morale with great thoroughness and in a remarkably short time. Afterward he handled it in all its actions against the enemy with marked success, displaying courage, resourcefulness, tactical skill, and military leadership of the highest order.

DAVIS, CHARLES G.
R—Geneseo, Ill.
B—Geneseo, Ill.
G. O. No. 53, W. D., 1921.

Colonel, Field Artillery, U. S. Army.
For services while in command of the 123d Field Artillery, 33d Division, during the St. Mihiel and Meuse-Argonne offensives.

DAVIS, EDWARD
R—Chicago, Ill.
B—Litchfield, Ill.
G. O. No. 124, W. D., 1919.

Colonel, Cavalry, U. S. Army.
For services while serving as military attaché at The Hague, Netherlands.

DAVIS, EDWIN G.
R—Samaria, Idaho.
B—Samaria, Idaho.
G. O. No. 111, W. D., 1919.

Colonel, Judge Advocate General's Department, U. S. Army.
As chief of the disciplinary division of the office of the Judge Advocate General of the Army he contributed a most helpful means of avoiding serious errors in in the administration of military justice during the war.

DAVIS, JOSEPH R.
R—Lowell, Ark.
B—Springdale, Ark.
G. O. No. 95, W. D., 1919.

Colonel, Field Artillery, U. S. Army.
He commanded the 15th Field Artillery throughout all the major operations in which the Second Division participated, at all times proving himself an officer of unusual ability and sound judgment. Inspiring the members of his command by his aggressive spirit, he kept his regiment at all times in closest proximity to the infantry units which it supported, thereby contributing materially to the successes achieved by his division against the enemy.

DAVIS, MILTON F.
R—McCoy, Oreg.
B—Milton, Minn.
G. O. No. 15, W. D., 1923.

Colonel, Signal Corps, U. S. Army.
As chief of the schools section, division of military aeronautics, his work in perfecting a system of training was thorough and complete. His soundness of judgment, fairness in dealing with all the boards of officers and branches of the service, and unusual executive ability made his work a decisive factor in the successful production of trained air personnel. He rendered services of the highest order to the Government in a position of great responsibility.

DAVIS, RICHMOND P.
R—Statesville, N. C.
B—Statesville, N. C.
G. O. No. 19, W. D., 1922.

Brigadier general, U. S. Army.
For services as acting chief of the 9th Corps Artillery, in which position his direction of Artillery employment and his intelligent comment on its employment by subordinate commanders was conspicuous.

DAVIS, ROBERT C.
R—Lancaster, Pa.
B—Lancaster, Pa.
G. O. No. 12, W. D., 1919.

Brigadier general, U. S. Army.
As adjutant general of the American Expeditionary Forces he has performed his exacting duties with high professional skill and administrative ability. The exceptional efficiency of the Adjutant General's Department under his direction was a material factor in the success of the staff work at General Headquarters.

DAVIS, WILLIAM C.
R—McGrawville, N. Y.
B—McGrawville, N. Y.
G. O. No. 59, W. D., 1919.

Brigadier general, U. S. Army.
In command of the Artillery support of the Fifth Corps in November he rendered services of the highest order. Through his energy, intelligence, and skill his guns were ever ready for an emergency. The successes of the operations between the Meuse and the Argonne Forest were in a measure due to his strong support of the attacking Infantry.

DAVIS, WILLIAM D.
R—Neosho, Mo.
B—Duplain, Mich.
G. O. No. 98, W. D., 1919.
Distinguished service cross also awarded.

Colonel, Infantry, U. S. Army.
He served with marked success as commanding officer of the 361st Infantry, displaying military attainments of a high order. Inspiring his men by his faithful devotion to duty, he proved a potent factor in the achievements of the 91st Division. While ably directing his regiment in action during the early part of November he was killed by an enemy shell.
Posthumously awarded. Medal presented to widow, Mrs. Abbie Greene Davis.

DAVISON, HENRY P.
R—New York, N. Y.
B—Troy, Pa.
G. O. No. 95, W. D., 1919.

As chairman of the war council, American Red Cross, he assumed general direction of the war measures of that society, and by the exercise of rare tact and consummate powers of construction and direction brought it to a perfection of organization which made it possible to extend relief promptly and bountifully to our armies and to those of the allied nations. His dynamic qualities as a financier and his forceful personality assured to the soldier in the field and to the inhabitants of the devastated countries of Europe systematized measures of relief beyond the limits of specific statement.

DAWES, CHARLES G.
R—Evanston, Ill.
B—Marietta, Ohio.
G. O. No. 12, W. D., 1919.

Brigadier general, U. S. Army.
He rendered most conspicuous services in the organization of the general purchasing board as general purchasing agent of the American Expeditionary Forces and as the representative of the U. S. Army on the military board of allied supply. His rare abilities, sound business judgment, and aggressive energy were invaluable in securing needed supplies for the American armies in Europe.

DAY, LEE GARNET.
R—New York, N. Y.
B—New York, N. Y.
G. O. No. 59, W. D., 1919.

Major, Quartermaster Corps, U. S. Army.
In command of the regulating station at St. Dizier, France, he displayed extraordinary ability in the promptness with which he organized and assured a steady flow of supplies to the 1st Army in the advance against the St. Mihiel salient and in the Argonne offensive. It was largely due to his splendid efforts in a time of great emergency that our troops were provided with necessary ammunition and supplies.

DEAN, ELMER A.
R—Centerville, Tenn.
B—Centerville, Tenn.
G. O. No. 89, W. D., 1919.

Colonel, Medical Corps, U. S. Army.
He came to France with a base hospital unit, which he established. Later he organized and commanded the first large hospital center at Bazoilles. The success of this center in caring for a large number of sick and wounded was due in a large measure to his high professional attainments, zeal, and extraordinary executive ability.

DeARMOND, EDWARD H.
R—Butler, Mo.
B—Greenfield, Mo.
G. O. No. 31, W. D., 1922.

Brigadier general, U. S. Army.
As chief of staff of the 32d Division during the period of its training in the United States and in France, he contributed largely to the organization and efficiency of that division. As chief of the Field Artillery Section of the office of the Chief of Artillery, American Expeditionary Forces, he perfected the field artillery training system in France, and by his marked ability for organization and his able supervision of that training, rendered services of great value.

DeBEVOISE, CHARLES I.
R—Brooklyn, N. Y.
B—Brooklyn, N. Y.
G. O. No. 89, W. D., 1919.

Brigadier general, U. S. Army.
He served with credit as commander of trains and military police of the 27th Division. Later, in command of the 107th Infantry, 27th Division, he proved himself to be an energetic and resourceful leader during the operations against the Hindenburg line and those on the La Selle River. After being promoted to brigadier general he continued to render valuable services to the American Expeditionary Forces as commander of the 53d Infantry Brigade, rendering conspicuous services to the American Expeditionary Forces.

DEEMS, CLARENCE, Jr.
R—Baltimore, Md.
B—Charlottesville, Va.
G. O. No. 87, W. D., 1919.

Colonel, Field Artillery, U. S. Army.
He served creditably as commanding officer of the 321st Field Artillery, 82d Division, giving proof of conspicuous military attainments. Through his tireless energy and technical skill as an artillerist his regiment gave most effective assistance to the infantry which it supported, and at all times furnished whole-hearted cooperation to the infantry in the operations against the enemy.

DELAFIELD, JOHN ROSS.
R—New York, N. Y.
B—New York, N. Y.
G. O. No. 47, W. D., 1921.

Colonel, Ordnance Department, U. S. Army.
For services as chairman of the War Department board of contract adjustment.

DELANEY, JOHN T.
R—New York, N. Y.
B—New York, N. Y.
G. O. No. 56, W. D., 1922.

Lieutenant colonel, Field Artillery, U. S. Army.
As regimental commander of the 104th Field Artillery, 27th Division, he demonstrated professional attainments and ability of the highest order. By his sound tactical judgment and superior knowledge of artillery he most successfully directed his units in support of the 157th Infantry Brigade in the operations north of Verdun, Nov. 4 to 11, 1918. By keeping his elements close to the attacking Infantry he contributed in no small measure to the success of the Infantry brigade in these operations. He rendered services of conspicuous merit and signal worth to the American Expeditionary Forces.

DELANEY, MATTHEW A.
  R—Waymart, Pa.
  B—Waymart, Pa.
  G. O. No. 43, W. D., 1922.

Colonel, Medical Corps, U. S. Army.
As commanding officer, Base Hospital No. 10, at Le Treport, France, from May 6, 1917, to March, 1918, he displayed tireless energy and military attainments of a high order in the efficient operation of this important hospital on the western front. By his marked devotion to duty and splendid administrative abilities, great numbers of our own and allied sick and wounded were cared for, resulting in the saving of many lives. His services were of great value to the American Expeditionary Forces.

DELANO, FREDERIC ADRIAN
  R—Washington, D. C.
  B—China.
  G. O No. 47, W. D., 1921.

Colonel, Transportation Corps, U. S. Army.
For services as deputy of the Director General of Transportation, American Expeditionary Forces, in connection with the evacuation of American troops to the ports of embarkation, France, and later in connection with the work of the liquidation commission.

ᵈDELANO, JANE A.
  R—Washington, D. C.
  B—Watkins, N. Y.
  G. O. No. 61, W. D., 1919.

Director, Department of Nursing, American Red Cross.
She applied her great energy and used her powerful influence among the nurses of the country to secure enrollments in the American Red Cross. Through her great efforts and devotion to duty 18,732 nurses were secured and transferred to the Army Nurse Corps for service during the war. Thus she was a great factor in assisting the Medical Department in caring for the sick and wounded.
Posthumously awarded. Medal presented to American Red Cross.

DENGLER, FREDERICK L.
  R—Hot Springs, Ark.
  B—Hot Springs, Ark.
  G. O. No. 15, W. D., 1923.

Colonel (Coast Artillery Corps), General Staff Corps, U. S. Army.
As a member of the intelligence section of the General Staff, American Expeditionary Forces, he efficiently organized and directed with rare ability the operation of the subsection dealing with the enemy's man power, war material, economic conditions, prisoners and documents, including the enemy press. Later he was sent to the United States to expedite the organization and training of the intelligence personnel of the division selected for overseas service; and also to assist in the coordination of the work of the Intelligence Section of the American Expeditionary Forces with that of the War Department. In all these positions he displayed rare judgment, great initiative, and unremitting devotion to duty.

DENSON, ELEY P.
  R—High Point, N. C.
  B—Trinity, N C.
  G. O. No. 14, W. D., 1923.

Lieutenant colonel (Infantry), General Staff Corps, U. S. Army.
For services as assistant chief of staff, G-3, 28th Division, from Sept. 15 to Nov. 15, 1918. In this position he displayed untiring energy, sound professional judgment and tactical skill, and unremitting attention to duty, thus contributing extremely valuable services to the 28th Division in its operations against the enemy.

DERBY, RICHARD
  R—New York, N. Y.
  B—New York, N. Y.
  G. O. No. 39, W. D., 1920.

Lieutenant colonel, Medical Corps, U. S. Army.
As sanitary inspector of the 2d Division, and in charge of the front-line hospitalization and evacuation of the wounded during the active operations of the division in the Marne area, and later as division surgeon during the Meuse-Argonne offensive, he demonstrated high professional attainments, excellent judgment, and gallantry in the execution of his important duties. Due to his energy, the sanitary units under his control were amply provided with facilities for the proper care of the sick and wounded in the field. He has given services of significant worth to the American Expeditionary Forces.

DE TARNOWSKY, GEORGE
  R—Chicago, Ill.
  B—France.
  G. O. No. 27, W. D., 1922.

Lieutenant colonel, Medical Corps, U. S. Army.
In organizing and commanding American Red Cross Military Hospital No. 5, American Expeditionary Forces. Notwithstanding the serious deficiency in personnel and material, Colonel de Tarnowsky, by his energy and efficient administration, received and treated great numbers of wounded evacuated to the Paris district during June and July, 1918.

DEVOL, CARROLL A.
  R—Waterford, Ohio.
  B—Waterford, Ohio.
  G. O. No. 73, W. D., 1919.

Major general, U. S. Army, retired.
General Devol, first as depot quartermaster and later as zone supply officer in San Francisco, handled the service of supply and service of transportation on the Pacific coast during the war, being responsible for the supply of troops serving in the Philippines, Hawaii, Siberia, and Alaska and the camps on the Pacific coast. He handled this large responsibility with ability, good judgment, and conspicuous success.

DEWEY, BRADLEY
  R—Pittsburgh, Pa.
  B—Burlington, Vt.
  G. O. No. 47, W. D., 1919.

Colonel, Chemical Warfare Service, U. S. Army.
As chief of the gas defense production division in achieving under most trying circumstances remarkable results in supplying the American Expeditionary Forces with sufficient number of gas masks of high grade and of improved design.

DEWITT, JOHN L.
  R—Fort Monroe, Va.
  B—Sidney, Nebr.
  G. O. No. 59, W. D., 1919.

Colonel (Infantry), General Staff Corps, U. S. Army.
He organized the supply section of the General Staff of the 1st Army and successfully administered this important section during all the operations of that command. The results obtained by his untiring efforts and brilliant professional ability had a marked influence on the successes attained by the 1st Army.

DICKMAN, JOSEPH T.
  R—Wapakoneta, Ohio.
  B—Dayton, Ohio.
  G. O. No. 27, W. D., 1920.

Major general, U. S. Army.
For services as commander of the 3d Army, American Expeditionary Forces. He commanded the 3d Division and contributed in large measure to success in hurling back the final enemy general attack commencing July 14, 1918. He participated in the offensive operations northward to Vesle River; commanded the 4th Army Corps from Aug. 18 to Oct. 11, 1918, including the operation of the reduction of the St. Mihiel salient, and the 1st Army Corps during the Meuse-Argonne operations from Oct. 12 until after the armistice. Later he commanded the 3d Army of Occupation at Coblenz, Germany.

DICKSON, ROBERT A.
R—Waterford, N. Y.
B—Lansingburg, N. Y.
G. O. No. 4, W. D., 1923.

Lieutenant colonel, Sanitary Corps, U. S. Army.
As officer in charge of the administrative division in the office of the chief surgeon, American Expeditionary Forces, by his foresight, executive ability, and unusual knowledge of administrative details he successfully organized and directed a record system which functioned on a highly efficient basis throughout the World War, thereby rendering, in a position of great responsibility, conspicuous services to the American Expeditionary Forces.

DILLARD, JAMES B.
R—New Orleans, La.
B—Norfolk, Va.
G. O. No. 69, W. D., 1919.

Colonel, Ordnance Department, U. S. Army.
For services as chief of the Heavy Artillery section of the carriage division of the office of the Chief of Ordnance, in which capacity he was charged with the design and development of all railway and other heavy artillery; and later as chief of the engineering division of the office of the Chief of Ordnance, in which capacity he was charged with the design and development of all articles of ordnance supplied to the U. S. Army.

DILLON, THEODORE H.
R—Bedford, Ind.
B—Center Valley, Ind.
G. O. No. 59, W. D., 1921

Colonel, Corps of Engineers, U. S. Army.
For services as assistant chief engineer, 1st Army, a position of great responsibility. During the St. Mihiel and Meuse-Argonne operations he administered the first engineer troops and the organization, plans, and field work of the engineer plans, operations, and information section and other engineer enterprises conducted in the 1st Army. By his energy and keen application he rendered services of much worth.

DISQUE, BRICE PURSELL.
R—Cincinnati, Ohio.
B—California, Ohio.
G. O. No. 69, W. D., 1919.

Brigadier general, U. S. Army.
For services rendered in connection with the organization and administration of the spruce production activities of the Bureau of Aircraft Production while serving as officer in charge of the spruce production division and president of the United States Spruce Production Corporation.

DODD, TOWNSEND F.
R—Waukegan, Ill.
B—Anna, Ill.
G. O. No. 50, W. D., 1919.

Colonel, Air Service, U. S. Army.
He organized the aviation training school at Issoudon and successfully conducted the negotiations for the first purchase of aeroplanes from allied governments for the use of the American Expeditionary Forces. He later served with distinction as chief of the supply section, Air Service, American Expeditionary Forces, and as technical adviser and information officer of the Chief of the Air Service, 1st Army.

DODDS, WILLIAM H., jr.
R—Detroit, Mich.
B—Detroit, Mich.
G. O. No. 38, W. D., 1922.

Colonel, Field Artillery, U. S. Army.
As commander of the 6th Field Artillery, 1st Division, he handled the regiment so brilliantly under severe conditions throughout the St. Mihiel operation, Sept. 12-13, 1918, and the Meuse-Argonne operations, Sept. 30 to Nov. 11, 1918, that the regiment demonstrated an unusually high degree of efficiency and morale. He repeatedly displayed superior tactical judgment and knowledge of artillery, and by his exceptional ability, leadership, and devotion to duty rendered the maximum support to the Infantry of the 1st Division in effectively executing the most difficult missions assigned to him, thus rendering conspicuous services to the American Expeditionary Forces.

D'OLIER, FRANKLIN.
R—Riverton, N. J.
B—Burlington, N. J.
G. O. No. 62, W. D., 1919.

Lieutenant colonel, Quartermaster Corps, U. S. Army.
He displayed marked ability in the organization and efficient administration of the American salvage depot at St. Pierre des Corps, of which he was the commanding officer. To his untiring zeal and constant devotion to duty is due the success with which this plant, the largest industrial undertaking in the American Expeditionary Forces, was operated.

DONALDSON, THOMAS Q.
R—Greenville, S. C.
B—Greenville, S. C.
G. O. No. 89, W. D., 1919.

Brigadier general, U. S. Army.
As inspector general of the services of supply, by his energy, sound judgment, and able management he organized and brought to a state of marked efficiency the Inspector General's Department in the services of supply. He proved a most potent factor in raising the standard of discipline throughout the command, rendering services of conspicuous worth.

DONAVIN, KIRKWOOD HARRY.
R—Columbus, Ohio.
B—Delaware, Ohio.
G. O. No. 116, W. D., 1919.

Commander, U. S. Navy.
For services as chief of staff of the commander, Cruiser and Transport Fleet, Newport News division.

DONOVAN, WILLIAM J.
R—Buffalo, N. Y.
B—Buffalo, N. Y.
G. O. No. 43, W. D., 1922.
Medal of honor and distinguished-service cross also awarded.

Colonel, Infantry, U. S. Army.
As battalion commander 165th Infantry, 42d Division, during its operations in the Baccarat sector July 28-31, 1918, he demonstrated high professional attainments and marked ability. He displayed conspicuous energy and most efficient leadership in the advance of his battalion across the Ourcq River and the capture of strong enemy positions. In October, 1918, as lieutenant colonel he commanded the same regiment with marked success and distinction in the Meuse-Argonne offensive. His devotion to duty, heroism, and pronounced qualities of a commander enabled him to successfully accomplish all missions assigned to him in this important operation. From Jan. 3 to Mar. 3, 1919, as inspector instructor, Provost Marshall General's Department, he rendered services of great value to the American Expeditionary Forces.

DOREY, HALSTEAD.
R—St. Louis, Mo.
B—St. Louis, Mo.
G. O. No. 59, W. D., 1919.
Distinguished-service cross also awarded.

Colonel, Infantry, U. S. Army.
He commanded with distinction the 4th Infantry, 3d Division, during the battle of the Marne, the advance from the Marne to the Ourcq, and in the St. Mihiel and Argonne-Meuse offensives. It was his regiment that led the advance to the Ourcq, capturing Charmel, Charmel-Chateau, Villardelle Ferme, and Roncheres. The successes attained by his command were greatly influenced by the high qualities of leadership he continually displayed in all these operations.

DORR, GOLDTHWAITE H.
R—New York, N. Y.
B—Newark, N. J.
G. O. No. 2, W. D., 1920.

Assistant director of munitions.
His assistance to the director of munitions in procuring supplies and equipment for the Army and his subsequent activities in the settlement of the complex contractual relations resulting therefrom were of signal value to the United States and its Army.

DORSEY, FRANK M.
R—Cleveland Heights, Ohio.
B—Dresden, Ohio.
G. O. No. 69, W. D., 1919.

Colonel, Chemical Warfare Service, U. S. Army.
As a civilian and as chief of development division, Chemical Warfare Service, he has displayed fine technical skill and administrative ability in developing materials and processes which have contributed greatly to the achievements of the Chemical Warfare Service during the war.

DOWELL, CASSIUS M.
R—Terre Haute, Ind.
B—Landes, Ill.
G. O. No. 3, W. D., 1924.

Lieutenant colonel, Judge Advocate General's Department, U. S. Army.
As assistant to the provost marshal general, Apr. 10 to Sept. 3, 1917, he assisted in the administration of the selective service law, being charged with the responsibility of the appointments of all local and district draft boards. As judge advocate, 26th Division, Sept. 3, 1917, to Jan. 2, 1918; chief of staff, 26th Division, Jan. 2, 1918, to Apr. 17, 1918; lieutenant colonel, 102d Infantry, and for part of period commanding officer of that regiment, Apr. 17 to June 13, 1918; G–3, 26th Division, Sept. 15 to Dec. 31, 1918, he displayed sound tactical judgment, administrative and executive qualities of a high order, and unremitting attention and devotion to his duties, thus rendering services of great value to his Government in positions of great responsibility.

DOYLE, LUKE C.
R—Worcester, Mass.
B—Worcester, Mass.
G. O. No. 59, W. D., 1919.

Major, Sanitary Corps, U. S. Army.
As assistant regulating officer, G–4, General Headquarters, American Expeditionary Forces, he arranged the schedules of hospital and medical supply trains with marked ability and succeeded in maintaining those schedules, despite numerous difficulties. His aggressive action in time of emergency, whereby he surmounted unforeseen obstacles, together with the excellent performance of his duties, were material factors in the alleviation of much suffering and in the saving of many lives among the wounded sent from the front.

DRAIN, JAMES A.
R—Washington, D. C.
B—Warren County, Ill.
G. O. No. 31, W. D., 1922.

Lieutenant colonel, Ordnance Department, U. S. Army.
From his experience as ordnance officer of the 1st Division during its early months in France he rendered valuable service in assisting the Chief of Ordnance, American Expeditionary Forces, in formulating the policies for the supply and maintenance of ordnance for subsequent divisions. Later he performed important work in charge of the machine gun and small arms division in the office of the Chief of Ordnance, American Expeditionary Forces, and finally rendered service of a high order in representing his Government as American member of the Anglo-American Tank Commission.

DRAKE, CHARLES B.
R—Old Forge, Pa.
B—Old Forge, Pa.
G. O. No. 18, W. D., 1919.
Distinguished-service cross also awarded.

Brigadier general, U. S. Army.
For conspicuous service in the organization of the Motor Transport Corps.

DRAKE, FRANCIS E.
R—New York, N. Y.
B—Farmington, Mich.
G. O. No. 56, W. D., 1922.

Lieutenant colonel, Corps of Engineers, U. S. Army.
He served as chief of control bureau, office of general purchasing agent, American Expeditionary Forces. Through his office, for the purpose of coordination and approval, passed all the purchases of the American Expeditionary Forces in France, and due to his highly efficient organization in connection with the supply procurement work, close and harmonious relations were sustained at all times with the French Government. His unusual tact, marked ability, and great energy in a position of great responsibility were of incalculable benefit to the supply service of the American Expeditionary Forces.

DRAVO, RALPH M.
R—Edgeworth, Pa.
B—Pittsburgh, Pa.
G. O. No. 124, W. D., 1919.

As chief of the Pittsburgh ordnance district, in which capacity he maintained at all times the greatest degree of intelligent and enthusiastic cooperation between the Ordnance Department and manufacturers in his district, thereby attaining the maximum production of munitions in a minimum time; and also as chairman of the Pittsburgh ordnance district claims board, in which capacity his services have been invaluable to the Nation in adjusting equitably the $210,000,000 worth of outstanding contracts in his district in force at the signing of the armistice.

DRISCOLL, THOMAS A.
R—San Mateo, Calif.
B—Virginia City, Nev.
G. O. No. 56, W. D., 1922.

Lieutenant colonel, Infantry, U. S. Army.
As assistant chief of staff, G–2, 91st Division, during the operations of the division in the Meuse-Argonne offensive, and during the advance from the Lys to and beyond the Scheldt River, by his keen foresight, discriminating judgment, and organizing ability he was able at all times to give to his commanding officer accurate and complete information of the enemy dispositions and to furnish valuable assistance in the preparation of all attack orders, thereby rendering services of inestimable value to the success of the division.

DRUM, HUGH A.
R—Boston, Mass.
B—Sault Ste. Marie, Mich.
G. O. No. 12, W. D., 1919.

Brigadier general, U. S. Army.
Upon him as chief of staff of the 1st Army devolved the important duty of organizing the headquarters of this command and of coordinating the detailed staff work in its operations in the St. Mihiel and Argonne-Meuse offensives. His tact, zeal, and high professional attainments had a marked influence on the success that attended the operations of the 1st Army.

DUFFY, FRANCIS P.
  R—New York, N. Y.
  B—Canada.
  G. O. No. 62, W. D., 1919.
  Distinguished-service cross also awarded.

Captain, chaplain, 165th Infantry, 42d Division, U. S. Army.
He performed with distinction his combined duties as regimental and division chaplain, stimulating the work of all with whom he came in contact. When his division was in rest areas, he was tireless and devoted in his efforts to help all with whom he served. Whether in the front-line trenches or in an attack, he was with the troops, encouraging them to greater effort, an example of fearlessness and devotion to duty, helping to care for the sick and wounded, administering to the dying, and arranging for the burial of the dead.

DUGAN, THOMAS B.
  R—Baltimore, Md.
  B—Baltimore, Md.
  G. O. No. 59, W. D., 1919.

Brigadier general, U. S. Army.
He commanded the 70th Infantry Brigade, 35th Division, during a part of the Meuse-Argonne offensive with great distinction and marked ability. By his painstaking energy, zeal, and great initiative he proved to be a material factor in the successes of the division.

DUNCAN, GEORGE B.
  R—Lexington, Ky.
  B—Lexington, Ky.
  G. O. No. 12, W. D., 1919.

Major general, U. S. Army.
Arriving in France with the first contingent of American troops, he commanded in turn a regiment, brigade, and division with conspicuous success. In the command of the 77th Division in the Baccarat sector his sound military judgment, energy, and resolution were important factors in the successes gained. Later, in command of the 82d Division in the Argonne-Meuse offensive, he proved himself a brilliant leader, with great force and energy.

DUNLAP, WILLIAM R.
  R—Pittsburgh, Pa.
  B—Dubois, Pa.
  G. O. No. 15, W. D., 1923.

Colonel, Infantry, U. S. Army.
Promoted in turn captain, major, lieutenant colonel, and colonel, while serving with the Infantry regiments of the 28th Division, he displayed unusual administrative, executive, and organizing ability and sound judgment. His brilliant leadership of his battalion in action against the enemy, and later, that of the regiment, testified to the unusual skill and soldierly qualities of this officer. Colonel Dunlap's services contributed in a marked degree to the successful operations of the 28th Division and the American forces in France.

DUNN, JOHN M.
  R—Wilmington, Del.
  B—Wilmington, Del.
  G. O. No. 73, W. D., 1919.

Colonel (Coast Artillery Corps), General Staff Corps, U. S. Army.
For services as chief of the positive branch, military intelligence division, General Staff. To his untiring energy, zeal, and ability the efficiency of the service of gathering, collating, and distributing military information is largely due.

DUNWOODY, HALSEY.
  R—Scranton, Pa.
  B—Washington, D. C.
  G. O. No. 62, W. D., 1919.

Colonel (Coast Artillery Corps), Signal Corps, U. S. Army.
As chief of supply and assistant chief of Air Service, by his energy, tact, and executive ability, he built up an efficient supply service, capable of meeting the program for material, airplanes, motors, and equipment. He established and maintained excellent relations with the allied military authorities. His service was marked by exceptional administrative ability, comprehensive knowledge of the needs and conditions of the service, and whole-hearted devotion to his important tasks.

EARLE, RALPH.
  R—Worcester, Mass.
  B—Worcester, Mass.
  G. O. No. 2, W. D., 1920.

Admiral, U. S. Navy.
For services as chief of the Bureau of Ordnance, Navy Department, during the World War, in which position, by his close cooperation and energetic efforts, he greatly assisted the War Department in the arming of its troop and cargo transports.

EASBY-SMITH, JAMES S.
  R—Washington, D. C.
  B—Tuskaloosa, Ala.
  G. O. No. 10, W. D., 1920.

Colonel, Judge Advocate General's Department, U. S. Army.
For services to the Government in connection with the administration of the selective service law during the war. To all of the tasks assigned him he brought an indefatigable energy and rare unselfish devotion, without which their accomplishment would have been impossible.

ECKELS, CHARLES B.
  R—Washington, D. C.
  B—Harrisburg, Pa.
  G. O. No. 38, W. D., 1922.

Lieutenant colonel, Quartermaster Corps, U. S. Army.
As assistant to the chief quartermaster, American Expeditionary Forces, from July 23, 1917, to Jan. 31, 1919. During this period he served as chief of the finance division of the chief quartermaster's office and successfully handled the numerous complex financial questions arising in the American Expeditionary Forces in France, England, and Italy. He also performed the duty of chief disbursing officer for the American Expeditionary Forces in procuring and transferring funds for disbursement by the finance officers of the several services, until the establishment of the financial requisition officer. The success with which these offices of great responsibility functioned was due largely to his exceptional ability, devotion to duty, and conspicuous services.

*EDGAR, CHARLES.
  R—Essex Falls, N. J.
  B—Metuchen, N. J.
  G. O. No. 15, W. D., 1923.

Director of lumber, War Industries Board.
For services in connection with the operations of the war industries board during the World War. In his position as director of one of the sections of the board he rendered, through his broad vision, distinguished capacity, and business ability, services of inestimable value in marshaling the industrial forces of the Nation and mobilizing its economic resources—marked factors in assisting to make military success attainable. As director of lumber, he rendered through his untiring efforts and devotion to duty exceptionally valuable service to the War Department in connection with the procurement of lumber for the Army.
Posthumously awarded. Medal presented to widow, Mrs. Charles Edgar.

EDGAR, CLINTON G_____
 R—Detroit, Mich.
 B—Detroit, Mich.
 G. O. No. 15, W. D., 1923.

Colonel, Signal Corps, U. S. Army.
For services as officer in charge of the construction division, Signal Corps, aviation section. Being a trained engineer with exceptional capacity for organization and execution, he was placed in charge of the construction division, Signal Corps, and had complete charge and responsibility for all buildings constructed for the Air Service, which consisted of 27 aviation fields, 9 depots, 2 experimental stations, as well as many minor activities. He originated and installed the system of securing lands for all aviation fields, securing grants and gifts of all facilities, including water and electric lights. He was responsible for the conception, design, and production of the all-steel demountable standard hangar adopted by the Army and Navy for use in the Air Service. By his untiring energy, sound judgment, and large grasp of construction problems, he has rendered services of inestimable value to the Government in a position of great responsibility.

EDIE, ELLIOTT B_____
 R—Connellsville, Pa.
 B—Baltimore, Md.
 G. O. No. 56, W. D., 1921.

Lieutenant colonel, Medical Corps, U. S. Army.
As commander of the 305th Sanitary Train, consisting of the 317th, 318th, 319th, and 320th Field Hospitals and later as division surgeon, 80th Division, a position of great responsibility, he maintained suitable dressing stations and provided for the continuous evacuation of the wounded in an exceptionally efficient manner under conditions of almost constant fire.

EDIE, GUY L_____
 R—Christiansburg, Va.
 B—Christiansburg, Va.
 G. O. No. 59, W. D., 1919.

Colonel, Medical Corps, U. S. Army.
He was placed in charge of the medical service at Brest at the time when it became the chief port of debarkation for American troops and at a period when the arrival of troops in unprecedented numbers, and with many sick, overwhelmed all medical arrangements for their care. By his great resourcefulness he successfully overcame the many difficult problems that were presented.

EDWARDS, OLIVER_____
 R—Chesterfield, Mass.
 B—Chesterfield, Mass.
 G. O. No. 47, W. D., 1919.

Brigadier general, U. S. Army.
Due to his rare ability and high professional attainments he was selected to organize the machine-gun training center, the success of which was, in a large measure, due to his zealous and energetic administration.

EICHELBERGER, ROBERT L_____
 R—Urbana, Ohio.
 B—Urbana, Ohio.
 G. O. No. 56, W. D., 1922.
 Distinguished-service cross also
  awarded.

Lieutenant colonel (Infantry), General Staff Corps, U. S. Army.
As assistant chief of staff, G-2, with the American Expeditionary Forces in Siberia, he organized and directed the intelligence service of the American Expeditionary Forces in Siberia in a most able manner and under most trying circumstances. By his keen foresight, discriminating judgment, and brilliant professional attainments, exercised through his efficiently established organization, he was able to keep his commanding general well and fully informed at all times. His tireless energy and his keen insight into local conditions gave him a masterful grasp of the situation, which contributed materially to the success of the forces in Siberia. He rendered most conspicuous services of inestimable value to the Government in a position of great responsibility.

EISENHOWER, DWIGHT D_____
 R—Abilene, Kans.
 B—Greyson, Tex.
 G. O. No. 43, W. D., 1922.

Lieutenant colonel, Tank Corps, U. S. Army.
While commanding officer of the Tank Corps training center from Mar. 23, 1918, to Nov. 18, 1918, at Camp Colt, Gettysburg, Pa., he displayed unusual zeal, foresight, and marked administrative ability in the organization, training, and preparation for overseas service of technical troops of the Tank Corps.

EISENMAN, CHARLES_____
 R—Cleveland, Ohio.
 B—New York, N. Y.
 G. O. No. 47, W. D., 1919.

Vice chairman of committee on supplies, Council of National Defense.
His energy, courage, business ability, and foresight did much to enlist American industry in the service of our country and thus make possible the prompt and proper equipment of our armies with clothing and equipage.

ELLIOTT, WILLIAM_____
 R—San Rafael, Calif.
 B—Canada.
 G. O. No. 59, W. D., 1919.

Colonel, Quartermaster Corps, U. S. Army.
As quartermaster at Langres and at the regulating station at Is-sur-Tille, his energy and thorough knowledge of methods and standards of supply have been of the greatest value to the Government, particularly while depot quartermaster at Is-sur-Tille, during a period when the successful operations of some 20 divisions were dependent upon receiving supplies from that depot.

ELTINGE, LEROY_____
 R—Kingston, N. Y.
 B—South Woodstock, N. Y.
 G. O. No. 87, W. D., 1919.

Brigadier general, U. S. Army.
As a member of the operations section, General Staff, General Headquarters, American Expeditionary Forces, he exhibited sound military judgment and foresight in drafting important plans. Later as deputy chief of staff of the American Expeditionary Forces throughout the period of active operations and thereafter he discharged the important and complex duties of his position with admirable efficiency and by his untiring efforts and devotion to duty rendered conspicuous service to the Government.

ELY, HANSON E_____
 R—Iowa City, Iowa.
 B—Buffalo Grove, Iowa.
 G. O. No. 46, W. D., 1920.
 Distinguished service cross also
  awarded.

Major general, U. S. Army.
He commanded, with skill and marked ability, a regiment in trench operations north of Toul, west of Montdidier, and during the attack at Cantigny. Later as a brigade commander during the Aisne-Marne and St. Mihiel offensives and in the attack on the strong enemy position of Mont Blanc Ridge; and as a division commander during the latter phase of the Meuse-Argonne offensive, he demonstrated rare qualities of leadership.

EMBICK, STANLEY D_____
 R—Boiling Springs, Pa.
 B—Greencastle, Pa.
 G. O. No. 69, W. D., 1919.

Colonel, Signal Corps, U. S. Army.
As a member of the American section of the Supreme War Council, by his high professional qualifications, his breadth of vision, and his sound military judgment, he has rendered invaluable aid in solving the many complex problems that have come before the Supreme War Council.

EMERSON, HAVEN.
  R—New York, N. Y.
  B—New York, N. Y.
  G. O. No. 3, W. D., 1922.

Lieutenant colonel, Medical Corps, U. S. Army.
As chief epidemiologist of the office of the Chief surgeon, American Expeditionary Forces, where he perfected a system of keeping daily checks on all of the many contagious diseases which afflicted the American Expeditionary Forces. Much sickness and many lives were saved by this system, the practical working of which was due to the experience and tireless energy of this officer.

EMERSON, THOMAS HENRY.
  R—Arcata, Calif.
  B—Pittsburgh, Pa.
  G. O. No. 87, W. D., 1919.

Colonel, Corps of Engineers, U. S. Army.
As assistant chief of Staff, G-3, of the operations section of the 5th Army Corps, he performed his important duties with marked zeal. By his rare technical skill in originating and developing plans for operations against the enemy he rendered services of signal worth to the American Expeditionary Forces.

EMMONS, HAROLD H.
  R—Detroit, Mich.
  B—Detroit, Mich.
  G. O. No. 69, W. D., 1919.

Lieutenant, United States Naval Reserve Force.
For services as Chief of the Engine Production Department of the Air Service.

ENNIS, WILLIAM P.
  R—New York.
  B—San Francisco, Calif.
  G. O. No. 31, W. D., 1922.

Brigadier general, U. S. Army.
As director of the department of matériel, school of fire, for Field Artillery at Fort Sill, Okla., from May 9, 1918, to Aug. 30, 1918, his untiring energy, devotion to duty, exceptional qualifications for the task he had to perform, were directly responsible for the splendid organization developed.

ENOCHS, BERKELEY.
  R—Ironton, Ohio.
  B—Ironton, Ohio.
  G. O. No. 56, W. D., 1922.

Colonel, Infantry, U. S. Army.
As chief of staff, 39th Division, from August, 1917, until August, 1918, by his marked efficiency, loyal devotion to duty, and high military attainments, he played an important part in the successful organization and training of that division. Later, as assistant chief of staff, G-3, 4th Army Corps, from September, 1918, until April, 1919, he performed many duties of great responsibility in a highly meritorious manner, rendering services of signal worth to the American Expeditionary Forces.

ERICKSON, HJALMER.
  R—Brooklyn, N. Y.
  B—Norway.
  G. O. No. 59, W. D., 1919.

Colonel, Infantry, U. S. Army.
As commanding officer of the 26th Infantry, 1st Division, in all the operations east of the Aire River from Oct. 1 to 11, 1918, he rendered most meritorious service by displaying marked tactical ability, courage, and resourcefulness in the handling of numerous critical situations, thus enabling his regiment to advance steadily to all its objectives.

ERSKINE, JOHN.
  R—New York, N. Y.
  B—New York, N. Y.
  G. O. No. 89, W. D., 1919.

As chairman of the educational commission, he devoted himself with tireless energy to the problem of developing educational opportunities for the American soldiers in France while they were awaiting repatriation. To his rare educational ability, breadth of vision, and initiative is due, in a large measure, the success of the educational program of the American Expeditionary Forces.

ERWIN, JAMES B.
  R—Savannah, Ga.
  B—Savannah, Ga.
  G. O. No. 15, W. D., 1923.

Brigadier general, U. S. Army.
With sound technical skill, initiative, and untiring energy, he assisted in the organization and training of the 6th Division, and commanded with distinction the 12th Infantry Brigade during its operations in the Vosges sector, and during the Meuse-Argonne offensive, Nov. 1 to 11, 1918. His rare quality of leadership and unremitting devotion to duty were material factors in the successful operations of his division, contributing markedly to the accomplishments of the American Expeditionary Forces in France.

EVANS, ROBERT K.
  R—Mississippi.
  B—Jackson, Miss.
  G. O. No. 4, W. D., 1923.

Brigadier general, U. S. Army, retired.
As department commander, Philippine Department, between Aug. 5, 1917, and Aug. 5, 1918, he handled many difficult problems arising in that department with rare judgment, tact, and great skill.

EXTON, CHARLES W.
  R—Clinton, N. J.
  B—Clinton, N. J.
  G. O. No. 53, W. D., 1921.

Colonel, Infantry, U. S. Army.
As assistant G-5, General Headquarters, France, a position of great responsibility, he had complete responsibility for the control, supervision, and inspection of the detachments of American students in French Universities, which was part of the educational system of the American Forces in Germany. He performed all his duties with exceptional efficiency and rendered services of much value.

FAIR, JOHN S.
  R—Altoona, Pa.
  B—Dakota, Nebr.
  G. O. No. 81, W. D., 1919.

Colonel, Quartermaster Corps, U. S. Army.
He organized and operated the remount service, controlled the purchasing of fuel and forage for the Army, and organized and started into operation the conservation and reclamation division. By his enthusiasm and energy valuable results were obtained.

FAISON, SAMSON L.
  R—Faison, N. C.
  B—Faison, N. C.
  G. O. No. 59, W. D., 1919.

Brigadier general, U. S. Army.
He commanded with great credit the 60th Infantry Brigade, 30th Division, in the breaking of the enemy's Hindenburg line at Bellicourt, France, and in subsequent operations in which important captures were made, all marking him as a military commander of great energy and determination.

FARNSWORTH, CHARLES S.
  R—Clarion, Pa.
  B—Lycoming County, Pa.
  G. O. No. 111, W. D., 1919.

Major general, U. S. Army.
In command of the 37th Division, his efficient leadership and military ability were important factors in the successful operations in the Meuse-Argonne offensive, and later proved their worth when this division served with the French and Belgian forces in Belgium.

FASSETT, WILLIAM M.
  R—Nashua, N. H.
  B—Nashua, N. H.
  G. O. No. 59, W. D., 1919.

Brigadier general, U. S. Army.
In the forcing of the crossing of the Escault River, Belgium, in November, 1918, and the establishment of a bridgehead thereof, he demonstrated his ability as a leader. The successful operations of his brigade in this and in ensuing actions were greatly influenced by his efforts.

FELAND, LOGAN.
  R—Hopkinsville, Ky.
  B—Hopkinsville, Ky.
  G. O. No. 89, W. D., 1919.
  Distinguished-service cross also awarded.

Colonel, U. S. Marine Corps.
As lieutenant colonel and second in command of the 5th Regiment, United States Marine Corps, 2d Division, he had an important function in the training of that organization, and he participated creditably in its operations in the Aisne defensive and the fighting in the Chateau-Thierry section. Having taken command of his regiment as colonel shortly before the battle of Soissons, he led it with extraordinary skill throughout the remainder of its engagements, giving proof of the highest qualities of leadership and unceasing devotion to his important duties.

FELTON, SAMUEL M.
  R—Chicago, Ill.
  B—Philadelphia, Pa.
  G. O. No. 18, W. D., 1919.

Director general of military railways.
In supervising the supply of railway material and the organization of railway operation and construction troops, by his energetic and loyal service he has contributed materially to the success of the Army in the field.

FENTON, CHAUNCEY L.
  R—Lowellville, Ohio.
  B—Edinburg, Pa.
  G. O. No. 15, W. D., 1921.

Colonel (Coast Artillery Corps), General Staff Corps, U. S. Army.
As chief of a section of the General Staff during the period of demobilization and reorganization of the Army, he has rendered conspicuous service in the solution of intricate and important problems pertaining to the scientific utilization of the commissioned personnel of the Army.

FERGUSON, HARLEY B.
  R—Waynesville, N. C.
  B—Waynesville, N. C.
  G. O. No. 87, W. D., 1919.

Brigadier general, U. S. Army.
As chief engineer of the 2d Army Corps and later of the 2d Army, he demonstrated high professional attainments and marked initiative. Through his foresight and skill in directing important technical operations he was a notable factor in the successes of the combat troops, rendering invaluable services to the American Expeditionary Forces.

FERGUSSON, FRANK K.
  R—Riddleton, Tenn.
  B—Riddleton, Tenn.
  G. O. No. 47, W. D., 1919.

Brigadier general, U. S. Army.
As commandant of the Coast Artillery training center at Fort Monroe, Va., he rendered specially meritorious and conspicuous service in organizing and administering that center and in the preparation and execution of the plans for the organization, training, and equipment of the units of Coast Artillery for oversea service.

FIEBEGER, GUSTAV J.
  R—Akron, Ohio.
  B—Akron, Ohio.
  G. O. No. 19, W. D., 1922.

Colonel, professor of civil and military engineering, U. S. Military Academy.
As head of the department of civil and military engineering he for 26 years instructed, both personally and by textbook, the officers of the Army in the principles of warfare, principles later fruitfully applied by many of these officers as commanders in the World War.

FIFE, JAMES DOUGLAS.
  R—Charlottesville, Va.
  B—Charlottesville, Va.
  G. O. No. 59, W. D., 1919.

Colonel, Medical Corps, U. S. Army.
In command of Base Hospital No. 21, he served with distinction with the British Expeditionary Forces. He was later assigned to duty in the office of the chief surgeon in charge of hospital planning and construction, procurement of permanent buildings, establishment of hospitalization, liaison with the French authorities, the General Staff, and with the Engineers. In the performance of these multifarious duties he displayed conspicuous ability.

FINNEY, JOHN M. T.
  R—Baltimore, Md.
  B—Natchez, Miss.
  G. O. No. 50, W. D., 1919.

Brigadier general, U. S. Army.
He rendered distinguished services in the organization of surgical teams, for the purpose of affording expert surgical aid to the wounded in the immediate vicinity of the battle field. He has done much to standardize the practice of surgery in war, and giving so freely of his professional experience and skill he has in many ways rendered services of exceptional value to the Government.

FISHER, HENRY C.
  R—Washington, D. C.
  B—Montgomery County, Md.
  G. O. No. 10, W. D., 1922.

Colonel, Medical Corps, U. S. Army.
As chief sanitary inspector, American Expeditionary Forces, by the tireless application of his many years of military experience to correct abuses and secure uniformity and efficiency of hospital administration in the hospitals of the American Expeditionary Forces, he rendered services of great value to the Government.

FISHER, WILLIAM A., Jr.
  R—Baltimore, Md.
  B—Baltimore, Md.
  G. O. No. 59, W. D., 1921.

Lieutenant colonel, Medical Corps, U. S. Army.
Due to his great energy, excellent judgment, and unusual knowledge of the personnel capabilities of surgeons, he made it possible to organize surgical teams and distribute surgical personnel in a manner which made it possible for evacuation hospitals to function under the greatest difficulties in spite of personnel shortage.

FISKE, CHARLES NORMAN.
  R—Upton, Mass.
  B—Jaffrey, N. H.
  G. O. No. 116, W. D., 1919.

Captain, Medical Corps, U. S. Navy.
As force medical officer, his untiring energy, his foresight in sanitary inspection of ships, and his close cooperation with the Army authorities contributed greatly to the successful outcome of our oversea operations.

FISKE, HAROLD B.
  R—Salem, Oreg.
  B—Salem, Oreg.
  G. O. No. 12, W. D., 1919.

Brigadier general, U. S. Army.
In charge of the training section of the General Staff, this brilliant officer perfected and administered the efficient scheme of instruction through which the American Army in France was thoroughly trained for combat in the shortest possible time. By his great depth of vision, his foresight, and his clear conception of modern tactical training he has enabled our forces to enter each engagement with that preparedness and efficiency that have distinguished the American Army in each battle.

FITCH, ROGER S.
 R—Buffalo, N. Y.
 B—Buffalo, N. Y.
 G. O. No. 9, W. D., 1923.

Colonel (Cavalry), General Staff Corps, U. S. Army.
As chief of staff, 86th Division, Aug. 25, 1917, to Nov. 21, 1918, he displayed the highest professional qualifications; by his tireless energy and devotion to duty he rendered conspicuous service to the Government. While chief of staff, 86th Division, he voluntarily performed the duties of assistant chief of staff, G-3, with the 89th Division during the final advance of that division from the Bois-de-Bantheville during the Meuse-Argonne offensive. In November, 1918, he was assigned to duty as G-3, 7th Army Corps; in this position of great responsibility he rendered service of great value and contributed materially to the success of that corps.

FLAHERTY, JAMES A.
 R—Philadelphia, Pa.
 B—Philadelphia, Pa.
 G. O. No. 116, W. D., 1919.

Supreme Knight of the Knights of Columbus.
His high leadership and service rendered the Army of the United States were conspicuous.

FLASH, Mrs. ALICE H.
 R—Boston, Mass.
 B—Jefferson County, Ga.
 G. O. No. 9, W. D., 1923.

Chief nurse, Army Nurse Corps, U. S. Army.
As chief nurse of the Mesves hospital center, France, during the World War, she rendered invaluable assistance and made possible the efficient nursing of over 20,000 patients at one time. Her good judgment in dealing with very difficult personnel problems, her tact, and splendid example resulted in an unusually high standard of nursing efficiency at this center, in spite of the most trying physical conditions. She displayed marked executive ability and professional qualities in directing hundreds of nurses in the care of the sick and wounded.

FLEMING, ADRIAN S.
 R—Louisville, Ky.
 B—Midway, Ky.
 G. O. No. 87, W. D., 1919.

Brigadier General, U. S. Army.
He commanded with distinction the 158th Field Artillery Brigade, 83d Division, displaying aggressive leadership and the highest professional attainments. He contributed materially to the successful operations of the Infantry units to which his brigade was attached during the Meuse-Argonne offensive by the timely and accurate artillery support furnished by his regiments.

FLETCHER, FRANK F.
 R—Oskaloosa, Iowa.
 B—Oskaloosa, Iowa.
 G. O. No. 15, W. D., 1923.

Rear admiral, U. S. Navy, Navy representative, War Industries Board.
In connection with the operations of the War Industries Board during the World War as a member of the board he rendered, through his broad vision and distinguished capacity, services of inestimable value in marshaling the industrial forces of the Nation and mobilizing its economic resources—marked factors in assisting to make military success attainable. As one of the service representatives on the board his sound judgment and wide knowledge of military and naval matters contributed markedly to the successful prosecution of the war.

FLINT, JOSEPH M.
 R—New Haven, Conn.
 B—Chicago, Ill.
 G. O. No. 59, W. D., 1919.

Lieutenant colonel, Medical Corps, U. S. Army.
When placed in a position of great responsibility as commanding officer of Mobile Hospital No. 39 at Aulnois-sous-Vertuzey, France, he used extraordinary skill and sound judgment in the organization and operation of that unit, the first of its kind in the American Expeditionary Forces. In its formative period he was faced by great and unforeseen difficulties, but with untiring energy and genius he surmounted all obstacles, making his unit a model for all those subsequently organized.

FLOOD, BERNARD A.
 R—New York, N. Y.
 B—New York, N. Y.
 G. O. No. 43, W. D., 1922.

Captain, Provost Marshal General's Department, U. S. Army.
As chief inspector of the division of criminal investigation on the staff of the provost marshal general, American Expeditionary Forces, he organized, coordinated, and directed this important office in a highly efficient manner. The successes achieved by this section are largely due to his sound judgment, untiring efforts, and exceptional ability, and were of great value to the American Expeditionary Forces.

FOGG, OSCAR H.
 R—New York, N. Y.
 B—Philadelphia, Pa.
 G. O. No. 13, W. D., 1923.

Lieutenant colonel, Ordnance Department, U. S. Army.
For services as engineer in charge of the organization, establishment, and performance of the American Ordnance Repair Arsenal in France. In this great undertaking he displayed exceptional zeal, keen foresight, and sound judgment, and by his undaunted perseverance he established these extensive and highly important repair facilities, thereby rendering service of inestimable value, contributing materially to the success of the American Expeditionary Forces in France.

FOOTE, ALFRED F.
 R—Holyoke, Mass.
 B—Mooers Forks, N. Y.
 G. O. No. 4, W. D., 1923.

Lieutenant colonel, Infantry, U. S. Army.
As battalion commander during the operations of the 104th Infantry, 26th Division, at Apremont, France, Apr. 10-13, 1918, he demonstrated unusual initiative and marked efficiency and contributed materially to the successful stand of the regiment against the enemy's repeated attacks. As regimental commander during the Champagne-Marne offensive, by his devotion to duty, courage, and superior qualities as a commander he successfully accomplished all missions assigned to him. Later, as division inspector, 26th Division, by his tact, sound judgment, and ability he assisted materially in maintaining the high morale and discipline of the division.

FORBES, CHARLES R.
 R—Burton, Vashon Island, Wash.
 B—Scotland.
 G. O. No. 59, W. D., 1919.

Lieutenant colonel, Signal Corps, U. S. Army.
As division signal officer of the 33d Division he performed his duties with marked distinction, maintaining communication at all times within the division, with adjoining units, and with the higher command. His ability and untiring devotion to duty were great factors in insuring the successes achieved by the division.

FORD, JOSEPH H._____
   R—Washington, D. C.
   B—Washington, D. C.
   G. O. No. 38, W. D., 1922.

Colonel, Medical Corps, U. S. Army.
Colonel Ford organized and commanded a hospital center of 15,000 beds at Allerey, France. Due to his great force and ability, a hospital group was prepared for the care of the sick and wounded during the St. Mihiel and Meuse-Argonne offensives, when the need of hospital beds was critical. This adequate and efficient hospitalization contributed materially to the conservation of man power and to the subsequent success of our forces.

FORD, STANLEY H._____
   R—Columbus, Ohio.
   B—Columbus, Ohio.
   G. O. No. 78, W. D., 1919.

Colonel, Infantry, U. S. Army.
As chief of staff of the 27th Division he rendered valuable services in the operations of this division. By tireless energy, good judgment, and keen foresight he proved to be an important factor in the brilliant military operations of the 27th Division.

FOREMAN, ALBERT WATSON_____
   R—Wilmington, Del.
   B—Wilmington, Del.
   G. O. No. 87, W. D., 1919.

Colonel (Infantry), General Staff Corps, U. S. Army.
As assistant chief of staff, First Section of the 5th Army Corps, by his zealous application to his important duties he ably administered the service of supply, movement of troops, and the control of communication of the 5th Corps during the active operations of that unit against the enemy, rendering meritorious services to the American Expeditionary Forces.

FOREMAN, MILTON J._____
   R—Chicago, Ill.
   B—Chicago, Ill.
   G. O. No. 1, W. D., 1926.
   Distinguished-service cross also awarded.

Colonel, Field Artillery, U. S. Army.
Commanding the 122d Field Artillery, 33d Division, he gave proof of eminent technical attainments and assiduous zeal. Though handicapped by many adverse conditions due to difficult terrain and determined hostile resistance, he kept his batteries in close support of the Infantry and thereby rendered services of inestimable value during the St. Mihiel offensive and the advance to the Meuse in the last phase of the Meuse-Argonne offensive.

FOSDICK, RAYMOND B._____
   R—New York, N. Y.
   B—Buffalo, N. Y.
   G. O. No. 73, W. D., 1919.

For services as chairman of the commission on training camp activities.

FOSTER, CHARLES L._____
   R—Washington, D. C.
   B—Starkville, Miss.
   G. O. No. 56, W. D., 1921.

Colonel, Medical Corps, U. S. Army.
As surgeon at Base Section No. 1, services of supply, a position of great responsibility, due to his thorough training, energy, and ability, he rendered services of great value at this important base port.

FOSTER, CHARLES W._____
   R—Burlington, Vt.
   B—Michigan.
   G. O. No. 56, W. D., 1922.

Major (Cavalry), General Staff Corps, U. S. Army.
He served with the 3d Division as assistant G-3 during May and June, 1918, assistant chief of staff G-3 from June to September, 1918; as assistant G-3 1st and 7th Army Corps, from September to November, 1918; and assistant G-3, 3d Army, from November, 1918, to July, 1919. By his marked ability, devotion to duty, and high military attainments he contributed materially to the successes achieved by the commands with which he served

FOSTER, REGINALD L._____
   R—New York, N. Y.
   B—China.
   G. O. No. 38, W. D., 1922.

Colonel, Infantry, U. S. Army.
During the Meuse-Argonne offensive from Sept. 26, 1918, to Nov. 11, 1918, he commanded his regiment, the 52d Pioneer Infantry, with marked ability and success. By zealous application to his arduous duties, energetic efforts displayed in the construction of roads, care of enemy prisoners, police and salvage of the battle fields, and the handling of ammunition and supplies, he contributed materially to the success of the 5th Army Corps during that important operation, thereby rendering meritorious and distinguished services to the American Expeditionary Forces.

FOULOIS, BENJAMIN D._____
   R—Washington, Conn.
   B—Washington, Conn.
   G. O. No. 68, W. D., 1920.

Brigadier general, U. S. Army.
As Chief of Air Service of the American Expeditionary Forces during the early organization period, he displayed great ability and untiring energy in order to place that service on a firm and efficient basis. He conducted intricate negotiations with the French for the procurement of aircraft material, of sites for Air Service installations and of schools of instruction for the Air Service personnel. Similar negotiations were made by him with the English for the assembly of night bombing planes for our Air Service and for instructions of our personnel in English shops and in English aerodromes. Later he rendered valuable assistance in connection with maintenance of Air Service squadrons at the front.

FOWLER, HAROLD_____
   R—New York, N. Y.
   B—England.
   G. O. No. 108, W. D., 1919.

Colonel, Air Service, U. S. Army.
He rendered notable aid in planning the movements of the night bombing squads of the American Air Service. Later, appointed Air Service commander of the 3d Army, he assisted largely in the joint training of air and ground troops, at all times handling his troops well and establishing liaison between the air and ground forces.

FRANKLIN, BENJAMIN A._____
   R—Springfield, Mass.
   B—Northumberland County, Va.
   G. O. No. 124, W. D., 1919.

Lieutenant colonel, Ordnance Department, U. S. Army.
As production manager and assistant chief of the Bridgeport Ordnance District, in which capacity he maintained at all times the greatest degree of intelligent and enthusiastic cooperation between the Ordnance Department and manufacturers in his district, thereby attaining the maximum production of munitions in a minimum time, and also as chairman of the Bridgeport Ordnance District Claims Board, in which capacity his services have been invaluable to the Nation in adjusting equitably the $346,000,000 worth of outstanding contracts in his district in force at the signing of the armistice.

FRANKLIN, PHILIP A. S., Sr_____
R—New York, N. Y.
B—Ashland, Md.
G. O. No. 25, W. D., 1919.

Chairman, shipping control committee.
For services in connection with the embarkation service of the Army in the division of purchase, storage, and traffic. To his fine technical knowledge and energetic action is due, in a large measure, the efficient jurisdiction over dock facilities and floating equipment which has made possible the large movement of troops and supplies overseas.

FRANKLIN, WALTER S_____
R—Baltimore, Md.
B—Ashland, Md.
G. O. No. 56, W. D., 1921.

Lieutenant colonel, Army Transport Service, U. S. Army.
As officer in charge of embarkation service in the troop and cargo division of the Army Transport Service, a position of great responsibility, by his energy, tact, and initiative, arrangements were effected whereby a large share of the British ocean passenger tonnage was made available for the American use, and whereby additional allied and neutral tonnage for the return of the American Expeditionary Forces was procured. He has rendered service of much value.

FRASER, LEON_____
R—New York, N. Y.
B—Boston, Mass.
G. O. No. 31, W. D., 1922.

Major, Judge Advocate General's Department, U. S. Army.
For services as assistant judge advocate of the services of supply. In this capacity he bore an important part of the responsibility consequent upon deciding a wide variety of legal and administrative questions, including the reviewing of court-martial cases, the preparation of opinions, and the solution of the many problems which grew out of the interpretation of United States Statutes, Army Regulations, General Orders, and French and international law involved in the business operations of the American Expeditionary Forces.

FRAYNE, HUGH_____
R—Scranton, Pa.
B—Scranton, Pa.
G. O. No. 15, W. D., 1923.

Labor commissioner, War Industries Board.
In connection with the operations of the War Industries Board during the World War, as a member of the board he rendered, through his broad vision, distinguished capacity, and organizing ability, services of inestimable value in marshaling the industrial forces of the Nation and mobilizing its economic resources—marked factors in assisting to make military success attainable. As labor commissioner he contributed largely to the successful mobilization and conservation of man power for war industry. His untiring efforts and devotion to duty in this connection contributed markedly to the successful operations of the supply system of the Army.

FRIES, AMOS A_____
R—Central Point, Oreg.
B—Debello, Wis.
G. O. No. 95, W. D., 1919.

Brigadier general, U. S. Army.
As chief of the Chemical Warfare Service he was charged with the important task of training and equipping our troops for a form of warfare in which the American Army had had no experience prior to the present war. Both in securing proper defensive measures against gas and in developing new methods for its use as an offensive agency, he performed his arduous duties with marked success, thereby rendering valuable services to the American Expeditionary Forces.

FRINK, JAMES L_____
R—Springfield, Mo.
B—Ida Grove, Iowa.
G. O. No. 14, W. D., 1923.

Major (Infantry), General Staff Corps, U. S. Army.
From Oct. 19 to Nov. 11, 1918, during the Meuse-Argonne offensive, he rendered highly meritorious services as assistant chief of staff, G-3, 78th Division. In this position of great responsibility he displayed rare judgment in the selection of points in the line for attack and in the designation of forces to be used in the attack. As acting chief of staff of his division from November, 1918, until February, 1919, he displayed high professional attainments, unfailing energy, and devotion to duty, contributing in a material way to the success of his division.

FULMER, JOHN J_____
R—Reading, Pa.
B—Amityville, Pa.
G. O. No. 87, W. D., 1919.

Lieutenant colonel, Infantry, U. S. Army.
As director of the Infantry Specialists' School at Langres he achieved a notable success in the efficient training of thousands of officers. He also rendered invaluable service while a member of training section of the General Staff by establishing uniform and effective methods of instruction in musketry training throughout the American Expeditionary Forces.

FUQUA, STEPHEN O_____
R—Baton Rouge, La.
B—Baton Rouge, La.
G. O. No. 59, W. D., 1919.

Colonel (Infantry), General Staff Corps, U. S. Army.
In charge of the troop movement subsection of G-3, 1st Army, from its organization until he became chief of staff, 1st Division, he was responsible for and supervised the movements incident to the concentration of troops for the St. Mihiel and Meuse-Argonne offensives of the 1st Army, which involved many thousands of men and was accomplished with the greatest success. His untiring, painstaking, and energetic efforts had a marked effect on the success of these major operations.

FURLOW, JAMES WADSWORTH_____
R—Americus, Ga.
B—Americus, Ga.
G. O. No. 69, W. D., 1919.

Colonel, Quartermaster Corps, U. S. Army.
While on duty in the Motor Transport Corps his brilliant conception and able administration were largely responsible for the organization and highly successful operation of the plan for upkeep and maintenance of motor vehicles during the war.

*GALBRAITH, FREDERICK W., Jr_____
R—Cincinnati, Ohio.
B—Watertown, Mass
G. O. No. 59, W. D., 1921.
Distinguished-service cross also awarded.

Colonel, Infantry, U. S. Army.
As regimental commander of the 147th Infantry, 37th Division, by his energy and ability he organized, trained, and brought his regiment to a high state of efficiency and commanded it throughout its operations in the Meuse-Argonne and in the Flanders offensives of the Lys and of the Escaut River. He has rendered services of conspicuous worth.
Posthumously awarded. Medal presented to widow, Mrs. Esther G. Galbraith.

GALEN, ALBERT J.
R—Helena, Mont.
B—Broadwater County, Mont
G. O. No. 56, W. D., 1922.

Lieutenant colonel, Judge Advocate General's Department, U. S. Army.
As judge advocate of the American Expeditionary Forces in Siberia he organized this important department and administered its affairs with conspicuous efficiency. His marked legal ability, sound judgment, and untiring efforts were important factors in the splendid work of his department, and he at all times handled with great success the various military and international problems with which he was confronted. He contributed materially to the success of the forces in Siberia and rendered conspicuous services in a position of great responsibility.

GARBER, MAX B.
R—Marble Rock, Iowa.
B—Marble Rock, Iowa.
G. O. No. 35, W. D., 1920.

Lieutenant colonel, Infantry, U. S. Army.
He commanded with marked distinction the 59th Infantry during the attack on the Vesle. In this command he displayed those high qualities of ability, leadership, and personal courage that marked him as a determining factor in the successes achieved by the 4th Division.

*GARDNER, AUGUSTUS PEABODY.
R—Hamilton, Mass.
B—Boston, Mass.
G. O. No. 14, W. D., 1923.

Major, Infantry, U. S. Army.
Resigning as a Member of Congress to serve under his reserve commission as colonel, Adjutant General's Department, in 1917, he served first at headquarters, Eastern Department, Governors Island, N. Y., and later as adjutant, 31st Division. At his own urgent request he was appointed major, 121st Infantry, 31st Division, and commanded a battalion in that regiment until his death. His entire service was characterized by untiring zeal, devotion to duty, and marked success. His splendid example of patriotism will always serve as an inspiration to his countrymen.
Posthumously awarded. Medal presented to widow, formerly Mrs. Augustus P. Gardner, now Mrs. C. C. Williams.

GARDNER, FULTON Q. C.
R—Fort Smith, Ark.
B—Lafayette Springs, Miss.
G. O. No. 73, W. D., 1919.

Colonel (Coast Artillery Corps), General Staff Corps, U. S. Army.
For services as secretary of the General Staff.

GARFIELD, HARRY A.
R—Williamstown, Mass.
B—Hiram, Ohio.
G. O. No. 15, W. D., 1921.

Fuel administrator.
During the progress of the war, by his conduct of the Fuel Administration, he stimulated the production, conserved the use, and supervised the distribution of those supplies of fuel necessary for the support and transportation of the Armies of the United States, the maintenance of industry, the production of war supplies, and the health and well-being of the civil population upon which the successful prosecution of military activities depended.

GASKILL, CHARLES S.
R—Morristown, N. J.
B—Mount Holly, N. J.
G. O. No. 59, W. D., 1919.

Lieutenant colonel, Corps of Engineers, U. S. Army.
In charge of the locomotive and car repair shops at Nevers, he carried out the installation and operation of this plant, exhibiting rare executive ability and engineering qualifications of the highest order.

GASSER, LORENZO D.
R—Tiffin, Ohio.
B—Likens, Ohio.
G. O. No. 43, W. D., 1922.

Major, Infantry, U. S. Army.
As chief of motor transportation section, office of the assistant chief of staff, G–4, American Expeditionary Forces, he showed unusual ability, tireless energy, and a comprehensive grasp of details in preparing plans for the organization and operation of a general headquarters reserve of motor transportation. Later, as deputy assistant chief of staff, G–4, at advance general headquarters at Treves, Germany, in coordinating the plans for the reception and disposition of enemy war materials, he successfully handled a problem requiring great tact and high professional attainments, thereby rendering services of great value to the American Expeditionary Forces.

GAUCHE, EDWARD E.
R—New York, N. Y.
B—Lorado, Ohio.
G. O. No. 15, W. D., 1923.

Lieutenant colonel, Adjutant General's Department, U. S. Army.
Charged with the responsibility for the operation of the statistical and personnel divisions, services of supply, and for the work of all personnel adjutants, functioning under the statistical section, Adjutant General's Office, General Headquarters, stationed at all base ports during the repatriation of the American Army, he displayed administrative and executive ability of a high order, unusual resourcefulness, and sound professional judgment. He organized and maintained a service that covered the entire theater of operations, exclusive of the zone of the armies during the entire period of military activity. By his fitness and aptitude for the grave responsibilities placed upon him, his tireless energy and unceasing devotion to duty he contributed materially to the successful operations of the services of supply and the American forces in France.

GERLACH, FRANK C.
R—Wooster, Ohio.
B—Wooster, Ohio.
G. O. No. 43, W. D., 1922.

Colonel, Infantry, U. S. Army.
As lieutenant colonel, 146th Infantry, 37th Division, from May, 1917, until October, 1918, he organized and developed to a high state of efficiency a training school for commissioned candidates at Camp Sheridan, Ala., and the divisional training school in the Baccarat sector, France. Later, as colonel, 145th Infantry, 37th Division, he commanded his regiment with marked success during the Ypres-Lys offensive, showing at all times great energy, resourcefulness, and splendid leadership, thereby contributing materially to the successes achieved in this important operation.

GEROW, LEONARD T.
R—Petersburg, Va.
B—Dinwiddie County, Va.
G. O. No. 49, W. D., 1922.

Lieutenant colonel, Signal Corps, U. S. Army.
For services as officer in charge of the sales and disbursing division of the Signal Corps. With unusual ability and skill he conducted the financial affairs of the Signal Corps and handled the negotiations with such tact and energy that Signal Corps property urgently needed was secured, inspected, and delivered to depots with the minimum of delay. Later, in the negotiations connected with the disposal of Signal Corps plants and stocks, he again performed his exacting duties in a highly meritorious manner, thereby rendering services of great value to the American Expeditionary Forces in a position of great responsibility.

GHORMLEY, ROBERT LEE..............
   R—Moscow, Idaho.
   B—Portland, Oreg.
   G. O. No. 116, W. D., 1919.

Commander, U. S. Navy.
For services as assistant director of Overseas Division, Naval Overseas Transportation Service.

GIBBS, ELBERT A..............
   R—Pittsburgh, Pa.
   B—Kasson, Minn.
   G. O. No. 87, W. D., 1919.

Colonel, Corps of Engineers, U. S. Army.
He served with distinction as chief of the general construction section in the office of the director of construction and forestry. Charged with the supervision of important engineering construction projects, he gave proof of high professional attainments and keen foresight, rendering invaluable services to the American Expeditionary Forces.

GIBBS, GEORGE S..............
   R—Harlan, Iowa.
   B—Harlan, Iowa.
   G. O. No. 59, W. D., 1919.

Brigadier general, U. S. Army.
As assistant to the chief signal officer, American Expeditionary Forces, much of the efficiency of the Signal Service in the zone of advance was due to his splendid ability and to his skill in the handling of the tactical and technical operations of the Signal Corps organizations attached to the service at the front.

GIBSON, ADELNO..............
   R—Oskaloosa, Iowa.
   B—Marysville, Iowa.
   G. O. No. 9, W. D., 1923.

Lieutenant colonel (Coast Artillery Corps), General Staff Corps, U. S. Army.
As officer in charge of the personnel subsection of G–1, general headquarters from October, 1917, until August, 1918, he displayed unusual foresight, excellent judgment and resourcefulness in the conception, organization, and operation of the entire replacement system at general headquarters. By his constant devotion to duty, his great ability for original work, and his high professional attainments he materially contributed to the efficiency of the staff work at general headquarters, thereby rendering services of great value to the American Expeditionary Forces.

GIGNILLIAT, LEIGH R..............
   R—Culver, Ind.
   B—Savannah, Ga.
   G. O. No. 43, W. D., 1922.

Colonel, Infantry, U. S. Army.
As G–2 of the 84th Division from Oct. 6, 1917, until Nov. 9, 1918, and of the 37th Division from Nov. 9, 1918, until Mar. 15, 1919, he displayed an unusual devotion to duty and military attainments of a high order, which enabled him to place the intelligence sections of both divisions on a high plane of efficiency. From Mar. 15, 1919, to June 27, 1919, as the United States' representative on the Interallied Food Commission, by rare tact, great energy, and marked executive ability he solved with conspicuous success many perplexing problems of supply in our occupied area.

GILCHRIST, HARRY L..............
   R—Cleveland, Ohio.
   B—Waterloo, Iowa.
   G. O. No. 38, W. D., 1921.

Colonel, Medical Corps, U. S. Army.
As chief of the delousing and bathing services of the American Expeditionary Forces, by his superior administration and splendid efficiency he contributed materially to the success achieved by the Army at the ports of Brest, Bordeaux, and St. Nazaire in the return to the United States of the American Expeditionary Forces.

GIRL, CHRISTIAN..............
   R—Cleveland, Ohio.
   B—Elkhart, Ind.
   G. O. No. 3, W. D., 1922.

For services rendered in the organization of the production section, motors branch, Quartermaster Corps, which designed and produced the standardized motor truck.

GLASSFORD, PELHAM D..............
   R—Carthage, Mo.
   B—Las Vegas, N. Mex.
   G. O. No. 89, W. D., 1919.

Brigadier general, U. S. Army.
He served creditably at the Saumur Artillery School, at the First Corps Artillery School, and as commander of a regiment of field artillery during the Chateau-Thierry campaign. Subsequently, upon being promoted to the grade of brigadier general, he displayed high military attainments and unceasing energy as commander of the 51st Field Artillery Brigade, rendering invaluable services to the American Expeditionary Forces.

GLEAVES, ALBERT..............
   R—Nashville, Tenn.
   B—Nashville, Tenn.
   G. O. No. 116, W. D., 1919.

Vice admiral, U. S. Navy.
As commander of the cruiser and transport fleet, his untiring energy, close cooperation, and wide decisions contributed greatly to the successful oversea operations of the transport fleet, resulting in the successful transportation of the United States forces abroad.

GLEAVES, SAMUEL R..............
   R—Wytheville, Va.
   B—Independence, Va.
   G. O. No. 62, W. D., 1919.

Colonel (Cavalry), General Staff Corps, U. S. Army.
As G–3 of the 42d Division he displayed military attainments of a high order, being constant in devotion to his exacting duties. In the operations section, General Headquarters, American Expeditionary Forces, he handled all questions arising in that section pertaining to the arrival, location and issuance of orders for movements of units in the American Expeditionary Forces. In the solution of the perplexing problems which arose he brought to his task a high faculty for organization, coupled with sound judgment and a comprehensive grasp of service conditions.

GLENDINNING, ROBERT..............
   R—Philadelphia, Pa.
   B—Philadelphia, Pa.
   G. O. No. 38, W. D., 1922.

Lieutenant colonel, Air Service, U. S. Army.
As classification officer for the Air Service in the Casual Officers' Depot, Blois, France, through his untiring energy and sound judgment he developed an efficient system of classification, which was of great assistance to the American air forces. Later, as a representative of the Air Service, American Expeditionary Forces in Italy, he was charged with many and intricate negotiations with the Italian Government, which he conducted with rare intelligence, sound judgment, and successful results.

GLENNAN, JAMES D..............
   R—Washington, D. C.
   B—Rochester, N. Y.
   G. O. No. 89, W. D., 1919.

Brigadier general, U. S. Army.
In charge of the hospitalization division in the office of the chief surgeon, he directed the establishment, equipment and operation, as well as the evacuation service, of all the American hospitals in France. By his keen foresight, untiring energy, and administrative ability he solved successfully the numerous problems which confronted him, rendering services of the highest value to the American Expeditionary Forces.

GODSON, WILLIAM F. H.
R—New Bedford, Mass.
B—England.
G. O. No. 124, W. D., 1919.

Colonel, Cavalry, U. S. Army.
For services while serving as military attaché at Berne, Switzerland.

GOETHALS, GEORGE W.
R—New York, N. Y.
B—Brooklyn, N. Y.

G. O. No. 144, W. D., 1918.

Major general, U. S. Army.
For services in reorganizing the Quartermaster Department and in organizing and administering the Division of Purchase, Storage, and Traffic during the war.

GOLDTHWAIT, JOEL E.
R—Boston, Mass.
B—Marblehead, Mass.
G. O. No. 59, W. D., 1919.

Colonel, Medical Corps, U. S. Army.
As a member of the Medical Corps he has, by his unusual foresight and organizing ability, made it possible to reclaim for duty thousands of men suffering from physical defects. He has thereby materially conserved for combat service a great number of men who would have been lost to the service.

GOODALL, HARRY W.
R—Boston, Mass.
B—Wells Beach, Me.
G. O. No. 56, W. D., 1921.

Lieutenant colonel, Medical Corps, U. S. Army.
For services while in command of the gas hospital of the Justice Hospital group from the beginning of the St. Mihiel offensive until October, 1918.

GOODRICH, ANNIE W.
R—New York, N. Y.
B—New Brunswick, N. J.
G. O. No. 9, W. D., 1923.

Contract nurse, Army Nurse Corps, U. S. Army.
As organizer and first dean of the Army School of Nursing, by her individual energy, ability, and breadth of vision, she enrolled, trained, and placed at the disposal of the Medical Department of the Army 1,800 selected student nurses. These young women were of inestimable assistance to the Army Nurse Corps at a time when there was a shortage of nurses and an enormous influx of patients in Army hospitals because of the influenza epidemic. Through her efforts there was put into operation a system of training of student nurses which contributed in a remarkable degree to the success of the Medical Department in the saving of hundreds of lives during the World War.

GOODRICH, CHARLES C.
R—Orange, N. J.
B—Akron, Ohio.
G. O. No. 15, W. D., 1923.

Lieutenant colonel, Ordnance Department, U. S. Army.
As chief representative in England of the Liquidation Commission he displayed unusual professional ability, tact, resourcefulness and sound judgment. As executive officer in the office of the chief purchasing officer, Ordnance Department, American Expeditionary Forces, he coordinated and supervised the activities of that office and maintained cordial and effective relations with the French Ministry of Armament. The grave and important duties performed by Colonel Goodrich were of immeasurable value to the Government and contributed markedly to the successful operations of the American Expeditionary Forces.

GOODRICH, DAVID M.
R—New York, N. Y.
B—Akron, Ohio.
G. O. No. 116, W. D., 1919.

Lieutenant Colonel (Infantry), General Staff Corps, U. S. Army.
As assistant chief of staff, G-2, of the 78th Division, he rendered excellent services; as a student at the Army General Staff College at Langres he was eminently successful; as a member of the G-2 section at American Expeditionary Forces he performed duties of great importance; and as director of the liaison section of the interallied games committee he demonstrated superior executive ability by the satisfactory management of his many tasks which insured the success of the interallied games.

GOODWIN, ROBERT E.
R—Concord, Mass.
B—Cambridge, Mass.
G. O. No. 56, W. D., 1922.

Colonel, Field Artillery, U. S. Army.
As commanding officer of the 101st Field Artillery, 26th Division, from Sept. 11, 1918, to Apr. 28, 1919, by his high standards, exceptional ability, and unusual grasp of the principles of artillery, he rendered conspicuous service during the St. Mihiel offensive, and later during the operations north of Verdun. His sound judgment and tact, his unflagging energy coupled with the very close cooperation he maintained with the Infantry, were of very great assistance to the Infantry in these operations. His high technical attainments and extraordinary activity contributed materially to the success of the operations of his division.

GORDON, WALTER H.
R—St. Landry Parish, La.
B—Wilkinson County, Miss.
G. O. No. 70, W. D., 1919.

Major general, U. S. Army.
As brigade commander of the 10th Infantry Brigade, 5th Division, he showed great energy and zeal in the conduct of his brigade during the major part of its maneuvers. Later, as division commander of the 6th Division, by his painstaking efforts, he brought his division to a marked state of efficiency, rendering services of great value to the American Expeditionary Forces.

GORGAS, WILLIAM C.
R—Tuscaloosa, Ala.
B—Mobile, Ala.
G. O. No. 144, W. D., 1918.

Major general, U. S. Army.
For services as Surgeon General of the Army in organizing and administering the Medical Department during the war.

GORRELL, EDGAR S.
R—Baltimore, Md.
B—Baltimore, Md.
G. O. No. 59, W. D., 1919.

Colonel, Air Service, U. S. Army.
He rendered most excellent service as a member of the United States Aeronautical Commission charged with the selection of types of European aeronautical material to be manufactured in the United States and as the representative of the Air Service with the General Staff, American Expeditionary Forces. In the performance of his many important tasks he displayed good judgment, great energy, and showed that he possessed ability of a high order, which have been of invaluable service to the Government.

GOSMAN, GEORGE H. R.................
 R—New York, N. Y.
 B—Brooklyn, N. Y.
 G. O. No. 39, W. D., 1921.

Colonel, Medical Corps, U. S. Army.
As commanding officer of Evacuation Hospital No. 1, Colonel Gosman was confronted with the tremendous task of transporting a group of partly finished barracks into an efficient and systematic hospital. He displayed sound judgment and great organizing ability. As chief surgeon, 4th Army Corps, American Expeditionary Forces, during the St. Mihiel offensive he displayed high medical military attainments in the disposition of his units and the provision for evacuation of the wounded and for supplies.

GOSS, BYRON C...................
 R—Princeton, N. J.
 B—Rochester, Ind.
 G. O. No. 3, W. D., 1924.

Lieutenant colonel, Chemical Warfare Service, U. S. Army.
As chemical adviser in the office of the Chief of Chemical Warfare Service, later as chief gas officer of the 1st Army Corps, and finally of the 2d Army, by his untiring energy, exceptional ability, and wide knowledge of gases he rendered service of great value to the American Expeditionary Forces in practically every battle in which American troops were engaged, thereby contributing materially to our success.

GOTWALS, JOHN C...................
 R—Yorktown Heights, N. Y.
 B—Yerkes, Pa.
 G. O. No. 56, W. D., 1922.

Lieutenant colonel, Corps of Engineers, U. S. Army.
As chief searchlight officer, American Expeditionary Forces, he rendered conspicuous service in a position of great responsibility and in a field which was practically new to our service. By his unlimited energy, marked inventive faculties, and high technical skill, together with his ever present willingness to cooperate with our allies, he organized and directed with great success an exceedingly advanced technical service for the night protection of troop concentrations, communications, supply and manufacture establishments, in rear areas, thereby contributing materially to the success of the American Expeditionary Forces.

GOWENLOCK, THOMAS R...............
 R—Chicago, Ill.
 B—Clay Center, Kans.
 G. O. No. 14, W. D., 1923.

Major (Infantry), General Staff Corps, U. S. Army.
As assistant chief of staff, G-2, 1st Division, immediately before the St. Mihiel offensive he displayed resourcefulness, ability, and devotion to duty in securing information under the most adverse circumstances, upon which the attack of the division was based. Throughout the operations of the 1st Division in the Meuse-Argonne advance, Sept. 30 to Nov. 11, 1918, Major Gowenlock demonstrated an unusually high degree of efficiency, courage, and devotion, at all times displaying superior judgment, and seized every opportunity during critical situations to sustain the morale of the command, as well as to furnish division headquarters with indispensable information. By his exceptional ability and devotion to duty he executed the most difficult missions assigned to him, thus rendering important services to the American Expeditionary Forces in the operations against the enemy.

GRAHAM, ALDEN M..................
 R—San Francisco, Calif.
 B—Monmouth, Ill.
 G. O. No. 38, W. D., 1922.

Lieutenant colonel, Motor Transport Corps, U. S. Army.
As motor transport officer of the 3d Division and later of the 5th Army Corps during the entire Meuse-Argonne offensive, by his superior ability, skill, leadership, and capacity for organization he overcame most difficult conditions and maintained the transportation of the corps upon which the success of the operations so largely depended.

GRAHAM, GEORGE F..................
 R—Helena, Mont.
 B—Safe Harbor, Pa.
 G. O. No. 49, W. D., 1922.

Lieutenant colonel, Quartermaster Corps, U. S. Army.
As quartermaster, 42d Division, during the entire time of its operations from May until November, 1918, by his zeal, indefatigable efforts, and unusual ability as an organizer and administrator, he solved many difficult problems under most adverse circumstances and assured the supply of the division at all times with those articles for which he was responsible, thereby rendering services of great value to the American Expeditionary Forces.

GRAHAM, JAMES H..................
 R—Sound Beach, Conn.
 B—Louisville, Ky.
 G. O. No. 87, W. D., 1919.

Colonel, Corps of Engineers, U. S. Army.
In charge of all the engineer depots in France for more than six months during the initial period of our entry into the war, he performed his exacting duties with rare professional ability and unflagging zeal. Later, as supervisor of railroad and dock construction in the office of the chief engineer, he rendered services of great value in connection with development of dock facilities for the American Expeditionary Forces.

GRANT, ULYSSES S., 3d.............
 R—New York, N. Y
 B—Chicago, Ill.
 G. O. No. 69, W. D., 1919.

Colonel, Corps of Engineers, U. S. Army.
As secretary of the American section, Supreme War Council, he was entrusted with the important duty of coordinating the work of the joint secretariat of the Supreme War Council and of the joint secretariat of the Military Representatives of the Supreme War Council, and as a member of the War Prisoners' Commission, Berne, Switzerland, he has rendered conspicuous service to the Government.

GRANT, WALTER S..................
 R—Ithaca, N. Y.
 B—Ithaca, N. Y.
 G. O. No. 50, W. D., 1919.

Colonel (Cavalry), General Staff Corps, U. S. Army.
As deputy chief of staff of the 1st Army, by his high professional attainments and ability he rendered valuable assistance in the staff work preparatory to and during the St. Mihiel and Argonne-Meuse offensives. As chief of staff of the 1st Army Corps, he displayed the same tact, zeal, and energy which marked the previous character of his services.

GRAVES, ERNEST...................
 R—Chapel Hill, N. C.
 R—Chapel Hill, N. C.
 G. O. No. 59, W. D., 1919.

Colonel, Corps of Engineers, U. S. Army.
He was charged with the construction of the Gievres storage depot and later was appointed Engineer officer of the intermediate section, services of supply, where he was placed in charge of all construction projects west of Bourges. As Engineer officer of Base Section No. 2 and of the Advance Section, S. O. S., he performed the duties with which he was intrusted in a conspicuously meritorious manner. In the many responsible capacities in which he was employed the performance of his duty was characterized by sound judgment and untiring zeal.

GRAVES, WILLIAM S.
  R—Gatesville, Tex.
  B—Mount Calm, Tex.
  G. O. No. 18, W. D., 1919.

Major general, U. S. Army.
For services as an executive assistant to the Chief of Staff and as commanding general of the American Expeditionary Forces in Siberia.

GRAY, QUINN
  R—Waco, Tex.
  B—Plantersville, Tex.
  G. O. No. 108, W. D., 1919.

Colonel, Coast Artillery Corps, U. S. Army.
As an instructor at the Army General Staff College, he displayed high military attainments and unfailing energy, performing services of the greatest value in connection with the instruction and training of officers for general staff duty.

GREELY, JOHN N.
  R—Newburyport, Mass.
  B—Washington, D. C.
  G. O. No. 59, W. D., 1919.

Colonel, Field Artillery, U. S. Army.
As a member of the operations section of the General Staff, 1st Division, and later as chief of that section, he showed sound judgment in the tactical operations before Cantigny, Soissons, and St. Mihiel. As chief of staff of the 1st Division, he was a material factor in the success of the operations against the enemy in the Argonne-Meuse offensive, where he demonstrated ability of a high order.

GREELY, WILLIAM B.
  R—Washington, D. C.
  B—Oswego, N. Y.
  G. O. No. 3, W. D., 1924.

Lieutenant colonel, Corps of Engineers, U. S. Army.
In charge of the forestry section of the Division of Construction in Forestry from Sept. 1, 1918, to July 6, 1919, he supervised the operations of all forestry troops in France. He rendered highly important and valuable service to the Government, contributing markedly to the successes of the American forces in France.

GREEN, FREDERICK W.
  R—St. Louis, Mo.
  B—Rock Island, Ill.
  G. O. No. 59, W. D., 1919.

Lieutenant colonel, Army Transport Service, U. S. Army.
As superintendent of the port of Brest, he organized the task expeditiously and with great ability. Without previous organization or sufficient personnel to aid him, and confronted by many serious obstacles, he, by sheer force of will, supported by untiring energy, undertook a new work and created the organization which was competent to unload the largest ships in a surprisingly short period of time. His service was most valuable to the American Expeditionary Forces.

GRISCOM, LLOYD C.
  R—New York, N. Y
  B—Riverton, N. J
  G. O. No. 59, W. D., 1919.

Lieutenant colonel, Adjutant General's Department, U. S. Army.
He served with marked ability as adjutant of the 77th Division during the early days of its organization and training. As special representative of the Commander in Chief with the Minister of War of Great Britain, he fulfilled with great distinction and credit the duties of an office requiring ability, tact, and address.

GRISSINGER, JAY W.
  R—York, Pa.
  B—Mechanicsburg, Pa.
  G. O. No. 70, W. D., 1919.

Colonel, Medical Corps, U. S. Army.
As division surgeon of the 42d Division, and later as chief surgeon of the 1st Army Corps during its operations on the Marne and in the St. Mihiel and Meuse-Argonne offensives, he displayed qualities of leadership, high professional attainments, and rare judgment in energetically directing the work of the sanitary units under his control in providing front-line hospitalization and evacuation facilities for our sick and wounded in the field.

GROOME, JOHN C.
  R—Philadelphia, Pa.
  B—Philadelphia, Pa.
  G. O. No. 19, W. D., 1922.

Colonel, Signal Corps, U. S. Army.
For services as chief of the officers' leave bureau, France. By his untiring energy and devotion to duty Colonel Groome rendered a service of great value to the morale of the American Army at a difficult time.

GROVE, WILLIAM R.
  R—Denver, Colo.
  B—Montezuma, Iowa.
  G. O. No. 17, W. D., 1924.
  Medal of honor also awarded.

Colonel, Quartermaster Corps, U. S. Army.
Assistant to the chief of the supply division, Quartermaster General's office, a position of great responsibility, he was charged with the procurement of the subsistence supplies of the Army in the United States and in France, and to him is due the organization of the subsistence division of the office of the Quartermaster General. In cooperation with the Food Administration, he made arrangements for the procurement of all subsistence supplies required for the Army. He rendered services of much value.

GRUBER, EDMUND L.
  R—Cincinnati, Ohio.
  B—Cincinnati, Ohio.
  G. O. No. 69, W. D., 1919.

Colonel, Field Artillery, U. S. Army.
He displayed exceptional ability in planning the organization of Field Artillery brigade firing centers; in April, 1918, established such a center at Fort Sill, and during the remainder of the war displayed rare judgment and high professional attainments in the administration of this center.

GRUNERT, GEORGE
  R—White Haven, Pa.
  B—White Haven, Pa.
  G. O. No. 59, W. D., 1919.

Lieutenant colonel (Cavalry), General Staff Corps, U. S. Army.
With remarkable skill, constantly displaying zeal and high military attainments, he performed his exacting duties as assistant chief of staff, G-1, of the 1st Corps, during the successive operations at Chateau-Thierry, on the Ourcq and Vesle, and in the St. Mihiel and Argonne-Meuse offensives. By his untiring and painstaking efforts and unusual ability he performed the most difficult tasks, rendering services of great value to the Government.

GULICK, JOHN W.
  R—Goldsboro, N. C.
  B—Goldsboro, N. C.
  G. O. No. 19, W. D., 1920.

Colonel (Coast Artillery Corps), General Staff Corps, U. S. Army.
As assistant chief of the operations section and later as chief of staff of the Army Artillery of the 1st Army, he demonstrated a keen conception of all of the tactical situations which confronted the artillery of the 1st Army. By his high professional attainments and sound military judgment, he handled the many complex problems of the 1st Army Artillery with marked skill, and thereby contributed, in no small degree, to the success of this unit in the St. Mihiel and Meuse-Argonne offensives.

GULLION, ALLEN W_____
R—Carrollton, Ky.
B—Carrollton, Ky.
G. O. No. 9, W. D., 1923.

Lieutenant colonel, Judge Advocate General's Department, U. S. Army.
In the national administration of the selective service law from May 4, 1917, to Mar. 26, 1918. As chief of publicity and information under the provost marshal general he successfully conducted the campaign to popularize selective service. Later, as acting executive officer to the provost marshal general, he solved many intricate problems with firmness, promptness, and common sense. Finally, as the first chief of the mobilization division of the provost marshal general's office, he supervised all matters relating to the making and filling of calls and the accomplishment of individual inductions. To each of his varied and important duties he brought a high order of ability and remarkable powers of application. His services were of great value in raising our National Army.

GUNBY, FRANK M_____
R—Boston, Mass.
B—Charleston, S. C.
G. O. No. 103, W. D., 1919.

Colonel, Quartermaster Corps, U. S. Army.
As officer in charge of the engineering branch of the construction division of the Army. The success of the engineering features of the Army building program is in large measure due to Colonel Gunby's genius for organization, his ability to judge men and inspire in them a determination to succeed. The services he rendered are of signal worth.

GURNEY, SAMUEL C_____
R—Detroit, Mich.
B—England.
G. O. No. 53, W. D., 1921.

Lieutenant colonel, Medical Corps, U. S. Army.
For services in evacuating and personally caring for the wounded of the 3d Division under heavy shell and machine-gun fire daily during the defensive and offensive operations on the Marne, and great devotion under shell fire in personally establishing battalion aid stations at Chartreves, Jaulgonne, and Le Charmel between July 14 and Aug. 1, 1918.

HAAN, WILLIAM G_____
R—Crown Point, Ind.
B—Crown Point, Ind.
G. O. No. 12, W. D., 1919.

Major general, U. S. Army.
This officer, in command of the 32d Division, took a prominent part in the Argonne-Meuse offensive and in the brilliant and successful attack against the Cote Dame Marie, covering several days, which deprived the enemy of the key point of the position. His clear conception of the tactical situations involved showed him to be a military leader of superior order.

HACKETT, HORATIO B_____
R—Abilene, Kans.
B—Chicago, Ill.
G. O. No. 56, W. D., 1921.

Colonel, Field Artillery, U. S. Army.
For services while commanding the 124th Field Artillery, 33d Division, during the St. Mihiel and Meuse-Argonne offensives until he was severely wounded.

HAGERLING, SIDNEY A_____
R—Pittsburgh, Pa.
B—Pittsburgh, Pa.
G. O. No. 38, W. D., 1922.

Lieutenant colonel, Signal Corps, U. S. Army.
As division signal officer, 28th Division, during the operations of his division on the Vesle in September, 1918, and later during the Meuse-Argonne offensive, Colonel Hagerling by his complete knowledge of his duties, his ceaseless energy, his initiative, and his devotion to duty, regardless of extreme difficulties of terrain and heavy shell fire by the enemy in open-warfare situations, maintained the telephone net of the division intact at all times, thereby enabling the division commander to communicate with all elements of the division, even the front line, and assisting very materially in the control and conduct of the division and in bringing its operations to a very successful conclusion.

BAGOOD, JOHNSON_____
R—Columbia, S. C.
B—Orangeburg, S. C.
G. O. No. 12, W. D., 1919.

Brigadier general, U. S. Army.
As chief of staff of the services of supply of the American Expeditionary Forces in France his ability for organization, his energy, and his sound judgment were factors in the efficiency of this important branch. By his marked zeal and aggressiveness he greatly added to the successful administrations of the services of supply.

HALE, HARRY C_____
R—Galesburg, Ill.
B—Knoxville, Ill.
G. O. No. 38, W. D., 1922.

Major general, U. S. Army.
While in command of the 84th Division during its organization and training in the United States and after the armistice in command of the 26th Division in France, by his ceaseless energy and the closest personal supervision of the training, discipline, and supply of his commands, he displayed rare qualities of leadership, organization, tact, and judgment. His brilliant professional attainments, his steadfast devotion to duty, and his loyalty to superiors were reflected in the high standards maintained throughout the divisions under his command, and he thus rendered important services to the American Expeditionary Forces and contributed conspicuously to the success of the operations.

HALE, RICHARD K_____
R—Brookline, Mass.
B—Boston, Mass.
G. O. No. 56, W. D., 1922.

Colonel (Field Artillery), General Staff Corps, U. S. Army.
As assistant chief of staff, G-1, 2d Army Corps, from March, 1918, until April, 1919, he displayed exceptional ability in the organization and administration of that division of the corps staff. With rare tact he assisted in the establishment of most cordial relations with the British organizations with which the corps was serving. He showed excellent judgment and great administrative ability in the handling of important questions in the arrangements for the service of American troops with the British.

HALL, ELBERT J_____
R—Oakland, Calif.
B—San Jose, Calif.
G. O. No. 69, W. D., 1919.

Lieutenant colonel, Air Service, U. S. Army.
In the designing of the Liberty engine and subsequently in the adapting of the Le Rhone engine to the American methods of production and also in pushing to completion the American adaptation of the De Haviland plane.

102444°—27——46

HALL, HARRISON_____
 R—Dayton, Ohio.
 B—Dayton, Ohio.
 G. O. No. 31, W. D., 1922.

Colonel, Field Artillery, U. S. Army.
In organizing and assisting in the conduct of training camps in the United States. Later, as chief of staff of the American Embarkation Center at Le Mans and as chief of staff of the 1st Replacement Depot at St. Aignan, he assisted in the organization of a system of evacuation which insured a very satisfactory flow of personnel to ports of embarkation. In these capacities he displayed exceptional efficiency.

HALLORAN, PAUL S_____
 R—Dayton, Ohio.
 B—Camp Wright, Calif.
 G. O. No. 103, W. D., 1919.

Colonel, Medical Corps, U. S. Army.
He served with great credit as division surgeon of the 90th Division from the date of its organization throughout its service in the field, displaying sound judgment, marked professional skill, and untiring energy. By enforcing effective sanitary measures he maintained the combat strength of his division, and by his able direction of the medical services he was largely responsible for the proper care of the sick and wounded.

HALSEY, CHARLES W_____
 R—New York, N. Y.
 B—Newark, N. J.
 G. O. No. 53, W. D., 1921.

Major, Quartermaster Corps, U. S. Army.
As deputy of the chief quartermaster, in the service of the 4th section of the General Staff at General Headquarters, a position of great responsibility, in which he rendered services of marked worth in all details relative to the Quartermaster Corps.

HALSTEAD, ALEXANDER SEAMAN_____
 R—Philadelphia, Pa.
 B—Philadelphia, Pa.
 G. O. No. 30, W. D., 1921.

Rear admiral, U. S. Navy.
While in command of the United States naval forces at Brest, France, by his superior administration, sound judgment, and splendid cooperation he contributed materially to the success achieved by the Army at the port of Brest in the return to the United States of the American Expeditionary Forces.

HALSTEAD, LAURENCE_____
 R—Riverside, Ohio.
 B—Riverside, Ohio.
 G. O. No. 49, W. D., 1922.

Colonel, Infantry, U. S. Army.
As officer in charge of quartermaster schools and assistant to officer in charge of administrative division, office of the Quartermaster General, from April to August, 1917, he rendered valuable service in the organization and operation of these schools. As chief of staff, 84th Division, from August, 1917, to November, 1918, by his marked efficiency, loyal devotion to duty, and high military attainments, he played an important part in the successful organization, training, and operations of that division. Later as assistant chief of staff, G-3, 1st Army, from November, 1918, to April, 1919, he performed many tasks of great responsibility in a highly meritorious manner.

HALYBURTON, EDGAR M. (42848)_____
 R—Taylorsville, N. C.
 B—Stony Point, N. C.
 G. O. No. 72, W. D., 1920.

Sergeant, Company F, 16th Infantry, 1st Division, U. S. Army.
Sergeant Halyburton, while a prisoner in the hands of the German Government from November, 1917, to November, 1918, voluntarily took command of the different camps in which he was located and under difficult conditions established administrative and personnel headquarters, organized the men into units, billeted them systematically, established sanitary regulations and made equitable distribution of supplies; he established an intelligence service to prevent our men giving information to the enemy and prevent the enemy introducing propaganda. His patriotism and leadership under trying conditions were an inspiration to his fellow prisoners and contributed greatly to the amelioration of their hardships.

HAM, SAMUEL V_____
 R—Warrington, Ind.
 B—Markleville, Ind.
 G. O. No. 13, W. D., 1923.
 Distinguished-service cross also
  awarded.

Colonel, Infantry, U. S. Army.
As colonel, 109th Infantry, 28th Division, he displayed in combat rare qualities of leadership, unusual tactical judgment, and devotion to duty. Severely wounded in action, he returned to duty as commanding officer of troops at Is-sur-Tille, a position of great responsibility, in which he acquitted himself with great credit, contributing materially to the proper functioning of the services of supply, American Expeditionary Forces.

HAMBLETON, THOMAS EDWARD_____
 R—Lutherville, Md.
 B—Baltimore, Md.
 G. O. No. 59, W. D., 1919.

Colonel, Adjutant General's Department, U. S. Army.
He displayed unusual skill and untiring zeal in organizing and administering the statistical division of The Adjutant General's Office. With no precedent to guide or assist him, he showed marked initiative in this most difficult task, creating a wonderful record of achievement, which is a tribute to his ability and clear-sightedness amid a maze of details. Self-sacrificing in his devotion to duty, he achieved excellent results in all his endeavors.

HAMILTON, GEORGE L_____
 R—Boston, Mass.
 B—Covington, Ky.
 G. O. No. 47, W. D., 1921.

Lieutenant colonel, Postal Express Service, U. S. Army.
For services in organizing and administering the motor dispatch service of the American Expeditionary Forces.

HAMILTON, WESLEY W. K_____
 R—Dayton, Ohio.
 B—Cincinnati, Ohio.
 G. O. No. 27, W. D., 1922.

Colonel, Coast Artillery Corps, U. S. Army.
As adjutant, Base Section No. 3, in 1917–1918, in the initiation, development, extension, and administration of those headquarters, through whose agency vast supplies and over a million troops were forwarded to France, he rendered a service of great value.

HAMMOND, THOMAS W_____
 R—Ashland, Oreg.
 B—Ashland, Oreg.
 G. O. No. 15, W. D., 1921.

Colonel (Infantry), General Staff Corps, U. S. Army.
While a member of the General Staff in the early days of the war his judgment and ability were applied to the solution of intricate problems concerning the distribution of the draft. He rendered meritorious service in France both as a line and as a staff officer. The services rendered by him, pertaining to the preparation and development of the reorganization act of June 4, 1920, have been of great value to the Army.

HAND, DANIEL W_____
 R—Oakmont, Pa.
 B—St. Paul, Minn.
 G. O. No. 19, W. D., 1921.

Lieutenant colonel, Field Artillery, U. S. Army.
For services as assistant director and director of the department of firing in the School of Fire for Field Artillery, Fort Sill, Okla., from October, 1917, to October, 1918.

HANNER, JOHN W
R—Franklin, Tenn.
B—Franklin, Tenn.
G. O. No. 89, W. D., 1919.

Colonel, Medical Corps, U. S. Army.
As commanding officer of Evacuation Hospital No. 1 he displayed high professional attainments and loyal devotion to duty. Subsequently, as chief surgeon, 4th Army Corps, by his able supervision of the medical and sanitary units under his direction he rendered invaluable services in connection with the care of many sick and wounded.

HANNUM, WARREN T
R—Pottsville, Pa.
B—Pottsville, Pa.
G. O. No. 89, W. D., 1919.

Colonel, Corps of Engineers, U. S. Army.
As a member of the training section, General Staff, he efficiently supervised the technical and tactical training of engineer, gas, and tank troops and the operation of the schools for those services. In the performance of his manifold duties he displayed military attainments of a high order, rendering service of importance to the American Expeditionary Forces.

HANSON, MARCUS H
R—San Antonio, Tex.
B—Stow, Mass.
G. O. No. 56, W. D., 1921.

Lieutenant colonel, Quartermaster Corps, U. S. Army.
As assistant to the Chief Quartermaster, American Expeditionary Forces, in charge of the administrative division, a position of great responsibility, in which he rendered services of conspicuous worth.

HARBORD, JAMES G
R—Council Grove, Kans.
B—Bloomington, Ill.
G. O. No. 136, W. D., 1918.

Major general, U. S. Army.
As chief of staff of the American Expeditionary Forces, and later as commanding general, services of supply, in both of which important positions his great constructive ability and professional attainments have played an important part in the success obtained by our armies. Commanded Marine Brigade of 2d Division, Belleau Wood, and later ably commanded 2d Division during attack on Soissons, France, July 18, 1918.

HARDEMAN, LETCHER
R—Grays Summit, Mo.
B—Arrow Rock, Mo.
G. O. No. 38, W. D., 1922.

Colonel, Quartermaster Corps, U. S. Army.
From May 30, 1917, until Oct. 10, 1918, as principal assistant in the supply office of the Quartermaster General, he showed great executive ability, excellent judgment, and a rare understanding of supply problems. On Oct. 10, 1918, during the period of urgent demand for animals so necessary to our overseas forces, he became chief of the remount service, and by his knowledge, efficiency, and broad experience he was able to organize and perfect the system for the purchase, collection, and shipment of large numbers of animals to supply these demands. Immediately after the signing of the armistice, he again rendered conspicuous services by instituting the method by which 200,000 surplus animals were promptly and efficiently disposed of, resulting in the saving of an enormous sum to the Government.

HARDING, JOHN, Jr
R—Dayton, Ohio.
B—Nashville, Tenn.
G. O. No. 14, W. D., 1925.
Act of Congress Feb. 25, 1925.

Second lieutenant, Air Service Reserve, U. S. Army.
Lieutenant Harding, as assistant pilot of airplane No. 4, the New Orleans, and assistant engineer officer of the United States Army Air Service around-the-world flight from Apr. 6, 1924, to Sept. 28, 1924, displayed sound technical skill, a high spirit of cooperation, initiative, energy, and resourcefulness. His indefatigable energy, good judgment, and personal courage contributed largely to the success of this pioneer flight of airplanes around the world. In the performance of his arduous duties he aided in the accomplishment of an undertaking bringing great credit to himself and to the Military Establishment of the United States.

HARJES, HENRY H
R—New York, N. Y.
B—Paris, France.
G. O. No. 59, W. D., 1919.

Lieutenant colonel, Infantry, U. S. Army.
As chief liaison officer of the American Expeditionary Forces he rendered most valuable and important service in establishing and maintaining cordial relations between the French and American authorities. His efforts materially furthered that deep feeling of understanding which marks the association of the allied Armies.

HARMON, KENNETH B
R—Altoona, Pa.
B—San Francisco, Calif.
G. O. No. 74, W. D., 1919.

Lieutenant colonel, Ordnance Department, U. S. Army.
With exceptionally sound judgment and marked initiative, he displayed a wide comprehension of existing conditions, solving perplexing problems connected with the establishment and operation of the storage system of the Ordnance Department of the American Expeditionary Forces. He opened first a base, then an intermediate depot, and later an advance depot, accomplishing these tasks in spite of numerous obstacles. At all times tireless in energy he worked to insure an adequate supply of ordnance matériel for the troops at the front.

HARRELL, WILLIAM F
R—Marion, S. C.
B—Marion, S. C.
G. O. No. 59, W. D., 1919.
Distinguished-service cross also awarded.

Colonel, Infantry, U. S. Army.
He served through all operations of the 1st Division in this war, and at all times was conspicuous for his courage, judgment, and leadership. As battalion and regimental commander, he distinguished himself by his exceptionally energetic and efficient command of his units. During the rapid advance of the 1st Division upon Sedan he carried out a most difficult mission of the division in that he successfully covered its right flank in a night march of about 20 kilometers, across broken country, in the face of the enemy. Herein he exhibited the qualities of a most able commander.

HARRIES, GEORGE H
R—Washington, D. C.
B—South Wales.
G. O. No. 103, W. D., 1919.

Brigadier general, U. S. Army.
As commanding general of Base Section No. 5, he successfully directed the manifold activities at the port of Brest during the time when troop arrivals were at their maximum. He overcame seemingly insurmountable obstacles in coordinating and organizing his important task. Subsequently, upon being sent on a special mission to Berlin in connection with the repatriation of allied prisoners of war, he displayed commendable tact and energy.

HARRIS, CHARLES T., Jr ....
   R—Mexia, Tex.
   B—Mexia, Tex.
   G. O. No. 77, W. D., 1919.

Colonel, Ordnance Department, U. S. Army.
Chief of the American Mission of Powder and Explosive Manufacturers which visited England. Later, as chief of the powder and explosive section of the Engineering Division of the Office of the Chief of Ordnance, he applied foreign methods of manufacture to United States industry so successfully that not only were the needs of the United States fully met, but a considerable surplus of these materials was rendered available for the cobelligerents against Germany.

HARRIS, PETER C ....
   R—Cedartown, Ga.
   B—Kingston, Ga.
   G. O. No. 25, W. D., 1919.

Major general, The Adjutant General, U. S. Army.
During his service in the Adjutant General's Department, his zeal, energy and judgment have been made manifest by the reforms accomplished in record keeping systems in the War Department and in the Army.

HARRISON, CHARLES L ....
   R—Cincinnati, Ohio.
   B—Cincinnati, Ohio.
   G. O. No. 124, W. D., 1919.

As chief of the Cincinnati ordnance district, in which capacity he maintained at all times the greatest degree of intelligent and enthusiastic cooperation between the Ordnance Department and manufacturers in his district, thereby attaining the maximum production of munitions in a minimum time; and also as chairman of the Cincinnati ordnance district claims board, in which capacity his services have been invaluable to the Nation in adjusting equitably the $153,000,000 worth of outstanding contracts in his district in force at the signing of the armistice.

HART, WILLIAM H ....
   R—Bath, S. Dak.
   B—Winona, Minn.
   G. O. No. 59, W. D., 1919.

Colonel, Quartermaster Corps, U. S. Army.
While serving as quartermaster, Base Section No. 1, by his thorough knowledge of methods and standards of supplying troops, his resourcefulness, and comprehensive study of the innumerable details of the largest and most important supply bases in France, he executed the important duties with which he was intrusted in a highly satisfactory and especially efficient manner.

HARTE, RICHARD H ....
   R—Philadelphia, Pa.
   B—Rock Island, Ill.
   G. O. No. 9, W. D., 1923.

Colonel, Medical Corps, U. S. Army.
From May, 1917, until May, 1919, as director of professional services and later as commanding officer, Base Hospital No. 10, surgical consultant, American Expeditionary Forces, member of the Interallied Medical and Surgical Conference, and one of the pioneer instructors in the principles of battle surgery, by his high professional attainments, keen foresight, and untiring energy he solved successfully many problems which confronted him, rendering services of the greatest value to the American Expeditionary Forces.

HARTMAN, CHARLES D ....
   R—Brookhaven, Miss.
   B—Brookhaven, Miss.
   G. O. No. 14, W. D., 1923.

Colonel, Quartermaster Corps, U. S. Army.
As chief of the utilities section of the construction division, Quartermaster Corps, he was charged with the direction of the manifold activities involved in the maintenance of all fixed properties of the War Department, United States Army cantonments, including buildings and roads, the operation of all utilities, including water, sewage plants and systems, lighting, heating, refrigeration, and power plants throughout the United States. The untiring energy, sound professional judgment, administrative and executive ability displayed by Colonel Hartman contributed in a signal way to the successful operations of the Quartermaster Corps during the World War.

HARTMANN, EDWARD T ....
   R—Milwaukee, Wis.
   B—Chicago, Ill.
   G. O. No. 89, W. D., 1919.

Colonel, Infantry, U. S. Army.
He organized the 357th Infantry, 90th division, and commanded it with extraordinary ability during its training period and throughout its active operations. To his energy, zeal, and high qualities of leadership were largely due the consistently high standards of efficiency maintained in his regiment and the successes which it achieved in the St. Mihiel and Meuse-Argonne offensives.

HARTS, WILLIAM W ....
   R—Springfield, Ill.
   B—Springfield, Ill.
   G. O. No. 89, W. D., 1919.

Brigadier general, U. S. Army.
In command of the important District of Paris, by his painstaking efforts and able directorship he maintained a high standard of discipline and efficiency among his large command. By his tact and keen perception he handled numerous diplomatic affairs with great satisfaction, rendering services of a superior value to the American Expeditionary Forces.

HARTSHORN, EDWIN S ....
   R—New York, N. Y.
   B—Troy, N. Y.
   G. O. No. 98, W. D., 1919.

Colonel (Infantry), General Staff Corps, U. S. Army.
As chief of the coordination section, office of the executive assistant to the Chief of Staff, his energy, judgment, and foresight have been of exceptional value to the War Department and to the Army.

HARVEY, ALVA L. (6203842) ....
   R—Cleburne, Tex.
   B—Cleburne, Tex.
   G. O. No. 14, W. D., 1925.
   Act of Congress Feb. 25, 1925.

Staff sergeant, Air Service, U. S. Army.
Sergeant Harvey displayed unusual judgment, technical knowledge, and initiative in the preparation of the airplanes for the United States Army Air Service around-the-world flight and as mechanician and assistant pilot of airplane No. 1, the Seattle, from Apr. 6, 1924, until Apr. 30, 1924, when, due to an accident which resulted in the complete wreck of the airplane, he was forced to abandon the flight. His foresight, perseverance, and mechanical ability were very material factors in contributing to the successful accomplishment of this pioneer flight of airplanes around the world. In the performance of this great task he aided in bringing great credit to the military forces of the United States.

HASE, WILLIAM F ....
   R—Milwaukee, Wis.
   B—Milwaukee, Wis.
   G. O. No. 69, W. D., 1919.

Colonel, Coast Artillery Corps, U. S. Army.
For services as senior assistant to the Chief of Coast Artillery in the preparation and execution of plans for the effective accomplishment of the duties assigned to the Coast Artillery Corps in the operations in France.

HASKELL, WILLIAM N. R—Albany, N. Y. B—Bath-on-the-Hudson, N. Y. G. O. No. 59, W. D., 1919.	Colonel (Field Artillery), General Staff Corps, U. S. Army. He exhibited devotion, skill, and untiring energy as chief of the operations section, 4th Army Corps, during its organization and in the St. Mihiel offensive. As chief of the operations section, 2d Army, he rendered exceptionally meritorious service during the organization of that Army and in the operations north of Toul, October and November, 1918.
HATCH, HENRY J. R—Ionia, Mich. B—Charlotte, Mich. G. O. No. 19, W. D., 1922.	Brigadier general, U. S. Army. As Chief of Heavy Artillery Section, office of the Chief of Artillery of the American Expeditionary Forces in France, a position involving individual and independent responsibility, he performed services of inestimable value in connection with the organization, equipment, and training of the Artillery troops in France.
HATHAWAY, LEVY M. R—Owensboro, Ky. B—Owensboro, Ky. G. O. No. 56, W. D., 1922.	Colonel, Medical Corps, U. S. Army. As surgeon, 33d Division, throughout its organization, training, and combat operations, by his devotion to duty, untiring energy, and high professional attainments, he rendered conspicuous service, maintaining at all times a remarkable health record in the division. His handling of the wounded, involving personal exposure to heavy enemy fire during his daily inspections of the advanced dressing stations, was notable for its extraordinary efficiency. Later, as chief surgeon, 9th Army Corps, he again rendered highly meritorious service to the American Expeditionary Forces.
HAY, WILLIAM H. R—Drifton, Fla. B—Monticello, Fla. G. O. No. 89, W. D., 1919.	Major general, U. S. Army. As commander of the 184th Infantry Brigade he showed efficient leadership. Promoted to major generalship in the early part of October, 1918, he took command of the 28th Division, and by his marked ability and great energy he contributed to the successes attained by the division during the time in which he was in command. He rendered services of a high character to the American Expeditionary Forces.
HAYES, EDWARD S. R—Waterbury, Conn. B—Waterbury, Conn. G. O. No. 9, W. D., 1923. Distinguished-service cross also awarded.	Lieutenant colonel (Infantry), General Staff Corps, U. S. Army. As assistant chief of staff, G-3, 78th Division, in the St. Mihiel and Meuse-Argonne offensives, he was an important factor in the success of the division and rendered exceptional services by the persistence and courage of his front line reconnaissances; by his originality and the soundness of his plans of relief and attack; and by the clearness and correctness of all orders written by him for the operations of the division. His work was creative and constructive to a high degree and rendered in a highly meritorious manner in a position of great responsibility.
HAYES, GEORGE B. R—Paris, France. B—Canandaigua, N. Y. G. O. No. 72, W. D., 1920.	An eminent dental surgeon who placed freely the advantages of his professional attainments and the full facilities of his complete clinic at Paris at the services of the American medical personnel. By the markedly distinguished record made by him in jaw and facial surgery among the wounded of the American Expeditionary Forces, and his able directorate of the school for instruction of dental personnel in maxillo facial and prosthetic surgery, he has rendered services of preeminent worth.
HAYES, JAMES H. R—Atlantic City, N. J. B—Haddon Field, N. C. G. O. No. 49, W. D., 1922.	Lieutenant colonel, Judge Advocate General's Department, U. S. Army. As judge advocate of the 2d Division during its organization in October, 1917, and throughout all its combat operations, he handled with unusual merit many questions confronting him. By his high order of legal ability, broad vision, excellent judgment, and a most comprehensive knowledge of discipline and morale he rendered services of exceptional value to the American Expeditionary Forces.
HAYWARD, WILLIAM. R—New York, N. Y. B—Nebraska City, Nebr. G. O. No. 50, W. D., 1919.	Colonel, Infantry, U. S. Army. As commander of a regiment that was detached from the American Expeditionary Forces and served continuously with a French division, he was charged with particularly responsible and exacting duties, in the performance of which he at all times displayed commendable tact, personal bravery, and military leadership of a high order.
HEARN, CLINT C. R—Whitesboro, Tex. B—Weston, Tex. G. O. No. 59, W. D., 1921.	Brigadier general, U. S. Army. For services while in command of the 153d Field Artillery Brigade, 78th Division, during the Meuse-Argonne offensive.
HECKEL, EDWARD G. R—Detroit, Mich. B—Menasha, Wis. G. O. No. 56, W. D., 1922.	Colonel, Infantry, U. S. Army. He served with the 125th Infantry, 32d Division, as lieutenant colonel from the organization until September, 1918, and as colonel from then until June, 1919, displaying at all times military attainments of a high order. By his initiative, force, marked ability, and untiring devotion to duty he contributed materially to the successes of his division in four major operations against the enemy, thereby rendering services of inestimable value to the American Expeditionary Forces.
HECKMAN, JAMES C. R—Buffalo, N. Y. B—Phillipsburg, N. J. G. O. No. 9, W. D., 1923.	Colonel, Ordnance Department, U. S. Army. As chief, storage operations section, supply division, Office of the Chief of Ordnance, and later as chief of supply division, he assisted in formulating the general scheme of ordnance storage depots and planned and operated with marked success those assigned to the Ordnance Department. His sound judgment, wide business experience, and administrative ability contributed in a material way to the successful operations of the Ordnance Department.

HEGEMAN, HARRY A.
R—Brookings, S. D.
B—Sparta, Wis.
G. O. No. 59, W. D., 1919.

Colonel, Infantry, U. S. Army.
With technical skill and great energy he organized a large force of trained workmen for the repair of motor transports. He restored to service a great mass of accumulated dead transportation of all kinds and types, and kept in operation much transportation by timely repair. By his untiring efforts the motor transportation was maintained at such a standard as to become an important factor in the successes achieved by the American troops.

HEINTZELMAN, STUART.
R—Washington, D. C.
B—New York, N. Y.
G. O. No. 12, W. D., 1919.

Brigadier general, U. S. Army.
He organized the headquarters of the 4th Army Corps and later, as chief of staff of this corps, directed with great success the staff of this organization prior to and during the St. Mihiel offensive. As chief of staff of the 2d Army, he had a prominent part in organizing it as a fighting unit. His tact, energy, and military ability were important elements in the success of this command.

HELMICK, ELI A.
R—Weir City, Kans.
B—Quakers Point, Ind.
G. O. No. 95, W. D., 1919.

Major general, U. S. Army.
As commanding general, Base Section No. 5, he has displayed brilliant administrative ability in successfully directing the manifold activities under his supervision. By his energy in expediting the completion of the various engineering projects necessitated by the enlargement of Pontanezen Camp and the development of Brest as a foremost embarkation camp, he has rendered invaluable services to the American Expeditionary Forces.

HERBST, GEORGE A.
R—St. Paul, Minn.
B—St. Paul, Minn.
G. O. No. 15, W. D., 1923.

Colonel (Infantry), General Staff Corps, U. S. Army.
As assistant chief of staff, G-3, 2d Division, from Dec. 28, 1917, to Aug. 3, 1918, and assistant chief of staff, G-2, 2d Division, Aug. 4 to Sept. 19, 1918, G-2 section, General Headquarters, Sept. 20 to Oct. 2, 1918, assistant chief of staff, G-2, 7th Army Corps, Oct. 3, 1918, to Nov. 16, 1918, he displayed untiring energy, sound professional judgment, and devotion to duty. He rendered valuable services to the Government in positions of responsibility and contributed materially to the successful operations of the American Forces in France in actions against the enemy.

HERR, JOHN K.
R—Flemington, N. J.
B—White House Station, N. J.
G. O. No. 87, W. D., 1919.

Colonel, Cavalry, U. S. Army.
He showed marked ability as chief of staff of the 30th Division in the capture of Voormezeele and Lock Eight in the Ypres section in Belgium in September, 1918, and in the breaking of the Hindenburg line at Bellicourt, France, and the operations against the Selle River and the Sambre Canal, Sept. 29–Oct. 20, 1918. By his energy, zeal, and persistent efforts, coupled with sound tactical judgment, he materially contributed to the success of the operations.

HERRINGSHAW, WILLIAM F.
R—Cleveland, Ohio.
B—Cleveland, Ohio.
G. O. No. 43, W. D., 1922.

Colonel, Motor Transport Corps, U. S. Army.
As chief motor transport officer, 1st Army, American Expeditionary Forces, he at all times displayed initiative and marked ability. By his high military attainments, sound judgment, and untiring efforts in his tasks he successfully solved the many complex problems involving the supply, conservation, and repair of the motor transportation of the 1st Army during its operations in the St. Mihiel and Meuse-Argonne offensives. He has rendered services of signal worth to the American Expeditionary Forces.

HERRON, CHARLES D.
R—Crawfordsville, Ind.
B—Crawfordsville, Ind.
G. O. No. 59, W. D., 1921.

Colonel (Field Artillery), General Staff Corps, U. S. Army.
For services as chief of staff, 78th Division, during the Meuse-Argonne offensive.

HERSEY, MARK L.
R—East Corinth, Me.
B—Stetson, Me.
G. O. No. 62, W. D., 1919.

Major general, U. S. Army.
As a brigade commander during the latter part of the Meuse-Argonne operation he exhibited qualities of excellent leadership and sound judgment. His brigade attacked and penetrated the strong enemy position of Bois-des Loges and wrested this strong point from the enemy. The success of his brigade in this engagement was in a large measure due to his able leadership. Later he commanded with distinction the 4th Division during its operations in the occupied territory.

°HETRICK, HAROLD S.
R—Canterbury, Conn.
B—Kansas City, Mo.
G. O. No. 56, W. D., 1922.

Colonel (Corps of Engineers), General Staff Corps, U. S. Army.
As assistant chief of staff, G-4, 2d Army Corps, from Feb. 24, 1918, to Aug. 31, 1918, he displayed exceptional ability in the organization and administration of that division of the corps staff. In the supply and equipment of organizations of the corps under the British system and with British matériel he exhibited military attainments of a high order and contributed in a marked degree to the successful preparation for subsequent operations.
Posthumously awarded. Medal presented to uncle, Andrew J. Clarke.

HICKMAN, EDWIN A.
R—Lexington, Mo.
B—Independence, Mo.
G. O. No. 15, W. D., 1921.

Colonel (Infantry), General Staff Corps, U. S. Army.
For services as chief of a section of the General Staff which had charge of estimates and financial matters pertaining to the conduct of the war and the support of the Army.

HILDERBRAND, JOEL H.
R—Berkeley, Calif.
B—Camden, N. J.
G. O. No. 56, W. D., 1922.

Lieutenant colonel, Chemical Warfare Service, U. S. Army.
As commandant of the Chemical Warfare Service Experimental Field (Hanlen Field), American Expeditionary Forces, a position of great responsibility and also of considerable personal danger, his profound knowledge of chemistry, coupled with his rapid grasp of military problems, enabled him to render services of the utmost value in determining the best means for using gas and gas materials in the field.

HILGARD, MILOSH R. _____
R—Belleville, Ill.
B—New York, N. Y.
G. O. No. 62, W. D., 1919.

Colonel, Quartermaster Corps, U. S. Army.
He organized the operations of the Quartermaster Corps at the important bases of St. Nazaire and Bordeaux, and later established and operated the first American regulating station, through which he successfully supplied a great number of American troops serving in the zone of the Armies and operating at the front. The successful operation of this great station was due directly to his painstaking efforts, his zeal, and great energy.

HILL, ARTHUR D. _____
R—Boston, Mass.
B—France.
G. O. No. 59, W. D., 1921.

Lieutenant colonel, Judge Advocate General's Department, U. S. Army.
As assistant judge advocate, services of supply, and assistant chief finance officer, American Expeditionary Forces, positions of great responsibility, he demonstrated marked ability in performing the various tasks assigned to him. He performed important services for the rents, requisitions, and claims service, and rendered valuable assistance in dealings with the representatives of the French Government in all matters of claims in France.

HILL, JOHN PHILIP. _____
R—Baltimore, Md.
B—Annapolis, Md.
G. O. No. 43, W. D., 1922.

Lieutenant colonel, Judge Advocate General's Department, U. S. Army.
As judge advocate of the 29th Division from August, 1917, until December, 1918, and of the 8th Army Corps from December, 1918, until April, 1919, his marked legal ability, sound judgment, and tireless energy were important factors in the splendid work of his department. Representing his division as liaison officer at headquarters, 17th Army Corps (French), in October, 1918, during the Meuse-Argonne offensive, by his tact and constant devotion to duty he rendered conspicuous services in this important operation.

HINDS, ERNEST. _____
R—New Hope, Ala.
B—Red Hill, Ala.
G. O. No. 12, W. D., 1919.

Major general, U. S. Army.
As Chief of Artillery, 1st Army Corps, commanding general, Army Artillery, of the 1st American Army, and as Chief of Artillery, American Expeditionary Forces, he perfected and successfully directed the organization and training of the Artillery of the American Army in France.

HINES, FRANK T. _____
R—Salt Lake City, Utah.
B—Salt Lake City, Utah.
G. O. No. 144, W. D., 1918.

Brigadier general, U. S. Army.
For services as chief of embarkation in organizing and administering the embarkation service during the war.

HINES, JOHN FORE. _____
R—Bowling Green, Ky.
B—Kentucky.
G. O. No. 116, W. D., 1919.

Captain, U. S. Navy.
For services as chief of staff of the commander, cruiser and transport fleet, Newport News division.

HINES, JOHN L. _____
R—White Sulphur Springs, W. Va.
B—White Sulphur Springs, W. Va.
G. O. No. 12, W. D., 1919.
Distinguished-service cross also awarded.

Major general, U. S. Army.
As regimental, brigade, division, and corps commander he displayed marked ability in each of the important duties with which he was intrusted and exhibited in the operations near Montdidier and Soissons and in the St. Mihiel and Argonne-Meuse offensives his high attainments as a soldier and a commander.

HINKLE, CHARLES L. _____
R—Frankfort, Ind.
B—Pleasantville, Ind.
G. O. No. 47, W. D., 1921.

Lieutenant colonel, Transportation Corps, U. S. Army.
For services in connection with the railroad operations of the United States Transportation Corps in France.

HITT, PARKER. _____
R—Indianapolis, Ind.
B—Indianapolis, Ind.
G. O. No. 59, W. D., 1919.

Colonel, Signal Corps, U. S. Army.
By his sound judgment and untiring efforts he assisted in perfecting the satisfactory organization of the Signal Corps of the American Expeditionary Forces, and he displayed conspicuous merit in his capacity as signal officer of the 1st American Army.

HODGE, HENRY W. _____
R—New York, N. Y.
B—Washington, D. C.
G. O. No. 14, W. D., 1923.

Colonel, Corps of Engineers, U. S. Army.
As general manager of roadways, American Expeditionary Forces, he displayed sound professional judgment, technical skill, untiring energy, and devotion to duty, thus contributing in a marked degree to the successes of the American forces in France.

HODGE, S. H. _____
R—Chicago, Ill.
B—Murfreesboro, Tenn.
G. O. No. 2, W. D., 1920.

Placed in charge of storage procurement for the Army, he rendered invaluable service in obtaining for the use of the Army an enormous amount of storage space at a cost averaging less than the commercial rate, thereby effecting a very substantial saving and enabling the storage division to meet a most pressing need for space required to mobilize and preserve the immense quantities of supplies gathered for the use of the Army. By his technical knowledge, broad judgment, and energetic action valuable results were obtained.

HODGES, CAMPBELL B. _____
R—Ruston, La.
B—Youngs Point, La.
G. O. No. 27, W. D., 1922.

Colonel (Infantry), General Staff Corps, U. S. Army.
As acting chief of staff, 31st Division, he demonstrated the highest professional attainments, and through his zeal and never-failing tact he was responsible in a large measure for the development of the high efficiency of that division. Later, as a member of the War Department General Staff, he rendered most valuable service in the development of the present system of efficiency reporting and in the development of the system of classification of commissioned personnel now in use.

*HODGES, GEORGE. _____
R—Baltimore, Md.
B—Newark, N. J.
G. O. No. 70, W. D., 1919.

Manager of the troop movement section of the division of operations, U. S. Railroad Administration.
Mr. Hodges arranged all the details of the movement of troops from local draft boards to mobilization camps, between camps, or from mobilization camps to the ports of embarkation for shipment overseas. Troops in large numbers were moved on short notice, and he was responsible for the successful coordination and carrying out of these movements.
Posthumously awarded. Medal presented to widow, Mrs. George Hodges.

HODGES, HARRY F. R—Lincoln, Mass. B—Boston, Mass. G. O. No. 15, W. D., 1923.	Major general, U. S. Army. As commanding general, Camp Devens, Mass., he displayed unusual adminis- trative and executive ability, sound judgment, and high professional skill. He established a model system of schools and training, organized and trained the 76th Division, and in addition thereto trained for overseas service more than 40,000 men of other units. His untiring energy, devotion to duty, coupled with other outstanding soldierly qualities, contributed markedly to the successful operations of the American forces during the World War.
*HODGES, HARRY L. R—Norfolk, Va. B—Norfolk, Va. G. O. No. 3, W. D., 1922.	Lieutenant colonel (Field Artillery), General Staff Corps, U. S. Army. As chief of staff of the military board of allied supply Colonel Hodges provided important liaison between the American member of the board and the General Staff, General Headquarters. His duties, which were varied and complex, were ably performed, often under great difficulties and embarrassment, and they continued to include the responsibility of gathering the data and pre- senting the report on the supply systems of the allied armies and in the pre- sentation of this completed report, which finally received the approval of his own Government. Posthumously awarded. Medal presented to widow, Mrs. Harry L. Hodges.
HODGES, JOHN N. R—Baltimore, Md. B—Baltimore, Md. G. O. No. 9, W. D., 1923.	Lieutenant colonel, Corps of Engineers, U. S. Army. While in command of the 6th Engineers, 3d Division, attached to the British Expeditionary Forces, from Feb. 11 to Apr. 1, 1918, during which time the regiment was engaged in the construction of heavy bridges on the Somme and on fortification and combat duty, because of his high professional ability, his great energy, and devotion to duty he contributed materially to the accomplishments of the forces engaged in those operations.
°HODGSON, FREDERICK O. R—Athens, Ga. B—Athens, Ga. G. O. No. 38, W. D., 1922.	Colonel, Quartermaster Corps, U. S. Army. While serving as representative of the Quartermaster Corps on the general munitions board of the Council of National Defense from April, 1917, until his death, Aug. 5, 1917, he displayed most distinguished ability and performed his manifold duties in a most conspicuous manner in a position of great respon- sibility at a time of gravest importance. By his tact, foresight, and excellent judgment his services in connection with the development of the vast program of housing and supply for our Army were of material assistance to the success- ful prosecution of the war and were of signal worth to the Government. Posthumously awarded. Medal presented to widow, Mrs. Ida C. Hodgson.
HOF, SAMUEL. R—Boscobel, Wis. B—Boscobel, Wis. G. O. No. 15, W. D., 1923.	Colonel, Ordnance Department, U. S. Army. First, as commanding officer, Frankford Arsenal, from March 1918, to March, 1919, where, by his indefatigable energy, outstanding administrative ability, and thorough technical knowledge, he brought to a successful production basis tracer, incendiary, and armor-piercing small-arms ammunition, and supplied substantially all that was used by our troops; later, as acting chair- man of the ordnance claims board, where, by his energy, tact, and business ability, he secured the settlement of outstanding obligations, and later, as chief of field service, Ordnance Department, where he perfected the organiza- tion controlling the disposition of vast quantities of materials and plants left over from the war.
HOFFMAN, GEORGE M. R—Wilkes-Barre, Pa. B—Wilkes-Barre, Pa. G. O. No. 74, W. D., 1919.	Colonel, Corps of Engineers, U. S. Army. As chief engineer of the 1st Army Corps, by his great energy and marked tech- nical ability he built up a strongly efficient organization, which made itself felt in all operations of the 1st Army Corps and in a great measure contributed to the successes achieved during the active operations of the 1st Army Corps at St. Mihiel and in the Argonne.
HOGAN, JOHN P. R—New York, N. Y. B—Chicago, Ill. G. O. No. 56, W. D., 1922.	Lieutenant colonel, Corps of Engineers, U. S. Army. As officer in charge of the topographical sections, G–2, of the 5th Army Corps and the 2d Army, he contributed materially toward the success of the opera- tions. Joining the 2d Army staff upon organization, he exercised such energy, judgment, and foresight that remarkable results were obtained in preparing maps for an offensive action and the subsequent occupation of the enemy's territory.
HOLABIRD, JOHN A. R—Chicago, Ill. B—Evanston, Ill. G. O. No. 59, W. D., 1921.	Lieutenant colonel, Field Artillery, U. S. Army. As commander of the 12th Field Artillery, 2d Division, during the period from August to November, 1918, a position of great responsibility, he commanded his unit during the offensives of St. Mihiel, Mont Blanc, and Meuse-Argonne with much credit. He has rendered important services to the United States.
HOLBROOK, LUCIUS R. R—Sargent, Minn. B—Arkansaw, Wis. G. O. No. 19, W. D., 1926.	Brigadier general, U. S. Army. As commander of the 7th Field Artillery and the 1st Field Artillery Brigade he with great distinction directed the artillery support of the 1st Division in the attacks on Cantigny and the Soissons salient. His careful judgment and high military attainments were shown in the accuracy and timeliness of the fire from the batteries under his direction, which, despite the difficulties in- volved, contributed materially to the success of the operations.
HOLBROOK, WILLARD A. R—Arkansaw, Wis. B—Arkansaw, Wis. G. O. No. 47, W. D., 1919.	Major general, U. S. Army. As commanding general, Southern Department, where his firmness and tact in handling a threatening situation on the Mexican border materially im- proved the conditions between the United States and Mexico.

HOLMAN, JESSE R.
  R—Comanche, Tex.
  B—Fayette County, Tex.
  G. O. No. 59, W. D., 1919.

Colonel, Corps of Engineers, U. S. Army.
In charge of general construction in the vicinity of Bordeaux he displayed unusual judgment and great executive ability in the performance of the many duties assigned to him. In addition he rendered valuable service and advice to the other departments of Base Section No. 2 regarding construction.

HOMAN, CHARLES C.
  R—Cleveland, Ohio.
  B—Williamsport, Pa.
  G. O. No. 14, W. D., 1923.

Lieutenant colonel, Motor Transport Corps, U. S. Army.
As assistant to the director, Motor Transport Corps, American Expeditionary Forces, he was charged with the assignment and distribution of motor transport to division, corps, and Army troops, the general supervision of motor transport within the zone of the services of supply, maintaining therein the difficult task of a balanced and efficient motor transport service. In these positions of grave responsibility he displayed unusual administrative and executive ability, high technical skill, and unremitting attention to duty, thus contributing markedly to the successful accomplishments of the American Expeditionary Forces.

HONOR, WILLIAM H.
  R—Wyandotte, Mich.
  B—Canada.
  G. O. No. 53, W. D., 1921.

Lieutenant colonel, Medical Corps, U. S. Army.
For services in organizing and operating the medical section of the labor bureau, American Expeditionary Forces.

HOPKINS, JAY P.
  R—Cassopolis, Mich.
  B—Mattawan, Mich.
  G. O. No. 62, W. D., 1919.

Colonel, Coast Artillery Corps, U. S. Army.
As chief of the Antiaircraft Artillery Service in the American Expeditionary Forces he performed arduous tasks with distinction, at all times being ceaseless in devotion to his important duties. Displaying marked scientific attainments, he handled perplexing problems with which the service was continually confronted with sound judgment, untiring energy, and a wide comprehension of the needs to be supplied and the facilities available.

HOPWOOD, LUCIUS L.
  R—Des Moines, Iowa.
  B—Vinton, Iowa.
  G. O. No. 59, W. D., 1921.

Lieutenant colonel, Medical Corps, U. S. Army.
In the hospitalization division of the chief surgeon's office, a position of great responsibility, he had charge of the equipment and installation of hospitals throughout the services of supply in France, and it was due to his ability and energy that these units were so efficiently provided and maintained. He has rendered services of value to the United States.

HORSEY, HAMILTON R.
  R—Tampa, Fla.
  B—Tallahassee, Fla.
  G. O. No. 56, W. D., 1922.

Lieutenant colonel, Infantry, U. S. Army.
As assistant chief of staff, G-2, 26th Division, during the St. Mihiel and Meuse-Argonne offensives, Colonel Horsey, by his complete grasp and knowledge of his duties, by his unfailing devotion to duty, and by his untiring energy, kept division headquarters at all times accurately and promptly informed as to the location of our own front lines and enemy troops, thus assisting materially in bringing the operations of the division to a successful conclusion.

HORTON, WILLIAM E.
  R—Washington, D. C.
  B—Washington, D. C.
  G. O. No. 53, W. D., 1921.

Colonel, Quartermaster Corps, U. S. Army.
As chief quartermaster of the advance section, services of supply, a position of great responsibility, due to his untiring energy, the supplying of many thousands of troops in this section was successfully carried out and numerous Quartermaster Corps services and activities were organized and expeditiously administered. He has rendered services of great worth.

HOUGH, BENSON W.
  R—Delaware, Ohio.
  B—Delaware County, Ohio.
  G. O. No. 47, W. D., 1921.

Colonel, Infantry, U. S. Army.
As regimental commander in the military operations of the 42d Division in the Baccarat Sector, Mar. 24 to June 21, 1918; the second Battle of the Marne, in which the 42d Division participated in the defense of the line east of Chalons, June 28 to July 21, 1918; and in the offensive against Reims, Chateau-Thierry, Soissons salient, July 24 to Aug. 3, 1918.

HOWARD, CARRIE L.
  R—San Francisco, Calif.
  B—Colusa, Calif.
  G. O. No. 9, W. D., 1923

First lieutenant, chief nurse, Army Nurse Corps, U. S. Army.
As chief nurse at the port of embarkation, Hoboken, N. J., during the World War, she held a position of great responsibility. To her fell the duty of supervising the nursing departments of all the hospitals at the port of embarkation and the mobilization stations for nurses destined for overseas duty. By her efficiency, energy, and knowledge of administrative detail she added greatly to the proficiency of the Medical Department of the Army at those trying stations.

HOWARD, DEANE C.
  R—Coleraine, Mass.
  B—Coleraine, Mass.
  G. O. No. 69, W. D., 1919.

Colonel, Medical Corps, U. S. Army.
In organizing and administering the division of sanitation and the sanitary inspection service of the office of the Surgeon General of the Army he contributed greatly to the efficiency of the military service.

HOWE, THORNDYKE D.
  R—Lawrence, Mass.
  B—Lawrence, Mass.
  G. O. No. 59, W. D., 1919.

Colonel, Field Artillery, U. S. Army.
As chief of the Postal Express Service he organized and administered with marked ability the Postal Service of the American Expeditionary Forces. He displayed great breadth of vision and untiring zeal in overcoming the many obstacles that were encountered in the organization of the service of handling mail for our troops in Europe.

HOWELL, WILLEY.
  R—Fayetteville, Ark.
  B—Austin, Ark.
  G. O. No. 59, W. D., 1919.

Colonel (Infantry), General Staff Corps, U. S. Army.
As assistant chief of staff, G-2, of the 1st Army he organized and directed the operations of this section during the entire operations of the 1st Army. The results achieved by him during the St. Mihiel and Meuse-Argonne operations had a noted influence on the successes gained by the 1st Army and showed him to be an officer of sound judgment and marked ability.

HOWZE, ROBERT L.
R—Overton, Tex.
B—Overton, Tex.
G. O. No. 89, W. D., 1919.
Medal of honor also awarded.

Major general, U. S. Army.
As commander of the 3d Division on its march to the Rhine and during the occupation of the enemy territory he proved himself energetic and capable, exhibiting superb qualities of leadership. He maintained an unusually high standard of efficiency in his unit, rendering eminently conspicuous services as a division commander.

*HOYLE, ELI D.
R—Haynesville, Ala.
B—Canton, Ga.
G. O. No. 4, W. D., 1923.

Brigadier general, U. S. Army.
As department commander, Eastern Department, between Aug. 25, 1917, and Jan. 15, 1918, he handled many difficult problems arising in that department with rare judgment, tact, and great skill.
Posthumously awarded. Medal presented to widow, Mrs. Eli D. Hoyle.

HOYLE, RENE E. DE R.
R—Governors Island, N. Y.
B—West Point, N. Y.
G. O. No. 38, W. D., 1922.

Colonel, Field Artillery, U. S. Army.
As executive officer and later as assistant commandant of the School of Fire for Field Artillery, Fort Sill, Okla., during the period from November, 1917, to May, 1919, he displayed remarkable tact and excellent judgment, combined with executive and professional ability of a high order, in positions of great responsibility, thereby contributing materially toward bringing that school to a state of maximum efficiency in a time of great emergency.

HUBBARD, SAMUEL T.
R—New York, N. Y.
B—Greenville, N. J.
G. O. No. 13, W. D., 1923.

Major (Signal Corps), General Staff Corps, U. S. Army.
In the intelligence section, General Staff, General Headquarters, American Expeditionary Forces, he organized and directed the operation of the battle order subsection, the section responsible for determining the effectives and tactics of the enemy and the location of enemy divisions on the western front. As director of the intelligence school, American Expeditionary Forces, Langres, France, he displayed rare efficiency and grasp of his work. His zeal and administrative and technical ability enabled him to render his Government services of great value in positions of responsibility.

HUEBNER, CLARENCE R.
R—Bushton, Kans.
B—Bushton, Kans.
G. O. No. 56, W. D., 1922.
Distinguished-service cross and oak-leaf cluster also awarded.

Lieutenant colonel, Infantry, U. S. Army.
As captain, major, and lieutenant colonel of the 28th Infantry, 1st Division, throughout its training and active operations in France he successfully commanded all echelons of the regiment, participating with distinction in every engagement from Cantigny to Sedan, reorganizing his regiment after its heavy losses in the first phase of the Meuse-Argonne offensive, and inspiring it with the will and dash that carried it to the heights of Sedan. By his sound tactical judgment, his unusual leadership, and indefatigable energy he contributed in a marked manner to the various successes of his regiment and of the 1st Division and rendered to the American Expeditionary Forces most conspicuous services in a position of great responsibility.

HUGHES, JOHN C.
R—New York, N. Y.
B—Louisville, Ky.
G. O. No. 59, W. D., 1919.

Captain, Infantry, U. S. Army.
As aid-de-camp to the commander in chief, American Expeditionary Forces, he performed duties of an exacting nature with peculiar tact, ability, and untiring energy, proving himself sound in judgment and indefatigable in all tasks assigned to him. At all times he served with distinction, rendering exceptional service.

HUGHES, JOHN H.
R—New York, N. Y.
B—New York, N. Y.
G. O. No. 89, W. D., 1919.

Colonel, Inspector General's Department, U. S. Army.
As a member of the Inspector General's Department at the headquarters, services of supply, for an extended period of time, by his unflagging energy, sound judgment, and tact, he handled with conspicuous ability many difficult problems which constantly arose in the execution of his important office. He rendered services of signal worth to the American Expeditionary Forces.

HUGHES, WILLIAM N., JR.
R—Pittsburgh, Pa.
B—Columbia, Tenn.
G. O. No. 59, W. D., 1919.

Colonel (Infantry), General Staff Corps, U. S. Army.
While he was serving as G-3 and as chief of staff of the 42d Division, his efforts had an important bearing on the successes gained by the division in the Baccarat sector, at the second Battle of the Marne, the operations near Chalons, Chateau-Thierry, the St. Mihiel salient, and along the Meuse. His splendid judgment and tactical ability were of the greatest value and demonstrated military knowledge of a high order.

HUIDEKOPER, FREDERIC L.
R—Washington, D. C.
B—Meadville, Pa.
G. O. No. 59, W. D., 1921.

Lieutenant colonel, Adjutant General's Department, U. S. Army.
As liaison officer of the 33d Division, with the 18th French Division during the Meuse-Argonne offensive, Colonel Huidekoper did this liaison work as a volunteer and at times performed his duties in great danger of his life. This work was of exceedingly great responsibility.

HULEN, JOHN A.
R—Houston, Tex.
B—Centralia, Tex.
G. O. No. 56, W. D., 1921.

Brigadier general, U. S. Army.
For services while commanding the 72d Infantry Brigade of the 36th Division in the Meuse-Argonne offensive Oct. 8-28, 1918.

HULL, JOHN A.
R—Des Moines, Iowa.
B—Bloomfield, Iowa.
G. O. No. 59, W. D., 1919.

Colonel, Judge Advocate General's Department, U. S. Army.
As judge advocate of the service of supplies, he most creditably handled the questions brought before him. His sound legal training, his complete knowledge of military administration, and his clear conception of the new and difficult problems involved made his services of most exceptional value.

HUME, EDGAR ERSKINE.
R—Frankfort, Ky.
B—Frankfort, Ky.
G. O. No. 14, W. D., 1923.

Lieutenant colonel, Medical Corps, U. S. Army.
As chief medical officer and later as commissioner of the American Red Cross in Serbia, February, 1919, to June, 1920, with untiring energy, unremitting devotion to duty, and with rare administrative and professional skill he organized and operated an American sanitary service, reorganizing hospitals, dispensaries, and dressing stations for soldiers and civilians alike, and successfully combating an epidemic of typhus fever which had caused the death of 80 per cent of the Serbian doctors. From June, 1918, to February, 1919, in direct charge of an American base hospital which was later expanded by the addition of Italian hospitals into a composite hospital center in the Italian war zone, he rendered professional services of a highly conspicuous character

HUME, FRANK M.
  R—Houlton, Me.
  B—Bridgewater, Me.
  G. O. No. 56, W. D., 1921.

Colonel, Infantry, U. S. Army.
For services while in command of the 103d Infantry, 26th Division, during the St. Mihiel and Meuse-Argonne operations.

HUMPHREY, GILBERT EDWIN
  R—El Reno, Okla.
  B—Abilene, Kans.
  G. O. No. 77, W. D., 1919.

Colonel, Corps of Engineers, U. S. Army.
For services while in charge of the building and organizing of the engineer depot at Norfolk, Va., and later as director of storage, purchase, storage and traffic division, General Staff.

HUNT, CHARLES A.
  R—Nashua, N. H.
  B—Nashua, N. H.
  G. O. No. 59, W. D., 1919.

Colonel, Infantry, U. S. Army.
He commanded first a battalion of the 18th Infantry and later the regiment. He conducted his unit in every action with marked ability and skillful leadership, showing the finest qualities of good judgment, courage, and devotion to duty.

HUNT, IRVIN L.
  R—Point Arena, Calif.
  B—Boonville, Calif.
  G. O. No. 55, W. D., 1920.

Colonel, Infantry, U. S. Army.
He served with conspicuous success as the officer in charge of civil affairs in the occupied area with the 3d Army and with the American Forces in Germany. With excellent judgment and sound adherence to well-established policies in a field of intricate problems affecting the civil population, he perfected, through his wide comprehension of conditions, an effective organization, which contributed materially to the efficiency of those forces.

HUNT, ORA E.
  R—Point Arena, Calif.
  B—Berryessa, Calif.
  G. O. No. 95, W. D., 1919.

Brigadier general, U. S. Army.
As commander of the 6th Infantry Brigade, 3d Division, during the greater part of its active operations he achieved notable success, demonstrating high qualities of leadership. Through his exceptional tactical ability his brigade was enabled to overcome desperate hostile resistance during its participation in the Meuse-Argonne offensive. By his efforts he has contributed materially to the brilliant success of his brigade in that important operation.

HURLEY, EDWARD N.
  R—Wheaton, Ill.
  B—Galesburg, Ill.
  G. O. No. 46, W. D., 1919.

Chairman, U. S. Shipping Board.
With tireless energy, he surmounted extreme difficulties and increased trans-Atlantic tonnage to an extent to allow of a steady shipment, both of troops and necessary supplies. Unselfish in devotion to duty, sound in judgment, quick to act, he rendered a service to the world.

HURLEY, PATRICK J.
  R—Tulsa, Okla.
  B—Indian Territory.
  G. O. No. 68, W. D., 1920.

Lieutenant colonel, Judge Advocate General's Department, U. S. Army.
Assigned as judge advocate, Army Artillery, 1st Army, he rendered services of marked ability, performing, in addition to his manifold duties, the duties of adjutant general and of inspector general. Later, as judge advocate of the 6th Army Corps, he ably conducted the negotiations arising between the American Expeditionary Forces and the Grand Duchy of Luxemburg wherein he displayed sound judgment, marked zeal, and keen perception of existing conditions. He has rendered services of material worth to the American Expeditionary Forces.

HUTCHESON, GROTE.
  R—Newtown, Ohio.
  B—Cincinnati, Ohio.
  G. O. No. 18, W. D., 1919.

Major general, U. S. Army.
For services in the administration of the port of embarkation, Newport News, Va., in connection with the shipment of troops overseas.

HUTCHINSON, JAMES P.
  R—Philadelphia, Pa.
  B—Philadelphia, Pa.
  G. O. No. 56, W. D., 1922.

Colonel, Medical Corps, U. S. Army.
As commanding officer, American Red Cross Military Hospital No. 1, American Expeditionary Forces, he displayed exceptional ability in the organization and administration of that unit. By his devotion to duty, untiring energy, coupled with professional attainments of a high order, he rendered services of inestimable value in a position of great responsibility in the alleviation of the sufferings of our sick and wounded.

HUTTON, PAUL C.
  R—Goldsboro, N. C.
  B—Goldsboro, N. C.
  G. O. No. 59, W. D., 1919.

Colonel, Medical Corps, U. S. Army.
As chief surgeon of the Paris group from June 2 to July 26, 1918, during which period by his good judgment and untiring energy he promoted a hospitalization and evacuation system that insured prompt and excellent care and treatment of the wounded, he furnished the means of saving many lives and provided comfort for the wounded, thereby greatly adding to the morale of the combatant troops of both the American and the French engaged in the second Battle of the Marne.

HUTTON, WILLIAM H. H., Jr.
  R—Detroit, Mich.
  B—Fort Jefferson, Fla.
  G. O. No. 95, W. D., 1919.

Colonel, Air Service, U. S. Army.
As chief of personnel service and later assistant chief of the supply section, Air Service, he was charged with duties of a varied and difficult nature. He constantly displayed marked zeal and sound judgment in the solution of the important problems of supply and transportation of the Air Service, rendering services of inestimable value to the American Expeditionary Forces.

INGRAHAM, CHARLES NELSON
  R—Findlay, Ohio.
  B—Oil City, Pa.
  G. O. No. 116, W. D., 1919.

Lieutenant commander, U. S. Navy.
As force transport officer his untiring energy contributed greatly to the successful oversea movement of troops and supplies.

IRELAND, MARK L.
  R—Chesaning, Mich.
  B—Chesaning, Mich.
  G. O. No. 14, W. D., 1923.

Colonel, Motor Transport Corps, U. S. Army.
As chief of the repair division, office of the Director, Motor Transport Corps, American Expeditionary Forces, he displayed sound judgment, executive ability of a high order, and unremitting devotion to duty, thus contributing markedly to the successful operations of the Motor Transport Corps of the American Expeditionary Forces.

IRELAND, MERRITTE W. R—Columbia City, Ind. B—Columbia City, Ind. G. O. No. 12, W. D., 1919.	Major general, U. S. Army. As Chief Surgeon of the American Expeditionary Forces he supervised and perfected the organization of the medical department in France; and to his excellent judgment, untiring efforts, and high professional attainments are largely due the splendid efficiency with which the sick and wounded of the American Army have been cared for.
IRWIN, GEORGE LE R. R—Chicago, Ill. B—Fort Wayne, Mich. G. O. No. 19, W. D., 1920.	Brigadier general, U. S. Army. He commanded with ability the 57th Field Artillery Brigade, 32d Division, during the Marne-Aisne, Oise-Aisne, and Meuse-Argonne offensives. At all times he displayed keen judgment, high military attainments, and loyal devotion to duty. The success of the division whose advance he supported was due, in a large measure, to his eminent technical skill and ability as an artillerist.
JACKLING, DANIEL C. R—San Francisco, Calif. B—Appleton City, Mo. G. O. No. 118, W. D., 1919.	For services as director of United States Government explosive plants.
JACKSON, THOMAS H. R—Muskegon, Mich. B—Canada. G. O. No. 87, W. D., 1919.	Colonel, Corps of Engineers, U. S. Army. As a member of the division of construction and forestry he displayed untiring energy and marked ability in the performance of his important duties. His stupendous task was fraught with numerous difficulties, which he overcame with noteworthy success, rendering services of signal worth to the American Expeditionary Forces.
JACKSON, WILLIAM P. R—Palmyra, Mo. B—Palmyra, Mo. G. O. No. 28, W. D., 1921.	Brigadier general, U. S. Army. For services as brigade commander, 74th Infantry Brigade, 37th Division, American Expeditionary Forces, in operations against the enemy in France and Belgium.
JACOBSON, BENJAMIN L. R—Washington, D. C. B—Germany. G. O. No. 14, W. D., 1923.	Lieutenant colonel, Quartermaster Corps, U. S. Army. As executive officer in the office of the Quartermaster General of the Army, and in addition to the duties of that office, charged with the responsibility for the operations of the administrative, personnel, and regulations divisions of that office, he rendered highly important services to the Government. His sound judgment, rare administrative and executive ability, and his unremitting attention to duty were contributing factors in the success of the Quartermaster Corps in the supply of the Army.
JADWIN, EDGAR. R—Honesdale, Pa. B—Honesdale, Pa. G. O. No. 59, W. D., 1919.	Brigadier general, U. S. Army. As commanding officer of the 15th Engineers, he inaugurated the important project at Gievres. Later, in charge of the division of construction and forestry, he brought to this important task a splendidly trained mind and exceptionally high skill. His breadth of vision and sound judgment influenced greatly the successful completion of many vast construction projects undertaken by the American Expeditionary Forces.
JAMERSON, GEORGE H. R—Martinsville, Va. B—Martinsville, Va. G. O. No. 15, W. D., 1923.	Brigadier general, U. S. Army. As regimental commander, 317th Infantry, and later as brigade commander of the 159th Infantry Brigade of the 80th Division, he rendered conspicuous service in the organization and training of these units, and in the command thereof during the operations of his brigade in the Meuse-Argonne offensive. Displaying sound judgment, high professional skill, untiring energy, and devotion to duty, he contributed in a material way to the successful operations of his division and of the American forces in France.
*JANEWAY, THEODORE C. R—Baltimore, Md. B—New York, N. Y. G. O. No. 14, W. D., 1923.	Major, Medical Corps, U. S. Army. As chief of the division of internal medicine in the Surgeon General's Office from June 26 to Dec. 27, 1917, he distinguished himself by his conspicuous service in the organization and development of that division. Standing as one at the head of his profession in America, he was responsible for the selection of many prominent internists for war service, thereby rendering a service of inestimable value to the Government in caring for the sick. Posthumously awarded. Medal presented to widow, Mrs. Eleanor C. Janeway.
JARMAN, SANDERFORD. R—West Monroe, La. B—Boatner, La. G. O. No. 43, W. D., 1922.	Lieutenant colonel, Coast Artillery Corps, U. S. Army. From Sept. 14 to Nov. 26, 1918, as G-3, and later chief of staff of the commanding general, American Railway Artillery Reserve, with the 1st and 2d American Armies during the Meuse-Argonne offensive and the offensive planned to be launched by the 2d American Army on Nov. 11, 1918, he displayed the highest qualities as an organizer, and by his untiring energy and zeal, good judgment, and general excellence, as well as great technical ability, contributed largely to the success of the command.
JAY, NELSON D. R—Pelham Manor, N. Y. B—Elwood, Ill. G. O. No. 56, W. D., 1922.	Lieutenant colonel, Quartermaster Corps, U. S. Army. As assistant general purchasing agent, American Expeditionary Forces, by excellent judgment, untiring energy, and broad and comprehensive knowledge of business affairs, he solved in a highly meritorious manner many difficult and serious problems of supply which confronted our forces throughout the war. His services were rendered with conspicuous success in a position of great responsibility and were of inestimable value to the American Expeditionary Forces.

JENKS, GLEN F.
R—Clayville, N. Y.
B—Deansboro, N. Y.
G. O. No. 13, W. D., 1923.

Colonel, Ordnance Department, U. S. Army.
As chief of the Heavy Artillery Division, office of Chief Ordnance Officer American Expeditionary Forces, he displayed keen foresight and excellent judgment, combined with professional ability of the highest order, in the development of allied artillery programs and the supply and maintenance of heavy and railway artillery to the troops in France. Later, as chief ordnance inspector of artillery, American Expeditionary Forces, he organized and perfected to a high degree of efficiency the maintenance of the artillery of the 1st Army during the Meuse-Aronne operations, thereby contributing materially to the success of our forces.

JERVEY, HENRY.
R—Charleston, S. C.
B—Dublin, Va.
G. O. No. 144, W. D., 1918.

Major general, U. S. Army.
For services as director of operations, General Staff, and as assistant to the Chief of Staff in preparing and executing the plans involving the mobilization of personnel during the war.

JERVEY, JAMES P.
R—Atlanta, Ga.
B—Powhatan County, Va.
G. O. No. 87, W. D., 1919.

Colonel, Corps of Engineers, U. S. Army.
As commanding officer of the 304th Engineers, 79th Division, he performed his exacting duties with signal ability. His high technical skill and unflagging energy were largely responsible for keeping the roads in condition for the transportation of artillery and large quantities of supplies during the attack on Montfaucon and Nantillois in the latter part of September. By his great efforts he proved a potent factor in the successes achieved during these operations.

JETT, GEORGE HENRY.
R—Baker County, Oreg.
B—Baker County, Oreg.
G. O. No. 116, W. D., 1919.

Lieutenant, U. S. Naval Reserve Force.
For services as repair officer on the staff of the division commander, cruiser and transport force, Newport News, Va.

JEWELL, FRANK C.
R—Beloit, Wis.
B—Chicago, Ill.
G. O. No. 133, W. D., 1919.

Colonel, Coast Artillery Corps, U. S. Army.
As commander of the Railway Artillery Reserve attached to the 1st Army, American Expeditionary Forces, during the Meuse-Argonne operations, he performed his task with energy and marked ability, rendering valuable services to the American Expeditionary Forces.

JEWETT, FRANK B.
R—Wyoming, N. J.
B—Pasadena, Calif.
G. O. No. 47, W. D., 1919.

Lieutenant colonel, Signal Corps, U. S. Army.
For services in connection with the development of the radio telephone and the development and production of other technical apparatus for the Army.

JEWETT, HENRY C.
R—Buffalo, N. Y.
B—Buffalo, N. Y.
G. O. No. 74, W. D., 1919.
Distinguished-service cross also awarded.

Colonel, Corps of Engineers, U. S. Army.
In command of the 182d Infantry Brigade, 91st Division, in the Argonne he displayed exceptional qualities of leadership and tactical ability in important engagements. Later, as chief of staff of the 91st Division, he planned operations with sound judgment and a comprehensive understanding of existing conditions, showing military attainments and initiative of a high order. At all times he was untiring in energy and self-sacrificing in devotion to his exacting duties.

JOHNSON, ARTHUR.
R—St. Peter, Minn.
B—St. Peter, Minn.
G. O. No. 59, W. D., 1919.

Brigadier General, U. S. Army.
In command of the intermediate section, services of supply, he had the responsibility of forwarding to the front great quantities of supplies and thousands of replacements for the combatant units, in which important duty he displayed untiring zeal and exceptional executive ability.

JOHNSON, HUGH S.
R—Alva, Okla.
B—Fort Scott, Kans.
G. O. No. 1, W. D., 1926.

Colonel, Judge Advocate General's Department, U. S. Army.
For services in the Provost Marshal General's office in connection with the planning and execution of the draft laws.

JOHNSON, JACOB C.
R—Benton City, Mo.
B—Marietta, Ohio.
G. O. No. 62, W. D., 1919.

Colonel, Inspector General's Department, U. S. Army.
As inspector general of the 1st Army Corps and later of the 1st Army, he performed exacting tasks with distinction throughout the Marne-Chateau-Thierry, St. Mihiel, and Meuse-Argonne operations. Both during the months of actual fighting and the periods of training before and after the campaign, he displayed conspicuous devotion to duty, unfailing zeal, and loyalty, acting always with sound judgment.

JOHNSON, WAIT C.
R—Burlington, Vt.
B—Rutland, Vt.
G. O. No. 87, W. D., 1919.

Colonel (Infantry), General Staff Corps, U. S. Army
As athletic director, G-5, of the American Expeditionary Forces he was given the important and difficult task of planning and organizing an elaborate program of athletic training and competitions for American troops, embracing all branches of sport. By his zeal and sound judgment he carried this program to an eminently successful conclusion, thereby rendering an invaluable service in maintaining the morale and physical fitness of our troops during the trying period of repatriation.

JOHNSTON, EDWARD N.
R—Portland, Oreg.
B—St. Louis, Mo.
G. O. No. 56, W. D., 1922.

Colonel, Corps of Engineers, U. S. Army.
From September, 1917, until May, 1918, as commanding officer, 23d Highway Engineer Regiment, he displayed rare qualities of leadership in the organization and training of the regiment, which later performed excellent services throughout the war. As assistant to the chief of Chemical Warfare Service in France from June until December, 1918, in charge of the offensive division he showed ability of the highest order in the general supervision of operations of all gas troops. From December, 1918, until June, 1919, as acting chief of Chemical Warfare Service abroad, his keen business ability and sound judgment were important factors in the successful closing out of all chemical warfare activities in the American Expeditionary Forces.

JOHNSTON, GORDON.
R—Birmingham, Ala.
B—Charlotte, N. C.
G. O. No. 59, W. D., 1919.
Medal of honor and distinguished-service cross also awarded.

Colonel (Infantry), General Staff Corps, U. S. Army.
He showed great ability while chief of staff of the 82d Division in the operations in the Argonne area. The force of his energy and his masterful leadership manifested itself in the crowning successes of the division during the operations of this campaign.

JOHNSTON, JOHN A.
R—Washington, D. C.
B—Allegheny, Pa.
G. O. No. 4, W. D., 1923.

Brigadier general, U. S. Army.
As department commander, Northeastern Department, between Sept. 11 1917, and May 23, 1918, he handled many difficult problems arising in the department with rare judgment, tact, and great skill. Later, as commanding general of the 34th Division, which he took overseas, his marked efficiency, unusual initiative, and military attainments of a high order were important factors in the excellent standard of training attained by the division.

JOHNSTON, WILLIAM H.
R—Cincinnati, Ohio.
B—Cincinnati, Ohio.
G. O. No. 44, W. D., 1919.
Distinguished-service cross also awarded.

Major general, U. S. Army.
During the Argonne-Meuse offensive he commanded with skill and ability the 91st Division in the difficult advance that resulted in the taking of Epinonville. Later, in participation with the French, he led his division with marked distinction in the attack on and capture of the important city of Audenarde in the closing operations of the war in Belgium.

JOHNSTON, WILLIAM T.
R—Livingston, Mo.
B—Alexandria, Pa.
G. O. No. 126, W. D., 1919.

Colonel (Cavalry), General Staff Corps, U. S. Army.
He organized and administered the officers' training camps from the outbreak of the war until July 25, 1918, and thereafter rendered conspicuous service as chief of staff, Southern Department.

JOLY, CHARLES L.
R—New York, N. Y.
B—New Orleans, La.
G. O. No. 56, W. D., 1922.

Major, Chemical Warfare Service, U. S. Army.
By displaying untiring energy and enthusiasm in the performance of his important duties, he developed efficient gas discipline in the 32d Division, resulting in the prevention of gas fatalities. He was zealous and discerning in the training of troops, achieving brilliant successes. His service was marked by self-sacrificing devotion to the welfare and protection of the men who were with him.

JONES, CLIFFORD.
R—Norcross, Ga.
B—Cumming, Ga.
G. O. No. 56, W. D., 1922.

Colonel (Coast Artillery Corps), General Staff Corps, U. S. Army.
In the office of the executive assistant to the Chief of Staff during the World War and the following demobilization period, his tactfulness and initiative in meeting the varied situations presented and sound judgment in passing upon many matters of highest importance contributed materially to the successful functioning of that office during the war. During demobilization his conception and organization of the emergency discharge section of the office not only protected the War Department from impositions but served in a marked degree to preserve the morale of the civilian population during that trying period.

JONES, GLENN I.
R—Washington, D. C.
B—Washington, D. C.
G. O. No. 69, W. D., 1919.

Lieutenant colonel, Medical Corps, U. S. Army.
While surgeon of the 10th Division during the epidemic of Spanish influenza in that command his farsightedness in providing hospital facilities and his energetic and exceptionally efficient action in directing the care of patients resulted in a large reduction of mortality. His services show a rare devotion to duty in that, though himself a sufferer from the disease, his efforts were unabated.

JONES, HARVEY L.
R—Baltimore, Md.
B—Tuckerton, N. J.
C. O. No. 53, W. D., 1921.

Lieutenant colonel, Provost Marshal General's Department, U. S. Army.
In charge of the Military Police School at Autun, France, a position of great responsibility, due to his ability and energy, hundreds of officers and enlisted men were trained as military police for the Provost Marshal General's Department, and were afterwards distributed throughout the American Expeditionary Forces. His work was conspicuous for its thoroughness and important results. He rendered services of much value to the United States.

JONES, HILARY POLLARD.
R—Taylorsville, Va.
B—Hanover County, Va.
G. O. No. 116, W. D., 1919.

Rear admiral, U. S. Navy.
As commanding officer of the Newport News division of the cruiser and transport fleet, his successful administration and close cooperation with the Army authorities resulted in the efficient joint operations of the Army and Navy at the port of Hampton Roads.

JONES, JAMES S.
R—Wheeling, W. Va.
B—Wheeling, W. Va.
G. O. No. 59, W. D., 1919.

Lieutenant colonel, Adjutant General's Department, U. S. Army.
As assistant to the adjutant general at General Headquarters, American Expeditionary Forces, he displayed executive ability of the highest order in the efficient administration of the divisions successively assigned to him. Possessed of a keen mind for organization, with sound judgment, tact, and a thorough understanding of the intricate details of the office, he successfully surmounted innumerable obstacles, rendering service of signal worth to the American Expeditionary Forces and to the Government.

JONES, JOHN C.
R—Wynnewood, Pa.
B—New York, N. Y.
G. O. No. 124, W. D., 1919.

As chief of the Philadelphia ordnance district, in which capacity he maintained at all times the greatest degree of intelligent and enthusiastic cooperation between the Ordnance Department and manufacturers in his district, thereby attaining the maximum production of munitions in a minimum time; and also as chairman of the Philadelphia ordnance district claims board, in which capacity his services have been invaluable to the Nation in adjusting equitably the $271,000,000 worth of outstanding contracts in his district in force at the signing of the armistice.

JONES, PERCY L.
  R—Cleveland, Tenn.
  B—Bartow Co., Ga.
  G. O. No. 59, W. D., 1919.

Colonel, Medical Corps, U. S. Army.
He served with marked distinction as commander of the United States Ambulance Service with the French Armies. By the force of his energy, zeal, and ability he brought the units of that service to a high state of perfection. The splendid record held by this service is attributable to his great devotion and untiring efforts in accomplishing his tasks.

JONES, SAMUEL G.
  R—Montgomery, Ala.
  B—Montgomery, Ala.
  G. O. No. 87, W. D., 1919.

Colonel, Cavalry, U. S. Army.
As commanding officer of Winchester Camp, England, he was directly charged with the transportation of several hundred thousand American troops through England, a task of great magnitude and one involving many difficulties. By his tireless energy and keen application to his important duties he accomplished his task with marked success, rendering services of distinction to the American Expeditionary Forces.

JONES, WALTER C.
  R—Quincy, Mass.
  B—Quincy, Mass.
  G. O. No. 56, W. D., 1922.

Colonel, Quartermaster Corps, U. S. Army.
As quartermaster of the intermediate section, services of supply, American Expeditionary Forces, he performed his manifold duties with marked ability and outstanding successes. He was charged with the supervision of the Quartermaster Corps personnel of over 500 officers and 10,000 enlisted men, with the responsibility for the direction and control of the vast quantity of quartermaster stores required for the intermediate section, consisting of over 270,000 troops, and with the initial supply of four replacement (depot) divisions of over 300,000 troops. He fulfilled this tremendous responsibility with conspicuous and marked efficiency, handling with tact and keen judgment the many complex problems which constantly confronted him. His service was characterized with zealousness, resourcefulness, and farsightedness, and he proved himself equal to every emergency. He rendered services of signal worth to the American Expeditionary Forces.

JORDAN, CLARENCE L.
  R—Monticello, Ga.
  B—Monticello, Ga.
  G. O. No. 13, W. D., 1923.

First lieutenant, Ordnance Department, U. S. Army.
In active charge of the Army ammunition depot system of the 1st American Army, from its organization until the armistice, displaying great technical ability, sound judgment, exceptional zeal and energy, he successfully assured at all times efficient and adequate storage, protection, and issue of all classes of ammunition at the front, contributing materially to the success of the American Expeditionary Forces in France.

JORDAN, RICHARD H.
  R—Haymarket, Va.
  B—Haymarket, Va.
  G. O. No. 14, W. D., 1923.

Colonel, Quartermaster Corps, U. S. Army.
As senior assistant and executive assistant to the chief of transportation service from Jan. 11, 1918, to Apr. 15, 1919, with untiring energy, rare administrative and executive ability, and unremitting devotion to duty, he rendered service of great value to the Government in perfecting the necessary organization of the movement of troops from encampments throughout the United States to ports of embarkation and thence overseas.

JOY, BENJAMIN.
  R—Boston, Mass.
  B—Boston, Mass.
  G. O. No. 70, W. D., 1919.

Major, Infantry, U. S. Army.
As chief of the fiscal department in the office of the officer in charge of civil affairs in the occupied territory, he has handled problems of a delicate and complicated character with remarkable success, displaying marked administrative ability, breadth of vision, and a comprehensive knowledge of international financial questions.

JOYCE, KENYON A.
  R—Chicago, Ill.
  B—Brooklyn, N. Y.
  G. O. No. 3, W. D., 1921.

Colonel (Cavalry), General Staff Corps, U. S. Army.
As chief of staff of the 2d Depot Division, Le Mans, France, in the training and sending forward of replacements and in breaking up combat divisions for replacement he displayed broad professional attainments, a high degree of leadership, and unfailing tact. Later, as chief of the classification and data section, personnel branch, General Staff, he demonstrated rare executive and planning ability in developing a system for the scientific classification of officers with a view to their suitability and availability.

JUDAH, NOBLE B., Jr.
  R—Chicago, Ill.
  B—Chicago, Ill.
  G. O. No. 103, W. D., 1919.

Lieutenant colonel, Field Artillery, U. S. Army.
As assistant chief of staff, G-2, of the 42d Division during all its campaigns, by the skillful direction of the intelligence service he proved a material factor in the successes gained by his division. He at all times displayed assiduous application to his important task, rendering services of the utmost value.

JUDSON, WILLIAM V.
  R—Indianapolis, Ind.
  B—Indianapolis, Ind.
  G. O. No. 11, W. D., 1921.

Brigadier general, U. S. Army.
For services while serving as chief of the American military mission to Russia and military attaché to the American embassy at Petrograd, Russia.

JUNKERSFELD, PETER.
  R—Chicago, Ill.
  B—Sadorus, Ill.
  G. O. No. 89, W. D., 1919.

Colonel, Quartermaster Corps, U. S. Army.
As associate officer in charge of the building branch of the construction division of the Army, by his unremitting industry and energy, sound judgment, and knowledge of men he was of the most material assistance in the accomplishment of the construction program of the Army. He performed notable service as executive, organizer, and administrator.

KEAN, JEFFERSON R.
  R—Lynchburg, Va.
  B—Lynchburg, Va.
  G. O. No. 59, W. D., 1921.

Brigadier General, U. S. Army.
As chief of the department of military relief, American Red Cross, a position of great responsibility, by his foresight, marked efficiency, and energy he organized the base hospitals, which cared for many of our wounded, and administered the United States Ambulance Service for duty with the French Army, greatly assisting our ally. He rendered services of conspicuous worth to the United States.

KEECH, FRANK BROWNE_____
R—New York, N. Y.
B—Newport, Md.
G. O. No. 59, W. D., 1919.

Lieutenant colonel, Inspector General's Department, U. S. Army.
As inspector general of the port of embarkation, Newport News, Va., through his very able control and judgment in the management of his office the shipment of troops and supplies overseas was materially aided.

KELLER, CHARLES_____
R—Fort Sam Houston, Tex.
B—Coeur d'Alene, Idaho.
G. O. No. 59, W. D., 1919.

Colonel, Infantry, U. S. Army.
He took command of a regiment at a critical moment after two unsuccessful assaults had been made by the brigade. He reorganized the regiment under fire and made possible the taking and holding of the Bois-des-Ogons, thereby displaying the highest order of leadership and exhibiting the masterful qualities of a commander.

KELLER, CHARLES_____
R—New York, N. Y.
B—Rochester, N. Y.
G. O. No. 2, W. D., 1920.

Brigadier general, U. S. Army.
As assistant to the Chief of Engineers, U. S. Army, he was instrumental in initiating policies which protected the channels and anchorages of our important harbors from obstruction by enemy aliens. As power administrator on behalf of the United States he organized the power section of the War Industries Board and initiated measures as a result of which the war program of the country was successfully protected against serious delay due to power shortage.

KELLER, WILLIAM L._____
R—New York, N. Y.
B—Hartford, Conn.
G. O. No. 62, W. D., 1919.

Colonel, Medical Corps, U. S. Army.
As director of the professional services, Medical Department, American Expeditionary Forces, he displayed marked ability in the organization and assignment of the forces at his disposal for service in hospitals at the front and in the rear areas. He was discerning in his knowledge of conditions, using his insufficient personnel to the maximum advantage in relieving the suffering of our sick and wounded, and in obtaining prompt treatment for battle casualties. His comprehensive grasp of the problems which presented themselves resulted in the saving of many lives.

KELLEY, REGINALD H._____
R—Berkeley, Calif.
B—Fresno, Calif.
G. O. No. 39, W. D., 1920.

Colonel, Infantry, U. S. Army.
First as division machine-gun officer and later as commanding officer, 116th Infantry, 29th Division, he displayed sterling qualities of leadership. By his high military attainments, sound judgment, and self-sacrificing devotion to duty he proved to be a material factor in the successes achieved by the 29th Division in the offensive actions in which they participated.

KELLOND, FREDERIC G._____
R—Louisville, Ky.
B—Canada.
G. O. No. 103, W. D., 1919.

Colonel (Infantry), General Staff Corps, U. S. Army.
While in charge of the construction section of the equipment branch, General Staff, he was responsible for the work of that section regarding projects of great magnitude, and his services have been of great value.

KELLY, WILLIAM._____
R—West Superior, Wis.
B—New York, N. Y.
G. O. No. 59, W. D., 1919.

Colonel, Corps of Engineers, U. S. Army.
After serving with great credit in the field he took command of the important ports of La Rochelle and La Pallice. By his executive ability and great energy he promptly relieved congested conditions and made possible the uninterrupted flow of necessary supplies toward the front.

KELLY, WILLIAM, Jr._____
R—Brownsville, Tex.
B—Brownsville, Tex.
G. O. No. 69, W. D., 1919.

Colonel, Adjutant General's Department, U. S. Army.
For services in The Adjutant General's Department during the war. To his untiring energy and his sound and impartial judgment is due, in a large measure, the efficient action leading to the maintenance of the high standard of commissioned personnel during the war.

KENLY, WILLIAM L._____
R—Baltimore, Md.
B—Baltimore, Md.
G. O. No. 15, W. D., 1923.

Major general, U. S. Army.
As Chief of Air Service, American Expeditionary Forces, by his executive ability, clear conception, and broad mental grasp, he was able to overcome many obstacles and placed the training of Air Service personnel on an efficient basis. Later, as Director of Military Aeronautics in the United States, he successfully organized and accomplished the training of personnel for overseas service with a resulting high degree of morale throughout the Air Service and the efficient performance of duties at the front. Serving at the same time as a member of the Advisory Committee and Joint Army and Navy Airship Board, he rendered services of inestimable value to the Government in positions of great responsibility.

KENNEDY, JAMES M._____
R—Troy, S. C.
B—Abbeville, S. C.
G. O. No. 111, W. D., 1919.

Colonel, Medical Corps, U. S. Army.
As port surgeon, port of embarkation, Hoboken, N. J., he has organized, provided, and administered with conspicuous efficiency all of the hospitals required for the accommodation of our troops going overseas from that port, as well as for the large number of our sick and wounded soldiers returning home.

KENNEDY, JOHN T._____
R—Orangeburg, S. C.
B—Hendersonville, S. C.
G. O. No. 39, W. D., 1920.
Medal of honor also awarded.

Lieutenant colonel, Field Artillery, U. S. Army.
As a regimental commander during the St. Mihiel offensive and the Meuse-Argonne offensive he displayed conspicuous efficiency, marked aggressiveness, and leadership. By his exceptional technical and executive ability he solved many perplexing problems, although much handicapped by losses in men, material, and animals. He at all times rendered invaluable support to the attacking Infantry and proved to be a material factor in the result achieved.

KENNEDY, MOORHEAD C._____
R—Chambersburg, Pa.
B—Chambersburg, Pa.
G. O. No. 14, W. D., 1923.

Colonel, Transportation Corps, U. S. Army.
As Deputy Director General of Transportation, American Expeditionary Forces, in Paris, France, and later in London, England, he rendered services to the Government of an important and responsible character, his wide experience as a railroad executive, sound professional and technical ability, his untiring energy and devotion to duty contributing markedly to the successful operations of the Transportation Corps of the American Expeditionary Forces.

KERNAN, FRANCIS J............
R—Jacksonville, Fla.
B—Jacksonville, Fla.
G. O. No. 12, W. D., 1919.

Major general, U. S. Army.
He was intrusted with the important duty of organizing the services of supply of the American Expeditionary Forces in France, and the foundation then laid was later successfully carried to completion. As member of the War Prisoners' Commission, Berne, Switzerland, and of the American section of the Supreme War Council, he has rendered conspicuous services to the Government.

KERR, JAMES T............
R—Martins Ferry, Ohio.
B—Martins Ferry, Ohio.
G. O. No. 77, W. D., 1919.

Brigadier general, U. S. Army.
While in charge of the enlisted men's division of The Adjutant General's Office and of the recruitment of the Army, and later as executive assistant to The Adjutant General of the Army, his sound judgment and unremitting industry were important factors in the efficient administration of The Adjutant General's Department.

KERTH, MONROE C............
R—Cairo, Ill.
B—Cairo, Ill.
G. O. No. 56, W. D., 1922.

Colonel (Infantry), General Staff Corps, U. S. Army.
As military attaché at Petrograd, Russia, from Sept. 1, 1917, to Feb. 1, 1918, he performed his exacting duties with marked ability under most trying circumstances. As director of the Army General Staff College, American Expeditionary Forces, by his high professional attainments and unfailing energy, he rendered service of inestimable worth in connection with the instruction and training of officers for general staff duty. Later, as a member of the training and instruction branch, War Plans Division, General Staff, he demonstrated sound judgment, great breadth of vision, and keen foresight in the solution of the various difficult problems with which he was confronted.

KEVILLE, WILLIAM J............
R—Belmont, Mass.
B—Somerville, Mass.
G. O. No. 13, W. D., 1923.

Lieutenant colonel, Infantry, U. S. Army.
In command of the 101st Ammunition Train, 26th Division, throughout the period of organization, training, and operations in France. During the Aisne-Marne offensive he provided a continuous and adequate supply of ammunition for the 26th Division and to elements of the 28th, 42d, and 4th Divisions over an extensive territory and under all conditions of open warfare. Because of his high professional attainments, initiative, untiring energy, and devotion to duty he rendered extremely valuable services.

KEYSER, RALPH S............
R—Thoroughfare, Va.
B—Thoroughfare, Va.
G. O. No. 9, W. D., 1923.

Major, U. S. Marine Corps.
As Assistant Chief of Staff, 2d Division, G-2, from July 26, 1918, to July, 1919, with indefatigable zeal and excellent executive ability he so organized his section as to furnish prompt and comprehensive information of the enemy for the use of the division in its operations in the battles of St. Mihiel, Blanc Mont Ridge (Champagne), and the Meuse-Argonne, and in the march to the Rhine.

KILBOURNE, CHARLES E............
R—Portland, Oreg.
B—Fort Myer, Va.
G. O. No. 89, W. D., 1919.
Medal of honor and distinguished-service cross also awarded.

Brigadier general, U. S. Army.
As chief of staff of the 89th Division, he displayed military ability of the highest order, contributing to the successes achieved by that division during the St. Mihiel offensive. Later, upon his promotion to the grade of brigadier general, he continued to render valuable services in command of the 36th Artillery Brigade during the remainder of the campaign.

KILBRETH, JOHN W............
R—Southampton, Long Island, N. Y.
B—New York, N. Y.
G. O. No. 19, W. D., 1921.

Lieutenant colonel, Field Artillery, U. S. Army.
As director of the department of firing, School of Fire for Field Artillery, Fort Sill, Okla., from September, 1917, to May, 1918, he displayed professional attainments of the highest and most progressive order. He was primarily responsible for the excellent grounding received by thousands of officers in the principles of artillery firing, including those applicable to open warfare.

KILNER, WALTER G............
R—Syracuse, N. Y.
B—Shelby, N. Y.
G. O. No. 62, W. D., 1919.

Colonel, Signal Corps, U. S. Army.
By his personal efforts and efficient labors he organized the machinery necessary to train pilots, and successfully developed this branch of the Air Service. He overcame numerous difficulties inherent in the establishment of such an organization in a foreign country, and it was largely due to his efficiency that the Air Service was able to furnish well-trained personnel to the squadrons at the front. He at all times displayed marked devotion to duty, untiring energy, and sound judgment.

KILPATRICK, JOHN R............
R—New York, N. Y.
B—New York, N. Y.
G. O. No. 59, W. D., 1919.

Lieutenant colonel, Quartermaster Corps, U. S. Army.
In his capacity as a member of the 4th Section, General Staff, he exhibited exceptional tact and ability in promoting cooperation between the French and American services of transport and supply. He has by his energy, good judgment, and decisive action in the establishment, organization, and conduct of various regulating stations and railheads very materially assisted in insuring a steady and adequate flow of supplies to our armies in their operations.

KIMBALL, GORDON N............
R—Ogden, Utah.
B—Indianapolis, Ind.
G. O. No. 10, W. D., 1922.

Lieutenant colonel, Judge Advocate General's Department, U. S. Army.
As chief of the rents, requisitions, and claims service, advance section, services of supply, from November, 1918, to July, 1919, Colonel Kimball with exceptionally sound judgment, breadth of vision, and marked initiative organized and administered a work of great responsibility and magnitude.

KIMBALL, RICHARD H............
R—Cleburne, Tex.
B—Kimball, Tex.
G. O. No. 69, W. D., 1919.

Lieutenant colonel (Cavalry), General Staff Corps, U. S. Army.
Upon joining the operations division of the General Staff he assumed the responsibility of mobilization of the draft and classification and distribution of troops. His clear judgment, initiative, and energy have done much toward the successful accomplishment of the huge task involved in receiving and placing the drafted forces.

*KING, ALFRED K.
    R—Erie, Pa.
    B—Geneva, Ohio.
    G. O. No. 70, W. D., 1919.

Major, Field Artillery, U. S. Army.
As munitions officer of the 5th Army Corps he performed exacting duties with untiring energy, displaying high professional attainments and a complete understanding of the needs of the troops he supplied. He personally reconnoitered roads over which transportation was to be made in order that he might keep in touch with changing conditions and be prepared to meet sudden emergencies, in order that the steady flow of munitions to the front lines might be maintained. He rendered services of signal worth.
Posthumously awarded. Medal presented to widow, Mrs. Ruth W. King.

KING, CAMPBELL.
    R—Atlanta, Ga.
    B—Flat Rock, N. C.
    G. O. No. 59, W. D., 1919.

Brigadier general, U. S. Army.
He served with distinction as chief of staff of the 1st Division in the operations near Montdidier, the advance south of Soissons, and in the attack on the St. Mihiel salient. Later, as chief of staff of the 3d Army Corps during the Argonne-Meuse operations, by his splendid tactical judgment he rendered especially meritorious service.

KING, DAVID M.
    R—Smyrna, Ohio.
    B—Smyrna, Ohio.
    G. O. No. 78, W. D., 1919.

Colonel, Ordnance Department, U. S. Army.
Displaying exceptional technical knowledge and comprehension of existing conditions, he ably organized, installed, and operated in the services of supply and in the Army area an extensive chain of repair facilities for the maintenance of ordnance matériel. With tireless energy and unfailing devotion to his important duties he perfected a loyal and efficient organization, capable of meeting all demands made upon it.

KING, EDWARD L.
    R—Bridgewater, Mass.
    B—Bridgewater, Mass.
    G. O. No. 59, W. D., 1919.
    Distinguished-service cross also awarded.

Brigadier general, U. S. Army.
He served, with marked distinction as chief of staff of the 28th Division. Later as brigade commander, he planned and directed the operations resulting in the capture by the 65th Infantry Brigade of Chateau d'Aulnois and Marcheville, where he displayed great tactical skill and demonstrated his abilities as a commander.

KING, EDWARD P., Jr.
    R—Atlanta, Ga.
    B—Atlanta, Ga.
    G. O. No. 15, W. D., 1921.

Major, Field Artillery, U. S. Army.
As principal assistant to the Chief of Field Artillery, Mar. 23, 1918, to Nov. 11, 1918, he contributed largely to the successful solution of the difficult problems of expansion, organization, and training which then confronted the Field Artillery.

KING, THOMAS W.
    R—Redwood City, Calif.
    B—Sacramento, Calif.
    G. O. No. 56, W. D., 1922.

Lieutenant colonel, Adjutant General's Department, U. S. Army.
As Adjutant General of the American Expeditionary Forces in Siberia he performed his exacting duties with high professional skill and administrative ability. The exceptional efficiency of the Adjutant General's Department under his direction was a material factor in the success of the staff work at headquarters. Possessed of a keen mind for organization, with sound judgment, tact, and a thorough understanding of the intricate details of his office, he successfully surmounted innumerable obstacles and rendered service of signal worth to the Government

KING, VAN RENSSELAER C.
    R—Wilmington, N. C.
    B—New York, N. Y.
    G. O. No. 10, W. D., 1922.

Colonel, Transportation Corps, U. S. Army.
As general superintendent of transportation, American Expeditionary Forces, and as transportation representative with the Armistice Commission, he organized and successfully installed a central car record and a system for distributing rolling stock which were of incalculable value to the American Expeditionary Forces.

KINGMAN, JOHN J.
    R—Chattanooga, Tenn.
    B—Omaha, Nebr.
    G. O. No. 78, W. D., 1919.

Colonel (Corps of Engineers), General Staff Corps, U. S. Army.
As chief of staff of the 90th Division he displayed exceptional ability, planning important operations with sound judgment and wide comprehension of the conditions to be encountered. He was unflagging in energy and tireless in devotion to his exacting duties. Constantly confronted by perplexing military problems, he handled them with aggression and achieved brilliant successes.

KLEIN, HARRY T.
    R—Cincinnati, Ohio.
    B—Bellevue, Ky.
    G. O. No. 19, W. D., 1920.

Lieutenant colonel, Judge Advocate General's Department, U. S. Army.
By his initiative, intelligence, and devotion to the study of the French laws on the subject of requisitions and billeting, he has contributed in a marked degree to the highly satisfactory results obtained by the rents, requisitions, and claims service in the conduct of a task of great magnitude. Later, as chief requisition officer, by his high professional attainments and sound judgment, furthered by a keen analytical mind, solved many intricate problems which daily confronted him, in the acquisition of property by lease, and in the requisitioning of billets for our troops.

KLOEBER, ROYALL O.
    R—Washington, D. C.
    B—Lynchburg, Va.
    G. O. No. 108, W. D., 1919.

For services as Assistant Director of Finance. In this capacity he rendered most valuable assistance in the solution of the great financial problems which arose due to the war.

KNIGHT, GEORGE W.
    R—Newark, N. J.
    B—Newark, N. J.
    G. O. No. 47, W. D., 1921.

Lieutenant colonel, Corps of Engineers, U. S. Army.
Near Bethincourt, France, on Sept. 25, 1918, he was assigned the task of placing foot bridges over Forges River and cutting the wire in front of the enemy positions. In this his inspiring leadership and constant supervision were conspicuous. Later he organized the regiment for the Nov. 1 offensive, which entailed the building of nine bridges, every one of which was completed in time for the Artillery to keep pace with the Infantry. His organization of the work at hand enabled the advance to proceed without delay and also enabled the Infantry to have the support of the Artillery and to keep in close touch with their transport.

KNIGHT, JOHN T.
R—Farmville, Va.
B—Poplar Hill, Va.
G. O. No. 59, W. D., 1921.

Colonel, Quartermaster Corps, U. S. Army.
As chief quartermaster and superintendent of the Army Transport Service at the port of embarkation, Newport News, Va., a position of great responsibility, in which he prepared ships for convoy and executed the manifold duties of his office with conspicuous merit.

KNISKERN, ALBERT D.
R—Manistee, Mich.
B—Monee, Ill.
G. O. No. 77, W. D., 1919.

Brigadier general, U. S. Army.
For services in the organization and development of the supply system in the general supply depot, Chicago, Ill.

KOCH, STANLEY.
R—Bozeman, Mont.
B—Bozeman, Mont.
G. O. No. 27, W. D., 1922.

Lieutenant colonel, Infantry, U. S. Army.
As executive officer of the remount service, American Expeditionary Forces, by his zeal, efficiency, and devotion to duty, he has rendered great and beneficial service in the organization of the remount service in the American Expeditionary Forces.

KOEHLER, HERMAN J.
R—Milwaukee, Wis.
B—Milwaukee, Wis.
G. O. No. 34, W. D., 1919.

Lieutenant colonel, master of the sword, U. S. Army.
At the beginning of the war he was placed in charge of the physical training in officers' training camps. These and also four divisional camps were personally visited by him. He personally instructed 200,000 officers and enlisted men of the new Army.

KRAMER, HARRY C.
R—West Collingswood, N. J.
B—Philadelphia, Pa.
G. O. No. 16, W. D., 1923.

Lieutenant colonel, Judge Advocate General's Department, U. S. Army.
In the administration of the selective service law in various capacities in the office of the Provost Marshal General, and as executive officer in that office during the period of the great drafts in the early part of 1918 he displayed superior executive ability, sound judgment, and complete devotion to his duties. His service was of immeasureable value to his Government in positions of great responsibility.

KRAUTHOFF, CHARLES R.
R—Kansas City, Mo.
B—St. Louis, Mo.
G. O. No. 59, W. D., 1919.

Brigadier general, U. S. Army.
His energy and thorough knowledge of methods and standards of supply have been of exceptional value, particularly in directing European purchases for the Quartermaster Corps and in the difficult and complex transactions attending the payments to allied and other foreign creditors of the American Government.

KREGER, EDWARD A.
R—Cherokee, Iowa.
B—Keota, Iowa.
G. O. No. 47, W. D., 1919.
Distinguished-service cross also awarded.

Brigadier general, U. S. Army.
As Acting Judge Advocate General for the American Expeditionary Forces he organized and efficiently administered his office, performing exacting duties with marked distinction. His masterful knowledge of military law, his foresight, and practical comprehension of the complex problems involved in his work enabled him to perform it with noteworthy success. His counsel was wise; his decisions were just. His services to the American Expeditionary Forces have been of great value.

KROMER, LEON B.
R—Grand Rapids, Mich.
B—Grand Rapids, Mich.
G. O. No. 62, W. D., 1919.

Colonel (Field Artillery), General Staff Corps, U. S. Army.
As assistant chief of staff of the 82d Division during the St. Mihiel offensive he displayed military attainments of a high order in the planning of operations of great moment. Later as assistant chief of staff, G-3, 1st Corps, and assistant chief of staff, G-1, 1st Army, during the Meuse-Argonne operations, his initiative, sound judgment, and tireless energy solved difficult problems of traffic control and regulation, playing an important part in the successes achieved.

KRUEGER, WALTER.
R—Cincinnati, Ohio.
B—Germany.
G. O. No. 3, W. D., 1924.

Colonel (Infantry), General Staff Corps, U. S. Army.
He served as assistant in the Bureau of Militia Affairs; assistant chief of staff, G-3, and acting chief of staff, 84th Division; assistant chief of staff, G-3, 26th Division; chief of staff, Tank Corps; instructor, Line School, Langres, France; assistant chief of staff, G-3, 4th Corps; and assistant chief of staff, G-3, 6th Corps. By his high professional attainments, superior zeal, loyal devotion to duty, soldierly character, and his dominant leadership, he has exercised a determining influence upon the commands with which he has served, and has contributed in a marked degree to the success of the military operations of our forces.

KRUMM, LOUIS R.
R—Brooklyn, N. Y.
B—Columbus, Ohio.
G. O. No. 59, W. D., 1919.

Lieutenant colonel, Signal Corps, U. S. Army.
As supervisor of radio service of the Signal Corps in France, he organized and placed in satisfactory operation this important branch. The excellent results obtained by our telephonic interception stations are due to his masterful ability and exact scientific knowledge.

KUEGLE, ALBERT S.
R—Sturgis, S. Dak.
B—Columbiana, Ohio.
G. O. No. 89, W. D., 1919.

Lieutenant colonel (Infantry), General Staff Corps, U. S. Army.
As secretary of the General Staff and of the 3d Section thereof, at General Headquarters, American Expeditionary Forces, charged with executive duties of a responsible and exacting character, he performed these duties with merited success, displaying at all times a high degree of tact, zeal, and efficiency, rendering invaluable services to the American Expeditionary Forces.

KUMPE, GEORGE E.
R—White Sulphur Springs, Mont.
B—Leighton, Ala.
G. O. No. 38, W. D., 1922.

Colonel, Signal Corps, U. S. Army.
As chief signal officer of the 3d Army Corps, by his untiring energy, devotion to duty, and excellent technical attainments, Colonel Kumpe established and maintained under great difficulties of terrain and enemy shell fire such thorough signal arrangements as to enable his corps headquarters to maintain communication at all times with the divisions of the corps and adjacent units, thereby greatly contributing to the success of his corps in the Meuse-Argonne offensive.

LADD, EUGENE F. R—Thetford Center, Vt. B—Thetford Center, Vt. G. O. No. 47, W. D., 1919.	Brigadier general, U. S. Army. While in charge of the officers' division of The Adjutant General's Office his comprehensive grasp of the new situations developing and his technical ability enabled him to perform the duties of his office with rare distinction, thus contributing greatly to the rapid organization of our new Army.
LAHM, FRANK P. R—Mansfield, Ohio. B—Mansfield, Ohio. G. O. No. 70, W. D., 1919.	Colonel, Air Service, U. S. Army. A balloon pilot of marked ability and scientific attainments, he rendered valuable services to the American Expeditionary Forces by his untiring devotion to the innumerable problems which faced the Air Service during its organization in France. His broad experience in aeronautics played an important part in the formulation of policies of the Air Service and was reflected in its successes during the St. Mihiel offensive and subsequently in the operations of the 2d Army.
LAMONT, ROBERT P. R—Evanston, Ill. B—Detroit, Mich. G. O. No. 77, W. D., 1919.	Colonel, Ordnance Department, U. S. Army. As assistant to the chief of the procurement division, later as chief of the procurement division and as a member of the claims board of the Ordnance Department, he has rendered material assistance to the Nation's industry in adjusting equitably outstanding contracts with full justice to employers and employees alike.
*LAMPERT, JAMES G. B. R—Oshkosh, Wis. B—Oshkosh, Wis. G. O. No. 59, W. D., 1919.	Lieutenant colonel, Corps of Engineers, U. S. Army. He invented, developed, and superintended the production of the standard floating footbridge equipage, which was successfully used by the 1st Army in its attack east of the Meuse, near Dun. His services in connection with the organization and development of the bridge department of the Chief Engineer's office were of inestimable value. He showed ability, great foresight, and exact scientific knowledge, and his work had an important bearing on the successes achieved by our armies. Posthumously awarded. Medal presented to widow, Mrs. James G. B. Lampert.
LANGDON, RUSSELL C. R—Brooklyn, N. Y. B—Brooklyn, N. Y. G. O. No. 69, W. D., 1919. Distinguished-service cross also awarded.	Colonel, Infantry, U. S. Army. As commanding officer of the 127th Infantry, 32d Division, he demonstrated personal courage, marked tactical ability, and military leadership of a high order. The brilliant success he achieved in the capture of Fismes during the Aisne-Marne offensive, and in the taking of Juvigny and the subsequent advance to Terny-Sorny during the Oise-Aisne offensive, was repeated later during the operations of the Meuse-Argonne, when he was given the important task of conducting the attack on La Cote Dame Marie.
LANGFITT, WILLIAM C. R—Millersburgh, Ohio. B—Wellsburg, Va. G. O. No. 12, W. D., 1919.	Major general, U. S. Army. As director of light, railways, and roads, and later as chief of utilities, he displayed great ability and marked breadth of vision. As chief engineer of the American Expeditionary Forces his brilliant professional attainments, untiring energy, and devotion to duty placed his department in a state of efficiency and enabled it to perform its important function in the most satisfactory manner.
LANGSTON, JOHN D. R—Goldsboro, N. C. B—Aurora, N. C. G. O. No. 56, W. D., 1922.	Lieutenant colonel, Judge Advocate General's Department, U. S. Army. As executive officer in charge of the selective draft in North Carolina from December, 1917, until September, 1918, by his unusual executive ability, rare tact and skill, great initiative, and resourcefulness exercised at times under most trying and novel conditions which arose in connection with the administration of the selective service act, he achieved a pronounced and conspicuous success in the performance of difficult and highly responsible duties, thereby rendering services of great value to the Government.
LANZA, CONRAD H. R—Washington, D. C. B—New York, N. Y. G. O. No. 19, W. D., 1920.	Colonel, Field Artillery, U. S. Army. As chief of the operations section, Army Artillery, 1st Army, American Expeditionary Forces, he exhibited a high order of ability and judgment. By his clear tactical conception of complex situations, by his exercise of initiative, by his untiring energy, and by his self-sacrificing devotion to duty, he contributed in a marked degree to the successful employment of all of the artillery of the 1st Army.
LASSITER, WILLIAM R—Petersburg, Va. B—Petersburg, Va. G. O. No. 12, W. D , 1919.	Major general, U. S. Army. As commander of the 51st Field Artillery Brigade, as Chief of Artillery of the 1st and 4th Army Corps in turn, and as Chief of Artillery, 2d Army, he showed himself to be a leader of conspicuous ability. His energy and sound judgment influenced greatly the successful operations of his commands on the Vesle, at the St. Mihiel salient, and in the Toul sector. He later commanded with skill and marked success the 32d Infantry Division.
LATHBURY, BENJAMIN B. R—Philadelphia, Pa. B—Philadelphia, Pa. G. O. No. 11, W. D., 1921.	Colonel, Quartermaster Corps, U. S. Army. In the adjustment and settlement of many difficult and intricate claims problems his services were distinguished by zeal, excellent judgment, and a high comprehension of the principles involved and his success has been most marked.
LEA, LUKE R—Nashville, Tenn. B—Nashville, Tenn. G. O. No. 56, W. D., 1922.	Colonel, Field Artillery, U. S. Army. As commander, 114th Field Artillery, 30th Division, he organized, trained, and handled the regiment in a skillful manner during the St. Mihiel and Meuse-Argonne offensives. By his marked tactical judgment, knowledge of artillery, and loyal devotion to duty he rendered at all times the maximum of support to the Infantry in all tasks assigned his regiment, thereby rendering in a position of great responsibility services of great value to the American Expeditionary Forces.

LEACH, GEORGE E.
   R—Minneapolis, Minn.
   B—Cedar Rapids, Iowa.
   G. O. No. 89, W. D., 1919.
   Distinguished-service cross also
   awarded.

Colonel, Field Artillery, U. S. Army.
As commanding officer of the 151st Field Artillery, 42d Division, he displayed marked qualities of leadership. Maintaining a high standard of efficiency and morale, he constantly kept his regiment in close proximity to the attacking infantry, where he was able to furnish it accurate and timely assistance, which contributed materially to the successes gained.

LEARNARD, HENRY G.
   R—Napoleon, Mich.
   B—Wright City, Mo.
   G. O. No. 13, W. D., 1919.

Brigadier general, U. S. Army.
For services in the work of reorganization and administration within The Adjutant General's Department.

LEE, BURTON J.
   R—New York, N. Y.
   B—New Haven, Conn.
   G. O. No. 59, W. D., 1919.

Lieutenant colonel, Medical Corps, U. S. Army.
As surgical consultant attached to the 2d Division, he served continuously at the front, organizing his forces for the treatment and evacuation of the casualties with skill and marked success. He displayed unusual ability in the operations before Soissons, when in an emergency he organized, personally led, and directed surgical teams which cared for hundreds of wounded soldiers at a time when adequate hospitalization could not be established.

LEE, HARRY.
   R—Washington, D. C.
   B—Washington, D. C.
   G. O. No. 95, W. D., 1919.

Colonel, United States Marine Corps.
Having taken command of the 6th Regiment, United States Marine Corps, 2d Division, prior to the attack on the Bois-de-Belleau and Bouresches, he directed the operations of his regiment with remarkable success during all the major operations in which it participated. His ability as a tactical leader and his untiring energy were reflected in the brilliant achievements of his command.

LEE, JOHN C. H.
   R—Junction City, Kans.
   B—Junction City, Kans.
   G. O. No. 59, W. D., 1919.

Colonel (Corp of Engineers), General Staff Corps, U. S. Army.
In the preparations for the drive on the St. Mihiel salient in September, and for the Argonne-Meuse offensive in October, 1918, he had charge of the detailed arrangements for and the subsequent execution of the operations of the 89th Division. The successes attained by this division were largely due to his splendid staff coordination, marked tactical ability, and sound judgment.

LEE, JOSEPH.
   R—Boston, Mass.
   B—Brookline, Mass.
   G. O. No. 74, W. D., 1919.

For services as president of the War Camp Community Service.

LEE, RAYMOND E.
   R—Kansas City, Mo.
   B—St. Louis, Mo.
   G. O. No. 38, W. D., 1922.

Colonel, Field Artillery, U. S. Army.
As executive officer to the Chief of Field Artillery from Sept. 1, 1918, to the present time, Colonel Lee acted as Chief of Field Artillery for a period of about 10 days prior to the armistice during the absence of the Chief of Field Artillery. After the armistice he successfully planned and executed the demobilization of the commissioned personnel of his arm and also organized and conducted the Field Artillery section of the Officers' Reserve Corps in a most effective manner.

LEGGE, ALEX.
   R—Chicago, Ill.
   B—Dane County, Wis.
   G. O. No. 15, W. D., 1923.

Vice chairman, War Industries Board.
In connection with the operations of the War Industries Board during the World War. As a member of the board he rendered, through his broad vision, distinguished capacity, and business ability, services of inestimable value in marshaling the industrial forces of the Nation and mobilizing its economic resources—marked factors in assisting to make military success attainable. As vice chairman of the board itself and as chairman of its requirements division he rendered, through his untiring efforts and devotion to duty, exceptionally valuable service to the War Department in matters connected with the supply of raw materials and general munitions to the Army.

LEGGE, BARNWELL R.
   R—Charleston, S. C.
   B—Charleston, S. C.
   G. O. No. 56, W. D., 1922.
   Distinguished-service cross also
   awarded.

Lieutenant colonel, Infantry, U. S. Army.
As company, battalion, and regimental commander of the 26th Infantry throughout hostilities he successfully led his command through each of the offensives of the 1st Division. By his superior tactical judgment, manifest ability, and tireless energy, coupled with unusual leadership, he contributed in a brilliant manner to the success of the 1st Division. Later, as a division adjutant, he gave further proof of the highest qualities of military character, again demonstrating conspicuous service in a position of great responsibility.

LEHMAN, HERBERT H.
   R—Washington, D. C.
   B—New York, N. Y.
   G. O. No. 103, W. D., 1919.

Colonel, General Staff Corps, U. S. Army.
While with the purchase, storage, and traffic division of the General Staff as chief of the purchase branch, member of the board of contract adjustment, chairman of the advisory board on sales and contract termination, member of the War Department claims board, and assistant director of purchase, storage, and traffic, General Staff, his large business experience, breadth of vision, and sound judgment have been of inestimable value in formulating and in supervising the execution of the methods and policies followed in the cancellation of war contracts and obligations and in the settlement and adjustment of terminated obligations.

LEITCH, JOSEPH D.
   R—Clay Center, Nebr.
   B—Montague, Mich.
   G. O. No. 56, W. D., 1922.

Colonel (Infantry), General Staff Corps, U. S. Army.
As chief of staff of the American Expeditionary Forces in Siberia he gave proof of his great breadth of vision, keen foresight, sound judgment, and tact. By his brilliant professional attainments, coupled with great diplomacy, he handled most ably the many delicate situations with which he was confronted. His fine soldierly qualities were at all times outstanding, and by his masterful grasp of the situation he was able to meet successfully each new and difficult problem with which he was faced. He rendered most conspicuous services of inestimable value to the Government in a place of great responsibility and at a time of gravest importance.

LEJEUNE, JOHN A.
  R—Pointe Coupee Parish, La.
  B—Pointe Coupee Parish, La.
  G. O. No. 12, W. D., 1919.
Major general, U. S. Marine Corps.
He commanded the 2d Division in the successful operations of Thiaucourt, Massif Blanc Mont, St. Mihiel, and on the west bank of the Meuse. In the Argonne-Meuse offensive his division was directed with such sound military judgment and ability that it broke and held, by the vigor and rapidity of execution of its attack, enemy lines which had hitherto been considered impregnable.

LEONARD, GRACE E.
  R—New York, N. Y.
  B—Newark, N. J.
  G. O. No. 9, W. D., 1923.
First lieutenant, chief nurse, Army Nurse Corps, U. S. Army.
As assistant director of the nursing service, American Expeditionary Forces, during the World War, she rendered invaluable assistance at the headquarters of Base Section No. 3. She supervised all nursing activities of the United States Army hospitals established in England, and in many ways promoted the general welfare of the nursing staff and the sick and wounded under their care.

LEWIS, DEAN D.
  R—Chicago, Ill.
  B—Kewanee, Ill.
  G. O. No. 56, W. D., 1922.
Lieutenant colonel, Medical Corps, U. S. Army.
As chief of the surgical service of Evacuation Hospital No. 5 during the operations on the Marne, and the St. Mihiel, Meuse-Argonne, and Ypres-Lys offensives, by his tireless energy, organizing ability, and unusual surgical skill, he successfully demonstrated that war wounds could be operated upon in large numbers in front-line hospitals with limited personnel, thus conserving many lives among combat troops, thereby rendering in a position of great responsibility conspicuous service to the American Expeditionary Forces.

LEWIS, EDWARD M.
  R—New Albany, Ind.
  B—New Albany, Ind.
  G. O. No. 12, W. D., 1919.
Major general, U. S. Army.
He commanded with distinction the 30th American Division during its successful operations in Belgium with the 2d British Army, and later, with the 4th British Army in the offensive which resulted in the breaking of the enemy's Hindenburg line. During all these operations he exhibited great ability, determined energy, and marked devotion to duty.

LEWIS, FREDERICK W.
  R—Atlanta, Ga.
  B—Buffalo, N. Y.
  G. O. No. 105, W. D., 1919.
Colonel, Adjutant General's Department, U. S. Army.
As officer in charge of the publication division of The Adjutant General's Office. To his painstaking efforts, tact, energy, and zeal are due the accuracy with which publications issued to the Military Establishment through The Adjutant General of the Army were drawn and the promptness with which they were distributed.

LEWIS, GILBERT N.
  R—Berkeley, Calif.
  B—Weymouth, Mass.
  G. O. No. 56, W. D., 1922.
Lieutenant colonel, Chemical Warfare Service, U. S. Army.
By his unusual energy, marked ability, and high technical attainments he rendered extremely valuable service by securing first-hand data on the uses and effects of gas and submitting reports of such value that they became fundamentals upon which the gas-warfare policies of the American Expeditionary Forces were thereafter largely based. Later, as chief of the defense division, Chemical Warfare Service, he obtained a high state of efficiency in the protection of our officers and soldiers against enemy gas and furthered the successes of American arms by securing a better and more effective use of gas, especially mustard gas, against the enemy, thereby rendering services of great value to our Government.

LIGGETT, HUNTER.
  R—Birdsboro, Pa.
  B—Reading, Pa.
  G. O. No. 136, W. D., 1918.
Lieutenant general, U. S. Army.
As commander of the 1st Army of the American Expeditionary Forces, he commanded the 1st Army Corps and perfected its organization under difficult conditions of early service in France; engaged in active operations in reduction of the Marne salient and of the St. Mihiel salient, and participated in the actions in the Forest of Argonne; in command of 1st Army when German resistance was shattered west of the Meuse.

LINCOLN, CHARLES S.
  R—Ames, Iowa.
  B—Boonsboro, Iowa.
  G. O. No. 28, W. D., 1921.
Colonel (Infantry), General Staff Corps, U. S. Army.
For services while a member of the G-1 section of the General Staff, General Headquarters, American Expeditionary Forces, later as Deputy Chief of Staff, G-1, and later as Assistant Chief of Staff, G-1.

LINDSEY, JULIAN R.
  R—Irwinton, Ga.
  B—Irwinton, Ga.
  G. O. No. 59, W. D., 1919.
Brigadier general, U. S. Army.
The brilliant and successful attack of the 164th Infantry Brigade, 82d Division, commanded by him, in the Argonne Forest, showed a spirit of aggressiveness and leadership of a high order. The tactical advantage attained in this action, whereby St. Juvin and Grand Pre were laid open to attack, was largely due to his ability and energy.

LINDSLEY, HENRY D.
  R—New York, N. Y.
  B—Nashville, Tenn.
  G. O. No. 59, W. D., 1919.
Colonel, Adjutant General's Department, U. S. Army.
He conducted with extreme devotion to duty and marked zeal the many activities of the War Risk Insurance Bureau in France. Due to his executive ability he contributed very largely to the successful development, extension, and administration of that important service.

LININGER, CLARENCE.
  R—Wabash, Ind.
  B—Huntington, Ind.
  G. O. No. 44, W. D., 1919.
First lieutenant, 13th Cavalry, U. S. Army.
Lieutenant Lininger, while in action at Parral, Mexico, Apr. 12, 1916, proceeded under fire to the rescue of a dismounted man of his command who was in danger of falling into the hands of the enemy, and, taking him up behind him (Lieutenant Lininger) on his horse, carried him to safety.

LITTELL, ISAAC W.
  R—Elizabeth, N. J.
  B—Elizabeth, N. J.
  G. O. No. 105, W. D., 1919.
Brigadier general, U. S. Army.
As chief of the cantonment division of the Quartermaster General's Office he was charged with the task of building the camps and cantonments of the Army raised in summer of 1917 under conditions imposing almost insuperable obstacles. His completion of this task is a conspicuous example of the exercise of qualities of mind and character making up the highest type of officer.

LITTLE, BASCOM
R—Cleveland, Ohio.
B—Cleveland, Ohio.
G. O. No. 126, W. D., 1919.

Colonel, Ordnance Department, U. S. Army.
As chief of the production district in which were manufactured practically all the machine guns and automatic rifles supplied for the United States Army and later as special assistant to the Chief of Ordnance, in charge of the production of small arms, automatic rifles, machine guns, small-arms ammunition, etc., he successfully organized the industry of the country for the production of these items to meet the needs of the United States Army.

LIVERMORE, PHILIP W.
R—New York, N. Y.
B—New York, N. Y.
G. O. No. 62, W. D , 1919.

Captain, Ordnance Department, U. S. Army.
As director of regional and ministerial liaison and later as deputy for the chief liaison officer, he displayed unusual administrative ability and rare judgment. By untiring effort and devotion to duty, he was largely instrumental in placing American liaison on a sound footing. His forceful personality and keen intelligence contributed largely to the successes achieved by his department. At all times he showed marked initiative, unflagging energy, and zeal in the performance of exacting and delicate tasks.

LLEWELLYN, FRED W.
R—Seattle, Wash.
B—Hillsboro, Oreg.
G. O. No. 62, W. D., 1919.

Lieutenant colonel, Infantry, U. S. Army.
Assuming the responsibilities of the first section of the general staff of the 28th Division five days before the Meuse-Argonne offensive, he efficiently coordinated the several services. By his constant vigilance and ceaseless efforts the entire system of supply, traffic, and evacuation operated during the advance of more than 10 kilometers, in accordance with the plans he had arranged. He was tireless in his energy and devotion to important duties, displaying military attainments of high order.

LLOYD, CHARLES R.
R—San Francisco, Calif.
B—England.
G. O. No. 89, W. D., 1919.

Colonel, Field Artillery, U. S. Army.
He commanded with distinction the 10th Field Artillery, displaying marked ability as an artillerist. His unflagging zeal and sound judgment was revealed by the success achieved by his regiment in furthering the gains achieved by the 3d Division in its operations in the field.

LOCHRIDGE, P. D.
R—Rara Avis, Miss.
B—Bexar, Ala.
G. O. No. 75, W. D., 1919.

Brigadier general, U. S. Army.
For services to the allied and associated governments as Chief of Staff, American Section, Supreme War Council.

LOCKE, MORRIS E.
R—Cincinnati, Ohio.
B—Salt Lake City, Utah.
G. O. No. 95, W. D., 1919.

Colonel, Field Artillery, U. S. Army.
He commanded, with marked skill and initiative, the 102d Field Artillery, 26th Division, during the Chateau-Thierry campaign, where at all times he furnished valuable support to the advancing infantry. Later he served creditably as an instructor at the Army General Staff College at Langres, rendering important services to the American Expeditionary Forces.

LOGAN, JAMES A., Jr.
R—Bala, Pa.
B—Philadelphia, Pa.
G. O. No. 59, W. D., 1919.

Colonel (Quartermaster Corps), General Staff, U. S. Army.
His marked administrative ability enabled him to assist most ably in the direction of important operations while on duty at G-1, General Headquarters American Expeditionary Forces, as deputy chief of staff, 2d Army, and G-1, 3d Army. As American representative with the Franco-American War Affairs Commission, at Paris, he displayed unfailing tact, energy, and sound judgment in handling the intricate details of the relations between the French and American authorities, achieving signal success. His high military attainments were shown in the success with which he performed duties of vital moment.

LONERGAN, THOMAS C.
R—St. Louis, Mo.
B—St. Louis, Mo.
G. O. No. 14, W. D., 1923.

Lieutenant colonel (Infantry), General Staff Corps, U. S. Army.
As an instructor at the Army General Staff College, American Expeditionary Forces, he displayed high professional attainments and unfailing energy, performing services of inestimable worth in connection with the instruction and training of officers for General Staff duty. He prepared the "Notebook for the General Staff Officer" and the "Provisional Staff Manual," and was mainly responsible for the "Handbook of Division and Brigade Commanders," all of which proved to be most valuable books. As adjutant of the 159th Infantry Brigade, 80th Division, during the St. Mihiel offensive, and later as a member of the interallied games committee he gave further proof of his sterling ability, sound judgment, and keen foresight. In all of these positions he rendered most conspicuous services to the Government.

LONGAN, RUFUS E.
R—Sedalia, Mo.
B—Sedalia, Mo.
G. O. No. 124, W. D., 1919.

Brigadier general, U. S. Army.
For services as chief of staff, port of embarkation, Hoboken, N. J., from Dec. 15, 1917, to Dec. 16, 1918.

LONGLEY, FRANCIS F.
R—Dobbs Ferry, N. Y.
B—Chicago, Ill.
G. O. No. 59, W. D., 1919.

Colonel, Corps of Engineers, U. S. Army.
He has been in charge of the water supply service, and as commanding officer of the 26th Engineers, a water-supply regiment, since the fall of 1917. His untiring energy, unusual initiative, and good judgment have to a marked degree, been responsible for the plentiful supply of pure drinking water to the combatant troops, thereby materially assisting in maintaining the unusually low rates in sickness among our troops.

LORD, HERBERT M.
R—Rockland, Me.
B—Rockland, Me.
G. O. No. 47, W. D., 1919.

Brigadier general, U. S. Army.
As assistant to the Quartermaster General and later as Director of Finance. As such he was responsible for and had authority over the preparation of estimates, disbursements, money accounts, property accounts, finance reports, and pay and mileage of the Army. The success of the Finance Department was, in a large measure, due to his breadth of vision, executive ability, initiative, and energy.

LOREE, JAMES T.
  R—Albany, N. Y.
  B—Logansport, Ind.
  G. O. No. 116, W. D., 1919.

Colonel, Quartermaster Corps, U. S. Army.
He served in turn as assistant quartermaster of the 27th Division, as quartermaster of the 30th Division, and in the Provost Marshal General's Department, American Expeditionary Forces, in all of which capacities he displayed exceptional ability. His good judgment, combined with a knowledge of methods and high professional attainments, resulted in a superior standard of efficiency, reflecting the greatest credit upon himself and enabling him to render most valuable services.

LORENZ, WILLIAM F.
  R—Mendota, Wis.
  B—New York, N. Y.
  G. O. No. 59, W. D., 1921.

Major, Medical Corps, U. S. Army.
As commanding officer of Field Hospital No. 127, and while in personal charge of the Triage (sorting station for wounded) of the 32d Division during the combat activities of that division on the Marne, Oise-Aisne, and in the Meuse-Argonne, he so displayed indefatigable zeal and exceptionally good judgment in sorting, caring for, and evacuating thousands of wounded as to directly result in the saving of many lives.

LOTT, ABRAHAM G.
  R—Abilene, Kans.
  B—Gettysburg, Pa.
  G. O. No. 19, W. D., 1922.

Colonel, Adjutant General's Department, U. S. Army.
For services in The Adjutant General's office in connection with the procurement of commissioned personnel. Later, as chief of staff, 12th Division, and executive officer at Camp Devens, Mass., by his untiring energy, loyalty, and organizing ability he aided materially in the rapid demobilization of troops passing through this camp.

LOVE, JAMES M., Jr.
  R—Fairfax Courthouse, Va.
  B—Fairfax Courthouse, Va.
  G. O. No. 89, W. D., 1919.

Colonel, Infantry, U. S. Army.
As adjutant general, 2d Army Corps, and later as commanding officer of the 319th Infantry, 80th Division, he rendered services of great credit. By his marked tactical ability and unceasing energy he contributed materially to the successes achieved by the 80th Division in the Meuse-Argonne offensive.

LUBEROFF, GEORGE.
  R—Lexington, Ky.
  B—New York, N. Y.
  G. O. No 108, W. D., 1919.

Lieutenant colonel, Quartermaster Corps, U. S. Army.
As chief quartermaster of the 1st Army, by his great energy, complete experience, and loyal efforts, he maintained an efficient service and kept a steady flow of all necessary quartermaster supplies to the 1st Army, rendering services of great value to the American Expeditionary Forces.

LUTHER, WILLARD D.
  R—Milton, Mass.
  B—Attleboro, Mass.
  G. O. No. 13, W. D., 1923.

Lieutenant colonel, Field Artillery, U. S. Army.
As assistant chief of staff, G-3, and brigade adjutant, 51st Field Artillery Brigade, during the occupancy of the Boucq sector, the Aisne-Marne, and St. Mihiel offensives, and as assistant to assistant chief of staff, G-3, headquarters, 1st Army Artillery, during the Meuse-Argonne operations, he displayed outstanding executive ability, leadership, and technical skill of a high order, these qualities coupled with unremitting devotion to duty contributing in a material way to the successful operations of the organizations with which he served.

LYLE, HENRY H. M.
  R—New York, N. Y.
  B—Ireland.
  G. O. No. 59, W. D., 1921.

Colonel, Medical Corps, U. S. Army.
As director of the Army evacuation ambulance companies and sections of the 1st Army of the American Expeditionary Forces, a position of great responsibility. During the St. Mihiel and Meuse-Argonne offensives he so directed the functioning of the ambulances that, in spite of the great shortage of these, he was able at all times to transport the wounded expeditiously, thereby saving many lives and enhancing the morale of the combatant troops. By his eminent surgical skill he has devised a new practical method for the treatment of gunshot fractures. He has rendered service of much value.

LYNCH, CHARLES.
  R—Syracuse, N. Y.
  B—Syracuse, N. Y.
  G. O. No. 111, W. D., 1919.

Colonel, Medical Corps, U. S. Army.
As port surgeon, port of embarkation, Newport News, Va., his service in governing and controlling the agencies for caring for sick and wounded soldiers, protecting them against diseases, and safeguarding them prior to and during transport overseas were conspicuous.

LYNCH, GEORGE A.
  R—Blairstown, Iowa.
  B—Blairstown, Iowa.
  G. O. No. 55, W. D., 1920.

Lieutenant colonel (Infantry), General Staff Corps, U. S. Army.
As a member of the training section, General Headquarters, he was chiefly responsible for the revision of the Infantry Drill Regulations. In this important task he displayed a broad grasp of the tactical lessons of the war and showed sound judgment in adapting their principles to American needs, capabilities, and characteristics, thereby rendering services of signal worth to the American Expeditionary Forces.

LYNN, CLARK.
  R—Chicago, Ill.
  B—Hartford, Ind.
  G. O. No. 56, W. D., 1922.

Lieutenant colonel (Infantry), General Staff Corps, U. S. Army.
As assistant chief of staff, G-3, 91st Division, from July, 1918, to April, 1919, by his high professional attainments and marked ability he rendered valuable assistance in the staff work during the St. Mihiel, Meuse-Argonne, and Ypres-Lys offensives, his efficiency contributing materially to the success of his division in these operations.

*LYON, LEROY S.
  R—Richmond, Va.
  B—Petersburg, Va.
  G. O. No. 68, W. D., 1920.

Major general, U. S. Army.
As brigadier general commanding the 65th Field Artillery Brigade, 40th Division, he displayed splendid qualities of leadership and organizing ability and by his enthusiasm and energy he developed his brigade to a high state of efficiency. Later, as major general, commanding the 31st Division during its training, he exhibited marked tactical judgment, and his skill and leadership were largely responsible for the success achieved in perfecting the organization and training of his division.
Posthumously awarded. Medal presented to widow, Mrs. Leroy S. Lyon.

LYSTER, THEODORE C.
R—Detroit, Mich.
B—Fort Larned, Kans.
G. O. No. 34, W. D., 1919.

Colonel, Medical Corps, U. S. Army.
For duty rendered in the office of the Surgeon General as Chief of the Air Service Division.

McADAMS, JOHN P.
R—Hawesville, Ky.
B—Hawesville, Ky.
G. O. No. 59, W. D., 1919.

Colonel (Infantry), General Staff Corps, U. S. Army.
He served with marked distinction as chief of staff of the lines of communication and as deputy chief of staff of the Services of Supply. He administered the affairs with which he was intrusted with noteworthy and conspicuous efficiency, energy, and ability.

McAFEE, LARRY B.
R—Delphi, Ind.
B—Delphi, Ind.
G. O. No. 49, W. D., 1922.

Lieutenant colonel, Medical Corps, U. S. Army.
As chief surgeon, district of Paris, France, by his unusual executive ability, sound judgment, and high professional attainments he successfully handled a position of great responsibility presenting many difficult problems, thereby bringing comfort and health to thousands of our sick and wounded and contributing materially to the success of the American Expeditionary Forces.

McALEXANDER, ULYSSES G.
R—McPherson, Kans.
B—Dundas, Minn.
G. O. No. 59, W. D., 1919.
Distinguished-service cross also awarded.

Brigadier general, U. S. Army.
He commanded the 38th Infantry, 3d Division, with marked distinction in repelling the German attack at Mezy, south of the Marne, in July, 1918. He exhibited particular skill and energy as a brigade commander in the operations at the St. Mihiel salient and in the Argonne-Meuse offensive. The successful accomplishment of the missions of his brigade in all cases were in a large measure due to his sound judgment and leadership.

McANDREW, JAMES W.
R—Scranton, Pa.
B—Hawley, Pa.
G. O. No. 136, W. D., 1918.

Major general, U. S. Army.
As Chief of Staff of the American Expeditionary Forces, the development of the Army schools in France is largely due to his marked ability as an organizer and to his brilliant professional attainments. As Chief of Staff of the American Expeditionary Forces during the period of active operations, he has met every demand of his important position; by his advice and decisions he has materially contributed to the success of these forces; and he has at all times enjoyed in full the confidence of the Commander in Chief.

McANDREW, JOSEPH A.
R—Bentonville, Ark.
B—Osage Mills, Ark.
G. O. No. 87, W. D., 1919.

Lieutenant colonel, Infantry, U. S. Army.
He served with distinguished ability as an instructor in the use of infantry weapons at the 1st Corps School and also as director of the Infantry Specialists' School at Langres. Later, as a member of the training section of the General Staff, he supervised the instruction at the various corps schools and was directly responsible for the maintenance of sound tactical training, securing especially brilliant results in the training of infantry, rendering services of marked merit to the American Expeditionary Forces.

McANDREWS, JOSEPH R.
R—Chicago, Ill.
B—Chicago, Ill.
G. O. No. 47, W. D., 1919.

Colonel (Cavalry), General Staff, U. S. Army.
As senior officer in the small group of the operations division, General Staff, designated as the section in charge of priorities of equipment and shipment, he was charged with the handling of the whole matter of preparing units for movement to the ports for oversea service, and is now engaged in the reverse process of moving returning units from the ports to camps for demobilization, all of which has been marked by conspicuous ability and meritorious service to the Government.

MacARTHUR, DOUGLAS.
R—Milwaukee, Wis.
B—Little Rock Barracks, Ark.
G. O. No. 59, W. D., 1919.
Distinguished-service cross and oak-leaf cluster also awarded.

Brigadier general, U. S. Army.
He served with credit as chief of staff of the 42d Division in the operations at Chalons and at the Chateau-Thierry salient. In command of the 84th Infantry Brigade he showed himself to be a brilliant commander of skill and judgment. Later he served with distinction as commanding general of the 42d Division.

McARTHUR, JOHN C.
R—Aberdeen, S. Dak.
B—Plainview, Minn.
G. O. No. 15, W. D., 1923.

Colonel, Infantry, U. S. Army.
As commanding officer, 326th Infantry, 82d Division, from the date of its organization and during its occupancy of the Somme, Toul, Nancy, and Argonne sectors, May 10, 1918, to Nov. 19, 1918, he displayed sound tactical judgment, unremitting attention and devotion to duty, these qualities inspiring the officers and men of his command, bringing their morale to a high pitch and inciting them to a rare devotion to duty. His services as a regimental commander contributed in a very marked way to the successful operations of the 82d Division and of the American Expeditionary Forces in France.

McCAIN, HENRY P.
R—Jackson, Miss.
B—Carroll County, Miss.
G. O. No. 18, W. D., 1919.

Major general, The Adjutant General, U. S. Army.
In administering the Adjutant General's Department during the early period of the war, through his efficient management this department was able to meet the excessive burdens placed upon it.

McCAW, WALTER D.
R—Richmond, Va.
B—Richmond, Va.
G. O. No. 12, W. D., 1919.

Colonel, Medical Corps, U. S. Army.
His counsel and advice in the earlier stages of the operations of the American Expeditionary Forces were of particular benefit to the effective work of the Medical Department. As chief surgeon of the American Expeditionary Forces, in the later operations in the field, he maintained the splendid efficiency of that department at a critical time and solved each new problem presented with wisdom and marked ability.

McCLEAVE, EDWARD G_____
   R—Berkeley, Calif.
   B—St. Louis, Mo.
   G. O. No. 56, W. D., 1922.

Lieutenant colonel (Infantry), General Staff, U. S. Army.
Serving as assistant to the assistant chief of staff, G-4, Paris group and 1st Army, from January to December, 1918, he displayed great initiative, devotion to duty, and executive ability during the operations in and around Chateau-Thierry in June and July, while the divisions of the American Army were serving with the 6th French Army and during the St. Mihiel and Meuse-Argonne operations of the 1st Army, assisted in the organization of the section and in a position of great responsibility contributed materially to the successful accomplishments of the supply of the troops engaged in those operations.

McCLEAVE, ROBERT_____
   R—California.
   B—Fort Union, N. Mex.
   G. O. No. 116, W. D., 1919.

Colonel (Infantry), General Staff Corps, U. S. Army.
As G-3 of the 1st Army from July 25 to Oct. 15, 1918, during the Chateau-Thierry, St. Mihiel, and Meuse-Argonne operations, he displayed marked ability. Later, in the midst of operations, he was appointed chief of staff of the 3d Division. In gaining immediate and complete control of a difficult situation and in coordinating the work of the new staff, he showed conspicuous ability, and by his inspiring example of energy and zeal, he was largely responsible for the successes achieved by the 3d Division at Claire-Chenes Wood and the Bois-de-Foret.

McCLELLAN, BENJAMIN F_____
   R—Tallulah, La.
   B—Paulding Plantation, Miss
   G. O. No. 87, W. D., 1919

Lieutenant colonel, Infantry, U. S. Army.
Attached to the 5th Section of the General Staff, he displayed high professional attainments and marked executive ability in the general supervision of the entire group of army schools. As an inspector-instructor of infantry, his influence was an important factor in securing the correct tactical training of that arm, rendering creditable services to the American Expeditionary Forces.

McCLOSKEY, MANUS_____
   R—Pittsburgh, Pa.
   B—Pittsburgh, Pa.
   G. O. No. 53, W. D., 1920.

Colonel, Field Artillery, U. S. Army.
While in command of the 12th Field Artillery during all its operations with the 2d Division, until Aug. 16, 1918, he displayed marked ability and efficiency. He especially distinguished himself during the operations of the 2d Division. at the Bois-de-Belleau and Bouresches, when he commanded in addition to his own regiment, the 37th Field Artillery, French Army. By this service he contributed in no small measure to the success of the Infantry brigade in these operations. Later, as commanding general, 152d Field Artillery Brigade, he rendered able support to the attacking Infantry of the 77th Division.

McCORD, JAMES H_____
   R—St. Joseph, Mo.
   B—Savannah, Ga.
   G. O. No. 56, W. D., 1922.

Lieutenant colonel, Inspector General's Department, U. S. Army.
As executive officer in charge of the selective draft in Missouri, by his unusual executive ability, rare tact and skill, great initiative, and resourcefulness at times under most trying and novel conditions which arose in connection with the administration of the selective service act, he achieved a pronounced and conspicuous success in the performance of difficult and highly responsible duties, thereby rendering services of great value to the Government.

McCORMICK, CHESTER B_____
   R—Lansing, Mich.
   B—Petrolia, Pa.
   G. O. No. 3, W. D., 1922.

Colonel, Field Artillery, U. S. Army.
For services while commanding the 119th Field Artillery and at times the 57th Artillery Brigade during the Meuse-Argonne offensive. During these operations he displayed marked judgment and devotion to duty, and by the skillful handling of his command contributed materially to the success of the 57th Artillery Brigade in supporting the 32d and at other times five other divisions.

McCORMICK, ROBERT R_____
   R—Chicago, Ill.
   B—Chicago, Ill.
   G. O. No. 15, W. D., 1923.

Colonel, Field Artillery, U. S. Army.
As commander of the 1st Battalion, 5th Field Artillery, in the Ansauville sector and in the Cantigny sector, France, between Jan. 18 and May 28, 1918, and as lieutenant colonel, 122d Field Artillery, 33d Division, May 13 to July 29, 1918, and colonel, 61st Field Artillery, July 30 to Dec. 31, 1918, he displayed rare leadership and organizing ability, unusual executive ability, and sound technical judgment. By his ceaseless energy and his close supervision of training, discipline, and command in action against the enemy he contributed materially to the successful operations of the Artillery of the American Expeditionary Forces.

McCOY, FRANK R_____
   R—Lewistown, Pa.
   B—Lewistown, Pa.
   G. O. No. 12, W. D., 1919.

Brigadier general, U. S. Army.
As secretary of the General Staff, American Expeditionary Forces, his services were of particular value in the original organization of the forces in France. Later, in command of the 165th Infantry, 42d Division, in the Baccarat sector, and then in command of the 63d Infantry Brigade in the difficult fighting east of Reims, he had a prominent part in the successes achieved.

McCOY, JOHN C_____
   R—Paterson, N. J.
   B—New York, N. Y.
   G. O. No. 126, W. D., 1919.

Lieutenant colonel, Medical Corps, U. S. Army.
He served with conspicuous success as commanding officer of American Red Cross Hospital No. 111, at Jouy-sur-Morin and Chateau-Thierry from June to August, 1918. Though he was hampered by insufficient personnel and equipment, he nevertheless succeeded in caring for a large number of wounded from the Marne offensive, rendering invaluable services to the American Expeditionary Forces.

McCOY, ROBERT B_____
   R—Sparta, Wis.
   B—Kenosha, Wis.
   G. O. No. 69, W. D., 1919.

Colonel, Infantry, U. S. Army.
In command of the 128th Infantry throughout all the major operations in which the 32d Division participated, he proved himself a leader of sound judgment and exceptional ability. During the Oise-Aisne offensive he skillfully handled the delicate maneuver of straightening and changing the front on the left bank of his brigade during the attack on Terny-Sorny and later during the Meuse-Argonne offensive. In the attack on the Kremhilde-Stellung he performed another tactical operation of a high order in a flank movement which resulted in the taking of the town of Romagne.

McCREA, JAMES A.
  R—Woodmere, N. Y.
  B—Philadelphia, Pa.
  G. O. No. 59, W. D., 1919.

Colonel, Corps of Engineers, U. S. Army.
He rendered especially efficient services to the American Expeditionary Forces while acting as general manager of the transportation service and later as deputy director general of transportation in the advance section. He handled his duties in a most efficient manner, showing marked ability, great zeal, and energy.

McCULLAGH, SAMUEL.
  R—New York, N. Y.
  B—Philadelphia, Pa.
  G. O. No. 49, W. D., 1922.

Lieutenant colonel, Medical Corps, U. S. Army.
As chief of the evacuation branches, G-4, Paris group, and G-4, 1st Army, he was charged with the supervision, coordination, and control of evacuation and hospitalization of the sick and wounded. By his great foresight, initiative, judgment, devotion to duty, and the helpful cooperation exercised by him, he played an important part in the successful treatment and evacuation of many thousands of sick and wounded under the most adverse conditions, performing this important and responsible duty with conspicuous ability, thereby contributing materially to the success of the American Expeditionary Forces.

McDONALD, JOHN B.
  R—Athens, Ala.
  B—Athens, Ala.
  G. O. No. 59, W. D., 1919.
  Distinguished-service cross also awarded.

Brigadier general, U. S. Army.
While commanding the 181st Infantry Brigade during the advance of the 91st Division from Foret-de-Hesse, Argonne, France, in September, 1918, he was instrumental in the successes achieved. He directed the attack in person, and by his example of personal courage and by his sound tactical orders he so inspired his brigade that it was enabled to capture and hold a most important position.

MacDONNELL, JOHN G.
  R—Gordon, Wash.
  B—Spencer, Mass.
  G. O. No. 9, W. D., 1923.

Major, Cavalry, U. S. Army.
As provost marshal, 1st Army, during the Meuse-Argonne offensive, he was responsible for the maintenance of order in the extensive Army area in rear of corps areas, for the apprehension of stragglers and wrongdoers in that area, for the military police of areas relinquished by each corps as a consequence of the rapid advance of the corps of the 1st Army after Nov. 1, 1918. Due to his tact, resourcefulness, and marked efficiency he accomplished all of his missions in a highly meritorious manner, thereby contributing materially to the successes of the 1st Army.

*McDONOUGH, JOHN.
  R—Beech Grove, Ind.
  B—Plattsmouth, Nebr.
  G. O. No. 59, W. D., 1921.

Major, Transportation Corps, U. S. Army.
As general foreman of the St. Nazaire Locomotive Erection Shops, he managed this plant from its organization in September, 1917, until June, 1918, when he was appointed superintendent, and thence performed the important duties of this later office until his death in November, 1918. It is due to his energy, ability, and devotion to duty that such a great number of American locomotives erected in France were turned out with record speed. He has rendered services of much value.
Posthumously awarded. Medal presented to widow, Mrs. John McDonough.

MacDOWELL, CHARLES H.
  R—Chicago, Ill.
  B—Lewistown, Ill.
  G. O. No. 15, W. D., 1923.

Director of the Chemicals Division, War Industries Board.
In connection with the operations of the War Industries Board during the World War, in his position as director of one of the sections of the board he rendered, through his broad vision, distinguished capacity, and business ability, services of inestimable value in marshaling the industrial forces of the Nation and mobilizing its economic resources—marked factors in assisting to make military success attainable. As director of chemicals, he rendered, through his untiring efforts and devotion to duty, exceptionally valuable service to the War Department in connection with the procurement of chemicals, particularly those elements used in the manufacture of explosives for the Army.

McDOWELL, RALPH W.
  R—Altoona, Pa.
  B—Altoona, Pa.
  G. O. No. 38, W. D., 1921.

Lieutenant commander, Medical Corps, U. S. Navy.
For services as sanitary inspector and surgeon of the Arrondissement of Tours, France.

McFADDEN, GEORGE.
  R—Villanova, Pa.
  B—Philadelphia, Pa.
  G. O. No. 46, W. D., 1919.

Representative in France of the War Trade Board.
He represented in France with high ability the War Trade Board of the State Department. In close liaison with the General Purchasing Board of the American Expeditionary Forces, he had a guiding influence in determining the methods of the invaluable cooperation of the War Trade Board in the supply-procurement efforts of the Army in France and in neutral and other allied countries. With untiring energy, sound judgment, great ability, and devoted purpose he cooperated in many matters of vital importance to the American Expeditionary Forces.

McFADDEN, JAMES FRANKLIN.
  R—Rosemont, Pa.
  B—Philadelphia, Pa.
  G. O. No. 56, W. D., 1922.

Lieutenant colonel, Air Service, U. S. Army.
As officer in charge of the leave bureau at headquarters, services of supply, Tours, France, from Dec. 14, 1918, until June, 1919, by his energy, loyal devotion to duty, unusual executive and administrative ability he was largely responsible for the successful organization and most efficient functioning of that highly important bureau which provided transportation, quarters, subsistence, recreation, and amusement for over 400,000 men on leave from the front in 19 different leave areas in southern France. In a position of great responsibility he rendered conspicuous services to the American Expeditionary Forces.

McFarland, Earl.
R—Topeka, Kans.
B—Topeka, Kans.
G. O. No. 38, W. D., 1921.

Colonel, Ordnance Department, U. S. Army.
First in charge of the design, development, and production of all machine guns, automatic rifles, and accessories thereto for the Army of the United States, for service in organizing the industries of the country to meet the unprecedented demands for automatic arms created after the entrance of the United States into the World War, and later as special assistant to the Chief of Ordnance in charge of all matters pertaining to small arms, automatic arms, and equipment.

McGlachlin, Edward F., Jr.
R—Stevens Point, Wis.
B—Fond du Lac, Wis.
G. O. No. 12, W. D., 1919.

Major general, U. S. Army.
As commander of the Artillery of the 1st Army in its organization and subsequent operations he solved the difficult problems involved with rare military judgment. In the St. Mihiel and Argonne-Meuse offensives his qualities as a leader were demonstrated by the effective employment of Artillery that was planned and conducted under his direction. He later commanded with great ability and success the 1st Infantry Division of the American Expeditionary Forces.

McGuire, Stuart.
R—Richmond, Va.
B—Staunton, Va.
G. O. No. 59, W. D., 1921.

Lieutenant colonel, Medical Corps, U. S. Army.
As commanding officer of Base Hospital No. 45 he received and cared for a very large number of wounded evacuated to that hospital during the Battle of the Argonne, in September and October, 1918. By his administrative and professional skill he has reflected credit upon the Medical Department of the Army.

McIndoe, James F.
R—Lonaconing, Md.
B—Lonaconing, Md.
G. O. No. 56, W. D., 1921.

Colonel, Corps of Engineers, U. S. Army.
He rendered valuable services in bringing this unit to a high state of discipline and training which manifested itself later in all operations in which it participated. Later, as director of military engineers, a position of great responsibility, he rendered services of much worth.
Posthumously awarded. Medal presented to widow, Mrs. Irene McIndoe.

McIntyre, Frank.
R—Montgomery, Ala.
B—Montgomery, Ala.
G. O. No. 25, W. D., 1919.

Major general, U. S. Army.
As executive assistant to the Chief of Staff, his breadth of view and sound judgment have contributed materially to the formulation and carrying out of policies essential to the operation of the military establishment.

McKean, Josiah Slutts.
R—Canal Dover, Ohio.
B—Mount Hope, Ohio.
G. O. No. 116, W. D., 1919.

Rear admiral, U. S. Navy.
As acting Chief of Naval Operations, his advice and assistance greatly tended to the successful outcome of the many problems requiring the close cooperation of the Navy and Army.

McKernon, James F.
R—New York, N. Y.
B—Cambridge, N. Y.
G. O. No. 59, W. D., 1919.

Colonel, Medical Corps, U. S. Army.
He has, by his tireless devotion to duty and his willingness to work in any capacity, not only placed his remarkable ability freely and fully at the disposition of the wounded, but in addition he has set so high a standard of professional efficiency as to serve as an inspiration to all with whom he has come in contact.

McLean, Angus.
R—Detroit, Mich.
B—St. Clair, Mich.
G. O. No. 56, W. D., 1922.

Colonel, Medical Corps, U. S. Army.
As director of the professional services and later as commanding officer of Base Hospital No. 17 and surgical consultant in hospital formations at the front, by his tireless energy, great resourcefulness, and brilliant professional attainments he rendered services of inestimable value in the care of the sick and wounded of the American, British, and French Armies, thereby contributing materially to the success of the American Expeditionary Forces.

McLeer, Edward, Jr.
R—Brooklyn, N. Y.
B—Brooklyn, N. Y.
G. O. No. 16, W. D., 1923.

Lieutenant colonel, Infantry, U. S. Army.
With untiring energy and unremitting devotion to duty he supervised the instruction and training and commanded in action against the enemy the machine-gun organizations of the 27th Division. During the operations of the division against the Hindenburg line, Sept. 25–30, 1918, and in the Le Selle River campaign Oct. 15–20, 1918, in direct command of 14 machine-gun companies he displayed outstanding leadership, sound tactical judgment, and indomitable determination and courage, these qualities being reflected in the morale and high efficiency of the machine-gun organizations of this division. He rendered extremely valuable and important services to the Government in a position of great responsibility.

McMahon, Edmund J.
R—St. Louis, Mo.
B—St. Louis, Mo.
G. O. No. 56, W. D., 1921.

Colonel, Infantry, U. S. Army.
As commanding officer of the camp at St. Sulpice, France, a position of great responsibility, he demonstrated rare executive ability. The efficient operation and growth of this important base section was due, in a great measure, to his energy and capacity for handling a task of considerable magnitude.

McManus, George H.
R—Hudson, Iowa.
B—Hudson, Iowa.
G. O. No. 69, W. D., 1919.

Brigadier general, U. S. Army.
As executive officer and troop movement officer, port of embarkation, Hoboken, N. J., through his very able organization and administration of these important offices the transport of troops and supplies overseas was materially aided.

MacMillan, William T.
R—Mahanoy City, Pa.
B—Girard Manor, Pa.
G. O. No. 4, W. D., 1923.

Lieutenant colonel (Infantry), General Staff Corps, U. S. Army.
As assistant chief of staff, G-1, 78th Division, from July, 1918, until June, 1919, by his indefatigable zeal, keen foresight, exceptional executive and administrative ability he successfully organized and directed the supply and administrative services of the division so as to insure at all times an ample supply of rations and ammunition, overcoming many difficult obstacles under most trying circumstances, thereby rendering conspicuous services to the American Expeditionary Forces in a position of great responsibility.

MCNAIR, LESLEY J. R—Bemidji, Minn. B—Verndale, Minn. G. O. No. 59, W. D., 1919.	Brigadier general, U. S. Army. As the senior Artillery officer of the training section, General Staff, he displayed marked ability in correctly estimating the changing conditions and requirements of military tactics. He was largely responsible for impressing upon the American Army sound principles for the use of artillery and for improving methods for the support of Infantry, so necessary to the proper cooperation of the two arms.
MCNAIR, WILLIAM S. R—Tecumseh, Mich. B—Tecumseh, Mich. G. O. No. 49, W. D., 1922.	Major general, U. S. Army. Serving in turn as commander of the 1st Field Artillery Brigade, 1st Division, and the 151st Field Artillery Brigade, 76th Division, as chief of artillery, 1st Army Corps, during the latter part of the Meuse-Argonne offensive and as chief of artillery of the 1st Army from Nov. 18, 1918, until April, 1919, by his marked ability, sound judgment, and thorough knowledge of artillery he rendered conspicuous services in a position of great responsibility to the American Expeditionary Forces.
MCNEELY, JOHN D. R—St. Joseph, Mo. B—St. Joseph, Mo. G. O. No. 56, W. D., 1921.	Colonel, Infantry, U. S. Army. As deputy director, rents, requisitions, and claims service, a position of great responsibility, in which he discharged important administrative functions with noteworthy ability and displayed good judgment and tact while attached to the British Claims Commission.
MCNEIL, EDWIN C. R—Alexandria, Minn. B—Alexandria, Minn. G. O. No. 19, W. D., 1922.	Colonel, Judge Advocate General's Department, U. S. Army. As assistant judge advocate, American Expeditionary Forces, by his excellent administrative ability and sound judgment he rendered a service which enabled his department to expeditiously handle many questions of great moment.
MCRAE, DONALD MARION. R—Boston, Mass. B—Fort Snelling, Minn. G. O. No. 56, W. D., 1922.	Lieutenant colonel, Infantry, U. S. Army. As assistant chief of staff, G-2, 78th Division, from November, 1917, until December, 1918, he displayed military attainments of a high order. By his great energy, sound judgment, and efficient administration of his section he was able to secure complete and valuable information of the enemy, thereby contributing materially to the successful operations of his division.
MCRAE, JAMES H. R—Lumber City, Ga. B—Lumber City, Ga. G. O. No. 12, W. D., 1919.	Major general, U. S. Army. He commanded with great credit the 78th Division in the Argonne-Meuse offensive and had an important part in that operation which forced the enemy to abandon Grandpre. In this and other campaigns his personal influence on the result obtained showed a rich quality of military leadership.
MCROBERTS, SAMUEL. R—New York, N. Y. B—Malta Bend, Mo. G. O. No. 72, W. D., 1920.	Brigadier general, U. S. Army. As chief of the procurement division of the office of the Chief of Ordnance, in which capacity he was charged with the procurement, by purchase or manufacture, of all articles of ordnance supplied to the United States Army, and the execution of the necessary contracts in connection therewith.
MABEE, JAMES I. R—Rockwood, Mich. B—Gasport, N. Y. G. O. No. 9, W. D., 1923.	Colonel, Medical Corps, U. S. Army. He served with the 1st Division as sanitary inspector from June 6, 1917, to July, 1918, and as division surgeon from July 4, 1918, to Feb. 10, 1919, and then as chief surgeon, 3d Army Corps. By his high professional attainments, his ability for organization and for securing cooperation of his subordinates, and his tireless efforts, he effected the successful evacuation of many casualties suffered by the 1st Division in the Soissons, St. Mihiel, Meuse-Argonne, and Sedan attacks. At all times he rendered services of great value to the American Expeditionary Forces in positions of great responsibility.
MACNAB, ALEXANDER, J., Jr. R—Salmon, Idaho. B—Salmon, Idaho. G. O. No. 62, W. D., 1919.	Colonel, Infantry, U. S. Army. He installed an extensive system of target ranges in France and perfected methods for the training of marksmen, personally supervising the instruction of 200,000 Infantry replacements. As a member of the training section he applied his methods to the instruction of the Infantry of the American Expeditionary Forces with extraordinary success.
MACRAE, DONALD, Jr. R—Council Bluffs, Iowa. B—Council Bluffs, Iowa. G. O. No. 19, W. D., 1922.	Colonel, Medical Corps, U. S. Army. As commanding officer of Mobile Hospital No. 1, at Coulommiers and Chateau-Thierry, from June to August, 1918, Colonel Macrae promptly arranged his hospital under the most difficult conditions and inadequate equipment and personnel.
MADDOX, GEORGE W. R—Owenton, Ky. B—Rogers Gap, Ky. G. O. No. 56, W. D., 1922.	Lieutenant colonel (Infantry), General Staff Corps, U. S. Army. As assistant chief of staff, G-1, 82d Division, from June until September, 1918, and of the 81st Division from then until June, 1919, he rendered highly meritorious services. Due to his unusual foresight, great energy, and marked ability, he directed the supply and administrative services of those divisions so as to insure at all times a proper supply of food and munitions, often under great difficulties and lack of transportation, thereby rendering conspicuous services in a position of great responsibility to the American Expeditionary Forces.
MADDUX, HENRY C. R—Orange, Va. B—Harrisonburg, Va. G. O. No. 59, W. D., 1921.	Lieutenant colonel, Medical Corps, U. S. Army. In organizing and constructing the hospital center at Toul. His work in this showed a force, initiative, and character that contributed in an unusual way to the ability of the Medical Department and care for the sick and wounded of the American Expeditionary Forces.

MADDUX, RUFUS F._____
  R—Newport, Ky.
  B—Cincinnati, Ohio.
  G. O. No. 14, W. D., 1923.

Lieutenant colonel, Chemical Warfare Service, U. S. Army.
As commanding officer and organizer of the officers' reclassification depot at Blois, France, from April to July, 1918, by his tact, judgment, and untiring energy he made it a model of efficient organization. Later as gas officer of the 5th Army Corps from Aug. 5 to Oct. 13, 1918, he displayed a great resourcefulness, marked efficiency, exceptional initiative, tireless efforts in successfully meeting the unusual demands relating to gas defense of the troops in the St. Mihiel and Meuse-Argonne offensives, thereby rendering service of great value to the American Expeditionary Forces.

MAGRUDER, BRUCE_____
  R—Washington, D. C.
  B—Washington, D. C.
  G. O. No. 87, W. D., 1919.

Lieutenant colonel (Infantry), General Staff Corps, U. S. Army.
As executive officer of the intelligence section at General Headquarters, by his marked ability and zeal he performed duties of a most exacting nature, in connection with the administration and development of the section, with conspicuous merit, rendering services of great value to the American Expeditionary Forces.

MAGRUDER, LLOYD B_____
  R—Washington, D. C.
  B—Washington, D. C.
  G. O. No. 60, W. D., 1920.

Lieutenant colonel, Coast Artillery Corps, U. S. Army.
As inspector of the district of Paris, he conducted many intricate and delicate investigations with noteworthy ability and solved many involved problems arising among the American Expeditionary Forces with sound judgment. The zealous and able manner with which he pursued the manifold details of his office was an important factor in raising the morale of the American Expeditionary Forces in Paris. He has performed services of special significance for the American Expeditionary Forces.

MAJOR, DUNCAN K., Jr_____
  R—New York, N. Y.
  B—New York, N. Y.
  G. O. No. 89, W. D., 1919.

Colonel (Infantry), General Staff Corps, U. S. Army.
As chief of staff of the 26th Division he proved to be a capable and energetic staff officer of marked executive ability. At all times he exhibited rare qualities of military leadership. He rendered invaluable services to the American Expeditionary Forces.

MALONE, PAUL B_____
  R—New York, N. Y.
  B—Middletown, N. Y.
  G. O. No. 24, W. D., 1920.
  Distinguished-service cross also awarded.

Brigadier general, U. S. Army.
He demonstrated marked ability in the important duty of organizing the military training and educational system of the American Army in France. Later, in active operations against the enemy, he commanded with distinction a regiment in the trench operations of the Sommedieue sector, in the Aisne defensive, the operations near Chateau-Thierry, and in the Aisne-Marne offensive, and a brigade in the St. Mihiel and Meuse-Argonne offensives. In all of these capacities the merit he displayed was conspicuous.

MALONE, WILLIAM B_____
  R—Memphis, Tenn.
  B—Brownsville, Tenn.
  G. O. No. 38, W. D., 1922.

Major, Medical Corps, U. S. Army.
As chief of surgical teams in hospital formations at the front through all the combat activities of the American Expeditionary Forces from the Cantigny offensive to the close of the Meuse-Argonne offensive, with rare technical skill and high professional attainments he rendered service of a most conspicuous nature in a position of great responsibility and at a time of gravest importance.

MALONY, HARRY J_____
  R—Dundee, N. Y.
  B—Lakemont, N. Y.
  G. O. No. 78, W. D., 1919.

Lieutenant colonel, Ordnance Department, U. S. Army.
He successfully organized and administered the many complex and difficult operations connected with the arming and equipping of airplanes for services at the front, displaying sound judgment and acting with energy and initiative in times of emergency. He worked self-sacrificingly and devotedly that there might be no delays, overcoming serious obstacles by the exercise of good judgment and thorough understanding of conditions in the American Expeditionary Forces.

MANCHESTER, PERCIVAL_____
  R—Chicago, Ill.
  B—Chicago, Ill.
  G. O. No. 126, W. D., 1919.

Major, Ordnance Department, U. S. Army.
As base ordnance officer in Base Section No. 1, at St. Nazaire, France, he ably organized and administered important work with exceptional success. As commanding officer of Intermediate Ordnance Depot No. 2, at Gievres, he conducted important activities with sound judgment and marked devotion to duty, working with tireless energy for the improvement of the Ordnance Service.

MANLY, CLARENCE J_____
  R—New York, N. Y.
  B—Georgetown, Ky.
  G. O. No. 103, W. D., 1919.

Colonel, Medical Corps, U. S. Army.
He organized and commanded with signal ability the hospital center at Beaune, taking charge of it when it was in an unfinished state and at a time when increased facilities were urgently needed. Overcoming numerous adverse conditions, he expedited its completion, and rendered invaluable services in furnishing effective medical treatment for large numbers of sick and wounded of the American Expeditionary Forces.

MANN, CHARLES R_____
  R—Summitt, N. J.
  B—Orange, N. J.
  G. O. No. 124, W. D., 1919.

As chairman of the advisory board of the committee on education and special training, he gave invaluable service in the development of the training of technicians and mechanics for the Army, and in the organization of the Students' Army Training Corps.

MARCH, PEYTON C_____
  R—Easton, Pa.
  B—Easton, Pa.
  G. O. No. 4, W. D., 1919.
  Distinguished-service cross also awarded.

General, U. S. Army.
As commanding general of the Army Artillery of the 1st Army from Oct. 4, 1917, to Jan. 31, 1918, initiated and prepared the plans for the organization of the Artillery of the American Army in France; as Acting Chief of Staff of the U. S. Army from Mar. 2, 1918, to May 20, 1918, and as Chief of Staff of the U. S. Army (General, U. S. Army), after May 20, 1918, he performed with intelligence, zeal, and patriotic devotion, duties of inestimable value in the development, the direction and the carrying into effect of the military program of the United States.

MARKEY, DAVID JOHN_____
   R—Frederick, Md.
   B—Frederick, Md.
   G. O. No. 31, W. D., 1922.

Major, Infantry, U. S. Army.
While in action with the 58th Infantry Brigade, 29th Division, north of Verdun, 1918, in the absence of a regularly detailed brigade adjutant he performed the exacting duties of that office in addition to his duties as commander of the machine-gun battalion under circumstances requiring exceptional courage, tactical judgment, initiative, and endurance, all of which he displayed to a marked degree. By his brilliant professional attainments, his untiring energy and keen foresight he contributed largely to the subsequent successes of his brigade and division. He handled with rare skill and in a masterful manner the multifarious duties devolving upon him in a position of great responsibility.

MARSHALL, FRANCIS C_____
   R—Darlington, Wis.
   B—Galena, Ill.
   G. O. No. 38, W. D., 1922.

Brigadier general, U. S. Army.
In command of the 2d Infantry Brigade, 1st Division, during the Meuse-Argonne offensive from Oct. 20 to Nov. 11, 1918, when, by his energy, professional skill, and his pronounced qualities of leadership, especially in the attack of the 1st Division on the line of the Meuse, Nov. 6, 1918, and the subsequent operations against Sedan, Nov. 6–7, 1918, he contributed in large measure to the success of his division.

MARSHALL, GEORGE C., Jr_____
   R—Uniontown, Pa.
   B—Uniontown, Pa.
   G. O. No. 116, W. D., 1919.

Colonel (Infantry), General Staff Corps, U. S. Army.
He has performed the duties of assistant chief of staff, G–3, 1st Division, from June 26, 1917, to July 12, 1918. He served in the G–3 section, General Headquarters, American Expeditionary Forces, from July 13, 1918, to Aug. 19, 1918; in G–3 section, 1st Army, from Aug. 20, 1918, to Oct. 16, 1918; as assistant chief of staff, G–3, of the 1st Army, from Oct. 17 to Nov. 19, 1918; and as chief of staff of the 8th Army Corps from Nov. 20, 1918, to Jan. 15, 1919, during which period the 1st Division served in the Toul sector and at the Cantigny attack and the 1st Army operations in the St. Mihiel and Meuse-Argonne offensives. By untiring, painstaking, and energetic efforts he succeeded in all these undertakings. His efforts had a marked influence on the successes achieved by the units with which he served.

MARSHALL, RICHARD C., Jr_____
   R—Portsmouth, Va.
   B—Portsmouth, Va.
   G. O. No. 25, W. D., 1919.

Brigadier general, U. S. Army.
In the construction division of the Army, his zeal, judgment, and exceptional administrative ability have enabled serious difficulties to be overcome and the construction necessary for a great army to be provided.

MARSHALL, WALDO H_____
   R—New York, N. Y.
   B—Boston, Mass.
   G. O. No. 77, W. D., 1919.

First, as assistant to the chief of the production division of the Office of the Chief of Ordnance, in which capacity he was of material assistance in securing the production of all articles of ordnance supplied to the United States Army, and later as special assistant to the Chief of Ordnance, in which capacity he successfully organized the industry of the country for the production of artillery, artillery ammunition, etc.

MARTIN, CHARLES H_____
   R—Carmi, Ill.
   B—Albion, Ill.
   G. O. No. 87, W. D., 1919.

Major general, U. S. Army.
As commander of the 90th Division during the greater part of its service with the Army of Occupation, by his ceaseless energy he performed his duties with the utmost efficiency, giving the closest personal supervision to the training, discipline, and equipment of his division. His brilliant professional attainments and steadfast devotion to duty were reflected in the high standards maintained throughout the organizations under his command, rendering important services to the American Expeditionary Forces.

MARTIN, FRANKLIN H_____
   R—Kenilworth, Ill.
   B—Xenia, Wis.
   G. O. No. 3, W. D., 1922.

Colonel, Medical Corps, U. S. Army.
As chairman of the committee on medicine and sanitation of the Council of National Defense, a position of great responsibility, by his tireless energy and marked ability, he so coordinated the civil medical resources of the Nation as to meet the needs of the Government medical service. He rendered valuable assistance in solving the important medical problems of the war.

MARTIN, FREDERICK L_____
   R—Liberty, Ind.
   B—Liberty, Ind.
   G. O. No. 14, W. D., 1925.
   Act of Congress Feb. 25, 1925.

Major, Air Service, U. S. Army.
As commanding officer of the U. S. Army Air Service around-the-world flight and as pilot of Airplane No. 1, the Seattle, from Apr. 6, 1924, until Apr. 30, 1924, when, due to an accident which resulted in the complete wreck of his airplane, he was obliged to relinquish command of the expedition. Major Martin by his tireless energy, foresight, and thorough technical knowledge assisted materially in completing arrangements and developing the special equipment installed in the airplanes, and so perfected and organized the command that each unit would become self-sustaining and automatic in its operations in the event of separation from or disaster to the others. In the performance of this great task he aided in bringing great credit to the military forces of the United States.

MASON, CHARLES H_____
   R—St. Paul, Minn.
   B—Fort Sanders, Wyo.
   G. O. No. 9, W. D., 1923.

Colonel (Infantry) General Staff Corps, U. S. Army.
While serving as chief of MI–2, information section, military intelligence division, General Staff, he originated and put into practical application an exceedingly scientific and highly technical system of handling all positive intelligence received in the division, including its dissemination to all branches of the War Department and other coordinating branches of the Government. From Sept. 16 to Dec. 12, 1918, as assistant chief of staff, G–2, 36th Division, he displayed sound judgment and exceptional ability in the administration and operation of that section of the division staff. Later, when on duty with the American Peace Mission at Paris, France, he rendered services of great value to the Government.

MASTELLER, KENNETH C._____
  R—Bakersville, Calif.
  B—Pella, Iowa.
  G. O. No. 89, W. D., 1919.

Colonel (Coast Artillery Corps), General Staff Corps, U. S. Army.
As chief of the negative branch of the military intelligence division of the General Staff, in building up and developing the counterespionage service in this country, the plant protection service, the detection of fraud and graft, and the development battalion system.

MATTHEWS, HUGH_____
  R—Louisville, Tenn.
  B—Tennessee.
  G. O. No. 59, W. D., 1921.

Lieutenant colonel, U. S. Marine Corps.
As assistant chief of staff, G–1, 2d Division, a position of great responsibility in which he functioned with marked ability during the St. Mihiel, Blanc Mont, and Argonne offensives.

MAUBORGNE, JOSEPH O._____
  R—New York, N. Y.
  B—New York, N. Y.
  G. O. No. 81, W. D., 1919.

Lieutenant colonel, Signal Corps, U. S. Army.
As head of the engineering and research division of the Signal Corps he rendered conspicuous service in connection with coordinating the design and supply of new technical apparatus for the Signal Corps. He was largely responsible for the high type of radio equipment developed for our Army and rendered unusual service in connection with cipher telegraphy.

MAUS, LOUIS M._____
  R—Rockville, Md.
  B—Montgomery County, Md.
  G. O. No. 96, W. D., 1918.

First lieutenant, assistant surgeon, U. S. Army.
For services on the Belle Fourche River, N. Dak., on Nov. 5, 1877, in that while serving with a detachment suddenly surrounded by an overwhelming force of hostile Sioux Indians he succeeded in extricating the party from a most perilous position.

MAXFIELD, HOWARD HOYT_____
  R—Elizabeth, N. J.
  B—Bloomfield, N. J.
  G. O. No. 19, W. D., 1922.

Colonel, Corps of Engineers, U. S. Army.
As general superintendent of motor power of the Transportation Corps, American Expeditionary Forces, by his sound judgment and energy displayed in the organizing and administering of the activities of the mechanical department he rendered services of great value to the American Expeditionary Forces.

MAYER, BRANTZ_____
  R—Davenport, Iowa.
  B—Baltimore, Md.
  G. O. No. 116, W. D., 1919.

Lieutenant commander, U. S. Navy.
For services as supply officer, Newport News division, cruiser and transport force.

MAYES, JAMES J._____
  R—Springfield, Mo.
  B—Amsterdam, Ohio.
  G. O. No. 89, W. D., 1919.

Colonel, Judge Advocate General's Department, U. S. Army.
He served with marked ability as deputy judge advocate of the American Expeditionary Forces. Fitted for his important duties by wide experience and conspicuous legal attainments, he solved ably and expeditiously the many questions of great moment with which his department was called upon to deal.

MAYO, CHARLES H._____
  R—Rochester, Minn.
  B—Rochester, Minn.
  G. O. No. 27, W. D., 1920.

Colonel, Medical Corps, U. S. Army.
In addition to the manifold service to the Surgeon General by furnishing needed advice and counsel, he distinguished himself by exceptionally meritorious service to the Government in his work in the organization of surgical service and his invaluable assistance in the reorganization of the Medical Department on the scale demanded by the war.

MAYO, WILLIAM JAMES_____
  R—Rochester, Minn.
  B—Le Sueur, Minn.
  G. O. No. 69, W. D., 1919.

Colonel, Medical Corps, U. S. Army.
In addition to the manifold service to the Surgeon General by furnishing needed advice and counsel, he distinguished himself by exceptionally meritorious service to the Government in his work in the organization of surgical service and his invaluable assistance in the reorganization of the Medical Department on the scale demanded by the war.

MAYO-SMITH, RICHMOND_____
  R—Norwood, Mass.
  B—Easthampton, Long Island, N. Y.
  G. O. No. 56, W. D., 1922.

Lieutenant colonel, Chemical Warfare Service, U. S. Army.
As chief of the supply section, Chemical Warfare Service, by his tireless energy, foresight, and marked executive ability, he built and administered a supply organization consisting of 12 seaport, intermediate, and front-line depots, which fully met at all times the demands for chemical warfare supplies that active operations brought upon the service, thereby rendering services of great value to the American Expeditionary Forces.

MEARS, FREDERICK_____
  R—St. Paul, Minn.
  B—Fort Omaha, Nebr.
  G. O. No. 89, W. D., 1919.

Colonel, Corps of Engineers, U. S. Army.
He served with distinction as commanding officer of the 31st Railway Engineers, and later as assistant general manager and general manager railway department, Transportation Corps. Due to his remarkable executive ability and skill as an organizer the railways of the American Expeditionary Forces were operated with rare success, and the huge transportation problem, involving the carrying of tremendous quantities of supplies from the base ports to the front, was satisfactorily solved.

MENOHER, CHARLES T._____
  R—Johnstown, Pa.
  B—Johnstown, Pa.
  G. O. No. 12, W. D., 1919.

Major general, U. S. Army.
In command of the 42d Division from Chateau-Thierry to the conclusion of the Argonne-Meuse offensive, including the Baccarat sector, Reims, Vesles, and at the St. Mihiel salient, this officer with his division participated in all of those important engagements. The reputation as a fighting unit of the 42d Division is in no small measure due to the soldierly qualities and the military leadership of this officer.

MERCHANT, BERKELEY T._____
  R—Watervliet, N. Y.
  B—New York, N. Y.
  G. O. No. 108, W. D., 1919.

Lieutenant colonel, Quartermaster Corps, U. S. Army.
While on duty in the remount service, he performed his tasks efficiently. Later, appointed Chief Veterinarian of the American Expeditionary Forces, he administered, with marked success, the veterinary service, providing for effective means of evacuation of sick and wounded animals from the front and in placing the personnel of the Veterinary Corps on an efficiently functioning basis.

MERRILL, DANA T.
R—East Auburn, Me.
B—East Auburn, Me.
G. O. No. 43, W. D., 1922.

Colonel (Infantry), General Staff Corps, U. S. Army.
As chief of staff, 37th Division, during its organization, training, and entire combat period he displayed unflagging energy and marked ability. To his zeal, initiative, and military attainments was due in great measure the success of the division in both the Meuse-Argonne and Ypres-Lys offensives.

METCALFE, RAYMOND F.
R—Buffalo, N. Y.
B—West Salamanca, N. Y.
G. O. No. 15, W. D., 1921.

Colonel, Medical Corps, U. S. Army.
As division surgeon, 36th Division, in August and again in September, 1918, during serious epidemics of Spanish influenza, his farsightedness and his energetic and efficient action in personally directing the handling of these epidemics resulted in a very large reduction in the mortality.

METTS, JOHN VAN B.
R—Wilmington, N. C.
B—Wilmington, N. C.
G. O. No. 55, W. D., 1920.

Colonel, Infantry, U. S. Army.
He commanded with marked distinction the 119th Infantry, 30th Division, from the time of its organization and early training period to the completion of the active combat operations in the Ypres-Lys and Somme offensives. He especially distinguished himself while in command of his regiment on Sept. 29, 1918, during the assault on the Hindenburg line, near Bellicourt, France, where he displayed marked ability and sound judgment. He has rendered services of signal worth to the American Expeditionary Forces.

MEYER, VINCENT.
R—Brooklyn, N. Y.
B—New York, N. Y.
G. O. No. 39, W. D., 1920.

Major, Field Artillery, U. S. Army.
As a member of the supply section of the General Staff in charge of the munitions branch, he ably controlled and handled the details of ammunition supply for the 1st Army during the period of the St. Mihiel and Meuse-Argonne operations. In this important task he displayed marked initiative and energy, and assured by his skill and prompt action a steady supply of ammunition under most difficult conditions. His efforts were a contributing factor to the success of the supply service during these important operations.

MIDDLETON, TROY H.
R—Georgetown, Miss.
B—Hazlehurst, Miss.
G. O. No. 95, W. D., 1919.

Colonel, Infantry, U. S. Army.
As a battalion and a regimental commander of the 47th Infantry, 4th Division, he gave proof of conspicuous energy and marked tactical ability. He achieved notable successes in the operations near Sergy, along the Vesle River, and during the fierce fighting in the Bois-du-Fays and Bois-de-Foret of the Argonne-Meuse offensive, rendering invaluable services to the American Expeditionary Forces.

MILES, PERRY L.
R—Columbus, Ohio.
B—Westerville, Ohio.
G. O. No. 39, W. D., 1919.
Distinguished-service cross also awarded.

Colonel, Infantry, U. S. Army.
As commander of the 371st Infantry, 93d Division, which, during its active operations, was attached to the French forces, he conducted his regiment with conspicuous success, by his admirable tact and sound judgment he maintained at all times harmonious relationship with the allied forces to which his unit was attached, rendering valuable services to the American Expeditionary Forces.

MILLER, ORRIN DAVID.
R—Oakland, Calif.
B—Boonville, N. Y.
G. O. No. 56, W. D., 1921.

Lieutenant colonel, Transportation Corps, U. S. Army.
As executive officer of the Army Transportation Service, a position of great responsibility, throughout the whole period of existence of the service, by his untiring devotion to duty, loyalty, and marked ability, he proved a material factor in the success of the operation of the Army Transportation Service.

MILLER, REUBEN B.
R—Millington, Ill.
B—Canada.
G. O. No. 69, W. D., 1919.

Colonel, Medical Corps, U. S. Army.
In the reorganization and administration of the personnel branch of the office of the Surgeon General of the Army during the present war, he thereby contributed greatly to the proper care of the sick and wounded and thus increased the efficiency of the Army.

MILLER, TROUP.
R—Macon, Ga.
B—Perry, Ga.
G. O. No. 56, W. D., 1922.

Lieutenant colonel (Field Artillery), General Staff Corps, U. S. Army.
He served as adjutant, 82d Division, from December, 1917, until March, 1918; assistant chief of staff, G-1, 82d Division, from April to June and from September to December, 1918; assistant chief of staff, G-1, 1st Army Corps December, 1918, until February, 1919; and assistant chief of staff, G-4, 1st Army, from February to April, 1919, and then as G-4, intermediate section, services of supply. By his marked ability, sound judgment, and high military attainments he rendered meritorious services in positions of great responsibility, contributing in a large measure to the success attained by all the units with which he served.

MILLIKEN, SAYRES L.
R—Morgantown, W. Va.
B—Brownsville, Pa.
G. O. No. 9, W. D., 1923.

Captain, assistant superintendent, Army Nurse Corps, U. S. Army.
As chief nurse of the base hospital at Camp Sevier, S. C., during the early part of the World War, she was responsible for the nursing care of thousands of patients under most trying and difficult circumstances. Her tact, energy, and ability were greatly instrumental in saving many lives, particularly during the influenza epidemic. Her services were characterized by zeal and excellent judgment, and her achievements have been conspicuous. Later, while serving as assistant superintendent, Army Nurse Corps, in the office of the Surgeon General, her administrative ability and professional experience were of inestimable value to the Medical Department in providing proper nursing care for the sick and wounded.

MILLIKIN, JOHN.
R—Danville, Ind.
B—Danville, Ind.
G. O. No. 46, W. D., 1920.

Lieutenant colonel (Cavalry), General Staff Corps, U. S. Army.
As executive officer and assistant director of the Army General Staff College, at Langres he rendered conspicuous services. Later, as chief of the Military Police Corps Division of the Provost Marshal General's Department, American Expeditionary Forces, by his ability, untiring zeal, and sound judgment, he aided in a material way in producing an efficient organization. He has rendered services of great value to the American Expeditionary Forces.

MILLING, THOMAS DEW_____
   R—Franklin, La.
   B—Winnfield, La.
   G. O. No. 50, W. D., 1919.

Colonel, Air Service, U. S. Army.
First, as chief of staff and later as commander, he organized and conducted the operations of the Air Service of the 1st Army during the entire operations of that Army. By untiring, painstaking, and energetic efforts he succeeded in raising the efficiency of his command and insuring the proper cooperation with the land units. He exhibited professional attainments of the highest order and exercised a marked influence on the success of the 1st Army.

MINER, ASHER_____
   R—Wilkes-Barre, Pa.
   B—Wilkes-Barre, Pa.
   G. O. No. 89, W. D., 1919.
   Distinguished-service cross also awarded.

Colonel, Field Artillery, U. S. Army.
He served with notable success as commanding officer of the 109th Field Artillery, 28th Division, giving proof of high qualities of leadership. Inspiring his men by his self-sacrificing devotion to duty he maintained a creditable standard of efficiency in his regiment and constantly furnished the most effective artillery support to the attacking Infantry.

MINNIGERODE, FITZHUGH L._____
   R—Washington, D. C.
   B—Oatlands, Va.
   G. O. No. 120, W. D., 1919.
   Distinguished-service cross also awarded.

Second lieutenant, 8th Infantry, U. S. Army.
For bravery and presence of mind in going to the rescue of drowning men at Iloilo, Panay, Philippine Islands, Sept. 10, 1907.

MINOR, SIDNEY W._____
   R—Durham, N. C.
   B—Granville Co., N. C.
   G. O. No. 55, W. D., 1920.

Colonel, Infantry, U. S. Army.
As commander of the 120th Infantry, 30th Division, from the time of its organization and training to the completion of active combat operations in the Ypres-Lys and Somme offensives, he displayed at all times initiative and sound judgment. During the attack on the Hindenburg line, near Bellicourt, France, Sept. 29, 1918, and during the subsequent advance, he handled his regiment with distinction, capturing several towns, numerous cannon and many prisoners. He has rendered services of material worth to the American Expeditionary Forces.

MITCHELL, JAMES B._____
   R—Syracuse, N. Y.
   B—Syracuse, N. Y.
   G. O. No. 53, W. D., 1921.

Colonel, Inspector General's Department, U. S. Army.
As inspector general, and as chief of staff, Base Section No. 3, services of supply, American Expeditionary Forces, positions of great responsibility, he showed great energy and good judgment in the organization of Base Section No. 3 and in providing for the American troops at the base and for those passing through England. H rendered services of much value.

MITCHELL, WILLIAM_____
   R—Milwaukee, Wis.
   B—France.
   G. O. No. 87, W. D., 1919.
   Distinguished-service cross also awarded.

Brigadier general, U. S. Army.
As Air Service commander, first of the zone of advance and later of the 1st Army Corps, by his tireless energy and keen perception he performed duties of great importance with marked ability. Subsequently as commander, Air Service, of the 1st Army, and, in addition, after formation of the 2d Army as commander of Air Service of both armies, by his able direction of these vitally important services he proved to be a potent factor in the successes achieved during the operations of the American armies.

MITCHELL, WILLIAM A._____
   R—Columbus, Ga.
   B—Seale, Ala.
   G. O. No. 89, W. D., 1919.

Colonel, Corps of Engineers, U. S. Army.
Having taken command of the 2d Engineers, 2d Division, just prior to the battle of Soissons, he served with distinction as the leader of this regiment until the close of hostilities. Under his skillful direction his regiment successfully accomplished all the important technical missions assigned to it. His high military attainments were reflected by its efficiency in combat operations. Subsequent to the armistice he continued to render important services to the American Expeditionary Forces as chief of engineers of the 8th Army Corps.

MOLLOY, JANE G._____
   R—San Francisco, Calif.
   B—Kingston, N. Y.
   G. O. No. 9, W. D., 1923.

First lieutenant, chief nurse, Army Nurse Corps, U. S. Army.
As chief nurse of the base hospital at Camp Devens, Mass., during the World War, although greatly hampered by lack of personnel and equipment, she displayed unusual talent for organization. By her untiring, painstaking, and energetic efforts in the distribution of her staff and her personal care of the sick she rendered great service to the Army, particularly during the influenza epidemic, when the lives of many patients were saved through her efforts.

MONCRIEF, WILLIAM H._____
   R—Atlanta, Ga.
   B—Greensboro, Ga.
   G. O. No. 59, W. D., 1921.

Colonel, Medical Corps, U. S. Army.
In organizing and commanding the hospital center at Mesves, he took possession of the center in its unfinished condition and by great force and ability prepared a hospital group for the care of sick and wounded at a time when the need of hospital beds was critical.

MONTGOMERY, HENRY G._____
   R—New York, N. Y.
   B—Washington, D. C.
   G. O. No. 14, W. D., 1923.

Captain, Field Artillery, U. S. Army.
On his own initiative and with great vision, patriotism, and determination he developed, with funds provided by public-spirited citizens, the armored motor tank car which was adopted by the American Government, training at the same time a corps of technical experts who rendered important service in the Ordnance Department of the Army. Subsequently, as an officer of Field Artillery in France, he was charged with the responsibility of maintaining while in action the mechanical efficiency of the newly issued French material forming the heavy artillery of the 33d and 79th Divisions during the St. Mihiel and Meuse-Argonne operations. In this work he displayed technical skill of the highest order, reducing to a minimum the fire losses of his brigade, and contributed markedly to the success of his artillery brigade in these offensives.

MONTGOMERY, JOHN C............
  R—Elizabethtown, Ky.
  B—Elizabethtown, Ky.
  G. O. No. 70, W. D., 1919.

Colonel (Infantry), General Staff Corps, U. S. Army.
Serving successively as division inspector, 2d Division; assistant chief of staff, G-3, 1st Army Corps; and assistant chief of staff, G-3, 3d Army, he has been charged with duties of a most important nature, in the performance of which he has at all times manifested steadfast loyalty and military ability of a high order, rendering services of signal worth.

MONTGOMERY, WALTER C............
  R—New York, N. Y.
  B—New York, N. Y.
  G. O. No. 22, W. D., 1920.

Lieutenant colonel Medical Corps, U. S. Army.
He served with marked distinction as division surgeon of the 27th Division. When confronted with a shortage of personnel, he displayed marked initiative and resourcefulness in organizing additional sanitary personnel. During the action along the Hindenburg line, Sept. 25 and 30, 1918, by his high professional attainments, sound judgment, and loyal devotion to duty he so conducted the personnel at his disposal as to provide successfully for the evacuation of 4,000 casualties in four days.

MOODY, LUCIAN B............
  R—Huron, S. Dak.
  B—Huron, S. Dak.
  G. O. No. 30, W. D., 1921.

Colonel, Ordnance Department, U. S. Army.
For services as assistant to the chief ordnance officer, American Expeditionary Forces in France, and chief ordnance officer, Army of Occupation in Germany.

MOORE, HUGH B............
  R—Texas City, Tex.
  B—Huntland, Tenn.
  G. O. No. 89, W. D., 1919.

Lieutenant colonel, Quartermaster Corps, U. S. Army.
He served creditably as superintendent, Army Transport Service, at Brest, and later, director of the Army Transport Service, he successfully supervised the activities of this service in 40 ports. Actuated by self-sacrificing devotion to duty, he achieved marked success in expediting the movement of troops, rendering services of inestimable value to the American Expeditionary Forces.

MOORHEAD, JOHN J............
  R—New York, N. Y.
  B—New York, N. Y.
  G. O. No. 56, W. D., 1921.

Lieutenant colonel, Medical Corps, U. S. Army.
As commanding officer of American Red Cross Hospital No. 110, at Coincy, France, from June to November, 1918, he operated his hospital under the most difficult conditions and with inadequate personnel and equipment.

MOORMAN, FRANK............
  R—Edwardsville, Ill.
  B—Eureka, Mich.
  G. O. No. 59, W. D., 1919.

Colonel (Coast Artillery Corps), General Staff Corps, U. S. Army.
In a position of the greatest responsibility, he displayed peculiar genius, combined with exact scientific knowledge, in organizing, training, and operating the radio intelligence service of the intelligence section. Charged with the duty of intercepting and deciphering the radio messages of the enemy, he acted with initiative and foresight, achieving brilliant results.

MORENO, ARISTIDES............
  R—Birmingham, Ala.
  B—New York, N. Y.
  G. O. No. 59, W. D., 1919

Lieutenant colonel (Infantry), General Staff Corps, U. S. Army.
As a member of the intelligence section, he efficiently organized and directed the operations of the counter-espionage service in the American Expeditionary Forces, displaying marked talents in a position of great responsibility. His unusual powers of discernment, his tact, and sound judgment made possible effective cooperation with corresponding services of the allied armies. Due to his zeal and untiring devotion, the counter-espionage service attained exceptional proficiency.

MORGAN, CASEY BRUCE............
  R—New York.
  B—Augusta, Ga.
  G. O. No. 116, W. D., 1919.

Captain, U. S. Navy.
For services as force transport officer. His untiring energy contributed greatly to the successful movement of troops and supplies.

MORGAN, JOHN M............
  R—Minersville, Ohio.
  B—Minersville, Ohio.
  G. O. No. 89, W. D., 1919.

Colonel, Infantry, U. S. Army.
As commanding officer of the 309th Infantry, 78th Division, during the last two months of hostilities he displayed marked qualities of leadership and unflagging energy. By the skillful manner in which he conducted his regiment during the advance through the Bois-de-Loges in the first part of November he contributed materially to the successes of his division in its operations in the Meuse-Argonne offensive.

MORRISON, JOHN F............
  R—Schoharie, N. Y.
  B—Summit, N. Y
  G. O. No. 47, W. D., 1919.

Major general, U. S. Army.
For services as department commander, Western Department, in handling with great skill, tact, and sound judgment many difficult problems arising in his department.

MORROW, CHARLES H............
  R—Somerset, Ky.
  B—Somerset, Ky.
  G. O. No. 3, W. D., 1922.

Colonel, Infantry, U. S. Army.
In command of American forces in the Baikal sector, Siberia, Colonel Morrow with great energy, tact, and force handled a situation fraught with serious possibilities and rendered a service of great worth.

MORROW, DWIGHT W............
  R—Englewood, N. J.
  B—Huntington, W. Va.
  G. O. No. 46, W. D., 1919.

Member of the American Shipping Mission.
He was responsible for the first intelligent epitomization of the complete allied tonnage situation, and his able presentation of the situation to the allied countries materially affected the tonnage policy, resulting in all possible economy. By his tact and good judgment in matters affecting the establishment of the military board of allied supply he helped materially in the splendid results obtained by that organization.

MORROW, FRANK J............
  R—Omaha, Nebr.
  B—Fort Douglas, Utah.
  G. O. No. 49, W. D., 1922.

Colonel, Infantry, U. S. Army.
From September, 1917, to March, 1918, in the organization and operation of the 1st Division and 1st Corps Schools, he rendered valuable and unusual services under very great difficulties. From April to December, 1918, as a member of the training and instruction branch of the war-plans division, General Staff, and in direct supervision of the field officers' schools in the combat divisions, and the War College course for instruction of divisional staff officers, his zeal, marked ability, and high military attainments were important factors in successfully standardizing the training and instruction of a great many senior officers. Later, while in direct charge of the organization of the Reserve Officers' Training Corps, he again rendered invaluable services in building up this important branch of civilian military training along sound and enduring lines.

MORROW, HENRY M.
R—Omaha, Nebr.
B—Niles, Mich.
G. O. No. 49, W. D., 1922.

Colonel, Judge Advocate General's Department, U. S. Army.
As judge advocate, Philippine Department, by his legal ability and sound advice in the solution of unusual and highly important military and international problems, he contributed in a marked degree to the successful administration of military and civil affairs in the Philippines and China. Later as judge advocate of the 2d Army in France from October, 1918, until April, 1919, and then as judge advocate of the district of Paris until August, 1919, he displayed great zeal, marked legal training, and sound judgment in many difficult problems which confronted him and solved them with conspicuous success.

MORROW, WILLIAM M.
R—Fort Sidney, Nebr.
B—Niles, Mich.
G. O. No. 126, W. D., 1919.
Distinguished-service cross also awarded.

Colonel, Infantry, U. S. Army.
He served with conspicuous success as commanding officer of the 7th Infantry, 3d Division, succeeding in all of the difficult missions assigned to him. His sound judgment and untiring energy proved important factors in the successful operations of his division against the enemy.

MORSE, CHARLES F.
R—Montpelier, Vt.
B—Montpelier, Vt.
G. O. No. 14, W. D., 1920.

Colonel, Medical Corps, U. S. Army.
As director of the Veterinary Corps, by displaying exceptional energy, zeal, and good judgment he organized and administered with marked success a veterinary service capable of meeting every need in home territory and in the theater of operations. He provided effective means for the treatment of sick and wounded animals, for the prevention of disease among well animals, for the inspection of meat and dairy products used by the Army, and, through the establishment of schools of instruction, placed the personnel of the Veterinary Corps of the Army on a high plane of efficiency.

MORSE, ERNEST C.
R—Needham, Mass.
B—Lebanon, N. H.
G. O. No. 3, W. D., 1921.

Director of Sales, Supply Division, War Department, General Staff.
Charged with the very important duty of organizing and training a competent force for the entirely novel functions of supervising, coordinating, and directing the disposal, according to law, of the vast War Department surplus of supplies, materials, and properties of every description, and with the formulation and development of sales policies, he performed his manifold duties with marked ability, energy, and judgment, with the result that the United States disposed of great quantities of supplies at exceptionally advantageous prices.

MORTON, CHARLES G.
R—Brookline, Mass.
B—Cumberland, Me.
G. O. No. 12, W. D., 1919.

Major general, U. S. Army.
He commanded the 29th Division from the date of its organization until the end of hostilities, and led this division with skill and ability in the successful operations east and northeast of Verdun, which forced the enemy to maintain this front with strong forces, thus preventing an increase of hostile strength between the Argonne and the Meuse.

MOSELEY, GEORGE V. H.
R—Evanston, Ill.
B—Evanston, Ill.
G. O. No. 12, W. D., 1919.

Brigadier general, U. S. Army.
For services as Assistant Chief of Staff. He handled with great executive ability and rare understanding all problems of equipping and supplying the large number of American troops arriving and operating in France, and by his large grasp of supply problems and tireless energy he has conspicuously aided the successful administration of the supply department.

MOSES, ANDREW.
R—Strickling, Tex.
B—Strickling, Tex.
G. O. No. 38, W. D., 1922.

Brigadier general, U. S. Army.
Commanding the 316th Field Artillery, 81st Division, from August, 1917, and the 156th Field Artillery Brigade, 81st Division from June, 1918, until it was demobilized, he exhibited qualities of excellent leadership and military attainments of a high order. Later, as chairman of a joint board of review, he occupied a position of great responsibility, having full charge and control of the redelivery of all ships allocated to the War Department during the World War. By his administrative ability, excellent judgment, energy, and tact, he rendered conspicuous services in bringing about speedy and accurate settlements with the shipowners, which resulted in a large saving to the Government.

MOTT, JOHN R.
R—Montclair, N. J.
B—Livingston Manor, N. Y.
G. O. No. 73, W. D., 1919.

For services as general secretary of the national war work council of the Young Men's Christian Association of the United States.

MOTT, T. BENTLEY.
R—Leesburg, Va.
B—Leesburg, Va.
G. O. No. 59, W. D., 1919.

Colonel, Field Artillery, U. S. Army.
As chief liaison officer of the Commander in Chief, American Expeditionary Forces, at Allied General Headquarters, he performed the important duties with which he was charged with marked ability, and by his tact and sound judgment he materially assisted in insuring close cooperation between the French and American Armies.

MOUNT, JAMES R.
R—Kansas City, Kans.
B—Kansas City, Kans.
G. O. No. 59, W. D., 1919.

Colonel, Medical Corps, U. S. Army.
Arriving in France with the first American troops, he undertook the task of creating a medical supply depot and administering a medical supply service for the American Expeditionary Forces. Using his limited resources with great skill and judgment, he displayed unusual talent for organization and laid the foundation of an efficient medical supply service.

MOUNTFORD, FREDERICK A.
R—East Liverpool, Ohio.
B—England.
G. O. No. 56, W. D., 1922.

Lieutenant colonel, Coast Artillery Corps, U. S. Army.
As assistant to the Chief of Coast Artillery and in charge of the matériel section in that office he displayed excellent judgment and a thorough knowledge of the involved and intricate details in connection with matériel, thereby rendering highly meritorious service in the preparation and execution of plans for the effective accomplishment of the duties assigned to the Coast Artillery Corps in the operations in France.

MUDGETT, CHARLES F.
R—Valley City, N. Dak.
B—Ravenna, Mo.
G. O. No. 38, W. D., 1922.

Lieutenant colonel, Adjutant General's Department, U. S. Army.
As officer in charge of the enlisted division, Adjutant General's Office, General Headquarters, American Expeditionary Forces, he demonstrated the highest order of efficiency, and by his tact and sound judgment he handled in a masterful manner the many difficult problems that arose. During the demobilization he carried to successful completion the various projects relative to the discharge and return of enlisted personnel. By his splendid attainments and professional zeal he contributed materially to the success of the work of The Adjutant General's Office at General Headquarters, American Expeditionary Forces.

MUIR, CHARLES H.
R—Erie, Mich.
B—Erie, Mich.
G. O. No. 12, W. D., 1919.
Distinguished-service   cross   also
awarded.

Major general, U. S. Army.
As division and corps commander, commanding the 28th Division during the Argonne-Meuse offensive, and especially in the difficult operations which resulted in the clearing of the Argonne Forest, he proved himself to be an energetic leader of the highest professional attainments. As a corps commander he displayed the same fine qualities that characterized his service with a division.

MULLALLY, THORNWELL.
R—San Francisco, Calif.
B—Columbia, S. C.
G. O. No. 49, W. D., 1922.

Colonel, Field Artillery, U. S. Army.
He projected, recruited, organized, and mobilized the 144th Field Artillery, 40th Division, as an additional regiment of the California National Guard. Later, as colonel of that regiment, his sound judgment and marked ability as a leader were largely responsible for the successful training of his regiment both in the United States and France and in the development in that unit of over 168 officers for our Army.

MUNSON, EDWARD L.
R—New Haven, Conn.
B—New Haven, Conn.
G. O. No. 34, W. D., 1919.

Brigadier general, U. S. Army.
He developed the scheme of field training for officers and enlisted men of the Medical Department, directed the organization and administration of the medical officers' training camps, and organized and administered the Morale Branch of the General Staff.

MURPHY, FRED T.
R—St. Louis, Mo.
B—Detroit, Mich.
G. O. No. 59, W. D., 1919.

Colonel, Medical Corps, U. S. Army.
As director of Base Hospital No. 21, as supervisor of the evacuation of the sick and wounded of the 1st Army, and later as director of the Bureau of Medicine and Surgery of the American Red Cross he rendered most valuable assistance to the American Expeditionary Forces. Throughout his service he displayed unusual administrative ability and professional skill, combined with a genius for organization that contributed greatly to the efficiency of the Medical Service of the Army. Untiring in zeal and enthusiastic in his duty, he was an inspiration to those associated with him.

MURPHY, GRAYSON M. P.
R—New York, N. Y.
B—Philadelphia, Pa.
G. O. No. 59, W. D., 1919.

Lieutenant colonel (Infantry), General Staff Corps, U. S. Army.
He organized the work of the American Red Cross in Europe, and to his foresight, wisdom, and untiring efforts are largely due the splendid work performed for the American Expeditionary Forces by that institution. Later he displayed marked ability as assistant chief of staff of the 42d Division during the operations of that unit.

MURPHY, JOHN B.
R—Notre Dame, Ind.
B—Fort Robinson, Nebr.
G. O. No. 39, W. D., 1920.

Colonel, Coast Artillery Corps, U. S. Army.
As commanding officer of the 44th Artillery Regiment, Coast Artillery Corps, by his marked ability, energy, and resourcefulness, he organized, equipped, and trained a regiment of 8-inch howitzers in an extraordinarily brief period of time, thus enabling them to reach the front in time to be of valuable service during the critical days of April, 1918. Later, as a member of the operations section at General Headquarters, he rendered services of great value to the American Expeditionary Forces.

MURRAY, ARTHUR.
R—Bowling Green, Mo.
B—Bowling Green, Mo.
G. O. No. 4, W. D., 1923.

Major general, U. S. Army.
As department commander, Western Department, between August 29, 1917, and May 14, 1918, he handled many difficult problems arising in that department with rare judgment, tact, and great skill.

MURRAY, MAXWELL.
R—Fort Totten, N. Y.
B—West Point, N. Y.
G. O. No. 9, W. D., 1923.

Colonel, Field Artillery, U. S. Army.
As commander of the 5th Field Artillery, 1st Division, he handled the regiment so brilliantly under severe conditions during the assault and capture of Cantigny, May 28, 1918, and during the Aisne-Marne offensive, July 18-25, 1918, in the assault southeast of Soissons, that the regiment demonstrated an unusually high degree of efficiency and morale. He repeatedly displayed superior tactical judgment and knowledge of artillery and by his exceptional ability, leadership, and devotion to duty, he rendered the maximum of support to the Infantry of the 1st Division in effectively executing the most difficult missions assigned to him, thus rendering important services to the American Expeditionary Forces.

MURRAY, PETER.
R—Visalia, Calif.
B—Visalia, Calif.
G. O. No. 3, W. D., 1922.

Colonel (Quartermaster Corps), General Staff Corps, U. S. Army.
As chief, training and instruction branch, War Plans Division, General Staff, during 1919, under Colonel Murray's direction the foundation for the existing system of Army education and training was established and ways and means evolved to apply to peace training the lessons learned in the World War.

MURY, EDITH A. (now Mrs. EDITH A. KERSHAW).
R—Oakland, Calif.
B—Wadsworth, Nev.
G. O. 9, No. W. D., 1923.

Assistant superintendent, Army Nurse Corps, U. S. Army.
As chief nurse on duty at the nurses' mobilization station, Ellis Island, N. Y., during the early period of the World War, she rendered invaluable assistance to the Medical Department by instructing newly appointed chief nurses before embarkation for overseas, and through her untiring energy and unusual capabilities set a splendid example for all who came in contact with her. Later, as assistant superintendent, in the office of the Surgeon General, she contributed services of high value in supervising the demobilization of the Army Nurse Corps.

MYER, EDGAR A. R—Troy, N. Y. B—Fort Richardson, Tex. G. O. No. 89, W. D., 1919.	Colonel, Infantry, U. S. Army. Having taken command of the 129th Infantry, 33d Division, at a critical period during active operations, he displayed marked ability as a military leader by the successful manner in which he conducted his regiment. Constantly maintaining a high grade of morale among his command, he was able to accomplish with marked success all missions assigned to his unit.
MYERS, HU B. R—Shelbyville, Tenn. B—Shelbyville, Tenn. G. O. No. 95, W. D., 1919.	Colonel (Infantry), General Staff Corps, U. S. Army. As assistant chief of staff, G–3, and chief of staff, 2d Division, during its final operations, he performed his exacting staff duties with conspicuous ability. To his brilliant military attainments and untiring zeal were due, in a large measure, the successes achieved by his division in its attack on Massif du Mont at St. Etienne-a-Arnes, when the 2d Division served with the 4th French Army, and again during its advance from Landres-et-St. Georges to the Meuse during the final phase of the Meuse-Argonne offensive.
NAYLOR, WILLIAM K. R—St. Paul, Minn. B—Bloomington, Ill. G. O. No. 59, W. D., 1919.	Brigadier general, U. S. Army. While chief of staff of the 33d Division he exhibited conspicuous ability in the operations north of Verdun, France, in September and October, 1918. He frequently visited the front-line positions under heavy enemy artillery fire, and by his personal efforts and skillful dispositions was in a large measure responsible for the successes gained.
NELSON, ERIK H. R—New York, N. Y. B—Sweden. G. O. No. 14, W. D., 1925. Act of Congress, Feb. 25, 1925.	First lieutenant, Air Service, U. S. Army. Lieutenant Nelson, as pilot of Airplane No. 4, the *New Orleans*, and engineer officer of the U. S. Army Air Service around-the-world flight from Apr. 6, 1924, to Sept. 28, 1924, displayed sound technical skill, initiative, untiring energy, and resourcefulness and succeeded in piloting his airplane throughout the voyage. His sound judgment, indefatigable energy, and courageous conduct in the face of extraordinary perils contributed largely to the success of this pioneer flight of airplanes around the world. In the efficient performance of his arduous duties he aided in the accomplishment of an exploit which brought great credit to himself and to the Army of the United States.
NEVILLE, WENDELL C. R—Portsmouth, Va. B—Portsmouth, Va. G. O. No. 59, W. D., 1919.	Brigadier general, U. S. Marine Corps. While in command of the 5th Regiment, U. S. Marine Corps, and later of the 4th Infantry Brigade, 2d Division, he participated in the battles of Chateau-Thierry, the advance near Soissons, and the operations of St. Mihiel, Blanc Mont Ridge, and the Argonne-Meuse. In all of these he proved himself to be a leader of great skill and ability.
NICHOLSON, WILLIAM J. R—Washington, D. C. B—Washington, D. C. G. O. No. 50, W. D., 1919. Distinguished-service cross also awarded.	Brigadier general, U. S. Army. He commanded with distinction the 157th Infantry Brigade, 79th Division, from its organization to the time of the armistice—at all times with credit to himself and to his command.
NOBLE, ROBERT E. R—Anniston, Ala. B—Rome, Ga. G. O. No. 73, W. D., 1919.	Major general, U. S. Army. He had immediate charge of the personnel division of the Surgeon General's Office and solved the problem of getting medical officers into the Army during an increase from 1,500 at the beginning of the war to 30,000. He also had charge of the hospital division of the Surgeon General's Office, handling both of these large responsibilities with conspicuous success.
NOLAN, DENNIS E. R—Akron, N. Y. B—Akron, N. Y. G. O. No. 12, W. D., 1919. Distinguished-service cross also awarded.	Brigadier general, U. S. Army. He organized and administered with marked ability the intelligence section of the General Staff of the American Expeditionary Forces. His estimates of the complex and everchanging military and political situation, his sound judgment, and accurate discrimination were invaluable to the Government, and influenced greatly the success that attended the operations of the American armies in Europe.
NUTTMAN, LOUIS M. R—Newark, N. J. B—Newark, N. J. G. O. No. 89, W. D., 1919.	Brigadier general, U. S. Army. As a regimental commander of the 89th Division he displayed marked military ability, providing his regiment with efficient training which showed its effects by the excellent conduct of the regiment in combat. Later, when promoted to the grade of brigadier general, he demonstrated great executive ability in the organization and administration of the combat officers' replacement depot at Gondrecourt.
O'BRIEN, MICHAEL J. R—New York, N. Y. B—Ireland. G. O. No. 31, W. D., 1922.	Lieutenant colonel, Adjutant General's Department, U. S. Army. As officer in charge of the officers' division, Adjutant General's Office, General Headquarters, American Expeditionary Forces, he demonstrated the highest professional attainments in the discharge of his multifarious duties. By his zeal, good judgment, and never-failing tact he was responsible in a large measure for the development of the high efficiency of his division. During the demobilization of the American Expeditionary Forces, working without regard to hours, by his tireless energy and exceptional ability he carried to successful completion the various and difficult projects that arose. Later, in charge of the American Expeditionary Forces records in Washington, D. C., he maintained that same high standard of efficiency which marked his work in the Adjutant General's Office, General Headquarters, American Expeditionary Forces.

O'DONNELL, JOHN L.
R—Chicago, Ill.
B—Chicago, Ill.
G. O. No. 62, W. D., 1919.

First lieutenant, chaplain, 132d Infantry, 33d Division, U. S. Army.
As regimental chaplain he was ceaseless in his efforts to better the welfare of the men, and during the period of operations accompanied the attacking waves in every action in which the regiment took part. Exposing himself to artillery and machine-gun fire to care personally for the wounded, organizing parties of stretcher bearers, going without a thought of personal danger wherever he was needed, he set an example of courage and heroism, appreciably raising the morale of those with whom and for whom he worked.

OGDEN, HENRY H.
R—Woodville, Miss.
B—Woodville, Miss.
G. O. No. 14, W. D., 1925.
Act of Congress, Feb. 25, 1925.

Second lieutenant, Air Service, U. S. Army.
Lieutenant Ogden, as assistant pilot of Airplane No. 3, the *Boston*, and assistant supply officer of the U. S. Army Air Service around-the-world flight from Apr. 6, 1924, to Sept. 28, 1924, displayed to a marked degree technical skill, courage, energy, and resourcefulness in carrying out his supply duties, in addition to alternate piloting of Airplane No. 3 during the voyage. His foresight, perseverance, and mechanical ability were very material factors in contributing to the successful accomplishment of this pioneer flight of airplanes around the world. In the efficient performance of his arduous duties he aided in the accomplishment of an undertaking bringing great credit to himself and to the military forces of the United States.

OGDEN, HUGH WALKER.
R—Brookline, Mass.
B—Bath, Me.
G. O. No. 74, W. D., 1919.

Lieutenant colonel, Judge Advocate General's Department, U. S. Army.
As judge advocate and inspector of the 42d Division he rendered valuable services. He exhibited ability of a high order throughout the operations of the division. Later, assigned to the bureau of civil affairs for the Third Army, he performed his task with marked success.

OLIVER, LLEWELLYN W.
R—Escanaba, Mich.
B—Escanaba, Mich.
G. O. No. 59, W. D., 1921.

Colonel (Infantry), General Staff Corps, U. S. Army.
As chief of staff, port of embarkation, Hoboken, N. J., from Jan. 21, 1919, to Feb. 26, 1920, he occupied a position of great responsibility. The details of arrangements for demobilization and for handling the sick and wounded fell largely upon his shoulders. His energy, excellent judgment, and administrative ability were of greatest value to the Government.

OLIVER, ROBERT T.
R—Indianapolis, Ind.
B—Indianapolis, Ind.
G. O. No. 103, W. D., 1919.

Colonel, Dental Corps, U. S. Army.
As chief dental surgeon he displayed remarkable ability in the performance of his numerous and exacting duties. He directed the personnel, equipment, and operations of his department with sound judgment, showing resourcefulness in solving new problems which confronted him.

OLMSTED, EDWARD.
R—Elizabeth, N. J.
B—San Francisco, Calif.
G. O. 22, W. D., 1920.

Lieutenant colonel, Infantry, U. S. Army.
As assistant chief of staff, G-1, of the 27th Division, by his high military attainments, zeal, and keen perception of his manifold duties the technical services of the division were so promptly and effectively coordinated and supervised that the front-line units were at all times completely supplied with all necessities. He has rendered services of particular worth to the American Expeditionary Forces.

O'NEILL, JOHN.
R—Newburyport, Mass.
B—Newburyport, Mass.
G. O. No. 49, W. D., 1922.

Lieutenant colonel, Transportation Corps, U. S. Army.
From November, 1917, until November, 1919, at Base Section No. 5, Brest, France, while serving in succession as chief of stevedores, assistant general superintendent in charge of the operations and troop and cargo divisions, and finally as superintendent of that port, by virtue of his marked ability, indomitable energy, and his capacity for inspiring his men he overcame all difficulties and met every demand made upon his force. His remarkable achievements in the quick turn-around of large vessels carrying troops and supplies were of inestimable value to the American Expeditionary Forces.

ORD, JAMES B.
R—San Diego, Calif.
B—Mexico.
G. O. No. 96, W. D., 1918.

Second lieutenant, 6th Infantry, attached to 13th Cavalry, U. S. Army.
While in action at Parral, Mexico, Apr. 12, 1916, after being himself wounded, he dismounted from his horse under heavy fire, placed a wounded man on a horse, and assisted him from the field.

ORTON, EDWARD, Jr.
R—Columbus, Ohio.
B—Chester, N. Y.
G. O. No. 69, W. D., 1919.

Lieutenant colonel, Motor Transport Corps, U. S. Army.
His untiring energy and splendid judgment were displayed in the efficient organization of the Engineering Division of the Motor Transport Corps in bringing about standardization of equipment and supplies and in efficiently directing the forces of the motor industry to the mutual advantage of the Army and the industry itself.

O'RYAN, JOHN F.
R—New York, N. Y.
B—New York, N. Y.
G. O. No. 12, W. D., 1919.

Major general, U. S. Army.
As commander of the 27th Division in its successful operations with the British in France in the autumn of 1918 he displayed qualities of skill and aggressiveness which mark him as a leader of ability. In the breach of the Hindenburg line between St. Quentin and Cambrai the name of his division is linked with the British in adding new laurels to the allied forces in France.

OURY, WILLIAM H.
R—Lincoln, Nebr.
B—Smyth County, Va.
G. O. No. 78, W. D., 1919.

Colonel, Infantry, U. S. Army.
Placed in command of the 157th Infantry Brigade, 79th Division, during the Montfaucon drive, he displayed exceptional qualities of leadership and marked tactical skill. Continuing at the same time in command of his regiment, the 314th Infantry, he directed the men of his command, and by his dauntless determination carried them forward under heavy enemy fire. He proved himself untiring in energy and possessed of great initiative, sound judgment, and military attainments of high order.

OVENSHINE, ALEXANDER T. R—Philadelphia, Pa. B—Fort Leavenworth, Kans. G. O. No. 31, W. D., 1922.	Colonel, Inspector General's Department, U. S. Army. While serving as inspector general, 3d Army Corps, during the Meuse-Argonne offensive, Colonel Ovenshine was charged with many important reports and investigations, which service was performed with marked ability and good judgment, and aided his corps and his Army commanders in decisions culminating in the success of this offensive. Later, as inspector general, 2d Army, he rendered distinguished service of a similar nature in connection with the operations of the 2d Army from Oct. 12, 1918, to the day of the armistice.
PAEGELOW, JOHN A. R—Chicago, Ill. B—Germany. G. O. No. 95, W. D., 1919.	Lieutenant colonel, Air Service, U. S. Army. As commander of balloon service of the 1st Army Corps and 1st Army, he was well fitted for his important position both by long experience in aeronautics and by noted organizing ability. Through his untiring energy an efficient system of supply and transportation was developed in spite of the numerous difficulties which assailed him. The successes achieved by the balloon service in the second Battle of the Marne and in the St. Mihiel offensive are a tribute to the high character of services rendered by him to the American Expeditionary Forces.
PALMER, BRUCE. R—Harrison, Ill. B—Fort Wallace, Kans. G. O. No. 50, W. D., 1919.	Colonel (Cavalry), General Staff Corps, U. S. Army. As a member of the General Staff of the American Expeditionary Forces, on duty with the 1st section, first as chief of the tonnage division during a period of stress, befraught with difficulties, and later as deputy assistant chief of staff, he performed duties of great responsibility with marked ability, fidelity, and success, invariably displaying personal and professional attainments of a high order.
PALMER, FREDERICK. R—New York, N. Y. B—Pleasantville, Pa. G. O. No. 16, W. D., 1923.	Lieutenant colonel, Signal Corps, U. S. Army. Charged with the responsibility of drafting regulations covering the mail and press censorship, as well as formulating plans for the guidance of press correspondents with the American forces in France, he rendered conspicuously valuable services. Having served in all the zones occupied by the allied forces, his information, of far reaching value, was placed at the disposal of our Government. His broad experience with many armies, his outstanding skill as an observer, his untiring energy and devotion to duty contributed markedly to the successful operations of the American Expeditionary Forces.
PALMER, JOHN McA. R—Springfield, Ill. B—Carlinville, Ill. G. O. No. 12, W. D., 1919.	Colonel, Infantry, U. S. Army. In the organization of the operations section of the General Staff, American Expeditionary Forces, this officer displayed sound tactical judgment and breadth of vision, and the ultimate success of the American plan of campaign was largely due to his detailed plans. As commander of the 58th Infantry Brigade, 29th Division, during the severe fighting north of Verdun, in the Argonne-Meuse offensive, his services were conspicuous and his brigade successful.
PARK, RICHARD. R—Warren, N. H. B—Malden, Mass. G. O. No. 69, W. D., 1919.	Colonel, Corps of Engineers, U. S. Army. To his energy and good judgment may be largely attributed the rapid development and successful administration of the Engineer Training School and mobilization camp at Camp A. A. Humphreys, Va. His utilization of labor of troops to supplement construction forces is a fine example of initiative in meeting a critical situation.
PARKER, CORTLANDT. R—Newark, N. J. B—Fort Apache, Ariz. G. O. No. 49, W. D., 1922.	Colonel, Field Artillery, U. S. Army. He organized and conducted the training camp for Field Artillery at Camp Coetquidan and later, at the office of the Chief of Artillery, American Expeditionary Forces, by his superior professional attainments, his zeal, and keen foresight, he contributed in a marked manner to the successful conduct of Field Artillery training. As regimental commander of the 6th Field Artillery, 1st Division, in the Cantigny sector and in the Aisne-Marne offensive, he repeatedly displayed superior tactical judgment and knowledge of artillery, and by his exceptional ability, leadership, and devotion to duty he rendered the maximum support to the Infantry of the 1st Division in effectively executing the most difficult missions assigned to him, thus rendering in a position of great responsibility most important services to the American Expeditionary Forces.
PARKER, EDWIN B. R—Houston, Tex. B—Shelby County, Mo. G. O. No. 35, W. D., 1920.	As priority commissioner on the War Industries Board he formulated and directed policies for the fullest development of war industries with marked ability and foresight. Later and subsequent to the armistice he, as chairman of the United States Liquidation Commission, War Department, undertook the immense task of liquidating the mass of intricate transactions incurred between the War Department and the allied Governments during the war, and also the disposal of all surplus war stocks and property of the War Department in Europe. As directing head of this commission and by his patience, industry, and diplomacy he brought this immense task to a successful conclusion in less than one year, to the very great credit of the War Department and the United States and the continued and warm friendship of the nations involved.
PARKER, FRANK. R—Georgetown, S. C. B—Georgetown, S. C. G. O. No. 59, W. D., 1919.	Brigadier general, U. S. Army. He commanded with marked distinction the 18th United States Infantry. Later, as a brigade commander, he exhibited qualities of rare leadership, superb courage, and unusual initiative. Finally he commanded the 1st Division in the Argonne offensive in the autumn of 1918, where he showed himself to be a skilled leader of marked ability.

PARKER, HUGH A.
R—Royse City, Tex.
B—Hunt, Tex.
G. O. No. 49, W. D., 1922.

Colonel (Infantry), General Staff Corps, U. S. Army.
He served as assistant G-3, 1st Army Corps, during June and July, 1918; G-3, 4th Division, from August to October, 1918; chief of staff, 7th Division, from November, 1918, to February, 1919; and chief of staff, American Military Mission at Berlin. By his tact, sound judgment, marked ability, and loyal devotion to duty, he contributed materially to the success of all units with which he served.

PARKER, JAMES.
R—Newark, N. J.
B—Newark, N. J.
G. O. No. 15, W. D., 1923.
Medal of honor also awarded.

Major general, U. S. Army.
He served with great distinction as commander of the Southern Department, Fort Sam Houston, Tex., Mar. 31, 1917, to Aug. 25, 1917; and as division commander, 32d Division, from Aug. 25 to Dec. 11, 1917; division commander, 85th Division, Dec. 11, 1917, to Feb. 20, 1918, when having reached the statutory age he was retired from active service. In these positions of great responsibility he displayed rare and outstanding leadership, the organizations under command at all times showing the results of sound training, a high state of morale and discipline. His unusual professional attainments, sound judgment, and devotion to duty were material and important factors in the development of organizations of the American Army, and contributed in a signal way to their successful operations in action against the enemy.

PARKER, JOHN H.
R—Green Ridge, Mo.
B—Tipton, Mo.
G. O. No. 89, W. D., 1919.
Distinguished-service cross and three oak-leaf clusters also awarded.

Colonel, Infantry, U. S. Army.
As an instructor in the Army Machine Gun School at Langres, by his tireless efforts he secured the necessary equipment and ably instructed a large student body in the technical handling of one of the most important fire power weapons developed in the present war, rendering services of great value to the American Expeditionary Forces.

PARSONS, JAMES K.
R—Birmingham, Ala.
B—Rockford, Ala.
G. O. No. 59, W. D., 1921.
Distinguished-service cross also awarded.

Colonel, Infantry, U. S. Army.
He organized and commanded with energy and ability the embarkation camp at St. Nazaire, France, and handled with conspicuous success the reception, care, and departure of the large number of officers and soldiers passing through that camp en route to the United States. He demonstrated administrative abilities of a high order and performed services of great value to the American Expeditionary Forces.

PARSONS, WILLIAM BARCLAY.
R—New York, N. Y.
B—New York, N. Y.
G. O. No. 4, W. D., 1923.

Colonel, Corps of Engineers, U. S. Army.
He served as major, 11th Engineers (Railway), during its organization and training period; chairman of Engineering Railway Commission sent overseas to investigate and report upon railway conditions in France; lieutenant colonel and then colonel, 11th Engineers, during its combat operations. By his wide experience, sound judgment, and brilliant professional and technical attainments he handled many difficult problems which confronted him with conspicuous success, thereby rendering services of great value to the American Expeditionary Forces.

PATRICK, MASON M.
R—Lewisburg, W. Va.
B—Lewisburg, W. Va.
G. O. No. 12, W. D., 1919.

Major general, U. S. Army.
He displayed much ability and devotion to duty as director of construction and forestry, and later, as chief of the Air Service of the American Expeditionary Forces, he perfected and ably administered the organization of this important department.

PATTERSON, CHARLES H.
R—Harrisburg, Pa.
B—Harrisburg, Pa.
G. O. No. 15, W. D., 1921.

Colonel, Coast Artillery Corps, U. S. Army.
As an officer of the Inspector General's Department and as chief of the investigations division of that department his rare efficiency and good judgment in the investigation and treatment of difficult and intricate problems have materially facilitated the administration of the office of the Inspector General, and have been of great value to the War Department and to the Army.

PATTERSON, HANNAH J.
R—Pittsburgh, Pa.
B—Smithton, Pa.
G. O. No. 73, W. D., 1919.

Council of National Defense.
She devoted herself throughout the whole period of the war to executive work of the Women's Committee of the Council of National Defense, devoting herself with great ability and energy to the organization of the activities and interests of the women throughout the United States in the interest of the successful prosecution of the war and, by her efforts, contributed to the splendid cooperation on the part of the women of the country in the great national emergency.

PATTERSON, PAUL M.
R—Kansas City, Kans.
B—Anthony, Kans.
G. O. No. 9, W. D., 1923.

Captain, Medical Corps, U. S. Army.
As evacuation officer of the 1st Army in the St. Mihiel and Meuse-Argonne offensives he directed the routing of all battle casualties into Army hospitals at the front, supervised the daily supply of those hospitals, and so skillfully and energetically classified and coordinated the constant flow of evacuations to the rear by timely calls for and expeditious loading of hospital trains that beds were always available for thousands of sick and wounded of the 1st Army. He rendered services of inestimable value to the American Expeditionary Forces in a position of great responsibility.

PATTERSON, ROBERT U.
R—Baltimore, Md.
B—Canada.
G. O. No. 13, W. D., 1923.

Colonel, Medical Corps, U. S. Army.
As commanding officer, Base Hospital No. 5, United States Army, serving with the British Expeditionary Forces at Dannes-Camiers and Boulogne, France, he displayed tireless energy and military attainments of a high order in the efficient operation of this hospital. By his marked devotion to duty and administrative ability great numbers of our own and allied sick and wounded were treated, resulting in the saving of many lives. His services were of material value to the American Expeditionary Forces.

PATTON, GEORGE S., Jr
R—San Gabriel, Calif.
B—San Gabriel, Calif.
G. O. No. 103, W. D., 1919.
Distinguished-service cross also awarded.

Colonel, Tank Corps, U. S. Army.
By his energy and sound judgment he rendered very valuable services in his organization and direction of the tank center at the Army schools at Langres. In the employment of Tank Corps troops in combat he displayed high military attainments, zeal, and marked adaptability in a form of warfare comparatively new to the American Army.

PAULES, EARL G
R—Marietta, Pa.
B—Marietta, Pa.
G. O. No. 95, W. D., 1919.

Colonel, Corps of Engineers, U. S. Army.
As a member of the American Military Commission of Italy from April to July, 1918, he displayed tact and diplomacy in making a preliminary investigation of the Czechoslovak situation. Later, as commanding officer of the 7th Engineers, he participated creditably in the St. Mihiel and Meuse-Argonne offensives, materially aiding in the operations of the 5th Division by his skill in constructing bridges across the Meuse River in the face of desperate hostile resistance.

PEABODY, PAUL E
R—Los Angeles, Calif.
B—Chicago, Ill.
G. O. No. 22, W. D., 1920.

Major, Infantry, U. S. Army.
As an assistant chief of staff, G–1, 1st Division, during the attack on Soissons, he displayed marked ability. Later, as G–1, during the St. Mihiel and Meuse-Argonne offensives, by his extraordinary ability, his capacity for organization, and his brilliant execution of all details pertaining to administration of supply he overcame unusual difficulties and thereby contributed in a marked degree to the success of the operations of the 1st Division.

PEARCE, EARLE D'A
R—Thomaston, Ga.
B—Thomaston, Ga.
G. O. No. 56, W. D., 1922.

Colonel, Field Artillery, U. S. Army.
As commanding officer of the 319th Field Artillery, 82d Division in the organization and training of the regiment and in its very successful operations against the enemy in the Aisne-Marne, St. Mihiel, and Meuse-Argonne offensives, he displayed tireless energy, keen devotion to duty, and eminent technical skill as an artillerist, gave most effective support to the Infantry of the 82d and 80th Divisions, and very materially contributed to the successes attained by those units.

PECK, ALLEN STEELE
R—Stottville, N. Y.
B—West Barre, N. Y.
G. O. No. 78, W. D., 1919.

Lieutenant colonel, Corps of Engineers, U. S. Army.
As an officer of the Forestry Service he displayed exceptional tact and sound judgment in securing public and private grants for large quantities of timber from the French Government. At all times he pursued his task with great energy, achieving signal success. He organized and administered a project for the production of fuel wood, which proved entirely successful. In these endeavors he rendered valuable service to the American Expeditionary Forces.

PECK, CHARLES H
R—New York, N. Y.
B—Newton, Conn.
G. O. No. 59, W. D., 1919.

Lieutenant colonel, Medical Corps, U. S. Army.
As director of Base Hospital No. 15, which he had organized most efficiently, he displayed unusual skill and very marked ability in the conduct of that unit. Later, as senior consultant in general surgery for the American Expeditionary Forces, his professional attainments, wide experience, and sound advice proved of inestimable value in increasing the efficiency of the Medical Department of the United States Army.

PECK, ROBERT H
R—San Diego, Calif.
B—San Francisco, Calif.
G. O. No. 95, W. D., 1919.
Distinguished-service cross also awarded.

Colonel, Infantry, U. S. Army.
He rendered services of signal worth as commanding officer of the 47th Infantry, 4th Division, and subsequently of the 11th Infantry, 5th Division, successfully accomplishing all missions assigned to the regiment under his command during the active operations in which it participated. Constantly displaying unremitting zeal and tactical ability of a high order, he proved himself invaluable to the American Expeditionary Forces.

PEEK, ERNEST D
R—Oshkosh, Wis.
B—Oshkosh, Wis.
G. O. No. 72, W. D., 1920.

Colonel, Corps of Engineers, U. S. Army.
He organized and conducted the operations of the standard gauge and light railways of the 1st Army during its active operations, resulting in the reduction of the St. Mihiel salient and the recovery of the extensive Meuse-Argonne area. Although handicapped by lack of personnel and material, he pushed the enterprise to success. By untiring, painstaking, and energetic efforts in the use of the inadequate means at his disposal, he displayed unusual talent for organization and masterful execution.

PEEK, GEORGE N
R—Moline, Ill.
B—Polo, Ill.
G. O. No. 15, W. D., 1923.

Commissioner of Finished Products, War Industries Board.
In connection with the operations of the War Industries Board during the World War. As a member of the board he rendered, through his broad vision, distinguished capacity, and business ability, services of inestimable value in marshaling the industrial forces of the Nation and mobilizing its economic resources—marked factors in assisting to make military success attainable. As commissioner of finished products, it was largely through his untiring efforts and devotion to duty that the supply bureaus of the War Department were able to maintain a constant flow of munitions as well as supplies of a general character to the Army.

PEIRCE, WILLIAMS S
R—Burlington, Vt.
B—Burlington, Vt.
G. O. No. 25, W. D., 1919.

Brigadier general, U. S. Army.
While in charge of the Springfield Arsenal his exceptional ability contributed materially to increasing the output of small arms. As Assistant Chief of Ordnance he has rendered conspicuous service.

PENNELL, RALPH McT
R—Belton, S. C.
B—Belton, S. C.
G. O. No. 31, W. D., 1922.

Colonel, Field Artillery, U. S. Army.
As assistant to the Chief of Field Artillery from Apr. 16, 1918, to Sept. 4, 1918, he planned and executed those measures which provided a balanced production of different types of field artillery matériel and equipment and the selection of the types to be produced, and which determined the priorities of distribution of the same.

PENNER, CARL
  R—Milwaukee, Wis.
  B—Milwaukee, Wis.
  G. O. No. 47, W. D., 1921.

Colonel, 120th Field Artillery, 32d Division, U. S. Army.
He commanded his regiment during the Marne-Aisne, Oise-Aisne, and Meuse-Argonne offensives. His devotion to duty and skillful handling of his command proved a material factor in the success of the 57th Field Artillery Brigade while supporting the 32d Division, and at other times five other divisions.

PEPPER, SAMUEL D.
  R—Lansing, Mich.
  B—Canada.
  G. O. No. 56, W. D., 1922.

Lieutenant colonel, Judge Advocate General's Department, U. S. Army.
From August, 1917, until June, 1919, he served in turn as judge advocate of the 32d Division, 5th Army Corps, and advance section, Services of Supply. By his marked legal ability, excellent judgment, and thorough knowledge of discipline and morale, he rendered services of signal worth to all the units with which he served, thereby contributing materially to the success of the American Expeditionary Forces.

PERKINS, ALBERT T.
  R—St. Louis, Mo.
  B—Brunswick, Me.
  G. O. No. 59, W. D., 1919.

Colonel, Corps of Engineers, U. S. Army.
As deputy and later as manager of light railways, he undertook the task of organizing a light railway service for the American Expeditionary Forces. His long and complete railroad experience and knowledge assured the success of these lines. By his foresight in promptly gathering from the United States a generous supply of railway material he quickly brought the light railway service to a high degree of efficiency.

PERKINS, FRED MILTON
  R—Salem, Oreg.
  B—Salem, Oreg.
  G. O. No. 116, W. D., 1919.

Commander, U. S. Navy.
As flag secretary to the Commander, cruiser and transport fleet, his close cooperation with the Army authorities in the handling of troop ships contributed greatly to the successful outcome of our oversea operations.

PERKINS, JAMES H.
  R—Greenwich, Conn.
  B—Milton, Mass.
  G. O. No. 50, W. D., 1919.

Lieutenant colonel, Quartermaster Corps, U. S. Army.
He was in charge of the work of the American Red Cross in Europe for a period of time, and by his great energy and untiring efforts maintained that institution at a high state of excellence and rendered valuable assistance to the American Expeditionary Forces. While in the military service he displayed marked ability in the performance of the various duties with which he was intrusted.

PERSHING, JOHN J.
  R—Laclede, Mo.
  B—Linn County, Mo.
  G. O. No. 111, W. D., 1918.

General, U. S. Army, Commander in Chief, American Expeditionary Forces.
As a token of the gratitude of the American people to the commander of our armies in the field for his distinguished services, and in appreciation of the success which our armies have achieved under his leadership.

PERSONS, ELBERT E.
  R—Chicago, Ill.
  B—Prouts, Ohio.
  G. O. No. 49, W. D., 1922.

Colonel, Medical Corps, U. S. Army.
He organized the United States Army Ambulance Service Training School at Camp Crane, Allentown, Pa., and as its commanding officer from June, 1917, to May, 1918, by his great energy, marked executive ability, and wide professional knowledge, succeeded in rapidly training, equipping, and dispatching overseas ambulance units where they rendered excellent service with the French armies and later with our own forces. From June to December, 1918, he commanded the American Ambulance Service, serving on the Italian front, where he again rendered conspicuous service in a position of great responsibility.

PETERSON, VIRGIL L.
  R—Mannsville, Ky.
  B—Raywick, Ky.
  G. O. No. 56, W. D., 1922.

Colonel, Corps of Engineers, U. S. Army.
As commanding officer, Engineer Officers' Training Camp at Camp Lee, Va., from April to August, 1918, and director of training at Camp Humphreys, Va., until October, 1918, he displayed marked foresight, rare ability, and sound judgment in the reorganization and standardization of the instruction for engineer troops. By his organizing and training ability, indefatigable efforts, and high military attainments he successfully directed the training of 4,500 engineer officers and 20,000 enlisted men, thereby rendering services of great value to our Government in positions of great responsibility.

PEYTON, EPHRAIM G.
  R—Columbus, Miss.
  B—Gallatin, Miss.
  G. O. No. 59, W. D., 1919.

Colonel, Infantry, U. S. Army.
As the commanding officer of the 320th Infantry Regiment, 80th Division, in all its operations, by careful and painstaking preparations and skillful leadership, he enabled his regiment to carry always its tasks through to a successful end. At all times he displayed a high order of leadership and exhibited superb qualities as a commander.

PEYTON, PHILIP B.
  R—Charlottesville, Va.
  B—Nashville, Tenn.
  G. O. No. 59, W. D., 1919.

Colonel, 61st Infantry, 5th Division, U. S. Army.
He took command of a regiment which had undergone six days of shell fire and commanded it with such unusual skill as to enable the regiment to capture Aincreville, Bois-de-Babiemont, Doulcon, and, after crossing the Meuse, to capture Hill No. 292, Dun-sur-Meuse, Milly-devant-Dun, Lion-devant-Dun, Cote St. Germain, Chateau Charmois, and Mouzay, thereby displaying the highest order of leadership and exhibiting the masterful qualities of a commander.

PHILLIPS, ALBERT E.
  R—New Orleans, La.
  B—New Orleans, La.
  G. O. No. 19, W. D., 1922.

Colonel, Ordnance Department, U. S. Army.
In the development and operation of machine-gun centers, in the design and development of machine-gun equipment, and later he rendered invaluable service in the preparation of machine-gun manuals.

PICKERING, RICHARD R.
  R—Uniontown, Ala.
  B—Uniontown, Ala.
  G. O. No. 19, W. D., 1922.

Lieutenant colonel, Infantry, U. S. Army.
As commanding officer of the large embarkation camp at Camp Mills, Long Island, N. Y., from Apr. 5, 1918, to Sept. 13, 1918, and from Dec. 11, 1918, to May 8, 1919, he displayed executive ability of the highest order in a position of great responsibility. His work marked Colonel Pickering as an exceptional officer in his capacity as organizer and administrator, and by his energy and excellent judgment contributed in a notable degree to the success of troop movements at the port of embarkation, Hoboken, N. J.

PIERCE, CHARLES C_____
  R—Germantown, Pa.
  B—Salem, N. J.
  G. O. No. 59, W. D., 1919.

Lieutenant colonel, Quartermaster Corps, U. S. Army.
Serving as assistant to the chief quartermaster, American Expeditionary Forces, in the capacity of chief of Graves Registration Services since December, 1917, he displayed unusual ability and conscientious care in the performance of his exacting duties. Under his skillful administration the service functioned efficiently. He at all times showed great energy and performed his important task with exceptional success.

PIERCE, JUNNIUS_____
  R—Brooklyn, N. Y.
  B—Gainesville, Tex.
  G. O. No. 59, W. D., 1921.

Major, Quartermaster Corps, U. S. Army.
As chief quartermaster, adjutant general, chief of staff, and later commanding officer, Base Section No. 3, England, he served in these various capacities with great credit. As commanding officer of Base Section No. 3 he had charge of evacuation of troops and liquidation of American interests in England. By his energy, ability, and tact he conducted American Expeditionary Forces affairs in England to a successful conclusion.

PIERCE, PALMER E_____
  R—Traer, Iowa.
  B—Savanna, Ill.
  G. O. No. 56, W. D., 1922.

Brigadier general, U. S. Army.
His zeal, intelligence, and effective work in the preliminary organization of our industries for war contributed substantially to the progress made. From May, 1917, until March, 1919, he commanded the 54th Infantry Brigade, 27th Division, in a highly meritorious manner during all the operations of his division against the Hindenburg line. His sound judgment, marked ability, and skillful leadership were important factors in the successes attained by his division against the enemy.

PIERSON, ROBERT H_____
  R—Syracuse, N. Y.
  B—Fayettesville, N. Y.
  G. O. No. 95, W. D., 1919.

Colonel, Medical Corps, U. S. Army.
He served as division surgeon of the 5th Division from its organization until the close of hostilities, when he became chief surgeon of the 6th Army Corps. Due to his sound judgment and efficient direction of medical personnel, gas casualties in his division were reduced to a minimum. By his resourceful methods in combating disease he prevented the firing lines from being depleted at a critical time, maintaining a high standard of combat strength efficiency.

PIKE, SHEPARD L_____
  R—Plattsburg, N. Y.
  B—Plattsburg, N. Y.
  G. O. No. 103, W. D., 1919.

Lieutenant colonel, Infantry, U. S. Army.
As commandant of the Army Candidates' School at Langres, France, he organized an important institution and developed it to a high state of efficiency. His services in capably directing the training of more than 5,000 candidates for active duty at the front were of the utmost value to the American Expeditionary Forces.

PILLOW, JEROME G_____
  R—Helena, Ark.
  B—Columbia, Tenn.
  G. O. No. 9, W. D., 1923.

Lieutenant colonel (Cavalry), General Staff Corps, U. S. Army.
As assistant chief of staff, G–3, 32d Division, he displayed marked ability, high professional attainments, rendering immeasurably valuable services in the operations of his division in the offensives of Aisne-Marne, Oise-Aisne, the Meuse-Argonne, and during the march to the Rhine.

PILLSBURY, GEORGE B_____
  R—Tewksbury, Mass.
  B—Lowell, Mass.
  G. O. No. 56, W. D., 1922.

Colonel, Corps of Engineers, U. S. Army.
As corps engineer, 2d Army Corps, from October, 1918, to January, 1919, he displayed professional qualifications of an exceptionally high order. During operations which broke the Hindenburg line between Cambrai and St. Quentin his tireless energy and highly efficient work contributed in a marked degree to the success of the operations of the corps.

PLUNKETT, CHARLES PERSHALL_____
  R—Washington, D. C.
  B—Washington, D. C.
  G. O. No. 59, W. D., 1919.

Rear admiral, U. S. Navy.
He supervised the production, transportation to Europe, and the placing in action on the western front of the United States Naval Gun Battalion of five 14-inch guns on railway mounts, the most powerful artillery weapons brought into action against Germany and her allies during the war. In this stupendous undertaking the successful accomplishment of which had an important bearing on the outcome of the war, he displayed technical knowledge of a high order, combined with practical knowledge of the needs of the service and the difficulties to be encountered. He worked with unceasing zeal and devotion, rendering a service of rare distinction to the American Expeditionary Forces.

POILLON, ARTHUR_____
  R—New York, N. Y.
  B—New York, N. Y.
  G. O. No. 124, W. D., 1919.

Lieutenant colonel, Cavalry, U. S. Army.
For services while serving as military attaché at The Hague, Netherlands.

POOL, EUGENE H_____
  R—New York, N. Y.
  B—New York, N. Y.
  G. O. No. 49, W. D., 1922.

Lieutenant colonel, Medical Corps, U. S. Army.
As surgical consultant with the 4th Army Corps, 5th Army Corps, and then the 1st Army, he displayed unusual organizing ability, excellent judgment, and professional attainments of the highest order in directing the work of surgical teams in the care of large numbers of wounded in various hospitals at the front during the St. Mihiel and Meuse-Argonne offensives, thereby rendering services of great value to the American Expeditionary Forces.

POORE, BENJAMIN A_____
  R—Fitchburg, Mass.
  B—Center, Ala.
  G. O. No. 59, W. D., 1919.
  Distinguished-service cross also awarded.

Brigadier general, U. S. Army.
He commanded with distinction and ability the 7th Infantry Brigade, 4th Division, in the numerous engagements of the Argonne-Meuse campaign. By his energy and ability his brigade drove the enemy from Ruisseau-des-Forges and from the Bois-du-Fays. In these engagements important captures of many prisoners and much material were made by the troops of his command.

POPE, FRANCIS H.
R—St. Louis, Mo.
B—Fort Leavenworth, Kans.
G. O. No. 3, W. D., 1921.

Colonel, Quartermaster Corps, U. S. Army.
As chief of the Motor Transport Service, American Expeditionary Forces, he developed a rational organization for the operation and maintenance of the motor transport of the American Expeditionary Forces, and as Deputy Chief, Motor Transport Corps, American Expeditionary Forces, was largely instrumental in further developing and applying this organization, thereby rendering exceptionally meritorious and distinguished service to the United States.

POPE, WILLIAM R.
R—Pulaski, Tenn.
B—Pulaski, Tenn.
G. O. No. 4, W. D., 1923.

Colonel, Infantry, U. S. Army.
Having taken command of the 113th Infantry, 29th Division, shortly before the beginning of the Meuse-Argonne offensive, held it with signal ability throughout the period of its engagement in that operation. By his energy and resourcefulness in overcoming the numerous adverse conditions which confronted his regiment, he proved an inspiration to his men and an important factor in the successes of his division. As provost marshal general of the embarkation center at Le Mans, France, from December, 1918, until June, 1919, he displayed tact, marked efficiency, and executive and administrative ability of the highest order, thereby rendering highly conspicuous services in a position of great responsibility.

PORGES, GUSTAVE.
R—New York, N. Y.
B—Bohemia.
G. O. No. 39, W. D., 1920.

Colonel, Quartermaster Corps, U. S. Army.
He gave markedly able assistance in the establishment of the American Expeditionary Forces depot at Nevers and the Engineer camp at Vierzon. Later, when placed in charge of the purchase of subsistence, clothing, equipage, forage, and bedding, by his valuable knowledge of commercial conditions and by his persistent efforts and able negotiations, he secured much needed supplies from allied and neutral countries. He originated the manufacture in Europe of necessary items of food, clothing, and cloth for the use of the American Expeditionary Forces. Subsequently he rendered valuable assistance in the liquidation of the American Expeditionary Forces surplus supplies. At all times he showed remarkable energy in all matters affecting the welfare of the troops in the American Expeditionary Forces.

POSTON, ADELE S.
R—White Plains, N. Y.
B—Springdale, Ark.
G. O. No. 9, W. D., 1923.

Chief Nurse, Army Nurse Corps, U. S. Army.
As chief nurse of Base Hospital No. 117 (psychiatric unit), at La Fauche, France, during the World War, she performed very difficult and exacting duties with marked skill and distinction. By her professional efficiency, untiring energy, and tact, she made a large contribution to the success of this novel and highly important hospital of the American Expeditionary Forces.

POTTER, WILLIAM CHAPMAN.
R—Old Westbury, N. Y.
B—Chicago, Ill.
G. O. No. 118, W. D., 1919.

He reorganized the equipment division of the Signal Corps and organized and developed the Bureau of Aircraft Production.

POWER, NEAL.
R—San Francisco, Calif.
B—Washington, D. C.
G. O. No. 89, W. D., 1919.

Lieutenant colonel, Judge Advocate General's Department, U. S. Army.
As head of the special disciplinary division in the office of the judge advocate, American Expeditionary Forces, he was charged with duties of an exceptionally arduous and responsible nature, in the performance of which he displayed high professional attainments and notable devotion to duty.

*POWERS, CHARLES A.
R—Denver, Colo.
B—Lawrence, Mass.
G. O. No. 3, W. D., 1924.

Major, Medical Corps, U. S. Army.
As a surgeon, first with the French armies and later with the American Red Cross Military Hospital No. 1, he displayed untiring energy and surgical ability of the highest order. By his professional skill he revolutionized the surgical treatment of faces mutilated by war wounds, demonstrating to the world how to restore them to a normal condition, thereby rendering conspicuous service, by this great contribution, in saving the lives of many French and American soldiers.
Posthumously awarded. Medal presented to stepfather, Dr. A. J. Stevens.

PRATT, JOSEPH H.
R—Chapel Hill, N. C.
B—Hartford, Conn.
G. O. No. 36, W. D., 1922.

Colonel, Corps of Engineers, U. S. Army.
He commanded the 105th Engineers during its organization and training period in the entire operation of the 30th Division near Ypres, Belgium, and during the breaking of the Hindenburg line and the advance beyond. He displayed forceful energy, exceptional ability, and remarkable foresight in the solution of all engineer tasks, including the construction of railways and roads, as well as the location and destruction of mines and traps, thereby contributing materially to the success of the operations.

PRATT, WILLIAM VEAZIE.
R—Belfast, Me.
B—Belfast, Me.
G. O. No. 116, W. D., 1919.

Captain, U. S. Navy.
As assistant to the Chief of Naval Operations, his untiring energy and close cooperation with the Army in connection with its oversea movements of troops and supplies, and especially in the making up and routing of convoys, resulted in the successful movement of over 2,000,000 men without material loss of life.

PRENTISS, AUGUSTIN M.
R—Barnwell, S. C.
B—Chapel Hill, N. C.
G. O. No. 56, W. D., 1922.

Major, Chemical Warfare Service, U. S. Army.
As officer in charge of the ordnance section, Chemical Warfare Service, he displayed great energy, untiring devotion to duty, and high technical skill. Due to his thorough knowledge of ordnance supplies and material, especially shells and guns, he rendered invaluable assistance in the proper choice of gases, gas shells, and other materials used both by the gas troops and artillery throughout the war, thereby contributing materially to the successes of the American Expeditionary Forces.

PRICE, HOWARD C_____
R—Chester, Pa.
B—Chester, Pa.
G. O. No. 89, W. D., 1919.

Colonel, Infantry, U. S. Army.
He organized, trained, and commanded in active operations the 360th Infantry, 90th Division, which under his capable leadership was eminently successful as a combat unit. At all times he inspired a notable spirit among the members of his command. He displayed military attainments of a high order in the capture of the Foret-du-Bois le Pretre during the St. Mihiel offensive and the assault on the Freya Stellung in the Argonne-Meuse operations, rendering merited services to the American Expeditionary Forces.

PRICE, WILLIAM G., jr_____
R—Chester, Pa.
B—Chester, Pa,
G. O. No. 103, W. D., 1919.

Brigadier general, U. S. Army.
He commanded the 53d Field Artillery Brigade, 28th Division, with marked distinction, proving himself a tactical leader of extraordinary ability. Through the formidable assistance which his brigade furnished to the attacking Infantry during the engagement of the 91st Division from the Lys to the Scheldt, the rapid advance of the Infantry was insured and the success made more brilliant.

PRICE, XENOPHON H_____
R—Bay City, Mich.
B—Saginaw, Mich.
G. O. No. 87, W. D., 1919.

Lieutenant colonel, Corps of Engineers, U. S. Army.
He organized and was continuously in charge of all map-room data of the 3d Section, General Staff, at General Headquarters, American Expeditionary Forces. Through his energy, ability, and sound military judgment maps showing accurately the situation on the battle fronts were constantly available for outlying projected operations, and the data compiled by him is of incalculable historical value in preserving a record of the achievements of the American Expeditionary Forces.

PURINGTON, GEORGE A_____
R—Fort Sheridan, Ill.
B—Cleveland, Ohio.
G. O. No. 59, W. D., 1919.

Lieutenant colonel, Cavalry, U. S. Army.
He was engaged in keeping roads open and traffic moving in the advance of the 1st Army between the 26th of September and the 30th of Sepember, 1918, in the battle west of the Meuse. Due to his tireless effort and determination the supply of ammunition and food of the 3d and 5th Corps was insured. Although confronted with a most difficult task, he overcame all obstacles and crowned his efforts with great success.

PUSEY, FRED TAYLOR_____
R—Lima, Pa.
B—Philadelphia, Pa.
G. O. No. 38, W. D., 1922.

Lieutenant colonel, Quartermaster Corps, U. S. Army.
As division quartermaster, 28th Division, he displayed exceptional ability and zeal in developing the personnel and in handling the endless duties of his office, both in training and in combat, and constantly maintained the supply of food and equipment to the command and its various units under most difficult circumstances, thereby contributing materially to the success of this division during its active operations in France.

QUAKEMEYER, JOHN G_____
R—Yazoo City, Miss.
B—Yazoo City, Miss.
G. O. No. 59, W. D., 1919.

Lieutenant colonel, Cavalry, U. S. Army.
As chief of the American Mission at British General Headquarters, he administered the duties of the office with tact and ability, promoting cordial relations between members of the Allied Armies with whom he came in contact. As aide-de-camp to the Commander in Chief, he has performed his important duties with marked distinction and sound judgment.

RAFFERTY, WILLIAM A_____
R—Chicago, Ill.
B—Fort Wingate, N. Mex.
G. O. No. 56, W. D., 1922.

Major (Infantry), General Staff Corps, U. S. Army.
While in charge of the supply branch, 4th (Supply) Section, General Staff, 1st Army, American Expeditionary Forces, he displayed great initiative, zeal, and devotion to duty, and by his excellent judgment, forethought, and cooperation materially assisted in the solution of many difficult problems concerning the supply of the troops during the St. Mihiel and Meuse-Argonne offensives, thereby contributing in a marked degree to the successful accomplishments of the supply of the troops engaged in those operations.

RALSTON, FRANCIS W_____
R—Philadelphia, Pa.
B—Philadelphia, Pa.
G. O. No. 59, W. D., 1919.

Colonel, Coast Artillery Corps, U. S. Army.
His marked military attainments rendered his services most valuable while serving as adjutant of the 42d Division. As commandant of General Headquarters, American Expeditionary Forces, he performed exacting duties with distinction. By his unflagging energy, zeal, and sound judgment he solved difficult problems of administration, achieving most satisfactory results.

RAND, WILLIAM_____
R—New York, N. Y.
B—Chicago, Ill.
G. O. No. 15, W. D., 1923.

Colonel, Judge Advocate General's Department, U. S. Army.
As principal assistant in the office of the Acting Judge Advocate General for the American Expeditionary Forces in Europe and as a member of the board of review to which was referred records of trial by courts-martial in the American army operating on foreign soil, he performed duties of an important and far-reaching character and grave responsibility. His broad learning, his comprehensive grasp of the problems with which he had to deal, coupled with untiring industry and energy contributed markedly to the accomplishments of the office of the Judge Advocate General and of the American Expeditionary Forces.

READ, ALVAN C_____
R—Baton Rouge, La.
B—Lewisburg, Tenn.
G. O. No., 108, W. D., 1919.

Colonel, Infantry, U. S. Army.
As inspector general for the armies during their operations in the St. Mihiel and Meuse-Argonne offensives, by his keen observations of the conduct of units and leadership displayed by commanders he was able at all times to give valuable information as to the morale and efficiency of troops and their commanders. By the able handling of his important duties, prompt and adequate means were always provided for improving conditions as to these important factors in the conduct of operations. Later, as chief inspector of the Army of Occupation, he continued to render the same superior quality of service which marked that given by him prior to the armistice.

READ, GEORGE W.
R—Des Moines, Iowa.
B—Indianola, Iowa.
G. O. No. 12, W. D., 1919.

Major general, U. S. Army.
He commanded with distinction the 30th Division, and organized and commanded the 2d Army Corps in its operations with the British forces in France. He displayed qualities of leadership and professional attainments of a high order, and to his efforts are largely due the brilliant success achieved.

RECKORD, MILTON A.
R—Bel Air, Md.
B—Harford County, Md.
G. O. No. 89, W. D., 1919.

Colonel, Infantry, U. S. Army.
He served with distinction as commanding officer of the 115th Infantry, 29th Division at all times showing qualities of high military leadership and great tactical ability. Inspiring his men by his aggressive spirit and fervent devotion to his task, he led them with noted success through three weeks of constant action against the enemy during the operations north of Verdun.

REECE, B. CARROLL.
R—Butler, Tenn.
B—Butler, Tenn.
G. O. No. 59, W. D., 1919.

First lieutenant, Infantry, U. S. Army.
He showed energy, initiative, and military ability of a high order while serving as second lieutenant in the 102d Infantry, 26th Division, in command of a company and later a battalion. He led his company brilliantly in the attack upon the St. Mihiel salient and during the operations of the 26th Division north of Verdun. Confronted later by a task of great difficulty when placed in command of a battalion, which suffered heavy casualties and became badly disorganized, he displayed marked ability and determination in reorganizing his command and molding it into a good fighting unit, able under his leadership to achieve valuable results.

REED, DAVID A.
R—Pittsburgh, Pa.
B—Pittsburgh, Pa.
G. O. No. 56, W. D., 1922.

Major, Field Artillery, U. S. Army.
As battalion commander, 311th Field Artillery, 79th Division, he displayed exceptional ability as an organizer, instructor, and leader. By his professional attainments and tireless energy he was instrumental in bringing his command to a high state of efficiency. Later, as the Field Artillery member of the United States section of the Inter-Allied Armistice Commission at Spa and as acting chief of staff thereof, by his exceptional ability and insight into affairs requiring delicate and diplomatic treatment, he rendered conspicuous service to the American Expeditionary Forces.

REED, WALTER L.
R—Washington, D. C.
B—Fort Apache, Ariz.
G. O. No. 10, W. D., 1922.

Lieutenant colonel, Inspector General's Department U. S. Army.
In the organization and administration of the Inspector General's Department at Camp Pontanezen, Brest, France, thereby enabling that department to meet the excessive demands made upon it during the return of the American Expeditionary Forces through the port of Brest.

REES, ROBERT I.
R—Houghton, Mich.
B—Houghton, Mich.
G. O. No. 25, W. D., 1919.

Brigadier general, U. S. Army.
For services with the committee charged with education and special training in the Army. To his initiative and breadth of vision are largely due the successful measures for training of enlisted men for special services and the establishment of the Student Army Training Corps.

REEVES, IRA L.
R—Chicago, Ill.
B—Jefferson City, Mo.
G. O. No. 59, W. D., 1921.

Colonel, Infantry, U. S. Army.
As president of the American Expeditionary Forces University in France his fine ability as an organizer made possible the expeditious establishment of this institution, and to his initiative, energy, rare tact, and good judgment is due, in a large measure, the successful operation of the university.

REEVES, JAMES H.
R—Center, Ala.
B—Center, Ala.
G. O. No. 87, W. D., 1919.
Distinguished-service cross also awarded.

Colonel, Infantry, U. S. Army.
He organized the 353d Infantry, 89th Division, and commanded it with distinction during all but one month of its active service. The high qualities of leadership and unfailing devotion to duty displayed by him were responsible for the marked esprit and morale of his command. To his marked tactical ability and energy are largely due the brilliant successes achieved by his regiment during its operations against the enemy.

*REGISTER, EDWARD C.
R—Georgetown, S. C.
B—Rose Hill, N. C.
G. O. No. 9, W. D., 1923.

Lieutenant colonel, Medical Corps, U. S. Army.
While a member of the Polish Relief Expedition, volunteering for service at Tarnopol, Poland, the entire city being prostrate from the effects of typhus fever, 45 doctors having sacrificed their lives within the preceding two months. Upon arrival at Tarnopol he assumed entire charge of the situation, organized and established a 1,500-bed hospital equipped with supplies which had been concealed from enemy forces and found by him. Fifteen days after his arrival in the city he contracted typhus fever and died from its effects on January 3, 1920.
Posthumously awarded. Medal presented to widow, Mrs. Edward C. Register.

REILLY, HENRY J.
R—Winnetka, Ill.
B—Fort Barrancas, Fla.
G. O. No. 108, W. D., 1919.

Colonel, Field Artillery, U. S. Army.
In command of the 149th Field Artillery, 42d Division, he participated with credit in the operations of the 42d Division. Through his tireless energy and technical skill as an artillerist, his regiment gave most effective assistance to the Infantry which it supported.

REINHART, STANLEY E.
R—Polk, Ohio.
B—Polk, Ohio.
G. O. No. 70, W. D., 1919.

Major, Field Artillery, U. S. Army.
In command of a battery and subsequently a battalion of the 17th Field Artillery, 2d Division, he gave proof of high qualities of leadership and military attainments, notably during the operations near Soissons in July, 1918, when he skillfully maneuvered his battalion in front of the infantry under machine-gun fire from the enemy with but few casualties to his command. Later he rendered valuable and loyal service as chief of staff to the chief of artillery, 6th Army Corps.

REPLOGLE, J. LEONARD............... | Director of steel, War Industries Board.
R—New York, N. Y. | In connection with the operations of the War Industries Board during the
B—Bedford County, Pa. | World War. As a member of the board he rendered through his broad
G. O. No. 15, W. D., 1923. | vision, distinguished capacity, and business ability, services of inestimable value in marshaling the industrial forces of the Nation and mobilizing its economic resources—marked factors in assisting to make military success attainable. As director of steel, he rendered, through his untiring efforts and devotion to duty, exceptionally valuable service to the War Department in connection with the procurement and supply of steel and iron for the Army.

REPP, WILLIAM F.................... | Lieutenant colonel, Signal Corps, U. S. Army.
R—Philadelphia, Pa. | With his valuable assistance the Signal Corps was enabled originally to plan
B—Philadelphia, Pa. | for the immense network of the United States Army telegraph and telephone
G. O. No. 59, W. D., 1919. | lines now existing in France. To him is attributable the exceptionally high standard of efficiency attained by the telephone and telegraph service. As chief signal officer, advance section, services of supply, his services have been marked by a character of exceptional excellence.

RETHERS, HARRY F.................. | Colonel, Quartermaster Corps, U. S. Army.
R—San Francisco, Calif. | He distinguished himself by his extraordinary ability and exceptional skill in
B—San Francisco, Calif. | organizing the work of the Quartermaster Corps at Base Section No. 3. His
G. O. No. 59, W. D., 1919. | good judgment, combined with tact, knowledge of methods, and high professional attainments, resulted in a superior standard of efficiency, reflecting the greatest credit upon himself and enabling him to render most valuable services to the Government.

REYBOLD, EUGENE................... | Colonel, Coast Artillery Corps, U. S. Army.
R—Delaware City, Del. | As director of the department of enlisted specialists, Coast Artillery School,
B—Delaware City, Del. | where, by his excellent judgment, energy, and foresight, he enabled that
G. O. No. 3, W. D., 1921. | department to meet the demands made upon it in an effective manner.

REYNOLDS, CHARLES R............... | Colonel, Medical Corps, U. S. Army.
R—Elmira, N. Y. | As division surgeon of the 77th Division, as chief surgeon, 6th Army Corps,
B—Elmira, N. Y. | and later as chief surgeon, 2d Army, he displayed qualities of leadership, high
G. O. No. 89, W. D., 1919. | professional attainments, and rare judgment in energetically directing the work of the sanitary units under his control. By his foresight in providing front-line hospitalization and evacuation facilities for the sick and wounded in the field, he rendered services of signal merit to the American Expeditionary Forces.

REYNOLDS, FREDERICK P............. | Colonel, Medical Corps, U. S. Army.
R—Elmira, N. Y. | As surgeon of the advance section, services of supply, American Expeditionary
B—Elmira, N. Y. | Forces, he displayed rare judgment, unusual executive ability, and high
G. O. No. 30, W. D., 1921. | professional attainments in the institution of sanitary measures and in providing and supervising hospitalization and evacuation facilities for the sick and wounded flowing into the advanced areas from the principal centers of combat activity.

REYNOLDS, STEPHEN C.............. | Lieutenant colonel (Infantry), General Staff Corps, U. S. Army.
R—St. Louis, Mo. | As assistant chief of staff, G-1, of the 5th Division, by his keen application to his
B—Louisiana, Mo. | task he overcame almost insurmountable difficulties in maintaining communi-
G. O. No. 87, W. D., 1919. | cations and securing supplies for his division during the 27 days when it was advancing against the enemy north of Verdun. In the performance of his many duties he displayed indefatigable zeal and showed exceptional administrative ability, rendering valuable services to the American Expeditionary Forces.

RHEA, JAMES C.................... | Colonel (Cavalry), General Staff Corps, U. S. Army.
R—Strawn, Tex. | In charge of the operations section, and later as chief of staff and brigade com-
B—Hamburg, Iowa. | mander of the 2d Division, he played a conspicuous part in the successful en-
G. O. No. 59, W. D., 1919. | gagements at the St. Mihiel salient, Blanc Mont Ridge, and in the Argonne-
Distinguished-service cross also awarded. | Meuse, revealing traits of military knowledge and attainments of a high order.

RHOADS, THOMAS L................ | Colonel, Medical Corps, U. S. Army.
R—Fort Worth, Tex. | As division surgeon of the 80th Division, he had charge of the Medical Depart-
B—Boyertown, Pa. | ment's work of that unit throughout its combat activities. Due to his skillful
G. O. No. 62, W. D., 1919. | administration, it functioned smoothly and with precision at all times, caring properly for a large number of the sick and wounded. As chief surgeon of the 1st Army Corps, and later of the 1st Army, he displayed executive ability of high order, being constant and zealous in devotion to his arduous tasks.

RHODES, CHARLES D............... | Major general, U. S. Army.
R—Delaware, Ohio. | As commander of the Artillery brigade in support of the 82d Division, during
B—Delaware, Ohio. | the offensive operations of the St. Mihiel salient and again in command of an
G. O. No. 89, W. D., 1919. | Artillery brigade during the Meuse-Argonne offensive, by his marked ability
Distinguished-service cross also awarded. | shown in the conduct of his units he contributed in a noted degree to the successes attained. Later he served with distinction as a member of the Interallied Commission at Spa, rendering conspicuous services to the American Expeditionary Forces.

RHODES, MARIE B. (now Mrs. CLARENCE CASH). R—Pittsburgh, Pa. B—Pittsburgh, Pa. G. O. No. 9, W. D., 1923.	Nurse, Army Nurse Corps, U. S. Army. As chief of the nurses' equipment bureau of the military department, American Red Cross, in Paris, France, during the World War, she rendered invaluable service to the Army. She organized and developed a department which was able not only to supply and replace nurses' equipment but to transport the material all over France, even to rapidly moving units and teams at the front. By her remarkable business acumen and integrity, unusual resourcefulness, and initiative she made a contribution to the welfare, efficiency, and conduct of the American Expeditionary Forces nursing forces which can not be measured. In addition to her most arduous duties during the day she frequently spent part of the nights during the emergency giving her services as an anesthetist in the Army hospitals in Paris.
RICE, JOHN H. R—Webster Groves, Mo. B—St. Louis, Mo. G. O. No. 62, W. D., 1919.	Brigadier general, U. S. Army. As chief of the engineering division of the office of the Chief of Ordnance, he performed with peculiar ability his arduous duties in connection with the design and development of all articles of ordnance supplied to the United States Army. Later as Chief Ordnance Officer, American Expeditionary Forces, he was charged with the procurement and supply of all ordnance to our forces in France, which duties he performed with exceptional success, displaying untiring energy and zeal. He handled perplexing problems of supply with sound judgment, achieving most valuable results.
RICE, MERVYN A. R—Montclair, N. J. B—Rockland, Me. G. O. No. 56, W. D., 1922.	Lieutenant colonel, Ordnance Department, U. S. Army. As corps ordnance officer of the 2d Army Corps from February to October, 1918, he displayed exceptional ability in the organization and administration of the system used in equipping with ordnance the American troops serving on the British front. Later, his great tact and business ability assisted to a marked degree in the satisfactory settlement of important claims arising out of this service.
RICE, SEDGWICK R—St. Paul, Minn. B—St. Paul, Minn. G. O. No. 96, W. D., 1918.	Second lieutenant, 7th Cavalry, U. S. Army. For services against hostile Indians near the Catholic Mission on White Clay Creek, S. Dak., Dec. 30, 1890.
RICHARDSON, LORRAIN T. R—Janesville, Wis. B—Janesville, Wis. G. O. No. 56, W. D., 1922.	Colonel, Infantry, U. S. Army. As commander of the 322d Infantry, 81st Division, during its organization, training, and in all its combat operations he displayed marked efficiency, tireless energy, and military attainments of the highest order. By his sound judgment and skillful and energetic leadership he contributed materially to the successes achieved by the regiment against the enemy.
RICHARDSON, ROBERT C., Jr. R—Charleston, S. C. B—Charleston, S. C. G. O. No. 87, W. D., 1919.	Colonel (Cavalry), General Staff Corps, U. S. Army. He organized and conducted with great efficiency the important strategical and tactical liaison service of the 3d Section, General Staff, General Headquarters, American Expeditionary Forces. During the Meuse-Argonne operations he gave proof of notable military attainments and untiring devotion to duty by the efficient manner in which he organized and administered the advanced General Headquarters, rendering services of distinction to the American Expeditionary Forces.
RICHARDSON, WILDS P. R—Paris, Tex. B—Hunt County, Tex. G. O. No. 19, W. D., 1922.	Brigadier general, U. S. Army. As commanding general of the American Expeditionary Forces in North Russia, by his skillful handling of the many difficult situations which arose, he rendered a signal service to the United States Government.
RICKARDS, GEORGE C. R—Oil City, Pa. B—Philadelphia, Pa. G. O. No. 38, W. D., 1922.	Colonel, Infantry, U. S. Army. As commanding officer of the 112th Infantry, 28th Division, he proved himself a forceful and capable military leader. Maintaining at all times a high degree of efficiency in his regiment through his personal magnetism, heroism, zeal, and energy, he contributed materially to the successes achieved by the 28th Division in its operations against the enemy, rendering services of distinction to the American Expeditionary Forces.
RIGGS, KERR T. R—Cynthiana, Ky. B—Harrison County, Ky. G. O. No. 19, W. D., 1922.	Colonel (Cavalry), General Staff Corps, U. S. Army. As G-2, 2d Army Corps, he displayed exceptional ability in the organization and administration of that division of the corps staff. He also showed great ability and rare tact in his relations with the intelligence branch of the staffs of the British organizations with which the 2d Army Corps served. By his tireless energy and unceasing devotion to exacting duties, he contributed to a marked degree to the successes achieved by his organization.
RIVERS, WILLIAM C. R—Pulaski, Tenn. B—Pulaski, Tenn. G. O. No. 89, W. D., 1919.	Brigadier general, U. S. Army. As commander of the 76th Field Artillery, 3d Division, he was a material factor in stemming the tide of the enemy's advance during the second Battle of the Marne. Subsequently, upon being promoted to the grade of brigadier general, he displayed marked leadership and high military attainments in command of the 5th Field Artillery Brigade, 5th Division, in the Meuse-Argonne offensive.
ROBERTS, CHARLES D. R—Fort D. A. Russell, Wyo. B—Cheyenne Agency, S. Dak. G. O. No. 59, W. D., 1919. Medal of honor also awarded.	Colonel (Infantry), General Staff Corps, U. S. Army. He displayed unusual ability as chief of staff of the 81st Division in its organization, and in the conduct of its operations in the St. Die Sector, on Nov. 9, 10, and 11, 1918, near Verdum, where the division was enabled to advance some 5½ kilometers over marshy ground under heavy fire.

ROBERTS, GEORGE J.
R—East Orange, N. J.
B—Charlotte County, Va.
G. O. No. 124, W. D., 1919.

As chief of the New York ordnance district, in which capacity he maintained at all times the greatest degree of intelligent and enthusiastic cooperation between the Ordnance Department and manufacturers in his district, thereby attaining the maximum production of munitions in a minimum time; and also as chairman of the New York ordnance district claims board, in which capacity his services have been invaluable to the Nation in adjusting equitably the $525,000,000 worth of outstanding contracts in his district in force at the signing of the armistice.

ROBERTS, OSCAR E.
R—Taylor, Tex.
B—Milan County, Tex.
G. O. No. 15, W. D., 1923.

Colonel, Infantry, U. S. Army.
As colonel commanding the 144th Infantry, 36th Division, he displayed untiring energy, initiative, and resourcefulness both during the period of organization and training and during the combat operations of his regiment in France. His leadership, sound judgment, and devotion to duty were material factors in the successes of his regiment, brigade, and division in action against the enemy.

ROBERTS, THOMAS A.
R—Springfield, Ill.
B—Springfield, Ill.
G. O. No. 50, W. D., 1919.

Colonel, Cavalry, U. S. Army.
As commander of an American regiment on duty with the French Army, although confronted with many difficult situations, he handled all questions with marked success. His tasks were performed with ability, in a manner that reflected the greatest credit upon him; his preparations were careful, his leadership skillful. The excellent results achieved by his regiment are in a measure attributable to his sound judgment and military knowledge.

ROBERTSON, ASHLEY HERMAN.
R—Ashmore, Ill.
B—Ashmore, Ill.
G. O. No. 116, W. D., 1919.

Rear Admiral, U. S. Navy.
As force transport officer, his untiring energy contributed greatly to the successful oversea movement of troops and supplies.

ROBERTSON, SAMUEL ARTHUR.
R—San Benito, Tex.
B—Fort Benton, Mont.
G. O. No. 59, W. D., 1919.

Lieutenant colonel, Corps of Engineers, U. S. Army.
As general superintendent of construction of the light railways he managed all the intricate details of complex organization and classification of tasks with a master hand. With untrained personnel he established a record for speed in tracklaying of the 60-centimeter lines, exciting the admiration of our allies. During the advance of the 1st Army, by his ceaseless activity, tireless energy, and great knowledge he performed his duty with marked credit to the Government.

ROBINS, THOMAS MATTHEW.
R—Snow Hill, Md.
B—Snow Hill, Md.
G. O. No. 77, W. D., 1919.

Colonel, Corps of Engineers, U. S. Army.
For services while in charge of the engineer depot established in connection with the port of New York, and subsequently included in the port of embarkation at Hoboken, N. J.

ROBINSON, DONALD A.
R—Seattle, Wash.
B—Chippewa Falls, Wis.
G. O. No. 56, W. D., 1922.

Lieutenant colonel (Cavalry), General Staff Corps, U. S. Army.
As chief of the executive division, fourth section, General Staff, Headquarters, Services of Supply, American Expeditionary Forces, from Apr. 25, 1918, to Feb. 19, 1919, he was charged with the immediate coordination of major supply activities, including the difficult and gravely responsible task of adjusting priority of shipments and determining the order in which all movements of supplies from the base ports of the American Expeditionary Forces should proceed. In a position of great responsibility, in which large powers were delegated to him, he displayed to an unusual degree rare tact, excellent judgment, and the faculty of firm and prompt decision. By his successful handling of difficult supply situations of the most critical character directly affecting important operations in the field, he rendered services of the highest value to the American Expeditionary Forces.

ROBINSON, FRED J.
R—Miami Beach, Fla.
B—Detroit, Mich.
G. O. No. 124, W. D., 1919.

As chief of the Detroit Ordnance District, in which capacity he maintained at all times the greatest degree of intelligent and enthusiastic cooperation between the Ordnance Department and manufacturers in his district, thereby attaining the maximum production of munitions in a minimum time; and also as chairman of the Detroit Ordnance District Claims Board, in which capacity his services have been invaluable to the Nation in adjusting equitably the $271,000,000 worth of outstanding contracts in his district in force at the signing of the armistice.

ROCKENBACH, SAMUEL D.
R—Boonville, Mo.
B—Lynchburg, Va.
G. O. No. 76, W. D., 1919.

Brigadier general, U. S. Army.
As quartermaster of Base Section No. 1, St. Nazaire, from June to December, 1917, he rendered especially valuable services. Confronted with a problem of great magnitude, befraught with serious difficulties, he went about his task with keen determination, and by his energy and great zeal organized and efficiently operated the first American base in France. Later, as chief of the Tank Corps, by his tireless energy and keen determination he established schools of training for tank personnel and laid the foundation for the organization of the tank units. He ably directed the operations of the tanks with the First Army and contributed in a measure to the success attained.

ROGERS, HARRY L.
R—Orchard Lake, Mich.
B—Washington, D. C.
G. O. No. 12, W. D., 1919.

Major general, U. S. Army.
He has organized, perfected, and administered with great efficiency the quartermaster department in France. He was able to meet each emergency in times fraught with untold difficulties, and by his energy and untiring zeal he has insured to our troops a prompt and constant supply of quartermaster stores, without which the ultimate success of our Army could not have been obtained.

ROGERS, HENRY H. R—Tuxedo, N. Y. B—New York, N. Y. G. O. No. 47, W. D., 1921.	Lieutenant colonel, Field Artillery, U. S. Army. While commanding the 2d Corps Artillery Park which operated with the 5th Army Corps, in spite of great difficulties he delivered large quantities of ammunition to the troops, and by his resourcefulness, courage, and leadership he maintained his command in a high state of efficiency and morale, thereby contributing to the success of the operations of the corps.
ROGERS, JOSEPH A. R—Mullin, Tex. B—Cameron, Tex. G. O. No. 4, W. D., 1923.	Lieutenant colonel, Field Artillery, U. S. Army. He commanded the 1st Battalion, 13th Field Artillery, 3d Division, from July 1 to Oct. 3, 1918, and the 124th Field Artillery, 33d Division, from Oct. 4 until Nov. 17, 1918, at all times proving himself to be an officer of exceptional ability. By his sound tactical judgment, loyal devotion to duty, and great skill he supported the Infantry, to which attached, so effectively that he aided materially in the successful operations of several divisions in many important engagements.
ROHAYNE, JAMES. R—New York, N. Y. B—Ireland. G. O. No. 59, W. D., 1919.	Colonel, U. S. Army. He served as assistant commandant of the Army schools for eight months. By his energy, perseverance, and good judgment, in all matters connected with the Army schools, he exhibited high professional attainments and military qualities of a superior order.
ROOP, JAMES CLAWSON. R—Upland, Pa. B—Upland, Pa. G. O. No. 56, W. D., 1922.	Lieutenant colonel, Corps of Engineers, U. S. Army. He served as assistant to the general purchasing agent, American Expeditionary Forces. His marked ability and tact were important factors in numerous negotiations with the allied armies and governments, involving critical matters of supply to our Army. In the organization of the work of the general purchasing board and general purchasing agent, Colonel Roop throughout its existence was an indispensable factor in a position of great responsibility. He rendered most distinguished service in connection with important supplies of all kinds for the Army as well as in the organization of the general system of coordination between the supply services of the American Expeditionary Forces and the allied armies.
ROOSEVELT, THEODORE, Jr. R—New York, N. Y. B—Oyster Bay, Long Island, N. Y. G. O. No. 2, W. D., 1927. Distinguished-service cross also awarded.	Lieutenant colonel, Infantry, U. S. Army. As battalion and regimental commander, 26th Infantry, he displayed consistent gallantry, conspicuous energy, and marked efficiency in the operations around Cantigny, Soissons, and during the Meuse-Argonne offensive. By his devotion to duty, pronounced tactical ability, and brilliant qualities of leadership he contributed materially to the successes of his regiment and of the 1st Division. He rendered services of signal worth to the Government in a position of great responsibility at a time of gravest importance.
ROSE, WILLIAM H. R—Kelton, Pa. B—Safe Harbor, Pa. G. O. No. 25, W. D., 1919.	Brigadier general, U. S. Army. While in charge of the engineer depot he was charged with the system of purchase of supplies. His exceptional ability, judgment, and resourcefulness are apparent in the efficient solution of the many difficult problems involved and in the success attained in supplying the vast quantities of engineering supplies to the Army overseas.
RUCKER, WILLIAM H. R—Los Angeles, Calif. B—Fort Riley, Kans. G. O. No. 95, W. D., 1919.	Lieutenant colonel, Field Artillery, U. S. Army. As commander of the 107th Field Artillery, 28th Division, and of a French artillery regiment during the operations of the 32d Division on the Vesle River, he displayed consummate skill as an artillerist and showed notable qualities of leadership. Subsequently he commanded the 16th Field Artillery, 4th Division, and acted as group commander of French and American artillery units, where he furnished effective support to the Infantry during the St. Mihiel and Argonne-Meuse operations.
*RUCKMAN, JOHN W. R—Sidney, Ill. B—Sidney, Ill. G. O. No. 1, W. D., 1926.	Major general, U. S. Army. As department commander, Southern Department, between Aug. 30, 1917, and May 4, 1918, and department commander, Northeastern Department, between May 23, 1918, and July 20, 1918, he handled many difficult problems arising in these departments with rare judgment, tact, and great skill. Posthumously awarded. Medal presented to widow, Mrs. John W. Ruckman.
RUFFNER, ERNEST L. R—Buffalo, N. Y. B—Fort Leavenworth, Kans. G. O. No. 59, W. D., 1919.	Colonel, Medical Corps, U. S. Army. He served as surgeon of the intermediate section, Services of Supply, having under his supervision 30 base hospital units. He performed his strenuous and exacting duties in an unusually efficient manner, displaying rare judgment and professional attainments of the first order.
RUGGLES, COLDEN L'H. R—Poughkeepsie, N. Y. B—Omaha, Nebr. G. O. No. 73, W. D., 1919.	Brigadier general, U. S. Army. The conception and construction of the Aberdeen Proving Ground and its operation during the early and most difficult period of its history are a monument to his sagacity and unremitting labor.
RUGGLES, FRANCIS A. R—Washington, D. C. B—St. Paul, Minn. G. O. No. 49, W. D., 1922.	Colonel, Field Artillery, U. S. Army. As a battalion commander during the Aisne-Marne offensive, July 18-25, 1918, and as a regimental commander during the St. Mihiel offensive, Sept. 12-19, 1918, and the Meuse-Argonne offensive, Sept. 30, 1918, to Nov. 8, 1918. In all of these offensives he displayed conspicuous efficiency, marked aggressiveness, and unusual leadership. By his exceptional technical and executive ability he solved with sound judgment many perplexing problems, and although much handicapped by severe losses in men, materiel, and animals he at all times so commanded his regiment as to render invaluable support to the attacking Infantry, thus materially adding to the success of the operations of the 1st Division.

RULON, BLANCHE S.
  R—Pittsburgh, Pa.
  B—Waretown, N. J.
  G. O. No. 9, W. D., 1923.

Captain, Army Nurse Corps, U. S. Army.
As chief nurse of Base Hospital No. 27, at Angers, France, during the World War, and later as assistant to the director of the Nursing Service, American Expeditionary Forces, at Tours, France, she displayed qualities of leadership and organizing ability of the highest order. Through her skillful management and untiring energy she developed the nursing force at Base Hospital No. 27 to a high degree of proficiency and was of material assistance in establishing and maintaining a reputation for unusual efficiency for that hospital, which had the unique distinction of caring for the largest numer of patients of any single hospital at any one time in the American Expeditionary Forces. Upon her return to the United States she had charge of the claim department of the Army Nurse Corps in the office of the Surgeon General, and the efficiency, sound judgment, and knowledge of the principles involved displayed by her in that capacity made a signal contribution to the demobilization work of the Government.

RUMBOLD, FRANK M.
  R—St. Louis, Mo.
  B—Meeker Grove, Wis.
  G. O. No. 56, W. D., 1922.

Colonel, Field Artillery, U. S. Army.
As assistant to the chief, Militia Bureau, during the inception of the World War, in which office of great responsibility his genius and ability were applied to the organizing, training, and expansion of the National Guard, the success of which was due, in a large measure, to his zeal, devotion to duty, and unquestionable competency. For his marked ability in the organizing, training, and disciplining of the 128th Field Artillery Regiment, 35th Division, the successful functioning of which unit during the war may be attributed in large part to Colonel Rumbold's indefatigable efforts. For his untiring and successful efforts throughout his entire service to secure close cooperation between the National Guard and the Regular Army in order that these two elements for the national defense might function successfully as the Army of the United States.

RUSSEL, EDGAR
  R—Kingston, Mo.
  B—Pleasant Hill, Mo.
  G. O. No. 12, W. D., 1919.

Brigadier general, U. S. Army.
As Chief Signal Officer, American Expeditionary Forces, he has shown great ability in the organization and administration of his department and the results attained are largely due to his zeal and energy. The Signal Corps in France stands out as one of the masterful accomplishments of the American Expeditionary Forces, and to General Russell is due the credit for its foundation and organization.

RUSSELL, EDMUND A.
  R—Lake Forest, Ill.
  B—Athens, N. Y.
  G. O. No. 124, W. D., 1919.

As chief of the Chicago Ordnance District, in which capacity he maintained at all times the greatest degree of intelligent and enthusiastic cooperation between the Ordnance Department and the manufacturers in his district, thereby attaining the maximum production of munitions in a minimum time; and also as chairman of the Chicago Ordnance District Claims Board, in which capacity his services have been invaluable to the Nation in adjusting equitably the $325,000,000 worth of outstanding contracts in his district in force at the signing of the armistice.

RUSSELL, FREDERICK F.
  R—Brooklyn, N. Y.
  B—Auburn, N. Y.
  G. O. No. 69, W. D., 1919.

Colonel, Medical Corps, U. S. Army.
He organized and directed the division of laboratories and infectious diseases of the Surgeon General's Office during the present war and thereby contributed in great measure to the efficiency of the military forces.

RUSSELL, GEORGE M.
  R—Plymouth, N. H.
  B—Plymouth, N. H.
  G. O. No. 103, W. D., 1919.

Colonel (Field Artillery), General Staff Corps, U. S. Army.
As assistant chief of staff, G-2, of the 5th Army Corps, he directed the activities of the intelligence section with marked skill and untiring energy. By effecting the collection and dissemination of timely and accurate information, he was an important factor in the successes achieved by his corps.

RYAN, LILLIAN J.
  R—Denver, Colo.
  B—Leland.
  G. O. No. 9, W. D., 1923.

First lieutenant, Army Nurse Corps, U. S. Army.
As chief nurse of the base hospital at Camp Merritt, N. J., during the World War, she rendered signal service. By the display of excellent judgment, energy, and example under unusual difficulties she so directed her staff as to enable it to meet in an efficient manner all the demands made upon it, thus making a large contribution to the saving of lives. By her great organizing ability and untiring efforts she played an important part in the successful aftercare of thousands of tuberculous soldiers during the demobilization period.

RYAN, WILLIAM B.
  R—Greensburg, Pa.
  B—Fairfield, Vt.
  G. O. No. 103, W. D., 1919.

Lieutenant colonel, Corps of Engineers, U. S. Army.
In charge of the tonnage section of G-1, General Headquarters, he performed services of great value to the American Expeditionary Forces. Later, as supervisor of cargo and supplies at the port of Marseilles, by his zeal and energy he overcame all obstacles and successfully accomplished his important task.

SAFFARRANS, GEORGE C.
  R—Paducah, Ky.
  B—Memphis, Tenn.
  G. O. No. 56, W. D., 1922.

Colonel, Infantry, U. S. Army.
As provost marshal of the district of Paris from Jan. 3, 1918, to May 3, 1918, and subsequently in command of this important district during a period of gravest import and charged with most important duties, he labored unceasingly and succeeded in attaining excellent results. Aided by his superior tact and keen perception, he performd his difficult duties with sound judgment and handled numerous diplomatic affairs with great satisfaction, thereby rendering important service to the American Expeditionary Forces in positions of great responsibility.

ST. JOHN, FORDYCE B.\
R—Hackensack, N. J.\
B—Hackensack, N. J.\
G. O. No. 59, W. D., 1921.

Major, Medical Corps, U. S. Army.\
As commanding officer of Mobile Hospital No. 2, at Coincy, France, from June to August, 1918, he commanded his hospital under most difficult and trying conditions with inadequate equipment and personnel. Later he served continuously at the front, taking a most active part in the Medical Service of our military efforts. Due to his unusual organizing ability and his indomitable will, his unit was always close to the firing line.

SALMON, THOMAS W.\
R—West New Brighton, N. Y.\
B—Lansingburg, N. Y.\
G. O. No. 87, W. D., 1919.

Colonel, Medical Corps, U. S. Army.\
He has, by his constant, tireless, and conscientious work, as well as by his unusual judgment, done much to conserve man power for active front-line work. Of special value was his demonstration that war neurosis could be treated in advanced sanitary units with greater success than in base hospitals.

SALTZMAN, CHARLES McK.\
R—Des Moines, Iowa.\
B—Panora, Iowa.\
G. O. No. 47, W. D., 1919.

Colonel, Signal Corps, U. S. Army.\
While assigned to duty in the Air Service he voluntarily undertook and successfully accomplished the difficult task, in the face of many obstacles, of preparing an organization for the procurement and supply of Signal Corps equipment for the Army.

SAMPLE, WILLIAM R.\
R—Fort Smith, Ark.\
B—Memphis, Tenn.\
G. O. No. 39, W. D., 1920

Brigadier general, U. S. Army.\
In command of the advance section, Services of Supply, throughout the campaigns in France in 1918, by his skillful management of the supply of the Army, involving many emergency orders to be promptly met, he showed himself to be, under those difficult conditions, an officer of great ability and resource.

SANBORN, JOSEPH B.\
R—Chicago, Ill.\
B—Manchester, N. H.\
G. O. No. 89, W. D., 1919.\
Distinguished-service cross also awarded.

Colonel, Infantry, U. S. Army.\
He commanded the 131st Infantry, 33d Division, during all its campaigns against the enemy, displaying military leadership of a high order. His unremitting zeal and tactical skill were largely responsible for the success of his regiment in combat.

SANDS, ALFRED L. P.\
R—Pittsburgh, Pa.\
B—Pittsburgh, Pa.\
G. O. No. 56, W. D., 1922.

Colonel, Field Artillery, U. S. Army.\
Serving in turn as a battalion and regimental commander, he rendered exceptionally efficient services, and at all times was conspicuous for his courage, leadership, and high military attainments. As commanding officer of the 7th Field Artillery, 1st Division, and a group of attached French Artillery at Cantigny, in May, 1918, and later at Soissons in July, 1918, he displayed unusual ability, and by his high professional attainments and great tactical skill supported the Infantry, to which he was attached, so effectively that he aided materially in the successful operations of that brigade in those important engagements. Later, by his tireless energy, great resourcefulness, and efficiency, he organized and trained, in a highly satisfactory manner, the 67th Field Artillery.

SAUNDERS, EDWIN O.\
R—Sharpsburg, Ky.\
B—Sharpsburg, Ky.\
G. O. No. 49, W. D., 1922.

Lieutenant colonel, Judge Advocate General's Department, U. S. Army.\
As chief of the criminal investigation division of the Provost Marshal General's Department, American Expeditionary Forces, he had charge of and was responsible for the investigation of crime and bringing to trial of criminals. His remarkable skill as an organizer, his untiring efforts, and his sound judgment were mainly responsible for the reduction and prevention of crime in the American Expeditionary Forces, and the recovery of thousands of dollars of Government and private property. His services were of very great value to the American Expeditionary Forces.

SCHELLING, ERNEST H.\
R—Bar Harbor, Me.\
B—Belvidere, N. J.\
G. O. No. 14, W. D., 1923.

Major, Infantry, U. S. Army.\
While serving as assistant military attaché at the American Legation in Switzerland from September, 1917, to October, 1919, his great tact, initiative, resourcefulness, sound judgment and unremitting devotion to duty contributed markedly to the successful operations of the American and Allied forces during the World War.

SCHLEY, JULIAN L.\
R—Savannah, Ga.\
B—Savannah, Ga.\
G. O. No. 4, W. D., 1923.

Colonel, Corps of Engineers, U. S. Army.\
As commanding officer of the 307th Engineers and division engineer officer, 82d Division, and later as corps engineer, 5th Army Corps, during the St. Mihiel and Meuse-Argonne offensives, he displayed excellent qualities of leadership and command while serving with his regiment in the battle line, as well as superior technical attainments as an engineer, together with great zeal and devotion to duty. By the high degree of efficiency with which he performed his manifold duties, he contributed materially to the success of the operations of the commands with which he served. Later, as Director of Purchase, in the Purchase, Storage, and Traffic Division of the General Staff, and as a member of the War Department Claims Board, by his good judgment and keen foresight in undertakings of great difficulty and magnitude he rendered conspicuous service.

SCHMITT, WILLIAM J.\
R—St. Paul, Minn.\
B—St. Paul, Minn.\
G. O No. 19, W. D., 1921.

Second lieutenant, Quartermaster Corps, U. S. Army.\
While serving in the regulating stations at Creil, Noisy-le-Sec, St. Dizier, and Metz, France, he showed untiring application in his devotion to duty. He remained continually on duty during nightly bombardments at Creil, performing ably the important tasks assigned to him. During the battle of Chateau-Thierry and during the St. Mihiel and Argonne-Meuse offensives by his energetic action he aided materially in the maintenance of a steady flow of supplies to the troops at the front, at all times showing marked ability and initiative when faced with difficult problems of transportation arising from the evacuation of the wounded. He rendered valuable service to the Government.

SCHOEFFEL, FRANCIS H.
  R—Rochester, N. Y.
  B—Rochester, N. Y
  G. O. No. 14, W. D., 1923.

Lieutenant colonel, Inspector General's Department, U. S. Army.
As inspector general, port of embarkation, Hoboken, N. J., from Aug. 1, 1917, to Apr. 1, 1919, he was charged with inspection of troops, troop transports and trains, money accounts of disbursing officers, and duties of a similar nature, all of which he performed in a highly efficient and successful manner. With untiring energy, sound judgment, and unusual professional skill and with unremitting attention and devotion to duty he rendered extremely valuable services to the Government in a position of great responsibility.

SCHULL, HERMAN W.
  R—Watertown, S. Dak.
  B—England.
  G. O. No. 9, W. D., 1923.

Colonel, Ordnance Department, U. S. Army.
As assistant chief and acting chief of the inspection division, Ordnance Department. His broad-minded policy, zeal, and technical ability contributed in a conspicuous way to the success of the Ordnance Department in the procurement of munitions in the United States and Canada during the World War.

SCHULTZ, THEODORE.
  R—St. Louis, Mo.
  B—Orville, Ohio.
  G. O. No. 10, W. D., 1924.

Captain, 9th Cavalry U. S. Army.
At Naco, Ariz., Oct. 9–10, 1914, he commanded the most important and most exposed outposts of the American troops engaged in preserving order on the international boundary line. Although his troops were constantly exposed to fire from Mexican attacking forces, by his firmness, tact, and courageous actions he handled in a highly efficient manner a threatening situation, thereby preventing serious international complications and contributing materially to the preservation of law and order on the Mexican border.

SCHULZ, JOHN W. N.
  R—Wheeling, W. Va.
  B—Wheeling, W. Va.
  G. O. No. 56, W. D., 1922.

Colonel, Corps of Engineers, U. S. Army.
As representative of the Chemical Warfare Service at General Headquarters he rendered valuable services in the solution of many important problems relating to the offensive use of gas and also in planning more effective methods for the issue of equipment and for training in gas warfare. Later, as chief gas officer of the 1st Army, he was charged with the entire responsibility of the gas warfare in that army during the St. Mihiel and Meuse-Argonne offensives. By great ability and untiring energy, his efforts resulted in the prevention of large numbers of casualties and fatalities from enemy gases as well as increasing the use of gas against the enemy, thereby contributing to the success of the American arms.

SCOTT, ERNEST D.
  R—De Witt, Nebr.
  B—Canada.
  G. O. No. 15, W. D., 1923.

Colonel, Field Artillery, U. S. Army.
He commanded the American and allied light artillery in the Toul sector prior to and after its occupation by the 1st Division, and was largely responsible for the instruction and training of the 66th Field Artillery Brigade and of brigades of Coast Artillery in their training areas. As Heavy Artillery commander, 1st Army Corps, in the Champagne-Marne and the Aisne-Marne operations, and as brigade and grouping commander of the 66th Field Artillery Brigade and numerous French units in the operations of the 5th Army Corps in the St. Mihiel offensive, and in the operations of the Army artillery in the Meuse-Argonne offensive, he performed his duties with great efficiency and distinction.

SCOTT, FRANK A.
  R—West Mentor, Ohio.
  B—Cleveland, Ohio.
  G. O. No. 69, W. D., 1919.

In assisting in organizing and as chairman of the munitions standards board and the general munitions board. He was later first chairman of the War Industries Board. He thus contributed greatly in developing the War Department's programs.

SCOTT, HUGH L.
  R—Princeton, N. J.
  B—Danville, Ky.
  G. O. No. 47, W. D., 1919.

Major general, U. S. Army.
As Chief of Staff in advocating and persistently urging the adoption of the selective service law and as Commanding General, Camp Dix, N. J., in organizing and training the divisions and miscellaneous troops committed to his care during the war.

SCOTT, WALTER DILL.
  R—Evanston, Ill.
  B—Cookesville, Ill.
  G. O. No. 69, W. D., 1919.

Colonel, U. S. Army.
In originating, organizing, and putting into operation the system of classification of enlisted personnel now used in the United States Army.

SCOTT, WALTER J.
  R—Mount Morris, Ill.
  B—Brownsville, Tenn.
  G. O. No. 9, W. D., 1923.

Lieutenant colonel (Cavalry), General Staff Corps, U. S. Army.
As assistant chief of staff, G–1, 89th Division, due to his unusual foresight, indefatigable efforts, and great executive ability, he was able to keep a constant flow of supplies and ammunition for the troops under extremely difficult combat conditions, thereby contributing materially to the success of the division and the American Expeditionary Forces.

SCOWDEN, FRANK F.
  R—Albany, N. Y.
  B—Meadville, Pa.
  G. O. No. 14, W. D., 1923.

Lieutenant colonel, Motor Transport Corps, U. S. Army.
As executive officer in the office of the director, Motor Transport Corps, American Expeditionary Forces, he discharged with rare distinction duties of grave responsibility. His sound judgment, tact, administrative and technical ability contributed in a marked degree to the successful functioning of the Motor Transport Corps of the American Expeditionary Forces.

SCREWS, WILLIAM P.
  R—Montgomery, Ala.
  B—Montgomery, Ala.
  G. O. No. 19, W. D., 1920.

Colonel, Infantry, U. S. Army.
He commanded, with courage, resourcefulness, and great skill the 167th Infantry, 42d Division, from the time of its organization and early training throughout the successive phases of sector warfare, offensive combat in the battles of the Champagne and Ourcq, and in the St. Mihiel and Meuse-Argonne offensives. By his high military attainments, sound judgment, and devotion to duty he has contributed, in no small degree, to the successes achieved by the 42d Division.

SEAMAN, A. OWEN..................
    R—Greenville, Ill.
    B—Greenville, Ill.
    G. O. No. 77, W. D., 1919.

Colonel, Infantry, U. S. Army.
In the very efficient operation of the Motor Transport Corps, War Department General Staff, and in accomplishing the standardization of motor vehicles in the Army.

SEAMAN, GILBERT E..................
    R—Milwaukee, Wis.
    B—Alpena, Mich.
    G. O. No. 116, W. D., 1919.

Colonel, Medical Corps, U. S. Army.
After serving with conspicuous success as division surgeon of the 32d Division, he became chief surgeon, 6th Army Corps, and in this capacity was an important factor in the establishment of effective measures for treating numerous sick and wounded. Fitted for his exacting duties by wide experience and unusual ability, he rendered services of great value to the American Expeditionary Forces.

SEWELL, JOHN S..................
    R—Gantts Quarry, Ala.
    B—Butlers Landing, Tenn.
    G. O. No. 59, W. D., 1919.

Colonel, Corps of Engineers, U. S. Army.
In command of a regiment of Engineers and later as commander of the base port at St. Nazaire he displayed high engineering skill and long practical experience in the management of men. His genius together with his great energy and devotion to duty, contributed largely to the successful development and efficient operation of that base.

SHALLENBERGER, MARTIN C..................
    R—Alma, Nebr.
    B—Osceola, Nebr.
    G. O. No. 89, W. D., 1919.

Lieutenant colonel (Infantry), General Staff Corps, U. S. Army.
As assistant chief of staff, G-1, of the 3d Army Corps, during the Argonne-Meuse offensive, by his tireless efforts, marked organizing ability, and keen application to his numerous duties, he contributed in a large measure to the successes attained by his corps, rendering valuable services to the American Expeditionary Forces.

SHANKS, DAVID C..................
    R—Salem, Va.
    B—Salem, Va.
    G. O. No. 18, W. D., 1919.

Major general, U. S. Army.
For services in the administration of the port of embarkation, Hoboken, N. J., in connection with the shipment of troops overseas.

SHANNON, EDWARD C..................
    R—Columbia, Pa.
    B—Phoenixville, Pa.
    G. O. No. 10, W. D., 1920.

Colonel, Infantry, U. S. Army.
As commanding officer of the 111th Infantry, 28th Division, he proved himself a forceful and capable military leader. Maintaining at all times a high degree of efficiency in his regiment, he contributed materially to the successes achieved by the 28th Division in its operations against the enemy, rendering services of distinction to the American Expeditionary Forces.

SHARP, GEORGE A. (1623188)..................
    R—Breckenridge, Colo.
    B—Victor, Colo.
    G. O. No. 88, W. D., 1918.

Sergeant, Company A, 115th Engineers, 40th Division, U. S. Army.
For his bravery in entering a dangerous surf at Ocean Beach, Calif., on May 5, 1918, and rescuing three men and assisting in the rescue of Corporal Stein, Company B, 115th Engineers, at the risk of his own life.

SHAUGHNESSY, EDWARD H..................
    R—Chicago, Ill.
    B—Chicago, Ill.
    G. O. No. 95, W. D., 1919.

Lieutenant colonel, Transportation Corps, U. S. Army.
Serving successively as general superintendent, general manager, and acting deputy director general of transportation, by his energy, zeal, and able management he rendered services of the highest type to the Transportation Corps of the American Expeditionary Forces. In the performance of his manifold duties he constantly displayed marked enthusiasm, originality, and sound judgment.

SHAW, ANNA HOWARD..................
    R—Moylan, Pa.
    B—England.
    G. O. No. 69, W. D., 1919.

As chairman of the woman's committee of the Council of National Defense, she coordinated the mobilization and organization of women throughout the country in every phase of war work, including the securing of women for some of the various branches of the Army.

SHAW, HENRY A..................
    R—Worcester, Mass.
    B—Salem, Mass.
    G. O. No. 56, W. D., 1921.

Colonel, Medical Corps, U. S. Army.
As surgeon of Base Section No. 2, Services of Supply, a position of great responsibility. Due to his ability and energy, he brought the medical service of the port of Bordeaux and the hospitalization of the same to a high degree of efficiency. He has rendered services of much value.

SHEARMAN, LAWRENCE H..................
    R—Roslyn, N. Y.
    B—Washington, D. C.
    G. O. No. 46, W. D., 1919.

Member of American Inter-Allied Maritime Council.
As civilian member of the 1st Section of the General Staff, American Expeditionary Forces, he placed his mature experience and his extensive technical and business knowledge of the shipping industry at the disposal of the American Expeditionary Forces during a period of several months when tonnage and shipping problems were of the most vital importance. His clear vision, sound advice, and unfailing energy and loyalty were of the greatest value to his country and to the allied cause.

SHEDD, WILLIAM E., Jr..................
    R—Danville, Ill.
    B—Danville, Ill.
    G. O. No. 56, W. D., 1922.

Colonel, Coast Artillery Corps, U. S. Army.
He served with marked efficiency as instructor and then as director of the Heavy Artillery School in France. As assistant in the office of the Chief of Artillery, American Expeditionary Forces, and later chief of the Heavy Artillery section in that office, he rendered services of inestimable value in connection with the organization, equipment, and training of the Heavy Artillery troops in France.

SHEEHAN, MARY E..................
    R—Syracuse, N. Y.
    B—Cortland County, N. Y.
    G. O. No. 9, W. D., 1923.

First lieutenant, Army Nurse Corps, U. S. Army.
As chief nurse of the Vichy Hospital Center, France, during the World War, she organized the nursing service of that center, and by her tact, good judgment, energy, and personal devotion to duty contributed largely to the successful care and well-being of 11,000 sick and wounded.

SHEEN, HENRY H.
R—Norfolk, Va.
B—Quincy, Mass.
G. O. No. 56, W. D., 1922.
Distinguished-service cross also awarded.

Colonel, Infantry, U. S. Army.
As quartermaster, 39th Division, from September, 1917, until October, 1918, and acting chief of staff, same division, from October until December, 1918, he rendered highly meritorious services. As chief quartermaster, intermediate section, Services of Supply, he displayed sound judgment in the disposal of property valued at many millions of dollars. Later, as chief quartermaster and acting chief of staff, G–4, American Forces in Germany, he displayed marked ability and initiative in many large undertakings and perplexing problems confronting him, resulting in immense savings to the Government.

SHELBY, EVAN.
R—Washington, D. C.
B—Fayette County, Ky.
G. O. No. 89, W. D., 1919.

Colonel, Quartermaster Corps, U. S. Army.
As chief of the contracts branch of the Office of the Chief of Construction Division, in following up contracts, aiding in their interpretation, adjusting differences between the contractors and the Government, and advising on matters of procedure and the rights of the parties involved, he has displayed sound judgment, marked professional attainments, and extraordinary capacity for sustained and unremitting labor.

*SHELTON, GEORGE H.
R—Seymour, Conn.
B—Seymour, Conn.
G. O. No. 53, W. D., 1921.

Brigadier general, U. S. Army.
While commanding the 51st Infantry Brigade, 26th Division, during the St. Mihiel and Meuse-Argonne offensives.
Posthumously awarded. Medal presented to widow, Mrs. Bernice Shelton.

SHELTON, NENA.
R—Kansas City, Mo.
B—Lexington, Ky.
G. O. No. 9, W. D., 1923.

First lieutenant, Army Nurse Corps, U. S. Army.
As assistant to the director of nursing service, American Expeditionary Forces in Paris, France, during the World War, she contributed largely to the success of that force of over 10,000 nurses. Her zeal, good judgment, and energy added greatly to the efficiency with which the sick and wounded of the American Expeditionary Forces were cared for. Her faithfulness to detail and unfailing devotion to duty greatly facilitated the work of the Medical Department.

SHEPARD, JOHN L.
R—Galesburg, Ill.
B—Sheboygan, Wis.
G. O. No. 15, W. D., 1921.

Colonel, Medical Corps, U. S. Army.
In 1918, as surgeon of Camp Funston, Kans., and of the 89th Division, he displayed high administrative, technical, and constructive ability in preventive measures adopted against epidemics and in the conservation of physical defects by their segregation and development. Later he performed conspicuous services in connection with hospital demobilization in France and the return to the United States of the sick and wounded.

SHEPHERD, WILLIAM E., Jr.
R—New York, N. Y.
B—New York, N. Y.
G. O. No. 87, W. D., 1919.

Lieutenant colonel, Field Artillery, U. S. Army.
As assistant chief of staff, 3d Section, of the 5th Army Corps, and as chief of staff of Artillery of that corps, by his marked military attainments and devotion to his exacting duties, he ably planned the employment of the Corps of Artillery in its operations against the enemy, rendering services of great worth to the American Expeditionary Forces.

SHERMAN, WILLIAM C.
R—Augusta, Ga.
B—Augusta, Ga.
G. O. No. 56, W. D., 1922.

Lieutenant colonel (Corps of Engineers), General Staff Corps, U. S. Army.
As assistant chief of staff, G–2, 1st Division and 3d Army Corps, from February, 1918, until October, 1918, he skillfully organized and directed the service of information of the enemy which guided in the preparation of the orders under which his division and corps achieved their many victories. Later, as chief of staff, Air Service, 1st Army, from November 1 to November 11, 1918, he displayed great ability and by his rare tactical conceptions rendered exceptionally meritorious service, enabling the Air Service to function in a highly efficient manner at all times.

SHERRILL, CLARENCE O.
R—Raleigh, N. C.
B—Newton, N. C.
G. O. No. 53, W. D., 1921.

Colonel (Corps of Engineers), General Staff Corps, U. S. Army.
He organized the 302d Engineers and conducted their operations with the 77th Division until he became the division chief of staff. To his initiative, energy, and good judgment is due much of the success of the staff functioning of the division in its operations in the Argonne. He has rendered services of marked worth.

SHINKLE, EDWARD M.
R—Higginsport, Ohio.
B—Higginsport, Ohio.
G. O. No. 16, W. D., 1923.

Colonel, Ordnance Department, U. S. Army.
In charge of the ammunition section of the gun division, Ordnance Department, he was responsible for the design, development, and placing of orders and contracts for all ammunition and projectiles supplied to the American Army. Later, in charge of ammunition section, engineering division, in the office of Chief of Ordnance, his technical skill, executive ability, and sound judgment were highly important factors in the successful operations of the Ordnance Department during the World War.

SHIPLEY, WALTER V.
R—Arlington, Md.
B—Cockeysville, Md.
G. O. No. 56, W. D., 1922.

Lieutenant colonel, Quartermaster Corps, U. S. Army.
As quartermaster and assistant chief of staff, G–1, 29th Division, from July, 1917, until June, 1919, by his untiring zeal, great energy, tact, and sound judgment he was able to overcome many difficult problems of supply and assist to a marked degree in the operations of his division, thereby rendering services of great value to the American Expeditionary Forces.

SHOCKLEY, M. A. W.
R—Fort Scott, Kans.
B—Fort Scott, Kans.
G. O. No. 87, W. D., 1919.

Colonel (Medical Corps), General Staff Corps, U. S. Army.
As a member of the 5th Section, General Staff, he displayed sound judgment and administrative ability in organizing, supervising, and inspecting the various sanitary schools and in conducting the sanitary training of troops. He also initiated and planned the preliminary organization of schools for instruction in civil educational subjects, established after the cessation of hostilities, rendering invaluable services to the American Expeditionary Forces.

SHORT, WALTER CAMPBELL R—Columbus, Ohio. B—Columbus, Ohio. G. O. No. 70, W. D., 1919.	Colonel (Infantry), General Staff Corps, U. S. Army. Attached to the Fifth Section, General Staff, General Headquarters, American Expeditionary Forces, he rendered conspicuous service in inspecting and reporting upon front-line conditions pertaining to the work of his section. During the St. Mihiel and Meuse-Argonne operations of the 1st Army Corps he efficiently directed the instruction and training of machine-gun units at every available opportunity during rest periods. Later, as assistant chief of staff, G-5, 3d Army, he manifested the same assiduous devotion to duty in organizing schools, conducting necessary inspections, and carrying out the intensive training program.
SHREEVE, HERBERT E. R—Wyoming, N. J. B—England. G. O. No. 59, W. D., 1919.	Lieutenant colonel, Signal Corps, U. S. Army. As officer in charge of the Division of Research and Inspection of the Signal Corps, at Paris, he rendered exceptionally valuable service, resulting in marked improvement in the efficiency of Signal Corps equipment. By his exact scientific knowledge and inventive genius he assisted in solving problems arising both at the front and in the Services of Supply.
SHULER, GEORGE K. R—Lyons, N. Y. B—Lyons, N. Y. G. O. No. 59, W. D., 1919.	Major, U. S. Marine Corps. In command of the 3d Battalion, 6th Regiment, U. S. Marine Corps, 2d Division, he displayed leadership of the highest order and marked tactical ability, resulting in the capture by his command of large numbers of prisoners and machine guns in the battles at Blanc Mont and St. Etienne, France. In the advance to the north from Sommerance he showed rare judgment in maneuvering his battalion in a difficult position, making important captures of field artillery. Fearless, aggressive, and able, he twice accomplished missions of vital importance with brilliant success.
SHUMAN, JOHN B. R—La Crosse, Wis. B—Espey, Pa. G. O. No. 3, W. D., 1922.	Colonel (Infantry), Adjutant General's Department, U. S. Army. In The Adjutant General's Department during the war and the demobilization period, his unusual initiative and splendid judgment contributed in a large measure to the successful handling of the commissioned personnel of the Army. He rendered services of great worth.
SIBERT, WILLIAM L. R—Gadsden, Ala. B—Gadsden, Ala. G. O. No. 18, W. D., 1919.	Major General, U. S. Army. For services in the organization and administration of the Chemical Warfare Service, contributory to the successful prosecution of the war.
*SIGERFOOS, EDWARD R—Arcanum, Ohio. B—Potsdam, Ohio. G. O. No. 103, W. D., 1919.	Brigadier general, U. S. Army. He organized the Army School of the Line at Langres, and as its commandant displayed unceasing energy and marked military and executive ability in directing its activities. Through the thorough instruction furnished by this school, he contributed materially to the combat efficiency of line troops, thereby rendering services of inestimable value to the American Expeditionary Forces. Posthumously awarded. Medal presented to widow, Mrs. Edward Sigerfoos.
SILER, JOSEPH F. R—Opelika, Ala. B—Orion, Ala. G. O. No. 59, W. D., 1919.	Colonel, Medical Corps, U. S. Army. He has been in charge of the laboratory service of the American Expeditionary Forces. Due to his untiring zeal and high professional attainments, he has been able to render invaluable service in the prevention of the spread of infectious disease among our troops. Under his able instructions, medical officers were sent out equipped to handle the new medical and surgical problems of war in a manner not believed possible before the present war.
SIMMONS, GEORGE H. R—Chicago, Ill. B—England. G. O. No. 3, W. D., 1922.	Major, Medical Section, Officers' Reserve Corps, U. S. Army. By his thorough knowledge of the medical profession and by his great esteem therein, together with his whole-hearted devotion to his task, he rendered services of a signal worth in the procurement of physicians and surgeons for the Medical Corps of the Army and in his able advising of the War Department upon the qualifications of that great body of the medical profession who entered the Army.
SIMONDS, GEORGE S. R—Cresco, Iowa. B—Cresco, Iowa. G. O. No. 59, W. D., 1919.	Brigadier general, U. S. Army. He served with marked distinction as chief of staff of the 2d Army Corps during the important operations along the Hindenburg line in the region of the Sambre Canal. His great administrative ability was shown in the excellent manner in which he handled a large force of American soldiers serving with the British.
SIMPSON, JOHN R. R—Newton, Mass. B—Richmond, Ind. G. O. No. 56, W. D., 1921.	Colonel, Ordnance Department, U. S. Army. While on duty in the office of the Chief Ordnance Officer, American Expeditionary Forces, in connection with the requirements and procurement of ordnance supplies, by his accurate forecasting and energetic following up of deliveries, he secured an adequate and uninterrupted flow of ordnance material for the American Expeditionary Forces.
SIMPSON, WILLIAM H. R—Aledo, Tex. B—Weatherford, Tex. G. O. No. 59, W. D., 1921.	Lieutenant colonel (Infantry), General Staff Corps, U. S. Army. For services as assistant chief of staff, 33d Division, during the Meuse-Argonne offensive and later as chief of staff of this division.
SINGLETON, ASA L. R—Fort Valley, Ga. B—Taylor County, Ga. G. O. No. 126, W. D., 1919.	Colonel (Infantry), General Staff Corps, U. S. Army. As chief of staff of Base Section No. 5 he displayed exceptional administrative ability. The excellent results obtained in evacuating over 700,000 men through the port of Brest are due in no small measure to the efficient organization created by him. He has rendered services of signal worth to the American Expeditionary Forces.

SINGLETON, MARVIN E.
  R—St. Louis, Mo.
  B—Waxahachie, Tex.
  G. O. No. 124, W. D., 1919.

As chief of the St. Louis Ordnance District, in which capacity he maintained at all times the greatest degree of intelligent and enthusiastic cooperation between the Ordnance Department and manufacturers in his district, thereby attaining the maximum production of munitions in a minimum time; and also as chairman of the St. Louis Ordnance District Claims Board, in which capacity his services have been invaluable to the Nation in adjusting equitably the $122,000,000 worth of outstanding contracts in his district in force at the signing of the armistice.

SINNOTT, CATHERINE G.
  R—Nashville, Tenn.
  B—Middletown, Conn.
  G. O. No. 9, W. D., 1923.

Second lieutenant, Army Nurse Corps, U. S. Army.
As chief nurse of Camp Hospital No. 28, France, during the World War, this nurse exhibited marked efficiency and administrative ability. Later, as chief nurse of the Nurses' Concentration Camp at Savenay, France, she managed the affairs of nearly a thousand nurses with exceptional tact, industry, and good judgment. She performed the unusual duties assigned to her in a way not only to facilitate greatly the embarkation of nurses, but to maintain a high state of morale and efficient organization among them. She rendered a conspicuously worthy service to the American Expeditionary Forces. Her splendid leadership, tireless energy, and unselfish devotion to duty were an inspiration to all who came in contact with her.

SKINNER, GEORGE A.
  R—St. Paul, Minn.
  B—Osage, Iowa.
  G. O. No. 38, W. D., 1922.

Colonel, Medical Corps, U. S. Army.
Colonel Skinner organized and commanded a hospital center of 20,000 beds at Mars, France. Due to his great force and ability, a hospital group was prepared for the care of the sick and wounded during the St. Mihiel and Meuse-Argonne offensives, when the need of hospital beds was critical. This adequate and efficient hospitalization contributed materially to the conservation of man power and to the subsequent success of our forces.

SLADE, GEORGE T.
  R—St. Paul, Minn.
  B—New York, N. Y.
  G. O. No. 50, W. D., 1919.

Colonel, Transportation Corps, U. S. Army.
He served with marked distinction as deputy director general of transportation, first with the French ministry and later with the railroad department in the zone of the services of supply. Due to his tactful negotiations and zealous efforts, the Transportation Department secured efficient cooperation with the French railroads and was enabled to meet the tremendous demands imposed upon it by the rapid advance of our armies during the Argonne-Meuse battles.

SLADEN, FRED W.
  R—Omaha, Nebr.
  B—Lowell, Mass.
  G. O. No. 59, W. D., 1919.
  Distinguished-service cross also awarded.

Brigadier general, U. S. Army.
While commanding the 5th Infantry Brigade, 3d Division, in the Battle of the Marne in July and in the Argonne operations in France, in October, 1918, he demonstrated conspicuous qualities of ability and leadership. The successes that attended the operations of his brigade were influenced greatly by his energy, skill, and courage as a commander.

SLAUGHTER, NUGENT H.
  R—Washington, D. C.
  B—Danville, Va.
  G. O. No. 69, W. D., 1919.

Lieutenant colonel, Signal Corps, U. S. Army.
For services in the very successful development of the radio equipment of the United States Army.

SLAVENS, THOMAS H.
  R—Urbana, Mo.
  B—Portland Mills, Ind.
  G. O. No. 53, W. D., 1921.

Colonel, Infantry, U. S. Army.
As commander of the New York depot from July, 1917, to March, 1918, in which he demonstrated superb energy and marked executive ability in a position of great responsibility. Later, as commanding officer of the 51st Infantry, 6th Division, from October, 1918, he showed marked ability in the training of the regiment and during its operations in the Meuse-Argonne.

*SLIFER, HIRAM J.
  R—Chicago, Ill.
  B—Montgomery County, Pa.
  G. O. No. 59, W. D., 1919.

Lieutenant colonel, Corps of Engineers, U. S. Army.
He was charged with active field operations and the construction and operation of the light railways of the 1st Army during the St. Mihiel and Argonne-Meuse offensives. His efforts were unceasing, and, due to his resourcefulness and exceptional executive ability, he was an important factor in the successful operations of the light railways, assuring for the troops of the 1st Army a steady flow of munitions and supplies.
Posthumously awarded. Medal presented to widow, Mrs. Hiram J. Slifer.

SLOCUM, STEPHEN L'H.
  R—New York, N. Y.
  B—Cincinnati, Ohio.
  G. O. No. 124, W. D., 1919.

Lieutenant colonel, U. S. Army.
For services while serving as military attaché at London, England.

SMALLEY, HENRY R.
  R—Chicago, Ill.
  B—Chicago, Ill.
  G. O. No. 56, W. D., 1922.

Major, Cavalry, U. S. Army.
As adjutant and operations officer of the 5th Infantry Brigade of the 3d Division during its occupation of the Chateau-Thierry defensive sector, and during the Champagne-Marne defensive, the Aisne-Marne, St. Mihiel, and Meuse-Argonne offensives, by his military knowledge, devotion to duty, excellent judgment, and unhesitating assumption of responsibility he rendered invaluable services and materially assisted in the success attained by the brigade in its operations.

SMITH, EMERY T.
  R—San Francisco, Calif.
  B—Virginia City, Nev.
  G. O. No. 56, W. D., 1922.

Colonel, Field Artillery, U. S. Army.
As commander of the group of Artillery supporting the 33d Division in the attack on the east bank of the Meuse River, Oct. 8 to 13, 1918, he demonstrated professional attainments and ability of a high order. Later, as a regimental commader, by his sound tactical judgment and special knowledge of Artillery, he most successfully directed his units in the support of the attacking Infantry of the 79th Division in the operations north of Verdun from Nov. 4 to 11, 1918. By keeping his elements close to the attacking Infantry he contributed in no small measure to the success of these operations. He rendered services of conspicuous merit and signal worth to the American Expeditionary Forces in a position of great responsibility.

SMITH, ERNEST G.
  R—Wilkes-Barre, Pa.
  B—Martins Ferry, Ohio.
  G. O. No. 59, W. D., 1919.

Lieutenant colonel, Infantry, U. S. Army.
As chief of the casualty section, central records office, he performed with marked efficiency duties of a most exacting character. With untiring efforts, sound analytical ability, and masterful attention to detail he handled questions pertaining to casualties in the American Expeditionary Forces with noteworthy success.

SMITH, HARRY A.
  R—Atchison, Kans.
  B—Atchison, Kans.
  G. O. No. 12, W. D., 1919.

Brigadier general, U. S. Army.
He rendered most conspicuous service as commandant of the Army schools at Langres, France, the success of which was, in a large measure, due to his vision, zeal, and administrative ability. He later showed marked executive ability as officer in charge of the administration of civil affairs in the German territory occupied by the American Army.

SMITH, LOWELL H.
  R—Rockwell Field, Calif.
  B—Santa Barbara, Calif.
  G. O. No. 14, W. D., 1925.
  Act of Congress Feb. 25, 1925.

Captain, Air Service, U. S. Army.
Lieutenant Smith, as pilot of the Airplane No. 2, the *Chicago*, and later when placed in command of the United States Army Air Service around-the-world flight from Apr. 6, 1924, to Sept. 28, 1924, displayed untiring energy, courage, and resourcefulness during the entire period that the Air Service expedition was upon its hazardous undertaking. His leadership, sound judgment, and tenacity of purpose were material factors in the success of this pioneer flight of airplanes around the world. In the performance of his great task he brought to himself and to the military forces of the United States the signal honor of an achievement which is a testimonial to American thoroughness, courage, and resourcefulness.

SMITH, PERRIN L.
  R—Minneapolis, Minn.
  B—Henry, Ill.
  G. O. No. 105, W. D., 1919.

Colonel, Quartermaster Corps, U. S. Army.
To his great administrative ability, initiative, and tireless energy is due in great measure the very successful practical application of the regulations governing the payment of allotments made by officers and enlisted men, the handling of the various Liberty bond issues subscribed for through the Army allotment system, and other important financial matters connected with the Army during the emergency.

SMITH, WILLIAM R.
  R—Nashville, Tenn.
  B—Nashville, Tenn.
  G. O. No. 2, W. D. 1920.

Brigadier general, U. S. Army.
As commanding general, 36th Division, by his thorough and ceaseless efforts, coupled with a keen insight into the principles of military training, he brought his division to such a high standard of discipline and proficiency as to achieve conspicuous results in a major operation without previous service under fire. The excellent conduct of his division subsequent to the signing of the armistice reflects great credit on him. His services have been of great value to the American Expeditionary Forces.

SMITH, WINFORD HENRY.
  R—Baltimore, Md.
  B—Scarboro, Me.
  G. O. No. 3, W. D., 1922.

Colonel, Medical Corps, U. S. Army.
For services as a specialist in hospital management and construction.

SMITH, WRIGHT.
  R—Holly Oak, Del.
  B—New York, N. Y.
  G. O. No. 95, W. D., 1919.

Colonel, Field Artillery, U. S. Army.
As commander of the 13th Field Artillery, 4th Division, he proved himself an artillerist of extraordinary skill and ability. Due to his energy and determination, he overcame seemingly insurmountable obstacles, keeping his regiment at all times on the alert in order to take its positions promptly, and rendered most effective support to the advancing Infantry units.

SMITHER, HENRY C.
  R—Denver, Colo.
  B—Indian Territory.
  G. O. No. 50, W. D., 1919.

Colonel (Signal Corps), General Staff Corps, U. S. Army.
As assistant chief of staff, in charge of the supply section, of the General Staff of the Services of Supply, he demonstrated by his energy, zeal, and masterful efforts a high order of efficiency and ability. He organized the supply section and handled without friction the questions of priority and troop orders during the period of the arrival of American troops in France. With a rare gift of tact and address, he discharged most successfully his many important duties.

SNOW, WILLIAM J.
  R—Rivervale, N. J.
  B—Brooklyn, N. Y.
  G. O. No. 18, W. D., 1919.

Major general, U. S. Army.
For services in planning and executing those measures responsible for the efficiency of the Field Artillery during the war.

SNYDER, FREDERIC A.
  R—Williamsport, Pa.
  B—Williamsport, Pa.
  G. O. No. 87, W. D., 1919.

Colonel, Corps of Engineers, U. S. Army.
As division engineer officer of the 28th Division during its participation in the Aisne-Marne and the Meuse-Argonne offensives, he solved numerous and difficult problems with marked ability. By his tireless energy in the construction and maintenance of transportation routes and defensive positions he contributed in no small degree to the successes of the combat troops.

SNYDER, JOHN JACOB.
  R—New Oxford, Pa.
  B—Two Taverns, Pa.
  G. O. No. 116, W. D., 1919.

Commander, Medical Corps, U. S. Navy.
As force medical officer, his untiring energy and close cooperation with the Army authorities contributed greatly to the successful outcome of our oversea operations.

SOLBERT, OSCAR N.
  R—Worcester, Mass.
  B—Sweden.
  G. O. No. 124, W. D., 1919.

Colonel, Corps of Engineers, U. S. Army.
For services while serving as military attaché at Copenhagen, Denmark.

SOMERVELL, BREHON B.
  R—Little Rock, Ark.
  B—Little Rock, Ark.
  G. O. No. 14, W. D., 1923.
  Distinguished-service cross also awarded.

Lieutenant colonel, Corps of Engineers, U. S. Army.
As adjutant, 15th Engineers, during the period of organization and training; in charge of construction of the Mehun ammunition depot; in charge of the construction at 1s-sur-Tille depot, including the gas depot at Poinson and the Etain engine terminal; assistant chief of staff, G–3, and assistant chief of staff, G–1, 89th Division, from October, 1918, until the division returned to the United States, when he was assigned as assistant chief of staff, G–4, 3d Army. In all these positions he displayed unusual vision, initiative, sound judgment, and high professional skill, contributing in a conspicuous way to the successful operations of the American forces in France.

SPALDING, GEORGE R.
  R—Monroe, Mich.
  B—Monroe, Mich.
  G. O. No. 59, W. D., 1919.

Colonel, Corps of Engineers, U. S. Army.
He served with marked distinction as commanding officer of the 305th Engineers. 80th Division, as division engineer of the 80th Division, as chief engineer of the 5th Army Corps, and as chief engineer of the 1st and 3d Armies. At all times he exhibited professional attainments of the highest order in handling the difficult problems with which he was confronted.

SPAULDING, OLIVER L., Jr.
  R—St. Johns, Mich.
  B—St. Johns, Mich.
  G. O. No. 19, W. D., 1920.

Lieutenant colonel, Field Artillery, U. S. Army.
As assistant commandant, School of Fire for Field Artillery, Fort Sill, Okla., from December, 1917, to May, 1918. His constructive and administrative ability was of great value in the remarkable and successful expansion of that school to meet the war requirements of the Field Artillery. In especial, his work in connection with the coordination and development of the course of instruction contributed materially to the excellence of the Field Artillery education received by thousands of officers.

SPEAKS, CHARLES E.
  R—Akron, Ohio.
  B—Washington, D. C.
  G. O. No. 16, W. D., 1923.

Lieutenant colonel, Motor Transport Corps, U. S. Army.
He organized and operated with marked success the division for the procurement of motor transport supplies in the office of the director, Motor Transport Corps, American Expeditionary Forces, and by his intelligent application, wide experience, sound judgment, and devotion to duty successfully surmounted great difficulties in providing adequate motor transportation for the American Expeditionary Forces.

SPEAR, RAY.
  R—Spokane, Wash.
  B—Illinois.
  G. O. No. 116, W. D., 1919.

Captain, Supply Corps, U. S. Navy.
As force supply officer, the efficient performance of which duties contributed greatly to the successful provisioning of ships engaged in the transportation of troops and supplies overseas.

SPENCE, CARY F.
  R—Knoxville, Tenn.
  B—Knoxville, Tenn.
  G. O. No. 19, W. D., 1920.

Colonel, Infantry, U. S. Army.
He commanded with marked distinction the 117th Infantry, 30th Division, from the time of its organization and early training period to the completion of the active combat operations in the Ypres-Lys and Somme offensives. He especially distinguished himself while in command of his regiment on Oct. 8–9, 1918, when he advanced his line 2½ miles, capturing several towns, numerous cannon, and many prisoners.

SPENCER, EUGENE J.
  R—Webster Groves, Mo.
  B—St. Louis, Mo.
  G. O. No. 13, W. D., 1923.

Colonel, Corps of Engineers, U. S. Army.
As colonel, 32d Engineers, he constructed, in large part, the storage camp at St. Sulpice, Bordeaux, the receiving barracks of Genicourt, the new port at Talmont, the munitions depot at St. Loubes, France. As chief engineer of Base Section No. 2, Bordeaux, France, his duties included the construction and maintenance of roads and buildings, supervision of forests and posts. His high professional skill, unremitting energy, and devotion to duty contributed markedly to the success of the American Expeditionary Forces.

SPINKS, MARCELLUS G.
  R—Meridian, Miss.
  B—Meridian, Miss.
  G. O. No. 59, W. D., 1919.

Brigadier general, U. S. Army.
By his untiring efforts, zeal, and marked military efficiency in the performance of duties of responsibility as senior assistant of the Inspector General's Department in France he has rendered services of exceptional value to the Government.

SPRUANCE, WILLIAM C., Jr.
  R—Wilmington, Del.
  B—Wilmington, Del.
  G. O. No. 77, W. D., 1919.

Colonel, Ordnance Department, U. S. Army.
As chief of the powder section, production division, of the office of the Chief of Ordnance. Later as special assistant to the Chief of Ordnance in charge of chemicals, propellants, and explosives, and chief of the explosives, chemicals, and loading division, office of the Chief of Ordnance, in which capacities he successfully organized the industry of the country so as to yield at all times an ample supply of powder, not only for the needs of the U. S. Army, but to some extent for the needs of the cobelligerents against Germany.

SQUIER, GEORGE O.
  R—Dryden, Mich.
  B—Dryden, Mich.
  G. O. No. 103, W. D., 1919.

Major general, U. S. Army.
As Chief Signal Officer he has demonstrated scientific attainments of the highest order. His researches and contributions to the scientific equipment of the Signal Corps are noteworthy. The Signal Corps under him has been an extremely progressive and efficient organization.

STACKPOLE, PIERPONT L.
  R—Boston, Mass.
  B—Brookline, Mass.
  G. O. No. 126, W. D., 1919.

Lieutenant colonel, Field Artillery, U. S. Army.
As aide-de-camp to Lieut. Gen. Hunter Liggett, U. S. Army, he rendered exceptional services during the entire time that the latter commanded the 1st Army Corps, the 1st Army, and the 3d Army. By his military attainments and pronounced ability he proved to be a most important factor in the successes of the corps and also of the armies.

STANBERY, SANFORD B.
  R—California, Ohio.
  B—Millersburg, Ohio.
  G. O. No. 89, W. D., 1919.

Brigadier general, U. S. Army.
Having taken command of the 155th Infantry Brigade, 78th Division, prior to the attack of Nov. 1, he proved himself a forceful and capable military leader. With the tactical situation thoroughly in hand, by his zeal and good judgment he contributed to the brilliant results attained during the severe fighting in the advance toward Sedan in the final phase of the Meuse-Argonne offensive.

STANLEY, DAVID S.
R—Washington, D. C.
B—Dakota Territory.
G. O. No. 15, W. D., 1923.

Colonel, Quartermaster Corps, U. S. Army.
As chief quartermaster and chief of staff, Base Section No. 5, at Brest, France, by his great administrative ability, exceptional foresight, and tireless energy he handled numerous difficult problems of supply and transportation with unusual efficiency and success. In the performance of his great task he rendered services of conspicuous worth to the American Expeditionary Forces.

STANSFIELD, JAMES H.
R—Oak Park, Ill.
B—Bridgeport, Ill.
G. O. No. 56, W. D., 1921.

Lieutenant colonel, 132d Infantry, 33d Division, U. S. Army.
In the Bois-de-Chaume during Oct. 9-11, 1918, when, due to his rare presence of mind and courage, he prevented the disorganization of units of regiments that had suffered heavy casualties. He personally reorganized scattered groups and caused them to hold a line which appeared untenable. He has rendered services of great value.

STANTON, CHARLES E.
R—Salt Lake City. Utah.
B—Monticello, Ill.
G. O. No. 70, W. D., 1919.

Colonel, Quartermaster Corps, U. S. Army.
As chief disbursing officer in the office of the finance division, Quartermaster Corps, at Paris he performed his duties with unremitting zeal, displayed marked administrative ability and accurate judgment in solving problems of extraordinary difficulty, rendering services of marked worth.

STARBIRD, ALFRED A.
R—South Paris, Me.
B—South Paris, Me.
G. O. No. 19, W. D., 1922.

Brigadier general, U. S. Army.
For services in connection with the planning, organization, and administration of the post and subposts of Brest, Base Section No. 5, thereby contributing in a very great measure to the successful operation of this base during the return of the American Expeditionary Forces.

STARK, ALEXANDER N.
R—Norfolk, Va.
B—Norfolk, Va.
G. O. No. 59, W. D., 1919.

Colonel, Medical Corps, U. S. Army.
He served as chief surgeon of the 1st Army during all its offensives, charged with the organization and direction of the Medical Service, involving the treatment and evacuation of many thousands of sick and wounded under most adverse conditions. In this important capacity he performed his duties with marked ability. With good judgment, furthered by high professional attainments and tireless energy, he solved the difficult problems which arose, prevented much suffering, and saved the lives of many among the American and French wounded soldiers.

STARR, WILLIAM T.
R—New York, N. Y.
B—Indianapolis, Ind.
G. O. No. 59, W. D., 1921.

Lieutenant colonel, Military Police Corps, U. S. Army.
In the organization of the Provost Marshal General's Department, American Expeditionary Forces, from Nov. 18, 1918, to Feb. 1, 1919, in which capacity he displayed marked qualities of zeal and efficiency that were of great benefit to the American Expeditionary Forces.

STAYTON, WILLIAM H., jr.
R—Washington, D. C.
B—New York, N. Y.
G. O. No. 14, W. D., 1923.

Major, Adjutant General's Department, U. S. Army.
With rare vision, sound judgment, and signal ability, he organized and operated the office charged with the responsibility for the circulation of officers, soldiers, and civilians of the American Expeditionary Forces in France, and of the officers of the allied forces in the zones occupied by the American armies, maintaining active contact with British and French authorities and with the American military police throughout France with reference to such circulation. By his tact and initiative, without American precedent, he handled these activities with the utmost efficiency. Later, in charge of the appointment and promotion section of The Adjutant General's Office, American Expeditionary Forces, he displayed unusual executive and administrative ability. His services were of immeasurable value to the Government.

STEARNS, CUTHBERT P.
R—Denver, Colo.
B—Elizabeth, N. J.
G. O. No. 14, W. D., 1923.

Lieutenant colonel, Air Service, U. S. Army.
As chief of staff, spruce production division, Bureau of Aircraft Production, by his loyal devotion to duty, tact, industry, and resourcefulness, he handled with conspicuous success all the military problems of organizing the division and distributing the troops and supplies, an operation which involved constant and intimate relations with 150,000 civilian lumbermen and over 500 lumber mills, extending over an extensive territory. He rendered services of great value to the Government in a position of great responsibility.

STEBBINS, HORACE C.
R—New York, N. Y.
B—Boston, Mass.
G. O. No. 87, W. D., 1919.

Lieutenant colonel (Infantry), General Staff Corps, U. S. Army.
As assistant chief of staff, G-2, 3d Army Corps, he performed his important duties with merited ability and zeal. Through his efficient administration of the section, complete and timely intelligence of the enemy was promptly disseminated through the combatant troops, which aided materially in the successes of his corps.

STEESE, JAMES G.
R—Harrisburg, Pa.
B—Mount Holly Springs, Pa.
G. O. No. 47, W. D., 1919.

Colonel, Corps of Engineers, U. S. Army.
As assistant to the Chief of Engineers and in charge of the personnel, equipment, construction, and maps division of the office of the Chief of Engineers, he displayed exceptional ability in handling commissioned personnel matters and developed special apparatus and methods for the production of aerial navigation maps. Since September, 1918, the solution of the many difficult problems in the organization and operation of the personnel branch, General Staff, has been due largely to his special knowledge of personnel matters, his untiring zeal, good judgment, and exceptional administrative ability.

STEPHENS, JOHN E.
R—Brentwood, Tenn.
B—Brentwood, Tenn.
G. O. No. 56, W. D., 1922.

Brigadier general, U. S. Army.
As chief of the war plans branch of the war plans division of the General Staff from Oct. 4, 1917, to June 13, 1918, he was directly responsible for the preparation of plans for the organization of units of the Army and all special branches thereof. In this capacity he rendered most conspicuous services in a brilliant manner. By his keen foresight, great breadth of vision, and tireless energy he carried to successful completion the various plans and problems which confronted him. Later, as commander of the 61st Field Artillery Brigade, 36th Division, he again demonstrated these high professional attainments and splendid leadership, which at all times characterized his service.

STEENBERGER, HENRY S.
R—New York, N. Y.
B—New York, N. Y.
G. O. No. 33, W. D., 1922.

Lieutenant colonel, Quartermaster Corps, U. S. Army.
While serving as quartermaster of his division throughout the entire period of its operations in Belgium and France, by his untiring zeal, great energy, tact, and sound judgment he was able to overcome many difficult problems of supply, thereby rendering services of great value to the American Expeditionary Forces.

STETTINIUS, EDWARD R.
R—New York, N. Y.
B—St. Louis, Mo.
G. O. No. 25, W. D., 1919.

Who, as director general of purchases for the War Department, Second Assistant Secretary of War, and special representative in France of the Secretary of War in connection with the procurement of munitions for the American Expeditionary Forces, rendered conspicuous services. His broad vision and splendid judgment have been of the greatest value to the success of the military program.

STEVENS, JOHN F.
R—New York, N. Y.
B—West Gardiner, Me.
G. O. No. 95, W. D., 1919.

As head of the Railway Advisory Commission to Russia and special adviser of the Russian Ministry of Ways of Communication. In the midst of revolutionary conditions he has pursued his undertaking to rehabilitate Russia by the restoration of railway traffic. In a distant country, far from immediate support, he has maintained an unflagging devotion to duty which is now beginning to show the valuable results of his labor.

STEWART, MERCH B.
R—Glenn Falls, N. Y.
B—Mitchell Station, Va.
G. O. No. 49, W. D., 1922.

Brigadier general, U. S. Army.
As senior instructor at the Plattsburg Training Camp from May until August, 1917, he displayed organizing and training ability and talents of the highest order in successfully directing the training and selection of 6,000 officer candidates, thereby rendering services of inestimable value to our newly formed forces. As chief of staff, 76th Division, from August, 1917, until June, 1918, he again showed tireless energy, practical resourcefulness, and military attainments of the highest order. Later, as commander of the 175th Infantry Brigade, 88th Division, he performed his duties with marked ability and excellent judgment.

STEWART, REDMOND C.
R—Eccleston, Md.
B—Baltimore County, Md.
G. O. No. 56, W. D., 1922.

Major, Judge Advocate General's Department, U. S. Army.
He served with the 1st Division as division judge advocate throughout the entire hostilities. At all times, by his high-minded sense of duty, his personal example of energy, loyalty, and courage, he was a powerful and consistent influence for promoting the morale, harmony, spirit, and the high standards of the 1st Division. His superior professional attainments and his loyal devotion to duty contributed materially to the success of his division and made him a conspicuous figure in a position of great responsibility.

STILWELL, JOSEPH W.
R—Yonkers, N. Y.
B—Palatka, Fla.
G. O. No. 73, W. D., 1919.

Lieutenant colonel (Infantry), General Staff Corps, U. S. Army.
As assistant chief of staff, G–2, 4th Army Corps, during the St. Mihiel offensive and later during the operations in the Woevre, he displayed military attainments of a high order. With great energy and zeal he pursued the developments of the enemy activities on the corps front, securing invaluable information which assisted in a marked degree in the planning of the operations. He contributed by the excellent performance of his task to the success of these operations.

STIMSON, JULIA C.
R—St. Louis, Mo.
B—Worcester, Mass.
G. O. No. 79, W. D., 1919.

Chief nurse, Army Nurse Corps, U. S. Army.
As chief nurse of Base Hospital No. 21 she displayed marked organizing and administering ability while that unit was on active service with the British forces. Her devotion to duty was exceptional while she was chief nurse of the American Red Cross in France. Upon her appointment as director of nursing service of the American Expeditionary Forces, she performed exacting duties with conspicuous energy and achieved brilliant results. Thousands of sick and wounded were cared for properly through the efficient service she provided.

STIVERS, DANIEL G.
R—Butte, Mont.
B—Fort Davis, Tex.
G. O. No. 55, W. D., 1921.

Lieutenant colonel, Quartermaster Corps, U. S. Army.
As quartermaster of the 3d Division during the Aisne-Marne, St. Mihiel, and Meuse-Argonne offensives, in maintaining an excellent system of quartermaster supplies and utilities under the greatest difficulties of active service, his superb efforts contributed in a marked degree to the success of this division throughout its operations.

STOKES, MARCUS B.
R—Early Branch, S. C.
B—Colleton, S. C.
G. O. No. 49, W. D., 1922.

Colonel, Infantry, U. S. Army.
As commander of the 311th Infantry, 78th Division, during its organization, training, and in all of its operations, he handled all of his tasks with marked efficiency and in a manner that reflected great credit upon him. By his most careful and thorough preparations, sound judgment, skillful and energetic leadership, he contributed in a large measure to the successes achieved by his regiment against the enemy.

STONE, DAVID L.
R—Greenville, Miss.
B—Stoneville, Miss.
G. O. No. 59, W. D., 1919.

Colonel (Quartermaster Corps), General Staff Corps, U. S. Army.
As assistant chief of staff, G–1, 3d Division, as G–1 of that organization, and later as G–1, 2d Army, he performed with distinction his important duties. In the action from July 5 to Aug. 2, 1918, near Chateau-Thierry, and in the advance to the Ourcq River, he displayed tireless energy and ability of an unusually high order in supplying troops under most difficult conditions. Aggressive and resourceful, he proved equal to every emergency.

STONE, EDWARD R.
R—Spencer, Mass.
B—Spencer, Mass.
G. O. No. 89, W. D., 1919.
Distinguished-service cross also awarded.

Colonel, Infantry, U. S. Army.
As second in command of the 9th Infantry, 2d Division, he participated with credit in the Aisne defensive, the operations in the Chateau-Thierry sector, and in the Aisne-Marne offensive. Subsequently, upon being placed in command of the 23d Infantry, 2d Division, he led it with marked ability in the St. Mihiel offensive, and by his skillful leadership was largely responsible for the successes gained by this regiment in the Battle of Blanc Mont Ridge and the Meuse-Argonne offensive.

*STRAIGHT, WILLARD D_____
   R—New York, N. Y.
   B—Oswego, N. Y.
   G. O. No. 50, W. D., 1919.

Major, Adjutant General's Department, U. S. Army.
In the service of the organization, development, and administration of the War Risk Bureau his efforts resulted in marked efficiency in the handling of the large volume of insurance, as well as the numerous applications for allotments and allowances which covered almost the entire personnel of the American Expeditionary Forces. As an assistant in the first section of the general staff of the 1st Army he rendered particularly valuable services to the Government by his great energy and high ability.
Posthumously awarded. Medal presented to widow, Mrs. Donley Straight.

STRONG, GEORGE V_____

   R—Helena, Mont.
   B—Chicago, Ill.
   G. O. No. 38, W. D., 1922.

Lieutenant colonel (Judge Advocate General's Department), General Staff Corps, U. S. Army.
While on staff duty with Headquarters, 4th Army Corps and Headquarters, 2d Army, American Expeditionary Forces, he was in charge of all troop movements preparatory to the St. Mihiel attack and immediately following this attack, and was also in charge of all troop movements from the 4th Army Corps in the Toul sector to the Argonne front. By his tireless energy, keen foresight, and sound judgment he perfected the multifarious duties whereby all of these movements were carried to successful completion. During this period his services were conspicuously efficient and contributed materially to the success of these operations.

STRONG, RICHARD P_____
   R—Cambridge, Mass.
   B—Fortress Monroe, Va.
   G. O. No. 70, W. D., 1919.

Lieutenant colonel, Medical Corps, U. S. Army.
Possessed of the highest professional qualifications and actuated by zealous devotion to duty, he has rendered services of inestimable value to the American Expeditionary Forces, notably as president of a board appointed to investigate the cause of trench fever, a disease which had caused serious losses to the effectives of the allied armies. The scientific research of this board under his skillful direction led to the discovery of the means by which trench fever is transmitted and in the establishment of effective measures for its prevention.

SULTAN, DANIEL I_____
   R—Oxford, Miss.
   B—Oxford, Miss.
   G. O. No. 56, W. D., 1922.

Colonel (Corps of Engineers), General Staff Corps, U. S. Army.
As chief of the personnel section in the office of the executive assistant to the Chief of Staff of the Army during the war and the demobilization, he formulated policies covering commissioned personnel and handled with marked ability many complex questions of grave importance to the War Department and to the entire Army. His work was characterized by conspicuous breadth of vision and keen foresight. His splendid judgment and the sound policies initiated by him contributed in a large measure to the successful handling of the commissioned personnel of the Army. He rendered service of signal worth to the Government in a position of great responsibility.

SUMMERALL, CHARLES P_____
   R—Astatula, Fla.
   B—Lake City, Fla.
   G. O. No. 12, W. D., 1919.
   Distinguished-service cross also awarded.

Major general, U. S. Army.
He commanded in turn a brigade of the 1st Division in the operations near Montdidier, the 1st Division during the Soissons and St. Mihiel offensives and in the early battles of the Argonne-Meuse advance, and the 5th Army Corps in the later battles of this advance. In all of these important duties his calm courage, his clear judgment, and his soldierly character had a marked influence in the attainment of the successes of his commands.

SUMMERS, LELAND L_____
   R—Whitestone, N. Y.
   B—Cleves, Ohio.
   G. O. No. 15, W. D., 1923.

Technical advisor, War Industries Board.
In connection with the operations of the War Industries Board during the World War, as a member of the board he rendered through his broad vision, distinguished capacity, and business ability, services of inestimable value in marshaling the industrial forces of the Nation and mobilizing its economic resources—marked factors in assisting to make military success attainable. As technical advisor of the board and later as representative on the interallied munitions board he rendered, through his untiring efforts and devotion to duty, exceptionally valuable service to the War Department in connection with the procurement and supply of explosives for the Army.

°SUMNER, EDWIN VOSE, Jr_____
   R—Milton, Mass.
   B—Fort Niobrara, Nebr.
   G. O. No. 62, W. D., 1919.

Lieutenant colonel, Air Service, U. S. Army.
As commanding officer of the Air Service production and assembly center at Romorantin, he displayed peculiar administrative ability in coordinating the work of the many different elements at the largest Air Service project in the American Expeditionary Forces. The satisfactory results obtained at Romorantin were due largely to his tireless energy and skill in supervising and directing its operation. His example established a spirit of teamwork and accomplishment which were most marked.
Posthumously awarded. Medal presented to widow, Mrs. Edwin V. Sumner.

SUNDERLAND, ARCHIBALD H_____
   R—Delavan, Ill.
   B—Delavan, Ill.
   G. O. No. 47, W. D., 1919.

Brigadier general, U. S. Army.
As commandant of the Coast Artillery School and in the reorganization and administration of that institution, he thereby enabled it to meet effectively the demands made upon it for training candidates for commissions in the Coast Artillery Corps.

SWEENEY, WALTER C_____
   R—Wheeling, W. Va.
   B—Wheeling, W. Va.
   G. O. No. 59, W. D., 1919.

Colonel (Infantry), General Staff Corps, U. S. Army.
As chief of staff of the 28th Division he rendered conspicuously valuable services in the Argonne-Meuse offensive. In the capture of the strong enemy positions at le Chene Tondu, Apremont, Chatel-Chehery, and Hill No. 244, by his marked ability and tactical knowledge he proved a material factor in the successes achieved during these important operations.

SWEET, ETHEL E. (now MRS. THEODORE FALCONER). R—Canada. B—Detroit, Mich. G. O. No. 9, W. D., 1923.	Chief nurse, Army Nurse Corps, U. S. Army. As chief nurse of the nurses' mobilization stations in New York City, she contributed largely to the rapid and successful embarkation of over 10,000 nurses for overseas duty during the World War. She displayed exceptional zeal, foresight, and good judgment in organizing the separate staffs of this large group of nurses, and by her efficiency, industry, and tact made possible the transfer to Europe of this large contingent of women, an unusual and difficult feat performed under the most trying and perplexing conditions.
SWOPE, GERARD. R—New Brunswick, N. J. B—St. Louis, Mo. G. O. No. 10, W. D., 1920.	Assistant to the Director of Purchase, Storage, and Traffic, War Department. As one of the principal advisers and assistants to the Director of Purchase, Storage and Traffic he accomplished the task of working out the detailed plan for bringing under one head the direction and supervision of procurement, storage, and issue of all commodities and articles of equipment and supply needed for the Army. It was due to his foresight, ability, energy, and loyal cooperation that the procurement program for the great Army of 1918 was successfully planned, and he assisted materially in carrying it into effect, thereby contributing directly to the success of the military program.
SYMMONDS, CHARLES J. R—Kenosha, Wis. B—Holland, Mich. G. O. No. 59, W. D., 1919.	Colonel, Cavalry, U. S. Army. He commanded for many months the important intermediate storage depot at Gievres. He successfully administered a large personnel and supervised the growth of Gievres as a storage depot. He organized the system of supply from that station so efficiently that there were no shortages, either of food or material, at the regulating stations dependent upon Gievres for supply during all the active operations.
TAYLOR, BRAINERD. R—Newtonville, Mass. B—Malden, Mass. G. O. No. 89, W. D., 1919.	Colonel, Motor Transport Corps, U. S. Army. Serving as chief motor transport officer of the advance section, S. O. S., he gave proof of excellent judgment and untiring energy in the performance of his duties. By his success in overcoming numerous obstacles involved in the transportation of supplies and troops, he rendered conspicuous services to the American Expeditionary Forces.
TAYLOR, HARRY. R—Tilton, N. H. B—Tilton, N. H. G. O. No. 50, W. D, 1919.	Brigadier general, U. S. Army. Arriving in France June 11, 1917, as Chief Engineer, American Expeditionary Forces, he organized and administered the Engineer Department, which included the construction of wharves, depots, railways, barracks, and shelters throughout the theater of operations. He continued these duties with most marked and conspicuous ability, building a complete and efficiently functioning institution.
TAYLOR, JAMES D. R—Lake City, Fla. B—Lake City, Fla. G. O. No. 108, W. D., 1919.	First lieutenant, Infantry, U. S. Army. While commanding the station of Pantabangan, Luzon, P. I., in January, 1901, by his discretion and excellent judgment he obtained possession of the correspondence which made known the whereabouts of the insurgent chieftain, Aguinaldo, thus making possible the expedition resulting in his capture.
TEBBETTS, HARRY H. R—Haverhill, Mass. B—Great Falls, N. H. G. O. No. 53, W. D., 1921.	Colonel (Infantry), General Staff Corps, U. S. Army As assistant chief of staff, G-1, of the Services of Supply, he demonstrated marked energy and executive ability in the management of troop evacuation from France, and especially subsequent to the armistice in the repatriation of the American Expeditionary Forces, when several hundred thousand men were returned to the United States each month. He has rendered services of great value.
TEFFT, William H. R—Belmont, N. Y. B—Belmont, N. Y. G. O. No. 103, W. D., 1919.	Colonel, Medical Corps, U. S. Army. As commanding officer of Evacuation Hospital No. 7 at Chateau Montomglaust, he performed his exacting duties with unflagging energy and marked executive ability. Overcoming grave difficulties due to inadequate personnel and equipment, he succeeded in receiving, treating, and evacuating a large number of wounded from the Marne offensive with notable success, thereby rendering services of the utmost value to the American Expeditionary Forces.
TENNEY, CHARLES H. R—Longmeadow, Mass. B—Everett, Mass. G. O. No. 9, W. D., 1923.	Colonel, Ordnance Department, U. S. Army. First as chief of the financial and accounting divisions of the Ordnance Department, in which position he was charged with effecting the disbursement of all ordnance funds and with accounting for all ordnance property; next as special representative of the Chief of Ordnance to coordinate the financial and accounting operations of the Ordnance Department in the United States with those in France; and finally, as chairman and organizer of the ordnance salvage board, in which capacity he was charged with the duty of perfecting the organization and outlining the procedure of that board and with reviewing its recommendations with respect to the disposition of great quantities of surplus stores. He performed all his duties with zeal, sound judgment, and exceptional ability, thereby rendering services of great value to the Government.
TERRELL, JOHN P. R—Yonkers, N. Y. B—Yonkers, N. Y. G. O. No. 56, W. D., 1922.	Colonel (Coast Artillery Corps), General Staff Corps, U. S. Army. As assistant G-4, and later as G-4, 2d Army Corps, from July, 1918, to January, 1919, he displayed exceptional ability in the administration of that division of the corps staff. During the operations which broke the Hindenburg line between Cambrai and St. Quentin his great energy and able handling of matters of supply and transportation for the organizations of the corps contributed in a marked degree to the success of the operations.

THAYER, WILLIAM S.
  R—Baltimore, Md.
  B—Milton, Mass.
  G. O. No. 50, W. D., 1919.

Brigadier general, U. S. Army.
As chief consultant in medicine of the American Expeditionary Forces, with untiring zeal he devoted his time, energy, and high professional talents in promoting the organization of eminent medical officers for the prosecution of efficient treatment among the sick and wounded of the American Expeditionary Forces. Largely through his individual efforts, the treatment of the sick was so standardized, coordinated, and proficiently perfected as to result in a direct saving of many lives and a consequent conservation of man power and morale of these forces.

THELEN, MAX.
  R—Berkeley, Calif.
  B—Rising City, Nebr.
  G. O. No. 10, W. D., 1920.

Assistant to the Director of Purchase, Storage and Traffic, War Department.
As chief of the purchase branch of the Purchase, Storage, and Traffic Division he was responsible for the initiation of the purchase policy and the supervision of all contracts. His constructive ability and exceptional foresight are responsible for the results attained in the settlement of war contracts.

THOMAS, JOHN R., Jr.
  R—Chicago, Ill.
  B—Metropolis, Ill.
  G. O. No. 59, W. D., 1919.

Colonel (Infantry), General Staff Corps, U. S. Army.
As chief of the aviation division of the Intelligence Section, he displayed unusual energy and skill in the collection and dissemination of information regarding the enemy's air forces. During part of the period covered by the Argonne-Meuse offensive operations he acted as head of the Intelligence Section and performed the duties of that position with marked ability and sound judgment.

THOMPSON, CHARLES F.
  R—Jamestown, N. Dak.
  B—Jamestown, N. Dak.
  G. O. No. 59, W. D., 1919.

Lieutenant colonel (Infantry), General Staff Corps, U. S. Army.
As assistant chief of staff, G-2, of the 1st Army he aided in its organization by his skill and sound judgment, participating in the preliminary preparations and operations at the St. Mihiel salient. The successes achieved by his section are largely due to his high military attainments, his great energy, and painstaking devotion to duty. He served with equal ability as G-2 of the 2d Army in September, 1918, at all times showing great skill and accomplishing results of exceptional value.

THOMPSON, DORA E.
  R—Cold Spring, N. Y.
  B—New York, N. Y.
  G. O. No. 108, W. D., 1919.

Superintendent, Army Nurse Corps, U. S. Army.
To her accuracy, good judgment, and untiring devotion to duty is due the splendid management of the Army Nurse Corps during the emergency.

THOMPSON, JOHN T.
  R—Newport, Ky.
  B—Newport, Ky.
  G. O. No. 34, W. D., 1919.

Colonel, Ordnance Department, U. S. Army.
As chief of the small arms division of the office of the Chief of Ordnance, in which capacity he was charged with the design and production of all small arms and ammunition thereby supplied to the United States Army, which results he achieved with such signal success that serviceable rifles and ample ammunition therefor were at all times available for all troops ready to receive and use them.

THOMPSON, MELVILLE WITHINGTON.
  R—Spy Rock, N. Y.
  B—Washington, D. C.
  G. O. No. 14, W. D., 1920.

Lieutenant colonel, Air Service, U. S. Army.
As governor of the War Credits Board, he was untiring in his efforts, and his good judgment in solving the many complex questions confronting the board added materially to the proper discharge of its many responsibilities.

THORNE, ROBERT J.
  R—Lake Forest, Ill.
  B—Chicago, Ill.
  G. O. No. 18, W. D., 1919.

Assistant to Acting Quartermaster General, U. S. Army.
For services in the reorganization of the service of supply, thereby enabling the heavy demands due to an increased Army to be met.

TILLMAN, SAMUEL E.
  R—Shelbyville, Tenn.
  B—Shelbyville, Tenn.
  G. O. No. 77, W. D., 1919.

Brigadier general, U. S. Army.
For services as superintendent, U. S. Military Academy, during the period of the emergency.

TINLEY, MATHEW A.
  R—Council Bluffs, Iowa.
  B—Council Bluffs, Iowa.
  G. O. No. 78, W. D., 1919.

Colonel, Infantry, U. S. Army.
He displayed exceptional qualities of leadership in command of the 168th Infantry, 42d Division, which under his able leadership fulfilled every mission assigned to it. He was untiring in energy and devotion to his important duties, acting with sound judgment and initiative in times of emergency. His conduct was an inspiration to the men of his command, whom he led repeatedly in successful engagements.

TOBIN, WILLIAM H.
  R—San Francisco, Calif.
  B—Middleboro, Mass.
  G. O. No. 19, W. D., 1920.

Colonel, Coast Artillery Corps, U. S. Army.
As commander of the Army Artillery Park of the 1st Army, American Expeditionary Forces, by his broad grasp of the problems of ammunition supply he contributed in a marked degree to the solution of this most difficult problem and also to the success of the artillery of the 1st Army during the St. Mihiel and Meuse-Argonne offensives.

TOD, ROBERT E.
  R—New York, N. Y.
  B—Scotland.
  G. O. No. 49, W. D., 1922.

Commander, U. S. Navy, port officer, and later utilities officer at Base Section No. 5, Brest, France.
For conspicuously successful and cooperative industry in matters essential to the handling of transports and cargo ships of the United States; for the exercising of well-balanced business judgment in dealing with emergencies and for untiring and self-sacrificing effort to the end that ships be speedily turned around; for his subordination of all personal consideration in time of stress and for his helpfulness in all that concerned army troops, thereby rendering most conspicuous services to the American Expeditionary Forces in a position of great responsibility,

TODD, HENRY D., Jr. R—Philadelphia, Pa. B—Clauverick, N. Y. G. O. No. 24, W. D., 1920.	Brigadier general, U. S. Army. As commanding general of the 58th Field Artillery Brigade, 33d Division, he demonstrated marked skill as an artillery officer in the preparations for the attack of the 5th Corps on the Kriemhilde Stellung on Nov. 1, 1918, and in the support of the 89th Division in its further advance and crossing of the Meuse River from Nov. 6 to 11, 1918. The brigade which he commanded effectively supported the 1st, 91st, 32d, and 89th Divisions, during the period of the operations in which it served with them. His services have been of particular value to the American Expeditionary Forces.
TOLMAN, EDGAR B. R—Chicago, Ill. B—British India. G. O. No. 56, W. D., 1922.	Major, Infantry, U. S. Army. As executive officer in charge of the selective draft in Illinois, by his unusual executive ability, rare tact and skill, great initiative and resourcefulness exercised at times under most trying and novel conditions which arose in connection with the administration of the selective service act, he achieved a pronounced and conspicuous success in the performance of difficult and highly responsible duties, thereby rendering services of great value to the Government.
TOMPKINS, FRANK. R—Governors Island, N. Y. B—Washington, D. C. G. O. No. 96, W. D., 1918.	Major, 13th Cavalry, U. S. Army. At Columbus, N. Mex., Mar. 9, 1916, having requested and received authority to pursue a superior force of bandits into Mexico, carried on, after being wounded, a running fight with said bandits for several miles, inflicting heavy losses upon the bandits and only stopping the pursuit when men and horses were exhausted and ammunition reduced to a few rounds per man.
TOOMBS, LOUIS A. R—Meridian, Miss. B—Pickens, Miss. G. O. No. 59, W. D., 1921.	Lieutenant colonel, Adjutant General's Department, U. S. Army. As provost marshal in Italy, a position of great responsibility, during the critical situation arising from the decision of the peace conference on the Fiume question, by the means of good judgment and tact, he so managed the situation as to prevent all friction between the American Expeditionary Forces and the Italian populace. Due to his efficiency the military police of Italy were held to a high state of military training and discipline. He has rendered services of much value.
TOULMIN, HARRY A., Jr. R—Dayton, Ohio. B—Springfield, Ohio. G. O. No. 38, W. D., 1922.	Lieutenant colonel, Air Service, U. S. Army. As head of the coordination section of the staff of the Chief of Air Service, American Expeditionary Forces, he was charged with the responsibility for outlining and developing an organization to handle the many and grave problems of administration, mobilization, supply, and armament. In the performance of this duty he displayed rare intelligence, great initiative, broad vision, and an ability to obtain results, thus contributing materially to the success of the American Expeditionary Forces.
TOWNSHEND, ORVAL P. R—Shawneetown, Ill. B—Shawneetown, Ill. G. O. No. 11, W. D., 1921.	Lieutenant colonel, Infantry, U. S. Army. For services in connection with the mobilization, organization, and training of Porto Rico's quota of troops in the World War.
TRACY, EVARTS. R—Plainfield, N. J. B—New York, N. Y. G. O. No. 3, W. D., 1922.	Major, Corps of Engineers, U. S. Army. As the pioneer camouflage officer in the United States Army, by his marked ability he ably assisted in recruiting and organizing personnel for this important work and in preparing lists of equipment and necessary material for the carrying out of this enterprise. He served as chief instructor in camouflaging at the Army Engineer School at Langres from its organization until August, 1918. To him is due the success in developing a school course and a field exhibit that disseminated important knowledge among a large number of the combat personnel of the American Expeditionary Forces.
TRACY, JOSEPH P. R—Monroeton, Pa. B—Washington, D. C. G. O. No. 133, W. D., 1919.	Colonel, Coast Artillery Corps, U. S. Army. While in charge of the enlisted division of The Adjutant General's Office during the war, in which capacity his sound judgment and administrative ability were conspicuous.
TREAT, CHARLES G. R—Monroe, Wis. B—Dexter, Me. G. O. No. 55, W. D., 1920.	Brigadier general, U. S. Army. As chief of the American military mission to Italy and commanding Base Section No. 8, by his untiring devotion to duty, loyalty, and zeal, he performed his intricate duties with marked ability and sound judgment. By his cheerfulness and sound diplomatic ability he furthered those cordial relations which existed between the American and Italian troops, and was an important factor in maintaining the morale at a high state of efficiency during the trying days prior to the armistice.
TRIPP, GUY E. R—Washington, D. C. B—Wells, Me. G. O. No. 25, W. D., 1919.	Brigadier general, U. S. Army. Who, as chief of the production division of the Ordnance Department, and later as Assistant Chief of Ordnance, displayed fine technical ability and broad judgment in systematizing methods and practices resulting in the efficient cooperation of industries producing articles of ordnance for the Army.
TRIPPE, HARRY M. R—Whitewater, Wis. B—Whitewater, Wis. G. O. No. 56, W. D., 1921.	Lieutenant colonel, Corps of Engineers, U. S. Army. As commanding officer of the 308th Engineers, 83d Division, a position of great responsibility, much of the engineering success in facilitating the progress and supply of the 3d Army Corps during the Meuse-Argonne operation is due to his efforts and ability. He rendered important services to the United States.

TROTT, CLEMENT A.
   R—Milwaukee, Wis.
   B—Milwaukee, Wis.
   G. O. No. 59, W. D., 1919.

Colonel (Infantry) General Staff Corps, U. S. Army.
As chief of staff of the 5th Division, through his intimate knowledge of staff duties and the requirements of troops of the line he organized a staff which insured efficient cooperation in combat. His ability was shown in sound tactical directions to his division, which insured successes in four offensive operations.

TROWBRIDGE, AUGUSTUS.
   R—Princeton, N. J.
   B—New York, N. Y.
   G. O. No. 87, W. D., 1919.

Lieutenant colonel, General Staff Corps, U. S. Army.
As supervisor of the technique of flash and sound ranging, by his complete scientific knowledge and keen devotion to his important duties he rendered services of great value. Due to his good judgment and painstaking energy suitable personnel was selected and properly trained in the efficient operation of the flash and sound ranging service of the American Expeditionary Forces.

TRUESDELL, KARL.
   R—Washington, D. C.
   B—Moorhead, Minn.
   G. O. No. 103, W. D., 1919.

Lieutenant colonel, Signal Corps, U. S. Army.
As signal officer of the 1st Division and the 5th Army Corps he displayed high professional attainments and unflagging zeal. By his skill in directing the construction and maintenance of extensive telephone and wireless systems he contributed materially to the success of combat operations.

TURCK, RAYMOND C.
   R—Jacksonville, Fla.
   B—Gratiot County, Mich.
   G. O. No. 19, W. D., 1922.

Lieutenant colonel, Medical Corps, U. S. Army.
As division surgeon, 35th Division, during the Meuse-Argonne offensive, Colonel Turck organized the medical service of that division and provided hospitalization and evacuation facilities for the sick and wounded under conditions which rendered the service of the Medical Department unusually hazardous and difficult.

TURNBULL, SAMUEL J.
   R—Monticello, Fla.
   B—Monticello, Fla.
   G. O. No. 103, W. D., 1919.

Major, Medical Corps, U. S. Army.
As commanding officer of Evacuation Hospital No. 9, he performed his exacting duties with notable success. Overcoming numerous obstacles, by his keen foresight and administrative ability he was instrumental in securing the prompt evacuation and effective treatment of a large number of sick and wounded.

TUTTLE, ARNOLD D.
   R—Highland Falls, N. Y.
   B—Sturgis, S. Dak.
   G. O. No. 59, W. D., 1919.

Colonel (Medical Corps), General Staff Corps, U. S. Army.
In his capacity as assistant to the chief surgeon, and later as a member of the General Staff, American Expeditionary Forces, he supervised the preparation of hospitalization plans and their execution and assisted in the evacuation of sick and wounded from the battle fields in such manner as to greatly increase the efficiency of his department.

TWACHTMAN, JOHN ALDEN.
   R—Greenwich, Conn.
   B—Cincinnati, Ohio.
   G. O. No. 9, W. D., 1923.

Colonel, Field Artillery, U. S. Army.
As battalion and later regimental commander, 103d Field Artillery, 26th Division, in the Aisne-Marne, St. Mihiel, and Meuse-Argonne offensives, he was conspicuous for his courage, marked ability, and leadership qualities. At all times he displayed superior tactical judgment and knowledge of artillery, and by his devotion to duty, great resourcefulness, and high military attainments he rendered the maximum support to the Infantry to which he was attached, thereby contributing in a large measure to their successes.

TWELVETREE, HERBERT J.
   R—Cleveland, Ohio.
   B—Cleveland, Ohio.
   G. O. No. 13, W. D., 1923.

Lieutenant colonel (Infantry), General Staff Corps, U. S. Army.
As assistant chief of staff, G-1, 37th Division throughout its operations in France, he displayed unusual ability, leadership, resourcefulness, and high technical skill, thus contributing in a material way to the successful operations of his division.

TYDINGS, MILLARD E.
   R—Havre de Grace, Md.
   B—Havre de Grace, Md.
   G. O. No. 16, W. D., 1923.

Lieutenant colonel, Infantry, U. S. Army.
While commanding the 111th Machine Gun Battalion, 29th Division during the Meuse-Argonne operations, north of Verdun, Oct. 8 to 30, 1918, he distinguished himself by his energy, fearlessness, and high qualifications for the gravely responsible duties devolving upon him. The exceptionally effective use made by him of the weapons at his command rendered an advance possible against formidable hostile field works. His constant personal reconnaissance of frontline positions of the Infantry made possible an effective disposal of machine guns and artillery in the support of the efforts of the Infantry to advance and contributed in a large measure to the success of the brigade which his command was supporting.

TYLER, MAX CLAYTON.
   R—Fargo, N. Dak.
   B—Fargo, N. Dak.
   G. O. No. 69, W. D., 1919.

Colonel, Corps of Engineers, U. S. Army.
As executive officer and military advisor to the Director General of Military Railways, he has displayed high professional attainments and given valuable assistance in procuring personnel and equipment for the railway service abroad.

TYNDALL, ROBERT H.
   R—Indianapolis, Ind.
   B—Indianapolis, Ind.
   G. O. No. 13, W. D., 1923.

Colonel, Field Artillery, United States Army.
As commander, 150th Field Artillery, 42d Division, in the Baccarat, Champagne, Aisne-Marne, St. Mihiel, and Meuse-Argonne operations, part of which time he commanded one or more additional elements of the Artillery with which he was operating. His high technical attainments, his untiring energy and devotion to duty were important factors in the successful operations of the American Expeditionary Forces.

TYNER, GEORGE P.
   R—Chicago, Ill.
   B—Davenport, Iowa.
   G. O. No. 59, W. D., 1919.

Colonel (Cavalry), General Staff Corps, U. S. Army.
He served first as assistant G-4 of the 1st Army and later as G-4 of the 2d Army. He rendered devoted, skillful, and efficient service in the supply of the 1st and 2d Armies during the St. Mihiel offensive in the Forest of Argonne and in the Woevre. His painstaking and tireless energy contributed materially to the success of these operations.

Tyson, Lawrence D. R—Knoxville, Tenn. B—Greenville, N. C. G. O. No. 89, W. D., 1919.	Brigadier general, U. S. Army. He commanded with distinction the 59th Infantry Brigade, 30th Division, throughout its training period and during its active operations against the enemy. His determination and skill as a military leader were reflected in the successes of his brigade in the attack and capture of Brancourt and Premont, where a large number of prisoners and much material fell into our hands. He rendered services of great worth to the American Expeditionary Forces.
Ulio, James A. R—Fort Keogh, Mont. B—Fort Walla Walla, Wash. G. O. No. 89, W. D., 1919.	Lieutenant colonel, Infantry, U. S. Army. As assistant chief of staff, G–1, of the 4th Army Corps, he showed marked organizing and administrative ability. By his tireless efforts and ceaseless energy he contributed in a large degree to the successes achieved by the 4th Army Corps in the Toul sector and in the battles of the St. Mihiel salient. Later he handled with great success the evacuation and feeding of French civilians in the occupied territory recovered from the enemy, rendering invaluable services to the American Expeditionary Forces.
Upham, John S. R—Los Angeles, Calif. B—Fort Walla Walla, Wash. G. O. No. 14, W. D., 1923.	Lieutenant colonel (Infantry), General Staff Corps, U. S. Army. As assistant chief of staff, G–3, and acting chief of staff, 36th Division, during the organization and training of the division in the United States and in France. In these positions of great responsibility he displayed sound judgment, high professional skill, executive and administrative ability, and devotion to duty, his services contributing greatly to the successes of the 36th Division in its operations with the American Expeditionary Forces.
Upton, LaRoy S. R—Big Rapids, Mich. B—Decatur, Mich. G. O. No. 59, W. D., 1919. Distinguished-service cross also awarded.	Brigadier general, U. S. Army. He commanded with conspicuous ability the 9th Infantry in the trench sector south of Verdun and in all its operations before Chateau-Thierry. In the campaign north of Verdun, in October, as commander of the 57th Infantry Brigade, 29th Division, he exhibited qualities of brilliant leadership, successfully participating in the battles at Molleville Farm, Grand Montagne, Etraye, and in those east of the Meuse. At all times he remained near his front lines, personally directing the attacks and serving as a constant inspiration to his men.
Van Cise, Philip S. R—Denver, Colo. B—Deadwood, S. Dak. G. O. No. 56, W. D., 1922.	Lieutenant colonel (Infantry), General Staff Corps, U. S. Army. As assistant chief of staff, G–2, 81st Division, from September, 1918, until June, 1919, he displayed exceptional ability in the administration and operation of that section of the division staff. By his keen foresight, sound judgment, and military attainments of a high order he was able at all times to secure valuable information of the enemy and to keep his commanding general well informed, thereby contributing materially to the successful operations of his division.
Van Deman, Ralph H. R—Delaware, Ohio. B—Delaware, Ohio. G. O. No. 73, W. D., 1919.	Colonel (Infantry), General Staff Corps, U. S. Army. As chief of the military intelligence branch, General Staff, in organizing the Intelligence Service of the Army in the United States, to his ability, untiring zeal, and devotion to duty the building up of a very efficient Intelligence Service of the Army was largely due.
Vanderbilt, Cornelius. R—New York, N. Y. B—New York, N. Y. G. O. No. 118, W. D., 1919.	Brigadier general, U. S. Army. As commanding officer, 102d Engineers, and as Engineer officer of the 27th Division, his marked qualities of leadership and thorough training and instruction developed a high state of military efficiency in his command, as demonstrated throughout its entire service.
Vandervort, Lynnette L. R—Denver, Colo. B—La Salle, Ill. G. O. No. 9, W. D., 1923.	Chief nurse, Army Nurse Corps, U. S. Army. As chief nurse of the Mars Hospital Center, France, during the World War, she was largely responsible for the nursing care of thousands of sick and wounded at that center. Her work was characterized by great efficiency, tact, and good judgment. Later, as chief nurse of the Nurses' Embarkation Center at Vannes, France, she had under her care at one time as many as 1,100 nurses for whom she was responsible. Her efforts in this capacity to facilitate the work of demobilization, and to improve the general welfare of the nurses, contributed largely not only to the success of their concentration and organization for transfer to the United States but also to their high morale and physical well-being.
Van Horn, Robert O. R—Fort D. A. Russell, Wyo. B—Whipple Barracks, Ariz. G. O. No. 38, W. D., 1921.	Colonel, Signal Corps, U. S. Army. On the night of Nov. 3, 1918, he led his regiment, the 9th Infantry, 2d Division, against the enemy position in the edge of the Bois de Belval. The regiment passed through the woods and the enemy lines and took up a position 6 kilometers in rear of the enemy, capturing many prisoners and much war material. At daylight, Nov. 4, his regiment was heavily counterattacked but not dislodged. The effect of night penetration of the enemy lines caused the enemy on the right and left of the 2d Division sector to fall back to the east bank of the Meuse River.
Van Natta, Thomas F., Jr. R—St. Joseph, Mo. B—Atchison, Kans. G. O. No. 124, W. D., 1919.	Lieutenant colonel, Cavalry, U. S. Army. For services while serving as military attaché at Habana, Cuba.
Van Voorhis, Daniel. R—Zanesville, Ohio. B—Zanesville, Ohio. G. O. No. 69, W. D., 1919.	Colonel (Cavalry), General Staff Corps, U. S. Army. As chief of staff at the port of embarkation, Newport News, Va., his services in governing and controlling the troop-movement branch at the port of embarkation materially aided in the efficient transport of troops and supplies overseas.

VAUCLAIN, SAMUEL MATTHEWS
R—Rosemont, Pa.
B—Philadelphia, Pa.
G. O. No. 69, W. D., 1919.

He assisted in organizing the munitions standards board and was chairman of a subcommittee of that board which later became a subcommittee of the War Industries Board. He rendered valuable assistance in developing the War Department's program as to artillery and rifles.

VAUGHAN, VICTOR C.
R—Ann Arbor, Mich.
B—Mount Airy, Randolph County, Mo.
G. O. No. 69, W. D., 1919.

Colonel, Medical Corps, U. S. Army.
During his service in the office of the Surgeon General his contributions of advice and information have been of great value to the Army in connection with the control of communicable diseases. During the recent epidemic of influenza, in particular, his work was of extreme value.

VERDI, WILLIAM F.
R—New Haven, Conn.
B—Italy.
G. O. No. 27, W. D., 1922.

Major, Medical Corps, U. S. Army.
As surgical consultant and specialist in surgery of the chest in hospital formations at the front during the operations on the Marne and the St. Mihiel and Meuse-Argonne offensives.

VIDMER, GEORGE.
R—Mobile, Ala.
B—Mobile, Ala.
G. O. No. 16, W. D., 1920.
Distinguished-service cross also awarded.

Colonel, Infantry, U. S. Army.
As commander of the 306th Infantry, 77th Division, he demonstrated marked ability as a military leader. His sound judgment and tireless energy were largely responsible for the successes which his regiment gained in its operations against the enemy.

VINCETT, GEORGE H.
R—Butler, Pa.
B—Syracuse, N. Y.
G. O. No. 59, W. D., 1919.

Lieutenant colonel, Corps of Engineers, U. S. Army.
As chief of construction and operation of the car-erecting plant at La Rochelle he performed with credit a task of great magnitude. By his skill in organizing labor and ability in imbuing the men with enthusiasm, he was enabled to increase greatly the output of his plant. The persistent high quality of the duty performed by him greatly facilitated the major operations of the American armies in the field.

VORIS, ALVIN C.
R—Neoga, Ill.
B—Neoga, Ill.
G. O. No. 74, W. D., 1919.

Colonel, Signal Corps, U. S. Army.
As chief signal officer, successively, of the 1st Division, the 1st Army Corps, and the 3d Army, he rendered conspicuous services. With tireless energy and indefatigable zeal he performed a task of great magnitude, insuring at all times the installation and maintenance of communications throughout the Marne and Argonne-Meuse offensives, contributing in a marked degree to the successes attained.

WADE, LEIGH.
R—Cassopolis, Mich.
B—Cassopolis, Mich.
G. O. No. 14, W. D., 1925.
Act of Congress Feb. 25, 1925.

First lieutenant, Air Service, U. S. Army.
Lieutenant Wade, as pilot of airplane No. 3, the Boston, and supply officer of the United States Army Air Service around-the-world flight from Apr. 6, 1924, to Sept. 28, 1924, displayed to a remarkable degree courage, energy, and resourcefulness in carrying out these duties, in addition to actually piloting his airplane throughout the voyage. His sound judgment and foresight were material factors in contributing to the successful achievement of this pioneer flight of airplanes around the world. He has assisted materially in bringing a signal honor to himself and to the military forces of the United States.

WADHAMS, SANFORD H.
R—Torrington, Conn.
B—Torrington, Conn.
G. O. No. 59, W. D., 1919.

Colonel, Medical Corps, U. S. Army.
In his capacity as assistant to the Chief Surgeon, American Expeditionary Forces, and later as a member of the General Staff he ably supervised the hospitalization and evacuation activities of the Medical Corps in advanced areas. By his timely anticipation of requirements he assisted in a marked degree the support of our operations against the enemy.

WADSWORTH, ELIOT.
R—Boston, Mass.
B—Boston, Mass.
G. O. No. 95, W. D., 1919.

As vice chairman of the central committee, American Red Cross, he brought the great problem of systematized relief for our armies, those of the Allies, and for the stricken people of Europe to an eminently successful solution. By earnest, unselfish concentration of high faculties of organization and control he helped most materially to conserve life and reconstitute the wastage of war in the devastated areas, and made it possible to express the generosity of the American people in terms of substantial helpfulness.

WAHL, LUTZ.
R—Milwaukee, Wis.
B—Milwaukee, Wis.
G. O. No. 15, W. D., 1923.

Brigadier general, U. S. Army.
In command of the 58th Infantry, 4th Division, from Aug. 6, 1917, to Feb. 1, 1918, he demonstrated leadership of a high order, untiring energy, and sound judgment. As chief of the operations section, General Staff, War Department, from Feb. 4, 1918, to May 12, 1918, he displayed rare professional attainments, initiating and developing many valuable ideas in the organization of the operations section. As brigadier general commanding the 14th Infantry Brigade, 7th Division, from May 19, 1918, to Nov. 3, 1919, he again displayed unusual gifts of organization, leadership, and tactical judgment, both during the period of organization and training of his brigade, as well as in combat operations in France.

WAINER, MAX R.
R—Delaware City, Del.
B—Russia.
G. O. No. 59, W. D., 1919.

Lieutenant colonel, Quartermaster Corps, U. S. Army.
As assistant to the quartermaster at Nevers, by his zeal and rare talent for organization he contributed in a large measure to the prompt and efficient operation of the first advance supply depot of the American Expeditionary Forces. Later he proved himself sound in judgment and of exceptional ability when he organized and operated the classification depot at Blois. He showed marked discernment and determination in the reclassification and assignment of commissioned personnel, performing most exacting duties with brilliant success.

WAINWRIGHT, JONATHAN MAYHEW____
   R—Rye, N. Y.
   B—New York, N. Y.
   G. O. No. 55, W. D., 1920.

Lieutenant colonel, Inspector General's Department, U. S. Army.
As division inspector and more especially as an acting general staff officer of the 27th Division in the Dickebusch sector in Belgium, the Ypres-Lys offensive, and the battle of the La Selle River, in France, by his energy, efficient coordination of details, and persistent application to his task, he regulated all movements of the division, involving the evacuation of wounded, the relief of units of the line, the supplying of rations and ammunition, and the control of communications, with such marked success as incurred a minimum of loss in each operation.

WAINWRIGHT, JONATHAN M._____
   R—Chicago, Ill.
   B—Fort Walla Walla, Wash.
   G. O. No. 19, W· D., 1922.

Lieutenant colonel (Cavalry), General Staff Corps, U. S. Army.
As assistant chief of staff, 82d Division, first assistant to the assistant chief of staff, G-3, 3d Army, and later as assistant chief of staff, G-3, American Forces in Germany, by his untiring energy, devotion to duty, and exercise of initiative he contributed in a large measure to the success attained by the commands with which he served.

WAITE, HENRY M._____
   R—Dayton, Ohio.
   B—Toledo, Ohio.
   G. O. No. 22, W. D., 1920.

Colonel, Corps of Engineers, U. S. Army.
As deputy director general of transportation, headquarters, services of supply, later as constructing engineer of the Transportation Corps and deputy director general of transportation, Zone of the Armies, he displayed marked technical ability, initiative, and judgment of a high order. Subsequently, as a member of the bridgehead commission of the 3d Army, as chief motor transport officer of the 3d Army, and as advisor to the officer in charge of civil affairs at advanced general headquarters, he displayed those same high qualities which characterized his previous distinguished service.

WAITE, SUMNER_____
   R—Portland, Me.
   B—Highland Lake, Me.
   G. O. No. 56, W. D., 1922.

Major (Infantry), General Staff Corps, U. S. Army.
As assistant chief of staff, G-2, 37th Division, from Aug. 15, 1918, to Oct. 13, 1918, he organized and developed a splendid intelligence system by which he kept his division commander constantly well informed of the enemy on his front. By the skillful direction of the intelligence service he proved a material factor in the successes gained by his division. Aggressive and resourceful, he proved equal to every emergency. Later, as assistant chief of staff, G-3, of the same division, he demonstrated high professional attainments, sound tactical judgment, and keen farsightedness. He at all times displayed assiduous application to each important task, rendering services of signal worth and conspicuous merit in a position of great responsibility.

WALDRON, WILLIAM H._____
   R—Welch, W. Va.
   B—Huntington, W. Va.
   G. O. No. 19, W. D., 1922.
   Distinguished-service cross also awarded.

Colonel (Infantry), General Staff Corps, U. S. Army.
As chief of staff, 80th Division, during the Meuse-Argonne offensive his extraordinary energy, initiative, and ability contributed largely to the success of the operations of the division.

WALES, BOYD_____
   R—Howard, S. Dak.
   B—Brownville, Nebr.
   G. O. No. 59, W. D., 1921.

Colonel, Field Artillery, U. S. Army.
As commander of the 147th Regiment of Field Artillery of the 57th Field Artillery Brigade, 32d Division, he commanded his regiment with marked ability throughout the campaign of the Aisne-Marne, Oise-Aisne, and Meuse-Argonne. By his energy and devotion to duty he contributed materially to the success of the 57th Field Artillery Brigade during its support of the 32d Division, and at other times its support of five other divisions. He has rendered service of much value.

WALKE, WILLOUGHBY_____
   R—Norfolk, Va.
   B—Norfolk, Va.
   G. O. No. 133, W. D., 1919.

Colonel, Coast Artillery Corps, U. S. Army.
As commanding officer of the Middle Atlantic Coast Artillery District during the war, his services were conspicuous in the administration of that command and in the execution of all projects coming within his control for the organization and training of Coast Artillery Corps units for overseas service.

WALKER, GEORGE_____
   R—Baltimore, Md.
   B—York, S. C.
   G. O. No. 15, W. D., 1923.

Colonel, Medical Corps, U. S. Army.
As a member of Base Hospital No. 18, in the prevention of the spreading of diseases at the base ports, and later as clinical chief of genito-urinary section of the Medical Department in France, he rendered services of inestimable value to the Government. His untiring energy and unremitting devotion to duty, coupled with his technical knowledge as a professional urologist, were of material value and contributed markedly to the successful operations of the American forces in France.

WALKER, JOHN B._____
   R—New York, N. Y.
   B—Lodi, N. J.
   G. O. No. 31, W. D., 1922.

Colonel, Medical Corps, U. S. Army.
As commanding officer of Base Hospital No. 116, American Expeditionary Forces, and later as consultant in the United States during the period of demobilization. The services rendered by Colonel Walker in standardizing and supervising the treatment of the wounded suffering from gunshot fractures were of inestimable value to the Government and a material contribution to the rehabilitation of the disabled.

WALKER, KENZIE W._____
   R—Schulenburg, Tex.
   B—Pin Oak, Tex.
   G. O. No. 15, W. D., 1923.

Colonel, Finance Department, U. S. Army.
As assistant to the Chief of Finance, charged with the responsibility for the settlement of many thousands of claims of officers and men of the National Army, he displayed extraordinary administrative and executive ability, sound business judgment, unflagging energy, and devotion to duty. The services rendered the Government were of immeasurable value in a position of great responsibility.

WALKER, MERIWETHER L.
R—Lynchburg, Va.
B—Lynchburg, Va.
G. O. No. 78, W. D., 1919.

Brigadier general, U. S. Army.
As chief of the Motor Transport Service he rendered services of much value. With tireless energy he assailed an important task, and by his zealous efforts met all difficulties arising from irregular shipments and lack of adequate material, successfully organizing the Motor Transport Service, and brought it to a high state of efficiency, thereby materially assisting in the solution of the important problem of transportation in the American Expeditionary Forces.

WALKER, WILLIAM H.
R—Cambridge, Mass.
B—Pittsburgh, Pa.
G. O. No. 69, W. D., 1919.

Colonel, Chemical Warfare Service, U. S. Army.
His extraordinary technical ability, untiring industry, and great zeal have enabled remarkable results to be achieved in the Production Division of the Chemical Warfare Service in the face of many obstacles encountered.

WALLACE, FRED C.
R—McMinnville, Tenn.
B—McMinnville, Tenn.
G. O. No. 38, W. D., 1922.

Lieutenant colonel, Field Artillery, U. S. Army.
As inspector-instructor for the Chief of Field Artillery from Apr. 16, 1918, to Oct. 16, 1918, he rendered valuable service in raising the efficiency of Field Artillery brigades and in recommending the measures needful to be taken to prepare these brigades for service overseas.

WALLACE, WILLIAM B.
R—Marquette, Mich.
B—Canada.
G. O. No. 69, W. D., 1919.

Lieutenant colonel, Infantry, U. S. Army.
As a member of the American section, Supreme War Council, he has rendered invaluable service in handling with especial ability and good judgment matters of the greatest importance to all the allied and associated Governments.

WALSH, JAMES L.
R—Brookline, Mass.
B—Boston, Mass.
G. O. No. 38, W. D., 1922.

Colonel, Ordnance Department, U. S. Army.
First as chief of the personnel division, office of the Chief of Ordnance, which he organized and administered with conspicuous success during the first nine months of the war, a critical period during which all ordnance activities depended upon the successful handling of the personnel problem, and later as personal and executive assistant to the Chief of Ordnance, in which capacity his breadth of vision, tact, sound judgment, and loyalty were invaluable to the Government in the numerous highly confidential matters entrusted to his care. In each of these positions his services to the Government were exceptionally conspicuous and meritorious.

WALSH, ROBERT D.
R—Redwood City, Calif.
B—Alleghany, Calif.
G. O. No. 59, W. D., 1919.

Brigadier general, U. S. Army.
In command of the important base ports of St. Nazaire and Bordeaux, France, and as deputy director general of transportation, his services have been characterized by exceptional ability, energy, and devotion to duty.

WARBURTON, BARCLAY H.
R—Wyncote, Pa.
B—Philadelphia, Pa.
G. O. No. 38, W. D., 1922.

Major, Field Artillery, U. S. Army.
While serving as military attaché at Paris, France, by his devotion to duty, intelligent cooperation, and indefatigable efforts he rendered invaluable assistance and conspicuous service to the military representative of the United States on the Supreme War Council.

WARD, CABOT.
R—New York, N. Y.
B—New York, N. Y.
G. O. No. 59, W. D., 1919.

Lieutenant colonel (Air Service), General Staff Corps, U. S. Army.
As assistant chief of staff, in charge of the intelligence section of the Services of Supply, he has rendered services of the most valuable character. He has handled with great efficiency the important task of counterespionage throughout the American Expeditionary Forces and in the neighboring neutral countries. In this service he showed marked ability, combined with superior military knowledge.

WARD, FRANKLIN W.
R—Albany, N. Y.
B—Philadelphia, Pa.
G. O. No. 118, W. D., 1919.

Colonel, Infantry, U. S. Army.
For services as division adjutant and acting chief of staff of the 27th Division and as commanding officer of the 106th Infantry. As commanding officer, 106th Infantry, his personal courage, determination, and thoroughness in the handling of his regiment under heavy fire during the battle of the LeSelle River in the Somme offensive of October, 1918, were conspicuous.

WARD, RALPH T.
R—Denver, Colo.
B—Fayette, Mo.
G. O. No. 50, W. D., 1921.

Colonel (Corps of Engineers), General Staff Corps, U. S. Army.
As chief of the operations subsection G-3, 1st Army, Colonel Ward was given the responsibility of drawing up plans, preparing orders, making personal reconnaissances, and insuring mutual relations with adjacent armies. He fulfilled these functions with exceptional ability, and his work was largely responsible for the successes achieved during the St. Mihiel and Meuse-Argonne offensives.

WARFIELD, AUGUSTUS B.
R—Buffalo, N. Y.
B—Prattsburg, N. Y.
G. O. No. 9, W. D., 1923.

Colonel, Field Artillery, U. S. Army.
As commanding officer of the 322d Field Artillery, 83d Division, from August 22, 1917, until February 15, 1919, he displayed untiring energy, unusual administrative ability, and an unfailing dependability, these qualities being reflected in the excellence of his regiment. His outstanding ability as an organizer, his leadership and his devotion to duty were material factors in the successful operations of his division.

WARREN, CHARLES B.
R—Detroit, Mich.
B—Bay City, Mich.
G. O. No. 10, W. D., 1920.

Colonel, Judge Advocate General's Department, U. S. Army.
In connection with the administration of the selective-service law during the war, in all of his varied and important duties he displayed unselfish devotion, tireless energy, and extraordinary executive ability

WARREN, CHARLES ELLIOT _____ | Lieutenant colonel, Ordnance Department, U. S. Army.
   R—New York, N. Y. | While in charge of finances of the small arms division, Ordnance Office, where
   B—Brooklyn, N. Y. | his eminent ability as a financier and as an executive of large affairs, and also
   G. O. No. 27, W. D., 1922. | his previous military training, were invaluable in the early organization of
  | the division. Later, as chief of the small-arms section, procurement division,
  | Ordnance Office, and as one of the members and vice governor, the war credits
  | board, Office of the Secretary of War, he rendered conspicuous service in the
  | conduct of its immense affairs.

WASHBURN, FREDERIC A _____ | Lieutenant colonel, Medical Corps, U. S. Army.
   R—Boston, Mass. | As commanding officer of Base Hospital No. 6, American Expeditionary Forces,
   B—New Bedford, Mass. | and as surgeon of Base Section No. 3, positions of great responsibility, by his
   G. O. No. 59, W. D., 1921. | ability, energy, and whole-hearted devotion to duty he has rendered services
  | of great value.

WATKINS, LEWIS H _____ | Colonel (Corps of Engineers), General Staff Corps, U. S. Army.
   R—Franklin, Tenn. | As assistant chief of staff, G-5, 1st Army, he performed exacting duties with
   B—Nashville, Tenn. | marked energy and ability, achieving valuable results. Notwithstanding
   G. O. No. 62, W. D., 1919. | his many duties, he arranged to aid G-3, 1st Army, in the preparation of
  | plans for important operations. By his especial ability, military attainments,
  | and painstaking devotion to the tasks assigned to him he contributed in a
  | marked degree to the successes achieved by our troops.

WATSON, ERNEST E _____ | Major, Infantry, 341st Machine Gun Battalion, 89th Division, U. S. Army.
   R—St. Paul, Minn. | For services near Romagne, France, in October, 1918, in organizing the machine-
   B—Augusta, Ky. | gun defense of the 89th Division sector.
   G. O. No. 53, W. D., 1921. |

WATT, DAVID A _____ | Lieutenant colonel, Adjutant General's Department, U. S. Army.
   R—Hasbrouck Heights, N. J. | As adjutant, port of embarkation, Hoboken, N. J., Sept. 3, 1917, to Sept. 10,
   B—Sandusky, Ohio. | 1918, and Jan. 6, 1919, to Feb. 28, 1920, his untiring energy, resourcefulness,
   G. O. No. 16, W. D., 1923. | and devotion to duty, coupled with administrative and executive ability of
  | an exceptionally high order, were of immeasurable value to the Government
  | in a position of great responsibility.

WEBB, GEORGE H _____ | Colonel, Corps of Engineers, U.S. Army.
   R—Detroit, Mich. | He was intrusted with the execution of some of the largest construction enter-
   B—Dubuque, Iowa. | prises in France. Confronted by difficulties of labor, material, and equipment
   G. O. No. 59, W. D., 1919. | he set about his task with ceaseless energy, and by his resourcefulness, initia-
  | tive, and skill he overcame all obstacles and completed these difficult projects
  | with great success.

WEED, FRANK W _____ | Lieutenant colonel, Medical Corps, U. S. Army.
   R—Baltimore, Md. | In August, 1917, as sanitary inspector at Camp Funston, Kans., he initiated
   B—Baltimore, Md. | and perfected the organization and establishment of a standardized type of
   G. O. No. 9, W. D., 1923. | detention and quarantine camp, the successful operation of which resulted
  | in the installation of similar camps in all large cantonments throughout the
  | United States during the war. This original and constructive work of his
  | had a marked influence in controlling epidemic diseases, then prevalent, and
  | greatly facilitated the rapid mobilization and training of urgently needed
  | man power. From January until August, 1918, as general sanitary inspector,
  | Surgeon General's Office, he rendered services of the highest order. Later,
  | while on duty in the hospital division of the chief surgeon's office, American
  | Expeditionary Forces, as transportation officer in charge of hospital trains,
  | ambulances, and the movement of sick and wounded within the American
  | Expeditionary Forces to the United States during the period from January
  | to July, 1919, he directed the evacuation of over 100,000 sick and wounded to
  | the United States.

WEEKS, ALANSON _____ | Major, Medical Corps, U. S. Army.
   R—San Francisco, Calif. | During the World War, as surgical consultant and director of surgical teams in
   B—Allegan, Mich. | hospital formations at the front during the operations on the Marne, the St.
   G. O. No. 16, W. D. 1923. | Mihiel, and Meuse-Argonne offensives, and later in command of Base Hos-
  | pital No. 30, by his loyal devotion to duty, sound judgment, and brilliant
  | professional attainments, he rendered services of great value in the care of the
  | sick and wounded of the American troops, thereby contributing materially
  | to the success of the American Expeditionary Forces.

WEEMS, FONTAINE CARRINGTON _____ | Lieutenant colonel, General Staff Corps, U. S. Army.
   R—Washington, D. C. | As chief of the foreign relations section of the Purchase, Storage and Traffic
   B—Houston, Tex. | Division of the General Staff, he foresaw the necessity of the preservation of
   G. O. No. 14, W. D., 1920. | accurate data affecting the international relations of the War Department in
  | the matter of purchase of supplies by the allied Governments in the United
  | States, and by his prevision and care prevented the loss of information essential
  | to just and speedy liquidation. Thereafter, in association with those charged
  | with the settlement of widely ramifying, intricate, and involved business
  | relations, by his judgment, industry, and knowledge he made possible speedy
  | and just settlements, reflecting a high degree of credit upon the American
  | Army for its accuracy and fairness in business transactions with its allies.

WEIGEL, WILLIAM _____ | Major general, U. S. Army.
   R—New Brunswick, N. J. | As commander of a brigade of the 28th Division in the fighting on the Vesle of
   B—New Brunswick, N. J. | August, 1918, he inspired confidence by his constant activities and his aggres-
   G. O. No. 12, W. D., 1919. | sive pressing of the enemy at every opportunity, which resulted in driving
  | the hostile forces across the Vesle northward toward the Aisne.

WELBORN, IRA C. R—Mico, Miss. B—Mico, Miss. G. O. No. 18, W. D., 1919. Medal of honor also awarded.	Colonel, Tank Corps, U. S. Army. For services in the organization and administration of the Tank Corps.
WELCH, WILLIAM H. R—Baltimore, Md. B—Norfolk, Conn. G. O. No. 69, W. D., 1919.	Colonel, Medical Corps, U. S. Army. From his rich experience in scientific medicine, sanitation, public health, and medical education he helped materially in guiding the medical profession both in and out of the Army safely through the many difficulties of war.
WELD, DE WITT C., Jr. R—Brooklyn, N. Y. B—Brooklyn, N. Y. G. O. No. 56, W. D., 1922.	Colonel, Field Artillery, U. S. Army. As regimental commander of the 105th Field Artillery, 27th Division, he demonstrated professional attainments and ability of the highest order. By his sound tactical judgment and superior knowledge of artillery, he most successfully directed his units in support of the 158th Infantry Brigade in the operations north of Verdun, Nov. 4 to 11, 1918. By keeping his elements close to the attacking Infantry he contributed in no small measure to the success of the Infantry brigade in these operations. He rendered services of conspicuous merit and signal worth to the American Expeditionary Forces.
WELLES, EDWARD M., Jr. R—New York, N. Y. B—Addison, N. Y. G. O. No. 14, W. D., 1923.	Lieutenant colonel, Medical Corps, U. S. Army. In the office of the Chief Surgeon, American Expeditionary Forces, for nearly two years and embracing the entire period of combat activities, he was charged with all details concerning the reception and distribution and the classification and assignment of all officers, nurses, and enlisted men of the Medical Department serving overseas, a force aggregating approximately 250,000 individuals. In this position of great responsibility he displayed exceptional ability and rendered conspicuous service to the Government by directing with the greatest economy the distribution of all available personnel during periods of stress and threatened shortage, thereby materially contributing to the success of our forces in the field.
WELLS, BRIANT H. R—Salt Lake City, Utah. B—Salt Lake City, Utah. G. O. No. 62, W. D., 1919.	Brigadier general, U. S. Army. As chief of staff of the 4th Army Corps while it was in the front line in the Woevre he displayed military attainments of a high order in the planning of operations. Both then and subsequently, during the march to the Rhine and the occupation of German territory, his service was marked by tireless zeal, excellent judgment, and whole-hearted devotion to the performance of important tasks.
WELLS, FREDERICK B. R—Minneapolis, Minn. B—France. G. O. No. 69, W. D., 1919.	Colonel, Quartermaster Corps, U. S. Army. In the organization and operation of the entire storage system for the Army he has displayed marked ability, energy, and application, to which are due, in a large measure, the satisfactory results attained.
*WELSH, ROBERT S. R—Sault Ste. Marie, Mich. B—Canada. G. O. No. 50, W. D., 1919.	Colonel, Field Artillery, U. S. Army. He commanded the 314th Field Artillery, 80th Division, which later became part of the 3d Army Corps. He rendered exceptionally efficient service with the 80th Division, taking part in all operations of that division. He displayed a high order of leadership and exhibited those masterful qualities of a commander which insure success. Later, assigned to the 3d Army Corps, his devotion to duty and high professional attainments were again revealed. Posthumously awarded. Medal presented to widow, Mrs. Eleanor E. Welsh.
WELSH, WILLIAM E. R—Hanover, Pa. B—Hanover, Pa. G. O. No. 15, W. D., 1923.	Brigadier general, U. S. Army. As colonel, 346th Infantry, 87th Division, from September, 1917, to June, 1918, he demonstrated unusual leadership, organizing and training his regiment to a high state of efficiency and morale; as brigadier general, General Staff, and inspector-instructor of Infantry, training section, General Headquarters, American Expeditionary Forces, he displayed marked tactical ability and by his general supervision shared largely in the responsibility for the training in that arm. His duties of very great importance were carried out with conspicuous success.
WELSHIMER, ROBERT R. R—Neoga, Ill. B—Neoga, Ill. G. O. No. 3, W. D., 1921.	Colonel, Coast Artillery Corps, U. S. Army. As senior instructor at the Coast Artillery School, and later as commandant of that school in the organization and administration of that institution so as to result in effective accomplishment of its object.
WENTZ, DANIEL B. R—Philadelphia, Pa. B—Jeddo, Pa. G. O. No. 15, W. D., 1923.	Lieutenant colonel, Quartermaster Corps, U. S. Army. While in charge of the fuel branch, office of the chief quartermaster, American Expeditionary Forces, a position of great responsibility, he displayed extraordinary ability in the promptness with which he procured and forwarded a steady flow of fuel to the American Expeditionary Forces.
WESSON, CHARLES M. R—Centerville, Md. B—St. Louis, Mo. G. O. No. 49, W. D., 1922.	Colonel, Ordnance Department, U. S. Army. As commanding officer, Watertown Arsenal, Mass., from January to October, 1918, by his indefatigable energy, great administrative ability, and thorough technical knowledge he planned, erected, equipped, and brought to a highly efficient working basis a new factory for the manufacture of 240-mm. howitzer carriages, as well as a new forging plant for large-caliber guns—a definite contribution to the military power of the Nation. From November, 1918, until August, 1919, as commanding officer of the ordnance base repair shops at Mehun-sur-Yevre, France, he again rendered highly meritorious service in a position of great responsibility in salvaging ordnance matériel valued at millions of dollars and prepared it properly for shipment to the United States.

WESTERVELT, WILLIAM I.
R—Corpus Christi, Tex.
B—Corpus Christi, Tex.
G. O. No. 59, W. D., 1919.

Brigadier general, U. S. Army.
As assistant to the Chief of Artillery, through his initiative, organizing ability, and comprehensive knowledge of the technique and tactics of Artillery in all its branches, and particularly through his complete knowledge of Artillery material, he has rendered services of exceptional value to the Government.

*WESTNEDGE, JOSEPH B.
R—Kalamazoo, Mich.
B—Kalamazoo, Mich.
G. O. No. 70, W. D., 1919.

Colonel, Infantry, U. S. Army.
With signal ability he commanded the 126th Infantry, 32d Division, from the date of its organization to its final engagement during the Meuse-Argonne offensive, inspiring the members of his command by his personal courage and indefatigable zeal; he kept his regiment efficiently organized at all times, as demonstrated by the successful results obtained in its operations against the enemy. During his service at the front he contracted a disease which subsequently proved fatal.
Posthumously awarded. Medal presented to widow, Mrs. Eva M. Westnedge.

WESTOVER, OSCAR.
R—Bay City, Mich.
B—Bay City, Mich.
G. O. No. 14, W. D., 1923.

Lieutenant colonel, Air Service, U. S. Army.
He served in turn as signal officer, port of embarkation, Hoboken, N. J., chief of storage department, Signal Corps, and chief of storage and traffic division, Bureau of Aeronautical Production, Air Service. By his great initiative, painstaking attention to details, exceptional ability, and untiring efforts he installed and developed with conspicuous success at all ports of embarkation a complete system of keeping records of shipment of Signal Corps and Air Service property for overseas. His services were of inestimable value to the Government in a position of great responsibility.

WHALEY, ARTHUR M.
R—Sault Sainte Marie, Mich.
B—Canada.
G. O. No. 56, W. D., 1922.

Colonel, Medical Corps, U. S. Army.
He served with marked ability as surgeon, 30th Division, from the time of its organization and early training period to the completion of the Ypres-Lys and Somme offensives. The care and evacuation of the wounded during the active operations of the division were conducted with the greatest smoothness and efficiency, and it was due to the great energy and conspicuous ability displayed by him that his services were of the utmost value to the division.

WHEELER, CHARLES B.
R—Fergus Falls, Minn.
B—Mattison, Ill.
G. O. No. 56, W. D., 1922.

Brigadier general, U. S. Army.
He initiated, organized, and developed the plans for the successful operation of the supply division in the office of the Chief of Ordnance, which division received, transported, warehoused, issued, and maintained all items of ordnance stores and equipment manufactured and purchased for issue to the Army during the war. With fine business acumen and with the full conception of the magnitude and intricacy of the complex problems involved, he brought to full and complete fruition a well-balanced and successful working organization. By his wide vision and full comprehension of conditions and the needs of the service and by his unflagging energy to insure constant supply of ordnance materials, he rendered service of signal worth to the Government in a position of great responsibility.

WHEELER, RAYMOND A.
R—Peoria, Ill.
B—Orchard Mines, Ill.
G. O. No. 68, W. D., 1920.

Colonel, Corps of Engineers, U. S. Army.
As active regimental commander of the 4th Engineers, 4th Division, during the Aisne-Marne, the St. Mihiel, and the Meuse-Argonne offensives, he ably supported the 4th Division in these operations by the promptness and skill with which he constructed bridges across the Vesle, destroyed enemy wire, and built and maintained roads during the attacks in the Meuse-Argonne offensive. His able and expeditious support of the 3d and 5th Army Corps by constructing roads through the Argonne was a material factor in the rapid advance and ultimate success of the units of those corps during this important operation.

WHIPPLE, SHERBURNE.
R—Springfield, Mass.
B—Cold Spring, N. Y.
G. O. No. 49, W. D., 1922.

Lieutenant colonel, Infantry, U. S. Army.
As assistant chief of staff, G-1, of the 80th Division from June until December, 1918, he performed his duties with marked ability in connection with the service of supply and communications for his division. By his tireless energy, exceptional administrative ability, initiative, and sound judgment he successfully solved many perplexing problems, maintaining at all times an adequate supply of food and ammunition for the troops, thereby rendering valuable services to the American Expeditionary Forces.

WHITE, HERBERT A.
R—Plymouth, Iowa.
B—Worth County, Iowa.
G. O. No. 15, W. D., 1921.

Colonel, Judge Advocate General's Department, U. S. Army.
As acting Judge Advocate General for the American Expeditionary Forces and later for the American Forces in Germany and France he performed very difficult and exacting duties with marked skill and distinction. In connection with the vast civil business of the War Department which passed through his hands he displayed a singular force of decision and sound judgment.

WHITE, HERBERT H.
R—Boise, Idaho.
B—Boston, Mass.
G. O. No. 59, W. D., 1919.

Lieutenant colonel (Field Artillery), General Staff Corps, U. S. Army.
As executive officer of the 4th Section, General Staff, General Headquarters, American Expeditionary Forces, he was intimately associated with the organization of the Services of Supply and their direction. By his energy, ability, and good judgment in the discharge of important and arduous duties he greatly assisted in the successful operations of the Services of Supply in support of the forces in the field.

WHITEHEAD, HENRY C.
R—Hemphill, Tex.
B—Hemphill, Tex.
G. O. No. 59, W. D., 1919.

Colonel, Signal Corps, U. S. Army.
During the period of organization of the American Expeditionary Forces he rendered service of a superior order in the planning and the organization of the Air Service. As chief of staff, Air Service, he displayed sound judgment and great ability in solving the many problems with which he was confronted. Throughout the entire duration of the war his high professional attainments and untiring zeal have materially promoted the efficiency of the Air Service.

WHITFIELD, ROBERT.................
   R—Milledgeville, Ga.
   B—Milledgeville, Ga.
   G. O. No. 89, W. D., 1919.

Colonel (Infantry), General Staff Corps, U. S. Army.
While on duty with the operations branch of the operations division, General Staff, he was charged with a multitude of exacting and very responsible duties, all of which he performed with conspicuous accuracy and thoroughness.

WHITLEY, FRANKLIN L.................
   R—St. Louis, Mo.
   B—St. Louis, Mo.
   G. O. No. 27, W. D., 1922.

Lieutenant colonel, Infantry, U. S. Army.
In 1917, before instruction pamphlets were issued, Colonel Whitley prepared combat drill formations, suitable for war strength companies armed with new weapons. As battalion commander during the operations near Chateau-Thierry, he rendered valuable service. Due to his initiative and personal leadership, 39 days of constant contact with the enemy failed to break the morale of his organization. After the armistice, as chief of the decorations division, General Headquarters, American Expeditionary Forces, by his sound judgment, professional knowledge, and exceptional ability, this officer performed his manifold and responsible duties with the utmost efficiency. He has rendered services of material worth to the American Expeditionary Forces.

WHITMAN, WALTER M.................
   R—New York, N. Y.
   B—New York, N. Y.
   G. O. No. 89, W. D., 1919.
   Distinguished-service cross also awarded.

Colonel, Infantry, U. S. Army.
He commanded with marked distinction the 325th Infantry, 82d Division, throughout its period of service in France. An able and aggressive leader, he achieved eminent success in all the missions assigned to him, contributing materially to the achievements of his division.

WHITSON, MILTON J.................
   R—Seattle, Wash.
   B—Scott County, Iowa.
   G. O. No. 89, W. D., 1919.

Colonel, Quartermaster Corps, U. S. Army.
While officer in charge of the building branch of the construction division of the Army, Colonel Whitson's task was of staggering magnitude, and its successful accomplishment was in a great measure due to his qualities of organization, leadership, technical knowledge, and untiring energy.

WICKERSHAM, CORNELIUS W.................
   R—Cedarhurst, Long Island, N. Y.
   B—Greenwich, Conn.
   G. O. No. 38, W. D., 1921.

Major (Infantry), General Staff Corps, U. S. Army.
As acting assistant chief of staff, G-3, 4th Army Corps, and as assistant to the assistant chief of staff, G-3, 4th Army Corps, in the preparation and execution of the 4th Army Corps attack at St. Mihiel. Subsequently he was one of the principal officers to organize the 2d Army Headquarters.

WICKES, FORSYTH.................
   R—Tuxedo Park, N. Y.
   B—New York, N. Y.
   G. O. No. 62, W. D., 1919.

Major, Infantry, U. S. Army.
He showed rare ability in the preliminary organization of the American liaison service and wide comprehension of the importance of forward interallied liaison. While attached to French divisions in liaison with the 1st American Division he performed exacting duties of a delicate nature with energy and tact, achieving signal success. He aided materially in the maintenance of cordial relations between the French and American military authorities, his service being continuously marked by ability, sound judgment, and devotion to duty.

WIGMORE, JOHN H.................
   R—Chicago, Ill.
   B—San Francisco, Calif.
   G. O. No. 10, W. D., 1920.

Colonel, Judge Advocate General, U. S. Army.
In connection with the administration of the selective-service law during the war, he originated and put into execution an excellent system of classification of registrants and his sound judgment and ability for analysis contributed materially to the success of the department.

WILBY, FRANCIS B.................
   R—Arlington, Mass.
   B—Detroit, Mich.
   G. O. No. 14, W. D., 1923.

Colonel, Corps of Engineers, U. S. Army.
As assistant in charge of military engineering in the office of the Chief Engineer, American Expeditionary Forces, and later as division engineer of the 1st Division, he displayed unusual ability and professional attainments of a high order. As editor of the Engineer Field Notes, and as the author of a large number of them, his clear conception of the functions and duties of Engineer troops was most firmly impressed upon the combat engineers and contributed in a signal manner to their marked efficiency. By his rare technical skill and knowledge, keen adaptability to all conditions, he contributed materially to the success of the 1st Division in a position of great responsibility and in times and circumstances of the gravest importance.

WILGUS, WILLIAM J.................
   R—New York, N. Y.
   B—Buffalo, N. Y.
   G. O. No. 50, W. D., 1919.

Colonel, Corps of Engineers, U. S. Army.
As delegate of the special railway commission, Director General of Military Railways and Deputy Director General of Transportation. In all of these positions he has demonstrated exceptional ability and untiring energy. The foundation of the Army Transportation Service was largely due to his vision and remarkable judgment. He has shown a degree of devotion to duty far above any calls which would have been made upon him by military authority.

WILKINS, HARRY E.................
   R—Victor, Iowa.
   B—Genesee, Ill.
   G. O. No. 77, W. D., 1919.

Brigadier general, U. S. Army.
For services while in charge of the general supply depot, New York City.

WILLCUTT, JOSEPH N.................
   R—Cohasset, Mass.
   B—Cohasset, Mass.
   G. O. No. 95, W. D., 1919.

Colonel, Quartermaster Corps, U. S. Army.
As officer in charge of the construction of the National Guard camps he displayed qualities of leadership, energy, administrative ability, and devotion to duty which rendered possible the housing of the National Guard troops in an incredibly short space of time. Later he served with conspicuous success as chief of the procurement branch of the Construction Division of the Army.

WILLIAMS, ALEXANDER E.
  R—Little River Academy, N. C.
  B—Cumberland, N. C.
  G. O. No. 43, W. D., 1922.

Colonel, Quartermaster Corps, U. S. Army.
As chief quartermaster, Army of Occupation, he displayed untiring zeal and administrative ability of the highest order in the organization and operation of the supply system of the 3d Army. By his sound judgment, initiative, and resourcefulness he solved many perplexing problems of supply and finance in a most satisfactory manner, thereby effecting a great saving for the United States.

WILLIAMS, CLARENCE C.
  R—Nacoochee, Ga.
  B—Nacoochee, Ga.
  G. O. No. 12, W. D., 1919.

Major general, U. S. Army.
An officer of high professional attainments, who rendered particularly valuable services in the organization of the Ordnance Department of the American Expeditionary Forces and exhibited unusual ability in arranging for the procurement of ordnance material and ammunition for the American Army in Europe.

WILLIAMS, EZEKIEL J.
  R—Barnesville, Ga.
  B—Sparks, Ga.
  G. O. No. 89, W. D., 1919.

Colonel (Infantry), General Staff Corps, U. S. Army.
He served with distinction as chief of staff of the 36th Division from the date of its organization to the date of departure from France. He performed his manifold duties with unflagging energy and notable ability, rendering services of striking value to the American Expeditionary Forces.

WILLIAMS, HARRY C.
  R—New Town Landing, Miss.
  B—New Town Landing, Miss.
  G. O. No. 56, W. D., 1922.

Colonel, Field Artillery, U. S. Army.
As commanding officer, 320th Field Artillery, 82d Division, in the organization and training of the regiment and in its very successful operations against the enemy in the Aisne-Marne, St. Mihiel, and Meuse-Argonne offensives he displayed tireless energy, keen devotion to duty, and eminent technical skill as an artillerist, gave most effective support to the Infantry of the 82d and 80th Divisions, and very materially contributed to the successes attained by those units.

WILLIAMS, HERBERT O.
  R—Tupelo, Miss.
  B—Fulton, Miss
  G. O. No. 103, W. D., 1919.

Brigadier general, U. S. Army.
As an officer of the Inspector General's Department his rare efficiency, fearlessness, and good judgment in the inspection of large commands and in the investigation and solution of intricate problems presenting unusual difficulties have been of the greatest value and have materially facilitated the operations of the War Department and of the Army during the emergency.

WILLIAMS, RICHARD H.
  R—Jersey City Heights, N. J.
  B—Jersey City, N. J.
  G. O. No. 62, W. D., 1919.

Colonel (Coast Artillery Corps), General Staff Corps, U. S. Army.
As G-2, 1st Army Corps, he displayed rare ability in the organization and administration of that section, being tireless in the energy with which he handled each problem during successive offensives. Later as G-2, 3d Army, he achieved brilliant successes when confronted with duties of a most exacting and difficult nature, accomplishing all by his zeal and ability.

WILLIAMS, RICHARD H., Jr
  R—Mendham, N. J.
  B—New York, N. Y.
  G. O. No. 19, W. D., 1922.

Lieutenant colonel, Quartermaster Corps, U. S. Army.
In the remount service, American Expeditionary Forces, through his farsightedness he saw the necessity for and by his untiring effort succeeded in expediting the obtaining of animals for the American Expeditionary Forces that were of vital importance for our Army.

WILLIAMSON, SYDNEY B.
  R—New York, N. Y.
  B—Lexington, Va.
  G. O. No. 15, W. D., 1923.

Colonel, Corps of Engineers, U. S. Army.
As section engineer, intermediate section west, Services of Supply, France, he constructed hospitals, depots, camps, and miscellaneous structures. He displayed rare technical skill, broad vision and business experience, untiring energy and devotion to duty, contributing in a material way to the successful operations of the American forces in France.

WILLIFORD, FORREST E.
  R—Bayle City, Ill.
  B—Coffeen, Ill.
  G. O. No. 56, W. D., 1922.

Colonel, Coast Artillery Corps, U. S. Army.
He served with marked efficiency as director of the trench artillery school at Langres and commandant of the trench artillery center at Vitrey. Later, as chief of the trench artillery section in the office of the Chief of Artillery, American Expeditionary Forces, he initiated the plans of and controlled the training of this important branch of the Artillery arm with exceptional ability, rendering services of inestimable value to the American Expeditionary Forces.

WILLS, DAVIS B.
  R—Charlottesville, Va.
  B—Charlottesville, Va.
  G. O. No. 62, W. D., 1919.

Major, U. S. Marine Corps.
As chief paymaster of the United States Marine Corps, he performed arduous and complex duties under most trying conditions. Displaying rare initiative and administrative ability, he organized and conducted his department in such a manner as to relieve combat units of a mass of detail and administrative work. He was tireless in devotion to duty, able in its execution.

WILLS, VAN LEER
  R—Grand Rapids, Mich.
  B—Davidson County, Tenn.
  G. O. No. 15, W. D., 1923.

Colonel (Infantry), General Staff Corps, U. S. Army.
As assistant chief of staff, G-3, 92d Division, from Sept. 9 to Nov. 9, 1918; as deputy chief of staff, 1st Army, American Expeditionary Forces, from Nov. 11, 1918, to Apr. 20, 1919, and as acting chief of staff for various periods, his duties involving the direction of reequipment and supply of the 1st Army units which marched into Germany with the 3d American Army; the direction of the policing of the 1st Army's battlefields and the withdrawal of the 1st Army to rest areas; planning and supervising the training, recreation, and vocational training of the 1st Army while in rest areas awaiting transportation home; as assistant to chief of staff, G-4, Services of Supply, from April, 1919, to August, 1919, and assistant chief of staff, G-4, Services of Supply, from August, 1919, to October, 1919, he supervised the liquidation and disposal of the vast supplies involved in the dissolution of the Services of Supply. He displayed rare initiative, outstanding administrative and executive ability, and unremitting devotion to duty in these positions of grave responsibility, contributing signally to the successful repatriation of the American Army and the prompt and effective liquidation of the affairs of the American Expeditionary Forces.

WILMER, WILLIAM H.
R—Washington, D. C.
B—Powhatan County, Va.
G. O. No. 59, W. D., 1919.

Colonel, Medical Corps, U. S. Army.
As surgeon in charge of medical research laboratories, Air Service, American Expeditionary Forces, since September, 1918, he has rendered most distinguished service. His thorough knowledge of the psychology of flying officers and the expert tests applied efficiently and intelligently under his direction have done much to decrease the number of accidents at the flying schools in France and have established standards and furnished indications which will be of inestimable value in all future work to determine the qualifications of pilots and observers. The data collected by him is an evidence of his ability, his painstaking care, and of his thorough qualifications for the important work intrusted to him. The new methods, instruments, and appliances devised under his direction for testing candidates for pilots and observers have attracted the attention and been the subject of enthusiastic comment by officers of the allied services and will be of great importance in promoting the safety and more rapid development of aerial navigation.

WILSON, GEORGE K.
R—Pueblo, Colo.
B—Denver, Colo.
G. O. No. 50, W. D., 1919.

Colonel, Infantry, U. S. Army.
As assistant chief of staff in charge of the administrative section of the General Staff, first of a division, later a corps, and finally of an army, he displayed marked ability in every capacity in which he was employed. By his thorough knowledge and grasp of his duties he became a material factor in the successful operations of his several departments.

WILSON, HENRY B.
R—Camden, N. J.
B—Camden, N. J.
G. O. No. 56, W. D., 1921.

Vice admiral, U. S. Navy.
While stationed at Brest, in the capacity of commander of the United States Naval Forces in France, where he showed a keen appreciation of the necessity for the closest cooperation between the military and naval services, his valuable cooperation has to a great extent made possible the prompt functioning of the port of Brest.

WILSON, JAMES S.
R—Baltimore, Md.
B—San Francisco, Calif.
G. O. No. 56, W. D., 1922.

Colonel, Medical Corps, U. S. Army.
As Chief Surgeon of the American Expeditionary Forces in Siberia, he organized, supervised, and perfected the organization of the Medical Department in Siberia so as to meet successfully the complex sanitary conditions confronting the American troops. To his excellent judgment, untiring efforts, and high professional attainments is largely due the splendid efficiency which characterized the work of the Medical Department under his control. He handled in a masterful manner the organization of available sanitary forces to combat a threatened typhus epidemic in eastern Siberia. He rendered conspicuous service of signal worth to the Government in a position of great responsibility.

WILSON, LOUIS B.
R—Rochester, Minn.
B—Pittsburgh, Pa.
G. O. No. 27, W. D., 1922.

Colonel, Medical Corps, U. S. Army.
As assistant to the director of laboratories and infectious diseases, American Expeditionary Forces, he organized most efficiently a pathological service throughout the American Expeditionary Forces in France that was of inestimable value to the medical and surgical services.

WILSON, WALTER K.
R—Nashville, Tenn.
B—Nashville, Tenn.
G. O. No. 18, W. D., 1919

Colonel (Coast Artillery Corps), General Staff Corps, U. S. Army.
In the organization and administration of the cable service of the War Department in the United States, thereby enabling that service to meet the excessive demands made upon it during the war.

WILSON, WILLIAM H.
R—Cincinnati, Ohio.
B—Mount Vernon, N. Y.
G. O. No. 62, W. D., 1919.

Colonel (Coast Artillery Corps), General Staff Corps, U. S. Army.
He displayed extensive scientific knowledge, together with a keen practical grasp of conditions, as artillery inspector with the first battalion of American Railway Artillery in action against the enemy. As a member of the training section, he was at all times energetic and tactful in the supervision of training of railway, tractor, trench, and antiaircraft artillery. As its executive officer, he organized and conducted an item of the general system of the training section, being tireless in devotion to his important duties.

WINANS, EDWIN B.
R—Hamburg, Mich.
B—Hamburg, Mich.
G. O. No. 59, W. D., 1919.

Brigadier general, U. S. Army.
He showed marked efficiency and excellent judgment while commanding the 64th Infantry Brigade, 32d Division, in the actions at the second Battle of the Marne, in the attack and capture of Juvigny, and in the operations at Bois-de-la-Morine, Bois-de-Chene Sec, and Bantheville Woods. In these actions, by his tactical ability, he was always master of the situation and executed his plans with a confidence that was an inspiration to his troops.

WINGATE, GEORGE ALBERT.
R—Brooklyn, N. Y.
B—Brooklyn, N. Y.
G. O. No. 126, W. D., 1919.

Brigadier general, U. S. Army.
In command of the 52d Field Artillery Brigade, 27th Division, he served with marked distinction in the St. Mihiel operation, displaying military attainments of a high order. In the Meuse-Argonne offensive he proved himself possessed of exceptionally tactical ability, working with untiring energy that the Infantry might have all the advantages of Artillery support. With sound judgment, unusual foresight, and wide comprehension of conditions and facilities available, he conducted operations in that offensive with brilliant success, repeatedly solving the difficult problems incident thereto.

WINN, CHARLES D.
R—Paris, Ky.
B—Winchester, Ky.
G. O. No. 89, W. D., 1919.

Colonel, Field Artillery, U. S. Army.
As commanding officer of the 306th Field Artillery, 77th Division, he displayed high qualities of leadership. Maintaining a high standard of efficiency and morale in his regiment, he constantly kept his command in close proximity to the attacking infantry, furnishing it accurate and timely support, furthering its rapid advance, and contributing to the successes gained.

WINN, FRANK L.
R—Winchester, Ky.
B—Winchester, Ky.
G. O. No. 62, W. D., 1919.

Major general, U. S. Army.
As commander of the 177th Infantry Brigade and later of the 89th Division, he displayed military attainments of a high order and achieved signal successes. In the St. Mihiel and Meuse-Argonne offensives he accompanied the assaulting battalions and placed them on their objectives, inspiring all by his personal courage and gaining their confidence by his exceptional tactical skill and ability as a leader. At all times he was tireless in energy, showing keen judgment and initiative in handling difficult situations.

WINSHIP, BLANTON.
R—Macon, Ga.
B—Macon, Ga.
G. O. No. 19, W. D., 1920.
Distinguished-service cross also awarded.

Colonel, Judge Advocate General's Department, U. S. Army.
He served with distinction as judge advocate of the 42d Division and of the 1st Army. As commanding officer of the 110th Infantry, 28th Division, he displayed marked qualities of leadership. Later, as judge advocate of the Services of Supply, and as chief of the rents, requisitions, and claims service, he displayed professional attainments and judgment of a high order, contributing, in no small degree, to the success of the operations during the war and afterwards in the liquidation of our affairs in France.

WINSLOW, E. EVELETH.
R—Boston, Mass.
B—Washington, D. C.
G. O. No. 47, W. D., 1919.

Colonel, Corps of Engineers, U. S. Army.
While in charge of the military section of the office of the Chief of Engineers during the early period of the war his services were marked by the energy, zeal, and good judgment which were essential to the procurement of personnel and equipment and the organization and training of engineer organizations for oversea service.

WINTER, FRANCIS A.
R—St. Louis, Mo.
B—St. Francisville, La.
G. O. No. 59, W. D., 1919.

Brigadier general, U. S. Army.
As chief surgeon of the lines of communication, American Expeditionary Forces, from June to December, 1917, he organized medical units at the base ports and in camps in France. He established large supply depots, from which medical supplies were distributed to the American Expeditionary Forces, and by keen foresight and administrative ability made these supplies at all times available for our armies.

WISE, FREDERIC MAY.
R—Baltimore, Md.
B—Brooklyn, N. Y.
G. O. No. 39, W. D., 1920.

Colonel, U. S. Marine Corps.
He commanded with skill, ability, and gallantry the 59th Infantry, 4th Division, from Sept. 4, 1918, to Jan. 23, 1919. During the St. Mihiel offensive he personally directed the attack of his regiment against Manheulles and Fresnes-en-Woevre, which resulted in the capture of the enemy's line in this area. On Sept. 26, 1918, he directed the attack of his regiment which resulted in the capture of the Bois-de-Brieulles. From Sept. 26 to Oct. 21, 1918, his personal courage and aggressive attitude was an important factor in the successful operations of the 8th Infantry Brigade against Bois-de-Brieulles, Bois-du-Fays, Bois-de-Malaumont, Bois-de-Peut, and Bois-de-Foret. He has rendered services of signal worth to the American Expeditionary Forces.

WITTENMYER, EDMUND.
R—Dunbarton, Ohio.
B—Buford, Ohio.
G. O. No. 12, W. D., 1919.

Major general, U. S. Army.
He served with marked distinction as brigade commander in the Argonne-Meuse offensive and as division commander in the final operations in the Toul sector, and in both capacities, by his untiring efforts and breadth of vision, proved himself to be an able leader.

WOLF, PAUL A.
R—Kewanee, Ill.
B—Kewanee, Ill.
G. O. No. 59, W. D., 1919.

Brigadier general, U. S. Army.
In the attacks on Bois-de-Forges, St. Hilaire, Bois-de-Warville, and Bois-des Hautes-Epines, France, in September and October, 1918, the conspicuous success of the brigade was due to his splendid leadership and skill.

WOLFE, EDWIN P.
R—New York, N. Y.
B—Page County, Iowa.
G. O. No. 69, W. D., 1919.

Colonel, Medical Corps, U. S. Army.
He systematized and controlled the distribution of medical supplies with so much foresight and good judgment that his service was able to meet promptly all the emergencies in the United States as they occurred.

WOLFE, SAMUEL H.
R—New York, N. Y.
B—Baltimore, Md.
G. O. No. 43, W. D., 1922.

Colonel, Quartermaster Corps, U. S. Army.
As officer in charge of insurance matters, cantonment division, Quartermaster General's Office, by his unusual constructive ability, foresight, and familiarity with large financial problems he rendered conspicuous service resulting in the saving of large sums to the Government. As a member of the committee on labor of the advisory commission of the Council of National Defense, he again rendered invaluable services in the preparation of necessary legislation to provide for the dependents of enlisted personnel of the Army and Navy, which later became the war risk insurance act. In October, 1917, he demonstrated exceptional ability and resourcefulness in the organization and operation of the War Risk Insurance Bureau in France and England. Later, as assistant director and executive officer in the office of the Director of Finance, his thorough knowledge of financial problems proved of the greatest assistance to the Director of Finance and of inestimable value to the Government.

WOOD, LEONARD.
R—Boston, Mass.
B—Winchester, N. H.
G. O. No. 47, W. D., 1919.
Medal of honor also awarded.

Major general, U. S. Army.
As a department, division, and camp commander during the war, he has displayed qualities of leadership and professional attainments of a high order in the administration and training of his various commands, and has furthered in every way during the war the system of officers' training schools.

WOOD, ROBERT E.
R—Kansas City, Mo.
B—Kansas City, Mo.
G. O. No. 19, W. D., 1919.

Brigadier general, U. S. Army.
For services in connection with the reorganization and operation of the Services of Supply of the Army.

WOOD, WILLIAM T.
   R—Danville, Ill.
   B—Irving, Ill.
   G. O. No. 77, W. D., 1919.

Brigadier general, U. S. Army.
For services as senior assistant to the Inspector General of the Army.

WOOD, WINTHROP S.
   R—Farmington, Me.
   B—Washington, D. C.
   G. O. No. 38, W. D., 1922.

Colonel, Quartermaster Corps, U. S. Army.
In charge of the general supply depot, Jeffersonville, Ind., from April, 1917, to May, 1918, the successful organization, development, and administration of the system at that important depot for the supply of clothing and general equipment were largely due to his great energy, foresight, and marked executive ability. Later, as quartermaster, Base Section No. 6, American Expeditionary Forces, from August, 1918, to January, 1919, by his administrative ability and untiring zeal he rendered conspicuous services in the improvement and development of the organization of the supply system at this important depot.

WOODRUFF, JAMES A.
   R—Burke, Vt.
   B—Fort Shaw, Mont.
   G. O. No. 59, W. D., 1919.

Colonel, Corps of Engineers, U. S. Army.
He organized and commanded the 10th Forestry Engineers with marked ability. In spite of the difficult situations confronting him he developed the Forestry Service to a marked degree of excellence. By his great energy and devotion to duty he rendered service of the highest character to the Government.

WOODS, ARTHUR.
   R—New York, N. Y.
   B—Boston, Mass.
   G. O. No. 15, W. D., 1923.

Colonel, Air Service, U. S. Army.
As inspector of schools, Signal Corps, and then as chief of Personnel, Division of Military Aeronautics, Air Service, by his executive ability, clear conception, and broad mental grasp, he handled with conspicuous success many perplexing problems in the organization and administration of the system for assigning personnel. From November, 1918, until January, 1919, as assistant director of military aeronautics, in the solution of many new and intricate problems concerning demobilization and reorganization his work was characterized by sound judgment and untiring zeal. Later as special assistant to the Secretary of War in matters pertaining to securing employment for discharged soldiers he rendered valuable service in placing great numbers of these men in lucrative positions.

WOODS, GILBERT F.
   R—Chicago, Ill.
   B—Clarksville, Mo.
   G. O. No. 16, W. D., 1920.

Director of Real Estate Service.
He rendered invaluable service to the War Department in the acquisition, either by purchase, condemnation, requisition, donation, or lease, of all real estate required for the use of the Army during the World War, also in the disposal of such real estate as was no longer required. By his technical knowledge, broad judgment, and energetic action valuable results were obtained.

WOOLDRIDGE, JESSE W.
   R—San Francisco, Calif.
   B—Hopkinsville, Ky.
   G. O. No. 35, W. D., 1920.
   Distinguished-service cross also awarded.

Captain, Infantry, U. S. Army.
Near Mezy, France, July 15, 1918, when attacked by portions of three enemy regiments, Captain Wooldridge, by exceptional skill and ability, so inspired his company that he defeated these units and drove them by successive counter attacks from the sectors of his regiment and that of an adjoining regiment. After his company had suffered a loss of 70 per cent by casualties, he organized a platoon from cooks, mess attendants, runners, and Stokes mortar men, and led it in attack upon the last enemy assault wave, which he defeated. During these successive encounters his company captured over 400 of the enemy and broke the strong enemy attempt to cross the Marne in this sector.

WOOTEN, WILLIAM P.
   R—La Grange, N. C.
   B—La Grange, N. C.
   G. O. No. 95, W. D., 1919.

Colonel, Corps of Engineers, U. S. Army.
He served with credit as commanding officer of the 14th Railway Engineers during the operations of that regiment on the British front. Subsequently, while corps engineer of the 3d Army Corps, by his energy, foresight, and skill in accomplishing important engineering works, he contributed materially to the successful operations of his corps. Later, when appointed engineer of the 3d Army, he performed important duties in a most creditable manner.

WORCESTER, PHILIP H.
   R—Portland, Me.
   B—Norfolk, Va.
   G. O. No. 60, W. D., 1920.

Colonel, Coast Artillery Corps, U. S. Army.
As ordnance officer, Army Artillery, 1st Army, during the St. Mihiel and Meuse-Argonne offensives, by his untiring energy and loyal devotion to duty he organized and successfully administered the ammunition and ordnance supplies of the Army Artillery. He at all times displayed sound judgment and military attainments of a high order. He has rendered services of signal worth to the American Expeditionary Forces.

WRIGHT, JOHN W.
   R—Washington, D. C.
   B—Kirkwood, Mo.
   G. O. No. 103, W. D., 1919.

Colonel (Infantry), General Staff Corps, U. S. Army.
As assistant chief of staff, G-3, at Headquarters, Services of Supply, he was charged with the important duty of directing the movement of troop arrivals, billeting, and the supply of initial equipment to units. He at all times displayed indefatigable zeal and administrative ability of a high order, rendering services of inestimable value to the American Expeditionary Forces.

WRIGHT, WILLIAM M.
   R—Newark, N. J.
   B—Newark, N. J.
   G. O. No. 12, W. D., 1919.

Major general, U. S. Army.
He commanded in turn the 35th Division, the 3d, 5th, and 7th Army Corps, under the 8th French Army in the Vosges Mountains, and later commanded the 89th Division in the St. Mihiel offensive and in the final operations on the Meuse River, where he proved himself to be an energetic and aggressive leader.

WYLLIE, Robert E.
   R—Sanford, Fla.
   B—India.
   G. O. No. 47, W. D., 1919.

Colonel (Coast Artillery Corps), General Staff Corps, U. S. Army.
In assisting in organizing the first group of General Staff officers that ultimately developed into the operations branch and the equipment branch of the operations division of the General Staff. As chief assistant and later as head of the equipment branch his services were conspicuously useful to the Government and to the Army.

YARDLEY, HERBERT O. R—Washington, D. C. B—Worthington, Ind. G. O. No. 56, W. D., 1922.	Major (Signal Corps), General Staff Corps, U. S. Army. For services as chief of the communication section of the Military Intelligence Division, War Department General Staff, during the World War.
YEATMAN, POPE. R—Philadelphia, Pa. B—St. Louis, Mo. G. O. No. 15, W. D., 1923.	Director of the nonferrous metals section, War Industries Board. In connection with the operations of the War Industries Board during the World War, in his position as director of one of the sections of the board he rendered, through his broad vision, distinguished capacity, and business ability, services of inestimable value in marshaling the industrial forces of the Nation and mobilizing its economic resources—marked factors in assisting to make military success attainable. As director of nonferrous metals, he rendered, through his untiring efforts and devotion to duty, exceptionally valuable service to the War Department in connection with the procurement and supply of copper, lead, zinc, and other nonferrous metals for the Army.
YOUNG, HUGH HAMPTON. R—Baltimore, Md. B—San Antonio, Tex. G. O. No. 50, W. D., 1919.	Colonel, Medical Corps, U. S. Army. He has, by his constant application, tireless energy, and foresight, lowered the nonefficiency rate of combat organizations, due to certain contagious diseases far below prewar anticipations, and has thereby aided in the conservation of man power to a degree never before attainable.
YOUNGBERG, GILBERT A. R—Cannon Falls, Minn. B—Bellcreek, Minn. G. O. No. 59, W. D., 1919.	Colonel (Corps of Engineers), General Staff Corps, U. S. Army. He served as representative of the Engineer Department and later as the principal assistant to the chief of the 4th Section, General Staff, American Expeditionary Forces. He performed duties of the greatest importance in connection with construction projects of the Army. By his high professional attainments and tireless energy, his sound judgment and logical recommendations on questions of construction, supply, and transportation he materially assisted in the successes of our forces in the field. In all matters he displayed remarkable ability and rendered services of the highest character to the Government.
YOUNGER, JOHN. R—Columbus, Ohio. B—Scotland. G. O. No. 2, W. D., 1920.	For services as advisory engineer in the designing and production of standard motor vehicles adopted by the United States of America.
ZALINSKI, MOSES GRAY. R—Rochester, N. Y. B—Seneca Falls, N. Y. G. O. No. 56, W. D., 1921.	Colonel, Quartermaster Corps, U. S. Army. As quartermaster, Base Section No. 2, Bordeaux, a position of great responsibility, due to his long quartermaster experience, marked ability, and knowledge of the methods and standards of supply, he performed the numerous duties of his important office with great success. He has rendered service of much value to the United States.
ZANETTI, JOAQUIN E. R—New York, N. Y. B—San Domingo, West Indies. G. O. No. 56, W. D., 1922.	Lieutenant colonel, Chemical Warfare Service, U. S. Army. As chief liaison officer of the Chemical Warfare Service with the French forces, his untiring energy, thorough familiarity with the French language and methods, and his superior technical ability enabled him to gather an enormous amount of detailed information concerning the manufacture, handling, and use of gases, which were of inestimable value to our Government in the manufacture and supply of chemical warfare materials in the United States.
ZIEGAUS, IRVIN W. R—Olympia, Wash. B—Sharon, Wis. G. O. No. 56, W. D., 1922.	Captain, Infantry, U. S. Army. As executive officer in charge of the selective draft in Washington, by his unusual executive ability, rare tact and skill, great initiative and resourcefulness exercised at times under most trying and novel conditions which arose in connection with the administration of the selective service act, he achieved a pronounced and conspicuous success in the performance of difficult and highly responsible duties, thereby rendering services of great value to the Government.
ZINSSER, HANS. R—New York, N. Y. B—New York, N. Y. G. O. No. 10, W. D., 1922.	Lieutenant colonel, Medical Corps, U. S. Army. While acting as sanitary inspector of the 2d Army he organized, perfected, and administered with extraordinary and exceptional success a plan of military sanitation and epidemic-disease control.

BELGIANS

[Awarded for exceptionally meritorious and distinguished services in a position of great responsibility, under the provisions of the act of Congress approved July 9, 1918]

ALBERT_____
R—Belgium.
G. O. No. 121, W. D., 1919.

King of the Belgians.
To this distinguished soldier, Commander in Chief of the Belgian Army, this medal is presented as an expression of the high regard of the people of the United States and of their Army for the distinguished and patriotic service which he has rendered to the common cause on the battle fields of Europe.

ARNOULD, HENRY_____
R—Belgium.
G. O. No. 87, W. D., 1919.

Lieutenant general, Belgian Army.
He served with marked distinction as chief of artillery of the Belgian Army, rendering invaluable service in the conduct of operations against the enemy. At all times he showed zeal and devotion to duty, his high military attainments having marked effect in the successes achieved by the allied armies.

BARBIER, PHILIPPE_____
R—Belgium.
G. O. No. 29, W. D., 1919.

Lieutenant, Belgian Army.
For services rendered the United States Army while serving as acting military attaché to the Belgian Legation, Washington, D. C.

BERNHEIM, LOUIS_____
R—Belgium.
G. O. No. 87, W. D., 1919.

Lieutenant general, Belgian Army.
In command of the 1st Belgian Army Division he achieved most valuable results by his brilliant leadership. He prosecuted the operations against the enemy with judgment and vigor and his service was marked by signal success.

BIEBUYCK, A_____
R—Belgium.
G. O. No. 45, W. D., 1919.

Lieutenant general, commanding the 6th Belgian Army Corps, Belgian Army.
For services rendered to the American Expeditionary Forces and to the cause in which the United States has been engaged.

CABRA, ALPHONSE F. E_____
R—Belgium.
G. O. No. 3, W. D., 1922.

Lieutenant general, Belgian Army.
For services with the American forces in Germany.

CORNELLIE, EMILE F_____
R—Belgium.
G. O. No. 126, W. D., 1919.

Colonel, Belgian Army.
As commander of the Belgian naval base in Antwerp, Belgium, he has rendered conspicuous service to the United States. His great energy and sound judgment has been an important factor in the success of the operations of American Base Section No. 9, at Antwerp.

CUMONT, EUGENE F. M. H_____
R—Belgium.
G. O. No. 45, W. D., 1919.

Major, Belgian Army, Belgian representative on the Military Board of Allied Supply.
For services rendered to the American Expeditionary Forces and to the cause in which the United States has been engaged.

DE CEUNINCK, ARMAND_____
R—Belgium.
G. O. No. 87, W. D., 1919.

Lieutenant general, Belgian Army.
In command of the 4th Belgian Army Division he conducted operations against the enemy with signal success, displaying remarkable qualities of leadership and untiring devotion to his manifold duties. His services were of inestimable value.

DE GOLS, ISADORE_____
R—Belgium.
G. O. No. 126, W. D., 1919.

Major, Belgian Army.
As a member of the Interallied Commission on the Repatriation of Prisoners of War he has rendered highly meritorious service to the United States and allied Governments in connection with the repatriation of American and allied prisoners released by the armistice. He always displayed a cheerful and active interest in all that pertained to their welfare and rendered sympathetic and practical cooperation.

DELOBBE, H_____
R—Belgium.
G. O. No. 87, W. D., 1919.

General major, Belgian Army.
As chief of a staff section at Belgian General Headquarters he rendered most distinguished services. He displayed the highest military attainments and great zeal in the direction of operations against the enemy.

DE PAGE, ANTOINE_____
R—Belgium.
G. O. No. 72, W. D., 1920.

Colonel, surgeon, Belgian Army.
A distinguished surgeon and one of the pioneers in developing the modern treatment of battle casualties, he placed his eminent talents and extensive experience at the disposition of the medical department of the American Expeditionary Forces, and at all times lent his cooperation toward improving the treatment of the wounded. At his hospital at Le Panne, Belgium, he took an active personal interest in training medical officers of the American Army in the advances being made in battle surgery. Under his able supervision and guidance the observation and experience gained by these officers eventually resulted in saving the lives of many American wounded.

DEVEZE, ALBERT J. C_____
R—Belgium.
G. O. No. 19, W. D., 1922.

Captain, Belgian Army.
For services rendered to the allied cause and to the American forces in Germany.

D'OULTREMONT, GUY D R—Belgium. G. O. No. 126, W. D., 1919.	Commandant, Belgian Army. The same high character of services rendered by him from 1914 to 1918 with the Belgian Artillery in the field were again revealed during the period which he was attached to the Belgian Mission at American General Headquarters. As principal assistant to the chief of the Belgian Mission at American General Headquarters he rendered services of great value to the American Expeditionary Forces. By able advice and sound judgment, coupled with loyal support, he assisted us in all problems presented to him.
DRUBBEL, HONORE R—Belgium. G. O. No. 87, W. D., 1919.	Lieutenant general, Belgian Army. In command of the 2d Belgian Army Division he showed eminent qualities of leadership, at all times using his genius for military tactics to the best advantage in the operations against the enemy. His brilliant achievements had an important bearing upon the successful conduct of the war.
DU BOIS, AUGUSTO, D. J. A. M R—Belgium. G. O. No. 30, W. D., 1921.	Colonel, Belgian Army, Chief of Staff, Belgian Army of Occupation. For services rendered to the allied cause and to the American forces in Germany.
GILLAIN, C. C. V R—Belgium. G. O. No. 111, W. D., 1919.	Lieutenant general, Belgian Army, Chief of the General Staff of the Belgian Army. As an expression to him of the high regard of the people of the United States and of their Army for the distinguished and patriotic services which he has rendered to the common cause in which he has been associated on the battle fields of Europe.
GREINDL, LEON M R—Belgium. G. O. No. 87, W. D., 1919.	General major, Belgian Army. As chief of a staff section at Belgian General Headquarters he rendered invaluable service in the direction of the most important engineering operations. Confronted by stupendous tasks he performed all with distinction, showing exact scientific knowledge and great zeal in the performance of his arduous duties.
HEMELEERS-SHENLEY, LEON A. H R—Belgium. G. O. No. 126, W. D., 1919.	Captain, Belgian Army. As principal assistant to the chief of the Belgian Mission at American General Headquarters he rendered services of great value to the American Expeditionary Forces. By able advice and sound judgment, coupled with loyal support, he assisted us in all problems presented to him. Zealous in his efforts and wholehearted in cooperation, he at all times promoted the friendly relations between the Belgians and Americans.
JACQUES, JULES M. A R—Belgium. G. O. No. 17, W. D., 1919.	Lieutenant general, Belgian Army, in command of the 3d Belgian Army Division. For services rendered to the American Expeditionary Forces and to the cause, in which the United States has been engaged.
JUNGBLUTH, HARRY R—Belgium. G. O. No. 87, W. D., 1919.	Lieutenant general, Belgian Army. As Adjutant General, Chief of the Military Household of the King he occupied with distinction one of the most important offices in the Belgian Army. He displayed the highest military attainments and his sound advice was of inestimable value in the prosecution of the war against the enemy.
LEMAN, GEORGES R—Belgium. G. O. No. 39, W. D., 1920.	Lieutenant general, Belgian Army. To this distinguished officer this medal is presented as an expression of the high regard of the people of the United States and of their Army for the distinguished and patriotic services which he has rendered to the common cause on the battle fields of Europe.
MAGLINSE, HENRY H R—Belgium. G. O. No. 87, W. D., 1919.	General major, Belgian Army. As chief of a staff section at Belgian General Headquarters he rendered invaluable services in the direction of operations against the enemy. At all times he displayed the highest military attainments, untiring energy, and zeal in the performance of his distinguished duties.
MAHIEU, LOUIS R—Belgium. G. O. No. 126, W. D., 1919.	General major, Belgian Army. As Military Governor of the Province of Antwerp he has rendered conspicuous service to the United States. His hearty cooperation was of the greatest value to our forces in the establishment of American Base Section No. 9.
MERCHIE, SYLVIAN R—Belgium. G. O. No. 87, W. D., 1919.	General major, Belgian Army. As chief of staff in the office of the Belgian Secretary of War he rendered most important service in the prosecution of operations against the enemy. His high professional attainments furthered his rapid promotion in active service with troops, fitting him for the distinguished duties to which he was called later.
MICHEL, AUGUSTIN E R—Belgium. G. O. No. 87, W. D., 1919.	Lieutenant general, Belgian Army. In command of the Belgian Army of Occupation he performed his important duties with the greatest distinction, at all times displaying marked qualities of leadership and sound judgment at critical periods.
OSTERRIETH, LEON R—Belgium. G. O. No. 11, W. D., 1919.	Major, Belgian Army. For services rendered the United States Army while serving as chief of the Belgian Military Mission to the United States and acting military attaché to the Belgian Legation, Washington, D. C.

RUQUOY, L. H. R—Belgium. G. O. No. 17, W. D., 1919.	Lieutenant general, Belgian Army, in command of the 5th Belgian Army Division. For services rendered to the American Expeditionary Forces and to the cause in which the United States has been engaged.
SEGERS, PAUL. R—Belgium. G. O. No. 15, W. D., 1923.	Minister of Transportation, Belgium. Instrumental in the cession by the Belgian Government of steam motive power at a time when the lack of locomotives was a grave and serious handicap to the successful operations of the American Expeditionary Forces, he assisted in an important and conspicuous way in the rapid and efficient functioning of the lines of communication of the American Army.
TILKENS, AUGUST. R—Belgium. G. O. No. 87, W. D., 1919.	Colonel, Belgian Army. After serving with distinction in the command of troops he rendered most valuable services as aide-de-camp to the King of the Belgians. As member of the King's military household he showed high military attainments, and his advice proved uniformly sound.
TINANT, JULES T. A. E. L. R—Belgium. G. O. No. 45, W. D., 1919.	Major, Belgian Army, Chief, Belgian Mission, General Headquarters, American Expeditionary Forces. For services rendered to the American Expeditionary Forces and to the cause in which the United States has been engaged.
VAN DE VYVERE, A. R—Belgium. G. O. No. 9, W. D., 1923.	Minister of Finance, Belgium. A grave situation having arisen with reference to a lack of steam motive power for the American Expeditionary Forces, Mr. Van de Vyvere succeeded in providing the American forces through the Belgian Government with a large number of locomotives, thus contributing in a conspicuous way to the successful operations of the allied cause.
WARNANT, URSMAR A. R—Belgium. G. O. No. 59, W. D., 1921.	Captain, Belgian Army. For services as chief of the Belgian Mission attached to the American forces in France.

BRITISH

[Awarded for exceptionally meritorious and distinguished services in a position of great responsibility, under the provisions of the act of Congress approved July 9, 1918]

ADYE, JOHN. R—England. G. O. No. 14, W. D., 1920.	Major general, British Army. His services were of conspicuous merit as a member of the committee on return of prisoners of war. His large experience in the Near East, combined with rare judgment and quiet forcefulness, made his work especially valuable in securing the rapid return of American and allied prisoners and in alleviating their condition.
ALLENBY, EDMUND H. H. R—England. G. O. No. 126, W. D., 1919.	General, British Army. The American Army will ever remember his valiant services as commander of the 3d British Army in France from 1914 to 1917 and for his marvelous successes in Palestine, rendering services of distinction to the allied cause.
ANDERSON, STUART M. R—England. G. O. No. 126, W. D., 1919.	Brigadier general, British Army, Royal Artillery, 1st Australian Division, British Expeditionary Forces. The accurate and highly efficient support by the Artillery under his command contributed materially to the successful assaults on the Hindenburg line by the 30th U. S. Division, Sept. 29, 1918. The part played by him in achieving that success has won for him the deepest gratitude and admiration on the part of the American officers and soldiers with whom he was cooperating during the great advance.
ATKINSON, CHARLES F. R—England. G. O. No. 45, W. D., 1919.	Major, British Army, instructor, Army Intelligence School, American Expeditionary Forces. For services performed for the American Expeditionary Forces and to the cause in which the United States has been engaged.
AULD, SAMUEL J. M. R—England. G. O. No. 45, W. D., 1919.	Major, 4th Battalion, Royal Berkshire Regiment, British Army. For services rendered the U. S. Army while serving as liaison officer between the British and American Chemical Warfare Services.
BALFOUR, ALFRED G. R—England. G. O. No. 46, W. D., 1920.	Brigadier general, British Army. As commander of the embarkation area of the port of Southampton, by his earnest spirit of cooperation with the American authorities at the port of Southampton, he rendered valuable assistance, offering them every facility at his command for the proper handling of the important problem of troop shipments.
BEADON, ROGERS H. R—England. G. O. No. 45, W. D., 1919.	Colonel, British Army, British section, Supreme War Council. For services rendered to the American Expeditionary Forces and to the cause in which the United States has been engaged.
BESSELL-BROWNE, ALFRED J. R—England. G. O. No. 126, W. D., 1919.	Brigadier general, British Army, 5th Australian Division, British Expeditionary Forces. The relations of the American and Australian forces, cooperating with each other, were always marked with the utmost harmony, but never was this spirit more clearly manifested than during the period from Oct. 16 to 19, 1918, when the 30th U. S. Division went forward in attack against the enemy supported by artillery under his command. To his consummate technical skill and unflagging energy is due much credit for the successful results attained in these operations.
BETHELL, HUGH K. R—England. G. O. No. 126, W. D., 1919.	Brigadier general, British Army. He commanded with distinction the 66th British Division, with which the 27th American Division was affiliated during its training period. By his broad-minded policy, personal interest, and careful supervision he gave great assistance to the division, enabling the American troops to readily adapt themselves to the methods and conditions of service with a foreign army.
BIRCH, JAMES F. N. R—England. G. O. No. 45, W. D., 1919.	Lieutenant general, British Army, chief of artillery, British Expeditionary Forces. For services rendered to the American Expeditionary Forces and to the cause in which the United States has been engaged.
BIRDWOOD, WILLIAM R. R—England. G. O. No. 17, W. D., 1919.	General, British Army, commanding 5th Army, British Expeditionary Forces. For services rendered to the American Expeditionary Forces and to the cause in which the United States has been engaged.
BLAKE, DAVID V. J. R—England. G. O. No. 45, W. D., 1919.	Major, British Army, commanding officer, 3d Squadron, Australian Flying Corps, British Expeditionary Forces. For services performed for the American Expeditionary Forces and to the cause in which the United States has been engaged.
BOND, FRANCIS G. R—England. G. O. No. 45, W. D., 1919.	Major general, director of quarterings, British Army. For services performed for the American Expeditionary Forces and to the cause in which the United States has been engaged.

BONHAM-CARTER, CHARLES
R—England.
G. O. No. 87, W. D., 1919.

Brigadier general, British Army.
As general officer in charge of training at British General Headquarters during the period the 2d American Corps was in the British Expeditionary Forces, he rendered exceptional service to the U. S. Army. His knowledge of training methods was extensive, and with loyal cooperation he gave us the benefit of his experience.

BOWDLER, BASIL W. B.
R—England.
G. O. No. 126, W. D., 1919.

Lieutenant colonel, British Army.
While he was on duty in the intelligence section at British General Headquarters and as chief of the information section, his assistance and advice in inspecting and training American officers were of distinct advantage to the American Expeditionary Forces. He displayed military attainments of a high order, and at all times cooperated with the American military authorities whole-heartedly.

BOWLBY, ANTHONY A.
R—England.
G. O. No. 124, W. D., 1919.

Major general, British Army.
An eminent consulting surgeon, and while serving with the British Expeditionary Forces in France, with untiring zeal he devoted his time and energy toward cooperating with and unreservedly placing at the disposal of the American Expeditionary Forces his eminent talents, broad experience, and knowledge of general conditions in preventing wastage among our forces from wounds and disease. His research work in wound bacteriology and evacuation resulted in the saving of many lives among our sick and wounded.

BOYCE, CHARLES E.
R—England.
G. O. No. 45, W. D., 1919.

Major, British Army, attached as staff officer, units of Royal Field Artillery, serving with the 2d Army Corps, American Expeditionary Forces.
For services performed for the American Expeditionary Forces and to the cause in which the United States has been engaged.

BRIDGES, GEORGE T. M.
R—England.
G. O. No. 11, W. D., 1919.

Lieutenant general, British Army.
For services rendered the U. S. Army while serving as the military representative of the British Mission to the United States (April, 1917), and later (1918) as chief of British Military Missions to the United States.

BURGESS, WILLIAM L. H.
R—England.
G. O. No. 126, W. D., 1919.

Brigadier general, British Army.
He commanded with distinction the 4th Australian Artillery Division while it was in support of the 27th American Division during the operations near St. Souplet, east of the La Selle River. His consummate skill as an artillerist and forceful determination in keeping his batteries well to the front were most potent factors in the successes achieved. The services which he rendered to the American Expeditionary Forces were of inestimable value.

BURTCHAELL, CHARLES H.
R—England.
G. O. No. 87, W. D., 1919.

Lieutenant general, British Army.
As Director General of Medical Service, British Expeditionary Forces, he displayed untiring zeal, eminent talents, and broad experience in providing adequate hospitalization and evacuation facilities for the sick and wounded of the American troops serving with the British Armies. His individual efforts counted largely in enabling the American Medical Service to function efficiently.

BUSH, W. A.
R—England.
G. O. No. 87, W. D., 1919.

Captain, British Army, instructor, American Expeditionary Forces Gas Defense School, Hanlon Field, France.
At the request of the American Expeditionary Forces he was detailed as instructor in gas defense, and rendered services of extraordinary merit at the school at Hanlon Field. Largely as the result of his energy, skill, and exact knowledge the school accomplished a most important mission, providing a course of instruction which, when put in practice in the field, prevented many casualties.

BUTLER, RICHARD H. K.
R—England.
G. O. No. 45, W. D., 1919.

Lieutenant general, British Army, commanding 3d Army Corps, British Expeditionary Forces.
For services rendered to the American Expeditionary Forces and to the cause in which the United States has been engaged.

BYNG, JULIAN H. G.
R—England.
G. O. No. 17, W. D., 1919.

General, British Army, commanding 3d Army, British Expeditionary Forces.
For services rendered to the American Expeditionary Forces and to the cause in which the United States has been engaged.

CAMPBELL, RONALD B.
R—England.
G. O. No. 45, W. D., 1919.

Lieutenant colonel, British Army, deputy inspector of bayonet and physical Training, British Expeditionary Forces.
For services performed for the American Expeditionary Forces and to the cause in which the United States has been engaged.

CARTER, EVAN E.
R—England.
G. O. No. 45, W. D., 1919.

Major general, British Army, Director General of Forage and Supplies, British Expeditionary Forces.
For services performed for the American Expeditionary Forces and to the cause in which the United States has been engaged.

CASSEL, FELIX
R—England.
G. O. No. 46, W. D., 1920.

As Judge Advocate General of the British forces, he ably assisted the American authorities with whom he came in contact. By his sound advice and able counsel on all questions affecting military justice he rendered services of great value to the American Expeditionary Forces.

CHARTERIS, JOHN
R—England.
G. O. No. 126, W. D., 1919.

Colonel, British Army, bureau of intelligence, British Expeditionary Forces.
As brigadier general of intelligence at British General Headquarters he rendered most valuable service to the American Expeditionary Forces during the early period of the organization of the staff by placing every facility at the disposal of the American staff officers who were sent to British General Headquarters for the purpose of instruction. He displayed military attainments of a high order, tireless energy, and marked zeal in the performance of his exacting duties. He was at all times tactful and proved himself a loyal friend.

CHURCHILL, WINSTON_____
R—England.
G. O. No. 62, W. D., 1919.

He rendered the Allied cause service of inestimable value. As British Minister of Munitions, he was confronted with a task of great magnitude. With ability of a high order, energy, and marked devotion to duty, he handled with great success the trying problems with which he was constantly confronted. In the performance of his great task he rendered valuable service to the American Expeditionary Forces.

CLARKE, TRAVERS E_____
R—England.
G. O. No. 45, W. D., 1919.

Lieutenant general, British Army, quartermaster general, British Expeditionary Forces.
For services rendered to the American Expeditionary Forces and to the cause in which the United States has been engaged.

CORNWALL, JAMES H. M_____
R—England.
G. O. No. 126, W. D., 1919.

Lieutenant colonel, British Army.
While a member of the intelligence section at British General Headquarters he rendered us very valuable and distinguished services by his hearty cooperation in placing at the disposal of the American officers sent to British General Headquarters his experience and profound knowledge regarding the work of intelligence. He was at all times tactful and helpful, proving himself a loyal friend.

COVELL, FRANK C_____
R—England.
G. O. No. 126, W. D., 1919.

Captain, British Army.
As liaison officer at general headquarters, Base Section No. 3, in London, he performed his duties with great tact, skill, and indefatigable zeal. The harmony which constantly characterized the relations of the American authorities with the British War Office was due, in no small degree, to his faithful services.

COWANS, JOHN S_____
R—England.
G. O. No. 45, W. D., 1919.

General, quartermaster general to the forces, British Army.
For services performed for the American Expeditionary Forces and to the cause in which the United States has been engaged.

CRAVEN, FRANCIS W_____
R—England.
G. O. No. 27, W. D., 1919.

Lieutenant, British Navy, commanding His Majesty's destroyer Mounsey.
For rescuing 7 officers and 313 men of the American forces at sea on Oct. 16, 1918.

CROOKSHANK, SYDNEY D'A_____
R—England.
G. O. No. 2, W. D., 1920.

Major general, British Army.
The services General Crookshank rendered as Director General of Transportation of the British forces were of great merit. He heartily cooperated with the American authorities and gave much assistance to our Transportation Corps. He lent every effort to further those friendly relations which characterized the transportation services of the British and American Armies.

CUNINGHAME, THOMAS A. A. M_____
R—England.
G. O. No. 45, W. D., 1919.

Lieutenant colonel, British Army, instructor, Army Staff College, American Expeditionary Forces.
For services performed for the American Expeditionary Forces and to the cause in which the United States has been engaged.

CURRIE, ARTHUR W_____
R—England.
G. O. No. 17, W. D., 1919.

Lieutenant general, British Army, commanding Canadian Corps, British Expeditionary Forces.
For services rendered to the American Expeditionary Forces and to the cause in which the United States has been engaged.

CURRY, PHILIP A_____
R—England.
G. O. No. 69, W. D., 1919.

Major, British Army.
As director of transports for the British Ministry of Shipping at the port of New York during the movement of troops overseas.

DAVIDSON, GILBERT_____
R—England.
G. O. No. 87, W. D., 1919.

Lieutenant colonel, British Army, forage and supplies, British Expeditionary Forces.
He extended whole-hearted cooperation to our supply procurement agencies, rendering very valuable services to the American Expeditionary Forces. By his efforts and devotion to the American interests, great quantities of necessary supplies were made available for our troops.

DAVIDSON, JOHN H_____
R—England.
G. O. No. 45, W. D., 1919.

Major general, British Army, general staff operations, British Expeditionary Forces.
For services performed for the American Expeditionary Forces and to the cause in which the United States has been engaged.

DAVIES, FRANCIS J_____
R—England.
G. O. No. 46, W. D., 1920.

Lieutenant general, British Army.
As Secretary of the British War Office and member of the Army Council, by his great energy and marked military attainments, he rendered services of signal worth to the common cause.

DAWNAY, GUY_____
R—England.
G. O. No. 126, W. D., 1919.

Lieutenant colonel, British Army.
As head of the staff duties section of the General Staff at General Headquarters American Expeditionary Forces, he was an important factor in the victorious termination of the war by his brilliant military attainments and unremitting zeal. He constantly manifested the utmost consideration for the American units attached to the British Armies, aiding materially in their training and thereby rendering invaluable services to the American Expeditionary Forces.

DEEDES, CHARLES P_____
R—England.
G. O. No. 46, W. D., 1920.

Brigadier general, British Army.
As Deputy Director of Staff Duties of the British Army, by his tact, marked ability, and spirit of wholehearted cooperation with American officers on duty with the British Army staff, ably assisting them in every endeavor, he rendered services of great value to the American Expeditionary Forces.

DELANO-OSBORNE, OSBORNE H_____
R—England.
G. O. No. 45, W. D., 1919.

Brigadier general, director of movements (transportation), British Army.
For services performed for the American Expeditionary Forces and to the cause in which the United States has been engaged.

DRAKE, JOHN_____ R—England. G. O. No. 126, W. D., 1919.	Lieutenant colonel, British Army. As chief of the intelligence bureau section at British General Headquarters he at all times cooperated with the officers of the U. S. Army, and by his sound advice and cooperation rendered most valuable services to the American Expeditionary Forces. He was tireless in his devotion to his important duties, proving tactful and ready to come to our assistance at all times.
ELLES, HUGH J_____ R—England. G. O. No. 45, W. D., 1919.	Major general, British Army, commanding Tank Corps, British Expeditionary Forces. For services performed for the American Expeditionary Forces and to the cause in which the United States has been engaged.
EWART, RICHARD_____ R—England. G. O. No. 126, W. D., 1919.	Major general, British Army. As senior British member of the Interallied Commission on the Repatriation of Prisoners of War he has rendered conspicuous service to the United States and the allied Governments in connection with the sustenance, care, and homeward transportation of American and allied prisoners released by the armistice. He has at all times displayed a personal interest and supervising influence which resulted in speedy and comfortable repatriation of prisoners of war.
FORBES, ARTHUR W_____ R—England. G. O. No. 46, W. D., 1920.	Brigadier general, British Army. As commander of the embarkation area of the port of Liverpool he gave important assistance to the American Expeditionary Forces. At all times he exhibited a spirit of zealous cooperation with the American authorities, placing all the facilities of his important office at their disposal. He rendered service of conspicuous worth to the American Expeditionary Forces.
FORD, REGINALD_____ R—England. G. O. No. 45, W. D., 1919.	Major general, British Army, British representative, military board of allied supply. For services performed for the American Expeditionary Forces and to the cause in which the United States has been engaged.
FOULKES, CHARLES H_____ R—England. G. O. No. 45, W. D., 1919.	Brigadier general, British Army, director British Chemical Warfare Service. For services performed for the American Expeditionary Forces and to the cause in which the United States has been engaged.
FOWKE, GEORGE H_____ R—England. G. O. No. 45, W. D., 1919.	Lieutenant general, British Army, Adjutant General, British Expeditionary Forces. For services rendered to the American Expeditionary Forces and to the cause in which the United States has been engaged.
FOWLER, JOHN S_____ R—England. G. O. No. 126, W. D., 1919.	Major general, British Army. As Director of Signals of the British Expeditionary Forces he constantly rendered us valuable assistance in connection with the supply of needed material at critical times. He aided us very materially in promoting the efficiency of our electrical communications, displaying at all times military and scientific attainments of a high order. He was energetic in our behalf, proving himself a loyal friend of the American Expeditionary Forces.
FURSE, WILLIAM T_____ R—England. G. O. No. 45, W. D., 1919.	Lieutenant general, British Army, Master General of the Ordnance, British Expeditionary Forces. For services performed for the American Expeditionary Forces and to the cause in which the United States has been engaged.
GEIGER, GERALD J. P_____ R—England. G. O. No. 46, W. D., 1920_____	Major, British Army. As liaison officer of the British with the American Army, and later as chief of the British Mission at the Headquarters of the 3d Army, he has, by his tact, loyalty, and painstaking efforts, earnestly cooperated with the American authorities in handling the difficult problems which constantly arose, rendering services of great worth to the American Expeditionary Forces.
GELLIBRAND, JOHN_____ R—England. G. O. No. 126, W. D., 1919.	Major general, British Army. During the operations against the Hindenburg line near Ronssoy, in September, 1918, he commanded with brilliant leadership the 3d Australian Division, operating in close liaison with the 27th American Division. The fine spirit of comradeship prevailing between the officers and soldiers of these two divisions was in no small measure a reflection of the warm spirit of cooperation which he constantly manifested and his willingness to aid the American Expeditionary Forces in every way possible.
GILES, EDWARD D_____ R—England. G. O. No. 45, W. D., 1919.	Lieutenant colonel, British Army. As advisor to the training and instruction branch, war plans division, he rendered invaluable assistance in making possible the inception and successful conduct of war-time instruction at the Army War College.
GLYN, RALPH G_____ R—England. G. O. No. 45, W. D., 1919.	Major, British Army, instructor, Army Staff College, American Expeditionary Forces. For services performed for the American Expeditionary Forces and to the cause in which the United States has been engaged.
GOLIGHER, HUGH G_____ R—England. G. O. No. 126, W. D., 1919.	Brigadier general, British Army. Serving as a financial officer of the British Expeditionary Forces in France by his broad experience and extraordinary administrative ability he proved a notable factor in the success of the allied cause. He was at all times at the service of the American officers, who were charged with questions of supply and finance, and by his sound advice, unfailing courtesy, and loyal spirit of cooperation rendered services of the utmost value to the American Expeditionary Forces.

GOODWIN, THOMAS H. J. C.
R—England.
G. O. No. 87, W. D., 1919.

Lieutenant general, British Army.
As Surgeon General of the British Army, he placed at the disposal of the American divisions serving with the British forces all the evacuation and hospitalization facilities at his command. His eminent skill, ability, and broad experience enabled him to extend most useful cooperation.

GREEN, ARTHUR F. U.
R—England.
G. O. No. 14, W. D., 1920.

Colonel, British Army.
He has done conspicuous service as chief of staff of the British Mission. His sound judgment, sympathetic understanding, and steadfastness of purpose contributed in large measure to the furtherance of Anglo-American understanding and to the execution of the terms of the armistice.

GUTHRIE, CONNOP.
R—England.
G. O. No. 103, W. D., 1919.

British Ministry of Shipping.
As a member of the shipping control committee his services in connection with negotiations for British tonnage in the interchange of tonnage by the War Department and the British Government have been conspicuous.

HAIG, DOUGLAS.
R—England.
G. O. No. 111, W. D., 1918.

Field Marshal, British Army, commander in chief of the British Armies in France.
As an expression to him of the high regard of the people of the United States and of their Army, for the distinguished and patriotic services which he has rendered to the common cause in which he has been associated on the battlefields of Europe.

HAKING, RICHARD C. B.
R—England.
G. O. No. 14, W. D. 1920.

Lieutenant general, British Army.
He rendered conspicuously meritorious service as chief of the British Mission, standing shoulder to shoulder with the chief of the American Mission in all questions of policy, aiding with rare tact and a genial personality the cordial relations between the English-speaking missions. He saw 52 months' service in the field, commanded the XI Army Corps, and is the hero of the relief of Lille in 1918.

HARINGTON, CHARLES H.
R—England.
G. O. No. 45, W. D., 1919.

Major general, deputy chief of the Imperial General Staff, British Army.
For services performed for the American Expeditionary Forces and to the cause in which the United States has been engaged.

HEADLAM, JOHN E. W.
R—England.
G. O. No. 11, W. D., 1919.

Major general, British Army.
For services rendered the United States Army while serving as chief of the British Artillery Mission to the United States.

HEATH, GERARD M.
R—England.
G. O. No. 126, W. D., 1919.

Major general, British Army.
As Engineer in Chief of the British Expeditionary Forces he performed, with conspicuous success, highly responsible duties in the struggle against the common enemy. He gave unfailing support to the American units serving with the British Armies, and in spite of his numerous other tasks sought every opportunity to aid in developing their efficiency, thereby rendering services of the utmost value to the American Expeditionary Forces.

HOLLAND, HENRY W.
R—England.
G. O. No. 45, W. D., 1919.

Lieutenant colonel, censorship and publicity section, British Army.
For services performed for the American Expeditionary Forces and to the cause in which the United States has been engaged.

HONE, THOMAS N.
R—England.
G. O. No. 126, W. D., 1919.

Captain, British Army.
As assistant to the chief of the British Mission at American General Headquarters from December, 1917, to June, 1919, by his complete knowledge of the intelligence service, fostered by an earnest spirit of cooperation, he gave much valuable aid to our second section of the General Staff, offering them every facility at his command.

HORNE, HENRY S.
R—England.
G. O. No. 17, W. D., 1919.

General, British Army, commanding 1st Army, British Expeditionary Forces.
For services rendered to the American Expeditionary Forces and to the cause in which the United States has been engaged.

HUTCHINSON, HUGH M.
R—England.
G. O. No. 126, W. D., 1919.

Lieutenant colonel, British Army.
As instructor at the Gondrecourt schools, he rendered service of exceptional value in the training of a large number of officers of the American Expeditionary Forces. He was tireless in devotion to his important duties and displayed military attainments of high order. Tactful and forceful in the presentation of important subjects, he gave us very valuable assistance.

HUTCHISON, ROBERT.
R—England.
G. O. No. 45, W. D., 1919.

Major general, Director of Organization, British Army.
For services performed for the American Expeditionary Forces and to the cause in which the United States has been engaged.

JACK, EVAN M.
R—England.
G. O. No. 126, W. D., 1919.

Lieutenant colonel, British Army.
As officer in charge of maps in British General Headquarters he rendered valuable and distinguished services to the American Expeditionary Forces, making available to us all the information at his command. He provided for the instruction of American officers on the British front, and supplied us with special technical equipment that could not be obtained elsewhere.

JACOB, CLAUD W.
R—England.
G. O. No. 45, W. D., 1919.

Lieutenant general, British Army, commanding 2d Army Corps, British Expeditionary Forces.
For services performed for the American Expeditionary Forces and to the cause in which the United States has been engaged.

JONES, ROBERT R—England. G. O. No. 124, W. D., 1919.	Major general, British Army. An eminent orthopedic surgeon and chief of the division of orthopedic surgery in the British Army, he placed at the disposal of the medical service of the American Expeditionary Forces his eminent talents and broad experience in standardizing methods of treatment for the sick and wounded and took an active personal interest in class instruction of American medical officers in this very important branch of surgery.
JURY, EDWARD C. R—England. G. O. No. 126, W. D., 1919.	Lieutenant colonel, British Army. As principal assistant to the chief of the British Mission attached to the American General Headquarters of the period, January, 1918, to March, 1919, by his brilliant military ability and his earnest spirit of wholehearted cooperation with the personnel of our fourth and fifth sections of the General Staff, he gave much valuable assistance to these sections during the period of their organization and development, rendering services of inestimable value to the American Expeditionary Forces.
KNAPP, KEMPSTER K. R—England. G. O. No. 87, W. D., 1919.	Brigadier general, British Army. In command of the British artillery supporting the 2d American Corps during the operations from Sept. 27, 1918, to Oct. 21, 1918, he proved of invaluable assistance to our Infantry. He showed himself an indefatigable worker, a brilliant tactician, and a loyal friend.
LAMB, MALCOLM H. M. R—England. G. O. No. 126, W. D., 1919.	Major, British Army. As chief of intelligence, line communications, British Expeditionary Forces, he assisted most ably in the organization, instruction, and working of the American intelligence section in the Services of Supply. Due to his advice, experience, and aid many of the difficult problems confronting our Services of Supply during its preliminary organization were successfully overcome. The services which he rendered the American Expeditionary Forces were most valuable.
LAWRENCE, HERBERT A. R—England. G. O. No. 17, W. D., 1919.	Lieutenant general, British Army, Chief of Staff, British Expeditionary Forces. For services rendered to the American Expeditionary Forces and to the cause in which the United States has been engaged.
LAWRENCE, RICHARD C. B. R—England. G. O. No. 126, W. D., 1919.	Brigadier general, British Army, commanding the British base at Marseille, France. When the American base was established at Marseille he was of the utmost assistance in its development and maintenance by the practical aid and valuable advice he gave. Always ready, in case of need, to place at our disposition the valuable facilities acquired by the British base, during four years of operation, he made possible an increased supply to the American troops. By his fervent spirit of cooperation he rendered service of inestimable value to the American Expeditionary Forces.
LEISHMAN, WILLIAM B. R—England. G. O. No. 60, W. D., 1920.	Major general, Royal Army Medical Corps, British Army. By his marked energy and zealous cooperation with the Medical Service of the American Expeditionary Forces in important research work, he promoted the efficient treatment of the American sick and wounded. His remarkable achievements in the domain of preventive medicine and wound bacteriology resulted in the saving of many lives among our wounded. His services were of great consequence to the American Expeditionary Forces.
LIDBURY, CHARLES A. R—England. G. O. No. 45, W. D., 1919.	Major, British Army, attached to the 2d Army Corps, American Expeditionary Forces. For services performed for the American Expeditionary Forces and to the cause in which the United States has been engaged.
LIVESAY, ROBERT O'H. R—England. G. O. No. 126, W. D., 1919.	Brigadier general, British Army. As instructor at the American Army General Staff College, he rendered services of exceptional value to the American Expeditionary Forces in connection with the efficient training of our officers. He displayed military attainment of high order in the performance of his exacting duties, working always wholeheartedly in our interests. He proved himself sound in judgment, tactful, and a loyal friend.
LYNDEN-BELL, ARTHUR L. R—England. G. O. No. 126, W. D., 1919.	Major general, British Army. As Director of Staff Duties at British General Headquarters he performed with distinction the important duties of his high office. In the midst of his manifold and exacting tasks he interested himself repeatedly in behalf of the American Expeditionary Forces, rendering us service of exceptional value. Through his instrumentality masses of information requested by us of the British Mission at Washington, Versailles, and at American General Headquarters have been made available. Most helpful at all times, he has proved himself a loyal friend.
MACDONOGH, GEORGE M. W. R—England. G. O. No. 45, W. D., 1919.	Lieutenant general, adjutant general to the forces, British Army. For services performed for the American Expeditionary Forces and to the cause in which the United States has been engaged.
MCLACHLAN, JAMES D. R—England. G. O. No. 11, W. D., 1919.	Major general, British Army. For services rendered the U. S. Army while serving as military attaché to the British Embassy, Washington, D. C.

McNAMEE, JOHN T R—England. G. O. No. 126, W. D., 1919. Distinguished-service cross also awarded.	Captain, Royal Field Artillery, British Army, attached to the 1st Battalion, 1st Gas Regiment, American Expeditionary Forces. As instructor with the 1st Gas Regiment, American Expeditionary Forces, he worked unceasingly in developing aggressive forms of gas attack. He was tireless in his devotion to duty, showing particularly valuable ability in personally supervising the liaison and conduct of the gas operations in the Meuse-Argonne drive.
MacPHERSON, WILLIAM G R—England. G. O. No. 56, W. D., 1921.	Major general, British Army. As Deputy Director General of Medical Services of the British Armies in France he displayed exceptional energy, initiative, and good judgment in directing and supervising in the most expeditious manner possible the organization, equipment, and training of medical department units of the 2d Corps, American Expeditionary Forces, serving with the British Expeditionary Forces. It was largely through his cooperation and individual efforts that the Medical Service of the 2d Corps, American Expeditionary Forces, was prepared in time to proficiently discharge its duties to the sick and wounded in combat.
MAUD, HARRY R—England. G. O. No. 87, W. D., 1919.	Colonel, British Army, forage and supplies, British Expeditionary Forces. He gave most valuable assistance to the American Expeditionary Forces in the procurement of necessary supplies for our troops. He rendered tactful and most willing service, affording whole-hearted cooperation in his important duties.
MAY, REGINALD S R—England. G. O. No. 45, W. D., 1919.	Major general, British Army, Deputy Quartermaster General, British Expeditionary Forces. For services performed for the American Expeditionary Forces and to the cause in which the United States has been engaged.
MILLER, WALTER R—Canada. G. O. No. 29, W. D., 1919.	Major, Canadian forces. For services rendered the U. S. Army while serving as the liaison officer between the British Embassy, the Ministry of Militia and Defense, Dominion of Canada, and the War Department.
MILNE, GEORGE F R—England. G. O. No. 62, W. D., 1919.	Lieutenant general, British Army. As British Minister of War he displayed military attainments of a high order, achieving a brilliant success. Untiring in devotion to his important duties, he was aggressive and capable, rendering service of inestimable value to the American Expeditionary Forces and the allied cause.
MONASH, JOHN R—England G. O. No. 41, W. D., 1919.	Lieutenant general, British Army, commanding Australian Army Corps, British Expeditionary Forces. For services rendered to the American Expeditionary Forces and to the cause in which the United States has been engaged.
MONTGOMERY, ARCHIBALD A R—England. G. O. No. 126, W. D., 1919.	Major general, British Army. As chief of staff of the 4th British Army he directed the operations of the 2d American Army Corps with distinguished ability, displaying military attainments of the highest order. The officers and soldiers of the 27th, 30th, and 33d U. S. Divisions are justly proud of having served with their English comrades against the common foe and of having shared with them in the successes which were due, in no small degree, to his capable direction.
NASH, PHILIP A. M R—England. G. O. No. 87, W. D., 1919.	Major general, British Army. The service he rendered the American Expeditionary Forces as Inspector General of Transportation for the British Army and as a member of the Interallied Transportation Council was of the greatest value. He lent every possible assistance to the American military authorities, giving us sound advice and important information.
NEEDHAM, HENRY R—England. G. O. No. 45, W. D., 1919.	Lieutenant colonel, British Army, instructor at the Army General Staff College, American Expeditionary Forces. For services performed for the American Expeditionary Forces and to the cause in which the United States has been engaged.
OVERTON, GEORGE C. R R—England. G. O. No. 45, W. D., 1919.	Lieutenant colonel, British Army, liaison officer, American Rest Camp, Winchester, England. For services performed for the American Expeditionary Forces and to the cause in which the United States has been engaged.
PAKENHAM, HERCULES A R—England. G. O. No. 29, W. D., 1919.	Lieutenant colonel, General Staff, British Army. For services rendered the U. S. Army while serving as the liaison officer between the British and American Military Intelligence Services.
PARSONS, HAROLD D. E R—England. G. O. No. 87, W. D., 1919.	Major general, British Army. As Director of Equipment and Ordnance Stores of the British Expeditionary Forces he was able to render assistance of the greatest value to the American Expeditionary Forces. He aided us most markedly in the procurement of artillery material and ammunition from British sources, at all times giving loyal cooperation.
PEAL, EDWARD R R—England. G. O. No 45, W. D., 1919.	Lieutenant colonel, British Army, in charge of British Aviation Office at Paris, France. For services performed for the American Expeditionary Forces and to the cause in which the United States has been engaged.

PHILLIPS, OWEN F R—England. G. O. No. 126, W. D., 1919.	Brigadier general, British Army, 2d Australian Division, British Expeditionary Forces. Commanding the artillery in support of the 30th American Division in its operations of Oct. 8–11, 1918, he aided greatly in the sucesses achieved at that time during these operations. He displayed military ability of the highest order and a spirit of earnest cooperation which made the members of the American units proud to be associated with him in their operations against the common foe.
PLAYFAIR, PATRICK E. L R—England. G. O. No. 45, W. D., 1919.	Lieutenant colonel, British Army, commanding 13th Wing, Royal Air Forces. For services performed for the American Expeditionary Forces and to the cause in which the United States has been engaged.
PLUMER, HERBERT C. O R—England. G. O. No. 17, W. D., 1919.	General, British Army, commanding 2d Army, British Expeditionary Forces. For services rendered to the American Expeditionary Forces and to the cause in which the United States has been engaged.
PRITCHARD, CLIVE G R—England. G. O. No. 126, W. D., 1919	Brigadier general, British Army, chief of artillery, 4th Army, British Expeditionary Forces. The valuable services which he rendered to the American Expeditionary Forces in supplying our divisions with artillery while he was serving as deputy chief of artillery, 4th British Army, will ever be remembered. Giving us the benefit of his brilliant experience as an artillerist he made an extended tour of American Artillery camps and thereby aided materially in bringing this branch of our arms up to the standard required for effective combat.
PUCKLE, FREDERICK K R—England. G. O. No. 126, W. D., 1919.	Lieutenant colonel, British Army. With the British Mission at Washington and later in France, he rendered services of inestimable value in the training of officers for the U. S. Quartermaster Corps and in the preparation of that department for oversea service. He gave us great assistance also in connection with the preliminary arrangements for the supply of our forces. He was tireless in the performance of his exacting duties, tactful at all times, and a loyal friend.
RADCLIFFE, PERCY P. deB R—England. G. O. No. 45, W. D., 1919.	Major general, British Army, Director of Military Operations, British War Office. For services performed for the American Expeditionary Forces and to the cause in which the United States has been engaged.
RAWLINSON, HENRY S R—England. G. O. No. 17, W. D., 1919.	General, British Army, commanding 4th Army, British Expeditionary Forces. For services rendered to the American Expeditionary Forces and to the cause in which the United States has been engaged.
RENNISON, WILLIAM R—England. G. O. No. 45, W. D., 1919.	Major, British Army, instructor, Army School of the Line, American Expeditionary Forces. For services performed for the American Expeditionary Forces and to the cause in which the United States has been engaged.
ROBERTSON, WILLIAM R R—England. G. O. No. 45, W. D., 1919.	General, British Army, commander in chief, Great Britain. For services performed for the American Expeditionary Forces and to the cause in which the United States has been engaged.
RUDOLPH, FREDERICK R—England. G. O. No. 45, W. D., 1919.	Lieutenant general, British Army, commanding 10th British Army in Italy. For services performed for the American Expeditionary Forces and to the cause in which the United States has been engaged.
SACKVILLE-WEST, CHARLES J R—England. G. O. No. 75, W. D., 1919.	Major general, British Army. For services to the allied and associated Governments as permanent military representative, British section, Supreme War Council.
SALMON, GEOFFREY N R—England. G. O. No. 126, W. D., 1919	Lieutenant colonel, British Army. Detailed as an instructor to the school at Chatillon-sur-Seine, France, he displayed military attainments of a high order, and rendered very valuable services to the American Expeditionary Forces in connection with the training of its officers. He was tireless in devotion to his important duties, and at all times tactful and energetic in the presentation of important subjects.
SALMOND, JOHN M R—England. G. O. No. 87, W. D., 1919.	Major general, British Army. As general officer commanding the British Royal Air Forces in the field, he distinguished himself by the exceptionally valuable services he performed. He aided and furthered the training of the units of the U. S. Air Service attached to his command. He organized the training in the field of squadrons, pilots, ground officers, and mechanics of the American Air Service, rendering us most valuable assistance at all times.
SARGENT, HARRY N R—England. G. O. No. 87, W. D., 1919.	Brigadier general, British Army. As chief of the British Military Mission, Headquarters, Services of Supply, he provided for prompt and satisfactory procurement of enormous amounts of supplies from England, thereby contributing greatly to the successes achieved by the American Expeditionary Forces.
SINCLAIR-MACLAGAN, EWEN G R—England. G. O. No. 126, W. D., 1919.	Major general, British Army, Commanding 4th Australian Division, British Expeditionary Forces. The 27th and 30th U. S. Divisions profited by the wide experience and brilliant military attainments which he displayed as commander of the Australian division, when he served with the 2d American Army Corps and so ably assisted in the direction of the operations. The officers of these organizations count it a high privilege to have been associated with him during the final phases of the struggle with the common enemy, in which he and his forces played such a splendid part.

STUDD, HERBERT W R—England. G. O. No. 75, W. D., 1919.	Brigadier general, British Army. For services to the allied and associated Governments as chief of staff, British section, Supreme War Council.
SYKES, FREDERICK H R—England. G. O. No. 46, W. D., 1920.	Major general, British Army. As chief of staff of the British Air Forces, by his willing spirit of cooperation with the American authorities in all matters pertaining to the Air Service and its operation, he rendered service of great worth to the American Expeditionary Forces.
THOMAS, GEORGE P R—England. G. O. No. 126, W. D., 1919.	Captain, British Army. He rendered valuable services to the American Expeditionary Forces in connection with the training of American Artillery officers. He displayed high military attainments and a keen comprehension of conditions in the field. His service was continuously marked by devotion to duty, tireless energy and tact.
THOMPSON, HARRY N R—England. G. O. No. 124, W. D., 1919.	Major general, British Army. An eminent medical officer, and as Director of the Medical Service, 1st British Field Army in France, he placed his time and energy at the disposal of the American Expeditionary Forces. The sanitary school maintained in his army for teaching front-line medical requirements was utilized for the instruction of the American medical officers sent to him by classes. The observation and experience gained by these student officers under his able supervision and guidance eventually resulted in the saving of lives of many American wounded.
THORNTON, HENRY W R—England. G. O. No. 87, W. D., 1919.	Major general, British Army. As Paris representative of the Director General of Movements and Railways in London he rendered the greatest assistance to the American Expeditionary Forces in the procurement of hospital trains and supplies. He furnished us with information which proved most important in the development of the American Transportation Corps.
THWAITES, WILLIAM R—England. G. O. No. 45, W. D., 1919.	Major general, British Army, Director of Military Intelligence, British War Office. For services performed for the American Expeditionary Forces and to the cause in which the United States has been engaged.
TRENCHARD, HUGH M R—England. G. O. No. 45, W. D., 1919.	Major general, British Army, chief of Air Service, British Expeditionary Forces. For services performed for the American Expeditionary Forces and to the cause in which the United States has been engaged.
TROTTER, GERALD F R—England. G. O. No. 11, W. D., 1919.	Brigadier general, British Army. For services rendered the U. S. Army while serving as chief of the British Military (or Advisory) Mission to the United States.
TWINING, PHILIP G R—England. G. O. No. 126, W. D., 1919.	Major general, British Army. As Director of the Royal Engineers, by his active interest in behalf of the needs of the American services in both France and England, he has given valuable services to the American Expeditionary Forces. Always manifesting an aggressive spirit in furthering our efforts, he handled many important questions of supply with marked tact and diplomacy.
WAGSTAFF, CYRIL M R—England. G. O. No. 45, W. D., 1919.	Brigadier general, British Army, chief, British Mission, General Headquarters, American Expeditionary Forces. For services rendered to the American Expeditionary Forces and to the cause in which the United States has been engaged.
WALLACE, CUTHBERT S R—England. G. O. No. 124, W. D., 1919.	Major general, British Army. An eminent consulting surgeon, and while serving with the British Expeditionary Forces in France, with untiring zeal he devoted his time and energy toward promoting standard methods for efficient treatment of American sick and wounded, and unreservedly placed at the disposal of the medical service of the American Expeditionary Forces his eminent talents, broad experience, and knowledge of general conditions, with a view to assisting in the prevention of wastage among our forces from wounds and disease.
WATTS, HERBERT E R—England. G. O. No. 45, W. D., 1919.	Lieutenant general, British Army, commanding 19th Army Corps, British Expeditionary Forces. For services rendered to the American Expeditionary Forces and to the cause in which the United States has been engaged.
WELLESLEY, WILLIAM R R—England. G. O. No. 46, W. D., 1920.	Lieutenant colonel, British Army, Undersecretary of State for War. In his capacity of Undersecretary of State for War, he manifested a constant desire to assist the American Expeditionary Forces in every possible way. The cordial spirit of cooperation and harmony which has marked the relations between the British and American forces during their struggle against the common enemy has been due, in large measure, to his zealous efforts.
WILSON, HENRY H R—England. G. O. No. 45, W. D., 1919.	General, British Army, chief of the Imperial General Staff, British War Office. For services performed for the American Expeditionary Forces and to the cause in which the United States has been engaged.

WINTERBOTHAM, HAROLD ST. J. L. R—England. G. O. No. 126, W. D., 1919.	Lieutenant colonel, British Army. His practical advice, constructive criticism, and keen personal interest influenced greatly the organization and methods of training for topographic and ranging work in the American Expeditionary Forces. He personally furnished us with a large amount of important topographic data and information on survey and ranging work in the British and German Armies which could not have been obtained elsewhere and which proved of the greatest value to us.
WORTHINGTON, EDWARD S. R—England. G. O. No. 2, W. D., 1920.	Colonel, British Army. For services demanding the fullest of his time and energy. He exerted himself to such good purpose in behalf of medical officers of the U. S. Army that large numbers of them were sent for special and intensive instruction to the various special medical and hospital centers in the British Isles, where they received at the hands of Great Britain's most eminent instructors the lessons learned from three years of war, thereby enabling the American surgeons to apply the knowledge acquired, with the result that many lives and limbs were conserved in the American Expeditionary Forces.
WEIR, ANDREW. R—England. G. O. No. 62, W. D., 1919.	British Surveyor General of Supplies. He displayed ability of a high order, untiring devotion to duty, and zeal in the performance of his exacting duties. At all times he worked with singleness of purpose for the good of the allied cause, rendering service of exceptional value to the American Expeditionary Forces.
WEIR, WILLIAM. R—England. G. O. No. 62, W. D., 1919.	British Secretary of State for the Royal Air Forces. In the performance of his important duties he displayed great energy and ability of a high order. He handled difficult situations with tact and aggressiveness, achieving brilliant results. At all times he was zealous to the best interests of the American Expeditionary Forces.

[Awarded for exceptionally meritorious and distinguished services in a position of great responsibility, under the provisions of the act of Congress approved July 9, 1918]

ALBY, HENRI M. C. E. R—France. G. O. No. 45, W. D., 1919.	Major general, Chief of Staff, French Army. For services performed for the American Expeditionary Forces and to the cause in which the United States has been engaged.
ALCAN, ADRIEN H. R—France. G. O. No. 126, W. D., 1919.	Captain of infantry, French Military Mission at General Headquarters, American Expeditionary Forces. While on duty with the French Military Mission at General Headquarters, American Expeditionary Forces, he rendered service of exceptional value to the U. S. Army. His ability, tact, loyalty, and untiring efforts on our behalf proved of inestimable assistance in the successful execution of many important negotiations with the French Army. He went far beyond the bounds of duty to help us, proving himself a willing and devoted friend to our interests.
ALERME, MARIE M. E. R—France. G. O. No. 126, W. D., 1919.	Lieutenant colonel of Colonial Infantry, French Army. As assistant chief of the cabinet of the Minister of War, charged with the management of Franco-American affairs, by his tact, good judgment, and loyal spirit of cooperation with the American authorities he has rendered important service to the American Expeditionary Forces.
ALEXANDRE, GEORGES R. R—France. G. O. No. 126, W. D., 1919.	Brigadier general, French Army, chief of artillery, 5th Army Corps, American Expeditionary Forces. As chief of artillery of the 5th Army Corps he performed invaluable services to the American Expeditionary Forces. To his consummate skill as an artillerist was due in a large measure the success of the artillery in the operations of the 5th Army Corps in the St. Mihiel offensive and in the first phase of the Meuse-Argonne offensive. The loyal spirit of cooperation which he constantly manifested will ever be held firm in memory by his American colleagues.
ALEXANDRE, MARIE N. G. R—France. G. O. No. 126, W. D., 1919.	Lieutenant colonel of artillery, 3d Bureau, French Army. Possessed of military attainments of a high degree, he placed these at our disposal, rendering very valuable services to the American Expeditionary Forces by the advice he has given us in connection with artillery operations. At all times he was devoted to the best interests of the allied cause. He displayed tireless energy and ceaseless vigilance in our behalf, proving himself loyal to our service.
ALLAIN, EMILIEN. R—France. G. O. No. 126, W. D., 1919.	Captain, 315th Regiment of Infantry, French Army. By his able adaptation of the French combat formations to the American units and by the thorough instructions he gave the students of the Army Candidates' School in the technique of the combat formation, he has given valuable service to the American Expeditionary Forces. Later, as liaison officer between the marshal commanding the allied armies and American General Headquarters, by his courteous efforts and good judgment, he has rendered much valuable assistance to the American Army.
ALPHANDERY, LEVY. R—France. G. O. No. 126, W. D., 1919.	Captain of infantry, French Army, mayor of Chaumont, France. By his whole-hearted spirit of cooperation with the authorities at American General Headquarters in all matters concerning the American personnel at General Headquarters, he rendered services of great value to the American Expeditionary Forces.
ANDLAUER, JOSEPH L. M. R—France. G. O. No. 126, W. D., 1919	Brigadier general, commanding the 18th Division of Infantry, French Army. As commander of the 18th French Division he displayed conspicuous military attainments and the most loyal devotion to the allied cause. By his cordial spirit of cooperation and careful supervision of the training of the 1st, 2d, 26th, and 42d American Divisions during the time in which they were attached to his command, he rendered invaluable services to the American Expeditionary Forces.
ANDRIOT, MAURICE. R—France. G. O. No. 87, W. D., 1919.	Lieutenant colonel of infantry, French Army. As regulating officer for the French railroads at American Headquarters, Services of Supply, he displayed remarkable efficiency in meeting the needs of the American forces for railroad transportation. By his untiring energy and technical ability he satisfied all sudden calls made upon him for railroad facilities.
ARDON, PIERRE R. I. R—France. G. O. No. 45, W. D., 1919.	Lieutenant commander, French Army, instructor, Valdahon Training Camp, American Expeditionary Forces School of Fire Triangulation, Langres, France. For services performed for the American Expeditionary Forces and to the cause in which the United States has been engaged.

ARMENGAUD, PAUL F. M.
R—France.
G. O. No. 126, W. D., 1919.
Distinguished-service cross also awarded.

Major, Air Service, French Mission, General Headquarters, American Expeditionary Forces.
Throughout the entire organization and operation of the American Air Service in France, he rendered us services of exceptional value. He displayed most marked tactical conception of aerial activities, his advice proving of inestimable worth during the St. Mihiel and Meuse-Argonne offensives. Out of his wide experience he gave us his best, proving at all times devoted to our interests and going far beyond the bounds of his duties, as a member of the French Mission, to assist us.

BADRE, LOUIS H. J.
R—France.
G. O. No. 45, W. D., 1919.

Major of infantry, French Army, forestry officer, attached to the French Military Mission at General Headquarters, American Expeditionary Forces.
For services performed for the American Expeditionary Forces and to the cause in which the United States has been engaged.

BARRAUD, CLAUDIUS E.
R—France.
G. O. No. 45, W. D., 1919.

Interpreter officer of 1st class, French Army, instructor, Army Intelligence School, American Expeditionary Forces.
For services performed for the American Expeditionary Forces and to the cause in which the United States has been engaged.

BARRILLON, PAUL C.
R—France.
G. O. No. 45, W. D., 1919.

Major of engineers, French Army, chief of Service for Military Development of Ports, Minister of Public Works.
For services performed for the American Expeditionary Forces and to the cause in which the United States has been engaged.

BATAILLARD, FÉLICIEN F.
R—France.
G. O. No. 45, W. D., 1919.

Captain of gendarmes, French Army, provost marshal with the 1st Army Corps, American Expeditionary Forces.
For services performed for the American Expeditionary Forces and to the cause in which the United States has been engaged.

BEAUDEROM DE LAMAZE, LOUIS E. M. J. I.
R—France.
G. O. No. 126, W. D., 1919.

Brevet lieutenant colonel of infantry, French Army.
As chief of the personnel section of the Army General Staff, by his able executive ability and good judgment, he has rendered most distinguished service to the Allies. He has always shown a keen interest in the American Army and lent them every assistance which he could furnish. His services have been of great value to the American Expeditionary Forces.

BECQ, RAYMOND.
R—France.
G. O. No. 45, W. D., 1919.

Lieutenant colonel of engineers, chief of the telegraphic service of the 2d French Army.
For services performed for the American Expeditionary Forces and to the cause in which the United States has been engaged.

BELIN, EMILE E.
R—France.
G. O. No. 75, W. D., 1919.

Major general, French Army.
For services to the allied and associated Governments as permanent military representative, French section, Supreme War Council.

BELLOT, LEON H. A.
R—France.
G. O. No. 126, W. D., 1919.

Lieutenant colonel of artillery, Geographic Service of the French Army.
As a member of the Geographic Service of the army, he, by his sound practical advice and good judgment, proved of the utmost assistance to the American Expeditionary Forces. Due largely to his personal efforts, the closest cooperation existed between his army and our own Topographic Service. Just prior to the Meuse-Argonne he personally arranged for the Geographic Service to extend and enlarge its facilities, so as to assist the 1st American Army in every way possible. His services were of inestimable value.

BERDOULAT, PIERRE E.
R—France.
G. O. No. 126, W. D., 1919.

Major general, commanding the 20th Army Corps, French Army, with the 1st and 2d Divisions, American Expeditionary Forces.
His gallant and conspicuous services are connected with the deeds of the 1st and 2d United States Divisions in the counterattack south of Soissons, July, 1918.

BERGASSE, GASTON L. E. E.
R—France.
G. O. No. 3, W. D., 1922.

Medical inspector, director of medical service of the 20th Army Area, French Army.
For services in the organization of American hospital centers. In addition to his official cooperation in the larger affairs, his influence exerted in innumerable ways has been a material factor in the successful installation and operation of the activities of the medical department of Base Section No. 2.

BERGER, MICHEL D.
R—France.
G. O. No. 87, W. D., 1919.

Major, French Army, chief of D. T. M. A., detailed at French General Headquarters.
While serving in the D. T. M. A. he was in personal charge of practically all of the movements of troops and hospital trains. He was tireless and careful in supervising the transportation of our troops through France and in providing railroad facilities for the shipment of supplies. He accomplished with brilliant success a most arduous and important task.

BERNARD, FREDERIC.
R—France.
G. O. No. 87, W. D., 1919.

Quartermaster, French Navy.
With extraordinary heroism he achieved the seemingly impossible in rescuing 70 members of the crew of the U.S. Army transport Jinsen Maru, when that vessel foundered at Ile d'Yeu, France, Dec. 4, 1918. Braving the rocky surf, he got a line to the ship when other efforts had failed, and by that line sent all the members of the crew one by one to shore, being himself the last man to leave the ship. Displaying personal bravery of the highest order, he voluntarily jeopardized his life to perform this courageous service.

BERTHELOT, HENRI M. R—France. G. O. No. 17, W. D., 1919.	Major general, French Army, chief of the French Military Mission with the Rumanian Armies, commanding the French forces in the Orient. For services rendered to the American Expeditionary Forces and to the cause in which the United States has been engaged.
BERTHIER, EMILE J. R—France. G. O. No. 126, W. D., 1919.	Major of artillery, French Army . By his brilliant military attainments and steadfast devotion to duty he rendered invaluable services to the American Expeditionary Forces as an instructor in the army schools at Langres. Subsequently, upon being attached to the 3d American Army as liaison officer, he capably performed duties of the utmost importance in maintaining effective liaison between the French and American Armies of Occupation, doing much to promote the harmonious relations which prevailed among the two allied forces.
BERTIN, JEAN M. A. R—France. G. O. No. 126, W. D., 1919.	Second lieutenant of the 5th Regiment of Engineers, French Army, liaison. As liaison officer between the D. T. M. A. of the French Army and the Deputy Director General of Transportation at American General Headquarters, by his tactful, capable, and untiring efforts to assist the American Army, he has rendered services of great value to the American Expeditionary Forces. Loyal to the American Army, he at all times showed a friendly interest in all our needs.
BLETRY, CAMILLE L. C. R—France. G. O. No. 59, W. D., 1921.	Major of artillery, French Army. For services while conducting all coal negotiations between the French Government and the American Expeditionary Forces. His unfailing response to every demand made upon him and his friendship and energy have many times proved of inestimable value to our Armies.
BLONDLAT, ERNEST J. R—France. G. O. No. 126, W. D., 1919.	Major general, commanding the 2d Colonial Army Corps, French Army. It was under his command that the 2d Colonial Corps performed such glorious deeds as part of the 1st American Army during the St. Mihiel offensive.
BOQUET, ERNEST. G. O. No. 87, W. D., 1919.	Colonel of engineers, French Army. As director of Military Transports Service with the Armies he extended hearty cooperation to the American military authorities, aiding us greatly in solving transportation problems. He displayed marked ability in handling the intricate details of his important work and went far beyond the bounds of duty to render vital assistance to the American Expeditionary Forces.
BORDEAUX, JOSEPH P. E. R—France. G. O. No. 126, W. D., 1919.	Brigadier general, commanding the 18th Division of Infantry, French Army. In command of the 18th French Division he displayed superb military attainments and unfailing devotion to the allied cause. By his cordial spirit of cooperation and careful supervision of the instruction of the American units which were attached to his command during their training periods he rendered invaluable services to the American Expeditionary Forces.
BORELLI, GEORGES M. R—France. G. O. No. 45, W. D., 1919.	Captain of artillery, French Army, commissariat general of the Franco-American military affairs, liaison officer with the 158th Field Artillery Brigade, 83d Division, American Expeditionary Forces, and operations officer, 55th Field Artillery Brigade, 30th Division, American Expeditionary Forces. For services performed for the American Expeditionary Forces and to the cause in which the United States has been engaged.
BOUCHER, MARCEL A. F. R—France. G. O. No. 87, W. D., 1919.	Major of air service, French Army. As commander of the air service of the 4th French Army, he displayed a keen interest in the welfare and training of the American officers serving with the French squadrons. He imparted to them freely of his wide experience and knowledge of air-service offensive tactics. He showed himself at all times tactful, able, and possessed of complete technical attainments.
BOULANGER, EDOUARD. R—France. G. O. No. 24, W. D., 1920.	As Chief of Interministerial Service of Expenditures he rendered great service to the United States.
BOULANGER, PIERRE. R—France. G. O. No. 87, W. D., 1919.	Lieutenant of the air service, French Army. As chief of the American Bureau of the French Aeronautic Ministry, and secretary of the Interallied Aviation Committee, he displayed tact, executive ability, and wide experience in aviation matters, rendering invaluable assistance to the American Air Service. He evinced the keenest interest in our success and an exceptional understanding of our needs, proving himself at all times most loyal to our interests.
BOURGEOIS, MAURICE J. R—France. G. O. No. 126, W. D., 1919.	Captain of the air service, French Army. As aide-de-camp to Marshal Pétain, he, through his careful and prompt action, rendered valuable service to the allied cause. His constant and tireless efforts brought about the consolidation and thorough regulation of the Red Cross and other benevolent organizations, which proved to be of great worth to the comfort of the American and other allied soldiers.
BOURGEOIS, JOSEPH E. R. R—France. G. O. No. 87, W. D., 1919.	Major general, French Army. As Director of the Geographic Service of the Army he afforded most loyal cooperation to the American Expeditionary Forces. He rendered exceptional service by placing at our disposal the various departments under his control, including the use of maps and optical instruments that proved of inestimable value in the planning and execution of important military operations.

BOUSOUET, ACHILLE R. R—France. G. O. No. 87, W. D., 1919.	Captain of the General Staff, French Army. As assistant to the American regulating officers of Creil, Le Bourget, Noisy-le-Sec, St. Dizier, and Metz, France, he rendered invaluable services during successive military operations of the greatest importance. He labored unceasingly under most trying conditions in order that the supply of American combat troops might not be interrupted, thus ably furthering the American operations.
BOUVARD, HENRI. R—France. G. O. No. 46, W. D., 1920	Brevet major of infantry, French Army. As an aide to Marshal Pétain, he served with distinction in a position of great responsibility. He displayed military attainments of a high order, and was constant in his devotion to the best interests of the allied cause. In the midst of his exacting duties, he found time to render very valuable service to the American Expeditionary Forces, being ever ready to assist us by his sound judgment and wide knowledge of constantly changing conditions.
BREART DE BOISANGER, JOSEPH M. H. R—France. G. O. No. 126, W. D., 1919.	Colonel of infantry, French Army. As chief of the French Mission with the 2d American Army, he worked loyally in the interests of the allied cause, proving always both energetic and tactful in the performance of his exacting duties. He aided materially in the maintenance of cordial relations between the French and American military authorities. He at all times cooperated with us most helpfully, proving himself an able and loyal friend.
BREUCQ, HENRI A. R—France. G. O. No. 126, W. D., 1919.	Brevet major of infantry, French Army. As a member of the Franco-American Bureau, Army General Staff, by his sound judgment, tact, and keen interest in the American Army, he has handled many delicate problems affecting the American Expeditionary Forces with marked ability and in a most satisfactory manner. The services he has given us have been of great value.
BRUNON, LOUIS E. P. R—France. G. O. No. 126, W. D., 1919.	Brevet major of artillery, French Army. As assistant chief of the 2d Bureau at French General Headquarters, he, by his wholehearted cooperation and sound advice, was of the greatest assistance to the various officers of the American Expeditionary Forces with whom he came in contact. At all times he gladly extended to us every facility available to aid in the successful operations of our forces. He was tactful and loyal, proving himself a true friend.
BRUMOT DE ROUVRE, ANTOINE C. P. C. M. R—France. G. O. No. 87, W. D., 1919.	Lieutenant colonel of artillery, French Army. As regulating commissioner at St. Dizier, France, he labored unceasingly in the interests of the American Expeditionary Forces during the St. Mihiel and Argonne offensives. Notwithstanding that the responsibility of supplying two French armies rested upon him, the facilities at his command were put unreservedly at our service. Without his loyal cooperation the supply of our combat troops would have been most complicated.
BRUSSAUX, EDOUARD O. J. R—France. G. O. No. 126, W. D., 1919.	Brevet major of infantry, French Army. While he was on duty at the 2d Bureau, French General Headquarters, he cooperated in every way possible with American military authorities, placing at their disposal the facilities at his command. He thereby rendered service of great value to the American Expeditionary Forces, proving himself at all times a helpful, able, and loyal friend.
BUAT, EDOUARD A. L. R—France. G. O. No. 17, W. D., 1919.	Major general, French Army, chief of staff to Marshal Pétain. For services rendered to the American Expeditionary Forces and to the cause in which the United States has been engaged.
BUCAILLE-LITTINÈRE, HENRI. R—France. G. O. No. 87, W. D., 1919.	Captain of artillery, French Army. In connection with the supply and maintenance of automatic arms and machine guns he rendered service of inestimable value to the American Expeditionary Forces. When assigned as assistant to the chief inspector of machine guns and small arms, American Ordnance Department, he displayed high professional attainments, energy, and untiring devotion to duty, cooperating with us most loyally at all times.
CAMBON, JULES. R—France. G. O. No. 126, W. D., 1919.	As special adviser to the French Prime Minister for Franco-American Affairs he displayed the same spirit of cooperation and warm friendship for the American people which he had manifested during his distinguished service as the French Ambassador at Washington. Ever actuated by the desire to cement the cordial relations between the two Republics, he and his staff offered us every facility which his important post commanded, thereby rendering services of inestimable value to the American Expeditionary Forces.
CAOUOT, A. T. R—France. G. O. No. 87, W. D., 1919.	Major of air service, French Army. As chief of the technical section of French aeronautics, he displayed technical aviation knowledge of high order, and performed his exacting duties with untiring energy and devotion. He rendered very valuable services to the American Expeditionary Forces, gladly assisting us at all times in the solution of perplexing problems.
CARREL-BILLIARD, MARIE J. A. A. R—France. G. O. No. 59, W. D., 1921.	Surgeon major, 1st class, medical service, French Army. For services in important researches and in the application of bacteriological methods of control to the dressing and closing of wounds. To him is due the discovery of the Carrell-Dakin treatment for war wounds. By his important service he has shortened the period of hospitalization of wounded men to a marked degree, and to him is due in a great measure the saving of many lives and limbs and the prevention of many disabilities among the wounded of our Army,

CARTIER, FRANCOIS.......................
R—France.
G. O. No. 126, W. D., 1919.

Lieutenant colonel of engineers, French Army.
As chief of the wireless-section of the French Ministry of War he rendered most distinguished services to the American Expeditionary Forces by placing every facility at the disposal of the American Army and by assisting our officers repeatedly with his sound advice, helping them in the solution of the many technical problems which confronted our Intelligence Section.

CASTELLI, EMILE J. B..................
R—France.
G. O. No. 87, W. D., 1919.

Surgeon major, 1st class, French Army, medical section of the Central Bureau of Franco-American Relations.
As adviser to the Medical Service of the American Expeditionary Forces he assisted in the preparation of a comprehensive hospitalization program and was of the greatest assistance in putting this plan into effect. Realizing the importance of liaison between the French and American Medical Services, he organized the medical section of the Central Bureau of Franco-American Relations, which proved of the greatest mutual benefit to the French and American Armies.

CHARET, CHARLES E...................
R—France.
G. O. No. 14, W. D., 1920.

Colonel of artillery, French Army.
He rendered service of conspicuous merit as chairman of the committee on material. In securing the rapid, early, and complete delivery by the Germans of all the artillery and airplane material required by the armistice terms, the achievement of his committee was noteworthy.

CHARREYRE, EUGÈNE J..................
R—France.
G. O. No. 126, W. D., 1919.

Brevet major of infantry, French Army.
As chief of the 3d Bureau, General Staff of the group of French Armies of the East, he has rendered great assistance in the placing of the 1st American Division in the Camp de Lorraine and in furthering their rapid instruction in combat. During the American operations of St. Mihiel and the Meuse-Argonne he again ably assisted us in furnishing additional artillery support during the attacks. At all times he assured the best conditions of liaison between the staffs of the French and the American Armies.

CHEVALIER, LOUIS J. G...............
R—France.
G. O. No. 87, W. D., 1919.

Major general, French Army.
As inspector general of the Forest Service, Minister of Munitions and War Manufactures, he directed the French civil agencies in obtaining standing timber and manufactured forest products required by the allied armies. With great administrative ability he superintended the supply of lumber to the American Expeditionary Forces, at all times doing everything within his power to further our needs, affording most loyal cooperation.

CHICOYNEAU DE LAVALETTE DE COË-
  LOSOUET, CHARLES O. M.
R—France.
G. O. No. 126, W. D., 1919.

Lieutenant colonel of infantry, French Army.
In his capacity as chief of the French Geographic Bureau he went far beyond the bounds of duty to render valuable services to the American Expeditionary Forces. He placed a fund of valuable information at our disposal, being ever ready to give us the benefit of his wide knowledge. At all times he proved himself devoted to our interests and a most loyal friend.

CLAUDEL, HENRI E...................
R—France.
G. O. No. 45, W. D., 1919.

Major general, French Army, commander of the 26th and 79th divisions, American Expeditionary Forces.
For services rendered to the American Expeditionary Forces and to the cause in which the United States has been engaged.

CLAUDON, JOSEPH P. H...............
R—France.
G. O. No. 11, W. D., 1919.

Brevet colonel of infantry, French Army.
For services rendered the United States Army while serving as chief of the French Military Information Mission to the United States.

CLAVEILLE, ALBERT..................
R—France.
G. O. No. 62, W. D., 1919.

As French Minister of Public Works he rendered service of immense value to the Transportation Corps of the American Expeditionary Forces, showing himself at all times willing to go to any lengths to assist us. He displayed broad vision, a keen grasp of the essentials and great energy in overcoming difficulties in times of emergency.

CLAVEL, JEAN. M. G.................
R—France.
G. O. No. 126, W. D., 1919.

Colonel of engineers, French Army, at Base Section No. 2, American Expeditionary Forces, Bordeaux, France.
Through his wide experience and eminent technical ability he was of the utmost assistance in connection with the establishment of Base Section No. 2 at Bordeaux. Not only did he conduct the negotiations for placing at our disposal the indispensable Bassens dock, but he made possible the enlargement of our port facilities by the construction of railroad terminals. In numerous other ways he furnished us material aid by his sound advice and whole-hearted spirit of cooperation, rendering services of a high character to the American Expeditionary Forces.

CLEMENSON, LOUIS..................
R—France.
G. O. No. 87, W. D., 1919.

Lieutenant colonel, of infantry, French Army.
As regulating commissioner at Is-sur-Tille, France, he rendered invaluable assistance in the organization and operation of the American regulating station and depots at that point. His hearty cooperation and genius for organization aided materially in the achievement of most satisfactory results and the development of an efficient service.

COCHET, MAURICE D.................
R—France.
G. O. No. 126, W. D., 1919.

Lieutenant colonel of engineers, French Army.
In his capacity as a member of the French Mission attached to the American General Headquarters from June, 1917, to April, 1918, he gave us much assistance in all matters of organization of units, and especially those affecting the organization of artillery. By his good judgment and willing spirit of cooperation, coupled with his high military attainments, he rendered services of particular worth to the American Expeditionary Forces.

COFFEC, FREDERIC M. M. R—France. G. O. No. 126, W. D., 1919.	Brigadier general, French Army. As director of artillery, Ministry of War, during the formative period of our Artillery, he has given most valuable service to our Artillery personnel in the solution of the difficult problems of organization and training. He placed at our disposal the means of instruction and training for our Artillery. Through his efforts the Saumur Artillery School was opened to us, the instruction camps at Coetquidan, Meticon, Souge, and Valdahon, France, were prepared as centers of training, and the facilities of the Tractor Artillery School at Vincennes were opened to us. The services he has rendered to the American Expeditionary Forces have been of great value.
COLLARDET, LOUIS. R—France. G. O. No. 11, W. D., 1919.	Colonel of infantry, French Army, military attaché, French Embassy, Washington, D. C., United States of America. For services rendered the United States Army while serving as military attaché to the French Embassy and chief of French Military Missions to the United States.
COLONNA-CECCALDI, MARIE. R—France. G. O. No. 87, W. D., 1919.	Brevet lieutenant colonel of infantry, French Army. As chief of the French Mission in Base Section No. 2, American Expeditionary Forces, he furthered the combining of French and American resources and interests, aiding materially in the success of the American Expeditionary Forces and the allied cause. He displayed untiring energy, devotion to duty, and great tact in the performance of his manifold duties.
COMBY, LOUIS. R—France. G. O. No. 87, W. D., 1919.	Major general, French Army. In command of the 12th French Region, he showed unceasing devotion to the welfare and comfort of the American troops entering that territory, at all times affording the American military authorities loyal cooperation. He exercised extraordinary personal interest in Franco-American relations and rendered invaluable service to the American Expeditionary Forces.
CORVISART, CHARLES P. R. V. R—France G. O. No. 4, W. D., 1923.	Major general, commanding the 15th Army Corps, French Army. For services rendered to the American Expeditionary Forces and to the cause in which the United States has been engaged.
COUSSERGUE. R—France. G. O. No. 62, W. D., 1919.	As French regulating doctor at St. Dizier, France, he rendered the American Expeditionary Forces valuable service in making possible the evacuation of the forward areas during the St. Mihiel and Argonne offensives. When the number of American hospital trains available were found insufficient, he placed at our disposition the facilities of the French, laboring personally, day and night, in order that there might be no interruption of the service.
COUTANCEAU, MICHEL H. M. R—France. G. O. No. 87, W. D., 1919.	Major general, French Army. As commanding general of the 11th French Region, he rendered the American authorities most valuable assistance, meeting them always in a spirit of cordial cooperation. Through his willing help, billeting areas and many other facilities were placed promptly at the disposal of the American Expeditionary Forces.
CROCHET, EUGÈNE. R—France. G. O. No. 126, W. D., 1919.	Captain of infantry, French Army, liaison officer with the 1st Division, American Expeditionary Forces. As liaison officer with the 1st Division during all the active operations of the division from April to October, 1918, he showed exceptional ability and tireless energy in the performance of his exacting duties. Although his task was a difficult one, he labored incessantly to our best interests, manifesting at all times a spirit of loyalty and friendship for the American Expeditionary Forces.
CROLET, CHARLES. R—France. G. O. No. 24, W. D., 1920.	As chief, bureau of Franco-American Service, Administration Division, Ministry of War, he rendered great service to the United States in the adjustment of numerous and complicated accounts between the United States and France.
D'AMADE, ALBERT G. L. R—France. G. O. No. 126, W. D., 1919.	Major general, French Army, 16th French Region. As general commanding the 16th Region he has displayed the same brilliant military attainments and loyal devotion to the common cause which characterized his service with combat troops at the front. By his spirit of wholehearted cooperation and earnest desire to furnish every facility to the American troops in his region, he has rendered services of the utmost value to the American Expeditionary Forces.
DANTANT, GEORGES V. R—France. G. O. No. 126, W. D., 1919.	Major general, French Army, 13th French Region. After long and honorable service in active operations against the enemy he took command of the 13th Region, in which were subsequently located a large variety of American military activities. The unbroken harmony which characterized the relations between the French and American authorities was due in no small degree to his tact and breadth of vision, and by his constant desire to further our interests in every possible manner he rendered services of inestimable value to the American Expeditionary Forces.
D'ARMAU DE POUYDRAGUIN, LOUIS M. G. R—France. G. O. No. 17, W. D., 1924.	Major general, French Army. He commanded with distinction the 67th Chasseurs Division, with which a large number of American troops had the honor of serving. He continually showed every consideration for the Americans in his command, and by his skill in directing their training for effective combat he rendered services of the highest value to the American Expeditionary Forces.

D'AUVIN, FRANCOIS R.
R—France.
G. O. No. 126, W. D., 1919.

Major general, French Army.
His brilliant military attainments and forceful leadership were an important factor in the combat training of the American units which were attached to his command in the Chemin des Dames and Vosges Sectors. Later, as aide major general for personnel at French General Headquarters, he continued to display a desire to aid the American Army with all the important facilities at his disposal, thereby rendering services of the utmost value to the American Expeditionary Forces.

DEBAINS, FREDERIC H.
R—France.
G. O. No. 87, W. D., 1919.

Colonel of cavalry, French Army.
As chief of staff in the 18th French Region he gave invaluable assistance and cooperation in the operation of Base Section No. 2, American Expeditionary Forces. His efficient and painstaking efforts assisted most materially in the results achieved by the American military authorities. At all times he showed tact and a most valuable comprehension of existing conditions and our needs.

DE BARESCUT, MAURICE
R—France.
G. O. No. 17, W. D., 1924.

Brigadier general, French Army.
As aide major general for operations at French General Headquarters, he gave us information and advice of the greatest importance. At all times he extended whole-hearted cooperation to the American military authorities. He showed himself a tireless worker, brilliant tactician, and loyal friend.

DE BAZELAIRE, GEORGES.
R—France.
G. O. No. 45, W. D., 1919.

Major general, commanding the 7th Army Corps, French Army, with the 42d Division, American Expeditionary Forces.
For services rendered to the American Expeditionary Forces and to the cause in which the United States has been engaged.

DE BEAUMONT, MARIE J.
R—France.
G. O. No. 87, W. D., 1919.

Colonel of infantry, French Army.
As chief of the French Mission at Headquarters of Base Section No. 1, American Expeditionary Forces, he discharged his duties with tact, zeal, and distinguished ability. He contributed very largely by his personal efforts to creating conditions which made it possible for Base Section No. 1 to maintain a steady flow of supplies to the American troops.

DEBENEY, MARIE E.
R—France.
G. O. No. 17, W. D., 1919.

Major general, commanding the 1st French Army.
For services rendered to the American Expeditionary Forces and to the cause in which the United States has been engaged.

DE BOISSOUDY, ANTOINE P. T. J.
R—France.
G. O. No. 17, W. D., 1919.

Major general, commanding the French Army of Belgium.
For services rendered to the American Expeditionary Forces and to the cause in which the United States has been engaged.

DE CHAMBRUN, JACQUES A.
R—France.
G. O. No. 45, W. D., 1919.

Lieutenant colonel, 40th Regiment of Artillery, French Army.
For services rendered to the American Expeditionary Forces and to the cause in which the United States has been engaged.

DE COINTET, LÉON E.
R—France.
G. O. No. 126, W. D., 1919.

Brevet lieutenant colonel of artillery, French Army.
As head of the 2d Bureau, French General Headquarters, he placed every facility in his power at the disposal of officers of the American Expeditionary Forces. The cooperation he extended to us was hearty and sincere. He rendered us distinguished and valuable service and played an important part in our successes. At all times he acted with tact and energy and showed himself a loyal friend.

DE CURIERES DE CASTELNAU, NOËL M. J. E
R—France.
G. O. No. 17, W. D., 1919.

Major general, French Army, commanding the Group of Armies of the East.
For services rendered to the American Expeditionary Forces and to the cause in which the United States has been engaged.

DE FAUCIGNY-LUCINGE ET COLIGNY, AYMON J.-B.-M.
R—France.
G. O. No. 19, W. D., 1922.

Captain of cavalry, detached, French Army, head of Franco-American section, G. S., 15th Region.
For services rendered to the American Expeditionary Forces.

DEGOUTTE, JEAN M. J.
R—France.
G. O. No. 17, W. D., 1919.

Major general, French Army, commanding the group of the Armies of Flanders.
For services rendered to the American Expeditionary Forces and to the cause in which the United States has been engaged.

DE LA FERRONAYS.
R—France.
G. O. No. 87, W. D., 1919.

Major, French Army.
As personal liaison officer between the French Minister of War and the American Commander in Chief, he showed great tact and energy in the performance of his duties. He rendered service of exceptional value to the American Expeditionary Forces, laboring incessantly for the common cause, and doing much to promote the cordial relations between the French and American authorities.

DE LAGRANGE, AMAURY.
R—France.
G. O. No. 126, W. D., 1919.

Captain of air service, French Army.
His broad knowledge of the conditions of modern warfare, gained through creditable service at the front, coupled with his loyalty and friendship for the United States and the American Army, fitted him admirably for his important service as a member of the French Mission at Washington. Later upon his return to France, as chief of the aeronautical section of the French Mission at Tours, he assisted materially in the establishment of schools and training systems for our Air Service, thereby rendering services of the utmost value to the American Expeditionary Forces.

DELALAIN, JOSEPH L. P. R—France. G. O. No. 87, W. D., 1919.	Lieutenant colonel of infantry, French Army. As chief of staff of the communication zone of the French Army he was charged with the study and development of many of the most important projects affecting the American Expeditionary Forces. He displayed untiring energy in carrying out many delicate problems, and as supervising head of the whole work of the communication zone continually rendered valuable services to our armies.
DELASSUS, GEORGES A. R—France. G. O. No. 87, W. D., 1919.	Major of air service, French Army. As chief of the 1st Bureau, Office of the French Undersecretary of State for Aeronautics, he rendered services of especial value to the American Expeditionary Forces. Through his energetic efforts the American balloon companies in France were at all times supplied with the equipment necessary to efficient operation.
DELAUNAY, M. L. R—France. G. O. No. 27, W. D., 1922.	Captain of infantry, French Army. For services while serving in the capacity of liaison officer with the 78th Division, American Expeditionary Forces.
DE L'ESPÉE, JEAN F. M. H. R—France. G. O. No. 126, W. D., 1919.	Major general, French Army. In command of the 5th Region he performed his important duties with marked success. At all times he manifested the utmost consideration for American interests in the territory under his jurisdiction. By his generous action in placing at our disposal many French barracks and cantonments in the Department of Loire-et-Cher and at other places, he rendered services of the highest value to the American Expeditionary Forces.
DE L'ESTRADE, JACQUES E. R—France. G. O. No. 126, W. D., 1919.	Captain of air service, pilot, French Army, attached to Air Service headquarters, American Expeditionary Forces, Paris, France. During his connection with our Air Service headquarters in Paris he was of the greatest assistance in solving many problems relating to supply through his wide practical experience and intimate knowledge of French factory and industrial methods, particularly in connection with the repair of airplanes and the procurement of spare parts. Ever displaying a desire to further the interests of the Air Service of our Army in every possible way, he rendered services of high character to the American Expeditionary Forces.
DELIGNY, HENRI V. R—France. G. O. No. 87, W. D., 1919.	Major general, French Army. In command of the 3d French Region at the time of the formation of American Base Section No. 4, he gave most helpful attention to the needs of the American Expeditionary Forces, affording us his unhesitating support. His tireless efforts in our behalf assured the success of important projects in the base section, contributing largely to the achievements of our armies.
DE MARENCHES, CHARLES C. M. R—France. G. O. No. 45, W. D., 1919.	Captain of infantry, French Army, liaison officer between the commander in chief, American Expeditionary Forces, and Marshals Foch and Pétain. For services rendered to the American Expeditionary Forces and to the cause in which the United States has been engaged.
DE MARGUERY. R—France. G. O. No. 87, W. D., 1919.	Admiral, French Navy. As French naval representative at the port of Nantes, he cooperated at all times with the American authorities, showing ability of the highest order and tact in the performance of his duties. He gave us the greatest possible assistance in times of emergency, far exceeding the bounds of his duties to render important services to the American Expeditionary Forces.
DE MAUD'HUY, LOUIS E. R—France. G. O. No. 45, W. D., 1919.	Major general, French Army. For services rendered to the American Expeditionary Forces and to the cause in which the United States has been engaged.
DE MONTAL, LÉOPOLD P. R—France. G. O. No. 29, W. D., 1919.	Major, French Army. For services rendered the U. S. Army while serving as the liaison officer between the embassy, the High Commission of France, and the War Department.
DE NEUFLIZE, ANDRÉ P. R—France. G. O. No. 49, W. D., 1922.	Lieutenant, General Staff, French Army. As liaison officer between the French Ministry of Armament and the American services in France and later as chief of the armament section of the French High Commission in the United States, he rendered valuable services to the American Army. To his efforts may be attributed, in a great measure, the success which the American Government had in obtaining from French sources artillery matériel and ammunition needed to complete the equipment for American troops arriving in France.
DE NEUVILLE, SEBASTIEN H. R—France. G. O. No. 24, W. D., 1920.	Lieutenant of cavalry, French Army. As chief of the sales and purchases section, French Mission, with the American service at Paris, he rendered great service to the United States.
DENNERY. R—France. G. O. No. 62, W. D., 1919.	French inspector general of the telephone service of the Department of Posts, Telephones, and Telegraphs. He rendered the American Expeditionary Forces service of the greatest value in securing much-needed supplies and personnel for the Signal Corps. He was devoted in his efforts to secure all grants from the French Posts, Telephones, and Telegraphs administration to the best interests of our service.

DE PELACOT, JACQUES M. G. J.
R—France.
G. O. No. 126, W. D., 1919.

Lieutenant colonel of infantry, French Army.
As chief liaison officer with the 5th Army Corps from the date of its organization until a period subsequent to the suspension of hostilities, he performed his important duties with distinguished ability and a spirit of whole-hearted cooperation. At all times he assured efficient liaison between the American troops and adjacent French units. Subsequently, while he was a member of the French Mission attached to the 1st Army, he continued to render invaluable services to the American Expeditionary Forces.

DERCLE, CHARLES U.
R—France.
G. O. No. 59, W. D., 1921.

Surgeon colonel, service of military health, French Army.
For services as liaison officer and adviser to the Surgeon General. By his thorough knowledge he gave much needed and excellent advice upon medico-military subjects, which was of great assistance in the solution of important problems confronting the Surgeon General. He has rendered services of great value.

D'ESPEREY, LOUIS F. M. F.
R—France.
G. O. No. 17, W. D., 1919.

Major general, French Army, commander in chief of the Allied Armies of the Orient.
For services rendered to the American Expeditionary Forces and to the cause in which the United States has been engaged.

DE ST. QUENTIN, MARIE C. A. G.
R—France.
G. O. No. 87, W. D., 1919.

Major of artillery, French Army.
He took upon his own shoulders the complete reorganization of the French delivery services, that the American Air Service might be supplied in a time of great need with airplanes, motors, and a large number of spare parts. His foresight was marked, and he acted at all times with keen judgment and energy, rendering most valuable services to the American Expeditionary Forces.

DESTICKER, PIERRE H.
R—France.
G. O. No. 17, W. D., 1924.

Brigadier general, French Army.
As aide major general of Marshal Foch's staff, he collaborated with wholehearted interest with officers of the American Expeditionary Forces, and brought to their aid his sound judgment and military attainments in the solution of perplexing and intricate problems. At all times he was unswerving in devotion to his important and exacting duties, and proved a loyal friend. The services which he rendered us were of inestimable value.

DE VALDNER DE FREUNDSTEIN, MAURICE F.
R—France.
G. O. No. 59, W. D., 1921.

Reserve captain of cavalry, French Army.
For services rendered to the American Expeditionary Forces.

DHE, PAUL.
R—France.
G. O. No. 87, W. D., 1919.

Colonel of air service, French Army.
As Military Director of French Aeronautics he displayed untiring attention to the problems of the American Air Service, and by his earnest and wholehearted cooperation did much to render possible of execution the air program of the American Expeditionary Forces. In all of his relations with the American Air Service he acted in a broad-minded manner, showing a spirit of willing helpfulness at all times.

DIVE, GABRIEL A.
R—France.
G. O. No. 15, W. D., 1923.

Subintendant of the 3d class, French Army.
As a member of the staff of the inspector general of supply, French Army, he was charged with the responsibility for the cooperation of his section of the French General Staff with the American Expeditionary Forces with reference to supply problems. The rare technical skill, resourcefulness, tact and spirit of cooperation displayed by Colonel Dive, coupled with sound judgment and unremitting devotion to duty contributed markedly to the successful operations of the American forces in France.

DOMEJEAN, RAYMOND X.
R—France.
G. O. No. 87, W. D., 1919.

Major of artillery, French Army.
He rendered service of inestimable value to the American Expeditionary Forces in connection with the supply and maintenance of artillery matériel. Throughout the offensive of the 1st American Army, in the St. Mihiel and Meuse-Argonne operations, he commanded with great distinction the important artillery repair and supply establishment at Souhesmes.

DOUMENC, JOSEPH E. A.
R—France.
G. O. No. 87, W. D., 1919.

Major of artillery, French Army.
As chief of the French Automobile Service, he was intimately associated and uniformly helpful in the development of the American motor transportation system in France. With extraordinary foresight he organized the interallied motor reserve. His exceptional ability and personal efforts made possible the transport of large numbers of American troops at times when the success of operations depended on rapid transportation.

DUBAIL, AUGUSTIN Y. E.
R—France.
G. O. No. 87, W. D., 1919.

Major general, French Army, Military Governor of Paris, France.
Through his distinguished efforts one of the largest hospitals in Paris was made available for the use of the American Expeditionary Forces, rendering possible the hospitalization there of hundreds of the sick and wounded of the American armies.

DUCHENE, JOSEPH C.
R—France.
G. O. No. 126, W. D., 1919.

Lieutenant colonel of artillery, French Army.
In his capacity as chief of Marshal Pétain's cabinet he extended to us most helpful cooperation, far exceeding the bounds of his important duties to render us valuable assistance. Possessed of military attainments of a high order, he aided the allied cause greatly, and by the advice and information he so willingly furnished rendered very valuable services to the American Expeditionary Forces.

DUFFOUR, GASTON C. G. A. ....
R—France.
G. O. No. 126, W. D., 1919.

Lieutenant colonel of infantry, French Army.
In command of troops in the field and as chief of the French Operations Bureau he rendered services of inestimable value in the prosecution of the war. He gave us very material assistance by imparting to the chief of the American Military Mission at French General Headquarters information as to the plans of his bureau. At all times he afforded us whole-hearted cooperation, going far beyond the bounds of duty to render service to the American Expeditionary Forces.

DUFIEUX, JULIEN C. M. S. ....
R—France.
G. O. No. 17, W. D., 1924.

Brigadier general, French Army.
As assistant chief and chief of the French Operations Bureau he was always ready to cooperate with the American military authorities. He willingly furnished us with valuable information and wise advice. Able, tactful, and helpful, he rendered very valuable services to the American Expeditionary Forces.

DUMAS, CHARLES F. E. ....
R—France.
G. O. No. 126, W. D., 1919.

Lieutenant, French Army.
On duty with the French mission attached to the 4th American Army Corps, and later with the 2d American Army, he rendered most valuable assistance to both the G-1 and G-3 sections of the staff in the preparations for and during the St. Mihiel offensive. He prepared the orders for the movements of French troops operating in this offensive, and saw that they were properly executed. He also gave us great assistance in solving problems of traffic control and billeting, working cheerfully day and night performing arduous tasks and going far beyond the bounds of duty to render service to the American Expeditionary Forces.

DUMESNIL, M. J. L. ....
R—France.
G. O. No. 62, W. D., 1919.

French Undersecretary of State for Aviation, president of the interallied aviation committee.
With persistent effort and determination he supplied each American division going into the line with the same aviation equipment as was given to corresponding French units. He performed a task of tremendous magnitude with remarkable success, working at all times with great zeal for the American Expeditionary Forces.

DUMONT, GEORGES A. L. ....
R—France.
G. O. No. 19, W. D., 1922.

Colonel, French Army.
For services in connection with the supply and transportation of the American Expeditionary Forces.

DUPONT, CHARLES. ....
R—France.
G. O. No. 126, W. D., 1919.

Brigadier general, French Army.
As President of the Interallied Commission on the Repatriation of Prisoners of War he has rendered highly meritorious service to the United States and allied Governments in connection with the repatriation of American and allied prisoners released by the armistice. He always displayed a cheerful and active interest in all that pertained to their welfare and rendered sympathetic and practical cooperation.

DUPORT, PIERRE G. ....
R—France.
G. O. No. 126, W. D., 1919.

Major general, French Army, commanding the 37th and 77th Divisions, American Expeditionary Forces.
It was under his brilliant command that the 37th and 77th United States Divisions first entered the zone of operations and engaged the enemy at Baccarat and Badonvillers.

DURETTE, CLÉMENT. ....
R—France.
G. O. No. 19, W. D., 1922.

Major of artillery, French Army.
For services as head of the French Field Artillery Mission at the School of Fire for Field Artillery, Fort Sill, Okla., from September, 1917, to May, 1918. His knowledge of field artillery, tireless effort, unfailing tact, and loyal cooperation were largely responsible for the great value of the services of this mission to that school and to the Field Artillery.

DUTEY, HENRI. ....
R—France.
G. O. No. 49, W. D., 1922.

Lieutenant, French Army.
For services while serving as liaison officer with the 15th Field Artillery, 2d Division, American Expeditionary Forces.

DUVAL, MAURICE. ....
R—France.
G. O. No. 87, W. D., 1919.

Brigadier general, French Army.
As aide-major general of Air Service he cooperated in every way possible with the chief of the U. S. Air Service in the organization and development of the American air units at the front. He placed French units at our disposal during important offensive operations and at all times showed himself a most loyal friend. His service to the American Expeditionary Forces was most valuable.

EBENER, CHARLES. ....
R—France.
G. O. No. 87, W. D., 1919.

Major general, French Army.
As Military Governor of Lyon, France, and general commanding the 14th French Region, he served with distinction. Amid his manifold responsibilities and duties he at all times lent us his valuable assistance, greatly furthering the interests of the American Expeditionary Forces. He was in command of a region in which were located large American military centers, and the services rendered by him were of inestimable value.

ESTIENNE, JEAN B. E. ....
R—France.
G. O. No. 126, W. D., 1919.

Major general, French Army.
As chief of tank corps, French Army, he manifested a constant desire to aid our Tank Service in every possible way, assisting in the instruction of American officers at Recloses and giving us the benefit of his extensive experience by his valuable advice. He also came to our assistance at a critical time when he helped secure a large number of tanks from the French Government, thereby rendering invaluable services to the American Expeditionary Forces.

FAYOLLE, MARIE E.
R—France.
G. O. No. 17, W. D., 1919.

Major general, commanding Group of Armies of the Center, French Army.
For services rendered to the American Expeditionary Forces and to the cause in which the United States has been engaged.

FERRIÉ, GUSTAVE A.
R—France.
G. O. No. 87, W. D., 1919.

Colonel of engineers, French Army.
He rendered the American Expeditionary Forces services of exceptional value in his capacity as technical director of the military wireless. Without his counsel and unremitting devotion to our interests, it would have been exceedingly difficult to have equipped our forces with the indispensable radio apparatus which he placed so freely at our disposal.

FEVRIER, CHARLES.
R—France.
G. O. No. 59, W. D., 1921.

Medical inspector general, medical service, French Army.
For services as chief medical officer of the military government of Paris. During the period of the great offensives of Chateau-Thierry, Soissons, and the Meuse-Argonne, when the American sick and wounded were arriving in great numbers from the battle front and our hospital resources and accommodations were exhausted, he placed at the disposal of the American Medical Service the entire available space in the French military hospitals of Paris. During the development of the American hospital facilities, General Fevrier greatly assisted the American Medical Service in the location and obtaining of suitable buildings for hospitals. He has rendered service of great merit to the United States.

FILLONNEAU, ETIENNE.
R—France.
G. O. No. 87, W. D., 1919.

Brigadier general, French Army.
As chief of the French Mission at Headquarters, Services of Supply, American Expeditionary Forces, he proved assiduous, tactful, and efficient in the performance of his duties. As intermediary between the Services of Supply and the Bureau of Franco-American Relations, he was prompt and most helpful, rendering exceptional service to the American Expeditionary Forces.

FILLOUX, LOUIS J. F.
R—France.
G. O. No. 87, W. D., 1919.

Lieutenant colonel of artillery, French Army.
He designed and developed the 155-mm. Giant Power Filloux material, which proved indispensable to the American Expeditionary Forces. He rendered further valuable service by placing all his technical engineering ability and experience at our disposal for the manufacture of this material in America.

FOCH, FERDINAND.
R—France.
G. O. No. 111, W. D., 1918.

Marshal of France, commander in chief of the allied armies.
As an expression to him of the high regard of the people of the United States and of their Army, for the distinguished and patriotic services which he has rendered to the common cause in which he has been associated on the battle fields of Europe.

FORT, JULES E. L.
R—France.
G. O. No. 87, W. D., 1919.

Lieutenant colonel of infantry, French Army.
As chief of the French Mission of the 1st American Army throughout its operations he performed with distinction and success the task of coordinating the work of the American and French units. He was painstaking and untiring in his efficient efforts to maintain helpful cooperation.

FOURNIER, GASTON.
R—France.
G. O. No. 14, W. D., 1920.

Colonel, French Army.
He rendered specially meritorious service as chief of staff of the French Mission. He organized the business side of the mission, the arrangements for meetings, the editing, distributing, and filing of notes with skill, accuracy, and dispatch.

FOURNIER, PIERRE.
R—France.
G. O. No. 87, W. D., 1919.

Captain, French Army.
As chief of the 1st Bureau of the French Mission attached to General Headquarters, American Expeditionary Forces, his thorough knowledge of the French military service, unfailing tact, and spirit of cheerful cooperation greatly assisted the officers of the American General Staff, contributing to the success of the allied cause.

FRANTZ, PHILIPPE.
R—France.
G. O. No. 3, W. D., 1922.

Brigadier general, chief of staff, 2d French Army.
For services rendered the American Expeditionary Forces in connection with the Meuse-Argonne operation.

FRESOULS, EMILE S.
R—France.
G. O. No. 126, W. D., 1919.

Major of artillery, French Army.
As regulating officer at Dunkerque he occupied a particularly difficult position in being called upon to meet the needs of the British, French, Belgian, and American Armies. He, despite the complexities of situations with which he was confronted, rendered us exceptional service, lightening the task of supplying our troops in Belgium, and at all times extending whole-hearted cooperation. He brought his wide knowledge of existing conditions to bear upon difficult problems of transportation, handling them with peculiar success. At all times he displayed marked devotion to his exacting tasks. In times of need he proved himself a loyal friend of the American Expeditionary Forces.

FRID, GEORGE E.
R—France.
G. O. No. 30, W. D., 1921.

Brevet colonel, Deputy Chief of Staff, French Army of the Rhine and Allied Forces of Occupation.
For services to the allied cause and to the American forces in Germany.

GAMELIN, MAURICE G.
R—France.
G. O. No. 53, W. D., 1920

Brigadier general, French Army.
As commander of the 9th French Division of Infantry, he displayed conspicuous military attainments and most loyal devotion to the allied cause. He assigned a large number of French officers and noncommissioned officers as instructors, prepared complete program of final training, cooperated with our personnel, and carefully supervised the training of the 32d Division during its training period from May 16 to July 21, 1918. By his important efforts he has rendered invaluable services to the American Expeditionary Forces.

GANNE_____
R—France.
G. O. No. 62, W. D., 1919.

French delegate of the Commissioner General for the Franco-American War Affairs.
He displayed untiring energy and exceptional ability in handling relations between the French and American authorities. At all times tactful and courteous, he did much to cement the feelings of friendship between the two nations, rendering services of great value to the American Expeditionary Forces.

GASCOUIN, FIRMIN E._____
R.—France.
G. O. No. 19, W. D., 1922.

Brigadier general, French Army.
For services rendered to the American Expeditionary Forces.

GASSOUIN, JOSEPH M. G._____
R—France.
G. O. No. 126, W. D., 1919.

Brigadier general, French Army.
As director of French military railways he displayed the same administrative ability and devotion to duty which had characterized his service as a commander of a division in the field. He has at all times showed a broad-minded appreciation of the transportation needs of the American Army and by his cordial spirit of cooperation and valuable assistance in furnishing the necessary railway facilities has rendered a substantial service to the American Expeditionary Forces.

GAUCHER, LÉON M._____
R—France.
G. O. No. 126, W. D., 1919.

Brigadier general, French Army.
As commander of the 164th French Division of Infantry, he displayed brilliant leadership and military attainments of a high order. The splendid successes achieved by the 4th American Division are due in no small degree to the splendid initial training in actual warfare they received under his command when they first went into the trenches in the Toul sector.

GAUTHIER, ADOLPHE F. M._____
R—France.
G. O. No. 126, W. D., 1919.

Major of infantry, French Army, general staff of the 14th Region at Lyons, France.
As head of the section of the G-2, with American division, he displayed untiring zeal, energy, and devotion to his exacting duties. By his spirit of whole-hearted cooperation with our personnel in solving the many and complex problems with which we were confronted he proved himself a most efficient officer and valuable friend.

GAUTHIER, JEAN B._____
R—France.
G. O. No. 24, W. D., 1920.

Military intendant, French Army.
As Assistant Director, Bureau of the Undersecretary of State for the Liquidation of Stocks, he ably discharged a difficult and important task and rendered great service to the United States in the settlement of complicated claims and the liquidation of stocks involving millions of dollars.

GAY, ANTOINE_____
R—France.
G. O. No. 87, W. D., 1919.

Lieutenant, French Army, representative of the French railroads, G-4, Base Section No. 1, Services of Supply, American Expeditionary Forces.
As representative of the French railroads, he rendered the American Services of Supply invaluable assistance by obtaining for us the rolling stock necessary to forward vitally important shipments to the front. During the period of the Argonne offensive his services were of especial value.

GÉRARD, AUGUSTIN G. A._____
R—France.
G. O. No. 17, W. D., 1919.

Major general, commanding the 8th French Army.
For services rendered to the American Expeditionary Forces and to the cause in which the United States has been engaged.

*GÉROME, AUGUSTE C._____
R—France.
G. O. No. 87, W. D., 1919.

Major general, French Army.
In command of the 15th French Region, he rendered the American Expeditionary Forces most valuable assistance, showing himself resourceful and at all times willing to aid the American military authorities at the base port at Marseille. His tact was marked and he displayed ability of a high order, combined with energy and devotion to duty.
Posthumously awarded. Medal presented to wife, Madam Auguste C. Gérome.

GIGIDOT, JEAN R._____
R—France.
G. O. No. 126, W. D., 1919.

Major of artillery, French Army.
As chief of staff to the adviser of the Prime Minister for Franco-American affairs, by his cordial spirit of whole-hearted cooperation with the American authorities he has rendered services of marked distinction. He has ably assisted the American Expeditionary Forces in many important matters which came before his office.

GILLY, EUGÈNE L._____
R—France.
G. O. No. 87, W. D., 1919.

Captain, French Army.
As French Commandant of the port of St. Nazaire he cooperated whole-heartedly with the American authorities, extending them most valuable assistance. Due to his tireless efforts, facilities of the utmost importance were placed at our disposal. He showed marked ability and initiative in the performance of his arduous duties.

GIROD, LÉON A._____
R—France.
G. O. No. 87, W. D., 1919

Colonel of Air Service, French Army.
In his capacity as Chief of Training of the French Aviation Schools he opened those schools to our cadets at a time when our air program was seriously retarded by lack of trained pilots. He took a personal interest in the training of our pilots, and it is due, in a large measure, to his attention that we did not incur further delay in placing our squadrons at the front. His high military attainments enabled him to render us a very valuable service, which was enhanced by the spirit of friendship and cooperation he at all times manifested.

GODART, JUSTIN_____
R—France.
G. O. No. 126, W. D., 1919.

As Undersecretary of State, Chief of the Medical Department of the French Army, he exercised his influence and energy to assist in the initial hospitalization of the American Army, and by his cordial cooperation and untiring efforts expedited and facilitated that hospitalization, thus rendering valuable service to the American Expeditionary Forces.

GODEFROY, ANTOINE P. R—France. G. O. No. 87, W. D., 1919.	Lieutenant colonel of engineers, French Army. As chief of the French Mission attached to Base Section No. 6, American Expeditionary Forces, he rendered invaluable assistance in the development of Marseille as an American base port. He displayed tact, energy, and foresight in a position of great responsibility, and his assistance aided markedly in the prompt forwarding of supplies to the troops operating in the advanced zones.
GORJU, EUGÈNE. R—France. G. O. No. 126, W. D., 1919.	Captain of artillery, French Army. As delegate from the Director of the Automobile Service, French General Headquarters, to the 1st and 2d American Armies, he displayed great foresight and sound judgment, working ceaselessly and with untiring energy in our behalf. He aided us repeatedly in the solution of perplexing problems of transportation. At all times devoted to our interests, he proved himself a loyal friend.
GOUDCHAUX, JOSEPH M. R—France. G. O. No. 30, W. D., 1921.	Captain of General Staff, French Army. For services as commander of the French Mission with the 89th Division and the 9th Army Corps of the American Expeditionary Forces.
GOUIN, EDOUARD F. M. A. R—France. G. O. No. 126, W. D., 1919.	Lieutenant, 59th Regiment of Artillery, French Army. He rendered services of great worth to the American Expeditionary Forces while he was attached successively to the 1st Field Artillery Brigade and 1st Division, the 5th Army Corps, and the 9th Army Corps. His earnest devotion to duty and marked tactical ability displayed in the execution of important artillery missions have earned for him the lasting respect and high esteem of the American officers with whom he served.
GOURAUD, HENRI J. E. R—France. G. O. No. 17, W. D., 1919.	Major general, commanding the 4th French Army. For services rendered to the American Expeditionary Forces and to the cause in which the United States has been engaged.
GOURGUEN, LOUIS A. G. R—France. G. O. No. 126, W. D., 1919.	Lieutenant colonel of infantry, French Army. As chief of the 2d Bureau, French Ministry of War, he at all times effectively collaborated with the various missions in deciding questions of policy and work to be undertaken. By his tactful and wholehearted cooperation he rendered valuable and distinguished service to the American Expeditionary Forces. He was tireless in devotion to his important duties, and at all times showed himself a loyal friend, zealous in our behalf.
GOYBET, MARIANO F. J. R—France. G. O. No. 17, W. D., 1924.	Brigadier general, French Army. As commander of the 157th French Division of Infantry, he was an important factor in the successes of the Allies. By his valiant leadership and eminent tactical ability the officers and soldiers of the 371st and 372d American Infantry Regiments count it a great honor to have served as part of his command in the operations conducted by him in the Champagne and in the Vosges.
GRAZIANI, JEAN C. R—France. G. O. No. 3, W. D., 1922.	Major general, French Army. For services as commander in succession of the 28th Infantry Division, the 17th Army Corps, the 12th Army Corps, the French forces in Italy, and as the French representative on the Interallied Military Mission to Hungary, positions of great responsibility, in which he rendered valuable services in the furtherance of the allied cause.
GROUT, M. G. R—France. G. O. No. 87, W. D., 1919.	Rear admiral, French Navy, at Base Section No. 5, American Expeditionary Forces. He rendered services of marked distinction and value to the American Expeditionary Forces in lending his assistance during the formative period of American Base Section No. 5. Going far beyond the bounds of duty, he placed at the disposal of that base all of the facilities at his command.
GUERIN, ETIENNE F. M. R—France. G. O. No. 30, W. D., 1921.	Major general, French Army. During the assault of the 5th Army Corps in the St. Mihiel offensive, Sept. 12, 1918, General Guerin commanded the 15th Division of Colonial Infantry, French Army, as a part of the 5th Army Corps. By his superior skill and leadership, his resolute execution of orders, his loyal and aggressive devotion to the operations of the 1st American Army, and his high courage in combat, he contributed in a determining manner to the assault of the 5th Army Corps.
GUILLAUMAT, MARIE L. A. R—France. G. O. No. 17, W. D., 1919.	Major general commanding the 5th French Army. For services rendered to the American Expeditionary Forces and to the cause in which the United States has been engaged.
GUILLON, HENRI A. R—France. G. O. No. 87, W. D., 1919.	Lieutenant colonel of infantry, French Army. As chief of the French Mission at Headquarters, Services of Supply, American Expeditionary Forces, he was zealous, tactful, and energetic in the performance of his exacting duties. At all times he displayed tact and exceptional ability. His loyal cooperation proved of great assistance to the American Expeditionary Forces.
HALLIER, EUGÈNE H. R—France. G. O. No. 126, W. D., 1919.	Brigadier general, French Army. As Assistant Chief of the Army General Staff of the French Army and in his capacity as chief of the 2d Bureau of the General Staff, by his loyal cooperation in all matters concerning the American Army, he has rendered the American Expeditionary Forces eminent service. He at all times furthered those friendly relations which characterized all association of the French and American authorities.

HALLOUIN, LOUIS E. A _____
  R—France.
  G. O. No. 87, W. D., 1919.

Major general, French Army.
In command of the 18th French Region he gave the American military authorities earnest cooperation and sound advice on matters of great importance. His able assistance counted greatly in increasing the efficiency of Base Section No. 2. His tact, energy, and wide knowledge of conditions were most marked.

HALPHEN, HENRI J _____
  R—France.
  G. O. No. 126, W. D., 1919.

First lieutenant of artillery, French Army.
As a member of the artillery section of the French Mission, during the entire period of the American activities, by his energetic policy and loyal spirit of cooperation with our Artillery personnel, he rendered valuable assistance to the American Expeditionary Forces. His advice was sound; his judgment was good.

HANAUT, HENRI S. A _____
  R—France.
  G. O. No. 105, W. D., 1919.

Major of infantry, General Staff, French Army.
As a member of the French Military Commission he was on duty with the training and instruction branch of the war plans division of the General Staff. In this branch he was charged with teaching the higher phases of the military art, a course which only a talented and experienced officer could have conducted. His services to the United States were of inestimable value.

HANOTEAU, JEAN A _____
  R—France.
  G. O. No. 126, W. D., 1919.

Captain of infantry, French Army.
As a member of the staff on the cabinet of the Minister of War, charged with affairs concerning American and other allied nations, by his splendid tact, good judgment, and sympathetic understanding of the American people, he has promoted that friendly spirit which has marked all dealings of the French with the American authorities.

HANOTTE, MAURICE J. V _____
  R—France.
  G. O. No. 59, W. D., 1921.

Territorial surgeon, major of the 2d class, Medical Service, French Army.
For services as assistant to the director of the Medical Service, 9th Region, and as a member of the French Mission, Headquarters, Services of Supply. Due to his energy, devotion to his task, and keen judgment, he rendered valuable assistance to the American Expeditionary Forces in the procurement of buildings for hospitalization and in the placing of American soldiers in French hospitals. His cordial assistance and excellent advice have been of great service to the American medical officials.

HAVARD, VICTOR _____
  R—France.
  G. O. No. 87, W. D., 1919.

Colonel of infantry, French Army.
As regulating officer at Creil and Nantes he rendered great service to the American Expeditionary Forces in connection with the supply and transportation of the American units engaged at Cantigny and Chateau-Thierry. Later he rendered exceptionally valuable assistance in relation to the St. Mihiel and Argonne offensive, at all times displaying brilliant organizing ability and a keen spirit of cooperation.

HELLET, FRÉDÉRIC E. A _____
  R—France.
  G. O. No. 45, W. D., 1919.

Major general, French Army.
For services rendered to the American Expeditionary Forces and to the cause in which the United States has been engaged.

HENNOCQUE, EDMOND C. A _____
  R—France.
  G. O. No. 30, W. D., 1921.

Major general commanding the 4th Cavalry Division, French Army of the Rhine.
For services to the allied cause and to the American forces in Germany.

HERING, PIERRE _____
  R—France.
  G. O. No. 30, W. D., 1921.

Colonel of artillery, Deputy Chief of Staff, French Army of the Rhine and Allied Forces of Occupation.
For services to the allied cause and to the American forces in Germany.

HERR, FRÉDÉRIC G _____
  R—France.
  G. O. No. 87, W. D., 1919.

Major general, French Army.
In his capacity as Inspector General of Artillery of the French Army he rendered exceptionally valuable services to the American Expeditionary Forces in connection with the design and production of new artillery material. He gave wise advice and painstaking assistance in this task, placing all the information at his command at our disposal.

HIRSCHAUER, AUGUSTE E _____
  R—France.
  G. O. No. 17, W. D., 1919.

Major general, commanding the 2d French Army.
For services rendered to the American Expeditionary Forces and to the cause in which the United States has been engaged.

HUBERT, XAVIER, L. M. R _____
  R—France.
  G. O. No. 126, W. D., 1919.

Brigadier general, French Army.
As a commander of a unit at the front during a period of three years, he has given eminently distinguished service to the allied cause. Later, as commander of the 1st and 2d Subdivisions of the 18th Region, by his loyal cooperation with the American authorities and ever willing in spirit to assist them with all means at his disposal, he has rendered services of great worth to the American Expeditionary Forces.

HUE, EUGÈNE _____
  R—France.
  G. O. No. 126, W. D., 1919.

Major of infantry, French Army.
As chief of the 2d Bureau, French Mission, General Headquarters, American Expeditionary Forces, he displayed sympathetic intelligence and untiring ardor in interpreting and sustaining the American point of view in the many delicate problems that arose. Due to his zeal, a spirit of complete cooperation existed between the French services and the American military authorities. He rendered services of great distinction to the United States and France.

HUMBERT, GEORGES L _____
  R—France.
  G. O. No. 17, W. D., 1919.

Major general, commanding the 3d French Army.
For services rendered to the American Expeditionary Forces and to the cause in which the United States has been engaged.

JACQUEMIN, HENRI............................
  R—France.
  G. O. No. 126, W. D., 1919.

Major of cavalry, French Army.
As a member of the Presidency of the Council, Commissariat General of the Franco-American War Affairs, by his thorough application to the needs of the American troops and by his fervent cooperation with the American authorities he has rendered most distinguished services to the American Expeditionary Forces. The 28th Division will always remember his courteous and efficient efforts in their behalf in all their dealings with the French authorities under his able supervision.

JOFFRE, JOSEPH J. C............................
  R—France.
  G. O. No. 111, W. D., 1918.

Marshal of France.
As an expression to him of the high regard of the people of the United States and of their Army, for the distinguished and patriotic services which he has rendered to the common cause in which he has been associated on the battle fields of Europe.

JULLIEN, GEORGES L. E............................
  R—France.
  G. O. No. 87, W. D., 1919.

Major general, French Army.
As chief of engineers, he gave the American Expeditionary Forces hearty support at all times, rendering especially valuable service by supplying the 1st American Army with pontoon equipages and bridge material. He provided an ample training center for our troops and in solving the problems of water supply and barracks construction aided materially our engineering operations.

KLOTZ, MAURICE............................
  R—France.
  G. O. No. 46, W. D., 1920.

Captain of infantry, General Staff, French Army.
As liaison officer with the 1st American Depot Division and the 77th American Division, he rendered services of exceptional value to the American Expeditionary Forces. He acted with tact and energy in handling all questions between the American military authorities, with whom he served, and the French authorities, immeasurably facilitating our work. In the field, he assured efficient liaison with the French units operating on the flanks of the 77th American Division, rendering services of inestimable worth.

KOECHLIN-SCHWARTZ............................
  R—France.
  G. O. No. 126, W. D., 1919.

Colonel of cavalry, French Army.
As a lecturer at the American Army schools he displayed extraordinary enthusiasm and ability in the performance of his important duties. Energetic and always working for the best interests of the allied cause, he rendered very valuable services to the American Expeditionary Forces in connection with the training of a large number of its officers.

LABROSSE, HENRI............................
  R—France.
  G. O. No. 126, W. D., 1919.

Captain of cavalry, French Army.
As a member of the Army General Staff, by his excellent military ability, sound judgment, and tact, he has given most creditable service to the Allies. His interest in the American Army was shown by his ever willing spirit to assist us by all means at his disposal. His services have been of great value to the American Expeditionary Forces.

LACAZE, MARIE J. L............................
  R—France.
  G. O. No. 87, W. D., 1919.

Vice admiral, French Navy, commander in chief and prefect, 5th Maritime Region.
At a critical period of the war he performed invaluable service by giving assistance in the establishment of a supplementary American port at the French naval base of Toulon. He assigned docks, storage, and transport facilities to the American authorities, aiding them with his wise advice and experience, and thus assuring the rapid development of the American interests at this port.

LACOMBE, LOUIS F............................
  R—France.
  G. O. No. 75, W. D., 1919.

Major of infantry, French Army.
For service to the allied and associated Governments as chief of staff, French section, Supreme War Council.

LACOMBE DE LA TOUR, ALPHONSE E. E. E. X. J.
  R—France.
  G. O. No. 126, W. D., 1919.

Major general, French Army.
As commander of the 5th French Cavalry Division in its operations with the 1st American Corps, he handled his division with ability. Zealous in his effort to effectively use his force, he was severely wounded while making a reconnaissance of the terrain between Four de Paris and Varennes.

LASNET, ALEXANDRE B............................
  R—France.
  G. O. No. 39, W. D., 1920.

Medical inspector, French Army.
As chief surgeon of the 6th French Army, he placed all the facilities of his complete organization at the disposal of the American Expeditionary Forces for instruction purposes of our inexperienced medical personnel. In February, 1918, when the 26th Division became a part of the 6th French Army, he made the most complete and detailed arrangements for the care of our sick and wounded and assisted in every possible way in making their instruction period a success. During the Chateau-Thierry offensive, when several divisions of the American Expeditionary Forces were in the 6th French Army, he rendered significant aid to the American units. Later, when he became chief surgeon of the group of armies of the Center, he again manifested the same whole-hearted spirit of cooperation and helpfulness. No demands were ever too great to receive his consideration and no effort was left unmade by him to assist in meeting the many emergencies which confronted the medical department in the trying periods of 1918. His services were of great consequence to the American Expeditionary Forces.

LAVAL, EDOUARD C............................
  R—France.
  G. O. No. 87, W. D., 1919.

Surgeon principal, 2d class, French Army.
As a member of the 4th Bureau of French General Headquarters, he was closely associated with front-line medical tasks in the American Expeditionary Forces, and by his cooperation aided us in procuring hospital facilities. He displayed good judgment, broad experience, and unfailing courtesy, rendering service of inestimable value in the proper care of the sick and wounded.

LAVELLE, PAUL M. P. L.
R—France.
G. O. No. 87, W. D., 1919.

Major of infantry, French Army.
As a member of the 4th French Bureau, General Staff, he showed himself at all times willing to help the American Expeditionary Forces. To expedite the selection of sites for various American transportation projects, he traveled extensively, making a personal inspection. His wide experience, tireless energy, and loyal cooperation made his services of inestimable value.

LEBRUN, LÉONCE M.
R—France.
G. O. No. 45, W. D., 1919.

Major general, French Army, commanding the 3d French Region.
For services rendered to the American Expeditionary Forces and to the cause in which the United States has been engaged.

LECONTE, MARIE G. F.
R—France.
G. O. No. 45, W. D., 1919.

Major general, French Army, commanding the 33d French Army Corps.
For services rendered to the American Expeditionary Forces and to the cause in which the United States has been engaged.

LEFEVRE-PONTALIS, HENRI G.
R—France.
G. O. No. 126, W. D., 1919.

Lieutenant of infantry, French Army.
As a member of the special Franco-American section, Army General Staff, he has administered with marked ability many important matters concerning the American Army. Actuated at all times by a spirit of friendship for the American Forces, he has rendered them great assistance by the excellent manner in which he managed the many items which came before his bureau.

LEFORT, FERRÉOL F. G.
R—France.
G. O. No. 19, W. D., 1922.

Colonel of engineers, French Army, assistant chief of the D. M. T. A.
For services in connection with the transportation of American troops and supplies.

LEGRAND, ALBERT L.
R—France.
G. O. No. 87, W. D., 1919.

Captain, French Army.
As chief liaison officer with the American Rents, Requisitions, and Claims Service, he rendered most valuable service in conducting negotiations between the French and American authorities relative to the settlement of claims and in making of agreements relative to the occupancy of French Government property by the American Expeditionary Forces. He displayed tact and zeal at all times, working unreservedly for the good of the allied cause.

LEGRAND, EMILE E.
R—France.
G. O. No. 87, W. D., 1919.

Major general, French Army.
In command of the 15th French Region at the time of the creation of Marseille as an American base port, he gave most valuable assistance and advice to the American military authorities, rendering possible the rapid organization and development of the American base. His able cooperation assisted greatly in putting the port on an efficient basis capable of supplying the needs of the troops at the front.

LEGRAND, JACQUES G.
R—France.
G. O. No. 87, W. D., 1919.

Major of artillery, French Army.
As executive member of the artillery section of the French mission at American General Headquarters during the entire period of American activities he gave himself wholly to the varied details of organization, training, and equipment of the American artillery. His services were of great value to the American Expeditionary Forces.

LE HENAFF, JOSEPH H. F.
R—France.
G. O. No. 87, W. D., 1919.

Colonel of infantry, French Army.
As representative of the French Government on the Interallied Transportation Council he evinced great vision and excellent judgment in handling questions of interallied transportation. In helping to solve problems of supply he rendered assistance of the greatest value to the American Expeditionary Forces.

LEMERRE, LOUIS A.
R—France.
G. O. No. 126, W. D., 1919.

Lieutenant colonel of infantry, French Army.
As a member of the French Military Mission attached to General Headquarters, American Expeditionary Forces, he rendered great assistance in the selection of French officers and soldiers for duty with American divisions and other American units. He at all times displayed remarkable tact and sound judgment in furthering the important tasks under his charge. Equipped, as he was, by a long period of staff training and high professional ability furthered by a keen appreciation of the importance of his task, he contributed to the maintenance of these cordial relations which had been established between the French and American forces.

LEPELLETIER, LOUIS R. V.
R—France.
G. O. No. 46, W. D., 1920.

Colonel of artillery, French Army.
As chief liaison officer for the French War Office to American General Headquarters, he rendered services of a distinguished character. By his painstaking efforts and thorough understanding of conditions which the American Army had to meet, he gave valuable assistance, at all times manifesting a spirit of loyalty and friendship for the American Expeditionary Forces.

LE PELLETIER DE WOILLEMONT, BERNARD C. F. M. X. E. G.
R—France.
G. O. No. 49, W. D., 1922.
Distinguished-service cross also awarded.

Lieutenant of cavalry, French Army.
As liaison officer with the 2d Division he participated in all the battles in which the division was engaged. By his loyal devotion to duty, whole-hearted spirit of cooperation, and high military attainments he rendered invaluable services to the American Expeditionary Forces.

LE ROCH, JULIEN E.
R—France.
G. O. No. 24, W. D., 1920.

Captain of infantry, French Army.
As a member of the Liquidation Commission, American Expeditionary Forces, he rendered great service to the United States in the adjustment of numerous and complicated accounts between the United States and France.

Le Rond, Henri L. E. R—France. G. O. No. 126, W. D., 1919.	Brigadier general, General Staff, French Army. In his capacity as a member of the general staff attached to the Minister of War and supervisor of all Franco-American Missions in France, by his thorough military knowledge and fervent spirit of cooperation, he gave the American authorities invaluable assistance. He placed at the disposal of our officials all the facilities that his extensive office afforded. His services have been of inestimable worth to the American Expeditionary Forces.
Lescanne, Fernand L. J. R—France. G. O. No. 19, W. D., 1922.	Major of infantry, chief of staff of the D. A., French Army. For services rendered the American Expeditionary Forces.
L'Hopital, René M. M. R—France. G. O. No. 126, W. D., 1919.	Captain of artillery, French Army. As a member of the Cabinet of the Minister of War, he has at all times displayed a friendly interest in the American Army, rendering it courteous consideration and assistance in all matters which came before him. His services have been of great value to the American Expeditionary Forces.
Libaud, Emmanuel U. R—France. G. O. No. 19, W. D., 1922.	Major of infantry, staff officer, French Army. For services rendered to the American Expeditionary Forces.
Linard, Jean L. A. R—France. G. O. No. 45, W. D., 1919.	Colonel of artillery, French Army, chief, French mission, General Headquarters, American Expeditionary Forces. For services rendered to the American Expeditionary Forces and to the cause in which the United States has been engaged.
Lobez, Stanislas J. R—France. G. O. No. 126, W. D., 1919.	Colonel of cavalry, French Army. As chief of the French mission with the 32d American Division, he rendered us services of exceptional value in connection with the planning and operation of an efficient system of liaison. He displayed a wide comprehension of the needs and difficulties which would be encountered and his advice proved of the greatest assistance to us. He was energetic, tactful, and tireless in his devotion to our interests.
Lombard, Emmanuel E. R—France. G. O. No. 126, W. D., 1919.	Captain of artillery, French Army. As senior French instructor at the American Artillery Training Camp at Valdahon he rendered services of great value to the American Artillery. At all times he exhibited a spirit of whole-hearted cooperation with the American authorities, assisting them by all means at his command.
Lorain, —— R—France. G. O. No. 62, W. D., 1919.	French director of the Telephonic Exploitation of the Department of Posts, Telephones, and Telegraphs. He procured much-needed telephone material for the American Expeditionary Forces at critical times. His assistance was indispensable in obtaining leases for the long lines which formed the basis of our general telephone and telegraph system. Laboring unremittingly in our behalf, he rendered service of inestimable value.
Loucheur, Louis. R—France. G. O. No. 62, W. D., 1919.	French Minister of Armament. He displayed ability of high order in the performance of his important duties. In his relations with the American authorities he was tactful and zealous in our behalf, going far beyond the bounds of his duties to render valuable service and assistance to the American Expeditionary Forces.
Mabille, Marie J. H. R—France. G. O. No. 126, W. D., 1919.	Major, French Army. While he was a member of the 2d Bureau at French General Headquarters he manifested a warm spirit of cooperation by doing all in his power to aid the members of the American Military Mission at these headquarters, placing at their disposal all the important facilities of his office, rendering invaluable service to the American Expeditionary Forces. Subsequently, as chief of the 2d Bureau of the 2d French Army, he continued to perform valiant services in the common cause.
Maison, Léopold. R—France. G. O. No. 126, W. D., 1919.	Brevet colonel detached, French Army. As commanding officer of artillery, 132d French Division, he rendered us service of exceptional value by his accurate and rapid work in directing and controlling the fire of the divisional artillery which was supporting the 37th American Division. His efficiency resulted in our infantry receiving all the advantages of constant artillery support, and he was in a large measure responsible for the success achieved. He displayed military attainments of a high order and was constant and untiring in his efforts in our behalf.
Maistre, Paul A. M. R—France. G. O. No. 17, W. D., 1919.	Major general, French Army, commanding the Group of Armies of the Center. For services rendered to the American Expeditionary Forces and to the cause in which the United States has been engaged.
Maitre, Alphonse A. R—France. G. O. No. 87, W. D., 1919.	Colonel of artillery, French Army. As chief of the artillery section, French mission, during the entire period of American activities, he gave himself wholly to the varied details of the organization, training, and equipment of the American Artillery. His services were of inestimable value to the American Expeditionary Forces.
Mangin, Charles M. E. R—France. G. O. No. 17, W. D., 1919.	Major general, commanding the 10th French Army. For services rendered to the American Expeditionary Forces and to the cause in which the United States has been engaged.

MARCOMBE, MARIE J. P.
R—France.
G. O. No. 46, W. D., 1920.

Surgeon major, 1st class, French Army.
As regulating medical officer at Le Bourget, France, he performed highly responsible duties with conspicuous success. When the number of our wounded from the operations in the vicinity of Chateau-Thierry and Soissons became so large that our transportation facilities proved inadequate, he relieved the critical situation by placing French trains at our disposal and personally directed their operations. He also furnished supplies for American patients, cooperated in selecting advantageous evacuation points, and in numerous other ways made himself the instrumentality for saving many American lives and alleviating much suffering, thereby rendering services of incalculable value to the American Expeditionary Forces.

MARTIN-ZÉDÉ, HENRI.
R—France.
G. O. No. 126, W. D., 1919.

Lieutenant of cavalry, French Army.
As a member of the French mission attached to the American General Headquarters, he has constantly exerted himself in the interest of the American Army, rendering us every assistance which his office afforded. By his courtesy and good judgment he has done much to further the strengthening of those friendly relations which have characterized the services of the French and American forces.

MARTY, AUGUSTIN A.
R—France.
G. O. No. 46, W. D., 1920.

Paymaster general of 1st class, French Army, technical inspector, French Military Postal Service.
In charge of the French Military Postal Service, he gave us assistance of inestimable value in the planning and organization of our own military postal system. At all times he cooperated with us most whole-heartedly, giving us sound advice and proving at all times most helpful and tactful. The services which he rendered to the American Expeditionary Forces were of great worth to us.

MARZAC, A. JOSEPH.
R—France.
G. O. No. 87, W. D., 1919.

Major of artillery, French Army.
In command of the Aerial Gunnery school at Cazeaux he displayed exceptional zeal and technical knowledge, combined with keen interest in the training of American students detailed to this school. He enabled them to secure training which rendered them markedly efficient at the front, thus rendering most valuable services to the American Expeditionary Forces.

MATHIEU DE VIENNE, ALEXANDRE H. M.
R—France.
G. O. No. 24, W. D., 1920.

Major of cavalry, French Army.
As chief of the French mission attached to the American services in Paris, he rendered great service to the United States.

MAUGIN, JULES E.
R—France.
G. O. No. 19, W. D., 1922.

Major of infantry, French Army, assistant chief of staff of the communication zone.
For services in connection with the ammunition supply of the American Expeditionary Forces.

MAURIER, GEORGES T. P. H.
R—France.
G. O. No. 87, W. D., 1919.

Colonel of infantry detached, French Army.
As chief of the 4th French Bureau of the French General Staff he practically controlled transportation from the coast to the army zone. He rendered the American Expeditionary Forces service of great worth, assisting us most ably in handling the Army supply problem. He cooperated always most fully and unselfishly with the American authorities.

MAURIN, LOUIS F. T.
R—France.
G. O. No. 17, W. D., 1924.

Brigadier general, French Army.
As commander of the general reserve of artillery, by his earnest efforts and loyal cooperation with our forces the American Railway Artillery was always provided with suitable cantonments and equipment facilities and an ample supply of ammunition for every mission which was assigned to them. By the high quality of service he has rendered he has contributed materially to the successes achieved by our Railway Artillery during the operations against the enemy.

MAZEL, OLIVIER C. A. A.
R—France.
G. O. No. 126, W. D., 1919.

Major general, French Army.
After serving with eminent distinction at the front he was assigned to the command of the 4th Region at Le Mans, France, where he continued to perform important duties with conspicuous ability. By his warm spirit of cooperation and appreciation of American needs he was of material assistance in connection with the establishment and development of our embarkation and replacement center at Le Mans, thereby rendering valuable services to the American Expeditionary Forces.

MENARD, VICTOR R.
R—France.
G. O. No. 87, W. D., 1919.

Major of air service, French Army.
From the inception of the American Air Service he was its constant and reliable adviser, and rendered most important services in its training and development. He personally supervised the instruction of the first American pursuit squadrons. During the St. Mihiel attack the organization which he commanded was placed at our disposal, and in his personal direction of his group during the battle he showed military ability of a high order, rendering most distinguished service.

MEYER, MAURICE.
R—France.
G. O. No. 126, W. D., 1919.

Brevet lieutenant colonel, French Army.
He served with marked distinction as head of the 2d Bureau at French General Headquarters, performing duties of the utmost responsibility. At all times he manifested the most cordial spirit of cooperation and desire to assist our officers by all the means at his disposal, thereby rendering valuable services to the American Expeditionary Forces.

MICHEL, CAMILLE C. G. A.
R—France.
G. O. No. 30, W. D., 1921.

Brigadier general, chief of staff of the French Army of the Rhine and chief of staff of the Interallied Armies of Occupation in Germany.
For services to the allied cause and to the American forces in Germany.

MICHEL-LEVY, MARCEL J. B._____ R—France. G. O. No. 126, W. D., 1919.	Second lieutenant of infantry, French Army. He served with exceptional ability as chief of the administrative offices of the French Mission at General Headquarters, American Expeditionary Forces, displaying the same sound judgment and untiring zeal which had previously characterized his service in the field. By his admirable tact and broad sense of justice he aided materially in solving many delicate questions which arose between the American Army and the French civilians, thereby rendering services of great worth to the American Expeditionary Forces.
*MOINIER, CHARLES E._____ R—France. G. O. No. 87, W. D., 1919.	Major general, French Army. As military governor of Paris, France, he constantly rendered services of the greatest value to the American Expeditionary Forces, in whose interests he proved himself zealous and self-sacrificing. Occupying a position of high distinction and with a multitude of important duties claiming his attention, he yet found time to aid us with his wise advice and extended hearty cooperation to the American military authorities. Posthumously awarded. Medal presented to next of kin, Major Moinier.
MOLINIER, FRÉDÉRIC_____ R—France. G. O. No. 126, W. D., 1919.	Captain of cavalry, French Army. As an aide-de-camp on the staff of the marshal commanding the Armies of France, by his thorough military knowledge, tact, and keen judgment he has rendered valuable services to the allied cause. At all times he has shown an active interest in the American Army and has rendered them every assistance at his command.
MONIER, LÉON F. V._____ R—France. G. O. No. 126, W. D., 1919.	Major, French Army. When he became liaison officer between the military government of Paris and the office of the chief surgeon, district of Paris, he brought to this important position a broad professional experience and devoted loyalty. During the summer of 1918, when the shortage of our hospital facilities in Paris became acute, he worked with self-sacrificing energy and made available for us many additional beds in French hospitals, also securing numerous buildings which were subsequently converted into hospitals, thereby rendering invaluable services in securing proper care for the sick and wounded of the American Expeditionary Forces.
MONROE DIT ROE, MARIE L. J._____ R—France. G. O. No. 17, W. D., 1924.	Brigadier general, French Army. He commanded with great distinction the 69th French Infantry Division, which under his skillful leadership achieved brilliant successes in combat. The 1st American Division had a most valuable initial experience in actual warfare when it went into the trenches of the Toul sector under his supervision. He gave also equally valuable assistance in the training of other American divisions which were attached to his command.
MORDACQ, JEAN J. H._____ R—France. G. O. No. 87, W. D., 1919.	Major general, French Army. As chief of the military cabinet of the Minister of War, he at all times accorded most valuable assistance to the American Expeditionary Forces. In the discharge of his exacting duties he cooperated loyally with the American military authorities, and by his timely advice and wholehearted service greatly increased the efficiency of our forces.
MOREAU, FRÉDÉRIC P._____ R—France. G. O. No. 87, W. D., 1919.	Vice admiral, French Navy, at Base Section No. 5, American Expeditionary Forces, Brest, France. He rendered most valuable assistance to the American Expeditionary Forces in the solution of difficult problems arising in Base Section No. 5 at Brest. He unified the energies of the French and American authorities, working wholeheartedly in the interest of the allied cause.
MOREL, PAUL_____ R—France. G. O. No. 24, W. D., 1920.	As French assistant secretary of the board of finances, charged with the liquidation of stocks, he ably discharged difficult and important tasks and rendered great service to the United States in the settlement of complicated claims and the liquidation of stocks involving millions of dollars.
MORNET, CHARLES L. D._____ R—France. G. O. No. 87, W. D., 1919.	Rear admiral, French Navy. In command of the French Marine at Marseille he showed rare tact, judgment, and energy in the performance of his varied duties. He gave generously and untiringly of his services in furthering the interests of the American base, placing every facility at the disposal of the American authorities.
MORTIER, PIERRE F._____ R—France. G. O. No. 87, W. D., 1919.	Surgeon major of 1st class, French Army. As medical member of the French Mission at General Headquarters, American Expeditionary Forces, and later in the medical section of the Central Bureau of Franco-American Relations, he labored ceaselessly and with conspicuous success in the interests of the sick and wounded. He obtained for us hospital sites, hospital trains, and ambulances in times of emergency.
MOURIER, LOUIS_____ R—France. G. O. No. 62, W. D., 1919.	As undersecretary of state of the medical service, civilian chief of the medical department of the French Army, he placed all available resources of his great department, both in material and personnel, at the disposal of the American Expeditionary Forces. His advice was of great value, aiding us in the solution of many problems, and he rendered services of inestimable value in assisting us in securing proper evacuation and hospitalization for the sick and wounded.
MOYRAND, AUGUSTE E. M._____ R—France. G. O. No. 126, W. D., 1919.	Brevet lieutenant colonel, French Army. In his capacity as assistant chief, 3d Bureau, French Army, he proved himself at all times a loyal friend of the American Expeditionary Forces, being ever ready to aid us. He gave advice on important matters which proved of the greatest value in handling difficult problems which we were confronted with. At all times he afforded us most hearty cooperation in the performance of all his duties, rendering us valuable assistance.

MUTEAU, PAUL J. H. R—France. G. O. No. 126, W. D., 1919.	Major general, French Army. He served with notable success as commanding general of the 8th Region, in which was located our intermediate section, Services of Supply. By his sound judgment and unfailing cordiality he was of the utmost assistance in the solution of many problems which arose, thereby rendering services of distinction to the American Expeditionary Forces
NAULIN, STANISLAS. R—France. G. O. No. 45, W. D., 1919.	Major general, commanding the 21st Army Corps, French Army. For services rendered to the American Expeditionary Forces and to the cause in which the United States has been engaged.
NEVEGANS, PAUL E. R—France. G. O. No. 105, W. D., 1919.	Captain of artillery, French Army. As a member of the French Military Mission he was placed on duty with the training and instruction branch, war plans division of the General Staff. His brilliant mental and exceptional professional attainments, coupled with rare tact and tireless devotion to duty, caused his services to be of signal worth to the United States Army.
NIVELLE, ROBERT G. R—France. G. O. No. 3, W. D., 1921.	Major general, French Army, member of the Superior Council of War. To this very distinguished soldier of the French Army this medal is presented as an expression of the high regard of the people of the United States and their Army for the distinguished and patriotic service which he has rendered to the common cause on the battle fields of Europe.
NORMAND, ANDRÉ C. J. J. R—France. G. O. No. 46, W. D., 1920.	Major of infantry, 4th Bureau, General Staff of the Army, (transportation), French Army. As chief of section in charge of railroad transportation for the American Expeditionary Forces he rendered valuable services. He took an active interest in the transportation problems of our service, and by his energy, zeal, and good judgment aided us materially in solving the perplexing problems that confronted us.
NUDANT, PIERRE A. R—France. G. O. No. 14, W. D., 1920.	Major general, French Army. As a commander, he rendered notable service in the field in command of the 34th Army Corps. As a diplomat, his keen sense of the situation, his tact and firmness toward the enemy, sound judgment in questions of moment, and constant devotion to the cause of the Allies made his services as president of the Permanent International Armistice Commission conspicuously meritorious.
OLIVARI, CHARLES. R—France. G. O. No. 87, W. D., 1919.	Major of artillery, French Army. As chief of the French Military Mission with the 88th American Division he worked efficiently and tirelessly, both in the training area and in the front lines. His tactful and most capable direction of the efforts of the French officers assigned to the division met with exceptionally valuable results. He rendered efficient and valuable service to the American Expeditionary Forces.
OPPENHEIM, RENÉ. R—France. G. O. No. 24, W. D., 1920.	Major of engineers, French Army. As assistant to the chief, Bureau of the General Commission of France-American War Affairs, he rendered great service to the United States.
PAGÉZY, EUGÈNE H. J. R—France. G. O. No. 126, W. D., 1919.	Lieutenant colonel of artillery, French Army. As a member of Marshal Foch's staff he cooperated with us most efficiently in the solution of many problems which presented themselves in connection with the combined operations of the French and the American forces. With sound judgment, tact, and untiring energy he managed difficult situations, rendering us services of exceptional value.
PAGÉZY, JULES E. F. R—France. G. O. No. 87, W. D., 1919.	Lieutenant colonel of artillery, French Army. By his efforts in devising and developing the French system of fire control for antiaircraft artillery, adopted by our armies, and in command of the French Officers' Antiaircraft Artillery School at Arnouville les Genesse, he rendered most valuable service to the American Expeditionary Forces.
PAILLE, LOUIS J. R—France. G. O. No. 126, W. D., 1919	Lieutenant colonel of infantry, French Army. As Chief of Instruction Service he rendered very valuable assistance to the American Expeditionary Forces, aiding us continually by giving us the benefit of his able advice. He displayed military attainments of a high order, and in the midst of exacting duties he never failed to come to our assistance. He cooperated with us wholeheartedly, proving himself at all times a loyal friend.
PASSAGA, FÉNELON F. G. R—France. G. O. No. 45, W. D., 1919.	Major general, commanding the 32d Army Corps, French Army. For services rendered to the American Expeditionary Forces and to the cause in which the United States has been engaged.
PAULINIER, MARIE J. A. R—France. G. O. No. 45, W. D., 1919.	Major general, commanding the 40th Army Corps, French Army. For services rendered to the American Expeditionary Forces and to the cause in which the United States has been engaged.
PAYOT, CHARLES. R—France. G. O. No. 45, W. D., 1919.	Brigadier general, French Army, Director General of Communication and Supply for the Armies. For services rendered to the American Expeditionary Forces and to the cause in which the United States has been engaged.
PENET, HIPPOLYTE A. R—France. G. O. No. 45, W. D., 1919.	Major general, commanding the 12th Division of Infantry, French Army. For services rendered to the American Expeditionary Forces and to the cause in which the United States has been engaged.

PERIA, JEAN V. H.
R—France.
G. O. No. 126, W. D., 1919.

Controller of 1st class of the administration of the army, French Army.
As controller of the administration of the army and chief of the Franco-American Service of the Direction of Control, he rendered very valuable services to the American Expeditionary Forces in connection with the organization of the American Rents, Requisitions, and Claims Service. He aided us materially in the matter of negotiations for the installation of American projects throughout France, proving tactful, energetic, and wholeheartedly devoted to the interests of the allied cause.

PERRIN, CHARLES E. F.
R—France.
G. O. No. 30, W. D., 1921.

Brevet lieutenant colonel, French Army, chief of the Interallied Bureau of Operations, Interallied Armies of Occupation.
For services to the allied cause and to the American forces in Germany.

PESSON-DIDION, MAURICE.
R—France.
G. O. No. 126, W. D., 1919.

Captain of infantry, French Army.
As a member of the French Mission in Paris he gave invaluable aid and assistance to the American authorities on numerous occasions when it was necessary to obtain material quickly in order to further the operations of the American Army. At all times he showed earnest cooperation with all the American authorities in all the activities in which he was associated with them.

PÉTAIN, HENRI P. B. O. J.
R—France.
G. O. No. 114, W. D., 1918.

Marshal of France, commander in chief of the French Armies of the North and Northeast.
As an expression to him of the high regard of the people of the United States and of their Army, for the distinguished and patriotic services which he has rendered to the common cause in which he has been associated on the battle fields of Europe.

PETIT, PAUL A. J.
R—France.
G. O. No. 87, W. D., 1919.

Surgeon major of 2d class, Medical Service, French Army.
Attached to the medical section of the Central Bureau of Franco-American Relations and later at G-4, General Headquarters, American Expeditionary Forces, he gave whole-hearted assistance and cooperation at all times. His advice and distinguished ability materially aided us in the prompt evacuation and hospitalization of casualties.

PIARRON DE MONDESIR, JEAN F. L.
R—France.
G. O. No. 45, W. D., 1919.

Major general, commanding the 36th Army Corps, French Army.
For services rendered to the American Expeditionary Forces and to the cause in which the United States has been engaged.

PICARD, FRANCOIS.
R—France.
G. O. No. 46, W. D., 1920.

Captain of cavalry, French Army.
He served with marked distinction in his capacity as aide to Marshal Pétain, and rendered very valuable service to the allied cause. He was at all times energetic and tactful in behalf of the American Expeditionary Forces, and by the sound advice and accurate information with which he furnished us proved of very material assistance. At all times he cooperated with us wholeheartedly, and showed himself a loyal friend.

PIKETTY, PAUL.
R—France.
G. O. No. 126, W. D., 1919.

Major of artillery, French Army.
As Chief of the 3d Division of Artillery, by his able advice and assistance the vocation of the organization and training centers for the heavy artillery of the American Expeditionary Forces was decided upon and the equipment and organization of these training centers were accomplished. He displayed great personal interest in the solution of the details of his task, at all times exhibiting an energetic spirit in furthering the general plan and in securing grants from the Ministry of War. His services have been of a high order.

POMPÉ, DANIEL.
R—France.
G. O. No. 3, W. D., 1922.

Lieutenant colonel of artillery, French Army.
For services rendered to the American Expeditionary Forces.

PONT, FERDINAND A.
R—France.
G. O. No. 87, W. D., 1919.

Major general, French Army.
He served with marked distinction throughout the war and in positions of great responsibility rendered invaluable service to the American Expeditionary Forces. As Deputy Chief of Staff of the French Army he displayed brilliant military genius and was at all times ready to afford us most loyal cooperation.

POUPART, HARMAN.
R—France.
G. O. No. 126, W. D., 1919.

Major of infantry, 158th Regiment of Infantry, French Army.
In command of front-line troops he performed valiant and faithful service in the allied cause. When he became an assistant to the members of the American Military Mission at French General Headquarters his wide practical experience fitted him admirably for his important duties, and by his zeal in promoting the interests of our Army he rendered services of the highest value to the American Expeditionary Forces.

POUPINEL, RAYMOND J. E.
R—France.
G. O. No. 46, W. D., 1920.

Major of infantry, French Army.
As head of the allied armies section of the French General Staff and later in the same position under the supreme Interallied Commander, he labored unceasingly in the interests of the American Expeditionary Forces. Through his tact and unswerving devotion many problems of great difficulty were solved carefully and expeditiously. He rendered most distinguished service.

POUY, JEAN F.
R—France.
G. O. No. 39, W. D., 1920.

Medical inspector, French Army.
As chief of the hospitalization section of the undersecretary of state for the Medical Service, he collaborated with the Medical Service of the American Expeditionary Forces in the preparation of a comprehensive hospitalization program. Through his wide knowledge of the possible hospital resources of France and his persistent efforts to insure adequate care for American soldiers, the trying period preceding combat operations was successfully bridged. He met all demands made for the delivering of French hospitals to the American Expeditionary Forces, even to the extent of burdening his own service. By his broad-minded conception of our problem and his energetic assistance in a most important task, he rendered estimable services to the American Expeditionary Forces.

PRESTAUT, RENÉ C. J. R—France. G. O. No. 87, W. D., 1919.	Second lieutenant of engineers, French Army. By his exceptional enthusiasm, good judgment, and untiring energy in the performance of his arduous duties as instructor at the American Army signal schools at Langres he proved himself of great value in the instruction of the Signal Corps of the American Expeditionary Forces.
PUPIER, CLAUDE F. R—France. G. O. No. 126, W. D., 1919.	Interpreter officer of the 1st class, French Army. As military secretary and chief of cabinet to the marshal commanding the allied armies he was charged with duties of a most exacting and responsible character, which duties he performed with conspicuous success. In his official capacity he was called upon to treat matters of the utmost delicacy pertaining to American affairs, in the conduct of which he displayed admirable tact, comprehension, and marked ability, thereby rendering invaluable services to the American Expeditionary Forces.
PURNOT, ALBERT. R—France. G. O. No. 126, W. D., 1919.	Captain of infantry, French Army. As a member of the cabinet of the Minister of War by his good judgment and earnest cooperation with the American authorities he has rendered services of marked distinction to the American Expeditionary Forces.
RAGENEAU, CAMILLE M. R—France. G. O. No. 45, W. D., 1919.	Brigadier general, French Army, chief, French Mission, General Headquarters, American Expeditionary Forces. For services rendered to the American Expeditionary Forces and to the cause in which the United States has been engaged.
REBOUL, JACQUES F. R—France. G. O. No. 87, W. D., 1919.	Colonel of infantry, French Army. As Chief of the Franco-American Bureau of the Army General Staff, he performed eminent and important services to the U. S. Army in deciding many essential questions in our relations with the French. That he acted with exceptional success is proven by the cordial relations which were at all times maintained.
REILLE, G. R. C. F. XAVIER. R—France. G. O. No. 55, W. D., 1920.	Lieutenant colonel of artillery, French Army. As head of the artillery section of the French Mission to the United States during the World War by his untiring efforts he aided materially in the training of the Artillery in the United States.
REMOND, LOUIS. R—France. G. O. No. 29, W. D., 1919.	Colonel of artillery, French Army. For services rendered the U. S. Army while serving as chief of the French Artillery Mission to the United States.
RENIÉ, ANDRÉ J. R—France. G. O. No. 126, W. D., 1919.	Colonel of infantry, French Army. As chief of the Personnel Bureau and as head of the 1st Bureau of the General Staff of the French Army, he placed the facilities of his important office at the service of the American authorities, lending every assistance in his power to meet the needs of the American troops. His services, rendered with a loyal spirit of cooperation, have been of great value to the American Expeditionary Forces.
REQUICHOT, HENRY L. M. J. R—France. G. O. No. 87, W. D., 1919.	Major general, French Army. As commanding general of the 9th French Region, in which were located American Headquarters, Services of Supply, he showed himself uniformly helpful, giving us willing assistance and cooperation. He went far beyond the exacting duties of his office to aid the American Expeditionary Forces in securing necessary supplies.
REQUIN, EDOUARD J. R—France. G. O. No. 29, W. D., 1919.	Brevet lieutenant colonel, French Army. For services rendered the United States Army while serving as the personal representative of Marshals Joffre and Foch and as special delegate of the French General Staff to the United States.
RICHERT, AUGUSTIN X. R—France. G. O. No. 49, W. D., 1922.	Major of infantry, French Army. For services as an instructor of the 90th Division. During the training of the division at Aignay-le-Duc he was constantly in the field, and by his unceasing work and high military attainments he materially assisted the division to acquire as quickly as possible that proficiency which enabled it to join the fighting forces at the front. His services at this period were invaluable, and later at the front his courage and coolness were an inspiration to the men who saw him.
RIEGEL, GEORGES. R—France. G. O. No. 46, W. D., 1920.	Major of engineers, French Army. As representative of the 4th Bureau, French Armies of the North and Northeast, in the office of assistant chief of staff, G–4, he was charged with the supervision of the supply of French units and personnel placed at the disposition of the 1st American Army. By his high professional attainments and interest he assured the closest liaison between the French and American authorities, contributing largely to the success of the allied cause.
RIST, EDOUARD. R—France. G. O. No. 59, W. D., 1921.	Major of the 1st class, Medical Service, French Army. For services as an eminent scientist. By his untiring zeal, devotion, and energy he promoted the efficient treatment of the American sick and wounded. In his important research work he cooperated with the Medical Service of the American Expeditionary Forces in the fullest measure of devotion to duty. To him is due much credit for the arresting of the ravages of disease and injuries among our forces. His valuable research efforts in the domain of preventive medicine and wound bacteriology resulted in the saving of many lives among our wounded soldiers. He has rendered services of signal worth,

ROTTÉE, LUCIEN lt—France. G. O. No. 10, W. D., 1922.	Lieutenant of infantry, French Army. For services rendered the United States Army while serving as French liaison officer in the division of criminal investigation, American Expeditionary Forces.
ROUGET, JULES F. A. R—France. G. O. No. 39, W. D., 1920.	Medical inspector, French Army. As chief surgeon of a group of armies, and later as chief of the cabinet of the undersecretary of state for the medical service, by his broad-minded treatment of all important questions arising with regard to hospitalization, and by his earnest cooperations with our Medical Service in all matters affecting the furtherance of efficiency in the care of our sick and wounded, he has rendered services of an important character to the American Expeditionary Forces. He at all times gave ready assistance to the limit of his power, in all important medical matters which constantly came under his extensive jurisdictions. He rendered services of great consequence to the American Expeditionary Forces.
ROUX, P. R—France. G. O. No. 39, W. D., 1920.	Professor, medical inspector, French director of Pasteur Institute, Paris, France. By his untiring zeal and devotion to the promotion of efficient treatment of the American sick and wounded, and by his earnest cooperation with the medical service of the American Expeditionary Forces in important research work, the inroads of disease and infections from wounds among Americans were curtailed. By his eminent skill in scientific research in the field of preventive medicine and wound bacteriology, he achieved marvelous results in the saving of lives among our sick and wounded. He has rendered services of great consequence to the American Expeditionary Forces.
ROUX, PAUL L. R—France. G. O. No. 87, W. D., 1919.	Major of artillery, French Army. He organized schools for ordnance and artillery mechanics and inspectors in the American Field Artillery Training Camp at Valdahon and in the American Ordnance Training Center at Is-sur-Tille. In the training of officers and soldiers of our armies he showed high military attainments and achieved distinguished success.
ROZET, HENRI E. R—France. G. O. No. 126, W. D., 1919.	Lieutenant colonel of infantry, 3d Bureau, French Army. He was tireless in his efforts to keep the American Mission at French General Headquarters informed on all matters of importance to it, rendering very valuable services to the American Expeditionary Forces; he at all times afforded us most loyal cooperation, proving unflagging in his devotion to our interests. He displayed military attainments of a high order and gave us valuable advice in times of emergency.
SAINTE-CLAIRE-DEVILLE, CHARLES E. R—France. G. O. No. 87, W. D., 1919.	Major general, French Army. As inspector general of material and ammunition for the French Armies he placed at our disposal at all times the results of his wide experience and the facilities under his control. He rendered most valuable service to the American Expeditionary Forces in connection with the development of the American ordnance inspection and maintenance work.
SALAUN, H. R—France. G. O. No. 126, W. D., 1919.	Vice admiral, French Navy. As prefect maritime, governor of the place of Brest, by his splendid spirit of whole-hearted cooperation with the American authorities in all matters concerning the welfare and comfort of the large number of American troops which constantly passed through the important port of Brest, he has rendered services of signal worth to the American Expeditionary Forces. Every demand upon the facilities of his office received the same courteous consideration which characterized all the dealings of the Americans with him and his efficient personnel.
SCHMIDLIN, GEORGES A. R—France. G. O. No. 126, W. D., 1919.	Major of infantry, French Army. As a member of the Cabinet of the Minister of War he has always exhibited a sympathetic interest in the American Army, offering them every assistance which his office afforded. His services have been of a high order and of much value to the American Expeditionary Forces.
SCHMIDT, HENRI F. E. R—France. G. O. No. 39, W. D., 1921.	Brigadier general, French Army, commanding the garrison of Mayence, Germany. For services in the allied cause and to the American forces in Germany.
SELIGMAN, GERMAIN. R—France. G. O. No. 126, W. D., 1919.	Captain, 132d Regiment of Infantry, French Army, liaison officer, 1st Division, American Expeditionary Forces. By his high professional attainments and wide experience he rendered invaluable services to the American Expeditionary Forces as liaison officer of the 1st Division. Both during the training period of the division and subsequently during active operations he performed his task with excellent judgment, admirable tact, and earnest loyalty.
SISTERON, EUGÈNE P. J. R—France. G. O. No. 10, W. D., 1922.	Major of infantry, French Army, assistant chief of staff of the communications zone. For services rendered the American Expeditionary Forces.
SORDET, HENRI. R—France. G. O. No. 24, W. D., 1920.	Captain of infantry, French Army. As chief of a section of the Franco-American Mission in Paris, his duties required the daily handling of important matters with the Liquidation Commission. He rendered great service to the United States in the efficient performance of these difficult tasks.

SPITZ, MARIE C. L. R—France. G. O. No. 126, W. D., 1919.	Colonel of infantry, Chief of Staff, 2d Division of Infantry, French Army. He has been a most sincere and devoted friend to American interests, always willing to do more than his share to bring about a harmonious and friendly understanding between the French and American authorities. By his unfailing tact, sympathy, and understanding, and by his untiring efforts to successfully solve the many intricate problems that arose between the French and the Americans, he has rendered an invaluable service to the American Expeditionary Forces.
TARDIEU, ANDRÉ. R—France. G. O. No. 62, W. D., 1919.	He rendered service of great value for the American Expeditionary Forces as High Commissioner of the French Republic at Washington, D. C. Displaying tact, energy, and devotion to duties, he handled difficult problems with unswerving zeal for the good of the American Army.
THENAULT, GEORGES. R—France. G. O. No. 15, W. D., 1923.	Captain, Air Service, French Army. He commanded the Lafayette Escadrille from the time of its creation in April of 1916 until it was transferred to the Air Service, American Expeditionary Forces in January, 1918. He led the squadron throughout the operations in Alsace and the Verdun offensive in 1916, later participating in the allied offensive on the Somme and the Aisne. In 1917 he continued to command the squadron which was withdrawn from the line, and participated immediately afterwards in the operations on the Chemin des Dames. During this time more than 50 American pursuit pilots served in this unit. By his initiative, force, courage, and devotion to duty he contributed materially to the successful operations of the squadron and instilled in the American pilots under his command the principles of pursuit aviation which contributed materially to the success of other American pursuit organizations when these pilots were transferred. By his ability as a pilot and his frequent missions over the lines he set an admirable example to the American officers under his command.
THOMAS, JOSEPH C. A. R—France. G. O. No. 87, W. D., 1919.	Lieutenant colonel of artillery, French Army. As director of organization at the Saumur Artillery School, he rendered services of inestimable value to the American artillery. By his energy, enthusiasm, and devotion, he lightened the task involved in the training of young officers for the greatly expanded artillery establishment needed for the American Army.
TINARDON, M. A. R—France. G. O. No. 87, W. D., 1919.	Major of engineers, French Army. As a member of the 4th Bureau of the General Staff of the French Army he rendered invaluable service to the American Expeditionary Forces in connection with the selection of suitable sites for depots and hospitals. He gave us his time willingly, although his duties were pressing, aiding us most ably in the development of plans, and frequently furnishing us with labor and construction material in time of emergency.
TONGAS, GASTON. R—France. G. O. No. 87, W. D., 1919.	Colonel of engineers, French Army. As director of telegraphic service of the 2d line, he labored most zealously and efficiently in our interests at French General Headquarters. He willingly offered wise counsel and unfailing support in our applications for service and material.
TREMBLAY, FRANCOIS L. R—France. G. O. No. 49, W. D., 1922.	Major of infantry, French Army. As chief of the French officers assigned to the 90th Division during the training period and later as liaison officer of the division at the front, by his tireless efforts, willing spirit of cooperation, and his high military attainments he rendered services of great value to our troops and won the warmest admiration of both officers and men. On the front in the St. Mihiel and Meuse-Argonne offensives his conduct and courage were a marked example to the men of the 90th Division.
TUFFIER, T. R—France. G. O. No. 39, W. D., 1920.	Professor, French medical inspector, School of Medicine. By his untiring zeal and devotion to the promotion of efficient treatment of the American sick and wounded, and by his earnest cooperation with the medical service of the American Expeditionary Forces in important research work, the inroads of disease and infections from wounds among Americans were curtailed. By his eminent skill in scientific research in the field of preventive medicine and wound bacteriology, he achieved marvelous results in the saving of lives among our sick and wounded. He rendered services of great consequence to the American Expeditionary Forces.
VALDANT, HENRY C. R—France. G. O. No. 126, W. D., 1919.	Brigadier general, French Army. As chief of staff to the commanding general, Military Governor of Paris, he has at all times manifested a most cordial attitude in all Franco-American relations, offering us every assistance at the command of his far-reaching office. His services to the American Expeditionary Forces have been of great value.
VALLOTTE, PAUL C. A. R. R—France. G. O. No. 126, W. D., 1919.	Major of cavalry, French Army. As chief of the 2d-Bureau of Marshal Foch's staff he kept us constantly informed as to the general trend of operations as well as the situation of the enemy. He brought to bear on this important task an exact knowledge of the terrain, comprehensive understanding of the enemy movements, and the successive situations which confronted our troops. The services which he so willingly rendered to the American Expeditionary Forces were of exceptional value.
VANDENBURG, CHARLES A. R—France. G. O. No. 45, W. D., 1919.	Major general, commanding the 13th Army Corps, French Army. For services rendered to the American Expeditionary Forces and to the cause in which the United States has been engaged.

VAN HEEMS, ROGER A R—France. G. O. No. 19, W. D., 1922.	Major of infantry, French Army, regulating commissioner at Le Bourget, France. For services in connection with the supply and transportation of American troops during the operations at Chateau-Thierry.
VARAIGNE, HENRI A R—France. G. O. No. 126, W. D., 1919.	Major of infantry, French Army. As chairman of the French Mission attached to the Headquarters of the American General Purchasing Board he rendered services of inestimable importance in carrying on for us negotiations with the French Government for the procurement of au immense amount of material. He was indefatigable in his efforts in our behalf, proving himself able, tactful, and possessed of a wide comprehension of existing needs and conditions. Cooperating with us at all times most whole-heartedly, he proved a loyal friend.
VERLEY, EDOUARD R—France. G. O. No. 87, W. D., 1919.	Captain of artillery, French Army. As liaison officer with the troop movement bureau at G–4, General Headquarters, American Expeditionary Forces, he solved many difficult transportation problems incident to active operations. No task proved too large or too small for him to accept, and he accomplished with distinction and unfailing courtesy his many duties, rendering services of great value to us.
VERNAY, ROMAIN P. J. J R—France. G. O. No. 24, W. D., 1920.	As French military intendant in the bureau of the assistant secretary of the board of finances, liquidation of stocks, he ably discharged a difficult and important task and rendered great service to the United States in the settlement of complicated claims and the liquidation of stocks involving millions of dollars.
VIDAL, PAUL R—France. G. O. No. 87, W. D., 1919.	Major general, French Army. In command of the 7th French Region he was tireless in his devotion to our interests and to the success of our varied projects in the region of which he had charge. His relations with the American military authorities were always cordial and helpful. He always evinced personal interest in our plans and aided us materially by his sound advice.
VIDALON, JEAN R—France. G. O. No. 17, W. D., 1924.	Brigadier general, French Army. As Assistant Chief of Staff of the French Army and as head of the 2d section of the General Staff, particularly charged with relations with foreign armies and of the special Franco-American Bureau, he successfully and most satisfactorily managed and coordinated the relations of the ministry of war with the American authorities. To him is due much credit for the able direction of all the American units in French territory. His able assistance has contributed in a large measure to the friendly spirit that has existed between the French and the Americans.
VIGNAL, PAUL R—France. G. O. No. 11, W. D., 1919.	Brigadier general, French Army. For services rendered the United States Army while serving as military attaché to the French Embassy and chief of French Military Missions to the United States.
VINEL, LOUIS A R—France. G. O. No. 87, W. D., 1919.	Military intendant, French Army. As Quartermaster General of the French Army he showed ability of the highest order in handling his important duties and cooperated most loyally with the the American military authorities. In a position of great responsibility, he made special efforts to aid the American Expeditionary Forces.
WAHL, VICTOR E R—France. G. O. No. 126, W. D., 1919.	Colonel of infantry, French Army. As commander of the 1st Brigade, Artillery of Assault, by his marked ability, initiative, and sound judgment, he gave much valuable assistance to the American armies operating in the St. Mihiel and Meuse-Argonne offensives. His earnest cooperation in carrying out the details of the plans of operations had a marked influence on the success of the Tank Corps units in those important operations.
WALCH, CAMILLE R—France. G. O. No. 126, W. D., 1919.	Brigadier general, French Army. As chief of the French Artillery attached to the 1st Army Corps, American Expeditionary Forces, during the Meuse-Argonne offensive, he rendered us services of noteworthy distinction. His services were continuously marked by his qualities of leadership and military attainments, which were of a high order. Displaying tireless energy and devotion to his arduous duties, he strove diligently that our infantry might have all the benefits of efficient artillery support.
WEYGAND, MAXIME R—France. G. O. No. 17, W. D., 1919.	Brigadier general, French Army, Chief of Staff to Marshal Foch. For services rendered to the American Expeditionary Forces and to the cause in which the United States has been engaged.
WIRBEL, HENRI R—France. G. O. No. 45, W. D., 1919.	Major general, commanding the 21st Region, French Army. For services rendered to the American Expeditionary Forces and to the cause in which the United States has been engaged.
WISSEMANS, MAURICE M R—France. G. O. No. 39, W. D., 1920.	Medical inspector, Chief of the Medical Service, 2d French Army. When the 1st American Division went into the Verdun sector for training purposes, he collaborated with the medical service of the American Expeditionary Forces in every possible way to the end that our battle casualties should receive proper care. Later, as chief surgeon of the Group of Armies of the East, he actively participated in the preparation of the plans for the St. Mihiel and the Argonne offensives. His wide experience in battle conditions and his whole-hearted desire to be of assistance to his country's allies enabled him to render services of great value to the American Expeditionary Forces. He freely placed at the disposition of the American Expeditionary Forces all his resources as to hospitals, ambulance and train transportation, and medical material. His able advice and valuable assistance was a material factor in the provisions for the care and evacuation of our battle casualties. He rendered services of great consequence to the American Expeditionary Forces.

ITALIANS

[Awarded for exceptionally meritorious and distinguished services in a position of great responsibility, under the provisions of the act of Congress approved July 9, 1918]

ALBRICCI, CONTE ALBERICO_____
   R—Italy.
   G. O. No. 126, W. D., 1919.

Lieutenant general, Italian Army.
As commander of the 2d Italian Army Corps in France he rendered services of great distinction to the allied cause. Later as Minister of War he showed a keen spirit of cooperation with the American forces in Italy, assisting us by all means at the disposal of his extensive office. He has given services of great value to the American Expeditionary Forces.

ALLEGRETTI, LORENZO_____
   R—Italy.
   G. O. No. 45, W. D., 1919.

Major, commanding the Arditi Assault Battalion, Italian Army, while attached to the 332d Infantry, American Expeditionary Forces.
For services performed for the American Expeditionary Forces and to the cause in which the United States has been engaged.

ANGELOZZI, CAMILLO_____
   R—Italy.
   G. O. No. 126, W. D., 1919.

Lieutenant general, Italian Army.
As Director General of Engineers, Italian Army, he performed his exacting duties with eminent technical skill. At all times he was actuated by a desire to aid the American authorities by all the means at his command, thereby rendering services of the greatest value to the common cause.

APOLLONI, ENEAL_____
   R—Italy.
   G. O. No. 126, W. D., 1919.

Lieutenant colonel, Italian Army.
As liaison officer with the American Red Cross, by his admirable tact and helpful spirit of cooperation he aided materially in the effective relief work of this organization in caring for the sick and wounded of the American and Italian Armies.

AYMONINO, ALDO_____
   R—Italy.
   G. O. No. 45, W. D., 1919.

Colonel, Italian Army, chief of Group of Allied Missions of the Comando Supremo in Italy.
For services performed for the American Expeditionary Forces and to the cause in which the United States has been engaged.

BADOGLIO, PIETRO_____
   R—Italy.
   G. O. No. 17, W. D., 1919.

Lieutenant general, Deputy Chief of Staff, Italian Army.
For services rendered to the American Expeditionary Forces and to the cause in which the United States has been engaged.

BASSI, GUIDO_____
   R—Italy.
   G. O. No. 126, W. D., 1919.

Brigadier general, Italian Army.
As a member of the Interallied Commission on the Repatriation of Prisoners of War he has rendered highly meritorious service to the United States and allied Governments in connection with the repatriation of American and allied prisoners released by the armistice. He always displayed a cheerful and active interest in all that pertained to their welfare, and rendered sympathetic and practical cooperation.

BONGIOVANNI, LUIGI_____
   R—Italy.
   G. O. No. 87, W. D., 1919.

Major general, Italian Army, chief of the Italian Combat Air Forces.
In command of the Italian air forces during the time American pilots attached to Italian squadrons were on active duty at the Italian front, he exercised great ability, tact, and energy in his direction of the work of our officers. At all times he showed the highest military attainments, and his enthusiasm was an example to all. He rendered a most valuable service in his prosecution of operations against the enemy.

BUSINELLI, UGO_____
   R—Italy.
   G. O. No. 75, W. D., 1919.

Colonel, Italian Army.
For services to the allied and associated Governments as chief of staff, Italian section, Supreme War Council, Apr. 27, 1918, to Feb. 11, 1919.

CARPENTIERI, GIACOMO_____
   R—Italy.
   G. O. No. 126, W. D., 1919.

Colonel, Italian Army.
As chief of the Transportation Department, Italian Army, he displayed exceptional administrative ability and untiring energy in his important duties, thereby playing an important part in assuring the victorious termination of our struggles against the common enemy. Manifesting an earnest desire to aid the American forces in every possible way, he assisted us materially by furnishing adequate facilties for the transportation of our troops and supplies.

CAVALLERO, UGO_____
   R—Italy.
   G. O. No. 75, W. D., 1919.

General, Italian Army.
For services to the allied and associated Governments as permanent military representative, Italian section, Supreme War Council.

CAVIGLIA, ENRICO_____
   R—Italy.
   G. O. No. 126, W. D., 1919.

Lieutenant general, Italian Army.
In his capacity as Secretary of State for War, he was a potent factor in the victorious termination of our struggles against the common enemy. For his unfailing courtesy and constant desire to aid in every possible way the American forces serving in Italy, he will ever be held in enduring memory by the officers and soldiers of our forces serving on Italian soil.

CHIESA, EUGENIO_____
   R—Italy.
   G. O. No. 46, W. D., 1920.

As Minister of Aeronautics for Italy, he rendered very valuable services to the American Expeditionary Forces in connection with the training of our pilots and the supply of aviation material. At all times he displayed a helpful interest in our Air Service, going far beyond the bounds of duty to render us advice and assistance. He has proved himself possessed of a wide knowledge of all matters pertaining to aviation and was always a loyal friend.

DALLOLIO, ALFREDO R—Italy. G. O. No. 126, W. D., 1919.	Lieutenant general, Italian Army. As Inspector General of Artillery, Italian Army, he performed duties of the greatest importance with conspicuous ability. Actuated by a warm spirit of cooperation, he aided materially in maintaining the cordial relations which existed between the Italian and American military authorities.
DE ANGELIS, CIRO R—Italy. G. O. No. 45, W. D., 1919.	Major general, commanding 31st Infantry Division, 3d Italian Army. For services rendered to the American Expeditionary Forces and to the cause in which the United States has been engaged.
DELLA VALLE, FRANCESCO R—Italy. G. O. No. 46, W. D., 1920.	Major general, Italian Army. Serving as surgeon general of the Italian Army he performed duties of the greatest responsibility with conspicuous merit. By his sympathetic knowledge of American needs he was of the greatest assistance in securing proper care for our sick and wounded in Italy.
DE LUCA, MARCELLO R—Italy. G. O. No. 126, W. D., 1919.	General, Italian Army. As chief of staff of a division at the outbreak of the war, he gave excellent service; later as commander of an infantry brigade, he showed great initiative and marked tactical ability; and finally as chief of staff of the Army Corps of Genoa he rendered eminent distinguished service.
DE MARCORENGO, FABRIZIO ODETTI R—Italy. G. O. No. 126, W. D., 1919.	Lieutenant general, Italian Army. As remount inspector of the Italian Army he performed his exacting duties with conspicuous success. Through his earnest desire to give the American authorities the advantage of all facilities at his disposal, he rendered services of the highest value in promoting effective cooperation between the two armies.
DEVALLE, GIOVANNI R—Italy. G. O. No. 126, W. D., 1919.	Colonel, Italian Army. He served with distinction as chief of T. A. E. A. of the Italian Army, contributing materially to the success of the common cause. At all times manifesting a cordial spirit of cooperation, he did much toward promoting the harmonious relations existing between the Italian Army and the American forces serving in Italy.
DIAZ, ARMANDO R—Italy. G. O. No. 111, W. D., 1918.	Lieutenant general, commander in chief, Italian Armies. As an expression to him of the high regard of the people of the United States and of their Army, for the distinguished and patriotic services which he has rendered to the common cause in which he has been associated on the battle fields of Europe.
DI CAMPIGLIONE, ENRICO R—Italy. G. O. No. 29, W. D., 1919.	Captain, Italian Army. For services rendered the United States Army while serving as the liaison officer between the embassy, the High Commission of Italy, and the War Department.
DI ROBILLANT, MARIO NICOLIS R—Italy. G. O. No. 75, W. D., 1919.	Lieutenant general, Italian Army. For services to the allied and associated Governments as permanent military representative, Italian section, Supreme War Council, Apr. 27, 1918, to Feb. 11, 1919.
DI SAVOIA, EMANUELE FILIBERTO R—Italy. G. O. No. 126, W. D., 1919.	Lieutenant general, Italian Army, commanding the 3d Italian Army. For services rendered the American Expeditionary Forces and to the cause in which the United States has been engaged.
GIRALDI, GUGLIELMO PECORI R—Italy. G. O. No. 126, W. D., 1919.	Lieutenant general, Italian Army. As commander of the 1st Italian Army, he rendered eminent services to the 332d U. S. Infantry, whose good fortune it was to be attached to his command during its training period, and which owes much of its successful preparation for combat to his painstaking efforts in its behalf.
GRAZIOSI, EUGENIO R—Italy. G. O. No. 126, W. D., 1919.	Brigadier general, Italian Army. As Director of Military Transportation of the Army of Italy, he rendered valuable service to the allied cause. His energy and zealous efforts were a deciding factor in the successful movement of the American forces in Italy.
GUGLIELMOTTI, EMILIO R—Italy. G. O. No. 11, W. D., 1919.	Major general, Italian Army. For services rendered the U. S. Army while serving as military attaché to the Royal Italian Embassy, Washington, D. C.
GUIDONI, ALESSANDRO R—Italy. G. O. No. 87, W. D., 1919.	Major, Italian Army. As the Italian technical delegate to the Interallied Aviation Committee, he displayed unusual technical knowledge and perfect understanding of the various problems incident to aviation. He was most helpful in giving the American Air Service the benefit of his wide experience, and rendered most valuable assistance in solving technical problems with which we were often confronted.
HUNTINGTON, CARLO H R—Italy. G. O. No. 126, W. D., 1919.	Captain, Italian Army. As principal assistant to the chief of the Italian Mission attached to American headquarters he has rendered great assistance to the various staff services at general headquarters. At all times he has manifested a zealous spirit of full cooperation with our staff personnel, furthering these cordial relations which have always prevailed between the Italian and American services.
LANZA, PAOLO R—Italy. G. O. No. 126, W. D., 1919.	Colonel, Italian Army. As Assistant Director of Transportation by his whole-hearted cooperation he rendered valuable services to the American Expeditionary Forces.

LEVI, CAESER GIULIO_____
R—Italy.
G. O. No. 87, W. D., 1919.

General, Italian Army.
As the representative of the Italian Government on the Interallied Transportation Council he aided in the solution of difficult transportation problems, involving shipment of supplies from and to Italy. He proved an able executive and demonstrated the possession of broad vision and sound judgment.

MALLANDRA, GUISEPPE_____
R—Italy.
G. O. No. 126, W. D., 1919.

Major general, Italian Army.
In his capacity as general attached to the Italian War Ministry he showed a constant desire to aid the American forces in every possible way. By his helpful spirit of cooperation he did much toward the success of operations against the common enemy, rendering services of worth to the allied cause.

MARCHETTI, ODOARDO_____
R—Italy.
G. O. No. 126, W. D., 1919.

Colonel, Italian Army.
He performed his highly important duties as chief of the Intelligence Department of the Italian Army with conspicuous success, aiding thereby to a marked degree in the success of the operations against the common enemy. Actuated by a constant desire to assist the American authorities with all the facilities of his office, he contributed materially to the harmonious relations existing between the Italian Army and the American forces in Italy.

MARIENI, GIOVAN BATTISTA_____
R—Italy.
G. O. No. 126, W. D., 1919.

Lieutenant general, Italian Army.
He served with distinction as Inspector General of Engineers, Italian Army, displaying technical attainments and devoted loyalty to the common cause. Ever ready to aid the American authorities with all the facilities at his command, he was a potent factor in furthering the spirit of cooperation which marked the relations of the Italian and American forces.

MERRONE, ENRICO_____
R—Italy.
G. O. No. 45, W. D., 1919.

Major general, Italian Army, Italian representative, Military Board of Allied Supply.
For services rendered to the American Expeditionary Forces and to the cause in which the United States has been engaged.

MODENA, ANGELO_____
R—Italy.
G. O. No. 126, W. D., 1919.

Major general, Italian Army.
He served with distinction as Director General of Transport and Administration of the Italian Army. He was ever ready to aid the American forces by his sound advice and cordial spirit of cooperation, thereby aiding us materially in the solution of the many problems of transportation which arose during the continued operations against the enemy.

MOLTENI, FILIPPO_____
R—Italy.
G. O. No. 45, W. D., 1919.

Major, Italian Army, chief of Paris section, Italian Foreign Military Aeronautical Mission.
For services performed for the American Expeditionary Forces and to the cause in which the United States has been engaged.

MOMBELLI, ERNESTO_____
R—Italy.
G. O. No. 10, W. D., 1922.

Major general, Italian Army.
As corps commander of the Italian Oriental Expeditionary Force during the war and as the Italian representative on the Interallied Military Mission to Hungary, positions of great responsibility, in which he rendered valuable services in the furtherance of the Allied cause.

MONTUORI, LUCA_____
R—Italy.
G. O. No. 126, W. D., 1919.

Lieutenant general, Italian Army.
As commanding general of the 6th Italian Army he displayed eminent military attainments in the performance of his important duties. To his brilliant leadership was due, in no small degree, the success achieved by the Italian forces against the common foe.

PAOLINI, GIUSEPPE_____
R—Italy.
G. O. No. 45, W. D., 1919.

Lieutenant general, Italian Army, commanding 11th Army Corps, 3d Italian Army.
For services rendered to the American Expeditionary Forces and to the cause in which the United States has been engaged.

PERELLI, IPPOLITO_____
R—Italy.
G. O. No. 45, W. D., 1919.

Brigadier general, Italian Army, chief, Italian Mission, General Headquarters, American Expeditionary Forces.
For services rendered to the American Expeditionary Forces and to the cause in which the United States has been engaged.

PIRAINO, ANTONIO_____
R—Italy.
G. O. No. 46, W. D., 1920.

Brigadier general, Italian Army.
As Adjutant General of the Italian Army he performed his manifold duties with excellent success. At all times desirous of aiding the American forces in every possible way, he did much to further the spirit of harmony which marked the relations between the Italian and American Armies.

RAGIONI, RODOLFO_____
R—Italy.
G. O. No. 45, W. D., 1919.

Colonel, General Staff, Italian Army, Italian military delegate for British and American troops.
For services performed for the American Expeditionary Forces and to the cause in which the United States has been engaged.

RICALDONI, OTTAVIO_____
R—Italy.
G. O. No. 87, W. D., 1919.

Colonel, Air Service, Italian Army.
He furnished us with scientific and technical information needed in the development of the American Air Service in America, France, and in Italy. At all times he showed himself zealous in our behalf, going far beyond the bounds of his important duties to render us invaluable services. His judgment was sound, his advice helpful, his loyalty whole-hearted.

ROTA, ALFREDO_____
R—Italy.
G. O. No. 126, W. D., 1919.

Colonel, Italian Army.
By his efficient performance of his exacting duties as chief of the General Staff Department, Italian Army, he did much toward bringing about the victory which terminated our struggles against the common enemy. At all times he showed an ardent desire to give every consideration to the American forces serving in Italy, thereby furthering materially the effective cooperation of the Italian and American Armies.

RUISECCO, CLAUDIO R—Italy. G. O. No. 126, W. D., 1919.	Captain, Italian Army. In his capacity as assistant to the chief of the Italian Mission at American General Headquarters, he has constantly shown an earnest desire to furnish our staff services with all assistance at his command. With a loyal devotion to the common cause, he furthered the friendly relations between the two allied services.
SACHERO, GIACINTO R—Italy. G. O. No. 126, W. D., 1919.	Lieutenant general, Italian Army. As Director General of Artillery of the Italian Army he was charged with duties of a most important nature, which he performed with marked zeal and ability. He was ever ready to place the facilities of his office at the disposal of the American authorities and thereby rendered important service in assuring the efficient cooperation of the two armies.
SCIMECA, VITO R—Italy. G. O. No. 14, W. D., 1920.	Colonel, Italian Army. His services were of distinguished merit as chief of the Italian Mission. His conscientious attention to duty presented a united allied front to the enemy, while his diplomatic tact cemented the cordial relations among the allies.
SCIPIONI, SCIPIONE R—Italy. G. O. No. 17, W. D., 1919.	Major general, Italian Army. For services rendered to the American Expeditionary Forces and to the cause in which the United States has been engaged.
SOGNO, VITTORIO R—Italy. G. O. No. 126, W. D., 1919.	Lieutenant colonel, Italian Army. He served with conspicuous success as chief of the 3d section of the General Staff Department, Italian Army, performing services of the utmost importance in connection with the struggle against the common enemy. The warm spirit of cooperation which he constantly displayed in his dealings with the American authorities was a potent factor in cementing the cordial relations between the Italian Army and the American forces serving in Italy.
TASSONI, GIULIO R—Italy. G. O. No. 126, W. D., 1919.	Lieutenant general, Italian Army. He commanded with distinction the 4th Italian Army, performing duties of great importance with conspicuous success. His high military attainments and the able leadership were an important factor in the victorious termination of the struggle against the common enemy.
TOMMASI, DONATO ANTONIO R—Italy. G. O. No. 126, W. D., 1919.	Lieutenant general, Italian Army. As Judge Advocate General of the Italian Army he performed his exacting duties with high conceptions of justice and loyal devotion to the common cause. By his unfailing desire to aid the American forces in every possible way he did much to further the harmonious relations which existed between the two armies.
TONI, RENZO R—Italy. G. O. No. 75, W. D., 1919.	Lieutenant colonel, Italian Army. For services to the allied and associated Governments as chief of staff, Italian section, Supreme War Council.
TROIANI, ETTORE R—Italy. G. O. No. 126, W. D., 1919.	Lieutenant colonel, Italian Army. As chief of R. Section, Intelligence Department, he performed his important duties with rare skill and untiring energy. Always ready to aid the American authorities with all the facilities at his command, he did much toward making possible the efficient cooperation of the American forces with the Italian Army.
VACCHELLI, NICOLA R—Italy. G. O. No. 126, W. D., 1919.	Brigadier general, Italian Army. As Chief of General Staff Department, Italian Army, he played an important part in bringing about the successful termination of the common struggle against the Central Powers by his distinguished services in a duty of great responsibility. Ever ready to aid the American forces in Italy by all the means at his disposal, he was an important factor in maintaining the harmonious relations between the American and Italian authorities.
ZACCONE, VITTORIO R—Italy. G. O. No. 126, W. D., 1919.	Lieutenant general, Italian Army. He rendered invaluable services to the common cause as Quartermaster General, Italian Army, displaying keen foresight and notable executive ability. By his ardent desire to assist the American forces in every possible way he was a potent factor in fostering the spirit of cooperation which marked the relations of the Italian and American Armies.
ZANGHIERI, GIOVANNI R—Italy. G. O. No. 126, W. D., 1919.	Lieutenant colonel, Italian Army. As Chief of P. of W. Department, he performed highly exacting duties with unflagging energy and marked executive ability. At all times he aided the American authorities in every way possible, thereby assisting materially in furthering the cordial relations between the Italian Army and the American forces serving in Italy.
ZUGARO, FULVIO R—Italy. G. O. No. 126, W. D., 1919.	Colonel, Italian Army. Charged with highly responsible duties as Chief of Organization and Demobilization Department, Italian Army, he displayed eminent administrative ability and rendered services of great value to the common cause. By his broad-minded spirit of cooperation and unfailing courtesy he did much toward promoting the harmony which marked the relations between the Italian Army and the American forces serving in Italy.
ZUPELLI, VITTORIO R—Italy. G. O. No. 126, W. D., 1919.	Lieutenant general, Italian Army. As Secretary of State for War he performed highly responsible duties with conspicuous ability and devoted loyalty, thereby meriting a large share of the credit for the success of our common cause. The unbroken harmony which marked the relations of the American and Italian military authorities was in no small degree a reflection of the war spirit of cooperation which he constantly manifested.

JAPANESE

[Awarded for exceptionally meritorious and distinguished services in a position of great responsibility, under the provisions of the act of Congress approved July 9, 1918]

INOUYE, KAZUTSUGU R—Japan. G. O. No. 29, W. D., 1919.	Major general, Imperial Japanese Army. For services rendered the U. S. Army while serving as military attaché to the Imperial Japanese Embassy, Washington, D. C.
MIZUMACHI, T R—Japan. G. O. No. 11, W. D., 1919.	Lieutenant colonel, Imperial Japanese Army. For services rendered the U. S. Army while serving as military attaché to the Imperial Japanese Embassy, Washington, D. C.
OTANI, KIKUZO R—Japan. G. O. No. 45, W. D., 1919.	General, Imperial Japanese Army. For services as senior allied commander in Siberia.
TANAKA, GIICHI R—Japan. G. O. No. 19, W. D., 1922.	General, Imperial Japanese Army. For services rendered the United States during the World War and in his relations with the U. S. Army while serving as vice chief of the Japanese General Staff and later as Minister of War of the Empire of Japan.
UYEHARA, Y R—Japan. G. O. No. 11, W. D., 1919.	General, Imperial Japanese Army. For services rendered in the war against Germany and in his relations with the U. S. Army while serving as Chief of the General Staff, Imperial Japanese Army.
WATARI, HISAO R—Japan. G. O. No. 29, W. D., 1919.	Captain, Imperial Japanese Army. For services rendered the U. S. Army while serving as acting military attaché to the Imperial Japanese Embassy, Washington, D. C.

833

CITIZENS OF OTHER COUNTRIES

[Awarded for exceptionally meritorious and distinguished services in a position of great responsibility, under the provisions of the act of Congress approved July 9, 1918]

ARGENTINIAN

CHUTRO, PIETRO. R—Argentina. G. O. No. 2, W. D., 1920.	At his clinic in Paris, Doctor Chutro devoted himself unreservedly to teaching American medical officers the principles of the lessons learned through experience by the French and British surgeons in the first years of the war, with the result that the knowledge so imparted assisted in a great measure to conserve the life and limb of thousands of American and allied wounded.

RUMANIANS

RUDEANU, JOHN. R—Rumania. G. O. No. 10, W. D. 1922.	General, Rumanian Army. For services in connection with the American Military Mission in Hungary.
TEIUSANU, LIVIUS D. R—Rumania. G. O. No. 87, W. D., 1919.	Major, Rumanian Army. For services rendered the U. S. Army while serving as military attaché to the Rumanian Legation, Washington, D. C.
VASILESCU, CRISTEA. R—Rumania. G. O. No. 10, W. D., 1922.	Lieutenant colonel, Rumanian Army. For services in connection with the American Military Mission in Hungary.

RUSSIAN

MESTCHERINOFF, SERGE A. R—Russia. G. O. No. 53, W. D., 1921.	Lieutenant colonel, Russian Army. As deputy military attaché of Russia in France from March, 1917, to April, 1918, he demonstrated qualities of initiative, energy, and ability. Later, as attaché to the King of Montenegro, he showed marked devotion in furthering the friendly relation of the two Governments. He has rendered valuable services to the allied cause.

SERBIAN

°MISHICH, ZHIVOYIN. R—Serbia. G. O. No. 30, W. D., 1921.	Field marshal, Commander in Chief of Serbian Army. For services to the allied cause during the World War. Posthumously awarded. Medal forwarded to the military attaché, American Embassy, Belgrade, Serbia, for delivery to the next of kin.

MEMBERS OF THE ARMY, IN LIEU OF THE CERTIFICATE OF MERIT

[Distinguished-service medal issued in lieu of the certificate of merit under the provisions of the act of Congress approved July 9, 1918]

ABBOTT, GEORGE F. R—Worcester, Mass. B—Brandon, Vt.	Corporal, Company G, 9th Infantry, U. S. Army. For distinguished gallantry in action at Tientsin, China, July 13, 1900.
AKERS, THOMAS P. R—Lexington, Ky. B—Lexington, Ky.	Sergeant, 1st class, Signal Corps, U. S. Army. For distinguished service in the China relief expedition in China, Aug. 12, 1900.
ALBERTSON, EDWARD J. R—Sante Fe, N. Mex. B—Corpus Christi, Tex.	Private, Company F, 1st U. S. Volunteer Cavalry. For distinguished service in battle on July 1, 1898, at Santiago, Cuba.
ALDRIDGE, JOHN S. R—Watauga County, N. C. B—Mitchell County, N. C.	Corporal, Company D, 19th Infantry, U. S. Army. For gallant conduct in action at Mount Bud-Dajo, Jolo, P. I., Mar. 7, 1906.
ALEXANDER, JAMES R—Monroe, Ohio. B—Ohio.	First sergeant, Troop A, 4th Cavalry, U. S. Army. For distinguished service in engagement at Rio Chico Nueva Ecya, P. I., Dec. 6, 1900.
ALLEN, LUCIUS A. Near Donna, Tex., May 27, 1914. R—St. Elmo, Tenn. B—Buncombe County, N. C.	Private, Company D, 12th Cavalry, U. S. Army. For most distinguished courage and bravery beyond the call of duty in continuing his efforts to save a drowning comrade after having been seized and almost drowned himself in the Rio Grande River, near Donna, Tex., May 27, 1914.
ARNDT, ALVIN R—NR. B—Germany.	First sergeant, Troop I, 4th Cavalry, U. S. Army. For distinguished gallantry in action near Nozagaray, P. I., Feb. 15, 1899.
ASH, JOHN W. R—Chicago, Ill. B—Henderson, Ky.	Sergeant, Company E, 24th Infantry, U. S. Army. For conspicuous gallantry in action against Pulajanes at Tabon-Tabon, Leyte, P. I., July 24, 1906.
ASKEW, PRESTON R—Guthrie, Okla. B—Sulphur Springs, Tex.	Corporal, Company E, 24th Infantry, U. S. Army. For distinguished gallantry in action against Pulajanes at Tabon-Tabon, Leyte, P. I., July 24, 1906.
BAKER, WILLIAM B. R—New York, N. Y. B—New York, N. Y.	Corporal, Astor Battery, U. S. Army. For distinguished service in action at Manila, P. I., Aug. 13, 1898.
BANDIRA R—Mindanao, P. I. B—Mindanao, P. I.	Private, 51st Company, Philippine Scouts. For distinguished gallantry in action against hostile Moros, at Mount Bagsak, Jolo, P. I., June 15, 1913.
BARNES, WALTER K. R—Birmingham, Ala. B—Clay Center, Kans.	Acting hospital steward, Hospital Corps, U. S. Army. On June 3, 1900, at Province of Bulacan, P. I., during an engagement with insurgents he went forward to the firing line and applied first-aid dresssing to a soldier who had been wounded fatally.
BARNHOUSE, JOHN L. R—Fairmount, Ind. B—Fairmount, Ind.	Private, Company F, 17th Infantry, U. S. Army. For gallantry in action at Tempitan, Mindanao, P. I. on May 9, 1904.
BARRETT, MICHAEL R—Lynn, Mass. B—Ireland.	Sergeant, Company A, 7th Infantry, U. S. Army. For distinguished service in action at El Caney, Cuba, July 1, 1898.
BASSETT, DANIEL S. R—Philadelphia, Pa. B—Philadelphia, Pa.	Private, Company F, 21st Infantry, U. S. Army. For distinguished service at San Felipe Church, El Deposito, Manila Province, P. I., June 22, 1899.
BEASLEY, HALBERT M. R—Kinston, N. C. B—Granville County, N. C.	Sergeant, Hospital Corps, U. S. Army. Being injured himself he rendered aid to others injured in a wreck at Buckatunna, Miss., Oct. 19, 1913.
BELL, FRED R—NR. B—Ulster County, N. Y.	Sergeant, Battery H, 3d Artillery, U. S. Army. For distinguished service in the campaign near Malalos, P. I., Mar. 25-31, 1899.
BELMONT, JOHN R—Haverhill, Mass. B—Haverhill, Mass.	Private, Company B, 23d Infantry, U. S. Army. For heroic efforts to his own extreme exhaustion and danger, which resulted in saving a comrade from drowning at Galveston, Tex., June 18, 1914.
BERNHEIM, ALFRED A. R—Sacketts Harbor, N. Y. B—Little Big Horn River, Mont.	Sergeant, Company D, 9th Infantry, U. S. Army. For distinguished gallantry in battle of Tientsin, China, July 13, 1900.

BIEFER, ALBERT_____ R—New York, N. Y. B—Switzerland.	First sergeant, Company G, 13th Infantry, U. S. Army. For distinguished service in the battle of Santiago, Cuba, July 1, 1898.
BINCKLI, FREDERIC_____ R—New York, N. Y. B—Germany.	Private, Company H, 13th Infantry, U. S. Army. For distinguished service in action at Santiago, Cuba, July 1, 1898.
BOWDEN, WILLIAM H_____ R—Farmersville, Tex. B—Benton Co., Miss.	Corporal, Company C, 27th Infantry, U. S. Army. Rescued a fellow soldier from drowning in the Wisconsin River, Wis., on June 25, 1912.
BRADFORD, CLAUDE L_____ R—Port Hudson, La. B—Port Hudson, La.	Sergeant, Company B, 43d Infantry, U. S. Volunteers. For coolness, bravery, and good judgment in action against bolomen near Alang Alang, Leyte, P. I., Mar. 29, 1900.
BROADUS, LEWIS_____ R—Richmond, Va. B—Richmond, Va.	First sergeant, Company M, 25th Infantry, U. S. Army. For coolness, presence of mind and bravery in saving lives of others at Fort Niobrara, Nebr., on July 3, 1906.
BRYAN, WILLIAM_____ R—New York, N. Y. B—New York, N. Y.	Sergeant, 69th Company, Coast Artillery Corps, U. S. Army. For voluntarily entering a closed place and removing sacks of powder that were in close proximity to burning powder and smoldering débris, at the risk of his own life, thereby preventing further disaster after the explosion, at Fort Monroe, Va., July 21, 1910.
BUGBEE, FRED W_____ R—Tucson, Ariz. B—Oakland, Calif.	Private, Troop A, 1st U. S. Volunteer Cavalry. For distinguished service in battle at Santiago, Cuba, July 1, 1898.
CAPRON, HARRY W_____ R—NR. B—Waverly, Iowa.	Corporal, Troop B, 7th Cavalry, U. S. Army. For extraordinary gallantry in action at Wounded Knee Creek, S. Dak., Dec. 29, 1890.
CHICK, LEON H_____ R—Lynn, Mass. B—Rockport, Mass.	Sergeant, Battery H, 3d Artillery, U. S. Army. For risking his life in order to check a fire at Manila, P. I., Feb. 22, 1899.
CHISUM, JENNER Y_____ R—Jackson, Tenn. B—Jackson, Tenn.	First sergeant, Troop B, 6th Cavalry, U. S. Army. For distinguished service in twice attempting at the risk of his own life to rescue a fellow soldier from drowning in Wild Cat Creek at Fort Meade, S. Dak., May 25, 1907.
CLARK, JOHN J_____ R—White County, Ind. B—Tippecanoe County, Ind.	Quartermaster sergeant, 3d Infantry, U. S. Army. For distinguished service in battle at Santiago, Cuba, July 1, 1898.
CLARK, ORION L_____ R—Nemaha, Nebr. B—Nemaha, Nebr.	Private, Company B, 2d Infantry, U. S. Army. For distinguished service in battle at Santiago, Cuba, July 1, 1898.
CLARKE, HENRY N_____ R—NR. B—East Indies.	Private, Troop D, 3d Cavalry, U. S. Army. For distinguished service in a fire in the troop stables at Fort Sam Houston, Tex., on Aug. 12, 1892.
COMBS, KENRICK B_____ R—NR. B—Stafford County, Va.	Private, Troop F, 5th Cavalry, U. S. Army. For gallantry in action at Milk River, Colo., Sept. 29, 1879.
COOK, WILLIAM C_____ R—Owosso, Mich. B—Owosso, Mich.	Private, Company C, 13th Infantry, U. S. Army. For distinguished service in battle at Santiago, Cuba, July 1, 1898.
CORBETT, JOHN E_____ R—Brooklyn, N. Y. B—Long Island City, N. Y.	Recruit, Field Artillery, U. S. Army. For distinguished service on June 22, 1912, at Fort Slocum, N. Y., for rescuing a fellow soldier from drowning at the risk of his own life.
COX, ROY F_____ R—Junction City, Kans. B—Manhattan, Kans.	Corporal, Signal Corps, U. S. Army. For highly meritorious services in voluntarily traveling about 30 miles during a severe blizzard, rescuing a civilian from freezing near Lake Minto, and dragging by sled 65 miles to Fairbanks, Alaska, Feb. 26–29, 1908.
CRAVEN, RALPH G_____ R—Siler City, N. C. B—Beaufort, N. C.	Color sergeant, 6th Infantry, U. S. Army. For conspicuous gallantry in action at Mount Bud-Dajo, Jolo, P. I., Mar. 7, 1906.
CROSBY, SCOTT_____ R—NR. B—Marion County, Tenn.	Private, Company A, 24th Infantry, U. S. Army. For distinguished service in battle at Santiago, Cuba, July 1, 1898.
CROWELL, LEON_____ R—Monterey, Mexico. B—Portland, Ind.	Private, Company B, 4th Infantry, U. S. Army. For conspicuous gallantry in action near Dasmarinas, Luzon, P. I., June 19, 1899.
DAVIS, EDWARD_____ R—San Antonio, Tex. B—Nashville, Tenn.	Private, Troop H, 9th Cavalry, U. S. Army. For distinguished service in action at Santiago, Cuba, July 1, 1898.

DEAVEY, WILLIAM H_____ | Sergeant, Company H, 3d Infantry, U. S. Army.
R—New York, N. Y. | For distinguished service at Almacenes River, Luzon, P. I., Dec. 3, 1899.
B—Fort Edwards, N. Y. |

DELANEY, MICHAEL J_____ | Private, 24th Company, Coast Artillery Corps, U. S. Army.
R—Portland, Me. | For bravery in rescuing a comrade from drowning at Fort McKinley, Me., on
B—Portland, Me. | Apr. 30, 1913.

DILLMAN, WILLIAM_____ | Quartermaster sergeant, Company A, 13th Infantry, U. S. Army.
R—Terre Haute, Ind. | For distinguished service in action at Santiago, Cuba, on July 1, 1898.
B—Milton, Pa. |

DONALDSON, LORENZO D_____ | Private, Company F, 32d Infantry, U. S. Volunteers.
R—Fremont County, Iowa. | For distinguished gallantry in action at Abucay, Luzon, P. I., May 25, 1900.
B—Fremont County, Iowa. |

DONNELLY, JAMES_____ | Private, Company G, 18th Infantry, U. S. Army.
R—Cincinnati, Ohio. | For distinguished service at the fire at Fort Clark, Tex., Mar. 31, 1892.
B—Ireland. |

DOYLE, LAWRENCE_____ | Corporal, Company G, 11th Infantry, U. S. Army.
R—Newark, N. J. | For distinguished service at the Malabang River, Mindanao, P. I., Aug. 2, 1902.
B—New York, N. Y. |

DOZIER, ERNEST_____ | First-class sergeant, U. S. Volunteer Signal Corps.
R—San Francisco, Calif. | For distinguished service in action near Manila, P. I., Aug. 5, 1898.
B—East Oakland, Calif. |

ELDER, ROBERT W_____ | First sergeant, Troop L, 14th Cavalry, U. S. Army.
R—Boston, Mass. | For exceptional courage and daring in action at Cotto Pangpang, P. I., Feb.
B—Athol, Mass. | 14, 1904.

ELLIOTT, HENRY W_____ | Corporal, Troop H, 3d Cavalry, U. S. Army.
R—Indianapolis, Ind. | For distinguished service in battle at Santiago, Cuba, July 1, 1898.
B—Indianapolis, Ind. |

EVERSOLE, JOSEPH_____ | Corporal, Company H, 8th Infantry, U. S. Army.
R—Dutton, Ark. | For saving a fellow soldier from drowning at the risk of his own life at Nagasaki,
B—Perry County, Ky. | Japan, Sept. 7, 1917.

FARMER, LYLE G_____ | Corporal, Company I, 43d Infantry, U. S. Army.
R—Grand Meadow, Minn. | For saving Miss Eleanor Bourgeois from drowning in the artificial lake, dis-
B—Mears, Mich. | regarding personal risk and danger. Place: City Park, New Orleans, La.
 | Date: Apr. 27, 1918.

FEARINGTON, GEORGE W_____ | Private, Troop I, 9th Cavalry, U. S. Army.
R—Durham, N. C. | For excellent conduct and heroic service at Fort Duchesne, Utah, Dec. 13, 1899,
B—Durham County, N. C. | when the troop barracks were destroyed by fire, having taken post on the peak
 | of the seriously threatened building, remaining there in spite of the great heat,
 | applying water until the danger was over.

FEARNLEY, ARTHUR_____ | Corporal, 75th Company, Coast Artillery Corps, U. S. Army.
R—Oklahoma City, Okla. | For courage and tenacity of purpose in the face of dangers both known and
B—Wichita, Kans. | unknown while searching for a comrade who had been lost in the Koolau
 | Range, Oahu, Hawaii, Apr. 8, 1914.

FELDCAMP, GEORGE_____ | First sergeant, Company E, 12th Infantry, U. S. Army.
R—Cincinnati, Ohio. | For conspicuous gallantry in action against insurgents at Barrio of Mazambique,
B—Sedamsville, Ohio. | Ilocos Norte, P. I., Sept. 22, 1900.

FINERTY, WILLIAM M_____ | Corporal, Company E, 2d Infantry, U. S. Army.
R—Napoleon, Ohio. | For most distinguished conduct at Santiago, Cuba, July 2, 1898.
B—Napoleon, Ohio |

FISHER, ROBERT L_____ | Private, Battery A, 1st Field Artillery, U. S. Army.
R—Augusta, Ga. | For saving a comrade from drowning in Medicine Creek, near Fort Sill, Okla.,
B—Augusta, Ga. | June 29, 1907.

FLACH, JOHN_____ | Post commissary sergeant, U. S. Army.
R—Effingham County, Ill. | For distinguished service in Boston Harbor, Mass., Jan. 3, 1900.
B—Effingham County, Ill. |

FLANNERY, DAVID T_____ | First-class private, Company E, Signal Corps, U. S. Army.
R—Brooklyn, N. Y. | For distinguished gallantry in action at Big Bend, Luzon, P. I., Oct. 2, 1899.
B—Ireland. |

FLYNN, EDWARD_____ | Quartermaster sergeant, Company G, 21st Infantry, U. S. Army.
R—Cincinnati, Ohio. | For distinguished service in battle at Santiago, Cuba, July 1, 1898.
B—Ireland. |

FORTESCUE, GRANVILLE ROLAND_____ | Corporal, Troop E, 1st U. S. Volunteer Cavalry.
At Santiago, Cuba, July 1, 1898. | For distinguished service in battle at Santiago, Cuba, July 1, 1898.
R—New York, N. Y. |
B—New York, N. Y. |

GANNON, WILLIAM F_____ | Musician, Company L, 8th Infantry, U. S. Army.
R—Lowell, Mass. | For conspicuous gallantry in action against Pulajanes near La Paz, Leyte,
B—Lowell, Mass. | P. I., Dec. 5, 1906.

GOODE, BENJAMIN H.
  R—Abbeville County, S. C.
  B—Abbeville County, S. C.

Private, Company H, 24th Infantry, U. S. Army.
For most distinguished gallantry in action at Naguilian, Luzon, P. I., Dec. 7, 1899.

GOULD, CLARENCE S.
  R—Cone, Mich.
  B—Cone, Mich.

First sergeant, Troop C, 1st Cavalry, U. S. Army.
For gallant conduct in battle at Santiago, Cuba, July 1, 1898.

GOULD, FRED H.
  R—Los Gatos, Calif.
  B—Forest Grove, Oreg.

Corporal, Company H, 9th Infantry, U. S. Army.
For distinguished service in operations at Oras, Samar, P. I., January and February, 1902.

GROGAN, MICHAEL.
  R—Haverstraw, N. Y.
  B—England.

Corporal, Company B, 13th Infantry, U. S. Army.
For distinguished service in action at Santiago, Cuba, July 1, 1898.

GUNN, JAMES C.
  R—Eureka, Tex.
  B—Eureka, Tex.

Sergeant, 1st class, Hospital Corps, U. S. Army.
For distinguished service in action at Teruke-Utig, Jolo, P. I., on May 3, 1905.

HARRIS, JOHN C.
  R—Chicago, Ill.
  B—Canada.

Private, Company L, 4th Infantry, U. S. Army.
For distinguished service in the engagement near Dasmarinas, Luzon, P. I., June 19, 1899.

HARRISON, FRED A.
  R—La Junta, Colo.
  B—Kirwin, Kans.

Private, Troop C, 6th Cavalry, U. S. Army.
For distinguished service in engagement near San Nicolas, Cavite Province, P. I., Dec. 31, 1901.

HAWK, WILMER HARRISON.
  R—New Brighton, Pa.
  B—New Brighton, Pa.

Cook, 69th Company, Coast Artillery Corps, U. S. Army.
For voluntarily entering a closed place and removing sacks of powder that were in close proximity to burning powder and smoldering débris, at the risk of his own life, thereby preventing further disaster after the explosion, at Fort Monroe, Va., July 21, 1910.

HAWKINS, GEORGE P.
  R—Coldenham, N. Y.
  B—Greenwood, Nebr.

Private, Troop K, 14th Cavalry, U. S. Army.
For gallant conduct on Jan. 20, 1910, in rescuing at the risk of his own life an officer from drowning in the Quingua River, Luzon, P. I.

HECHT, HERMAN.
  R—Coryville, Ohio.
  B—Germany.

First sergeant, Company H, 4th Infantry, U. S. Army.
For distinguished service in battle at El Caney, Cuba, July 1, 1898.

HEINZE, JULIUS.
  R—Minneapolis, Minn.
  B—Germany.

Private, Hospital Corps, U. S. A.
For distinguished service in the fight on the Gandara River, Samar, P. I., Oct. 16, 1901.

HENNECKE, FRED.
  R—New York, N. Y.
  B—New York, N. Y.

Sergeant, Company L, 8th Infantry, U. S. Army.
For conspicuous gallantry in action against Pulajanes, near the village of La Paz, on the island of Leyte, P. I., Dec. 5, 1906.

HERBERT, THOMAS H.
  R—Washington, D. C.
  B—Montgomery County, Md.

Corporal, Troop E, 10th Cavalry, U. S. Army.
For distinguished service in battle at Santiago, Cuba, July 1, 1898.

HICKEY, EDWARD J.
  R—West Point, N. Y.
  B—West Point, N. Y.

Private, U. S. M. A., detachment Army Service men, Quartermaster department, U. S. Army.
For rescuing the body of a drowning boy at West Point, N. Y., Nov. 28, 1904.

HICKMAN, TAYLOR B.
  R—Knoxville, Tenn.
  B—Sevier County, Tenn.

Private, Company C, 9th Infantry, U. S. Army.
For distinguished service in battle at Tientsin, China, July 13, 1900.

HILYARD, SAMUEL W.
  R—Louisville, Ky.
  B—Louisville, Ky.

Artificer, Company E, 13th Infantry, U. S. Army.
For distinguished service in battle at Santiago, Cuba, July 1, 1898.

HOGAN, JAMES E.
  R—Chicago, Ill.
  B—Boston, Mass.

First-class sergeant, Signal Corps, U. S. Army.
For energy and good judgment displayed in administration of relief to, and safe transportation of three badly frozen enlisted men of Signal Corps from Summit to North Fork, Alaska, Jan. 3, 4, 5, 6, and 7, 1906.

HOGAN, THOMAS.
  R—N.E.
  B—Wayne County, Mich.

Private, Troop E, 3d Cavalry, U. S. Army.
For bravery in action with hostile Utes at Milk Creek, Colo., Sept. 29, 1879.

HOUSTON, ADAM.
  R—Pulaski County, Va.
  B—Pulaski County, Va.

First sergeant, Troop C, 10th Cavalry, U. S. Army.
For distinguished service in battle at Santiago, Cuba, July 1, 1898.

HOWE, CHARLES S.
  R—Chicago, Ill.
  B—Brooklyn, N. Y.

Sergeant, Company D, 17th Infantry, U. S. Army.
For distinguished service in the engagement near San Isidro, Luzon, P. I., July 29, 1900.

HUGHES, DAVID L.
  R—Tucson, Ariz.
  B—Tucson, Ariz.

Sergeant, Troop B, 1st U. S. Volunteer Cavalry.
For distinguished service in battle at Santiago, Cuba, on July 1, 1898.

HUMPHREY, CHARLES L.
  R—W. Boylston, Mass.
  B—Holden, Mass.

Private, Company H, 9th Infantry, U. S. Army.
For distinguished service in the battle of Tientsin, China, July 13, 1900.

HUNSAKER, IRVIN L. R—Cobden, Ill. B—Cobden, Ill.	First sergeant, Company H, 18th Infantry, U. S. Army. For distinguished gallantry in battle at Jaro, Panay, P. I., Feb. 12, 1899.
HUNTER, CHARLES. R—Mansfield, Ohio. B—Mansfield, Ohio.	Sergeant, Company E, 12th Infantry, U. S. Army. For conspicuous gallantry in action against insurgents at Barrio of Mazambique, Ilocos Norte, P. I., Sept. 22, 1900.
HYATT, THADDEUS R. R—Quallatown, N. C. B—Jackson County, N. C.	Corporal, Company L, 19th Infantry, U. S. Army. For capturing two of the worst criminals in Porto Rico, Nov. 8, 1898, near Guayanilla, P. R.
JACKSON, PERRY B. R—Ozark, Ala. B—Jonesburg, Mo.	Corporal, Company E, 6th Infantry, U. S. Army. For exceptional bravery in action at Mount Bud-Dajo, Jolo, P. I., Mar. 7, 1906.
JACKSON, PETER. R—New York, N. Y. B—New York, N. Y.	Corporal, Company G, 24th Infantry, U. S. Army. For distinguished service in battle at Santiago, Cuba, July 1, 1898.
JANOWSKI, FRANK. R—Chicago, Ill. B—Germany.	Artificer, Company D, 13th Infantry, U. S. Army. For distinguished service in battle at Santiago, Cuba, July 1, 1898.
JENSEN, JULIUS. R—New York, N. Y. B—Germany.	Sergeant major, 21st Infantry, U. S. Army. For distinguished service in battle at Santiago, Cuba, July 1, 1898.
JOHNSON, SANT. R—Jackson, Ky. B—Kentucky.	Corporal, Troop G, 3d Cavalry, U. S. Army. For distinguished service at a fire at Fort Apache, Ariz., Apr. 12, 1904.
JOHNSON, THOMAS. R—Lexington, Ky. B—Woodford County, Ky.	First sergeant, Company I, 24th Infantry, U. S. Army. For pursuing and disarming an enlisted man bent on murdering his first sergeant at Camp McGrath, Batangas, P. I., Aug. 22, 1912.
JONES, FRED B. R—Nough, Tenn. B—Cocke County, Tenn.	Sergeant, 109th Company, Coast Artillery Corps, U. S. Army. For attempting to rescue a civilian from drowning at Fort Greble, R. I., May 28, 1907.
JONES, WILLIE R. R—Atlanta, Ga. B—Atlanta, Ga.	Private, Battery A, 4th Field Artillery, U. S. Army. For saving a comrade from drowning at the risk of his own life at Vera Cruz, Mexico, on Sept. 1, 1914.
JORDAN, JOSEPH K. R—New York, N. Y. B—New York N, Y.	Private, Company H, 43d Infantry, U. S. Volunteers. For distinguished service in the attack by insurgents at Calbayog, Samar, P. I., Mar. 26, 1900.
KAINE, PATRICK. R—NR. B—Ireland.	First sergeant, Company D, 3d Infantry, U. S. Army. For saving a comrade from drowning at the risk of his own life at Leech Lake, Minn., June 16, 1898.
KALBER, LOUIS. R—NR. B—Germany.	Corporal, Company G, 9th Infantry, U. S. Army. For distinguished service at Singalan, P. I., Aug. 9, 1899.
KARSTEN, CHARLES. R—NR. B—Germany.	First sergeant, Troop D, 1st Cavalry, U. S. Army. For distinguished service in battle at Santiago, Cuba, July 1, 1898.
KEISTER, GUY A. R—Marion, Ind. B—Tipton County, Ind.	Private, Company F, 8th Infantry, U. S. Army. For saving a comrade from drowning at Lakeside, Calif., May 7, 1911.
KELLY, MICHAEL. R—Lewiston, Me. B—West Minot, Me.	First sergeant, Company C, 21st Infantry, U. S. Army. For distinguished service in battle at Santiago, Cuba, July 1, 1898.
KENNEDY, CECIL W. R—Central Lake, Mich. B—Wetzel, Mich.	Private, Troop B, 12th Cavalry, U. S. Army. For distinguished conduct in passing through a zone infested with bandits and carrying a message to his troop commander after having had two horses shot from under him at Progreso, Tex., Sept. 24, 1915, while serving as private, Troop B, 12th Cavalry, U. S. Army.
KLINGENSMITH, SAMUEL. R—Greensburg, Pa. B—West Newton, Pa.	Private, Troop F, 5th Cavalry, U. S. Army. For gallantry in action at Milk River, Colo., Sept. 29, 1879.
KOENIG, EDWARD. R—New York, N. Y. B—New York, N. Y.	Private, Company G, 8th Infantry, U. S. Army. For distinguished service on July 13, 1905, at the Governors Island Dock, New York, N. Y., in saving from drowning, at the risk of his own life, a civilian who had jumped into the bay with the intention of committing suicide.
LEABACK, CHARLES C. R—Chicago, Ill. B—Reading, Pa.	First sergeant, Company F, 9th Infantry, U. S. Army. For distinguished service near Tarlac, P. I., Nov. 19, 1899.

LEAKINS, JOHN ADAM R—Taneytown, Md. B—Libertytown, Md.	Private, Company C, 13th Infantry, U. S. Army. For distinguished service in battle at Santiago, Cuba, July 1, 1898.
LEONARD, CHARLES L. R—Lancaster, Pa. B—Lancaster, Pa.	Hospital steward, U. S. Army. For distinguished gallantry in the assault on Fort Bacolod, Lake Lanao, Mindanao, P. I., Apr. 8, 1903.
LEWIS, OLIVER. R—Nebo, N. C. B—Nebo, N. C.	Sergeant, 73d Company, Coast Artillery Corps, U. S. Army. For saving a girl from drowning at the risk of his own life at Fort Monroe, Va., July 27, 1915.
LEWIS, THOMAS R—NR. B—Burlington, Vt.	Private, Troop E, 3d Cavalry, U. S. Army. For conspicuous bravery before hostile Utes at Milk Creek, Colo, Sept. 29, 1879.
LIESMANN, FREDERICK J R—Dixon, Mo. B—Pierce City, Mo.	Corporal, Company B, 16th Infantry, U. S. Army. For distinguished service in battle at Santiago, Cuba, July 1, 1898. Oak-leaf cluster. Sergeant, Company M, 38th Infantry, U. S. Volunteers. For most conspicuous gallantry in action near San Juan de Bocboc, Luzon, P. I., July 1, 1900.
LIPSCOMB, SPENCER K R—East Seattle, Wash. B—St. George, W. Va.	Corporal, Company G, 14th Infantry, U. S. Army. For distinguished gallantry in action near Manila, P. I., on Feb. 5, 1899.
LOFTUS, JOHN R—Philadelphia, Pa. B—Ireland.	Private, Company C, 13th Infantry, U. S. Army. For distinguished service in battle at Santiago, Cuba, July 1, 1898.
LONG, JOHN R—Baltimore, Md. B—Baltimore, Md.	Artificer, Battery F, 3d Artillery, U. S. Army. For services rendered during fire at Fort Sam Houston, Tex., on Aug. 12, 1892.
LONGACRE, BENJAMIN F R—Lewistown, Pa. B—Mifflintown, Pa.	First sergeant, Troop F, 12th Cavalry, U. S. Army. For saving a comrade from drowning at the risk of his own life at Louisville, Colo., July 8, 1914.
LOOMIS, WILLIAM E R—Montrose, Pa. B—Forest Lake, Pa.	Corporal, Company B, 21st Infantry, U. S. Army. For distinguished service in battle at Santiago, Cuba, July 1, 1898.
MCBRIDE, JOHN R—Boston, Mass. B—Boston, Mass.	Sergeant, Troop A, 3d Cavalry, U. S. Army. For distinguished service in battle at Manimani River, Cuba, July 23, 1898.
MCCOMYN, RICHARD H R—NR. B—Ireland.	Private, Troop A, 6th Cavalry, U. S. Army. For distinguished service in engagement at San Juan, Santiago, Cuba, July 1, 1898.
MCCUTCHEON, JAMES R—Washington County, Minn. B—Washington County, Minn.	Private, Company G, 13th Infantry, U. S. Army. For distinguished service in battle at Santiago, Cuba, July 1, 1898.
MCDONALD, FRANK R R—Oologah, Ind. T. B—Keysport, Ill.	Trumpeter, Troop L, 1st U. S. Volunteer Cavalry. For distinguished service in battle at Santiago, Cuba, July 1, 1898.
MCDONALD, JOHN R—Sugar Notch, Pa. B—Shenandoah, Pa.	First sergeant, 77th Company, Coast Artillery Corps, U. S. Army. For saving from drowning an employee of the United States Engineer Department, at Otis Wharf, Boston, Mass., June 24, 1905.
MCDONALD, WILBERT L R—Huntsville, Tenn. B—Huntsville, Tenn.	Sergeant, Company D, 19th Infantry, U. S. Army. For gallant conduct in action at Mount Bud-Dajo, Jolo, P. I., Mar. 7, 1906.
MCGURTY, FRANK P R—New York, N. Y. B—Worcester, Mass.	Private, Company E, 22d Infantry, U. S. Army. For distinguished service during the San Francisco fire and earthquake, April, 1906.
MCMILLEN, JAMES L R—Fountain County, Ind. B—Fountain County, Ind.	Private, Company H, 12th Infantry, U. S. Army. For distinguished service at El Caney, Cuba, July 1, 1898.
MCNARNEY, FRANK T R—Lock Haven, Pa. B—McElhattan, Pa.	Sergeant, Company H, 10th Infantry, U. S. Army. For distinguished service in battle at Santiago, Cuba, July 1, 1898.
MAHONEY, DENIS R—New York, N. Y. B—Ireland.	Private, Company I, 36th Infantry, U. S. Volunteers. For gallantry in action near Guagua, P. I., Aug. 16, 1899.
MALONE, JOHN R—Anacostia, D. C. B—New York, N. Y.	Corporal, 47th Company, Coast Artillery Corps, U. S. Army. For rescuing with assistance, a comrade from drowning at Fort Washington, Md., Feb. 6, 1910.
MERDINGER, GEORGE R—NR. B—Germany.	First sergeant, Company H, 21st Infantry, U. S. Army. For distinguished service in battle at Santiago, Cuba, July 1, 1898.

MILLER, RICHARD. R—Louisville, Ky. B—Madison County, Ky.	Sergeant, Troop F, 9th Cavalry, U. S. Army. For distinguished service in the engagement at Tagbac, Albay, P. I., Dec. 17 1900.
MOLL, JULIUS. R—NR. B—Germany.	Sergeant major, 6th Cavalry, U. S. Army. For distinguished service in battle at Santiago, Cuba, July 1, 1898.
MOORE, JOHN I. R—Columbia, S. C. B—Hagood, S. C.	Sergeant, Company C, 27th Infantry, U. S. Army. For distinguished service in the assault on Fort Bacolod, Lake Lanao, Mindanao, P. I., Apr. 8, 1903.
MORRISON, EDWARD W. R—Colfax, Ind. B—Boone County, Ind.	Private, 1st class, Hospital Corps, U. S. Army. For rendering first aid to wounded comrades under fire of hostile Moros at Mamaya Peak, P. I., Dec. 15, 1913.
MORROW, BISHOP L. R—Mark, La. B—Osage, Mo.	Sergeant, Company I, 28th Infantry, U. S. Army. For bravery in saving the life of a wounded comrade by defending him against the treacherous attack of three Moros, all of whom he killed or wounded, near Pantar on the Iligan-Lake Lanao Military Road, P. I., May 1, 1903.
MOSELY, THOMAS. R—Brandenburg, Ky. B—Brandenburg, Ky.	Private, Hospital Corps, U. S. Army. For distinguished gallantry in action against hostile Moros at Mount Bagsak, Jolo, P. I., June 11, 1913.
MULHERN, BARTHOLOMEW. R—NR. B—Ireland.	Sergeant, Troop E, 3d Cavalry, U. S. Army. For distinguished service in battle at Santiago, Cuba, July 1, 1898.
MURPHY, MICHAEL J. R—Boston, Mass. B—Ireland.	First sergeant, Company D, 13th Infantry, U. S. Army. For distinguished service in battle at Santiago, Cuba, July 1, 1898.
MYERS, EDWIN. R—Brooklyn, N. Y. B—Bayard, Ohio.	Corporal, Signal Corps, U. S. Army. For distinguished service at Siassi, P. I., May 11, 1905.
NAGEL, THEODORE. R—Brooklyn, N. Y. B—Germany.	First sergeant, Company A, 13th Infantry, U. S. Army. For distinguished service in action at Santiago, Cuba, July 1, 1898.
NEAL, LOYD. R—Darlington, S. C. B—Darlington, S. C.	Musician, Battery H, 3d Artillery, U. S. Army. For distinguished service in action near Manila, P. I., July 31, 1898.
NICHOLS, EDWARD T. R—Boston, Mass. B—Bangor, Me.	First sergeant, Company D, 21st Infantry, U. S. Army. For distinguished service in action at Santiago, Cuba, July 1, 1898.
O'CONNOR, JOHN C. R—Jersey City, N. J. B—Ireland.	First sergeant, Battery G, 3d Artillery, U. S. Army. For distinguished services in the campaign in the Philippine Islands, July 31, 1898, to Jan. 1, 1901.
ODEN, GEORGE J. R—Los Angeles, Calif. R—Germany.	Sergeant major, 36th Infantry, U. S. Volunteers. For distinguished gallantry in action near Mangaperen, Luzon, P. I., Nov. 28, 1899.
ODEN, OSCAR N. R—San Francisco, Calif. B—Eaton County, Mich.	Trumpeter, Troop I, 10th Cavalry, U. S. Army. For distinguished service in action at Santiago, Cuba, July 1, 1898.
ODIN, ARTHUR S. R—Syracuse, N. Y. B—Syracuse, N. Y.	Private, Company I, 9th Infantry, U. S. Army. For distinguished service in the action near Zapote River, P. I., June 13, 1899.
O'KEEFFE, DANIEL. R—New York, N. Y. B—Albany, N. Y.	Private, Battery B, 7th Artillery, U. S. Army. For heroic conduct in rescuing from drowning a corporal of his battery who fell overboard from the tugboat Alert in Long Island Sound, Apr. 13, 1900, while the boat was plying between New London, Conn., and the islands.
O'KEEFE, JOHN P. R—New York, N. Y. B—New York, N. Y.	Corporal, Company M, 2d Infantry, U. S. Army. For imperiling his life in rescuing a drowning militiaman at American Lake, Wash., Aug. 19, 1908.
OLSEN, STANLEY R. R—Schenectady, N. Y. B—Frankfort, N. Y.	Private, 1st class, Troop H, 6th Cavalry, U. S. Army. For distinguished service in saving from death by drowning Private Emil A. Saboslay, Troop B, 6th Cavalry, at Espia, Chihuahua, Mexico, May 17, 1916.
OLSON, ROY C. R—Kansas City, Mo. B—Kansas City, Mo.	Corporal, Company D, 19th Infantry, U. S. Army. For gallant conduct in action at Mount Bud-Dajo, Jolo, P. I., Mar. 7, 1906, in willingly and unhesitatingly responding to a call for a few men to occupy an exposed point in advance of the line in plain view and within 25 or 30 yards of the enemy, from which point he kept up a most effective fire until the assault was made.
OLSSON, HENRY. R—NR. B—Springfield, Me.	Corporal, Battery E, 1st Artillery, U. S. Army. For distinguished service in the action at San Cristobal Bridge, P. I., July 30, 1899.

O'REILLY, JAMES F.
  R—Westville, N. Y.
  B—Westville, N. Y.

Corporal, Company B, 9th Infantry, U. S. Army.
For distinguished service in action at Santiago, Cuba, July 1, 1898.

O'ROURKE, JOHN.
  R—NR.
  B—Ireland.

First sergeant, Company C, 17th Infantry, U. S. Army.
For distinguished service in action at El Caney, Cuba, July 1, 1898.

PAREER, JESSE E.
  R—Charleston, W. Va.
  B—Charleston, W. Va.

Artificer, Company D, 24th Infantry, U. S. Army.
For brave and faithful conduct (assisting a wounded officer to a place of safety) while exposed to a severe fire at Santiago, Cuba, July 1, 1898.

PARKS, CHARLES E.
  R—Gas City, Ind.
  B—Grant County, Ind.

Private, 69th Company, Coast Artillery Corps, U. S. Army.
For voluntarily entering a closed place and removing sacks of powder that were in close proximity to burning powder and smoldering débris, at the risk of his own life, thereby preventing further disaster after the explosion at Fort Monroe, Va., July 21, 1910.

PASCHAL, JESSE J.
  R—Louisiana, Mo.
  B—Trenton, Mo.

Private, 14th Company, Coast Artillery Corps, U. S. Army.
For voluntarily imperiling his own life in the rescue of fellow soldiers at Fort Screven, Ga., July 4, 1906.

PAYNE, WILLIAM.
  R—Chattanooga, Tenn.
  B—Nashville, Tenn.

Sergeant, Troop E, 10th Cavalry, U. S. Army.
For distinguished service in battle at Santiago, Cuba, July 1, 1898.

PHILBIN, PATRICK.
  R—Dunmore, Pa.
  B—Ireland.

Sergeant, Company L, 15th Infantry, U. S. Army.
For distinguished service in the action at the barrio of San Juan, P. I., Dec. 31, 1900.

PITTS, WILLIAM C.
  R—Mineral Point, Wis.
  B—Milwaukee, Wis.

Private, Company F, 21st Infantry, U. S. Army.
For saving from drowning, near Florence, Ariz., a citizen of Florence, Ariz., on Mar. 7, 1908.

POTTER, HARRY E.
  R—Monticello, Me.
  B—Penobscot, Me.

First sergeant, Company A, 37th Infantry, U. S. Volunteers.
For distinguished conduct June 6, 1900, near Majajay, Laguna Province, Luzon, P. I., in exposing himself to the fire of the enemy at close range in order to remove a wounded comrade to shelter.

PRICE, WILSON C.
  R—W. Philadelphia, Pa.
  B—Philadelphia, Pa.

Private, Company F, 9th Infantry, U. S. Army.
For distinguished gallantry in battle of Tientsin, China, July 13, 1900.

PUMPHREY, GEORGE W.
  R—Baltimore, Md.
  B—Anne Arundel County, Md.

Corporal, Troop H, 9th Cavalry, U. S. Army.
For distinguished service in battle at Santiago, Cuba, July 1, 1898.

REED, JOSEPH E.
  R—Huntingdon, Pa.
  B—Drafton, Pa.

Private, Company L, 47th Infantry, U. S. Volunteers.
For distinguished gallantry in action at Bulusan, Luzon, P. I., Nov. 11, 1900.

RICHARDSON, HURLEY O.
  R—Lorley, Md.
  B—Berlin, Md.

Sergeant, Troop C, 6th Cavalry, U. S. Army.
For conspicuous bravery and efficiency in action against hostile Moros on Patian Island, P. I., July 4, 1909.

RICHMOND, CHARLES C.
  R—Toledo, Ohio.
  B—White House, Ohio.

First sergeant, Company L, 30th Infantry, U. S. Volunteers.
For distinguished gallantry in action on Arabaon Mountain, Luzon, P. I., Mar. 26, 1900.

ROSE, FRANKLIN.
  R—St. Louis, Mo.
  B—Fulton County, Ohio.

Post commissary sergeant, U. S. Army.
For distinguished service during the cyclone at Galveston, Tex., Sept. 8, 1900.

ROSSER, HENRY.
  R—Stork, N. C.
  B—Chatham County, N. C.

Corporal, Company A, 17th Infantry, U. S. Army.
For distinguished service in the attack on Angeles, P. I., Oct. 16, 1899.

ROWBOTTOM, HENRY T.
  R—Valdez, Alaska.
  B—England.

First-class private, Signal Corps, U. S. Army.
For rescuing two soldiers clinging to a capsized canoe in the Yukon River, Alaska, on July 3, 1913.

RYAN, THOMAS.
  R—Baltimore, Md.
  B—Ireland.

First sergeant, Troop K, 1st Cavalry, U. S. Army.
For distinguished service in battle at Las Guasimas, Cuba, June 24, 1898.

RYDER, WILLIAM.
  R—Chicago, Ill.
  B—Canada.

Sergeant, Company G, 13th Infantry, U. S. Army.
For distinguished service in the battle at Santiago, Cuba, July 1, 1898.

SACKNUS, HENRY W.
  R—Cleveland, Ohio.
  B—Cleveland, Ohio.

Private, Troop C, 6th Cavalry, U. S. Army.
For exceptional gallantry in an engagement at Talisay, P. I., Mar. 29, 1901.

SATCHELL, JAMES.
  R—Eastville, Va.
  B—Eastville, Va.

Corporal, Company A, 24th Infantry, U. S. Army.
For distinguished service in battle at Santiago, Cuba, July 1, 1898.

SCALETTA, PAUL.
  R—San Jose, Calif.
  B—San Jose, Calif.

Corporal, Company B, 159th Infantry, 40th Division, U. S. Army.
For repeatedly entering a dangerous surf and saving the lives of other soldiers at the risk of his own life at Ocean Beach, Calif., May 5, 1918.

SCHUCK, WILLIAM J R—Williamsville, N. Y. B—Williamsville, N. Y.	Sergeant, Company K, 6th Infantry, U. S. Army. For distinguished service in action at Bobong, Negros, P. I., July 19, 1899.
SCHWARZ, ADOLPHUS A R—Edwardsville, Ill. B—Edwardsville, Ill.	Private, Troop K, 3d Cavalry, U. S. Army. For excellent conduct and heroic service during the burning of a troop barracks at Jefferson Barracks, Mo., Apr. 22, 1896.
SELMIRE, GEORGE R—Philadelphia, Pa. B—Philadelphia, Pa.	Sergeant, Company H, 7th Infantry, U. S. Army. For distinguished service in battle at El Caney, Cuba, July 1, 1898.
SEUFERT, LOUIS P R—Buffalo, N. Y. B—Germany.	Corporal, Company H, 13th Infantry, U. S. Army. For distinguished service in battle at Santiago, Cuba, July 1, 1898.
SHADDEAU, HENRY D R—Swan Creek, Mich. B—Owosso, Mich.	Corporal, Company F, 17th Infantry, U. S. Army. For exceptional and conspicuous bravery in action at Simpiton, Mindanao, P. I., on May 8, 1904.
SHAFFER, SAMUEL W R—Collington, Md. B—Baltimore, Md.	Sergeant major, 7th Infantry, U. S. Army. For distinguished service in battle of El Caney, Cuba, July 1, 1898.
SHEWBRIDGE, SMITH M R—Cincinnati, Ohio. B—Charlestown, W. Va.	First sergeant, Company B, 17th Infantry, U. S. Army. For gallantry in action at Tamparan, Mindanao, P. I., Apr. 5, 1904.
SILLITO, LOUIS A R—Fort Myers, Fla. B—Nashville, Tenn.	Private, Company C, 3d Infantry, U. S. Volunteers. For volunteering to nurse and nursing yellow fever patients at Guantanamo, Cuba, Sept. 1, 1898.
SIMPSON, ALFRED G R—Columbia, Ind. B—Fayette County, Ind.	Private, Company C, 3d Infantry, U. S. Army. For distinguished service in battle at Santiago, Cuba, July 2, 1898.
SLOAN, JAMES R—Salt Lake City, Utah. B—Meadville, Pa.	Private, Company G, 16th Infantry, U. S. Army. For distinguished service in battle at Santiago, Cuba, July 1, 1898.
SMITH, GEORGE R—Edgefield, Tenn. B—De Kalb, Tenn.	Corporal, Company A, 7th Infantry, U. S. Army. For distinguished service in battle at El Caney, Cuba, July 1, 1898.
SMITH, JACK R—Marksville, La. B—Marksville, La.	Lance Corporal, Company E, 21st Infantry, U. S. Army. For conspicuous bravery in action at Arroyo de Lanao, Samar, P. I., June 4, 1905.
SMITH, LUCIIIOUS R—Pittsburgh, Pa. B—Etowah County, Ala.	Private, Troop D, 10th Cavalry, U. S. Army. For distinguished service in battle at Santiago, Cuba, July 1, 1898.
SMITH, ROBERT M R—NR. B—Detroit, Mich.	Corporal, Company B, 7th Infantry, U. S. Army. For distinguished service in battle at El Caney, Cuba, July 1, 1898.
SMITH, WESLEY W R—New York, N. Y. B—New York, N. Y.	Private, Company D, 21st Infantry, U. S. Army. For distinguished service in battle at Santiago, Cuba, July 1, 1898.
SMITH, WILLIE B R—Lockesburg, Ark. B—Union County, Ark.	Corporal, Battery O, 1st Artillery, U. S. Army. For distinguished service during the cyclone at Galveston, Tex., Sept. 8, 1900.
SMYTH, CHARLES E R—Chicago, Ill. B—New York, N. Y.	Sergeant, Troop A, 14th Cavalry, U. S. Army. For manner in which he handled his detachment of 8 men while surrounded by and under fire of almost 100 Mexican bandits at Glen Springs, Tex., May 5, 1916.
STAPLES, FRANK R—Brooklyn, N. Y. B—Brooklyn, N. Y.	Private, Company F, 27th Infantry, U. S. Army. For distinguished gallantry in the assault on Fort Pitacus, P. I., May 4, 1903.
STEGER, EDWARD R—Detroit, Mich. B—Germany.	Private, Company C, 1st Infantry, U. S. Army. Assisted in rescuing a fellow soldier from drowning at the risk of his own life in the sea near Haleiwa, Hawaii, Oct. 22, 1914.
STEVENS, JACOB W R—Baltimore, Md. B—Franktown, Va.	First sergeant, Company K, 24th Infantry, U. S. Army. For distinguished service in engagement near Santa Ana, P. I., Oct. 6, 1899.
STEWART, CLYDE H R—Pawnee, Okla. B—Seneca, Kans.	Corporal, Company F, 7th Infantry, U. S. Army. For saving a child from drowning at Fort Wayne, Mich., Sept. 7, 1908.
STOCKFLETH, HENRY R—NR. B—Germany.	Corporal, Battery H, 3d Artillery, U. S. Army. For distinguished service in action near Manila, P. I., July 31, 1898.

STOKES, EARNEST R—Wartrace, Tenn. B—Wartrace, Tenn.	Private, Company F, 24th Infantry, U. S. Army. For most distinguished gallantry in action at Naguilian, Luzon, P. I., Dec. 7, 1899.
STOKES, GEORGE P. R—Rocky Ford, Colo. B—Rogers, Ark.	Private, Company I, 21st Infantry, U. S. Army. For marked courage and intelligence in saving a comrade from drowning at Lake Lanao, P. I., Dec. 12, 1909.
TEETER, MILO C. R—New Albany, Pa. B—New Albany, Pa.	Private, Company E, 19th Infantry, U. S. Army. Rescued a comrade from drowning at Penaranda, P. I., on Feb. 22, 1912.
THORNTON, WILLIAM R—Great Falls, Mont. B—Washington Court House, Ohio.	Corporal, Company G, 24th Infantry, U. S. Army. For distinguished service in battle at Santiago, Cuba, July 1, 1898.
TOBIN, WILLIAM J. R—NR. B—Fort Warren, Mass.	First sergeant, Battery G, 7th Artillery, U. S. Army. For distinguished service in Boston Harbor, Mass., Jan. 3, 1900.
TOM, JACOB W. R—Paulding, Ohio. B—Hancock County, Ohio.	Quartermaster sergeant, Company D, 19th Infantry, U. S. Army. For gallant conduct in action at Mount Bud-Dajo, Jolo, P. I., Mar. 7, 1906.
TOMLINSON, FREDERICK. R—Philadelphia, Pa. B—Memphis, Tenn.	Corporal, 138th Company, Coast Artillery Corps, U. S. Army. Rescued the body of a comrade from a burning launch on the Pasig River, Manila, P. I., Sept. 9, 1912.
TURNER, CLEMON. R—Cincinnati, Ohio. B—Gainesville, Ala.	Private, Company K, 24th Infantry, U. S. Army. For rescuing a fellow soldier from drowning at the risk of his own life, near Camp McGrath, P. I., Nov. 12, 1914.
TURNER, VICTOR. R—Homestead, Pa. B—Homestead, Pa.	Musician, Company C, 17th Infantry, U. S. Army. For meritorious conduct in saving the life of a drowning comrade at the risk of his own at Sweetwater Creek, Ga., June 29, 1909.
VAN CAMPEN, HIEL. R—Jeffersonville, Ind. B—Jeffersonville, Ind.	Private, Company E, 1st Infantry, U. S. Army. For his assistance in saving a comrade from drowning at great risk of his own life near Haleiwa, Hawaiian Islands, Oct. 22, 1914.
VENUS, CHARLES. R—Lawrenceville, Ill. B—Lawrenceville, Ill.	Sergeant, Company I, 23d Infantry, U. S. Army. For exceptional courage in swimming and wading to shore from a capsized boat and guiding a rescue boat to the relief of three comrades whom he had left on the upturned hull in Galveston Bay, Tex., June 29, 1914.
VILLUMSEN, HANS. R—Philadelphia, Pa. B—Denmark.	Sergeant, Company D, 10th Infantry, U. S. Army. For distinguished service in battle at Santiago, Cuba, July 1, 1898.
VOLKMAR, WALTER S. R—NR. B—Philadelphia, Pa.	Sergeant, Signal Corps, U. S. Army. For distinguished services in subduing a fire which threatened to destroy public property, at Fort Sam Houston, Tex., Mar. 15, 1898.
WALKER, ARTHUR L., Jr. R—Brookline, Mass. B—Yonkers, N. Y.	Private, Headquarters Company, 301st Infantry, 76th Division, U. S. Army For saving the life of a fellow soldier when the latter was thrown into the water by the capsizing of a canoe, at the risk of his own life, at Robbins Pond, Camp Devens, Mass., May 4, 1918.
WALKER, JOHN. R—Thomhill, Va. B—Orange County, Va.	Corporal, Troop D, 10th Cavalry, U. S. Army. For distinguished service in action at Santiago, Cuba, July 1, 1898.
WARNER, JAMES H. R—NR. B—Tioga County, N. Y.	First sergeant, Company D, 4th Infantry, U. S. Army. For distinguished service in action near Dasmarinas, P. I., June 19, 1899.
WEBER, ANTON. R—New York, N. Y. B—Switzerland.	Quartermaster sergeant, Company H, 13th Infantry, U. S. Army. For distinguished service in action at Santiago, Cuba, July 1, 1898.
WEISS, GEORGE. R—St. Louis, Mo. B—Prussia.	Sergeant, Troop F, 4th Cavalry, U. S. Army. For gallant conduct in action at Mount Bud-Dajo, Jolo, P. I., Mar. 7, 1906.
WHELAN, JOHN. R—New York, N. Y. B—Ireland.	Sergeant, Company F, 27th Infantry, U. S. Army. For conspicuous gallantry in action at Bayang, P. I., May 2, 1902.
WHITE, JOSEPH. R—NR. B—New Orleans, La.	Musician, Company B, 24th Infantry, U. S. Army. For distinguished service at the Rio Grande River, Cabanatuan, P. I., Nov. 8, 1900.
WHITNEY, PERCY M. R—Washington, D. C. B—Washington, D. C.	Private, Troop E, 6th Cavalry, U. S. Army. For distinguished service in unhesitatingly plunging into the Palico River, P. I., Sept. 2, 1901, and at the risk of his own life attempting to rescue from drowning Sergt. John W. Harris, of said troop.

WIGLEY, DORPHIN C. R—Pine Creek, Pa. B—Pine Creek, Pa.	Private, 39th Company, Coast Artillery Corps, U. S. Army. For voluntarily entering a closed place and removing sacks of powder that were in close proximity to burning powder and smoldering débris, at the risk of his own life, thereby preventing further disaster after the explosion, at Fort Monroe, Va., July 21, 1910.
WILKINS, GEORGE W. R—Sioux City, Iowa. B—Sioux City, Iowa.	Sergeant, Company G, 39th Infantry, U. S. Volunteers. For distinguished gallantry in action at San Cristobal River, Luzon, P. I., on Jan. 1, 1900.
WILLFORD, JAMES W. R—Urbana, Ohio. B—Urbana, Ohio.	Sergeant, Company G, 9th Infantry, U. S. Army. For great coolness and good judgment displayed in commanding a battery when ambushed by natives near Basey, Samar, P. I., Sept. 1, 1901.
WILLIAMS, RICHARD R—Dayton, Ohio. B—Cincinnati, Ohio.	Corporal, Company B, 24th Infantry, U. S. Army. For distinguished service in action at Santiago, Cuba, July 1, 1898.
WILLIAMS, THOMAS S. R—NR. B—Glasgow, Ky.	Private, Company E, 2d Infantry, U. S. Army. For most distinguished conduct at Santiago, Cuba, July 2, 1898.
WINTER, JOHN G. R—San Antonio, Tex. B—Waco, Tex.	Private, Troop F, 1st U. S. Volunteer Cavalry. For distinguished service in battle at Santiago, Cuba, July 1, 1898.
WINTERS, OSCAR F. R—NR. B—Flintstone, Md.	Corporal, Company F, 9th Infantry, U. S. Army. For distinguished service in action at Santiago, Cuba, July 2, 1898.
YOUNG, JOHN C. R—NR. B—Canada.	Sergeant major, 3d Infantry, U. S. Army. For rescuing a comrade from drowning in Minnesota River near Fort Snelling, Minn., Nov. 19, 1896. Oak-leaf cluster. For distinguished service in battle at El Caney, Cuba, July 1, 1898.

www.ingramcontent.com/pod-product-compliance
Lightning Source LLC
Chambersburg PA
CBHW080334220326
41598CB00030B/4499